The Best of the Best

IEEE Press
445 Hoes Lane
Piscataway, NJ 08854

The Best of the Best

Fifty Years of Communications and Networking Research

Edited by

William H. Tranter
Desmond P. Taylor
Rodger E. Ziemer
Nicholas F. Maxemchuk
Jon W. Mark

A JOHN WILEY & SONS, INC., PUBLICATION

Published by John Wiley & Sons, Inc. Published simultaneously in Canada.

For general information on our other products and services or for technical support, please contact our Customer Care Department within the United States at (800) 762-2974, outside the United States at (317) 572-3993 or fax (317) 572-4002.

Wiley also publishes its books in a variety of electronic formats. Some content that appears in print may not be available in electronic format. For information about Wiley products, visit our web site at www.wiley.com.

Library of Congress Cataloging-in-Publication Data is available.

ISBN-13 978-0-470-11268-7
ISBN-10 0-470-11268-9

Printed in the United States of America.

10 9 8 7 6 5 4 3 2 1

Contents

* This paper selected for both Physical and Link Layer Aspects of Communications and Networking Sections.

Preface

The year 2002 marked the fiftieth anniversary of the founding of the IEEE Communications Society. Various special events (two Grand Reunions at IEEE ICC 2002 in New York and IEEE GLOBEOM 2002 in Taipei) took place to celebrate this significant milestone. A number of publications were part of this celebration. For example, the *IEEE Communications Magazine, 50th Anniversary Issue,* was published in May 2002. The Society issued a 2 DVD compilation of 28,000 papers, *Communications Engineering Technology; A Comprehensive Collection of Papers 1953-2001,* and a 128-page volume *A Brief History of Communications.* Another activity, focusing on archival research papers was a compilation of outstanding papers that have appeared in the Society's transactions and journals during the past five decades. These papers represent a selection of various key research papers that have been published by the Communications Society. Moreover, they provide a snapshot history of the evolution of communications systems over this five-decade period. This compilation was known as the "Best of the Best" and was distributed via the web page of the IEEE Communications Society, and was also distributed at Communications Society conferences and other appropriate events during 2002. This volume is at long last a hard copy edition of this compilation.

The editors fully realize that there will not be unanimous agreement that the chosen papers represent the best possible selection and that a list of "best papers," like beauty, is in the eye of the beholder. We certainly agree that many excellent papers have appeared in the publications of the Communications Society that are not included in this publication. In order to select the papers included here the editors attempted to put in place a process that would result in a useful reference, represent a good cross section of the best papers, and provide a historical perspective of the development of communications over the past fifty years.

It should be pointed out that the papers included herein represent the best of "our" best with our referring to the IEEE Communications Society. The journals represented include the *IRE Transactions on Communications Systems,* the *IEEE Transactions on Communications Technology,* the *IEEE Transactions on Communications,* the *IEEE Journal on Selected Areas in Communications,* and the *IEEE/ACM Transactions on Networking.* We should point out that two additional publications focusing on research results have recently joined the suite of Communications Society publications: *IEEE Communications Letters* and the *IEEE Transactions on Wireless Communications.* There are a number of journals that publish research-level papers in communications that are not part of the IEEE Communications Society and are therefore not represented here. The *IEEE Transactions on Information Theory* and special issues of the *Proceedings of the IEEE* devoted to communications are of particular note as are several publications of the IEE (Institution of Electrical Engineers, now the IET).

The volume has two major sections. The first section contains 41 papers dealing with physical and link layer aspects of communications. These papers are arranged chronologically within the section. The second section consists of 16 network papers that deal with the higher layers in communications systems. This second set of papers was compiled by a second set of guest editors, whose expertise lies in the networks and protocols area. They, too, are arranged in chronological order within the section.

Selection Process for the Physical and Link Layer Communications Papers

As the first step in the process of choosing the physical and link layer papers, the editors compiled a preliminary list of 26 papers dealing with physical and link layer communications. The initial list was developed from a list of those papers receiving the most citations and those papers receiving "best paper" awards. Also included on the list were candidate papers based on our experience as Transactions and Journal editors and as members of the research community. The next step was to circulate this preliminary list to 43 members of the Communications Society who had extensive service as editors of various IEEE Communications Society publications or as leaders in the research community. This group was asked to respond to the preliminary list and suggest additions or deletions along with the reasons for their suggestions. After considering the inputs received, the list was expanded from 26 to the 41 papers appearing in this book. It is significant that only 17 of the initially selected 26 papers survived this review process. It is clear from this statistic that the reviewers took their jobs seriously and we thank them for their efforts and for their support of this project. The names of these reviewers appear in Supplement A of this introduction.

Selection Process for the Networking Papers

The selection of papers in the networking area was a bit more free-form than in the physical layer communications area. In May 2001, invitations were sent to 159 members of the networking community to nominate two or three papers, other than their own, that most affected their own work. The nominators were asked to briefly indicate why they considered the papers important. While a large number of very good papers received nominations, a set of 16 papers dominated. Those papers are presented in this issue.

The only constraint that we placed on the nominations was that they appeared in one of the IEEE Communications Society's three archival journals, the *IEEE Transactions on Communications,* the *IEEE Journal on Selected Areas in Communications,* or the *IEEE/ACM Transactions on Networking.* Networking, to a much greater extent than physical communications, is also in the domain of several other societies, and many important publications in networking are not in this list because they appeared in a journal of one of our sister societies or in one of the many conferences in this area.

If you count the number of papers you will note that 56 papers appear, not 57. This is because one paper "The Throughput of Packet Broadcasting Channels" by Norman Abramson was selected for both sections.

The Past Few Years

The initial plan was to publish this volume in 2002. However, due to a number of reasons this did not prove practical. Now, through a partnership of the IEEE Press and John Wiley, publication has been made possible.

Since several years have passed between the original submissions of the most recent papers appearing here, the editors considered a method for updating this volume. All of our research publications sponsor a best paper award and it was decided to list the recipients of these awards given in 2000-2005 within this introduction. These papers appear in Supplement B of this introduction. A complete list of recent (1994-2005 for most journals) awards can be found on the Communications Society home page appearing at

http://www.comsoc.org/socstr/org/operation/awards/paperawards.html/

A Brief Perspective on Communications

In order to place the papers in this volume in perspective we provide a brief outline of the development of communications here and of the IEEE Communications Society in the next section. Much of what follows in this section and the next is based on *A Brief History of Communications,* which was published by the IEEE Communications Society as a part of the 50th anniversary events. The material in this booklet is available on the web site of the IEEE History Center along with lists of publications and extensive related material. The URL for the IEEE History Center is

http://www.ieee.org/web/aboutus/history_center/

The interested reader is encouraged to consult this site for additional perspectives on the history of electrical engineering, including communications, and on the history of the IEEE Communications Society.

Although the first 50 years of the IEEE Communications Society spans the period (1952-2002) many agree that the period of "modern" communications has its roots in 1947. The year 1947 was a benchmark year for the communications industry and was marked by two events which, at the time, were not viewed as closely related. The first of these events was the publication of the seminal paper *A Mathematical Theory of Communications,* by Claude Shannon. This paper quantitatively defined information in the communications context and established fundamental limits on the performance of communication systems. The concepts of source coding and channel (error-correction) coding have their roots in this paper, as does the field of information theory. Without the contributions of information theory, and error-correction coding in particular, modern communication systems, which are often required to perform well in heavy interference and noise environments, would simply not exist. The second event of 1947 was the invention of the point-contact transistor by Shockley, Bardeen, and Brattain. The transistor led to the integrated circuit and microelectronics. Microelectronics enabled the development of many communication devices ranging from CD and DVD players, to cell phones and satellite communication systems. Microelectronics also made possible the development of high-speed desktop computers at reasonable cost and this enabled the development of computer communication networks culminating in today's internet. It should be pointed out that both of these events were products of research performed at Bell Laboratories.

This is not to say that important strides in the fields of communications were not made prior to 1947. Indeed the early days of communications (pre 1947 in this context) saw the development of much of the basic theory of electrical circuits, electronics, and electromagnetics that led to the invention of the telegraph, telephone, AM and FM radio, and television. Few today realize that the first working transatlantic telegraph cable was laid in 1857-1858, approximately 150 years ago. The now famous 1928 paper by Nyquist developed many of the concepts required to design modern, high-speed data communications systems for band-limited channels. The late 1930's and early 1940's also saw the invention of radar initially by the British. Radar systems were among the first communication systems to utilize the "modern" tools of statistical communication theory. The time between World War II and 1952 was dominated by the further development of both radio and television.

Plans for an undersea telephone cable were initially formulated in 1952 between Bell Telephone and the British Post Office. This cable, connecting the US and the UK, was installed in 1955 and 1956. Known as TAT1, this cable successfully operated from 1956 to 1979. A number of other telephone cables followed, notably cables from France and Newfoundland (TAT-2), Alaska to Washington state, and Hawaii to California. It should be recalled that undersea telephone cables were preceded by undersea telegraph cables by 100 years!

The next revolution in transcontinental communications was provided by the communications satellite. The first artificial earth satellite, Sputnik, was successfully launched in October 1957, by the Soviet Union. This was followed a few months later by the first successful US satellite, Explorer I. By the end of the 1950's, the enabling technologies required for commercial satellite communications were in place. The first communications satellite, Echo I, was launched in August, 1960. Echo I was a passive satellite in which signals were bounced off the surface of the satellite. While Echo I demonstrated the viability of satellite communications, very high transmitter power was required to overcome path losses. The answer to this problem was, of course, active satellites capable of receiving and retransmitting signals using high-gain antennas.

The first active communications satellites, Telstar I and Telstar II, were launched in 1962 and 1963, respectively. In addition to telephone channels, Telstar II provided one television channel. These were not commercial systems but were developed to demonstrate the viability of satellite communications systems. In 1964 INTELSAT, an organization involving the communications agencies of over 100 countries, was formed as an international body to develop and operate the global satellite communications system. INTELSAT deployed four generations of satellite systems operating in geosynchronous orbit. Telstar and INTELSAT provided the basic building blocks for modern satellite-based communications providing intercontinental telephone and television relaying, subscription television and radio, global positioning systems (GPS), and a host of other services.

Another significant advance in high-speed communications was the development of optical fiber systems in the late 1960s. The two enabling technologies were the invention of the laser in 1959-60 and the development of the glass fiber waveguide by Kao and Hockham in 1966. The first field installation of an optical fiber system using a semiconductor laser took place in the mid 1970s.

Two additional developments of the past three decades changed the way we use and view communications. These two developments were the computer network and the wireless cellular system. These two developments gave rise to intense research activity. This research activity continues to this day and gave rise to two research journals published by the IEEE Communications Society. These are the *IEEE/ACM Transactions on Networking,* focusing on computer communications, and the *IEEE Transactions on Wireless Communication,* focusing on wireless systems.

Although the ARPANET was not the first computer network to be demonstrated, the ARPANET was the first network to use packet switching and is considered by most to be the first network having long-term significance. The development of the ARPANET was funded by the Advanced Research Projects Agency (ARPA), an arm of the U.S. Department of Defense. The ARPANET was based on the seminal research of Kleinrock, Baran, and Davies and connected a number of universities and research laboratories in the U.S. ARPANET, despite its capabilities, was not widely used until Robert Kahn demonstrated ARPANET capabilities in 1972 at the first International Conference on Computer Communications. Also in 1972, Packet Communications, Inc., was formed to market communications services along the lines of the ARPANET model. The first major result of this initiative was Telnet, which served seven U.S. cities in 1975.

The work of Robert Kahn and Vinton Cerf, along with many other researchers, was instrumental in transforming the ARPANET into what we know today as the Internet, which is a network of networks. Cerf and Kahn proposed the Internet architecture which, because of its flexibility and decentralized nature, was able to accommodate a wide range of users and applications. They developed the host-to-host protocol and the Internet protocol (TCP/IP) which was fundamental in supporting the growth of the Internet. In addition, the 1980s saw the personal computer gaining a place both in the home and on the on the office desktop and this rapid expansion of computer access gave the masses network accessibility. Today the Internet, coupled with a wide variety of supporting software, such as web browsers and search engines, supplies a variety of services including email, information processing, voice and video transmission, as well as applications to education and entertainment.

Wireless cellular communications, like the Internet, has witnessed recent explosive growth. Much of the fundamental research leading to wireless cellular communications was performed at Bell Laboratories in the 1960s and a series of papers focused on this research was published in the *Bell Systems Technical Journal.* The cellular concept was proposed to the FCC in 1968. At this time much of the theory was in place and by 1980, the hardware necessary for the practical implementation of wireless cellular systems was available. In 1983 the FCC allocated spectrum for wireless services in the 800 MHz frequency band. Later that year the first wireless cellular system was deployed by Ameritech in Chicago, IL. From this beginning the wireless industry grew at an exponential rate worldwide. Wireless communications has brought telephone service to many people living in environments where wired telephone service is simply impractical and in many countries the number of cell phones significantly exceeds the number of wired phones. Much of the current research is aimed at improving the quality of service, increasing bandwidth, and developing new services for the user.

Hopefully this brief perspective helps place the papers in this book in historical perspective and provides insight into the exciting nature of the communications industry.

A Brief Perspective on the IEEE Communications Society

The IEEE Communications Society, like all of the IEEE, has its roots in the AIEE (American Institute of Electrical Engineers) and the IRE (Institute of Radio Engineers). The older of these two institutes, the AIEE, dates back to 1884 and, at least early in the history of the AIEE, had a focus on telegraph and telephone. This focus, however, soon shifted to electric power generation and distribution. In order to maintain technical diversity a number of Special Committees were organized. A Committee on Telegraphy and Telephony was formed in 1903. In 1915 the Special Committees were renamed Technical Committees. This put in place the technical committee structure that lives to this day.

The Institute of Radio Engineers (IRE) was formed in 1912 by engineers wishing to have a technical institute focusing on wireless radio transmission and the related electronics. In 1937, the IRE formed a technical committee structure with the first six committees named broadcast, electroacoustics, radio receiving, television and facsimile, transmitting and antennas, and wave propagation. The IRE grew rapidly and there was considerable competition between the AIEE and the IRE for members. As a result, between 1950 and 1960 the AIEE formed a number of technical committees focusing on communications, including technical committees on television broadcasting, communication theory, data communication, and space communication. In 1952, the IRE Professional Group on Radio Communications was formed and later that year was renamed the IRE Professional Group on Communications Systems. This semi-autonomous group grew rapidly, formed a series of technical conferences, and initiated publication of the *IRE Transactions on Communications Systems.* This publication was the forerunner of the *IEEE Transactions on Communications.*

The formation of the IRE Professional Group on Communications Systems in 1952 marks Year 1 of the 50-year history spanned by this collection of technical papers. It should be noted that three papers in this collection by Costas (March 1957), Hancock and Lucky (December 1960), Arthurs and Dym (December 1962), and Prosser (December 1962) were originally published in the *IRE Transactions on Communications Systems.* When the AIEE and the IRE formally merged in 1963 the *IRE Transactions on Communications Systems* was renamed the *IEEE Transactions on Communications Systems* and two papers in this volume by Bello (December 1963), and Baran (March 1964) were originally published in the *IEEE Transactions on Communications Systems.* In 1964 the IEEE Group on Communications Technology was formed, basically from the IRE Professional Group Communications Systems. The *IEEE Transactions* was renamed the *IEEE Transactions on Communications Technology,* as the papers in this volume published between 1964 and 1971 indicate. When the IEEE Communications Society was formed in 1972 the *IEEE Transactions on Communications Technology* became the *IEEE Transactions on Communications* and retains this name to this day.

Thus, in a real sense, the *IRE Transactions on Communications Systems,* the *IEEE Transactions on Communications Systems,* the *IEEE Transactions on Communications Technology,* and finally the *IEEE Transactions on Communications* all represent the same journal, which changed names from time-to-time because of a merger and several reorganizations. The organizational structure and many of the volunteers which led these publications changed little from year-to-year although the number of papers published experienced tremendous growth. The influence of the four technical journals on the field of communications has been tremendous and represents a history of research performed in the field of communications. This is evidenced by the fact that 45 of the 56 papers in this volume come from these four journals.

As the IEEE Communications Society matured and the field became broader several new journals were added. The *IEEE Journal on Selected Areas in Communications* began publication in 1983 to serve as a journal focusing on single topics of emerging research interest. Prior to 1983, the *Transactions on Communications* published special issues when the need was apparent. Note that 10 papers in this volume were originally published in the *IEEE Journal on Selected Areas in Communica-*

tions. In 1993, publication of the *IEEE/ACM Transactions on Networking* was initiated to serve the rapidly advancing networking field. Two papers in this volume are from that publication. Two new research journals were recently added to the ComSoc's suite of archival research journals, namely *IEEE Communications Letters* and *IEEE Transactions on Wireless Communications,* but they are too new for papers published in these two journals to be represented here.

Dedication

We trust that you, the reader, will find this volume to be a useful reference in your further work in communications and that it will serve also to provide you with a historical perspective on the development of modern communications systems. We hope that students who browse this volume will be inspired by the work represented by the papers contained herein, will receive a sense of the excitement and opportunities represented within the communications industry, and will be motivated to make their own contributions to the field. Finally, we hope that you will find the volume to be a useful reminder of the fiftieth anniversary of the Communications Society and that it will serve as an indication of the many problems that remain to be solved in the worldwide communications systems research and development scene. We dedicate this volume to those who have advanced the field of communications over the past half century. The editors thank the many people who contributed to the development of this volume and to the authors of the outstanding papers contained herein. Without these contributions this volume would not have been possible.

Acknowledgement

Planning for the the 50th anniversary of the IEEE Communications Society began in 1998 with The Fiftieth Anniversary Advisory Board (FAAB) chaired by Jack McDonald. Their recommendations were accepted by the BOG in 1999 and the 50th Anniversary Implementation Committee was formed. Members included Celia Desmond, Roberto de Marca, Harvey Freeman, Jack McDonald, Tom Plevyak, Curtis Siller, and Jack Howell.

The IEEE History Center at Rutgers University researched and wrote a short history of the technology which was subsequently published by the IEEE Communications Society in a small paperback "A Brief History of Communications Technology." The editors want to acknowledge the efforts of David Hochfelder, David Morton, William Aspray, Andrew Goldstein, and Robert Colburn of the History Center and Chip Larkin of AT&T. Additional thanks go to Amos Joel for preparing and compiling much of the information regarding the history of the society. Substantial parts of the Preface of "The Best of the Best" have been based on "A Brief History of Communications Technology."

The editors also wish to thank Communications Society President Nim Cheung for his efforts to support the print version of "The Best of the Best."

For Physical and Link Layer Communications
William H. Tranter
Desmond P. Taylor
Rodger E. Ziemer

For Networking
Nick Maxemchuk
Jon Mark

Preface Supplement A—Reviewers for Physical and Link Layer Communications

Bob Aaron
Zeke Bar-Ness
Vijay K. Bhargava
David G. Daut
David Falconer
David Goodman
Joachim Hagennauer
Joe L. LoCicero
Peter J. McLane
Raymond L. Pickholtz
Steve S. Rappaport
K. Sam Shanmugan
Gordon Stuber
Sergio Verdu
Stephen B. Wicker
Ian F. Akyildiz
Norm C. Beaulieu
Ezio Biglieri
Anthony Epheremides
Costas N. Georghiades
Paul E. Green, Jr.
Isreal Korn
Robert W. Lucky
Larry B. Milstein
John J. Proakis
Donald L. Schilling
Nelson Sollenberger
Gottfried Ungerboeck
Andrew J. Viterbi
Fred T. Andrews
Sergio Benedetto
Rob Calderbank
Joseph B. Evans
Jerry Gibson
Larry J. Greenstein
Khaled B. Letaief
James L. Massey
James W. Modestino
Michael B. Persley
Mischa Schwartz
Ray Steele
Reinaldo A. Valenzuela
Steven B. Weinstein

Preface Supplement B—Recent Award Winning Papers Appearing in Research Journals Published by the IEEE Communications Society

The Leonard G. Abraham Prize Paper Award is presented annually to the best paper published in the *IEEE Journal on Selected Areas in Communications.*

Tai-Ann Chen, M. P. Fitz, Wen-Yi Kuo, M. D. Zoltowski, and H. Grimm, "A Space-Time Model for Frequency Nonselective Rayleigh Fading Channels with Applications to Space-Time Modems," *IEEE Journal on Selected Areas in Communications,* Vol. 18, No. 7, pp. 1175-1190, July 2000.

R. R. Müller, and S. Verdù, "Design and Analysis of Low-Complexity Interference Mitigation on Vector Channels," *IEEE Journal on Selected Areas in Communications,* Vol. 19, No. 8, pp 1429-1441, August 2001.

G. Cherubini, E. Eleftheriou, and S. Olcer, "Filtered Multitone Modulation for Very High-Speed Digital Subscriber Lines," *IEEE Journal on Selected Areas in Communications,* Vol. 20, No. 5, pp. 1016-1028, June 2002.

Shiwen Mao, Shunan Lin, Shivendra S. Panwar, Yao Wang, Emre Celebi, "Video Transport Over Ad Hoc Networks: Multistream Coding With Multipath Transport," *IEEE Journal on Selected Areas in Communications,* Vol. 21, No. 10, pp. 1721-1737, December 2003.

Parvathinathan Venkitasubramaniam, Srihari Adireddy, Lang Tong, "Sensor Networks With Mobile Access: Optimal Random Access and Coding," *IEEE Journal on Selected Areas in Communications,* Vol. 22, No. 6, pp. 1058-1068, August 2004.

Moritz Borgmann, Helmut Bölcskei, "Noncoherent Space-Frequency Coded MIMO-OFDM," *IEEE Journal on Selected Areas in Communications,* Vol. 23, No.9, pp. 1799 -1810, September 2005.

The Stephen O. Rice Prize Paper Award is presented annually to the best paper published in the *IEEE Transactions on Communications.* (Note: No award was given in 2002.)

S. L. Ariyavisitakul, "Turbo Space-Time Processing to Improve Wireless Channel Capacity," *IEEE Transactions on Communications,* Vol. 48, No. 8, pp. 1347-1359, August 2000.

C. L. Miller, D. P. Taylor, P. T. Gough, "Estimation of Co-Channel Signals with Linear Complexity," *IEEE Transactions on Communications,* Vol. 29, No. 11, pp. 1997-2005, November 2001.

A. Sendonaris, E. Erkip, B. Aazhang, "User Cooperation Diversity - Part I: System Description" and "User Cooperation Diversity - Part II: Implementation Aspects and Performance Analysis," *IEEE Transactions on Communications,* Vol. 51, No. 11, pp. 1927-1948, November 2003

S. ten Brink, G. Kramer, and A. Ashikhmin, "Design of Low-Density Parity-Check Codes for Modulation and Detection," *IEEE Transactions on Communications,* Vol. 52, No. 4, pp. 670-678, April 2004

B. M. Hochwald, C. B. Peel, and A. L. Swindlehurst, " A Vector-Perturbation Technique for Near-Capacity Multiantenna Multiuser Communication - Part I: Channel Inversion and Regularization," *IEEE Transactions on Communications,* Vol. 53, No.1, pp. 195-202, January 2005, and "A Vector-Perturbation Technique for Near-Capacity Multiantenna Multiuser Communication - Part II: Perturbation," *IEEE Transactions on Communications,* Vol. 53, No.3, pp. 537-544, March 2005.

The William R. Bennett Prize Paper Award is presented annually to the best paper published in the *IEEE/ACM Transactions on Networking.*

C. K. Wong, M. G. Gouda, and S. S. Lam, "Secure Group Communications Using Key Graphs," *IEEE/ACM Transactions on Networking,* Vol. 8, No. 1, pp. 16-30, February 2000.

S. Floyd and V. Paxson, "Difficulties in Simulating the Internet," *IEEE/ACM Transactions on Networking,* Vol. 9, No. 4, pp. 392-403, August 2001.

Wu-chang Feng, K. G. Shin, D. D. Kandlur, and D. Saha, "The BLUE Active Queue Management Algorithms," *IEEE/ACM Transactions on Networking,* Vol. 10, No. 4, pp. 513-528, August 2002.

I. Stoica, R. Morris, D. Liben-Nowell, D. R. Karger, M. F. Kaashoek, F. Dabek, and H. Balakrishnan, "Chord: A Scalable Peer-to-Peer Lookup Protocol for Internet Applications," *IEEE/ACM Transactions on Networking,* Vol. 11, No. 1, pp. 17-32, February 2003

N.Spring, R. Mahajan, D. Wetherall, and T. Anderson," Measuring ISP Topologies With Rocketfuel " *IEEE/ACM Transactions on Networking,* Vol. 12, No. 1, pp. 2-16, February 2004.

J. L. Sobrinho. "An Algebraic Theory of Dynamic Network Routing," *IEEE/ACM Transactions on Networking,* Vol. 13, No. 5, pp. 1160-1173, October 2005.

The IEEE Marconi Prize Paper Award is given annually to the best paper published in the *IEEE Transactions on Wireless Communications.* (Note: This journal began publication in 2002 and no award was given in 2004.)

C. Rose, S. Ulukus, and R. D. Yates, "Wireless Systems and Interference Avoidance," *IEEE Transactions on Wireless Communications,* Vol. 1, No. 3, pp. 415-428, July 2002.

Y. Xin, Z. Wang, and G. B. Giannakis, "Space-Time Diversity Systems Based on Linear Constellation Precoding," *IEEE Transactions on Wireless Communications,* Vol. 2, No. 2, pp. 294-309, March 2003.

Y. Zhu and H. Jafarkhani, "Differential Modulation Based On Quasi-Orthogonal Codes," *IEEE Transactions on Wireless Communications,* Vol. 4, No. 6, pp. 3018-3030, November 2005.

PHYSICAL AND LINK LAYER ASPECTS OF COMMUNICATIONS

Turbo Space–Time Processing to Improve Wireless Channel Capacity

Sirikiat Lek Ariyavisitakul, *Senior Member, IEEE*

***Abstract*—By deriving a generalized Shannon capacity formula for multiple-input, multiple-output Rayleigh fading channels, and by suggesting a layered space–time architecture concept that attains a tight lower bound on the capacity achievable, Foschini has shown a potential enormous increase in the information capacity of a wireless system employing multiple-element antenna arrays at both the transmitter and receiver. The layered space–time architecture allows signal processing complexity to grow linearly, rather than exponentially, with the promised capacity increase. This paper includes two important contributions: First, we show that *Foschini's lower bound is, in fact, the Shannon bound* when the output signal-to-noise ratio (SNR) of the space–time processing in each layer is represented by the corresponding "matched filter" bound. This proves the optimality of the layered space–time concept. Second, we present an embodiment of this concept for a coded system operating at a low average SNR and in the presence of possible intersymbol interference. This embodiment utilizes the already advanced space–time filtering, coding and turbo processing techniques to provide yet a practical solution to the processing needed. Performance results are provided for quasi-static Rayleigh fading channels with no channel estimation errors. We see for the first time that the Shannon capacity for wireless communications can be both *increased* by N times (where N is the number of the antenna elements at the transmitter and receiver) and *achieved* within about 3 dB in average SNR, about 2 dB of which is a loss due to the practical coding scheme we assume—the layered space–time processing itself is nearly information-lossless!**

***Index Terms*—Equalization, interference suppression, space–time processing, turbo processing.**

I. Introduction

"TURBO" and "space–time" are two of the most explored concepts in modern-day communication theory and wireless research. From a communication theorist's viewpoint, "turbo" coding/processing is a way to approach the Shannon limit on channel capacity, while "space–time" processing is a way to increase the possible capacity by exploiting the rich multipath nature of fading wireless environments. We will see through a specific embodiment in this paper that combining the two concepts provides even a practical way to both increase and approach the possible wireless channel capacity.

With growing bit rate demand in wireless communications, it is especially important to use the spectral resource efficiently. The basic information theory results reported by Foschini and Gans [1] have promised extremely high spectral efficiencies possible through multiple-element antenna array technology. In high scattering wireless environments (e.g., troposcatter, cellular, and indoor radio), the use of multiple spatially separated and/or differently polarized antennas at the receiver has been very effective in providing diversity against fading [2], [3]. Receiver diversity techniques also create signal processing opportunities for interference suppression and equalization (e.g., [4]–[6]). However, using multiple antennas at either the transmitter or the receiver does not enable a significant gain in the possible channel capacity. According to [1], the Shannon capacity for a system with 1 transmit and N receive antennas scales only logarithmically with N, as $N \to \infty$. For a system using N transmit and 1 receive antennas, asymptotically there is no additional capacity to be gained, assuming that the transmit power is divided equally among the N antennas.

Foschini and Gans [1] have shown that the asymptotic capacity of multiple-input, multiple-output (MIMO) Rayleigh fading channels grows, instead, linearly with N when N antennas are used at *both* the transmitter *and* the receiver. Furthermore, in [7], Foschini suggested a layered space–time architecture concept that can attain a tight lower bound on the capacity achievable. In this layered space–time architecture, N information bit streams are transmitted simultaneously (in the same frequency band) using N diversity antennas. The receiver uses another N diversity antennas to decouple and detect the N transmitted signals, one signal at a time. The decoupling process in each of the N processing "layers" involves a combination of *nulling out* the interference from yet undetected signals (N diversity antennas can null up to $N-1$ interferers, regardless of the angles-of-arrival [5]) and *canceling out* the interference from already detected signals. One very significant aspect of this architecture is that it allows an N-dimensional signal processing problem—which would otherwise be solvable only through multiuser detection methods [8] with m^N complexity (m is the signal constellation size)—to be solved with only N similar 1-D processing steps. Namely, the processing complexity grows only linearly with the promised capacity.

This paper includes two important contributions. First, we show that Foschini's lower bound is, in fact, the Shannon bound when the output SNR of the space–time processing in each layer is represented by the corresponding "matched filter" bound [6], i.e., the maximum SNR achievable in a hypothetical situation where the array processing weights to suppress the remaining interference in each layer are chosen to maximize the output signal-to-interference-plus-noise ratio and any possible

Paper approved by K. B. Letaief, the Editor for Wireless Systems of the IEEE Communications Society. Manuscript received September 15, 1999; revised December 3, 1999. This paper was presented at the IEEE International Conference on Communications, New Orleans, LA, June 2000.

The author is with the Home Wireless Networks, Norcross, GA 30071 USA (e-mail: lek@homewireless.com).

Publisher Item Identifier S 0090-6778(00)07111-7.

Reprinted from *IEEE Transactions on Communications,* vol. 48, no. 8, August 2000.

intersymbol interference (ISI) is assumed to be completely eliminated by some means of equalization. The "matched filter" bound has been shown to be approachable using minimum mean-square error (MMSE) space–time filtering techniques [6].[1] By showing the equivalence of the generalized Foschini's bound and the Shannon bound, we essentially prove the optimality of the layered space–time concept.

Second, we present an embodiment of Foschini's layered space–time concept for a coded system operating at a low average SNR and in the presence of unavoidable ISI. Previously, a different embodiment has been provided in [9] for an uncoded system with variable signal constellation sizes, operating at a high average SNR without ISI. Adding coding redundancy might, at first, seem conflicting with the desire to increase the channel bit rate. Our justification is as follows: First, we seek to enhance the channel capacity from a system perspective. We use "noise" in SNR to represent all system impairments, including thermal noise and multiuser interference. The ability to operate at low SNR's means that more users per unit area can occupy the same bandwidth simultaneously. Second, we anticipate the use of adaptive-rate coding schemes to permit different degrees of error protection according to the channel SNR's. Incremental redundancy transmission [10], currently being considered for the Enhance Data Services for GSM Evolution (EDGE; GSM stands for Global System for Mobile Communications) standard, is an efficient way to implement adaptive code rates without requiring channel SNR monitoring. With such adaptive-rate coding, the system does not "waste" spectral resources under good channel conditions.

Meanwhile, the iterative processing principle used in turbo and serial concatenated coding [11]–[15] has been successfully applied to a wide variety of joint detection and decoding problems. One such application is the so-called "turbo equalization" [16]–[19], where successive maximum *a posteriori* (MAP) processing is performed by the equalizer and channel decoder to provide *a priori* information about the transmit sequence to one another. Similar to the layered space–time concept, turbo processing allows a multi-dimensional (*two*-dimensional in this case) problem to be optimally solved with successive 1-D processing steps without much performance penalty. In this paper, we apply the turbo principle to layered space–time processing in order to prevent decision errors produced in each layer from catastrophically affecting the signal detection in subsequent layers.

We consider two possible coded layered space–time structures: one applying coding across the multiple signal processing layers, and the other assuming independent coding within each layer. Similar to [1], we assume a *quasi-static* random Rayleigh channel model, where the channel characteristics are stationary within each data block, but statistically independent between different data blocks, different antennas, and, in the case of dispersive multipath channels, different paths. The system is assumed to have similar ISI situations as in EDGE and GSM, where multipath dispersions may last up to several symbol periods [20]. We show that near-capacity performance is achievable using 1-D processing and coding techniques that are already practical and "legacy-compatible" with the EDGE standard, e.g., the use of bit-interleaved 8-ary phase-shift keying (8-PSK) with rate-1/3 convolutional coding and an equalizer with a similar length and structure.

A slightly different layered space–time approach based on *space–time coding* [23], [24] has been studied in [25]. Although it is difficult to make a general comparison, we will see later that our coded layered space–time approach does by far outperform the results reported in [25] for $N = 4$ and $N = 8$. On the other hand, for $N = 2$, space–time coded quaternary phase-shift keying (QPSK) without layered processing appears to be the best known technique for achieving a spectral efficiency of 2 bps/Hz.

This paper is organized as follows. Section II provides a brief review of Foschini's layered space–time concept. Section III describes the two coded layered space–time architectures and presents a capacity analysis which reveals the equivalence of a generalized Foschini's lower bound formula and the true capacity bound. Section IV provides details on the array processing, equalization, and iterative MAP techniques. Section V presents performance results. A summary and conclusions are given in Section VI.

II. Background Theory

We briefly review the theory behind Foschini's layered space–time concept. The generalized Shannon capacity for a MIMO Rayleigh fading system with N transmit and M receive antennas is given in [1] as

$$C = \log_2 \left[\det\left(\boldsymbol{I} + \frac{\rho}{N}\boldsymbol{H}\boldsymbol{H}^{\dagger}\right)\right] \tag{1}$$

where $\boldsymbol{H}$ is an $M \times N$ matrix, the (i, j)th element of which is the normalized channel transfer function of the transmission link between the jth transmit antenna and the ith receive antenna, $\boldsymbol{I}$ is the $M \times M$ identity matrix, ρ is the average SNR per receive antenna, and $\det(\cdot)$and superscript $\dagger$ denote determinant and conjugate transpose. It is assumed that the transmit power is equally divided among the N transmit antennas. The normalization of the channel transfer function is done such that the average (over Rayleigh fading) of its squared magnitude is equal to unity.

The lower bound on capacity is provided in [1] as

$$C > \sum_{k=N-M+1}^{N} \log_2 \left[1 + \frac{\rho}{N}\chi_{2k}^2\right] \triangleq C_F \tag{2}$$

where is a chi-squared random variable with degrees of freedom. For $M = N$

$$C_F = \sum_{k=1}^{N} \log_2 \left[1 + \frac{\rho}{N}\chi_{2k}^2\right]. \tag{3}$$

Since χ_{2k}^2 represents a fading channel with a diversity order of k, the lower-bound capacity in (3) can be viewed as the sum of the capacities of N independent channels with increasing diversity orders from 1 to N. This suggests a layered space–time approach [7] for detecting the N transmitted signals as follows:

[1]In a flat fading case, MMSE array processing achieves exactly the "matched filter" bound performance.

In the first layer, the receiver detects a first transmitted signal by nulling out interference from $N-1$ other transmitted signals through array processing. Assuming a "zero forcing" (ZF) constraint, one receive antenna is needed to completely correlate and subtract each interference [5]. Thus, the overall process of nulling $N-1$ interferences leaves the receiver with $N-(N-1)=1$ degree of freedom to provide diversity for detecting the first signal, i.e., a diversity order of 1 (or simply no diversity). Once detected, the first signal is subtracted out from the received signals on all N antennas.

In the second layer, the receiver performs similar interference nulling to detect a second transmitted signal. This time, since there are only $N-2$ remaining interferences, the receiver affords a diversity order of 2. The detected signal is again subtracted out from the received signals provided by the first layer.

Repeating the above interference nulling/canceling step through N layers, we see that the receiver affords an increasing order of diversity from 1 to N. If the capacities achieved in individual layers can be combined in some manner, then the layered space–time approach just mentioned will achieve the capacity lower bound expressed in (3). We will explore two capacity combining possibilities in the next section.

Note that the capacity and capacity low bound given in (1)–(3) are actually frequency-dependent. We here provide an explicit capacity formula for *band-limited, frequency-selective* channels (some variables are redefined to be consistent with later analytical development).

$$C = \langle \log_2[\det(\mathfrak{R}\aleph^{-1})] \rangle \tag{4}$$

where, as shown in equations (5)–(8) at the bottom of the page, $\mathfrak{R}$ is the frequency-domain correlation matrix of the signals on M receive antennas, $\aleph_j(f)$ is the noise power density at frequency f on the jth receive antenna, T is the symbol period, $H_{ij}(f)$ is the channel transfer function (not normalized) of the transmission link between the ith transmit antenna and the jth receive antenna, and superscripts $*$ and T denote complex conjugate and transpose. Note in (7) and (8) that we consider the folded spectra $H_{ij}(f-(m/T))$ and $\aleph_j(f-(m/T))$ of the channel transfer function and noise power density, where $m=-J,\ldots,J$ (J is finite because the signal sources are assumed to be band-limited). This is to take into account the effect of excess bandwidth and symbol-rate sampling when the frequency selectivity of the channel is not symmetrical around the Nyquist band edges. Even though we assume white Gaussian noise, the noise power density near and outside the Nyquist band edges actually attenuates with the receive filter transfer function. From our experiment (assuming a square-root Nyquist filter with a 50% rolloff factor), the computed capacity can be underestimated by as much as 0.5 dB if this attenuation is not taken into account.

III. Coded Layered Space–Time Architectures

A. Basic Concepts

We consider two coded layered space–time approaches as shown in Fig. 1(a) and (b). In the first approach, named "LST-I" (LST stands for "layered space–time"), the coded information bits are interleaved across the N parallel data streams $\boldsymbol{x}_1, \boldsymbol{x}_2, \cdots, \boldsymbol{x}_N$, where $\boldsymbol{x}_i$ denotes a sequence of complex-valued, transmit data symbols (e.g., 8-PSK symbols). The receiver first decouples the N data streams through interference nulling/cancellation, as described in Section II, then deinterleaves and decodes all the data streams as one information block. In the second approach, "LST-II," the information is first divided into N uncoded bit sequences $\boldsymbol{u}_1, \boldsymbol{u}_2, \ldots, \boldsymbol{u}_N$, each of which is independently encoded, interleaved, and symbol-mapped to generate one of the N parallel data streams. At the receiver, the

$$\langle\cdot\rangle \triangleq T\int_{-(1/2T)}^{(1/2T)} [\cdot]\, df \tag{5}$$

$$\mathfrak{R} \triangleq \sum_{i=1}^{N} \boldsymbol{H}_i^* \boldsymbol{H}_i^T + \aleph \tag{6}$$

$$\aleph \triangleq \begin{bmatrix} \aleph_1\left(f-\frac{J}{T}\right) & & & & & & \mathbf{0} \\ & \ddots & & & & & \\ & & \aleph_M\left(f-\frac{J}{T}\right) & & & & \\ & & & \ddots & & & \\ & & & & \aleph_1\left(f+\frac{J}{T}\right) & & \\ & & & & & \ddots & \\ \mathbf{0} & & & & & & \aleph_M\left(f+\frac{J}{T}\right) \end{bmatrix} \tag{7}$$

$$\boldsymbol{H}_i \triangleq \left[H_{i1}\left(f-\frac{J}{T}\right) \cdots H_{i,M}\left(f-\frac{J}{T}\right) \cdots\cdots H_{i1}\left(f+\frac{J}{T}\right) \cdots H_{i,M}\left(f+\frac{J}{T}\right)\right]^T \tag{8}$$

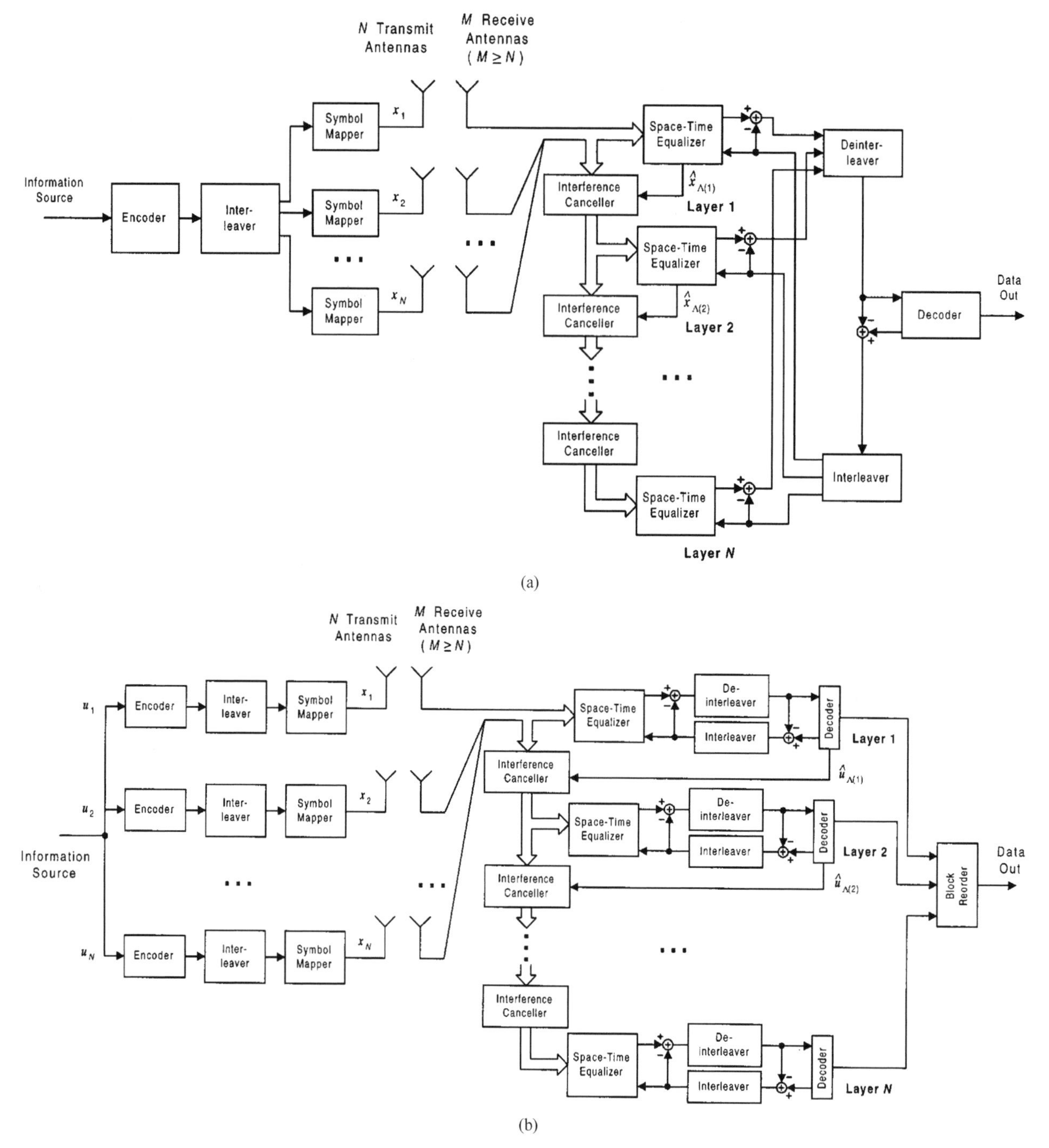

Fig. 1. Coded layered space–time architecture: (a) LST-I and (b) LST-II.

N data streams are decoupled and independently deinterleaved and decoded. The output of LST-II produces N information blocks at a rate of $1/N$ times the output rate of LST-I.

In Fig. 1(a) and (b), "space–time equalizer" refers to a combined array processing (for interference nulling) and equalization function. Instead of the ZF criterion, we assume that the optimization of the antenna/equalizer weights is based on a MMSE criterion, which in general provides better performance than a ZF approach. Foschini [7] has also indicated a potential performance benefit of using MMSE (or "maximum SNR") rather than ZF in a layered space–time architecture. Although we show M receive antennas in Fig. 1(a) and (b) ($M \geq N$ is the sufficient condition for nulling $N-1$ interference), we only consider $M = N$ in this study.

Similar to [9], the underlying assumption of our layered space–time architecture is that the receiver can order the detections of N data streams such that an undetected layer always has the strongest received SNR. In LST-I, the space–time equalizer in each layer must provide data decisions $\hat{\boldsymbol{x}}_{\Lambda(i)}$ (Λ denotes the permutation due to layer ordering) to the interference canceller, since decoding cannot be done until all the layers are processed. In LST-II, the interference cancellation in each layer can use more reliable data decisions $\hat{\boldsymbol{u}}_{\Lambda(i)}$ provided by the decoder. Thus, LST-I is more prone to decision errors than

LST-II. In order to minimize the effects of decision errors, and also to improve the joint detection/decoding performance in general, we assume the use of turbo processing in our layered space–time architecture. As shown in Fig. 1(a) and (b), the space–time equalizers and the decoders provide *extrinsic* soft information to one another by subtracting the received soft information from the newly computed soft information. Details on MMSE space–time equalization and turbo processing will be provided in the Section IV.

B. Capacity Analysis

Without getting into the detail of all the processing functions, we first discuss the general differences between the two coded layered space–time approaches. In particular, we are most interested in the *capacity combining* aspects of the two approaches.

Let SNR_k denote the output SNR of the array processing in the kth layer. First, we note that, in LST-II, the capacity of each processing layer is bounded by the spectral efficiency R of the modulation and coding in each layer, e.g., $R = 1$ for 8-PSK with rate 1/3 coding. Thus, the total capacity of LST-II is given by (similar to (1)–(3), we write capacity without showing the frequency dependence)

$$C_{\mathrm{LST-II}} = \sum_{k=1}^{N} \min\{R, \log_2[1 + \mathrm{SNR}_k]\}. \quad (9)$$

Without layer ordering, it is most likely that the overall performance of LST-II will be largely influenced by the error probability of the first processing layer with a diversity order of only 1. In contrast, our simulation results in Section V will show that LST-II with layer ordering can actually achieve a diversity order of approximately N.

Since coding is performed across all the processing layers in LST-I, the achieved SNR in each layer will contribute to the overall layer processing performance. As Foschini [7] indicated, such a coding scheme should be able to achieve the capacity lower bound in (3). Here, we provide a generalized formula of Foschini's lower bound by removing the ZF constraint and instead using SNR_k as the generalized output SNR.

$$C_F = \sum_{k=1}^{N} \log_2[1 + \mathrm{SNR}_k]. \quad (10)$$

Reference [6] provides output SNR formulas for different types of optimum space–time processors. Here, it is of great interest to express the capacity lower bound using the best performance achievable. In the following equation, we represent SNR_k in (10) by the "matched filter" bound-the maximum achievable SNR by any space–time processing receiver:

$$C_{F,MF} = \left\langle \sum_{k=1}^{N} \log_2[1 + \Gamma_k(f)] \right\rangle \quad (11)$$

where $\Gamma_k(f)$ is the "matched filter" bound[2] given by equation (15) in Section IV-A (simply a rewriting of the result in [6]).

[2]Note that the "matched filter" bound usually refers to the integrated SNR $\langle\Gamma_k(f)\rangle$ over the signal bandwidth (e.g., [6]). However, in the capacity context, we assume the best possible way to exploit the SNR's in all frequency components

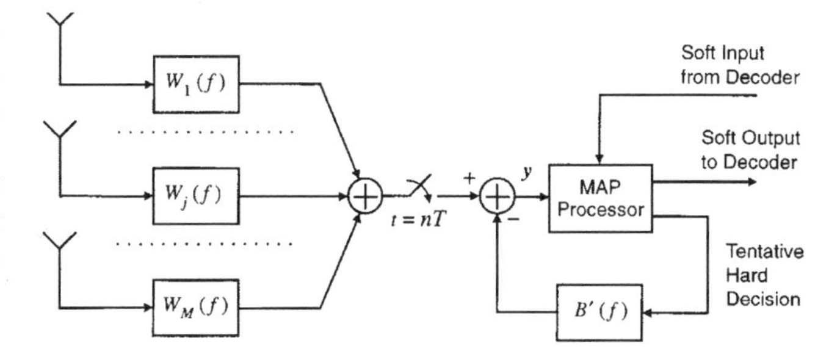

Fig. 2. Space-time DDFSE with MAP processing

Note that (11) is an explicit formula similar to (4); it shows the frequency dependence of the output SNR and the integration of capacity over the signal bandwidth. Also, we assume that the kth layer has $k - 1$ interferences.

In the process of analyzing the meaning of (11), we discovered an identical relationship between (11) and (4) regardless of how the layers are ordered. We show the proof in the Appendix (this proof is valid even when $M \neq N$). *Thus, Foschini's lower bound (3) is actually the true Shannon capacity bound when the output SNR of the space–time processing in each layer is represented by the corresponding "matched filter" bound.* This proves the optimality of the layered space–time concept.

The capacity analysis presented above is based on the assumption of perfect layer detection, i.e., no decision errors affecting the detection in subsequent layers. In reality, LST-I is more prone to decision errors than LST-II and layer ordering becomes important for both schemes. Our simulation results in Section V will demonstrate how decision errors affect the actual performance of the two coded layered space–time approaches.

IV. Signal Processing Functions

A. Space–Time Equalization

We consider combined array processing and equalization in order to cope with dispersive channels. A space–time equalizer, consisting of a spatial/temporal whitening filter, followed by a decision-feedback equalizer (DFE) or maximum-likelihood sequence estimator, can suppress both ISI and dispersive interference [6]. The space–time equalizer used in this study is shown in Fig. 2. It consists of a linear feedforward filter $W_j(f)$, $j = 1, \ldots, M$, on each diversity branch, a combiner, symbol-rate sampler, soft-input, soft-output (SISO) MAP sequence estimator, and synchronous linear feedback filter $B'(f)$. The feedforward filters $\{W_j(f)\}$ are shown as continuous-time filters, but they can be implemented in practice using fractionally-spaced tapped delay lines. The combined use of a sequence estimator and feedback filter after diversity combining is similar to the structure of a delayed decision-feedback sequence estimator (DDFSE) [27]. Thus, we refer to the space–time equalizer in Fig. 2 as a "space–time DDFSE." A "space–time DFE" is a structure where the sequence estimator is replaced by a memoryless hard slicer.

It has been shown in [20] that a space–time DDFSE with a sequence estimator memory of μ and a feedback filter of length $L_B - \mu$ can be optimized in a MMSE manner as if it was a space–time DFE with a feedback filter of length L_B. In fact, numerical results in [6] showed that an optimum space–time DFE (with unconstrained filter lengths and no feedback decision errors) can perform within only 1–2 dB of the ideal "matched filter" bound performance. Thus, in order to have a practical receiver structure for layered space–time processing, we consider a space–time DDFSE with a minimum sequence estimator memory, i.e., $\mu = 1$. The sequence estimator is used only to provide a trellis structure needed for turbo processing, and presumably more reliable feedback decisions than the slicer used in a space–time DFE. Details on MAP processing will be given in Section IV-B.

We first provide a brief review of the space–time filtering theory. Based on the space–time DFE equivalent model described above, the MMSE solution for the feedforward filters $\{W_j(f)\}$ with unconstrained length can be given using the results of [6] (see also [28])

$$\boldsymbol{W} = \mathfrak{R}_k^{-1}\boldsymbol{H}_k^*(1+B(f)) = \mathfrak{R}_{k-1}^{-1}\boldsymbol{H}_k^*\frac{1+B(f)}{1+\Gamma_k(f)} \tag{12}$$

where

$$\mathfrak{R}_k = \sum_{i=1}^{k}\boldsymbol{H}_i^*\boldsymbol{H}_i^T + \aleph \tag{13}$$

$$\boldsymbol{W} \triangleq \left[W_1\left(f-\frac{J}{T}\right)\cdots W_M\left(f-\frac{J}{T}\right)\cdots \right.$$
$$\left.\cdots W_1\left(f+\frac{J}{T}\right)\cdots W_M\left(f+\frac{J}{T}\right)\right]^T \tag{14}$$

$$\Gamma_k(f) \triangleq \boldsymbol{H}_k^T\mathfrak{R}_{k-1}^{-1}\boldsymbol{H}_k^*. \tag{15}$$

In the above equations, we assume that there are a total of k signal sources and we use $\boldsymbol{H}_k$ to indicate the channel vector [see (8)] of the *desired* signal. The remaining $k-1$ signals are interferences. $B(f)$ in (12) is the feedback filter of the space–time DFE, which, from our assumption of $\mu = 1$, is only "1 tap" longer than $B'(f)$. $\Gamma_k(f)$ is the signal-to-interference-plus-noise power density ratio at frequency f, i.e., the "matched filter" bound. $B(f)$ can be determined through spectral factorization of $1+\Gamma_k(f)$.

Equation (12) indicates that the optimum feedforward filter consists of a spatial/temporal filter $\mathfrak{R}_{k-1}^{-1}\boldsymbol{H}_1^*$, which performs prewhitening ($\mathfrak{R}_{k-1}^{-1/2}$ is the whitening filter of interference and noise) and matching to the desired channel, followed by a temporal filter $(1+B(f))/(1+\Gamma_k(f))$, which is an anticausal post-whitening filter for suppressing precursor ISI.

A filter length analysis of the optimum space–time DFE described above is provided in [6]. We will first consider a finite-length realization of the space–time DFE based on the results presented there. We assume that the system has similar ISI situations as in EDGE and GSM. Namely, using the multipath delay profiles specified for EDGE and GSM (see Tables I and II), and assuming the same symbol rate of 270.833 kbaud ($T = 3.692$ μs) with Nyquist filtering (partial response signaling is used in EDGE and GSM), the ISI lasts up to five symbol periods for the hilly terrain (HT) profile in Table II. According to the empirical filter length formulas in [6], the feedforward filter on each branch should have the following causal and anticausal lengths to achieve near-optimum performance

TABLE I
GSM TYPICAL URBAN (TU) CHANNEL MODEL

Path Delay (μs)	0.0	0.2	0.5	1.6	2.3	5.0
Path Power (dB)	−3.0	0.0	−2.0	−6.0	−8.0	−10.0

TABLE II
GSM HILLY TERRAIN (HT) CHANNEL MODEL

Path Delay (μs)	0.0	0.2	0.4	0.6	15.0	17.2
Path Power (dB)	0.0	−2.0	−4.0	−7.0	−6.0	−12.0

$$L_C \approx K(N-1)(\rho_{\text{dB}}/10)$$
$$L_A \approx K + KN(\rho_{\text{dB}}/10) \tag{16}$$

where K is the channel memory, N is used here to indicate the total number of signals, including the desired and interference signals, and ρ_{dB} is the average SNR in decibels. In our case, $K = 5$, and assume for example that the system has four transmit and four receive antennas ($N = 4$) and the operating range of average SNR is around 5 dB ($\rho_{\text{dB}} = 5$). The required filter length, including the *center* tap, will be $L_C + L_A + 1 \approx 23.5$. This is a highly impractical number, considering that four such filters are required, one per each receive antenna. Furthermore, as mentioned earlier, the optimum feedforward filters should be implemented using fractionally-spaced tapped delay lines. If a $T/2$-spaced filters are used, the total number of taps will be doubled. Such a space–time system with about 200 coefficients would be nearly impossible to compute in any radio link design.

Faced with such impracticality of an ideal signal processing arrangement, we proceed to consider a *suboptimum* option. First, we will use *symbol-spaced* instead of fractionally-spaced feedforward filters. In order to avoid significant performance penalties, a channel estimation-based timing recovery algorithm described in [29] will be used to optimize the symbol timing and the decision delay of the center tap relative to the measured channel impulse response. In principle, such timing optimization also allows the DFE to use a feedforward filter with a shorter span than the channel memory while achieving a reasonable performance [29], [30]. After experimenting with a number of significantly reduced filter length options, we decided on the following suboptimum space–time equalizer structure. The feedforward filter on each branch has a total of nine symbol-spaced taps, which are positioned such that $L_C = L_A = 4$. The feedback filter has a length of 8, i.e., $L_B = 9$ with the MAP processor memory $\mu = 1$ included (in order to completely cancel postcursor ISI, L_B must be at least as large as the channel memory plus the number of causal taps in the feedforward filter). The method in [29] is used to optimize the symbol timing and the decision delay of the center tap as described above. Direct matrix inversion is used to set all the filter coefficients in a standard MMSE linear processing fashion [4], [26], [31], assuming perfect channel estimation.

B. Turbo Processing

The turbo processing technique used in this study is also based on a standard approach—the reader is referred to the rich literature [11]–[19], [21]–[22], [32]–[35] for a thorough treatment of this subject. The space–time equalizer and the decoder both performs SISO sequence estimation to compute the *a posteriori* probability (APP) of the transmit data symbols. This sequence estimation is done using the Bahl–Cocke–Jelinek–Raviv (BCJR) forward/backward algorithm. In the following, we describe the basic principle of the iterative detection/decoding process.

Using the BCJR algorithm, the MAP processor in the space–time equalizer with m^{μ} states (m is the signal constellation size, e.g., $m = 8$ for 8-PSK, and $\mu = 1$ in our case) computes the APP $P[c_k|\boldsymbol{y}]$ of the kth coded bit c_k based on the observation $\boldsymbol{y}$, where $\boldsymbol{y}$ is the equalizer output sequence corresponding to all the data symbols in a received block (see Fig. 2), and the *a priori* information provided by the decoder (this is not available in the first "turbo" iteration). The logarithm $\lambda(c_k) \triangleq \log_e(P[c_k|\boldsymbol{y}])$ of this APP can be regarded as the sum of two terms

$$\lambda(c_k) = \lambda^p(c_k) + \lambda^e(c_k) \tag{17}$$

where $\lambda^p(c_k) \triangleq \log_e(P[c_k])$ is the logarithm of the *a prior* information provided by the decoder, and $\lambda^e(c_k)$ is called the "extrinsic" information. In each "turbo" iteration, the space–time equalizer subtracts $\lambda^p(c_k)$ from the newly computed value of $\lambda(c_k)$ to obtain the extrinsic information $\lambda^e(c_k)$ [see Fig. 1(a) and (b)]. The entire sequence $\{\lambda^e(c_k)\}$ is deinterleaved and forwarded to the decoder.

Similarly, the decoder computes the log-APP $v(c_k) \triangleq \log_e(P[c_k|\{\lambda^e(c_k)\}])$ based on the deinterleaved extrinsic information provided by the space–time equalizer, and subtract $\lambda^e(c_k)$ from it to obtain extrinsic information $v^e(c_k)$. The extrinsic information is then interleaved and forwarded to the equalizer as the new *a priori* information $\lambda^p(c_k)$ for the next "turbo" iteration.

The interleaver considered in this study is a *pseudo-random* interleaver, i.e., we generate a pseudo-random permutation of numbers from 1 to l, where l is the block length, and then use this permutation as a *fixed* interleaver.

In combining the branch metric obtained from the equalizer output with the branch metric obtained from the soft input provided by the decoder, the MAP processor in the space–time equalizer must compute the *a priori* information for each 8-PSK symbol x_k from the three soft inputs ($\lambda^p(c_{3k})$, $\lambda^p(c_{3k+1})$, $\lambda^p(c_{3k+2})$). We assume that this is done by way of summing the three soft inputs as if the three coded bits were transmitted from independent sources (these soft inputs are actually not independent when conditioned on the observed waveform of the entire data burst). This is a suboptimal method, which is known to cause a "random modulation" performance degradation effect in bit-interleaved coded modulation [36]–[38]. However, this effect can be overcome by iterative decoding [38], which is implicit in our turbo space–time processing approach.

As noted earlier, in LST-I, the space–time equalizer in each layer must provide immediate data decisions to be used for interference cancellation. Since these decisions are not "protected" by coding, they are prone to errors. In this study, we explore a soft decision technique to minimize the effect of decision errors. The optimum soft decision can be computed by averaging all the possible transmit symbols weighted by their APP's [39]

$$\hat{x}_k = \sum_{x \in X} sP[x_k = x|\boldsymbol{y}] \tag{18}$$

where X includes all the complex-valued 8-PSK constellation points. Since $P[x_k = x|\boldsymbol{y}]$ can be obtained along with the computation of the APP $P[c_k|\boldsymbol{y}]$, this soft decision approach can be implemented with nearly no additional cost in complexity. Similarly, we apply the same technique to compute soft decision outputs in LST-II.

V. Performance Results

A. Performance Criteria and System Assumptions

We now present performance results of the layered space–time concepts described so far. The performance measure is the block-error rate (BLER) over Rayleigh fading. The results are obtained through Monte Carlo simulation. The BLER is averaged over up to 40 000 channel realizations. Each block contains 400 information bits (before coding).

In comparing the performance results to channel capacity, we follow the convention of a number of previous works (e.g., [9], [23]) to compare the computed BLER with the "outage capacity" [1], i.e., the probability that a specified bit rate is not supported by the channel capacity. This is a vague comparison, since the Shannon limit refers to the highest error-free bit rate possible for long encoded blocks but it does not specify how long the blocks should be. Nevertheless, such a comparison should still be meaningful as long as the block length and BLER are specified. This is similar to the way a bit-error rate of 10^{-5} is commonly used as the "error free" reference for an additive white Gaussian noise (AWGN) channel.

In order to assess the best performance achievable, we assume that the channel characteristics can be perfectly estimated at the receiver. Similarly, the choices of 1-D processing and coding techniques are important to deliver the best possible performance. We try to optimize these choices while keeping them as practical as possible. Except for the use of array processing and iterative MAP algorithms, all the radio link techniques assumed in this study are "legacy-compatible" with the EDGE standard (note also that vast research interest in turbo coding has made simplified MAP algorithms available [21], [22] that are not much more complex than the conventional Viterbi algorithm). None of these techniques are claimed to be optimum. Yet, our results indicate that near-capacity performance is achievable when combining them through the coded layered space–time architectures.

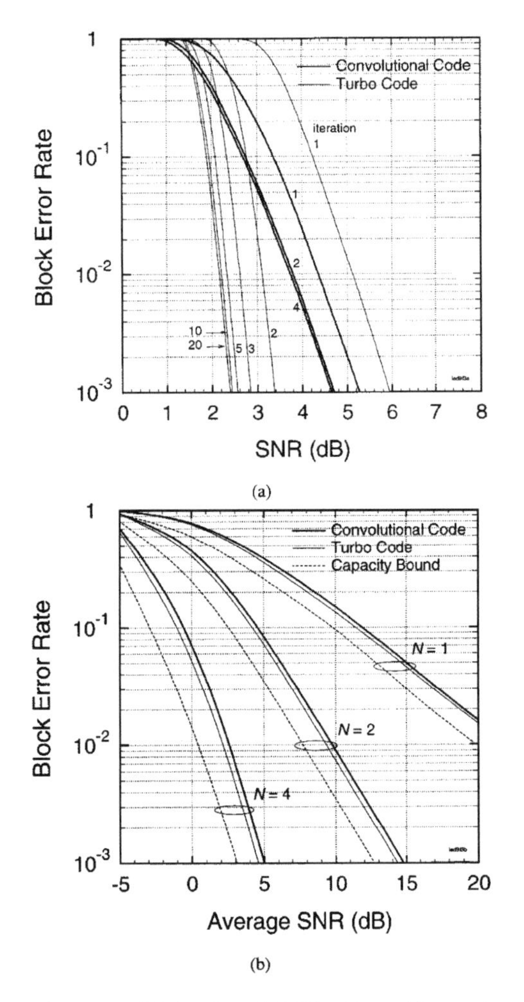

Fig. 3. Performance of bit-interleaved 8-PSK with rate-1/3 convolutional and turbo coding over (a) AWGN channel, and (b) quasi-static flat Rayleigh fading channels with N receive diversity antennas.

B. Choosing the Coding Scheme

We consider a bit-interleaved coded modulation scheme using 8-PSK with Gray mapping and rate-1/3 coding. Square-root Nyquist filtering with 30% rolloff is assumed at the transmitter and receiver. Bit-interleaved coded modulation has been shown [36], [37] to outperform traditional trellis-coded modulation in fast fading channels (where *time* diversity can be exploited through sufficient interleaving) and it can be improved upon by considering a better mapping technique that permits a large Euclidean distance without sacrificing the maximum Hamming distance of the baseline coding scheme [38]. In this paper, though, since quasi-static fading is our basic assumption, the code by itself must be able to withstand deep fades. In principle, any code that performs well in an AWGN channel is considered a good candidate—turbo codes are among the strongest candidates that come to mind.

Fig. 3(a) and (b) provide a performance comparison between two rate-1/3 coding schemes: one using a 64-state convolutional code with (octal) generators $(G_1, G_2, G_3) = (155, 117, 123)$ (the same code as proposed for EDGE [20]) and the other using a turbo code with two identical 16-state recursive encoders similar to the scheme originally proposed by Berrou and Glavieux [12] (the results here assume generators $(G_1, G_2) = (23, 31)$, which appeared to perform slightly better than other generators we tested). Both schemes assume the use of bit-interleaved 8-PSK with Gray mapping. The turbo coding scheme has an additional interleaver within the encoder, which uses another pseudo-randomly generated permutation. The receiver structure is consistent with what we have described so far. Note, however, that we assume a minimum number of filter taps (only one feedforward tap per branch and no feedback filter) whenever there is no delay spread assumed, although the MAP processor in the DDFSE is always used for iterative detection/decoding as described in Section IV-B. For the turbo coding scheme, "one iteration" means a full cycle of three processes: 1) MAP processing in DDFSE; 2) turbo decoding by the first decoder; and 3) turbo decoding by the second decoder.

Fig. 3(a) shows the performance of the two coding schemes in an AWGN channel. First, we note that the performance of convolutional coding also benefits from iterative processing. This is due to the suboptimal nature of the decoding scheme, i.e., the "random modulation" effect described earlier, which can be improved through iterative decoding. Fig. 3(a) shows that most of the improvement is achieved within two decoding cycles. For turbo coding, the performance still improves even after five iterations, but saturates quickly after ten iterations. At 10^{-3} BLER (approximately equivalent to 10^{-5} bit-error rate), turbo coding outperforms convolutional coding by about 2.2 dB, and the required SNR is within only 2.4 dB of the 0-dB Shannon limit for a spectral efficiency of 1 bps/Hz (8-PSK with rate-1/3 coding).

However, when we look at the average BLER performance over quasi-static flat fading channels in Fig. 3(b), the benefit of turbo coding (with ten iterations) over convolutional coding (with two iterations) is reduced to only about 0.5 dB at any value of the average SNR and for all the assumed numbers of receive diversity antennas. This is not surprising for two reasons: First, it is well known that the average BLER is determined mostly by the probability of fading events that results in high BLER's. If we look at the relative performance at a BLER of, say, above 10% in Fig. 3(a), the difference between the two coding schemes is indeed less than 1 dB. Second, the performance of convolutional coding over fading channels is already within about 2 dB of the capacity bound—the capacity bound in this case is defined as the probability that the combined output SNR of all diversity branches is below the 0-dB Shannon limit. Thus, there is not much room for further improvement.

Based on the fact that the performances of the two coding schemes are quite similar in quasi-static fading channels, we will only consider convolutional coding in the remainder of this paper.

C. Layered Space–Time Performance

We first look at the performance of the two coded layered space–time approaches over a flat Rayleigh fading channel. Fig. 4 shows the different capacity bounds for this channel, assuming $N = 2$, 4, and 8, where N is the number of transmit and receive antennas. Again, although we plot the results as "block error rate," the capacity bound is defined as the probability that the specified spectral efficiency R ($R = N$ in this case) is not supported by each of the differently defined channel capacities. C denotes the Shannon capacity bound given by (4)

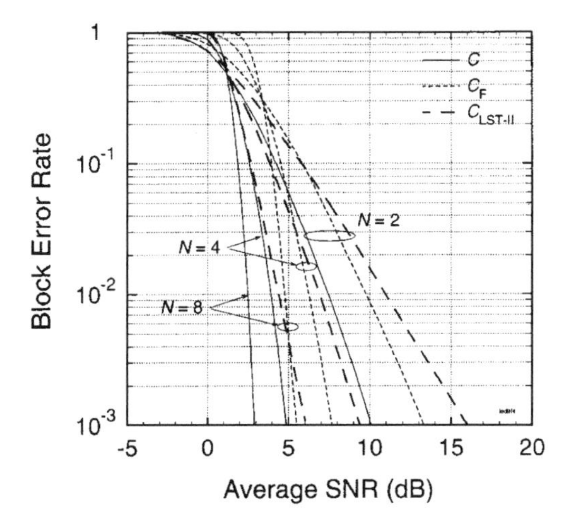

Fig. 4. Capacity bounds for quasi-static flat Rayleigh fading channels with N transmit and N receive antennas. C: Shannon capacity, C_F: Foschini (original) bound, and $C_{\text{LST-II}}$: capacity bound for LST-II.

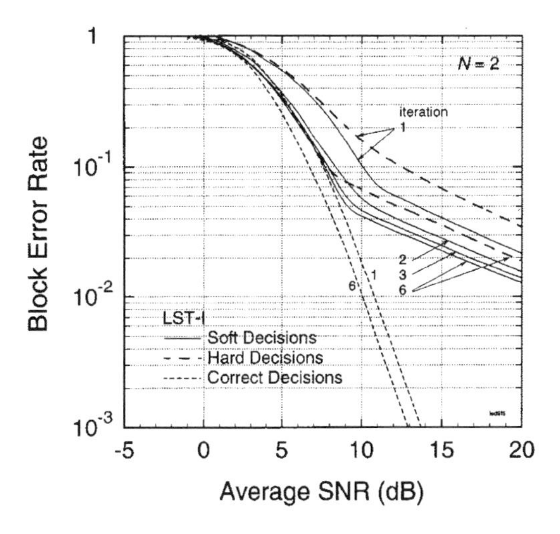

Fig. 5. Layered space–time performance of LST-I with two transmit and two receive antennas ($N = 2$). Quasi-static flat Rayleigh fading channel.

[which is equivalent to the generalized Foschini bound $C_{F,MF}$ in (11)], C_F denotes the original Foschini bound (with the ZF constraint) in (3), and $C_{\text{LST-II}}$ is the capacity bound for LST-II in (9) (however, the results for $C_{\text{LST-II}}$ are obtained simply by averaging the probability that $R = 1$ is not supported by each processing layer). Note that $C_{\text{LST-II}}$ can indeed provide approximately a diversity order of N; this is attributed largely to the use of layer ordering as discussed earlier. Note also that all the bounds show an improvement with increasing N. This means that the capacity actually increases more than linearly with the number of transmit and receive antennas. However, there is a diminishing improvement as N increases to a much larger number.

Fig. 5 shows the simulation results for LST-I with 2 transmit and 2 receive antennas (i.e., $N = 2$). Three sets of results are provided, assuming: 1) soft decisions; 2) hard decisions; and 3) correct decisions in each layer (note that the DDFSE always uses tentative decisions and provides soft outputs to the decoder). We see that, although soft decisions offer some improvement over hard decisions, the impact of decision errors is still quite noticeable. Fig. 6 shows similar results for $N = 4$. Here, the impact of decision errors is not as significant as the previous results, and turbo processing and soft decisions help to reduce much of this impact. With three iterations, the effect of decision errors almost completely disappears when using soft decisions. Decision errors have a lesser effect for a larger N because of the greater diversity order available through array processing and layer ordering.

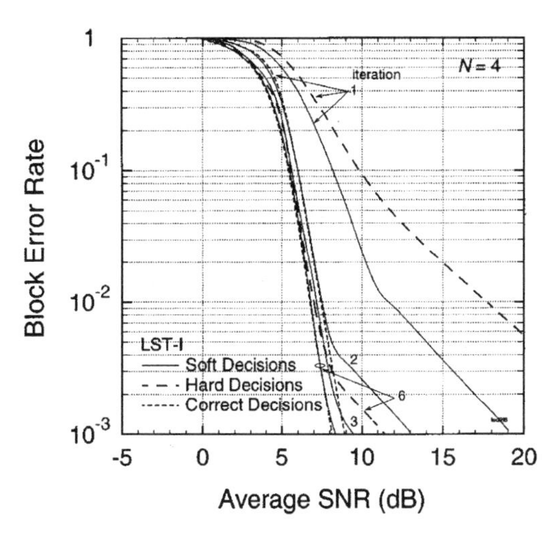

Fig. 6. Layered space–time performance of LST-I with four transmit and four receive antennas ($N = 4$). Quasi-static flat Rayleigh fading channel.

In Fig. 7, we compare the results using soft decisions and six "turbo" iterations with the Shannon capacity bound. For $N = 4$ and 8, the performance of LST-I is within 2.5–3 dB of the capacity bound at 10% BLER (and about 3–3.5 dB at 1% BLER). Since the BLER may vary as a function of the block size,[3] it is also important to consider the *processing* loss by discounting the loss due to the inefficiency of modulation and coding. As shown in Fig. 3(b), there is already a gap of about 2 dB between the performance of our coding scheme and the Shannon limit.

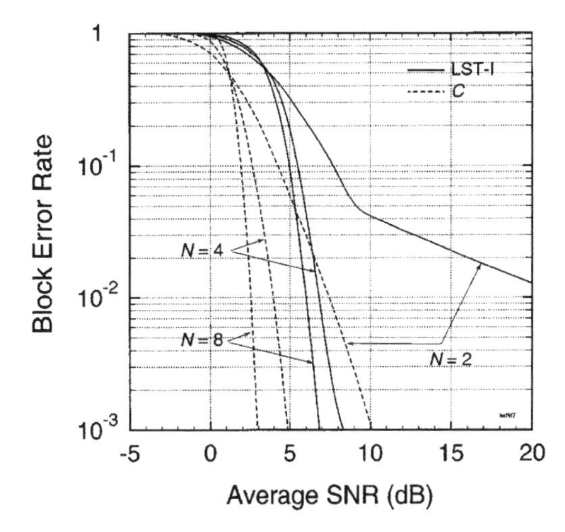

Fig. 7. Layered space–time performance of LST-I for $N = 2$, 4, and 8. Soft decisions. 6 iterations. Quasi-static flat Rayleigh fading channel.

[3]As an example, when we double the block size, the required average SNR is 0.2–0.4 dB greater than the results shown here. However, this difference in average SNR applies uniformly to all results, with or without layered space–time processing.

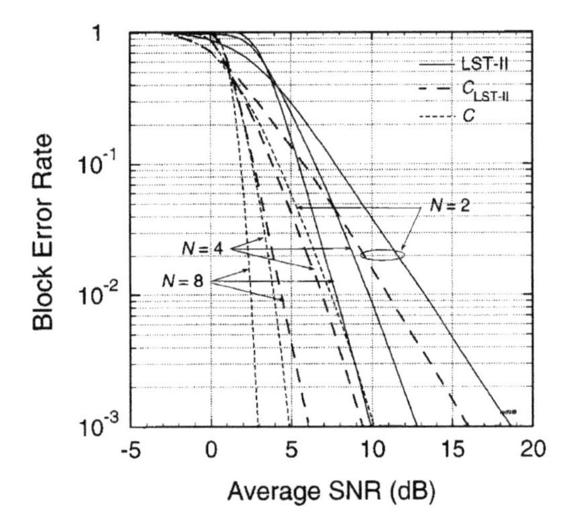

Fig. 8. Layered space–time performance of LST-II for $N = 2$, 4, and 8. Soft decisions. 2 iterations. Quasi-static flat Rayleigh fading channel.

Thus, the actual loss due to layered space–time processing is only about 0.5–1 dB at 10% BLER.

Fig. 8 shows the performance of LST-II for $N = 2$, 4, and 8, compared to the Shannon and $C_{\text{LST-II}}$ bounds. Soft decisions and two "turbo" iterations are assumed. (In this case, we found the effect of decision errors to be marginal, i.e., the results with hard and correct decisions were generally within 1 dB of the results shown here. Also, we found turbo processing with more than two iterations to provide little improvement.) At 10% BLER, the performance of LST-II is 2.5 dB from the $C_{\text{LST-II}}$ bound, and 3.5 dB from the Shannon bound. At 1% BLER, however, the loss compared to the Shannon bound can be as much as 6 dB.

From the above results, we conclude that, for $N = 4$ and 8, LST-I outperforms LST-II by a margin of 0.5 dB (at 10% BLER) to 3 dB (at 1% BLER). For $N = 2$, the performance of LST-I is greatly affected by decision errors (note that, even in this case, LST-I still performs as well as LST-II at 10% BLER), whereas LST-II can reach a lower BLER at high average SNR. Based on these results, the layered space–time approach is not highly recommended for $N = 2$. As mentioned earlier, space–time coding is a better alternative to achieve a spectral efficiency of 2 bps/Hz. For instance, a 64-state space–time coded QPSK can perform to within 2 dB of the Shannon capacity bound [23].

D. Frequency-Selective Channels

Finally, we present an example of performance results for frequency-selective fading channels. This example assumes $N = 4$ and the use of soft decisions for both LST-I and LST-II. Fig. 9(a) and (b) show the results for the TU and HT profiles, defined in Tables I and II. Again, we only show results with two "turbo" iterations for LST-II because little improvement can be achieved with more iterations in this case. For both delay profiles, the performance at 10% BLER is within 3 dB of the Shannon bound for LST-I with six iterations, and within 4 dB for LST-II with two iterations. At a lower BLER, the loss relative to the bound is greater for HT than for TU. This is due to the limitation of

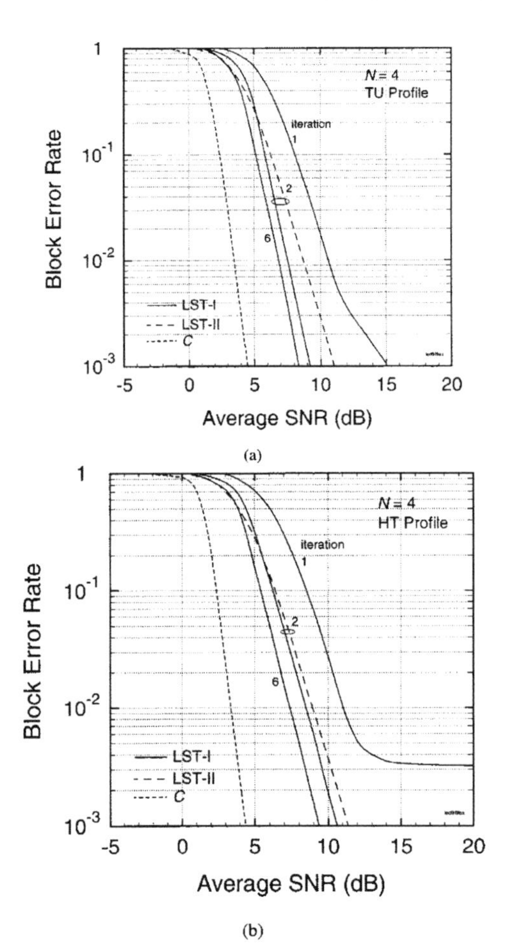

Fig. 9. Layered space–time performance of LST-I and LST-II over frequency-selective channels: (a) TU profile. (b) HT profile. $N = 4$. Soft decisions.

the suboptimum space–time equalizer structure we assume, as already discussed in Section IV-A.

VI. Conclusion

By deriving the generalized Shannon capacity formula and suggesting a layered space–time architecture that attains a tight lower bound on the capacity achievable, Foschini has laid a significant theoretical foundation for improving the wireless channel capacity through multiple-element array technology. We have shown that Foschini's lower bound is actually the true Shannon bound when the output SNR of the space–time processing in each layer is represented by the corresponding "matched filter" bound. We then provided two coded layered space–time approaches as an embodiment of this concept. For a large number of transmit and receive antennas, coding across the layers provides a better performance than independent coding within each layer. However, with two transmit and two receive antennas, the former is heavily affected by decision errors and, therefore, provides a poorer performance than the latter.

The underlying coding and signal processing techniques used in this study are based on practical but suboptimal approaches. Yet, such suboptimality can be greatly compensated for by iterative processing. Overall, our coded layered space–time approaches can achieve a performance within about 3 dB of the Shannon bound at 10% BLER, about 2 dB of which is a loss due to the practical coding scheme we assume. Thus, not only is the layered space–time architecture exactly what the Shannon limit has prescribed in a theoretical sense, but it also provides an attractive general methodology for improving and achieving the wireless channel capacity.

APPENDIX

PROVING THE EQUIVALENCE OF FOSCHINI BOUND AND SHANNON CAPACITY

Using the mathematical induction method, we will prove that (4) and (11) are identical. In order to do so, we must show that

$$\det(\mathfrak{R}\aleph^{-1}) = \prod_{k=1}^{N}(1+\Gamma_k(f)) \tag{19}$$

where $\mathfrak{R}$ in the above equation is equivalent to $\mathfrak{R}_N$ defined in (13). Again, we assume that the kth layer has $k-1$ interferences. Note also that the proof provided here is independent of the number of receive antennas M (the dimension of $\mathfrak{R}$) and the way the layers are ordered.

We start by assuming 1 signal source and M receive antennas. It can be easily shown [1] that

$$\det(\mathfrak{R}_1\aleph^{-1}) = 1 + \boldsymbol{H}_1^T\aleph^{-1}\boldsymbol{H}_1^* = 1+\Gamma_1(f). \tag{20}$$

Next, we assume that (19) is true for the case of $n-1$ signal sources, i.e.,

$$\det(\mathfrak{R}_{n-1}\aleph^{-1}) = \prod_{k=1}^{n-1}(1+\Gamma_k(f)). \tag{21}$$

We then show in the following that, given (21), (19) is also true for n signal sources.

First, we note from (13) that

$$\mathfrak{R}_n = \mathfrak{R}_{n-1} + \boldsymbol{H}_n^*\boldsymbol{H}_n^T. \tag{22}$$

Using the matrix inversion lemma [4, Appendix D], we can show that [similar to (12)]

$$\mathfrak{R}_n^{-1}\boldsymbol{H}_n^* = \frac{\mathfrak{R}_{n-1}^{-1}\boldsymbol{H}_n^*}{1+\Gamma_n(f)}. \tag{23}$$

It follows that

$$(\mathfrak{R}_n\aleph^{-1})\boldsymbol{H}_n^* = \aleph\mathfrak{R}_n^{-1}\boldsymbol{H}_n^* = \frac{\aleph\mathfrak{R}_{n-1}^{-1}\boldsymbol{H}_n^*}{1+\Gamma_n(f)} = \frac{(\mathfrak{R}_{n-1}\aleph^{-1})^{-1}\boldsymbol{H}_n^*}{1+\Gamma_n(f)}. \tag{24}$$

For convenience, let

$$\boldsymbol{A} \triangleq \mathfrak{R}_n\aleph^{-1} \text{ and } \boldsymbol{B} \triangleq \mathfrak{R}_{n-1}\aleph^{-1}. \tag{25}$$

We can rewrite (24) as

$$\boldsymbol{A}^{-1}\boldsymbol{H}_n^* = \frac{\boldsymbol{B}^{-1}\boldsymbol{H}_n^*}{1+\Gamma_n(f)}. \tag{26}$$

Furthermore, using the matrix identity [40]

$$\boldsymbol{A}^{-1} = \frac{\boldsymbol{A}_{\text{adj}}}{\det(\boldsymbol{A})} \tag{27}$$

where $\boldsymbol{A}_{\text{adj}}$ is called the *adjugate matrix* of matrix $\boldsymbol{A}$, we can rewrite (26) as

$$\frac{\boldsymbol{A}_{\text{adj}}}{\det(\boldsymbol{A})}\boldsymbol{H}_n^* = \frac{\boldsymbol{B}_{\text{adj}}}{\det(\boldsymbol{B})}\frac{\boldsymbol{H}_n^*}{1+\Gamma_n(f)}. \tag{28}$$

By replacing $\det(\boldsymbol{B})$ in the above equation using (21), we obtain

$$\frac{\boldsymbol{A}_{\text{adj}}\boldsymbol{H}_n^*}{\det(\boldsymbol{A})} = \frac{\boldsymbol{B}_{\text{adj}}\boldsymbol{H}_n^*}{\prod_{k\ 1}^{n}(1+\Gamma_k(f))}. \tag{29}$$

Our goal is to prove that

$$\det(\boldsymbol{A}) = \prod_{k=1}^{n}(1+\Gamma_k(f)). \tag{30}$$

Thus, given (29), we must show that

$$\boldsymbol{A}_{\text{adj}}\boldsymbol{H}_n^* = \boldsymbol{B}_{\text{adj}}\boldsymbol{H}_n^*. \tag{31}$$

From (22) and (25), we have

$$\begin{aligned}\boldsymbol{A} &= \boldsymbol{B} + \boldsymbol{H}_n^*\boldsymbol{H}_n^T\aleph^{-1}\\ &= \left[\boldsymbol{b}_1 + \boldsymbol{H}_n^*\frac{H_{n1}(f)}{\aleph_1(f)},\ \boldsymbol{b}_2 + \boldsymbol{H}_n^*\frac{H_{n2}(f)}{\aleph_2(f)},\ \cdots,\right.\\ &\qquad \left.\boldsymbol{b}_M + \boldsymbol{H}_n^*\frac{H_{nM}(f)}{\aleph_M(f)}\right]\end{aligned} \tag{32}$$

where $\boldsymbol{b}_j$ is the jth column vector of $\boldsymbol{B}$, for $j = 1, \cdots, M$; for convenience, we use M here to indicate the *overall* receive diversity order, including the effects of both multiple antennas and excess bandwidth.

We now prove (31) by showing that the jth element of $\boldsymbol{A}_{\text{adj}}\boldsymbol{H}_n^*$ is equal to the jth element of $\boldsymbol{B}_{\text{adj}}\boldsymbol{H}_n^*$, for $j = 1, \ldots, M$. Note that the jth element of $\boldsymbol{A}_{\text{adj}}\boldsymbol{H}_n^*$ is given by $\det(\overline{\boldsymbol{A}}_j)$, where $\overline{\boldsymbol{A}}_j$ is obtained by replacing the jth column of $\boldsymbol{A}$ by $\boldsymbol{H}_n^*$. Similarly, the jth element of $\boldsymbol{B}_{\text{adj}}\boldsymbol{H}_n^*$ is given by $\det(\overline{\boldsymbol{B}}_j)$, where $\overline{\boldsymbol{B}}_j$ is obtained by replacing the jth column of $\boldsymbol{B}$ by $\boldsymbol{H}_n^*$.

Using (32) and the linear properties of determinants, we can show that

$$\det(\overline{\boldsymbol{A}}_j) = \det\left[\boldsymbol{b}_1 + \boldsymbol{H}_n^*\frac{H_{n1}(f)}{\aleph_1(f)},\ \cdots,\ \underset{\boldsymbol{H}_n^*}{j\text{th column}},\ \cdots\cdots\right] \tag{33}$$

$$\begin{aligned}&= \det[\boldsymbol{b}_1, \cdots, \boldsymbol{H}_n^*, \cdots\cdots]\\ &\quad + \det\left[\boldsymbol{H}_n^*\frac{H_{n1}(f)}{\aleph_1(f)}, \cdots, \boldsymbol{H}_n^*, \cdots\cdots\right]\\ &= \det[\boldsymbol{b}_1, \cdots, \boldsymbol{H}_n^*, \cdots\cdots]\\ &\quad + \frac{H_{n1}(f)}{\aleph_1(f)}\det[\boldsymbol{H}_n^*, \cdots, \boldsymbol{H}_n^*, \cdots\cdots]\\ &= \det[\boldsymbol{b}_1, \cdots, \boldsymbol{H}_n^*, \cdots\cdots].\end{aligned} \tag{34}$$

Similarly, we can expand the result of (34) with respect to the second column, the third column, and so on (except for the jth column). Eventually, we obtain

$$\det(\overline{\boldsymbol{A}}_j) = \det[\boldsymbol{b}_1, \boldsymbol{b}_2, \cdots, \underset{\text{column}}{\overset{j\text{th}}{\boldsymbol{H}_n^*}}, \cdots, \boldsymbol{b}_M] = \det(\overline{\boldsymbol{B}}_j) \tag{35}$$

which proves (31). The proof of (19) is therefore complete.

ACKNOWLEDGMENT

The author wishes to thank G. J. Foschini for reviewing the information theory part of the original manuscript of this paper and suggesting several improvements, and for the enormous inspiration he provided by inventing the layered space–time concept. The author also benefited from discussions with K. R. Narayanan, I. Lee, and X. Li. Finally, the author thanks the anonymous reviewers for valuable comments and suggestions.

REFERENCES

[1] G. J. Foschini and M. J. Gans, "On limits of wireless communication in a fading environment when using multiple antennas," *Wireless Pers. Commun.*, vol. 6, no. 3, pp. 311–335, Mar. 1998.

[2] W. C. Jakes Jr., Ed., *Microwave Mobile Communications*. New York: Wiley, 1974.

[3] D. C. Cox, "Universal digital portable radio communications," *Proc. IEEE*, vol. 75, pp. 436–477, Apr. 1987.

[4] R. A. Monzingo and T. W. Miller, *Introduction to Adaptive Arrays*. New York: Wiley, 1980.

[5] J. H. Winters, J. Salz, and R. D. Gitlin, "The impact of antenna diversity on the capacity of wireless communications systems," *IEEE Trans. Commun.*, vol. 42, pp. 1740–1751, Apr. 1994.

[6] S. L. Ariyavisitakul, J. H. Winters, and I. Lee, "Optimum space–time processors with dispersive interference: Unified analysis and required filter span," *IEEE Trans. Commun.*, vol. 47, pp. 1073–1083, July 1999.

[7] G. J. Foschini, "Layered space–time architecture for wireless communication in a fading environment when using multiple antennas," *Bell Labs Tech. J.*, vol. 1, no. 2, pp. 41–59, Autumn 1996.

[8] S. Verdu, *Multiuser Detection*. Cambridge, U.K.: Cambridge Univ. Press, 1998.

[9] G. J. Foschini, G. D. Golden, R. A. Valenzuela, and P. W. Wolaniansky, "Simplified processing for high spectral efficiency wireless communication employing multi-element arrays," *IEEE J. Select. Areas Commun.*, vol. 17, pp. 1841–1852, Nov. 1999.

[10] R. van Nobelen, N. Seshadri, J. Whitehead, and S. Timiri, "An adaptive radio link protocol with enhanced data rates for GSM evolution," *IEEE Pers. Commun.*, vol. 6, pp. 54–63, Feb. 1999.

[11] C. Berrou, A. Glavieux, and P. Thitimajshima, "Near Shannon limit error-correction coding and decoding: Turbo codes," in *Proc. IEEE ICC'93*, Geneva, Switzerland, May 1993, pp. 1064–1070.

[12] C. Berrou and A. Glavieux, "Near optimum error correcting coding and decoding: Turbo codes," *IEEE Trans. Commun.*, vol. 44, pp. 1261–1271, Oct. 1996.

[13] S. Benedetto and G. Montorsi, "Unveiling turbo codes: Some results on parallel concatenated coding schemes," *IEEE Trans. Inform. Theory*, vol. 42, pp. 409–428, Mar. 1996.

[14] S. Benedetto, D. Divsalar, G. Montorsi, and F. Pollara, "Serial concatenation of interleaved codes: Design and performance analysis," *IEEE Trans. Inform. Theory*, vol. 44, pp. 409–429, Apr. 1998.

[15] K. R. Narayanan and G. L. Stuber, "A serial concatenation approach to iterative demodulation and decoding," *IEEE Trans. Commun.*, vol. 47, pp. 956–961, July 1999.

[16] C. Douillard, C. B. M. Jezequel, A. Picart, P. Didier, and A. Glavieux, "Iterative correction of intersymbol interference: Turbo equalization," *Eur. Trans. Telecommun.*, vol. 6, pp. 507–511, Sept. 1995.

[17] A. Picart, P. Didier, and A. Glavieux, "Turbo-detection: A new approach to combat channel frequency selectivity," in *Proc. IEEE ICC'97*, Montreal, Quebec, Canada, June 1997, pp. 1498–1502.

[18] D. Raphaeli and Y. Zarai, "Combined turbo equalization and turbo decoding," *IEEE Commun. Lett.*, vol. 2, pp. 107–109, Apr. 1998.

[19] J. Garcia-Frias and J. D. Villasenor, "Combined blind equalization and turbo decoding," in *Proc. IEEE ICC'99, Communication Theory Mini-Conference*, Vancouver, BC, Canada, June 1999, pp. 52–57.

[20] S. L. Ariyavisitakul, J. H. Winters, and N. R. Sollenberger, "Joint equalization and interference suppression for high data rate wireless systems," in *Proc. IEEE VTC'99*, Houston, TX, May 1999, pp. 700–706.

[21] P. Jung, "Novel low complexity decoder for turbo codes," *Electron. Lett.*, vol. 31, no. 2, pp. 86–87, Jan. 1995.

[22] A. J. Viterbi, "An intuitive justification and a simplified implementation of the MAP decoder for convolutional codes," *IEEE J. Select. Areas Commun.*, vol. 16, pp. 260–264, Feb. 1998.

[23] V. Tarokh, N. Seshadri, and A. R. Calderbank, "Space-time codes for high data rate wireless communications: Performance analysis and code construction," *IEEE Trans. Inform. Theory*, vol. 44, pp. 744–765, Mar. 1998.

[24] A. F. Naguib, V. Tarokh, N. Seshadri, and A. R. Calderbank, "A space–time coding modem for high-data-rate wireless communications," *IEEE J. Select. Areas Commun.*, vol. 16, pp. 1459–1478, Oct. 1998.

[25] V. Tarokh, A. F. Naguib, N. Seshadri, and A. R. Calderbank, "Combined array processing and space–time coding," *IEEE Trans. Inform. Theory*, vol. 45, pp. 1121–1128, May 1999.

[26] J. G. Proakis, *Digital Communications*, 2nd ed. New York: McGraw-Hill, 1989.

[27] A. Duel-Hallen and C. Heegard, "Delayed decision-feedback sequence estimation," *IEEE Trans. Commun.*, vol. 37, pp. 428–436, May 1989.

[28] S. L. Ariyavisitakul and I. Lee, "The equivalence of two unified solutions for optimum space–time processing," *IEEE Trans. Commun.*, to be published.

[29] S. Ariyavisitakul and L. J. Greenstein, "Reduced-complexity equalization techniques for broadband wireless channels," *IEEE J. Select. Areas Commun.*, vol. 15, pp. 5–15, Jan. 1997.

[30] P. A. Voois, I. Lee, and J. M. Cioffi, "The effect of decision delay in finite-length decision feedback equalization," *IEEE Trans. Inform. Theory*, vol. 42, pp. 618–621, Mar. 1996.

[31] S. Haykin, *Adaptive Filter Theory*. Englewood Cliffs, NJ: Prentice-Hall, 1991.

[32] L. R. Bahl, J. Cocke, F. Jelinek, and J. Raviv, "Optimal decoding of linear codes for minimizing symbol error rate," *IEEE Trans. Inform. Theory*, vol. IT-20, pp. 284–287, Mar. 1974.

[33] J. Hagenauer, E. Offer, and L. Papke, "Iterative decoding of binary block and convolutional codes," *IEEE Trans. Inform. Theory*, vol. 42, pp. 429–445, Mar. 1996.

[34] S. Benedetto, D. Divsalar, G. Montorsi, and F. Pollara, "Soft-output decoding algorithms for continuous decoding of parallel concatenated convolutional codes," in *Proc. IEEE ICC'96*, Dallas, TX, June 1996, pp. 112–117.

[35] R. J. McEliece, D. J. C. MacKay, and J.-F. Cheng, "Turbo decoding as an instance of Pearl's "belief propagation" algorithm," *IEEE J. Select. Areas Commun.*, vol. 16, pp. 140–152, Feb. 1998.

[36] E. Zehavi, "8-PSK trellis codes for a Rayleigh channel," *IEEE Trans. Commun.*, vol. 40, pp. 873–883, May 1992.

[37] G. Caire, G. Taricco, and E. Biglieri, "Bit-interleaved coded modulation," *IEEE Trans. Inform. Theory*, vol. 44, pp. 927–946, May 1998.

[38] X. Li and J. Ritcey, "Trellis-coded modulation with bit interleaving and iterative decoding," *IEEE Trans. Commun.*, vol. 17, pp. 715–724, Apr. 1999.

[39] S. Ariyavisitakul and Y. Li, "Joint coding and decision feedback equalization for broadband wireless channels," *IEEE J. Select. Areas Commun.*, vol. 16, pp. 1670–1678, Dec. 1998.

[40] G. Strang, *Linear Algebra and Its Applications*. Orlando, FL: Harcourt Brace Jovanovich, 1980.

PHOTO
NOT
AVAILABLE

Sirikat Lek Ariyavisitakul (S'85–M'88–SM'93) received the B.S., M.S., and Ph.D. degrees in electrical engineering from Kyoto University, Kyoto, Japan, in 1983, 1985, and 1988, respectively.

He is currently Director of Research at Home Wireless Networks (HWN), Norcross, GA. Prior to HWN, he was with Bellcore (now Telcordia), Red Bank, NJ, from 1988 to 1994, and with AT&T Bell Laboratories (later AT&T Laboratories) in Holdmel, NJ, and Red Bank, NJ, from 1994 to 1998. His research interests include communication theory, signal processing, and coding techniques for wireless communications. He has published about 30 journal articles and 40 conference papers in the areas of equalization, interference suppression, synchronization, modulation, CDMA and power control, coding and frequency hopping, and wireless system architectures and infrastructures. He holds 14 U.S. patents with several pending in these areas.

Dr. Ariyavisitakul has served as Editor of the IEEE TRANSACTIONS ON COMMUNICATIONS since 1995, and the Secretary of the Communication Theory Technical Committee of the IEEE Communications Society since 1997. He has organized and chaired technical sessions in a number of IEEE conferences. He received the 1988 Niwa Memorial Award in Tokyo, Japan, for outstanding research and publication. He is a member of the Institute of Electronics, Information, and Communication Engineers of Japan.

A Simple Transmit Diversity Technique for Wireless Communications

Siavash M. Alamouti

Abstract— **This paper presents a simple two-branch transmit diversity scheme. Using two transmit antennas and one receive antenna the scheme provides the same diversity order as maximal-ratio receiver combining (MRRC) with one transmit antenna, and two receive antennas. It is also shown that the scheme may easily be generalized to two transmit antennas and receive antennas to provide a diversity order of 2 . The new scheme does not require any bandwidth expansion any feedback from the receiver to the transmitter and its computation complexity is similar to MRRC.**

Index Terms—**Antenna array processing, baseband processing, diversity, estimation and detection, fade mitigation, maximal-ratio combining, Rayleigh fading, smart antennas, space block coding, space–time coding, transmit diversity, wireless communications.**

I. Introduction

THE NEXT-generation wireless systems are required to have high voice quality as compared to current cellular mobile radio standards and provide high bit rate data services (up to 2 Mbits/s). At the same time, the remote units are supposed to be small lightweight pocket communicators. Furthermore, they are to operate reliably in different types of environments: macro, micro, and picocellular; urban, suburban, and rural; indoor and outdoor. In other words, the next generation systems are supposed to have better quality and coverage, be more power and bandwidth efficient, and be deployed in diverse environments. Yet the services must remain affordable for widespread market acceptance. Inevitably, the new pocket communicators must remain relatively simple. Fortunately, however, the economy of scale may allow more complex base stations. In fact, it appears that base station complexity may be the only plausible trade space for achieving the requirements of next generation wireless systems.

The fundamental phenomenon which makes reliable wireless transmission difficult is time-varying multipath fading [1]. It is this phenomenon which makes tetherless transmission a challenge when compared to fiber, coaxial cable, line-of-sight microwave or even satellite transmissions.

Increasing the quality or reducing the effective error rate in a multipath fading channel is extremely difficult. In additive white Gaussian noise (AWGN), using typical modulation and coding schemes, reducing the effective bit error rate (BER) from 10^{-2} to 10^{-3} may require only 1- or 2-dB higher signal-to-noise ratio (SNR). Achieving the same in a multipath fading environment, however, may require up to 10 dB improvement in SNR. The improvement in SNR may not be achieved by higher transmit power or additional bandwidth, as it is contrary to the requirements of next generation systems. It is therefore crucial to effectively combat or reduce the effect of fading at both the remote units and the base stations, without additional power or any sacrifice in bandwidth.

Theoretically, the most effective technique to mitigate multipath fading in a wireless channel is transmitter power control. If channel conditions as experienced by the receiver on one side of the link are known at the transmitter on the other side, the transmitter can predistort the signal in order to overcome the effect of the channel at the receiver. There are two fundamental problems with this approach. The major problem is the required transmitter dynamic range. For the transmitter to overcome a certain level of fading, it must increase its power by that same level, which in most cases is not practical because of radiation power limitations and the size and cost of the amplifiers. The second problem is that the transmitter does not have any knowledge of the channel experienced by the receiver except in systems where the uplink (remote to base) and downlink (base to remote) transmissions are carried over the same frequency. Hence, the channel information has to be fed back from the receiver to the transmitter, which results in throughput degradation and considerable added complexity to both the transmitter and the receiver. Moreover, in some applications there may not be a link to feed back the channel information.

Other effective techniques are time and frequency diversity. Time interleaving, together with error correction coding, can provide diversity improvement. The same holds for spread spectrum. However, time interleaving results in large delays when the channel is slowly varying. Equivalently, spread spectrum techniques are ineffective when the coherence bandwidth of the channel is larger than the spreading bandwidth or, equivalently, where there is relatively small delay spread in the channel.

In most scattering environments, antenna diversity is a practical, effective and, hence, a widely applied technique for reducing the effect of multipath fading [1]. The classical approach is to use multiple antennas at the receiver and perform combining or selection and switching in order to improve the quality of the received signal. The major problem with using the receive diversity approach is the cost, size, and power of the remote units. The use of multiple antennas and radio frequency (RF) chains (or selection and switching circuits) makes the remote units larger and more expensive. As a result, diversity techniques have almost exclusively been

Manuscript received September 1, 1997; revised February 1, 1998.
The author was with AT&T Wireless Services, Redmond, WA, USA. He is currently with Cadence Design Systems, Alta Business Unit, Bellevue, WA 98005-3016 USA (e-mail: siavash@cadence.com).
Publisher Item Identifier S 0733-8716(98)07885-8.

Reprinted from *IEEE Journal on Select Areas in Communications*, vol. 16, no. 8, October 1998.

applied to base stations to improve their reception quality. A base station often serves hundreds to thousands of remote units. It is therefore more economical to add equipment to base stations rather than the remote units. For this reason, transmit diversity schemes are very attractive. For instance, one antenna and one transmit chain may be added to a base station to improve the reception quality of all the remote units in that base station's coverage area.[1] The alternative is to add more antennas and receivers to all the remote units. The first solution is definitely more economical.

Recently, some interesting approaches for transmit diversity have been suggested. A delay diversity scheme was proposed by Wittneben [2], [3] for base station simulcasting and later, independently, a similar scheme was suggested by Seshadri and Winters [4], [5] for a single base station in which copies of the same symbol are transmitted through multiple antennas at different times, hence creating an artificial multipath distortion. A maximum likelihood sequence estimator (MLSE) or a minimum mean squared error (MMSE) equalizer is then used to resolve multipath distortion and obtain diversity gain. Another interesting approach is space–time trellis coding, introduced in [6], where symbols are encoded according to the antennas through which they are simultaneously transmitted and are decoded using a maximum likelihood decoder. This scheme is very effective, as it combines the benefits of forward error correction (FEC) coding and diversity transmission to provide considerable performance gains. The cost for this scheme is additional processing, which increases exponentially as a function of bandwidth efficiency (bits/s/Hz) and the required diversity order. Therefore, for some applications it may not be practical or cost-effective.

The technique proposed in this paper is a simple transmit diversity scheme which improves the signal quality at the receiver on one side of the link by simple processing across two transmit antennas on the opposite side. The obtained diversity order is equal to applying maximal-ratio receiver combining (MRRC) with two antennas at the receiver. The scheme may easily be generalized to two transmit antennas and M receive antennas to provide a diversity order of $2M$. This is done without any feedback from the receiver to the transmitter and with small computation complexity. The scheme requires no bandwidth expansion, as redundancy is applied in space across multiple antennas, not in time or frequency.

The new transmit diversity scheme can improve the error performance, data rate, or capacity of wireless communications systems. The decreased sensitivity to fading may allow the use of higher level modulation schemes to increase the effective data rate, or smaller reuse factors in a multicell environment to increase system capacity. The scheme may also be used to increase the range or the coverage area of wireless systems. In other words, the new scheme is effective in all of the applications where system capacity is limited by multipath fading and, hence, may be a simple and cost-effective way to address the market demands for quality and efficiency without a complete redesign of existing systems. Furthermore, the scheme seems to be a superb candidate for next-generation wireless systems, as it effectively reduces the effect of fading at the remote units using multiple transmit antennas at the base stations.

In Section II, the classical maximal ratio receive diversity combining is discussed and simple mathematical descriptions are given. In Section III, the new two-branch transmit diversity schemes with one and with two receive antennas are discussed. In Section IV, the bit-error performance of the new scheme with coherent binary phase-shift keying (BPSK) modulation is presented and is compared with MRRC. There are cost and performance differences between the practical implementations of the proposed scheme and the classical MRRC. These differences are discussed in detail in Section V.

II. Classical Maximal-Ratio Receive Combining (MRRC) Scheme

Fig. 1 shows the baseband representation of the classical two-branch MRRC.

At a given time, a signal s_0 is sent from the transmitter. The channel including the effects of the transmit chain, the airlink, and the receive chain may be modeled by a complex multiplicative distortion composed of a magnitude response and a phase response. The channel between the transmit antenna and the receive antenna zero is denoted by $\boldsymbol{h_0}$ and between the transmit antenna and the receive antenna one is denoted by $\boldsymbol{h_1}$ where

$$\begin{aligned} \boldsymbol{h_0} &= \alpha_0 e^{j\theta_0} \\ \boldsymbol{h_1} &= \alpha_1 e^{j\theta_1}. \end{aligned} \tag{1}$$

Noise and interference are added at the two receivers. The resulting received baseband signals are

$$\begin{aligned} \boldsymbol{r_0} &= \boldsymbol{h_0 s_0} + \boldsymbol{n_0} \\ \boldsymbol{r_1} &= \boldsymbol{h_1 s_0} + \boldsymbol{n_1} \end{aligned} \tag{2}$$

where $\boldsymbol{n_0}$ and $\boldsymbol{n_1}$ represent complex noise and interference.

Assuming $\boldsymbol{n_0}$ and $\boldsymbol{n_1}$ are Gaussian distributed, the maximum likelihood decision rule at the receiver for these received signals is to choose signal $\boldsymbol{s_i}$ if and only if (iff)

$$\begin{aligned} d^2(\boldsymbol{r_0}, \boldsymbol{h_0 s_i}) + d^2(\boldsymbol{r_1}, \boldsymbol{h_1 s_i}) &\le d^2(\boldsymbol{r_0}, \boldsymbol{h_0 s_k}) \\ &+ d^2(\boldsymbol{r_1}, \boldsymbol{h_1 s_k}), \qquad \forall \boldsymbol{i} \neq \boldsymbol{k} \end{aligned} \tag{3}$$

where $d^2(\mathrm{x}, \mathrm{y})$ is the squared Euclidean distance between signals $\boldsymbol{x}$ and $\boldsymbol{y}$ calculated by the following expression:

$$d^2(\boldsymbol{x}, \boldsymbol{y}) = (\boldsymbol{x} - \boldsymbol{y})(\boldsymbol{x}^* - \boldsymbol{y}^*). \tag{4}$$

The receiver combining scheme for two-branch MRRC is as follows:

$$\begin{aligned} \tilde{\boldsymbol{s}}_0 &= \boldsymbol{h}_0^* \boldsymbol{r_0} + \boldsymbol{h}_1^* \boldsymbol{r_1} \\ &= \boldsymbol{h}_0^*(\boldsymbol{h_0 s_0} + \boldsymbol{n_0}) + \boldsymbol{h}_1^*(\boldsymbol{h_1 s_0} + \boldsymbol{n_1}) \\ &= (\alpha_0^2 + \alpha_1^2)\boldsymbol{s_0} + \boldsymbol{h}_0^* \boldsymbol{n_0} + \boldsymbol{h}_1^* \boldsymbol{n_1}. \end{aligned} \tag{5}$$

Expanding (3) and using (4) and (5) we get

choose $\boldsymbol{s_i}$ iff

$$\begin{aligned} &(\alpha_0^2 + \alpha_1^2)|\boldsymbol{s_i}|^2 - \tilde{\boldsymbol{s}}_0 \boldsymbol{s}_i^* - \tilde{\boldsymbol{s}}_0^* \boldsymbol{s_i} \\ &\quad \le (\alpha_0^2 + \alpha_1^2)|\boldsymbol{s_k}|^2 - \tilde{\boldsymbol{s}}_0 \boldsymbol{s}_k^* - \tilde{\boldsymbol{s}}_0^* \boldsymbol{s_k}, \qquad \forall \boldsymbol{i} \neq \boldsymbol{k} \end{aligned} \tag{6}$$

[1] In fact, many cellular base stations already have two receive antennas for receive diversity. The same antennas may be used for transmit diversity.

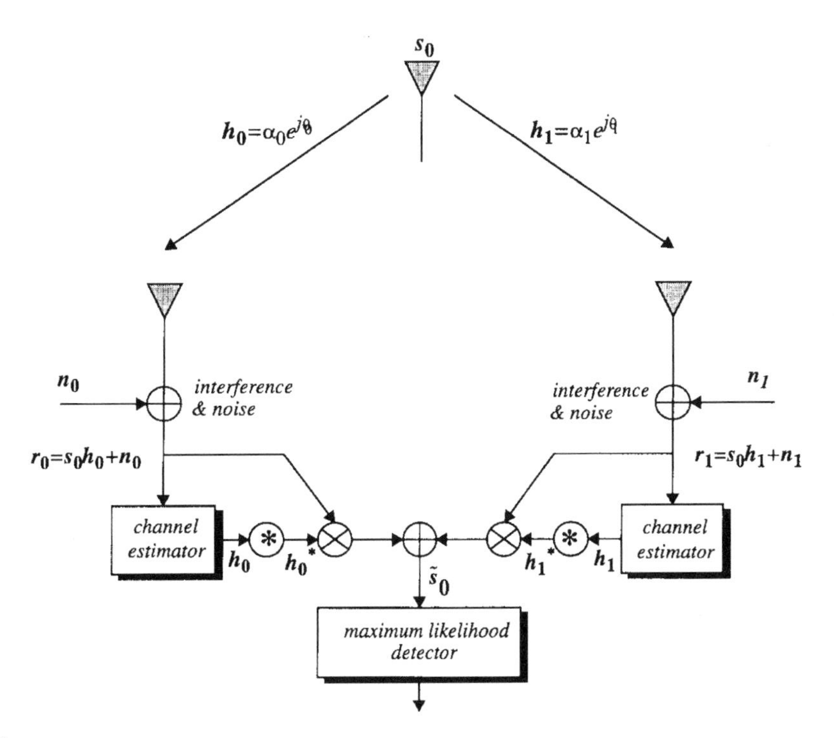

Fig. 1. Two-branch MRRC.

or equivalently

choose s_i iff

$$(\alpha_0^2+\alpha_1^2-1)|s_i|^2+d^2(\tilde{s}_0, s_i) \leq (\alpha_0^2+\alpha_1^2-1)|s_k|^2+d^2(\tilde{s}_0, s_k), \qquad \forall i \neq k. \tag{7}$$

For PSK signals (equal energy constellations)

$$|s_i|^2=|s_k|^2=E_s, \qquad \forall i, k \tag{8}$$

where E_s is the energy of the signal. Therefore, for PSK signals, the decision rule in (7) may be simplified to

choose s_i iff

$$d^2(\tilde{s}_0, s_i) \leq d^2(\tilde{s}_0, s_k), \qquad \forall i \neq k. \tag{9}$$

The maximal-ratio combiner may then construct the signal ~ , as shown in Fig. 1, so that the maximum likelihood detector may produce $\hat{s}_0$, which is a maximum likelihood estimate of s_0.

III. The New Transmit Diversity Scheme

A. Two-Branch Transmit Diversity with One Receiver

Fig. 2 shows the baseband representation of the new two-branch transmit diversity scheme.

The scheme uses two transmit antennas and one receive antenna and may be defined by the following three functions:

- the encoding and transmission sequence of information symbols at the transmitter;
- the combining scheme at the receiver;
- the decision rule for maximum likelihood detection.

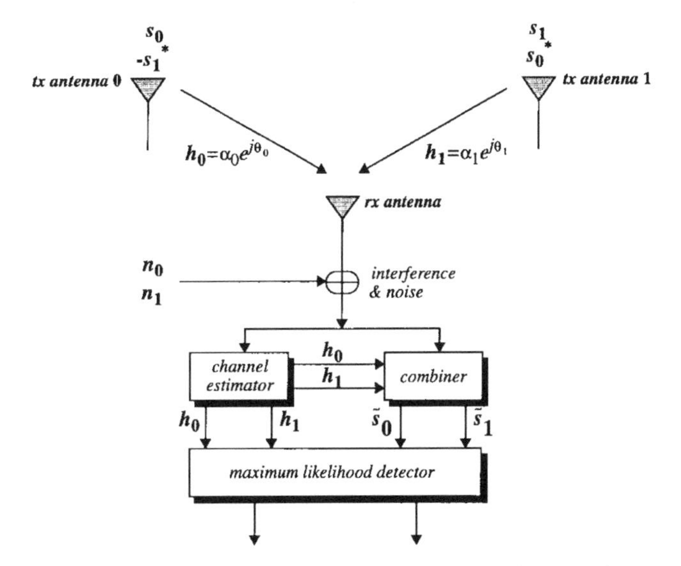

Fig. 2. The new two-branch transmit diversity scheme with one receiver.

1) The Encoding and Transmission Sequence: At a given symbol period, two signals are simultaneously transmitted from the two antennas. The signal transmitted from antenna zero is denoted by s_0 and from antenna one by s_1. During the next symbol period signal $(-s_1^*)$ is transmitted from antenna zero, and signal s_0^* is transmitted from antenna one where $*$ is the complex conjugate operation. This sequence is shown in Table I.

In Table I, the encoding is done in space and time (space–time coding). The encoding, however, may also be done in space and frequency. Instead of two adjacent symbol periods, two adjacent carriers may be used (space–frequency coding).

TABLE I
THE ENCODING AND TRANSMISSION SEQUENCE FOR THE TWO-BRANCH TRANSMIT DIVERSITY SCHEME

	antenna 0	antenna 1
time t	s_0	s_1
time $t + T$	$-s_1^*$	s_0^*

The channel at time t may be modeled by a complex multiplicative distortion $h_0(t)$ for transmit antenna zero and $h_1(t)$ for transmit antenna one. Assuming that fading is constant across two consecutive symbols, we can write

$$\begin{aligned} h_0(t) &= h_0(t+T) = h_0 = \alpha_0 e^{j\theta_0} \\ h_1(t) &= h_1(t+T) = h_1 = \alpha_1 e^{j\theta_1} \end{aligned} \tag{10}$$

where T is the symbol duration. The received signals can then be expressed as

$$\begin{aligned} r_0 &= r(t) = h_0 s_0 + h_1 s_1 + n_0 \\ r_1 &= r(t+T) = -h_0 s_1^* + h_1 s_0^* + n_1 \end{aligned} \tag{11}$$

where r_0 and r_1 are the received signals at time t and $t+T$ and n_0 and n_1 are complex random variables representing receiver noise and interference.

2) The Combining Scheme: The combiner shown in Fig. 2 builds the following two combined signals that are sent to the maximum likelihood detector:

$$\begin{aligned} \tilde{s}_0 &= h_0^* r_0 + h_1 r_1^* \\ \tilde{s}_1 &= h_1^* r_0 - h_0 r_1^*. \end{aligned} \tag{12}$$

It is important to note that this combining scheme is different from the MRRC in (5). Substituting (10) and (11) into (12) we get

$$\begin{aligned} \tilde{s}_0 &= (\alpha_0^2 + \alpha_1^2) s_0 + h_0^* n_0 + h_1 n_1^* \\ \tilde{s}_1 &= (\alpha_0^2 + \alpha_1^2) s_1 - h_0 n_1^* + h *_1 n_0. \end{aligned} \tag{13}$$

3) The Maximum Likelihood Decision Rule: These combined signals are then sent to the maximum likelihood detector which, for each of the signals s_0 and s_1, uses the decision rule expressed in (7) or (9) for PSK signals.

The resulting combined signals in (13) are equivalent to that obtained from two-branch MRRC in (5). The only difference is phase rotations on the noise components which do not degrade the effective SNR. Therefore, the resulting diversity order from the new two-branch transmit diversity scheme with one receiver is equal to that of two-branch MRRC.

B. Two-Branch Transmit Diversity with M Receivers

There may be applications where a higher order of diversity is needed and multiple receive antennas at the remote units are feasible. In such cases, it is possible to provide a diversity order of $2M$ with two transmit and M receive antennas. For illustration, we discuss the special case of two transmit and two receive antennas in detail. The generalization to M receive antennas is trivial.

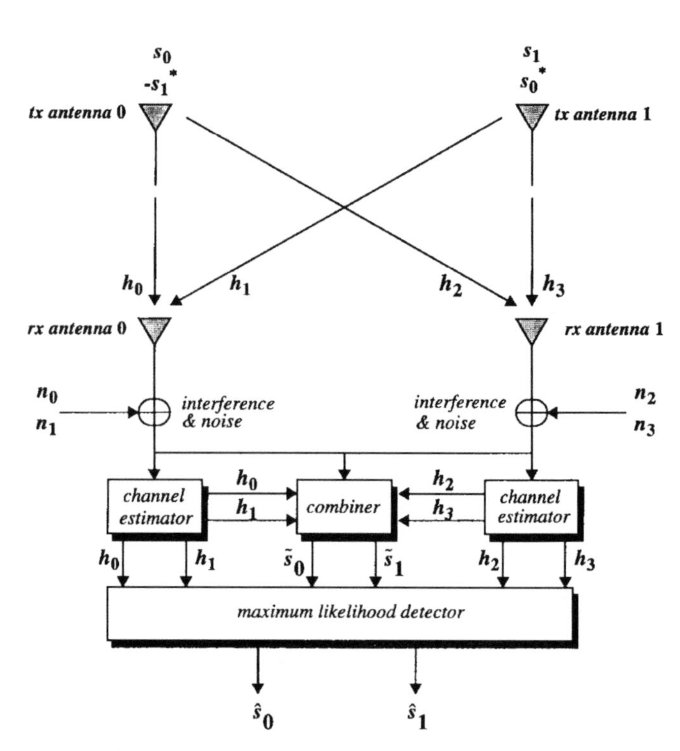

Fig. 3. The new two-branch transmit diversity scheme with two receivers.

TABLE II
THE DEFINITION OF CHANNELS BETWEEN THE TRANSMIT AND RECEIVE ANTENNAS

	rx antenna 0	rx antenna 1
tx antenna 0	h_0	h_2
tx antenna 1	h_1	h_3

TABLE III
THE NOTATION FOR THE RECEIVED SIGNALS AT THE TWO RECEIVE ANTENNAS

	rx antenna 0	rx antenna 1
time t	r_0	r_2
time $t + T$	r_1	r_3

Fig. 3 shows the baseband representation of the new scheme with two transmit and two receive antennas.

The encoding and transmission sequence of the information symbols for this configuration is identical to the case of a single receiver, shown in Table I. Table II defines the channels between the transmit and receive antennas, and Table III defines the notation for the received signal at the two receive antennas.

Where

$$\begin{aligned} r_0 &= h_0 s_0 + h_1 s_1 + n_0 \\ r_1 &= -h_0 s_1^* + h_1 s_0^* + n_1 \\ r_2 &= h_2 s_0 + h_3 s_1 + n_2 \\ r_3 &= -h_2 s_1^* + h_3 s_0^* + n_3 \end{aligned} \tag{14}$$

n_0, n_1, n_2, and n_3 are complex random variables representing receiver thermal noise and interference. The combiner in Fig. 3 builds the following two signals that are sent to the maximum

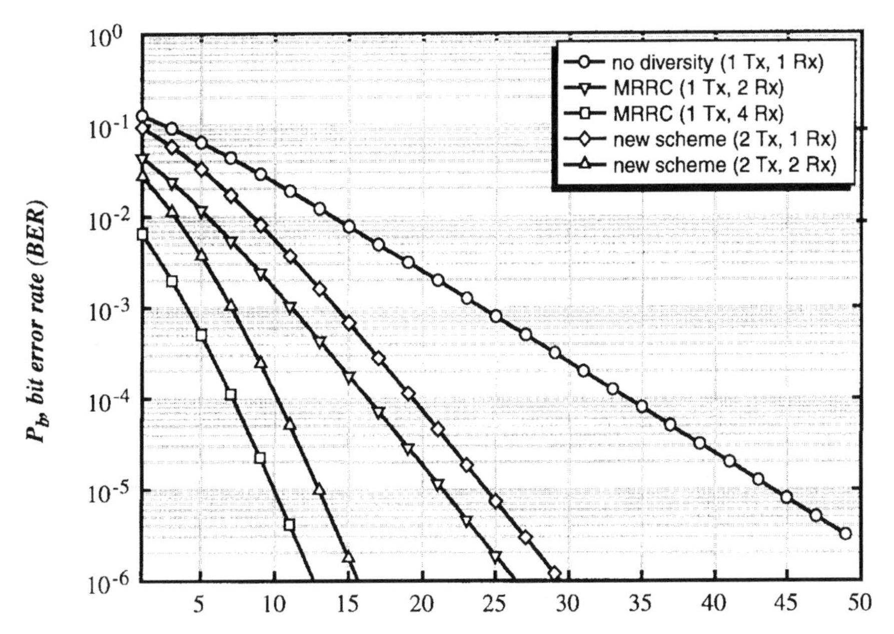

Fig. 4. The BER performance comparison of coherent BPSK with MRRC and two-branch transmit diversity in Rayleigh fading.

likelihood detector:

$$\begin{aligned} \tilde{s}_0 &= h_0^* r_0 + h_1 r_1^* + h_2^* r_2 + h_3 r_3^* \\ \tilde{s}_1 &= h_1^* r_0 - h_0 r_1^* + h_3^* r_2 - h_2 r_3^*. \end{aligned} \tag{15}$$

Substituting the appropriate equations we have

$$\begin{aligned} \tilde{s}_0 &= (\alpha_0^2 + \alpha_1^2 + \alpha_2^2 + \alpha_3^2) s_0 + h_0^* n_0 + h_1 n_1^* \\ &\quad + h_2^* n_2 + h_3 n_3^* \\ s_1 &= (\alpha_0 + \alpha_1 + \alpha_2 + \alpha_3) s_1 - h_0 n_1 + h_1 n_0 \\ &\quad - h_2 n_3^* + h_3^* n_2. \end{aligned} \tag{16}$$

These combined signals are then sent to the maximum likelihood decoder which for signal s_0 uses the decision criteria expressed in (17) or (18) for PSK signals.

Choose s_i iff

$$\begin{aligned} &(\alpha_0^2 + \alpha_1^2 + \alpha_2^2 + \alpha_3^2 - 1)|s_i|^2 + d^2(\tilde{s}_0, s_i) \\ &\le (\alpha_0^2 + \alpha_1^2 + \alpha_2^2 + \alpha_3^2 - 1|s_k|^2 + d^2(\tilde{s}_0, s_k). \end{aligned} \tag{17}$$

Choose s_i iff

$$d^2(\tilde{s}_0, s_i) \le d^2(\tilde{s}_0, s_k), \qquad \forall i \neq k. \tag{18}$$

Similarly, for s_1 using the decision rule is to choose signal s_i iff

$$\begin{aligned} &(\alpha_0^2 + \alpha_1^2 + \alpha_2^2 + \alpha_3^2 - 1)|s_i|^2 + d^2(\tilde{s}_1, s_i) \\ &\le (\alpha_0^2 + \alpha_1^2 + \alpha_2^2 + \alpha_3^2 - 1)|s_k|^2 + d^2(\tilde{s}_1, s_k) \end{aligned} \tag{19}$$

or, for PSK signals,

choose s_i iff

$$d^2(\tilde{s}_1, s_i) \le d^2(\tilde{s}_1, s_k), \qquad \forall i \neq k. \tag{20}$$

The combined signals in (16) are equivalent to that of four-branch MRRC, not shown in the paper. Therefore, the resulting diversity order from the new two-branch transmit diversity scheme with two receivers is equal to that of the four-branch MRRC scheme.

It is interesting to note that the combined signals from the two receive antennas are the simple addition of the combined signals from each receive antenna, i.e., the combining scheme is identical to the case with a single receive antenna. We may hence conclude that, using two transmit and M receive antennas, we can use the combiner for each receive antenna and then simply add the combined signals from all the receive antennas to obtain the same diversity order as $2M$-branch MRRC. In other words, using two antennas at the transmitter, the scheme doubles the diversity order of systems with one transmit and multiple receive antennas.

An interesting configuration may be to employ two antennas at each side of the link, with a transmitter and receiver chain connected to each antenna to obtain a diversity order of four at both sides of the link.

IV. Error Performance Simulations

The diversity gain is a function of many parameters, including the modulation scheme and FEC coding. Fig. 4 shows the BER performance of uncoded coherent BPSK for MRRC and the new transmit diversity scheme in Rayleigh fading.

It is assumed that the total transmit power from the two antennas for the new scheme is the same as the transmit power from the single transmit antenna for MRRC. It is also assumed that the amplitudes of fading from each transmit antenna to each receive antenna are mutually uncorrelated Rayleigh distributed and that the average signal powers at each receive antenna from each transmit antenna are the same. Further, we assume that the receiver has perfect knowledge of the channel.

Although the assumptions in the simulations may seem highly unrealistic, they provide reference performance curves for comparison with known techniques. An important issue is whether the new scheme is any more sensitive to real-world sources of degradation. This issue is addressed in Section V.

As shown in Fig. 4, the performance of the new scheme with two transmitters and a single receiver is 3 dB worse than two-branch MRRC. As explained in more detail later in Section V-A, the 3-dB penalty is incurred because the simulations assume that each transmit antenna radiates half the energy in order to ensure the same total radiated power as with one transmit antenna. If each transmit antenna in the new scheme was to radiate the same energy as the single transmit antenna for MRRC, however, the performance would be identical. In other words, if the BER was drawn against the average SNR per transmit antenna, then the performance curves for the new scheme would shift 3 dB to the left and overlap with the MRRC curves. Nevertheless, even with the equal total radiated power assumption, the diversity gain for the new scheme with one receive antenna at a BER of 10^{-4} is about 15 dB. Similarly, assuming equal total radiated power, the diversity gain of the new scheme with two receive antennas at a BER of 10^{-4} is about 24 dB, which is 3 dB worse than MRRC with one transmit antenna and four receive antennas.

As stated before, these performance curves are simple reference illustrations. The important conclusion is that the new scheme provides similar performance to MRRC, regardless of the employed coding and modulation schemes. Many publications have reported the performance of various coding and modulation schemes with MRRC. The results from these publications may be used to predict the performance of the new scheme with these coding and modulation techniques.

V. Implementation Issues

So far in this report, we have shown, mathematically, that the new transmit diversity scheme with two transmit and M receive antennas is equivalent to MRRC with one transmit antenna and $2M$ receive antennas. From practical implementation aspects, however, the two systems may differ. This section discusses some of the observed difference between the two schemes.

A. Power Requirements

The new scheme requires the simultaneous transmission of two different symbols out of two antennas. If the system is radiation power limited, in order to have the same total radiated power from two transmit antennas the energy allocated to each symbol should be halved. This results in a 3-dB penalty in the error performance. However, the 3-dB reduction of power in each transmit chain translates to cheaper, smaller, or less linear power amplifiers. A 3-dB reduction in amplifiers power handling is very significant and may be desirable in some cases. It is often less expensive (or more desirable from intermodulation distortion effects) to employ two half-power amplifiers rather than a single full power amplifier. Moreover, if the limitation is only due to RF power handling (amplifier sizing, linearity, etc.), then the total radiated power may be doubled and no performance penalty is incurred.

B. Sensitivity to Channel Estimation Errors

Throughout this paper, it is assumed that the receiver has perfect knowledge of the channel. The channel information may be derived by pilot symbol insertion and extraction [7], [8]. Known symbols are transmitted periodically from the transmitter to the receiver. The receiver extracts the samples and interpolates them to construct an estimate of the channel for every data symbol transmitted.

There are many factors that may degrade the performance of pilot insertion and extraction techniques, such as mismatched interpolation coefficients and quantization effects. The dominant source of estimation errors for narrowband systems, however, is time variance of the channel. The channel estimation error is minimized when the pilot insertion frequency is greater or equal to the channel Nyquist sampling rate, which is two times the maximum Doppler frequency. Therefore, as long as the channel is sampled at a sufficient rate, there is little degradation due to channel estimation errors. For receive diversity combining schemes with M antennas, at a given time, M independent samples of the M channels are available. With M transmitters and a single receiver, however, the estimates of the M channels must be derived from a single received signal. The channel estimation task is therefore different. To estimate the channel from one transmit antenna to the receive antenna the pilot symbols must be transmitted only from the corresponding transmit antenna. To estimate all the channels, the pilots must alternate between the antennas (or orthogonal pilot symbols have to be transmitted from the antennas). In either case, M times as many pilots are needed. This means that for the two-branch transmit diversity schemes discussed in this report, twice as many pilots as in the two-branch receiver combining scheme are needed.

C. The Delay Effects

With N branch transmit diversity, if the transformed copies of the signals are transmitted at N distinct intervals from all the antennas, the decoding delay is N symbol periods. That is, for the two-branch diversity scheme, the delay is two symbol periods. For a multicarrier system, however, if the copies are sent at the same time and on different carrier frequencies, then the decoding delay is only one symbol period.

D. Antenna Configurations

For all practical purposes, the primary requirement for diversity improvement is that the signals transmitted from the different antennas be sufficiently uncorrelated (less than 0.7 correlation) and that they have almost equal average power (less than 3-dB difference). Since the wireless medium is reciprocal, the guidelines for transmit antenna configurations are the same as receive antenna configurations. For instance, there have been many measurements and experimental results indicating that if two receive antennas are used to provide diversity at the base station receiver, they must be on the order of ten wavelengths apart to provide sufficient decorrelation. Similarly, measurements show that to get the same diversity improvement at the remote units it is sufficient to separate the antennas at the remote station by about three wavelengths.[2] This is due to the difference in the nature of the scattering environment in the proximity of the remote and base stations. The remote stations are usually surrounded by nearby scatterers, while the base station is often placed at a higher altitude, with no nearby scatterers.

Now assume that two transmit antennas are used at the base station to provide diversity at the remote station on the other side of the link. The important question is how far apart should the transmit antennas be to provide diversity at the remote receiver. The answer is that the separation requirements for receive diversity on one side of the link are identical to the requirements for transmit diversity on the other side of link. This is because the propagation medium between the transmitter and receiver in either direction are identical. In other words, to provide sufficient decorrelation between the signals transmitted from the two transmit antennas at the base station, we must have on the order of ten wavelengths of separation between the two transmit antennas. Equivalently, the transmit antennas at the remote units must be separated by about three wavelengths to provide diversity at the base station.

It is worth noting that this property allows the use of existing receive diversity antennas at the base stations for transmit diversity. Also, where possible, two antennas may be used for both transmit and receive at the base and the remote units, to provide a diversity order of four at both sides of the link.

E. Soft Failure

One of the advantages of receive diversity combining schemes is the added reliability due to multiple receive chains. Should one of the receive chains fail, and the other receive chain is operational, then the performance loss is on the order of the diversity gain. In other words, the signal may still be detected, but with inferior quality. This is commonly referred to as soft failure. Fortunately, the new transmit diversity scheme provides the same soft failure. To illustrate this, we can assume that the transmit chain for antenna one in Fig. 2 is disabled, i.e., $\boldsymbol{h}_1 = 0$. Therefore, the received signals may be described as [see (11)]

$$\begin{aligned} \boldsymbol{r}_0 &= \boldsymbol{h}_0 \boldsymbol{s}_0 + \boldsymbol{n}_0 \\ \boldsymbol{r}_1 &= -\boldsymbol{h}_0 \boldsymbol{s}_1^* + \boldsymbol{n}_1. \end{aligned} \tag{21}$$

The combiner shown in Fig. 2 builds the following two combined signals according to (12):

$$\begin{aligned} \tilde{s}_0 &= \boldsymbol{h}_0^* \boldsymbol{r}_0 = \boldsymbol{h}_0^*(\boldsymbol{h}_0 \boldsymbol{s}_0 + \boldsymbol{n}_0) = \alpha_0^2 \boldsymbol{s}_0 + \boldsymbol{h}_0^* \boldsymbol{n}_0 \\ \tilde{s}_1 &= -\boldsymbol{h}_0 \boldsymbol{r}_1^* = -\boldsymbol{h}_0(-\boldsymbol{h}_0^* \boldsymbol{s}_1 + \boldsymbol{n}_1^*) = \alpha_0^2 \boldsymbol{s}_1 - \boldsymbol{h}_0 \boldsymbol{n}_1^*. \end{aligned} \tag{22}$$

These combined signals are the same as if there was no diversity. Therefore, the diversity gain is lost but the signal may still be detected. For the scheme with two transmit and two receive antennas, both the transmit and receive chains are protected by this redundancy scheme.

[2] The separation required depends on many factors such as antenna heights and the scattering environment. The figures given apply mostly to macrocell urban and suburban environments with relatively large base station antenna heights.

F. Impact on Interference

The new scheme requires the simultaneous transmission of signals from two antennas. Although half the power is transmitted from each antenna, it appears that the number of potential interferers is doubled, i.e., we have twice the number of interferers, each with half the interference power. It is often assumed that in the presence of many interferers, the overall interference is Gaussian distributed. Depending on the application, if this assumption holds, the new scheme results in the same distribution and power of interference within the system. If interference has properties where interference cancellation schemes (array processing techniques) may be effectively used, however, the scheme may have impact on the system design. It is not clear whether the impact is positive or negative. The use of transmit diversity schemes (for fade mitigation) in conjunction with array processing techniques for interference mitigation has been studied for space-time trellis codes [9]. Similar efforts are under way to extend these techniques to the new transmit diversity scheme.

VI. Conclusions and Discussions

A new transmit diversity scheme has been presented. It is shown that, using two transmit antennas and one receive antenna, the new scheme provides the same diversity order as MRRC with one transmit and two receive antennas. It is further shown that the scheme may easily be generalized to two transmit antennas and M receive antennas to provide a diversity order of $2M$. An obvious application of the scheme is to provide diversity improvement at all the remote units in a wireless system, using two transmit antennas at the base stations instead of two receive antennas at all the remote terminals. The scheme does not require any feedback from the receiver to the transmitter and its computation complexity is similar to MRRC. When compared with MRRC, if the total radiated power is to remain the same, the transmit diversity scheme has a 3-dB disadvantage because of the simultaneous transmission of two distinct symbols from two antennas. Otherwise, if the total radiated power is doubled, then its performance is identical to MRRC. Moreover, assuming equal radiated power, the scheme requires two half-power amplifiers compared to one full power amplifier for MRRC, which may be advantageous for system implementation. The new scheme also requires twice the number of pilot symbols for channel estimation when pilot insertion and extraction is used.

Acknowledgment

The author would like to thank V. Tarokh, N. Seshadri, A. Naguib, and R. Calderbank of AT&T Labs Research for their critical feedback and encouragement; T. Lo of AT&T Wireless Services for the investigation of the antenna pattern generated by this scheme (not included in the paper but important for validation purposes); and T. Alberty of Bosch Telecom for his insightful comments.

References

[1] W. C. Jakes, Ed., *Microwave Mobile Communications.* New York: Wiley, 1974.

[2] A. Wittneben, "Base station modulation diversity for digital SIMULCAST," in *Proc. 1991 IEEE Vehicular Technology Conf. (VTC 41st)*, May 1991, pp. 848–853.
[3] A. Wittneben, "A new bandwidth efficient transmit antenna modulation diversity scheme for linear digital modulation," in *Proc. 1993 IEEE International Conf. Communications (ICC'93)*, May 1993, pp. 1630–1634.
[4] N. Seshadri and J. H. Winters, "Two signaling schemes for improving the error performance of FDD transmission systems using transmitter antenna diversity," in *Proc. 1993 IEEE Vehicular Technology Conf. (VTC 43rd)*, May 1993, pp. 508–511.
[5] J. H. Winters, "The diversity gain of transmit diversity in wireless systems with Rayleigh fading," in *Proc. 1994 ICC/SUPERCOMM*, New Orleans, LA, May 1994, vol. 2, pp. 1121–1125.
[6] V. Tarokh, N. Seshadri, and A. R. Calderbank, "Space-time codes for high data rate wireless communication: Performance criteria and code construction," *IEEE Trans. Inform. Theory*, Mar. 1998.
[7] J. K. Cavers, "An analysis of pilot symbol assisted modulation for Rayleigh fading channels," *IEEE Trans. Veh. Technol.*, vol. 40, pp. 686–693, 1991.
[8] S. Sampei and T. Sunaga, "Rayleigh fading compensation method for 16 QAM in digital land mobile radio channels," in *Proc. IEEE Vehicular Technology Conf.*, San Francisco, CA, 1989, pp. 640–646.
[9] V. Tarokh, A. Naguib, N. Seshadri, and A. R. Calderbank, "Space-time codes for wireless communication: Combined array processing and space time coding," *IEEE Trans. Inform. Theory*, Mar. 1998.

Siavash M. Alamouti received the B.S. and the M.Sc. degrees in electrical engineering from the University of British Columbia, Vancouver, Canada, in 1989 and 1991, respectively.

He has been involved in research and development activities in wireless communications since 1989. He is currently with the Alta Business Unit of Cadence Design Systems, Sunnyvale, CA, where he is a Senior Technical Leader involved in the specification and design of electronic design automation (EDA) tools for next-generation wireless communications systems. From 1995 to 1998, he was a Senior Scientist at the Strategic Technology Group of AT&T Wireless Services, Redmond, WA, where he was involved in air-link physical layer design of wireless systems. Prior to that, he was with Mobile Data Solutions, Inc. (MDSI), Richmond, Canada, and for three years was at MPR Teltech, Vancouver, Canada, where he was a Member of the Technical Staff involved in the design of physical and MAC layers for proprietary mobile data systems. His areas of interest include smart antenna techniques, coding and modulation, and physical and MAC layer design of narrowband and wideband wireless communication systems.

A Space–Time Coding Modem for High-Data-Rate Wireless Communications

Ayman F. Naguib, *Member, IEEE,* Vahid Tarokh, *Member, IEEE,*
Nambirajan Seshadri, *Senior Member, IEEE,* and A. Robert Calderbank, *Fellow, IEEE*

***Abstract*— This paper presents the theory and practice of a new advanced modem technology suitable for high-data-rate wireless communications and presents its performance over a frequency-flat Rayleigh fading channel. The new technology is based on *space–time coded modulation* (STCM) [1]–[5] with multiple transmit and/or multiple receive antennas and *orthogonal pilot sequence insertion* (O-PSI). In this approach, data is encoded by a space–time (ST) channel encoder and the output of the encoder is split into N streams to be simultaneously transmitted using N transmit antennas. The transmitter inserts periodic orthogonal pilot sequences in each of the simultaneously transmitted bursts. The receiver uses those pilot sequences to estimate the fading channel. When combined with an appropriately designed interpolation filter, accurate channel state information (CSI) can be estimated for the decoding process. Simulation results of the proposed modem, as applied to the IS-136 cellular standard, are presented. We present the *frame error rate* (FER) performance results as a function of the signal-to-noise ratio (SNR) and the maximum Doppler frequency, in the presence of timing and frequency offset errors. Simulation results show that for 10% FER, a 32-state eight-phase-shift keyed (8-PSK) ST code with two transmit and two receive antennas can support data rates up to 55.8 kb/s on a 30-kHz channel, at an SNR of 11.7 dB and a maximum Doppler frequency of 180 Hz. Simulation results for other codes and other channel conditions are also provided. We also compare the performance of the proposed STCM scheme with delay-diversity schemes and conclude that STCM can provide significant SNR improvement over simple delay diversity.**

***Index Terms*— Coded modulation, space–time (ST) coding, space–time processing, wireless communications.**

I. Introduction

THE realization of wireless communications, providing high data rate and high quality information exchange between two portable terminals that may be located anywhere in the world, and the vision of a new telephone service based on a single phone that acts as a traditional cellular phone when used outdoors and as a conventional high-quality phone when used indoors [6] has been the new communication challenge in recent years and will continue to be for years to come. The great popularity of cordless phones, cellular phones, radio paging, portable computing, and other personal communication services (PCS's) demonstrates the rising demand for these services. Rapid growth in mobile computing and other wireless data services is inspiring many proposals for high-speed data services in the range of 64–144 kb/s for a microcellular-wide area and high-mobility applications and up to 2 Mb/s for indoor applications [7]. Research challenges in this area include the development of efficient coding and modulation and signal processing techniques to improve the quality and spectral efficiency of wireless communications and better techniques for sharing the limited spectrum among different high-capacity users.

The physical limitations of the wireless channel presents a fundamental technical challenge for reliable communications. The channel is susceptible to time-varying impairments such as noise, interference, and multipath. Limitations on the power and size of the communications and computing devices in a mobile handset are a second major design consideration. Most personal communications and wireless services portables are meant to be carried in a briefcase and/or pocket and must, therefore, be small and lightweight, which translates to a low power requirement since small batteries must be used. Many of the signal processing techniques which may be used for reliable communications and efficient spectral utilization, however, demand significant processing power, precluding the use of low-power devices. Continuing advances in very large scale integration (VLSI) and integrated circuit technology for low power applications will provide a partial solution to this problem. Hence, placing a higher signal processing burden on fixed locations (base stations), with relatively larger power resources than the portables, makes good engineering sense.

Perhaps the single most important factor in providing reliable communications over wireless channels is diversity. Diversity techniques which may be used include time, frequency, and space diversity.

- *Time diversity:* Channel coding in combination with limited interleaving is used to provide time diversity. However, while channel coding is extremely effective in fast-fading environments (high mobility), it offers very little protection under slow fading (low mobility) unless significant interleaving delays can be tolerated.
- *Frequency diversity:* The fact that signals transmitted over different frequencies induce different multipath structures and independent fading is exploited to provide frequency diversity (sometimes referred to as path diversity). In time division multiple access (TDMA) systems, frequency diversity is obtained by the use of equalizers [8] when the multipath delay spread is a significant fraction of a symbol period. The global system for mobile communications (GSM) uses frequency hopping to provide frequency

Manuscript received October 30, 1997; revised March 30, 1998. This paper was presented in part at IEEE GLOBECOM'97, Phoenix, AZ.
The authors are with AT&T Labs-Research, Florham Park, NJ 07932 USA.
Publisher Item Identifier S 0733-8716(98)07895-0.

Reprinted from *IEEE Journal on Selected Areas in Communications,* vol. 16, no. 8, October 1998.

diversity. In direct sequence code division multiple access (DS-CDMA) systems, RAKE receivers [9], [10] are used to obtain path diversity. When the multipath delay spread is small, as compared to the symbol period, however, frequency or path diversity does not exist.

- *Space diversity:* The receiver/transmitter uses multiple antennas that are separated for reception/transmission and/or differently polarized antennas to create independent fading channels. Currently, multiple antennas at base stations are used for receive diversity at the base. It is difficult, however, to have more than one or two antennas at the portable unit due to the size limitations and cost of multiple chains of RF down conversion.

In this paper we present the theory and practice of a new advanced modem technology suitable for high-data-rate wireless communications based on *space–time coded modulation* (STCM) with multiple transmit antennas [1]–[5] and *orthogonal pilot sequences insertion* (O-PSI). At the transmitter, each block of data is first optionally encoded using a high-rate Reed Solomon (RS) block encoder followed by a *space–time* (ST) channel encoder. The spatial and temporal properties of STCM guarantee that diversity is achieved at the transmitter, while maintaining optional receive diversity, without any sacrifice in transmission rate. The output of the ST encoder is split into N streams that are simultaneously transmitted using N transmit antennas. Each stream of encoded symbols is then independently interleaved, using a block symbol-by-symbol interleaver. The transmitter inserts periodic orthogonal pilot sequences in each one of the simultaneously transmitted blocks. Each block is then pulse-shaped and transmitted from a different antenna. Since the signal at each receive antenna is a linear superposition of the N transmitted signals, the receiver uses the orthogonal pilot sequences to estimate the different fading channels. The receiver then uses an appropriately designed interpolation filter to interpolate those estimates and obtain accurate channel state information (CSI). The interpolated channel estimates, along with the received samples, are then deinterleaved using a block symbol-by-symbol deinterleaver and passed to a vector maximum likelihood sequence decoder, followed by an RS decoder.

The information theoretic aspects of transmit diversity were addressed in [13]–[16]. Previous work on transmit diversity can be classified into three broad categories: schemes using feedback; schemes with feedforward or training information but no feedback; and blind schemes. The first category uses feedback, either explicitly or implicitly, from the receiver to the transmitter to train the transmitter. For instance, in time division duplex (TDD) systems [11], the same antenna weights are used for reception and transmission so that feedback is implicit in the exploitation of channel symmetry. These weights are chosen during reception to maximize the received signal-to-noise ratio (SNR) and, during transmission, to weight the amplitudes of the transmitted signals. Therefore, this will also maximize the SNR at the portable receiver. Explicit feedback includes switched diversity systems with feedback [12]. In practice, however, vehicle movement and interference dynamics cause a mismatch between the channel perceived by the transmitter and that perceived by the receiver.

Transmit diversity schemes mentioned in the second category use linear processing at the transmitter to spread the information across antennas. At the receiver, information is recovered by an optimal receiver. Feedforward information is required to estimate the channel from the transmitter to the receiver. These estimates are used to compensate for the channel response at the receiver. The first scheme of this type was proposed by Wittneben [17] and it includes the delay-diversity scheme of [18] as a special case. The linear processing techniques were also studied in [19] and [20]. It was shown in [21] and [22] that delay-diversity schemes are indeed optimal in providing diversity, in the sense that the diversity gain experienced at the receiver (which is assumed to be optimal) is equal to the diversity gain obtained with receive diversity. The linear filtering used at the transmitter can be viewed as a channel code that takes binary or integer input and creates real valued output. This paper shows that there is a significant gain to be realized by viewing this problem from a coding perspective, rather than from a purely signal processing point of view.

The third category does not require feedback or feedforward information. Instead, it uses multiple transmit antennas combined with channel coding to provide diversity. An example of this approach is the use of channel coding along with phase sweeping [23] or frequency offset [24] with multiple transmit antennas to simulate fast fading. An appropriately designed channel code/interleaver pair is used to provide the diversity benefit. Another approach in this category is to encode information by a channel code and transmit the code symbols, using different antennas, in an orthogonal manner. This can be done by either time multiplexing [23], or by using orthogonal spreading sequences for different antennas [24]. The disadvantage of these schemes, as compared to the previous two categories, is the loss in bandwidth efficiency due to the use of the channel code. Using appropriate coding it is possible to relax the orthogonality requirement needed in these schemes and to obtain the diversity, as well as a coding gain, without sacrificing bandwidth. This will be possible if one views the whole system as a multiple input/multiple output system and uses channel codes that are designed with that view in mind.

Pilot symbol insertion (PSI) has been used to obtain channel estimates for coherent detection and for decoding channel codes over fast flat-fading channels [26]–[32]. The advantage of the PSI technique is that it neither requires complex signal processing nor does it increase the peak factor of the modulated carrier. In [27] through [29] applications and implementations of PS-aided coherent modems are presented. In [26] and [31], the performance of PS-aided coherent modems is studied by theoretical analysis.

The organization of this paper is as follows. In Section II we briefly review the theory of STCM. The reader is referred to [1]–[5] for a detailed treatment of the theory. We present two specific ST codes based on eight-phase-shift keyed (8-PSK) and 16-QAM signaling constellations. We also present an ST code representation for the delay-diversity scheme based on the 8-PSK constellation. These ST codes, as well as the delay-diversity code, will be used in the simulations.

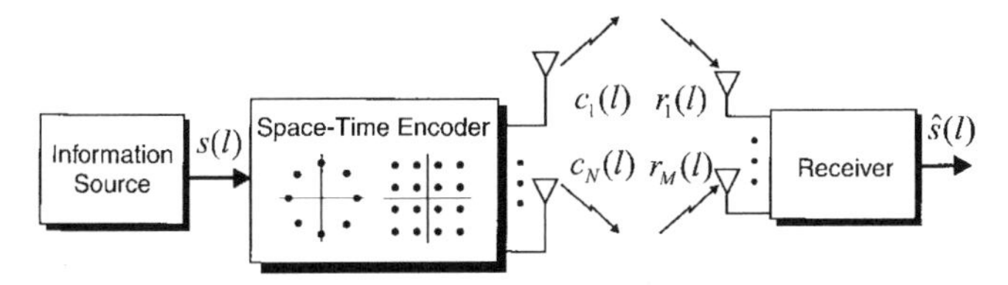

Fig. 1. ST coding.

In Section III, an STCM-based modem architecture and its different signal processing blocks is described. Simulation results for the proposed modem based on 32-state 8-PSK and 16-state 16-quadrature amplitude modulation ST (QAM ST) codes are presented in Section IV. The frame error rate (FER) performance as a function of SNR and maximum Doppler frequency, as well as the effects of antenna correlation and interpolation filter on the FER performance, are examined. In addition, the performance of the 32-state 8-PSK ST code is compared to the performance of the delay-diversity scheme with an 8-PSK constellation. Finally, Section V includes our conclusions and remarks.

II. Space-Time (ST) Coding

In this section we will describe a basic model for a communication system that employs ST coding with N transmit antennas and M receive antennas. As shown in Fig. 1, the information symbol $s(l)$ at time l is encoded by the ST encoder as N code symbols $c_1(l), c_2(l), \quad , c_N(l)$. Each code symbol is transmitted *simultaneously* from a different antenna. The encoder chooses the N code symbols to transmit, so that both the coding gain and diversity gain are maximized.

Signals arriving at different receive antennas undergo independent fading. The signal at each receive antenna is a noisy superposition of the faded versions of the N transmitted signals. A flat-fading channel is assumed. Let E_s be the average energy of the signal constellation. The constellation points are scaled by a factor of $\sqrt{E_s}$ such that the average energy of the constellation points is 1. Let $r_j(l), j = 1 \cdots M$ be the received signal at antenna j after matched filtering. Assuming ideal timing and frequency information, we have

$$r_j(l) = \sqrt{E_s} \cdot \sum_{i=1}^{N} \alpha_{ij}(l) c_i(l) + \eta_j(l), \quad j = 1, \cdots, M \quad (1)$$

where $\eta_j(l)$ are independent samples of a zero-mean complex white Gaussian process with two-sided power spectral density $N_0/2$ per dimension. It is also assumed that $\eta_j(l)$ and $\eta_k(l)$ are independent for $j \neq k, 1 \leq j, k \leq M$. The gain $\alpha_{ij}(l)$ models the complex fading channel gain from transmit antenna i to receive antenna j. The channel gain α_{ij} is modeled as a low-pass filtered complex Gaussian random process with zero-mean, variance one, and autocorrelation function $R_\alpha(\tau) = J_0(2\pi f_d \tau)$, where $J_0(\cdot)$ is the zeroth-order bessel function of the first kind and f_d is the maximum Doppler frequency [33]. It is also assumed that $\alpha_{ij}(l)$ and $\alpha_{qk}(l)$ are independent for $i \neq q$ or $j \neq k, 1 \leq i, q \leq N, 1 \leq j, k \leq M$. This condition is satisfied if the transmit antennas are well separated (by more than $\lambda/2$) or by using antennas with different polarization.

Let $\boldsymbol{c}_l = [c_1(l), c_2(l), \cdots, c_N(l)]^T$ be the $N \times 1$ code vector transmitted from the N antennas at time l, $\boldsymbol{\alpha}_j(l) = [\alpha_{1j}(l), \alpha_{2j}(l), \cdots, \alpha_{Nj}(l)]^T$ be the corresponding $N \times 1$ channel vector from the N transmit antennas to the jth receive antenna, and $\boldsymbol{r}(l) = [r_1(l), r_2(l), \cdots, r_M(l)]^T$ be the $M \times 1$ received signal vector. Also, let $\boldsymbol{\eta}(l) = [\eta_1(l), \eta_2(l), \cdots, \eta_M(l)]^T$ be the $M \times 1$ noise vector at the receive antennas. Let us define the $M \times N$ channel matrix $\mathcal{H}_l$ from the N transmit to the M receive antennas as $\mathcal{H}(l) = [\boldsymbol{\alpha}_1(l), \boldsymbol{\alpha}_2(l), \cdots, \boldsymbol{\alpha}_M(l)]^T$. Equation (1) can be rewritten in a matrix form as

$$\boldsymbol{r}(l) = \sqrt{E_s} \cdot \mathcal{H}(l) \cdot \boldsymbol{c}_l + \boldsymbol{\eta}(l). \quad (2)$$

We can easily see that the SNR *per receive antenna* is given by

$$\text{SNR} = \frac{N \cdot E_s}{N_o}. \quad (3)$$

A. Performance Criterion

Suppose that the *code vector* sequence

$$\mathcal{C} = \boldsymbol{c}_1, \boldsymbol{c}_2, \cdots, \boldsymbol{c}_L$$

was transmitted. We consider the probability that the decoder decides erroneously in favor of the legitimate code vector sequence

$$\tilde{\mathcal{C}} = \tilde{\boldsymbol{c}}_1, \tilde{\boldsymbol{c}}_2, \cdots, \tilde{\boldsymbol{c}}_L.$$

Assuming that for each frame or block of data of length the ideal CSI $\mathcal{H}(l), l = 1, \cdots, L$ are available at the receiver, the probability of transmitting $\mathcal{C}$ and deciding in favor of $\tilde{\mathcal{C}}$ is well upper bounded by [34]

$$P(\mathcal{C} \rightarrow \tilde{\mathcal{C}} | \mathcal{H}(l), l = 1, \cdots, L) = Q\left(\sqrt{\frac{\mathcal{D}^2(\mathcal{C}, \tilde{\mathcal{C}}) E_s}{2N_o}}\right) \quad (4)$$

$$\leq \exp\left(-\mathcal{D}^2(\mathcal{C}, \tilde{\mathcal{C}}) \cdot E_s / 4N_o\right) \quad (5)$$

where $Q(x) = (1/\sqrt{2\pi}) \int_x^\infty \exp(-x^2/2)\, dx$ and

$$\mathcal{D}^2(\mathcal{C}\ \tilde{\mathcal{C}}) \quad \sum_{l=1}^{L} \|\mathcal{H}(l)(\quad \tilde{} \quad)\|^2$$

It is clear that in order to minimize the pairwise error probability we need to maximize $\mathcal{D}^2(\mathcal{C}, \tilde{\mathcal{C}})$ (with the proper design of the ST code). It is clear, however, that $\mathcal{D}^2(\mathcal{C}, \tilde{\mathcal{C}})$ is a function of the maximum Doppler frequency. Therefore, we will derive the performance criterion for designing the ST code, assuming that the fading is static over the block. In this case $\mathcal{H}(l) = \mathcal{H} = [\boldsymbol{\alpha}_1, \boldsymbol{\alpha}_2, \cdots, \boldsymbol{\alpha}_M]^T, l = 1, \cdots, L$ and we can easily verify that

$$\mathcal{D}^2(\mathcal{C}, \tilde{\mathcal{C}}) = \sum_{j=1}^{M} \boldsymbol{\alpha}_j^* \mathcal{A}(\mathcal{C}, \tilde{\mathcal{C}}) \boldsymbol{\alpha}_j \quad (6)$$

where

$$\mathcal{A}(\mathcal{C}, \tilde{\mathcal{C}}) = \sum_{l=1}^{L} (\boldsymbol{c}_l - \tilde{\boldsymbol{c}}_l)(\boldsymbol{c}_l - \tilde{\boldsymbol{c}}_l)^*. \quad (7)$$

We can also verify that the $N \times N$ matrix $\mathcal{A}(\mathcal{C},\tilde{\mathcal{C}})$ is Hermitian and is equal to $\mathcal{B}(\mathcal{C},\tilde{\mathcal{C}})\mathcal{B}^*(\mathcal{C},\tilde{\mathcal{C}})$ where $\mathcal{B}(\mathcal{C},\tilde{\mathcal{C}})$ is $N \times L$ and represents the error sequence $\mathcal{C} - \tilde{\mathcal{C}}$. The matrix $\mathcal{B}$ is a square root of $\mathcal{A}$. Since $\mathcal{A}$ is Hermitian we can write $\mathcal{A}$ as $\boldsymbol{U}\boldsymbol{\Lambda}\boldsymbol{U}^*$ [35] where $\boldsymbol{U}$ is unitary[1] and $\boldsymbol{\Lambda}$ is a diagonal matrix where the diagonal elements $\lambda_n, n = 1, \cdots, N$ are nonnegative and are the eigenvalues of $\mathcal{A}$. Therefore, we can write $\mathcal{D}^2(\mathcal{C},\tilde{\mathcal{C}})$ as

$$\mathcal{D}^2(\mathcal{C},\tilde{\mathcal{C}}) = \sum_{j=1}^{M} \boldsymbol{\beta}_j^* \boldsymbol{\Lambda} \boldsymbol{\beta}_j \tag{8}$$

where $\boldsymbol{\beta}_j = \boldsymbol{U}^* \boldsymbol{\alpha}_j$. Since $\boldsymbol{U}$ is unitary and $\boldsymbol{\alpha}_j$ is a complex Gaussian random vector with zero mean and covariance $\boldsymbol{I}$, then $\boldsymbol{\beta}_j$ will be also a complex Gaussian random vector with zero mean and covariance $\boldsymbol{I}$. Hence, we will have

$$\mathcal{D}^2(\mathcal{C},\tilde{\mathcal{C}}) = \sum_{j=1}^{M} \sum_{i=1}^{N} \lambda_i |\beta_{ij}|^2. \tag{9}$$

The random variable $\nu_{ij} = |\beta_{ij}|^2$ has a χ^2 distribution with two degrees of freedom, that is

$$\nu_{ij} \sim f_\nu(\nu) = e^{-\nu} \quad \text{for} \quad \nu > 0 \quad \text{and 0 otherwise.} \tag{10}$$

Thus, to compute an upper bound on the average pairwise error probability we simply average the right-hand side of (5) to arrive at

$$P(\mathcal{C} \to \tilde{\mathcal{C}}) \leq \left(\prod_{i=1}^{N} \frac{1}{1 + \lambda_i \cdot (E_s/4N_o)} \right)^M. \tag{11}$$

Let r denote the rank of the matrix $\mathcal{A}$ (which is also equal to the rank of $\mathcal{B}$). Then $\mathcal{A}$ has exactly $N - r$ zero eigenvalues. Without loss of generality, let us assume that $\lambda_1, \lambda_2, \cdots, \lambda_r$ are the nonzero eigenvalues, then it follows from (11) that

$$P(\mathcal{C} \to \tilde{\mathcal{C}}) \leq \left(\prod_{i=1}^{r} \lambda_i \right)^{-M} \cdot (E_s/4N_o)^{-rM}. \tag{12}$$

We can easily see that the probability of error bound in (12) is similar to the probability of error bound for trellis coded modulation for fading channels and, thus, a diversity gain of rM and a coding gain of $g_r = (\lambda_1 \lambda_2 \cdots \lambda_r)^{1/r}$ are achieved [36]. From the above analysis, we arrive at the following design criteria.

- *The Rank Criterion:* In order to achieve the maximum diversity NM, the rank of the matrix $\mathcal{B}(\mathcal{C},\tilde{\mathcal{C}})$ has to be full rank for any two code vector sequences $\mathcal{C}$ and $\tilde{\mathcal{C}}$. If $\mathcal{B}(\mathcal{C},\tilde{\mathcal{C}})$ has a minimum rank r over the set of two tuples of distinct code vector sequences, then a diversity of rM is achieved.
- *The Determinant Criterion:* Suppose that a diversity benet rM is our target. The minimum of $g_r = (\lambda_1 \lambda_2 \cdots \lambda_r)^{1/r}$ taken over all pairs of distinct code vector sequences $\mathcal{C}$ and $\tilde{\mathcal{C}}$ is the coding gain. The design target is to maximize g_r.

[1] An $n \times n$ matrix $\boldsymbol{U}$ is unitary if and only if $\boldsymbol{U}\boldsymbol{U}^* = \boldsymbol{I}$.

B. Maximum Likelihood Vector Decoder

As before, we assume that the ideal CSI $\mathcal{H}(l), l = 1, \cdots, L$ are available at the receiver. We can derive the maximum likelihood decoding rule for the ST code as follows. Suppose that a code vector sequence

$$\mathcal{C} = \boldsymbol{c}_1, \boldsymbol{c}_2, \cdots, \boldsymbol{c}_L$$

has been transmitted, and

$$\mathcal{R} = \boldsymbol{r}_1, \boldsymbol{r}_2, \cdots, \boldsymbol{r}_L$$

has been received, where $\boldsymbol{r}_l$ is given by (2). At the receiver, optimum decoding amounts to choosing a vector code sequence

$$\tilde{\mathcal{C}} = \tilde{\boldsymbol{c}}_1, \tilde{\boldsymbol{c}}_2, \cdots, \tilde{\boldsymbol{c}}_L$$

for which the *a posteriori* probability

$$\Pr(\tilde{\mathcal{C}} | \mathcal{R}, \mathcal{H}(l), l = 1, \cdots, L)$$

is maximized. Assuming that all the code words are equiprobable, and since the noise vector is assumed to be a multivariate allitive white Gaussian noise (AWGN), it can be easily shown that the optimum decoder is [34]

$$\tilde{\mathcal{C}} = \arg \min_{\mathcal{C} = \tilde{\boldsymbol{c}}_1, \cdots, \tilde{\boldsymbol{c}}_L} \sum_{l=1}^{L} ||\boldsymbol{r}(l) - \sqrt{E_s} \cdot \mathcal{H}(l) \cdot \tilde{\boldsymbol{c}}_l||^2. \tag{13}$$

It is obvious that the optimum decoder in (13) can be implemented using the Viterbi algorithm when the ST code has a trellis representation. In practice, the receiver has to estimate the CSI, and techniques to accurately estimate the multichannel CSI for STCM will be discussed later. CSI estimation errors, however, will limit the performance of STCM. In this case, let $\hat{\mathcal{H}}(l)$ denote the CSI estimate at time l such that

$$\hat{\mathcal{H}}(l) = \mathcal{H}(l) + \boldsymbol{\Delta}_{\mathcal{H}}(l) \tag{14}$$

where the error matrix $\boldsymbol{\Delta}_{\mathcal{H}}(l)$ represents the error in the CSI estimates. The (i, j) element of $\boldsymbol{\Delta}_{\mathcal{H}}(l), e_{ij}(l)$ represents the error in the estimate of the channel gain $\alpha_{ij}(l)$. Since these channels are assumed to be independent, the $e_{ij}(l)$'s are also independent and are modeled as identically distributed Gaussian random variables with zero mean and variance σ_e^2. Moreover, we will also assume that $\{\boldsymbol{\Delta}_{\mathcal{H}}(l)\}_{l=1,\cdots,L}$ are independent. This is true if we assume innite interleaving depth. For a nite block length, however, these errors will be correlated. In this case we have

$$\boldsymbol{r}_l = \sqrt{E_s} \hat{\mathcal{H}}(l) \boldsymbol{c}_l + \tilde{\boldsymbol{\eta}}(l) \tag{15}$$

where

$$\tilde{\boldsymbol{\eta}}(l) = \boldsymbol{\eta}(l) - \sqrt{E_s} \boldsymbol{\Delta}_{\mathcal{H}}(l) \boldsymbol{c}_l.$$

We can easily verify that $\tilde{\boldsymbol{\eta}}(l)$ is a zero-mean Gaussian random vector with covariance $\Sigma(l) \cdot \boldsymbol{I}$ where

$$\Sigma(l) = N_o + \sqrt{E_s} \sigma_e^2 ||\boldsymbol{c}_l||^2. \tag{16}$$

In this case, and conditioned on $\hat{\mathcal{H}}(l)$, the log likelihood to be minimized for optimum decoding is given by

$$\sum_{l=1}^{L} \{\Sigma^{-1}(l) \cdot ||\boldsymbol{r}(l) - \sqrt{E_s} \cdot \hat{\mathcal{H}}(l) \cdot \boldsymbol{c}_l||^2 + \log \Sigma(l)\}. \tag{17}$$

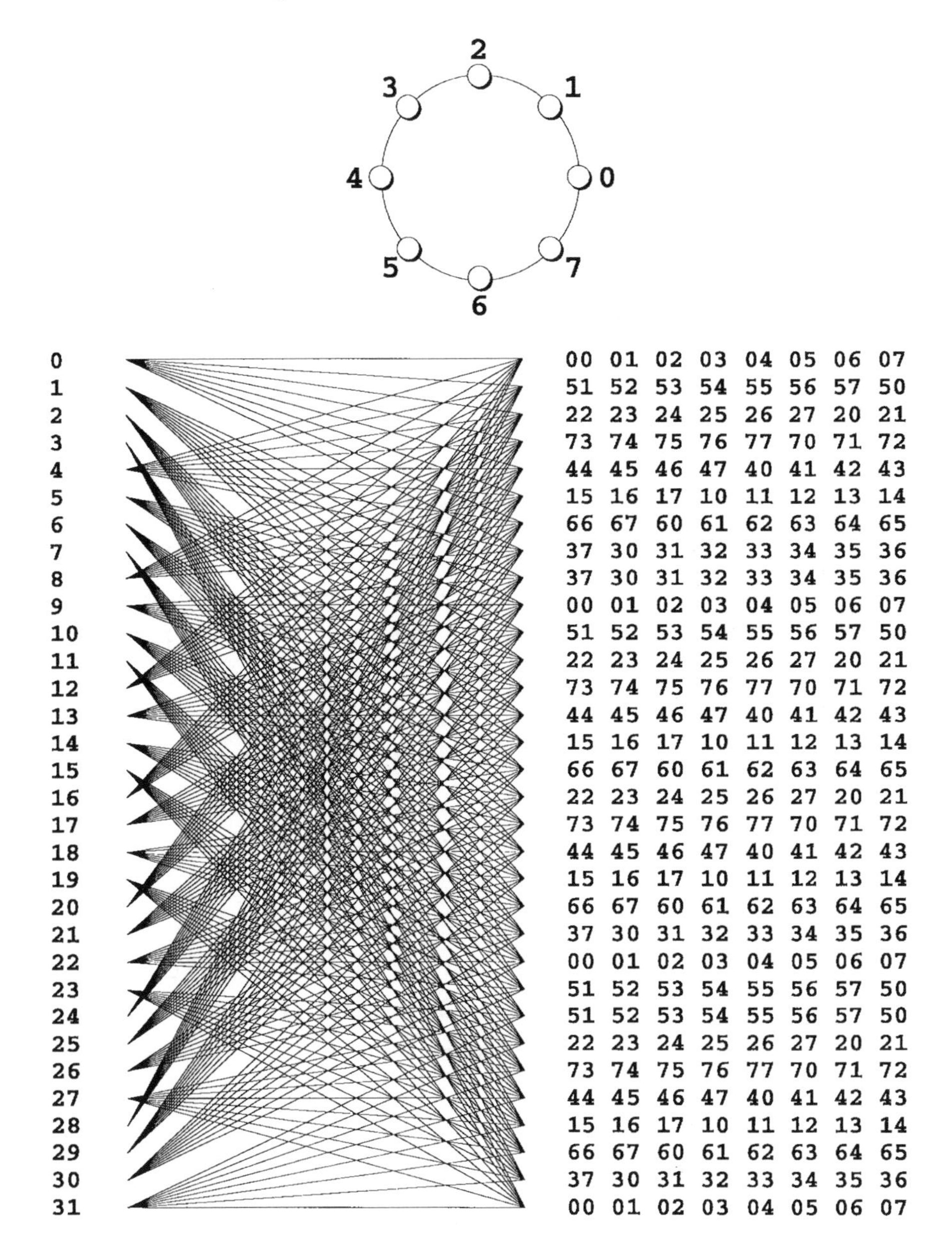

Fig. 2. 8-PSK 32-state ST code with two transmit antennas.

For the case of constant envelope signals such as PSK, $\Sigma(l)$ does not depend on the transmitted code vector c_l and, therefore, the metric in (17) reduces to that in (13), replacing $\mathcal{H}(l)$ with the CSI estimate $\hat{\mathcal{H}}(l)$. This means that the decoding rule in (13) is still optimum for equal energy constellation, e.g., PSK [5], even in the presence of channel estimation errors. For QAM signals, however, this will be true only if we have ideal CSI, or when the channel estimation error is negligible compared to the channel noise, i.e., $E_S\sigma_e^2||c_l||^2 \ll N_o$.

C. Examples of ST Codes

Here we give two examples of ST codes that were designed, using the above criteria, for two transmit antennas. The reader is referred to [1] for further examples of ST codes.

- *Example 1:* Here we provide an 8-PSK 32-state ST code designed for two transmit antennas. Consider the 8-PSK constellation as labeled in Fig. 2. Fig. 2 also shows the trellis description for this code. Each row in the matrix represents the edge labels for transitions from the corresponding state. The edge label s_1s_2 indicates that symbol s_1 is transmitted over the rst antenna and that symbol s_2 is transmitted over the second antenna. The input bit stream to the ST encoder is divided into groups of three bits, and each group is mapped into one of eight constellation points. This code has a bandwidth ef ciency of 3 bits/channel use.
- *Example 2:* Here we provide a 16-QAM 16-State ST code designed for two transmit antennas. Consider the 16-QAM constellation, as labeled in Fig. 3, using hexadecimal notation. Fig. 3 also shows the trellis description for this code. The input bit stream to the ST encoder is divided into groups of four bits and each group is mapped into one of 16 constellation points. This code has a bandwidth ef ciency of 4 bits/channel use.

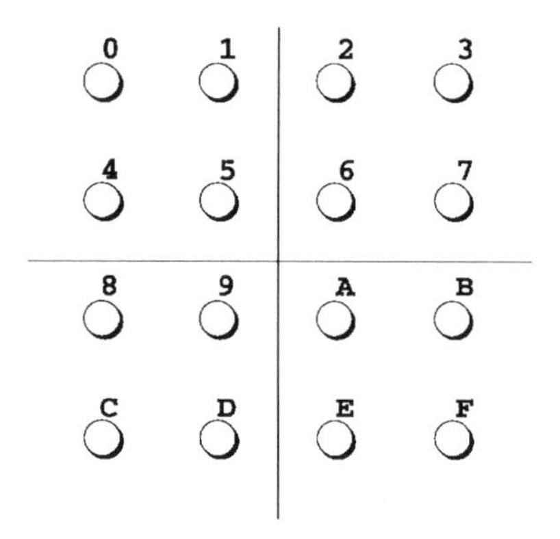

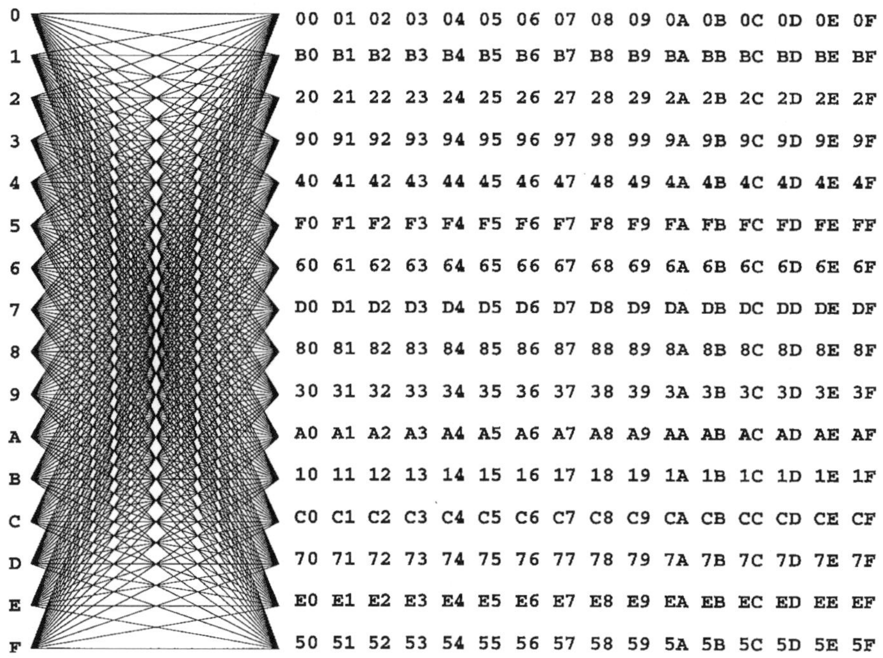

Fig. 3. Sixteen-QAM 16-state ST code with two transmit antennas.

D. Comparison with Delay Diversity

We observe that the delay-diversity scheme of [18] and [19] can be viewed as an ST code and, therefore, the performance analysis presented above applies to it. Consider the delay-diversity scheme of [18] and [19] where the channel encoder is a rate 1/2 block repetition code defined over some signal alphabet. Let $\bar{c}_1(l)\bar{c}_2(l)$ be the output of the channel encoder where $\bar{c}_1(l)$ is to be transmitted from antenna 1 and $\bar{c}_2(l)$ is to be transmitted from antenna two, one symbol later. This can be viewed as an ST code by defining the *code vector* $\mathbf{c}(l)$ as

$$c_l = \begin{pmatrix} \bar{c}_1(l) \\ \bar{c}_2(l-1) \end{pmatrix}. \tag{18}$$

Now, let us consider the 8-PSK constellation in Fig. 4. It is easy to show that the ST code realization of this delay-diversity scheme has the trellis representation in Fig. 4. The minimum determinant of this code is $(2-\sqrt{2})^2$.

Next, consider the block code

$$\mathcal{C} = \{00, 15, 22, 37, 44, 51, 66, 73\} \tag{19}$$

of length two defined over the 8-PSK alphabet instead of the repetition code. This block code is the best, in the sense of product distance [18], among all the codes of cardinality eight and of length two, defined over the 8-PSK alpha-

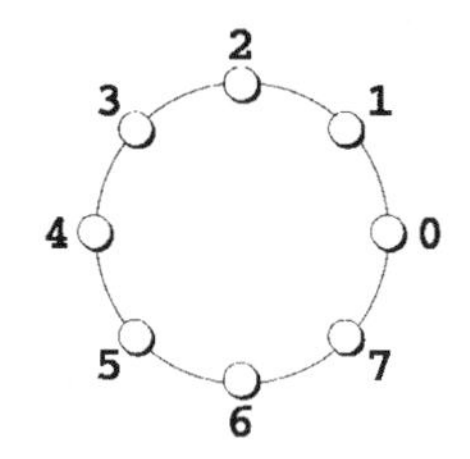

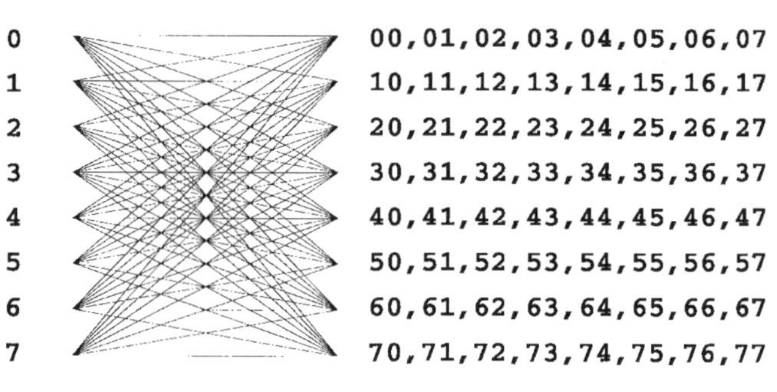

Fig. 4. ST coding realization of a delay-diversity 8-PSK eight-state code with two transmit antennas.

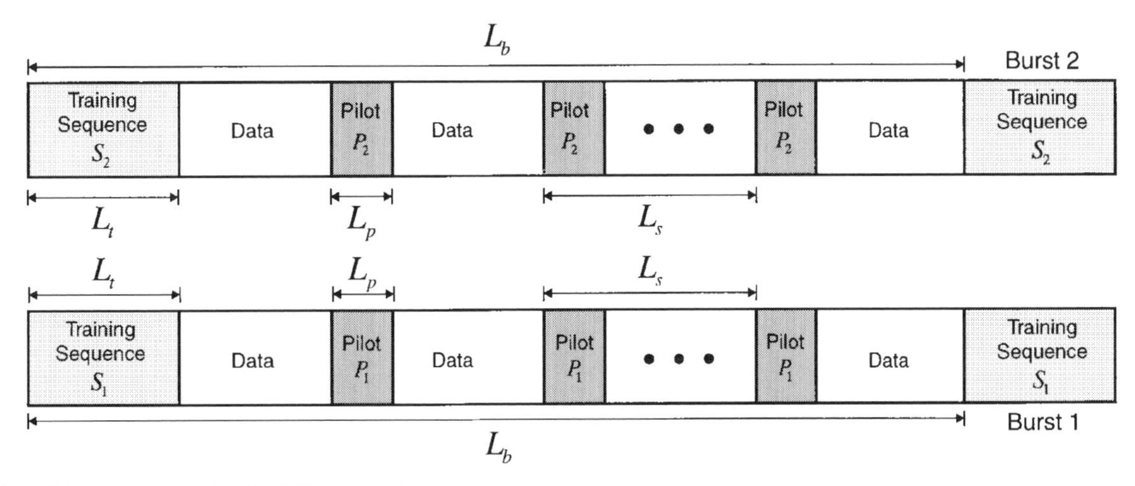

Fig. 5. Downlink slot structure for STCM-based modem.

bet. This means that the minimum of the product distance $|c_1 - \tilde{c}_1||c_2 - \tilde{c}_2|$ between pairs of distinct code words $\boldsymbol{c} = c_1 c_2 \in \boldsymbol{C}$ and $\tilde{\boldsymbol{c}} = \tilde{c}_1\tilde{c}_2 \in \boldsymbol{C}$ is the maximum among all such codes. The delay-diversity code constructed from this repetition code is identical to the 8-PSK eight-state ST code [1]. The minimum determinant of this code is two. Simulation results in Section IV will show an advantage of up to 9 dB for the proposed ST coded modulation scheme with the 8-PSK 32-state ST code, over the delay-diversity code with 8-PSK (obtained by the use of repetition code).

III. System Architecture

In this section, we will present a general architecture for a narrowband TDMA/STCM-based modem with N transmit antennas suitable for wireless communications. Without loss of generality, we will assume that $N = 2$. For brevity, we will also present the modem architecture for the downlink only. The uplink modem will have a similar architecture, except that the framing and timing structure will be different and must allow for a guard time between different asynchronous (due to difference in propagation delay) bursts from different users. The transmit antennas are assumed to be placed far enough apart so that each transmit signal will experience independent fading. Independent fading may be also obtained by the use of two dually polarized transmit antennas.

A. Timing and Framing Structure

The system architecture that we propose is similar, but not identical, to that of the IS-136 US cellular standard. Let W be the bandwidth of each of the frequency channels, R_s be the raw symbol rate, N_f be the number of TDMA frames per second for each frequency channel, and N_s be the number of time slots per TDMA frame. Fig. 5 shows the basic TDMA time slot structure. A signaling format which interleaves training and synchronization sequences, pilot sequences, and data is used. In each TDMA slot two bursts are transmitted, one from each antenna. Each burst is L_b symbols long and begins with a training sequence of length L_t symbols. The training sequences $\boldsymbol{S}_1$ and $\boldsymbol{S}_2$ will be used for timing and frequency synchronization at the receiver. In addition, every L_s symbols, the transmitter inserts N_p pilot sequences $\boldsymbol{P}_1$ and $\boldsymbol{P}_2$, each, of length L_p symbols. The length of the pilot sequences L_p should be at least equal to the number of transmit antennas N. In addition, we may note that the symbols used for pilots do not necessarily belong to the same symbol alphabet used for sending the information symbols. Without loss of generality, we will assume that the pilot and synchronization symbols are taken from a constant envelope constellation (8-PSK or $\pi/4$-shifted differential PSK (DPSK), for example). The pilot sequences $\boldsymbol{P}_1$ and $\boldsymbol{P}_2$, along with the training sequences $\boldsymbol{S}_1$ and $\boldsymbol{S}_2$, will be used at the receiver to estimate the channel from each of the transmit antennas to the receiver. *In general, with N transmit antennas we will have N different synchronization sequences* $\boldsymbol{S}_1, \boldsymbol{S}_2, \cdots, \boldsymbol{S}_N$, *and N different pilot sequences* $\boldsymbol{P}_1, \boldsymbol{P}_2, \cdots, \boldsymbol{P}_N$. Since signals at the receiver antennas will be linear superpositions of all transmitted signals, we choose the training sequences $\boldsymbol{S}_1$ and $\boldsymbol{S}_2$ and the pilot sequences $\boldsymbol{P}_1$ and $\boldsymbol{P}_2$ to be orthogonal sequences. Thus, the number of data symbols L_d in each burst is

$$L_d = L_b - L_t - N_p \cdot L_p. \tag{20}$$

B. Transmitter Model

Fig. 6 shows a block diagram for the transmitter where, in addition to the ST encoder, a high-rate RS block encoder is used as an outer code. The reason for using an outer block code is that, as it will be seen later from the simulation, at reasonable values of SNR, when only the ST code is used most of the frame errors are due to very few symbol errors per frame, most of which can be recovered by the use of an outer block code. The overall coding strategy of the modem is called *concatenated ST coding*. Depending on the desired error correction capability of the RS code, its rate, and the signal constellation used, the dimensions of the RS code should be chosen so that we have an integer number of RS code words per one TDMA slot. In this case we will be able to decode each slot immediately, without the need to wait for other bursts, thereby minimizing the decoding delay.

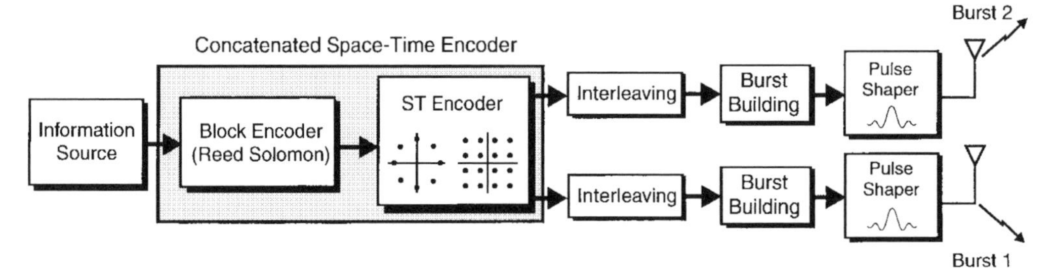

Fig. 6. Base station transmitter with STCM and two transmit antennas.

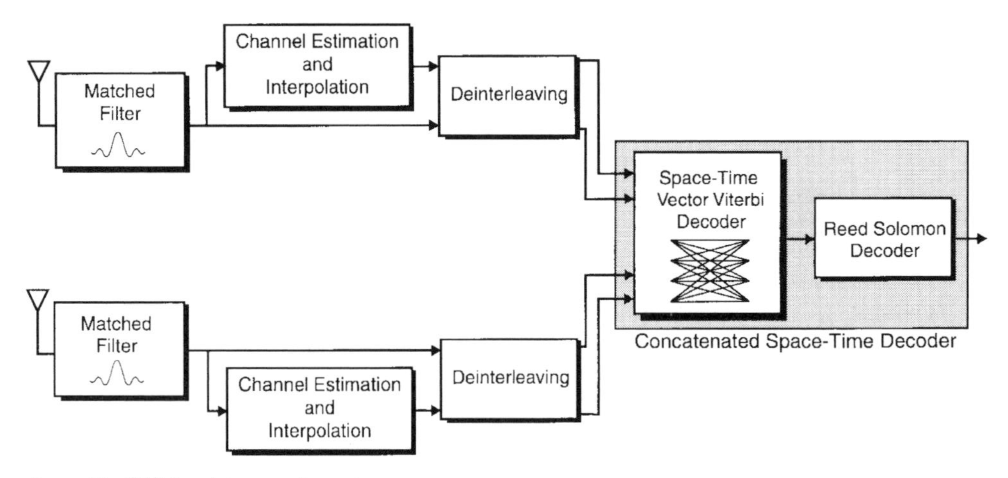

Fig. 7. Mobile receiver with STCM and two receive antennas.

Let B be the number of information bits/modulation symbols. We assume that the RS code used is a (N_c, K_c) code over $GF(2^q)$. The K_c $GF(2^q)$ symbols are rst created by partitioning a block of $q \cdot K_c$ information bits into K_c groups of q bits, each. Similarly, the N_c $GF(2^q)$ output symbols are split into $L_d = (q \cdot N_c/B)$ modulation symbols. Thus, the data throughput of the system is

$$\rho_d = \frac{K_c}{N_c} \cdot \left(1 - \frac{L_t + N_p \cdot L_p}{L_b}\right). \tag{21}$$

The output of the RS encoder is then encoded by an ST channel encoder and the output of the ST encoder is split into two streams of encoded modulation symbols. Each stream of encoded symbols is then independently interleaved using a block symbol-by-symbol interleaver. The transmitter inserts the corresponding training and periodic pilot sequences in each of the two bursts. Each burst is then pulse-shaped and transmitted from the corresponding antenna. In this case, we can write the signal transmitted from the ith antenna, $i = 1, 2$, as

$$s_i(t) = \sqrt{E_s} \cdot \sum_l c_i(l) p(t - lT_s) \tag{22}$$

where $T_s = 1/R_s$ is the symbol period and $p(t)$ is the transmit lter pulse shaping function. Without loss of generality, we will assume that $p(t)$ is a square root raised-cosine ($\sqrt{\mathrm{RC}}$) pulse shape given by [34]

$$p(t) = \frac{4\epsilon}{\pi\sqrt{T_s}} \cdot \frac{\cos\left((1+\epsilon)\pi t/T_s\right) + \dfrac{\sin\left((1-\epsilon)\pi t/T_s\right)}{(4\epsilon t/T_s)}}{(4\epsilon t/T_s)^2 - 1} \tag{23}$$

where ϵ is the bandwidth expansion or roll-off factor. Since $p(t)$ is noncausal, we truncate $p(t)$ to $\pm 3T_s$ around $t = 0$.

C. Receiver Model

Fig. 7 shows the corresponding block diagram of a mobile receiver equipped with two receive antennas. After down conversion to baseband, the received signal at each antenna element is ltered using a receive lter with impulse response $\overline{p}(t)$ that is matched to the transmit pulse shape $p(t)$. In the case of a $\sqrt{\mathrm{RC}}$ transmit lter, $\overline{p}(t) = p(t)$. The output of the matched lters is over sampled at a rate f_{AD} that is Q times faster than the symbol rate R_s, that is, $f_{AD} = Q \cdot R_s$. Received samples corresponding to the training sequences $\boldsymbol{S}_1$ and $\boldsymbol{S}_2$ are used for timing and frequency synchronization. The received samples at the optimum sampling instant are then split into two streams. The rst one contains the received samples corresponding to the pilot and training symbols. These are used to estimate the corresponding CSI $\hat{\mathcal{H}}(l)$ at the pilot and training sequence symbols. The receiver then uses an appropriately designed interpolation lter to interpolate those trained CSI estimates and obtain accurate interpolated CSI estimates for the whole burst. The second stream contains the received samples corresponding to the superimposed information symbols. The interpolated CSI estimates, along with the received samples corresponding to the information symbols, are then deinterleaved using a block symbol-by-symbol deinterleaver and passed to a vector maximum likelihood sequence decoder, followed by an RS decoder.

D. Received Signal Model

We can write the received signal at the jth antenna, $r_j(t)$, as

$$r_j(t) = \sqrt{E_s} \cdot \sum_{i=1}^{N} \tilde{\alpha}_{ij}(t) \sum_{n} c_i(n) p(t - nT_s) + \eta_j(t) \tag{24}$$

where the overall complex channel variable $\tilde{\alpha}_{ij}(t)$ incorporates both the channel gain and the effect of the residual frequency offset and is given by

$$\tilde{\alpha}_{ij}(t) = \alpha_{ij}(t) e^{j2\pi f_o t} \tag{25}$$

where f_o is the residual frequency offset after automatic frequency control (AFC). We can easily see that $\tilde{\alpha}_{ij}(t)$ is bandlimited to $f_o + f_d$. The autocorrelation function of $\tilde{\alpha}_{ij}(t)$ is given by

$$R_{\tilde{\alpha}}(\tau) = J_o(2\pi f_d \tau) e^{j2\pi f_o \tau}. \tag{26}$$

De ne $y_j(t)$ as the matched lter output at the jth antenna, which is given by[2]

$$\begin{aligned} y_j(t) &= \int r_j(\tau) p(t - \tau)\, d\tau \\ &= \sqrt{E_s} \cdot \sum_{i=1}^{N} \tilde{\alpha}_{ij}(t) \sum_{n} c_i(n) \tilde{p}(t - nT_s) + \overline{\eta}_j(t) \end{aligned} \tag{27}$$

where $\tilde{p}(t) = p(t) * p(t)$ is the raised-cosine pulse shape and the colored noise $\overline{\eta}_j(t)$ is given by

$$\overline{\eta}_j(t) = \int \eta_j(\tau) p(t - \tau)\, d\tau. \tag{28}$$

We note that $\overline{\eta}_j(t)$ has a zero mean and an autocorrelation of

$$R_\eta(t_1 - t_2) = N_o \cdot \tilde{p}(t_1 - t_2). \tag{29}$$

For $|t_1 - t_2| = lT_s$ where l is an integer, we have $R_\eta(lT_s) = N_o$ for $l = 0$ and 0 otherwise. Thus, the noise at the output of the matched lter will be uncorrelated when sampled at the symbol rate.

[2]Here we have ignored the intersymbol interference (ISI) due to the time varying nature of the fading, which is a reasonable assumption to make when the fading bandwidth is much smaller than the pulse shape bandwidth.

As we mentioned earlier, the output of the matched lter will be sampled at a rate f_{AD} that is Q times faster than the symbol rate R_s, that is $f_{AD} = Q \cdot R_s$. Let us assume that the sampling time for A/D conversion is

$$t_{l,k} = lT_s + \frac{k}{Q} T_s + \delta T_s \tag{30a}$$

$$l = 1, 2, \cdots \tag{30b}$$

$$k = 0, 1, 2, \cdots, Q - 1 \tag{30c}$$

where δT_s is the timing error. Therefore, we can write the samples at the jth antenna matched lter output as

$$\begin{aligned} y_j(t_{l,k}) = {} & \sqrt{E_s} \cdot \sum_{i=1}^{N} \tilde{\alpha}_{ij}(t_{l,k}) c_i(l) \tilde{p}(t_{l,k} - lT_s) \\ & + v_j(t_{l,k}) + \overline{\eta}_j(t_{l,k}) \end{aligned} \tag{31}$$

where $v_j(t_{l,k})$ represents the intersymbol interference (ISI) due to other transmitted symbols and is given by

$$v_j(t_{l,k}) = \sqrt{E_s} \cdot \sum_{i=1}^{N} \tilde{\alpha}_{ij}(t_{l,k}) \sum_{n \neq l} c_i(n) \cdot \tilde{p}(t_{l,k} - nT_s). \tag{32}$$

In addition, when the fading bandwidth is much smaller than the reciprocal of the symbol period it is reasonable to assume that the fading is constant over one symbol period, that is

$$\tilde{\alpha}_{ij}(t_{l,k}) \approx \tilde{\alpha}_{ij}(t_{l,0}) = \tilde{\alpha}_{ij}(lT_s) = \tilde{\alpha}_{ij}(l).$$

In this case, we can rewrite $y_j(t_{l,k})$ as

$$\begin{aligned} y_j(l,k) &= y_j(t_{l,k}) \\ &= \sqrt{E_s} \cdot \tilde{\alpha}_{ij}(l) \cdot \sum_{i=1}^{N} c_i(l) \tilde{p}(t_{l,k} - lT_s) \\ &\quad + v_j(t_{l,k}) + \overline{\eta}_j(t_{l,k}). \end{aligned} \tag{33}$$

E. Timing and Frequency Synchronization

A frequency offset in the order of 1 ppm, which corresponds to 1.9 kHz at a carrier frequency of 1.9 GHz, will exist in the baseband signal. This frequency offset can be coarsely compensated for using an AFC circuit [37]. After the coarse frequency offset compensation, a residual frequency offset f_o in the order of 0.1 ppm will still exist in the baseband signal and can be compensated for, as we will see later, as part of the channel estimation [38].

For symbol timing synchronization the receiver uses the $Q \times L_t$ samples, corresponding to the training symbols as follows. Consider the jth antenna signal at the output of the matched lter, as given by (33). First, we make the reasonable assumption that the channel is almost constant over the duration of the training sequence and is equal to the value of the channel in the middle of the training period, that is, $\tilde{\alpha}_{ij}(0) \approx \tilde{\alpha}_{ij}(L_t - 1) = \tilde{\alpha}_{ij}(L_t/2)$. Also, let us de ne the overall noise term as $z_j(l,k) = z_j(t_{k,l}) = v_j(t_{k,l}) + \overline{\eta}_j(t_{k,l})$. Let us consider the received samples $y_j(0,k), y_j(1,k), \cdots, y_j(L_t - 1, k)$. Let $\tilde{p}(l,k) = \tilde{p}(t_{l,k} - lT_s)$. Then, we may note that $\tilde{p}(0,k) = \tilde{p}(1,k) = \cdots = \tilde{p}(L_t - 1, k) = \tilde{p}(kT_s/Q)$. De ne $\boldsymbol{X}_j(k)$ as

$$\begin{aligned}\boldsymbol{X}_j(k) &= [y_j(0,k,0) y_j(1,k) \cdots y_j(L_t-1,k)]^T \\ &= A_s(k)\begin{bmatrix} c_1(0) & c_2(0) & \cdots & c_N(0) \\ c_1(1) & c_2(1) & \cdots & c_N(1) \\ \vdots & \vdots & \vdots & \vdots \\ c_1(L_t-1) & c_2(L_t-1) & \cdots & c_N(L_t-1)\end{bmatrix} \\ &\cdot \begin{bmatrix}\tilde{\alpha}_{1j}(L_t/2) \\ \tilde{\alpha}_{2j}(L_t/2) \\ \vdots \\ \tilde{\alpha}_{Nj}(L_t/2)\end{bmatrix} + \begin{bmatrix} z_j(0,k) \\ z_j(1,k) \\ \vdots \\ z_j(L_t-1,k)\end{bmatrix} \\ &= A_s(k)[\boldsymbol{S}_1 \quad \boldsymbol{S}_2 \quad \cdots \quad \boldsymbol{S}_N]\tilde{\boldsymbol{\alpha}}_j(L_t/2) + \boldsymbol{z}_j(k) \end{aligned} \tag{34}$$

where $A_s(k) = \sqrt{E_s}\tilde{p}(kT_s/Q)$. We assume (erroneously[3]) that the noise term due to ISI $v_j(l,k) = v_j(lT_s + kT_s/Q)$ is modeled as uncorrelated Gaussian noise with zero mean and variance $\sigma_v^2(k)$. Therefore, the overall noise vector $\boldsymbol{z}_k$ is zero mean with covariance $\sigma_z^2(k)\boldsymbol{I}$ where $\sigma_z^2(k) = \sigma_v^2(k) + N_o$. De ne $\Psi_{ij}(k) = \boldsymbol{S}_i^* \boldsymbol{X}_j(k)$. Since the training sequences are orthogonal, it is easy to verify that

$$\Psi_{ij}(k) = A_s(k)\tilde{\alpha}_{ij}(L_t/2)||\boldsymbol{S}_i||^2 + \overline{z}_{ij}(k) \tag{35}$$

where $\overline{z}_{ij}(k) = \boldsymbol{S}_i^* \boldsymbol{z}_j(k)$. The symbol timing synchronization algorithm estimates which k is closest to the optimum sampling instant in each frame. This value of k can be estimated using maximum likelihood estimation. Similar to the development in [40], we can shows that the log likelihood function for the symbol timing synchronization can be approximated by

$$\Lambda_{\text{ML}}(k) = \sum_{j=1}^{M}\sum_{i=1}^{N}\Lambda_{ij}(k) = \sum_{j=1}^{M}\sum_{i=1}^{N}|\Psi_{ij}(k)|^2, \quad k = 0, 1, \cdots, Q-1. \tag{36}$$

The optimum sampling instant k^* is obtained by searching for the value k that gives the maximum value of $\Lambda_{ML}(k)$. Because in (36) we only use the envelope of $\Psi_{ij}(k)$, this method will be robust against phase distortion due to the Rayleigh fading, especially in deep fades.

F. Channel Estimation

Consider the jth receive antenna output after matched ltering. We can write the received signal samples for the lth symbol within the burst at the optimum sampling instant as

$$y_j(l) = \sqrt{E_s}\tilde{p}(\delta T_s')\sum_{i=1}^{N}\tilde{\alpha}_{ij}(l)c_i(l) + v_j(l) + \overline{\eta}_j(l) \tag{37}$$

where $\delta T_s'$ is the timing error after timing synchronization, $\overline{\eta}_j(l)$ is the AWGN with zero mean and variance $N_o/2$ per dimension, and $v_j(l)$ is the ISI due to the timing error which is modeled as uncorrelated Gaussian noise with zero mean and variance σ_v^2.

[3]This approximation is only valid if the code symbols are Gaussian [39] and if the timing error is constant, but otherwise unknown.

Consider the output samples corresponding to the nth pilot sequence within the burst $y_j((n-1)L_s+1), y_j((n-1)L_s+2), \cdots, y_j((n-1)L_s+L_p)$. As before, a reasonable assumption to make is that the channel is almost constant over the duration of the pilot sequence and is equal to the value of the channel in the middle of the pilot sequence, that is, $\tilde{\alpha}_{ij}(n) = \tilde{\alpha}_{ij}((n-1)L_s+1) \approx \tilde{\alpha}_{ij}((n-1)L_s+L_p) = \tilde{\alpha}_{ij}((n-1)L_s+L_p/2)$. As before, we de ne the overall noise term as $z_j(n) = \overline{\eta}_j(n) + v_j(n)$. De ne $\boldsymbol{Y}_j(n)$ as

$$\begin{aligned}\boldsymbol{Y}_j(n) &= [y_j((n-1)L_s+1) y_j((n-1)L_s+2) \\ &\quad \cdots y_j((n-1)L_s+L_p)]^T \\ &= A_s\begin{bmatrix} c_1(1) & c_2(1) & \cdots & c_N(1) \\ c_1(2) & c_2(2) & \cdots & c_N(2) \\ \vdots & \vdots & \cdots & \vdots \\ c_1(L_p) & c_2(L_p) & \cdots & c_N(L_p)\end{bmatrix} \\ &\cdot \begin{bmatrix}\tilde{\alpha}_{1j}(n) \\ \tilde{\alpha}_{2j}(n) \\ \vdots \\ \tilde{\alpha}_{Nj}(n)\end{bmatrix} + \begin{bmatrix} z_j((n-1)Ls+1) \\ z_j((n-1)Ls+2) \\ \vdots \\ z_j((n-1)L_s+L_p)\end{bmatrix} \\ &= A_s[\boldsymbol{P}_1 \quad \boldsymbol{P}_2 \quad \cdots \quad \boldsymbol{P}_N]\tilde{\boldsymbol{\alpha}}_j(n) + \boldsymbol{z}_j(n) \end{aligned} \tag{38}$$

where $A_s = \sqrt{E_s}\tilde{p}(\delta T_s')$. Using the fact that $\boldsymbol{P}_1, \boldsymbol{P}_2, \cdots, \boldsymbol{P}_N$ are orthogonal, we can immediately see that the minimum mean square error (MMSE) estimate of $\tilde{\alpha}_{ij}(l)$ is given by

$$\hat{\tilde{\alpha}}_{ij}(l) = \frac{\boldsymbol{P}_i^* \boldsymbol{Y}_j(n)}{A_s \cdot ||\boldsymbol{P}_i||^2}, \qquad i = 1, 2, \cdots, N. \tag{39}$$

It is easy to show that

$$\hat{\tilde{\alpha}}_{ij}(n) = \tilde{\alpha}_{ij}(n) + e_{ij}(n) \tag{40}$$

where $e_{ij}(n)$ is the estimation error due to the noise and ISI and is given by

$$e_{ij}(n) = \frac{\boldsymbol{P}_i^* \boldsymbol{z}_j(l)}{A_s \cdot ||\boldsymbol{P}_i||^2}, \qquad i = 1, 2, \cdots, N. \tag{41}$$

Since $\boldsymbol{z}_j(n)$ is assumed to be a zero-mean Gaussian random vector it easy to see that $e_{ij}(n)$ will be also Gaussian with zero mean and variance

$$\sigma_e^2(n) = \frac{\sigma_v^2 + N_o}{A_s^2 \cdot ||\boldsymbol{P}_i||^2}. \tag{42}$$

Note that, in the case of pilot symbols with constant envelope, $||\boldsymbol{P}_i||^2 = L_p$ and in this case we will have

$$\sigma_e^2(n) = \frac{\sigma_v^2 + N_o}{A_s^2 \cdot L_p}. \tag{43}$$

At nominal SNR's, and when the timing error is relatively small, the variance of the term due to ISI will be very small as compared to the thermal noise variance, that is, $\sigma_v^2 \ll N_o$. In addition, we will have $A_s^2 \approx E_s$. In this case, the variance of the estimation error will be given by

$$\sigma_e^2 \approx \frac{1}{(E_s/N_o) \cdot L_p}. \tag{44}$$

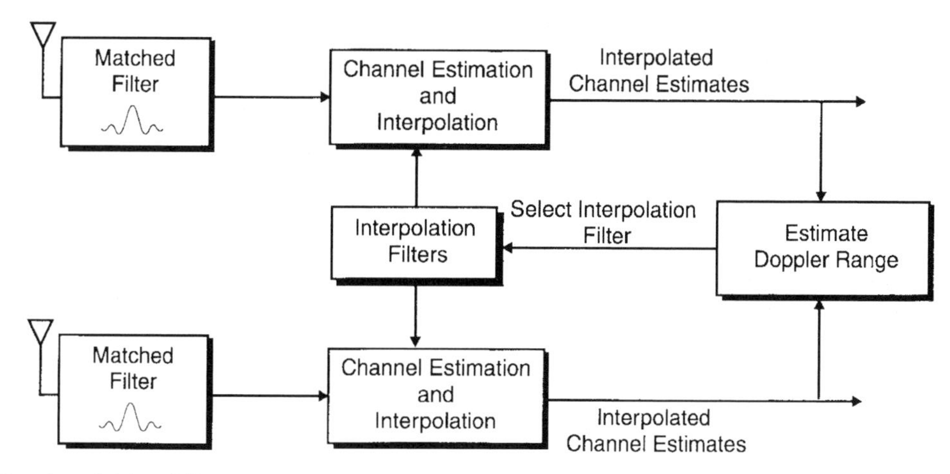

Fig. 8. Quasi-adaptive channel interpolation.

In order to minimize the overall system delay we will assume that the receiver estimates the CSI in any given time slot using the pilot and training sequences in that slot only. Therefore, we will avoid the need to wait for future bursts in order to be able to estimate the channel and perform the decoding of the slot. In addition, other time slots may be carrying bursts that correspond to old IS-136 wireless channels or any other bursts with different slot structure and, thus, no pilot symbols will exist in those slots.

G. Channel Interpolation

Without loss of generality, let us assume that the number of trained channel estimates in each time slot is N_e. For clarity of notation, let $h_{ij}(n)$ denote the trained channel estimate $\{\hat{\alpha}_{ij}(n), n = 1, \cdots, N_e\}$. These trained channel estimates need to be interpolated to obtain a complete CSI for the whole time slot. To satisfy the Nyquist criterion, the normalized sampling rate of the trained channel estimates must satisfy

$$\overline{f}'_s = \frac{f'_s}{2(f_d + f_o)} = \frac{1}{(L_s \cdot T_s) \cdot 2(f_o + f_d)} \geq 1. \quad (45)$$

Moreover, in order to compensate for the fact that, in estimating the channel over any time slot, we are using the pilot and training symbols in that time slot only, $\overline{f}'_s$ should be slightly higher than one. Here, we will briey consider two different approaches for interpolating the channel estimates.

1) Wiener Interpolation Filter (WIF): In this approach, we use a multichannel generalization of the WIF proposed in [26]. In this case, the receiver estimates the channel gain $\tilde{\alpha}_{ij}(l), 1 \leq i \leq N, 1 \leq j \leq M$ for the lth symbol position in the burst as a linear combination of the trained channel estimates

$$\hat{\tilde{\alpha}}_{ij}(l) = \sum_{n=1}^{N_e} w^*(n,l) h_{ij}(n) = \boldsymbol{w}^*(l)\boldsymbol{h}_{ij} \quad (46)$$

where $\boldsymbol{w}(l) = [w(1,l) \quad w(2,l) \quad \cdots \quad w(N_e,l)]^T$ and $\boldsymbol{h}_{ij} = [h_{ij}(1) \quad h_{ij}(2) \quad \cdots \quad h_{ij}(N_e)]^T$. Note that the interpolator coef cients $\boldsymbol{w}(l)$ are different for every symbol position in the burst. These coef cients are chosen such that the mean squared error (MSE) between the channel gain $\tilde{\alpha}_{ij}(l)$ and its *interpolated* estimate $\hat{\tilde{\alpha}}_{ij}(l)$ is minimized. In this case, it is known that the optimum interpolator coef cients are given by the Wiener solution [41]

$$\boldsymbol{w}(l) = \boldsymbol{R}_h^{-1}\boldsymbol{a}(l) \quad (47)$$

where $\boldsymbol{R}_h$ is an $N_e \times N_e$ matrix and $\boldsymbol{a}(l)$ is an $N_e \times 1$ vector and are given by

$$\boldsymbol{R}_h = E\{\boldsymbol{h}_{ij}\boldsymbol{h}_{ij}^*\} \quad \text{and}$$
$$\boldsymbol{a}(l) = E\{\tilde{\alpha}_{ij}^*(l)\boldsymbol{h}_{ij}\}.$$

From (26) and (40) we can easily see that the (m,n) element of $\boldsymbol{R}_h$ and the nth element of $\boldsymbol{a}(l)$ are given by

$$R_{h,mn} = \sigma_e^2 + J_o(2\pi f_d(m-n)T_s)e^{j2\pi f_o(m-n)T_s} \quad \text{and}$$
$$a_n(l) = J_o(2\pi f_d(l-n)T_s)e^{j2\pi f_o(l-n)T_s}$$

respectively. In this case, we can easily show that the MMSE in the interpolated estimates $\sigma_i^2(l)$ is given by

$$\sigma_{ij}^2(l) = \sigma_\alpha^2 - \boldsymbol{a}^*(l)\boldsymbol{w}(l) \quad (48)$$

where σ_α^2 is the variance of $\alpha_{ij}(l)$ which was assumed to be one.

As in [26], however, the optimum WIF assumes knowledge of the SNR (or more speci cally the estimation error variance σ_e^2 which can be related to the SNR, as shown above), the maximum Doppler frequency f_d, and the residual frequency offset f_o and therefore will be different for different values of the SNR, f_d and f_o. This would be very complex for practical implementation. In a practical scenario, the lter will be optimized for the worst case f_d and f_o such that it will have a bandwidth that is wide enough for all possible time variations of the channel. In this case, however, the performance at low f_d and f_o will be the same as that of the worst case (noise in the interpolated estimates will have a larger variance due to the large bandwidth of the interpolation lter). In addition, if $f_d + f_o$ exceeds the lter bandwidth, aliasing in the interpolated CSI will occur and we will have a signi cant mismatch between the interpolated CSI and the ideal one. This will lead to a signi cant error oor. Here, we consider a quasi-adaptive approach to remedy this problem. This approach is shown in Fig. 8. In this approach, we divide the range of all possible f_d into different nonoverlapping ranges. For every range of Doppler frequencies, we design an optimum WIF for the maximum Doppler frequency in that range and use it for the whole range. By observing the correlations of the interpolated channel estimates from the previous time slots, or by observing its frame error rate (FER), the receiver selects which lter to use.

2) Low-Pass Interpolation Filter (LPIF) In this approach, a time invariant nite impulse response (FIR) digital low pass lter is used to interpolate the channel estimates in every time slot. This approach is similar to that in [42] where an FIR low-pass lter with unit sample response equal to a truncated raised-cosine pulse is used for interpolation. Here, however, we use the approach described in [43] to design an optimum FIR low-pass lter for interpolation that will minimize the error between the interpolated channel estimates and its true value. This approach will be briey described below. For full mathematical treatment, however, the reader is referred to [43].

We are given a sequence $x(l)$, the values of which are possibly nonzero only at $l = kL_s$ where $k = 0, \pm 1, \pm 2, \cdots$. The sequence $x(l)$ is considered as being a sampled version of an unknown, but bandlimited sequence $u(l)$

$$x(l) = \begin{cases} u(l), & l = kL_s, k = 0, \pm 1, \pm 2, \cdots, \\ 0, & \text{otherwise.} \end{cases} \quad (49)$$

The sequence $u(l)$ is assumed to be bandlimited with

$$U(\omega) = 0 \quad \text{for} \quad |\omega| \geq \delta \cdot \pi / L_s, 0 < \delta \leq 1. \quad (50)$$

The sequence $x(l)$ here corresponds to the channel samples at the pilot positions. Let us assume that $g(l)$ is the unit sample response of the FIR interpolating lter, which, given every L_s sample of the sequence $u(l)$, interpolates the remaining $L_s - 1$ samples using K past and K future samples. It is easy to verify that the length of the lter will be then $2KL_s + 1$. The unit sample response $g(l)$ is designed such that the error

$$\varepsilon^2 = \sum_l ||(x * g)(l) - u(l)||^2 \quad (51)$$

is minimized. The method in [43] divides $g(l)$ into L_s subsequences $g_\mu(k) = g(kL_s + \mu), \mu = 0, 1, \cdots, L_s - 1, k = 0, \pm 1, \pm 2, \cdots$. The minimization of (51) results in the following set of linear equations for each μ

$$\sum_{m=-K}^{K-1} g_\mu(m)\phi((m-k)L_s) = \phi(kL_s + \mu) k = -K, \cdots, K-1$$
$$\text{and } \mu = 0, 1, \cdots, L_s - 1 \quad (52)$$

where $\phi(n)$ is the autocorrelation function of $u(l)$ and is given by

$$\phi(n) = \int_{-\delta\pi/L_s}^{\delta\pi/L_s} |U(\omega)|^2 e^{-j\omega n} \, d\omega. \quad (53)$$

Equation (52) can be put in a matrix form as

$$\boldsymbol{\Phi} \boldsymbol{g}_\mu = \boldsymbol{\phi}_\mu \quad (54)$$

where

$$\boldsymbol{\Phi} = \begin{bmatrix} \phi(0) & \phi(L_s) & \cdots & \phi((2K-1)L_s) \\ \phi(L_s) & \phi(0) & \cdots & \phi((2K-2)L_s) \\ \vdots & \vdots & \ddots & \vdots \\ \phi((2K-1)L_s) & \phi((2K-2)L_s) & \cdots & \phi(0) \end{bmatrix}$$

is a Toeplitz matrix that does not depend on μ. Furthermore

$$\boldsymbol{g}_\mu = [g(-KL_s + \mu) g(-(K-1)L_s + \mu), \cdots, g((K-1)L_s + \mu)]^T$$
$$\boldsymbol{\phi}_\mu = [\phi(-KL_s + \mu) \phi(-(K-1)L_s + \mu), \cdots, \phi((K-1)L_s + \mu)]^T.$$

This, in some sense, resembles the WIF approach described above, except that the same lter is used for all points in the slot as compared to the WIF, in which a different lter (or set of weights) is used for each point in the slot. Also, since we require the lter to be time invariant, the lter bandwidth should satisfy the condition in (45) for the worst case maximum Doppler frequency and frequency offset. As explained above, this will degrade the performance at low f_d and f_d. In addition, since the receiver estimates the CSI in any given time slot using the pilot and training sequences in that slot only, interpolated CSI near the ends of the slot will exhibit a larger MSE than those near the middle of the slot.

IV. Simulation Results

In this section, we present the simulation results for the STCM-based modem architecture described above. These results will be presented for both the 8-PSK 32-state and the 16-QAM 16-state ST codes presented in Section II. We will briey describe the simulation scenario in Sections IV-A and IV-B. The results of these simulations are presented in Sections IV-C through IV-G.

A. Time Slot Structure and Signaling Format

In all of the simulations, we assume IS-136 *basic* channelization and framing, except that the slot structure of the STCM-based modem will be different. For the purpose of comparison, we will briey describe the channelization and framing structure in IS-136. On each 30-kHz channel, the IS-136 standard de nes 25 frames of data per second ($N_f = 25$), each of which is then further subdivided into six time slots ($N_s = 6$). Each time slot is of 6.667-ms duration and carries 162 modulation symbols (the raw symbol rate R_s is 24 300 symbols/s). These symbols, in turn, include, 130 symbols for data or speech and 32 symbols for synchronization and control overhead. Under normal operating conditions, a single user is provided with exactly two time slots per frame, which guarantees the user a symbol rate of $2 \times 130 \times 25 = 6500$ symbols per second. The IS-136 uses $\pi/4$-DQPSK for modulation, which supports two bits per symbol. This means that the net (*uncoded*) bit rate over a 30-kHz channel is 39 kb/s.

Fig. 9 shows the slot structure for the STCM-based modem with two transmit antennas, using IS-136 basic channelization and framing. As with the IS-136 standard, we also assume

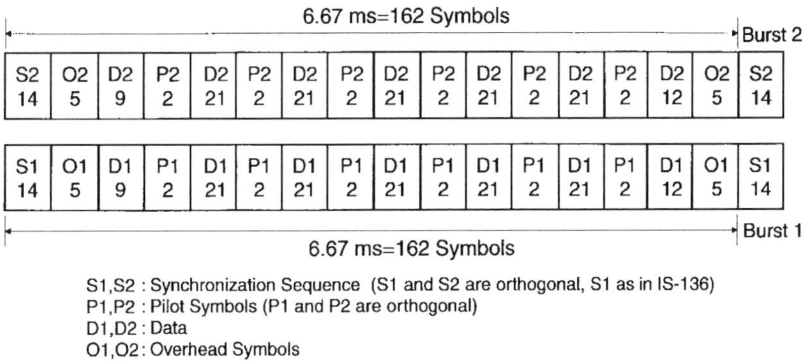

6.67 ms=162 Symbols — Burst 2

S2	O2	D2	P2	D2	P2	D2	P2	D2	P2	D2	P2	D2	P2	D2	O2	S2
14	5	9	2	21	2	21	2	21	2	21	2	21	2	12	5	14

S1	O1	D1	P1	D1	P1	D1	P1	D1	P1	D1	P1	D1	P1	D1	O1	S1
14	5	9	2	21	2	21	2	21	2	21	2	21	2	12	5	14

6.67 ms=162 Symbols — Burst 1

S1,S2 : Synchronization Sequence (S1 and S2 are orthogonal, S1 as in IS-136)
P1,P2 : Pilot Symbols (P1 and P2 are orthogonal)
D1,D2 : Data
O1,O2 : Overhead Symbols

Fig. 9. Slot structure for STCM-based modem based on IS-136 timing and framing structure.

TABLE I
SNR_w(dB) USED FOR DESIGNING THE WIF'S FOR 8-PSK

1 Tx 1 Rx No ST code	2 Tx 1 Rx 32-State Code	2 Tx 2 Rx 32-State Code	2 Tx 1 Rx DD Code	2 Tx 2 Rx DD Code
25	20	12	30	17.5

TABLE II
SNR_w(dB) USED FOR DESIGNING THE WIF'S FOR 16-QAM

1 Tx 1 Rx No ST code	2 Tx 1 Rx 16-State Code	2 Tx 2 Rx 16-State Code
35	22.5	15

the same symbol rate R_s of 24 300 symbols/s. Each burst of 6.667 ms is 162 symbols long ($L_b = 162$) and starts with a 14 symbols training sequence ($L_t = 14$) that will be used for timing and frequency synchronization. The training sequence is also used to estimate the channel at the middle of the training sequence. In addition, the transmitter inserts six two-symbol ($N_p = 6, L_p = 2$) pilot sequences $\boldsymbol{P}_1$ and $\boldsymbol{P}_2$ that are used for channel estimation. Thus, in each burst we are left with 136 symbols, ten of which will be reserved for control overhead and 126 of which will be used for information. The 126 symbols in each burst are interleaved by a 14×9 symbol-by-symbol block interleaver. The $\sqrt{\text{RC}}$ pulse shape $p(t)$ has a roll-off factor of 0.35.

B. Channel Estimation and Interpolation

As we pointed out before, signals at the receive antennas will be a linear superposition of the two transmitted bursts and we need the two training sequences $\boldsymbol{S}_1$ and $\boldsymbol{S}_2$, as well as the pilot sequences $\boldsymbol{P}_1$ and $\boldsymbol{P}_2$, to be orthogonal sequences. We use the same $\pi/4$-DQPSK synchronization sequence speci ed in the IS-136 standard for $\boldsymbol{S}_1$. This will allow the STCM-based service to coexist with old IS-136 services and, at the same time, ensure backward compatibility with IS-136.[4] We assume that the synchronization and pilot symbols have the same energy per symbol as the information symbols. As we mentioned before, in estimating the channel over any burst, the receiver uses the training and pilot sequences in that burst only since other time slots may be carrying bursts that correspond to old IS-136 wireless channels or other bursts with different structure. In addition, this will minimize the overall system delay. Thus, the receiver will use $\boldsymbol{S}_1$ and $\boldsymbol{S}_2$ at the beginning

[4] Some of today's IS-136 mobile phones use the synchronization sequence in other time slots to update their equalizer and maintain timing and frequency synchronization.

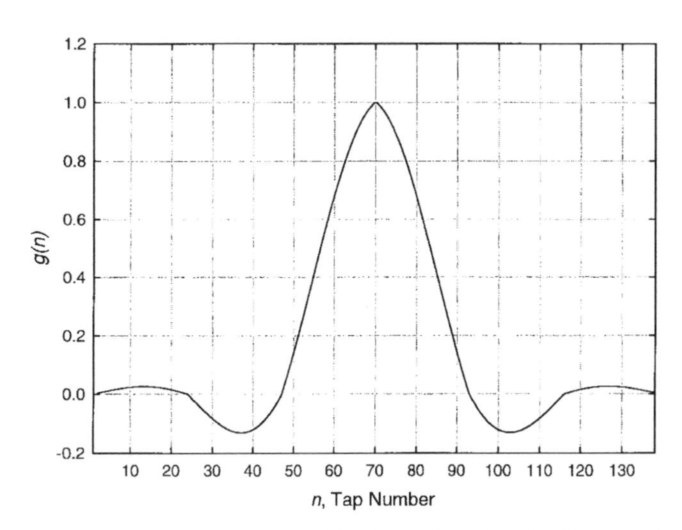

Fig. 10. Unit sample response of LPIF designed for $L_s = 23, K = 3$, and $\delta = 0.5$.

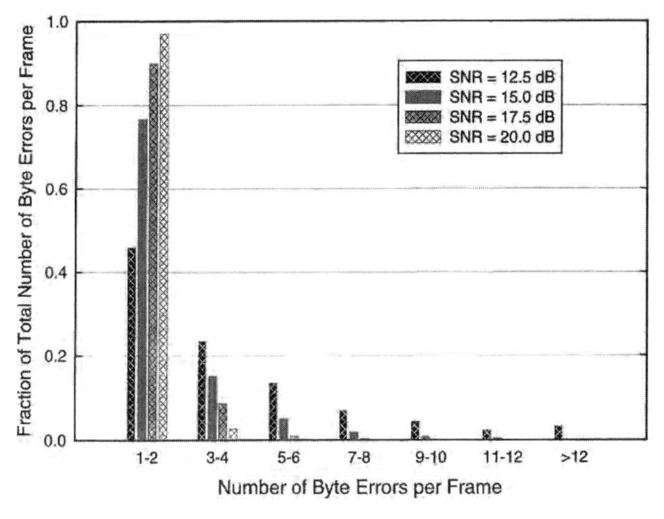

Fig. 11. Error histogram of the 16-QAM 16-state ST code without an outer RS with two transmit and two receive antennas and optimized WIF at $f_d = 180$ Hz.

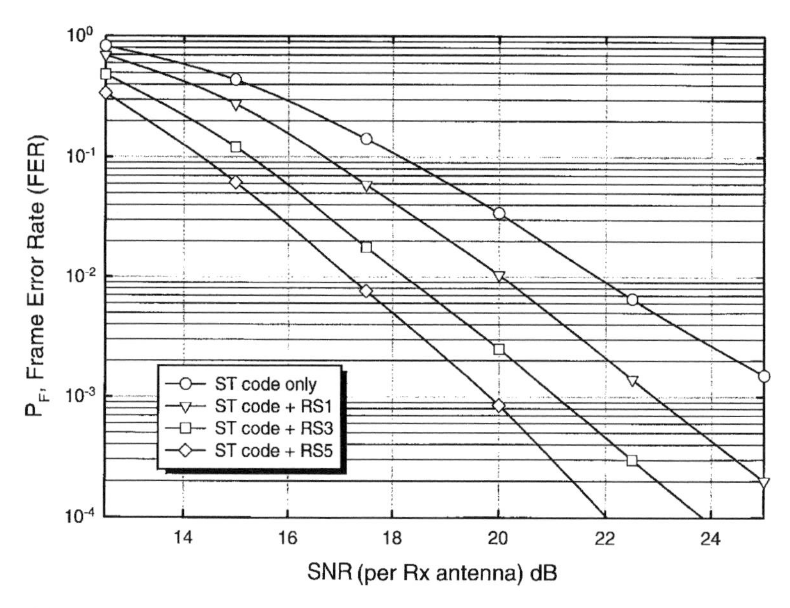

Fig. 12. Peformance of the 16-QAM 16-state ST code with two transmit and two receive antennas and $f_d = 180$ Hz.

of the current time slot, the six pilot sequences $\boldsymbol{P}_1$ and $\boldsymbol{P}_2$, as well as $\boldsymbol{S}_1$ and $\boldsymbol{S}_2$, at the beginning of the next time slot (which may belong to a different user) to obtain eight estimates per TDMA time slot ($N_e = 8$) for the channel from each of the transmitting antennas to the receiver. The sampling period of these channel estimates $T_s' = L_s \cdot T_s$ where $L_s = 23$, in our case, can be easily seen to be 23/24 300 which corresponds to a sampling frequency $f_s' \approx 1056$Hz . In all of the simulations we will consider maximum Doppler frequencies up to 180 Hz, a residual frequency offset f_o of 200 Hz, an over sampling factor $Q = 8$, and a timing error δT_s which is uniformly distributed over $\pm T_s/16$. Thus, an f_s' of 1056 Hz will satisfy the requirement in (45).

For the WIF, we assumed that the 200-Hz Doppler range is divided into four subranges: 0–20, 20–80, 80–140, and 140–200 Hz. Four different WIF's were designed, one for each subrange. These lters were optimized at a frequency offset f_o of 200 Hz, maximum Doppler frequencies of 20, 80, 140, and 200 Hz, respectively, and an SNR_w that will depend on the ST code used and the number of transmit and receive antennas used. Tables I and II list the SNR_w used for designing the WIF's for both the 8-PSK and 16-QAM cases we considered.

For the LPIF, the approach described in Section III-G2 was used to design a time-invariant low-pass lter with $L_s = 23, K = 23, \delta = 0.5$. Fig. 10 shows the unit sample response $g(n)$ of the LPIF. The low-pass lter was designed such that it will have its 3-dB cutoff frequency at 528 Hz.

C. 16-QAM Results

For the 16-QAM 16-state space time code, shown in Fig. 3, we simulated the STCM-based modem without an outer RS code. Fig. 11 shows the number of errors per frame as a fraction of the total number of errors per frame for two transmit and two receive antennas at a maximum Doppler frequency $f_d = 180$ Hz. From this gure we can easily see that for SNR's of more than 15 dB, more than 80% of the frame errors are due

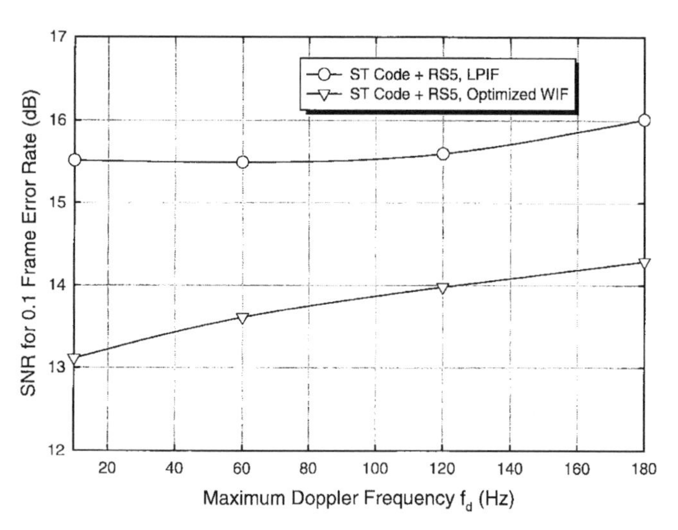

Fig. 13. SNR performance at 10% FER as a function of f_d of the 16-QAM 16-state ST code with two transmit and two receive antennas.

TABLE III
SNR (dB) REQUIRED FOR 10% FER FOR THE 16-QAM 16-STATE ST CODE FOR DIFFERENT BIT RATES

Maximum Doppler Frequency f_d	Diversity	ST Code (74.4 kbps)	ST Code + RS3 (67.2 kbps)	ST Code + RS5 (62.4 kbps)
10 Hz	2Tx 1Rx	22.4	21.1	20.5
	2Tx 2Rx	15.7	13.8	13.2
	1Tx 1Rx	**28.0**	**25.6**	**24.6**
	1Tx 2Rx	**20.7**	**18.1**	**17.5**
180 Hz	2Tx 1Rx	30.6	25.0	23.0
	2Tx 2Rx	18.1	15.3	14.3
	1Tx 1Rx	**> 50**	**32.4**	**28.9**
	1Tx 2Rx	**24.5**	**20.3**	**18.9**

to one or two symbol errors and 90% of them are due to four or fewer symbol errors. These errors can be corrected by using a high-rate outer code. Therefore, we considered three different shortened RS codes over $GF(2^8)$ for the outer code. The rst

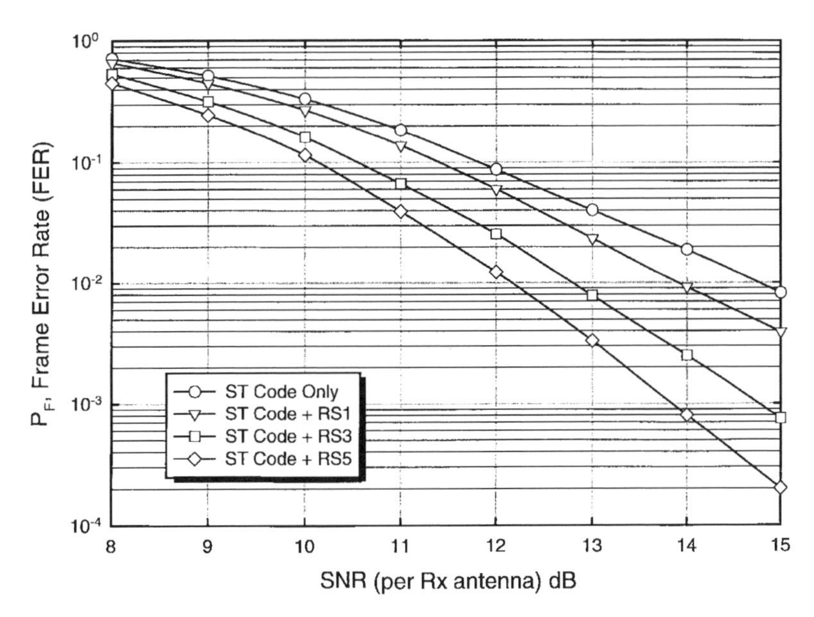

Fig. 14. Performance of the 8-PSK 32-state ST code with two transmit and two receive antennas and $f_d = 180$ Hz.

code, referred to as RS1, is a shortened RS(62, 60) code that corrects single-byte errors. The second RS code, referred to as RS3, is a shortened RS(62, 56) code that corrects three-byte errors, and the third RS code, referred to as RS5, is a shortened RS(62, 52) code that corrects Þve-byte errors. For RS1, for example, the $62GF(2^8)$ symbols are Þrst created by partitioning a block of 480 information bits into 60 groups of eight bits each. These 60 bytes are then encoded by RS1 to give 62 bytes or RS symbols. The output $62GF(2^8)$ symbols ($62 \times 8 = 496$ bits) are then partitioned into 124 16-QAM symbols, two modulation symbols per one RS symbol. The 124 16-QAM symbols are then padded with two 16-QAM zero symbols to force the ST encoder to go back to the zero state.[5] The 126 16-QAM symbols are then encoded using the ST encoder.

Fig. 12 shows the FER performance of the 16-QAM 16-state ST code with two transmit and two receive antennas and a maximum Doppler frequency $f_d = 180$ using optimized WIF. From this Þgure we can see that the ST code alone needs an SNR of 18 dB to achieve 10% FER. However, when the ST code is concatenated with RS5, for example, the required SNR is 14.5 dB, which is a 3.5-dB gain over the ST code alone. In this case, however, the net bit rate (over a 30-kHz channel) at 10% FER will be reduced from 74.4 kb/s to 62.4 kb/s. Fig. 13 shows the SNR required for 10% FER versus the maximum Doppler frequency f_d for the 16-QAM 16-state ST code, concatenated with RS5 and two transmit and two receive antennas. We plot the results for both the LPIF and WIF. We can easily see a 2.5Ð3.5 dB advantage for the WIF over the LPIF for maximum Doppler frequencies up to 180 Hz. As expected, the WIF will have a better performance than the LPIF (since we are optimizing the Þlter coefÞcients for every point in the slot), although it would be more computationally expensive and would require more memory to store the coefÞcients.

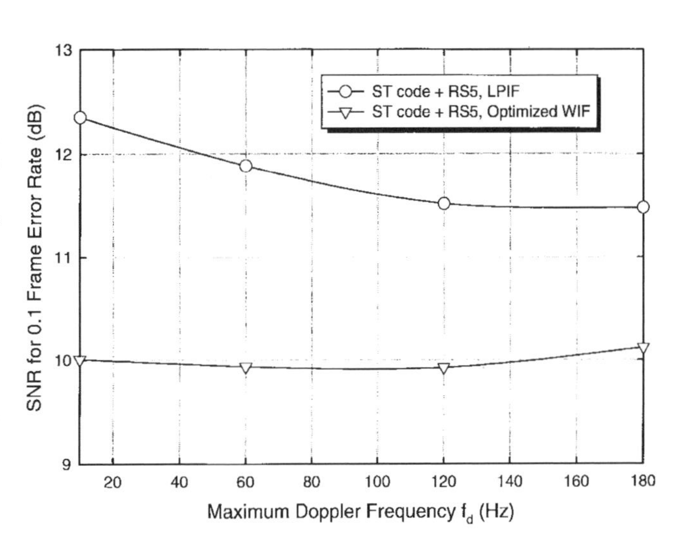

Fig. 15. SNR performance at 10% FER as a function of f_d of the 8-PSK 32-state ST code with two transmit and two receive antennas.

TABLE IV
SNR (dB) REQUIRED FOR 10% FER FOR THE EIGHT-PSK 32-STATE ST CODE FOR DIFFERENT BIT RATES

Maximum Doppler Frequency	Diversity	ST Code (55.8 kbps)	ST Code + RS3 (50.4 kbps)	ST Code + RS5 (46.8 kbps)
10 Hz	2Tx 1Rx	18.3	17.3	16.9
	2Tx 2Rx	11.2	10.3	10.1
	1Tx 1Rx	**25.5**	**23.1**	**22.0**
	1Tx 2Rx	**18.4**	**15.8**	**14.9**
180 Hz	2Tx 1Rx	20.1	17.6	17.1
	2Tx 2Rx	11.8	10.6	10.2
	1Tx 1Rx	**41.0**	**28.5**	**25.4**
	1Tx 2Rx	**21.9**	**17.6**	**16.2**

Table III summarizes the SNR performance at 10% FER for the 16-QAM 16-state ST code case. It shows the required

[5] For the 16-QAM 16-state ST code only one 16-QAM zero symbol is needed to terminate the trellis, the other symbol is merely a dummy symbol.

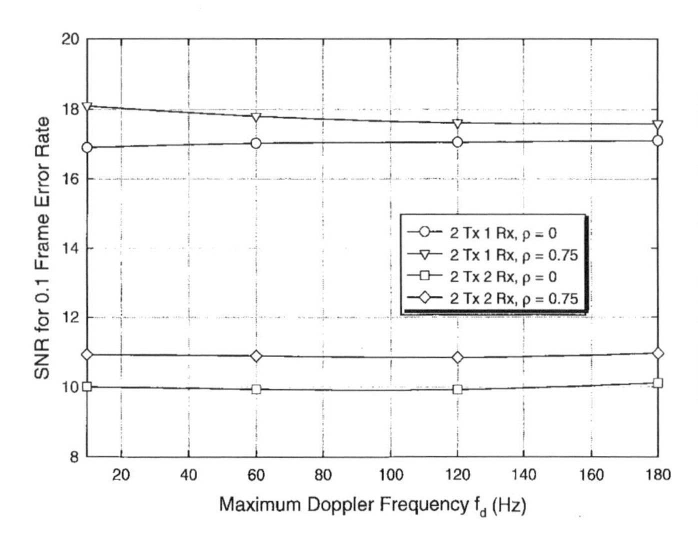

Fig. 16. Effect of transmit antenna correlation on the SNR performance at 10% FER for the 8-PSK 32-state ST code as a function of f_d.

SNR for different number of transmit and receive antennas and different bit rates, for maximum Doppler frequencies of 10 and 180 Hz. For all of these cases we assumed that the WIF is used for to obtain the interpolated CSI. In addition, we also included the case when there is only one transmit antenna as a reference at the transmitter, which corresponds to the case where no ST coding is used. From these numbers, one can easily see the improvement in the SNR performance due to the use of the ST code with transmit antennas. For example, when using the space time code alone with two transmit and one receive antennas, at a maximum Doppler frequency of 10 Hz, an improvements of 5.6 dB (over the system with one transmit and one receive antenna) is achieved. For the same case, at a maximum Doppler frequency of 180 Hz, this improvement is even larger at more than 20 dB.

D. 8-PSK Results

For the 8-PSK constellation, we considered the same three RS codes used for the 16-QAM case, except that the code polynomial is now deÞned over $GF(2^6)$ and each symbol is 6 bits long. For RS1, in this case, the 62 $GF(2^6)$ symbols are Þrst created by partitioning a block of 360 information bits into 60 groups of 6 bits each. The output 62 $GF(2^6)$ symbols are then partitioned into 124 8-PSK symbols, two modulation symbols per one RS symbol. The 124 8-PSK symbols are then padded with two 8-PSK zero symbols to force the ST encoder to go back to the zero state. The 126 8-PSK symbols are then encoded using the ST encoder.

Fig. 14 shows the FER performance of the 8-PSK 32-state ST code with two transmit and two receive antennas and $f_d = 180$ with WIF. From this Þgure, we can see that a 10% FER can be achieved at 11.7-dB SNR and 10-dB SNR with an 8-PSK 32-state ST code, concatenated with RS1 and RS5, respectively. This corresponds to a net bit rate (over a 30-kHz channel) of 54 kb/s and 46.8 kb/s, respectively. Fig. 15 shows the SNR required for 10% FER versus f_d for the 8-PSK 32-state ST code, concatenated with RS5 and two receive antennas. As in the 16-QAM 16-state ST code case,

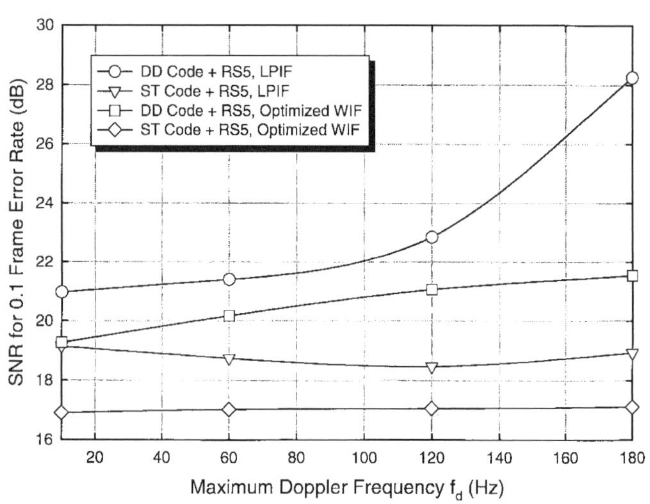

Fig. 17. SNR performance at 10% FER. Performance of delay diversity versus 8-PSK 32-state ST code with two transmit and one receive antennas as a function of f_d.

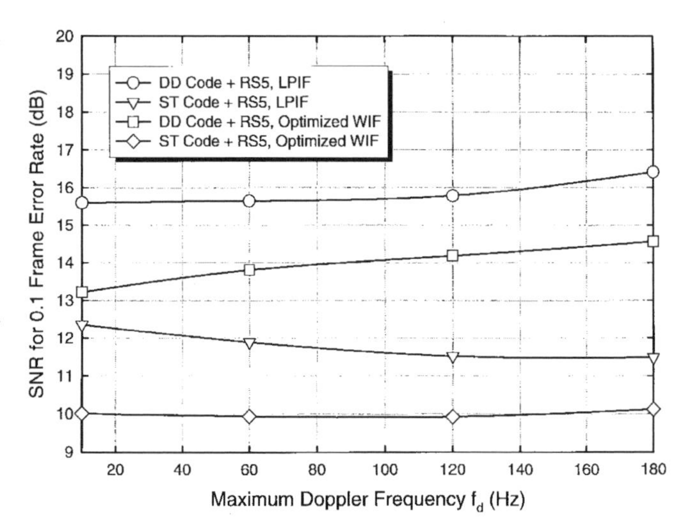

Fig. 18. SNR performance at 10% FER. Performance of delay diversity versus 8-PSK 32-state ST code with two transmit and two receive antennas as a function of f_d.

we also plot the results for both the LPIF and the WIF. We can see a 2.5-dB advantage for the WIF over the LPIF at 10 Hz. At 180 Hz, the WIF advantage over the LPIF is only 1.5 dB.

Table IV summarizes the SNR performance at 10% FER for the 8-PSK 32-state ST code case. It also shows the required SNR for different numbers of transmit and receive antennas and different bit rates, for maximum Doppler frequencies of 10 and 180 Hz. As before, we assume that the WIF is used to obtain the interpolated CSI. Similar to the 16-QAM case, we can easily see the improvement in the SNR, due to the use of the ST code, as compared to the case when only one transmit antenna is used (no ST coding).

E. Effect of Transmit Antenna Correlation

Next, we study the effect of transmit antenna correlation on the STCM-based modem performance. In this case, we

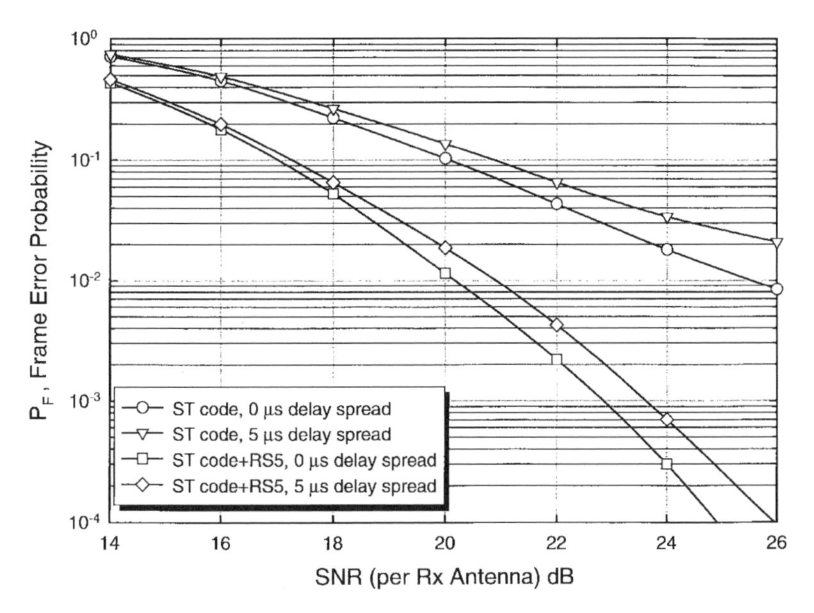

Fig. 19. Performance of the 8-PSK 32-state ST code with two transmit and one receive antennas at $f_d = 180$ Hz in a TU environment with delay spread of 5 μs (GSM TU channel model).

assumed that the channel gains from the two transmit antennas to the jth receive antenna are correlated such that

$$E\{\boldsymbol{\alpha}_j(l)\boldsymbol{\alpha}_j^*(l)\} = \begin{bmatrix} 1 & \rho \\ \rho & 1 \end{bmatrix}$$

where $\boldsymbol{\alpha}_j(l) = [\alpha_{1j}(l)\alpha_{2j}(l)]^T$. Fig. 16 also shows the SNR required for 10% FER as a function of the maximum Doppler frequency for $\rho = 0.75$ and $\rho = 0$ (*uncorrelated channel gains*) for the 8-PSK 32-state ST code with two transmit antennas. We can easily see that, even though the channels from the two transmit antennas are highly correlated, the performance was degraded by less than 1 dB for both the one- and two-receive antennas cases.

TABLE V
THE GSM TU CHANNEL MODEL: DELAY SPREAD = 5 μs

Delay (μs)	0.0	0.2	0.5	1.6	2.3	5.0
Strength (dB)	-3.0	0.0	-2.0	-6.0	-8.0	-10.0

TABLE VI
THE GSM HT CHANNEL MODEL: DELAY SPREAD = 17 μs

Delay (μs)	0.0	0.1	0.3	0.5	15.0	17.0
Strength (dB)	0.0	-1.5	-4.5	-7.5	-6.0	-12.0

F. Performance of ST Coding Versus Delay Diversity

Here, we compare the performance of the STCM scheme versus the simple delay-diversity scheme. For that purpose, we consider the delay-diversity scheme with 8-PSK constellation. In this case the delay-diversity scheme will have the ST coding representation shown in Fig. 4. We simulated the STCM-based modem with the delay-diversity code shown in Fig. 4 as its ST code. Figs. 17 and 18 show the SNR required for 10% FER as a function of the maximum Doppler frequency f_d for the cases with one and two receive antennas, respectively. We show the results for both the WIF and the LPIF. We also show the corresponding results for the 8-PSK 32-state ST code. We can immediately see that the STCM scheme has an approximately 4-dB SNR advantage over simple delay diversity with two receive antennas for both the LPIF and the WIF. For one receive antenna and the WIF, the SNR advantage of STCM over simple delay diversity is about 2.5 and 4.5 dB for $f_d = 10$ and 180 Hz, respectively. For the one-receive antenna case and LPIF, this advantage goes up 9 dB at a maximum Doppler frequency of 180 Hz. The superior performance of the ST code over the delay-diversity scheme is due to the extra coding gain provided by the code.

G. Performance in Delay Spread Channels

In all of our discussions and simulations so far we have assumed that $h_{ij}(t)$, the channel impulse response (CIR) from ith transmitting antenna to jth receiving antenna, is a frequency-ßat channel. That is, the channel impulse response is assumed to be

$$h_{ij}(\tau; t) = \alpha_{ij}(t)\delta(\tau - \tau_o)$$

where $\alpha_{ij}(t)$ is the channel gain deÞned earlier and τ_o is the propagation delay. This model is generally valid as long as the delay spread of the channel is *much* less than the symbol period. Measurements for typical urban (TU) and hilly terrain (HT) propagation environments, however, show delay spreads of up to 5 and 17 μs [44], respectively. In this case, the CIR will be

$$h_{ij}(\tau; t) = \sum_{u=1}^{U} \alpha_{ij,u}(t)\delta(\tau - \tau_u)$$

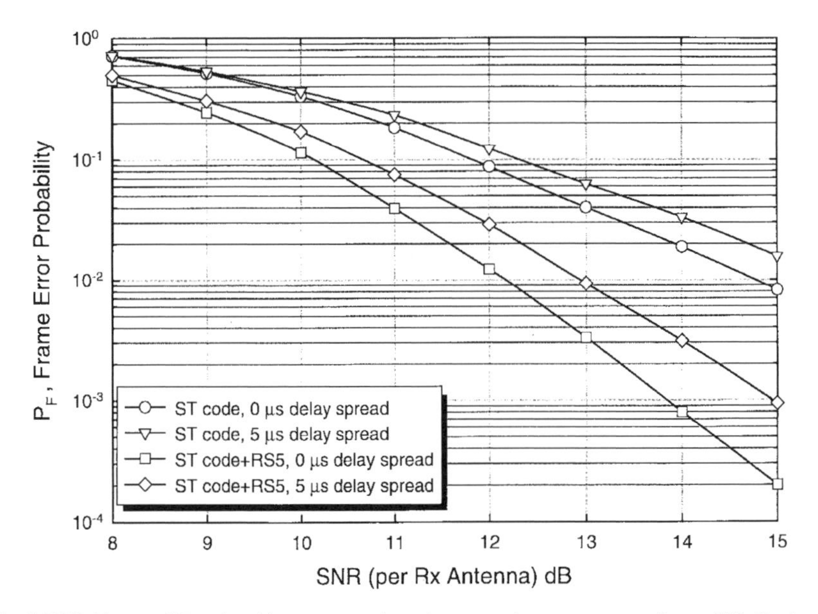

Fig. 20. Performance of the 8-PSK 32-state ST code with two transmit and two receive antennas at $f_d = 180$ Hz in a TU environment with delay spread of 5 μs (GSM TU channel model).

where $\alpha_{ij,u}(t)$ and τ_u are the complex channel gain and propagation delay for the uth multipath component.

Tables V and VI show the GSM measurement-based channel models for TU and HT channels [44]. We simulated the STCM-based modem described above with the 8-PSK 32-state space time code for both the TU and HT channel models. Figs. 19 and 20 show the FER performance for the STCM model with one and two receive antennas at a maximum Doppler frequency of 180 Hz and a TU channel model with 5 μs delay spread, respectively. From these two Þgures, we can easily see that for the TU channel model, and at $f_d = 180$ Hz, the performance degradation due to the multipath is 0.5 dB or less at an FER of 10%. Fig. 21 shows the SNR performance at 10% FER for the STCM-based modem using the TU channel model, which shows a performance degradation (at 10% FER) of less than 0.5 dB for the one receive antenna case and less than 1 dB for the two receive antenna case. For the HT channel model, however, the results showed a substantial error ßooring due to the severe ISI caused by the channel. In this case it is very clear that an equalizer must be used, which is currently under investigation.

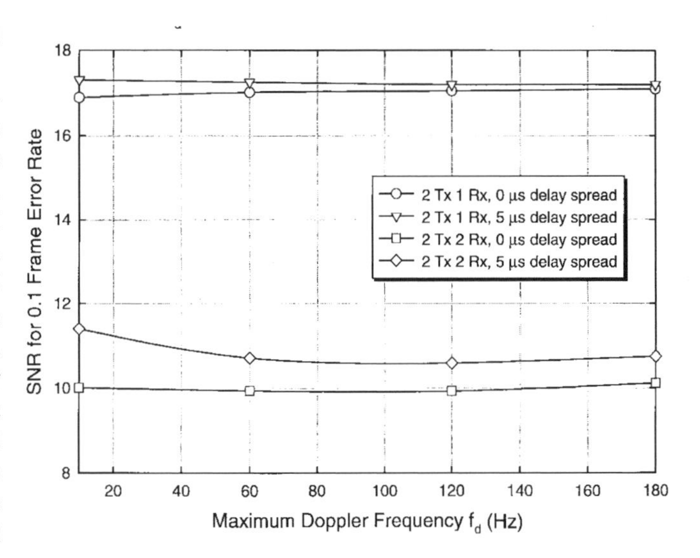

Fig. 21. Effect of delay spread on the SNR performance at 10% FER, or the 8-PSK 32-state ST code as a function of f_d.

V. Discussion and Conclusion

We have proposed a new advanced modem technology for high-data-rate wireless communications. This technology is based on the use of concatenated STCM with multiple transmit and/or multiple receive antennas. The spatial and temporal properties of STCM guarantee that, unlike with other transmit diversity techniques, diversity is achieved at the transmitter while maintaining optional receive diversity, without any sacriÞce in transmission rate. The multichannel CSI required at the receiver for decoding is estimated using O-PSI techniques. A detailed design for a narrowband TDMA/STCM-based modem has been presented. Simulation results for the proposed STCM-based modem show great promise for STCM techniques as a powerful channel coding method for high-data-rate wireless applications. For example, the 16-QAM 16-state STCM-based modem with two transmit and two receive antennas, presented earlier, can achieve a net bit rate of 74.4 kb/s with 10% FER at SNR of 16 dB which is a 2.6 times increase in data rate, as compared to the net bit rate of 28.8 kb/s offered by the current IS-136 [45]. In addition, the STCM modem achieves these bit rates at an SNR that is lower than that required by current systems [45]. Simulation results also showed that the STCM-based modem performs well even when there is a correlation between the two transmit antennas, which suggests that the same concept can be easily applied at both base stations and handsets. In addition, the performance of the STCM-based modem in a typical urban propagation environment was very close to that under a frequency-ßat channel response. For propagation environments with large delay spreads ($\geq T_s/4$), however, multichannel equalization is necessary in order to maintain the performance of the STCM-based modem at acceptable levels. Efforts to design good multichannel equalizers for STCM are now under investigation. Research on the interaction and combination of STCM with other techniques, such as orthogonal frequency division multiplexing (OFDM), maximum likelihood (ML) decoding and interference cancellation, and beamforming is now being pursued [46]–[48].

ACKNOWLEDGMENT

The authors would like to thank the reviewers for their helpful comments and their thorough review. Their remarks greatly improved the presentation of the paper.

REFERENCES

[1] V. Tarokh, N. Seshadri, and A. R. Calderbank, "Space-time codes for high data rate wireless communications: Performance criterion and code construction," *IEEE Trans. Inform. Theory*, vol. 44, pp. 744–765, Mar. 1998.
[2] ——, "Space-time codes for high data rate wireless communications: Performance criterion and code construction," in *Proc. IEEE ICC'97*, Montreal, Canada, 1997, pp. 299–303.
[3] N. Seshadri, V. Tarokh, and A. R. Calderbank, "Space-time codes for high data rate wireless communications: Code construction," in *Proc. IEEE VTC'97*, Phoenix, AZ, 1997, pp. 637–641.
[4] V. Tarokh, A. F. Naguib, N. Seshadri, and A. R. Calderbank, "Space-time codes for high data rate wireless communications: Mismatch analysis," in *Proc. IEEE ICC'97*, Montreal, Canada, 1997, pp. 309–313.
[5] V. Tarokh, A. F. Naguib, N. Seshadri, and A. Calderbank, "Space-time codes for high data rate wireless communications: Performance criteria in the presence of channel estimation errors, mobility, and multiple paths," *IEEE Trans. Commun.*, to be published.
[6] N. Sollenberger and S. Kasturia, "Evolution of TDMA (Is-54/IS-136) to foster further growth of PCS," in *Proc. ICUPC Int. Conf. Universal Personal Communications, 1996*, Boston, MA, 1996.
[7] "Special issue on the European path toward UMTS," *IEEE Personal Commun. Mag.*, vol. 2, Feb. 1995.
[8] J. G. Proakis, "Adaptive equalization for TDMA digital mobile radio," *IEEE Trans. Veh. Technol.*, vol. 40, pp. 333–341, May 1991.
[9] R. Price and J. P. E. Green, "A communication technique for multipath channels," *Proc. IRE*, vol. 46, pp. 555–570, Mar. 1958.
[10] G. Turin, "Introduction to spread-spectrum antimultipath techniques and their application to urban digital radio," *Proc. IEEE*, vol. 68, pp. 328–353, Mar. 1980.
[11] P. S. Henry and B. S. Glance, "A new approach to high capacity digital mobile radio," *Bell Syst. Tech. J.*, vol. 51, no. 8, pp. 1611–1630, Sept. 1972.
[12] J. H. Winters, "Switched diversity with feedback for DPSK mobile radio systems," *IEEE Trans. Veh. Technol.*, vol. 32, pp. 134–150, Feb. 1983.
[13] G. J. Foschini and M. J. Gans, submitted for publication.
[14] G. Foscini, "Layered space-time architecture for wireless communication in a fading environment when using multi-element antennas," *AT&T Tech. J.*, vol. 1, Autumn 1996.
[15] A. Narula, M. Trott, and G. Wornell, submitted for publication.
[16] E. Teletar, "Capacity of multi-antenna Gaussian channels," AT&T Bell Laboratories, Murray Hill, NJ, Tech. Memo., 1995.
[17] A. Wittneben, "Base station modulation diversity for digital SIMULCAST," in *Proc. IEEE VTC'91*, St. Louis, MO, 1991, vol. 1, pp. 848–853.
[18] N. Seshadri and J. H. Winters, "Two schemes for improving the performance of frequency-division duplex (FDD) transmission systems using transmitter antenna diversity," *Int. J. Wireless Inform. Networks*, vol. 1, no. 1, pp. 49–60, Jan. 1994.
[19] A. Wittneben, "A new bandwidth efficient transmit antenna modulation diversity scheme for linear digital modulation," in *Proc. IEEE ICC'93*, Geneva, Switzerland, vol. 3, pp. 1630–1634.
[20] J.-C. Guey, M. P. Fitz, M. R. Bell, and W.-Y. Kuo, "Signal design for transmitter diversity wireless communication systems over Rayleigh fading channels," in *Proc. IEEE VTC'96*, Atlanta, GA, vol. 1, pp. 136–140.
[21] J. H. Winters, "Diversity gain of transmit diversity in wireless systems with Rayleigh fading," in *Proc. IEEE ICC'94*, New Orleans, LA, vol. 2, pp. 1121–1125.
[22] ——, "Diversity gain of transmit diversity in wireless systems with Rayleigh fading," *IEEE Trans. Veh. Technol.*, vol. 47, pp. 119–123, Feb. 1998.
[23] A. Hiroike, F. Adachi, and N. Nakajima, "Combined effects of phase sweeping transmitter diversity and channel coding," *IEEE Trans. Veh. Technol.*, vol. 41, pp. 170–176, May 92.
[24] T. Hattori and K. Hirade, "Multitransmitter simulcast digital signal transmission by using frequency offset strategy in land mobile radiotelephone," *IEEE. Trans. Veh. Technol.*, vol. VT-27, pp. 231–238, 1978.
[25] V. Weerackody, "Diversity for the direct-sequence spread spectrum system using multiple transmit antennas," in *Proc. ICC'93*, Geneva, Switzerland, May, 1993, vol. 3, pp. 1503–1506.
[26] J. K. Cavers, "An analysis of pilot symbol assisted modulation for Rayleigh faded channels," *IEEE Trans. Veh. Technol.*, vol. 40, pp. 683–693, Nov. 1991.
[27] S. Sampei and T. Sunaga, "Rayleigh fading compensation method for 16 QAM in digital land mobile radio channels," in *Proc. IEEE VTC'89*, San Francisco, CA, May 1989, vol. I, pp. 640–646.
[28] M. L. Moher and J. H. Lodge, "TCMP-A modulation and coding strategy for Rician fading channels," *IEEE J. Select. Areas Commun.*, vol. 7, pp. 1347–1355, Dec. 1989.
[29] R. J. Young, J. H. Lodge, and L. C. Pacola, "An implementation of a reference symbol approach to generic modulation in fading channels," in *Proc. Int. Mobile Satellite Conf.*, Ottawa, Canada, June 1990, pp. 182–187.
[30] J. Yang and K. Feher, "A digital Rayleigh fade compensation technology for coherent OQPSK System," in *Proc. IEEE VTC'90*, Orlando, FL, May 1990, pp. 732–737.
[31] C. L. Liu and K. Feher, "A new generation of Rayleigh fade compensated $\pi/4$-QPSK coherent modem," in *Proc. IEEE VTC'90*, Orlando, FL, May 1990, pp. 482–486.
[32] A. Aghamohammadi, H. Meyr, and G. Asheid, "A new method for phase synchronization and automatic gain control of linearly modulated signals on frequency-flat fading channel," *IEEE Trans. Commun.*, vol. 39, pp. 25–29, Jan. 1991.
[33] W. C. Jakes, *Microwave Mobile Communications*. New York: Wiley, 1974.
[34] J. G. Proakis, *Digital Communications*, 2nd ed. New York: McGraw-Hill, 1989.
[35] R. A. Horn and C. R. Johnson, *Matrix Analysis*. Cambridge: Cambridge Univ. Press, 1985.
[36] E. Biglieri, D. Divsalar, P. J. McLane, and M. K. Simon, *Introduction to Trellis Coded Modulation with Applications*. New York: Maxwell Macmillan, 1991.
[37] F. D. Natali, "AFC tracking algorithms," *IEEE Trans. Commun.*, vol. COM-32, pp. 935–947, Aug. 1984.
[38] S. Sampei and T. Sunaga, "Rayleigh fading compensation method for QAM in digital land mobile radio channels," *IEEE Trans. Veh. Technol.*, vol. 42, pp. 137–147, May 1993.
[39] J. Cioffi, "Digital communication," class notes, Stanford University, 1996.
[40] K. Feher, *Digital Communications Sattelite/Earth Station Engineering*. Englewood Cliffs, NJ: Prentice-Hall, 1981.
[41] G. H. Golub and C. F. V. Loan, *Matrix Computations*, 2nd ed. Baltimore, MD: Johns Hopkins Press, 1989.
[42] N. W. K. Lo, D. D. Falconer, and A. U. H. Sheikh, "Adaptive equalization and diversity combining for mobile radio using interpolated channel estimates," *IEEE Trans. Veh. Technol.*, vol. 40, pp. 636–645, Aug. 1991.
[43] G. Oetken and T. W. Parks, "New results in the design of digital interpolators," *IEEE Trans. Acoustics, Speech, Signal Processing*, June 1975, pp. 301–309.
[44] K. Pahlavan and A. H. Levesque, *Wireless Information Networks*. New York: Wiley, 1995.
[45] E. Gelblum and N. Seshadri, "High-rate coded modulation schemes for 16 Kbps speech in wireless systems," in *Proc. 1997 47th IEEE Vehicular Technology Conf.*, Phoenix, AZ, May 1997, pp. 349–353.
[46] V. Tarokh, A. F. Naguib, N. Seshadri, and A. R. Calderbank, "Array signal processing and space–time coding for very high data rate wireless communications," submitted for publication.
[47] A. F. Naguib, N. Seshadri, V. Tarokh, and S. Alamouti, "Combined interference cancellation and ML decoding of block space–time codes," submitted for publication.
[48] D. Agrawal, V. Tarokh, A. F. Naguib, and N. Seshadri, "Space-time coded OFDM for high data rate wireless communications over wideband channels," in *Proc. IEEE VTC'98*, Ottawa, Canada, May 1998.

Ayman F. Naguib (S'91–M'96) received the B.Sc. degree (with honors) and the M.S. degree in electrical engineering from Cairo University, Cairo, Egypt, in 1987 and 1990, respectively, and the M.S. degree in statistics and the Ph.D. degree in electrical engineering from Stanford University, Stanford, CA, in 1993 and 1996, respectively.

From 1987 to 1989, he served at the Signal Processing Laboratory, The Military Technical College, Cairo, Egypt. From 1989 to 1990, he was employed with Cairo University as a Research and Teaching Assistant in the Communication Theory Group, Department of Electrical Engineering. From 1990 to 1995, he was a Research and Teaching Assistant in the Information Systems Laboratory, Stanford University. In 1996, he joined AT&T Labs-Research, Florham Park, NJ, as a Senior Member of the Technical Staff. His current research interests include signal processing and coding for high-data-rate wireless and digital communications and modem design for broadband systems.

Vahid Tarokh (M'97) received the Ph.D. degree in electrical engineering from the University of Waterloo, Waterloo, Ontario, Canada, in 1995.

He is currently a Senior Member of the Technical Staff at AT&T Labs-Research, Florham Park, NJ.

Nambirajan Seshadri (S'81–M'82–SM'95) received the B.S. degree in electronics and communications engineering from the University of Madras, Madaras, India, in 1982 and the M.S. and Ph.D. degrees in electrical and computer engineering from Rensselaer Polytechnic Institute, Troy, NY, in 1984 and 1986, respectively.

He was a Distinguished Member of the Technical Staff at AT&T Bell Laboratories, Murray Hill, NJ, and is now Head of the Communications Research Department at AT&T Labs-Research, Florham Park, NJ. His technical interests include coding and modulation, diversity techniques, and reliable transmission of audio-visual signals over wireless channels.

Dr. Seshadri has just completed his term as Associate Editor for Coding Techniques for the IEEE TRANSACTIONS ON INFORMATION THEORY.

A. Robert Calderbank (M'89–SM'97–F'98) received the B.S. degree in mathematics from Warwick University, Coventry, U.K., in 1975, the M.S. degree in mathematics from Oxford University, Oxford, U.K., in 1976, and the Ph.D. degree in mathematics from the California Institute of Technology, Pasadena, in 1980.

He joined AT&T Bell Laboratories in 1980, and was a Department Head in the Mathematical Sciences Research Center at Murray Hill. He is currently Director of the Information Sciences Research Center at AT&T Labs-Research, Florham Park, NJ. His research interests include algebraic coding theory, wireless data transmission, and quantum computing. At the University of Michigan, and at Princeton University he developed and taught an innovative course on bandwidth-efficient communication.

Dr. Calderbank was the Associate Editor for Coding Techniques for the IEEE TRANSACTIONS ON INFORMATION THEORY from 1986 to 1989. He was also a Guest Editor for a Special Issue of the IEEE TRANSACTIONS ON INFORMATION THEORY dedicated to coding for storage devices. He served on the Board of Governors of the IEEE Information Theory Society from 1990 to 1996. He was a corecipient of the 1995 Prize Paper Award from the Information Theory Society for his work on the Z4 linearity of the Kerdock and Preparata codes. He is currently Editor-in-Chief of the IEEE TRANSACTIONS ON INFORMATION THEORY.

Near Optimum Error Correcting Coding And Decoding: Turbo-Codes

Claude Berrou, *Member, IEEE,* and Alain Glavieux

***Abstract*— This paper presents a new family of convolutional codes, nicknamed turbo-codes, built from a particular concatenation of two recursive systematic codes, linked together by nonuniform interleaving. Decoding calls on iterative processing in which each component decoder takes advantage of the work of the other at the previous step, with the aid of the original concept of extrinsic information. For sufficiently large interleaving sizes, the correcting performance of turbo-codes, investigated by simulation, appears to be close to the theoretical limit predicted by Shannon.**

I. Introduction

CONVOLUTIONAL error correcting or channel coding has become widespread in the design of digital transmission systems. One major reason for this is the possibility of achieving real-time decoding without noticeable information losses thanks to the well-known soft-input Viterbi algorithm [1]. Moreover, the same decoder may serve for various coding rates by means of puncturing [2], allowing the same silicon product to be used in different applications. Two kinds of convolutional codes are of practical interest: nonsystematic convolutional (NSC) and recursive systematic convolutional (RSC) codes. Though RSC codes have the same free distance d_f as NSC codes and exhibit better performance at low signal to noise ratios (SNR's) and/or when punctured, only NSC codes have actually been considered for channel coding, except in Trellis-coded modulation (TCM) [3]. Section II presents the principle and the performance of RSC codes, which are at the root of the study expounded in this article.

For a given rate, the error-correcting power of convolutional codes, measured as the coding gain at a certain binary error rate (BER) in comparison with the uncoded transmission, grows more or less linearly with code memory ν. Fig. 1 (from [4]) shows the achievable coding gains for different rates, and corresponding bandwidth expansion rates, by using classical NSC codes with $\nu = 2, 4, 6$ and 8, for a BER of 10^{-6}. For instance, with $R = 1/2$, each unit added to ν adds about 0.5 dB more to the coding gain, up to $\nu = 6$; for $\nu = 8$, the additional gain is lower. Unfortunately, the complexity of the decoder is not a linear function of ν and it grows exponentially as $\nu \cdot 2^{\nu}$. Factor 2 represents the number of states processed by the decoder and the multiplying factor ν accounts for the complexity of the memory part (metrics and survivor memory). Other technical limitations like the interconnection constraint in the silicon decoder lay-out, make the value of six a practical upper limit for ν for most applications, especially for high data rates.

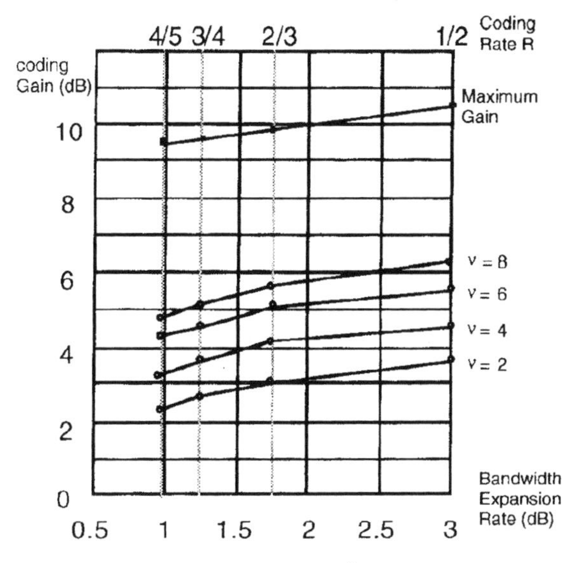

Fig. 1. Coding gains at BER equal to 10^{-6}, achievable with NSC codes (three-bit quantization, from [4], and maximum possible gains for 1/2, 2/3, 3/4, and 4/5 rates in a Gaussian channel, with quaternary phase shift keying (QPSK) modulation.

In order to obtain high coding gains with moderate decoding complexity, concatenation has proved to be an attractive scheme. Classically, concatenation has consisted in cascading a block code (the outer code, typically a Reed-Solomon code) and a convolutional code (the inner code) in a serial structure. Another concatenated code, which has been given the familiar name of *turbo-code,* with an original parallel organization of two RSC elementary codes, is described in Section III. Some comments about the distance properties of this composite

Paper approved by D. Divsalar, the Editor for Coding Theory and Applications of the IEEE Communications Society. Manuscript received March 3, 1995; revised December 6, 1995 and February 26, 1996. This paper was presented at ICC'93, Geneva, Switzerland, May 1993.

The authors are with Ecole Nationale Supérieure des Télécommunications de Bretagne, BP 832 29285 BREST CEDEX, France.

Publisher Item Identifier S 0090-6778(96)07373-4.

Reprinted from *IEEE Transactions on Communications,* vol. 44, no. 10, October 1996.

code, are propounded. When decoded by an iterative process, turbo-codes offer near optimum performance. The way to achieve this decoding with the *Maximum A Posteriori* (MAP) algorithm is detailed in Sections IV, V, VI, and some basic results are given in Section VII.

II. Recursive Systematic Convolutional Codes

A. Introduction

Consider a binary rate $R = 1/2$ convolutional encoder with constraint length K and memory $\nu = K - 1$. The input to the encoder at time k is a bit d_k and the corresponding binary couple (X_k, Y_k) is equal to

$$X_k = \sum_{i=0}^{\nu} g_{1i} d_{k-i} \qquad g_{1i} = 0, 1 \tag{1a}$$

$$Y_k = \sum_{i=0}^{\nu} g_{2i} d_{k-i} \qquad g_{2i} = 0, 1 \tag{1b}$$

where $G_1: \{g_{1i}\}, G_2: \{g_{2i}\}$ are the two encoder generators, expressed in octal form. It is well known that the BER of a classical NSC code is lower than that of a classical nonrecursive systematic convolutional code with the same memory ν at large SNR's, since its free distance is smaller than that of a NSC code [5]. At low SNR's, it is in general the other way round. The RSC code, presented below, combines the properties of NSC and systematic codes. In particular, it can be better than the equivalent NSC code, at any SNR, for code rates larger than 2/3.

A binary rate RSC code is obtained from a NSC code by using a feedback loop and setting one of the two outputs X_k or Y_k equal to the input bit d_k. The shift register (memory) input is no longer the bit d_k but is a new binary variable a_k. If $X_k = d_k$ (respectively, $Y_k = d_k$), the output Y_k (resp. X_k) is defined by (1b) [respectively, (1a)] by substituting d_k for a_k and variable a_k is recursively calculated as

$$a_k = d_k + \sum_{i=1}^{\nu} \gamma_i a_{k-i} \tag{2}$$

where γ_i is respectively equal to g_{1i} if $X_k = d_k$ and to g_{2i} if $Y_k = d_k$. Equation (2) can be rewritten as

$$d_k = \sum_{i=0}^{\nu} \gamma_i a_{k-i}. \tag{3}$$

Taking into account $X_k = d_k$ or $Y_k = d_k$, the RSC encoder output $C_k = (X_k, Y_k)$ has exactly the same expression as the NSC encoder outputs if $g_{10} = g_{20} = 1$ and by substituting d_k for a_k in (1a) or (1b).

Two RSC encoders with memory $\nu = 2$ and rate $R = 1/2$, obtained from a NSC encoder defined by generators $G_1 = 7, G_2 = 5$, are depicted in Fig. 2.

Generally, we assume that the input bit d_k takes values zero or one with the same probability. From (2), we can show that variable a_k exhibits the same statistical property

$$\Pr\{a_k = 0/a_{k-\nu} = \varepsilon_\nu, \cdots a_{k-i} = \varepsilon_i, \cdots a_{k-1} = \varepsilon_1\} = \Pr\{d_k = \varepsilon\} = \tfrac{1}{2} \tag{4}$$

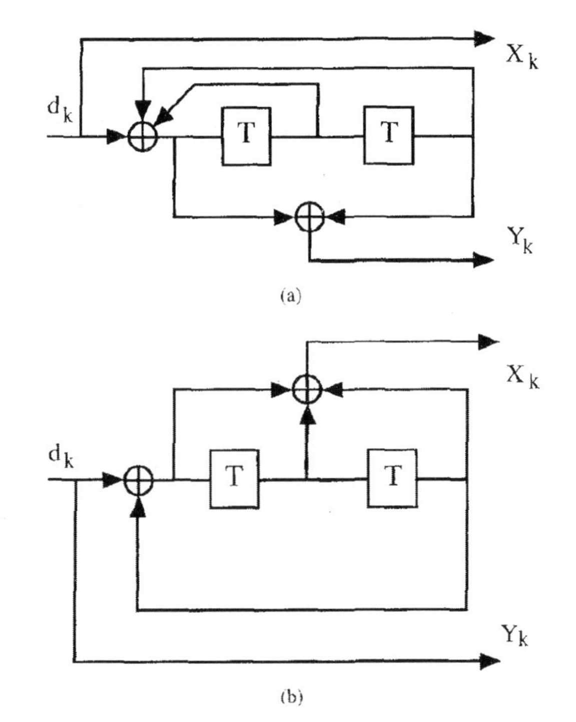

Fig. 2. Two associated Recursive systematic convolutional (RSC) encoders with memory $\nu = 2$, rate $R = 1/2$ and generators $G_1 = 7, G_2 = 5$.

where ε is equal to

$$\varepsilon = \sum_{I=1}^{\nu} \gamma_i \varepsilon_i; \quad \varepsilon_i = 0, 1. \tag{5}$$

Thus, the transition state probabilities $\pi(\boldsymbol{S}_k = \boldsymbol{m}/\boldsymbol{S}_{k-1} = \boldsymbol{m}')$, where $\boldsymbol{S}_k = \boldsymbol{m}$ and $\boldsymbol{S}_{k-1} = \boldsymbol{m}'$ are, respectively, the encoder state at time k and at time $(k-1)$, are identical for the equivalent RSC and NSC codes; moreover these two codes have the same free distance d_f. However, for a same input sequence $\{d_k\}$, the two output sequences $\{X_k\}$ and $\{Y_k\}$ are different for RSC and NSC codes.

When puncturing is considered, some output bits X_k or Y_k are deleted according to a chosen perforation pattern defined by a matrix $\boldsymbol{P}$. For instance, starting from a rate $R = 1/2$ code, the matrix $\boldsymbol{P}$ of rate 2/3 punctured code can be equal to

$$\boldsymbol{P} = \begin{bmatrix} 1 & 1 \\ 1 & 0 \end{bmatrix}. \tag{6}$$

For the punctured RSC code, bit d_k must be emitted at each time k. This is obviously done if all the elements belonging to the first or second row of matrix $\boldsymbol{P}$ are equal to one. When the best perforation pattern for a punctured NSC code is such that matrix $\boldsymbol{P}$ has null elements in the first and the second rows, the punctured recursive convolutional code is no longer systematic. In order to use the same matrix $\boldsymbol{P}$ for both RSC and NSC codes, the RSC encoder is now defined by (3), (7), and (8)

$$Y_k = \sum_{i=0}^{\nu} \lambda_i a_{k-i} \tag{7}$$

$$X_k = d_k \tag{8}$$

where coefficients γ_i [see (3)] and λ_i are, respectively, equal to g_{1i} and g_{2i} when element $p_{1j}; 1 \leq j \leq n$ of matrix $\boldsymbol{P}$ is equal to one and to g_{2i} and g_{1i} when p_{1j} is equal to zero.

B. Recursive Systematic Code Performance

In order to compare the performance of RSC and NSC codes, we determined their weight spectrum and their BER. The weight spectrum of a code is made up of two sets of coefficients $a(d)$ and $W(d)$ obtained from two series expansions related to the code transfer function $T(D, N)$ [6]

$$T(D,N)\Big|_{N=1} = \sum_{d=d_f}^{\infty} a(d)D^d \tag{9}$$

$$\frac{\partial T(D,N)}{\partial N}\Big|_{N=1} = \sum_{d=d_f}^{\infty} W(d)D^d \tag{10}$$

where d_f is the free distance of the code, $a(d)$ is the number of paths at Hamming distance d from the "null" path and $W(d)$ is the total Hamming weight of input sequences $\{d_k\}$ used to generate all paths at distance d from the "null" path. In general, it is not easy to calculate the transfer function of a punctured code, that is why the first coefficients $a(d)$ and $W(d)$ are directly obtained from the trellis by using an algorithm derived from [7]. From coefficients $W(d)$, a tight upper bound of error probability can be calculated for large SNR's [6]

$$P_e \leq \sum_{d=d_f}^{\infty} W(d)P(d). \tag{11}$$

For a memoryless Gaussian channel with binary modulation (PSK, QPSK), probability $P(d)$ is equal to

$$P(d) = \frac{1}{2}\operatorname{erfc}\left[\sqrt{dR\frac{E_b}{N_0}}\right] \tag{12}$$

where E_b/N_0 is the energy per information bit to noise power spectral density ratio and R is the code rate.

In [8], a large number of RSC codes have been investigated and their performance was compared to that of NSC codes, in term of weight spectrum and of BER. Coefficients $a(d)$ are the same for RSC and NSC codes but the coefficients $\{W_{\mathrm{RSC}}(d)\}$ of RSC codes have a tendency to increase more slowly in function of d than the coefficients $\{W_{\mathrm{NSC}}(d)\}$ of NSC codes, whatever the rate R and whatever the memory ν. Thus, at low SNR's, the BER of the RSC code is always smaller than the BER of the equivalent NSC code.

In general, for rates $R \leq 2/3$, the first coefficients $(W_{\mathrm{RSC}}(d_f)), W_{\mathrm{RSC}}(d_f+1))$ are larger than those of NSC codes, therefore, at large SNR's, the performance of NSC codes is a little better than that of RSC codes. When the rate is larger than 2/3, it is easy to find RSC codes whose performance is better than that of NSC codes at any SNR.

In order to illustrate the performance of RSC codes, the BER of RSC and NSC codes are plotted in Fig. 3 for different values of R and for an encoder with memory $\nu = 6$ and generators 133, 171. For instance, at $P_e = 10^{-1}$, the coding gain with this RSC code, relative to the equivalent NSC code,

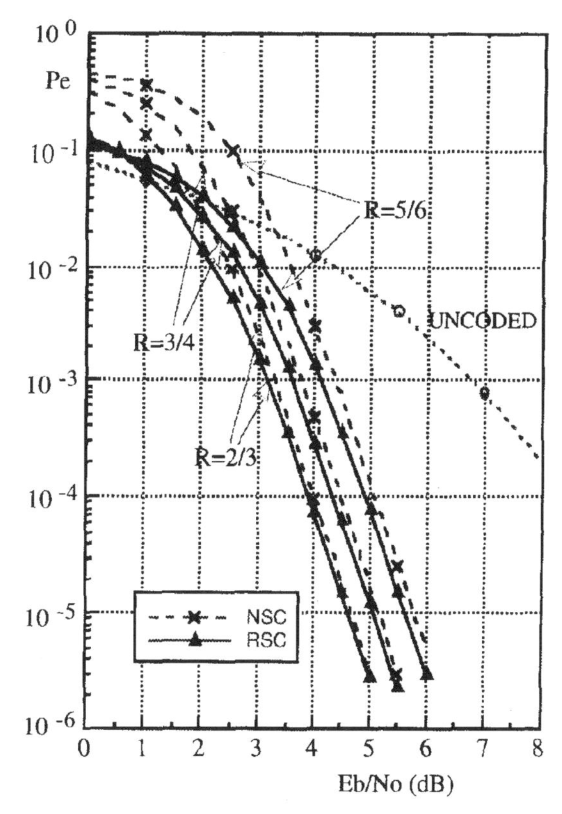

Fig. 3. P_e of punctured RSC and NSC codes for different values of rate R and memory $\nu = 6$, generators $G_1 = 133, G_2 = 171$.

is approximately 0.8 dB at $R = 2/3$, whereas at $R = 3/4$, it reaches 1.75 dB.

III. Parallel Concatenation with Non Uniform Interleaving: Turbo-Code

A. Construction of the Code

The use of systematic codes enables the construction of a concatenated encoder in the form given in Fig. 4, called *parallel concatenation*. The data flow (d_k at time k) goes directly to a first elementary RSC encoder C_1 and after interleaving, it feeds (d_n at time k) a second elementary RSC encoder C_2. These two encoders are not necessarily identical. Data d_k is systematically transmitted as symbol X_k and redundancies Y_{1k} and Y_{2k} produced by C_1 and C_2 may be completely transmitted for an $R = 1/3$ encoding or punctured for higher rates. The two elementary coding rates R_1 and R_2 associated with C_1 and C_2, after puncturing, may be different, but for the best decoding performance, they will satisfy$R_1 \leq R_2$. The global rate R of the composite code, R_1 and R_2 are linked by

$$\frac{1}{R} = \frac{1}{R_1} + \frac{1}{R_2} - 1. \tag{13}$$

Unlike the classical (serial) concatenation, parallel concatenation enables the elementary encoders, and therefore the associated elementary decoders, to run with the same clock. This point constitutes an important simplification for the design of the associated circuits, in a concatenated scheme.

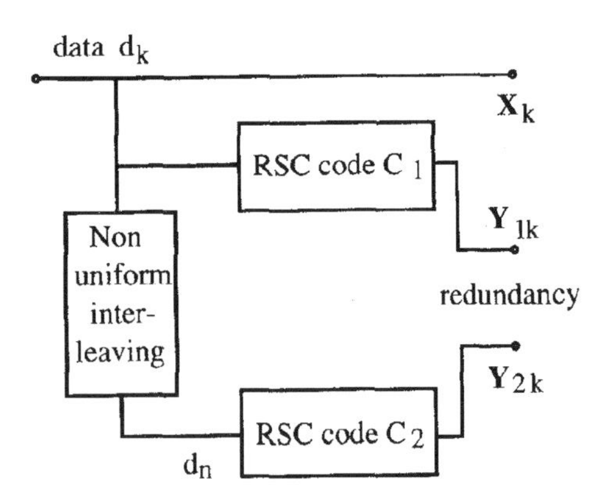

Fig. 4. Basic turbo-encoder (rate 1/3).

B. Distance Properties

Consider for instance elementary codes C_1 and C_2 with memory $\nu = 4$ and encoder polynomials $G_1 = 23, G_2 = 35$. Redundancy Y_k is one time every second time, either Y_{1k} or Y_{2k}. Then the global rate is $R = 1/2$ with $R_1 = R_2 = 2/3$. The code being linear, the distance properties will be considered relatively to the "all zero" or "null" sequence. Both encoders C_1, C_2 and the interleaver are initialized to the "all zero" state and a sequence $\{d_k\}$ containing w "1"s, is fed to the turbo-encoder. w is called the *input weight*.

Some definitions: let us call a *Finite Codeword* (FC) of an elementary RSC encoder an output sequence with a finite distance from the "all zero" sequence (i.e., a limited number of "1"s in output sequences $\{X_k\}$ and $\{Y_k\}$). Because of recursivity, only some input sequences, fitting with the linear feedback register (LFR) generation of Y_k, give FC's. These particular input sequences are named *FC (input) patterns*. Let us also call a *global FC* an output sequence of the turbo-code, with a finite distance from the "all zero" sequence (i.e., a limited number of "1"s in output sequences $\{X_k\}, \{Y_{1k}\}$ and $\{Y_{2k}\}$). The distance $d_q(w)$ of an elementary FC ($q = 1$ for $C_1, q = 2$ for C_2), associated with an input sequence $\{d_k\}$ with weight w, is the sum of the two contributions of $\{X_k\}$ and $\{Y_{qk}\}$

$$d_q(w) = d_{Xq}(w) + d_{Yq}(w). \tag{14}$$

Since the codes are systematic, $d_{Xq}(w) = w$

$$d_q(w) = w + d_{Yq}(w). \tag{15}$$

The distance $d(w)$ of a global FC is given likewise by

$$d(w) = w + d_{Y1}(w) + d_{Y2}(w). \tag{16}$$

1) Uniform Interleaving: Consider a uniform block interleaving using an $M \cdot M$ square matrix with M large enough (i.e., ≥ 32), and generally a power of 2. Data are written linewise and read columnwise. As explained above, the matrix is filled with "0"s except for some "1"s and we are now going to state some of the possible patterns of "1"s leading to global FC's and evaluate their associated distances. Beforehand, note that, for each elementary code, the minimal value for input weight w is two, because of their recursive structure. For the particular case of $w = 2$, the delay between the two data at "1" is 15 or a multiple of 15, since the LFR associated with the redundancy generation, with polynomial $G = 23$, is a maximum length LFR. If the delay is not a multiple of 15, then the associated sequence $\{Y_k\}$ will contain "0"s and "1"s indefinitely.

Global FC's with Input Weight $w = 2$

Global FC's with $w = 2$ have to be FC's with input weight $w = 2$ for each of the constituent codes. Then the two data at "1" in the interleaving memory must be located at places which are distant by a multiple of 15, when both writing and reading. For lack of an extensive analysis of the possible patterns which satisfy this double constraint, let us consider only the case where the two "1"s in $\{d_k\}$ are time-separated by the lowest value (i. e. 15). It is also a FC pattern for code C_2 if the two data at "1" are memorized on a single line, when writing in the interleaving memory, and thus the span between the two "1"s is $15 \cdot M$, when reading. From (16), the distance associated with this input pattern is approximately equal to

$$\begin{aligned} d(2) &\approx 2 + \min\{d_{Y1}(2)\} + \mathrm{INT}((15 \cdot M + 1)/4) \\ &= 2 + 4 + \mathrm{INT}((15 \cdot M + 1)/4) \end{aligned} \tag{17}$$

where $\mathrm{INT}(\cdot)$ stands for integer part of $(\cdot)$. The first term represents input weight w, the second term the distance added by redundancy Y_1 of C_1 (rate 2/3), the third term the distance added by redundancy Y_2. The latter is given assuming that, for this second 2/3 rate code, the $(15 \cdot M + 1)/2$ Y_2 symbols are at "1", one out of two times statistically, which explains the additional division by 2. With $M = 64$ for instance, the distance is $d(2) \approx 246$. This value is for a particular case of a pattern of two "1"s but we imagine it is a realistic example for all FC's with input weight 2. If the two "1"s are not located on a single line when writing in the interleaving matrix, and if M is larger than 30, the span between the two "1"s, when reading from the interleaving matrix, is higher than $15 \cdot M$ and the last term in (17) is increased.

Global FC's with $w = 3$

Global FC's with input weight $w = 3$ have to be elementary FC's with input weight $w = 3$ for each of the constituent codes. The inventory of the patterns with three "1"s which satisfy this double constraint is not easy to make. It is not easier to give the slightest example. Nevertheless, we can consider that associated distances are similar to the case of input weight two codewords, because the FC for C_2 is still several times M long.

Global FC's with $w = 4$

Here is the first pattern of a global FC which can be viewed as the separate combination of two minimal (input weight $w = 2$) elementary FC patterns both for C_1 and C_2. When writing in the interleaving memory, the four data at "1" are located at the four corners of a square or a rectangle, the sides of which have lengths equal to 15 or a multiple of 15 [Fig. 5(a)]. The minimum distance is given by a square pattern, with side length equal to 15, corresponding to minimum values

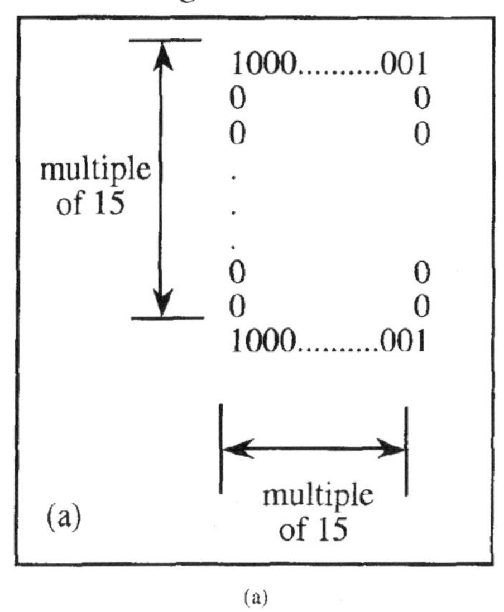

(a)

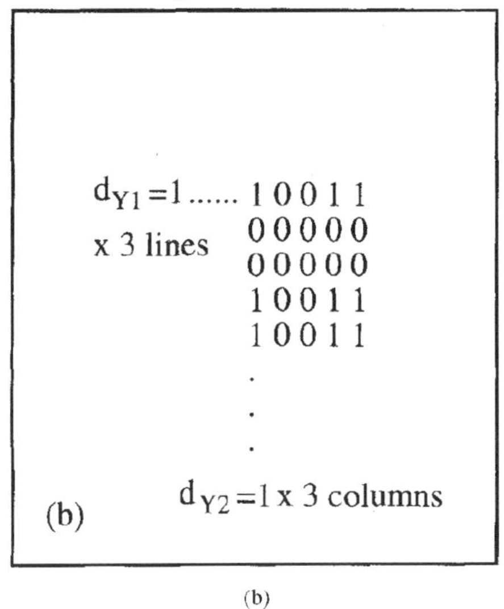

(b)

Fig. 5. (a) Input pattern for separable global FC's with input weight $w = 4$. (b) Input pattern for a global FC with $w = 9$ and total distance of 15.

of $d_{Y1}(2)$ and $d_{Y2}(2)$

$$\begin{aligned} d(4) &= 4 + 2.\min\{d_{Y1}(2)\} + 2.\min\{d_{Y2}(2)\} \\ &= 4 + 8 + 8 = 20. \end{aligned} \tag{18}$$

Global FC's with $w > 4$

As previously, one has to distinguish between cases where the global FC is separable into two, or more, elementary FC's for both codes, or not. The shortest distances are given by separable codewords (this occurs for weights 6, 8, 9, 10, $12, \cdots$). We made an exhaustive research of possible patterns up to $w = 10$ and found a minimum distance of 15 for the pattern given in Fig. 5(b), with $w = 9$, which corresponds to three separable FC's for each of the two codes, and the six FC's having d_{Y1} or d_{Y2} equal to 1.

To conclude this review of the global FC's with the lowest values of we can retain the result obtained for the case in Fig. 5(b) as the minimum distance d_m of this particular turbo-code ($\nu = 4$, polynomials 23, 35, $R_1 = R_2 = 2/3$). Other codes were considered with different values of ν and various polynomials; in all cases, the minimum distance seems to correspond to input sequences with weights 4, 6, or 9. The values of d_m are within 10 and 20, which is not a remarkable property. So as to obtain larger values of d_m, turbo-coding employs nonuniform interleaving.

2) Nonuniform interleaving: It is obvious that patterns giving the shortest distances, such as those represented in Fig. 5, can be "broken" by appropriate nonuniform interleaving, in order to transform a separable FC pattern into either a nonseparable or a non FC. Nonuniform interleaving must satisfy two main conditions: the maximum scattering of data, as in usual interleaving, and the maximum disorder in the interleaved data sequence. The latter, which may be in conflict with the former, is to make redundancy generation by the two encoders as diverse as possible. In this case, if the decision by the decoder associated with C_1 about particular data implies a few items of redundancy Y_1, then the corresponding decision by the decoder associated with C_2 will rely on a large number of values Y_2, and vice-versa. Then, the minimum distance of the turbo-code may be increased to much larger values than that given by uniform interleaving.

The mathematical aspects of nonuniform interleaving are not trivial and have still to be studied. For instance, how can it be ensured that an interleaver, able to "break" patterns corresponding to $w = 4, 6$, or 9, will not drastically lower the distance for a particular case of a $w = 2$ input sequence? On the other hand, the interleaving equations have to be limited in complexity for silicon applications, because several interleavers and de-interleavers have to be employed in a real-time turbo-decoder (see Section VI-A). However that may be, as we have tried to explain up to now, one important property of the turbo-code is that its minimum distance d_m is not fixed, chiefly, by the constituent RSC codes but by the interleaving function; and finding out the optimum interleaver for turbo-codes remains a real challenge.

For the results which are presented in Section VII, we used an empirical approach and chose an interleaving procedure in which, for reading, the column index is a function of the line index. Let i and j be the addresses of line and column for writing, and i_r and j_r the addresses of line and column for reading. For an $M \cdot M$ memory (M being a power of two), i, j, i_r and j_r have values between 0 and $M - 1$. Nonuniform interleaving may be described by

$$\begin{aligned} i_r &= (M/2 + 1)(i + j) && \mod \cdot M \\ \xi &= (i + j) && \mod \cdot 8 \\ j_r &= [P(\xi) \cdot (j + 1)] - 1 && \mod \cdot M \end{aligned} \tag{19}$$

$P(\cdot)$ is a number, relatively prime with M, which is a function of line address $(i + j) \bmod \cdot 8$. Note that reading is performed diagonally in order to avoid possible effects of a relation between M and the period of puncturing. A multiplying factor $(M/2 + 1)$ is used to prevent two neighboring data written on two consecutive lines from remaining neighbors in reading, in a similar way as given in [9].

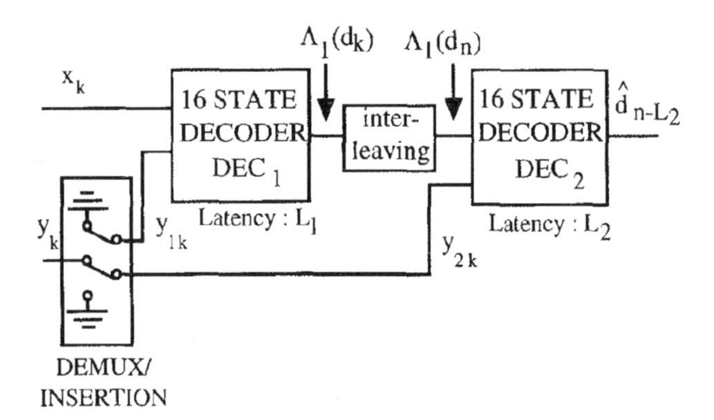

Fig. 6. Principle of the decoder in accordance with a serial concatenation scheme.

IV. Decoding Turbo-Codes

The decoder depicted in Fig. 6, is made up of two elementary decoders (DEC_1 and DEC_2) in a serial concatenation scheme. The first elementary decoder DEC_1 is associated with the lower rate R_1 encoder C_1 and yields a weighted decision.

For a discrete memoryless Gaussian channel and a binary modulation, the decoder input is made up of a couple R_k of two random variables x_k and y_k, at time k

$$\begin{aligned} x_k &= (2X_k - 1) + i_k \\ y_k &= (2Y_k - 1) + q_k \end{aligned} \tag{20}$$

where i_k and q_k are two independent noises with the same variance σ^2. The redundant information y_k is demultiplexed and sent to decoder DEC_1 when $Y_k = Y_{1k}$ and toward decoder DEC_2 when $Y_k = Y_{2k}$. When the redundant information of a given encoder (C_1 or C_2) is not emitted, the corresponding decoder input is set to analog zero. This is performed by the DEMUX/INSERTION block.

The logarithm of likelihood ratio (LLR), $\Lambda_1(d_k)$ associated with each decoded bit d_k by the first decoder DEC_1 can be used as a relevant piece of information for the second decoder DEC_2

$$\Lambda_1(d_k) = \log \frac{\Pr\{d_k = 1/\text{observation}\}}{\Pr\{d_k = 0/\text{observation}\}} \tag{21}$$

where $\Pr\{d_k = i/\text{observation}\}, i = 0, 1$ is the *a posteriori* probability (APP) of the bit d_k.

A. Optimal Decoding of RSC Codes with Weighted Decision

The VITERBI algorithm is an optimal decoding method which minimizes the probability of sequence error for convolutional codes. Unfortunately, this algorithm is not able to yield directly the APP for each decoded bit. A relevant algorithm for this purpose has been proposed by BAHL *et al.* [10]. This algorithm minimizes the bit error probability in decoding linear block and convolutional codes and yields the APP for each decoded bit. For RSC codes, the BAHL *et al.* algorithm must be modified in order to take into account their recursive character.

Modified BAHL et al. Algorithm for RSC Codes: Consider an RSC code with constraint length K; at time k the encoder state S_k is represented by a K-uple

$$S_k = (a_k, a_{k-1} \cdots\cdots a_{k-K+2}). \tag{22}$$

Let us also suppose that the information bit sequence $\{d_k\}$ is made up of N independent bits d_k, taking values zero and one with equal probability and that the encoder initial state S_0 and final state S_N are both equal to zero, *i.e*

$$S_0 = S_N = (0, 0 \cdots\cdots 0) = 0. \tag{23}$$

The encoder output codeword sequence, noted $C_1^N = \{C_1 \cdots\cdots C_k \cdots\cdots C_N\}$ where $C_k = (X_k, Y_k)$ is the input to a discrete Gaussian memoryless channel whose output is the sequence $R_1^N = \{R_1 \cdots\cdots R_k \cdots\cdots R_N\}$ where $R_k = (x_k, y_k)$ is defined by (20).

The APP of a decoded bit d_k can be derived from the joint probability $\lambda_k^i(m)$ defined by

$$\lambda_k^i(m) = \Pr\{d_k = i, S_k = m/R_1^N\} \tag{24}$$

and thus, the APP of a decoded bit d_k is equal to

$$\Pr\{d_k = i/R_1^N\} = \sum_m \lambda_k^i(m), \qquad i = 0, 1. \tag{25}$$

From (21) and (24), the LLR $\Lambda(d_k)$ associated with a decoded bit d_k can be written as

$$\Lambda(d_k) = \log \frac{\sum_m \lambda_k^1(m)}{\sum_m \lambda_k^0(m)}. \tag{26}$$

Finally the decoder can make a decision by comparing $\Lambda(d_k)$ to a threshold equal to zero

$$\begin{aligned} \hat{d}_k &= 1 \quad \text{if} \quad \Lambda(d_k) \geq 0 \\ \hat{d}_k &= 0 \quad \text{if} \quad \Lambda(d_k) < 0. \end{aligned} \tag{27}$$

From the definition (24) of $\lambda_k^i(m)$, the LLR $\Lambda(d_k)$ can be written as shown at the bottom of the page.

Using the BAYES rule and taking into account that events after time k are not influenced by observation R_1^k and bit d_k

$$\Lambda(d_k) = \log \frac{\sum_m \sum_{m'} \Pr\{d_k = 1, S_k = m, S_{k-1} = m', R_1^{k-1}, R_k, R_{k+1}^N\}}{\sum_m \sum_{m'} \Pr\{d_k = 0, S_k = m, S_{k-1} = m', R_1^{k-1}, R_k, R_{k+1}^N\}} \tag{28}$$

if state S_k is known, the LLR $\Lambda(d_k)$ is equal

$$\Lambda(d_k) = \log \frac{\sum_m \sum_{m'} \Pr\{R_{k-1}^N/S_k = m\} \Pr\{S_{k-1} = m'/R_1^{k-1}\}}{\sum_m \sum_{m'} \Pr\{R_{k+1}^N/S_k = m\} \Pr\{S_{k-1} = m'/R_1^{k-1}\}} \cdot \frac{\Pr\{d_k = 1, S_k = m, R_k/S_{k-1} = m'\}}{\Pr\{d_k = 0, S_k = m, R_k/S_{k-1} = m'\}}. \quad (29)$$

In order to compute the LLR $\Lambda(d_k)$, as in [11] let us introduce the probability functions $\alpha_k(m)$, $\beta_k(m)$ and $\gamma_i(R_k, m', m)$ defined by

$$\alpha_k(m) = \Pr\{S_k = m/R_1^k\} \quad (30a)$$

$$\beta_k(m) = \frac{\Pr\{R_{k+1}^N/S_k = m\}}{\Pr\{R_{k+1}^N/R_1^k\}} \quad (30b)$$

$$\gamma_i(R_k, m', m) = \Pr\{d_k = i, S_k = m, R_k/S_{k-1} = m'\}. \quad (30c)$$

Using the LLR definition (29) and (30a), (30b) and (30c), $\Lambda(d_k)$ is equal to

$$\Lambda(d_k) = \log \frac{\sum_m \sum_{m'} \gamma_1(R_k, m', m)\alpha_{k-1}(m')\beta_k(m)}{\sum_m \sum_{m'} \gamma_0(R_k, m', m)\alpha_{k-1}(m')\beta_k(m)} \quad (31)$$

where probabilities $\alpha_k(m)$ and $\beta_k(m)$ can be recursively calculated from probability $\gamma_i(R_k, m', m)$ as in [12]

$$\alpha_k(m) = \frac{\sum_{m'} \sum_{i=0}^{1} \gamma_i(R_k, m', m)\alpha_{k-1}(m')}{\sum_m \sum_{m'} \sum_{i=0}^{1} \gamma_i(R_k, m', m)\alpha_{k-1}(m')} \quad (32)$$

$$\beta_k(m) = \frac{\sum_{m'} \sum_{i=0}^{1} \gamma_i(R_{k+1}, m', m)\beta_{k+1}(m')}{\sum_m \sum_{m'} \sum_{i=0}^{1} \gamma_i(R_{k+1}, m, m')\alpha_k(m')}. \quad (33)$$

The probability $\gamma_i(R_k, m', m)$ can be determined from transition probabilities of the discrete Gaussian memoryless channel and transition probabilities of the encoder trellis. From (30c), $\gamma_i(R_k, m', m)$ is given by

$$\begin{aligned}\gamma_i(R_k, m', m) &= p(R_k/d_k = i, S_k = m, S_{k-1} = m') \\ &\cdot q(d_k = i/S_k = m, S_{k-1} = m') \\ &\cdot \pi(S_k = m/S_{k-1} = m')\end{aligned} \quad (34)$$

where $p(\cdot/\cdot)$ is the transition probability of the discrete Gaussian memoryless channel. Conditionally to $(d_k = i, S_k = m, S_{k-1} = m')$, x_k and y_k $(R_k = (x_k, y_k))$ are two uncorrelated Gaussian variables and thus we obtain

$$\begin{aligned}p(R_k/d_k = i, S_k = m, S_{k-1} = m') &= p(x_k/d_k = i, S_k = m, S_{k-1} = m') \\ &\cdot p(y_k/d_k = i, S_k = m, S_{k-1} = m').\end{aligned} \quad (35)$$

Since the convolutional encoder is a deterministic machine, $q(d_k = i/S_k = m, S_{k-1} = m')$ is equal to 0 or 1. The transition state probabilities $\pi(S_k = m/S_{k-1} = m')$ of the trellis are defined by the encoder input statistic. Generally, $\Pr\{d_k = 1\} = \Pr\{d_k = 0\} = 1/2$ and since there are two possible transitions from each state, $\pi(S_k = m/S_{k-1} = m') = 1/2$ for each of these transitions.

Modified BAHL et al. Algorithm:

Step 0: Probabilities $\alpha_0(m)$ are initialized according to condition (23)

$$\alpha_0(0) = 1; \quad \alpha_0(m) = 0 \quad \forall m \neq 0. \quad (36a)$$

Concerning the probabilities $\beta_N(m)$, the following conditions were used

$$\beta_N(m) = \frac{1}{N} \forall m \quad (36b)$$

since it is very difficult to put the turbo encoder at state 0, at a given time. Obviously with this condition, the last bits of each block are not taken into account for evaluating the BER.

Step 1: For each observation R_k, probabilities $\alpha_k(m)$ and $\gamma_i(R_k, m', m)$ are computed using (32) and (34), respectively.

Step 2: When sequence R_1^N has been completely received, probabilities $\beta_k(m)$ are computed using (33), and the LLR associated with each decoded bit d_k is computed from (31).

V. Extrinsic Information from the Decoder

In this chapter, we will show that the LLR $\Lambda(d_k)$ associated with each decoded bit d_k, is the sum of the LLR's of d_k at the decoder input and of other information called *extrinsic information,* generated by the decoder.

Since the encoder is systematic $(X_k = d_k)$, the transition probability $p(x_k/d_k = i, S_k = m, S_{k-1} = m')$ in expression $\gamma_i(R_k, m', m)$ is independent of state values S_k and S_{k-1}. Therefore we can factorize this transition probability in the numerator and in the denominator of (31)

$$\Lambda(d_k) = \log \frac{p(x_k/d_k = 1)}{p(x_k/d_k = 0)} + \log \frac{\sum_m \sum_{m'} \gamma_1(y_k, m', m)\alpha_{k-1}(m')\beta_k(m)}{\sum_m \sum_{m'} \gamma_0(y_k, m', m)\alpha_{k-1}(m')\beta_k(m)}. \quad (37)$$

Conditionally to $d_k = 1$ (resp. $d_k = 0$), variables x_k are Gaussian with mean 1 (resp. -1) and variance σ^2, thus the LLR $\Lambda(d_k)$ is still equal to

$$\Lambda(d_k)\frac{2}{\sigma^2} x_k + W_k \quad (38)$$

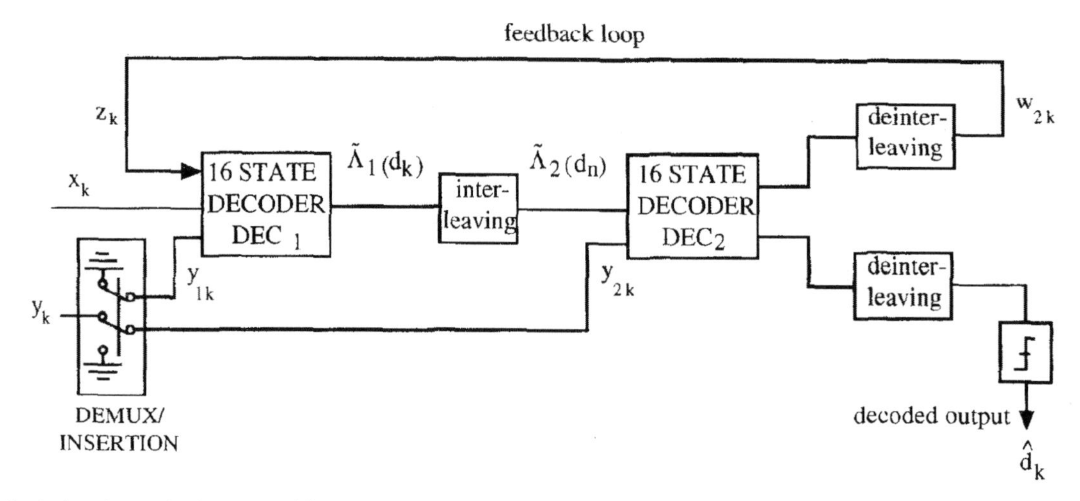

Fig. 7. Feedback decoder (under 0 internal delay assumption).

where

$$W_k = \Lambda(d_k)\,|_{x_k=0} = \log \frac{\sum_{m}\sum_{m'} \gamma_1(y_k, m', m)\alpha_{k-1}(m')\beta_k(m)}{\sum_{m}\sum_{m'} \gamma_0(y_k, m', m)\alpha_{k-1}(m')\beta_k(m)}. \tag{39}$$

W_k is a function of the redundant information introduced by the encoder. In general, W_k has the same sign as d_k; therefore W_k may improve the LLR associated with each decoded bit d_k. This quantity represents the extrinsic information supplied by the decoder and does not depend on decoder input x_k. This property will be used for decoding the two parallel concatenated encoders.

VI. Decoding Scheme of Parallel Concatenation Codes

In the decoding scheme represented in Fig. 6, decoder DEC_1 computes LLR $\Lambda_1(d_k)$ for each transmitted bit d_k from sequences $\{x_k\}$ and $\{y_k\}$, then decoder DEC_2 performs the decoding of sequence $\{d_k\}$ from sequences $\{\Lambda_1(d_k)\}$ and $\{y_k\}$. Decoder DEC_1 uses the modified BAHL *et al.* algorithm and decoder DEC_2 may use the VITERBI algorithm. The global decoding rule is not optimal because the first decoder uses only a fraction of the available redundant information. Therefore it is possible to improve the performance of this serial decoder by using a feedback loop.

A. Decoding with a Feedback Loop

Both decoders DEC_1 and DEC_2 now use the modified BAHL *et al.* algorithm. We have seen in section V that the LLR at the decoder output can be expressed as a sum of two terms if the noises at the decoder inputs are independent at each time k. Hence, if the noise at the decoder DEC_2 inputs are independent, the LLR $\Lambda_2(d_k)$ at the decoder DEC_2 output can be written as

$$\Lambda_2(d_k) = f(\Lambda_1(d_k)) + W_{2k} \tag{40}$$

with

$$\Lambda_1(d_k) = \frac{2}{\sigma^2} x_k + W_{1k}. \tag{41}$$

From (39), we can see that the decoder DEC_2 extrinsic information W_{2k} is a function of the sequence $\{\Lambda_1(d_n)\}_{n\neq k}$. Since $\Lambda_1(d_n)$ depends on observation R_1^N, extrinsic information W_{2k} is correlated with observations x_k and y_{1k}, regarding the noise. Nevertheless, from (39), the greater $|n-k|$ is, the less correlated are $\Lambda_1(d_n)$ and observations x_k, y_k. Thus, due to the presence of interleaving between decoders DEC_1 and DEC_2, extrinsic information W_{2k} and observations x_k, y_{1k} are weakly correlated. Therefore extrinsic information W_{2k} and observations x_k, y_{1k} can be used jointly for carrying out a new decoding of bit d_k, the extrinsic information $z_k = W_{2k}$ acting as a diversity effect.

In Fig. 7, we have depicted a new decoding scheme using the extrinsic information W_{2k} generated by decoder DEC_2 in a feedback loop. For simplicity this drawing does not take into account the different delays introduced by decoder DEC_1 and DEC_2, and interleaving.

The first decoder DEC_1 now has three data inputs: (x_k, y_k^1, z_k), and probabilities $\alpha_{1k}(m)$ and $\beta_{1k}(m)$ are computed by substituting $R_k = (x_k, y_{1k}, z_k)$ for $R_k = (x_k, y_{1k})$ in (32) and (33). Taking into account that z_k is weakly correlated with x_k and y_{1k} and supposing that z_k can be approximated by a Gaussian variable with variance $\sigma_z^2 \neq \sigma^2$, the transition probability of the discrete gaussian memoryless channel can be now factorized in three terms

$$\begin{aligned} p(R_k/d_k &= i, S_k = m, S_{k-1} = m') \\ &= p(x_k/\cdot)p(y_k^1/\cdot)p(z_k/\cdot). \end{aligned} \tag{42}$$

Encoder C_1 with initial rate R_1, through the feedback loop, is now equivalent to a rate R_1' encoder with

$$R_1' = \frac{R_1}{1+R_1}. \tag{43}$$

The first decoder obtains additional redundant information with Z_k that may significantly improve its performance ; the term turbo-code is given for this feedback decoder scheme with reference to the principle of the turbo engine.

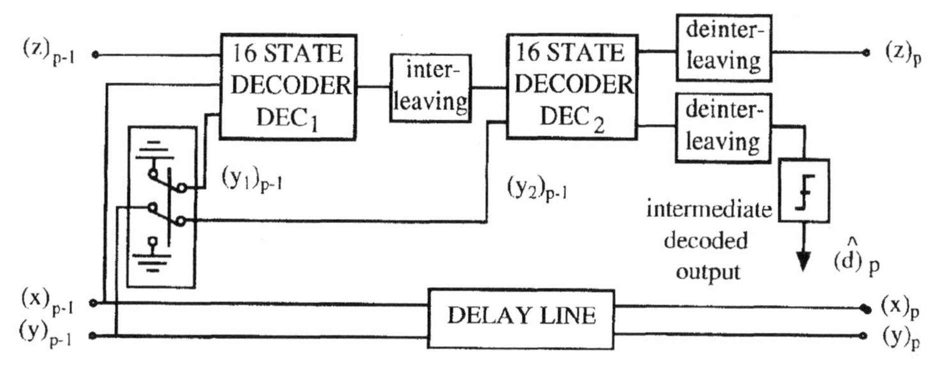

Fig. 8. Decoding module (level p).

With the feedback decoder, the LLR $\Lambda_1(d_k)$ generated by decoder DEC_1 is now equal to

$$\Lambda_1(d_k) = \frac{2}{\sigma^2} x_k + \frac{2}{\sigma_z^2} z_k + W_{1k} \tag{44}$$

where W_{1k} depends on sequence $\{z_n\}_{n \neq k}$. As indicated above, information z_k has been built by decoder DEC_2. Therefore z_k must not be used as input information for decoder DEC_2. Thus decoder DEC_2 input sequences will be sequences $\{\tilde{\Lambda}_1(d_n)\}$ and $\{y_{2k}\}$ with

$$\tilde{\Lambda}_1(d_n) = \Lambda_1(d_n)_{z_n=0}. \tag{45}$$

Finally, from (40), decoder DEC_2 extrinsic information $z_k = W_{2k}$ after deinterleaving can be written as

$$z_k = W_{2k} = \Lambda_2(d_k)|_{\tilde{\Lambda}_1(d_k)=0} \tag{46}$$

and the decision at the decoder DEC output is

$$\hat{d}_k = \mathrm{sign}[\Lambda_2(d_k)]. \tag{47}$$

The decoding delays introduced by the component decoders, the interleaver and the deinterleaver imply that the feedback piece of information z_k must be used through an iterative process.

The global decoder circuit is made up of P pipelined identical elementary decoders. The pth decoder DEC (Fig. 8) input, is made up of demodulator output sequences $(x)_p$ and $(y)_p$ through a delay line and of extrinsic information $(z)_p$ generated by the $(p-1)$th decoder DEC. Note that the variance σ_z^2 of $(z)_p$ and the variance of $\tilde{\Lambda}_1(d_k)$ must be estimated at each decoding step p.

For example, the variance σ_z^2 is estimated for each M^2 interleaving matrix by the following:

$$\sigma_z^2 = \frac{1}{M^2} \sum_{k=1}^{M^2} (|z_k| - m_z)^2 \tag{48a}$$

where m_z is equal to

$$m_z = \frac{1}{M^2} \sum_{k=1}^{M^2} |z_k|. \tag{48b}$$

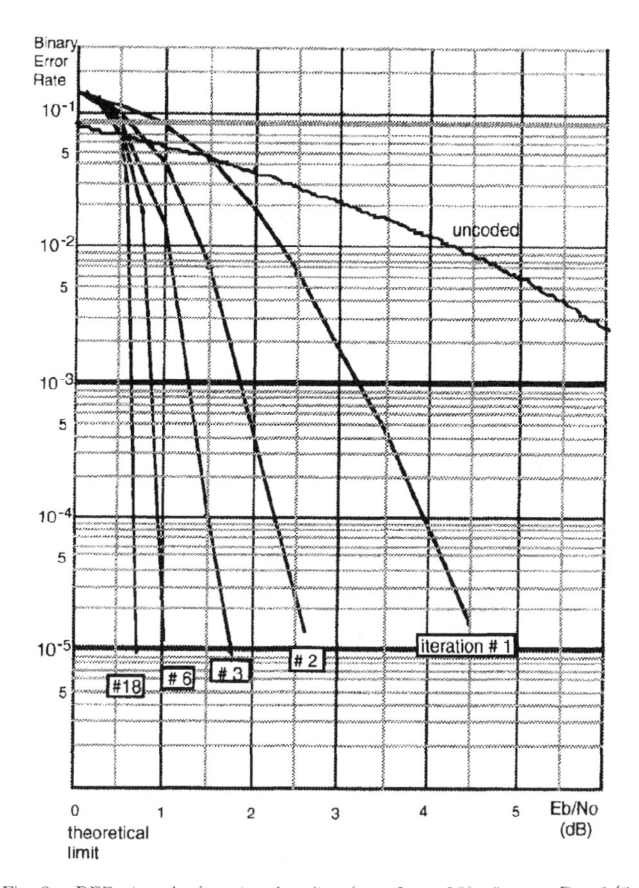

Fig. 9. BER given by iterative decoding ($p = 1, \cdots 18$) of a rate $R = 1/2$ encoder, memory $\nu = 4$, generators $G_1 = 37, G_2 = 21$, with interleaving 256×256.

B. Interleaving

The interleaver is made up of an $M \cdot M$ matrix and bits $\{d_k\}$ are written row by row and read following the nonuniform rule given in Section III-B2. This nonuniform reading procedure is able to spread the residual error blocks of rectangular form, that may set up in the interleaver located behind the first decoder DEC_1, and to give a large free distance to the concatenated (parallel) code.

For the simulations, a 256.256 matrix has ben used and from (19), the addresses of line i_r and column j_r for reading are the following:

$$i_r = 129(i + j) \qquad \mathrm{mod} \cdot 256$$

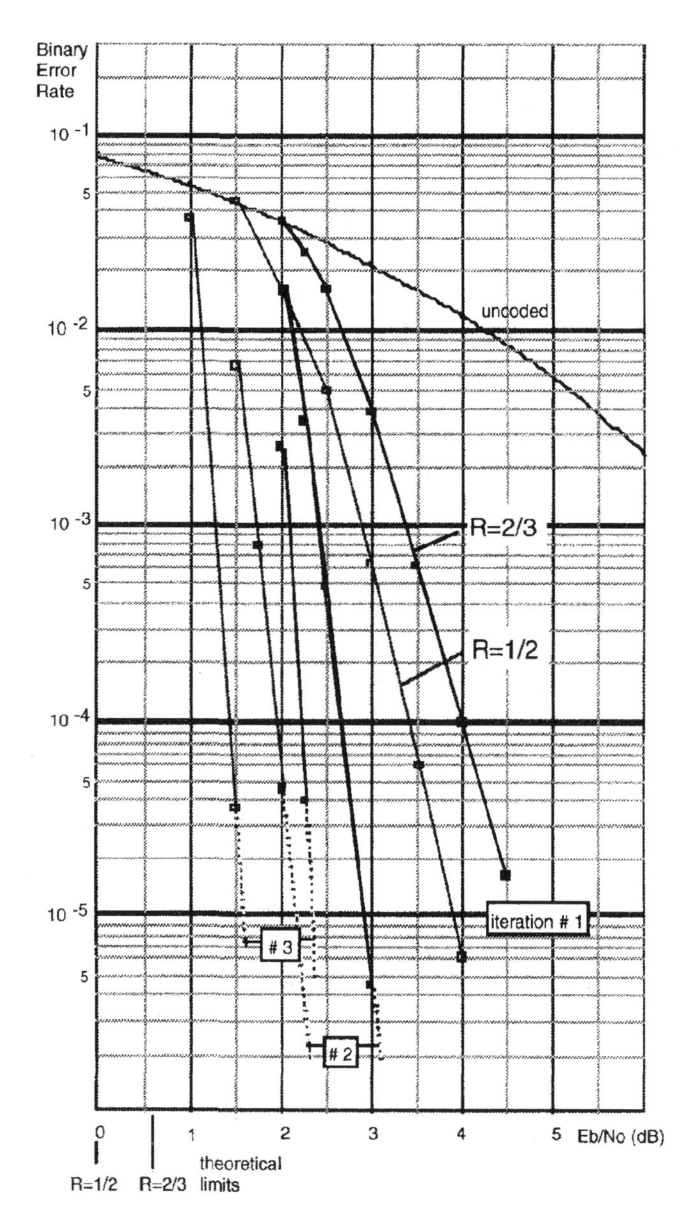

Fig. 10. BER given by iterative decoding ($p = 1, 2, 3$) of code with memory $\nu = 4$, generators $G_1 = 23, G_2 = 35$, rate $R = 1/2$ and $R = 2/3$ and interleaving 256×256 decoding.

$$\begin{aligned} \xi &= (i + j) && \text{mod} \cdot 8 \\ j_r &= [P(\xi) \cdot (j + 1)] - 1 && \text{mod} \cdot 256 \end{aligned} \tag{49a}$$

with

$$\begin{aligned} &P(0) = 17; \quad P(1) = 37; \quad P(2) = 19; \quad P(3) = 29 \\ &P(4) = 41; \quad P(5) = 23; \quad P(6) = 13; \quad P(7) = 7. \end{aligned} \tag{49b}$$

VII. Results

For a rate $R = 1/2$ encoder with memory $\nu = 4$, and for generators $G_1 = 37, G_2 = 21$ and parallel concatenation 2/3//2/3 ($R_1 = 2/3, R_2 = 2/3$), the BER has been computed after each decoding step using the Monte Carlo method, as a function of signal to noise ratio E_b/N_0. The interleaver consists of a 256×256 matrix and the modified BAHL *et al.* algorithm has been used with a data block length of $N = 65\,536$ bits. In order to evaluate a BER equal to 10^{-5}, we have considered 256 data blocks *i.e.* approximately 16×10^6 b d_k. The BER versus E_b/N_0, for different values of p is plotted in Fig. 9. For any given SNR greater than 0 dB, the BER decreases as a function of the decoding step p. The coding gain is fairly high for the first values of $p (p = 1, 2, 3)$ and carries on increasing for the subsequent values of p. For $p = 18$ for instance, the BER is lower than 10^{-5} at $E_b/N_0 = 0.7$ dB with generators $G_1 = 37, G_2 = 21$. Remember that the Shannon limit for a binary modulation is $P_e = 0$ (several authors take $P_e = 10^{-5}$ as a reference) for $E_b/N_0 = 0$ dB. With the parallel concatenation of RSC convolutional codes and feedback decoding, the performance is 0.7 dB from Shannon's limit. The influence of the memory ν on the BER has also been examined. For memory greater than 4, at $E_b/N_0 = 0.7$ dB, the BER is slightly worse at the first ($p = 1$) decoding step and the feedback decoding is inefficient to improve the final BER. For ν smaller than 4, at $E_b/N_0 = 0.7$ dB, the BER is slightly better at the first decoding step than for ν equal to four, but the correction capacity of encoders C_1 and C_2 is too weak to improve the BER with feedback decoding. For $\nu = 3$ (i.e., with only eight-states decoders) and after iteration 18, a BER of 10^{-5} is achieved at $E_b/N_0 = 0.9$ dB.

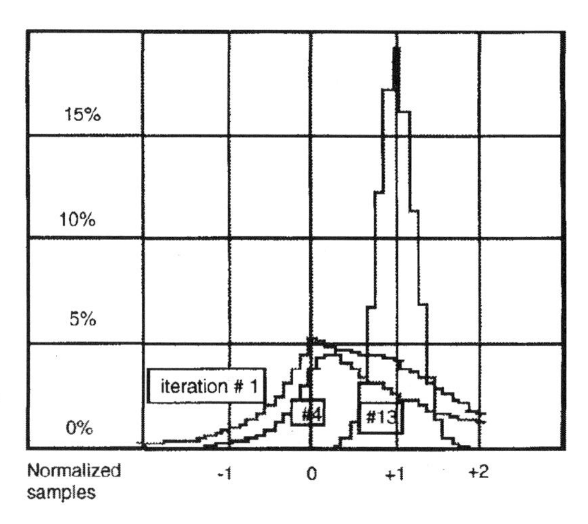

Fig. 11. Histograms of extrinsic information z_k after iterations #1, 4, 13 at $E_b/N_0 = 0.8$ dB; all information bits $d_k = 1$.

For ν equal to 4, we have tested several generators (G_1, G_2) and the best results were achieved with $G_1 = 37, G_2 = 21$ and at least six iterations ($p \geq 6$). For p smaller than four, generators $G_1 = 23, G_2 = 35$ lead to better performance than generators $G_1 = 37, G_2 = 21$, at large SNR's (Fig. 10). This result can be understood since, with generators $G_1 = 23, G_2 = 35$, the first coefficients $W_{RSC}(d)$ are smaller than with generators $G_1 = 37, G_2 = 21$. For very low SNR's, the BER can sometimes increase during the iterative decoding process. In order to overcome this effect the extrinsic information z_k has been divided by $[1 + \theta|\tilde{\Lambda}_1(d_k)|]$.$\theta$ acts as a stability factor and its value of 0.15 was adopted after several simulation tests at $E_b/N_0 = 0.7$ dB.

We have also plotted the BER for a rate $R = 2/3$ encoder with memory $\nu = 4$, generators $G_1 = 23, G_2 = 35$ and parallel concatenation 4/5//4/5 in Fig. 10. For a nonuniform interleaver consisting of a 256×256 matrix and for $p = 3$, the BER is equal to 10^{-5} at $E_b/N_0 = 1.6$ dB and thus the performance is only 1 dB from Shannon's limit.

In Fig. 11, the histogram of extrinsic information $(z)_p$ (generators $G_1 = 37, G_2 = 21$.) has been drawn for several values of iteration p, with all bits equal to one and for a low SNR ($E_b/N_0 = 0.8$ dB). For $p = 1$ (first iteration), extrinsic information $(z)_p$ is very poor about bit d_k, and furthermore, the gaussian hypothesis made above for extrinsic information $(z)_p$, is not satisfied! Nevertheless, when iteration p increases, the histogram merges toward a Gaussian law with a mean equal to one, after normalization. For instance, for $p = 13$, extrinsic information $(z)_p$ becomes relevant information concerning bits d_k.

Note that the concept of turbo-codes has been recently extended to block codes, in particular by Pyndiah *et al.* [13].

Acknowledgment

The authors wish to express their grateful acknowledgment to D. Divsalar for his constructive remarks about the organization of this paper.

References

[1] G. D. Forney, "The Viterbi algorithm," *Proc. IEEE*, vol. 61, no. 3, pp. 268–278, Mar. 1973.
[2] J. B. Cain, G. C. Clark, and J. M. Geist, "Punctured convolutional codes of rate (n-1)/n and simplified maximum likelihood decoding," *IEEE Trans. Inform. Theory*, vol. IT-25, pp. 97–100, Jan 1979.
[3] G. Ungerboeck, "Channel coding with multilevel/phase signals," *IEEE Trans. Inform. Theory*, vol. IT-28, no. 1, pp. 55–67, Jan. 1982.
[4] Y. Yasuda, K. Kashiki, and Y. Hirata, "High-rate punctured convolutional codes for soft-decision Viterbi decoding," *IEEE Trans. Commun.*, vol. COM-32, no. 3, Mar. 1984.
[5] A. J. Viterbi and J. K. Omura. *Principles of digital communication and coding*. New York: MacGraw-Hill, 1979.
[6] A. J. Viterbi, "Convolutional codes and their performance in communications systems," *IEEE Trans. Commun.*, vol. COM-19, Oct. 1971.
[7] M. Cedervall and R. Johannesson, "A fast algorithm for computing distance spectrum of convolutional codes," *IEEE Trans. Inform. Theory*, vol. 35, no. 6, Nov. 1989.
[8] P. Thitimajshima, "Les codes convolutifs récursifs systématiques et leur application à la concaténation parallèle," (in French), Ph.D. no. 284, Université de Bretagne Occidentale, Brest, France, Dec. 1993.
[9] E. Dunscombe and F. C. Piper, "Optimal interleaving scheme for convolutional coding," *Electron. Lett.*, vol. 25, no. 22, pp. 1517–1518, Oct. 1989.
[10] L. R. Bahl, J. Cocke, F. Jelinek, and J. Raviv, "Optimal decoding of linear codes for minimizing symbol error rate," *IEEE Trans. Inform. Theory*, vol. IT-20, pp. 248–287, Mar. 1974.
[11] J. Hagenauer, P. Robertson, and L. Papke, "Iterative ("TURBO") decoding of systematic convolutional codes with the MAP and SOVA algorithms," in *Proc. ITG'94*, 1994
[12] C. Berrou, A. Glavieux, and P. Thitimajshima, "Near Shannon limit error-correcting coding and decoding: Turbo-codes," in *ICC'93*, Geneva, Switzerland, May 93, pp. 1064–1070.
[13] R. Pyndiah, A. Glavieux, A. Picart, and S. Jacq, "Near optimum decoding of products codes," in *GLOBECOM'94*, San Francisco, CA, Dec. 94, pp. 339–343.

PHOTO NOT AVAILABLE

Claude Berrou (M'84) was born in Penmarc'h, France in 1951. He received the electrical engineering degree from the "Institut National Polytechnique", Grenoble, France, in 1975.

In 1978, he joined the Ecole Nationale Supérieure des Télécommunications de Bretagne (France Télécom University), where he is currently a Professor of electronic engineering. His research focuses on joint algorithms and VLSI implementations for digital communications, especially error-correcting codes, and synchronization techniques.

PHOTO NOT AVAILABLE

Alain Glavieux was born in France in 1949. He received the engineering degree from the Ecole Nationale Supérieure des Télécommunications, Paris, France in 1978.

He joined the Ecole Nationale Supérieure des Télécommunications de Bretagne, Brest, France, in 1979, where he is currently Head of Department Signal and Communications. His research interests include channel coding, communications over fading channels, and digital modulation.

Erlang Capacity of a Power Controlled CDMA System

Audrey M. Viterbi and Andrew J. Viterbi

Abstract—**This paper presents an approach to the evaluation of the reverse link capacity of a CDMA cellular voice system which employs power control and a variable rate vocoder based on voice activity. It is shown that the Erlang capacity of CDMA is many times that of conventional analog systems and several times that of other digital multiple access systems.**

I. Introduction

FOR any multiuser communication system, the measure of its economic usefulness is not the maximum number of users which can be serviced at one time, but rather the peak load that can be supported with a given quality and with availability of service as measured by the blocking probability (the probability that a new user will find all channels busy and hence be denied service, generally accompanied by a busy signal). Adequate service is usually associated with a blocking probability of 2 percent or less. The average traffic load in terms of average number of users requesting service resulting in this blocking probability is called the *Erlang capacity* of the system.

In virtually all existing multiuser circuit-switched systems, blocking occurs when all frequency slots or time slots have been assigned to a voice conversation or message. In code division multiple access (CDMA) systems, in contrast, users all share a common spectral frequency allocation over the time that they are active. Hence, new users can be accepted as long as there are receiver-processors to service them, independent of time and frequency allocations. We shall assume that a sufficient number of such processors is provided in the common base station such that the probability of a new arrival finding them all busy is negligible. Rather, blocking in CDMA systems will be defined to occur when the interference level, due primarily to other user activity, reaches a predetermined level above the background noise level of mainly thermal origin. While this interference-to-noise ratio could, in principle, be made arbitrarily large, when the ratio exceeds a given level (about 10 dB nominally), the interference increase per additional user grows very rapidly, yielding diminishing returns and potentially leading to instability. Consequently, we shall establish blocking in CDMA as the event that the total interference-to-background noise level exceeds $1/\eta$ (where $\eta = 0.1$ corresponding to 10 dB), and we determine the Erlang capacity which results in a 1 percent blocking probability. We emphasize, however, that this is a "soft blocking" condition, which can be relaxed as will be shown, as contrasted to the "hard blocking" condition wherein channels are all occupied.

Also, in conventional systems, a fraction of the time or frequency slots must be set aside for users to transmit requests for initiating service, and a protocol must be established for multiple requests when two or more users collide in simultaneously requesting service. In CDMA systems, even the users seeking to initiate access can share the common medium. Of course, they add to the total interference and hence lower the Erlang capacity to some degree. We shall demonstrate that this reduction is very small for initial access requests whose signaling time is on the order of a few percent of the average duration of a call or message.

Conclusions will be drawn regarding the relative increase in Erlang capacity of a direct sequence spread-spectrum CDMA system over existing FDMA and TDMA systems.

Manuscript received March 16, 1992; revised November 16, 1992. Part of this paper was presented at the IEEE International Symposium on Information Theory, San Antonio, TX, January 17–22, 1993.

The authors are with Qualcomm Inc., San Diego, CA 92121-1617.

IEEE Log Number 9208658.

II. Conventional Blocking

In conventional multiple access systems, such as FDMA and TDMA, traffic channels are allocated to users as long as there are channels available, after which all incoming traffic is blocked until a channel becomes free at the end of a call. The blocking probability is obtained from the classical Erlang analysis of the $M/M/S/S$ queue, where the first M refers to a Poisson arrival rate of λ calls/s; the second M refers to exponential service time with mean $1/\mu$ s/call; the first S refers to the number of servers (channels); and the second S refers to the maximum number of users supported before blockage occurs.

The Erlang-B formula [1] gives the blocking probability under these conditions

$$P_{\text{blocking}} = \frac{(\lambda/\mu)^S/S!}{\sum_{k=0}^{S} (\lambda/\mu)^k/k!}$$

where λ/μ is the offered average traffic measured in Erlangs. Of course, the average active number of users equals $(\lambda/\mu)(1 - P_{\text{blocking}})$.

Thus, for the conventional *AMPS system* with 30 kHz channels $K = 7$ frequency reuse factor[1] and 3 sectors, the number

[1] It should be noted that the reuse factor of 7 applies strictly only to North American analog systems. Digital TDMA systems have been proposed for other regions with lower reuse factors (4 or even 3). Field experience is not yet available, however, to support general use of such lowered reuse factors.

Reprinted from *IEEE Journal on Selected Areas in Communications,* vol. 11, No. 6, August 1993.

The Best of the Best. Edited by W. H. Tranter, D. P. Taylor, R. E. Ziemer, N. F. Maxemchuk, and J. W. Mark.

of channels (servers) in 12.5 MHz is

$$S_{\text{AMPS}} = \frac{12.5\text{ MHz}}{(30\text{ kHz})(7)(3)} = 19\text{ channel / sector}.$$

Similarly, for a *3-slot TDMA system*, which otherwise uses the same channelization, sectorization, and reuse factor as AMPS, and is generally *called D-AMPS,*

$$S_{D-\text{AMPS}} = 57\text{ channel / sector}.$$

The blocking probabilities *P*(blocking) as a function of the average number of calls offered at any instant λ/μ are obtained from the Erlang-B formula. This establishes that the offered traffic *Erlang capacity* per sector for a *blocking probability equal to 2 percent* is respectively,

$$(\lambda/\mu)_{\text{AMPS}} = 12.34\text{ Erlangs},$$
$$(\lambda/\mu)_{D-\text{AMPS}} = 46.8\text{ Erlangs}.$$

III. CDMA Reverse Link Erlang Capacity Bounds and Approximations

For the CDMA reverse link (or uplink), which is the limiting direction, blocking is defined to occur when the total collection of users both within the given sector (cell) and in other cells introduce an amount of interference density I_0 so great that it exceeds the background noise level N_0 by an amount $1/\eta$, taken to be 10 dB.

If there were always

1) a constant number of users N_u in every sector,
2) each (perfectly power controlled) user were transmitting continually, and
3) required the same E_b/I_0 (under all propagation conditions),

then, as established in [2] and [3], the number of users N_u would be determined by equating

$$N_u(\text{Signal Power / User}) + \text{Other Cell Interference} + \text{Thermal Noise} = \text{Total Interference}.$$

Taking

W = spread-spectrum bandwidth
R = data rate,
E_b = bit energy,
N_0 = thermal (or background) noise density,
I_0 = maximum total acceptable interference density (interference power normalized by W), and
f = ratio of other cell interference (at base station for given sector)-to-own sector interference[2] then the condition for nonblocking is

$$N_u E_b R(1+f) + N_0 W \le I_0 W \tag{1}$$

whence it follows that

$$N_u \le \frac{W/R}{E_b/I_0} \cdot \frac{1-\eta}{1+f} \tag{2}$$

[2]In [7], a tight upper bound on other-cell interference is derived, which gives $f = 0.55$ for the propagation parameters of interest.

where

$$\eta = N_0/I_0 = 0.1\text{ (nominally)}. \tag{3}$$

In fact, however, none of the three assumptions above holds since

a) the number of active calls is a Poisson random variable with mean λ/μ (there is no hard limit on servers);
b) each user is gated on with probability ρ and off with $1-\rho$;
c) each user's required energy-to-interference E_b/I_0 ratio is varied according to propagation conditions to achieve the desired frame error rate ($\approx$1 percent).

For simplicity, we continue to assume that all cells are equally loaded (with the same number of users per cell and sector which are uniformly distributed over each sector).

With assumptions a), b), and c) replacing 1), 2), and 3), the condition for nonblocking, replacing (1) becomes

$$\sum_{i=1}^{k} \nu_i E_{bi} R + \sum_{j}^{\text{other cells}} \sum_{i=1}^{k} \nu_{i(j)} E_{b_{i(j)}} R + N_0 W \le I_0 W \tag{4}$$

with k, the number of users/sector, being a Poisson random variable with mean λ/μ; and ν being the binary random variable taking values 0 and 1, which represents voice activity, with

$$P(\nu = 1) = \rho. \tag{5}$$

Dividing by $I_0 R$ and defining

$$\epsilon = E_b/I_0, \tag{6}$$

the nonblocking condition (4) becomes

$$Z \triangleq \sum_{i=1}^{k} \nu_i \epsilon_i + \sum_{j}^{\text{other cells}} \sum_{i=1}^{k} \nu_i^{(j)} \epsilon_i^{(j)} \le (W/R)(1-\eta) \tag{7}$$

where η is given by (3). Hence, the blocking probability for CDMA becomes

$$P_{\text{blocking}} = \Pr[Z > (W/R)(1-\eta)]. \tag{8}$$

Setting this equal to a given value (nominally 1 percent) establishes the Erlang capacity of a CDMA cellular system. Again, we note that this is a "soft blocking" phenomenon which can be occasionally relaxed by allowing I_0/N_0, and consequently $1-\eta$, to increase.

Naturally, when condition (8) is exceeded, call quality will suffer. Thus, this probability is kept sufficiently low so that we can ensure high availability of good quality service. Conventional multiple access systems also are limited to providing good quality service only on the order of 90–99 percent of the time because of the variability of interference from just one or a few other users, in contrast with the CDMA case where quality depends on an average over the entire user population.

To evaluate this blocking probability, we must determine the distribution function of the random variable Z which in

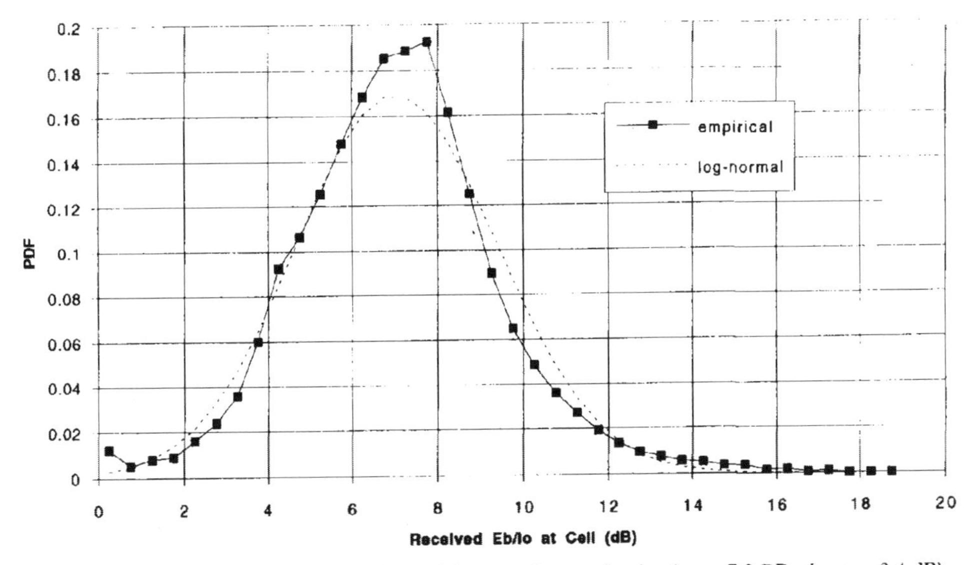

Fig. 1. Empirical E_b/I_0 probability density and log-normal approximation (m = 7.0 DB; sigma = 2.4 dB).

turn depends on the random variables ν, k, and ϵ, representing voice activity, number of users in a sector, and E_b/I_0 of any user, respectively. The distribution of ν is given by (5). Since k is Poisson, its distribution is given by

$$p_k = \Pr(k \text{ active users / sector}) = \frac{(\lambda/\mu)^k}{k!} e^{-\lambda/\mu} \tag{9}$$

where λ and μ are the arrival and service rates as defined in the previous section.

On the other hand, ϵ, the E_b/I_0 ratio of a single user, depends on the power control mechanism which attempts to equalize the performance of all users. It has been demonstrated that inaccuracy in power control loops are approximately log-normally distributed with a standard deviation between 1 and 2 dB [4]. However, since under some propagation conditions (e.g., with excessive multipath) higher than normal E_b/I_0 will be required to achieve the desired low error rates, the overall distribution, also log-normally distributed, will have a larger standard derivation. An example drawn from field trials with all cells fully loaded, in which E_b/I_0 is varied in order to maintain frame error rate below 1 percent is shown by the histogram of Fig. 1. As demonstrated by the dotted curve, the histogram is closely approximated by a log-normal probability density, with the mean and standard deviation of the normal exponent equal to 7 and 2.4 dB, respectively. Hence, we shall use the log-normal approximation[3]

$$\epsilon = 10^{x/10} \tag{10}$$

where x is a Gaussian variable with mean $m \approx 7$ dB and standard deviation $\sigma \approx 2.5$ dB. Note then that the first and second moments of ϵ are given by

$$E(\epsilon) = E\left(e^{\beta x}\right) = \exp\left[(\beta\sigma)^2/2\right] \exp(\beta m) \tag{11}$$

$$E(\epsilon^2) = E\left(e^{2\beta x}\right) = \exp\left[2(\beta\sigma)^2\right] \exp(2\beta m),$$
$$\beta = (\ln 10)/10. \tag{12}$$

Although all moments exist, the moment generating function of ϵ does not converge; hence the ordinary Chernoff bound for the blocking probability (8) cannot be obtained. In Appendix I, we derive a modified Chernoff upper bound obtained by treating the upper end of the distribution of ϵ separately. The result for a single sector (no interference from other cells) is

$$P_{\text{blocking}} < \underset{\substack{s>0 \\ \tau>0}}{\text{Min}} \exp\left\{\rho(\lambda/\mu)\left[E\left(e^{s\epsilon'_T}\right) - 1\right] - sA\right\} + \rho(\lambda/\mu)Q(\tau/\sigma) \tag{13}$$

where[4]

$$A \triangleq \frac{(W/R)(1-\eta)}{\exp(\beta m)} = \frac{(W/R)(1-\eta)}{E_b/I_{0_{\text{median}}}} \tag{14}$$

$$E\left(e^{s\epsilon'_T}\right) = \int_{-\infty}^{\tau/\sigma} \exp\left[s e^{\beta\sigma\zeta}\right] e^{-\zeta^2/2}\, d\zeta/\sqrt{2\pi} \tag{15}$$

$$Q(\tau/\sigma) = \int_{\tau/\sigma}^{\infty} e^{-\zeta^2/2}\, d\zeta/\sqrt{2\pi}\,. \tag{16}$$

This bound, obtained through numerical integration of (15), is plotted, for the single sector case, as the upper curve of Fig. 2.

[3] In the following, we could use the exact empirical distribution in numerically computing the Chernoff bound (13). Using the log-normal approximation gives us more flexibility in obtaining a general analytical result. Actually, the approximation is slightly pessimistic (an upper bound) since it appears from Fig. 1 that the empirical histogram is slightly skewed to the left of the log-normal approximation with the same mean and standard deviation. Fig. 1 is typical of data from a large number of field tests conducted in widely varying terrain in a large number of cities and in several countries.

[4] Note that m is the mean of the exponent of the log-normal variable. Since $e^{\beta m} = 10^{m/10}$, this is the value of E_b/I_0 corresponding to the mean (dB) value, but it is not the true mean. It can be shown by symmetry that it is the median, while the true mean is related to it by (11).

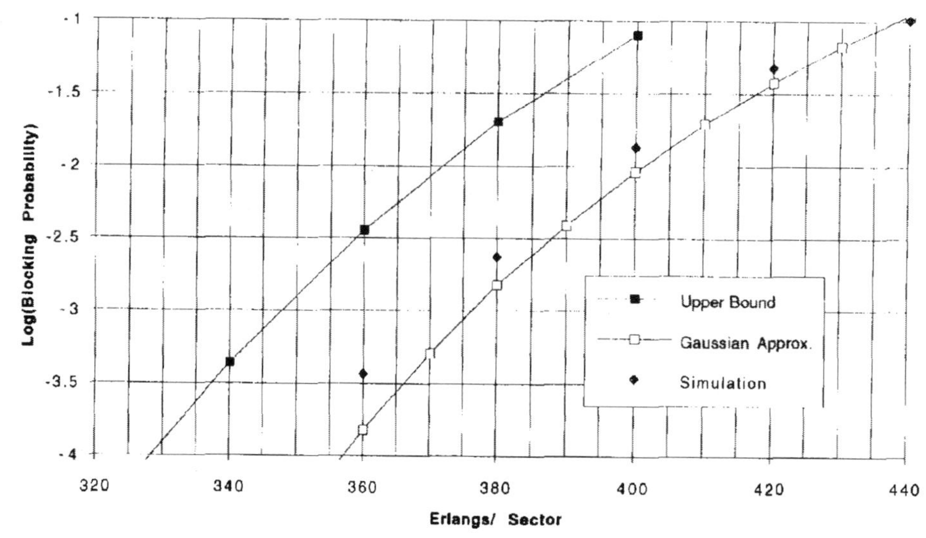

Fig. 2. Blocking probabilities for single cell interference (CDMA parameters: $W/R = 1280$; voice act. = 0.4; $I_0/N_0 = 10$ dB; median $E_b/I_0 = 7$ dB; sigma = 2.5 dB).

A much simpler approach is to assume a central limit theorem approximation for Z and to compute its mean and variance. Then

$$P_{\text{blocking}} \approx Q\left[\frac{A - E(Z')}{\sqrt{\text{Var } Z'}}\right] \quad (17)$$

where $Z' = Z/\exp(\beta m) = Z/(E_b/I_0)_{\text{median}}$.

Now since Z is the sum of k random variables, where k is itself a random variable, we have from [5], letting $\epsilon' = \epsilon/\exp(\beta m)$,

$$E(Z') = E(k)E(\nu\epsilon') = (\lambda/\mu)\rho \exp\left[(\beta\sigma)^2/2\right] \quad (18)$$

$$\text{Var}(Z') = E(k)\,\text{Var}(\nu\epsilon') + \text{Var}(k)[E(\nu\epsilon')]^2. \quad (19)$$

But since k is a Poisson variable, $E(k) = \text{Var}(k) = \lambda/\mu$, so that

$$\text{Var}(Z') = \lambda/\mu\left[E(\nu\epsilon')^2\right] = \lambda/\mu E(\nu^2)E(\epsilon'^2) = (\lambda/\mu)\rho \exp\left[2(\beta\sigma)^2\right]. \quad (20)$$

Using (18) and (20) in (17) yields the lower curve of Fig. 2.

It is commonly accepted that a Chernoff upper bound overestimates the probability by almost an order of magnitude. Moreover, a simulation involving on the order of one million frames was run to evaluate the tightness of the bound and the accuracy of the approximation. Results shown in Fig. 2 indicate excellent agreement with the approximation, with at most *1 percent discrepancy* in Erlang capacity for the nominal 1 percent blocking probability. Thus, we shall henceforth use only central limit approximations, given the much greater ease of computation, which will allow us to obtain more general and more easily employed results. We note that for any specific case, a strict upper bound can always be obtained by numerically computing the Chernoff bound based on the empirical distribution of E_b/I_0 as given in (A.3) of Appendix I. We note, however, that the approximation underestimates this upper bound by at most a few percent.

TABLE I
FACTOR f FOR SEVERAL VALUES OF SIGMA (FOURTH POWER LAW)

δ (dB)	f
0	0.44
2	0.43
4	0.45
6	0.49
8	0.55
10	0.66
12	0.91

IV. General Erlang Capacity Formula Including Other Cell Interference

Before deriving the general formula, we consider the effect of users in other cells which are power controlled by other base stations. It has been shown both analytically [3] and by simulation that the interference from surrounding cells increases the average level at the base station under consideration by a fraction between $1/2$ and $2/3$ of that of the desired cell's users when the propagation attenuation is proportional to the fourth power of the distance times a log-normally distributed component whose differential standard deviation is 8 dB. The results of an improved upper bound on the mean outer-cell interference fraction f as derived in [7] and shown in Table I.

We note also that each user which is controlled by an other-cell base station will also have an E_b/I_0 which is distributed according to the histogram of Fig. 1. Hence, this other-user interference can also be modeled by the same log-normal distribution as assumed for users of the desired cell. We can now modify (18) to accommodate other-cell users as well. The total number of other-cell users is generally much larger, but their average power is equivalent to that of kf users. We shall therefore model them as such, recognizing that this is somewhat pessimistic since, with a larger number of smaller received power users, the mean power will remain the same but the variance will be reduced. In any case, accepting this approach as an overbound, we find that the mean and variance (18) and (20) are simply increased by the factor $1 + f$. Thus, we may restate (17)–(20), when other-cell (power controlled) interfering users are included, as follows:

$$P_{\text{blocking}} \approx Q\left[\frac{A - E(Z')}{\sqrt{\text{Var } Z'}}\right] \quad (21)$$

$$E(Z') = (\lambda/\mu)\,\rho\,(1+f)\exp\left[(\beta\sigma)^2/2\right] \quad (22)$$

$$\text{Var}(Z') = (\lambda/\mu)\,\rho\,(1+f)\exp\left[2(\beta\sigma)^2\right]. \quad (23)$$

Inverting (21) yields the quadratic equation

$$x\alpha^4\left[Q^{-1}(P_{\text{blocking}})\right]^2 = [A - x\alpha]^2 \quad (24)$$

where $x = (\lambda/\mu)\,\rho\,(1+f)$, and $\alpha = \exp[(\beta\sigma)^2/2]$, while A is given by (14). Its solution is

$$x = \frac{A}{\alpha}\left[1 + \frac{\alpha^3 B}{2}\left(1 - \sqrt{1 + \frac{4}{\alpha^3 B}}\right)\right] \quad (25)$$

where

$$B = \frac{\left[Q^{-1}(P_{\text{blocking}})\right]}{A} = \frac{(E_b/I_0)_{\text{median}}\left[Q^{-1}(P_{\text{blocking}})\right]^2}{(W/R)(1-\eta)}. \quad (26)$$

Using (24) and (25), this can be expressed as a formula for Erlang capacity

$$\frac{\lambda}{\mu} = \frac{(1-\eta)(W/R)F(B,\sigma)}{\rho(1+f)(E_b/I_0)_{\text{median}}} \quad \text{Erlangs/Sector} \quad (27)$$

where

$$F(B,\sigma) = \exp\left[-(\beta\sigma)^2/2\right] \cdot \left\{1 + (B/2)\exp\left[3(\beta\sigma)^2/2\right] \cdot \left(1 - \sqrt{1 + 4\exp\left[-3(\beta\sigma)^2/2\right]/B}\right)\right\} \quad (28)$$

with B given by (26). Fig. 3 is a plot of $(Q^{-1})^2$—the factor of B which depends on P_{blocking}. Fig. 4 shows plots of $F(B,\sigma)$ as a function of B for several values of σ, the standard deviation in decibels of the power-controlled E_b/I_0. We note finally that the Erlang capacity, as given by (27), is the same as the capacity predicted under ideal assumptions (2) but increased by the inverse of the voice activity factor $1/\rho$ and decreased by the reduction factor $F(B,\sigma)$ given in Fig. 4. We note from Fig. 4 that a power control inaccuracy of $\sigma = 2.5$ dB causes only about as much capacity reduction as does the variability in arrival times and voice activity. Numerically, for $E_b/I_0 = 7$ dB, $\eta = 0.1$, $W/R = 31$ dB and $P_{\text{blocking}} = 1$ percent we obtain $B = 0.024$. Then, from Fig. 4, we note that $F(B,\sigma = 2.5) = 0.695$ while $F(B,\sigma = 0) = 0.86$. Thus, it follows that the incremental loss due to power control inaccuracy $F(B,\sigma = 2.5)/F(B,\sigma = 0) = 0.8$. This means that in this case a standard deviation of 2.5 dB in power control causes a reduction of only 1 dB in capacity. Fig. 3 and 4 also show that results are not very sensitive to the level of blocking probability. Doubling the latter only increases capacity by about 2 percent. Note finally that substituting $F(B,\sigma = 2.5) = 0.695$ in the general formula (27), we obtain, for the parameters noted above along with voice activity $\rho = 0.4$ and other-cell factor $f = 0.55$ (as obtained from Table I), an Erlang capacity $\lambda/\mu = 258$ Erlangs/sector. Thus, comparison to Section II gives the ratio

$$\frac{C_{\text{CDMA}}}{C_{\text{AMPS}}} \approx \frac{258}{12.34} \approx 20.9.$$

Based on the simulation results shown in Fig. 2, which suggest reduction of the approximate results by one to two percent, an estimate in excess of a twentyfold increase in Erlang capacity is thus justified.

V. Designing for Minimum Transmitted Power

In the preceding, the total received interference-to-noise I_0/N_0 was assumed fixed, and blocking probability was evaluated as a function of average user loading λ/μ. Now suppose the blocking probability is fixed at 1 percent, but for each user loading λ/μ, the minimum value of I_0/N_0 is determined to achieve this. This can be established by varying $\eta = N_0/I_0 < 1$ and determining from (27) the λ/μ value for which blocking probability equals 1 percent for each I_0/N_0 level.

More importantly, from I_0/N_0, we may determine the minimum value of received signal power-to-background noise for each user, and consequently the minimum transmitted power per user, given the link attenuation, the receiver sensitivity, and antenna gains. The minimum required received signal-to-background noise per user $S/(N_0W)$ is obtained by equating the average sum of the per-user ratios weighted by the average voice activity factor to the total other user interference-to-noise ratio

$$\rho\frac{\lambda}{\mu}\,\frac{S}{N_0W} = \left(\frac{I_0 - N_0}{N_0}\right).$$

Thus,

$$\frac{S}{N_0W} = \frac{1}{\rho(\lambda/\mu)}\left(\frac{I_0}{N_0} - 1\right) = \frac{(1/\eta) - 1}{\rho(\lambda/\mu)} \quad (29)$$

where $1/\eta = I_0/N_0$ is fixed, as noted above, and λ/μ is the resulting loading to achieve $P_{\text{blocking}} = 0.01$ for this value of η, as given by (26)–(28). Fig. 5 shows the per-user signal-to-background noise ratio as a function of the relative Erlang capacity for the same parameters used in the last section. The inverse of η, the interference-to-noise density I_0/N_0 ranges from 1 to 10 dB as the capacity varies over a factor of 8. The important point is that the per-user power can be reduced by about 8 dB for lightly loaded cells. It also helps to justify the choice of $I_0/N_0 = 10$ dB for blocking when all cells are heavily loaded, since per-user power requirements increase rapidly above this point.

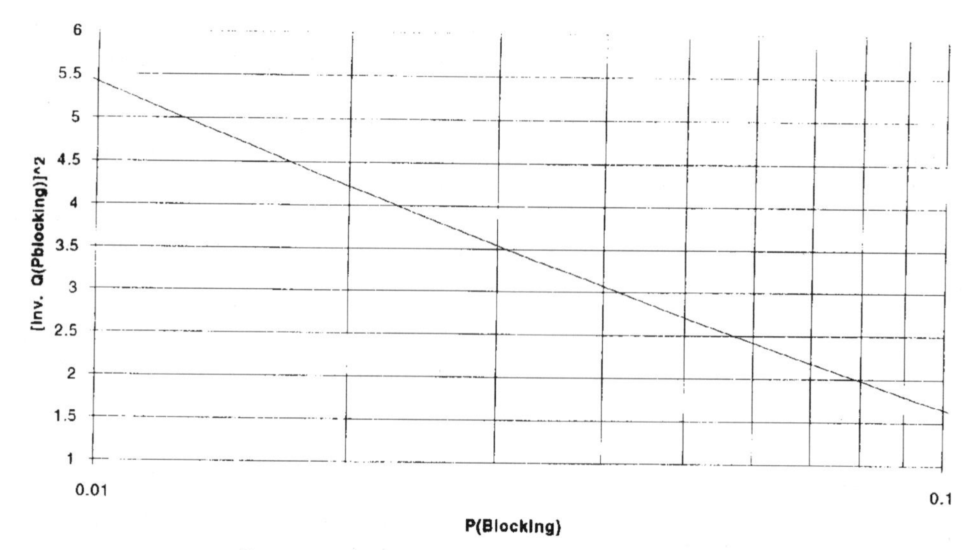

Fig. 3. Variable factor in B as a function of P (blocking).

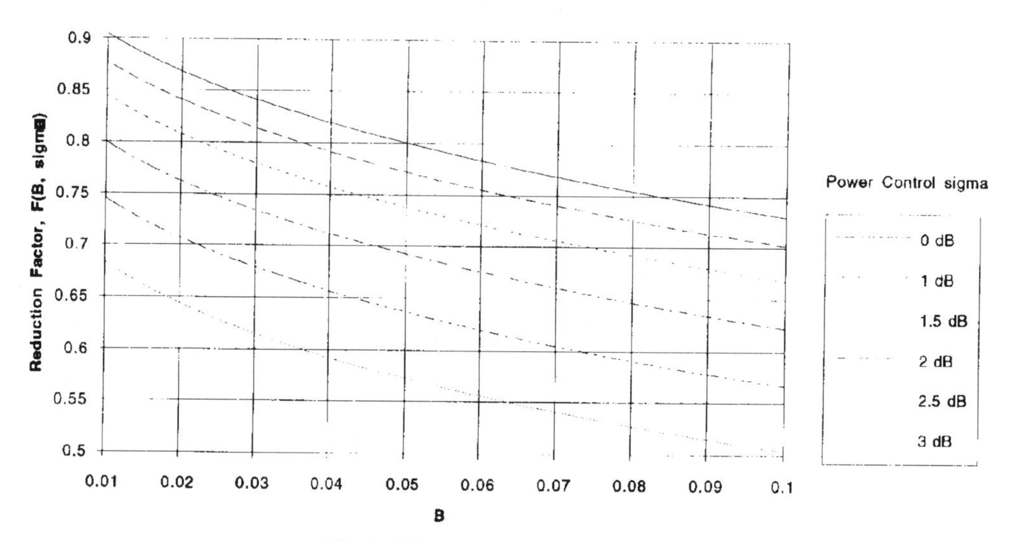

Fig. 4. Erlang capacity reduction factors.

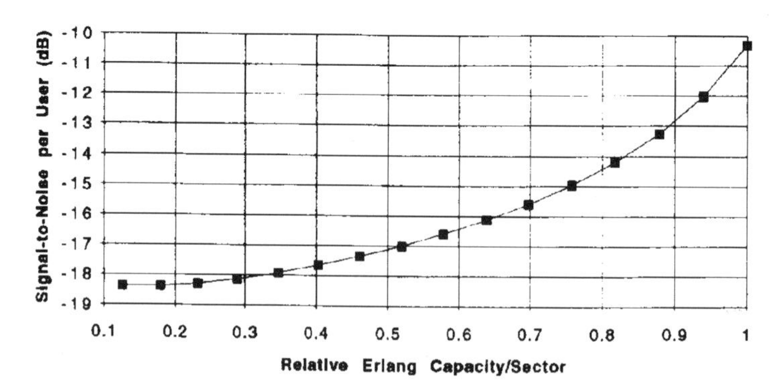

Fig. 5. Signal-to-noise per user as a function of relative sector capacity.

VI. Initial Access

Prior to initiating a call on the reverse link, a user must signal a request to the base station. In conventional systems, a time or frequency slot is allocated for the purpose of access request, and a protocol is provided for recovering from collisions which occur when two or more new users send access requests simultaneously. In CDMA, in which the allocated commodity is energy rather than time or frequency, access requests can share the common channel with ongoing users. The arrival rate of user requests is taken to be λ calls/s, which is the same as that for ongoing calls under the assumption that all requests are eventually served. Since newly arriving users are not power controlled until their requests are recognized, the initial power level will be taken to be a random variable uniformly distributed from 0 to a maximum value, corresponding to a (bit) energy level of E_M. Thus, the initial access energy level is γE_M where γ is a random variable with probability density function

$$p(\gamma) = \begin{cases} 1, & 0 < \gamma < 1 \\ 0, & \text{otherwise}. \end{cases} \quad (30)$$

If this initial power level is not sufficient for detection,[5] and hence acknowledgment is not received, the user increases his power in constant decibel steps every frame until his request is acknowledged. Thus, the initial access user's power grows exponentially (i.e., linearly in decibels) with time, taken to be continuous since the frame time is only tens of milliseconds long. Hence, the energy as a function of time for initial access requests is

$$E(t) = \gamma E_M e^{\delta t} \quad (31)$$

where δ is fixed. Letting τ denote the time required for an initial access user to be detected, it follows from (30) to (31) that

$$\begin{aligned} \Pr(\tau > T) &= \Pr(\gamma E_M e^{\delta T} < E_M) \\ &= \Pr(\gamma < e^{-\delta T}) = e^{-\delta T}, \qquad T > 0 \end{aligned} \quad (32)$$

where the last equation follows from the fact that γ is uniformly distributed on the unit interval.

Thus, the "service time" for any initial access requests (time required for acceptance) for all users is exponentially distributed with mean $1/\delta$ s/message. The consequences of this observation are very significant. We have thus shown that with Poisson distributed arrivals and exponentially distributed service time, *the output distribution is also Poisson*. This then guarantees that the distribution for users initiating service is the same as for newly accessing users. Furthermore, at any given time, the power distribution for users which have not yet been accepted is also uniform. As for the service time distribution, according to (32), this is exponential with mean service time equal to $1/\delta$.

Thus, the total interference of (4) is augmented by the interference from initial accesses. When normalized as in (18), this introduces an additional term in the mean and variance of Z', (18) and (20), respectively. Calling this Z'', we have, ignoring for the moment surrounding cells,

$$E(Z'') = E(Z') + (\lambda/\delta)E(\gamma E_M/I_0)/e^{\beta m} \quad (33)$$

$$\mathrm{Var}(Z'') = \mathrm{Var}(Z') + (\lambda/\delta)E(\gamma^2 E_M/I_0)/e^{2\beta m} \quad (34)$$

where γ is uniformly distributed on the unit interval and is independent of E_M/I_0. Thus $E(\gamma) = 1/2$ and $E(\gamma^2) = 1/3$. We take $E_M/I_0 = \theta(E_b/I_0)$, and hence log-normally distributed with the same σ as E_b/I_0. Then $E(E_M/I_0) = \theta E(\epsilon), E[(E_M/I_0)^2] = \theta^2 E(\epsilon^2)$.

If surrounding cells support the same coverage load as the given cell, initial accesses will also produce the same relative interference. Introducing this term as well, we obtain, from (33) and (34),

$$\begin{aligned} E(Z'') &= \Big\{(\lambda/\mu)\,\rho\,\exp\big[(\beta\sigma)^2/2\big] \\ &\quad + (\lambda/\delta)(1/2)\,\theta \exp\big[(\beta\sigma)^2/2\big]\Big\}[1+f] \\ &= \rho(\lambda/\mu)\left(1 + \frac{\theta\mu}{2\rho\delta}\right)\exp\big[(\beta\sigma)^2/2\big][1+f] \end{aligned} \quad (35)$$

$$\begin{aligned} \mathrm{Var}(Z'') &= \Big\{(\lambda/\mu)\,\rho\,\exp\big[2(\beta\sigma)^2\big] \\ &\quad + (\lambda/\delta)(1/3)\,\theta^2 \exp\big[2(\beta\sigma)^2\big]\Big\}[1+f] \\ &= \rho\,(\lambda/\mu)\left(1 + \frac{\theta^2\mu}{3\rho\delta}\right)\exp\big[2(\beta\sigma)^2\big][1+f]. \end{aligned} \quad (36)$$

If we take the arbitrary scale factor $\theta = 3/2$—a reasonable choice since this places the detection bit energy level at 50 percent higher than the operating bit energy level—we find that the effect is to increase the Erlang level (λ/μ) by the factor

$$F = 1 + \left(\frac{3}{4\rho}\right)\left(\frac{\mu}{\delta}\right) = 1 + 1.88(\mu/\delta) \quad \text{for } \rho = 0.4. \quad (37)$$

Note that μ/δ is the ratio of mean detection time-to-mean message duration, which should be very small. If, for example, $\mu/\delta = 0.0055$, corresponding to a mean access time of 1 s (50 frame)[6] for a 3-min mean call duration, this means that the effect of initial accesses is to reduce the Erlang traffic by about 1 percent (a negligible cost considering the notable advantages).

VII. Conclusions

The foregoing analysis represents a conservative estimate of the Erlang capacity of direct sequence spread-spectrum CDMA. Classical blocking was replaced by the condition that the total interference exceeds the background noise by 10 dB, and Erlang capacity was defined as the traffic load corresponding to a 1 percent probability that this event occurs. The three random variables which contribute to this event were modeled as follows:

1) Poisson traffic arrival, exponentially distributed message length, and arbitrarily many servers ($M/M/\infty$ in the terminology of queuing theory [1]);

2) voice activity factor ρ equal to 40 percent, as established by extensive experimental measurements on ordinary speech conversations [6];

[5] We assume initially that the access request is detected (immediately) when the initial access user's energy reaches E_M, but it remains undetected until that point. Below, we shall take E_M to be itself a random variable, with distribution similar to E_b. It should also be noted that initial access detection is performed on an unmodulated signal, other than for the user's pseudorandom code which is known to the base station. It should thus be more robust than the demodulator performance at the same power level.

[6] Note that at a data rate of 9600 bits/s, each 20 ms voice frame contains 192 bits. For $E_b/I_0 \approx 7$ dB, the frame energy-to-interference level is about 32 dB—an ample value to virtually assure detection in one frame.

3) individual user's received bit energy-to-interference density E_b/I_0 whose distribution was determined empirically from field measurements and supported by analytical results [4].

The main conclusion is that the Erlang capacity of CDMA is about 20 times that of AMPS. This is based on establishing the blocking condition as the event that the total interference-to-noise-ratio exceeds 10 dB. This "soft" blocking condition can be relaxed to allow a maximum ratio of 13 dB. for example, for particularly heavily loaded sectors. When traffic is light, much lower interference-to-noise ratios can be imposed which translate into much lower mobile transmitted powers, a particularly valuable feature in prolonging battery life for portable subscriber units.

Finally, a significant byproduct of spectrum sharing is the very small Erlang capacity overhead required to accommodate initial access requests along with ongoing traffic, as demonstrated in the last section.

Appendix
Modified Chernoff Bound for a Single Sector

Starting with (8), without considering outer cells, we may bound (8) by introducing an arbitrary threshold on each term $\nu_i\epsilon_i'$, where $\epsilon' = \epsilon/e^{\beta m}$. Thus,

$$\begin{aligned} P_{\text{blocking}} &= \Pr\left(\sum_{i=1}^{k} \nu_i\epsilon_i > (W/R)(1-\eta)\right) \\ &= \Pr\left(\sum_{i=1}^{k} \nu_i\epsilon_i' > A\right) \\ &< \Pr\left(\sum_{i=1}^{k} \nu_i\epsilon_i' > A \,|\, \nu_i\epsilon_i' < T \quad \text{for all } i\right) \\ &\quad + \Pr(\nu_i\epsilon_i' > T \quad \text{for any } i) \end{aligned} \tag{A.1}$$

where A is given by (14).

Bounding the first term by its (conditional) Chernoff bound and the second by a union bound, we have, using (9),

$$\begin{aligned} P_{\text{blocking}} &< E_k E_{\epsilon'} E_\nu\left([\exp(s\nu\epsilon')]^k \,|\, \epsilon' < T\right)e^{-sA} \\ &\quad + E_k\left[\sum_{i=1}^{k} \Pr(\epsilon_i' > T)\Pr(\nu_i = 1)\right] \\ &= E_k\left(\left[\rho E_{\epsilon}^{s\epsilon'} + (1-\rho)\right]^k \,\middle|\, \epsilon' < T\right)e^{-sA} \\ &\quad + \rho(\lambda/\mu)\Pr(\epsilon' > T) \\ &= \exp\left[\rho(\lambda/\mu)E\left(e^{s\epsilon'}t\right) - 1\right]e^{-sA} \\ &\quad + \rho(\lambda/\mu)\Pr(\epsilon' > T) \end{aligned} \tag{A.2}$$

$$\begin{aligned} Ee^{s\epsilon'_T} &= \int_{-\infty}^{(\ln T)/\beta} \exp(se^{\beta\xi}) \frac{e^{-\xi^2/(2\sigma^2)}}{\sqrt{2\pi}\sigma}\, d\xi \\ &= \int_{-\infty}^{\tau/\delta} \exp(se^{\beta\sigma\zeta})e^{-\zeta^2/2}\, d\zeta/\sqrt{2\pi} \end{aligned} \tag{A.3}$$

where $\tau = (\ln T)/\beta$, and

$$\begin{aligned} \Pr(\epsilon' > T) &= \Pr(\xi > (\ln T)/\beta) \\ &= \int_{(\ln T)/\beta}^{\infty} e\frac{-\xi^2/2\sigma^2}{\sqrt{2\pi}\sigma} = Q(\tau/\sigma). \end{aligned} \tag{A.4}$$

Inserting (A.4) and (A.3) into (A.2), and minimizing $s > 0$ and $\tau > 0$, yields (13)–(16).

The total interference is then the sum of (B.3) and (B.7), where the latter is integrated over all regions not including the six central region pairs A and B. The result equals $N_u(1+f)$. By performing this numerical integration, the other-cell interference factor f was obtained for various values of μ and σ_0, as shown in Table I.

Acknowledgment

The authors are grateful to E. Zehavi for suggesting the approach to the modified Chernoff bound of the Appendix, and to the highly motivated QUALCOMM team whose tireless efforts proved the feasibility of CDMA technology, and particularly of the power control methods which led to the empirical results shown in Fig. 1.

References

[1] D. Bertsekas and R. Gallager, *Data Networks*. Englewood Cliffs, NJ: Prentice-Hall, 1987.
[2] G. R. Cooper and R. W. Nettleton, "A spread spectrum technique for high capacity mobile communications," *IEEE Trans. Veh. Technol.*, vol. VT-27, pp. 264–275, Nov. 1978.
[3] K. S. Gilhousen, I. M. Jacobs, R. Padovani, A. J. Viterbi, L. A. Weaver, Jr., and C. E. Wheatley, III, "On the capacity of a cellular CDMA system," *IEEE Trans. Veh. Technol.*, vol. 40, May 1991.
[4] A. J. Viterbi, A. M. Viterbi, and E. Zehavi, "Performance of power-controlled wideband terrestrial digital communications," to appear in *IEEE Trans. Commun.*, 1993.
[5] W. Feller, *An Introduction to Probability Theory and Its Applications, Vol. I, 2nd ed.* New York: Wiley, 1957.
[6] P. T. Brady, "A statistical analysis of on–off patterns in 16 conversations," *Bell Syst. Tech. J.*, vol. 47, pp. 73–91, Jan. 1968.
[7] A. J. Viterbi, A. M. Viterbi, and E. Zehavi, "Other-cell interference in cellular power-controlled CDMA," *IEEE Trans. Commun.*, to appear.

PHOTO NOT AVAILABLE

Audrey M. Viterbi received the B.S. degree in electrical engineering and computer science and the B.A. degree in mathematics from the University of California, San Diego, in 1979. She received the M.S. and Ph.D. degrees in electrical engineering from the University of California, Berkeley, in 1981 and 1985, respectively.

She was an Assistant Professor of Electrical and Computer Engineering at the University of California, Irvine, from 1985 to 1990. Since 1990 she has been a Staff Engineer at Qualcomm, Inc., San Diego, where she is currently involved in the development of the CDMA cellular telephone system. Her professional interests are in the area of modeling and performance evaluation of computer-communication networks and communications systems.

PHOTO
NOT
AVAILABLE

Andrew J. Viterbi received the S.B. and S.M. degrees from the Massachusetts Institute of Technology, Cambridge, in 1957, and the Ph.D. degree from the University of Southern California in 1962.

In his first employment after graduating from M.I.T., he was a member of the project team at C.I.T. Jet Propulsion Laboratory which designed and implemented the telemetry equipment on the first successful U.S. satellite, *Explorer I*. In the early 1960's at the same laboratory, he was one of the first communication engineers to recognize the potential and propose digital transmission techniques for space and satellite telecommunication systems. As a Professor in the UCLA School of Engineering and Applied Science from 1963 to 1973, he did fundamental work in digital communication theory and wrote two books on the subject, for which he received numerous professional society awards and international recognition. In 1968 he co-founded LINKABIT Corporation and served as its Executive Vice President from 1974 to 1982, and as President from 1982 to 1984. On July 1, 1985, he co-founded and became Vice Chairman and Chief Technical Officer of QUALCOMM, Inc., a company specializing in mobile satellite and terrestrial communication and signal processing technology. Since 1975 he has been associated with the University of California, San Diego, and since 1985, as Professor (quarter time) of Electrical and Computer Engineering.

Dr. Viterbi is a member of the U.S. National Academy of Engineering. He is past Chairman of the Visiting Committee for the Electrical Engineering Department of Technion, Israel Institute of Technology, past Distinguished Lecturer at the University of Illinois and the University of British Columbia, and he is presently a member of the M.I.T. Visiting Committee for Electrical Engineering and Computer Science. In 1986 he was recognized with the Annual Outstanding Engineering Graduate Award by the University of Southern California, and in 1990 he received an honorary Doctor of Engineering Degree from the University of Waterloo, Ont. Canada. He presented the Shannon Lecture at the 1991 International Symposium on Information Theory. He has received three paper awards, culminating in the 1968 IEEE Information Theory Group Outstanding Paper Award. He has also received four major international awards: the 1975 Christopher Columbus International Award (from the Italian National Research Council sponsored by the City of Genoa); the 1984 Alexander Graham Bell Medal (from IEEE, sponsored by AT&T) "for exceptional contributions to the advancement of telecommunications;" the 1990 Marconi International Fellowship Award; and was co-recipient of the 1992 NEC C&C Foundation Award.

Decorrelating Decision-Feedback Multiuser Detector for Synchronous Code-Divison Multiple-Access Channel

Alexandra Duel-Hallen

Abstract— **Several multiuser detectors for code-division multiple-access (CDMA) systems have been studied recently. We propose a new decorrelating decision-feedback detector (DF) for synchronous CDMA which utilizes decisions of the stronger users when forming decisions for the weaker ones. The complexity of DF is linear in the number of users, and it requires only one decision per user. Performance gains with respect to the linear decorrelating detector are more significant for relatively weak users, and the error probability of the weakest user approaches the single-user bound as interferers grow stronger. The error rate of DF is compared to those of the decorrelator and the two-stage detector.**

I. Introduction

We consider a code-division multiple-access (CDMA) system where the receiver observes the sum of synchronously transmitted signals from several users in additive noise. The synchronous assumption holds in several important practical systems [2], [3] and provides a simple model for studying detection algorithms which in their general form can be applied to the asynchronous case and to channels with fading and multipath. Since the conventional detector often fails to produce reliable decisions for the CDMA channel, several new multiuser detectors have been proposed [1]–[4]. The optimum receiver achieves low error probability at the expense of high computational complexity [1]. When the number of users is large, it is desirable to use a simple but reliable suboptimum detector. The linear decorrelating detector, e.g., [2] (the decorrelator) can significantly outperform the conventional receiver. If a synchronous CDMA system is modeled as a time-varying single-user channel with intersymbol interference (ISI), the decorrelator becomes analogous to the zero-forcing equalizer [6, p. 358], since it eliminates multiuser interference at the expense of increased noise power. Decision-feedback equalizers often have significantly lower error rates than linear detectors in single-user channels [6, p. 383], and have been shown superior to linear detectors in several multi-user systems [7], [8]. This observation motivates our study of decision-feedback detection for synchronous CDMA. Other recent approaches to multiuser detection include multistage detectors [3], [5]. They improve the performance of the decorrelator under certain operating conditions (e.g., when interfering users are stronger than the user under consideration), and their error rate approaches that of the optimum detector as energies of the interfering users grow. However, several stages of decisions result in higher complexity relative to the linear and decision-feedback detectors.

The decorrelating decision-feedback detector proposed in this paper is suitable for synchronous CDMA systems, since it utilizes the difference in users' energies. Its forward and feedback filters are chosen to eliminate all multiuser interference provided that the feedback data are correct. Decisions for users are made in the order for decreasing received energies. The receiver for each user linearly combines sampled outputs of the matrix matched filter with decisions of all stronger interfering users. Thus, the receiver for the strongest user does not involve feedback, and is equivalent to the decorrelator. On the other hand, the detector for the weakest user utilizes decisions of all other users and ideally (i.e., when feedback symbols are correct) achieves the performance of the single-user system.

In [4], a successive cancellation method was proposed for a coded CDMA system. Although the current paper was originally written without the knowledge of that technique, the ideas are quite similar. However, the emphasis of this work and its extension to the asynchronous case [9] is on the receiver filter optimization in addition to the cancellation of stronger users.

With the exception of the decorrelator [2], all of the multiuser detectors studied in [1]–[5] and the present approach require the knowledge of users' energies. Thus, although these detectors offer theoretical performance advantages over the conventional detector, they rely on the ability to obtain accurate estimates of users' energies in time-variant CDMA channels. These have to be estimated using the conventional tracking methods. In the coded case, provided that the users are ranked in the order of decreasing strength, more accurate estimates of the users' energies can be calculated using the re-encoded signal [4]. We briefly outline a technique for estimating energies in coded and uncoded systems for the simple synchronous CDMA model studied in the current paper.

In Section II, the discrete-time vector model for synchronous CDMA and the linear decorrelating detector are considered. In Section III, we derive a white-noise discrete-time model, the decorrelating decision-feedback detector and describe a method for estimating energies. Performance estimates, simulation results and a comparison with the decorrelator and the two-stage detector are presented in Section III.

Paper approved by the Editor for Adaptive Signal Processing of the IEEE Communications Society. Manuscript received October 12, 1990; revised August 15, 1991. This work was supported in part by NSF under Grant ECS8352220.

The author is with the Department of Electrical and Computer Engineering, North Carolina State University, Raleigh, NC 27695.

IEEE Log Number 9207340.

Reprinted from *IEEE Transactions on Communications,* vol. 41, no. 2, February 1993.

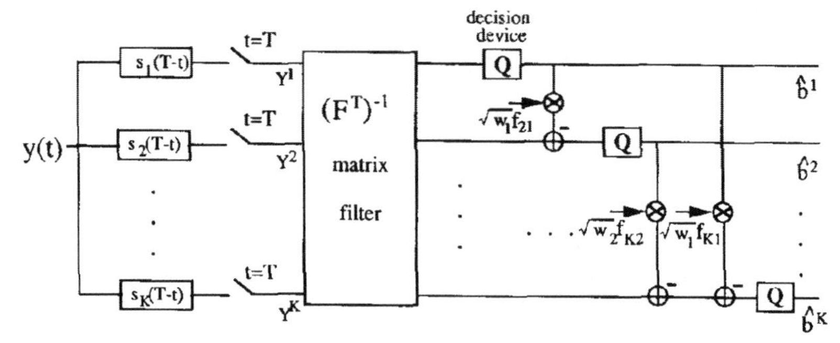

Fig. 1. Matched-filter receiver and the decorrelating decision-feedbackdetector for synchronous CDMA.

II. The System Model and the Decorrelator

Consider a synchronous CDMA system with K users and a set of preassigned normalized signature waveforms $s_k(t)$, $k = 1, \cdots, K$ where each waveform is restricted to a symbol interval of duration T, $\int_0^T s_i^2(t)\,dt = 1$, and signature waveforms are linearly independent. The input alphabet of each user is antipodal binary $\boldsymbol{X} = \{-1, +1\}$. The input bit sequence for each user is i.i.d., equiprobable, and independent of other users. It is sufficient to consider a single transmission. Suppose the input vector is given by $\boldsymbol{b} = (b_1, \cdots, b_K)^T$, $b_k \in \boldsymbol{X}$, and assume that input energies w_k are nondecreasing, i.e., $w_1 \geq w_2 \geq \cdots \geq w_K$. (We initially assume that the receiver knows the energies.) When a bank of filters matched to the set of signature waveforms is followed by samplers at time T, the following discrete-time output vector arises (Fig. 1) [2]

$$\boldsymbol{y} = \boldsymbol{RWb} + \boldsymbol{z}\,, \tag{1}$$

where $\boldsymbol{R}$ is a $K \times K$ positive definite matrix of signature waveform cross-correlations, and $\boldsymbol{W}$ is a diagonal matrix with $W_{i,i} = \sqrt{w_i}$, $i = 1, \cdots, K$. The term $\boldsymbol{z}$ is a Gaussian noise vector with the $K \times K$ autocorrelation matrix $\boldsymbol{R}(\boldsymbol{z}) = \sigma^2 \boldsymbol{R}$ where $R(\boldsymbol{z})_{i,j} = E(z_i, z_j)$.

The objective of a multiuser detector is to recover the input data vector $\boldsymbol{b}$ given the output vector $\boldsymbol{y}$. In the linear decorrelating detector, the matrix filter $\boldsymbol{R}^{-1}$ followed by a set of decision devices (sign detectors) is applied to $\boldsymbol{y}$. The output of the matrix filter is

$$\hat{\boldsymbol{y}} = \boldsymbol{Wb} + \tilde{z} \tag{2}$$

where $\tilde{\boldsymbol{z}}$ is a Gaussian noise vector with the autocorrelation matrix $\boldsymbol{R}(\tilde{\boldsymbol{z}}) = \sigma^2 \boldsymbol{R}^{-1}$. Thus, the probability that the kth input is recovered incorrectly is

$$Pe_k(\text{dec}) = Q\left(\sqrt{w_k / \left[\sigma^2 (R^{-1})_{k,k}\right]}\right) \tag{3}$$

where the Q-function is $Q(x) = \frac{1}{\sqrt{2\pi}} \int_x^\infty e^{-\frac{y^2}{2}}\,dy$.

III. White Noise Discrete-Time Model and the Decorrelating Decision-Feedback Detector

In synchronous CDMA, a white noise model can be obtained by factoring the positive definite matrix of cross-correlations as $\boldsymbol{R} = \boldsymbol{F}^T \boldsymbol{F}$ where $\boldsymbol{F}$ is a lower triangular matrix (see Cholesky decomposition algorithm [11, p. 88]). If the filter with response $\left(\boldsymbol{F}^T\right)^{-1}$ is applied to the sampled output of the matched filter (1), the resulting output vector is

$$\check{\boldsymbol{y}} = \boldsymbol{FWb} + \boldsymbol{n} \tag{4}$$

where $\boldsymbol{n}$ is a white Gaussian noise vector with the autocorrelation matrix $\boldsymbol{R}(\boldsymbol{n}) = \sigma^2 \boldsymbol{I}$. ($\boldsymbol{I}$ is the $(K \times K)$ identity matrix.) The discrete-time models (4) and (1) correspond to the outputs of the standard and whitened matched filters, respectively, in single-user channels with ISI [6, p. 353].

Since the components of the noise vector $\boldsymbol{n}$ in (4) are uncorrelated, the optimum (maximum likelihood) detector for synchronous CDMA [3] has the Euclidean metric $\|\check{\boldsymbol{y}} - \hat{\boldsymbol{t}}\|^2 = \sum_{k=1}^K \left(\check{y}_k - \hat{t}_k\right)^2$ where $\hat{\boldsymbol{t}}$ is the signal associated with an input $\hat{\boldsymbol{b}}$, i.e., $\hat{\boldsymbol{t}} = \boldsymbol{FW}\hat{\boldsymbol{b}}$. Both the metric and the expression for the probability of error of the optimal detector have a simpler derivation when the model (4) is used instead of (1).

The model (4) also gives rise to the decorrelating decision-feedback detector (DF). The kth component of $\check{\boldsymbol{y}}$ is given by $\check{y}_k = F_{k,k}\sqrt{w_k}b_k + \sum_{i=1}^{k-1} F_{k,i}\sqrt{w_i}b_i + n_k$. Since this expression does not contain a multiuser interference term for the strongest user ($k = 1$), a decision for this user is made first: $\hat{b}_1 = \text{sgn}(\check{y}_1)$. Multiuser interference for the second user is $F_{2,1}\sqrt{w_1}b_1$. Since a decision for the first user is available, we can use feedback in estimating the second symbol. Thus, the second decision is $\check{y}_2 - F_{2,1}\sqrt{w_1}\hat{b}_1$. Similarly, for the $k$the user, multiuser interference depends on stronger users ($i = 1, \cdots, k-1$). Decisions for these users have already been made, and they can be used to form a feedback term (Fig. 1), i.e.,

$$\begin{aligned} \hat{b}_k &= \text{sgn}\left(\check{y}_k - \sum_{i=1}^{k-1} F_{k,i}\sqrt{w_i}\hat{b}_i\right) \\ &= \text{sgn}\left(F_{k,k}\sqrt{w_k}b_k + \sum_{i=1}^{k-1} F_{k,i}\sqrt{w_i}\right. \\ &\qquad \left. \cdot \left(b_i - \hat{b}_i\right) + n_k\right). \end{aligned} \tag{5}$$

To summarize, the decorrelating decision-feedback detector is characterized by a feedback filter $\boldsymbol{B} = \left(\boldsymbol{F} - \boldsymbol{F}^d\right)\boldsymbol{W}$, where $\boldsymbol{F}^d$ is a diagonal matrix obtained from $\boldsymbol{F}$ by setting all off-diagonal elements to zero. The filter is fed by the vector of decisions $\hat{b}$. The vector input to the set of decision devices is $\tilde{y}-B\hat{b} = F^dWb+\left(F - F^d\right)W\left(b - \hat{b}\right)+n$. Since B is lower triangular with zeros along the diagonal, only previously made decisions (i.e., $\hat{b}_{k-1}, \hat{b}_{k-2}, \cdots, \hat{b}_1$) are required for forming the input to the kth quantizer. DF corresponds to the zero-forcing decision-feedback equalizer for ISI channels [6, p. 383] since it attempts to cancel all multiuser interference. The strictly lower triangular B corresponds to the purely causal feedback filter used in single-user systems.

Of course, it is possible to remove the requirement that received energies are nondecreasing. However, the multistage method studied in [3] and our analysis indicate that feedback is primarily beneficial when interfering users are stronger. This observation motivated our choice of the receiver structure.

An important measure of performance for a decision-feedback detector is the signal-to-noise ratio at the input to the decision device under the assumption of correct previous decisions. From (5), the signal-to-noise ratio for the DF is

$$\mathrm{SNR}_k = F_{k,k}^2 w_k/\sigma^2 . \tag{6}$$

Given the same order of making decisions, this SNR is the largest achievable by any decision-feedback detector which attempts to cancel all multi-user interference. To show this, consider a $(K \times K)$ matrix filter M applied to the white-noise output (4) which result in a discrete-time model $MFWb + \hat{z}$, where the interference matrix MF is lower triangular. Note that this requires M to be lower triangular, and the autocorrelation of the noise is $R(\hat{z}) = \sigma^2 MM^T$. The lower triangular structure of the interference matrix allows to construct a decision-feedback detector with the kth signal-to-noise ratio $M_{k,k}^2 F_{k,k}^2 w_k/(\sigma^2 \sum_{i=1}^k M_{k,i}^2)$. We observe that this signal-to-noise ratio is maximized for each k if $M = I$, and the optimality of DF results.

It can be shown that DF is equivalent to a noise-canceling detector obtained from the discrete-time model (2). Since the inverse of the autocorrelation matrix R is given by $R^{-1} = F^{-1}\left(F^T\right)^{-1}$, (2) can be rewritten as

$$\tilde{y} = Wb + F^{-1}n \tag{7}$$

where n is the white noise vector in (4). Since F^{-1} is lower triangular, i.e., $\tilde{y}_k = \sqrt{w_k}b_k + \sum_{i=1}^k \left(F^{-1}\right)_{k,i} n_i$, we can construct a detector which uses earlier decisions of the noise sequence $\hat{n}_1, \cdots, \hat{n}_{k-1}$ to reduce the variance of the noise in the kth component of (2) (or (7)). The kth decision of this detector is given by $\hat{b}_k = \mathrm{sgn}\left(\tilde{b}_k\right)$ where

$$\tilde{b}_k = \tilde{y}_k - \sum_{i=1}^{k-1} \left(F^{-1}\right)_{k,i} \hat{n}_i , \tag{8}$$

and the kth noise-decisions is $\hat{n}_k = F_{k,k}\left(\tilde{b}_k - \sqrt{w_k}\hat{b}_k\right)$. This noise-canceling detector is related to a partial feedback multistage detector studied in [5]. In the second stage of that detector, a subset of previous noise estimates is used to reduce the noise variance in (2). We observe that if this subset is restricted to stronger users (i.e., users $1, \cdots, k-1$ when making a decision for the kth user), the second stage of the detector in [5] and DF (or the noise-cancelling detector described above) have the same inputs to the decisions devices provided previous decisions are correct.

In time-varying CDMA channels (e.g., mobile radio), the energies of users need to be estimated and updated for proper cancellation of multiuser interference. Consider the noise-cancelling detector described in the previous paragraph. Suppose the receiver does not know the energies, and a single transmission occurs. Given the output of the decorrelating filter (7), it is reasonable to estimate $\sqrt{w_k}$ as $\widehat{\sqrt{w_k}} = |\tilde{y}_k|$. This implies that the kth noise-decision is $\hat{n}_k = 0$. Thus, the resulting detector does not cancel the noise, and is equivalent to the decorrelator. If the energies vary slowly, it is possible to improve upon this estimation method by averaging the absolute values of previous N outputs, i.e., the estimate at time n is $\widehat{\sqrt{w_k}}(n) = \sum_{i=0}^{N-1} \alpha_i |\tilde{y}_k(n-i)|$ where α_i is a nonnegative nonincreasing sequence of length N, i.e., $\alpha_i \geq \alpha_{i+1}$, and $\sum_{i=0}^{N-1} \alpha_i = 1$. If $N = \infty$, the estimator can be implemented as an IIR filter. (If the receiver knows the ranking of energies, but not their values, the estimate of energy of the kth user can be computed after cancellation, i.e., the components of the sequence $\tilde{b}_k(i)$ (8) can be averaged to obtain less noisy estimates.) Of course, the weights α_i have to be chosen appropriately. If the energy varies slowly, an accurate estimate is obtained by choosing a large N and a slowly decreasing sequence α_i. However, if the variations are fast, more emphasis should be placed on more recent components of the sequence $\tilde{y}_k$. When the energies jump unpredictably, we obtain the decorrelator as shown above. Note that the decorrelator is the maximum-likelihood detector when the energies are unknown, e.g., [2]. It also achieves the optimum near-far resistance [2]. If there are "jumps" and "quiet periods" in the energy values, the receiver can detect these and switch between the decorrelating and the feedback modes.

The above discussion applies to relatively high signal-to-noise ratio systems. When the channel is very noisy, coding has to be employed. In [4], a cancellation method was proposed for a coded system. Similarly, DF can be applied to encoded signals. In this case, received sequences of stronger users (extend (4) or (7) to sequences) need to be decoded first, and then re-encoded and fed back to the receivers of weaker users. To estimate the energy of the kth user, the re-encoded kth sequence can be correlated with the sequence $\tilde{b}_k(i)$ (8) (see [4]). In this case, ranking of energies has to be performed *a priori* based on previous estimates and other tracking methods.

IV. Analysis and Examples

First, assume that the energies of users are estimated correctly. The signal-to-noise ratio (6) gives rise to the probability of error of DF for the kth user under the assumption of correct previous decisions

$$\tilde{P}e_k(\mathrm{DF}) = Q(F_{k,k}\sqrt{w_k}/\sigma) . \tag{9}$$

It is easy to show that $F_{k,k}^2 \geq 1/\left(R^{-1}\right)_{k,k}$. Note that for the strongest user $(k=1)$, $F_{1,1}^2 = 1/\left(R^{-1}\right)_{1,1}$ and the estimate (9) gives the probability of error since the receiver for this user does not utilize feedback. Therefore, the error rates of the DF (9) and the decorrelator (3) are the same for the strongest user. For $k > 0$, the last inequality is tight provided that multiuser interference affects the kth user. Thus, an improvement over the performance of the decorrelator is suggested by the comparison of (9) and (3). (When DF and the other methods [1]–[3] are compared to the decorrelator, it should be taken into account that the decorrelator has the following advantage: it does not require the knowledge of energies.) Finally, for the weakest user, $F_{K,K}^2 = 1$, and the ideal performance of DF (9) agrees with the error probability of the single-user system given by

$$Pe_k(\mathrm{SU}) = Q(\sqrt{w_k}/\sigma). \tag{10}$$

To find the exact error rate of DF for the kth user, one has to average the conditional error probability given a particular error pattern for users $1, \cdots, k-1$ over all such error patterns:

$$P_k = \frac{1}{2} E_{\Delta b_1, \cdots, \Delta b_{k-1}} \cdot Q\left(\frac{F_{k,k}\sqrt{w_k} + \sum_{i=1}^{k-1} F_{k,i}\sqrt{w_i}\Delta b_i}{\sigma}\right) \tag{11}$$

where the error pattern for the ith user is $\Delta b_i = (b_i - \hat{b}_i)$. For example, consider a two-user system with $R_{1,2} = r$. Then $F_{1,1} = \sqrt{1 - r^2}$ and $F_{2,1} = r$. The error rate for the stronger user ($k = 1$) is given by the error rate of the decorrelator: $P_1 = Q(\sqrt{w_1(1 - r^2)}/\sigma)$, and the input to the decision device of the second user is $\sqrt{w_2}b_2 + \sqrt{w_1}r(b_1 - \hat{b}_1) + n_2$. By averaging over all possible values taken on by $b_1 - \hat{b}_1$, we derive the error rate for the weaker user

$$Pe_2(\mathrm{DF}) = (1 - P_1)Q\left(\frac{\sqrt{w_2}}{\sigma}\right) + \frac{P_1}{2}\left[Q\left(\frac{\sqrt{w_2} - 2r\sqrt{w_1}}{\sigma}\right) + Q\left(\frac{\sqrt{w_2} + 2r\sqrt{w_1}}{\sigma}\right)\right]. \tag{12}$$

This expression also gives the error rate of the second stage detector in [3]. We use (12) and numerical results obtained in [3] to analyze performance of DF in this two-user system. Suppose the energy of the second user w_2 is fixed. As w_1 grows (P_1 becomes smaller), the first Q-function dominates (12), and the error rate of the weaker user approaches the single-user bound (10). Thus, DF is attractive in near–far situations. The expression (12) also gives insight into the behavior of DF with the order of users interchanged, i.e., when $w_1 < w_2$. This analysis is of interest if the users are not ranked properly. In this case, the second and third terms in (12) (error propagation) are more significant, and feedback is not as useful. In fact, in a certain region of values taken by w_1, the error probability of DF is higher than that of the decorrelator (in the Figs. 2, 3 of [3], they differ at most by a factor of 20). Finally, as w_1 decreases, the DF detector again performs better than the decorrelator and finally approaches the error rate of the single-user system. However, the gain is not significant, and for small w_1, the conventional detector is a better choice than either DF or the decorrelator.

For a large number of users, exact computation of error probability becomes complex (see, for example, the analysis for the multistage detector in [3]). The problem is similar to computing the exact probability of a decision-feedback equalizer for a single-user system. Although this probability can be bounded analytically, computer simulations are most commonly used in determining the error rate and the effects of error propagation [6]. We show results of simulations for a four-user system in Example 2. It is possible to study general performance trends by considering dominating terms in the expression for the error rate (11). Suppose the energy w_k is fixed. Then the error rate of DF for the kth user does not depend on energies of users $k + 1, \cdots, K$, and approaches the ideal error probability (9) as energies of users $1, \cdots, k - 1$ grow. In particular, the error rate of the weakest user tends to the error probability of the single-user system (10). It is interesting to note that if the energies of users $1, \cdots, i$ grow, but $w_{i+1}, \cdots, w_k$ remain fixed, the limiting error rates of DF for users $i + 1, \cdots, k$ are found by studying the $(k - i)$-user white-noise model with a lower triangular interference matrix (a submatrix of $\boldsymbol{F}$) whose m, nth component ($m \geq n$, $m = 1, \cdots, k - i$) is given by $F_{i+m,i+n}$. We can construct a decision-feedback detector for this model and find its error rate for the mth user. This error probability is approached by the error rate of DF for the $(i + m)$th user in the original K-user system as energies of users $1, \cdots, i$ grow. Example 2 illustrates the situation for a four-user system.

Finally, we investigate the effect of errors in estimating energies on the performance of DF. (Note that incorrect ranking was discussed following (12).) Suppose the ith energy estimate is $\sqrt{\hat{w}_i}$, and define $\Delta w_i = \sqrt{w_i} - \sqrt{\hat{w}_i}$. The input to the kth decision device is then $F_{k,k}\sqrt{w_k}b_k + \sum_{i=1}^{k-1} F_{k,i}(\sqrt{w_i}(b_i - \hat{b}_i) + \Delta w_i \hat{b}_i) + n_k$. Assuming correct previous decisions, the remaining interference is $\sum_{i=1}^{k-1} F_{k,i}\Delta w_i \hat{b}_i$. If Δw_i are large relative to w_k, DF suffers from multiuser interference similarly to the conventional detector. However, using estimation techniques described in the previous section, we can keep $\sqrt{\hat{w}_i}$ close to the output value $\tilde{y}_i$ (7) in the case of rapidly varying energies, and obtain more accurate estimates when energies vary slowly. Then, as discussed in Section III (for the equivalent noise-canceling detector), the performance of DF is at least as good as that of the decorrelator.

Example 1: We consider the bandwidth-efficient two-user synchronous CDMA system studied in [3] with signal cross-correlation $R_{1,2} = r = 0.7$. The error probability of DF for the second user is given by (12). Fig. 2 depicts error probabilities for the second user versus the difference between input signal-to-noise ratios where $\mathrm{SNR}(2) = w_2^2/\sigma^2 = 11$ dB. We observe that DF has lower error rate than the decorrelator (3). As the first user grows stronger, the performance of DF approaches that of the single-user system (10).

As was pointed out earlier, the error probabilities of the two-stage detector and DF are the same for the weaker user in a two-user system. In addition to the data in Fig. 2, Fig. 2 of [3] also shows error rates of the conventional, decorrelator, optimum, optimum linear [2], and three-stage detectors for this

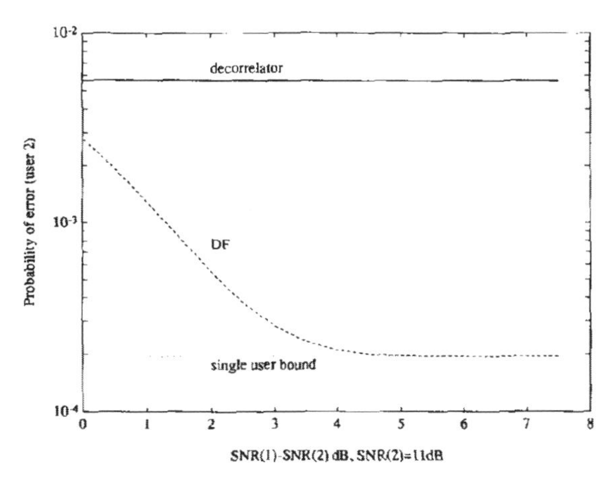

Fig. 2. Error probabilities for two-user channel, $r = 0.7$. (Example 1).

example. We observe that performance of DF is close to that of the optimal detector, and other methods have higher error rates.

Example 2: In [3], multistage detectors are evaluated for a multiuser system based on a set of signature waveforms derived from Gold sequences of length seven. The objective of this study is to gain insight into the performance of multistage detectors in asynchronous bandwidth-efficient CDMA with many users. It is of interest to conduct a similar study here. We consider the same set of signature waveforms, and assume that four users are active. The corresponding matrix of cross-correlations is given in Fig. 3. The probability of error of the decorrelator (3), the single-user bound (10) and the simulated performance of DF are shown for the weakest user in Fig. 3. The input signal-to-noise ratio is $\mathrm{SNR}(4) = 11$ dB for the weakest user, and varies from 11 to 17 dB for other users. We observe that DF has significantly lower error rate than the decorrelator, and approaches the single-user bound as interfering users grow stronger. We also find that the performance of DF is close to that of the two-stage detector (see [3, Fig. 5(c)]), although in this case the two detectors are not the same. The second-stage detector uses outputs of the decorrelator to cancel interference terms, whereas DF feeds back its decisions for stronger users, which, in general, are better than those of the decorrelator (see below). Since both DF and two-stage detectors of [3], [5] are difficult to analyze, it would be desirable to compare these detectors for various examples using simulations for the asynchronous system, this is accomplished in [9], [10]. We also note that DF is simpler to implement than two-stage detectors, since it does not require a first stage.

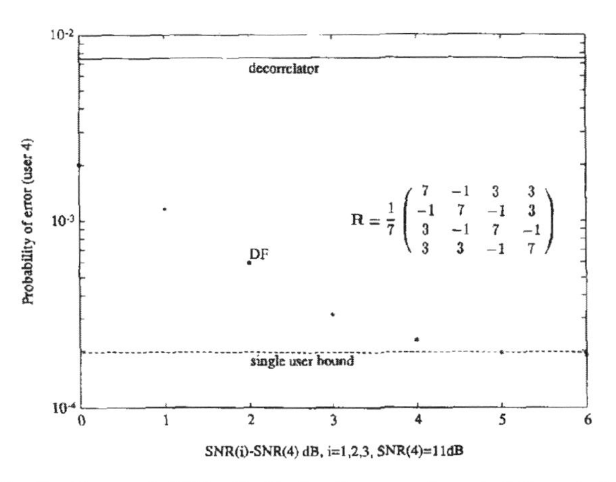

Fig. 3. Error probabilities for the first four-user channel of Example 2.

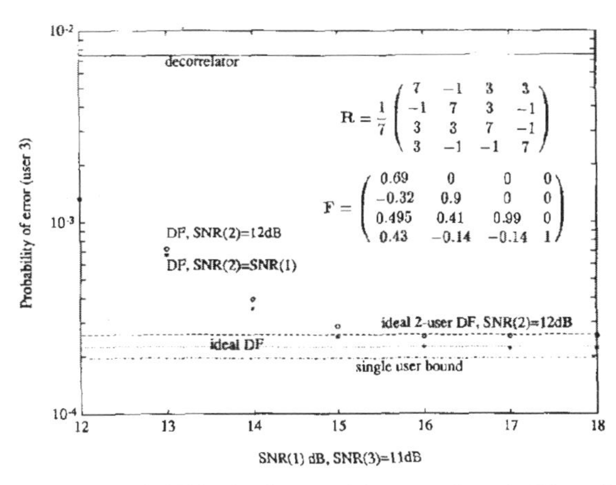

Fig. 4. Error probabilities for the second four-user channel of Example 2.

Next, we evaluate the error rate of DF for user 3 with $\mathrm{SNR}(3) = 11$ dB in the same four-user system, but with a different order of users. The matrix of cross-correlation and the corresponding matrix $\boldsymbol{F}$ are shown in Fig. 4 along with simulation results (individual points) and analytical performance estimates (lines). In the first simulation study, we allowed the energies of the first two users grow. We observe that the error rate of DF converges to the ideal performance (9) which is close to the single-user bound (10), since $F_{3,3} \approx 1$. In the second study, we fixed $\mathrm{SNR}(2) = 12$ dB, while the energy of the first user was allowed to grow. We find that the error rate of user 3 approaches the error rate of the decision-feedback detector for the weaker user in a white-noise two-user system with the interference matrix $\begin{pmatrix} F_{2,2} & 0 \\ F_{3,2} & F_{3,3} \end{pmatrix}$. The exact expression for this "ideal two-user DF" error probability is derived similarly to (12). We also observe that the decorrelator performs worse than DF for all operating points we have considered.

REFERENCES

[1] S. Verdu, "Minimum probability of error for asynchronous Gaussian multiple-access channel," *IEEE Trans. Inform. Theory,* vol. IT-32, pp. 85–96, Jan. 1986.
[2] R. Lupas and S. Verdu, "Linear multiuser detectors for synchronous code-division multiple-access channel," *IEEE Trans. Inform. Theory,* vol. IT-35, pp. 123–136, Jan. 1989.
[3] M. K. Varanasi and B. Aazhang, "Near-optimum detection in synchronous code-division multiple access systems," *IEEE Trans. Commun.,* vol. 39, pp. 725–736, May 1991.
[4] A. J. Viterbi, "Very low rate convolutional codes for maximum theoretical performance of spread-spectrum multiple-access channel," *IEEE J. Select. Areas Commun.,* vol. 8, pp. 641–649, May 1990.
[5] R. Lupas, "Near-far resistant linear multiuser detection," Ph.D. dissertation, Princeton Univ., Jan. 1989.
[6] J. G. Proakis, *Digital Communications.* New York: McGraw-Hill, 1983.
[7] M. Kavehrad and J. Salz, "Cross-polarization cancellation and equalization in digital data transmission over dually polarized fading radio channels," *Bell Syst. Tech. J.,* vol. 64, no. 10, pp. 2211–2245, Dec. 1985.
[8] A. Duel-Hallen, "Equalizers for multiple input/multiple output channels and PAM systems with cyclostationary input sequences," *IEEE J. Select. Areas Commun.,* to appear.
[9] A. Duel-Hallen, "On suboptimal detection for asynchronous CDMA channels,"in *Proc. 1992 CISS,* Princeton, NJ, pp. 838–843.
[10] ——, "Decision-feedback detector for asynchronous CDMA," *IEEE Trans. Commun.,* to be published.
[11] G. Golub and C. Van Loan, *MATRIX Computations.* Baltimore, MD: The Johns Hopkins University Press, 1983.

Dense Wavelength Division Multiplexing Networks: Principles and Applications

CHARLES A. BRACKETT, SENIOR MEMBER, IEEE

(Invited Paper)

***Abstract*—The very broad bandwidth of low-loss optical transmission in a single-mode fiber and the recent improvements in single-frequency tunable lasers have stimulated significant advances in dense wavelength division multiplexed optical networks. This technology, including wavelength-sensitive optical switching and routing elements and passive optical elements, has made it possible to consider the use of wavelength as another dimension, in addition to time and space, in network and switch design. The independence of optical signals at different wavelengths makes this a natural choice for multiple-access networks, for applications which benefit from shared transmission media, and for networks in which very large throughputs are required.**

In this paper, we review recent progress on multiwavelength networks, some of the limitations which affect the performance of such networks, and present examples of several network and switch proposals based on these ideas. We also discuss the critical technologies that are essential to progress in this field.

I. Introduction

THE advent of a single-mode optical fiber has presented communications engineers with the exciting dilemma of a transmission medium which has a bandwidth that exceeds both the speeds at which it can be accessed by conventional means and the aggregate information rates for which it is likely to be used. The low-loss region of a single-mode fiber extends over wavelengths from roughly 1.2 to 1.6 μm, which is an optical bandwidth of more than 30 THz. An information capacity of 30 Tb/s would be enough to deliver a channel of 100 Mb/s to hundreds of thousands of destinations on a single fiber. To utilize such a bandwidth fully would require individual optical pulse widths of a few tens of femtoseconds. Technologies do exist for generating such ultrashort pulses in research laboratories but systems employing them are not yet practical.

These examples serve to illustrate two principles upon which modern work on dense wavelength division multiplexing (WDM) is based. First, the bandwidth of the fiber is most easily accessed in the wavelength domain directly, rather than in the time domain. Second, it is possible to take advantage of the enormous bandwidth by using wavelength to perform such network- and system-oriented functions as routing, switching, and service segregation.

Manuscript received November 16, 1989; revised March 30, 1990.

The author is with Bell Communications Research, Morristown, NJ 07960.

IEEE Log Number 9036369.

In this paper, we review recent work on dense WDM as it is applied to networks. We take the view that the most important applications are likely to be other than simple high-capacity point-to-point transmission links, in which wavelength becomes an integral part of the interconnection fabric. Our interests here, then, are in the new network architectures which are made possible by this multiwavelength technology and in the directions and capabilities of the technologies themselves.

A few definitions of terms are in order. It is common to refer to dense WDM for systems where the wavelength spacing is on the order of 1 nm and to optical frequency division multiplexing (FDM) for systems where the optical frequency spacings are on the order of the signal bandwidth or bit rate. Such a distinction is useful in indicating the frequency and wavelength scales of interest and the degree of complexity required in the control and selection of wavelengths, but makes no fundamental architectural difference. We, therefore, make no distinction here between WDM and FDM except when we consider the technologies used for their implementation and the performance limits achievable.

In contrast with dense WDM, conventional WDM (not dense) technology was directed at upgrading the capacity of installed point-to-point transmission systems, typically by the addition of two, three, or four additional wavelengths usually separated by several tens, or even hundreds, of nanometers in wavelength.

Early attempts at introducing conventional WDM were not successful because WDM requires separate transmitter/receiver components at each wavelength and this proved to be more expensive than adding additional stages of time-division multiplexing to achieve increased capacity by running at higher speeds. Three things have now changed, which have influenced WDM greatly and serve as a point of departure for this paper. First, the technology has improved greatly, with the introduction of distributed feedback (DFB), distributed Bragg reflector (DBR), and other narrow linewidth lasers, and reasonably inexpensive passive components such as star couplers and fused-fiber multiplexers. Second, it has been realized that WDM has network applications (referred to above as the main subject of this paper) beyond the simple increase of link capacity. The third change is that bit rates have now

Reprinted from *IEEE Journal on Selected Areas in Communications,* vol. 8, no. 6, August 1990.

increased to the neighborhood of 2 Gb/s in commercial systems and to 10–20 Gb/s in research laboratories. It is getting harder to increase these rates beyond these levels with time-division multiplexing, although further progress will be undoubtedly made. At the same time, it is getting easier to increase capacity with WDM.

A key feature of dense WDM is that the discrete wavelengths form an orthogonal set of carriers which can be separated, routed, and switched without interfering with each other, as long as the total light intensity is kept sufficiently low. It is this use of wavelength and its processing in passive network elements which distinguish optical networks, in general, from other network technologies.

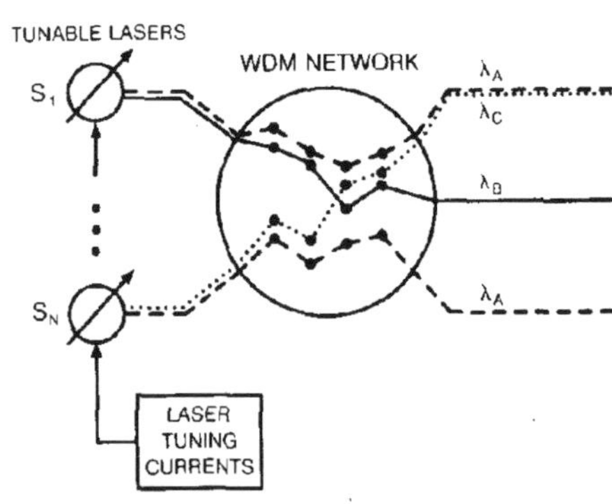

Fig. 1. A wavelength routing network illustrating the features of non-blocking wavelength addressing.

II. WDM Architectures

The two general architectural forms that have been most commonly used in WDM networks are wavelength routing networks and broadcast-and-select networks. These are illustrated in Figs. 1 and 2, respectively. Wavelength routing networks are composed of one or more wavelength-selective elements and have the property that the path that the signal takes through the network is uniquely determined by the wavelength of the signal and the port through which it enters the network. So, for example, in Fig. 1 an $N \times N$ network is shown in which N tunable laser sources are interconnected with N (wavelength independent) receivers through a network consisting of perhaps several WDM elements. By tuning to a selected wavelength, the signal from a given laser can be routed to a selected output port on the network. Since there are N inputs and N outputs, one might expect N^2 wavelengths would be required to form a complete interconnection. It turns out, however, that it can always be arranged so that with only N wavelengths, N inputs can be interconnected with N outputs in a completely noninterfering way. That this is so can be seen in Fig. 3, drawn for a 4×4 network, which shows one possible interconnection pattern and arrangement of WDM units and the associated wavelength assignment table. For example, in Fig. 3, the wavelength to go from input port $S1$ to output port $R3$ is λ_2. It is possible to address each output port uniquely by choice of λ and no output port can receive any given wavelength from more than one input. This is extendible to any size network with just N wavelengths but it does require N^2 interconnection fibers between the WDM stages. It is possible to eliminate the interconnection between the source and receiver of any given port, thereby eliminating the need for λ_0 and, therefore, requiring only N-1 wavelengths. Whether this is desirable depends on the system design, which may, for example, use the wavelength λ_0 as transmitted through the WDM system to provide information on the health of the system or for wavelength registration measurements.

The second major architectural type is the broadcast-and-select network illustrated in Fig. 2. In this network, all inputs are combined in a star coupler and broadcast to all outputs. Several different possibilities exist, depending

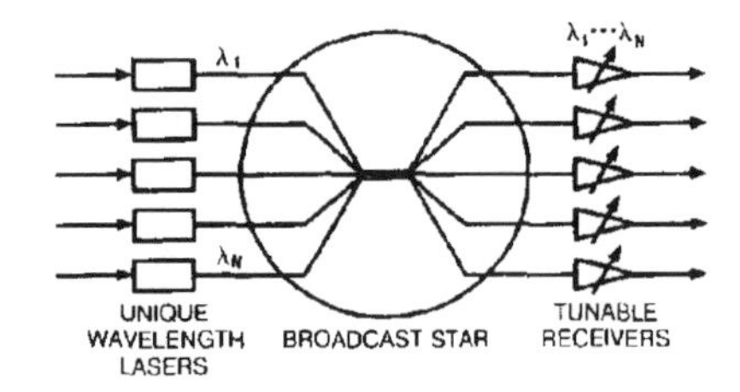

Fig. 2. An example of one broadcast-and-select network with fixed-wavelength lasers and tunable receivers. The schematic diagram for the star illustrates the combining and splitting functions typically achieved through interconnection of 2×2 directional couplers.

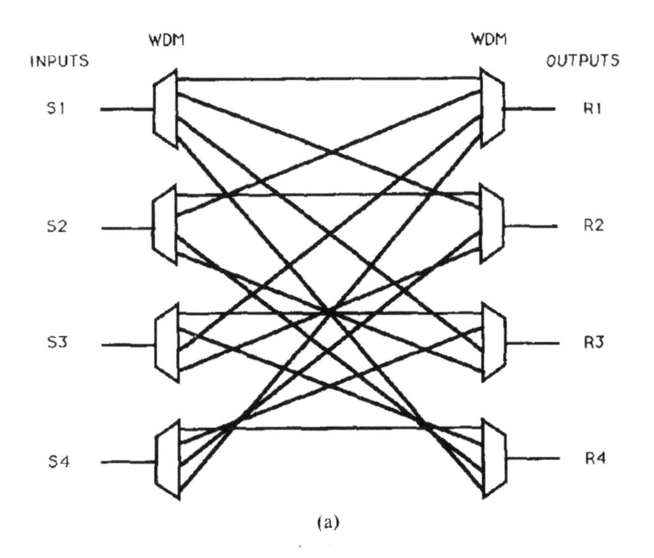

INPUT	OUTPUT R1	R2	R3	R4
S1	λ0	λ1	λ2	λ3
S2	λ1	λ0	λ3	λ2
S3	λ2	λ3	λ0	λ1
S4	λ3	λ2	λ1	λ0

(b)

Fig. 3. An example of: (a) a 4×4 WDM interconnection network; and (b) the associated wavelength assignment table. N wavelengths can always be used to completely interconnect an $N \times N$ network.

on whether the input lasers, the output receivers, or both, are made tunable.

If the input lasers are tunable and the output receivers are tuned to fixed wavelengths, the architecture is basically a space-division switch in function. Output-port contention exists in such a network, so that a means must be provided for contention resolution. The properties of this network are that it uses wavelength addressing of the output ports, but that with only a single wavelength selectable at each output, only point-to-point connections are possible and multicast connections cannot be achieved. It is possible with acoustooptic technology to make a filter which can simultaneously select more than one wavelength (see Sections III and VI-B). With such a filter, it is possible to make a many-to-one connection.

If the output receivers are made tunable but the input lasers are tuned to fixed unique wavelengths, the broadcast-and-select architecture supports multicast connections. This is achieved by arranging to have more than one receiver tuned to the same source wavelength at the same time. Output-port contention exists in this mode also and is exacerbated by the multicast function. The performance of such a network can be severely limited by the means for resolution of the contention and the means for communicating the tuning information to the output port receivers. Several proposed schemes for doing this will be presented in Section V.

If both the transmitters and receivers are made tunable, the possibility exists for reducing the number of wavelengths required but with the result that there are not enough wavelengths available to support simultaneous $N \times N$ interconnection. This introduces network or switch blocking in the wavelength domain which introduces the need for more complicated protocols.

It is obvious that the above network types may be combined with each other and with more traditional space-division photonic switching to generate a very broad range of network architectures. The above network types do, however, represent two classical elements. Any particular architectural arrangement must be chosen to suit the application. And, it will be seen below that the telecommunications applications of these techniques are of a wide range, running from circuit and packet switching to local optical distribution and local-area interconnection networks.

III. Wavelength Switching

In addition to the broadcast-and-select and wavelength routing functions of optical networks as discussed above, another possible network function is "wavelength switching." There are actually two types of wavelength switching, one of which dynamically switches signals from one path to another by changing the WDM routing in the network. The other type of wavelength switching is really wavelength conversion, where the information on a signal is transferred from an optical carrier at one wavelength to another. In this section, we will describe examples of both.

A wavelength-selective space-division switch is one which will select an arbitrary subset of the wavelengths $\lambda_1 \cdots \lambda_N$ on one fiber and redirect them to another fiber. This selection must be rearrangeable, so that the subset selected can be changed whenever desired, and the larger the subset which can be selected simultaneously, the better. It is also important to have the switching action fast enough to meet the application need.

A very interesting device of this type, which has been recently demonstrated [1], [2], is an acoustooptic tunable filter which utilizes the acoustooptic coupling between the TM and TE polarization modes in a birefringent optical waveguide in, for example, $LiNbO_3$. The principle of operation is illustrated, for a bulk device, in Fig. 4. Here, three wavelengths are shown propagating through a polarizing beam splitter (PBS), an acoustooptic narrowband polarization converter, and a second polarizing beam splitter. After passing through the first PBS, all three wavelengths have the same polarization. In the acoustooptic crystal, an acoustic wave introduces a diffraction grating which couples together the two polarization modes but only over a narrow wavelength range on the order of 1 or 2 nm wide. For that selected wavelength, λ_2 in the figure, the polarization is converted from TE to TM, or vice versa. The second PBS then resolves the two polarizations into two orthogonal output beams, thus performing the function of tunable narrowband optical filtering. The conversion efficiency can exceed 98% and is a function of the acoustic drive power and the interaction length. To change the wavelength selected, only the acoustic drive frequency need be changed and this can be done, typically, in a few microseconds.

Moreover, by injecting more than one RF drive signal into the acoustic transducer, more than one optical wavelength can be simultaneously selected. The limit to the number of simultaneously selectable wavelengths is set by the maximum power dissipation allowable in the acoustic transducer divided by the drive power required for complete conversion. Cheung *et al.* [3] have recently demonstrated an integrated version of such a tunable filter with simultaneous selection of five wavelengths. The total number of resolvable wavelengths runs into the hundreds. They have also demonstrated a polarization-independent version, illustrated in Fig. 5, which takes the form of a four-port device that has the remarkable ability to interchange any one or more of the wavelengths on one input with the same subset of wavelengths on the other input [4]. One very useful application of such a device would be as a dynamically tunable ADD–DROP multiplexer for WDM systems.

The second type of wavelength-switching component is one which effectively transfers information from one wavelength to another wavelength. This is most easily done by detecting the first signal and using the detected current to modulate a laser at the desired second wavelength. By using a tunable laser for the second wavelength, it is possible to switch information from one wavelength to any of the set of output wavelengths, and

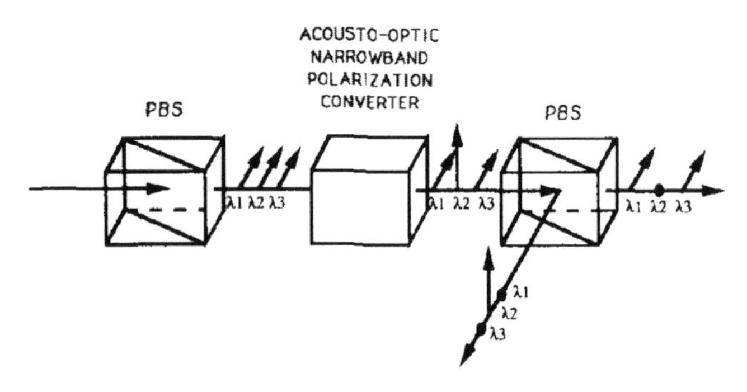

Fig. 4. The principle of the acoustooptic tunable filter.

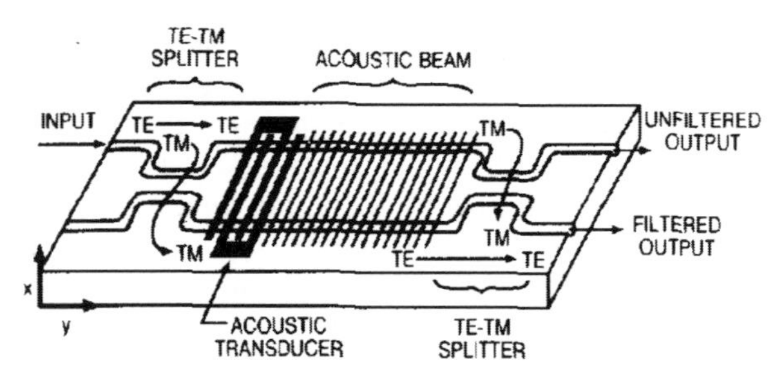

Fig. 5. An integrated-waveguide, polarization-independent acoustooptic tunable filter. Actually a four-port device, this filter can act as a wavelength exchanger in which some selected set of wavelengths on the input are exchanged in the output signal.

a switch has been proposed based upon this principle [5]. In a second type of wavelength conversion device [6], a four-section bistable laser is triggered by an optical input signal at one wavelength but can be tuned to oscillate over a range of wavelengths. In this way, wavelength conversion was demonstrated over a range of about 3 nm.

Applications for wavelength conversion range from switching via wavelength routing (much like a time-slot interchanger in a digital switch) to wavelength-based interconnection gateways between local networks.

Together with the principles of wavelength routing and broadcast-and-select WDM networks, the ability to switch and convert wavelengths will make possible a network in which wavelength takes its place alongside time and space as another network dimension [7]-[9]. It is not expected that the complexity of WDM functions will be nearly as great as in electronic circuits but the addition of dynamic wavelength assignment and manipulation, in addition to time and space multiplexing, will produce networks with significantly increased flexibility and capacity.

IV. Star Network Bandwidth Limits

It is useful to consider the various limits to the bandwidth of fiber-star networks. Consider the star of Fig. 6, assumed to be a passive $N \times N$ fiber star, comprised of several stages of 2×2 couplers. We assume that each of the independent N sources transmits on a different wavelength. Our concern here is, in some measure, to estimate the quantity of data that may be transmitted through such

Fig. 6. The general $N \times N$ passive star network, having a total bandwidth of $B \cdot N$.

a network. We make the distinction here, however, between the *bandwidth* of such a fabric and the *throughput* the network would achieve in real operation. Throughput is a very difficult quantity to estimate because it depends on the details of the traffic in nontrivial ways. It also depends on the algorithm chosen to control the flow of information in the network. A simpler concept, which is useful here, is that of the bandwidth of the network. We define the network bandwidth to be the product of the bit rate on each source or wavelength B and the number of sources or wavelengths N (N also being equal to the dimension of the star).

$$BW = B \cdot N.$$

There are several limits to this bandwidth that can be considered and we desire to calculate these limits as a function of the number of wavelengths N.

A. Bit Rate

First, there is the speed of the driving circuits for the transmission and reception of signals. In Fig. 7, the network bandwidth is plotted versus N. The straight lines with unity slope indicate network bandwidth limits imposed by electronics running at the indicated bit rates. Clearly, there is nothing very fundamental about these limits as they are limited only by the ever-increasing speeds of the electronics technology.

B. Power Budget

Second is the power-budget limit imposed by the finite available source power and minimum detectable receiver power, along with the $1/N$ power splitting ratio of the star and the star's losses. Henry [10], [11] has considered the fundamental limits of such a star network. The available power at each output of the star P_N is given by

$$P_N = \frac{P_T \cdot \beta^{\log_2 N}}{N}$$

where P_T is the transmitted power from each source and β is the transmission factor which accounts for the loss of each 2×2 coupler element.

When the network is operated at maximum bandwidth, the available power P_N will be equal to the receiver sensitivity, defined by $P_R = h\nu n_p \cdot B$, where h is Planck's constant, ν is the optical frequency, n_p is the receiver sensitivity in photons per bit, and B is the bit rate. Thus,

$$B \cdot N = \frac{P_T \cdot \beta^{\log_2 N}}{h\nu \cdot n_p}.$$

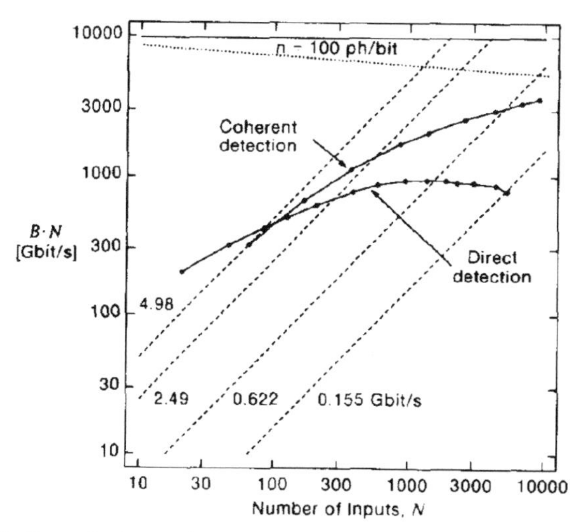

Fig. 7. Power budget limits to the bandwidth of star networks for both ideal receivers (n = 100 photons per bit sensitivity independent of bit rate) and for laboratory best-results for coherent and direct detection. The dotted line illustrates the effect of 0.2 dB attenuation per coupler-stage on the ideal case.

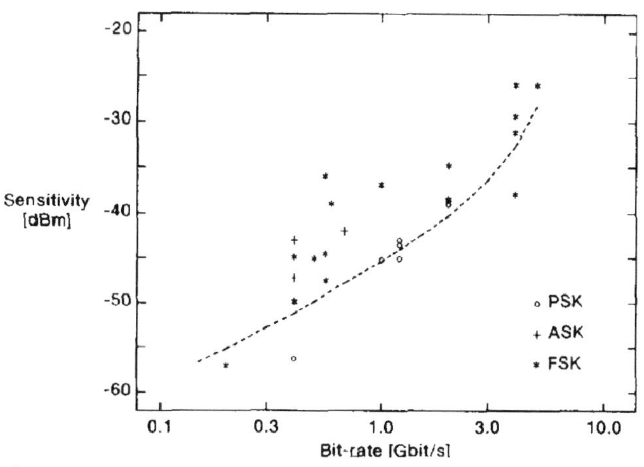

Fig. 8. Reported receiver sensitivities for coherent detection, with the empirical locus of best results used in the power budget limits of Fig. 7. See Table I.

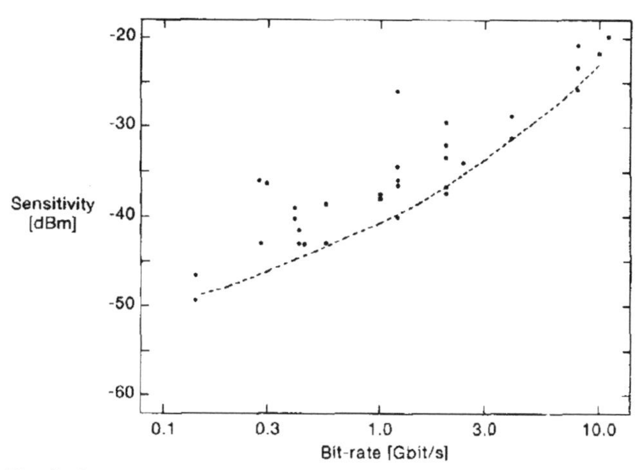

Fig. 9. Reported receiver sensitivities for direct detection with the empirical locus of best results used in the power budget limits of Fig. 7. See Table II.

For ideal receivers, where performance is limited only by photon-counting fluctuation in the incident laser light, n_p is a constant, independent of bit rate, whose value is typically between 10 and 20, depending upon the details of the modulation and detection schemes. A value more representative of practical heterodyne receivers is n_p = 100. We also assume P_T = 1 mW, an additional 3 dB for margin, 1 dB for connector losses, and 5 dB for filter and component losses. For lossless couplers, $\beta = 1$, resulting in $B \cdot N$ = 10 Tb/s at a wavelength of 1.5 μm, independent of N, as shown by the 100 photon per bit line at the top of Fig. 7. For lossy couplers, β is still close to 1 and $B \cdot N$ decreases slowly with increasing N, as shown by the dotted line in Fig. 7, for a loss of 0.2 dB per coupler. This corresponds to $\beta = 0.955$ and a decrease of $B \cdot N$ by 37% for N = 1024. It is seen that network bandwidths of several tens of terabits per second can be supported in fiber star networks under these rather idealized conditions, providing that such large stars can actually be made and that the bit rate can be high enough.

The fundamental limiting factor here was taken to be the sensitivity n_p of the detection process which was assumed constant, as is theoretically expected, for both heterodyne and direct-detection idealized receivers. In more realistic receivers, electronic noise, limited local oscillator power (in the heterodyne case), and bandwidth limitations further restrict the $B \cdot N$ product and make it dependent on the bit rate B. We have plotted recent research results for heterodyne-receiver sensitivities in Fig. 8 and for direct-detection receivers in Fig. 9 as a function of the bit rate. These results include both FSK and PSK modulation schemes for the heterodyne case, and p-i-n and APD receivers for the direct-detection case. Details are given in Tables I and II. We have drawn an empirical locus fitting the lower edge of these sensitivities for both cases (the dotted lines in Figs. 8 and 9). These loci were then used to construct the power-budget limit curves shown in Fig. 7. The same assumptions regarding transmitted power and excess losses as made above pertain here.

We see from Fig. 7 that the decrease in receiver sensitivities at high bit rates places additional restrictions on the total capacity $B \cdot N$ that become increasingly severe at high bit rates. At lower bit rates, the heterodyne receivers achieve near-ideal n_p, yielding $B \cdot N$ values that approach the 100 photon/b bound. For the direct detection case, $B \cdot N$ becomes nearly bit-rate independent for bit rates between 155 and 2500 Mb/s. The maximum value of $B \cdot N$ approaches 1 Tb/s for direct detection at approximately 600 Mb/s for a star size exceeding N = 1000.

To increase the bandwidth $B \cdot N$ of such a star network would require fundamental improvements in the receiver sensitivities at high bit rates or an increase in transmitter power. Increases in transmitter powers and receiver sen-

TABLE I
RECEIVER SENSITIVITIES FOR COHERENT DETECTION

Modulation	Wavel. [μm]	Bit-rate [Mbit/s]	Sensit. [dBm]	Sensit. [phot./bit]	Reference
FSK	1.5	5000	-27(ηP)	3012	Emura, Vodhanel.. (NEC, BELLCORE)
CPFSK	1.55	4000	-38	310	Iwashita.. (NTT), Elect. Letter V24 N12, 88
FSK	1.55	4000	-31.3	1445	Iwashita.. (NTT)(88) [2]
FSK	1.5	4000	-30.5(ηP)	1682	Emura, Vodhanel.. (NEC, BELLCORE)
FSK		4000	-26		Iwashita (NTT), OFC'88
FSK	1.55	2000	-38.6	535	Iwashita.. (NTT)(87) [2]
FSK		2000	-38.5		Iwashita (NTT), OFC'88
FSK	1.55	2000	-34.9	1260	Iwashita.. (NTT)(87) [2]
FSK	1.50	1000	-37.0	1500	Vodhanel.. (BELLCORE)(86) [2]
FSK	1.55	600	-39.1	1600	Chikama.. (FUJITSU)(87) [2]
FSK	1.55	560	-36.0	3380	Vodhanel.. (BELLCORE)(86) [2]
FSK	1.5	500	-45.1	467	Noe, Gimlett.. (BELLCORE), OFC'89
FSK	1.5	400	-50.0	190	Olsson.. (AT&T)(88) [2]
FSK	1.55	400	-49.9	195	Iwashita.. (NTT)(86) [2]
FSK	1.54	400	-45	613	Shibutani.. (NEC), OFC'89
FSK	1.53	200	-57.2	74	Glana.. (AT&T), p.d. OFC'89
DPSK	1.53	2000	-39.0	480	Gnauck.. (AT&T)(87) [2]
DPSK	1.54	1200	-45.0	200	Yamazaki.. (NEC)(87) [2]
DPSK	1.54	1200	-43.5	290	Kuwahara.. (FUJITSU)(88) [2]
DPSK	1.54	1200	-43.0	315	Yamazaki.. (NEC)(87) [2]
DPSK	1.53	1000	-45.2	230	Linke.. (AT&T)(86) [2]
DPSK	1.53	565	-47.6	237	Creaner.. (BRIT. TELECOM), OFC'89
DPSK	1.54	560	-44.6	480	Naito.. (FUJITSU), OFC'89
DPSK	1.53	400	-56.3	45	Linke.. (AT&T)(86) [2]
ASK	1.53	680	-42.0	710	Davis.. (STL)(87) [2]
ASK	1.53	400	-47.2	365	Linke.. (AT&T)(86) [2]
ASK	1.5	400	-43.0	960	Linke.. (AT&T)(86) [2]

All numbers for BER $\leq 10^{-9}$.

[2] In R. L. Linke and A. H. Gnauck "High Capacity Lightwave Systems", Journal of Lightwave Technology, V6, N11, November 1988.

TABLE II
RECEIVER SENSITIVITIES FOR DIRECT DETECTION

Detector	Wavel. [μm]	Bit-rate [Mbit/s]	Sensit. [dBm]	Sensit. [phot./bit]	Reference
pin-InGaAs	1.5	11000	-19.8	7185	Gimlett (89), subm. JLT
APD-InGaAs	1.55	10000	-21.7	5273	Fujita.. (NEC), p.d. OFC'88
pin (preamp.)	1.307	8000	-29.5	923	Gnauck.. (AT&T), OFC'89
APD-InGaAs	1.3	8000	-25.8	2150	Kasper..(87) [1]
APD-InGaAs	1.3	8000	-23.3	3825	Gnauck.. (AT&T), Globecom'88
pin	1.307	8000	-20.8	6838	In Gnauck.. , OFC'89
APD	1.32	4000	-31.2	1260	Eisenstein, Tucker.. (AT&T), OFC'89
APD-InGaAs	1.51	4000	-31.2	1440	Kasper..(85) [1]
APD		4000	-28.8		Tucker.. (AT&T), OFC'88
APD	1.3/1.5	2400	-34	1085/1252	Henmi.. (NEC), OFC'89
APD-InGaAs	1.54	2000	-37.4	705	Shikada.. (85) [1]
APD-InGaAs	1.51	2000	-36.6	847	Kasper..(85) [1]
APD-Ge	1.3	2000	-33.4	1494	Yamada.. (82) [1]
APD-Ge	1.55	2000	-32.0	2460	Yamada.. (82) [1]
APD-InGaAs	1.548	2000	-29.5	4370	Yamamoto.. (KDD), OFC'88
APD-InGaAs	1.54	1200	-40.0	646	Shikada.. (85) [1]
pin-InGaAs	1.53	1200	-36.5	1436	Brain.. (84) [1]
APD-Ge	1.3	1200	-35.9	1401	Yamada.. (82) [1]
APD-Ge	1.55	1200	-34.4	2359	Yamada.. (82) [1]
pin-InGaAs/InP(OEIC)	1.32	1200	-26	13903	Inomoto.. (NEC), p.d. OFC'88
APD-InGaAs	1.55	1000	-38.0	1235	Campbell.. (83) [1]
APD-InGaAs	1.3	1000	-37.5	1162	Campbell.. (83) [1]
APD-InGaAs	1.54	565	-42.9	703	Shikada.. (85) [1]
pin-InGaAs	1.3	565	-38.5	1635	Smith.. (82) [1]
APD-Ge	1.52	446	-43.1	840	Toba.. (84) [1]
APD-InGaAs	1.55	420	-43.0	930	Campbell.. (83) [1]
APD-InGaAs	1.3	420	-41.5	1102	Campbell.. (83) [1]
APD-Ge	1.3	400	-40.2	1561	Yamada.. (82) [1]
APD-Ge	1.55	400	-39.0	2454	Yamada.. (82) [1]
pin-InGaAs	1.3	296	-36.3	5179	Snodgrass.. (84) [1]
pin-InGaAs	1.3	280	-43.0	1170	Smith.. (82) [1]
pin-InGaAs	1.3	274	-36.0	5995	Lee.. (80) [1]
APD-Ge	1.52	140	-49.3	642	Walker.. (84) [1]
pin-InGaAs	1.3	140	-46.5	1046	Smith.. (82) [1]

All numbers for BER $\leq 10^{-9}$

[1] In S. E. Miller and I. P. Kaminow "Optical Fiber Telecommunications II", Chapter 18: B. L. Kasper "Receiver Design", Academic Press, 1988.

sitivities are certainly possible. These examples illustrate, however, that the power budget is a severe restriction and results in much lower maximum capacities than the theoretical ideal. Essentially, a 10 dB reduction in power budget would reduce the star bandwidth by a factor of 10.

C. *Spectrum Access*

The third limiting factor of some practical significance is the amount of the fiber bandwidth which can be accessed by the attendant multiwavelength lasers and receivers. Again, Henry has examined the idealized case

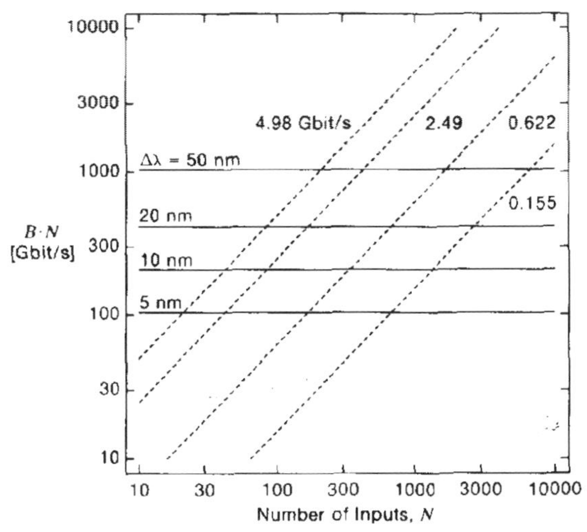

Fig. 10. Bandwidth limits of star networks for limited optical-spectrum access, as with tunable lasers or receivers.

[10] in which it is assumed that the entire 1.2–1.6 μm low-loss region of the spectrum can be accessed. The present state of the art is, however, limited to much less than that.

Assume that the channel spacing is ρ times the bit rate B (it has been shown [12], [13] that a minimum value for ρ of about 6 is required to minimize crosstalk and is equally applicable to various kinds of direct and heterodyne detection)

$$\Delta\nu = \rho B.$$

If we take $\Delta\Omega$ to be the optical frequency range accessible to a tunable laser or receiver, then

$$\Delta\Omega = N\Delta\nu$$

where N is the number of optical channels. These two equations imply that

$$B \cdot N = \Delta\Omega/\rho.$$

Thus, the maximum $B \cdot N$ that the network can support is limited by the accessible optical spectrum. The current state of the art for $\Delta\Omega$ is a sensitive function of the speed with which one needs to tune from one wavelength to another, which is dictated by the intended application. Note that this limit applies to the total accessible wavelength range and could be achieved with the use of more than one tunable laser or receiver, if necessary. More will be said later about the values of $\Delta\Omega$ for various laser and filter technologies and their tuning speeds. For now, let it suffice to say that electronically tunable semiconductor lasers have been demonstrated with approximately 10 nm tuning range and external cavity lasers have been demonstrated with tuning ranges of approximately 100 nm. Using $\rho = 6$, the $B \cdot N$ limit corresponding to various tuning ranges $\Delta\Omega$ are plotted in Fig. 10.

The conclusion from Fig. 10 is clear. If the spectral-access limit is not to be more restrictive than the power-budget limit, a tuning range of at least 50 nm is necessary for the direct detection case. For coherent detection, $\Delta\Lambda$ even greater than 50 nm would be required (at an increased capacity, of course).

TABLE III
MULTIWAVELENGTH NETWORK EXPERIMENTS

YEAR	SYSTEM	Nλ	CAPACITY (PER λ)	CHANNEL SPACING	TYPE
1985	BTRL	7	280Mb/s	15nm	BROADCAST STAR TUNABLE RECEIVERS
1985	AT&T	10	2Gb/s	1.3nm	POINT-POINT TRANSMISSION
1986	HHI	10(128)	70Mb/s	6GHz	COHERENT BROADCAST STAR
1987	BCR (LAMBDANET)	18	2Gb/s	2nm	BROADCAST STAR INTERCONNECT
1987	BCR(FOX)	8	1Gb/s	0.5Å	BROADCAST STAR INTERCONNECT
1987	AT&T	3(30-1000)	45Mb/s	300MHz	FDM BROADCAST STAR
1987	AT&T	3(~10^5)	45Mb/s	300MHz	FDM COHERENT BROADCAST STAR
1988	BCR(PPL)	20	1.2Gb/s	2nm	BIDIRECTIONAL WAVELENGTH ROUTING
1988	NTT	8	-	5GHz	FDM BROADCAST STAR
1988	BCR	16	2Gb/s	2nm	ETALON TUNED BROADCAST STAR
1988	AT&T	2(128)	250Mb/s	1.5GHz	BROADCAST STAR: TUNABLE FIBER AMPLIFIER
1988	BCR	2	1Gb/s	0.23nm	TUNABLE LASER AMP. 1 ns PACKET SWITCHING

This qualitative description of the bandwidth properties of optical star networks neglects many factors. One such missing factor, which will be mentioned in conjunction with some of the later system proposals, is that there are additional limits to the information throughput of such networks which depend on the network control algorithm, the network architecture, the traffic statistics, and other factors. Most of this is beyond the scope of this paper but these effects should be recognized as important and have traditionally been the limiting factors in the throughput of real switches and networks.

V. MULTIWAVELENGTH NETWORK EXPERIMENTS

In this section, we present a brief history of dense WDM network experiments and then discuss several of them in more detail as they relate to specific new technology or new system and application concepts.

The early history of dense WDM is sketched in Table III. The field really had its beginnings in the experiments of BTRL [14] and AT&T Bell Labs [15] in 1985. The BTRL experiment introduced the concept of the multiwavelength star network, operating in the broadcast mode and using mechanically tunable filters at each receiver. The filters were continuously tunable holographic reflection filters in dichromated gelatin, having a tunable range of 400 nm with a bandwidth of approximately 10 nm. The star size was 8 × 8 and the closest spacing of wavelengths in the experiment was 15 nm, limited by both the filter and laser linewidths. This experiment demonstrated the use of multiple wavelengths for simultaneous transmission of several different services and traffic types over the same network. The filters were tuned by hand and, therefore, were very slow, even though of very broad tuning range.

The Bell Labs experiment was the first to demonstrate channel spacings on the order of 1 nm. Using a diffraction-grating multiplexer and demultiplexer, 10 DFB lasers were multiplexed onto the same fiber in what was really the first dense-WDM experiment. The channel

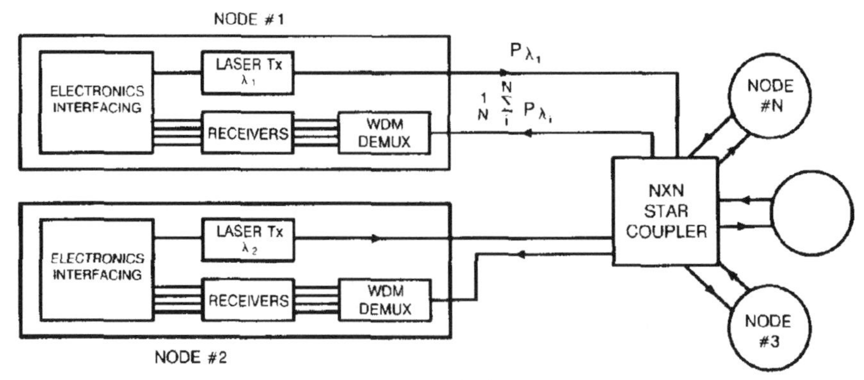

Fig. 11. LAMBDANET.

spacing was 1.3 nm and the operating bit rate was 2 Gb/s per channel. This was a point-to-point WDM demonstration, achieving a 1.37 Tb · km/s bandwidth-distance product.

Taken together, these demonstrations gave birth to the idea of using dense WDM in network applications.

In 1986, the Heinrich Hertz Institute [16] reported the first broadcast-star demonstration of video distribution using coherent technology. They demonstrated 10 optical channels, spaced by 6 GHz in the optical frequency domain, each carrying a 70 Mb/s encoded video channel. The experiment was conducted on a partial 128 × 128 star in order to demonstrate a power budget sufficient to serve a large number of subscribers.

With these experiments as a background, we will now discuss the rest of the demonstrations listed in Table III in detail, trying to point out the significance of each of these contributions.

A. LAMBDANET

The LAMBDANET demonstration [17], whose architecture is shown in Fig. 11, is similar in several respects to the BTRL demonstration mentioned above. As in the BTRL network, each transmitter is equipped with a laser emitting at a unique wavelength; the wavelength, in a sense, is an identifier of the source. Also, at the center of the network is a broadcast star, so that each of the wavelengths in the network is broadcast to every receiving node.

What is new in the LAMBDANET is the use of an array of N receivers at each receiving node in the network, using a grating demultiplexer to separate the different optical channels. Each transmitter time-division multiplexes its traffic destined for all other network nodes into a single high-speed data stream to modulate its own laser. Each receiving node simultaneously receives all of the traffic of the entire network, with a subsequent selection, by electronic circuits, of the traffic destined for that node. This creates an internally passive, nonblocking, and completely connected mesh-equivalent network which can be rearranged in capacity and connectivity under electronic control at its periphery. The passive nature of the fiber, the grating demultiplexer, and the star make it possible to consider transporting signals of differing bit rates or signal format, or both analog and digital, at different wavelengths.

The LAMBDANET demonstration also introduced a new level of technology. The basic star at the center of the network consisted of a 16 × 16 fused single-mode fiber star made of wavelength-flattened components. The 18 DFB lasers were selected for 2 nm wavelength spacing and were thermally tuned ± 10°C to achieve ± 1 nm fine tuning. They were packaged in rack-mountable, self-controlling plug-in units containing stabilized bias and temperature control. They could be modulated with either analog or digital signals to over 2 Gb/s. The demultiplexer grating was a commercial WDM component of the Stimax design [18], with one single-mode input fiber and 20 multimode output fibers all connectorized.

The capabilities of LAMBDANET were demonstrated in three experiments, summarized in Table IV. The first was using 18 wavelengths (the input side of the star was expanded to make an equivalent 18 × 16 star), each running at 1.5 Gb/s with a transmission distance of 57.8 km. This represented the transmission of 27 Gb/s over 57 km on each of 16 output fibers for a point-to-point figure of merit of 1.56 Tb · km/s and an aggregate point-to-multipoint figure of merit of 21.5 Tb · km · node/s. In the second experiment, the bit rate was increased to 2 Gb/s but, due to laser chirping, fiber dispersion, and limited power, only 16 wavelengths could be used and the transmission distance was reduced to 40 km, resulting in lower figures of merit. For the third experiment, labeled "WDM" in Table IV, the star was replaced by a second grating multiplexer, increasing the available power at the receiver, with the result that 18 wavelengths were successfully transmitted at 2 Gb/s over 57.5 km, with the results shown in the table.

There were several new results from the LAMBDANET demonstration. The first was the architecture itself with its nonblocking, fully connected properties. The second was the demonstration of dense WDM with prototype-packaged components and lasers and the removal of the optical bench. The third was the new transmission records noted above. The LAMBDANET concepts served as a foundation for several switching and routing architectures. A detailed discussion of the LAMBDANET archi-

TABLE IV
LAMBDANET EXPERIMENTAL RESULTS

	POINT-TO-POINT Tbit-km/sec	POINT-TO-MULTIPOINT (Tbit-km/sec) • node
LAMBDANET 1.5 Gbit/s, 18 λ 's, 57.8 km	1.56	21.5
LAMBDANET 2.0 Gbit/s, 16 λ 's, 40.0 km	1.28	18.0
WDM 2.0 Gbit/s, 18 λ 's, 57.5 km	2.07	****

tecture and its applications is given elsewhere in this issue [19].

B. *Fast Optical Cross Connect, FOX*

In an experimental architecture dubbed FOX, Arthurs *et al.* [20] and Cooper *et al.* [21] investigated the potential of using fast-tunable lasers in an application to parallel-processing computers. The architecture is shown in Fig. 12. The object of the design is to provide a high-speed (low latency) cross connect which allows any processor to talk to any of the shared memories. The cross connect includes two star networks: one for the signals sent from the processors to the memories and the other for the signals sent from the memories to the processors. Both of these are broadcast-and-select networks, the selection being done by the fixed-tuned filters at the receivers in both directions. The transmitters, however, are tunable over the set of receiver filter wavelengths. To address a particular memory, for example, the processor tunes its wavelength to that memory's filter and vice versa in the opposite direction.

In such a network, the possibility of contention exists, and in general will affect the overall throughput or response time. For this application, where the utilization of the memory access is relatively low [20], a binary exponential-backoff algorithm was shown to be sufficient to obtain good performance.

The technology issue here was to demonstrate the fast tunability of lasers so that the network is not slowed down waiting for the transmitter to tune its desired wavelength. There is no magic rule as to how quickly lasers must be tuned. It depends critically on the application. For a packet switch or a FOX-like application with data packets ranging from 100 ns to 1 μs, tuning times less than a few tens of nanoseconds will ensure reasonable efficiency. A summary of experiments on fast-tunable lasers demonstrating tuning times in the range of 1–10 ns is given in Section VI-A and is treated in more detail in the paper by Kobrinski in this issue [22].

Prior to these experiments and the proposals to use tunable lasers in fast-switching applications, the interest in tunable lasers had been largely confined to coherent transmission research and the need for high-speed tunability had not been envisioned. The FOX experiment demonstrated the need for very high-speed tunable WDM devices and that the highest speeds are currently achievable

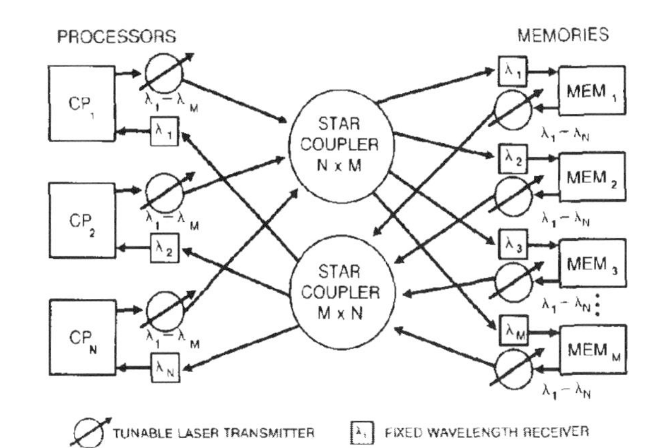

Fig. 12. FOX (fast optical cross connect): illustrating the use of tunable lasers in wavelength-addressable network structures.

with injection-current tuned devices. This experiment also emphasized that not only the speed of tuning, but the ability to attain a given frequency and hold for an extended time, is of paramount interest. In such a system as FOX, one would desire to have a table of absolute injection currents to be used to generate each specific wavelength. Thus, heating effects must somehow be overcome, either by locking to some reference or, as in FOX, by using a tuning current schedule which maintained constant power dissipation in the laser as it is tuned, or perhaps both.

C. *Hybrid Packet Switching System, HYPASS*

In a further extension [23] of the above systems, by using both high-speed tunable lasers and tunable receivers, a proposal has been made for a packet switch that may be capable of several hundreds of Gb/s throughput. The basic idea for HYPASS is shown in Fig. 13. It is called a hybrid electronic/optical switch because it follows the basic philosophy of using electronics to do the signal processing and optics to do the signal transport. In the center of the switch, there are two $N \times N$ star couplers. But in HYPASS, one is dedicated to the transport of the information data from the switch inputs to the switch outputs and the other is for carrying the output-port status information back to the input as part of the control algorithm. In the transport path, the tunable laser is tuned to address the desired output port. In the control path, the input-port tunable receiver is tuned to listen to a polling signal from the desired output port. An incoming packet on a fiber trunk is detected and stored in an input-port buffer. The header is then read and the appropriate output port determined. At the appropriate time, the input-port laser and receiver are then both tuned to the wavelength corresponding to the destination output port and the input port awaits a polling signal from the destination port for a packet transmission. When the output port is ready to receive another packet, the output-port status monitor generates a signal which is then broadcast to all input ports on a wavelength specific to the particular destination port. Those input ports which have packets destined for the polling port will then transmit. If there is

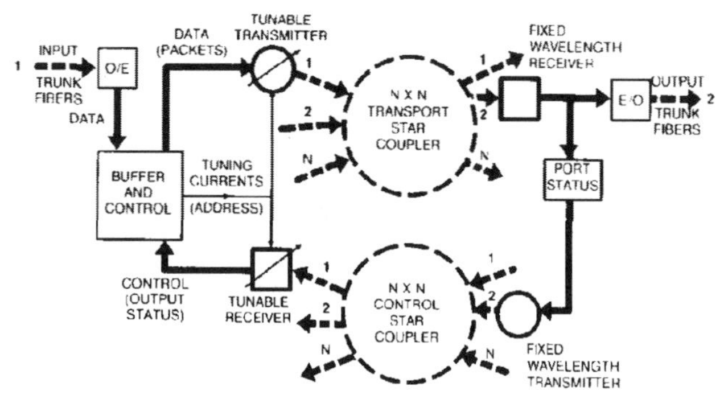

Fig. 13. HYPASS (hybrid packet switching system): illustrating the use of electronics for signal processing and tunable transmitters and receivers in the signal transport portions of a high-capacity packet switch.

more than one such packet, a collision will occur and a tree-polling algorithm is used to resolve the contention.

This structure has the advantage that each input port is being processed in parallel with the others, so the processing can be done without increasing the overhead. The performance analysis of the switch for very idealized traffic assumptions has been performed [23] and it is found that the throughput saturates at 0.31 packets per time-slot per port.

This architecture requires, in addition to high-speed tunable lasers, high-speed tunable receivers. A brief review of tunable filters is given in the section on critical technologies. The basic conclusion is that, for switching applications which require tuning in less than 10 ns, there are two likely candidates. One is the use of a permanently tuned array of receivers, one for each wavelength on the system, with the desired wavelength being selected by an electronic gate [24]. Such an electronic gate can be made very fast, the wavelength tuning is very robust, and the only drawback is the number of receivers which can be integrated together economically in an array. The second candidate is to use some form of injection-current tunable amplifier. This has been done in the form of a tunable DFB laser-diode amplifier [25], [26] with demonstrated switching times of 1–10 ns and up to as many as eight wavelength channels. The principal limitations here are the relatively small tuning ranges such devices have so far shown (0.6 nm) and the need to compensate for thermal drifts.

In a variation on the HYPASS architecture, the use of totally electronic contention resolution has been considered in a structure called BHYPASS [27]. In principle, contention resolution requires processing only the output-port address information which is contained in the packet headers. This can be done in a variety of electronic circuits, one of the most appealing being the self-routing Batcher-banyan structures proposed for electronic packet switching for broad-band ISDN [28]. The output-port headers would be processed through the Batcher-banyan circuit. Those packets winning the contention would then be allowed to be transported through a multiwavelength star network resembling the transport portion of the HYPASS.

D. STAR-TRACK

One of the potential disadvantages of the above networks is that they do not easily lend themselves to selective multicast service requests, i.e., the ability to switch one input to more than one output at the same time and to resolve the contention that may develop with other packets from other inputs. One approach to this for electronic Batcher-banyan switches is to make copies of the original packet to be multicast and to route each of those copies through the switch separately.

In a proposal intended to take advantage of the inherent broadcast nature of the optical star, Lee *et al.* [29] have proposed a network they call STAR-TRACK. This network consists of a broadcast star with fixed-wavelength inputs ($\lambda_1 \cdots \lambda_N$) and tunable receivers. The contention-resolution and network-control functions are handled by an electronic ring-reservation scheme as shown in Fig. 14. A token, which consists of a subtoken for each of the N output ports, is generated and passed through each of the input ports. Each input port then examines the subtoken field for the desired output port. If the subtoken is empty, that is if the output port has not yet been reserved, then the input port may write its own address into that subtoken field, thereby reserving the output port in question for the next packet slot. If the subtoken is not empty, then that input port must wait to transport its token until the next packet slot. Selective multicast is done very simply by writing one input port (or source) address into several output port subtoken fields.

The entire token circulates through all of the inputs in the write phase, setting up the reservations for the entire network, and then it circulates through all of the output ports in the read phase. Each output port examines its own subtoken field to determine if it has been reserved and the input port to which it should tune its optical receiver or filter. It is envisioned that the entire contention-resolution and reservation cycle would take place during the transmission cycle for the previous packet, thereby minimizing the processing overhead.

In addition to the selective multicast feature, STAR-TRACK allows a straightforward implementation of prioritized and reserved traffic and call splitting to enhance the throughput of the switch in the multicast case.

There are many factors which limit the performance of such a switch, such as the bandwidth limits of star networks as listed in a previous section. A significant consideration is the time it takes to process the token through the entire ring. Such a reservation scheme is sequential in nature and each input port must be processed, whether it has traffic to be switched or not. This is quite different from the situation for HYPASS and BHYPASS, in which the input ports are processed in parallel. The need for the ring to complete its cycle within the transmission time of the previous packet means that the electronics of the ring must run at a high rate and this rate of ring processing may, in fact, determine the maximum number of switch ports as a function of such parameters as packet length, input-port bit rate, and tuning times.

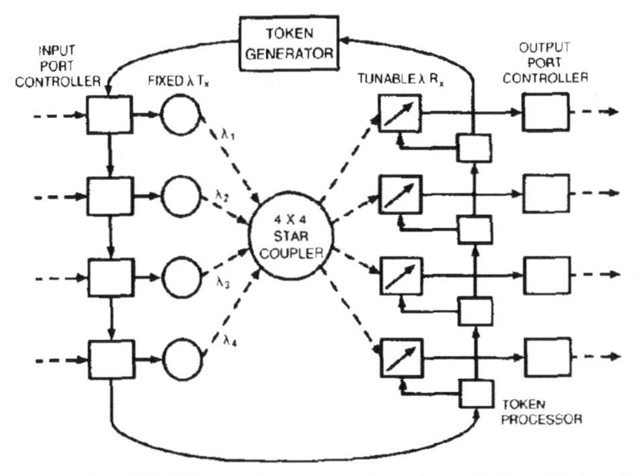

Fig. 14. STAR-TRACK, a multiwavelength packet-switching fabric using fixed-wavelength transmitters and tunable receivers to achieve selective multicast and prioritized traffic handling in a packet switch. A reservation scheme is implemented in the track to resolve contention, schedule transmission, and carry tuning information to the receivers.

E. *Wavelength Routing*

Another potential application of dense WDM, besides the switching functions mentioned above, is the use of wavelength routing. As discussed earlier, this is where the wavelength determines the path which a signal takes through the network. A proposal known as the passive photonic loop (PPL) [30] has been suggested for routing signals on fiber in the local telephone loop. In the customary plans for fiber to the home, a feeder fiber carries the signals for many subscribers in a time-division multiplexed (TDM) format out to a remote electronics site where they are demultiplexed and then distributed over fiber to each individual subscriber. This remote site requires significant electronic power, maintenance, and transmission equipment. In the PPL, as shown in Fig. 15, the signals for each customer (in both directions) are carried on separate wavelengths and separated for subscriber distribution at the remote site by a passive WDM component such as a grating demultiplexer. The remote site, being passive, would then dissipate no power, private channel assignments to each subscriber would be maintained, the economics would be improved by maintenance and reliability considerations, and all the remote electronic functions are moved back to the central-office location. An experiment has been performed, to demonstrate the feasibility of these concepts, with 10 wavelengths in each direction at 1.2 Gb/s on each laser.

Another application of wavelength routing is to nearly-fixed-assignment cross-connect applications, where circuit connections and capacity are slow to change and do not require active switches. The general scheme of such a cross connect is shown in Fig. 16. Here, the N-node network is shown in a two-sided model with the transmit side on the left and the receive side on the right. Each of the transmitting nodes assigns a wavelength λ_{ij} for the traffic from node i to node j and then multiplexes these wavelengths onto a fiber to a central WDM cross-connect location. At the cross connect, the signals are demultiplexed, cross-connected to a multiplexer associated with their intended destination, and remultiplexed and transmitted to the receive side of each network node. This is exactly the WDM routing scheme discussed earlier in this paper, where it was shown that the λ_{ij} need not be distinct and that only N distinct wavelengths are required to form the complete cross-connect function.

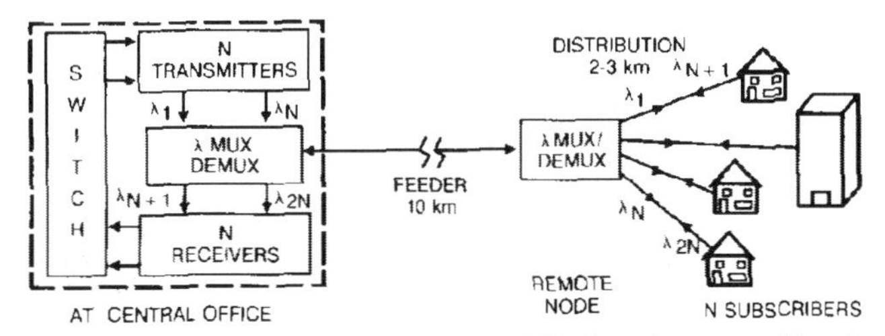

Fig. 15. The passive photonic loop, using WDM routing to provide private channels to subscribers with a passive remote-node double-star architecture.

These principles have been discussed by Kobrinski [31] in regard to telephone central-office trunk circuits and by Hill in regard to more general interconnect structures [32] and long-haul networks [33]. The significance of these cross-connect structures is that they again are passive, that the achievable total traffic of such a network is very high, and that the principal topological function they offer is that of pair-gain. One potential drawback is the amount of traffic that is multiplexed onto a single fiber, which would be lost and a node isolated if there were a fiber fault.

F. *Broadcast Video Distribution*

A specialized area of application for dense WDM techniques is in networks for the distribution of broadcast services, particularly video, to the subscriber premises. Many experiments using a wide variety of technologies have been performed, all using the same basic concept. Multiple broad-band channels may be time-division multiplexed onto a high-speed digital data stream, with each such stream modulating a laser at a unique wavelength. Several of these different wavelength signals may then be combined in a broadcast star to both mix them together and distribute them to several outgoing fibers for delivery to subscribers in a passive broadcast fashion. This appears to be a physically natural application for dense WDM, taking account of the high-speed nature of the optical channel, the noninterference between different wavelength carriers, and the ease and efficiency of passive-star architectures. Channel selection at the subscriber would be done with a tunable optical receiver which could be either a coherent receiver with a tunable-laser local oscillator or with a tunable optical filter. This application does not require very fast tuning speeds but it does require one optical receiver at each subscriber terminal (e.g., TV set), which may be economically unattractive.

For this kind of architecture, nonsymmetrical architectures are likely where there are N sources and M destinations, with N much less than M. In this particular case,

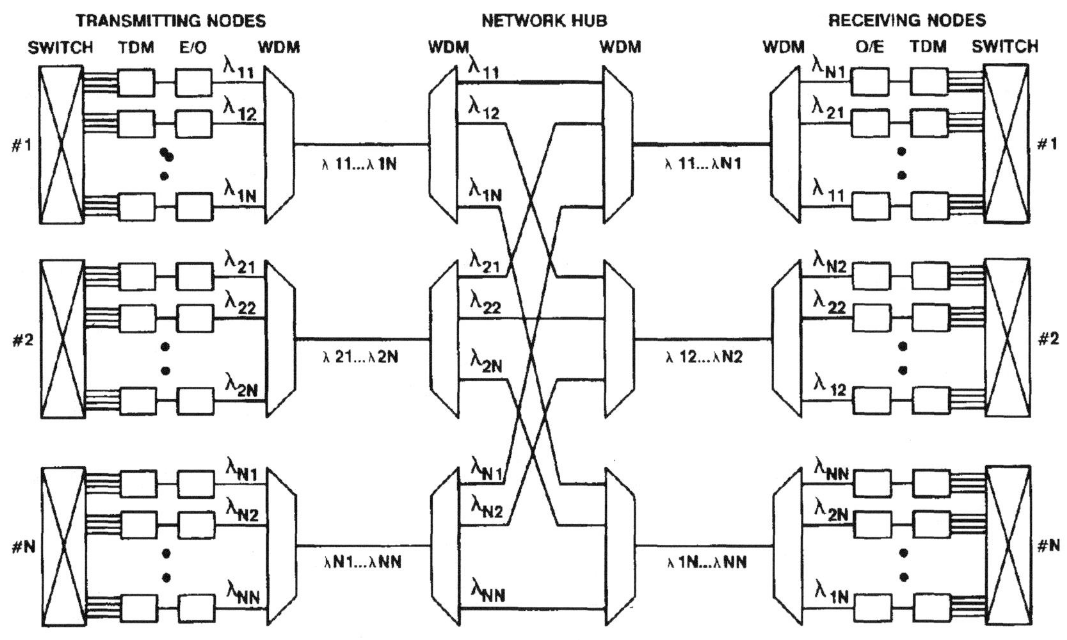

Fig. 16. A WDM cross-connect proposal, using wavelength routing to cross-connect central office switches. Wavelength reuse allows use of only N independent wavelengths for an $N \times N$ cross-connect.

the splitting loss of each channel is a factor of $1/M$ and the total power on each output fiber becomes $(N/M) \cdot P_0$. Here, optical amplifiers can be of significant advantage in overcoming the splitting loss N/M, as well as any additional transmission losses, in order to dramatically increase the number of subscriber nodes that may be served.

The first such experiment using multiwavelength technology was the HHI experiment with 10-channel coherent video demonstration [16]. It was later updated to include nine channels of 70 Mb/s video and one channel of high-definition TV. In their experiment, they demonstrated a sufficient power budget to serve 128 customers from a single star with no amplifier.

Kaminow [12] and Glance [34], [35], in a pair of related experiments, demonstrated three-channel distribution of 45 Mb/s data using highly stabilized external-cavity lasers with optical channel spacings of 300 MHz in a broadcast-star configuration with FSK modulation.

In the experiment reported by Kaminow, direct-detection was used, the key factor being a highly selective tunable fiber Fabry-Perot filter developed by Stone [36]. In such an arrangement, the number of channels that may be resolved without interference from neighboring channels is estimated to be the finesse F divided by 6, and the channel separation must be at least six times the bit rate of the signal. The finesse of the fiber Fabry-Perot filters reported by Stone have been in the range of 200, allowing perhaps 30 channels. It may be possible, at an increased loss, to use a two-stage filter [37] to increase the effective overall finesse to about 3000, thus increasing the number of channels to $\approx$ 1000, since the number of channels for a two-stage filter can be increased to $F/3$.

One very important feature this experiment demonstrated is the ability to space direct-detection channels with about the same density that one would expect for a coherent system. The difference between direct and coherent detection (other than the sensitivity) is essentially the cost and stability of filtering in the optical domain for direct detection or at microwave intermediate frequencies for the heterodyne case.

In the experiment reported by Glance [34], a coherent receiver employing a frequency-tracking tunable local oscillator was used to select the desired optical channel. The number of channels would, in this case, be limited by the tuning of the local-oscillator laser and could, in principle, be very large.

Another means of tuning an optical receiver [38] used an angle-tuned Fabry-Perot filter. In this experiment, 2 nm wavelength separations were used and 16 channels were demonstrated at 2 Gb/s per channel. Because of the slow tuning speed of such a device, the primary application would be toward systems which need to select a single channel out of many and to hold it for a reasonably long time. Subscriber video broadcast selection is one such application. Using 16 wavelengths and 12 digital channels per wavelength, such a system could distribute 192 video channels to a very large number of subscribers.

A novel tunable-filtering technique for such broadcast applications involved the use of stimulated Brillouin scattering (SBS) to produce a tunable narrowband amplifier [39]. This experiment actually demonstrated only two channels, but in a simulated 1×128 star, and used inexpensive off-the-shelf GaAs lasers. At the receiving end, a laser is used to pump the SBS amplifier which consisted of a length of fiber at the subscriber site.

In more recent experiments, several coherent broadcast-star demonstrations have been reported [40]–[42]. The issues these experiments have addressed have included: 1) increasing the number of channels; 2) wavelength stabilization; 3) absolute and relative wavelength registration; and 4) demonstrating network wavelength-control procedures and cold-start operation.

It is not clear whether economic factors will favor the use of dense WDM to provide broadcast video services to the subscriber. Analog and digital distribution of CATV-like services using subcarrier multiplexing techniques [43]-[45] may be less costly to implement in the near future. But, the progress in heterodyne systems and devices has been much more rapid than most people expected.

VI. Critical Technologies

In the dense WDM network examples given above, there are two critical technologies which stand out: tunable lasers and tunable filters or optical receivers. The parameters of importance are: 1) the number of resolvable channels; and 2) the speed with which these channels can be accessed.

Each application has different requirements on these two parameters. A tunable filter to select channels carrying home entertainment video may not require as short a tuning time as a filter operating as an information gateway, and an even shorter tuning time may be required for filters operating in the internal part of a high-speed switch. The number of channels required will also be sensitive to the application. If achieving the maximum throughput of a star-based network is the goal, then a very large tuning range/number of channels will be required. If wavelength routing is being used to provide fiber to the subscriber, then a smaller number of channels may be desired.

There are other critical technologies which we will not deal with in this paper, one of which is optical amplifiers. Optical amplifiers are particularly important to WDM systems because the cost of regenerating individual wavelength signals would be prohibitive, whereas broad-band amplification of many channels in a single device is both possible and highly desirable. The reader is referred to the literature for more information on this rapidly expanding subject [46].

In this section, we give a very brief review of the state of the art of tunable lasers and filters and take a look at a new type of optical star which may be very important. This review is by no means complete, nor is it meant to explain the details of operation of any of these devices, but is meant simply to indicate the overall characteristics of the devices in these general families.

A. Tunable Lasers

A brief status of tunable lasers is shown in Fig. 17, in which the tuning range of the laser is plotted versus the tuning speed (inverse tuning time). Four families of tunable lasers are shown: thermal tuning, mechanical tuning, injection-current tuning, and acoustooptic tuning.

Thermal tuning of single-frequency DFB lasers, for example, is limited to about $\pm$ 1 nm and is limited to very slow changes, on the order of milliseconds or longer. This sort of tuning is very useful for stabilizing lasers in the laboratory or for fine tuning their frequency. The LAMBDANET experiment used temperature stabilization and tuning very successfully to demonstrate that wavelength stabilization is not as critical to that sort of system as had been thought.

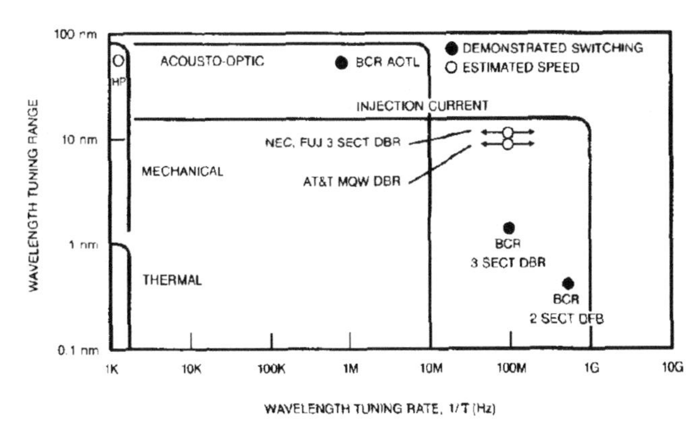

Fig. 17. Four families of tunable lasers, with tuning range plotted versus tuning rate (inverse tuning time). The fastest tuning is achieved in injection-current tuned lasers. The broadest tuning range is achieved for mechanically and acoustooptically tuned lasers. Solid data points are for measurements of wavelength switching times and open data points are for estimated switching times.

External cavity, mechanically tuned lasers have a much broader tuning range, equal, essentially, to the useful spectral region of optical gain of the laser diode. The principal drawback of these lasers (besides their bulk and mechanical stability requirements) is the very limited tuning speed, due to the mechanical movements required for tuning. An actual model of such a laser is on the market for use in laboratory research [47], with piezoelectric tuning over a narrow (GHz) tuning range and manual tuning over 40 nm.

There are a variety of semiconductor lasers which are tuned by adjusting the injection current in one or more sections of the laser. These include the ordinary single-section DFB laser, the dual-section DFB laser, and the three-section DBR laser. These lasers have, by far, the fastest tuning times of any laser class, on the order of a few nanoseconds having been measured, but the tuning range is limited to 10 nm or so. In these devices, the injected carriers act to change the effective index of refraction within the optical cavity and the change of index is proportional to the square root of the current density. In practice, the fractional change in wavelength is equal to the fractional change in the effective index n

$$\frac{\Delta\lambda}{\lambda} = \frac{\Delta n}{n}.$$

Since the maximum index change is about 1%, a maximum tuning range of about 10-15 nm is expected.

As shown in Fig. 17, an experiment on a dual-section DFB (DS-DFB) laser demonstrated a tuning range of 3.2 Å, with a switching time between any two wavelengths of less than 5 ns [21]. In this experiment, eight independent wavelengths separated by 0.45 Å were demonstrated. A particular feature of the experiment was to maintain constant total current through the diode as the tuning currents in the two sections were changed and, in that way, the tuning/current characteristics of the laser were maintained independent of tuning history.

Another experiment on a three-section DBR laser [48] demonstrated 20 wavelength tuning while simultaneously amplitude modulating the laser at 200 Mb/s. Tuning times increased to about 15 ns and significant thermal effects were observed. A unique FSK modulation scheme [49] was used on the same laser to demonstrate the capability of ≈ 50 channels, each carrying 50 Mb/s of information, in a manner which avoided the thermal effects. Tuning times were still limited to about 15 ns. Details of these experiments are given in another paper in this issue [22].

A tunable three-section DBR laser has been used to make a fast-tuning coherent receiver [50]. Wavelength switching times of 1.8 ns were measured over a wavelength range of up to 12.5 nm.

Another very interesting result is the report of Koren [51], [52] on the integration of three tunable DBR lasers (individual tuning ranges ≈ 7 nm), an amplifier, and a waveguide combiner on a single chip. The laser center frequencies (at 1.56 μm) were staggered so that the total tuning range of the output was approximately 20 nm with an output power of about 1 mW. Tuning speeds were not measured.

Another form of a tunable laser is an external-cavity semiconductor laser with an electronically tunable filter within the cavity. Such lasers have been demonstrated with both acoustooptic [53] and electrooptic [54] tunable filters. The tuning ranges are limited by either the tuning range of the filters or by the gain spectrum of the semiconductor laser, whichever is smaller. Tuning ranges of 7 and 83 nm have been demonstrated for the electrooptic and acoustooptic cases, respectively. The tuning time of the acoustooptic tunable laser was limited by the acoustic velocity in the filter to a few microseconds. The tuning time of the electrooptic tunable laser was not measured. These lasers are not continuously tunable, however, because they hop from mode to mode of the semiconductor laser and oscillate simultaneously in several modes of the external cavity. Tighter coupling of the laser diode to the external cavity and better antireflection coatings on the diode may eliminate the mode hopping.

We have already discussed the bandwidth limits to optical star networks due to the limitations of the available optical spectrum or the tuning range of the optical devices. Current tuning ranges of semiconductor tunable lasers are in the 4–10 nm range, depending on whether continuous tunability is needed or not. We showed earlier that a range of approximately 50 nm is needed before the optical power budget becomes the limiting factor and the ultimate bandwidth of the network can be realized. There is, therefore, need for considerable improvement in the laser tuning range.

B. Tunable Filters

For the purposes of this paper, we classify tunable filters into three categories: passive, active, and tunable optical amplifiers, as shown in Table V.

The passive category is composed of those basically passive wavelength-selective components which can be made tunable by varying some mechanical element of the filter, such as mirror position or etalon angle. This includes Fabry–Perot etalons, tunable fiber Fabry–Perot filters, and tunable Mach–Zehnder filters. For the Fabry–Perot filters, as discussed earlier, the number of resolvable wavelengths is related to the value of the finesse F of the filter. Normal Fabry–Perot filters and etalons have finesses of up to 200 and, in FSK systems, this implies a number of channels of $F/6$ or about 30. This can be increased by raising the finesse or by operating two filters in tandem with suitable isolation, for which the finesse may reach 3000 and the number of channels 1000. The advantages of such filters are their ready availability and the very fine frequency resolution that can be achieved. The disadvantages are, primarily, their tuning speed and losses.

The Mach–Zehnder integrated-optic-interferometer [55] tunable filter, listed in Table V, is a waveguide device with $\log_2(N)$ stages, each of which passes every other incoming wavelength in a divide-by-two fashion until only the desired wavelength remains. This filter has recently been demonstrated with 100 wavelengths [56] separated by 10 GHz in optical frequency, and with thermal control of the exact tuning. The measured tuning times were on the order of milliseconds. The number of simultaneously resolvable wavelengths is limited by the number of stages required and the loss incurred in each stage.

Several recent reviews of tunable filter technologies provide more detail [22], [57]–[59].

In the active category, there are two filters based upon wavelength-selective polarization transformation by either electrooptic [60], [61] or acoustooptic means [62]. In both cases, the orthogonal polarizations of the waveguide are coupled together at a specific tunable wavelength. In the electrooptic case, the wavelength selected is tuned by changing the dc voltage on the electrodes; in the acoustooptic case, the wavelength is tuned by changing the frequency of the acoustic drive frequency [58].

A filter bandwidth (full-width half-maximum) of ≈ 1 nm has been achieved by both filters [62], [63]. However, the acoustooptic tunable filter has a much broader tuning range than the electrooptic type, namely the entire 1.3–1.56 μm range versus 16 nm. This is because the effective grating in the acoustooptic case is dynamic and its pitch changes with the acoustic-drive frequency, whereas for the electrooptic case, the grating is fixed by the metalization pattern.

Both types of filter are controlled electronically and can be tuned reasonably fast. The acoustooptic filter is limited to tuning times of a few microseconds by the acoustic propagation velocity, whereas the electrooptic filter can

TABLE V
TUNABLE FILTER TYPES AND STATUS

PASSIVE	RESOLUTION	RANGE	NO CHANNELS	SPEED	
ETALON (F ~ 200)			~ 30*	ms*	BCR
FIBER FABRY-PEROT (F ~ 200)			~ 30*	ms*	AT&T
TWO-STAGE (F ~ 3000)			~ 1000*	ms*	
WAVEGUIDE MACH-ZEHNDER	0.38Å (5 GHz)	45Å (600 GHz)	128*	ms	NTT
ACTIVE					
ELECTRO OPTIC TE/TM	6Å	160Å	~ 10	ns*	AT&T
ACOUSTO OPTIC TE/TM	10Å	400 nm	~ 100	~ 10μs	BCR
LASER DIODE AMPLIFIERS					
DFB FILTER AMPLIFIER	1-2Å	4-5Å	2-3	1 ns	BCR
2-SECTION DFB AMPLIFIER	0.85Å	6Å	8	ns*	NEC

*ESTIMATE

be tuned typically in the nanosecond range. Fairly large voltages are required for the electrooptic case (± 80 V), which may determine its actual achievable tuning speed. The acoustooptic drive power for an integrated acoustooptic case has been demonstrated to be on the order of 100 to 200 mW.

Another significant difference is that the acoustooptic filter can be driven with drive signals of many frequencies simultaneously, thereby selecting more than one wavelength simultaneously [2]. Such a filter opens the possibility of switching and rearranging wavelengths just as if they were bits in a register.

The third category of filter in Table V is that of laser-diode amplifiers as tunable filters. Operation of a resonant laser structure, such as a DFB or DBR laser, below threshold results in narrowband amplification. In a simple DFB structure, this is tunable by simply adjusting the injection current [25], however, the gain changes as well as the bandwidth. An improved version of the DFB tunable filter includes a phase-control section [64]. This phase-control section allows separate adjustment of the gain and wavelength so that a constant gain and bandwidth were achieved across 6 Å tuning, allowing eight channels to be selected [65]. The resolution of this laser-diode tunable filter was about 0.25 Å. Switching times on the order of several nanoseconds have been measured [26]. This experiment also demonstrated an interesting wavelength-division switching scheme in which information can be circuit switched from one wavelength to any other [66].

It is also possible to make a tunable receiver by using a tunable local oscillator in a coherent receiver. A great deal of work has been done on designing such receivers, including their frequency registration and stability, but very little has been reported on the fast tunability of such receivers. One such report, however, demonstrates a tuning time of less than 2 ns for a limited tuning range [50]. It is entirely possible that such receivers are the only key to fully maximizing the $B \cdot N$ product of optical networks but it will require local oscillator tuning over 50 nm or more to reach that maximum.

The research on fast tunable filters is much less advanced than that on fast tunable lasers but it is important, for many network applications, that these filters be advanced to a status commensurate with that of tunable lasers. The significance of these results is that tunable lasers and receivers have many properties and capabilities sufficient to make high-speed packet switches, but much more work needs to be done to increase the total number of channels, or tuning range, of these devices if even a small fraction of the usable fiber bandwidth is to be utilized for such switching applications.

C. Star Couplers

A central piece of most of the network architectures discussed in this paper, and indeed in the research work done in this area to date, is the optical star coupler. The advantages of this sort of network have been duly pointed out but without much mention of how these couplers are made and what prospects we have for future developments in reducing the costs and improving the performance of such components. The most prevalent form of a single-mode star coupler is composed of many stages of individual 2×2 fused fiber couplers. For an $N \times N$ star, where N is a power of 2, the number of such stages required is $\log_2 N$ and the number of 2×2 couplers required is $N/2 \log_2 N$. These stars have low excess loss and good coupling uniformity, they are compatible with the transmission fiber and, therefore, have low coupling losses, and they are reasonably easy to manufacture and are, therefore, readily available.

However, they are still expensive in large sizes. This is because the number of required 2×2 stars grows faster than N. A new approach, based on an integrated optics technology, offers the promise of large, inexpensive stars [67]. This approach consists of two arrays of strip waveguides separated by a "free-space" planar-waveguide-propagating region. Each input waveguide acts as an antenna radiating into the unguided region and each output waveguide acts as a receiving antenna. Dragone [68] has demonstrated a device with $N = 19$ on a glass substrate, the excess losses are about 3.5 dB and are, in principle, independent of N, and there is very little wavelength dependence. Arrays of 100×100 are thought to be possible.

If manufacturable, this technology represents a significant advance in the technology for optical networks.

VII. IMPLICATIONS

The claim is made in this paper that the technology for incorporating wavelength as an additional dimension in telecommunications networks is almost at hand and that the feasibility of such networks has been demonstrated. It is clear that there is much remaining work to be done. Examples of this would be the stabilization and registration of wavelengths in the fast-switching mode, further improvements in the tuning ranges and resolution of tunable sources and filters, particularly for fast tunability and, perhaps most importantly, studies of network protocols, control structures, and architectures which make the best use of the extremely broad-band systems which are now conceivable.

The overall implication, however, is that wavelength is a network dimension which cannot be neglected. Not only does it unlock more bandwidth, it also offers new network forms. The passivity of optical components adds function while maintaining processing simplicity. Not only can wavelength be used to further increase the total signal-multiplexing capability, but it can also be used for functions related to routing, switching, and service segregation. Whether the technology used is direct detection or heterodyne detection, the basic network forms and architectures are much the same. The issues will ultimately be decided on two criteria: service requirements and cost effectiveness.

ACKNOWLEDGMENT

The author wishes to acknowledge the substantial contributions of T. Renner to the work presented in Section IV and the critical reading of this manuscript by E. Goldstein and T. Chapuran.

REFERENCES

[1] K. W. Cheung, M. M. Choy, and H. Kobrinski, "Electronic wavelength tuning using acousto-optic tunable filter with broad continuous tuning range and narrow channel spacing," *IEEE Photon. Technol. Lett.*, vol. 1, pp. 38-40, 1989.

[2] K. W. Cheung, D. A. Smith, J. E. Baran, and B. L. Heffner, "Multiple channel operation of an integrated acousto-optic tunable filter," *Electron. Lett.*, vol. 25, pp. 375-376, 1989.

[3] K. W. Cheung, S. C. Liew, D. A. Smith, C. N. Lo, J. E. Baran, and J. J. Johnson, "Simultaneous five-wavelength filtering at 2.2 nm wavelength separation using an integrated-optic tunable filter with subcarrier detection," in *Proc. ECOC '89*, Gothenburg, Sweden, Sept. 1989, vol. 1, pp. 312-315.

[4] D. A. Smith, J. E. Baran, K. W. Cheung, and J. J. Johnson, "Polarization-independent acoustically-tunable optical filter," in *Proc. ECOC '89*, Gothenburg, Sweden, Sept. 1989, vol. 3, postdeadline paper, pp. 70-73.

[5] M. Fujiwara *et al.*, "A coherent photonic wavelength division switching system for broadband networks," in *Proc. ECOC '88*, Brighton, England, Sept. 1988, pp. 139-141.

[6] S. Yamakoshi, K. Kondo, M. Kuno, Y. Kotaki, and H. Imai, "An optical-wavelength conversion laser with tunable range of 30 Å," in *Tech. Dig. OFC '88*, New Orleans, LA, 1988, paper PD10.

[7] C. A. Brackett, "Dense WDM networks," in *Proc. ECOC '88*, Brighton, England, Sept. 1988, pp. 533-540.

[8] S. Suzuki, M. Fujiwara, and S. Murata, "Photonic wavelength-division and time-division hybrid switching networks for large line-capacity broadband switching systems," in *Conf. Rec., ICC '88*, Hollywood, FL, 1988, paper 29.2.

[9] D. W. Smith, P. Healey, and S. A. Cassidy, "Multidimensional optical switching networks," in *Conf. Rec., Globecom '89*, Dallas, TX, 1989, paper 1.2.

[10] P. S. Henry, R. A. Linke, and A. H. Gnauck, "Introduction to lightwave systems," in *Optical Fiber Telecommunications II*, S. E. Miller and I. P. Kaminow, Eds. New York: Academic, 1988, ch. 21.

[11] P. S. Henry, "High-capacity lightwave local area networks," *IEEE Commun. Mag.*, pp. 20-26, Oct. 1989.

[12] I. P. Kaminow, P. P. Iannone, J. Stone, and L. W. Stulz, "FDMA-FSK star network with a tunable optical filter demultiplexer," *J. Lightwave Technol.*, vol. 6, pp. 1406-1414, 1988.

[13] L. G. Kazovsky, "Multichannel coherent optical communications systems," *J. Lightwave Technol.*, vol. LT-5, pp. 1095-1102, 1987.

[14] D. B. Payne and J. R. Stern, "Single mode optical local networks," in *Conf. Proc., GLOBECOM '85*, Houston, TX, 1985, paper 39.5; and D. B. Payne and J. R. Stern, "Transparent single mode fiber optical networks," *J. Lightwave Technol.*, vol. LT-4, pp. 864-869, 1986.

[15] N. A. Olsson, J. Hegarty, R. A. Logan, L. F. Johnson, K. L. Walker, L. G. Cohen, B. L. Kasper, and J. C. Campbell, "68.3 km transmission with 1.37 Tbit km/s capacity using wavelength division multiplexing of ten single-frequency lasers at 1.5 μm," *Electron. Lett.*, vol. 21, pp. 105-106, 1985.

[16] E.-J. Bachus, R.-P. Braun, C. Caspar, E. Grossman, H. Foisel, K. Hermes, H. Lamping, B. Strebel, and F. J. Westphal, "Ten-channel coherent optical fiber transmission," *Electron. Lett.*, vol. 22, pp. 1002-1003, 1986.

[17] M. S. Goodman, J. L. Gimlett, H. Kobrinski, M. P. Vecchi, and R. M. Bulley, "The LAMBDANET multiwavelength network: Architecture, applications, and demonstrations," *IEEE J. Select. Areas Commun.*, this issue, pp. 995-1004; and H. Kobrinski, R. M. Bulley, M. S. Goodman, M. P. Vecchi, C. A. Brackett, L. Curtis, and J. L. Gimlett, "Demonstration of high capacity in the LAMBDANET architecture: A multiwavelength optical network," *Electron. Lett.*, vol. 23, p. 284, 1987.

[18] J. P. Laude and J. M. Lerner, "Wavelength division multiplexing/demultiplexing (WDM) using diffraction gratings," *Application, Theory, and Fabrication of Periodic Structures, SPIE*, vol. 503, 1984; and J. P. Laude *et al.*, "STIMAX, a grating multiplexer for monomode or multimode fibers," in *Proc. ECOC '83*, 1983, pp. 417-420.

[19] M. S. Goodman, J. L. Gimlett, H. Kobrinski, M. P. Vecchi, and R. M. Bulley, "The LAMBDANET multiwavelength network: Architecture, applications, and demonstrations," *IEEE J. Select. Areas Commun.*, this issue, pp. 995-1004.

[20] E. Arthurs, J. M. Cooper, M. S. Goodman, H. Kobrinski, M. Tur, and M. P. Vecchi, "Multiwavelength optical crossconnect for parallel-processing computers," *Electron. Lett.*, vol. 24, pp. 119-120, 1986.

[21] J. Cooper, J. Dixon, M. S. Goodman, H. Kobrinski, M. P. Vecchi, E. Arthurs, S. G. Menocal, M. Tur, and S. Tsuji, "Nanosecond wavelength switching with a double-section distributed feedback laser," in *Conf. Proc., CLEO '88*, Anaheim, CA, 1988, paper WA4.

[22] H. Kobrinski, M. P. Vecchi, M. S. Goodman, E. L. Goldstein, T. E. Chapuran, J. M. Cooper, C. E. Zah, and S. G. Menocal, "Fast wavelength-switching of laser transmitters and amplifiers," *IEEE J. Select. Areas Commun.*, this issue, pp. 1190-1202.

[23] E. Arthurs, M. S. Goodman, H. Kobrinski, and M. P. Vecchi, "HYPASS: An optoelectronic hybrid packet-switching system," *IEEE J. Select. Areas Commun.*, vol. 6, pp. 1500-1510, 1988.

[24] P. A. Kirkby, N. Baker, W. S. Lee, Y. Kanabar, and R. Worthington, "Multichannel grating demultiplexer receivers for high density wavelength multiplexed systems," in *Conf. Proc., IOOC'89*, Kobe, Japan, 1989.

[25] H. Kobrinski, M. P. Vecchi, E. L. Goldstein, and R. M. Bulley, "Wavelength selection with nanosecond switching times using DFB laser amplifiers," *Electron. Lett.*, vol. 24, pp. 969-970, 1988.

[26] M. Nishio, S. Suzuki, N. Shimosaka, T. Numai, T. Miyakawa, M. Fujiwara, and M. Itoh, "An experiment on photonic wavelength-division and time division hybrid switching," in *Tech. Dig. Photon. Switch. Top. Meet.*, Opt. Soc. Amer., Salt Lake City, UT, 1989.

[27] M. S. Goodman, "Multiwavelength networks and new approaches to packet switching," *IEEE Commun. Mag.*, vol. 27, pp. 27-35, Oct. 1989.

[28] J. Y. Hui and E. Arthurs, "A broadband packet switch for integrated transport," *IEEE J. Select. Areas Commun.*, vol. SAC-5, pp. 1264-1273, 1987.

[29] T. T. Lee, M. S. Goodman, and E. Arthurs, "A broadband optical multicast switch," to be presented at the *ISS '90*, 1990.

[30] S. S. Wagner, H. Kobrinski, T. J. Robe, H. L. Lemberg, and L. S. Smoot, "A passive photonic loop architecture employing wavelength-division multiplexing," in *Conf. Proc., GLOBECOM '88*, 1988, pp. 1569-1573.

[31] H. Kobrinski, "Crossconnection of WDM high-speed channels," *Electron. Lett.*, vol. 23, p. 975, 1987.

[32] G. R. Hill, "A wavelength routing approach to optical communications networks," presented at *IEEE INFOCOM '88*, 1988.

[33] G. R. Hill, P. J. Chidgey, and J. Davidson, "Wavelength routing for long haul networks," in *Conf. Proc., Int. Conf. Commun., ICC '89*, Boston, MA, 1989, vol. 2, pp. 734-738.

[34] B. Glance, K. Pollock, C. A. Burrus, B. L. Kasper, G. Eisenstein, and L. W. Stulz, "Densely spaced WDM coherent optical star network," *Electron. Lett.*, vol. 23, no. 17, pp. 875-876, 1987.

[35] B. Glance, J. Stone, K. J. Pollock, P. J. Fitzgerald, C. A. Burrus, B. L. Kasper, and L. W. Stulz, "Densely spaced FDM coherent star

network with optical signals confined to equally spaced frequencies," *J. Lightwave Technol.*, vol. 6, pp. 1770–1781, 1988.
[36] J. Stone and L. W. Stulz, "Pigtailed high-finesse tunable fiber Fabry–Perot interferometers with large, medium, and small free spectral ranges," *Electron. Lett.*, vol. 23, no. 15, pp. 781–783, 1987.
[37] I. P. Kaminow, P. P. Iannone, J. Stone, and L. W. Stulz, "Frequency division multiple access network demultiplexing with a tunable Vernier fiber Fabry–Perot filter," in *Conf. Proc., CLEO '88*, Anaheim, CA, Apr. 1988.
[38] C. Lin, H. Kobrinski, A. Frenkel, and C. A. Brackett, "Wavelength-tunable 16 optical channel transmission experiment at 2 Gb/s and 600 Mb/s for broadband subscriber distribution," *Electron. Lett.*, vol. 24, pp. 1215–1217, 1988.
[39] A. R. Chraplyvy and R. W. Tkach, "Narrowband tunable optical filter for channel selection in densely packed WDM systems," *Electron. Lett.*, vol. 22, pp. 1084–1085, 1986.
[40] M. Shibutani *et al.*, "Ten-channel coherent optical FDM broadcasting system," in *Tech. Dig. OFC '89*, Houston, TX, 1989, paper ThC2.
[41] B. Glance, T. L. Koch, O. Scaramucci, K. C. Reichmann, L. D. Tzeng, U. Koren, and C. A. Burrus, "Densely spaced FDM coherent system with near quantum-limited sensitivity and computer-controlled random access channel selection," in *Tech. Dig. OFC '89*, Houston, TX, Feb. 1989, postdeadline paper.
[42] R. E. Wagner *et al.*, "16-channel coherent broadcast network at 155 Mb/s," in *Tech. Dig. OFC '89*, Houston, TX, 1989, paper PD12.
[43] W. I. Way, C. E. Zah, S. G. Menocal, C. Caneau, F. Favire, F. K. Shokoohi, T. P. Lee, and N. K. Cheung, "90-channel FM video transmission to 2048 terminals using two inline traveling-wave laser amplifiers in a 1300 nm subcarrier multiplexed optical system," in *Proc. ECOC '88*, Brighton, UK, Sept. 1988, postdeadline paper, pp. 37–40.
[44] R. Olshansky, V. A. Lanzisera, and P. Hill, "Simultaneous transmission of 100 Mb/s at baseband and 60 FM video channels for a wideband optical communication network," *Electron Lett.*, vol. 24, pp. 1234–1235, Sept. 1988.
[45] T. E. Darcie *et al.*, "Bidirectional multichannel 1.44 Gb/s lightwave distribution system using subcarrier multiplexing," *Electron. Lett.*, vol. 24, no. 11, pp. 649–650, 1988.
[46] See, for example, *Proc. IOOC '89*, Kobe, Japan, 1989.
[47] BT&D Technol., Wilmington, DE.
[48] H. Kobrinski *et al.*, "Simultaneous fast wavelength switching and direct data modulation using a 3-sectioned DBR laser with 2.2 nm continuous tuning range," in *Tech. Dig. OFC '89*, Houston, TX, 1989, paper PD3.
[49] H. Kobrinski, M. P. Vecchi, T. E. Chapuran, J. B. Georges, C. E. Zah, C. Caneau, S. G. Menocal, P. S. D. Lin, A. S. Gozdz, and F. J. Favire, "Fast multiwavelength switching and simultaneous FSK modulation using a 3-section DBR laser," in *Conf. Proc., ECOC '89*, Gothenburg, Sweden, Sept. 1989, pp. 292–295.
[50] N. Shimosaka, M. Fujiwara, S. Murata, N. Henmi, K. Emura, and S. Suzuki, "Photonic wavelength-division and time-division hybrid switching system utilizing coherent optical detection," in *Proc. ECOC '89*, Gothenburg, Sweden, Sept. 1989, vol. 1, pp. 268–271.
[51] U. Koren, T. L. Koch, B. I. Miller, G. Eisenstein, and B. H. Bosworth, "Wavelength division multiplexing light source with integrated quantum well tunable lasers and optical amplifiers," *Appl. Phys. Lett.*, vol. 54, pp. 2056–2058, 1989.
[52] U. Koren *et al.*, "An integrated tunable light source with extended tunability range," in *Proc. Int. Opt. Opt. Commun. Conf. (IOOC)*, Kobe, Japan, Aug. 1989, paper 19A2-3.
[53] G. Coquin, K. W. Cheung, and M. Choy, "Single- and multiple-wavelength operation of acousto-optically tuned lasers at 1.3 microns," in *Proc. 11th IEEE Int. Semiconductor Laser Conf.*, Boston, MA, 1988, pp. 130–131.
[54] F. Heismann *et al.*, "Narrow-linewidth, electro-optically tunable InGaAsP-Ti:$LiNbO_3$ extended cavity laser," *Appl. Phys. Lett.*, vol. 51, pp. 164–165, 1987.
[55] H. Toba, K. Oda, N. Takato, and K. Nosu, "5 GHz-spaced, eight-channel guided-wave tunable multi/demultiplexer for optical FDM transmission systems," *Electron. Lett.*, vol. 23, pp. 788–789, 1987.
[56] H. Toba, K. Oda, K. Nakanishi, N. Shibata, K. Nosu, N. Takato, and M. Fukuda, "100-channel optical FDM transmission/distribution at 622 Mb/s over 50 km," *Opt. Fiber Commun. Conf., OFC '90*, San Francisco, CA, 1990, postdeadline paper PD1.
[57] I. P. Kaminow, "Non-coherent photonic frequency-multiplexed multiple access networks," *IEEE Network*, vol. 3, pp. 4–12, Mar. 1989.
[58] H. Kobrinski and K. W. Cheung, "Wavelength-tunable optical filters: Applications and technologies," *IEEE Commun. Mag.*, vol. 27, pp. 53–63, Oct. 1989.
[59] J. M. Senior and S. D. Cusworth, "Devices for wavelength multiplexing and demultiplexing," *IEE Proc.*, vol. 136, pt. J, no. 3, pp. 183–202, June 1989.
[60] F. Heismann, L. L. Buhl, and R. C. Alferness, "Electro-optically tunable, narrowband Ti:$LiNbO_3$ wavelength filter," *Electron. Lett.*, vol. 23, no. 11, pp. 572–573, 1987.
[61] F. Heismann, W. Warzanskyj, R. C. Alferness, and L. L. Buhl, "Narrowband double-pass wavelength filter with broad tuning range," in *Conf. Dig., IGWO '88*, Santa Fe, NM, 1988.
[62] B. L. Heffner, D. A. Smith, J. E. Baran, A. Yi-Yan, and K. W. Cheung, "Improved acoustically-tunable optical filter on X-cut LiNbO3," *Electron. Lett.*, vol. 24, pp. 1562–1563, 1988.
[63] W. Warzanskyj, F. Heismann, and R. C. Alferness, "Polarization-independent electro-optically tunable narrow-band wavelength filter," *Appl. Phys. Lett.*, vol. 53, pp. 13–15, 1988.
[64] T. Numai, S. Murata, T. Sasaki, and I. Mito, "1.5 μm tunable wavelength filter using phase-shift controllable DFB LD with wide tuning range and high constant gain," in *Conf. Proc., ECOC '88*, Brighton, UK, 1988, pp. 243–246.
[65] M. Nishio, T. Numai, S. Suzuki, M. Fujiwara, M. Itoh, and S. Murata, "Eight-channel wavelength-division switching experiment using wide-tuning-range DFB LD filters," in *Conf. Proc., ECOC '88*, Brighton, UK, 1988, pp. 49–52.
[66] S. Suzuki, M. Nishio, T. Numai, M. Fujiwara, M. Itoh, S. Murata, and N. Shimosaka, "A photonic wavelength-division switching system using tunable laser diode filters," in *Conf. Rec., ICC '89*, Boston, MA, 1989, paper 23.1.
[67] C. Dragone, "Efficient $N \times N$ star couplers using Fourier optics," *J. Lightwave Technol.*, vol. 7, pp. 479–489, 1989.
[68] C. Dragone, C. H. Henry, I. P. Kaminow, and R. C. Kistler, "Efficient multichannel integrated optics star coupler on silicon," *IEEE Photon. Technol. Lett.*, vol. 1, pp. 241–243, Aug. 1989.

PHOTO NOT AVAILABLE

Charles A. Brackett (S'60–M'64–SM'89) received the Ph.D. degree in electrical engineering from the University of Michigan, Ann Arbor, in 1968.

From 1968 to 1981, he was employed at AT&T Bell Laboratories, Murray Hill, NJ, where he worked on semiconductor microwave oscillators, optical receivers and laser transmitters, and optical data links. From 1981 to 1984, he was Supervisor of the Lightwave Receivers Group, Allentown, PA, where he was responsible for the final development and introduction into manufacture of all optical receivers for AT&T's transmission systems. In 1984, he joined Bell Communications Research, Morristown, NJ, as District Manager of Exploratory Optical Networks Research. In this capacity, he initiated work on multiwavelength optical networks and on optical-code division multiple-access systems. His interests include photonic device technology, optical networks, and optical switching.

Dr. Brackett is a member of the Optical Society of America and has been elected to Tau Beta Pi, Eta Kappa Nu, Sigma Xi, and Phi Kappa Phi.

Near–Far Resistance of Multiuser Detectors in Asynchronous Channels

RUXANDRA LUPAS, STUDENT MEMBER, IEEE, AND SERGIO VERDÚ, SENIOR MEMBER, IEEE

Abstract—**We consider an asynchronous code-division multiple-access environment in which the receiver has knowledge of the signature waveforms of all the users. Under the assumption of white Gaussian background noise, we compare detectors by their worst case bit error rate in a low background noise near–far environment where the received energies of the users are unknown to the receiver and are not necessarily similar.**

Conventional single-user detection in a multiuser channel is not near–far resistant, while the substantially higher performance of the optimum multiuser detector requires exponential complexity in the number of users. We explore suboptimal demodulation schemes which exhibit a low order of complexity while not exhibiting the impairment of the conventional single-user detector. Attention is focused on linear detectors, and it is shown that there exists a linear detector whose bit-error-rate is independent of the energy of the interfering users. Moreover it is shown that the near–far resistance of optimum multiuser detection can be achieved by a linear detector. The optimum linear detector for worst-case energies is found, along with existence conditions, which are always satisfied in the models of practical interest.

I. Introduction

THE near–far problem is the principal shortcoming of current radio networks using direct-sequence spread-spectrum multiple-access (DS/SSMA) communication systems. Those systems achieve multiple-access capability by assigning a distinct signature waveform to each user from a set of waveforms with low mutual crosscorrelations. Then, when the sum of the signals modulated by several asynchronous users is received, it is possible to recover the information transmitted by correlating the received process with replicas of the assigned signature waveforms. This demodulation scheme is conventionally used in practice, and its performance is satisfactory if two conditions are satisfied: first, the assigned signals need to have low crosscorrelations for all possible relative delays between the data streams transmitted by the asynchronous users, and second the powers of the received signals cannot be very dissimilar. If either of these conditions is not fulfilled, then the bit-error-rate and the antijamming capability of the conventional detector are degraded substantially. The reason why system performance is unacceptable when the received energies are dissimilar even with good (i.e., quasiorthogonal) signal constellations, is that the output of each correlator or matched filter contains a spurious component which is linear in the amplitude of each of the interfering users. Thus, as the multiuser interference grows, the bit-error-rate increases until the conventional detector is unable to recover the messages transmitted by the weak users.

Due to the severe reduction of the multiple-access capability and the increase of vulnerability to hostile sources caused by the near–far problem and its ubiquity in networks with dynamically changing topologies (such as mobile radio), its alleviation has been a target of researchers in the area for several years. However, success has been very limited and the only remedies implemented in practice have been to use power control or to design signals with more stringent crosscorrelation properties, which as we have noted, does not eliminate the near–far problem.

The viewpoint of this paper is that the near–far problem is not an inherent shortcoming of DS/SSMA systems, but of the conventional single-user detector. The optimum multiuser detector was obtained in [1] and was shown to be near–far resistant in the sense that a (very good) performance level can be guaranteed regardless of the relative energy of the transmitters. The optimum multiuser detector consists of a bank of matched filters and a Viterbi algorithm whose complexity is exponential in the number of users. In decentralized applications (where each receiver is only interested in demodulating the data sent by one transmitter), it is possible to drastically reduce the complexity of the optimum receiver (without compromising performance) by neglecting all but the comparatively powerful interferers. However, in this paper we propose a receiver (which we refer to as the decorrelating receiver) whose complexity is only linear in the number of users, and whose bit-error-rate is independent of the powers of the interferers at the receiver. Moreover, the decorrelating receiver achieves optimum near–far resistance (in a sense to be defined precisely in the sequel). The only requirement is the knowledge of the signature waveforms of the interfering users, and, in particular, no knowledge of the received energies is required, in contrast to the optimum receiver.

This paper generalizes the results obtained in [7] in the case of synchronous code-division multiple-access channels. Other recent attempts to derive detectors for multiuser channels include [9]–[11].

The multiple-access channel model considered in this paper is spelled out in Section II, as well as the general structure of the proposed detector. In Section III, we present the performance measure of interest, the near–far resistance and we show that the near–far resistance of the optimum multiuser detector can be achieved by a linear detector (the decorrelating detector), which is explicitly obtained in Section IV, as well as its implementable version as a linear time-invariant system. Section V gives a numerical comparison of the error probabilities of the decorrelating receiver and the conventional receiver in a scenario of practical interest.

II. Multiuser Communication Model

Let the receiver input signal be

$$r(t) = S(t, b) + n(t) \tag{2.1}$$

where $n(t)$ is white Gaussian noise with power spectral density σ^2 and

$$S(t, b) = \sum_{i=-M}^{M} \sum_{k=1}^{K} b_k(i)\sqrt{w_k(i)}\tilde{s}_k(t - iT - \tau_k) \tag{2.2}$$

is the element of $\mathcal{L}_2$ (the Hilbert space of square-integrable functions) which contains the information sequence $b = \{b(i) = [b_1(i), \cdots, b_K(i)], b_k(i) \in \{-1, 1\}, k = 1, \cdots, K;\ i =$

Paper approved by the Editor for Spread Spectrum of the IEEE Communications Society. Manuscript received March 24, 1988; revised September 14, 1988. This work was supported in part by the U.S. Army Research Office under Contract DAAL03-87-K-0062. This paper was presented in part at the 1988 IEEE International Symposium on Information Theory, Kobe, Japan, June 1988.

The authors are with the Department of Electrical Engineering, Princeton University, Princeton, NJ 08544.

IEEE Log Number 9034846.

Reprinted from *IEEE Transactions on Communications,* vol. 38, no. 4, April 1990.

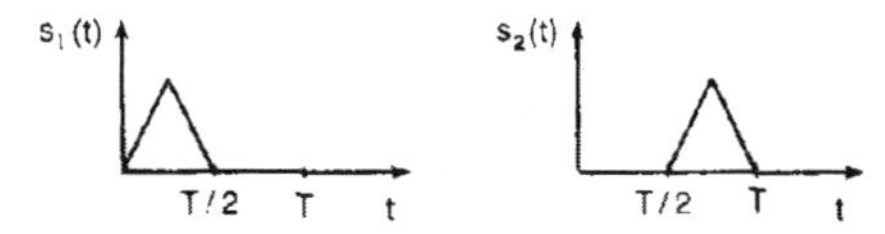

Fig. 1. Example of signature waveforms which can violate the LIA.

$-M,\cdots,M\}$, $\tilde{s}_k(t)$ is the normalized signature waveform of user k and is zero outside the interval $[0,T]$, and $w_k(i)$ is the received energy of user k in the ith time slot. Let $N = 2M+1$ be the length of the transmitted sequence. Without loss of generality it is assumed that the users are numbered such that their delays satisfy $0 \le \tau_1 \le \cdots \le \tau_K < T$. The normalized signal $\tilde{S}(t,\boldsymbol{b})$ is the receiver input signal corresponding to unit energies.

Define the vector space $L = \{\boldsymbol{x} = [\boldsymbol{x}(-M),\cdots,\boldsymbol{x}(M)] = [[x_1(-M),\cdots,x_K(-M)],\cdots,[x_1(M),\cdots,x_K(M)]]^T,\ x_k(i)\in\mathbb{R},\ k=1,\cdots,K,\ i=-M,\cdots,M\}$, (each element of which can be equivalently viewed as a sequence of N $(K*1)$-vectors or as one single $(NK*1)$-vector), and define the (k,i)th unit vector $\boldsymbol{u}^{k,i}$ in L with components $u_j^{k,i}(l) = \delta_{kj}\delta_{li}$. Note that the set of possible transmitted sequences $\boldsymbol{b}$ is a subset of L, obtained by restricting the components of the vector $\boldsymbol{x}$ to take on the values ± 1. Let $\langle\cdot,\cdot\rangle$ denote the usual inner product on $\mathcal{L}_2$, i.e., the integral of the product over the region of support, with induced norm $\|\cdot\|$. Henceforth, we make the following assumption on $\tilde{S}(t,\boldsymbol{b})$.

1) Linear Independence Assumption (LIA):

$$\forall\, \boldsymbol{v}\in L,\ \boldsymbol{v}\ne 0 \Rightarrow \left\|\tilde{S}(t,\boldsymbol{v})\right\| \ne 0. \tag{2.3}$$

In other words, no matter what the received energies are, the received signal does not vanish everywhere if at least one of the users has transmitted a symbol. This condition fails to hold only in pathological nonpractical cases with very heavy crosscorrelation between the signals, such as the two-user example in Fig. 1. There if the delay between the users is $T/2$, the received signal can be identically zero although transmissions have been made [this happens if, for all i, $b_2(i) = -b_1(i)$]. It is shown in Appendix II that such a situation will arise with probability zero if the *a priori* unknown delays are uniformly distributed, which is the case in the asynchronous channel used by noncooperating users. Basically, in order to violate the LIA, a subset of the users must be effectively synchronous and the modulating signals of this subset have to be heavily correlated. The LIA will be in effect in the rest of the paper. If it is removed all the given results can be generalized in a manner analogous to the treatment of the synchronous transmission case [7].

The sampled output of the normalized matched filter for the ith bit of the kth user, $i=-M,\cdots,M$, is

$$y_k(i) = \int_{iT+\tau_k}^{iT+T+\tau_k} r(t)\tilde{s}_k(t-iT-\tau_k)\,dt \tag{2.4}$$

$$= \int_{-\infty}^{\infty} S(t,\boldsymbol{b})\tilde{s}_k(t-iT-\tau_k)\,dt + \int_{-\infty}^{\infty} n(t)\tilde{s}_k(t-iT-\tau_k)\,dt \tag{2.5}$$

where the second equality is valid since the signals are zero outside $[0,T]$. It is well established (e.g., [1]) that the whole sequence $\boldsymbol{y}$ of outputs of the bank of K matched filters, with components $y_k(i)$ given by (2.5), for $k=1,\cdots,K$, $i=-M,\cdots,M$, is a sufficient statistic for decision on the most likely transmitted information sequence $\boldsymbol{b}$. The multiuser demodulation problem which needs to be solved at the receiver is to recover the transmitted sequence $\boldsymbol{b}\in L$ from the sequence $\boldsymbol{y}\in L$. Motivated by the state of the art—where the choice lies between the optimum multiuser detector, which is of exponential complexity and the ad hoc single user detector whose performance degrades to zero for sufficiently high interference energy—we define a class of simple detectors and optimize performance within this class, to obtain an acceptable error probability versus complexity tradeoff.

A *linear detector* for bit i of user k is characterized by $\boldsymbol{v}^{k,i}\in L$. The decision of the detector is given by the polarity of the inner product of $\boldsymbol{v}^{k,i}$ and the vector $\boldsymbol{y}$ of matched filter outputs, which is equal to

$$\sum_{l=-M}^{M}\sum_{j=1}^{K} v_j^{k,i}(l)y_j(l) = \int_{-\infty}^{\infty} \tilde{S}(t,\boldsymbol{wb})\tilde{S}(t,\boldsymbol{v}^{k,i})\,dt + n_{k,i} \tag{2.6}$$

$$= \langle \tilde{S}(t,\boldsymbol{wb}), \tilde{S}(t,\boldsymbol{v}^{k,i})\rangle + n_{k,i} \tag{2.7}$$

where for any information sequence $\boldsymbol{b}$, $\boldsymbol{wb}$ will denote the sequence of amplitudes $\boldsymbol{wb} = \{[\sqrt{w_1(i)}b_1(i),\cdots,\sqrt{w_K(i)}b_K(i)],\ i=-M,\cdots,M\}$. $n_{k,i}$ is the noise component at the output of the cascade of matched filter, sampler and detector, hence is a Gaussian zero-mean random variable with variance given by

$$E[n_{k,j}^2] = \sum_{l,i} v_k(l)v_j(i)\int_{-\infty}^{\infty}\sigma^2\tilde{s}_k(t-lT-\tau_k)\tilde{s}_j(t-iT-\tau_j)\,dt$$

$$= \sigma^2\left\|\tilde{S}(t,\boldsymbol{v}^{k,i})\right\|^2. \tag{2.8}$$

The receiver decides on the ith bit of the kth user according to the rule

$$\hat{b}_k(i) = \operatorname{sgn}\sum_{l=-M}^{M}\sum_{j=1}^{K} v_j^{k,i}(l)y_j(l) \tag{2.9}$$

$$= \operatorname{sgn}(\langle\tilde{S}(t,\boldsymbol{wb}),\tilde{S}(t,\boldsymbol{v}^{k,i})\rangle + n_{k,i}). \tag{2.10}$$

Wherever it is clear from the context, the superscripts k,i will be omitted.

2) Matrix Notation: It is convenient to introduce the following compact notation. Define the $K*K$ normalized signal crosscorrelation matrices $\boldsymbol{R}(l)$ whose entries are given by

$$R_{kj}(l) = \int_{-\infty}^{\infty}\tilde{s}_k(t-\tau_k)\tilde{s}_j(t+lT-\tau_j)\,dt. \tag{2.11}$$

Then, since the modulating signals are zero outside $[0,T]$

$$\boldsymbol{R}(l) = 0\ \forall\ |l|>1, \tag{2.12}$$

$$\boldsymbol{R}(-l) = \boldsymbol{R}^T(l), \tag{2.13}$$

and, if the users are numbered according to increasing delays, $\boldsymbol{R}(1)$ is an upper triangular matrix with zero diagonal. Also let $\boldsymbol{W}(l) = \operatorname{diag}([\sqrt{w_1(l)},\cdots,\sqrt{w_K(l)}])$. With this notation the matched filter outputs for $l=\{-M,\cdots,M\}$ can be written in vector form as (cf., [8])

$$\boldsymbol{y}(l) = \boldsymbol{R}(-1)\boldsymbol{W}(l+1)\boldsymbol{b}(l+1) + \boldsymbol{R}(0)\boldsymbol{W}(l)\boldsymbol{b}(l) + \boldsymbol{R}(1)\boldsymbol{W}(l-1)\boldsymbol{b}(l-1) + \boldsymbol{n}(l), \tag{2.14}$$

as can be seen for each component by inserting (2.1) into (2.4). We adopt the convention that $\boldsymbol{b}(-M-1) = \boldsymbol{b}(M+1) = 0$. $\boldsymbol{n}(l)$ is the matched filter output noise vector, with autocorrelation matrix given by

$$E[\boldsymbol{n}(i)\boldsymbol{n}^T(j)] = \sigma^2\boldsymbol{R}(i-j). \tag{2.15}$$

The entries of the matrices $\boldsymbol{R}(i)$, $i=-1,0,1$ are obtained at the receiver by cross-correlating appropriately delayed replicas of the normalized signature waveforms according to (2.11). Note that no additional complexity is hereby required of the receiver, since knowledge of the normalized signature waveforms and the capability to lock onto the respective delays are necessary for matched filtering and sampling at the instant of maximal signal-to-noise ratio.

In contrast to (2.5) the asynchronous nature of the problem is clearly transparent in (2.14). To make this notation more compact

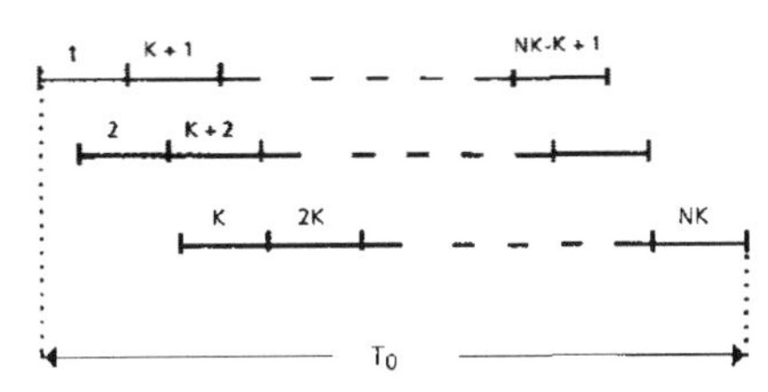

Fig. 2. Equivalent synchronous transmitted sequence.

we define the $NK*NK$ symmetric block-Toeplitz matrix $\mathcal{R}$ and the $NK*NK$ diagonal matrix $\mathcal{W}$, as follows:

$$\mathcal{R} = \begin{pmatrix} R(0) & R(-1) & 0 & \cdots & 0 \\ R(1) & R(0) & R(-1) & & \vdots \\ 0 & R(1) & R(0) & \ddots & 0 \\ \vdots & & \ddots & \ddots & R(-1) \\ 0 & \cdots & 0 & R(1) & R(0) \end{pmatrix}, \quad (2.16)$$

$$\mathcal{W} = \operatorname{diag}([\sqrt{w_1(-M)}, \cdots, \sqrt{w_K(-M)}, \cdots, \sqrt{w_1(M)}, \cdots, \sqrt{w_K(M)}]). \quad (2.17)$$

In this notation the matched filter output vector y depends on b via, from (2.14)

$$y = \mathcal{R}\mathcal{W}b + n. \quad (2.19)$$

The matrix $\mathcal{R}$ can be interpreted as the cross-correlation matrix for an equivalent synchronous problem where the whole transmitted sequence is considered to result from $N*K$ users, labeled as shown in Fig. 2, during one transmission interval of duration $T_e = N*T + \tau_K - \tau_1$. Then the results presented here for finite transmission length can be derived via analysis of synchronous multiuser communication, as done in [7]. However, the approach taken in this paper is more general and gives more insight into the nature of the problem. The limit $N \to \infty$ is considered in Section IV-B.

The decision made on the ith bit of the kth user at the output of the detector v is:

$$\hat{b}_k(i) = \operatorname{sgn} v^T y = \operatorname{sgn} v^T(\mathcal{R}\mathcal{W}b + n). \quad (2.20)$$

As for the inner product, for all x, y in L

$$\langle \tilde{S}(t, x), \tilde{S}(t, y) \rangle = x^T \mathcal{R} y. \quad (2.21)$$

It can be seen from (2.21) and from (2.3) that $\mathcal{R}$ is positive definite.

III. Near-Far Resistance

The main performance measure we are interested in is the bit-error-rate in the high signal-to-background noise region. Thus, even though the background thermal noise is not neglected, our main focus will be on the underlying performance degradation due to multiple-access interference. This performance degradation is conveniently quantified by the *asymptotic efficiency* which was introduced in [1]-[2], and is defined as follows. Let $P_k(\sigma)$ denote the bit-error-rate of the kth user when the spectral level of the background white Gaussian noise is σ^2, and let $e_k(\sigma)$ be such that $P_k(\sigma) = Q(\sqrt{e_k(\sigma)}/\sigma)$.[1]

Then, $e_k(\sigma)$ is actually the energy that the kth user would require to achieve bit-error-rate $P_k(\sigma)$ in the same white Gaussian channel but without interfering users. Hence, we refer to $e_k(\sigma)$ as the *effective energy* of the kth user, and the *efficiency* or ratio between the effective and actual energies $e_k(\sigma)/w_k$ is a number between 0 and 1 which characterizes the performance loss due to the existence of other users in the channel. Thus, the *asymptotic efficiency* (for high SNR) of a transmitter whose bit-error-rate curve and energy are given by $P_k(\sigma)$ and w_k, respectively, is

$$\eta_k = \lim_{\sigma \to 0} \frac{e_k(\sigma)}{w_k} = \sup\left\{0 \le r \le 1;\ \lim_{\sigma \to 0} P_k(\sigma)/Q\left(\frac{\sqrt{r w_k}}{\sigma}\right) < \infty\right\} \quad (3.1)$$

where the last equation follows immediately upon substitution of $P_k(\sigma)$ by its expression in terms of the effective energy. In order to visualize intuitively the asymptotic efficiency, note that the logarithm of the bit-error-rate $P_k(\sigma)$ decays asymptotically with the same slope as the logarithm of the bit-error-rate of a single-user with energy $\eta_k w_k$. Therefore, if $\lim_{\sigma \to 0} P_k(\sigma) > 0$, (i.e., there is an irreducible probability of error even in the absence of background noise), then the asymptotic efficiency is zero. Conversely, nonzero asymptotic efficiency implies that the bit-error-rate goes to zero (as $\sigma \to 0$) exponentially in $1/\sigma^2$.

While asymptotic efficiency and low-noise bit-error-rate are equivalent performance measures, asymptotic efficiency has the advantage of being analytically tractable and of resulting in explicit expressions for the detectors we are interested in. For example, while the probability of error of the optimum multiuser detector does not admit an explicit expression, its asymptotic efficiency is given by [2]

$$\eta_{k,i} = \frac{1}{w_k(i)} \min_{\epsilon \in Z_k} \|\tilde{S}(t, w\epsilon)\|^2 \quad (3.2)$$

where Z_k is the set of error-sequences $\epsilon = \{\epsilon(i) \in \{-1, 0, 1\}^K,\ i = -M, \cdots, M,\ \epsilon_k(i) = 1\}$ that affect the ith bit of the kth user. It was shown in [3] (see also [15]) that the numerical computation of the asymptotic efficiency of optimum multiuser detection given by (3.2) is an NP-complete combinatorial optimization problem.

In an environment where the transmission energies change in time, e.g., if the transmitters are mobile, a performance measure of interest for any detector is its kth *user near-far resistance*, $\overline{\eta_{k,i}}$, which is defined for each detector as its worst case asymptotic efficiency for bit i of user k over all possible energies of the other (interfering and noninterfering) bits, i.e.,

$$\overline{\eta_{k,i}} = \inf_{\substack{w_j(l) \ge 0 \\ (j,l) \ne (k,i)}} \eta_{k,i}. \quad (3.3)$$

In our definition of near-far resistance we model the most general case where the energies of the users are allowed to be time-dependent. This captures the worst case operating conditions of the detector, which are, for example, encountered in mobile radio communication, due to positioning and tracking variations. In the case where the energies are constrained to be arbitrary but nonvarying the present near-far resistance is a lower bound. That case is not amenable to closed-form analysis, since one has to deal with a combinatorial optimization problem.

For illustration consider the two-user case. If the user energies are constant over time, i.e., $w_1(i) = w_1$, $w_2(i) = w_2$, the asymptotic efficiency of the optimal multiuser detector given by (3.2) reduces to [2]:

$$\eta_1 = \min\left\{1,\ 1 + \frac{w_2}{w_1} - 2\max\{|\rho_{12}|, |\rho_{21}|\}\frac{\sqrt{w_2}}{\sqrt{w_1}},\ 1 + 2\frac{w_2}{w_1} - 2(|\rho_{12}| + |\rho_{21}|)\frac{\sqrt{w_2}}{\sqrt{w_1}}\right\}$$

and hence

$$\eta_{\min} \triangleq \min_{\substack{w_2 \\ w_1 \text{ const.}}} \eta_1 = \min\left\{1 - \rho_{12}^2,\ 1 - \rho_{21}^2,\ 1 - \rho_{12}^2 - \rho_{21}^2 + \frac{(|\rho_{12}| - |\rho_{21}|)^2}{2}\right\}, \quad (3.4)$$

[1] $Q(x) = \int_x^\infty (1/\sqrt{2\pi}) e^{-v^2/2}\, dv$.

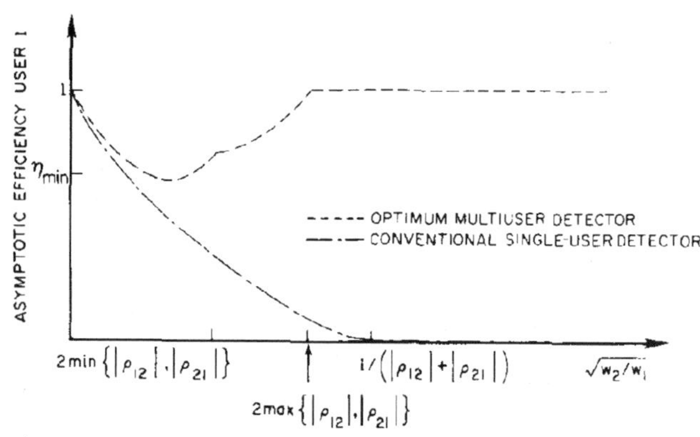

Fig. 3. Asymptotic efficiencies in the two-user case for infinite transmitted sequence length, when the user energies are constant over time (here we chose $|\rho_{12}|, |\rho_{21}| = 0.3, 0.5$).

and analogously for user 2 where $\rho_{12} = R_{12}(0)$ and $\rho_{21} = R_{12}(1)$. The dependence of η_1 for constant energies on the energy ratio is shown in Fig. 3. Note that the optimal multiuser detector is near-far resistant, and in fact has an asymptotic efficiency of unity for sufficiently powerful interference ([2]). Note also that in this case three different error-sequences minimize (3.2) for different values of w_2/w_1, as can be seen from the discontinuity points of the derivative of η. The minimum of η over constant energies, $\eta_{\min}$, is an upper bound on the near-far resistance of optimum multiuser detection $\bar{\eta}$, which is the minimum asymptotic efficiency over unconstrained energies.

The near-far resistance of the optimal multiuser detector is important since it is the least upper bound on the near-far resistance of any detector, and a measure of the relative performance of any suboptimal detector. From (3.2) and the definition of near-far resistance it is equal to

$$\overline{\eta_{k,i}} = \inf_{\substack{w_j(l)\geq 0\\ (j,l)\neq(k,i)}} \frac{1}{w_k(i)} \min_{\epsilon\in Z_k} \|\tilde{S}(t, w\epsilon)\|^2 \tag{3.5}$$

$$= \inf_{\substack{w_j(l)\geq 0\\ (j,l)\neq(k,i)}} \min_{\epsilon\in Z_k} \left\|\tilde{S}\left(t, \frac{1}{\sqrt{w_k(i)}} w\epsilon\right)\right\|^2 \tag{3.6}$$

$$= \inf_{\substack{y\in L\\ y_k(i)=1}} \|\tilde{S}(t,y)\|^2. \tag{3.7}$$

In Section IV, we obtain a closed-form expression for (3.7) as the reciprocal of the (k, i)th diagonal element (see footnote 2) of the inverse of $\mathcal{R}$. Hence, the near-far resistance of optimum multiuser resistance is guaranteed to be nonzero because of the linear independence assumption of (2.3), which ensures that $\mathcal{R}$ is invertible.

We now turn to the performance analysis of the linear detectors introduced above. The probability of error at decision upon $b_k(i)$ of the linear detector v is, from (2.10):

$$P_k(i) = P(\hat{b}_k(i) \neq b_k(i)) \tag{3.8}$$

$$= P(\langle\tilde{S}(t, wb), \tilde{S}(t, v)\rangle + n_{k,i} < 0 | b_k(i) = 1). \tag{3.9}$$

The equality follows since the hypotheses $+1, -1$ are assumed equally likely. Let B be the set of possible transmitted sequences. From (2.8) $n_{k,i}$ is a zero-mean Gaussian random variable with variance $\sigma^2\|\tilde{S}(t, v)\|^2$, hence the probability of error in (3.9) is a sum of Q-functions, one for each possible interfering bit-combination. For $\sigma \to 0$ the Q-function with the smallest argument dominates the error probability, hence from (3.1), since the expression below can be shown (cf. [15]) to be upper bounded by 1, the asymptotic efficiency achieved by the linear detector v for the ith bit of the kth user is

$$\eta_{k,i}(v) = \frac{1}{w_k(i)} \max{}^2 \left\{0, \min_{\substack{b\in B\\ b_k(i)=1}} \frac{\langle\tilde{S}(t, wb), \tilde{S}(t, v)\rangle}{\|\tilde{S}(t, v)\|}\right\}. \tag{3.10}$$

Knowledge of the asymptotic efficiency of a linear detector is equivalent to knowledge of the worst case probability of error over the bit sequences of the interfering users, since this error probability, which is a Q-function, is set equal to $Q(\sqrt{\eta_{k,i}(v)w_k(i)}/\sigma)$ to obtain (3.10).

For illustration consider the conventional single-user detector in the two-user case. We have $v = u^{k,i}$ (recall that $u^{k,i}$ is the (k, i)th unit vector in the space L of linear detectors). If the user energies are constant over time, i.e., $w_1(i) = w_1$, $w_2(i) = w_2$, the asymptotic efficiency of the conventional single-user detector is found from (3.10) to be

$$\eta_1^c = \max{}^2 \left\{0, 1 - (|\rho_{12}| - |\rho_{21}|)\frac{\sqrt{w_2}}{\sqrt{w_1}}\right\} \tag{3.11}$$

and analogously for user 2. The dependence of η_1^c for constant energies on the energy ratio is shown in Fig. 3. Note that the asymptotic efficiency of the conventional single-user detector is zero for sufficiently high interference energy ($\sqrt{w_2}/\sqrt{w_1} > 1/(|\rho_{12}| + |\rho_{21}|)$). This implies that its near-far resistance is zero, which is what we want to remedy.

There are three quantities of interest in this communication environment, on the one hand the transmitted bit-sequence and the set of energies, both of which depend only on the transmitters and determine the *operating points* for the receiver, and on the other hand the data-processing detector v at the receiver, which we called a *linear detector*. In determining which linear detector to choose at the receiver a useful procedure is the *minimax* approach, in which the design goal is to optimize the worst case performance of the receiver over the class of operating points. Thus we are interested in finding the *maximin linear detector*, whose worst case performance over all allowable input sequences is the highest in the class of linear detectors. The following result quantifies the performance of the maximin detector, in the sequel denoted by v^*.

Proposition 1: There exists a linear detector (which is independent of the received energies) that achieves optimum near-far resistance (i.e., the near-far resistance of the optimum multiuser detector). ●

Proof: From (3.10) the asymptotic efficiency of the linear detector v is

$$\eta_{k,i}(v) = \frac{1}{w_k(i)} \max{}^2 \left\{0, \min_{\substack{b\in B\\ b_k(i)=1}} \frac{\langle\tilde{S}(t, wb), \tilde{S}(t, v)\rangle}{\|\tilde{S}(t, v)\|}\right\} \tag{3.12}$$

$$\times \min_{\substack{b\in B\\ b_k(i)=1}} \frac{1}{w_k(i)} \max{}^2 \left\{0, \frac{\langle\tilde{S}(t, wb), \tilde{S}(t, v)\rangle}{\|\tilde{S}(t, v)\|}\right\} \tag{3.13}$$

$$= \min_{\substack{b\in B\\ b_k(i)=1}} \frac{1}{w_k(i)} \max{}^2 \left\{0, \frac{b^T \mathcal{W}\mathcal{R}v}{\sqrt{v^T\mathcal{R}v}}\right\} \tag{3.14}$$

where in the last equality we have used the compact matrix notation of (2.21) for simplicity. We are interested in the linear detector with the highest worst case asymptotic efficiency, i.e., whose near-far resistance is

$$\overline{\eta_{k,i}(v^*)} = \sup_{\substack{v\in L\\ \|\tilde{S}(t,v)\|\neq 0}} \inf_{\substack{w_j(l)\geq 0\\ (j,l)\neq(k,i)}} \eta_{k,i}(v) \tag{3.15}$$

$$= \sup_{\substack{v\in L\\ v^T\mathcal{R}v\neq 0}} \underbrace{\inf_{\substack{w_j(l)\geq 0\\ (j,l)\neq(k,i)}} \min_{\substack{b\in B\\ b_k(i)=1}} \frac{1}{w_k(i)} \max{}^2 \left\{0, \frac{b^T \mathcal{W}\mathcal{R}v}{\sqrt{v^T\mathcal{R}v}}\right\}} \tag{3.16}$$

$$= \sup_{\substack{v \in L \\ y^T \mathcal{R} v \neq 0}} \inf_{\substack{y \in L \\ y_k(i)=1}} \max^2 \left\{ 0, \frac{y^T \mathcal{R} v}{\sqrt{v^T \mathcal{R} v}} \right\} \tag{3.17}$$

$$= \max^2 \left\{ 0, \sup_{\substack{v \in L \\ v^T \mathcal{R} v \neq 0}} \inf_{\substack{y \in L \\ y_k(i)=1}} \frac{y^T \mathcal{R} y}{\sqrt{v^T \mathcal{R} v}} \right\} \tag{3.18}$$

where we have set $y_j(l) = b_j(l)\sqrt{w_j(l)}/\sqrt{w_k(i)}$ for the third equality. Let $M(v, y)$ denote the penalty function $y^T \mathcal{R} v / \sqrt{v^T \mathcal{R} v}$ where the first argument is from the set of detectors and the second from the set of operating points, both specified in (3.18). We show in Appendix 1 that $M(v, y)$ has a saddle point, i.e.,

$$\sup_{\substack{v \in L \\ v^T \mathcal{R} v \neq 0}} \inf_{\substack{v \in L \\ y_k(i)=1}} \frac{y^T \mathcal{R} v}{\sqrt{v^T \mathcal{R} v}} = \inf_{\substack{y \in L \\ y_k(i)=1}} \sup_{\substack{v \in L \\ v^T \mathcal{R} v \neq 0}} \frac{y^T \mathcal{R} v}{\sqrt{v^T \mathcal{R} v}}, \tag{3.19}$$

which establishes the existence of v^* and hence

$$\overline{\eta_{k,i}(v^*)} = \max^2 \left\{ 0, \inf_{\substack{y \in L \\ y_k(i)=1}} \sup_{\substack{v \in L \\ v^T \mathcal{R} v \neq 0}} \frac{y^T \mathcal{R} v}{\sqrt{v^T \mathcal{R} v}} \right\} \tag{3.20}$$

$$= \max^2 \left\{ 0, \inf_{\substack{y \in L \\ y_k(i)=1}} \sqrt{y^T \mathcal{R} y} \right\} \tag{3.21}$$

$$= \inf_{\substack{y \in L \\ y_k(i)=1}} \|\tilde{S}(t, y)\|^2 \tag{3.22}$$

$$= \overline{\eta_{k,i}} \tag{3.23}$$

where the second equality is obtained in (A.1), the third line follows since $\mathcal{R}$ is nonnegative definite and the last equality was obtained in (3.7). ◇

The reason why the near–far optimum linear receiver achieves the same near–far resistance as the optimum receiver can be understood as follows. Let Ω be the set of multiuser signals modulated by all positive amplitudes, i.e., $\Omega = \{\tilde{S}(t, y), y \in L\}$ and let Ξ denote the subset of Ω such that the amplitude of the ith symbol of the kth user is fixed to 1, i.e., $\Xi = \{\tilde{S}(t, y), y \in L, y_k(i) = 1\}$ (note that Ξ is a convex set, and because of the LIA it does not include the origin). Since the penalty function in (3.18) is invariant to scaling of v and the operator $\mathcal{R}$ is positive definite, (3.18) can be rewritten as

$$\overline{\eta_{k,i}(v^*)} = \max^2 \left\{ 0, \sup_{\substack{v \in L \\ |\tilde{S}(t,v)|=1}} \inf_{\substack{y \in L \\ y_k(i)=1}} \langle \tilde{S}(t, y), \tilde{S}(t, v) \rangle \right\} \tag{3.24}$$

$$= \max^2 \left\{ 0, \sup_{\substack{v \in \Omega \\ |v|=1}} \inf_{y \in \Xi} \langle y, v \rangle \right\}. \tag{3.25}$$

Therefore the kth user decorrelating filter can be viewed as the unit-norm multiuser waveform whose minimum inner product with the elements of Ξ is highest. But since Ξ is a convex set, that signal is a scaled version of the closest vector in Ξ to the origin (Fig. 4), and its near-far resistance [cf. (3.22)] is the norm squared of that vector. But, as (3.7) indicates, the square of the distance from Ξ to the origin is precisely the near–far resistance of the optimum detector.

Equation (3.7) leads to a nice intuitive interpretation of near–far resistance. Rewrite this equation, using the definition of $\tilde{S}(t, \cdot)$, as

$$\overline{\eta_{k,i}} = \inf_{\substack{y_j(l) \in R \\ (j,l) \neq (k,i)}} \left\| \tilde{s}_k(t - iT - \tau_k) + \sum_{(j,l) \neq (k,i)} y_j(l) \tilde{s}_j(t - lT - \tau_j) \right\|^2 \tag{3.26}$$

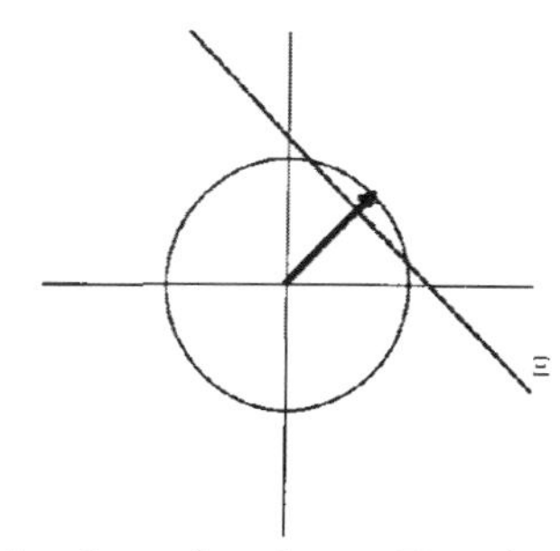

Fig. 4. Interpretation of near–far resistance. Vector in boldface corresponds to decorrelating filter.

Letting $\{y_j(l)\}$ vary over the admissible set, the second term above generates all points of a linear subspace which includes the origin, therefore the infimum in (3.26) is the distance of $\tilde{s}_k(t - iT - \tau_k)$ to this space, i.e.,

$$\overline{\eta_{k,i}} = d^2(\tilde{s}_k(t - iT - \tau_k), \text{ span } \{\tilde{s}_j(t - lT - \tau_j), (j, l) \neq (k, i)\}) \tag{3.27}$$

where $d(a, b)$ denotes the Euclidean distance between the $\mathcal{L}_2$ elements a and b. In the synchronous case because the time-support is disjoint, the infimum in (3.26) is achieved when $y_j(l) = 0, l \neq i$, and (3.27) reduces to

$$\overline{\eta_k} = d^2(\tilde{s}_k(t), \text{ span } \{\tilde{s}_j(t), j \neq k\}), \tag{3.28}$$

i.e., the kth user near–far resistance in a synchronous channel is the square of the *distance of the kth user signal to the space spanned by the signals of the interfering users*. Viewing the asynchronous problem in terms of the equivalent synchronous system with $N*K$ users and period NT, the near–far resistance of asynchronous communication allows for the same interpretation. Note, however, that the shifted versions $s_k(t - lT - \tau_k), l \neq i$ of the kth user signal affect the near–far resistance of the ith symbol of user k.

The following section characterizes the linear detector that achieves the optimum near–far resistance anticipated by Proposition 1.

IV. The Decorrelating Detector

We first assume N to be finite, as in the case in all communication environments, and characterize the linear filter which achieves the near–far resistance of optimum multiuser detection. This filter is nonstationary for finite N. The limit as $N \to \infty$ is then considered, yielding a stationary noncausal limiting filter, and hence, after appropriate truncation of the noncausal part, an approximation of the near–far optimal linear filter which can be implemented easily.

A. The Finite Sequence Length Case

Definition: A decorrelating detector $d^{k,i}$ for the ith bit of the kth user is a linear detector for which

$$\mathcal{R} d^{k,i} = u^{k,i} \tag{4.1}$$

or equivalently, from (2.21), $\langle \tilde{S}(t, v), \tilde{S}(t, d^{k,i}) \rangle = v_k(i)$, for all v in L.

Existence: By the LIA, statement (4.2) below holds for all k, i. Hence, the following equivalences show the existence of the decorrelating detectors for each bit of each user.

$$\forall v \in L \text{ with } v_k(i) \neq 0\text{: } \|\tilde{S}(t, v)\| \neq 0 \tag{4.2}$$

$$\Leftrightarrow \forall v \in L \text{ with } v_k(i) \neq 0\text{: } v^T \mathcal{R} v \neq 0 \tag{4.3}$$

$$\Leftrightarrow \nexists v \in L \text{ with } v_k(i) \neq 0 \text{ s.t. } \mathcal{R} v = 0 \tag{4.4}$$

$\Leftrightarrow$ the $(k, i)^{th}$ column[2] *of* $\mathcal{R}$ is

[2] We refer to the (k, i)th row (or column) of a matrix of the dimension of $\mathcal{R}$ when we want to name the kth row (or column) within the ith block in vertical (horizontal) direction. This notation was adopted since $\mathcal{R}$ is block-Toeplitz.

$$\times \quad \text{linearly independent of the others} \tag{4.5}$$

$$\Leftrightarrow \exists \boldsymbol{d} \text{ s.t. } \mathcal{R}\boldsymbol{d} = \boldsymbol{u}^{k,i}. \tag{4.6}$$

Properties:

i) The decorrelating detector for each bit of each user is invariant with respect to received energies and does not require knowledge thereof.

Proof: Since the elements of the matrix $\mathcal{R}$ are normalized crosscorrelation coefficients, the defining equation (4.1) is energy independent.

ii) The decorrelating detector eliminates the multiuser interference present in the respective matched filter output. (Hence its name).

Proof: From (2.20) the decision made on the ith bit of the kth user at the output of the decorrelating filter $\boldsymbol{d}$ is,

$$\begin{aligned}\hat{b}_k(i) &= \text{sgn}(\boldsymbol{d}^T \mathcal{R} \mathcal{W} \boldsymbol{b} + \boldsymbol{d}^T \boldsymbol{n}) \\ &= \text{sgn}(\sqrt{w_k(i)} b_k(i) + \boldsymbol{d}^T \boldsymbol{n}).\end{aligned} \tag{4.7}$$

Interestingly, this natural strategy, though not necessarily optimal for specific user-energies, is optimal with respect to the worst possible distribution of energies.

iii) The kth-user bit-error-rate of the decorrelating detector is independent of the energies of the interfering users $w_j(i)$, $j \neq k$, $i = -M, \cdots, M$.

Proof: It follows from (4.7) that the decision statistic that is compared to a zero threshold is independent of the energies of the interfering users.

iv) The efficiency of the decorrelating detector is independent of the energies and is given by

$$\eta^d_{k,i} = \max{}^2 \left\{ 0, \min_{\substack{b \in B \\ b_k(i)=1}} \frac{1}{\sqrt{w_k(i)}} \frac{\langle \tilde{S}(t, \mathcal{W}\boldsymbol{b}), \tilde{S}(t, \boldsymbol{d}) \rangle}{\|\tilde{S}(t, \boldsymbol{d})\|} \right\} \tag{4.8}$$

$$= \max{}^2 \left\{ 0, \min_{\substack{b \in B \\ b_k(i)=1}} \frac{1}{\sqrt{w_k(i)}} \frac{\sqrt{w_k(i)} b_k(i)}{\sqrt{d_k(i)}} \right\} \tag{4.9}$$

$$= \frac{1}{d_k(i)}, \tag{4.10}$$

which by i) is energy-independent.

v) The decorrelating detector is the worst case optimal linear detector, and achieves the near-far resistance of optimum multiuser detection.

Proof: The proof of Proposition 1 is constructive, hence the first part of v) was obtained as a byproduct in Appendix 1. Here is a shorter proof, using the following fact. Any single linear strategy which is not decorrelating has a near-far resistance of zero. This is shown as follows. The near-far resistance of a linear filter is (cf. (3.18)):

$$\overline{\eta_{k,i}(\boldsymbol{v})} = \max{}^2 \left\{ 0, \inf_{\substack{v \in L \\ y_k(i)=1}} \frac{\boldsymbol{v}^T \mathcal{R} \boldsymbol{y}}{\sqrt{\boldsymbol{v}^T \mathcal{R} \boldsymbol{v}}} \right\}. \tag{4.11}$$

Unless $\mathcal{R}\boldsymbol{v} = \boldsymbol{u}^{k,i}$ (note invariance of η to scaling of $\boldsymbol{v}$) the value of the inf-term is $-\infty$. Hence any linear filter which is not decorrelating has a near-far resistance $\bar{\eta} = 0$. This fact together with the nonzero asymptotic efficiency (4.10) of the decorrelating detector establish optimality of the decorrelating detector within the class of linear filters. Therefore the second part of v) results from Proposition 1.

Note that since the asymptotic efficiency of the decorrelating detector is independent of energies (Property iv) it equals the near-far resistance. This gives us an explicit solution for the Hilbert space optimization problem we obtained for the near-far resistance of optimal multiuser detection in (3.7), namely,

$$\overline{\eta_{k,i}} = \eta^d_{k,i} = \frac{1}{d_k(i)} \tag{4.12}$$

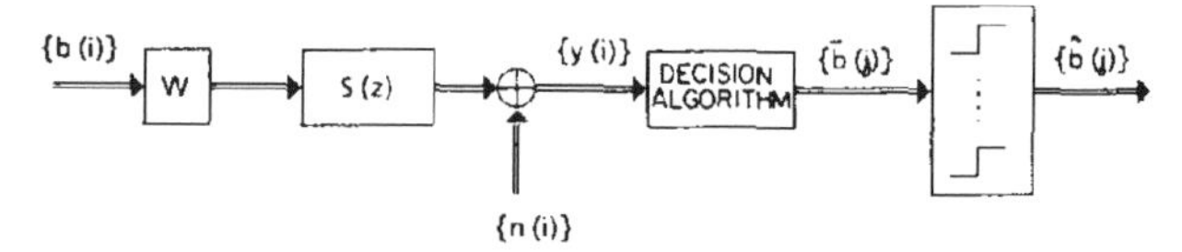

Fig. 5. Equivalent communication system.

and outlines an alternative proof for Proposition 1: we could have explicitly solved the above optimization problem by proceeding along the same lines as in Appendix 1, postulated the decorrelating detector by reasoning as in Fact under v), and shown that the asymptotic efficiency of the decorrelating detector and the near-far resistance of optimal multiuser detection are equal (see [7]). However, the game theoretic proof provides more insight into the nature of the solution.

Property iii) is of special importance. By this property the decorrelating detector does not become multiple-access limited, no matter how strong the multiple-access interference is. Also the decorrelating detector demodulates the data perfectly in the absence of noise, as can be seen from (4.7).

Characterization: We would now like to find an explicit expression for the decorrelating detector which we have up to now defined implicitly. It follows immediately from (4.1) and the uniqueness of the inverse of an invertible matrix that the decorrelating detector for the ith bit of user k is the (k, i)th row of the inverse of $\mathcal{R}$.

From the above and (4.10) the asymptotic efficiency of the decorrelating detector for the ith bit of user k is given by the (k, i)th diagonal element of the inverse of $\mathcal{R}$:

$$\eta^d_{k,i} = \frac{1}{\mathcal{R}^{-1}_{(k,i),(k,i)}}. \tag{4.13}$$

For the values of N encountered in practical applications, inverting a $NK*NK$ matrix is not possible. This issue is addressed in Section IV-B where we represent the decorrelating detector as a K-input K-output time-varying linear filter, and then show that in the limit as N tends to infinity the filter becomes time-invariant.

B. The Limiting Case $N \to \infty$

Proposition 2: As the length of the transmitted sequence increases ($N \to \infty$) the decorrelating detector approaches the K-input K-output linear time-invariant filter with transfer function

$$\boldsymbol{G}(z) = [\boldsymbol{R}^T(1)z + \boldsymbol{R}(0) + \boldsymbol{R}(1)z^{-1}]^{-1}. \tag{4.14}$$

●

Proof: From (2.14) and (2.13) the matched filter outputs for $l = \{-M, \cdots, M\}$ are

$$\begin{aligned}\boldsymbol{y}(l) = &\boldsymbol{R}^T(1)\boldsymbol{W}(l+1)\boldsymbol{b}(l+1) + \boldsymbol{R}(0)\boldsymbol{W}(l)\boldsymbol{b}(l) \\ &+ \boldsymbol{R}(1)\boldsymbol{W}(l-1)\boldsymbol{b}(l-1) + \boldsymbol{n}(l)\end{aligned} \tag{4.15}$$

where $\boldsymbol{b}(-M-1) = \boldsymbol{b}(M+1) = 0$. Taking z-transforms and letting N go to infinity we have

$$Y(z) = S(z)[WB](z) + N(z) \tag{4.16}$$

where $[WB](z)$ is the z-transform of the sequence $\boldsymbol{wb} = \{[\sqrt{w_1(i)} b_1(i), \cdots, \sqrt{w_K(i)} b_K(i)]\}$, the matrix $\boldsymbol{S}(z)$ is

$$\boldsymbol{S}(z) = \boldsymbol{R}^T(1)z + \boldsymbol{R}(0) + \boldsymbol{R}(1)z^{-1} \tag{4.17}$$

and $Y(z)$, $B(z)$ and $N(z)$ are, respectively, the vector-valued z-transforms of the matched filter output sequence, the transmitted sequence, and the noise sequence at the output of the matched filters. $\boldsymbol{S}(z)$ can be interpreted as the equivalent transfer function of the multiuser communication system between transmitter and decision algorithm, as illustrated in Fig. 5. In this setting the optimal receiver problem is to find the transfer function matrix $\boldsymbol{G}(z)$ of a K-input K-output linear time-invariant filter, at the output of which

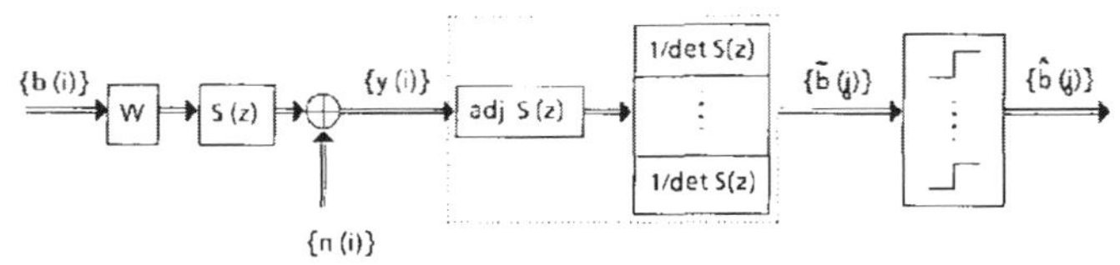

Fig. 6. Interpretation of the decorrelating detector.

a sign-decision yields estimates of the transmitted sequence which are optimal in a certain sense. In our case the optimality criterion is the near–far resistance, and we have demonstrated that the optimal filter is the decorrelating filter, which is the filter that eliminates the multiuser interference, i.e., is the K-input K-output time invariant linear filter which recovers the transmitted data in the absence of noise. Its transfer function is therefore the inverse of the equivalent transfer function $S(z)$:

$$G(z) = [S(z)]^{-1}. \tag{4.18}$$

◇

The effect of the inverse filter $[S(z)]^{-1}$ can be interpreted as illustrated in Fig. 6. The decorrelating filter can be viewed as the cascade of a finite impulse response filter with transfer function *adjoint* $S(z)$, which decorrelates the users, but introduces intersymbol interference among the previously noninterfering symbols of the same user, and of a second filter, consisting of a bank of K identical filters with transfer function $[\det S(z)]^{-1}$, which removes this intersymbol interference. Whereas the region of convergence of the z-transform can always be chosen so as to make $S(z)$ invertible, attention has to be paid to the issue of stability.

Proposition 3: There is a stable, noncausal realization of the decorrelating detector, if and only if the signal cross-correlations are such that

$$\det S(e^{j\omega}) = \det[R^T(1)e^{j\omega} + R(0) + R(1)e^{-j\omega}] \neq 0, \ \forall\, \omega \in [0, 2\pi]. \tag{4.19}$$

●

Proof: As long as $\det S(z)$ has no zeros on the unit circle, a nonempty convergence region of $S^{-1}(z)$ can be chosen which includes the unit circle. Thus, stability can be achieved. But, since $R(0)$ is symmetric,

$$\det S(z) = \det S^T(z) = \det S(z^{-1}).$$

Hence, the stable version of the decorrelating detector will be noncausal. (As a side remark, the matrix $S(e^{j\omega})$ is nonnegative define for all ω, cf. [15]). ◇

Condition (4.19) is equivalent to the limit of the LIA as $N \to \infty$. Both are necessary and sufficient conditions for system invertibility. The LIA requires that the output of a system (the system between the user bit-streams and the matched filter outputs) not be identically zero if the input is nonzero. Hence different inputs generate different outputs, i.e., the system is invertible. For a linear system the requirement that nonzero input produce nonzero output is equivalent to requiring that the transfer matrix be nonsingular on the unit circle. Assume the transfer matrix is singular at the angular frequency ω_0. Necessity follows since otherwise the input sequence consisting of a complex exponential at ω_0 times a vector in the nullspace of the transfer matrix evaluated at ω_0 yields zero output, since the transfer function on the unit circle gives the magnitude and phase of the system response to complex exponentials. On the other hand, sufficiency can be established by using Parseval's relation extended to multivariable systems:

$$\sum_{n=-\infty}^{\infty} \|y_n\|^2 = \frac{1}{2\pi}\int_0^{2\pi} \|Y(e^{j\omega})\|^2\, d\omega$$

$$= \frac{1}{2\pi}\int_0^{2\pi} \|H(e^{j\omega})X(e^{j\omega})\|^2\, d\omega.$$

Hence, for a zero output sequence y_n the vector $H(e^{j\omega})X(e^{j\omega})$ has to vanish for all ω, which implies that $H(e^{j\omega})$ is singular whenever $X(e^{j\omega})$ is nonzero. This establishes the claimed equivalence.

Proposition 4: Condition (4.19) of Proposition 3 is equivalent to

$$\min_{\substack{x^*x=1\\ x\in\mathbb{C}}} \left(x^*R(0)x - \sqrt{(x^*R_+x)^2 + (x^*R_-x)^2}\right) > 0 \tag{4.20}$$

where $R_+ = R^T(1)+R(1)$ and $R_- = j(R^T(1)-R(1))$. The * denotes the complex conjugate. ●

Note that both R_+ and R_- are Hermitian. The proof of Proposition 4 is given in [15], together with the following two results.

— A necessary condition for (4.20) is that the matrices $R(0) + R(1) + R^T(1)$ and $R(0) - R(1) - R^T(1)$ be nonsingular.

— A sufficient condition for (4.20) is that

$$\lambda^2_{\min}(R(0)) > \max\{\lambda^2_{\max}(R_+), \lambda^2_{\min}(R_+)\} + \lambda^2_{\max}(R_-).$$

The following results quantify the asymptotic efficiency achieved by the limiting decorrelating detector.

Proposition 5: Let

$$[S(z)]^{-1} = \sum_{m=-\infty}^{\infty} D(m)z^{-m}. \tag{4.21}$$

Then the asymptotic efficiency of the limiting decorrelating detector for the kth user is given by

$$\eta_k^d = \frac{1}{D_{kk}(0)} \tag{4.22}$$

$$= \left[\frac{1}{2\pi}\int_0^{2\pi} [R^T(1)e^{j\omega} + R(0) + R(1)e^{-j\omega}]^{-1}_{kk}\, d\omega\right]^{-1}. \tag{4.23}$$

●

Proof: From Proposition 2 the z-transform of the decision statistic at the output of the limiting decorrelating detector is given by

$$G(z)Y(z) = [WB](z) + [S(z)]^{-1}N(z) = [WB](z) + N'(z)$$

where $N'(z)$ is the z-transform of the (stationary) filtered Gaussian background noise vector sequence. The z-transform of its covariance matrix sequence $E[n'(\cdot)n'^T(\cdot + i)]$ is equal to $\sigma^2[S(z)]^{-1}$, hence with (4.21) n'_k is a zero-mean Gaussian random variable with variance $\sigma^2 D_{kk}(0)$. Therefore, the probability of error for the kth user equals

$$P_k = P(n'_k > \sqrt{w_k}) = Q\left(\frac{\sqrt{w_k}}{\sigma\sqrt{D_{kk}(0)}}\right). \tag{4.24}$$

From here, using the definition of asymptotic efficiency, the first equality follows. For the second, note that applying the inverse z-transform and definition (4.21), we obtain

$$D_{kk}(0) = \frac{1}{2\pi}\int_0^{2\pi} [S(e^{j\omega})]^{-1}_{kk}\, d\omega,$$

and the result follows using (4.17). ◇

Proposition 6: The asymptotic efficiency of the limiting decorrelating detector for the kth user is strictly positive, and lower bounded by

$$\eta_k^d \geq \left[\max_{\omega\in[0,2\pi]} \left|[R^T(1)e^{j\omega} + R(0) + R(1)e^{-j\omega}]^{-1}_{kk}\right|\right]^{-1} > 0. \tag{4.25}$$

●

Proof: From (4.22), (4.23)

$$D_{kk}(0) \le \max_{\omega \in [0,2\pi]} \left| [R^T(1)e^{j\omega} + R(0) + R(1)e^{-j\omega}]^{-1}_{kk} \right|. \quad (4.26)$$

Hence,

$$\eta_k^d = \frac{1}{D_{kk}(0)} \ge \left[\max_{\omega \in [0,2\pi]} \left| [R^T(1)e^{j\omega} + R(0) + R(1)e^{-j\omega}]^{-1}_{kk} \right| \right]^{-1}$$

$$\ge \frac{\min_\omega |\det[R^T(1)e^{j\omega} + R(0) + R(1)e^{-j\omega}]|}{\max_\omega |\mathrm{adj}_k [R^T(1)e^{j\omega} + R(0) + R(1)e^{-j\omega}]|}, \quad (4.27)$$

which is positive by Proposition 3. ◇

Proposition 7: In the two-user case let $R_{12}(0) = \rho_{12}$ and $R_{12}(1) = \rho_{21}$. Then the asymptotic efficiency of the decorrelating detector for infinite sequence length is given by

$$\eta_1^d = \eta_2^d = \sqrt{(1-\rho_{12}^2-\rho_{21}^2)^2 - 4\rho_{12}^2\rho_{21}^2}$$

$$= \sqrt{[1-(\rho_{12}+\rho_{21})^2][1-(\rho_{12}-\rho_{21})^2]}. \quad (4.28)$$

●

Proof: This formula can be obtained by particularizing Proposition 5 or by minimizing the asymptotic efficiency of optimal multiuser detection in the two-user case with respect to energies. Alternatively, we will prove (4.28) by taking the limit as $N \to \infty$ of the asymptotic efficiency of the decorrelating filter for the central bits in a length N sequence. We will then have proved that in the two-user case the limit of the asymptotic efficiency of the finite-length decorrelating detector as $N \to \infty$ is indeed the asymptotic efficiency of the limiting decorrelating detector.

Recall that the asymptotic efficiency of the decorrelating detector is given by the reciprocal of the corresponding diagonal element of $\mathcal{R}^{-1}$. We need to find explicit expressions for the central diagonal elements of the inverse of the matrix $\mathcal{R}$ as a function of N. We have

$$\mathcal{R} = \begin{pmatrix} 1 & \rho_{12} & 0 & 0 & \\ \rho_{12} & 1 & \rho_{21} & 0 & \\ 0 & \rho_{21} & 1 & \rho_{12} & \ddots \\ 0 & 0 & \rho_{12} & 1 & \ddots \\ & \ddots & \ddots & \ddots & \ddots \end{pmatrix}. \quad (4.29)$$

Denote by Δ_n the determinant of the above $n*n$ matrix. It is easy to see from the structure of $\mathcal{R}$ that Δ_n satisfies the recursion

$$\Delta_n = \Delta_{n-1} - \begin{cases} \rho_{12}^2 \Delta_{n-2}, & n \text{ even} \\ \rho_{21}^2 \Delta_{n-2}, & n \text{ odd.} \end{cases} \quad (4.30)$$

Hence, we can write

$$\begin{bmatrix} \Delta_{2n} \\ \Delta_{2n-1} \end{bmatrix} = \begin{bmatrix} 1-\rho_{12}^2 & -\rho_{21}^2 \\ 1 & -\rho_{21}^2 \end{bmatrix} \begin{bmatrix} \Delta_{2n-2} \\ \Delta_{2n-3} \end{bmatrix}. \quad (4.31)$$

If we consider the sequence of $4n*4n$ matrices for simplicity, the central diagonal element of the inverse of $\mathcal{R}$ is $\Delta_{4n}/(\Delta_{2n-1}\Delta_{2n})$. Hence, after introducing the state vector

$$x_n = \begin{bmatrix} \Delta_{2n} \\ \Delta_{2n-1} \end{bmatrix}, \quad (4.32)$$

we see that finding Δ_{2n}, Δ_{2n-1} requires finding the trajectory of the unforced linear dynamic system

$$x_n = \begin{bmatrix} 1-\rho_{12}^2 & -\rho_{21}^2 \\ 1 & -\rho_{21}^2 \end{bmatrix} x_{n-1}, \; x_1 = \begin{bmatrix} 1-\rho_{12}^2 \\ 1 \end{bmatrix},$$

i.e.,

$$x = \begin{bmatrix} 1-\rho_{12}^2 & -\rho_{21}^2 \\ 1 & -\rho_{21}^2 \end{bmatrix}^n \begin{bmatrix} 1 \\ 0 \end{bmatrix}. \quad (4.33)$$

The eigenvalues of this system are found to be

$$\lambda_{1,2} = \frac{1-\rho_{12}^2-\rho_{21}^2 \pm \sqrt{(1-\rho_{12}^2-\rho_{21}^2)^2 - 4\rho_{12}^2\rho_{21}^2}}{2}.$$

We see $0 < \lambda_1 < \lambda_2 < 1$. After finding the corresponding eigenvectors it follows that:

$$x_n = \begin{bmatrix} \lambda_1+\rho_{12}^2 & \lambda_2+\rho_{21}^2 \\ 1 & 1 \end{bmatrix} \begin{bmatrix} \lambda_1^n & 0 \\ 0 & \lambda_2^n \end{bmatrix} \cdot \begin{bmatrix} 1 & -(\lambda_2+\rho_{21}^2) \\ -1 & \lambda_1+\rho_{21}^2 \end{bmatrix} \begin{bmatrix} 1 \\ 0 \end{bmatrix} \frac{1}{\lambda_1-\lambda_2}$$

$$= \begin{bmatrix} \lambda_1+\rho_{12}^2 & \lambda_2+\rho_{21}^2 \\ 1 & 1 \end{bmatrix} \begin{bmatrix} \lambda_1^n \\ -\lambda_2^n \end{bmatrix} \frac{1}{\lambda_1-\lambda_2}. \quad (4.34)$$

Hence the central diagonal element of the inverse of $\mathcal{R}$ is

$$\frac{\Delta_{4n}}{\Delta_{2n-1}\Delta_{2n}} = \frac{[1 \;\; 0]x_{2n}}{[0 \;\; 1]x_n[1 \;\; 0]x_n}$$

$$= \frac{(\lambda_1/\lambda_2)^{2n}(\lambda_1+\rho_{21}^2) - (\lambda_2+\rho_{21}^2)}{[(\lambda_1/\lambda_2)^n - 1][(\lambda_1/\lambda_2)^n(\lambda_1+\rho_{21}^2) - (\lambda_2+\rho_{21}^2)]} \cdot (\lambda_1-\lambda_2). \quad (4.35)$$

So finally

$$\eta^d = \lim_{n\to\infty} \frac{\Delta_{4n}}{\Delta_{2n-1}\Delta_{2n}} = \lambda_2 - \lambda_1 = \sqrt{(1-\rho_{12}^2-\rho_{21}^2)^2 - 4\rho_{12}^2\rho_{21}^2}.$$

◇

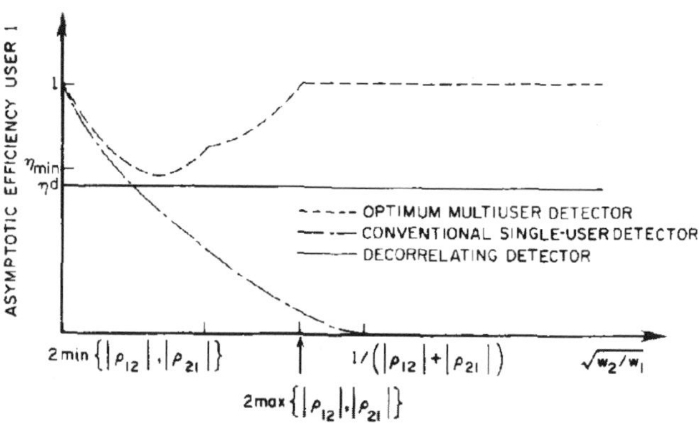

Fig. 7. Asymptotic efficiencies in the two-user case for infinite transmitted sequence length, when the user energies are constant over time (here we chose $|\rho_{12}|$, $|\rho_{21}| = 0.3, 0.5$ which yields $\eta_{\min} = 0.68$, $\eta^d = 0.59$).

Fig. 7 shows the asymptotic efficiency of the decorrelating detector for infinite transmitted sequence length in the two user case. Note its invariance with respect to energies. The discrepancy between η^d and $\eta_{\min}$, defined in (3.4), is due to the fact that $\eta_{\min}$ is higher than the near-far resistance of optimum multiuser detection, since for $\eta_{\min}$ the energies are constrained to be constant over time.

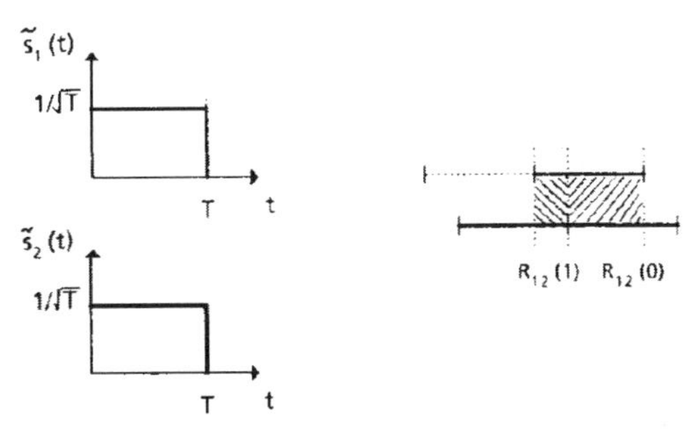

Fig. 8. Signals and crosscorrelations of example (4.42).

The fact that the stable version of the decorrelating filter turns out to be noncausal is not surprising. Due to the lack of synchronism among the users any decision based on less than the entire received waveform is suboptimal. In practice, since the filter is stable, the more remote symbols will count less heavily, and truncation of the noncausal part will be performed after a suitable delay without affecting performance appreciably. For illustration consider the two-user case where we let $R_{12}(0) = \rho_{12}$ and $R_{12}(1) = \rho_{21}$. Then

$$S(z) = \begin{pmatrix} 1 & \rho_{12} + \rho_{21} z^{-1} \\ \rho_{12} + \rho_{21} z & 1 \end{pmatrix}$$

and the transfer function of the decorrelating detector as given by (4.27) is

$$S^{-1}(z) = \frac{1}{1 - \rho_{12}^2 - \rho_{21}^2 - \rho_{12}\rho_{21} z - \rho_{12}\rho_{21} z^{-1}} \cdot \begin{pmatrix} 1 & -(\rho_{12} + \rho_{21} z^{-1}) \\ -(\rho_{12} + \rho_{21} z) & 1 \end{pmatrix}. \quad (4.36)$$

We are interested in the impulse response $f(n)$ of the IIR part of the above filter. Taking the inverse z-transform it is found to be

$$f(n) = Z^{-1}\left[\frac{1}{1 - \rho_{12}^2 - \rho_{21}^2 - \rho_{12}\rho_{21} z - \rho_{12}\rho_{21} z^{-1}}\right] = \frac{\xi^{|n|}}{\eta} \quad (4.37)$$

where $\xi = (1 - \rho_{12}^2 - \rho_{21}^2 - \eta)/(2\rho_{12}\rho_{21})$ and η is the asymptotic efficiency which is given by Proposition 7. It can be checked that $|\xi| \leq 1$, with equality if $|\rho_{12}| + |\rho_{21}| = 1$, which can be shown to coincide with the condition imposed by Proposition 3 for the two-user case. In the latter case the asymptotic efficiency is zero, which follows from Proposition 7. Otherwise, since $|\xi| < 1$ the limiting filter is stable, with symmetric coefficients which decay with rate ξ. In practical applications the filter will be approximated up to any desired precision by truncation of the noncausal part to a finite number of filter coefficients. For illustration the decay rate ξ of the filter coefficients and the achievable asymptotic efficiency η are plotted in Fig. 9 as functions of ρ_{12} and ρ_{21}.

Poor cross-correlation properties among the signature waveforms could imply that the limiting filter $G(z)$ does not exist, although the decorrelating detector exists for finite-length transmitted sequences. We give an example to illustrate this fact. For $K = 2$ it is straightforward to show that the condition of Proposition 3 is satisfied for all signal constellations for which $|R_{12}(0)| = |R_{12}(1)| \neq 1$. This is the case unless the normalized waveforms coincide modulo circular shifts and sign changes.

Consider the trivial signal case where both users are assigned the same rectangular waveform, as shown in Fig. 8. Abbreviate $R_{12}(0)$, which is the crosscorrelation between bits in the same signaling interval, by $r = \tau/T \in [0, 1)$, then in this case $R_{12}(1)$, which is the crosscorrelation between bits in adjacent intervals, is $1 - r$. Then,

$$S(z) = \begin{pmatrix} 1 & r + (1-r)z^{-1} \\ r + (1-r)z & 1 \end{pmatrix} \quad (4.38)$$

becomes singular for $z = 1$, hence there is no stable limiting inverse filter. And if it existed its asymptotic efficiency, as given by (4.28), would be zero. This is not surprising, for an infinite sequence of transmitted bits where both users use the same waveform. However, for finite length sequences advantage can be taken of the marginal effects of having bits which are not affected by either past or future bits. For finite N the decorrelating detector exists unless $r = 0$, i.e., when the transmissions are not synchronous. This is in accord with the multiarrival condition given in Appendix 2, and with the results obtained in the synchronous case [7].

V. Error Probabilities: Numerical Examples

In the sequel, we compare the performances of the conventional and of the decorrelating detector. Without loss of generality we consider the error probability of user 1 in a channel shared by several active users. The conventional detector decides for the sign of the kth component of the matched filter output vector, given by (2.14). Therefore its average error probability over the bit sequences of the interfering users equals

$$\frac{1}{2^2(K-1)} \sum_{b_j(0),\, b_j(-1),\, j\neq 1} Q\left(\frac{\sqrt{w_1} - \sum_{j=2}^{K} [R_{1j}(0)b_j(0) + R_{1j}(1)b_j(-1)]\sqrt{w_j}}{\sigma}\right), \quad (5.1)$$

whereas its worst case error probability over the interfering bit sequences equals

$$Q\left(\frac{\sqrt{w_1} - \sum_{j=2}^{K} [|R_{1j}(0)| + |R_{1j}(1)|]\sqrt{w_j}}{\sigma}\right). \quad (5.2)$$

The probability of error of the decorrelating detector equals, from (4.24),

$$Q\left(\frac{\sqrt{w_1}}{\sigma\sqrt{D_{11}(0)}}\right),$$

with (the equivalence with (4.23) is easy to show, cf. [15])

$$D_{11}(0) = \frac{1}{\pi}\int_0^{\pi} [R(1)^T e^{j\omega} + R(0) + R(1)e^{-j\omega}]_{11}^{-1}\, d\omega. \quad (5.3)$$

The delays enter the above formulas implicitly via the crosscorrelation matrices, which are functions thereof and of the chosen signature sequences. In the following examples, we have chosen a set of spread-spectrum m-sequences of length 31.

In Fig. 10 we use, for comparison purposes to previous works ([14], [1]), the set of 3 sequences reported in [12, Table V] to be optimal with respect to a signal-to-multiple-access interference parameter when the conventional detector is used. We consider a baseband environment with $K - 1$ active equal energy interferers, whose delay relative to each other is fixed. Fig. 10, for $K = 3$, shows the 1st user error probability of the conventional receiver versus SNR_1, the signal-to-background-noise ratio of user 1, for different values

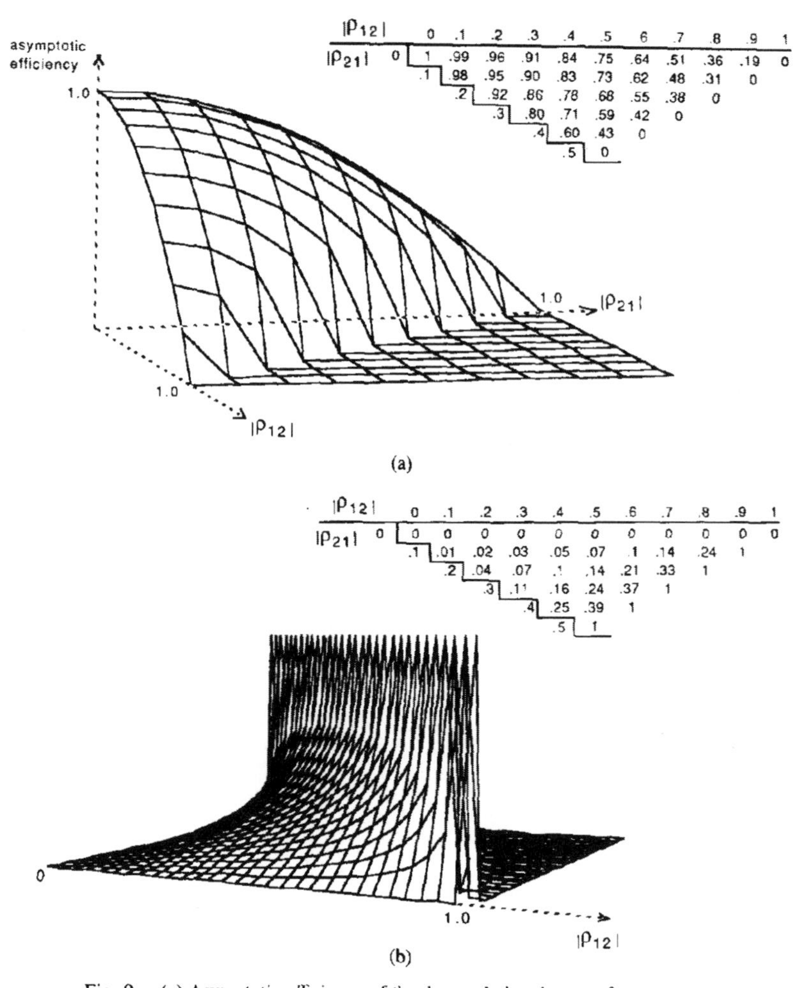

$\|\rho_{21}\|$ \ $\|\rho_{12}\|$	0	.1	.2	.3	.4	.5	6	.7	.8	.9	1
0	1	.99	.96	.91	.84	.75	.64	.51	.36	.19	0
.1		.98	.95	.90	.83	.73	.62	.48	.31	0	
.2			.92	.86	.78	.68	.55	.38	0		
.3				.80	.71	.59	.42	0			
.4					.60	.43	0				
.5						0					

$\|\rho_{21}\|$ \ $\|\rho_{12}\|$	0	.1	.2	.3	.4	.5	.6	.7	.8	.9	1
0	0	0	0	0	0	0	0	0	0	0	0
.1		.01	.02	.03	.05	.07	.1	.14	.24	1	
.2			.04	.07	.1	.14	.21	.33	1		
.3				.11	.16	.24	.37	1			
.4					.25	.39	1				
.5						1					

Fig. 9. (a) Asymptotic efficiency of the decorrelating detector for two users as a function of the partial crosscorrelations of their signature waveforms. (b) Decay rates of the coefficients of the IIR part of the decorrelating detector for two users, symmetric in ρ_{12} and ρ_{21}.

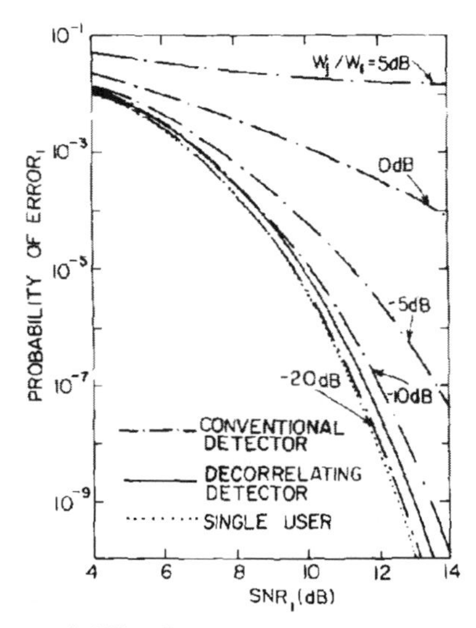

Fig. 10. Error probability of user 1 with 2 active equal energy interferers, each of energy w_j, averaged over the interfering bit sequences and over the delay of user 1, for the decorrelating and conventional receiver versus the SNR of user 1, for m-sequences of length 31 and different interference levels.

of the energy ratio $\mathrm{SNR}_j/\mathrm{SNR}_1$, averaged over the bit sequences of the two interferers and over the delay of user 1. Also shown are the 1st user error probability of the decorrelating detector and the error probability of the single user channel. From Fig. 10 we see the strong dependence of the performance of the conventional receiver on the relative energies of the active users. While the error probability of the decorrelating detector is invariant to the energy of interfering users, the performance of the conventional receiver deteriorates rapidly for increasing interference, till for an energy ratio above 5 dB the conventional receiver becomes practically multiple-access limited. (For a sufficiently high level of nonorthogonal interference the error probability of the conventional receiver can be seen to become irreducible. E.g., in the two-user synchronous case, for $\sqrt{w_2}/\sqrt{w_1} = (1+\Delta)/\rho$ where ρ is the normalized crosscorrelation coefficient between the two signature signals and $\Delta \geq 0$, the error probability of the conventional receiver tends to 1/4 if $\Delta = 0$ and to 1/2 if $\Delta > 0$ for increasing SNR of user 1). Note that if the energies of all the users are equal the decorrelating detector is around two orders of magnitude better than the conventional receiver at 10 dB. Only if the multiple-access interference level plays a subordinate role compared to the background noise does the conventional detector outperform the decorrelating detector, which pays a penalty for combatting the interference instead of ignoring it. Similar results were obtained regardless of the actual value of the relative delay between the two interfering users.

Fig. 11 shows the same setting as above, in the case $K = 6$. We have used the set of autooptimal m-sequences of length 31 found in

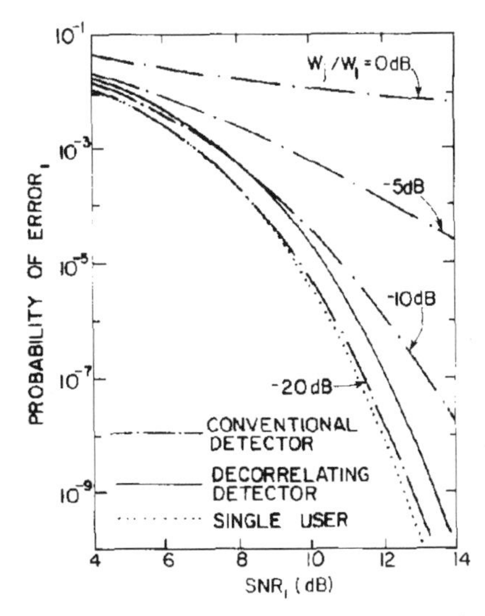

Fig. 11. Same as Fig. 10, with 5 active equal energy interferers.

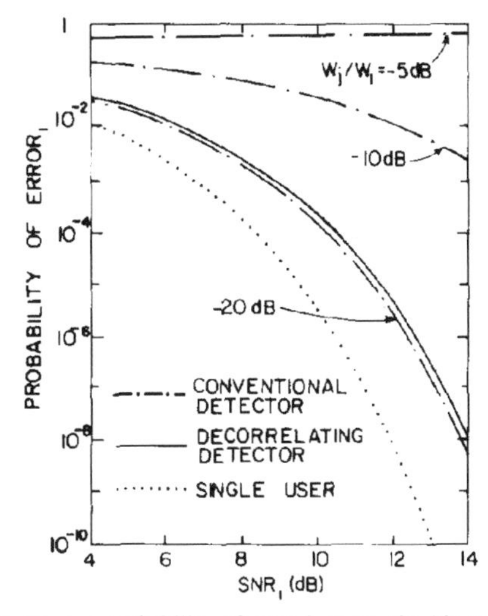

Fig. 12. Worst case probability of user 1 w.r.t. the bit sequences of the interfering users, with 9 equal energy interferers.

[13, Fig. A.1] to be optimal with respect to certain peak and mean-square correlation parameters which play an important role in the error probability analysis of the conventional detector. Comparing Fig. 11 to Fig. 10 we see the same qualitative error probability relation between the two detectors, and again the strong near-far limitation of the conventional receiver. Since there are more active interferers the performance advantage of the decorrelating detector in a near-far environment is even more pronounced: if the energies of all the users are equal the decorrelating detector is almost three orders of magnitude better than the conventional receiver at 10 dB.

Finally, Fig. 12 shows the worst case probability of the conventional detector over the sequences of interfering users, as given by (5.2), for $K = 10$. The signature sequence set used for $K = 6$ has been expanded—without trying to optimize, as before, with respect to the performance of the conventional detector. The shown error probabilities are typical, varying very little if different sets of delays are used because of the good crosscorrelation properties of m-sequences.

Overall the generated error probability curves show the pronounced superiority of the decorrelating receiver in a near-far environment, and whenever sufficiently many users are active even if their energies are well below the energy of the desired user. Note, finally, that the selected signature sequences were optimal with respect to the performance of the conventional receiver. It would be interesting to investigate the possible performance gain of using the decorrelating detector in conjunction with a set of signature sequences optimized for its use.

VI. Conclusions

In this paper, we have obtained a linear multiuser detector, the decorrelating detector, for demodulation of asynchronous code-division multiplexed signals in white Gaussian channels. The bit-error-rate of this detector is independent of the energy of the interfering users and exhibits the same degree of near-far resistance as the optimum multiuser detector obtained in [1]. Since the decorrelating detector does not require knowledge of the received energies and its complexity is only linear in the number of users, it emerges as the solution of choice in near-far environments with a large number of users.

In applications where each receiver is interested in demodulating the information transmitted by only one user, it is easy to decentralize the K-user decorrelating receiver since it can be implemented as K separate (continuous-time) single-input (discrete-time) single-output filters. Each of those filters can be viewed as a modification of the conventional single-user matched filter where instead of correlating the channel output with the signature waveform of the user of interest, we use its projection on the subspace orthogonal to the space spanned by the interfering signals.

Note, finally, that if the filter is actually an approximation to the decorrelating receiver, due to, for example, finite accuracy in the computation of the crosscorrelations or truncation of the impulse response (Section IV-B), it will no longer be orthogonal to the subspace of the interfering signals and therefore it will not be near-far resistant in the worst case sense adopted in this paper. However, the effect on the bit-error-rate will be arbitrarily small with a good enough approximation to the decorrelating receiver, and therefore the bit-error-rate will be very insensitive to the energy level of the interferers. Hence, the resistance to the near-far problem can be preserved within any desired energy range.

Appendix 1

Saddle-Point Property in (3.19)

Though the penalty function of (3.18) looks similar to the signal-to-noise ratio functional encountered in the robust matched filtering problem [5], $|\langle h, s\rangle|^2/\langle h, \sum h\rangle$, the problem is different here because the numerator can be negative. Thus we have to establish the result "from scratch." In order to show that $M(v, y)$ has a saddle point, i.e., satisfies (3.19), we show that it satisfies the requirements of the following theorem.

Theorem [4, Thm. 2.1]: Suppose Q is a convex set and $M(v, \cdot)$ is convex on Q for every $v \in H$. Then if (v_L, y_L) is a *regular pair*[3] for (H, Q, M), the following are equivalent:

a) $y_L \in \arg\min_{y \in Q} \sup_{v \in H} M(v, y)$,

b) (v_L, y_L) is a saddle point solution for (H, Q, M).

This theorem establishes that if we exhibit a regular pair whose second argument satisfies a), the game (H, Q, M) has a saddle point, which means that the sequence of max and min in (3.18) can be interchanged. In the following, we find a suitable regular pair, thereby proving (3.19).

Clearly the convexity conditions are satisfied (the set of detectors is not required to be topologized). We need to find a candidate regular pair. Note that the value of inf term in (3.18) is $-\infty$ (which gives a near-far resistance of zero) unless v is picked such that $\Re v = u^{k,i}$ (η is invariant with respect to scaling of v). $u^{k,i}$ is the (k, i)th unit vector in the Hilbert space L, defined as $u_j^{k,i}(l) = \delta_{kj}\delta_{li}$. This gives us a candidate for an optimal detector v_L: d, with $\Re d = u^{k,i}$. Existence of such a vector is shown in (4.6) to follow from the LIA of (2.3).

[3] $(v_L, y_L) \in H \times Q$ is a regular pair for (H, Q, M) if, for every $y \in Q$ such that $y_\alpha = (1-\alpha)y_L + \alpha y \in Q$ for $\alpha \in [0, 1]$, we have

$$\sup_{v \in H} M(v, y_\alpha) - M(v_L, y_\alpha) = o(\alpha).$$

(If this detector is indeed optimal, which follows if the candidate pair is regular and satisfies a), and coincides with v^*).

Next we find a y_L which meets the requirement of point a) of the theorem. Using the Cauchy–Schwarz inequality, we find that

$$\sup_{v \in H} M(v, y) = \sup_{\substack{v \in L \\ v^T \mathcal{R} v \neq 0}} \frac{y^T \mathcal{R} v}{\sqrt{v^T \mathcal{R} v}} = \sqrt{y^T \mathcal{R} y} \tag{A.1}$$

where the inner product is maximized for $v = ky + \{x \in L: \mathcal{R}x = 0\}$.

We now need to solve the Hilbert space optimization problem

$$\inf y^T \mathcal{R} y \tag{A.2}$$

$$\text{subject to } y_k(i) = 1.$$

Using (2.21) and the definition of d we can rewrite the minimization problem under consideration as

$$\inf \|y\|_R \tag{A.3}$$

$$\text{subject to } \langle d, y \rangle_R = 1.$$

$\|\cdot\|_R$ is a norm since $\mathcal{R}$ is positive definite. We have obtained a minimum-norm optimization problem in Hilbert space. To prove existence of a solution we need to show that constraint set to be closed, which holds since the Hilbert space is finite dimensional. (Even for $N \to \infty$, when we have an infinite dimensional optimization problem, we could use the fact that the codimension is finite. The problem there is that the signals are no longer square integrable.) The constraint, $y_k(i) = 1$, is equivalent to $y = u^{k,i} + \{x: \langle x, d \rangle_R = 0\}$. $\mathcal{A} = [d]$, the subspace generated by d, is a closed subspace of dimension 1. Hence the constraint set $\{x: \langle x, d \rangle_R = 0\} = \mathcal{A}^\perp$ is closed. We now have a minimum-norm optimization problem in Hilbert space over a closed subspace. Hence, the Projection Theorem, [6], guarantees existence [so we can replace the inf by a min, as required in a)] and uniqueness of a minimizing equivalence class y^*, with

$$y^* \in \{\mathcal{A}^\perp + u^{k,i}\} \cap \mathcal{A}^{\perp\perp} = \{\mathcal{A}^\perp + u^{k,i}\} \cap \mathcal{A} \tag{A.4}$$

where equality holds since A is closed. Hence $y_k^*(i) = 1$ and $y^* = kd$, which implies

$$y^* = \frac{1}{d_k(i)} d. \tag{A.5}$$

We now have a candidate regular pair which satisfies a): $(v_L, y_L) = (d, (d_k(i))^{-1} d)$. From (A.1) and the definition of regularity we have to check the dependence on α of

$$\begin{aligned}\sqrt{y_\alpha^T \mathcal{R} y_\alpha} - \frac{y_\alpha^T \mathcal{R} v_L}{\sqrt{v_L^T \mathcal{R} v_L}} &= \sqrt{d^T \mathcal{R} d + 2\alpha (y - d)^T \mathcal{R} d + \alpha^2 (y - d)^T \mathcal{R} (y - d)} \\ &\quad - \sqrt{\frac{1}{d_k(i)}} \\ &= \sqrt{\frac{1}{d_k(i)} + \alpha^2 (y - d)^T \mathcal{R} (y - d)} - \sqrt{\frac{1}{d_k(i)}}.\end{aligned} \tag{A.6}$$

We have repeatedly used the decorrelating property of d. Since $\sqrt{1 + x} \le 1 + 1/2x$, the above quantity lies in the interval $[0, (y - d)^T \mathcal{R} (y - d) \sqrt{d_k(i)}/2 \alpha^2]$, hence divided by α goes to 0 when $\alpha \downarrow 0$. Thus $(d, (d_k(i))^{-1} d)$ is a regular pair which satisfies point a) of the theorem. Hence it follows from the theorem that the penalty function $y^T \mathcal{R} v / \sqrt{v^T \mathcal{R} v}$ has a saddle point, i.e.,

$$\sup_{\substack{v \in L \\ v^T \mathcal{R} v \neq 1}} \inf_{\substack{y \in L \\ y_k(i) = 1}} \frac{y^T \mathcal{R} v}{\sqrt{v^T \mathcal{R} v}} = \inf_{\substack{y \in L \\ y_k(i) = 1}} \inf_{\substack{y \in L \\ v^T \mathcal{R} v \neq 1}} \frac{y^T \mathcal{R} v}{\sqrt{v^T \mathcal{R} v}}. \tag{A.7}$$

◇

Appendix 2

Sufficient Conditions for Linear Independence

Suppose that, for a fixed signal set,

i) $\{\tau_1, \cdots, \tau_K\}$ are continuous random variables,

ii) $\{\tau_1, \cdots, \tau_K\}$ are independent random variables,

iii) $w_k(i) \neq 0$.

Then almost surely there is no $v \in L$, $v_k(i) \neq 0$ such that $\tilde{S}(t, v) = 0$.

Proof: Define the times of effective arrival and departure of the ith signal of the kth user [1], as

$$\lambda_{i,k}^a = \tau_k + iT + \sup \left\{ \tau \in [0, T), \int_0^\tau s_k^2(t)\, dt = 0 \right\} \tag{A.8}$$

and

$$\lambda_{i,k}^d = \tau_k + iT + \inf \left\{ \tau \in (0, T], \int_\tau^T s_k^2(t)\, dt = 0 \right\}, \tag{A.9}$$

respectively.

Since $v_k(i) \neq 0$ there is a first and a last symbol that differs from zero. It is readily apparent that in order to have $\tilde{S}(t, v) = 0$, the effective arrival of the first (and the effective departure of the last) symbol that differs from zero must be a point of effective multiarrival (respectively multideparture). Note that this property does not depend on the particular v chosen, but only on the set of delays. From (A.8), (A.9), the effective times of arrival and departure inherit from the delays the properties of being continuously valued and mutually independent. Therefore, the result follows, since the set of delays $\{\tau_1, \cdots, \tau_K\}$ for which multiarrival points result has measure zero. ◇

References

[1] S. Verdú, "Minimum probability of error for asynchronous Gaussian multiple-access channels," *IEEE Trans. Inform. Theory*, vol. IT-32, pp. 85–96, Jan. 1986.

[2] ——, "Optimum multi-user asymptotic efficiency," *IEEE Trans. Commun.*, vol. COM-34, pp. 890–897, Sept. 1986.

[3] ——, "Computational complexity of optimum multi-user detection," *Algorithmica*, vol. 4, pp. 303–312, 1989.

[4] S. Verdú and H. V. Poor, "On minimax robustness: A general approach and applications," *IEEE Trans. Information Theory*, vol. IT-30, pp. 328–340, Mar. 1984.

[5] ——, "Minimax robust discrete-time matched filters," *IEEE Trans. Commun.*, vol. COM-31, pp. 208–215, Feb. 1983.

[6] D. Luenberger, *Optimization by Vector Space Methods*. New York: Wiley, 1969.

[7] R. Lupas and S. Verdú, "Linear multiuser detectors for synchronous code-division multiple-access channels," *IEEE Trans. Inform. Theory*, vol. IT-34 1988.

[8] S. Verdú, "Minimum probability of error for asynchronous multiple access communication systems," in *Proc. 1983 IEEE Military Commun. Conf.*, Washington, DC, Nov. 1983, pp. 213–219.

[9] H. V. Poor and S. Verdú, "Single-user detectors for multiuser channels," *IEEE Trans. Commun.*, vol. COM-36, pp. 50–60, Jan. 1988.

[10] M. K. Varanasi and B. Aazhang, "Multistage detection in asynchronous code-division multiple access communications," *IEEE Trans. Commun.*, see this issue, pp. 509–519.

[11] C. Rushforth and Z. Xie, "Multiuser signal detection using sequential decoding," *1988 IEEE Military Commun. Conf.*, San Diego, CA, Oct. 1988.

[12] F. D. Garber and M. B. Pursley, "Optimal phases of maximal-length sequences for asynchronous spread-spectrum multiplexing," *Electron. Lett.*, vol. 16, pp. 756–757, Sept. 1980.

[13] M. B. Pursley and H. F. A. Roefs, "Numerical evaluation of correlation parameters for optimal phases of binary shift-register sequences," *IEEE Trans. Commun.*, vol. COM-27, pp. 1597–1604, Oct. 1979.

[14] E. A. Geraniotis and M. B. Pursley, "Error probability for direct-sequence spread-spectrum multiple-access communications—Part II: Approximations," *IEEE Trans. Commun.*, vol. COM-30, pp. 985–995, May 1982.
[15] R. Lupas, "Near-far resistant linear multiuser detection," Ph.D. dissertation, Princeton Univ., Jan. 1989.

PHOTO NOT AVAILABLE

Ruxandra Lupas (S'86) was born in Bucharest, Romania, on March 13, 1962. She received the Vordiplom degree in electrical engineering from the Technical University of Munich, West Germany, in 1983, and the Ph.D. degree in electrical engineering from Princeton University in 1989.

She was a Research Associate at DFVLR (German Aerospace Research Establishment), Munich, in the Summer of 1987 and she is currently a member of the Technical Staff at Siemens, Munich, Germany. Her research interests are in communication systems and artificial intelligence.

Ms. Lupas is a member of the German National Scholarship Foundation for having been a winner of the German High-School Competition in Mathematics in 1979, 1980, and 1981. She received an IBM Graduate Fellowship in 1986.

PHOTO NOT AVAILABLE

Sergio Verdú (S'80–M'84–SM'88) received the Ph.D. degree in electrical engineering from the University of Illinois at Urbana-Champaign in 1984.

Upon completion of his doctorate he joined the Faculty of Princeton University, Princeton, NJ, where he is an Associate Professor of Electrical Engineering. His current research interests are in the areas of multiuser communication, information theory, and statistical signal processing.

Dr. Verdú is a recipient of the National University Prize of Spain, the Rheinstein Outstanding Junior Faculty Award of the School of Engineering and Applied Science in Princeton University, and the NSF Presidential Young Investigator Award. He is currently serving as Associate Editor of the IEEE TRANSACTIONS ON AUTOMATIC CONTROL and is a member of the Board of Governors of the IEEE Information Theory Society.

Multiple-Symbol Differential Detection of MPSK

DARIUSH DIVSALAR, MEMBER, IEEE, AND MARVIN K. SIMON, FELLOW, IEEE

Abstract—**A differential detection technique for MPSK, which uses multiple-symbol observation interval, is presented and its performance analyzed and simulated. The technique makes use of maximum-likelihood sequence estimation of the transmitted phases rather than symbol-by-symbol detection as in conventional differential detection. As such the performance of this multiple-symbol detection scheme fills the gap between conventional (two-symbol observation) differentially coherent detection of MPSK and ideal coherent of MPSK with differential encoding. The amount of improvement gained over conventional differential detection depends on the number of phases M and the number of additional symbol intervals added to the observation. What is particularly interesting is that substantial performance improvement can be obtained for only one or two additional symbol intervals of observation. The analysis and simulation results presented are for uncoded MPSK.**

I. Introduction

It is well known that, in applications where simplicity and robustness of implementation take precedence over achieving the best system performance, differential detection is an attractive alternative to coherent detection. Aside from implementation considerations, it is also possible that the transmission environment may be sufficiently degraded, e.g., a multipath fading channel, that acquiring and tracking a coherent demodulation reference signal is difficult if not impossible. Here again, differential detection is a possible, and perhaps the only, solution.

In the past, differential detection of multiple-phase shift keying (MPSK) has been accomplished by comparing the received phase in a given symbol interval to that in the previous symbol interval and making a multilevel decision on the difference between these two phases [1]. An implementation of such a receiver and the analysis of its error rate performance on an additive white Gaussian noise (AWGN) channel may also be found in [2, ch. 5]. In arriving at the results in [1], [2], the assumption was made that the received carrier reference phase is constant over at least two symbol intervals and thus has no effect on the decision process when the above-mentioned phase difference is taken. This assumption is crucial to the analysis but is also realistic in many practical applications. Also, since the information is carried in the difference between adjacent received phases, the input information must be differentially encoded before transmission over the channel.

Although differential detection eliminates the need for carrier acquisition and tracking in the receiver, it suffers from a performance penalty (additional required SNR at a given bit error rate) when compared to ideal (perfect carrier phase reference) coherent detection. The amount of this performance penalty increases with the number of phases M and is significant for $M \geq 4$. For example, at a bit error probability $P_b = 10^{-5}$, differentially detected BPSK (often abbreviated as DPSK) requires about 0.75 dB more bit energy-to-noise ratio (E_b/N_0) than coherently detected BPSK. For QPSK ($M = 4$), the difference in E_b/N_0 between differential detection and ideal coherent detection at $P_b = 10^{-5}$ is about 2.2 dB. Finally for 8PSK, the corresponding difference in E_b/N_0 performance between the two is greater than 2.5 dB.

Thus, it is natural to ask: is there a way of enhancing the conventional (two symbol observation) differential detection technique so as to recover a portion of the performance lost relative to that of coherent detection and yet, still maintain a simple and robust implementation? Furthermore, if this is possible, what is the tradeoff between the amount of performance recovered and the additional complexity added to the conventional differential detection implementation. The answers to these questions stem from the idea of allowing the observation interval over which symbol decisions are made to be longer than two symbol intervals while at the same time making a *joint* decision on several symbols simultaneously as opposed to symbol-by-symbol detection. As such one must extend the previous assumption on the duration of time over which the carrier phase is constant to be commensurate with the extended observation interval. For observations on the order of three or four symbol intervals, this is still a reasonable assumption in many applications.

The theoretical framework in which we shall develop this so-called *multiple-bit differential detection* technique is the maximum-likelihood approach to statistical detection. In the next section, we derive the appropriate maximum-likelihood algorithm for differential detection of MPSK and show how the conventional technique is a special case of this more general model. Since, as mentioned above, we will be making joint symbol decisions in this new configuration, the technique is a form of *maximum-likelihood sequence estimation* although no coding of the input information is implied. We hasten to add, however, that the theory developed here for uncoded M-ary DPSK (MDPSK) is easily extended to include, for example, trellis-coded MDPSK. This subject will be reported on by the authors in a forthcoming paper.

II. Maximum-Likelihood Detection of MPSK over an AWGN Channel

Consider the transmission of MPSK signals over an AWGN channel. The transmitted signal in the interval $kT \leq t \leq (k+1)T$ has the complex form

$$s_k = \sqrt{2P}e^{j\phi_k} \tag{1}$$

where P denotes the constant signal power, T denotes the MPSK symbol interval, and ϕ_k the transmitted phase which takes on one of M uniformly distributed values $\beta_m = 2\pi m/M$; $m = 0, 1, \cdots, M-1$ around the unit circle. The corresponding received signal is then

$$r_k = s_k e^{j\theta_k} + n_k \tag{2}$$

where n_k is a sample of zero-mean complex Gaussian noise with variance

$$\sigma_n^2 = \frac{2N_0}{T} \tag{3}$$

and θ_k is an arbitrary phase introduced by the channel which, in the absence of any side information, is assumed to be uniformly distributed in the interval $(-\pi, \pi)$.

Consider now a received sequence of length N and assume that θ_k is independent of k over the length of this sequence, i.e., $\theta_k = \theta$. Analogous to (2), the received sequence r is expressed as

$$r = se^{j\theta} + n \tag{4}$$

Paper approved by the Editor for Modulation Theory and Nonlinear Channels of the IEEE Communications Society. Manuscript received March 10, 1989; revised May 31, 1989. This work was performed at the Jet Propulsion Laboratory, California Institute of Technology under a contract from NASA.

The authors are with the Jet Propulsion Laboratory, California Institute of Technology, Pasadena, CA 91109.

IEEE Log Number 8933656.

Reprinted from *IEEE Transactions on Communications*, vol. 38, no. 3, March 1990.

where r_k, s_k, and n_k are, respectively, the kth components of the N-length sequences $\boldsymbol{r}$, $\boldsymbol{s}$, and $\boldsymbol{n}$. For the assumed AWGN model, the *a posteriori* probability of $\boldsymbol{r}$ given $\boldsymbol{s}$ and θ is

$$p(\boldsymbol{r}|\boldsymbol{s}, \theta) = \frac{1}{(2\pi\sigma_n^2)^N} \exp\left\{-\frac{\|\boldsymbol{r} - \boldsymbol{s}e^{j\theta}\|^2}{2\sigma_n^2}\right\} \quad (5)$$

where

$$\|\boldsymbol{r} - \boldsymbol{s}e^{j\theta}\|^2 = \sum_{i=0}^{N-1} |r_{k-i} - s_{k-i}e^{j\theta}|^2. \quad (6)$$

Simplifying the right-hand side of (6) results in

$$\begin{aligned}\|\boldsymbol{r} - \boldsymbol{s}e^{j\theta}\|^2 &= \sum_{i=0}^{N-1} [|r_{k-i}|^2 + |s_{k-i}|^2] \\ &\quad - 2\,\mathrm{Re}\left\{\sum_{i=0}^{N-1} r_{k-i}s_{k-i}^*\right\}\cos\theta \\ &\quad - 2\,\mathrm{Im}\left\{\sum_{i=0}^{N-1} r_{k-i}s_{k-i}^*\right\}\sin\theta \\ &= \sum_{i=0}^{N-1}[|r_{k-i}|^2 + |s_{k-i}|^2] \\ &\quad - 2\left|\sum_{i=0}^{N-1} r_{k-i}s_{k-i}^*\right|\cos(\theta - \alpha)\end{aligned} \quad (7)$$

where

$$\alpha = \tan^{-1}\frac{\mathrm{Im}\left\{\sum_{i=0}^{N-1} j r_{k-i}s_{k-i}^*\right\}}{\mathrm{Re}\left\{\sum_{i=0}^{N-1} r_{k-i}s_{k-i}^*\right\}}. \quad (8)$$

Since θ has been assumed to be uniformly distributed, then the *a posteriori* probability of $\boldsymbol{r}$ given $\boldsymbol{s}$ is simply

$$\begin{aligned}p(\boldsymbol{r}|\boldsymbol{s}) &= \int_{-\pi}^{\pi} p(\boldsymbol{r}|\boldsymbol{s}, \theta)p(\theta)\,d\theta \\ &= \frac{1}{(2\pi\sigma_n^2)^N}\exp\left\{-\frac{1}{2\sigma_n^2}\sum_{i=0}^{N-1}[|r_{k-i}|^2 + |s_{k-i}|^2]\right\} \\ &\quad \times I_0\left(\frac{1}{\sigma_n^2}\left|\sum_{i=0}^{N-1} r_{k-i}s_{k-i}^*\right|\right)\end{aligned} \quad (9)$$

where $I_0(x)$ is the zeroth order modified Bessel function of the first kind. Note that for MPSK, $|s_k|^2$ is constant for all phases. Thus, since $I_0(x)$ is a monotonically increasing function of its argument, maximizing $p(\boldsymbol{r}|\boldsymbol{s})$ over $\boldsymbol{s}$ is equivalent to finding

$$\max_i \left|\sum_{i=0}^{N-1} r_{k-i}s_{k-i}^*\right|^2 \quad (10)$$

which, using (1), results in the decision rule

$$\text{choose } \hat{\boldsymbol{\phi}} \text{ if } \left|\sum_{i=0}^{N-1} r_{k-i}e^{-j\hat{\phi}_{k-i}}\right|^2 \text{ is maximum} \quad (11)$$

where $\hat{\boldsymbol{\phi}}$ is a particular sequence of the β_m's. Note that this decision rule has a phase ambiguity associated with it since the addition of an arbitrary fixed phase, say ϕ_a, to all N estimated phases $\hat{\phi}_k, \hat{\phi}_{k-1}, \cdots, \hat{\phi}_{k-N+1}$ results in the same decision for $\hat{\boldsymbol{\phi}}$. Thus, letting $\phi_a = \hat{\phi}_{k-N+1}$, the above decision rule can be alternately expressed as choosing the sequence $\hat{\boldsymbol{\phi}}$ that maximizes the statistic

$$\eta \triangleq \left|\sum_{i=0}^{N-1} r_{k-i}e^{-j(\hat{\phi}_{k-i} - \hat{\phi}_{k-N+1})}\right|^2. \quad (12)$$

To resolve the above phase ambiguity, one should differentially encode the phase information at the transmitter. Letting

$$\phi_k = \phi_{k-1} + \Delta\phi_k \quad (13)$$

where now $\Delta\phi_k$ denotes the input data phase corresponding to the kth transmission interval and ϕ_k the differentially encoded version of it, then

$$\phi_{k-i} - \phi_{k-N+1} = \sum_{m=0}^{N-i-2} \Delta\phi_{k-i-m} \quad (14)$$

and the above decision statistic becomes

$$\eta = \left|r_{k-N+1} + \sum_{i=0}^{N-2} r_{k-i}e^{-j\sum_{m=0}^{N-i-2}\Delta\hat{\phi}_{k-i-m}}\right|^2. \quad (15)$$

This statistic implies that we observe the received signal over N symbol time intervals and from this observation make a simultaneous decision on $N - 1$ data phases.

Some special cases of (15) are of interest. For $N = 1$, i.e., an observation of the received signal over one symbol interval, (15) simplifies to

$$\eta = |r_k|^2 \quad (16)$$

which is completely independent of the input data phases and thus cannot be used for making decisions on differentially encoded MPSK modulation. In fact, the statistic of (16) corresponds to the classical case of noncoherent detection which is not applicable to phase modulation.

Next, let $N = 2$, in which case (15) becomes

$$\eta = |r_{k-1} + r_k e^{-j\Delta\hat{\phi}_k}|^2 = |r_{k-1}|^2 + |r_k|^2 + 2\,\mathrm{Re}\{r_k r_{k-1}^* e^{-j\Delta\hat{\phi}_k}\}. \quad (17)$$

This results in the well-known decision rule for conventional MDPSK, namely,

$$\text{choose } \Delta\hat{\phi}_k \text{ if } \mathrm{Re}\{r_k r_{k-1}^* e^{-j\Delta\hat{\phi}_k}\} \text{ is maximum} \quad (18)$$

which is implemented in complex form as in Fig. 1. Thus, we see from this approach that conventional differential detection of MPSK is the optimum receiver in the sense of minimizing the symbol error probability given that the unknown carrier phase is constant over two symbol times. This result is not new other than, perhaps, the approach taken to demonstrate it.

Now, to see a new structure, we consider (15) for $N = 3$. Here we have

$$\begin{aligned}\eta &= |r_{k-2} + r_k e^{-j(\Delta\hat{\phi}_k + \Delta\hat{\phi}_{k-1})} + r_{k-1}e^{-j\Delta\hat{\phi}_{k-1}}|^2 \\ &= |r_{k-2}|^2 + |r_{k-1}|^2 + |r_k|^2 + 2\,\mathrm{Re}\{r_k r_{k-2}^* e^{-j(\Delta\hat{\phi}_k + \Delta\hat{\phi}_{k-1})}\} \\ &\quad + 2\,\mathrm{Re}\{r_{k-1}r_{k-2}^* e^{-j\Delta\hat{\phi}_{k-1}}\} + 2\,\mathrm{Re}\{r_k r_{k-1}^* e^{-j\Delta\hat{\phi}_k}\}.\end{aligned} \quad (19)$$

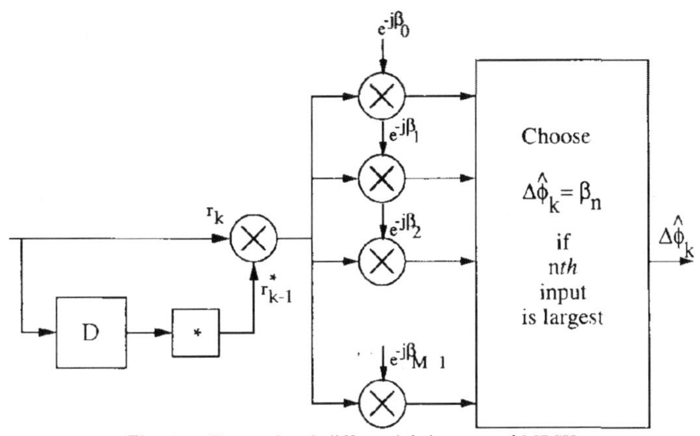

Fig. 1. Conventional differential detector of MPSK.

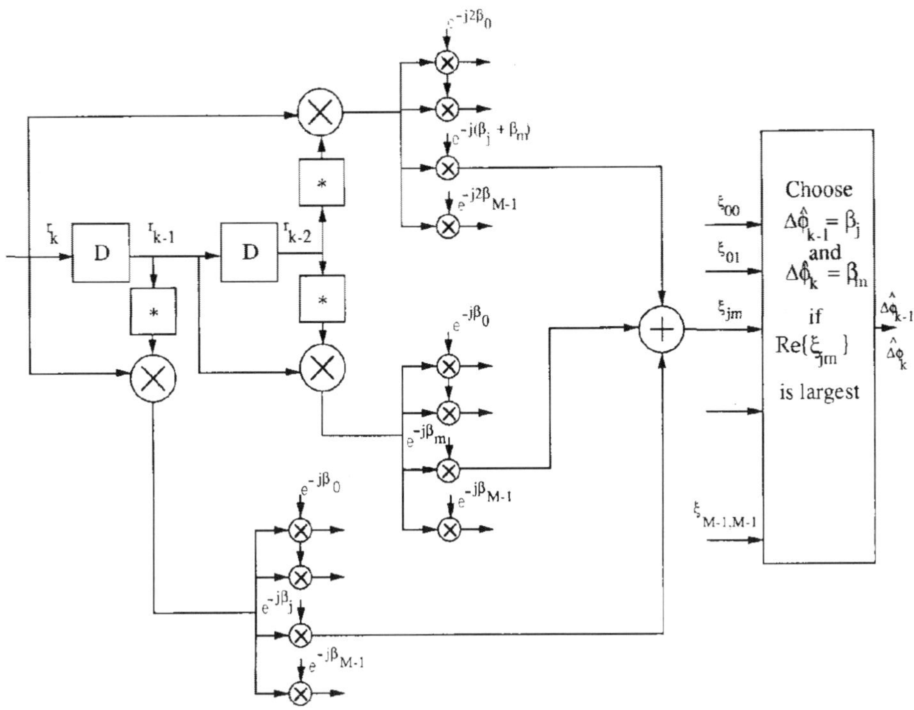

Fig. 2. Parallel implementation of multiple bit differential detector; $N = 3$.

Thus, the decision rule becomes

$$\text{choose } \Delta\hat{\phi}_k \text{ and } \Delta\hat{\phi}_{k-1} \text{ if}$$
$$\mathrm{Re}\left\{r_k r^*_{k-1} e^{-j\Delta\hat{\phi}_k} + r_{k-1} r^*_{k-2} e^{-j\Delta\hat{\phi}_{k-1}} + r_k r^*_{k-2} e^{-j(\Delta\hat{\phi}_k + \Delta\hat{\phi}_{k-1})}\right\} \text{ is maximum.} \tag{20}$$

Note that the first and second terms of the metric used in the decision rule of (20) are identical to those used to make successive and independent decisions on $\Delta\phi_k$ and $\Delta\phi_{k-1}$, respectively, in conventional MDPSK. The third term in the optimum metric is a combination of the first two and is required to make an optimum *joint* decision on $\Delta\phi_k$ and $\Delta\phi_{k-1}$.

Clearly, a receiver implemented on the basis of (20) will outperform conventional MDPSK. Before demonstrating the amount of this performance improvement as a function of the number of phases M we first discuss the implementation of the optimum $N = 3$ receiver. Fig. 2 is a parallel implementation of the decision rule of (20). It should be noted that the M^2 phasors[1] needed to perform the phase rotations of the output $r_k(r_{k-2})^*$ can be obtained using a matrix which performs all possible multiplications of the M phasors $e^{-j\beta_0}, e^{-j\beta_1}, \cdots, e^{-j\beta_{M-1}}$ with themselves. Fig. 3 is a series implementation of the same decision rule which, although simpler in appearance than Fig. 2, requires envelope normalization and additional delay elements.

III. Bit Error Probability Performance

To obtain a simple upper bound on the average bit error probability P_b of the proposed N-bit detection scheme, we use a union bound analogous to that used for upper bounding the performance of error correction coded systems. In particular, the upper bound on P_b is the sum of the pairwise error probabilities associated with each $(N-1)$-bit error sequence. Each pairwise error probability is then either evaluated directly or itself upper bounded. Mathematically speaking, let $\boldsymbol{\Delta\phi} = (\Delta\phi_k, \Delta\phi_{k-1}, \cdots, \Delta\phi_{k-N+2})$ denote the sequence of $N-1$ information phases and $\boldsymbol{\Delta\hat{\phi}} = (\Delta\hat{\phi}_k, \Delta\hat{\phi}_{k-1}, \cdots, \Delta\hat{\phi}_{k-N+2})$ be the corresponding sequence of detected phases. Let $\boldsymbol{u}$ be the sequence of $b = (N-1)\log_2 M$ information bits that produces $\boldsymbol{\Delta\phi}$ at the transmitter and $\hat{\boldsymbol{u}}$ the sequence of b bits that result from the detection of

[1]In reality, only M phasors are needed since the sum angle $\Delta\phi_k + \Delta\phi_{k-1}$ when taken modulo 2π ranges over the set $\beta_0, \beta_1, \cdots, \beta_{M-1}$.

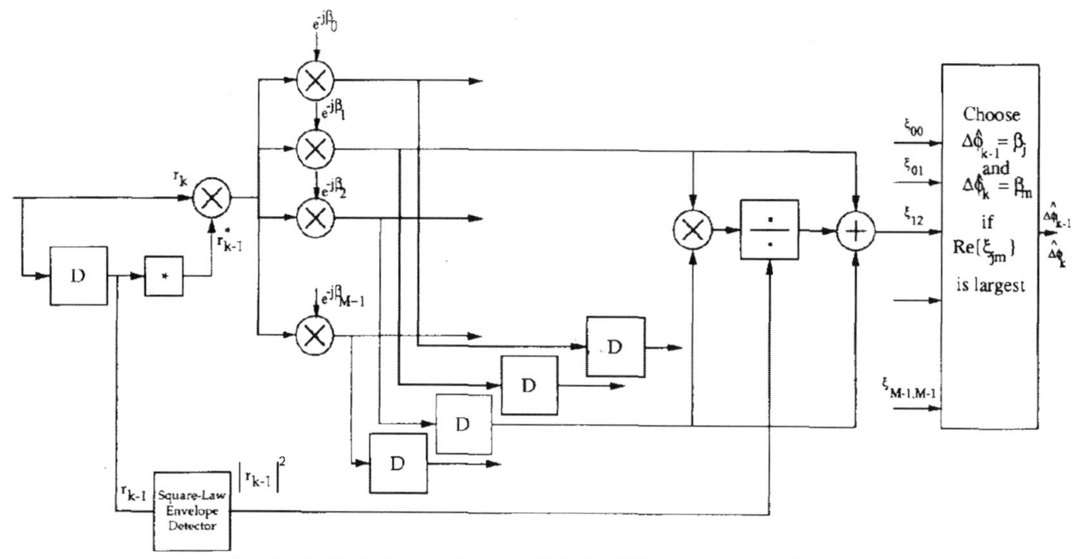

Fig. 3. Serial implementation of multiple bit differential detector; $N = 3$.

$\Delta\hat{\boldsymbol{\phi}}$. Then,

$$P_b \le \frac{1}{b}\frac{1}{M^{N-1}}\sum\sum_{\Delta\boldsymbol{\phi}\ne\Delta\hat{\boldsymbol{\phi}}} w(\boldsymbol{u},\hat{\boldsymbol{u}})\Pr\{\hat{\eta}>\eta|\Delta\boldsymbol{\phi}\} \tag{21}$$

where $w(\boldsymbol{u},\hat{\boldsymbol{u}})$ denotes the Hamming distance between $\boldsymbol{u}$ and $\hat{\boldsymbol{u}}$ and $\Pr\{\hat{\eta}>\eta|\Delta\boldsymbol{\phi}\}$ denotes the pairwise probability that $\Delta\hat{\boldsymbol{\phi}}$ is incorrectly chosen when indeed $\Delta\boldsymbol{\phi}$ was sent. The decision statistic η is defined in (15) and the corresponding error statistic $\hat{\eta}$ is identical to (15) with each $\Delta\phi_k$ replaced by $\Delta\hat{\phi}_k$. For symmetric signalling sets (such as MPSK), (21) satisfies a uniform error probability (UEP) criterion, i.e., the probability of error is independent of which input phase sequence $\Delta\boldsymbol{\phi}$ is chosen as the correct sequence. Under these conditions, (21) simplifies to

$$P_b \le \frac{1}{(N-1)\log_2 M}\sum_{\Delta\hat{\boldsymbol{\phi}}\ne\Delta\boldsymbol{\phi}} w(\boldsymbol{u},\hat{\boldsymbol{u}})\Pr\{\hat{\eta}>\eta|\Delta\boldsymbol{\phi}\} \tag{22}$$

where $\Delta\boldsymbol{\phi}$ is any input sequence, (e.g., the null sequence $(0, 0, \cdots, 0) = \mathbf{0}$.)

A. Evaluation of the Pairwise Error Probability

To compute $\Pr\{\hat{\eta}>\eta|\Delta\boldsymbol{\phi}\}$, we use the approach taken in [3] for evaluating the performance of noncoherent FSK. It is convenient to define

$$w(\Delta\boldsymbol{\phi}) = \sum_{i=0}^{N-1} r_{k-i}e^{-j\sum_{m=0}^{N-i-2}\Delta\phi_{k-i-m}};$$

$$w(\Delta\hat{\boldsymbol{\phi}}) = \sum_{i=0}^{N-1} r_{k-i}e^{-j\sum_{m=0}^{N-i-2}\Delta\hat{\phi}_{k-i-m}} \tag{23}$$

in which case,

$$\eta = |w(\Delta\boldsymbol{\phi})|^2;\ \hat{\eta} = |w(\Delta\hat{\boldsymbol{\phi}})|^2. \tag{24}$$

Then, the pairwise error probability $\Pr\{\hat{\eta}>\eta|\Delta\boldsymbol{\phi}\}$ is derived in Appendix A as

$$\Pr\{\hat{\eta}>\eta|\Delta\boldsymbol{\phi}\} = \frac{1}{2}[1 - Q(\sqrt{b}, \sqrt{a}) + Q(\sqrt{a}, \sqrt{b})] \tag{25}$$

where $Q(\alpha, \beta)$ is Marcum's Q-function [4] and

$$\begin{Bmatrix} b \\ a \end{Bmatrix} = \frac{E_s}{2N_0}[N \pm \sqrt{N^2 - |\delta|^2}] \tag{26}$$

with $E_s = PT$ denoting the energy per data symbol and

$$\delta \triangleq \sum_{i=0}^{N-1} e^{j\sum_{m=0}^{N-i-2}(\Delta\phi_{k-i-m} - \Delta\hat{\phi}_{k-i-m})}$$

$$\triangleq \sum_{i=0}^{N-1} e^{j\sum_{m=0}^{N-i-2}\delta\phi_{k-i-m}} \tag{27}$$

In (27), it is understood that the summation in the exponent evaluates to zero if the upper index is negative.

Note that for any given N, M, and input data sequence $\Delta\boldsymbol{\phi}$, δ can be evaluated for each error sequence $\Delta\hat{\boldsymbol{\phi}}$. We now consider the evaluation of (22) and (25) for some special cases.

C. Case 1: Conventional DPSK (N=2, M=2)

From (27), we immediately get $\delta = 0$ and thus from (26)

$$\begin{Bmatrix} b \\ a \end{Bmatrix} = \begin{Bmatrix} \frac{2E_s}{N_0} \\ 0 \end{Bmatrix}. \tag{28}$$

Substituting (28) into (25) gives

$$\Pr\{\hat{\eta}>\eta|\Delta\boldsymbol{\phi}\} = \frac{1}{2}\left[1 - Q\left(\sqrt{\frac{2E_s}{N_0}}, 0\right) + Q\left(0, \sqrt{\frac{2E_s}{N_0}}\right)\right]. \tag{29}$$

From the definition of the Q-function,

$$Q(\alpha, 0) = 1;\ Q(0, \beta) = \exp\left(-\frac{\beta^2}{2}\right). \tag{30}$$

Since for the binary case the pairwise error probability is indeed equal to the bit error probability, we have from (29) and (30) that

$$P_b = \frac{1}{2}\exp\left(-\frac{E_s}{N_0}\right) \tag{31}$$

which is the well-known result for DPSK.

D. Case 2: $N = 3$, $M = 2$

Here there are three possible error sequences of length 2. The pertinent results related to the evaluation of (26) and (27) are given below

$\Delta\phi_k - \Delta\hat{\phi}_k$	$\Delta\phi_{k-1} - \Delta\hat{\phi}_{k-1}$	δ	$\begin{Bmatrix} b \\ a \end{Bmatrix}$
0	π	-1	$\frac{E_s}{N_0}\left[\frac{3}{2} \pm \sqrt{2}\right]$. (32)
π	0	$+1$	$\frac{E_s}{N_0}\left[\frac{3}{2} \pm \sqrt{2}\right]$
π	π	$+1$	$\frac{E_s}{N_0}\left[\frac{3}{2} \pm \sqrt{2}\right]$

Since the Hamming distance $w(\boldsymbol{u}, \hat{\boldsymbol{u}})$ is equal to 1 for the first two error sequences and is equal to 2 for the third sequence, then using (32) in (25) and (26), the upper bound on bit error probability as given by (22) is evaluated as

$$P_b \le 1 - Q\left(\sqrt{\frac{E_s}{N_0}\left(\frac{3}{2} + \sqrt{2}\right)}, \sqrt{\frac{E_s}{N_0}\left(\frac{3}{2} - \sqrt{2}\right)}\right) + Q\left(\sqrt{\frac{E_s}{N_0}\left(\frac{3}{2} - \sqrt{2}\right)}, \sqrt{\frac{E_s}{N_0}\left(\frac{3}{2} + \sqrt{2}\right)}\right). \quad (33)$$

To see how much performance is gained by extending the observation interval for differential detection from $N = 2$ (conventional) to $N = 3$, we must compare (33) to (31). Due to the complex form of (33) this comparison is not readily obvious without resorting to numerical evaluation. On the other hand, by examining the asymptotic (large E_s/N_0) behavior of (33) we can get an immediate fix on this gain.

When both arguments of the Q-function are large, the following asymptotic approximations are valid [6]:

$$Q(\alpha, \beta) \cong 1 - \frac{1}{\alpha - \beta}\sqrt{\frac{\beta}{2\pi\alpha}} \exp\left\{-\frac{(\alpha-\beta)^2}{2}\right\}; \alpha \gg \beta \gg 1$$

$$Q(\alpha, \beta) \cong \frac{1}{\beta - \alpha}\sqrt{\frac{\beta}{2\pi\alpha}} \exp\left\{-\frac{(\beta-\alpha)^2}{2}\right\}; \beta \gg \alpha \gg 1. \quad (34)$$

Using these approximations, (25) becomes

$$\Pr\{\hat{\eta} > \eta | \boldsymbol{\Delta\phi}\} = \frac{1}{2}\left[\frac{1}{\sqrt{b} - \sqrt{a}}\left(\sqrt{\frac{\sqrt{a/b}}{2\pi}} + \sqrt{\frac{\sqrt{b/a}}{2\pi}}\right) \times \exp\left\{-\frac{(\sqrt{b} - \sqrt{a})^2}{2}\right\}\right] \quad (35)$$

or, from (26),

$$\begin{aligned}\Pr\{\hat{\eta} > \eta | \boldsymbol{\Delta\phi}\} &= \frac{1}{2\sqrt{2\pi\frac{E_s}{N_0}(N - |\delta|)}} \times \left(\left[\frac{N - \sqrt{N^2 - |\delta|^2}}{N + \sqrt{N^2 - |\delta|^2}}\right]^{1/4} + \left[\frac{N + \sqrt{N^2 - |\delta|^2}}{N - \sqrt{N^2 - |\delta|^2}}\right]^{1/4}\right) \times \exp\left\{-\frac{E_s}{2N_0}(N - |\delta|)\right\} \\ &= \frac{1}{2\sqrt{\pi\frac{E_s}{N_0}}}\left(\sqrt{\frac{N + |\delta|}{|\delta|(N - |\delta|)}}\right) \times \exp\left\{-\frac{E_s}{2N_0}(N - |\delta|)\right\}. \end{aligned} \quad (36)$$

For $N = 3$ and $M = 2$, $|\delta| = 1$ from (32). Thus, (33) becomes

$$P_b \lesssim \frac{2\sqrt{2}}{\sqrt{\pi\frac{E_s}{N_0}}}\left[\frac{1}{2}\exp\left\{-\frac{E_s}{N_0}\right\}\right]. \quad (37)$$

Comparing (37) to (31), we observe that the factor in front of the term in brackets in (37) represents a bound on the improvement in performance obtained by increasing the memory of the decision by one symbol interval from $N = 2$ to $N = 3$.

E. General Asymptotic Results

In the general case for arbitrary N, the dominant terms in the bit error probability occur for the sequences that result in the minimum value of $N - |\delta|$. One can easily show that this minimum value will certainly occur for the error sequence $\boldsymbol{\Delta\hat{\phi}}$ having $N - 1$ elements equal to the correct sequence $\boldsymbol{\Delta\phi}$ and one element with the smallest error. Thus,

$$\begin{aligned}\min_{\boldsymbol{\Delta\phi}, \boldsymbol{\Delta\hat{\phi}}} (N - |\delta|) &= N - |N - 1 + e^{j(\Delta\phi_k - \Delta\hat{\phi}_k)_{\min}}| \\ &= N - \sqrt{(N-1)^2 + (N-1)(2 - d_{\min}^2) + 1} \\ &= N - |\delta|_{\max} \end{aligned} \quad (38)$$

where

$$d_{\min}^2 = 4\sin^2\frac{(\Delta\phi_k - \Delta\hat{\phi}_k)_{\min}}{2} = 4\sin^2\frac{\pi}{M}. \quad (39)$$

Also note that for $|\delta| = |\delta|_{\max}$, (26) reduces to

$$\begin{Bmatrix} b \\ a \end{Bmatrix} = \frac{E_s}{2N_0}\left[N \pm 2\sqrt{N-1}\sin\frac{\pi}{M}\right]. \quad (40)$$

Thus, the average bit error probability is approximately upper bounded by

$$P_b \le \frac{1}{(N-1)\log_2 M}\left(\sum_{\boldsymbol{\Delta\hat{\phi}} \ne \boldsymbol{\Delta\phi}} w(\boldsymbol{u}, \hat{\boldsymbol{u}})\right) \times \frac{1}{2\sqrt{\pi\frac{E_s}{N_0}}}\left(\sqrt{\frac{N + |\delta|_{\max}}{|\delta|_{\max}(N - |\delta|_{\max})}}\right) \times \exp\left\{-\frac{E_s}{2N_0}(N - |\delta|_{\max})\right\} \quad (41)$$

where $w(\boldsymbol{u}, \hat{\boldsymbol{u}})$ corresponds only to those error sequences that result in $|\delta|_{\max}$.

For the binary case ($M = 2$), we have from (39) that $d_{\min}^2 = 4$ and hence $N - |\delta|_{\max} = 2$. Similarly, it is straightforward to show (see Appendix B) that the sum of Hamming distances required in (41) is given by

$$\sum_{\boldsymbol{\Delta\hat{\phi}} \ne \boldsymbol{\Delta\phi}} w(\boldsymbol{u}, \hat{\boldsymbol{u}}) = \begin{cases} 2(N-1); & N > 2 \\ 1; & N = 2 \end{cases}. \quad (42)$$

Thus, (41) simplifies to

$$P_b \lesssim \frac{2}{\sqrt{\pi\frac{E_s}{N_0}}}\left(\sqrt{\frac{N-1}{N-2}}\right)\left[\frac{1}{2}\exp\left\{-\frac{E_s}{N_0}\right\}\right] \quad (43)$$

which is the generalization of (37) for arbitrary N[2].

[2]Note that (43) is not valid for $N = 2$ since in that case $|\delta|_{\max} = 0$ [see (38) and (39)] and thus the inequalities in (34) are not satisfied.

Equation (43) has an interesting interpretation as N gets large. Taking the limit of (43) as $N \to \infty$, we get

$$P_b \lesssim \frac{1}{\sqrt{\pi \frac{E_s}{N_0}}} \exp\left\{-\frac{E_s}{N_0}\right\} \tag{44}$$

which can be expressed in terms of the asymptotic expansion of the complementary error function,

$$\operatorname{erfc} x \triangleq \frac{2}{\sqrt{\pi}} \int_x^{\infty} \exp(-y^2)\,dy \cong \frac{1}{\sqrt{\pi}x} \exp(-x^2) \tag{45}$$

by

$$P_b \lesssim \operatorname{erfc}\sqrt{\frac{E_s}{N_0}}. \tag{46}$$

For *coherent* detection of binary PSK (BPSK) with differential encoding and decoding, the bit error probability performance is given by [2; ch. 5]

$$P_b = \left(\operatorname{erfc}\sqrt{\frac{E_s}{N_0}}\right)\left(1 - \frac{1}{2}\operatorname{erfc}\sqrt{\frac{E_s}{N_0}}\right) \tag{47}$$

which has an asymptotic upper bound identical to (46). Thus, as one might expect, *the performance of multiple-symbol differentially detected BPSK approaches that of ideal coherent detection BPSK with differential encoding in the limit as the observation interval (decision memory) approaches infinity.*

A similar limiting behavior as the above may be observed for other values of M. In particular, it can be shown (see Appendix B) that for $M > 2$ and a Gray code [2] mapping of bits to symbols, the sum of Hamming distances corresponding to $|\delta|_{\max}$ is given by

$$\sum_{\Delta\hat{\phi} \neq \Delta\phi} w(u, \hat{u}) = \begin{cases} 4(N-1); & N > 2 \\ 2; & N = 2 \end{cases}. \tag{48}$$

Using (48) in (41), we get (for $N > 2$)

$$P_b \le \frac{2}{(\log_2 M)\sqrt{\pi \frac{E_s}{N_0}}} \left(\sqrt{\frac{N + |\delta|_{\max}}{|\delta|_{\max}(N - |\delta|_{\max})}}\right) \times \exp\left\{-\frac{E_s}{2N_0}(N - |\delta|_{\max})\right\} \tag{49}$$

where, from (38) and (39),

$$|\delta|_{\max} = \sqrt{(N-1)^2 + 2(N-1)\left(1 - 2\sin^2\frac{\pi}{M}\right) + 1}. \tag{50}$$

For $N = 2$, the upper bound on bit error probability becomes

$$P_b \le \frac{1}{(\log_2 M)\sqrt{2\pi \frac{E_s}{N_0}}} \left(\frac{\cos\frac{\pi}{2M}}{\sin\frac{\pi}{2M}\sqrt{\cos\frac{\pi}{M}}}\right) \times \exp\left\{-\frac{2E_s}{N_0}\sin^2\frac{\pi}{2M}\right\}. \tag{51}$$

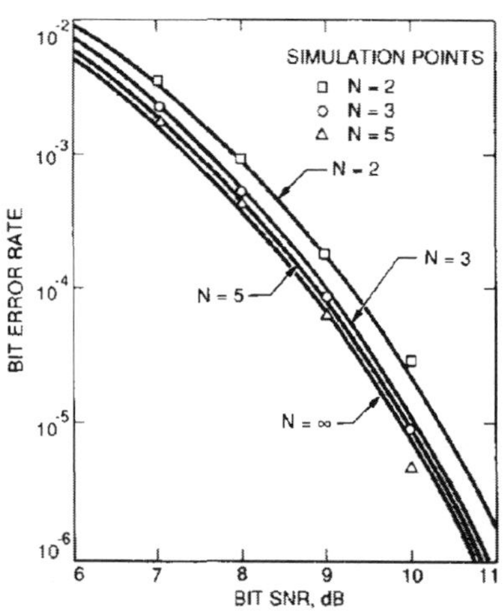

Fig. 4. Bit error probability versus E_b/N_0 for multiple differential detection of MPSK; $M = 2$.

As N gets large, $|\delta|_{\max} \to N - 2\sin^2 \pi/M$ and (49) reduces to

$$P_b \le \frac{1}{(\log_2 M)\sqrt{\pi \frac{E_s}{N_0}\sin^2\frac{\pi}{M}}} \exp\left\{-\frac{E_s}{N_0}\sin^2\frac{\pi}{M}\right\}$$

$$\cong \frac{1}{\log_2 M}\operatorname{erfc}\left(\sqrt{\frac{E_s}{N_0}}\sin\frac{\pi}{M}\right) \tag{52}$$

which is identical to the asymptotic bit error probability for *coherent* detection of MPSK with differential encoding and decoding (see [2, eqs. (5-91), (5-92) and (5-113)])[3]

For example, for QPSK ($M = 4$), the symbol error probability is given by [2; Eq. (5-115)]

$$P_s = 2\operatorname{erfc}\left(\sqrt{\frac{E_s}{2N_0}}\right) - 2\operatorname{erfc}^2\left(\sqrt{\frac{E_s}{2N_0}}\right) + \operatorname{erfc}^3\left(\sqrt{\frac{E_s}{2N_0}}\right) - \frac{1}{4}\operatorname{erfc}^4\left(\sqrt{\frac{E_s}{2N_0}}\right). \tag{53}$$

Since for a Gray code bit to symbol mapping

$$P_b \cong \frac{P_s}{\log_2 M} = \frac{P_s}{2} \tag{54}$$

then (54) together with (53) has an asymptotic upper bound identical to (52).

Figs. 4, 5, and 6 are illustrations of the upper bounds of (49) and (51) for $M = 2$, 4, and 8, respectively. In each figure, the length (in MPSK symbols) of the observation interval N is a parameter varying from $N = 2$ (conventional MDPSK) to $N = \infty$ (ideal coherent detection). Also indicated on the figures are computer simulation results corresponding to the exact performance. We observe that, for example, for binary DPSK, extending the observation interval from $N = 2$ to $N = 3$ recovers more than half of the E_b/N_0 loss of differential

[3]It should be noted that the result in (52) can be obtained by observing that, for large N, (40) satisfies $\sqrt{b} \gg \sqrt{b} - \sqrt{a} > 0$. In this case, (25) can be approximated by [5; Appendix A]

$$\Pr\{\hat{\eta} > \eta | \Delta\phi\} = \frac{1}{2}[1 - Q(\sqrt{b}, \sqrt{a}) + Q(\sqrt{a}, \sqrt{b})]$$

$$\cong \frac{1}{2}\operatorname{erfc}\left(\frac{\sqrt{b} - \sqrt{a}}{\sqrt{2}}\right).$$

Using this relation in (22) gives the asymptotic bit error probability in (52).

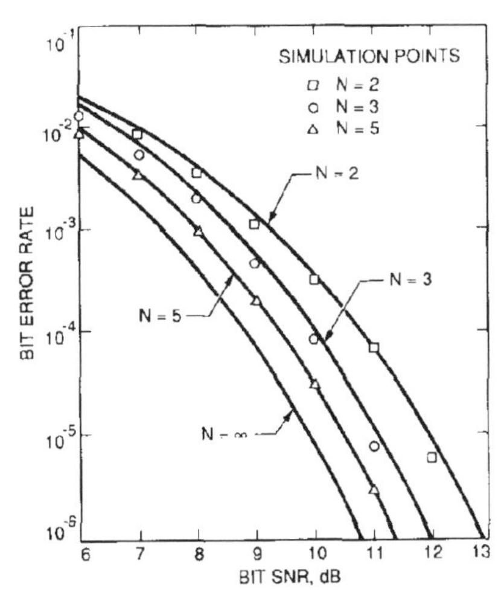

Fig. 5. Bit error probability versus E_b/N_0 for multiple differential detection of MPSK; $M = 4$.

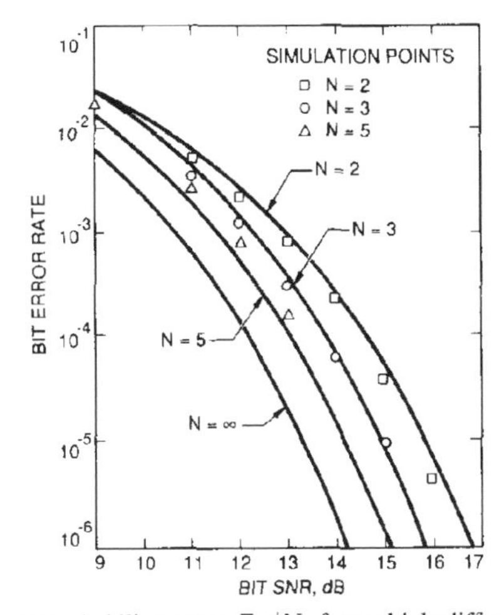

Fig. 6. Bit error probability versus E_b/N_0 for multiple differential detection of MPSK; $M = 8$.

detection versus coherent detection with differential encoding. For $M = 4$, the improvement in E_b/N_0 performance of $N = 3$ relative to $N = 2$ is more than 1 dB which is slightly less than half of the total difference between differential detection and coherent detection with differential encoding.

IV. Conclusions

We have demonstrated a multiple-symbol differential detection technique for MPSK which is based on maximum-likelihood sequence estimation (MLSE) of the transmitted phases rather than symbol-by-symbol detection. The performance of this multiple-symbol scheme fills the gap between ideal coherent and differentially coherent detection of MPSK. The amount of improvement gained over differentially coherent detection (two symbol observation interval per decision) depends on the number of additional symbol intervals added to the observation. In the limit as the observation interval approaches infinity, the performance approaches that of ideal coherent detection with different encoding. Practically, this limiting performance is approached with observation times (decision memory) only on the order of a few additional symbols intervals. Thus, even in situations (e.g., benign environments) where one would ordinarily not turn to differential detection, it might now be desirable to employ multiple-symbol differential detection for reasons related to simplicity of implementation. For example, the acquisition and maintenance of a locked carrier tracking loop as required in a coherent detection system is not needed here.

Although the analysis and results have been derived for an uncoded MPSK modulation, it should be obvious that this scheme applies to other forms of modulation as well as coded systems. The results of these investigations are available in forthcoming papers by the authors.

Appendix A

Evaluation of the Pairwise Error Probability

In [3], it is shown that for complex Gaussian random variables z_1 and z_2 with identical variances and arbitrary means and covariance, the pairwise probability of error $\Pr\{|z_2|^2 > |z_1|^2\}$ is given by

$$\Pr\{|z_2|^2 > |z_1|^2\} = \frac{1}{2}[1 - Q(\sqrt{b}, \sqrt{a}) + Q(\sqrt{a}, \sqrt{b})] \quad \text{(A-1)}$$

where $Q(\alpha, \beta)$ is Marcum's Q-function [4] and

$$\begin{Bmatrix} b \\ a \end{Bmatrix} = \frac{1}{2N_s} \times \left\{ \frac{S_1 + S_2 - 2|\rho|\sqrt{S_1 S_2}\cos(\theta_1 - \theta_2 + \phi)}{1 - |\rho|^2} \pm \frac{S_1 - S_2}{\sqrt{1 - |\rho|^2}} \right\} \quad \text{(A-2)}$$

with

$$S_1 \triangleq \frac{1}{2}|\bar{z}_1|^2; \; S_2 \triangleq \frac{1}{2}|\bar{z}_2|^2$$

$$N_s \triangleq \frac{1}{2}\overline{|z_1 - \bar{z}_1|^2} = \frac{1}{2}\overline{|z_2 - \bar{z}_2|^2}$$

$$\rho \triangleq \frac{1}{2N_s}\overline{(z_1 - \bar{z}_1)^*(z_2 - \bar{z}_2)}$$

$$\phi = \arg\{\rho\}; \; \theta_1 = \arg\{\bar{z}_1\}; \; \theta_2 = \arg\{\bar{z}_2\}. \quad \text{(A-3)}$$

Here we associate z_1 and z_2 with $w(\Delta\boldsymbol{\phi})$ and $w(\Delta\hat{\boldsymbol{\phi}})$, respectively, of (23).

Using (1) and (2) in (A-3), we get

$$\bar{z}_1 = \sqrt{2P}\sum_{i=0}^{N-1} e^{j\phi_{k-i}} e^{-j\sum_{m=0}^{N-i-2}\Delta\phi_{k-i-m}} = \sqrt{2P}e^{j\phi_{k-N+1}}N$$

$$\bar{z}_2 = \sqrt{2P}\sum_{i=0}^{N-1} e^{j\phi_{k-i}} e^{-j\sum_{m=0}^{N-i-2}\Delta\hat{\phi}_{k-i-m}} = \sqrt{2P}e^{j\phi_{k-N+1}}\delta \quad \text{(A-4)}$$

where

$$\delta \triangleq \sum_{i=0}^{N-1} e^{j\sum_{m=0}^{N-i-2}(\Delta\phi_{k-i-m} - \Delta\hat{\phi}_{k-i-m})} \quad \text{(A-5)}$$

where it is understood that the summation equals zero if the upper summation index is negative. Substituting (A-4) into (A-3), gives

$$S_1 = PN^2; S_2 = P|\delta|^2. \quad \text{(A-6)}$$

Also, using (3)

$$N_s = \frac{1}{2}E\left\{\left|\sum_{i=0}^{N-1} n_{k-i} e^{-j\sum_{m=0}^{N-i-2}\Delta\phi_{k-i-m}}\right|^2\right\} = \frac{NN_0}{T} \quad \text{(A-7)}$$

and

$$\rho = \frac{1}{2N_s}\sum_{i=0}^{N-1}\sum_{n=0}^{N-1}E\{n_{k-i}n_{k-n}^*\} \times e^{j(\sum_{m=0}^{N-i-2}\Delta\phi_{k-i-m}-\sum_{m=0}^{N-n-2}\Delta\hat{\phi}_{k-n-m})}$$

$$= \frac{1}{2N_s}\sum_{i=0}^{N-1}\frac{2N_0}{T}e^{j(\sum_{m=0}^{N-i-2}\Delta\phi_{k-i-m}-\Delta\hat{\phi}_{k-i-m})} = \frac{\delta}{N}. \quad \text{(A-8)}$$

Finally, substituting (A-6)-(A-8) into (A-1) gives the desired result, namely,

$$\Pr\{\hat{\eta} > \eta|\Delta\phi\} = \frac{1}{2}[1 - Q(\sqrt{b}, \sqrt{a}) + Q(\sqrt{a}, \sqrt{b})] \quad \text{(A-9)}$$

where

$$\begin{Bmatrix} b \\ a \end{Bmatrix} = \frac{E_s}{2N_0}[N \pm \sqrt{N^2 - |\delta|^2}] \quad \text{(A-10)}$$

and $E_s = PT$ is the energy per data symbol.

Appendix B

Proof of (42) and (48)

Starting with the definition of δ in (27), we now write it in the form

$$\delta = 1 + \sum_{i=1}^{N-1}e^{j\alpha_i} \quad \text{(B-1)}$$

where

$$\alpha_i \triangleq \sum_{n=1}^{i}\delta\phi_{k-N+1+n}. \quad \text{(B-2)}$$

Thus, δ is the sum of N unit vectors the first of whose arguments is zero and the rest of whose arguments are increasingly larger sums of the phase errors in accordance with (B-2). Note that the values of the accumulated phase errors α_i's also range over the set $\pm 2\pi m/M$; $m = 0, 1, \cdots, M/2-1$. We are interested in determining the various possible solutions for the $\delta\phi$'s such that the maximum value of the magnitude of δ, namely, $|\delta|_{\max}$ of (49) is achieved.

For arbitrary M, there are four situations that achieve $|\delta|_{\max}$. We shall refer to these as Cases 1, 2, 3, and 4 which are described as follows.

Case 1: All $N-1$ vectors $e^{j\alpha_i}$; $i = 1, 2, \cdots, N-1$ must be collinear and equal to $e^{j2\pi/M}$. Thus, $\alpha_i = 2\pi/M$; $i = 1, 2, \cdots, N-1$ which, in accordance with (B-2), has the single solution

$$\delta\phi_{k-N+2} = \frac{2\pi}{M};\ \delta\phi_{k-N+i} = 0;\ i = 3, 4, \cdots, N. \quad \text{(B-3)}$$

Case 2: All $N-1$ vectors $e^{j\alpha_i}$; $i = 1, 2, \cdots, N-1$ must be collinear and equal to $e^{-j2\pi/M}$. Thus, $\alpha_i = -2\pi/M$; $i = 1, 2, \cdots, N-1$ which, in accordance with (B-2), has the single solution

$$\delta\phi_{k-N+2} = -\frac{2\pi}{M};\ \delta\phi_{k-N+i} = 0;\ i = 3, 4, \cdots, N. \quad \text{(B-4)}$$

Case 3: Any $N-2$ vectors $e^{j\alpha_i}$ must be collinear and equal to $e^{j0} = 1$ and the remaining vector must be equal to $e^{j2\pi/M}$. For this case there are $N-1$ different solutions. For example, suppose first that $\alpha_1 = 2\pi/M$, and $\alpha_i = 0$; $i = 2, 3, \cdots, N-1$. Then,

$$\delta\phi_{k-N+2} = \frac{2\pi}{M};\ \delta\phi_{k-N+3} = -\frac{2\pi}{M};\ \delta\phi_{k-N+i} = 0; i = 4, 5, \cdots, N. \quad \text{(B-5)}$$

Next, let $\alpha_2 = 2\pi/M$, and $\alpha_i = 0$; $i = 1, 3, 4, \cdots, N-1$. Then,

$$\delta\phi_{k-N+3} = \frac{2\pi}{M};\ \delta\phi_{k-N+4} = -\frac{2\pi}{M};$$
$$\delta\phi_{k-N+i} = 0; i = 2, 5, 6, \cdots, N. \quad \text{(B-6)}$$

In general for $\alpha_l = 2\pi/M$, $l = 1, 2, \cdots, N-2$ and $\alpha_i = 0$; $i = 1, 2, \cdots, N-1$; $i \neq l$, we have the solution

$$\delta\phi_{k-N+l+1} = \frac{2\pi}{M};\ \delta\phi_{k-N+l+2} = -\frac{2\pi}{M};$$
$$\delta\phi_{k-N+i} = 0;\ i = 1, 2, \cdots, l, l+3, \cdots, N. \quad \text{(B-7)}$$

Finally, for $\alpha_{N-1} = 2\pi/M$ and $\alpha_i = 0$; $i = 1, 2, \cdots, N-2$, the solution is

$$\delta\phi_k = \frac{2\pi}{M};\ \delta\phi_{k-N+i} = 0; i = 1, 2, \cdots, N-1. \quad \text{(B-8)}$$

Case 4: Any $N-2$ vectors $e^{j\alpha_i}$ must be collinear and equal to $e^{j0} = 1$ and the remaining vector must be equal to $e^{-j2\pi/M}$. For this case there are again $N-1$ different solutions which are identical to those described by (B-5) to (B-8) with $2\pi/M$ replaced by $-2\pi/M$ and vice versa.

We note that for Cases 3 and 4, $N-2$ of the solutions are characterized by having one $\delta\phi = 2\pi/M$, one $\delta\phi = -2\pi/M$, and the rest of the $\delta\phi$'s equal to zero. The remaining solution has one $\delta\phi = 2\pi/M$ and the rest of the $\delta\phi$'s equal to zero.

To compute the accumulated Hamming distance of (47) where $w(u, \hat{u})$ corresponds only to those error sequences that result in $|\delta|_{\max}$, we proceed as follows. We assume a Gray code bit to symbol assignment where $\delta\phi = \pm 2\pi/M$ corresponds to an adjacent phase symbol error and thus a single bit error or a Hamming distance equal to 1. Also, a value $\delta\phi = 0$ implies no symbol error or a Hamming distance equal to zero. Thus, the following accumulated Hamming distances occur for each of the four cases.

Case 1:

$$w(u, \hat{u}) = 1. \quad \text{(B-9)}$$

Case 2:

$$w(u, \hat{u}) = 1. \quad \text{(B-10)}$$

Case 3:

$$w(u, \hat{u}) = (2)(N-2) + (1)(1) = 2(N-2) + 1. \quad \text{(B-11)}$$

Case 4:

$$w(u, \hat{u}) = (2)(N-2) + (1)(1) = 2(N-2) + 1. \quad \text{(B-12)}$$

Finally, the accumulated Hamming distance is obtained by summing (B-9) through (B-12) which yields (for $N > 2$)

$$\sum_{\Delta\hat{\phi}\neq\Delta\phi} w(u, \hat{u}) = 4(N-1) \quad \text{(B-13)}$$

which agrees with (48).

For $N = 2$, Cases 3 and 4 do not occur since $N-2 = 0$. Thus, the accumulated Hamming distance is merely the sum of (B-9) and (B-10) which yields

$$\sum_{\Delta\hat{\phi}\neq\Delta\phi} w(u, \hat{u}) = 2 \quad \text{(B-14)}$$

in agreement with (47).

For $M = 2$, $e^{j2\pi/M} = e^{j\pi} = e^{-j2\pi/M}$ and thus Cases 1 and 2 are one and the same and similarly for Cases 3 and 4. Thus, for binary multiple bit DPSK, we have only *half* the solutions in which case

(B-13) becomes

$$\sum_{\Delta\phi \neq \Delta\hat{\phi}} w(u, \hat{u}) = \begin{cases} 2(N-1); & N > 2 \\ 1; & N = 2 \end{cases} \quad \text{(B-15)}$$

which agrees with (42). Q.E.D.

REFERENCES

[1] J. G. Lawton, "Investigation of digital data communication systems," Rep. no. UA-1420-S-1, Cornell Aeronautical Laboratory, Inc., Buffalo, NY, Jan. 3, 1961. Also available as ASTIA Document no. 256 584.

[2] W. C. Lindsey and M. K. Simon, *Telecommunication Systems Engineering*. Englewood Cliffs, NJ: Prentice-Hall, 1973.

[3] S. Stein, "Unified analysis of certain coherent and noncoherent binary communication communications systems," *IEEE Trans. Inform. Theory*, pp. 43–51, Jan. 1964.

[4] J. Marcum, "Tables of Q Functions," RAND Corp. Rep. M-339, Jan. 1950.

[5] M. Schwartz, W. R. Bennett, and S. Stein, *Communication Systems and Techniques*. New York: McGraw-Hill, 1966.

[6] C. W. Helstrom, *Statistical Theory of Signal Detection*. Pergamon, p. 154.

PHOTO NOT AVAILABLE

Dariush Divsalar (S'76–M'79) was born in Anzali, Iran, on July 18, 1947. He received the B.Sc. degree in electrical engineering from the University of Tehran, Iran, in 1970, the M.S., Engineer, and Ph.D. degrees, in electrical engineering from the University of California, Los Angeles, in 1975, 1977, and 1978, respectively.

He has been with the Jet Propulsion Laboratory, California Institute of Technology, Pasadena, since 1978, where he has been working on advanced deep space communication systems techniques for future space exploration and was leading the modulation coding research work for the NASA Mobile Satellite Experiment (MSAT-X). His area of interest is coding and modulation. He has taught coding and digital communication theory courses at UCLA and Northrop University for the past five years.

Dr. Divsalar has coauthored a chapter on telemetry systems, modulation, and coding in *Deep-Space Telecommunication Systems Engineering* (New York: Plenum, 1983). He is also a member of Sigma Xi and is Editor for Coding Theory and Applications for IEEE TRANSACTIONS ON COMMUNICATIONS.

PHOTO NOT AVAILABLE

Marvin K. Simon (S'60–F'78) received the B.S. degree in electrical engineering from the City College of New York, NY, in 1960, the M.S. degree in electrical engineering from Princeton University, Princeton, NJ, in 1961, and the Ph.D. degree from New York University, New York, NY, in 1966.

During the years 1961 to 1963, and again from 1966 to 1968, he was employed at the Bell Telephone Laboratories, Holmdel, NJ, where he was involved in theoretical studies of digital communication systems. Concurrent with his doctoral studies between 1963 and 1966, he was employed as a full-time Instructor of Electrical Engineering at New York University. Since 1968, he has been with the Jet Propulsion Laboratory, where he is presently a Senior Research Engineer. His research has explored all aspects of the synchronization problem associated with digital communication systems, in particular, the application of phase-locked loops. More recently, his interests have turned toward the study of modulation and coding techniques for efficient spectrum utilization and spread-spectrum communications.

Dr. Simon is a winner of the 1984 IEEE Centennial Medal, and a member of Eta Kappa Nu, Tau Beta Pi, and Sigma Xi. He is a winner of NASA Exceptional Service Medal in recognition of outstanding contributions in analysis and design of space communication systems. He is also listed in American Men and Women of Science, Who's Who in Technology, Who's Who in Engineering, and Who's Who in the West. in the past he has served as Editor of Communication Theory for the IEEE TRANSACTIONS ON COMMUNICATIONS, and Technical Program Chairman of the 1977 National Telecommunications Conference. In addition to his technical contributions in other areas, Dr. Simon has published over 90 journal papers on the above subjects and is coauthor with W. C. Lindsey of the textbook *Telecommunication Systems Engineering* (Prentice-Hall, Englewood Cliffs, NJ 1973) and *Phase-Locked Loops and Their Application* (IEEE Press). His work has also appeared in the textbook *Deep Space Telecommunication Systems Engineering*, by J. H. Yuen (Plenum 1983) and *Spread Spectrum Communications*, by C. E. Cook, F. W. Ellersick, L. B. Milstein, and D. L. Schilling (IEEE Press). Recently, he has coauthored with J. K. Omura, R. A. Scholtz, and B. K. Levitt, the 3-volume textbook *Spread Spectrum Communications* (Computer Science, Rockville, MD).

The Effects of Time Delay Spread on Portable Radio Communications Channels with Digital Modulation

JUSTIN C-I CHUANG

Abstract—**Frequency-selective fading caused by multipath time delay spread degrades digital communication channels by causing intersymbol interference, thus resulting in an irreducible BER and imposing a upper limit on the data symbol rate. In this paper, a frequency-selective, slowly fading channel is studied by computer simulation. The unfiltered BPSK, QPSK, OQPSK, and MSK modulations are considered first to illustrate the physical insights and the error mechanisms. Two classes of modulation with spectral-shaping filtering are studied next to assess the tradeoff between spectral occupancy and the performance under the influence of time delay spread. The simulation is very flexible so that different channel parameters can be studied and optimized either individually or collectively. The irreducible BER averaged over fading samples with a given delay profile is used to compare different modulation/detection methods, while the cumulative distribution of short-term BER is employed to show allowable data symbol rates for given values of delay spread. It is found that both GMSK and QPSK with a raised-cosine Nyquist pulse are suitable for a TDM/TDMA digital portable communications channel.**

I. Introduction

The emergence of a demand for personal portable communications, along with a need for integrated voice and data in our mobile society, has made universal digital portable communications [1] highly desirable. A candidate for providing a flexible, high-capacity digital service is the TDM/TDMA (time-division multiplexing/time-division multiple-access) channel allocation scheme [1], [2]. TDM architecture has several advantages over the more conventional FDM (frequency-division multiplexing) such as simpler radio hardware and variable-rate voice communications [3]; however, the signaling rate required for such a system is higher. As a result, the influence of frequency-selective fading caused by delay spread becomes a crucial issue [4]–[6].

A flexible computer simulation is presented in this paper to assess the effects of delay spread and to optimize system parameters such as modulation/detection scheme and data symbol rate. Adaptive equalization of the fading channel is not assumed for this study.

Section II illustrates the channel model, and Section III describes the simulation. Numerical results for some unfiltered modulations are shown in Section IV. In Section V modulations employing spectral-shaping filters, as would be used in a practical system are studied. Section VI summarizes the results of the paper.

Manuscript received January 14, 1987; revised February 17, 1987.
The author is with Bell Communications Research Incorporated, Red Bank, NJ 07701.
IEEE Log Number 8714459.

II. Model

A multipath radio propagation channel can be described mathematically by its impulse response $h(t)$. A baseband model of a multipath radio channel is shown in Fig. 1. A quasi-static channel is considered in this study, that is, the channel is assumed to be time-invariant over many symbol periods. This is a reasonable assumption for a portable communications system with around 1 GHz carrier frequency and several hundred kbits/s signaling rate, since the maximum Doppler frequency is only on the order of 10^{-5} times the bit rate.

Let $u(t)$ be the baseband representation of the modulated waveform. The total received waveform is then

$$z(t) = r(t) + n(t) \tag{1}$$

where $n(t)$ is the additive Gaussian noise and

$$r(t) = u(t) * h(t). \tag{2}$$

Intersymbol interference (ISI) caused by the delay spread of $h(t)$ results in the frequency-selective fading. With increasing signal-to-noise ratio (SNR), an irreducible "floor" of bit error rate (BER) is approached because ISI increases in proportion to the signal level.

In this paper, the short-term, small-scale ("microscopic") signal variations due to multipath fading are studied; therefore, the "macroscopic" effects such as shadowing and distance-related attenuation are normalized as unity. When an overall system design is considered, proper scaling factors should be incorporated in a power budget to account for these long-term, large-scale variations [1].

The impulse response of the channel can be expressed as [7]

$$h(t) = \sum_m A_m e^{j\phi_m} \delta(t - \tau_m). \tag{3}$$

A general form of $h(t)$ can be expressed as a continuous-time function. Both A_m and ϕ_m are slowly varying random quantities that introduce a small Doppler shift for the communications channel. A commonly accepted model [8], [9] suggests that A_m be a random variable with a Rayleigh distribution and ϕ_m be a random variable uniformly distributed from 0 to 2π; therefore, $h(t)$ is a zero-mean complex Gaussian random variable. Physically, at a specific time delay, the received signal approaches a complex Gaussian distribution according to the central limit theo-

Reprinted from *IEEE Journal on Selected Areas in Communications,* vol. SAC-5, no. 5, June 1987.

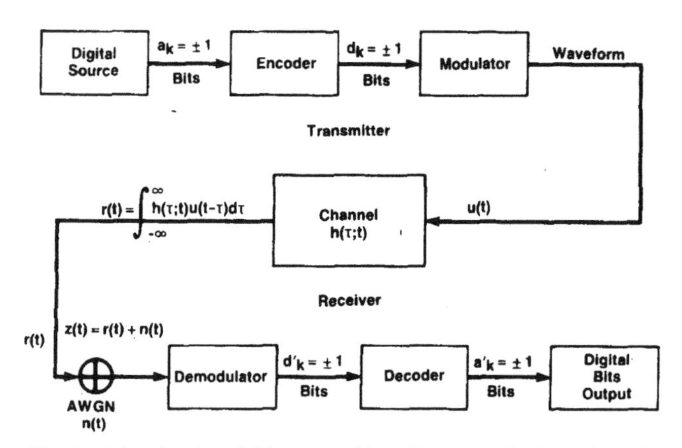

Fig. 1. A baseband model for a portable radio communications channel.

rem because it is a combination of signals from a large number of unresolved paths.

For $t \neq t'$, it is reasonable to assume that $h(t)$ and $h(t')$ are uncorrelated since they are composed of signals from independent sets of paths with different Doppler spectra ("uncorrelated scattering" [9]). Consequently,

$$\langle h^*(t)\, h(t') \rangle = p(t)\, \delta(t - t') \tag{4}$$

where $\langle\ \rangle$ denotes the ensemble average, and

$$p(t) = \langle |h(t)|^2 \rangle. \tag{5}$$

The function $p(t)$ is a measurable profile [10] called the "power delay profile." Some measured results for $p(t)$ in the portable radio environment are documented by Devasirvatham [11], [12].

A measure of the width of $p(t)$ is the root mean-square (rms) delay spread τ defined as the square root of the second central moment [9]. That is

$$\tau = \left[\frac{\int (t - D)^2 p(t)\, dt}{\int p(t)\, dt} \right]^{1/2} \tag{6}$$

where the average delay D, i.e., the centroid of $p(t)$, is

$$D = \frac{\int tp(t)\, dt}{\int p(t)\, dt}. \tag{7}$$

The error rates for transmission through a channel with a small delay spread are most strongly dependent on the normalized rms delay spread [4], [5], [13]. Normalized delay spread is defined as

$$d = \frac{\tau}{T} \tag{8}$$

where T is the symbol period. In this study, $d \leq 0.2$ is considered.

III. Simulation

Fig. 2 is the flowchart for coherent detection of the modulations considered except GMSK. The simulations for GMSK and differential detection involve minor modifications to what is shown.

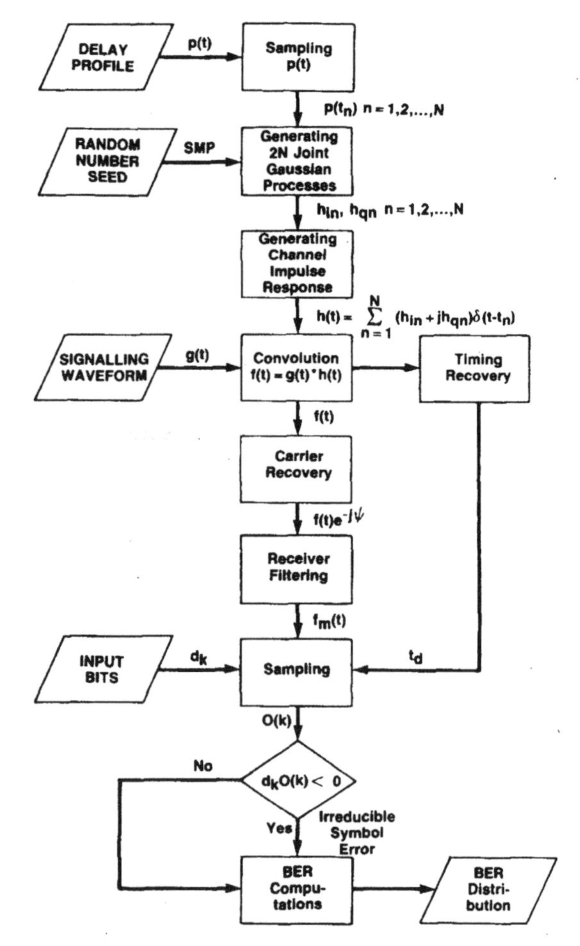

Fig. 2. The flowchart of a computer simulation of a frequency-selective fading channel.

A. Sampling $p(t)$

The power delay profile $p(t)$ is sampled at $t = t_n$ where $n = 1, 2, \cdots, N$. The spacing between time instants is chosen according to the accuracy needed.

In addition to a measured profile [11] as shown in Fig. 3, three idealized profiles are considered:

a) Gaussian profile

$$p(t) = \frac{1}{\sqrt{2\pi}\tau} \exp\left\{ -\frac{1}{2}(t/\tau)^2 \right\}. \tag{9a}$$

b) One-sided exponential profile

$$p(t) = \begin{cases} \frac{1}{\tau} \exp\{-t/\tau\} & \text{for } t \geq 0 \\ 0 & \text{for } t < 0. \end{cases} \tag{9b}$$

c) Equal amplitude two-ray profile

$$p(t) = \tfrac{1}{2}[\delta(t - \tau) + \delta(t + \tau)]. \tag{9c}$$

Both the Gaussian and the two-ray profiles are represented

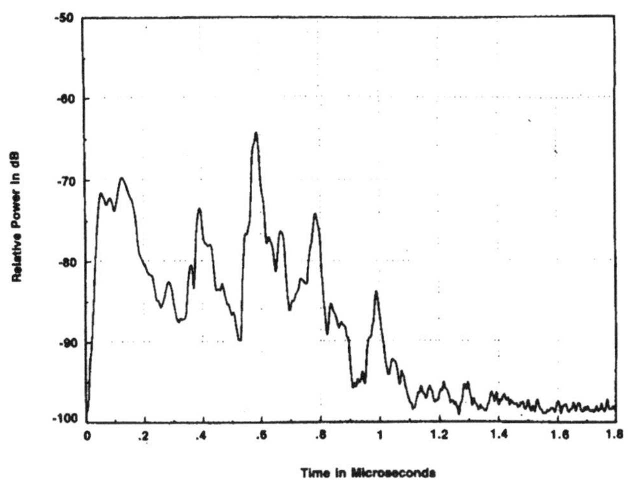

Fig. 3. A power delay profile obtained from measurements done in an office building; the rms delay spread is approximately 250 ns.

by noncausal functions because the average delays are removed for mathematical convenience.

The channel impulse response at t_n, denoted by $h(t_n) = h_{in} + jh_{qn}$, is a complex zero-mean Gaussian random process with variance $p(t_n)$; therefore, h_{in} and h_{qn} are uncorrelated Gaussian processes each with variance $p(t_n)/2$.

B. Generating Gaussian Random Processes and h(t)

From (4), the covariance matrix for the $2N$-dimension Gaussian random "vector" $\{h_{i1}, h_{q1}, \cdots, h_{iN}, h_{qN}\}$ can be expressed as

$$M = \begin{bmatrix} \frac{p(t_1)}{2} & & & & \\ & \frac{p(t_1)}{2} & & 0 & \\ & 0 & \ddots & & \\ & & & \frac{p(t_N)}{2} & \\ & & & & \frac{p(t_N)}{2} \end{bmatrix}. \quad (10)$$

A subroutine is called to generate the $2N$-deviate Gaussian random vectors with a covariance matrix M. A sample of the channel impulse response can then be constructed by the following formula:

$$h(t) = \sum_{n=1}^{N} (h_{in} + jh_{qn})\,\delta(t - t_n). \quad (11)$$

C. Convolving the Modulated Waveform with h(t)

All the modulations treated here except GMSK can be simulated as quadrature amplitude modulation (QAM) with a signaling waveform $g(t)$ of symbol period T. GMSK is simulated as a form of frequency modulation (FM).

1) Unfiltered BPSK, QPSK, OQPSK: Signaling with a rectangular pulse,

$$g(t) = g_p(t) \equiv \begin{cases} 1 & -T/2 \le t \le T/2 \\ 0 & \text{otherwise.} \end{cases} \quad (12a)$$

Note that 1) BPSK is simulated by one input bit sequence; 2) QPSK and OQPSK are simulated by two independent input bit sequences; and 3) a time offset of $T/2$ offset is introduced between the two sequences for OQPSK.

2) MSK: Signaling with a sinusoidal pulse,

$$g(t) = g_m(t) \equiv \begin{cases} \cos\frac{\pi t}{T} & -T/2 \le t \le T/2 \\ 0 & \text{otherwise.} \end{cases} \quad (12b)$$

Note that MSK is simulated in the same way as OQPSK except with a different signaling waveform.

3) QPSK with raised-cosine Nyquist signaling pulse [7] (RC-QPSK): Signaling with a pulse with the following form.

$$g(t) = g_c(t) \quad (12c)$$

where $g_c(t)$ can be realized by a filter that has the square root of a raised-cosine spectrum with a roll-off factor α. This signaling waveform and its matched filter form an ISI-free Nyquist pulse in the absence of delay spread.

4) GMSK: Signaling with a pulse with the following form.

$$g(t) = g_p(t) * g_g(t) \quad (12d)$$

where $g_p(t)$ is a rectangular pulse expressed by (12a) and $g_g(t)$ is the impulse response of a Gaussian filter whose width is controlled by the BT_b (3 dB bandwidth normalized by bit rate) product [14].

The linear combination of binary pulses is used to frequency-modulate the carrier, with the total phase change for each pulse being $\pm\pi/2$.

With the exception of GMSK, the received waveform is a linear combination of $g(t) * h(t)$.

D. Carrier Recovery for Coherent Detection

For a portable communications system using around 1 GHz for the carrier and several hundred Kbits/s for the signaling rate, the following assumptions are reasonable.

a) The bandwidth of the carrier recovery circuit is assumed to be *much higher* than the channel fading rate; therefore, the carrier phase can be tracked as if $h(t)$ is time-invariant.

b) The bandwidth of the carrier recovery loop is *much lower* than the symbol rate; therefore, for an idealized carrier recovery circuit that removes the modulation, the carrier phase is extracted from the average of $h(t)$ over several symbol periods.

Phase jitter on the recovered carrier caused by Gaussian noise is neglected as we focus on the effects of delay spread; therefore, the low SNR performance should be viewed as the idealized case, which is reasonable because of b), above.

For $d \leq 0.2$, the average of $h(t)$ over several symbols can be approximated by the average over the whole time axis; as a result, the recovered carrier phase is then

$$\psi \approx \text{phase of} \int h(t)\, dt = \text{phase of } H(f)|_{f=0} \quad (13)$$

where $H(f)$ is the Fourier transform of $h(t)$.

The recovered carrier can be represented by

$$e^{j\psi} = \frac{H(0)}{|H(0)|}. \quad (14)$$

Carrier recovery is then treated by replacing $h(t)$ by $h(t)e^{-j\psi}$ in the simulation.

E. Receiver Filtering

For all the modulations considered except GMSK, the optimal receiver filter in the absence of delay spread is the matched filter with impulse response $g^*(-t)$; this is the receiver filter assumed in the simulation. For GMSK, we simulate the type of receiver described by Murota and Hirada [14]; this is a parallel implementation of the MSK receiver, which is a suboptimal receiver for GMSK in the absence of delay spread [15].

F. Timing Recovery and Sampling

A squaring timing loop [16] is used in the simulation to recover the timing clock. The timing delay t_d caused by a multipath channel was analyzed in an earlier paper [17]. The timing jitter caused by Gaussian noise is neglected; therefore, the BER computed under the low SNR condition is a lower bound.

For small d, the recovered timing tracks the centroid of $|h(t)|^2$. Two cases of timing recovery are considered: If the fading rate is much higher than the bandwidth of the timing recovery loop, $p(t)$ is used to generate detection timing; whereas the short-term $|h(t)|^2$ is used when the fading rate is much lower than the bandwidth of the timing recovery loop. It will be seen later that the BER performances are about the same for both cases if d is small.

Once t_d is computed, the demodulated waveform $r_m(t)$ is sampled at $t = kT + t_d$ for the kth bit; the result, $O(k)$, is used for symbol detection.

G. Symbol Detection and BER Computation

The quantity $O(k)$ is normalized in such a way that it has a 0 dB rms value in the absence of delay spread. The effect of AWGN is then included in the following ways for the coherent detection:

a) If $O(k)d_k \geq 0$, errors occur if AWGN causes errors; the resulting short-term BER, $P_e(k)$, is

$$P_e(k) = \frac{1}{2} \operatorname{erfc}\left(\sqrt{O^2(k)\frac{E_b}{N_0}}\right) \quad (15a)$$

where E_b/N_0 is average signal-to-noise ratio per bit at the input of the receiver and "erfc" is the complementary error function.

b) If $O(k)d_k < 0$, errors occur if AWGN causes no errors; therefore,

$$P_e(k) = 1 - \frac{1}{2} \operatorname{erfc}\left(\sqrt{O^2(k)\frac{E_b}{N_0}}\right). \quad (15b)$$

For differential detection of BPSK, sampled values of $r_m(t)\, r_m^*(t - T)$ are computed to determine the irreducible BER. To compute the short-term BER when AWGN is present, the BER formula, $(1/2)e^{-E_b/N_0}$, is not exact because differential detection is not a linear operation. However, we can get a reasonable approximation for the conditions assumed in this paper by multiplying the signal-to-noise ratio by $|O(k)\, O(k - 1)|$ and then using the above BER formula. The reasons are as follows:

a) This formula is exact for $E_b/N_0 \rightarrow \infty$, i.e., when AWGN is not present; therefore, the irreducible BER is accurate and the short-term BER at high SNR is a good approximation.

b) This formula is exact if $|O(k)| = |O(k - 1)|$ because the noise processes are effectively changed by the same ratio for the two bits. It will be seen from the next section that ISI is much smaller than the signal component unless the signal is in a deep fade as a result of multipath cancellation for the range $0.02 \leq d \leq 0.2$ simulated; therefore, $|O(k)| \approx |O(k - 1)|$ in most cases. When the signal is in a deep fade, $|O(k)\, O(k - 1)|$ is very small, the BER is very large, the channel is unusable, and the exact value of the BER is of less interest.

Since the comparison of the delay-spread effects on the two detection methods is based on the high SNR (irreducible BER) performance, the low SNR region where the approximation is less accurate does not effect the conclusions reached in the paper. The approximate BER calculation for differential detection should be used with caution, especially when SNR is low and the value of d is large.

For the simulation of GMSK, only the irreducible BER is computed.

For all the modulations treated except GMSK, $O(k)$ is a weighted sum of the samples of $f_m(t)$ (coherent detection) or $f_m(t)\, f_m^*(t - T)$ (differential detection) where $f_m(t) = g(t) * h(t) * g^*(-t)$. For example, $O(k)$ can be expressed in the following form for coherent QPSK,

$$O(k) = \operatorname{Re}\left\{\sum_{n=-\infty}^{\infty} (d_{n1} + jd_{n2}) f_m[(k - n)T + t_d]\right\} \quad (16)$$

where Re denotes the real part and $d_{n1}, d_{n2} = \pm 1$. It is clear that $d_{k1} f_m(t_d)$ is the desired signal term in the kth interval, while other terms cause intersymbol interference. Interference between I and Q rails in general exists ("cross-rail interference").

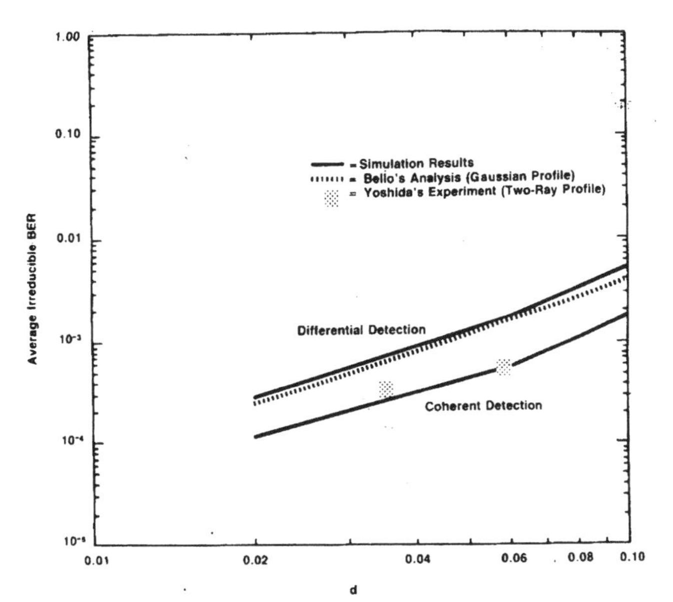

Fig. 4. Comparison of results of the simulation with those obtained by analysis and experiment.

IV. Results for the Unfiltered Modulations

In this section, results for the unfiltered modulations are presented. Major parameters in the simulation are 1) type of modulation: BPSK, QPSK, OQPSK, and MSK; 2) type of detection: coherent and differential; 3) delay profile and its normalized rms delay spread; 4) ratio of the bandwidth B of the timing recovery loop to the fading rate F; and 5) the signal-to-noise ratio: E_b/N_0.

Simulations using different profiles indicate that the BER performance is not sensitive to the shape of the delay profile, for the range of d simulated. Examples involving different profiles are shown in the following; however, the results apply regardless of the profile used.

A. *Verification of the Simulation with Results in the Existing Literature*

Bello and Nelin [4] calculated the irreducible BER averaged over the multipath fading samples for differentially detected BPSK in a channel with a Gaussian delay profile. Yoshida *et al.* [18] performed microscopic BER measurements for coherent BPSK with a two-ray profile. Fig. 4 indicates that the results of the present simulations compare well with the available analytical and experimental results.

B. *The Error Mechanisms and the Effects of the Timing Recovery Circuit*

In Figs. 5 and 6, two examples are shown to illustrate the error mechanisms for small d; a rectangular signaling pulse with a matched receiver filter is considered. Both figures indicate how three consecutive symbols combine and interfere with one another to form the output $O(k)$ at the kth detection timing. The channel impulse response in each example has two equal-amplitude "rays" which are in-phase for Fig. 5 and out-of-phase for Fig. 6.

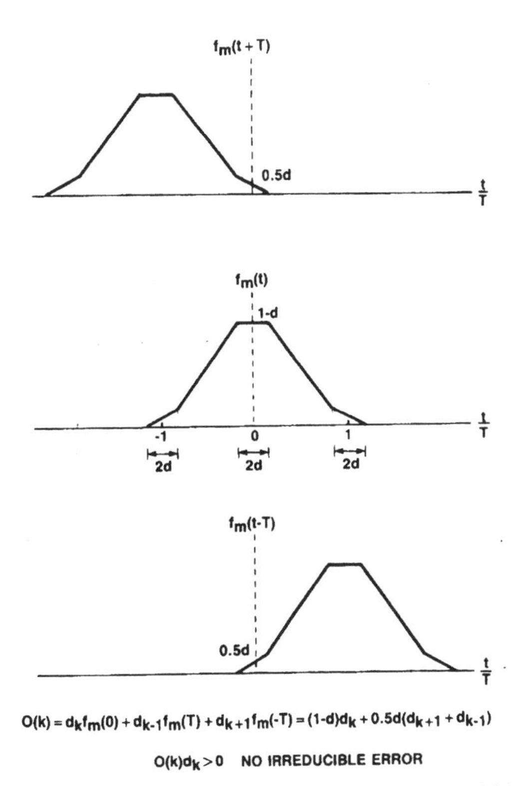

Fig. 5. Illustration of the error mechanism by using a two-ray model (equal-amplitude and in-phase rays).

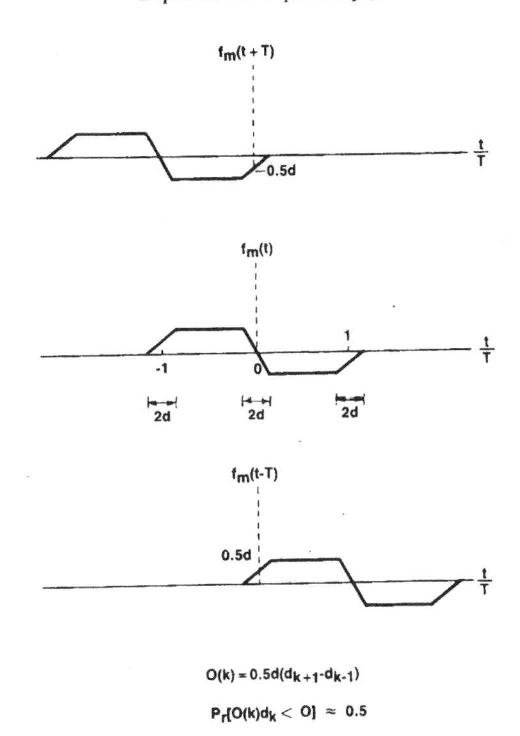

Fig. 6. Illustration of the error mechanism by using a two-ray model (equal-amplitude and out-of-phase rays).

The "irreducible" errors (i.e., those made at very high SNR) cannot occur unless ISI outweighs the signal component at the sampling instant; therefore, there are three error mechanisms: 1) A faded signal component caused

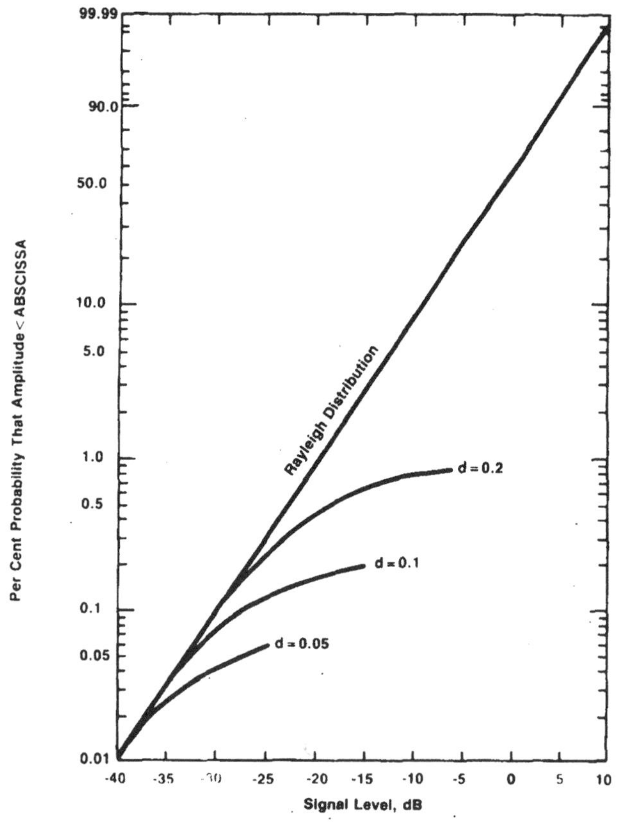

Fig. 7. Amplitude distributions of the detected waveform when irreducible symbol errors occur for coherent BPSK in a channel with a two-ray profile. The Rayleigh distribution results for the entire set of $h(t)$.

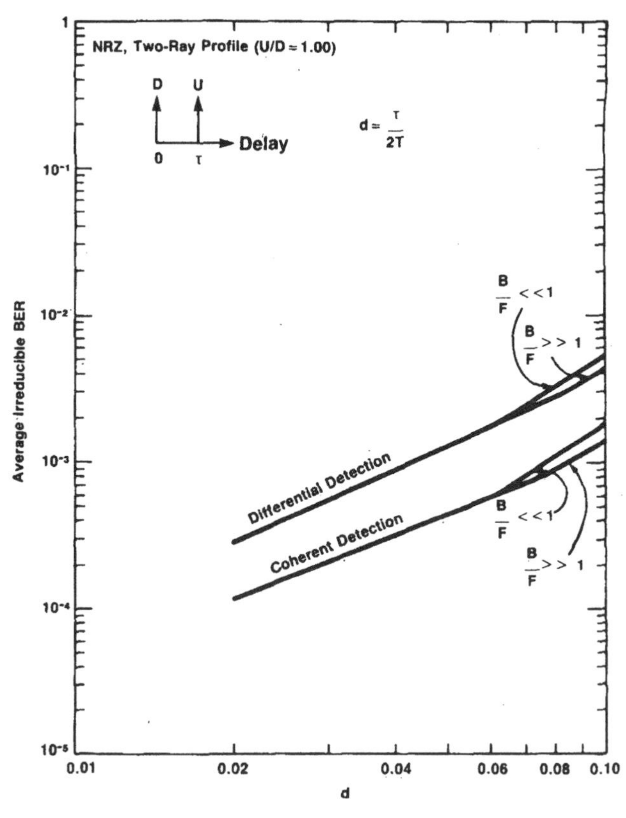

Fig. 8. The irreducible BER performance for BPSK with two kinds of detection and two B/F ratios. (two-ray profile)

by multipath cancellation; 2) ISI caused by nonzero d; and 3) shift of sampling timing as a result of delay spread. It is clear from both figures that, for small d, 1) ISI causes a very small perturbation at the tail end; and 2) a small shift of sampling timing has a negligible effect on the irreducible BER. Therefore, the major error mechanism is signal fading in this case. For example, the signal component is much stronger than ISI in Fig. 5; as a result, irreducible errors will not occur. On the contrary, the signal component for Fig. 6 is in a deep fade and $O(k)$ depends only one two adjacent symbols; a BER of about 0.5 is expected because the output bit is uncorrelated with the corresponding input bit.

It is indicated by the simulation that the occurrence of irreducible symbol errors is very bursty. For example, for a channel with a Gaussian profile and $d = 0.05$, a simulation for differential detection of BPSK yields a 1.5×10^{-3} average irreducible BER. In this simulation, only 10 out of 2000 samples of $h(t)$ result in irreducible symbol errors; in these 10, the BER is very high.

To determine whether the burst of errors is a result of envelope fading, cumulative distributions of the amplitude of $O(k)$ when the irreducible symbol error occurs are computed. The distributions shown in Fig. 7 correspond to two-ray channels with $d = 0.05$, 0.1 and 0.2 for coherent detection of BPSK, along with a Rayleigh distribution, which is the amplitude distribution of the entire sample space of $h(t)$ (with or without symbol errors). The rms signal level for the Rayleigh-faded signal is 0 dB. All cumulative distributions are computed as probabilities in the entire sample space. It is indicated that $O(k)$ is always in a fade when irreducible symbol errors occur; for example, the depth of fading is at least 15 dB with respect to the rms signal level for $d \leq 0.1$.

To determine the influence of timing error on the irreducible BER, more simulations were performed. In the case of a two-ray propagation environment, a comparison of the average irreducible BER for a system using a very fast squaring timing recovery circuit (i.e., loop bandwidth B is much higher than the fading rate F) with one using a very slow timing circuit (i.e., $B \ll F$) for BPSK is shown in Fig. 8. It shows that if delay spread is not severe, a very fast timing loop improves the irreducible BER performance only slightly. This suggests that timing error is not the major mechanism for the bursty irreducible symbol errors. Simulations using other profiles indicate the same result. In the following, only results for $B \ll F$ are shown.

In summary: 1) For small delay spread, envelope fading is the most important mechanism causing error bursts; and 2) for severe delay spread, extrapolation of Fig. 8 suggests that timing error could be a significant factor if the timing recovery circuit is not fast enough.

The most significant implication of this result is that diversity selection can be effective in this case of small

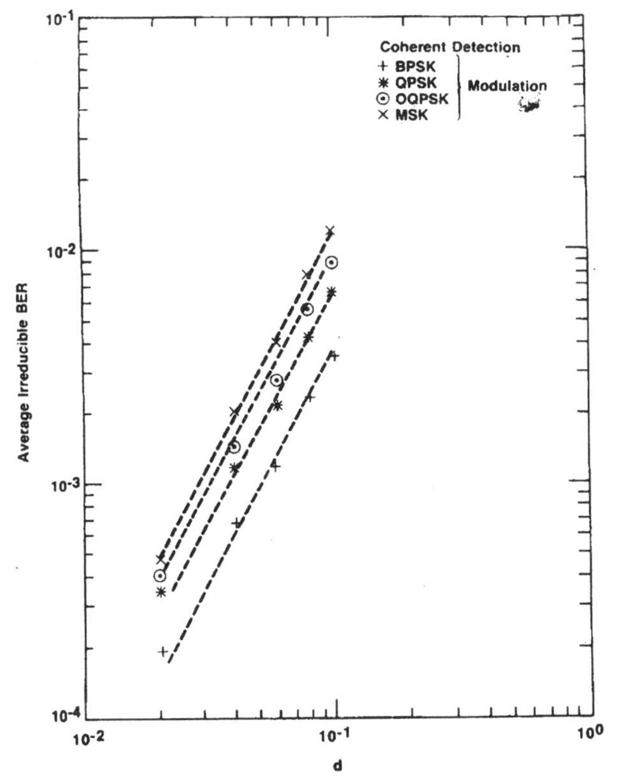

Fig. 9. The irreducible BER performance for different modulations with coherent detection for a channel with a Gaussian profile. The parameter d is the rms delay spread normalized by symbol period.

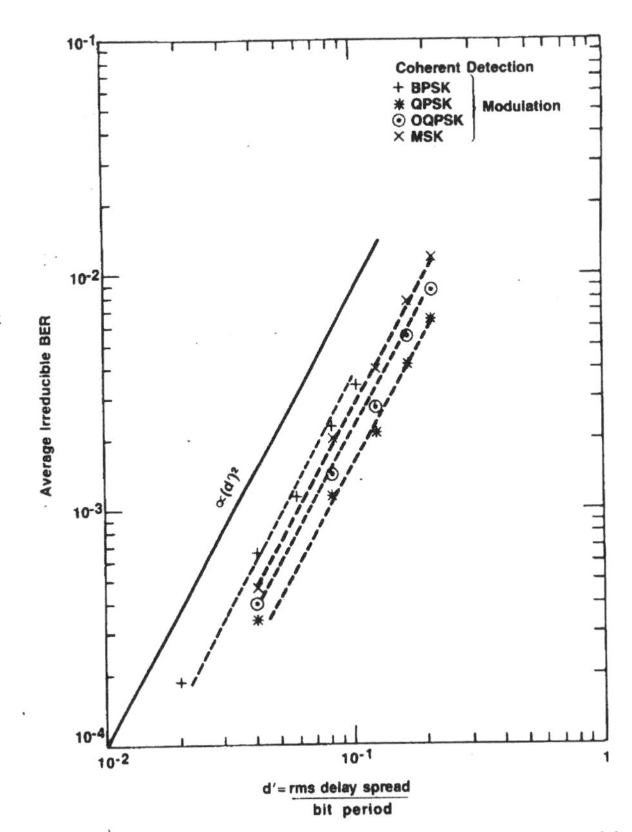

Fig. 10. The same set of curves as in Fig. 9, plotted against rms delay spread normalized by bit period.

delay spread because an irreducible symbol error rarely occurs for a diversity branch that has a high received power.

C. *Comparison of Modulation and Detection Methods*

Fig. 8 indicates that coherent detection performs better than differential detection; we shall focus the discussion on coherent detection.

Fig. 9 shows the average irreducible BER as functions of d for different unfiltered modulation methods with coherent detection; the multipath channel is simulated by using a Gaussian delay profile. This figure indicates that the delay spread performance of various unfiltered modulations when normalized to the same symbol period is ranked in the following order: 1) BPSK, 2) QPSK, 3) OQPSK, 4) MSK. The performance of BPSK is the best because cross-rail interference does not exist. Both OQPSK and MSK have a $T/2$ timing offset between two bit sequences, hence the cross-rail ISI is more severe; therefore, their performances are inferior to that of QPSK.

In Fig. 9, the normalization factor for parameter d is the symbol period T, during which two bits are transmitted over the channel for QPSK, OQPSK, and MSK, while only one bit is sent for BPSK. A fairer comparison of performance for the same information capacity should be based on $d' = \tau / T_b$ where τ is the rms delay spread and T_b is the bit period. Fig. 10 is the same set of functions as those in Fig. 9 plotted against d'. When this normalization is applied, it is clear that 4-level modulations (QPSK, OQPSK, and MSK) are more resistant to delay spread than BPSK for constant information throughput.

Higher level modulations were also considered. For example, Fig. 11 indicates that the performance of 8-PSK as SNR approaches infinity is not superior to that of QPSK even though it derives 3 bits per symbol. Since higher level modulations are less efficient than 4-level modulations at low SNR, we shall concentrate on 4-level modulations in the next section.

It is also interesting to note that all the curves shown in Figs. 9, 10, and 11 are nearly parallel to a straight line of slope 2; that is, an order of magnitude increase in delay spread results in about two orders of magnitude increase in the irreducible BER within the range of d simulated. By showing that the group delay is a Student's t distribution with two degrees of freedom, Andersen *et al.* [13] have proved that the irreducible BER caused by frequency-selective fading is proportional to d^2 when d is small and that the proportionality constant depends on the method of modulation and detection. The results of the simulation are consistent with this earlier result.

D. *Cumulative Distribution of BER*

Because symbol errors are very bursty, it is important to predict the cumulative distribution functions (cdf) of BER, namely, the probability that short-term BER performance is worse than a certain value, say, 10^{-3}. Fig.

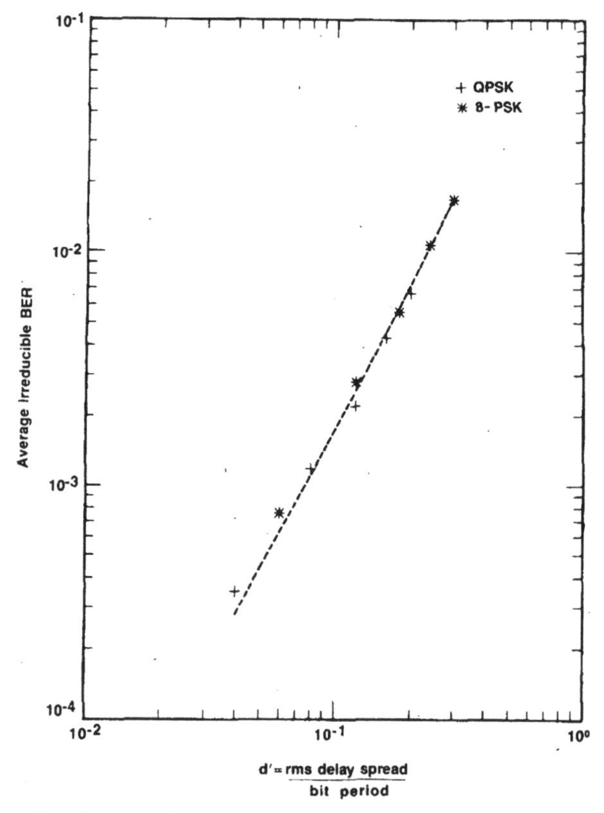

Fig. 11. The irreducible BER performance for QPSK and 8-PSK. The parameter d' is the rms delay spread normalized by bit period.

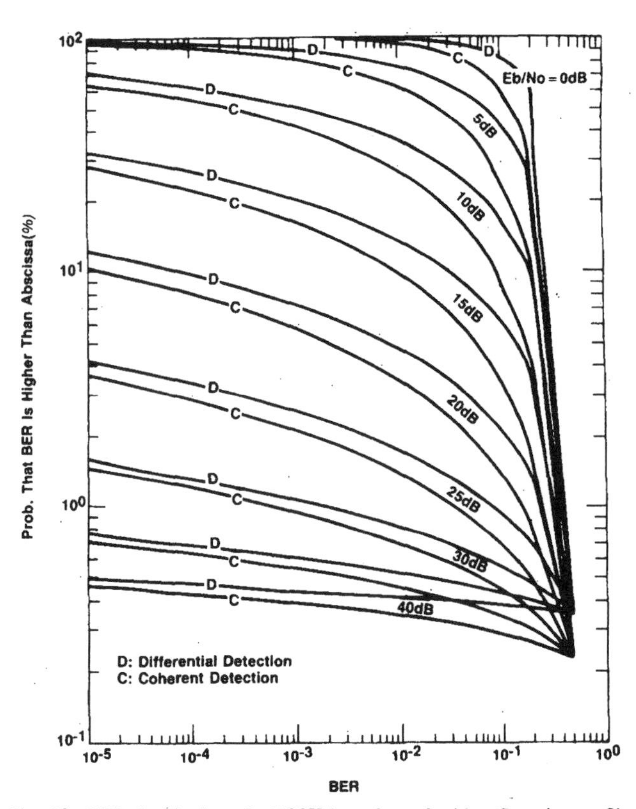

Fig. 12. BER distributions for BPSK in a channel with a Gaussian profile (d = 0.08).

12 is a typical cdf plot which includes a set of distribution curves for BPSK with two detections (coherent and differential) simulated by using a Gaussian profile with d = 0.08. In the example set of curves, the average E_b/N_0 (sometimes called the "local mean") is varied from 0 dB to 40 dB in 5 dB steps so that a change of signal level due to shadow fading and other large-scale variations can be accounted for.

For each cdf plot, the abscissa indicates a better BER performance in the left-hand part of the figure while the ordinate indicates a more reliable coverage in the lower part of the figure; therefore, the curve appearing in the left-most and lowest position represents the "best" combination. Both axes are expressed in a log scale. It is easy to see from Fig. 12 that, as expected, 1) coherent detection is better than differential detection and 2) as SNR increases, the performance gets better; however, diminishing returns are observed as the "irreducible BER" is approached.

V. Results for Modulations with Spectral-Shaping Filters

Two classes of spectrally efficient modulation methods that have been considered for practical applications in mobile and satellite communications are considered: 1) Gaussian-filtered minimum shift keying (GMSK) [14] and 2) QPSK with a raised-cosine Nyquist pulse (RC-QPSK) [19].

A. General

Similar to the unfiltered modulations in Section IV, we find that: 1) Timing jitter caused by small delay spread is not crucial; 2) coherent detection is the more desired choice; 3) the BER performance is relatively insensitive to the shape of the delay profiles, but this sensitivity increases with d; and 4) the irreducible BER increases about two orders of magnitude as d increases from 0.02 to 0.2.

A measured power delay profile with about 250 ns rms delay spread as shown in Fig. 3 was used to simulate a real-world portable communications channel. Only results for coherent detection will be shown.

B. RC-QPSK

Fig. 13 shows the average irreducible BER performance of RC-QPSK as a function of roll-off factor α, along with that for unfiltered QPSK, for different values of rms delay spread. As α increases, the irreducible BER for a given value of d decreases monotonically due to decreasing ISI; however, the spectral occupancy also increases. It is interesting to note that RC-QPSK with $\alpha \geq$ 0.75 is more resistant to delay spread than the unfiltered NRZ-QPSK.

C. GMSK

Fig. 14 shows the BER performance as a function of the BT_b product of the GMSK premodulation filter, along with that for unfiltered MSK modulation. It is found that the best BER performance is achieved by choosing BT_b =

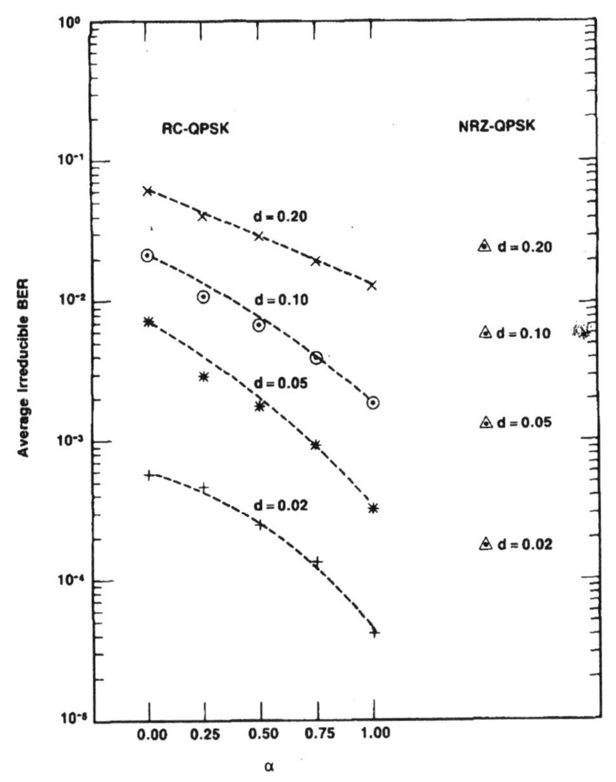

Fig. 13. The irreducible BER performance for RC-QPSK with coherent detection for a channel with a delay profile as shown in Fig. 3. Results for the unfiltered QPSK are also shown for comparison. The parameter α is the roll-off factor in raised-cosine filter. The parameter d is the rms delay spread normalized by symbol period.

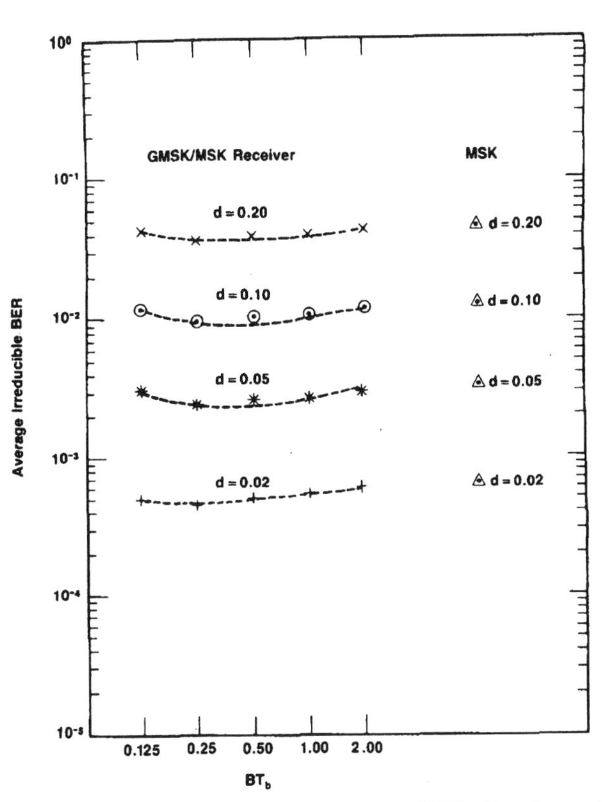

Fig. 14. The irreducible BER performance for GMSK with coherent detection for a channel with a delay profile as shown in Fig. 3. Results for the unfiltered MSK are also shown for comparison. The parameter BT_b is the 3-dB bandwidth of the premodulation Gaussian filter normalized by bit rate. The parameter d is the rms delay spread normalized by sumbol period.

0.25 for the premodulation filter; however the performance is not very sensitive to BT_b. This result is somewhat surprising at the first glance because it suggests that we get no penalty in performance for better spectral efficiency until the spectrum is compressed too much. The reason becomes clear if we examine how the limiting case, i.e., MSK, can be generated. For MSK, two bit sequences (I and Q rails) are offset by T_b; therefore, cross-rail coupling is the dominant impairment when ISI is not severe. GMSK cannot be generated in the same way since no equivalent linear filters can be found for the two rails; however, similar orthogonal decomposition can be used to explain the results. Although ISI from the inphase rail is higher for a smaller BT_b product, the cross-rail interference is reduced by a smoother signaling pulse as a result of a sharper filter. The balance between I rail ISI and Q rail ISI makes the BER performance relatively insensitive to the change of BT_b product until it is lower than 0.25. In the absence of delay spread, no cross-rail interference is present because I and Q rails are completely decoupled after carrier recovery; therefore, GMSK performance suffers some degradation as compared to MSK for better spectral efficiency [14].

D. Design Considerations and Example

The maximum allowable d for a channel *without adaptive equalization and diversity combining* can be determined from cdf curves for a given performance criterion. Fig. 15 shows the cdf curves for RC-QPSK with $\alpha = 0.25$ at 30 dB average E_b/N_0 for different values of rms delay spread. It can be seen that, for example, to guarantee a better than 10^{-3} BER performance at 30 dB average E_b/N_0 for 99 percent of the samples of $h(t)$ for a given delay profile, the value of d' should not exceed 0.1; this corresponds to a 2.5 μs minimum bit period for the 250 ns rms delay spread profile shown in Fig. 3. Therefore, the maximum digital bit rate for these conditions would be about 400 Kbits/s.

By employing coherent GMSK modulation with BT_b = 0.25, about the same performance is achieved; these modulations are also comparable in terms of spectral occupancy [14], [19]. RC-QPSK can trade bandwidth occupancy for increased resistance to delay spread but not constant-envelope, like GMSK is.

Different criteria can be used to determine the proper signaling rate for a system. If the reliability requirement is relaxed to be 90 percent of the samples of $h(t)$, the value of d can be higher than 0.2 even with $\alpha = 0.0$; however, other impairments could be even more serious than delay spread if α is too small.

Diversity can be used to reduce the BER floor for the same average SNR and d or to achieve the same BER with a large d or a smaller average SNR [5], [6]; therefore, a

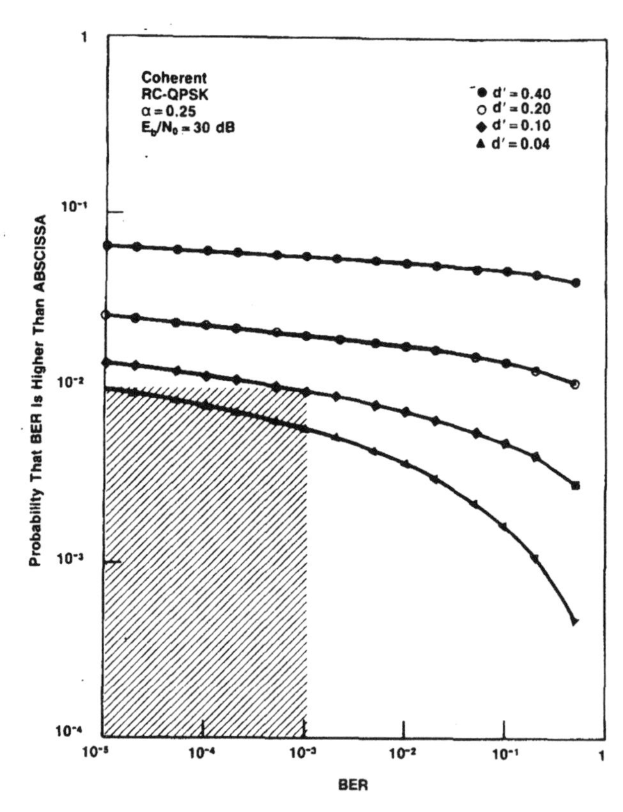

Fig. 15. Determining the maximum allowable signaling rate by using a cdf plot. The parameter d' is the rms delay spread normalized by bit period.

higher signaling rate is possible with even better performance if diversity is implemented.

VI. Conclusion

A flexible simulation for evaluating the BER performance of a frequency-selective, slowly fading digital radio channel has been described. Results from the simulation for normalized delay spread in the range of $0.02 \leq d \leq 0.2$ are consistent with the following performance characteristics:

a) Only a small fraction of the multipath channel impulse responses encountered will exhibit "irreducible" bit errors; however, once channel conditions which cause errors occur, the resulting short-term BER is very high.

b) The major error mechanism is envelope fading; the degradation due to timing error caused by a small delay spread is not significant.

c) Since the irreducible errors occur only when the signal is in a deep fade, diversity will be an effective way to lower the irreducible BER or to permit higher transmission rates for a given delay spread.

d) The BER performance is more sensitive to the rms value of the delay spread than to the shape of the delay profile. By measuring the rms delay spread, both symbol rate and the modulation/detection scheme can be chosen using the simulation. A power-of-two dependence on the rms delay spread for the irreducible BER is found.

e) The BER averaged over fading samples of a given delay profile provides a comparison among different modulation/detection schemes; cumulative distributions can be used to determine the allowable signaling rate.

f) Coherent detection is more resistant to delay spread than is differential detection.

g) 4-level modulation yields greater information rates for a given delay spread than does 2-level modulation.

h) GMSK with $BT_b = 0.25$ is near optimum for resistance to delay spread.

i) RC-QPSK performance degrades monotonically as the roll-off factor is reduced from 1 to 0.

j) It is possible to transmit a few hundred kbits/s using a TDM/TDMA architecture in a typical portable radio environment without diversity or equalization.

Acknowledgments

The author would like to thank D. C. Cox, P. T. Porter, and H. W. Arnold, for their discussions and guidance during the course of this work. Special thanks are due to D. Devasirvatham for providing a delay profile measured by him to be used in the simulation. Finally, the author is grateful to the anonymous reviewers for their comments that improve the quality of this paper.

References

[1] D. C. Cox, "Universal portable radio communications," *IEEE Trans. Veh. Technol.*, vol. VT-34, pp. 117-121, Aug. 1985.
[2] K. Raith, J. Stjernvall, and J. Uddenfeldt, "Multipath equalization for digital cellular radio operating at 300 Kbit/s," in *Proc. 1986 IEEE VT Conf.*, May 20-22, 1986.
[3] D. C. Cox, H. W. Arnold, and P. T. Porter, "Universal digital portable communications: An applied research perspective," in *Proc. ICC'86*, June 22-25, 1986.
[4] P. A. Bello and B. D. Nelin, "The effects of frequency selective fading on the binary error probabilities of incoherent and differentially coherent matched filter receivers," *IEEE Trans. Commun. Syst.*, vol. CS-11, pp. 170-186, June 1963.
[5] B. Glance and L. J. Greenstein, "Frequency-selective fading effects in digital mobile radio with diversity combining," *IEEE Trans. Commun.*, vol. COM-31, pp. 1085-1094, Sept. 1983.
[6] J. H. Winters and Y. S. Yeh, "On the performance of wideband digital radio transmission within buildings using diversity," in *Proc. GLOBECOM'85 Conf.*, Dec. 1985.
[7] J. G. Proakis, *Digital Communications*. New York: McGraw-Hill, 1983.
[8] W. C. Jakes, Ed., *Microwave Mobile Communications*. New York: Wiley, 1974.
[9] P. A. Bello, "Characterization of randomly time-variant linear channels," *IEEE Trans. Commun. Syst.*, vol. CS-11, pp. 360-393, Dec. 1963.
[10] D. C. Cox and R. P. Leck, "Correlation bandwidth and delay spread multipath propagation statistics for 910 MHz urban mobile radio channels," *IEEE Trans. Commun.*, vol. COM-23, pp. 1271-1280, Nov. 1975.
[11] D. M. J. Devasirvatham, "Time delay spread measurements of wideband radio signals within a building," *Electron. Lett.*, vol. 20, pp. 950-951, Nov. 1984.
[12] D. M. J. Devasirvatham, "Time delay spread and signal level measurements of 850 MHz radio waves in building environments," *IEEE Trans. Antennas Propagat.*, vol. 34, pp. 1300-1305, Nov. 1986.
[13] J. B. Andersen, S. L. Lauritzen, and C. Thommesen, "Statistics of

phase derivatives in mobile communications," in *Proc. 1986 IEEE VT Conf.*, May 20–22, 1986.
[14] K. Murota and K. Hirade, "GMSK modulation for digital mobile radio telephony," *IEEE Trans. Commun.*, vol. COM-29, pp. 1044–1050, July 1981.
[15] J. B. Anderson, T. Aulin, and C-E Sundberg, *Digital Phase Modulation*. New York: Plenum, 1986.
[16] J. K. Holms, "Tracking performance of the filter and square bit synchronizer," *IEEE Trans. Commun.*, vol. COM-28, pp. 1154–1158, Aug. 1980.
[17] J. C-I Chuang, "The effects of multipath delay spread on timing recovery," in *Proc. ICC'86*, June 23–25, 1986.
[18] S. Yoshida, F. Ikegami, and T. Takeuchi, "A mechanism of burst error occurrence due to multipath propagation in digital mobile radio," in *Proc. ISAP'85 Conf.*, Aug. 1985, pp. 561–564.
[19] K. Feher, *Digital Communications: Satellite/Earth Station Engineering*. Englewood Cliffs, NJ: Prentice-Hall, 1983.

PHOTO NOT AVAILABLE

Justin C-I Chuang (S'80–M'83) was born in Taiwan on December 2, 1954. He received the B.S.E.E. degree from National Taiwan University, Taipei, Taiwan, in 1977, and the M.S. and Ph.D. degrees in electrical engineering from Michigan State University, East Lasing, in 1980 and 1983, respectively.

From 1979 to 1982, he was a research assistant at MSU doing transient electromagnetics research on radar target discrimination. From 1982 to 1984, he was with GE Corporate Research and Development, Schenectady, NY, where he studied personal and mobile communications. Since 1984, he has been with Bell Communications Research, Red Bank, NJ, where he is a member of Technical Staff in a group conducting research on techniques for future systems to provide tetherless communications.

Dr. Chuang is a member of Phi Kappa Phi.

A Statistical Model for Indoor Multipath Propagation

ADEL A. M. SALEH, FELLOW, IEEE, AND REINALDO A. VALENZUELA, MEMBER, IEEE

***Abstract*—The results of indoor multipath propagation measurements using 10 ns, 1.5 GHz, radarlike pulses are presented for a medium-size office building. The observed channel was very slowly time varying, with the delay spread extending over a range up to about 200 ns and rms values of up to about 50 ns. The attenuation varied over a 60 dB dynamic range. A simple statistical multipath model of the indoor radio channel is also presented, which fits our measurements well, and more importantly, appears to be extendable to other buildings. With this model, the received signal rays arrive in clusters. The rays have independent uniform phases, and independent Rayleigh amplitudes with variances that decay exponentially with cluster and ray delays. The clusters, and the rays within the cluster, form Poisson arrival processes with different, but fixed, rates. The clusters are formed by the building superstructure, while the individual rays are formed by objects in the vicinities of the transmitter and the receiver.**

I. Introduction

The use of radio for indoor data or voice communications, e.g., within an office building, a warehouse, a factory, a hospital, a convention center, or an apartment building, is an attractive proposition. It would free the users from the cords tying them to particular locations within these buildings, thus offering true mobility, which is convenient and sometimes even necessary. It would also drastically reduce wiring in a new building, and would provide the flexibility of changing or creating various communications services in existing buildings without the need for expensive, time-consuming rewiring. The challenge is to offer such services to the majority of the people in a building, not just a selected few. This would most certainly involve a sophisticated local radio communications system whose engineering would require the knowledge of the spatial and temporal statistics of the signal attenuation, the multipath delay spread, and even the impulse response of the indoor radio channel involved.

Various measurements have been reported in the literature for the attenuation of microwave CW signals propagating within buildings [1], [2] or into buildings [3]–[6]. The work of Alexander given in [2], which was done at 900 MHz, is particularly useful since it gives the power law representing the signal attenuation as a function of distance for various types of building.

The first multipath delay spread measurements within a building were reported recently by Devasirvatham [7]. He used a carrier at 850 MHz, biphase-modulated with a 40 Mbit/s maximal-length pseudonoise code, resulting in a 25 ns time resolution. The measurements were made in a large office building (AT&T Bell Laboratories main location in Holmdel, NJ), which occupies an area of 315 × 110 m and is 6 stories high.

In the present paper, we present the results of multipath delay spread and attenuation measurements within a medium-size two-story office building (AT&T Bell Laboratories, Crawford Hill location in Holmdel, NJ), whose first floor plan is sketched in Fig. 1. The external walls of this building are made of steel beams and glass, while the internal walls are almost entirely made of wood studs covered with plasterboard. The rooms contained typical metal office furniture and/or laboratory equipment. The measurements were made by using low-power 1.5 GHz radarlike pulses to obtain a large ensemble of the channel's impulse responses with a time resolution of about 5 ns. Based on these measurements, as well as some of the others cited above, we propose a statistical model of the indoor radio channel, which 1) has enough flexibility to permit reasonably accurate fitting of the measured channel responses, 2) is simple enough to use in simulation and analysis of various indoor communications schemes, and 3) appears to be extendable (by adjusting its parameters) to represent the channel within other buildings.

II. Measurement Setup

A schematic diagram of the measurement setup is shown in Fig. 2. An RF sweep oscillator was used to generate a 1.5 GHz CW signal, which was then modulated by a train of 10 ns pulses with 600 ns repetition period, which is longer than any delay observed in the test building. This radarlike signal was amplified and transmitted via a vertically polarized discone antenna [8] whose radiation pattern is omnidirectional in the horizontal plane. The discone was chosen over a vertical dipole, which has an almost identical radiation pattern, because of its superior bandwidth, which provided the flexibility of changing or sweeping the signaling frequency when desired. The average transmitted power could be adjusted with a step attenuator over a range from a fraction of a nanowatt to several milliwatts.

At the receiver, a second vertically polarized discone antenna was used followed by a low-noise (3 dB noise figure) FET amplifier chain with a 60-dB gain over a frequency range of 1 to 2 GHz. (Actually, the state of polarization of the receiving antenna did not matter from a statistical viewpoint.) The signal was then detected with a sensitive square-law envelope detector whose output was displayed on a computer-controlled digital storage oscil-

Manuscript received January 23, 1986; revised September 24, 1986.
The authors are with AT&T Bell Laboratories, Crawford Hill Laboratory, Holmdel, NJ 07733.
IEEE Log Number 8612162.

Reprinted from *IEEE Journal on Selected Areas in Communications,* vol. SAC-5, no. 2, February 1987.

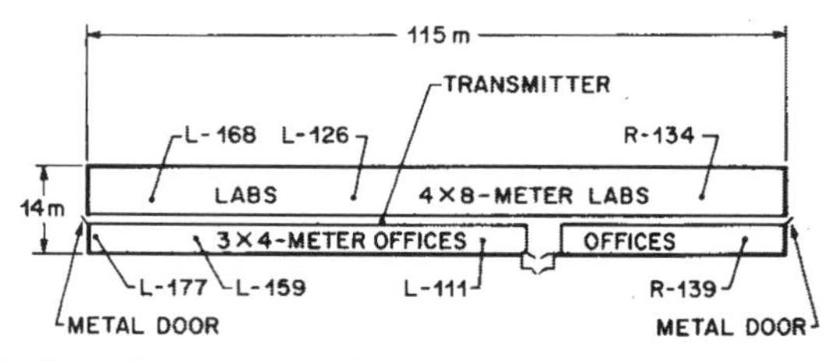

Fig. 1. Plan of the first floor of AT&T Bell Laboratories' Crawford Hill building in Holmdel, NJ.

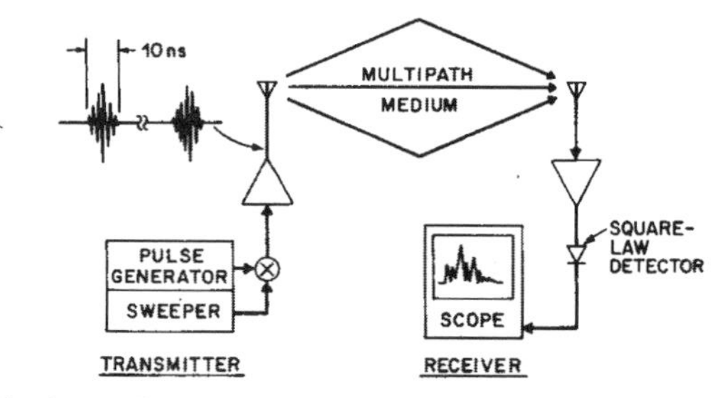

Fig. 2. A schematic representation of the measurements setup.

loscope (Tektronix model 7854). The dynamic range of our setup was more than 90 dB, which was achieved by manually adjusting the step attenuator at the transmitter as well as the oscilloscope's vertical gain.

The transmitter was fixed in the hallway near the center of the first floor of the building (see Fig. 1) with its antenna located at a height of about 2 m. The receiver, with the antenna at the same height, was moved to collect measurements in the hallway and in several rooms throughout the same floor. (No floor-to-floor penetration measurements are presented since the signal level in that case was generally too small to detect with our setup.) A total of 8 rooms were measured with about 25 pulse responses at various locations within each room. Both the transmitter and the receiver were stationary during the acquisition of each pulse response. For reasons to be explained in the next section, the transmitted signal frequency was swept at a rate of about 100 Hz over $\pm$ 100 MHz around the 1.5-GHz center frequency. The waveform processing capability of the oscilloscope was used to average the received frequency-swept pulse responses, and to transfer that average to a computer for storage. The averaging process took about 15 s to complete. To freeze the channel during that time, we made sure that people in the vicinity of the transmitting and receiving antennas, or those in the hallway between the antennas, stopped moving. (People moving within their offices, or in other locations in and around the building, had a negligible effect on the measurements.) A coaxial cable was used to trigger the oscilloscope from the transmitter's pulse generator to guarantee a stable timing reference under our swept conditions.

III. Mathematical Formulation of the Method of Measurements

Each transmitted RF radarlike pulse has the complex time domain representation

$$x(t) = p(t)\, e^{j(\omega t + \phi)}, \tag{1}$$

where $p(t)$ is the baseband pulse shape (which in our case has a width of about 10 ns), ω is the RF angular frequency (which is nominally $2\pi \times 1.5$ GHz in our experiment), and ϕ is an arbitrary phase.

The channel is represented by multiple paths or rays having real positive gains $\{\beta_k\}$, propagation delays $\{\tau_k\}$, and associated phase shifts $\{\theta_k\}$, where k is the path index; in principle, k extends from 0 to ∞. Thus, the complex, low-pass channel impulse response is given by [9]

$$h(t) = \sum_k \beta_k e^{j\theta_k} \delta(t - \tau_k), \tag{2}$$

where $\delta(\cdot)$ is the Dirac delta function.

Because of the motion of people and equipment in and around the building, the parameters β_k, τ_k, and θ_k are randomly time-varying functions. However, the rate of their variations is very slow compared to any useful signaling rates that are likely to be considered, e.g., higher than tens of kbit/s. Thus, these parameters can be treated as virtually time-invariant random variables.

It follows from (1) and (2) that the received signal, which is the time convolution of $x(t)$ and $h(t)$, is given by

$$y(t) = \sum_k \beta_k p(t - \tau_k)\, e^{j[\omega(t-\tau_k) + \phi + \theta_k]}. \tag{3}$$

Upon passing through the square-law envelope detector, as indicated in Fig. 2, the power profile displayed on the oscilloscope becomes

$$|y(t)|^2 = \sum_k \sum_l \{\beta_k \beta_l p(t - \tau_k)\, p(t - \tau_l) \cdot e^{j[\theta_k - \theta_l + \omega(\tau_l - \tau_k)]}\}. \tag{4}$$

If there were no overlap of pulses, i.e., in our experiments, if $|\tau_k - \tau_l| \geq 10$ ns when $k \neq l$, then (4) would have reduced to

$$|y(t)|^2 = \sum_k \beta_k^2 p^2(t - \tau_k), \quad \text{no overlap}, \tag{5}$$

which is much simpler to use than (4) in estimating the β's and the τ's from the measured $|y(t)|^2$ waveform.

To obtain the simplification of (5) even with pulse overlap, we make the reasonable assumption that the θ's are statistically independent uniform random variables over $[0, 2\pi)$. In this case, the mathematical expectation of (4) with respect to the θ's yields

$$E_\theta\{|y(t)|^2\} = \sum_k \beta_k^2 p^2(t - \tau_k), \tag{6}$$

which is identical to (5) but allows for pulse overlap.

The mathematical expectation with respect to the θ's was, in effect, accomplished in our experiment by averaging the oscilloscope's power waveform while sweeping the transmitted RF frequency. Indeed, if the total swept bandwidth Δf is sufficiently large, it can be shown using

(4) that the frequency-average power profile is given by

$$s(t) \equiv \frac{1}{\Delta f} \int_{-\Delta f/2}^{\Delta f/2} |y(t)|^2 \, df \approx \sum_k \beta_k^2 p^2(t - \tau_k). \tag{7}$$

In our case, $\Delta f = 200$ MHz, which is sufficient to satisfy (7) for pulses that are separated by more than about 4 ns. Thus, overlapping pulses of closer separation could not be resolved and were considered to belong to "one path."

Given a frequency-averaged power profile $s(t)$, and knowing the shape of the transmitted pulse $p(t)$, we need to find pairs $\{(\tau_k, \beta_k^2), k = 1, 2, 3, \cdots\}$ that fit (7) under some minimum error criterion. A reasonable way of doing this is by picking $\{\tau_k\}$ to represent discrete points on the time axis, say, every 1-5 ns, then proceeding to find the corresponding $\{\beta_k^2\}$ to minimize the mean-square error involved in fitting (7) under the constraint that $\beta_k^2 \geq 0$, all k. This is a classic problem in *quadratic programming*, whose rigorous automated solution is somewhat involved. Because of the limited number of waveforms that we had to process (8 rooms times 25 profiles per room), we actually performed an interactive heuristic computer optimization. The resulting errors were still within the measurement uncertainty.

IV. Review of Measured Responses

A typical sequence of pulse responses involved in measuring a room is shown in Fig. 3. Actually, the responses shown are frequency-averaged power profiles, $s(t)$, as defined in (7). In all cases, the horizontal time axis covers 10 ns per division, the vertical axis is linear in power (not decibels), and the vertical sensitivity of the oscilloscope is left unchanged. The dB numbers shown in the various figures corresponds to the setting of the transmitter's step attenuator.

Fig. 3(a) shows the squared envelope of the transmitted pulse obtained by a direct coaxial connection between transmitter and receiver. In Fig. 3(b), the receiving and transmitting antennas are both located in the hallway at a 1-m separation. Some pulse broadening is observed, and the peak signal level has dropped by about 34 dB. This is only 2 dB more than the theoretical free-space power transmission ratio [10]

$$P_{\text{rec}}/P_{\text{trans}} = G_t G_r [\lambda_0/4\pi r]^2, \tag{8}$$

where G_t and G_r (≈ 1.6) are, respectively, the gains of the transmitting and receiving discone antennas, λ_0 ($= 0.2m$) is the RF wavelength, and r ($= 1m$) is the antenna separation. Note that, within a building, this relation is expected to hold only for small r, where the direct line-of-sight ray dominates all other rays.

Fig. 3(c) is obtained by moving the receiver about 60 m, still in the hallway directly in front of the room in question (R-139 in Fig. 1). A strong 60 ns echo now exists, which was found to occur as a result of a reflected

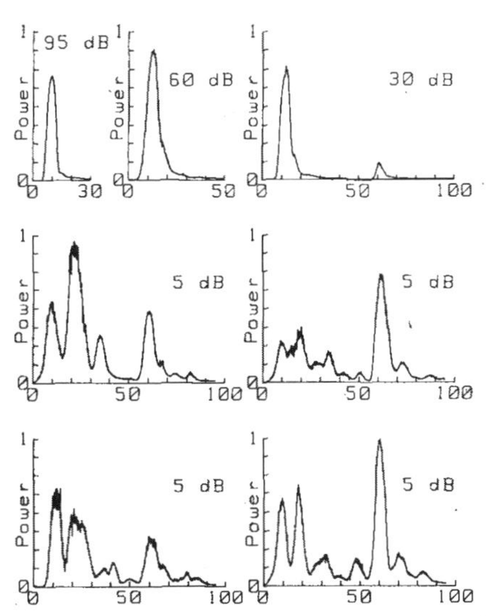

Fig. 3. A sample sequence of frequency-averaged power-profile measurements. The dB figures represent settings of the transmitter's attenuator.

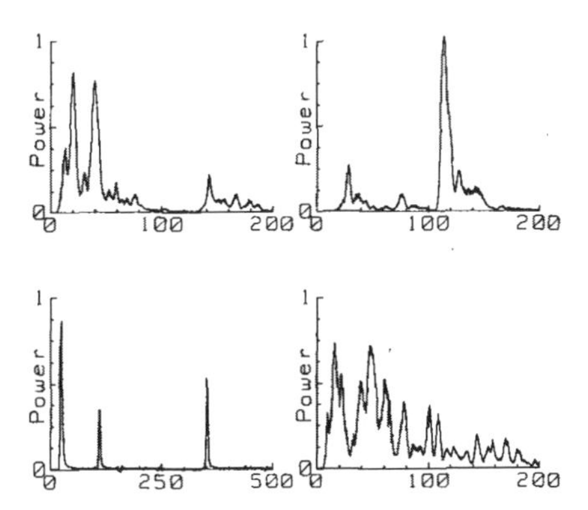

Fig. 4. Four interesting examples of frequency-averaged power profiles.

wave from the metal door in the right-hand side of Fig. 1. The signal level dropped from the previous 1-m location by about 30 dB.

Fig. 3(d)-(g) corresponds to four measurements inside the room on the vertices of a 0.23-m square (9-in tile). The received echos now extend over about 100 ns, and the signal level has dropped by an additional 25 dB relative to the previous location in the hallway. Notice some spatial correlation among the four profiles. Notice also that the groups of rays starting at about 60 ns in Fig. 3(d)-(g) coincide with the small echo in Fig. 3(c), which is taken in the hallway just outside the room.

Fig. 4 shows four other measured pulse responses in different locations within the building. Fig. 4(a) (Room L-159 in Fig. 1) shows two clearly separated clusters of arriving rays covering a 200 ns time span. Fig. 4(b) (Room R-134 in Fig. 1) shows a 100 ns delayed echo that is much stronger than the first arriving rays. Fig. 4(c) (hallway in front of Room R-134 in Fig. 1) shows a strong echo that is delayed about 325 ns, which is the largest delay of a relevant echo that we observed. Fig. 4(d) corresponds to measurements done where both transmitter and receiver were on the second floor of the building, with the receiver located in a large ($\approx 13 \times 17$ m) open machine shop with cinder-block surrounding walls. Notice the high density of the received rays over the entire 200 ns time axis.

V. The Multipath Power Gain and rms Delay Spread

A. Definitions

Getting the set of pairs $\{(\tau_k, \beta_k)\}$, as defined in (2), and deducing their joint statistics is, of course, the ultimate desired result of modeling. However, there are two simple parameters that are useful in describing the overall characteristics of the multipath profile. These are the total *multipath power gain*

$$G \equiv \sum_k \beta_k^2, \tag{9}$$

which is actually less than unity, and the *rms delay spread* [11]

$$\sigma_\tau \equiv \sqrt{\overline{\tau^2} - (\overline{\tau})^2}, \tag{10}$$

where

$$\overline{\tau^n} \equiv \frac{\sum_k \tau_k^n \beta_k^2}{\sum_k \beta_k^2}, \quad n = 1, 2. \tag{11}$$

The power gain is useful in estimating such things as the signal-to-noise ratio of a communications system. The rms delay spread is a measure of the temporal extent of the multipath delay profile, which relates to performance degradation caused by intersymbol interference. In fact, in some communications systems with data rates small compared to $1/\sigma_\tau$, σ_τ is found to be the most important single parameter determining the performance [12]. This, however, is not true in general [12].

The above parameters can actually be obtained directly from the received frequency-averaged power profile, $s(t)$ defined in (7), without the need to know the individual β's and τ's. Define the received power profile moments

$$M_n \equiv \int_{-\infty}^{\infty} t^n s(t)\, dt, \quad n = 0, 1, 2, \tag{12}$$

the transmitted pulse moments,

$$m_n \equiv \int_{-\infty}^{\infty} t^n p^2(t)\, dt, \quad n = 0, 1, 2, \tag{13}$$

the corresponding averages

$$\overline{t_s^n} = M_n/M_0, \quad n = 1,2, \tag{14}$$

$$\overline{t_p^n} = m_n/m_0, \quad n = 1,2, \tag{15}$$

and the variances

$$\sigma_s^2 = \overline{t_s^2} - (\overline{t}_s)^2, \tag{16}$$

$$\sigma_p^2 = \overline{t_p^2} - (\overline{t}_p)^2. \tag{17}$$

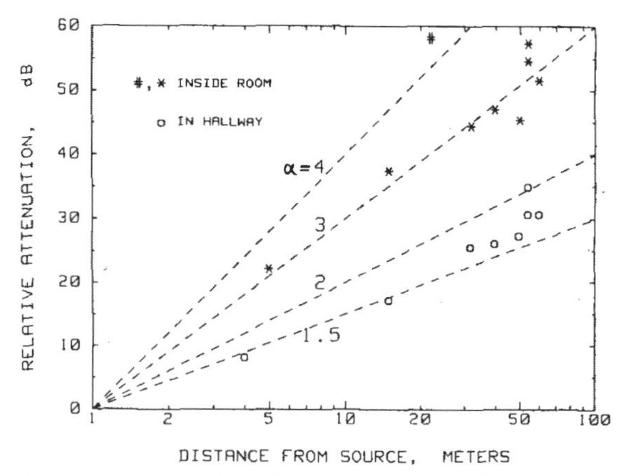

Fig. 5. Measured signal attenuation versus distance relative to 1-m separation. The dashed lines are for different values of the distance-power law exponent.

It can be shown that

$$G = M_0/m_0, \tag{18}$$

$$\overline{\tau} = \overline{t}_s - \overline{t}_p, \tag{19}$$

$$\sigma_\tau^2 = \sigma_s^2 - \sigma_p^2. \tag{20}$$

Note that G and σ_τ, but not $\overline{\tau}$, are independent of the choice of the time origin.

B. Distance-Power Law

The spatial average value $\overline{G}$ of the multipath power gain in the neighborhood of a point at a distance r from the transmitter is, in general, a decreasing function of r. Usually, this function is represented by a distance-power law of the form.

$$\overline{G}(r) \sim r^{-\alpha}. \tag{21}$$

In free space, $\alpha = 2$, and the power gain obeys an inverse-square law as given by (8).

A logarithmic plot of our measured attenuation

$$L(r) = -10 \log_{10} [G(r)/G(1 \text{ m})], \tag{22}$$

in decibels, versus r, in meters, is given in Fig. 5. Note that $G(1 \text{ m})$ can be computed reasonably accurately from (8).

Each rounded square in Fig. 5 represents one measurement with the receiver located in the same hallway as the transmitter. These measurements obey a power-law relation with $\alpha < 2$, which is the result of a waveguiding effect. Each asterisk in Fig. 5 represents the average of about 20 points within a room at the indicated distance from the transmitter. Recall, from Fig. 1, that all the rooms are located off the same hallway containing the transmitter. The value $\alpha = 3$ fits our room measurements reasonably well.

The single point indicated by the sharp sign in Fig. 5 represents the average of about 20 measurements where both the transmitter and the receiver were in the second floor of the building, with the receiver located in a large

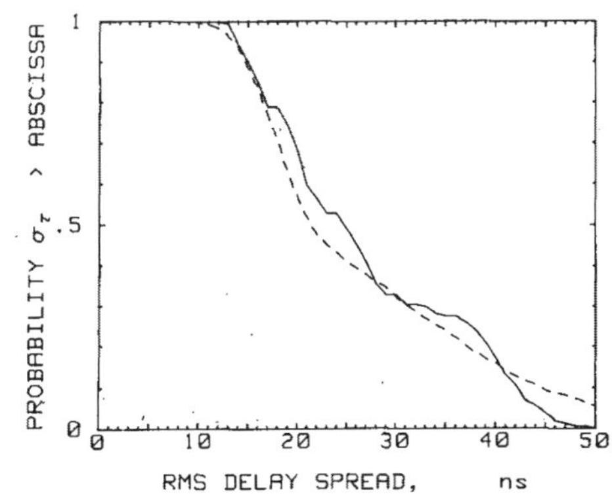

Fig. 6. Measured cumulative probability distribution of the rms delay spread for all rooms (solid line), and the result of a simulation using our model (dashed line).

open machine shop with cinder-block surrounding walls [same room as that for Fig. 4(d)]. Unlike the one-dimensional hallway geometry of Fig. 1, the shop in this case was located off a second hallway that is perpendicular to the transmitter's hallway. This resulted in a value of α slightly larger than 4. This is in more general agreement with Alexander's results [2] for this type of building. According to him, α can even be as large as 6 for office buildings with metalized partitions. (In a recent conference, Murray, Arnold, and Cox [19] reported on values of α between 3 and 4 measured in our building at 815 MHz.)

C. Statistics of the rms Delay Spread

Our measured rms delay spread, σ_τ, within the individual rooms was not generally correlated with the distance to the transmitter as had been previously predicted by Schmid [13]. Rather, it was somewhat related to the local surroundings of the transmitter and the receiver, e.g., to their proximity to such large reflectors as the end metal doors in Fig. 1. This is more consistent with the result of Turin *et al.* [14] of the urban mobile-radio channel.

The combined cumulative distribution of σ_τ resulting from a total of about 150 measurements within 7 rooms on the first floor of the building (see Fig. 1) is plotted by the solid line in Fig. 6. The results indicate that the rms delay spread has a median value of about 25 ns and a maximum value of about 50 ns. These numbers are a factor of five less than those measured by Devasirvatham in a much larger office building [7]. (In a recent conference, Devasirvatham [20] reported on rms delay spread of up to 200 ns measured in our building at 850 MHz. This large value, however, occurred only when the received signal levels were extremely low, e.g., 100–140 dB of path loss. When he truncated the data to a subset having path loss less than 100 dB, as is the situation in our case, he obtained results comparable to ours.)

We did not include in Fig. 6 the handful of measurement taken in the hallway, which occasionally yielded values of σ_τ much larger than 50 ns. For example, the measurement of Fig. 4(c) corresponds to $\sigma_\tau \approx 150$ ns. In fact, hallway measurements were excluded from our model development process altogether because, unlike room measurements, they include a direct line-of-sight ray from transmitter to receiver as well as strong rays with long delays, which is not the typical situation in most office buildings.

VI. Proposed Model

A. Basic Conjectures

We employ the discrete representation of the channel's impulse response given by (2). As mentioned earlier, the phase angles $\{\theta_k\}$ will be assumed *a priori* to be statistically independent random variables with a uniform distribution over $[0, 2\pi)$. We believe that this is self-evident and needs no experimental justification. Our goal now is to find the joint statistics of the paths gains $\{\beta_k\}$ and arrival times $\{\tau_k\}$.

We start with the attractive conjecture, which was pioneered by Turin *et al.* [14] in modeling urban mobile-radio channels, that $\{\tau_k\}$ forms a Poisson arrival-time sequence with some mean arrival rate λ. The path gain β_k associated with every τ_k is then picked from some probability distribution whose moments (e.g., mean and mean square) are functions of τ_k that eventually vanishes for large values of τ_k.

For simplicity, it would be desirable to have a model with λ being a constant and the β's statistically independent from one another. A straightforward implementation of these features, however, can lead to inconsistency with our experimental observations, as will be discussed next.

B. Clustering of Rays

Observations of measured pulse responses, such as some of those given in Figs. 3 and 4, indicate that rays generally arrive in clusters. Similar observations were found in the experimental data of Turin *et al.* [14]. This is clearly not consistent with a simple Poisson arrival-time model with constant λ. Indeed, Turin *et al.* as well as subsequent researchers [15], [16] who further refined their work, abandoned this model in favor of a Markov-type model in which the time axis is, in effect, divided into "bins," with the probability of a ray arriving within a given bin related to whether or not a ray actually arrived in the previous bin. Furthermore, in their model, the path gains in successive bins were also correlated. Although these features of their model fit the experimental data well, including the ray clustering effect, the Markovian nature of their model makes its use in analysis quite complex.

We now present an alternative multipath model, which is flexible enough to fit the experimental data reasonably well, while retaining the basic features of a constant-rate Poisson arrival-time process and mutually independent path gains, thus making its use in analysis relatively simple. In addition, and perhaps more importantly, the model can be explained from a physical viewpoint, thus making it more readily extendable to other types of buildings.

C. General Description of the Model

Our model starts with the physical realization that rays arrive in clusters. The cluster arrival times, i.e., the arrival times of the first rays of the clusters, are modeled as a Poisson arrival process with some fixed rate Λ. Within each cluster, subsequent rays also arrive according to a Poisson process with another fixed rate λ. Typically, each cluster consists of many rays, i.e., $\lambda >> \Lambda$.

Let the arrival time of the lth cluster be denoted by T_l, $l = 0, 1, 2, \cdots$. Moreover, let the arrival time of the kth ray measured from the beginning of the lth cluster be denoted by τ_{kl}, $k = 0, 1, 2, \cdots$. By definition, for the first cluster, $T_0 = 0$, and for the first ray within the lth cluster, $\tau_{0l} = 0$. Thus, according to our model, T_l and τ_{kl} are described by the independent interarrival exponential probability density functions

$$p(T_l \mid T_{l-1}) = \Lambda \exp\left[-\Lambda(T_l - T_{l-1})\right], \qquad l > 0, \tag{23}$$

$$p(\tau_{kl} \mid \tau_{(k-1)l}) = \lambda \exp\left[-\lambda(\tau_{kl} - \tau_{(k-1)l})\right], \quad k > 0. \tag{24}$$

Let the gain of the kth ray of the lth cluster be denoted by β_{kl} and its phase by θ_{kl}. Thus, instead of (2), our complex, low-pass impulse response of the channel is given by

$$h(t) = \sum_{l=0}^{\infty} \sum_{k=0}^{\infty} \beta_{kl} e^{j\theta_{kl}} \delta(t - T_l - \tau_{kl}). \tag{25}$$

Recall that $\{\theta_{kl}\}$ are statistically independent uniform random variables over $[0, 2\pi)$. Furthermore, the $\{\beta_{kl}\}$ are statistically independent positive random variables whose probability distributions are to be discussed in Section VII and whose mean square values $\{\overline{\beta_{kl}^2}\}$ are monotonically decreasing functions of $\{T_l\}$ and $\{\tau_{kl}\}$. In our model,

$$\begin{aligned} \overline{\beta_{kl}^2} &\equiv \overline{\beta^2(T_l, \tau_{kl})} \\ &= \overline{\beta^2(0,0)}\, e^{-T_l/\Gamma} e^{-\tau_{kl}/\gamma}, \end{aligned} \tag{26}$$

where $\overline{\beta^2(0,0)} = \overline{\beta_{00}^2}$ is the average power gain of the first ray of the first cluster, and Γ and γ are power-delay time constants for the clusters and the rays, respectively. A sketch that clarifies our model up to this point is given in Fig. 7.

Note that clusters generally overlap. For example, if for some k, $\tau_{kl} \geq T_{l+1} - T_l$, then the lth and the $(l+1)$th clusters overlap for all subsequent values of k. Typically, however, $\Gamma > \gamma$ and the expected power of the rays in a cluster decay faster than the expected power of the first ray of the next cluster. Thus, if $\Delta T \equiv T_{l+1} - T_l$ is sufficiently large such that $\exp[-\Delta T/\gamma] << \exp[-\Delta T/\Gamma]$, then the lth and $(l+1)$th clusters will appear disjoint.

Note also that, in principle, rays and clusters extend over an infinite time, as signified by the double infinite sum in (25). However, practically speaking, the sum over l stops when $\exp(-T_l/\Gamma) << 1$, and that over k stops when $\exp(-\tau_{kl}/\gamma) << 1$. In our room measurements, rays and clusters outside a roughly 200 ns observation window, although they exist, were in general too small to be detected.

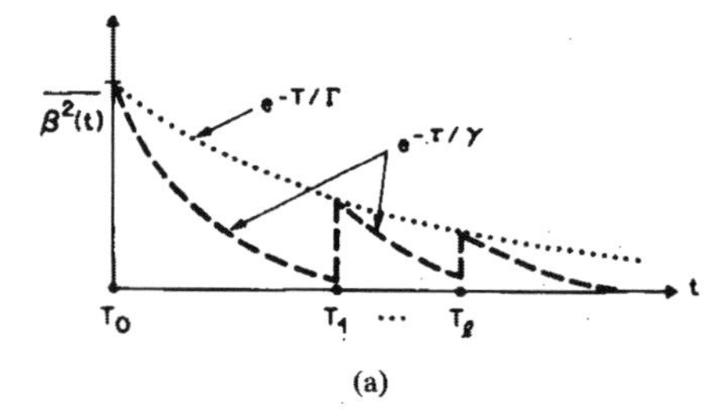

(a)

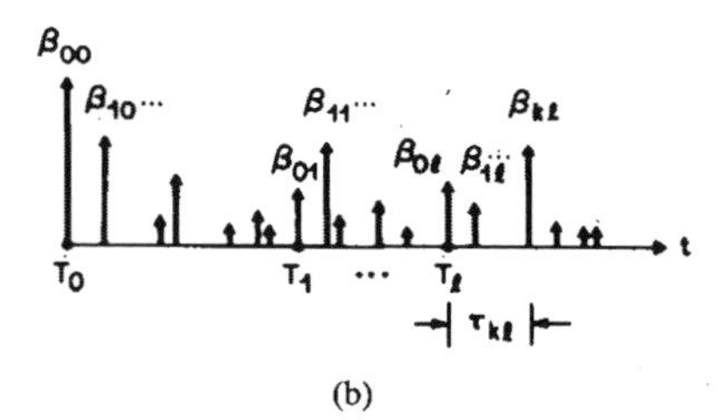

(b)

Fig. 7. A schematical representation of our model. (a) Exponentially decaying ray and cluster average powers. (b) A realization of the impulse response.

VII. Model Parameters and Physical Interpretations

A. The Cluster Arrival Rate, Λ

The first arriving cluster of rays is formed by the transmitted wave following a more-or-less "direct" path to the receiver. Such a path, which is not usually a straight line, comprises mostly open spaces (e.g., hallways), and goes through a few, but not too many, walls. Subsequent clusters result from reflections from the building superstructure (e.g., large metalized external or internal walls and doors). In our building, the two end metal doors (see Fig. 1) were the main cause of the additional clusters. Note that with this physical picture, the number and arrival times of clusters should be the same for all locations within any given room, which is what we observed experimentally.

One-half of the rooms we measured showed no evidence of additional clusters within the 200 ns observation window. (The first cluster, of course, was always present.) The other half showed, essentially, one additional relevant cluster. Using the Poisson distribution associated with (23) with a cluster arrival rate Λ, we obtain the probability of having n additional clusters in a 200 ns period as $P(n) = (200\Lambda)^n \exp(-200\Lambda)/n!$, $n = 0, 1, 2, \cdots$. We thus find that to fit our cluster observations, $1/\Lambda$ needs to be roughly in the range from 200 to 300 ns, in which case $P(0) = 0.37$ to 0.51, $P(1) = 0.37$ to 0.34, and $P(n > 1) = 0.26$ to 0.15. With only eight rooms measured, we cannot at this point be more precise in picking Λ. In Section VIII-A we will see, from another point of view, that $1/\Lambda \approx 300$ ns.

B. The Ray Arrival Rate, λ

By resolving the individual rays in about 200 power profile measurements similar to those in Figs. 3 and 4, we estimate $1/\lambda$ to be in the range of 5–10 ns. The range uncertainty comes from the fact that our ray-resolving algorithm, coupled with our measurements sensitivity, is unable to detect many weak rays, in particular, those falling near strong rays. The higher the sensitivity, the more (weak) rays we would find, and hence, the larger the value of λ. At the same time, the probability distribution of the path gains $\{\beta_k\}$ would be increased for small values of β's. Thus, the appropriate choice of λ is strongly coupled to the probability distribution of the β's. We find that a consistent choice is $1/\lambda = 5$ ns coupled with Rayleigh-distributed β's having the appropriate mean-square value (see Section VII-D). Actually, as will be discussed later (see Section VIII-B), smaller values of $1/\lambda$ could have been employed, even down to the limiting value of zero (i.e., continuous ray arrival process). This, however, is beyond our measurement time resolution.

C. The Ray and the Cluster Power-Decay Time Constants, γ and Γ

As mentioned in Section VII-A, the number of arrival times of clusters, say, $T_0, T_1, \cdots, T_L$, are the same for all locations within a given room. It follows from (26) that, within that room, the expected value of the ray power as a function of time, measured from the arrival point of the first ray of the first cluster, is given by

$$\overline{\beta^2(t)} = \overline{\beta^2(0, 0)} \sum_{l=0}^{L} e^{-T_l/\Gamma} e^{-(t-T_l)/\gamma} \,\mathrm{U}\,(t - T_l), \tag{27}$$

where $\mathrm{U}(t)$ is the unit step function, which equals one for $t \geq 0$, and zero for $t < 0$. A sketch of (27) is shown in Fig. 7(a).

An estimate of $\overline{\beta^2(t)}$, and hence of γ and Γ, was obtained for a given room by aligning the time origin and taking the average of many measured power profiles, $s(t)$, within that room. Actually, what one obtains from this process is an estimate of the time convolution of $\overline{\beta^2(t)}$ and $p^2(t)$, the square of the transmitted pulse. However, since this pulse is narrow, the effect of the convolution is negligible. Four different examples of such space-averaged $s(t)$ are shown in Fig. 8. More than 20 power profiles were averaged in each case. Fig. 8(a) indicates that only one cluster reached the receiver in that case. Each of the remaining cases shows two arriving clusters. The time spread evident in the leading edges of a few clusters in Fig. 8 is mainly due to fundamental uncertainty in aligning the time origins of the various measured power profiles.

By fitting decaying exponentials to each cluster in Fig. 8, as well as to similar measurements taken in other

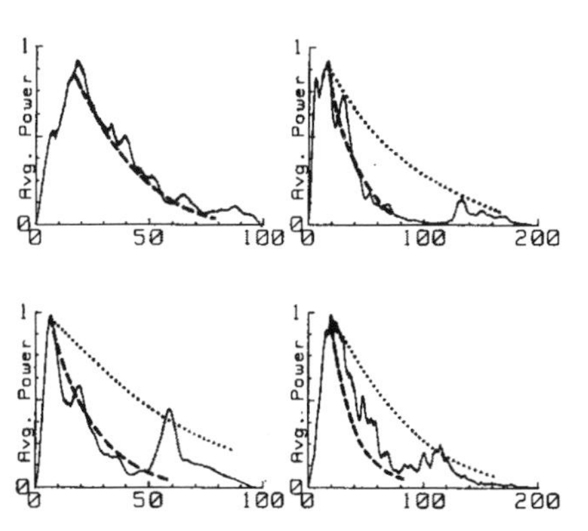

Fig. 8. Four spatially averaged power profiles within various rooms. The dashed lines correspond to exponential power decay profile of the rays and the clusters.

rooms, we find that, on the average, rays within a cluster decay with an approximate time constant of $\gamma = 20$ ns (see dashed lines in Fig. 8). Similarly, by fitting decaying exponentials through the leading peaks of successive clusters, we find that the clusters themselves decay, on the average, with an approximate time constant of $\Gamma = 60$ ns (see dashed lines in Fig. 8).

The use of exponentials in (26) and (27) to represent the decays of the powers of the rays and clusters as functions of time has an intuitively appealing interpretation. Consider, for example, our physical picture of the rays bouncing back and forth in the vicinity of the receiver and/or the transmitter to form a cluster. On the average, with each bounce, the wave suffers some average delay (say, equivalent to the width of a room), and some average decibels of attenuation (which depends on the surrounding materials of the walls, furnitures, etc.). In this case, the power level *in decibels* of each successive ray would be proportional to the time delay of that ray, which results in our exponential power decay characteristics. Note that from this picture, γ and Γ would be increased if the building walls were more reflective and/or if the sizes of the rooms and the building itself were increased.

D. The Probability Distribution of the Path Gains, $\{\beta_{kl}\}$

So far, our model gives the expected value of the path power gain $\overline{\beta_{kl}^2}$ as a function of the associated cluster and ray delays T_l and τ_{kl}. We now make the assumption, which is reasonably supported by our observations, that the probability distribution of the normalized power gain $\beta_{kl}^2/\overline{\beta_{kl}^2}$ is independent of the associated delays, or for that matter, of the location within the building. Under this assumption, we can put the measured power gains of *all* our resolved paths into one, supposedly homogeneous, data pool by simply normalizing by the appropriate (i.e., measured within the same room and having the same delay) spatially averaged power profiles, such as those shown in Fig. 8.

The cumulative distribution of $\beta_{kl}^2/\overline{\beta_{kl}^2}$, obtained as described above, is shown by the solid line in Fig. 9. The dashed line in that figure is the unity-mean exponential

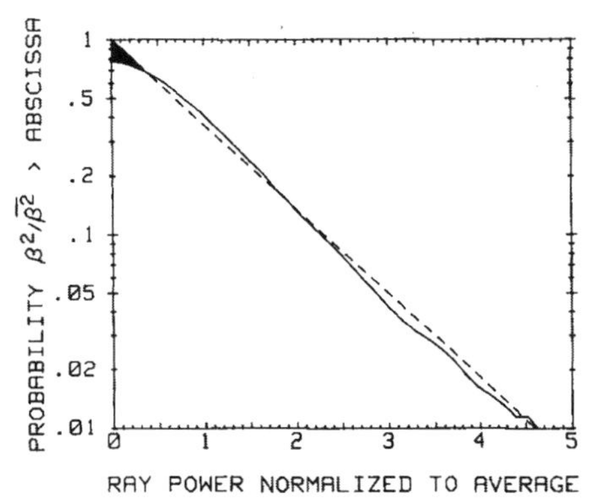

Fig. 9. Measured cumulative probability distribution of the normalized ray power for all rooms (solid line), and the best fitting exponential distribution (dashed line).

cumulative distribution, $P[\beta_{kl}^2/\overline{\beta_{kl}^2} > X] = \exp(-X)$, which appears as a straight line on the semilog plot. This results in the exponential probability density function,

$$p(\beta_{kl}^2) = (\overline{\beta_{kl}^2})^{-1} \exp(-\beta_{kl}^2/\overline{\beta_{kl}^2}), \tag{29}$$

for the path power gain, or, equivalently, the Rayleigh probability density function,

$$p(\beta_{kl}) = (2\beta_{kl}/\overline{\beta_{kl}^2}) \exp(-\beta_{kl}^2/\overline{\beta_{kl}^2}), \tag{30}$$

for the path voltage gain.

As mentioned earlier, our ray-resolving algorithm, coupled with our measurement sensitivity, is unable to detect many weak rays, in particular, those falling near strong rays. Thus, in fitting the unity-mean exponential cumulative distribution to the data in Fig. 9, we let the total number of resolved rays be a floating optimization parameter, instead of fixing it to the actual number observed. In effect, we let the optimization procedure estimate the number of weak rays that we could have missed. As indicated from Fig. 9, with the solid line intercepting the ordinate axis at a value of about 0.8 instead of 1, a 25 percent increase in the total number of rays is needed to achieve the best fit. This increase is reflected in the choice of λ given in Section VII-B. As can be observed from the shaded area in Fig. 9, the power of the needed rays is less than about 25 percent of the local average. Such weak rays could indeed have been missed by our ray-resolving algorithm.

It is appropriate to mention at this point that various other distributions could have fit our path gain data as well as, or even better than, the Rayleigh distribution of (30). For example, a log-normal distribution, with a standard deviation of about 4 dB, fits our data well, even without inflating the number of weak rays as was done above. However, we believe that those undetected weak rays actually exist, and that the Rayleigh distribution, which is much simpler to work with in analysis than the log-normal distribution, is quite adequate for our model.

Having rays with Rayleigh amplitudes and uniform phases, which corresponds to a complex Gaussian process, can physically occur if what we call a "ray" is actually the sum of many independent rays arriving within our time resolution. Invoking the central limit theorem then leads to the aforementioned complex Gaussian process. This is basically the same logic leading to the continuous Rayleigh model, or more precisely, the Gaussian wide-sense stationary uncorrelated scattering (GWSSUS) model of the urban mobile radio channel advocated by Bello, Cox, and others [17], [18], [12]. On the other hand, Turin and others [14]–[16] model the same channel as having discrete correlated rays with log-normal amplitudes. By comparison, our model of the indoor radio channel consists of discrete uncorrelated rays with Rayleigh amplitudes. As will be seen shortly, our model can easily be extended to a continuous model of the GWSSUS type.

VIII. Using the Model

A. Simulation Procedure

Suppose that we wish to simulate the indoor channel between a centrally located transmitter/receiver and some room at a distance r. The first step is to generate the cluster arrival times, $T_1, T_2, \cdots$, through the use of the exponential distribution of (23) with $T_0 = 0$. Before proceeding with the generation of the rays within the clusters, we need to estimate $\overline{\beta^2(0, 0)}$, the average power of the first ray of the first cluster. This quantity is directly related to the average multipath power gain $\overline{G}(r)$ for the room in question. It follows from (21), (23), and Fig. 5 that $\overline{G}(r) = G(1\text{ m})\, r^{-\alpha}$, where $G(1\text{ m})$ can be approximated by (8), and $\alpha = 3$ to 4 in our building, but can be up to 6 in other builds [2].

The relation between $\overline{G}(r)$ and $\overline{\beta^2(0, 0)}$ can be found from the definition of G in Section V-A, and from $\overline{\beta^2(t)}$, the expected value of the ray power as a function of time, given in (27). With λ being the average number of rays per unit time, it follows that[1]

$$\overline{G}(r) = \lambda \int_0^\infty \overline{\beta^2(t)}\, dt = \gamma\lambda\, \overline{\beta^2(0, 0)} \sum_{l=0}^{L} e^{-T_l/\Gamma}. \tag{31}$$

The summation term in (31) is usually dominated by the first term, i.e., the first cluster. For example, with T_l statistically described in (23), it can be shown that the average value of this summation is $1 + \Gamma\Lambda$. The 1 accounts for the first cluster and $\Gamma\Lambda$ accounts for subsequent clusters. In our case, with $\Gamma = 60$ ns and $1/\Lambda$ between 200 and 300 ns, this average value is between 1.3 and 1.2. Considering the ± 7 dB deviation in fitting $\overline{G}(r)$ to the power law $r^{-\alpha}$ shown in Fig. 5, the small increase (≈ 1 dB) due to subsequent clusters can safely be neglected. (Note, however, that these delayed clusters have a major

[1] Actually, more complete computation would yield (31) with $\gamma\lambda$ replaced by $(1 + \gamma\lambda)$. The added 1 accounts for the first ray of each cluster. However, usually $\gamma\lambda \gg 1$, and (31) is a valid approximation.

effect on the value of the rms delay spread.) We can finally compute $\beta^2(0, 0)$ from

$$\overline{\beta^2(0, 0)} \approx (\gamma\lambda)^{-1} G(1 \text{ m}) r^{-\alpha} \qquad (32)$$

where $G(1 \text{ m})$ is given approximately by (8), with $r = 1$.

Now we can proceed to generate the rays within the clusters. First, the relative ray arrival times $\{\tau_{kl}\}$ are generated through the use of the exponential distribution of (24). Next, the ray amplitudes $\{\beta_{kl}\}$ are generated from the Rayleigh distribution of (30), with $\overline{\beta_{kl}^2}$ given by (26). To complete the picture, the associated phase angles $\{\theta_{kl}\}$ are chosen from a uniform distribution. Depending on the application, the process of ray and cluster generation would stop when the average power of the generated rays drops below some threshold.

As mentioned in Section VII-A, we estimate the mean time between clusters in our building to be in the range $1/\Lambda$ = 200–300 ns. To resolve this uncertainty, we simulated the channel as described above, computed the rms delay spread σ_τ for each simulation point, and plotted its cumulative distribution. With our other parameters being γ = 20 ns, Γ = 60 ns, and $1/\lambda$ = 5 ns, the best fit to the cumulative distribution obtained directly from measurements, the solid line in Fig. 6, was for $1/\Lambda$ = 300 ns. The corresponding distribution obtained from simulations is given by the dashed line in Fig. 6.

B. Discrete versus Continuous Model

Thus far, our model is presented as a discrete ray arrival process. This would be more convenient for use in studies using computer simulations. We now examine the possibility of having a continuous model.

We note that, by letting the ray arrival rate λ approach infinity, while having $\overline{\beta^2(0, 0)}$ approach zero such that their product is finite and satisfies (32), and by noting that the complex ray gains form a Gaussian process, our model mathematically approaches the continuous GWSSUS model mentioned in Section VII-D. Such a model would be more suitable for analysis [12], [17]. Some measured power delay profiles, such as the one shown in Fig. 4(d), suggest that such a continuous model might indeed represent the physical reality. On the other hand, other measurements, such as those of Fig. 3(d)–(g), which were taken on the vertices of an 0.23-m square, suggest a discrete model. This follows from the correlation evident between the various profiles, which could be attributed to three or four dominant waves traversing the observation points.

With our time resolution of about 5 ns, and with our finding of a mean time between rays of $1/\lambda \approx 5$ ns, we are unable to determine whether the indoor radio channel is best described by a discrete or a continuous model. In many studies, the answer to this question may not be relevant, while in others, the answer may be important. For example, a truly discrete model would yield a strong spatial correlation of the amplitudes of the rays received in some vicinity. This would yield some space diversity systems useless. A continuous model, on the other hand, results in the opposite conclusion. We believe from our measurements that, in most situations within the indoor channel, a continuous model or, equivalently, a high-density discrete model, is closer to the physical reality. However, as mentioned earlier, some exceptions do exist. A definitive settlement of this issue awaits future experiments involving either a better time resolution or simultaneous measurements with two or more closely spaced receiving antennas.

IX. Summary and Conclusions

We presented the results of 1.5 GHz, pulsed, multipath propagation measurements between two vertically polarized omnidirectional antennas located on the same floor within a medium-size building (AT&T Bell Laboratories, Crawford Hill location in Holmdel, NJ). A novel method of measurement was used, which involves averaging the square-law-detected received pulse response while sweeping the frequency of the transmitted pulse. This enabled us to resolve the multipath channel impulse response within about 5 ns.

Our results show the following. 1) The indoor channel is quasi-static, or *very slowly* time varying (related to people's movements). 2) The nature and statistics of the channel's impulse response is virtually independent of the states of polarization of the transmitting and receiving antennas, provided that there is no line-of-sight path between them. 3) The maximum observed delay spread in the building was 100–200 ns within rooms, with occasional delays of more than 300 ns within hallways. 4) The measured rms delay spread within rooms had a median value of 25 ns, and a maximum value of 50 ns, both being a factor of five less than those measured by Devasirvatham in a much larger building [7]. 5) The signal attenuation with no line-of-sight path varied over a 60 dB range and seems to obey an inverse distance-power law with an exponent between 3 and 4, which is in general agreement with Alexander's results [2] for this type of building.

We have also developed a simple statistical multipath model of the indoor radio channel, which fits our measurements well and, more importantly, appears to be extendable (by adjusting the values of its parameters) to other buildings. In our model, the rays of the received signal arrive in clusters. The received ray amplitudes are independent Rayleigh random variables with variances that decay exponentially with cluster delay as well as with ray delay within a cluster. The corresponding phase angles are independent uniform random variables over $[0, 2\pi)$. The clusters, as well as the rays within a cluster, form Poisson arrival processes with different, but fixed, rates. Equivalently, the clusters and the rays have exponentially distributed interarrival times. The formation of the clusters is related to the building superstructure (e.g., large metalized external or internal walls and doors). The rays within a cluster are formed by multiple reflections from objects in the vicinities of the transmitter and the receiver (e.g., room walls, furnishings, and people).

A detailed summary of the method of using the model was given in Section VIII. Both a discrete and a continuous version of the model are possible. The former is more suitable for computer simulations, while the latter would be more desirable in analysis.

ACKNOWLEDGMENT

We thank Prof. P. J. McLane for many stimulating discussions, especially about the discrete versus continuous multipath models. We also thank L. J. Greenstein, D. J. Goodman, and M. M. Kavehrad for their useful suggestions and comments, and R. A. Semplak and G. J. Owens for their design of the discone antennas used in the experiment. We also acknowledge the valuable contributions of R. D. Nash in the early phase of the experiment, and the able help of M. F. Wazowicz in maintaining the equipment. Last, but not least, we thank T. A. Saleh and O. A. Saleh for their help in the tedious task of gathering the valuable experimental data.

REFERENCES

[1] S. E. Alexander, "Radio propagation within buildings at 900 MHz," *Electron Lett.*, vol. 18, no. 21, pp. 913-914, Oct. 14, 1982.

[2] S. E. Alexander, "Characterizing buildings for propagation at 900 MHz," *Electron Lett.*, vol. 19, no. 20, p. 860, Sept. 29, 1983.

[3] H. H. Hoffman and D. C. Cox, "Attenuation of 900 MHz radio waves propagating into a metal building," *IEEE Trans. Antennas Propagat.*, vol. AP-30, pp. 808-811, July 1982.

[4] D. C. Cox, R. R. Murray, and A. W. Norris, "Measurements of 800 MHz radio transmission into buildings with metallic walls," *Bell Syst. Tech. J.*, vol. 62, no. 9, pp. 2695-2717, Nov. 1983.

[5] —, "800 MHz attenuation measured in and around suburban houses," *AT&T Bell Lab. Tech. J.*, vol. 63, no. 6, pp. 921-954, July-Aug. 1984.

[6] D. M. J. Devasirvatham, "Time delay speed measurements of 850 MHz radio waves in building environments," in *GLOBECOM '85 Conf. Rec.*, vol. 2, Dec. 1985, pp. 970-973.

[7] —, "The delay spread measurements of wideband radio signals within a building," *Electron. Lett.*, vol. 20, no. 23, pp. 950-951, Nov. 8, 1984.

[8] A. G. Kandoian, "Three new antenna types and their applications," *Proc. IRE*, vol. 34, no. 2, pp. 70W-75W, Feb. 1946.

[9] J. G. Proakis, *Digital Communications.* New York: McGraw-Hill, 1983, ch. 7.

[10] S. Ramo, J. R. Whinnery, and T. Van Duzer, *Fields and Waves in Communication Electronics.* New York: Wiley, 1965, p. 717, eq. (7).

[11] D. C. Cox, "Delay Doppler characteristics of multipath propagation at 910 MHz in a suburban mobile radio environment," *IEEE Trans. Antennas Propagat.*, vol. AP-20, pp. 625-635, Sept. 1972.

[12] B. Glance and L. J. Greenstein, "Frequency selective fading effects in digital mobile radio with diversity combining," *IEEE Trans. Commun.*, vol. COM-31, pp. 1085-1094, Sept. 1983.

[13] H. F. Schmid, "A prediction model for multipath propagation of pulse signals at VHF and UHF over irregular terrain," *IEEE Trans. Antenna Propagat.*, vol. AP-18, pp. 253-258, Mar. 1970.

[14] G. L. Turin, F. D. Clapp, T. L. Johnston, S. B. Fine, and D. Lavry, "A statistical model of urban multipath propagation," *IEEE Trans. Veh. Technol.*, vol. VT-21, pp. 1-9, Feb. 1972.

[15] H. Suzuki, "A statistical model for urban radio propagation," *IEEE Trans. Commun.*, vol. COM-25, pp. 673-680, July 1977.

[16] H. Hashemi, "Simulation of the urban radio propagation channel," *IEEE Trans. Veh. Technol.*, vol. VT-28, Aug. 1979.

[17] A. P. Bello and B. D. Nelin, "The effect of frequency selective fading on the binary error probability of incoherent and differentially coherent matched filter receivers," *IEEE Trans. Commun. Syst.*, vol. CS-11, pp. 170-186, June 1963.

[18] D. C. Cox and R. P. Leck, "Correlation bandwidth and delay spread multipath propagation statistics for 910 MHz urban mobile radio channels," *IEEE Trans. Commun.*, vol. COM-23, pp. 1271-1280, Nov. 1975.

[19] R. R. Murray, H. W. Arnold, and D. C. Cox, "815 MHz radio attenuation measured within a commercial building," in *Dig. 1986 IEEE Int. Symp. Antennas Propagat.*, vol. 1, June 1986, pp. 209-212.

[20] D. M. J. Devasirvatham, "A comparison of time delay spread measurements within two dissimilar office buildings," in *ICC'86 Conf. Rec.*, vol. 2, June 1986, pp. 852-857.

PHOTO NOT AVAILABLE

Adel A. M. Saleh (M'70-SM'76-F'87) was born in Alexandria, Egypt, on July 8, 1942. He received the B.Sc. degree in electrical engineering from the University of Alexandria, Alexandria, Egypt, in 1963, and the M.S. and Ph.D. degrees in electrical engineering from the Massachusetts Institute of Technology, Cambridge, in 1967 and 1970, respectively.

From 1963 to 1965 he worked as an Instructor at the University of Alexandria. In 1970 he joined AT&T Bell Laboratories, Holmdel, NJ, where he is engaged in research on microwave and optical components and communications systems.

Dr. Saleh is a member of Sigma Xi.

PHOTO NOT AVAILABLE

Reinaldo A. Valenzuela (M'85) received the B.Sc.(Eng.) degree from the School of Engineering, University of Chile, in 1977. He received the Ph.D. degree in 1982 through the study of "techniques for transmultiplexer design."

From 1975 to 1977 he worked for Thomson-CSF (Chile) in the setting-up and commissioning of large telecommunication networks. In 1978 he joined the Department of Electrical Engineering, Imperial College, London, England, where, in 1981, he was appointed a Research Assistant, sponsored by British Telecom for the study of digital filters for CODEC applications. He then joined DATABIT Ltd. (U.K.) where he participated in the analysis and design of a full duplex, echo cancelling system for the transmission of 88 kbits/s data and analog voice signals over the subscriber loop. Since 1984 he has been a member of the Communications Methods Research Department, AT&T Bell Laboratories, Holmdel, NJ, where he has been studying local area networks issues such as indoor microwave propagation and the integration of voice and data in multiple access packet networks.

Standardized Fiber Optic Transmission Systems—A Synchronous Optical Network View

RODNEY J. BOEHM, YAU-CHAU CHING, MEMBER, IEEE, C. GEORGE GRIFFITH, SENIOR MEMBER, IEEE, AND FREDERICK A. SAAL, MEMBER, IEEE

Abstract—**The deployment of fiber optic systems has drastically altered the complexion of the digital network, and standardization of optical interface parameters has assumed paramount importance. This paper reports on the activities in various standards organizations, with emphasis on a synchronous network proposal which is currently being discussed in the T1 Committee.**

STANDARDIZATION work on various fiber optical components has been active in the EIA and the CCITT standards arenas for several years. EIA has generated numerous FOTP's (fiber optic test procedures) which deal with standardizing definitions and test procedures for fiber optical components, while CCITT Recommendations G.651 and G.652 address similar topics. Standardization of fiber optical transmission systems is, however, still in its infant stages. There are three main reasons for this late start of system standardization process. First, systems are more complex than their components and therefore require more time to mature. Second, there are a number of manufacturers, both domestic and foreign, producting fiber optic transmission systems, but none seems to be dominant enough to establish a de facto standard. Third, the seven regional companies, the numerous independent exchange and interexchange carriers have diverse interests and preferences and have not been able to dictate a unified standard. With the rapid advances of fiber optic technology, new systems are being introduced to the marketplace steadily, and the consequences of these nonstandardized systems are a fragmented network and vast confusion to users.

For this rapidly changing technology and fragmented network, one might ask,'' Do we really need a standard for fiber optic systems?'' From the perspective of an operating telephone company, his goals are to generate new revenues and to reduce costs. New revenues can be generated either by introducing new services or by providing more cost-effective services to a wider customer base. Cost reduction can be achieved either through savings on operational expenses or through purchase of more competitively priced equipment. From a manufacturer's viewpoint, his goals are also to generate new revenues and to reduce costs. New revenues can be generated by introducing new products and by expanding market shares. Costs can be reduced by using advanced technology, ingenious system design, and cost-effective manufacturing processes. Cost can also be reduced by expanding the market and thereby achieving the economics of scale. Standards are beneficial to both groups in their pursuance of revenue enhancement and cost reduction.

Standards are valuable to operating telephone companies because they lead to the possibility of procuring compatible equipment from multiple vendors (which tends to imply more competitive pricing). Standardization also reduces operational costs by unifying operation and maintenance procedures, simplifying training, and lowering sparing requirements (because the equipment has a common maintenance interface and can be interconnected without special interfacing assemblies). For manufacturers, standards lead to economics of scale in production and engineering because multiple customers for the manufacturer's product have the same or nearly the same requirements.

The introduction of new services can benefit both the manufacturers and the operating companies alike by providing more product opportunities to manufacturers and more revenue to operating companies, and standardization is essential in introducing new services. A telecommunications network by its very nature requires coordination for any services. A national standard is therefore necessary for an orderly introduction of new services. Many of the new services introduced into the network in past years can be attributed to the vertically integrated Bell System. After the divestiture of AT&T, the need for such telecommunication standards was well recognized and Committee T1 was established by ECSA to formally set these standards so that a network can evolve and progress gracefully. The structure and missions of Committee T1 have been documented in [1] and will not be repeated here. However, much of this paper concentrates on the activities of T1, particularly that of the T1X1.2 working group (carrier-to-carrier interfaces) and T1X1.4 working group (digital hierarchical rates and formats).

Manuscript received March 21, 1986.
R. J. Boehm is with Rockwell International Corporation, Dallas, TX 75207.
Y.-C. Ching, C. G. Griffith, and F. A. Saal are with Bell Communications Research, Inc., Red Bank, NJ 07701.
IEEE Log Number 8610775.

I. What Needs Standardization?

There are two separate categories in a fiber optic system that need standardization: optical parameters and signal

Reprinted from *IEEE Journal on Selected Areas in Communications,* vol. SAC-4, no. 9, December 1986.

format. To start, let us consider the simple point-to-point transmission systems that use DS3 bundles either as multiplexer inputs or outputs. For optical transmission around the DS3 rate, a system usually starts with an M13 multiplexer. Additional overhead capacity is then added so that the transmission system can be maintained. For a higher speed system, several DS3 bit streams are fed into a multiplexer, maintenance overhead is added, and the resultant signal is then converted to an optical form. Due to the variations in the overhead channels and the stuffing mechanisms, most of these systems have unique rates and formats. Additionally, technological advances in fibers and transducers have resulted in vast variations in optical parameters such as optical wavelengths, spectral widths, attenuations, waveshapes, and dispersion characteristics. These different rates, formats, and variations in optical parameters lead to incompatibilities in fiber optic transmission systems and confusion to users.

II. Optical Parameters

Standardization of optical parameters is currently being addressed in three areas: standardization of optical parameter definitions and measurement techniques, standardization of fiber optic transmission system regeneration section design methodology, and standardization of the values of optical parameters for transmission systems.

The definitions and measurement techniques are least controversial in that there is consensus that such standards need be established. Difficulties arise from the rapid technological advances and the fact that many properties are not well understood. However, great progress has been made in this direction in IEEE, EIA, and CCITT [2]–[4]. Although these activities do not address the questions of fiber systems per se, they are nonetheless an integral part of the system standardization work.

The regenerator section design methodology is somewhat more controversial in that there is no agreement as to how a standard should be used. Regenerator section design methodology also encounters difficulties from rapid technological advances and the fact that many parameters are still not well understood. To date, the designs are nominally loss limited, with dispersion and bandwidth limitations serving only as checking mechanisms. Dispersion penalties are devised for regions where dispersion does not drastically affect the performance of the system. As system speed increases, dispersion becomes the limiting factor, and still there is no accurate and verifiable measure of dispersion limitation of a fiber optic system that a span designer can use as a guide [5].

There is also disagreement on whether a worst case approach or a statistical approach should be preferred. With the demise of the Bell System, each operating company has begun engineering its own fiber routes; formerly Western Electric usually served as the system integrator. Operating companies also frequently purchase fiber and terminal equipment separately from numerous suppliers. It is therefore imperative that the specifications of the products be consistent and complete to allow the telco personnel to design regenerator sections. A standard design methodology and the associated set of parameter specifications would allow such regenerator section design. In assuming the design responsibilities, however, the telco personnel also assume the responsibilities of guaranteeing the overall end-to-end transmission performance. A conservative approach to this is to apply a worst case design, using the worst case parameters. Since the worst case parameter values are generally guaranteed by the suppliers, the discrepancy between the design and the actual performance can be easily traced to individual components. The worst case design, suffers from overdesigning. The final architecture often contains too many regenerators and excessive loss and dispersion margins. A statistical method alleviates some of the overdesign problems, but introduces uncertainty in component performance as well as a small percentage of statistical failures.

Regenerator section design methodology is currently being addressed in the EIA FO2.1 working group. A multimode fiber system design methodology has been included in a document entitled "Optical fiber digital transmission systems—Consideration for users and suppliers" [6] and a methodology for single-mode systems is under development.

Standardization of the actual values of optical parameters is the most controversial. Since these parameters are changing drastically with technological advances, unless standards are implemented carefully, they might include unnecessary constraints which might limit innovation on the part of the manufacturer. Therefore, it is important to standardize only what is absolutely necessary to allow various manufacturers' products to interface and to interoperate properly. T1X1 has initiated a project in this direction, and a fiber optical subworking group in T1X1.2 has been established to address the questions. The original purpose of this group is to establish a standard so that a multiowner fiber system (also known as midfiber meet) can be installed without lengthy contract negotiations [7]. This standard, when established, may also have more far-reaching implications in that it would allow optical cross-connect capabilities in an operating company's own network, thus greatly simplifying the design restrictions and partially realizing the flexibilities of a fiber network.

Despite the rapid changes in technology, there are indications that optical parameters standards are beneficial at some low speeds (i.e., < 150 Mbits/s) in a short-loop environment. For these applications, regenerators can be avoided. In fact, improvements of fiber and transducer manufacturing processes and quality control will make it possible to design a regenerator section not by detailed computation of the parameters, but by a simple table look-up procedure, in much the same manner as voice circuits are designed today. Because of the low-loss fibers, the high-power transmitters, and the high-sensitivity receivers, there will be enough flexibility in fiber optic transmission systems to allow the introduction of optical cross-connect systems without unduly limiting the applications of these systems. To ensure compatibility, only the signal rate and format and a minimum set of optical parameters need to be standardized.

The question of a standardized signal rate and format is being addressed by another subworking group in T1X1—subworking group on rates and formats for fiber optic network interfaces of the T1X1.4 working group [8]. Rates and formats also depend on how the fiber optic systems are deployed: as a point-to-point transmission system or as part of a fiber optical network. Presently, transmission systems are designed and deployed as point-to-point systems, without much consideration of the services they can carry or the network to which they belong. Transmission parameters and the signal format are largely determined by the characteristics of the transmission media. The introduction of fiber optic systems and the Integrated Service Digital Network (ISDN) may have changed the nature of the network, and the standardization process should be changed accordingly.

III. Point-to-Point Transmission Systems

Existing digital networks are delineated by digital cross-connect systems. A digital hierarchy is defined at these digital cross-connect points. In North America, there are DS1, DS2, and DS3 signals. Typically, the cross-connect serves as a maintenance span boundary. Span performance specifications need to be satisfied between the two cross-connect points. The means by which these performance specifications are satisfied is usually the function of the line-terminating equipment, which often also serves as a multiplexer. When the T1 carrier was first introduced, the maintenance features were few and simple. There were no error checks or operations channels. The order wire was provided via a separate copper wire pair. Fault location was performed off line by injecting a bipolar test signal into the line, and looping the signal back at the appropriate regenerator through a second separate pair. As technology advanced, i.e., as higher capacity systems were deployed, more automated features for maintenance were included. Error checking and fault location became part of the line-terminating equipment and the operations channel became part of signal overhead. Often, the order wire was also included in the overhead channels. Furthermore, the operations channels were expanded to include administration and provisioning. Since none of these functions has any standard features, each manufacturer was forced to choose its own interpretation and implementation. Often, the choice of overhead functions is dictated by the coding technique and the bit rates.

For a fiber optic system, standardization of these overhead functions need not be influenced by technological limitations. There is enough bandwidth in almost all fibers to allow as many overhead functions as necessary. The difficulty lies in the fact that none of these overheads has been standardized, and they will not be standardized in the near future. In the T1X1.4 subworking group, an approach was adopted to standardize only the channel capacity and the locations of overhead channels. The details of overhead have been left for future work, either in T1X1 or in T1M1 (internetwork operations, administrations, maintenance (OAM), and provisioning). The tentative list of the overheads as proposed by the subworking group is listed in Appendix A [9]. Note that sufficient overhead channels are being reserved for future expansion or for manufacturer and user proprietary uses. Note also that a channel called Span Channel Identification (SCID) is included in the list. The function of such a channel will become apparent in the section of synchronous multiplexing (Section VII).

Unlike other transmission media, the specific rate of a fiber optic system also does not depend on technology. A system at 140 Mbits/s and a system at 150 Mbits/s can both provide three DS3 signals as their information payload, with slightly different overhead capabilities. These two systems probably perform in the same manner and cost the same amount. Thus, the choice between these two options is almost arbitrary. Performance differences become apparent only when the rates are high (above 500 Mbits/s) and far apart. However, there are still great difficulties in arriving at a standard digital rate for the fiber optic systems, mainly due to political and business considerations. Any standard rate and format which do not agree with an existing product will introduce development and revision costs for respective manufacturers.

While the present arrangement for point-to-point systems is workable without a standard, the continued proliferation of incompatible systems has drawbacks. All high-speed signals must be demultiplexed down to the standard DS3 level for interconnection. All overhead signals are terminated at the line-terminating equipment and routed via some low-speed lines to their appropriate destinations. There are also possibilities of accepting the existing or future digital hierarchical signal rates and formats and concentrating only on arriving at a set of standard optical parameters. For example, the current DS3 format and the proposed SYNTRAN and DS4E formats can all be used for fiber optic systems [10], [11]. SYNTRAN and DS4E are particularly promising since the current activities in T1X1.4 also stress a set of standard operations channels for these signals. This approach is similar to work by CCITT on fiber optic systems [12]. However, this approach has its own limitations. First, DS3 signals do not have enough overhead capacity, even with the SYNTRAN modifications, to make them full line signals, so their application may be limited. Second, there are numerous asynchronous DS2 signals and some DS4E signals in the current network. Even if new standards can be adopted at these two rates, it is only the beginning. It would be difficult to carry these existing signals with the new standard systems. Third, newer systems at higher rates are being introduced rapidly and there is no standardization work at these high rates.

In addition, there is some apprehension about this laissez-faire approach. As more fiber systems are introduced into the network, there will be more incompatible systems that can be interconnected only via DS3 cross connects. Current requirements call for a distance limitation of 450 ft between the cross connect and the line-termination equipment and lead to congested Central Offices. At the DS4E rate, the distance limitation is even more restrictive (270 ft). There will also be more incompatible operations systems associated with these fiber systems. The overhead signals will still be carried around the network via outdated low-speed copper wires, not utilizing the enormous flexibility that a fiber can provide. Most of all, this approach is inconsistent with the philosophy of ISDN. Proliferation of incompatible fiber optic systems, which do not integrate ISDN services, will lead to a fragmented network and definitely stifle the development of ISDN. For these reasons, we will also examine standardization from a network approach. That is, instead of always interfacing at the standard electrical cross connect of a DS3 signal, we will examine the prospect of interfacing optically at some standard optical cross-connect systems. Because the maintenance and other signal associated overheads are embedded in the signal when it traverses the cross connect, the signal standards will be considered not only for the integrity of a transmission span, but also for the integrity of the network, the fiber network.

IV. A Network Approach

What is a fiber network? The deployment of fiber systems has drastically altered the complexion of the digital network. The two most outstanding features of these transmission systems are low loss and large bandwidth. The low loss allows us to build fiber transmission systems with few or no regenerators. The large bandwidth allows us to build very high-speed links. In addition, low loss and large bandwidth are obtained without special treatment to the fiber media, so that a low-speed link can be easily upgraded to a higher speed system without modifications to the media. Only the electronics need to be changed. To date, most fiber transmission systems are being deployed in trunks and long-haul links. However, there is a trend towards the penetration of the loop distribution plants. A most interesting and fruitful development of fiber systems will occur in the distribution plant. With fiber, wide-band services can be introduced into distribution plant economically. The bundle size of these wide-band services can be varied drastically and requires a flexible network. If these wide-band services are fed to hubbing points where they can be switched either manually or electronically, the hubbing configuration allows flexible capacity. The hubbing configuration may also be economical because it minimizes the construction cost, although using more fiber.

The cost of providing electrical to optical conversion is still relatively high, and will be for many years to come. Thus, it may be economical to provide an optical cross-connect system at the hub such that hubbing can be performed in the optical domain without incurring the back-to-back conversion cost. Because of the low loss of the fiber and therefore the long regenerator span length, an optical cross-connect system can cover a wide geographical area. Such wide coverage may lead to the deployment of optical cross-connect systems in areas with only medium densities of customers.

Whether or not the hubbing is performed at an optical cross-connect or with an electronic digital cross-connect system, the hub should provide flexible capacity. This capacity should be easily rearranged or reassigned. Standardization is essential in providing this flexible capacity. Using the existing fiber optic systems, all signals would have to be decomposed down to the DS3 level in the electrical form for interconnection. These DS3 signals, however, lack sufficient overhead functions to allow them to become line signals. Thus, a new standard signal may be necessary. To allow flexible rearrangement of the capacity, the high-speed signals should have the same format or they should have the same channel rearrangement capability, which calls for a standard overhead function. To have flexible capacity, it is advisable to have a number of bundle sizes available. However, under the existing hierarchical structure, these bundle sizes may be overly restrictive. Thus, a new hierarchical structure may also be beneficial. To simplify the administration procedures, it may also be advisable to devise a new basic bundle size and to provide services and to transport signals in multiples of this basic bundle. To simplify reassigning channel capacity, it is also proposed to use a synchronous multiplexing technique, which will be described in a later section. At this moment, the signal in a basic bundle size will be referred to as synchronous transport signal, level 1 (STS-1).

The STS-1 is a basic signal which transports 64 kbit/s voice services and connects to the existing voice-oriented network. Two requirements should be considered for this feature. First, the existing network is heavily DS1 and DS3 oriented. These DS1 and DS3 signals could be either synchronous or asynchronous. Regardless of the origins of its component signals, STS-1 should be easily decomposed into DS1's and DS3's. Second, savings can be realized through direct fiber termination to a voice switching system. STS-1 should be easily decomposed into DS0's at the switches. As mentioned earlier, this STS-1 should also incorporate features of an ISDN for easy and economical access for a variety of new services.

All of these arguments seem to call for a new fiber optic system rate and format standard. The T1X1.4 subworking group on rates and formats for fiber optic network interfaces has been working on such a project since April 1985. The basic concept came from a contribution by Bellcore in November 1984, which proposed a network called Synchronous Optical Network (SONET) [13], [14]. In the next section, we will briefly describe SONET. For identification purposes, we have retained the name of SONET for this discussion. It should be acknowledged that the present status of the project is the result of the joint ongoing efforts of the members of the T1X1.4 subworking group and the features described may be significantly different from both the original contribution and the final form.

V. Synchronous Optical Network (SONET) Components

SONET is a fiber network that consists of four standardized products families that could be easily interconnected. Of course, standardization only refers to the interface. The actual implementation should be left to the manufacturers to maximize the benefit of technological advances and designer's ingenuity.

To start, SONET offers a standard family of fiber optic transmission system product possibilities. These transmission systems perform the same functions as the existing transmission systems, with the exception that their interface specifications are standardized. It is recognized that the purpose of the standardization is to allow flexible network interconnectibility. Therefore, the rate and format are devised to allow such flexibility. The inputs to these transmission systems are DS3 signals, both synchronous and asynchronous. The output is a family of optical signals called optical carrier, level M (OC-M) supporting M signals at the DS3 level. Corresponding to each OC-M, there is an electrical signal called synchronous transport signal, level M(STS-M), which has the same rate, format, and functionalities as that of OC-M. The basic electrical signal, STS-1, is approximately 50 Mbits/s and can accept one DS3 signal. An STS-1 also has an abundance of overhead capacity for network administration and operations support. Therefore, the optical signal OC-1 can be transmitted as a line signal. A high-speed signal, OC-M, is generated by, first, bit-interleaved multiplexing of M STS-1's to form an STS-M and, then converting STS-M to an optical signal. With the capabilities of the overhead channels and the simple bit-interleaved multiplexing, these transmission systems can be easily interconnected.

The second family of SONET products are the OC-1/OC-M converters, the optical add–drop multiplexers, the optical cross-connect systems, and the electronic digital cross-connect systems. These are the devices by which the transmission systems are interconnected. OC-1/OC-M converters and optical add–drop multiplexers are currently perceived as electrical equipments that accept optical signals OC-1's and multiplex them to OC-M's. In the future, they may use optical processing as well as electrical processing. With these four types of equipment, an OC-M signal can be easily rearranged at the hub and an STS-1 signal can be easily reassigned to different routes. It should be noted that the equipment to function properly, the optical network should be synchronized; thus, we have the acronym SONET. Synchronization is necessary because a simple synchronous bit-interleaved multiplexing technique is employed to all multiplexing equipments.

A third family of products are those that convert services at sub-DS3 rates to OC-M transport directly without first converting to DS3's. They include the fiber optic transmission systems that accept DS1 signals (both line-terminating multiplexers and add–drop multiplexers) and DS0 signals. Should ISDN services be transmitted as a primary service ($23B + D$ channels at 1.536 Mbits/s), they will be carried by the fiber network as DS1-rate signals. For ISDN services at rates between that of DS1 and DS3, they will probably be multiplexed up to a DS3 signal first.

The fourth family of products are those that provide broad-band services that cannot be carried by one STS-1. A transparent DS4E signal or an enhanced quality video signal may require the capacity of an STS-3. A high-definition video signal may require the capacity of an STS-12. There are debates concerning how these services can be efficiently provided in SONET. Some suggest that the basic bundle size should be increased to around 140 Mbits/s so that both DS4E and enhanced quality video signal can be carried by STS-1. They also argue that equipment at approximately 140 Mbits/s is both technologically feasible and economically attractive. Defining STS-1 at this high speed may also simplify network administration procedure if these high-speed services are as widespread as someone has anticipated.

Another suggestion for these high-speed services is to devise a new signal called single service STS-3 (SSTS-3) and maintain STS-1 at 50 Mbits/s. An SSTS-3 retains the simplicity of an STS-3 in that it appears as three bit-interleaved multiplexed STS-1 signals as far as span maintenance is concerned. However, it retains the integrity of a single signal by not allowing the bit stream to be broken in the process of transmission. This last restriction is imposed so that elaborate and expensive timing recovery circuits can be avoided. The advantage of this approach is that both voiceband services and broad-band services can be provided with the same family of signals. Should a new wide-band service be introduced into the network at another rate, say STS-6, this approach can be easily extended to accept such a new service. The relative merits of choosing STS-1 at 140 Mbits/s or at 50 Mbits/s with SSTS-3 modifications is still under discussion in the T1X1.4 working group at the time of this writing.

Because of the network approach, the products necessary to form SONET include not only the transmission systems, but also distribution and switching systems. Although the SONET transmission systems per se may not have many economical advantages over that of existing fiber optic transmission systems when used point-to-point, the capability of easily interconnecting between them and the deployment of an associated interconnect and access system will lead to a simple, flexible, and economical network. In addition, there is also a consensus in the T1X1.4 subworking group that this is a golden opportunity to redirect the network development for future growth, and the opportunity may be lost forever if we do not act promptly.

With so many new products to be introduced, an easy deployment scheme which does not ignore the existing network is crucial. Initially, only point-to-point fiber optic transmission systems are deployed. These equipments can operate with an internal crystal-controlled clock. As soon as it becomes desirable to deploy some interconnecting systems such as OC-1/OC *M* converters, optical add-drop multiplexers, optical cross-connect systems, and electronic digital cross-connect systems, all SONET equipments would have to be synchronized. This second step probably starts with the deployment of OC-1/OC-*M* converters and optical add-drop multiplexers, coupled with some simple optical splicing and terminating devices. Only when the network is sufficiently full would an optical cross-connect system or an electronic digital cross-connect system be deployed. This completes the core work of SONET. However, not many new services have been introduced at this stage. Both voice and data signals are probably converted to DS3 for optical transmission. SYNTRAN terminals can be used efficiently to convert DS0 and DS1 signals to DS3.

The third step involves deploying new equipment that converts services directly to OC-*M*'s to take full advantage of SONET capabilities. Switching system interfaces to an optical signal could be produced cheaply if the rate and format are chosen carefully. Additionally, the standard can provide an optical ISDN interface to CPE which may be requested by customers in the future. Broad-band ISDN services can also be provided with optical lines. It is difficult to enumerate all new services that could be provided by the optical network. However, it should be emphasized that, once the interface to the customers is defined and the mechanisms of transporting the signal through the network is established, new services will proliferate rapidly.

VI. The SONET Rates and Formats

The goals which were established to guide the selection of STS-1 rate and format are

- the ability to transport DS3 signals,
- the ability to transport DS1 signals,
- the rate to be not much higher than 50 Mbits/s to take advantage of inexpensive VLSI processing for framing and related operations,
- the frame repetition rate of 125 μs to facilitate easy definition of 64 kbit/s channels, and
- a regular frame structure to separate overhead channels from information payload.

It was also desirable to select a frame format such that switch termination and broad-band ISDN access can be implemented simply and economically. However, it was realized that with the diverse requirements of switch termination and uncertainties in broad-band services, there is probably no perfect rate and format to satisfy all of these conditions. The signal currently being favored by the T1X1.4 subworking group is at 49.920 Mbits/s which can be separated into 26 overhead channels and 754 information channels. This STS-1 signal can accept one DS3 signal or 29 asynchronous DS1 signals. For signals other than the existing digital hierarchy, the information payload can be flexibly rechannelized.

An important feature of the overhead channels is the inclusion of a span channel identification marker (SCID). SCID is a bit sequence assigned to each STS-1 just prior to multiplexing and is used to quickly identify constituent STS-1's of a high rate signal. Each time an STS-1 is multiplexed, this SCID is rewritten. The inclusion of SCID allows an arbitrary numer of STS-1's to be multiplexed with a simple bit-interleaved multiplexing operation. The resultant high-speed signal does not require any additional overhead channels. Hence, the entire family of STS-*M* signals is defined once a standard rate and format for the STS-1 signal are adopted.

VII. SONET Optical Parameters

The development of optical parameter standards and the introduction of fiber optic equipments are intimately related. Since SONET products are deployed in three steps and in many different applications, the requirements of a standard on optical parameters also evolves in stages. In any case, because of the rapidly advancing technology, these standards should be minimal, flexible, and upgradable. Appendix B shows a minimum set of optical parameters which were discussed in the T1X1.2 subworking group. For a point-to-point fiber optic transmission system, each manufacturer could use parameters of his choice to fully exploit technology. However, the set of parameters in Appendix B could be used as a guideline since they may eventually be interconnected to other systems. This set is especially important for lower speed transmission systems since the technology has matured and stabilized at these speeds.

When an optical cross-connect system is deployed, it is also necessary to define the power level at the cross-connect points. Assuming that the optical cross-connect system is at the midpoint of a transmission system, we might arbitrarily set the power at, for example, -20 dBm. Since most transmitters have a power of higher than -5 dBm and most receivers have a sensitivity of better than -35 dBm, the optical cross-connect system could still find wide usage. Assuming further an average fiber loss of 0.5 dB/km, the optical cross-connect system could cover an area of up to 30 km in radius without using regenerators.

For high-speed systems, standardization still involves definitions, test procedures, and span design methodology. As new fibers and transducers are introduced, new definitions, test procedures, and span design methodology would have to be established. Fortunately for SONET, these high-speed transmission systems can still be used to interconnect two SONET hubs as long as they conform to the SONET format standards.

VIII. Conclusion

In this paper, we have reported various standards issues of fiber optic systems. In particular, we have addressed these issues from a synchronous optical network's viewpoint. The flexibility of such a network can only be realized if a standard exists, both in terms of signal format and optical parameters. However, such a standard also leads to a number of new products and new services. We have acknowledged the difficulties in adopting any standards, but we have also stressed the opportunities of such a standard.

Appendix A

Recommendation: Melding contributions T1X1.4/85-054 and T1X1.4/86-012 into the original T1X1.4/85-053 contribution resulted in the following recommended standard optical interface overhead structure (as of December 1985).

1) Framing: Two framing channels ($A1$, $A2$) are recommended:

Subframe	Pattern
$A1$	1111 0110
$A2$	0010 1000

The reframe is accomplished by examining the 8 bit blocks until one of the two framing patterns is identified. Then the other pattern is tracked. This framing pattern supports the DS3 and SYNTRAN requirements.

2) Channel Error Monitoring: This function consists of two 8 bit channels ($B1$ and $B2$), a cyclic redundancy check code-8 (CRC-8) for span error check, and a CRC-8 code for network error check.

3) Channel Identification: Two channels are recommended for user channel identification: one channel ($C1$) as a span information payload identification number—a unique number assigned just prior to multiplexing that stays with that payload until demultiplexing—and a network signal label channel ($C2$) that defines how the information payload is organized in the frame and how it is constructed.

This latter channel will be used to identify DS3 asynchronous, SYNTRAN, digital video, etc., payload—to include the need to group multiple payloads if such a need exists.

4) User Facility Maintenance and Control: Three channels for facility maintenance requirements are recommended. Two channels ($D1$ and $D2$) are allocated for span maintenance activities, e.g., parity, automatic protection switching, fault locating, terminal and switching office surveillance, and other span-to-span measurements. The T1M1 Working Group has a task (see attached T1M1.3 Task Proposal, dated June 12, 1985) to identify the specific channel definition and usage of these two channels, but the T1M1 representative in the Ad Hoc Group recommended that two channels be allocated to these functions, after the intricacies of protection switching were reviewed. A third channel ($D3$) is assigned for the monitoring, control, and analysis of the multispan networks. This latter channel is made available for packet switching and centralized maintenance applications.

5) Order Wire: Two order wire channels were allocated: local ($E1$) and express ($E2$). These are reserved for communication between central offices and, where required, repeater, hubs, and remote terminal locations. This reservation is in line with current industry practices.

6) User Proprietary Channels: Again, three channels have been allocated to this requirement: one channel ($F1$) is allocated to the EC/IEC user for his input of span information such as data communication for use in maintenance activities and remoting the alarms external to the span equipment, and two channels ($F2$ and $F3$) for the EC/IEC use as end-to-end VF communications channels for themselves and/or their end users.

7) Manufacturer's Proprietary Channel: Two channels ($G1$ and $G2$) are recommended to be reserved for manufacturer's use for unique enhancements he might introduce.

8) Frame Alignment Control: Three channels are recommended to allow for asynchronous impairments in an otherwise synchronous network. The first two ($H1$ and $H2$) are redundant (reducing error possibility) channels that are used for indicating the action to be performed, i.e., clocks in instantaneous synchronization (no action taken), add a sync bit, delete a bit, etc. The third channel ($H3$) is the bit to be acted upon. (Note: This channel will contain 7, 8, or 9 bits—depending on $H1/H2$. Seven bits of this channel are available for other assignments.)

9) Unreserved Channels: Seven channels ($J1$, $J2$, $J3$, $J4$, $J5$, $J6$, $J7$) allocated to future growth needs for the industry. Additional growth channels may become available as some of the above requirements are refined. The final format was designed with this latter factor in mind.

The final format of the recommended overhead is

$A1$	$D3$	$H1$	$H2$	$H3$	$J1$	$D1$	$D2$	$B1$	$B2$	$C1$	$F1$	J
1	2	3	4	5	6	7	8	9	10	11	12	13
$A2$	$F2$	$F3$	$J3$	$C2$	$E1$	$E2$	$J4$	$J5$	$J6$	$J7$	$G1$	$G2$
14	15	16	17	18	19	20	21	22	23	24	25	26

Appendix B
Transmitter Optical Parameters (Minimum Set)

	Ranges Under Consideration in T1X1.2		
Wavelength	Nominal	1310	nm
	Minimum	1280	nm
	Maximum	1340	nm
Spectral Width	Maximum	9	nm
Extinction Ratio	Minimum	10:1	
Coding	Nonreturn-to-Zero (NRZ)		
Pulse Shape	To be determined		

References

[1] I. Lifchus, "Committee T1 now ready to address interconnection standards," *Telephony*, 1984.
[2] "IEEE standard definitions of terms relating to fiber optics," IEEE STD 812-1984.
[3] J. R. Neigh, "Voluntary standards for fiber optics from EIA 7 IEC," FOC/LAN 84.
[4] G. Bonaventura and U. Rossi, "Standardization within CCITT of optical fibres for telecommunication systems," presented at GLOBECOM, 1984.

[5] K. Ogawa, "Considerations for single-mode fiber systems," *Bell Syst. Tech. J.*, vol. 61, pp. 1919–1931, Oct. 1982.
[6] EIA-522, Electron. Industry Association.
[7] T1X1.2 contributions, T1X1.2/84-017.
[8] T1X1.4 contribution, "Rates and formats for optical network interfaces," T1X1.4/85-017.
[9] T1X1.4 contribution, "Recommendations of ad hoc group on overhead structure," T1X1.4/85-23.
[10] T1X1.4 contribution, "Interface specification for synchronous DS3 format," T1X1.4/85-020.
[11] T1X1.4 contribution, "Fourth hierarchy level at 139264 kbits/s," T1X1.4/85-021.
[12] CCITT Recommendations G.951 and G.952.
[13] T1X1.4 contribution, "Synchronous optical network (SONET)," T1X1.4/85-005.
[14] Bell Commun. Res., Inc., "Synchronous optical network (SONET)," TA-TSY-000253, Apr. 1985.

PHOTO NOT AVAILABLE

Rodney J. Boehm received the B.S. and M.E. degrees from Texas A&M University, College Station, both in electrical engineering.

He worked at Bell Telephone Laboratories designing high-speed multiplexing equipment for digital microwave radio. In 1983 he joined Bell Communications Research in the fiber optic systems area where he helped develop the Synchronous Optical Network (SONET) concept. During this time, he helped introduce SONET to the T1 standards body to achieve a standardized optical interface. He is now a Product Line Administrator with the Collins Transmission Systems Division of Rockwell International, Dallas, TX.

PHOTO NOT AVAILABLE

Yau-Chau Ching (S'64–M'69) received the B.E.E.E. degree from City College of New York, New York, NY, in 1966 and the Ph.D. degree from New York University, NY, in 1969 under a National Science Foundation Fellowship.

From 1969 to 1984 he was with Bell Laboratories where he did a series of exploratory development work in data compression, such as interframe video coding, digital speech interpolation, embedded ADPCM coding, and fast packet network. Since the divestiture of the Bell System, he has been with Bell Communications Research where he is District Manager, Fiber Optic Systems Engineering, responsible for technical requirements of fiber optical systems and the SONET project.

Dr. Ching is a member of Tau Beta Pi and Eta Kappa Nu.

PHOTO NOT AVAILABLE

C. George Griffith (M'59–SM'67) graduated from the University of Missouri, Columbia, and received the M.S.E.E. degree from San Jose State University, San Jose, CA.

He joined GTE Lenkurt, San Carlos, CA, after service as a Fire Control Officer in the U.S. Navy, was assigned in 1976 as Vice President and Product Manager of GTE Communications Products, Stamford, CT, and in 1984, joined Bell Communications Research. He is Assistant Vice President in charge of interoffice technology. He has spent 30 years in research and development, manufacturing, and marketing in the field of telecommunications.

Mr. Griffith has organized International Conferences on Communications sponsored by the IEEE Communications Society.

PHOTO NOT AVAILABLE

Frederick A. Saal (M'84) received the B.S.E.E. and B.S.E.P. degrees from Lehigh University, Easton, PA, in 1954.

From 1954 to 1984 he was with Bell Laboratories where he was first associated with the development of the first commercial TASI system. From 1961 to 1984 he was responsible for development of various digital transmission equipments and transmission peripherals for digital switches. Since the divestiture of the Bell System, he has been with Bell Communications Research where he is Division Manager, Interoffice Transport Requirements, responsible for technical requirements of interoffice transmission systems.

Mr. Saal is a member of Tau Beta Pi, Eta Kappa Nu, and Phi Beta Kappa.

Analysis and Simulation of a Digital Mobile Channel Using Orthogonal Frequency Division Multiplexing

LEONARD J. CIMINI, JR., MEMBER, IEEE

Abstract—**This paper discusses the analysis and simulation of a technique for combating the effects of multipath propagation and cochannel interference on a narrow-band digital mobile channel. This system uses the discrete Fourier transform to orthogonally frequency multiplex many narrow subchannels, each signaling at a very low rate, into one high-rate channel. When this technique is used with pilot-based correction, the effects of flat Rayleigh fading can be reduced significantly. An improvement in signal-to-interference ratio of 6 dB can be obtained over the bursty Rayleigh channel. In addition, with each subchannel signaling at a low rate, this technique can provide added protection against delay spread. To enhance the behavior of the technique in a heavily frequency-selective environment, interpolated pilots are used. A frequency offset reference scheme is employed for the pilots to improve protection against cochannel interference.**

I. Introduction

SEVERE multipath propagation, arising from multiple scattering by buildings and other structures in the vicinity of a mobile unit, makes the design of a mobile communication channel very challenging [1]. This scattering produces rapid random amplitude and phase variations in the received signal as the vehicle moves in the multipath field. In addition, the vehicle motion introduces a Doppler shift, which causes a broadening of the signal spectrum. Measurements confirm that the short-term statistics of the resultant signal envelope approximate a Rayleigh distribution.

Multipath fading may also be frequency selective, that is, the complex fading envelope of the received signal at one frequency may be only partially correlated with the received envelope at a different frequency. This decorrelation is due to the difference in propagation time delays associated with the various scattered waves making up the total signal. The spread in arrival times, known as delay spread, causes transmitted data pulses to overlap, resulting in intersymbol interference. In a typical urban environment, a spread of several microseconds and greater can be occasionally expected.

There is an additional impairment in a *cellular* mobile system. The available radio channels are reused at different locations within the overall cellular service area in order to use the assigned spectrum more efficiently. Thus, mobiles simultaneously using the same channel in different locations interfere with each other. This is termed cochannel interference and is often the dominant impairment.

In addition, there is a long-term variation of the local mean of the received signal, called shadow fading. Shadow fading in a mobile radio environment is caused by large obstacles blocking the transmission path. This impairment is alleviated in cellular systems by using transmitted and received base-station signals at two different geographical locations [1], and will not be discussed in this paper.

Given the harsh mobile environment and the scarcity of available spectrum, it is desirable to look for channel designs which provide good performance for both speech and data transmission, and which are also bandwidth efficient. The channel designs presented in this paper could accommodate speech or data transmission. For the *narrow* channel assumed, a low-bit-rate speech coder would be required. For example, a 7.5 kHz channel using the system proposed in this paper can support 8.6 kbits/s. In what follows, the channel will be assumed to be transmitting data symbols.

In a conventional serial data system, the symbols are transmitted sequentially, with the frequency spectrum of each data symbol allowed to occupy the entire available bandwidth. Due to the bursty nature of the Rayleigh channel, several adjacent symbols may be completely destroyed during a fade. To illustrate the severity of the problem, consider the following example. Assume that there is a cochannel interferer with an average power level 17 dB below that of the desired signal. This condition occurs approximately 10 percent of the time in a cellular mobile system. A fade 17 dB below the average level will bury the desired signal in the interference. At a carrier frequency of 850 MHz and a vehicle speed of 60 mph, the average fade duration for a fade 17 dB below the local mean of the desired signal is 0.75 ms [1]. For a data rate of 10 kbits/s, 7 or 8 adjacent bits would be destroyed during such a fade.

In a serial system, higher data rates can be achieved, at the expense of a degradation in performance, by using higher order modulations or, at the expense of increased channel bandwidth, by decreasing the symbol interval. However, delay spread imposes a waiting period that determines when the next pulse can be transmitted. This waiting period requires that the signaling be reduced to a rate much less than the reciprocal of the delay spread to prevent intersymbol interference. Decreasing the symbol interval makes the system more susceptible to delay spread impairments.

A parallel or multiplexed data system offers possibilities for alleviating many of the problems encountered with serial systems. A parallel system is one in which several sequential streams of data are transmitted simultaneously, so that at any instant many data elements are being transmitted. In such a system, the spectrum of an individual data element normally occupies only a small part of the available bandwidth. In a classical parallel data system, the total signal frequency band is divided into N *nonoverlapping* frequency subchannels. Each subchannel is modulated with a separate symbol and, then, the N subchannels are frequency multiplexed. A more efficient use of bandwidth can be obtained with a parallel system if the spectra of the individual subchannels are permitted to overlap, with specific orthogonality constraints imposed to facilitate separation of the subchannels at the receiver.

A parallel approach has the advantage of spreading out a fade over many symbols. This effectively randomizes the burst errors caused by the Rayleigh fading, so that instead of several adjacent symbols being completely destroyed, many symbols are only slightly distorted. This allows precise reconstruction of a majority of them. A parallel approach has the additional advantage of spreading out the total signaling interval, thereby reducing the sensitivity of the system to delay spread.

Several systems have previously used orthogonal frequency

Paper approved by the Editor for Radio Communication of the IEEE Communications Society for publication without oral presentation. Manuscript received June 18, 1984; revised January 14, 1985.
The author is with AT&T Bell Laboratories, Holmdel, NJ 07733.

Reprinted from *IEEE Transactions on Communications,* vol. COM-33, no. 7, July 1985.

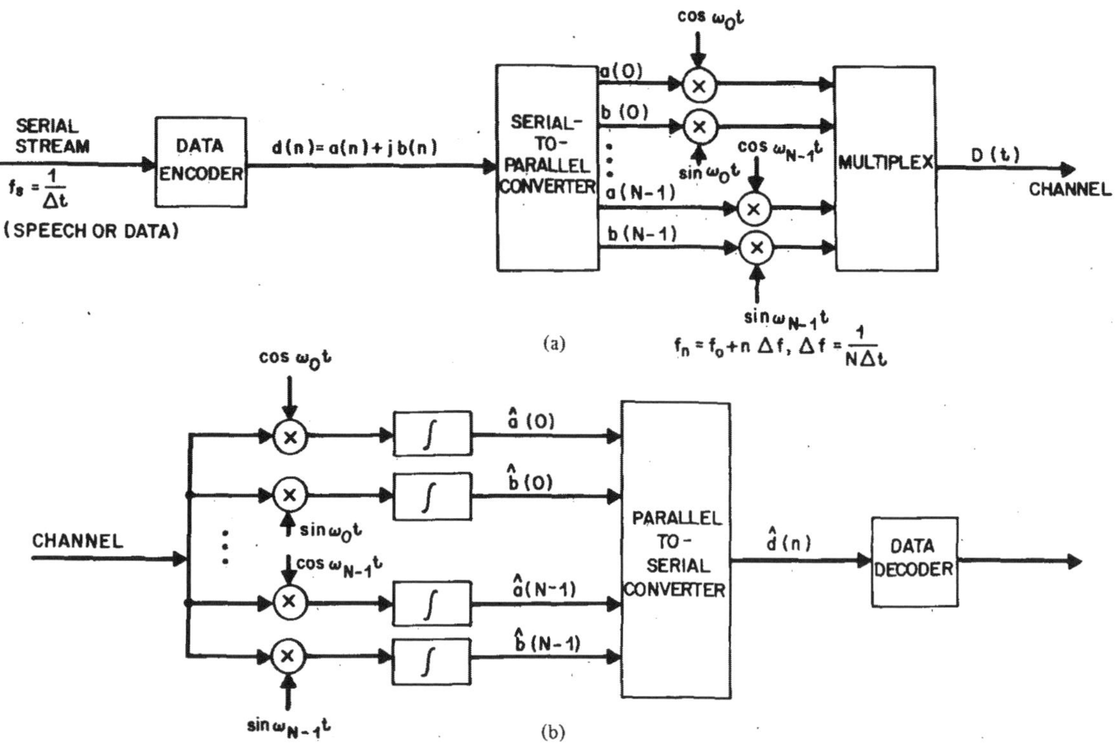

Fig. 1. Basic OFDM system. (a) Transmitter. (b) Receiver.

division multiplexing (OFDM) [3]-[8]. In particular, in the early 1960's, this technique was used in several high-frequency military systems (for example, KINEPLEX [9], ANDEFT [10], KATHRYN [11], [12]), where fast fading was not a problem. Similar modems have found applications in voice bandwidth data communications (for example, [13]) to alleviate the degradations caused by an impulsive noise environment.

In this paper, a parallel system which uses the OFDM technique is described. In Section II an analysis and simulation of the basic system, using pilot-based correction, is presented. In Section III a practical 7.5 kHz channel design is presented, along with a discussion of several of the problems encountered in reliably retrieving the pilots used in the data correction process. Several solutions to these problems are also presented.

This investigation is simplified by the assumption that the sole source of additive signal degradation is cochannel interference—thermal noise is assumed negligible. Man-made environmental noise, such as that caused by automotive ignitions or neon lights, is also ignored. However, these impairments are basically impulsive and their effect should be greatly reduced by this technique.

II. Basic Principles of Operation

A. Orthogonal Frequency Division Multiplexing (OFDM)

When an efficient use of bandwidth is not required, the most effective parallel system uses conventional frequency division multiplexing where the spectra of the different subchannels do not overlap. In such a system, there is sufficient guard space between adjacent subchannels to isolate them at the receiver using conventional filters. A much more efficient use of bandwidth can be obtained with a parallel system if the spectra of the individual subchannels are permitted to overlap. With the addition of coherent detection and the use of subcarrier tones separated by the reciprocal of the signaling element duration (orthogonal tones), independent separation of the multiplexed tones is possible.

Consider the system shown in Fig. 1. The transmitted spectral shape is chosen so that interchannel interference does not occur; that is, the spectra of the individual subchannels are zero at the other subcarrier frequencies. The N serial data elements (spaced by $\Delta t = 1/f_s$ where f_s is the symbol rate) modulate N subcarrier frequencies, which are then frequency division multiplexed. The signaling interval T has been increased to $N\Delta t$, which makes the system less susceptible to delay spread impairments. In addition, the subcarrier frequencies are separated by multiples of $1/T$ so that, with no signal distortion in transmission, the coherent detection of a signal element in any one subchannel of the parallel system gives no output for a received element in any other subchannel. Using a two-dimensional digital modulation format, the data symbols $d(n)$ can be represented as $a(n) + jb(n)$ (where $a(n)$ and $b(n)$ are real sequences representing the in-phase and quadrature components, respectively) and the transmitted waveform can be represented as

$$D(t) = \sum_{n=0}^{N-1} \{a(n)\cos(\omega_n t) + b(n)\sin(\omega_n t)\} \tag{1}$$

where $f_n = f_0 + n\Delta f$ and $\Delta f = 1/N\Delta t$. This expression and the following analyses can be easily extended to include pulse shaping other than the assumed rectangular shape.

Theoretically, M-ary digital modulation schemes using OFDM can achieve a bandwidth efficiency, defined as bit rate per unit bandwidth, of $\log_2 M$ bits/s/Hz. This is easily shown as follows. Given that the symbol rate of the serial data stream is $1/\Delta t$, the bit rate for a corresponding M-ary system is $\log_2 M/\Delta t$. Each subchannel, however, transmits at a much lower rate, $\log_2 M/(N\Delta t)$. The total bandwidth of the OFDM system is

$$B = f_{N-1} - f_0 + 2\delta \tag{2}$$

where f_n is the nth subcarrier and δ is the one-sided bandwidth of the subchannel (where the bandwidth is considered as the

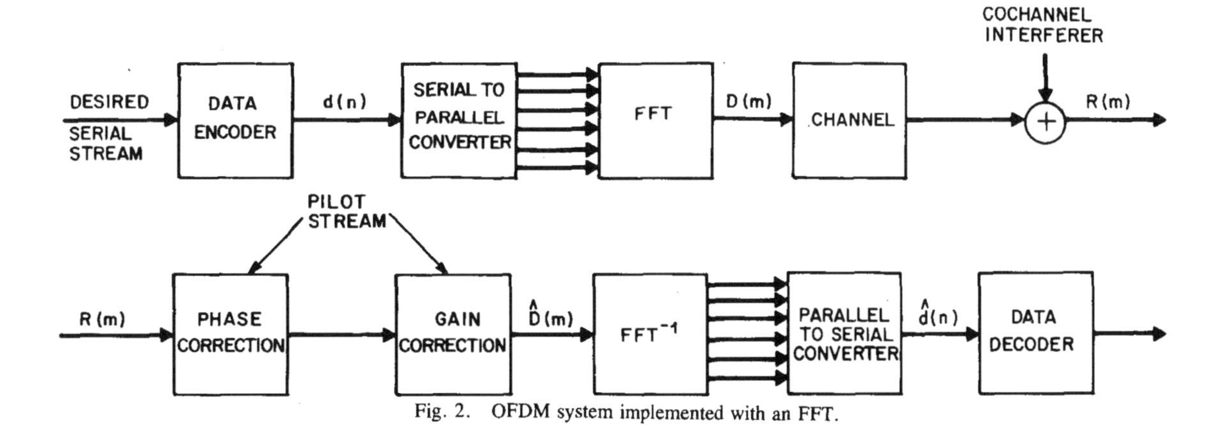

Fig. 2. OFDM system implemented with an FFT.

distance to the first null). The subcarriers are uniformly spaced so that $f_{N-1} - f_0 = (N - 1)\Delta f$. Since $\Delta f = 1/N\Delta t$ due to the orthogonality constraint, $f_{N-1} - f_0 = (1 - (1/N))(1/\Delta t)$. Therefore, the bandwidth efficiency β becomes

$$\beta = \frac{\log_2 M}{\left(1 - \frac{1}{N}\right) + 2\delta\Delta t}. \tag{3}$$

For orthogonal frequency spacing and strictly band-limited spectra (bandwidth Δf) with $\delta = \frac{1}{2}\Delta f = 1/2N\Delta t$, $\beta = \log_2 M$ bits/s/Hz. In reality, however, the spectra overflow this minimum bandwidth by some factor α so that $\delta = (1 + \alpha)(1/2N\Delta t)$ and the efficiency (3) becomes

$$\beta = \frac{\log_2 M}{1 + \frac{\alpha}{N}} < \log_2 M. \tag{4}$$

To obtain the highest bandwidth efficiency in an OFDM system, N must be large and α must be small.

B. Implementation of OFDM Using the Discrete Fourier Transform

The principal objections to the use of parallel systems are the complexity of the equipment required to implement the system, and the possibility of severe mutual interference among subchannels when the transmission medium distorts the signal. The equipment complexity (filters, modulators, etc.) can be greatly reduced by eliminating any pulse shaping, and by using the discrete Fourier transform (DFT) to implement the modulation processes, as shown in [7], [8]. There it is shown that a multitone data signal is effectively the Fourier transform of the original data stream, and that a bank of coherent demodulators is effectively an inverse Fourier transform. This can be seen by writing (1) as

$$D(t) = \mathrm{Re}\left[\sum_{n=0}^{N-1} d(n)e^{-j\omega_n t}\right]. \tag{5}$$

Letting $t = m\Delta t$, the resulting sampled sequence $D(m)$ is seen as the real part of the DFT of the sequence $d(n)$.[1] The act of truncating the signal to the interval $(0, N\Delta t)$ imposes a $\sin x/x$ frequency response on each subchannel with zeros at multiples of $1/T$. This spectral shape has large sidelobes, and gives rise to significant interchannel interference in the presence of multipath. This point will be discussed in more detail in Section III.

[1] It is convenient in this paper to think of $d(n)$ as being in the frequency domain and $D(m)$ as being in the time domain, contrary to the usual engineering interpretation [8].

Further reductions in complexity are possible by using the fast Fourier transform (FFT) algorithm to implement the DFT when N is large.

C. Pilot-Based Correction

If the transmission channel is distortionless, the orthogonality of the subcarriers allows the transmitted signals to be received without error at the receiver. Consider the system in Fig. 2 with the block of data represented by the sequence of N complex numbers $\{d(0), d(1), \cdots, d(N-1)\}$. These complex numbers are generated by the data encoder from a binary data sequence. A DFT is performed on this block of data, giving the transmitted symbols[2]

$$D(m) = \mathrm{DFT}\{d(n)\} = \sum_{n=0}^{N-1} d(n)e^{-j(2\pi/N)nm}. \tag{6}$$

Notice that this is a sampled version of (5) where the complex notation has been retained. All future analyses will be done in the complex domain. Under the assumption of a distortionless channel, the received data sequence (the output of the inverse DFT) will be exactly the transmitted sequence due to the orthogonality of the subcarrier tones (exponentials).

If the transmission channel distorts the signal, this orthogonality is impaired. In a flat Rayleigh fading environment (i.e., the environment is not frequency selective), the effects of the Rayleigh channel can be represented as a multiplicative noise process on the transmitted signal. This multiplicative process is characterized by a complex fading envelope with samples $Z(m) = A(m)e^{j\theta(m)}$ where the $A(m)$ are samples from a Rayleigh distribution and the $\theta(m)$ are samples from a uniform distribution [1]. These samples multiply the sequence of (6) to give

$$R(m) = Z(m)D(m). \tag{7}$$

The output data sequence $\hat{d}(k)$ is the inverse DFT of (7),

$$\begin{aligned}\hat{d}(k) &= \frac{1}{N}\sum_{m=0}^{N-1} Z(m)D(m)e^{j(2\pi/N)km}\\ &= \sum_{n=0}^{N-1} d(n)\left[\frac{1}{N}\sum_{m=0}^{N-1} Z(m)e^{j(2\pi/N)m(k-n)}\right]\\ &= \sum_{n=0}^{N-1} d(n)z(k-n)\end{aligned} \tag{8}$$

[2] Throughout this paper, all indexes will be assumed to belong to the set $\{0, 1, 2, \cdots, N-1\}$.

where $z(n)$ is the inverse DFT of $Z(m)$. It can be seen from (8) that there is a complex-weighted averaging of the samples of the complex fading envelope. If $Z(m) = 1$ for all m (the distortionless channel), $z(k-n)$ is simply the Kronecker delta function δ_{kn} and $\hat{d}(k) = d(k)$. In the presence of fading, $z(k-n) \neq \delta_{kn}$ and

$$\hat{d}(k) = d(k)z(0) + \sum_{\substack{n=0 \\ n \neq k}}^{N-1} d(n)z(k-n). \tag{9}$$

The second term on the right represents the interchannel (intersymbol) interference caused by the loss of orthogonality. Without correction for the fading, the output sequence is corrupted by intersymbol interference *even if there is no cochannel interferer.*

Pilot-based correction provides an amplitude and phase reference which can be used to counteract the unwanted effects of multipath propagation. Similar considerations have been analyzed for single-sideband mobile radio systems [14], [15]. Coherent detection, by definition, requires a phase reference; however, gain correction is also needed in an OFDM system in a fading environment to remove intersymbol interference. If phase and gain correction is employed in the absence of cochannel interference, it is easily shown, in (9), that $\hat{d}(k) = d(k)$.

In a cellular mobile system, the dominant transmission impairment often comes from other users using the same carrier frequency. It is assumed here that the desired signal and a *single* undesired cochannel interferer are received simultaneously, and that both are digital signals modulated by different data sequences with identical signaling rates. It is also assumed that they are subject to mutually independent Rayleigh fading.

When a cochannel interferer is present in the received signal, it is not advantageous to do unlimited gain correction, due to the possibility of enhancing the energy of the interferer during deep fades of the desired signal. The detrimental effects of unlimited gain correction in the presence of a cochannel interferer can be seen as follows. Let $D(m)$ be the desired transmitted signal sequence and let $I(m)$ be the corresponding cochannel interferer sequence. With $Z_d(m) = A_d(m)e^{j\theta_d(m)}$ and $Z_i(m) = A_i(m)e^{j\theta_i(m)}$ the desired and interferer complex fading sequences, respectively, the sequence present at the receiver can be represented as

$$R(m) = Z_d(m)D(m) + \sqrt{\gamma}Z_i(m)I(m) \tag{10}$$

where γ is the interference-to-signal power ratio (SIR^{-1}). $R(m)$ is corrected by a complex correction sequence $Z_c(m) = Z_p(m)$, the complex pilot fading envelope, giving

$$\hat{D}(m) = \frac{R(m)}{Z_c(m)} = \frac{Z_d(m)}{Z_p(m)} D(m) + \sqrt{\gamma}\frac{Z_i(m)}{Z_p(m)} I(m). \tag{11}$$

Taking the inverse DFT of (11), the received data sequence becomes

$$\hat{d}(k) = \sum_{n=0}^{N-1} d(n)z(k-n) + \sqrt{\gamma} \sum_{m=0}^{N-1} \frac{1}{N} \frac{Z_i(m)}{Z_p(m)} \cdot I(m)e^{j(2\pi/N)mk} \tag{12}$$

where

$$z(k-n) = \frac{1}{N} \sum_{m=0}^{N-1} \frac{Z_d(m)}{Z_p(m)} e^{j(2\pi/N)m(k-n)}.$$

If unlimited gain and phase correction is used [i.e., $Z_p(m) = Z_d(m)$], $z(k-n) = \delta_{kn}$, there is no intersymbol interference, and (12) becomes

$$\hat{d}(k) = d(k) + \sqrt{\gamma}\, \frac{1}{N} \sum_{m=0}^{N-1} I(m) \frac{Z_i(m)}{Z_d(m)} e^{j(2\pi/N)mk}. \tag{13}$$

The only distortion is caused by the cochannel interferer. However, since $Z_i(m)$ and $Z_d(m)$ are statistically independent, the desired signal may be in a fade when the interferer is not, and unlimited gain correction may boost the interferer average energy above that of the desired signal.

One alternative to unlimited gain and phase correction is to have a limit on the gain correction, so as not to follow the desired signal into deep fades [1]. This is done at the expense of increased intersymbol interference, due to imperfect correction of the desired signal. In this situation, the correction signal is of the form

$$Z_c(m) = \begin{cases} A_d(m)e^{j\theta_d(m)} & \text{when } A_d(m) > \epsilon \\ \epsilon e^{j\theta_d(m)} & \text{when } A_d(m) \leqslant \epsilon \end{cases} \tag{14}$$

where ϵ is the gain limit and is defined relative to the average value of the local field strength. Therefore, in (12), $z(k-n) \neq \delta_{kn}$, resulting in intersymbol interference. Consequently, there is a tradeoff between increasing the intersymbol interference and boosting the cochannel interference energy.

Another alternative is to develop an optimum gain correction factor which takes both distortion effects into account. An optimum gain correction factor $F(m)$ has been derived by minimizing the mean-square distortion between $D(m)$ and $\hat{D}(m)$. The derivation of $F(m)$ has been omitted for the sake of brevity. The correction sequence then becomes

$$\begin{aligned} Z_c(m) &= Z_p(m)F(m) \\ &= Z_d(m)\left[1 + \gamma\left(\frac{A_i(m)}{A_d(m)}\right)^2\right]. \end{aligned} \tag{15}$$

This correction procedure would be more difficult to implement than the gain limiting procedure described above.

In addition to the impairments caused by intersymbol and cochannel interference, frequency-selective fading may also be present. This phenomenon causes a decorrelation of the received signal envelopes at different frequencies, lessening the effectiveness of the pilot-correction procedure, since a data point which is being corrected may be decorrelated from the corresponding pilot complex fading envelope.

Finally, one of the major advantages of the OFDM technique is its ability to "average" out impairments, making the bursty Rayleigh channel appear much less bursty. The extent to which this averaging approaches a Gaussian channel depends on the correlation between samples of the complex fading envelope. It can be seen that as N increases, more independent fades are averaged. This enables burst errors to be randomized and thereby aids in bit error correction. This property will be more evident in the simulation results, which indicate that the curves for the bit error rate fall between the linear Rayleigh channel curves and the exponential Gaussian channel curves. For large N and high vehicle speeds, the bit error curve approaches that for a Gaussian channel.

D. Distortion Analyses

Several mechanisms contribute to the overall distortion of the desired signal. In this section, emphasis is on the contributions due to gain limiting, evident in increased intersymbol interference, and due to cochannel interference. The distortion resulting from decorrelation of the pilot due to frequency-selective fading or due to interference on the pilot is considered in Section II-F.

First, consider the case of gain-limited correction, where the amplitude correction is bounded to follow fades only as deep as ϵ. Assume that the random processes which produce the random sequences are ergodic, thereby permitting the equivalence of time and ensemble averages. The pilot complex fading envelope at a particular instant in time is $Z_p(m) = Z_d(m)$ and the correction sequence is

$$Z_c(m) = \max (A_d(m), \epsilon) e^{j\theta_d(m)}. \tag{16}$$

The corrected output samples become

$$\hat{D}(m) = \frac{R(m)}{Z_c(m)} = D(m)\frac{Z_d(m)}{Z_c(m)} + \sqrt{\gamma} I(m) \frac{Z_i(m)}{Z_c(m)}$$
$$= D(m)A_d(m) \min\left(\frac{1}{A_d(m)}, \frac{1}{\epsilon}\right)$$
$$+ \sqrt{\gamma} I(m) A_i(m) e^{j(\theta_i(m) - \theta_d(m))}$$
$$\cdot \min\left(\frac{1}{A_d(m)}, \frac{1}{\epsilon}\right). \tag{17}$$

The signal-to-distortion ratio (SDR) can be defined as in [14],

$$\text{SDR} = \frac{\overline{|D(m)|^2}}{\overline{|\hat{D}(m) - D(m)|^2}} \tag{18}$$

where $\overline{X}$ denotes a time average of X. Assuming $\overline{|D(m)|^2} = \overline{|I(m)|^2} = 1$, the denominator in (18) reduces to

$$\overline{|\hat{D}(m) - D(m)|^2}$$
$$= \overline{A_d{}^2(m) \min^2\left(\frac{1}{A_d(m)}, \frac{1}{\epsilon}\right)}$$
$$+ \gamma \overline{A_i{}^2(m) \min^2\left(\frac{1}{A_d(m)}, \frac{1}{\epsilon}\right)}. \tag{19}$$

Assuming time averages can be replaced by expected values and assuming $A_d(m)$ and $A_i(m)$ are statistically independent and Rayleigh distributed, (18) becomes, after some manipulations,

$$\text{SDR} = \left\{\left(\frac{\text{SIR} + 1}{\text{SIR}}\right) \frac{1}{\epsilon^2} [1 - e^{-\epsilon^2}] + 1 - \frac{\sqrt{\pi}}{\epsilon} \operatorname{erf}(\epsilon) + \frac{E_1(\epsilon^2)}{\text{SIR}}\right\}^{-1} \tag{20}$$

where $E_1(x) = -[\Psi + \ln(x) + (\sum_{n=1}^{\infty} (-1)^n x^n / nn!)]$ and Ψ is Euler's constant (=0.57721566 ···). The SDR in (20) is plotted in Fig. 3 for several values of SIR. Obviously, if SIR $= \infty$ (no cochannel interference), the results reduce to that in [14] and no gain limit should be used. However, for SIR $< \infty$ the curves clearly indicate the tradeoff between intersymbol interference, caused by gain limiting, and boosting of the cochannel interference average energy, caused by unlimited gain correction. If unlimited gain correction is used, SDR $= -\infty$, indicating that the interferer completely distorts the desired signal. Notice, there is a definite maximum which is fairly flat. Although SDR as defined here is an analog transmission quality measure, it does indicate the degree to which intersymbol interference, caused by imperfect gain correction, and cochannel interference are problems. These factors are critically important in digital transmission. The SDR also clearly shows the tradeoffs which must be made when choosing the appropriate gain limit. This, in turn, directly affects the bit error rate (BER), as shown in the next section.

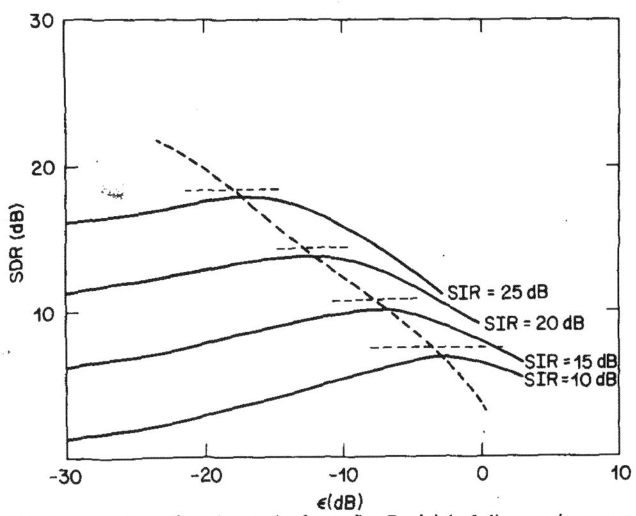

Fig. 3. Signal-to-distortion ratio for a flat Rayleigh fading environment when gain-limited correction is used.

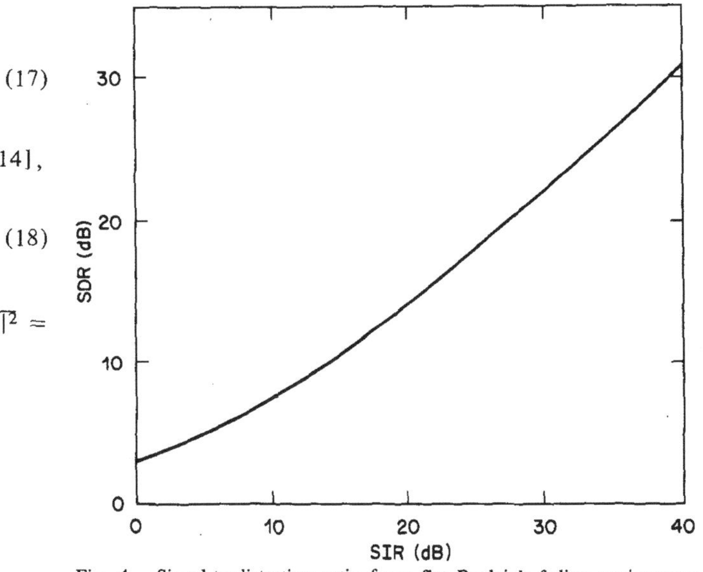

Fig. 4. Signal-to-distortion ratio for a flat Rayleigh fading environment when the optimum gain correction factor is used.

Similar results can be derived for optimum gain correction, as in (15), and the SDR can be shown to be

$$\text{SDR} = \left[\left(\frac{1}{\text{SIR} - 1}\right)\left[\frac{\text{SIR}}{\text{SIR} - 1} \ln(\text{SIR}) - 1\right]\right]^{-1} \tag{21}$$

which is plotted versus SIR in Fig. 4. This curve indicates the *best* performance for a given SIR. Notice, by comparing Figs. 3 and 4, that using gain-limited correction does not sacrifice much if the gain limit is in the vicinity of the maximum. Both of these results could be used as an aid in determining the appropriate level for gain limiting for a given SIR.

E. Simulation Results

Initial simulations were performed assuming flat Rayleigh fading. In these simulations, it was also assumed that the pilots could be recovered perfectly; that is, there is no interference or distortion on the pilots. A symbol rate and channel bandwidth of 7.5 kHz[3] were used and the BER was determined for several values of SIR (100 000 bits were used in the simulation to provide statistical significance). Based on this choice for the bandwidth, the maximum bit rate is 7.5 $\log_2 M$ kbits/s. Several parameters were varied in this initial investigation, the most important being N, the number of subchannels, and v, the vehicle speed. Both quantities are very important factors in determining the ability of this system to effectively randomize the burst errors created by the Rayleigh fading. The fading rate is directly proportional to the vehicle speed. In particular, at a carrier frequency of 850 MHz, independent fades are about 7 in apart, giving a fade every 6.6 ms at 60 mph. Therefore, for a given value of N, higher vehicle speeds should result in better performance because more fades are included in the averaging process. Similarly, for a given vehicle speed, if N is large, the total signaling interval is also large and more fades are again used in the averaging process.

The results shown in Fig. 5, where quadrature phase shift keying (QPSK) has been employed, indicate the improvement possible if OFDM is used with gain correction under the assumptions of ideal pilot recovery and a flat Rayleigh fading environment. Results for both optimum gain correction, as in (15), and gain-limited correction, as in (14), are given. These results clearly indicate the effects of vehicle speed and the number of subchannels. At a carrier frequency of 850 MHz and with a vehicle speed of 60 mph, with gain-limited correction, improvements in SIR of 6-7 dB[4] have been obtained using 512 subchannels ($T = 68$ ms). This is in comparison to a flat Rayleigh channel using coherent detection ($N = 1$) with QPSK. A reduction in speed to 30 mph results in a loss in performance of less than 1 dB. A reduction of N to 128 subchannels ($T = 17$ ms) results in an additional 2 dB loss, because fewer independent fades are included in the averaging process. For the cases where gain-limited correction is used, the BER curves shown are for the "best" absolute gain limit. If the optimum gain correction factor can be determined, an additional improvement of 1 dB can be obtained. The sensitivity of the BER on the gain limit, shown in Fig. 6, indicates that adaptive gain limiting, or some "intelligent" guess at the gain limit based on the distortion curves, may be required.

F. Effects of Frequency-Selective Fading

When good correlation exists between the fading statistics of the pilot tone and those of the fading information signal, almost total suppression of the unwanted amplitude and phase fluctuations is possible. The simulation results given in Section II-E were obtained under the assumption that the fading on the pilot and the desired signal were totally correlated. This is a valid consideration when there is no interference on the pilot and when the fading is not frequency selective.

In general, however, the mobile environment is frequency selective, due to the existence of a spread in arrival times of the various multipath components. In this case, the correlation in phase and amplitude between two pilots separated in frequency is high for small frequency separation, and falls essentially to zero as the separation substantially exceeds the correlation bandwidth [1]. The gain correction process, as will be seen, requires a high degree of correlation between the phase and

[3] Such a channel allows a factor of 4 improvement in spectral efficiency over the current 30 kHz cellular mobile telephone service channel.

[4] All comparisons in this paper will be made at a BER level of 10^{-2}.

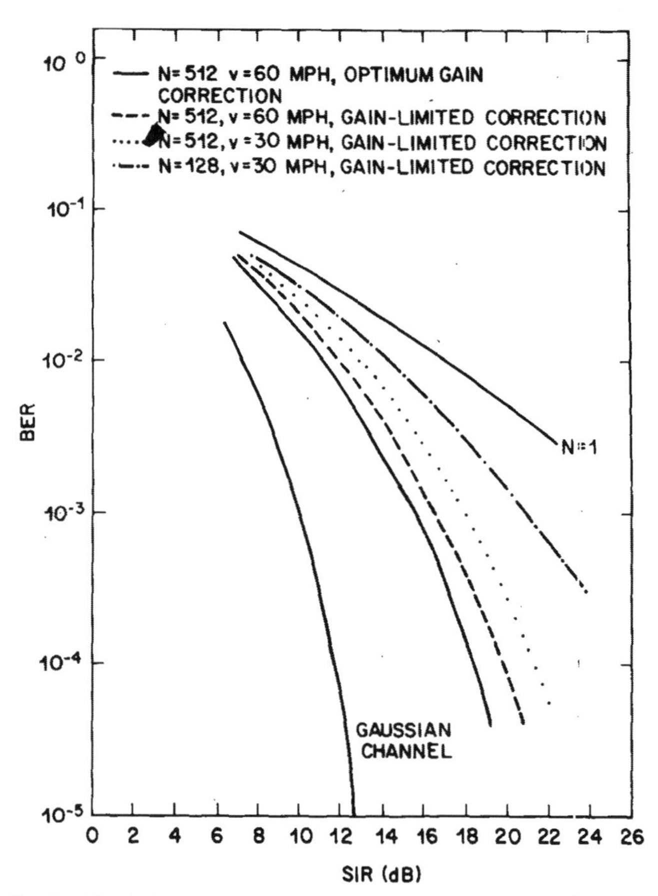

Fig. 5. Simulation results assuming perfect pilot recovery in a flat Rayleigh fading environment (QPSK, $f_s = 7.5$ kHz).

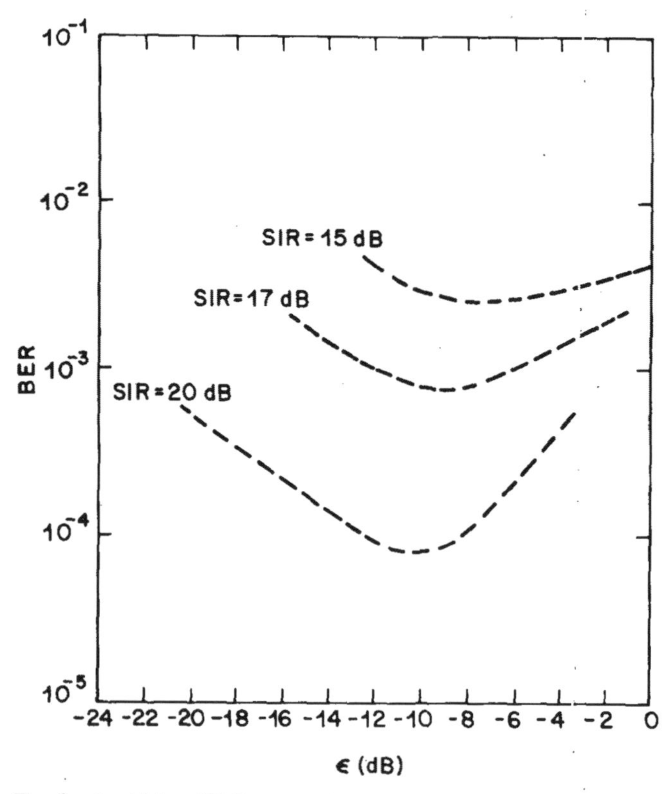

Fig. 6. Sensitivity of BER to variations in gain limit ($N = 512$, $v = 60$ mph, $f_s = 7.5$ kHz).

amplitude variations of the pilot and that of the phase and amplitude variations imposed on the data.

A simple way to estimate the effects of a delay spread environment is to compute the equivalent decorrelation between the pilot and the data caused by the frequency-selective fading (for example, see [12]). These calculations depend on the model used to describe the dispersive channel. In the simulations, the delay spread channel is simply modeled as a two-impulse channel response with equal-amplitude signal and echo separated by some time Δ. As in [2, sect. 9.8], an approximation model will be assumed for the delay distribution. In this model, the probability density function of the delay is represented as two equal-amplitude, equally likely impulses separated by some delay Δ. This is the model which was employed in the simulation and is a sufficient approximation for $s < \pi/2$. The rms delay spread in this case is simply $\Delta/2$. The corresponding complex correlation coefficient can be shown to be

$$\rho(s, \tau, \sigma) \approx \frac{J_0(\omega_m \tau)}{\sqrt{2}} \, [1 + \cos(s\Delta)]^{1/2} \tag{22}$$

where s is the frequency separation in rad/s, τ is the separation in time between two samples, and ω_m is the maximum Doppler shift in rad/s. Letting $\tau = 0$ (without loss of generality)

$$\rho(s, \sigma) = \frac{1}{\sqrt{2}} \, [1 + \cos(s\Delta)]^{1/2}. \tag{23}$$

The coherence bandwidth, defined as the frequency separation when the envelope correlation (ρ^2) is 0.5, is easily computed to be $1/\Delta$. According to published data for New York City [16], the rms delay spread is usually less than 3 μs ($\Delta = 6$ μs). However, other measurements in Newark [17] suggest that values of Δ as high as 50 μs are possible. Values of $\Delta = 25$ μs were found to occur often enough and with sufficient repeatability in Newark to be regarded seriously. The usefulness of the OFDM technique has been illustrated for a flat Rayleigh fading environment. It remains to show how much the system degrades in a delay spread environment where the pilots and the data become decorrelated.

In the simulation, a bound on the effects of delay spread has been obtained by appropriately decorrelating the pilot fading sequence and the fading data sequence. At a specific instant in time, the correlation between the complex fading envelope of the pilot, $Z_p(m)$, and the complex fading envelope of the data, $Z_d(m)$, is some value $\rho < 1$. It is easily shown that this will be true if the pilot complex fading envelope is chosen as

$$Z_p(m) = \rho Z_d(m) + \sqrt{1 - \rho^2}\, Z_x(m) \tag{24}$$

where $Z_x(m)$ is statistically independent from $Z_d(m)$. In the previous section, ρ was assumed to be unity. Obviously, this process causes additional distortion. Using limited gain correction, it can be shown that in a frequency-selective environment the SDR is

$$\mathrm{SDR} = \left[\left(\frac{\mathrm{SIR}+1}{\mathrm{SIR}}\right)\frac{1}{\epsilon^2}\,(1 - e^{-\epsilon^2}) + 1 - \frac{\rho\sqrt{\pi}\cos\phi}{\epsilon}\,\mathrm{erf}(\epsilon) + \left(1 - \rho^2 + \frac{1}{\mathrm{SIR}}\right)E_1(\epsilon^2)\right]^{-1} \tag{25}$$

where $\phi = \arctan[-s\sigma]$. This reduces to (20) for $\rho = 1$ and reduces to results in [14] when $\mathrm{SIR} = \infty$ and $\phi = 0$.

Simulation results have been obtained assuming $\Delta = 50$ μs (a particularly bad case measured in Newark). This value has been chosen to determine what degradations occur in an approximately worst-case dispersive environment. Frequency separations between the pilot and the data of 500 Hz and 1 kHz have been used. The results are shown in Fig. 7 and indicate that, for severe frequency-selective fading, a loss of 5 or 6 dB in BER is obtained over the flat fading case for 1 kHz separation, and less than 2 dB for 500 Hz separation. Similar degradations occur when there is interference on the pilot, either from the data spreading into the pilot due to the Doppler shift induced by the motion of the mobile, or by a cochannel pilot. The interference from a cochannel pilot decorrelates the received pilot and data signal envelopes. Thus, cochannel pilot interference induces correction distortion in a manner very similar to that induced by operating in a high delay spread environment. Obviously, in a practical implementation of this modem, some techniques must be devised to overcome the detrimental effects of delay spread and pilot interference. These will be treated in the next section.

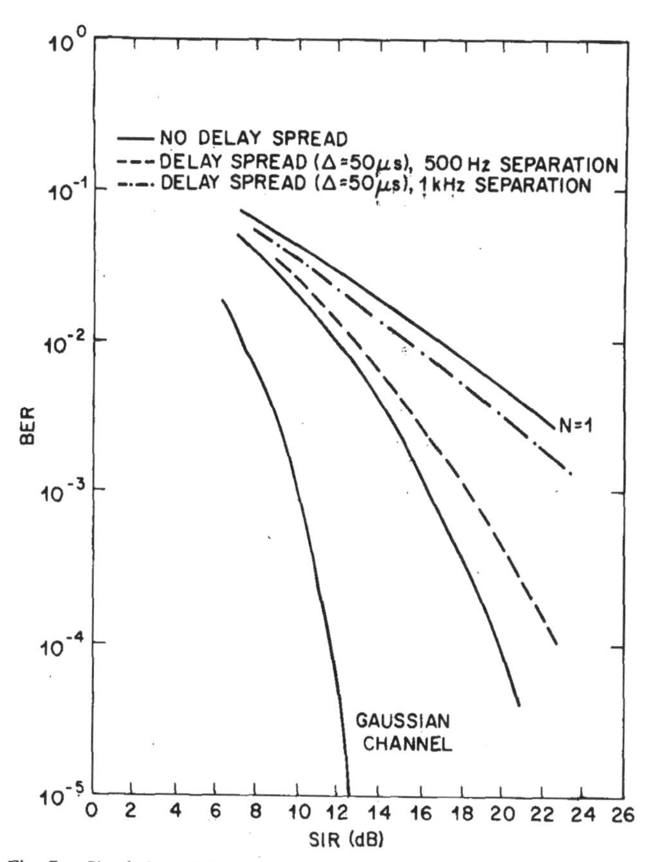

Fig. 7. Simulation results in a frequency-selective environment (QPSK, f_s = 7.5 kHz, N = 512, v = 60 mph, gain-limited correction).

III. Practical Considerations

A. Problems in Pilot Retrieval

In a practical OFDM system, a method for accurately correcting the data is required, which in turn requires a method for reliably retrieving the pilot signals. In addition, the complex fading envelope of the pilot signal must be highly correlated with the complex fading envelope of the data. A pilot signal, located somewhere within the transmission band, must be sent with the data. This pilot will be distorted in transmission by cochannel interference, by adjacent data symbols spreading into the pilot due to the motion of the vehicle, by filtering processes, and by the decorrelation between the pilot fading envelope and the signal fading envelope caused by a frequency-selective environment. Techniques will be presented which alleviate these distortions.

B. Techniques for Reliable Pilot Retrieval

When there is adequate separation between the pilot tone and its neighboring information components, it is possible to separate the spread pilot from the spread signal components, without fear of overlap, by suitable filtering in the receiver. Assuming a top vehicle speed of 60 mph and transmission at 850 MHz, each subchannel spreads ±80 Hz (due to the Doppler shift). Therefore, a 200 Hz spacing between the pilot frequency and the nearest data subcarrier provides more than enough protection against overlap. However, due to the use of the DFT, each subchannel possesses a sin x/x spectrum which has fairly large sidelobes, and which may cause problems if the orthogonality of the subcarriers is impaired. The sidelobes can be reduced either by filtering the individual subchannels before transmission (for example, see [6]), or by extending the frame of data in time and by requiring gradual rather than abrupt rolloffs of the transmitted waveform [8]. In this paper, the latter technique is employed to avoid the complexity created by prefiltering. In particular, the data block is modulo extended in time, and then the extended sections are shaped with a raised cosine and added as shown in Fig. 8. This reduces the sidelobes but widens the main lobe. The length of the extension determines the width of the main lobe beyond that of the sin x/x spectrum.

When the effects of the data on the pilot signal have been sufficiently attenuated, cochannel pilot interference can be attacked. A frequency offset reference transmission scheme works well to minimize the effects of cochannel pilot interference. In this scheme, the slot for pilots in the total band is further divided into several slots for cochannel interferer pilots. Specifically, in this paper, three adjacent slots are made available, increasing the total pilot allocation to 1000 Hz. This technique effectively increases frequency reuse by a factor of 3.

In a frequency-selective environment, a high correlation between the complex fading envelopes of the pilot and the data must be ensured. Based on the results in Section II-F, this depends on the severity of the delay spread and on the frequency separation between the pilot and the data subcarrier. Assuming a harsh delay spread environment (for example, $\Delta = 50\ \mu s$), the pilot and data must be very close in frequency (<500 Hz) to constrain the loss due to decorrelation to 2 dB. If a 7.5 kHz channel is used, placing the pilot in the center of the band results in data symbols at the edges that are 3.75 kHz away from the pilot. This large separation causes a large degradation in BER. With smaller amounts of delay spread (for example, 5 μs) this is not a problem. This large degradation can be avoided if two pilots, separated in frequency, are used and the appropriate complex correction signal is obtained by interpolation *in frequency* between these two pilots.

In general, separate amplitude (A) and phase (θ) interpolation is not appropriate. For the two-impulse channel model for delay spread which is being used here, the amplitude response is as shown in Fig. 9. It is possible for both pilot fading envelopes to be at peaks, due to the randomness of the initial position of the 7.5 kHz data window. The resulting amplitude and phase interpolation will be inadequate to follow the null in frequency caused by the fading. Interpolation separately the real (in-phase, I) and imaginary (quadrature, Q) parts of the complex fading envelopes of the pilots will enable such a correction to occur. Simulation results, presented in Fig. 12, indicate that I-Q interpolation gives 8 dB improvement over A-θ interpolation for 50 μs delay spread. Similar results will be obtained for other models for the dispersive channel. The technique depends only on nulls occurring in the channel frequency response.

Finally, the design of a mobile telephone system must also include measures to limit adjacent channel interference. This is accomplished by leaving 250 Hz gaps at each end of the band. The final data setup used in the simulations is shown in Fig. 10. The total bandwidth used by the data has now been reduced to 5 kHz.

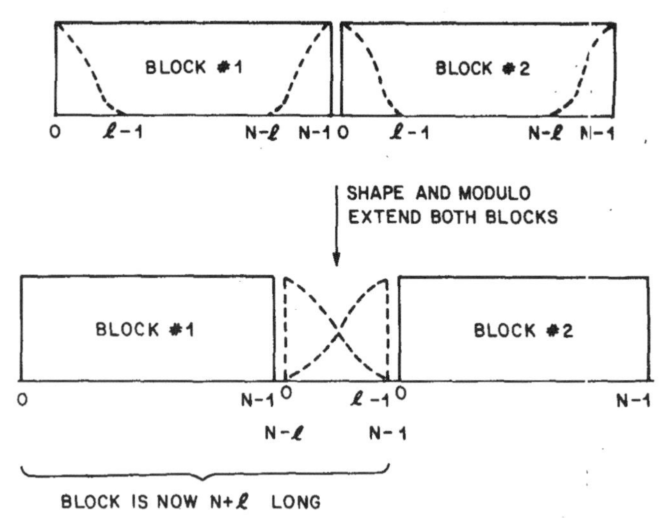

Fig. 8. Modulo extension and shaping of data block to reduce interchannel interference caused by impairment of orthogonality.

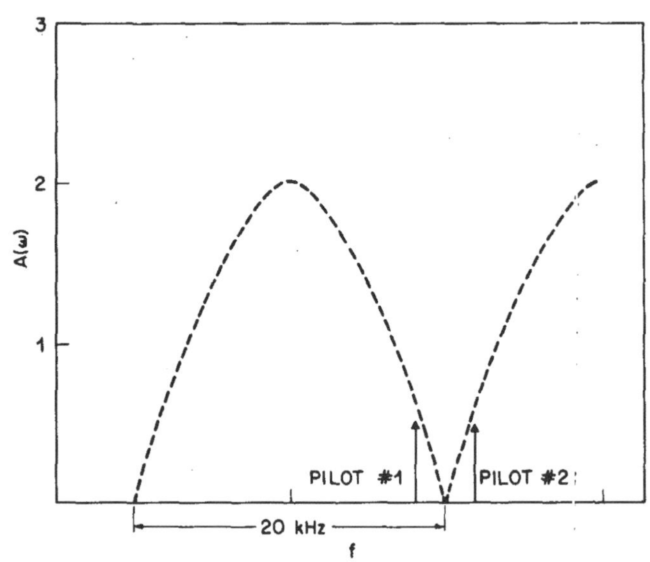

Fig. 9. Possible amplitude response for two-impulse channel model for delay spread, $\Delta = 50\ \mu s$, $A(\omega) = 2\ |\cos \omega\Delta/2|$.

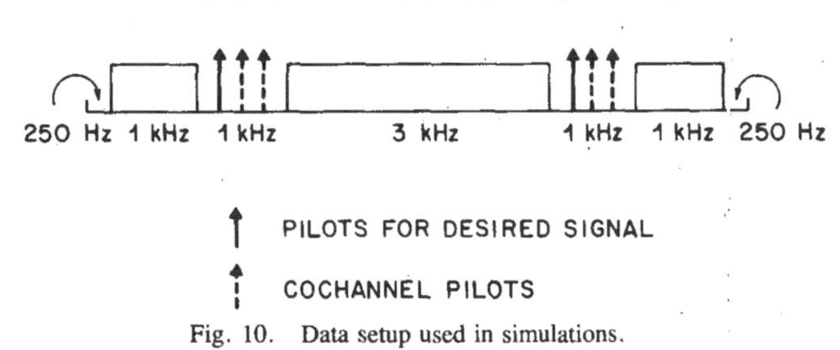

Fig. 10. Data setup used in simulations.

C. Basic System Setup and Pertinent Parameters for Simulation

Simulations were performed incorporating all of the design features presented in the previous section. The pertinent system parameters were carefully selected, based on an initial assumption of a 7.5 kHz channel. T, the total signaling interval,

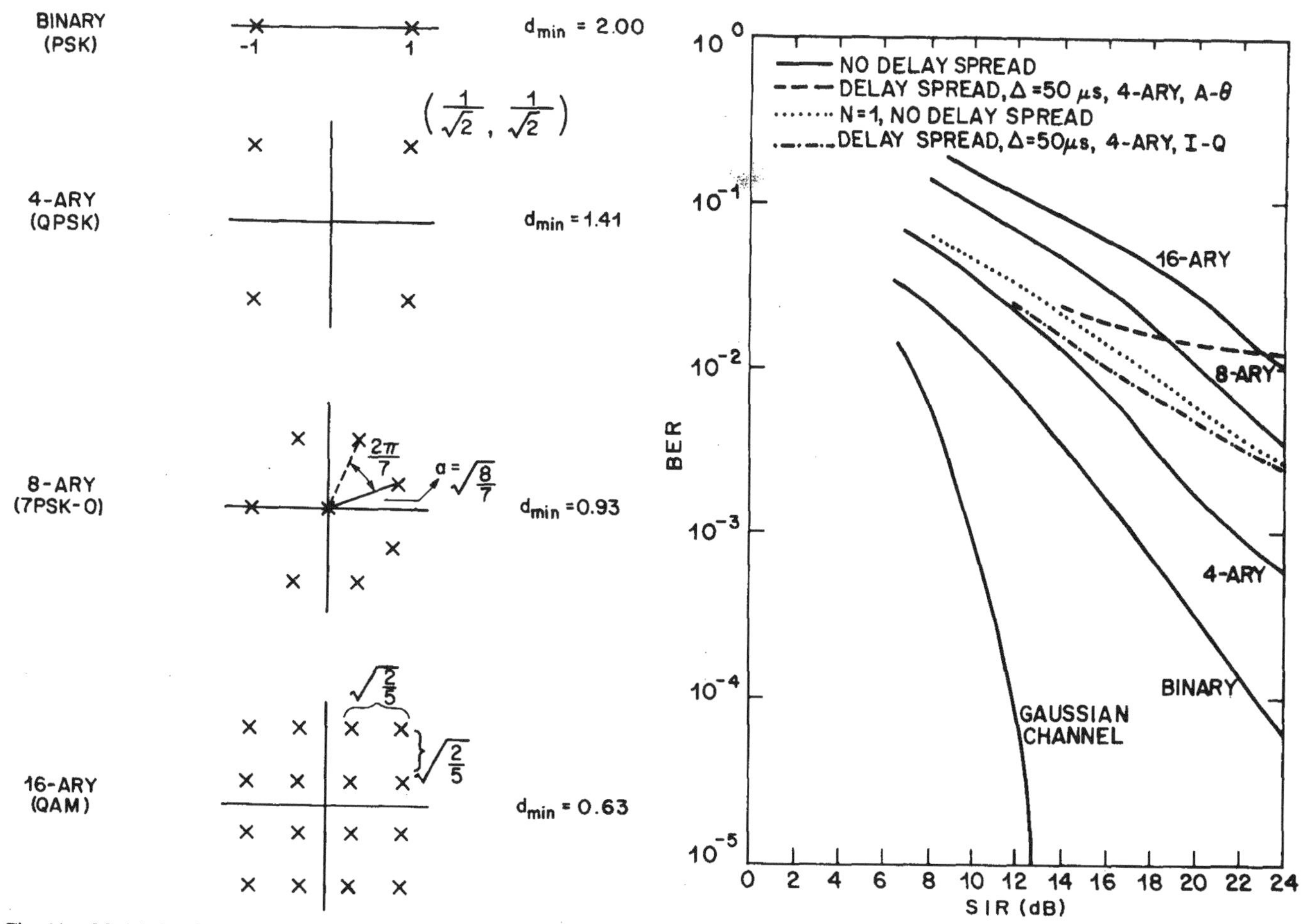

Fig. 11. Modulation formats used in simulations (average energy = 1).

Fig. 12. Simulation results (N = 128, v = 30 mph, f_s = 7.5 kHz).

should be chosen to be much greater than the average duration between fades, so that many fades are averaged in the receiver. In the previous simulation, T was 68 ms and good results were obtained. However, based on subjective opinions, a signaling interval of 68 ms was shown to result in a noticeable round trip delay for speech transmission when buffering and processing times are included. Also, by making T too large, the subchannels are closer together and the interchannel interference due to spreading may become unacceptable. An interval on the order of 20 ms (N = 128) is more satisfactory. This decrease in T results in a degradation of about 2 dB in performance compared to T = 68 ms.

Note that the average interval between fades tends to infinity as the speed of the mobile approaches zero. Therefore, T must be very large at low speeds to obtain any averaging. A solution to this problem is to make the vehicles always appear to move at some substantial speed, such as 30 mph. One way this can be done is by forcing an antenna to oscillate when the vehicle is moving at low speeds, as in [18].

Assuming an initial sample rate of 7.5 kHz and letting N = 128 gives T = 17.06 ms and Δf = 58.59 Hz. Assuming a total interval of 20 ms, these parameters allow 2.93 ms of extension. Sacrificing two bands of 1000 Hz each for pilot protection and 250 Hz at either end for adjacent channel interference protection leaves space for 86 data channels and a bandwidth efficiency of $\beta = 0.57 \log_2 M$ bits/s/Hz. For QPSK, this corresponds to a maximum data rate of 8.6 kbits/s.

D. Simulation Results

In this simulation, several different two-dimensional modulation formats were considered. These are shown in Fig. 11, along with the corresponding minimum distances. Results were obtained for binary (BPSK), 4-ary (QPSK), 8-ary (7PSK-0), and 16-ary (QAM) modulation schemes. The 8-ary constellation used a seven-point phase shift keying pattern with an additional data point at the origin. This requires a two-step detection procedure, but gives 1.7 dB improvement in performance over conventional 8-PSK, based on comparison of the minimum distances in each constellation. These patterns could form the basis of a multimode system.

Simulation results for a flat Rayleigh fading environment are shown in Fig. 12. Gain-limited correction has been used in the data correction process. These results are computed for a vehicle speed of 30 mph. For a vehicle speed of 60 mph, there is an improvement of 1 dB. Each pilot was also run 10 dB higher in energy than any individual data point, for additional protection of the pilots. By using the frequency offset reference technique to combat cochannel pilot interference, the total pilot-to-distortion ratio is 25 dB, a reasonable level of distortion. The pilot filters were implemented, in the simulation, as low-pass FIR filters with 100 Hz passbands. These filters added no significant distortion to the pilots. The results in Fig. 12 indicate the improvements which are attainable with this technique. Gains of 2-3 dB were obtained for OFDM-QPSK over coherent QPSK. An additional 3 dB is gained over differential QPSK (DQPSK).

Fig. 12 also indicates the BER for QPSK when there is significant delay spread. Delay spread was implemented as a statistically independent delayed echo added to the desired signal. At a BER = 10^{-2}, there is a loss of only 2 dB in SIR for Δ = 50 μs. For Δ = 10 μs, the loss is about 0.5 dB. The losses due to delay spread have been minimized by interpolating *in frequency* ten appropriately spaced pilots, which are then each used to correct the data sequence. The appropriate sym-

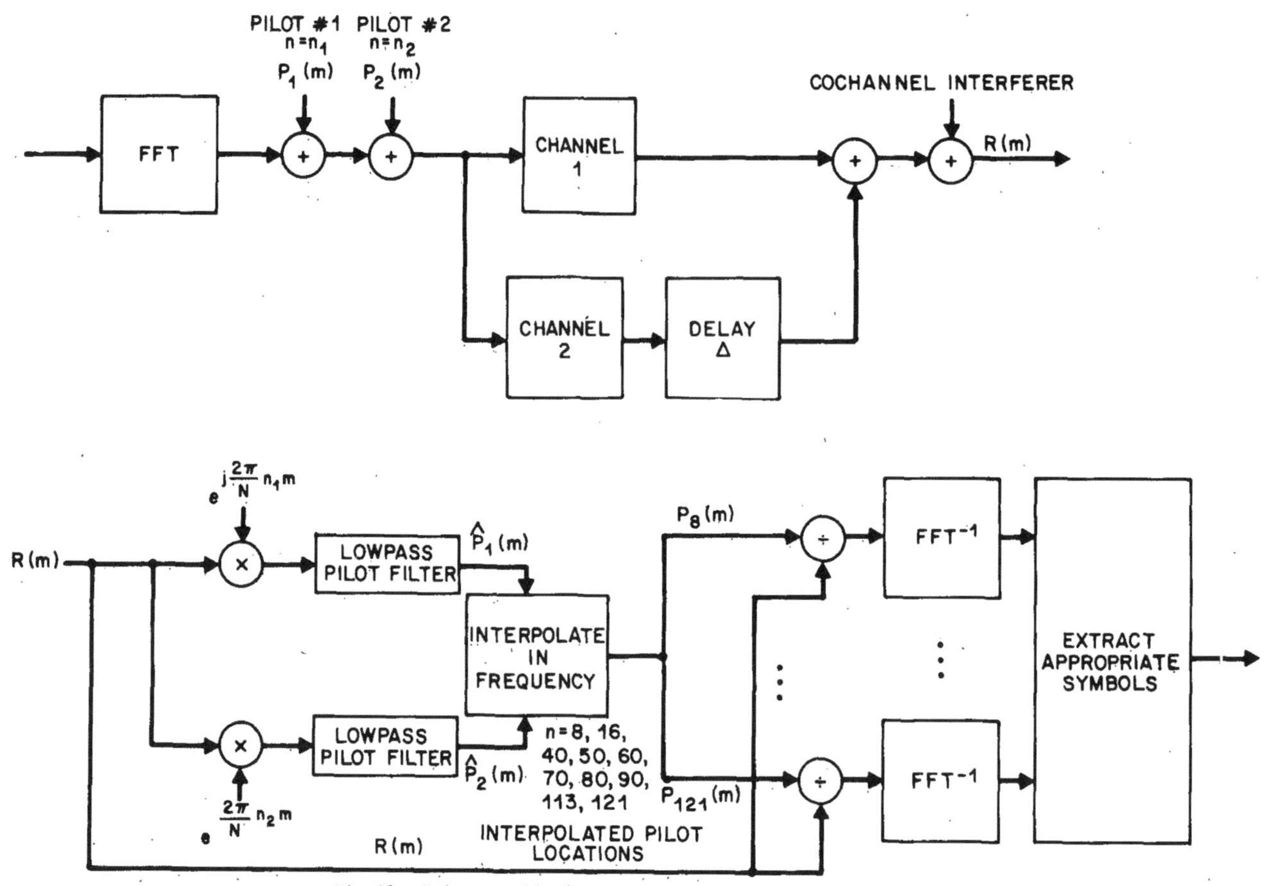

Fig. 13. Delay spread implementation and pilot correction procedure.

bols are then chosen at the output. The correction procedure is indicated in Fig. 13 and the symbols corrected by each pilot are given in Table I.

Variations such as changing the pilot locations, changing the number and locations of the interpolated pilots, and changing the length of the time extension have a large effect on the BER. Based upon numerous trials, it has been determined that for ten interpolated pilots, the system simulation used in this section is close to an optimum.

TABLE I
DATA SYMBOLS CORRECTED BY EACH INTERPOLATED PILOT

INTERPOLATED PILOT POSITION	DATA SYMBOLS CORRECTED
8	5,11
16	12,21
40	38,44
50	45,54
60	55,64
70	65,74
80	75,84
90	85,91
113	109,117
121	118,123

IV. Conclusions and Further Investigations

In this paper, a cellular mobile radio system based on orthogonally frequency division multiplexing many low-rate subchannels into one higher rate channel was analyzed and simulated. This technique, when used with pilot-based correction, was shown to provide large improvements in BER performance in a *flat Rayleigh fading* environment. Degradations due to *severe* delay spread were kept to a minimum by frequency interpolation of two pilots. Using a frequency offset scheme for the pilots limited the effects of cochannel pilot interference. These considerations provided very good BER performance, at the expense of a decrease in the overall bandwidth efficiency. The averaging ability of the OFDM system, which makes the bursty Rayleigh channel appear nearly Gaussian, provides this large BER improvement. Additional improvements are possible if larger signaling intervals are permissible.

Other factors which also impair the orthogonality of the subcarriers, such as phase and gain hits, phase jitter, and frequency offset, have not been considered. In [13], these impairments were considered for a voice-bandwidth data modem. There it is shown that some form of frequency domain equalization can be used to alleviate the effects of these impairments.

Some alternative coding schemes for the OFDM system have been briefly considered. Two possibilities for coding in this system are analog and Ungerboeck codes. Analog codes have recently gained quite a bit of interest (for example, see [19]). One approach is to use fixed transmitted symbols to correct symbols in error. In initial trials for the OFDM system, this technique was not very effective, because the large noise term on each symbol caused by cochannel interference was enough to offset the effectiveness of the algorithm. Further study in this area may prove fruitful. Ungerboeck codes [20] map data symbols into a higher order constellation and, in a Gaussian, white noise environment, have been shown to provide as much as 3 dB coding gain without sacrificing data rate or requiring more bandwidth. These codes, although untested in a fading environment, present a possibility for our system.

Acknowledgment

The author would like to acknowledge gratefully N. Sollenberger and K. Leland for their helpful suggestions throughout the course of this project.

References

[1] W. C. Jakes, Jr., Ed., *Microwave Mobile Communications*. New York: Wiley, 1974.
[2] W. C. Y. Lee, *Mobile Communications Engineering*. New York: McGraw-Hill, 1982.
[3] R. W. Chang, "Synthesis of band-limited orthogonal signals for multichannel data transmission," *Bell Syst. Tech. J.*, vol. 45, pp. 1775–1796, Dec. 1966.
[4] ——, "Orthogonal frequency division multiplexing," U.S. Patent 3 488 445, filed Nov. 14, 1966, issued Jan. 6, 1970.
[5] B. R. Saltzberg, "Performance of an efficient data transmission system," *IEEE Trans. Commun. Technol.*, vol. COM-15, pp. 805–813, Dec. 1967.
[6] R. W. Chang and R. A. Gibbey, "A theoretical study of performance of an orthogonal multiplexing data transmission scheme," *IEEE Trans. Commun. Technol.*, vol. COM-16, pp. 529–540, Aug. 1968.
[7] J. Salz and S. B. Weinstein, "Fourier transform communication system," in *Proc. ACM Symp. Probl. Optimiz. Data Commun. Syst.*, Pine Mountain, GA, October 13–16, 1969, pp. 99–128.
[8] S. B. Weinstein and P. M. Ebert, "Data transmission by frequency-division multiplexing using the discrete Fourier transform," *IEEE Trans. Commun. Technol.*, vol. COM-19, pp. 628–634, Oct. 1971.
[9] M. L. Doelz, E. T. Heald, and D. L. Martin, "Binary data transmission techniques for linear systems," *Proc. IRE*, vol. 45, pp. 656–661, May 1957.
[10] G. C. Porter, "Error distribution and diversity performance of a frequency-differential PSK HF modem," *IEEE Trans. Commun. Technol.*, vol. COM-16, pp. 567–575, Aug. 1968.
[11] M. S. Zimmerman and A. L. Kirsch, "The AN/GSC-10 (KATHRYN) variable rate data modem for HF radio," *IEEE Trans. Comrıun. Technol.*, vol. COM-15, pp. 197–205, Apr. 1967.
[12] P. A. Bello, "Selective fading limitations of the Kathryn modem and some system design considerations," *IEEE Trans. Commun. Technol.*, vol. COM-13, pp. 320–333, Sept. 1965.
[13] W. E. Keasler, Jr., "Reliabile data communications over the voice bandwidth telephone channel using orthogonal frequency division multiplexing," Ph.D. dissertation, Univ. Illinois, Urbana-Champaign, 1982.
[14] K. W. Leland and N. R. Sollenberger, "Impairment mechanisms for SSB mobile communications at UHF with pilot-based Doppler/fading correction," *Bell Syst. Tech. J.*, vol. 59, pp. 1923–1942, Dec. 1980.
[15] J. P. McGeehan and A. J. Bateman, "Theoretical and experimental investigation of feedforward signal regeneration as a means of combating multipath propagation effects in pilot-based SSB mobile radio systems," *IEEE Trans. Veh. Technol.*, vol. VT-32, pp. 106–120, Feb. 1983.
[16] D. C. Cox and R. P. Leck, "Distributions of multipath delay spread and average excess delay for 910 MHz urban mobile radio path," *IEEE Trans. Antennas Propagat.*, vol. AP-23, pp. 206–213, Mar. 1975.
[17] K. W. Leland and N. R. Sollenberger, private communication.
[18] W. C. Wong, R. Steele, B. Glance, and D. Horn, "Time diversity with adaptive error detection to combat Rayleigh fading in digital mobile radio," *IEEE Trans. Commun.*, vol. COM-31, pp. 378–387, Mar. 1983.
[19] J. K. Wolf, "Redundancy, the discrete Fourier transform, and impulse noise cancellation," *IEEE Trans. Commun.*, vol. COM-31, pp. 458–461, Mar. 1983.
[20] G. Ungerboeck, "Channel coding with multilevel/phase signals," *IEEE Trans. Inform. Theory*, vol. IT-28, pp. 55–67, Jan. 1982.

Leonard J. Cimini, Jr. (S'77–M'82) was born in Philadelphia, PA, on April 19, 1956. He received the B.S.E., M.S.E., and Ph.D. degrees in electrical engineering from the University of Pennsylvania, Philadelphia, in 1978, 1979, and 1982, respectively. During his graduate work he was supported by a National Science Foundation Fellowship.

He has been employed by AT&T Bell Laboratories, Holmdel, NJ, since 1982, working in the area of mobile radio systems. His main research interests are in the general areas of signal processing and communications systems. He also teaches courses in communication systems at Monmouth College, West Long Branch, NJ.

Dr. Cimini is a member of Tau Beta Pi and Eta Kappa Nu.

Efficient Modulation for Band-Limited Channels

G. DAVID FORNEY, JR., FELLOW, IEEE, ROBERT G. GALLAGER, FELLOW, IEEE, GORDON R. LANG, FRED M. LONGSTAFF, AND SHAHID U. QURESHI, SENIOR MEMBER, IEEE

Abstract—This paper attempts to present a comprehensive tutorial survey of the development of efficient modulation techniques for band-limited channels, such as telephone channels. After a history of advances in commercial high-speed modems and a discussion of theoretical limits, it reviews efforts to optimize two-dimensional signal constellations and presents further elaborations of uncoded modulation. Its principal emphasis, however, is on coded modulation techniques, in which there is an explosion of current interest, both for research and for practical application. Both block-coded and trellis-coded modulation are covered, in a common framework. A few new techniques are presented.

I. Historical Introduction

BAND-LIMITED channels (as opposed to power-limited) are those on which the signal-to-noise ratio is high enough so that the channel can support a number of bits/Hz of bandwidth. The telephone channel (particularly the dedicated private line) has historically been the scene of the earliest application of the most efficient modulation techniques for band-limited channels. The reasons have to do both with the commercial importance of such channels and with the fact that they can be modeled to first order as linear time-invariant channels, sharply band-limited between typically 300–3000 Hz, with high signal-to-noise ratios, typically 28 dB or greater. Their relatively low bandwidth permits a great deal of signal processing per transmission element, and therefore early application of the most advanced techniques, which have often then been applied several years later to broader-band channels (e.g., radio).

The earliest commercially important telephone-line modems appeared in the 1950's and used frequency shift keying to achieve speeds of 300 bits/s (Bell 103), or 1200 bits/s on private lines (Bell 202). The earliest commercially important synchronous modem was the Bell 201, introduced in about 1962, which used 4-phase modulation in a nominal 1200 Hz bandwidth to achieve 2400 bits/s on private lines. This remained the state of the art for most of the decade. (It was not unknown in this period to encounter users who thought that Nyquist or Shannon or someone else had proved that 2400 bits/s was about the maximum rate theoretically possible on phone lines.)

The first commercially important 4800 bit/s modem was the Milgo 4400/48, introduced in about 1967, which included a manually adjustable equalizer to allow use of a nominal 1600 Hz bandwidth in conjunction with 8-phase modulation to send 3 bits/Hz. The development of digital adaptive equalization soon allowed expansion of the nominal bandwidth to 2400 Hz, essentially the full telephone line bandwidth. Following a first generation of single-sideband 9600 bit/s modems in the late 1960's, which were only marginally successful, broad success was achieved by the Codex 9600C, introduced in 1971, which used quadrature amplitude modulation (QAM) with a 16-point signal constellation to send 4 bits/Hz in a nominal 2400 Hz bandwidth. (16-point QAM had been used at 4800 bits/s by ESE and ADS about 1970.) Modems with these characteristics remained the state of the art for another decade, and it was thought by many (including some of the present authors, who should have known better) that higher-speed modems would never be widely used commercially.

In 1980, a first generation of 14 400 bit/s modems was introduced by Paradyne (MP 14400), followed in 1981 by the Codex/ESE SP14.4 and by others. These modems simply extended 2400 Hz QAM modulation to 6 bits/Hz by using 64-point signal constellations, and by using advanced implementation techniques and exploiting the gradual upgrading of the telephone network, proved to operate successfully over a high percentage of circuits. In a second generation of 14.4 kbit/s modems that will begin appearing in 1984, coded QAM modulation is being introduced to provide greater performance margins. The principal focus of this paper will be on these new coded modulation techniques.

This evolution of high-speed modems to ever higher bit rates using successively more complicated modulation schemes is summarized in Table I, along with the designation and year of final adoption of CCITT international standards embodying these schemes. How far can this evolution go? History would suggest caution in stipulating any ultimate ceiling. However, without any dramatic general upgrading of the telephone network, we venture to say that 19.2 kbits/s is the maximum conceivable rate for a

Manuscript received February 14, 1984; revised May 3, 1984.
G. D. Forney, Jr. is with the Motorola Information Systems Group, Mansfield, MA 02048.
R. G. Gallager is with the Department of Electrical Engineering and Computer Science, Massachusetts Institute of Technology, Cambridge, MA 02139.
G. R. Lang and F. M. Longstaff are with Motorola Information Systems Ltd., (ESE) Rexdale, Ont., Canada M9V 1C1.
S. U. Qureshi is with the Codex Corporation, Canton, MA 02021.

Reprinted from *IEEE Journal on Selected Areas in Communications,* vol. SAC-2, no. 5, September 1984.

TABLE I
MODEM MILESTONES

YEAR	MODEL	SPEED	BANDWIDTH	n	CONSTELLATION	COMMENTS	CCITT STANDARD
1962	Bell 201	2400	1200 Hz	2	4-phase	fixed eq.	V.26 (1968)
1967	Milgo 4400/48	4800	1600 Hz	3	8-phase	manual eq.	V.27 (1972)
1971	Codex 9600 C	9600	2400 Hz	4	16-QAM	adaptive eq.	V.29 (1976)
1980	Paradyne MP14400	14,400	2400 Hz	6	64-QAM	rectangular	
1981	Codex/ESE SP14.4	14,400	2400 Hz	6	64-QAM	hexagonal	
1984		14,400	2400 Hz	6	128-QAM	8-state trellis	

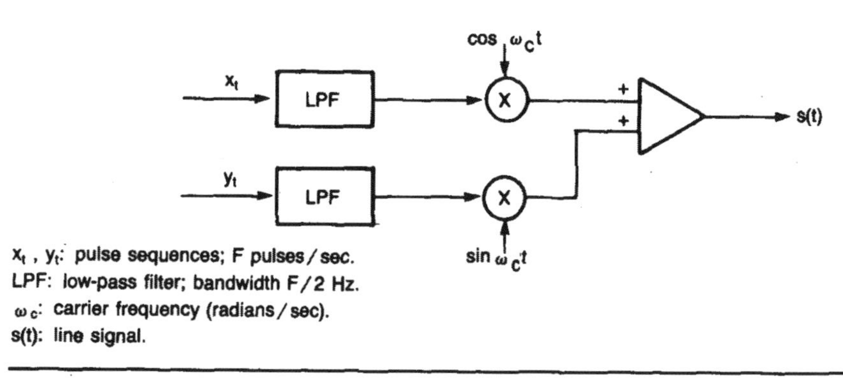

Fig. 1. Canonical QAM modulator.

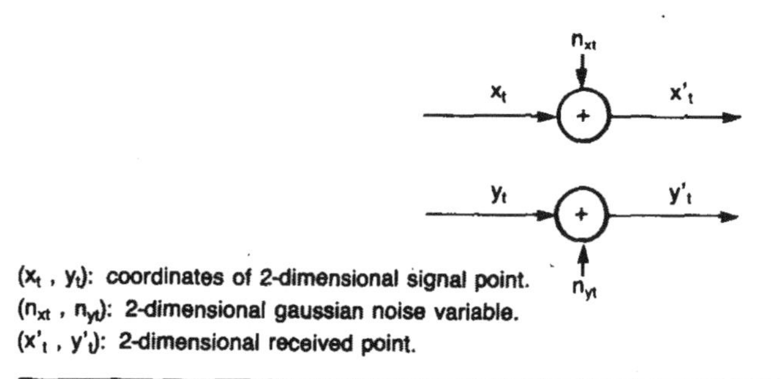

Fig. 2. QAM channel model.

telephone-line modem for general use, even with all-out use of the most powerful coded modulation. We shall see.

II. Channel Model and Basic Limits

A quadrature amplitude modulator can be used to generate any standard linear double-sideband modulated carrier signal (including forms of phase modulation and phase/amplitude modulation), which includes all types of modulation in general use in synchronous modems. A canonical QAM modulator is shown in Fig. 1.

Assuming that the only channel impairment is Gaussian noise and that the receiver achieves perfect carrier phase tracking, the simple model of Fig. 2 applies. Signals are sent in pairs (x_t, y_t); the channel is essentially two-dimensional. We shall call such a pair a "signal point," imagined to lie on a two-dimensional plane. Signal points are sent at a regular rate of F points/s where F Hz is the nominal (Nyquist) bandwidth of the channel. A signal point is also called a "symbol," and the symbol interval is $1/Fs$. The model indicates that the two signal point coordinates (x_t, y_t) are independently transmitted over decoupled parallel channels and perturbed by Gaussian noise variables (n_{xt}, n_{yt}), each with zero mean and variance N. Alternatively, we could regard the two-dimensional signal point as being perturbed by a two-dimensional Gaussian noise variable. If the average energy (the mean squared value) of each signal coordinate is S, then the signal-to-noise ratio is S/N.

The simplest method of digital signaling through such a system is to use one-dimensional pulse amplitude modulation (PAM) independently for each signal coordinate. (This is sometimes called narrow-sense QAM.) In PAM, to send m bits/dimension, each signal point coordinate takes on one of 2^m equally likely equispaced levels, conventionally $\pm 1, \pm 3, \pm 5, \cdots, \pm(2^m - 1)$. The average energy of each coordinate is then

$$S_m = (4^m - 1)/3,$$

and it follows that

$$S_{m+1} = 4S_m + 1.$$

That is, it takes approximately (asymptotically, exactly) 4 times as much energy (or 6 dB) to send an additional 1 bit/dimension or 2 bits/symbol. The probability $P(E)$ that either x_t or y_t is in error is upperbounded and closely approximated by the probability that the two-dimensional Gaussian noise vector (n_{xt}, n_{yt}) lies outside a circle of radius 1, which is easily calculated to be $P(E) = \exp(-1/2N)$. A noise variance N of about 1/24 therefore yields an error probability per symbol in the range of about 6×10^{-6}.

The channel capacity of the Gaussian channel was calculated in Shannon's original papers [1]. Subject to an energy constraint $\overline{x_t^2} \leqslant S$, the capacity is

$$C = (1/2)\log_2(1 + S/N) \text{ bits/dimension},$$

achieved when x_t is a zero-mean Gaussian variable with variance S. Note that when S becomes large with N constant, it takes approximately (asymptotically, exactly) 4 times as much power to increase the capacity by an additional 1 bit/dimension. Therefore the ratio of bits/dimension achieved by PAM to channel capacity approaches 1 as S becomes large. This fact was used in [2] to argue that coding has little to offer on highly band-limited channels.

We can make a more quantitative estimate of the potential gains from coding as follows. Using narrow-sense QAM (two-dimensional PAM) to send m bits/dimension or $n = 2m$ bits/symbol at an acceptable error rate (of the order of 10^{-5}–10^{-6}) requires an average energy in each dimension of

$$S = (2^n - 1)/3$$

when $N = 1/24$, or $S/N = 8 \times 2^n$ for n moderately large. If channel capacity could be achieved, we could send about $n' = \log_2(S/N)$ bits/symbol, or about $n + 3$ bits/symbol at the same signal-to-noise ratio. Thus, the potential gain is about 3 bits/symbol or, alternatively, about a factor of 8 (9 dB) of power savings.

Many authors (see, e.g., [3], [4]) regard the parameter R_o as a better estimate than C of the maximum rate that is practically achievable using coding. On the Gaussian channel, R_o is [5]

$$R_o = (1/2)\log_2(1 + S/2N) \text{ bits/dimension}.$$

It thus takes a factor of 2 (3 dB) more power to signal at R_o than to signal at C. The maximum practical improvement obtainable by coding might therefore be estimated as of the order of 6 dB, or 2 bits/symbol (although the R_o estimate is not universally accepted).

In what follows we shall show that simple coding techniques gain about 3 dB or 1 bit/symbol, while the most elaborate techniques described have theoretical gains of the order of 6 dB or about 2 bits/symbol. This is entirely consistent with the R_o estimate given above, and suggests that little can be gained by seeking still more elaborate schemes. (In [6] the capacity of the telephone channel was estimated as of the order of 23 500 bits/s, roughly consistent with what we are saying here.)

III. Uncoded Modulation Systems

Digital QAM signaling schemes are conventionally and usefully represented by two-dimensional constellations of all possible signal points. A 2^n-point constellation can be used to send n bits/symbol. A fair amount of effort has gone into finding "optimum" constellations. We shall shortly see that the payoff for this effort on purely Gaussian-noise channels is relatively slight, although the schemes found are helpful precursors for more elaborate schemes.

A. Rectangular Constellations

A brief flurry of theoretical papers in the early 1960's [7]–[10] developed two-dimensional signal constellations from various viewpoints. The most interesting for our present purposes are the family of constellations developed by Campopiano and Glazer [9], reproduced in Fig. 3. (We have taken the liberty of substituting a "cross constellation" for theirs at $n = 7$; the two are equally good.) For even integer numbers of bits/symbol, the constellations are simply representations of two independent PAM channels, so the constellations are square and have points drawn from the rectangular lattice of points with odd-integer coordinates. It takes about 6 dB more power to send 2 more bits/symbol, as expected. For odd integer numbers of bits/symbol, the constellations lie within an envelope in the form of a cross (and have hence come to be called "cross constellations") and the points are drawn from the same rectangular lattice (except for the 8-point constellation, where the outer points are put on the axes for symmetry and energy savings). With the figures scaled so that the minimum distance between any two points is equal to 2, the average signal energy in absolute terms and in dB is as given in Table II. We see that the "cross constellations" require about 3 dB more or less than the next lower or higher square constellation, respectively, as we would expect.

The Campopiano–Glazer construction can be generalized as follows: from an infinite array of points closely packed in a regular array or lattice, select a closely packed subset of 2^n points as a signal constellation. This important principle is at the root of much recent work. We shall explore applications of this principle, working up from the simpler to the more sophisticated.

When constellations are drawn from a regular lattice within some enclosing boundary, the following asymptotic

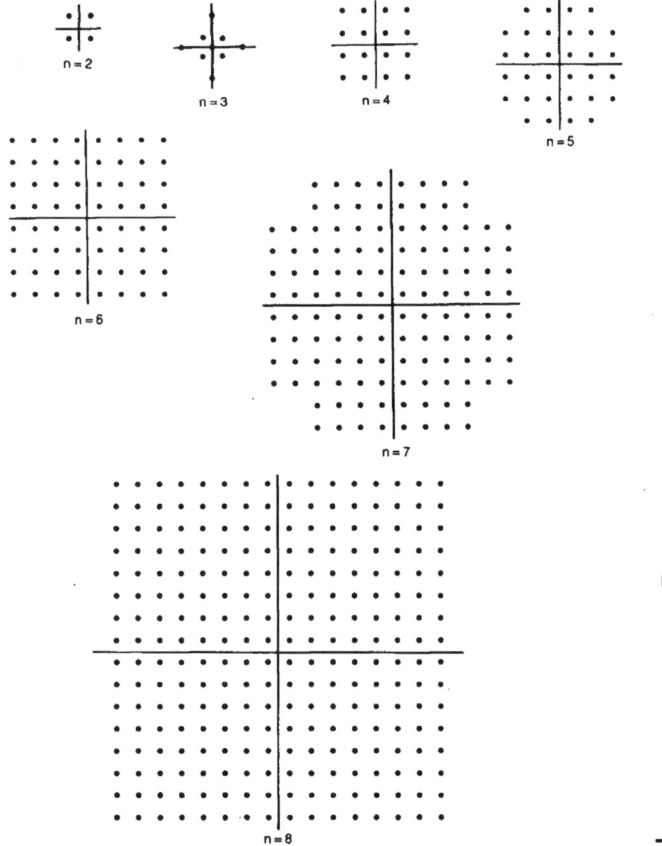

Fig. 3. Rectangular signal constellations (after Campopiano and Glazer [9]).

TABLE II
CAMPOPIANO–GLAZER CONSTELLATIONS

No. Points	S	(dB)	$\hat{S}$	(dB)
4	2	3.0	2.7	4.3
8	5.5	7.4	-	-
16	10	10.0	10.7	10.3
32	20	13.0	20.7	13.2
64	42	16.2	42.7	16.3
128	82	19.1	82.7	19.2
256	170	22.3	170.7	22.3

approximation is useful and remarkably accurate for constellations of even moderate size (16 points and up). Let A be the area of the region consisting of points that are closer to a given lattice point than to any other (i.e., the decision region belonging to that point, or the Voronoi (or Dirichlet) region [11]). If we take some boundary that encloses a region in space of area B, then the number of lattice points in that region will be approximately B/A, and the average energy of these lattice points will be approximately equal to the average energy $(x^2 + y^2)$ of all points within the boundary.

For example, the Campopiano–Glazer constellations are drawn from a rectangular lattice with $A = 4$, and are bounded by a square of side $2 \times 2^{n/2}$ and area $B_s = 4 \times 2^n$ for n even, and by a cross of area $B_c = 4 \times 2^n$ for n odd, with dimensions as shown in Fig. 4. The energy calculated by the integral approximation is

$$\hat{S} = (2/3)2^n \text{ (square)}$$
$$= (31/48)2^n \text{ (cross)}.$$

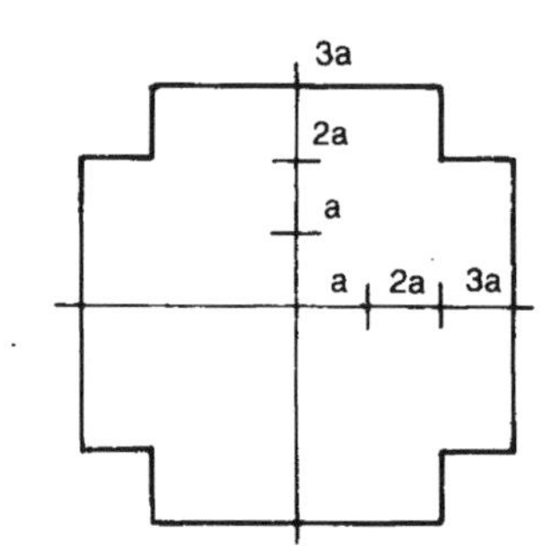

[$a = 2^{(n-1)/2}$ for Campopiano/Glazer constellations, n odd].

Fig. 4. Cross constellation boundary.

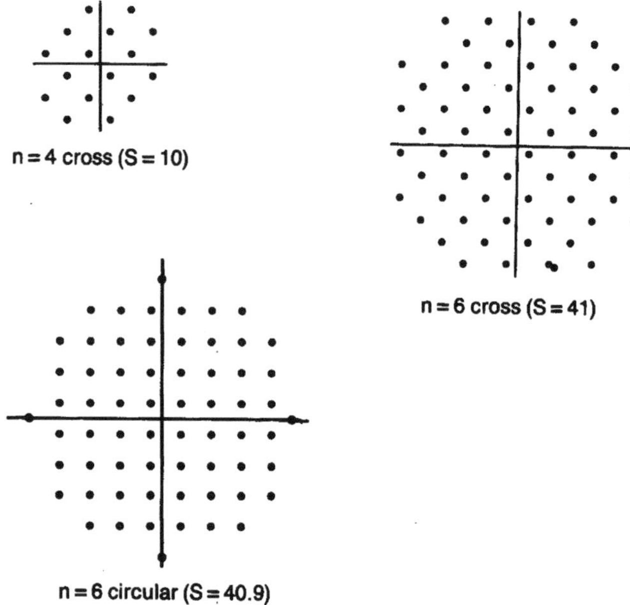

Fig. 5. Improved rectangular constellations.

Comparisons between S and $\hat{S}$ are given in Table II; the approximations are good ones. Furthermore, note that the cross is slightly more efficient than the square, by a factor of $31/32$ or 0.14 dB (because it is more like a circle); this suggests that the cross would be the better shape even for n even, and indeed this is the case and can be easily achieved by taking alternate points from the next higher cross constellation, as shown in Fig. 5 for $n = 4$ and $n = 6$. The $n = 4$ cross constellation is as good as the conventional 4×4 square constellation, and the $n = 6$ cross constellation is 0.1 dB better than the 8×8 square constellation, about as predicted.

Of course, the best enclosing boundary would be a circle, the geometrical figure of least average energy for a given area. A circle with radius R has area πR^2 and average energy equal to $R^2/2$; setting $R^2 = 4 \times 2^n$, we find that the average energy for a 2^n-point circular constellation ought to be about

$$\hat{S} = (2/\pi)2^n \text{ (circle)}$$

which would be only about $\pi/3$ or 0.20 dB better than the square, or 0.06 dB better than the cross. Fig. 5 also shows a

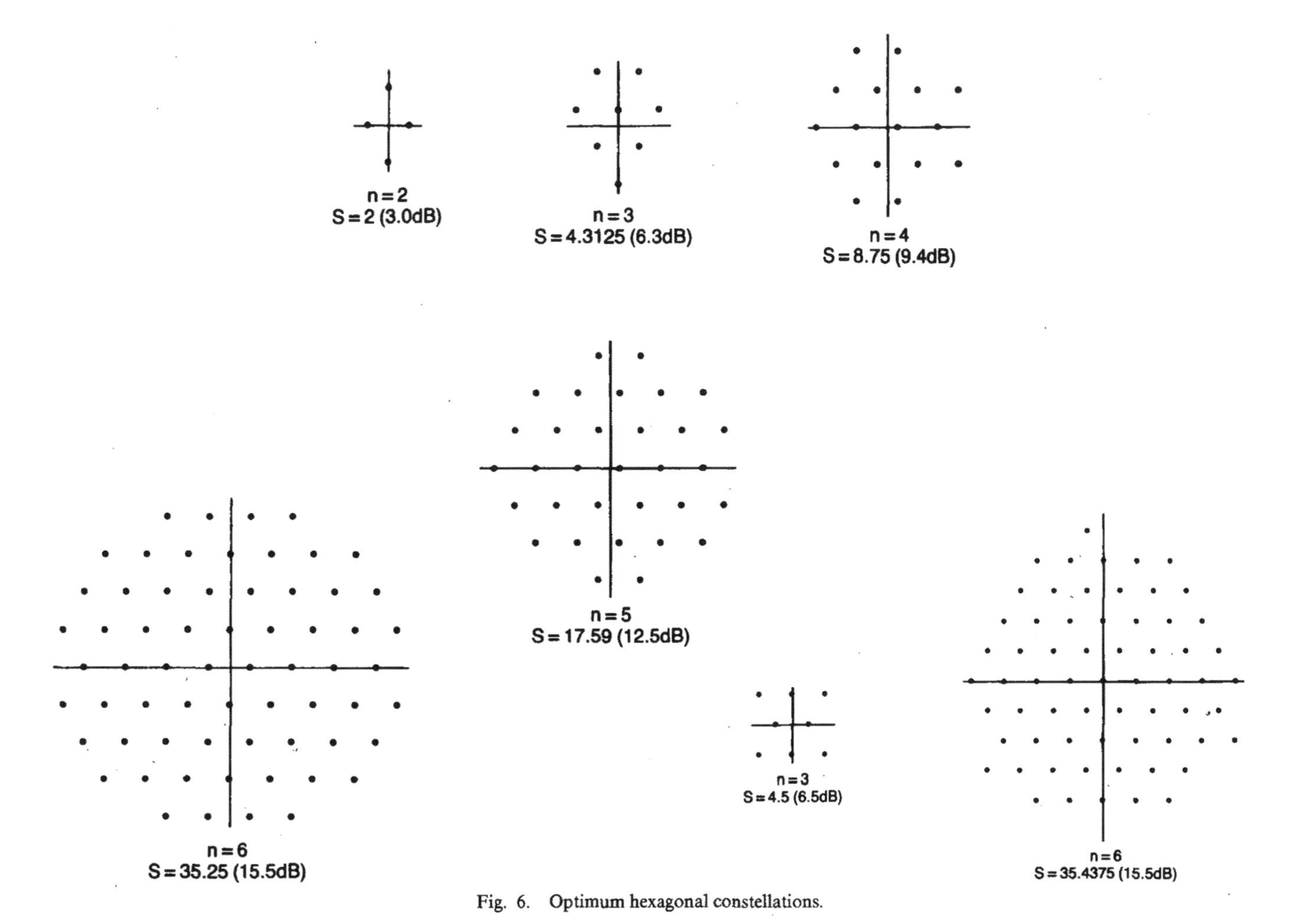

Fig. 6. Optimum hexagonal constellations.

more circular constellation for $n = 6$, with the outer points moved to the axes as in Campopiano–Glazer's 8-point constellation; this constellation has been used in the Paradyne 14.4 kbit/s modem and, like the $n = 6$ cross, is about 0.1 dB better than the $n = 6$ square.

B. Hexagonal Constellations

As the densest lattice in two dimensions is the hexagonal lattice (try penny packing), constellations using points from a hexagonal lattice ought to be the most efficient. Indeed, the area of the hexagonal Voronoi region for a hexagonal lattice with minimum distance 2 is $2\sqrt{3} = 3.464$, or 0.866 the size of the square region, which according to our approximation principle should translate to a 0.6 dB gain for a hexagonal constellation over a rectangular one with the same boundary. (A hexagonal boundary, as suggested in [12], has an energy efficiency within 0.03 dB of the circular boundary, or 0.03 dB better than the cross.)

Fig. 6 shows the best hexagonal packings for $n = 2$ through 6. For $n \geq 4$, the predicted 0.6 dB gain is effectively obtained over the best rectangular packings. (Historical notes: suggestions that the hexagonal lattice would asymptotically be the best were made very early; see, e.g., [13]. The suboptimal $n = 3$ "double diamond" structure was actually used in a 4800 bit/s Hycom modem in the mid-1970's. There was a great deal of attention to $n = 4$ structures in the early 1970's because of their importance in 9600 bit/s modems; the rather strange-looking one shown here was apparently first discovered by Foschini *et al.* [14], and is still the best 16-point constellation known. The $n = 6$ suboptimal structure is used in the Codex/ESE SP14.4 modem.)

IV. Elaborations of Uncoded Modulation

In this section we shall discuss further variants of uncoded modulation: constellations with nonuniform probabilities, higher-dimensional uncoded constellations, and constellations for nonintegral numbers of bits/symbol.

A. Nonuniform Probabilities

Attainment of the channel capacity bound requires that the signal points have a Gaussian probability distribution, whereas with all the constellations of the previous section it is implicit that points are to be used with equal probabilities. A uniform circular distribution of radius R has average energy $S_c = R^2/2$ and entropy $H_c = \log_2 \pi R^2$; a

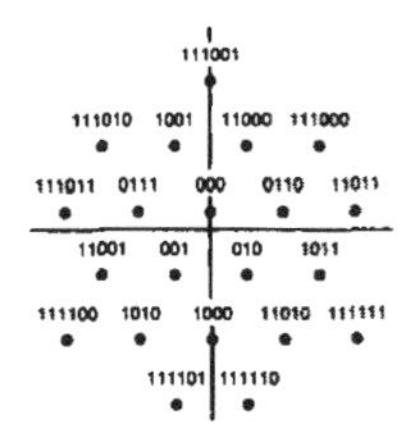

Fig. 7. Source-coded constellation.

TABLE III
ENERGY SAVINGS FROM N-SPHERE MAPPING

N	Gain	dB
2	1.05	.20
4	1.11	.45
8	1.18	.73
16	1.25	.98
24	1.29	1.10
32	1.31	1.17
48	1.34	1.26
64	1.35	1.31

Gain $= \frac{\pi(N+2)}{12}[(N/2)!]^{-2/N}$ (for N even)

two-dimensional Gaussian distribution of variance σ^2 in each dimension has average energy $S_g = 2\sigma^2$ and entropy $H_g = \log_2 2\pi e\sigma^2$. Thus,

$$H_c = \log_2 2\pi S_c$$
$$H_g = \log_2 \pi e S_g.$$

To yield the same entropy, the Gaussian distribution requires a factor of $e/2 = 1.36$ (or 1.33 dB) less average energy than the circular distribution.

Implementation of a constellation with nonuniform probabilities presents a number of practical problems. One possible way of achieving some of the potential gain is to divide the incoming data bits into words of nonuniform length according to a prefix code, and then to map the prefix code words into signal points drawn as before from a regular two-dimensional lattice. The probability associated with a prefix code word of length t bits is then 2^{-t}. For example, Fig. 7 gives a set of prefix code words and a mapping onto the hexagonal lattice that yields an average energy of $S = 7.02$ while transmitting an average of 4 bits/symbol, an improvement of close to 1 dB over the best $n = 4$ uniform code known. Of course, the fact that the number of data bits transmitted per unit time is a random variable leads to system problems (e.g., buffering, delay) that may outweigh any possible improvement in signal-to-noise margin.

B. Higher-Dimensional Constellations

It is possible to achieve the same gain in another way by coding blocks of data into higher-dimensional constellations without going to the true block coding to be described in later sections. (By "true coding," we refer to schemes in which the distance between sequences in a higher number of dimensions is greater than that between points in two dimensions.) We have already seen in Section III-A that a small (0.2 dB) gain is possible by going from one-dimensional PAM to two-dimensional QAM and choosing points on a two-dimensional rectangular lattice from within a circular rather than a square boundary. In the same way, by going to a higher number N of dimensions and choosing points on an N-dimensional rectangular lattice from within an N-sphere rather than an N-cube, further modest savings are possible. Table III gives the energy savings possible in N dimensions, based on the difference between average energy of an N-sphere versus an N-cube of the same volume. Note that as N goes to infinity, the gain goes to $\pi e/6$ (by the Stirling approximation, $(n!)^{-1/n}$ goes to e/n), or 1.53 dB; the improvement over $N = 2$ goes to $e/2$ or 1.33 dB, as computed above. This is because for large N the probabilities of points in any two dimensions become nonuniform and ultimately Gaussian. (It seems remarkable that a purely geometric fact like the asymptotic ratio of the second moment of an N-sphere to that of an N-cube can be derived from an information-theoretic entropy calculation in 2-space, but so it can.)

Implementation of such a scheme also involves added complexity that may outweigh the performance gain. To send n bits/symbol in N dimensions (assuming N even), incoming bits must be grouped in blocks of $Nn/2$. Some sort of mapping must then be made into the $2^{Nn/2}$ N-dimensional vectors with odd-integer coordinates (assuming a rectangular lattice) which have least energy among all such vectors. This can rapidly become a huge task; and a corresponding inverse mapping must be made at the receiver. Compromises can be made to simplify the mapping, at the cost of some suboptimality in energy efficiency; e.g., the cross is an effective compromise between the square and the circle in two dimensions.

C. Nonintegral Number of Bits/Symbol

It is sometimes desirable (as we shall see in Section VI) to transmit a nonintegral number of bits/symbol. Since in general an additional 1 bit/symbol costs about an additional 3 dB, it ought to be possible to send an additional 1/2 bit/symbol for about 1.5 dB. In this section we give a simple method that effectively achieves such performance. The method can be generalized to other simple binary fractions at the expected costs, but we shall omit the generalization here.

To send $n + 1/2$ bits/symbol, we proceed as follows. Use a signal constellation comprising 2^n "inner points" drawn from a regular grid, such as any of those of Section III, and an additional 2^{n-1} "outer points" drawn from the same grid and of as little average energy as possible, subject to whatever symmetry constraints may be imposed. Incoming bits are then grouped into blocks of $2n + 1$ bits and sent in two successive symbol intervals as follows. One bit in the block determines whether any outer point is to be used. If not, the remaining $2n$ bits are used, n at a time, to select two inner points. If so, then one additional bit selects which of the two signals is to be an outer point, $n - 1$ bits select which outer point, and the remaining n bits select which inner point for the other signal. (That is, at most one outer point is sent.) With random data, the average

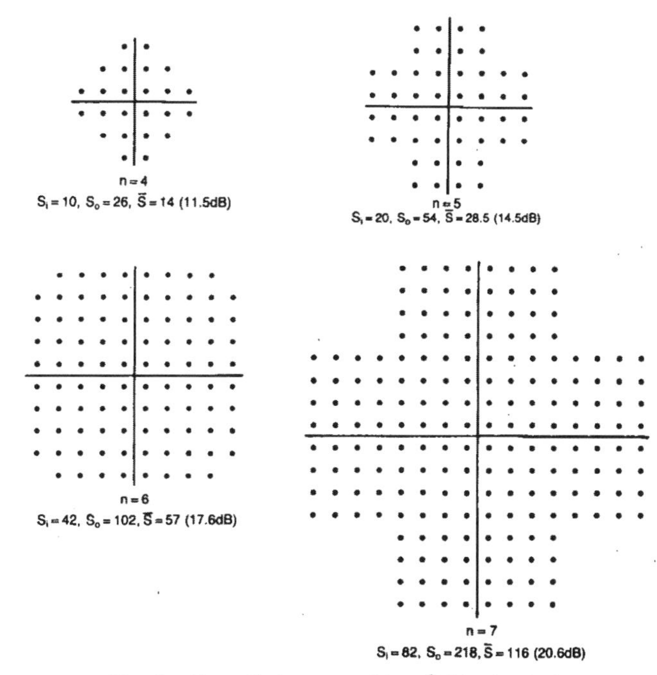

Fig. 8. Constellations to send $(n+\frac{1}{2})$ bits/symbol.

energy is 3/4 the average energy of inner points plus 1/4 the average energy of outer points. Fig. 8 shows constellations of 24, 48, 96, and 192 points that can be used in such schemes for $4 \leqslant n \leqslant 7$; the average energy in all cases for $n+1/2$ bits/symbol is approximately halfway between that needed for n and that for $n+1$ bits/symbol. Thus, these constellations are intermediate between the Campopiano/Glazer constellations in the same way that the cross constellations are intermediate between the squares. (In fact, it can be shown that the 2-dimensional cross constellations can be derived from 1-dimensional PAM constellations with "inner" and "outer" points in an analogous way.)

V. Coding Fundamentals

Heretofore we have been concerned with methods of mapping input bits to signal point constellations in two or more dimensions, where in higher dimensions the bits simply lie on the lattice that is the Cartesian product of two-dimensional rectangular lattices, so that the distance between points in N-space is no different from that in two dimensions. Now we shall begin to discuss methods of coding of sequences of signal points, where for the purposes of this paper we mean by coding (or "channel coding") the introduction of interdependencies between sequences of signal points such that not all sequences are possible; as a consequence, perhaps surprisingly, the minimum distance $d_{\min}$ in N-space between two possible sequences is greater than the minimum distance d_o in 2-space between two signal points in the constellation from which signal points are drawn. Use of maximum likelihood sequence detection at the receiver yields a "coding gain" of a factor of $d_{\min}^2/d_o^2$ in energy efficiency, less whatever additional energy is needed for signaling. (In practice, some of the "coding gain" may be lost due to there being a large number of sequences at distance $d_{\min}$ from the correct sequence and therefore a large number of possibilities for error, called the "error coefficient" effect. We shall not be able to discuss the "error coefficient" much in this paper, but offer some general remarks at the end of Section VIII.)

Conventional coding techniques cannot be directly applied in conjunction with band-limited modulation techniques, at least with significant gain. (In 1970–1971, at least four companies prototyped conventional coding schemes for use in high-speed modems; two of the companies failed, and two shortly withdrew their products from the market.) In recent years, however, a number of effective coding techniques have been developed for such applications. The most important point to be made in this paper is that all of these coding schemes can be developed from a common conceptual principle. This principle was set forth clearly by Ungerboeck [15], who called it "mapping by set partitioning," although its roots may perhaps be found elsewhere as well. We describe it in this section, and in succeeding sections then use it to develop all known and some new coding schemes, both block and trellis.

We shall consider only 2-dimensional constellations with points drawn from a 2-dimensional rectangular grid. (From research to date, we cannot find any advantage to starting with hexagonal grids when higher orders of coding are to be used.)

Such a constellation can be divided into two subsets by assigning alternate points to each subset; i.e., according to the pattern

The resulting two subsets (A and B) have the following properties.

a) The points in each subset lie on a rectangular grid (rotated 45° with respect to the original grid).

b) The minimum squared distance between points within a subset is twice the minimum squared distance $[d_o^2]$ between points in the original constellation.

Furthermore, because of the first property, the partitioning can be repeated to yield 4, 8, 16, $\cdots$ subsets with similar properties, and in particular within-subset squared distances of 4, 8, 16, $\cdots$ times d_o^2. Fig. 9 shows the 64-point square constellation divided into two subsets of 32 points, 4 subsets of 16 points, 8 subsets of 8 points, and

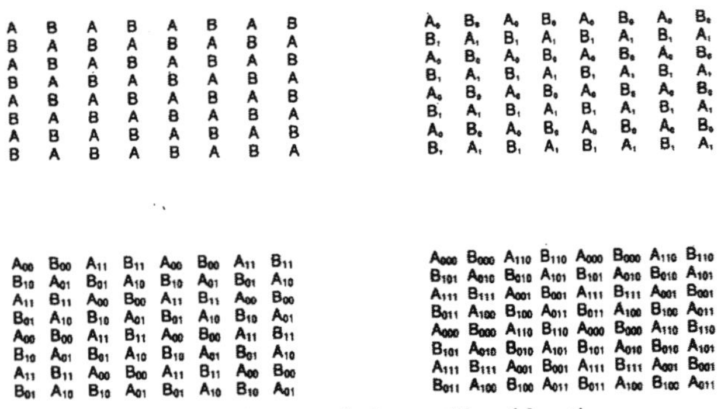

Fig. 9. 64-point constellation partitioned four times.

16 subsets of 4 points. The subset nomenclature is according to the following binary tree:

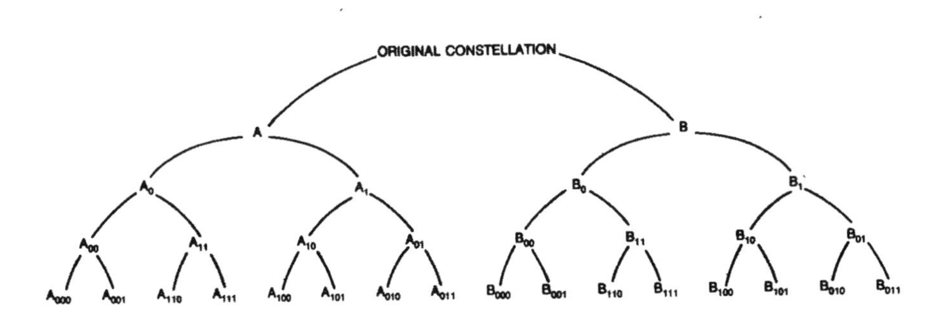

This nomenclature is different from that of Ungerboeck [15], who uses a more natural subscript notation that reflects the successive binary partitions; our nomenclature will be useful in the next section, where we shall use the fact (evident by inspection) that at both the 4-subset and 8-subset levels, the minimum squared distance between two A subsets, say, is $2d_o^2$ times the Hamming distance between their subscripts: e.g., $d^2(A_0, A_1) = 2d_o^2$; $d^2(A_{00}, A_{01}) = 2d_o^2$; $d^2(A_{00}, A_{11}) = 4d_o^2$; and so forth.

These subsets may then be used to implement relatively simple but effective coding schemes, illustrated in general in Fig. 10. Certain incoming data bits are encoded in a binary encoder, resulting in a larger number of coded bits. The coded bits are then used to select which subsets are to be used for each symbol. The remaining incoming bits are not coded, but merely select points from the selected subsets, with the signal constellation chosen large enough to accommodate all incoming bits. The coding scheme is thus more or less decoupled from the choice of constellation, as long as it is of the rectangular grid type. The coding gain is effectively determined by the distance properties of the subsets combined with those of the binary code, regardless of the size of the constellation. On the other hand, the constellation size, boundary, symmetries, and other "uncoded" properties such as were investigated in earlier sections are more or less independent of the coded bits and are determined by the mapping of the remaining bits. This may be regarded as effectively decou-

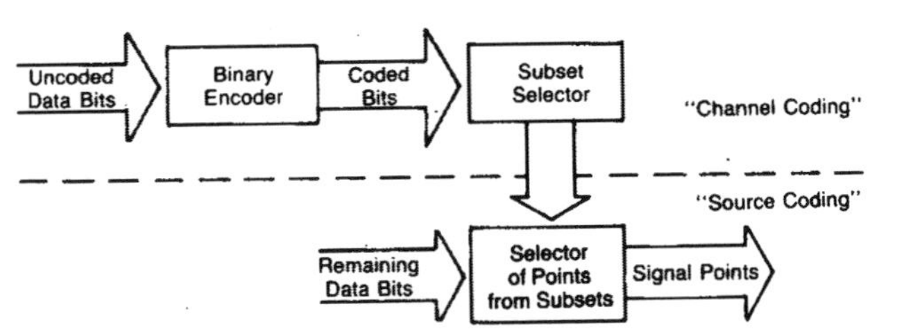

Fig. 10. General coding scheme.

pling "source coding" from "channel coding" and is important both conceptually and in implementation.

We now show how this general scheme can be applied to both block and trellis codes, with performance approaching the R_o estimate.

VI. Block Codes

In Section IV-B we saw what could be achieved by using points from an N-dimensional rectangular lattice and using an N-sphere, rather than an N-cube, as a boundary.

The rectangular lattice comprising all N-dimensional vectors with odd-integer coordinates is not the most densely packed for any N greater than 1; for example, for $N = 2$, the hexagonal lattice is 0.6 dB denser, as we have seen. Finding the densest lattice in N dimensions is an old and well-studied problem in the mathematical literature. Table IV gives the densest packings currently known for all N up to 24 and selected larger N, with the improvement in packing density over the rectangular lattice given in absolute and in dB terms [16]. The dimensions $N = 4$, 8, 16, and 24 are locally particularly good and are known to be optimum in the sense of being the densest possible lattice packing in these dimensions.

A body of recent work [11], [12], [18]–[21] generalizes the Campopiano–Glazer construction to N dimensions by taking all points on the densest lattice in N-space that lie within an N-sphere, where the radius of the sphere is chosen just large enough to enclose 2^{mN} points, to send m bits per dimension. These codes obtain a "coding gain" over PAM which is a combination of both the lattice packing density gain of Table IV ("channel coding") and the N-sphere/N-cube boundary gain of Table III ("source coding"). The resulting coding gains achievable for $N = 2$,

TABLE IV
ENERGY SAVINGS FROM DENSE LATTICES

N	Density	Gain	dB
1	.500	1.00	.00
2	.289	1.15	.62
3	.177	1.26	1.00
4	.125	1.41	1.51
5	.088	1.52	1.81
6	.072	1.67	2.21
7	.063	1.81	2.58
8	.063	2.00	3.01
9	.044	2.00	3.01
10	.037	2.07	3.16
11	.035	2.18	3.38
12	.037	2.31	3.63
13	.035	2.39	3.78
14	.036	2.49	3.96
15	.044	2.64	4.21
16	.063	2.83	4.52
17	.063	2.89	4.60
18	.072	2.99	4.75
19	.088	3.10	4.91
20	.125	3.25	5.11
21	.177	3.39	5.30
22	.289	3.57	5.53
23	.500	3.77	5.76
24	1.000	4.00	6.02
32	1.000	4.00	6.02
36	2.000	4.16	6.19
40	16.000	4.59	6.62
48	16832.947	6.00	7.78
64	4194304.000	6.44	8.09

TABLE V
COMBINED ENERGY SAVINGS

N	Gain	dB
2	1.21	.82
4	1.57	1.96
8	2.36	3.74
16	3.54	5.50
24	5.16	7.12
32	5.24	7.19
48	8.04	9.04
64	8.69	9.40

4, 8, 16, 24, 32, 48 and 64 are shown in Table V. Because the number of near neighbors in these densely packed lattices becomes very large, the total number of error events ("error coefficient") becomes large, which reduces the coding gain realized in practice. Also, the mapping of all mN bit combinations to their corresponding signal points can be an enormous task, even if all possible symmetries and simplifications are cleverly exploited [20], [21].

We will now show that certain of these dense N-dimensional lattices can be constructed using 2-dimensional rectangular lattices, the subset partitioning idea, and simple binary block codes. In particular, we shall give constructions for $N=4$, 8, 16 and 24 that form a natural sequence both in complexity and in nominal coding gain (respectively 1.5, 3.0, 4.5, and 6.0 dB, using the simplest implementations). (Cusack [21a] has recently shown how to construct dense 2^n-dimensional lattices from 2-dimensional lattices using Reed–Muller codes, for any n; for $N=4$, 8, and 16, the lattices obtained are the same as those we obtain here.)

To generate the optimum N-dimensional lattices for $N=4$, 8, 16, and 24, we shall use sequences of 2, 4, 8, and 12 points from the 2-dimensional rectangular lattice, partitioned as shown in Fig. 9, into 2, 4, 8, and 16 subsets, respectively.

The 4-dimensional lattice is generated by taking all sequences of two points in which both points come from the same subset, i.e., sequences of the form (A, A) or (B, B).

The 8-dimensional lattice consists of all sequences of four points in which all points are either A points or B points and further in which the 4 subset subscripts satisfy an overall parity check, $i_1+i_2+i_3+i_4=0$; e.g., sequences of the form (A_0, A_0, A_0, A_0), (B_0, B_1, B_0, B_1), and so forth. (In other words, the subscripts must be codewords in the (4,3) single-parity-check block code, whose minimum Hamming distance between codewords is 2.)

The 16-dimensional lattice consists of all sequences of eight points in which all points are either A points or B points, and further, in which the 16 subset subscripts (each subset now having two subscripts) are codewords in the (16,11) extended Hamming code, whose minimum Hamming distance between codewords is 4.

The 24-dimensional lattice consists of all sequences of 12 points in which all points are either A points or B points; the 24 (i, j) subscripts are codewords in the (24,12) Golay code, known to have minimum Hamming distance 8, and further, in which the third subscripts k are constrained to satisfy an overall parity check in the following way: if the sequence is of all A points, then overall k parity is even (an even number are equal to 1), while if the sequence is of all B points, overall k parity is odd.

If the minimum squared distance between points in the 2-dimensional constellation is d_o^2, then the minimum squared distance between points (sequences) in these higher-dimensional lattices can be shown to be $2d_o^2$, $4d_o^2$, $8d_o^2$, and $16d_o^2$, respectively, as follows.

a) A sequence of A points and a sequence of B points differ from each other by squared distance at least d_o^2 in every point and therefore by at least $2d_o^2$, $4d_o^2$, $8d_o^2$, and $12d_o^2$ in total. In fact, in the 24-dimensional case there is a distance of at least $5d_o^2$ in at least one symbol, so the minimum squared distance between A sequences and B sequences is at least $16d_o^2$. The proof depends on the properties of the Golay code as well as the particular partitioning shown in Fig. 9 and is in the Appendix.

b) Two different sequences with points all from the same sequence of subsets must differ in at least one point by the minimum within-subset squared distance, which is $2d_o^2$, $4d_o^2$, $8d_o^2$, or $16d_o^2$, respectively. This is all we need to establish $2d_o^2$ as the minimum squared distance between sequences in the 4-dimensional case.

c) For $N=8$, 16, and 24, the i or (i, j) subset subscripts are drawn from (4,3), (16,11), or (24,12) codes with minimum Hamming distances 2, 4, and 8, respectively. By the relation between subscript Hamming distance d_H and subset squared distance $d_S^2 = 2d_H d_o^2$ given in Section V, two sequences with points drawn from subsets of the same type (A or B) but different i or (i, j) subscripts must differ by squared distance at least $4d_o^2$, $8d_o^2$, or $16d_o^2$, respectively. This is all we need for the 8- and 16-dimensional cases.

d) For $N=24$, two sequences of points from subsets of the same type and with the same (i, j) subscripts but different k subscripts must differ by at least $8d_o^2$ in at least two symbols because of the overall k parity check, and the fact that the minimum squared distance between points of the same type and with the same (i, j) subscripts is $8d_o^2$. This concludes the 24-dimensional proof.

To send m bits/symbol using these lattices, we need to encode a block of mN bits into one of 2^{mN} lattice points. To maximize coding gain, the 2^{mN} lattice points of least energy should be chosen; however, implementation of the mappings from bits to points and vice versa becomes complex. Simpler methods will now be given, using the binary codes used to construct the lattices, and the constellations either of Fig. 3 (for $N=8$ and 24) or of Fig. 8, along with the method of sending half-integral numbers of bits/symbol given in Section IV-C (for $N=4$ and 16). The cost in coding gain is relatively small, ranging from a few tenths of a decibel for $N=4$ or 8, up to about 1 dB for $N=24$; it is upperbounded by the N-sphere/N-cube gain given in Table III.

Of the block of mN bits, we always use one bit to specify whether A or B points will be used. For $N=8$, 16, and 24, a further set of bits is used as input to a binary block coder, which produces appropriate codewords to be used as subset subscript designators: 3 bits to produce 4 for $N=8$, 11 bits to produce 16 for $N=16$, and $12+11=23$ bits to produce $24+12=36$ for $N=24$. Thus, a total of 1, 4, 12, or 24 incoming bits are used as in Fig. 10 to select the subsets, or $\frac{1}{2}$, 1, $1\frac{1}{2}$, or 2 bits/symbol, respectively.

The remaining bits are used to select points from the selected subsets. We use the rectangular constellations of Figs. 3 and 8 as follows: for $N=4$, constellations of 1.5×2^m points as in Fig. 8, divided into two $1.5\times 2^{m-1}$-point subsets A and B; for $N=8$, constellations of 2^{m+1} points as in Fig. 3, divided into four 2^{m-1}-point subsets; for $N=16$, constellations of $1.5\times 2^{m+1}$ points as in Fig. 8, divided into eight $1.5\times 2^{m-2}$-point subsets; and for $N=24$, constellations of 2^{m+2} points as in Fig. 3, divided into 16 2^{m-2}-point subsets. (In all cases m must be large enough so that the subsets resulting from the partitioning have equal size and other desired properties, e.g., symmetries.) These can be used to send $m-\frac{1}{2}$, $m-1$, $m-1\frac{1}{2}$, or $m-2$ bits/symbol, where for $N=4$ and 16, the method of sending half-integral bits/symbol of Section IV-C may be used.

Since the minimum squared distance in N-space is 2, 4, 8, or 16 times the minimum squared distance for an uncoded 2^m-point constellation, there is a distance gain of 3, 6, 9, or 12 dB, respectively. However, the expanded constellations required with coded modulation cost 1.5, 3, 4.5, and 6 dB, respectively, yielding a net coding gain of 1.5, 3, 4.5, and 6 dB for $N=4$, 8, 16, and 24. The family relationship of this progression of codes is apparent.

Maximum likelihood sequence detection of the lattice point closest to a sequence of received points is easy for $N=4$ and 8. General methods are given in [22]. For $N=8$, given four received points, assume first that A points were sent. Find the closest A point to each received point, and check subscript parity of the four subsets tentatively decided. If the parity check fails, change the least reliable decision to the next closest A point (which must be in the other A subset). This gives the best A sequence satisfying the parity-check constraint. Repeat, assuming that B points were sent, to get the best B sequence. Compare the best A and B sequences, and choose the better as the final decision.

Detection for $N=16$ or 24 is harder, involving either generalization of soft-decision decoding of the (16,11) or (24,12) block codes to perform error correction on tentative decision subscripts (as above; the method of changing the least reliable decision if an overall parity check fails was used in the 1950's as a soft-decision error-correction method for single-parity-check codes by Wagner), or exhaustive search of a neighborhood of the received sequence in N-space. Note that as its first step the decoder can always choose the closest point in each subset to each received point as representative of that subset, there being no reason to prefer any more distant point, and then proceed to determine the best sequence of subsets using those points (with their distances from the received point) as proxies for the corresponding subsets; the decoding task may thus be partitioned in the same way as coding is partitioned in Fig. 10.

VII. Trellis Codes

On power-limited channels (such as the satellite channel), convolutional coding techniques have more or less become the standard (although there are some who continue to champion block codes [23]). Generally, anything that can be achieved with a block code can be achieved with somewhat greater simplicity with a convolutional code. We have just seen that relatively simple ($N=8$) block codes can achieve of the order of 3 dB coding gain on band-limited channels, and relatively complex ($N=24$) block codes can achieve of the order of 6 dB. We shall now see that trellis codes can do the same, perhaps a bit more simply.

A. Ungerboeck Codes

For band-limited channels, the trellis codes came first, in the work of Ungerboeck [15]. In Ungerboeck's paper, to send n bits/symbol with two-dimensional modulation, a constellation of 2^{n+1} points is used, partitioned into 4 or 8 subsets. 1 or 2 incoming bits/symbol enter a rate-$\frac{1}{2}$ or rate-$\frac{2}{3}$ binary convolutional encoder, and the resulting 2 or 3 coded bits/symbol specify which subset is to be used. The remaining incoming bits specify which point from the selected subset is to be used.

The coding gain obtainable increases with the number M of states in the convolutional encoder. Ungerboeck's simplest scheme uses a 4-state encoder and achieves a nominal 3 dB coding gain (a factor of 4, or 6 dB, in increased sequence distance, less 3 dB due to use of the larger 2^{n+1}-point constellation). His most complex scheme uses a 128-state encoder and gains 6 dB (the limit with 8 subsets and a 2^{n+1}-point constellation since the within-subset distance is $8d_o^2$ for a 9 dB gain, less the 3 dB due to the larger constellation). Table VI gives the coding gains obtained by

TABLE VI
CODING GAINS FOR UNGERBOECK CODES

States	Gain	dB
4	2.0	3.0
8	2.5	4.0
16	3.0	4.8
32	3.0	4.8
64	3.5	5.4
128	4.0	6.0

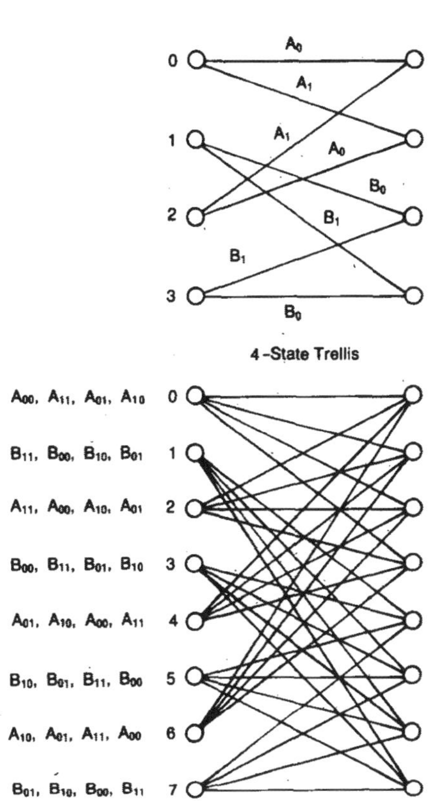

Fig. 11. Ungerboeck 4- and 8-state trellis codes.

Ungerboeck for these and intermediate numbers of states.

Decoding is assumed to be by the Viterbi algorithm [24], a maximum likelihood sequence estimation procedure for any trellis code. The complexity of such a decoder is roughly proportional to the number of encoder states. With these codes, each branch in the trellis corresponds to a subset rather than to an individual signal point; but if the first step in decoding is to determine the best signal point within each subset (the one closest to the received point), then that point and its metric (squared distance from the received point) can be used thereafter for that branch, and Viterbi decoding can proceed in a conventional manner. Fig. 11 gives trellises with branches labeled by subset for Ungerboeck's 4-state and 8-state codes.

(Note: the block codes of the previous section can be represented as trellises; Fig. 12 shows the trellises corresponding to the $N=4$ and $N=8$ codes. Viterbi decoding could therefore be used for them as well. It is interesting that the block code with 3 dB coding gain is also associated with a 4-state trellis, albeit decomposable into two parallel

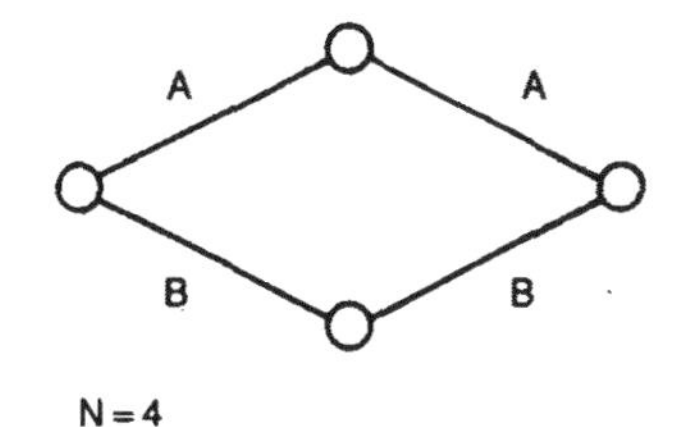

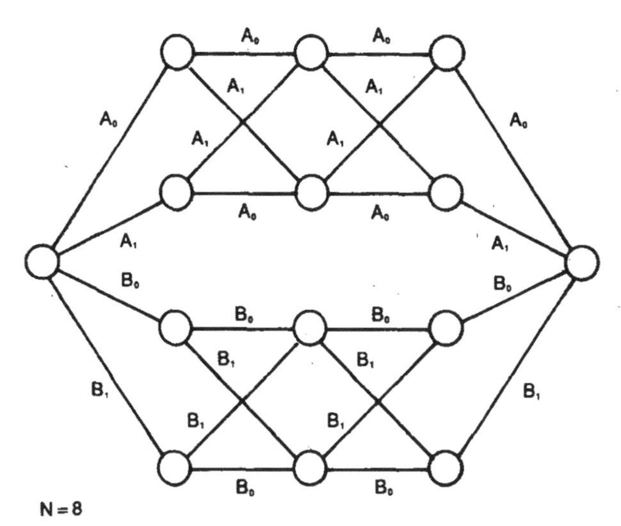

Fig. 12. Trellises corresponding to $N=4$ and $N=8$ block codes.

2-state trellises. The $N=16$ block code can similarly be associated with a 64-state trellis decomposable into two parallel 32-state trellises, and the $N=24$ block code can be associated with a 2×4096-state trellis.)

8-state trellis codes with nominal 4 dB coding gain are in the process of being adopted as international CCITT standards for 9600 bit/s transmission over the switched (dial) telephone network [25] and potentially for 14.4 kbit/s transmission over private lines as well [26]. A slight variant [27] of the Ungerboeck scheme involving a nonlinear convolutional encoder is being used in these standards; with this variant, whose trellis is shown in Fig. 13, a 90° rotation of a coded sequence is another coded sequence, so that differential coding techniques may be used. The distance properties and therefore coding gain of the variant are apparently identical to those of Ungerboeck's 8-state scheme.

B. Other Trellis Codes

The Ungerboeck codes seem to cover the range of possible coding gains with complexity of the order of what we might expect, and may therefore be taken as benchmarks of how much complexity is needed to achieve different coding gains in the 3–6 dB range. Can they be improved upon? From our research, the answer seems to be: yes, but not very much. In this section we shall describe two schemes that exhibit modest improvements and some new ideas: a 2-state code that has a nominal coding gain of almost 3 dB, and an 8-state trellis code with a coding gain

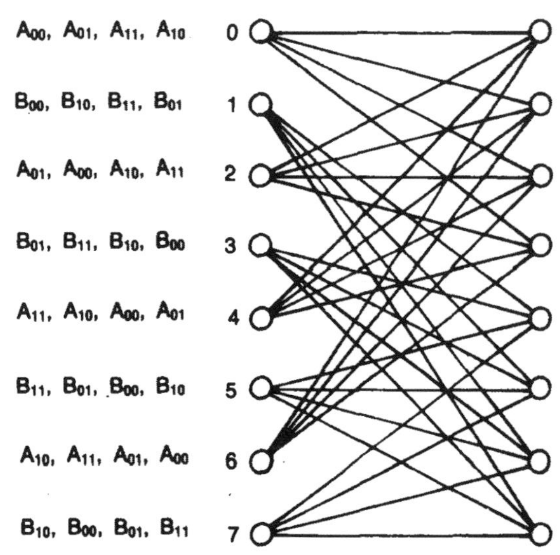

Fig. 13. Nonlinear 8-state trellis code with 90° symmetry.

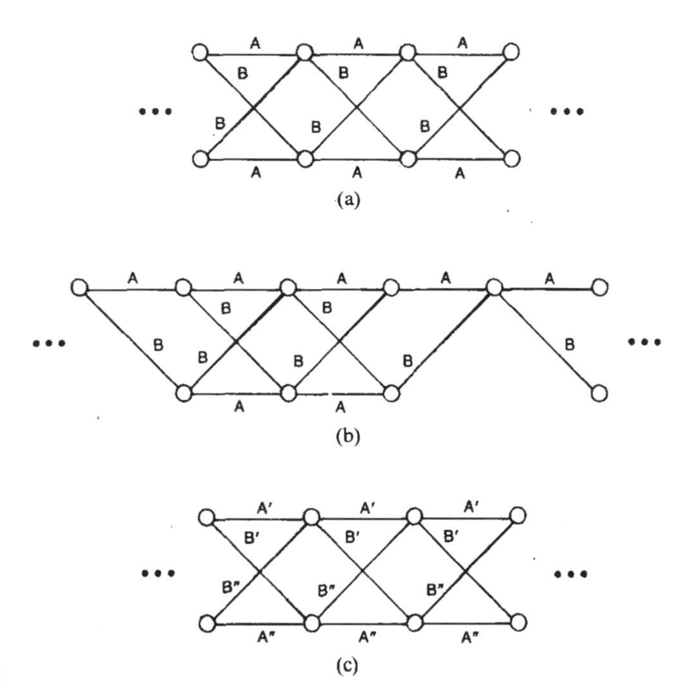

Fig. 14. 2-state trellises. (a) Infinite nonredundant 2-state trellis. (b) 2-state trellis terminated every 4 symbols. (c) Time-invariant 2-state trellis with $A' \neq A''$, $B' \neq B''$.

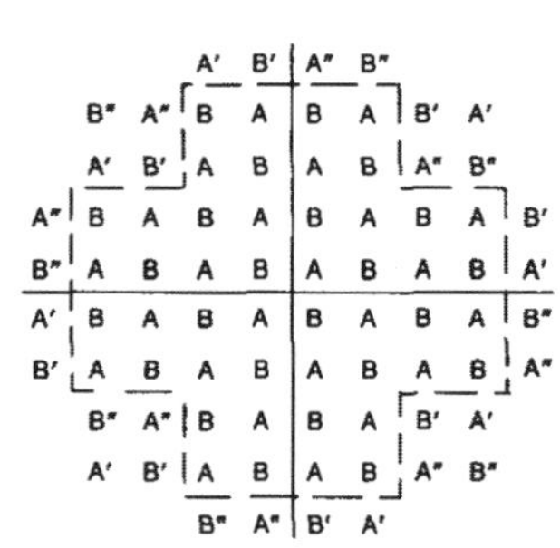

A′ B′ A″ B″
B″ A″ B A B A B′ A′
A′ B′ A B A B A″ B″
A″ B A B A B A B A B′
B″ A B A B A B A B A′
A′ B A B A B A B A B″
B′ A B A B A B A B A″
B″ A″ B A B A B′ A′
A′ B′ A B A B A″ B″
B″ A″ B′ A′

Fig. 15. 80-point signal constellation. 32-point subsets A', A'', B', B'' have 24 common points, 8 unique points; e.g., $A' = 24$ A inner points plus 8 A' outer points. $\overline{S_i} = 31.3$; $\overline{S_o} = 80$; $S = 43.5$ (16.4 dB).

of 4.5 dB. The new idea in the 2-state scheme is to use subsets that are partially overlapping and partially distinct. The new idea in the 8-state scheme is to use a 4-dimensional constellation rather than 2-dimensional as the basic constellation. (4-dimensional trellis codes have also been studied by Wilson [28] and Fang *et al.* [28a].)

1) 2-state Trellis Code: If a 2^n-point signal constellation is partitioned into two subsets of A points and B points, and channel coding is done using the (infinite) 2-state trellis shown at the top of Fig. 14, it would appear at first glance that a 3 dB gain is obtained at no cost. This scheme sends n bits/symbol with no signal constellation expansion; further, any two sequences that start at one common node and end at another differ by a squared distance of at least $2d_o^2$, because the paths differ by at least d_o^2 when they diverge and another d_o^2 when they merge, so the nominal coding gain is apparently a factor of 2 or 3 dB.

Of course, we cannot get something for nothing, and the fallacy in this scheme (called "catastrophic error propagation" in the convolutional coding literature) is that there are paths of infinite length starting from a common node that never remerge and have squared distance only d_o^2, namely any two paths of the form $AXYZ\cdots$ and $BXYZ\cdots$. (The "error coefficient" is infinite.)

One way of curing this problem is to terminate the trellis every b symbols by forcing it to a single node, illustrated in Fig. 14(b). In other words, at the bth symbol, the subset is constrained to be A or B, as necessary to reach the designated node. Only $n-1$ bits can be used to determine the bth symbol, so there is a cost of 1 bit per b symbols of transmission capacity, but now a legitimate coding gain of 3 dB is obtained minus $(1/b)\times 3$ dB for the rate loss. The 8-space block code would operate in just this way if it used only A points, partitioned into A_0 and A_1 (see the top half of its trellis in Fig. 12); happily it is possible to insert a similar code made up of B points into the interstices of the A code lattice without compromising distance, and the additional bit involved in specifying A or B compensates for the bit lost at the fourth symbol, and allows a full 3 dB gain. This terminated trellis code may be regarded as a generalization of a single-parity-check block code.

Another way of gaining almost 3 dB while using a time-invariant trellis code is as follows. The signal constellation is modestly expanded to include $(1+p)2^n$ points, arranged on a rectangular grid and divided as usual into A and B subsets of $(1+p)2^{n-1}$ points each. The A and B subsets are further divided into $(1-p)2^{n-1}$ "inner" points and $2p$ "outer" points. Finally, sets A', A'', B', and B'', each of 2^{n-1} points, are created as follows: A' and A'' both include the $(1-p)2^{n-1}$ inner A points, but each includes a different half of the $2p$ outer points; and similarly with B' and B''. Thus A' and A'' are partially overlapping and partially disjoint, and so are B' and B''. The probability that a random choice from A' will not be a member of A'' is p. Such a construction is illustrated for $n = 6$ and $p = 0.25$ in Fig. 15.

Now the sets A', A'', B', and B'' may be used in a 2-state trellis as illustrated in the last part of Fig. 14. The squared distance between two paths beginning and ending at common nodes remains $2d_o^2$ since the basic distance properties between A subsets and B subsets remain. But now, although there are still pairs of infinite sequences comprising only inner points that start from a common node and never accumulate distance of more than d_o^2, their probability is zero. The squared distance between any branch that uses an outer point and its counterpart branch must be at least $2d_o^2$ since the counterpart branch cannot use the same outer point and the distance between different points in the same subset is at least $2d_o^2$. For practical purposes, this means that a sequence containing such a branch cannot be confused with a sequence containing the counterpart branch, their squared distance being at least $3d_o^2$, so that in effect, whenever an outer point is sent, reconvergence to a common node is forced, as in the trellis termination method. Here, however, the reconvergence is probabilistic and happens on average every $1/p$ symbols, e.g., every 4 symbols if $p = 0.25$.

There is a slight reduction in power due to the increased constellation size; e.g., the average power using the constellation of Fig. 15 is 43.5 or 16.4 dB, versus 42 or 16.2 dB for the 8×8 constellation, or 41 (16.1 dB) for the Fig. 5 $n = 6$ cross constellation. Use of an integral approximation gives an estimate of additional power required of a factor of $1 + p^2$, or 1.0625 (0.26 dB) for $p = 0.25$. This can be made as small as desired by reducing p; the cost, however, is a greater average time to converge and a greater average number of near-neighbor sequences ("error coefficient") increasing inversely proportional to p; for this code the "error coefficient" is rather large and must be taken into account. Any p greater than zero in principle avoids catastrophic error propagation; a p of about 0.25 seems a good choice in practice.

2) 8-State Trellis Code: For the 8-state four-dimensional scheme, we shall use a two-dimensional rectangular grid divided into four subsets as before. The binary convolutional encoder for this scheme, however, operates on pairs of symbols rather than single symbols. An appropriate encoder is shown in Fig. 16. During each pair of symbol intervals, three bits enter the encoder and four coded bits are produced. The first two coded bits select the subset for the first symbol and the second two bits select the subset for the second symbol.

If the Hamming distance between two encoded sequences is K, then the squared distance between the mappings onto grid points is at least Kd_o^2. We now show that the minimum free Hamming distance of this convolutional code is 4. First note that the response of the encoder to a single 1 on any input line is a sequence with even weight, from which it follows that all encoded sequences have even weight and the minimum free distance is even. By inspection, it is easy to verify that there is no encoded sequence of weight 2, and a simultaneous 1 on all inputs yields an encoded sequence of weight 4. Thus, the squared distance between any two sequences corresponding to different encoded outputs is at least $4d_o^2$. If the encoded outputs are the same, but different elements are chosen from the same

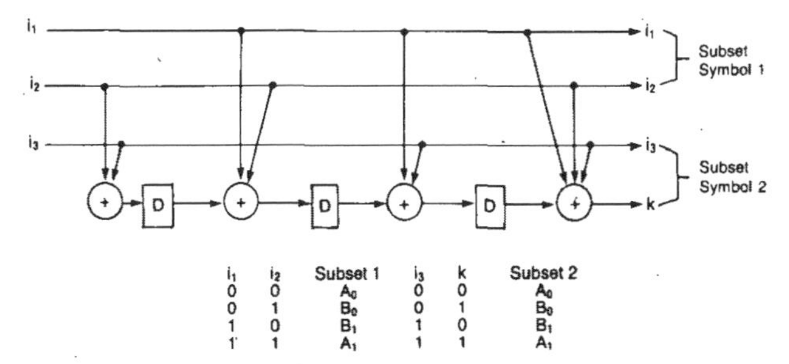

i_1	i_2	Subset 1	i_3	k	Subset 2
0	0	A_0	0	0	A_0
0	1	B_0	0	1	B_0
1	0	B_1	1	0	B_1
1	1	A_1	1	1	A_1

Fig. 16. Encoder for 8-state four-dimensional code.

subsets, then again the squared distance is at least $4d_o^2$. Suppose now that the signal constellation contains 2^n points. Then, over two symbol intervals, $2n-1$ bits enter the modem and one parity check is generated, giving $2n$ bits to select the two signal points. Since $n - \frac{1}{2}$ bits/symbol enter the modem, there is a loss of 1.5 dB due to the larger signal constellation and a gain of 6 dB in distance for a net nominal coding gain of 4.5 dB.

Section IV discussed encoding for a half-integral number of bits/symbol, and that method can be applied here. An alternative which is somewhat more attractive is to have an integer number n of bits/symbol enter the modem, with three bits entering the convolutional encoder each pair of symbols and $2n-3$ bits entering a prefix-code source coder as described in Section IV. This both yields an integer number of bits/symbol and also gains some of the possible 1.33 dB for nonuniform probabilities.

VIII. Concluding Remarks

It has not been possible in this paper to cover a number of topics that are of importance in practice.

The only channel disturbance considered has been white Gaussian noise. Other disturbances are usually controlling on telephone channels. There is some accumulating experience that the coded modulation schemes are often more robust relative to uncoded schemes than would be predicted by Gaussian noise calculations against some important disturbances, such as nonlinear distortion and phase jitter, perhaps due to the memory inherent in coded modulation and sequence estimation over multiple symbols.

Because of the symmetries of attractive constellations, e.g., the 90° symmetry of most of our rectangular constellations, there may be an ambiguity in phase at the receiver. In general, there are two ways to handle this ambiguity with coded modulation. If the code is such that, on a sequence basis, every 90° rotation of a code sequence is another legitimate code sequence, then it will be possible by differential quadrantal coding to make end-to-end transmission transparent to 90° rotations. Alternatively, if 90° rotations do not give valid code sequences in general, then it will be possible eventually to detect this and to force receiver phase to a valid setting. The former technique is generally preferred. Of the codes we have discussed, the block codes generally are differentially codable and the trellis codes are generally not, although they can often be modified to be; e.g., the modification of Fig. 11 shown in Fig. 13.

Finally, we have mostly used nominal coding gain as a figure of merit of coding schemes. In fact, error probabilities for coded systems on Gaussian channels are typically [24] of the form $P(E) = K\exp(-E)$, where the exponent E is governed by the nominal coding gain and the "error coefficient" K is of the order of the number of coded sequences at minimum distance from an average transmitted sequence. In general, the error coefficient

a) increases with the complexity of coding;

b) can cost a significant fraction of a dB for coding schemes with moderate (3–4 dB) gain, for error probabilities in the 10^{-5}–10^{-6} range;

c) can become very large for schemes with large (6+dB) gain, such as the block codes with $N = 24$, or the most complex trellis codes; and

d) is generally significantly larger for block codes than for trellis codes with comparable nominal coding gain.

Thus the error coefficient cannot be ignored in a more detailed assessment of coded systems.

VIII. Summary

On the band-limited channel, dense packing of 2-dimensional constellations with optimal (circular) boundaries yields less than 1 dB improvement over simple pulse amplitude modulation. Uncoded schemes in higher dimensions or, alternatively, source coding can gain somewhat more than 1 dB by using signal points with nonuniform probabilities. These gains pale by comparison with what can be obtained with (channel) coding, where relatively simple block or trellis codes easily yield coding gains of the order of 3 dB, or 1 bit/symbol. Relatively complex block and trellis codes have been constructed that yield of the order of 6 dB, or 2 bits/symbol. Because this is as much gain as would be predicted using the R_o estimate and is only 3 dB below the capacity limit, it seems unlikely that further major improvements are possible. However, within the spectrum of performance of already known schemes, there will likely be some further embellishments that will reduce implementation complexity or have other desirable properties, such as the differentially coded variant of Ungerboeck's 8-state trellis code that is likely to become an international standard.

Appendix
Proof That 24-Space Lattice Has $d_{\min}^2 = 16d_o^2$

If the grid of triple-subscripted signal points illustrated in Fig. 9 is rotated 45° with a point A_{000} at the origin, and scaled so that the coordinates (x, y) of all points are integers, then the following hold true.

a) For A points, both x and y are even; for B points, both x and y are odd.

b) For A points, $i = 0$ iff $y = 0 \bmod 4$, and $j = 0$ iff $x = 0 \bmod 4$; for B points, $i = 0$ iff $y = 1 \bmod 4$, and $j = 0$ iff $x = 1 \bmod 4$.

c) For A points, $x + y = 2i + 2j + 4k \bmod 8$; for B points, $x + y = 2i + 2j + 4k + 2 \bmod 8$.

Two sequences with points from different groups differ by at least 1 in every one of the 24 (x, y) coordinates. They cannot all differ only by 1, however, because of the following. The sum S of all coordinates satisfies

$$S = 2w_{ij} + 4w_k \text{ modulo } 8, \quad \text{for } A \text{ sequences;}$$
$$2w_{ij} + 4w_k \text{ modulo } 8, \quad \text{for } B \text{ sequences}$$

where w_{ij} is the number of (i, j) subscripts equal to 1, and w_k the number of k subscripts equal to 1. But since the (i, j) subscripts form a Golay code word and all such words have weights equal to integer multiples of 4, and since w_k is even for A sequences and odd for B sequences by construction,

$$S = 0 \text{ modulo } 8, \quad \text{for } A \text{ sequences;}$$
$$4 \text{ modulo } 8, \quad \text{for } B \text{ sequences.}$$

Now suppose that there were a B sequence that differed from an A sequence by ± 1 in every coordinate, and let m be the number of coordinates in which the difference was $+1$. Since the sum S_A of the A coordinates is $0 \bmod 8$ and the sum S_B of the B coordinates is $4 \bmod 8$, and $S_A - S_B = m - (24 - m) = 2m \bmod 8$, it follows that $2m = 4 \bmod 8$, or $m = 2 \bmod 4$. Now, the construction of the array is such that if a B point has an x coordinate 1 larger than the x coordinate of an A point, then the j subscript is the same, whereas if it is 1 smaller, then the j subscript is different; similarly a difference of $+1$ in y gives the same i subscript, while a difference of -1 gives the opposite one. Thus, $m = w_{ij}$. But, $w_{ij} = 0 \bmod 4$, so $m = 0 \bmod 4$; contradiction. Hence, any B sequence must differ from every A sequence by at least ± 3 in one coordinate, Q.E.D.

This lattice and its distance properties were originally discovered by Leech [29].

References

[1] C. E. Shannon and W. Weaver, *A Mathematical Theory of Communication*. Urbana, IL: Univ. Illinois Press, 1949.

[2] G. D. Forney, Jr., "Coding and its application in space communications," *IEEE Spectrum*, vol. 7, pp. 47–58, 1970.

[3] J. M. Wozencraft and R. S. Kennedy, "Modulation and demodulation for probabilistic coding," *IEEE Trans. Inform. Theory*, vol. IT-12, pp. 291–297, 1966.

[4] J. L. Massey, "Coding and modulation in digital communications," in *Proc. 1974 Int. Zurich Seminar Digital Commun.*, Zurich, Switzerland, pp. E2(1)–E2(4), 1974.

[5] J. M. Wozencraft and I. M. Jacobs, *Principles of Communication Engineering*. New York: Wiley, 1965.

[6] R. W. Lucky, J. Salz, and E. J. Weldon, Jr., *Principles of Data Communications*. New York: McGraw-Hill, 1968, ch. 3.

[7] C. R. Cahn, "Combined digital phase and amplitude modulation communication systems," *IRE Trans. Commun. Syst.*, vol. CS-8, pp. 150–154, 1960.

[8] J. C. Hancock and R. W. Lucky, "Performance of combined amplitude and phase-modulated communication systems," *IRE Trans. Commun. Syst.*, vol. CS-8, pp. 232–237, 1960.

[9] C. N. Campopiano and B. G. Glazer, "A coherent digital amplitude and phase modulation scheme," *IRE Trans. Commun. Syst.*, vol. CS-10, pp. 90–95, 1962.

[10] R. W. Lucky and J. C. Hancock, "On the optimum performance of N-ary systems having two degrees of freedom," *IRE Trans. Commun. Syst.*, vol. CS-10, pp. 185-192, 1962.
[11] J. H. Conway and N. J. A. Sloane, "Voronoi regions of lattices, second moments of polytopes, and quantization," *IEEE Trans. Inform. Theory*, vol. IT-28, pp. 211-226, 1982.
[12] J. H. Conway and N. J. A. Sloane, "A fast encoding method for lattice codes and quantizers," *IEEE Trans. Inform. Theory*, vol. IT-29, pp. 820-824, 1983.
[13] R. W. Lucky, "Digital phase and amplitude modulated communication systems," Ph.D. dissertation, Purdue Univ., Lafayette, IN, 1961 p. 89 (attributed to Cahn).
[14] G. J. Foschini, R. D. Gitlin, and S. B. Weinstein, "Optimization of two-dimensional signal constellations in the presence of Gaussian noise," *IEEE Trans. Commun.*, vol. COM-22, pp. 28-38, 1974.
[15] G. Ungerboeck, "Channel coding with multilevel/phase signals," *IEEE Trans. Inform. Theory*, vol. IT-28, pp. 55-67, 1982.
[16] J. Leech and N. J. A. Sloane, "Sphere packings and error-correcting codes," *Can. J. Math.*, vol. 23, pp. 718-745, 1971.
[17] G. R. Welti and J. S. Lee, "Digital transmission with coherent four-dimensional modulation," *IEEE Trans. Inform. Theory*, vol. IT-20, pp. 497-502, 1974.
[18] Canada, "An aspect of future modem development," CCITT Contrib. COM-Sp.A No. 142, 1974.
[19] N. J. A. Sloane, "Tables of sphere packings and spherical codes," *IEEE Trans. Inform. Theory*, vol. IT-27, pp. 327-338, 1981.
[20] P. deBuda, "Encoding and decoding algorithms for an optimal lattice-based code," in *ICC Conv. Rec.*, pp. 65.3. 1-65.3.5, 1981.
[21] Canada, "Block coding for improved modem performance," CCITT Contrib. COM-XVII No. 112, 1983.
[21a] E. L. Cusack, "Error control codes for QAM signalling," *Electron. Lett.*, vol. 20, pp. 62-63, 1984.
[22] J. H. Conway and N. J. A. Sloane, "Fast quantizing and decoding algorithms for lattice quantizers and codes," *IEEE Trans. Inform. Theory*, vol. IT-28, pp. 227-232, 1982.
[23] E. R. Berlekamp, "The technology of error-correcting codes," *Proc. IEEE*, vol. 68, pp. 564-593, 1980.
[24] G. D. Forney, Jr., "The Viterbi algorithm," *Proc. IEEE*, vol. 61, pp. 268-278, 1973.
[25] Rapporteur on 9600 bit/s duplex family modem, "Draft recommendation V.32 for a family of 2-wire, duplex modems operating at data signalling rates of up to 9600 bit/s for use on the general switched telephone network and on leased telephone-type circuits," CCITT Contrib. COM-XVII No. T30, Geneva, Switzerland, Mar. 1984; also see J. D. Brownlie and E. L. Cusack, "Duplex transmission at 4800 and 9600 bit/s on the general switched telephone network and the use of channel coding with a partitioned signal constellation," in *Proc. Zurich Int. Sem. Digital Commun.*, 1984.
[26] Rapporteur on 14 400 bit/s modem, "Working draft recommendation V.CC for a 14 400 bit/s modem standardized for use on point-to-point 4-wire leased telephone-type circuits," CCITT Study Group XVII, 1984.
[27] L. F. Wei, "Rotationally invariant convolutional channel coding with expanded signal space—Part II: Nonlinear codes," *IEEE J. Select. Areas Commun.*, vol. SAC-2, pp. xxx-xxx, Sept. 1984.
[28] S. G. Wilson and H. A. Sleeper, "Four-dimensional modulation and coding: an alternate to frequency reuse," Univ. Virginia, Rep. UVA/528200/EE83/107, Sept. 1983.
[28a] R. Fang and W. Lee, "Four-dimensionally coded PSK systems for combatting effects of severe ISI and CCI," in *Proc. 1983 IEEE Globecom Conv. Rec.*, pp. 30.4. 1-30.4.7, 1983.
[29] J. Leech, "Notes on sphere packings," *Can. J. Math.*, vol. 19, pp. 251-267, 1967.

PHOTO NOT AVAILABLE

G. David Forney, Jr. (S'58-M'61-F'73) was born in New York, NY, on March 6, 1940. He received the B.S.E. degree in electrical engineering from Princeton University, Princeton, NJ, in 1961, and the M.S. and Sc.D. degrees in electrical engineering from the Massachusetts Institute of Technology, Cambridge, MA, in 1963 and 1965, respectively.

In 1965, he joined the Codex Corporation, and became a Vice President and Director in 1970. Since 1982 he has been Vice President and Director of Technology and Planning of the Motorola Information Systems Group, Mansfield, MA. During the 1971-1972 academic year he was a Visiting Scientist at Stanford University, Stanford, CA, and he has also served as an Adjunct Professor at the Massachusetts Institute of Technology.

Dr. Forney was Editor of the IEEE TRANSACTIONS ON INFORMATION THEORY from 1970 to 1973 and has held numerous other IEEE positions. He was winner of the 1970 Information Theory Group Prize Paper Award and the 1972 Browder J. Thompson Memorial Prize Paper Award. He was elected to membership in the National Academy of Engineering in 1983.

PHOTO NOT AVAILABLE

Robert G. Gallager (S'58-M'61-F'68) received the S.B. degree in electrical engineering from the University of Pennsylvania, Philadelphia, PA, in 1953, and the S.M. and Sc.D. degrees in electrical engineering from the Massachusetts Institute of Technology, Cambridge, MA, in 1957 and 1960, respectively.

From 1953 to 1954 he was a Member of the Technical Staff at Bell Telephone Laboratories, and from 1954 to 1956 was in the Signal Corps of the U.S. Army. He has been with the Massachusetts Institute of Technology since 1956 and was Associate Chairman of the faculty from 1973 to 1975. He is currently a Professor in the Department of Electrical Engineering and Computer Science, Associate Director of the Laboratory for Information and Decision Systems, and Co-chairman of the Department Area I (Control, Communication, and Operations Research). He is a consultant to Codex Corporation, Western Union, and CNR. He is the author of the textbook *Information Theory and Reliable Communication* (New York: Wiley, 1968). His major research interests are data communication networks, information theory, and communication engineering.

Dr. Gallager was awarded the IEEE Baker Prize Paper Award in 1966 for the paper "A simple derivation of the coding theorem and some applications." He has been a member of the Administrative Committee of the IEEE Group on Information Theory, from 1965 to 1970 and from 1979 to the present, was Chairman of the Group in 1971, and was elected a member of the National Academy of Engineering in 1979.

PHOTO NOT AVAILABLE

Gordon R. Lang was born on September 7, 1922, in London, Ont., Canada. He received the B.Sc. degree (Hons.) in mathematics and physics, radio physics option, from the University of Western Ontario, London, in 1949.

He was with Ferranti Electric Ltd. from 1949 to 1965. He was a founding member of ESE Limited (now Motorola Information Systems Ltd.) in January 1965 and became a director of the company. He was appointed Associate Professor with the Department of Electrical Engineering, University of Toronto, Canada, in January 1965. He increased activity at ESE Limited as Director of Research in 1969 and became an Adjunct Professor with the Department of Electrical Engineering, University of Toronto. He is author or co-author of over 60 published papers plus a variety of conference papers. He has obtained approximately 16 patents. Presently, he is Vice-President of Research and Development at Motorola Information Systems Ltd.

Dr. Lang is a member of the Association of Professional Engineers of Ontario.

PHOTO NOT AVAILABLE

Fred M. Longstaff was born March 3, 1926, in Thornloe, Ont., Canada. He received the M.A.Sc. degree in engineering physics from the University of Toronto, Canada, in 1951.

He was with Ferranti Electric Ltd. from 1949 to 1965. He was a founding member of ESE Limited in January 1965 and became a Director of the company. He has had extensive experience since the early 1950's in the architecture of computer systems and in computer software, leading to the position since 1965 of Vice-President of Systems and Software Development with ESE Limited (now Motorola Information Systems Ltd.), Rexdale, Ont.

Dr. Longstaff is a member of the Association of Professional Engineers of Ontario and the Association for Computing Machinery.

PHOTO NOT AVAILABLE

Shahid V. Qureshi (S'68–M'73–SM'81) was born in Peshawar, Pakistan, on September 22, 1945. He received the B.Sc. degree from the University of Engineering and Technology, Lahore, Pakistan, in 1967, the M.Sc. degree from the University of Alberta, Edmonton, Canada, in 1970 and the Ph.D. degree from the University of Toronto, Toronto, Canada, in 1973, all in electrical engineering.

From 1967 to 1968 he was a Lecturer with the Department of Electrical Engineering, University of Engineering and Technology, Lahore, Pakistan. He held the Canadian Commonwealth Scholarship from 1968 to 1972.

Since 1973 he has been with Codex Corporation, where he is currently Senior Director of Modulation Products Research. His interests include data communication, computer architecture for signal processing and the application of digital signal processing to communication.

Optimum Combining in Digital Mobile Radio with Cochannel Interference

JACK H. WINTERS, MEMBER, IEEE

Abstract —This paper studies optimum signal combining for space diversity reception in cellular mobile radio systems. With optimum combining, the signals received by the antennas are weighted and combined to maximize the output signal-to-interference-plus-noise ratio. Thus, with cochannel interference, space diversity is used not only to combat Rayleigh fading of the desired signal (as with maximal ratio combining) but also to reduce the power of interfering signals at the receiver. We use analytical and computer simulation techniques to determine the performance of optimum combining when the received desired and interfering signals are subject to Rayleigh fading. Results show that optimum combining is significantly better than maximal ratio combining even when the number of interferers is greater than the number of antennas. Results for typical cellular mobile radio systems show that optimum combining increases the output signal-to-interference ratio at the receiver by several decibels. Thus, systems can require fewer base station antennas and/or achieve increased channel capacity through greater frequency reuse. We also describe techniques for implementing optimum combining with least mean square (LMS) adaptive arrays.

I. Introduction

SPACE diversity provides an attractive means for improving the performance of mobile radio systems. With space diversity, the signals from the receiving antennas can be combined to combat multipath fading of the desired signal and reduce the relative power of interfering signals.

Previous studies of mobile radio systems (e.g., [1]) have considered space diversity only for combating multipath fading of the desired signal. Interference at each receiving antenna is assumed to be independent in these studies. Under this condition, maximal ratio combining[1] [1, p. 316] produces the highest output signal-to-interference-plus-noise ratio (SINR) at the receiver. However, in most systems (in particular, cellular mobile radio systems [1]) the same interfering signals are present at each of the receiving antennas. Thus, the received signals can be combined to suppress these interfering signals in addition to combating desired signal fading and thereby achieve higher output SINR than maximal ratio combining.

The output SINR can be maximized by using adaptive array techniques at the receiver (e.g., [2]–[4]). We will not analyze the performance of the various adaptive array techniques in this paper, but only study the performance of the optimum combiner that maximizes the output SINR. Although the adaptive array (i.e., optimum combiner) has been studied extensively, it has not been previously analyzed with the fading conditions of digital mobile radio.

This paper studies the performance of the optimum combiner in digital mobile radio systems. We assume flat Rayleigh fading across the signal channel and independent fading between antennas. The average bit error rate (BER) of the optimum combiner is studied for coherent detection of phase shift keyed (PSK) signals. Analytical and computer simulation results show that optimum combining is significantly better than maximal ratio combining even when there are more interferers than receive antennas. Results for typical cellular mobile radio systems show that the optimum combiner can increase the output SINR several decibels more than maximal ratio combining.

In Section II we describe the optimum combiner. Section III studies the BER of the optimum combiner when the desired and interfering signals are subject to Rayleigh fading. We discuss analytical results for one interferer and Monte Carlo simulation results for multiple interferers. In Section IV we consider the optimum combiner performance in cellular mobile radio systems. Section V discusses the possible methods for implementing the optimum combiner in mobile radio with a least mean square (LMS) [3] adaptive array. A summary and conclusions are presented in Section VI.

II. Optimum Combiner

A. Description and Weight Equation

Fig. 1 shows a block diagram of an M element space diversity combiner. The signal received by the ith element $y_i(t)$ is split with a quadrature hybrid into an in-phase signal $x_{I_i}(t)$ and a quadrature signal $x_{Q_i}(t)$. These signals are then multiplied by a controllable weight $w_{I_i}(t)$ or $w_{Q_i}(t)$. The weighted signals are then summed to form the array output $s_o(t)$.

The space diversity combiner can be described mathematically using complex notation [5]. Let the weight vector $\mathbf{w}$ be given by

Manuscript received February 25, 1983; revised November 23, 1983. This paper was presented at the International Conference on Communications, Boston, MA, June 1983.

The author is with the Radio Research Laboratory, AT&T Bell Laboratories, Holmdel, NJ 07733.

[1] In maximal ratio combining, the received signals are weighted proportionately to their signal-voltage-to-noise-power ratios and combined in phase.

Reprinted from *IEEE Journal on Selected Areas in Communications,* vol. SAC-2, no. 4, July 1984.

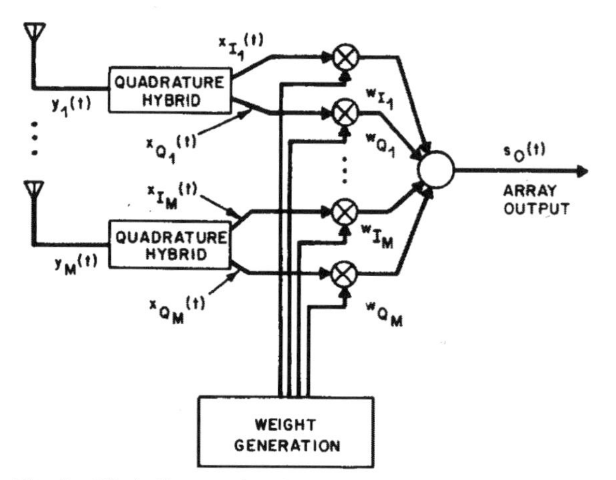

Fig. 1. Block diagram of an M element space diversity combiner.

$$\mathbf{w} = \begin{bmatrix} w_{I_1} \\ \vdots \\ w_{I_M} \end{bmatrix} - j \begin{bmatrix} w_{Q_1} \\ \vdots \\ w_{Q_M} \end{bmatrix} \tag{1}$$

and the received signal vector $\mathbf{x}$ be given by

$$\mathbf{x} = \begin{bmatrix} x_{I_1} \\ \vdots \\ x_{I_M} \end{bmatrix} + j \begin{bmatrix} x_{Q_1} \\ \vdots \\ x_{Q_M} \end{bmatrix}. \tag{2}$$

The received signal consists of the desired signal, thermal noise, and interference and, therefore, can be expressed as

$$\mathbf{x} = \mathbf{x}_d + \mathbf{x}_n + \sum_{j=1}^{L} \mathbf{x}_j \tag{3}$$

where $\mathbf{x}_d$, $\mathbf{x}_n$, and $\mathbf{x}_j$ are the received desired signal, noise, and jth interfering signal vectors, respectively, and L is the number of interferers. Furthermore, let $s_d(t)$ and $s_j(t)$ be the desired and jth interfering signals as they are transmitted, respectively, with

$$E\left[s_d^2(t)\right] = 1 \tag{4}$$

and

$$E\left[s_j^2(t)\right] = 1 \qquad \text{for } 1 \leqslant j \leqslant L. \tag{5}$$

Then $\mathbf{x}$ can be expressed as

$$\mathbf{x} = \mathbf{u}_d s_d(t) + \mathbf{x}_n + \sum_{j=1}^{L} \mathbf{u}_j s_j(t) \tag{6}$$

where $\mathbf{u}_d$ and $\mathbf{u}_j$ are the desired and jth interfering signal propagation vectors, respectively.

The received interference-plus-noise correlation matrix is given by

$$\mathbf{R}_{nn} = E\left[\left(\mathbf{x}_n + \sum_{j=1}^{L} \mathbf{x}_j\right)^* \left(\mathbf{x}_n + \sum_{j=1}^{L} \mathbf{x}_j\right)^T\right] \tag{7}$$

where the superscripts * and T denote conjugate and transpose, respectively. Assuming the noise and interfering signals are uncorrelated, we can show that

$$\mathbf{R}_{nn} = \sigma^2 \mathbf{I} + \sum_{j=1}^{L} E\left[\mathbf{u}_j^* \mathbf{u}_j^T\right] \tag{8}$$

where σ^2 is the noise power and $\mathbf{I}$ is the identity matrix. In (8) the expected value is taken over a period much less than the reciprocal of the fading rate (e.g., several bit intervals). Note that we have assumed that the fading rate is much less than the bit rate.

Finally, the equation for the weights that maximize the output SINR is (from [6])

$$\mathbf{w} = \alpha \mathbf{R}_{nn}^{-1} \mathbf{u}_d^* \tag{9}$$

where α is a constant,[2] and the superscript -1 denotes the inverse of the matrix.

B. Discussion

The mobile radio environment is quite different from the signal environment in which adaptive arrays (i.e., optimum combiners) are usually employed. In a typical adaptive array application in a nonfading environment, at the receiver there are only a few interfering signals, and their power is much greater than that of the desired signal. The adaptive array places nulls in the antenna pattern in the direction of these interferers, greatly suppressing these signals in the array output. In general, an M element array can null up to $M-2$ interfering signals and still optimize desired signal reception. The output SINR is, therefore, substantially increased by the array.

In mobile radio systems, on the other hand, at the receiver there can be several interfering signals whose power is close to that of the desired signal, and numerous interfering signals whose power is much less than that of the desired signal. Therefore, the number of interfering signals may be much greater than M, and the array may not be able to greatly suppress every interfering signal. Thus, the array output SINR may not be markedly increased by the array.

To be useful in mobile radio systems, however, the adaptive array does not have to greatly suppress interfering signals or vastly increase the output SINR. Interfering signals need only to be reduced in power by a few decibels so that their power is below the sum of the power of other interferers. Furthermore, a substantial output SINR improvement is not required because a several decibel increase in output SINR can make possible large increases in the channel capacity of the system. Thus, although the signal environment in mobile radio systems is quite different from that of the typical adaptive array system, adaptive array techniques can still offer significant advantages.

One other major difference is the ability of the array to resolve closely spaced transmitters. In a nonfading environment the adaptive array cannot suppress an interfering

[2] Note that α does not affect the array output SINR (i.e., the array performance) and, therefore, we will not consider its value.

signal if the angular separation between the interfering and desired transmiters is too small. In this case, the desired signal phase difference between receive antennas is nearly the same as that for the interfering signal. Therefore, the array cannot both null one signal and enhance reception of the other. The use of additional antennas results in a small decrease in the required angular separation, but the problem remains.

In mobile radio, because of multipath, the signal phase at one antenna is independent of the signal phase at another antenna when the antenna separation is greater than half of a wavelength (several inches at 800 MHz) [1, p. 311].[3] Therefore, the adaptive array antenna pattern is meaningless. Similarly, at the receive antennas, the signal phases from two different transmit antennas are independent when the transmit antennas are more than half of a wavelength apart. Therefore, for all practical purposes, the received signal phases are independent of vehicle location. Thus, the resolution of interfering and desired signals does not depend on how closely the vehicles are located. Instead, for all locations, there is a small probability that the array cannot resolve the two signals. This occurs when the phase differences between the antennas are nearly the same for both the desired and interfering signals. With moving mobiles,[4] however, the period of time with unresolved signals is very brief, and the performance of the adaptive array can be averaged over the fading. Furthermore, since the signal phase differences between the antennas are independent, the probability of unresolved signals with M antennas is approximately equal to the probability of unresolved signals with two antennas raised to the $M-1$ power. Thus, each additional antenna greatly decreases the probability of unresolved signals, and this probability becomes negligible with only a few antennas.

In summary, in mobile radio the adaptive array cannot resolve desired and interfering signals a small percentage of the time, rather than over a given angular separation. Therefore, we need not be concerned with vehicle location for the resolving of the signals. In the following analysis we study the optimum combiner performance averaged over the Rayleigh fading.

III. Optimum Combiner Performance with Fading

This section studies the performance of the optimum combiner when the received signals are subject to fading. For the analysis we assume flat fading across the channel, with the fading independent between antennas. We assume that the received signal has an envelope with a Rayleigh distribution and a phase with a uniform distribution. We study the steady state performance of the optimum combiner, first determining the output SINR distribution and then the BER for coherent detection of PSK. All results are compared to those for maximal ratio combining.

In cellular mobile radio, the interference plus noise at the receiver consists primarily of cochannel interference. In a typical system, there are numerous cochannel interfering signals, each of which affects the performance of the optimum combiner. An exact analysis of the performance is, therefore, quite complicated, especially since, with fading, each of these signals has a random amplitude. Therefore, in the analysis in this paper, we consider only the strongest interferers individually. The remaining interfering signals are combined and considered as lumped interference that is uncorrelated between antennas. Since with Rayleigh fading the in-phase and quadrature components of each of the received interfering signals have a Gaussian distribution, the components of the sum of these signals also have a Gaussian distribution, and the sum can be considered as thermal noise. Thus, under this assumption the combiner cannot suppress the lumped interference, and we are therefore analyzing a worse situation since the actual combiner performance will be better. Therefore, although this analysis does not show the maximum improvement (over maximal ratio combining) with optimum combining, it does show most of it. Note that if we consider all the interference as lumped interference, the results are identical to those for maximal ratio combining.

The analysis involves several parameters which are defined as follows:

$$\Gamma = \frac{\text{mean received desired signal power per antenna}}{\text{mean received noise plus interference power per antenna}} \tag{10}$$

$$\Gamma_d = \frac{\text{mean received desired signal power per antenna}}{\text{mean received noise power per antenna}} \tag{11}$$

$$\Gamma_j = \frac{\text{mean received } j\text{th interferer signal power per antenna}}{\text{mean received noise power per antenna}} \tag{12}$$

$$\gamma_R = \frac{\text{local mean desired signal power at the array output}}{\text{mean noise plus interference power at the array output}} \tag{13}$$

and

$$\gamma = \frac{\text{local mean desired signal power at the array output}}{\text{local mean noise plus interference power at the array output}} \tag{14}$$

In the above definitions, mean is the average over the Rayleigh fading, and local mean is the average over a period less than the reciprocal of the fading rate (e.g., several bit durations). It is useful to note that

[3] In our analysis we assume the antennas are spaced far enough apart so that the received signal phases are independent.

[4] If all mobiles are stationary, channel reassignment can be used to eliminate the problem.

$$\Gamma = \frac{\Gamma_d}{1 + \sum_{j=1}^{L} \Gamma_j}. \tag{15}$$

A. Analytical Results with One Interferer

We first consider γ_R with one interferer. γ_R can be determined from [6, weight equation (9)]

$$\gamma_R = \mathbf{u}_d^T \mathbf{R}_{nn}^{-1} \mathbf{u}_d^* \tag{16}$$

where from (8)

$$\mathbf{R}_{nn} = \sigma^2 \mathbf{I} + E\left[\mathbf{u}_1^* \mathbf{u}_1^T\right]. \tag{17}$$

We note that with our fading model the components of $\mathbf{u}_d$ and $\mathbf{u}_1$ are complex Gaussian random variables that vary at the fading rate. Thus, to determine γ_R, the expected value in (17) must be averaged over the Rayleigh fading. It can also be seen that γ_R will vary at the fading rate.

The probability density function of γ_R can be calculated to be given by [7]

$$p(\gamma_R) = \frac{e^{-\gamma_R/\Gamma_d}\left(\frac{\gamma_R}{\Gamma_d}\right)^{M-1}(1+M\Gamma_1)}{\Gamma_d(M-2)!} \cdot \int_0^1 e^{-((\gamma_R/\Gamma_d)M\Gamma_1)t}(1-t)^{M-2}\,dt. \tag{18}$$

Thus, the cumulative distribution function is given by

$$P(\gamma_R) = \int_0^{\gamma_R/\Gamma_d} \frac{e^{-x}x^{M-1}(1+M\Gamma_1)}{(M-2)!} \cdot \int_0^1 e^{-xM\Gamma_1 t}(1-t)^{M-2}\,dt\,dx. \tag{19}$$

In the above equations it is seen that γ_R can be normalized by Γ_d. Therefore, from (15) we can also normalize γ_R by Γ and compare the performance of optimum combining to that of maximal ratio combining for fixed average received SINR.

Fig. 2 shows the cumulative distribution function of γ_R versus γ_R/Γ with optimum combining for several values of M and Γ_1. The $\Gamma_1 = 0$ distribution curve is also the distribution curve for maximal ratio combining. Fig. 2 shows that for fixed average received SINR the distribution function decreases as the interference power becomes a larger proportion of the total noise-plus-interference power. The decrease becomes even greater as M increases. Thus, optimum combining improves the receiver's performance the most when the interferer's power is large compared to the thermal noise power and there are several antennas.

As in [7] let us now consider the effect of a very high power interferer on optimum combining. Without interference (or with maximal ratio combining) $\Gamma_1 = 0$, and the cumulative distribution function is given by

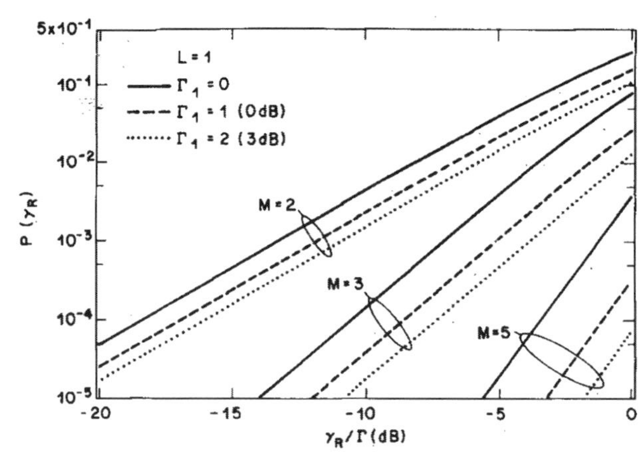

Fig. 2. The cumulative distribution function of γ_R versus γ_R/Γ for optimum combining when the desired and interfering signals are subject to fading. Results are shown for one interferer with several values of M and Γ_1. The distribution function for γ_R with fixed average received SINR is shown to decrease as the power of the interferer becomes a larger proportion of the total noise plus interference power. The decrease is even larger as M increases.

$$P(\gamma_R) = \int_0^{\gamma_R/\Gamma_d} \frac{e^{-x}x^{M-1}}{(M-1)!}\,dx \tag{20}$$

or

$$P(\gamma_R) = 1 - e^{-\gamma_R/\Gamma_d} \sum_{k=1}^{M} \frac{\left(\frac{\gamma_R}{\Gamma_d}\right)^{k-1}}{(k-1)!} \tag{21}$$

which agrees with [1, p. 319]. With a high power interferer $\Gamma_1 = \infty$, and the cumulative distribution function is given by

$$P(\gamma_R) = \int_0^{\gamma_R/\Gamma_d} \frac{e^{-x}x^{M-2}}{(M-2)!}\,dx \tag{22}$$

or

$$P(\gamma_R) = 1 - e^{-\gamma_R/\Gamma_d} \sum_{k=1}^{M-1} \frac{\left(\frac{\gamma_R}{\Gamma_d}\right)^{k-1}}{(k-1)!}. \tag{23}$$

Thus, optimum combining with an infinitely strong interferer gives the same results as maximal ratio combining without the interferer and with one less antenna. In other words, optimum combining with a strong interferer and an additional antenna will always do better than maximal ratio combining without the interferer.

For digital mobile radio this has the following impact. Let us consider a system where the required system performance can be achieved with maximal ratio combining at the base station receiver and adaptive retransmission (see Section V-B) for base-to-mobile transmission. Then, another mobile can be added per channel per cell by using optimum combining and adding one antenna at the base station. Thus, optimum combining provides a relatively simple means for growth in a system.

The BER for coherent detection of PSK is given by

$$\mathrm{BER} = \int_0^{\infty} p(\gamma_R) \frac{1}{2} \operatorname{erfc}\left(\sqrt{\Gamma \gamma_R}\right) d\gamma_R. \tag{24}$$

From (18) and (24) the BER for optimum combining with one interferer can be calculated to be given by [7]

$$\mathrm{BER} = \frac{(-1)^{M-1}(1+M\Gamma_1)}{2(M\Gamma_1)^{M-1}} \left\{ -\frac{M\Gamma_1}{1+M\Gamma_1} + \sqrt{\frac{\Gamma_d}{1+\Gamma_d}} - \frac{1}{1+M\Gamma_1}\sqrt{\frac{\Gamma_d}{1+M\Gamma_1+\Gamma_d}} - \sum_{k=1}^{M-2} (-M\Gamma_1)^k \cdot \left[1 - \sqrt{\frac{\Gamma_d}{1+\Gamma_d}}\left(1 + \sum_{i=1}^{k} \frac{(2i-1)!!}{i!(2+2\Gamma_d)^i}\right)\right]\right\} \tag{25}$$

where

$$(2i-1)!! = 1 \cdot 3 \cdot 5 \cdot \cdots \cdot (2i-1). \tag{26}$$

Similarly, for maximal ratio combining the BER can be determined from (21) and (24) as [8]

$$\mathrm{BER} = 2^{-M}\left(1 - \sqrt{\frac{\Gamma}{1+\Gamma}}\right)^M \sum_{k=0}^{M-1} \binom{M-1+k}{k} \cdot 2^{-k}\left(1 + \sqrt{\frac{\Gamma}{1+\Gamma}}\right)^k. \tag{27}$$

Fig. 3 shows the BER versus the average received SINR (Γ) for optimum combining with one interferer. Results are shown for several values of M and Γ_1. The results for $\Gamma_1 = 0$ are the same as those for maximal ratio combining.

Let us define the optimum combining improvement as the decrease (in decibels) in the average received SINR required for a given BER as compared to maximal ratio combining. Fig. 3 shows that the improvement is nearly independent of Γ for BER's less than 10^{-2}. Thus, for most systems of interest, the improvement is independent of Γ_d and depends only on Γ_1 and M. That is, the improvement depends on the interferer's power relative to the combined power of the other interferers and not the power of the desired signal.

In Fig. 4 the optimum combining improvement is plotted versus Γ_1 for one interferer and several values of M. The results are shown for a BER of 10^{-3}, but as discussed above, similar results can be obtained for other BER values (less than 10^{-2}). Results show that as Γ_1 and M increase, the improvement also increases.

Fig. 4 also shows the maximum improvement that can be achieved if the interferer is completely nulled in the array output. The difference between the maximum and the actual improvement for a given M ($M > 2$) is shown to be constant for large Γ_1. From (23) it can be seen that this

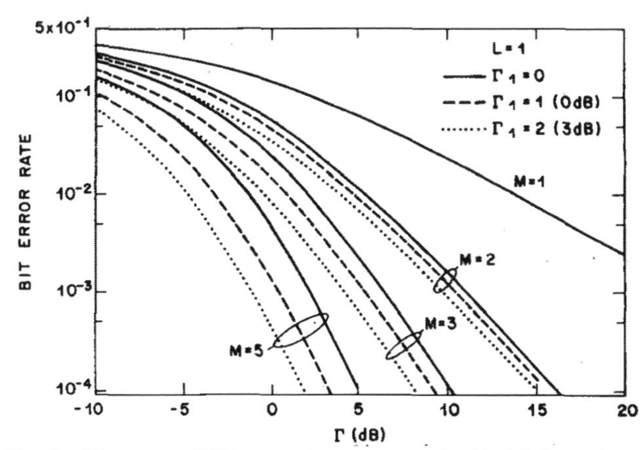

Fig. 3. The average BER versus the average received SINR for optimum combining with one interferer. Results are shown for several values of M and Γ_1. The improvement with optimum combining (in decibels) is shown to be nearly independent of the average received SINR for BER's less than 10^{-2}.

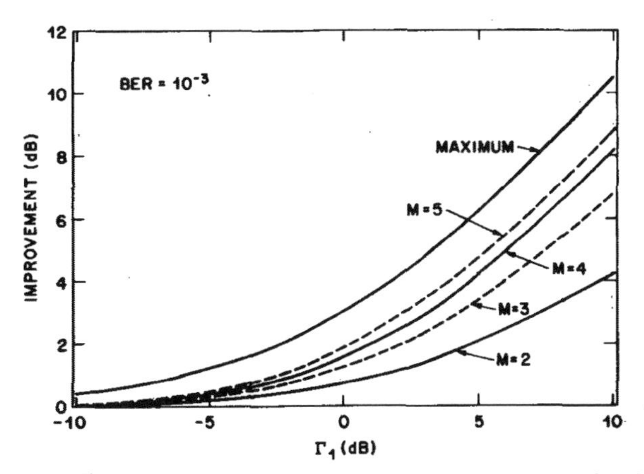

Fig. 4. The improvement of optimum combining over maximal ratio combining versus Γ_1 with one interferer for several values of M and a 10^{-3} BER. Results show that as Γ_1 and M increase, the improvement becomes significant.

difference is the increase in the required received SINR with the loss of one antenna with maximal ratio combining. For example, the required received SINR with maximal ratio combining is 2.3 dB for $M = 5$ and 4.0 dB for $M = 4$. Thus, for $M = 5$ the optimum combining improvement is 1.7 dB less than maximum for large Γ_1.

B. Simulation Results with Multiple Interferers

With two or more interferers, it is extremely difficult to determine analytically the optimum combiner performance. Therefore, in this section we use Monte Carlo simulation to determine the performance of optimum combining. For the simulation we consider γ rather than γ_R as in the previous section.

For a given bit duration, the array output SINR is given by

$$\gamma = \frac{P_d}{P_{i+n}} \tag{28}$$

where P_d is the power of the desired signal and P_{i+n} is the power of the interference plus noise. The desired signal power is given by

$$P_d = \frac{1}{2}|\mathbf{w}^\dagger \mathbf{u}_d^*|^2 \tag{29}$$

where the superscript † denotes complex conjugate transpose. The power of the interference plus noise is given by

$$P_{i+n} = \frac{1}{2}\left|\mathbf{w}^\dagger\left[\mathbf{u}_n + \sum_{j=1}^{L}\mathbf{u}_j\right]^*\right|^2 \tag{30}$$

where $\mathbf{u}_n$ is the vector of the thermal noise vectors at the receiver. The components of $\mathbf{u}_n$ are independent complex Gaussian random variables with a variance corresponding to the noise power. The array output SINR is then given by

$$\gamma = \frac{|\mathbf{w}^\dagger \mathbf{u}_d^*|^2}{\left|\mathbf{w}^\dagger\left(\mathbf{u}_n + \sum_{j=1}^{L}\mathbf{u}_j\right)^*\right|^2}. \tag{31}$$

The weight vector for the optimum combiner is given in (9). As can be seen from (8), $\mathbf{R}_{nn}$ is Hermitian, and therefore, from (9)

$$\mathbf{w}^\dagger = \alpha \mathbf{u}_d^T \mathbf{R}_{nn}^{-1}. \tag{32}$$

Thus, the SINR can be expressed as

$$\gamma = \frac{|\mathbf{u}_d^T \mathbf{R}_{nn}^{-1} \mathbf{u}_d^*|^2}{\left|\mathbf{u}_d^T \mathbf{R}_{nn}^{-1}\left(\mathbf{u}_n + \sum_{j=1}^{L}\mathbf{u}_j\right)^*\right|^2}. \tag{33}$$

With Rayleigh fading, the components of $\mathbf{u}_d$, $\mathbf{u}_n$, and $\mathbf{u}_j$ are complex Gaussian random variables with zero mean and variance Γ_d, σ^2, and Γ_j, respectively. Therefore, through Monte Carlo simulation, the probability distribution of γ can be determined.

We now discuss the distribution of γ with maximal ratio combining so that a comparison to optimum combining can be made. For maximal ratio combining, the weights are given by

$$\mathbf{w} = \mathbf{u}_d^* \tag{34}$$

or

$$\mathbf{w}^\dagger = \mathbf{u}_d^T. \tag{35}$$

Therefore, from (31) the SINR is given by

$$\gamma = \frac{|\mathbf{u}_d^T \mathbf{u}_d^*|^2}{\left|\mathbf{u}_d^T\left(\mathbf{u}_n + \sum_{j=1}^{L}\mathbf{u}_j\right)^*\right|^2}. \tag{36}$$

Since the components of $\mathbf{u}_n + \Sigma_{j=1}^{L}\mathbf{u}_j$ are the sum of independent complex Gaussian random variables, the components are also independent complex Gaussian random variables. Thus, the probability density function of γ can be determined analytically to be given by [1, p. 367]

$$p(\gamma) = \frac{M\left(\frac{\gamma}{\Gamma}\right)^{M-1}}{\Gamma\left(1+\frac{\gamma}{\Gamma}\right)^{M+1}} \tag{37}$$

and the cumulative distribution function is given by

$$P(\gamma) = \left(\frac{\frac{\gamma}{\Gamma}}{1+\frac{\gamma}{\Gamma}}\right)^M. \tag{38}$$

The cumulative distribution of γ is plotted versus γ/Γ in Fig. 5. Simulation results with 100 000 samples are shown for optimum combining with two interferers that have 3 dB higher power than the noise, and analytical results are shown for maximal ratio combining. With 100 000 samples there are small deviations in the simulation results only for very small values of the distribution function. Fig. 5 shows that optimum combining significantly decreases the value of the distribution function as compared to maximal ratio combining. This decrease becomes even greater as M increases.

The BER can be determined from the cumulative distribution function by the equation

$$\text{BER} = \frac{1}{\pi}\int_0^1 \cos^{-1}(\sqrt{\gamma})\, p(\gamma)\, d\gamma \tag{39}$$

or

$$\text{BER} = \frac{1}{\pi}\int_0^1 \cos^{-1}(\sqrt{\gamma})\left(\frac{dP(\gamma)}{d\gamma}\right) d\gamma. \tag{40}$$

Thus, the BER for optimum combining was determined from the simulation results using the above equation. Since the cumulative distribution function can be determined for γ normalized by Γ, from one simulation run we can determine the BER over a wide range of Γ's. Similarly, the BER for maximal ratio combining is seen from (37) and (39) to be given by

$$\text{BER} = \frac{M}{\pi}\int_0^{1/\Gamma} \cos^{-1}(\sqrt{\Gamma x})\frac{x^{M-1}}{(1+x)^{M+1}}\, dx \tag{41}$$

which is numerically equivalent to (27).

For one interferer, the BER results obtained using the above equations (with a simulation using 100 000 samples) agree with the analytical results shown in Fig. 3.

For two interferers, the BER results are shown in Fig. 6. The simulation used 100 000 samples per data point. Fig. 6 shows that there is a marked improvement with optimum combining as the number of antennas increases. For example, for a BER of 10^{-3} and M equal to 5, optimum combining requires 4.2 dB less SINR than maximal ratio combining. Thus, in this case, optimum combining with five antennas (which requires -1.9 dB for a 10^{-3} BER) is

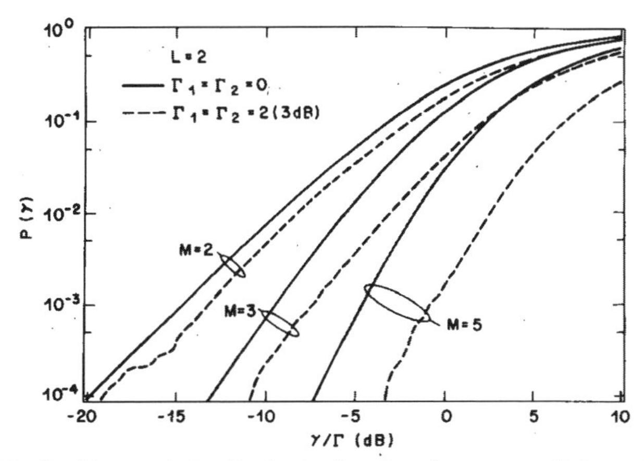

Fig. 5. The cumulative distribution function of γ versus γ/Γ for optimum combining with two interferers that have 3 dB higher power than the noise. Analytical results for maximal ratio combining ($\Gamma_1 = \Gamma_1 = 0$) are also shown. Optimum combining is seen to significantly decrease the distribution function.

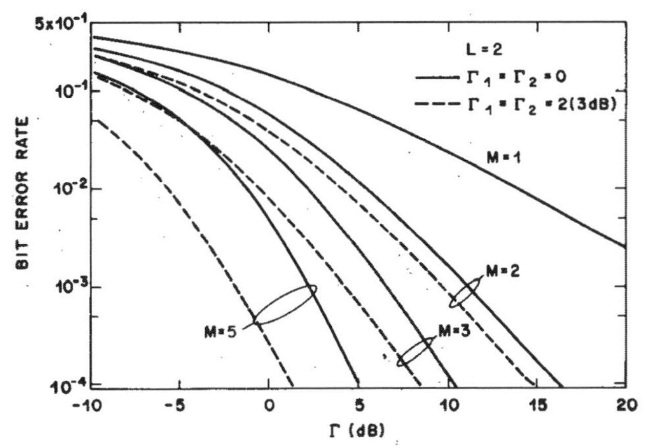

Fig. 6. The average BER versus the average received SINR for optimum combining with two interferers. Results are shown for optimum combining with two equal power interferers ($\Gamma_1 = \Gamma_2 = 2$) and for maximal ratio combining ($\Gamma_1 = \Gamma_2 = 0$). There is a marked improvement with optimum combining as the number of antennas increases.

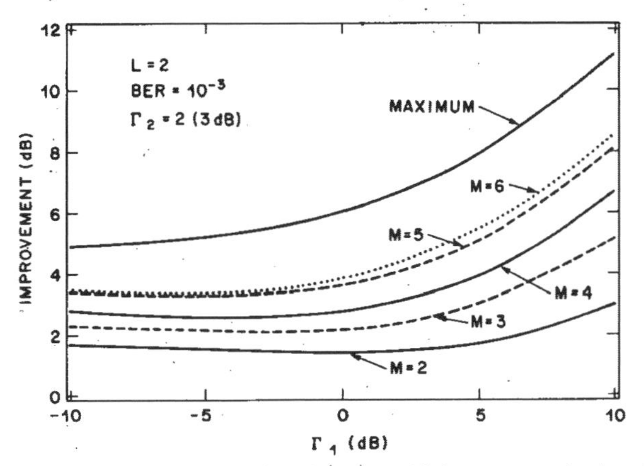

Fig. 7. The improvement of optimum combining over maximal ratio combining versus the signal-to-noise ratio of one interferer when there is also a second interferer with a 3 dB signal-to-noise ratio. The improvement is within about 2 dB of the maximum improvement with six or more antennas.

better than maximal ratio combining with nine antennas (which, from (27), requires -1.7 dB for a 10^{-3} BER).

Fig. 7 shows the optimum combiner improvement over maximal ratio combining versus the signal-to-noise ratio of one interferer when there is also a second interferer with a 3 dB signal-to-noise ratio. The simulation used 100 000 samples per data point. The results are shown for a 10^{-3} BER, but as seen in Fig. 6, these results are similar to the results for other BER's less than 10^{-2}. Fig. 7 also shows the maximum improvement possible if both interfering signals are completely nulled in the receiver output (i.e., the difference between the maximal ratio combiner performance with and without interference). The improvement is within about 2 dB of the maximum with six or more antennas.

Fig. 8 shows the improvement versus the number of antennas with one to six equal power ($\Gamma_j = 3$ dB) interferers. Again, 100 000 samples per data point were used. The improvement is shown to be between 1–6 dB as M varies from 2 to 8. Thus, optimum combining has some improvement over maximal ratio combining even with a few antennas, and the improvement greatly increases with the number of antennas.

Although the results of Fig. 8 are for equal power interferers with a particular value of Γ_j, they demonstrate the following characteristics of optimum combining that apply to other interference cases as well. First, when the number of antennas is much greater than the number of interferers, the improvement is limited. That is, in this case there is little improvement (relative to maximal ratio combining) with additional antennas. This can also be seen from Figs. 4 and 7. Second, except for the above case, the increase in the improvement (in decibels) with each additional antenna is approximately constant (about 0.6 dB for $\Gamma_j = 3$ dB). Finally, the most interesting characteristic is that there is a large improvement even when the number of interferers is greater than the number of antennas. This implies that in analyzing systems we must consider many interferers individually even if there are only a few antennas. For example, consider the case of five antennas with six interferers, each with Γ_j equal to 3 dB. From Fig. 8, the improvement is 2.7 dB. However, if only five interferers are considered individually, and the power of the sixth one is combined with the thermal noise, Γ_j is -1.8 dB and the improvement is only 1.6 dB. Thus, we must consider individually as many interferers as possible to determine accurately the actual optimum combining improvement.

IV. Performance in Typical Systems

This section studies the performance of optimum combining in typical cellular mobile radio systems. Using the techniques of Section III, we study optimum combining when the signals are subject to Rayleigh fading.[5] Optimum combining is studied only at the base station receiver because multiple antennas and the associated signal processing for optimum combining are less costly to implement at the base station than on numerous mobiles. (Adaptive retransmission with time division [1], [9] can be used to

[5] In an actual mobile radio system, the signals are also subject to shadow fading [9] which greatly complicates analysis. We therefore only consider Rayleigh fading so that system comparisons can easily be made.

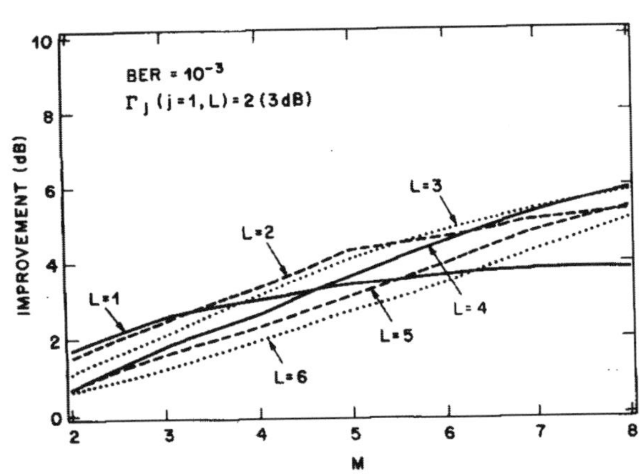

Fig. 8. The improvement of optimum combining versus the number of antennas with one to six equal power interferers. The improvement is between 1–6 dB as M varies from 2 to 8.

improve reception at the mobile with multiple base station antennas only (see Section V-B).) As before, all results for optimum combining are compared to maximal ratio combining.

Analysis of optimum combining with numerous interferers requires a substantial amount of computer time. It is therefore nearly impossible to determine the average performance of the adaptive array in the typical cellular system with random mobile locations. Therefore, in this section we consider a worst case scenario only, i.e., the mobile transmitting the desired signal is at the point in the cell farthest from the base station, and the interfering mobiles in the surrounding cells are as close as possible to the base station of the desired mobile. Furthermore, in the analysis we consider only the six strongest interferers individually. The power of the other interferers is combined and considered as thermal noise.

The systems studied involve two different cell geometries with hexagonal cells. In one geometry the base stations are located at the cell center, and in the other geometry the base stations are at the three alternate corners of the cell and are equipped with sectoral horns. In the latter geometry, each of the base station's three antennas has a 120° beamwidth and serves the three adjoining cells. We also consider both frequency reuse in every cell and the use of three channel sets. Furthermore, because in the typical system the signal strength falls with the inverse of the distance raised to between the third and fourth power, we also consider these two extremes.[6]

The performance of optimum combining and maximal ratio combining in typical mobile radio systems is shown in Table I. For each of the systems described above, Table I lists the number of antennas required to achieve a 10^{-3} BER and the average output SINR margin. We also show the margin with an additional antenna.

The results show that with three-corner base station geometry and frequency reuse in every cell, optimum combining more than halves the required number of antennas. Furthermore, the increase in margin with an additional

[6]The calculation of the power of the signals in these cellular systems will not be described here. The method is similar to that described in [10].

TABLE I
COMPARISON OF OPTIMUM AND MAXIMAL RATIO COMBINING IN TYPICAL MOBILE RADIO SYSTEMS—THE NUMBER OF ANTENNAS REQUIRED AND THE SINR MARGIN FOR A 10^{-3} BER

Case			Maximal Ratio Combining		Optimum Combining	
Base Station Geometry	Frequency Reuse	Decay Exponential	Number of Antennas	SINR Margins[a] (dB)	Number of antennas	SINR Margin[a] (dB)
3-corner	1	3	14	0.5	6	0.0
			15	1.1	7	1.7
		4	12	0.1	5	1.9
			13	0.7	6	4.5
	3	3	3	0.3	3	2.6
			4	2.9	4	5.8
		4	2	1.0	2	3.0
			3	5.5	3	8.6
Centrally Located	1	3	55	0.0	17	0.4
			69	1.0	18	0.9
		4	55	0.0	11	0.5
			69	1.0	12	1.1
	3	3	7	0.3	5	0.8
			8	1.2	6	2.4
		4	5	1.6	4	2.7
			6	2.9	5	5.3

[a]Margins are accurate to within a few tenths of a decibel and were determined from simulation results using 100 000 samples.

antenna is much greater. With the same geometry and three channel sets, even though only a few antennas are required with maximal ratio combining, optimum combining increases the margin by 2–3 dB. With centrally located base stations and frequency reuse in every cell, optimum combining substantially reduces the number of antennas. As few as 11 antennas are required with optimum combining as compared to more than 50 with maximal ratio combining. Finally, with three channel sets, optimum combining requires one less antenna and has higher margins.

Thus, the improvement with optimum combining is the largest in systems where a large number of antennas is required because of low received SINR. However, even with high SINR and few antennas, the improvement is 2 dB or more. Therefore, the results for typical cellular systems agree with those of Section III (i.e., Fig. 8).

In an actual system we would expect the optimum combining improvement to be even greater than that shown in Table I because of the following three reasons. First, all the channels in all the cells may not always be occupied. Thus, the total interference power will be less, and the power of the strongest interferers (when transmitting) relative to the power of the sum of the other interferers Γ_j will be higher. As shown in Section III, as Γ_j increases, the optimum combining improvement increases. Second, with random mobile locations rather than the worst case, the total interference power will be lower. Thus, Γ_j for the strongest interferers (those closest to the desired mobile's base station) will be higher, and therefore, so will the improvement. Third, for the results in Table I only the six strongest interferers were considered individually, and thus the results are somewhat pessimistic.

Finally, we note that in actual systems the fading can be non-Rayleigh with direct paths existing between an interfering mobile and a base station (i.e., the fading might not be independent at each antenna). Under these conditions, the performance of maximal ratio combining can be significantly degraded while optimum combining can still achieve the maximum output SINR.

V. Implementation

In this section we discuss the implementation of optimum combining in mobile radio. We consider the use of an LMS [3] adaptive array at the base station receiver and adaptive retransmission with time division for base-to-mobile transmission. For the LMS adaptive array, we discuss the dynamic range, reference signal generation, and modulation technique.

A. The LMS Adaptive Array

1) Description: Of the various adaptive array techniques [2]–[4] that can be used in mobile radio, the LMS technique appears to be the most practical one for mobile radio because it is not too complex to implement and it does not require that the desired signal phase difference between antennas be known *a priori* at the receiver.

Fig. 9 shows a block diagram of an M element LMS adaptive array. It is similar to the optimum combiner of Fig. 1 except for the addition of a reference signal $r(t)$ and an error signal $e(t)$. As shown in Fig. 9, the array output is subtracted from a reference signal (described below) $r(t)$ to form the error signal $e(t)$. The element weights are generated from the error signal and the $x_{I_i}(t)$ and $x_{Q_i}(t)$ signals by using the LMS algorithm which minimizes the power of the error signal.

The reference signal is used by the array to distinguish between the desired and interfering signals at the receiver. It must be correlated with the desired signal and uncorrelated with any interference. Under these conditions the minimization of the power of the error signal suppresses interfering signals and enhances the desired signal in the array output. Generation of the reference signal in digital mobile radio systems is described in Section V-A3).

We now consider the weight equation for the LMS adaptive array in a mobile radio system. In the typical system the bit rate is 32 kbits/s, and the carrier frequency is about 840 MHz. With the signal bandwidth 1.5 times the bit rate, the relative bandwidth of the mobile radio channel is only 0.006 percent, and we can consider the signal as narrow band. For narrow-band signals, the weight equation for the LMS array is given by [6, eq. (9)], i.e., the LMS adaptive array maximizes the output SINR. However, these are the steady state weights, and in mobile radio the signal environment is continuously changing. Therefore, we must consider the transient performance of the array. That is, because the weights are constantly changing, the performance will be degraded somewhat from that of the optimum combiner. (Analysis of the transient performance is not considered in this paper.) Also, we must consider the dynamic range of the LMS adaptive array.

2) Dynamic Range: One limitation of the LMS adaptive array technique is the dynamic range over which it can operate. In an LMS adaptive array, the speed of response to the weights is proportional to the strength of the signals at the array input. For the array to operate properly, the weights must change fast enough to track the fading of the

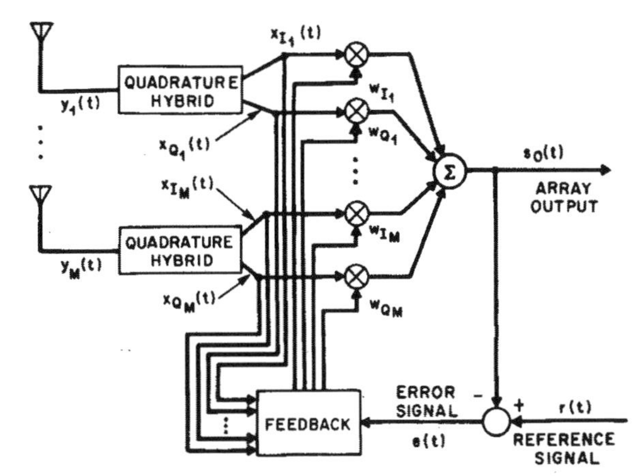

Fig. 9. Block diagram of an M element LMS adaptive array.

desired and interfering signals. However, the weights must also change much more slowly than the data rate so that the data modulation is not altered. It has been shown [11] that for PSK signals the maximum rate of change in the weights without significant data distortion is about 0.2 times the data rate. For the typical mobile radio system, the maximum fading rate is about 70 Hz (for a carrier frequency of 840 MHz and a vehicle speed of 55 mi/h), and the code rate is 32 kbits/s. Thus, the permissible range in signal power at the array input is given by

$$\text{Dynamic Range} = \frac{0.2\times 32\times 10^3}{70} \doteq 20 \text{ dB}. \tag{42}$$

The received signals in a mobile radio system vary by more than 20 dB, however, and therefore automatic transmitter power control (which could add significantly to the cost of the mobile radio) is required to control the power of the strongest signals at the receiver. With this power reasonably fixed, the dynamic range determines the power ratio of the strongest to the weakest received signal that the array can track. A 20 dB dynamic range is certainly not large, but it is more than adequate for mobile radio for the reasons described below.

In the mobile radio systems studied in this paper (see Section IV), the average received SINR at each antenna is relatively small. This is because an adaptive array is not needed when the received SINR is large. For example, for maximal ratio combining with two antennas, an average received SINR at each antenna of 11 dB [1] is required for coherent detection of PSK with a 10^{-3} BER. For optimum combining the required SINR is less with two antennas and, of course, even lower with more antennas. Thus, the received SINR is much less than 20 dB for all cases of interest. (It is typically between -5 and 5 dB.)

A small received SINR affects array operation as follows. First, if the power of an interfering signal is more than 20 dB below the desired signal's power at an antenna, the array need not track the interfering signal at that antenna because it has a negligible effect on the output SINR. Second, if the power of an interfering signal is more than 20 dB higher than the desired signal's power at an antenna, the array need not track the desired signal at that antenna because the resulting weight for the antenna will be almost zero. Thus, because the received SINR is small in the systems where the LMS adaptive array is practical, a 20 dB dynamic range is adequate. Note that if the received SINR is large (e.g., greater than 20 dB, as in a lightly loaded system), the LMS adaptive array will have the same performance as maximal ratio combining.

3) Reference Signal Generation and Modulation Technique: The LMS adaptive array must be able to distinguish between the desired signal and any interfering signals. This is accomplished through the use of a reference signal as discussed in Section V-A1). The reference signal must be correlated with the desired signal and uncorrelated with any interference.

A reference signal generation technique that allows for signal discrimination is described in [12] and involves the use of pseudonoise codes with spread-spectrum techniques. To generate the spread-spectrum signal the pseudonoise code symbols, generated from a maximal length feedback shift register, are mixed with lower speed voice (data) bits, and the resulting bits are used to generate a PSK signal. The code modulation frequency is an integer multiple of the voice bit rate, and this multiple is defined as the spreading ratio k.

The reference signal is generated from the biphase spread-spectrum signal using the loop shown in Fig. 10. The array output is first mixed with a locally generated signal modulated by the pseudonoise code. When the codes of the locally generated signal and the desired signal in the array output are synchronized, the desired signal's spectrum is collapsed to the data bandwidth. The mixer output is then passed through a filter with this bandwidth. The biphase desired signal is therefore unchanged by the filter. The filter output is then hard limited so that the reference signal will have constant amplitude. The hard-limiter output is mixed with the locally generated signal to produce a biphase reference signal. The reference signal is therefore an amplitude scaled replica of the desired signal. Any interference signal without the proper code has its waveform drastically altered by the reference loop. When the coded locally generated signal is mixed with the interference, the interference spectrum is spread by the code bandwidth. The bandpass filter further changes the interference component out of the mixer. As a result, the interference at the array output is uncorrelated with the reference signal. Thus, with spread spectrum, a reference signal is continuously generated that is correlated with the desired signal and uncorrelated with any interference. Furthermore, since pseudonoise codes are used, every mobile can be distinguished by a unique code.

Unfortunately, spread spectrum increases the biphase signal bandwidth by a factor of k and therefore increases both the total cochannel interference power and the number of interferers in cellular mobile radio. For example, with frequency reuse in every cell, the cochannel interference power and the number of interferers from surrounding cells are increased by factors of k and $2k-1$, respectively. This increase in interference power is canceled by the processing gain of spread spectrum, but the increased number of interferers degrades the performance of the LMS adaptive array. Furthermore, $2(k-1)$ cochannel interferers are now present within the desired mobile's cell. Thus, even with a small spreading ratio (e.g., 5 or less) the performance of the LMS adaptive array with the biphase spread-spectrum signal can be worse than that of maximal ratio combining, making the LMS system impractical.

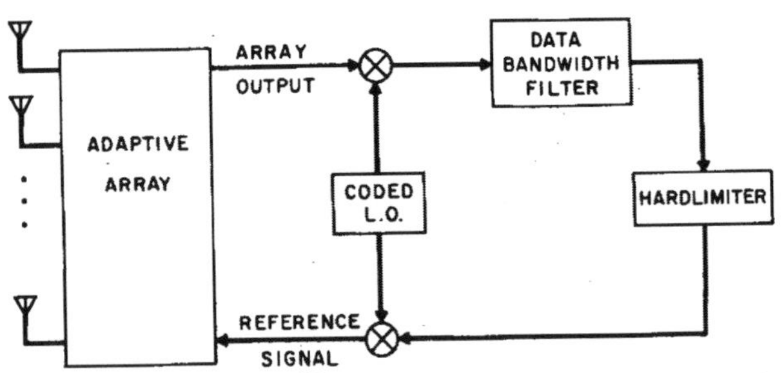

Fig. 10. Reference signal generation loop with the adaptive array. When the desired signal is a biphase spread-spectrum signal, the reference signal is correlated with it but not with any interference.

The bandwidth increase with spread spectrum and its associated problems can be overcome in the following way. The biphase spread-spectrum signal is combined with an orthogonal biphase signal modulated by the voice bits only (see [13]). The data modulation rate of the orthogonal biphase signal is the same as the code modulation rate of the biphase spread-spectrum signal. The resulting four-phase signal therefore has a bandwidth determined by the data rate only, i.e., the bandwidth is not increased by the spreading ratio. Furthermore, a reference signal for the four-phase signal can be generated from its biphase spread-spectrum signal component using the loop described earlier. As shown in [14], the performance of the LMS adaptive array with the four-phase signal is close to that with the biphase signal. Therefore, with this system, we can generate a reference signal without any increase in interference power or the number of interferers and achieve an improvement with an LMS adaptive array close to that for optimum combining which is shown in Sections III and IV.

We now describe the modulation technique in detail by describing three possible ways to modulate the four-phase signal. The simplest technique is for the voice bits to modulate only the orthogonal biphase signal. The biphase spread-spectrum signal then contains the code plus data bits for transferring information from the mobile to the base station. With this first technique, the signal bandwidth corresponds to the voice bit rate r (e.g., 32 kbits/s). However, the energy-per-bit-to-noise (interference) density ratio E_b/N_o is half that of a biphase signal. Thus, the improvement with an LMS adaptive array is 3 dB less than that shown in Sections III and IV. A data channel is also available, however, with an r/k data rate. Furthermore, since the E_b/N_o for the data bits is k times that for the voice bits (because of the spread spectrum), the BER for the data bits is very low.

If a data channel is not required, then voice bits can

replace the data bits. With this second technique, the voice bits are split into two channels, one modulating the biphase spread-spectrum signal and the other modulating the orthogonal biphase signal. The bit rate for the latter channel is k times that for the biphase spread-spectrum signal. The signal bandwidth is reduced by $k/(k+1)$ as compared to the first technique. However, the E_b/N_o of the voice bits on the biphase spread-spectrum signal is k times that on the orthogonal biphase signal. Through appropriate coding techniques, this difference can be used to improve the overall BER.

We can equalize the BER for both channels by decreasing the power of the biphase spread-spectrum signal by $1/k$. With this third technique the E_b/N_o for the voice bits is just $k/(k+1)$ times that for a biphase signal. For example, with k equal to 5, the improvement with an LMS adaptive array is 0.8 dB less than that shown in Sections III and IV. Table II summarizes the above results for the three modulation techniques.

A block diagram of the four-phase signal generation circuitry for the three modulation techniques is shown in Fig. 11. The code symbols of duration Δ are mixed with either voice or data bits of duration $k\Delta$. The resulting symbols modulate a local oscillator to generate a biphase spread-spectrum signal. As shown in the lower portion of Fig. 11, voice bits, also of duration Δ, modulate the local oscillator signal shifted by 90° to generate the orthogonal biphase signal. This signal is then combined with the biphase spread-spectrum signal to obtain the four-phase signal. By adjusting the biphase spread-spectrum signal level with β and modulating this signal with either voice or data bits, we can generate any of the three four-phase signals listed in Table II.

TABLE II
FOUR-PHASE SIGNAL PARAMETERS FOR THREE MODULATION TECHNIQUES IN AN LMS ADAPTIVE ARRAY SYSTEM

Technique No.	Relative Biphase Signal Powers	Spread-Spectrum Biphase Signal			Orthogonal Biphase Signal		
		Information Bits	Bit Rate [a]	E_b/N_o [b]	Information Bits	Bit Rate	E_b/N_o [b]
1	1:1	Data	r/k	$k/2$	Voice	r	0.5
2	1:1	Voice	$r/(k+1)$	$k/2$	Voice	$(\frac{k}{k+1})r$	0.5
3	$1/k$:1	Voice	$r/(k+1)$	$k/(k+1)$	Voice	$(\frac{k}{k+1})r$	$k/(k+1)$

[a] The code modulation rate is k times the bit rate.
[b] Relative to biphase signals.

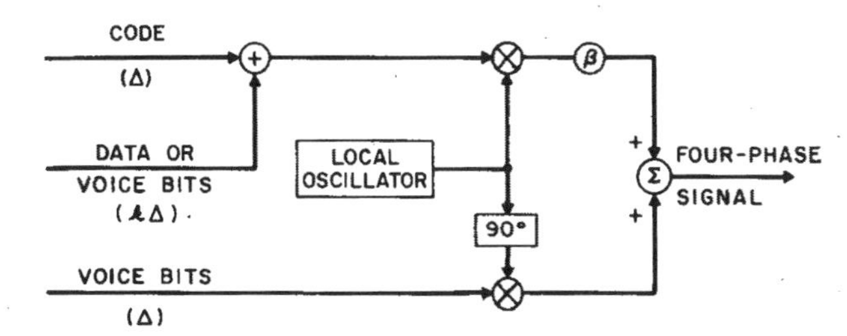

Fig. 11. Block diagram of the four-phase signal generation circuitry for the LMS adaptive array. A biphase spread-spectrum signal, modulated by code symbols plus data or voice bits, is combined with an orthogonal biphase signal, modulated by voice bits, to generate the four-phase signal.

B. Base-to-Mobile Transmission

As we have shown, the LMS technique can significantly improve signal reception at the base station. This improvement is, of course, also desired at the mobile. However, since there are many more mobiles than base stations, it is economically desirable to add the complexity of the LMS technique (particularly multiple antennas) only to the base stations.

Adaptive retransmission with time division [1], [9] can be used to improve reception at the mobile with multiple base station antennas only. With adaptive retransmission, the base station transmits at the same frequency as it receives, using the complex conjugate of the receiving weights. With time division, a single channel is time shared by both directions of transmission. Thus, with the LMS technique, during mobile-to-base transmission the antenna element weights are adjusted to maximize the signal-to-noise ratio at the receiver output. During base-to-mobile transmission, the complex conjugate of the receiving weights are used so that the signals from the base station antennas combine to enhance reception of the signal at the desired mobile and to suppress this signal at other mobiles. Therefore, by keeping the time intervals for transmitting and receiving much shorter than the fading rate (e.g., transmitting in 10 bit blocks), we can achieve the advantages of the LMS technique at both the mobile and the base station.

With adaptive retransmission using the LMS technique, each base station transmits in a way that maximizes the power of the signal received by the desired mobile relative to the total power of the signal received by all other mobiles. Thus, at the mobiles, interfering base station signals are suppressed and the improvement in the performance with the LMS technique as compared to maximal ratio combining should be similar to that at the base stations. The actual improvement for a given mobile, however, depends on the interference environment of every base station. Because of the complexity of the analysis, we will not study this improvement in detail. It should be noted, though, that for base-to-mobile transmission, spread spectrum on the signal is not required because a reference signal is not generated at the mobile. Therefore, without the degradation with the modulation scheme in the mobile-to-base transmission (see Section V-A-3), the BER at the mobile may be lower than that at the base station.

VI. SUMMARY AND CONCLUSIONS

In this paper we have studied optimum combining for digital mobile radio systems. The combining technique is optimum in that it maximizes the output SINR at the receiver even with cochannel interference. We determined the BER performance of optimum combining in a Rayleigh fading environment and compared the performance to that of maximal ratio combining. Results showed that with cochannel interference there is some improvement over maximal ratio combining with only a few receiving antennas, but there is significant improvement with several antennas. With optimum combining, the typical cellular system was seen to have greater margins and require fewer antennas than with maximal ratio combining. Finally, we described how optimum combining can be implemented in mobile radio with LMS adaptive arrays. Thus, we have shown that optimum combining is a practical means for increasing the channel capacity and performance of digital mobile radio systems.

REFERENCES

[1] W. C. Jakes Jr. *et al.*, *Microwave Mobile Communications.* New York: Wiley, 1974.

[2] R. A. Monzingo and T. W. Miller, *Introduction to Adaptive Arrays.* New York: Wiley, 1980.

[3] B. Widrow, P. E. Mantey, L. J. Griffiths, and B. B. Goode, "Adaptive antenna systems," *Proc. IEEE*, vol. 55, p. 2143, Dec. 1967.

[4] S. R. Applebaum, "Adaptive arrays," *IEEE Trans. Antennas Propagat.*, vol. AP-24, p. 585, Sept. 1976.

[5] B. Widrow, J. McCool, and M. Ball, "The complex LMS algorithm," *Proc. IEEE*, vol. 63, p. 719, Apr. 1975.

[6] C. A. Baird, Jr. and C. L. Zahm, "Performance criteria for narrowband array processing," *1971 Conf. Decision Contr.*, Miami Beach, FL, Dec. 15–17, 1971, p. 564.

[7] V. M. Bogachev and I. G. Kiselev, "Optimum combining of signals in space-diversity reception," *Telecommun. Radio Eng.*, vol. 34/35, p. 83, Oct. 1980.

[8] P. Bello and B. D. Nelin, "Predetection diversity combining with selectively fading channels," *IRE Trans. Commun. Syst.*, vol. CS-10, p. 32, Mar. 1962.

[9] P. S. Henry and B. S. Glance, "A new approach to high-capacity digital mobile radio," *Bell Syst. Tech. J.*, vol. 60, no. 8, p. 1891, Oct. 1981.

[10] Y. S. Yeh and D. O. Reudink, "Efficient spectrum utilization for mobile radio systems using space diversity," *IEEE Trans. Commun.*, vol. COM-30, p. 447, Mar. 1982.

[11] T. W. Miller, "The transient response of adaptive arrays in TDMA systems," Electrosci. Lab., Dep. Elec. Eng., Ohio State Univ., Columbus, OH, Rep. 4116-1, p. 287, June 1976.

[12] R. T. Compton, Jr., "An adaptive array in a spread-spectrum communication system," *Proc. IEEE*, vol. 66, p. 289, Mar. 1978.

[13] J. H. Winters, "Increased data rates for communication systems with adaptive antennas," in *Proc. IEEE Inter. Conf. Commun.*, June 1982.

[14] —, "A four-phase modulation system for use with an adaptive array," Ph.D. dissertation, Ohio State Univ., Columbus, OH, July 1981.

PHOTO NOT AVAILABLE

Jack H. Winters (S'77–S'78–S'80–M'81) was born in Canton, OH on September 17, 1954. He received the B.S.E.E. degree from the University of Cincinnati, Cincinnati, OH, in 1977, and the M.S. and Ph.D. degrees in electrical engineering from The Ohio State University, Columbus, in 1978 and 1981, respectively.

From 1973 to 1976 he was a Professional Practice Student at the Communications Satellite Corporation, Washington, DC. He was a Graduate Research Associate with the Electro-Science Laboratory, The Ohio State University, from 1977 to 1981. He is now with the Radio Research Laboratory, AT&T Bell Laboratories, Holmdel, NJ, where he is studying digital satellite and mobile communication systems.

Dr. Winters is a member of Sigma Xi.

MMSE Equalization of Interference on Fading Diversity Channels

PETER MONSEN, MEMBER, IEEE

Abstract—**Adaptive equalization is used in digital transmission systems with parallel fading channels. The equalization combines the diversity channels and reduces intersymbol interference due to multipath returns. When interference is present and correlated from channel to channel, the equalizer can also reduce its effect on the quality of information transfer. Important applications for interference cancellation occur in diversity troposcatter systems in the presence of jamming, diversity high frequency (HF) systems which must cope with interfering skywaves, and space diversity line-of-sight (LOS) radio systems where adjacent channel interference is a problem.**

In this paper we develop the general formulation for minimum mean square error (MMSE) equalization of interference in digital transmission diversity systems. The problem formulation includes the use of available receiver decisions to assist in MMSE processing. The effects of intersymbol interference are included in the analysis through a critical approximation which assumes sufficient processor capability to reduce ISI effects to levels small enough for satisfactory communication. The analysis also develops the concept of additional implicit or intrinsic diversity which results from channel multipath dispersion. It shows how the MMSE processor sacrifices diversity to suppress interference even when the interference arrives in the main beams of the receiver antenna patterns. The condition of near synchronous same-path interference is also addressed. Because the spatial angle of arrival of the interference may result in delay differences between interference signals in different antenna channels, interference delay compensation may be required. We show that this effect is compensated for with a small number of appropriately spaced equalizer taps.

I. Introduction

In a digital transmission system, the criterion of most importance is usually bit error rate or some function of its distribution if the channel is fading. The use of a mean square error (MSE) criterion for processor optimization rather than bit error rate results in inferior bit error rate performance but provides a generally simple processor structure. In many channels with intersymbol interference, the MSE criterion and the restriction to *linear* processing results in only a small performance inferiority. Linear MMSE equalizers were initially suggested for removal of intersymbol interference in telephone channel applications [1], [2]. This concept was extended [3] to include parallel channels in a fading environment. The equalizer in this latter application not only removed intersymbol interference but also performed channel Doppler compensation and maximal ratio diversity combining. An improved performance capability was realized when the multichannel linear equalizer was augmented with decision-feedback [4] for cancellation of previous receiver decisions. The application emphasis for these MMSE multichannel processors has centered on intersymbol interference removal and the provision of additional implicit diversity through coherent recombining of multipath returns [5], [6]. Although it has been historically recognized that these processors could also cancel correlated interference, this feature has not been exploited for the following reasons.

1) Cancellation of correlated interference signals results in loss of signal diversity, which was needed for fading protection. System design typically provided enough diversity for signal fading with the premise or hope that interference would not be a problem.

2) The cancellation of large interferers such as jammers presents a tracking problem to the processor because the signal component is "hidden" in the interference signal.

3) In many communication applications such as LOS microwave radio, the use of multiple antenna aperture (space diversity) channels, which are required for interference cancellation, has only recently become popular.

Diversity concepts such as angle diversity [7] in troposcatter and LOS radio systems[1] and polarization diversity in HF systems can be used to augment existing diversity techniques to compensate for the diversity of loss inherent in interference cancellation. Also we will show in this paper that the diversity loss associated with interference cancellation is ameliorated by the multipath or implicit diversity. The tracking problem can be solved by a preprocessor which adaptively orthogonalizes the channel outputs, thus separating the large interference from the signal component. This preprocessor concept has been widely used in null steering applications [8] for jammer cancellation. The null steerer without the equalizer is inferior, however, because it does not optimize the *desired signal* characteristics, and because under certain conditions it may null out the desired signal.

Evaluation of the performance of an MMSE processor is complicated by the presence of fading signals and intersymbol interference. In this paper an approach to determine this performance is presented along with some example results.

II. Problem Description

Consider a general D-diversity channel with arbitrary noise inputs. The additive noise may represent adjacent channel interference, hostile jamming signals, atmospheric noise, thermal noise, or a combination of disturbance signals. The MMSE equalizer will exploit the correlation between these noise sources to reduce their effects. In multisource systems, i.e., where more than one independent information source is transmitted over the same radio route, we also will consider using receiver decisions on adjacent channels for interference cancellation. Because of the good quality of communications required in most applications, it is reasonable to assume that these adjacent channel receiver decisions are correct. The bandpass nature of radio channels allows us to define the communication system in terms of the in-phase and quadrature com-

Paper approved by the Editor for Communication Theory of the IEEE Communications Society for publication after presentation at the International Conference on Communications, Denver, CO, June 1981. Manuscript received July 25, 1981; revised April 11, 1983.

The author is with SIGNATRON, Inc., Lexington, MA 02173.

[1] Although existing LOS systems do not use angle diversity, recent propagation analysis shows promise for this technique in this application.

Reprinted from *IEEE Transactions on Communications,* vol. COM-32, no. 1, January 1984.

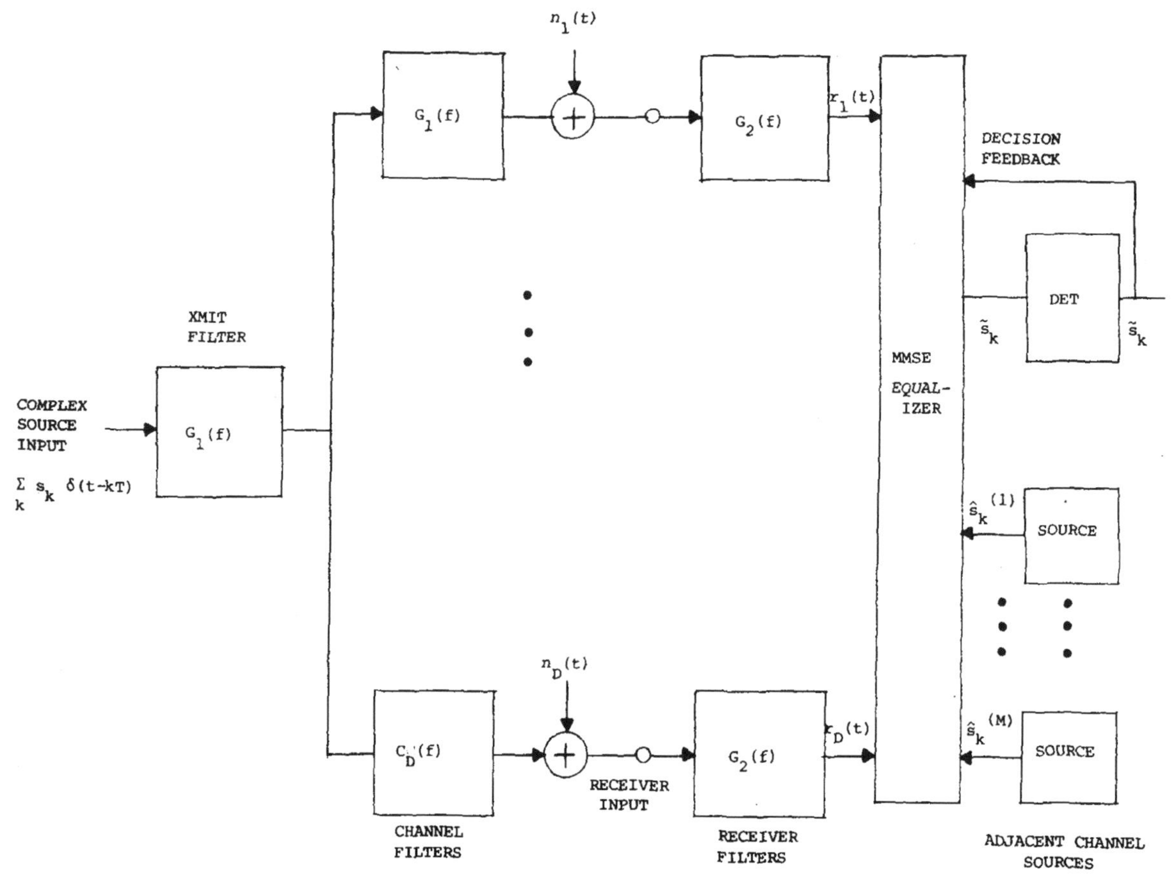

Fig. 1. MMSE equalizer, system model.

ponents of the signals and channel impulse responses. A complex notation which assigns the in-phase components to a real part representation and the quadrature components to the imaginary part eliminates the carrier frequency from the signal analysis. The complex signal (or impulse response) is converted into the real bandpass representation by the operation

$$x_B(t) = \text{Re}\,(x(t)e^{j2\pi f_0 t}) \tag{1}$$

where f_0 is the carrier frequency and $x(t)$ is the complex notation signal. Using this notation the general D-diversity MMSE equalization problem is shown in Fig. 1. The source inputs s_k are complex samples selected from a set which defines the particular modulation used. The most common example is QPSK which selects s_k from the set $(\pm 1 \pm j)$. The transfer functions of the transmit filter $G_1(f)$, the channel filters $C_I(f)$, $I = 1, 2, \cdots, D$, and the receive filter $G_2(f)$ are the Fourier transforms of the respective complex impulse responses. The MMSE equalizer combines the D-diversity signals $r_I(t)$, $I = 1, 2, \cdots, D$, the past receiver decisions $\hat{s}_{k-i}$, $i > 0$, and the M adjacent channel receiver decisions $\hat{s}_{k-j}^{(i)}$, $i = 1, 2, \cdots, M$, $j > 0$.

The criterion of goodness in the MMSE equalizer is that the mean square value of the error signal $s_k - \tilde{s}_k$ is minimized. The error signal is the difference between the transmitted source symbol and the symbol output of the equalizer.

The channel filters in a radio application change with time. Our assumption in the analysis is that the time variation is slow compared to the information rate such that either some fraction of the information can be used to track the time variations or a decision-directed approach can be used to adapt the MMSE equalizer to these time variations. Our interest here is to derive the fading channel performance characteristics of the MMSE equalizer under conditions of perfect channel tracking. The tracking problem can be solved by using least-mean-squares or Kalman adaptation algorithms [9], [10].

III. MMSE Equalizer Structure

The MMSE equalizer under consideration uses a tapped delay line filter for each of the D-diversity channels and combnes the outputs of each of these filters with a linear combination of the receiver decision sources. [The multiple channel tapped delay line filter corresponds to a linear equalizer MMSE processor. For the decision-feedback equalizer MMSE processor, available receiver decision sources are used for cancellation of past receiver decisions on the desired signal and cancellation of possible present and past receiver decisions of adjacent channel signals. In some applications it may be possible to introduce delay into the desired signal channel to allow time for formation of these adjacent channel decisions.] The use of decisions rather than the adjacent channel sample itself derives from our desire to cancel adjacent channel *signals* which are interfering with our desired signal and an assumption of relatively low error rates even for these tentative decisions. Error rates on the order of 10^{-2} are excellent from a decision-feedback viewpoint, yet quality communications usually demands 10^{-4} or better.

The cancellation of adjacent channel interference with decision-feedback requires that these channels be near synchronous, digital, and have receivers at the same receive terminal. The near synchronous requirement assumes that the data rates of the interference and desired signal are close enough that the relative change is a slow variation with respect to the adaptation rate of the processor. With the high-stability, high-accuracy clocks required for digital transmission, the slow variation requirement will be satisfied in many practical applications. Future military digital LOS microwave systems are an

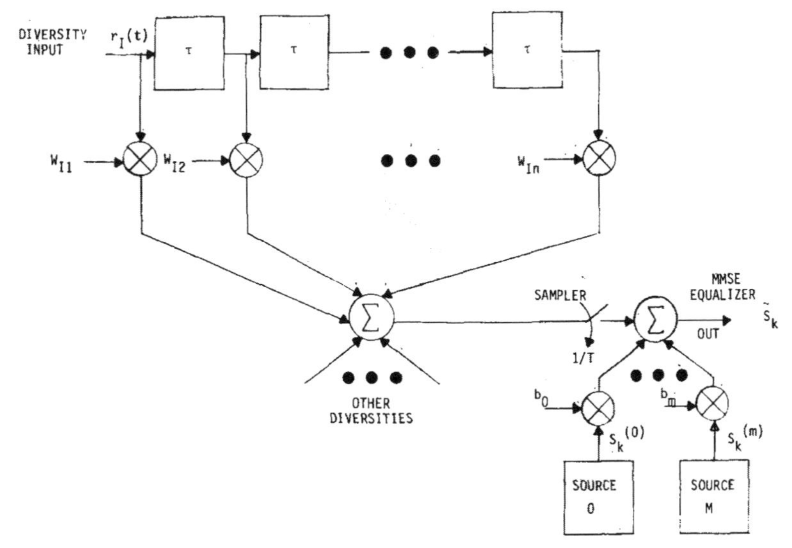

Fig. 2. MMSE processor.

example where decision-feedback cancellation of adjacent channel interference may be applicable because of standardized data rate requirements and path convergence at a single terminal in congested networks. Of course, when receiver decisions are not available, interference must be processed and cancelled by the linear equalizer portion of the MMSE processor.

The receiver decisions to be used in the cancellation are past symbols $\hat{s}_{k-j}$, $j > 0$, from the desired signal channel and both present and past adjacent channel symbols $\hat{s}_{k-j}^{(i)}$, $i = 1, 2, \cdots, M$, $j \geqslant 0$. For convenience we number these symbols $\hat{s}_k^{(l)}$, $l = 0, 1, 2, \cdots, M$. With this notational simplification the linear combining accomplished by the MMSE equalizer is shown in Fig. 2. The equalizer uses the optimum set of complex weights w_{Ij}, $I = 1, 2, \cdots, D$, $j = 1, 2, \cdots, n$, and b_l, $l = 0, 1, \cdots, M$ to minimize the mean square error between the equalizer output $\hat{s}_k$ and the transmitted symbol s_k. In practice this is accomplished either by using a time division multiplexed reference that is known to the receiver and adapting only during the reference receptions or by using receiver decisions as the "known" transmitted symbol. Combinations of these approaches are also possible.

Since we are interested in the performance of the equalizer and not in its adaptation, we can achieve one further notational simplification by considering each of the $r_I(t - j_\tau)$, $j = 0, 1, n - 1$, as a separate input. Furthermore, we can move the weight past the symbol sampler on each of these inputs and consider a pure sampled data system. We rename the complete set of $d = nD$ diversity signals beginning with $r_1(t_0 + kT)$, $r_1(t_0 + kT - \tau), \cdots, r_2(t_0 + kT)$, $r_2(t_0 + kT - \tau), \cdots$ as $\tilde{r}_1(k)$, $\tilde{r}_2(k), \cdots, \tilde{r}_d(k)$. The complex weights are renumbered $w_1, w_2, \cdots, w_d$, respectively. The MMSE equalizer can then be represented by the summation over the d diversity channels and the $M + 1$ interference sources.

$$\tilde{s}_k = \sum_{i=1}^{d} w_i \tilde{r}_i(k) + \sum_{i=0}^{M} b_i \hat{s}_k^{(i)} \tag{2a}$$

which suggests an obvious vector notation

$$\tilde{s}_k = \boldsymbol{w} \cdot \tilde{\boldsymbol{r}}_k + \boldsymbol{b} \cdot \boldsymbol{s}_k. \tag{2b}$$

Note that these d diversity channels are composed of the product of D *explicit*, e.g., frequency, space, angle, diversity channels and n *implicit* diversity channels. The latter derive from a coherent combining of the channel multipath structure by the tapped delay lines of the equalizer.

The first dot product term in the equalizer output $\tilde{s}_k$ is a linear equalizer component which provides for diversity protection, reduction of intersymbol interference (ISI) from future symbols, s_{k+j}, $j > 0$, and reduction of interference sources which are correlated between diversity branches. The linear equalizer component does not cancel adjacent channel interference for which receiver decisions are available, as this can be done more efficiently by the second dot product term in the equalizer output equation. This decision-feedback equalizer component cancels ISI from past symbols whether they are in the desired or an adjacent channel. In our analysis we assume that the $\hat{s}_k^{(i)}$ are correct. From this assumption it follows that any cancellation which can be done by the decision-feedback equalizer will be done there and not in the linear equalizer, because in the former such cancellation is noise free, whereas it is not noise free in the latter. A mathematical proof illustrating this result for a single channel application without adjacent channel interference is given in [4].

We can take advantage of this result and further simplify the problem at hand by defining a set of purged input signal samples which have all the available desired receiver decisions and available adjacent channel interference receiver decisions removed. Since interference due to these available decisions is cancelled by the decision feedback-equalizer with no noise enhancement, their removal from the signal cannot affect the calculation of the equalizer performance. In this calculation, we need only consider the signal combiner and a purged signal input set.

IV. Purged Signal Characteristics

For signal transmission of complex discrete source symbols, the ith diversity signal at the sampling time kT is the convolution of the source symbols and the channel impulse response $h(t)$ plus an additive noise term and adjacent channel interference. For the $d = nD$ diversity channels, this diversity signal has the representation

$$\tilde{r}_i(k) = \sum_{p=-\infty}^{\infty} s_{k-p} h_{I+1}(t_0 + pT - i\tau + In\tau) + u_i(k) + \tilde{u}_i(k)$$

$$i = In + 1, \cdots, In + n$$

$$I = 0, 1, \cdots, D - 1 \tag{3}$$

where we define $u_i(k)$ as the additive noise component which includes no available adjacent channel decisions and $\tilde{u}_i(k)$ as the filtered version of the available adjacent channel decisions. The purged signal results from subtracting the adjacent channel interference $\tilde{u}_i(k)$ and past ISI to obtain

$$x_i(k) = \sum_{p=-\infty}^{0} h_{ip} s_{k-p} + u_i(k), \quad i = 1, 2, \cdots, d \tag{4}$$

when the sampled data channel h_{ip} has been defined as

$$h_{ip} = h_{I+1}(t_0 + pT - i\tau + In\tau)$$
$$i = In + 1, \cdots, In + n$$
$$I = 0, 1, \cdots, D - 1. \tag{5}$$

We subtract the interference terms for which receiver decisions are available from the equalizer input for convenience of analysis, as they will be cancelled out in subsequent operations by the equalizer.

The solution for the MMSE equalizer performance will clearly depend on the statistics of the sample data channel h_{ip} and the noise sample $u_i(k)$. We need to relate these quantities and their statistics to channel parameters prior to the sampling operation.

The channel impulse response $h_L(t)$ which is related to the sampled data response through (5) is the convolution of the transmit, receive, and Lth diversity channel filters.

$$h_L(t) = \int G_1(f) G_2(f) C_L(f) e^{j2\pi ft}\, dt, \quad L = 1, 2, \cdots, D. \tag{6}$$

The noise term in the purged signal represents the additive receiver noise and all interference which cannot be cancelled using receiver decisions. In our model we assume stationary zero mean Gaussian noise so a complete description is given by the noise covariance matrix

$$E[u_i(k) u_l^*(k + p)] \equiv U_{il}(p).$$

This matrix can be related to purged cross-channel noise spectra $N_0 S_n^{PQ}(f)$, $P, Q = 1, 2, \cdots, D$ for the D explicit diversity channels by

$$U_{il}(p) = N_0 \int_{-\infty}^{\infty} |G_2(f)|^2 S_n^{I+1, L+1} \cdot (f) e^{j2\pi[(l-LD)-(i-ID)]ft}\, df, \tag{7}$$
$$i = In + 1, \cdots, In + n$$
$$l = Ln + 1, \cdots, Ln + n$$
$$I, L = 0, 1, \cdots, D - 1.$$

The purged cross-channel noise spectra are found by cross correlation and Fourier transform operations on all noise sources for which receiver decisions are not available for cancellation. The normalized constant N_0 is selected as the average noise spectral density over the frequency band of interest of the weakest diversity noise source.

Two different time scale averages are needed for the analysis. Since the equalizer can track the channel variations, from its viewpoint the channel gains h_{il} are fixed and known and the optimum weights are a function of these gains and the noise statistics. The "short-term" time average for the noise statistics is denoted by the expected value notation $E(\cdot)$. Since the channel gains vary with time, the signal-to-noise ratio after optimum equalization must be considered a random variable with respect to the channel variations. "Long-term" averages with respect to the channel are denoted by an overbar. For most fading radio channels, the channel gain h_{ip} is a zero mean complex Gaussian process which is completely characterized by the set of long-term covariance matrices

$$H_{il}(k) = \overline{h_{ip} h_{l,p-k}^*} \quad i, l = 1, 2, \cdots, d. \tag{8}$$

These covariance matrices can be related back to cross-diversity delay power spectra through the relationship between the channel impulse response and the sampled data responses; cf. (6) and (5). The long-term cross-diversity delay power spectra measure the scattered power in a particular delay cell and have the following form:

$$P_{LP}(\tau) = \int_{-\infty}^{\infty} \overline{C_L(f) C_P^*(f + \nu)} e^{j2\pi\nu\tau}\, d\nu, \quad L, P = 1, 2, \cdots, D. \tag{9}$$

V. MMSE Equalizer Performance

The problem to be solved is to find the statistical performance of an MMSE equalizer in the face of multipath distorion, indirect route interference, and direct route interference. In our model the direct route interference from adjacent channels can be cancelled, and thus, its statistical description is not relevant to the performance results. The multipath distortion and indirect route interference are summarized, respectively, by the cross-diversity delay power spectra (9) and the indirect route cross-diversity noise spectra $N_0 S_n^{PQ}(f)$, $P, Q = 1, 2, \cdots, D$. We would like to find the bit error rate (BER) long-term statistical distribution as a function of these two quantities. We concentrate on the BER distribution because most communication applications are more sensitive to error rates above some threshold rather than the average BER.

The output of a combiner that operates on a purged signal input can be written

$$\tilde{s}_k = \sum_{i=1}^{d} w_i x_i(k) \tag{10}$$

or as the dot product in vector form

$$\tilde{s}_k = \mathbf{w} \cdot \mathbf{x}_k \equiv w' x_k, \; w = \{w\}, \; _i x = \{x_k(k)_i\} \tag{11}$$

where "$'$" denotes vector transpose.

A. ISI Approximation

It is easy to show that the optimum weight vector under the criterion of minimum mean square value of $s_k - \tilde{s}_k$ is

$$W_0 = [E(\mathbf{x}_k \mathbf{x}_k')]^{-1} E(\mathbf{x}_k s_k^*). \tag{12}$$

This solution is not very helpful in a fading channel situation, since the above short-term statistics are a function of the channel gains which, in turn, are random processes due to the channel fading. The major stumbling block to the analysis is the ISI term, which results in random channel gains in the denominator of the equalizer output signal-to-noise ratio. By approximating the ISI term by a nonfading additive noise contribution, the output SNR has only random channel terms in the numerator and analysis can proceed. This approximation provides meaningful results for two reasons. First, the error rate statistics are affected more by the fading of the channel vector which multiplies the signal term rather than the interference terms. Second, interference in many applications dominates the ISI effects such that the ISI could even be neglected. This approximation is valid when the equalizer has sufficient length to reduce the ISI resulting from filters and channel multipath to a level small compared with the decision thresholds. For example, if the equalizer length is short compared to the channel multipath spread, ISI effects will dominate performance and a nonfading, Gaussian noise representation is a poor representation. Of course, this example is of small practical interest because communications is not satisfactory under these conditions. The approximation of the fading ISI by a nonfading component is accomplished by replacing $x_i(k)$ in (4) with the expression

$$x_i(k) \doteq s_k h_{i0} + v_i(k). \tag{13}$$

In vector notation

$$\boldsymbol{x}_k = s_k \boldsymbol{h}_0 + \boldsymbol{v}_k \tag{14}$$

where $v_i(k)$ is a complex zero mean Gaussian noise which includes both the channel noise and a noise component which approximates the effects of ISI. Assuming independent source digits, the short-term covariance matrix for the ISI component in (4) is a function of the channel gains and is given by

$$\sum_{p<0} \boldsymbol{h}_p \boldsymbol{h}_p{}' \equiv \sum_{p<0} h_{ip} h_{jp}{}^* \quad i, j = 1, 2, \cdots, d. \tag{15}$$

This covariance matrix shows up in the denominator of the equalizer signal-to-noise ratio when (12) is used. The approximation is to replace (15) with a scaled average value with respect to the channel gains. The noise term in (14) then has the short-term covariance matrix

$$Q \equiv E(\boldsymbol{v}_k \boldsymbol{v}_k{}') = E(\boldsymbol{u}_k \boldsymbol{u}_k{}') + \alpha \sum_{l<0} \overline{\boldsymbol{h}_l \boldsymbol{h}_l{}'} \tag{16}$$

and α is an attenuation constant which reflects the larger tail of a Gaussian noise distribution relative to a source sequence distribution. This constant may be selected from theoretical considerations or empirically by comparing measured and predicted values.

B. Equivalent Uncorrelated Diversity/White Noise System

The MMSE equalizer combines the $d = nD$ diversity signals $x_i(k)$, $i = 1, 2, \cdots, d$, which are a function of the channel gain vectors $\boldsymbol{h}_i$ and the equivalent noise vector $\boldsymbol{v}_k$ which includes the average fading effect of ISI. The diversity signals $x_i(k)$ in (13) are long-term correlated (with respect to the channel gains) so that it is difficult to assess the effective diversity protection. Also, the noise components exhibit short-term correlation which the equalizer can exploit in order to cancel interference. These effects can be quantitatively assessed by means of a linear transformation of the input process. This transformation will yield an identical problem but with long-term uncorrelated diversity channels and short-term uncorrelated noise sources. The transformation is

$$\boldsymbol{z}_k = T'\boldsymbol{x}_k = s_k T'\boldsymbol{h}_0 + T'\boldsymbol{v}_k \tag{17}$$

and we require

$$\text{uncorrelated diversity: } T'\overline{\boldsymbol{h}_0 \boldsymbol{h}_0{}'}T = T'HT \equiv \frac{\overline{\epsilon_b}}{N_0}\Gamma,$$

$$\Gamma_{ij} = 0, \quad i = j; \quad \Gamma_{ii} = \lambda_i \tag{18}$$

$$\text{uncorrelated noise: } T'E(\boldsymbol{v}_k \boldsymbol{v}_k{}')T' = T'QT = I. \tag{19}$$

The transformation T which satisfies these requirements is

$$T = MA^{-1/2}N \tag{20}$$

where M is the modal matrix for the noise matrix Q, i.e.,

$$Q = MAM', \quad A = M'QM, \quad M' = M^{-1}, \quad A_{ij} = 0, \quad i = j \tag{21}$$

and N is the modal matrix for the whitened channel gain matrix

$$W = A^{-1/2}M'HMA^{-1/2}. \tag{22}$$

The diversity matrix Γ is conveniently normalized by the ratio of average bit energy $\overline{\epsilon}_b$ to the noise spectral density constant N_0. We will see subsequently that this fixes the sum of the diversity eigenvalues λ_i to unity or less.

We can define new channel and noise vectors as follows:

$$\boldsymbol{q} \equiv T'\boldsymbol{h}_0 \tag{23}$$

$$\boldsymbol{y}_k \equiv T'\boldsymbol{v}_k \tag{24}$$

and note that $\boldsymbol{y}_k$ is a unit variance white noise process.

The equalizer output has the form

$$\tilde{s}_k = s_k \boldsymbol{w}'\boldsymbol{g} + \boldsymbol{w}'\boldsymbol{y}_k. \tag{25}$$

Since $\boldsymbol{y}_k$ is a white noise process, the optimum solution to the minimization problem is the matched filter [11],

$$\boldsymbol{w}_0 = c\boldsymbol{g}. \tag{26}$$

Since the constant c multiplies both signal and noise, its value is of no importance.

The equivalent uncorrelated diversity/white noise system with optimum combining is shown in Fig. 3. In this equivalent system, performance is completely determined by the set of d explicit/implicit diversity gains λ_i, $i = 1, 2, \cdots, d$ and the average bit energy-to-noise power ratio.

C. Bit Error Rate (BER) Statistics

Compare the detector input $\tilde{s}_k$ (25) with an ideal QPSK transmission through a white noise channel. In the ideal QPSK system, if the received energy per bit is ϵ_b watts and the noise spectral diversity is N_0 W/Hz, the detector input is

$$\tilde{s}_k = \epsilon_b s_k + \sqrt{\epsilon_b N_0}\,\eta_k \tag{27}$$

where η_k is a ZMCG noise with unity covariance. The error rate for this ideal system is

$$p = \tfrac{1}{2} \operatorname{erfc} \sqrt{\epsilon_b / N_0} \tag{28}$$

which leads to a signal-to-noise ratio (SNR) definition of bit energy at the detector divided by the noise power in both the in-phase (real) and quadrature (imaginary) channels.

For the MMSE equalizer the detector SNR can then be computed from (25) and (26) as

$$\gamma = \frac{(\boldsymbol{g}'\boldsymbol{g})^2}{\boldsymbol{g}'\boldsymbol{g}} = \boldsymbol{g}'\boldsymbol{g}. \tag{29}$$

Since $\boldsymbol{h}_0$ is a random Gaussian channel vector, the vector $\boldsymbol{g}$, which is a linear transformation of $\boldsymbol{h}_0$, is also a Gaussian vector.

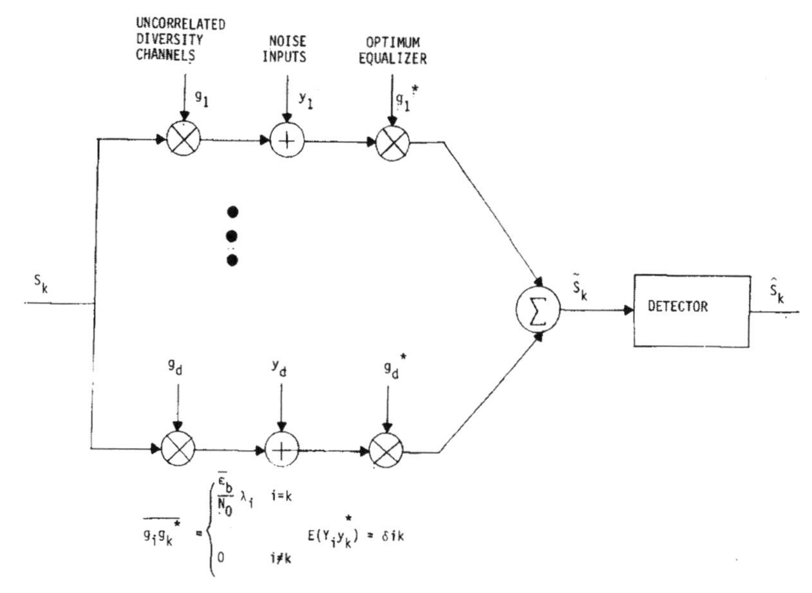

Fig. 3. Uncorrelated diversity equivalent system.

The bit error rate for an MMSE system is also a random variable. Depending on the detection system used, we have

$$p = \tfrac{1}{2} \operatorname{erfc} \sqrt{\gamma} \qquad \text{coherent detection} \tag{30}$$

$$p = \tfrac{1}{2} e^{-\gamma} \qquad \text{differential detection} \tag{31}$$[2]

In order to obtain the statistics for the bit error rate, it is necessary to obtain the probability density function for the detector SNR γ.

Because the components of g are uncorrelated, the transform of the pdf of γ is simply the product

$$P_\gamma(s) = \int_0^\infty p_\gamma(x) e^{-sx}\, dx = \prod_{i=1}^{D} \frac{1}{1 + s\lambda_i \dfrac{\bar{\epsilon}_b}{N_0}}. \tag{32}$$

By taking the inverse transform, the pdf for γ can be determined and all the BER statistics obtained.

In a nonfading ideal channel application, the equalizer SNR is ϵ_b/N_0. Thus, by a conservation of energy argument, the average equalizer SNR cannot exceed $\bar{\epsilon}_b/N_0$ for a nonideal system, i.e.,

$$\bar{\gamma} = \overline{g'g} = \frac{\bar{\epsilon}_b}{N_0} \sum_{i=1}^{d} \lambda_i \leqslant \frac{\bar{\epsilon}_b}{N_0}. \tag{33}$$

The trace of the diagonalized covariance matrix is upper bounded by unity,

$$\sum_{i=1}^{d} \lambda_i \leqslant 1 \tag{34}$$

and the λ_i are the normalized diversity gains for each of the d diversity channels. Note that it is not necessary to perform all the matrix multiplications indicated in (18) in order to obtain the diversity eigenvalues. It is easy to show that the diversity eigenvalues solve the generally non-Hermitian matrix equation

$$HQ^{-1}u = \lambda u \tag{35}$$

where H is the channel covariance matrix and Q is the noise matrix augmented to include ISI effects. The non-Hermitian property results because the product of two Hermitian matrices is not necessarily Hermitian. The eigenvalues correspond to a transformation of a nonsingular covariance matrix, and are thus real and positive definite.

The general procedure for determining the BER statistics can then be summarized as follows.

1) Determine the channel covariance matrix $H_{il}(0) \equiv H = \overline{h_0 h_0'}$ [cf. (8)] where the ith diversity component of the channel vector h_{i0} is the sampled data version [cf. (5)] of the channel impulse response.

2) Determine the effective noise matrix Q [cf. (16)] which is the sum of the additive noise covariance matrix and a covariance matrix of future ISI contribution. The additive noise matrix is the sum of the thermal noise matrix and adjacent and cochannel signal interference matrices for incoherent sources. When interference results from near synchronous digital systems on the same path, it is assumed that they can be cancelled out.

3) Perform the matrix multiplication HQ^{-1} and determine the $d = nD$ diversity eigenvalues of the product matrix using a non-Hermitian procedure.

4) Obtain the SNR pdf by computing the inverse Laplace transform of the d-fold product of single pole $(1 + s\lambda_i\bar{\epsilon}_b/N_0)^{-1}$ diversity contributions.

5) Determine the required BER statistics from operations on the SNR pdf and the BER equation as a function of SNR. For example, the average BER for differential detection does not require inverse Laplace transformation, viz.,

$$\bar{p} = P_\gamma(1).$$

The outage probability is the probability that a particular BER is exceeded, i.e., the cumulative BER distribution:

$$P_0(p_c) = \text{prob}\,(p \geqslant p_c) = \text{prob}\,(\gamma \leqslant \gamma_c) = \int_0^{\gamma_c} p_\gamma(x)\, dx$$

[2] Although most MMSE systems are coherent, the differential detection characteristic is a useful approximation to practical modem performance curves on the Gaussian noise channel.

where the critical value of SNR γ_c is obtained by inverting the functional relationship given a critical BER value p_c.

VI. Performance Results, an Example

The potential for interference reduction in a high-speed digital troposcatter system is illustrated by computing the BER error rate and outage probability for a typical diversity configuration and a range of multipath and interference conditions. The results show that the interference suppression is a function of the spatial angle of arrival of the interference relative to the center line direction (antenna boresight) of the antenna beam.

For our example we assume a fourth-order diversity system composed of two antennas (dual space) and two frequencies (dual frequency). This diversity configuration is denoted $2S/2F$. The MMSE processor considered is a decision-feedback equalizer with a tapped-delay line (TDL) filter on each of the four diversity branches. The TDL filter consists of three taps spaced at approximately a reciprocal bandwidth. This spacing provides for good ISI removal, which, in general, is more important than delay compensation of off-axis spatial interference. The interference delay is equal to $(d_{sep}/c) \sin \theta$ where d_{sep} is the antenna separation, c is the speed of light, and θ is the arrival angle relative to the antenna boresight. The example calculation used a worse case interference delay d_{sep}/c of 60 ns. The tap spacing selected for the TDL filter was 100 ns.

For this model, there are $D = 4$ explicit diversity channels and $n = 3$ implicit diversity channels which require the determination of 12 eigenvalues. The channel model corresponds to a 168 mi, 4.5 GHz troposcatter link. The average bit error rate and outage probability (BER $> 10^{-4}$) performance of a 10 Mbit/s QPSK system in the absence of interference ($J/S = -\infty$ dB) and for a QPSK jammer with signal power ratios of 10 and 30 dB are shown in Fig. 4. Performance depends on the normalized multipath spread $2S/T$ where T is the QPSK symbol interval and S is the rms multipath spread. Generally, performance initially improves with increasing $2S/T$ due to enhanced implicit diversity and then falls off for large $2S/T$ due to ISI limitations.

When interference is present, diversity is sacrificed in order to cancel the interference. When there is no implicit diversity, one order of explicit diversity is lost for each independent jammer which is cancelled. The implicit diversity plays a larger role because of this effect. The implicit diversity "gain" due to coherent combining of the multipath is the dB difference between the predicted no interference curves and the flat fading (no multipath delayed echoes, $2S/T = 0$) reference. This gain is seen in Fig. 4 to be modest, maybe 1/2 dB at 10^{-5} average BER. With two jammers and flat fading conditions, two orders of diversity must be sacrificed to cancel the jammers. The E_b/N_0 required in a flat fading dual diversity system for a 10^{-5} BER is 21.3 dB [13], yet from Fig. 4 only 17.7 dB is required to achieve a 10^{-5} BER with two 30 dB stronger jammers. Thus, the implicit diversity gain when interference is present is 3.6 dB. In other words, multipath delayed echoes increase the diversity potential, and this diversity potential is more significant when some diversity has to be sacrificed for jammer cancellation. Multiple taps are required per diversity to realize this implicit diversity gain.

When the antenna is pointed right at the interference, the *best* suppression is possible. As the interference angle of arrival increases from zero, suppression is reduced because of relative delay between the interfering signals on each antenna port. For larger arrival angles, this effect is compensated for by the increased attenuation of the sidelobe pattern of the antenna. The results in Fig. 4 are for an interference arrival angle of 0.15°. The beamwidth of this example system is 0.55° so that the interference is almost at the antenna center line. Table I indicates how the suppression due to the antenna sidelobe pattern helps to reduce the interference effect. Note that the MMSE processor does a little better for closer-in interference angles because the interference delay limits suppression at larger angles. In general, the MMSE processor requires multiple taps per diversity if large suppression ratios are to be achieved. The choice of three taps per diversity in this example results in good performance over the entire range of possible interference arrival angles.

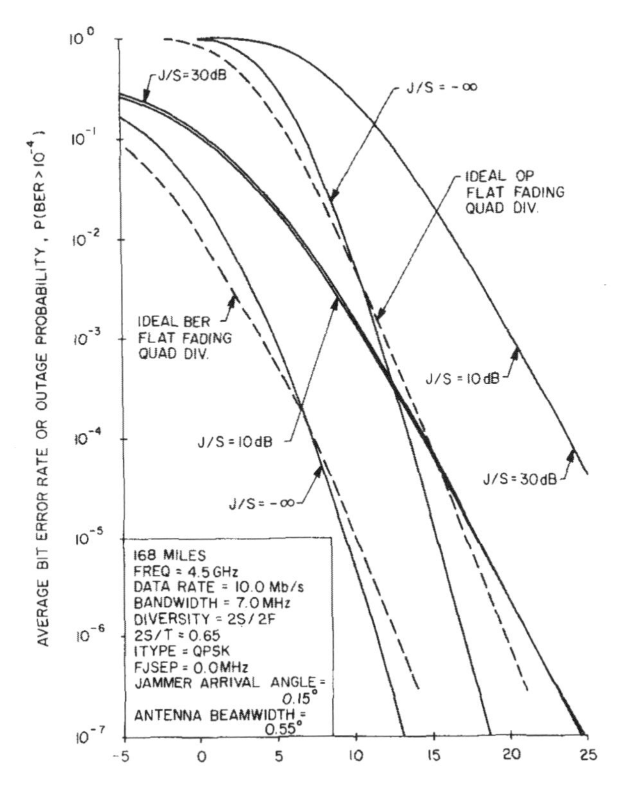

Fig. 4. Predicted MMSE performance with in-band interference.

TABLE I
PERFORMANCE VERSUS ARRIVAL ANGLE 30 dB J/S

JAMMER ANGLE OFF BORESIGHT	ANTENNA SIDELOBE SUPPRESSION (dB)	J/S AFTER ANTENNA (dB)	JAMMER RELATIVE DELAY (ns)	10^{-5} BER DEGRADATION (dB)
0.15°	0.8	29.2	0.16	8.2
0.5°	7.3	22.7	0.52	8.4
1.0°	14.1	15.9	1.0	8.4
2.0°	21.5	8.5	2.1	8.3
10.0°	38.9	-8.9	10.4	4.5

The model can also be used to predict performance when the interference is out of band, i.e., adjacent channel interference. Performance is shown in Fig. 5 when the interference is another QPSK signal but offset by two channel allocations (21 MHz). Here the degradations are not as large as the in-band interference example because of receiver filtering of some of the adjacent channel interference. The receiver filtering provides about 50 dB of rejection in this example. The dashed lines indicate experimental results on a troposcatter model [6] under the adjacent channel conditions defined in Fig. 5. A troposcatter simulator was used to synthesize the quadruple diversity fading channel. Good agreement between predicted and measured results was obtained.

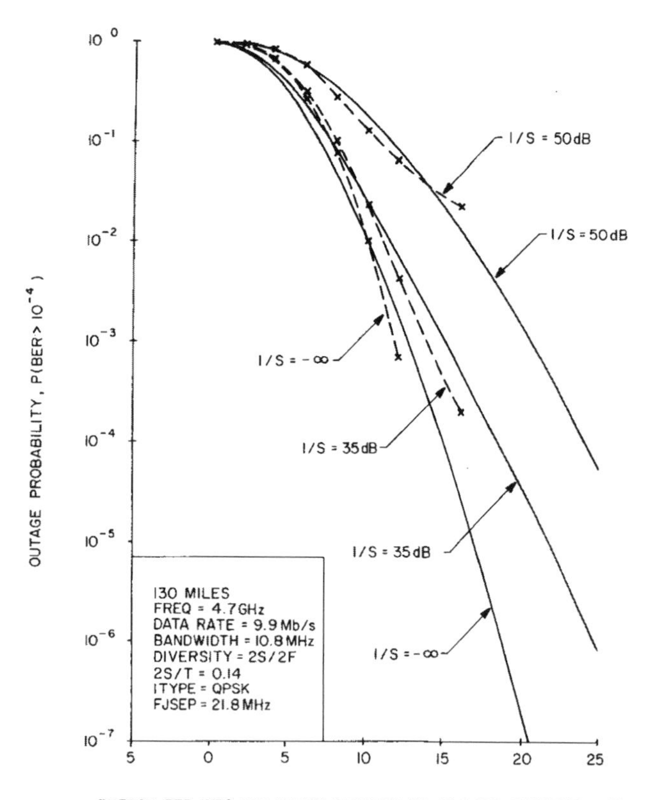

Fig. 5. Predicted MMSE performance with out-of-band interference.

VII. Summary

A procedure for calculation of BER statistics for an MMSE processor in a fading diversity digital transmission system with interference sources has been presented. The procedure includes the effects of both near-synchronous and incoherent interference sources. It also includes the signal enhancement resulting from implicit diversity and the penalty due to multipath induced intersymbol interference effects. The method of calculation requires the determination of a set of eigenvalues from the product of a channel covariance and noise covariance matrix. Results from these calculations illustrate the following qualitative features.

- Diversity must be sacrificed for interference suppression.
- Best cancellation occurs for spatial interference in the main beams of the antennas.
- A TDL filter for delay compensation of off-axis spatial interference is required for good suppression results.
- Antenna sidelobes contribute significantly to the total suppression of off-axis interference and compensate for the delay distortion effect.
- Multipath *helps* reduce the diversity sacrifice for interference suppression provided enough equalizer taps are available.

The quantitative results developed for the example system are significant because they illustrate the large suppression capability of the MMSE processor. Since many actual communication applications do not have large excess bandwidth available for spread-spectrum protection against hostile interference threats, the suppression achievable with multiple antenna processing is the primary resource available.

This paper has not dealt with the adaptation of the MMSE processor because of the additional scope of this problem. We would mention here, though, that it is generally necessary to utilize an approach which compensates for the slow tracking modes associated with following the fading signal in the presence of large interference. This can be accomplished by Kalman filter techniques [10] or by the cascade of Gram-Schmidt [12] orthogonalizing filters with an MMSE processor.

Acknowledgment

The author would like to acknowledge the many helpful technical discussions with his colleagues at SIGNATRON, in particular Dr. S. Parl and Dr. S. Mui. Experimental data used in Fig. 5 were provided by J. Gadoury of GTE Sylvania.

References

[1] R. W. Lucky, "Techniques for adaptive equalization of digital communications systems," *Bell Syst. Tech. J.*, vol. 45, pp. 255–286, Feb. 1966.

[2] J. G. Proakis, "Adaptive receivers for digital signaling over random or unknown channels," Ph.D. dissertation, Harvard Univ., Cambridge, MA, 1966.

[3] D. M. Brady, "Adaptive coherent diversity receiver for data transmission through dispersive media," in *Conf. Rec., IEEE Int. Conf. Commun.*, San Francisco, CA, June 1970.

[4] P. Monsen, "Feedback equalization for fading dispersive channels," *IEEE Trans. Inform. Theory*, vol. IT-17, pp. 56–64, Jan. 1971.

[5] ——, "Theoretical and measured performance of a DFE modem on a fading multipath channel," *IEEE Trans. Commun.*, vol. COM-25, pp. 1144–1153, Oct. 1977.

[6] D. R. Kern and P. Monsen, "Megabit digital troposcatter subsystem (MDTS)," GTE Sylvania, Needham, MA, and SIGNATRON, Lexington, MA, Final Rep. ECOM-74-0040-F.

[7] G. Krause and P. Monsen, "Results of an angle diversity field test experiment," in *Nat. Telecommun. Conf. Rec.*, 1978.

[8] J. N. Pierce and S. Parl, "Digital tropo ECCM techiques study," RADC Final Rep. Contr. F30602-76-C-0225, by SIGNATRON, Inc., Lexington, MA, 1977.

[9] B. Widrow, "Adaptive sampled-data systems," in *Proc. 1st Int. Cong. IFAC, Automat. Remote Contr.*, Moscow, USSR, pp. 423–429, 1960.

[10] R. D. Gitlin and F. R. Magee, Jr., "Self-orthogonalizing adaptive equalization algorithm," *IEEE Trans. Commun.*, vol. COM-25, pp. 666–672, July 1977.

[11] L. A. Wainstein and V. D. Zubakov, *Extraction of Signals from Noise*. Englewood Cliffs, NJ: Prentice-Hall, 1962.

[12] F. B. Hildebrand, *Methods of Applied Mathematics*. Englewood Cliffs, NJ: Prentice-Hall, 1952.

[13] M. Schwartz, W. R. Bennett, and S. Stein, *Communications Systems and Techniques*. New York: McGraw-Hill, 1966, ch. 10.

PHOTO NOT AVAILABLE

Peter Monsen (S'60–M'64) received the B.S. degree in electrical engineering from Northeastern University, Boston, MA, in 1962, the M.S. degree in operation research from the Massachusetts Institute of Technology, Cambridge, in 1963, and the Eng.Sc.D. degree in electrical engineering from Columbia University, New York, NY, in 1970.

During a two-year period beginning in January 1964, he served as a Lieutenant in the U.S. Army at the Defense Communication Agency, Arlington, VA, where he was concerned with communication systems engineering. From 1966 to 1972 he was with Bell Laboratories, Holmdel, NJ, where he was Supervisor of a Transmission Studies Group whose work included fading-channel characterization and adaptive equalization. He is currently with SIGNATRON, Inc., Lexington, MA, where he was responsible for the development of a 12.6 Mbit/s troposcatter modem. His current areas of interest at SIGNATRON include adaptive signal processors, antijam techniques, and error correction coding. He has four patents on adaptive equalization modems.

The SL Undersea Lightguide System

PETER K. RUNGE AND PATRICK R. TRISCHITTA, MEMBER, IEEE

Abstract—**A digital optical fiber undersea cable system targeted for transatlantic service in 1988 is now under development at Bell Laboratories. The system uses single-mode fibers to carry data at a bit rate of 280 Mbits/s. Using digital speech compression techniques, a total system capacity of over 35 000 two-way voice channels can be realized. With laser transmitters at 1.3 μm, repeater spacings are expected to exceed 35 km. This paper discusses system parameters, repeaters, fiber and cable design, terminal equipment, and system measurements.**

INTRODUCTION

THE inherent advantages of fiber optic transmission can be truly exploited in the development of long distance, large capacity digital systems. A transoceanic cable system is a prime example of such a system which can directly benefit from the use of fiber optics. The requirements of a transoceanic cable system are such that the low loss, large bandwidth, and small size of a lightguide cable make a digital fiber optic undersea cable system both technically and economically attractive. By reviewing the history of Bell System deployment of transatlantic telephone cable systems (TAT-1–TAT-7) and by showing the growth in demand for transatlantic voice circuits, we can see how the development of a digital fiber optic undersea cable system can radically change the trends in submarine cable deployment.

The first transatlantic systems developed by Bell Laboratories were the Submarine B systems (SB) and were installed across the Atlantic Ocean in 1956 and 1959 (TAT-1 and TAT-2). (The Submarine A system (SA) was a prototype trial between Florida and Cuba.) By using frequency-division multiplexing, these analog systems had a capacity of 48, 3 kHz two-way voice circuits with a repeater spacing of 70 km. Using a larger transmission bandwidth and a larger diameter cable with a shorter repeater spacing, the SD systems followed in 1963 and 1965 each with a capacity of 138 voice circuits and a repeater spacing of 36 km. In 1970, the SF system was installed across the Atlantic, providing an additional 845 circuits at a shorter repeater spacing of 18 km. Finally, in 1976 (TAT-6) and in an installation planned for 1983, (TAT-7) the SG systems each provide an additional 4200 voice channels with a repeater spacing of 9.4 km.

With the demand for international voice circuits growing at an annual rate of 20–30 percent, a need exists to place into service many additional voice channels. (See Fig. 1.) In addition, the demand for a mix of voice, TV, and data signals along a transmission medium which is free from long transmission delay, constantly changing environmental interference, and possible political intervention is increasing, making under-

Manuscript received May 15, 1982; revised November 15, 1982.
The authors are with Bell Laboratories, Holmdel, NJ 07733.

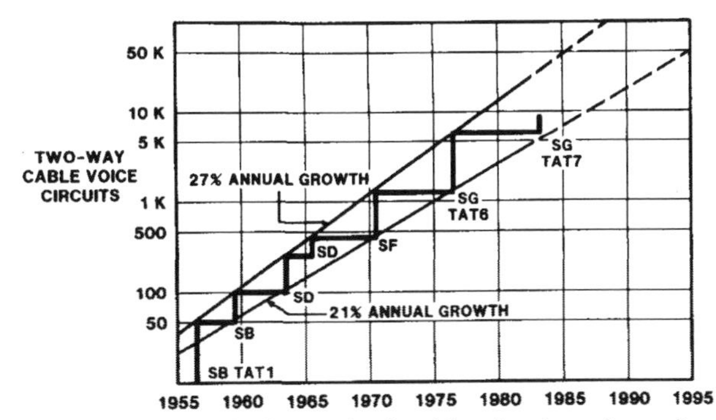

Fig. 1. The growth of transatlantic cable voice channel capacity.

sea cable systems a very valuable commodity. In order to meet this estimated demand with a larger capacity analog coaxial cable system, still larger bandwidths would require larger diameter coaxial cable and even shorter repeater spacings. A next-generation analog coaxial cable system SH which would have a capacity of 16 000 voice channels at a repeater spacing of only 4.3 km was constructed, but the maturing of the single-mode optical fiber technology has provided an opportunity to develop a new family of digital optical undersea cable systems with an expected economic saving over existing technologies [1].

System SL, under development and proposed for installation in 1988 as the TAT-8 undersea cable system, is a state-of-the-art digital fiber optic undersea cable system with a cable diameter and a repeater spacing similar (1/2) to the first SB systems, but with a voice channel capacity almost 1000 times that of the SB systems. Figs. 2 and 3 graphically show how this fiber optic system will change the trends in submarine cable deployment.

THE DIGITAL OPTICAL UNDERSEA CABLE SYSTEM—SYSTEM SL

Fig. 4 shows a schematic diagram of the SL system. The system consists of on-shore terminal equipment, the undersea cable, and the undersea repeaters which house the opto-electronic regenerators.

The terminal equipment contains a digital mutliplexer which accepts data signals directly and video signals through high-speed A/D converters and bandwidth compressors. The analog voice signals will be converted to digital signals using a combination of ADPCM (adaptive differential pulse code modulation) codecs and digital TASI (time assignment speech interpolation) equipment. The average digital bit rate per voice channel under heavy load conditions should be approximately 16 kbits/s, a four-fold improvement over standard PCM. The on-shore terminal also contains a high voltage power sup-

Reprinted from *IEEE Journal on Selected Areas in Communications,* vol. SAC-1, no. 3, April 1983.

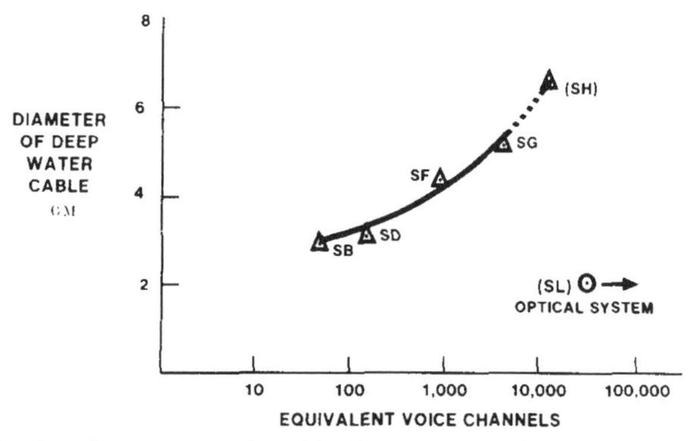

Fig. 2. Comparison of the cable diameters of previous cable systems.

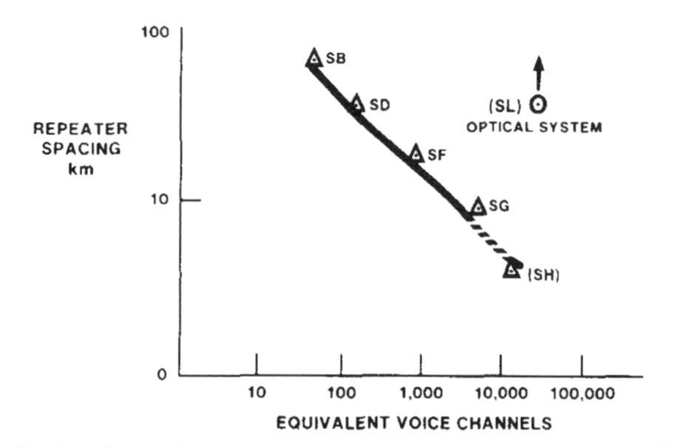

Fig. 3. Comparison of repeater spacing of previous cable systems.

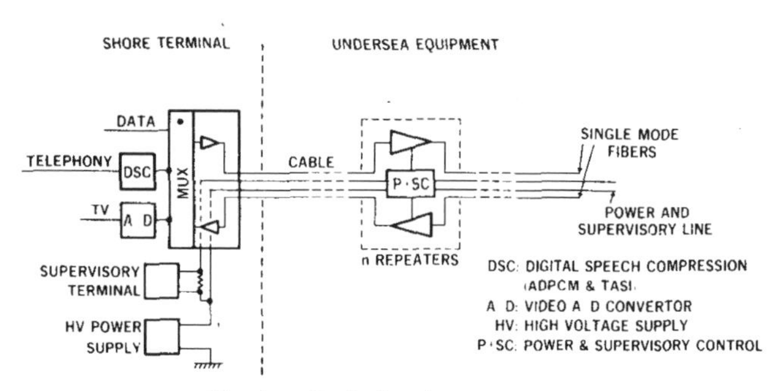

Fig. 4. Optical undersea system.

ply which provides a constant current for repeater powering, and a terminal for supervisory control.

The undersea system length is 6500 km for a transatlantic crossing with a maximum ocean depth of 5.5 km. The system also contains provisions for possible undersea branching of the cable system. Up to five undersea *Y* type branches could provide cable terminals to six different countries (Fig. 5). The system will operate at a bit rate of 280 Mbits/s over each of four fibers with a repeater spacing of more than 35 km making the capacity of the SL system over 35 000 two-way voice channels. As in all undersea systems, reliability is very important as the system's goal is a 25 year life with not more than three repairs by a cable ship during that time [2].

THE UNDERSEA CABLE

A cross-sectional view of the lightguide cable [3] is shown in Fig. 6. The cable consists of a central core with a 2.6 mm outside diameter containing up to 12 fibers. The fibers are embedded in an elastomer and are helically wound around a central copper clad steel wire called the kingwire. The elastomer is used as a cushion for the fiber and to reduce cabling induced microbending losses. The elastomer core is surrounded by a thin covering of nylon and then a series of steel strands which provide cable strength. Surrounding these strength members is a continuously welded copper cylinder which acts as a hermetic seal for the fibers and as a power conductor for the undersea repeaters. Surrounding the conductor is a layer of low density polyethylene providing cable insulation and abrasion resistance. The outside diameter for the complete SL undersea cable is 21 mm compared to a 53 mm OD for the SG analog cable. The small size of the fiber optic cable results in material cost savings, and in easier development and storage aboard the cable laying ship.

THE FIBER

In order to support a bit rate of 280 Mbits/s over the longest possible distances, single-mode fiber must be used. Fig. 7 shows how bit rate, dispersion, and fiber loss will limit the repeater spacing of an optical system. At a bit rate of 280 Mbits/s, the transmission distance through multimode fiber is limited by modal dispersion. With a fiber bandwidth-distance product of 2500 MHz/km, repeater spacing is limited to under 10 km. In contrast, single-mode fiber supports just two modes of propagation, one of each polarization. The modal dispersion of the two polarization modes is small with little or no reported transmission penalty. The transmission properties of single-mode fiber are limited by chromatic dispersion and fiber loss. Chromatic dispersion is the sum of waveguide and material dispersions and arises from the variations of propagation delay with wavelength. By using a depressed-cladding refractive-index profile as shown in Fig. 8 [4], chromatic dispersion can be reduced to zero at the 1.3 μm wavelength while maintaining good mode confinement. This maximizes the fiber's transmission bandwidth at $\lambda = 1.3$ μm. Fig. 9 shows the measured values of group delay versus wavelength and the subsequent dispersion for a 101 km length of fiber. With a narrow laser source line width (2-4 nm) matched to the zero-dispersion wavelength of the single-mode fiber, bit rate and repeater spacing are limited only by fiber loss. This is shown in Fig. 7 for a single-mode fiber with a bandwidth-distance product of 25 GHz/km and a loss of 0.7 dB/km [5].

At $\lambda = 1.3$ μm, the fiber loss is at a local minimum. At this wavelength, single-mode fiber can support high bit rates over long distances, making it the most attractive medium for transoceanic cable systems. Fig. 10 shows the fiber loss versus wavelength for a 101 km length of single-mode fiber. At $\lambda = 1.3$ the loss is 0.38 dB/km and at 1.5 μm the fiber loss is 0.29 dB/km including 11 splices. Although at $\lambda = 1.5$ μm the fiber loss is at the absolute minimum, difficulty in achieving both minimum loss and zero chromatic dispersion at 1.5 μm has made the 1.3 μm wavelength the most logical choice for current system development. Future systems will most likely be deployed at the 1.5 μm wavelength when reliable single frequency lasers or single-mode fiber optimized the low loss and zero chromatic dispersion at $\lambda = 1.5$ μm become available [6], [7].

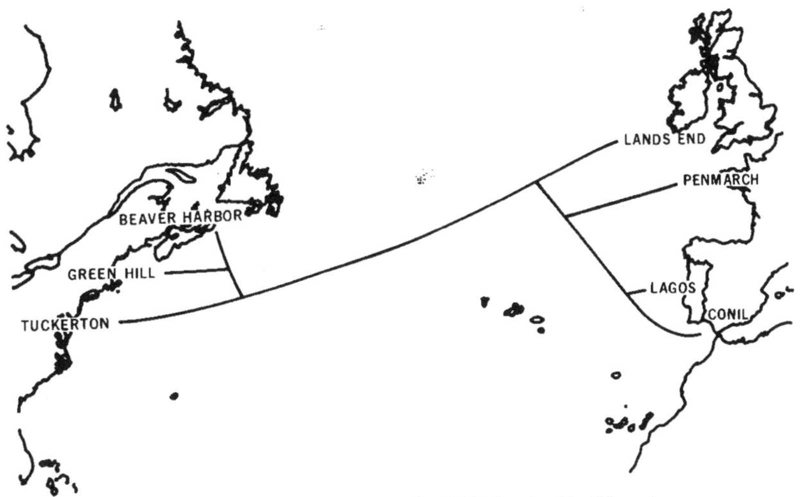

Fig. 5. A possible configuration for TAT-8 using the SL system.

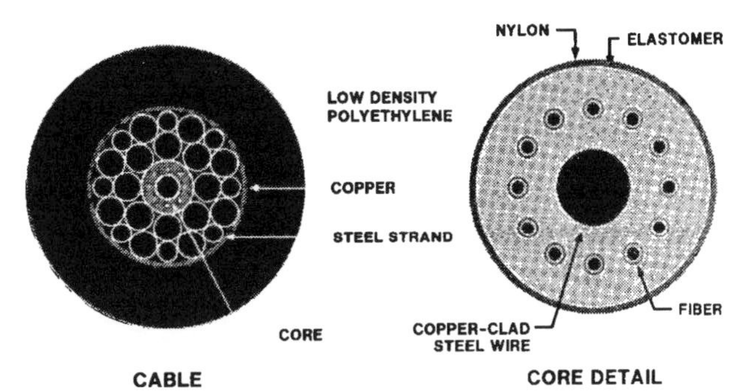

Fig. 6. Embedded-fiber core cable.

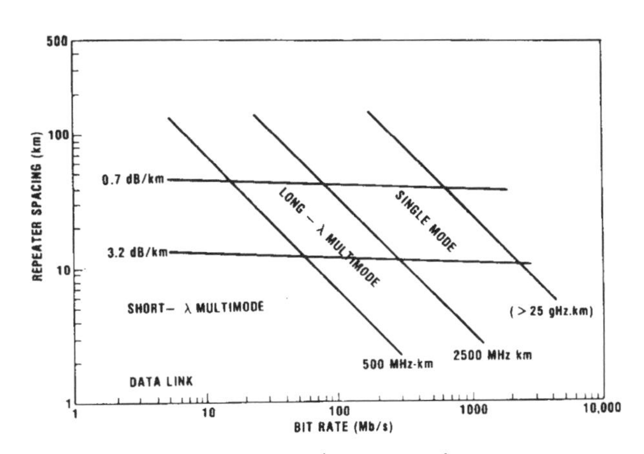

Fig. 7. Bit rate/repeater spacing.

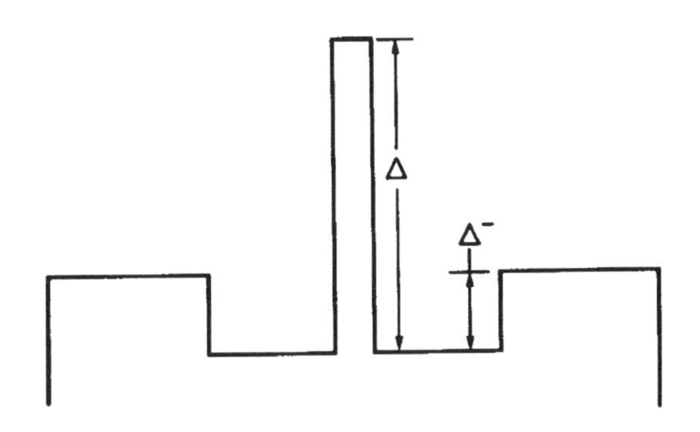

Fig. 8. Refractive-index profile of a depressed-cladding single-mode fiber.

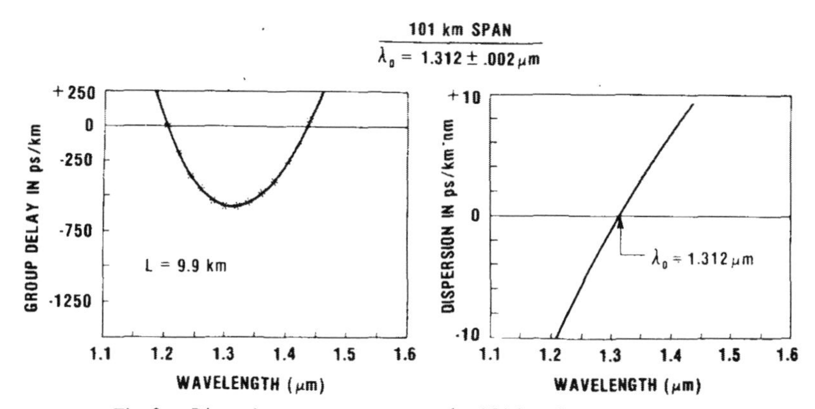

Fig. 9. Dispersion measurements on the 101 km single-mode fiber.

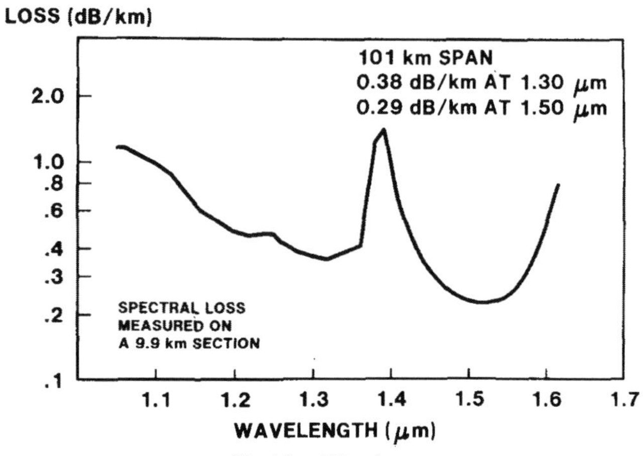

Fig. 10. Fiber loss.

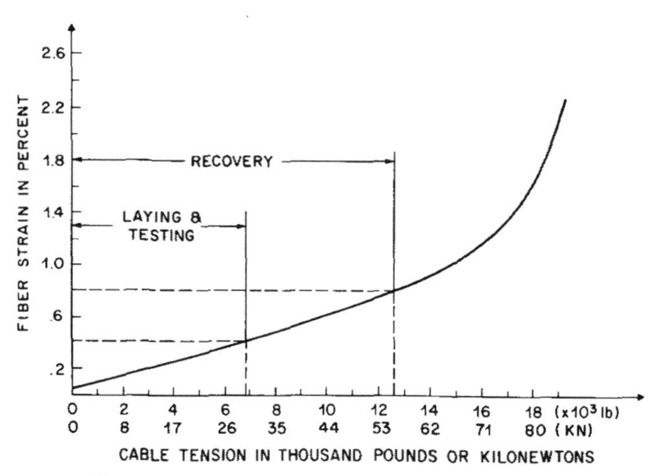

Fig. 11. The fiber strain during deep sea operations.

The single-mode fiber preform is made by depositing cladding and core material into starting tubes by the modified chemical vapor deposition process (MCVD) [8]. These preforms when drawn produce low-loss, large bandwidth, and high strength single-mode fibers. High strength fibers are necessary to survive the tension placed on the fibers during cable laying and recovery (Fig. 11). The most severe recovery operation in deep water may subject the fiber to a strain of approximately 0.8 percent which imposes a stringent proof-test requirement on the fiber. To survive the long-time tensile load during recovery operations, the fibers must be proof tested at stresses of approximately 200 kpsi as shown in Fig. 12 [9].

Fig. 13 shows the fiber loss for a recent 19 km cable manufacturing run at each manufacturing step. The loss of the fibers increases when they are rewound at high tension on cable machine bobbins to start the cabling process but return to their original value after the elastomer is added. The results at λ = 1.3 μm show that the cable manufacturing process does not significantly increase the loss in the fibers.

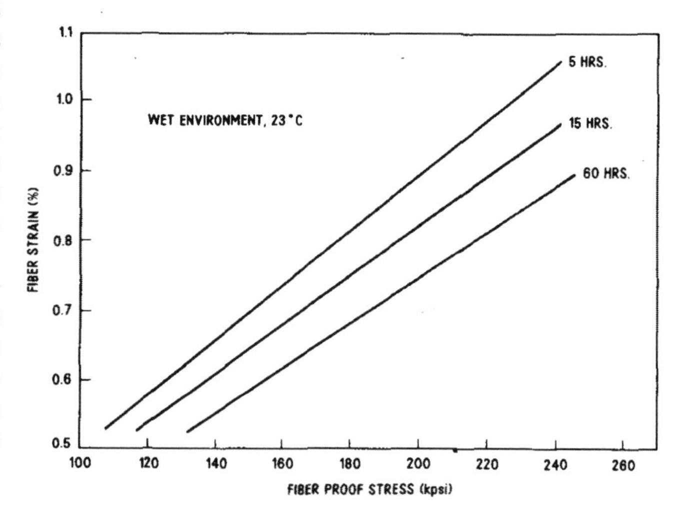

Fig. 12. Proof test requirements on undersea fiber.

THE UNDERSEA REPEATER

A rendering of a possible physical layout of a repeater is shown in Fig. 14 and a functional schematic is shown in Fig. 15. Powering for the repeater is provided over the copper

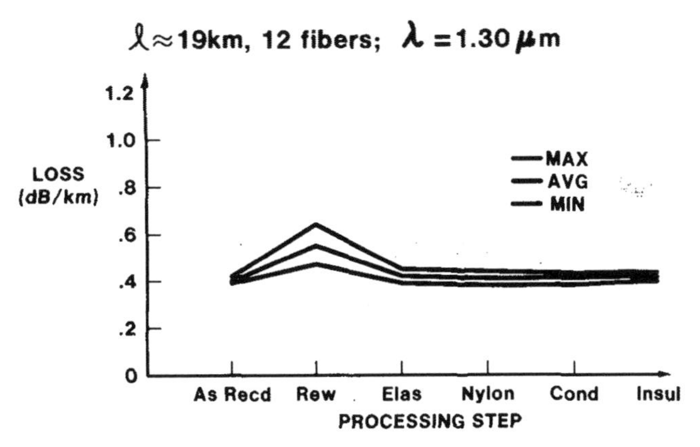

Fig. 13. Cabled fiber loss.

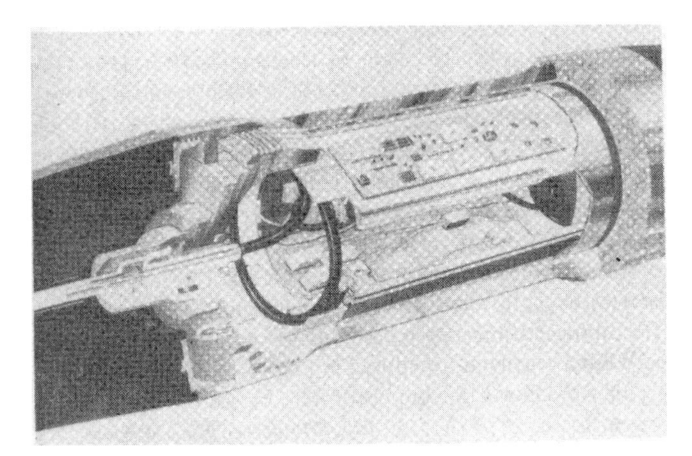

Fig. 14. Physical model of SL repeater.

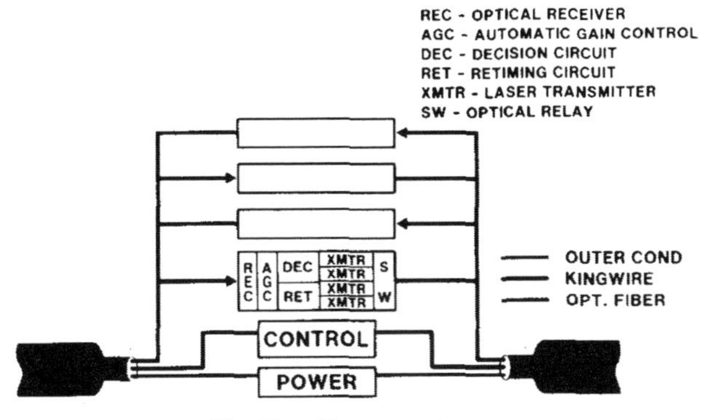

Fig. 15. SL repeater layout.

conductor of the cable with the sea acting as the ground conductor. Each optoelectronic regenerator contains a receiver, integrated circuits, and up to four laser transmitters connected to the output fiber through a 1 X 4 optical fiber relay [10]. By using one active laser with up to three cold standby lasers, the laser reliability requirement is reduced by several orders of magnitude. The laser reliability model assumes a log-normal distribution of laser failures with time ($\sigma = 2$) and an MTTF (mean time to failure) equal to 7.4 X median.) If the number of cold standby transmitters or spares is zero, the required MTTF of the lasers is 444 million hours; if 1, MTTF is 37; if 2, MTTF is 6.6; if 3, MTTF is 3.7. By using a low rate digital supervisory signal over the kingwire of the cable, continuous monitoring of all laser diode currents can take place, and the fiber relay can be activated to switch to a fresh laser when the shore terminal determines that a standby laser is needed. Laser diode lifetests are in progress to verify this model and establish the number of spare lasers required in each regenerator in order to meet the overall system reliability requirements.

THE OPTOELECTRONIC REGENERATORS

A block diagram of the optoelectronic regenerator is shown in Fig. 16 [11]. The optical pulses from the incoming fiber are coupled into an InGaAs p-i-n diode receiver followed by a silicon bipolar transimpedance amplifier. Although an APD would have a better sensitivity than a p-i-n, the high voltage supply needed for an APD would require a dc-dc converter. This extra device would add complexity and decrease reliability. A silicon bipolar transimpedance front-end amplifier is used for reasons of reliability, dynamic range, and ease of equalization. Silicon IC technology is a more proven technology than GaAs IC technology. A bipolar transimpedance amplifier will have a larger dynamic range than an FET integrating amplifier without requiring special line coding. A dynamic range of 15–20 dB (optical) is needed of the receiver to allow for manufacturing variation and aging of system components over the 25 year life of the system.

The NRZ signal from the receiver is passively equalized to a raised cosine shape and passed to an automatic gain control amplifier. This amplifier provides over 40 dB of variable gain (electrical) on a highly integrated monolythic Si IC. Integrated circuits are used to increase reliability and decrease cost. The constant amplitude signal is then split in two, feeding the decision circuit IC and the retiming circuit to recover a timing signal.

The timing signal is recovered by passing the data signal through a nonlinear device (a squarer) to obtain a discrete frequency component at the baud rate. This timing component is then filtered by a surface acoustic wave (SAW) filter [12]. The high Q of the filter limits the rms jitter accumulation of a transatlantic number of cascaded regenerators [13]. The output of the SAW filter is amplified by another IC amplifier and is used as the timing signal for the decision circuit.

The output of the decision circuit modulates a single transverse mode buried-heterostructure laser diode. The operating wavelength of the laser is near the zero-dispersion wavelength of the fiber ($\lambda = 1.3$ μm) and has an average light output power of 1 mW (0 dBm). By monitoring the back face of the laser with a p-i-n diode, the laser diode current is adjusted to keep the output light power constant. The laser diode current is then monitored by the supervisory circuit, and up to three spare lasers are available per regenerator and can be activated and switched in when appropriate. The light is coupled into the output single-mode fiber by a spherical lens formed on the

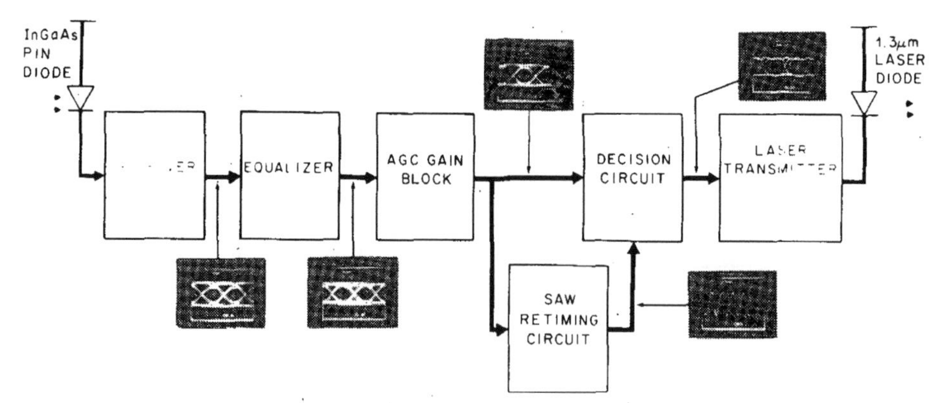

Fig. 16. Block diagram of SL regenerator.

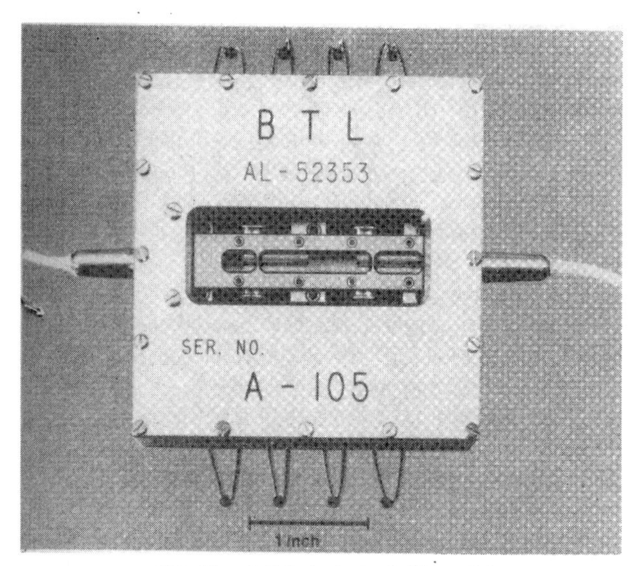

Fig. 17. 1 X 4 single-mode fiber relay.

fiber end. Coupling efficiencies of 40 percent have been reported [14].

THE SINGLE-MODE FIBER RELAY

The 1 X 4 single-mode optical fiber relay under consideration for laser sparing is shown in Fig. 17. It uses a movable array of precision etched silicon *V*-groove chips that accurately align the ~10 μm core fibers. Models have been built with insertion losses of less than 1 dB through each of the four input ports. The switching time is ~3 ms. These characteristics have been maintained over more than 250 000 switching cycles. The switch design is such that it limits the amount of light reflected back into the laser diode. The optical feedback bit error rate penalty is limited to less than 0.1 dB [15].

SYSTEM MEASUREMENTS–BIT ERROR RATE AND JITTER ACCUMULATION

Two system experiments have recently been conducted in order to measure the bit error rate in a long fiber and the jitter accumulation for a large number of cascaded regenerators.

The first experiment demonstrated error-free transmission at 274 Mbits/s over 101 km of single-mode fiber [16]. The fiber met all system requirements for deployment in a transatlantic crossing. The total loss at 1.3 μm of the 101 km span was 38 dB, or 0.38 dB/km including 11 splices. Fibers and splices had a minimum tensile proof-test strength of 200 kpsi thus meeting the necessary long term tensile strength requirement for recovery operations from deep water. The zero-dispersion wavelength for the fiber was 1.312 μm, and the laser wavelength was 1.301 μm, minimizing dispersion effects. The optoelectronic regenerator used was a highly integrated SAW filter regenerator pictured in Fig. 18.

Fig. 19 shows the bit error rate versus received optical power for a $2^{15} - 1$ bit pseudorandom NRZ data at 274 Mbits/s for various lengths of fiber. The overlapping curves demonstrate that dispersion effects were unmeasurable for fiber lengths up to 101 km. Also shown are measurements at 420 Mbits/s.

The second experiment was a circulating loop jitter accumulation experiment. By looping the regenerator's output back onto its input with fiber precisely cut to the length of a $2^{15} - 1$ bit pseudorandom word at 274 Mbits/s (approximately 25 km), a simulation of the accumlated jitter through a chain of cascaded optoelectronic regenerators can be made. Fig. 20 shows the experimental setup and Fig. 21 shows the accumulated jitter of 33 regenerator-fiber spans using the same type of laser, fiber, and regenerator as in the long fiber experiment. This experiment can eventually be used to simulate the jitter accumulation in over 200 cascaded regenerators, in order to confirm experimentally and prior to system installation that the system jitter specification will be met.

CONCLUSIONS

We have described elements of the digital optical fiber undersea cable system now under development for the TAT-8 transatlantic cable system at Bell Laboratories. The system will be a technological advance for undersea cable systems at expected cost savings over existing technology. This system, when installed, will set a new milestone in the history of fiber optic system design, introducing the first optical system to an application which demands extraordinary reliability.

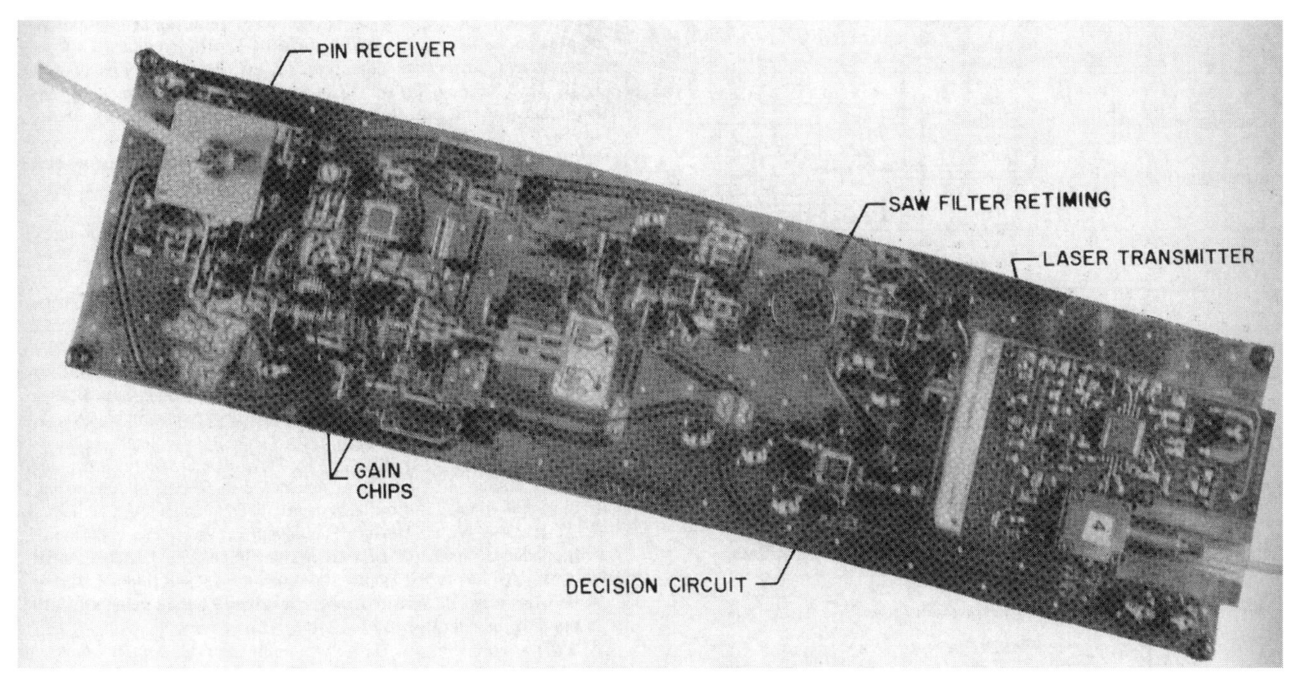

Fig. 18. A highly integrated SL regenerator.

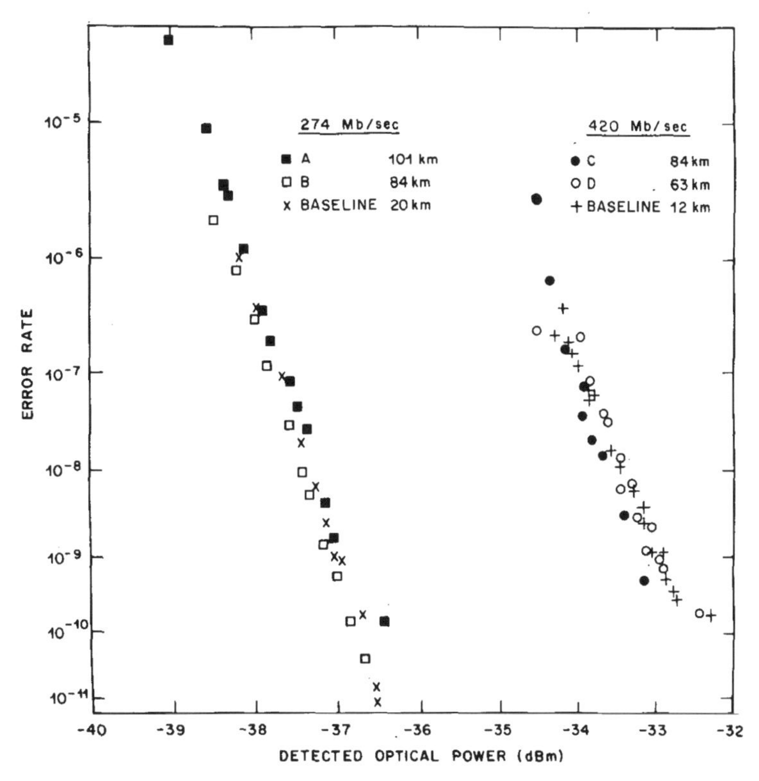

Fig. 19. Bit error rate measurements on 101 km span of fiber.

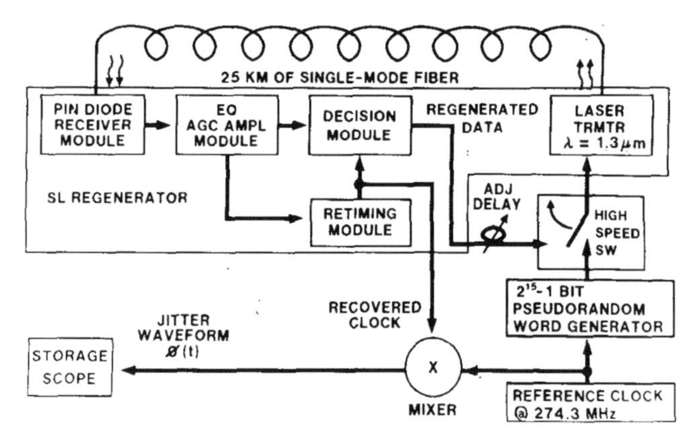

Fig. 20. Circulating loop jitter accumulation experiment.

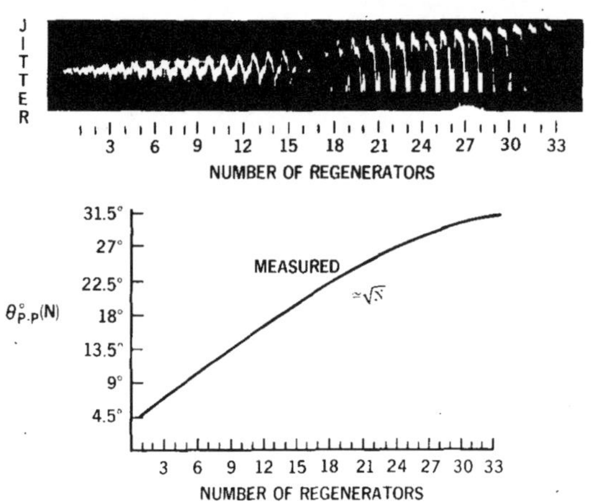

Fig. 21. Simulated jitter accumulation of a chain of identical regenerators.

ACKNOWLEDGMENT

The work and accomplishments of many people, both inside and outside the Bell System, are described briefly in this paper. We are indebted to them for their generous contributions of material for inclusion in this paper.

REFERENCES

[1] C. O. Anderson, R. F. Gleason, P. T. Hutchinson, and P. K. Runge, "An undersea communication system using fiberguide cable," *Proc. IEEE*, vol. 68, Oct. 1980.

[2] P. K. Runge, "A high capacity optical-fiber undersea cable system," in *Electron. Conf. Rec.*, New York, Apr. 1981, paper 8B/2.

[3] R. F. Gleason, R. C. Mondello, B. W. Fellows, and D. A. Hadfield, "Design and manufacture of an experimental lightwave cable for undersea transmission systems," in *Proc. Int. Wire Cable Symp.*, Nov. 14–16, 1978, p. 385.

[4] P. D. Lazay, A. D. Pearson, W. A. Reed, and P. J. Lamaire, "An improved single mode fiber design," in *Proc. CLEO 1981*, June 10–12, 1981, Washington, DC, paper W6 6-1.

[5] P. D. Lazay and A. D. Pearson, "Developments in single mode fiber design, materials and performance at Bell Laboratories," *IEEE J. Quantum Electron.*, Special Issue on Guided Wave Technology, Apr. 1982.

[6] K. R. Preston, K. C. Woollard, and K. H. Cameron, "External cavity controlled single longitudinal mode laser transmitter module," *Electron. Lett.*, vol. 17, Nov. 1981.

[7] L. G. Cohen, C. Lin, and W. G. French, "Tailoring zero chromatic dispersion into the 1.5-1.6 μm low-loss spectral region of single-mode fibers," *Electron. Lett.*, vol. 15, pp. 334–335, 1979.

[8] J. B. MacChesney, P. B. O'Connor, and H. M. Presby, "A new technique for the preparation of low-loss and graded-index optical fibers," *Proc. IEEE*, vol. 62, p. 1280, 1974.

[9] D. Kalish and B. K. Tariyal, "Static and dynamic fatigue of a polymer-coated fused silica optical fiber," *J. Amer. Ceram. Soc.*, vol. 61, Nov. 1978.

[10] W. C. Young and L. Curtis, "Cascaded multipole switch for single mode and multimode optical fibers," *Electron. Lett.*, vol. 17, Nov. 1981.

[11] R. E. Wagner, S. M. Abbott, R. F. Gleason, R. M. Paski, A. G. Richardson, D. G. Ross, and R. D. Tuminaro, "Lightwave undersea cable system" in *Proc. Int. Conf. Commun.*, June 1982.

[12] R. L. Rosenberg, D. G. Ross, and P. R. Trischitta, "SAW filter requirements for clock recovery in digital long haul optical fiber communication systems," in *Proc. Ultrasonic Symp.*, Chicago, IL, 1981.

[13] R. L. Rosenberg, D. G. Ross, P. R. Trischitta, D. A. Fishman, and C. B. Armitage, "Optical fiber repeatered transmission systems utilizing SAW filter," in *Proc. Ultrason. Symp.*, San Diego, CA, 1982.

[14] R. T. Ku and W. H. Dufft, "Hemispherical microlens coupling of semiconductor laser to single mode fiber," presented at OFC 1982, Phoenix, AZ, Apr. 15, 1982.

[15] V. J. Mazurczyk, "Sensitivity of single mode buried heterostructure lasers to reflected power at 274 Mb/s," *Electron. Lett.*, vol. 17, Feb. 5, 1981.

[16] M. M. Boenke, R. E. Wagner, and D. J. Will, "Transmission experiments through 101 km and 84 km of single mode fiber at 274 Mb/s and 420 Mb/s," *Electron. Lett.*, Nov. 1982.

Peter K. Runge received the M.S.E.E. degree in 1963 and the Ph.D.E.E. degree in 1967 from the Technical University of Braunschweig, Germany.

He joined Bell Laboratories, Holmdel, NJ, in 1967. He has been engaged in research of He-Ne, and organic dye lasers, exploratory development of fiber optic repeaters, and single fiber optic connectors. In 1976, he became Supervisor, responsible for the development of single fiber optic connectors. He was also engaged in exploratory development of active and passive fiber optic components. In 1979, he lead a group responsible for research on undersea lightwave systems. In 1980, he assumed his present position, Head, Undersea Lightwave Systems Development Department.

Patrick R. Trischitta (S'77–M'83) received the B.E.E. (with honors) degree in 1979 and the M.S.E.E. degree in 1980 from the Georgia Institute of Technology, Atlanta, GA. He is presently working toward the Ph.D. degree in electrical engineering at Rutgers University, New Brunswick, NJ.

Since 1980, he has been a member of the Technical Staff in the Undersea Lightwave Regenerator Development Group at Bell Laboratories, Holmdel, NJ.

Mr. Trischitta is a member of Eta Kappa Nu. He serves as the Associate Editor for Fiber Optics for the IEEE TRANSACTIONS ON COMMUNICATIONS.

Distribution of the Phase Angle Between Two Vectors Perturbed by Gaussian Noise

R. F. PAWULA, S. O. RICE, AND J. H. ROBERTS

Abstract—**The probability distribution of the phase angle between two vectors perturbed by correlated Gaussian noises is studied in detail. Definite integral expressions are derived for the distribution function, and its asymptotic behavior for large signal-to-noise is found for "small," "near $\pi/2$," and "large" angles. The results are applied to obtain new formulas for the symbol error rate in MDPSK, to calculate the distribution of instantaneous frequency, to study the error rate in digital FM with partial-bit integration in the post-detection filter, and to obtain a simplified expresion for the error rate in DPSK with a phase error in the reference signal. In the degenerate case in which one of the vectors is noise free, the results lead to the symbol error rate in MPSK.**

I. INTRODUCTION

THE distribution of the phase angle between two vectors arises in a variety of applications in communications systems [6], [11], and common cases are those in which information is transmitted in the phase or frequency of a sinusoid. Typically, the noise-corrupted version of such a sinusoid has the form

$$z(t) = A(t) \cos [\omega_c t + \phi(t)] + n(t) \tag{1}$$

in which ω_c is the carrier radian frequency, $A(t)$ and $\phi(t)$ are the time-varying amplitude and phase of the sinusoid, and $n(t)$ is a narrow-band Gaussian noise. In dealing with signals of this form, the vector (or phasor) representation is convenient to employ. With subscripts to denote different time instants, vector forms of $z(t)$ for the two times $t_1 = t$ and $t_2 = t + \tau$ are

$$\left.\begin{aligned} z_1 &= \text{Re}\{A_1 e^{i\omega_c t + i\phi_1} + (x_1 + iy_1)e^{i\omega_c t}\} \\ &= \text{Re}\{R_1 e^{i\omega_c t + i\theta_1}\} \\ z_2 &= \text{Re}\{A_2 e^{i\omega_c (t+\tau) + i\phi_2} + (x_2 + iy_2)e^{i\omega_c t}\} \\ &= \text{Re}\{R_2 e^{i\omega_c t + i\theta_2}\}. \end{aligned}\right\} \tag{2}$$

Graphical representations of these two vectors are shown in Fig. 1. The quantities x_j and y_j; $j = 1, 2$, are Gaussian variables related to the in-phase and quadrature components in the narrow-band representation of the noise at time t_j with assumed statistical properties $\bar{x}_j = \bar{y}_j = 0$ and var x_j = var $y_j = \sigma_j^2$. x_1 and y_1 are, in fact, the I and Q components at time t_1, whereas these components at time t_2 are the real and imaginary parts of $(x_2 + iy_2) \exp(-i\omega_c t)$.

Here θ_1 and θ_2 are assumed to lie between $-\pi$ and π. Consequently, $\theta_2 - \theta_1$ can range from -2π to 2π. However, we shall be concerned with the angle ψ defined as the modulo 2π difference between θ_2 and θ_1, viz.

$$\psi = (\theta_2 - \theta_1) \bmod 2\pi; \qquad -\pi \leqslant \theta_1, \theta_2 \leqslant \pi. \tag{3}$$

The modulo 2π appearing in this definition will be understood to mean "over *some* 2π interval" where the appropriate choice of the 2π interval may vary from application to application. We shall consider the probability $P\{\psi_1 \leqslant \psi \leqslant \psi_2\}$ where $\psi_1 \leqslant \psi_2$ and the interval (ψ_1, ψ_2) is assumed to be contained entirely within the particular 2π interval of definition being used.

Previous works which have considered the distribution of the phase angle between two vectors perturbed by Gaussian noise have been concerned primarily with problems in angle modulation and demodulation. References [8] and [11] contain extensive results in this area along with many references to related studies. Although the present paper may be considered as extensions and generalizations of some of the results found in these references, along with some new applications, we have attempted to make the paper self-contained. Another area in which the phase angle between two noisy vectors arises is that of phase comparison monopulse radar [14].

Expressions for the probability $P\{\psi_1 \leqslant \psi \leqslant \psi_2\}$ are presented in Section II for three different cases of signal parameter conditions and noise correlation. These results are derived by making use of an unconventional approach which relates the desired probability to a functional of the joint characteristic function of the narrow-band waveform. A second derivation in one of the simpler cases follows traditional lines and is given in an Appendix. Using the results, various asymptotic formulas for the probability $P\{\psi \geqslant \psi_0\}$ that ψ lies above some value ψ_0 are obtained in Section III. The results of Sections II and III are then applied in Section IV to the study of the calculation of error probabilities in MPSK, MDPSK, and digital FM communications systems, and to the distribution of instantaneous frequency. In particular, new formulas for the symbol error rate in MDPSK are obtained and compared with previously known results. The performance of narrow-band digital FM is considered for

Paper approved by the Editor for Communication Theory of the IEEE Communications Society for publication without oral presentation. Manuscript received September 29, 1981; revised March 25, 1982. The work of R. F. Pawula was performed under subcontract to Random Applications, Inc., San Diego, CA.

R. F. Pawula is at 552 Savoy St., San Diego, CA 92106.

S. O. Rice is with the Department of Electrical Engineering and Computer Science, University of California at San Diego, La Jolla, CA 92093.

J. H. Roberts is with Plessey Electronic Systems Ltd., Roke Manor, Romsey, Hants., England.

Reprinted from *IEEE Transactions on Communications,* vol. COM-30, no. 8, August 1982.

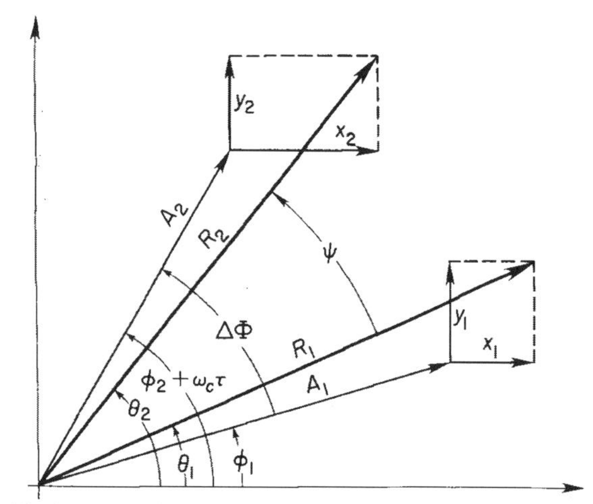

Fig. 1. Angle between two vectors perturbed by Gaussian noise.

limiter-discriminator detection in which the postdetection filter integrates for only a fraction of the bit time. In Section V, the Fourier series approach is briefly compared, by way of several examples, with the preceding treatments. The final section summarizes and discusses the results.

II. THE DISTRIBUTION $P\{\psi_1 \leqslant \psi \leqslant \psi_2\}$

A. Quantities Used in the Analysis

It is convenient to first define several parameters used in the analysis:

$$\Delta\Phi = (\omega_c\tau + \phi_2 - \phi_1) \bmod 2\pi^1 \tag{4a}$$

$$U = \tfrac{1}{2}(\rho_2 + \rho_1) \tag{4b}$$

$$V = \tfrac{1}{2}(\rho_2 - \rho_1) \tag{4c}$$

$$W = \sqrt{\rho_1\rho_2} = \sqrt{U^2 - V^2} \tag{4d}$$

where $\rho_j = A_j^2/2\sigma_j^2$ is the instantaneous signal-to-noise ratio. $\Delta\Phi$ is the signal phase change in the absence of noise and will be referred to simply as the "signal phase change." Note that $\Delta\Phi$ includes the carrier phase shift $\omega_c\tau$, for generality, although this term is frequently eliminated by receiver processing. The parameter U is the average signal-to-noise ratio of the two vectors and W is their geometric mean. $|V|$ is a measure of the "spread" in the SNR values.

When the noises perturbing the ends of the two vectors are correlated, the correlations between the x and y components will be assumed to take on the specialized forms

$$r = \frac{\overline{x_1x_2}}{\sigma_1\sigma_2} = \frac{\overline{y_1y_2}}{\sigma_1\sigma_2}$$

$$\lambda = \frac{\overline{x_1y_2}}{\sigma_1\sigma_2} = -\frac{\overline{y_1x_2}}{\sigma_1\sigma_2}. \tag{5}$$

As is known, when the noise is a wide-sense stationary random process, r and λ can be expressed in terms of the noise power spectrum. In this case, if the two-sided spectrum is $[S(f - f_c) + S(-f - f_c)]/2$, then

$$r = r_c(\tau)\cos\omega_c\tau - r_s(\tau)\sin\omega_c\tau$$

$$\lambda = r_c(\tau)\sin\omega_c\tau + r_s(\tau)\cos\omega_c\tau$$

where $r_c(\tau) = \int_{-\infty}^{\infty} S(f)\cos(2\pi f\tau)df$ and $r_s(\tau) = \int_{-\infty}^{\infty} S(f)\sin(2\pi f\tau)df$ and the spectral density has been assumed to be normalized to contain unity power.

Certain special cases of the above parameters will be considered in the following. In the case of "equal signal conditions," the two vectors will be meant to have equal SNR's with noise, and to lie in the same direction without noise, so that $\rho_1 = \rho_2$ and $\phi_1 = \phi_2$. Consequently $\Delta\Phi = 0$ and $V = 0$ (in this case, $W = U$ also). In the case of "uncorrelated noises," both $r = 0$ and $\lambda = 0$.

B. Distribution of a Modulo 2π Difference

The probability density function $p(\psi)$ of the modulo 2π difference $\psi = (\theta_2 - \theta_1) \bmod 2\pi$; $-\pi \leqslant \theta_1, \theta_2 \leqslant \pi$, can be found by integrating the joint probability density $p(\theta_1, \theta_2)$ of the phase angles θ_1 and θ_2 over the regions of the θ_1, θ_2 plane in which $(\theta_2 - \theta_1) \bmod 2\pi$ lies between ψ and $\psi + d\psi$ and θ_1 and θ_2 lie between $\pm\pi$. These regions are indicated in Fig. 2 for a particular value of ψ and, in general, are determined by the intersections of two of the three lines $\theta_2 = \theta_1 + \psi + 2\pi k; k = 0, \pm 1$, with the square of side 2π centered at the origin. Since the joint density $p(\theta_1, \theta_2)$ is of 2π periodicity[2] in each of its arguments, the small shaded region in the lower right-hand corner of the square can be raised by an amount 2π so that the resulting shaded region will be one continuous strip. Consequently, $p(\psi)$ can be determined from $p(\theta_1, \theta_2)$ by the integration

$$p(\psi) = \int_{-\pi}^{\pi} p(\theta_1, \theta_1 + \psi)\, d\theta_1. \tag{6}$$

In the more general case when θ_1 and θ_2 are not restricted to values in $(-\pi, \pi)$, the density of $\psi = (\theta_2 - \theta_1) \bmod 2\pi$ can be expressed in terms of the density of $\psi_a = \theta_2 - \theta_1$, the "actual" phase difference without the modulo 2π condition. The relation is

$$p(\psi) = \sum_{k=-\infty}^{\infty} p_a(\psi + 2\pi k) \tag{7}$$

where $p_a(\psi_a)$ is the density of ψ_a. In either case, the probability $P\{\psi_1 \leqslant \psi \leqslant \psi_2\}$ is then given by

$$P\{\psi_1 \leqslant \psi \leqslant \psi_2\} = \int_{\psi_1}^{\psi_2} p(\psi)\, d\psi. \tag{8}$$

[1] Again, the modulo 2π is to be interpreted so that the resulting $\Delta\Phi$ lies within the 2π interval of interest. However, in practical cases, $\Delta\Phi$ will usually be determined *first*, and the interval of definition then chosen. Frequently, this interval will be $(\Delta\Phi - \pi, \Delta\Phi + \pi)$. It is sufficient for our purposes here that $-\pi \leqslant \Delta\Phi \leqslant \pi$. Extensions to other values of $\Delta\Phi$ require that the intervals of definition of θ_1 and θ_2 be appropriately modified.

[2] Strictly speaking, it is not the joint density that is periodic, but rather a new function defined as the periodic extension of the joint density function that is periodic. A distinction will be drawn only when necessary.

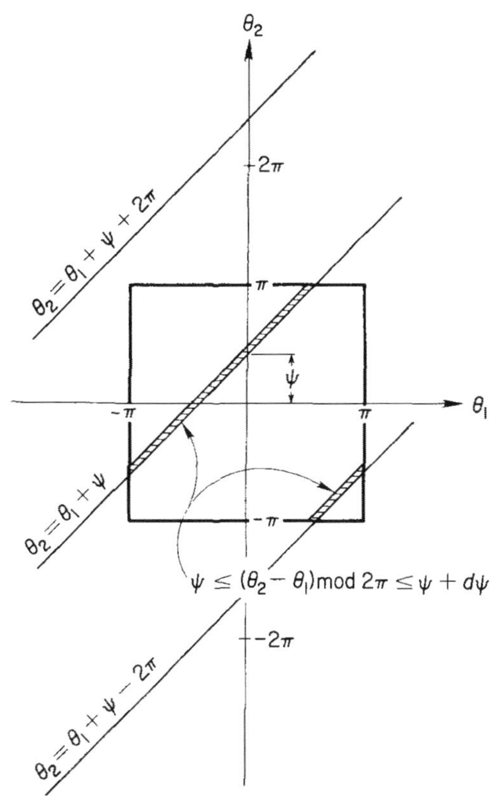

Fig. 2. Regions of integration for probability density of $\psi = (\theta_2 - \theta_1) \bmod 2\pi$.

As will later be seen, the probability $P\{\psi_1 \leqslant \psi \leqslant \psi_2\}$ can also be calculated without recourse to the intermediate step of first obtaining $p(\psi)$; however, (6) will indeed be employed in an alternative derivation which follows the more traditional and conventional approach.

C. Expressions for $P\{\psi_1 \leqslant \psi \leqslant \psi_2\}$

Let ψ_1 and ψ_2, with $\psi_1 < \psi_2$, be angles lying within the particular 2π interval of interest. Then the probability $P\{\psi_1 \leqslant \psi \leqslant \psi_2\}$ can be expressed in terms of an auxiliary function $F(\psi)$:

$$P\{\psi_1 \leqslant \psi \leqslant \psi_2\} = \begin{cases} F(\psi_2) - F(\psi_1) + 1; & \psi_1 < \Delta\Phi < \psi_2, \\ F(\psi_2) - F(\psi_1); & \psi_1 > \Delta\Phi \text{ or } \psi_2 < \Delta\Phi. \end{cases} \tag{9}$$

By its definition, $F(\psi)$ is periodic with period 2π, and is discontinuous at the "singular point" $\psi = \Delta\Phi$ with a jump $F(\Delta\Phi+) - F(\Delta\Phi-) = -1$. When either ψ_1 or ψ_2 equals $\Delta\Phi$, the value of $F(\psi)$ is to be interpreted as a right-hand or a left-hand limit; i.e., $F(\psi_1) = F(\Delta\Phi+)$ if $\psi_1 = \Delta\Phi$, and $F(\psi_2) = F(\Delta\Phi-)$ if $\psi_2 = \Delta\Phi$.

$F(\psi)$ and the classical distribution function of probability theory, say $\hat{F}(\psi)$, are related through[3]

$$\hat{F}(\psi) = F(\psi) - F(\psi_{\min}) + \tfrac{1}{2}\,[1 + \operatorname{sgn}(\psi - \Delta\Phi)]; \qquad \psi_{\min} \leqslant \psi \leqslant \psi_{\max}$$

[3] $\operatorname{sgn} x = +1$ if $x \geqslant 0$ and $= -1$ if $x < 0$.

where $(\psi_{\min}, \psi_{\max})$ is the interval of definition. Here, the signum fuction cancels the discontinuity in $F(\psi)$, and the $-F(\psi_{\min})$ term makes $\hat{F}(\psi_{\min}) = 0$, $\hat{F}(\psi_{\max}) = 1$. At all points of continuity of $F(\psi)$, $p(\psi) = \partial F(\psi)/\partial\psi = \partial\hat{F}(\psi)/\partial\psi$. We shall find it more convenient to deal with $F(\psi)$ than with $\hat{F}(\psi)$. In this section $F(\psi)$ is stated for three[4] different combinations of signal conditions and noise correlation. A derivation of the results will be given in Section II-E. Particular forms of $F(\psi)$ are as follows.

Case I—Equal Signal Conditions ($\Delta\Phi = 0$, $V = 0$), Uncorrelated Noises ($r = \lambda = 0$):

$$F(\psi) = \frac{-\sin\psi}{4\pi} \int_{-\pi/2}^{\pi/2} dt\, \frac{e^{-U(1-\cos\psi\cos t)}}{1 - \cos\psi\cos t} \tag{10}$$

Case II—Unequal Signal Conditions, Uncorrelated Noises:

$$F(\psi) = \frac{W\sin(\Delta\Phi - \psi)}{4\pi} \int_{-\pi/2}^{\pi/2} dt \cdot \frac{e^{-[U - V\sin t - W\cos(\Delta\Phi - \psi)\cos t]}}{U - V\sin t - W\cos(\Delta\Phi - \psi)\cos t} \tag{11}$$

Case III—Unequal Signal Conditions, Correlated Noises:

$$F(\psi) = \int_{-\pi/2}^{\pi/2} dt\, \frac{e^{-E}}{4\pi} \left[\frac{W\sin(\Delta\Phi - \psi)}{U - V\sin t - W\cos(\Delta\Phi - \psi)\cos t} + \frac{r\sin\psi - \lambda\cos\psi}{1 - (r\cos\psi + \lambda\sin\psi)\cos t} \right] \tag{12}$$

where

$$E = \frac{U - V\sin t - W\cos(\Delta\Phi - \psi)\cos t}{1 - (r\cos\psi + \lambda\sin\psi)\cos t}.$$

Note that in Case II $F(\psi)$ is a function of $\psi - \Delta\Phi$ only; however, this simple relation no longer holds in Case III because of the noise correlations. The behavior of $F(\psi)$ in Case I is illustrated in Fig. 3.

Equations (10)-(12) are written in a manner which illustrates how increasingly complex signal and noise conditions affect the results. Some simplification of the integrands, at the expense of complicating the limits, is possible in the most general case through a change of variables by a transformation due to Euler and Legendre (see (31) in Section II-E) which results in

Case III—Simplified Integrand Representation:[5]

$$F(\psi_i) = \frac{W\sin(\Delta\Phi - \psi_i)}{4\pi\sin\gamma_i} \int_{-\gamma_i-\delta_i}^{\gamma_i-\delta_i} d\theta\, \frac{e^{-(\alpha_i - \beta_i\cos\theta)}}{\alpha_i - \beta_i\cos\theta} + \frac{r\sin\psi_i - \lambda\cos\psi_i}{4\pi\sin\gamma_i} \int_{-\gamma_i-\delta_i}^{\gamma_i-\delta_i} d\theta\, e^{-(\alpha_i - \beta_i\cos\theta)} \tag{13}$$

[4] The corresponding probability density functions are given in Appendix B [cf. (B-1) and (B-6)].

[5] A subscript "i" will be used to indicate explicitly ψ dependence whenever other quantities in the same expression are ψ dependent. Also, because angles will be restricted to their principal values, the signum function is used in (14c) to give the correct sign of the square root.

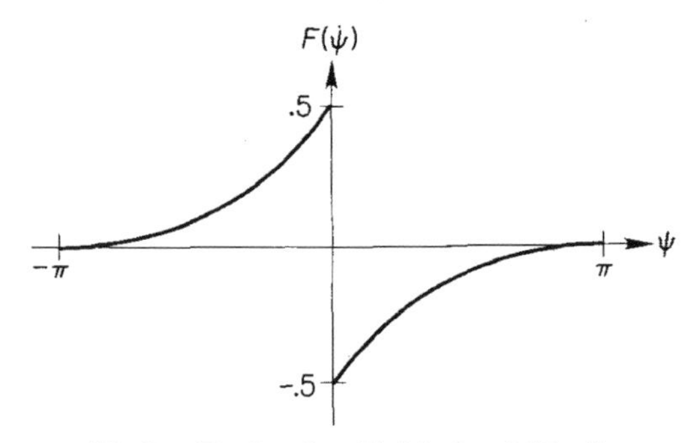

Fig. 3. The function $F(\psi)$ in Case I ($U = 1$).

where

$$\gamma_i = \cos^{-1}[-(r\cos\psi_i + \lambda\sin\psi_i)] \quad (14a)$$

$$\alpha_i = \frac{U + W\cos(\Delta\Phi - \psi_i)\cos\gamma_i}{\sin^2\gamma_i} \quad (14b)$$

$$\beta_i = \operatorname{sgn}[W\cos(\Delta\Phi - \psi_i) + U\cos\gamma_i] \cdot \sqrt{\alpha_i^2 - \frac{W^2\sin^2(\Delta\Phi - \psi_i)}{\sin^2\gamma_i}} \quad (14c)$$

$$\delta_i = \tan^{-1}\frac{V\sin\gamma_i}{W\cos(\Delta\Phi - \psi_i) + U\cos\gamma_i} = \sin^{-1}\frac{V}{\beta_i\sin\gamma_i} \quad (14d)$$

and the principal values of the arccosine on $(0, \pi)$, and of the arctangent and arcsine on $(-\pi/2, \pi/2)$, are to be taken. This form is not only helpful for numerical work, but will be found to be useful in Section III in obtaining asymptotic formulas, and in Section IV in deriving the distribution of instantaneous frequency.

D. Some Special Forms for $F(\psi)$

In this section, certain special situations are identified in which $F(\psi)$ and the difference $F(\psi_2) - F(\psi_1)$ assume simple forms.

First, in Case I, $F(\psi)$ has the special values [see (10)]

$$F(0\pm) = \mp\tfrac{1}{2}, \quad F(\pm\pi/2) = \mp\tfrac{1}{4}e^{-U}, \quad F(\pm\pi) = 0$$

and if the interval of definition of ψ is $(-\pi, \pi)$, from $F(\pi) = 0$ we get

$$P\{\psi \geq \psi_0\} = \begin{cases} 1 - F(\psi_0); & \text{if } \psi_0 < 0 \\ -F(\psi_0); & \text{if } \psi_0 > 0. \end{cases} \quad (15)$$

In particular, $P\{\psi \geq 0\} = 1/2$ and $P\{\psi \geq \pi/2\} = \exp(-U)/4$.

In Case II, from (11)

$$F(\Delta\Phi - \pi/2) = \tfrac{1}{4}[1 - (W/U)Ie(V/U, U)] = \frac{W}{4\pi}\int_0^\pi d\theta\,\frac{e^{-(U - V\cos\theta)}}{U - V\cos\theta} \quad (16)$$

where the $Ie(\cdot)$ function is defined in Appendix C. Also in Case II, $F(\Delta\Phi \pm \pi) = 0$, so that if the interval of definition of ψ is $(\Delta\Phi - \pi, \Delta\Phi + \pi)$, then

$$P\{\psi \geq \psi_0\} = \begin{cases} 1 - F(\psi_0); & \text{if } \psi_0 < \Delta\Phi \\ -F(\psi_0); & \text{if } \psi_0 > \Delta\Phi. \end{cases} \quad (17)$$

In all three cases, $F(\psi)$ is of 2π-periodicity in ψ so that $F(\psi) - F(\psi \pm 2\pi) = 0$. This guarantees that the probability density function $p(\psi)$ integrates to unity over any 2π interval (containing $\Delta\Phi$).

In Case III, the difference $F(\psi) - F(\psi - \pi)$ is expressible in terms of tabulated functions. Upon forming the difference, and using (13) and the definitions (14a)–(14d), the difference terms can be combined into integrals with limits ranging over 2π intervals, and there follows

$$F(\psi_i) - F(\psi_i - \pi) = \tfrac{1}{2}\operatorname{sgn}[\sin(\Delta\Phi - \psi_i)][1 - \sqrt{1 - \beta_i^2/\alpha_i^2}\, Ie(\beta_i/\alpha_i, \alpha_i)] + \frac{r\sin\psi_i - \lambda\cos\psi_i}{2\sin\gamma_i}\, e^{-\alpha_i}I_0(\beta_i) \quad (18)$$

where $I_0(\cdot)$ is a modified Bessel function. We note that this last equation can also be expressed in terms of the Marcum Q-function ([7] and [11, p. 85]).

E. Derivation of $P\{\psi_1 \leq \psi \leq \psi_2\}$

A derivation of $P\{\psi_1 \leq \psi \leq \psi_2\}$ for the most general case, Case III, of a signal with unequal parameters and correlated noises will be given using the following unconventional approach which makes use of the joint characteristic function of the narrow-band waveform under consideration. This approach has found wide application to problems related to noise zero crossings, instantaneous frequency, and FM and PM demodulation [11]. An alternate derivation for Case II is given in Appendix B.

For a general, noise-corrupted, narrow-band waveform with samples $z_1 = R_1\cos(\omega_c t + \theta_1)$, $z_2 = R_2\cos(\omega_c t + \theta_2)$, the joint characteristic function $\Phi(u, v)$ is [11, ch. 2]

$$\Phi(u, v) = E[e^{iuz_1 + ivz_2}] = E[J_0(\sqrt{u^2R_1^2 + v^2R_2^2 + 2uvR_1R_2\cos(\theta_2 - \theta_1)})] + \text{HF terms} \quad (19)$$

in which the expectation in the final form is over the random variables R_1, R_2, θ_1, and θ_2 and, also, the high-frequency terms will be assumed to vanish by virture of the expectation.[6] By subtracting some angle, say ξ, from the argument of the cosine prior to doing the expectation, a closely related *conditional* characteristic function $\Phi(u, v \mid \xi)$ can be defined as

$$\Phi(u, v \mid \xi) = E[J_0(\sqrt{u^2R_1^2 + v^2R_2^2 + 2uvR_1R_2\cos(\theta_2 - \theta_1 - \xi)}) \mid \xi] \quad (20)$$

so that $\Phi(u, v) = \Phi(u, v \mid \xi = 0)$. This conditional characteristic function may be regarded as the joint characteristic function of the two random variables $z_1' = R_1\cos(\omega_c t + \theta_1 + \xi)$

[6] Without the loss of generality, the oscillator which produces ω_c can be assumed to have a uniformly distributed random phase; and, as a consequence, the high frequency terms will vanish when averaged over this phase.

and $z_2 = R_2 \cos(\omega_c t + \theta_2)$. Then, using this definition, Roberts' general result for the distribution of the modulo 2π reduced phase angle $\psi = (\theta_2 - \theta_1) \bmod 2\pi$ can be stated as [11, sect. 2.4]

$$P\{\psi_1 \leqslant \psi \leqslant \psi_2\} = G(\psi_2) - G(\psi_1) \tag{21}$$

where

$$G(\psi_i) = \frac{\psi_i}{2\pi} - \frac{1}{2\pi} \int_0^\infty \int_0^\infty \frac{\partial}{\partial \psi_i} \Phi(-u, v \mid \psi_i) \frac{du\, dv}{uv}. \tag{22}$$

A derivation of this relation is given in Appendix A. Physically, $G(\psi)$ is related to the classical distribution function $\hat{F}(\psi)$ by $\hat{F}(\psi) = G(\psi) - G(\psi_{\min})$.

In order to apply this result, it is necessary first to calculate the joint conditional characteristic function which, in the case of independent signal and noise, can be written as the product $\Phi_{S+N} = \Phi_S \times \Phi_N$, where the subscripts "$S$" and "$N$" refer to the signal and noise components, respectively. In the case of the particular narrow-band waveform of interest [cf. (2)], these signal and noise components are readily shown to be (cf. [11, eq. (2.10)])

$$\Phi_S(u, v \mid \psi_i) = J_0\left(\sqrt{u^2 A_1^2 + v^2 A_2^2 + 2uvA_1A_2 \cos(\Delta\Phi - \psi_i)}\right) \tag{23}$$

$$\Phi_N(u, v \mid \psi_i) = e^{-(\sigma_1^2 u^2 + \sigma_2^2 v^2 - 2\sigma_1 \sigma_2 uv \cos \gamma_i)/2} \tag{24}$$

where γ_i is defined by (14a), and where[7] $\Delta\Phi$ is the signal phase change as given by (4a). Now, it is a matter of using these in (22) and performing the differentiation and integrations. These operations are simplified somewhat by first making a transformation to polar coordinates by means of

$$u = \frac{\sqrt{2}R}{\sigma_1} \cos\frac{\theta}{2}; \quad v = \frac{\sqrt{2}R}{\sigma_2} \sin\frac{\theta}{2} \tag{25}$$

and then the change $\theta' = \pi/2 - \theta$. Dropping the primes, we find

$$G(\psi_i) = \frac{\psi_i}{2\pi} - \int_{-\pi/2}^{\pi/2} \frac{d\theta}{2\pi \cos\theta} \int_0^\infty \frac{dR}{R} \frac{\partial}{\partial \psi_i} \cdot \left[J_0\left(2R\sqrt{U - V\sin\theta - W\cos(\Delta\Phi - \psi_i)\cos\theta}\right) \times e^{-R^2(1+\cos\gamma_i \cos\theta)}\right]. \tag{26}$$

Now, performing the differentiation and employing Hankel's exponential integrals (Appendix C) to do the remaining integrals over R gives

$$G(\psi_i) = F(\psi_i) + g(\psi_i) \tag{27}$$

where $F(\psi_i)$ is given by (12) and

$$g(\psi_i) = \frac{\psi_i}{2\pi} - \frac{W \sin(\Delta\Phi - \psi_i)}{4\pi} \cdot \int_{-\pi/2}^{\pi/2} \frac{d\theta}{U - V\sin\theta - W\cos(\Delta\Phi - \psi_i)\cos\theta}. \tag{28}$$

With the changes of variable $\theta' = \pi/2 - \theta$, $z = \tan(\theta'/2)$, the integral in $g(\psi_i)$ can be evaluated straightforwardly and yields[8]

$$g(\psi_i) = \begin{cases} \dfrac{\Delta\Phi + \pi}{2\pi}, & \text{if } \Delta\Phi < \psi_i \leqslant \Delta\Phi + 2\pi, \\[2ex] \dfrac{\Delta\Phi - \pi}{2\pi}, & \text{if } \Delta\Phi - 2\pi \leqslant \psi_i \leqslant \Delta\Phi. \end{cases} \tag{29}$$

Hence, for $\psi_1 < \psi_2$,

$$g(\psi_2) - g(\psi_1) = \begin{cases} 1, & \text{if } \psi_1 < \Delta\Phi \leqslant \psi_2, \\ 0, & \text{otherwise,} \end{cases} \tag{30}$$

which completes the derivation of (9) and (12). Equation (13) follows from (12) by use of a change of variables due to Euler and Legendre [4, p. 316]

$$(1 - d\cos t)(1 + d\cos\theta) = 1 - d^2, \tag{31}$$

where the transformation is from the variable t to the variable θ, and where $d = r\cos\psi_i + \lambda \sin\psi_i = -\cos\gamma_i$.

III. ASYMPTOTIC FORMULAS FOR $P\{\psi \geqslant \psi_0\}$

The probability $P\{\psi \geqslant \psi_0\}$ that ψ lies above some value ψ_0 is frequently of interest in certain communications problems since this probability can often be related to the probability of making an error in detecting an information symbol. In many cases this probability is not expressible in terms of tabulated functions and, in addition, values of this probability for a large parameter, the signal-to-noise ratio, are of primary concern. Hence, in this section, asymptotic formulas for $P\{\psi \geqslant \psi_0\}$ will be considered for large U. For the sake of definiteness, the interval of definition of ψ will be taken to be $\Delta\Phi - \pi \leqslant \psi \leqslant \Delta\Phi + \pi$, since $\psi = \Delta\Phi$ in the absence of noise. Then, $P\{\psi \geqslant \psi_0\} = P\{\psi_0 \leqslant \psi \leqslant \Delta\Phi + \pi\}$ so that (9)

[7] The symbol "Φ" is being used both to denote a characteristic function and in the definition of the signal phase change. The arguments of the characteristic function identify it and there should be no danger of confusion.

[8] A similar integral is evaluated in Appendix A. Cf. (A-4). The value of $g(\psi_i)$ at the jump $\psi_i = \Delta\Phi$ has been chosen as the limit from the left. The actual values of $g(\psi_2) - g(\psi_1)$ at the jumps $\Delta\Phi = \psi_1$ and $\Delta\Phi = \psi_2$ are of no consequence in computing $P\{\psi_1 \leqslant \psi \leqslant \psi_2\}$ if the discontinuity in $g(\psi_i)$ is chosen to cancel that in $F(\psi_i)$, so that the sum $F(\psi_i) + g(\psi_i)$ is continuous.

applies with $\psi_1 = \psi_0$ and $\psi_2 = \Delta\Phi + \pi$. In the following, we take $\psi_0 > \Delta\Phi$ so that $P\{\psi \geqslant \psi_0\} = F(\Delta\Phi + \pi) - F(\psi_0)$. Furthermore, in Cases I and II, $F(\Delta\Phi + \pi) = 0$; and in Case III, $F(\Delta\Phi + \pi)$ will be negligible in comparison to $-F(\psi_0)$ unless ψ_0 is near $\Delta\Phi + \pi$. Thus, $P\{\psi \geqslant \psi_0\} = -F(\psi_0)$ in Cases I and II, and in Case III, when ψ_0 is not near $\Delta\Phi + \pi$, $P\{\psi \geqslant \psi_0\} \sim -F(\psi_0)$.

A. Small $\psi_0 - \Delta\Phi$

The term $\alpha_i - \beta_i \cos\theta$ appearing in the exponents of the integrands in (13) will have a minimum at $\theta = 0$ provided that $\beta_i > 0$. This minimum will lie within the range of integration whenever $0 < \gamma_i - \delta_i < 2\gamma_i$, and the condition $\beta_i > 0$ will be satisfied when [cf. (14c)]

$$W \cos(\Delta\Phi - \psi_i) > U(r \cos\psi_i + \lambda \sin\psi_i). \quad (32)$$

The left-hand side of this inequality is largest when $\Delta\Phi - \psi_i = 0$. Furthermore, at $\psi_i = \Delta\Phi$, the minimum of $\alpha_i - \beta_i$ is zero as can be seen from (14c), viz.

$$\alpha_i - \beta_i = \frac{W^2 \sin^2(\Delta\Phi - \psi_i)}{(\alpha_i + \beta_i)\sin^2\gamma_i} \quad (33)$$

Assuming the conditions for a minimum within the range of integration to be fulfilled, $\cos\theta$ in (13) can be approximated by $1 - \theta^2/2$, the range of integration extended to $\pm\infty$, and use made of the identity[9] [1, eq. (7.4.11)]

$$\int_{-\infty}^{\infty} \frac{e^{-at^2}}{x^2 + t^2}\, dt = \frac{\pi}{x} e^{ax^2} \operatorname{erfc}(x\sqrt{a}) \quad (34)$$

which then leads to the asymptotic formulas[10]

i) Case I:

$$P\{\psi \geqslant \psi_0\} \sim \sqrt{\frac{1 + \cos\psi_0}{8\cos\psi_0}} \operatorname{erfc}\sqrt{U(1 - \cos\psi_0)} \quad (35a)$$

ii) Case II:

$$P\{\psi \geqslant \Delta\Phi + \psi_0\} \sim \sqrt{\frac{U + \beta_0}{8\beta_0}} \operatorname{erfc}\sqrt{U - \beta_0} \quad (35b)$$

iii) Case III:

$$P\{\psi \geqslant \psi_0\} \sim \sqrt{\frac{\alpha_i + \beta_i}{8\beta_i}} \operatorname{erfc}\sqrt{\alpha_i - \beta_i} - \frac{r \sin\psi_0 - \lambda\cos\psi_0}{\sin\gamma_i} \frac{e^{-(\alpha_i - \beta_i)}}{\sqrt{8\pi\beta_i}} \quad (35c)$$

[9] $\operatorname{erfc} z = (2/\sqrt{\pi})\int_z^\infty \exp(-t^2)\, dt$.

[10] Equations (35a), (35b), and (35c) can be further simplified with, however, loss of some accuracy, by use of the asymptotic behavior of the complementary error function; i.e., $\operatorname{erfc} z \sim (1/z\sqrt{\pi}) \exp(-z^2)$ for large z.

where

$$\beta_0 = \sqrt{U^2 \cos^2\psi_0 + V^2 \sin^2\psi_0}$$

and, in (35c), γ_i, α_i, and β_i are defined by (14a)-(14c) with $\psi_i = \psi_0$.

B. $\psi_0 - \Delta\Phi = \pi/2$

The point $\psi_0 = \Delta\Phi + \pi/2$ is of special importance, for this point is a "turning point" in the vicinity of which the character of the asymptotic behavior of $P\{\psi \geqslant \psi_0\}$ changes. In Cases I and II, closed-form expressions can be obtained and are

i) Case I:

$$P\{\psi \geqslant \pi/2\} = \tfrac{1}{4} e^{-U} \quad (36)$$

ii) Case II:

$$P\{\psi \geqslant \Delta\Phi + \pi/2\} = \tfrac{1}{4}[1 - (W/U)Ie(V/U, U)]. \quad (37)$$

These follow directly from (15) and (16).

C. $\psi_0 - \Delta\Phi$ Near $\pi/2$

Simple asymptotic formulas in the vicinity of the point $\psi_0 = \Delta\Phi + \pi/2$ evidently cannot be obtained, except in Case I, without further assumptions about the relative sizes of the system parameters. For Cases I and II, the asymptotic formulas are

i) Case I:

$$P\{\psi \geqslant \psi_0\} \sim \tfrac{1}{4} e^{-U} \sin\psi_0 [I_0(U\cos\psi_0) + \mathbf{L}_0(U\cos\psi_0)] \quad (38)$$

ii) Case II:

$$P\{\psi \geqslant \Delta\Phi + \psi_0\} \sim \frac{We^{-U}\sin\psi_0}{4\pi U} \int_{-\pi/2-\delta_0}^{\pi/2-\delta_0} d\theta\, e^{\beta_0\cos\theta} \quad (39)$$

where $\beta_0 = \operatorname{sgn}(\cos\psi_0)\sqrt{U^2\cos^2\psi_0 + V^2\sin^2\psi_0}$, $\delta_0 = \sin^{-1}(V/\beta_0)$, and $V/U \ll 1$. $\mathbf{L}_0(\cdot)$ is a modified Struve function [1, ch. 12]. These last two equations follow directly from (10) and (11) by approximating the denominators by their values at $\psi = \Delta\Phi + \pi/2$. Equation (39) is not much simpler than the exact expression (13) and, hence, for computational purposes (13) is to be preferred.

D. Large $\psi_0 - \Delta\Phi$

Since the left-hand side of (32) is negative for $\psi_i - \Delta\Phi$ near π, β_i will have a negative sign in Cases I and II and also in Case III for small correlations r and λ. Asymptotic formulas in these cases can be obtained directly from (13) by noting that the major contributions to the integrals come from the endpoints; i.e., the minima of the exponent with a negative β_i. Expanding about these points, we are led to the asymptotic formulas (cf. [8, Appendix]):

i) Case I:

$$P\{\psi \geqslant \psi_0\} \sim \frac{e^{-U}(-\tan\psi_0)}{2\pi U} \quad (40)$$

ii) Case II:

$$P\{\psi \geqslant \psi_0\} \sim \frac{e^{-U}\tan(\Delta\Phi - \psi_0)}{2\pi W^2}(U\cosh V + V\sinh V) \quad (41)$$

iii) Case III: Although the asymptotic formula can be worked out in this case, the end result is long and cumbersome and will not be given here in detail. It can be shown that

$$P\{\psi \geqslant \psi_0\} \sim F(\Delta\Phi + \pi) - F(\psi_0) \tag{42}$$

where the asymptotic form of $F(\Delta\Phi + \pi)$ is the sum of two nonzero terms and that of $F(\psi_0)$ is composed of the sum of four.

IV. APPLICATIONS

This section presents some illustrative applications of the preceding results. In the first of these, new formulas for the symbol error probability in MDPSK are given, and are compared with previously known formulas. The second example shows how the distribution of the instantaneous frequency of a narrow-band wave can be obtained from the Case III simplified representation. In the third application, use of the Case III distribution is illustrated in the calculation of the bit error rate performance of narrow-band digital FM, with limiter-discriminator detection and partial-bit integration postdetection filtering. These results are also compared with those for an ideal frequency detector by making use of the results for the distribution of instantaneous frequency obtained in the second application. In a fourth application, a simplified expression is obtained for the BER in DPSK with a phase error in the reference signal. The final example considers the error probability in MPSK.

A. Symbol Error Rate in MDPSK

In M-ary differential phase shift keying (MDPSK), information is transmitted as the phase difference between two consecutive sinusoidal pulses, and thus the receiver decision variable can be viewed as being the phase difference between two vectors perturbed by Gaussian noise. For a maximum *a posteriori* probability receiver employing differentially coherent detection of equiprobable, equal energy MDPSK signals, the probability of a symbol error is [6, ch. 5]

$$P_E(M) = 2 \int_{\pi/M}^{\pi} p(\psi)\, d\psi \tag{43}$$

in which the noise has been taken to be uncorrelated at the sampling instants. Using the results of Section II, for the Case I distribution with $\psi_2 = \pi$ and $\psi_1 = \pi/M$, we find the symbol error probability[11] to be

$$P_E(M) = \frac{\sin(\pi/M)}{2\pi} \int_{-\pi/2}^{\pi/2} dt\, \frac{e^{-\rho[1-\cos(\pi/M)\cos t]}}{1 - \cos(\pi/M)\cos t} \tag{44}$$

[11] We are indebted to M. K. Simon for pointing out the following relation between the bit and symbol error probabilities in the case of quaternary DPSK, $M = 4$. In DQPSK, a symbol error results in a *single* bit error whenever $\pi/4 \leqslant |\psi| \leqslant 3\pi/4$, and results in *two* bit errors when $3\pi/4 \leqslant |\psi| \leqslant \pi$. Using these, we can show that the *bit* error probability $P_B(4)$ is related to the *symbol* error probability $P_E(4)$ by

$$P_B(4) = (1/2)P_E(4) + P\{3\pi/4 \leqslant \psi \leqslant \pi\}.$$

The probability $P\{3\pi/4 \leqslant \psi \leqslant \pi\}$ can be evaluated exactly from (9) and (10); however, it is sufficient for our purposes to use the asymptotic form given by (40), namely,

$$P\{3\pi/4 \leqslant \psi \leqslant \pi\} \sim e^{-\rho}/(2\pi\rho).$$

Consequently, for the region of interest (ρ greater than 10 dB), $P_B(4) = (1/2)P_E(4)$ with little error. This approximation is then better than the commonly used $P_B(4) = (2/3)P_E(4)$, which corresponds to orthogonal signals; i.e., $P_B(M) = MP_E(M)/2(M-1)$ for $M = 4$.

where ρ is the symbol signal-to-noise ratio, related to the signal energy per bit-to-noise spectral density ratio by

$$\frac{E_b}{N_0} = \frac{\rho}{\log_2 M}\,. \tag{45}$$

When $M = 2$, (44) gives, upon inspection, $P_E(2) = \exp(-\rho)/2$.

For large ρ, the asymptotic formula[12] for $P_E(M)$ follows from the results of Section III-A for Case I and small ψ_0, and is [cf. (35a)]

$$P_E(M) \sim \sqrt{\frac{1 + \cos(\pi/M)}{2\cos(\pi/M)}}\ \operatorname{erfc}\sqrt{\rho[1 - \cos(\pi/M)]}\,; \qquad M \geqslant 3. \tag{46}$$

Several other approximations have been developed for $P_E(M)$ and are listed here for comparison with the above result. For large ρ, Fleck and Trabka [5] derived the approximation

$$P_E(M) \cong \operatorname{erfc} X + \frac{Xe^{-X^2}}{4\sqrt{\pi}(\rho + 1/8)} \tag{47}$$

where $X = \sqrt{\rho[1 - \cos(\pi/M)]}$. Arthurs and Dym [2] later gave the estimate

$$P_E(M) \cong \operatorname{erfc}\left[\sqrt{\rho}\,\sin\frac{\pi}{\sqrt{2}M}\right] \tag{48}$$

and, finally, Bussgang and Leiter [3] obtained the upper bound

$$P_E(M) \cong \operatorname{erfc}\left[\frac{\rho}{\sqrt{1 + 2\rho}}\,\sin\frac{\pi}{M}\right] \tag{49}$$

which is also a good approximation for large ρ. Although all of these three approximations are good for large signal-to-noise ratios, none is asymptotic in the sense that the error in using it becomes vanishingly small as ρ becomes infinitely large.

The Fleck and Trabka approximation is closest in form to the asymptotic formula, (46). Neglecting the 1/8 in the denominator of (47), employing the asymptotic expansion for the erfc function in both (46) and (47), and then forming the quotient of these two gives (for $M \geqslant 3$)

$$\frac{P_E(M),\ \text{Fleck \& Trabka}}{P_E(M),\ \text{eq. (46)}} \sim \frac{\sqrt{\cos(\pi/M)}[1 + \frac{1}{2}\sin^2(\pi/2M)]}{\cos(\pi/2M)} \equiv Q \tag{50}$$

[12] Another asymptotic formula can be obtained from (44) by noting that the limits of integration can be extended to $\pm\pi$ with an error that becomes vanishingly small as ρ gets infinitely large. This gives the asymptotic formula

$$P_E(M) \sim 1 - \left(\sin\frac{\pi}{M}\right) Ie\left(\cos\frac{\pi}{M}, \rho\right); \qquad M \geqslant 3$$

which has been obtained previously when $M = 4$ [11, p. 247].

which is independent of ρ. For $M = 3, 4, 8$, and 16, values of this ratio are

M	Q
3	0.9186
4	0.9768
8	0.9987
16	0.9999

Hence, although Fleck and Trabka's result is not asymptotic in the strict sense of the term, it differs in its leading term from the asymptotic formulas by less than 9 percent in the worst case when $M = 3$, and by less than 3 percent in the more ordinary situation when M is a power of 2 [$M > 2$ is tacitly assumed, since an exact expression is known for $P_E(2)$].

B. Distribution of Instantaneous Frequency

The distribution of the instantaneous frequency of a narrow-band signal may be regarded as a limiting case of the distribution of the phase angle between two vectors perturbed by Gaussian noise, if the vectors are appropriately regarded as samples of a narrow-band waveform with vanishingly small time separation. The Case III simplified representation of Section II-C for $P\{\psi_1 \leqslant \psi \leqslant \psi_2\}$ provides a convenient mechanism for readily determining the probability $P\{\omega \leqslant \omega_0\}$, that the instantaneous radian frequency lies below some value ω_0. In the remainder of this section, we will consider only the case that the noise power is not time-dependent so that $\sigma_1 = \sigma_2$.

The instantaneous radian frequency of a narrow-band signal is the time derivative of its instantaneous phase; i.e.

$$\omega = \lim_{\tau \to 0} \frac{\theta_2 - \theta_1}{\tau} = \lim_{\tau \to 0} \frac{\psi}{\tau} \tag{51}$$

in which $\theta_1 = \theta(t)$, $\theta_2 = \theta(t + \tau)$ and, in the final limit, $\theta_2 - \theta_1$ has been replaced by its modulo 2π difference $\psi = (\theta_2 - \theta_1) \bmod 2\pi$ (cf. [11, ch. 6]). Then, formally, the probability $P\{\omega \leqslant \omega_0\}$ can be determined using (a) as the limit

$$\begin{aligned} P\{\omega \leqslant \omega_0\} &= 1 - P\{\omega > \omega_0\}, \\ &= 1 - \lim_{\tau \to 0} P\{\psi > \omega_0 \tau\}, \\ &= \begin{cases} \lim_{\tau \to 0} F(\omega_0 \tau) - \lim_{\tau \to 0} F(\Delta\Phi + \pi); & \text{if } \omega_0 < \dot{\Phi}, \\ 1 + \lim_{\tau \to 0} F(\omega_0 \tau) - \lim_{\tau \to 0} F(\Delta\Phi + \pi); & \text{if } \omega_0 > \dot{\Phi}, \end{cases} \end{aligned} \tag{52}$$

where, by (4a), $\dot{\Phi} = \omega_c + \dot{\phi}$ and $\dot{\phi} = \lim_{\tau \to 0} \tau^{-1}(\phi_2 - \phi_1)$.

The limits in (52) can be evaluated using (13) by first regarding all of the parameters U, V, W, $\Delta\Phi$, r, and λ as functions of τ, and then letting $\tau \to 0$. For small τ, $r(\tau) \approx 1 + \ddot{r}\tau^2/2$, where $\ddot{r} = [d^2 r(\tau)/d\tau^2]_{\tau=0}$; $\lambda(\tau) \approx \dot{\lambda}\tau$, where $\dot{\lambda} = [d\lambda(\tau)/d\tau]_{\tau=0}$; and (13) shows immediately that $F(\Delta\Phi + \pi) \to 0$ as $\tau \to 0$. Hence, only $F(\omega_0\tau)$ needs to be further considered. The key to the evaluation of the limit of $F(\omega_0\tau)$ as $\tau \to 0$ is to note first from (14a) that $\gamma_i \to \pi$ as $\tau \to 0$. Then the limits of integration in the integrals of (13) are over 2π ranges and, consequently,

1) the values of δ_i in (13) are unimportant,
2) the algebraic sign of β_i is unimportant, and
3) the integrals in (13) are expressible in terms of I_0 and Ie functions as in (18).

Thus, employing (13) to evaluate the limit of $F(\omega_0\tau)$ as $\tau \to 0$, and using the result in (52), we find (cf. [11, eq. (6.6)])

$$\begin{aligned} P\{\omega \leqslant \omega_0\} &= \tfrac{1}{2}[1 - \operatorname{sgn}(\dot{\Phi} - \omega_0)\sqrt{1 - \beta^2/\alpha^2}\, Ie(\beta/\alpha, \alpha)] \\ &\quad + \frac{\omega_0 - \dot{\lambda}}{2(\omega_0^2 - \ddot{r} - 2\omega_0\dot{\lambda})^{1/2}} e^{-\alpha} I_0(\beta) \end{aligned} \tag{53}$$

in which α and β are the limits as $\tau \to 0$ of (14b) and (14c), and are

$$\alpha = \frac{U}{2}\left[1 + \frac{(\omega_0 - \dot{\Phi})^2 + (\dot{A}/A)^2}{\omega_0^2 - \ddot{r} - 2\omega_0\dot{\lambda}}\right] \tag{54}$$

$$\beta = \left[\alpha^2 - \frac{U^2(\omega_0 - \dot{\Phi})^2}{\omega_0^2 - \ddot{r} - 2\omega_0\dot{\lambda}}\right]^{1/2} \tag{55}$$

and $\dot{A} = dA(t)/dt$ and $U = U(t)$ [cf. (1)]. Equation (53) can also be written as

$$P\{\omega - \omega_c \leqslant \nu\} = \frac{1}{2}\left[1 + \frac{\alpha^{-1}U(\nu - \dot{\phi})\, Ie(\beta/\alpha, \alpha) + (\nu - \overline{\omega}_S)e^{-\alpha}I_0(\beta)}{\sqrt{\sigma_S^2 + (\nu - \overline{\omega}_S)^2}}\right] \tag{53a}$$

where $\nu = \omega_0 - \omega_c$ and $\dot{\lambda}$, $\ddot{r}$ have been expressed in terms of $\overline{\omega}_S$, σ_S^2, and ω_c. Here $\dot{\lambda} = \omega_c + \overline{\omega}_S$ and $\ddot{r} = -\dot{\lambda}^2 - \sigma_S^2$. The quantities $\overline{\omega}_S$ and σ_S^2 depend on the normalized baseband power spectrum $S(f)$ through $\overline{\omega}_S = \int_{-\infty}^{\infty} \omega S(f)\, df$ and $\sigma_S^2 = \int_{-\infty}^{\infty} (\omega - \overline{\omega}_S)^2 S(f)\, df$. The expressions for α and β corresponding to (53a) can be obtained by replacing $\omega_0 - \dot{\Phi}$ by $\nu - \dot{\phi}$ in (54) and $\omega_0 - \ddot{r} - 2\omega_0\dot{\lambda}$ by $\sigma_S^2 + (\nu - \overline{\omega}_S)^2$ in (54) and (55). Equivalent results, expressed in terms of the Marcum Q-function, were obtained by Salz and Stein [12].[13]

C. Narrow-Band Digital FM with Partial-Bit Integration

The bit error probability for narrow-band digital FM with limiter–discriminator detection and integrate and dump (I&D) output filtering is considered in detail in [8]; however, this reference exclusively treats the uncorrelated noises case of an output filter (see Fig. 4) which integrates over the *entire* bit time. The more involved problem of *partial-bit* integration, i.e., that in which the output filter integrates over less than the full bit time, requires additional analytical tools that are

[13] The probability density function of the instantaneous frequency, which can be obtained by differentiating (53a), has been given by Gatkin *et al.* [21].

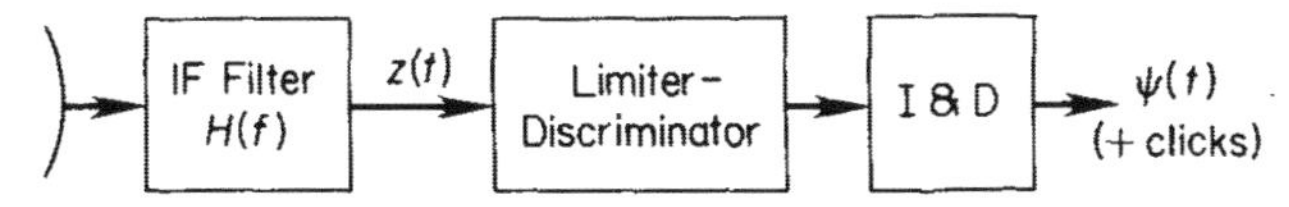

Fig. 4. Narrow-band digital FM.

provided by our results in Case III of the present paper. In this example, we briefly indicate how the Case III results can be used in the BER calculation in the partial-bit case.

In the "worst case" situation of an alternating one-zero data stream, the operation of narrow-band digital FM can be partly modeled in terms of the phase angle between two vectors perturbed by Gaussian noise, when the underlying sinusoid [cf. (1)] has an amplitude and phase of the forms (cf. [8, eqs. (12) and (13)] for a linear phase IF)

$$A(t) = \sqrt{\left(a_0 + a_2 \cos 2\pi \frac{t}{T}\right)^2 + \left(a_1 \sin \pi \frac{t}{T}\right)^2} \tag{56a}$$

and

$$\phi(t) = \tan^{-1} \frac{a_1 \sin \pi \frac{t}{T}}{a_0 + a_2 \cos 2\pi \frac{t}{T}} \tag{56b}$$

where a_0, a_1, and a_2 are constants ($a_0 > 0$, $a_0 \to a_2$) and T is the bit time. Letting τ, $0 \leq \tau \leq T$, denote the partial-bit integration time, we can express the phase angle of interest between the noisy vectors as

$$\psi = \left[\theta\left(\frac{\tau}{2}\right) - \theta\left(-\frac{\tau}{2}\right)\right] \bmod 2\pi \tag{57}$$

and, consequently, the parameters of (4) can be identified as

$$\Delta\Phi(\tau) = \phi\left(\frac{\tau}{2}\right) - \phi\left(-\frac{\tau}{2}\right), \quad U(\tau) = \frac{A^2\left(\frac{\tau}{2}\right)}{2\sigma^2}, \quad V = 0, \; W = U \tag{58}$$

where σ^2 is the noise power after IF filtering. The noise correlations of (5) are determined by the system "narrow-band" IF filter; and for $(\omega_c\tau) \bmod 2\pi = 0$ and for a filter with a symmetric low-pass equivalent $H(f)$, $\lambda = 0$ and $r = r(\tau)$ is given by

$$r(\tau) = \frac{F^{-1}\{|H(f)|^2\}}{\int_{-\infty}^{\infty} |H(f)|^2 \, df}. \tag{59}$$

The overall bit error probability is composed of averages over various bit patterns, and contains contributions from both the continuous and the "click" components of the discriminator output noise [8]. Part of this calculation, the only part to be considered here, requires the evaluation of $P\{\psi \leq 0\}$ for $\Delta\Phi > 0$. The desired result for this probability follows from $P\{\psi \leq 0\} = F(0) - F(\Delta\Phi - \pi)$ and (12), and is

$$P\{\psi \leq 0\} = \frac{\sin \Delta\Phi}{4\pi} \int_{-\pi/2}^{\pi/2} dt \, \frac{\exp\left[-U \dfrac{1 - \cos \Delta\Phi \cos t}{1 - r \cos t}\right]}{1 - \cos \Delta\Phi \cos t} + \frac{r \sin \Delta\Phi}{4\pi} \int_{-\pi/2}^{\pi/2} dt \, \frac{\exp\left[-U \dfrac{1 + \cos t}{1 + r \cos \Delta\Phi \cos t}\right]}{1 + r \cos \Delta\Phi \cos t} \tag{60}$$

In the limit $\tau \to 0$, the output filter may be regarded as a sampler which samples the instantaneous radian frequency, say ω, of the discriminator output. The above probability then becomes $P\{\omega \leq 0\}$, which can be found in the limit of the above equation, but is more directly given by (53) with $\omega_0 = 0$, $\lambda = 0$, $\dot{A} = 0$, viz.

$$P\{\omega \leq 0\} = \tfrac{1}{2}\left[1 - \sqrt{1 - \beta^2/\alpha^2}\, Ie(\beta/\alpha, \alpha)\right] \tag{61}$$

where

$$\alpha = \frac{U(0)}{2}\left(1 - \frac{\dot{\Phi}^2}{\ddot{r}}\right), \quad \beta = \frac{U(0)}{2}\left(1 + \frac{\dot{\Phi}^2}{\ddot{r}}\right),$$

$$\dot{\Phi} = \frac{\pi a_1}{T(a_0 + a_2)};$$

and

$$\ddot{r} = [d^2 r(\tau)/d\tau^2]_{\tau=0}.$$

D. The Effect of Phase Error on DPSK BER

In a recent paper, Blachman [13] considers the effect of an error in the delay of the preceding signal, in calculating the error probability in decoding the present signal, in differential phase-shift keying (DPSK). By recasting this situation in terms of the phase angle between two vectors perturbed by Gaussian noise, a simplification results for the error probability.

If we let θ denote a constant phase error in the reference signal, the error probability P_B can be written as an infinite series, with terms involving modified Bessel functions as [13, eq. (7)]

$$P_B = \tfrac{1}{2} - \frac{\sqrt{\rho_1 \rho_2}}{2} e^{-(\rho_1 + \rho_2)/2} \sum_{n=0}^{\infty} \frac{(-1)^n}{2n+1} \cdot \left[I_n\left(\frac{\rho_1}{2}\right) + I_{n+1}\left(\frac{\rho_1}{2}\right)\right] \times \left[I_n\left(\frac{\rho_2}{2}\right) + I_{n+1}\left(\frac{\rho_2}{2}\right)\right] \cos(2n+1)\theta \tag{62}$$

where ρ_1 and ρ_2 are the instantaneous signal-to-noise ratios of the two vectors. Using (18) with $\psi_i = -\pi/2, r = \lambda = 0$, and $\Delta\Phi = \theta$, for $|\theta| \leqslant \pi/2$, the above equation admits to the simplification

$$P_B = \tfrac{1}{2}[1 - \sqrt{1-Y^2} Ie(Y, U)] \tag{63}$$

where

$$Y = \sqrt{1-(W/U)^2 \cos^2\theta},$$

and U and W are given by (4).

The equivalence between (62) and (63)[14] can be shown directly by the following steps. From (4), (C-1), and Neumann's addition theorem (cf. [1, eq. (9.1.79)]), $U^{-1} Ie(Y, U)$ has the representations

$$U^{-1} Ie(Y, U)$$

$$= \int_0^1 d\xi e^{-(x+y)\xi} I_0(\xi\sqrt{x^2+y^2-2xy\cos 2\theta}) \tag{64}$$

$$= \sum_{m=0}^{\infty} (-1)^m \epsilon_m \cos 2m\theta \int_0^1 d\xi$$

$$\cdot e^{-(x+y)\xi} I_m(x\xi) I_m(y\xi) \tag{65}$$

in which $x = \rho_1/2$, $y = \rho_2/2$, $\epsilon_0 = 1$, and $\epsilon_m = 2$ for $m > 0$. When (65) is used in (63), after some rearrangement (63) becomes a Fourier series in $\cos(2n+1)\theta$, as is (62). Equality follows by equating coefficients and use of the identity (which can be verified by differentiation)

$$\frac{\partial}{\partial\xi}\{\xi e^{-(x+y)\xi}[I_n(x\xi) + I_{n+1}(x\xi)][I_n(y\xi) + I_{n+1}(y\xi)]\}$$

$$= (2n+1)e^{-(x+y)\xi}[I_n(x\xi)I_n(y\xi) - I_{n+1}(x\xi)I_{n+1}(y\xi)] \tag{66}$$

to evaluate the remaining integral over ξ.

E. Symbol Error Rate in MPSK

A degenerate case of the distribution of the phase angle between two vectors perturbed by Gaussian noise is that in which one of the vectors is noise free. This situation is representative of MPSK; hence, appropriate limits of our previous results can be used to obtain the symbol error probability in MPSK.

When the second vector, say, has zero noise, the probability density function of the phase angle between the two vectors has the well-known forms [15]

$$p(\psi) = \int_0^\infty dx \frac{x}{\pi} e^{-(x^2+\rho-2x\sqrt{\rho}\cos\psi)} \tag{67}$$

$$= \frac{e^{-\rho}}{2\pi} + \sqrt{\frac{\rho}{4\pi}} e^{-\rho\sin^2\psi} \cos\psi \operatorname{erfc}(-\sqrt{\rho}\cos\psi) \tag{68}$$

in which $\rho = \rho_1$ and $-\pi \leqslant \psi \leqslant \pi$ ($\Delta\Phi = 0$ since MPSK is implicitly coherent signaling). The limit as $\rho_2 \to \infty$ (or $\sigma_2 \to 0$) of the function $F(\psi)$ will first be shown to be consistent with this $p(\psi)$, and will then be applied to the MPSK symbol error probability calculation.

Taking the limit $\rho_2 \to \infty$ of (11) (with $\Delta\Phi = 0$) leads to

$$F(\psi) = \frac{-\operatorname{sgn}\psi}{2\pi} \int_{-\infty}^{\cot|\psi|} \frac{dx}{1+x^2} e^{-\rho(1+x^2)\sin^2\psi} \tag{69}$$

which can also be written as

$$F(\psi) = \frac{-\operatorname{sgn}\psi}{2\pi} \int_{-\pi/2}^{\pi/2-|\psi|} d\theta e^{-\rho\sin^2\psi\sec^2\theta}. \tag{70}$$

In taking the limit of (11) to get (69), it is to be noted that both U and $V \to \infty$ as $\rho_2 \to \infty$ and that the integrand of (11) becomes peaked in the vicinity of $t = t_0$, $t_0 = \tan^{-1}\{V/(W\cos\psi)\}$. Expanding the integrand in powers of $(t - t_0)$ and passing to the limit gives (69). Equation (70) then follows by the change of variable $x = \tan\theta$. That (68) and (69) are consistent is readily verified by showing that these satisfy $p(\psi) = \partial F(\psi)/\partial\psi$.

Turning now to MPSK: for equiprobable, equal energy MPSK signals, the symbol error probability for the optimum receiver is $P_E(M) = 2P\{\psi > \pi/M\}$. Consequently, employing (70) to evaluate $F(\pi) - F(\pi/M)$ gives

$$P_E(M) = \frac{1}{\pi} \int_{-\pi/2}^{\pi/2-\pi/M} d\theta e^{-\rho\sin^2(\pi/M)\sec^2\theta} \tag{71}$$

which is essentially the same as the result of Weinstein [17]. When $M = 2$, (71), with the aid of (69) and (34), reduces to the well-known form $P_E(2) = (1/2)\operatorname{erfc}(\sqrt{\rho})$. Also, from (71), it is straightforward to show that $P_E(M) \sim \operatorname{erfc}[\sqrt{\rho}\sin(\pi/M)]$ for $M > 2$ (cf. [6, p. 231]).

V. RELATION TO THE FOURIER SERIES APPROACH

In most of the preceding, definite integral expressions have been presented for the probability densities, distributions, and related quantities of interest. However, some of these densities and distributions can be obtained in a quite general way as Fourier series [22]. It is the goal of this section to briefly explore the series approach and show its relation to our previous work. Primarily the Fourier series method appears to be limited to Cases I and II in which there is no correlation between the noises perturbing the endpoints of the vectors; however, the

14 We are grateful to N. M. Blachman for the interesting bit of history that a special case of (63) was obtained, but apparently not published, by L. Lewandowski in 1961.

method will be found to have some application in Case III. Additionally, the Fourier series method quite naturally allows for extensions to situations which fall outside the scope of the main part of this paper, situations in which the phase angle of interest is the result of the modulo 2π combination of more than two components.

The starting point for the Fourier series approach is the Fourier series of the periodic extension of the Bennett probability density, of Section IV-E, for the phase noise of the angle of a single vector; i.e.,

$$p_1(\theta) = \int_0^\infty dx \, \frac{x}{\pi} \, e^{-(x^2+\rho^2-2x\sqrt{\rho}\cos\theta)}; \qquad -\pi \leqslant \theta \leqslant \pi. \tag{72}$$

The Fourier series corresponding to this $p_1(\theta)$ can be shown to be (cf. Prabhu [22], Middleton [18, p. 417] and Lindsey [19, p. 190])

$$p_1(\theta) = \frac{1}{2\pi} \sum_{k=0}^{\infty} \epsilon_k \tilde{\alpha}_k(\rho) \cos k\theta; \qquad \epsilon_0 = 1, \qquad \epsilon_k = 2 \text{ for } k > 0 \tag{73}$$

where[15]

$$\tilde{\alpha}_k(\rho) = \sqrt{\frac{\pi\rho}{4}} \, e^{-\rho/2} \left[I_{k/2-1/2}\left(\frac{\rho}{2}\right) + I_{k/2+1/2}\left(\frac{\rho}{2}\right) \right] \tag{74}$$

From this series, the series for the density of $\psi = (2\theta)$ mod 2π follows by noting that $p(\psi) = [p_1(\psi/2) + p_1(\psi/2 + \pi)]/2$, which leads to

$$p(\psi) = \frac{1}{2\pi} \sum_{k=0}^{\infty} \epsilon_k \tilde{\alpha}_{2k}(\rho) \cos k\psi; \qquad \psi = (2\theta) \bmod 2\pi. \tag{75}$$

The second ingredient in the Fourier series approach is the observation, made from (6), that the Fourier series for the combination of two independent θ's, with densities of the form of (73) or (75), has $\tilde{\alpha}$-coefficients which are just products of the $\tilde{\alpha}$-coefficients of the component pdf's. Then, series for the densities of combinations such as $\psi = (\theta_2 - \theta_1)$ mod 2π and $\psi = (2\theta_2 - \theta_1 - \theta_3)$ mod 2π[16] [all the θ's are independent with densities given by (73)] can be written by inspection, viz.

$$p(\psi) = \frac{1}{2\pi} \sum_{k=0}^{\infty} \epsilon_k \tilde{\alpha}_k(\rho_1) \tilde{\alpha}_k(\rho_2) \cos k\psi; \qquad \psi = (\theta_2 - \theta_1) \bmod 2\pi \tag{76}$$

$$p(\psi) = \frac{1}{2\pi} \sum_{k=0}^{\infty} \epsilon_k \tilde{\alpha}_k(\rho_1) \tilde{\alpha}_{2k}(\rho_2) \tilde{\alpha}_k(\rho_3) \cos k\psi; \qquad \psi = (2\theta_2 - \theta_1 - \theta_3) \bmod 2\pi. \tag{77}$$

Other combinations can be handled similarly.

Probabilities stemming from the series representations of such densities can be determined by termwise integration. For example, Blachman's error probability P_B given by (62) is just the termwise integral of (76) over the intervals $(-\pi, -\pi/2 - \theta)$ and $(\pi/2 - \theta, \pi)$, viz.

$$P_B = \tfrac{1}{2} - \frac{2}{\pi} \sum_{k \text{ odd}}^{\infty} \tilde{\alpha}_k(\rho_1) \tilde{\alpha}_k(\rho_2) \frac{\sin k(\pi/2 - \theta)}{k}. \tag{78}$$

Further, distributions like $P\{\psi \geqslant \psi_0\}$ in Cases I and II, of concern in previous parts of this paper, can be obtained as definite integral expressions by directly summing their corresponding Fourier series. However, the $\tilde{\alpha}$-coefficient of the combination density becomes increasingly complicated as the modulo 2π combination gets more involved, so that it becomes more difficult to sum the Fourier series. For example, we have not as yet succeeded in summing either the series for $p(\psi)$ or its counterpart for $P\{\psi \geqslant \psi_0\}$ in the case of $\psi = (2\theta_2 - \theta_1 - \theta_3)$ mod 2π.

VI. SUMMARY AND CONCLUSIONS

The probability distribution of the modulo 2π phase angle between two vectors perturbed by correlated Gaussian noise has been studied in detail. For various types of signal conditions and noise correlation, the distribution and its asymptotic behavior have been presented. Some of the results were applied to problems in angle modulation in communication systems, and in finding the distribution of instantaneous frequency.

Even in the most complicated case, the expressions for the distribution require only the numerical evaluation of single integrals, and these are easily done on a digital computer or hand-held programmable calculator to ten digit accuracy using the methods described in [10]. Such methods were used to get numerical comparisons of the asymptotic formulas with exact results and, in general, the asymptotic results were found to be quite good for $U > 10$ dB and over the entire range of ψ_0.

By following the techniques used, the results can be extended to other special situations of interest in phase and frequency modulation (cf. [6, ch. 5]).

APPENDIX A

CHARACTERISTIC FUNCTION METHOD

Here we sketch the derivation of (21) and (22). The original development in [11] has been changed somewhat and rearranged into a step-by-step procedure.

1) Define a periodic sawtooth function $s(\xi)$ of period 2π such that

$$s(\xi) = \xi; \qquad -\pi < \xi \leqslant \pi \tag{A-1}$$

[15] $\tilde{\alpha}_1(\rho)$ is the well-known *signal suppression factor* [6, p. 50].

[16] This combination arises in Manchester encoded digital FM [20].

and use it to construct the function

$$\frac{1}{2\pi}\left[(\psi_2-\psi_1)+s(\psi-\psi_2+\pi)-s(\psi-\psi_1+\pi)\right].$$

This function of ψ is discontinuous at $\psi=\psi_1$ and $\psi=\psi_2$. It is equal to 1 in $\psi_1<\psi\leqslant\psi_2$ and to 0 elsewhere in $-\pi\leqslant\psi\leqslant\pi$. Then (cf. [11, eq. (2.26)])

$$P\{\psi_1\leqslant\psi\leqslant\psi_2\}=\frac{\psi_2-\psi_1}{2\pi}+\frac{1}{2\pi}\mathbf{E}[s(\theta_2-\theta_1-\psi_2+\pi)-s(\theta_2-\theta_1-\psi_1+\pi)] \qquad \text{(A-2)}$$

where the expectation extends over the random variables θ_1 and θ_2. By the periodicity of $s(\cdot)$, the mod 2π in $\psi=(\theta_2-\theta_1)$ mod 2π is accounted for.

2) Note the basic relation [11, eq. (2.20)]

$$s(\xi)=\int_0^\infty\int_0^\infty\frac{\partial}{\partial\xi}J_0(\sqrt{x^2+y^2+2xy\cos\xi})\frac{dx\,dy}{xy} \qquad \text{(A-3)}$$

which can be verified by changing to polar coordinates $x=R\cos\phi$, $y=R\sin\phi$. Then the integration with respect to R can be performed immediately (since it is equivalent to integrating $J_0'(cR)$, c being a constant). The integration with respect to ϕ can be done by setting $2\phi=\pi/2-t$ and using

$$\int_0^{\pi/2}\frac{dt}{1+\cos\xi\cos t}=\frac{\xi}{\sin\xi};\quad -\pi\leqslant\xi\leqslant\pi. \qquad \text{(A-4)}$$

3) Replace ξ in (A-3) by $\theta_2-\theta_1-\xi$ and change the variables of integration to u, v where $x=uR_1$, $y=vR_2$ and $R_1, R_2>0$. This carries (A-3) into

$$s(\theta_2-\theta_1-\xi)=-\int_0^\infty\int_0^\infty\frac{du\,dv}{uv}\frac{\partial}{\partial\xi}\cdot J_0(\sqrt{u^2R_1{}^2+v^2R_2{}^2+2uvR_1R_2\cos(\theta_2-\theta_1-\xi)}). \qquad \text{(A-5)}$$

Taking the expected value of both sides of (A-5) and setting $\xi=\psi_i-\pi$ (so $\partial/\partial\xi=\partial/\partial\psi_i$) gives a form[17] for $\mathbf{E}[s(\theta_2-\theta_1-\psi_i+\pi)]$ that can be used in (A-2) to get an expression for $P\{\psi_1\leqslant\psi\leqslant\psi_2\}$. The desired equations (21) and (22) now follow when (20) is used to express $\mathbf{E}[J_0(\cdot)]$ as $\Phi(-u, v\,|\,\psi_i)$, the conditional characteristic function of Section II-E.

[17] Interchanging the order of expectation and integration can be justified by standard methods of analysis (see [16]). If the double integral is first expressed in the polar coordinates (R,ϕ) then the most difficult step to justify is the interchange of the expectation with the R-integration. The validity of this step can be shown by first writing the R-integral as one from 0 to M plus a second from M to ∞. Then the interchange can be made in the first part with the finite limits and, as $M\to\infty$, the contribution from the second part can be shown to be nil by use of the Riemann–Lebesgue lemma (or the asymptotic behavior of the Bessel function).

APPENDIX B

ALTERNATE DERIVATION OF $P\{\psi_1\leqslant\psi\leqslant\psi_2\}$

This Appendix presents an alternate derivation of $P\{\psi_1\leqslant\psi\leqslant\psi_2\}$ which follows more along classical lines than the derivation given in Section II-E. For the sake of simplicity, only Case II is considered in some detail. The derivation can be extended to Case III, although the algebra involved becomes tedious.

Case II: Initially, the derivation proceeds along the lines followed by Fleck and Trabka [5], who start with the four-dimensional joint Gaussian probability density function $p(x_1, y_1, x_2, y_2)$ of the four noise variables in (2). A change is made to polar coordinates, and the resulting radius vectors integrated over, leading to the joint density $p(\theta_1,\theta_2)$. Employing (6), the result for $p(\psi)$, $\psi=(\theta_2-\theta_1)$ mod 2π, can be written as (cf. [8, Appendix])

$$p(\psi)=\frac{1}{4\pi}\int_{-\pi/2}^{\pi/2}dt e^{-E}(1+2U-E)\cos t \qquad \text{(B-1)}$$

where $E=U-V\sin t-W\cos(\Delta\Phi-\psi)\cos t$. The observation is made that the negative exponent E satisfies the identity

$$E(2U-E)=\left(\frac{\partial E}{\partial t}\right)^2+\sec^2 t\left(\frac{\partial E}{\partial\psi}\right)^2 \qquad \text{(B-2)}$$

so that multiplying and dividing the integrand of (B-1) by E, using (B-2), and integrating over (ψ_1,ψ_2) gives

$$\begin{aligned}P\{\psi_1\leqslant\psi\leqslant\psi_2\}&=\frac{1}{4\pi}\int_{\psi_1}^{\psi_2}d\psi\int_{-\pi/2}^{\pi/2}dt e^{-E}\cos t\\&+\frac{1}{4\pi}\int_{\psi_1}^{\psi_2}d\psi\int_{-\pi/2}^{\pi/2}\left(\frac{-\cos t}{E}\frac{\partial E}{\partial t}\right)\left(\frac{\partial e^{-E}}{\partial t}dt\right)\\&+\frac{1}{4\pi}\int_{-\pi/2}^{\pi/2}dt\int_{\psi_1}^{\psi_2}\left(\frac{-1}{E\cos t}\frac{\partial E}{\partial\psi}\right)\left(\frac{\partial e^{-E}}{\partial\psi}d\psi\right)\end{aligned} \qquad \text{(B-3)}$$

Upon integrating the second integral by parts with respect to t, and the third integral by parts with respect to ψ, it will be found formally that all of the remaining double integrals cancel, and that there results the single integral

$$P\{\psi_1\leqslant\psi\leqslant\psi_2\}=\frac{1}{4\pi}\int_{-\pi/2}^{\pi/2}dt\left[\frac{-e^{-E}}{E\cos t}\frac{\partial E}{\partial\psi}\right]_{\psi_1}^{\psi_2}. \qquad \text{(B-4)}$$

However, care must be taken if the singularity $E = 0$ lies within the range of integration, for then the steps performed in going between the last two equations would not be valid. This singularity occurs when

$$\psi = \Delta\Phi, \quad \text{and} \quad t = \tan^{-1} \frac{V}{W} \tag{B-5}$$

and will be avoided if $\Delta\Phi$ lies outside the range of integration. If $\Delta\Phi$ lies within the range of integration, then the above derivation must be repeated for the complementary probability $P\{\psi < \psi_1 \text{ or } \psi > \psi_2\}$. Evaluating the integrand of (B-4) then leads to (9) and (11), which completes the proof.

Case III: In extending this derivation to Case III, the probability density function $p(\psi)$ becomes

$$p(\psi) = \frac{1 - r^2 - \lambda^2}{4\pi} \int_{-\pi/2}^{\pi/2} dt \cdot \frac{e^{-E} \cos t}{[1 - (r \cos \psi + \lambda \sin \psi) \cos t]^2} \cdot \left[1 - E + 2 \frac{U - W(r \cos \Delta\Phi + \lambda \sin \Delta\Phi)}{1 - r^2 - \lambda^2}\right] \tag{B-6}$$

where

$$E = \frac{U - V \sin t - W \cos (\Delta\Phi - \psi) \cos t}{1 - (r \cos \psi + \lambda \sin \psi) \cos t}$$

and the analog of (B-2) is

$$\left(\frac{\partial E}{\partial t}\right)^2 + \sec^2 t \left(\frac{\partial E}{\partial \psi}\right)^2 = \frac{E[2U - 2W(r \cos \Delta\Phi + \lambda \sin \Delta\Phi) - E(1 - r^2 - \lambda^2)]}{[1 - (r \cos \psi + \lambda \sin \psi) \cos t]^2} \tag{B-7}$$

APPENDIX C

SOME INTEGRAL RESULTS

This Appendix summarizes certain integral results needed in the main part of the paper. Also listed are two interesting definite integrals which are related to the analysis, and which do not appear to be given in most tables.

1) *The* $Ie(k, x)$ *Function*: This function is defined and tabulated in [9], and [11] contains additional tabulations. For $|k| < 1$

$$Ie(k, x) = \int_0^x e^{-t} I_0(kt)\, dt \tag{C-1}$$

$$= \frac{1}{\sqrt{1 - k^2}} - \frac{1}{\pi} \int_0^\pi \frac{e^{-x(1 - k \cos \theta)}}{1 - k \cos \theta}\, d\theta \tag{C-2}$$

where $I_0(\cdot)$ is a modified Bessel function [see [9, eq. (A3-5)] for (C-2)].

2) *Hankel Exponential Integrals ([1, eq. (11.4.28)]*:

$$\int_0^\infty t J_0(at) e^{-bt^2}\, dt = \frac{1}{2b} e^{-a^2/4b} \tag{C-3}$$

$$\int_0^\infty J_1(at) e^{-bt^2}\, dt = \frac{1}{a}(1 - e^{-a^2/4b}). \tag{C-4}$$

3) The following two integrals can be obtained by use of the Euler-Legendre transformation, (31):

$$\frac{\sqrt{1 - d^2}}{2\pi} \int_{-\pi}^{\pi} dt \frac{\exp\left[\dfrac{a + b \sin t + c \cos t}{1 + d \cos t}\right]}{1 + d \cos t} = e^{-\alpha} I_0(\beta) \tag{C-5}$$

$$\frac{\sqrt{1 - d^2}}{2\pi} \int_{-\pi}^{\pi} dt \frac{\exp\left[\dfrac{a + b \sin t + c \cos t}{1 + d \cos t}\right]}{a + b \sin t + c \cos t} = \frac{1}{\alpha} Ie(\beta/\alpha, \alpha) - \frac{1}{\sqrt{\alpha^2 - \beta^2}} \tag{C-6}$$

where

$$\alpha = \frac{cd - a}{1 - d^2}, \quad \beta = \sqrt{\alpha^2 - \frac{a^2 - b^2 - c^2}{1 - d^2}}. \tag{C-7}$$

REFERENCES

[1] M. Abramowitz and I. A. Stegun, *Handbook of Mathematical Functions*. New York: Dover, 1972.

[2] E. Arthurs and H. Dym, "On the optimum detection of digital signals in the presence of white Gaussian noise—A geometric interpretation and a study of three basic data transmission systems," *IRE Trans. Commun. Syst.*, vol. CS-10, pp. 336–372, Dec. 1962.

[3] J. J. Bussgang and M. Leiter, "Error rate approximations for differential phase-shift keying," *IEEE Trans. Commun. Syst.*, vol. CS-12, pp. 18–27, Mar. 1964.

[4] J. Edwards, *A Treatise on the Integral Calculus, Vol. II*. London, England: Macmillan, 1922.

[5] J. T. Fleck and E. A. Trabka, "Error probabilities of multiple-state differentially coherent phase-shift keyed systems in the presence of white, Gaussian noise," in *Investigation of Digital Data Communication Systems*, Rep. UA-1420-S-1, J. G. Lawton, Ed., Cornell Aeronaut. Lab., Inc., Buffalo, NY, Jan. 1961, Detect Memo 2A; available as NTIS Doc. AD256584.

[6] W. C. Lindsey and M. K. Simon, *Telecommunication Systems Engineering*. Englewood Cliffs, NJ: Prentice-Hall, 1973.

[7] J. I. Marcum, "A statistical theory of target detection by pulsed radar," *IRE Trans. Inform. Theory*, vol. IT-6, Apr. 1960, pp. 59–267.

[8] R. F. Pawula, "On the theory of error rates for narrow-band digital FM," *IEEE Trans. Commun.*, vol. COM-29, pp. 1634–1643, Nov. 1981.

[9] S. O. Rice, "Statistical properties of a sine wave plus random noise," *Bell Syst. Tech. J.*, vol. 27, pp. 109–157, Jan. 1948.

[10] ——, "Efficient evaluation of integrals of analytic functions by the trapezoidal rule," *Bell Syst. Tech. J.*, vol. 52, pp. 707–722, May–June 1973.

[11] J. H. Roberts, *Angle Modulation*. England: Peregrinus, 1977.
[12] J. Salz and S. Stein, "Distribution of instantaneous frequency for signal plus noise," *IEEE Trans. Inform. Theory*, vol. IT-10, pp. 272–274, Oct. 1964.
[13] N. M. Blachman, "The effect of phase error on DPSK error probability," *IEEE Trans. Commun.*, vol. COM-29, pp. 364–365, Mar. 1981.
[14] A. J. Rainal, "Monopulse radars excited by Gaussian signals," *IEEE Trans. Aerosp. Electron. Syst.*, vol. AES-2, pp. 337–345, May 1966.
[15] W. R. Bennett, "Methods of solving noise problems," *Proc. IRE*, vol. 44, pp. 609–638, May 1956.
[16] E. C. Titchmarsh, *The Theory of Functions*. London, England: Oxford Univ. Press, 1939.
[17] F. S. Weinstein, "A table of the cumulative probability distribution of the phase of a sine wave in narrow-band normal noise," *IEEE Trans. Inform. Theory*, vol. IT-23, pp. 640–643, Sept. 1977.
[18] D. Middleton, *An Introduction to Statistical Communication Theory*. New York: McGraw-Hill, 1960.
[19] W. C. Lindsey, *Synchronization Systems in Communication and Control*. Englewood Cliffs, NJ: Prentice-Hall, 1972.
[20] I. Korn, "Comments on 'Limiter discriminator detection performance of Manchester and NRZ coded FSK,'" *IEEE Trans. Aerosp. Electron. Syst.*, vol. AES-16, p. 415, May 1980.
[21] N. G. Gatkin, V. A. Geranin, M. I. Karnovskiy, L. G. Krasnyy, and N. I. Cherney, "Probability density of phase derivative of the sum of a modulated signal and Gaussian noise," *Radio Eng. Electron. Phys. (USSR)*, No. 8, pp. 1223–1229, 1965.
[22] V. K. Prabhu, "Error-rate considerations for digital phase-modulation systems," *IEEE Trans. Commun. Technol.*, vol. COM-17, pp. 33–42, Feb. 1969.

★

PHOTO
NOT
AVAILABLE

R. F. Pawula was born in Chicago, IL, on May 17, 1936. He attended the Illinois Institute of Technology, Chicago, and the Massachusetts Institute of Technology, Cambridge, and received the Ph.D. degree in electrical engineering from the California Institute of Technology, Pasadena, in 1965.

He served two years in the U.S. Army prior to his college enrollment. He was a tenured faculty member of the University of California, San Diego, when he retired in 1975. Since that time he has been a communications and radar Consultant and a commercial pilot. His special fields of interest include spread-spectrum communications, communication satellite system analysis, precision DME system design, intermodulation distortion, multipath, jamming, modulation and demodulation, and error correction coding and decoding. He is also interested in long-range avionics development.

Dr. Pawula is a member of Sigma Xi, Tau Beta Pi, Eta Kappa Nu, Phi Eta Sigma, Rho Epsilon, and Alpha Eta Rho. He was an Alfred Sloan Fellow at M.I.T. and a Howard Hughes Fellow at Caltech. He also belongs to the Aircraft Owners and Pilots Association, the Experimental Aircraft Association, and the Baja Bush Pilots.

PHOTO
NOT
AVAILABLE

S. O. Rice was born in Shedds, OR. He received the B.S. degree in electrical engineering at Oregon State College, Corvallis, in 1929, did graduate work in 1930 at the California Institute of Technology, Pasadena, and received the D.Sc. (Hon.) degree from Oregon State College, in 1961.

From 1930 to 1972 he was employed by Bell Telephone Laboratories, where he was a Consultant on mathematical problems and Head of the Communications Analysis Research Department. During this time he was concerned with various aspects of communication theory, particularly those areas involving random phenomena and noise. Since 1973 he has been a Research Physicist in the Department of Electrical Engineering and Computer Sciences at the University of California, San Diego.

PHOTO
NOT
AVAILABLE

J. H. Roberts received the degree (with honours) in mathematics from Reading University, Reading, Berks., England, in 1954.

Following military service, he joined the General Electric Company at their Wembley, England, research laboratories. Here he worked in close association with R. G. Medhurst on a variety of problems in trunk radio links and FM systems. In 1966 he joined the Plessey Company, Roke Manor, Romsey, Hants., England. His book *Angle Modulation* (IEE Series on Telecommunications, No. 5) was published in 1977.

Mr. Roberts received the IEE Electronics Premium from 1968 for two papers on FM threshold entension techniques.

Cochannel Interference Considerations in Frequency Reuse Small-Coverage-Area Radio Systems

DONALD C. COX, FELLOW, IEEE

Abstract—**Frequency reuse small-coverage-area radio systems having hexagonal and square coverage areas are compared. Comparison is made on the basis of average signal to average interference ($\bar{S}/\bar{I}$) in the corners of the areas and on the basis of the expected probability of S/I exceeding some system threshold for at least one base station that is eligible to provide service. The difference in performance between square and hexogonal systems is small, smaller than the usual uncertainties in the propagation parameters needed in the performance estimates.**

Results suggest that, even if the signal strength decreases as slowly as the inverse cube of the distance and the standard deviation of the large scale signal variation is as large as 10 dB, good service probabilities (on the order of 99 percent) can be provided in small-coverage-area radio systems using 30–40 channel sets.

I. INTRODUCTION

THE reuse of radio frequencies makes cochannel interference a fundamental consideration in small-coverage-area radio systems. A parameter sometimes used to evaluate the performance of such systems or to compare similar systems is the average signal-to-average interference ratio ($\bar{S}/\bar{I}$) produced at the corner of a coverage area by the first tier of cochannel interferers. Since both signal and interference vary randomly, a more definitive parameter is the expected probability that the signal-to-interference ratio is below some acceptable level [1]. Calculation of this more definitive parameter is more involved than calculation of $\bar{S}/\bar{I}$. Values of $\bar{S}/\bar{I}$ are obtained as an intermediate step in determining the expected probability.

Coverage areas in frequency reuse systems usually are idealized in the form of hexagons or squares for the purpose of analysis [1]-[4]. The square geometry is simpler but under some conditions the hexagonal coverage areas are expected to provide a better system configuration [4].

An early paper [1] considered some aspects of this problem for square and hexagonal coverage areas. These early comparisons lumped the effects of small scale Rayleigh distributed signal fluctuations caused by multipath [2] and the large scale signal variations caused by shadowing [2] into one Rayleigh distributed random variable.

It is well known that several branches of diversity are quite effective in mitigating small scale multipath fluctuations [2]. Thus, the cochannel interference considerations for frequency reuse systems employing diversity are more strongly dependent on the large scale signal variations. At a fixed radius from an omni-directional base station antenna, large scale signal variation can be represented by a log-normally distributed random variable [2] (i.e., the attenuation in decibels is normally distributed). The mean of the signal power $\bar{S}$ decreases with radial distance r from the base station and is inversely proportional to some power m of the distance, i.e., $\bar{S} = k/r^m$ where k is a proportionality constant [2].

Paper approved by the Editor for Communication Theory of the IEEE Communications Society for publication without oral presentation. Manuscript received November 10, 1980; revised July 13, 1981.

The author is with Bell Laboratories, Holmdel, NJ 07733.

In Section II of this paper, values of $\bar{S}/\bar{I}$ are estimated at coverage area corners. Estimates are made for different numbers of radio channel sets that correspond to different frequency reuse factors for hexagonal and square coverage areas. The values of $\bar{S}/\bar{I}$ are used in Section III to estimate the expected probability that the signal-to-interference ratio is less than a threshold appropriate for differential phase-shift-keyed (DPSK) digital signals. The values of propagation and system parameters used for the comparisons are based on preliminary estimates for a portable radio telephone environment. Interference from more than one tier of cochannel base stations is included.

II. AVERAGE SIGNAL-TO-AVERAGE INTERFERENCE ESTIMATES FOR SQUARE AND HEXAGONAL COVERAGE AREAS

Distances between base stations and portable radio telephones (portaphones) will be short (on the order of 1000 ft) because of limited transmitter power (on the order of 10 mW). Therefore, the value for the exponent m in the distance dependence of average signal power is expected to be smaller than that for the greater distances involved in mobile radio [2]. Estimates made by extrapolating mobile radio measurements suggest that $m \gtrsim 3$. This propagation parameter and the geometrical distance from the coverage area layout establish the ratio of average signal power $\bar{S}$ to average interference power $\bar{I}$ as

$$\bar{S}/\bar{I} = 1\Big/\left(r^m \sum_j \frac{1}{d_j^{\,m}}\right) \tag{1}$$

where r is the distance to the portaphone from the base station providing service, d_j is the distance to the portaphone from the jth cochannel interfering base station, and j ranges over all the interfering base stations. This assumes that the interference from all base stations is incoherent and independent so that addition of interference powers is appropriate.

The signal and interference from the first tier of interfering base stations for a layout of square coverage areas is de-

Reprinted from *IEEE Transactions on Communications,* vol. COM-30, no. 1, January 1982.

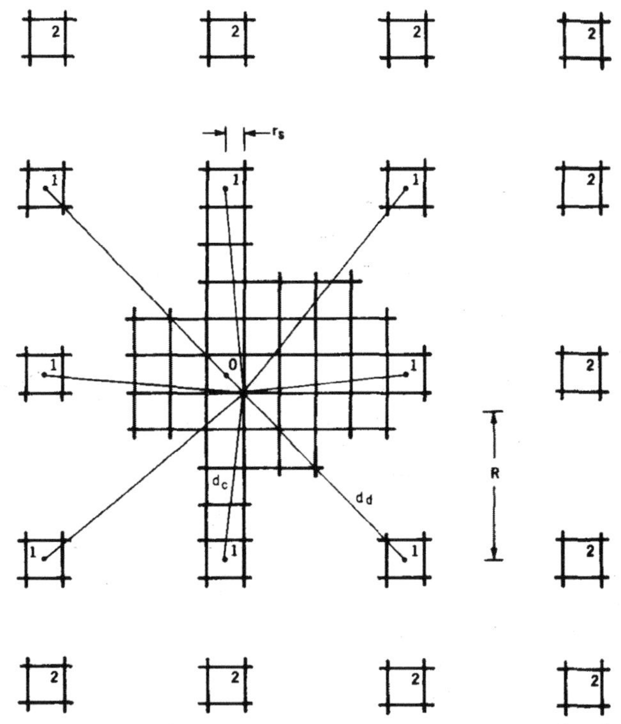

Fig. 1. Layout of square coverage areas. Coverage areas numbered 1 are in the first cochannel interference tier from area 0. Areas numbered 2 are in the second tier.

picted in Fig. 1. The squares numbered 1 are in the first tier of cochannel interferers surrounding the coverage area being considered and labeled 0. Squares labeled 2 are in the second tier, etc. A similar situation for hexagonal areas is depicted in Fig. 2. An approximate relationship for $\bar{S}/\bar{I}$ in the corner for square areas as a function of the number of tiers of interfering base stations is derived in Appendix A as

$$\bar{S}/\bar{I} \approx \frac{2^{m-2} R^m}{1 + (2)^{m/2}} \bigg/ \sum_{t=1}^{T} \frac{1}{t^{m-1}} \tag{2}$$

where t is the tier number and T is the number of tiers of interfering base stations in the radio system. The reuse interval R measures the number of coverage areas between areas using the same radio channels.[1] This approximate relationship is compared with exact calculations for $T = 1$ in Appendix A and is shown to be within $\frac{1}{2}$ dB for $R > 3$ and $m < 4$. The accuracy is much better for lower values of m, for larger values of R, and for larger values of T.

A similar approximate relationship for $\bar{S}/\bar{I}$ in corners for hexagonal coverage areas is derived in Appendix B and compared with the exact relationship for $T = 1$. For hexagonal areas

$$\bar{S}/\bar{I} \approx (3n)^m \bigg/ \left(6 \sum_{t=1}^{T} \frac{1}{t^{m-1}} \right) \tag{3}$$

where n is a parameter related to reuse of radio channels. The

[1] For example, $R = 5$ coverage areas in Fig. 1.

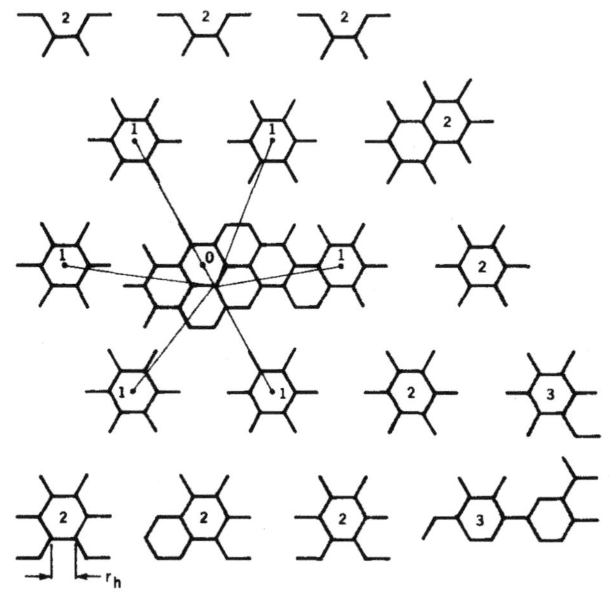

Fig. 2. Layout of hexagonal coverage areas. Coverage areas numbered 1 are in the first cochannel interference tier from area 0. Areas 2 and 3 are in the second and third tiers.

center-to-center separation between coverage areas using the same channel is $3nr_h$, where r_h is the "radius" of the hexagonal area measured from the center to a corner. Accuracy for this approximation is better than that for the square coverage areas.

Reuse of channels in a fixed channel assignment plan [2]-[4] implies that the total number of channels available to the system are divided into a number of separate channel sets. Only certain patterns of channel sets repeat over a plane surface for hexagonal or square coverage areas. Thus, only a restricted group of numbers of channel sets are realizable. For square areas, the allowable numbers of channels sets, C_s, are given by

$$C_s = k^2 + l^2 \tag{4}$$

where k and l range over the positive integers [1], [2]. For hexagonal areas, the allowable numbers of sets, C_h, are given by

$$C_h = k^2 + l^2 + kl \tag{5}$$

where again k and l range over the positive intergers [1], [2], [4].

The reuse parameters R and n, used earlier, have integer values and are strictly applicable for only a subgroup of the allowable number of channel sets. For example, with R ranging over the integers for squares, the subgroup of numbers of channel sets ranges over 1, 4, 9, 16, 25, ···, $(R)^2$, ··· where the total numbers of channel sets ranges over 1, 2, 4, 5, 8, 9, 10, 13, 16, 17, 20, 25, ···, $k^2 + l^2$, ···. Curves based on calculations from the R^2 subgroup and interpolated between for the intermediate numbers of channel sets are sufficiently accurate for system comparisons for numbers of channel sets greater than 9 or 10.

systems infinite in extent involves large sums of random variables that have different means; however, from Fig. 3 it is evident that more than half of the interference is contributed by the first interference tier. Therefore, a reasonable approximation for σ_i can be obtained by considering a random process that is the sum of eight identical log-normal random variables for square coverage areas and the sum of six random variables for hexagonal areas. Monte Carlo calculations of up to 100 000 sums and using three different normal random number generators resulted in $\sigma_i = 5$ dB for squares and $\sigma_i = 5.5$ dB for hexagons. This assumes a σ of 10 dB for the basic process being summed. These values are in reasonable agreement with [6].[4] Summing six random processes for hexagons with $\sigma = 7$ dB resulted in $\sigma_i = 3.7$ dB.

Equation (8) is plotted as the upper five curves in Fig. 4 for hexagonal and square areas, for $\sigma_s = 10$ dB and 7 dB, for a system margin of $\gamma = 8$ dB, and for $m = 3$ and 4. The calculations are based on $\overline{S}/\overline{I}$ from Fig. 3. The probability P_1 of not having adequate S/I from a single base station to a coverage area corner is quite large, even for $m = 4$ and for large numbers of channel sets. If there were only one possible serving base station the situation would be quite grim. However, as illustrated in Figs. 5 and 6, there are four base stations at the same minimum distance from a corner for square areas and three for hexagonal areas. These different base stations use different radio channels and are associated with a different set of interferers. Therefore, fluctuations on the signals and interferences should be independent and the probability P_p of $s - i < \gamma$ for the set of q minimum distance "primary server" base stations is $P(s - i < \gamma)$ for one station raised to the qth power, that is, $P_{ps} = P_1{}^4 = P^4(s - i < \gamma)$ for squares or $P_{ph} = P_1{}^3 = P^3(s - i < \gamma)$ for hexagons. These probabilities[5] are also plotted in Fig. 4.

Inspection of Fig. 5 shows that there are eight base stations surrounding the corner being considered that are at a distance only $\sqrt{5}$ greater than that of the primary servers. These eight secondary stations have a lower $\overline{S}/\overline{I}$ but still have a probability below 1.0 that $s - i < \gamma$. Since the direct paths between the eight secondary servers and the corner are separated by only a small angle from the paths to the primary servers, the signals are probably not all independent. A reasonable estimate might be to say that the eight secondary servers are equivalent to four independent secondary servers. Then $P(s - i < \gamma)$ can be calculated using (6) and $\overline{S}/\overline{I}$ for the secondary servers. The contribution P_{ss} of the four equivalent independent secondary servers is again this single station probability raised to the 4th power. The overall probability of $s - i < \gamma$ for the four primary and four secondary servers is then $P_{ps}P_{ss}$ which is also plotted in Fig. 4 for $m = 3$ and $\sigma_s = 10$ dB. The $\overline{S}/\overline{I}$ for the secondary servers was determined by adjusting $\overline{S}$ for the increased distance and assuming that $\overline{I}$ was the same as for the primary servers.

[4] The Monte Carlo results were not quite as sensitive to truncation of the normal distribution as was expected from [6].

[5] The probabilities P_{ps} and P_{ph} can be viewed as the probabilities of service that would result for these system parameters for a system using the four or three nearest base stations in a selection diversity configuration to combat the large-scale log-normal signal variation if the different base stations use different radio channels as assumed here.

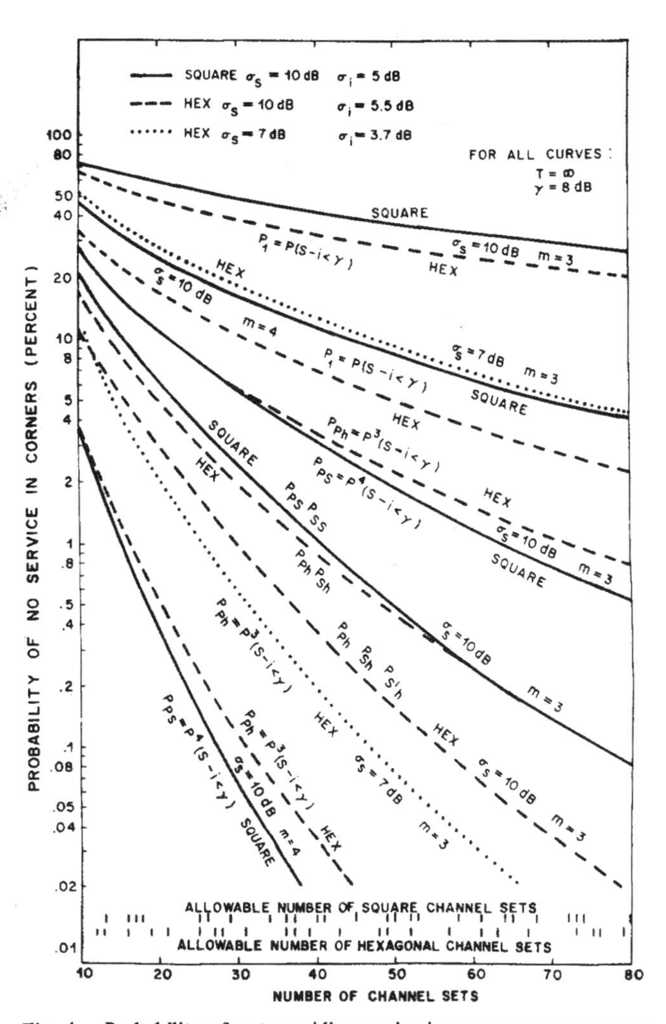

Fig. 4. Probability of not providing service in a coverage area corner because of excessive cochannel interference for systems with different numbers of channel sets. Curves are discussed in text.

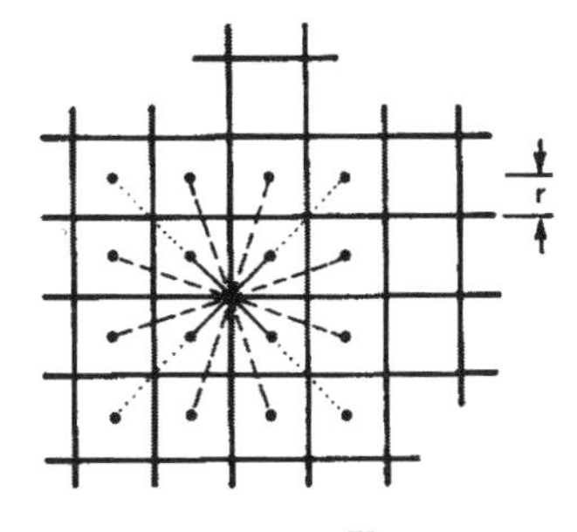

Fig. 5. Square coverage areas illustrating possible center-located base stations that could provide service at a corner of four coverage areas.

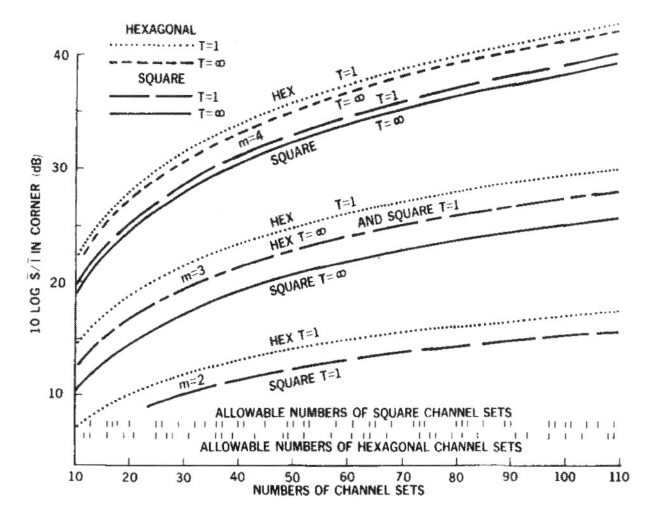

Fig. 3. Average signal to average cochannel interference in a coverage area corner for systems with different numbers of channel sets. Curve parameters are defined in the text.

The $\bar{S}/\bar{I}$ at corners of areas are shown in Fig. 3 as functions of numbers of channel sets for hexagonal and square coverage areas. The curves are interpolated between points calculated from (2) and (3). The tic marks along the horizontal axis indicate the allowable numbers of channel sets for squares and hexagons as indicated. The curves are for $T = 1$ and $T = \infty$, i.e., for a system with only one tier of cochannel base stations and for an infinite system.

From Fig. 3, it is obvious that $\bar{S}/\bar{I}$ in corners is greater for hexagonal cells for all m, for any number of tiers, and for numbers of channel sets greater than 9 or 10. It is also evident in Fig. 3 that $\bar{S}/\bar{I}$ in corners is more sensitive to the propagation parameter m than to any other factor. This is unfortunate since the greatest uncertainty is in this empirically derived parameter. The knee in the curves between 10 and 30 channel sets suggests that systems requiring more than 30 channel sets may experience diminishing returns in numbers of channels invested in the system.

III. COMPARISON OF HEXAGONAL AND SQUARE COVERAGE AREA SYSTEMS IN TERMS OF SERVICE PROBABILITY

Since both S and I are random variables, the fraction P_1 of corners (or alternately the fraction of area closely surrounding a corner) having an S/I less than some system threshold is the quantity that measures the quality of service provided.[2]

As discussed earlier, S can be modeled as a log-normal random variable.[3] The sum of log-normal variables, I, is also approximately log-normal [6], [7]. Then s and i can be defined as

$$s = 10 \log S$$

$$i = 10 \log I \tag{6}$$

[2] This is based on interference considerations alone. Of course, absolute signal level must be considered also.

[3] Recall that diversity is assumed to mitigate the small scale signal variation so this analysis considers only the large scale variation.

where log indicates a logarithm to the base 10. The variables s and i are normally distributed with probability densities

$$p(s) = \frac{1}{\sigma_s\sqrt{2\pi}} \exp\left[-\frac{(s-\bar{s})^2}{2\sigma_s^2}\right] \tag{7a}$$

and

$$p(i) = \frac{1}{\sigma_i\sqrt{2\pi}} \exp\left[-\frac{(i-\bar{i})^2}{2\sigma_i^2}\right] \tag{7b}$$

where $\bar{s}$ and $\bar{i}$ are the dB means and σ_s and σ_i are the dB standard deviations of the normal processes.

The fraction P_1 is then given by

$$P_1 = P(s - i < \gamma) = \frac{1}{\sqrt{2\pi}} \int_{z=-\infty}^{-V} \exp[-z^2/2]\, dz \tag{8}$$

where γ is the system signal-to-interference threshold in decibels, and

$$V = \frac{\bar{s} - \bar{i} - \gamma}{\sqrt{\sigma_s^2 + \sigma_i^2}}. \tag{9}$$

It remains then to determine $\bar{s} - \bar{i}$ in terms of $\bar{S}/\bar{I}$ and to select appropriate values for γ, σ_s, and σ_i.

A reasonable estimate for σ_s obtained by combining standard deviations for mobile radio propagation [2] and for building penetration attenuation [8] is $\sigma_s \approx 10$ dB. As indicated in Section II, interference power I is the sum of interference power from many sources and may be expected to be noise-like. From [9], a signal-to-noise ratio of 8 dB results in an error probability of 10^{-3} for differential phase-shift-keyed (DPSK) signals. Since error rates of several parts in 10^{-3} are acceptable for voice transmission, a value of $\gamma = 8$ dB will be used for the system signal-to-interference threshold.

If $u = \ln U$ and u is normally distributed with mean $\bar{u}$ and standard deviation σ_u, then [5], [7]

$$\bar{U} = \exp[\bar{u}] \exp[\sigma_u^2/2] \tag{10a}$$

and

$$\sigma_U^2 = [\exp(\sigma_u^2) - 1] \exp(2\bar{u}) \exp(\sigma_u^2) \tag{10b}$$

where $\bar{U}$ is the mean of U and σ_U is the standard deviation of U. From (6) and (10a) it follows that

$$\bar{s} - \bar{i} = 10 \log(\bar{S}/\bar{I}) - (\sigma_s^2 - \sigma_i^2)\,(\ln 10)/20. \tag{11}$$

Estimating σ_i is made difficult because of the inherent intractability of the problem of sums of log-normal processes [6], [7]. The sum of a number of log-normal random variables is a random variable that is also approximately log-normal [6], [7]. This fact has been verified by this author by Monte Carlo computer calculations. The total interference for radio

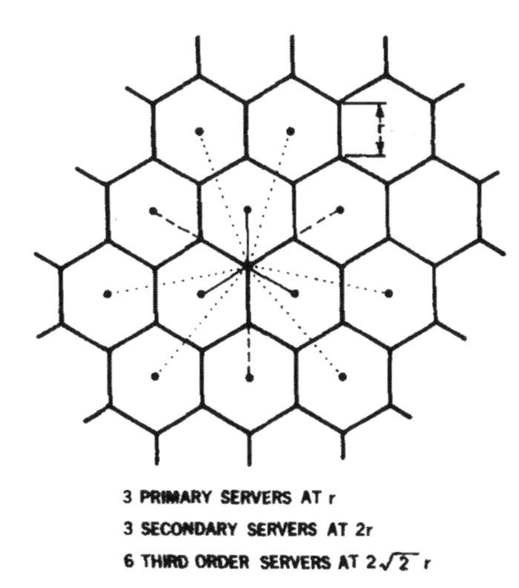

Fig. 6. Hexagonal coverage areas illustrating possible center-located base stations that could provide service at a corner of three coverage areas.

For the hexagonal arrangement in Fig. 6, it is evident that the three base stations twice as far from the corner to be served as the primary servers could also serve the corner. Their contribution to the probability that $s - i < \gamma$ is $P_{sh} = P^3(s - i < \gamma)$ where, in this case, $P(s - i < \gamma)$ is computed based on $\bar{S}/\bar{I}$ at the corner for one secondary server. The overall probability that $s - i < \gamma$ for the three primary and three secondary servers is then $P_{ph}P_{sh}$ which is plotted in Fig. 3 for $m = 3$ and $\sigma_s = 10$ dB.

The third-order set of servers in Fig. 5 are at a distance from the corner to be served that is three times the distance of the primary servers. These third-order base stations lie along the same paths from the corners as the primary servers and are likely to have signal variations that are highly correlated with the variations of the primary servers; thus, they cannot be expected to improve service to the corner significantly. Any farther out base stations would also be of questionable help and so will not be considered.

There are six third-order servers in the hexagonal arrangement in Fig. 6. The same argument about path proximity applies here as was used for the second-order servers for the square arrangement. Thus, these six third-order servers will be assumed to have the effect of three independent ones, with a probability that $s - i < \gamma$ of $P_{s'h} = P^3(s - i < \gamma)$. The overall probability of $s - i < \gamma$ for the hexagonal areas is then the product of the probabilities for first-, second- and third-order servers of $P_{ph}P_{sh}P_{s'h}$, which is also plotted in Fig. 4 for $m = 3$ and $\sigma_s = 10$ dB.

Looking at Fig. 4 it is evident that, in considering first-order servers only, the square coverage area arrangement provides lower probability of no service, i.e., a higher probability of service, for the equivalent number of channel sets. However, the hexagonal arrangement appears to provide approximately equal service when second-order servers are considered and better service when third-order servers are included. The independence assumptions weigh heavily on these conclusions and since no data exist to confirm or refute the assumptions, this apparent superiority of the hexagonal arrangement must be viewed cautiously. In any event, it is obvious that the second-order serving base stations, for both squares and hexagons, contribute significantly to the service probabilities for the corners. For comparison in terms of probability of service, the parameters that matter most, i.e., m and σ_s in the propagation law, are the parameters that are least well known.

IV. SERVICE PROBABILITY AT LOCATIONS OTHER THAN CORNERS

In this section, the probability that $s - i < \gamma$ discussed in the last section for coverage area corners will be considered at interior points in a coverage area. Consideration will be confined to the following set of conditions: hexagonal coverage areas, 27 channel sets, propagation exponent $m = 3$, system threshold $\gamma = 8$ dB, signal standard deviation $\sigma_s = 10$ dB, and interference standard deviation $\sigma_i = 5$ dB. The details will change somewhat for other parameter values but the overall trends are expected to be the same over the range of parameters that can be reasonably expected.

Fig. 7 illustrates the shortening of some paths to potential serving base stations and lengthening of other paths as the location needing service shifts from the corner of a coverage area toward the center. New values of $\bar{S}/\bar{I}$ can be calculated for each new set of path lengths and $P(s - i < \gamma)$ determined from each $\bar{S}/\bar{I}$. This has been done for the conditions stated on the figure. These results are plotted in Fig. 8 as a function of normalized distance from the center of the closest base station (see Fig. 7) along the line between the center and a corner, i.e., 1 is at the corner and 0 is at the center.

The upper two curves in Fig. 8 are associated with the closest base station and the next two base stations as labeled. The combined probability that $s - i < \gamma$ for these three is the third curve down from the top.[6] These three base stations were the set of primary servers for the corner and remain that for locations between the corner and the center. The fourth curve down from the top is the combined probability for the three primary and three secondary servers. The probability for the three primary, three secondary, and three of the six third-order servers discussed in the previous section is plotted as the bottom curve. For the curves that included the three closest base stations, the decrease in probability associated with the closest base station tends to be offset by increases in other probabilities until the normalized distance decreases to about 0.5 or 0.6. Then the rapid decrease associated with the closest base station overwhelms the other factors. The single data point at 0.866 is for the midpoint of a coverage area edge, lying midway between two corners.

The curves in Fig. 8 are replotted in Fig. 9 as functions of the square of the distance from the center, i.e., the square of the horizontal axis values in Fig. 8. This parameter is related to the fraction of the area experiencing the corresponding probability value or lower. The single data point at 0.75 illustrates the very small difference in probability for the midpoint along a coverage area edge as compared to a point at the same distance from the center along the line between the center and a corner. This single point in Figs. 8 and 9 indicates that

[6] As mentioned earlier, this probability also represents the performance of selection diversity against the large scale signal variations.

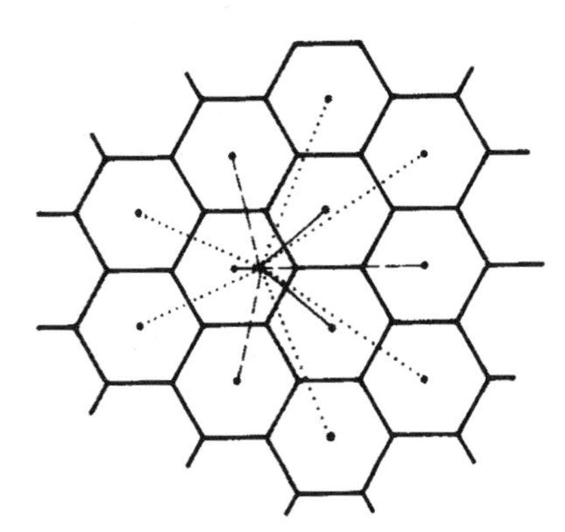

Fig. 7. Hexagonal coverage areas similar to Fig. 6 illustrating possible service to a location interior to a coverage area.

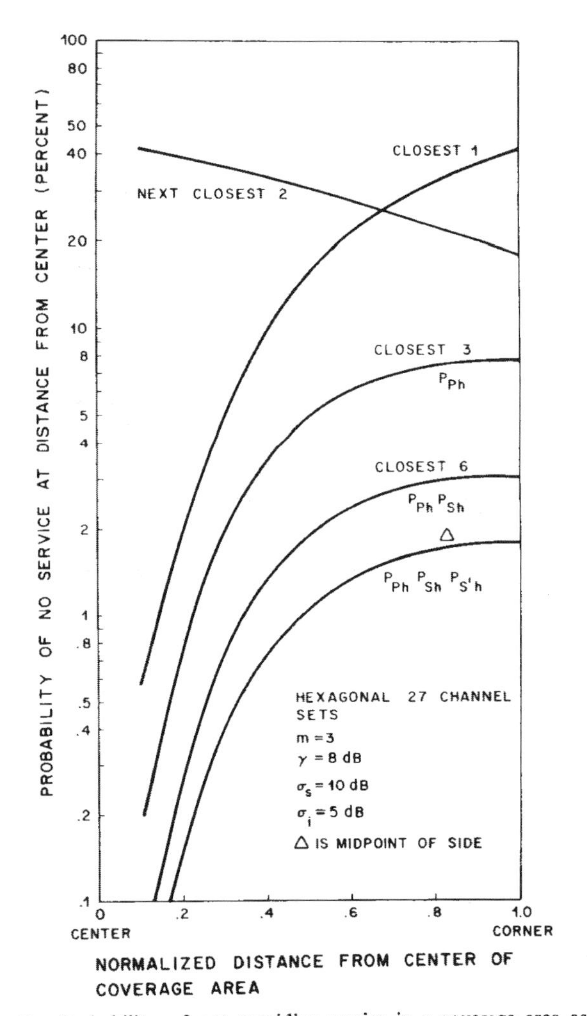

Fig. 8. Probability of not providing service in a coverage area as a function of distance from the base station located in the center of the area.

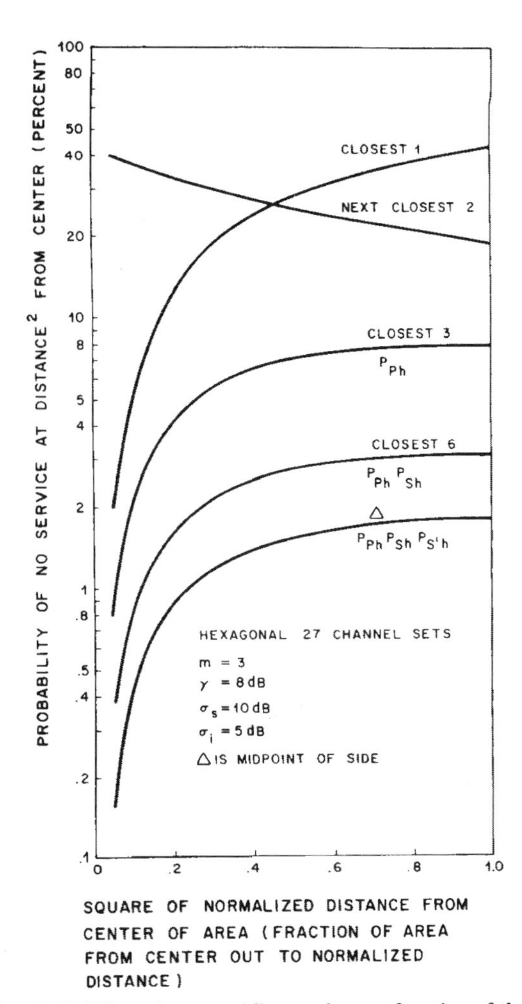

Fig. 9. Probability of not providing service as a function of the square of the distance from a center-located base station. The square of the distance is proportional to the area up to that distance.

the dominant factor is distance from the center, i.e., that direction is not important. Therefore, the distance squared is a measure of the fraction of the area affected. The curves in Fig. 9 illustrate, then, that the probability of $s - i < \gamma$, i.e., the probability of not receiving service, is nearly the value at the corner for 60–70 percent of the coverage area.

V. CONCLUSIONS

Comparisons were made between hexagonal and square shaped coverage areas for frequency reuse small-coverage-area radio systems. In general, the differences in performance between systems with the two different shaped areas are smaller than differences due to uncertainties in other system parameters. When the systems are compared only on the basis of average signal-to-average interference $\overline{S}/\overline{I}$ at the coverage area corners, the hexagons are better; however, when compared on the basis of probability of receiving service from any one of the closest possible serving base stations, the squares are better. Comparison on the basis of service probability using nearly all possible serving base stations suggests that, using this criterion, the hexagons may be somewhat better than the squares. Probabilities calculated for these service comparisons also indicate the effectiveness of selection diversity in combating large-scale signal variations.

Radio systems are compared assuming an infinite set of cochannel frequency reuses, a required S/I of at least 8 dB, and log-normal large-scale signal variations with a 10 dB standard deviation. These systems provide good service probabilities (on the order of 99 percent) with either hexagonal or square coverage areas if the available radio channels are subdivided into 30–40 channel sets, *even* when the propagation law changes as slowly as inverse distance cubed and has a standard deviation as large as 10 dB. Significantly fewer channel sets are required or, alternately, a significantly higher service probability is provided as the propagation power law increases from third toward fourth power and as the standard deviation decreases from 10 dB. Service probabilities remain relatively constant over 60–70 percent of the coverage area that is furthest from the central base station.

APPENDIX A

Consider an arrangement of square coverage areas as depicted in Fig. 1 with omnidirectional antennas located at area centers. Let R be the reuse interval defined in Section II, r_s the half width of an area, and t the tier number. The number of interfering cochannel base stations in a tier is $8t$. The distance from the center of a coverage area boundary to the center of the closest interfering base station is $(2Rt - 1)r_s$. Also, the distance d_c between a coverage area corner closest to the closest interfering base station and that interfering station is approximately $(2Rt - 1)r_s$ for large R. For large R, $d_c \approx 2Rtr_s$.

The shortest distance from a corner of a coverage area to an interfering base station at the adjacent corner of the interference tier (along the diagonal of the tier) is $d_d = (2Rt - 1)\sqrt{2}r_s$. Again, for large R this becomes $d_d \approx 2Rtr_s\sqrt{2}$.

The level of interference power or signal power at a point is k/d^m where d is distance between the point and a base station, m is the distance dependence of the attenuation law, and k is a constant depending on antennas, frequency, etc. The average interference from a tier is

$$\bar{I}_t \approx \frac{4k}{(2Rr_s)^m t^{m-1}} \cdot \frac{1 + (2)^{m/2}}{(2)^{m/2}}$$

where half of the distances from a coverage area to the interference tier are approximated by d_c and half by d_d. The total interference $\bar{I}$ contributed by the first T tiers is then

$$\bar{I} \approx \frac{4k[1 + (2)^{m/2}]}{(2Rr_s\sqrt{2})^m} \sum_{t=1}^{T} \frac{1}{t^{m-1}}.$$

At the corner of a coverage area the average signal is $\bar{S} = k/r_s{}^m(2)^{m/2}$. Then at the corner the $\bar{S}/\bar{I}$ contributed by the first T tiers is

$$\bar{S}/\bar{I} \approx \frac{2^{m-2}R^m}{1 + (2)^{m/2}} \Bigg/ \sum_{t+1}^{T} \frac{1}{t^{m-1}}. \quad (2)$$

The approximate relationship in (2) slightly overestimates values of $\bar{S}/\bar{I}$. The approximation is best for small m and for large R and deteriorates as m increases or R decreases. These trends are indicated in Table I which contains entries of the quantity (approx. $\bar{S}/\bar{I}$–exact $\bar{S}/\bar{I}$) in decibels for the least accurate case of one interference tier and ranging over values of m and R.

TABLE I
DECIBEL DIFFERENCES BETWEEN APPROXIMATE FORMULA $\bar{S}/\bar{I}$ AND ACTUAL $\bar{S}/\bar{I}$ FOR CORNERS OF SQUARE COVERAGE AREAS. INTERFERENCE IS FROM THE FIRST TIER ONLY ($t = 1$).

Channel Sets	R \ m	2	3	3.5	4
9	3	.2	.4	.6	.8
16	4	.1	.25	.35	.5
25	5	.07	.2	.2	.3
36	6	.05	.1	.1	.2
64	8	.02	.06	.1	.1

TABLE II
DECIBEL DIFFERENCE BETWEEN APPROXIMATE FORMULA $\bar{S}/\bar{I}$ AND ACTUAL $\bar{S}/\bar{I}$ FOR CORNERS OF HEXAGONAL COVERAGE AREAS. INTERFERENCE IS FROM THE FIRST TIER ONLY ($t = 1$).

Channel Sets	n \ m	2	3	3.5	4
12	2	0.10	.2	.3	.4
27	3	0.04	.1	.15	.2
48	4	0.02	.06	.1	.1

APPENDIX B

Consider an arrangement of hexagonal coverage areas as depicted in Fig. 2. The distances d_i from the centers of interfering coverage areas to the center of the serving area (area 0 in Fig. 2) are $d_i = 3ntr_h$ where n is a parameter related to the reuse of radio channels, r_h is the radius of the hexagonal areas (see Fig. 2), and t is the tier number. For only a subset of the allowable hexagonal reuse patterns (channel sets), n is an integer. The approximate $\bar{S}/\bar{I}$ at a coverage area corner from a set of interferers is obtained by substituting into (1) the center-to-center distance, $d_i = 3ntr_h$, for d_j and letting $r = r_h$.

By noting that there are $6t$ interferers for each tier t, numbered as in Fig. 2, and considering only complete tiers, the sum over j can be replaced by a sum over t, yielding (3). Some of the distances from interferers to the corner are overestimated and some are underestimated, but these effects balance out quite well as illustrated in Table II. The entires in Table II are (approx. $\bar{S}/\bar{I}$–exact $\bar{S}/\bar{I}$) in decibels for the least accurate case of one interference tier and ranging over values of m and n. The approximate formula slightly overestimates $\bar{S}/\bar{I}$.

REFERENCES

[1] J. S. Engel, "The effects of cochannel interference on the parameters of a small-cell mobile telephone system," *IEEE Veh. Technol.*, vol. VT-18, pp. 110–116, Nov. 1969.

[2] W. C. Jakes, Jr., *Microwave Mobile Communications*. New York: Wiley, 1974, chs. 1, 2, 5–7.
[3] W. R. Young, "Introduction, background, and objectives," *Bell Syst. Tech. J.*, Special Issue on Advanced Mobile Phone Service, vol. 58, pp. 1–14, Jan. 1979.
[4] V. H. MacDonald, "The cellular concept," *Bell Syst. Tech. J.*, Special Issue on Advanced Mobile Phone Service, vol. 58, pp. 15–42, Jan. 1979.
[5] J. Aitchison and J. A. C. Brown, *The Lognormal Distribution*. Cambridge, England: Cambridge Univ. Press, 1957.
[6] I. Nasell, "Some properties of power sums of truncated normal random variables," *Bell Syst. Tech. J.*, vol. 46, pp. 2091–2110, Nov. 1967.
[7] L. F. Fenton, "The sum of log-normal probability distributions in scatter transmission systems," *IRE Trans. Commun. Syst.*, vol. CS-9, pp. 57–67, Mar. 1960.
[8] P. I. Wells, "The attenuation of UHF radio signals by houses," *IEEE Trans. Veh. Technol.*, vol. VT-26, pp. 358–362, Nov. 1977.
[9] M. Schwartz, W. R. Bennett, and S. Stein, *Communication Systems and Techniques*. New York: McGraw-Hill, 1966, p. 299.

Donald C. Cox (S'57–M'60–SM'71–F'79) received the B.S. and M.S. degrees from the University of Nebraska, Lincoln, in 1959 and 1960, respectively, and the Ph.D. degree from Stanford University, Stanford, CA, in 1968, all in electrical engineering.

From 1960 to 1963 he did microwave communications system design for the U.S.A.F. Dyna-Soar at Wright-Patterson AFB, OH. From 1963 to 1968 he was at Stanford University doing tunnel diode amplifier design and research on microwave propagation in the troposhere. From 1968 to 1973 he was a member of the Technical Staff of Bell Laboratories, Holmdel, NJ, doing research in mobile radio propagation and on high capacity mobile radio systems. He is now Supervisor of a group at Bell Laboratories doing propagation and systems research for portable-radio telephony and for millimeter-wave satellite communications.

Dr. Cox is a member of Commissions B, C, and F of USNC/URSI, Sigma Xi, Sigma Tau, Eta Kappa Nu, and Pi Mu Epsilon, and is a Registered Professional Engineer in Ohio and Nebraska.

GMSK Modulation for Digital Mobile Radio Telephony

KAZUAKI MUROTA, Member, IEEE, and
KENKICHI HIRADI, Member, IEEE

Abstract—This paper is concerned with digital modulation for future mobile radio telephone services. First, the specific requirements on the digital modulation for mobile radio use are described. Then, premodulation Gaussian filtered minimum shift keying (GMSK) with coherent detection is proposed as an effective digital modulation for the present purpose, and its fundamental properties are clarified with the aid of machine computation. The constitution of modulator and demodulator is then discussed from the viewpoints of mobile radio applications. The superiority of this modulation is supported by some experimental test results.

I. INTRODUCTION

It is well known that voice transmission in many VHF and UHF mobile radio telephone systems has usually been made by using a single-channel-per-carrier (SCPC) analog FM transmission technique. However, in order to provide highly secure voice and/or high-speed data transmission by the use of large-scale integrated (LSI) transceivers, digital mobile radio transmission is currently being studied [1]-[7]. While digital transmission can surely bring many advantages, some technical problems must be solved. This paper is concerned with a digital modulation for future mobile radio communications.

From the viewpoint of mobile radio use, the out-of-band radiation power in the adjacent channel should be generally suppressed 60-80 dB below that in the desired channel. So as to satisfy this severe requirement, it is necessary to manipulate the RF output signal spectrum. Such a spectrum manipulation cannot usually be performed at the final RF stage in the multichannel SCPC transceivers because the transmitted RF frequency is variable. Therefore, intermediate-frequency (IF) or baseband filtering with frequency up conversion is mostly used. However, when such a spectrum-manipulated signal is translated up and passed through a nonlinear class-C power amplifier, the required spectrum manipulation should not be violated by the nonlinearities. In order to mitigate the impairments, some narrow-band digital modulation schemes with constant or less fluctuated envelope property have been researched [8]-[10].

In this paper, premodulation Gaussian filtered minimum shift keying (GMSK) with coherent detection is proposed as an effective digital modulation for the present purpose, and its fundamental properties are analyzed with the aid of machine computation. The relationship between out-of-band radiation suppression and bit-error-rate (BER) performance is made clear. Constitution of the modulator and demodulator is then discussed. The superiority of this modulation is supported by some experimental test results.

Paper approved by the Editor for Communication Theory of the IEEE Communications Society for publication after presentation at 29th IEEE Vehicular Technology Conference, Chicago, IL, March 1979. Manuscript received May 28, 1980; revised January 5, 1981.

The authors are with the Yokosuka Electrical Communication Laboratory, Nippon Telegraph and Telephone Public Corporation, Kanagawa-Ken, Japan.

II. GMSK MODULATION

A. Spectrum Manipulation of MSK

Minimum shift keying (MSK), which is binary digital FM with a modulation index of 0.5, has the following good properties: constant envelope, relatively narrow bandwidth, and coherent detection capability [11]-[13]. However, it does not satisfy the severe requirements with respect to out-of-band radiation for SCPC mobile radio. MSK can be generated by direct FM modulation. As is easily found, the output power spectrum of MSK can be manipulated by using a premodulation low-pass filter (LPF), keeping the constant envelope property, as shown in Fig. 1. To make the output power spectrum compact, the premodulation LPF should have the following properties:

1) narrow bandwidth and sharp cutoff
2) lower overshoot impulse response
3) preservation of the filter output pulse area which corresponds to a phase shift $\pi/2$.

Condition 1) is needed to suppress the high-frequency components, 2) is to protect against excessive instantaneous frequency deviation, and 3) is for coherent detection to be applicable as simple MSK.

Generally, the introduction of the premodulation LPF violates the minimum frequency spacing constraint and the fixed-phase constraint of MSK. However, the above two constraints are not intrinsic requirements for effective coherent binary FM with modulation index 0.5. Such a premodulation-filtered MSK signal can be detected coherently because its

Reprinted from *IEEE Transactions on Communications,* vol. COM-29, no. 7, July 1981.

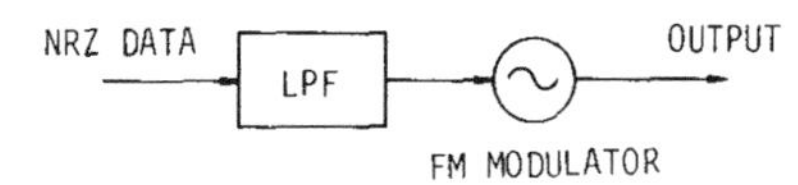

Fig. 1. Premodulation baseband-filtered MSK.

pattern-averaged phase-transition trajectory does not deviate from that of simple MSK.

B. Fundamental Properties of GMSK

A Gaussian LPF satisfies all the above-described characteristics. Consequently, the modified MSK modulation using a premodulation Gaussian LPF can be expected to be an excellent digital modulation technique for the present purpose. Such a modified MSK is named Gaussian MSK or GMSK in connection with Gaussian low-pass filtering. Let us now investigate the GMSK modulation from various aspects.

Output Power Spectrum: Fig. 2 shows the machine-computed results of the output power spectrum of the GMSK signal versus the normalized frequency difference from the carrier center frequency $(f - f_c)T$ where the normalized 3 dB-down bandwidth of the premodulation Gaussian LPF B_bT is a parameter. The spectrum for GMSK with $B_bT = 0.2$ is nearly equal to that of TFM.

The effective variable parameter B_bT can be selected by the system designer considering overall spectrum efficiency of the cellular zone structure.

Fig. 3 shows the machine-computed results of the fractional power in the desired channel versus the normalized bandwidth of the predetection rectangular bandpass filter (BPF) B_iT. Table I shows the occupied bandwidth for the prescribed percentage of power where B_bT is also a variable parameter. For comparison, the occupied bandwidth of TFM is also shown in Table I.

Fig. 4 shows the machine-computed results of the ratio of the out-of-band radiation power in the adjacent channel to the total power in the desired channel where the normalized channel spacing f_sT is taken as the abscissa and both channels are assumed to have the ideal rectangular bandpass characteristics with $B_iT = 1$. The situation of $f_sT = 1.5$ and $B_iT = 1$ corresponds to the case of $f_s \cong 25$ kHz and $B_i = 16$ kHz when $f_b = 1/T = 16$ kbits/s. From Fig. 4, it is found that the GMSK with $B_bT = 0.28$ can be adopted as the digital modulation for conventional VHF and UHF SCPC mobile radio communications without carrier frequency drift where the ratio of out-of-band radiation power in the adjacent channel to the total power in the desired channel must be lower than −60 dB. When a certain amount of carrier frequency drift (for example $\Delta f = \pm 1.5$ kHz) exists, $B_bT = 0.2$ is needed.

BER Performance: Let us now consider the theoretical BER performance of GMSK modulation using coherent detection in the presence of additive white Gaussian noise.

Since the GMSK modulation of interest is a certain kind of binary digital modulation, its BER performance bound in the high SNR condition is approximately represented as

$$P_e = \frac{1}{2} \operatorname{erfc}\left(\frac{d_{\min}}{2\sqrt{N_0}}\right) \tag{1}$$

where N_0 is the power spectrum density of the additive white Gaussian noise and erfc( ) is the complementary error func-

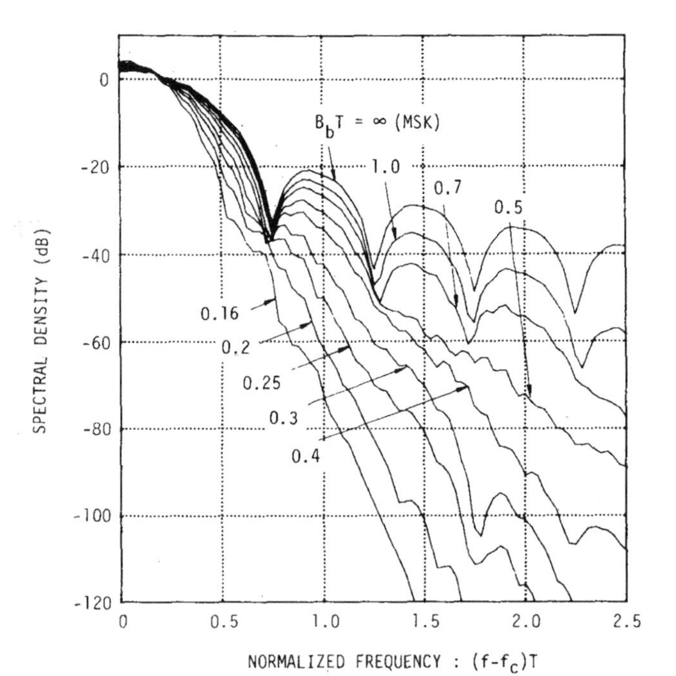

Fig. 2. Power spectra of GMSK.

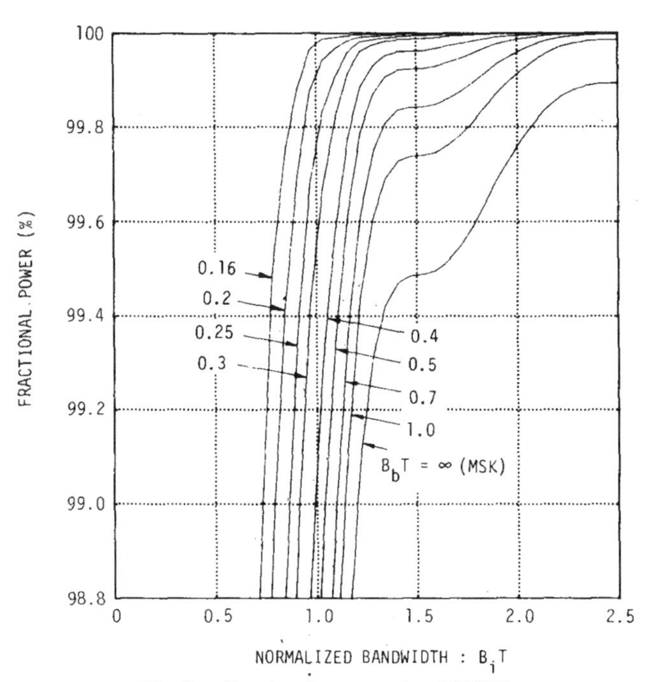

Fig. 3. Fractional power ratio of GMSK.

TABLE I
OCCUPIED BANDWIDTH CONTAINING A GIVEN PERCENTAGE POWER

B_bT \ %	90	99	99.9	99.99
0.2	0.52	0.79	0.99	1.22
0.25	0.57	0.86	1.09	1.37
0.5	0.69	1.04	1.33	2.08
MSK	0.78	1.20	2.76	6.00
TFM	0.52	0.79	1.02	1.37

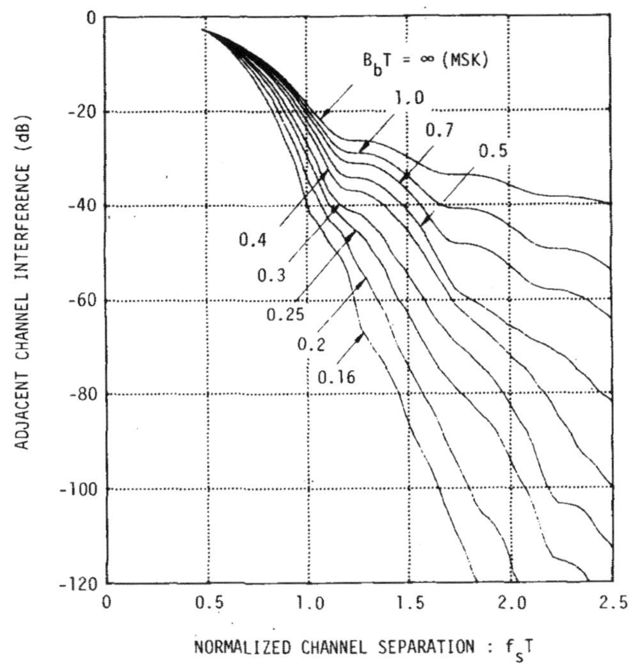

Fig. 4. Adjacent channel interference of GMSK.

tion given by

$$\mathrm{erfc}\,(x) = \frac{2}{\sqrt{\pi}} \int_x^{\infty} \exp(-u^2)\, du. \tag{2}$$

Furthermore, $d_{\min}$ is the minimum value of the signal distance d between mark and space in Hilbert space observed during the time interval from t_1 to t_2 and d is defined by

$$d^2 = \frac{1}{2} \int_{t_1}^{t_2} |u_m(t) - u_s(t)|^2\, dt \tag{3}$$

where $u_m(t)$ and $u_s(t)$ are the complex signal waveforms corresponding to the mark and the space transmissions, respectively.

While the BER performance bound given by (1) is attained only when the ideal maximum likelihood detection is adopted, it gives an approximate solution for the ideal BER performance of GMSK modulation with coherent detection.

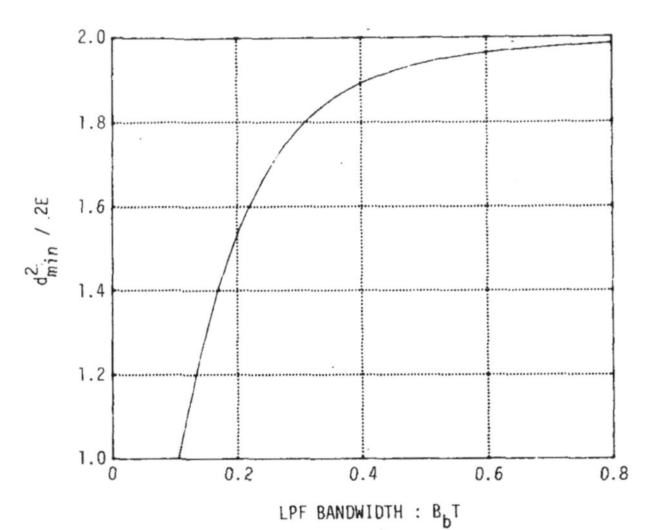

Fig. 5. Normalized minimum signal distance of GMSK.

Fig. 5 shows the machine-computed results for $d_{\min}$ of the GMSK signal versus B_bT where E_b denotes the signal energy per bit defined by

$$E_b = \frac{1}{2} \int_0^T |u_m(t)|^2\, dt = \frac{1}{2} \int_0^T |u_s(t)|^2\, dt. \tag{4}$$

In the case $B_bT \to \infty$, which corresponds to the simple MSK signal, Fig. 5 yields $d_{\min} = 2\sqrt{E_b}$, which is that of antipodal transmission. It is noticed that the meaningful observation time interval for the GMSK signal $t_2 - t_1$ may be made longer than $2T$, which corresponds to that for the simple MSK signal, due to the intersymbol interference (ISI) effect on the phase transitions.

Substituting the machine-computed results of $d_{\min}$ into (1), the BER performance of the GMSK modulation with coherent detection is obtained. Fig. 6 shows the performance degradation of GMSK from antipodal transmission due to the ISI effect of the premodulation LPF. This figure shows that the performance degradation is small and that the required E_b/N_0 of GMSK with $B_bT = 0.25$ does not exceed more than 0.7 dB compared to that of antipodal transmission.

III. IMPLEMENTATION

A. Modulator

The simple and easy method is to modulate the frequency of VCO directly by the use of baseband Gaussian pulse stream, as shown in Fig. 1. However, this modulator has the weak point that it is difficult to keep the center frequency within the allowable value under the restriction of maintaining the linearity and the sensitivity for the required FM modulation. Such a weak point can be removed by the use of an elaborate PLL modulator with a precisely designed transfer characteristics or an orthogonal modulator with digital waveform generators [14]. Instead of such a modulator, a $\pi/2$-shift binary PSK (BPSK) modulator followed by a suitable PLL phase smoother, as shown in Fig. 7, is considered to be a prominent alternative where the transfer characteristics of this PLL are also designed for the output power spectrum to satisfy the required condition.

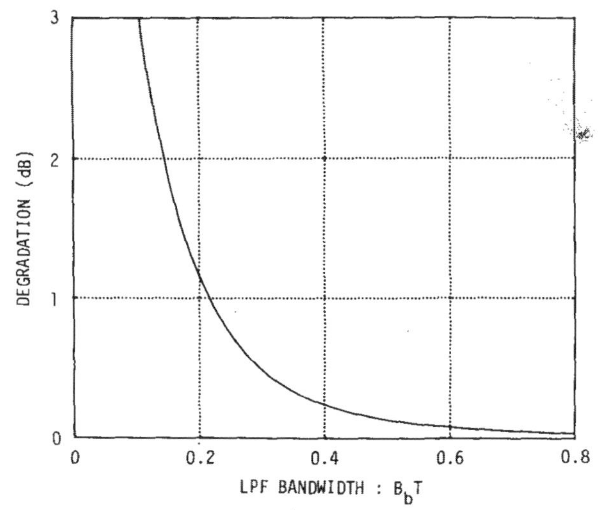

Fig. 6. Theoretical E_b/N_0 degradation of GMSK.

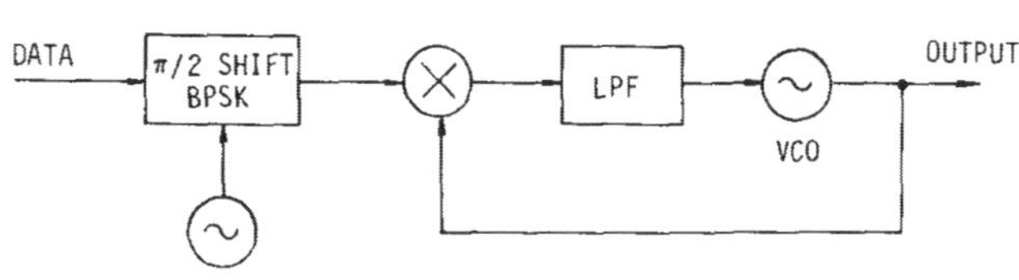

Fig. 7. PLL-type GMSK modulator.

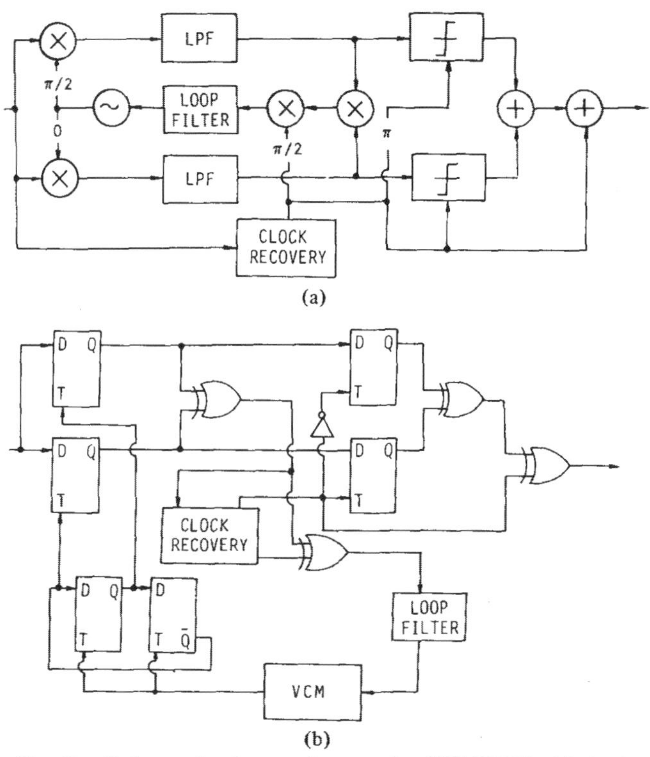

Fig. 8. Orthogonal coherent detector for MSK/GMSK. (a) Analog type. (b) Digital type.

B. Demodulator

Similar to the simple MSK or TFM system, the orthogonal coherent detector is also applicable for the GMSK system. When realizing such an orthogonal coherent detector, one of the most important and difficult problems is how to recover the reference carrier and the timing clock. The most typical method is de Buda's one [12]. In his method, the reference carrier is recovered by dividing by four the sum of the two discrete frequencies contained in the frequency doubler output and the timing clock is directly recovered by their difference. Remembering that the action of the well-known Costas loop as a carrier recovery circuit for BPSK systems is equivalent to that of a PLL with a frequency doubler [15], de Buda's method is realized by the equivalent one shown in Fig. 8(a). This modified method can easily be implemented by conventional digital logic circuits and its configuration is also shown in Fig. 8(b). In this configuration, two D flip-flops act as the quadrature product demodulators and both of the Exclusive-Or logic circuits are used for the baseband multipliers. Furthermore, the mutually orthogonal reference carriers are generated by the use of two D flip-flops, and the VCO center frequency is then set equal to the four times carrier center frequency. This configuration is considered to be especially suitable for the mobile radio unit which must be simplified, miniaturized, and economized.

IV. EXPERIMENTS

A. Test System

Fig. 9 shows the block diagram of the experimental test system where the carrier frequency and the bit rate are $f_c = 70$ MHz and $f_b = 16$ kbits/s, respectively. A pseudonoise (PN) pulse sequence with a repetition period of $N = (2^{15} - 1)$ bits is generated by the 15-stage feedback shift register (FSR) and is used as a test pattern signal. After passing through a pre-

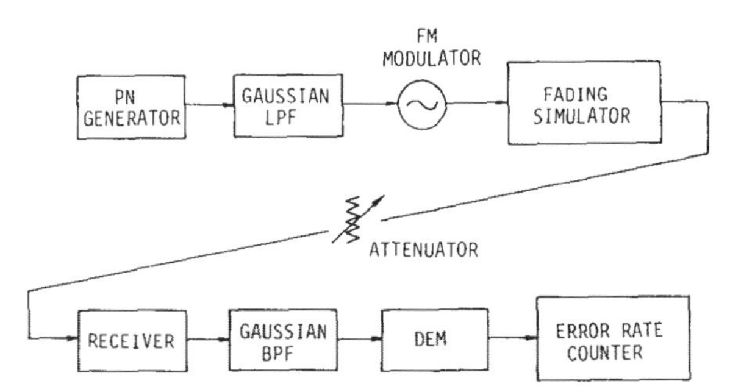

Fig. 9. Block diagram of experimental test system.

modulation Gaussian LPF having a variable bandwidth B_b, the PN sequence is put into the synthesized RF signal generator having an external FM modulation capability. The frequency deviation of the RF signal generator is set equal to $\Delta f_d = \pm 4$ kHz, which corresponds to the MSK condition for the 16 kbits/s transmission. Then the GMSK signal of our choice is obtained as the RF signal generator output, and is transmitted into the receiver via the Rayleigh fading simulator [16]. Predetection bandpass filtering in the receiver is performed by the precisely designed Gaussian bandpass crystal filter. The bandpass-filtered output is demodulated by the digital orthogonal coherent detector shown in Fig. 8. The regenerated output is fed into the error-rate counter for the BER measurement.

B. Power Spectrum and Eye Pattern

Fig. 10 shows the measured power spectra of the RF signal generator output when B_bT is a variable parameter. It is clearly seen that the measured results agree well with the machine-computed ones shown in Fig. 2. Moreover, GMSK with $B_bT = 0.25$ is shown to satisfy the severe requirements

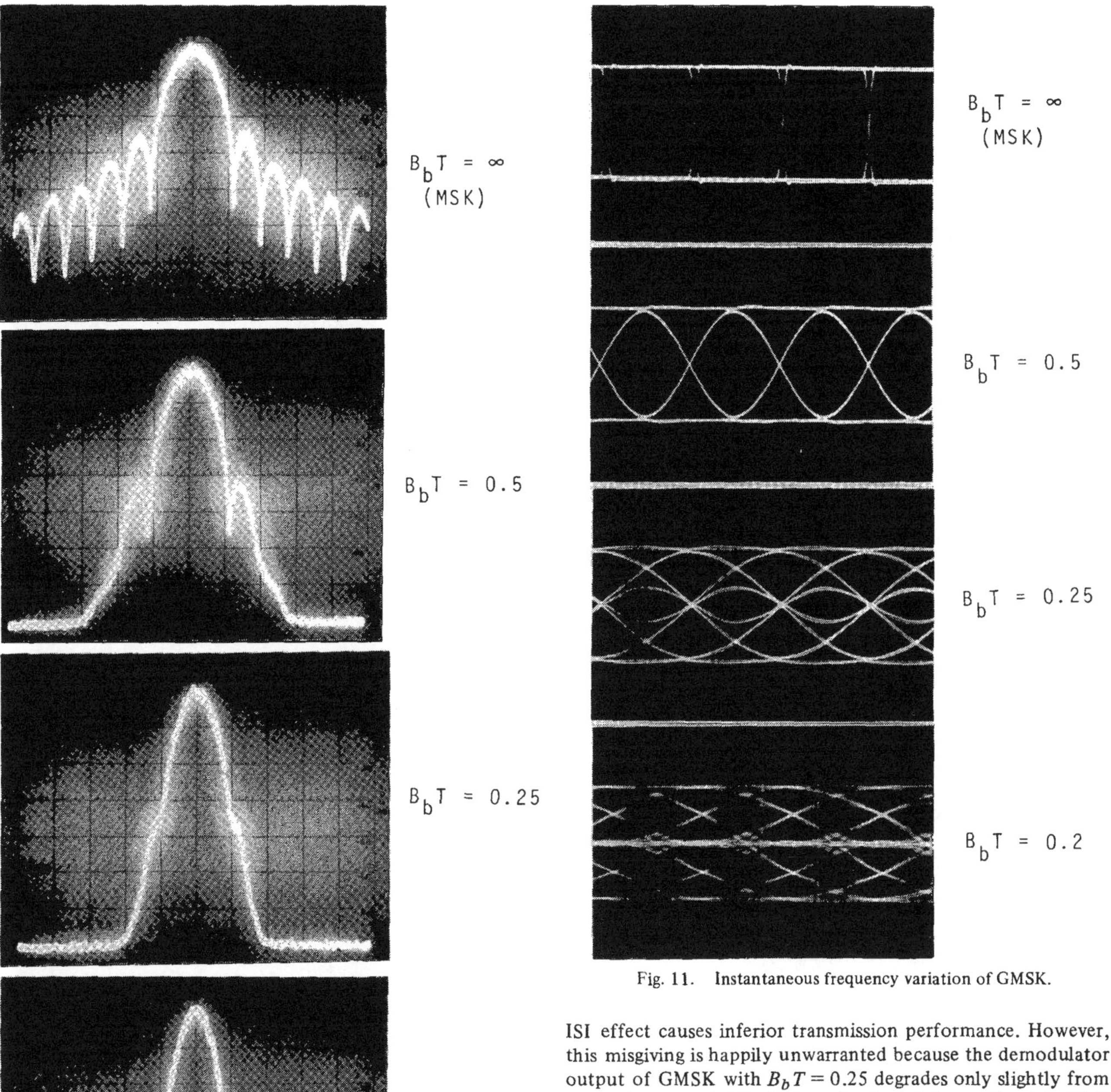

Fig. 10. Measured power spectra of GMSK (*V*: 10 dB/div., *H*: 10 kHz/div.).

Fig. 11. Instantaneous frequency variation of GMSK.

of the out-of-band radiation of SCPC mobile radio communications. The corresponding eye pattern measured at the premodulation Gaussian LPF output is shown in Fig. 11. This figure shows that the above satisfactory performance of the out-of-band radiation can only be attained by the sacrifice of introducing severe ISI effects into the baseband waveform of the FM modulator input. It might be feared that such a severe ISI effect causes inferior transmission performance. However, this misgiving is happily unwarranted because the demodulator output of GMSK with $B_bT = 0.25$ degrades only slightly from that of simple MSK. It is easily found from Fig. 12 which shows the respective eye patterns measured by the analog-type orthogonal coherent detector shown in Fig. 8(a). It is also certified from the BER performance test results described later.

C. Static BER Performance

Fig. 13 shows experimental test results for static BER performance in the nonfading environment where the normalized 3 dB-down bandwidth of the premodulation Gaussian LPF, B_bT, is a variable parameter and the normalized 3 dB-down bandwidth of the predetection Gaussian BPF is $B_iT \cong 0.63$, i.e., $B_i = 10$ kHz for $f_b = 1/T = 16$ kbits/s. The condition $B_iT \cong 0.63$ is nearly optimum, as shown in Fig. 14. From Fig. 13, performance degradation of GMSK with $B_bT = 0.25$ relative to simple MSK is found to be only 1.0 dB. Moreover, the measured static BER performance of simple MSK degrades by 0.7 dB from the theoretical one of ideal antipodal binary

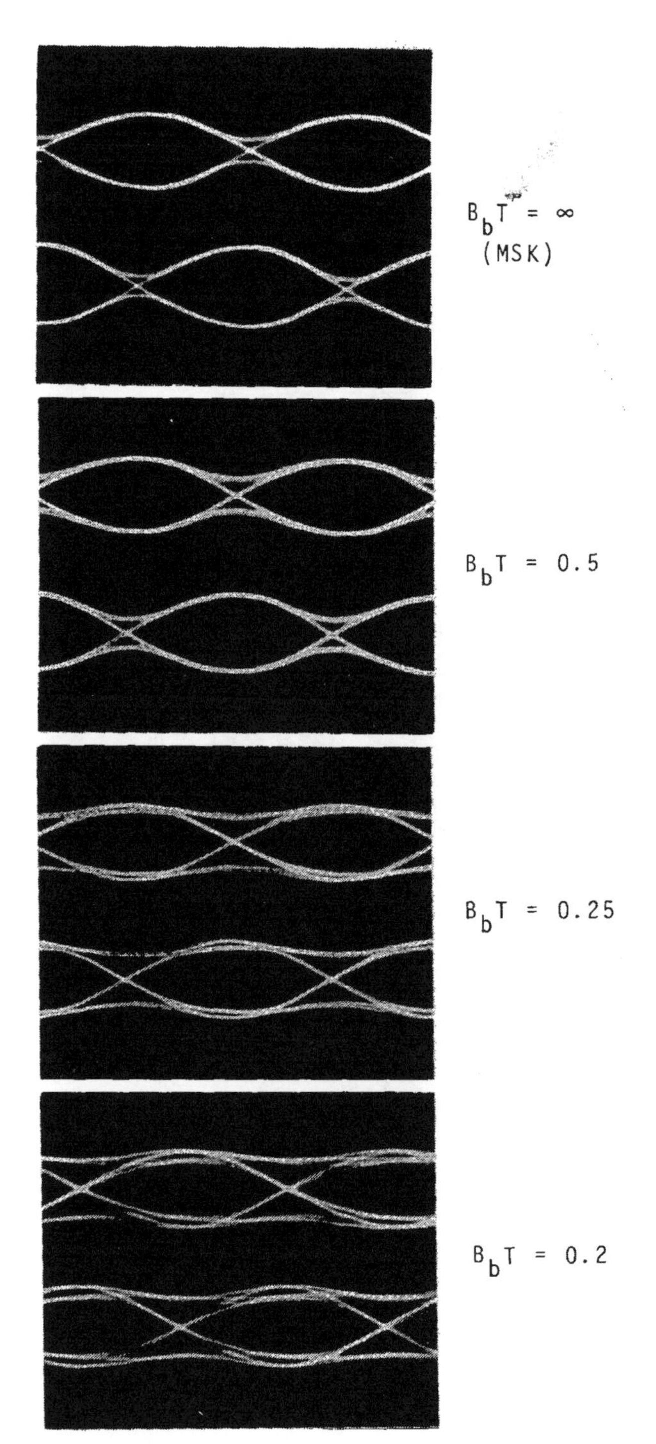

Fig. 12. GMSK eye patterns demodulated by orthogonal coherent detector.

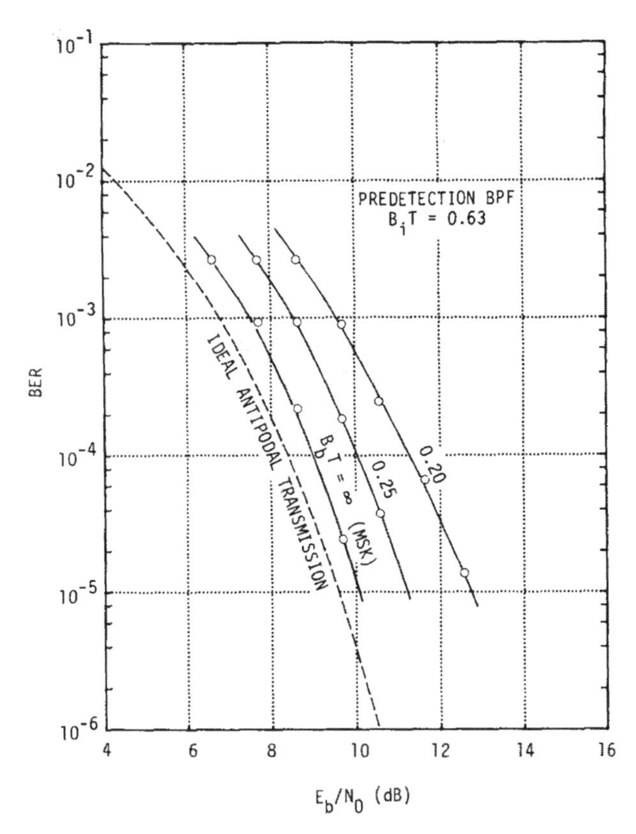

Fig. 13. Static BER performance.

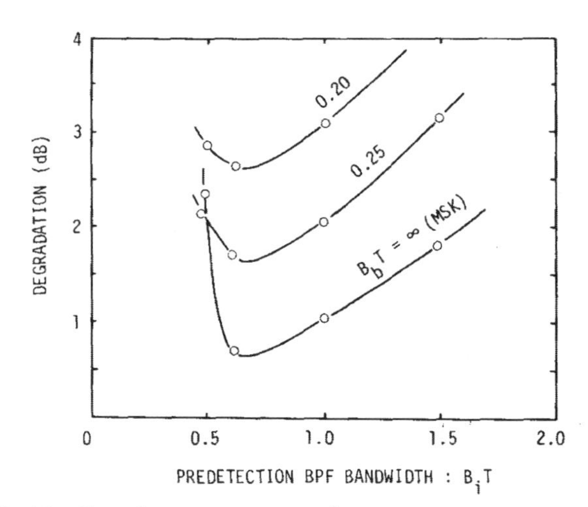

Fig. 14. Degradation of required E_b/N_0 for obtaining BER of 10^{-3}.

transmission system. If γ denotes the received signal energy-to-noise density ratio, i.e., E_b/N_0, the measured static BER performance in the nonfading environment can be approximated as

$$P_e(\gamma) \cong \frac{1}{2} \operatorname{erfc}(\sqrt{\alpha\gamma}) \tag{5}$$

where erfc( ) is the complementary error function given by (2) and α is a constant parameter determined as

$$\alpha \cong \begin{cases} 0.68 & \text{for GMSK with } B_bT = 0.25 \\ 0.85 & \text{for simple MSK } (B_bT \to \infty). \end{cases} \tag{6}$$

The above-obtained results can be estimated by the degradation of the minimum signal distance shown in Figs. 5 and 6.

D. Dynamic BER Performance

In the practical V/UHF land mobile radio environment, signal transmission between a fixed base station and a moving vehicle is usually performed via random multiple propagation routes. Consequently, fast and deep multipath fading, which can generally be treated by the well-known Rayleigh fading model, appears on the received signals of both stations and degrades the signal transmission performance severely.

In particular, when a quasi-stationary slow Rayleigh fading

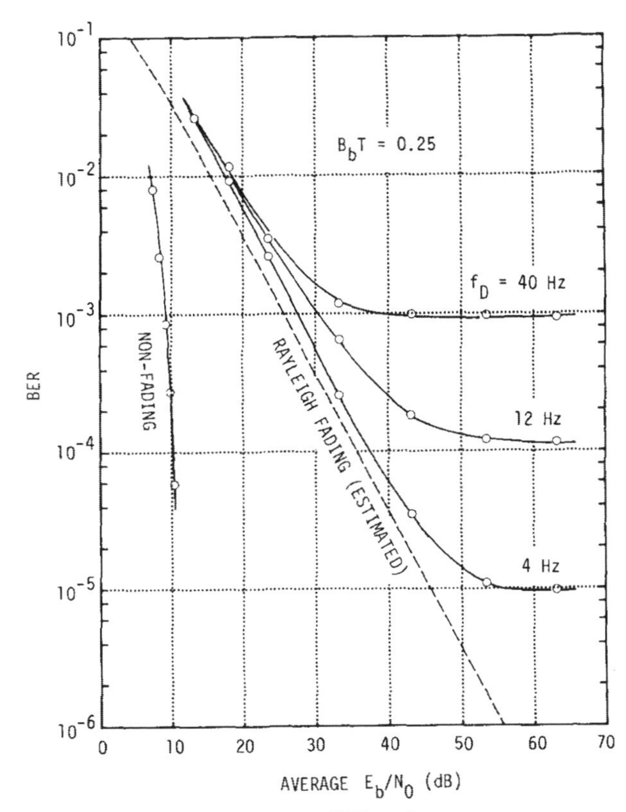

Fig. 15. Dynamic BER performance.

model is assumed, dynamic BER performance is given by

$$P_e(\Gamma) = \int_0^\infty P_e(\gamma)p(\gamma)\,d\gamma \tag{7}$$

where Γ is the average E_b/N_0 and $p(\gamma)$ is the probability density function (pdf) of γ given by

$$p(\gamma) = \frac{1}{\Gamma}\exp\left(-\frac{\gamma}{\Gamma}\right). \tag{8}$$

Substituting (5) and (8) into (7) yields

$$P_e(\Gamma) \cong \frac{1}{2}\left(1 - \sqrt{\frac{\alpha\Gamma}{\alpha\Gamma + 1}}\right) \cong \frac{1}{4\alpha\Gamma} \tag{9}$$

where α is the constant parameter given by (6).

However, the dynamic BER performance in the fast Rayleigh fading environment, where the temporal variation effect of the fading cannot be neglected, has not yet been theoretically estimated because the tracking performance of the carrier recovery circuit in such environment cannot be analyzed. Fig. 15 shows the experimental test results of dynamic BER performance of the GMSK with $B_bT = 0.25$ in the simulated fast Rayleigh fading environment where the maximum Doppler frequency, i.e., the fading rate f_D, is a variable parameter. For comparison, theoretically estimated dynamic BER performance in the quasi-stationary slow Rayleigh fading environment, i.e., $f_DT \to 0$, is also shown by the dashed line in the same figure.

V. CONCLUSION

As an effective digital modulation for mobile radio use, premodulation Gaussian-filtered minimum shift keying (GMSK) modulation with coherent detection has been proposed. The fundamental properties have been analyzed with the aid of machine computation. The constitution of modulator and demodulator has also been discussed. The superiority of this modulation has been supported by experimental results.

ACKNOWLEDGMENT

The authors wish to thank Dr. K. Miyauchi, S. Ito, K. Izumi, and Dr. S. Seki for their helpful guidance. They also are grateful to Dr. M. Ishizuka and H. Suzuki for their fruitful discussions.

REFERENCES

[1] O. Bettinger, "Digital speech transmission for mobile radio service," *Elec. Commun.*, vol. 47, pp. 224–230, 1972.

[2] J. S. Bitler and C. O. Stevens, "A UHF mobile telephone system using digital modulation: Preliminary study," *IEEE Trans. Vehic. Technol.*, vol.VT-22, pp. 78–81, Aug. 1973.

[3] N. S. Jayant, R. W. Schafer, and M. R. Karim, "Step-size-transmitting differential coders for mobile telephony," in *Proc. IEEE Int. Conf. Commun.*, June 1975, pp. 30/6–30/10.

[4] D. L. Duttweiler and D. G. Messerschmitt, "Nearly instantaneous companding and time diversity as applied to mobile radio transmission," in *Proc. IEEE Int. Conf. Commun.*, June 1975, pp. 40/12–40/15.

[5] J. C. Feggeler, "A study of digitized speech in mobile telephony," presented at the Symp. on Microwave Mobile Commun., session V-3, Boulder, CO, Sept.–Oct. 1976.

[6] H. M. Sachs, "Digital voice considerations for the land mobile radio services," in *Proc. IEEE 27th Vehic. Technol. Conf.*, Mar. 1977, pp. 207–219.

[7] K. Hirade and M. Ishizuka, "Feasibility of digital voice transmission in mobile radio communications," *Paper Tech. Group. IECE Japan*, vol. CS78-2, Apr. 1978.

[8] F. G. Jenks, P. D. Morgan, and C. S. Warren, "Use of four-level phase modulation for digital mobile radio," *IEEE Trans. Electromagn. Compat.*, vol. EMC-14, pp. 113–128, Nov. 1972.

[9] P. K. Kwan, "The effects of filtering and limiting a double-binary PSK signal," *IEEE Trans. Aerosp. Electron. Syst.*, vol. AES-5, pp. 589–594, July 1969.

[10] S. A. Rhodes, "Effects of hardlimiting on bandlimited transmission with conventional and offset QPSK modulation," in *Proc. IEEE Nat. Telecommun. Conf.*, 1972, pp. 20F/1–20F/7.

[11] H. C. van den Elzen and P. van der Wurf, "A simple method of calculating the characteristics of FSK signals with modulation index 0.5," *IEEE Trans. Commun.*, vol. COM-20, pp. 139–147, Apr. 1972.

[12] R. de Buda, "Coherent demodulation of frequency shift keying with low deviation ratio," *IEEE Trans. Commun.*, vol. COM-20, pp. 466–470, June 1972.

[13] H. Miyakawa *et al.*, "Digital phase modulation scheme using continuous-phase waveform," *Trans. IECE Japan*, vol. 58-A, pp. 767–774, Dec. 1975.

[14] F. de Jager and C. B. Dekker, "Tamed frequency modulation, a novel method to achieve spectrum economy in digital transmission," *IEEE Trans. Commun.*, vol. COM-20, pp. 534–542, May 1978.

[15] R. L. Didday and W. C. Lindsey, "Subcarrier tracking methods and communication system design," *IEEE Trans. Commun. Technol.*, vol. COM-16, pp. 541–550, Aug. 1968.

[16] K. Hirade *et al.* "Fading simulator for land mobile radio communications," *Trans. IECE Japan*, vol. 58-B, pp. 449–459, Sept. 1975.

Continuous Phase Modulation—Part I: Full Response Signaling

TOR AULIN, MEMBER, IEEE, AND CARL-ERIK W. SUNDBERG, MEMBER, IEEE

Abstract—**The continuous phase modulation (CPM) signaling scheme has gained interest in recent years because of its attractive spectral properties. Data symbol pulse shaping has previously been studied with regard to spectra, for binary data and modulation index 0.5. In this paper these results have been extended to the M-ary case, where the pulse shaping is over a one symbol interval, the so-called full response systems. Results are given for modulation indexes of practical interest, concerning both performance and spectrum. Comparisons are made with minimum shift keying (MSK) and systems have been found which are significantly better in E_b/N_0 for a large signal-to-noise ratio (SNR) without expanded bandwidth. Schemes with the same bit error probability as MSK but with considerably smaller bandwidth have also been found. Significant improvement in both power and bandwidth are obtained by increasing the number of levels M from 2 to 4.**

I. INTRODUCTION

FOR digital transmission over bandlimited channels, the demand for bandwidth efficient constant envelope signaling schemes with good reliability has increased in recent years. A system often used in practice is multilevel phase shift keying, M-ary PSK, which has the drawback that, although, for M equal 2 or 4, the receiver sensitivity is acceptable, the signal is too wide-band because of discontinuous phase. Thus, RF-filtering has to be performed before transmission causing a nonconstant envelope signal and a decreased receiver sensitivity. The so-called minimum shift keying (MSK), or fast frequency shift keying (FFSK), binary signaling schemes opened new prospects since the error probability performance is the same as coherent 2- or 4-ary PSK but the spectrum is narrower for large frequencies. Choosing an M larger than 4 (e.g., $M = 8$ or $M = 16$) in the MPSK system makes the main lobe of the spectrum narrower, but the sensitivity to noise is considerably increased.

A general definition of continuous phase modulation (CPM) systems is given in the next section. Assume that each data symbol only affects the instantaneous frequency of the transmitted signal in one symbol interval and that the phase is a continuous function of time. This defines the subclass full response CPM systems considered in this paper. In Part II more general CPM schemes are considered. In some cases the phase is allowed to be discontinuous while maintaining the coupling between the phase in successive symbol intervals.

By increasing M, an interesting tradeoff between symbol error probability at large signal-to-noise ratio (SNR) and spectrum is achieved. This trade off is studied for modulation indexes of practical interest and also for systems where the instantaneous frequency is not constant over each symbol interval.

Manuscript received March 19, 1980; revised September 19, 1980. This work was supported by the Swedish Board of Technical Development under Grant 79-3594.

The authors are with the Department of Telecommunication Theory, University of Lund, Fack, S-220 07 Lund, Sweden.

The channel noise is assumed to be additive, white Gaussian throughout the paper. The symbol error probability for an optimum detector at large SNR is calculated using the minimum Euclidean distance between any two signals in the signal space [3]. The optimum detector operates coherently, and due to the continuous phase, the detector must observe the received signal for more than one symbol interval to make a decision about a specific symbol [3].

II. GENERAL SYSTEM DESCRIPTION

For CPM systems, the transmitted signal is

$$s(t, \alpha) = \sqrt{\frac{2E}{T}} \cos (2\pi f_0 t + \varphi(t, \alpha) + \varphi_0) \tag{1}$$

where the information carrying phase is

$$\varphi(t, \alpha) = 2\pi h \int_{-\infty}^{t} \sum_{i=-\infty}^{\infty} \alpha_i g(\tau - iT)\, d\tau; \qquad -\infty < t < \infty \tag{2}$$

and $\boldsymbol{\alpha} = \cdots \alpha_{-2}\ \alpha_{-1}\ \alpha_0\ \alpha_1 \cdots$ is an infinitely long sequence of uncorrelated M-ary data symbols, each taking one of the values

$$\alpha_i = \pm 1, \pm 3, \ldots, \pm(M-1); \qquad i = 0, \pm 1, \pm 2, \cdots \tag{3}$$

with equal probability $1/M$. (M is assumed even.)

E is the symbol energy, T is the symbol time, f_0 is the carrier frequency, and φ_0 is an arbitrary constant phase shift which without loss of generality can be set to zero in the case of coherent transmission.

The variable h is referred to as the modulation index, and the amplitude of the baseband pulse $g(t)$ is chosen to give the maximum phase change $\alpha h \pi$ radians over each symbol interval when all the data symbols in the sequence $\boldsymbol{\alpha}$ take the same value α. For implementation reasons, rational values of the modulation index h are used. This is discussed in some detail in Part II.

To have a CPM signal, the information carrying phase $\varphi(t, \alpha)$ is a continuous function of time t, which implies that the frequency baseband pulse $g(t)$ does not contain any im-

Reprinted from *IEEE Transactions on Communications*, vol. COM-29, no. 3, March 1981.

The Best of the Best. Edited by W. H. Tranter, D. P. Taylor, R. E. Ziemer, N. F. Maxemchuk, and J. W. Mark.

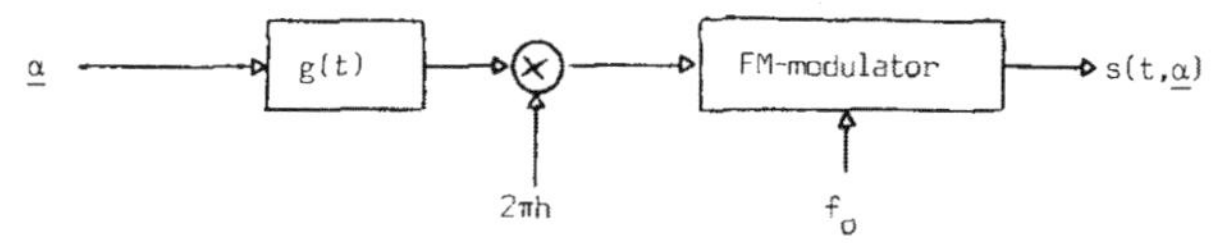

Fig. 1. Schematic modulator for CPM.

pulses. A schematic modulator is shown in Fig. 1. Note that a CPM signal always has a constant envelope.

Defining the baseband phase response (phase pulse)

$$q(t) = \int_{-\infty}^{t} g(\tau)\, d\tau; \qquad -\infty < t < \infty \tag{4}$$

it is seen that the phase of the CPM signal is formed by

$$\varphi(t, \alpha) = 2\pi h \sum_{i=-\infty}^{\infty} \alpha_i q(t - iT); \qquad -\infty < t < \infty. \tag{5}$$

A causal CPM system is obtained if the frequency pulse $g(t)$ satisfies

$$\begin{cases} g(t) \equiv 0; & t < 0, \quad t > LT \\ g(t) \neq 0; & 0 \leq t \leq LT \end{cases} \tag{6}$$

where the pulse length L measured in symbol intervals T may be infinite. $L = 1$ yields full response schemes considered in this part. The normalizing constraint for the frequency pulse $g(t)$ can be expressed as

$$q(LT) = \tfrac{1}{2}. \tag{7}$$

The CPFSK modulation schemes [5], [7], [13] are a subclass of the CPM signaling scheme where the instantaneous frequency is constant over each symbol interval. Thus, for a full response CPFSK modulation scheme, we have

$$q(t) = \begin{cases} 0; & t \leq 0 \\ \dfrac{t}{2T}; & 0 \leq t \leq T \\ \tfrac{1}{2}; & t \geq T \end{cases} \tag{8}$$

which corresponds to linear phase trajectories over each symbol interval (see Fig. 2). Note that although the scheme is full response, the actual phase in any specific symbol interval depends upon the previous data symbols.

The CPM signal is assumed to be transmitted over an additive, white, and Gaussian channel having a one-sided noise power spectral density N_0. Thus, the signal available for observation is

$$r(t) = s(t, \alpha) + n(t); \qquad -\infty < t < \infty \tag{9}$$

where $n(t)$ is a Gaussian random process having zero mean and one-sided power spectral density N_0. A detector which minimizes the probability of erroneous decisions must observe the received signal $r(t)$ over the entire time axis and choose the

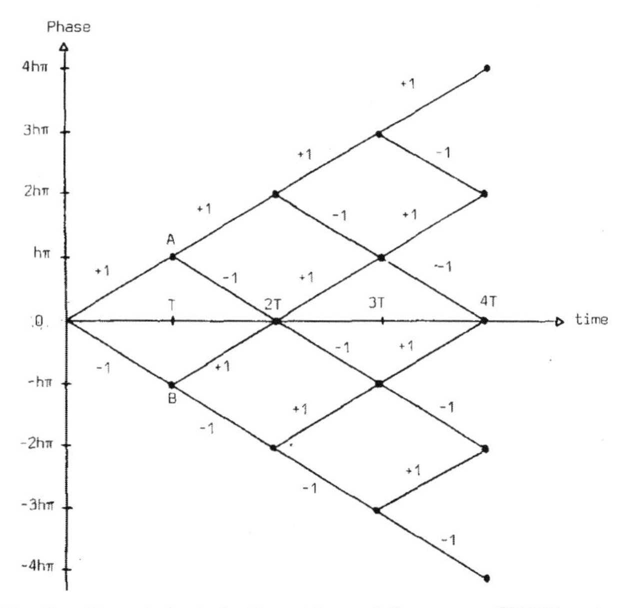

Fig. 2. Phase trajectories for a binary full response CPFSK system. Four bit time intervals are shown.

infinitely long sequence $\tilde{\alpha}$ which minimizes the error probability. This is referred to as maximum likelihood sequence estimation (MLSE). In order to be able to study the performance of an optimum MLSE detector, a suboptimum detector is studied instead. The limiting case of this suboptimum detector is the MLSE detector [7], [11].

This suboptimum detector observes the received signal $r(t)$ for N symbol intervals to make a decision about a specific data symbol, say α_0. Thus, the receiver observes the signal

$$r(t) = s(t, \alpha) + n(t); \qquad 0 \leq t \leq NT \tag{10}$$

and if we let $N \to \infty$, an MLSE detector is obtained. An optimum detector maximizes the likelihood function [3], [5]

$$\Lambda_N[r(t)] = e^{-\frac{2}{N_0}\int_0^{NT} [(r(t) - s(t,\tilde{\alpha})]^2 dt} \tag{11}$$

and since the quantities $\int_0^{NT} r^2(t)dt$ and $\int_0^{NT} s^2(t, \tilde{\alpha})dt$ are independent of $\tilde{\alpha}_N = \tilde{\alpha}_0, \tilde{\alpha}_1, \cdots, \tilde{\alpha}_{N-1}$, one can, as well, maximize

$$\Lambda_N'[r(t)] = e^{\frac{2}{N_0}\int_0^{NT} r(t)s(t,\tilde{\alpha})\, dt} \tag{12}$$

or the log likelihood function

$$\log\, (\Lambda_N'[r(t)] = \int_0^{NT} r(t)s(t, \tilde{\alpha})\, dt. \tag{13}$$

Since there are M^N sequences, $\tilde{\alpha}_N = \tilde{\alpha}_0, \tilde{\alpha}_1, \cdots, \tilde{\alpha}_{N-1}$, but the detector is only interested in finding an estimate $\tilde{\alpha}_0$ of α_0, the M^N sequences can be formed into M groups:

$$\begin{cases} \tilde{\alpha}_{1,N}, \tilde{\alpha}_{3,N}, \cdots, \tilde{\alpha}_{(M-1),N} \\ \tilde{\alpha}_{-1,N}, \tilde{\alpha}_{-3,N}, \cdots, \tilde{\alpha}_{-(M-1),N} \end{cases} \tag{14}$$

where

$$\begin{cases} \tilde{\alpha}_{k,N} = k, \tilde{\alpha}_1, \tilde{\alpha}_2, \cdots, \tilde{\alpha}_{N-1} \\ k = \pm 1, \pm 3, \cdots, \pm(M-1) \end{cases} \tag{15}$$

and it is not necessary for the detector to find the specific sequence $\tilde{\boldsymbol{\alpha}}_N$ which maximizes (13) and to choose $\tilde{\alpha}_0$ as an estimate of α_0. Instead, the detector must find which group of sequences $\tilde{\boldsymbol{\alpha}}_{k,N}$, $k = \pm 1, \pm 3, \cdots, \pm(M-1)$ jointly maximizes (13), and take $\tilde{\alpha}_0$ as the group belonging. It is believed, however, that for large SNR the two detectors have the same performance [7].

The probability of an erroneous decision can be upper bounded by using the union bound [3], [5], [10]

$$P_e \leqslant \frac{1}{M^{N-1}} \sum_{k} \sum_{\substack{l \\ k \neq l}} Q\left[\frac{D(\boldsymbol{\alpha}_{k,N}, \boldsymbol{\alpha}_{l,N})}{\sqrt{2N_0}}\right] \tag{16}$$

where

$$Q(x) = \frac{1}{\sqrt{2\pi}} \int_x^{\infty} e^{-\frac{t^2}{2}} \, dt \tag{17}$$

and the summation is taken over all pairs of sequences defined by (15), with the restriction that $k \neq l$, $k, l = \pm 1, \pm 3, \cdots, \pm(M-1)$. $D[\boldsymbol{\alpha}_{k,N}, \boldsymbol{\alpha}_{l,N}]$ is the Euclidean distance between the signals $s(t, \boldsymbol{\alpha}_{k,N})$ and $s(t, \boldsymbol{\alpha}_{l,N})$.

The squared Euclidean distance can be written

$$\begin{aligned} D^2(\boldsymbol{\alpha}_{k,N}, \boldsymbol{\alpha}_{l,N}) &= \int_0^{NT} [s(t, \boldsymbol{\alpha}_{k,N}) - s(t, \boldsymbol{\alpha}_{l,N})]^2 \, dt \\ &= \sum_{i=0}^{N-1} \int_{iT}^{(i+1)T} [s(t, \boldsymbol{\alpha}_{k,N}) - s(t, \boldsymbol{\alpha}_{l,N})]^2 \, dt. \end{aligned} \tag{18}$$

Assuming $2\pi f_0 T \gg 1$, this can be written

$$\begin{aligned} D^2(\boldsymbol{\alpha}_{k,N}, \boldsymbol{\alpha}_{l,N}) &= 2E\left(N - \frac{1}{T} \int_0^{NT} \cos\left[2\pi h \sum_{i=0}^{N-1} (\alpha_i{}^k - \alpha_i{}^l) \right.\right. \\ &\quad \left.\left. \cdot\, q(t - iT)\right] dt\right). \end{aligned} \tag{19}$$

The superscript denotes the value of the first symbol in a sequence of N symbols, i.e.,

$$\begin{cases} \alpha_0{}^k = k \\ \alpha_i{}^k = \alpha_i; \qquad i = 1, 2, \cdots, N-1. \end{cases} \tag{20}$$

Equation (19) can be written

$$\begin{aligned} D^2(\boldsymbol{\alpha}_{k,N}, \boldsymbol{\alpha}_{l,N}) \\ = 2E\left(N - \frac{1}{T} \int_0^{NT} \cos\left[\varphi(t, \boldsymbol{\alpha}_{k,N} - \boldsymbol{\alpha}_{l,N})\right] dt\right) \end{aligned} \tag{21}$$

Thus, it is sufficient to consider the difference sequence

$$\boldsymbol{\gamma}_N = \boldsymbol{\alpha}_{k,N} - \boldsymbol{\alpha}_{l,N} \tag{22}$$

instead of the pair of sequences $\boldsymbol{\alpha}_{k,N}$ and $\boldsymbol{\alpha}_{l,N}$.

The approximation

$$P_e \approx \Gamma_0 \cdot Q\left[\frac{D_{\min,N}}{\sqrt{2N_0}}\right] \tag{23}$$

of (16) is good for large E/N_0. It is now assumed that E/N_0 is sufficiently large for this approximation to be valid. The limitations of this assumption is considered in detail in [20]. Γ_0 is a positive constant, independent of E/N_0, and $D_{\min,N}$ is the minimum of $D(\boldsymbol{\alpha}_{k,N}, \boldsymbol{\alpha}_{l,N})$ with respect to the pair of sequences $\boldsymbol{\alpha}_{k,N}$ and $\boldsymbol{\alpha}_{l,N}$ with the restriction that $k \neq l$. This quantity can also be calculated using the difference sequence $\boldsymbol{\gamma}_N$ through

$$D_{\min,N}^2 = 2E \cdot \min_{\boldsymbol{\gamma}_N} \left\{N - \frac{1}{T} \int_0^{NT} \cos\left[\varphi(t, \boldsymbol{\gamma}_N)\right] dt\right\} \tag{24}$$

with the restriction that

$$\begin{cases} \gamma_0 = 2, 4, 6, \cdots, 2(M-1) \\ \gamma_i = 0, \pm 2, \pm 4, \cdots, \pm 2(M-1); \quad i = 1, 2, \cdots, N-1. \end{cases} \tag{25}$$

In Part II, we will only deal with squared Euclidean distances normalized by bit energy

$$d^2 = \frac{D^2}{2E_b} \tag{26}$$

Note that

$$E = E_b \cdot \log_2(M) \tag{27}$$

where E_b is the bit energy. Thus, error probability comparisons for large SNR can be made directly in E_b/N_0 between systems even if they have different M. Only values that are powers of two ($M = 2, 4, 8, \cdots$) will be considered. As a reference point in the following, note that $d_{\min}^2 = 2$ for MSK, binary PSK (BPSK), and quaternary phase shift keying (QPSK). For the case of full response CPFSK systems, calculations of $d_{\min,N}^2$ has been considered for both the binary case [1], [5]-[7], [13] and also for $M = 4$ and $M = 8$ to some extent [10], [11], [13].

III. BOUNDS ON THE MINIMUM EUCLIDEAN DISTANCE

An important tool for the analysis of CPM systems is the so-called phase tree. This tree is formed by all phase trajectories $\varphi(t, \boldsymbol{\alpha})$ having a common start value zero at $t = 0$. The ensemble is over the sequence $\boldsymbol{\alpha}$ and Fig. 2 shows a part of the phase tree for a binary full response CPFSK system. A more general case is shown in Fig. 3 where two phase trees for a quaternary CPM system having different frequency baseband pulses $g(t)$ are shown.

To calculate the minimum squared Euclidean distance for an observation length of N symbols, all pairs of phase trajectories in the phase tree over N symbol intervals must be considered. The phase trajectories must not coincide over the first symbol interval however. The Euclidean distance is calculated according to (21) for all these pairs, and the minimum of these Euclidean distances is the desired result. It is of great importance to remember that the phase must always be viewed modulo 2π in conjunction with distance calculations. A practical method to do this is to form a cylinder by folding the phase tree [16], [17]. Trajectories which seem to be far apart in the phase tree might actually be very close or even coincide when viewed modulo 2π.

It is clear from (18) that, for a fixed pair of phase trajectories, the Euclidean distance is a nondecreasing function of the observation length N. If just a few pairs of infinitely long sequences are chosen, an upper bound on the minimum Euclidean distance at all values of the observation interval N is obtained. Good candidates for these infinitely long pairs are pairs that merge as soon as possible. Two phase trajectories merge at a certain time if they coincide all the time thereafter. These merges are called *inevitable* if they occur independently of h. Thus, an upper bound on the minimum Euclidean distance is obtained as a function of the modulation index h for all N.

Applying this method to the scheme in Fig. 2, it is seen that if a pair of sequences is chosen as

$$\begin{cases} \alpha_{+1} = +1, -1, \alpha_2, \alpha_3, \cdots \\ \alpha_{-1} = -1, +1, \alpha_2, \alpha_3, \cdots \end{cases} \tag{28}$$

the two phase trajectories coincide for all $t \geqslant 2T$. Thus, the upper bound on the normalized minimum squared Euclidean distance is

$$d_B{}^2(h) = 2 - \frac{1}{T} \int_0^{2T} \cos\,[2\pi h(2q(t) - 2q(t-T))]\,dt \tag{29}$$

where (21) was used. For the binary CPFSK system, i.e., linear phase trajectories, the result is

$$d_B{}^2(h) = 2\left(1 - \frac{\sin 2\pi h}{2\pi h}\right). \tag{30}$$

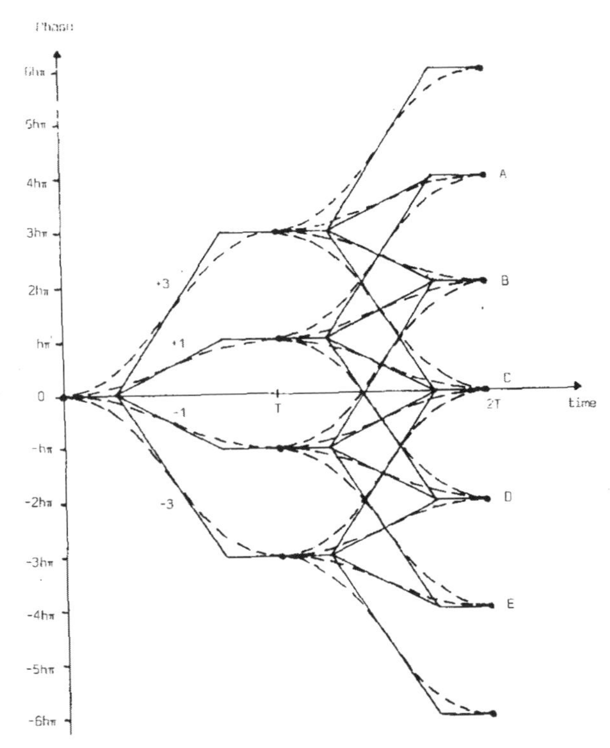

Fig. 3. Phase trees for $M = 4$ CPM schemes with two different baseband pulses $g(t)$. The α, β function [see (56)] with $\alpha = 0.25$, is $\beta = 0$ is shown by a solid line and the HCS, half cycle sinusoid [see (58)] is shown by a dashed line.

It can be noted that instead of using the pair of sequences (28), the single difference sequence $\gamma = +2, -2, 0, 0, \cdots$ could be used together with (24) for calculation of the normalized squared Euclidean distance.

Turning to the quaternary case, we can find pairs of phase trajectories merging at $t = 2T$. Fig. 3 shows two examples of phase trees for this case. These merges occur at the points labeled A, B, C, D, and E in the phase tree. Unlike the binary case, there is more than one merge point, and two different pairs of phase trajectories can have the same merge point. There are only three phase differences however, namely, those having phase difference $+2h\pi$, $+4h\pi$, and $+6h\pi$ at $t = T$.

It is easily seen that an upper bound on the minimum Euclidean squared distance for the M-ary case is obtained by using the difference sequences

$$\gamma = \gamma_0, -\gamma_0, 0, 0, 0; \qquad \gamma_0 = 2, 4, 6, \cdots, 2(M-1) \tag{31}$$

and taking the minimum of the resulting Euclidean distances, i.e.,

$$d_B{}^2(h) = \log_2(M) \cdot \min_{1 \leqslant k \leqslant M-1} \left\{ 2 - \frac{1}{T} \int_0^{2T} \cos\,[2\pi h \cdot 2k(q(t) - q(t-T))]\,dt \right\} \tag{32}$$

which, for the M-ary CPFSK system, specializes to

$$d_B{}^2(h) = \log_2 (M) \cdot \min_{1 \leqslant k \leqslant M-1} 2\left(1 - \frac{\sin k2\pi h}{k2\pi h}\right). \quad (33)$$

Fig. 4 shows the minimum distance $d^2(h)$ for binary CPFSK as a function of h and N. The upper bound $d_B{}^2$ is shown dashed. Note the peculiar behavior at $h = 1$. This will be discussed later.

It can be noticed that for all full response CPM systems with the property

$$\int_0^T g(\tau)\, d\tau = 0 \quad (34)$$

a merge can occur at $t = T$ and thus the difference sequences

$$\gamma = \gamma_0, 0, 0, 0, \cdots; \qquad \gamma_0 = 2, 4, 6, \cdots, 2(M-1) \quad (35)$$

yield the upper bound

$$d_B{}^2(h) = \log_2 (M) \cdot \min_{1 \leqslant k \leqslant M-1} \left\{1 - \frac{1}{T}\int_0^T \cos\, [2\pi h \cdot 2q(t)]\, dt\right\}. \quad (36)$$

This class of pulses is called weak [20], and is not considered because distance properties are poor. Only positive pulses $g(t)$ will be considered below. Furthermore, they are assumed to be symmetric with

$$g(t) = g(T-t); \qquad 0 \leqslant t \leqslant T. \quad (37)$$

Weak Modulation Indexes, h_c

For the construction of the upper bound $d_B{}^2(h)$ on the minimum squared Euclidean distance, pairs of phase trajectories giving merges at $t = 2T$ were used. These merges occur independently of the value of h. For all pulses $g(t)$ except weak ones the first inevitable (for all h), merge occurs at $t = 2T$. For specific values of the modulation index h, however, other merges are also possible. In the binary case (see Fig. 2), a merge can occur at $t = T$ if the difference sequence is chosen to be $\gamma = +2, 0, 0, \cdots$ and the modulation index h is an integer. This is because the two points labeled A and B are a multiple of 2π apart, and thus coincide modulo 2π.

For an M-ary full response system, the phase trajectories take the values $\pm 2\pi hq(T), \pm 6\pi hq(T), \cdots, \pm 2(M-1)h\pi q(T)$, which for positive $g(t)$ pulses reduces to $\pm h\pi, \pm 3h\pi, \cdots, \pm(M-1)h\pi$. Thus, there are $M - 1$ phase differences between the nodes in the phase tree at $t = T$, and merges occur at $t = T$ for h-values given by

$$\pi h \cdot \gamma_0 = 2\pi \cdot n; \quad \gamma_0 = 2, 4, 6, \cdots, 2(M-1)$$
$$n = 1, 2, \cdots \quad (38)$$

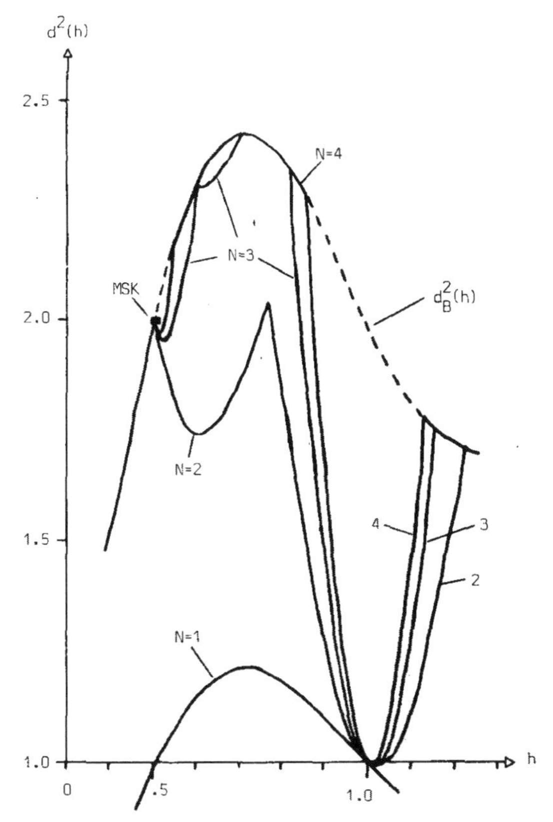

Fig. 4. Normalized squared Euclidean distance versus modulation index for binary CPFSK: upper bound (dashed line) and $d^2(h)$ for N = 1, 2, 3 and 4 bit decision intervals.

The modulation indexes [defined by (38)]

$$h_c = \frac{n}{k}; \qquad k = 1, 2, \cdots, M-1$$
$$n = 1, 2, \cdots \quad (39)$$

are called *weak* modulation indexes of the first order. For these modulation indexes, the minimum Euclidean distance is normally below the upper bound for all values of N; see Fig. 4 for $h = 1$. Sometimes the minimum distance for weak modulation index values is considerably below $d_B{}^2(h_c)$. For such cases, h_c is called a *catastrophic* modulation index value [17]-[20].

Merges of the weak (catastrophic) type occur at any observation length. Thus, weak h-values of the second order are defined by

$$\pi h(\gamma_0 + \gamma_1) = 2\pi \cdot n \qquad \gamma_0 = 2, 4, 6, \cdots, 2(M-1) \quad (40)$$
$$\gamma_1 = \pm 2, \pm 4, \pm 6, \cdots, \pm 2(M-1)$$
$$n = 1, 2, \cdots$$

As we will see later, the effect of weak modulation indexes of a higher order than one are of minor importance, since the corresponding Euclidean distance for the pairs of phase trajectories causing the merge is above the upper bound $d_B{}^2(h)$ [20]. For weak modulation indexes of the second order, the corresponding Euclidean distance might be on the upper bound $d_B^2(h)$; for the third order and higher it is strictly above. But from (39) it can be seen that the number of first-order weak modulation indexes grows rapidly with M.

For weak modulation indexes of the first order, it is sufficient to calculate the corresponding Euclidean distance over the first symbol interval. Since the calculation of the upper bound uses the two first symbol intervals, it can be concluded that $d^2(h_c)$ can be, and normally is, smaller than $d_B^2(h_c)$. Thus, for weak modulation indexes of the first order, $d_{\min}^2(h_c)$ might be smaller than $d_B^2(h_c)$ (for details, see [20]). Furthermore, it is shown in detail [17], [20] that only h_c of the first order can influence the minimum distance calculation for full response CPM systems.

Tightness of the Upper Bound $d_B^2(h)$

A powerful property of the upper bound on the minimum Euclidean distance is that except for weak modulation indexes of the first order, the minimum Euclidean distance itself equals this bound if the observation interval is long enough [18], [20]. Denoting the minimum normalized squared Euclidean distance for an N symbol observation interval $d_{\min,N}^2(h)$, we have that

$$d_{\min,N}^2(h) = d_B^2(h) \tag{41}$$

if

$$\begin{cases} N \geqslant N_B(h) \\ h \neq h_c: \quad \text{(first order).} \end{cases} \tag{42}$$

$N_B(h)$ is the number of symbol intervals required to reach the upper bound for the specific modulation index h.

If a specific pair of phase trajectories never merge, the Euclidean distance will grow without limit. This is true because the Euclidean distance calculated over each symbol interval is positive. Since the minimum distance was previously shown to be upper bounded, the pair of phase trajectories giving the minimum distance must eventually merge.

For a modulation index near a first-order weak modulation index, the difference sequence (35) gives the smallest growing Euclidean distance with N:

$$d^2(h) = \log_2(M) \min_{1 \leqslant k \leqslant M-1} \left\{ 1 - \frac{1}{T}\int_0^T \cos\,[2\pi hk\gamma_0 q(t)]\, dt + (N-1)(1 - \cos\,[\pi hk\gamma_0]) \right\} \tag{43}$$

and the smallest growth rate per symbol interval is

$$\log_2 M \cdot \min_{1 \leqslant k \leqslant M-1} \{1 - \cos\,[\pi hk\gamma_0]\}. \tag{44}$$

For sufficiently large N in (43), the minimum distance for the considered h is given by $d_B^2(h)$ since $d^2(h)$ will exceed $d_B^2(h)$.

The tightness of the upper bound and the behavior of (43) is illustrated in all the minimum distance figures.

Optimization of the Upper Bound $d_B^2(h)$

For frequency pulses $g(t)$ having the symmetry property (37), the expression for the upper bound can be written

$$d_B^2(h) = \log_2 M \min_{1 \leqslant k \leqslant M-1} \left\{ 2\left(1 - \frac{1}{T}\int_0^T \cos\,[4\pi hkq(t)]\, dt\right)\right\} \tag{45}$$

and since $\cos(\cdot) \geqslant -1$, $d_B^2(h)$ can never exceed $4 \cdot \log_2 M$. Thus, at most, an improvement of 3 dB in E/N_0 for a large SNR might be obtained for the binary case compared to MSK.

Furthermore, (45) can also be written

$$d_B^2(h) = \log_2 M \cdot \min_{1 \leqslant k \leqslant M-1} \left\{ 2\left(1 - 2\cos\,[\pi hk] \cdot \frac{1}{T}\int_0^{T/2} \cos\,[4\pi hkq_0(t) - \pi hk]\, dt\right)\right\} \tag{46}$$

where

$$q_0(t) = q(t + T/2). \tag{47}$$

The binary case ($M = 2$) will be considered first. In this case,

$$d_B^2(h) = 2\left[1 - 2\cos \pi h \cdot \frac{1}{T}\int_0^{T/2} \cos\,[4\pi hq_0(t) - \pi h]\, dt\right]. \tag{48}$$

It can at once be observed that $\cos \pi h = 0$ for $h = 1/2, 3/2, 5/2$, etc. Thus, the upper bound equals two, independent of the shape of the frequency pulse $g(t)$. Actually, this is true for all pulses $g(t)$ (except weak ones) [16]. This is of particular interest since much attention has been devoted to the case of $h = 1/2$ [6], [12], [14], [15].

To maximize $d_B^2(h)$ for the binary case, the last term in (48) must be minimized. Two different cases can be distinguished:

Case I:

$$0 \leqslant h \leqslant \tfrac{1}{2} \text{ where } \cos \pi h \geqslant 0. \tag{49}$$

Case II:

$$\tfrac{1}{2} \leqslant h \leqslant 1 \text{ where } \cos \pi h \leqslant 0. \tag{50}$$

To maximize $d_B^2(h)$, the integral in (48) must be minimized for case I and maximized for case II. It is also clear that the pulse, which maximized $d_B^2(h)$ for case I, minimizes $d_B^2(h)$ for case II and vice versa.

To make the integral in (48) as small as possible, the argument inside the cosine must be as close to π as possible. This yields for the interval $0 \leqslant h \leqslant 1$

$$\begin{cases} q_0(t) = \frac{1}{2}; \text{ case I} \\ q_0(t) = \frac{1}{4}; \text{ case II} \end{cases} \tag{51}$$

with the resulting phase responses

$$q_1(t) = \begin{cases} 0; & 0 \leqslant t < \frac{T}{2} \\ \frac{1}{2}; & t > \frac{T}{2} \end{cases} \quad \text{case I} \tag{52}$$

and

$$q_2(t) = \begin{cases} 0; & t = 0 \\ \frac{1}{4}; & 0 < t < T \\ \frac{1}{2}; & t \geqslant T \end{cases} \quad \text{case II.} \tag{53}$$

From the sign symmetry, to minimize $d_B^2(h)$ the phase responses above have to be interchanged with each other for the respective cases.

The upper bound $d_B^2(h)$ for the two phase responses is now found as

$$\begin{cases} d_B^2(h) = 1 - \cos 2\pi h; & \text{using } q_1(t) \\ d_B^2(h) = 2(1 - \cos \pi h); & \text{using } q_2(t) \end{cases} \quad 0 \leqslant h \leqslant 1. \tag{54}$$

Thus, for any binary scheme, $d_B^2(h)$ is bounded by

$$\begin{cases} 1 - \cos 2\pi h \leqslant d_B^2(h) \leqslant 2(1 - \cos \pi h); & 0 \leqslant h \leqslant \frac{1}{2} \\ 2(1 - \cos \pi h) \leqslant d_B^2(h) \leqslant 1 - \cos 2\pi h; & \frac{1}{2} \leqslant h \leqslant 1. \end{cases} \tag{55}$$

An analogous technique can be used to derive bounds for $h \geqslant 1$.

The two phase responses $q_1(t)$ and $q_2(t)$ are members of the class called α, β functions [16], [17] defind by

$$q(t) = \begin{cases} \frac{\beta}{2\alpha}\frac{t}{T}; & 0 \leqslant t \leqslant \alpha T \\ \frac{\alpha}{2}\frac{t}{T}\frac{1-2\beta}{1-2\alpha} + \frac{\beta}{2}; & \alpha T \leqslant t \leqslant (1-\alpha)T \\ \frac{\beta}{2\alpha}\left(\frac{t}{T} - 1\right) + \frac{1}{2}; & (1-\alpha)T \leqslant t \leqslant T \\ \frac{1}{2}; & t \geqslant T. \end{cases} \tag{56}$$

Hence, when $h = 1/2$, $q_1(t)$ corresponds to binary PSK but with a $T/2$ time offset. The phase response $q_2(t)$ corresponds to a so-called plateau function; i.e., the phase changes only in the beginning and the end of the symbol interval, but remains constant (forms a plateau) in the middle of the symbol interval. Note that $q_1(t)$ and $q_2(t)$ give systems with discontinuous phase. However, the ensemble of possible phase trajectories is completely known to the receiver just as for all CPM systems. This gives a slight generalization of the considered class of systems.

It is interesting to note that the plateau function with $\beta = 1/2$ gives the upper bound [17]

$$d_B^2(h) = 2\left\{1 - (1 - 2\alpha)\cos \pi h - \frac{\alpha}{h\pi}\sin 2\pi h\right\} \tag{57}$$

which, for small values of α, approaches a value of 4 near $h = 1$. Since $h = 1$ is a first-order weak modulation index, it is concluded that large values of N are required to make $d^2_{\min}(h)$ equal $d_B^2(h)$ near $h = 1$.

In practice, the phase responses $q_1(t)$ and $q_2(t)$ are not attractive because of their spectra. This will be discussed more in Section VI. From a spectral point of view, the phase during a symbol interval should change slowly and smoothly, and the following frequency pulses with corresponding phase responses are of interest. The first one is

$$g(t) = \begin{cases} 0; & t \leqslant 0, t \geqslant T \\ \frac{\pi}{4T}\sin\frac{\pi t}{T}; & 0 \leqslant t \leqslant T \end{cases} \tag{58}$$

with the corresponding phase response

$$q(t) = \begin{cases} 0; & t \leqslant 0 \\ \frac{1}{4}\left(1 - \cos\frac{\pi t}{T}\right); & 0 \leqslant t \leqslant T \\ \frac{1}{2}; & t \geqslant T \end{cases} \tag{59}$$

and like CPFSK this pulse $g(t)$ has no continuous derivatives at the end points [14]. The pulse in itself is continuous, however, unlike CPFSK. Since the frequency pulse is a half cycle sinusoid, this scheme will be referred to as half cycle sinusoid (HCS) [see Fig. 3 (dashed tree) for the quaternary case].

Another pulse of interest is

$$g(t) = \begin{cases} 0; & t \leqslant 0, t \geqslant T \\ \frac{1}{2T}\left(1 - \cos\frac{2\pi t}{T}\right); & 0 \leqslant t \leqslant T \end{cases} \tag{60}$$

with the corresponding phase response

$$q(t) = \begin{cases} 0; & t \leqslant 0 \\ \frac{1}{2}\left(\frac{t}{T} - \frac{1}{2\pi}\sin\frac{2\pi t}{T}\right); & 0 \leqslant t \leqslant T \\ \frac{1}{2}; & t \geqslant T. \end{cases} \tag{61}$$

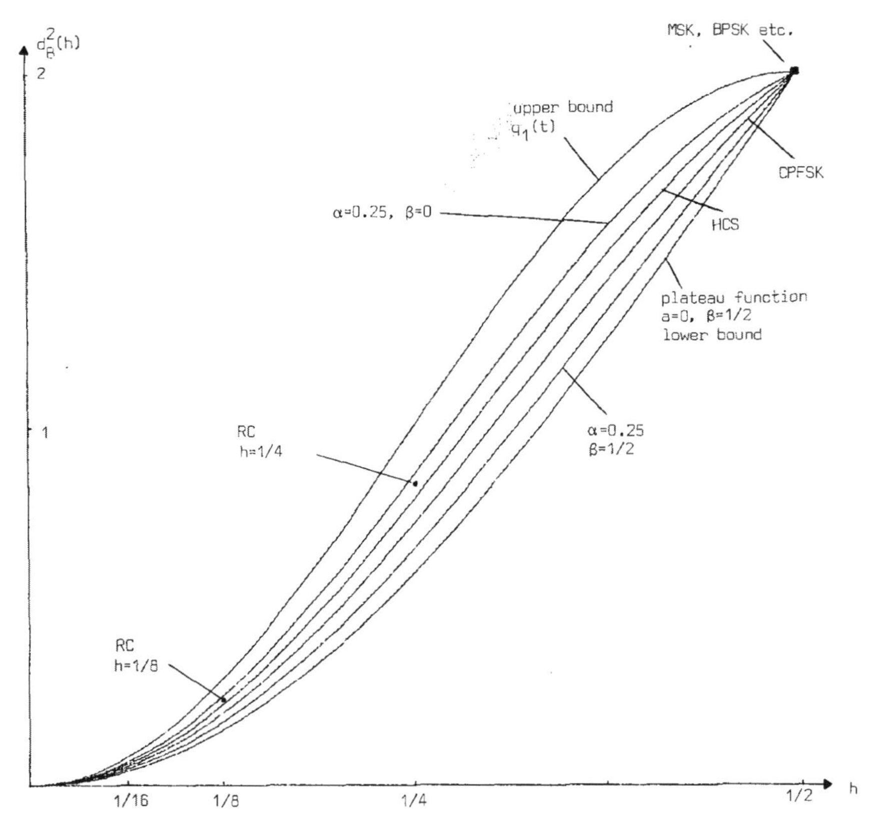

Fig. 5. Upper bound comparison for $M = 2$, $0 \leqslant h \leqslant 0.5$. The bound for $q_1(t)$ is the upper bound on all upper bounds $d_B{}^2(h)$, and the $\alpha = 0$ plateau function is the lower bound on all the upper bounds $d_B{}^2(h)$ in the considered interval.

This pulse $g(t)$ has one continuous derivative at the end points $t = 0$ and $t = T$. Since this frequency pulse is a raised cosine function, it will be referred to as raised cosine (RC). When $h = 1/2$, this scheme has previously been referred to as SFSK [12].

Fig. 5 shows the upper and lower bounds on $d_B{}^2(h)$ for all binary schemes in the region $0 \leqslant h \leqslant 1/2$, computed with (55). Fig. 5 also shows $d_B{}^2(h)$ for CPFSK, HCS (with formula in [18]), and α, β functions with $\alpha = 0.25$, $\beta = 0.5$ and $\alpha = 0.25$, $\beta = 0$. The bound for the RC scheme is in between that of $\alpha = 0.25, \beta = 0$ and HCS.

The problem of finding frequency pulses $g(t)$ that optimize $d_B{}^2(h)$, given h and M, is far more complicated in the general M-ary case than for the binary case. This general problem has not been solved. The reason is that the upper bound is constructed from the minimum of more than one function, and h varies with fixed M, different functions take the minimum value. This is also true for fixed h and M when the frequency pulse $g(t)$ is varied. For an M-ary scheme with $0 \leqslant h \leqslant 1/M$ the binary bounds (55) on $d_B{}^2(h)$ still apply after multiplication with $\log_2 M$.

Of course, also for M-ary schemes $d_B{}^2(h) \leqslant 4\log_2 M$. However, due to the fact that the bound $d_B{}^2(h)$ in this case is formed by taking the minimum of several component functions (32), this maximum value can never be reached [17], [20].

It is previously known [5], [7], that the h-value maximizing the minimum Euclidean distance ($N \geqslant 3$) for binary CPFSK is $h = 0.715$. This value of h also maximizes $d_B{}^2(h)$ for this scheme. The same h was also shown by Kotelnikov [1] to maximize the Euclidean distance when $N = 1$. It is possible to find the values of h which maximize $d_B{}^2(h)$ for M-ary CPFSK [see (33)] and they are given in Table I together with the maximum value of $d_B{}^2(h)$ for $M = 2, 4, 8, 16$, and 32.

The optimum occurs for h-values slightly below $h = 1$. Unfortunately, $h = 1$ is a first-order weak modulation index for all M, but if N is made large enough $(N \geqslant N_B) d^2_{\min}(h_0)$ equals $d_B{}^2(h_0)$.

For the quaternary case, the CPFSK scheme gives a maximum of $d_B{}^2(h_0) = 4.232$. In [17] it is shown that a scheme based on the α, β function with $\alpha = 0$, $\beta = 0.17$ for $h = 0.62$ gives a minimum distance of 4.62. This value came out of a nonexhaustive search for quaternary schemes yielding large distance values. However, better schemes may exist.

IV. NUMERICAL RESULTS ON THE MINIMUM EUCLIDEAN DISTANCE

In this section numerical results on the minimum normalized squared Euclidean distance will be given in form of graphs. These graphs present the minimum Euclidean distance versus the modulation index h for specific schemes and different values of N, the number of received signal intervals observed. Thus, these graphs will show what is below the upper bound $d_B{}^2(h)$, and also how large N has to be made in a

TABLE I
OPTIMUM h-VALUES AND CORRESPONDING NORMALIZED EUCLIDEAN DISTANCES FOR M-ARY CPFSK SCHEMES

M	Optimum h h_o	$d_B^2(h_o)$	N_B
2	.715	2.434	3
4	.914	4.232	9
8	.964	6.141	41
16	.983	8.088	178
32	.992	10.050	777

specific situation, to make the minimum Euclidean distance $d^2_{\min}(h)$ equal to $d_B{}^2(h)$.

Plateau Functions

As an example of a plateau function, a binary scheme with a phase response very similar to that of $\alpha = 0.05$ and $\beta = 1/2$ will be chosen. The difference between the chosen phase function and the α, β function is that the phase does not vary linearly in the intervals $0 \leqslant t \leqslant \alpha T$ and $(1 - \alpha) T \leqslant t \leqslant T$. Instead, the phase varies like a raised cosine (for an exact definition, see b-functions, $b = 0.05$ in [16]).

Fig. 6 shows $d^2_{\min}(h)$ for this binary system. The number of observed bit intervals is $N = 1, 2, \cdots, 7$. The upper bound on the minimum Euclidean distance is also shown by a dashed line where the minimum Euclidean distance still does not equal $d_B{}^2(h)$.

CPFSK and M-ary PSK

Fig. 4 shows the well-known minimum distance for binary CPFSK for $N = 1, 2, 3$, and 4 observed bit intervals. Also shown in Fig. 4 is the upper bound $d_B{}^2(h)$. It can be noted that $h = 1/2$ corresponds to MSK and gives $d^2_{\min}(1/2) = 2$, which is the same antipodal signaling, e.g., BPSK. The required observation interval for PSK is one bit interval, and for detectors making bit by bit decisions, PSK is optimum [3]. The required observation interval for MSK is two bit intervals, and the asymptotic performance in terms of error probability is the same as that for PSK. The optimum modulation index for CPFSK is $h = 0.715$ when the number of observed symbol intervals is 3. This gives the minimum Euclidean distance $d^2_{\min}(0.715) = 2.43$ and thus a gain of 0.85 dB in terms of E_b/N_0 is obtained compared to MSK or PSK.

The minimum normalized squared Euclidean distance versus the modulation index h is shown in Fig. 7 for the quaternary CPFSK system (see also [10]). Note that the upper bound $d_B{}^2(h)$ (shown by a dashed line where it is not reached) is twice the minimum distance for a receiver observation interval of the $N = 1$ symbol. This is because the rectangular frequency pulse $g(t)$ has the symmetry property (37). The maximum value of $d_B{}^2(h)$ is approximately reached for $N = 8$ observed symbol intervals (compare to Table I).

It is clear from (39) that the first-order weak modulation indexes in the interval $0 \leqslant h \leqslant 2$ are $h_c = 1/3, 1/2, 2/3, 1, 4/3, 3/2, 5/3$, and 2, and the effect of some of these early merges can be clearly seen in Fig. 7. Note that most of these weak

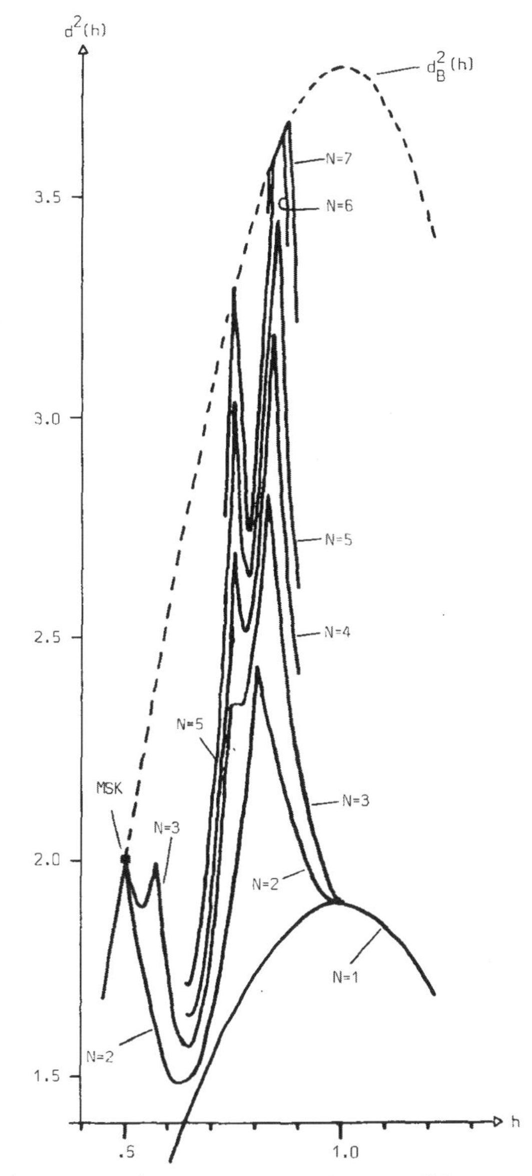

Fig. 6. Normalized squared minimum distances $d^2(h)$ versus modulation index for a b-function with $b = 0.5$. This phase function is very similar to the α-β function with $\alpha = 0.5$, $\beta = 1/2$ (see [16] or [18]).

indexes are catastrophic. The minimum Euclidean distance for these is no better than 2.

It is interesting to compare the minimum distance for the quaternary CPFSK system to QPSK (phase response $q_1(t)$, $h = 1/4$). As indicated in Fig. 7, the minimum squared distance for QPSK is $d^2_{\min} = 2$, and for the quaternary CPFSK system it is slightly below this value for $h = 1/4$. This is a different relative performance level than that for $M = 2$. For $M = 2$, $h = 1/2$ all schemes have the minimum squared distance $d^2_{\min} = 2$, CPFSK and PSK included.

The minimum distance for the octal ($M = 8$) CPFSK system is given in Fig. 8 $N = 1, 2, 3$ and in some interval for $N = 4$ and 5. The upper bound $d_B{}^2(h)$, which, as usual, is shown by a dashed line where it is not reached, is like the quaternary case reached with $N = 2$ observed symbol intervals for low

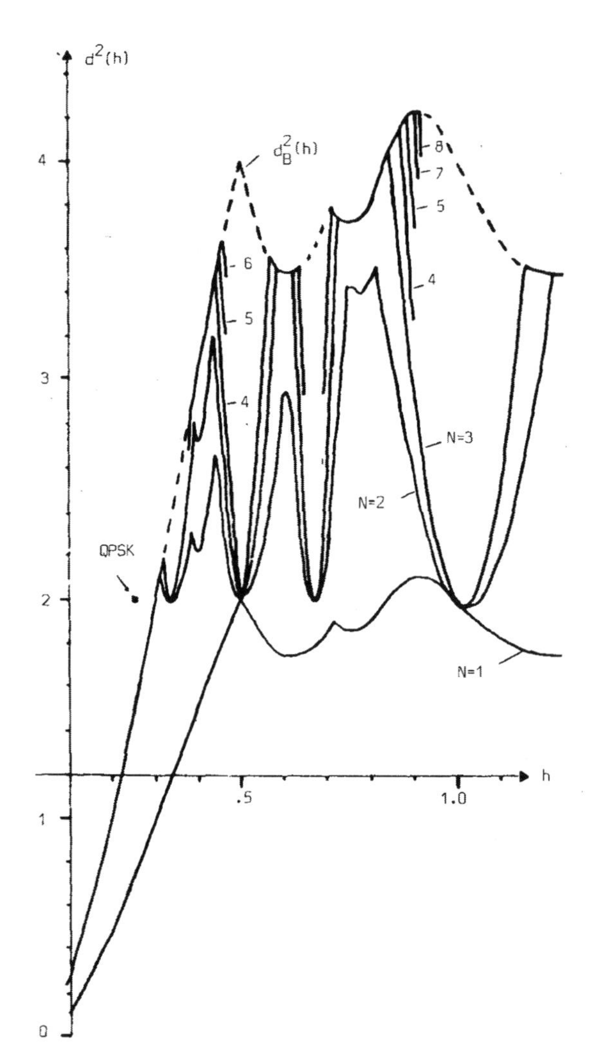

Fig. 7. Minimum normalized squared distance versus modulation index for $M = 4$ CPFSK.

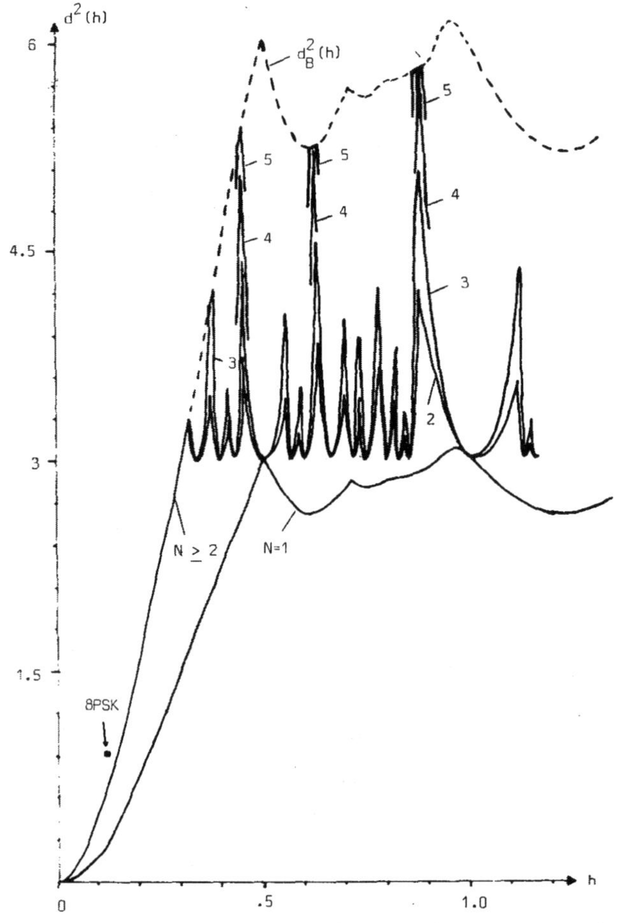

Fig. 8. Minimum normalized squared Euclidean distance versus modulation index for $M = 8$, CPFSK.

modulation indexes. Compared to the quaternary case, the number of first-order weak (catastrophic) modulation indexes has increased in the interval of $0.3 \leqslant h \leqslant 1$. Larger values of N are required to reach $d_B{}^2(h)$, compared to the quaternary and especially the binary case.

The scheme 8PSK ($q_1(t)$, $h = 1/8$), previously shown to maximize $d_B{}^2(1/8)$ is also indicated in Fig. 8, and it is seen that octal CPFSK yields the same minimum distance if h is chosen slightly larger than $h = 1/8$ and if $N = 2$. Much larger distances can be obtained for the CPFSK system by choosing, for instance, $h \cong 0.45$ and $N \geqslant 5$.

Distance properties of CPFSK schemes with larger values of M have been investigated in [17], [20]. The maximum attainable minimum distance value grows with M, but the number of first-order weak (castastrophic) modulation index values also grows with M, as does the length of the observation interval necessary for reaching the upper bound $d_B{}^2(h)$. However, for $h \lesssim 0.3$, $N = 2$ is sufficient for all M. As an illustration to the behavior discussed above, Table I shows N_B which is a lower bound on the observation interval for reaching the upper bound.

HCS, RC, and M-ary PSK

The HCS system yields a phase tree where the phase trajectories are always raised cosine shaped over each symbol interval. Fig. 3 shows the quaternary case.

The upper bound $d_B{}^2(h)$ for HCS, $M = 2$ is given by [18]

$$d_B{}^2(h) = 2(1 - \cos \pi h \cdot J_0(\pi h)) \tag{62}$$

where $J_0(\cdot)$ is the Bessel function of the first kind and zero order. The maximum value of $d_B{}^2(h)$ is smaller than that for binary CPFSK, but this maximum value is still reached with $N = 3$ observed symbols ($d^2_{\min} = 2.187$ for $h = 0.626$, [18]). For $h = 1/2$, $d^2_{\min}(h)$ equals that of MSK and PSK, as in all binary full response CPM systems. In the region of $0 < h < 1/2$, the upper bound is reached with $N = 2$ observed symbols as in binary CPFSK. Fig. 5 shows that HCS gives a larger minimum distance than binary CPFSK in this region.

The minimum Euclidean distances for the quaternary HSC system are given in Fig. 9 when $N = 1, 2, 3,$ and 4 observed symbol intervals, and QPSK is also indicated. $d_B{}^2(1/4)$ is still smaller than the minimum distance for QPSK of course, but since the upper bound for HCS is reached with $N = 2$ observed symbol intervals in the region of $0 < h \lesssim 0.3$, HCS is better than CPFSK. Note that the value of the minimum Euclidean distance at $h = h_c$ no longer equals 2 as for CPFSK. This is due to

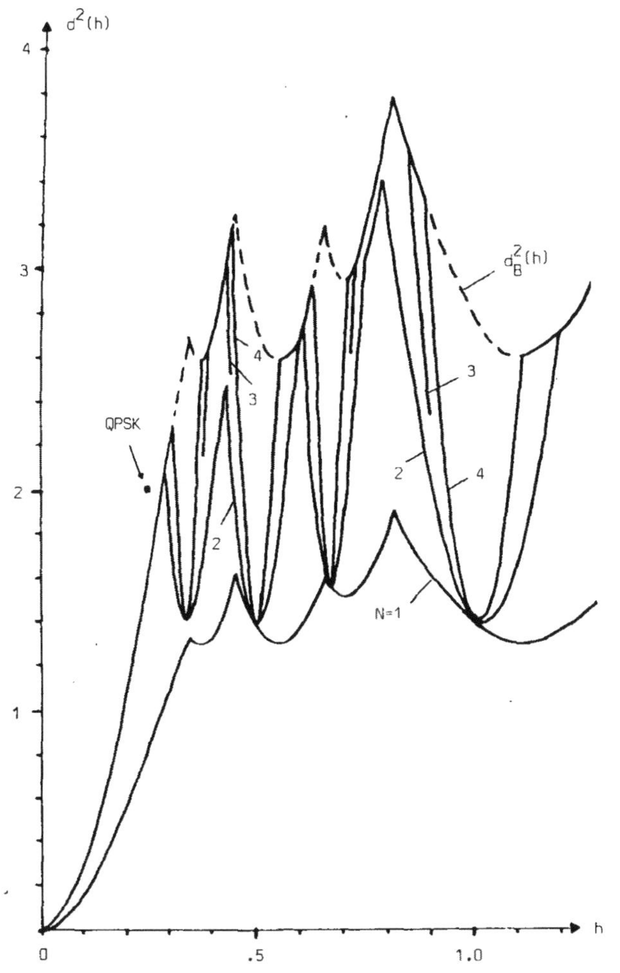

Fig. 9. Minimum normalized squared distance versus modulation index h for $M = 4$, HCS system.

the pulse shaping. The results for the octal case (see Fig. 10) follow the same trend as for the CPFSK system; i.e., the upper bound is reached with $N = 2$ for low h-values and the number of catastrophic modulation indexes has increased compared to the quaternary case.

V. POWER SPECTRUM

The power spectral density for the full response CPM schemes considered in this paper can be calculated with formulas given in [2]. The data symbols are assumed to be independent and identically distributed. For the case of full response CPFSK systems, the spectrum can be expressed directly in terms of elementary functions [2] and for RC systems in terms of Bessel functions [12].

Spectra for systems with different values of M should be compared at the same bit rate. The bit rate normalized variable $f \cdot T_b$ is used where

$$T_b = \frac{T}{\log_2 M}. \tag{63}$$

Hence, the power spectra $R_0(f)$ are plotted against the bit rate normalized frequency separation from carrier. Plots of the commonly used function [14], [15]

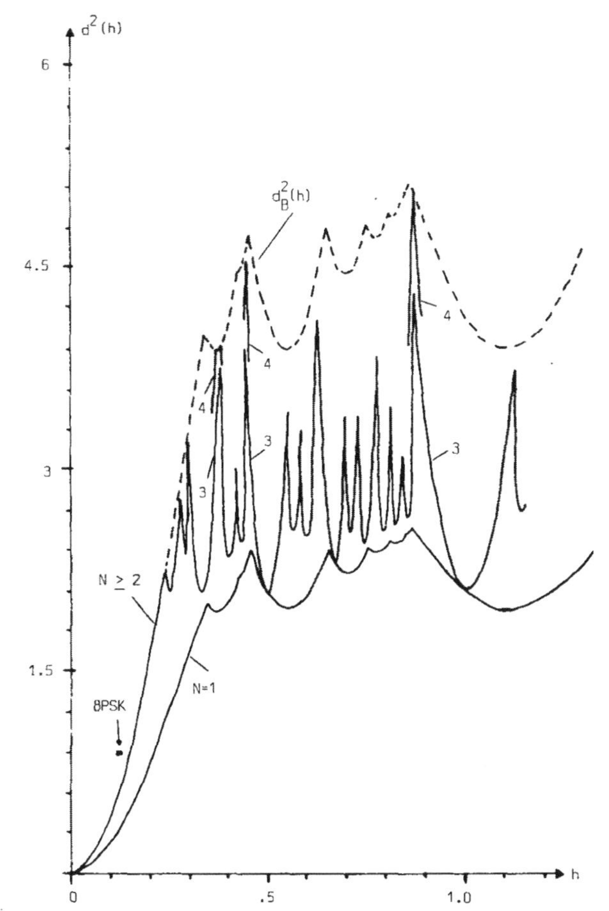

Fig. 10. Minimum normalized squared Euclidean distance versus modulation index h for $M = 8$, HCS system.

$$P_{ob}(B) = \frac{\int_B^\infty R_0(f)\,df}{\int_0^\infty R_0(f)\,df} \tag{64}$$

which gives the fractional out of band power at the one-sided bandwidth B, will also be given.

Fig. 11 shows the power spectra (double-sided) for M-ary CPFSK with $h = 1/M$, $M = 2, 4, 8$. The corresponding fractional out of band power plots are shown in Fig. 12 (for other values of h, see [17], [20]). It is well known that for fixed M and $g(t)$, the spectrum widens for increasing h. For certain h values discrete components occur. Fig. 11 and the distance figures illustrate the fact that for a roughly fixed distance, the spectral main lobe is decreasing with increasing M. The behavior of the spectra for large frequencies (i.e., the spectral tails) depends only on the number c of continuous derivatives of the instantaneous phase. It is shown in [8] that the tails decrease with f as $|f|^{-2(c+2)}$. For CPFSK, c equals 0.

Fig. 13 shows the spectra for the quaternary CPFSK, HCS, and RC schemes for $h = 1/4$. Note that for increased values of c the main lobe becomes larger. The main lobe widens intuitively due to the presence of higher phase slopes over a portion

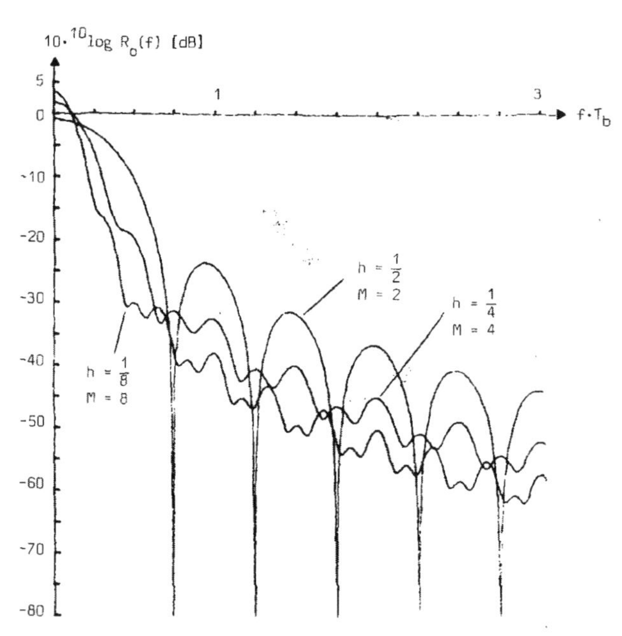

Fig. 11. Normalized power spectral density in decibels for M-ary (M = 2, 4, and 8) CPFSK with modulation indexes H = 1/2, 1/4, and 1/8, respectively.

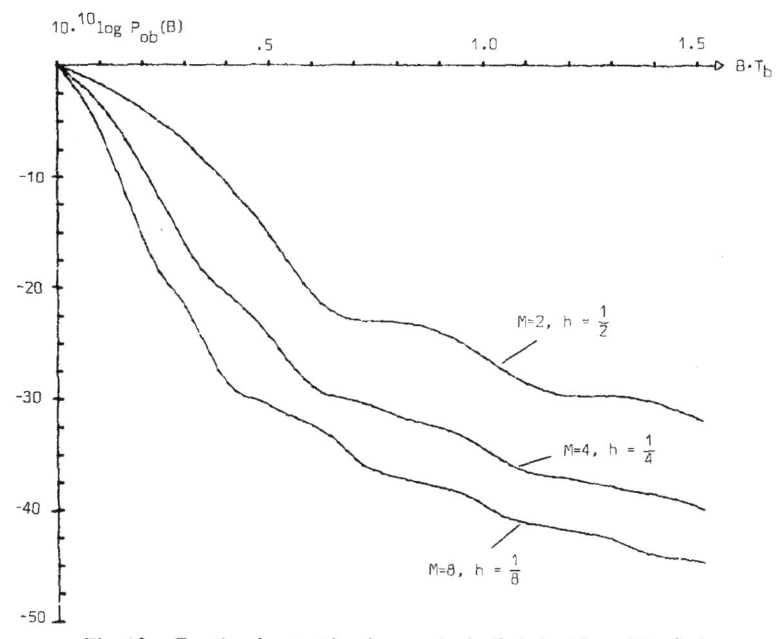

Fig. 12. Fractional out of band power in decibels for M-ary (M = 2, 4, and 8) CPFSK with modulation indexes h = 1/2, 1/4, 1/8, respectively.

of the pulse for non-CPFSK schemes. The spectral tail of HCS behaves like f^{-6} and like f^{-8} for RC. Further spectra for these schemes are plotted in [17], [19], [20].

The spectra of schemes with plateau functions are investigated in [16], [18]. As might be expected, the rapid phase change in the beginning and the end of each symbol interval gives wide spectra. The previously mentioned spectral tail behavior versus c is also applicabe in this case. However, f must be impractically large before this asymptotic behavior is dominating. Furthermore, it was concluded above that large $d^2_{\min}$ (close to 4) are reached with plateau functions with $M = 2$ and h close to 1. For $h = 1$, the power spectrum contains spectral lines however.

VI. DISCUSSION AND CONCLUSIONS

From the distance and spectrum results above and in [17], [19], [20], it is evident that M-ary full response CPM schemes have both bandwidth compaction properties and yield gain in E_b/N_0 as compared to MSK. Schemes within this class of CPM systems can also be designed to give a large gain in E_b/N_0 with

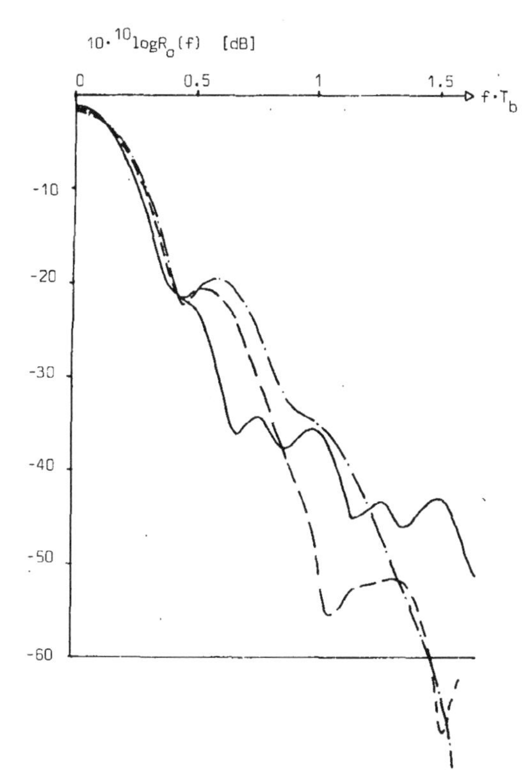

Fig. 13. Normalized power spectral densities in decibles for quaternary CPFSK with modulation index h = 1/4. The schemes are CPFSK (solid line), HCS (dashed line), and RC (dash-dotted line).

TABLE II
BANDWIDTH/DISTANCE TRADEOFF FOR SOME M-ARY CPFSK SYSTEMS

CPFSK scheme	Bandwidth $2B\cdot T_b$ 90%	99%	99.9%	$D^2_{min}/2E_b$	Gain over MSK, dB	N_B symbols
M=2 h=.5	0.78	1.20	2.78	2.0	0	2
M=4 h=.25	0.42	0.80	1.42	1.45	-1.38	2
M=8 h=.125	0.30	0.54	0.96	.60	-5.23	2
M=4 h=.40	0.68	1.08	2.08	3.04	1.82	4
M=4 h=.45	0.76	1.18	2.20	3.60	2.56	5
M=8 h=.30	0.70	1.00	1.76	3.0	1.76	2
M=8 h=.45	1.04	1.40	2.36	5.40	4.31	5

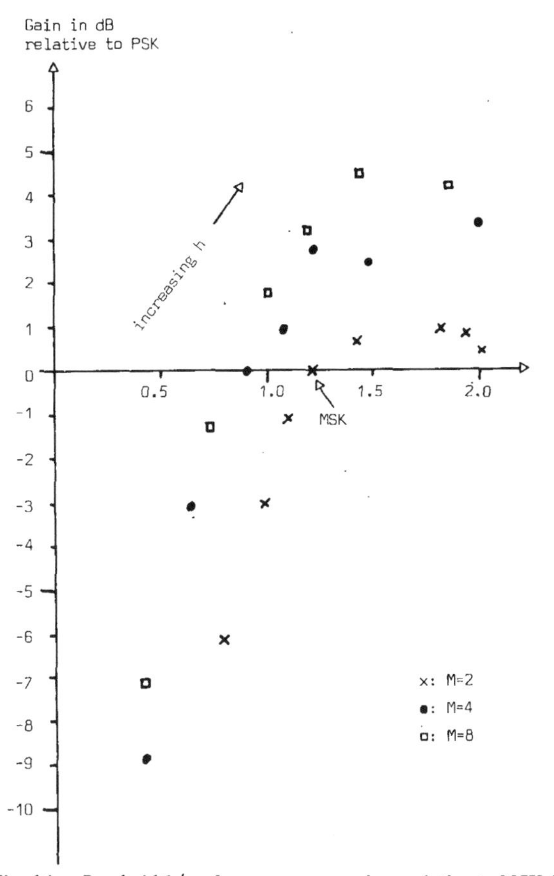

Fig. 14. Bandwidth/performance comparison relative to MSK for various CPFSK systems (99 percent fractional out of band power bandwidth, see Table II).

the same bandwidth as MSK, or considerably smaller bandwidth at the expense of an increased E_b/N_0. This holds, for example, for M-ary CPFSK. The same also holds for systems like HCS and RC, and in these cases the tradeoff between bandwidth and gain in E_b/N_0 at large SNR is even more attractive.

In the binary case, plateau functions are a way to achieve considerable gains in terms of E_b/N_0, which unfortunately gives poor spectra.

In Table II comparisons between various CPFSK schemes are made, both concerning bandwidth and gain in terms of E_b/N_0 (dB) at large SNR. The reference system is MSK.

Three different definitions of bandwidth will be used. The normalized bandwidth (double-sided) is defined at $2BT_b$, for which 90, 99, or 99.9 percent of the total signal power is within the frequency band $|f - f_0| \leq B$.

Table II also gives the number of observed symbols N_B required to reach the given minimum squared Euclidean distance value. The quaternary scheme with h = 0.45 has approximately the same bandwidth as MSK (99 percent bandwidth) and yields a gain in E_b/N_0 of 2.56 dB. The octal scheme with h = 0.45 gives a slight bandwidth expansion when compared to MSK (at 99 percent bandwidth), but gives the gain 4.31 dB in terms of E_b/N_0.

A more exhaustive comparison between different M-ary CPFSK systems can be found in Fig. 14. In this figure the gain in decibles of various schemes is shown versus the 99 percent bandwidth. The schems are binary (indicated by x), quaternary (indicated by •) and octal (indicated by □). Note the superior performance of quaternary and octal schemes. In the binary case it was possible to find the frequency pulses $g(t)$ maximizing the upper bound on the minimum Euclidean distance. This was not the case for multilevel systems, and it is believed that the optimum frequency pulse depends on M and h. In the interval $0 < h < 1/M$, M-ary PSK was shown to yield the largest minimum Euclidean distance, but the HCS and RC schemes are not far from this optimum. However, the two latter schemes have much smaller spectral tails than M-ary PSK.

It was shown that the number of first-order weak modulation indexes grows with M, thus putting a practical limit on how large an M should be chosen.

It is interesting to note that for the schemes considered in this paper, a gain in terms of E_b/N_0 is obtained without expanded bandwidth, compared to MSK. This is different from the case with a channel coded MSK system, where the spectrum must be expanded by a factor of $1/R$ where R is the code rate [3], [9]. For the CPM systems, no parity symbols are transmitted, and the total signal energy is devoted to the information symbols.

This paper explores the distance and bandwidth properties of full response CPM systems. In spite of the restriction that the schemes must be a full response type (i.e., the instantaneous frequency only depends on one data symbol), we have found considerable improvements. However, larger improvements are obtainable with partial response systems (the instantaneous frequency depends on more than one data symbol). This class of system is considered in part II. We have intentionally omitted all problems dealing with transmitter and receiver considerations. These problems will be treated in a unified manner in Part II.

REFERENCES

[1] V. A. Kotelnikov, *The Theory of Optimum Noise Immunity*. New York: Dover, 1960.

[2] R. R. Anderson and J. Salz, "Spectra of digital FM," *Bell Syst. Tech. J.*, vol. 44, pp. 1165–1189, July–Aug. 1965.

[3] J. M. Wozencraft and I. M. Jacobs, *Principles of Communication Engineering*. New York: Wiley, 1965.

[4] R. W. Lucky, J. Salz, and E. J. Weldon, Jr. *Principles of Data Communication*. New York: McGraw-Hill, 1968.

[5] M. G. Pelchat, R. C. Davis, and M. B. Luntz, "Coherent demodulation of continuous phase binary FSK signals," in *Proc. Int. Telemetering Conf.*, Washington, DC, 1971, pp. 181–190.

[6] R. deBuda, "Coherent demodulation of frequency-shift keying with low deviation ratio," *IEEE Trans. Commun.*, vol. COM-20, pp. 429–436, June 1972.

[7] W. P. Osborne and M. B. Luntz, "Coherent and noncoherent detection of CPFSK," *IEEE Trans. Commun.*, vol. COM-22, pp. 1023–1036, Aug. 1974.

[8] T. J. Baker, "Asymptotic behaviour of digital FM spectra," *IEEE Trans. Commun.*, vol. COM-22, pp. 1585–1594, Oct. 1974.

[9] W. C. Lindsey and M. K. Simon, *Telecommunication Systems Engineering*. Englewood Cliffs, NJ: Prentice-Hall, 1974.

[10] T. A. Schonhoff, "Symbol error probabilities for M-ary coherent continuous phase frequency-shift keying (CPFSK)," in *Proc. IEEE Int. Conf. Commun. Conf. Record*, San Francisco, CA, 1975, pp. 34.5–34.8.

[11] T. A. Schonhoff, "Bandwidth vs performance considerations for CPFSK," in *Proc. IEEE National Telecommun. Conf. Record*, 1975, pp. 38.1–38.5.

[12] F. Amoroso, "Pulse and spectrum manipulation in the minimum (frequency) shift keying (MSK) format," *IEEE Trans. Commun.*, vol. COM-24, pp. 381–384, Mar. 1976.

[13] T. A. Schonhoff, "Symbol error probabilities for M-ary CPFSK: Coherent and noncoherent detection," *IEEE Trans. Commun.*, vol. COM-24, pp. 644–652, June 1976.

[14] M. K. Simon, "A generalization of the minimum-shift-keying (MSK)-type signaling based upon input data symbol pulse shaping," *IEEE Trans. Commun.*, vol. COM-24, pp. 845–856, Aug. 1976.

[15] M. Rabzel and S. Pasupathy, "Spectral shaping in minimum shift keying (MSK) type signals," *IEEE Trans. Commun.*, vol. COM-26, pp. 189–195, Jan. 1978.

[16] T. Aulin and C-E. Sundberg, "Binary CPFSK type of signaling with input data symbol pulse shaping—Error probability and spectrum," Telecommunication Theory, Techn. Rep. TR-99, Univ. Lund, Lund, Sweden, July 1978.

[17] ——, "M-ary CPFSK type of signaling with input data symbol pulse shaping—Minimum distance and spectrum," Telecommunication Theory, Techn. Rep. TR-111, Univ. Lund, Lund, Sweden, Aug. 1978.

[18] ——, "Bounds on the performance of binary CPFSK type of signaling with input data symbol pulse shaping," in *Proc. IEEE Nat. Telecommun. Conf. Record*, Birmingham, AL, 1978, pp. 6.5.1–6.5.5.

[19] ——, "M-ary CPFSK type of signaling with input data symbol pulse shaping—Minimum distance and spectrum," in *Proc. IEEE Int. Conf. Commun. Conf. Record*, Boston, MA, 1979, pp. 42.3.1–42.3.6.

[20] T. Aulin, "CPM—A power and bandwidth efficient digital constant envelope modulation scheme," Dr. Techn. dissertation, Telecommunication Theory, Univ. Lund, Lund, Sweden, Nov. 1979.

[21] T. Aulin, N. Rydbeck, and C.-E. W. Sundberg, "Continuous phase modulation—Part II: Partial response signaling," this issue, pp. 210–225.

Continuous Phase Modulation—Part II: Partial Response Signaling

TOR AULIN, MEMBER, IEEE, NILS RYDBECK, AND CARL-ERIK W. SUNDBERG, MEMBER, IEEE

Abstract—**An analysis of constant envelope digital partial response continuous phase modulation (CPM) systems is reported. Coherent detection is assumed and the channel is Gaussian. The receiver observes the received signal over more than one symbol interval to make use of the correlative properties of the transmitted signal. The systems are *M*-ary, and baseband pulse shaping over several symbol intervals is considered. An optimum receiver based on the Viterbi algorithm is presented. Constant envelope digital modulation schemes with excellent spectral tail properties are given. The spectra have extremely low sidelobes. It is concluded that partial response CPM systems have spectrum compaction properties. Furthermore, at equal or even smaller bandwidth than minimum shift keying (MSK), a considerable gain in transmitter power can be obtained. This gain increases with *M*. Receiver and transmitter configurations are presented.**

I. INTRODUCTION

THIS paper presents constant envelope digital modulation systems having both good symbol error and spectral properties. The systems developed and studied are called partial response continuous phase modulation (CPM) systems. This is because the data symbols modulate the instantaneous phase of the transmitted signal and this phase is a continuous function of time. One single data symbol affects this phase over more than one symbol interval, an approach called *partial response signaling* [1], [3], [8]. The general class of signals is defined in Part I.

Previously, power spectra for systems of this type have been analyzed, for example, in [5], [9], and [14]. In this paper the analysis of the performance of the optimum detector for an additive white Gaussian channel is given. Some work on a suboptimum detector for a very special case of the CPM signaling scheme appears in [13].

As was treated in Part I of this paper, several attempts have been made to improve the asymptotic spectral properties for binary *full response* CPM systems with modulation index $h = 1/2$, compared to minimum shift keying (MSK). Considerable asymptotic improvement can be obtained by making the phase smoother at the bit transition instants, but even though the spectral properties with these schemes are asymptotically good, they are sometimes inferior to MSK for low and intermediate frequencies. But a result of Part I is that if full response CPM systems are to be used, a more attractive tradeoff between error probability performance and spectrum is achieved by using more levels than two and moderate smoothing of the phase at the symbol transition instants.

As will be shown in this paper, the use of partial response CPM systems yields a more attractive tradeoff between error probability and spectrum than does the full response systems. The spectral properties are improved at almost all frequencies and the first side-lobes are considerably lower. This spectral improvement takes place without increase of the probability of symbol error at practical signal to noise ratios. The improvement is obtained by introducing memory in the modulation process. The price for the improvements is system complexity, especially with respect to the optimum receiver. In practice, only rational values of the modulation index are useful, but it will be convenient here to imagine that h is real.

Manuscript received March 19, 1980; revised September 19, 1980. This work was supported by the Swedish Board of Technical Development under Contract 79-3594.

T. Aulin and C.-E. W. Sundberg are with the Department of Telecommunication Theory, University of Lund, Fack, S-220 07, Lund, Sweden.

N. Rydbeck is with SRA Communications AB, S 163 00, Spånga, Sweden.

II. PROPERTIES OF THE MINIMUM EUCLIDEAN DISTANCE

In this section general properties of the minimum Euclidean distance as a function of the real valued modulation index h are given for partial response CPM systems. This distance is defined in Part I.

The Phase Tree and the Phase Difference Tree

An important tool for calculation of the minimum Euclidian distance is the so-called phase tree. This tree is formed by the ensemble of phase trajectories having a common start phase (root), say zero, at time $t = 0$. The data symbols for all the phase trajectories in the tree before this time are all equal. The value of these previous data symbols (the pre-history) can be chosen arbitrarily, but for unifying purposes they will be chosen to be $M - 1$; i.e., $\alpha_i = M - 1$; $i = -1, -2, \cdots$. An example of a phase tree for a binary partial response CPFSK system is shown in Fig. 1 and the frequency pulse $g(t)$ is

$$g(t) = \begin{cases} \dfrac{1}{6T}; & 0 < t < 3T \\ 0; & \text{otherwise.} \end{cases} \tag{1}$$

When a calculation of the minimum Euclidean distance is to be performed, the first pair of data symbols by definition must be different, and hence the phase trajectories are always different over this symbol interval. Going deeper into the tree however, it is always possible to find a pair of phase trajectories which coincide (modulo 2π) at a specific time, and do so ever after. This is called a merge. Denoting the time where the two phase trajectories merge by t_m, it can be seen that the calculation of the Euclidean distance only has to be performed over the interval $0 \leqslant t \leqslant t_m$; see (21), Part I.

Reprinted from *IEEE Transactions on Communications*, vol. COM-29, no. 3, March 1981.

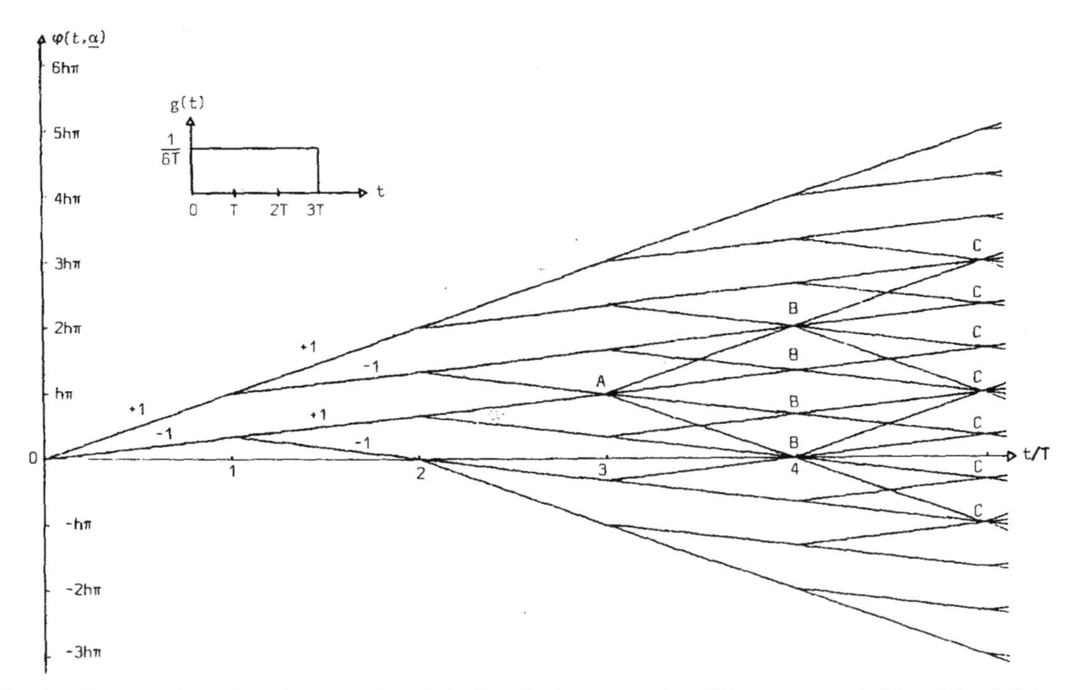

Fig. 1. Phase tree (ensemble of phase trajectories) when the frequency pulse $g(t)$ is constant and of length $L = 3$. It is assumed that all the binary ($M = 2$) data symbols prior to $t = 0$ are all +1. Note that A is a crossing and not a merge. B and C denotes first and second merges, respectively. Note: characters with underbars appear boldface in text.

The phase difference trajectories are defined by

$$\varphi(t, \gamma) = 2\pi h \sum_{i=-\infty}^{\infty} \gamma_i q(t - iT);$$

$$\gamma_i = 0, \pm 2, \pm 4, \cdots, \pm 2(M-1). \quad (2)$$

The phase difference tree is now obtained from the ensemble of phase difference trajectories. Since the first pair of data symbols must be different in the phase tree, the first difference symbol γ_0 must not equal zero in the phase difference tree. In the phase tree the prehistory is the same for all phase trajectories, and thus the difference symbols forming the prehistory in the phase difference tree must all be zero. This means that the phase difference tree does not depend on the prehistory and hence, neither do the Euclidean distances (see (24), Part I).

Since there is a sign symmetry among the difference sequences and thus also among the phase difference trajectories, this can be removed for calculation of Euclidean distances. This is clear from I-(24) and the fact that cosine is an even function. For calculations of the minimum Euclidean distance, when the receiver observation interval equals N symbol intervals, it is sufficient to consider the phase difference tree, using the difference sequences γ_N defined by

$$\begin{cases} \gamma_i = 0 & i < 0 \\ \gamma_0 = 2, 4, 6, \cdots, 2(M-1) & \\ \gamma_i = 0, \pm 2, \pm 4, \cdots, \pm 2(M-1); & i = 1, 2, \cdots, N-1 \end{cases} \quad (3)$$

The merges are easily identified in the phase difference tree, since this corresponds to a phase difference trajectory which is identically equal to zero for all $t \geqslant t_m$.

In Fig. 2 the phase difference tree with sign symmetry removed is given for the frequency pulse defined by (1). Merges are also shown. Note that in Fig. 1 pairs of phase trajectories must be considered in calculating Euclidean distance, but in Fig. 2 only single phase difference trajectories need to be.

Weak Modulation Indices h_c

As was mentioned earlier, a phase difference trajectory identically equal to zero for all $t \geqslant t_m$ defines a merge. Since the phase difference tree always must be viewed modulo 2π in conjunction with distance calculations, a merge that depends upon the modulation index h is obtained if there exists a phase difference trajectory which equals a nonzero multiple of 2π for all $t \geqslant t_c$. The value of a phase difference trajectory depends on the modulation index h, and a situation like this occurs if

1) there exists a phase difference trajectory which is a constant not equal to zero for all $t \geqslant t_c$;

2) the modulation index h is chosen so that this constant is a multiple of 2π.

Since any phase difference trajectory is achieved by feeding the difference sequence into a filter having the impulse response $q(t)$ and multiplying the output of this filter by $2\pi h$, the first requirement is met for $t_c = LT$ by choosing

$$\gamma_i = \begin{cases} 0; & i < 0 \\ 2, 4, 6, \cdots, 2(M-1); & i = 0 \\ 0; & i > 0 \end{cases} \quad (4)$$

since for frequency pulses $g(t)$ of length L symbol intervals $q(t) = q(LT)$, $t \geqslant LT$. Sometimes this can occur also for $t_c =$

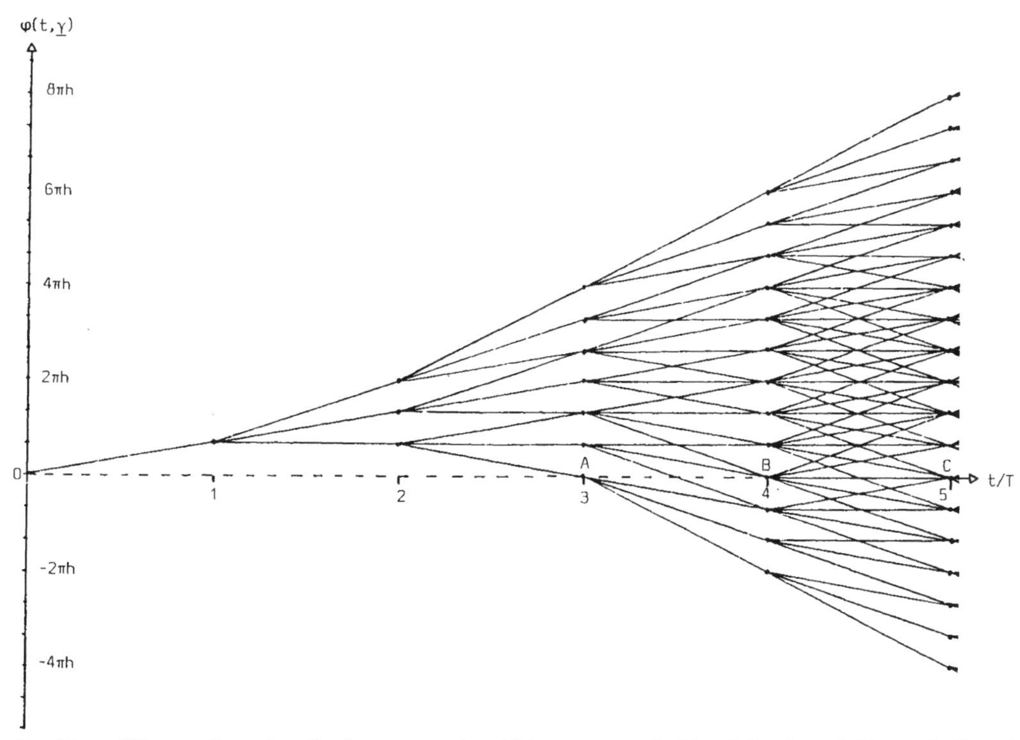

Fig. 2. Phase difference tree when the frequency pulse $g(t)$ is constant and of length $L = 3$ symbol intervals. Note that A is not a merge since the trajectory cannot be identically zero in the future. B and C denotes merges. Characters with underbars appear boldface in text.

$(L-1)T$, $(L-2)T$ and so on, but only for specific phase responses $q(t)$.

The second requirement above is satisfied if

$$2\pi h_c \gamma_0 q(LT) = k \cdot 2\pi; \qquad k = 1, 2, \cdots$$
$$\gamma_0 = 2, 4, 6, \cdots, 2(M-1) \qquad (5)$$

where it is assumed that $q(LT) \neq 0$. Thus, for these modulation indices, in the sequel called *weak*, h_c satisfies

$$h_c = \frac{k}{\gamma_0 q(LT)}; \qquad k = 1, 2, \cdots$$
$$\gamma_0 = 2, 4, 6, \cdots, 2(M-1). \qquad (6)$$

A merge, modulo 2π, occurs at $t = LT$.

In a similar way, the phase difference trajectory can be made constant for all $t \geqslant t_c = (L + \Delta L)T$ by choosing

$$\gamma_i = \begin{cases} 0; & i < 0 \\ 2, 4, 6, \cdots, 2(M-1); & i = 0 \\ 0, \pm 2, \pm 4, \cdots, \pm 2(M-1); & i = 1, 2, \cdots, \Delta L \\ 0; & i > \Delta L \end{cases} \qquad (7)$$

and the weak modulation indices are

$$h_c = \frac{k}{q(LT) \sum_{i=0}^{\Delta L} \gamma_i}; \qquad k = 1, 2, 3, \cdots \quad q(LT) \neq 0 \quad \sum_{i=0}^{\Delta L} \gamma_i \neq 0 \qquad (8)$$

It can be noted that for a given finite interval of the modulation index, the number of weak modulation indices within this interval increases with M as in Part I.

In Fig. 3 the minimum normalized squared Euclidean distance is shown versus h when $N = 1, 2, \cdots, 10$ observed symbol intervals for the binary CPM system given by (1). It is seen that for most h the minimum Euclidean distance increases to an upper bound $d_B{}^2(h)$ with N, but not for the weak modulation indices

$$h_c = \tfrac{3}{4}, 1, \tfrac{9}{8}, \tfrac{6}{5}, \tfrac{9}{7}, \tfrac{3}{2}. \qquad (9)$$

It can also be seen that some of the "weak" modulation indices, e.g., $h_c = 4/5$, $7/8$ do not affect the growth of the minimum Euclidean distance with N. This will be discussed in more detail later. However, indices such as $h_c = 3/2$ are catastrophic.

An Upper Bound on the Minimum Euclidean Distance

A phase difference trajectory identically equal to zero for all $t \geqslant t_m$ does not increase the Euclidean distance if the observation interval is made longer than $N = t_m/T$ symbol intervals. By choosing any fixed difference sequence so that a merge is obtained, a limited upper bound on the minimum Euclidean distance is obtained.

The first time instant for which any phase difference trajectory can be made identically zero ever after is in general $t = (L + 1)T$, where L is the length of the causal frequency pulse $g(t)$. This is called the first merge, and the phase difference trajectories giving this merge are obtained by choosing the difference sequences

$$\gamma_i = \begin{cases} 0 & i < 0 \\ \gamma_0 = 2, 4, 6, \cdots, 2(M-1); & i = 0 \\ -\gamma_0 & i = 1 \\ 0 & i > 1 \end{cases} \qquad (10)$$

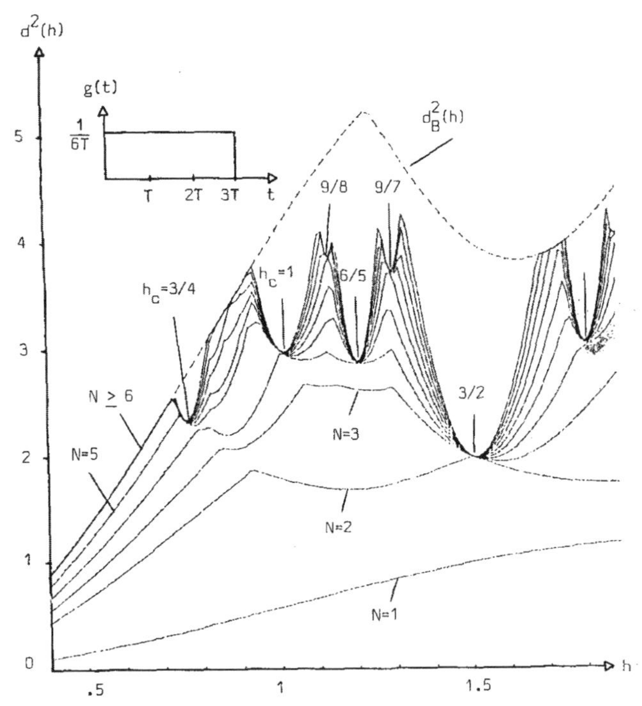

Fig. 3. The upper bound on the minimum Euclidean distance (dashed) and minimum distance curves (solid) for $1 \leqslant N \leqslant 10$ observed symbol intervals. The pulse $g(t)$ is constant and of duration $3T$ [equation (1)] and $M = 2$.

and thus the phase difference trajectories are

$$\varphi(t, \gamma) = \begin{cases} 0; & t \leqslant 0 \\ \gamma_0 \cdot 2\pi h q(t); & 0 \leqslant t \leqslant T \\ \gamma_0 \cdot 2\pi h[q(t) - q(t-T)]; & T \leqslant t \leqslant (L+1)T \\ \gamma_0 \cdot 2\pi h[q((L+1)T) - q(LT)] = 0; & t \geqslant (L+1)T \end{cases} \tag{11}$$

By taking the minimum of the Euclidean distances between the signals having the phase difference trajectories above, an upper bound on the minimum Euclidean distance as a function of h, for fix M and $g(t)$, is achieved just as in Part I. The resulting bound is more complex, however.

One can also consider second merges, that is phase difference trajectories which merge at $t = (L + 2)T$. These phase difference trajectories are obtained from (2) by using

$$\gamma_i = \begin{cases} 0; & i < 0 \\ 2, 4, 6, \cdots, 2(M-1); & i = 0 \\ 0, \pm 2, \pm 4, \cdots, \pm 2(M-1); & i = 1, 2 \\ 0; & i > 2 \end{cases} \tag{12}$$

satisfying

$$\sum_{i=0}^{2} \gamma_i = 0. \tag{13}$$

It might happen that the Euclidean distance associated with these phase difference trajectories for a specific h is smaller than those giving the first merge. Hence, the former upper bound can be tightened by also taking the minimum of the Euclidean distances associated with the second merge. This new upper bound might also be further tightened by taking third merges into account, and so on.

The exact number of merges needed to give an upper bound on the minimum Euclidean distance which cannot be further tightened is not known in the general case, but in no case treated below were merges later than the Lth needed [16]-[18], [21]. For the full response case ($L = 1$) it was shown in Part I that first merges give the tight bound.

Fig. 3 shows the resulting upper bound dashed for the CPM scheme defined by (1) together with actual minimum normalized Euclidean distances when $N = 1, 2, \cdots, 10$ observed symbol intervals. The upper bound is the minimum of three functions corresponding to the first three merges [21].

As was mentioned earlier, not all weak modulation indices affect the minimum Euclidean distance. In this case the phase difference trajectories associated with (7) give larger Euclidean distances than the upper bound. Only those weak modulation indices which have phase difference trajectories having distance smaller than the upper bound affect the minimum Euclidean distance [21].

Weak Systems

In general, the first inevitable (independent of h) merge occurs at $t = (L + 1)T$. However, there are partial response CPM systems which have earlier merges, depending on the shape of the frequency pulse $g(t)$ and the number of levels M, but not the modulation index h. A partial response system is said to be *weak of order* L_c, if the first inevitable merge occurs at $t = (L + 1 - L_c)T$. L_c cannot be larger than L. A class of first-order weak systems are those where the frequency pulse $g(t)$ integrates to zero, i.e., $q(LT) = 0$. The sequence (10) of course still gives a merge, but not the earliest one. By choosing

$$\gamma_i = \begin{cases} 0; & i \neq 0 \\ \gamma_0 = 2, 4, 6, \cdots, 2(M-1); & i = 0 \end{cases} \tag{14}$$

the phase difference trajectory now equals zero for all $t \geqslant LT$; see (2). Examples of weak systems are given in [18] and [21]. The partial response CPFSK system having the frequency pulse

$$g(t) = \begin{cases} \dfrac{1}{4T}; & 0 < t < T \\ -\dfrac{1}{4T}; & T < t < 2T \\ 0; & \text{otherwise} \end{cases} \tag{15}$$

is a weak system of the first order [21].

The main conclusion concerning weak systems is that they should be avoided, since the potential maximum value of the minimum Euclidean distance, indicated by the upper bound $d_B{}^2(h)$, will never be reached due to the early merges. More results for weak systems can be found in [18]. We have found that there is little difficulty choosing pulses in such a way that the resulting schemes does not have the weak property considered above. This will be clearly illustrated in the following.

III. A SEQUENTIAL ALGORITHM FOR COMPUTATION OF THE MINIMUM EUCLIDEAN DISTANCE

A fast sequential algorithm for computation of the minimum Euclidean distance, for any real valued modulation index h and large numbers N of observed symbol intervals, has been developed. The algorithm is sequential in N and uses the basic properties of the minimum Euclidean distance, namely that it is upper bounded and is a nondecreasing function of N given $g(t)$, h, and M. The sequential property is obtained from the fact that squared Euclidean distances for coherent CPM systems are additive. This means that if the squared Euclidean distance has been calculated for N symbol intervals, the squared Euclidean distance for $N+1$ observed symbol intervals is obtained by just adding an increment to the previously calculated squared Euclidean distance. This holds for fixed difference sequences γ, and can be seen from the expression for the normalized squared Euclidean distance

$$d^2(\gamma_N, h) = N - \frac{1}{T}\sum_{i=0}^{N-1}\int_{iT}^{(i+1)T}\cos\left[2\pi h\left(\sum_{j=i-L+1}^{i}\gamma_j \cdot q(t-jT) + q(LT)\sum_{j=0}^{i-L}\gamma_j\right)\right]dt \tag{16}$$

Thus,

$$d^2(\gamma_{N+1}, h) = d^2(\gamma_N, h) + 1 - \frac{1}{T}\int_{NT}^{(N+1)T}\cos\left[2\pi h\left(\sum_{j=N-L+1}^{N}\gamma_j \cdot q(t-jT) + q(LT)\sum_{j=0}^{N-L}\gamma_j\right)\right]dt. \tag{17}$$

A brute force method for calculating the minimum Euclidean distance for a given N is to compute $d^2(\gamma_N, h)$ for all sequences γ_N defined by (3) and take the minimum of all the achieved quantities. The number of calculations required using this method grows exponentially with N since the number of difference sequences of length N is $(M-1)(2M-1)^{N-1}$ and even for $M = 2$ it is unrealistic to calculate the minimum Euclidean distance for moderate N-values.

A flowchart for a more efficient algorithm can be seen in Fig. 4. It is assumed that the minimum Euclidean distance is to be calculated for a CPM system when the modulation index is $h = h_{\min}, h_{\min+\Delta h}, \cdots, h_{\max-\Delta h}, h_{\max}$ and for $1 \leqslant N \leqslant N_{\max}$ observed symbol intervals. The first step is to compute the upper bound $d_B{}^2(h)$ on the squared minimum Euclidean distance for the given h-values. Now all the Euclidean distances for $N = 1$ observed symbol intervals are computed, i.e., the difference sequence if $\gamma_0 = 2, 4, \cdots, 2(M-1)$. If any of these distances is larger than the upper bound for that specific h-value, the entire subtree having the corresponding γ_0–value will never be used. This can be done since the Euclidean distance is a nondecreasing function of N for fixed h. The minimum of these Euclidean distances gives the minimum Euclidean distance over $N = 1$ observed symbol intervals.

When the minimum Euclidean distance for $N = 2$ observed symbol intervals is to be computed, only those γ_0-values whose corresponding distances when $N = 1$ did not exceed the upper bound are used. The second component of the difference sequence is chosen to be $\gamma_1 = -2(M-1), -2(M-2), \cdots, 2(M-1)$, and the Euclidean distance is computed through (17). Again, only those sequences whose corresponding Euclidean distances are not above the upper bound will be stored as possible difference sequences for $N = 3$. The algorithm continues like this up to the maximum N-value, $N_{\max}$. Through this procedure subtrees are continuously deleted.

It is clear that any finite upper bound on the minimum Euclidean distance can be used for cutting subtrees. If the upper bound is too loose, however, the number of possible difference sequences will grow. The algorithm will naturally be as fast as possible if the smallest possible upper bound is used.

This algorithm has empirically been found to increase only linearly in computational complexity with N, and not exponentially as the brute force method. Without this algorithm the results on the minimum Euclidean distances presented in the next chapter would be impossible to achieve. Using this algorithm allows observation intervals up to a couple of hundred symbol intervals, although $N \lesssim 100$ is usually sufficient.

IV. NUMERICAL RESULTS ON THE MINIMUM EUCLIDEAN DISTANCE

In the previous sections tools have been developed for evaluating the minimum Euclidean distance of any CPM system, and thus also the performance in terms of symbol error probability at large SNR. These tools will now be applied to selected classes of CPM systems. The chosen schemes are in no specific sense optimum. Systems are achieved, which are reasonably easy to describe and analyze and which have attractive properties.

The first class of partial response CPM systems analyzed in this chapter is defined by the frequency pulse

$$g(t) = \begin{cases} \dfrac{1}{2LT}\left(1 - \cos\left[\dfrac{2\pi t}{LT}\right]\right); & 0 \leqslant t \leqslant LT \\ 0; & \text{otherwise,} \end{cases} \tag{18}$$

i.e., the frequency pulse $g(t)$ is a raised cosine over L symbol intervals. The class is obtained by varying M, L, and the modulation index h.

Another class of CPM systems which also will be considered are those having the Fourier transform of the frequency pulse $g(t)$

$$G(f) = F\{g(t)\} = \begin{cases} \dfrac{1}{4}\left(1 + \cos\left[\dfrac{\pi f LT}{2}\right]\right); & |f| \leqslant \dfrac{2}{LT} \\ 0; & \text{otherwise,} \end{cases} \tag{19}$$

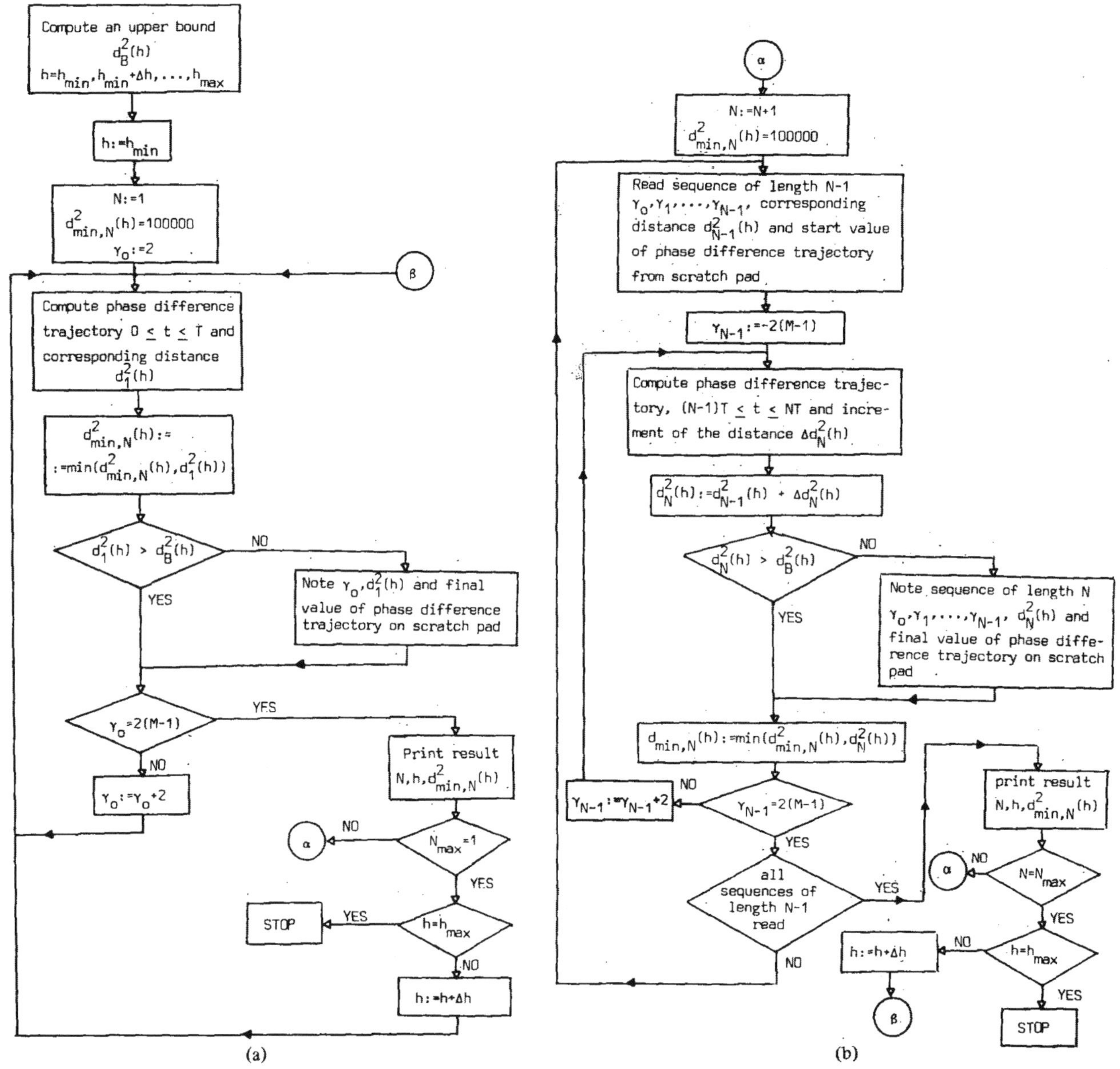

Fig. 4. (a) Flowchart for the sequential algorithm for computation of minimum distances $d^2_{min,N}(h)$, $1 \le N \le N_{max}$, $h_{min} \le h \le h_{max}$. (b) Continued.

i.e., the Fourier transform of the frequency pulse $g(t)$ is raised cosine shaped [4]. The frequency pulse is

$$g(t) = \frac{1}{LT} \frac{\sin\left[\frac{2\pi t}{LT}\right]}{\frac{2\pi t}{LT}} \cdot \frac{\cos\left[\frac{2\pi t}{LT}\right]}{1-4\left(\frac{2t}{LT}\right)^2} \tag{20}$$

which has an infinite duration. The third class of CPM systems has previously been considered [13] for the specific case $h = 1/2$, $M = 2$, and an approximation of the frequency pulse used is [17], [21]

$$g(t) = \tfrac{1}{8}\left[g_0(t-T) + 2g_0(t) + g_0(t+T)\right] \tag{21}$$

where

$$g_0(t) \approx \frac{1}{T}\left[\frac{\sin(\pi t/T)}{\pi t/T} - \frac{\pi^2}{24} \cdot \frac{2\sin(\pi t/T) - (2\pi t/T)\cos(\pi t/T) - (\pi t/T)^2 \sin(\pi t/T)}{(\pi t/T)^3}\right] \tag{22}$$

This scheme will be referred to as the TFM (tamed frequency modulation) system as in [13]. It will be analyzed only for the binary case, but the modulation index will not be constrained to 1/2 as in the approximate analysis presented in [13].

The schemes given by (18) will be denoted $1RC$ $(= RC)$,

$2RC$; $3RC$, etc. since the frequency pulse $g(t)$ is a raised cosine of length $L = 1, 2, 3$ etc. The schemes defined by (20) will be denoted $1SRC$, $2SRC$ etc. since the frequency pulse $g(t)$ has a Fourier transform which is raised cosine shaped (Spectral Raised Cosine) and the width of the main-lobe of this pulse is $L = 1, 2, 3$ etc.

Binary Systems

The RC class will now be analyzed by means of the minimum Euclidean distance. The simplest of these systems is naturally the $1RC$ scheme, which was considered in Part I. The minimum normalized squared Euclidean distance for the $2RC$ scheme is shown in Fig. 5 when the receiver observation interval is $N = 1, 2, 3, 4$, and 5 symbol intervals. The upper bound on the minimum Euclidean distance is shown dashed. For $h = 1/2$, all binary full response CPM systems including MSK have the minimum squared Euclidean distance 2 (see Part I). For this h-value the $2RC$ system yields almost the same distance when $N = 3$; the exact figure is 1.97. Hence, MSK and $2RC$ with $h = 1/2$ have almost the same performance in terms of symbol error probability for large SNR, but the optimum detector for the $2RC$ system must observe one symbol interval more than that for the MSK system. It can be expected, however, that the $2RC$, $h = 1/2$ system has a more compact spectrum due to its smooth phase tree, and this will be discussed in Section V.

For the binary full response CPFSK system, the maximum value of the minimum Euclidean distance is 2.43 when $h = 0.715$ and $N = 3$ [23]. Fig. 5 shows that larger values of the minimum Euclidean distance can be obtained for the $2RC$ system. When $h \approx 0.8$ the minimum Euclidean distance is 2.65 for $N = 4$. From a spectral point of view, however, low modulation indices should be used. It can be seen that in the region $0 < h \leq 1/2$, the upper bound is reached with $N = 3$ observed symbol intervals.

The phase tree for the binary $3RC$ system is shown in Fig. 6 and the specific shape of a phase depends upon the present and the two preceeding data symbols. This makes the phase tree yet smoother than for the $2RC$ case [20]. The first merge occurs at $t = 4T$. Note the straight lines in the phase tree. These occur for all RC-schemes and they can be used for synchronization [24], [28]. Note also that the slope in the phase tree is never larger than $h\pi/T$ (i.e., the slope of the full response CPFSK tree, Fig. 2 in Part I). This is true for all binary RC-schemes. The minimum normalized squared Euclidean distances for this scheme are shown in Fig. 7 when $N = 1, 2, \cdots, 6$ and $N = 15$ observed symbol intervals. The upper bound $d_B{}^2(h)$ has been calculated through the method described in Section III, using the first, second, and third merges.

For $N = 1$ observed symbol interval, the minimum Euclidean distance is very poor for almost every modulation index, but already with $N = 2$ it is increased significantly and when $N = 4$ the upper bound is reached in the region $0 < h \leq 1/2$. The upper bound in this region is lower than for the $2RC$ system, however. Outside this region larger minimum Euclidean distances can now be obtained, and for $h \approx 0.85$ the upper bound is reached with $N = 6$ observed symbol intervals and takes values around 3.35. The maximum value of the upper bound occurs for an h-value slightly smaller than $h = 1$.

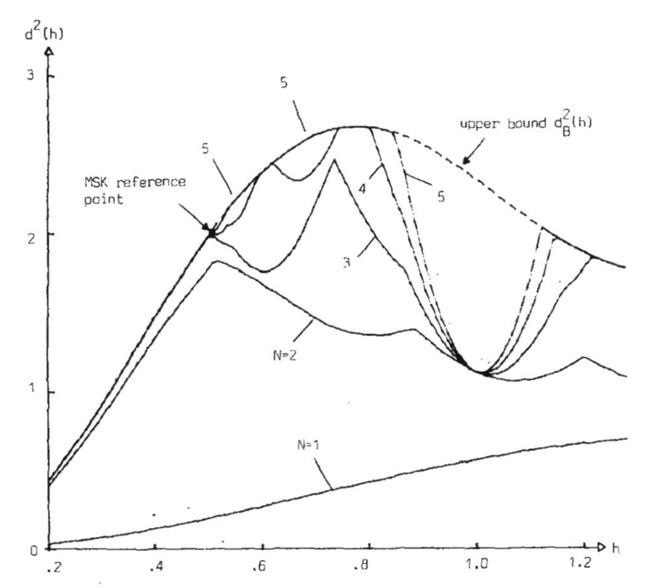

Fig. 5. Minimum squared Euclidean distance versus modulation index h for the binary $2RC$ scheme.

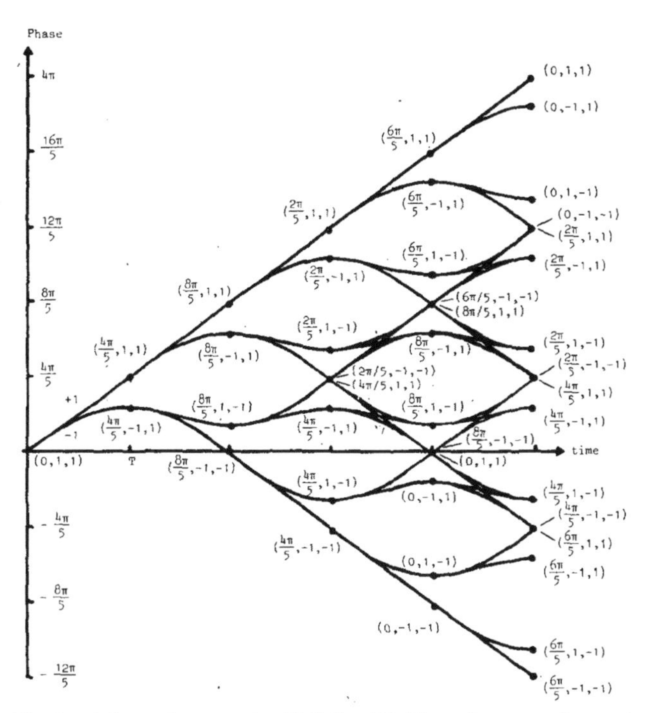

Fig. 6. Binary phase tree for $3RC$, $h = 4/5$. The assignment of states is used in Section VI.

There is an apparent weak modulation index when $h = 2/3$, but this is not a true weak modulation index since the minimun Euclidean distance in fact increases (by a small amount) when N is increased. This behavior is caused by a crossing and not a merge, and the phase trajectories are close after this crossing. The effect can be seen in the phase tree in Fig. 6, for this specific modulation index, if the phase is viewed modulo 2π.

Phase trees and minimum distance graphs for $4RC$, $5RC$ are found in [17], [20], and [21]. The distance/modulation index plot of the binary $6RC$ scheme is shown in Fig. 8. Note that the peak $d^2(h)$ value is larger than those for the shorter

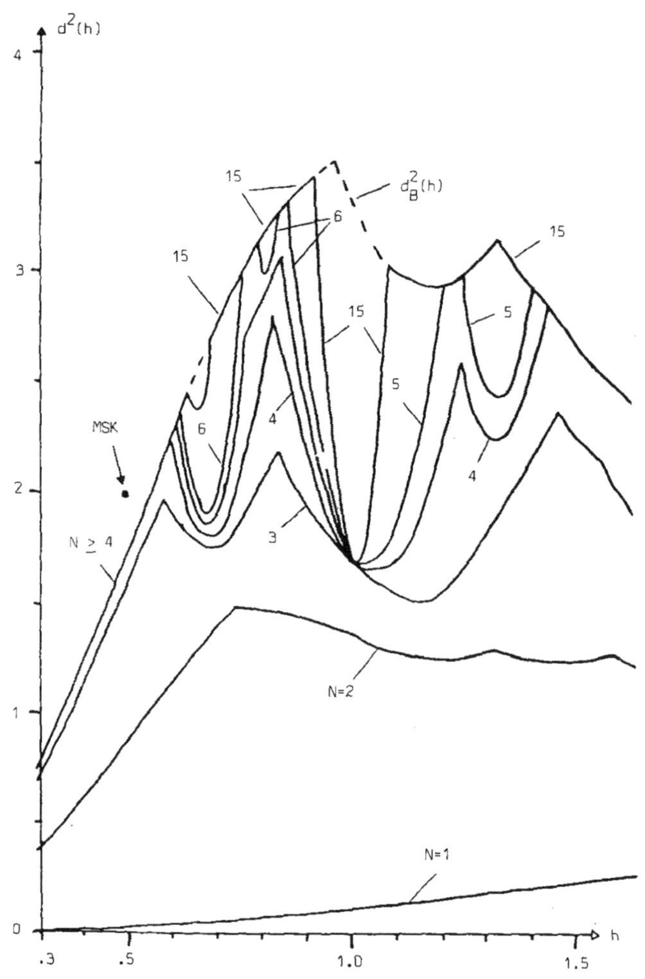

Fig. 7. Minimum squared Euclidean distance versus modulation index h for binary, 3*RC*. N is the receiver observation interval. The upper bound is shown dashed.

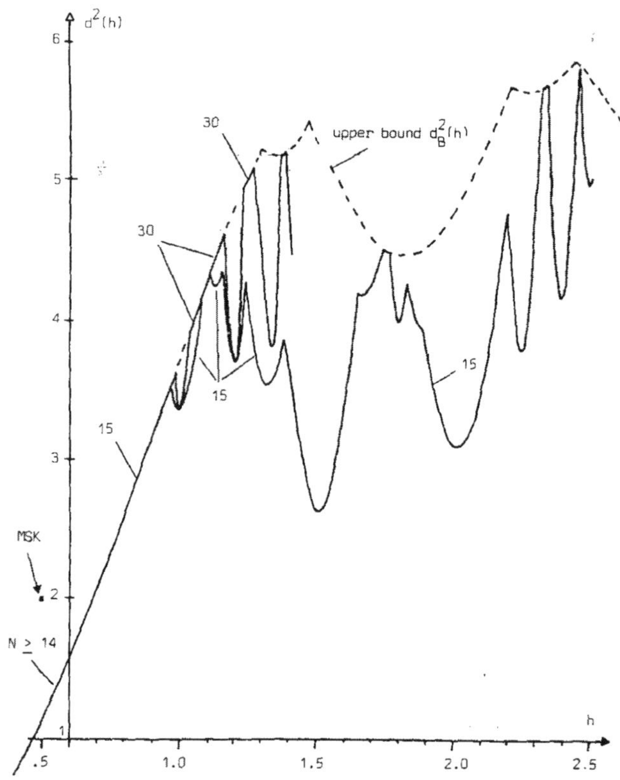

Fig. 8. Minimum squared Euclidean distance versus h for $M = 2$, 6*RC*.

pulses and it occurs for a larger h-value. Also note that the distance value at h-values below 0.5 decreases with L. Fig. 8 also shows that longer pulses require longer observation intervals. For the binary 6*RC* system, minimum Euclidean distance values above 5 can be obtained when $h \approx 1.25$ and $N = 30$ observed symbol intervals. This corresponds to an asymptotic gain in SNR of at least 4 dB compared to MSK.

These trends are summarized in Fig. 9 using data in [17] and [21], where upper bounds for the *RC* family $3 \leqslant L \leqslant 6$ are shown. Except for a few weak h-values, the minimum distance equals $d_B{}^2(h)$ for sufficiently large N-values.

The binary system 8*RC* is second-order weak, i.e., the upper bound on the minimum Euclidean distance calculated by using the merge difference sequences can be further tightened by also taking into account the sequence giving the early merge at $t = 7T$. The two systems 7*RC* and 9*RC* are not weak according to the above definition, but since they are very close to the 8*RC* system they are "nearly weak." Thus, for 7*RC* and 9*RC* N must be chosen very large for all modulation indices to make the minimum Euclidean distance equal to the upper bound $d_B{}^2(h)$, even for very low h-values. For further details, see [17] and [21].

From a spectral point of view it is intuitively appealing to use strictly bandlimited frequency pulses $g(t)$. A frequently used pulse for digital AM-systems is the *SRC* pulse [4]. This pulse is of course of infinite duration.

The phase tree for the binary 3*SRC* system with the frequency pulse $g(t)$ given by (20) with $L = 3$ truncated symmetrically to a total length of 7 symbol intervals is very similar to the 3*RC* tree in Fig. 6. It is believed that an optimum detector for the 3*RC* system works well if the transmitted signal is 3*SRC*, but the distortion in the 3*SRC* phase tree will appear as a slightly increased noise level. This assumption must of course be properly shown. Fig. 10 shows the minimum normalized squared Euclidean distances for the binary 3*SRC* scheme. The pulse $g(t)$ has been truncated to a total length of $7T$. By comparing these minimum Euclidean distances to those for the binary 3*RC* system (see Fig. 7), it is indeed seen that they are very similar. The number of observed symbol intervals required to reach the upper bound $d_B{}^2(h)$ must be increased for the 3*SRC* system, however. This is due to the tails of the 3*SRC* pulse.

A bandwidth efficient digital CPM system was recently developed and given the name tamed frequency modulation [13], or TFM. This binary system uses the bandlimited frequency pulse given by (21). This system has modulation index 1/2, and the detector is constructed with the assumption that the transmitted signal is linearly modulated, and is therefore suboptimum. The detector is simple, however, and simulations have indicated that the performance in terms of symbol error probability for a given SNR is very close to the optimum detector [13]. This system will now be analyzed by means of the minimum Euclidean distance; it has been generalized to include all real valued modulation indices. The result can be seen

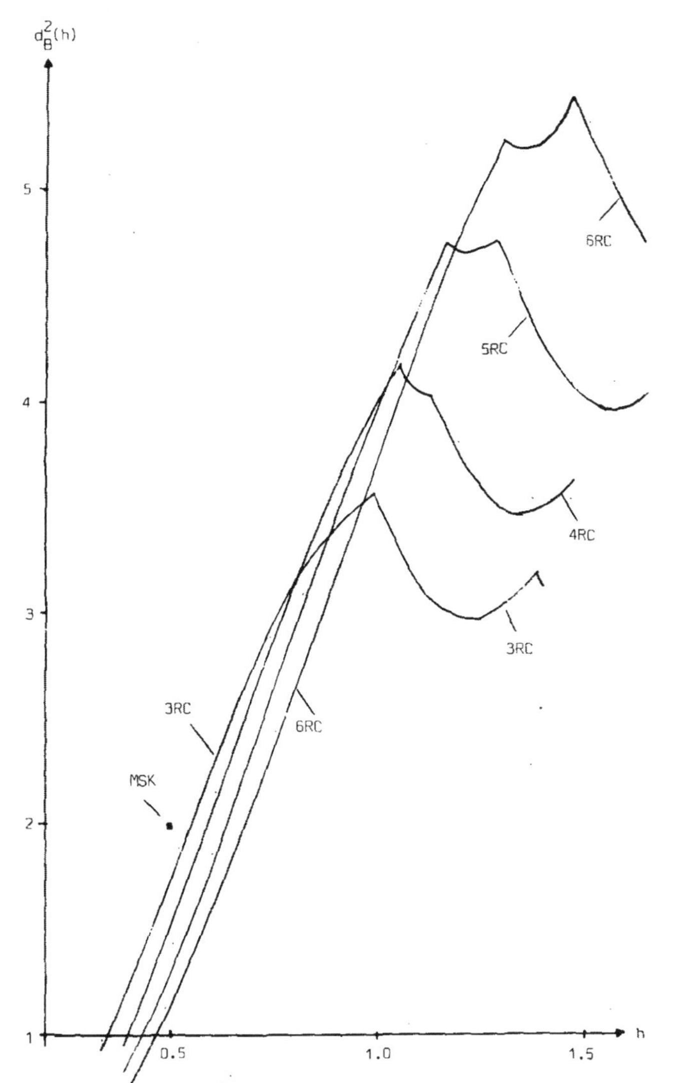

Fig. 9. Summary of upper bounds for $3RC$–$6RC$. MSK is shown for comparison.

in Fig. 11. The subpulse $g_0(t)$ given by (22) has been truncated to a total length of five symbol intervals.

The main-lobe of the frequency pulse $g(t)$ used for the TFM system is approximatively $3.7T$. For $h = 1/2$ the upper bound on the minimum Euclidean distance is reached with $N = 10$ observed symbol intervals and equals 1.58. The value of the upper bound for $3RC$ and $4RC$ when $h = 1/2$ is 1.76 and 1.51, respectively.

Quaternary and Octal Systems

In Part I it was found that multilevel systems ($M = 4, 8, \cdots$) yield larger minimum Euclidean distances than binary systems. The first quaternary partial response CPM system within the RC class is $2RC$. Now four phase branches leave every node in the phase tree, depending on the data symbols $\pm 1, \pm 3$. The phase tree for the binary $2RC$ system forms a subtree of the quaternary system. The first merges occur at $t = 3T$.

The result of the distance calculations, using the algorithm described in Section III, are shown in Fig. 12 for the $M = 4$, $3RC$ scheme. Distance calculations for various M-ary RC

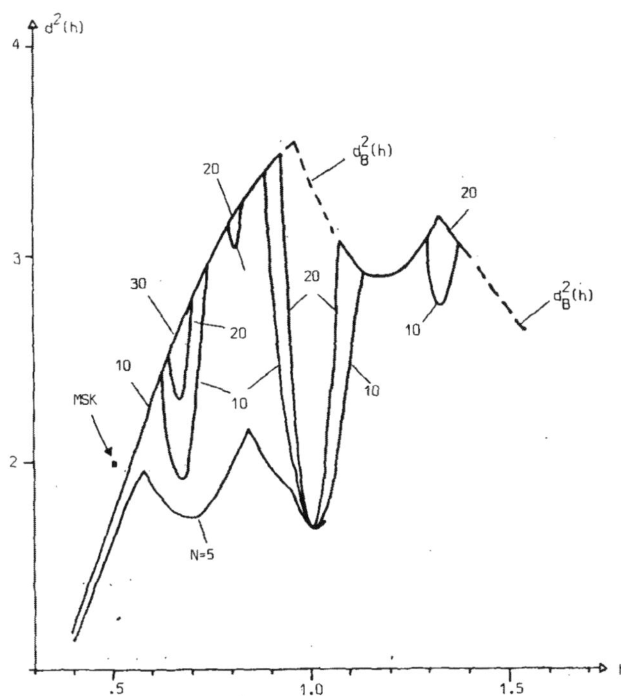

Fig. 10. Minimum squared Euclidean distances versus h for a binary modulation system with a spectral raised cosine pulse $3SRC$, truncated to a total length of $7T$.

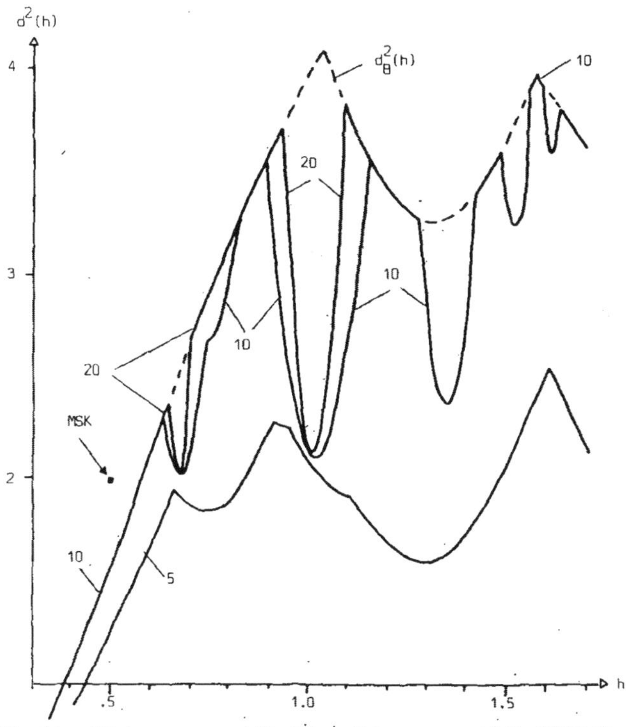

Fig. 11. Minimum squared Euclidean distances versus h for TFM. The subpulse $g_0(t)$ is truncated to a length of $5T$.

schemes are reported in [17] and [21]. The trends already observed for the binary system again appear as the length of the frequency pulse $g(t)$ is increased. The upper bound decreases for low modulation indices and increases for large modulation indices. Actually, very large values of the minimum

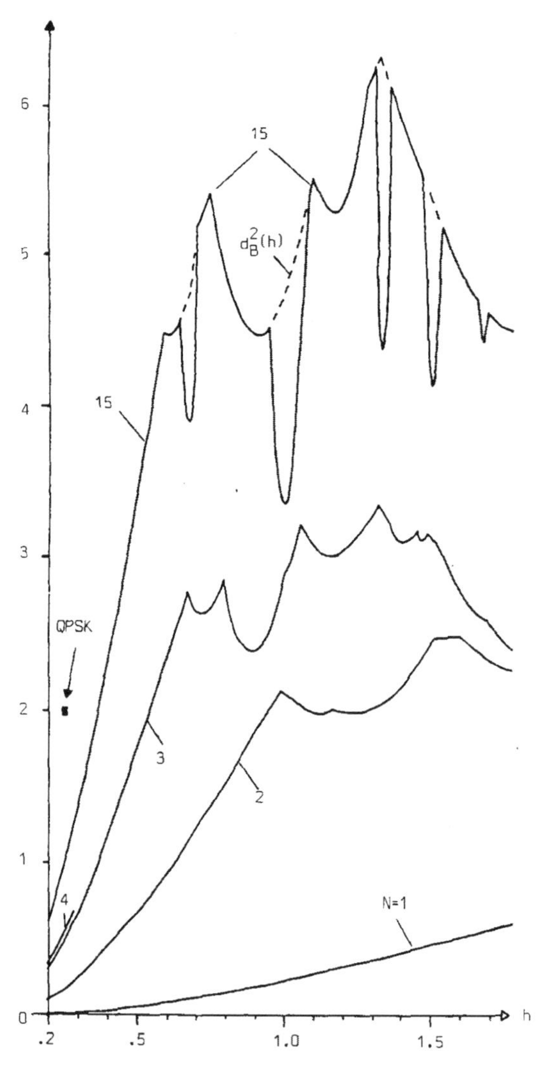

Fig. 12. Minimum squared normalized Euclidean distance versus h for $M = 4$, $3RC$. The upper bound is reached with $N = 12$ symbols in the interval $0.25 \leq h \leq 0.5$.

Euclidean distance are obtained in the region $h \approx 1.3$ and a receiver observation interval of $N = 15$ symbol intervals is enough to make the minimum Euclidean distance coincide with the upper bound. For these h-values the minimum Euclidean distance takes values around 6.28. This corresponds to an asymptotic gain of 5 dB compared to MSK or QPSK.

Minimum Euclidean distance calculations have also been performed for the octal ($M = 8$) and hexadecimal ($M = 16$) systems using the frequency pulses $2RC$ and $3RC$. The results can be found in [17] and [21]. The main conclusion concerning these results is that the trend for increasing L holds, i.e., lower values of $d_B^2(h)$ for low modulation indices, larger values of $d_B^2(h)$ for large modulation indices and an increased number of symbol intervals required in order to make the actual minimum Euclidean distance equal to the upper bound $d_B^2(h)$. The octal $3RC$ system yields a minimum Euclidean distance of 9.03 when $h = 1.38$ [21] with $N = 12$ symbol intervals required. The asymptotic gain in terms of E_b/N_0 compared to MSK or QPSK is 6.5 dB.

V. POWER SPECTRA

The power spectra (double-sided) can be obtained by numerical calculations, simulations in software, and hardware measurements. Numerical calculations have been carried out by using formulas in [25]. These computer calculations are very time consuming, especially for large M and L values and for large fT values. The bulk of the spectra in this paper have been obtained by means of simulations, a much faster method. For each pulse shape, M, h, and L value, the simulated spectra have been compared to numerically calculated spectra and a close fit was observed [26]. Comparisons have also been made to previously published numerically calculated spectra in [5], [9], and [14].

The simulated spectra have been calculated by the use of a well-tested simulation program [17]. In these simulations, which are made in discrete time, the complex envelope $e^{j\varphi(t,\alpha)}$ has been sampled four times per symbol interval. The data symbol sequence is a randomly generated 128 symbols long M-ary sequence, and the estimated power spectrum is the squared magnitude of the discrete Fourier transform (DFT) applied to the complex envelope $e^{j\varphi(t,\alpha)}$. These limitations introduce some distortion in the spectra and in some cases simulations have been performed using eight samples per symbol interval of the complex envelope. No significant change in the result has been observed.

The behavior of the power spectra for large frequencies is also of interest. It is not feasible to use the above-mentioned methods to obtain the spectral tail behavior due to numerical inaccuracy. In [5] the asymptotic behavior of partial response CPM systems is given. The number of continuous derivatives of $g(t)$ determines the fall-off rate for large frequencies [5], [26], as discussed below.

Fig. 13 shows the power spectra for the binary CPM systems $2RC$, $3RC$, and $4RC$ when $h = 1/2$. For comparison the spectrum for MSK is also shown. The frequency is normalized with the bit rate $1/T_b$, where $T = T_b \cdot \log_2 M$ and the spectra are shown in a logarithmic scale (dB). The spectrum for MSK falls off as $|f|^{-4}$ and for the RC-systems as $|f|^{-8}$ for large frequencies [5]. For full response CPM systems the improved fall-off rate has to be paid for by increased first side-lobes. It can be seen from Fig. 13, that by also increasing the length of the frequency pulse $g(t)$, the side-lobes can be made to decrease. The power spectra are thus compact and the first side-lobe very small with long pulses.

The effect of changing the modulation index h for the binary RC systems have been studied in detail in [17], [21], and [26]. The general trend is that the spectra become wider for increasing h-values, but normally maintain their smooth shape. As an example, see Fig. 14. For h-values equal to integer values, spectral lines occur [5], [25].

As can be expected, even more compact spectra can be obtained by using strictly bandlimited pulses $g(t)$ of the SRC-type. Fig. 15 shows the power spectra for the binary system $6SRC$, when the modulation index is $h = 0.4$, 0.5, 0.8, and 1.2. In this simulation the infinitely long frequency pulse given by (20) has not been significantly truncated [17]. It can be seen that these spectra are more compact than those for the

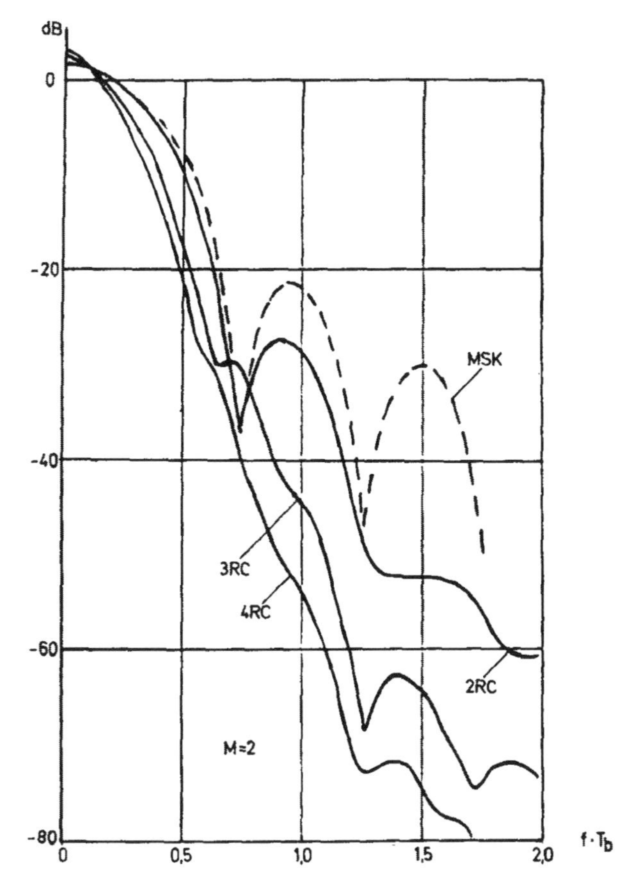

Fig. 13. Power spectra for binary CPM schemes with various baseband pulses. $h = 0.5$.

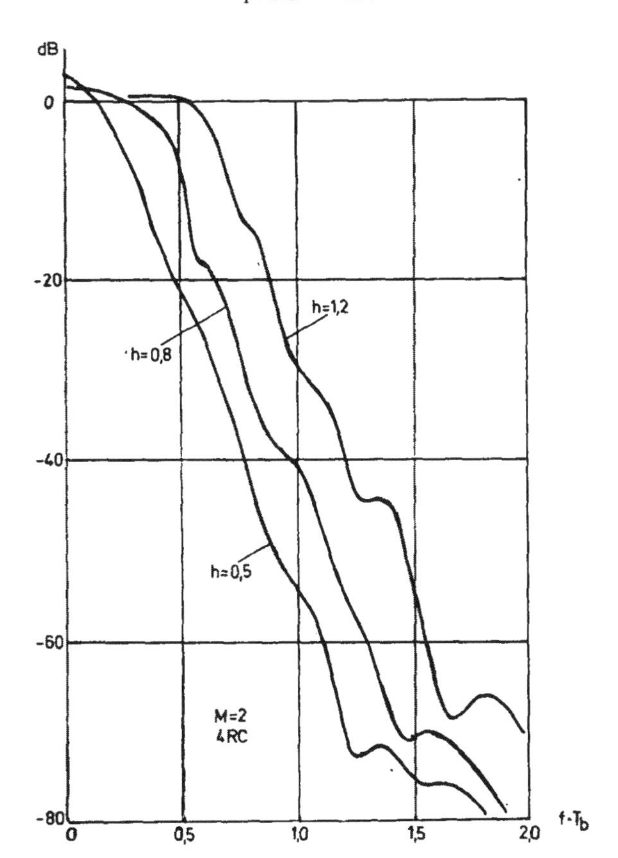

Fig. 14. Power spectra for $M = 2$, $4RC$. $h = 0.5$, 0.8, and 1.2.

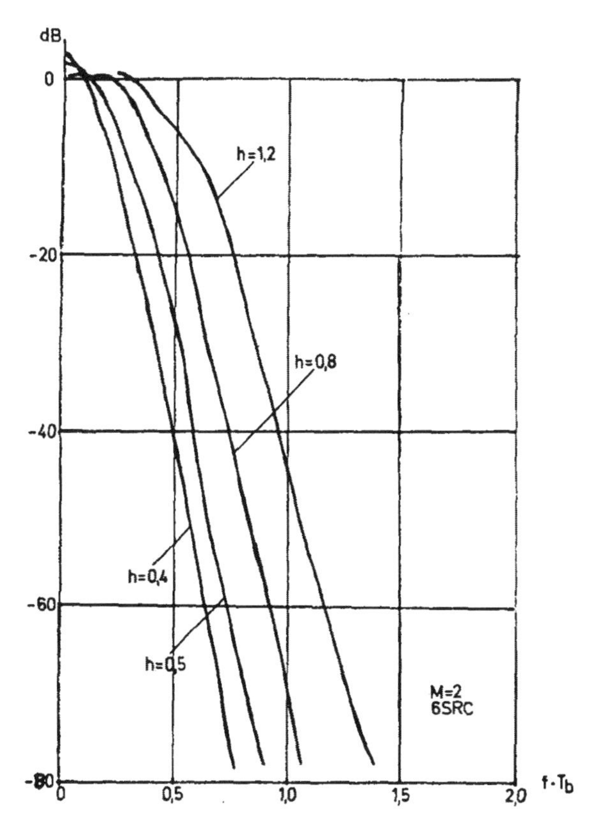

Fig. 15. Power spectra for $M = 2$, $6SRC$. No pulse truncation.

RC system for large frequencies. No side-lobes at all can be distinguished for the $6SRC$ spectra shown.

Bandwidth is not a quantity which is precisely defined for signals that are not strictly bandlimited. A common way of defining bandwidth for such signals is by means of the fractional out-of-band power (see Part I). Naturally, this definition of bandwidth can also be used for the CPM signals considered in this paper. However, since the power spectra are extremely compact and the out-of-band computation is not simple, bandwidth will instead be defined directly from the power spectra. Thus, the bandwidth $2BT_b$ is the bit rate normalized frequency for which the spectrum itself equals −60 dB when $f = B$. The level −60 dB has been chosen in order to illustrate constant envelope modulation schemes which have spectra sufficiently well-behaved to make bandlimiting radio-frequency filtering unnecessary.

Just like full response CPM systems multilevel partial response CPM systems yield more compact spectra than binary systems for fixed frequency pulse and distance. An example of this behavior is shown in Fig. 16, where the power spectra for the quaternary system ($M = 4, 3RC$) is given for $h = 0.25, 0.4$, 0.5, and 0.6. By comparing to the corresponding binary systems, it can be seen that the spectra have been compressed by using four levels instead of two [17], [21]. Note that the frequency is normalized with the bit rate $1/T_b$ and not the symbol rate $1/T$.

Power spectra for multilevel systems having strictly bandlimited frequency pulses $g(t)$ are shown in [17] and [21]. The systems there are quaternary $3SRC$ and $4SRC$, and octal $3SRC$, respectively. For each of these systems spectra are shown for

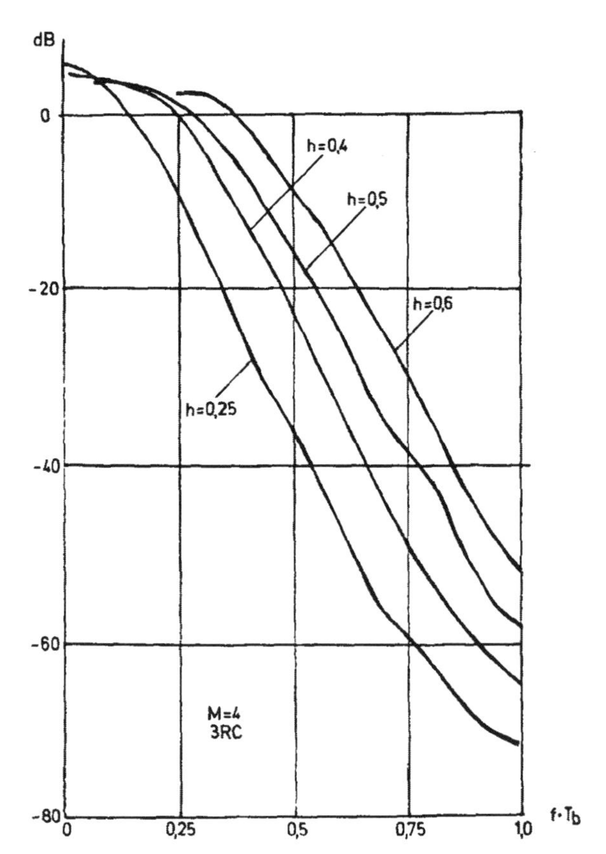

Fig. 16. Power spectra for $M = 4$, $3RC$. Note that the frequency is normalized with the bit rate $1/T_b$ in all spectrum figures.

various h-values. The same comparisons between the RC and the SRC schemes hold for the M-ary as for the binary case.

VI. TRANSMITTER AND RECEIVER STRUCTURES

It is assumed that the pulse $g(t)$ has finite length LT, i.e., $g(t) \equiv 0$ for $t < 0$ and $t > LT$. Since $g(t)$ is time limited, $q(t)$ is 0 for $t \leqslant 0$ and constant at $q(LT)$ for $t \geqslant LT$. For positive pulses $g(t)$, e.g., the raised cosine pulses (18), $q(LT) = 1/2$. Thus, the information carrying phase I-(5) can be written as

$$\begin{aligned}\varphi(t, \alpha) &= 2\pi h \sum_{i=-\infty}^{n} \alpha_i q(t - iT)\\ &= 2\pi h \sum_{i=n-L+1}^{n} \alpha_i q(t - iT) + h\pi \sum_{i=-\infty}^{n-L} \alpha_i,\\ &\qquad nT \leqslant t \leqslant (n+1)T. \end{aligned} \tag{23}$$

Hence, for given h and $g(t)$ and for any symbol interval n, the phase $\varphi(t, \alpha)$ is defined by α_n, the *correlative state vector* $(\alpha_{n-1}, \alpha_{n-2}, \cdots, \alpha_{n-L+1})$ and the *phase state* θ_n, where

$$\theta_n = h\pi \sum_{i=-\infty}^{n-L} \alpha_i \bmod 2\pi. \tag{24}$$

The number of correlative states is finite and equal to $M^{(L-1)}$. For rational modulation indices the phase tree is reduced to a phase trellis [10], [12], [22], [28]. For $h = 2k/p$ (k, p integers) there are p different phase states with values $0, 2\pi/p, 2\cdot 2\pi/p, \cdots, (p-1)2\pi/p$. The state is defined by the L-tuple $\sigma_n = (\theta_n, \alpha_{n-1}, \alpha_{n-2}, \cdots, \alpha_{n-L+1})$. The total number of states is $S = pM^{(L-1)}$. Transmitter and receiver structures utilizing these properties are described below.

Fig. 6 shows the phase tree for a binary system with a raised cosine pulse of length $L = 3$ ($3RC$) for $h = 4/5$. Phase states and correlative states are assigned to the nodes in the phase tree. The root node is arbitrarily given phase state 0. Each node in the tree is labeled with the state $(\theta_n, \alpha_{n-1}, \alpha_{n-2})$. The state trellis diagram can be derived from Fig. 6.

The transmitted signal can always be written

$$s(t, \alpha) = \sqrt{\frac{2E}{T}}\,[I(t)\cos(2\pi f_0 t) - Q(t)\sin(2\pi f_0 t)] \tag{25}$$

where

$$\begin{cases} I(t) = \cos[\varphi(t, \alpha)]\\ Q(t) = \sin[\varphi(t, \alpha)]. \end{cases} \tag{26}$$

From (23) we have

$$\varphi(t, \alpha) = \theta(t, \alpha) + \theta_n; \qquad nT \leqslant t \leqslant (n+1)T \tag{27}$$

where

$$\theta(t, \alpha) = 2\pi h \sum_{i=n-L+1}^{n} \alpha_i q(t - iT). \tag{28}$$

Hence, for $nT \leqslant t \leqslant (n+1)T$

$$\begin{cases} I(t) = \cos[\theta(t, \alpha)]\cos\theta_n - \sin[\theta(t, \alpha)]\sin\theta_n\\ Q(t) = \cos[\theta(t, \alpha)]\sin\theta_n + \sin[\theta(t, \alpha)]\cos\theta_n. \end{cases} \tag{29}$$

The basic transmitter structure is given by (25) and (26). The I- and Q-generators can be implemented in different ways. Fig. 17 shows an example based on (27) and (28). By also using (29), the ROM (read only memory) size can be reduced by a factor of p, but adders and multipliers must be used as in Fig. 18. Alternative structures are presented in [24]. These transmitters work for all rational h-values and time limited pulses $g(t)$. Only the ROM contents change. An exact rational relationship between h and $1/T$ is obtained. Both ROM's in Fig. 18 (phase branch ROM's) are of size $N_s \cdot N_q \cdot M^L$ bits, where N_s is the number of samples per symbol interval and N_q is the number of bits per sample. The size of the $\cos(\theta_n)$ and $\sin(\theta_n)$ ROM's is $p \cdot N_q$. The phase states are accessed sequentially [24].

From a spectral point of view it is desirable to use strictly bandlimited pulses $g(t)$, e.g., the SRC pulse given by (20). This pulse must be truncated to some length L_T, and hence the phase branch ROM's become a factor $M^{(L_T - L)}$ larger, compared to an RC pulse of length L.

Receiver Structures

The receiver observes the signal $r(t) = s(t, \alpha) + n(t)$, where the noise $n(t)$ is Gaussian and white (see Part I). The MLSE

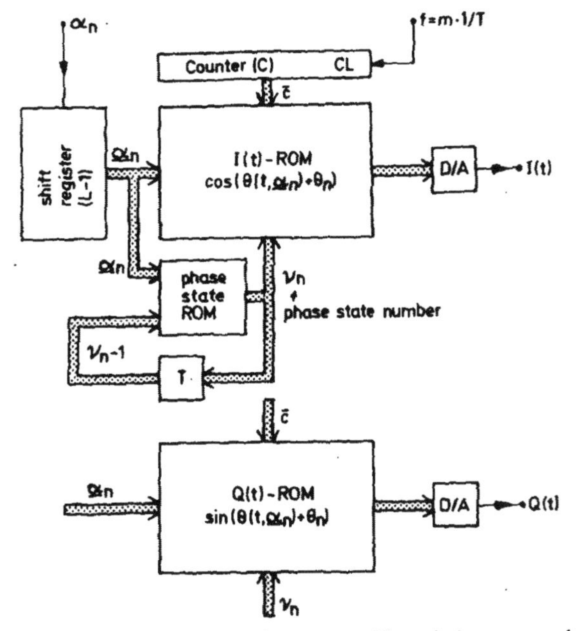

Fig. 17. *I*- and *Q*-generators. Characters with underbars appear boldface in text.

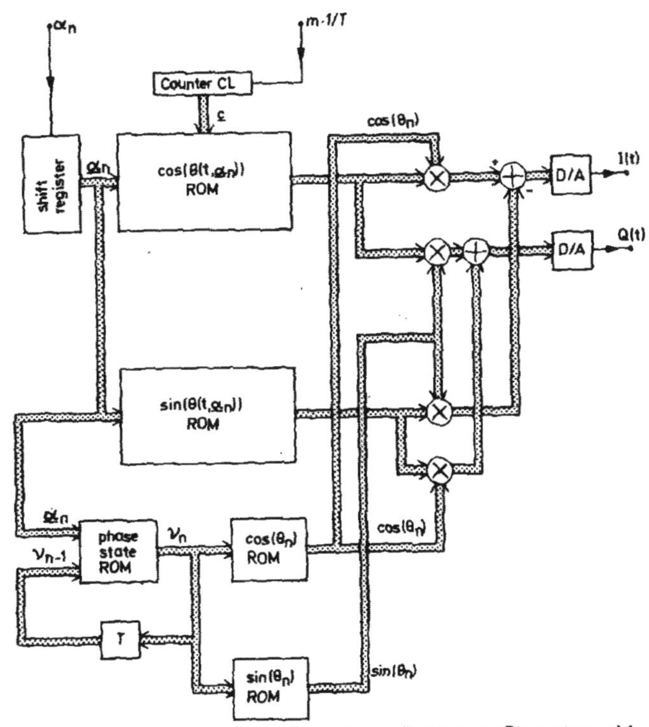

Fig. 18. *I*- and *Q*-generators with reduced ROM size. Characters with underbars appear boldface in text.

receiver maximizes the log likelihood function [2]

$$\log_e [p_{r(t)|\tilde{\alpha}}(r(t)\,|\,\tilde{\alpha})] \sim -\int_{-\infty}^{\infty} [r(t) - s(t, \tilde{\alpha})]^2\, dt \qquad (30)$$

with respect to the infinitely long estimated sequence $\tilde{\alpha}$. The maximizing sequence $\tilde{\alpha}$ is the maximum likelihood sequence estimate and $p_{r(t)|\tilde{\alpha}}$ is the probability density function for the observed signal $r(t)$ conditioned on the infinitely long sequence $\tilde{\alpha}$. It is equivalent to maximize the correlation

$$J(\tilde{\alpha}) = \int_{-\infty}^{\infty} r(t) \cdot s(t, \tilde{\alpha})\, dt. \qquad (31)$$

Now define

$$J_n(\tilde{\alpha}) = \int_{-\infty}^{(n+1)T} r(t) \cdot s(t, \tilde{\alpha})\, dt. \qquad (32)$$

Thus, it is possible to write

$$J_n(\tilde{\alpha}) = J_{n-1}(\tilde{\alpha}) + Z_n(\tilde{\alpha}) \qquad (33)$$

where

$$Z_n(\tilde{\alpha}) = \int_{nT}^{(n+1)T} r(t) \cdot \cos(\omega_0 t + \varphi(t, \tilde{\alpha}))\, dt. \qquad (34)$$

Using the above formulas it is possible to calculate the function $\boldsymbol{J}(\tilde{\alpha})$ recursively through (33) and the metric $\boldsymbol{Z}_n(\tilde{\alpha})$. This metric is recognized as a correlation between the received signal over the nth symbol interval and an estimated signal from the receiver.

The Viterbi algorithm [7] is a recursive procedure to choose those sequences that maximizes the log likelihood function up to the nth symbol interval. A trellis is used to choose among possible extensions of those sequences. The receiver computes $\boldsymbol{Z}_n(\tilde{\alpha}_n, \tilde{\theta}_n)$ for all M^L possible sequences $\tilde{\alpha}_n = \{\tilde{\alpha}_n, \tilde{\alpha}_{n-1}, \cdots, \tilde{\alpha}_{n-L+1}\}$ and all p possible $\tilde{\theta}_n$. This makes $p \cdot M^L$ different $\boldsymbol{Z}_n$ values. Rewriting (34) using (27) yields

$$Z_n(\tilde{\alpha}_n, \tilde{\theta}_n) = \int_{nT}^{(n+1)T} r(t) \cdot \cos(\omega_0 t + \theta(t, \tilde{\alpha}_n) + \tilde{\theta}_n)\, dt. \qquad (35)$$

It is seen that $\boldsymbol{Z}_n(\tilde{\alpha}_n, \tilde{\theta}_n)$ is obtained by feeding the signal $r(t)$ into a filter and sampling the output of the filter at $t = (n + 1)T$. In this case a bank of bandpass filters must be used.

The noise $n(t)$ is written

$$n(t) = x(t) \cdot \cos \omega_0 t - y(t) \cdot \sin \omega_0 t. \qquad (36)$$

Using the basic quadrature receiver the received quadrature components are

$$\begin{cases} \hat{I}(t) = \dfrac{1}{2}\left[\sqrt{\dfrac{2E}{T}}\, I(t) + x(t)\right] \\ \hat{Q}(t) = \dfrac{1}{2}\left[\sqrt{\dfrac{2E}{T}}\, Q(t) + y(t)\right]. \end{cases} \qquad (37)$$

By inserting these components in (35) and omitting double

frequency terms, we have

$$\begin{aligned} Z_n(\tilde{\alpha}_n, \tilde{\theta}_n) &= \cos(\tilde{\theta}_n) \int_{nT}^{(n+1)T} \hat{I}(t) \cdot \cos(\theta(t, \tilde{\alpha}_n))\, dt \\ &+ \cos(\tilde{\theta}_n) \int_{nT}^{(n+1)T} \hat{Q}(t) \cdot \sin(\theta(t, \tilde{\alpha}_n))\, dt \\ &+ \sin(\tilde{\theta}_n) \int_{nT}^{(n+1)T} \hat{Q}(t) \cdot \cos(\theta(t, \tilde{\alpha}_n))\, dt \\ &- \sin(\tilde{\theta}_n) \int_{nT}^{(n+1)T} \hat{I}(t) \cdot \sin(\theta(t, \tilde{\alpha}_n))\, dt. \end{aligned} \tag{38}$$

This can be interpreted as $4M^L$ baseband filters with the impulse response

$$h_c(t, \tilde{\alpha}_n) = \begin{cases} \cos\left[2\pi h \sum_{j=-L+1}^{0} \tilde{\alpha}_j q((1-j)T - t)\right] \\ 0 \quad \text{for } t \text{ outside } [0, T] \end{cases} \tag{39}$$

and

$$h_s(t, \tilde{\alpha}_n) = \begin{cases} \sin\left[2\pi h \sum_{j=-L+1}^{0} \tilde{\alpha}_j q((1-j)T - t)\right] \\ 0 \quad \text{for } t \text{ outside } [0, T]. \end{cases} \tag{40}$$

The number of filters required can be reduced by a factor of two by observing that every $\tilde{\alpha}$-sequence has a corresponding sequence with reversed sign. Fig. 19 shows an optimum receiver with $F = 2M^L$ matched filters. The outputs of these filters are sampled once every symbol interval. The metrics $Z_n(\tilde{\alpha}_n, \tilde{\theta}_n)$ are obtained by the use of (38) and used by the Viterbi algorithm. A delay of N_T symbol intervals is introduced. Alternative receiver structures are given in [11], [12] and [24].

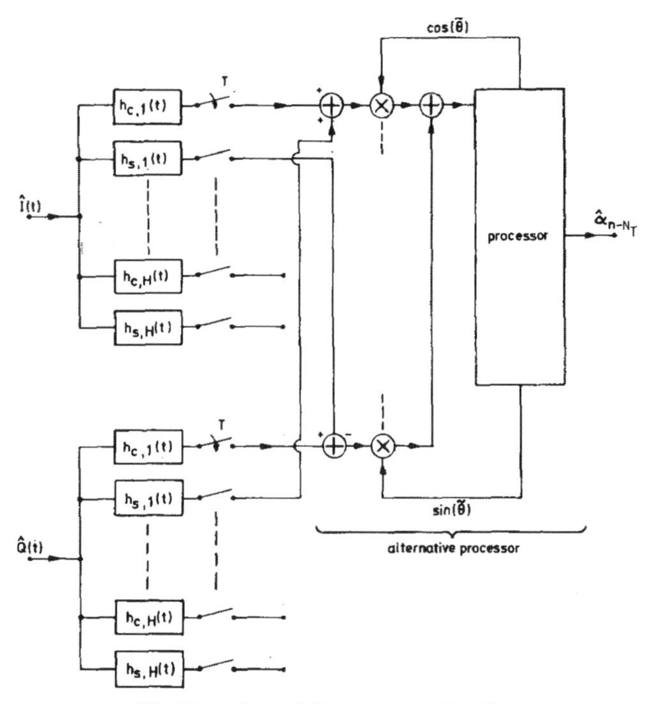

Fig. 19. Matched filter receiver. $H = F/4$.

A method has been developed for calculation of upper bounds on the symbol error probability for a Viterbi detector having a path memory of length N_T symbol intervals [21], [27]. Fig. 20 shows the result of this calculation for the binary $3RC$ scheme with $h = 4/5$ when $N_T = 1, 2, \cdots, 20$ and $N_T = \infty$. A lower bound is also shown. As a reference the bit error probability for QPSK (BPSK) is shown dashed. The asymptotic behavior cannot be improved by making $N_T > N_B$ [27], where N_B is the value of N where $d^2_{\min} = d_B{}^2$. Compare the improvements of the upper bounds with N_T in Fig. 20 to the growth of $d^2_{\min}$ with N for $H = 4/5$ in Fig. 7.

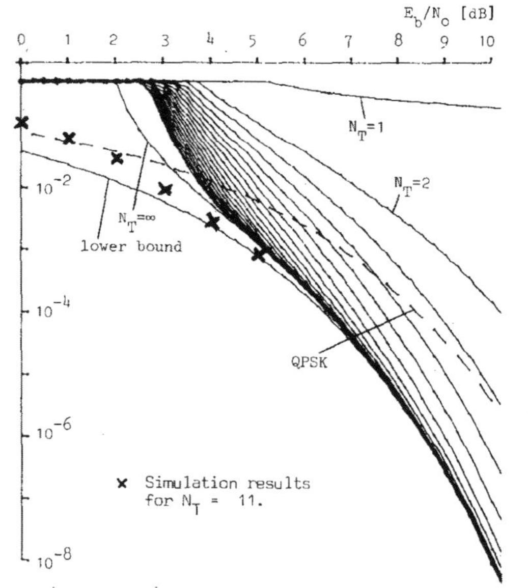

Fig. 20. Upper bounds on the bit error probability for $3RC$, $M = 2$, $h = 4/5$. A lower bound is also shown. Compare QPSK (dashed).

The lower bound shown in Fig. 20 is based on the minimum Euclidean distance and it is seen that if $N_T \gtrsim N_B$, the upper and lower bounds are close even for fairly large error probabilities. By further increasing N_T, improvement is achieved for low SNR. For every low SNR the upper bounds are loose. From the calculation of $d^2_{\min}$ yielding the value 3.17 (see Fig. 7), the asymptotic gain in E_b/N_0 is 2.00 dB compared to QPSK(MSK). From Fig. 20, it is seen that this gain is achieved even at fairly low SNR. This conclusion holds for a large variety of schemes and it can thus be concluded that the minimum Euclidean distance is sufficient for the characterization of performance in terms of symbol error probability [21], [27]. Even where the upper bound is loose the minimum distance appears to be accurate. For the scheme considered in Fig. 20 some simulation results are shown. In this simulation a coherent Viterbi detector with $N_T = 11$ was used. We have obtained results similar in appearance for a wide variety of modulation systems [21].

VII. DISCUSSION AND CONCLUSIONS

In this paper classes of M-ary partial response CPM systems have been analyzed, with respect to their minimum Euclidean distance, error probability, and spectral properties. Naturally it is desirable to have a system which yields both large values of the minimum Euclidean distance (small error probability) and a compact spectrum. To start with binary systems, the RC schemes give lower values of the upper bound on the minimum Euclidean distance for low h-values and larger for large h-values. To have an RC system with main-lobe width L bit intervals with the same performance as MSK for large SNR, the modulation index must be chosen larger and larger as L is increased. For the $3RC$ system h must be 0.56 roughly, and for the $4RC$ system, about .60. Although the modulation index is slightly increased, the spectra for the two partial response systems are far more attractive than that for MSK. This is true also for the asymptotic spectral behavior. At $f{\cdot}T_b = 1$ the spectrum for $3RC$ with $h = 0.56$ is 23 dB lower than the MSK spectrum and the corresponding figure for $4RC$ with $h = 0.6$ is 33 dB. For the $5RC$ system, the modulation index must be $h \approx .63$ to have the minimum Euclidean distance 2, i.e., the same as MSK. The spectral behavior of the system $5RC$ with $h \approx 0.63$ is far more compact than for the MSK system and at $f{\cdot}T_b = 1$ there is a 40 dB difference in favor for the RC system. This is a clear trend: Using RC systems with longer frequency pulses and with the modulation index chosen to yield the same performance in terms of error probability for large SNR as MSK, spectrally more efficient systems are obtained. The number of observed symbol intervals must grow however.

Binary RC systems can also be chosen which yield large gains in E_b/N_0. Take the $6RC$ with $h = 1.28$. When $N = 30$ observed bit intervals a gain of 4 dB in E_b/N_0 is obtained compared to MSK. When $|f{\cdot}T_b| \gtrsim 0.8$ the spectrum for $6RC$ with $h = 1.28$ is below the MSK spectrum, and the asymptotic properties of the $6RC$ spectrum are of course more attractive.

If a smaller minimum Euclidean distance than two can be accepted, which is sometimes the case, still narrower band binary RC schemes can be considered. For example, the binary RC systems with minimum Euclidean distance 1/2 have a loss of 6 dB in E_b/N_0. Systems having this performance are $2RC$, $h \approx 0.21$; $3RC$, $h \approx 0.25$; $6RC$, $h \approx 0.34$. These systems are very narrow band.

The favorable minimum Euclidean distance versus bandwidth tradeoff pointed out for a few cases of binary partial response CPM is even more pronounced for systems using more levels than two. Quaternary RC systems having the same minimum Euclidean distance as MSK are $2RC$, $h \approx 0.32$, and $3RC$, $h \approx 0.37$. From [21], the quaternary systems give more compact spectra for the same distance. The octal system $2RC$, $h \approx 0.25$ and the hexadecimal system $2RC$, $h \approx 0.21$ yield the same minimum Euclidean distance as MSK. The modulation indices are very low and thus good spectral properties can be expected.

Large gains in terms of E_b/N_0 can be achieved especially for multilevel systems, and as an example the system $M = 8$, $3RC$, $h \approx 1.38$ is chosen. With $N = 12$ observed symbol intervals the minimum normalized square Euclidean distance equals 9.03, and thus the gain in E_b/N_0 for large SNR is 6.5 dB.

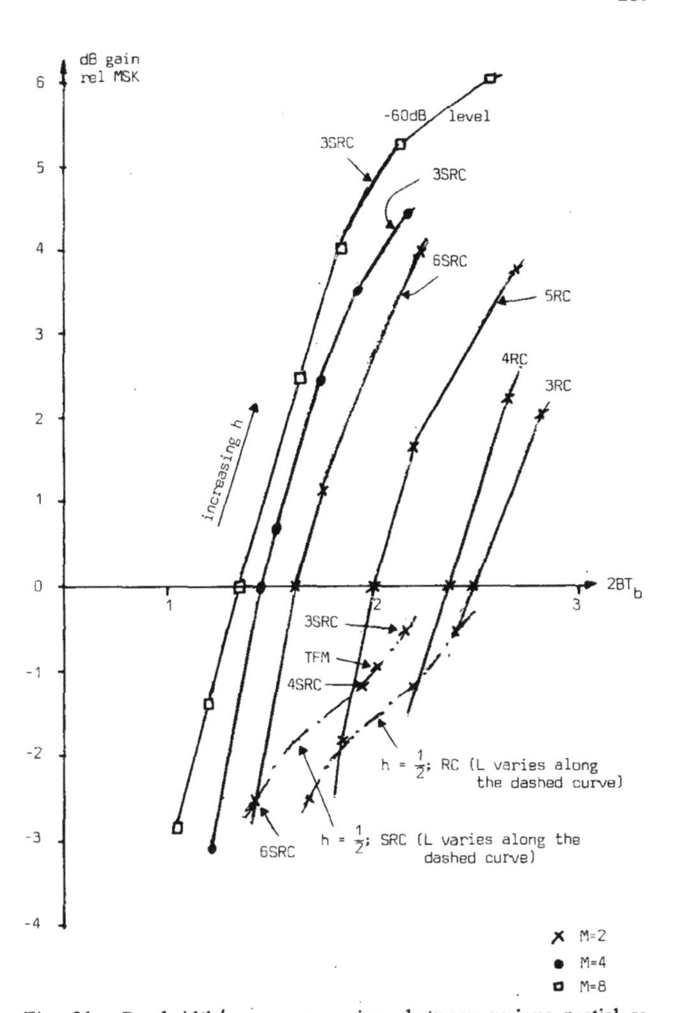

Fig. 21. Bandwidth/power comparison between various partial response CPM systems.

A summary of results concerning the bandwidth–minimum Euclidean distance tradeoff is given in Fig. 21. The bandwidth has been calculated using the definition in Section V, i.e., the double-sided bit rate normalized frequency where the double-sided power spectrum takes the value –60 dB. This figure shows the asymptotic gain in E_b/N_0 compared to MSK (and QPSK) versus the corresponding bandwidth for a number of M-ary partial response CPM schemes. For example, for the scheme $M = 8$, $3SRC$ a number of bandwidth and distance values are shown as □ in the figure. To underline the fact that only h varies, but that the pulse shape and the number of levels are the same, these points are connected. The interaction between distance and bandwidth for a fixed scheme when h varies is clearly demonstrated. The figure also shows the clear improvement which is obtained by increasing L and/or M. Another trend shown in Fig. 21 is the connection between binary RC and SRC schemes for $h = 1/2$ and increasing L. Both bandwidth and distance change with L for fixed h. Plots like figure 21 can be drawn for alternative definitions of bandwidth, depending on the application. The plots will be different, but the relative positions of the different schemes will remain unchanged.

The main conclusion concerning the systems considered in this paper is that digital constant envelope modulation systems can be found which are both power and bandwidth efficient. Nonbinary (e.g., $M = 4$) systems are especially attractive. A specific system is achieved by specifying the number of levels M, the modulation index h, and the frequency pulse $g(t)$. The modulation index should be rational for implementation reasons. The price that has to be paid for systems with optimum receivers is complexity.

REFERENCES

[1] A. Lender, "The duobinary technique for high speed data transmission," *IEEE Trans. Commun. Electron.*, vol. COM-11, pp. 214–218, May 1963.

[2] J. M. Wozencraft and I. M. Jacobs, *Principles of Communication Engineering*. New York: Wiley, 1965.

[3] A. Lender, "Correlative level coding for binary data transmission," *IEEE Spectrum*, vol. 3, pp. 104–115, Feb. 1966.

[4] R. W. Lucky, J. Salz, and E. J. Weldon, *Principles of Data Communication*. New York: McGraw-Hill, 1968.

[5] T. J. Baker, "Asymptotic behavior of digital FM spectra," *IEEE Trans. Commun.*, vol. COM-22, pp. 1585–1594, Oct. 1974.

[6] W. C. Lindsey and M. K. Simon, *Telecommunication Systems Engineering*. Englewood Cliffs, NJ: Prentice-Hall, 1974.

[7] G. D. Forney, "The Viterbi algorithm," *Proc. IEEE*, vol. 61, pp. 268–278, Mar. 1973.

[8] P. Kabal and S. Pasupathy, "Partial response signaling," *IEEE Trans. Commun.*, vol. COM-23, pp. 921–934, Sept. 1975.

[9] G. J. Garrison, "A power spectral density analysis for digital FM," *IEEE Trans. Commun.*, vol. COM-23, pp. 1228–1243, Nov. 1975.

[10] J. B. Anderson and R. de Buda, "Better phase-modulation error performance using trellis phase codes," *Electron. Lett.*, vol. 12, pp. 587–588, Oct. 1976.

[11] R. C. Davis, "An experimental 4-ary CPFSK modem for line-of-sight microwave digital data transmission," in *EASCON Conf. Rec.*, Washington, DC, 1978, pp. 674–682.

[12] T. A. Schonhoff, H. E. Nichols, and H. M. Gibbons, "Use of the MLSE algorithm to demodulate CPFSK," in *Proc. Int. Conf. Commun.*, Toronto, Canada, 1978, pp. 25.4.1–25.4.5.

[13] F. deJager and C. B. Dekker, "Tamed frequency modulation, a novel method to achieve spectrum economy in digital transmission," *IEEE Trans. Commun.*, vol. COM-26, pp. 534–542, May 1978.

[14] G. S. Deshpande and P. H. Wittke, "The spectrum of correlative encoded FSK," in *Proc. Int. Conf. Commun.*, Toronto, Canada 1978, pp. 25.3.1–25.3.5.

[15] T. Aulin, N. Rydbeck, and C-E. Sundberg, "Bandwidth efficient digital FM with coherent phase tree demodulation," Telecommun. Theory, Univ. of Lund, Lund, Sweden, Tech. Rep. TR-102, May 1978.

[16] ——, "Bandwidth efficient constant-envelope digital signalling with phase-tree demodulation," *Electron. Lett.*, vol. 14, pp. 487–489, July 1978.

[17] ——, "Further results on digital FM with coherent phase tree demodulation–minimum distance and spectrum," Telecommun. Theory, Univ. of Lund, Lund, Sweden, Tech. Rep. TR-119, Nov. 1978.

[18] T. Aulin and C-E. Sundberg, "Minimum distance properties of M-ary correlative encoded CPFSK," Telecommun. Theory, Univ. of Lund, Lund, Sweden, Tech. Rep. TR-120, Nov. 1978.

[19] N. Rydbeck and C-E. Sundberg, "Recent results on spectrally efficient constant envelope digital modulation methods," in *Proc. IEEE Int. Conf. Commun.*, Boston, MA, 1979, pp. 42.1.1–42.1.6.

[20] T. Aulin, N. Rydbeck, and C-E. Sundberg, "Bandwidth efficient digital FM with coherent phase tree demodulation," in *Proc. IEEE Int. Conf. Commun.*, Boston, MA, 1979, pp. 42.4.1–42.4.6.

[21] T. Aulin, "CPM—A power and bandwidth efficient digital constant envelope modulation scheme," Ph.D. dissertation, Telecommun. Theory, Univ. of Lund, Lund, Sweden, Nov. 1979.

[22] T. Aulin, N. Rydbeck, and C-E. Sundberg, "Performance of constant envelope M-ary digital FM-systems and their implementation," in *Proc. Nat. Telecommun. Conf.*, Washington, DC, 1979, pp. 55.1.1–55.1.6.

[23] T. Aulin and C-E. Sundberg, "Continuous phase modulation—Part I: Full response signaling," this issue, pp. 196–209.

[24] T. Aulin, N. Rydbeck, and C-E. Sundberg, "Transmitter and receiver structures for M-ary partial response FM. Synchronization considerations," Telecommun. Theory, Univ. of Lund, Lund, Sweden, Tech. Rep. TR-121, June 1979.

[25] R. R. Anderson and J. Salz, "Spectra of digital FM," *Bell Syst. Tech. J.*, vol. 44, pp. 1165–1189, July–Aug. 1965.

[26] T. Aulin and C-E. Sundberg, "Digital FM spectra–Numerical calculations and asymptotic behaviour," Telecommun. Theory, Univ. of Lund, Lund, Sweden, Tech. Rep. TR-141, May 1980.

[27] T. Aulin, "Symbol error probability bounds for coherently Viterbi detected digital FM," Telecommun. Theory, Univ. of Lund, Lund, Sweden, Tech. Rep. TR-131, Oct. 1979.

[28] T. Aulin, N. Rydbeck, and C-E. Sundberg, "Transmitter and receiver structures for M-ary partial response FM," in *Proc. 1980 Int. Zürich Seminar on Digital Commun.*, Mar. 4–6, 1980, pp. A2.1–A2.6.

Carrier and Bit Synchronization in Data Communication—A Tutorial Review

L. E. FRANKS, FELLOW, IEEE

Abstract—This paper examines the problems of carrier phase estimation and symbol timing estimation for carrier-type synchronous digital data signals, with tutorial objectives foremost. Carrier phase recovery for suppressed-carrier versions of double sideband (DSB), vestigial sideband (VSB), and quadrature amplitude modulation (QAM) signal formats is considered first. Then the problem of symbol timing recovery for a baseband pulse-amplitude modulation (PAM) signal is examined. Timing recovery circuits based on elementary statistical properties are discussed as well as timing recovery based on maximum-likelihood estimation theory. A relatively simple approach to evaluation of timing recovery circuit performance in terms of rms jitter of the timing parameters is presented.

I. INTRODUCTION

IN digital data communication there is a hierarchy of synchronization problems to be considered. First, assuming that a carrier-type system is involved, there is the problem of *carrier synchronization* which concerns the generation of a reference carrier with a phase closely matching that of the data signal. This reference carrier is used at the data receiver to perform a coherent demodulation operation, creating a baseband data signal. Next comes the problem of synchronizing a receiver clock with the baseband data-symbol sequence. This is commonly called *bit synchronization*, even when the symbol alphabet happens not to be binary.

Depending on the type of system under consideration, problems of *word-*, *frame-*, and *packet-synchronization* will be encountered further down the hierarchy. A feature that distinguishes the latter problems from those of carrier and bit synchronization is that they are usually solved by means of special design of the message format, involving the repetitive insertion of bits or words into the data sequence solely for synchronization purposes. On the other hand, it is desirable that carrier and bit synchronization be effected without multiplexing special timing signals onto the data signal, which would use up a portion of the available channel capacity. Only timing recovery problems of this type are discussed in this paper. This excludes those systems wherein the transmitted signal contains an unmodulated component of sinusoidal carrier (such as with "on-off" keying). When an unmodulated component or pilot is present, the standard approach to carrier synchronization is to use a phase-locked loop (PLL) which locks onto the carrier component, and has a narrow enough loop bandwidth so as not to be excessively perturbed by the sideband components of the signal. There is a vast literature on the performance and design of the PLL and there are several textbooks dealing with synchronous communication systems which treat the PLL in great detail [1]-[5]. Although we consider only suppressed-carrier signal formats here, the PLL material is still relevant since these devices are often used as component parts of the overall phase recovery system.

For modulation formats which exhibit a high bandwidth efficiency, i.e., which have a large "bits per cycle" figure of merit, we find the accuracy requirements on carrier and bit synchronization increasingly severe. Unfortunately, it is also in these high-efficiency systems that we find it most difficult to extract accurate carrier phase and symbol timing information by means of simple operations performed on the received signal. The pressure to develop higher efficiency data transmission has led to a dramatically increased interest in timing recovery problems and, in particular, in the ultimate performance that can be achieved with optimal recovery schemes.

We begin our review of carrier synchronization problems with a brief discussion of the major types of modulation format. In each case (DSB, VSB, or QAM), we assume coherent demodulation whereby the received signal is multiplied by a locally generated reference carrier and the product is passed through a low-pass filter. We can get some idea of the phase accuracy, or degree of coherency, requirements for the various modulation formats by examining the expressions for the coherent detector output, assuming a noise-free input. Let us assume that the message signal, say, $a(t)$, is incorporated by the modulation scheme into the complex envelope $\beta(t)$ of the carrier signal.[1]

$$y(t) = \text{Re}\,[\beta(t) \exp(j\theta) \exp(j2\pi f_0 t)] \tag{1}$$

and the reference carrier $r(t)$ is characterized by a constant complex envelope

$$r(t) = \text{Re}\,[\exp(j\hat{\theta}) \exp(j2\pi f_0 t)]. \tag{2}$$

From (A-8), the output of the coherent detector is

$$z_1(t) = \tfrac{1}{2}\,\text{Re}\,[\beta(t) \exp(j\theta - j\hat{\theta})]. \tag{3}$$

For the case of DSB modulation, we have $\beta(t) = a(t) + j\theta$, so $z_1(t)$ is simply proportional to $a(t)$. The phase error $\theta - \hat{\theta}$ in the reference carrier has only a second-order effect

Manuscript received June 28, 1979; revised March 26, 1980.
The author is with the Department of Electrical and Computer Engineering, University of Massachusetts, Amherst, MA 01003.

[1] See the Appendix for definitions and basic relations concerning complex envelope representation of signals.

Reprinted from *IEEE Transactions on Communications*, vol. COM-28, no. 8, August 1980.

on detector performance. The only loss is that phase error causes a reduction, proportional to $\cos^2 (\theta - \hat{\theta})$, in signal-to-noise ratio at the detector output when additive noise is present on the received signal.

For VSB modulation, however, phase error produces a more severe distortion. In this case $\beta(t) = a(t) + j\tilde{a}(t)$, where $\tilde{a}(t)$ is related to $a(t)$ by a time-invariant filtering operation which causes a cancellation of a major portion of one of the sidebands. In the limiting case of complete cancellation of a sideband (SSB), we have $\tilde{a}(t) = \hat{a}(t)$, the Hilbert transform of $a(t)$ [6]. The coherent detector output (3) for the VSB signal is

$$z_1(t) = \tfrac{1}{2}\, a(t) \cos (\theta - \hat{\theta}) - \tfrac{1}{2}\, \tilde{a}(t) \sin (\theta - \hat{\theta}) \tag{4}$$

and the second term in (4) introduces an interference called *quadrature distortion* when $\hat{\theta} \neq \theta$. As $\tilde{a}(t)$ has roughly the same power level as $a(t)$, a relatively small phase error must be maintained for low distortion, e.g., about 0.032 radian error for a 30 dB signal-to-distortion ratio.

In the QAM case, two superimposed DSB signals at the same carrier frequency are employed by making $\beta(t) = a(t) + jb(t)$, where $a(t)$ and $b(t)$ are two separate, possibly independent, message signals. A dual coherent detector, using a reference carrier and its $\pi/2$ phase-shifted version, separates the received signal into its in-phase (I) and quadrature (Q) components. Again considering only the noise-free case, these components are

$$\begin{aligned} c_I(t) &= \tfrac{1}{2}\, a(t) \cos (\theta - \hat{\theta}) - \tfrac{1}{2}\, b(t) \sin (\theta - \hat{\theta}) \\ c_Q(t) &= \tfrac{1}{2}\, b(t) \cos (\theta - \hat{\theta}) + \tfrac{1}{2}\, a(t) \sin (\theta - \hat{\theta}). \end{aligned} \tag{5}$$

From (5) it is clear that $\hat{\theta} \neq \theta$ introduces a *crosstalk* interference into the I and Q channels. As $a(t)$ and $b(t)$ can be expected to be at similar power levels, the phase accuracy requirements for QAM are high compared to straight DSB modulation.

From the previous discussion we see that the price for the approximate doubling of bandwidth efficiency in VSB or QAM, relative to DSB, is a greatly increased sensitivity to phase error. The problem is compounded by the fact that carrier phase recovery is much more difficult for VSB and QAM, compared to DSB.

II. CARRIER PHASE RECOVERY

Before examining specific carrier recovery circuits for the suppressed-carrier format, it is helpful to ask, "What properties must the carrier signal $y(t)$ possess in order that operations on $y(t)$ will produce a good estimate of the phase parameter θ?" A general answer to this question lies in the *cyclostationary* nature of the $y(t)$ process.[2] A cyclostationary process has statistical moments which are periodic in time, rather than constant as in the case of stationary processes [2], [6], [7]. To a large extent, synchronization capability can be characterized by the lowest-order moments of the process, such as the mean and autocorrelation. The $y(t)$ process is said to be cyclostationary in the wide sense if $E[y(t)]$ and $k_{yy}(t + \tau, t) = E[y(t + \tau)y(t)]$ are both periodic functions of t. A process modeled by (1) is typically cyclostationary with a period of $1/f_0$ or $1/2f_0$. The statistical moments of this process depend upon the value of the phase parameter θ and it is not surprising that efficient phase estimation procedures are similar to moment estimation procedures. It is important to note here that we are regarding θ as an unknown but nonrandom parameter. If instead we regarded θ as a random parameter uniformly distributed over a 2π interval, then the $y(t)$ process would typically be stationary, not cyclostationary.

A general property of cyclostationary processes is that there may be a correlation between components in different frequency bands, in contrast to the situation for stationary processes [8]. For carrier-type signals, the significance lies in the correlation between message components centered around the carrier frequency ($+f_0$) and the image components around ($-f_0$). This correlation is characterized by the cross-correlation function $k_{\beta\beta^*}(\tau) = E[\beta(t + \tau)\beta(t)]$ for a $y(t)$ process as in (1) when $\beta(t)$ is a stationary process.[3]

Considering first the DSB case with $\beta(t) = a(t) + j\theta$, and using (A-10) we have

$$\begin{aligned} k_{yy}(t + \tau, t) = {} & \tfrac{1}{2} \operatorname{Re} [k_{aa}(\tau) \exp (j2\pi f_0 \tau)] \\ & + \tfrac{1}{2} \operatorname{Re} [k_{aa}(\tau) \exp (j4\pi f_0 t + j2\pi f_0 \tau + j2\theta)] \end{aligned} \tag{6}$$

where the second term in (6) exhibits the periodicity in t that makes $y(t)$ a cyclostationary process.

We are assuming that $y(t)$ contains no periodic components. Consider what happens, however, when $y(t)$ is passed through a square-law device. We see immediately from (6) that the output of the squarer has a periodic mean value, since

$$\begin{aligned} E[y^2(t)] &= k_{yy}(t, t) \\ &= \tfrac{1}{2}\, k_{aa}(0) + \tfrac{1}{2}\, k_{aa}(0) \operatorname{Re} [\exp (j2\theta + j4\pi f_0 t)]. \end{aligned} \tag{7}$$

If the squarer output is passed through a bandpass filter with transfer function $H(f)$ as shown in Fig. 1, and if $H(f)$ has a unity-gain passband in the vicinity of $f = 2f_0$, then the mean value of the filter output is a sinusoid with frequency $2f_0$, phase 2θ, and amplitude $\tfrac{1}{2}E[a^2(t)]$. In this sense, the squarer has produced a periodic component from the $y(t)$ signal.

It is often stated that the effect of the squarer is to produce a discrete component (a line at $2f_0$) in the spectrum of its output signal. This statement lacks precision and can lead to serious misinterpretations because $y^2(t)$ is not a stationary process, so the usual spectral density concept has no meaning. A stationary process can be derived from $y^2(t)$ by phase randomizing [6], but then the relevance to carrier phase recovery is lost because the discrete component has a completely indeterminate phase.

[2] In [2], these processes are called *periodic nonstationary*.

[3] Despite its appearance, this is not an autocorrelation function, due to the definition of autocorrelation for complex processes; see (A-11).

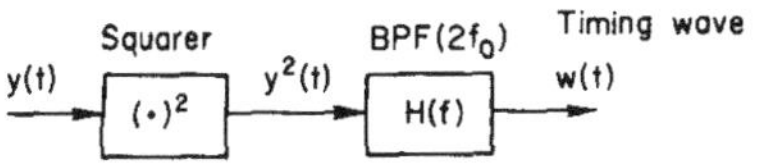

Fig. 1. Timing recovery circuit.

The output of the bandpass filter in Fig. 1 can be used directly to generate a reference carrier. Assuming that $H(f)$ completely suppresses the low-frequency terms [see (A-8)] the filter output is the *reference waveform*

$$w(t) = \tfrac{1}{2}\ \mathrm{Re}\,\{[\omega \otimes \beta^2](t) \exp(j2\theta) \exp(j4\pi f_0 t)\} \tag{8}$$

where the convolution product $[\omega \otimes \beta^2]$ represents the filtering action of $H(f)$ in terms of its low-pass equivalent $\Omega(f)$ in (A-5). For the DSB case, $\beta^2(t) = a^2(t)$ is real and $\omega(t)$ is real[4] if $H(f)$ has a symmetric response about $2f_0$. Then the phase of the reference waveform is 2θ and the amplitude of the reference waveform fluctuates slowly [depending on the bandwidth of $H(f)$]. The reference carrier can be obtained by passing $w(t)$ through an infinite-gain clipper which removes the amplitude fluctuations. The square wave from the clipper can drive a frequency divider circuit which halves the frequency and phase. Alternatively, the bandpass filter output can be tracked by a PLL and the PLL oscillator output passed through the frequency-divider circuit.

There is another tracking loop arrangement, called the Costas loop, where the voltage-controlled oscillator (VCO) operates directly at f_0. We digress momentarily to describe the Costas loop and to point out that it is equivalent to the squarer followed by a PLL [1]-[3]. The equivalence is established by noting that the inputs to the loop filters in the two configurations shown in Fig. 2 are identical. In the PLL quiescent lock condition, the VCO output is in quadrature with the input signal so we introduce a $\pi/2$ phase shift into the VCO in the configurations of Fig. 2. Then using (A-8) to get the output of the multiplier/low-pass filter combinations, we see that the input to the loop filter is

$$v(t) = \tfrac{1}{8}\ \mathrm{Re}\,[A^2\beta^2(t) \exp(j2\theta - j2\hat{\theta} - j\pi/2)] \tag{9}$$

in both configurations if the amplitude of the VCO output is taken as $\frac{1}{2}A^2$ in the squarer/PLL configuration, and taken as A in the Costas loop.

Going back to (8), we see that phase recovery is perfect if $[\omega \otimes \beta^2]$ is real. Assuming $\omega(t)$ real, a phase error will result only if a quadrature component [relative to $\beta^2(t)$] appears at the output of the squarer. This points out the error, from a different viewpoint, of using the phase randomized spectrum of the squarer output to analyze the phase recovery performance because the spectrum approach obliterates the distinction between I and Q components. For the DSB case, a quadrature component will appear at the squarer output only if there is a quadrature component of interference added to the input signal $y(t)$. We can demonstrate this effect by considering the

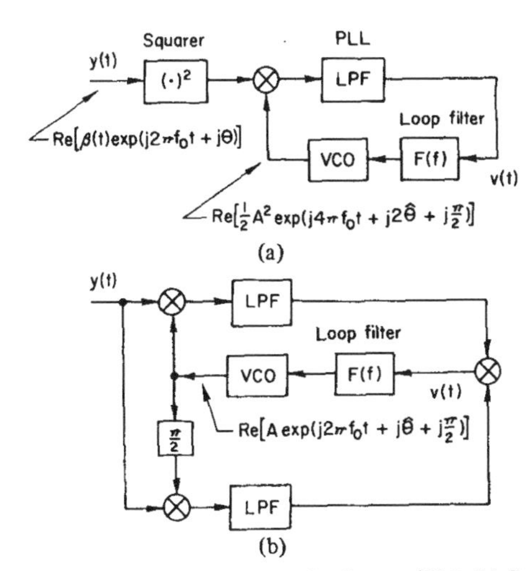

Fig. 2. Carrier phase tracking loops. (a) Squarer/PLL (b) Costas loop.

input signal to be $z(t) = y(t) + n(t)$ where $n(t)$ is white noise with a double-sided spectral density of N_0 W/Hz. We can represent $n(t)$ by the complex envelope, $[u_I(t) + ju_Q(t)] \exp(j\theta)$ where, from (A-15) the I and Q noise components relative to a phase θ are uncorrelated and have a spectral density of $2N_0$. The resulting phase of the reference waveform (8) is

$$2\hat{\theta} = 2\theta + \tan^{-1}\left[\frac{2\omega \otimes (\beta u_Q + u_I u_Q)}{\omega \otimes (\beta^2 + 2\beta u_I + u_I{}^2 - u_Q{}^2)}\right]. \tag{10}$$

We can approximate the phase error $\phi = \hat{\theta} - \theta$ (also called *phase jitter* because $\hat{\theta}$ is a quantity that fluctuates with time) by neglecting the noise $\times$ noise term in the numerator and both signal $\times$ noise and noise $\times$ noise terms in the denominator in (10). Furthermore, we replace $\omega \otimes \beta^2$ by its expected value (averaging over the message process) and use the $\tan^{-1} x \cong x$ approximation. With all these simplifications, which are valid at sufficiently high signal-to-noise ratio and with sufficiently narrow-band $H(f)$, it is easy to derive an expression for the variance of the phase jitter.

$$\mathrm{var}\ \phi = (2N_0 B)S^{-1} \tag{11a}$$

$$= \left(\frac{S}{N}\right)^{-1}\left(\frac{B}{W}\right) \tag{11b}$$

where

$$B \triangleq \int_{-\infty}^{\infty} |\Omega(f)|^2\, df = \int_0^{\infty} |H(f)|^2\, df$$

is the *noise bandwidth* of the bandpass filter, recalling that we have set $\Omega(0) = 1$. The message signal power is $S = E[a^2(t)]$ and for the second version of the jitter formula (11b) we have assumed a signal bandwidth of W Hz and have defined a noise power over this band of $N = 2N_0 W$. This allows the satisfying physical interpretation of jitter variance being inversely proportional to signal-to-noise ratio and directly proportional to the bandwidth ratio of the phase recovery circuit and the message signal. For the smaller signal-to-noise ratios, the accuracy and convenience of the expression can be maintained by incorporating a correction factor known as the *squaring loss* [3].

[4] A real $\omega(t)$ corresponds to the case where the cross-coupling paths between input and output I and Q components in Fig. 10 are absent. If the bandpass function $H(f)$ does not exhibit the symmetrical amplitude response and antisymmetrical phase response about $2f_0$ for a real $\omega(t)$, then there simply is a fixed phase offset introduced by the bandpass filter.

When the signal itself carries a significant quadrature component, as in the case of the VSB signal, there will be a quadrature component at the squarer output that interferes with the phase recovery operation even at high signal-to-noise ratios. Let us suppose that the VSB signal is obtained by filtering a DSB signal with a bandpass filter with a real transfer function (no phase shift) and with a cutoff in the vicinity of f_0. The resulting quadrature component for the VSB signal is $\tilde{a}(t) = [p_Q \otimes a](t)$ and $p_Q(t)$ is derived from the low-pass equivalent transfer function for the bandpass filter in accordance with (A-7). The real transfer function condition makes $p_Q(t)$ an odd function of time, which also makes the cross-correlation function for $a(t)$ and $\tilde{a}(t)$ an odd function.

The result is that, for $\beta(t) = a(t) + j\tilde{a}(t)$, the autocorrelation for the VSB signal is

$$k_{yy}(t+\tau, t) = \tfrac{1}{2} \operatorname{Re} [\{k_{aa}(\tau) + k_{\tilde{a}\tilde{a}}(\tau) + j2k_{\tilde{a}\tilde{a}}(\tau)\} \cdot \exp(j2\pi f_0 \tau)] + \tfrac{1}{2} \operatorname{Re} [\{k_{aa}(\tau) - k_{\tilde{a}\tilde{a}}(\tau)\} \cdot \exp(j4\pi f_0 t + j2\pi f_0 \tau + j2\theta)]. \quad (12)$$

Comparing (12) with (6), we see that the second, cyclostationary, term is much smaller for the VSB case than the DSB case since the autocorrelation functions for $a(t)$ and $\tilde{a}(t)$ differ only to the extent that some of the low-frequency components in $\tilde{a}(t)$ are missing because of the VSB rolloff characteristic. Although the jitter performance will be poorer, the phase recovery circuit in Fig. 1 can still be used since the mean value of the reference waveform is a sinusoid exhibiting the desired phase, but with an amplitude which is proportional to the difference in power levels in $a(t)$ and $\tilde{a}(t)$.

$$E[w(t)] = \tfrac{1}{2} [k_{aa}(0) - k_{\tilde{a}\tilde{a}}(0)] \operatorname{Re} [\exp(j4\pi f_0 t + j2\theta)]. \quad (13)$$

However, it is not possible to get a very simple formula for the variance of phase jitter, as in (11), because the power spectral density of the quadrature component of $\beta^2(t)$, which is proportional to $a(t)\,\tilde{a}(t)$, vanishes at $f = 0$, unlike in the additive noise case. An accurate variance expression must take into account the particular shape of the $\Omega(f)$ filtering function as well as the shape of the VSB rolloff characteristic.

Our examination of phase recovery for DSB (with additive noise) and VSB modulation formats has indicated that rms phase jitter can be made as small as desired by making the width of $\Omega(f)$ sufficiently small. The corresponding parameter in the case of the tracking loop configuration is called the loop bandwidth [3]. These results, however, are for *steady-state* phase jitter since the signals at the receiver input were presumed to extend into the remote past. The difficulty with a very narrow phase recovery bandwidth is that excessive time is taken to get to the steady-state condition when a new signal process begins. This time interval is referred to as the *acquisition time* of the recovery circuit and in switched communication networks or polling systems it is usually very important to keep this interval small, even at the expense of the larger steady-state phase jitter. One way to accommodate the conflicting objectives in designing a carrier recovery circuit is to specify a minimum phase-recovery bandwidth and then adjust other parameters of the system to minimize the steady-state phase jitter.

Another problem with a very narrow-band bandpass filter is in the inherent mistuning sensitivity, where mistuning is a result of inaccuracies in filter element values or a result of small inaccuracies or drift in the carrier frequency. This problem is avoided with tracking loop configurations since they lock onto the carrier frequency. One the other hand, tracking loops have some problems also, one of the more serious being the "hang-up" problem [9] whereby the nonlinear nature of the loop can produce some greatly prolonged acquisition times.

Although we have modeled the phase recovery problem in terms of a constant unknown carrier phase, it may be important in some situations to consider the presence of fairly rapid fluctuations in carrier phase (independent of the message process). Such fluctuations are often called *phase noise* and if the spectral density of these fluctuations has a greater bandwidth than that of the phase recovery circuits, there is a phase error due to the inability to track the carrier phase. Phase error of this type, even in steady state, becomes *larger* as the bandwidth of the recovery circuits decreases.

Another practical consideration is a π-radian phase ambiguity in the phase recovery circuits we have been discussing. The result is a polarity ambiguity in the coherently demodulated signal. In many cases this polarity ambiguity is unimportant, but otherwise some *a priori* knowledge about the message signal will have to be used to resolve the ambiguity.

For a QAM signal with $\beta(t) = a(t) + jb(t)$, where $a(t)$ and $b(t)$ are independent zero-mean stationary processes, we get

$$k_{yy}(t+\tau, t) = \tfrac{1}{2} \operatorname{Re} [\{k_{aa}(\tau) + k_{bb}(\tau)\} \exp(j2\pi f_0 \tau)] + \tfrac{1}{2} \operatorname{Re} [\{k_{aa}(\tau) - k_{bb}(\tau)\} \cdot \exp(j4\pi f_0 t + j2\pi f_0 \tau + j2\theta)] \quad (14)$$

and the situation is very similar to the VSB case (12). In this case where $a(t)$ and $b(t)$ are uncorrelated, the mean reference waveform has the correct phase, but the amplitude vanishes if the power levels in the I and Q channels are the same.

$$E[w(t)] = \tfrac{1}{2} [k_{aa}(0) - k_{bb}(0)] \operatorname{Re} [\exp(j4\pi f_0 t + j2\theta)]. \quad (15)$$

Hence, unless the QAM format is intentionally unbalanced, the squaring approach in Fig. 1 does not work. We briefly examine what happens when the squarer is replaced by a fourth-power device in the recovery schemes we have been considering. From (1), we can obtain

$$y^4(t) = \tfrac{1}{8} \operatorname{Re} [\beta^4(t) \exp(j8\pi f_0 t + j4\theta)] + \tfrac{1}{2} \operatorname{Re} [\,|\beta(t)|^2 \beta^2(t) \exp(j4\pi f_0 t + j2\theta)] + \tfrac{3}{8} |\beta(t)|^4. \quad (16)$$

Now if we use a bandpass filter tuned to $4f_0$ which passes only the first term in (16), then the mean reference waveform at the filter output is

$$E[w(t)] = \frac{1}{4} \operatorname{Re} [\{\overline{a^4} - 3(\overline{a^2})^2\} \exp (j8\pi f_0 t + j4\theta)] \quad (17)$$

still assuming independent $a(t)$ and $b(t)$ and a balanced QAM format, i.e., $k_{aa}(0) = k_{bb}(0) = \overline{a^2}$. Hence, a mean reference waveform exists even in the balanced QAM case if a fourth-power device is used.[5]

One very popular QAM format is quadriphase-shift keying (QPSK) where the standard carrier recovery technique is to use a fourth-power device followed by a PLL or to use an equivalent "double" Costas loop configuration [3]. The QPSK format, with independent data symbols, can be regarded as two independent binary phase-shift-keyed (BPSK) signals in phase quadrature. In a nonbandlimited situation each BPSK signal can be regarded as DSB-AM where the message waveform has a rectangular shape characterized by $a(t) = \pm 1$. In this case, the complex envelope of the QPSK signal is characterized by $\beta(t) = (\pm 1 \pm j)/\sqrt{2}$ or $\beta(t) = \exp\ (j(\pi/4) + j(\pi/2)k)$ with $k = 0, 1, 2,$ or 3. The result is that $\beta^4(t) = -1$ and the $4f_0$ component in (16) is a pure sinusoid with no fluctuations in either phase or amplitude. For PSK systems with a larger alphabet of phase positions, the result of (17) cannot generally be used as the I and Q components are no longer independent. Analysis of the larger alphabet cases shows that higher-order nonlinearities are required for successful phase recovery [3], [10]. For any balanced QAM format, such as QPSK, the phase recovery circuits discussed here give a $\pi/2$-radian phase ambiguity. This problem is often handled by use of a differential PSK scheme, whereby the information is transmitted as a sequence of phase changes rather than absolute values of phase.

III. PAM TIMING RECOVERY

The receiver synchronization problem in baseband PAM transmission is to find the correct sampling instants for extracting a sequence of numerical values from the received signal. For a synchronous pulse sequence with a pulse rate of $1/T$, the sampler operates synchronously at the same rate and the problem is to determine the correct sampling phase within a T-second interval. The model for the baseband PAM signal is

$$x(t) = \sum_{k=-\infty}^{\infty} a_k g(t - kT - \tau) \quad (18)$$

where $\{a_k\}$ is the message sequence and $g(t)$ is the signaling pulse. We want to make an accurate determination of τ, from operations performed on $x(t)$. We assume that $g(t)$ is so defined that the best sampling instants are at $t = kT + \tau$; $k = 0, \pm 1, \pm 2, \cdots$. The objective is to recover a close replica of the message sequence $\{a_k\}$ in terms of the sequence $\{\hat{a}_k = x(kT + \hat{\tau})\}$, assuming a normalization of $g(0) = 1$. In the noise-free case, the difference between a_k and $\hat{a}_k$ is due to intersymbol interference which can be minimized by proper shaping of the data pulse $g(t)$. With perfect timing ($\hat{\tau} = \tau$), the intersymbol interference is

$$\hat{a}_k - a_k = \sum_{n \neq k} a_n g(kT - nT) \quad (19)$$

and this term can be made to vanish for pulses satisfying the Nyquist criterion, i.e., $g(nT) = 0$ for $n \neq 0$. For bandlimited Nyquist pulses, the intersymbol interference will not be zero when $\hat{\tau} \neq \tau$, and if the bandwidth is not significantly greater than the Nyquist bandwidth ($1/2T$) the intersymbol interference can be quite severe even for small values of timing error. The problem is especially acute for multilevel (nonbinary) data sequences where timing accuracy of only a few percent of the symbol period is often required.

Symbol timing recovery is remarkably similar in most respects to carrier phase recovery and we find that similar signal processing will yield suitable estimates of the parameter τ. In the discussion to follow, we assume that $\{a_k\}$ is a zero-mean stationary sequence with independent elements. The resulting PAM signal (18) is a zero-mean cyclostationary process, although there are no periodic components present [6]. The square of the PAM signal does, however, possess a periodic mean value.

$$E[x^2(t)] = \overline{a^2} \sum_k g^2(t - kT - \tau). \quad (20)$$

Using the Poisson Sum Formula [6], we can express (20) in the more convenient form of a Fourier series whose coefficients are given by the Fourier transform of $g^2(t)$.

$$E[x^2(t)] = \frac{\overline{a^2}}{T} \sum_{\ell} A_{\ell} \exp\left(\frac{j2\pi\ell}{T}(t - \tau)\right) \quad (21)$$

where

$$A_{\ell} \triangleq \int_{-\infty}^{\infty} G\left(\frac{\ell}{T} - f\right) G(f)\, df.$$

For high bandwidth efficiency, we are often concerned with data pulses whose bandwidth is at most equal to twice the Nyquist bandwidth. Then $|G(f)| = 0$ for $|f| > 1/T$ and there are only three nonzero terms ($\ell = 0, \pm 1$) in (21).

This result suggests the use of a timing recovery circuit of the same form as shown in Fig. 1, where now the bandpass filter is tuned to the symbol rate, $1/T$. Alternate zero crossings of $w(t)$, a *timing wave* analogous to the reference waveform in Section II, are used as indications of the correct sampling instants. Letting $H(1/T) = 1$, the mean timing wave is a sinusoid with a phase of $-2\pi\tau/T$, for a real $G(f)$.

$$E[w(t)] = \frac{\overline{a^2}}{T} \operatorname{Re}\left[A_1 \exp\left(j\frac{2\pi t}{T} - j\frac{2\pi\tau}{T}\right)\right]. \quad (22)$$

We see that the zero crossings of the mean timing wave are at a fixed time offset ($T/4$) relative to the desired sampling instants.

[5] Unless $a(t)$ and $b(t)$ are Gaussian processes, for then $\overline{a^4} = 3(\overline{a^2})^2$.

This timing offset can be handled by counting logic in the clock circuitry, or by designing $H(f)$ to incorporate a $\pi/2$ phase shift at $f = 1/T$.

The actual zero crossings of $w(t)$ fluctuate about the desired sampling instants because the timing wave depends on the actual realization of the entire data sequence. Different zero crossings result for different data sequences and for this reason the fluctuation in zero crossings is sometimes called *pattern-dependent* jitter to distinguish it from jitter produced by additive noise on the PAM signal. To evaluate the statistical nature of the pattern-dependent jitter, we need to calculate the variance of the timing wave. This is a fairly complicated expression in terms of $H(f)$ and $G(f)$ but it can be evaluated numerically to study the effects of a variety of parameters (bandwidth, mistuning, rolloff shape, etc.) relating to data pulse shape and the bandpass filter transfer function [11]. For a relatively narrow-band real $H(f)$ and real $G(f)$ bandlimited as mentioned previously, the variance expression has the form

$$\operatorname{var} w(t) = C_0 + C_1 \cos \frac{4\pi}{T}(t - \tau) \tag{23}$$

where $C_0 \geqslant C_1 > 0$ are constants depending on $G(f)$ and $H(f)$. The cyclostationarity of the timing wave is apparent from this expression. As the bandwidth of $H(f)$ approaches zero, the value of C_1 approaches C_0 so that the variance has a great fluctuation over one symbol period. Note that the minimum variance occurs just at the instant of the mean zero crossings, hence the fluctuations in zero crossings are much less than would be expected from a consideration of the average variance of the timing wave over a symbol period. This again points out the error in disregarding the cyclostationary nature of the timing wave process as, for example, in using the power spectral density of the squarer output to analyze the jitter phenomenon.

The mean timing wave (22) can be regarded as a kind of *discriminator characteristic* or *S-curve* for measuring the parameter τ. For the bandlimited case we are discussing here, this S-curve is just a sinusoid, with a zero crossing at the true value of the parameter. Discrimination is enhanced by increasing the slope at the zero crossing. As this slope is proportional to A_1, we can see how the shape of the data pulse $g(t)$ affects timing recovery. From (21) we see that the value of A_1 depends on the amount of overlap of the functions $G(f)$ and $G(1/T - f)$, and hence it depends on the amount by which the bandwidth of $G(f)$ exceeds the $1/2T$ Nyquist bandwidth. With no excess bandwidth, $A_1 = 0$ and this method of timing recovery fails. The situation improves rapidly as the excess bandwidth factor increases from 0 to 100 percent. With very large increases in bandwidth there are more harmonic components in the mean timing wave, and its zero crossing slope can be further increased without increasing signal level by proper phasing of these components. On the other hand, there are systems where the spectral distribution is such that the fractional amount of energy above $1/2T$ is very small. An important case is that of (class IV) partial response signaling where the pulse shape is chosen to produce a spectral null at $1/2T$. This spectral null in combination with a sharp baseband rolloff characteristic can result in very small values of A_1. Although the class IV partial response format exhibits a relatively high tolerance to timing error [12], it is likely that some other recovery scheme may have to be used. Some of the proposed schemes [13], [14] closely resemble the data-aided approach discussed in Section IV.

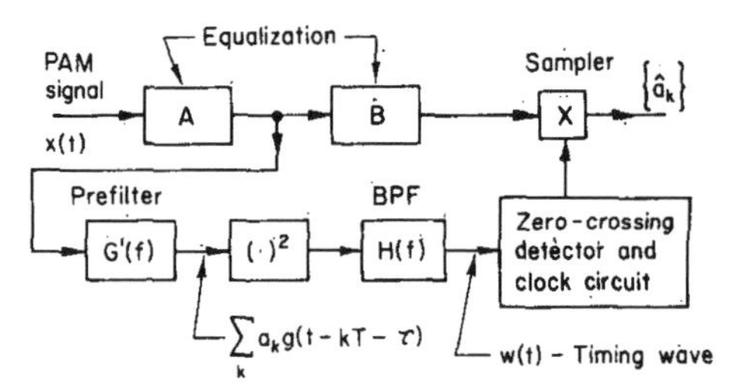

Fig. 3. Baseband PAM receiver with timing recovery.

Calculation of the statistical properties of the actual zero crossings of the timing wave is difficult. A useful approximation can be obtained by locating the zero crossings by linear extrapolation using the mean slope at the mean zero crossing. When this approach is used, the expression for timing jitter variance becomes [11]

$$\operatorname{var} \hat{\tau}/T = (\overline{a^2})^{-1} \left(\frac{T}{2\pi A_1} \right)^2 (C_0 - C_1). \tag{24}$$

In order to reduce this pattern-dependent jitter, there is fortunately an attractive alternative to making the bandwidth of $H(f)$ very small, which increases acquisition time in the same manner as for carrier phase recovery circuits, or to making the bandwidth of $G(f)$ very large. There are symmetry conditions that can be imposed upon $H(f)$ and $G(f)$ that make $C_1 = C_0$ in (24), resulting in nonfluctuating zero crossings. These conditions are simply that $G(f)$ be a bandpass characteristic symmetric about $1/2T$, with a bandwidth not exceeding $1/2T$, and $H(f)$ be symmetric about $1/T$. The symmetry in $G(f)$ can be accomplished by prefiltering the PAM signal before it enters the squarer [11], [15]. Since the timing recovery path is distinct from the data signal path, the prefiltering can be performed without influencing the data signal equalization, as shown in the baseband receiver configuration of Fig. 3.

Although we are dealing with a baseband signal process, it is interesting to observe that the timing jitter problem can be studied by means of complex envelopes and decomposition into I and Q components, as in the carrier phase recovery case [16]. One way to do this is to let $\gamma(t)$ be the complex envelope of $g(t)$, relative to a frequency $f_0 = 1/2T$. This makes $\Gamma(f)$ bandlimited to $|f| < 1/2T$. Then, taking $\tau = 0$ for convenience, the output of the squarer is

$$x^2(t) = \tfrac{1}{2} \operatorname{Re} \left[\left\{ \sum_k (-1)^k a_k \gamma(t - kT) \right\}^2 \exp(j2\pi t/T) \right]$$

$$+ \tfrac{1}{2} \left| \sum_k (-1)^k a_k \gamma(t - kT) \right|^2. \tag{25}$$

The second term in (25) can be disregarded as not being passed by $H(f)$. The first term is expressed by a complex envelope relative to $f_0 = 1/T$. It is the quadrature component (imaginary part) of this complex envelope that produces timing jitter. This component is

$$b_Q(t) = \sum_k \sum_m a_k a_m (-1)^{k+m} c_I(t - kT) c_Q(t - mT) \tag{26}$$

where $c_I(t)$ and $c_Q(t)$ are the real and imaginary parts of $\gamma(t)$. The Fourier transform of (26), evaluated at $f = 0$, is

$$B_Q(0) = \int_{-\infty}^{\infty} \left| M\left(\nu - \frac{1}{2T}\right) \right|^2 C_I(\nu) C_Q(-\nu)\, d\nu = 0 \tag{27}$$

where $M(f) = \Sigma\, a_k \exp(-j2\pi kTf)$ is the transform of the data sequence. The integral (27) vanishes because, for a real $G(f)$ i.e., a real $\Gamma(f)$, the integrand is an odd function. The situation is similar to the VSB carrier signal case, where the spectrum of the quadrature component at the squarer output vanished at $f = 0$. We see here also that the particular shape of $H(f)$ will have a major influence when calculating the jitter variance because the spectrum of the jitter-producing component goes to zero just at the center of its passband.

IV. MAXIMUM-LIKELIHOOD PARAMETER ESTIMATION

The foregoing carrier- and bit-synchronization circuits were developed on a rather heuristic basis and a natural question arises as to how much improvement in parameter estimation could result from the choice of other circuit configurations or circuit parameters. It seems natural to regard θ and τ as unknown but nonrandom parameters which suggests the maximum-likelihood (ML) estimator as the preferred strategy [17]. Some authors have used the maximum *a posteriori* probability (MAP) receiver by modeling θ and τ as random parameters with specified *a priori* probability density functions. However, in most situations the *a priori* knowledge about θ is only accurate to within many carrier cycles, or in the case of τ, to within many symbol periods. As our concern is with estimation modulo 2π for θ or modulo T for τ, we would use a "folded" version of the *a priori* density functions, resulting in a nearly uniform distribution over the interval. In this case, the ML approach estimates and MAP estimates would be essentially identical. We find that the phase and timing-recovery circuits based on the ML approach may not be drastically different from the circuits already considered. In fact, under the proper conditions, the circuits we have examined can be close approximations to ML estimators. One of the main advantages of the ML approach, in addition to suggesting appropriate circuit configurations, is that simple lower bounds on jitter performance can be developed to serve as benchmarks for evaluating performance of the actual recovery circuits employed.

In this section we begin with discussion of ML carrier phase recovery with a rather general specification of the message signal process. We show that the Costas loop, or the equivalent squarer/BPF, can be designed to closely approximate the ML phase estimator. Then we present a similar development for ML estimation of symbol timing for a baseband PAM signal. We introduce the idea of using information about the data sequence to aid the timing recovery process and we later make comparisons to show the effectiveness of such data-aided schemes. Extension of the idea to joint recovery of both carrier phase and symbol timing parameters is discussed in Section V.

To formulate the problem in terms of ML estimation, we require that the receiver perform operations on a T_0-second record of the received signal, $z(t) = y(t, \theta) + n(t)$, to estimate the parameter θ, assumed essentially constant over the T_0-second interval. This interval is called the *observation interval* and the T_0 parameter would be selected in accordance with acquisition time requirements. Estimation procedures based on data from a single observation interval will be referred to as *one-shot estimation.* We find that the one-shot ML estimators lead to the simplest methods for evaluating jitter performance. On the other hand, the preferred implementation of recovery circuits is usually in the form of tracking loops where the parameter estimates are being continuously updated. Fortunately, it is a relatively simple matter to relate the rms one-shot estimation error to the steady-state error of the tracking loop and the loop bandwidth is directly related to the T_0 parameter.

We shall assume that the additive noise $n(t)$ is Gaussian and white with a double-sided spectral density of N_0 W/Hz. Initially we consider the situation where $y(t, \theta)$ is completely known except for the parameter θ. The resulting likelihood function, with argument $\hat{\theta}$ which can be regarded as a trial estimate of the parameter, is given by

$$L(\hat{\theta}) = \exp\left\{ -\frac{1}{2N_0} \int_{T_0} [z(t) - y(t, \hat{\theta})]^2\, dt \right\}. \tag{28}$$

The ML estimate is the value of θ which minimizes the integral in (28). This integral expresses the signal space distance between the functions $z(t)$ and $y(t, \hat{\theta})$ defined on the interval T_0 [6], [17]. Expanding the binomial term in (28), we see that

$$\Lambda(\hat{\theta}) = \ln L(\hat{\theta}) = \frac{1}{N_0} \int_{T_0} z(t) y(t, \hat{\theta})\, dt + \text{constant} \tag{29}$$

since $z^2(t)$ is independent of $\hat{\theta}$, and if $\hat{\theta}$ is a time shift or phase shift parameter, then the integral of $y^2(t, \hat{\theta})$ over a relatively long T_0 interval would have only a small variation with $\hat{\theta}$. The first term in (29) is often called the correlation between the received signal $z(t)$ and the reference signal $y(t, \hat{\theta})$ so that in this "known-signal" case, the ML receiver is a correlator, and $\hat{\theta}$ is varied so as to maximize the correlation.

When $y(t, \hat{\theta})$ contains random message parameters, the appropriate likelihood function for estimating θ is obtained by averaging $L(\hat{\theta})$—not $\ln L(\hat{\theta})$—over these message parameters. We shall illustrate the method using the example of carrier phase estimation on a DSB signal where the modulating signal $a(t)$ is a zero-mean, Gaussian random process with a substantially flat spectrum bandlimited to W Hz. Finding the expectation of $L(\hat{\theta})$ with respect to the Gaussian message process can be done without great difficulty by making a Karhunen-Loeve expansion of the process to give a series representation with independent coefficients [18]. The result of this averaging gives a log-likelihood function closely approximated by

$$\Lambda(\hat{\theta}) = \int_{T_0} [\operatorname{Re} \alpha(t) \exp(-j\hat{\theta})]^2 \, dt$$
$$= \tfrac{1}{2} \operatorname{Re}\left[\exp(-j2\hat{\theta}) \int_{T_0} \alpha^2(t)\, dt\right] + \frac{1}{2} \int_{T_0} |\alpha(t)|^2 \, dt \tag{30}$$

where $\alpha(t)$ is the complex envelope, relative to f_0, of the received signal. We ignore the second integral in (30) as it is independent of $\hat{\theta}$. (30) suggests a practical implementation of the ML phase estimator. Consider a receiver structure which produces the complex signal

$$\lambda(t) = \int_{t-T_0}^{t} \alpha^2(s)\, ds \triangleq \rho(t) \exp(j2\tilde{\theta}(t)). \tag{31}$$

The integral in (31) is the convolution product of $\alpha^2(t)$ and a T_0-second rectangle, hence $\lambda(t)$ can be regarded as the complex envelope of the output of the squarer/bandpass filter configuration of Fig. 1. In this case, $H(f)$ corresponds to a sinc $(T_0 f)$ shape centered at $2f_0$. Writing $\lambda(t)$ in polar form as shown in (31), we see that the corresponding term in (30) is maximized, at any t, by choosing $\hat{\theta} = \tilde{\theta}(t)$. In other words, by suitably designing the shape of the bandpass transfer function, the simple structure of Fig. 1 is a ML phase estimator, in the sense that the instantaneous phase of the timing wave output is the best estimate of the DSB carrier phase based on observations over only the past T_0 seconds. The phase jitter can be evaluated approximately by the same method leading to (11), with the result that

$$\operatorname{var} \phi = \frac{2N_0}{ST_0} = \frac{1}{WT_0}\left(\frac{S}{N}\right)^{-1} \tag{32}$$

where $S = k_{aa}(0)$ is the signal power and $N = 2N_0 W$ is noise power over the signal band.

The tracking loop version of this phase estimator is developed by forming a loop error signal proportional to the derivative of Λ with respect to $\hat{\theta}$. Then, as the loop action tends to drive the error signal to zero, the resulting value of $\hat{\theta}$ should correspond to a maximum of Λ. Since the control voltage $v(t)$ for a VCO normally controls frequency, rather than phase, we suppress the integration in (30) and let

$$v(t) = \frac{1}{8} \frac{\partial}{\partial \hat{\theta}} [\operatorname{Re} \alpha(t) \exp(-j\hat{\theta})]^2$$
$$= \tfrac{1}{8} \operatorname{Re} [\alpha^2(t) \exp(-j2\hat{\theta} - j\pi/2)] \tag{33}$$

which is the same control voltage that appears in the Costas loop (9) and Fig. 2(b), with the normalization, $A = 1$. If the VCO had a voltage-controlled phase, rather than frequency, we would include the T_0-second integration effect by means of the loop filter. However, in this case we simply let $F(f) = 1$ and rely on the integration inherent in the VCO. The parameter T_0 is related to loop performance by adjusting the loop gain factor M, which is proportional to loop bandwidth, so that the steady-state jitter variance for the loop is identical to (32) for the nontracking implementation.

For the DSB signal with additive noise, we have

$$\alpha(t) = [a(t) + u_I(t) + ju_Q(t)] \exp(j\theta) \tag{34}$$

where u_I and u_Q are the I and Q components of noise relative to the carrier phase θ. Letting the VCO gain constant be M (hertz/volt) so that $\dot{\hat{\theta}}(t) = 2\pi M v(t)$ and assuming a high signal-to-noise ratio so that the second-order noise effects can be neglected, a linearized loop equation for phase error, $\phi = \hat{\theta} - \theta$, appears as

$$\dot{\phi}(t) + 2\pi B_L \phi(t) = \frac{2\pi B_L}{S} u_Q(t) a(t) - \frac{2\pi B_L}{S} b(t)\phi(t). \tag{35}$$

The difficult part of solving this equation to get the steady-state variance of ϕ is the second driving term where $b(t) \triangleq a^2(t) - S$ and $\phi(t)$ are clearly not independent. It turns out however, that if the loop bandwidth parameter $B_L = \frac{1}{4} MS$ is sufficiently small compared to signal bandwidth W, then this term can be neglected. The other excitation term $u_Q a$ can be treated as white noise with a spectral density of $2N_0 S$, and the steady-state variance of ϕ can be determined by conventional frequency-domain techniques. The result is

$$\operatorname{var} \phi = 2\pi B_L N_0 / S \tag{36}$$

and, equating (36) and (32) we find that $B_L = 1/\pi T_0$ is the relation sought between observation interval and tracking loop bandwidth.

Turning now to ML timing recovery for the baseband PAM signal, with $\hat{\tau}$ replacing $\hat{\theta}$, and using (18) for $y(t, \hat{\tau})$, the log-likelihood function for the case of a known signal (29) becomes

$$\Lambda(\hat{\tau}) = \frac{1}{N_0} \sum_{k=-\infty}^{\infty} a_k q_k \tag{37}$$

where

$$q_k(\hat{\tau}) = \int_{T_0} z(t) g(t - kT - \hat{\tau})\, dt.$$

It is possible to use this expression directly for timing recovery in a situation where a relatively long sequence, say K, of known symbols is transmitted as a preamble to the actual message sequence. The receiver would store the K-symbol sequence and attempt to establish the correct timing before the end of the preamble. The idea can also be used during message transmission if the symbols are digitized, so that the receiver makes decisions as to which of the finite number of possible symbols have been transmitted. The receiver decisions are then assumed to be correct, at least for the purposes of timing recovery. The bootstrap type of operation is referred to as *decision-directed* or *data-aided* timing recovery and it has received extensive study for both symbol timing and carrier phase recovery [19]-[22]. In the following, we shall use the term "data-aided" to refer to both modes of operation, i.e., the start-up mode where a known data sequence is being transmitted and the tracking mode where the symbol detector output sequence is used.

For recovery strategies which are not data aided, we need to average the likelihood function (37) over the random data variables. If we assume that the $\{a_k\}$ are independent Gaussian random variables and also that the data pulses have unit energy and are orthogonal over the T_0 interval, i.e.,

$$\int_{T_0} g(t-kT)g(t-mT)\,dt = \delta_{km} \tag{38}$$

then the log-likelihood function is given by

$$\Lambda(\hat{\tau}) = \frac{1}{2N_0} \sum_{k=-\infty}^{\infty} q_k^2(\hat{\tau}) \tag{39}$$

where the q_k are the same quantities defined in (37). Although the Gaussian density is obviously not an accurate model for digital data signals, we want to consider it here because it provides the link between the ML estimators and the estimators of Sections II and III based on statistical moment properties. It is the Gaussian assumption that leads to the square-law type of nonlinearity. If we consider equiprobable binary data, for example, the corresponding log-likelihood function is [23]-[25]

$$\Lambda(\hat{\tau}) = \sum_k \ln\cosh \frac{1}{N_0} q_k(\hat{\tau}) \tag{40}$$

and since $\ln\cosh x \cong \frac{1}{2}x^2$ for small x, the square-law nonlinearity is near optimum at the lower signal-to-noise ratios. The log-likelihood function for equiprobable independent multilevel data has also been derived [25], [26]. When the Gaussian assumption is used for the data, it is also possible to consider correlated data as well as nonorthogonal pulses, i.e., when (38) does not hold. Both of these effects can be dealt with by replacing the q_k-sequence in (39) by a linear discrete-time filtered version of this sequence [27]. In summary, we find that recovery circuits based on the Gaussian-distributed data assumption are somewhat simpler than the optimum circuits and in most situations the jitter performance is not appreciably worse. We note that the method for evaluating rms jitter, presented in Section VI, does not depend on the particular kind of density function used to characterize the data.

When it comes to implementation of receivers based on (37) and (39) for the data-aided (DA) or nondata-aided (NDA) strategies, we usually resort to an approximation which involves replacing the infinite sum by a K-term sum, where $KT = T_0$, and replacing the finite integration interval by an infinite interval. Then the approximate implementable, log-likelihood function in the DA case is taken as

$$\tilde{\Lambda}(\hat{\tau}) = \frac{1}{N_0} \sum_{k=0}^{K-1} a_k \tilde{q}_k \tag{41}$$

where

$$\tilde{q}_k(\hat{\tau}) = \int_{-\infty}^{\infty} z(t)g(t-kT-\hat{\tau})\,dt.$$

With this approximation, the integral is a convolution integral, and $\tilde{q}_k$ can be interpreted as the sampled (at $t = kT + \hat{\tau}$) output of a matched filter having the impulse response $g(-t)$. The same approximation is used for the NDA case (39) and the orthogonality condition (38) can be interpreted to mean that the matched filter response to a single data pulse is a pulse satisfying the Nyquist criterion. This approximation, which leads to relatively simple implementations for the recovery circuits, does introduce a degradation from the idealized ML performance. An interesting interpretation of the effect of the approximation is that it introduces a pattern-dependent component of jitter, as discussed in Section VI.

For tracking loop implementation of these timing recovery strategies, we use a voltage-controlled clock (VCC) driven by

$$v(t) = \frac{\partial}{\partial \hat{\tau}} [a_k \tilde{q}_k] = -a_k \int_{-\infty}^{\infty} z(t)\dot{g}(t-kT-\hat{\tau})\,dt \tag{42}$$

for the DA case, and

$$v(t) = \frac{\partial}{\partial \hat{\tau}} [\tfrac{1}{2} \tilde{q}_k^2] = -\left[\int_{-\infty}^{\infty} z(t)g(t-kT-\hat{\tau})\,dt\right] \cdot \left[\int_{-\infty}^{\infty} z(t)\dot{g}(t-kT-\hat{\tau})\,dt\right] \tag{43}$$

for the NDA case. The K-term summation is suppressed, being replaced by the integration action of the VCC as in the case of the Costas loop phase recovery circuit discussed earlier. Similar also is the relation between T_0 and the loop bandwidth, the loop gain being adjusted so that the steady-state variance of timing jitter is the same as for one-shot estimation in a single observation interval. The result is also $B_L = 1/\pi T_0$ [26]. The tracking loop configurations are evident from inspection of (42) and (43).

One structure will serve for both strategies by incorporating a DA/NDA mode switch as shown in Fig. 4. This could be quite useful in a system that uses the DA strategy on a message preamble, then switches to the NDA strategy when the message symbols begin. Notice that the NDA configuration is remarkably like a Costas loop, which suggests the existence of an equivalent realization using a square-law device. This alternative and equivalent form is shown in Fig. 5. The corresponding implementation of (40) for NDA recovery with binary data involves the same structure as shown in Fig. 4, except that a tanh(·) nonlinearity is incorporated into the upper path of the NDA loop [25].

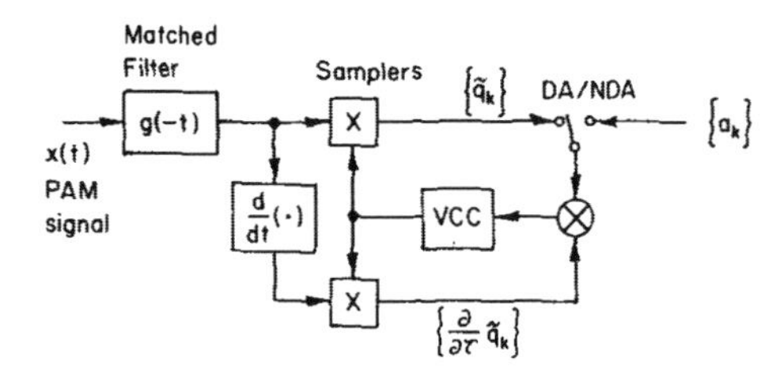

Fig. 4. ML baseband PAM timing recovery circuit.

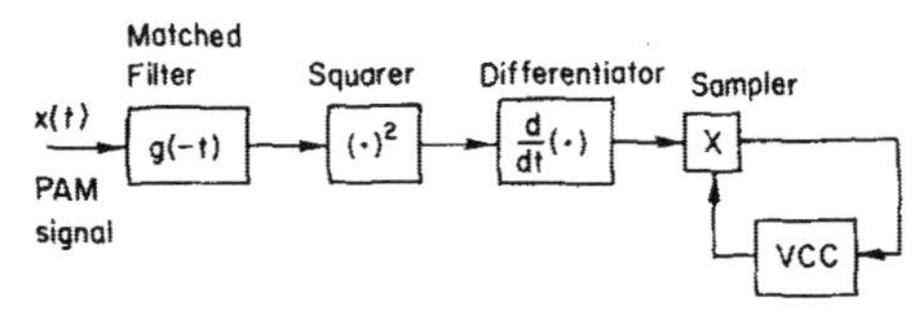

Fig. 5. Alternative implementation of NDA baseband timing recovery.

V. JOINT RECOVERY OF CARRIER PHASE AND SYMBOL TIMING

When a carrier system, such as VSB/PAM or QAM/PAM is used to transmit a digital data signal, we have the possibility of jointly estimating the carrier phase and symbol timing parameters. Such a strategy certainly cannot be worse than estimating the parameters individually and, in some cases, join estimation gives remarkable improvements. Some authors have extended the idea to joint estimation of the data sequence and the two timing parameters [28]-[30]. We shall not consider this latter possibility here, but shall consider both DA and NDA joint parameter estimation. Our conjecture is that, in the majority of applications, DA recovery performance differs little from that of joint estimation of data and timing parameters.

We consider first the QAM/PAM data signal case where we want to estimate θ and τ in

$$y(t;\theta,\tau) = \mathrm{Re}\left[\left\{\sum_k a_k g(t-kT-\tau) + jb_k h(t-kT-\tau)\right\} \cdot \exp(j\theta)\exp(j2\pi f_0 t)\right] \tag{44}$$

from receiver measurements on $z(t) = y(t) + n(t)$ over a T_0-second observation interval. The implementable version of the log-likelihood function for the DA case is

$$\tilde{\Lambda}(\hat{\theta},\hat{\tau}) = \frac{1}{N_0}\sum_{k=0}^{K-1} a_k\tilde{q}_k + b_k\tilde{p}_k \tag{45}$$

where

$$\tilde{q}_k(\hat{\theta},\hat{\tau}) = \mathrm{Re}\left[\exp(-j\hat{\theta})\int_{-\infty}^{\infty}\alpha(t)g(t-kT-\hat{\tau})\,dt\right]$$

$$\tilde{p}_k(\hat{\theta},\hat{\tau}) = \mathrm{Re}\left[-j\exp(-j\hat{\theta})\int_{-\infty}^{\infty}\alpha(t)h(t-kT-\hat{\tau})\,dt\right]$$

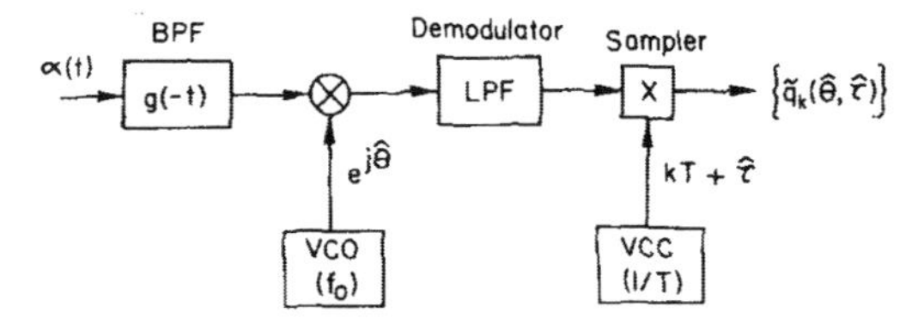

Fig. 6. Receiver implementation of the $\tilde{q}_k(\hat{\theta},\hat{\tau})$ test statistic.

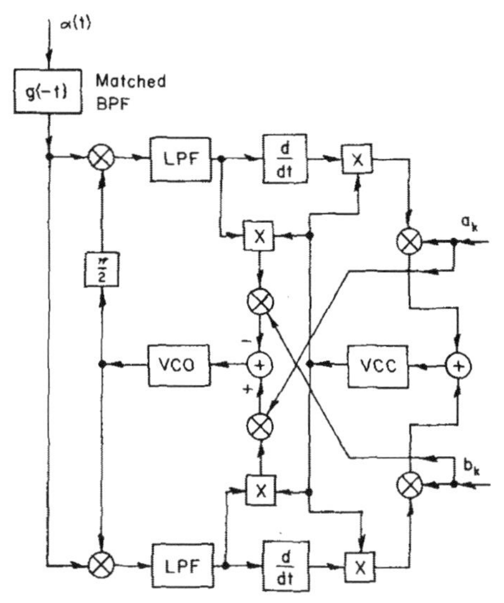

Fig. 7. Data-aided QAM joint tracking loop for carrier phase and symbol timing.

and $\alpha(t)$ is the complex envelope of the received signal. The $\tilde{q}_k$ quantities are interpreted as the sampled (at $t = kT + \hat{\tau}$) output of a coherent demodulator (operating at a phase $\hat{\theta}$) whose input is a bandpass filtered version of the received signal. The receiver implementation for these quantities is shown in Fig. 6, and a similar implementation would provide the $\tilde{p}_k$ quantities.

For the joint tracking loop the partial derivatives of $\tilde{\Lambda}$ with respect to $\hat{\theta}$ and $\hat{\tau}$, without the K-term summation, are used to update the VCO and VCC frequencies once every T seconds. For the normal QAM case we let $h(t) = g(t)$ and some simplifications result, for then $\partial\tilde{q}_k/\partial\hat{\theta} = \tilde{p}_k$ and $\partial\tilde{p}_k/\partial\hat{\theta} = -\tilde{q}_k$. The $\partial\tilde{q}_k/\partial\hat{\tau}$ and $\partial\tilde{p}_k/\partial\hat{\tau}$ quantities are obtained by differentiating the I and Q baseband signals before sampling. The complete tracking loop implementation for the DA case is shown in Fig. 7.

For balanced QAM/PAM with identical pulse shapes and statistically identical independent data in the I and Q channels, the NDA mode of recovery fails [27] because the NDA log-likelihood function is

$$\tilde{\Lambda}(\hat{\theta},\hat{\tau}) = \frac{1}{2N_0}\sum_{k=0}^{K-1} \tilde{q}_k^2 + \tilde{p}_k^2 \tag{46}$$

and this is independent of $\hat{\theta}$ under the previous assumptions. Fortunately, a simple modification makes the NDA mode effective. This modification is $h(t) = g(t \pm T/2)$ and the format is called *staggered* QAM (SQAM). The implementation is similar, but somewhat more complex, to that shown in Fig. 7

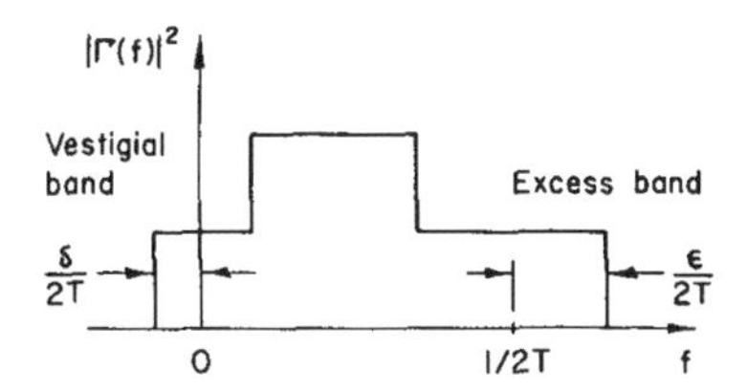

Fig. 8. Energy spectral density for VSB data pulse.

because additional samplers are needed for sampling at both $kT + \hat{\tau}$ and $kT + \hat{\tau} \pm T/2$.

Considering now the one-dimensional VSB/PAM case, the log-likelihood functions are the same as (45) and (46) with the b_k and $\tilde{p}_k$ quantities omitted, and with

$$\tilde{q}_k(\hat{\theta}, \hat{\tau}) = \mathrm{Re}\left[\exp(-j\hat{\theta}) \int_{-\infty}^{\infty} \alpha(t)\gamma * (t - kT - \hat{\tau})\, dt\right]. \tag{47}$$

In (47), $\gamma(t) = g(t) + j\,\tilde{g}(t)$ is the complex envelope of a single, unit-amplitude carrier data pulse. The orthogonality condition corresponding to (38) for the NDA case can be satisfied by a pulse whose energy spectrum $|\Gamma(f)|^2$ has a shape of the form shown in Fig. 8, exhibiting a Nyquist-type of symmetry in both of its rolloff regions.

In contrast to the QAM case, we find that the VCO and VCC frequency-control voltages should be derived as linear combinations of the partial derivatives of $\tilde{\Lambda}$ with respect to $\hat{\theta}$ and $\hat{\tau}$, i.e., there is a coupling between the parameter estimates. To show this, we consider the approximate solution for the one-shot estimator based on a Taylor series expansion of $\tilde{\Lambda}$ about trial values of θ_0 and τ_0. If these values are sufficiently close to the true values, then we can take as refined estimates, θ_1 and τ_1, the solutions of

$$\begin{bmatrix} \tilde{\Lambda}_{\theta\theta}(\theta_0, \tau_0) & \tilde{\Lambda}_{\theta\tau}(\theta_0, \tau_0) \\ \tilde{\Lambda}_{\theta\tau}(\theta_0, \tau_0) & \tilde{\Lambda}_{\tau\tau}(\theta_0, \tau_0) \end{bmatrix} \begin{bmatrix} \theta_1 - \theta_0 \\ \tau_1 - \tau_0 \end{bmatrix} = \begin{bmatrix} -\tilde{\Lambda}_{\theta}(\theta_0, \tau_0) \\ -\tilde{\Lambda}_{\tau}(\theta_0, \tau_0) \end{bmatrix} \tag{48}$$

where the subscripts in (48) denote partial derivatives. The solution of (48) is greatly simplified if the 2 X 2 matrix is replaced by its mean value, and this is valid at moderately high signal-to-noise ratio and moderately long observation intervals. Then we have a simple form of estimation given by

$$\begin{bmatrix} \theta_1 - \theta_0 \\ \tau_1 - \tau_0 \end{bmatrix} = -A^{-1} \begin{bmatrix} \tilde{\Lambda}_{\theta}(\theta_0, \tau_0) \\ \tilde{\Lambda}_{\tau}(\theta_0, \tau_0) \end{bmatrix} \tag{49}$$

where the 2 X 2 matrix A is the expected value of the matrix in (48). The A matrix can be regarded as a generalization of the A_1 quantity in (22), (24) for the single-parameter recovery problem. For the joint tracking loop, the VCO and VCC control voltages are linear combinations of the Λ_θ and Λ_τ quantities (without the K-term summation) as characterized by the inverse of the A matrix. In the QAM and SQAM cases, the A matrix is diagonal so that no coupling is required, but for VSB there are strong off-diagonal terms. It has been shown [30]-[32] that loop convergence rates can be substantially improved by incorporating this coupling on the control signals.

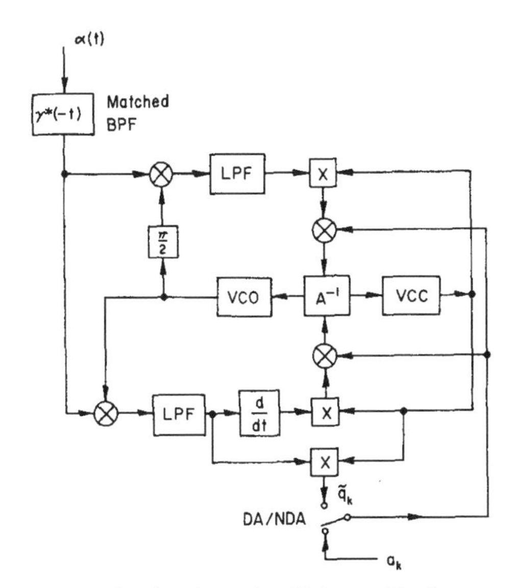

Fig. 9. One-dimensional joint tracking loop.

A block diagram for the one-dimensional joint tracking loop is shown in Fig. 9. A DA/NDA mode switch is shown in the diagram, but it must be recognized that the A^{-1} coupling matrix is a compromise value in either one or both modes because the A matrix is quite different for the DA and NDA cases, as discussed in the next section. Another practical consideration is that the configuration of Fig. 9 can also be used for QAM and SQAM with some loss in performance. Here we just eliminate the $\tilde{p}_k$ quantities in (45) or (46) and use the matched BPF for the I-channel pulse, i.e., let $\gamma^*(-t) = g(-t)$. The loss in performance will be about 3 dB or greater, depending on signal-to-noise ratio and on which parameter is considered, as there are some cancellations of pattern-dependent jitter in the configuration of Fig. 7 which are not possible in that of Fig. 9.

VI. PERFORMANCE OF TIMING RECOVERY SCHEMES

A convenient approach to evaluating timing recovery circuit performance, is to derive expressions for rms phase-and timing-jitter directly from (49), or its one-dimensional counterpart for individual estimation of θ or τ. For these calculations we assume θ_0 and τ_0 are the true values of the parameters, so that the left-hand side of (49) gives the jitter variables. The equation is linearized in the sense that the jitter variables depend linearly on the receiver measurements $\tilde{\Lambda}_\theta$ and $\tilde{\Lambda}_\tau$. As in all analyses of this type, the results are accurate if the jitter is relatively small, which in this case generally means a moderately high signal-to-noise ratio and a moderately long observation interval. This approach affords an effective means to study jitter performance with respect to the values of all system parameters, such as signal-to-noise ratio, T_0 (or K), excess bandwidth, and pulse shape. It also allows comparison of jitter performance of the various modulation formats and evaluation of the DA strategy relative to the NDA strategy.

Another important aspect that can be examined from the rms jitter calculations is the effect of the implementation approximations $\Lambda \rightarrow \tilde{\Lambda}$ and $q_k \rightarrow \tilde{q}_k$. Let us take as an illustrative example the one-dimensional case of baseband timing recovery. For the implementable DA case we get

$$\text{var}\ \hat{\tau} = \text{var}\ \tilde{\Lambda}_\tau(\tau)[E\{\tilde{\Lambda}_{\tau\tau}(\tau)\}]^{-2}$$

$$= \frac{N_0}{a^2 KD} + \frac{\sum_{k=0}^{K-1} \sum_{m \in K'} \dot{r}^2(mT - kT)}{K^2 D^2} \tag{50}$$

where

$$r(t) = \int_{-\infty}^{\infty} g(s+t)g(s)\, ds$$

and

$$D = -\ddot{r}(0) = \int_{-\infty}^{\infty} \dot{g}^2(t)\, dt$$

is the energy in the time derivative of a single data pulse. The notation $m \in K'$ in (50) means that the sum is taken for all $k < 0$ and $\geqslant K$, i.e., just the terms not used in the other sum. The first term in (50) is seen to vary inversely with K, signal-to-noise ratio, and the energy in $\dot{g}(t)$. From this it is obvious that "sharp-edged" pulses can give excellent timing recovery performance. In fact, the entire denominator in the first term can be taken approximately as the expected value of the energy of the time-derivative of the received PAM signal over the T_0 -second observation interval.

Another significant aspect of the first term in (50) is that it gives the entire jitter variance if Λ and q_k are used instead of $\tilde{\Lambda}$ and $\tilde{q}_k$. In other words, the second term in (50) gives the additional jitter variance resulting from the practical implementation considerations. This term does not depend on the noise level and it can be regarded as the effect of the pattern-dependent component of jitter. It varies inversely with K^2 since the numerator is essentially a constant even for moderately small values of K. Thus we see that if severe requirements are placed on acquisition time (small K), then the effect of this pattern-dependent term is apt to dominate. Otherwise, for larger K, the first term may be dominant and the difference between true ML estimation and its implementable approximation may be negligible.

Although the variance expressions are somewhat different, the same general conclusions about the two types of jitter terms hold for the NDA timing recovery case, and for joint timing and carrier phase recovery [26]. Another physical interpretation, in the case of carrier phase recovery, is as follows. There are two random interference components producing jitter in the carrier phase tracking loop; one is due to the additive noise on the input signal, and the other due to the message sidebands of the carrier signal. It is primarily the quadrature components of these interferences that cause the jitter. The quandrature noise has a flat spectrum about the carrier frequency, so the jitter variance due to this interference should vary in direct proportion to the loop bandwidth. The quadrature data dependent interference has a spectral null in the vicinity of the carrier frequency, so we would expect jitter variance due to this effect to increase faster than linearly with loop bandwidth. Hence, for a given signal-to-noise ratio, and for a large enough loop bandwidth (rapid acquisition) we would expect the pattern-dependent term to dominate.

The relative performance of the different modulation formats is governed primarily by the size of the elements of the A matrix. Also, the A matrix almost completely characterizes the difference in performance of the DA and NDA strategies. For example, in the NDA/VSB case, the A matrix elements contain terms proportional to the integral of the product of $|\Gamma(f)|^2$ and $|\Gamma(1/T - f)|^2$ [26], [27]. Thus the size of the terms depends on the amount of overlap of the pulse energy spectrum and its frequency-translated version, and this depends on the amount of excess bandwidth available. For the staircase shape shown in Fig. 8, the term is directly proportional to the excess bandwidth factor ϵ. As a result, the rms jitter has a $1/\epsilon$ behavior and performance is unacceptable at very small excess bandwidth. On the other hand, the A matrix for DA/VSB has a completely different dependence on $\Gamma(f)$ and it results in a finite jitter variance at $\epsilon = 0$ and a much slower rate of decrease for increasing ϵ. In fact, with excess bandwidths over about 30 percent, the difference between DA and NDA jitter is usually small enough so that it may not be worth the additional circuit complexity to implement the DA strategy [26].

Finally, the variance expressions can be used to compare performance with different pulse shapes or to solve for optimum pulse shapes. There is no universally optimal pulse shape to cover the variety of cases discussed here. For one thing, we can see from (50) that the optimal pulse shape can depend on the signal-to-noise ratio. It is found, however, that the staircase, or "double-jump," rolloff pictured in Fig. 8 is optimal in certain cases and tends to be desirable in all cases. For example, it is better than the familiar "raised-cosine" rolloff, by a factor of approximately 2 in jitter variance [26]. It is interesting to note that such pulse shaping is also optimal from the standpoint of providing maximal immunity to timing or phase offsets [33], [34]. This fact accentuates the importance of proper pulse shaping for overall system performance.

APPENDIX

COMPLEX ENVELOPE REPRESENTATION OF SIGNALS

A straightforward extension of the familiar two-dimensional phasor representation for sinusoidal signals has proven to be a great convenience for dealing with carrier-type data signals where properties of amplitude and phase shift are of special significance. As a supplement to this paper only the most basic relationships are presented. More details and the derivations of the formulas can be found in some texts on communication systems or in [6, chs. 4 and 7].

An arbitrary signal $x(t)$ can be represented exactly by a complex envelope $\gamma(t)$ relative to a "center" frequency f_0, which for modulated-carrier signals is usually, but not necessarily, taken as the frequency of the unmodulated carrier.

$$x(t) = \text{Re}\ [\gamma(t) \exp(j2\pi f_0 t)]. \tag{A-1}$$

Expressing the complex number $\gamma(t)$ in polar form reveals directly the instantaneous *amplitude* $\rho(t)$ and *phase* $\theta(t)$ of the signal.

$$\gamma(t) = \rho(t) \exp [j\theta(t)] = c_I(t) + jc_Q(t). \tag{A-2}$$

In some situations, the rectangular form of $\gamma(t)$ in (A-2) has a more direct bearing on the problem as it decomposes the signal into its *in-phase* and *quadrature* (I and Q) components.

$$x(t) = c_I(t) \cos 2\pi f_0 t - c_Q(t) \sin 2\pi f_0 t. \tag{A-3}$$

Equation (A-1) might be regarded as one part of a transform pair. The other equation, i.e., how to get $\gamma(t)$, given $x(t)$, presents a small problem. Due to the nature of the "real part of" operator Re, there is not a unique $\gamma(t)$ for a given $x(t)$. We solve this problem by making the definition

$$\gamma(t) = [x(t) + j\hat{x}(t)] \exp(-j2\pi f_0 t) \tag{A-4}$$

where $\hat{x}(t)$ is the Hilbert transform of $x(t)$. The prescription for getting $\gamma(t)$ from $x(t)$ is especially simple in the frequency domain. The Fourier transform $\Gamma(f)$ is obtained by doubling $X(f)$, suppressing all negative-frequency values, and frequency-translating the result downward by an amount f_0. Incidentally, using this approach no narrow-band approximations concerning $x(t)$ are necessary, and an arbitrary value of f_0 can be selected.

We now characterize the two most important signal processing operations, filtering and multiplication, in terms of equivalent operations on complex envelopes. Consider first the time-invariant bandpass filtering operation in Fig. 10. We express the bandpass transfer function $H(f)$ in terms of an equivalent low-pass transfer function $\Omega(f)$, according to

$$H(f) = \Omega(f - f_0) + \Omega^*(-f - f_0) \tag{A-5}$$

$\Omega(f)$ is not necessarily a physical transfer function. If $H(f)$ exhibits asymmetry about f_0, then $\Omega(f)$ is asymmetric about $f = 0$ and the corresponding impulse response $\omega(t)$ is complex. In fact, $\omega(t)$ is precisely the complex envelope of $2h(t)$, where $h(t)$ is the real impulse response of the bandpass filter.

Straightforward manipulation shows that the input-output relation for complex envelopes is also a time-domain convolution

$$\beta(t) = [\omega \otimes \gamma](t) \tag{A-6}$$

and this result is general because of our particular method for defining the complex envelope in (A-4). If we express $\omega(t)$ in terms of its real and imaginary parts, $\omega(t) = p_I(t) + j\,p_Q(t)$, then the two-port bandpass filtering operation can be represented by a real four-port filter with separate ports for the I and Q input and output. The four-port filter is a lattice configuration involving the transfer functions $P_I(f)$ and $P_Q(f)$ as shown in Fig. 10.

$$P_I(f) = \tfrac{1}{2}\,\Omega(f) + \tfrac{1}{2}\,\Omega^*(-f)$$

$$P_Q(f) = \frac{1}{2j}\,\Omega(f) - \frac{1}{2j}\,\Omega^*(-f). \tag{A-7}$$

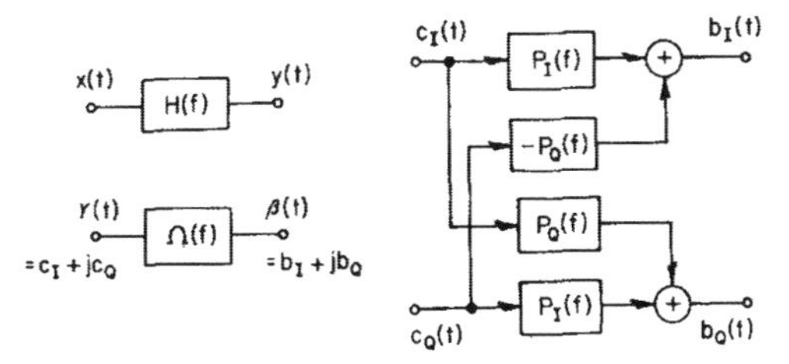

Fig. 10. Bandpass filtering and low-pass equivalent operation on complex envelope signals.

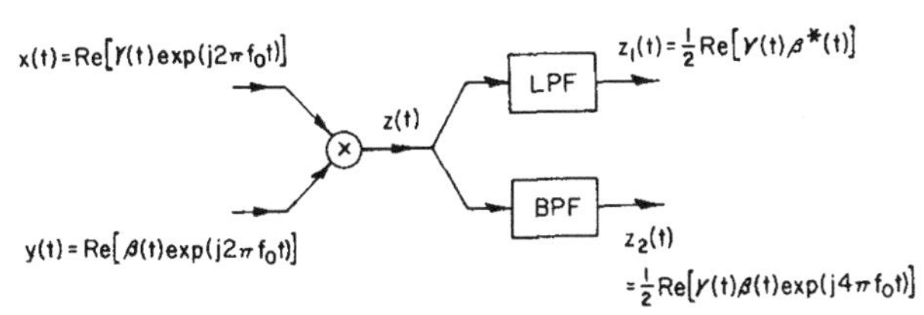

Fig. 11. Low-frequency and $2f_0$ terms of product of two bandpass signals.

Notice that if $H(f)$ is symmetric about f_0, then $P_Q(f) = 0$ (this is the definition of symmetry for a bandpass filter) and there is no cross coupling of the I and Q components in the filtering operation.

Next we consider the output of a multiplier circuit, $z(t) = x(t)y(t)$, when the two inputs are expressed in complex envelope notation. From (A-8), the multiplier output consists of two terms, one representing low-frequency components and the other representing components around $2f_0$.

$$z(t) = \text{Re}\,[\gamma(t) \exp(j2\pi f_0 t)]\ \text{Re}\,[\beta(t) \exp(j2\pi f_0 t)]$$

$$= \tfrac{1}{2}\,\text{Re}\,[\gamma(t)\beta^*(t)] + \tfrac{1}{2}\,\text{Re}\,[\gamma(t)\beta(t) \exp(j4\pi f_0 t)]. \tag{A-8}$$

In most applications a multiplier is followed by either a low-pass filter (LPF) or a bandpass filter (BPF), as shown in Fig. 11, in order to select either the first or second term in (A-8) and completely reject the other term. In our application we may regard $y(t)$ as the reference carrier; then the LPF output $z_1(t)$ is the response of a coherent demodulator to $x(t)$. If $y(t) = x(t)$, so that the multiplier is really a squarer circuit, the BPF output $z_2(t)$ can be used for carrier phase recovery. Its complex envelope, relative to $2f_0$, is proportional to $\gamma^2\,(t)$.

Finally, when the bandpass signal is modeled as a random process, we use the same correspondence, (A-1) and (A-4), between the real process $x(t)$ and the complex envelope process $\gamma(t)$. It is of interest to relate the statistical properties of $x(t)$ to those of its in-phase and quadrature components, relative to some f_0. First we note that $E[x(t)] = \text{Re}\,\{E[\gamma(t)] \exp(j2\pi f_0 t)\}$; hence for a wide-sense stationary (WSS) $x(t)$ process, $\gamma(t)$ must be a zero-mean process, in order that $E[x(t)]$ be independent of t. Proceeding to an examination of second-order moments, it is a simple matter to show that $\gamma(t)$ must be a WSS process if $x(t)$ is to be a WSS process. The converse is not true. A WSS $\gamma(t)$ may produce a nonstationary $x(t)$, indicated as follows. Rewriting (A-1) as

$$x(t) = \tfrac{1}{2}\,\gamma(t) \exp(j2\pi f_0 t) + \tfrac{1}{2}\,\gamma^*(t) \exp(-j2\pi f_0 t) \tag{A-9}$$

the autocorrelation for $x(t)$ can be expressed as

$$k_{xx}(t+\tau, t) = E[x(t+\tau)x(t)] = \tfrac{1}{2}\,\mathrm{Re}\,[k_{\gamma\gamma}(\tau) \cdot \exp(j2\pi f_0\tau)] + \tfrac{1}{2}\,\mathrm{Re}\,[k_{\gamma\gamma^*}(\tau) \cdot \exp(j4\pi f_0 t + j2\pi f_0\tau)] \quad \text{(A-10)}$$

where, for complex WSS processes, we define the autocorrelation of $\gamma(t)$ as

$$k_{\gamma\gamma}(\tau) = E[\gamma(t+\tau)\gamma^*(t)]. \quad \text{(A-11)}$$

The quantity $k_{\gamma\gamma^*}(\tau) = E[\gamma(t+\tau)\,\gamma\,(t)]$ in (A-10) can be regarded as the cross correlation between signal components centered at $+f_0$ and at $-f_0$. If $x(t)$ is WSS, then this cross correlation must vanish in order that the t-dependent term in (A-10) vanish. Otherwise $x(t)$ is a cyclostationary process.

If we let $\gamma(t) = u(t) + j\,v(t)$, where the I and Q processes, $u(t)$ and $v(t)$, are jointly WSS, then we have

$$k_{\gamma\gamma^*}(\tau) = k_{uu}(\tau) - k_{vv}(\tau) + j[k_{vu}(\tau) + k_{uv}(\tau)] \quad \text{(A-12)}$$

and the condition for stationarity of $x(t)$ requires that

$$k_{uu}(\tau) = k_{vv}(\tau) \quad \text{and } k_{vu}(\tau) = -k_{uv}(\tau). \quad \text{(A-13)}$$

Thus for a WSS bandpass process, the I and Q components are balanced in the sense that they have the same autocorrelation function. Also, the cross correlation of the I and Q components must be an odd function, since $k_{vu}(\tau) = k_{uv}(-\tau)$ for any pair of WSS processes. For example, $u(t) = v(t)$ would satisfy the autocorrelation condition in (A-13), but not the cross correlation condition. The size of $k_{\gamma\gamma^*}(\tau)$ indicates the degree of cyclostationarity of a bandpass process. In the extreme case where either the I or Q component is missing, as in DSB-AM, we would have $k_{\gamma\gamma^*}(\tau) = \pm\, k_{\gamma\gamma}(\tau)$, e.g., for $v(t) = 0$,

$$k_{xx}(t+\tau, t) = k_{uu}(\tau)\cos(2\pi f_0\tau)\cos(2\pi f_0 t + 2\pi f_0\tau). \quad \text{(A-14)}$$

In modeling an additive noise process $n(t)$ on received signals, we often use the white-noise assumption wherein $k_{nn}(\tau) = N_0\delta(\tau)$. If we let $r(t) + j\,s(t)$ be the complex envelope of the process relative to any f_0 which is significantly larger than the passband width of the signals, then the white-noise process is equivalently modeled by I and Q processes whose correlation functions are given by

$$k_{rr}(\tau) = k_{ss}(\tau) = 2N_0\delta(\tau); \qquad k_{rs}(\tau) = 0. \quad \text{(A-15)}$$

REFERENCES

[1] J. J. Stiffler, *Theory of Synchronous Communications.* Englewood Cliffs, NJ: Prentice-Hall, 1971.

[2] W. C. Lindsey, *Synchronization Systems in Communication and Control.* Englewood Cliffs, NJ: Prentice-Hall, 1972.

[3] W. C. Lindsey and M. K. Simon, *Telecommunication Systems Engineering.* Englewood Cliffs, NJ: Prentice-Hall, 1973.

[4] F. M. Gardner, *Phaselock Techniques*, 2nd ed. New York: Wiley, 1979.

[5] A. J. Viterbi, *Principles of Coherent Communication.* New York: McGraw-Hill, 1966.

[6] L. E. Franks, *Signal Theory.* Englewood Cliffs, NJ: Prentice-Hall, 1969.

[7] W. R. Bennett, "Statistics of regenerative digital transmission," *Bell Syst. Tech. J.*, vol. 37, pp. 1501–1542, Nov. 1958.

[8] W. A. Gardner and L. E. Franks, "Characterization of cyclostationary random signal processes," *IEEE Trans. Inform. Theory*, vol. IT-21, pp. 4–14, Jan. 1975.

[9] F. M. Gardner, "Hangup in phase-lock loops," *IEEE Trans. Commun.*, vol. COM-25, pp. 1210–1214, Oct. 1977.

[10] M. K. Simon, "Further results on optimum receiver structures for digital phase and amplitude modulated signals," presented at 1978 Int. Conf. Commun., Toronto, Canada, 1978.

[11] L. E. Franks and J. P. Bubrouski, "Statistical properties of timing jitter in a PAM timing recovery scheme," *IEEE Trans. Commun.*, vol. COM-22, pp. 913–920, July 1974.

[12] P. Kabal and S. Pasupathy, "Partial response signaling," *IEEE Trans. Commun.*, vol. COM-23, pp. 921–934, Sept. 1975.

[13] H. Sailer, "Timing recovery in data transmission systems using multilevel partial response signaling," presented at 1975 Int. Conf. Commun., San Francisco, CA, 1975.

[14] S. U. H. Qureshi, "Timing recovery for equalized partial response systems," *IEEE Trans. Commun.*, vol. COM-24, pp. 1326–1330, Dec. 1976.

[15] E. Roza, "Analysis of phase-locked timing extraction circuits for pulse code transmission," *IEEE Trans. Commun.*, vol. COM-22, pp. 1236–1249, Sept. 1974.

[16] F. M. Gardner, "Self-noise in synchronizers," this issue, pp. 1159–1163.

[17] H. L. Van Trees, *Detection, Estimation and Modulation Theory, Part I.* New York: Wiley, 1968.

[18] L. E. Franks, "Acquisition of carrier and timing data-I," in *Signal Processing in Communication and Control.* Groningen, The Netherlands: Noordhoff, 1975, pp. 429–447.

[19] W. C. Lindsey and M. K. Simon, "Data-aided carrier tracking loop," *IEEE Trans. Commun.*, vol. COM-19, pp. 157–168, Apr. 1971.

[20] R. Matyas and P. J. McLane, "Decision-aided tracking loops for channels with phase jitter and intersymbol interference," *IEEE Trans. Commun.*, vol. COM-22, pp. 1014–1023, Aug. 1974.

[21] M. K. Simon and J. G. Smith, "Offset quadrature communications with decision-feedback carrier synchronization," *IEEE Trans. Commun.*, vol. COM-22, pp. 1576–1584, Oct. 1974.

[22] U. Mengali, "Synchronization of QAM signals in the presence of ISI," *IEEE Trans. Aerosp. Electron. Syst.*, vol. AES-12, pp. 556–560, Sept. 1976.

[23] A. L. McBride and A. P. Sage, "Optimum estimation of bit synchronization," *IEEE Trans. Aerosp. Electron. Syst.*, vol. AES-5, pp. 525–536, May 1969.

[24] P. A. Wintz and E. J. Luecke, "Performance of optimum and suboptimum synchronizers," *IEEE Trans. Commun.*, vol. COM-17, pp. 380–389, June 1969.

[25] R. D. Gitlin and J. Salz, "Timing recovery in PAM systems," *Bell Syst. Tech. J.*, vol. 50, pp. 1645–1669, May–June 1971.

[26] M. H. Meyers and L. E. Franks, "Joint carrier phase and symbol timing for PAM systems," this issue, pp. 1121–1129.

[27] L. E. Franks, "Timing recovery problems in data communication," in *Communication Systems and Random Process Theory.* Groningen, The Netherlands: Sijthoff and Noordhoff, 1978, pp. 111–127.

[28] H. Kobayashi, "Simultaneous adaptive estimation and decision algorithm for carrier modulated data transmission systems," *IEEE Trans. Commun.*, vol. COM-19, pp. 268–280, June 1971.

[29] G. Ungerboeck, "Adaptive maximum likelihood receiver for carrier-modulated data transmission systems," *IEEE Trans. Commun.*, vol. COM-22, pp. 624–636, May 1974.

[30] D. D. Falconer and J. Salz, "Optimal reception of digital data over the Gaussian channel with unknown delay and phase jitter," *IEEE Trans. Inform. Theory*, vol. IT-23, pp. 117–126, Jan. 1977.

[31] U. Mengali, "Joint phase and timing acquisition in data transmission," *IEEE Trans. Commun.*, vol. COM-25, pp. 1174–1185, Oct. 1977.

[32] M. Mancianti, U. Mengali, and R. Reggiannini, "A fast start-up algorithm for channel parameter acquisition in SSB-AM data transmission," presented at 1979 Int. Conf. Commun., Boston, MA, 1979.

[33] L. E. Franks, "Further results on Nyquist's problem in pulse transmission," *IEEE Trans. Commun.*, vol. COM-16, pp. 337–340, Apr. 1968.

[34] F. S. Hill, "Optimum pulse shapes for PAM data transmission using VSB modulation," *IEEE Trans. Commun.*, vol. COM-23, pp. 352–361, Mar. 1975.

PHOTO
NOT
AVAILABLE

L. E. Franks (S'48–M'61–SM'71–F'77) was born in San Mateo, CA, on November 8, 1931. He received the B.S. degree in electrial engineering from Oregon State University, Corvallis, in 1952, and the M.S. and Ph.D. degrees in electrical engineering from Stanford University, Stanford, CA, in 1953 and 1957, respectively.

In 1958 he joined Bell Laboratories, Murray Hill, NJ, working on filter design and signal analysis problems. He moved to Bell Laboratories, North Andover, MA, in 1962 to serve as Supervisor of the Data Systems Analysis Group. In 1969 he became a Faculty Member at the University of Massachusetts, Amherst, where he is currently Professor of Electrical and Computer Engineering and is engaged in research and teaching in signal processing and communication systems. He served as Chairman of the Department between 1975 and 1978. He was Academic Visitor at Imperial College, London, England, in 1979.

Dr. Franks is the author of the textbook, *Signal Theory* (Englewood Cliffs, NJ: Prentice-Hall, 1969). He is a member of the URSI Commission C, the Communication Theory and Data Communication Systems Committees of the IEEE Communications Society, and currently is the Associate Editor for Communications for the IEEE TRANSACTIONS ON INFORMATION THEORY.

Tamed Frequency Modulation, A Novel Method to Achieve Spectrum Economy in Digital Transmission

FRANK de JAGER AND CORNELIS B. DEKKER, MEMBER, IEEE

Abstract—This paper describes a new type of frequency modulation, called Tamed Frequency Modulation (TFM), for digital transmission. The desired constraint of a constant envelope signal is combined with a maximum of spectrum economy which is of great importance, particularly in radio channels. The out-of-band radiation is substantially less as compared with other known constant envelope modulation techniques. With synchronous detection, a penalty of only 1 dB in error performance is encountered as compared with four-phase modulation. The idea behind TFM is the proper control of the frequency of the transmitter oscillator, such that the phase of the modulated signal becomes a smooth function of time with correlative properties. Simple and flexible implementation schemes are described.

I. INTRODUCTION

IN the last fifteen years numerous modulation systems for efficient digital transmission via telephone lines have been introduced. In almost all cases the resultant modulated signals exhibit amplitude variations. Such systems are implemented with linear amplifiers and linear modulators. For radio communication constant-envelope modulated signals are preferable due to existing system constraints in power economy and the consequent use of non-linear power amplifiers. Quite naturally this leads to the use of frequency modulation.

However, the spectrum of an FM signal is relatively wide. In order to narrow the spectrum, a channel filter with a precisely prescribed attenuation and phase characteristic may be used but this is not attractive in radio equipment. Besides, a filter of this kind cannot be used if the transmitted center frequency has to be changed.

Another method for narrowing the spectrum is to shape the data at the input of the frequency modulator by means of a filter. The requirements to be met by this premodulation filter are manifold. The use of the filter has to result in a transmitted spectrum which is as narrow as possible in order to be efficient with regard to spectrum economy. We may, in addition, require that the receiver has to be able to detect the signal reliably, even with an unknown frequency shift between the transmitter and the receiver. Moreover, if an orthogonal coherent demodulator is used, the detection should in principle be independent of the level of the received signal.

A solution to these problems, called Tamed Frequency Modulation (TFM), is described in the present paper. The concept of TFM is explained in the first section. In the next section we calculate the necessary transfer characteristic of the premodulation filter. In the following section the influence of some parameters of the transfer characteristic on the TFM spectrum will be investigated. The description of the receiver and the calculation of the receiving filters follows next. It will then be shown that the implementation can be simple and flexible. The last section presents a further evaluation to obtain other types of synchronous frequency modulation.

Manuscript received October 7, 1977.
The authors are with Philips Research Laboratories, Eindhoven, The Netherlands.

II. THE CONCEPT OF TFM

To start with, we will consider four-phase modulation (4ϕ), well known from digital transmission via telephone lines. With this modulation method information can be transmitted efficiently in a relatively narrow bandwidth. In the two-dimensional representation of its signal space diagram in Fig. 1(a), four signal points are defined on a circle. Every two bits of the incoming binary data stream are encoded into one of these points. Between two sampling moments, the phasor $v(t)$, representing the modulation, moves from one point to another. The way in which this takes place determines the spectrum width of the modulated signal. The magnitude of the phasor is often deliberately not kept constant between the sampling moments, in order to reduce the spectrum width. However, as mentioned in the introduction, we prefer a constant amplitude for our communication applications.

The coherent orthogonal demodulator is the optimum receiver structure for 4ϕ-modulation [1]. A certain frequency shift between the transmitter and the receiver can be tolerated because a carrier recovery system is used. Detection depends only on the polarity of the X and Y components of the incoming signal. Some of the conditions mentioned in the introduction are thus automatically fulfilled.

In order to avoid amplitude variations the phasor $v(t)$ must remain on a circle during the transitions from one signal point to another. An example of a system with this performance is frequency shift keying with a modulation index $m = 0.5$ [2], also called Fast Frequency Shift Keying (FFSK) [3] or Minimum Shift Keying (MSK) [4], whose signal space diagram is given in Fig. 1(b). During a sampling interval the phasor moves from a signal point to one of the two neighboring ones. The corresponding variation of the phase $\phi(t)$ of the modulated signal is indicated by the dashed line in Fig. 2. However, the power spectral density function (PSDF) [5] of MSK, represented by the dashed line in Fig. 3, is somewhat wide.

This is mainly due to the sharp edges in the phase path. The spectrum will be narrowed if the edges are smoothed and coherent detection remains possible if the phase values remain the same at the sampling moments, $t = mT$ (m integer and $T = 1/f_{\text{bit}} = 2\pi/\omega_{\text{bit}}$).

Reprinted from *IEEE Transactions on Communications,* vol. COM-26, no. 5, May 1978.

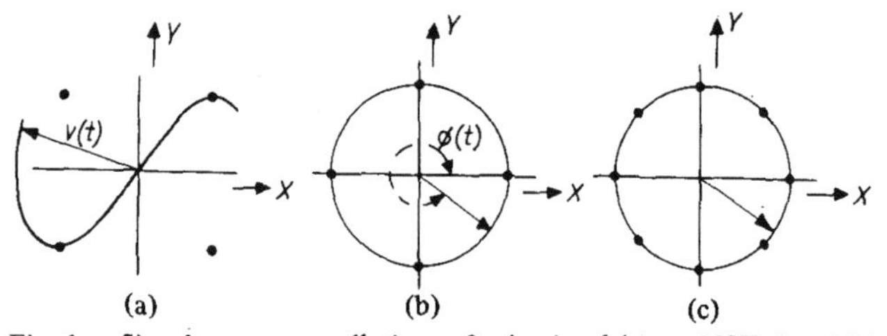

Fig. 1. Signal space constellations of a 4ϕ-signal (a), an MSK signal (b) and a TFM signal (c).

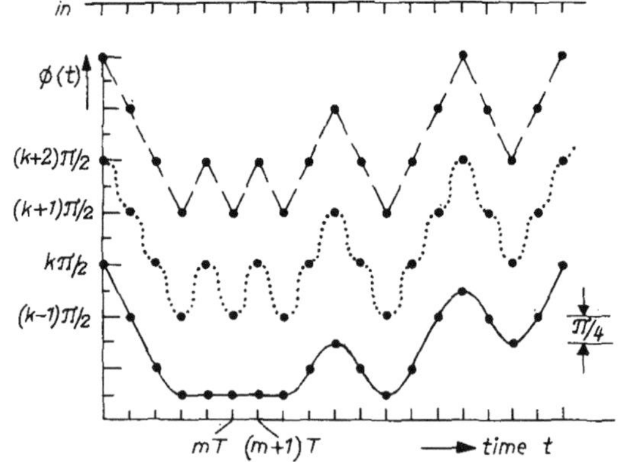

Fig. 2. Phase behavior of MSK (- - -), MSK with sinusoidal smoothing (· · · ·) and TFM (——).

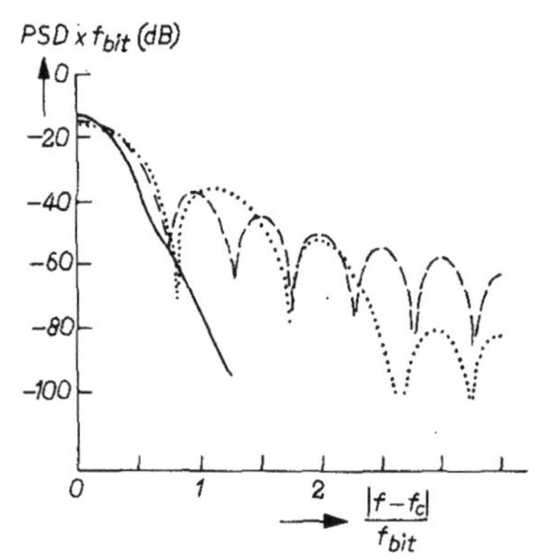

Fig. 3. Power spectral density functions of MSK (- - -), MSK with sinusoidal smoothing (· · · ·) and TFM (——).

In the literature considerable attention has been devoted to finding the optimal smoothing, in the sense of giving a PSDF which is as narrow as possible. Amoroso [4] used the sinusoidal smoothing represented by the dotted line in Fig. 2. The sharp edges have disappeared but the maximum slope of the phase path has considerably increased. The corresponding PSDF, given in Fig. 3 by the dotted line, therefore shows but little improvement.

We found that it is possible to obtain a considerably better spectrum efficiency by prescribing a phase path as depicted by the solid line in Fig. 2. We named this modulation method Tamed Frequency Modulation (TFM). The phase path is characterized by the fact that the values of the phase at the successive sampling moments $t = (m-1)T, mT, (m+1)T, \cdots$ are obtained from the data via a different code rule from that for MSK. If the data signal $a(t)$ is defined by

$$a(t) = \sum_{n=-\infty}^{\infty} a_n \cdot \delta(t - nT), \qquad \text{with } a_n = +1 \text{ or } -1, \quad (1)$$

then for MSK

$$\phi(mT + T) - \phi(mT) = (\pi/2) \cdot a_m \quad (2)$$

and for TFM

$$\phi(mT + T) - \phi(mT) = (\pi/2) \cdot (a_{m-1}/4 + a_m/2 + a_{m+1}/4) \quad (3)$$

with $\phi(0) = 0$ if $a_0 \cdot a_1 = 1$ and $\phi(0) = \pi/4$ if $a_0 \cdot a_1 = -1$.

From this code rule it is easily established that phase-changes of $\pi/2$ are obtained if three succeeding bits have the same polarity, and the phase remains constant for three bits of alternating polarity. Phase changes of $\pi/4$ are connected with the bit configurations $++-$, $+--$, $-++$ and $--+$.

In addition the phase should vary as smoothly as possible in going through the values obtained with (3), as shown in Fig. 2. The signal space diagram of TFM is shown in Fig. 1(c). The Tamed Frequency Modulation should not be mixed up with eight-phase modulation, because the phase values of the TFM signal at successive sampling moments are dependent. For example, it is possible only to obtain a constant phase for phase values which are odd multiples of $\pi/4$.

III. SPECIFICATION OF THE PREMODULATION FILTER

The wanted fluctuations of the phase as a function of time are obtained when the signal is applied to a frequency modulator as shown in Fig. 4. The premodulation filter $G(\omega)$ has to shape the data signal $a(t)$ so as to obtain the wanted smooth phase fluctuations. Let the deviation sensitivity of the modulator be K_0 radians/volt/second. With the definition of $a(t)$ in (1) and the impulse response $g(t)$ of the filter, the phase $\phi(t)$ can be written as

$$\phi(t) = K_0 \cdot \int_{-\infty}^{t} \left[\sum_{n=-\infty}^{\infty} a_n \cdot g(\tau - nT) \right] \cdot d\tau + C,$$

or

$$\phi(t) = K_0 \cdot \sum_{n=-\infty}^{\infty} a_n \cdot x(t - nT) + C, \quad (4)$$

where

$$x(t) = \int_{-\infty}^{t} g(\tau) \cdot d\tau$$

and C is a constant. At the sampling moment $t = mT$ the phase becomes

$$\phi(mT) = K_0 \cdot \sum_{n=-\infty}^{\infty} a_n \cdot x(mT - nT) + C$$

and

$$\phi(mT + T) - \phi(mT)$$

$$= K_0 \cdot \sum_{n=-\infty}^{\infty} a_n \cdot [x(mT + T - nT) - x(mT - nT)]$$

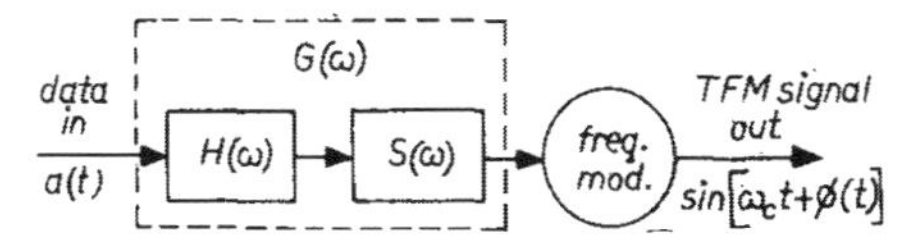

Fig. 4. Basic circuit for the generation of a TFM signal.

or

$$\phi(mT+T)-\phi(mT)$$
$$=K_0 \sum_{l=-\infty}^{\infty} a_{m-l}[x(lT+T)-x(lT)]. \tag{5}$$

From eq. 3 it can be seen that the right-hand side of (5) can be equal to $\pm\pi/2$, $\pm\pi/4$ or 0 radians.

Writing the code rule in (3) in more detail as

$$\phi(mT+T)-\phi(mT)$$
$$=(\pi/2)\cdot(\cdots+a_{m-2}\cdot 0+a_{m-1}/4+a_m/2+a_{m+1}/4$$
$$+a_{m+2}\cdot 0+\cdots), \tag{6}$$

then, with (5), we obtain

$$x(lT+T)-x(lT)=\begin{cases}\pi/(8K_0) & \text{for } l=1,\\ \pi/(4K_0) & \text{for } l=0,\\ \pi/(8K_0) & \text{for } l=-1,\\ 0 & \text{otherwise.}\end{cases} \tag{7}$$

Combining this with the definition of $x(t)$ in (4) we obtain

$$\int_{lT}^{(l+1)T} g(t)\cdot dt=\begin{cases}\pi/(8K_0) & \text{for } |l|=1,\\ \pi/(4K_0) & \text{for } l=0,\\ 0 & \text{otherwise.}\end{cases} \tag{8}$$

The relation in (8) gives the condition for the impulse response $g(t)$ needed to ensure that the phase goes through the values at the sampling moments shown in Fig. 2 for TFM. This condition is certainly fulfilled if $g(t)$ is derived from a single pulse $h(t)$, satisfying the third Nyquist criterion* [6], [7] by simple scaling and delay operations with a network $S(\omega)$ as is shown in Fig. 5. The transfer characteristic of this network is given by

$$S(\omega)=[\pi/(8K_0)]\cdot e^{-j\omega T}+[\pi/(4K_0)]$$
$$+[\pi/(8K_0)]\cdot e^{j\omega T}$$
$$=[\pi/(2K_0)]\cdot\cos^2(\omega T/2). \tag{9}$$

*An impulse response $h(t)$ satisfies the third Nyquist criterion if, for any integer l:

$$\int_{(2l-1)\cdot T/2}^{(2l+1)\cdot T/2} h(t)\,dt=\begin{cases}1 & \text{for } l=0,\\ 0 & \text{otherwise.}\end{cases}$$

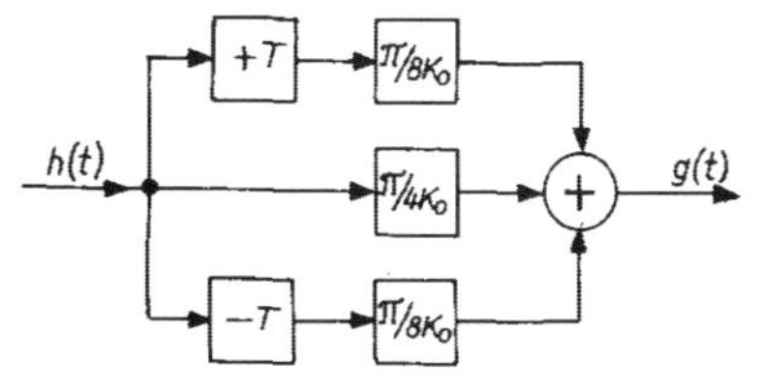

Fig. 5. Network with a transfer characteristic $S(\omega)=[\pi\cos^2(\omega T/2)]/(2K_0)$.

The overall shape factor $G(\omega)$ of the premodulation filter can now be written as

$$G(\omega)=H(\omega)\cdot S(\omega)$$
$$=[\pi/(2K_0)]\cdot H(\omega)\cdot\cos^2(\omega T/2) \tag{10}$$

where $H(\omega)$ applies to any filter giving a pulse satisfying the third Nyquist criterion.

IV. DIFFERENT TFM SYSTEMS

In the previous sections we have seen that the combination of a premodulation filter $G(\omega)$, defined in (10), and a frequency modulator with deviation sensitivity K_0, gives rise to a phase behavior as given in (5), thus yielding a TFM system. The filter $S(\omega)$, being a part of $G(\omega)$, determines the total amount of phase increase or decrease during a sampling interval, while the other part of $G(\omega)$, namely $H(\omega)$, prescribes the phase path from $\phi(mT)$ to $\phi(mT+T)$. In this section we will look at the influence of different functions for $H(\omega)$ on the TFM spectrum.

Generally, $H(\omega)$ can be written as [7]:

$$H(\omega)=[(\omega T)/(2\sin(\omega T/2))]\cdot N_1(\omega), \tag{11}$$

where $N_1(\omega)$ is the Fourier spectrum of a function satisfying the first Nyquist criterion. A class of Nyquist characteristics which has been extensively used and studied is the raised-cosine characteristic [1]:

$$N_1(\omega)=\begin{cases}1 & \text{for } 0\leqslant|\omega|\leqslant\pi(1-\alpha)/T,\\ [1-\sin((T\omega-\pi)/2\alpha)]/2 & \\ \quad\text{for } \pi(1-\alpha)/T\leqslant|\omega|\leqslant\pi(1+\alpha)/T, & \\ 0 & \text{otherwise.}\end{cases} \tag{12}$$

The only variable left is the roll-off factor α ($0\leqslant\alpha\leqslant 1$). For various values of α the PSDF's of the corresponding TFM systems will be calculated.

In following the calculation method of Garrison [5], we truncate the length of the impulse response $g(t)$ of the network $G(\omega)$ to five symbol intervals ($5T$), and we henceforth approximate the impulse response with eight samples per bit interval T. If Garrison's method is applied to a system in which $G(\omega)$ is implemented by analog means, then the results incorporate an approximation error. However, for a digital implementation this error does not occur.

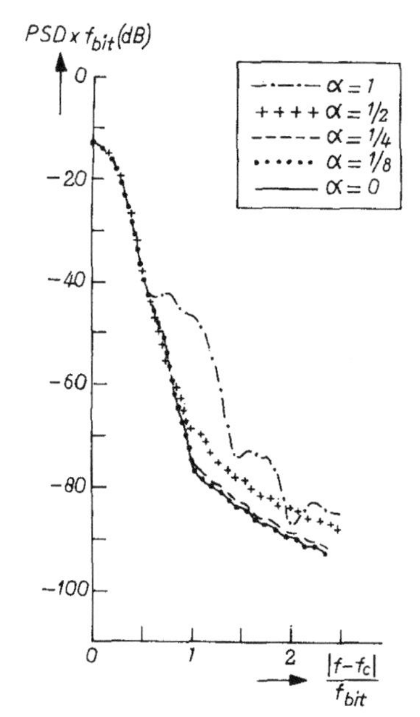

Fig. 6. PSDF's of the TFM signal with different roll-off factors in the transfer characteristic of the premodulation filter.

The resulting power spectral density functions**, plotted in Fig. 6, show no great improvement of the out-of-band radiation when the roll-off factor α is made smaller than 0.25.

A certain truncation of the length of the impulse response $g(t)$ also has to be accepted, so as to keep the amount of hardware for the implementation of $G(\omega)$ small. In the following, the PSDF of TFM is calculated three times for different truncation lengths ($3T$, $5T$, $7T$). For $H(\omega)$ we have chosen a filter with the smallest bandwidth ($\alpha = 0$) [8]:

$$H(\omega) = \begin{cases} \omega T/(2 \sin(\omega T/2)) & \text{for } |\omega| \leq \pi/T, \\ 0 & \text{otherwise.} \end{cases} \tag{13}$$

The shape factor of $G(\omega)$ with this $H(\omega)$ is

$$G(\omega) = \begin{cases} [\pi\omega T/(4K_0 \cdot \sin(\omega T/2))] \cdot \cos^2(\omega T/2) \\ \quad \text{for } |\omega| \leq \pi/T, \\ 0 \quad \text{otherwise} \end{cases} \tag{14}$$

and the corresponding $g(t)$ is shown in Fig. 7. The PSDF's of TFM systems with different truncation lengths of $g(t)$, depicted in Fig. 8, show a considerable reduction of the out-of-band radiation if the length is increased. When we draw the PSDF of the $7T$-version of TFM in Fig. 3 (solid line), we see finally the great improvement obtained by TFM in comparison with MSK and MSK with sinusoidal smoothing.

V. RECEIVER STRUCTURE

From the signal space diagram of TFM in Fig. 1(c) it can be seen that an orthogonal coherent demodulator can be used as

** The 0 dB reference is a constant but arbitrary value throughout the paper.

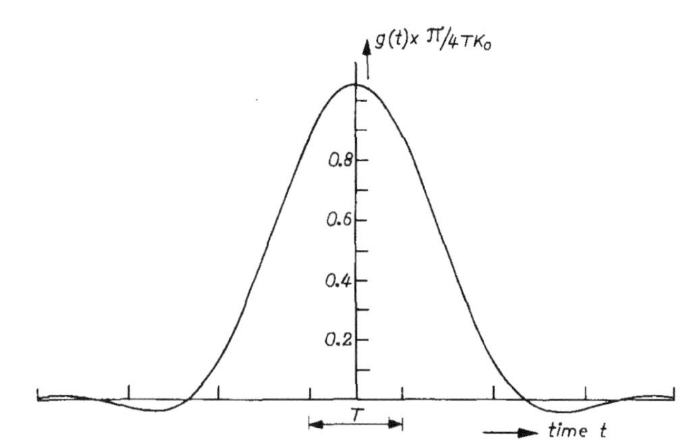

Fig. 7. Impulse response $g(t)$ of the premodulation filter.

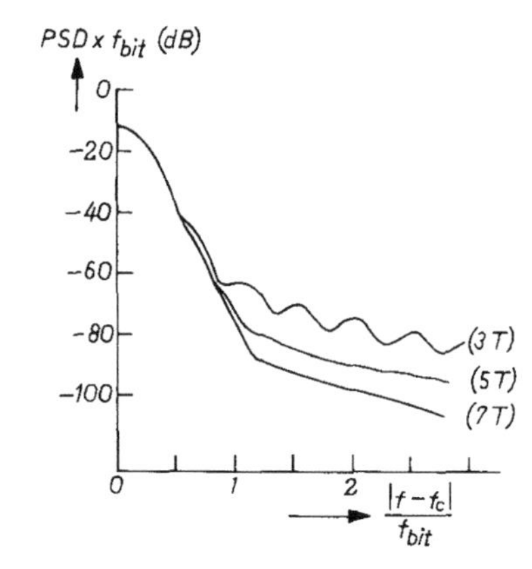

Fig. 8. PSDF's of the TFM signal with different truncations of the length of the impulse response of the premodulation filter.

receiver (Fig. 9). The incoming TFM signal $\sin(\omega_c t + \phi(t))$ is multiplied by respectively $\sin(\omega_c t)$ and $\cos(\omega_c t)$. This results in baseband signals $\cos[\phi(t)]$ and $\sin[\phi(t)]$ which follow from the phase function in Fig. 2. The observed eye patterns are shown in Fig. 10(a) and these signals arrive at the input of the low-pass filters $A_1(\omega)$ and $A_2(\omega)$, together denoted as $A_{1,2}(\omega)$. These filters provide for minimization of the error probability. Finally, a decoder of the same configuration as used by de Buda [3] in his MSK demodulator, can produce the output data signal. In the MSK case, the demodulated signals in de Buda's decoder can only have the amplitude 1 at the sampling moments, while in TFM the amplitude can be 1 or 0.707. This is due to the smoothed phase path occurring in TFM, which however does not affect the polarity of the demodulated signals. This can be shown in a straight-forward way by comparing the phase paths of both FFSK and TFM (Fig. 2).

The deterioration of the eye opening of TFM will cause a somewhat higher error probability, but in this section we will see that the penalty is only 1 dB as compared with 4ϕ-modulation. The question of error propagation and the need for differential encoding for TFM are the same as for MSK [3].

Also, carrier recovery and clock recovery for TFM and MSK are quite similar. We will come back to this in the following section.

The low-pass filters $A_{1,2}(\omega)$ need some special attention. The filters have to minimize the error probability. This can be

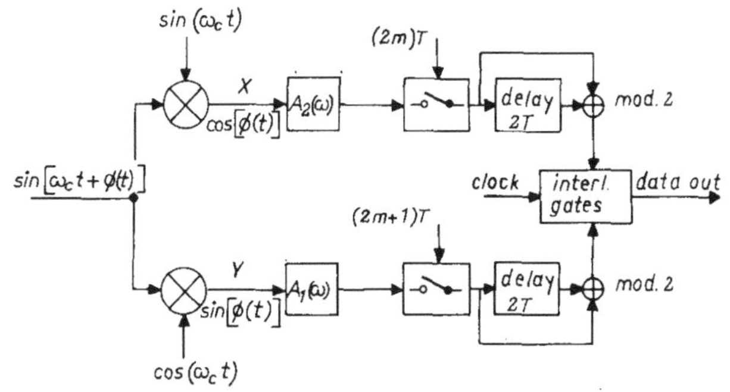

Fig. 9. Orthogonal coherent receiver structure.

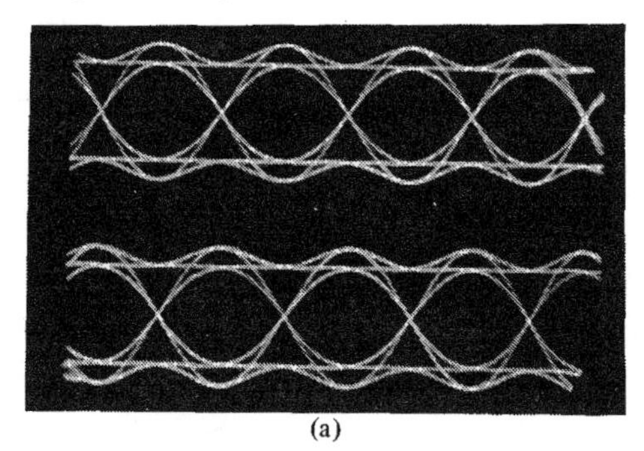

(a)

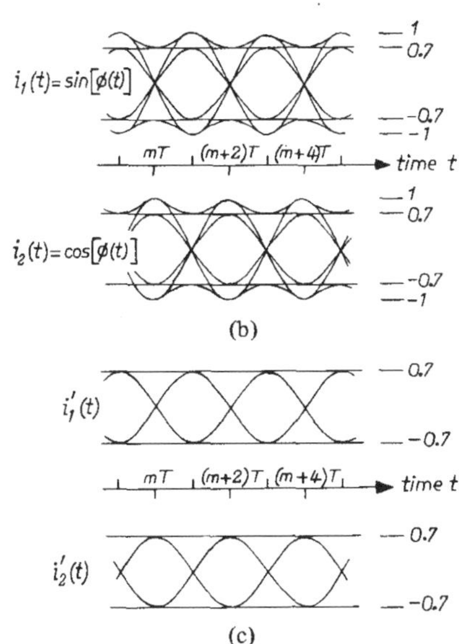

Fig. 10. Observed eye patterns of the demodulated signals (a, b) and eye patterns of the approximating signals (c).

done by simply minimizing the noise variance σ^2 in the output signal, if the intersymbol interference is taken to be zero. The solution to this problem is described in the following.

If we suppose for the moment that the transfer characteristics from the data input of the transmitter to the inputs of the filters can be regarded to be the transfer characteristics of two linear networks $I_1(\omega)$ and $I_2(\omega)$, then the optimum receiver filters can be found according to Lucky et al. [1]. Since it is not at all obvious that $I_1(\omega)$ and $I_2(\omega)$ for TFM exist and even so, how they should be derived analytically, we make approximating descriptions, which considerably simplifies this problem.

The observed eye patterns of Fig. 10(a) are schematically redrawn in Fig. 10(b). The eye opening at the sampling moment can see to be $\sqrt{2}(\approx 1.4)$. Two slightly different signals $i_1'(t)$ and $i_2'(t)$ with the same opening are shown in Fig. 10(c). These signals can be constructed by superimposing impulse responses $z(t)$, which can be written as:

$$z(t)=\begin{cases}\sqrt{2}\,[1+\cos(\pi t/(2T))/4] & \text{for } |t|\leqslant 2T,\\ 0 & \text{otherwise.}\end{cases} \tag{15}$$

Now the linear expressions for $i_1'(t)$ and $i_2'(t)$ can be given as

$$i_1'(t)=\sum_{p=-\infty}^{\infty} v_p\cdot\{z(t-2Tp)\}, \tag{16a}$$

where $v_p=+1$ or -1, and

$$i_2'(t)=\sum_{p=-\infty}^{\infty} w_p\cdot\{z(t-T-2Tp)\}, \tag{16b}$$

where $w_p=+1$ or -1.

The corresponding shape factors $I_1'(\omega)$ and $I_2'(\omega)$ are

$$I_1'(\omega)=I_2'(\omega)=I'(\omega)=[\sin(2\omega T)]/[2\omega T(1-(2\omega T/\pi)^2)]. \tag{17}$$

The power spectral density function $[I'(\omega)]^2$ of the signals in (16a) and (16b) is shown in Fig. 11 and approximates the PSDF, $[I(\omega)]^2$, derived from Fig. 8, of the real input signals. With this $I'(\omega)$ rather than $I_1(\omega)$ and $I_2(\omega)$ a useful, nearly optimum filter $A_1(\omega)=A_2(\omega)=A(\omega)$ can easily be found, leading to the transfer characteristic which is shown in Fig. 12. Its equivalent noise bandwidth ω_r is found to be:

$$\omega_r=\int_0^{\infty} A^2(\omega)\cdot d\omega\approx 0.7\,\pi/T. \tag{18}$$

With this approximate filter the calculated bit error probability is only 1 dB worse for TFM (Fig. 13) in comparison with 4ϕ modulation as calculated by Lucky et al. [1].

In practice the difference will be even smaller because ideal rectangular low-pass filters used by Lucky et al. in their description, are not practically feasible. One final observation has to be made. The foregoing calculation of the filters is based on the assumption of white noise, but in practice the out-of-band radiation of transmitters in neighboring channels has to be added and the total disturbing signal received cannot be interpreted as white noise. If this effect is non-negligible the filter $A(\omega)$ has to be re-optimized.

VI. IMPLEMENTATION

a) Transmitter

In this section we give two different implementation diagrams each having its own advantages and disadvantages. The diagram of the transmitter in Fig. 4 is the basis for the first implementation. In the description of the TFM signal we have

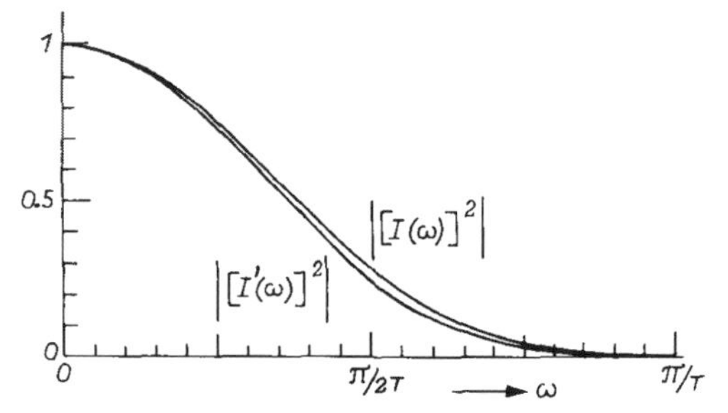

Fig. 11. PSDF's of the demodulated signals and the approximate signals.

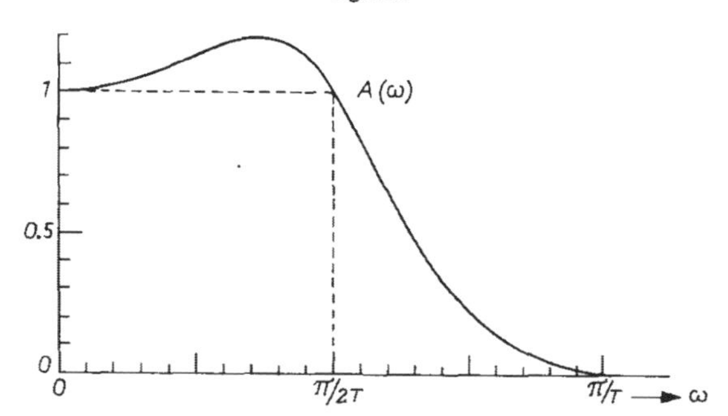

Fig. 12. Transfer characteristic of the low-pass filter $A(\omega)$.

$$P_{4\phi} = \frac{1}{2}\operatorname{erfc}(\sqrt{\eta})$$

$$P_{TFM} = \frac{1}{4}\operatorname{erfc}(\sqrt{\eta}) + \frac{1}{4}\operatorname{erfc}(\sqrt{.69\eta})$$

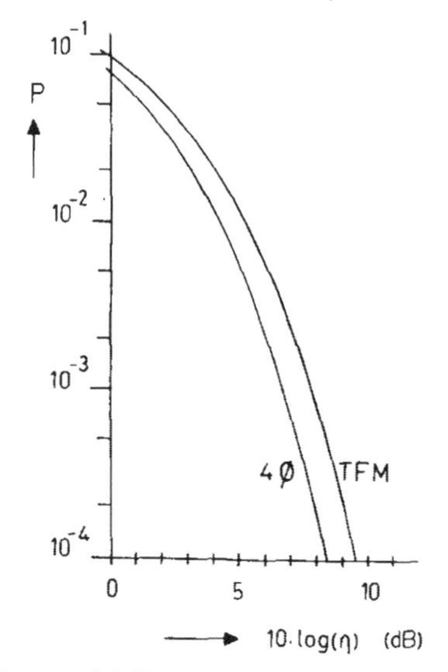

Fig. 13. Error probability curves for 4ϕ modulation and TFM with ideal recovered carrier and clock. The variable η is the signal-to-noise ratio at the input of the receiver in a bandwidth $1/T$.

assumed the center frequency ω_c and the deviation sensitivity K_0 to be invariant. In practice, however, they are insufficiently constant. Extra measures have to be taken, as shown in Fig. 14(a), to keep these parameters at the prescribed values. An adder is inserted for control of the center frequency, while a multiplier can be used to keep the deviation of the frequency modulator at the correct value when K_0 varies. Both are controlled by means of a detector. This circuit has to generate the two control signals to make the phase $\phi(t)$ in the output TFM signal sin $[\omega_c t + \phi(t)]$ pass through the prescribed values at the sampling moments, as shown in Fig. 2. In addition, the

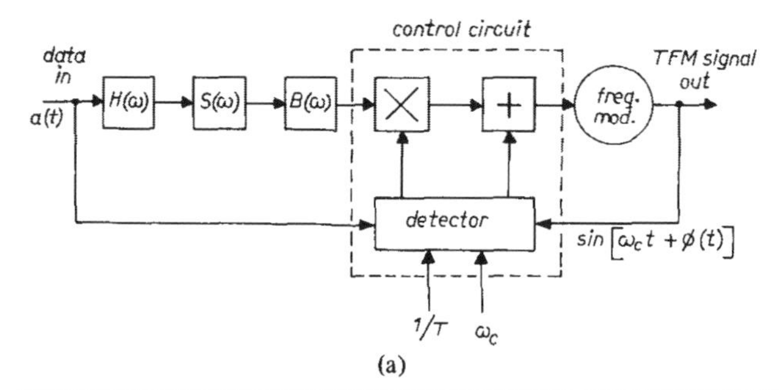

(a)

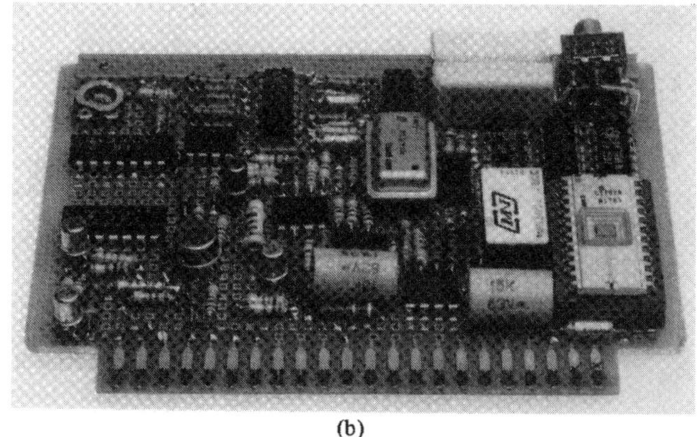

(b)

Fig. 14. Diagram of the TFM transmitter with control circuit (a) and the implementation (b) without the oscillator.

center frequency has to be kept at the specified value. The detector therefore receives as input parameters the input data signal $a(t)$, the sampling moments and the value of the center frequency. These parameters give the information about the center frequency needed and the increase or decrease of the phase per bit interval. The control circuit can be thought of as consisting of two cooperating phase-locked loops. The analytical optimization of this system is difficult since it is a two-dimensional control process.

By combining the filters $H(\omega)$ and $S(\omega)$, the premodulation filter can be easily implemented by means of a digital filter [9]. An extra low-pass filter $B(\omega)$ has to be added to reject spurious signals around multiples of the sampling frequency f_s of the digital filter. Fig. 14(b) shows the implementation of the control circuit and the filters. The advantage of this type of TFM transmitter is that the output TFM signal can exactly meet the constant amplitude condition. A disadvantage is the presence of a feedback system which might cause instabilities.

The other type of transmitter, which is shown in Fig. 15(a), is based on a quadrature modulator. Two signals sin $[\phi(t)]$ and cos $[\phi(t)]$ are fed from a network E to two product modulators operating in quadrature. It will be seen that the output signal is sin $[\omega_c t + \phi(t)]$, i.e., the wanted TFM signal. This signal can thus be applied to a class-C power amplifier, without introducing extra out-of-band radiation. A more detailed implementation diagram is given in Fig. 15(b). The data a_{n+3} enter a shift register with a length of q bits. The value q corresponds with the number of bit intervals to which the length of the impulse response $g(t)$ is truncated ($q \geqslant 3$), as described in the previous section. For the moment we have taken $q = 5$. From equation 3 it can be seen that the difference in phase

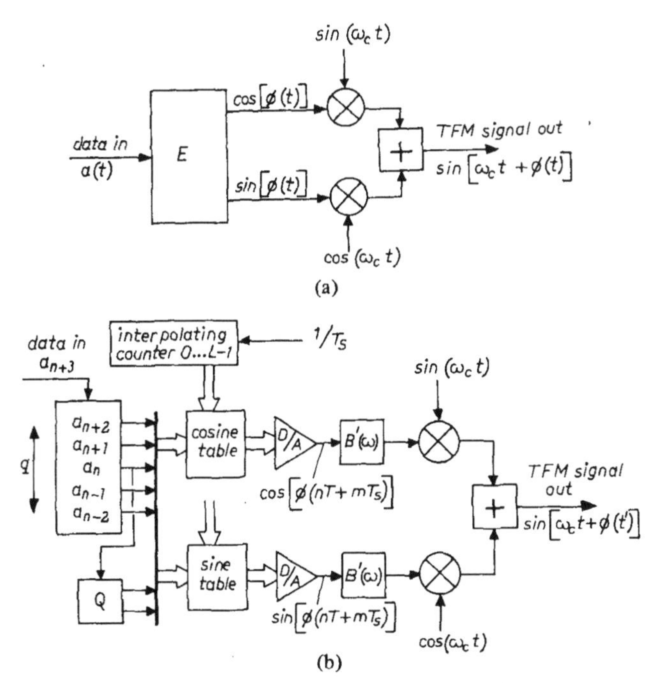

Fig. 15. Basic diagram of the TFM transmitter without a feedback system (a) and a more detailed diagram (b).

between two sampling moments does not exceed $\pm\pi/2$ radians. The cross-over to another quadrant takes place at the sampling moments. Within each quadrant, the phase path is completely determined by the impulse response $g(t)$, truncated over $5T$, and the values a_{n-2}, a_{n-1}, a_n, a_{n+1} and a_{n+2} which are present in the shift register. Moreover, it is necessary to remember in which quadrant the phase path is located. From equation 3 it can be deduced that the phase shifts to a following quadrant in the phase diagram if two successive data symbols have the same value +1. It shifts in the opposite direction if the value is −1. It remains in the original quadrant if the two data symbols do not have the same value. A modified up/down counter, here called quadrant counter Q, can perform this task. The number of the quadrant is represented by two output bits.

The 7 bits, 2 for the quadrant information and 5 for the phase path, form the address for two digital memories called the sine table and the cosine table. The sin $[\phi(t)]$ and cos $[\phi(t)]$, corresponding to the 7 bits, are stored in these memories. The size of the memories increases with q. These "tables" are read out with a sampling frequency f_s. Generally speaking, $f_s = 1/T_s = L \cdot f_{\text{bit}}$, where L is an interpolation factor [10]. The sampled values sin $[\phi(nT + mT_s)]$ and cos $[\phi(nT + mT_s)]$, with $m = 0, 1, 2, \cdots, L-1$, are supplied via digital-to-analog converters and low-pass filters $B'(\omega)$ to the modulators. The accuracy of the converters is limited. This means that we get some distortion which can be considered as noise. At the moment, the accuracy required of the D/A converters for a certain permissible amount of out-of-band radiation in a neighboring channel has been determined heuristically. The implementation of the digital signal processing part described above is shown in Fig. 16. The accuracy of the D/A converters is eight bits and the interpolation factor $L = 8$.

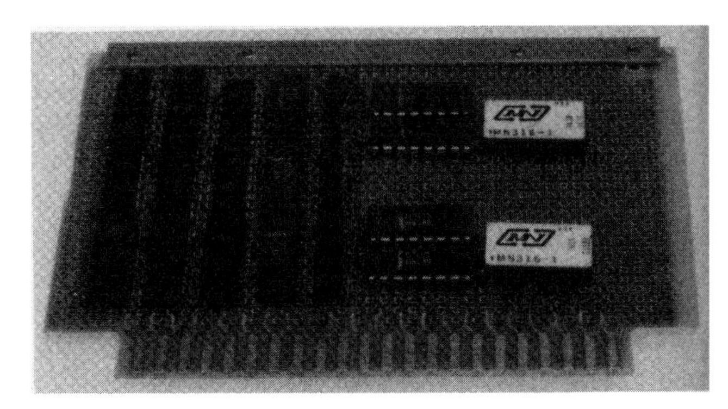

Fig. 16. Implementation of the digital signal processing part E of the TFM transmitter in Fig. 15.

In order to suppress the spurious signals around $\omega_c \pm 2\pi f_s \cdot r$ (r integer) in the modulated signal, low-pass filters $B'(\omega)$ are used. The group delays of these filters should be frequency-invariant and equal in the pass-band. If the cut-off frequency is too low, unwanted variations of the amplitude and phase of the TFM signal will occur. If the interpolation factor L is taken to be large enough, e.g., eight or sixteen, an acceptable cut-off frequency may be $4 \cdot f_{\text{bit}}$ or $8 \cdot f_{\text{bit}}$.

The two parts of the quadrature modulator should have the same flat amplitude and phase characteristic in the frequency band concerned. If they are not the same, amplitude variations and unwanted phase variations will occur, which cannot be eliminated by linear means. For 70 MHz, at which we implemented our TFM system, inequalities of 0.5% have to be taken into account. The power spectral density function measured at the output of the TFM transmitter, shown in Fig. 15 and 16, is given in Fig. 17 and shows good similarity to the corresponding calculated power spectral density function in Fig. 8 ($5T$ version). The flat part in the measured spectrum for $|(f-f_c)/f_{\text{bit}}| > 1$ is caused by the distortion of the 8-bit D/A converters.

The advantage of this type of transmitter is the absence of a feedback system. Its disadvantage is the occurrence of small amplitude variations in practice.

b) Receiver

In the implementation of the receiver according to Fig. 9 a carrier recovery and a clock recovery system have to be provided. These can be similar to the ones suggested by de Buda [3]. The generation of clock and phase reference is based on the fact that the maximal amount of phase change per symbol interval T for both TFM and MSK equals $\pm\pi/2$ radians. Using a frequency doubling circuit, two discrete frequencies are generated. The difference of the frequencies corresponds to the clock frequency and the sum corresponds to four times the carrier frequency.

VII. FURTHER EVALUATION

In the previous sections we have seen that the use of a system according to Fig. 4, where the premodulation filter is defined by $G(\omega) = H(\omega) \cdot S(\omega) = [\pi/(2K_0)] \cdot H(\omega) \cdot \cos^2(\omega T/2)$, gives rise to a TFM signal. The influence of dif-

TABLE 1

$H(\omega)$	$S(\omega)$: $\pi/(2K_0)$	$S(\omega)$: $[\pi \cos(\omega T/2)]/(2K_0)$	$S(\omega)$: $[\pi \cos^2(\omega T/2)]/(2K_0)$
$\dfrac{\sin(\omega T/2)}{\omega T/2}$	(a) Fast Freq. Shift Keying	(b) Duobinary Freq. Shift Keying $m = 0.5$	(c) Tamed Freq. Shift Keying
$\dfrac{\omega T/2}{\sin(\omega T/2)}$ for $\lvert\omega\rvert \leqslant \dfrac{\pi}{T}$; 0 otherwise	(a') Fast Freq. Modulation	(b') Duobinary Freq. Modulation $m = 0.5$	(c') Tamed Freq. Modulation

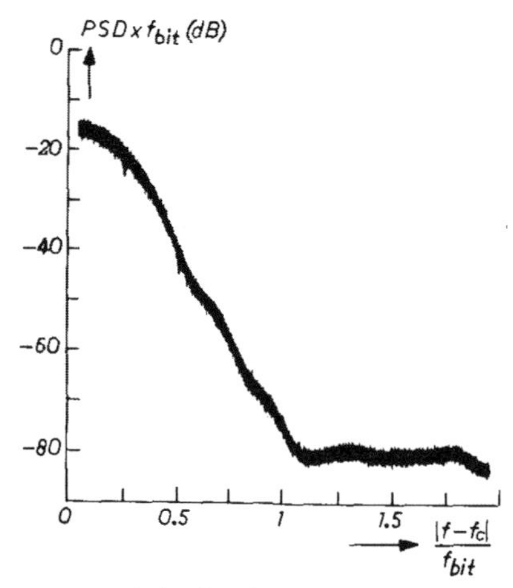

Fig. 17. Power spectral density function measured at the output of the TFM transmitter shown in Figs. 15 and 16, with 8-bit D/A converters, $L = 8$, truncation to $5T$ and $\alpha = 0$.

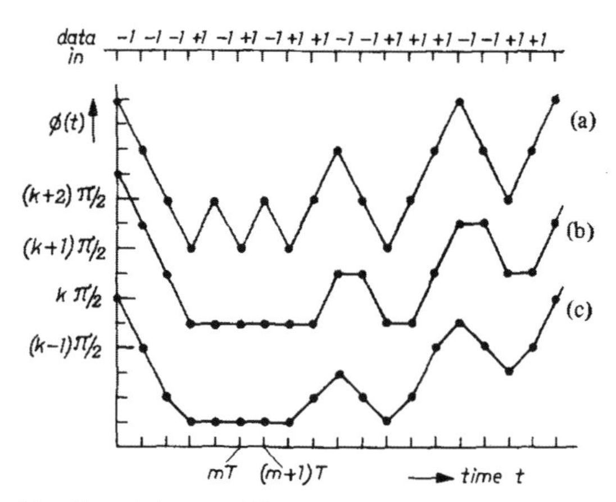

Fig. 18. Phase behavior of FFSK or MSK (a), Duobinary FSK with $m = 0.5$ (b) and Tamed FSK (c).

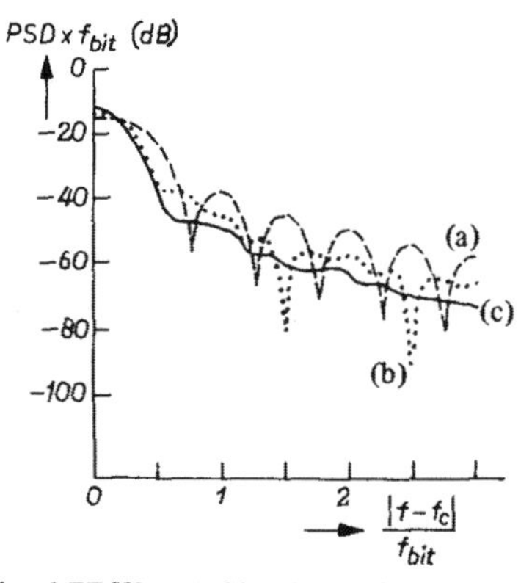

Fig. 19. PSDF's of FFSK or MSK (a), Duobinary FSK with $m = 0.5$ (b), and Tamed FSK (c).

ferent functions for $H(\omega)$ on the outgoing TFM spectrum has been investigated. In this section we compare the spectrum of TFM with spectra of other types of synchronous frequency modulation, which also exhibit a constant amplitude. Originally for $H(\omega)$ the function of equation (13) was used, giving rise to a continuous modulating signal. Now, for $H(\omega)$ a different function is chosen,

$$H(\omega) = [2 \sin(\omega T/2)]/(\omega T), \qquad -\infty < \omega < \infty \tag{19}$$

producing a non-continuous modulating signal. In this case the impulse response $h(t)$ is of rectangular form and obviously satisfies the third Nyquist criterion. To clearly distinguish both cases, we talk about keying versus modulation (Table 1).

Several of these systems have been described in literature before, e.g., Fast Frequency Shift Keying [2, 3], Duobinary Frequency Shift Keying $m = 0.5$ [5, 11] and Duobinary Frequency Modulation $m = 0.5$ [5]***.

In Fig. 18 the phase behaviors for the three frequency shift keying systems are depicted for a certain data stream. In each of these systems the phase changes linearly with time between the transition points. The outgoing PSDF's of these systems are given in Fig. 19, showing a slight improvement in spectrum economy when a high-order correlative coding is used. According to the classification of Kretzmer [12], several different functions for $S(\omega)$ can be taken, but only the three functions used give rise to a two-level eye opening if an orthogonal coherent demodulator is used as the receiver. The spectra of these three systems can be made much narrower, however, by the use of a filter $H(\omega)$ defined in relation (13). This is shown in Fig. 20.

CONCLUSION

In this paper we have described a novel and promising type of frequency modulation, named Tamed Frequency Modu-

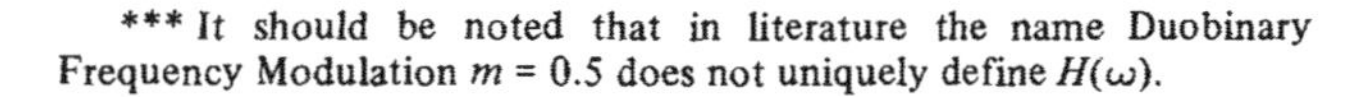

*** It should be noted that in literature the name Duobinary Frequency Modulation $m = 0.5$ does not uniquely define $H(\omega)$.

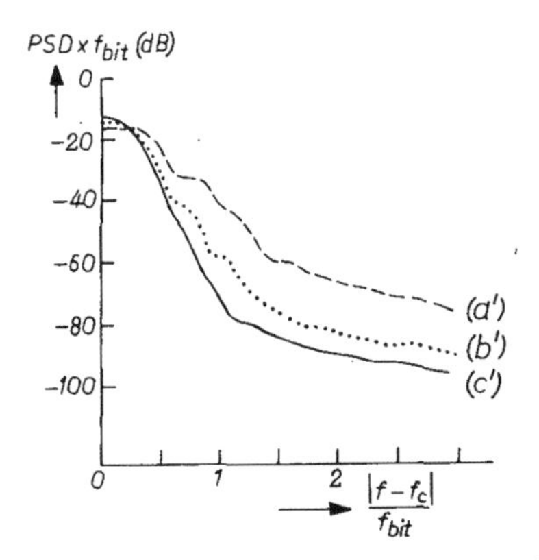

Fig. 20. PSDF's of Fast Frequency Modulation (a'), Duobinary Frequency Modulation with $m = 0.5$ (b') and Tamed Frequency Modulation (c'). In each system the truncation interval equals $5T$.

lation, for digital transmission. A very low out-of-band radiation is obtained as compared with other constant-envelope modulation techniques. In this way the severe constraints of the radio field can be met with the receiving filter shown in Fig. 12 and the PSDF of TFM in Fig. 3. For example, with a radio channel spacing of 25 kHz and a required data rate of 16 kbits/s the power radiated into the adjacent channel can be 85 dB lower than the power radiated into the wanted channel.

The detection quality is almost the same as for four-phase modulation.

Finally, it is shown that the implementation of the TFM transmitter and receiver can be relatively simple.

ACKNOWLEDGMENT

The authors would like to thank D. Muilwijk and B. van de Ham of Philips Telecommunication Industries for their contributions to the investigation.

REFERENCES

[1] R. W. Lucky, J. Salz, E. J. Weldon Jr., *Principles of Data Communication*, McGraw-Hill Book Company, New York, 1968.

[2] H. C. van den Elzen, P. van de Wurf, A Simple Method of Calculating the Characteristics of FSK Signals with Modulation Index 0.5, *IEEE Transactions on Communications*, vol. COM-20, No. 2, pp. 139-147, April 1972.

[3] R. de Buda, Coherent Demodulation of Frequency Shift Keying with Low Deviation Ratio, *IEEE Transactions on Communications*, vol. COM-20, No. 3, pp. 429-435, June 1972.

[4] F. Amoroso, Pulse and Spectrum Manipulation in the Minimum (Frequency) Shift Keying (MSK) Format, *IEEE Transactions on Communications*, vol. COM-24, No. 3, pp. 381-384, March 1976.

[5] G. J. Garrison, A Power Spectral Density Analysis for Digital FM, *IEEE Transactions on Communications*, vol. COM-23, No. 11, pp. 1228-1243, Nov. 1975.

[6] H. Nyquist, Certain Topics in Telegraph Transmission Theory, *AIEE Trans.*, vol. 47, pp. 617-644, April 1928.

[7] S. Pasupathy, Nyquist's Third Criterion, *Proceedings of the IEEE*, vol. 62, No. 6, pp. 860-861, June 1974.

[8] W. R. Bennett, J. R. Davey, *Data Transmission*, McGraw-Hill Book Company, 1965.

[9] A. D. Sypherd, Design of Digital Filters Using Read-Only Memories, *Proceedings of the NEC*, Chicago, vol. 25, 8-10 Dec. 1969, pp. 691-693.

[10] F. A. M. Snijders, N. A. M. Verhoeckx, H. A. van Essen, P. J. van Gerwen, Digital Generation of Linearly Modulated Data Waveforms, *IEEE Transactions on Communications*, vol. COM-23, No. 11, pp. 1259-1270, Nov. 1975.

[11] A. Lender, A Synchronous Signal with Dual Properties for Digital Communications, *IEEE Transactions on Communication Technology*, vol. COM-13, No. 2, pp. 202-208, June 1965.

[12] E. R. Kretzmer, Generalization of a Technique for Binary Data Communication, *IEEE Transactions on Communication Technology*, vol. COM-14, No. 1, pp. 67-68, Febr. 1966.

Frank de Jager was born in Amsterdam, The Netherlands, on June 13, 1919. He was graduated from the Technical University of Delft in 1946, when he joined the Philips Research Laboratories in Eindhoven and was assigned to the Telecommunications Department where he worked in the fields of carrier telephony, deltamodulation, vocoders, companders, data transmission, automatic equalization and, finally, in radio communication. In 1958 he received, together with Johannes A. Greefkes the Veder-Award for work on speech transmission with low signal-to-noise ratios. In 1972, together again with Johannes A. Greefkes, he received the 1972 IEEE Award in International Communication in honor of Hernand and Sosthênes Behn for contributions to communications systems research, in particular, for inventions in the delta-modulation area.

Cornelis B. Dekker (M'76) was born in The Netherlands, on May 23, 1950. He received the degree in electrical engineering from the Technical University, Delft, The Netherlands, in 1973.

After his military service he joined the Telecommunications Department of the Philips Research Laboratories where he is now working on digital communication via radio channels.

Mr. Dekker is a member of the Netherlands Electronics and Radio Society.

Performance Evaluation for Phase-Coded Spread-Spectrum Multiple-Access Communication—Part I: System Analysis

MICHAEL B. PURSLEY, MEMBER, IEEE

Abstract—An analysis of an asynchronous phase-coded spread-spectrum multiple-access communication system is presented. The results of this analysis reveal which code parameters have the greatest impact on communication performance and provide analytical tools for use in preliminary system design. Emphasis is placed on average performance rather than worst-case performance and on code parameters which can be computed easily.

I. INTRODUCTION

IN RECENT YEARS there has been increased interest in a class of multiple-access techniques known as code-division multiple access (CDMA). The CDMA techniques are those multiple-access methods in which the multiple-access capability is due primarily to coding and in which—unlike traditional time- and frequency-division multiple access—there is no requirement for precise time or frequency coordination between the transmitters in the system. CDMA techniques have been considered for a variety of satellite systems including the NASA tracking and data-relay system [17], systems to provide communication to aircraft and other mobile users [11], air traffic control systems [18], and military satellite communication systems. In certain satellite communication systems, CDMA techniques can be designed to provide multiple-access capability and, simultaneously, to reduce the effects of multipath distortion [12].

The most common form of CDMA is spread-spectrum multiple access (SSMA) in which each user is assigned a particular code sequence which is modulated on the carrier along with the digital data. The SSMA techniques are characterized by the use of a high-rate code (i.e., many code symbols per data symbol) which has the effect of spreading the bandwidth of the data signal. The two most common forms of SSMA are frequency-hopped SSMA and phase-coded SSMA. The first of these two was used in the TATS modulation system for the Lincoln Experimental Satellites and is described in detail in [11]. Phase-coded SSMA (also known as direct-sequence spread spectrum [5], [6]) utilizes the most common form of spread-spectrum modulation: the carrier is phase modulated by the digital data sequence and the code sequence.

Manuscript received January 4, 1976; revised March 17, 1977. This work was supported in part by the National Science Foundation under Grant ENG75-22621 and in part by the Joint Services Electronics Program under Contract DAAB-07-72-C-0259. A portion of this paper was presented at the International Telemetering Conference, Los Angeles, CA, September 28-30, 1976.

The author is with the Coordinated Science Laboratory and the Department of Electrical Engineering, University of Illinois, Urbana, IL 61801.

Although phase-coded spread-spectrum modulation has been considered for a wide variety of purposes [5], [9], [12], we are concerned in this paper only with its use in achieving multiple-access capability.

The main topic of this paper is phase-coded SSMA system analysis. We concentrate on communication performance rather than on acquisition and tracking performance, so that the performance measures of interest are error rate and signal-to-noise ratio. Although various aspects of phase-coded SSMA communication were discussed in a number of publications which appeared in the mid-1960's (e.g., [1], [2], [4], [10], [16]), there were very few analytical results on *asynchronous* phase-coded systems and little had been done to identify the important code parameters for asynchronous phase-coded SSMA applications. Most of this work implicitly or explicitly assumed a synchronous model and therefore dealt only with the *periodic* cross-correlation properties of the code sequence. Further, nearly all of the results on cross-correlation properties of sequences dealt with only the periodic correlation (e.g., [8]).

One of the first detailed investigations of asynchronous phase-coded SSMA system performance which dealt with *aperiodic* cross-correlation effects was published in 1969 by Anderson and Wintz [3]. They obtained a bound on the signal-to-noise ratio at the output of the correlation receiver for a SSMA system with a hard-limiter in the channel. The need for considering the aperiodic cross-correlation properties of the code sequences is clearly demonstrated in their paper [3, pp. 286]. Since that time, many additional results have been obtained (e.g., [12], [13], [19]) which help clarify the role of aperiodic correlation in asynchronous phase-coded SSMA communication. In this paper, we present some of these results and their implications.

II. THE PHASE-CODED SSMA SYSTEM MODEL

The SSMA system model that we will consider is shown in Figure 1 for K users. The k-th user's data signal $b_k(t)$ is a sequence of unit amplitude, positive and negative, rectangular pulses of duration T. This signal represents the k-th user's binary information sequence. The k-th user is assigned a code waveform $a_k(t)$ which consists of a periodic sequence of unit amplitude, positive and negative, rectangular pulses of duration T_c. If $(a_j^{(k)})$ is the corresponding sequence of elements of $\{+1, -1\}$ then we write $a_k(t)$ as

$$a_k(t) = \sum_{j=-\infty}^{\infty} a_j^{(k)} p_{T_c}(t - jT_c)$$

Reprinted from *IEEE Transactions on Communications,* vol. COM-25, no. 8, August 1977.

The Best of the Best. Edited by W. H. Tranter, D. P. Taylor, R. E. Ziemer, N. F. Maxemchuk, and J. W. Mark.

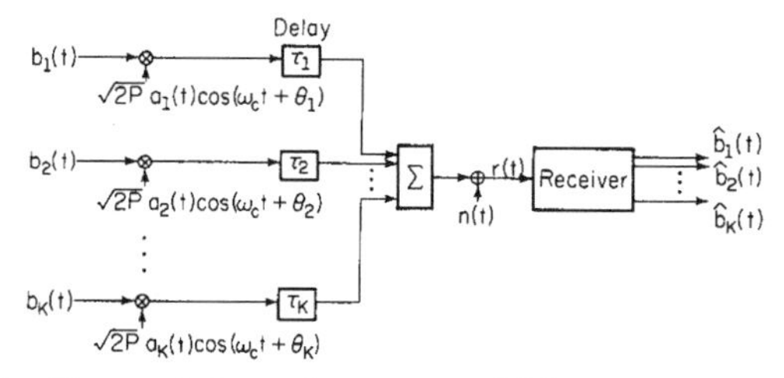

Fig. 1. Phase-coded spread-spectrum multiple-access system model.

where $p_\tau(t) = 1$ for $0 \leqslant t < \tau$ and $p_\tau(t) = 0$ otherwise. We assume that the k-th user's code sequence $(a_j^{(k)})$ has period $N = T/T_c$ so that there is one code period $a_0^{(k)}, a_1^{(k)}, \cdots, a_{N-1}^{(k)}$ per data symbol. The results presented can easily be generalized to multiple code periods per data symbol.

The data signal $b_k(t)$ is modulated onto the phase-coded carrier $c_k(t)$, which is given by

$$c_k(t) = \sqrt{2P}\sin(\omega_c t + \theta_k + (\pi/2)a_k(t))$$
$$= \sqrt{2P}\,a_k(t)\cos(\omega_c t + \theta_k).$$

Thus, the transmitted signal for the k-th user is

$$s_k(t) = \sqrt{2P}\sin(\omega_c t + \theta_k + (\pi/2)a_k(t)b_k(t))$$
$$= \sqrt{2P}\,a_k(t)b_k(t)\cos(\omega_c t + \theta_k).$$

In the above expressions θ_k represents the phase of the k-th carrier, ω_c represents the common center frequency, and P represents the common signal power. The results that follow can easily be modified for unequal center frequencies and power levels.

If the SSMA system is completely synchronized, then the time delays τ_k shown in the model of Figure 1 can be ignored (i.e., $\tau_k = 0$ for $k = 1, 2, \cdots, K$). This would require a common timing reference for the K transmitters and it would necessitate compensation for delays in the various transmission paths. This is generally not feasible and hence the transmitters are not time-synchronous. For asynchronous systems the received signal $r(t)$ in Figure 1 is given by

$$r(t) = n(t) + \sum_{k=1}^{K} \sqrt{2P}\,a_k(t-\tau_k)b_k(t-\tau_k)\cos(\omega_c t + \phi_k)$$

where $\phi_k = \theta_k - \omega_c\tau_k$ and $n(t)$ is the channel noise process which we assume to be a white Gaussian process with two-sided spectral density $N_0/2$. Since we are concerned with relative phase shifts modulo 2π and relative time delays modulo T, there is no loss in generality in assuming $\theta_i = 0$ and $\tau_i = 0$ and considering only $0 \leqslant \tau_k < T$ and $0 \leqslant \theta_k < 2\pi$ for $k \neq i$.

If the received signal $r(t)$ is the input to a correlation receiver matched to $s_i(t)$, the output is

$$Z_i = \int_0^T r(t)a_i(t)\cos\omega_c t\,dt.$$

In all that follows we assume $\omega_c \gg T^{-1}$ since the frequency response of a realistic hardware implementation of the correlation receiver is such that we can then ignore the double frequency component of $r(t)\cos\omega_c t$. The condition $\omega_c \gg T^{-1}$ is always satisfied in a practical SSMA communication system. The data signal $b_k(t)$ can be expressed as

$$b_k(t) = \sum_{l=-\infty}^{\infty} b_{k,l}\,p_T(t - lT)$$

where $b_{k,l} \in \{+1, -1\}$. The output of the correlation receiver at $t = T$ is given by

$$Z_i = \sqrt{P/2}\left\{b_{i,0}T + \sum_{\substack{k=1\\k\neq i}}^{K} [b_{k,-1}R_{k,i}(\tau_k) + b_{k,0}\hat{R}_{k,i}(\tau_k)] \cdot \cos\phi_k\right\} + \int_0^T n(t)a_i(t)\cos\omega_c t\,dt \tag{1}$$

where $R_{k,i}$ and $\hat{R}_{k,i}$ are the continuous-time partial cross-correlation functions defined by

$$R_{k,i}(\tau) = \int_0^{\tau} a_k(t-\tau)a_i(t)\,dt,$$

$$\hat{R}_{k,i}(\tau) = \int_\tau^T a_k(t-\tau)a_i(t)\,dt \tag{2}$$

for $0 \leqslant \tau \leqslant T$. It is easy to see that for $0 \leqslant lT_c \leqslant \tau \leqslant (l+1)T_c \leqslant T$, these two cross-correlation functions can be written as

$$R_{k,i}(\tau) = C_{k,i}(l-N)T_c + [C_{k,i}(l+1-N) - C_{k,i}(l-N)] \cdot (\tau - lT_c) \tag{3}$$

and

$$\hat{R}_{k,i}(\tau) = C_{k,i}(l)T_c + [C_{k,i}(l+1) - C_{k,i}(l)](\tau - lT_c) \tag{4}$$

where the discrete aperiodic cross-correlation function $C_{k,i}$ for the sequences $(a_j^{(k)})$ and $(a_j^{(i)})$ is defined by

$$C_{k,i}(l) = \begin{cases} \sum_{j=0}^{N-1-l} a_j^{(k)} a_{j+l}^{(i)}, & 0 \leqslant l \leqslant N-1 \\ \sum_{j=0}^{N-1+l} a_{j-l}^{(k)} a_j^{(i)}, & 1-N \leqslant l < 0 \\ 0, & |l| \geqslant N. \end{cases}$$

The periodic cross-correlation function $\theta_{k,i}$ is given by

$$\theta_{k,i}(l) = \sum_{j=0}^{N-1} a_j^{(k)} a_{j+l}^{(i)}$$

for any integer l. Notice that $\theta_{k,i}(l) = C_{k,i}(l) + C_{k,i}(l-N)$ for $0 \leqslant l < N$. We also define $\hat{\theta}_{k,i}(l) = C_{k,i}(l) - C_{k,i}(l-N)$

for $0 \leqslant l < N$. The function $\hat{\theta}_{k,i}$ is called the odd cross-correlation function by Massey and Uhran [13] since it has the property $\hat{\theta}_{i,k}(l) = -\hat{\theta}_{k,i}(N-l)$, whereas the periodic (or even) cross-correlation function satisfies $\theta_{i,k}(l) = \theta_{k,i}(N-l)$. Both of these relationships follow from the observation that $C_{i,k}(l) = C_{k,i}(-l)$.

If $v_{k,i}(\tau_k)$ is defined by

$$v_{k,i}(\tau_k) = [b_{k,-1}R_{k,i}(\tau_k) + b_{k,0}\hat{R}_{k,i}(\tau_k)] \cos \phi_k$$

then $\sqrt{P/2}\, v_{k,i}(\tau_k)$ is the contribution of the k-th signal to the output Z_i of the correlation receiver matched to $s_i(t)$. For fixed τ_k, $v_{k,i}(\tau_k)$ depends only on ϕ_k, the data symbols $b_{k,-1}$ and $b_{k,0}$, and the aperiodic cross-correlation function (or, the periodic and odd cross-correlation functions). Specifically, if l_k is an integer for which $l_k T_c \leqslant \tau_k \leqslant (l_k + 1)T_c$ and $b_{k,0} = b_{k,-1}$, then

$$v_{k,i}(\tau_k) = b_{k,0}\{\theta_{k,i}(l_k)T_c + [\theta_{k,i}(l_k+1) - \theta_{k,i}(l_k)](\tau_k - l_k T_c)\} \cos \phi_k . \quad (5)$$

On the other hand, if $b_{k,0} \neq b_{k,-1}$, then

$$v_{k,i}(\tau_k) = b_{k,0}\{\hat{\theta}_{k,i}(l_k)T_c + [\hat{\theta}_{k,i}(l_k+1) - \hat{\theta}_{k,i}(l_k)](\tau_k - l_k T_c)\} \cos \phi_k . \quad (6)$$

III. SYSTEM ANALYSIS: WORST-CASE PERFORMANCE

Up to this point we have not explicitly indicated which code sequence parameters should be optimized. The ideal situation would be to find a code for which the error probabilities $\Pr(Z_i > 0 \mid b_{i,0} = -1)$ and $\Pr(Z_i < 0 \mid b_{i,0} = +1)$ are small for all values of the parameters $\tau_k, \phi_k, b_{k,-1}$, and $b_{k,0}$. It is clear from symmetry considerations that, for any code, the set of values that one of the two probabilities takes on as the parameters are varied is the same as the corresponding set for the other probability. In particular, the two probabilities have the same *maximum* value $P_{\max}{}^{(i)}$ for any given code. One code selection procedure that is often suggested is to choose the code that gives the smallest value of $P_{\max}{}^{(i)}$; that is, the maximum value of the error probability is minimized. This approach is open to the usual criticism of minimax methods which is that too much emphasis is placed on the worst-case parameter values. However, the minimax approach is warranted for certain systems so we will pursue it further before suggesting an alternative.

If $b_{i,0} = -1$, $P_{\max}{}^{(i)}$ depends on the maximum value of the sum of the $v_{k,i}(\tau_k)$ over all $k \neq i$. From (5) and (6) it is clear that the maximum value of $v_{k,i}(\tau_k)$ is achieved when τ_k is an integer multiple of T_c and when $\phi_k = 0$. That is, the maximum value of $u_{k,i}(\tau_k) \triangleq v_{k,i}(\tau_k)/T_c$ is of the form

$$[b_{k,-1}C_{k,i}(l-N) + b_{k,0}C_{k,i}(l)]$$

for $l \in \{0, 1, 2, \cdots, N-1\}$, $b_{k,-1} \in \{+1, -1\}$, and $b_{k,0} \in \{+1, -1\}$. For a fixed l, this quantity has four possible values, $\pm\theta_{k,i}(l)$ and $\pm\hat{\theta}_{k,i}(l)$. The maximum of these values over all values of l is $\lambda_{k,i} = \max\{\gamma_{k,i}, \hat{\gamma}_{k,i}\}$ where $\gamma_{k,i} = \max|\theta_{k,i}(l)|$ and $\hat{\gamma}_{k,i} = \max|\hat{\theta}_{k,i}(l)|$.

From the above discussion we conclude that if $b_{i,0} = -1$, the maximum error probability for the i-th receiver corresponds to the maximum value of $u_{k,i}(\tau_k)$ for each $k \neq i$ and that this maximum value is $\lambda_{k,i}$. The same argument can be applied for $b_{i,0} = +1$ in which case the maximum error probability corresponds to the minimum value of $u_{k,i}(\tau_k)$ for each $k \neq i$ and $\min u_{k,i}(\tau_k) = -\lambda_{k,i}$. Thus, $P_{\max}{}^{(i)}$ is minimized if the quantity $\Lambda_i = \Sigma_{k \neq i} \lambda_{k,i}$ is minimized. In fact

$$P_{\max}{}^{(i)} = 1 - \Phi([1 - (\Lambda_i/N)]\sqrt{2E/N_0}) \quad (7)$$

where Φ is the standard (i.e., zero mean, unit variance) Gaussian cumulative distribution function and $E = PT$ is the energy per data bit. We define $P_{\max} = \max P_{\max}{}^{(i)}$ and $\Lambda = \max \Lambda_i$, where the maximization is over i, and notice that

$$P_{\max} = 1 - \Phi([1 - (\Lambda/N)]\sqrt{2E/N_0}) \quad (8)$$

$$\leqslant 1 - \Phi([1 - (K-1)(\lambda/N)]\sqrt{2E/N_0}) \quad (9)$$

where λ is the maximum of $\lambda_{k,i}$ over all i and k such that $1 \leqslant i < k \leqslant K$. The property $\lambda_{k,i} = \lambda_{i,k}$ was used to obtain (9). An additional code parameter that is of interest is the maximum magnitude of the aperiodic cross-correlation

$$C_c = \max\{|C_{k,i}(l)|: \ 1-N \leqslant l \leqslant N-1, 1 \leqslant i < k \leqslant K\}.$$

Notice that $\lambda \leqslant 2C_c$ and hence

$$P_{\max} \leqslant 1 - \Phi([1 - (K-1)(2C_c/N)]\sqrt{2E/N_0}) . \quad (10)$$

Expressions (8)-(10) provide upper bounds on the worst-case error probability which may be useful for certain SSMA systems. However, unless the period N of the code sequences is much larger than the number of users K, the terms Λ/N, $\lambda(K-1)/N$, and $2C_c(K-1)/N$ which appear in these bounds will often be greater than unity. For this situation, not only are the bounds of no value, but also the maximum error probability itself is not a useful performance parameter. If in such situations, the large cross-correlation values arise for only a few values of the delay parameters, $\tau_1, \tau_2, \cdots, \tau_K$, it is more meaningful to consider the average performance rather than the worst case. Two important measures of average performance are the average error probability, which is discussed in [19], and the average signal-to-noise ratio, which is discussed in the next section.

IV. SYSTEM ANALYSIS: AVERAGE SIGNAL-TO-NOISE RATIO

In this section we present an alternative approach to phase-coded SSMA system analysis which leads to a new parameter upon which to base code-sequence selection and evaluation. In this approach we treat the phase shifts, time delays, and data symbols as mutually independent random variables. The interference terms appearing in (1) are random and are treated as additional noise. The signal-to-noise ratio,

SNR_i, at the output of the i-th correlation receiver is one of the most important performance measures that can be obtained with a reasonable amount of computation. We should point out that this signal-to-noise ratio is computed by means of probabilistic averages (expectations) with respect to the phase shifts, time delays, and data symbols. However, such averages can also be interpreted as time averages since, in practice, these variables are actually slowly varying time functions which can be modeled as stationary ergodic random processes.

As in the previous section, there is no loss of generality in assuming $\phi_i = 0$ and $\tau_i = 0$ when considering Z_i, the output of the i-th correlation receiver. Also, because of the symmetry involved we need consider only $b_{i,0} = +1$. The desired signal component of Z_i is then $\sqrt{P/2}\,T$ while the variance of the noise component of Z_i is

$$\text{Var}\{Z_i\} = \left(\frac{P}{4T}\right) \sum_{\substack{k=1 \\ k\neq i}}^{K} \int_0^T R_{k,i}^2(\tau) + \hat{R}_{k,i}^2(\tau)\, d\tau + \tfrac{1}{4}N_0 T$$

$$= \left(\frac{P}{4T}\right) \sum_{\substack{k=1 \\ k\neq i}}^{K} \sum_{l=0}^{N-1} \int_{lT_c}^{(l+1)T_c} R_{k,i}^2(\tau) + \hat{R}_{k,i}^2(\tau)\, d\tau + \tfrac{1}{4}N_0 T \tag{11}$$

where the expectation has been computed with respect to the mutually independent random variables ϕ_k, τ_k, $b_{k,-1}$, and $b_{k,0}$ for $1 \leqslant k \leqslant K$ and $k \neq i$. We have assumed that ϕ_k is uniformly distributed on the interval $[0, 2\pi]$ and τ_k is uniformly distributed on the interval $[0, T]$ for $k \neq i$. Also, the data symbols $b_{k,l}$ are assumed to take values +1 or −1 with equal probability for $k \neq i$.

We next substitute for $R_{k,i}(\tau)$ and $\hat{R}_{k,i}(\tau)$ from (3) and (4) into (11). Upon evaluating the resulting integral we find that

$$\text{Var}\{Z_i\} = \frac{PT^2}{12N^3}\left(\sum_{\substack{k=1 \\ k\neq i}}^{K} r_{k,i}\right) + \tfrac{1}{4}N_0 T \tag{12}$$

where

$$r_{k,i} = \sum_{l=0}^{N-1} \{C_{k,i}^2(l-N) + C_{k,i}(l-N)C_{k,i}(l-N+1) + C_{k,i}^2(l-N+1) + C_{k,i}^2(l) + C_{k,i}(l)C_{k,i}(l+1) + C_{k,i}^2(l+1)\}.$$

This last expression can be written in terms of the cross-correlation parameters $\mu_{k,i}(n)$ which are defined by

$$\mu_{k,i}(n) = \sum_{l=1-N}^{N-1} C_{k,i}(l)C_{k,i}(l+n). \tag{13}$$

Notice that

$$\mu_{k,i}(0) = \sum_{l=1-N}^{N-1} C_{k,i}^2(l) = \sum_{l=0}^{N-1} C_{k,i}^2(l-N) + C_{k,i}^2(l) = \sum_{l=0}^{N-1} C_{k,i}^2(l-N+1) + C_{k,i}^2(l+1)$$

and

$$\mu_{k,i}(1) = \sum_{l=1-N}^{N-1} C_{k,i}(l)C_{k,i}(l+1) = \sum_{l=0}^{N-1} C_{k,i}(l-N)C_{k,i}(l-N+1) + C_{k,i}(l)C_{k,i}(l+1).$$

Therefore,

$$r_{k,i} = 2\mu_{k,i}(0) + \mu_{k,i}(1). \tag{14}$$

The signal-to-noise ratio is $\sqrt{\tfrac{1}{2}P}\,T$ divided by the rms noise $\sqrt{\text{Var}\,Z_i}$, which is

$$\text{SNR}_i = \left\{(6N^3)^{-1} \sum_{\substack{k=1 \\ k\neq i}}^{K} [2\mu_{k,i}(0) + \mu_{k,i}(1)] + \frac{N_0}{2E}\right\}^{-1/2}. \tag{15}$$

It is shown in [14] that $\mu_{k,i}(n)$ can be computed directly from the aperiodic autocorrelation functions for $(a_j^{(k)})$ and $(a_j^{(i)})$. Thus, the signal-to-noise ratio can be evaluated without knowledge of the cross-correlation functions.

Note that for $K = 1$, (15) reduces to $\text{SNR}_i = \sqrt{2E/N_0}$ which has associated error probability $P_e = 1 - \Phi(\sqrt{2E/N_0})$. In general for $K > 1$ the error probability will not be exactly $1 - \Phi(\text{SNR}_i)$, but this is typically a very good approximation for values of N and K of interest in practical systems. Quantitative results on the accuracy of this approximation have been obtained by Yao [19]. Numerical results on the evaluation of the signal-to-noise ratio for sequences of period $N = 511$ and for $K = 10, 20, 30$ and 40 are given in [13].

Finally, we should mention that for preliminary system design it is useful to be able to carry out a tradeoff between the parameters K, N, and E/N_0. Such a tradeoff can be based on the approximation

$$(6N^3)^{-1} \sum_{\substack{k=1 \\ k\neq i}}^{K} r_{k,i} \approx (K-1)/3N \tag{16}$$

which yields

$$\text{SNR}_i \approx \left\{\frac{K-1}{3N} + \frac{N_0}{2E}\right\}^{-1/2}. \tag{17}$$

In [15], it is shown that the right-hand side of (16) is actually the expectation of the left-hand side when random sequences are employed. The main use of (17) would be to first determine roughly what code-sequence length N, bit energy E, and noise density $N_0/2$ are required to achieve a given signal-to-noise ratio for a given number of users K. A more detailed investigation of the performance can then be carried out using (15) for specific code sequences.

REFERENCES

[1] J. M. Aein, "Multiple access to a hard-limiting communication-satellite repeater," *IEEE Transactions on Space Electronics and Telemetry,* vol. SET-10, pp. 159-167, December 1964.

[2] J. M. Aein and J. W. Schwartz (editors), "Multiple access to a communication satellite with a hard-limiting repeater–Volume II: Proceedings of the IDA multiple access summer study," Institute for Defense Analysis, Report R-108, 1965.

[3] D. R. Anderson and P. A. Wintz, "Analysis of a spread-spectrum multiple-access system with a hard limiter," *IEEE Transactions on Communication Technology,* vol. COM-17, pp. 285-290, April 1969.

[4] H. Blasbalg, "A comparison of pseudo-noise and conventional modulation for multiple-access satellite communications," *IBM Journal,* vol. 9, pp. 241-255, July 1965.

[5] R. C. Dixon, *Spread Spectrum Systems.* New York: Wiley, 1976.

[6] R. C. Dixon (editor), *Spread Spectrum Techniques.* New York: IEEE Press, 1976.

[7] L. A. Gerhardt (lecture series director), "Spread Spectrum Communications," AGARD Lecture Series No. 58, NATO, July 1973.

[8] R. Gold, "Optimal binary sequences for spread spectrum multiplexing," *IEEE Transactions on Information Theory,* vol. IT-13, pp. 619-621, October 1967.

[9] S. W. Golomb (editor), *Digital Communications with Space Applications,* Englewood Cliffs, N. J.: Prentice-Hall, 1964.

[10] J. Kaiser, J. W. Schwartz, and J. M. Aein, "Multiple access to a communication satellite with a hard-limiting repeater.--Volume I: Modulation techniques and their applications," Institute for Defense Analysis, Report R-108, January 1965.

[11] I. L. Lebow, K. L. Jordan, and P. R. Drouilhet, Jr., "Satellite communications to mobile platforms," *Proceedings of the IEEE,* vol. 59, pp. 139-159, February 1971.

[12] J. L. Massey and J. J. Uhran, "Sub-baud coding," *Proceedings of the Thirteenth Annual Allerton Conference on Circuit and System Theory,* pp. 539-547, October 1975 (see also "Final report for multipath study," Department of Electrical Engineering, University of Notre Dame, 1969).

[13] M. B. Pursley, "Evaluating performance of codes for spread spectrum multiple access communications," *Proceedings of the Twelfth Annual Allerton Conference on Circuit and System Theory,* pp. 765-774, October 1974 (see also "Tracking and data relay satellite system configuration and tradeoff study," Volume 4, Appendix D, Hughes Aircraft Company, Space and Communications Group, El Segundo, California, Report 20642R, September 1972).

[14] M. B. Pursley and D. V. Sarwate, "Performance evaluation for phase-coded spread-spectrum multiple-access communication–Part II: Code sequence analysis," this issue, pp. 800-803.

[15] H. F. A. Roefs and M. B. Pursley, "Correlation parameters of random sequences and maximal length sequences for spread-spectrum multiple-access communication," *1976 IEEE Canadian Communications and Power Conference,* pp. 141-143, October 1976.

[16] J. W. Schwartz, J. M. Aein, and J. Kaiser, "Modulation techniques for multiple access to a hard-limiting satellite repeater," *Proceedings of the IEEE,* vol. 54, pp. 763-777, May 1966.

[17] R. A. Stampfl and A. E. Jones, "Tracking and data relay satellites," *IEEE Transactions on Aerospace and Electronic Systems,* vol. AES-6, pp. 276-289, May 1970.

[18] I. G. Stiglitz, "Multiple-access considerations–A satellite example," *IEEE Transactions on Communications,* vol. COM-21, pp. 577-582, May 1973.

[19] K. Yao, "Error probability of asynchronous spread spectrum multiple access communication systems," this issue, pp. 803-809.

★

Michael B. Pursley (S'68–M'68–S'72–M'74) was born in Winchester, Indiana on August 10, 1945. He studied electrical engineering at Purdue University where he received the B.S. degree with highest distinction in 1967 and the M.S. degree in 1968. In 1974 he received the Ph.D. degree in electrical engineering from the University of Southern California.

He held a summer position in the Laser and Radar Electronics Section of the Hughes Aircraft Company, Los Angeles, California, in 1967 and a position in the Systems Analysis Section of the Nortronics Division of Northop Corporation in 1968. In December of 1968 he rejoined the Hughes Aircraft Company as a Member of the Technical Staff and was involved in satellite communication systems design and analysis; he was promoted to Staff Engineer in 1973. He was a Hughes Doctoral Fellow at the University of Southern California from 1971 until 1974 and a Research Assistant during 1973. From January through June of 1974 he was an Acting Assistant Professor at the University of California, Los Angeles. Since June, 1974 he has been an Assistant Professor in the Department of Electrical Engineering and the Coordinated Science Laboratory at the University of Illinois, Urbana, Illinois where his research work has been in the general area of information theory and stochastic processes with applications to source coding and communication systems. His current interests are in universal source coding, spread-spectrum multiple-access communication systems, and multiple-user information theory.

Dr. Pursley is a member of Phi Eta Sigma, Tau Beta Pi, and the Institute of Mathematical Statistics. He was treasurer and is presently secretary of the Joint Chapter of the Chicago, Central Illinois, Central Indiana, and South Bend Sections of the IEEE Information Theory Group.

The Throughput of Packet Broadcasting Channels

NORMAN ABRAMSON, FELLOW, IEEE

Abstract—Packet broadcasting is a form of data communications architecture which can combine the features of packet switching with those of broadcast channels for data communication networks. Much of the basic theory of packet broadcasting has been presented as a byproduct in a sequence of papers with a distinctly practical emphasis. In this paper we provide a unified presentation of packet broadcasting theory.

In Section II we introduce the theory of packet broadcasting data networks. In Section III we provide some theoretical results dealing with the performance of a packet broadcasting network when the users of the network have a variety of data rates. In Section IV we deal with packet broadcasting networks distributed in space, and in Section V we derive some properties of power-limited packet broadcasting channels, showing that the throughput of such channels can approach that of equivalent point-to-point channels.

I. INTRODUCTION

A. Packet Switching and Packet Broadcasting

THE transition of packet-switched computer networks from experimental [1] to operational [2] status during 1975 provides convincing evidence of the value of this form of communications architecture. Packet switching, or statistical multiplexing [3], can provide a powerful means of sharing communications resources among large number of data communications users when those users can be characterized by a high ratio of peak to average data rates. Under such circumstances, data from each user are buffered, address and control information is added in a "header," and the resulting bit sequence, or "packet," is routed through a shared communications resource by a sequence of node switches [4], [5].

Packet-switched networks, however, still employ point-to-point communication channels and large multiplexing switches for routing and flow control in a fashion similar to conventional circuit switched networks. In some situations [6]-[10] it is desirable to combine the efficiencies achievable by a packet communications architecture with other advantages obtained by use of broadcast communication channels. Among these advantages are elimination of routing and network switches, system modularity, and overall system simplicity. In addition, certain kinds of channels available to the communications systems designer, notably satellite channels, are basically broadcast in their structure. In such cases use of these channels in their natural broadcast mode can lead to significant system performance advantages [11], [12].

Manuscript received January 19, 1976; revised June 11, 1976. This work was supported by The ALOHA System, a research project at the University of Hawaii which is supported by the Advanced Research Projects Agency of the Department of Defense and monitored by NASA Ames Research Center under Contract NAS2-8590. The views and conclusions contained in this paper are those of the author and should not be interpreted as necessarily representing the official policies, either expressed or implied, of the Advanced Research Projects Agency of the United States Government.

The author is with The ALOHA System, University of Hawaii, Honolulu, HI 96822.

B. Outline of Results

Packet broadcasting is a form of data communications architecture which can combine the features of packet switching with those of broadcast channels for data communication networks. Much of the basic theory of packet broadcasting has been presented as a byproduct in a sequence of papers with a distinctly practical emphasis. In this paper we provide a unified presentation of packet broadcasting theory.

In Section II we introduce the theory of packet broadcasting as implemented in the ALOHA System at the University of Hawaii; also in Section II we explain a modification of the basic ALOHA method, called slotting. In Section III we provide some theoretical results dealing with the performance of a packet broadcasting channel when the users of the channel have a variety of data rates. In Section IV we deal with packet broadcasting networks distributed in space, and present some incomplete results on the theoretical properties of such networks. Finally, in Section V we derive some properties of power limited packet broadcasting channels showing that the throughput of such channels can approach that of equivalent point-to-point channels. This result is of importance in satellite systems using small earth stations since it implies that the multiple access capability and the complete connectivity (in the topological sense) of packet broadcasting channels can be obtained at no price in average throughput.

II. PACKET BROADCASTING CHANNELS

A. Operation of a Packet Broadcasting Channel

Consider a number of widely separated users, each wanting to transmit short packets over a common high-speed channel. Assume that the rate at which users generate packets is such that the average time between packets from a single user is much greater than the time needed to transmit a single packet. In Fig. 1 we indicate a sequence of packets transmitted by a typical user.

Conventional time or frequency multiplexing methods (TDMA or FDMA) or some kind of polling scheme could be employed to share the channel among the users. Some of the disadvantages of these methods for users with high peak-to-average data rates are discussed by Carleial and Hellman [13]. In addition, under certain conditions polling may require unacceptable system complexity and extra delay.

In a packet broadcasting system the simplest possible solution to this multiplexing problem is employed. Each user transmits its packets over the common broadcast channel in a completely unsynchronized (from one user to another) manner. If each individual user of a packet broadcasting chan-

Reprinted from *IEEE Transactions on Communications*, vol. COM-25, no. 1, January 1977.

Fig. 1. Packets from a typical user.

Fig. 2. Packets from several users on an ALOHA channel.

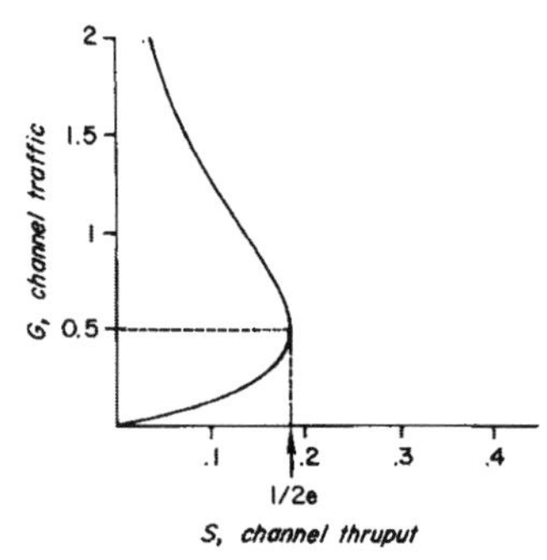

Fig. 3. Channel throughput versus channel traffic for an ALOHA channel.

nel is required to have a low duty cycle, the probability of a packet from one user interfering with a packet from another user is small as long as the total number of users on the common channel is not too large. As the number of users increases, however, the number of packet overlaps increases and the probability that a packet will be lost due to an overlap also increases. The question of how many users can share such a channel and the analysis of various methods of dealing with packets lost due to overlap are the primary concerns of this paper. In Fig. 2 we show a packet broadcasting channel with two overlapping packets. Since the first packet broadcasting channel was put into operation in the ALOHA System radio-linked computer network at the University of Hawaii [6], they have been referred to as ALOHA channels.

B. ALOHA Capacity

A transmitted packet can be received incorrectly or lost completely because of two different types of errors: 1) random noise errors and 2) errors caused by packet overlap. In this paper we assume that the first type of error can be ignored, and we shall be concerned only with errors caused by packet overlap. In Section II-D we describe several methods of dealing with the problem of packets lost due to overlap, but first we derive the basic results which tell us how many packets can be transmitted with no overlap.

Assume that the start times of packets in the channel comprise a Poisson point process with parameter λ packets/second. If each packet lasts τ seconds, we can define the normalized channel traffic G where

$$G = \lambda\tau. \tag{1}$$

If we assume that only those packets which do not overlap with any other packet are received correctly, we may define $\lambda' < \lambda$ as the rate of occurrence of those packets which are received correctly. Then we define the normalized channel thruput S by

$$S = \lambda'\tau. \tag{2}$$

The probability that a packet will not overlap a given packet is just the probability that no packet starts τ seconds before or τ seconds after the start time of the given packet. Then, since the point process formed from the start times of all packets in the channel was assumed Poisson, the probability that a packet will not overlap any other packet is $e^{-2\lambda\tau}$, or e^{-2G}. Therefore

$$S = Ge^{-2G} \tag{3}$$

and we may plot the channel throughput versus channel traffic for an ALOHA channel (Fig. 3).

From Fig. 3 we see that as the channel traffic increases, the throughput also increases until it reaches its maximum at $S = 1/2e = 0.184$. This value of throughput is known as the capacity of an ALOHA channel, and it occurs for a value of channel traffic equal to 0.5. If we increase the channel traffic above 0.5, the throughput of the channel will decrease.

C. Application of an ALOHA Channel

In order to indicate the capabilities of such a channel for use in an interactive network of alphanumeric computer terminals, consider the 9600 bits/s packet broadcasting channel used in the ALOHA System. From the results of Section II-B we see that the maximum average throughput of this channel is 9600 bits/s times $1/2e$, or about 1600 bits/s. If we assume the conservative [14] figure of 5 bits/s as the average data rate (including overhead) from each active[1] terminal in the network, this channel can handle the traffic of over 300 active terminals and each terminal will operate at a peak data rate of 9600 bits/s. Of course, the total number of terminals in such a network can be much larger than 300 since only a fraction of all terminals will be active and a terminal consumes no channel resources when it is not active.

D. Recovery of Lost Packets

Since the packet broadcasting technique we have described will result in some packets being lost due to packet overlaps, it is necessary to introduce some technique to compensate for this loss. We may list four different packet recovery techniques for dealing with the problem of lost packets. The first three make use of a feedback channel to the packet transmitter and the repetition of lost packets, while the fourth is based on coding.

1) Positive Acknowledgments (POSACKS): Perhaps the most direct way to handle lost packets is to require the receiver of the packet to acknowledge correct receipt of the packet. Each packet is transmitted and then stored in the transmitter's buffer until a POSACK is received from the receiver. If a POSACK is not received in a given amount of time, the transmitter can repeat the transmission and continue to repeat until a POSACK is received or until some other criterion is met. The POSACK can be transmitted on a sepa-

[1] A terminal is defined as active from the time a user transmits an attempt to log on until he transmits a log off message.

rate channel (as in the ALOHANET [6]) or transmitted on the same channel as the original packets (as in the ARPA packet radio system [15]). An error detection code and a packet numbering system can be used to increase the reliability of this technique.

2) Transponder Packet Broadcasting: Certain communication channels—notably communication satellite channels—transmit packets on one frequency to a transponder which retransmits the packets on a second frequency. In such cases all units in a packet broadcasting network can receive their own packet retransmissions, determine whether a packet overlap has occurred, and repeat the packet if necessary. This technique has been employed in ATS-1 satellite experiments in the Pacific Educational Computer Network (PACNET) [16] and in the ARPA Atlantic INTELSAT IV packet broadcasting experiments [17].

3) Carrier Sense Packet Broadcasting: For ground-based packet broadcasting networks where the signal propagation time over the furthest transmission path is much less than the packet duration, it is feasible to provide each transmission unit with a device to inhibit packet transmission while another unit is detected transmitting. A carrier sense capability can increase the channel throughput, even if these conditions are not met, when used in conjunction with other packet recovery methods. Carrier sense systems have been analyzed by Tobagi [18] and by Kleinrock and Tobagi [19]. A comprehensive yet compact analysis of such systems is provided in [42].

4) Packet Recovery Codes: When a user employs a packet broadcasting channel to transmit long files by breaking them into large numbers of packets, it is possible to encode the files so that packets lost due to broadcasting overlap can be recovered. It is clear that some of the existing classes of multiple burst error-correcting codes [20] and cyclic product codes [21] can be used for packet recovery in transmissions of long files. It is also clear that these codes are not as efficient as possible for packet recovery and that considerable work remains to be done in this area.

E. Slotted Channels

It is possible to modify the completely unsynchronized use of the ALOHA channel described above in order to increase the maximum throughput of the channel. In the pure ALOHA channel each user simply transmits a packet when ready without any attempt to coordinate his transmission with those of other users. While this strategy has a certain elegance, it does lead to somewhat inefficient channel utilization. If we establish a time base and require each user to start his packet only at certain fixed instants, it is possible to increase the maximum value of the channel thruput. In this kind of channel, called a slotted ALOHA channel, a central clock establishes a time base for a sequence of "slots" of the same duration as a packet transmission [41]. Then when a user has a packet to transmit, he synchronizes the start of his transmission to the start of a slot. In this fashion, if two messages conflict they will overlap completely, rather than partially.

To analyze the slotted ALOHA channel, define G_i as the probability that the ith user will transmit a packet in some slot. Assume that each user operates independently of all other users, and that whether or not a user transmits a packet in a given slot does not depend upon the state of any previous slot. If we have n users, we can define the normalized channel traffic for the slotted channel G where

$$G = \sum_{i=1}^{n} G_i. \tag{4}$$

Note that G may be greater than 1.

As before, we can also consider the rate at which a user sends packets which do not experience an overlap with other user packets. Define $S_i \leqslant G_i$ as the probability that a user sends a packet and that this packet is the only packet in its slot. If we have n users, then we define the normalized channel throughput for the slotted channel S where

$$S = \sum_{i=1}^{n} S_i. \tag{5}$$

Note that S is less than or equal to 1 and $S \leqslant G$.

For the slotted ALOHA channel with n independent users, the probability that a packet from the ith user will not experience an interference from one of the other users is

$$\prod_{\substack{j=1 \\ j\neq i}}^{n} (1 - G_j).$$

Therefore we may write the following relationship between the message rate and the traffic rate of the ith user:

$$S_i = G_i \prod_{\substack{j=1 \\ j\neq i}}^{n} (1 - G_j). \tag{6}$$

If all users are identical, we have

$$S_i = \frac{S}{n} \tag{7}$$

and

$$G_i = \frac{G}{n} \tag{8}$$

so that (6) can be written

$$S = G\left(1 - \frac{G}{n}\right)^{n-1} \tag{9}$$

and in the limit as $n \to \infty$, we have

$$S = Ge^{-G}. \tag{10}$$

Equation (10) is plotted in Fig. 4 (curve labeled "slotted ALOHA"). Note that the message rate of the slotted ALOHA channel reaches a maximum value of $1/e = 0.368$, twice the capacity of the pure ALOHA channel.

This result for slotted ALOHA channels was first derived by Roberts [41] using a different method.

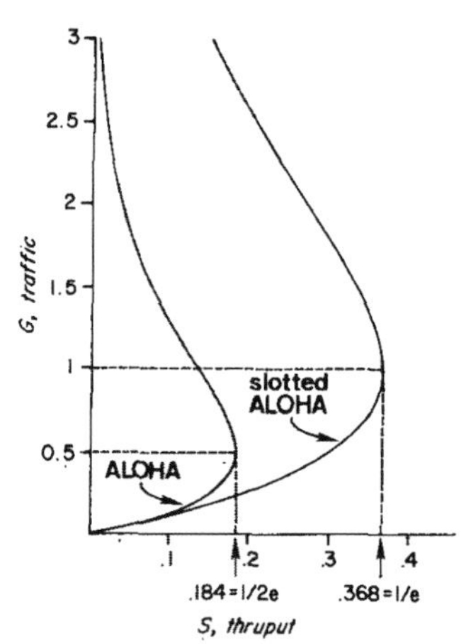

Fig. 4. Traffic versus throughput for an ALOHA channel and a slotted ALOHA channel.

III. PACKET BROADCASTING WITH MIXED DATA RATES

A. Unslotted Case: Variable Packet Lengths

In Section II we were concerned with the analysis of ALOHA channels carrying a homogeneous mix of packets. If some channel users have a higher average data rate than others, however, the high rate users must either transmit packets more frequently or transmit longer packets. In this section we shall analyze the unslotted ALOHA channel when carrying packets of different lengths, and we shall analyze the slotted ALOHA channel when the probability of transmitting in a given slot varies from user to user.

Let us assume an unslotted ALOHA channel with two different possible packet durations, τ_2 and τ_1. Assume $\tau_2 \geqslant \tau_1$, and therefore we refer to the two different length packets as long packets and short packets, respectively. Assume also the start times of the long packets and short packets form two Poisson point processes with parameters λ_2 and λ_1 packets/second, and that the two Poisson point processes are mutually independent. Then we can define the normalized channel traffic for those packets of duration τ_i:

$$G_i = \lambda_i \tau_i, \qquad i = 1, 2. \tag{11}$$

Again assume that only those packets which do not overlap with any other packet are received correctly and define $\lambda_i' < \lambda_i$ as the rate of occurrence of those packets of duration τ_i which are received correctly. Define the normalized throughput of packets of duration τ_i as

$$S_i = \lambda_i' \tau_i, \qquad i = 1, 2. \tag{12}$$

Since we assumed two independent Poisson point processes, the probability that a short packet will be received correctly is

$$\exp\left[-\lambda_1(2\tau_1) - \lambda_2(\tau_1 + \tau_2)\right] \tag{13}$$

and if we define

$$G_{12} \triangleq \lambda_1 \tau_2 \tag{14a}$$

$$G_{21} \triangleq \lambda_2 \tau_1 \tag{14b}$$

(13) becomes

$$\exp\left[-2G_1 - G_{21} - G_2\right]. \tag{15}$$

Therefore

$$S_1 = G_1 \exp\left[-2G_1 - G_{21} - G_2\right] \tag{16a}$$

and, by a similar argument, the throughput of long packets is

$$S_2 = G_2 \exp\left[-G_{12} - G_1 - 2G_2\right]. \tag{16b}$$

For any given values of λ_1 and λ_2 we may calculate G_1, G_2, G_{12}, and G_{21}; substitution of these values into (16a) and (16b) will allow calculation of the throughputs S_1 and S_2. Therefore (16a) and (16b) may be used to define an allowable set of throughput pairs (S_1, S_2) in the (S_1, S_2) plane.

To determine the boundary of this region we define

$$\alpha \triangleq \frac{\tau_2}{\tau_1}. \tag{17}$$

Note that $\alpha \geqslant 1$. We may rewrite (16a) and (16b) in terms of α, the ratio of long packet duration to short packet duration:

$$S_1 = G_1 \exp\left[-2G_1 - \left(1 + \frac{1}{\alpha}\right)G_2\right] \tag{18a}$$

$$S_2 = G_2 \exp\left[-(1+\alpha)G_1 - 2G_2\right]. \tag{18b}$$

The boundary of the set of allowable (S_1, S_2) pairs in the (S_1, S_2) plane is defined by setting the Jacobian

$$J = \left|\frac{\partial S_i}{\partial G_j}\right|, \qquad i,j = 1, 2 \tag{19}$$

equal to zero. A simple calculation shows that the Jacobian is zero when

$$G_2 = \frac{1 - 2G_1}{\left(\dfrac{(\alpha-1)^2}{\alpha} G_1 + 2\right)}. \tag{20}$$

Note that this checks for $G_1 = 0$ and for $\alpha = 1$.

We need only substitute this expression for G_2 into (18a) and (18b) to obtain two equations for S_1, the short packet throughput, and S_2, the long packet throughput, in terms of the single parameter G_1; and as G_1 varies from 0 (all long packets) to 1/2 (all short packets), we will trace out the boundary of the achievable values of throughput in the (S_1, S_2) plane. These achievable throughput regions are indicated for several values of α in Fig. 5.

The basic conclusion of this analysis is that the total channel throughput can undergo a significant decrease if all packets are not of the same length. Thus if the two different

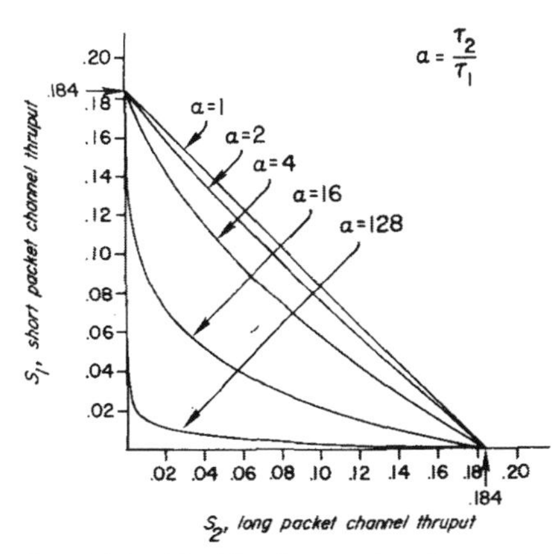

Fig. 5. Achievable throughput regions in an unslotted ALOHA channel.

packet lengths differ by a large factor, it is often preferable to break up long packets into many shorter packets as long as the overhead necessary to transmit the text in each packet is small. Ferguson [23] has generalized these results to show that channel throughput is maximized over all possible packet length distributions with fixed length packets.

In view of this discouraging result, we might conclude that an inhomogeneous mix of users inevitably leads to a decrease in the maximum value of channel throughput. Surprisingly, this conclusion is not warranted, and we shall show in Section III-B that a mix of users of varied data rates can lead to an increase in the maximum values of channel throughput.

B. Slotted Case: Variable Packet Rates

In the section we shall consider a slotted ALOHA channel used by n users, possibly with different values of channel traffic G_i. From (6) we have a set of n nonlinear equations relating the channel traffics and the channel throughputs for these n users:

$$S_i = G_i \prod_{\substack{j=1 \\ j \neq i}}^{n} (1 - G_j), \qquad i = 1, 2, \cdots, n. \tag{21}$$

Define

$$\alpha = \prod_{j=1}^{n} (1 - G_j); \tag{22}$$

then (21) can be written

$$S_i = \frac{G_i}{1 - G_i}\, \alpha, \qquad i = 1, 2, \cdots, n. \tag{23}$$

For any set of n acceptable traffic rates $G_1, G_2, \cdots, G_n$, these n equations define a set of channel throughputs $S_1, S_2, \cdots, S_n$ or a region in an n-dimensional space whose coordinates are the S_i. In order to find the boundary of this region, we calculate the Jacobian:

$$J = \left| \frac{\partial S_j}{\partial G_k} \right|, \qquad j, k = 1, 2, \cdots, n. \tag{24}$$

Since

$$\frac{\partial S_j}{\partial G_k} = \begin{cases} \prod_{\substack{i=1 \\ i \neq j}}^{n} (1 - G_i), & j = k \\ -G_j \prod_{\substack{i=1 \\ i \neq j,k}}^{n} (1 - G_i), & j \neq k \end{cases} \tag{25}$$

after some algebra we may write the Jacobian as

$$J = \alpha^{n-2} \begin{vmatrix} (1 - G_1) & -G_1 & -G_1 & \cdots \\ -G_2 & (1 - G_2) & -G_2 & \\ -G_3 & -G_3 & (1 - G_3) & \\ & \vdots & & \end{vmatrix}$$

$$= \alpha^{n-2} [1 - G_1 - G_2 - \cdots - G_n]. \tag{26}$$

Thus the condition for maximum channel throughputs is

$$\sum_i G_i = 1. \tag{27}$$

This condition can then be used to define a boundary to the n-dimensional region of allowable throughputs $S_1, S_2, \cdots, S_n$.

Consider the special case of two classes of users with n_1 users in class 1 and n_2 users in class 2:

$$n_1 + n_2 = n. \tag{28}$$

Let S_1 and G_1 be the throughputs and traffic rates for users in class 1, and let S_2 and G_2 be the throughputs and traffic rates for users in class 2. Then the n equations (21) can be written as the two equations

$$S_1 = G_1 (1 - G_1)^{n_1 - 1} (1 - G_2)^{n_2} \tag{29a}$$

$$S_2 = G_2 (1 - G_2)^{n_2 - 1} (1 - G_1)^{n_1}. \tag{29b}$$

For any pair of acceptable traffic rates G_1 and G_2, these two equations define a pair of channel throughputs S_1 and S_2 or a region in the (S_1, S_2) plane.

From (27) we know that the boundary of this region is defined by the condition

$$n_1 G_1 + n_2 G_2 = 1. \tag{30}$$

We can use (30) to substitute for G_1 in (29a) and (29b) and obtain two equations for S_1 and S_2 in terms of a single parameter G_2. Then as G_2 varies from 0 to 1, the resulting (S_1, S_2) pairs define the boundary of the region we seek. These achievable regions are indicated for various values of n_1 and n_2 in Figs. 6 and 7.

The important point to notice from Figs. 6 and 7 is that in a lightly loaded slotted ALOHA channel, a single large user can transmit data at a significant percentage of the total channel data rate, thus allowing use of the channel at rates well above the limit of $1/e$ or 37 percent obtained when all users have the same message rate. A throughput data rate above the $1/e$ limit

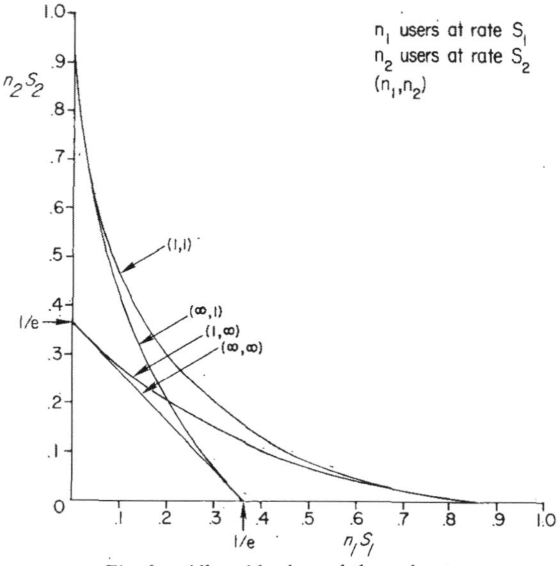

Fig. 6. Allowable channel throughputs.

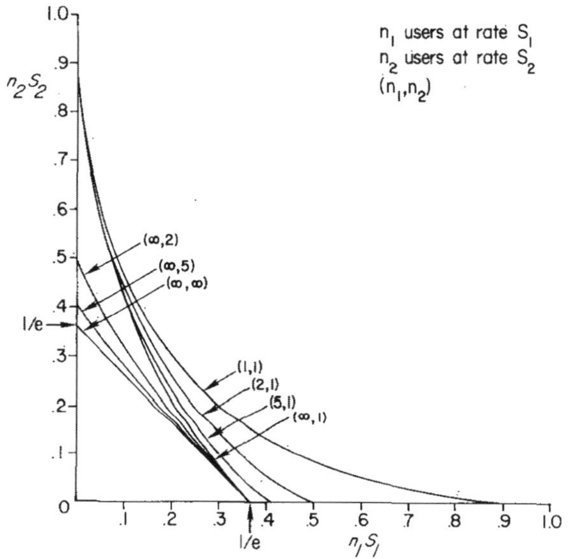

Fig. 7. Allowable channel throughputs.

has been referred to as "excess capacity" [24]. Excess capacity is important for a lightly loaded packet broadcasting network consisting of many interactive terminal users and a small number of users who send large but infrequent files over the channel. Operation of the channel in a lightly loaded condition, of course, may not be desirable in a bandwidth-limited channel. For a communications satellite where the average power in the satellite transponder limits the channel, however, operation in a lightly loaded packet-switched mode is an attractive alternative. Since the satellite will transmit power only when it is relaying a packet, the duty cycle in the transponder will be small and the average power used will be low (see Section V).

Finally, we note that it is possible to deal with certain limiting cases in more detail, to obtain equations for the boundary of the allowable (S_1,S_2) region.

1) For $n_1 = n_2 = 1$:

Upon using (30) in (29), we obtain

$$S_1 = G_1^2 \tag{31a}$$

$$S_2 = (1 - G_1)^2. \tag{31b}$$

2) For $n_2 \to \infty$:

$$S_1 = G_1(1 - G_1)^{n_1 - 1} \cdot \exp\left[-(1 - n_1 G_1)\right] \tag{32a}$$

$$S_2 = (1 - n_1 G_1)(1 - G_1)^{n_1 - 1} \cdot \exp\left[-(1 - n_1 G_1)\right]. \tag{32b}$$

3) For $n_1 = n_2 \to \infty$:

$$S_1 = \frac{G_1}{e} \tag{33a}$$

$$S_2 = \frac{1 - G_1}{e} \tag{33b}$$

Additional details dealing with excess capacity and the delay experienced with this kind of use of a slotted ALOHA channel may be found in [11] and [25]. A different view of the use of a slotted packet broadcasting for different sources may be found in [43].

IV. SPATIAL PROPERTIES OF PACKET BROADCASTING NETWORKS

A. Packet Repeaters

In this section we deal with certain spatial properties of packet broadcasting networks. Not long after the initial units of the ALOHA System went into operation, it was realized that the range of the network could be extended beyond the range of a single radio link in the network (about 200 km) by the use of packet repeaters. A packet repeater operates in much the same manner as a conventional radio repeater with one major exception. Since radio transmission in a packet broadcasting network is intermittent, a packet repeater can receive a packet and retransmit that packet in the same frequency band by turning off its receiver during a retransmission burst. Thus a packet repeater can sidestep many of the frequency allocation and spatial cell problems [26] of conventional land-based repeater networks.

The use of packet repeaters leads to the consideration of packet broadcasting networks with more than one central station distributed over very large areas. Users transmit a packet, and if the packet cannot be received directly by its destination, it is forwarded to its destination by one or more packet repeaters according to some routing algorithm [27]. The study of such networks has led to the analysis of two communication theory issues related to the performance of the networks: 1) capture effect and 2) the distribution of packet traffic and packet throughput in space.

B. Capture Effect

Up to this point we have analyzed packet broadcasting channels under the pessimistic assumption that if two packets overlap at the receiver, both packets are lost. In fact, this assumption provides a lower bound to the performance of real packet broadcasting channels, since in many receivers the stronger of two overlapping packets may capture the receiver and may be received without error. Metzner [40] has used this fact to derive an interesting result, showing that by dividing users into two groups—one transmitting at high power and the other at low power—the maximum throughput can be increased by about 50 percent. This result is of importance for packet broadcasting networks with a mixture of data and packetized speech traffic.

In order to include the effect of capture in a packet broadcasting network, we consider a distribution of packet generators over a two-dimensional plane and a single packet broadcasting receiver which receives packets from these generators [41]. The receiver then may be viewed as a "packet sink" and the packet generators as a distribution of "packet sources" in the plane. We assume that the rate of generation of packets in a given area depends only on r, the distance from the packet sink, and is independent of direction θ.

Then we may define a traffic density and a throughput density analogous to the normalized traffic G and normalized thoughput S defined in Section II-B.

$G(r)$ = normalized packet traffic per unit area at a distance r.

$S(r)$ = normalized packet thruput per unit area at a distance r.

The traffic due to all packet generators in a differential ring of width dr at a radius r is

$$G(r)\, 2\pi r\, dr. \tag{34}$$

We assume that packets from different users are generated so that the packet starting times of all packets generated in the differential ring constitute a Poisson point process. Then since the sum of two independent Poisson processes is a Poisson point process, if users in different rings are independent, the start times of all packets generated in a circle of radius r also constitute a Poisson point process, and the total traffic generated by all users within a distance r of the center is

$$\int_0^r G(x) 2\pi x\, dx. \tag{35}$$

If we assume that a packet from a user at a distance r from the center will be received correctly unless it is overlapped by a packet sent from a user at a distance ar or less ($a \geqslant 1$), then using the results of Section II-B the probability that such a packet will be received correctly is

$$\exp\left[-4\pi \int_0^{ar} G(x) x\, dx\right]. \tag{36}$$

Any packet generated from a packet source in the circle of radius ar shown in Fig. 8 will interfere with packets generated from a source in the circle of radius r. A packet generated outside the circle of radius ar will not interfere with packets generated from a source in the circle of radius r.

We can relate the normalized packet throughput to the normalized packet traffic in the usual way:

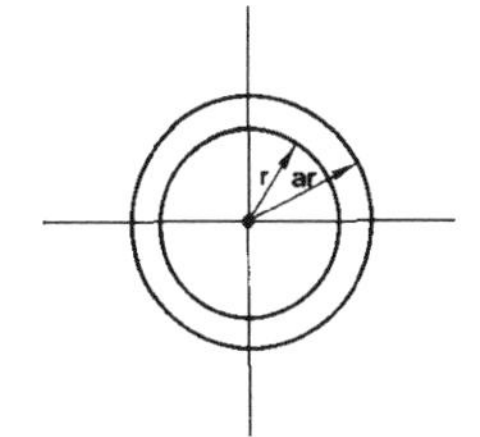

Fig. 8. Regions of interfering packets.

$$2\pi r S(r)\, dr = 2\pi r G(r) \exp\left[-4\pi \int_0^{ar} G(x) x\, dx\right] dr$$

or

$$S(r) = G(r) \exp\left[-4\pi \int_0^{ar} G(x) x\, dx\right]. \tag{37}$$

If we take a derivative of (37) with respect to r and use (37) to substitute for the exponential, we get

$$S'(r)G(r) = G'(r)S(r) - 4\pi r a^2 S(r)G(r)G(ar). \tag{38}$$

We have not found a general solution of (38) for relating $S(r)$ to $G(r)$ in the presence of capture. We have been able to analyze two special cases, however.

C. Two Solutions

In the first of these special cases we assume a constant traffic density $G(r)$. We can then show that the throughput density $S(r)$ has a Gaussian form, due to the fact that those packets generated further from the receiver will be received correctly less frequently than those packets generated close to the receiver.

In the second special case analyzed we assume a constant packet throughput density $S(r)$ and perfect capture ($a = 1$). Under these assumptions, the packet traffic density will increase as the distance from the receiver increases. We show that there exists a radius r_0 such that the packet traffic density is finite within a circle of radius r_0 around the receiver, while the packet traffic density becomes unbounded on the circle of radius r_0.

For the important case of a packet broadcasting channel distributed over some geographical area and using a packet retransmission policy (Section II-D), this result has an interesting interpretation. In such a situation any packet transmitted from a terminal located within the circle of radius r_0 will be received correctly with probability one (after a finite number of retransmissions), while the expected number of retransmissions required for a packet transmitted from a terminal further from the center than r_0 will be unbounded. Thus there exists a circle of radius r_0 such that terminals transmitting from within this circle can get their packets into the central receiver, while terminals transmitting from outside this circle spend all their time retransmitting their packets in vain. We call r_0 the Sisyphus distance of the ALOHA channel.

1) Constant Packet Traffic Density: Assume the density of normalized packet traffic is constant over the plane

$$G(r) = G_0 \tag{39}$$

and define the distance r_1 as the radius of a circle within which the total packet traffic is unity:

$$\pi r_1{}^2 G_0 \triangleq 1. \tag{40}$$

Then (38) reduces to

$$S'(r) = -\frac{4ra^2}{r_1{}^2} S(r) \tag{41a}$$

with the boundary condition

$$S_0 = G_0 \tag{41b}$$

so that the packet throughput density is

$$S(r) = G_0 \exp\left[-2a^2\left(\frac{r}{r_1}\right)^2\right] \tag{42}$$

and the total normalized packet thruput from a circle of radius r is

$$S = \int_0^r S(r')2\pi r'\, dr'$$
$$= \frac{1}{2a^2}\left\{1 - \exp\left[-2\left(\frac{ar}{r_1}\right)^2\right]\right\} \tag{43}$$

and

$$\lim_{r\to\infty} S = \frac{1}{2a^2}. \tag{44}$$

Note that a total throughput which can be supported by a single packet sink with "perfect capture" ($a = 1$) is equal to one half.

2) Constant Packet Throughput Density: Another case of interest where we have found a solution for (38) is that of constant packet throughput density in the plane. Assume

$$S(r) = S_0 \tag{45}$$

over the region in the plane where $S(r)$ and $G(r)$ are bounded.

Then (38) becomes

$$G'(r) = 4\pi ra^2 G(r)G(ar). \tag{46}$$

For the case of $a = 1$ (perfect capture), (46) becomes

$$G'(r) = 4\pi rG^2(r) \tag{47}$$

with the boundary condition

$$G(0) = S_0 \tag{48}$$

so that

$$G(r) = \frac{S_0}{1 - 2\pi r^2 S_0}. \tag{49}$$

Note that the normalized packet traffic per unit area is finite

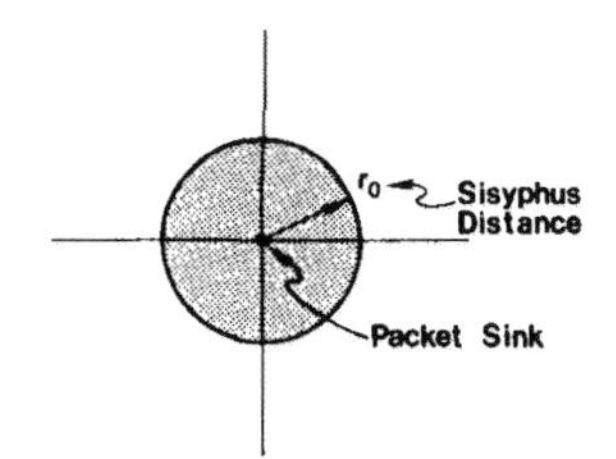

Fig. 9. Region of constant packet throughput S_0 for a single packet sink.

for

$$0 \leqslant r < r_0 \tag{50}$$

where

$$r_0 \triangleq [2\pi S_0]^{-1/2} \tag{51}$$

and r_0 is the Sisyphus distance mentioned in Section IV-C. Note that the Sisyphus distance also has the property that

$$\pi r_0{}^2 S_0 = \frac{1}{2}. \tag{52}$$

As in the previous case, the total packet throughput which can be supported by a single packet sink operating with perfect capture is one half.

V. PACKET BROADCASTING WITH AVERAGE POWER LIMITATIONS

A. Satellite Packet Broadcasting

In previous sections we have analyzed the performance of packet broadcasting channels and compared the performance of these channels to that of conventional point-to-point channels operating at the same peak data rate. Such a comparison is of interest in the case of channels limited by multiple access interference rather than noise, since an increase in the transmitted power of such channels will not lead to improved performance. But just as the average data rate of a packet broadcasting channel can be well below its peak data rate when it is operated at a low duty cycle, the average transmitted power of a packet broadcasting channel can be well below its peak transmitted power.

In this section we analyze the throughput of a packet broadcasting channel when compared to that of a conventional point-to-point channel of the same average power. This analysis is of interest in the case of satellite information systems employing thousands of small earth stations. For a satellite system the fundamental limitation in the downlink is the average power available in the satellite transponder rather than the peak power. Our results show that in the limit of large numbers of small earth stations, the packet throughput approaches 100 percent of the point-to-point capacity. Thus the multiple access capability and the complete connectivity (in the topological sense) of an ALOHA channel can be obtained at no price in average throughput. Furthermore, since our results suggest the use of higher peak power in the satellite transponder (while the average power is kept constant), the small earth stations may use smaller antennas and simpler receivers and modems than would be necessary in a conventional system.

In existing satellite systems the TWT output power in each transponder cannot be varied dynamically. In such systems the advantages implied by our analysis may be realized by frequency-division sharing a single transponder among several voice users and a single channel, operating in an ALOHA mode or some other burst mode, and occupying a frequency band equivalent to one or more voice users. The type of operation implied by our analysis also suggests investigation of high peak power satellite burst transponders (perhaps employing power devices similar to those used in radar systems) for use in information systems composed of large numbers of ultra-small earth stations.

B. Burst Power and Average Power

The capacity of a satellite channel can be calculated by the classical Shannon equation

$$C = W \log \left(1 + \frac{P}{N}\right) \tag{53}$$

where C is the capacity in bits (if the log is a base two logarithm), W is the channel bandwidth, P is the average received signal power at the earth station, and N is the average noise power at the earth station. Equation (53) expresses the capacity of the satellite channel under the assumption that the transponder transmits continuously.

If the channel is used in burst mode the transponder will emit power only when a data burst occurs, and the average power out of the transponder will be less than the burst power. Let D be the ratio of the average power transmitted to the power transmitted during a data burst. For a linear transponder D will equal the channel traffic G, and for a hard-limiting transponder D will equal the duty cycle of the channel. For both the unslotted and slotted ALOHA channel the duty cycle is $1 - e^{-G}$. Thus for a linear transponder[2]

$$D = G, \tag{54a}$$

while for a hard-limiting transponder

$$D = 1 - e^{-G}. \tag{54b}$$

Note that in the case of a hard-limiting transponder with small values of channel traffic, the duty cycle approaches that of a linear transponder.

If we retain P as the notation for the average signal power received at the earth station, the power received during a data burst will be P/D. Thus (53) should be modified in two ways.

[2] Our analysis is of significance only for $G < 1$. The analysis is formally correct, however, for all G, even though the designation of the power transmitted during bursts as "peak power" becomes inappropriate for the linear transponder case when $G > 1$. (In such a situation the "peak power" is less than the average power.)

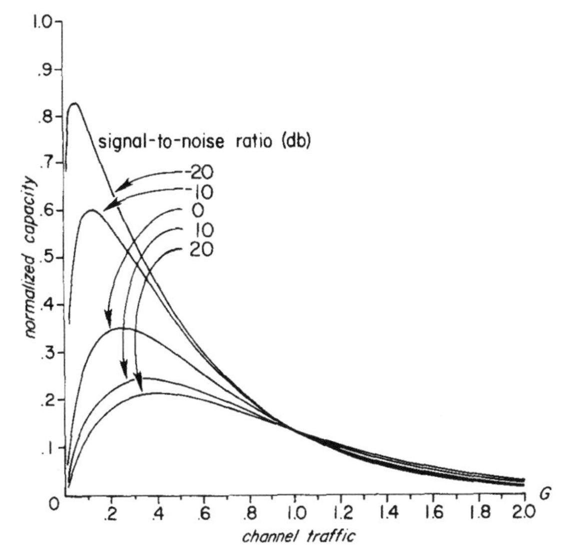

Fig. 10. Linear transponder; unslotted channel.

1) We replace W by SW to account for the fact that the channel is only used intermittently.

2) We replace P in (53) by P/D to keep the average power of the channel fixed at P.

We should note that when we make these changes, we are assuming that the packet length of the system is long enough so that the asymptotic assumptions which are used to derive (53) still apply. In practice, this is not a problem.

With these two changes then, we have four different cases.

1) Unslotted channel, linear transponder:

$$C_1 = Ge^{-2G} W \log \left(1 + \frac{P}{GN}\right). \tag{55a}$$

2) Unslotted channel, limiting transponder:

$$C_2 = Ge^{-2G} W \log \left(1 + \frac{P}{(1 - e^{-g})N}\right). \tag{55b}$$

3) Slotted channel, linear transponder:

$$C_3 = Ge^{-G} W \log \left(1 + \frac{P}{GN}\right). \tag{55c}$$

4) Slotted channel, limiting transponder:

$$C_4 = Ge^{-G} W \log \left(1 + \frac{P}{(1 - e^{-G})N}\right). \tag{55d}$$

We have calculated the normalized capacities C_i/C for $i = 1, 2, 3, 4$ for different values of P/N, the signal-to-noise ratio of the earth station when the transponder operates continuously. The normalized capacities are plotted in Figs. 10, 11, 12, and 13 for P/N equal to -20, -10, 0, 10, and 20 dB. Of particular interest in these curves is the fact that the highest values of C_i/C occur just where we would want them to occur—for small values of channel traffic (G) and for small earth stations (low P/N). In the limit we have (for a fixed value of G)

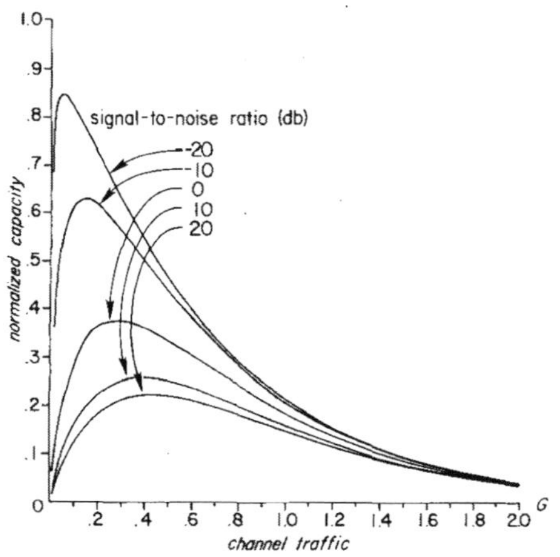

Fig. 11. Limiting transponder; unslotted channel.

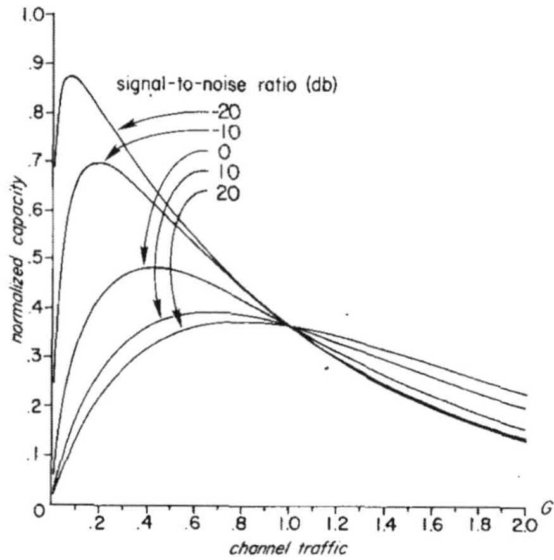

Fig. 12. Linear transponder; slotted channel.

$$\lim_{\frac{P}{N}\to 0} \frac{C_i}{C} = \frac{S}{D}, \qquad i = 1, 2, 3, 4 \tag{56}$$

so that

1) unslotted channels, linear transponder

$$\lim_{\frac{P}{N}\to 0} \frac{C_1}{C} = e^{-2G} \tag{57a}$$

2) unslotted channels, limiting transponder

$$\lim_{\frac{P}{N}\to 0} \frac{C_2}{C} = \frac{Ge^{-2G}}{(1-e^{-G})} \tag{57b}$$

3) slotted channel, linear transponder

$$\lim_{\frac{P}{N}\to 0} \frac{C_3}{C} = e^{-G}$$

4) slotted channel, limiting transponder

$$\lim_{\frac{P}{N}\to 0} \frac{C_4}{C} = \frac{Ge^{-G}}{(1-e^{-G})} \tag{57d}$$

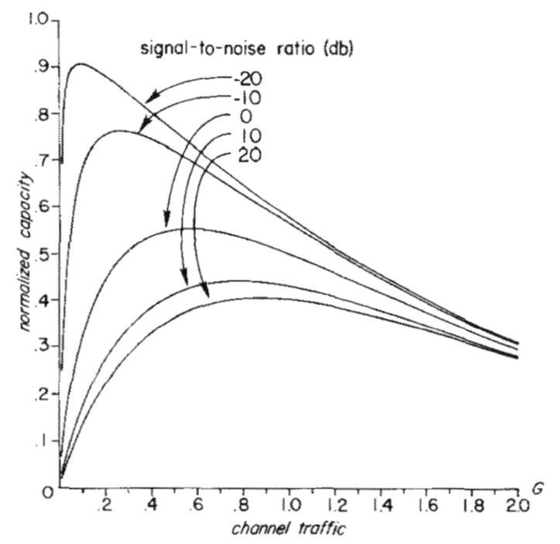

Fig. 13. Limiting transponder; slotted channel.

and in all cases

$$\lim_{G\to 0} \lim_{\frac{P}{N}\to 0} \frac{C_i}{C} = 1. \tag{58}$$

Thus this multiplexing technique allows a network of small inexpensive earth stations to achieve the maximum value of channel capacity, at the same time providing complete connectivity and multiple access capability.

VI. BACKGROUND AND ACKNOWLEDGMENT

The term "packet broadcasting" was first coined by Robert Metcalfe in his Ph.D. dissertation [28]. As is often the case with simple ideas, the concept of combining burst transmission and Poisson user statistics to provide random access to a channel has occurred independently to a number of investigators. The first attempt at an analysis of such a system of which I am aware is contained in an internal Bell Laboratories memorandum by Schroeder [29], suggested by an earlier paper by Pierce and Hopper [30]. Two other early related papers were written by Costas [31] and Fulton [32]. Of course, a theoretical analysis is not necessary in order to build such a system, and anyone who has sat in a taxi listening to the staccato voice bursts of a radio dispatcher and a set of taxi drivers sharing a single voice channel will recognize the operation of a voice packet broadcasting channel using a carrier sense protocol. And even after an analysis is available, the concept of packet broadcasting may be suggested without reference to the theory [33].

The first papers analyzing packet broadcasting in the form implemented in the ALOHA System [6] assumed fixed packet throughput and a retransmission protocol as described in Section II-D-1). This approach leads to a number of questions involving optimum retransmission policy [28], the behavior of the channel with a finite number of users [39], stability of the channel [13], and transmission of long files by means of various reservation schemes [34], [44]. A comprehensive treatment of these as well as other interesting packet broadcasting questions may be found in Kleinrock [42]. In this paper we have taken a different approach by assuming a given packet traffic rather than throughput. With such a starting point, the questions mentioned above do not assume key importance in the theory, although their practical importance is not diminished.

Much of the theory of packet broadcasting was developed in two working groups sponsored by the Advanced Research Projects Agency of the Department of Defense. These groups circulated a private series of working papers—the ARPANET Satellite System notes (ASS notes) and the Packet Radio Temporary notes (PRT notes)—where many of the theoretical results described or referenced in this paper appeared for the first time. Unfortunately, the several references to ASS notes in papers subsequently published in the open literature may have produced some confusion in the minds of those trying to trace the references. Among the most significant of the ASS note and PRT note results was the first derivation of the capacity of a slotted ALOHA channel and the first analysis of the use of the capture effect in packet broadcasting, both by Larry Roberts. That note has since been republished in the open literature [41].

The results of Section III-A dealing with two different packet lengths were suggested by an ASS note written by Tom Gaarder, and the results of Section III-B dealing with the excess capacity of a slotted channel were suggested by an ASS note written by Randy Rettberg. Other problems which were first analyzed in ASS notes or PRT notes but not emphasized in this paper include various packet broadcasting reservation systems [22], [35], [36], carrier sense packet broadcasting [18], [19], and questions dealing with packet routing and protocol issues in a network of repeaters [37]. The reader interested in theoretical network protocol questions should also see Gallagher [38], although this work did not originate in an ASS note or PRT note.

The first system to employ packet broadcasting techniques was the ALOHA System computer network at the University of Hawaii in 1970. Subsequently, packet repeaters were added to the network and packet broadcasting by satellite was demonstrated in the system. Some of the people involved in the implementation and development of the system were Richard Binder, Chris Harrison, Alan Okinaka, and David Wax.

The historical relevance of [29] and [32] was pointed out to me by Joe Aein, to whom I am indebted, in spite of my embarassment at having forgotten I was thesis supervisor on the second of these papers.

REFERENCES

[1] L. G. Roberts and B. D. Wessler, "Computer network development to achieve resource sharing," in *1970 Spring Joint Comput. Conf., AFIPS Conf. Proc.*, vol. 36. Montvale, NJ: AFIPS Press, 1970, pp. 543–549.

[2] L. G. Roberts, "Data by the packet," *IEEE Spectrum*, vol. 11, pp. 46–51, Feb. 1974.

[3] W. W. Chu, "A study of asynchronous time division multiplexing for time-sharing computer systems," in *1969 Spring Joint Comput. Conf., AFIPS Conf. Proc.*, vol. 35. Montvale, NJ: AFIPS Press, 1969, pp. 669–678.

[4] F. Heart, R. Kahn, S. Ornstein, W. Crowther, and D. Walden, "The interface message processor for the ARPA computer network," in *1970 Spring Joint Comput. Conf., AFIPS Conf. Proc.*, vol. 36. Montvale, NJ: AFIPS Press, 1970, pp. 551–567.

[5] H. Frank, I. Frisch, and W. Chou, "Topological considerations in the design of the ARPA computer network," in *1970 Spring Joint Comput. Conf., AFIPS Conf. Proc.*, vol. 36. Montvale, NJ: AFIPS Press, 1970, pp. 581–587.

[6] N. Abramson, "The ALOHA system—Another alternative for computer communications," in *1970 Fall Joint Comput. Conf., AFIPS Conf. Proc.*, vol. 37. Montvale, NJ: AFIPS Press, 1970, pp. 281–285.

[7] R. E. Kahn, "The organization of computer resources into a packet radio network," in *Nat. Comput. Conf., AFIPS Conf. Proc.*, vol. 44, May 1975, pp. 177–186.

[8] N. Abramson and E. R. Cacciamani, Jr., "Satellites: Not just a big cable in the sky," *IEEE Spectrum*, vol. 12, pp. 36–40, Sept. 1975.

[9] I. T. Frisch, "Technical problems in nationwide networking and interconnection," *IEEE Trans. Commun.*, vol. COM-23, pp. 78–88, Jan. 1975.

[10] R. M. Metcalfe and D. R. Boggs, "ETHERNET: Distributed packet switching for local computer networks," *Commun. Ass. Comput. Mach.*, to be published.

[11] N. Abramson, "Packet switching with satellites," in *Nat. Comput. Conf., AFIPS Conf. Proc.*, vol. 42, 1973, pp. 695–702.

[12] R. D. Rosner, "Optimization of the number of ground stations in a domestic satellite system," in *EASCON'75 Rec.*, Sept. 29–Oct. 1, 1975, pp. 64A–64J.

[13] A. B. Carleial and M. E. Hellman, "Bistable behavior of ALOHA-type systems," *IEEE Trans. Commun.*, vol. COM-23, pp. 401–410, Apr. 1975.

[14] P. E. Jackson and C. D. Stubbs, "A study of multiaccess computer communications," in *1969 Spring Joint Comput. Conf., AFIPS Conf. Proc.*, vol. 34. Montvale, NJ: AFIPS Press, 1969, pp. 491–504.

[15] T. J. Klein, "A tactical packet radio system," in *Proc. Nat. Telecommun. Conf.*, New Orleans, LA, Dec. 1975.

[16] K. Ah Mai, "Organizational alternatives for a Pacific educational computer-communication network," ALOHA Syst. Tech. Rep. CN74-27, Univ. Hawaii, Honolulu, May 1974.

[17] R. Binder, R. Rettberg, and D. Walden, *The Atlantic Satellite Packet Broadcast and Gateway Experiments*. Cambridge, MA: Bolt Beranek and Newman, 1975.

[18] L. Kleinrock and F. A. Tobagi, "Packet switching in radio channels: Part I—Carrier sense multiple-access modes and their throughput-delay characteristics," *IEEE Trans. Commun.*, vol. COM-23, pp. 1400–1416, Dec. 1975.

[19] F. A. Tobagi and L. Kleinrock, "Packet switching in radio channels: Part II—The hidden terminal problem in carrier sense multiple-access and the busy-tone solution," *IEEE Trans. Commun.*, vol. COM-23, pp. 1417–1433, Dec. 1975.

[20] R. T. Chien, L. R. Bahl, and D. T. Tang, "Correction of two erasure bursts," *IEEE Trans. Inform. Theory*, vol. IT-15, pp. 186–187, Jan. 1969.

[21] N. Abramson, "Cyclic code groups," *Problems of Inform. Transmission*, Acad. Sci. USSR, Moscow, vol. 6, no. 2, 1970.

[22] L. G. Roberts, "Dynamic allocation of satellite capacity through packet reservation," in *Nat. Comput. Conf. AFIPS Conf. Proc.*, vol. 42, June 1973, pp. 711–716.

[23] M. J. Ferguson, "A study of unslotted ALOHA with arbitrary message lengths," in *Proc. 4th Data Commun. Symp.*, Quebec, Canada, Oct. 7–9, 1975, pp. 5-20–5-25.

[24] R. Rettberg, "Random ALOHA with slots-excess capacity," ARPANET Satellite Syst. Note 18, NIC Document 11865, Stanford Res. Inst., Menlo Park, CA, Oct. 11, 1972.

[25] N. Abramson, "Excess capacity of a slotted ALOHA channel (continued)," ARPANET Satellite Syst. Note 30, NIC Document 13044, Stanford Res. Inst., Menlo Park, CA, Dec. 6, 1972.

[26] L. Schiff, "Random-access digital communication for mobile radio in a cellular environment," *IEEE Trans. Commun.*, vol. COM-22, pp. 688–692, May 1974.

[27] H. Frank, R. M. Van Slyke, and I. Gitman, "Packet radio network design—System considerations," in *Nat. Comput. Conf., AFIPS Conf. Proc.*, vol. 44, May 1975, pp. 217–232.

[28] R. M. Metcalfe, "Packet communication," Rep. MAC TR-114, Project MAC, Massachussetts Inst. Technol., Cambridge, July 1973.

[29] M. R. Schroeder, "Nonsynchronous time multiplex system for speech transmission," Bell Lab. Memo., Jan. 19, 1959.

[30] J. R. Pierce and A. L. Hopper, "Nonsynchronous time division with holding and random sampling," *Proc. IRE*, vol. 40, pp. 1079–1088, Sept. 1952.

[31] J. P. Costas, "Poisson, Shannon, and the radio amateur," *Proc. IRE*, vol. 47, p. 2058, Dec. 1959.

[32] F. F. Fulton, Jr., "Channel utilization by intermittent transmitters," Tech. Rep. 2004-2, Stanford Electron. Lab., Stanford Univ., Stanford, CA, May 12, 1961.

[33] K. D. Levin, "The overlapping problem and performance degradation of mobile digital communication systems," *IEEE Trans. Commun.*, vol. COM-23, pp. 1342-1347, Nov. 1975.

[34] R. Binder, "A dynamic packet-switching system for satellite broadcast channels," in *Conf. Rec., Int. Conf. Commun.*, vol. III, June 1975, pp. 41-1-41-5.

[35] S. S. Lam and L. Kleinrock, "Packet switching in a multiaccess broadcast channel: Dynamic control procedures," *IEEE Trans. Commun.*, vol. COM-23, pp. 891-904, Sept. 1975.

[36] L. Kleinrock and S. S. Lam, "Packet switching in a multiaccess broadcast channel: Performance evaluation," *IEEE Trans. Commun.*, vol. COM-23, pp. 410-423, Apr. 1975.

[37] I. Gitman, "On the capacity of slotted ALOHA networks and some design problems," *IEEE Trans. Commun.*, vol. COM-23, pp. 305-317, Mar. 1975.

[38] R. G. Gallager, "Basic limits on protocol information in data communication networks," *IEEE Trans. Inform. Theory*, vol. IT-22, pp. 385-398, July 1976.

[39] M. J. Ferguson, "On the control, stability, and waiting time in a slotted ALOHA random-access system," *IEEE Trans. Commun.*, vol. COM-23, pp. 1306-1311, Nov. 1975.

[40] J. J. Metzner, "On improving utilization in ALOHA networks," *IEEE Trans. Commun.*, vol. COM-24, pp. 447-448, Apr. 1976.

[41] L. G. Roberts, "ALOHA packet system with and without slots and capture," *Comput. Commun. Rev.*, vol. 5, pp. 28-42, Apr. 1975.

[42] L. Kleinrock, *Queueing Systems, Volume 2: Computer Applications.* New York: Wiley, 1976, pp. 360-407.

[43] I. Gitman, R. M. Van Slyke, and H. Frank, "On splitting random accessed broadcast communication channels," in *Proc. 7th Hawaii Int. Conf. Syst. Sci.–Suppl. on Comput. Nets*, Western Periodicals, Jan. 1974, pp. 81-85.

[44] W. Crowther, R. Rettberg, D. Walden, S. Ornstein, and F. Heart, "A system for broadcast communication: Reservation ALOHA," in *Proc. 6th Hawaii Int. Conf. Syst. Sci.*, Western Periodicals, Jan. 1973, pp. 371-374.

PHOTO NOT AVAILABLE

Norman Abramson (S'55-M'59-F'73) received the A.B. degree in physics from Harvard University, Cambridge, MA, the M.A. degree in physics from the University of California, Los Angeles, and the Ph.D. degree in electrical engineering from Stanford University, Stanford, CA.

In 1966 he was appointed Professor of Electrical Engineering and Professor of Information Sciences at the University of Hawaii, Honolulu. Before moving to Hawaii he taught communication theory at Stanford, Berkeley, and Harvard. From 1968 to 1970 he served as Chairman of the Information Sciences Program at the University of Hawaii. He is now Director of The ALOHA System–a university research project concerned with new forms of computer communication networks. He has served as a consultant in communication theory, satellite data transmission, and computer networks to several government and industrial laboratories in the United States. He has also served as a UN expert in computer networks at the International Computer Education Center, Budapest, Hungary.

Dr. Abramson is the author of *Information Theory and Coding* (New York: McGraw-Hill) and co-editor of *Computer-Communication Networks* (Englewood Cliffs, NJ: Prentice-Hall). He is a past Chairman of the Administrative Committee of the IEEE Information Theory Group and a member of the Editorial Board of IEEE SPECTRUM. In 1972 he was the recipient of the IEEE Sixth Region Achievement Award.

Maximum Likelihood Receiver for Multiple Channel Transmission Systems

W. VAN ETTEN

Abstract—A maximum likelihood (ML) estimator for digital sequences disturbed by Gaussian noise, intersymbol interference (ISI) and interchannel interference (ICI) is derived. It is shown that the sampled outputs of the multiple matched filter (MMF) form a set of sufficient statistics for estimating the input vector sequence. Two ML vector sequence estimation algorithms are presented. One makes use of the sampled output data of the multiple whitened matched filter and is called the vector Viterbi algorithm. The other one is a modification of the vector Viterbi algorithm and uses directly the sampled output of the MMF. It appears that, under a certain condition, the error performance is asymptotically as good as if both ISI and ICI were absent.

I. INTRODUCTION

It has first been pointed out by Shnidman [1] that intersymbol interference (ISI) and crosstalk between multiplexed signals are essentially identical phenomena. Kaye and George have worked out this idea by investigating the transmission of multiplexed signals over multiple channel and diversity systems [2]. The author of the underlying concise paper has presented a unified theory for treating intersymbol interference and interchannel interference (ICI) as one type of disturbance [3]. We will call the combined effect of these disturbances multidimensional interference (MDI). In the following the essentials of [3] are summarized.

The generalized Nyquist criterion formulated by Shnidman is restated in matrix notation. Furthermore an optimal linear receiver is derived, consisting of a multiple matched filter (MMF) followed by a multiple tapped delay line (MTDL). As optimization criterion is used minimum error probability and it appears that this optimum linear receiver has the same structure as the receiver derived by Kaye and George under the minimum mean-square error criterion. For a suboptimum criterion (minimum Pr (e) under the constraint that the multidimensional Nyquist criterion is satisfied) a theorem is given to calculate the tap coefficients for this case.

Up to this point is appears that several concepts known from ISI literature can be generalized for MDI. Recently maximum likelihood (ML) sequence estimation of data distrubed by noise and ISI received considerable attention [4]-[6]. Now the question arises whether these concepts can also be generalized for sequences transmitted over multiple channel systems where the output data are disturbed by noise and MDI. This concise paper gives a positive answer to this question.

II. THE MULTIPLE CHANNEL COMMUNICATION MODEL

The transmission system, to be considered in this concise paper, has M inputs and M outputs. To each input j a data sequence $\Sigma_l\, a_{l,j}\delta(t - lT)$ is applied which we want to estimate at the receiving end of the communication system. The symbols $a_{l,j}$ are elements of the alphabet $\{0,1,\cdots, L-1\}$. Symbols that are applied to the several inputs of the system at the same instant lT are ordered systematically in the input vector

$$x_l \triangleq \begin{bmatrix} a_{l,1} \\ a_{l,2} \\ \cdot \\ \cdot \\ \cdot \\ a_{l,M} \end{bmatrix}. \tag{1}$$

With the input vector sequence we associate the vector D transform

$$x(D) \triangleq \sum_l x_l D^l \tag{2}$$

where D is the delay operator. In our investigations a linear, dispersive, and time-invariant multiple channel model is assumed (Fig. 1). This means that a linear relation exists between each input and each output signal and that the output signal due to the excitation of more than one input is the sum of the individual responses to the inputs in question. The relation between input j and output i is denoted by the impulse response $c_{ij}(t)$. All these responses are assumed to be square-integrable and of finite duration. Further we assume that the output signals are disturbed by MDI and additive, zero-mean, white Gaussian noise. Each output i is corrupted by a different noise signal $n_i(t)$, but it is assumed that these noise signals are uncorrelated and all have the same double-sided spectral density N_0. These assumptions are not a restriction of the generality as is shown in [3].

III. THE STATISTICAL SUFFICIENCY OF THE MMF OUTPUT

In this section we shall show that if the MMF, as defined in [3], is used as multiple linear receiving filter, then the sampled outputs of this MMF form a set of sufficient statistics for estimating the vector input sequence $x(D)$.

The impulse response $c_{ij}(t)$ is considered as an element of a matrix

$$C(t) \triangleq \begin{bmatrix} c_{11}(t) & c_{12}(t) & \cdot & \cdot & c_{1M}(t) \\ c_{21}(t) & c_{22}(t) & \cdot & \cdot & c_{2M}(t) \\ \cdot & \cdot & \cdot & \cdot & \cdot \\ \cdot & \cdot & \cdot & \cdot & \cdot \\ \cdot & \cdot & \cdot & \cdot & \cdot \\ c_{M1}(t) & \cdot & \cdot & \cdot & c_{MM}(t) \end{bmatrix} \tag{3}$$

which defines the behavior of the multiple channel system. If the MMF is described in an analog way, it will be clear this its response is denoted by $C^T(-t)$. Assume that the multiple channel system is excited by an arbitrary, single-input vector x. Defining in this case the signal at output i of the multiple channel system by $s_i(t)$, we can write the total system output as a vector as follows:

$$s(t) \triangleq \begin{bmatrix} s_1(t) \\ s_2(t) \\ \cdot \\ \cdot \\ \cdot \\ s_M(t) \end{bmatrix} \tag{4}$$

called the vector output signal. The noise is also given as a vector

Paper approved by the Associate Editor for European Contributions of the IEEE Communications Society for publication without oral presentation. Manuscript received April 14, 1975; revised July 16, 1975.

The author is with the Department of Electrical Engineering, Eindhoven Institute of Technology, Eindhoven, The Netherlands.

Reprinted from *IEEE Transactions on Communications,* February 1976.

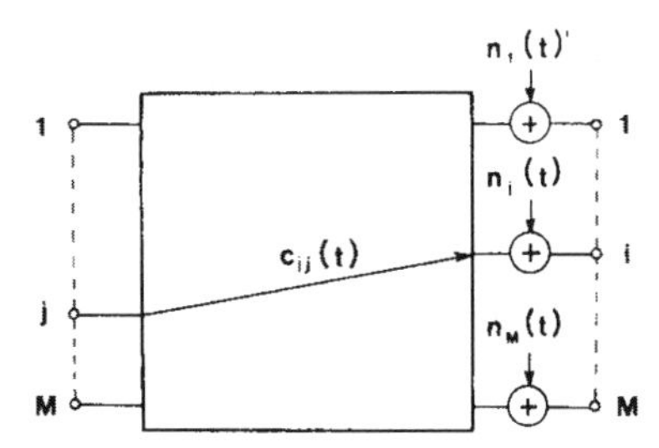

Fig. 1. Multiple channel communication model.

$$n(t) \triangleq \begin{bmatrix} n_1(t) \\ n_2(t) \\ \cdot \\ \cdot \\ \cdot \\ n_M(t) \end{bmatrix} \tag{5}$$

called the vector noise.

In the following we shall use several times the inner product of matrices, the components of which consist of time functions. Such a product is denoted as $\langle A(t),B(t)\rangle$ and defined by

$$\langle A(t),B(t)\rangle_{ij} \triangleq \sum_n \int_{-\infty}^{\infty} a_{in}(t)b_{nj}(t)\,dt. \tag{6}$$

The sampled output of the MMF, in the absence of noise, is given by the signal vector

$$\mathbf{s} = \langle C^T(t),s(t)\rangle \tag{7}$$

whereas the inverse transformation from signal vector to output vector signal is

$$s(t) = [C^T(t)]^T G\mathbf{s} = C(t)G\mathbf{s} \tag{8}$$

where G is a matrix to be determined. Substituting (8) in (7) gives

$$G = [\langle C^T(t), C(t)\rangle]^{-1}. \tag{9}$$

So the systems to be treated must have the property that the matrix G exists. This requirement, however, is quite trivial, because at systems not possessing this property it is impossible to recover even a single-input vector from the sampled MMF outputs. The sampled output noise, if the signal is absent, can be written as

$$\mathbf{n} = \langle C^T(t), n(t)\rangle. \tag{10}$$

According to (10) the relevant vector noise, being that part of the input vector noise that is left after projection of $n(t)$ at the signal space, is denoted by

$$n_r(t) = [C^T(t)]^T G\mathbf{n} = C(t)G\mathbf{n}. \tag{11}$$

By means of the definition

$$\mathbf{v} \triangleq \mathbf{s} + \mathbf{n}. \tag{12}$$

The equivalent received vector signal is written as

$$\nu(t) = C(t)G\mathbf{v} \tag{13}$$

which menas that for the sampled output there is no difference whether the true received vector signal $s(t) + n(t)$ or the vector signal $\nu(t)$ is supplied to the input of the MMF. Writing out (13) yields

$$\nu(t) = C(t)G\mathbf{v} = C(t)G\mathbf{s} + C(t)G\mathbf{n} = s(t) + n_r(t). \tag{14}$$

Thus $C(t)$ is a basis for the signal space spanned by both $s(t)$ and $n_r(t)$ [7, ch. 4], which proves that the sampled MMF output is a sufficient statistic for estimation of a single input vector x. Now the following theorem can be stated.

Theorem 1: If at each instant lT a vector x_l is transmitted, then the vector output sequence

$$\nu(D) \triangleq \sum_l \nu_l D^l \tag{15}$$

forms a set of sufficient statistics for estimating the vector input sequence $x(D)$ (see [4] and [7]).

IV. THE MULTIPLE WHITENED MATCHED FILTER

Now consider the system consisting of the channel in cascade with the MMF as a multiple channel system with M inputs and M outputs. The impulse response from input j to output n of this system is called $v_{nj}(t)$ and can be written as

$$v_{nj}(t) = \sum_{i=1}^{M} c_{in}(-t) * c_{ij}(t) = \sum_{i=1}^{M} \int_{-\infty}^{\infty} c_{in}(\tau - t)c_{ij}(\tau)\,d\tau. \tag{16}$$

where $*$ means convolution. Define

$$V_l \triangleq \begin{bmatrix} v_{11}(lT) & v_{12}(lT) & \cdot & \cdot & v_{1M}(lT) \\ v_{21}(lT) & v_{22}(lT) & \cdot & \cdot & v_{2M}(lT) \\ \vdots & \vdots & \vdots & \vdots & \vdots \\ v_{M1}(lT) & v_{M2}(lT) & \cdot & \cdot & v_{MM}(lT) \end{bmatrix} \tag{17}$$

and

$$V(D) \triangleq \sum_l V_l D^l. \tag{18}$$

By means of (16) it is easy to see that (18) is equivalent to

$$V(D) = \langle C^T(D^{-1},t), C(D,t)\rangle \tag{19}$$

where $C(D,t)$ is a matrix with components consisting of the chip D transforms [4] of the components of $C(t)$. The crosscorrelation of the output noise signals at outputs n and m is given by

$$\begin{aligned} \phi_{nm}(\rho) &= \sum_{i=1}^{M} N_0 \int_{-\infty}^{\infty} c_{in}(-t)c_{im}(-t-\rho)\,dt \\ &= \sum_{i=1}^{M} N_0 \int_{-\infty}^{\infty} c_{in}(t)c_{im}(t-\rho)\,dt. \end{aligned} \tag{20}$$

Sampling this function we define its D transform as follows:

$$\phi_{nm}(D) \triangleq \sum_l \phi_{nm}(lT)D^l. \tag{21}$$

If all $\phi_{nm}(D)$ are collected in a matrix we get the spectral matrix

$$\Phi(D) = N_0 \langle C^T(D,t), C(D^{-1},t) \rangle. \tag{22}$$

Relation (22) can easily be verified by means of (20). In [8] and [9] it is shown that a matrix $H(D^{-1})$ can be found such that

$$\Phi(D) = N_0 H(D) H^T(D^{-1}) \tag{23}$$

with both $H(D^{-1})$ and $H^{-1}(D^{-1})$ stable and nonanticipatory. Comparing (19) and (22) it is obvious that

$$V(D) = H(D^{-1}) H^T(D). \tag{24}$$

Now we conclude that the sampled output of the MMF can be written as

$$\nu(D) = H(D^{-1}) H^T(D) x(D) + H(D^{-1}) n(D) \tag{25}$$

where $n(D)$ is the sampled input noise vector sequence. The output noise

$$n'(D) = H(D^{-1}) n(D) \tag{26}$$

is colored Gaussian with spectral matrix $\Phi(D)$. This follows from

$$E[H(D)n(D^{-1})\{H(D^{-1})n(D)\}^T]$$
$$= E[H(D)n(D^{-1})n^T(D)H^T(D^{-1})] = N_0 H(D) H^T(D^{-1}). \tag{27}$$

From (25) it is seen that the output noise is whitened by the following operation:

$$z(D) \triangleq H^{-1}(D^{-1}) \nu(D)$$
$$= H^T(D) x(D) + n(D) = y(D) + n(D) \tag{28}$$

which means physically that an MTDL [3] with transfer $H^{-1}(D^{-1})$ is placed after the MMF. It has been mentioned in the foregoing that $H^{-1}(D^{-1})$ is stable and nonanticipatory and thus realizable. The MMF followed by the MTDL is called multiple whitened matched filter and is characterized by its chip D transform

$$W(D,t) \triangleq H^{-1}(D^{-1}) C^T(D^{-1},t). \tag{29}$$

If the impulse response from input n to output m is denoted by $w_{mn}(t)$, the set of functions $w_{mn}(t - kT)$ is orthonormal in both the time and space dimension as is seen from

$$\Phi_{ww}(D) = \langle W(D^{-1},t), W^T(D,t) \rangle$$
$$= H^{-1}(D) \langle C^T(D,t), C(D^{-1},t) \rangle \{H^{-1}(D^{-1})\}^T$$
$$= H^{-1}(D) V(D^{-1}) \{H^T(D^{-1})\}^{-1}$$
$$= H^{-1}(D) H(D) H^T(D^{-1}) \{H^T(D^{-1})\}^{-1} = I. \tag{30}$$

In the foregoing section we concluded that $\nu(D)$ forms a set of sufficient statistics for estimation $x(D)$, but $z(D)$ is found by the reversible linear transformation $H^{-1}(D^{-1})$ on $\nu(D)$. Thus $z(D)$ forms a set of sufficient statistics for estimating $x(D)$ also. This section is resumed in the following theorem.

Theorem 2: Let $C(t)$ be the matrix of impulse responses of the multiple channel transmission system and $H(D^{-1})H^T(D)$ a factorization of

$$V(D) = \langle C^T(D^{-1},t), C(D,t) \rangle \tag{31}$$

such that both $H(D^{-1})$ and $H^{-1}(D^{-1})$ are stable and nonanticipatory. Then the multiple filter whose chip D transform is

$$W(D,t) = H^{-1}(D^{-1}) C^T(D,t) \tag{32}$$

is realizable and is called a multiple whitened matched filter and its sampled outputs give a vector sequence

$$z(D) = H^T(D) x(D) + n(D) \tag{33}$$

in which $n(D)$ is a white Gaussian noise vector sequence, and which is a set of sufficient statistics for estimation of the vector input sequence $x(D)$ where $n(D)$ white Gaussian is to be interpreted in both the discrete time and space dimension.

The multiple whitened matched filter found in this section is a generalized version of the whitened matched filter derived in [4]. This generalized filter is capable of optimizing the signal-to-noise ratio of the outputs of a multiple channel transmission system in which both ISI and ICI together with noise contribute to the disturbance, under the constraint that the output noise must be white in the two-dimensional sense.

V. THE VECTOR VITERBI ALGORITHM

In the preceding sections we have derived a structure giving a set of sufficient statistics for estimating the input vector sequence of a multiple channel transmission system from the observations of the output. This output is disturbed by MDI and noise. The noisy part of the multiple whitened matched filter output samples are shown to be uncorrelated and thus independent, since we have assumed that the noise is Gaussian. From this it follows that the Viterbi algorithm is a powerful tool to perform ML estimation of the input vector sequence $x(D)$. The vector Viterbi algorithm is a vector version of the algorithm used to make ML estimations on digital sequences and which is extensively described in [4] and [5]. The vector sequence $y(D)$ may be considered to be generated by a multiple finite state machine, driven by an input vector sequence $x(D)$ (see Fig. 2). As the state of this finite state machine we define

$$s_l \triangleq \{x_{l-1}, x_{l-2}, \cdots, x_{l-N}\} \tag{34}$$

where N is the degree of the matrix polynomial $H^T(D)$. The state s_l can take on L^{NM} different values. We can depict the successive states of the multiple finite state machine, together with all allowable transitions, in a trellis diagram. Each transition T_k in this trellis diagram is associated with an input vector x_{kl} and a certain value of the output signal y_{kl}. Given the observation z_l, the log likelihood of transition T_k is given by

$$\ln p(z_l - y_{kl}) = -\ln (\sqrt{2\pi N_0})^M$$
$$- \frac{1}{2N_0} \sum_{i=1}^{M} (z_{l,i} - y_{kl,i})^2 \tag{35}$$

where $z_{l,i}$ and $y_{kl,i}$ are the ith components of, respectively, z_l and y_{kl}. In ML sequence estimation the first term of the right member of (35), being independent of l, can be omitted and the same holds for the factor $1/2N_0$ in the second term. The squared distance of an observation at instant lT to a certain allowable transition T_k, characterized by y_{kl}, is defined by

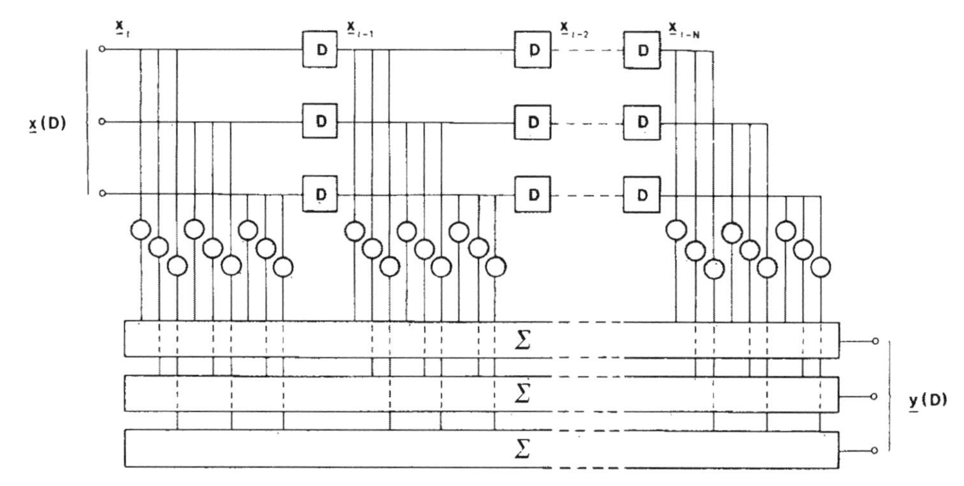

Fig. 2. Model of a multiple finite state machine.

$$D_{kl}^2 \triangleq \sum_{i=1}^{M} (z_{l,i} - y_{kl,i})^2. \tag{36}$$

The updating of the metrics, belonging to the several states, together with the updating of the corresponding path-registers proceeds as follows.

1) The metrics of all the states that belong to transitions, terminating in the same state, are increased with the corresponding squared distance D_{kl}^2.

2) Select the smallest increased metric as survivor-metric for the new state. The path-register of this new state is to be filled with the content of the path-register of the old state of the selected transition. This new path-register content is then updated with the elements of the input alphabet that belong to the selected transition.

The vector Viterbi algorithm does not differ fundamentally from the scalar version; the only differences, which are in fact generalizations, are as follows.

1) The operation "squared distance computation" is a computation in the vector sense defined by the Euclidean squared distance (36).

2) Each element of the path-register consists of M components and must be shifted and updated parallel in the path-register.

At this point the vector Viterbi algorithm is in fact reduced to the scalar version and we refer to [4] and [5] for further details. It will be clear that in a multiple channel transmission system the number of states is growing exponentially with the number of channels.

VI. THE VECTOR UNGERBOECK ALGORITHM

Ungerboeck has given an alternative recursive algorithm for making ML sequence estimations on data that are disturbed by ISI and white Gaussian noise [6]. Using this algorithm, the tapped delay line is omitted and the sampled output of the matched filter is directly used as input for the algorithm. In the following we shall generalize the Ungerboeck algorithm for ML vector sequence estimation of data that are disturbed by MDI and white Gaussian noise.

If a vector sequence $x(D)$ is transmitted the corresponding received vector signal is defined as follows:

$$u(t) \triangleq \sum_l C(t - lT)x_l + n(t). \tag{37}$$

Among all possible input sequences $\xi(D)$ we choose as estimate $\hat{x}(D)$ for $x(D)$ that vector sequence which maximizes $\ln p[u(t) \mid \xi(D)]$, which means minimizing over all allowable $\xi(D)$

$$J = \left\| u(t) - \sum_l C(t - lT)\,\xi_l \right\|_2^2$$
$$= \left\langle \left[u(t) - \sum_l C(t - lT)\,\xi_l \right]^T, \right.$$
$$\left. \cdot \left[u(t) - \sum_k C(t - kT)\,\xi_k \right] \right\rangle. \tag{38}$$

Writing out (38) the expression for J becomes

$$J = \langle u^T(t), u(t)\rangle - \left\langle u^T(t), \sum_k C(t - kT)\,\xi_k \right\rangle$$
$$- \left\langle \sum_l \xi_l^T C^T(t - lT), u(t) \right\rangle$$
$$+ \left\langle \sum_l \xi_l^T C^T(t - lT), \sum_k C(t - kT)\xi_k \right\rangle. \tag{39}$$

Define

$$v_l \triangleq \langle C^T(t - lT), u(t)\rangle. \tag{40}$$

This vector is interpreted as the sampled output of the MMF. By means of definition (40) J is written as

$$J = \langle u^T(t), u(t)\rangle$$
$$- 2\sum_l \xi_l^T v_l + \sum_l \sum_k \xi_l^T V_{l-k}\,\xi_k. \tag{41}$$

The first term of (41) is independent of ξ_l and thus may be ignored during the minimization process. The metric $J(\xi(D))$ can be calculated in a recursive manner.

$$J_l(\cdots, \xi_{l-1}, \xi_l) = J_{l-1}(\cdots, \xi_{l-1}) + F(v_l;\ \xi_{l-N}, \cdots, \xi_l) \tag{42}$$

with

$$F(\nu_l;\ \xi_{l-N},\ \cdots,\ \xi_l) = \xi_l^T\left[2\nu_l - V_0\,\xi_l - 2\sum_{k=1}^{N} V_k\,\xi_{l-k}\right] \quad (43)$$

where N is determined by the length of the $V(D)$ matrix sequence, according to

$$V(D) = \sum_{l=-N}^{N} V_l D^l. \quad (44)$$

Here the survivor-metric $\tilde{J}_l$ is introduced, which is defined as follows:

$$\tilde{J}_l(s_l) \triangleq \tilde{J}_l\{\xi_{l+1-N},\ \cdots,\ \xi_l\} \triangleq \min_{\{\cdots,\ \xi_{l-N}\}} \{J_l(\cdots,\ \xi_{l-N},\ \xi_{l-N+1},\ \cdots,\ \xi_l)\}. \quad (45)$$

The sequence $(\cdots, \xi_{l-N})$, which results in a minimum of (45) is called the path-history of the survivor-state

$$s_l \triangleq (\xi_{l+1-N},\ \cdots,\ \xi_l). \quad (46)$$

It is easy to see that there are again L^{NM} different survivor-states. One can imagine that these survivor-states correspond to the states of a finite state machine. From this point of view the principles of the Ungerboeck algorithm coincide from now on those of the Viterbi algorithm. For further details see [6].

Expression (43) is now to be used for the calculation of the squared distance of an observation to an allowable transition, and the finite state machine has as much different states as the finite state machine of the Viterbi algorithm.

Although at first glance the metric calculation of the Ungerboeck algorithm seems more complicated than that of the Viterbi algorithm, a second inspection of (43) shows that the metric up-dating is a rather simple operation from a programming point of view. Namely the quantity

$$\xi_l^T\left[V_0\,\xi_l + 2\sum_{k=1}^{N} V_k\,\xi_{l-k}\right]$$

only depends on the channel response, which is assumed to be fixed, and on the transitions to be considered. So this value can be stored in a memory and need not be calculated in real time.

VIII. THE ERROR PERFORMANCE OF THE ML RECEIVER

The investigations, given in this section, are closely related to the methods given in [4] and [6]. Remember that $x(D)$ represents the transmitted vector sequence and that the vector sequence estimated by the ML receiver is denoted by $\hat{x}(D)$. Then

$$e(D) \triangleq \hat{x}(D) - x(D) \quad (47)$$

defines the error vector sequence. Assuming stationarity, the starting point of an error event ϵ can be associated with $t = 0$:

$$\epsilon\colon e(D) = e_0 + e_1 D + \cdots + e_H D^H \quad \text{with } \|e_i\|_2 \geq \delta_0, \quad (48)$$

where δ_0 denotes the minimum nonzero value of the Euclidean norm of the error vector e_i. This value equals the minimum distance

$$\delta_0 = \min_{i \neq j}\{\|a_{l,j} - a_{l,i}\|\}. \quad (49)$$

From [6] we know that the error event probability is written as

$$\Pr(\epsilon) = \Pr(\epsilon_1)\Pr(\epsilon_2 \mid \epsilon_1) \leqslant \Pr(\epsilon_1)\Pr(\epsilon_2' \mid \epsilon_1) \quad (50)$$

where the subevents ϵ_1, ϵ_2, and ϵ_2' are defined as follows.

- ϵ_1 $x(D)$ is such that $x(D) + e(D)$ is an allowable data vector sequence;
- ϵ_2 noise vector sequence is such that $x(D) + e(D)$ has ML (within the observation interval); and
- ϵ_2' noise vector sequence is such that $x(D) + e(D)$ has greater likelihood than $x(D)$, but not necessarily ML.

From the preceding section it is concluded that $\Pr(\epsilon_2' \mid \epsilon_1)$ is the probability that

$$J(x(D)) > J(x(D) + e(D)). \quad (51)$$

It can be proven that inequality (51) is identical with

$$\delta^2(\epsilon) \triangleq \|V_0^{-1}\|_2 \sum_{l=0}^{H}\sum_{k=0}^{H} e_l^T V_{l-k} e_k < 2\,\|V_0^{-1}\|_2 \sum_{l=0}^{H} e_l^T n_l' \quad (52)$$

where n_l' are the sample values of the noise at the output of the MMF. The quantity $\delta(\epsilon)$ is called the distance of the error event ϵ. Consider the random variable α given by the right member of (52).

$$\alpha \triangleq 2\,\|V_0^{-1}\|_2 \sum_{l=0}^{H} e_l^T n_l'. \quad (53)$$

This random variable is Gaussian distributed with zero-mean and variance

$$E[\alpha^2] = 4N_0\,\|V_0^{-1}\|_2\,\delta^2(\epsilon). \quad (54)$$

From this it follows that

$$\Pr(\epsilon_2' \mid \epsilon_1) = \Pr(\alpha > \delta^2(\epsilon)) = Q\left(\frac{\delta(\epsilon)}{2\sqrt{N_0}\,\|V_0^{-1}\|_2^{1/2}}\right) \quad (55)$$

where the well-known $Q(\cdot)$ function is defined in [7].

Let E be the set of all possible error events ϵ. Then the probability that any error event occurs becomes

$$\Pr(E) = \sum_{\epsilon \in E} \Pr(\epsilon). \quad (56)$$

Let Δ be the set of all possible $\delta(\epsilon)$ and E_δ the subset of error

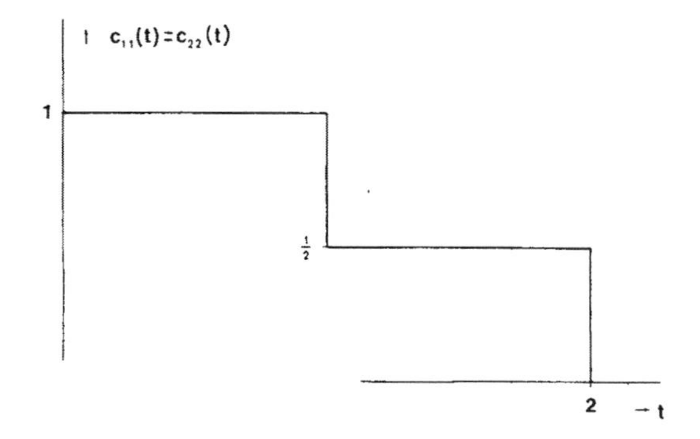

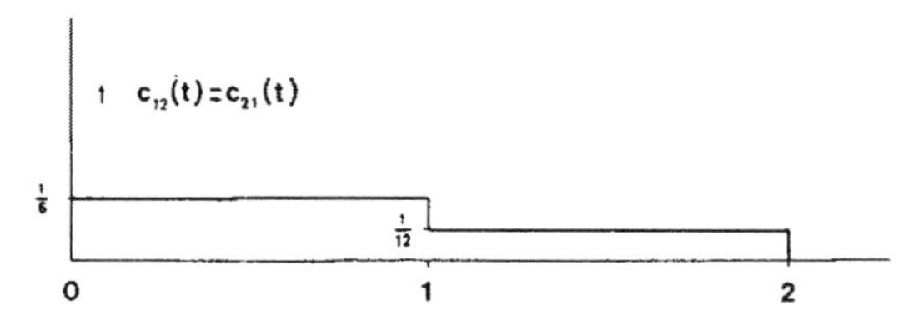

Fig. 3. Received signal set for the example.

events for which $\delta(\epsilon) = \delta$. Then from (50) the error event probability is bounded by

$$\Pr(E) \leqslant \sum_{\delta\in\Delta} Q\left(\frac{\delta}{2\sqrt{N_0}\,\|V_0^{-1}\|_2^{1/2}}\right) \sum_{\epsilon\in E_\delta} \Pr(\epsilon_1). \tag{57}$$

Because of the exponential behavior of the $Q(\cdot)$ function for large argument values, this expression will already at moderate signal-to-noise ratios be dominated by the term involving the minimum value $\delta_{\min}$ out of the set Δ. At moderate and large signal-to-noise ratios ϵ_2' implies ϵ_2 with a probability almost equal to one. For these SNR values Pr (E) is approximated by

$$\Pr(E) \simeq Q\left(\frac{\delta_{\min}}{2\sqrt{N_0}\,\|V_0^{-1}\|_2^{1/2}}\right) \sum_{\epsilon\in E_{\delta_{\min}}} \Pr(\epsilon_1). \tag{58}$$

Assuming the input symbols $a_{l,j}$ to be independent of each other and equiprobable, the probability of ϵ_1 is written as

$$\Pr(\epsilon_1) = \prod_{i=1}^{M} \prod_{l=0}^{H} \frac{L - |e_{l,i}|}{L} \tag{59}$$

with $e_{l,i}$ the ith component of e_l.

In the Appendix it is shown that under the constraint

$$\|V_0^{-1}\|_2 \sum_{l=-\infty}^{\infty}{}' \|V_l\|_2 \leqslant 1 \tag{60}$$

not any error event has smaller distance than the single error events with distance δ_0. With a single error event we mean an error sequence that consists of one error vector ($e(D) = e_0$) and from this vector only one component differs from zero. In this situation the single error events with distance δ_0 dominate the expression for the error event probability and the error event probability equals the symbol error probability

$$\Pr(e) \simeq Q\left(\frac{\delta_0}{2\sqrt{N_0}\,\|V_0^{-1}\|_2^{1/2}}\right) \sum_{\epsilon\in E_{\delta_0}} \frac{L-1}{L}. \tag{61}$$

Since $\delta_0^2/\|E_0^{-1}\|_2$ is the total amount of energy that is measured as the receiving end at transmission of a single symbol out of the set E_{δ_0}, the symbol error probability is not increased by MDI.

VIII. AN EXAMPLE

As an example we take a multiple channel with $M = 2$. The components of the transmission matrix $C(t)$ are as given in Fig. 3. We take $T = 1$ and for this system the $V(D)$ matrix polynomial is as follows:

$$\left.\begin{aligned} V_0 &= \frac{5}{144}\begin{bmatrix} 37 & 12 \\ 12 & 37 \end{bmatrix} \\ V_1 = V_{-1} &= \frac{1}{72}\begin{bmatrix} 37 & 12 \\ 12 & 37 \end{bmatrix} \end{aligned}\right\}. \tag{62}$$

One can easily verify that this $V(D)$ satisfies condition (60). Decomposition of $V(D)$ according to (24) yields

$$H^T(D) = \frac{1}{12}\begin{bmatrix} 6 & 1 \\ 1 & 6 \end{bmatrix}(2 + D). \tag{63}$$

By means of this matrix sequence the given system is simulated on a minicomputer. In Fig. 4 the error probability for a binary alphabet $\{+1, -1\}$ is plotted as function of the signal-to-noise ratio, together with the Pr (e) for isolated pulses. The two curves merge at a Pr (e) of about 10^{-4}. So, for error probabilities smaller than 10^{-4}, the performance of the ML receiver is as good as if MDI were absent. In the case of larger error probabilities the difference between the two curves is

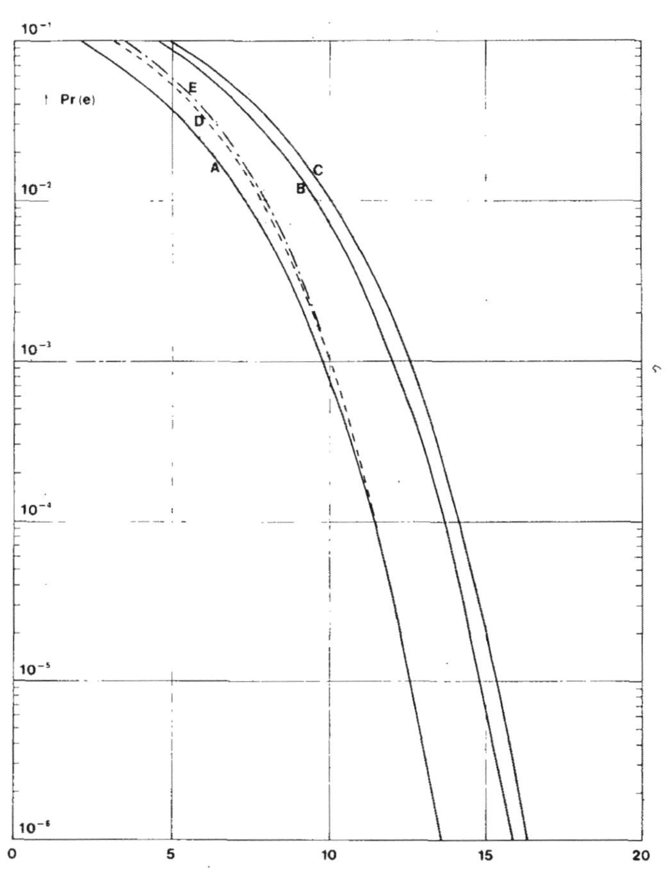

Fig. 4. Symbol error probability versus signal-to-noise ratio.

Curve A Single pulse;
Curve B monochannel with linear correction and bit-by-bit detection;
Curve C multiple channel with $M = 2$, linear correction and bit-by-bit detection;
Curve D monochannel with ML sequence estimation; and
Curve E multiple channel with $M = 2$ and ML vector sequence estimation.

maximal 1.2 dB. These results are compared with those of an optimum constrained linear receiver [3]. The difference between the linear receiver and the single-pulse performance is 2.7 dB, showing the superiority of the vector ML receiver. We also simulated a ML receiver for a monochannel with impulse response $c_{11}(t)$. Now the maximum difference with the single-pulse performance appears to be 1 dB, whereas the two curves also merge at a Pr (e) value of about 10^{-4}. Linear correction with bit-by-bit detection gives an increase of 2.2 dB in this case.

N.B.: At the simulations the path-register length was 16 bits in all cases. The number of transmissions was chosen such that the real error probability lies, with a probability 0.9, within an interval of 10 percent around the plotted value.

IX. SUMMARY AND CONCLUSIONS

It is shown that the MMF outputs form a set of sufficient statistics for estimating the transmitted vector sequence over a multiple channel system. A multiple whitened matched filter is derived, the output of which is used to perform ML vector sequence estimation by means of the vector version of the Viterbi algorithm. A modified algorithm, pointed out by Ungerboeck, is also generalized to combat the noise and MDI. If this algorithm is used MTDL is omitted and the sampled output of the MMF is directly used as input data for the algorithm. Finally, the error performance of the ML receiver for a multiple channel system disturbed by noise and MDI is calculated. From the latter investigations it follows that, under a certain constraint, for moderate and large SNR's the error performance is not substantially influenced by MDI, i.e., the symbol error probability is approximated by the value found if a single pulse is transmitted. It is concluded from this concise paper that ICI plays the same role as ISI. If these two disturbances are simultaneously considered, then MDI can, under the given constraints, be treated as a generalization of ISI and the concepts of ML sequence estimation on data disturbed by noise and ISI are also generalized for noise and MDI.

APPENDIX

Let

$$\| e_0 \|_2 \geqslant \delta_0 \tag{64}$$

and let $\{V_l\}_{l=-\infty}^{\infty}$ be given and assume

$$\| V_0^{-1} \|_2 \sum_{l=-\infty}^{\infty}{}' \| V_l \|_2 \leqslant 1. \tag{65}$$

The matrix V_0 equals $\langle C^T(t), C(t)\rangle$ and it is easy to show that this matrix is positive definite, under the condition derived in Section III.

$$\delta^2(\epsilon) = \| V_0^{-1} \|_2 \sum_{l=-H}^{H} \sum_{k=0}^{H} e_{l+k}^T V_l e_k$$
$$= \| V_0^{-1} \|_2 \sum_{k=0}^{H} e_k^T V_0 e_k$$
$$+ \| V_0^{-1} \|_2 \sum_{l=-H}^{H}{}' \sum_{k=0}^{H} e_{l+k}^T V_l e_k. \tag{66}$$

Consider the first term of (66). Because V_0 is positive definite we have the inequality

$$e_k^T V_0 e_k \geqslant \lambda_{\min}(V_0) e_k^T e_k \tag{67}$$

where $\lambda_{\min}(V_0)$ is the smallest eigenvalue of V_0. Moreover,

$$\| V_0^{-1} \|_2 = \frac{1}{\lambda_{\min}(V_0)}. \tag{68}$$

From (67) and (68) it follows

$$\| V_0^{-1} \|_2 \sum_{k=0}^{H} e_k^T V_0 e_k \geqslant \sum_{k=0}^{H} e_k^T e_k$$
$$= \sum_{k=0}^{H} \| e_k \|_2^2. \tag{69}$$

Consider now the second term of (66). Due to the Schwarz inequality and from what is given we have

$$\left| \| V_0^{-1} \|_2 \sum_{l=-H}^{H}{}' \sum_{k=0}^{H} e_{l+k}^T V_l e_k \right|$$
$$\leqslant \| V_0^{-1} \|_2 \sum_{l=-H}^{H}{}' \| V_l \|_2 \sum_{k=0}^{H} \| e_{l+k}^T \|_2 \cdot \| e_k \|_2$$
$$\leqslant \left\{ \| V_0^{-1} \|_2 \sum_{l=-H}^{H}{}' \| V_l \|_2 \right\} \left\{ \sum_{k=0}^{H} \| e_k \|_2^2 - \delta_0^2 \right\}. \tag{70}$$

From (69) and (70) it follows

$$\delta^2(\epsilon) \geqslant \left(\sum_{k=0}^{H} \| e_k \|_2^2 - \delta_0^2 \right)$$
$$- \left\{ \| V_0^{-1} \|_2 \sum_{l=-H}^{H}{}' \| V_l \|_2 \right\}$$
$$\cdot \left\{ \sum_{k=0}^{H} \| e_k \|_2^2 - \delta_0^2 \right\} + \delta_0^2$$
$$\geqslant \delta_0^2. \tag{71}$$

This last inequality holds if (65) is satisfied.

ACKNOWLEDGMENT

The author would like to thank Prof. J. van der Plaats and Prof. J. P. M. Schalkwijk for the stimulating discussions on the subject. He also wants to acknowledge L. S. de Jong for giving the proof of the Appendix.

REFERENCES

[1] D. A. Shnidman, "A generalized Nyquist criterion and optimum linear receiver for a pulse modulation system," *Bell Syst. Tech. J.*, vol. 46, pp. 2163-2177, Nov. 1967.

[2] A. R. Kaye and D. A. George, "Transmission of multiplexed PAM signals over multiple channel and diversity systems," *IEEE Trans. Commun. Technol.*, vol. COM-18, pp. 520-526, Oct. 1970.

[3] W. van Etten, "An optimum linear receiver for multiple channel digital transmission systems," *IEEE Trans. Commun.* (Concise Papers), vol. COM-23, pp. 828-834, Aug. 1975.

[4] G. D. Forney, Jr., "Maximum-likelihood sequence estimation of digital sequences in the presence of intersymbol interference," *IEEE Trans. Inform. Theory*, vol. IT-18, pp. 363-378, May 1972.

[5] J. K. Omura, "Optimal receiver design for convolutional codes and channels with memory via control theoretical concepts," *Inform. Sci.*, vol. 3, pp. 243-266, 1971.

[6] G. Ungerboeck, "Adaptive maximum-likelihood receiver for carrier-modulated data-transmission systems," *IEEE Trans. Commun.*, vol. COM-22, pp. 624-636, May 1974.

[7] J. M. Wozencraft and I. M. Jacobs, *Principles of Communication Engineering.* New York: Wiley, 1965.

[8] D. N. Prabhakar Murthy, "Factorization of discrete-process spectral matrices," *IEEE Trans. Inform. Theory* (Corresp.), vol. IT-19, pp. 693-696, Sept. 1973.

[9] P. R. Motyka and J. A. Cadzow, "The factorization of discrete-process spectral matrices," *IEEE Trans. Automat. Contr.*, vol. AC-12, pp. 698-707, Dec. 1967.

[10] F. B. Hildebrand, *Methods of Applied Mathematics.* Englewood Cliffs, NJ: Prentice-Hall, 1952.

An Optimum Linear Receiver for Multiple Channel Digital Transmission Systems

W. VAN ETTEN

Abstract—**An optimum linear receiver for multiple channel digital transmission systems is developed for the minimum P_e and for the zero-forcing criterion. A multidimensional Nyquist criterion is defined together with a theorem on the optimality of a finite lenght multiple tapped delay line. Furthermore an algorithm is given to calculate the tap settings of this multiple tapped delay line. This algorithm simplifies in those cases where the noise is so small that it can be neglected. Finally as an example the transmission of binary data over a cable, consisting of four identical wires, symmetrically situated inside a cylindrical shield, is considered.**

I. INTRODUCTION

In this paper we shall investigate the transmission of digital signals over a multiple channel system, where each channel is used to transmit a data sequence. This configuration is included in the more general structure considered by Kaye and George [1]. We, however, use a technique that leads to an optimum structure for both the zero-forcing and minimum error probability criterion, instead of the minimum mean-square error criterion Kaye and George used.

Besides the intersymbol interference (ISI), interchannel interference (ICI) can be one of the major problems in such a multiple channel digital transmission system. ISI is disturbance of an output signal by symbols that originate from the corresponding input but that are shifted in time with respect to the symbol of interest. ICI is disturbance of an output signal by symbols that do not originate from the corresponding input but from input symbols that belong to neighboring channels. We introduce the name multidimensional interference (MDI) for the combined effect of ISI and ICI. Because the equalization of ISI also changes the ICI at the output, and the other way round, only a simultaneous treatment of these two phenomena can be successful in combating the overall degradation. In the following we generalize for MDI some techniques known from the ISI literature. As examples of systems where these methods can be applied, we mention multiwire cables and multichannel radio systems that make use of perpendicular polarized waves in a common frequency band.

Paper approved by the Associate Editor for European Contributions of the IEEE Communications Society for publication without oral presentation. Manuscript received November 27, 1974; revised April 3, 1975.

The author is with the Eindhoven University of Technology, Eindhoven, The Netherlands.

II. THE MULTIPLE CHANNEL COMMUNICATION MODEL

The multiple channel transmission system, to be treated in this paper, has M inputs and M outputs, where to each input j a data sequence $\sum_l a_j{}^l \delta(t - lT)$ is applied which we want to detect at output j. The symbols $a_j{}^l$ are elements of the alphabet $\{0,1,\cdots, L-1\}$ and are chosen equiprobable and independent of each other. In our investigations a linear, dispersive, and time-invariant channel model is assumed (Fig. 1); this means that a linear relation exists between each input and each output signal and that the output signal due to the excitation of more than one input is the sum of the individual responses to the several inputs. The relation between input j and output i is denoted by the impulse response $r_{ij}(t)$. It is assumed that the output signals are disturbed by MDI and that zero-mean white Gaussian noise is added to them. Each output is corrupted by a different noise signal $n_i(t)$.

III. THE OPTIMUM LINEAR RECEIVER

By means of an optimum linear receiver and bit by bit detection on each channel output we make an estimate of the several input sequences. The receiving filter is assumed to be linear in the sense described in the preceding section. The linear relation between input i and output n of this filter is denoted by the impulse response $h_{ni}(t)$ (see Fig. 2). The following method yields an optimum solution for the linear multiple channel receiving filter for both the zero-forcing (zero MDI) and minimum bit error probability criterion. Assuming that the several noise sample functions $n_i(t)$ are independent of each other, then the noise variance at output n of the receiving filter can be written as follows:

Reprinted from *IEEE Transactions on Communications,* August 1975.

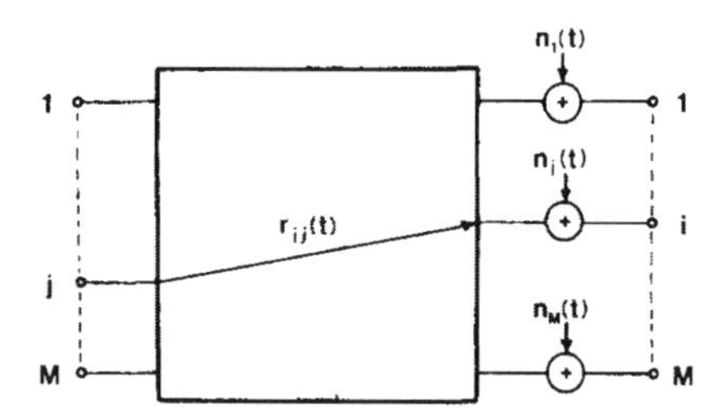

Fig. 1. Multiple channel communication model.

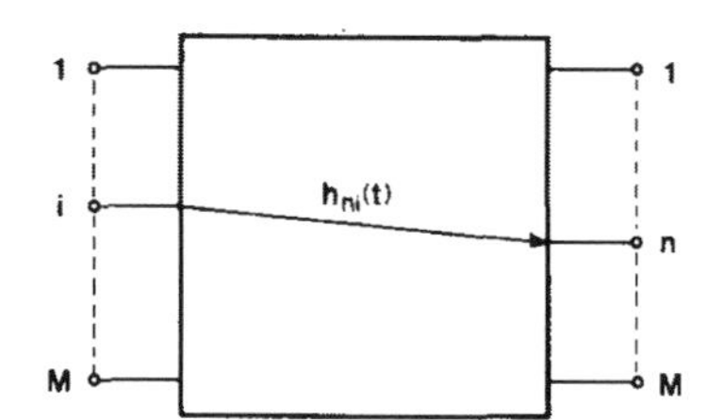

Fig. 2. Multiple linear receiving filter.

$$\sigma_n^2 = \sum_{i=1}^{M} N_i \int_0^\infty h_{ni}^2(\tau)\, d\tau \tag{1}$$

where N_i is the density of the noise spectrum of $n_i(t)$. Investigating the optimum structure of the linear receiving filter a technique is used that is presented in [4] and [5]. This means that all signal values that contribute to the possible sample values of the signal at output n are fixed. Then the noise variance σ_n^2 is minimized subject to these constraints. Defining the input vector

$$x^k \triangleq \begin{bmatrix} a_1^k \\ a_2^k \\ \cdot \\ \cdot \\ \cdot \\ a_M^k \end{bmatrix} \tag{2}$$

the constraints are found by considering the sample values of the signals at output n due to the L^M possible input vectors x^k. The latter sample values are found in the following way.

Assuming that at time $t = 0$ the vector x^k is applied to the inputs of the channel, then the response at output n of the receiving filter is given by

$$s_n^k(t) = \sum_{j=1}^{M} a_j^k \sum_{i=1}^{M} \int_0^\infty h_{ni}(\tau) r_{ij}(t-\tau)\, d\tau. \tag{3}$$

At the instant $t_s + lT$, this response has the value

$$s_n^k(t_s + lT) = \sum_{j=1}^{M} a_j^k \sum_{i=1}^{M} \int_0^\infty h_{ni}(\tau) r_{ij}(t_s + lT - \tau)\, d\tau. \tag{4}$$

In the minimization process these values for all k and l must be kept constant, therefore we have to minimize the functional

$$J_n = \sum_{i=1}^{M} N_i \int_0^\infty h_{ni}^2(\tau)\, d\tau - 2 \sum_{k=1}^{L^M} \sum_l \lambda_{nkl} \sum_{j=1}^{M} a_j^k \cdot \sum_{i=1}^{M} \int_0^\infty h_{ni}(\tau) r_{ij}(t_s + lT - \tau)\, d\tau. \tag{5}$$

Applying the calculus of variations to expression (5) yields

$$h_{ni}(t) = \frac{1}{N_i} \sum_{k=1}^{L^M} \sum_l \lambda_{nkl} \sum_{j=1}^{M} a_j^k r_{ij}(t_s + lT - t). \tag{6}$$

For the sake of simplicity we take $N_i = N$ for all i. This assumption and the assumption that the noise functions $n_i(t)$ are uncorrelated are not a restriction of the generality, as is shown in Appendix II.

With

$$c_{njl} = \frac{1}{N} \sum_{k=1}^{L^M} a_j^k \lambda_{nkl} \tag{7}$$

(6) reduces to

$$h_{ni}(t) = \sum_{j=1}^{M} \sum_l c_{njl} r_{ij}(t_s + lT - t). \tag{8}$$

The structure of the entire receiving filter follows from this equation. Each $h_{ni}(t)$ consists of a bank of matched filters, the outputs of which are added and the output signals of all $h_{ni}(t)$, which belong to the same receiving filter output n, are added again. Assuming that t_s is larger than the largest duration of all $s_n^k(t)$, then a reduction of the receiving filter is possible and Fig. 3 depicts the result for $M = 3$, for instance. For ease of notation the time axis is shifted such that $t_s = 0$. At each filter input i we see an array of filters matched to the particular responses at channel output i due to the individual excitation of the several inputs. Then all the outputs of the filters matched to the responses due to the same input are summed to form the primed outputs 1′-2′-3′. This part of the filter we call the multiple matched filter (MMF) (inputs 1-2-3 and outputs 1′-2′-3′). Each primed output is followed by a delay line with elements D giving a delay T. The rest of the receiving filter consists of M summing circuits and from each delayed primed output there is a weighted connection (with weighting coefficient c_{njl}) to each adder. This part of the filter we call the multiple tapped delay line (MTDL) (inputs 1′-2′-3′ and outputs 1″-2″-3″). The weighting coefficients c_{njl} have to be chosen such as to meet the optimization criterion. In the case of the minimum P_e criterion it is impossible to find an analytical solution for the set $\{c_{njl}\}$. By means of a steepest descent method one can find an approximation. Zero MDI offers the possibility to calculate the tap coefficients in a rather easy way as will be shown in Section V and to check the practical realization by means of the eye pattern, while the error probability P_e is of the same order of magnitude as when the minimum P_e criterion is used especially for large signal-to-noise ratios.

Considering the cascade connection of the channel, the MMF and the MTDL, the impulse responses of this overall system evaluated at the discrete instants lT are denoted by

$$F_l \triangleq \begin{bmatrix} f_{11}(lT) & f_{12}(lT) & \cdot\ \cdot & f_{1M}(lT) \\ f_{21}(lT) & f_{22}(lT) & \cdot\ \cdot & f_{2M}(lT) \\ \cdot & \cdot & & \cdot \\ \cdot & \cdot & & \cdot \\ \cdot & \cdot & & \cdot \\ f_{M1}(lT) & f_{M2}(lT) & \cdot\ \cdot & f_{MM}(lT) \end{bmatrix} \tag{9}$$

with $f_{nj}(t)$ the response at output n of this system as result of a delta excitation on input j.

Further we define

$$F(D) \triangleq \sum_l F_l D^l \tag{10}$$

where D is the delay operator.

A measure for MDI is now defined as follows:

$$I_n \triangleq \frac{\sum_l \sum_{i=1}^{M} |f_{ni}(lT)| - |f_{nn}(0)|}{|f_{nn}(0)|} \tag{11}$$

which is the worst case distortion due to MDI on output n. The overall worst case MDI distortion is given by

$$I_0 = \max_n (I_n). \tag{12}$$

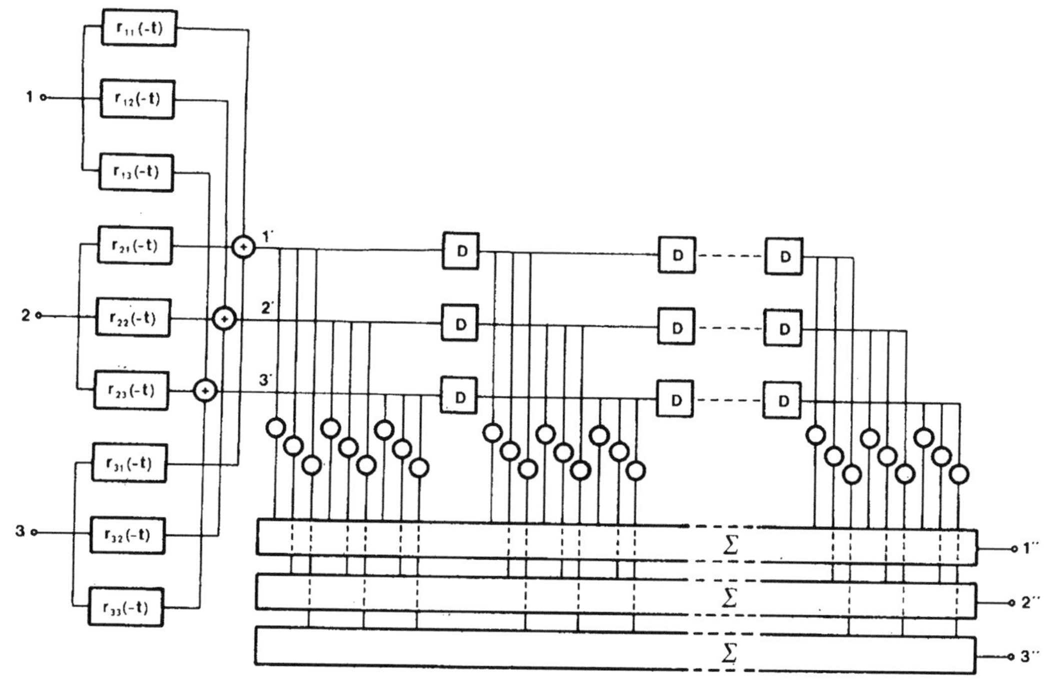

Fig. 3. Structure of the multiple linear receiving filter.

The terms "zero MDI" and "zero-forcing" are used here if $I_0 = 0$. By means of (10) and (12) we formulate a multidimensional Nyquist criterion which fits Shnidman's generalized Nyquist criterion [2].

Theorem 1: A multiple channel transmission system described by (10) satisfies the multidimensional Nyquist criterion if

$$F(D) = I \tag{13}$$

where I is the $M \times M$ identity matrix.

It will be clear from the foregoing that for a system satisfying the multidimensional Nyquist criterion the MDI will be zero.

Now consider the channel in cascade with the MMF as a multiple channel system with M inputs and M outputs. The impulse response from input j to output m of this system is called $v_{mj}(t)$ and can be written as

$$v_{mj}(t) = \sum_{i=1}^{M} r_{ij}(t) * r_{im}(-t) \tag{14}$$

where $*$ means convolution. Define

$$V_l \triangleq \begin{bmatrix} v_{11}(lT) & v_{12}(lT) & \cdot\ \cdot & v_{1M}(lT) \\ v_{21}(lT) & v_{22}(lT) & \cdot\ \cdot & v_{2M}(lT) \\ \cdot & \cdot & & \cdot \\ \cdot & \cdot & & \cdot \\ \cdot & \cdot & & \cdot \\ v_{M1}(lT) & v_{M2}(lT) & \cdot\ \cdot & v_{MM}(lT) \end{bmatrix} \tag{15}$$

and

$$V(D) \triangleq \sum_l V_l D^l. \tag{16}$$

The MTDL is also a multiple linear filter. For this system we define

$$C_l \triangleq \begin{bmatrix} c_{11l} & c_{12l} & \cdot\ \cdot & c_{1Ml} \\ c_{21l} & c_{22l} & \cdot\ \cdot & c_{2Ml} \\ \cdot & \cdot & & \cdot \\ \cdot & \cdot & & \cdot \\ \cdot & \cdot & & \cdot \\ c_{M1l} & c_{M2l} & \cdot\ \cdot & c_{MMl} \end{bmatrix} \tag{17}$$

and

$$C(D) \triangleq \sum_l C_l D^l. \tag{18}$$

From the definitions (10), (16), and (18) it follows

$$F(D) = C(D) \cdot V(D). \tag{19}$$

In Section V we shall give a procedure to calculate the tap coefficients described by $C(D)$.

IV. THE ERROR PROBABILITY OF THE EQUALIZED SYSTEM

If in a multiple channel transmission system it is possible to satisfy the multidimensional Nyquist criterion and the system has an optimum constraint receiver as described in the foregoing, the mean error probability of channel n of such a system is denoted by

$$P_{en} = 2\frac{L-1}{L} Q\left(\frac{d}{2\sigma_n}\right) \tag{20}$$

where the well-known $Q(\cdot)$ function is defined in [6, p. 82] and d is the smallest difference between two output levels. As the smallest difference between two elements of the input alphabet is taken unity and because of (13), d equals one. The noise variance at output n is calculated from (1) and (8)

$$\sigma_n^2 = \sum_m \sum_l \sum_{i=1}^{M} \sum_{j=1}^{M} \sum_{k=1}^{M} N c_{njm} c_{nkl} \int_{-\infty}^{\infty} r_{ik}(lT - \tau) r_{ij}(mT - \tau)\, d\tau. \tag{21}$$

For the equalized system the impulse response from input j to output n, evaluated at the instant mT, can be written as

$$f_{nj}(mT) = \sum_l \sum_{i=1}^{M} \sum_{k=1}^{M} c_{nkl} \cdot \int_{-\infty}^{\infty} r_{ik}(lT - \tau) r_{ij}(mT - \tau)\, d\tau = \delta_{m}\delta_{nj} \tag{22}$$

as is derived in [2]. Substituting (22) reduces (21) to the simple form

$$\sigma_n^2 = Nc_{nn0} \tag{23}$$

which, if substituted in (20), gives for the error probability of channel n

$$P_{en} = 2\frac{L-1}{L} Q\left(\frac{1}{2(Nc_{nn0})^{1/2}}\right). \tag{24}$$

V. THE OPTIMUM REALIZABLE MTDL

The index l of the $C(D)$ sequence runs from minus infinite to plus infinite and as a consequence the MTDL becomes infinitely long. In practice we have to make it of finite length and in this case (13) cannot be satisfied exactly. If the MTDL is of length $2K$ the optimum tap settings are given by the following theorem.

Theorem 2: If $V_0 = I$ and $\sum_l' \| V_l \| < 1$, then an upper bound of the MDI distortion I_0 is minimal for those tap setting matrices which cause $F_l = 0$, $|l| \leq K$, $l \neq 0$; where the primed summation excludes the term with $l = 0$ and the infinite norm is taken (which is the maximum over all rows of the sum of the absolute values of the components of the rows).

This theorem will be proven in Appendix I. In the special case that $\sum_l' \| F_l \|$ represents the worst case MDI I_0, this distortion itself is minimized. Under these constraints this theorem is a generalization of a theorem derived by Lucky for ISI [3, p. 138]. If $V_0 \neq I$ we can force V_0 to equal the identity matrix by placing between the MMF and the MTDL a multiple channel system with matrix D-transform V_0^{-1}. To apply the theorem all V_l matrices must then be replaced by $V_0^{-1}V_l$. In the case that $\sum_l' \| V_l \|$ represents the worst case MDI at the MMF outputs, a sufficient condition to satisfy the requirement $\sum_l' \| V_l \| < 1$ is that at none of the MMF outputs, the eye pattern is closed if $a_j^l \in \{+1, -1\}$.

The tap settings as stated in Theorem 2 are calculated as follows. Define the composite matrices

$$C \triangleq \begin{bmatrix} C_{-K} \\ C_{-K+1} \\ \cdot \\ \cdot \\ \cdot \\ C_K \end{bmatrix}, \tag{25}$$

$$V \triangleq \begin{bmatrix} V_0 & V_1 & \cdot & \cdot & V_{2K} \\ V_{-1} & V_0 & \cdot & & V_{2K-1} \\ V_{-2} & V_{-1} & V_0 & \cdot & V_{2K-2} \\ \cdot & \cdot & \cdot & & \cdot \\ \cdot & \cdot & \cdot & & \cdot \\ \cdot & \cdot & \cdot & & \cdot \\ V_{-2K} & V_{-2K+1} & \cdot & \cdot & V_0 \end{bmatrix} \tag{26}$$

and

$$E \triangleq \begin{bmatrix} 0 \\ 0 \\ \cdot \\ \cdot \\ \cdot \\ 0 \\ I \\ 0 \\ \cdot \\ \cdot \\ \cdot \\ 0 \\ 0 \end{bmatrix} \tag{27}$$

where 0 is the all zero matrix. To satisfy Theorem 2 we have the relation

$$C^T V = E^T. \tag{28}$$

This equation is further simplified if we look at (14), (15), and (26). It is easy to see that

$$V^T = V \tag{29}$$

so that

$$VC = E. \tag{30}$$

The solution of this equation decomposes into M times the solution of a set of linear equations, one time for each column of C as wanted vector and the corresponding column of E as known vector.

In systems where the noise does not play an important role, MDI distortion correction can directly be applied to the channel response. In this situation (29) is not true in general, but it is sometimes possible to choose $t_s < T$ giving a simplification of the expression for C_l. It is easy to see that the matrix sequence C_l starts now at $l = 0$ and runs to plus infinite. Analogous to (15) and (16) we define

$$R_l \triangleq \begin{bmatrix} r_{11}(lT) & r_{12}(lT) & \cdot & \cdot & r_{1M}(lT) \\ r_{21}(lT) & r_{22}(lT) & & & r_{2M}(lT) \\ \cdot & \cdot & \cdot & \cdot & \cdot \\ \cdot & \cdot & & & \cdot \\ \cdot & \cdot & & & \cdot \\ r_{M1}(lT) & r_{M2}(lT) & \cdot & \cdot & r_{MM}(lT) \end{bmatrix} \tag{31}$$

and

$$R(D) \triangleq \sum_{l=0}^{\infty} R_l D^l. \tag{32}$$

By applying Theorem 1 to this system it follows that the tap coefficients are determined by the recurrence relation

$$C_0 = R_0^{-1}$$
$$C_l = -R_0^{-1} \sum_{i=0}^{l-1} R_{l-i} C_i \qquad l \geq 1. \tag{33}$$

It will be clear that Theorem 2 is also valid now with the restriction that l has only positive values and the length of the MTDL is K. In this case the MTDL is also realizable as M shift registers with resistance matrices at the sending end. One can derive that, in doing so, the expression (33) for the C_l matrices stay unchanged.

Decision feedback is another possibility to eliminate MDI [7]. Then the MMF is followed by a "forward" MTDL and a "feedback" MTDL.

VI. AN EXAMPLE

As an example we implemented the transmission of binary data over a multiwire cable, consisting of four identical wires which are symmetrically situated within a cylindrical shield (see Fig. 4). The cable has a length of 1 km and the bit rate is taken 5 Mbit/s. In this example the length of the cable, the bit rate and the sending pulses are such that the noise can be neglected, thus the relations (33) are used for calculating the tap coefficients. We have measured the following matrices:

$$R_0 = \begin{bmatrix} 1 & 0.24 & 0.24 & 0.13 \\ 0.24 & 1 & 0.13 & 0.24 \\ 0.24 & 0.13 & 1 & 0.24 \\ 0.13 & 0.24 & 0.24 & 1 \end{bmatrix}$$

$$R_1 = 0.26I, \qquad R_2 = 0.11I, \qquad R_3 = 0.07I, \qquad R_4 = 0.04I. \tag{34}$$

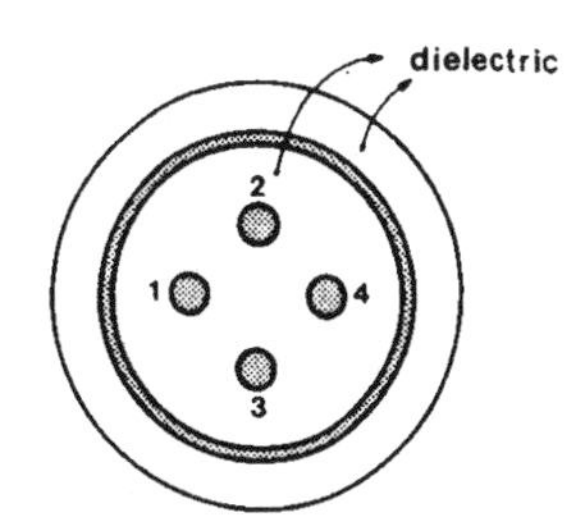

Fig. 4. Cross section of the 4-wire cable within cylindrical shield.

One can verify that $\sum_l' \| R_l R_0^{-1} \| < 1$, thus Theorem 2 can be applied. The calculated C_l matrices are

$$C_0 = \begin{bmatrix} 1 & -0.21 & -0.21 & -0.03 \\ -0.21 & 1 & -0.03 & -0.21 \\ -0.21 & -0.03 & 1 & -0.21 \\ -0.03 & -0.21 & -0.21 & 1 \end{bmatrix}$$

$$C_1 = \begin{bmatrix} -0.31 & 0.12 & 0.12 & -0.01 \\ 0.12 & -0.31 & -0.01 & 0.12 \\ 0.12 & -0.01 & -0.31 & 0.12 \\ -0.01 & 0.12 & 0.12 & -0.31 \end{bmatrix}. \tag{35}$$

Because of the several kinds of symmetry in both the R_l and C_l matrices, $\sum_l' \| R_l R_0^{-1} \|$ represents the worst case MDI before the MTDL. Moreover, the output matrices F_l show the same symmetry and thus $\sum_l' \| F_l \|$ represents the worst case MDI at the output, so that Theorem 2 is valid in its full consequence.

At the realization of the C_l matrices, tap coefficients equal or smaller than 3 percent are omitted because these values do not give a substantial improvement of the eye opening. All components of C_2, C_3, etc., are smaller than 3 percent, that is why they are not given at (35). Only C_0 and C_1 are realized and at these matrices the connections between a certain wire and the diagonal opposed one are omitted too. This MTDL is implemented as 4 shift registers at the sending end which are connected to the cable by means of resistance matrices forming the tap coefficients. Fig. 5 shows the eye pattern at the receiving end of the cable if all wires are excited and it is seen that the unequalized system has a fully closed eye as is calculated from (34). Fig. 6 shows the eye pattern of the system characterized by $R(D)R_0^{-1}$ which means that a multiple channel system with matrix D-transform R_0^{-1} is placed between the transmitter and the sending end of the cable. The eye pattern of this system is not closed, which shows that $\sum_l' \| R_l R_0^{-1} \| < 1$. Finally Fig. 7 shows the eye pattern of the equalized system and it appears that the multidimensional Nyquist criterion is satisfied rather well.

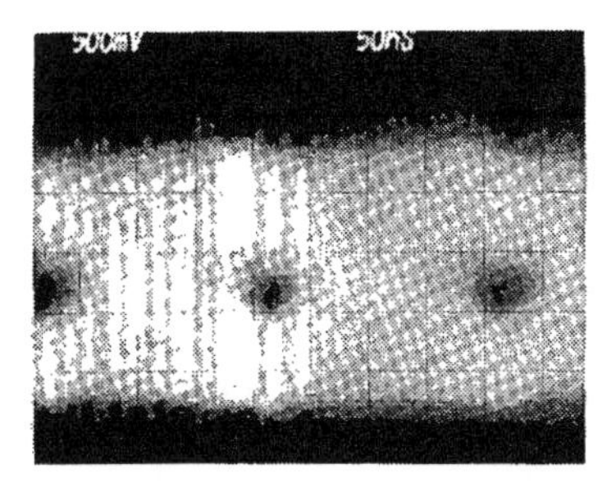

Fig. 5. Eye pattern of the unequalized system.

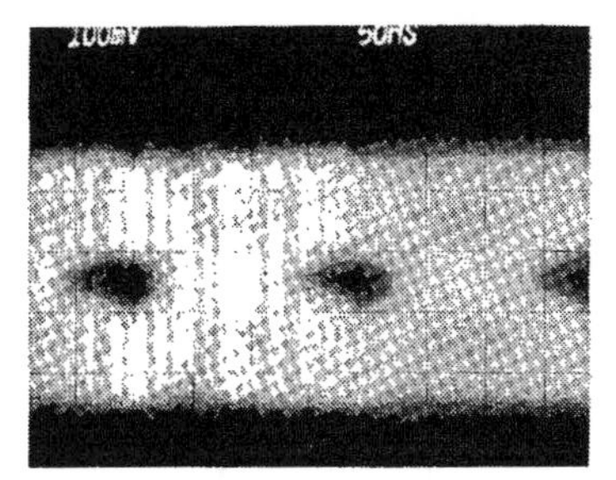

Fig. 6. Eye pattern of the system $R(D)R_0^{-1}$.

VII. CONCLUSIONS

It is shown that for a multiple channel transmission system both the optimum linear receiver (minimum P_e) and the optimum linear constraint receiver (minimum P_e under zero-forcing condition) have the same structure as the optimum linear receiver found by Kaye and George applying the minimum mean-square error criterion. Moreover it appears that by means of the multidimensional Nyquist criterion and the generalization for MDI of a theorem by Lucky for ISI it is rather easy to find the optimum tap settings for a finite length MTDL. The algorithm to calculate the tap settings is further simplified in the case that the noise is unimportant and the sampling instant is smaller than the bit time.

By means of the methods developed in this paper it is shown that MDI is the generalization of ISI.

APPENDIX I

PROOF OF THEOREM 2

Let $\{V_i\}_{i=-\infty}^{\infty}$ be given with $V_0 = I$ and let

$$M_0 = \sum_{i=-\infty}^{\infty}{}' \| V_i \| < 1. \tag{36}$$

Let

$$A = \sum_{n=-\infty}^{\infty}{}' \| \sum_{j=-N}^{N} C_j V_{n-j} \| \tag{37}$$

under the constraint

$$\sum_{j=-N}^{N} C_j V_{-j} = I. \tag{38}$$

We shall prove that a minimum for A exists and that this minimum occurs if

$$\sum_{j=-N}^{N} C_j V_{n-j} = 0, \qquad n = -N,\cdots,-1,1,\cdots,N. \tag{39}$$

Proof: Due to (38), (37) can be written as follows:

$$A = \sum_{n=-\infty}^{\infty}{}' \| \sum_{j=-N}^{N} C_j (V_{n-j} - V_{-j} V_n) + V_n \|. \tag{40}$$

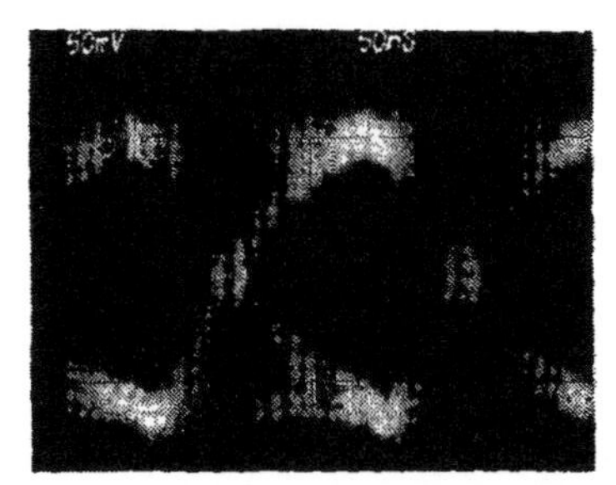

Fig. 7. Eye pattern of the equalized system.

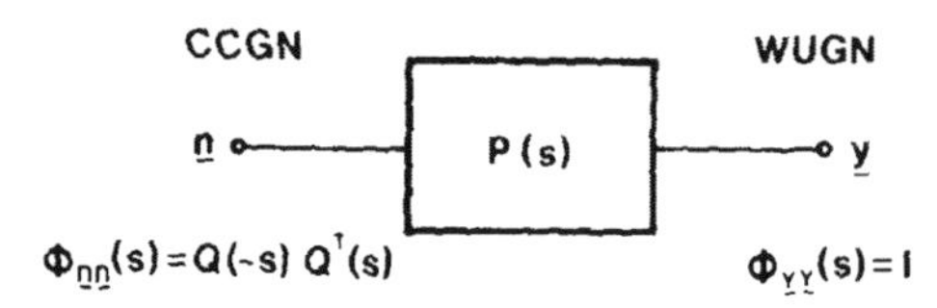

Fig. 8. Multiple noise whitening filter.

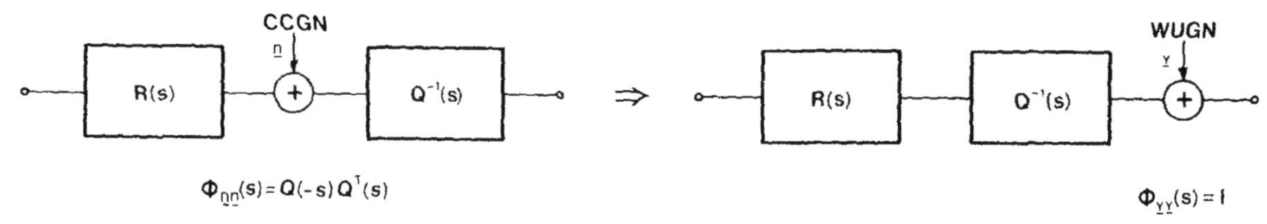

Fig. 9. System $R(s)$ disturbed by CCGN is replaced by the system $Q^{-1}(s)R(s)$ disturbed by WUGN.

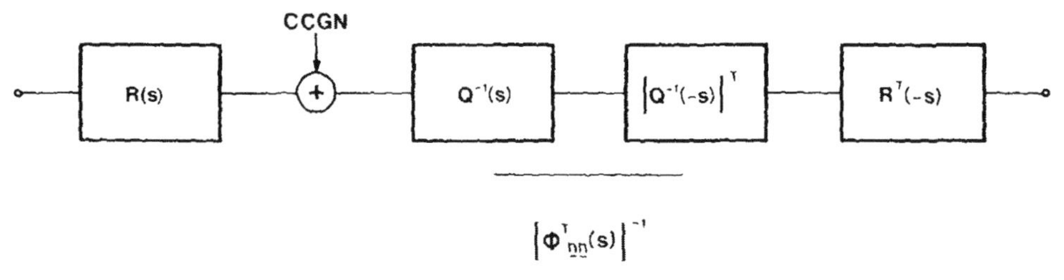

Fig. 10. Multiple channel system disturbed by CCGN in cascade with the multiple whitening matched filter.

Let A be minimal in $(C_{-N}^*,\cdots,C_N^*)$ and let its value there be A^*. Consider A in the point $(C_{-N}^*,\cdots,C_k^*+E_k,\cdots,C_N^*)$ and let its value there be $\tilde{A}$. Now we must have

$$A^* \leq \tilde{A}. \tag{41}$$

From (40) it follows

$$\begin{aligned}\tilde{A} &= \sum_{n=-\infty}^{\infty}{}' \left\| \sum_{j=-N}^{N} C_j^*(V_{n-j} - V_{-j}V_n) + V_n + E_k(V_{n-k} - V_{-k}V_n) \right\| \\ &\leq A^* - \|H_k^*\| + \sum_{n=-\infty;n\neq k}^{\infty}{}' \|V_{n-k}\|\cdot\|E_k\| \\ &\quad + \sum_{n=-\infty;n\neq k}^{\infty}{}' \|V_n\|\cdot\|E_k\|\cdot\|V_{-k}\| \\ &\quad + \|H_k^* + E_k(I - V_{-k}V_k)\|\end{aligned} \tag{42}$$

where

$$H_k^* \triangleq \sum_{j=-N}^{N} C_j^* V_{k-j}. \tag{43}$$

Choose

$$E_k = -\delta H_k^*(I - V_{-k}V_k)^{-1} \qquad 0 < \delta < 1. \tag{44}$$

This is possible because the inverse of $(I - V_{-k}V_k)$ exists and besides that

$$\|(I - V_{-k}V_k)^{-1}\| \leq \frac{1}{1 - \|V_{-k}\|\cdot\|V_k\|}. \tag{45}$$

By means of (44), (42) becomes

$$\begin{aligned}\tilde{A} - A^* &\leq \|H_k^*\| \left[-1 + \frac{1}{1 - \|V_{-k}\|\cdot\|V_k\|} \{M_0 - \|V_{-k}\| + (M_0 - \|V_k\|)\|V_{-k}\|\} + 1 - \delta \right] \\ &\leq \frac{\delta\|H_k^*\|}{1 - \|V_{-k}\|\cdot\|V_k\|} [M_0 - \|V_{-k}\| + M_0\|V_{-k}\| - \|V_k\|\cdot\|V_{-k}\| - 1 + \|V_k\|\cdot\|V_{-k}\|] \\ &\leq \frac{\delta\|H_k^*\|}{1 - \|V_{-k}\|\cdot\|V_k\|} [(M_0 - 1)(1 + \|V_{-k}\|)].\end{aligned} \tag{46}$$

From (46) it follows that

$$\|H_k^*\| = 0, \tag{47}$$

because otherwise there is a contradiction with (41).

APPENDIX II

In this Appendix we prove that the assumptions that the noise functions $n_i(t)$ are white and uncorrelated are not a restriction of the generality; i.e., a system not satisfying these assumptions can be transformed into a system that meets these requirements. The proof starts with the remark that the spectral matrix (which is the Laplace transform of the correlation matrix) of the input noise can be factored, according to [8], in the following way:

$$\Phi_{nn}(s) = Q(-s)Q^T(s) \tag{48}$$

where s is the bilateral Laplace variable. Assume that we have a system with transfer matrix $P(s)$ such that the spectral matrix of the output noise is the identity matrix if the input spectral matrix is given by (48). Then the spectral matrix of the output y of $P(s)$ is written as follows [8]:

$$\Phi_{yy}(s) = P(-s)Q(-s)Q^T(s)P^T(s) \tag{49}$$

(see Fig. 8). From this it follows that

$$P(s) = Q^{-1}(s) \tag{50}$$

satisfies the requirement of white, uncorrelated output noise. A procedure for finding a $Q(s)$ such that both $Q(s)$ and $Q^{-1}(s)$ are stable is also given in [8]. Now we shall further investigate the MMF for colored, correlated Gaussian noise (CCGN). The several impulse responses $r_{ij}(t)$ of the multiple channel system are written in a matrix $R(t)$. From (50) it follows that the multiple channel transmission system with transfer matrix $R(s)$ disturbed by CCGN with spectral matrix $\Phi_{nn}(s)$ can be replaced by a multiple channel transmission system with transfer matrix $Q^{-1}(s)R(s)$ disturbed by white, uncorrelated, Gaussian noise (WUGN) (see Fig. 9). The MMF for this latter system is given by

$$[Q^{-1}(-s)R(-s)]^T = R^T(-s)[Q^T(-s)]^{-1}. \tag{51}$$

Note that the MMF for the system with impulse response matrix $R(t)$ disturbed by WUGN is given by $R^T(-t)$.

So that the MMF for the original system can be written as

$$R^T(-s)[Q^T(-s)]^{-1}Q^{-1}(s) = R^T(-s)[\Phi_{nn}{}^T(s)]^{-1} \tag{52}$$

(see Fig. 10). This MMF we call multiple whitening matched filter (MWMF).

ACKNOWLEDGMENT

The author wishes to thank J. van der Plaats for initiating and stimulating the work on interchannel interference and L. S. de Jong for giving the proof of Theorem 2.

REFERENCES

[1] A. R. Kaye and D. A. George, "Transmission of multiplexed PAM signals over multiple channel and diversity systems," *IEEE Trans. Commun. Technol.*, vol. COM-18, pp. 520–525, Oct. 1970.
[2] D. A. Shnidman, "A generalized Nyquist criterion and optimum linear receiver for a pulse modulation system," *Bell Syst. Tech. J.*, pp. 2163–2177, Nov. 1967.
[3] R. W. Lucky, J. Salz, and E. J. Weldon, Jr., *Principles of Data Communication*. New York: McGraw-Hill, 1968.
[4] M. R. Aaron and D. W. Tufts, "Intersymbol interference and error probability," *IEEE Trans. Inform. Theory*, vol. IT-12, pp. 26–34, Jan. 1966.
[5] D. W. Tufts, "Nyquist's problem: The joint optimization of transmitter and receiver in pulse amplitude modulation," *Proc. IEEE*, vol. 53, pp. 248–259, Mar. 1965.
[6] J. M. Wozencraft and I. M. Jacobs, *Principles of Communication Engineering*. New York: Wiley, 1965.
[7] M. E. Austin, "Equalization of dispersive channels using decision feedback," Quarterly Progress Rep., M.I.T. Res. Lab. Electron., Cambridge, Mass., no. 84, pp. 227–243, 1967.
[8] M. C. Davis, "Factoring the spectral matrix," *IEEE Trans. Automat. Contr.*, vol. AC-8, pp. 296–305, Oct. 1963.

Adaptive Maximum-Likelihood Receiver for Carrier-Modulated Data-Transmission Systems

GOTTFRIED UNGERBOECK

Abstract—A new look is taken at maximum-likelihood sequence estimation in the presence of intersymbol interference. A uniform receiver structure for linear carrier-modulated data-transmission systems is derived which for decision making uses a modified version of the Viterbi algorithm. The algorithm operates directly on the output signal of a complex matched filter and, in contrast to the original algorithm, requires no squaring operations; only multiplications by discrete pulse-amplitude values are needed. Decoding of redundantly coded sequences is included in the consideration. The reason and limits for the superior error performance of the receiver over a conventional receiver employing zero-forcing equalization and symbol-by-symbol decision making are explained. An adjustment algorithm for jointly approximating the matched filter by a transversal filter, estimating intersymbol interference present at the transversal filter output, and controlling the demodulating carrier phase and the sample timing, is presented.

I. INTRODUCTION

THE design of an optimum receiver for synchronous data-transmission systems that employ linear carrier-modulation techniques [1] is discussed. All forms of digita amplitude modulation (AM), phase modulation (PM) and combinations thereof, are covered in a unified manner Without further mention in this paper, the results are als applicable to baseband transmission.

In synchronous data-transmission systems intersymbo interference (ISI) and noise, along with errors in th demodulating carrier phase and the sample timing, ar the primary impediments to reliable data reception [1] The goal of this paper is to present a receiver structur that deals with all these effects in an optimum way an an adaptive manner. In deriving the receiver the concep of maximum-likelihood (ML) sequence estimation [2], [3 will be applied. This assures that the receiver is optimun in the sense of sequence-error probability, provided tha data sequences have equal *a priori* probability.

The modulation schemes considered in this paper ca be viewed in the framework of digital quadrature ampli tude modulation (QAM) [1]. They can therefore be repre sented by an equivalent linear baseband model that differ from a real baseband system only by the fact that signal and channel responses are complex functions [2], [4], [5]

Conventional receivers for synchronous data signal

Paper approved by the Associate Editor for Communication Theory of the IEEE Communications Society for publication without oral presentation. Manuscript received May 18, 1973; revised December 21, 1973.

The author is with the IBM Zurich Research Laboratory, Rüschlikon, Switzerland.

Reprinted from *IEEE Transactions on Communications,* vol. COM-22, no. 5, May 1974.

comprise a linear receiver filter or equalizer, a symbol-rate sampler, and a quantizer for establishing symbol-by-symbol decisions. A decoder, possibly with error-detection and/or error-correction capability, may follow. The purpose of the receiver filter is to eliminate intersymbol interference while maintaining a high signal-to-noise ratio (SNR). It has been observed by several authors [1], [6]–[9] that for various performance criteria the optimum linear receiver filter can be factored as a matched filter (MF) and a transversal filter with tap spacings equal to the symbol interval. The MF establishes an optimum SNR irrespective of the residual ISI at its output. The transversal filter then eliminates or at least reduces intersymbol interference at the expense of diminishing the SNR.

If symbols of a data sequence are correlated by some coding law, a better way than making symbol-by-symbol decisions is to base decisions on the entire sequence received. The same argument holds true if data sequences are disturbed by ISI. The correlation introduced by ISI between successive sample values is of discrete nature as in the case of coding, in the sense that a data symbol can be disturbed by adjacent data symbols only in a finite number of ways. ISI can even be viewed as an unintended form of partial response coding [1]. Receivers that perform sequence decisions or in some other way exploit the discreteness of ISI exhibit highly nonlinear structures. Decision feedback equalization [10], [11] represents the earliest step in this direction. Later, several other nonlinear receiver structures were described [12]–[17]. In view of the present state of the art, many of these approaches can be regarded as attempts to avoid, by nonlinear processing methods, noise enhancement which would otherwise occur if ISI were eliminated by linear filtering.

A new nonlinear receiver structure was introduced by Forney [18]. The receiver consists of a "whitened" MF (i.e., an MF followed by a transversal filter that whitens the noise), a symbol-rate sampler, and a recursive nonlinear processor that employs the Viterbi algorithm in order to perform ML sequence decisions. The Viterbi algorithm was originally invented for decoding of convolutional codes [19]. Soon thereafter the algorithm was shown to yield ML sequence decisions and that it could be regarded as a specific form of dynamic programming [20]–[22]. Its applicability to receivers for channels with intersymbol interference and correlative level coding was noticed by Omura [23] and Kobayashi [24]–[26]. A survey on the Viterbi algorithm is given by Forney [27]. Very recently, adaptive versions of Forney's receiver have been proposed [28], [29], and its combination with decision-feedback equalization has been suggested [30].

In Forney's [18] receiver, whitening of the noise is essential because the Viterbi algorithm requires that noise components of successive samples be statistically independent. In this paper a receiver similar to that of Forney will be described. The receiver employs a modified Viterbi algorithm that operates directly on the MF output without whitening the noise. In a different form the algorithm has already been used for estimating and subtracting ISI terms from disturbed binary data sequences [17]. Here the algorithm is restated and extended so that it performs ML decisions for complex-valued multilevel sequences. Apparently, Mackechnie [31] has independently found the same algorithm.

In Section II we define the modulation scheme. The general structure of the maximum likelihood receiver (MLR) is outlined in Section III. In Section IV we derive the modified Viterbi algorithm. The error performance of the MLR is discussed in Section V and compared with the error performance of the conventional receiver. Finally, a fully adaptive version of the MLR is presented in Section VI.

II. MODULATION SCHEME

We consider a synchronous linear carrier-modulated data-transmission system with coherent demodulation of the general form shown in Fig. 1. By combining in-phase and quadrature components into complex-valued signals (indicated by heavy lines in Fig. 1), all linear carrier-modulation schemes can be treated in a concise and uniform manner [2], [4], [5]. Prior to modulation with carrier frequency ω_c, the receiver establishes a pulse-amplitude modulated (PAM) signal of the form

$$x(t) = \sum_n a_n f(t - nT) \tag{1}$$

where the sequence $\{a_n\}$ represents the data symbols, T is the symbol spacing, and $f(t)$ denotes the transmitted baseband signal element. Generally, $\{a_n\}$ and $f(t)$ may be complex (but usually only one of them is; see Table I).[1] The data symbols are selected from a finite alphabet and may possibly succeed one another only in accordance with some redundant coding rule.

Assuming a linear dispersive transmission medium with impulse response $g_c(t)$ and additive noise $w_c(t)$, the receiver will observe the real signal

$$y_c(t) = g_c(t) * \mathrm{Re}\,\{\sqrt{2}x(t) \exp(j\omega_c t)\} + w_c(t) \tag{2}$$

where $*$ denotes convolution. One side of the spectrum of $y_c(t)$ is redundant and can therefore be eliminated without loss of information; the remaining part must be transposed back in the baseband. In Fig. 1 we adhere to the conventional approach of demodulating by transposing first and then eliminating components around twice the carrier frequency. The demodulated signal thus becomes

$$y(t) = \sum_n a_n h(t - nT) + w(t) \tag{3}$$

where

$$\begin{aligned} h(t) &= [g_c(t) \exp(-j\omega_c t - j\varphi_c)] * f(t) \\ &= g(t) * f(t) \end{aligned} \tag{4}$$

[1] A more general class of PAM signals is conceivable where $f(t)$ depends on n or a_n.

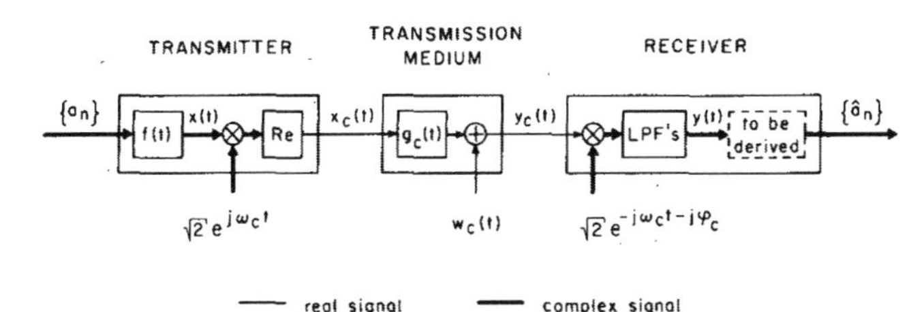

Fig. 1. General linear carrier-modulated data-transmission system.

TABLE I

Modulation Scheme	$\{a_n\}$	$f(t)$
DSB-AM	real	real
SSB, VSB-AM	real	complex[a]
PM	$\mid$ complex $\mid = 1$	real
AM-PM	complex	real

[a] SBB: $f(t) = f_1(t) \pm j\mathcal{H}\{f_1(t)\}$, where $\mathcal{H}$ is the Hilbert transform [1].

and

$$w(t) = \sqrt{2}w_c(t) \exp(-j\omega_c t - j\varphi_c). \tag{5}$$

In (5) the effect of low-pass filtering the transposed noise is neglected since it affects only noise components outside the signal bandwidth of interest. Our channel model does not include frequency offset and phase jitter. It is understood that the demodulating carrier phase φ_c accounts for these effects.

III. STRUCTURE OF THE MAXIMUM-LIKELIHOOD RECEIVER

The objective of the receiver is to estimate $\{a_n\}$ from a given signal $y(t)$. Let the receiver observe $y(t)$ within a time interval I which is supposed to be long enough so that the precise conditions at the boundaries of I are insignificant for the total observation. Let $\{\alpha_n\}$ be a hypothetical sequence of pulse amplitudes transmitted during I. The MLR by its definition [2], [3] determines as the best estimate of $\{a_n\}$ the sequence $\{\alpha_n\} = \{\hat{a}_n\}$ that maximizes the likelihood function $p[y(t), t \in I \mid \{\alpha_n\}]$.

In the following paragraphs the shape of the signal element $h(t)$ and the exact timing of received signal elements are assumed to be known. The noise of the transmission medium is supposed to be stationary Gaussian noise with zero mean and autocorrelation function $W_c(\tau)$. From (5) the autocorrelation function of $w(t)$ is obtained as

$$W(\tau) = E[\bar{w}(t)w(t+\tau)] = \bar{W}(-\tau)$$

$$= 2W_c(\tau) \exp(-j\omega_c\tau). \tag{6}$$

For example, if $w_c(t)$ is white Gaussian noise (WGN) with double-sided spectral density N_0, then $W_c(\tau) = N_0\delta(\tau)$ and $W(\tau) = 2N_0\delta(\tau)$. In view of this important case it is appropriate to introduce

$$W(\tau) = 2N_0K(\tau), \quad \text{WGN}: K(\tau) = \delta(\tau). \tag{7}$$

If $\{\alpha_n\}$ were the actual sequence of the pulse amplitudes transmitted during I, then

$$w(t \mid \{\alpha_n\}) = y(t) - \sum_{nT \in I} \alpha_n h(t - nT), \qquad t \in I \tag{8}$$

must be the realization of the noise signal $w(t)$. Hence, owing to the Gaussian-noise assumption, the likelihood function becomes (apart from a constant of proportionality) [32]

$$p[y(t), t \in I \mid \{\alpha_n\}] = p[w(t \mid \{\alpha_n\})]$$

$$\sim \exp\left\{-\frac{1}{4N_0}\int_I\int_I \bar{w}(t_1 \mid \{\alpha_n\})K^{-1}(t_1 - t_2) \cdot w(t_2 \mid \{\alpha_n\})\, dt_1\, dt_2\right\} \tag{9}$$

where $K^{-1}(\tau)$ is the inverse of $K(\tau)$

$$K(\tau) * K^{-1}(\tau) = \delta(\tau). \tag{10}$$

The correctness of (9) for the complex-signal case is proven in Appendix I. Substituting (8) into (9) and considering only terms that depend on $\{\alpha_n\}$, yields

$$p[y(t), t \in I \mid \{\alpha_n\}]$$

$$\sim \exp\left\{\frac{1}{4N_0}\left[\sum_{nT\in I} 2\,\mathrm{Re}\,(\bar{\alpha}_n z_n) - \sum_{iT\in I}\sum_{kT\in I} \bar{\alpha}_i s_{i-k} \alpha_k\right]\right\} \tag{11}$$

where

$$z_n = \int_I\int_I \bar{h}(t_1 - nT)K^{-1}(t_1 - t_2)y(t_2)\, dt_1\, dt_2 \tag{12}$$

$$s_l = \int_I\int_I \bar{h}(t_1 - iT)K^{-1}(t_1 - t_2)h(t_2 - kT)\, dt_1\, dt_2$$

$$= \bar{s}_{-l}, \quad l = k - i. \tag{13}$$

The quantities z_n and s_l can be interpreted as sample values taken at the output of a complex MF with impulse response function[2]

$$g_{\mathrm{MF}}(t) = \bar{h}(-t) * K^{-1}(t). \tag{14}$$

The derivation presented is mathematically weak in that it assumes $K^{-1}(t)$ exists. This is not the case if the spectral power density of the noise becomes zero somewhere along the frequency axis. The difficulty can be avoided by defining z_n and s_l in terms of the reproducing-kernel Hilbert space (RKHS) approach [33], [34]. Here it is sufficient to consider the frequency-domain equivalent of (14) given by

$$G_{\mathrm{MF}}(f) = \bar{H}(f)/K(f) \tag{15}$$

where $G_{\mathrm{MF}}(f)$, $H(f)$, and $K(f)$ are the Fourier transforms of $g_{\mathrm{MF}}(t)$, $h(t)$, and $K(t)$, respectively. It follows from (15) that $g_{\mathrm{MF}}(t)$ exists if the spectral power density

[2] Here we are not concerned with realizability.

of the noise does not vanish within the frequency band where signal energy is received. Only in this case have we a truly probabilistic (nonsingular) receiver problem. It can easily be shown that

$$z_n = g_{MF}(t) * y(t) \mid_{t=nT} = \sum_l a_{n-l} s_l + r_n \quad (16)$$

$$s_l = g_{MF}(t) * h(t) \mid_{t=lT} = \bar{s}_{-l} \quad (17)$$

and that the covariance of the noise samples r_n reads

$$R_l = E(\bar{r}_n r_{n+l}) = 2N_0 s_l. \quad (18)$$

Since the noise of the transmission medium does not exhibit distinct properties relative to the demodulating carrier phase, the following relations must hold:

$$E[\mathrm{Re}\,(r_n)\,\mathrm{Re}\,(r_{n+l})] = E[\mathrm{Im}\,(r_n)\,\mathrm{Im}\,(r_{n+l})] = N_0\,\mathrm{Re}\,(s_l) \quad (19)$$

$$E[\mathrm{Re}\,(r_n)\,\mathrm{Im}\,(r_{n+l})] = -E[\mathrm{Im}\,(r_n)\,\mathrm{Re}\,(r_{n+l})] = N_0\,\mathrm{Im}\,(s_l). \quad (20)$$

The similarity of s_l and R_l expressed by (18) implies that $s_0 \geq |s_l|$ and that the Fourier transform of the sampled signal element $\{s_l\}$ is a real nonnegative function

$$S^*(f) = \sum_l s_l \exp(-j2\pi flT) \geq 0 \quad (21)$$

with period $1/T$. Clearly, the MF performs a complete phase equalization, but does not necessarily eliminate ISI (ISI: $s_l \neq 0$ for $l \neq 0$). The main effect of the MF is that it maximizes the SNR, which we define as

$$S/N_{MF} = \frac{\text{instantaneous peak power of a single signal element}}{\text{average power of the real part of the noise}} = \frac{s_0^2}{E\{\mathrm{Re}\,(r_0)^2\}} = \frac{s_0}{N_0}. \quad (22)$$

The part of the receiver which in Fig. 1 was left open can now be specified, as indicated in Fig. 2. From (16) it comprises a MF and a symbol-rate sampling device sampling at times nT. It follows a processor, called maximum-likelihood sequence estimator (MLSE), that determines as the most likely sequence transmitted the sequence $\{\alpha_n\} = \{\hat{a}_n\}$ that maximizes the likelihood function given by (11), or equivalently, that assigns the maximum value to the metric

$$J_I(\{\alpha_l\}) = \sum_{nT\in I} 2\,\mathrm{Re}\,(\bar{\alpha}_n z_n) - \sum_{iT\in I}\sum_{kT\in I} \bar{\alpha}_i s_{i-k} \alpha_k. \quad (23)$$

The values of s_l are assumed to be known. The sequence $\{z_n\}$ contains all relevant information available about $\{a_n\}$ and hence forms a so-called set of sufficient statistics [2], [3]. The main difficulty in finding $\{\hat{a}_n\}$ lies in the fact that $\{\hat{a}_n\}$ must only be sought among discrete sequences

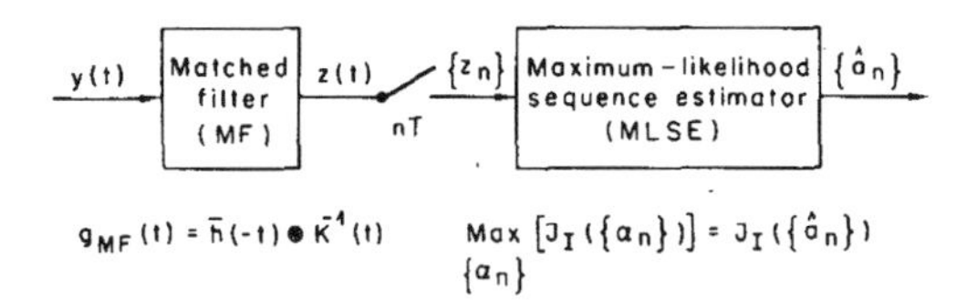

Fig. 2. MLR structure.

$\{\alpha_n\}$ which comply with the coding rule. The exact solution to this discrete maximization problem is presented in Section IV.

Solving the problem approximately by first determining the nondiscrete sequence $\{\alpha_n\} = \{z_{Ln}\}$ that maximizes (23) and then quantizing the elements of $\{z_{Ln}\}$ in independent symbol-by-symbol fashion, leads to the optimum conventional receiver [5]. Applying the familiar calculus of variation to (23), one finds that $\{z_{Ln}\}$ is obtained from $\{z_n\}$ by a linear transversal filter which, having the transfer function $1/S^*(f)$, eliminates ISI. The series arrangement of the MF and the transversal filter is known as the optimum linear equalizer, which for zero ISI yields maximum SNR [6], [8]. Calculating the SNR at the transversal filter output in a manner equivalent to (22), gives

$$S/N_L = \frac{S/N_{MF}}{T^2 \int_0^{1/T} S^*(f)\,df \int_0^{1/T} 1/S^*(f)\,df} \leq S/N_{MF}. \quad (24)$$

This reflects the obvious fact that, since the MF provides the absolutely largest SNR, elimination of ISI by a subsequent filter must diminish the SNR. Equation (24) indicates, however, that a significant loss will occur only if somewhere along the frequency axis $S^*(f)$ dips considerably below the average value.

For systems that transmit only real pulse amplitudes, i.e., double-sideband amplitude modulation (DSB-AM), vestigial-sideband amplitude modulation (VSB-AM), and single-sideband amplitude modulation (SSB-AM), it follows from (23) that only the real output of the MF is relevant. In those cases $S^*(f)$ should be replaced by $\frac{1}{2}[S^*(-f) + S^*(f)]$ without further mention throughout the paper.

IV. MAXIMUM-LIKELIHOOD SEQUENCE ESTIMATION

In this section the exact solution to the discrete maximization problem of (23) is presented. The MLSE algorithm that will be derived determines the most likely sequence $\{\hat{a}_n\}$ among sequences $\{\alpha_n\}$ that satisfy the coding rule. Clearly, the straightforward approach of computing $J_I(\{\alpha_n\})$ for all sequences allowed, and selecting the sequence that yields the maximum value, is impracticable in view of the length and number of possible messages. Instead, by applying the principles of dynamic programing [22], we shall be able to conceive a nonlinear recursive algorithm that performs the same selection with greatly reduced computational effort. The MLSE algorithm thus obtained represents a modified version of the well-known Viterbi algorithm [18]–[21], [23]–[30].

The algorithm is obtained, observing $s_l = \bar{s}_{-l}$, by first realizing from (23) that $J_I(\{\alpha_n\})$ can iteratively be computed by the recursive relation

$$J_n(\cdots,\alpha_{n-1},\alpha_n) = J_{n-1}(\cdots,\alpha_{n-1}) + \mathrm{Re}\,[\bar{\alpha}_n(2z_n - s_0\alpha_n - 2\sum_{k\leq n-1} s_{n-k}\alpha_k)]. \quad (25)$$

Conditions concerning the boundaries of I are not needed since once we have a recursive relationship, the length of I becomes unimportant. We now assume that at the MF output ISI from a particular signal element is limited to L preceding and L following sampling instants:

$$s_l = 0, \qquad |l| > L. \quad (26)$$

Changing indices we obtain from (25) and (26)

$$J_n(\cdots,\alpha_{n-1},\alpha_n) = J_{n-1}(\cdots,\alpha_{n-1}) + \mathrm{Re}\,[\bar{\alpha}_n(2z_n - s_0\alpha_n - 2\sum_{l=1}^{L} s_l\alpha_{n-l})]. \quad (27)$$

We recall that sequences may be coded. For most transmission codes a state representation is appropriate. Let μ_j be the state of the coder after α_j has been transmitted. The coding state μ_j determines which sequences can be further transmitted. Given μ_j and an allowable sequence $\alpha_{j+1},\alpha_{j+2},\cdots,\alpha_{j+k}$, the state μ_{j+k} is uniquely determined as

$$\mu_j:\alpha_{j+1},\alpha_{j+2},\cdots,\alpha_{j+k} \longrightarrow \mu_{j+k}. \quad (28)$$

The sequence of states $\{\mu_j\}$ is Markovian, in the sense that $\Pr(\mu_j \mid \mu_{j-1},\mu_{j-2},\cdots) = \Pr(\mu_j \mid \mu_{j-1})$.

Let us now consider the metric

$$\tilde{J}_n(\mu_{n-L}:\alpha_{n-L+1},\cdots,\alpha_n) = \tilde{J}_n(\sigma_n) = \max_{\{\cdots,\alpha_{n-L-1},\alpha_{n-L}\}\to\mu_{n-L}} \{J_n(\cdots,\alpha_{n-L},\alpha_{n-L+1},\cdots,\alpha_n)\} \quad (29)$$

where the maximum is taken over all allowable sequences $\{\cdots,\alpha_{n-L-1},\alpha_{n-L}\}$ that put the coder into the state μ_{n-L}. In accordance with the VA literature, $\tilde{J}_n$ is called survivor metric. There exist as many survivor metrics as there are survivor states

$$\sigma_n \triangleq \mu_{n-L}:\alpha_{n-L+1},\cdots,\alpha_n. \quad (30)$$

Clearly, the succession of states $\{\sigma_n\}$ is again Markovian. Associated with each σ_n is a unique path history, namely, the sequence $\{\cdots,\alpha_{n-L-1},\alpha_{n-L}\}$, which in (29) yields the maximum. With respect to σ_n this sequence is credited ML among all other sequences. It is not difficult to see that the further one looks back from time $n - L$, the less will a path history depend on the specific σ_n to which it belongs. One can therefore expect that all σ_n will have a common path history up to some time $n - L - m$; m being a nonnegative random variable. Obviously, the common portion of the path histories concurs with the most likely sequence $\{\hat{\alpha}_n\}$ for which we are looking.

The final step in deriving the MLSE algorithm is to apply the maximum operation defined by (29) to (27). Introducing the notation of a survivor metric also on the right-hand side, we obtain

$$\tilde{J}_n(\sigma_n) = 2\,\mathrm{Re}\,(\bar{\alpha}_n z_n) + \max_{\{\sigma_{n-1}\}\to\sigma_n} \{\tilde{J}_{n-1}(\sigma_{n-1}) - F(\sigma_{n-1},\sigma_n)\} \quad (31)$$

where the maximum is taken over all states $\{\sigma_{n-1}\}$ that have σ_n as a possible successor state, and

$$F(\sigma_{n-1},\sigma_n) = \bar{\alpha}_n s_0 \alpha_n + 2\,\mathrm{Re}\,(\bar{\alpha}_n \sum_{l=1}^{L} s_l\alpha_{n-l}). \quad (32)$$

Verifying (31), the reader will observe that L is just the minimum number of pulse amplitudes that must be associated with σ_n. Thus, L takes on the role of a constraint length inherent to ISI. Equation (31) enables us to calculate survivor metrics and path histories in recursive fashion. The path history of a particular σ_n is obtained by extending the path history of the σ_{n-1}, which in (31) yields the maximum by the α_{n-L} associated with the selected σ_{n-1}. At each sampling instant n, survivor metrics and path histories must be calculated for all possible states σ_n. Instead of expressing path histories in terms of pulse amplitudes α_{n-L}, they could also be represented in any other one-to-one related terms.

This concludes the essential part in the derivation of the MLSE algorithm. The algorithm can be extended to provide maximum *a posteriori* probability (MAP) decisions, as is shown in Appendix II. However, for reasons given there, the performance improvement which thereby can be attained will usually be insignificant.

The algorithm is identical to the original Viterbi algorithm if there is no ISI at the MF output, i.e., $F(\sigma_{n-1},\sigma_n) = \bar{\alpha}_n s_0 \alpha_n$. In the presence of ISI the algorithm differs from the original Viterbi algorithm in that it operates directly on the MF output where noise samples are correlated according to (18). For the original Viterbi algorithm statistical independence of the noise samples is essential. Forney [18] proposed to decorrelate the MF output noise by a transversal filter which thereby reduces the number of nonzero sample values of the signal element from $2L + 1$ to $L + 1$. The constraint length inherent to ISI is therefore the same, namely, L, for both the Viterbi algorithm as used by Forney and its modified version presented in this section.[3] Clearly, this indicates some fundamental lower

[3] G. D. Forney, private communication.

limit to the complexity of MLSE. Yet the modified Viterbi algorithm offers computational advantages in that the large number of squaring operations needed for the original Viterbi algorithm [29] are no longer required. Only the simple multiplications by discrete pulse amplitude values occurring in Re $(\bar{a}_n z_n)$ must be executed in real time. It was observed by Price [35] that Forney's whitened MF is identical to the optimum linear input section of a decision-feedback receiver. In Section VI we shall see that basically the same principle can be applied in order to realize the MF and the whitened MF in adaptive form. Hence in this respect the two algorithms are about equal.

We shall now illustrate the algorithm by a specific example. Let us consider a simple binary run-length-limited code with $a_n \in \{0,1\}$ and runs no longer than two of the same symbol. The state-transition diagram of this code is shown in Fig. 3(a). According to (30), with $L = 1$ the following survivor states and allowed transitions between them are obtained:

$$\sigma^1 \triangleq (\mu^1:1) \rightarrow (\mu^3:\{0,1\}) \triangleq \{\sigma^4,\sigma^5\}$$
$$\sigma^2 \triangleq (\mu^2:0) \rightarrow (\mu^1:1) \triangleq \sigma^1$$
$$\sigma^3 \triangleq (\mu^2:1) \rightarrow (\mu^3:\{0,1\}) \triangleq \{\sigma^4,\sigma^5\}$$
$$\sigma^4 \triangleq (\mu^3:0) \rightarrow (\mu^2:\{0,1\} \triangleq \{\sigma^2,\sigma^3\}$$
$$\sigma^5 \triangleq (\mu^3:1) \rightarrow (\mu^4:0) \triangleq \sigma^6$$
$$\sigma^6 \triangleq (\mu^4:0) \rightarrow (\mu^2:\{0,1\}) \triangleq \{\sigma^2,\sigma^3\}.$$

The corresponding state-transition diagram is depicted in Fig. 3(b). By introducing the time parameter explicitly, we obtain Fig. 4, in which allowed transitions are indicated by dashed lines (the so-called trellis picture [20]). The solid lines, as an example, represent the path histories of the six possible states σ_n and demonstrate their tendency to merge at some time $n - L - m$ into a common path. As the algorithm is used to compute the path histories of the states σ_{n+i}, $i = 1,2,\cdots$, new path-history branches appear on the right, whereas certain existing branches are not continued further and disappear. In this way, with some random time lag $L + m$, $m \gtrsim 0$, a common path history develops from left to right.

In order to obtain the output sequence $\{\hat{a}_n\}$ only the last, say, M, pulse amplitudes of each path history have to be stored. M should be chosen such that the probability for $m > M$ is negligible compared with the projected error probability of the ideal system (infinite M). Then, at time n, the α_{n-L-M} of all path histories will with high probability be identical; hence any one of them can be taken as $\hat{a}_{n-L-M}$. The path histories can now be shortened by the α_{n-L-M}. Thus the path histories are kept at length M, and a constant delay through the MLSE of $L + M$ symbol intervals results. For decoding of convolutional codes the value of M has been discussed in the literature [20], [36], [37]. At the present time no such results are available for the ISI case.

A refined way of selecting $\hat{a}_{n-L-M}$, which allows for a

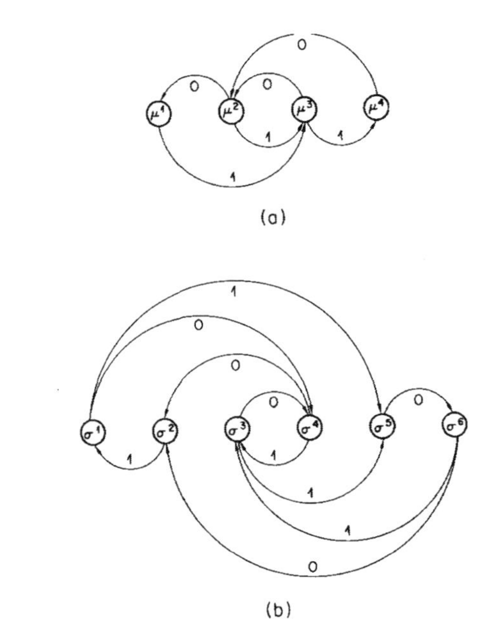

Fig. 3. (a) State-transition diagram of binary ($a_n = \{0,1\}$) run-length-limited code with runs ≤ 2. (b) State-transition diagram of corresponding survivor states for $L = 1$.

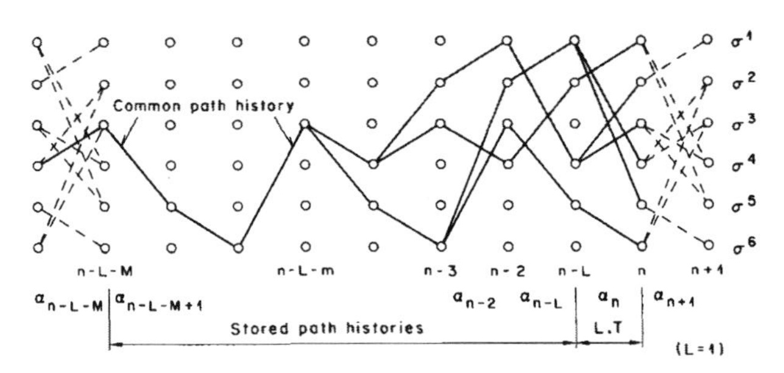

Fig. 4. Time-explicit representation of Fig. 3(b) (dashed lines) and illustration of path histories at time n (solid lines).

reduction of M, is to take α_{n-L-M} for the path history corresponding to the largest survivor metric. From (31) it is clear that without countermeasures the survivor metrics would steadily increase in value. A suitable method of confining the survivor metrics to a finite range is to subtract the largest $\tilde{J}_{n-1}$ from all $\tilde{J}_n$ after each iteration.

V. ERROR PERFORMANCE

Since the receiver of this paper realizes the same decision rule as Forney's receiver [18], it is not surprising that identical error performance will be found. In this section, following closely Forney's approach, we present a short derivation of the error-event probability for the modified Viterbi-algorithm case. The influence of ISI present at the MF output on the error performance of the MLR is discussed, and bounds for essentially no influence are given in explicit form. The results are compared with the error performance of the optimum conventional receiver.

We recall that $\{a_n\}$ represents the data sequence transmitted, whereas $\{\hat{a}_n\}$ denotes the sequence estimated by the receiver. Then

$$\{e_n\} = \{\hat{a}_n\} - \{a_n\} \tag{33}$$

is the error sequence. Since consecutive symbol errors are generally not independent of each other, the concept of sequence error events must be used. Hence, as error events we consider short sequences of symbol errors that intuitively are short compared with the mean time between them and that occur independently of each other. Presuming stationarity, the beginning of a specific error event $\mathcal{E}$ can arbitrarily be aligned with time 0:

$$\mathcal{E}: \{e_n\} = \cdots,0,0,e_0,e_1,\cdots,e_H,0,0,\cdots; \qquad |e_0|, |e_H| \geq \delta_0, \quad H \geq 0. \tag{34}$$

Here δ_0 denotes the minimum symbol error distance

$$\delta_0 = \min_{i \neq k} \{|a^{(i)} - a^{(k)}|\}. \tag{35}$$

We are not concerned with the meaning of error events in terms of erroneous bits of information.

Let E be the set of events $\mathcal{E}$ permitted by the transmission code. For a distinct event $\mathcal{E}$ to happen, two subevents must occur:

$\mathcal{E}_1$: $\{a_n\}$ is such that $\{a_n\} + \{e_n\}$ is an allowable data sequence;

$\mathcal{E}_2$: the noise terms are such that $\{a_n\} + \{e_n\}$ has ML (within the observation interval).

It is useful to define beyond that the subevent

$\mathcal{E}_2'$: the noise terms are such that $\{a_n\} + \{e_n\}$ has greater likelihood than $\{a_n\}$, but not necessarily ML.

Then we have

$$\Pr(\mathcal{E}) = \Pr(\mathcal{E}_1)\Pr(\mathcal{E}_2 \mid \mathcal{E}_1) \leq \Pr(\mathcal{E}_1)\Pr(\mathcal{E}_2' \mid \mathcal{E}_1) \tag{36}$$

where $\Pr(\mathcal{E}_1)$ depends only on the coding scheme. Events $\mathcal{E}_1$ are generally not mutually exclusive. Note that in (36) conditioning of $\mathcal{E}_2$ and $\mathcal{E}_2'$ on $\mathcal{E}_1$ tightens the given bound, since prescribing $\mathcal{E}_1$ reduces the number of other events that could, when $\mathcal{E}_2'$ occurs, still have greater likelihood, so that $\mathcal{E}_2$ would not be satisfied. From (23) we conclude that $\Pr(\mathcal{E}_2' \mid \mathcal{E}_1)$ is the probability that

$$J_I(\{a_n\}) < J_I(\{a_n\} + \{e_n\}). \tag{37}$$

By substituting (16) into (23), and observing (33) and $s_l = \bar{s}_{-l}$, (37) becomes

$$\delta^2(\mathcal{E}) \triangleq \frac{1}{s_0} \sum_{i=0}^{H} \sum_{k=0}^{H} \bar{e}_i s_{i-k} e_k < \frac{2}{s_0} \operatorname{Re}\left[\sum_{i=0}^{H} \bar{e}_i r_i\right]. \tag{38}$$

We call $\delta(\mathcal{E})$ the distance of $\mathcal{E}$. The right-hand side of (38) is a normally distributed random variable with zero mean and, from (18), (19), and (20), variance

$$\operatorname{var}\left\{\frac{2}{s_0} \operatorname{Re}\left[\sum_{i=0}^{H} \bar{e}_i r_i\right]\right\} = \frac{4N_0}{s_0} \delta^2(\mathcal{E}). \tag{39}$$

Hence, observing (22), the probability of (37) being satisfied is given by

$$\Pr(\mathcal{E}_2' \mid \mathcal{E}_1) = Q[(S/N_{\mathrm{MF}})^{1/2}\delta(\mathcal{E})/2] \tag{40}$$

where

$$Q(x) = \frac{1}{(2\pi)^{1/2}} \int_x^\infty \exp(-y^2/2)\, dy$$

$$\simeq \frac{1}{x(2\pi)^{1/2}} \exp(-x^2/2), \qquad x > 3.5. \tag{41}$$

We continue as indicated by Forney [18]. Let $E(\delta)$ be the subset of E containing all events $\mathcal{E}$ with distance $\delta(\mathcal{E}) = \delta$. Let Δ be the set of the possible values of δ. From (36) and (40) the probability that any error event $\mathcal{E}$ occurs becomes and is upper bounded by

$$\Pr(E) = \sum_{\mathcal{E} \in E} \Pr(\mathcal{E}) \leq \sum_{\delta \in \Delta} Q((S/N_{\mathrm{MF}})^{1/2}\delta/2) \sum_{\mathcal{E} \in E(\delta)} \Pr(\mathcal{E}_1). \tag{42}$$

Owing to the steep decrease of $Q(x)$, the right-hand side of (42) will already at moderate SNR be dominated by the term involving the smallest value in Δ, denoted by $\delta_{\min}$. Likewise, the bound given by (36) becomes tight for all $\mathcal{E} \in E(\delta_{\min})$, as then $\mathcal{E}_2'$ very likely implies $\mathcal{E}_2$. Consequently, as the SNR is increased, $\Pr(E)$ approaches asymptotically

$$\Pr(E) \simeq Q((S/N_{\mathrm{MF}})^{1/2}\delta_{\min}/2) \sum_{\mathcal{E} \in E(\delta_{\min})} \Pr(\mathcal{E}_1) \tag{43}$$

where

$$\delta_{\min}^2 = \min_{\mathcal{E} \in E} \left\{\frac{1}{s_0} \sum_{i=0}^{H} \sum_{k=0}^{H} \bar{e}_i s_{i-k} e_k\right\}. \tag{44}$$

Equations (43) and (44) differ only in notation from Forney's original finding.

In the following this result should be discussed in more detail. Specifically, we are interested in the influence of ISI on the value of $\delta_{\min}$. Using Parseval's theorem, from (21) and (38), $\delta^2(\mathcal{E})$ can be rewritten in the form

$$\delta^2(\mathcal{E}) = \frac{T}{s_0} \int_0^{1/T} S^*(f) E^*(f)\, df \tag{45}$$

where $E^*(f)$ is the energy density spectrum of the error sequence $\{e_n\}$,

$$E^*(f) = \sum_{i=0}^{H} \sum_{k=0}^{H} \bar{e}_i \exp[j2\pi f(i-k)T] e_k \geq 0. \tag{46}$$

If $S^*(f)$ were constant (no ISI at the MF output), (45) becomes

$$\delta^2(\mathcal{E}) = T \int_0^{1/T} E^*(f)\, df = \sum_{i=0}^{H} |e_i|^2 \geq w_H(\mathcal{E})\delta_0^2 \tag{47}$$

where $w_H \geq 1$ denotes the number of nonzero symbol errors of $\mathcal{E}$. In this case $\delta_{\min}$ would simply be the smallest Euclidian distance between any two allowed data sequences. In a noncoded system, where pulse amplitudes of a given alphabet may occur in arbitrary succession, single error events ($w_H = 1$) with minimum distance $\delta_{\min} = \delta_0$ would be the dominating error events. The value of $\delta_{\min}$ can be increased by redundant sequence coding, e.g., by convolutional encoding [36]. If $S^*(f)$ is not constant, ISI at the MF output distorts the space in which error-event distances are measured. Depending on the weighting of $E^*(f)$ by $S^*(f)$, error-event distances can become smaller or larger. By sequence coding one can prevent that error sequences are allowed which have spectral peaks where $S^*(f)$ is small. This is precisely what is accomplished by correlative-level (partial-response) coding [25]. Clearly, if $S^*(f)$ vanishes on one side of $f = 0$, as in SSB or VSB systems, only (real) data sequences with symmetric error-sequence spectra $E^*(f) = E^*(-f)$ can be transmitted.

We now limit our attention to noncoded systems. As long as ISI does not exceed limits discussed further in the following paragraphs, we have $\delta_{\min} = \delta_0$. From (43) the probability of occurrence of the then dominating single error events becomes

$$\Pr(E) \simeq Q((S/N_{\text{MF}})^{1/2}\delta_0/2) \cdot \sum_{|e_0|=\delta_0} \Pr(e_0 \text{ allowed}), \qquad \delta_{\min} = \delta_0. \tag{48}$$

For comparison, the error performance of the optimum conventional receiver is given by

$$\Pr(E) \simeq Q((S/N_L)^{1/2}\delta_0/2) \sum_{|e_0|=\delta_0} \Pr(e_0 \text{ allowed}), \qquad S/N_L \leq S/N_{\text{MF}}. \tag{49}$$

We note that in (48) ISI at the MF output has essentially no influence on the error performance of the MLR, whereas in (49) ISI affects the error performance of the conventional receiver through the loss of SNR expressed by (24). The evaluation of Pr (E) is shown by Fig. 5 for a specific octal AM–PM scheme.

In order to determine the degree of ISI up to which (48) holds, we must look for multiple error events ($w_H \geq 2$) with distance smaller than δ_0. Such error events would then be more probable than the minimum single error events. A first condition for the nonexistence of such events can be derived from (45) and the inequality expressed in (47). Noting that

$$\delta^2(\mathcal{E}) \geq \frac{1}{s_0} \min\{S^*(f)\} w_H(\mathcal{E})\delta_0^2 \tag{50}$$

it follows that if

$$\min\{S^*(f)\} w_H(\mathcal{E}) \geq s_0 \triangleq T \int_0^{1/T} S^*(f)\, df \tag{51}$$

is satisfied, no event $\mathcal{E}$ can have smaller distance than δ_0. Hence, a sufficient but not necessary condition for the nonexistence of multiple error events with distance smaller

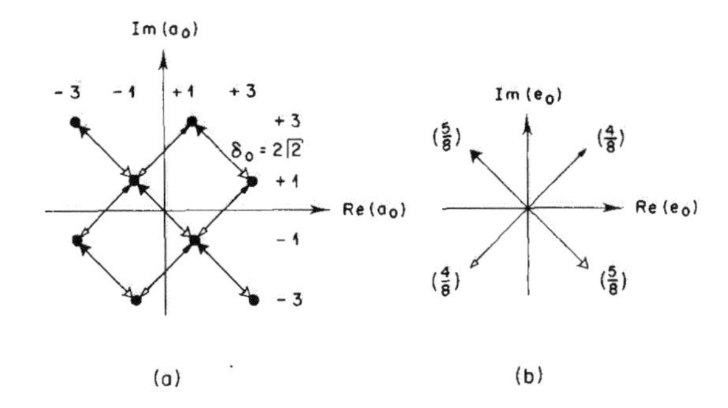

Fig. 5. (a) Minimum symbol error distance in a specific octal pulse-amplitude alphabet. (b) Minimum symbol errors and probabilities that they may occur (assuming that pulse amplitudes are transmitted with equal probability). $\sum_{|e0|=\delta 0} \Pr(e_0 \text{ allowed}) = 2(\frac{4}{8} + \frac{5}{8}) = \frac{9}{4}$, $\Pr(E) \simeq \frac{9}{4} Q((2 \cdot S/N)^{1/2})$.

than δ_0 is that $S^*(f)$ dips nowhere more than 6 dB [20 log ($w_H = 2$)] below average value.

A second generally less restrictive condition is that

$$\sum_{l \neq 0} |s_l| \leq s_0 \tag{52}$$

which is the familiar condition for peak distortion at the MF output being smaller than unity. In order to prove this sufficient but again not necessary condition for the nonexistence of error events $\mathcal{E}$ with distance smaller than δ_0, one should first realize from (34) and the Schwarz inequality that

$$\left| \sum_{k=0}^{H} \bar{e}_{l+k} e_k \right| \leq \sum_{k=0}^{H} |e_k|^2 - \delta_0^2, \qquad l \neq 0. \tag{53}$$

Condition (52) can then be verified by transforming and bounding $\delta^2(\mathcal{E})$ as follows:

$$\begin{aligned} \delta^2(\mathcal{E}) &= \frac{1}{s_0} \sum_l s_l \sum_{k=0}^{H} \bar{e}_{l+k} e_k \\ &\geq \sum_{k=0}^{H} |e_k|^2 - \frac{1}{s_0} \sum_{l \neq 0} |s_l| \left| \sum_{k=0}^{H} \bar{e}_{l+k} e_k \right| \\ &\geq \delta_0^2 + \left(\sum_{k=0}^{H} |e_k|^2 - \delta_0^2\right)\left(1 - \frac{1}{s_0} \sum_{l \neq 0} |s_l|\right) \\ &\geq \delta_0^2 \qquad \text{[holds, if (52) is true].} \end{aligned} \tag{54}$$

Comparing (52) with the definition of $S^*(f)$ in (21) reveals that at distinct frequencies, $S^*(f)$ may approach zero level without this significantly affecting the error performance of the MLR. This was first observed by Kobayashi [25] for ML decoding of $(2m - 1)$-ary correlative-level encoded signals. A signal of this kind can just as well be interpreted as a noncoded m-ary signal with intentionally introduced ISI, which causes $S^*(f)$ to become zero (usually) at $f = 0$ and/or $1/2T$. For the MLR the two concepts are equivalent. A conventional receiver, however, can interpret such signals only as $(2m - 1)$-ary coded sequences and thereby loses in the limit 3 dB, unless error correcting schemes are used. But even then the loss can only partly be compensated, since the hard decisions made by the symbol-by-symbol decision circuit of the conventional receiver cause an irreversible loss of information.

VI. AUTOMATIC RECEIVER ADAPTATION

So far the exact signal and timing characteristics have been assumed to be known. However, in a realistic case the MLR must at least be able to extract the carrier phase and sample timing from the signal received. Beyond that, automatic adjustment of the MF will often be desirable or necessary. In this section we present an algorithm that simultaneously adjusts the demodulating carrier phase and the sample timing, approximates the MF by a transversal filter, and estimates ISI present at the approximated MF output. The algorithm works in decision-directed mode in much the same way as described by Kobayashi [5] and Qureshi and Newhall [29].

In the proposed fully adaptive MLR the MF is approximated by a transversal filter, similar to the familiar adaptive equalizers described by Lucky *et al.* [1], [38], [39] and others [5], [40]–[45]. An analog implementation will be assumed. Assuming $N + 1$ taps equally spaced by τ_p seconds with tap gains g_i, $0 \le i \le N$, the output signal of the transversal filter at the nth sampling instant $nT + \tau_s$, where τ_s denotes the sampling phase, becomes

$$\hat{z}_n = \sum_{i=0}^{N} g_i y(nT + \tau_s - i\tau_p, \varphi_c) \triangleq \sum_{i=0}^{N} g_i y_{ni}(\tau_s, \varphi_c). \quad (55)$$

Note that according to (4) and (5) we have $y(t,\varphi_c) \triangleq y(t,\varphi_c = 0) \exp(-j\varphi_c)$, where φ_c is the demodulating carrier phase. The ideal MF characteristic can be accomplished if $1/2\tau_p$ exceeds or at least is equal to the bandwidth of $y(t)$, and $N\tau_p$ corresponds to the duration of the signal element $h(t)$.

For the derivation of the adjustment algorithm we make the usual assumption that the transmitted data sequence $\{a_n\}$ is known. The decision delay of the Viterbi algorithm will be taken into account later when we devise the final adaptive MLR structure. In Section III we have seen that the MF is rigorously defined by the fact that it minimizes the noise power relative to the instantaneous peak power of the signal element. Let $\hat{s}_l$, $|l| \le L$, be estimated values of the sample values of the signal element at the transversal filter output. Then

$$\hat{r}_n = \hat{z}_n - \sum_{l=-L}^{L} \hat{s}_l a_{n-l} \quad (56)$$

represents the estimated noise component of $\hat{z}_n$. In order to adjust the parameters g_i, $\hat{s}_l$, τ_s, and φ_c, the variance of $\hat{r}_n$,

$$\begin{aligned} \operatorname{var}(\hat{r}_n) &= \sum_{i=0}^{N}\sum_{k=0}^{N} \bar{g}_i E[\bar{y}_{ni}(\tau_s,\varphi_c) y_{nk}(\tau_s,\varphi_c)] g_k \\ &\quad - 2\operatorname{Re}\{\sum_{i=0}^{N}\sum_{l=-L}^{L} \bar{g}_i E[\bar{y}_{ni}(\tau_s,\varphi_c) a_{n-l}] \hat{s}_l\} \\ &\quad + \sum_{i=-L}^{L}\sum_{k=-L}^{L} \bar{\hat{s}}_i E[\bar{a}_{n-i} a_{n-k}] \hat{s}_k \end{aligned} \quad (57)$$

must be minimized as a function of these parameters, with $\hat{s}_0$ held constant. Differentiating (57) and applying the Robbins–Monro stochastic approximation method [46] leads to the stochastic steepest-descent algorithm comprising the following recursive relations:

$$g_i^{(n+1)} = g_i^{(n)} - \alpha_g^{(n)} \hat{r}_n \bar{y}_{ni}, \qquad 0 \le i \le N \quad (58)$$

$$\hat{s}_l^{(n+1)} = \hat{s}_l^{(n)} + \alpha_s^{(n)} \hat{r}_n \bar{a}_{n-l}, \qquad 1 \le |l| \le L \quad (59)$$

$$\tau_s^{(n+1)} = \tau_s^{(n)} - \alpha_\tau^{(n)} \operatorname{Re}(\hat{r}_n \bar{\dot{\hat{z}}}_n), \quad (60)$$

$$\varphi_c^{(n+1)} = \varphi_c^{(n)} + \alpha_\varphi^{(n)} \operatorname{Im}(\hat{r}_n \bar{\hat{z}}_n). \quad (61)$$

The step-size gains α_g, α_s, α_τ, and α_φ must be positive and may depend on n. In (60) $\dot{\hat{z}}_n$ denotes the time derivative of the transversal filter output at the nth sampling instant. As the algorithm adjusts the transversal filter as MF, the values $\hat{s}_l$ approach the values s_l required by the MLSE algorithm.

Equations (58) and (59) differ from the corresponding equations of an adaptive decision-feedback equalizer, or whitened MF, only by the fact that here at the transversal filter output L preceding *and* L trailing ISI terms are considered. The algorithm will force ISI outside this interval to zero. The true MF characteristic may often require a large value of L. However, since the complexity of the MLSE algorithm increases exponentially with L, for the choice of L a compromise suggests itself [29]. In many practical cases already with values $L = 1$ or $L = 2$, a good approximation of the ideal MF characteristic will be obtained. The potential advantages of MLSE can thus be exploited to a commensurate degree at a still manageable receiver complexity.

Introducing the symmetry condition $\hat{s}_{-l} = \bar{\hat{s}}_l$ into (56), we obtain instead of (59)

$$\hat{s}_l^{(n+1)} = \hat{s}_l^{(n)} + \alpha_s^{(n)} (\bar{\hat{r}}_n a_{n+l} + \hat{r}_n \bar{a}_{n-l}), \qquad 1 \le l \le L. \quad (62)$$

This modification has the desirable effect of forcing the transversal filter to produce at its output a symmetric signal element even if L and the transversal filter parameters are not fully adequate to achieve therewith the ideal MF characteristic.

Equations (60) and (61) have been reported by Kobayashi [5]. They describe the operation of two first-order phase-locked loops. Theoretically, if by (58) the (complex) tap gains are adapted, the adjustment of τ_s and φ_c appears to be not really necessary. In practice, however, these phases must be controlled in order to compensate carrier and sampling frequency offsets. In case of considerable offset one might even add second-order terms to (60) and (61).

The structure of the proposed MLR is seen in Fig. 6. It is basically a combination of the approaches of Kobayashi [5] and Qureshi and Newhall [29], except that here the transversal filter approximates a true MF with ISI at the transversal filter not being predefined. The receiver operates in decision-directed mode with the MLSE exhibiting a decision delay of $M + L$ symbol intervals (see

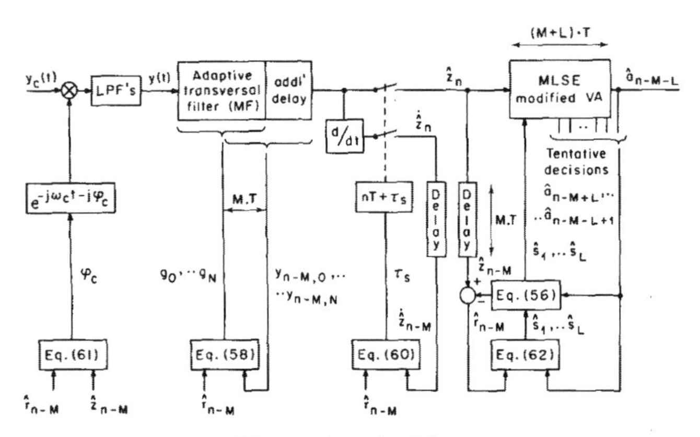

Fig. 6. Adaptive MLR.

Section IV). In order to shorten the feedback delay, tentative decisions taken from the path history with largest survivor metric are employed in the feedback paths as suggested by Qureshi [30]. With this approach, delays of M symbol intervals must be used in the forward paths. Not shown in Fig. 6 is the possibility of incorporating decision-feedback cancellation of further trailing ISI in the receiver [30].

In the remainder of this section we discuss topics related to convergence and convexity of the adjustment algorithm. It must be assumed that already safe enough decisions are available. To begin with, τ_s and φ_c are considered as given constant values. With sufficiently small step-size gains α_g and α_s, convergence from arbitrary initial settings $g_i^{(0)}$ and $\hat{s}_l^{(0)}$ towards globally optimum settings $g_i^*(\tau_s,\varphi_c)$ and $\hat{s}_l^*(\tau_s)$ is assured by the quadratic positive-definite nature of var $(\hat{r}_n)$ in g_i and $\hat{s}_l$. For adaptive zero-forcing equalizers (equivalent to $L = 0$) it has been shown [45] that fastest convergence takes place if α_g corresponds to $1/[N \times$ average power of $y(t)]$. The additional variability introduced by (59) and eventually (60) and (61), and the delay of M symbol intervals necessary for decision-directed operation, suggest that α_g must here be somewhat smaller than the preceding value.

In Appendix III we calculate $g_i^*(\tau_s,\varphi_c)$ and $\hat{s}_l^*(\tau_s)$, thus showing that with adequate values of N, L, τ_p, and τ_s the transversal filter indeed assumes the desired MF characteristic. We further evaluate the minimum value of var $(\hat{r}_n)$ as a function of τ_s. On the whole, convexity is found within an interval comparable to the length of the transversal filter delay line; yet, depending on the value of τ_p, there can be some small ripple. If $1/2\tau_p$ does not exceed the bandwidth of $y(t)$ this ripple can even become quite pronounced. In any case, the form of var $(\hat{r}_n)$ will guarantee that τ_s does not drift away when all quantities g_i, $\hat{s}_l$, τ_s, and φ_c are adjusted simultaneously.

Another case of interest relates to carrier and timing synchronization in an otherwise already optimized and fixed receiver. In Appendix III var $(\hat{r}_n)$ is calculated as a function of τ_s and φ_c for fixed values $g_i^*(\tau_s^*,\varphi_c^*)$ and $\hat{s}_l^*(\tau_s^*)$, τ_s^* and φ_c^* being the timing phase and carrier phase to which the optimum receiver settings correspond. It is shown that var $(\hat{r}_n)$ is convex in a region around τ_s^* and φ_c^*, which depends mainly on the form of the signal element at the MF output. With the receiver working in decision-directed mode and disregarding phase ambiguity, τ_s and φ_c must in principle be close to the optimum settings. Convergence towards τ_s^* and φ_c^* should therefore generally not be a problem.

VII. SUMMARY AND CONCLUSIONS

A uniform fully adaptive receiver structure has been derived for synchronous data-transmission systems that employ linear carrier-modulation techniques. The structure realizes the ML sequence rule. In the receiver, first an information reduction to a set of sufficient statistics takes place by the demodulation, matched filtering, and symbol-rate sampling process. Sequence estimation is performed by a modified Viterbi algorithm that exhibits the same performance characteristics as the original scheme. The algorithm represents an attractive design alternative due to the fact that squaring operations are no longer needed. Besides add and compare operations, only a few simple multiplications by discrete pulse-amplitude values must be performed in real time. In addition to performance gains realized by the MLSE principle, one may expect that the approximation of the MF will generally require fewer filter taps than are needed for the zero-forcing equalizer of a conventional receiver. The proposed adaptation scheme permits compromise solutions between the conventional receiver and the ideal ML receiver.

The choice of the decoding delay of MLSE in the presence of ISI and the dynamics of the presented adjustment algorithm have not been discussed in detail. Also the issues of effective QAM coding and joint transmitter-receiver design have not been addressed. These could be fruitful areas for further research. For example, how should the transmitter filter be designed for a given channel characteristic in order to attain with (52) as secondary condition maximum SNR at the matched filter output? The specific implementation of ML receivers will be another interesting topic. Recent progress in circuit technology will allow here for much more complex designs than we are still used to.

APPENDIX I

PROOF OF (9)

Owing to the one-to-one relation between $w_c(t)$ and $w(t)$ expressed by (5) we have [32]

$$p[w_c(t), t \in I] = p[w(t), t \in I] \sim \exp\left\{-\frac{1}{2}\int_I\int_I w_c(t_1)W_c^{-1}(t_1 - t_2)w_c(t_2)\,dt_1\,dt_2\right\}. \tag{A1}$$

It follows from (6), (7), and (10) that

$$W_c^{-1}(\tau) = \frac{1}{N_0}K^{-1}(\tau)\exp(+j\omega_c\tau). \tag{A2}$$

Substituting (A2) into (A1) and observing (5) we obtain

$$p[w(t),t \in I] \sim \exp\left\{-\frac{1}{4N_0}\int_I\int_I \bar{w}(t_1)K^{-1}(t_1-t_2)w(t_2)\,dt_1\,dt_2\right\}. \tag{A3}$$

APPENDIX II

EXTENSION OF THE MLSE ALGORITHM TO THE MAP RULE

We proceed as indicated by Forney [27]. To satisfy the MAP rule the algorithm has to determine the sequence $\{\alpha_n\}$ which maximizes

$$\Pr[\{\alpha_n\} \mid y(t),t \in I] \sim p[y(t),t \in I \mid \{\alpha_n\}]\Pr[\{\alpha_n\}]. \tag{A4}$$

Since there is a one-to-one correspondence between $\{\alpha_n\}$ and the sequence of survivor states $\{\sigma_n\}$, and since $\{\sigma_n\}$ is Markovian, we have

$$\Pr[\{\alpha_n\}] = \Pr[\{\sigma_n\}] \sim \prod_{nT\epsilon I}\Pr(\sigma_n \mid \sigma_{n-1}). \tag{A5}$$

It follows from (30) that

$$\Pr(\sigma_n \mid \sigma_{n-1}) = \Pr(\mu_n \mid u_{n-1}). \tag{A6}$$

Taking the logarithm of (A4) and observing (A5), it is seen that transitions (σ_{n-1},σ_n) are to be weighted by $\ln[\Pr(\sigma_n \mid \sigma_{n-1})]$. In this way the MAP version of (31) becomes

$$\tilde{J}_n(\sigma_n) = 2\,\mathrm{Re}\,(\bar{\alpha}_n z_n) + \max_{\{\sigma_{n-1}\}\to\sigma_n}\{\tilde{J}_{n-1}(\sigma_{n-1}) - F(\sigma_{n-1},\sigma_n) + 4N_0\ln[\Pr(\sigma_n \mid \sigma_{n-1})]\}. \tag{A7}$$

The factor $4N_0$ follows from (11) and, according to (22), is inversely proportional to the SNR at the MF output. The ML rule and the MAP rule are therefore equivalent for infinite SNR. The MAP rule can offer a significant advantage only at very low SNR's and when a code is used that leads to considerable differences among the conditional probabilities $\Pr(\sigma_n \mid \sigma_{n-1})$.

APPENDIX III

First, we show that minimizing var $(\hat{r}_n)$ with $\hat{s}_0$ held constant indeed adjusts the transversal filter as MF, and that thereby the values of s_l are provided to a sufficient degree of approximation. Second, we study the convexity of var $(\hat{r}_n)$ relative to the sampling phase τ_s. Assuming sampling instant $n = 0$, we drop the index n in the following calculations.

To begin with, the sampling phase τ_s and the demodulating carrier phase φ_c are considered as given constant values. It follows from (57) that the optimum tap gains g_i^* and values $\hat{s}_l^*$ which minimize var $(\hat{r})$ must satisfy

$$\sum_{k=0}^{N} E(\bar{y}_i y_k)g_k^* - \sum_{l=-L}^{L} E(\bar{y}_i a_{-l})\hat{s}_l^* = 0, \qquad 0 \le i \le N \tag{A8}$$

and

$$\sum_{i=0}^{N} E(\bar{a}_{-l}y_i)g_i^* - \sum_{k=-L}^{L} E(\bar{a}_{-l}a_{-k})\hat{s}_k^* = \begin{cases}0, & 1 \le |l| \le L\\ -\lambda, & l = 0\end{cases} \tag{A9}$$

where λ acts as Lagrangian multiplier. Substitution o (A8) and (A9) into (57) yields

$$\mathrm{var}\,[\hat{r} \mid \{g_i^*(\tau_s,\varphi_c)\},\{\hat{s}_l^*(\tau_s)\}] = \mathrm{var}\,(\hat{r} \mid \tau_s) = \lambda. \tag{A10}$$

From (3) and (55) we have

$$y_i(\tau_s) = \sum_k a_k h(\tau_s - i\tau_p - kT) + w(\tau_s - i\tau_p). \tag{A11}$$

With power series notations

$$h(t,D) = \sum_k h(t+kT)D^k \tag{A12}$$

$$A(D) = \sum_k E(\bar{a}_0 a_k)D^k \tag{A13}$$

$$\hat{s}^*(D) = \sum_{l=-L}^{L}\hat{s}_l^* D^l \tag{A14}$$

(A8) and (A9) can be rewritten in the form

$$\{\bar{h}(\tau_s - i\tau_p,D^{-1})\sum_{k=0}^{N} A(D)h(\tau_s - k\tau_p,D)g_k^*\}_{D0} + \sum_{k=0}^{N} W[(i-k)\tau_p]g_k^* = \{\bar{h}(\tau_s - i\tau_p,D^{-1})A(D)\hat{s}^*(D)\}_{D0}, \qquad 0 \le i \le N \tag{A15}$$

and

$$\{\sum_{k=0}^{N} A(D)h(\tau_s - k\tau_p,D)g_k^*\}_{Dl} = \{A(D)\hat{s}^*(D) - \lambda\}_{Dl}, \qquad |l| \le L. \tag{A16}$$

Here $\{\cdot\}_{Dl}$ indicates the coefficient belonging to D^l. Witl L not being too restricted, and $w_{ik} = W[(i-k)\tau_p]$ substitution of (A16) into (A15) yields in good approx imation

$$\sum_{k=0}^{N} w_{ik}g_k^* \simeq \lambda\bar{h}(\tau_s - i\tau_p), \qquad 0 \le i \le N. \tag{A17}$$

Let w_{ik}^{-1} be the elements of the inverse of the $(N+1) \times$

$(N+1)$ matrix with elements w_{ik}. Then from (A17),

$$g_k^* \simeq \lambda \sum_{i=0}^{N} \bar{h}(\tau_s - i\tau_p) w_{ki}^{-1}, \qquad 0 \leq k \leq N. \quad \text{(A18)}$$

Substituting (A18) into (A16), and observing (A12) and (A14), we obtain

$$\hat{s}_l^* \simeq \lambda \sum_{i=0}^{N} \sum_{k=0}^{N} \bar{h}(\tau_s - i\tau_p) w_{ki}^{-1} h(\tau_s - k\tau_p + lT) + \lambda \{A^{-1}(D)\}_{Dl}, \qquad |l| \leq L. \quad \text{(A19)}$$

For moderate SNR we can neglect the last term in (A19) which involves minor approximation. Comparison of (A18) with (14) and of (A19) with (17) exhibits that with adequate values of N, τ_p, L, and τ_s, the desired adjustment of the adaptive MLR will be achieved.

We investigate now the convexity of var $(\hat{r})$ relative to τ_s. Considering (A10) and determining λ from (A19) with $l = 0$, we find

$$\text{var}\ (\hat{r} \mid \tau_s) \simeq \frac{\hat{s}_0}{\sum_{i=0}^{N} \sum_{k=0}^{N} \bar{h}(\tau_s - i\tau_p) w_{ki}^{-1} h(\tau_s - k\tau_p)}. \quad \text{(A20)}$$

The denominator of (A20) expresses the weighted energy of the values of $h(t)$ seen at time $t = \tau_s$ at the transversal filter taps. We assume that the length of the transversal filter delay line $N\tau_p$ exceeds or at least corresponds to the duration of $h(t)$. Var $(\hat{r} \mid \tau_s)$ will then, on the whole, be convex within an interval comparable to $N\tau_p$. Unless the tap spacing τ_p is very small, however, there will be some ripple within this region.

Suppose now that for some τ_s^* and φ_c^* the optimum values g_i^* and $\hat{s}_l^*$ are given. In attempting to resynchronize the receiver, only τ_s and φ_c should be readjusted. Let $\Delta\tau_s = \tau_s - \tau_s^*$ and $\Delta\varphi_c = \varphi_c - \varphi_c^*$. In (57) only the second term depends noticeably on $\Delta\tau_s$ and $\Delta\varphi_c$. Using (A11) and

$$\sum_{i=0}^{N} g_i^*(\tau_s^*, \varphi_c^*) h(\tau_s - i\tau_p + lT, \varphi_c) \simeq s(\Delta\tau_s + lT) \exp(-j\Delta\varphi_c) \quad \text{(A21)}$$

where $s(t)$ denotes the time-continuous signal element at the MF output $[s(lT) = s_l]$, (57) becomes

$$\text{var}\ (\hat{r} \mid \Delta\tau_s, \Delta\varphi_c) \simeq \text{first term} + \text{third term} - 2\,\text{Re} \left\{ \sum_{l=-L}^{L} \sum_{k} \bar{s}(lT) E(\bar{a}_k a_l) \cdot s(\Delta\tau_s + kT) \exp(-j\Delta\varphi_c) \right\}. \quad \text{(A22)}$$

Minimizing (A22) with respect to $\Delta\varphi_c$ yields

$$\text{var}\ (\hat{r} \mid \Delta\tau_s) \simeq \text{first term} + \text{third term} - 2 \left| \sum_{l=-L}^{L} \sum_{k} \bar{s}(lT) E(\bar{a}_k a_l) s(\Delta\tau_s + kT) \right|. \quad \text{(A23)}$$

The form of (A23) permits local minima to occur within short distance from τ_s^*. However, with the receiver working in decision-directed mode, $|\Delta\tau_s| < T/2$ can be assumed, and hence convergence towards τ_s^* and φ_c^* can hardly be a problem.

REFERENCES

[1] R. W. Lucky, J. Salz, and E. J. Weldon, Jr., *Principles of Data Communication.* New York: McGraw-Hill, 1968.
[2] C. W. Helstrom, *Statistical Theory of Signal Detection,* revised ed. New York: Pergamon, 1968.
[3] H. L. Van Trees, *Detection, Estimation and Modulation Theory,* pt. I. New York: Wiley, 1968.
[4] a) H. B. Voelcker, "Toward a unified theory of modulation,—Part I: Phase-envelope relationships," *Proc. IEEE,* vol. 54, pp. 340–353, Mar. 1966.
b) ——, "Toward a unified theory of modulation—Part II: Zero manipulation," *Proc. IEEE,* vol. 54, pp. 735–755; May 1966.
[5] H. Kobayashi, "Simultaneous adaptive estimation and decision algorithm for carrier modulated data transmission systems," *IEEE Trans. Commun. Technol.,* vol. COM-19, pp. 268–280, June 1971.
[6] D. W. Tufts, "Nyquist's problem—The joint optimization of transmitter and receiver in pulse amplitude modulation," *Proc. IEEE,* vol. 53, pp. 248–259, Mar. 1965.
[7] M. R. Aaron and D. W. Tufts, "Intersymbol interference and error probability," *IEEE Trans. Inform. Theory,* vol. IT-12, pp. 26–34, Jan. 1966.
[8] T. Berger and D. W. Tufts, "Optimum pulse amplitude modulation, Part I: Transmitter-receiver design and bounds from information theory," *IEEE Trans. Inform. Theory,* vol. IT-13, pp. 196–208, Apr. 1967.
[9] T. Ericson, "Structure of optimum receiving filters in data transmission systems," *IEEE Trans. Inform. Theory* (Corresp.), vol. IT-17, pp. 352–353, May 1971.
[10] M. E. Austin, "Decision-feedback equalization for digital communication over dispersive channels," Sc.D. thesis, Mass. Inst. Technol., Cambridge, May 1967.
[11] D. A. George, R. R. Bowen, and J. R. Storey, "An adaptive decision-feedback equalizer," *IEEE Trans. Commun. Technol.,* vol. COM-19, pp. 281–293, June 1971.
[12] R. W. Chang and J. C. Hancock, "On receiver structures for channels having memory," *IEEE Trans. Inform. Theory,* vol. IT-12, pp. 463–468, Oct. 1966.
[13] K. Abend, T. J. Harley, B. D. Frichman, and C. Gumacos, "On optimum receivers for channels having memory," *IEEE Trans. Inform. Theory* (Corresp.), vol. IT-14, pp. 819–820, Nov. 1968.
[14] R. R. Bowen, "Bayesian decision procedure for interfering digital signals," *IEEE Trans. Inform. Theory* (Corresp.), vol. IT-15, pp. 506–507, July 1969.
[15] R. A. Gonsalves, "Maximum-likelihood receiver for digital data transmission," *IEEE Trans. Commun. Technol.,* vol. COM-16, pp. 392–398, June 1968.
[16] K. Abend and B. D. Fritchman, "Statistical detection for communication channels with intersymbol interference," *Proc. IEEE,* vol. 58, pp. 779–785, May 1970.
[17] G. Ungerboeck, "Nonlinear equalization of binary signals in Gaussian noise," *IEEE Trans. Commun. Technol.,* vol. COM-19, pp. 1128–1137, Dec. 1971.
[18] G. D. Forney, "Maximum-likelihood sequence estimation of digital sequences in the presence of intersymbol interference," *IEEE Trans. Inform. Theory,* vol. IT-18, pp. 363–378, May 1972.
[19] A. J. Viterbi, "Error bounds for convolutional codes and an asymptotically optimum decoding algorithm," *IEEE Trans. Inform. Theory,* vol. IT-13, pp. 260–269, Apr. 1967.
[20] G. D. Forney, "Review of random tree codes," NASA Amer. Res. Cen., Moffett Field, Calif., Contract NAS2-3637, NASA CR 73176, Final Rep., Appendix A, Dec. 1967.

[21] J. K. Omura, "On the Viterbi decoding algorithm," *IEEE Trans. Inform. Theory* (Corresp.), vol. IT-15, pp. 177–179, Jan. 1969.
[22] R. Bellman, *Dynamic Programming.* Princeton, N. J.: Princeton Univ. Press, 1957.
[23] J. K. Omura, "On optimum receivers for channels with intersymbol interference" (Abstract), presented at the IEEE Int. Symp. Information Theory, Noordwijk, Holland, 1970.
[24] H. Kobayashi, "Application of probabilistic decoding to digital magnetic recording systems," *IBM J. Res. Develop.*, vol. 15, pp. 64–74, Jan. 1971.
[25] ——, "Correlative level coding and maximum-likelihood decoding," *IEEE Trans. Inform. Theory*, vol. IT-17, pp. 586–594, Sept. 1971.
[26] ——, "A survey of coding schemes for transmission or recording of digital data," *IEEE Trans. Commun. Technol.*, vol. COM-19, pp. 1087–1100, Dec. 1971.
[27] G. D. Forney, "The Viterbi algorithm," *Proc. IEEE*, vol. 61, pp. 268–278, Mar. 1973.
[28] F. R. Magee, Jr., and J. G. Proakis, "Adaptive maximum-likelihood sequence estimation for digital signaling in the presence of intersymbol interference," *IEEE Trans. Inform. Theory* (Corresp.), vol. IT-19, pp. 120–124, Jan. 1973.
[29] S. U. H. Qureshi and E. E. Newhall, "An adaptive receiver for data transmission over time-dispersive channels," *IEEE Trans. Inform. Theory*, vol. IT-19, pp. 448–457, July 1973.
[30] S. U. H. Qureshi, "An adaptive decision-feedback receiver using maximum-likelihood sequence estimation," presented at the 1973 Int. Communications Conf., Seattle, Wash.
[31] L. K. Mackechnie, "Receivers for channels with intersymbol interference" (Abstract), presented at the IEEE Int. Symp. Information Theory, 1972, p. 82.
[32] M. Schwartz, "Abstract vector spaces applied to problems in detection and estimation theory," *IEEE Trans. Inform. Theory*, vol. IT-12, pp. 327–336, July 1966.
[33] J. Capon, "Hilbert space methods for detection theory and pattern recognition," *IEEE Trans. Inform. Theory*, vol. IT-11, pp. 247–259, Apr. 1965.
[34] T. Kailath, "RKHS approach to detection and estimation problems—Part I: Deterministic signals in Gaussian noise," *IEEE Trans. Inform. Theory*, vol. IT-17, pp. 530–549, Sept. 1971.
[35] R. Price, "Nonlinearly feedback-equalized PAM versus capacity for noise filter channels," presented at the 1972 Int. Conf. Communications, Philadelphia, Pa.
[36] A. J. Viterbi, "Convolutional codes and their performance in communication systems," *IEEE Trans. Commun. Technol.*, vol. COM-19, pp. 751–772, Oct. 1971.
[37] J. A. Heller and I. M. Jacobs, "Viterbi decoding for satellite and space communications," *IEEE Trans. Commun. Technol.*, vol. COM-19, pp. 835–848, Oct. 1971.
[38] R. W. Lucky, "Automatic equalization for digital communication," *Bell Syst. Tech. J.*, vol. 44, pp. 547–588, Apr. 1965.
[39] ——, "Techniques for adaptive equalization of digital communication systems," *Bell Syst. Tech. J.*, vol. 45, pp. 255–286, Feb. 1966.
[40] M. J. DiToro, "Communication in time-frequency spread media using adaptive equalization," *Proc. IEEE*, vol. 56, pp. 1653–1679, Oct. 1968.
[41] A. Gersho, "Adaptive equalization of highly dispersive channels for data transmission," *Bell Syst. Tech. J.*, vol. 48, pp. 55–70, Jan. 1969.
[42] J. G. Proakis and J. H. Miller, "An adaptive receiver for digital signaling through channels with intersymbol interference," *IEEE Trans. Inform. Theory*, vol. IT-15, pp. 484–497, July 1969.
[43] D. Hirsch and W. J. Wolf, "A simple adaptive equalizer for efficient data transmission," *IEEE Trans. Commun. Technol.*, vol. COM-18, pp. 5–11, Feb. 1970.
[44] K. Möhrmann, "Einige Verfahren zur adaptiven Einstellung von Entzerrern für die schnelle Datenübertragung," *Nachrichtentech. Z.*, vol. 24, pp. 18–24, Jan. 1971.
[45] G. Ungerboeck, "Theory on the speed of convergence in adaptive equalizers for digital communication," *IBM J. Res. Develop.*, vol. 16, pp. 546–555, Nov. 1972.
[46] H. Robbins and S. Monro, "A stochastic approximation method," *Ann. Math. Stat.*, pp. 400–407, 1951.

★

PHOTO NOT AVAILABLE

Gottfried Ungerboeck was born in Vienna, Austria, in 1940. He received the Dipl.Ing. degree in telecommunications from the Technische Hochschule, Vienna, in 1964 and the Ph.D. degree from the Swiss Federal Institute of Technology, Zurich, Switzerland, in 1970.

After graduating from the Technische Hochschule, he joined the technical staff of the Wiener Schwachstromwerke, Vienna. In 1967 he became a member of the IBM Zurich Research Laboratory, Rüschlikon, Switzerland. He worked in speech processing, data switching, and data transmission theory. His present interests cover signal processing and detection and estimation theory.

Error Probability in the Presence of Intersymbol Interference and Additive Noise for Multilevel Digital Signals

SERGIO BENEDETTO, GIROLAMO DE VINCENTIIS, AND ANGELO LUVISON

Abstract—A new method is presented to compute the average probability of error in the presence of intersymbol interference and additive noise for multilevel pulse-amplitude-modulation (PAM) and partial-response-coded (PRC) signaling schemes.

The method is based upon nonclassical Gauss quadrature rules (GQR) and suffers no limitation on noise statistics, so that it applies also for non-Gaussian noise.

Moreover it yields some remarkable advantages as compared with other methods, in particular, with the series expansion method that has recently received considerable attention.

Expressions for the truncation error are also given and their derivation is reported in the Appendix.

Finally, examples of applications are presented and comparisons with other methods are carried out.

I. Introduction

THIS paper deals with the evaluation of the average probability of error in multilevel pulse-amplitude-modulation (PAM) and partial-response-coded (PRC) data transmission systems in the presence of both intersymbol interference and additive noise.

As this problem is one of the most important in the determination of the system performance, it has received considerable attention by the researchers in the area of digital communication theory.

The first obvious approach (exhaustive method) is to consider a truncated M-pulse-train approximation of the real channel. The evaluation of the probability of error is performed by computing the conditional error probability for each of the L^M possible sequences of data, L being the number of levels, and then averaging over all the sequences. However the total number of such sequences is limited by the computer time needed to perform the average. As a consequence only few interfering samples can be taken into account and the approximation of the true channel becomes very poor, especially when dealing with multilevel systems.

When the additive noise is Gaussian, some authors evaluate an upper bound to the error probability by either the worst case sequence [1] or the Chernoff inequality [2], [3]; other authors [4], [5], [7] compute the error probability for binary and multilevel systems by means of a Hermite polynomials series expansion, or, only for binary systems, by a Gram–Charlier expansion [6], [8].

All these methods suffer some disadvantages: 1) the former gives bounds that are often too loose; 2) the latter provides oscillating results when the channel distortion, or the signal-to-noise ratio, or the number of levels increase, in spite of the absolute convergence of the series that has been theoretically proved.

In this paper we present a new method to evaluate the average probability of error within any desired accuracy. The proposed procedure is based on nonclassical Gauss quadrature rules (GQR) and suffers no limitation on noise statistics, so that it applies as well to non-Gaussian noise.

Moreover it always assures accurate results and very satisfactory performances even when other methods fail [4], [7]. The computer time required in the numerical evaluation is shorter, especially when the error probability is computed for many values of the signal-to-noise ratio.

In Section II the problem is stated and a general expression for the error probability is derived. In Section III the method to compute the probability of error is explained, together with a brief description of GQR's. In Section IV the truncation error due to the finite number of points in the quadrature rule is analyzed and an upper bound for the error is found. A discussion about the roundoff errors is also given. Finally, in Section V, the method is applied to some particular examples of additive Gaussian noise channels.

II. Statement of the Problem

The received signal at the input to the decision device for an L-level digital PAM system is well known to be

$$y(t) = \sum_{h=-\infty}^{+\infty} a_h r(t - hT) + n(t), \tag{1}$$

where $n(t)$ is a noise process, $r(t)$ the impulse response of the overall time-invariant linear system, T the signaling period, and the sequence of a_h the random stream of symbols that can assume the values

$$\pm d, \pm 3d, \pm 5d, \cdots, \pm(L-1)\,d \qquad (L \text{ even}) \tag{2}$$

with probabilities

$$p_k \triangleq K\{a_h = [2K - \operatorname{sgn}(k)]\,d\}, \quad -L/2 \leqslant k \leqslant L/2,\ k \neq 0 \tag{3}$$

Paper approved by the Data Communications Committee of the IEEE Communications Society for publication without oral presentation. Manuscript received February 3, 1972.

S. Benedetto is with the Istituto di Elettronica e Telecomunicazioni del Politecnico di Torino, Turin, Italy.

G. DeVincentiis and A. Luvison are with the Centro Studi e Laboratori Telecomunicazioni (CSELT), Turin, Italy.

Reprinted from *IEEE Transactions on Communications*, vol. COM-21, no. 3, March 1973.

The Best of the Best. Edited by W. H. Tranter, D. P. Taylor, R. E. Ziemer, N. F. Maxemchuk, and J. W. Mark.

such that

$$p_k = p_{-k}. \tag{4}$$

At the detector, $y(t)$ is sampled every T seconds to determine the amplitude of the transmitted symbols. At the sampling time t_0, the signal can be written in the form

$$y_0 = a_0 r_0 + \sum_{h=-\infty}^{+\infty}{}' a_h r_h + n_0, \tag{5}$$

where

$$\begin{aligned} y_0 &= y(t_0) \\ r_0 &= r(t_0) \\ r_h &= r(t_0 - hT) \\ n_0 &= n(t_0) \end{aligned} \tag{6}$$

and Σ' does not include the term $h = 0$.

The first term in (5) is the desired signal, whereas the second and the third terms represent the intersymbol interference and the noise, respectively. The set of the slicing levels at the detector are

$$-(L-2)\,dr_0, \cdots, -2dr_0, 0, 2dr_0, \cdots, (L-2)\,dr_0. \tag{7}$$

Let us assume that

$$X = \sum_{h=-\infty}^{+\infty}{}' a_h r_h \tag{8}$$

is a random variable; so the average error probability has the following expression [7]:

$$\begin{aligned} P(e) &= \sum_{k=-L/2}^{L/2}{}' p_k P\{|X+n_0| > dr_0\} - p_{L/2} P\{X+n_0 > dr_0\} \\ &\qquad - p_{-L/2} P\{X+n_0 < -dr_0\} \\ &= P\{|X+n_0| > dr_0\} - p_{L/2} P\{X+n_0 > dr_0\} \\ &\qquad - p_{-L/2} P\{X+n_0 < -dr_0\}. \end{aligned} \tag{9}$$

Because of the hypothesis in (4), the probability density function (pdf) of X is an even function. Also, if the pdf of the noise sample n_0 is even, (9) becomes

$$P(e) = 2(1 - p_{L/2}) P\{X + n_0 > dr_0\}. \tag{10}$$

Denoting with $D(\cdot)$ the distribution function of the random variable (RV) n_0, and observing that

$$\begin{aligned} P\{X+n_0 > dr_0\} &= E[P\{x+n_0 > dr_0 \mid X = x\}] \\ &= E[1 - D(dr_0 - x)] \end{aligned} \tag{11}$$

the average error probability assumes the following form

$$P(e) = 2(1 - p_{L/2}) E[1 - D(dr_0 - x)]. \tag{12}$$

We suppose henceforth that the impulse response has only a finite number M of interfering samples, that is

$$r_h = 0, \quad h < -p, \; h > q, \; p + q = M. \tag{13}$$

This is a reasonable approximation for any actual channel; moreover it allows us to avoid the mathematical difficulties arising from the problem of the convergence of a series of RV's to an RV (see [4] or [6] for a discussion about this problem).

As a matter of fact, under this hypothesis, the pdf $w(\cdot)$ of the RV X always exists, being a finite sum of the kind

$$w(x) = \sum_{i=1}^{L^M} q_i \delta(x - x_i) \tag{14}$$

where $\delta(\cdot)$ is the Dirac delta function and x_i the discrete values taken by X, and $q_i = P\{X = x_i\}$. Thus (12) can be represented formally in two ways: with the aid of generalized functions, which are involved in (14), we have

$$P(e) = 2(1 - p_{L/2}) \int_{\mathfrak{X}} [1 - D(dr_0 - x)]\, w(x)\, dx, \tag{15}$$

where $[\mathfrak{X}]$ is the range of the RV X; or by introducing the Stieltjes integral we are led to write

$$P(e) = 2(1 - p_{L/2}) \int_{\mathfrak{X}} [1 - D(dr_0 - x)]\, dF(x). \tag{16}$$

The last representation is based on the fact that $F(x)$ is the distribution function of the RV X, thus having bounded variations in $[\mathfrak{X}]$.

In Section III we shall give a method to compute $P(e)$, through (15) or (16), overcoming the difficulties due to the fact that the pdf of X cannot be known explicitly unless a direct enumeration of all possible sequences is performed.

III. Computation of the Probability of Error

Equation (15) or (16) shows that the problem to be solved is the averaging of $[1 - D(dr_0 - x)]$ with respect to the RV X whose pdf is unknown.

In the case of Gaussian noise, several authors expanded the error probability into a series approximation in terms of the moments of X and of Hermite polynomials for multilevel signaling schemes [5], [7], or Hermite functions for binary signaling schemes [6], [8].

These techniques provide values of the probability of error more accurate than the Chernoff bound. Moreover the computation is much faster than using the exhaustive method. Unfortunately, the series expansions behave critically, that is, they provide oscillating results, when either the signal-to-noise ratio or the intersymbol interference increases.

Following a different approach, we have computed the average in (15) or (16) by means of GQR's [9]-[11], which guarantee, in the area of approximate integration, the highest degree of precision.

We are concerned with an integral of the form

$$\int_a^b f(x)\, \omega(x)\, dx, \tag{17}$$

where $\omega(x)$ is a "weight" function (maybe unknown), $f(x)$ an arbitrary function of some wide class, and $[a, b]$ any finite or infinite interval. In order to compute numerically the integral

(17), it is sufficient to assume that the weight function satisfies these conditions: 1) $\omega(x)$ is nonnegative, integrable in $[a, b]$, with

$$\int_a^b \omega(x)\, dx > 0; \tag{18}$$

2) the products $x^k \omega(x)$, for any nonnegative integer k, are such that the integrals

$$\int_a^b |x|^k \omega(x)\, dx \tag{19}$$

are definite and finite.

Though approximation theory makes good use of several wider classes of functions, we shall only consider the class of functions $f(x)$ that have $2m$ continuous derivatives in the interval $[a, b]$, that is $f(x) \in C^{2m}[a, b]$.

The pdf $w(x)$ in (14) satisfies the previously stated conditions 1) and 2), so we shall use in the following a representation of the integrals as in (15), though the Stieltjes integral notation could also be used [9, pp. 15-16] for a completely equivalent approach.

A. Gauss Quadrature Rules Approach

The most widely investigated method to approximate the integral (17) uses a linear combination of values of the function $f(x)$, i.e.,

$$\int_a^b f(x)\, \omega(x)\, dx \simeq \sum_{i=1}^{m} w_i f(x_i). \tag{20}$$

It is usually said that this quadrature rule has degree of precision n if it is exact whenever $f(x)$ is a polynomial of degree $\leqslant n$.

The x_i are called the abscissas of the formula and the w_i the coefficients or weights, so the set $\{w_i, x_i\}_{i=1}^m$ is called a quadrature rule corresponding to the weight function $\omega(x)$.

If $\omega(x)$ satisfies conditions 1) and 2), previously stated, then m abscissas and weights can be found to make (20) exact for all polynomials of degree $n \leqslant 2m - 1$; this is the highest degree of precision that can be obtained using m points.

As such formulas were originally considered by Gauss, a quadrature rule like (20) is called a GQR if it is an exact equality whenever $f(x)$ is a polynomial of degree $2m - 1$ or lower.

Beyond the advantages of accuracy, GQR's do not require an explicit knowledge of the weight function. In fact, as it will be shown later on, a rule $\{w_i, x_i\}_{i=1}^m$ can be obtained from the moments of $\omega(x)$ so defined

$$\int_a^b x^k \omega(x)\, dx, \qquad k = 0, 1, \cdots, 2m \tag{21}$$

and this fact is of great practical usefulness. Therefore, the moments' accuracy represents the key point of this computational approach.

A systematic introduction to the theory of Gauss quadrature formulas is given in Krylov [10]; in the following, we shall sketch the essential properties that are used in the construction of GQR's, with particular regard to the case of unknown weight functions.

Let $\omega(x)$ satisfy conditions 1) and 2). For $\omega(x)$ it is possible to define a sequence of polynomials $\{P_n(x)\}$ that are orthonormal with respect to $\omega(x)$ and in which $P_n(x)$ is of exact degree n so that

$$\int_a^b P_n(x) P_j(x)\, \omega(x)\, dx = \delta_{nj}. \tag{22}$$

The polynomial

$$P_n(x) = K_n \prod_{i=1}^{n} (x - x_i), \qquad K_n > 0$$

has n real roots x_i located in the interval $[a, b]$.

The roots of the orthogonal polynomials play an important role in GQR's. In fact, the following theorem holds [11, p. 35].

Theorem: Let $f(x) \in C^{2m}[a, b]$, then

$$\int_a^b f(x)\, \omega(x)\, dx = \sum_{i=1}^{m} w_i f(x_i) + \frac{f^{(2m)}(\xi)}{K_m^2 (2m)!}, \qquad (a \leqslant \xi \leqslant b) \tag{23}$$

where

$$w_i = -\frac{K_{m+1}}{K_m} \cdot \frac{1}{P_{m+1}(x_i) P'_m(x_i)}, \qquad P'_m(x_i) = \left. \frac{dP_m(x)}{dx} \right|_{x=x_i}, \qquad i = 1, \cdots, m. \tag{24}$$

Moreover it can be proved that each set of orthonormal polynomials satisfies the following three-term recurrence relationship

$$x P_{n-1}(x) = \beta_{n-1} P_{n-2}(x) + \alpha_n P_{n-1}(x) + \beta_n P_n(x), \qquad n = 1, 2, \cdots, m, \tag{25}$$

where $P_{-1}(x) \equiv 0$ and $P_0(x) \equiv 1$.

By observing that the remainder of (23) is zero for all polynomials of degree $\leqslant 2m - 1$, we conclude that (23) is a GQR.

Quadrature rules $\{w_i, x_i\}_{i=1}^m$ have several important convergence properties; in fact these formulas converge to the true value of the integral for almost any conceivable function that can be met in practice [9, p. 15], even if $f(x) \notin C^{2m}[a, b]$; namely

$$\lim_{m \to \infty} \sum_{i=1}^{m} w_i f(x_i) = \int_a^b f(x)\, \omega(x)\, dx. \tag{26}$$

Hopefully, the quadrature rule $\{w_i, x_i\}_{i=1}^m$ corresponding to the weight function $\omega(x)$ is available in tabulated form, but more likely it is not; so the constructive aspect of the formulas becomes very important. Several algorithms have been proposed in order to compute $\{w_i, x_i\}_{i=1}^m$. The procedure generally recommended consists in generating the set of orthonormal polynomials associated with the weight function, thus obtaining x_i as the zeros of $P_m(x)$, and w_i starting from these orthonormal polynomials.

An alternative approach, suggested by Golub and Welsch [12], performs the computation of $\{w_i, x_i\}_{i=1}^m$ using the first $2m + 1$ moments of $\omega(x)$. This approach seemed particularly suitable for practical purposes, because it leaves out of consideration the knowledge of $\omega(x)$. In the rest of the subsection we shall briefly outline this method; the development of the algorithm, the mathematical details and proofs are reported in [12].

Basically, the algorithm consists of two steps: 1) evaluation of the coefficients $\{\alpha_n\}$ and $\{\beta_n\}$ of the three-term recurrence relationship (25) by means of the first $2m + 1$ moments of $\omega(x)$ defined in (21) and 2) generation of a symmetric tridiagonal matrix, whose elements depend on the coefficients $\{\alpha_n\}$ and $\{\beta_n\}$. The weights w_i are found as the first components of the eigenvectors of this matrix, whereas the abscissas x_i are the corresponding eigenvalues.

B. Application to the Evaluation of the Error Probability

The previously described procedure fits well into our problem, if we identify the weight function $\omega(x)$ with the pdf $w(x)$ of the RV X. In fact the average error probability (15) can be computed as

$$P(e) \simeq 2(1 - p_{L/2}) \sum_{i=1}^{N} w_i [1 - D(dr_0 - x_i)] \tag{27}$$

$\{w_i, x_i\}_{i=1}^N$ being an N-point GQR. A sequence of formulas similar to (27) does converge to the exact value of $P(e)$, when $N \to \infty$, except for the roundoff errors.

For the particular case of Gaussian noise with variance σ_n^2, (27) becomes

$$P(e) \simeq (1 - p_{L/2}) \sum_{i=1}^{N} w_i \operatorname{erfc}\left(\frac{dr_0}{\sqrt{2}\,\sigma_n} - \frac{x_i}{\sqrt{2}\,\sigma_n}\right). \tag{28}$$

We must emphasize that this technique is independent of the noise statistics, whereas the methods up to the presently known [4]-[8] are strictly based on the "Gaussianity" of the noise process.

So the proposed approach seems to be very powerful, even in the presence of other types of disturbances, e.g., phase jitter [13] or channels with fading, and when dealing with different systems of data transmission, for which the conditional error probability is easily written but cannot be expanded in series as in [4]-[8].

Moreover the method just described for multilevel PAM transmission systems, applies as well in the case of PRC signaling schemes [14], [15].

Some cautions must be used in the interpretation of (12), as it was explained in [7] for the series expansion method, namely: 1) the term $p_{L/2}$ is now the probability of the outermost level we can have before the detector; for instance, in the case of class-4 PRC systems, we have $p_{L/2} = 1/L^2$ if the source symbols are equally likely and 2) the vector r of the interfering impulse response samples

$$\{r_i\}_{\substack{i=-p\\ i\neq 0}}^{q}$$

must be modified according to the class of PRC, because we must take into account only the undesired intersymbol interference.

In Table I we give the expressions of $p_{L/2}$ and r for all the classes of PRC when the source symbols are equally likely. With the aid of this table one can evaluate the error probability for both PAM and PRC systems by means of the same computational procedure.

C. Evaluation of the Moments

In the preceding section we showed how the error probability can be evaluated by an N-point GQR if the first $2N + 1$ moments of the RV X, which represents the intersymbol interference, are known.

Then our aim is the accurate evaluation of the moments

$$M_k = E[X^k] = \int_{\mathfrak{X}} x^k w(x)\, dx \tag{29}$$

without resorting to the trivial direct enumeration, which requires too much computer time.

We shall describe two methods that allow a recurrent evaluation of the moments. The first, described in detail in [7], leads to the following expression

$$M_{2k} = \sum_{j=1}^{k} \binom{2k-1}{2j-1} (-1)^j M_{2(k-j)} f^{(2j-1)}(0), \tag{30}$$

where $M_0 = 1$, $k = 1, 2, \cdots, N$ and

$$f^{(2j-1)}(0) = (2j-1)! \sum_{i=-p}^{q}{}' C_{2j-1}^{(i)}, \tag{31}$$

where

$$C_{2j-1}^{(i)} = (-1)^j (dr_i)^{2j} \frac{1}{(2j-1)!} \sum_{n=1}^{L/2} 2p_n(2n-1)^{2j} - \sum_{k=1}^{j-1} \cdot C_{2j-2k-1}^{(i)} \cdot (-1)^k \cdot (dr_i)^{2k} \cdot \frac{1}{(2k)!} \sum_{m=1}^{L/2} 2p_m(2m-1)^{2k}. \tag{32}$$

Since the source symbols a_h are independently distributed RV's with zero mean, the odd order moments are all equal to zero.

When the source symbols are equally likely, (30)-(32) reduce easily to the following formula [5]

$$M_{2k} = -\sum_{i=1}^{k} \binom{2k-1}{2i-1} M_{2(k-i)} \cdot (-1)^i \frac{2^{2i}[L^{2i}-1]}{2i} |B_{2i}| \cdot \sum_{h=-p}^{q}{}' (dr_h)^{2i}, \tag{33}$$

where B_{2i} are the Bernoulli numbers.

TABLE I

Signal	$P_{L/2}$	Intersymbol vector $\underline{r}$
CLASS I PRC	$\frac{1}{L^2}$	$\underline{r} = \{r_{-p}, \ldots, r_{-1}, r_1 - r_0, r_2, \ldots, r_q\}$
CLASS II PRC	$\frac{1}{L(2L-1)}$	$\underline{r} = \{r_{-p}, \ldots, r_{-1} - r_0/2, r_1 - r_0/2, \ldots, r_q\}$
CLASS III PRC	$\frac{1}{L(2L-1)}$	$\underline{r} = \{r_{-p}, \ldots, r_{-1} - 2r_0, r_1 + r_0, \ldots, r_q\}$
CLASS IV PRC	$\frac{1}{L^2}$	$\underline{r} = \{r_{-p}, \ldots, r_{-1}, r_1, r_2 + r_0, \ldots, r_q\}$
CLASS V PRC	$\frac{1}{L(2L-1)}$	$\underline{r} = \{r_{-p}, \ldots, r_{-1} + r_0/2, r_1 + r_0/2, \ldots, r_q\}$
PAM	$\frac{1}{L}$	$\underline{r} = \{r_{-p}, \ldots, r_{-1}, r_1, \ldots, r_q\}$

Intersymbol vector according to the coded signal class of partial response.

The second method was proposed by Prabhu [16]. Let us number from 1 to M the M interfering samples that are significantly different from zero, so that

$$X = \sum_{h=1}^{M} a_h r_h = \sum_{h=1}^{M} X_h. \tag{34}$$

If we let

$$Y_n = \sum_{h=1}^{n} X_h \tag{35}$$

the even order moments of the RV X are given by

$$M_{2k} = \sum_{i=0}^{k} \binom{2k}{2i} E[Y_{M-1}^{2i}]\, E[X_M^{2k-2i}], \quad k = 0, 1, \cdots, N \tag{36}$$

because of the statistical independence of Y_{M-1} and X_M. Equation (36) allows us to evaluate, by recurrence, the moments of X through the knowledge of the moments

$$E[X_h^{2k}], \quad h = 1, 2, \cdots, M, \quad k = 0, 1, \cdots, N \tag{37}$$

and these are given by [see (3)]

$$E[X_h^{2k}] = (dr_h)^{2k} \sum_{i=-L/2}^{L/2}{}' \; p_i [2i - \operatorname{sgn}(i)]^{2k}. \tag{38}$$

IV. Estimation of Computational Errors

In the numerical evaluation of the probability of error using (27) three different types of error occur, two of them due to truncation and one to roundoff.

In this section we give an upper bound to the error depending on truncation of the quadrature rule, and briefly discuss the other two types of errors.

A. Truncation Errors

Two different truncation errors would have to be taken into account: the first is the error due to the truncation of the impulse response and the second is the error involved in the quadrature formula (27).

An analysis of the first error is given in [8]. Moreover it is possible to determine how many samples significantly contribute to the error probability within a preassigned amount by a preliminary computer run. Another way is to increase the number of the interfering samples until $P(e)$ stops changing.

The second truncation error is analyzed in the Appendix for the case of Gaussian noise. Therefore the error probability can be evaluated by taking a finite number of terms

$$P(e) = (1 - P_{L/2}) \sum_{i=1}^{N} w_i \operatorname{erfc}\left(\frac{dr_0}{\sqrt{2}\,\sigma_n} - \frac{x_i}{\sqrt{2}\,\sigma_n}\right) + R_N, \tag{39}$$

where R_N represents the truncation error and is upper bounded by

$$\begin{cases} |R_N| < A \exp\left[-\dfrac{(d|r_0| - \max\{\xi\})^2}{4\sigma_n^2}\right], & \max\{\xi\} < d|r_0|;\ \xi \in [\mathfrak{X}] \\ |R_N| < A, & \max\{\xi\} \geqslant d|r_0|, \end{cases} \tag{40}$$

where

$$A = \frac{B[(2N-1)!]^{1/2}\, 2^{N+\frac{1}{2}}\, (1 - p_{L/2})}{(\pi)^{1/2}\, (2N)!\, K_N^2 (2\sigma_n^2)^N}, \quad B \simeq 1.086\,435.$$

It must be noticed that this bound is useful only if it is sufficiently tight, which in some cases does not happen. In these cases, however, there is a very effective way usually followed in numerical computation to judge the accuracy of the obtained result. In fact, since convergence is assured to the true value of $P(e)$, it is sufficient to check how many significant digits remain unchanged as N increases and to continue the iteration until the desired accuracy. In all cases we tried, no oscillation was observed and the convergence was obtained, at least in the first three significant digits.

Also the bound to the truncation error in the series expansion method sometimes suffers the same drawback, but in that case the result can oscillate because of roundoff errors, when the convergence to the true value is not reached with a low number of terms; thus any reasonable stopping criterion is prevented.

Examples of the truncation error bound and of the other way to check the accuracy will be given in Section V.

B. Roundoff Error

In the evaluation of the right-hand side of (27) the limited accuracy of the digital computer introduces also a roundoff error. This error is usually negligible in Gauss quadrature formulas of low order, especially with regard to the evaluation of the probability of error, whose accuracy is satisfactory when 2 or 3 digits are correctly known.

On the other hand the high speed of the present computers allows one to use high-order formulas, thus making significant the roundoff error. In order to reduce this drawback, one may resort to double precision computation, evaluating integer quantities in integer arithmetic, as far as possible.

The knowledge of the moments with high precision is an important condition to further improve the accuracy in the computation of w_i and x_i. To this purpose, the computational method proposed by Prabhu seems to be quite satisfactory, since relationship (36) involves the summation of positive quantities only.

A mathematical discussion of the roundoff error effects is given in [17]; it is shown that, roughly speaking, the lower is the quantity $\min_i \{w_i\}$, the more the GQR is sensitive to the loss of accuracy with increasing N. So the values of the coefficients give a significant insight into the overall precision.

Summarizing, the main limiting factor is due to the values of the moments that are affected by roundoff errors because they are computed by recurrence formulas.

Due to this fact, the coefficients $\{\alpha_i\}$ and $\{\beta_i\}$ of (25) may fail to be computed from the moments, when N becomes large. Nevertheless our algorithm, though applied in many cases different from those analyzed in Section V [13], [18], [19], never failed. In fact, GQR's always converged for N ranging between 6 and 15.

It may be noticed that the roundoff error does not happen only in our method of computing $P(e)$. In fact, the same arguments apply to any series expansion method, where each term of the approximation is obtained by multiplying a value of a Hermite polynomial (or function) obtained recurrently by a term (such as a moment) that is computed iteratively as well. Moreover the terms in the series expansion may be very large and have opposite signs, whereas the approximation (28) uses only positive and very regular quantities.

When particular attention should be devoted to the roundoff error in special applications, the proposed procedure could be modified following two ways: 1) the first lies in the field of numerical differentiation of analytic functions [20], basing the kth moment on the corresponding derivative of the characteristic function of the RV X; 2) the second one consists in applying a recent procedure [21], which allows one to construct GQR's starting from "modified moments," such as

$$m_k = \int_{\mathcal{X}} P_k(x)\, w(x)\, dx,$$

where $P_k(x)$ belongs to a set of orthogonal polynomials suitably chosen. Extensions of the proposed method to evaluate $P(e)$ following these ways seem possible; however such possibilities have not yet been pursued.

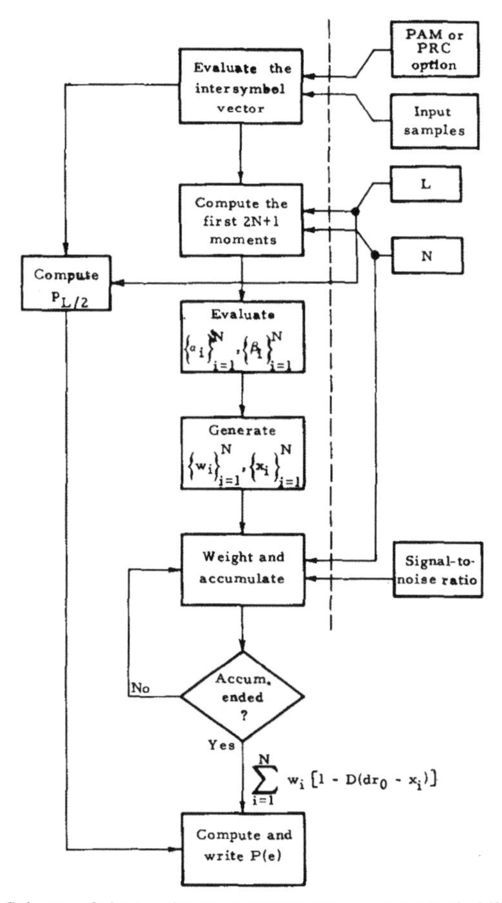

Fig. 1. Scheme of the routine to compute average error probability $P(e)$.

V. Examples and Comparisons

The examples reported hereafter have been chosen to show the differences between the method of GQR's and the series expansion rules [4], [5], [7]. When it was possible, the exhaustive method was also used as a basis of comparison. The noise process was supposed to be Gaussian. The computational procedure requires the following steps: 1) evaluation of the first $2N+1$ moments $\{M_k\}_{k=0}^{2N}$ using (30)–(33) or (36) and (38) starting from the vector $\boldsymbol{r}$ of the interfering samples; 2) determination of the three-term recurrence relationship through the coefficients $\{\alpha_i\}_{i=1}^N$ and $\{\beta_i\}_{i=1}^N$ that appear in (25); 3) generation of the weights $\{w_i\}_{i=1}^N$ and abscissas $\{x_i\}_{i=1}^N$ of the GQR by means of the numerical algorithm described in Section III-A; and 4) evaluation of the average error probability through (28). A block diagram of the preceding outlined procedure is drawn in Fig. 1.

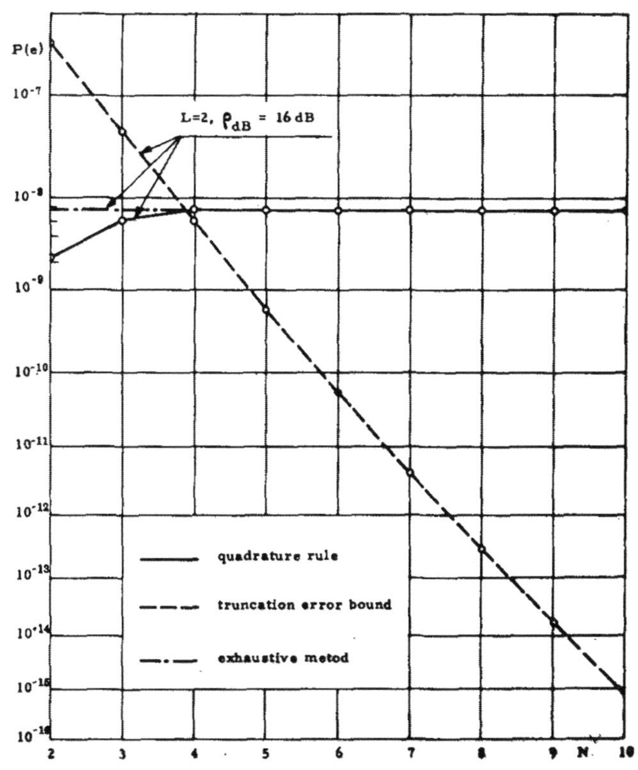

Fig. 2. Error probability and upper bound to truncation error versus number N of points of the quadrature formula. 11-pulse truncation approximation. Ideal band-limited signal of (41). Sampling time deviation $0.05T$. Exhaustive-method probability of error is also reported.

A. Ideal Band-Limited Pulse

We consider the case of binary PAM transmission when the received pulse is assumed to have the form

$$r(t) = \frac{\sin(\pi t/T)}{\pi t/T}. \tag{41}$$

For a truncated 11-pulse train approximation the exhaustive-method error probability, the error probability computed with our method and the bound (40) to the truncation error are shown in Fig. 2 as a function of the number N of points of the quadrature formula. The sampling time deviation from the nominal value is $0.05T$ and the signal-to-noise ratio is taken to be 16 dB. In this subsection and in the following, the number of interfering samples has been chosen small enough to allow the exhaustive method to compute the exact probability of error, for the sake of comparison.

In Fig. 3 the exact error probability, the error probability evaluated either with the series expansion method or with GQR are reported for a sampling time deviation of $0.2T$. N_i is the number of terms of the series expansion and N is the order of the GQR according to (28). The curve giving the results of the series expansion method ends with the eighth term because the successive approximation of nine terms gives a negative value for $P(e)$.

In order to check the accuracy and the convergence, the numerical values of $P(e)$ are reported in Table II.

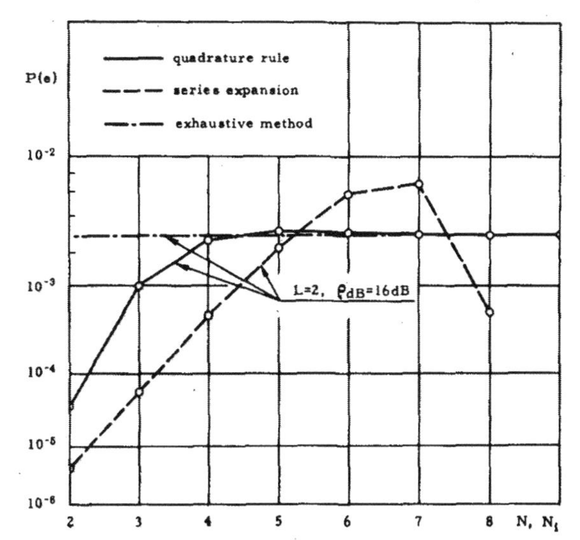

Fig. 3. Comparison of error probabilities versus number N of quadrature points or number N_i of terms of the series expansion. 11-pulse truncation approximation of ideal band-limited signal. Sampling time deviation $0.2T$. Exhaustive-method error probability is also reported.

TABLE II

Exhaustive Method P(e) = 2.7614 (-3)					
N	Quadrature Rule P(e)		Series Expansion P(e)		N_i
2	3.77	(-5)	4.20	(-6)	2
3	8.86	(-4)	5.98	(-5)	3
4	2.4	(-3)	4.71	(-4)	4
5	2.9	(-3)	2.14	(-3)	5
6	2.8	(-3)	5.46	(-3)	6
7	2.74	(-3)	6.56	(-3)	7
8	2.75	(-3)	5.06	(-4)	8
9	2.766	(-3)	—		9
10	2.7617	(-3)	—		10
11	2.7615	(-3)	—		11
12	2.76164	(-3)	—		12
13	2.761639	(-3)	—		13
14	2.761636	(-3)	—		14

Numerical values of the probabilities of error obtained with the Gauss quadrature rule, the series expansion, and the exhaustive methods. 11-pulse truncation approximation of the ideal band-limited signal. Sampling time deviation $0.2T$.

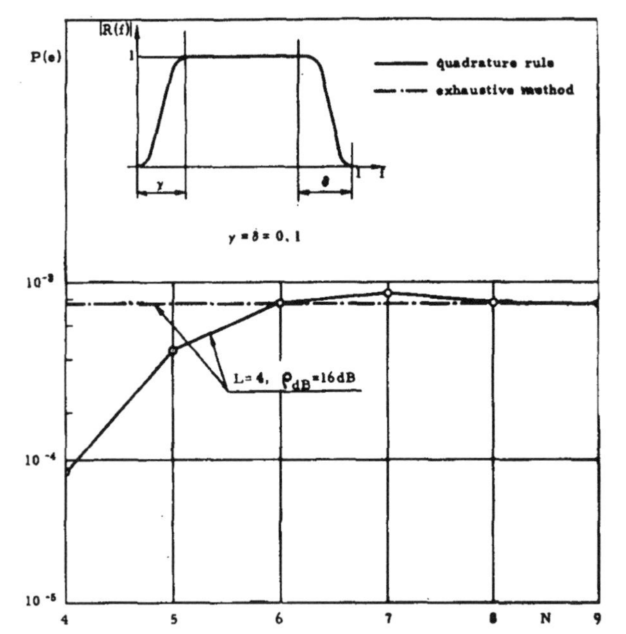

Fig. 4. Error probability versus number N of quadrature points. 9-pulse truncation approximation of PRC signal of (42). Exhaustive-method error probability is also reported.

B. A 4-Level PRC System

In Fig. 4 the results of the computation of $P(e)$ in a class-4 PRC system with four levels by both exhaustive and GQR methods are presented for a truncated 9-pulse train approximation of the following impulse response:

$$r(t) = \frac{\sin(2\pi t)\cos(2\pi\gamma t)}{2\pi t[1-(4\gamma t)^2]} - \frac{\sin(2\pi\delta t)\cos(2\pi\delta t)}{2\pi t[1-(4\delta t)^2]}, \qquad (42)$$

where the parameters γ and δ have the meaning described in the same figure. The signal-to-noise ratio is chosen equal to 19 dB.

The curves are very significant, and show that the GQR method behaves very well; in fact the convergence to the exact value is fast and accurate (5 correct significant digits).

The exhaustive method, used to evaluate the exact value of $P(e)$, required about 2 h of computer time, while the GQR method needed just a few seconds. In this case we found that the series expansion method [5], [7] behaved critically giving very inaccurate results.

C. An Equalized Channel

The method described in this paper has been used extensively to evaluate the performance of multilevel class-4 PRC data transmission systems with a 16-tap zero forcing automatic equalizer.

In this case, it was necessary to consider a truncated impulse response of 60 samples, so the exhaustive method was inapplicable because of the required computer time. Thus, we restricted our comparisons to the series approximation and GQR methods.

In Fig. 5 the case of an 8-level system is reported, having a

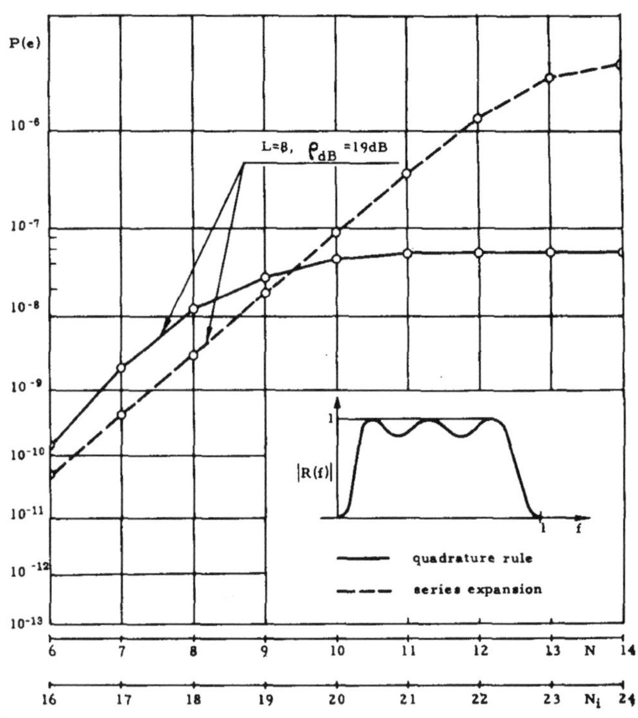

Fig. 5. Comparison of error probabilities versus number N of quadrature points and of terms N_i of series expansion. 8-level source symbols are class-4 PRC and passed through the channel of the figure.

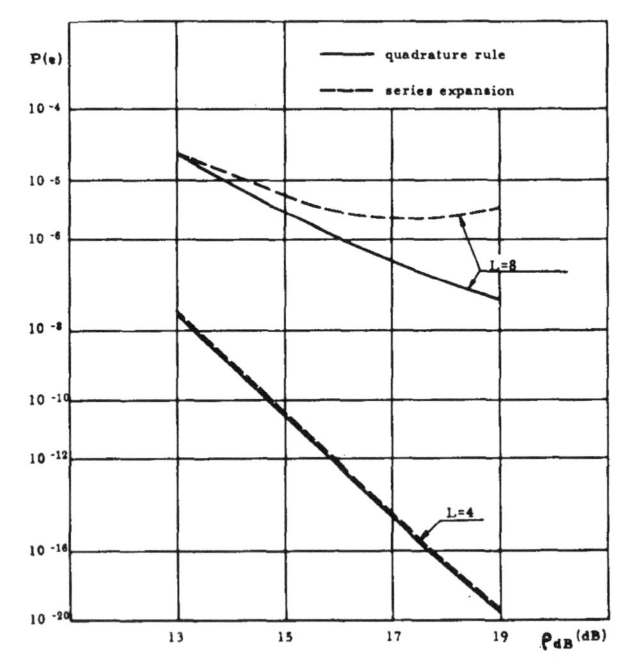

Fig. 6. Comparison of error probabilities versus signal-to-noise ratio, with number of levels as a parameter. Channel is the same as Fig. 4.

signal-to-noise ratio equal to 19 dB. The curves show that the GQR method converges to a final value without oscillations, while the series approximation does not, and it gives a negative value for $P(e)$ at the 25th iteration.

The channel we are considering presents an amplitude characteristic similar to that drawn in the same figure, and a parabolic group delay. This kind of channel model is described in detail in [18].

In Fig. 6 an example of the curves of $P(e)$ versus the signal-to-noise ratio is presented, in the case of 4- and 8-level class-4 PRC systems. It is interesting to observe that the curves of $P(e)$ given by these two methods coincide in the case of $L = 4$. On the opposite, with $L = 8$, the values of $P(e)$ are almost the same only for small values of the signal-to-noise ratio, while they diverge as the signal-to-noise ratio increases. These results agree with the statements reported in the Introduction.

It must be noticed also that a remarkable saving of computer time may be achieved using the GQR method, when the error probability has to be computed for many values of the signal-to-noise ratio (as in this case). In fact the first three steps in the computational procedure are the same for all the values of the signal-to-noise ratio and therefore are performed only once.

VI. Conclusions

A new method has been presented to compute the average probability of error for multilevel PAM and PRC data transmission systems with intersymbol interference and additive noise.

The method is based upon nonclassical GQR's and suffers no limitation on noise statistics.

After the explanation of the computational technique and the extension to the PRC case, some examples of application have been given either for binary PAM or multilevel PRC signaling schemes.

The results were quite satisfactory even when the other known methods were found to fail, either for the required computer time (exhaustive method) or because the obtained results were oscillating (series expansion method).

The GQR algorithm applied to the computation of $P(e)$ has been proved to be rapidly convergent in all practical cases that were considered.

Finally it must be noticed that the proposed algorithm can be applied iteratively to construct GQR's. So the integration with respect to other random variables, whose moments are known, may be performed. This technique has been used, for instance, to compute the error probability in the presence of a random phase jitter [13] where a further integration over the corresponding random variable had to be carried out. Moreover the method does not apply only to baseband pulse transmission, but it can be extended to deal with modulation systems, such as the M-phase PSK.

Many of these problems are being investigated and satisfactory results are hopefully expected.

Acknowledgment

The authors gratefully acknowledge the help of L. Lambarelli in organizing the computer work and results.

Appendix

In this Appendix, we shall find an upper bound to the error that occurs in the truncation of GQR's, when the noise process is Gaussian. Rewriting (39) according to (23), we have

$$P(e) = (1 - p_{L/2}) \sum_{i=1}^{N} w_i \operatorname{erfc}\left(\frac{dr_0}{\sqrt{2}\,\sigma_n} - \frac{x_i}{\sqrt{2}\,\sigma_n}\right) + (1 - p_{L/2}) \cdot \frac{\operatorname{erfc}^{(2N)}\left(\frac{dr_0}{\sqrt{2}\,\sigma_n} - \frac{x}{\sqrt{2}\,\sigma_n}\right)\Big|_{x=\xi}}{(2N)!\,K_N^2}, \quad \xi \in [\mathfrak{X}], \tag{43}$$

where K_N is shown to be

$$K_N = \prod_{i=1}^{N} (\beta_i)^{-1} \tag{44}$$

and is easily computed during step 2) of the computational procedure described in Section V.

The error involved in truncating the quadrature rule to the first N terms is thus

$$R_N(\xi) = \frac{\operatorname{erfc}^{(2N)}\left(\frac{dr_0}{\sqrt{2}\,\sigma_n} - \frac{x}{\sqrt{2}\,\sigma_n}\right)\Big|_{x=\xi}}{(2N)!\,K_N^2} (1 - p_{L/2}). \tag{45}$$

But [22, p. 298]

$$\operatorname{erfc}^{(n)}(z) = (-1)^{n-1}\, 2(\pi)^{-1/2} H_{n-1}(z) \exp(-z^2), \tag{46}$$

where $H_{n-1}(\cdot)$ is the Hermite polynomial of order $n - 1$. Therefore

$$|R_N(\xi)| = \frac{2(1 - p_{L/2})}{(\pi)^{1/2}\,(2N)!\,K_N^2\,(\sqrt{2}\,\sigma_n)^{2N}} \cdot \left|H_{2N-1}\left(\frac{dr_0}{\sqrt{2}\,\sigma_n} - \frac{\xi}{\sqrt{2}\,\sigma_n}\right)\right| \cdot \exp\left[-\frac{(dr_0 - \xi)^2}{2\sigma_n^2}\right]. \tag{47}$$

We know that [22, p. 787]

$$|H_n(z)| < B \exp(z^2/2)\, 2^{n/2}\, (n!)^{1/2}, \qquad B \simeq 1.086\,435. \tag{48}$$

Hence, substituting (48) in (47) and letting $n = 2N - 1$, one obtains

$$|R_N(\xi)| < A \exp\left[-\frac{(dr_0 - \xi)^2}{4\sigma_n^2}\right], \tag{49}$$

where the coefficient A is given by

$$A = \frac{B[(2N-1)!]^{1/2}\, 2^{N+\frac{1}{2}}\,(1 - p_{L/2})}{(\pi)^{1/2}\,(2N)!\,K_N^2\,(2\sigma_n^2)^N} \tag{50}$$

and does not depend on ξ.

If we choose ξ in order to maximize the right-hand side of (49), we have

$$|R_N| < A \max_{\xi} \left\{\exp\left[-\frac{(dr_0 - \xi)^2}{4\sigma_n^2}\right]\right\} \tag{51}$$

with

$$\max_{\xi} \left\{\exp\left[-\frac{(dr_0 - \xi)^2}{4\sigma_n^2}\right]\right\} = \begin{cases} \exp\left[-\frac{(d\,|r_0| - \max\{\xi\})^2}{4\sigma_n^2}\right], & \max\{\xi\} < d\,|r_0|, \\ 1, & \max\{\xi\} \geq d\,|r_0| \end{cases} \tag{52}$$

and

$$\max\{\xi\} = \sum_{h=-p}^{q}{}' (L-1)\, d\, |r_h|. \qquad (53)$$

Equation (51) with (50), (52), and (53) gives the desired bound to the truncation error.

REFERENCES

[1] R. W. Lucky, J. Salz, and E. J. Weldon, *Principles of Data Communication.* New York: McGraw-Hill, 1968, p. 44.

[2] B. R. Saltzberg, "Intersymbol interference error bounds with application to ideal bandlimited signaling," *IEEE Trans. Inform. Theory*, vol. IT-14, pp. 563-568, July 1968.

[3] R. Lugannani, "Intersymbol interference and probability of error in digital systems," *IEEE Trans. Inform. Theory*, vol. IT-15, pp. 682-688, Nov. 1969.

[4] E. Y. Ho and Y. S. Yeh, "A new approach for evaluating the error probability in the presence of intersymbol interference and additive Gaussian noise," *Bell Syst. Tech. J.*, vol. 49, pp. 2249-2265, Nov. 1970.

[5] —, "Error probability of a multilevel digital system with intersymbol interference and Gaussian noise," *Bell Syst. Tech. J.*, vol. 50, pp. 1017-1023, Mar. 1971.

[6] M. I. Celebiler and O. Shimbo, "The probability of error due to intersymbol interference and Gaussian noise in digital communication systems," COMSAT, Tech. Memo, May 5, 1970.

[7] S. Benedetto, E. Biglieri, and R. Dogliotti, "Probabilità di errore per trasmissione numerica a più livelli e codificazione lineare," *Alta Freq.*, vol. 40, pp. 725-732, Sept. 1971.

[8] O. Shimbo and M. I. Celebiler, "The probability of error due to intersymbol interference and Gaussian noise in digital communication systems," *IEEE Trans. Commun. Technol.*, vol. COM-19, pp. 113-119, Apr. 1971.

[9] A. H. Stroud and D. Secrest, *Gaussian Quadrature Formulas.* Englewood Cliffs, N.J.: Prentice-Hall, 1966.

[10] V. I. Krylov, *Approximate Calculation of Integrals.* New York: Macmillan, 1962.

[11] P. J. Davis and P. Rabinowitz, *Numerical Integration.* Waltham, Mass.: Blaisdell, 1967.

[12] G. H. Golub and J. H. Welsch, "Calculation of Gauss quadrature rules," *Math. Comput.*, vol. 23, pp. 221-230, Apr. 1969.

[13] S. Benedetto, G. De Vincentiis, and A. Luvison, "The effect of phase jitter on the performances of automatic equalizers," in *Conf. Rec. 1972 IEEE Int. Conf. Communications*, Philadelphia, Pa., June 1972.

[14] A. Lender, "The duobinary technique for high-speed data transmission," *IEEE Trans. Commun. Electron.*, vol. 82, pp. 214-218, May 1963.

[15] E. R. Kretzmer, "Binary data communication by partial response transmission," in *IEEE Conf. Rec. Ann. Communication Conv.*, pp. 451-455, Boulder, Colo., June 1965.

[16] V. K. Prabhu, "Some considerations of error bounds in digital systems," *Bell Syst. Tech. J.*, vol. 50, pp. 3127-3151, Dec. 1971.

[17] W. Gautschi, "Construction of Gauss-Christoffel quadrature formulas," *Math. Comput.*, vol. 22, pp. 251-270, Apr. 1968.

[18] S. Benedetto, V. Castellani, C. Cianci, and U. Mazzei, "On the efficient bandwidth utilization in digital transmission," *Advis. Group Aerosp. Res. Dev. (NATO), 23rd Meeting Avionics Panel*, London, England, May 1972.

[19] S. Benedetto, V. Castellani, and G. De Vincentiis, "Error probability in the presence of intersymbol interference and additive noise for correlated digital signals," to be published.

[20] J. N. Lyness, "Differentiation formulas for analytic functions," *Math. Comput.*, vol. 22, pp. 352-362, Apr. 1968.

[21] W. Gautschi, "On the construction of Gaussian quadrature rules from modified moments," *Math. Comput.*, vol. 24, pp. 245-260, Apr. 1970.

[22] M. Abramowitz and I. A. Stegun, Eds. *Handbook of Mathematical Functions.* Washington, D.C.: Nat. Bur. Stand., 1967.

PHOTO NOT AVAILABLE

Sergio Benedetto was born in Turin, Italy, on January 18, 1945. He received the Dr.Ing. degree in electronic engineering from the Politecnico of Torino, Turin, Italy, where he is Associate Professor in Electrical Communications. His current interests involve statistical communication theory and data communication systems.

PHOTO NOT AVAILABLE

Girolamo DeVincentiis was born in Taranto, Italy, on June 26, 1946. He received the Dr.Ing. degree in electronic engineering from the Politecnico di Torino, Turin, Italy, on December 23, 1969.

He joined the Centro Studi e Laboratori Telecomunicazioni (CSELT), Turin, Italy, in 1970 and is currently a Research Engineer in the Scientific Section of the Scientific Secretariat. At CSELT he worked in the field of digital filtering and data transmission. Other areas of work include nonlinear coding theory and computer-aided design of digital circuits.

PHOTO NOT AVAILABLE

Angelo Luvison was born in Turin, Italy, on November 30, 1944. He received the Dr.Ing. degree in electronic engineering from the Politecnico di Torino, Turin, Italy, on January 30, 1969.

In 1969 he joined the Communication and Information Theory Section in the Transmission Department of the Centro Studi e Laboratori Telecomunicazioni (CSELT), Turin, Italy. He is a Research Engineer in the Scientific Section of the Scientific Secretariat of CSELT. He has been engaged in PCM transmission on radio links, and specifically in computerized system analysis and optimization. He has also worked on a variety of problems of high-speed data transmission, digital signal processing, and numerical analysis.

Mr. Luvison is a member of the Associazione Elettrotecnica ed Elettronica Italiana (AEI).

Coherent Demodulation of Frequency-Shift Keying with Low Deviation Ratio

RUDI de BUDA
Member, IEEE

Abstract—A coherent binary FSK modulation system is discussed, that has the following properties: 1) it is phase coherent; 2) it has a low deviation ratio, $h = \frac{1}{2}$; 3) it occupies a small RF bandwidth, typically only 0.75 times the bit rate, without need for intersymbol interference correction; 4) it uses as receiver a self-synchronizing circuit and a phase detector, which together achieve optimal decisions; and 5) its error performance is about 3 dB better than that of conventional FSK.

I. Introduction

This paper describes a scheme to transmit binary signals over an RF link.

Usually such digital signals are encoded either into frequency-shift keying (FSK) or phase-shift keying (PSK). While FSK has the advantage of simpler circuitry, it is generally conceded that PSK has the advantage of better utilization of bandwidth and signal-to-noise ratio.

It will, however, be shown in this paper that the coherent FSK with the deviation ratio 0.5 has an optimal detector, which is also simple to construct. Using this detector, one can receive faster pulse trains than with any other binary FSK or PSK of equal bandwidth and signal-to-noise ratio.

II. Phase Coherence

We consider a phase-coherent binary frequency-shift-keyed (FSK) signal and call a pulse at frequency f_1 a "mark" and at $f_2 > f_1$, a "space." Each pulse has duration T.

If the signal is narrow band, e.g., at IF or RF, then we may describe it by its preenvelope [1] $u(t)$;

$$u(t) = \exp j\{\pi(f_1 + f_2)t + \phi(t)\}, \qquad (1)$$

where the phase function $\phi(t)$ is continuous if the FSK is coherent.

Comparison of (1) with the preenvelope of a space

$$u_{\text{space}} = \exp j\{2\pi f_2 t + \phi(0)\}, \qquad 0 \leq t \leq T$$

gives

$$\phi(T) - \phi(0) = \pi(f_2 - f_1)T. \qquad (2)$$

Paper approved by the Communications Theory Committee of the IEEE Communications Society after presentation at the 1971 International Conference on Communications, Montreal, Que., Canada, June 14–16, 1971. Manuscript received July 21, 1971; revised December 10, 1971.
The author is with the Canadian General Electric Company, Ltd., Toronto, Ont., and McMaster University, Hamilton, Ont., Canada.

Reprinted from *IEEE Transactions on Communications,* June 1972.

For a mark, the phase decreases by this same amount. This suggests introducing, by definition, the dimensionless parameter h;

$$h = (f_2 - f_1)T, \tag{3}$$

which is called modulation index [2] or frequency deviation ratio.

A space increases the phase by πh and a mark reduces it by πh. After s spaces and m marks, i.e., at time $(s + m)T$, we have then the phase shift [3, fig. 1]

$$\phi\{(s + m)T\} - \phi(0) = (s - m)h\pi, \tag{4}$$

which is an even (odd) multiple of πh when $(s + m)$ is even (odd). The phase at all other times is obtained by linear interpolation. The possible values of $\phi(t)$ are shown in Fig. 1.

Since the phase $\phi(t)$ is a continuous function of time, the signal itself is continuous and has thus lower spectral sidelobes than a pulsed signal. Consequently, a coherent FSK with low deviation ratio h occupies less bandwidth than other comparable systems. This has been noticed for some time [2]–[6] and will be utilized here.

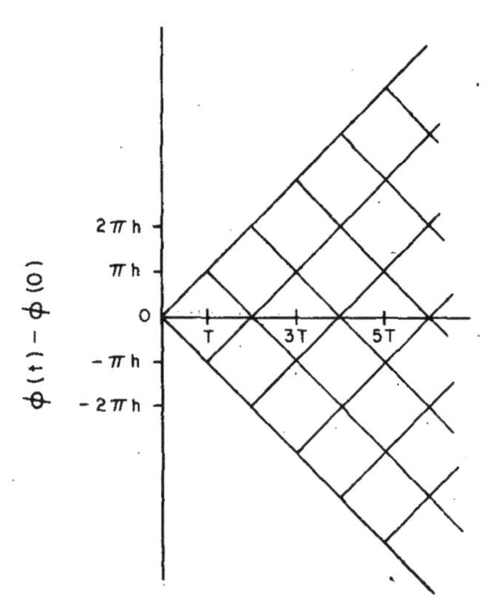

Fig. 1. Possible values $\phi(t) - \phi(0)$.

III. FSK With Low Modulation Index h

From here on we will restrict our attention to coherent FSK with small modulation index h and in particular to the three cases

$$h = 1.0$$

$$h = 0.71$$

$$h = 0.5.$$

The FSK with $h = 1.0$ has been described in detail by Sunde [6]; it has not found wide acceptance. It uses signal power inefficiently because it transmits some power at the spectral lines f_1 and f_2 [7]. More on this topic will be found in Section VII.

The case $h = 0.71$ has been claimed to be optimum [8], [9] under certain conditions, which will be discussed later. This system has been extensively studied by Thjung, who shows in particular that, in a very narrow band coherent FSK with $h = 0.7$ is better than phase-shift keying PSK, because the FSK has lower intersymbol interference [10].

The system that will occupy most of this report is the coherent FSK with deviation ratio of exactly $h = \frac{1}{2}$. It has a spectrum that falls off smoothly but rapidly [3, fig. 2], [4, fig. 4] and it has been singled out as a good example of band conservation [5], (11). Apart from this, the superior qualities of this modulation seem to have attracted little attention until very recently [12]–[14] and one finds often that PSK is accepted as superior to FSK [15], [16].

After some comments on the so-called optimum at $h = 0.71$, the coherent FSK with deviation ratio $h = 0.5$ will be taken up again. First its optimal decision structure will be given, which requires a coherent demodulation; and self-synchronizing circuits will then be described, which are essential to obtain the optimal performance. With such a receiver, the signal requires less bandwidth than binary PSK and less power than any other binary FSK and comes close to four-phase PSK, which has much more complex receiver synchronization requirements. These advantages might be indicated by dubbing this FSK modulation the "fast FSK."

IV. Discussion of the Optimization of h

The statement that this fast FSK is better than all other FSK seems to be at variance with the accepted value of $h = 0.715 \cdots$ for optimum error probability. Therefore, it will be instructive to repeat here the derivation of the optimum at $h = 0.715$, so that we can see more readily where a false restriction has been applied [17].

We assume, that at time $t = 0$ the phase of the received signal is given, say $\phi(o)$, and that for $0 \leq t \leq T$ either space or mark is transmitted.

We assume further, that the IF center frequency is sufficiently high. Then the narrow-band assumption is valid, so that the signals can be expressed by their preenvelopes;

$$u_{\text{mark}} = \exp j[2\pi f_1 t + \phi(0)]$$

or

$$u_{\text{space}} = \exp j[2\pi f_2 t + \phi(0)].$$

Next, white Gaussian noise is added to the signal. Then a decision must be made at the receiver as to whether mark or space has been transmitted (see Fig. 2).

The lowest error of this decision is obtained when the distance between the two signals in Hilbert space is largest [18]. Since mark or space are transmitted only for $0 \leq t \leq T$, one evaluates thus first the integral

$$D^2(h) = \int_0^T |u_s - u_m|^2 \, dt \tag{5}$$

and then maximizes $D(h)$ with respect to h.

Simple substitution gives, with (3),

$$|u_s - u_m| = 2 \sin \frac{\pi h t}{T}$$

and

$$D^2(h) = 4 \int_0^T \sin^2 \frac{\pi h t}{T} \, dt = 2T[1 - \operatorname{sinc} 2h]. \tag{6}$$

Woodward's sinc function [1, p. 29] is known to have its first and lowest minimum (−0.217) when

$$\tan 2\pi h = 2\pi h. \tag{7}$$

The maximum of $D(h)$ occurs therefore when $h = 0.715$ so that this is assumed to be the optimum value for h.

The fallacy in the preceding derivation lies in the arbitrary selection of the limits $0 \leq t \leq T$ in the integral that defines the distance $D(h)$, rather than integrating over $(-\infty, \infty)$. The same applies to the experimental work, which seems to confirm the theory, but bases the decision also only upon the

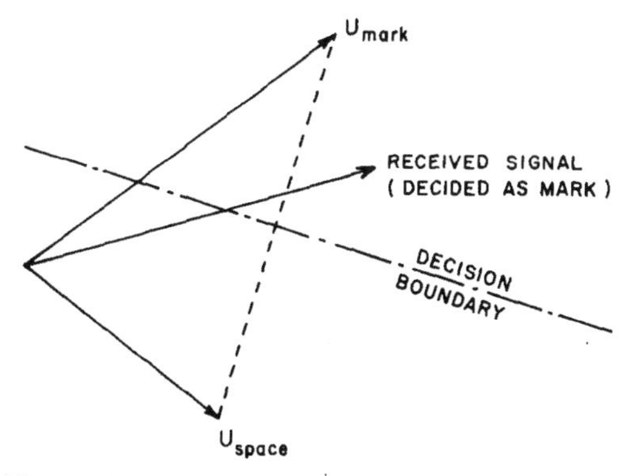

Fig. 2. Hilbert space representation of the mark-space decision.

interval $(0, T)$. This is inadequate, because the interval $(0, T)$ is not the complete time that is available to make the decision. It is true that T is the duration of the transmission of one bit and we must make one decision per transmitted bit; but the bit phases are interrelated and it is a basic tenet of information theory that better performance is obtained when several signals are jointly investigated over a longer period. Now low modulation index coherent FSK retains useful phase information beyond the time of the bit transmission [15]. In the binary case this extends over two time intervals. (For suitably selected m-ary FSK, the phase coherence can extend over several time intervals, but this will not be pursued here.) In particular, the phase in the succeeding time interval $T \leq t \leq 2T$ contains unused information about the transmitted frequency of the earlier time interval $0 \leq t \leq T$; this information increases when h decreases (below 0.71) and becomes particularly important at $h = 0.5$. A receiver that fails to use this information will therefore achieve at best a local maximum, but not an optimum, and thus we should expect a receiver with lower h and a decision interval $2T$ to perform better than the so-called optimum at $h = 0.715$.

V. Optimum Reception of Coherent FSK

If we inspect Fig. 1, or (4), then it is evident that the phase at times that are odd (even) multiples of T will be odd (even) multiples of πh. Since all phases are modulo 2π, the case $h = \frac{1}{2}$ stands out, because then the phase can take only the two values $\pm \pi/2$ at odd times and only the two values 0 and π at even times.

Thus one may design a receiver to make a decision between the two permitted phase values. It will be shown that this is an optimal decision scheme, provided that the decision is made after observing not the signal in one time interval, but the real (imaginary) part of its preenvelope in two time intervals: the one before and the one after the phase that we wish to estimate. This can be seen by elementary methods.

Let $\phi(0) = 0$ or π be the phase at $t = 0$ and let $\omega = \pi(f_1 + f_2)$. Then

$$u(t) = \begin{cases} +\exp j\left(\omega t \pm \dfrac{\pi t}{2T}\right), & \phi(0) = 0, \\ -\exp j\left(\omega t \pm \dfrac{\pi t}{2T}\right), & \phi(0) = \pi, \end{cases} \qquad 0 \leq t \leq T. \tag{8}$$

The sign within the exponent is positive for space, negative for mark.

A similar result holds for $-T \leq t \leq 0$, except that the sign within the exponent is not necessarily the same in both intervals; but this is immaterial, because we can form

$$\mathrm{Re}\left\{\exp j\left(\pm\frac{\pi t}{2T}\right)\right\} = \cos\frac{\pi t}{2T}$$

regardless of whether mark or space is being sent in either interval. Consequently, in the given interval of duration $2T$

$$\mathrm{Re}\,\{u(t)\exp -i\omega t\} = \begin{cases} +\cos\dfrac{\pi t}{2T}, & \phi(0) = 0, \\ -\cos\dfrac{\pi t}{2T}, & \phi(0) = \pi, \end{cases} \qquad -T \leq t \leq T. \tag{9}$$

The function $\mathrm{Re}\,\{u(t)\exp - j\omega t\}$ depends only on $\phi(0)$, not on any other $\phi(nT)$, thus all unwanted information has been removed. The decision between $\phi(0) = 0$ and $\phi(0) = \pi$ can be optimally made, based on the waveform received in $(-T, T)$, since the signals $\pm \cos(\pi t/2T)$, $(-T \leq t \leq T)$ are antipodal and contain each the energy of one bit [18]. The probability of error is therefore the same as that of a coherent binary receiver (or same energy per bit), i.e., as function of the SNR γ

$$P_{\mathrm{FFSK}} = \mathrm{erfc}\,\sqrt{2\gamma} \doteq \frac{1}{\sqrt{2\pi}\,\gamma}\exp -\gamma.$$

This is over 3 dB better than the error probability of conventional FSK [15],

$$P_{\mathrm{FSK}} = \tfrac{1}{2}\exp -\gamma/2.$$

The optimal receiver structure is a clocked matched filter, for instance implemented by an integrate-dump circuit in baseband, which performs

$$J(0) = \int_{-T}^{T} \cos\frac{\pi t}{2T}\,\mathrm{Re}\,\{u(t)\exp -j\omega t\}\,dt \tag{10}$$

and decides

$$\phi(0) = 0, \qquad J(0) > 0$$
$$\phi(0) = \pi, \qquad J(0) < 0.$$

Corresponding decisions in every second interval pair give $\phi(2nT)$ and this yields half the number of transmitted bits.

The other half comes from the imaginary part of u, where a similar relation holds for $\phi(T)$ as decided during the interval $(0, 2T)$,

$$J(T) = \int_0^{2T} \sin\frac{\pi t}{2T}\,\mathrm{Im}\,\{u(t)\exp -j\omega t\}\,dt$$
$$\phi(T) = \frac{\pi}{2}, \qquad J(T) > 0 \tag{11}$$
$$\phi(T) = -\frac{\pi}{2}, \qquad J(T) < 0$$

and so on. Finally, the transmitted bit stream can be recovered from the resulting $\phi(nT)$.

No further effort to optimize the receiver is required; we can concentrate on implementing the mathematical receiver structure with practical circuits.

VI. Shifting Into Baseband

$\mathrm{Re}\,\{u(t)\exp - j\omega t\}$ and $\mathrm{Im}\,\{u(t)\exp - j\omega t\}$ are obtained from the received signal

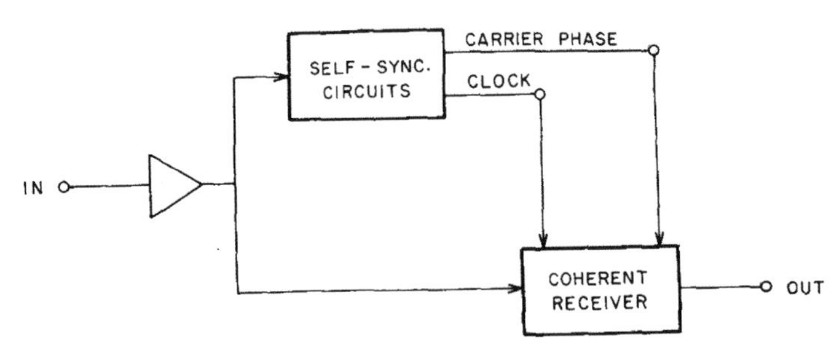

Fig. 3. Self-synchronizing receiver.

$$s(t) = \text{Re}\,\{u(t)\}$$

by beating it into baseband. This requires generating a phase-locked carrier at frequency $\omega/2\pi$, multiplying $s(t)$ with cos ωt and sin ωt, respectively, in two balanced modulators, and then removing the second-harmonic terms from the two outputs, which by then are usually called I channel and Q channel.

The other operations of (10) and (11) are straightforward. They require at the receiver two auxiliary signals: the clock that is needed for integrating over an interval of length $2T$ and the precisely phase-locked carrier signals cos ωt and sin ωt.

There are various ways to transport both signals, but the most elegant method is to regenerate them at the receiver by a self-synchronizing scheme, rather than wasting power and bandwidth for transmitting phase and clock reference signals.

Fig. 3 illustrates the overall schematic of a self-synchronizing receiver; after a preamplifier, the signals are split into two paths, one extracts the timing and phasing information, necessarily at a much slower rate than the signaling information [19]; and the other path contains a coherent detector of the signals, using the reference signals from the self-synchronizing subsystem.

VII. Self-Synchronization of the Receiver

The self-synchronization of the fast FSK is most easily explained by reference to Sunde's FSK [6], [7], which has $h = 1.0$. Since h is a (normalized) frequency difference, a frequency-doubling circuit doubles h and thus transforms the fast FSK into Sunde's FSK.

Now Sunde's FSK has the stated disadvantage, that part of its power is in the spectral lines of the two carriers and thus not available for signaling. But we do not need the signaling—this is done at $h = \frac{1}{2}$; on the contrary we turn the spectral lines of Sunde's FSK into an advantage, since they are at $2f_1$ and $2f_2$ and therefore ideally suited to generate for the fast FSK the phase reference.

$$f_c = (2f_1 + 2f_2)/4$$

and the clock

$$1/T = 2f_2 - 2f_1.$$

In principle, this could be done with the circuit in Fig. 4.

A frequency doubler feeds two phase-locked loops (PLL), which extract $2f_1$ and $2f_2$, and send them to a balanced modulator.

The low-frequency output $2f_2 - 2f_1$ of the modulator is the clock, the high-frequency output $2f_1 + 2f_2$ is divided by 4 to generate the two phase reference signals. All the reference signals can be generated practically error free, by narrowing the PLL bandwidth, provided that sufficient time is available for initial synchronization.

The phase and clock signals are then used for a conventional coherent demodulation and detection of the signals in the main path, which do not pass through the doubler stage, but remain at $h = \frac{1}{2}$; and this, in principle, is the receiver.

If certain periodic signals are used as modulation, then the frequency doubler may generate a line at $2f_c = f_1 + f_2$. This frequency should not be used in the self-synchronizing circuit, since a spectral line at $2f_c$ is in general not available. Some care is necessary to prevent false locking of the PLL if they should find $2f_c$ in the presence of such a periodic modulation.

A different problem is the possible loss of lock if a long string of only mark (or only space) is received. Some improvement can be obtained by modulating the remaining line with the clock to regenerate the missing line, but this works only as long as the clock itself does not drift out of sync. This problem is common to all FSK. The best remedy, if available, is suitable source encoding to prevent such long strings from occurring.

VIII. 90° Phase Ambiguity

The divide-by-four circuit generates a phase ambiguity of the reference phases by some multiple of 90°. Unlike the 4-phase PSK, the fast FSK can resolve this ambiguity, for instance by comparing the coherent output with a bit stream from the frequency doubler. This bit stream is a conventional FSK, except with some noise added (by the doubling circuit) and it is not desirable to use it directly in the receiver, but it still contains enough good bits to resolve the ambiguity if comparison is made over several time intervals.

An alternative method is easier to implement. We treat the ambiguity in two ways. The 90° ambiguity can be removed by a more careful layout of the self-synchronizing circuit, keeping in mind that we really have a binary system and should not allow fourth-order effects to cause phase ambiguities. Fig. 5 illustrates the circuitry.

The circuit of Fig. 5 has the same doubler, PLL, and master clock as the circuit of Fig. 4, but it generates the phase references differently. The outputs of the PLL are divided by two and then added or subtracted. Since the phase at the divider output is ambiguous by 180°, we do not know which adder gives the sum and which the difference, but this does not matter since the outputs are, respectively, $\pm$cos ωt cos $\pi t/2T$ and $\pm$sin ωt sin $\pi t/2T$ and both are the required reference signals, when they are gated between the nulls of their respective modulation.

IX. 180° Phase Ambiguity

The only ambiguity left is the $\pm$ sign in front of the generated reference and this can be taken care of by the method of differential encoding, as is usually done in coherent binary PSK.

Therefore we relate the transmitted bit stream $a_n = \pm 1$, $n = 1, 2 \cdots$ to the received signals in the I and Q channels. First we notice that because of the 180° ambiguities only the phase differences $\phi[(n-1)T] - \phi\,[(n+1)T]$ can be used. Inspection of Fig. 1 shows that a sequence mark-space or space-mark leaves ϕ unchanged, while a sequence mark-mark

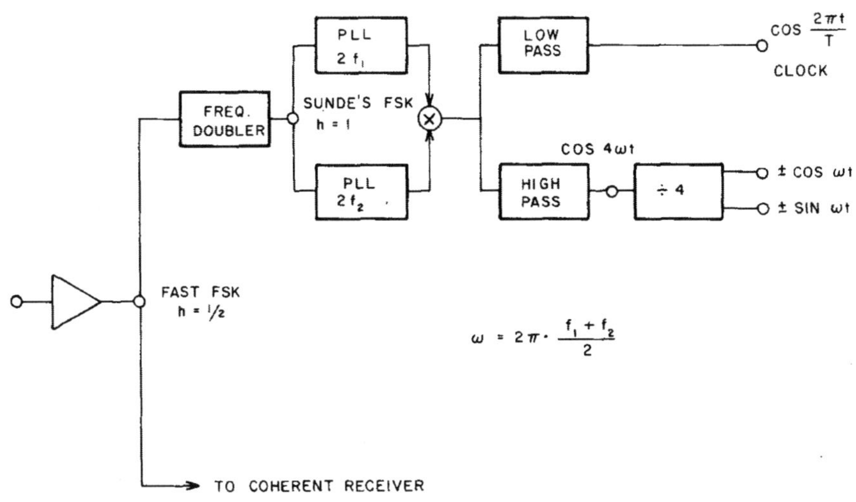

Fig. 4. Generation of clock and phase reference.

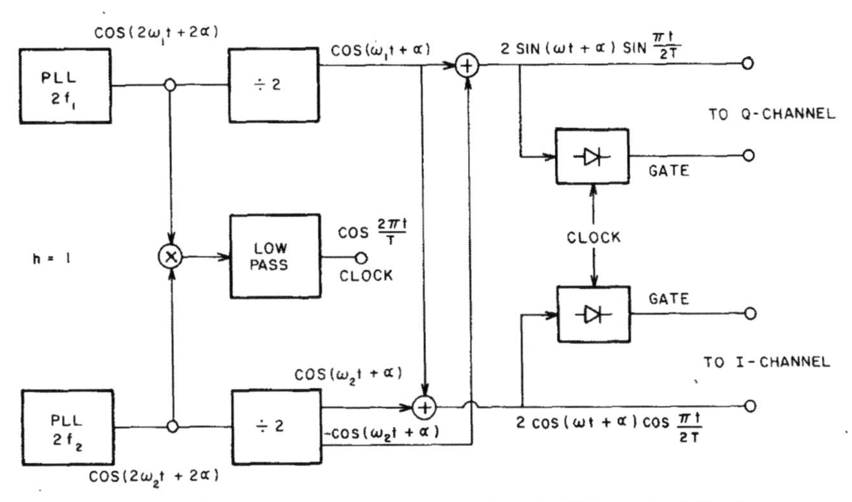

Fig. 5. Alternative generation of references from the PLL outputs of Fig. 4

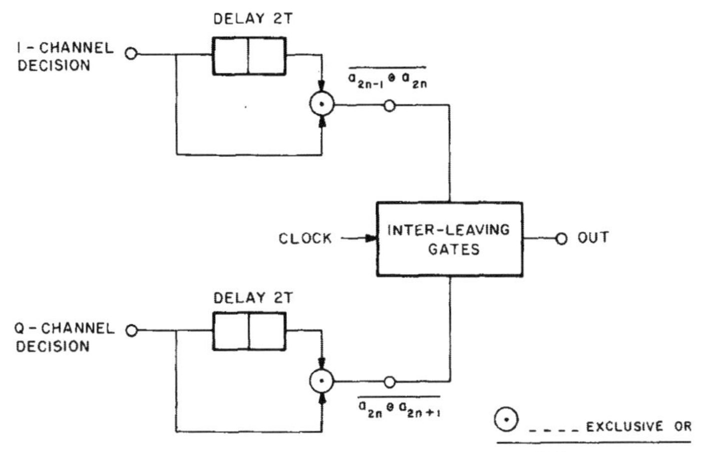

Fig. 6. Digital outputs when $\{a_n\}$ was transmitted.

or space-space changes ϕ by 180° after 2 bit intervals. Thus, at time $t = 2nT$, a transition at the input corresponds to no transition at the output of the I channels and at time $t = (2n + 1)T$, a transition at the input corresponds to no transition in the Q channel, and vice versa.

The circuit of Fig. 6 shows the various logic operations that recover the differentially encoded input from the outputs of the I and Q channels.

The circuit decodes the differentially encoded input $a_n \cdot a_{n+1}$ when a_n was sent.

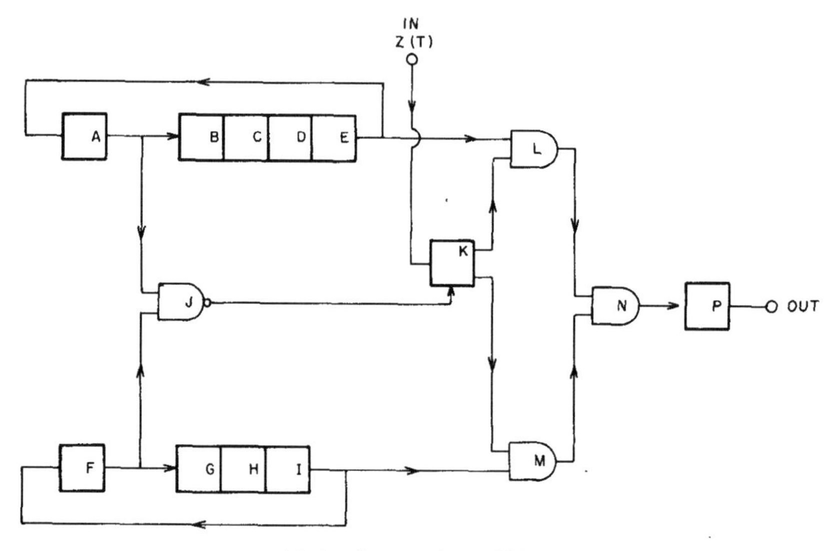

Fig. 7. Generator for fast FSK.

X. Effect of Filtering and Intersymbol Interference

If the fast FSK signals are passed through a narrow-band filter, at the transmitter and/or at the receiver front end, then the signals available for the phase decisions will no longer be independent. Two types of interference may occur: 1) within each channel and 2) between the I and the Q channel.

The second type can be avoided by careful circuit designs. There will be no such interference if the overall filter action is that of a symmetrical bandpass,

$$G(f_c + f) = G^*(f_c - f)$$

because such a bandpass has a weighting function

$$g(t) \cdot \exp - 2\pi j f_c t,$$

where $g(t)$ is a real function.

The bandpass action is then the correlation of $\exp j\phi(t)$ with $g(t)$ and this leaves the real part of $\exp j\phi(t)$ real and the imaginary part imaginary, so that no interference between the I channel and the Q channel occurs.

The intersymbol interference within the I (or Q) channel cannot be removed so easily. It limits the speed with which we can transmit bits through a given system. But, in each of the two baseband channels, bits occur at half-speed ($1/2T$) and the signaling function is continuous, since switching occurs only at the nulls of $\cos \pi t/2T$. Thus the intersymbol interference will be considerably smaller than when rectangular pulses are transmitted in an IF bandwidth of $1.2/T$.

When the IF bandwidth is $0.75/T$, then the intersymbol interference of the fast FSK is still negligible. Higher bit rates are possible by making the intersymbol correction by a transversal filter [20], [21] at baseband, which is by now a well-established technique. Using only an intersymbol correction before decomposition into I and Q channel seems to give poorer results [22].

The output of the narrow-band filter will show amplitude fluctuations, mainly increases near the mark-space transition. This is unavoidable. If one tries to remove the amplitude by a limiter, then the sidebands are regenerated and the effective filter bandwidth is widened [12]. There is also the usual small reduction of SNR due to hard limiter action, but the probability of error does not increase significantly. The decision is based upon the quadrant in which the received signal is in a given interval. If we now make no intersymbol interference correction, but ideally investigate per interval only one sample of signal plus noise plus interference, then the decision made from this sample, and hence the error probability, is not affected by hard limiting after the filter.

XI. Digital Generation of the Fast FSK Signal

When generating the fast FSK, for good performance it is required not only to generate a coherent FSK, but also to hold the normalized frequency excursion h to precisely the value of $h = \frac{1}{2}$.

This can be achieved by a suitable feedback circuit around a conventional coherent FSK; or by a circuit that generates two sine waves at f_1 and $f_2 = f_1 + h/T$, each with positive and negative polarity, and a logic circuit, which, on frequency shifting, selects that polarity of the new frequency that maintains phase coherence.

A more elegant generator of the fast FSK can be constructed as follows.

First Sunde's FSK [6] is generated at IF, then sent through a flip-flop, which divides by 2; then the harmonics of the resulting square wave are removed by a zonal filter and the resulting signal is the fast FSK, since Sunde's FSK has a normalized frequency shift of $h = 1$, which, after the frequency divider stage, becomes $h = \frac{1}{2}$.

If the desired frequency range allows it, then Sunde's FSK can be obtained in an all-digital circuit as follows.

Let n be some small integer. Generate a master clock with frequency $n(+ 1)/T$.

Count down to derive three pulse trains, the first with n/T; the other with $(n + 1)/T$ pulses, and the third train with $1/T$ pulses. This third train serves as clock for the bit-stream encoder.

In each time interval T we select n pulses from the first train if we transmit mark or $(n + 1)$ pulses from the second train, if we transmit space. This can be done by a digital circuit under control of the master clock. The fundamental of the resulting pulse train is then an FSK with $h = 1$; but it is not necessary to obtain the fundamental because the train itself is ideally suited to drive a divide-by-two flip-flop. The fundamental at the flip-flop output is then the fast FSK.

Fig. 7 illustrates a digital circuit that implements this scheme for $n = 4$.

XII. Conclusion

The fast FSK is a binary FSK with these priorities.

1) It is phase coherent.
2) It has a low deviation ratio, $h = \frac{1}{2}$.
3) It occupies a small bandwidth.

4) It uses a phase detector as demodulator. First, with the help of a modification of the Costas receiver [23], self-synchronization of phase and clock is achieved; then the signals are coherently demodulated and detected.

Such a receiver shows a distinct improvement over the bandwidth and SNR required for conventional reception of binary FSK signals. For instance, when no correction for intersymbol interference is used, then the fast FSK can be transmitted in an RF bandwidth of only 0.75 times the bit rate, yet received with about 3 dB less power than conventional FSK at equal bit rate and error probability.

The theory of the scheme is fairly simple, once it is recognized where the information, which makes this improvement possible, is to be found; and the implementation is straightforward.

References

[1] P. M. Woodward, *Probability and Information Theory, With Applications to Radar.* New York: Pergamon, 1953, pp. 40–42.

[2] W. Postl, "Die spektrale Leistungsdichte bei Frequenzmodulation eines Traegers mit einem stochastischen Telegraphie-signal," *Frequenz*, vol. 17, pp. 107–110, Mar. 1963.

[3] M. G. Pelchat, "The autocorrelation function and power spectrum of PCM/FM with random binary modulating waveforms," *IEEE Trans. Space Electron. Telem.*, vol. SET-10, pp. 39–44, Mar. 1964.

[4] W. R. Bennett and S. O. Rice, "Spectral density and autocorrelation functions associated with binary frequency-shift keying," *Bell Syst. Tech. J.*, vol. 42, pp. 2355–2385, Sept. 1963.

[5] H. J. von Baeyer, "Band limitation and error rate in digital UHF-FM transmission," *IEEE Trans. Commun. Syst.*, vol. CS-11, pp. 110–117, Mar. 1963.

[6] E. D. Sunde, "Ideal binary pulse transmission by AM and FM," *Bell Syst. Tech. J.*, vol. 38, pp. 1357–1426, Nov. 1959.

[7] W. R. Bennett and J. Salz, "Binary data transmission by FM over a real channel," *Bell Syst. Tech. J.*, vol. 42, pp. 2387–2426, Sept. 1963.

[8] V. A. Kotel'nikov, *The Theory of Optimum Noise Immunity* (in Russian, 1947). New York: McGraw-Hill, 1959, p. 40.

[9] E. F. Smith, "Attainable error probabilities in demodulation of random binary PCM/FM waveforms," *IRE Trans. Space Electron. Telem.*, vol. SET-8, pp. 290–297, Dec. 1962.

[10] T. T. Tjhung and P. H. Wittke, "Carrier transmission of binary data in a restricted band," *IEEE Trans. Commun. Technol.*, vol. COM-18, pp. 295–304, Aug. 1970.

[11] M. L. Doelz *et al.*, "Minimum-shift data communication system," U. S. Patent 2 977 417, Mar. 1961.

[12] J. R. Boykin, "Spectrum economy for filtered and limited FSK signals," *Conf. Rec., 1970 IEEE Conf. Communications*, pt. 1, pp. 21.29–21.34.

[13] W. A. Sullivan, "High-capacity microwave system for digital data transmission," *Conf. Rec., Int. Conf. Communications*, Montreal, Que., Canada, June 1971, pp. 23.4–23.8.

[14] H. C. van den Elzen and P. van der Wurf, "A simple method of calculating the characteristics of FSK signals with modulation index 0.5," *IEEE Trans. Communications*, vol. COM-20, pp. 139–147, Apr. 1972.

[15] M. Schwartz, W. R. Bennett, and S. Stein, *Communication Systems and Techniques.* New York: McGraw-Hill, 1966, pp. 298–302, eq. 7.5.8; pp. 341–342.

[16] B. P. Lathi, *Communications Systems.* New York: Wiley, 1968, p. 419.

[17] R. de Buda and H. Anto, "About FSK with low modulation index," Can. Gen. Elec. Co., Ltd., Tech. Inf. Ser. Rep. RQ69EE11, Dec. 1969.

[18] J. M. Wozencraft and I. M. Jacobs, *Principles of Communication Engineering.* New York: Wiley, 1965, pp. 248–250.

[19] R. de Buda, "The phaselock to a suppressed carrier in additive Gaussian noise," Can. Gen. Elec. Co., Ltd., Rep. TIS RQ70EE7, Sept. 1970.

[20] R. W. Lucky, "Techniques for adaptive equalization of digital communication systems," *Bell Syst. Tech. J.*, vol. 45, pp. 255–286, Feb. 1966.

[21] D. A. George, D. C. Coll, A. R. Kaye, and R. R. Bowen, "Channel equalization for data transmission," *Eng. J.*, vol. 53, pp. 20–31, May 1970.

[22] J. L. Pearce and P. H. Wittke, "Optimum reception of digital FM signals," *Dig. 1970 IEEE Symp. Communications*, Montreal, Que., Canada, Nov. 12–13, 1970.

[23] J. P. Costas, "Synchronous communications," *Proc. IRE*, vol. 44, pp. 1713–1718, Dec. 1956.

Data Transmission by Frequency-Division Multiplexing Using the Discrete Fourier Transform

S. B. WEINSTEIN, MEMBER, IEEE, AND PAUL M. EBERT, MEMBER, IEEE

Abstract—The Fourier transform data communication system is a realization of frequency-division multiplexing (FDM) in which discrete Fourier transforms are computed as part of the modulation and demodulation processes. In addition to eliminating the banks of subcarrier oscillators and coherent demodulators usually required in FDM systems, a completely digital implementation can be built around a special-purpose computer performing the fast Fourier transform. In this paper, the system is described and the effects of linear channel distortion are investigated. Signal design criteria and equalization algorithms are derived and explained. A differential phase modulation scheme is presented that obviates any equalization.

I. Introduction

DATA ARE usually sent as a serial pulse train, but there has long been interest in frequency-division multiplexing with overlapping subchannels as a means of avoiding equalization, combating impulsive noise, and making fuller use of the available bandwidth. These "parallel data" systems, in which each member of a sequence of N digits modulates a subcarrier, have been studied in [2] and [4]. Multitone systems are widely used and have proved to be effective in [3], [8], and [9]. Fig. 1 compares the transmissions of a serial and a parallel system.

For a large number of channels, the arrays of sinusoidal generators and coherent demodulators required in parallel systems become unreasonably expensive and complex. However, it can be shown [1] that a multitone data signal is effectively the Fourier transform of the original serial data train, and that the bank of coherent demodulators is effectively an inverse Fourier transform generator. This point of view suggests a completely digital modem built around a special-purpose computer performing the fast Fourier transform (FFT). Fourier transform techniques, although not necessarily the signal format described in this paper, have been incorporated into several military data communication systems [5]–[7].

Because each subchannel covers only a small fraction of the original bandwidth, equalization is potentially simpler than for a serial system. In particular, for very narrow subchannels, soundings made at the centers of the subchannels may be used in simple transformations of the receiver output data to produce excellent estimates of the original data. Further, a simple equalization algorithm will minimize mean-square distortion on each subchannel, and differential encoding of the original data may make it possible to avoid equalization altogether.

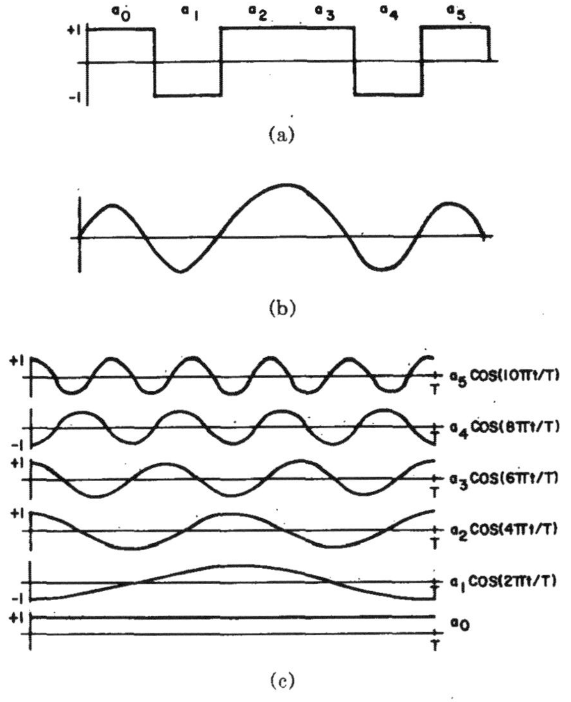

Fig. 1. Comparison of waveforms in serial and parallel data transmission systems. (a) Serial stream of six binary digits. (b) Typical appearance of baseband serial transmission. (c) Typical appearance of waveforms that are summed to create parallel data signals.

II. Frequency-Division Multiplexing as a Discrete Transformation

Consider a data sequence $(d_0, d_1, \cdots, d_{n-1})$, where each d_n is a complex number $d_n = a_n + jb_n$.

If a discrete Fourier transform (DFT) is performed on the vector $\{2d_n\}_{n=0}^{N-1}$, the result is a vector $S = (S_0, S_1, \cdots, S_{n-1})$ of N complex numbers, with

$$S_m = \sum_{n=0}^{N-1} 2d_n e^{-j(2\pi nm/N)} = 2\sum_{n=0}^{N-1} d_n e^{-j2\pi f_n t_m}$$

$$m = 0, 1, \cdots, N-1, \tag{1}$$

where

$$f_n \triangleq \frac{n}{n\,\Delta t} \tag{2}$$

Paper approved by the Data Communications Committee of the IEEE Communication Technology Group for publication after presentation at the 1971 IEEE International Conference on Communications, Montreal, Que., Canada, June 14–16. Manuscript received January 15, 1971; revised March 29, 1971.

The authors are with the Advanced Data Communications Department, Bell Telephone Laboratories, Holmdel, N. J.

Reprinted from *IEEE Transactions on Communication Technology,* vol. COM-19, no. 5, October 1971.

$$t_m \triangleq m\,\Delta t \tag{3}$$

and Δt is an arbitrarily chosen interval. The real part of the vector S has components

$$Y_m = 2\sum_{n=0}^{N-1}(a_n \cos 2\pi f_n t_m + b_n \sin 2\pi f_n t_m), \quad m = 0, 1, \cdots, N-1. \tag{4}$$

If these components are applied to a low-pass filter at time intervals Δt, a signal is obtained that closely approximates the frequency-division multiplexed signal

$$y(t) = 2\sum_{n=0}^{N-1}(a_n \cos 2\pi f_n t + b_n \sin 2\pi f_n t), \quad 0 \le t \le N\,\Delta t. \tag{5}$$

A block diagram of the communication system in which $y(t)$ is the transmitted signal appears in Fig. 2.

Demodulation at the receiver is carried out via a discrete Fourier transformation of a vector of samples of the received signal. Because only the real part of the Fourier transform has been transmitted, it is necessary to sample twice as fast as expected, i.e., at intervals $\Delta t/2$. When there is no channel distortion, the receiver DFT operates on the $2N$ samples

$$Y_k = y\left(k\frac{\Delta t}{2}\right) = 2\sum_{n=0}^{N-1}\left(a_n \cos\frac{2\pi nk}{2N} + b_n \sin\frac{2\pi nk}{2N}\right), \quad K = 0, 1, \cdots, 2N-1, \tag{6}$$

where definitions (2) and (3) have been substituted into (4). The DFT yields

$$z_l = \frac{1}{2N}\sum_{k=0}^{2N-1} Y_k e^{-j(2\pi lk/2N)} = \begin{cases} 2a_0, & l = 0 \\ a_l - jb_l, & l = 1, 2, \cdots, N-1 \\ \text{irrelevant}, & l > N-1, \end{cases} \tag{7}$$

where the equality

$$\frac{1}{2N}\sum_{k=0}^{2N-1} e^{j(2\pi mk/2N)} = \begin{cases} 1, & m = 0, \pm 2N, \pm 4N, \cdots \\ 0, & \text{otherwise} \end{cases} \tag{8}$$

has been employed. The original data a_l and b_l are available (except for $l = 0$) as the real and imaginary components, respectively, of z_l, as indicated in Fig. 2. A synchronizing signal is required, but one or several channels of the transmitted signal can readily be utilized for this purpose.

Because the sinusoidal components of the parallel data signal $y(t)$ are truncated in time, the power density spectrum of $y(t)$ consists of $[\sin(f)/f]^2$-shaped spectra, as sketched in Fig. 3. Nevertheless, the data on the different subchannels can be completely separated by the DFT operation of (7). This will not be exactly true when linear channel distortion affects the received signal, but it will be shown later that a modest reduction in transmission rate eliminates most interferences.

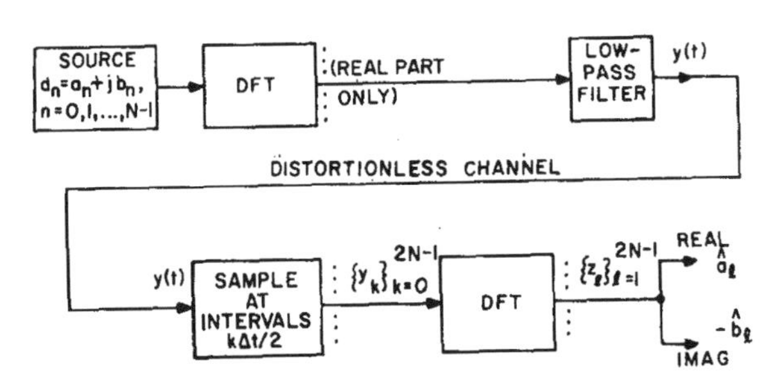

Fig. 2. Fourier transform communication system in absence of channel distortion.

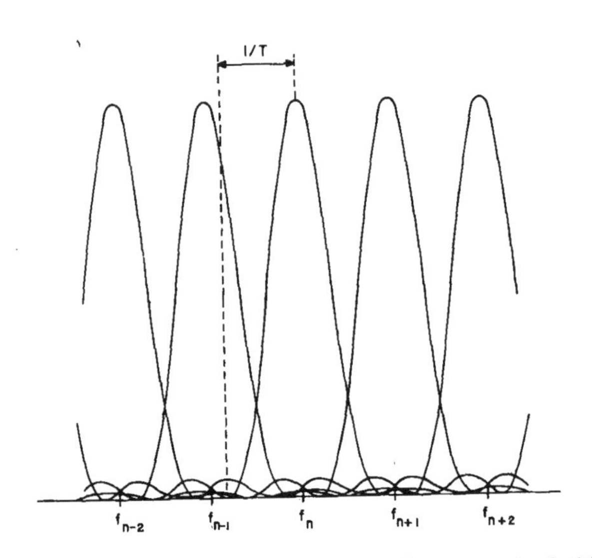

Fig. 3. Power density spectra of subchannel components of $y(t)$.

III. Equalization by Use of Channel Soundings

Except for the added linear channel distortion and final equalizer, the Fourier transform data communication system shown in Fig. 4 is identical to that of Fig. 2. Ideally, the discrete Fourier transformation in the receiver should be replaced by another linear transformation, derived in [1], which minimizes the error in the receiver output. However, it is preferable, if possible, to retain the DFT with its "fast" implementations and carry out suboptimal but adequate correctional transformations at the receiver output. The system of Fig. 4 performs this approximate equalization.

Consider the waveform at the receiver input,

$$r(t) = y(t) * h(t), \tag{9}$$

where the asterisk denotes convolution. This waveform is a collection of truncated sinusoids modified by a linear filter. If the sinusoid $\cos 2\pi f_n t$ were not truncated, then the result of passing it through a channel with transfer function $H(f)$ would be $H_n \cos(2\pi f_n t + \phi_n)$, where

$$H_n = |H(f_n)| \qquad \phi_n = \tan^{-1}\left(\frac{\operatorname{Im} H(f_n)}{\operatorname{Re} H(f_n)}\right). \tag{10}$$

The sinusoids in the transmitted signal $y(t)$ [see (5)] are truncated to the interval $(0, N\Delta t)$, so that the nth subchannel must accommodate a $[\sin N\pi(f - f_n)\Delta t]/$

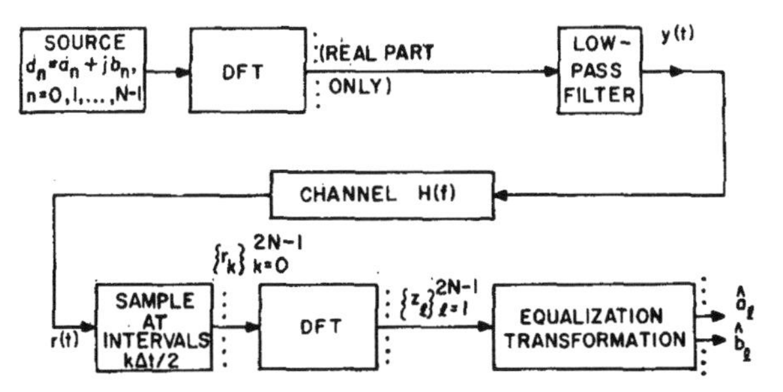

Fig. 4. Fourier transform communication system including linear channel distortion and final equalization.

$[N\pi(f - f_n)\Delta t]$ spectrum instead of the impulse at f_n, which would correspond to a pure sinusoid. However, if $1/(N\Delta t)$ is small compared with the total transmission bandwidth, then $H(f)$ does not change significantly over the subchannel and an approximate expression for the received signal $r(t)$ is

$$\begin{aligned} r(t) &\cong 2 \sum_{n=1}^{N-1} H_n[a_n \cos(2\pi f_n t + \phi_n) \\ &\qquad + b_n \sin(2\pi f_n t + \phi_n)] + 2H_0 a_0 \\ &= 2 \sum_{n=1}^{N-1} H_n[(a_n \cos\phi_n + b_n \sin\phi_n) \cos 2\pi f_n t \\ &\qquad + (b_n \cos\phi_n - a_n \sin\phi_n) \sin 2\pi f_n t] \\ &\qquad + 2H_0 a_0, \qquad 0 \le t \le N\,\Delta t. \end{aligned} \tag{11}$$

As indicated in Fig. 4, $r(t)$ is sampled at times $k(\Delta t/2)$, $k = 0, 1, \cdots, 2N - 1$, and the samples $\{r_k\}$ are applied to a discrete Fourier transformer. The output of the DFT is

$$z_l = \frac{1}{2N} \sum_{k=0}^{2N-1} r_k e^{-j(2\pi lk/2N)}$$

$$\cong \begin{cases} 2H_0 a_0, & l = 0 \\ H_l[a_l \cos\phi_l + b_l \sin\phi_l] - H_l[b_l \cos\phi_l \\ \quad - a_l \sin\phi_l], & l = 1, 2, \cdots, N-1 \\ \text{irrelevant}, & l > N - 1. \end{cases} \tag{12}$$

Estimates of a_l and b_l are obtained from the computations

$$\left.\begin{aligned} \hat{a}_l &= \frac{1}{H_l} [\mathrm{Re}(z_l) \cos\phi_l + \mathrm{Im}(z_l) \sin\phi_l] \\ \hat{b}_l &= \frac{1}{H_l} [\mathrm{Re}(z_l) \sin\phi_l - \mathrm{Im}(z_l) \cos\phi_l] \end{aligned}\right\},$$

$$l = 1, 2, \cdots, N - 1. \tag{13a}$$

In complex notation, the appropriate computation is

$$\hat{a}_l - j\hat{b}_l = w_l z_l, \tag{13b}$$

where

$$w_l = \frac{1}{H_l} [\cos\phi_l - j \sin\phi_l]. \tag{13c}$$

Equations (13a–c) describe a 2×2 transformation to be performed on each of the DFT outputs z_l, $l = 1, 2, \cdots, N - 1$. For a reasonably large N and a typical communication channel, the approximation of $H(f)$ by a constant over each subchannel, which leads to (13a–c), may be adequate. However, linear rather than constant approximations to the amplitude and phase of the channel transfer function as it affects each subchannel waveform are much closer to reality. The following section examines the consequences of these approximations. It is shown that the truncated subchannel sinusoids are delayed by differing amounts, and that distortion is concentrated at the on–off transitions of these waveforms. Further, the magnitude of the distortion is proportional to the abruptness of the transitions. Hence a "guard space," consisting of a modest increase in the signal duration together with a smoothing of the on–off transitions, will eliminate most interference among channels and between adjacent transmission blocks. The individual channels can then be equalized in accord with (13a–c).

IV. Approximate Analysis of the Effects of Channel Distortion

The transmitted signal $y(t)$ as given by (5) exists only on the interval $(0, N\Delta t)$, so that each subchannel must, as noted earlier, accommodate a $\sin f/f$ type spectrum. As suggested in the last section, let this spectrum be narrowed by increasing the signal duration to some $T > N\,\Delta\,t$ and requiring gradual rather than abrupt roll-offs of the transmitted waveform. Specifically, the transmitted signal will be redefined as

$$y(t) = 2g_a(t) \sum_{n=0}^{N-1} [a_n \cos(2\pi f_n t) + b_n \sin(2\pi f_n t)], \tag{14}$$

where an appropriate $g_a(t)$ is

$$g_a(t) = \begin{cases} \frac{1}{2}\left[1 + \cos \frac{\pi t}{2aT}\right], & -2aT \le t < 0 \\ 1, & 0 \le t < T \\ \frac{1}{2}\left[1 + \cos \frac{\pi(t - T)}{2aT}\right], & T \le t < (1 + 2a)T \\ 0 \quad \text{elsewhere}. \end{cases} \tag{15}$$

The "window function" $g_a(t)$ is sketched in Fig. 5.

When $y(t)$ is passed through the channel filter with impulse response $h(t)$, the received signal is

$$\begin{aligned} r(t) &= \sum_{n=0}^{N-1} \{a_n[2h(t)*g_a(t) \cos 2\pi f_n t] \\ &\qquad + b_n[2h(t)*g_a(t) \sin 2\pi f_n t]\} \\ &= \sum_{n=0}^{N-1} \{a_n q_a^{(n)}(t) + b_n q_b^{(n)}(t)\}, \end{aligned} \tag{16}$$

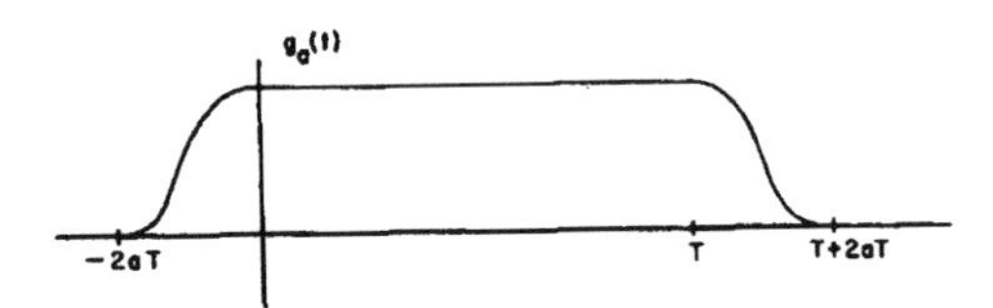

Fig. 5. Window $g_a(t)$ multiplying all subchannel sinusoids in transmitted signal.

where

$$q_a^{(n)}(t) = 2h(t) * g_a(t) \cos 2\pi f_n t$$
$$q_b^{(n)}(t) = 2h(t) * g_a(t) \sin 2\pi f_n t. \qquad (17)$$

In Appendix I, linear approximations to the amplitude and phase of $H(f)$ around $f = \pm f_n$ result in the following approximate expression for $q_a^{(n)}(t)$.

$$q_a^{(n)}(t) \cong 2H_n \cos[2\pi f_n t + \phi_n] g_a(t - \beta_n) + \frac{\alpha_n}{\pi} \sin[2\pi f_n t + \phi_n] \frac{d}{dt} g_a(t - \beta_n), \qquad (18)$$

where (H_n, ϕ_n) is the channel sounding at frequency f_n, and α_n and β_n are the slopes at $f = f_n$ of the linear approximations to amplitude and phase of $H(f)$, as shown in Fig. 6. A similar expression results for $q_b^{(n)}(t)$. The first term on the right-hand side of (18) is the nth cosine element in the transmitted signal (14), except that it is modified by a channel sounding (H_n, ϕ_n) and subjected to an envelope delay β_n. Interblock interference can result if a delayed sinusoid from a previous block impinges on the current sampling period. The second term is distortion arising from the amplitude variations of $H(f)$, and it is a potential source of interchannel interference. Suppose, however, that T is large enough so that

$$T > (2N - 1)\frac{\Delta t}{2} + \max_n (\beta_n) - \min_n (\beta_n), \qquad (19)$$

as pictured in Fig. 7. Then for all n there exists a time $t_0 > \max \beta_n$ such that

$$\frac{d}{dt} g_a(t - \beta_n) = 0, \qquad t_0 \le t \le (2N - 1)\frac{\Delta t}{2} + t_0. \qquad (20)$$

Fig. 7 shows where the interval $(t_0, t_0 + (2N - 1)(\Delta t/2))$ is located with respect to the minimum and maximum values of the time shift β_n. Thus,

$$q_a^{(n)}(t) \cong 2H_n \cos(2\pi f_n t + \phi_n), \qquad t_0 \le t \le t_0 + \frac{(2N-1)\Delta t}{2}. \qquad (21)$$

By a similar derivation,

$$q_b^{(n)}(t) \cong 2H_n \sin(2\pi f_n t + \phi_n), \qquad t_0 \le t \le t_0 + \frac{(2N-1)\Delta t}{2}. \qquad (22)$$

Therefore, substituting (21) and (22) into (16) (except for $n = 0$),

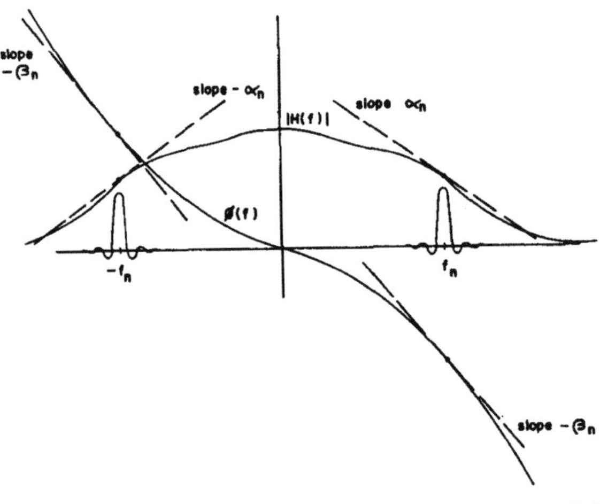

Fig. 6. Linear approximations to amplitude and phase of $H(f)$ in relation to spectra $G_a(f - fn)$ and $G_a(f + fn)$.

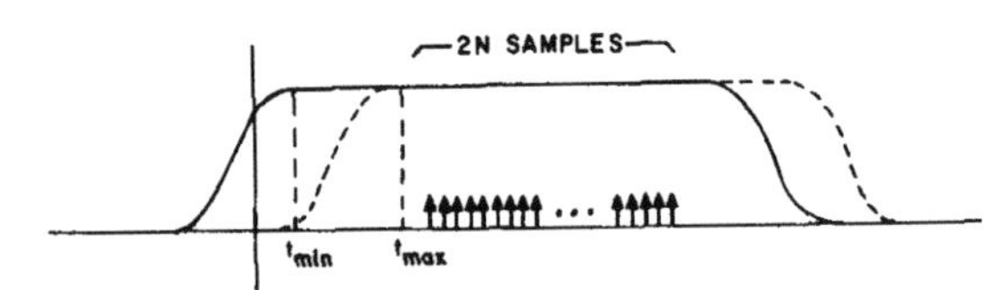

Fig. 7. Shifted versions of $g_a(t)$ corresponding to subchannels with minimum and maximum delay and locations of samples taken by receiver. Here $t_{min} = \min_n \beta_n$, $t_{max} = \max_n \beta_n$.

$$r(t) \cong 2\sum_{n=1}^{N-1} H_n[a_n \cos(2\pi f_n t + \phi_n) + b_n \sin(2\pi f_n t + \phi_n)] + 2H_0 a_0, \qquad t_0 \le t \le t_0 + (2N - 1)\frac{\Delta t}{2}. \qquad (23)$$

Except for the shifted domain of definition, (23) is identical to (11), which led to (13) for retrieval of the data. It can be shown that initiating sampling of $r(t)$ at $t = t_0$ instead of at $t = 0$ is equivalent to incrementing each phase ϕ_l by $f_l\, t_0$ rad in the equalization equations (13).

Intuitively, $r(t)$ reduces to (23) because linear distortion delays different spectral components by different amounts and responds to abrupt transitions with "ringing." If the subchannel waveforms are examined during an interval when they are all present, and if all the on-off transitions lie well outside this interval, then the waveforms will look like the collection of sinusoids described by (5). The linear approximations to amplitude and phase of $H(f)$ restrict the ringing from the transition periods to those periods themselves. In practice, the ringing can be expected to die out very rapidly outside of the transition periods. This use of a guard space is a common technique, as, for example, described in [3].

V. Algorithm for Minimizing Mean-Square Distortion

Under the assumption supported by the results of the last section that interchannel interference is negligible, a simple algorithm can be devised for determination of the parameters $\cos\phi_l/H_l$ and $\sin\phi_l/H_l$ on each channel, which minimize mean-square distortion when used in the transformation of (13a–c). Because of propagation delay and the presence of noise, these parameters may not exactly correspond to a sounding of the lth subcarrier channel. The use of an automatic equalization procedure based on the minimum mean-square distortion algorithm leads to accurate demodulation without having to make precise channel soundings. Further, the algorithm can work adaptively after an initial coarse adjustment.

The receiver produces estimates $\hat{a}_l$ and $\hat{b}_l$ according to the formulas

$$\hat{a}_l = T_{1l}\,\mathrm{Re}\,Z_l + T_{2l}\,\mathrm{Im}\,Z_l$$
$$\hat{b}_l = T_{2l}\,\mathrm{Re}\,Z_l - T_{1l}\,\mathrm{Im}\,Z_l, \tag{24}$$

which resemble (13a–c), except that the T coefficients are to be chosen to minimize the estimation error.[1] Mean-square distortion is defined by

$$\begin{aligned}\epsilon &= E\sum_{l=1}^{N-1}[e_{la}^2 + e_{lb}^2]\\ &= E\sum_{l=1}^{N-1}[(T_{1l}\cdot\mathrm{Re}\,Z_l + T_{2l}\cdot\mathrm{Im}\,Z_l - a_l)^2\\ &\qquad + (T_{2l}\cdot\mathrm{Re}\,Z_l - T_{1l}\cdot\mathrm{Im}\,Z_l - b_l)^2],\end{aligned} \tag{25}$$

where

$$e_{la} \triangleq \hat{a}_l - a_l$$
$$e_{lb} \triangleq \hat{b}_l - b_l. \tag{26}$$

It is shown in Appendix II that ϵ is a convex function of the vector $\bar{T}$, where

$$\bar{T} = (T_{11}, T_{21}, T_{12}, T_{22}, \cdots, T_{1(n-1)}, T_{2(n-1)}). \tag{27}$$

Thus a steepest descent algorithm is sure to converge to the vector $\bar{T}_0$ yielding the minimum mean-square distortion.

The $l_{1,2}$th components of the gradient of ϵ with respect to $\bar{T}$ are

$$(\nabla\epsilon)_{l1} = \frac{\partial\epsilon}{\partial T_{1l}} = E[2e_{la}\,\mathrm{Re}\,Z_l - 2e_{lb}\,\mathrm{Im}\,Z_l]$$
$$(\nabla\epsilon)_{l2} = \frac{\partial\epsilon}{\partial T_{2l}} = E[2e_{la}\,\mathrm{Im}\,Z_l + 2e_{lb}\,\mathrm{Re}\,Z_l]. \tag{28}$$

The steepest descent algorithm makes changes at the end of each block transmission in a direction opposite to the gradient:

$$\Delta T_{1l} = -k[e_{la}\,\mathrm{Re}\,Z_l - e_{lb}\,\mathrm{Im}\,Z_l]$$
$$\Delta T_{2l} = -k[e_{la}\,\mathrm{Im}\,Z_l + e_{lb}\,\mathrm{Re}\,Z_l], \tag{29}$$

where k controls the step size. A block diagram of the implementation of (30) is shown in Fig. 8. The initial value of $\bar{T}$ is probably best obtained from crude channel soundings, or specified as some "typical" vector quantity. It is expected that the first round of adjustments, made at the end of the first block transmission, will suffice to reduce the error to a low level, if it is not already low with the initial value of $\bar{T}$. At $\Delta t = 0.5$ ms, the length of one block before addition of a guard space will vary from about 8 ms (16 subchannels) to about 64 ms (128 subchannels). This block length, plus the guard space necessary to minimize interference, is a transmission delay that cannot be avoided.

The equalization algorithm given here only equalizes distortion due to cochannel interference (channels on the same frequency) and completely ignores interchannel interference. An unpublished analysis by the authors shows that for this equalizer, the interchannel interference becomes arbitrarily small as the number of subchannels increases. This is true even without any of the signal modification described in Section IV.

[1] The vector $\bar{T}$ and its subscripted components should not be confused with the signal duration T used earlier.

VI. Transmission Without Equalization

We have shown that for narrow subchannels the channel can be equalized by multiplying z_l, the receiver output for the lth subcarrier channel, by a number w_l [see (13c)]. This is simply compensation for attenuation and phase shift in that particular subchannel. The interference among subchannels is made small by using a guard time and smooth transitions between blocks.

For binary transmission, the attenuation need not be compensated, and if differential phase transmission between subchannels is used, no phase equalization is needed. In order for this technique to work, the difference in phase of the transmission channel transfer function $H(f)$ between adjacent subchannels should be small. Assume this is the case and let

$$d_n = (a_n - jb_n)d_{n-1}, \tag{30}$$

where a_n and b_n are binary information digits on the nth subchannel. For the first block transmission, d_0 is necessarily a fixed reference. At the output of the DFT in the receiver, form the product

$$\begin{aligned}z_n z_{n-1}^* &= h_n d_n h_{n-1}^* d_{n-1}^*\\ &= (a_n - jb_n)\,|h_n|^2\,|d_{n-1}|^2\\ &\quad + (a_n - jb_n)h_n(h_{n-1}^* - h_n^*)\,|d_{n-1}|^2,\\ &\qquad n = 1, 2, \cdots, N-1,\end{aligned} \tag{31}$$

where $h_n = H(f_n)$ is the complex channel transfer function at the center frequency of the nth subchannel. The last part of (31) is the information signal times an unknown amplitude term, plus an error term depending on $h_n - h_{n-1}$. For binary transmission, the information signal can be reliably recovered if the second term is less than half of the first term. This is equivalent to saying that the phase of h does not change by more than

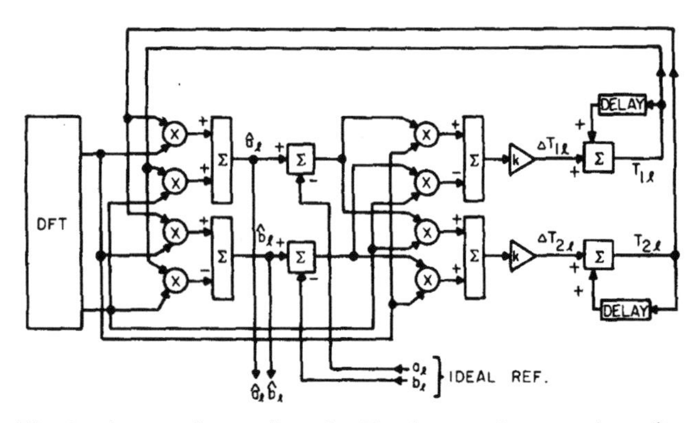

Fig. 8. Automatic equalizer for Fourier transform receiver. One such apparatus is required for each of the $(n - 1)$ usable outputs of the discrete Fourier transformer.

$30°$ between the centers of adjacent subchannels. Because a_n and b_n are binary, $(a_n - jb_n)$ may be recovered by determining which quadrant of the complex plane contains $z_n z_{n-1}^*$, even though h_n is unknown.

For the second and all subsequent block transmissions, d_0 can carry information by comparing it with d_0 from the previous T-second transmission, i.e.,

$$d_{0\ \text{current}} = (a_0 - jb_0) d_{0\ \text{previous}}. \tag{32}$$

Appendix I

Received Signal Under Linear Approximations to Amplitude and Phase of the Channel Transfer Function

The received signal is given by (16) as

$$r(t) = \sum_{n=0}^{N-1} [a_n q_a^{(n)}(t) + b_n q_b^{(n)}(t)], \tag{33}$$

where

$$q_a^{(n)}(t) = 2h(t) * [g_a(t) \cos (2\pi f_n t)]$$

$$q_b^{(n)}(t) = 2h(t) * [g_a(t) \sin (2\pi f_n t)]. \tag{34}$$

Restricting attention for the time being to $q_a^{(n)}(t)$, the Fourier transform of this function is

$$Q_a^{(n)}(f) = H(f)[G_a(f - f_n) + G_a(f + f_n)], \tag{35}$$

where $G_a(f)$ is the Fourier transform of the "window function" $g_a(t)$. We now make linear approximations to the amplitude and phase of $H(f)$ as they affect the separate spectra $G_a(f - f_n)$ and $G(f + f_n)$ in (35). For $G_a(f - f_n)$, the approximation is

$$H(f) \cong [H_n + \alpha_n(f - f_n)] e^{j[\phi_n - \beta_n(f - f_n)]}, \tag{36}$$

and for $G_a(f + f_n)$ the approximation is

$$H(f) \cong [H_n - \alpha_n(f - f_n)] e^{-j[\phi_n + \beta_n(f + f_n)]}. \tag{37}$$

The validity of these approximations depends on the narrowness of $G_a(f)$ and the size of f_n, as illustrated in Fig. 6. One can always select a window function $g_a(t)$ for which the approximations of (36) and (37) lead to arbitrarily accurate expressions for $q_a^{(n)}(t)$ and $q_b^{(n)}(t)$ except for the lowest values of n.

Substituting approximations (36) and (37) into (35),

$$Q_a^{(n)}(f) \cong [H_n + \alpha_n(f - f_n)] e^{j[\phi_n - \beta_n(f - f_n)]} + [H_n - \alpha_n(f - f_n)] e^{-j[\phi_n + \beta_n(f + f_n)]} G_a(f + f_n). \tag{38}$$

Thus

$$\begin{aligned} q_a^{(n)}(t) &= \int_{-\infty}^{\infty} Q_a^{(n)}(f) e^{j2\pi f t}\, df \\ &\cong \int_{-\infty}^{\infty} [H_n + \alpha_n(f - f_n)] e^{j[\phi_n - \beta_n(f - f_n)]} \\ &\quad \cdot G_a(f - f_n) e^{j2\pi f t}\, df + \int_{-\infty}^{\infty} [H_n - \alpha_n(f + f_n)] \\ &\quad \cdot e^{-j[\phi_n + \beta_n(f + f_n)]} G_a(f + f_n) e^{j2\pi f t}\, df. \end{aligned} \tag{39}$$

After appropriate changes of variable, these Fourier transforms can be evaluated by recalling the following Fourier transform pairs

$$\begin{aligned} x(t - t_0) &\Leftrightarrow e^{-jft_0} X(f) \\ \frac{d}{dt} x(t - t_0) &\Leftrightarrow j2\pi f e^{-jft_0} X(f). \end{aligned} \tag{40}$$

Thus

$$q_a^{(n)}(t) \cong 2H_n \cos [2\pi f_n t + \phi_n] g_a(t - \beta_n) + \frac{\alpha_n}{\pi} \sin [2\pi f_n t + \phi_n] \frac{d}{dt} g_a(t - \beta_n). \tag{41}$$

This approximation appears as (19) in the main body of this paper. A similar derivation yields an analogous approximation to $q_b^{(n)}(t)$.

Appendix II

Convexity of Mean-Square Error

The mean-square error has been defined as

$$\epsilon(\bar{T}) = E \sum_{l=1}^{N-1} \{[T_{1l} \operatorname{Re} z_l + T_{2l} \operatorname{Im} z_l - a_l]^2 + [T_{2l} \operatorname{Re} z_l - T_{1l} \operatorname{Im} z_l - b_l]^2\}, \tag{42}$$

where the vector $\bar{T}$ is defined as

$$\bar{T} = (T_{11}, T_{21}, T_{22}, T_{22}, \cdots, T_{1(N-1)}, T_{2(N-1)}). \tag{43}$$

Taking a single term of the sum, we have

$$\begin{aligned} &T_{1l}^2(\operatorname{Re} z_l)^2 + T_{2l}^2(\operatorname{Im} z_l)^2 + 2 \operatorname{Re} Z_l \operatorname{Im} z_l T_{1l} T_{2l} \\ &\quad - 2a_l[T_{1l} \operatorname{Re} z_l + T_{2l} \operatorname{Im} z_l] + a_l^2 + T_{2l}^2(\operatorname{Re} z_l)^2 \\ &\quad + T_{2l}^2(\operatorname{Im} z_l)^2 - 2 \operatorname{Re} z_l \operatorname{Im} z_l T_{1l} T_{2l} \\ &\quad - 2b_l[T_{2l} \operatorname{Re} z_l - T_{1l} \operatorname{Im} z_l] + b_l^2 \\ &= (T_{1l}^2 + T_{2l}^2) |z_l|^2 + 2T_{1l}(b_l \operatorname{Im} z_l - a_l \operatorname{Re} z_l) \\ &\quad - 2T_{2l}(a_l \operatorname{Im} z_l + b_l \operatorname{Re} z_l) + a_l^2 + b_l^2, \end{aligned}$$

which is the sum of convex functions and is thus convex itself. Since $\epsilon(\bar{T})$ is the sum of convex functions, it too is convex.

ACKNOWLEDGMENT

The authors are indebted to J. E. Mazo for comments on approximation techniques, and to J. Salz for earlier work on this project.

REFERENCES

[1] J. Salz and S. B. Weinstein, "Fourier transform communication system," presented at the Ass. Comput. Machinery Conf. Computers and Communication, Pine Mountain, Ga., Oct. 1969.
[2] B. R. Salzberg, "Performance of an efficient parallel data transmission system," *IEEE Trans. Commun. Technol.*, vol. COM-15, Dec. 1967, pp. 805–811.
[3] M. L. Doelz, E. T. Heald, and D. L. Martin, "Binary data transmission techniques for linear systems," *Proc. IRE*, vol. 45, May 1957, pp. 656–661.
[4] R. W. Chang and R. A. Gibby, "A theoretical study of performance of an orthogonal multiplexing data transmission scheme," *IEEE Trans. Commun. Technol.*, vol. COM-16, Aug. 1968, pp. 529–540.
[5] E. N. Powers and M. S. Zimmerman, "TADIM—A digital implementation of a multichannel data modem," presented at the 1968 IEEE Int. Conf. Communications, Philadelphia, Pa.
[6] W. W. Abbott, R. C. Benoit, Jr., and R. A. Northrup, "An all-digital adaptive data modem," presented at the IEEE Computers and Communication Conf., Rome, N. Y., 1969.
[7] W. W. Abbott, L. W. Blocker, and G. A. Bailey, "Adaptive data modem," Page Commun. Eng., Inc., RADC—TR-69-296, Sept. 1969.
[8] M. S. Zimmerman and A. L. Kirsch, "The AN/GSC-10 (KATHRYN) variable rate data modem for HF radio," *IEEE Trans. Commun. Technol.*, vol. COM-15, Apr. 1967, pp. 197–204.
[9] J. L. Holsinger, "Digital communication over fixed time-continuous channels with memory, with special applications to telephone channels," M.I.T. Res. Lab. Electron., Cambridge, Rep. 430, 1964.

PHOTO NOT AVAILABLE

S. B. Weinstein (S'59–M'66) was born in New York, N. Y., on November 25, 1938. He received the B.S. degree from the Massachusetts Institute of Technology, Cambridge, in 1960, the M.S. degree from the University of Michigan, Ann Arbor, in 1962, and the Ph.D. degree from the University of California, Berkeley, in 1966.

Upon finishing his graduate studies he worked for approximately one year with the Philips Research Laboratories in Eindhoven, the Netherlands, before joining the Advanced Data Communications Department at Bell Telephone Laboratories, Holmdel, N. J., in 1968. He works in the areas of statistical communication theory and data communications.

PHOTO NOT AVAILABLE

Paul M. Ebert (M'60) was born in Madison, Wis., on December 30, 1935. He received the B.S. degree from the University of Wisconsin, Madison, in 1958, and the M.S. and Sc D. degrees from the Massachusetts Institute of Technology, Cambridge, in 1962 and 1965, respectively.

From 1958 to 1960 he was a member of the Airborne Communications Division at the Radio Corporation of America, Camden, N. J. Since 1965 he has been a member of the Advanced Data Communications Department, Bell Telephone Laboratories, Holmdel, N. J. He has worked in the fields of information theory, coding, and digital filtering.

Viterbi Decoding for Satellite and Space Communication

JERROLD A. HELLER
Member, IEEE

IRWIN MARK JACOBS
Member, IEEE

Abstract—Convolutional coding and Viterbi decoding, along with binary phase-shift keyed modulation, is presented as an efficient system for reliable communication on power limited satellite and space channels. Performance results, obtained theoretically and through computer simulation, are given for optimum short constraint length codes for a range of code constraint lengths and code rates. System efficiency is compared for hard receiver quantization and 4 and 8 level soft quantization. The effects on performance of varying of certain parameters relevant to decoder complexity and cost is examined. Quantitative performance degradation due to imperfect carrier phase coherence is evaluated and compared to that of an uncoded system. As an example of decoder performance versus complexity, a recently implemented 2-Mbit/s constraint length 7 Viterbi decoder is discussed. Finally a comparison is made between Viterbi and sequential decoding in terms of suitability to various system requirements.

Paper approved by the Communication Theory Committee of the IEEE Communication Technology Group for publication without oral presentation. This work was supported in part by NASA Ames Research Center under Contract NAS2-6024. Manuscript received June 10, 1971.
The authors are with the Linkabit Corporation, San Diego, Calif.

I. Introduction

THE SATELLITE and space communication channels are likely candidates for the cost-effective use of coding to improve communication efficiency. The primary additive disturbance on these channels can usually be accurately modeled by Gaussian noise which is "white" enough to be essentially independent from one bit time interval to the next, and, particularly on the space channel but also in many instances on satellite channels, sufficient bandwidth is available to permit moderate bandwidth expansion. Two effective decoding algorithms for independent noise (memoryless) channels have been developed and refined, namely sequential and Viterbi decoding of convolutional codes. These theoretical accomplishments, combined with real communication

Reprinted from *IEEE Transactions on Communication Technology,* vol. COM-19, no. 5, October 1971.

needs and the availability of low-cost complex digital integrated circuits, make possible practical and powerful high-speed decoders for satellite and space communication.

Communication from a distant and isolated object in space to a ground-based station presents certain system problems which are not nearly as critical in earth-based communication systems. The most obvious among these is the high cost of space-platform power. It is desirable to design a system which is as efficient as practical in order to minimize the spacecraft weight necessary to generate power.

The modulated signal power at a ground station receiver front end P depends upon the transmitted power, the transmitting and receiving antenna gains, and propagation path losses. Primarily due to thermal activity at the receiver front end, wideband noise is added to the received signal, resulting in a received signal power-to-noise ratio (P/N_0), where N_0 is the single-sided noise spectral density. The noise is usually accurately modeled as being both white and Gaussian. Other perturbations caused by uncertainty in carrier phase at the demodulator and inaccuracies in receiver AGC are treated in Sections IV and V.

The efficiency of a communication system is usefully measured by the received energy per bit to noise ratio (E_b/N_0) required to achieve a specified system bit error rate. The E_b/N_0 is expressable in terms of the modulating signal power by the relationship

$$\frac{E_b}{N_0} = \frac{P}{N_0} \cdot \frac{1}{R} \tag{1}$$

where R is the information rate in bits per second. Alternatively, (1) can be written as

$$R = \frac{P/N_0}{E_b/N_0}. \tag{2}$$

The payoff for using modulation and/or coding techniques which reduce the E_b/N_0 required for a given bit error probability is an increase in allowable data rate and/or a decrease in necessary received P/N_0.

As a point of reference, it is traditional to compare the efficiency of modulation-coding schemes with that of a hypothetical system operating at channel capacity. Channel capacity for an infinite bandwidth white Gaussian noise channel with average power P is [1]

$$C_\infty = \frac{P}{N_0 \ln 2} \text{ bit/s}. \tag{3}$$

From (1), when $R = C_\infty$,

$$E_b/N_0 = \ln 2 \triangleq \frac{E_{b,\min}}{N_0}. \tag{4}$$

Thus, the lower bound on achievable E_b/N_0 is about -1.6 dB.

Without coding, required E_b/N_0 can be minimized by selecting an efficient modulation technique. For example, 180° binary phase-shift keying (BPSK) is more efficient than binary frequency shift keying (BFSK). For a desired bit error rate of 10^{-5}, an E_b/N_0 of 9.6 dB is required using BPSK (antipodal) modulation, whereas, 12.6 dB is required with BFSK (orthogonal) modulation. Quadraphase-shift keying (QPSK) is often used to conserve bandwith. Under the assumption of perfect phase coherence, QPSK has the same performance as BPSK.

In designing a communication system to operate at a specified data rate, the improvement in efficiency to be realized using coding must be weighed against the relative costs. Potential alternatives include increasing the transmitted power, increasing the transmitting antenna gain, and/or the receiving antenna area, and accepting a higher probability of bit error. In many applications, a minimum P_e is required and the incremental cost per decibel increase in P/N_0 is now greater (often much greater) than the cost of reducing the needed E_b/N_0 through coding. Soft decision Viterbi and hard decision sequential decoding can provide a relatively inexpensive 4–6-dB improvement in required E_b/N_0 (at a 10^{-5} bit error rate), even at multimegabit data rates. Sequential decoding is extensively discussed in [5]. In Section VII, we compare these techniques. Sections II and III examine various aspects of Viterbi decoding and present curves permitting system tradeoffs. In Section VI, a particular implementation of a Viterbi decoder is discussed to provide one benchmark for cost-complexity discussions.

In the discussion that follows, we assume that the channel is power limited rather than bandwidth limited. This assumption is realistic for many present day and future systems; however, the trend, especially in satellite repeaters, is to larger P/N_0 without a proportional increase in available bandwidth. For this reason, we will limit consideration to codes which involve a "bandwidth expansion" of 3 or less; that is, we assume that from 1 to 3 binary symbols can be transmitted over the channel for each bit of information communicated without appreciable intersymbol interference.

II. System

A. Convolutional Encoder

Fig. 1 shows a general binary-input binary-output convolutional coder. The encoder consists of a kK stage binary shift register and v mod-2 adders. Each of the mod-2 adders is connected to certain of the shift register stages. The pattern of connections specifies the code. Information bits are shifted into the encoder shift register k bits at a time. After each k bit shift, the outputs of the mod-2 adders are sampled sequentially yielding the code symbols. These code symbols are then used by the modulator to specify the waveforms to be sent over the channel. Since v code symbols are generated for each set of k information bits, the code rate R_N is k/v information bits per code symbol, where $k < v$. The constraint length of the code is K, since that is the number of k bit shifts over which a single information bit can

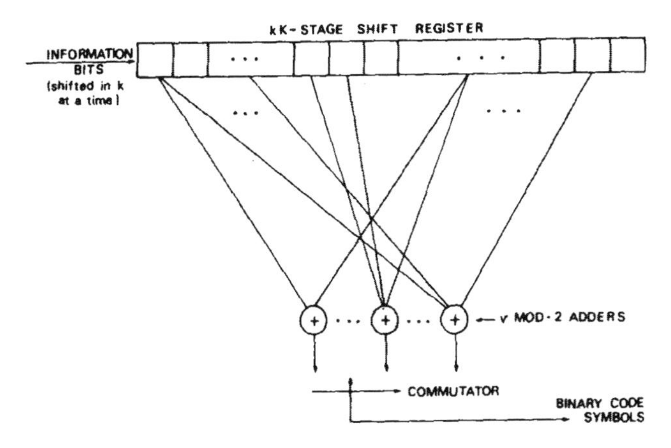

Fig. 1. Rate k/n convolutional encoder.

influence the encoder output. The state of the convolutional encoder is the contents of the first $k(K - 1)$ shift register stages. The encoder state together with the next k input bits uniquely specify the v output symbols.

As an example, a $K = 3$, $k = 1$, $v = 2$ encoder is shown in Fig. 2(a). The first two coder stages specify the state of the encoder; thus, there are 4 possible states. The code words, or sequences of code symbols, generated by the encoder for various input information bit sequences is shown in the code "trellis" [2] of Fig. 2(b). The code trellis is really just a state diagram for the encoder of Fig. 2(a). The four states are represented by circled binary numbers corresponding to the contents of the first two stages of the encoder. The lines or "branches" joining states indicate state transitions due to the input of single information bits. Dashed and solid lines correspond to "1" and "0" input information bits, respectively. The trellis is drawn under the assumption that the encoder is in state 00 at time 0. If the first information bit were a 1, the encoder would go to state 10 and would output the code symbols 11. Code symbols generated are shown adjacent to the trellis branches. As an example, the input data sequence 101 $\cdots$ generates the code symbol sequence 111000 $\cdots$. Further interpretations of the encoder state diagram and a discussion of "good" convolutional codes is presented in [3].

B. Modulation

The binary symbols output by the encoder are used to modulate an RF carrier sinusoid. Here we restrict our attention to the case of 180° BPSK modulation. Each code symbol results in the transmission of a pulse of carrier at either of two 180° separated phases. A squence of code symbols produces a uniformly spaced sequence of biphase pulses. The signal component of the received waveform thus has the form

$$s(t) = \sum_i \sqrt{2E_s}\, p(t - iT_s) \cos(2\pi f_c t + x_i \pi/2 + \theta)$$
$$= \sqrt{2E_s} \cos(2\pi f_c t + \theta) \sum_i \cdot x_i p(t - iT_s). \quad (5)$$

Here x_i is ± 1 depending on whether the ith code symbol is 0 or 1. The function $p(t)$ is a convenient unit energy low-pass pulse waveform, f_c is the carrier frequency, E_s is the energy per pulse, and T_s is the time between successive code symbols. E_s and T_s are defined by the relationships

$$E_s = R_N E_b = kE_b/v \quad (6)$$

and

$$T_s = R_N/R. \quad (7)$$

There are several reasons for restricting attention to BPSK modulation. Three important ones are as follows.

1) BPSK signals are convenient to generate and amplify. Traveling wave tube amplifiers operate most efficiently at or near saturation. This nonlinear amplification would degrade performance with multilevel amplitude modulated waveforms.

2) It can be shown that antipodal (BPSK) modulation results in little increase in required E_b/N_0 compared to optimum signaling when E_s/N_0 is low [4].

3) BPSK modulation of quadrature carriers is equivalent to quadraphase (QPSK) modulation of one carrier. Thus, QPSK need not be separately treated except for synchronization and phase error requirements.

C. Demodulation and Quantization

At the receiver, the signal $s(t)$ of (5), is observed added to white Gaussian noise. When the carrier phase θ is known, the optimum demodulator consists of an integrate and dump filter matched to $p(t)$ cos $(2\pi f_c t + \theta)$. At time jT_s, the demodulator outputs data r_j relevant to the jth code symbol. Normalizing the matched filter output by dividing by $\sqrt{N_0/2}$ yields

$$r_j = x_j \sqrt{2E_s/N_0} + n_j \quad (8)$$

when n_j is a zero-mean unit variance Gaussian random variable. Each n_j is independent of all others.

To facilitate digital processing by the decoder, the continuous r_j must be quantized. The simplest quantization is a hard decision with 0 output if r_j is greater than zero and 1 output otherwise. Here, the received data are represented by only one bit per code symbol. Without coding, the matched filter sampler hard quantizer is an optimum receiver.

When coding is used, hard quantization of the received data usually entails a loss of about 2 dB in E_b/N_0 compared with infinitely fine quantization [4], [5]. Much of this loss can be recouped by quantizing r_j to 4 or 8 levels instead of merely 2. Adding additional levels of quantization necessitates a 2- or 3-bit representation of each r_j. Fig. 3(a) and (b) shows two quantization schemes with 4 and 8 levels, respectively. Here the quantization level thresholds are spaced evenly. The spacing is 1.0 for 4 levels and 0.5 for 8 levels. Uniform quantization threshold spacings of 1.0 and 0.5 can be shown by analytical means and through simulation to be very close to optimum for 4- and 8-level quantiza-

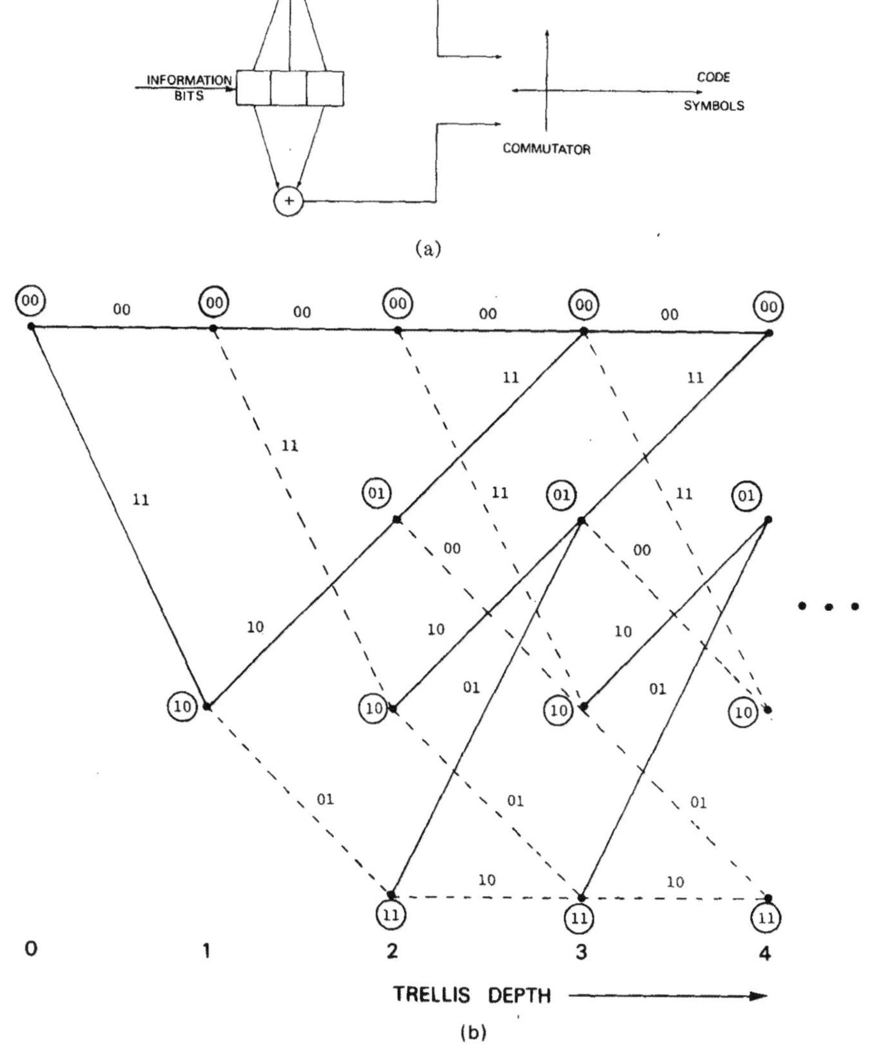

Fig. 2. (a) $K = 3$, $R_N = 1/2$ convolutional encoder. (b) Code trellis diagram.

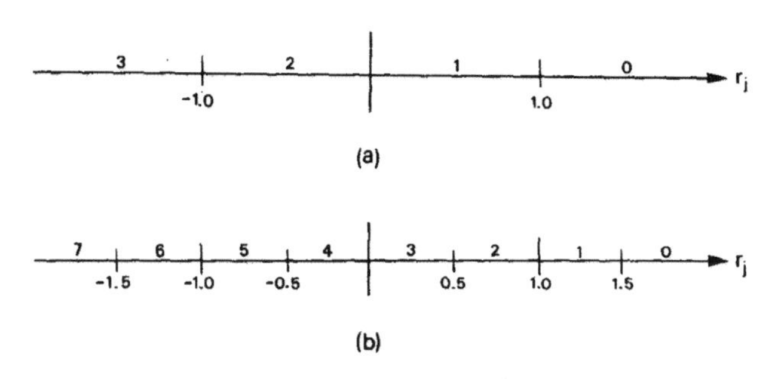

Fig. 3. Receiver quantization thresholds and intervals for (a) 4-level and (b) 8-level quantization.

tion. Furthermore, 8-level quantization results in a loss of less than 0.25 dB compared to infinitely fine quantization; therefore, quantization to more than 8 levels can yield little performance improvement. We confine our attention to hard decision quantization and the 4 and 8 level schemes shown in Fig. 3.

Receiver quantization converts the modulator, Gaussian channel, and demodulator into a discrete channel with two inputs (0 or 1, the code symbols) and 2, 4, or 8 outputs. The channel transition probabilities are a function only of the symbol signal-to-noise ratio E_s/N_0. For example, with 8-level quantization, the probability of receiving interval 6 given that a 0-code symbol is sent is the probability that a unit-variance Gaussian random variable with mean $\sqrt{2E_s/N_0}$ lies between -1.0 and -1.5 in value.

III. Viterbi Decoding

A. Basic Algorithm

The maximum likelihood or Viterbi decoding algorithm was discovered and analyzed by Viterbi [6] in 1967. Viterbi decoding was first shown to be an efficient and practical decoding technique for short constraint length codes by Heller [7], [8]. Forney [2] and Omura [12] demonstrated that the algorithm was in fact maximum likelihood.

A thorough discussion of the Viterbi decoding algorithm is presented by Viterbi [3]. Here, it will suffice to briefly review the algorithm and elaborate on those features and parameters which bear on decoder performance and complexity on satellite and space communication channels.

Referring to the code trellis diagram of Fig. 2(b), a brute-force maximum likelihood decoder would calculate the likelihood of the received data for code symbol sequences on all paths through the trellis. The path with the largest likelihood would then be selected, and the information bits corresponding to that path would form the decoder output. Unfortunately, the number of paths for an L bit information sequence is 2^L; thus, this brute force decoding quickly becomes impractical as L increases.

With Viterbi decoding, it is possible to greatly reduce the effort required for maximum likelihood decoding by taking advantage of the special structure of the code trellis. Referring to Fig. 2(b), it is clear that the trellis assumes a fixed periodic structure after trellis depth 3 (in general, K) is reached. After this point, each of the 4 states can be entered from either of two preceding states. At depth 3, for instance, there are 8 code paths, 2 entering each state. For example, state 00 at level 3 has the two paths entering it corresponding to the information sequences 000 and 100. These paths are said to have diverged at state 00, depth 0 and remerged at state 00, depth 3. Paths remerge after 2 [in general $k(K - 1)$] consecutive identical information bits. A Viterbi decoder calculates the likelihood of each of the 2^k paths entering a given state and eliminates from further consideration all but the most likely path that leads to that state. This is done for each of the $2^{k(K-1)}$ states at a given trellis depth; after each decoding operation only one path remains leading to each state. The decoder then proceeds one level deeper into the trellis and repeats the process.

For the $K = 3$ code trellis of Fig. 2(b), there are 8 paths at depth 3. Decoding at depth 3 eliminates 1 path entering each state. The result is that 4 paths are left. Going on to depth 4, the decoder is again faced with 8 paths. Decoding again eliminates 4 of these paths, and so on. Note that in eliminating the less likely paths entering each state, the Viterbi decoder will not reject any path which would have been selected by the brute force maximum likelihood decoder.

The decoder as described thus far never actually decides upon one most likely path. It always retains a set of $2^{k(K-1)}$ paths after each decoding step. Each retained path is the most likely path to have entered a given encoder state. One way of selecting a single most likely path is to periodically force the encoder into a prearranged state by inputting a $K - k$ bit fixed information sequence to the encoder after each set of L information bits. The decoder can then select that path leading to the known encoder state as its (1 bit) output.

The great advantage of the Viterbi maximum likelihood decoder is that the number of decoder operations performed in decoding L bits is only $L2^{k(K-1)}$, which is linear in L. Of course, Viterbi decoding as a practical technique is limited to relatively short constraint length codes due to the exponential dependence of decoder operations per bit decoded on K. Fortunately, as will be shown, excellent decoder performance is possible with good short constraint length codes.

B. Path Memory

In order to make the Viterbi algorithm a practical decoding technique, certain refinements on the basic algorithm are desirable. First of all, periodically forcing the encoder into a known state by using preset sequences multiplexed into the data stream is neither operationally desirable nor necessary. It can be shown [2], [9] that with high probability, the $2^{k(K-1)}$ decoder selected paths will not be mutually disjoint very far back from the present decoding depth. All of the $2^{k(K-1)}$ paths tend to have a common stem which eventually branches off to the various states. This suggests that if the decoder stores enough of the past information bit history of each of the $2^{k(K-1)}$ paths, then the oldest bits on all paths will be identical. If a fixed amount of path history storage is provided, the decoder can output the oldest bit on an arbitrary path each time it steps one level deeper into the trellis. The amount of path storage required u is equal to the number of states, $2^{k(K-1)}$ multiplied by the length of the information bit path history per state h,

$$u = h2^{k(K-1)}. \tag{9}$$

Since the path memory represents a significant portion of the total cost of a Viterbi decoder, it is desirable to minimize the required path history length h. One refinement which allows for a smaller value of h is to use the oldest bit on the most likely of the $2^{k(K-1)}$ paths as the decoder output, rather than the oldest bit on an arbitrary path. It has been demonstrated theoretically [2] and through simulation [9] that a value of h of 4 or 5 times the code constraint length is sufficient for negligible degradation from optimum decoder performance. Simulation results showing performance degradation incurred with smaller path history lengths are presented and discussed in Section IV.

C. State and Branch Metric Quantization

The path comparisons made for paths entering each state require the calculation of the likelihood of each path involved for the particular received information. Since the channel is memoryless, the path likelihood function is the product of the likelihoods of the individual code symbols [3]

$$P(\mathbf{r}^*/\mathbf{x}^l) = \prod_i P(r_i^*/x_i^l) \tag{10}$$

where $\mathbf{r}^* = (r_1^*, r_2^*, \cdots, r_i^*, \cdots)$ is the vector of quantized receiver outputs and $\mathbf{x}^l = (x_1^l, x_2^l, \cdots, x_i^l, \cdots)$ is the code symbol vector for the lth trellis path. In order to avoid multiplication, the logarithm of the likelihood is a preferable path metric

$$\begin{aligned} M_l &= \log p(\mathbf{r}^*/\mathbf{x}^l) \\ &= \sum_j \log p(r_j^*/x_j^l) \triangleq \sum_j m_j^l \end{aligned} \tag{11}$$

where M_l is the metric of the lth path and m_j^l is the metric of the jth code symbol on the lth path. With this type of additive metric, when a path is extended by one branch, the metric of the new path is the sum of the new branch symbol metrics and the old path metric. To facilitate this calculation, the path metric for the best path leading to each state must be stored by the decoder as a state metric. This is an addition to the path information bit history storage required.

Viterbi decoder operation can then be summarized as follows, taking the $K = 3$ case of Fig. 2 as an example.

1) The metric for the 2 paths entering state 00 are calculated by adding the previous state metrics of states 00 and 01 to the branch metrics of the upper and lower branches entering state 00, respectively.

2) The largest of the two new path metrics is stored as the new state metric for state 00. The new path history for state 00 is the path history of the state on the winning path augmented by a 0 or 1 depending on whether state 00 or 01 was on the winning path.

3) This add-compare-select (ACS) operation is performed for the paths entering each of the other 3 states.

4) The oldest bit on the path with the largest new path metric forms the decoder output.

Since the code symbol metrics must be represented in digital form in the decoder, the effects of metric quantization come into question. Simulation has shown that decoder performance is quite insensitive to symbol metric quantization. In fact, use of the integers as symbol metrics instead of log likelihoods results in a negligible performance degradation with 2-, 4-, or 8-level receiver quantization [7], [9]. Fig. 4 shows such a set of metrics for the 8-level quantized channel. Use of these symbol metrics implies that symbol metrics as well as the received symbols themselves may be represented by 1, 2, or 3 bits for 2-, 4-, and 8-level receiver quantization, respectively.

D. Unknown Starting State

It has been assumed thus far that a Viterbi decoder has knowledge of the encoder starting state before decoding begins. Thus, in Fig. 2(b), the starting state is assumed to be 00. A known starting state may be operationally undesirable since it requires that the decoder know when transmission commences. In reality, it has been found through simulation that a Viterbi decoder may start decoding at any arbitrary point in a transmission, if all state metrics are initially reset to zero. The first 3–4 constraint lengths worth of data output by the decoder will be more or less unreliable because of the unknown encoder starting state. However, after about 4 constraint lengths, the state metrics with high probability have values independent of the starting values and steady-state reliable operation results.

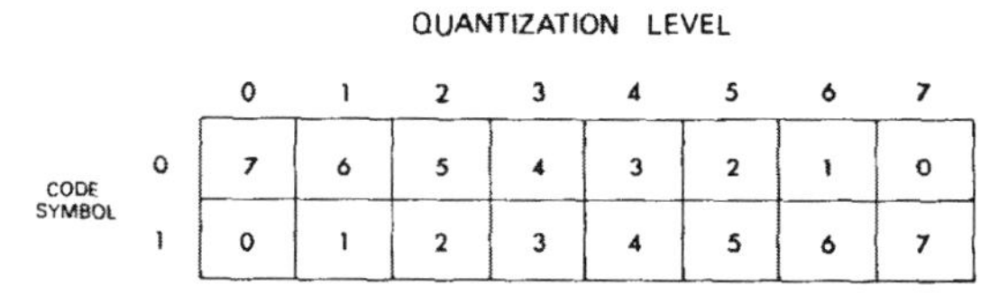

CODE SYMBOL \ QUANTIZATION LEVEL	0	1	2	3	4	5	6	7
0	7	6	5	4	3	2	1	0
1	0	1	2	3	4	5	6	7

Fig. 4. Integer code symbol metrics for 8-level receiver quantization.

IV. Simulation and Error Probability Bound Results

A. Tradeoffs Between Bit Error Probability and E_b/N_0 for Rate 1/2 Codes

Viterbi has derived tight upper bounds to bit error probability for Viterbi decoding based on the convolutional code transfer function [3]. These bounds are particularly tight for the white Gaussian noise channel for error probabilities less than about 10^{-4}. This bound has been numerically evaluated over a range of E_b/N_0 for a variety of codes. The upper bound is presented along with some of the 8-level receiver quantized simulation results for comparison. The upper bound also provides performance data at very-low bit error rates, where simulation results are not available due to excessive computer time required. In comparing the upper bounds to the simulation results, it is important to keep in mind that the upper bound was derived for an infinitely finely quantized receiver output.

The convolutional codes used in the simulations were found through exhaustive computer search [9], [10]. The search criterion was maximization of the minimum free distance for a given code constraint length [3]. Where two codes had the same minimum free distance, the number of codewords at that distance and the higher order free distances were used for code selection. Simulations have consistently shown that the free distance criterion yields codes with the minimum error probability. The principal results of the simulations and code transfer function bounds are shown in Figs. 5, 6, and 7. All of these figures show bit error rate versus E_b/N_0 for Viterbi decoders using optimum rate 1/2 convolutional codes. In all cases, the decoder path history length was 32 bits. In all simulation runs, at least 25 error events contributed to the compiled statistics.

B. Performance Depending on Quantization, Path History, and Receiver Automatic Gain Control

The simulation results in Figs. 5 and 6 are for soft (8-level) receiver quantization. Equally spaced demodulation thresholds are used as shown in Fig. 3(b). This choice of 8-level quantizer thresholds is within a broad range of near optimum values, as will be shown presently. The transfer function bound is for infinitely finely quantized received data, although tight bounds for any degree of quantization can be obtained. Allowing for the 0.20–0.25 dB loss usually associated with 8-level receiver quantization compared with infinite quantization, the transfer function bound curves are in excellent

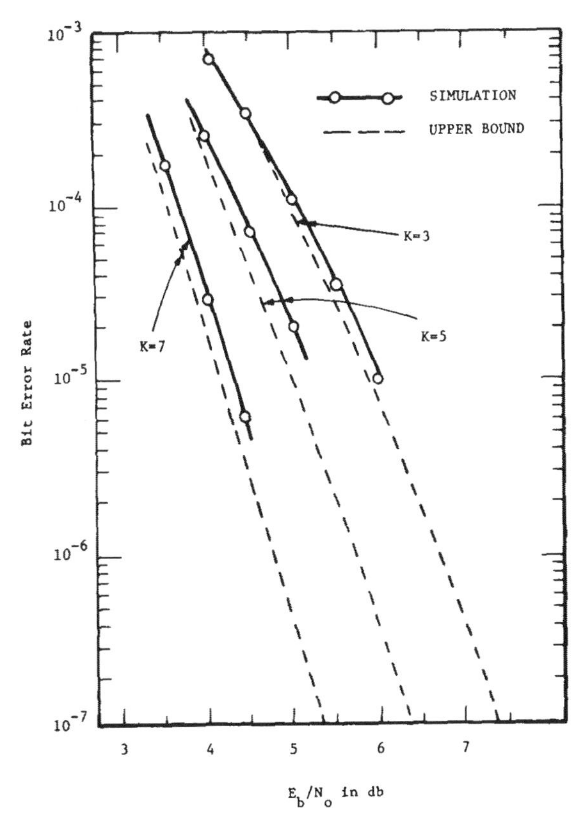

Fig. 5. Bit error rate versus E_b/N_0 for rate 1/2 Viterbi decoding. 8-level quantized simulations with 32-bit paths, and infinitely finely quantized transfer function bound, $K = 3, 5, 7$.

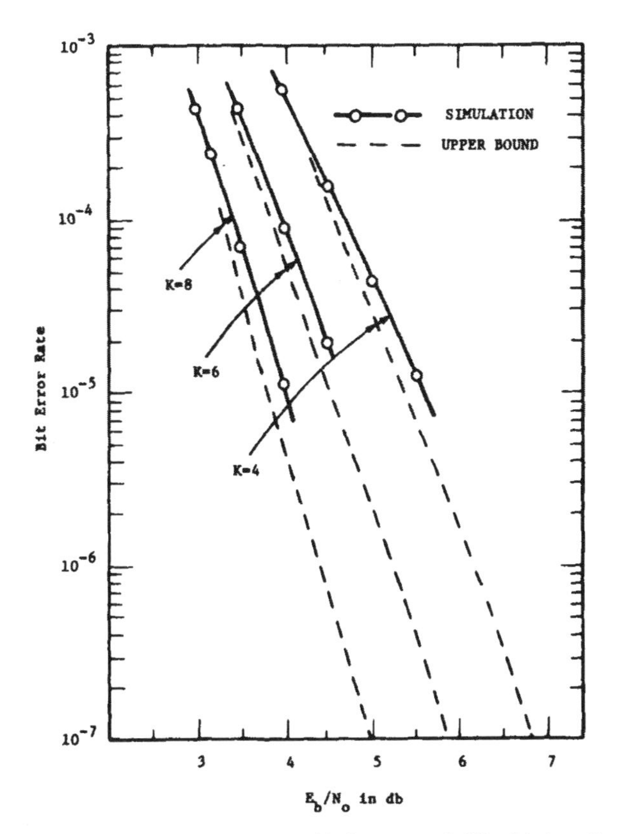

Fig. 6. Bit error rate versus E_b/N_0 for rate 1/2 Viterbi decoding 8-level quantized simulations with 32-bit paths, and infinitely finely quantized transfer function bound, $K = 4, 6, 8$.

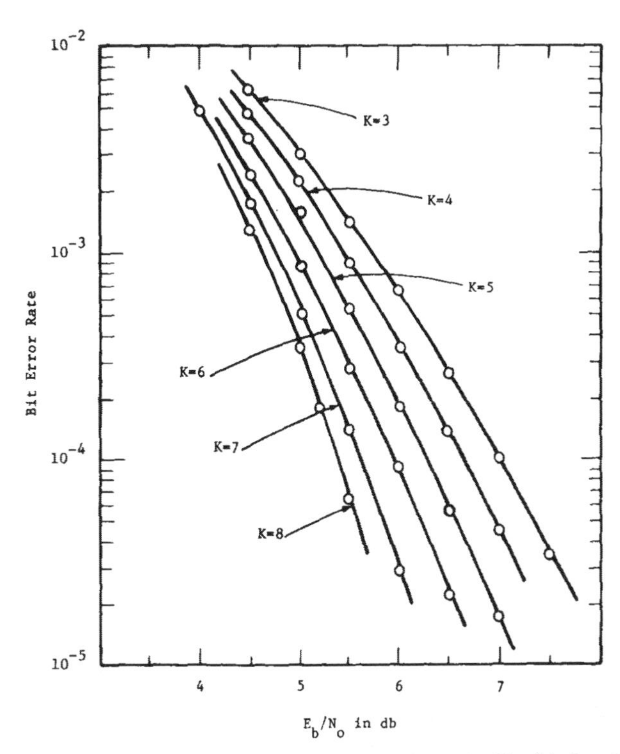

Fig. 7. Bit error rate versus E_b/N_0 for rate 1/2 Viterbi decoding. Hard quantized received data with 32-bit paths; $K = 3$ through 8.

agreement with simulation results in the 10^{-4} to 10^{-5} bit error rate range.

Since the accuracy of the transfer function bound increases with E_b/N_0, decoder performance can be ascertained accurately in the 10^{-5} to 10^{-8} region even in the absence of simulation. The symbol metrics used in the simulation were the equally spaced integers as shown in Fig. 4.

Fig. 7 gives the simulation results for Viterbi decoding with hard receiver quantization. The same optimum rate 1/2, $K = 3$ through $K = 8$ codes were used here as in the 8-level quantized simulations.

The following points are obvious from the performance curves.

1) 2-level quantization is everywhere close to 2-dB inferior to 8-level quantization.

2) Each increment in K provides an improvement in efficiency of something less than 0.5 dB at a bit error rate of 10^{-5}.

3) Performance improvement versus K increases with decreasing bit error rate.

To observe the effects of varying receiver quantization more closely, simulation performance data are presented in Fig. 8 for the $K = 5$, rate 1/2 code, with 2-, 4-, and 8-level receiver quantization. The $Q = 8$- and $Q = 4$-level thresholds are those of Fig. 3.

Fig. 9 shows bit error rate performance versus E_b/N_0 for three values of path history length (8, 16, and 32) using the rate 1/2, $K = 5$ code, for both 2- and 8-level received data quantization. (The length 32 path curve is identical to the $K = 5$ curve in Fig. 5.) Performance with

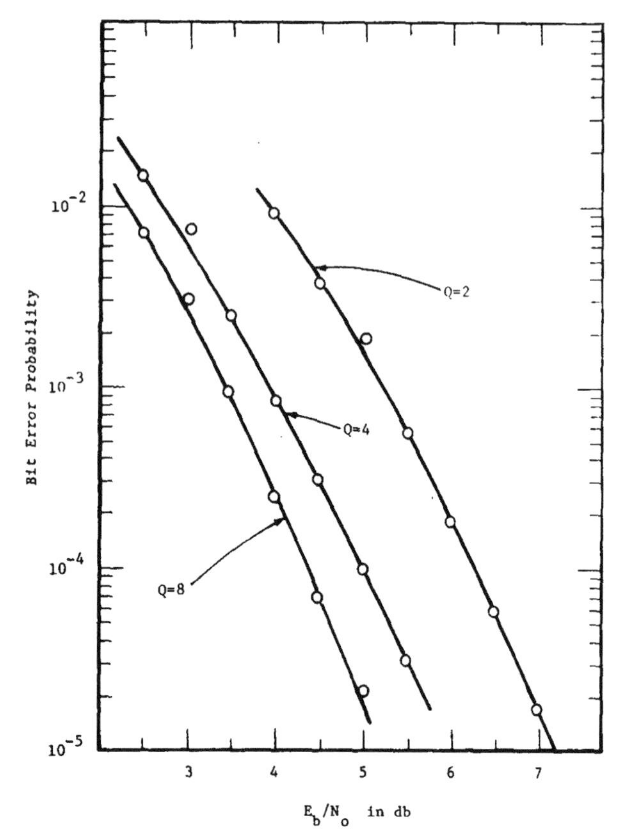

Fig. 8. Performance comparison of Viterbi decoding using rate 1/2, $K = 5$ code with 2-, 4-, and 8-level quantization. Path length = 32 bits.

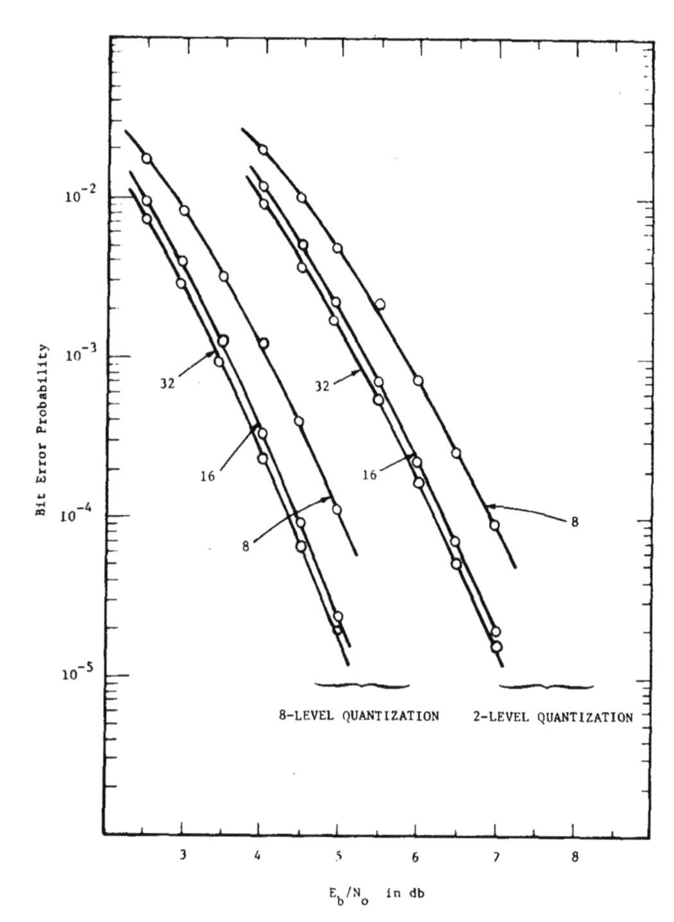

Fig. 9. Performance comparison of Viterbi decoding using rate 1/2, $K = 5$ code with 8-, 16-, and 32-bit path lengths and 2- and 8-level quantization.

length 32 paths is essentially identical to that of an infinite path decoder. Even for a path length of only 16, there is only a small degradation in performance. As previously mentioned, other simulations have shown that a path length of 4–5 constraint lengths is sufficient for other constraint lengths as well.

Coded systems that make use of receiver outputs quantized to more than two levels require an analog-to-digital converter at the modem matched filter output, with thresholds that depend on correct measurement of the noise variance. Since the level settings are effectively controlled by the automatic gain control (AGC) circuitry in the modem, it is of interest to investigate the sensitivity of decoder performance to an inaccurate or drifting AGC signal. Fig. 10 shows the decoder performance variation as a function of A–D converter level threshold spacing. In all cases, the thresholds are uniformly spaced. These simulations use the $K = 5$ rate 1/2 code with $E_b/N_0 = 3.5$ dB. It is evident that Viterbi decoding performance is quite insensitive to wide variations in AGC gain. In fact, performance is essentially constant over a range of spacing from 0.5 to 0.7. This allows for a variation in AGC gain of better than ±20 percent with no significant performance degradation.

C. Performance of Codes of Other Rates

The preceding simulation results have concentrated on Viterbi decoding of rate 1/2 convolutional codes. The

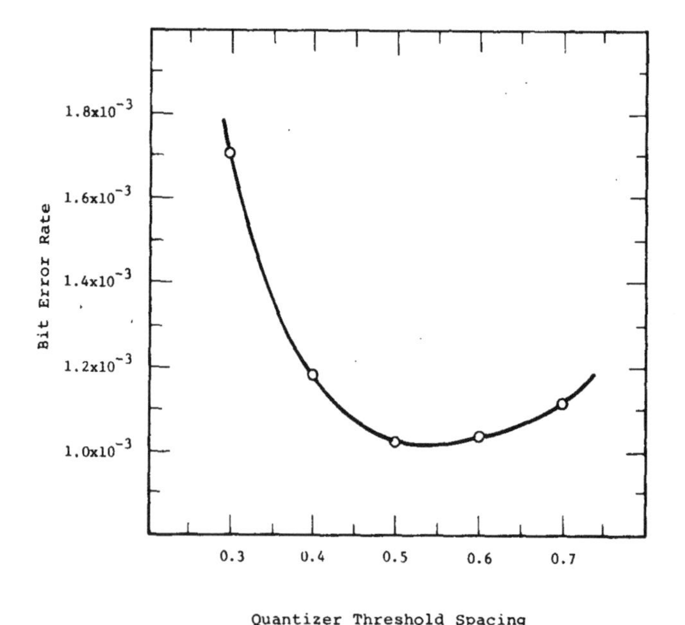

Fig. 10. Viterbi decoder bit error rate performance as function of quantizer threshold level spacing; $K = 5$, rate 1/2, $E_b/N_0 = 3.5$ dB, 8-level quantization with equally spaced thresholds.

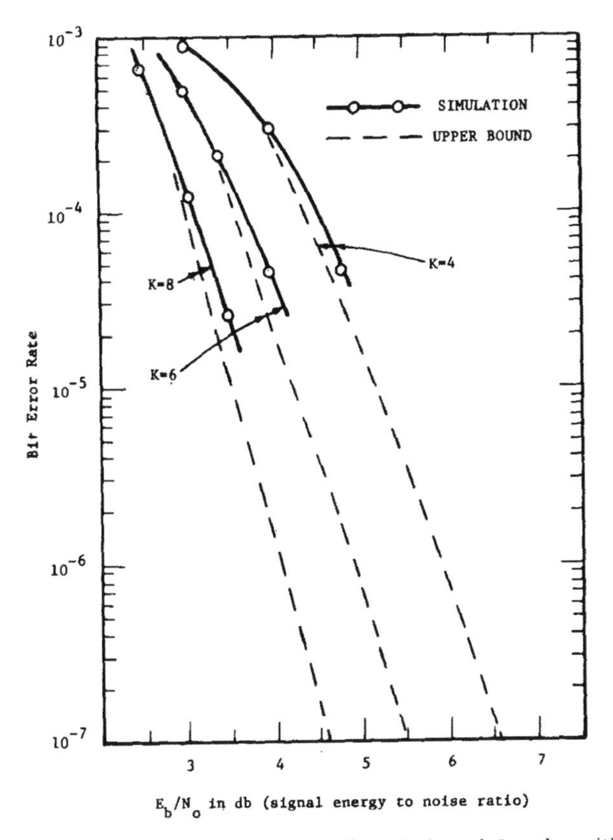

Fig. 11. Performance of rate 1/3, $K = 4$, 6, and 8 codes with Viterbi decoding.

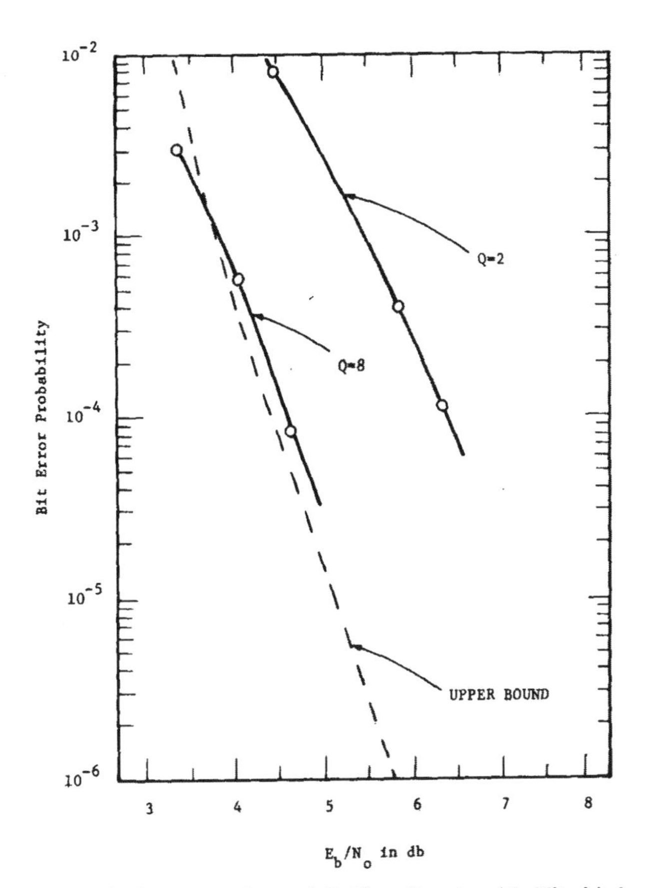

Fig. 12. Performance of rate 2/3 $K = 3$ code with Viterbi decoding. Numerical bound and simulation results.

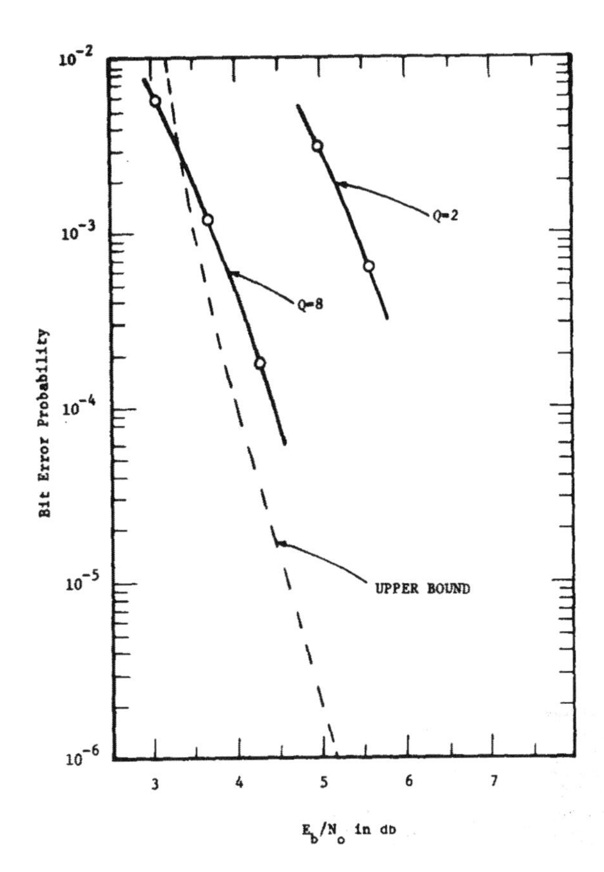

Fig. 13. Performance of rate 2/3 $K = 4$ code with Viterbi decoding. Numerical bound and simulation results.

results on performance fluctuation due to decoder parameter variation carry over to other code rates with minor changes.

Code rates less than 1/2 buy improved performance at the expense of increased bandwith expansion and more difficult symbol tracking due to decreased symbol energy-to-noise ratios. Rates above 1/2 conserve bandwidth but are less efficient in energy.

Fig. 11 shows bit error rate versus E_b/N_0 performance obtained from simulations of Viterbi decoding with optimum rate 1/3, $K = 4$, 6, and 8 codes, and 8-level quantization. Figs. 12 and 13 show numerical bound and simulation performance results for rate 2/3 $K = 3$ and $K = 4$ codes, respectively. Simulation curves are for 2- and 8-level quantization, while the numerical bound curves are for infinitely fine receiver quantization.

Comparing the performance data obtained through simulations of Viterbi decoders with rate 1/2 (Figs. 5, 6, and 7), and rate 1/3 codes, it is apparent that the latter offers a 0.3-to-0.5-dB improvement over the former for fixed K, in the range reported. This is close to the improvement in efficiency of a channel with capacity 1/3 compared with one of capacity 1/2, and is therefore expected.

Comparison of the higher rate codes with the rate 1/2 codes may also be made over the range spanned by the simulation and analytical data. The fairest comparison is probably between decoders with similar number of states, and hence similar decoder complexity. Thus, the $K = 3$ rate 2/3 data should be compared with the $K = 5$, rate 1/2 data.

Fig. 14 shows the union bounds on performances for the rate 2/3, $K = 3$, and rate 1/2, $K = 5$ codes. Both encoders have 16 states. The free distance d_f equals 7 for the rate 1/2 code and 5 for the rate 2/3 codes. At very high E_b/N_0, the rate 1/2 must be superior. This is because asymptotically, at high E_b/N_0, the error probability varies as

$$P_e \sim n_e \exp(-d_f E_s/N_0) = n_e \exp(-d_f R_N E_b/N_0)$$

where n_e is the number of bit errors contributed by codewords at distance d_f. This gives the rate 1/2 code an advantage of about 0.2 dB in the limit.

In Fig. 14, the difference between the two curves is about 0.1 dB in the error probability range of 10^{-6} to 10^{-9}. This small difference is due to the fact that the rate 2/3 code used happens to be a particularly good code; the value of n_e is smaller for it than for the rate 1/2 code and this difference is significant even for P_e as small as 10^{-9}.

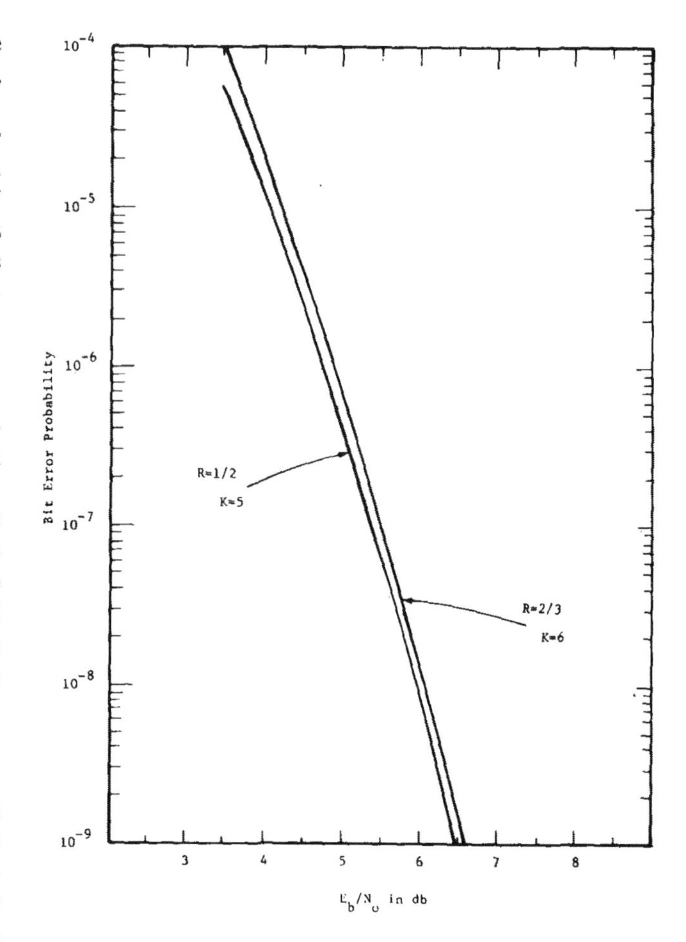

Fig. 14. Bit error probability bound for rate 1/2, $K = 5$, and rate 2/3, $K = 3$ code.

V. Imperfect Carrier Phase Coherence

Thus far it has been assumd that carrier phase is known exactly at the receiver. In real systems this is usually not the case. Oscillator instabilities and uncompensated doppler shifts necessitate closed loop carrier phase tracking at the receiver. Since the carrier loop tracks a noisy received signal, the phase reference it provides for demodulation will not be perfect.

An inaccurate carrier phase reference at the demodulator will degrade system performance. In particular a constant error ϕ in the demodulator phase will cause the signal component of the matched filter output to be suppressed by the factor $\cos\phi$ (see [4, ch. 7]).

$$r_i = \pm\sqrt{\frac{2E_s}{N_0}} \cos\phi + n_i. \qquad (12)$$

The effect of an imperfect carrier phase reference on performance is always worse for coded than uncoded systems. This is because coded systems are characterized by steeper error probability versus E_b/N_0 curves than uncoded systems. An imperfect carrier phase reference causes an apparent loss in received energy-to-noise ratio. Since the coded curve is steeper, the loss in E_b/N_0 degrades error probability to a greater extent. Furthermore, unless care is taken in the design of the phase-tracking loop, the phase error might be higher for the coded system than for an uncoded system, since loop performance may depend upon E_s/N_0, which is significantly smaller for coded than uncoded systems.

For convolutional coding with phase coherent demodulation and Viterbi decoding, exact analytical expressions for bit error rate P_e versus E_b/N_0 are not attainable. The simulation results of the preceding section, however, define a relationship between P_e and E_b/N_0 that can be written formally as

$$P_e = f\left(\frac{E_b}{N_0}\right) \qquad (13)$$

for a given code, receiver quantization, and Viterbi decoder. Since the carrier phase is being tracked in the presence of noise the phase error ϕ will vary with time. To simplify analysis, assume that the data rate is large compared to the carrier loop bandwidth so that the phase error does not vary significantly during perhaps 20–30 information bit times. Viterbi decoder output errors are typically several bits in length and are very rarely longer than 10–20 bits when the overall decoder bit error probability is less than 10^{-3}. Therefore, the phase error is assumed to be constant over the length of almost any decoder error. This being the case, the bit error probability for a constant phase error ϕ, can be written as

$$P_e(\phi) = f\left(\frac{E_b}{N_0}\cos^2\phi\right) \qquad (14)$$

from (12) and (13). This result uses the fact that received signal energy is degraded by $\cos^2\phi$. If ϕ is a random variable with distribution $p(\phi)$, the resulting error probability averaged on ϕ is

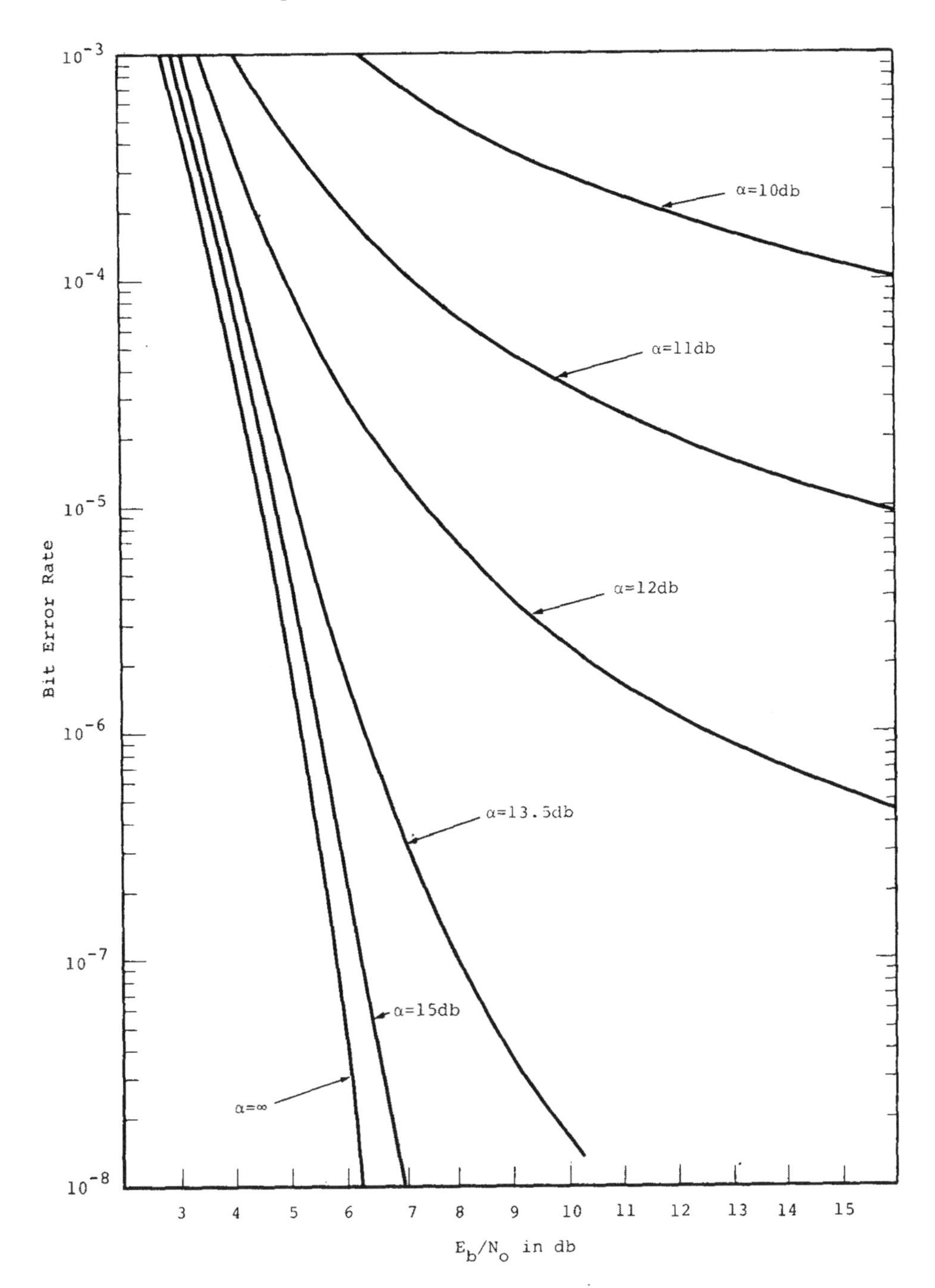

Fig. 15. Performance curves for rate 1/2; $K = 7$ Viterbi decoder with 8-level quantization as a function of carrier phase tracking loop signal-to-noise ratio α.

$$P_e' = \int_{-\pi}^{\pi} p(\phi) P_e(\phi)\, d\phi. \tag{15}$$

For the second-order phase-locked loop

$$p(\phi) = \frac{e^{\alpha \cos \phi}}{2\pi I_0(\alpha)}, \qquad \alpha \gg 1 \tag{16}$$

where $I_0(\cdot)$ is the zeroth order modified Bessel function and α is the loop signal-to-noise ratio [11]. Using this distribution and the P_e versus E_b/N_0 curve for the $K = 7$, rate 1/2 code of Fig. 5, the P_e integral of (15) has been evaluated for several values of α. The results are shown in Fig. 15 as curves of P_e' versus E_b/N_0 with α as a parameter (the $K = 7$, rate 1/2 simulation curve of Fig. 5 was extrapolated to get the high E_b/N_0 results shown in this figure). These curves exhibit the same general shape as those for uncoded binary PSK modulation with phase coherence provided by a carrier tracking loop. As expected, the losses due to imperfect coherence are somewhat greater with than without coding. Fig. 16 shows the additional E_b/N_0 required to maintain a 10^{-5} bit error rate as a function of loop signal to noise ratio α. Curves are shown for the case of uncoded BPSK and rate 1/2, $K = 7$ convolutional encoding—Viterbi decoding.

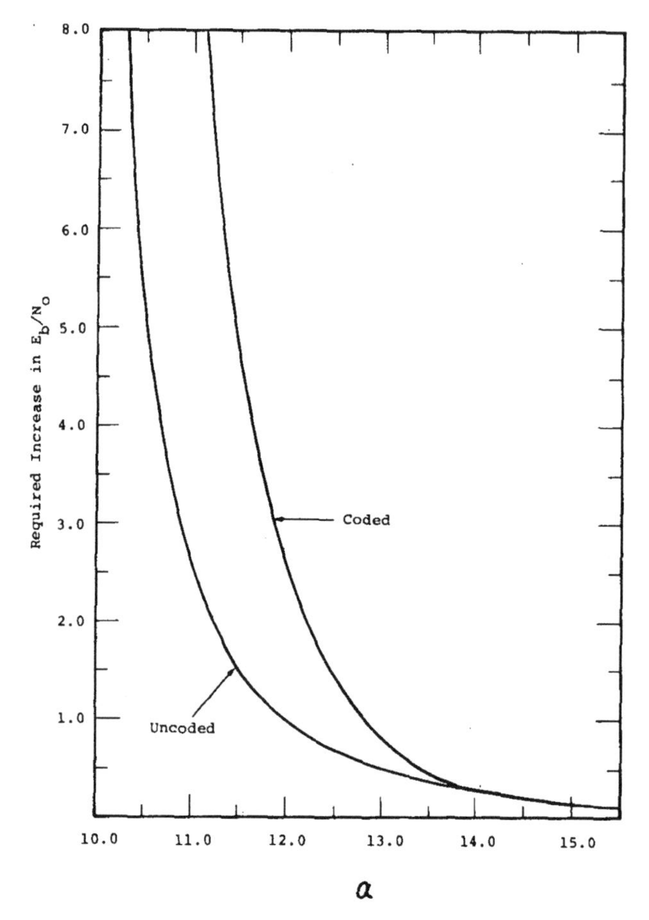

Fig. 16. Comparison to increase in E_b/N_0 due to imperfect phase coherence necessary to maintain 10^{-5} bit error rate for uncoded BPSK and $K = 7$; rate 1/2, $Q = 8$ Viterbi decoding.

VI. Implementation of a Viterbi Decoder

It is convenient to break the basic Viterbi decoder into five functional units; an input or branch metric calculation section, an ACS arithmetic section, and a path memory and output section. Information can be thought of as passing successively from one section to the next.

The branch metric calculation section accepts the input data and calculates (or looks up) the metric for each distinct branch. For a rate 1/2 code, four branches are possible corresponding to transmission of 00, 01, 10, and 11. For a rate 1/3 or rate 2/3 code, eight distinct branch metrics are possible. Note that this is the only section of the decoder that is directly concerned with the number of bits of quantization of the received data, and hence, the only section whose complexity is directly dependent on quantization. (The complexity of the ACS section also depends on quantization indirectly, in that the number of bits required for storing state metrics increases with the number of bits of quantization.) The input section is generally not critical in terms of either complexity or speed limitations. Its complexity does double, however, for each increase of the denominator of the rate R_N by one.

The ACS sections perform the basic arithmetic calculations of the decoder. For a rate $1/v$ decoder, an ACS is used to add the state metrics for two states to the appropriate branch metrics, to compare the resulting two sums, and to select the larger. The decision is transmitted to the path memory section and the larger of the two sums becomes a new state metric. One ACS function must be performed for each of the 2^{K-1} states. In a fully parallel very-high-speed decoder, 2^{K-1} ACS units are required. In general, the speed of the ACS unit places an upper bound on the speed of the decoder. For slower decoders, e.g., R less than several megabits per second for T²L logic, ACS units may be time shared, decreasing decoder cost significantly. Complexity of the ACS unit is strongly dependent upon required decoder speed. It should be noted that implementation of Viterbi decoders is greatly simplified by the fact that all ACS units perform identical functions and can be realized by a set of identical circuits.

The path memory section must store about a 4 constraint length history of decisions for each state. The memory requirements are thus nontrivial. Considerable advantage can be taken of new integrated-circuits memories to keep the equipment cost small. However, the complexity of the path memory and the ACS units both increase by a factor slightly larger than 2 for each increase in constraint length of 1. Thus, an increase in system performance of about 0.4 dB at a bit error rate of 10^{-5}, which can be achieved by increasing K by 1, comes at a cost of slightly more than doubling decoder complexity.

A complete decoder also must include interface circuits, synchronization circuits, timing circuits, and generally an encoder. A recent implementation[1] of a $K = 7$, rate 1/2 self-synchronized Viterbi decoder capable of operating at up to $R = 2$ Mbit/s with 2-, 4-, or 8-level quantized data required a total of 356 TTL integrated circuits for all functions. As noted in Fig. 5, this relatively simple decoder provides over 5-dB E_b/N_0 advantage over an uncoded BPSK system at $P_e = 10^{-5}$, and 6-dB advantage at $P^e = 10^{-8}$, when soft quantization is used.

VII. Comparison of Sequential and Viterbi Decoding

Both sequential and Viterbi decoding offer practical alternatives to a communications engineer designing a high-performance efficient communication system. The two decoders have significant differences which are noted below. Both are capable of very-high-speed operation.[2]

[1] The Linkabit LV7026 decoder is designed for use with differentially encoded BPSK or QPSK systems. It automatically resolves demodulator phase ambiguities and establishes node synchronization without manual intervention.

[2] The Linkabit LS4157 sequential decoder is capable of operation at data rates up to $R = 50$ Mbit/s. It uses a constraint length $K = 41$, rate 1/2 code and accepts only hard quantized data. The decoder is fully self-synchronizing. The coding advantage over uncoded data is 4.4 dB at $P_e = 10^{-5}$ at $R = 50$ Mbit/s and greater than 6 dB at $P_e = 10^{-8}$. The coding advantage is larger at lower data rates.

A. Error Probability

It should be recalled that, since the complexity of sequential decoders is relatively independent of constraint length, the constraint length is typically made quite large to provide a very small probability of undetected error. Usually the important contributor of errors is received data buffer overflow due to a computational overload. Such an event causes a long burst of rather noisy output data until the decoder reestablishes code synchronization. During this burst, the probability of bit error is that of the raw channel, perhaps $P_c = 3 \times 10^{-2}$.

Error from a Viterbi decoder occurs in short bursts of length at most 10 to 20. Systems that are sensitive to long bursts of errors should thus use Viterbi decoding. Systems that can tolerate occasional long bursts, with an error indication provided if desired by the decoder, should consider sequential decoding.

The curve of error probability versus E_b/N_0 tends to be much steeper for a sequential decoder than for a Viterbi decoder because of the difference in K. Thus, the sequential decoding advantage tends to increase as lower probabilities of bit error are demanded, although, as before, many errors tend to come in widely separated noisy bursts.

B. Decoder Delay

Sequential decoders tend to require long buffers of at least 200 bits and as much as several thousand bits to smooth out the variations in computational load. Viterbi decoders require a path memory of at most 64 bits. Thus the decoding delay differs by up to two orders of magnitude.

C. Long Tail Required to Terminate Sequences

In time-division multiplexed systems, bursts of separately encoded data may be received at the same decoder from different sources. In these instances, it may be desirable to time share the decoder. As noted in Section III, termination of encoding can be achieved by transmitting a known sequence of length $K - 1$, thus causing the encoder to enter a known state. Since K is typically larger for sequential decoding, the "tailing off" of the encoded sequence can cause a significant degradation in system efficiency. The tailing off of the short constraint length codes for Viterbi decoding causes a much smaller degradation.

If time and implementation permit the storage of the decoder state without code termination, then the cost of tailing off can be ignored. The design of such a time-shared sequential decoder remains for future work.

D. Rates Other than 1/2 and Soft Quantization

Viterbi decoders for rate 1/3 and 8-level quantization are not significantly more complex than those for rate 1/2 and 4- or 2-level quantization. The chief costs occur in the input section of the decoder as discussed in Section VI. In particular, the soft quantized data are processed in the input section and then incorporated in the branch and state metrics. No storage is required. A sequential decoder, on the other hand, must store several thousand branches of received data, each branch containing $\log_2 Q/R_N$ bits for rate R_N and Q level quantization. Although the possibility exists of gaining 0.4 dB by using rate 1/3 rather than rate 1/2 and of gaining 2 dB by using soft decisions rather than hard, these advantages are bought in sequential decoding at a formidable storage and processing cost. In general, then, practical high-rate sequential decoders are limited to rate 1/2 and hard decisions. (It is conceivable that this cost could be minimized by operating the decoder at a very high ratio of computation rate to average bit rate, thereby minimizing the number of branches required in the buffer.)

A second argument against soft quantization with sequential decoding involves the sensitivity of the probability of buffer overflow to channel variations. In Fig. 10, it was demonstrated that changes in receiver AGC of ±20 percent had negligible effect on the performance of a Viterbi decoder. The degradation is much more pronounced for sequential decoding, since the computational load is very sensitive to changes in channel parameters. Thus, part of the 2-dB gain anticipated for soft decisions might be lost unless great care was exercised in controlling receiver AGC precisely.

In comparing sequential decoding and Viterbi decoding, it thus appears fair to consider soft decisions only for the Viterbi decoder. Under these conditions, the efficiency advantage of a long constraint length sequential decoder is considerably diluted. Consequently, performance of a rate 1/2, $K = 41$ sequential decoder is no better than a rate 1/2 Viterbi decoder of constraint length 5 to 7 (depending on the speed factor, that is, the ratio of computation rate to bit rate) at a P_e of 10^{-5}. The sequential decoder does show a distinct advantage for P_e of 10^{-8} or smaller.

On the other hand, building a system without receiver quantization lowers system costs, since a considerably more crude AGC may be used.

E. Sensitivity to Phase Error and Bursty Conditions on the Channel

The performance of Viterbi decoding under slowly fluctuating phase error was presented in Fig. 15. A similar calculation would indicate much greater degradation in the case of sequential decoding, since the error probability curve is much steeper. Furthermore, this estimate would be optimistic in the case of sequential decoding, since the assumption that the phase varied so slowly that errors occurred independently would probably not hold for sequential decoding. Thus, more careful design of the phase-tracking loop is indicated for a system utilizing sequential decoding rather than Viterbi decoding.

VIII. Conclusions

Viterbi decoding has been shown to be a practical method for improving satellite and space communication efficiency by 4–6 dB, at a bit error rate of 10^{-5}. The successful implementation of 2-Mbit/s constraint-length-7 Viterbi decoders effectively demonstrates that the technique is well beyond the stage of being a theoretical curiosity. In fact, a major effort has been under way for the past 2–3 years with the aim of modifying and adapting the algorithm for minimum complexity implementation without sacrificing performance significantly.

In addition, Viterbi decoding has been shown to "degrade gracefully" in the presence of adverse channel or receiver conditions. In particular, the error probability does not change precipitously with E_b/N_0 as is the case with coding techniques that use longer codes and/or require variable decoding effort, such as sequential decoding. This ensures that performance degradation due to an imperfect phase or bit timing reference, or a slight correlation between noise samples, will be minimal. Requirements on AGC accuracy, even for soft decisions, were shown to be quite loose.

Finally the results presented here should provide the communication engineer with the information necessary to evaluate the applicability of Viterbi decoding to space and satellite communication systems with a wide range of requirements and constraints.

References

[1] C. E. Shannon, "Communication is the presence of noise," *Proc. IRE*, vol, 37, Jan. 1949, pp. 10–21.
[2] Codex Corp., Final Rep. "Coding system design for advanced solar missions," Contract NAS 2-3637, NASA Ames Res. Cent., Moffett Field, Calif.
[3] A. J. Viterbi, "Convolutional codes and their performance in communication systems," this issue, pp. 751–772.
[4] J. M. Wozencraft and I. M. Jacobs, *Principles of Communication Engineering*. New York: Wiley, 1965.
[5] I. M. Jacobs. "Sequential decoding for efficient communication from deep space," *IEEE Trans. Commun. Technol.*, vol. COM-15, Aug. 1967, pp. 492–501.
[6] A. J. Viterbi, "Error bounds for convolutional codes and an asymptotically optimum decoding algorithm," *IEEE Trans. Inform. Theory*, vol. IT-13, Apr. 1967, pp. 260–269.
[7] J. A. Heller, "Short constraint length convolutional codes," Jet Propulsion Lab., California Inst. Technol., Space Programs Summary 37–54, vol. III, Oct./Nov., 1968, pp. 171–177.
[8] ——, "Improved performance of short constraint length convolutional codes," Jet Propulsion Lab., California Inst. Technol., Space Programs Summary 37–56, vol. III, Feb./Mar. 1969, pp. 83–84.
[9] Linkabit Corp., Final Rep., "Coding systems study for high data rate telemetry links," Contract NAS2-6024, NASA Ames Res. Ctr. Rep. CR-114278, Moffett Field, Calif.
[10] J. P. Odenwalder, "Optimum decoding of convolutional codes," Ph.D. dissertation, Syst. Sci. Dep., Univ. California, Los Angeles, 1970.
[11] A. J. Viterbi, *Principles of Coherent Communication*. New York: McGraw-Hill, 1966.
[12] J. K. Omura, "On the Viterbi decoding algorithm," *IEEE Trans. Inform. Theory* (Corresp.), vol. IT-15, Jan. 1969, pp. 177–179.

PHOTO NOT AVAILABLE

Jerrold A. Heller (M'68) was born in New York, N. Y., on June 30, 1941. He received the B.E.E. degree in 1963 from Syracuse University, Syracuse, N. Y., and the M.S. and Ph.D. degrees in electrical engineering from the Massachusetts Institute of Technology, Cambridge, in 1964 and 1967, respectively. During his first year at M.I.T. he was a National Science Foundation Fellow.

In 1965 he joined the M.I.T. Research Laboratory of Electronics where he was a Xerox Fellow for the following two years. He held summer positions in 1962 at the Bell Telephone Laboratories, New York, N. Y., and in 1963 at the IBM Research Center, Yorktown Heights, N. Y., where he worked on the logical design of digital systems. From 1967 to 1969 he was with the Communications Research Section of the Jet Propulsion Laboratory, Pasadena, Calif., where his work centered on the application of coding to deep-space communication. He is presently Director of Technical Operations for the Linkabit Corporation, San Diego, Calif. Currently, his work is concerned with coding for space, HF, and communication satellite channels.

Dr. Heller is a member of Tau Beta Pi, Eta Kappa Nu, and Sigma Xi.

PHOTO NOT AVAILABLE

Irwin Mark Jacobs (S'55–M'60) was born in New Bedford, Mass., on October 18, 1933. He received the B.E.E. degree from Cornell University, Ithaca, N. Y., in 1956, and the S.M. and Sc.D. degrees from the Massachusetts Institute of Technology, Cambridge, in 1957 and 1959, respectively. He was the recipient of a McMullin Regional Scholarship and a General Electric Teachers Conference Scholarship at Cornell and participated in the engineering cooperative program in association with the Cornell Aeronautical Laboratory, Buffalo, N. Y. In graduate school, he was a General Electric Fellow and an Industrial Fellow of Electronics.

In 1959, he was appointed Assistant Professor of Electrical Engineering at M.I.T. and was a Member of the staff of the Research Laboratory of Electronics. He was promoted to Associate Professor in 1964. On leave from M.I.T., he spent the academic year 1964–1965 as a NASA Resident Research Fellow at the Jet Propulsion Laboratory, Pasadena, Calif., and was concerned principally with coding for deep-space communications. In 1966, he accepted an appointment as Associate Professor of Applied Physics and Information Science at the University of California, San Diego. In 1970 he was promoted to full Professor. In 1968 he cofounded Linkabit Corporation, of which he is now President. He is presently on leave from the University of California and devoting full time to Linkabit Corporation. He is currently working in the area of information and computer science.

Dr. Jacobs is a member of Phi Kappa Phi, Sigma Xi, Eta Kappa Nu, Tau Beta Pi, and the Association for Computing Machinery.

Convolutional Codes and Their Performance in Communication Systems

ANDREW J. VITERBI
Senior Member, IEEE

Abstract—**This tutorial paper begins with an elementary presentation of the fundamental properties and structure of convolutional codes and proceeds with the development of the maximum likelihood decoder. The powerful tool of generating function analysis is demonstrated to yield for arbitrary codes both the distance properties and upper bounds on the bit error probability for communication over any memoryless channel. Previous results on code ensemble average error probabilities are also derived and extended by these techniques. Finally, practical considerations concerning finite decoding memory, metric representation, and synchronization are discussed.**

I. Introduction

ALTHOUGH convolutional codes, first introduced by Elias [1], have been applied over the past decade to increase the efficiency of numerous communication systems, where they invariably outperform block codes of the same order of complexity, there remains to date a lack of acceptance of convolutional coding and decoding techniques on the part of many communication technologists. In most cases, this is due to an incomplete understanding of convolutional codes, whose cause can be traced primarily to the sizable literature in this field, composed largely of papers which emphasize details of the decoding algorithms rather than the more fundamental unifying concepts, and which, until recently, have been divided into two nearly disjoint subsets. This malady is shared by the block-coding literature, wherein the algebraic decoders and probabilistic decoders have been at odds for a considerably longer period.

The convolutional code dichotomy owes its origins to the development of sequential (probabilistic) decoding by Wozencraft [2] and of threshold (feedback, algebraic) decoding by Massey [3]. Until recently the two disciplines flourished almost independently, each with its own literature, applications, and enthusiasts. The Fano sequential decoding algorithm [4] was soon found to

Paper approved by the Communication Theory Committee of the IEEE Communication Technology Group for publication without oral presentation. Manuscript received January 7, 1971; revised June 11, 1971.

The author is with the School of Engineering and Applied Science, University of California, Los Angeles, Calif. 90024, and the Linkabit Corporation, San Diego, Calif.

Reprinted from *IEEE Transactions on Communications Technology*, vol. COM-19, no. 5, October 1971.

greatly outperform earlier versions of sequential decoders both in theory and practice. Meanwhile the feedback decoding advocates were encouraged by the burst-error correcting capabilities of the codes which render them quite useful for channels with memory.

To add to the confusion, yet a third decoding technique emerged with the Viterbi decoding algorithm [9], which was soon thereafter shown to yield maximum likelihood decisions (Forney [12], Omura [17]). Although this approach is probabilistic and emerged primarily from the sequential-decoding oriented discipline, it leads naturally to a more fundamental approach to convolutional code representation and performance analysis. Furthermore, by emphasizing the decoding-invariant properties of convolutional codes, one arrives directly to the maximum likelihood decoding algorithm and from it to the alternate approaches which lead to sequential decoding on the one hand and feedback decoding on the other. This decoding algorithm has recently found numerous applications in communication systems, two of which are covered in this issue (Heller and Jacobs [24], Cohen *et al.* [25]). It is particularly desirable for efficient communication at very high data rates, where very low error rates are not required, or where large decoding delays are intolerable.

Foremost among the recent works which seek to unify these various branches of convolutional coding theory is that of Forney [12], [21], [22], *et seq.*, which includes a three-part contribution devoted, respectively, to algebraic structure, maximum likelihood decoding, and sequential decoding. This paper, which began as an attempt to present the author's original paper [9] to a broader audience,[1] is another such effort at consolidating this discipline.

It begins with an elementary presentation of the fundamental properties and structure of convolutional codes and proceeds to a natural development of the maximum likelihood decoder. The relative distances among codewords are then determined by means of the generating function (or transfer function) of the code state diagram. This in turn leads to the evaluation of coded communication system performance on any memoryless channel. Performance is first evaluated for the specific cases of the binary symmetric channel (BSC) and the additive white Gaussian noise (AWGN) channel with biphase (or quadriphase) modulation, and finally generalized to other memoryless channels. New results are obtained for the evaluation of specific codes (by the generating function technique), rather than the ensemble average of a class of codes, as had been done previously, and for bit error probability, as distinguished from event error probability.

The previous ensemble average results are then extended to bit error probability bounds for the class of time-varying convolutional codes by means of a generalized generating function approach; explicit results are obtained for the limiting case of a very noisy channel and compared with the corresponding results for block codes. Finally, practical considerations concerning finite memory, metric representation, and synchronization are discussed. Further and more explicit details on these problems and detailed results of performance analysis and simulation are given in the paper by Heller and Jacobs [24].

While sequential decoding is not treated explicitly in this paper, the fundamentals and techniques presented here lead naturally to an elegant tutorial presentation of this subject, particularly if, following Jelinek [18], one begins with the recently proposed stack sequential decoding algorithm proposed independently by Jelinek and Zigangirov [7], which is far simpler to describe and understand then the original sequential algorithms. Such a development, which proceeds from maximum likelihood decoding to sequential decoding, exploiting the similarities in performance and analysis has been undertaken by Forney [22]. Similarly, the potentials and limitations of feedback decoders can be better understood with the background of the fundamental decoding-invariant convolutional code properties previously mentioned, as demonstrated, for example, by the recent work of Morrissey [15].

II. Code Representation

A convolutional encoder is a linear finite-state machine consisting of a K-stage shift register and n linear algebraic function generators. The input data, which is usually, though not necessarily, binary, is shifted along the register b bits at a time. An example with $K = 3$, $n = 2, b = 1$ is shown in Fig. 1.

The binary input data and output code sequences are indicated on Fig. 1. The first three input bits, 0, 1, and 1, generate the code outputs 00, 11, and 01, respectively. We shall pursue this example to develop various representations of convolutional codes and their properties. The techniques thus developed will then be shown to generalize directly to any convolutional code.

It is traditional and instructive to exhibit a convolutional code by means of a tree diagram as shown in Fig. 2.

If the first input bit is a zero, the code symbols are those shown on the first upper branch, while if it is a one, the output code symbols are those shown on the first lower branch. Similarly, if the second input bit is a zero, we trace the tree diagram to the next upper branch, while if it is a one, we trace the diagram downward. In this manner all 32 possible outputs for the first five inputs may be traced.

From the diagram it also becomes clear that after the first three branches the structure becomes repetitive. In fact, we readily recognize that beyond the third branch the code symbols on branches emanating from the two nodes labeled a are identical, and similarly for all the

[1] This material first appeared in unpublished form as the notes for the Linkabit Corp., "Seminar on convolutional codes," Jan. 1970.

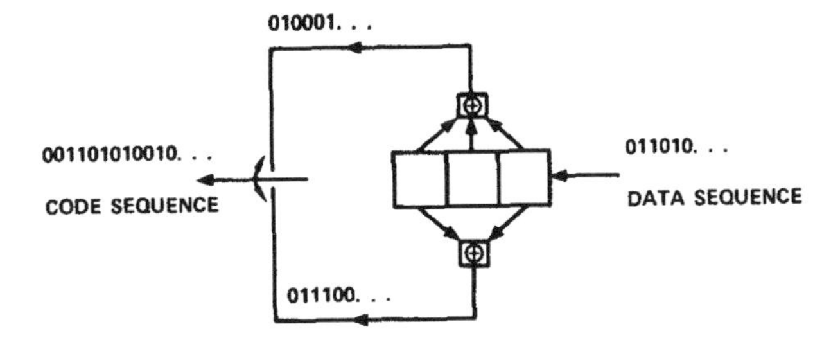

Fig. 1. Convolutional coder for $K = 3$, $n = 2$, $b = 1$.

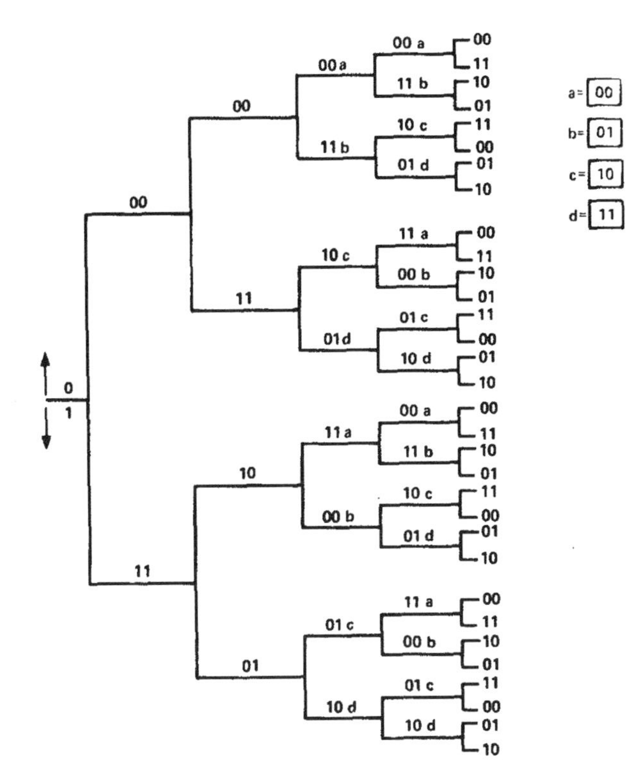

Fig. 2. Tree-code representation for coder of Fig. 1.

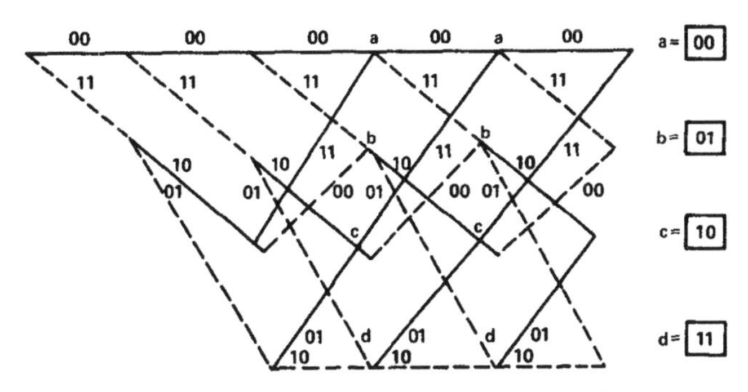

Fig. 3. Trellis-code representation for coder of Fig. 1.

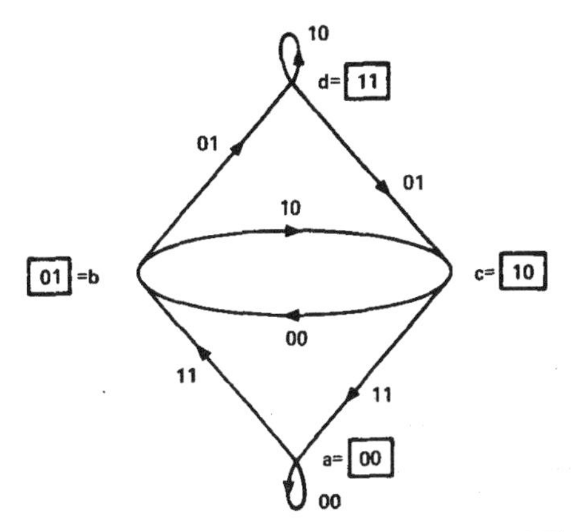

Fig 4. State-diagram representation for coder of Fig. 1.

identically labeled pairs of nodes. The reason for this is obvious from examination of the encoder. As the fourth input bit enters the coder at the right, the first data bit falls off on the left end and no longer influences the output code symbols. Consequently, the data sequences $100xy\cdots$ and $000xy\cdots$ generate the same code symbols after the third branch and, as is shown in the tree diagram, both nodes labeled a can be joined together.

This leads to redrawing the tree diagram as shown in Fig. 3. This has been called a trellis diagram [12], since a trellis is a tree-like structure with remerging branches. We adopt the convention here that code branches produced by a "zero" input bit are shown as solid lines and code branches produced by a "one" input bit are shown dashed.

The completely repetitive structure of the trellis diagram suggests a further reduction in the representation of the code to the state diagram of Fig. 4. The "states" of the state diagram are labeled according to the nodes of the trellis diagram. However, since the states correspond merely to the last two input bits to the coder we may use these bits to denote the nodes or states of this diagram.

We observe finally that the state diagram can be drawn directly by observing the finite-state machine properties of the encoder and particularly the fact that a four-state directed graph can be used to represent uniquely the input-output relation of the eight-state machine. For the nodes represent the previous two bits while the present bit is indicated by the transition branch; for example, if the encoder (machine) contains 011, this is represented in the diagram by the transition from state $b = 01$ to state $d = 11$ and the corresponding branch indicates the code symbol outputs 01.

III. Minimum Distance Decoder for Binary Symmetric Channel

On a BSC, errors which transform a channel code symbol 0 to 1 or 1 to 0 are assumed to occur independently from symbol to symbol with probability p. If all input (message) sequences are equally likely, the decoder which minimizes the overall error probability for any code, block or convolutional, is one which examines the error-corrupted received sequence $y_1y_2\cdots y_j\cdots$ and chooses the data sequence corresponding to the transmitted code sequence $x_1x_2\cdots x_j\cdots$, which is closest to the received sequence in the sense of Hamming distance; that is, the transmitted sequence which differs from the received sequence in the minimum number of symbols.

Referring first to the tree diagram, this implies that we should choose that path in the tree whose code sequence differs in the minimum number of symbols from the received sequence. However, recognizing that the transmitted code branches remerge continually, we may equally limit our choice to the possible paths in the trellis diagram of Fig. 3. Examination of this diagram indicates that it is unnecessary to consider the entire received sequence (which conceivably could be thousands or millions of symbols in length) at one time in deciding upon the most likely (minimum distance) transmitted sequence. In particular, immediately after the third branch we may determine which of the two paths leading to node or state a is more likely to have been sent. For example, if 010001 is received, it is clear that this is at distance 2 from 000000 while it is at distance 3 from 111011 and consequently we may exclude the lower path into node a. For, no matter what the subsequent received symbols will be, they will effect the distances only over subsequent branches after these two paths have remerged and consequently in exactly the same way. The same can be said for pairs of paths merging at the other three nodes after the third branch. We shall refer to the minimum distance path of the two paths merging at a given node as the "survivor." Thus it is necessary only to remember which was the minimum distance path from the received sequence (or survivor) at each node, as well as the value of that minimum distance. This is necessary because at the next node level we must compare the two branches merging at each node level, which were survivors at the previous level for different nodes; e.g., the comparison at node a after the fourth branch is among the survivors of comparisons at nodes a and c after the third branch. For example, if the received sequence over the first four branches is 01000111, the survivor at the third node level for node a is 000000 with distance 2 and at node c it is 110101, also with distance 2. In going from the third node level to the fourth the received sequence agrees precisely with the survivor from c but has distance 2 from the survivor from a. Hence the survivor at node a of the fourth level is the data sequence 1100 which produced the code sequence 11010111 which is at (minimum) distance 2 from the received sequence.

In this way we may proceed through the received sequence and at each step for each state preserve one surviving path and its distance from the received sequence, which is more generally called *metric*. The only difficulty which may arise is the possibility that in a given comparison between merging paths, the distances or metrics are identical. Then we may simply flip a coin as is done for block codewords at equal distances from the received sequence. For even if we preserved both of the equally valid contenders, further received symbols would affect both metrics in exactly the same way and thus not further influence our choice.

This decoding algorithm was first proposed by Viterbi [9] in the more general context of arbitrary memoryless channels. Another description of the algorithm can be obtained from the state-diagram representation of Fig. 4. Suppose we sought that path around the directed state diagram, arriving at node a after the kth transition, whose code symbols are at a minimum distance from the received sequence. But clearly this minimum distance path to node a at time k can be only one of two candidates: the miminum distance path to node a at time $k - 1$ and the minimum distance path to node c at time $k - 1$. The comparison is performed by adding the new distance accumulated in the kth transition by each of these paths to their minimum distances (metrics) at time $k - 1$.

It appears thus that the state diagram also represents a system diagram for this decoder. With each node or state we associate a storage register which remembers the minimum distance path into the state after each transition as well as a metric register which remembers its (minimum) distance from the received sequence. Furthermore, comparisons are made at each step between the two paths which lead into each node. Thus four comparators must also be provided.

There remains only the question of truncating the algorithm and ultimately deciding on one path rather than four. This is easily done by forcing the last two input bits to the coder to be 00. Then the final state of the code must be $a = 00$ and consequently the ultimate survivor is the survivor at node a, after the insertion into the coder of the two dummy zeros and transmission of the corresponding four code symbols. In terms of the trellis diagram this means that the number of states is reduced from four to two by the insertion of the first zero and to a single state by the insertion of the second. The diagram is thus truncated in the same way as it was begun.

We shall proceed to generalize these code representations and optimal decoding algorithm to general convolutional codes and arbitrary memoryless channels, including the Gaussian channel, in Sections V and VI. However, first we shall exploit the state diagram further to determine the relative distance properties of binary convolutional codes.

IV. Distance Properties of Convolutional Codes

We continue to pursue the example of Fig. 1 for the sake of clarity; in the next section we shall easily generalize results. It is well known that convolutional codes are group codes. Thus there is no loss in generality in computing the distance from the all zeros codeword to all the other codewords, for this set of distances is the same as the set of distances from any specific codeword to all the others.

For this purpose we may again use either the trellis diagram or the state diagram. We first of all redraw the trellis diagram in Fig. 5 labeling the branches according to their distances from the all zeros path. Now consider all the paths that merge with the all zeros for the first time at some arbitrary node j.

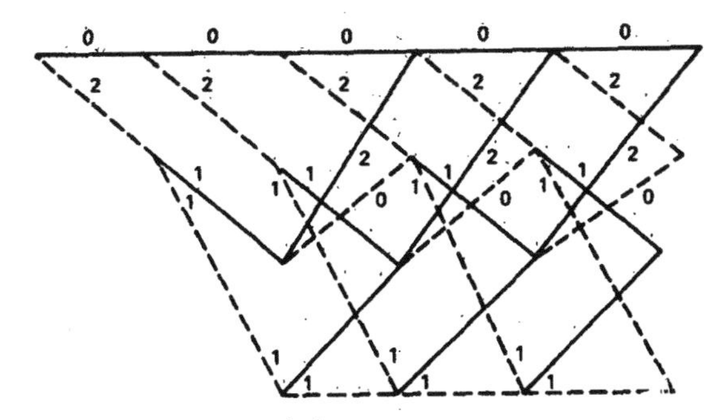

Fig. 5. Trellis diagram labeled with distances from all zeros path.

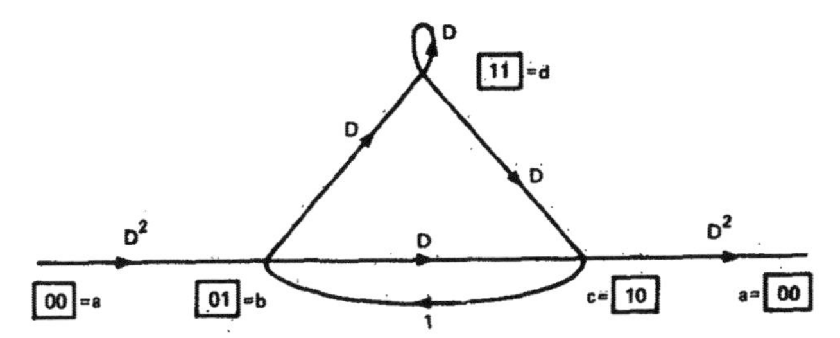

Fig. 6. State diagram labeled according to distance from all zeros path.

It is seen from the diagram that of these paths there will be just one path at distance 5 from the all zeros path and this diverged from it three branches back. Similarly there are two at distance 6 from it, one which diverged 4 branches back and the other which diverged 5 branches back, and so forth. We note also that the input bits for distance 5 path are 00 ⋯ 0100 and thus differ in only one input bit from the all zeros, while the distance 6 paths are 00 ⋯ 01100 and 00 ⋯ 010100 and thus each differs in 2 input bits from the all zeros path. The minimum distance, sometimes called the minimum "free" distance, among all paths is thus seen to be 5. This implies that any pair of channel errors can be corrected, for two errors will cause the received sequence to be at distance 2 from the transmitted (correct) sequence but it will be at least at distance 3 from any other possible code sequence. It appears that with enough patience the distance of all paths from the all zeros (or any arbitrary) path can be so determined from the trellis diagram.

However, by examining instead the state diagram we can readily obtain a closed form expression whose expansion yields directly and effortlessly all the distance information. We begin by labeling the branches of the state diagram of Fig. 4 either D^2, D, or $D^0 = 1$, where the exponent corresponds to the distance of the particular branch from the corresponding branch of the all zeros path. Also we split open the node $a = 00$, since circulation around this self-loop simply corresponds to branches of the all zeros path whose distance from itself is obviously zero. The result is Fig. 6. Now as is clear from examination of the trellis diagram, every path which arrives at state $a = 00$ at node level j, must have at some previous node level (possibly the first) originated at this same state $a = 00$. All such paths can be traced on the modified state diagram. Adding branch exponents we see that path $a\ b\ c\ a$ is at distance 5 from the correct path, paths $a\ b\ d\ c\ a$ and $a\ b\ c\ b\ c\ a$ are both at distance 6, and so forth, for the *generating functions* of the output sequence weights of these paths are D^5 and D^6, respectively

Now we may evaluate the generating function of all paths merging with the all zeros at the jth node level simply by evaluating the generating function of all the weights of the output sequences of the finite-state machine.[2] The result in this case is

$$T(D) = \frac{D^5}{1 - 2D}$$
$$= D^5 + 2D^6 + 4D^7 + \cdots + 2^k D^{k+5} + \cdots . \quad (1)$$

This verifies our previous observation and in fact shows that among the paths which merge with the all zeros at a given node there are 2^k paths at distance $k + 5$ from the all zeros.

Of course, (1) holds for an infinitely long code sequence; if we are dealing with the jth node level, we must truncate the series at some point. This is most easily done by considering the additional information indicated in the modified state diagram of Fig. 7.

The L terms will be used to determine the length of a given path; since each branch has an L, the exponent of the L factor will be augmented by one every time a branch is passed through. The N term is included only if that branch transition was caused by an input data "one," corresponding to a dotted branch in the trellis diagram. The generating function of this augmented state diagram is then

$$T(D, L, N)$$
$$= \frac{D^5L^3N}{1 - DL(1 + L)N}$$
$$= D^5L^3N + D^6L^4(1 + L)N^2 + D^7L^5(1 + L)^2N^3$$
$$+ \cdots + D^{5+k}L^{3+k}(1 + L)^kN^{1+k} + \cdots . \quad (2)$$

Thus we have verified that of the two distance 6 paths one is of length 4 and the other is of length 5 and both differ in 2 input bits from the all zeros.[3] Also, of the distance 7 paths, one is of length 5, two are of length 6, and one is of length 7; all four paths correspond to input sequences with three ones. If we are interested in the jth node level, clearly we should truncate the series such that no terms of power greater than L^j are included.

We have thus fully determined the properties of all paths in the convolutional code. This will be useful later in evaluating error probability performance of codes used over arbitrary memoryless channels.

[2] Alternatively, this can be regarded as the transfer function of the diagram regarded as a signal flow graph.

[3] Thus if the all zeros was the correct path and the noise causes us to choose one of the incorrect paths, two bit errors will be made.

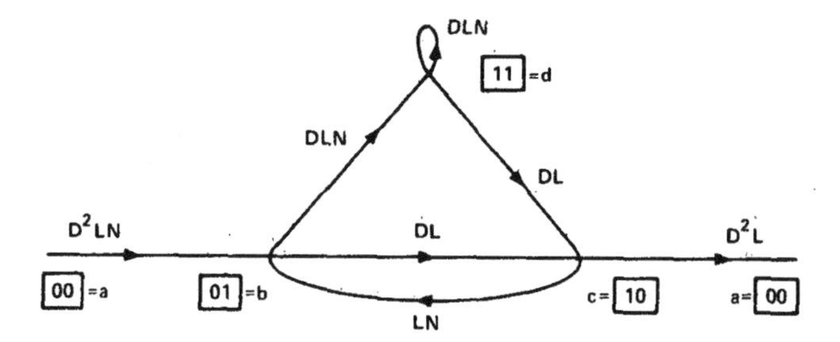

Fig. 7. State diagram labeled according to distance, length, and number of input ones.

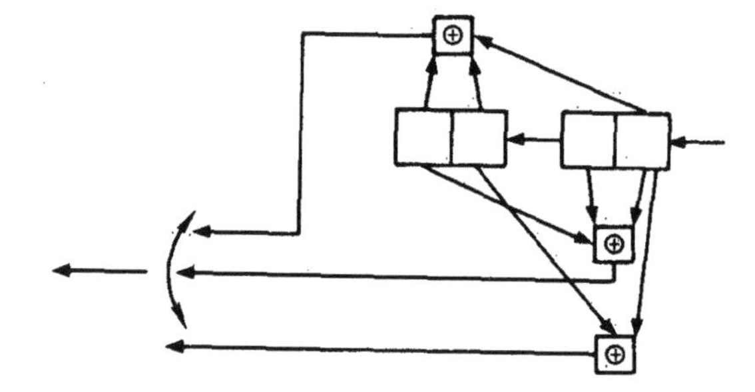

Fig. 8. Coder for $K = 2$, $b = 2$, $n = 3$, and $R = 2/3$.

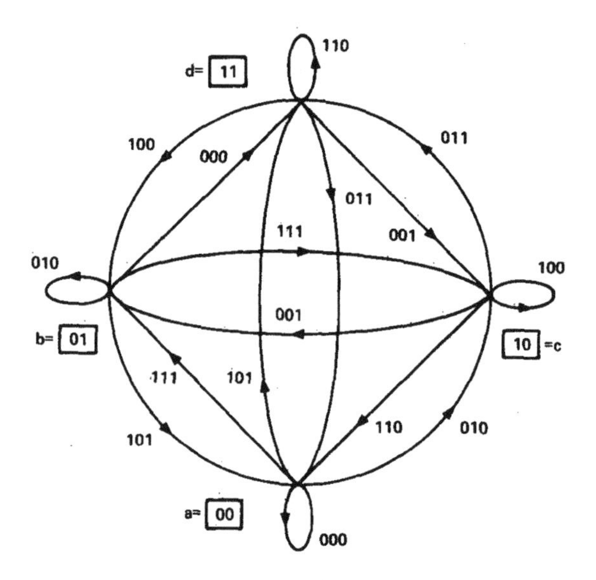

Fig. 9. State diagram for code of Fig. 8.

V. Generalization to Arbitrary Convolutional Codes

The generalization of these techniques to arbitrary binary-tree ($b = 1$) convolutional codes is immediate. That is, a coder with a K-stage shift register and n mod-2 adders will produce a trellis or state diagram with 2^{K-1} nodes or states and each branch will contain n code symbols. The rate of this code is then

$$R = \frac{1}{n} \text{ bits/code symbol.}$$

The example pursued in the previous sections had rate $R = 1/2$. The primary characteristic of the binary-tree codes is that only two branches exit from and enter each node.

If rates other than $1/n$ are desired we must make $b > 1$, where b is the number of bits shifted into the register at one time. An example for $K = 2$, $b = 2$, $n = 3$, and consequently rate $R = 2/3$ is shown in Fig. 8 and its state diagram is shown in Fig. 9. It differs from the binary-tree codes only in that each node is connected to four other nodes, and for general b it will be connected to 2^b nodes. Still all the preceding techniques including the trellis and state-diagram generating function analysis are still applicable. It must be noted, however, that the minimum distance decoder must make comparisons among all the paths entering each node at each level of the trellis and select one survivor out of four (or out of 2^b in general).

VI. Generalization of Optimal Decoder to Arbitrary Memoryless Channels

Fig. 10 exhibits a communication system employing a convolutional code. The convolutional encoder is precisely the device studied in the preceding sections. The data sequence is generally binary ($a_i = 0$ or 1) and the code sequence is divided into subsequences where $\mathbf{x}_i$ represents the n code symbols generated just after the input bit a_i enters the coder: that is, the symbols of the jth branch. In terms of the example of Fig. 1, $a_3 = 1$ and $\mathbf{x}_3 = 01$. The channel output or received sequence is similarly denoted. $\mathbf{y}_i$ represents the n symbols received when the n code symbols of $\mathbf{x}_i$ were transmitted. This model includes the BSC wherein the $\mathbf{y}_i$ are binary n vectors each of whose symbols differs from the corresponding symbol of $\mathbf{x}_i$ with probability p and is identical to it with probability $1 - p$.

For completely general channels it is readily shown [6], [14] that if all input data sequences are equally likely, the decoder which minimizes the error probability is one which compares the conditional probabilities, also called likelihood functions, $P(\mathbf{y} \mid \mathbf{x}^{(m)})$, where $\mathbf{y}$ is the overall received sequence and $\mathbf{x}^{(m)}$ is one of the possible transmitted sequences, and decides in favor of the maximum. This is called a maximum likelihood decoder. The likelihood functions are given or computed from the specifications of the channel. Generally it is more convenient to compare the quantities $\log P(\mathbf{y} \mid \mathbf{x}^{(m)})$ called the log-likelihood functions and the result is unaltered since the logarithm is a monotonic function of its (always positive) argument.

To illustrate, let us consider again the BSC. Here each transmitted symbol is altered with probability $p < 1/2$. Now suppose we have received a particular N-dimensional binary sequence $\mathbf{y}$ and are considering a possible transmitted N-dimensional code sequence $\mathbf{x}^{(m)}$ which differs in d_m symbols from $\mathbf{y}$ (that is, the Hamming distance between $\mathbf{x}^{(m)}$ and $\mathbf{y}$ is d_m). Then since the channel is memoryless (i.e., it affects each symbol independently of all the others), the probability

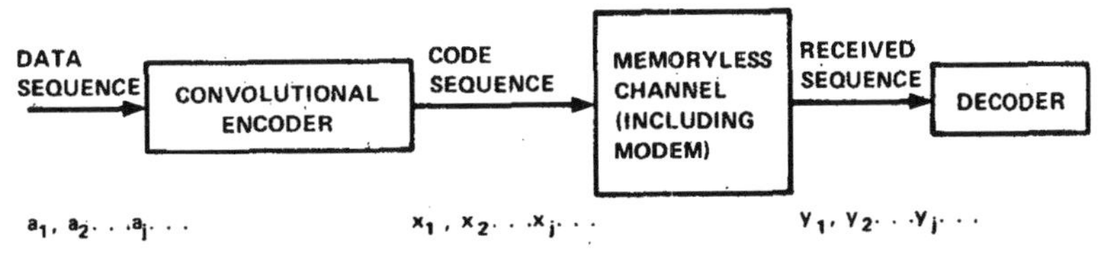

Fig. 10. Communication system employing convolutional codes.

that this $\mathbf{x}^{(m)}$ was transformed to the specific received $\mathbf{y}$ at distance d_m from it is

$$P(\mathbf{y} \mid \mathbf{x}^{(m)}) = p^{d_m}(1 - p)^{N-d_m}$$

and the log-likelihood function is thus

$$\log P(\mathbf{y} \mid \mathbf{x}^{(m)}) = -d_m \log (1 - p/p) + N \log (1 - p)$$

Now if we compute this quantity for each possible transmitted sequence, it is clear that the second term is constant in each case. Furthermore, since we may assume $p < 1/2$ (otherwise the role of 0 and 1 is simply interchanged at the receiver), we may express this as

$$\log P(\mathbf{y} \mid \mathbf{x}^{(m)}) = -\alpha d_m - \beta \tag{3}$$

where α and β are positive constants and d_m is the (positive) distance. Consequently, it is clear that maximizing the log-likelihood function is equivalent to minimizing the Hamming distance d_m. Thus for the BSC to minimize the error probability we should choose that code sequence at minimum distance from the received sequence, as we have indicated and done in preceding sections.

We now consider a more physical practical channel: the AWGN channel with biphase[4] phase-shift keying (PSK) modulation. The modulator and optimum demodulator (correlator or integrate-and dump filter) for this channel are shown in Fig. 11.

We use the notation that x_{jk} is the kth code symbol for the jth branch. Each binary symbol (which we take here for convenience to be ± 1) modulates the carrier by $\pm \Pi/2$ radians for T seconds. The transmission rate is, therefore, $1/T$ symbols/second or $b/nT = R/T$ bit/s. The function ϵ_s is the energy transmitted for each symbol. The energy per bit is, therefore $\epsilon_b = \epsilon_s/R$. The white Gaussian noise is a zero-mean random process of one-sided spectral density N_0 W/Hz, which affects each symbol independently. It then follows directly that the channel output symbol y_{jk} is a Gaussian random variable whose mean is $\sqrt{\epsilon_s}x_{jk}$ (i.e., $+\sqrt{\epsilon_s}$ if $x_{jk} = 1$ and $-\sqrt{\epsilon_s}$ if $x_{jk} = -1$) and whose variance is $N_0/2$. Thus the conditional probability density (or likelihood) function of y_{jk} given x_{jk} is

$$p(y_{jk} \mid x_{jk}) = \frac{\exp[-(y_{jk} - \sqrt{\epsilon_s}\, x_{jk})^2/N_0]}{\sqrt{\Pi N_0}}. \tag{4}$$

The likelihood function for the jth branch of a particular

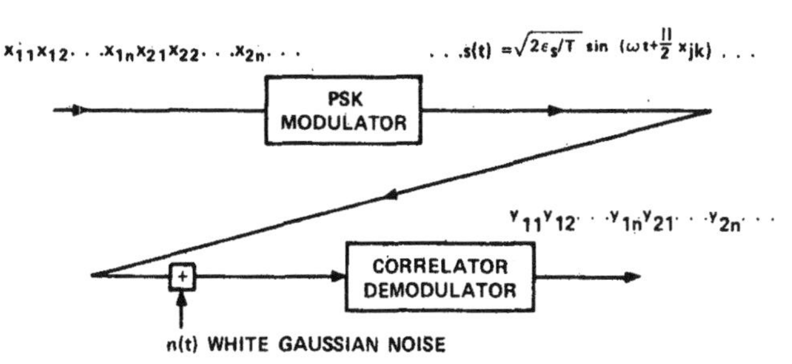

Fig. 11. Modem for additive white Gaussian noise PSK modulated memoryless channel.

code path $\mathbf{x}_j^{(m)}$

$$p(\mathbf{y}_j \mid \mathbf{x}_j^{(m)}) = \prod_{k=1}^{n} p(y_{jk} \mid x_{jk}^{(m)})$$

since each symbol is affected independently by the white Gaussian noise, and thus the log-likelihood function for the jth branch is

$$\begin{aligned}
\ln p(\mathbf{y}_j \mid \mathbf{x}_j^{(m)}) &= \sum_{k=1}^{n} \ln p(y_{jk} \mid x_{jk}^{(m)}) \\
&= -\frac{1}{N_0} \sum_{k=1}^{n} [y_{jk} - \sqrt{\epsilon_s}\, x_{jk}^{(m)}]^2 - \tfrac{1}{2} \ln \frac{\Pi}{N_0} \\
&= \frac{2\sqrt{\epsilon_s}}{N_0} \sum_{k=1}^{n} y_{jk}x_{jk}^{(m)} - \frac{\epsilon_s}{N_0} \sum_{k=1}^{n} [x_{jk}^{(m)}]^2 \\
&\quad - \frac{1}{N_0} \sum_{k=1}^{n} y_{jk}^2 - \tfrac{1}{2} \ln \frac{\Pi}{N_0} \\
&= C \sum_{k=1}^{n} y_{jk}x_{jk}^{(m)} - D
\end{aligned} \tag{5}$$

where C and D are independent of m, and we have used the fact that $[x_{jk}^{(m)}]^2 = 1$. Similarly, the log-likelihood[5] function for any path is the sum of the log-likelihood functions for each of its branches.

We have thus shown that the maximum likelihood decoder for the memoryless AWGN biphase (or quadriphase) modulated channel is one which forms the inner product between the received (real number) sequence and the code sequence (consisting of ± 1) and chooses the path corresponding to the greatest. Thus the *metric* for this channel is the inner product (5) as contrasted with the distance[6] metric used for the BSC.

[4] The results are the same for quadriphase PSK with coherent reception. The analysis proceeds in the same way, if we treat quadriphase PSK as two parallel independent biphase PSK channels.

[5] We have used the natural logarithm here, but obviously a change of base results merely in a scale factor.

[6] Actually it is easily shown that maximizing an inner product is equivalent to minimizing the Euclidean distance between the corresponding vectors.

For convolutional codes the structure of the code paths was described in Sections II–V. In Section III the optimum decoder was derived for the BSC. It now becomes clear that if we substitute the inner product metric $\Sigma y_{jk}x_{jk}^{(m)}$ for the distance metric $\Sigma d_{jk}^{(m)}$, used for the BSC, all the arguments used in Section III for the latter apply equally to this Gaussian channel. In particular the optimum decoder has a block diagram represented by the code state diagram. At step j the stored metric for each state (which is the maximum of the metrics of all the paths leading to this state at this time) is augmented by the branch metrics for branches emanating from this state. The comparisons are performed among all pairs of (or in general sets of 2^b) branches entering each state and the *maxima* are selected as the new most likely paths. The history (input data) of each new survivor must again be stored and the decoder is now ready for step $j+1$.

Clearly, this argument generalizes to any memoryless channel and we must simply use the appropriate metric $\ln P(\mathbf{y} \mid \mathbf{x}^{(m)})$, which may always be determined from the statistical description of the channel. This includes, among others, AWGN channels employing other forms of modulation.[7]

In the next section, we apply the analysis of convolutional code distance properties of Section IV to determine the error probabilities of specific codes on more general memoryless channels.

VII. Performance of Convolutional Codes on Memoryless Channels

In Section IV we analyzed the distance properties of convolutional codes employing a state-diagram generating function technique. We now extend this approach to obtain tight upper bounds on the error probability of such codes. We shall consider the BSC, the AWGN channel and more general memoryless channels, in that order. We shall obtain both the first-event error probability, which is the probability that the correct path is excluded (not a survivor) for the first time at the jth step, and the bit error probability which is the expected ratio of bit errors to total number of bits transmitted.

A. Binary Symmetric Channel

The first-event error probability is readily obtained from the generating function $T(D)$ [(5) for the code of Fig. 1, which we shall again pursue for demonstrative purposes]. We may assume, without loss of generality, since we are dealing with group codes, that the all zeros path was transmitted. Then a first-event error is made at the jth step if this path is excluded by selecting another path merging with the all zeros at node a at the jth level.

Now suppose that the previous-level survivors were such that the path compared with the all zeros at step j is the path whose data sequence is $00 \cdots 0100$ corresponding to nodes $a \cdots a\ a\ b\ c\ a$ (see Fig. 4.). This differs from the correct (all zeros) path in five symbols. Consequently an error will be made in this comparison if the BSC caused three or more errors in these particular five symbols. Hence the probability of an error in this specific comparison is

$$P_5 = \sum_{e=3}^{5} \binom{5}{e} p^e (1-p)^{5-e}. \tag{6}$$

On the other hand, there is no assurance that this particular distance five path will have previously survived so as to be compared with the correct path at the jth step. If either of the distance 6 paths were compared instead, then four or more errors in the six different symbols will definitely cause an error in the survivor decision, while three errors will cause a tie which, if resolved by coin flipping, will result in an error only half the time. Then the probability if this comparison is made is

$$P_6 = \frac{1}{2}\binom{6}{3} p^3 (1-p)^3 + \sum_{e=4}^{6} \binom{6}{e} p^e (1-p)^{6-e}. \tag{7}$$

Similarly, if the previously surviving paths were such that a distance d path is compared with the correct path at the jth step, the resulting error probability is

$$P_k = \begin{cases} \displaystyle\sum_{e=(k+1)/2}^{k} \binom{k}{e} p^e (1-p)^{k-e}, & k \text{ odd} \\ \displaystyle\frac{1}{2}\binom{k}{k/2} p^{k/2} (1-p)^{k/2} + \sum_{e=k/2+1}^{k} \binom{k}{e} p^e (1-p)^{k-e}, & k \text{ even.} \end{cases} \tag{8}$$

Now at step j, since there is no simple way of determining previous survivors, we may overbound the probability of a first-event error by the sum of the error probabilities for all possible paths which merge with the correct path at this point. Note this *union bound* is indeed an upper bound because two or more such paths may both have distance closer to the received sequence than the correct path (even though only one has survived to this point) and thus the events are not disjoint. For the example with generating function (1) it follows that the first-event error probability[8] is bounded by

$$P_E < P_5 + 2P_6 + 4P_7 + \cdots + 2^k P_{k+5} + \cdots \tag{9}$$

where P_k is given by (8).

In Section VII-C it will be shown that (8) can be upper bounded by (see (39)).

$$P_k < 2^k p (1-p)^{k/2}. \tag{10}$$

Using this, the first-event error probability bound (9)

[7] Although more elaborate modulators, such as multiple FSK or multiphase modulators, might be employed, Jacobs [11] has shown that the most effective as well as the simplest system for wide-band space and satellite channels is the binary PSK modulator considered in the example of this section. We note again that the performance of quadriphase modulation is the same as for biphase modulation, when both are coherently demodulated.

[8] We are ignoring the finite length of the path, but the expression is still valid since it is an upper bound.

can be more loosely bounded by

$$P_E < \sum_{k=5}^{\infty} 2^{k-5} 2^k p(1-p)^{k/2}$$
$$= \frac{[2\sqrt{p(1-p)}]^5}{1-4\sqrt{p(1-p)}} = T(D)\,|_{D=2\sqrt{p(1-p)}} \quad (11)$$

where $T(D)$ is just the generating function of (1)

It follows easily that for a general binary-tree ($b = 1$) convolutional code with generating function

$$T(D) = \sum_{k=d}^{\infty} a_k D^k \quad (12)$$

the first-event error probability is bounded by the generalization of (9).

$$P_E < \sum_{k=d}^{\infty} a_k P_k \quad (13)$$

where P_k is given by (8) and more loosely upper bounded by the generalization of (11)

$$P_E < T(D)\,|_{D=2\sqrt{p(1-p)}}. \quad (14)$$

Whenever a decision error occurs, one or more bits will be incorrectly decoded. Specifically, those bits in which the path selected differs from the correct path will be incorrect. If only one error were ever made in decoding an arbitrary long code path, the number of bits in error in this incorrect path could easily be obtained from the augmented generating function $T(D, N)$ (such as given by (2) with factors in L deleted). For the exponents of the N factors indicate the number of bit errors for the given incorrect path arriving at node a at the jth level.

After the first error has been made, the incorrect paths no longer will be compared with a path which is overall correct, but rather with a path which has diverged from the correct path over some span of branches (see Fig. 12). If the correct path $\mathbf{x}$ has been excluded by a decision error at step j in favor of path $\mathbf{x}'$, the decision at step $j + 1$ will be between $\mathbf{x}'$ and $\mathbf{x}''$. Now the (first-event) error probability of (13) or (14) is for a comparison, at any step, between path $\mathbf{x}$ and any other path merging with it at that step, including path $\mathbf{x}''$ in this case. However, since the metric[9] for path $\mathbf{x}'$ is greater than the metric for $\mathbf{x}$, for on this basis the correct path was excluded at step j, the probability that path $\mathbf{x}''$ metric exceeds path $\mathbf{x}'$ metric at step $j + 1$ is less than the probability that path $\mathbf{x}''$ exceeds the (correct) path $\mathbf{x}$ metric at this point. Consequently, the probability of a new incorrect path being selected after a previous error has occurred is upper bounded by the first-event error probability at that step.

Moreover, when a second error follows closely after a first error, it often occurs (as in Fig. 12) that the erroneous bit(s) of path $\mathbf{x}''$ overlap the erroneous bit(s) of path $\mathbf{x}'$. With this in mind, we now show that for a

[9] Negative distance from the received sequence for the BSC, but clearly this argument generalizes to any memoryless channel.

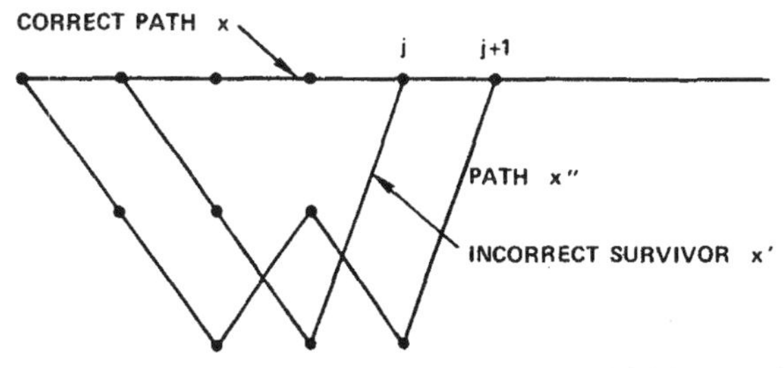

Fig. 12. Example of decoding decision after initial error has occurred.

binary-tree code if we weight each term of the first-event error probability bound at any step by the number of erroneous bits for each possible erroneous path merging with the correct path at that node level, we upper bound the bit error probability. For, a given step decision corresponds to decoder action on one more bit of the transmitted data sequence; the first-event error probability union bound with each term weighted by the corresponding number of bit errors is an upper bound on the expected number of bit errors caused by this action. Summing the expected number of bit errors over L steps, which as was just shown may result in overestimating through double counting, gives an upper bound on the expected number of bit errors in L branches for arbitrary L. But since the upper bound on expected number of bit errors is the same at each step, it follows, upon dividing the sum of L equal terms by L, that this expected number of bit errors per step is just the bit error probability P_B, for a binary-tree code ($b = 1$). If $b > 1$, then we must divide this expression by b, the number of bits encoded and decoded per step.

To illustrate the calculation of P_B for a convolutional code, let us consider again the example of Fig. 1. Its transfer function in D and N is obtained from (2), letting $L = 1$, since we are not now interested in the lengths of incorrect paths, to be

$$T(D, N) = \frac{D^5 N}{1 - 2DN}$$
$$= D^5 N + 2D^6 N^2 = \cdots + 2^k D^{k+5} N^{k+1} + \cdots. \quad (15)$$

The exponents of the factors in N in each term determine the number of bit errors for the path(s) corresponding to that term. Since $T(D) = T(D, N)\,|_{N=1}$ yields the first-event error probability P_E, each of whose terms must be weighted by the exponent of N to obtain P_B, it follows that we should first differentiate $T(D, N)$ at $N = 1$ to obtain

$$\left.\frac{dT(D, N)}{dN}\right|_{N=1}$$
$$= D^5 + 2 \cdot 2D^6 + 3 \cdot 4D^7 + \cdots + (k+1)2^k D^{k+5} + \cdots$$
$$= \frac{D^5}{(1-2D)^2}. \quad (16)$$

Then from this we obtain, as in (9), that for the BSC

$$P_B < P_5 + 2\cdot 2P_6 + 3\cdot 4P_7 + \cdots + (k+1)2^k P_{k+5} + \cdots \quad (17)$$

where P_k is given by (8).

If for P_k we use the upper bound (10) we obtain the weaker but simpler bound

$$P_B < \sum_{k=5}^{\infty} (k-4)2^{k-5}[4p(1-p)]^{k/2} = \left.\frac{dT(D,N)}{dN}\right|_{N=1, D=2\sqrt{p(1-p)}} = \frac{|2\sqrt{p(1-p)}|^5}{[1-4\sqrt{p(1-p)}]^2}. \quad (18)$$

More generally for any binary-tree ($b = 1$) code used on the BSC if

$$\left.\frac{dT(D,N)}{dN}\right|_{N=1} = \sum_{k=d}^{\infty} c_k D^k \quad (19)$$

then corresponding to (17)

$$P_B < \sum_{k=d}^{\infty} c_k P_k \quad (20)$$

and corresponding to (18) we have the weaker bound

$$P_B < \left.\frac{dT(D,N)}{dN}\right|_{N=1, D=2\sqrt{p(1-p)}}. \quad (21)$$

For a nonbinary-tree code ($b \neq 1$), all these expressions must be divided by b.

The results of (14) and (18) will be extended to more general memoryless channels, but first we shall consider one more specific channel of particular interest.

B. AWGN Biphase-Modulated Channel

As was shown in Section VI the decoder for this channel operates in exactly the same way as for the BSC, except that instead of Hamming distance it uses the metric

$$\sum_i \sum_{j=1}^{n} x_{ij} y_{ij}$$

where $x_{ij} = \pm 1$ are the transmitted code symbols, y_{ij} the corresponding received (demodulated) symbols, and j runs over the n symbols of each branch while i runs over all the branches in a particular path. Hence, to analyze its performance we may proceed exactly as in Section VII-A except that the appropriate pairwise-decision errors P_k must be substituted for those of (6) to (8).

As before we assume, without loss of generality, that the correct (transmitted) path $\mathbf{x}$ has $x_{ij} = +1$ for all i and j (corresponding to the all zeros if the input symbols were 0 and 1). Let us consider an incorrect path $\mathbf{x}'$ merging with the correct path at a particular step, which has k negative symbols ($x_{ij}' = -1$) and the remainder positive. Such a path may be incorrectly chosen only if it has a higher metric than the correct path, i.e.,

$$\sum_i \sum_{j=1}^{n} x_{ij}' y_{ij} \geq \sum_i \sum_{j=1}^{n} x_{ij} y_{ij}$$

or

$$\sum_i \sum_{j=1}^{n} (x_{ij}' - x_{ij}) y_{ij} \geq 0$$

where i runs over all branches in the two paths. But since, as we have assumed, the paths $\mathbf{x}$ and $\mathbf{x}'$ differ in exactly k symbols, wherein $x_{ij} = 1$ and $x_{ij}' = -1$, the pairwise error probability is just

$$P_k = \Pr\left\{\sum_i \sum_{j=1}^{n} (x_{ij}' - x_{ij}) y_{ij} \geq 0\right\} = \Pr\left\{\sum_{r=1}^{k} (x_r' - x_r) y_r \geq 0\right\} = \Pr\left\{-2\sum_{r=1}^{k} y_r \geq 0\right\} = \Pr\left\{\sum_{r=1}^{k} y_r \leq 0\right\} \quad (22)$$

where r runs over the k symbols wherein the two paths differ. Now it was shown in Section VI that the y_{ij} are independent Gaussian random variables of variance $N_0/2$ and mean $\sqrt{\epsilon_s} x_{ij}$, where x_{ij} is the actually transmitted code symbol. Since we are assuming that the (correct) transmitted path has $x_{ij} = +1$ for all i and j, it follows that y_{ij} or y_r has mean $\sqrt{\epsilon_s}$ and variance $N_0/2$. Therefore, since the k variables y_r are independent and Gaussian, the sum $Z = \sum_{r=1}^{k} y_r$ is also Gaussian with mean $k\sqrt{\epsilon_s}$ and variance $kN_0/2$.

Consequently,

$$P_k = \Pr(Z < 0) = \int_{-\infty}^{0} \frac{\exp(-Z - k\sqrt{\epsilon_s})^2/kN_0}{\sqrt{\Pi k N_0}} dZ = \int_{\sqrt{2k\epsilon_s/N_0}}^{\infty} \left[\frac{\exp(-x^2/2)}{\sqrt{2\Pi}}\right] dx \triangleq \operatorname{erfc}\sqrt{\frac{2k\epsilon_s}{N_0}}. \quad (23)$$

We recall from Section VI that ϵ_s is the symbol energy, which is related to the bit energy by $\epsilon_s = R\epsilon_b$, where $R = b/n$. The bound on P_E then follows exactly as in Section VII-A and we obtain the same general bound as (13)

$$P_E < \sum_{k=d}^{\infty} a_k P_k \quad (24)$$

where a_k are the coefficients of

$$T(D) = \sum_{k=d}^{\infty} a_k D^k \quad (25)$$

and where d is the minimum distance between any two paths in the code. We may simplify this procedure considerably while loosening the bound only slightly for this channel by observing that for $x \geq 0$, $y \geq 0$,

$$\operatorname{erfc}\sqrt{x+y} \leq \exp\left(\frac{-y}{2}\right) \operatorname{erfc}\sqrt{x}. \quad (26)$$

Consequently, for $k \geq d$, letting $l = k - d$, we have from (23)

$$P_k = \operatorname{erfc} \sqrt{\frac{2k\epsilon_s}{N_0}} = \operatorname{erfc} \sqrt{\frac{2(d+l)\epsilon_s}{N_0}}$$
$$\leq \exp\left(\frac{-l\epsilon_s}{N_0}\right) \operatorname{erfc} \sqrt{\frac{2d\epsilon_s}{N_0}} \qquad (27)$$

whence the bound of (24), using (27), becomes

$$P_E < \sum_{k=d}^{\infty} a_k P_k \leq \operatorname{erfc} \sqrt{\frac{2d\epsilon_s}{N_0}} \sum_{k=d}^{\infty} a_k \exp\left[\frac{-(k-d)\epsilon_s}{N_0}\right]$$

or

$$P_E < \operatorname{erfc} \sqrt{\frac{2d\epsilon_s}{N_0}} \exp\left(\frac{d\epsilon_s}{N_0}\right) T(D) \,\Big|_{D=\exp(-\epsilon_s/N_0)}. \qquad (28)$$

The bit error probability can be obtained in exactly the same way. Just as for the BSC [(19) and (20)] we have that for a binary-tree code

$$P_B < \sum_{k=d}^{\infty} c_k P_k \qquad (29)$$

where c_k are the coefficients of

$$\left.\frac{dT(D, N)}{dN}\right|_{N=1} = \sum_{k=d}^{\infty} c_k\, D^k. \qquad (30)$$

Thus following the came arguments which led from (24) to (28) we have for a binary-tree code

$$P_B < \operatorname{erfc} \sqrt{\frac{2d\epsilon_s}{N_0}} \exp\left(\frac{d\epsilon_s}{N_0}\right) \left.\frac{dT(D, N)}{dN}\right|_{N=1,\, D=\exp(-\epsilon_s/N_0)} \qquad (31)$$

For $b > 1$, this expression must be divided by b.

To illustrate the application of this result we consider the code of Fig. 1 with parameters $K = 3$, $R = 1/2$, whose transfer function is given by (15). For this case since $R = 1/2$ and $\epsilon_s = 1/2\ \epsilon_b$, we obtain

$$P_B < \frac{\operatorname{erfc} \sqrt{5\epsilon_b/N_0}}{(1 - 2e^{-\epsilon_b/2N_0}}. \qquad (32)$$

Since the number of states in the state diagram grows exponentially with K, direct calculation of the generating function becomes unmanageable for $K > 4$. On the other hand, a generating function calculation is basically just a matrix inversion (see Appendix I), which can be performed numerically for a given value of D. The derivative at $N = 1$ can be upper bounded by evaluating the first difference $[T(D, 1 + \epsilon) - T(D, 1)]/\epsilon$, for small ϵ. A computer program has been written to evaluate (31) for any constraint length up to $K = 10$ and all rates $R = 1/n$ as well as $R = 2/3$ and $R = 3/4$. Extensive results of these calculations are given in the paper by Heller and Jacobs [24], along with the results of simulations of the corresponding codes and channels. The simulations verify the tightness of the bounds.

In the next section, these bounding techniques will be extended to more general memoryless channels, from which (28) and (31) can be obtained directly, but without the first two factors. Since the product of the first two factors is always less than one, the more general bound is somewhat weaker.

C. General Memoryless Channels

As was indicated in Section VI, for equally likely input data sequences, the minimum error probability decoder chooses the path which maximizes the log-likelihood function (metric)

$$\ln P(\mathbf{y} \mid \mathbf{x}^{(m)})$$

over all possible paths $\mathbf{x}^{(m)}$. If each symbol is transmitted (or modulates the transmitter) independent of all preceding and succeeding symbols, and the interference corrupts each symbol independently of all the others, then the channel, which includes the modem, is said to be memoryless[10] and the log-likelihood function

$$\ln P(\mathbf{y} \mid \mathbf{x}^{(m)}) = \sum_i \sum_{j=1}^{n} \ln P(y_{ij} \mid x_{ij}^{(m)})$$

where $x_{ij}^{(m)}$ is a code symbol of the mth path, y_{ij} is the corresponding received (demodulated) symbol, j runs over the n symbols of each branch, and i runs over the branches in the given path. This includes the special cases considered in Sections VII-A and -B.

The decoder is the same as for the BSC except for using this more general metric. Decisions are made after each set of new branch metrics have been added to the previously stored metrics. To analyze performance, we must merely evaluate P_k, the pairwise error probability for an incorrect path which differs in k symbols from the correct path, as was done for the special channels of Sections VII-A and -B. Proceeding as in (22), letting x_{ij} and x_{ij}' denote symbols of the correct and incorrect paths, respectively, we obtain

$$P_k(\mathbf{x}, \mathbf{x}')$$
$$= \Pr\left[\sum_i \sum_{j=1}^{n} \ln P(y_{ij} \mid x_{ij}') > \sum_i \sum_{j=1}^{n} \ln P(y_{ij} \mid x_{ij})\right]$$
$$= \Pr\left\{\sum_{r=1}^{k} \ln \frac{P(y_r \mid x_r')}{P(y_r \mid x_r)} > 0\right\}$$
$$= \Pr\left\{\prod_{r=1}^{k} \frac{P(y_r \mid x_r')}{P(y_r \mid x_r)} > 1\right\} \qquad (33)$$

where r runs over the k code symbols in which the paths differ. This probability can be rewritten as

$$P_k(\mathbf{x}, \mathbf{x}') = \sum_{\mathbf{y} \in Y_k} \prod_{r=1}^{k} P(y_r \mid x_r) \qquad (34)$$

where Y_k is the set of all vectors $\mathbf{y} = (y_1, y_2, \cdots, y_r, \cdots, y_k)$ for which

[10] Often more than one code symbol in a given branch is used to modulate the transmitter at one time. In this case, provided the interference still affects succeeding branches independently, the channel can still be treated as memoryless but now the symbol likelihood functions are replaced by branch likelihood functions and (33) is replaced by a single sum over i.

$$\prod_{r=1}^{k} \frac{P(y_r \mid x_r')}{P(y_r \mid x_r)} > 1. \tag{35}$$

But if this is the case, then

$$P_k(\mathbf{x}, \mathbf{x}') < \sum_{\mathbf{y} \in Y_k} \prod_{r=1}^{k} P(y_r \mid x_r) \left[\frac{P(y_r \mid x_r')}{P(y_r \mid x_r)}\right]^{1/2}$$
$$< \sum_{\text{all } \mathbf{y} \in Y} \prod_{r=1}^{k} P(y_r \mid x_r)^{1/2} P(y_r \mid x_r')^{1/2} \tag{36}$$

where Y is the entire space of received vectors.[11] The first inequality is valid because we are multiplying the summand by a quantity greater than unity,[12] and the second because we are merely extending the sum of positive terms over a larger set. Finally we may break up the k-dimensional sum over $\mathbf{y}$ into k one-dimensional summations over $y_1, y_2, \cdots, y_k$, respectively, and this yields

$$P_k(\mathbf{x}, \mathbf{x}') \le \sum_{y_1} \sum_{y_2} \cdots \sum_{y_k} \prod_{r=1}^{k} P(y_r \mid x_r)^{1/2} P(y_r \mid x_r')^{1/2}$$
$$= \prod_{r=1}^{k} \sum_{y_r} P(y_r \mid x_r)^{1/2} P(y_r \mid x_r')^{1/2} \tag{37}$$

To illustrate the use of this bound we consider the two specific channels treated above. For the BSC, y_r is either equal to x_r, the transmitted symbol, or to $\bar{x}_r$, its complement. Now y_r depends on x_r through the channel statistics. Thus

$$P(y_r = x_r) = 1 - p$$
$$P(y_r = \bar{x}_r) = p. \tag{38}$$

For each symbol in the set $r = 1, 2, \cdots, k$ by definition $x_r \neq x_r'$. Hence for each term in the sum if $x_r = 0$, $x_r' = 1$ or vice versa. Hence, whatever x_r and x_r' may be

$$\sum_{y_r=0}^{1} P(y_r \mid x_r)^{1/2} P(y_r \mid x_r')^{1/2} = 2p^{1/2}(1 - p)^{1/2}$$

and the product (37) of k identical factors is

$$P_k = 2^k p^{k/2} (1 - p)^{k/2} \tag{39}$$

for all pairs of correct and incorrect paths. This was used in Section VII-A to obtain the bounds (11) and (21).

For the AWGN channel of Section VII-B we showed that the likelihood functions (probability densities) were

$$p(y_r \mid x_r) = \frac{\exp[-(y_r - \sqrt{\epsilon_s}\, x_r)^2 / N_0]}{\sqrt{\Pi N_0}} \tag{40}$$

where $x_r = +1$ or -1 and

$$x_r + x_r' = 0. \tag{41}$$

Since y_r is a real variable, the space of y_r is the real line and the sum in (37) becomes the integral

$$\int_{-\infty}^{\infty} p(y_r \mid x_r)^{1/2} p(y_r \mid x_r')^{1/2}\, dy_r = \frac{1}{\sqrt{\Pi N_0}} \int_{-\infty}^{\infty} \exp\left\{-\frac{[(y_r - \sqrt{\epsilon_s}\, x_r)^2 + (y_r - \sqrt{\epsilon_s}\, x_r')^2]}{2N_0}\right\} dy_r$$
$$= \frac{1}{\sqrt{\Pi N_0}} \int_{-\infty}^{\infty} \exp\left[\frac{-(y_r^2 + \epsilon_s)}{N_0}\right] dy_r = \exp\left(\frac{-\epsilon_s}{N_0}\right)$$

where we have used (41) and $x_r^2 = x_r'^2 = 1$. The product of these k identical terms is, therefore,

$$P_k < \exp\left(\frac{-k\epsilon_s}{N_0}\right) \tag{42}$$

for all pairs of correct and incorrect paths. Inserting these bounds in the general expressions (24) and (29), and using (25) and (30) yields the bound on first-event error probability and bit error probability.

$$P_E < T(D)\big|_{D=\exp(-\epsilon_s/N_0)} \tag{43}$$

$$P_B < \left.\frac{dT(D, N)}{dN}\right|_{N=1,\, D=\exp(-\epsilon_s/N_0)} \tag{44}$$

which are somewhat (though not exponentially) weaker than (28) and (31).

A characteristic feature of both the BSC and the AWGN channel is that they affect each symbol in the same way independent of its location in the sequence. Any memoryless channel has this property provided it is stationary (statistically time invariant). For a stationary memoryless channel (37) reduces to

$$P_k(\mathbf{x}, \mathbf{x}') < [\sum_{y_r} P(y_r \mid x_r)^{1/2} P(y_r \mid x_r')^{1/2}]^k \triangleq D_0^k \tag{45}$$

where[13]

$$D_0 \triangleq \sum_{y_r} P(y_r \mid x_r)^{1/2} P(y_r \mid x_r')^{1/2} < 1. \tag{46}$$

While this bound on P_k is valid for all such channels, clearly it depends on the actual values assumed by the symbols x_r and x_r', of the correct and incorrect path, and these will generally vary according to the pairs of paths $\mathbf{x}$ and $\mathbf{x}'$ in question. However, if the input symbols are binary, x and $\bar{x}$, whenever $x_r = x$, then $x_r' = \bar{x}$,

[11] This would be the set of all 2^k k-dimensional binary vectors for the BSC, and Euclidean k space for the AWGN channel. Note also that the bound of (36) may be improved for asymmetric channels by changing the two exponents of ½ to s and $1 - s$, respectively, where $0 < s < 1$.

[12] The square root of a quantity greater than one is also greater than one.

[13] For an asymmetric channel this bound may be improved by changing the two exponents 1/2 to s and $1 - s$, respectively, where $0 < s < 1$.

so that for any input-binary memoryless channel (46) becomes

$$D_0 = \sum_y P(y \mid x)^{1/2} P(y \mid \bar{x})^{1/2} \tag{47}$$

and consequently

$$P_E < T(D) \big|_{D=D_0} \tag{48}$$

$$P_B < \left. \frac{dT(D, N)}{dN} \right|_{N=1, D=D_0} \tag{49}$$

where D_0 is given by (47). Other examples of channels of this type are FSK modulation over the AWGN (both coherent and noncoherent) and Rayleigh fading channels.

VIII. Systematic and Nonsystematic Convolutional Codes

The term *systematic* convolutional code refers to a code on each of whose branches one of the code symbols is just the data bit generating that branch. Thus a systematic coder will have its stages connected to only $n - 1$ adders, the nth being replaced by a direct line from the first stage to the commutator. Fig. 13 shows an $R = 1/2$ systematic coder for $K = 3$.

It is well known that for group block codes, any nonsystematic code can be transformed into a systematic code which performs exactly as well. This is not the case for convolutional codes. The reason for this is that, as was shown in Section VII, the performance of a code on any channel depends largely on the relative distances between codewords and particularly on the minimum free distance d, which is the exponent of D in the leading term of the generating function. Eliminating one of the adders results in a reduction of d. For example, the maximum free distance code for $K = 3$ is that of Fig. 13 and this has $d = 4$, while the nonsystematic $K = 3$ code of Fig. 1 has minimum free distance $d = 5$. Table I shows the maximum minimum free distance for systematic and nonsystematic codes for $K = 2$ through 5. For large constraint lengths the results are even more widely separated. In fact, Bucher and Heller [19] have shown that for asymptotically large K, the performance of a systematic code of constraint length K is approximately the same as that of a nonsystematic code of constraint length $K(1 - R)$. Thus for $R = 1/2$ and very large K, systematic codes have the performance of nonsystematic codes of half the constraint length, while requiring exactly the same optimal decoder complexity. For $R = 3/4$, the constraint length is effectively divided by 4.

IX. Catastrophic Error Propagation in Convolutional Codes

Massey and Sain [13] have defined a catastrophic error as the event that a finite number of channel symbol errors causes an infinite number of data bit errors to be decoded. Furthermore, they showed that a necessary and sufficient condition for a convolutional code to produce

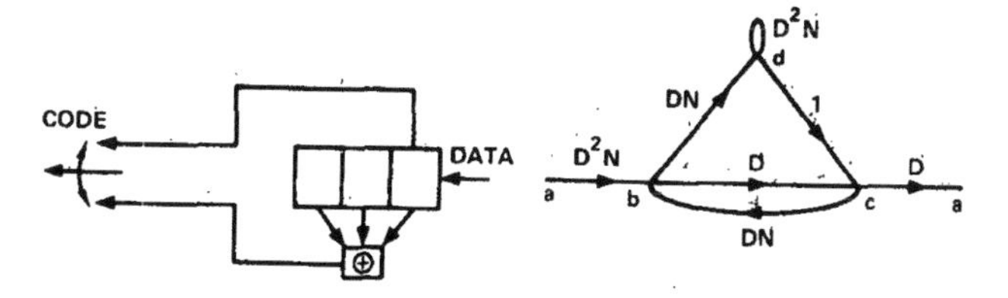

Fig. 13. Systematic convolution coder for $K = 3$ and $r = 1/2$.

TABLE I
Maximum–Minimum Free Distance

K	Systematic	Nonsystematic[a]
2	3	3
3	4	5
4	4	6
5	5	7

[a] We have excluded catastrophic codes (see Section IX); $R = \frac{1}{2}$.

catastrophic errors is that all of the adders have tap sequences, represented as polynomials, with a common factor.

In terms of the state diagram it is easily seen that catastrophic errors can occur if and only if any closed loop path in the diagram has a zero weight (i.e, the exponent of D for the loop path is zero). To illustrate this, we consider the example of Fig. 14.

Assuming that the all zeros is the correct path, the incorrect path $a\ b\ d\ d \cdots d\ c\ a$ has exactly 6 ones, no matter how many times we go around the self loop d. Thus for a BSC, for example, four-channel errors may cause us to choose this incorrect path or consequently make an arbitrarily large number of bit errors (equal to two plus the number of times the self loop is traversed). Similarly for the AWGN channel this incorrect path with arbitrarily many corresponding bit errors will be chosen with probability erfc $\sqrt{6\epsilon_b/N_0}$.

Another necessary and sufficient condition for catastrophic error propagation, recently found by Odenwalder [20] is that any nonzero data path in the trellis or state diagram produces $K - 1$ consecutive branches with all zero code symbols.

We observe also that for binary-tree ($R = 1/n$) codes, if each adder of the coder has an even number of connections, then the self loop corresponding to the all ones (data) state will have zero weight and consequently the code will be catastrophic.

The main advantage of a systematic code is that it can never be catastrophic, since each closed loop must contain at least one branch generated by a nonzero data bit and thus having a nonzero code symbol. Still it can be shown [23] that only a small fraction of nonsystematic codes is catastrophic (in fact, $1/(2^n - 1)$ for binary-tree $R = 1/n$ codes. We note further that if catastrophic errors are ignored, nonsystematic codes with even larger free distance than those of Table I exist.

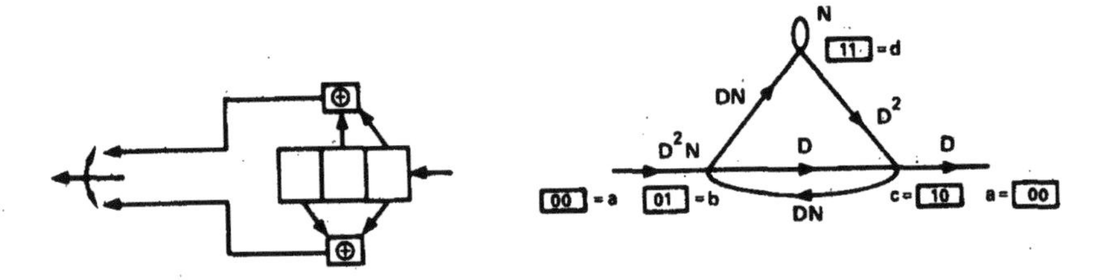

Fig. 14. Coder displaying catastrophic error propagation.

X. Performance Bounds for Best Convolutional Codes for General Memoryless Channels and Comparison with Block Codes

We begin by considering the path structure of a binary-tree[14] ($b = 1$) convolutional code of any constraint K, independent of the specific coder used. For this purpose we need only determine $T(L)$ the generating function for the state diagram with each branch labeled merely by L so that the exponent of each term of the infinite series expansion of $T(L)$ determines the length over which an incorrect path differs from the correct path before merging with it at a given node level. (See Fig. 7 and (2) with $D = N = 1$).

After some manipulation of the state-transition matrix of the state diagram of a binary-tree convolutional code of constraint length K, it is shown in Appendix I[15] that

$$T(L) = \frac{L^K(1 - L)}{1 - 2L + L^K} < \frac{L^K}{1 - 2L} = L^K(1 + 2L + 4L^2 + \cdots + 2^k L^k + \cdots) \qquad (50)$$

where the inequality indicates that more paths are being counted than actually exist. The expression (50) indicates that of the paths merging with the correct path at a given node level there is no more than one of length K, no more than two of length $K + 1$, no more than three of length $K + 2$, etc.

We have purposely avoided considering the actual code or coder configuration so that the preceding expressions are valid for all binary-tree codes of constraint length K. We now extend our class of codes to include time-varying convolutional codes. A time-varying coder is one in which the tap positions may be changed after each shift of the bits in the register. We consider the ensemble of all possible time-varying codes, which includes as a subset the ensemble of all fixed codes, for a given constraint length K. We further impose a uniform probabilistic measure on all codes in this ensemble by randomly reselecting each tap position after each shift of the register. This can be done by hypothetically flipping a coin nK times after each shift, once for each stage of the register and for each of the n adders. If the outcome is a head we connect the particular stage to the particular adder; if it is a tail we do not. Since this is repeated for each new branch, the result is that for each branch of the trellis the code sequence is a random binary n-dimensional vector. Furthermore, it can be shown that the distribution of these random code sequences is the same for each branch at each node level except for the all zeros path, which must necessarily produce the all zeros code sequence on each branch. To avoid treating the all zeros path differently, we ensure statistical uniformity by requiring further that after each shift a random binary n-dimensional vector be added to each branch[16] and that this also be reselected after each shift. (This additional artificiality is unnecessary for input-binary channels but is required to prove our result for general memoryless channels). Further details of this procedure are given in Viterbi [9].

We now seek a bound on the average error probability of this ensemble of codes relative to the measure (random-selection process) imposed. We begin by considering the probability that after transmission over a memoryless channel the metric of one of the fewer than 2^k paths merging with the correct path after differing in $K + k$ branches, is greater than the correct metric. Let $\mathbf{x}_i$ be the correct (transmitted) sequence and $\mathbf{x}_i'$ an incorrect sequence for the ith branch of the two paths. Then following the argument which led to (37) we have that the probability that the given incorrect path may cause an error is bounded by

$$P_{K+k}(\mathbf{x}, \mathbf{x}') < \prod_{i=1}^{K+k} \sum_{\mathbf{y}_i} P(\mathbf{y}_i \mid \mathbf{x}_i)^{1/2} P(\mathbf{y}_i \mid \mathbf{x}_i')^{1/2} \qquad (51)$$

where the product is over all $K + k$ branches in the path. If we now average over the ensemble of codes constructed above we obtain

$$\bar{P}_{K+k} < \prod_{i=1}^{K+k} \sum_{\mathbf{x}_i} \sum_{\mathbf{x}_i'} \sum_{\mathbf{y}_i} q(\mathbf{x}_i) P(\mathbf{y}_i \mid \mathbf{x}_i)^{1/2} q(\mathbf{x}_i') P(\mathbf{y}_i \mid \mathbf{x}_i')^{1/2} \qquad (52)$$

where $q(\mathbf{x})$ is the measure imposed on the code symbols of each branch by the random selection, and because of the statistical uniformity of all branches we have

$$\bar{P}_{K+k} < \{\sum_{\mathbf{y}} [\sum_{\mathbf{x}} q(\mathbf{x}) P(\mathbf{y} \mid \mathbf{x})^{1/2}]^2\}^{K+k} = 2^{-(K+k)nR_0} \qquad (53)$$

[14] Although for clarity all results will be derived for $b = 1$, the extension to $b > 1$ is direct and the results will be indicated at the end of this Section.

[15] This generating function can also be used to obtain error bounds for orthogonal convolutional codes all of whose branches have the same weight, as is shown in Appendix I.

[16] The same vector is added to all branches at a given node level.

where

$$R_0 \triangleq -\frac{1}{n}\log_2\{\sum_{\mathbf{y}}[\sum_{\mathbf{x}} q(\mathbf{x})P(\mathbf{y}\mid\mathbf{x})^{1/2}]^2\}. \quad (54)$$

Note that the random vectors $\mathbf{x}$ and $\mathbf{y}$ are n dimensional. If each symbol is transmitted independently on a memoryless channel, such as was the case in the channels of Sections VII-A and -B, (54) is reduced further to

$$R_0 = -\log_2\{\sum_{y}[\sum_{x} q(x)P(y\mid x)^{1/2}]^2\} \quad (55)$$

where x and y are now scalar random variables associated with each code symbol. Note also that because of the statistical uniformity of the code, the results are independent of which path was transmitted and which incorrect path we are considering.

Proceeding as in Section VII, it follows that a union bound on the ensemble average of the first-event error probability is obtained by substituting $\bar{P}_{K+k}$ for L^{K+k} in (50). Thus

$$\bar{P}_E < \sum_{k=0}^{\infty} 2^k\bar{P}_{K+k} < \sum_{k=0}^{\infty} 2^k 2^{-(K+k)R_0/R} = \frac{2^{-KR_0/R}}{1-2^{-(R_0/R-1)}} \quad (56)$$

where we have used the fact that since $b = 1$, $R = 1/n$ bits/symbol.

To bound the bit error probability we must weight each term of (56) by the number of bit errors for the corresponding incorrect path. This could be done by evaluating the transfer function $T(L, N)$ as in Section VII (see also Appendix I), but a simpler approach, which yields a simpler bound which is nearly as tight, is to recognize that an incorrectly chosen path which merges with the correct path after $K + k$ branches can produce no more $k + 1$ bit errors. For, any path which merges with the correct path at a given level must be generated by data which coincides with the correct path data over the last $K - 1$ branches prior to merging, since only in this way can the coder register be filled with the same bits as the correct path, which is the condition for merging. Hence the number of incorrect bits due to a path which differs from the correct path in $K + k$ branches can be no greater than $K + k - (K - 1) = k + 1$.

Hence we may overbound $\bar{P}_B$ by weighting the kth term of (56) by $k + 1$, which results in

$$\bar{P}_B < \sum_{k=0}^{\infty}(k+1)2^{-k(R_0/R-1)}2^{-KR_0/R} = \frac{2^{-KR_0/R}}{[1-2^{-(R_0/R-1)}]^2}. \quad (57)$$

The bounds of (56) and (57) are finite only for rates $R < R_0$, and R_0 can be shown to be always less than the channel capacity.

To improve on these bounds when $R > R_0$, we must improve on the union bound approach by obtaining a single bound on the probability that any one of the fewer than 2^k paths which differ from the correct path in $K + k$ branches has a metric higher than the correct path at a given node level. This bound, first derived by Gallager [5] for block codes, is always less than 2^k times the bound for each individual path. Letting $Q_{K+k} \triangleq$ Pr (any one of 2^k incorrect path metrics > correct path metric), Gallager [5] has shown that its ensemble average for the code ensemble is bounded by

$$\bar{Q}_{K+k} < 2^{k\rho}2^{-(K+k)nE_0(\rho)} \quad (58)$$

where

$$E_0(\rho) = -\frac{1}{n}\log_2\sum_{\mathbf{y}}[\sum_{\mathbf{x}} q(\mathbf{x})p(\mathbf{y}\mid\mathbf{x})^{1/1+\rho}]^{1+\rho}, \quad 0 < \rho \leq 1 \quad (59)$$

where ρ is an arbitrary parameter which we shall choose to minimize the bound. It is easily seen that $E_0(0) = 0$, while $E_0(1) = R_0$, in which case $\bar{Q}_{K+k} = 2^k\bar{P}_{K+k}$, the ordinary union bound of (56). We bound the overall ensemble first-event error probability by the probability of the union of these composite events given by (58). Thus we find

$$\bar{P}_E < \sum_{k=0}^{\infty}\bar{Q}_{K+k} < \frac{2^{-KE_0(\rho)/R}}{1-2^{-(E_0(\rho)/R-\rho)}}. \quad (60)$$

Clearly (60) reduces to (56) when $\rho = 1$.

To determine the bit error probability using this approach, we must recognize that $\bar{Q}_{K+k}$ refers to 2^k different incorrect paths, each with a different number of incorrect bits. However, just as was observed in deriving (57), an incorrect path which differs from the correct path in $K + k$ branches prior to merging can produce at most $k + 1$ bit errors. Hence weighting the kth term of (60) by $k + 1$, we obtain

$$\bar{P}_B < \sum_{k=0}^{\infty}(k+1)\bar{Q}_{K+k} < \sum_{k=0}^{\infty}(k+1)2^{-k(E_0(\rho)/R-\rho)}2^{-KE_0(\rho)/R} = \frac{2^{-KE_0(\rho)/R}}{[1-2^{-(E_0(\rho)/R-\rho)}]^2}, \quad 0 < \rho \leq 1. \quad (61)$$

Clearly (61) reduces to (57) when $\rho = 1$.

Before we can interpret the results of (56), (57), (60), and (61) it is essential that we establish some of the properties of $E_0(\rho)$ $(0 < \rho \leq 1)$ defined by (59). It can be shown [5], [14] that for any memoryless channel, $E_0(\rho)$ is a concave monotonic nondecreasing function as shown in Fig. 15 with $E_0(0) = 0$ and $E_0(1) = R_0$.

Where the derivative $E_0'(\rho)$ exists, it decreases with ρ and it follows easily from the definition that

$$\lim_{\rho\to 0} E_0'(\rho) = \frac{1}{n}\sum_{\mathbf{y}}\sum_{\mathbf{x}} q(\mathbf{x})\log_2\frac{P(\mathbf{y}\mid\mathbf{x})}{\sum_{\mathbf{x}'} q(\mathbf{x}')P(\mathbf{y}\mid\mathbf{x}')} = \frac{1}{n}I(X^n, Y^n) \triangleq C \quad (62)$$

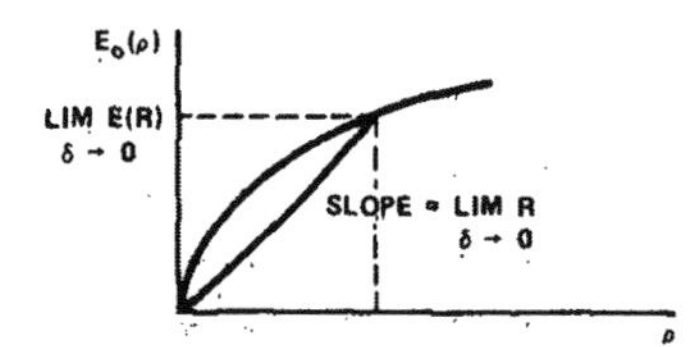

Fig. 15. Example of $E_0(\rho)$ function for general memoryless channel.

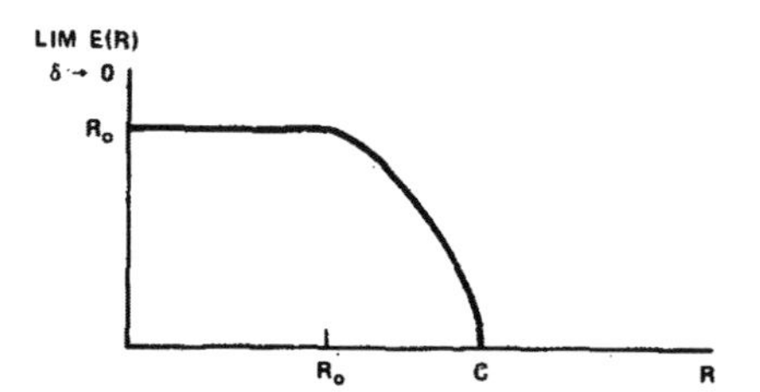

Fig. 16. Typical limiting value of exponent of (67).

the mutual information of the channel[17] where X^n and Y^n are the channel input and output spaces, respectively, for each branch sequence. Consequently, it follows that to minimize the bounds (60) and (61), we must make $\rho \leq 1$ as large as possible to maximize the exponent of the numerator, but at the same time we must ensure that

$$R < \frac{E_0(\rho)}{\rho}$$

in order to keep the denominator positive. Thus since $E_0(1) = R_0$ and $E_0(\rho) < R_0$, for $\rho < 1$, it follows that for $R < R_0$ and sufficiently large K we should choose $\rho = 1$, or equivalently use the bounds (56) and (57). We may thus combine all the above bounds into the expressions

$$\bar{P}_E < \frac{2^{-KE(R)/R}}{1 - 2^{-\delta(R)}} \tag{63}$$

$$\bar{P}_B < \frac{2^{-KE(R)/R}}{[1 - 2^{-\delta(R)}]^2} \tag{64}$$

where

$$E(R) = \begin{cases} R_0, & 0 \leq R < R_0 \\ E_0(\rho), & R_0 < R < C, \quad 0 < \rho \leq 1 \end{cases} \tag{65}$$

$$\delta(R) = \begin{cases} R_0/R - 1, & 0 < R < R_0 \\ E_0(\rho)/R - \rho, & R_0 \leq R < C, \quad 0 < \rho \leq 1. \end{cases} \tag{66}$$

To minimize the numerators of (63) and (64) for $R > R_0$ we should choose ρ as large as possible, since $E_0(\rho)$ is a nondecreasing function of ρ. However, we are limited by the necessity of making $\delta(R) > 0$ to keep the denominator from becoming zero. On the other hand, as the constraint length K becomes very large we may choose $\delta(R) = \delta$ very small. In particular, as δ approaches 0, (65) approaches

$$\lim_{\delta \to 0} E(R) = \begin{cases} R_0, & 0 < R < R_0 \\ E_0(\rho), & R_0 \leq R = E_0(\rho)/\rho < C, \\ & 0 < \rho \leq 1. \end{cases} \tag{67}$$

[17] C can be made equal to the channel capacity by properly choosing the ensemble measure $q(\mathbf{x})$. For an input-binary channel the random binary convolutional coder described above achieves this. Otherwise a further transformation of the branch sequence into a smaller set of nonbinary sequences is required [9].

Fig. 15 demonstrates the graphical determination of $\lim_{\delta \to 0} E(R)$ from $E_0(\rho)$.

It follows from the properties of $E_0(\rho)$ described, that for $R > R_0$, $\lim_{\delta \to 0} E(R)$ decreases from R_0 to 0 as R increases from R_0 to C, but that it remains positive for all rates less than C. The function is shown for a typical channel in Fig. 16.

It is particularly instructive to obtain specific bounds, in the limiting case, for the class of "very noisy" channels, which includes the BSC with $p = 1/2 - \gamma$ where $|\gamma| \ll 1$ and the biphase modulated AWGN with $\epsilon_s/N_0 \ll 1$. For this class of channels it can be shown [5] that

$$E_0(\rho) = \frac{\rho C}{1 + \rho} \tag{68}$$

and consequently $R_0 = E_0(1) = C/2$. (For the BSC, $C = \gamma^2/2 \ln 2$ while for the AWGN, $C = \epsilon_s/N_0 \ln 2$.)

For the very noisy channel, suppose we let $\rho = C/R - 1$, so that using (68) we obtain $E_0(\rho) = C - R$. Then in the limit as $\delta \to 0$ (65) becomes for a very noisy channel

$$\lim_{\delta \to 0} E(R) = \begin{cases} C/2, & 0 \leq R \leq C/2 \\ C - R, & C/2 \leq R \leq C. \end{cases} \tag{69}$$

This limiting form of $E(R)$ is shown in Fig. 17.

The bounds (63) and (64) are for the average error probabilities of the ensemble of codes relative to the measure induced by random selection of the time-varying coder tap sequences. At least one code in the ensemble must perform better than the average. Thus the bounds (63) and (64) hold for the best time-varying binary-tree convolutional coder of constraint length K. Whether there exists a fixed convolutional code with this performance is an unsolved problem. However, for small K the results of Section VII seem to indicate that these bounds are valid also for fixed codes.

To determine the tightness of the upper bounds, it is useful to have lower bounds for convolutional code error probabilities. It can be shown [9] that for all $R < C$

$$P_B \geq P_E > 2^{-K[E_L(R)/R - o(K)]} \tag{70}$$

where

$$\begin{aligned} E_L(R) &= E_0(\rho) \\ R &= E_0(\rho)/\rho \end{aligned}, \quad 0 \leq \rho < \infty, \quad 0 \leq R \leq C \tag{71}$$

and $o(K) \to 0$ as $K \to \infty$. Comparison of the parametric

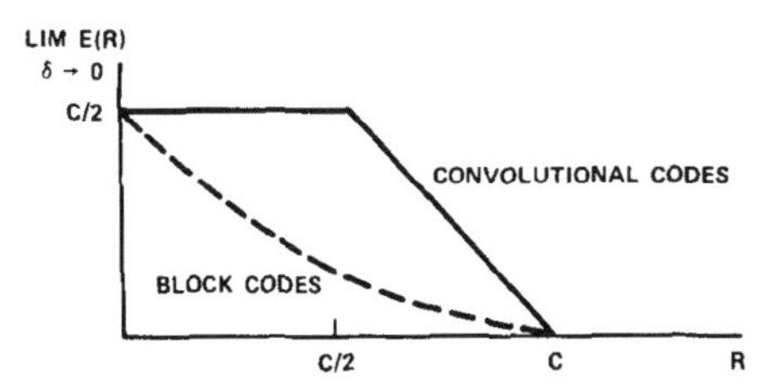

Fig. 17. Limiting values of $E(R)$ for very noisy channels.

equations (67) with (71), shows that

$$E_L(R) = \lim_{\delta\to 0} E(R)$$

for $R > R_0$ but is greater for low rates.

For very noisy channels, it follows easily from (71) and (68) that

$$E_L(R) = C - R, \qquad 0 \le R \le C.$$

Actually, however, tighter lower bounds for $R < C/2$ (Viterbi [9]) show that for very noisy channels

$$E_L(R) = \begin{cases} C/2, & 0 \le R \le C/2 \\ C - R, & C/2 \le R < C, \end{cases} \tag{72}$$

which is precisely the result of (69) or of Fig. 17. It follows that, at least for very noisy channels, the exponential bounds are asymptotically exact.

All the results derived in this section can be extended directly to nonbinary ($b > 1$) codes. It is easily shown (Viterbi [9]) that the same results hold with $R = b/n$, R_0 and $E_0(\rho)$ multiplied by b, and all event probability upper bounds multiplied by $2^b - 1$, and bit probability upper bounds multiplied by $(2^b - 1)/b$.

Clearly, the ensemble of codes considered here is nonsystematic. However, by a modification of the arguments used here, Bucher and Heller [19] restricted the ensemble to systematic time-varying convolutional codes (i.e., codes for which b code symbols of each branch correspond to the data which generates the branch) and obtained all the above results modified only to the extent that the exponents $E(R)$ and $E_L(R)$ are multiplied by $1 - R$. (See also Section VIII.)

Finally, it is most revealing to compare the asymptotic results for the best convolutional codes of a given constraint length with the corresponding asymptotic results for the best block codes of a given block length. Suppose that K bits are coded into a block code of length N so that $R = K/N$ bits/code symbol. Then it can be shown (Gallager [5], Shannon *et al.* [8]) that for the best block code, the bit error probability is bounded above and below by

$$2^{-K[E_{Lb}(R)/R + o(K)]} < P_B < 2^{-KE_b(R)/R} \tag{73}$$

where

$$E_b(R) = \max_{0\le\rho\le 1} [E_0(\rho) - \rho R]$$

$$E_{Lb}(R) \le \max_{0\le\rho} [E_0(\rho) - \rho R].$$

Both $E_b(R)$ and $E_{Lb}(R)$ are functions of R which for all $R > 0$ are less than the exponents $E(R)$ and $E_L(R)$ for convolutional codes [9]. In particular, for very noisy channels they both become [5]

$$E_b(R) = E_{Lb}(R) = \begin{cases} C/2 - R \\ (\sqrt{C} - \sqrt{R})^2. \end{cases} \tag{74}$$

This is plotted as a dotted curve in Fig. 17.

Thus it is clear by comparing the magnitudes of the negative exponents of (73) and (64) that, at least for very noisy channels, a convolutional code performs much better asymptotically than the corresponding block code of the same order of complexity. In particular at $R = C/2$, the ratio of exponents is 5.8, indicating that to achieve equivalent performance asymptotically the block length must be over five times the constraint length of the convolutional code. Similar degrees of relative performance can be shown for more general memoryless channels [9].

More significant from a practical viewpoint, for short constraint lengths also, convolutional codes considerably outperform block codes of the same order of complexity.

XI. Path Memory Truncation Metric Quantization and Synchronization

A major problem which arises in the implementation of a maximum likelihood decoder is the length of the path history which must be stored. In our previous discussion we ignored this important point and therefore implicitly assumed that all past data would be stored. A final decision was made by forcing the coder into a known (all zeros) state. We now remove this impractical condition. Suppose we truncate the path memories after M bits (branches) have been accumulated, by comparing all 2^K metrics for a maximum and deciding on the bit corresponding to that path (out of 2^K) with the highest metric M branches forward. If M is several times as large as K, the additional bit errors introduced in this way are very few, as we shall now demonstrate using the asymptotic results of the last section.

An additional bit error may occur due to memory truncation after M branches, if the bit selected is from an incorrect path which differed from the correct path M branches back and which has a higher metric, but which would ultimately be eliminated by the maximum likelihood decoder. But for a binary-tree code there can be no more than 2^M distinct paths which differ from the correct path M branches back. Of these we need concern ourselves only with those which have not merged with the correct path in the intervening nodes. As was originally shown by Forney [12], using the ensemble arguments of Section X we may bound the average probability of this event by [see (58)]

$$\bar{P}_t < 2^{M\rho} 2^{-ME_0(\rho)/R}, \qquad 0 < \rho \le 1. \tag{75}$$

To minimize this bound we should maximize the exponent $E_0(\rho)/R - \rho$ with respect to ρ on the unit interval. But this yields exactly $E_b(R)$, the upper bound exponent of (73) for block codes. Thus

$$\bar{P}_t < 2^{-ME_b(R)/R} \tag{76}$$

where $E_b(R)$ is the block coding exponent.

We conclude therefore that the memory truncation error is less than the bit error probability bound without truncation, provided the bound of (76) is less than the bound of (64). This will certainly be assured if

$$ME_b(R) > KE(R). \tag{77}$$

For very noisy channels we have from (69) and (74) or Fig. 17, that

$$\frac{M}{K} > \begin{cases} \dfrac{1}{1 - 2R/C}, & 0 \leq R \leq C/4 \\ \dfrac{1}{2(1 - \sqrt{R/C})^2}, & C/4 \leq R \leq C/2 \\ \dfrac{1 - R/C}{(1 - \sqrt{R/C})^2}, & C/2 < R < C. \end{cases}$$

For example, at $R = C/2$ this indicates that it suffices to take $M > (5.8)K$.

Another problem faced by a system designer is the amount of storage required by the metrics (or log-likelihood functions) for each of the 2^K paths. For a BSC this poses no difficulty since the metric is just the Hamming distance which is at most n, the number of code symbols, per branch. For the AWGN, on the other hand, the optimum metric is a real number, the analog output of a correlator, matched filter, or integrate-and-dump circuit. Since digital storage is generally required, it is necessary to quantize this analog metric. However, once the components y_{jk} of the optimum metric of (5), which are the correlator outputs, have been quantized to Q levels, the channel is no longer an AWGN channel. For biphase modulation, for example, it becomes a binary input Q-ary output discrete memoryless channel, whose transition probabilities are readily calculated as a function of the energy-to-noise density and the quantization levels. The optimum metric is not obtained by replacing y_{jk} by its quantized value $Q(y_{jk})$ in (5) but rather it is the log-likelihood function $\log P(\mathbf{y} \mid \mathbf{x}^{(m)})$ for the binary-input Q-ary-output channel.

Nevertheless, extensive simulation [24] indicates that for 8-level quantization even use of the suboptimal metric $\sum_k Q(y_{jk})x_{jk}^{(m)}$ results in a degradation of no more than 0.25 dB relative to the maximum likelihood decoder for the unquantized AWGN, and that use of the optimum metric is only negligibly superior to this. However, this is not the case for sequential decoding, where the difference in performance between optimal and suboptimal metrics is significant [11].

In a practical system other considerations than error performance for a given degree of decoder complexity often dictate the selection of a coding system. Chief among these are often the synchronization requirements. Convolutional codes utilizing maximum likelihood decoding are particularly advantageous in that no block synchronization is ever required. For block codes, decoding cannot begin until the initial point of each block has been located. Practical systems often require more complexity in the synchronization system than in the decoder. On the other hand, as we have by now amply illustrated, a maximum likelihood decoder for a convolutional code does not require any block synchronization because the coder is free running (i.e., it performs identical operations for each successive input bit and does not require that K bits be input before generating an output). Furthermore, the decoder does not require knowledge of past inputs to start decoding; it may as well assume that all previous bits were zeros. This is not to say that initially the decoder will operate as well, in the sense of error performance, as if the preceding bits of the correct path were known. On the other hand, consider a decoder which starts with an initially known path but makes an error at some point and excludes the correct path. Immediately thereafter it will be operating as if it had just been turned on with an unknown and incorrectly chosen previous path history. That this decoder will recover and stop making errors within a finite number of branches follows from our previous discussions in which it was shown that, other than for catastrophic codes, error sequences are always finite. Hence our initially unsynchronized decoder will operate just like a decoder which has just made an error and will thus always achieve synchronization and generally will produce correct decisions after a limited number of initial errors. Simulations have demonstrated that synchronization generally takes no more than four or five constraint lengths of received symbols.

Although, as we have just shown, branch synchronization is not required, code symbol synchronization within a branch is necessary. Thus, for example, for a binary-tree rate $R = 1/2$ code, we must resolve the two-way ambiguity as to where each two code-symbol branch begins. This is called node synchronization. Clearly if we make the wrong decisions, errors will constantly be made thereafter. However, this situation can easily be detected because the mismatch will cause *all* the path metrics to be small, since in fact there will not be any correct path in this case. We can thus detect this event and change our decision as to node synchronization (cf. Heller and Jacobs [24]). Of course, for an $R = 1/n$ code, we may have to repeat our choice n times, once for each of the symbols on a branch, but since n represents the redundancy factor or bandwidth expansion, practical systems rarely use $n > 4$.

XII. Other Decoding Algorithms for Convolutional Codes

This paper has treated primarily maximum likelihood decoding of convolutional codes. The reason for this was two-fold: 1) maximum likelihood decoding is closely related to the structure of convolutional codes and its consideration enhances our understanding of the ultimate capabilities, performance, and limitation of these codes; 2) for reasonably short constraint lengths ($K < 10$) its implementation is quite feasible[18] and worthwhile because of its optimality. Furthermore for $K \leq 6$, the complexity of maximum likelihood decoding is sufficiently limited that a completely parallel implementation (separate metric calculators) is possible. This minimizes the decoding time per bit and affords the possibility of extremely high decoding speeds [24].

Longer constraint lengths are required for extremely low error probabilities at high rates. Since the storage and computational complexity are proportional to 2^K, maximum likelihood decoders become impractical for $K > 10$. At this point *sequential decoding* [2], [4], [6] becomes attractive. This is an algorithm which sequentially searches the code tree in an attempt to find a path whose metric rises faster than some predetermined, but variable, threshold. Since the difference between the correct path metric and any incorrect path metric increases with constraint length, for large K generally the correct path will be found by this algorithm. The main drawback is that the number of incorrect path branches, and consequently the computation complexity, is a random variable depending on the channel noise. For $R < R_0$, it is shown that the average number of incorrect branches searched per decoded bit is bounded [6], while for $R > R_0$ it is not; hence R_0 is called the computational cutoff rate. To make storage requirements reasonable, it is necessary to make the decoding speed (branches/s) somewhat larger than the bit rate, thus somewhat limiting the maximum bit rate capability. Also, even though the average number of branches searched per bit is finite, it may sometimes become very large, resulting in a storage overflow and consequently relatively long sequences being erased. The stack sequential decoding algorithm [7], [18] provides a very simple and elegant presentation of the key concepts in sequential decoding, although the Fano algorithm [4] is generally preferable practically.

For a number of reasons, including buffer size requirements, computation speed, and metric sensitivity, sequential decoding of data transmitted at rates above about 100 K bits/s is practical only for hard-quantized binary received data (that is, for channels in which a hard decision —0 or 1— is made for each demodulated symbol). For the biphase modulated AWGN channel, of course, hard quantization (2 levels or 1 bit) results in an efficiency loss of approximately 2 dB compared with soft quantization (8 or more levels—3 or more bits). On the other hand, with maximum likelihood decoding, by employing a parallel implementation, short constraint length codes ($K \leq 6$) can be decoded at very high data rates (10 to 100 Mbits/s) even with soft quantization. In addition, the insensitivity to metric accuracy and simplicity of synchronization render maximum likelihood decoding generally preferable when moderate error probabilities are sufficient. In particular, since sequential decoding is limited by the overflow problem to operate at code rates somewhat below R_0, it appears that for the AWGN the crossover point above which maximum likelihood decoding is preferable to sequential decoding occurs at values of P_B somewhere between 10^{-3} and 10^{-5}, depending on the transmitted data rate. As the data rate increases the P_B crossover point decreases.

A third technique for decoding convolutional codes is known as *feedback decoding,* with threshold decoding [3] as a subclass. A feedback decoder basically makes a decision on a particular bit or branch in the decoding tree or trellis based on the received symbols for a limited number of branches beyond this point. Even though the decision is irrevocable, for limited constraint lengths (which are appropriate considering the limited number of branches involved in a decision) errors will propagate only for moderate lengths. When transmission is over a binary symmetric channel, by employing only codes with certain algebraic (orthogonal) properties, the decision on a given branch can be based on a linear function of the received symbols, called the *syndrome,* whose dimensionality is equal to the number of branches involved in the decision. One particularly simple decision criterion based on this syndrome, referred to as *threshold decoding,* is mechanizable in a very inexpensive manner. However, feedback decoders in general, and threshold decoders in particular, have an error-correcting capability equivalent to very short constraint length codes and consequently do not compare favorably with the performance of maximum likelihood or sequential decoding.

However, feedback decoders are particularly well suited to correcting error bursts which may occur in fading channels. Burst errors are generally best handled by using interleaved codes: that is, employing L convolutional codes so that the jth, $(L + j)$th $(2L + j)$th, etc., bits are encoded into one code for each $j = 0, 1, \cdots, L - 1$. This will cause any burst of length less than L to be broken up into random errors for the L independently operating decoders. Interleaving can be achieved by simply inserting $L - 1$ stage delay lines between stages of the convolutional encoder; the resulting single encoder then generates the L interleaved codes. The significant advantage of a feedback or threshold decoder is that the same technique can be employed in the decoder resulting in a single (time-shared) decoder rather than L decoders, providing feasible implementations for hard-quantized channels, even for protection against error bursts of thousands of bits. Details of feedback decoding

[18] Performing metric calculations and comparisons serially.

are treated extensively in Massey [3], Gallager [14], and Lucky *et al.* [16].

Appendix I

Generating Function for Structure of a Binary-Tree Convolutional Code for Arbitrary K and Error Bounds for Orthogonal Codes

We derive here the distance-invariant ($D = 1$) generating function $T(L, N)$ for any binary tree ($b = 1$) convolutional code of arbitrary constraint length K. It is most convenient in the general case to begin with the finite-state machine state-transition matrix for the linear equations among the state (node) variables. We exhibit this in terms of N and L for a $K = 4$ code as follows:

$$\begin{bmatrix} 1 & 0 & 0 & -NL & 0 & 0 & 0 \\ -L & 1 & 0 & 0 & -L & 0 & 0 \\ -NL & 0 & 1 & 0 & -NL & 0 & 0 \\ 0 & -L & 0 & 1 & 0 & -L & 0 \\ 0 & -NL & 0 & 0 & 1 & -NL & 0 \\ 0 & 0 & -L & 0 & 0 & 1 & -L \\ 0 & 0 & -NL & 0 & 0 & 0 & 1-NL \end{bmatrix} \cdot \begin{bmatrix} X_{001} \\ X_{010} \\ X_{011} \\ X_{100} \\ X_{101} \\ X_{110} \\ X_{111} \end{bmatrix} = \begin{bmatrix} NL \\ 0 \\ 0 \\ 0 \\ 0 \\ 0 \\ 0 \end{bmatrix}. \tag{78}$$

This pattern can be easily seen to generalize to a $2^{K-1} - 1$ dimensional square matrix of this form for any binary-tree code of constraint length K, and in general the generating function

$$T(L, N) = LX_{100\cdots 0},$$

$$\text{where } 100 \cdots 0 \text{ contains } (K-2) \text{ zeros.} \tag{79}$$

From this general pattern it is easily shown that the matrix can be reduced to a dimension of 2^{K-2}. First combining adjacent rows, from the second to the last, pairwise, one obtains the set of $2^{K-2} - 1$ relations

$$NX_{j_1 j_2 \cdots j_{K-2} 0} = X_{j_1 j_2 \cdots j_{K-2} 1} \tag{80}$$

where $j_1, j_2, \cdots, j_{K-2}$ runs over all binary vectors except for the all zeros. Substitution of (80) into (78) yields a 2^{K-2}-dimensional matrix equation. The result for $K = 4$ is

$$\begin{bmatrix} 1 & 0 & -L & 0 \\ -NL & 1 & -NL & 0 \\ 0 & -L & 1 & -L \\ 0 & -NL & 0 & 1-NL \end{bmatrix} \cdot \begin{bmatrix} X_{001} \\ X_{011} \\ X_{101} \\ X_{111} \end{bmatrix} = \begin{bmatrix} NL \\ 0 \\ 0 \\ 0 \end{bmatrix}. \tag{81}$$

Defining the new variable

$$X'_{00\cdots 01} = NL\, X_{00\cdots 01} + X_{00\cdots 11} \tag{82}$$

(which corresponds to adding the second row to NL times the first), we obtain finally a $2^{K-2} - 1$ dimensional matrix equation, which for $K = 4$ is

$$\begin{bmatrix} 1 & -N(L+L^2) & 0 \\ -L & 1 & -L \\ -NL & 0 & 1-NL \end{bmatrix} \cdot \begin{bmatrix} X'_{001} \\ X_{101} \\ X_{111} \end{bmatrix} = \begin{bmatrix} N^2L^2 \\ 0 \\ 0 \end{bmatrix}. \tag{83}$$

Note that (83) is the same as (78) for K reduced by unity, but with modifications in two places, both in the first row; namely, the first component on the right side is squared, and the middle term of the first row is reduced by an amount NL^2. Although we have given the explicit result only for $K = 4$, it is easily seen to be valid for any K.

Since in all respects, except these two, the matrix after this sequence of reductions is the same as the original but with its dimension reduced corresponding to a reduction of K by unity, we may proceed to perform this sequence of reductions again. The steps will be the same except that now in place of (80), we have

$$NX_{j_1 j_2 \cdots j_{K-3} 01} = X_{j_1 j_2 \cdots j_{K-3} 11} \tag{80'}$$

and in place of (82)

$$X''_{00\cdots 01} = NL\, X'_{00\cdots 01} + X_{00\cdots 111} \tag{82'}$$

while in place of (81) the right of center term of the first row is $-(L + L^2)$ and the first component on the right side is N^2L^2. Similarly in place of (83) the center term of the first row is $-N(L + L^2 + L^3)$ and the first component on the right side is N^3L^3.

Performing this sequence of reductions $K - 2$ times in all, but omitting the last step—leading from (81) to (83)—in the last reduction, the original $2^{K-1} - 1$ equations are reduced in the general case to the two equations

$$\begin{bmatrix} 1 & -(L + L^2 + \cdots L^{K-2}) \\ -NL & 1 - NL \end{bmatrix} \cdot \begin{bmatrix} X_{00-01}^{(K-3)} \\ X_{11\cdots 1} \end{bmatrix} = \begin{bmatrix} (NL)^{K-2} \\ 0 \end{bmatrix} \tag{84}$$

whence it follows that

$$X_{11\cdots 1} = \frac{(NL)^{K-1}}{1 - N(L + L^2 + \cdots + L^{K-1})} \tag{85}$$

Applying (79) and the $K - 2$ extensions of (80) and (80′) we find

$$\begin{aligned} T(L, N) &= LX_{100\cdots 00} = LN^{-1}X_{100\cdots 01} \\ &= LN^{-2}X_{100\cdots 011} = \cdots = LN^{-(K-2)}X_{11\cdots 1} \\ &= \frac{NL^K}{1 - N(L + L^2 + \cdots + L^{K-1})} \\ &= \frac{NL^K(1 - L)}{1 - L(1 + N) + NL^K} \end{aligned} \tag{86}$$

If we require only the path length structure, and not the number of bit errors corresponding to any incorrect path, we may set $N = 1$ in (86) and obtain

$$T(L) = \frac{L^K}{1 - (L + L^2 + \cdots + L^{K-1})} = \frac{L^K(1 - L)}{1 - 2L + L^K}. \tag{87}$$

If we denote as an upper bound an expression which is the generating function of more paths than exist in our state diagram, we have

$$T(L) < \frac{L^K}{1 - 2L}. \tag{88}$$

As an additional application of this generating function technique, we now obtain bounds on P_E and P_B for the class of orthogonal convolutional (tree) codes introduced by Viterbi [10]. For this class of codes, to each of the 2^K branches of the K-state diagram there corresponds one of 2^K orthogonal signals. Given that each signal is orthogonal to all others in $n \geq 1$ dimensions, corresponding to n channel symbols or transmission times (as, for example, if each signal consists of n different pulses out of $2^K n$ possible positions), then the weight of each branch is n. Consequently, if we replace L, the path length enumerator, by D^n in (86) we obtain for orthogonal codes

$$T(D, N) = \frac{ND^{nK}(1 - D^n)}{1 - D^n(1 + N) + ND^{nK}}. \tag{89}$$

Then using (48) and (49), the first-event error probability for orthogonal codes is bounded by

$$P_E < \frac{D_0^{nK}(1 - D_0^n)}{1 - 2D_0^n + D_0^{nK}} < \frac{D_0^{nK}(1 - D_0^n)}{1 - 2D_0^n} \tag{90}$$

and the bit error probability bound is

$$\begin{aligned} P_B &< \left.\frac{dT(N, D)}{dN}\right|_{N=1, D=D_0} \\ &= \frac{D_0^{nK}(1 - D_0^n)^2}{(1 - 2D_0^n + D_0^{nK})^2} < \frac{D_0^{nK}(1 - D_0^n)^2}{(1 - 2D_0^n)^2} \end{aligned} \tag{91}$$

where D_0 is a function of the channel transition probabilities or energy-to-noise ratio and is given by (46).

Acknowledgment

The author gratefully acknowledges the considerable stimulation he has received over the course of writing the several versions of this paper from Dr. J. A. Heller, whose recent work strongly complements and enhances this effort, for numerous discussions and suggestions and for his assistance in its presentation at the Linkabit Corporation "Seminars on Convolutional Codes." This tutorial approach owes part of its origin to Dr. G. D. Forney, Jr., whose imaginative and perceptive reinterpretation of my original work has aided immeasurably in rendering it more comprehensible. Also, thanks are due to Dr. J. K. Omura for his careful and detailed reading and correction of the manuscript during his presentation of this material in the UCLA graduate course on information theory.

References

[1] P. Elias, "Coding for noisy channels," in *1955 IRE Nat. Conv. Rec.*, vol. 3, pt. 4, pp. 37–46.

[2] J. M. Wozencraft, "Sequential decoding for reliable communication," in *1957 IRE Nat. Conv. Record*, vol. 5, pt. 2, pp. 11–25.

[3] J. L. Massey, *Threshold Decoding*. Cambridge, Mass.: M.I.T. Press, 1963.

[4] R. M. Fano, "A heuristic discussion of probabilistic decoding," *IEEE Trans. Inform. Theory*, vol. IT-9, Apr. 1963, pp. 64–74.

[5] R. G. Gallager, "A simple derivation of the coding theorem and some applications," *IEEE Trans. Inform. Theory*, vol. IT-11, Jan. 1965, pp. 3–18.

[6] J. M. Wozencraft and I. M. Jacobs, *Principles of Communication Engineering*. New York: Wiley, 1965.

[7] K. S. Zigangirov, "Some sequential decoding procedures," *Probl. Peredach Inform.*, vol. 2, no. 4, 1966, pp. 13–25.

[8] C. E. Shannon, R. G. Gallager, and E. R. Berlekamp, "Lower bounds to error probability for coding on discrete memoryless channels," *Inform. Contr.*, vol. 10, 1967, pt. I, pp. 65–103, pt. II, pp. 522–552.

[9] A. J. Viterbi, "Error bounds for convolutional codes and an asymptotically optimum decoding algorithm," *IEEE Trans. Inform. Theory*, vol. IT-13, Apr. 1967, pp. 260–269.

[10] ——, "Orthogonal tree codes for communication in the presence of white Gaussian noise," *IEEE Trans. Commun. Technol.*, vol. COM-15, April 1967, pp. 238–242.

[11] I. M. Jacobs, "Sequential decoding for efficient communication from deep space," *IEEE Trans. Commun. Technol.*, vol. COM-15, Aug. 1968, pp. 492–501.

[12] G. D. Forney, Jr., "Coding system design for advanced solar missions," submitted to NASA Ames Res. Ctr. by Codex Corp., Watertown, Mass., Final Rep., Contract NAS2-3637, Dec. 1967.

[13] J. L. Massey and M. K. Sain, "Inverses of linear sequential circuits," *IEEE Trans. Comput.*, vol. C-17, Apr. 1968, pp. 330–337.

[14] R. G. Gallager, *Information Theory and Reliable Communication*. New York: Wiley, 1968.

[15] T. N. Morrissey, "Analysis of decoders for convolutional codes by stochastic sequential machine methods," Univ. Notre Dame, Notre Dame, Ind., Tech. Rep. EE-682, May 1968.

[16] R. W. Lucky, J. Salz, and E. J. Weldon, *Principles of Data Communication*. New York: McGraw-Hill, 1968.

[17] J. K. Omura, "On the Viterbi decoding algorithm," *IEEE Trans. Inform. Theory*, vol. IT-15, Jan. 1969, pp. 177–179.
[18] F. Jelinek, "Fast sequential decoding algorithm using a stack," *IBM J. Res. Dev.*, vol. 13, no. 6, Nov. 1969, pp. 675–685.
[19] E. A. Bucher and J. A. Heller, "Error probability bounds for systematic convolutional codes," *IEEE Trans. Inform. Theory*, vol. IT-16, Mar. 1970, pp. 219–224.
[20] J. P. Odenwalder, "Optimal decoding of convolutional codes," Ph.D. dissertation, Dep. Syst. Sci., Sch. Eng. Appl. Sci., Univ. California, Los Angeles, 1970.
[21] G. D. Forney, Jr., "Coding and its application in space communications," *IEEE Spectrum*, vol. 7, June 1970, pp. 47–58.
[22] ——, "Convolutional codes I: Algebraic structure," *IEEE Trans. Inform. Theory*, vol. IT-16, Nov. 1970, pp. 720–738; "II: Maximum likelihood decoding," and "III: Sequential decoding," *IEEE Trans. Inform. Theory*, to be published.
[23] W. J. Rosenberg, "Structural properties of convolutional codes," Ph.D. dissertation, Dep. Syst. Sci., Sch. Eng. Appl. Sci., Univ. California, Los Angeles, 1971.
[24] J. A. Heller and I. M. Jacobs, "Viterbi decoding for satellite and space communication," this issue, pp. 835–848.
[25] A. R. Cohen, J. A. Heller, and A. J. Viterbi, "A new coding technique for asynchronous multiple access communication," this issue, pp. 849–855.

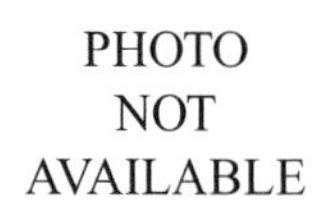

PHOTO NOT AVAILABLE

Andrew J. Viterbi (S'54–M'58–SM'63) was born in Bergamo, Italy, on March 9, 1935. He received the B.S. and M.S. degrees in electrical engineering from the Massachusetts Institute of Technology, Cambridge, in 1957, and the Ph.D. degree in electrical engineering from the University of Southern California, Los Angeles, in 1962.

While attending M.I.T., he participated in the cooperative program at the Raytheon Company. In 1957 he joined the Jet Propulsion Laboratory where he became a Research Group Supervisor in the Communications Systems Research Section. In 1963 he joined the faculty of the University of California, Los Angeles, as an Assistant Professor. In 1965 he was promoted to Associate Professor and in 1969 to Professor of Engineering and Applied Science. He was a cofounder in 1968 of Linkabit Corporation of which he is presently Vice President.

Dr. Viterbi is a member of the Editorial Boards of the PROCEEDINGS OF THE IEEE and of the journal *Information and Control*. He is a member of Sigma Xi, Tau Beta Pi, and Eta Kappa Nu and has served on several governmental advisory committees and panels. He is the coauthor of a book on digital communication and author of another on coherent communication, and he has received three awards for his journal publications.

An Adaptive Decision Feedback Equalizer

DONALD A. GEORGE, MEMBER, IEEE, ROBERT R. BOWEN, MEMBER, IEEE, AND JOHN R. STOREY, MEMBER, IEEE

Abstract—**An adaptive decision feedback equalizer to detect digital information transmitted by pulse-amplitude modulation (PAM) through a noisy dispersive linear channel is described, and its performance through several channels is evaluated by means of analysis, computer simulation, and hardware simulation. For the channels considered, the performance of both the fixed and the adaptive decision feedback equalizers are found to be notably better than that obtained with a similar linear equalizer.**

The fixed equalizer, which may be used when the channel characteristics are known, exhibits performance which is close to that of the optimum, but impractical, Bayesian receiver and is considerably superior to that of the linear equalizer. The adaptive decision feedback equalizer, which is used when the channel impulse response is unknown or time varying, has a better transient and steady-state performance than the adaptive linear equalizer. The sensitivity of the receiver structure to adjustment and quantization errors is not pronounced.

I. Introduction

An adaptive decision feedback equalizer is described, and its performance is discussed in this paper. The equalizer is used to recover a sequence of digits that has been transmitted at a high rate over a noisy dispersive linear communications channel by some linear modulation process. The channel is used efficiently by sending the digital information at such a high rate that there is intersymbol interference at the receiver input between several successive digits. The receiver is able to combat both the additive noise and the intersymbol interference, and also to adapt itself to an unknown or slowly varying channel without the aid of a training digit sequence. Thus it can "track" a continual slow drift in channel characteristics without interrupting the message transmission. Past decisions about the digits are used in minimizing the intersymbol interference by coherently subtracting the interference from previously detected digits, and also are used in adapting the equalizer parameters to a change in channel characteristics. It is shown that this receiver is insensitive to quantization of the input signal and quantization and adjustment of its own parameters, and so can be constructed at reasonable cost.

Paper approved by the Data Communication Systems Committee of the IEEE Communication Technology Group for publication after presentation at the 1970 IEEE International Conference on Communications, San Francisco, Calif., June 8–10. Manuscript received September 25, 1970; revised December 23, 1970.

D. A. George is with the Department of Electrical Engineering, Carleton University, Ottawa, Ont., Canada.

R. R. Bowen and J. R. Storey are with Communications Research Center, Department of Communications, Ottawa, Ont., Canada.

The adaptive linear transversal equalizer [1]–[3] has been developed in recent years to accomplish the task outlined previously. With that receiver it has been possible to utilize unknown or slowly varying dispersive channels much more effectively than was possible with fixed lumped-parameter equalizers. Concurrently, however, it has been shown [4]–[7] that the statistically optimum receiver for the recovery of the digit sequence, when the dispersive channel is known, is nonlinear. At high data rates the performance of this receiver is much better than that possible with the transversal equalizer, which is the optimum linear receiver. Unfortunately, the statistically optimum receiver is very complex when there is a large amount of intersymbol interference, and is not practical with today's technology. This suggests that one seek a statistically suboptimum receiver that is practical and has a performance that is significantly better than that of any receiver that is constrained to be linear. A decision feedback equalizer, described by Austin [8], is such a receiver. It is shown in Fig. 1. This equalizer is not adaptive but an adaptive version may readily be obtained, as is shown in this paper.

The decision feedback equalizer is similar to the transversal equalizer in that both have a filter matched to the isolated received pulse, followed by a baud-rate tapped delay line. However, it makes use of the fact that at the transversal equalizer output there is intersymbol interference caused by both undetected digits and previously detected digits. If the previous decisions are correct, they can be used to coherently substract the intersymbol interference caused by the previously detected digits. This is done by passing the past decisions through the feedback tapped delay line. The feedback delay line tap values are chosen on the assumption that these past decisions are all correct. The matched filter and the forward tapped delay line are used to minimize the effects of the additive noise and the intersymbol interference from undetected digits. Errors at the output of this equalizer occur in bursts, of course, because a decision error in the feedback delay line tends to cause yet more incorrect decisions. However, the equalizer is able to recover spontaneously from this condition. Simulation studies show that the performance of the decision feedback equalizer can be considerably better than that of the linear equalizer even though its output errors occur in bursts.

In Section II of this paper the decision feedback equalizer is described, and its performance is compared with the performance of a number of other receivers. This com-

Reprinted from *IEEE Transactions on Communication Technology*, vol. COM-19, no. 3, June 1971.

The Best of the Best. Edited by W. H. Tranter, D. P. Taylor, R. E. Ziemer, N. F. Maxemchuk, and J. W. Mark.

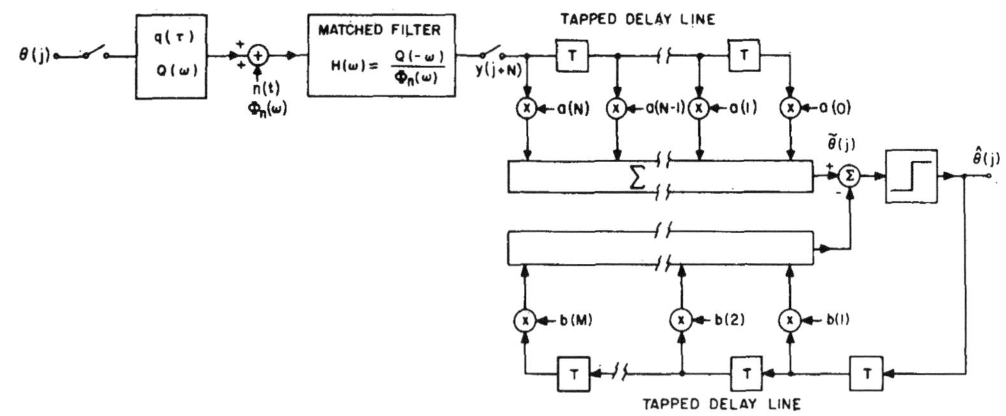

Fig. 1. Decision feedback equalizer.

parison is done both analytically and by digital computer simulation. It is shown by example that the decision feedback equalizer is an attractive compromise between what is theoretically possible and what is now in use. Next, in Section III, it is demonstrated that the decision feedback equalizer may be made adaptive to an unknown channel. Two different adaptation algorithms are described and compared by hardware simulation studies. In the cases considered, the adaptive decision feedback equalizer significantly outperformed the adaptive linear equalizer, and a training sequence was not required for adaptation. Rather, the decisions can be used for adaptation as well as to coherently substract the intersymbol interference. Finally, in Section IV, the practical nature of the equalizer is demonstrated by showing that the number of delay line taps that it requires is modest, and that its digital implementation requires no finer quantization than the linear equalizer.

II. Fixed Equalizer

In this section the fixed nonadaptive decision feedback equalizer for a known dispersive channel will be examined. The error rate of this equalizer is a lower bound on the error rate of an equalizer that must also adapt to random changes in the channel characteristics. The basic assumption made in deriving the receiver is that the decisions made by the receiver as to the transmitted signal samples are essentially correct. Given the bit error rate requirements in modern communication systems, this is a valid assumption. It is furthermore assumed that the analog signal from the communications channel has been demodulated, filtered, and sampled at the digit baud rate with the appropriate phase. Previous work [1], [8] has indicated the desirability of a matched filter before the sampler, as shown in Fig. 1. In the approach taken in this paper, any suitable band-limiting filter may be used, at the price of some loss in performance. It remains to determine the tap gains $\{a(k);\ k = 0,1,\cdots,N\}$ and $\{b(m);\ m = 1,2,3,\cdots,M\}$, as illustrated in Fig. 1. In the adaptive form of the equalizer these taps are automatically adjusted; for the fixed equalizer the tap gains must be calculated and manually adjusted after the channel characteristics are determined.

The equalizer makes the estimate

$$\tilde{\theta}(j) = \sum_{k=0}^{N} a(k)y(j+k) - \sum_{l=1}^{M} b(l)\hat{\theta}(j-l) \qquad (1)$$

about $\theta(j)$, the digit that is sent at time $t = jT$, and then converts this estimate to a final decision $\hat{\theta}(j)$ with a nonlinear memoryless circuit. (If the digits $\{\theta(j)\}$ are binary this circuit is a clipping circuit with zero bias. If the digits are m-ary the circuit is an m-output quantizer.) In (1) $y(j+k)$ is the output of the initial filter at time $t = (j+k)T$. One method of choosing the tap values is to adjust for the minimization of the probability that $\hat{\theta}(j) \neq \theta(j)$. However, this direct optimization is difficult because an analytical expression for the error probability in terms of the equalizer tap values is not known. A practicable way to "optimize" the tap values is to choose them such that the output mean-square error $E[e^2(j)]$ is minimized, where

$$e(j) \triangleq \tilde{\theta}(j) - \theta(j). \qquad (2)$$

As shown later, this leads to a set of linear equations that specify the tap values. This method of optimization is also attractive because it can be used to make the equalizer self-optimizing or adaptive to an unknown or a slowly varying channel. While this optimization does not minimize the digit error probability directly, computer simulation studies [6] have shown that the probability density function of the error $e(j)$ is close to Gaussian, and so the two performance criteria are similar.

The process of determining the tap values starts with the evaluation of the mean-square error:

$$\begin{aligned} E[e^2(j)] &= E[\{\tilde{\theta}(j) - \theta(j)\}^2] \\ &= E[\{\sum_{k=0}^{N} a(k)y(j+k) \\ &\quad - \sum_{l=1}^{M} b(l)\hat{\theta}(j-l) - \theta(j)\}^2] \end{aligned} \qquad (3)$$

where the signal sample $y(j+k)$ is

$$y(j+k) = \sum_{i=0}^{M} \theta(j+k-i)x(i) + n(j+k) \quad (4)$$

and $x(i)$ is the value of the impulse response of the linear modulator, the channel, and the initial receiver filter after a delay iT, and $n(j+k)$ is the additive noise at the output of the initial filter at time $t = (j+k)T$. The forward tap $a(n)$ is optimum when

$$\frac{\partial E[e^2(j)]}{\partial a(n)} = 2E[\{\sum_{k=0}^{N} a(k)y(j+k) - \sum_{l=1}^{M} b(l)\hat{\theta}(j-l) - \theta(j)\}\cdot y(j+n)]$$
$$= 2E[e(j)y(j+n)]$$
$$= 0, \quad n = 0,1,\cdots,N. \quad (5)$$

Similarly, the feedback tap $b(m)$ is optimum when

$$\frac{\partial E[e^2(j)]}{\partial b(m)} = 2E[\{\sum_{k=0}^{N} a(k)y(j+k) - \sum_{l=1}^{M} b(l)\hat{\theta}(j-l) - \theta(j)\}\cdot\hat{\theta}(j-m)]$$
$$= -2E[e(j)\hat{\theta}(j-m)]$$
$$= 0, \quad m = 1,2,\cdots,M. \quad (6)$$

The $M+N+1$ (5) and (6) can be written as a set of $M+N+1$ linear equations with the unknowns $a(i)$, $i = 0,1,\cdots,N$ and $b(k)$, $k = 1,2,\cdots,M$. This is done by reversing the order of the summation and the averaging in (5) and (6) and by assuming that $\hat{\theta}(j) = \theta(j)$ and that $E[\theta(j)\cdot\theta(j+k)]$ is zero if $k \neq 0$. The resulting equations are

$$\sum_{i=0}^{N} a(i)\{\Psi(i,k) + \phi_n(k-i)\} = x(k), \quad k = 0,1,\cdots,N \quad (7)$$

and

$$b(m) = \sum_{i=0}^{N} a(i)x(m+i), \quad m = 1,2,\cdots,M \quad (8)$$

where $\phi_n(k-i)$ is the autocorrelation function of the noise $n(t)$ at delay $\tau = (k-i)T$, and $\Psi(i,k)$ is defined by the equation

$$\Psi(i,k) \triangleq \sum_{l=0}^{i} x(l)x(l+k-i). \quad (9)$$

In the particular case where a matched filter is used ahead of the tapped delay lines, the $x(t)$ is the autocorrelation function of the impulse response of the modulator and channel in cascade, and (7)–(9) become equivalent to those given by Austin [8].

In general, it is quite difficult to calculate the digit error probability at the equalizer output. The calculation is particularly difficult because the assumption that all past digits are strictly correct will, of course, be violated, and the errors may tend to occur in bursts. Nonetheless, some idea of the improved performance of the decision feedback equalizer over the transversal equalizer can be obtained by assuming an ideal equalizer with an infinite number of taps and a matched filter in an environment of white additive noise with spectral density N_0. In this case the equations for the optimum tap values of a transversal equalizer are

$$\sum_{j=-\infty}^{\infty} c(j)\{\phi_q(j-k) + N_0\delta(j-k)\} = \delta(k), \quad \text{for all } k \quad (10)$$

where $\delta(\cdot)$ is the Kronecker delta function and

$$\phi_q(j-k) \triangleq \int_0^{\infty} q(\tau)q(\tau-(j-k)T)\,d\tau \quad (11)$$

where $q(\tau)$ is the impulse response of the modulator and the channel in cascade. The mean-square error of the output of this equalizer is

$$E[e^2(j)] = N_0 c(0). \quad (12)$$

In the same situation the forward taps of the decision feedback equalizer are given by

$$\sum_{j=0}^{\infty} a(j)\{\phi_q(j-m) + N_0\delta(j-m)\} = \delta(m), \quad \text{for } m \geq 0. \quad (13)$$

As before, the feedback tap values are given by (8). If the past decisions $\{\hat{\theta}(j-m)\}$ are all correct, the decision feedback equalizer output mean-square error is

$$E[e^2(j)] = N_0 a(0). \quad (14)$$

The performances of the two equalizers can now be compared by comparing (12) with (14) through the medium of an "equivalent received pulse" which has samples $p(i)$ at the sampling instants $t = iT$. Use of sampled data notation and the z transform is convenient here. The transformed pulse shape $P(z)$ is defined to be

$$P(z) \triangleq \sum_{i=0}^{\infty} p(i)z^{-i}. \quad (15)$$

With this notation, the equivalent received pulse for the problem at hand is given by

$$P(z)P(z^{-1}) = N_0 + \Phi_q(z) \quad (16)$$

where

$$\Phi_q(z) = \sum_{i=-\infty}^{\infty} \phi_q(i)z^{-i}. \quad (17)$$

The quantity of significance here is the inverse of the "equivalent received pulse", which is defined by

$$R(z) = \frac{1}{P(z)} \tag{18}$$

and the $r(i)$ can be determined by simple long division.

Substituting the results of this series of definitions back into (12) and (14), gives the mean-square error

$$E[e^2(j)] = N_0 \sum_{i=0}^{\infty} r^2(i) \tag{19}$$

for the transversal equalizer and

$$E[e^2(j)] = N_0 r^2(0) \tag{20}$$

for the decision feedback equalizer. It should be noted that these results are for an idealized situation, and as such form a lower bound on the actual mean-square error; however, they do allow ready comparisons. For example, for the simple case where $p(i) = P \exp(-i\gamma)$, $i \geq 0$, the advantage of the decision feedback equalizer is limited to 3 dB, since

$$\sum_{i=0}^{\infty} r^2(i) = P\{1 + \exp(-2\gamma)\} \tag{21}$$

and

$$r^2(0) = P. \tag{22}$$

In contrast, if the background noise spectral density N_0 is small and the actual pulse $q(\tau)$ is a rectangular pulse of length L, where the interpulse period T is αL, $1/2 \leq \alpha \leq 1$, then the ratio of the mean-square errors is

$$\frac{\sum_{i=0}^{\infty} r^2(i)}{r^2(0)} = \frac{1}{1 - \beta^2} \tag{23}$$

where

$$\frac{\beta^2}{1 + \beta^2} \triangleq 1 - \alpha. \tag{24}$$

The ratio becomes very large as $\alpha \rightarrow 1/2$. Thus the advantage achieved by using the decision feedback equalizer depends on the channel impulse response, and can, in some cases, be quite large.

The performance of the equalizers were compared also by Monte Carlo simulation of the two equalizers on a digital computer. In addition, the statistically optimum or Bayesian demodulator [6] was simulated to determine how close to the optimum performance were the performances of the statistically suboptimum, but much simpler, equalizers. In one series of simulation tests the isolated received pulse was $A\tau e^{-\alpha\tau}$ and the additive noise was Gaussian and white with power spectral density N_0. The message was a sequence of independent binary digits. The measured error probabilities at the output of the decision feedback equalizer, the linear equalizer, the Bayesian

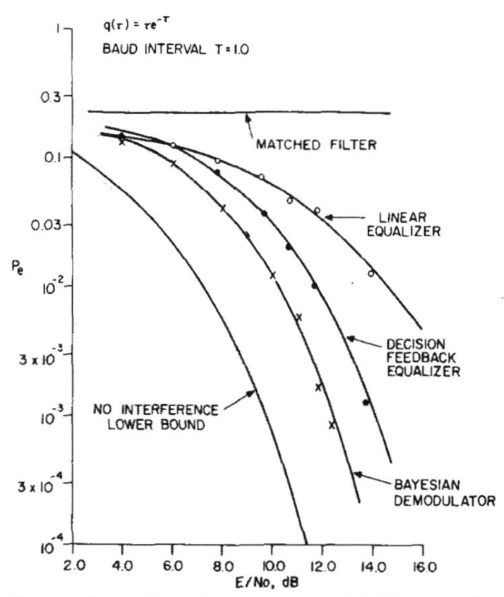

Fig. 2. Comparison of receiver error probabilities as functions of signal to background noise ratio.

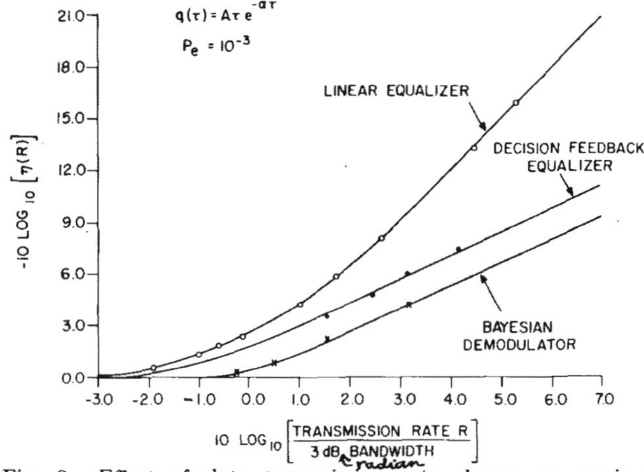

Fig. 3. Effect of data transmission rate change on receiver performances.

demodulator, and a matched filter with clipper are shown in Fig. 2, as a function of the signal to background noise level E/N_0. (E is the isolated pulse energy, equal to $\phi_q(0)$ of (11).) Also shown is the error function curve, the performance that one could achieve if there was no intersymbol interference. In this series of tests the channel parameter α was unity. A filter matched to $A\tau e^{-\tau}$ was used as part of both the linear and the decision feedback equalizer. Approximately one hundred errors were observed in making each error probability measurement. A sequence of tests was done with different numbers of forward and feedback equalizer taps. It was found that the linear equalizer performance improved as the number of taps was increased to 5, but a further increase did not appreciably improve the performance. In the same way, 3 forward taps and 6 feedback taps were found for the decision feedback equalizer. (The question of the number

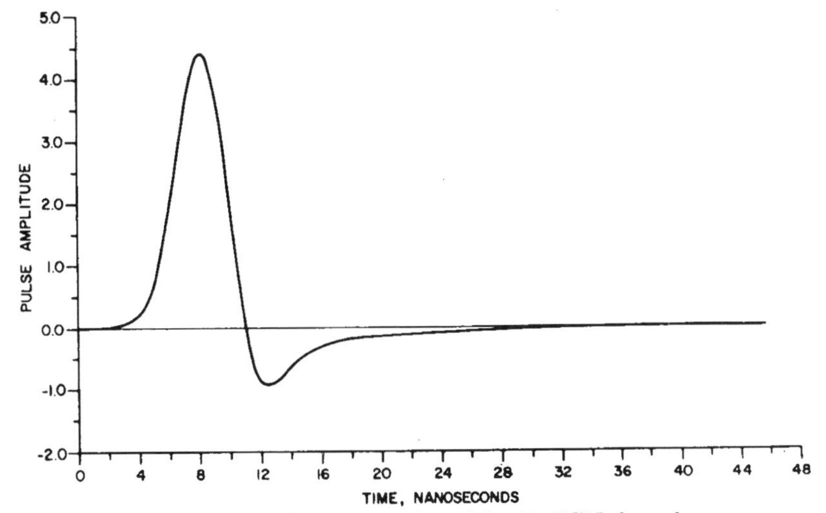

Fig. 4. Impulse response of coaxial cable PCM channel.

of taps is discussed in more detail in Section IV.) As expected, it was observed that the errors at the decision feedback equalizer output occurred in short bursts. At low signal to background noise ratios, ≤6 dB in this case, the bursts occur so frequently that the linear equalizer performance is better. However, at higher signal-to-noise ratios (SNRs) the improved ability of the decision feedback equalizer to reduce the intersymbol interference is more important than the tendency to produce errors in bursts.

It was found that at low error probabilities, wherever the digit error probability was less than 10^{-2}, the performance of the decision feedback equalizer, the linear equalizer, and the Bayesian demodulator could all be accurately described by the empirical formula

$$P[e] = 0.5 \left\{1 - \operatorname{erf}\left[\frac{\eta(R)E}{2N_0}\right]\right\} \tag{25}$$

where R is the data transmission rate, defined for this example to be equal to $(\alpha T)^{-1}$. $\eta(R)$ may be considered to be the efficiency of the modem, and must be in the range $0 \leq \eta(R) \leq 1.0$. In all cases $\eta \rightarrow 0$ as $R \rightarrow \infty$ and $\eta \rightarrow 1.0$ as $R \rightarrow 0$. $\eta(R)$ for the three demodulators, measured at 10^{-3} digit error probability, is shown in Fig. 3 as a function of R. At all transmission rates the efficiency of the decision feedback equalizer was greater than that of the linear equalizer and less than that of the Bayesian demodulator. At high transmission rates the efficiency of both nonlinear receivers decreased by 4.0 dB when the rate was doubled. The efficiency of the Bayesian demodulator was 2.0 dB better than that of the decision feedback equalizer at all high rates. In contrast, the efficiency of the linear equalizer decreased by 9.0 dB when the rate was doubled. This difference in rate of efficiency decrease becomes very important, of course, if the channel is used at very high rates and at high SNRs.

The usefulness of the decision feedback equalizer in a coaxial cable pulse-code modulation (PCM) link was also

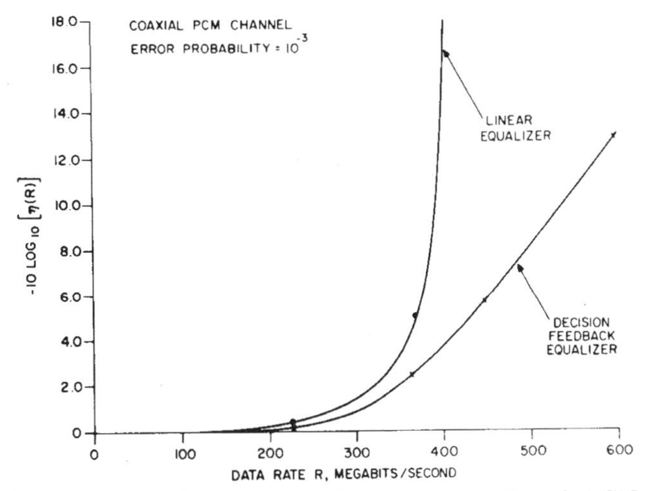

Fig. 5. Effect of data transmission rate change through PCM channel.

evaluated with a computer Monte Carlo simulation program. In this series of tests the "channel" included the transmitter, a solid coaxial cable with an air dielectric, and a fixed lumped-parameter equalizer. This channel has no dc response, a 100-MHz bandwidth at the −3.0-dB points, a 240-MHz bandwidth at the −20.0-dB points, and a 70.0-dB per octave roll-off at higher frequencies. (This contrasts with the previous example, where the roll-off was 12.0 dB per octave.) The channel impulse response is shown in Fig. 4. The nominal data rate through this channel without further equalization is 225 Mbit/s. Simulation tests showed that inclusion of a linear transversal equalizer with a matched filter would allow one to increase the data rate to 400 Mbit/s, but not beyond. In contrast, the decision feedback equalizer with a matched filter can be used at 450 Mbit/s with only 6.0 dB more signal strength than that required at low data rates, and even higher rates if the signal strength is increased further. The efficiency $\eta(R)$ of the linear and the decision feedback equalizer for this channel example, measured at 10^{-3} digit error probability, is shown in Fig. 5.

Thus both the performance analysis and the simulation results indicate that the decision feedback equalizer performance is considerably better than that of the linear equalizer. Moreover, in the examples in which the much more complex statistically optimum demodulator was also evaluated, the decision feedback equalizer performance was quite close to this limiting performance. However, these results cannot be extended to other channels without either a simulation study in each case or the development of an appropriate analysis technique.

III. Adaptive Equalizer

In this section it is shown that the decision feedback equalizer can be made adaptive to unknown or slowly varying channels, i.e., channels in which the impulse response does not change appreciably during the transmission of several hundred digits. The dynamic performance of the decision feedback equalizer, that is, the performance of the equalizer while it is adapting, is described.

A method by which the decision feedback equalizer can be made adaptive can be seen from (2), (6), and (7). For any set of tap values

$$\frac{\partial E[e^2(j)]}{\partial a(n)} = 2E[e(j)y(j+n)] \tag{26}$$

and

$$\frac{\partial E[e^2(j)]}{\partial b(m)} = -2E[e(j)\hat{\theta}(j-m)]. \tag{27}$$

If $e(j)$ of (2) is replaced by

$$\hat{e}(j) = \tilde{\theta}(j) - \hat{\theta}(j) \tag{28}$$

by assuming that the decisions are correct, then all the signals in (26) and (27) are available at the receiver. When the error probability is low this substitution does not change the value of (26) or (27) appreciably. By changing the forward tap values by amounts approximately proportional to $-E[\hat{e}(j)y(j+k)]$, and the feedback tap values by amounts approximately proportional to $E[\hat{e}(j)\hat{\theta}(j-m)]$, the taps are automatically adjusted to near their optimum values. Thus the forward taps are adjusted by means of measurement of the cross correlation between the error and the input signals, just as for the linear transversal equalizer. On the other hand, adjustment of the feedback taps makes use of the cross correlation between the error and the output signal, i.e., the decisions.

The potential of this type of algorithm can be seen by observing how the mean-square error $E[e^2(j)]$ depends on tap value errors. Let $\{a(k);\ k = 0,1,\cdots,N\}$ and $\{b(l);\ l = 1,2,\cdots,M\}$ be the actual tap values, and $\{a_0(k); k = 0,1,\cdots,N\}$ and $\{b_0(l); l = 1,2,\cdots,M\}$ be the optimum tap values, specified by (7) and (8). Then the

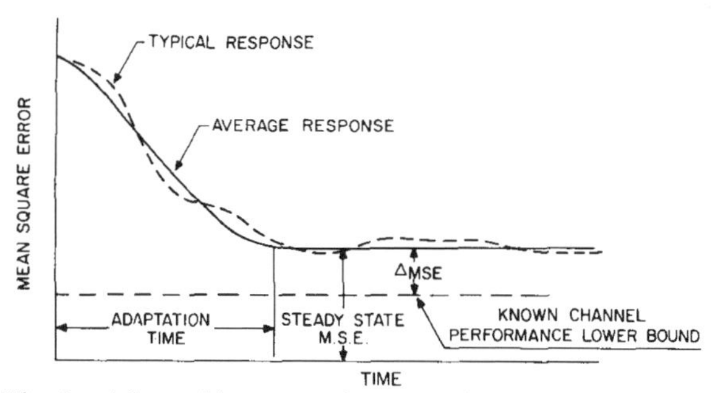

Fig. 6. Adaptation to step change in channel impulse response.

tap value error is

$$\begin{aligned} d(k) &\triangleq a(k) - a_0(k), \qquad k = 0,1,\cdots,N \\ &\triangleq b(-k) - b_0(-k), \qquad k = -1,\cdots,-M. \end{aligned} \tag{29}$$

Let us also define a set of signals $\{z(j+i)\}$ by

$$\begin{aligned} z(j+i) &= y(j+i), \qquad i = 0,1,\cdots,N \\ &= \hat{\theta}(j+i), \qquad i = -1,\cdots,-M. \end{aligned} \tag{30}$$

Then it can be shown that if the tap values are in error the mean-square error is

$$E[e^2(j)] = E[e_0^2(j)] + \sum_{i=-M}^{M}\sum_{k=-M}^{N} d(i)d(k)E[z(j+i)z(j+k)] \tag{31}$$

where $e_0(j)$ would be the error if the tap values were all correct. (The assumption was made that $\theta(j-m) = \hat{\theta}(j-m)$, $m = 1,2,\cdots,M$, to derive (31).) Thus the mean-square error is a quadratic function of the tap gain errors, in the same way that the mean-square error of the linear equalizer is a quadratic function of its tap gain errors. Because of this, there are no "locally optimum" tap gain settings, and a hill-climbing adaptation can be made to readily converge close to the correct set of tap values given by (7) and (8).

Of course, the cross correlations $E[\hat{e}(j)y(j+k)]$ and $E[\hat{e}(j)\hat{\theta}(j-m)]$ cannot be measured exactly in a finite time, so any particular sequence of tap adjustments is a sample function of a random process. A Robbins–Monro procedure [9] would be applicable if the channel were unknown but time invariant. However, if the channel is slowly varying then the adaptive algorithm must be capable of "tracking" a slowly varying channel and of "learning" the optimum tap values for an unknown channel. In that case a procedure such as the Robbins–Monro procedure is not applicable, and a compromise between a smaller steady-state mean-square error and a shorter adaptation time to channel changes is necessary.

There are a number of adaptation algorithms available, in which the exact details of the algorithm are somewhat different. A typical response of an adaptive receiver to a step change in the channel characteristics when any of these algorithms is used is shown in Fig. 6. The mean-square error at the receiver output is plotted as a function of the number of baud intervals after a step change in the channel characteristics. It is a nonstationary random variable, of course, since it is a function of the equalizer tap values, which in turn are nonstationary random variables. Also shown in Fig. 6 is the average of many such adaptation curves. The two most important characteristics of such a curve are the "steady-state" mean-square error, and the "adaptation time," the time required to reach the steady-state mean-square error after a specific change in the channel characteristics. The adaptation curves that were obtained in a series of simulation studies are discussed later in the paper, but some general observations may be made now.

First, since a cross correlation is being measured, the variance of the square of the signals involved contribute significantly to the measurement error. Particularly when binary signals are involved, there is a notable difference between the input samples $\{y(j)\}$ and the output decisions $\{\hat{\theta}(j)\}$ in this regard. In particular,

$$E[\{\hat{\theta}^2(j) - E[\hat{\theta}^2(j)]\}^2] = 0 \tag{32}$$

when

$$\theta(j) = \pm 1$$

and

$$E[\{y^2(j) - E[\{y^2(j)]\}^2] = 2\{\sum_i q^2(i))^2 + E^2[n^2(j)]\}. \tag{33}$$

(The additive noise samples $n(j)$ are assumed to be Gaussian in the calculation.) Certainly then in the case of strong intersymbol interference, $\sum q^2(i)$ is large and so $\hat{\theta}^2(j)$ is much less "noisy" than $y^2(j)$. Thus an estimate of $E[e(j)\hat{\theta}(j-m)]$ would usually involve less error than an estimate of $E[e(j)y(j+k)]$.

Whether this implies that the decision feedback equalizer can adapt more rapidly or more closely to the performance possible from a fixed optimum equalizer than can the linear transversal equalizer, and that the feedback taps can adapt more easily than the forward taps, depends on the sensitivity of the tap adjustments. The partial derivatives of (5) and (6) indicate this sensitivity. Since the absolute signal levels are, of course, arbitrary, the decisions $\hat{\theta}(j)$ are taken to be ± 1 and the average signal power $E[y^2(j)]$ is taken as unity. This effectively means that the taps $\{a(k)\}$ and $\{b(l)\}$ are of the same order of magnitude. This done, the sensitivity of the tap values can be evaluated, giving:

$$\frac{\partial^2 E[e^2(j)]}{\partial a^2(n)} = 2E[y^2(j+n)] = 2 \tag{34}$$

$$\frac{\partial^2 E[e^2(j)]}{\partial b^2(m)} = 2E[\hat{\theta}^2(j-m)] = 2. \tag{35}$$

The tap values are thereby shown to be equally sensitive to adjustment, and this point combined with that of the previous paragraph implies that the feedback taps can be adjusted more quickly and/or more accurately than the forward taps or the taps of a transversal equalizer. Thus the decision feedback equalizer would be expected to have a better adaptation performance than the transversal equalizer. This has also been observed experimentally, as will be described.

The advantages of using the cross correlation between the error and the decisions suggest the use of this same measurement for adjustment of the transversal or forward tap values. Of course, the taps would not converge to the correct value to minimize the error due to the additive noise and the intersymbol interference, but in some cases at least the taps converge to a value close to the correct value. As these taps may be subject to less error due to the cross correlation measurement, improved adaptive behavior can result.

Suppose we specify the taps $\{a(i)\}$ by

$$E[e(j)\hat{\theta}(j+l)] = 0, \qquad l = 0,1,\cdots,N \tag{36}$$

rather than by (5). Then the tap values are the solution to the equations

$$\sum_{i=0}^{N} a(i)x(i-l) = \delta(l), \qquad l = 0,1,\cdots,N \tag{37}$$

rather than to (7). Thus adjustment of the forward taps by cross correlation between the error and the decisions "forces zeros" in the overall transmission characteristic. If the equalizer had an infinite number of correctly spaced taps, specified by (37), the result would be an inverse filter. In the limiting situation of no additive noise and a similarly ideal equalizer, an equalizer adjusted by the minimum mean-square error approach would also yield an inverse filter. Consequently, it is not surprising that in some high SNR situations where effective equalization is being obtained, the two methods give similar results.

A potential difficulty with this "zero-forcing" algorithm is that only as many system impulse response zeros can be forced as there are taps in the delay line, with one additional tap reserved to force a unit response at the desired time. The overall system impulse response can become large both before and after this interval over which the response is forced to zero. In contrast, when the equalizer is adjusted by the minimum output mean-square error approach, the mean-square contribution of the total system impulse response is minimized, not just the response over an interval as large as the equalizer delay line. Note, however, that if the taps of the transversal filter of the decision feedback equalizer are adjusted by the zero-forcing algorithm, adjustment of the taps of the feedback filter will automatically cancel any large impulse response after the main pulse, without causing a large impulse response at an even greater delay. This is the basic idea behind the decision feedback equalizer, based on the assumption that the decisions in the feed-

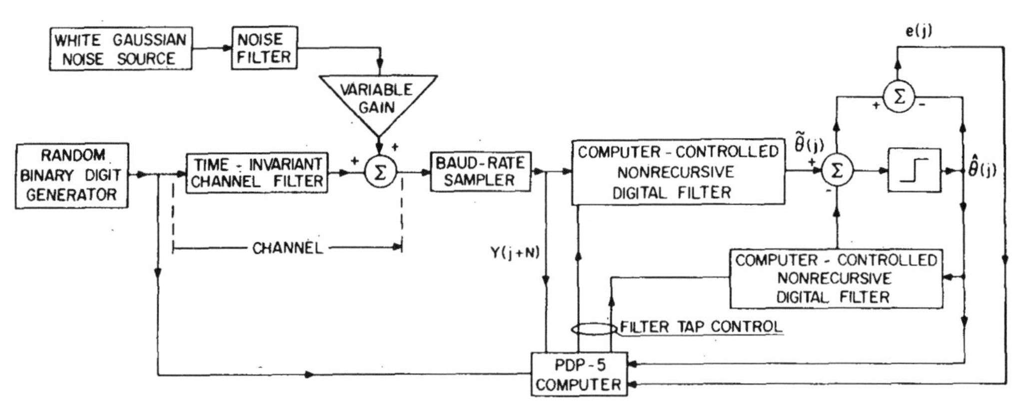

Fig. 7. Hardware simulator.

back delay line are correct. Thus only the system impulse response before the area in which zeros are forced can contribute to the output error. From this and from considerations of the errors associated with measuring $E[e(j)\hat{\theta}(j+k)]$ and $E[e(j)y(j+k)]$, it is hypothesized that the decision feedback equalizer can use the cross correlation between the error and the decisions to advantage for adjustment of both forward and feedback taps. The experimental results that are described later substantiate this hypothesis.

The actual adaptation algorithm that was used in the experimental investigation will now be discussed. As shown, the transversal or forward tap gain should be changed by an amount proportional to a measure of $-E[e(j)y(j+k)]$ or $-E[e(j)\hat{\theta}(j+k)]$, and the feedback tap by an amount proportional to a measure of $E[e(j)\hat{\theta}(j-m)]$. Actually, rather than taking a fixed finite time average of these products and then changing the tap values, the adaptation is done indirectly with an algorithm similar to that developed by Lucky [2] for adaptation of the linear equalizer. The tap values are changed in the following way.

1) An accumulator for the tap is set equal to zero.

2) Each time a digit is processed, the product $\hat{e}(j)y(j+k)$ for the forward tap $a(k)$, or $-\hat{e}(j)\hat{\theta}(j-m)$ for the feedback tap $b(m)$, is added to the accumulator. Only the signs of $\hat{e}(j)$, $y(j+k)$, and $\hat{\theta}(j-m)$ are used in this calculation to simplify the equalizer synthesis.

3) If the accumulator contents exceed a threshold $+V$, then the tap value is decreased by Δ and step 1 is repeated. If the contents become less than $-V$, the tap value is increased by Δ and step 1 is repeated. If the contents remain between $+V$ and $-V$, then step 2 is repeated.

In the alternate procedure previously discussed, the $\hat{e}(j)y(j+k)$ of step 2 are replaced by $\hat{e}(j)\hat{\theta}(j+k)$. Both the transversal and decision feedback equalizer were tested using each of these adaptation algorithms.

The adaptive equalizers were evaluated by observing their ability to adapt to an unknown but fixed channel, rather than to a time-varying channel. This was done by measuring the mean-square error at the equalizer output as a function of adaptation time. A hardware simulator under the control of a PDP-5 digital computer was used to do this. A block diagram of the simulator is shown in Fig. 7. The transmitted message was the pseudorandom output of the m-sequence generator corresponding to the polynomial $x^{31}+x^{28}+x^{27}+x^{24}+x^{17}+x^{16}+x^{9}+x^{8}+1$. In some cases a time-invariant analog filter was used to simulate the channel. The filter output was sampled at the baud rate after the addition of filtered Gaussian noise. In other tests, a 32 tap 12-bit baud rate nonrecursive digital filter was used to simulate the channel filter. In this case the additive noise was sampled at the baud rate and then added to the dispersed signal. In both cases, the composite sampled signal was processed with a 7-bit baud rate digital filter, as shown in Fig. 7. The sum of the number of taps in the two nonrecursive filters could be as great as eleven, with any division of taps between the two. These filter tap values were under direct computer control.

At the beginning of each adaptation test, all taps of the decision feedback equalizer except the last forward tap were set to zero, so that the output would be θ if there were no noise or intersymbol interference. The adaptive transversal equalizer was tested in a similar way.

The digital computer was used to change the tap values, and took the sequences $\{y(j)\}$, $\{\hat{e}(j)\}$, and $\{\hat{\theta}(j)\}$ directly as inputs. This method was used to avoid construction of adaptation circuitry for each tap. As a result, only a few of the binary digits that were processed by the equalizer were used for adaptation processes. The digits that were used are called "independent digits", because the time between successive observations is long compared to the times over which the autocorrelation functions of $e(j)$ and $y(j)$ are significant. A specified number of these independent digits, usually 100, were processed according to the preceding algorithm to change the taps. Then 2000 digits were used to estimate $E[e^2(j)]$, without changing either the tap values or the accumulator contents. Then 100 more samples were used for adaptation purposes, followed by another measurement of $E[e^2(j)]$. This sequence continues until it is evident that the equalizer has reached a "steady-state" mode of operation where the trend in mean-square error is no longer

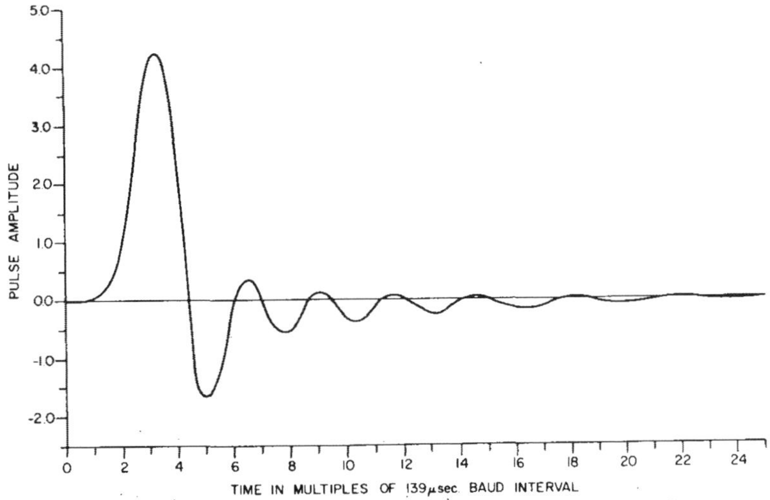

Fig. 8. Impulse response of simulated telephone cable channel.

changing with time. The results of 50 such adaptation runs are then averaged to give a mean adaptation curve.

Both the signal samples $\{y(j)\}$ and the equalizer tap values were quantized with a maximum accuracy of 7 bits. (More will be said about quantization accuracy requirements in Section IV.) The least significant bit of the tap gain values was changed during adaptation each time the threshold $+V$ or $-V$ was exceeded. Thus the adaptation parameter Δ in these tests is $2^{-6} \simeq 0.016$.

Tests were carried out to determine whether the decision feedback equalizer can adapt better than the linear equalizer to an unknown channel, whether the results are valid for a variety of channels, whether use of a learning sequence is necessary or even advantageous, and whether or not use of an estimate of $E[e(j)\hat{\theta}(j+k)]$ results in better adaptation than an estimate of $E[e(j)y(j+k)]$ for the forward taps $\{a(k)\}$. The channels that were simulated in these tests included a channel with an exponential impulse response, a coaxial cable PCM channel, and an audio-loaded telephone line.

One series of tests was made to compare the performances of the adaptive linear and decision feedback equalizers in an audio-loaded telephone cable system, and to determine the advantage that could be gained by using a training sequence. The channel filter was a lumped-parameter filter designed by Bell Canada to simulate a 15 000 ft audio-loaded telephone cable and was terminated in 600 ohms. The impulse response of the filter is shown in Fig. 8. Binary information was transmitted through this channel at 7200 bit/s. Neither any intentional additive Gaussian noise nor a filter matched to the isolated received pulse were used in this series of tests. Because of the resulting mismatch, choice of the third of 11 taps as the "main" tap minimized the linear equalizer output mean-square error. Similarly, it was found that the decision feedback equalizer with 4 forward taps and 7 feedback taps made the best use of the 11 available taps. The adaptation thresholds were set at ± 20 during this series of tests. It

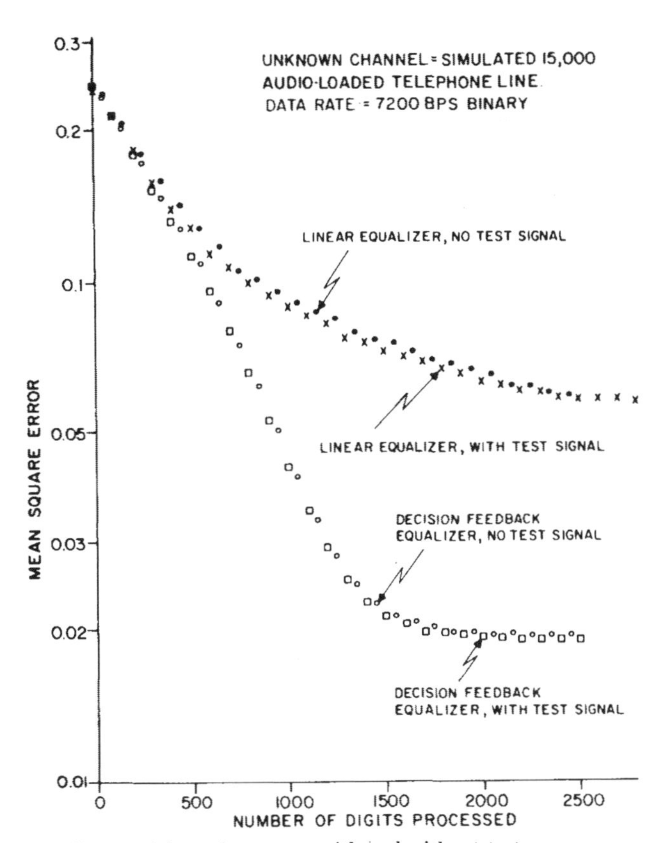

Fig. 9. Adaptation curves with and without test sequence.

was found that adaptation based on the measurement of $E[e(j)\hat{\theta}(j+k)]$ resulted in a better performance than measurement of $E[e(j)y(j+k)]$. The mean-square errors at the equalizer outputs, using the former measurement, are shown in Fig. 9 as a function of the number of independent digits that were processed. As shown, the steady-state mean-square error of the decision feedback equalizer is 5 dB better than that of the linear equalizer. This is consistent with the fixed equalizer tests that were

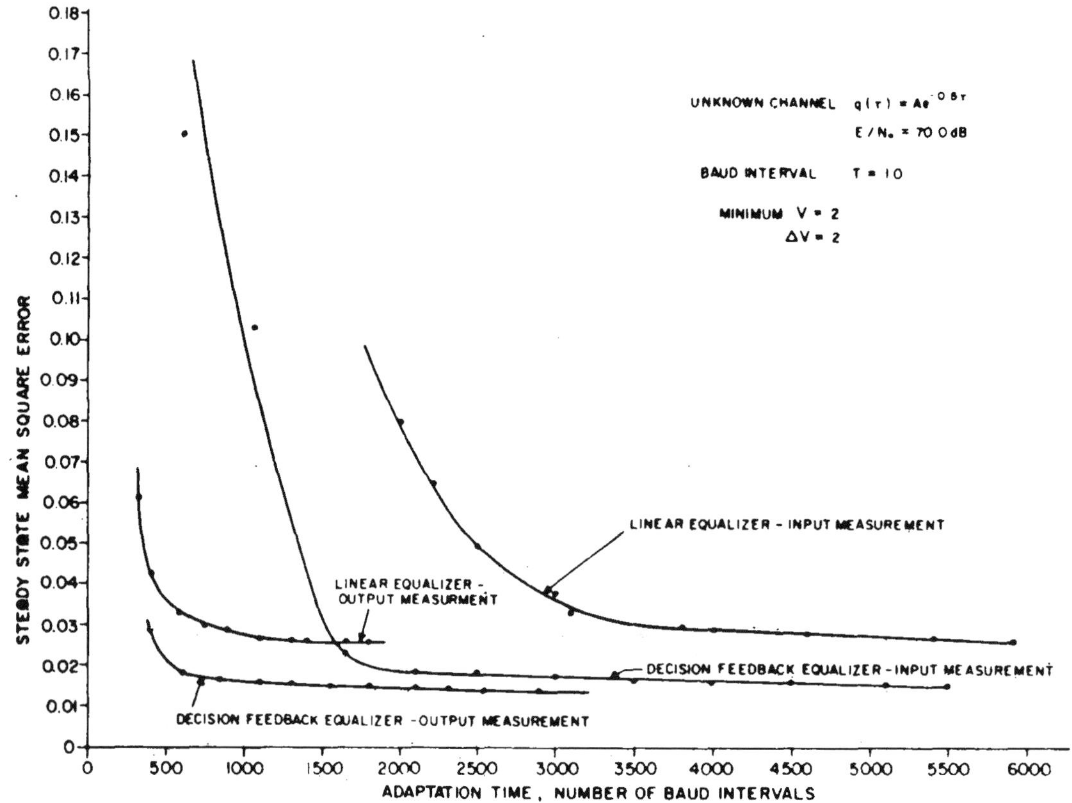

Fig. 10. Adaptation performances at various threshold values, high signal-to-background noise ratio example.

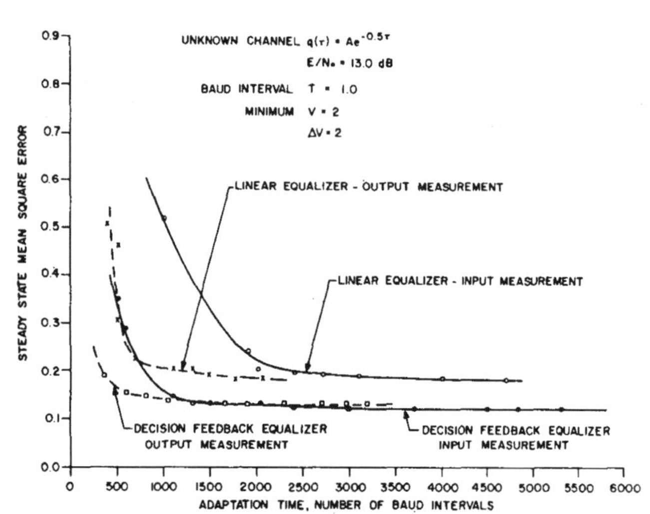

Fig. 11. Adaptation performances at various threshold values, low signal to background noise ratio example.

described in the previous section. (The results shown in Fig. 9, and the results to follow, are normalized in amplitude such that $|\theta(j)| = 1$.) The decision feedback equalizer converged to its steady-state performance after 1500 digits were processed, or in about 0.2 s if every digit was used for adaptation.

It is notable that knowledge of the correct digit sequence did not improve the adaptation of either equalizer significantly; use of the decisions gave almost the same performance. However, the digit sequence was the output of a pseudorandom generator. It is likely that the transmission of certain redundant sequences could cause a significant increase in adaptation time, especially if the "eye" at the equalizer output were initially closed.

Two important characteristics that can be taken from a graph such as Fig. 9 are the "steady-state" mean-square error and the "adaptation time". Unfortunately, it is not possible to improve both these characteristics simultaneously by changing the adaption parameters V and Δ. By increasing V, or decreasing Δ, or doing both, the mean-square error can be reduced, but the adaptation time is increased. A series of tests was made to determine how the mean-square error and the adaptation time are dependent on the threshold V. In these tests the channel impulse response was $Ae^{-0.5\tau/T}$. A filter matched to this pulse was used as part of the equalizer in this series of tests. Measurements were made at high SNR, approximately 70 dB, and at 13.0 dB. The effect of using both "input" measurements $\hat{e}(j)y(j+k)$ and "output" measurements $\hat{e}(j)\hat{\theta}(j+k)$ were tested with both the decision feedback equalizer and the linear equalizer. The effect of changing V from 2 to 24 in increments of 2 is shown in Figs. 10 and 11. As is shown, the performance of the decision feedback equalizer is better than that of the linear equalizer when either tap-error measurement is used. Also, the performance of both equalizers when adaptation is based on $\hat{e}(j)\hat{\theta}(j+k)$ is better than when it is based on $\hat{e}(j)y(j+k)$, except for the decision feedback case when $E/N_0 = 13$ dB and a large threshold was used. In all cases a much quicker adaptation could be achieved by using $\hat{e}(j)\hat{\theta}(j+k)$. In all cases a minimum product of adaptation time and mean-square error was achieved when V was 4.

Thus both the experimental results and the analysis indicate that the decision feedback equalizer can be made adaptive, and that its adaptive performance is better than that of the linear equalizer. Estimates of either $E[\hat{e}(j)y(j+k)]$ or $E[\hat{e}(j)\hat{\theta}(j+k)]$ can be used to modify the forward taps. The experimental results described previously show that the latter measurement is better in many cases. However, the work of Hirsch and Wolf [10] indicates that when other channels are used measurement of $E[\hat{e}(j)\hat{\theta}(j+k)]$ cannot be used to make the linear equalizer adaptive. Further investigation is required to determine which is the better measurement to adapt the decision feedback equalizer to such channels.

IV. Practical Considerations

It has been shown previously that the performance of the decision feedback equalizer is considerably better than that of the linear equalizer, both when the channel impulse response is known and when it is not. This result is particularly important when one realizes that the decision feedback equalizer is no more complex than the linear equalizer. For instance, in the telephone cable example the decision feedback equalizer with 4 forward taps and 7 feedback taps outperformed the linear equalizer with 11 taps. In both cases the input signal and the tap values were quantized to 7 bits. In fact, the decision feedback equalizer is potentially simpler, since the feedback delay line can be a single bit shift register if the data is a binary one.

It was determined from computer simulation studies that the decision feedback equalizer is no more sensitive to signal quantization errors or tap quantization errors than the linear equalizer, even though it achieves its superior performance by coherent subtraction of the intersymbol interference. This is consistent with the tap sensitivity analysis of the previous section, (34) and (35). In the computer simulation study the matched filter output $y(j)$ and the equalizer tap values could be independently quantized to any specified accuracy. The results of a typical test are shown in Fig. 12. In this test the channel impulse response was $A\tau e^{-0.8\tau/T}$, the signal-to-background noise ratio was 16.0 dB, the linear equalizer had 5 taps, and the decision feedback equalizer had 3 forward taps with 8 feedback taps. Two curves of Fig. 12, one for each equalizer, show the error probability as a function of the number of quantization bits when all quantities are quantized with the same accuracy. As shown, the error probability starts to increase when the number of quantization bits is reduced to 8. In contrast, when the quantization of the signal was held at 10 bits, the tap gain quantization of both equalizers could be reduced to 6 bits before any significant increase in error probability was observed.

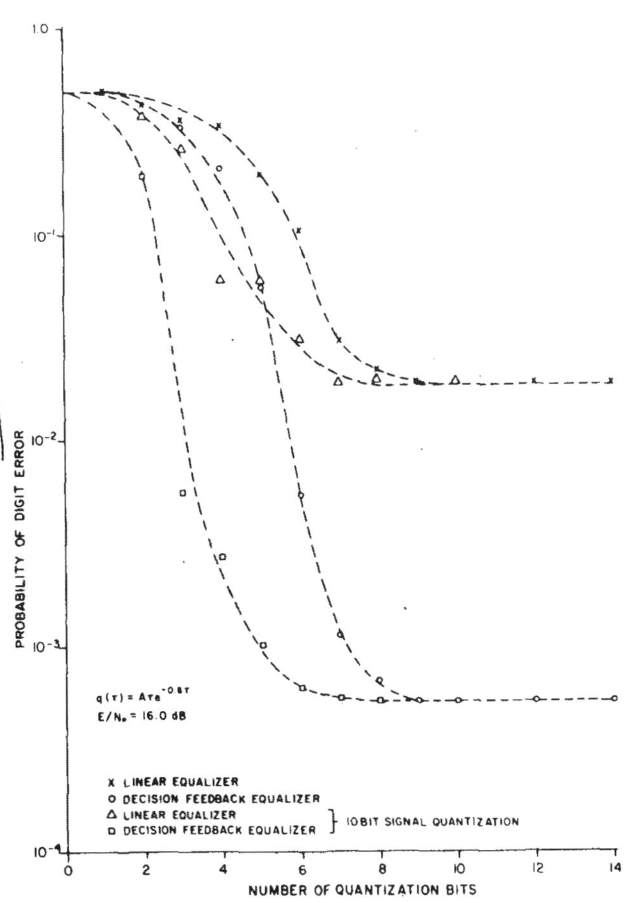

Fig. 12. Effect of signal and tap quantization on receiver performance.

In a separate experiment, quantization of only the signal values $\{y(j)\}$ resulted in a performance very similar to that achieved when all quantities were quantized equally. It is believed that the reason for the 2-bit difference in quantization requirements is that the signal includes the additive noise and the intersymbol interference, and so the quantization error is a larger percentage of the desired signal than of the total signal $y(t)$.

Similar results were observed when the coaxial cable PCM channel was simulated. Over that channel at 360 Mbit/s it was necessary to use 7-bit accuracy for the signal and 6-bit accuracy for the taps. At 450 Mbit/s the linear equalizer could not effectively equalize the channel, even with as many as 21 taps and with no quantization error in either the signal or the tap values. The decision feedback equalizer with 5 forward taps and 6 feedback taps required 9-bit signal quantization accuracy and 7-bit tap accuracy.

These results are directly applicable if a digital synthesis method is used. They show that the decision feedback equalizer is no more sensitive to quantization inaccuracies than the linear equalizer, and so is no more expensive to construct. If an analog synthesis method is used, these results indicate that the decision feedback equalizer is no more sensitive to component inaccuracies or signal distortion than the linear equalizer.

TABLE I
STEADY-STATE OUTPUT SNRS AS FUNCTION OF NUMBER OF TAPS

Number of Feedback Taps	Number of Forward Taps			
	2	3	4	5
1	9.6	10.7	11.1	11.0
2	11.4	13.3	14.9	14.2
3	11.9	14.5	15.7	15.4
4	11.9	14.3	15.9	16.0
5	11.8	14.7	16.0	16.1
6	11.8	14.9	16.1	16.2
7	11.7	14.9		
8	11.7	14.8		

The required number of equalizer taps was also examined. Computer simulation tests showed that, when the channel had a simple impulse response such as $Ae^{-\alpha\tau}$ or $A^{\tau}e^{-\alpha\tau}$, both equalizers equalized the channel as well with a few transversal taps as with many, but that the decision feedback equalizer required several feedback taps to realize its full potential, and as many as 10 or 12 at very high rates. Note however, from Figs. 2 and 3, that the more complex decision feedback equalizer could attain a performance not possible with a linear equalizer with any number of taps. In the more complex channel examples the linear equalizer did not retain this advantage in simplicity. Computer simulation tests of the coaxial cable PCM channel at 360 Mbit/s, with a 20.0-dB signal to background noise ratio, showed that the linear equalizer required 9 taps to realize its full potential, and the decision feedback equalizer required 4 forward taps and 5 feedback taps.

A series of hardware simulation tests was carried out to determine the number of taps required by the adaptive decision feedback equalizer. In these tests 450-Mbit/s transmission over the PCM channel at high signal to background noise ratio was simulated. The threshold V was held at 16 during these tests. The steady-state output SNR in dB is shown as a function of the number of taps in Table I. As shown, 4 forward taps and 4 to 6 feedback taps is a good compromise between better performance and higher cost. Use of more than 6 feedback taps degrades the performance, because the amount of quantization noise and adaptation noise that the tap introduces is more than the amount of intersymbol interference that is removed.

From these experimental results it is concluded that the decision feedback equalizer is a practical as well as a very high performance receiver.

V. CONCLUSIONS

It has been shown that the decision feedback equalizer can be used to detect digital information that has been sent at high rates over an unknown or slowly varying dispersive channel. The equalizer's ability to combat intersymbol interference caused by several channels was investigated experimentally. In each of the examples that were investigated the digit error probability of the decision feedback equalizer was considerably less than that of the linear equalizer. As well, at any specified low error probability the decision feedback equalizer allowed data transmission at rates beyond that possible with the linear equalizer. Two practical algorithms are described that make the decision feedback equalizer adaptive to unknown or slowly varying channels. It was found experimentally that the algorithm based on the cross correlations between the estimated error and the estimated digits gave the better performance, and that a training sequence was not necessary for adaptation. Also, it was found that the decision feedback equalizer is no more sensitive to quantization errors than the linear equalizer. Because of these advantages, and because its performance is close to that of the much more complex Bayesian demodulator when the channel is known, the decision feedback equalizer should be considered whenever linear modulation techniques are used to transmit digital information over dispersive linear channels at a high rate.

This is additionally verified in the theoretical portions of the paper where it is shown that the idealized decision feedback equalizer will always yield smaller mean-square error than the transversal equalizer. As well, theoretical considerations indicate that the adaptive properties of the decision feedback equalizer will tend to be superior.

REFERENCES

[1] D. C. Coll and D. A. George, "A receiver for time-dispersed pulses," in *Conf. Rec., 1965 IEEE Int. Conf. Communications*, pp. 753–758.
[2] R. W. Lucky, "Techniques for adaptive equalization of digital communication systems," *Bell Syst. Tech. J.*, vol. 45, Feb. 1966, pp. 255–268.
[3] J. G. Proakis and J. H. Miller, "An adaptive receiver for digital signaling through channels with intersymbol interference," *IEEE Trans. Inform. Theory*, vol. IT-15, July 1969, pp. 484–497.
[4] R. A. Gonsalves, "Maximum-likelihood receiver for digital transmission," *IEEE Trans. Commun. Technol.*, vol. COM-16, June 1968, pp. 392–398.
[5] R. R. Bowen, "Bayesian decision procedure for interfering digital signals," *IEEE Trans. Inform. Theory* (Corresp.), vol. IT-15, July 1969, pp. 506–507.
[6] ——, "Bayesian detection of noisy time-dispersed pulse sequences," Ph.D. dissertation, Carleton Univ., Ottawa, Ont., Canada, Sept. 1969.
[7] K. Abend and B. D. Fritchman, "Statistical detection for communication channels with intersymbol interference," *Proc. IEEE*, vol. 58, May 1970, pp. 779–785.
[8] M. E. Austin, "Decision feedback equalization for digital communication over dispersive channels," M.I.T./R.L.E. Tech. Rep. 461, Aug. 11, 1967.
[9] H. Robbins and S. Monro, "A stochastic approximation method," *Ann. Math. Statist.*, vol. 22, 1951, pp. 400–407.
[10] D. Hirsch and W. J. Wolf, "A simple adaptive equalizer for efficient data transmission," *IEEE Trans. Commun. Technol.*, vol. COM-18, Feb. 1970, pp. 5–12.

PHOTO NOT AVAILABLE

Donald A. George (S'54–M'59) was born in Galt, Ont., Canada, on April 24, 1932. He received the B.Eng. degree in engineering physics from McGill University, Montreal, P. Q., Canada, in 1955, the M.S. degree in electrical engineering from Stanford University, Stanford, Calif., in 1956, and the Sc.D. degree in electrical engineering from Massachusetts Institute of Technology, Cambridge, in 1959.

From 1959 to 1962 he was an Assistant Professor of Electrical Engineering, University of New Brunswick,

Fredericton, Canada. Since then he has been a member of the Faculty of Engineering, Carleton University, Ottawa, Ont., Canada. While teaching, he has been a Consultant to a number of organizations principally the Defence Research Telecommunications Establishment (now the Communications Research Center) and Canadian Westinghouse Company, Ltd. Also, he spent three summers and a nine-month sabbatical period with the Communications Research Center. At Carleton University, while being concerned with the development of the engineering program in general, he has been particularly involved in building up graduate activity in communications. His recent research activity has been in the area of optimum and adaptive PAM systems and in signal processing with small computers. At present, he is a Professor of Engineering and Dean of Engineering at Carleton University. His research interests are in communication and information theory, cybernetics and systems, and signal processing.

PHOTO NOT AVAILABLE

Robert R. Bowen (S'57–M'61) was born in Peterborough, Ont., Canada, on June 10, 1935. He received the B.Sc. degree in engineering physics and the M.Sc. degree in electrical engineering from Queen's University, Kingston, Ont., in 1958 and 1960, and the Ph.D. degree in electrical engineering from Carleton University, Ottawa, Ont., in 1970.

In 1960 he joined the Defence Research Telecommunications Establishment, which has since become the Communications Research Center of the Department of Communications, Ottawa, Ont., and worked on radar signal processing problems. In 1967 he returned to the Communications Research Center, where he continued his work on the transmission of data through dispersive media. More recently, he and two colleagues have developed a computer language for simulation of adaptive communication systems.

PHOTO NOT AVAILABLE

John R. Storey (M'68) was born in Trelewis, Wales, on September 19, 1926. He graduated from Lewis School, Pengam, Wales.

After graduation he joined the Royal Navy for a period of three years. In 1952 he joined Decca Radar, Ltd., Surrey, England, working primarily on the development of a baseband communication system. He emigrated to Canada in 1955 and joined the Ferranti Packard Company where his main interest was in HF communications using meteor trial reflections. In 1957 he joined the Avro Aircraft Company, Malton, and was involved in data processing during the flight trials of the Avro Arrow aircraft. He is now with the Communications Research Center, Ottawa, Ont., Canada, (formerly the Defence Research Telecommunications Establishment) having joined them in 1959. He has worked on projects involved with pulse compression techniques for ionospheric sounding and on research in the field of adaptive receivers for data transmission. He now heads a communication engineering group currently working on an automated system to measure noise in the HF communication spectrum.

Performance of Optimum and Suboptimum Synchronizers

PAUL A. WINTZ, MEMBER, IEEE, AND EDGAR J. LUECKE, MEMBER, IEEE

REFERENCE: Wintz, P. A., and Luecke, E. J.: PERFORMANCE OF OPTIMUM AND SUBOPTIMUM SYNCHRONIZERS,[1] School of Electrical Engineering, Purdue University, Lafayette, Ind. 47907 and School of Electrical Engineering, Valparaiso University, Valparaiso, Ind. 46383. Rec'd 12/16/68. Paper 69TP22-COM, approved by the IEEE Communication Theory Committee for publication without oral presentation. IEEE TRANS. ON COMMUNICATION TECHNOLOGY, 17-3, June 1969, pp. 380–389.

ABSTRACT: The optimum (maximum-likelihood) synchronizer for extracting bit synchronization directly from a binary data stream is presented along with some simple suboptimum synchronizers that perform almost as well. The manner in which the performances of these systems depend on the pertinent system parameters as determined by a combined program of analysis, simulation, and laboratory experimentation are reported. Both synchronizer jitter and the degradation in error rate due to jitter are considered.

I. Introduction

DIGITAL transmission systems require receivers that are synchronized to the incoming signal format. Three levels of synchronization are usually required. Bit synchronization is required since optimum (matched-filter) detectors require knowledge of the start and stop times of each incoming symbol in order to make bit decisions. Word synchronization is also required in order to sort out bits into their appropriate words. Finally, frame synchronization is required by the data user if he is to distinguish between word groups, or frames. Further comments on these synchronization problems are given in [1].

This paper is concerned with the problem of bit synchronization. There are, of course, a number of strategies for obtaining bit synchronization by transmitting extra power and/or bandwidth. For example, the standard teletype format uses start and stop signals to synchronize the encoder and decoder. Approximately 33 percent of the transmission time is alloted to symbol synchronization. In other systems, synchronization words are inserted with the data words at regular intervals, and the receiver first obtains word synchronization by locking to these synchronization sequences. If word synchronization is obtained accurately enough, it can be divided down to obtain bit synchronization. Sometimes a separate subchannel is used exclusively for transmitting synchronization information.

On the other hand, self-bit synchronization systems extract bit synchronization directly from the data stream with no additional power, bandwidth, etc., added for synchronization purposes. We restrict our comments to this class of synchronization strategies. The ease of extracting timing information directly from the data stream depends on the amount of energy in the incoming waveform at the frequency corresponding to the bit rate. Since a random sequence of equally likely antipodal signals contains no energy at the bit rate frequency, some type of nonlinear processing is required. The analysis of nonlinear systems with noisy inputs is not a trivial problem. However, some progress has been made. Van Horn [2] suggested a synchronizer consisting of a bank of correlators similar to some radar systems, but made no claims about its performance or practicality. Stiffler [3] proposed an optimum (maximum-likelihood) strategy that operates on the infinite past or enough of the past to make the truncation error negligible. Van Trees [4] analyzed a system consisting of a data channel and a synchronization channel and concluded that the optimum strategy for dividing the transmitter power between these two channels devotes all available transmitter power to the data channel and obtains synchronization from it. (An analogous result relative to word synchronization was established by Chase [5] who showed that the channel capacity for the discrete memoryless unsynchronized channel is the same as the capacity for the synchronized channel.) Wintz and Hancock [6] evaluated the performance of an adaptive estimator-correlator detection system for signals of unknown arrival times. The error rate was given in terms of the signal-to-noise ratio (SNR) and the rms timing error, but no synchronizer structure was suggested.

In this paper we report the results of an investigation aimed at determining the important parameters associated with self-bit synchronizing systems and the physical relationships between these parameters. Toward this end we present the structure of the optimum (maximum-likelihood) self-bit synchronizer and also some suboptimum structures that perform nearly as well as the optimum but are much easier to implement. The manner in which the performances of these optimum and suboptimum systems depend on the pertinent system parameters are determined and the performances compared. The important parameters were found to be the input SNR, the shape of the signaling waveform, the memory time of the synchronizer, the synchronizer nonlinearity,

[1] This research was supported by NASA under Grant NsG-553.

Reprinted from *IEEE Transactions on Communication Technology*, vol. COM-17, no. 3, June 1969.

and the bandwidth of the low-pass filter preceding the synchronizer. The performance of some suboptimum systems (properly designed) are quite close to the performance of the optimum system and, therefore, represent a reasonable engineering solution to the synchronization problem. Synchronizer performance is very dependent on signal waveform. In particular, rounded pulses result in significantly better performance (approximately 1 order of magnitude) than square pulses; the best nonlinearity is "log cosh", but "square law" and "magnitude" nonlinearities perform nearly as well for rounded pulses while the hard limiter is better for square pulses. We present only a discussion of the results of this study. Detailed derivations, etc., can be found in [7]. The problem of bit slippage is not considered.

II. Model

A mathematical model for the synchronization problem is presented in Fig. 1. A random sequence of positive and negative pulses is transmitted. The equation for the positive symbol waveform is $s(t)t \in (0,T)$ and its energy is E as illustrated in Fig. 1(a). For noiseless reception the received signal sequence would appear as illustrated in Fig. 1(b), $S(t;\zeta,\tau)$, and for the more realistic case (with noise) as in Fig. 1(b), $X(t)$. The received (noiseless) signal is characterized by the random process $S(t;\zeta,\tau)$ with two sources of randomness indicated by the two parameters ζ and τ. ζ is a parameter that governs the particular sequence transmitted. We assume that all possible sequences are equally likely. τ is a parameter that indicates epoch [the first zero crossing of $S(t;\zeta,\tau)$] relative to some point (chosen arbitrarily and labeled $t = 0$) on the time scale. We assume that τ is equally likely to be any number between $-T/2$ and $+T/2$. The observed random process $X(t;\tau)$ is the sum of the signal process and a noise process, i e.,

$$X(t;\tau) = S(t;\zeta,\tau) + N(t) \tag{1}$$

where the noise process $N(t)$ is assumed to be a Gaussian noise process with noise power N_0 W/Hz.

A synchronizer can be modeled as a transformation G of the random process $X(t;\tau)$ into the random variable $\hat{\tau}$, i e.,

$$\hat{\tau} = G[X(t;\tau)]. \tag{2}$$

We define the synchronizer error (jitter) to be

$$\epsilon = \hat{\tau} - \tau \tag{3}$$

where ϵ is a random variable whose range is $(-T/2,+T/2)$.

III. Performance Measures

On what basis should we say that one synchronizer is better (or worse) than another? Two different performance measures for synchronizers are illustrated in Fig. 2.

In the first, illustrated in Fig. 2(a), the synchronizer is considered a system by itself and some measure of

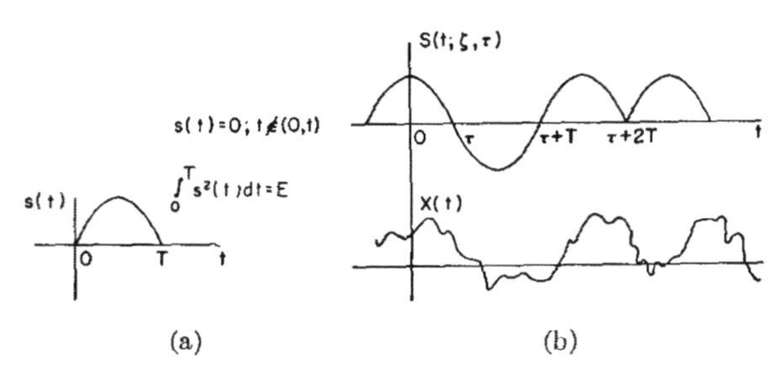

Fig. 1. (a) Positive symbol waveform. (b) Random processes $S(t;\zeta,\tau)$ and $X(t)$.

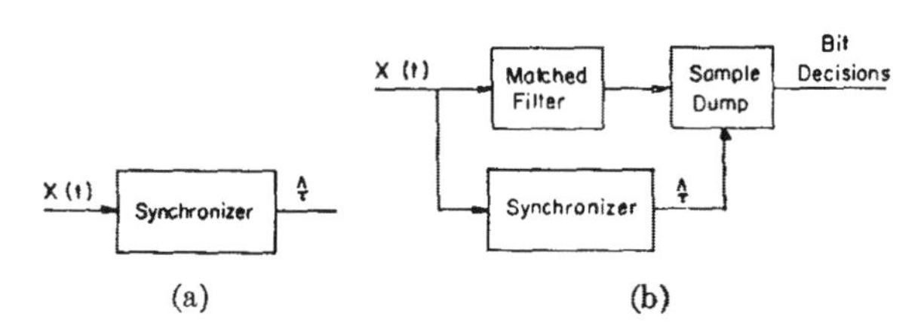

Fig. 2. (a) Synchronizer. (b) Detection system.

the jitter at the synchronizer output is the performance measure. One possibility is the rms error $(\overline{\epsilon^2})^{1/2}$. However, because we found it easier to measure in the laboratory, we have chosen the mean magnitude error $\overline{|\epsilon|}$ to be the performance measure. For typical unimodal probability density functions encountered in practice $\overline{|\epsilon|} \approx 0.8(\overline{\epsilon^2})^{1/2}$.

A complete digital detection system consists of a matched filter (or correlator), a sample-hold-dump circuit, and a synchronizer, as illustrated in Fig. 2(b). In this system the synchronizer has no utility of its own. Rather, its sole purpose is to provide timing information (the start and stop times of each incoming signal) to the sample and dump switches. If the bit timing information is not exact, the detector performance is degraded, i.e., the error rate is increased relative to what it would be for perfect synchronization. Hence a reasonable performance measure for this system is the fractional increase in transmitter power required to achieve the same probability of error that would be obtained with perfect synchronization. That is, suppose that with perfect synchronization, the digital detection system operates at an average SNR E/N_0 so that the corresponding probability of detection error is given by [14]

$$P_E = \frac{1}{2}\left(1 - \operatorname{erf}\sqrt{\frac{E}{N_0}}\right). \tag{4}$$

Then with synchronization jitter it can be shown (after considerable computation) that the probability of detection error is given by [7]

$$P_E{}^* = \frac{1}{2}\int_0^{T/2}\left\{1 - \frac{1}{2}\operatorname{erf}\left[\sqrt{\frac{E}{N_0}}\left(\frac{R(\epsilon) + R(T-\epsilon)}{R(0)}\right)\right]\right.$$
$$\left. - \frac{1}{2}\operatorname{erf}\left[\sqrt{\frac{E}{N_0}}\left(\frac{R(\epsilon) - R(T-\epsilon)}{R(0)}\right)\right]\right\} p(\epsilon)\, d\epsilon \tag{5}$$

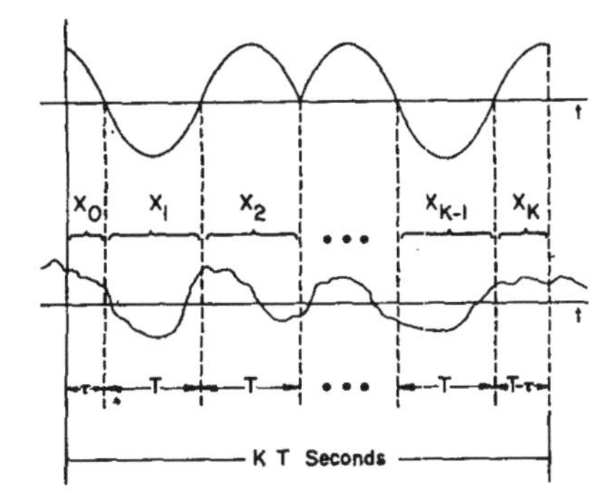

Fig. 3. Received signal without and with noise.

where $p(\epsilon)$ is the probability density function of the synchronizer error and $R(\epsilon)$ is the autocorrelation function of the positive symbol waveform, i.e.,

$$R(\epsilon) = \int_{-0}^{T} s(t)s(t+\epsilon)\, dt. \tag{6}$$

Equation (5) takes into account the intersymbol interference effect due to timing errors, i.e., a timing error of ϵ seconds results in the decision being based on the last ϵ seconds of one signal and the first $T - \epsilon$ seconds of the following signal. It is easy to show that $P_E^* \geq P_E$ with $P_E^* = P_E$ only for $p(\epsilon) = \delta(\epsilon)$ (perfect synchronization). We now define the degradation due to jitter to be the fractional increase in SNR required at the input to the system of Fig. 2(b) in order to decrease P_E^* to P_E. That is, in (5) replace E/N_0 by $E/N_0 + \Delta$ and find Δ such that $P_E^* = P_E$. Then the fractional increase in SNR $\Delta/(E/N_0)$ is called the degradation. Note that the degradation depends on the input SNR E/N_0, the signal waveform through its autocorrelation function $R(\epsilon)$, and the synchronizer through $p(\epsilon)$. $p(\epsilon)$, in turn, depends on the input SNR E/N_0 and the signal waveform $s(t)$ as well as the synchronizer structure G.

IV. Optimum Synchronizer

The synchronization problem was presented graphically in Fig. 1(b), $X(t)$. A random sequence of anticorrelated signaling waveforms (in noise) is observed; the problem is to determine the epoch τ. We use the maximum-likelihood parameter estimation technique to derive the structure of the optimum self-bit synchronizer, i.e., the maximum likelihood estimator of the epoch parameter τ.

The basic assumptions and notation are illustrated in Fig. 3. We assume a record of received data precisely KT seconds long, where T is the known symbol duration (T^{-1} is the bit rate in bits per second) and K is an integer called the memory (K is the number of bits over which the estimate is made). Even though the record is exactly K symbols long, it contains parts of $K + 1$ consecutive symbols since the last τ seconds of one symbol are included at the beginning of the record and the first $T - \tau$ seconds of another symbol are included at the end of the record. Let the received data (signal plus noise) in the $K + 1$ time intervals be represented by the $K + 1$ column matrices $X_0, X_1, \cdots, X_K$ as in Fig. 3. (The representation of waveforms by vectors is discussed in Appendix B of [14].) Hence we write

$$X_i = S_i + N_i, \quad i = 0,1,\cdots,K \tag{7}$$

where S_i is a column matrix representing the signal waveform in the ith interval ($S_i = \pm S$, where S represents the positive symbol waveform except that S_0 and S_K represent the last τ seconds and the first $T - \tau$ seconds, respectively, of the positive symbol waveform) and N_i is a column matrix representing the noise waveform in the ith interval. We assume that the noises in the ith and jth time intervals are uncorrelated for $i \neq j$, and that in each interval the noise has zero mean and known covariance ϕ_i, i.e.,

$$\begin{aligned} &\overline{N_i} = 0 \\ &\overline{N_i N_i^t} = \phi_i, \quad i = 0,1,\cdots,K, \quad N_i^t = N_i \text{ transpose} \\ &\overline{N_i N_j^t} = 0, \quad ij = 0,1,\cdots,K, \quad i \neq j. \end{aligned} \tag{8}$$

Under these assumptions (and those stated in Section II) it can be shown that the a posteriori probability density function $p(\tau \mid X_0, X_1, \cdots, X_K)$ can be written in the form [7]

$$p(\tau \mid X_0,\cdots,X_K) = C \frac{p(\tau)}{p(X_0,\cdots,X_K)} \prod_{j=0}^{K} \cosh\left[X_j^t \phi_j^{-1} S_j(\tau)\right] \tag{9}$$

where C is a constant, $p(\tau)$ is the a priori probability density on τ, and $S(\tau)$ is a column matrix representing the positive symbol waveform shifted by τ seconds. The maximum likelihood estimate of τ is the value of τ that maximizes $p(\tau \mid X_0,\cdots,X_K)$. Assuming no a priori knowledge of τ [$p(\tau) = T^{-1}; -T/2 \leq \tau \leq T/2$] the maximum likelihood estimate is the value of τ that maximizes $\prod_{j=1}^{K} \cosh[X_j^t \phi_j^{-1} S(\tau)]$. Or, since this quantity is always positive and the logarithm is a monotonically increasing function for positive arguments, an equivalent expression for the maximum likelihood estimator is $\hat{\tau}$ such that

$$\max_{\tau} \left\{ \sum_{j=0}^{K} \log\cosh\left[X_j \phi_j^{-1} S_j(\tau)\right] \right\} = \sum_{j=0}^{K} \log\cosh\left[X_j \phi_j^{-1} S_j(\hat{\tau})\right]. \tag{10}$$

For white noise (10) reduces to

$$\max_{\tau} \left\{ \sum_{j=0}^{K} \log\cosh\left[X_j S_j(\tau)\right] \right\} = \sum_{j=0}^{K} \log\cosh\left[X_j S_j(\hat{\tau})\right]. \tag{11}$$

To implement the optimum synchronizer for the white noise case we need a device that chooses a value for τ, say $\tau = \tau^*$, and then correlates the first τ^* seconds of the record with the last τ^* seconds of the positive symbol waveform, computes the log hyperbolic cosine of this quantity, and stores the resulting number. Next, the

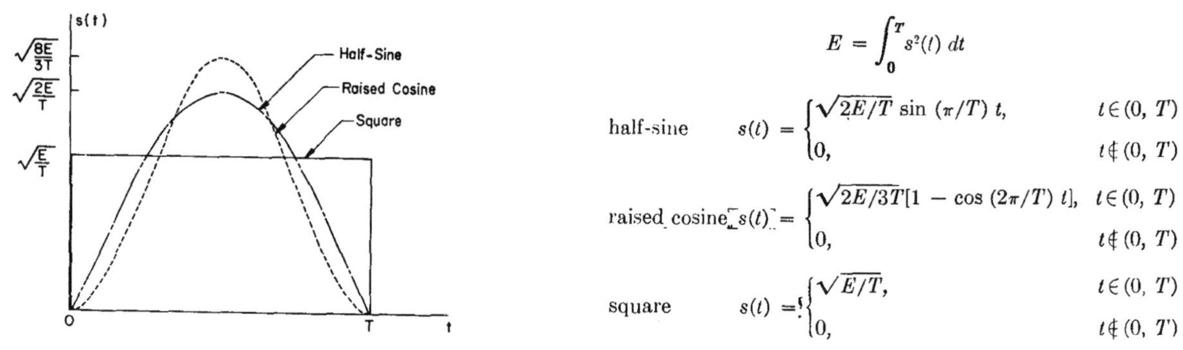

s(t)
$\sqrt{\frac{8E}{3T}}$
$\sqrt{\frac{2E}{T}}$
$\sqrt{\frac{E}{T}}$
Half-Sine
Raised Cosine
Square
0
T
t
$E = \int_0^T s^2(t)\, dt$
half-sine $s(t) = \begin{cases} \sqrt{2E/T} \sin(\pi/T)\, t, & t \in (0, T) \\ 0, & t \notin (0, T) \end{cases}$
raised cosine $s(t) = \begin{cases} \sqrt{2E/3T}[1 - \cos(2\pi/T)\, t], & t \in (0, T) \\ 0, & t \notin (0, T) \end{cases}$
square $s(t) = \begin{cases} \sqrt{E/T}, & t \in (0, T) \\ 0, & t \notin (0, T) \end{cases}$

Fig. 4. Signaling waveforms; all three waveforms have energy.

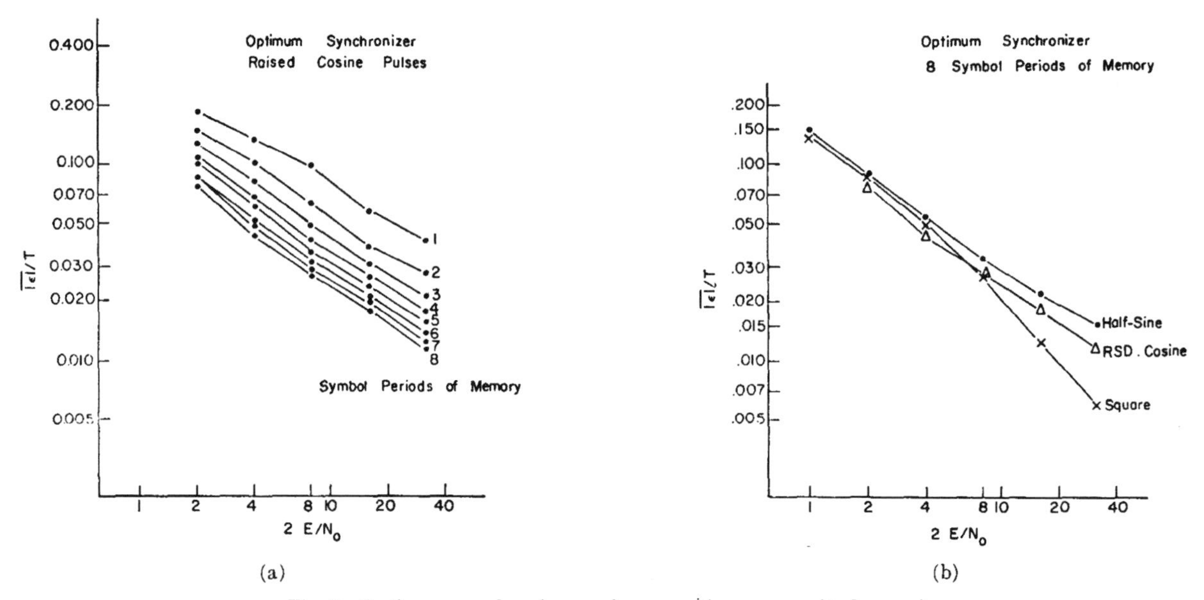

Fig. 5. Optimum synchronizer performance (mean magnitude error).

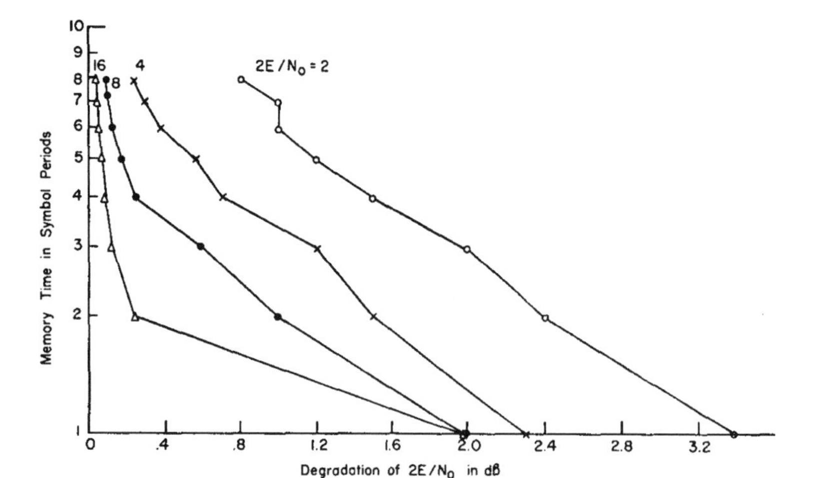

Fig. 6. Optimum synchronizer performance (degradation).

T-second portion of the record from τ^* to $\tau^* + T$ is correlated with the positive symbol waveform, the log hyperbolic cosine of this number computed, and the result added to the number previously obtained. The same procedure is followed for each of the remaining $K - 1$ T-second intervals. For the last interval, the last $T - \tau^*$ seconds of the record is correlated with the first $T - \tau^*$ seconds of the positive symbol waveform, the log hyperbolic cosine is taken, and this number is added to the sum of the first K numbers. The device has now evaluated (11) for $\tau = \tau^*$ and obtained the number

$$\sum_{j=0}^{K} \log \cosh [X_j S_j(\tau^*)].$$

This procedure must now be repeated for all $-T/2 \leq \tau \leq T/2$ to obtain $\max_\tau \{\sum_{j=0}^K \log \cosh [X_j S_j(\tau)]\}$ and the corresponding estimate $\hat{\tau}$. For colored noise the procedure is the same except that the data X_j are correlated with the matrix $Q(\tau) = \phi_j^{-1} S_j(\tau)$ rather than $S_j(\tau)$.

Because of its complexity, the optimum (maximum-likelihood) synchronizer does not appear to represent a realistic engineering solution to the synchronization problem. However, its performance is of interest. Even though we would not build the optimum synchronizer, knowledge of its performance would give us a standard against which to judge the performances of various suboptimum systems which are easier to implement.

It is not, of course, feasible to evaluate (11) for all $-T/2 \leq \tau \leq T/2$. However, an approximate solution can be obtained by evaluating (11) for a large but finite number of values for τ. Using this procedure and the Monte Carlo technique, the performance of the optimum synchronizer was evaluated on a digital computer for the 3 symbol waveforms shown in Fig. 4, various memory times K, and SNRs E/N_0.

Results for the raised cosine symbol are presented in Fig. 5(a) for memory times of $K = 1,2,\cdots,8$ symbol durations. The curves indicate that $|\epsilon|$ varies inversely with the square root of the SNR E/N_0 and with the square root of memory time K. This result is typical of adaptive systems (see, for example, [14]). Results for the raised cosine, half-sine, and square pulses are compared in Fig. 5(b) for $K = 8$.

Fig. 6 shows the amount of degradation that the optimum synchronizer introduces in the error performance of the matched filter detection system of Fig. 2(b). The abcissa gives the fractional amount of SNR that must be added to achieve the same error performance that would be achieved with perfect synchronization. The ordinate gives the memory required to achieve the specified degradation. All of the curves increase without bound as the degradation approaches zero. This is to be expected since for no degradation in error performance perfect synchronization is required; this can be obtained only with an infinite memory. Hence we are forced to accept some degradation. Fig. 6 gives the memory required by the optimum synchronizer to achieve any specified degradation for input SNRs of 2,4,8,16.

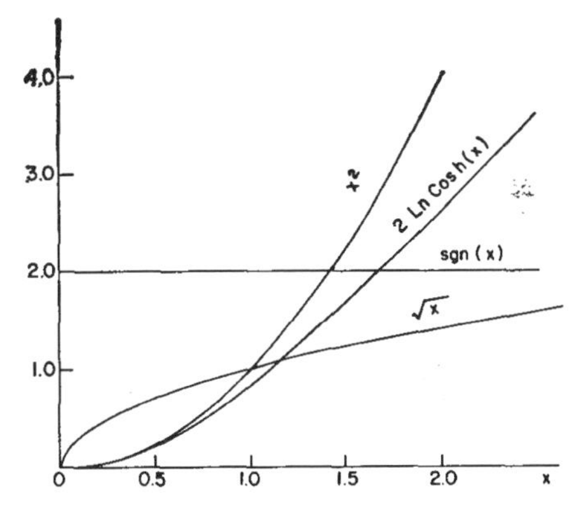

Fig. 7. Synchronizer nonlinearities.

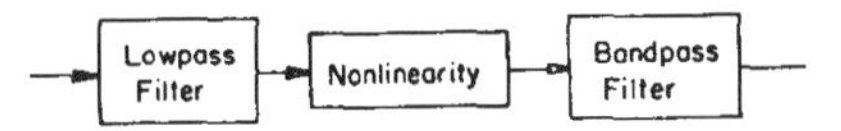

Fig. 8. Suboptimum synchronizer.

V. Suboptimum Synchronizers

In Section IV we showed the structure of the optimum (maximum-likelihood) synchronizer and interpreted it in terms of physical operations on the received data. In this section we use the optimum structure as a guide to a (suboptimum) synchronizer structure that is easier to implement.

Referring to (11), we recall that the first operation performed on the received data is a correlation or matched filter operation. For the pulse type signals of Fig. 4 a single-pole low-pass RC filter with time constant approximately equal to the symbol duration $RC \approx T$ is a reasonably good approximation to the matched filters. The log cosh nonlinearity is nearly a square law device for low-level inputs and a magnitude device for high-level inputs, as illustrated in Fig. 7. We choose to use a square law device rather than a log cosh device because it is easier to analyze. (We later investigate the effect of the nonlinearity on system performance.) Finally, the summation in (11) can be interpreted as an averaging operation on the nonlinearity outputs. In our analog system a bandpass filter can serve to time average the nonlinearity outputs due to all past inputs. Hence we arrive at the suboptimum system presented in Fig. 8. The input sequence of positive and negative symbols (and noise) is passed into the low-pass filter which eliminates some of the noise thereby increasing the SNR. Unfortunately, it also introduces some intersymbol interference because of its memory. (Each input pulse is smeared into the subsequent pulses.) The filter output is passed into the nonlinearity which flips the negative pulses (and noise) positive. The resulting waveform is not periodic (it would not be periodic even in the noiseless case because of the smearing effects of the low-pass filter on the random input sequence), but has some energy at the bit rate frequency. The bandpass filter is centered at the frequency

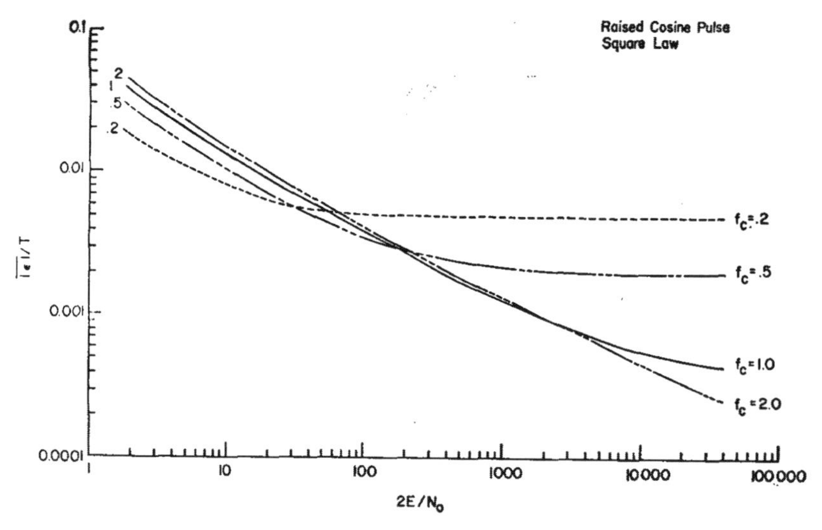

Fig. 9. Effect of bandwidth of low-pass filter.

corresponding to the bit rate and passes only the energy near the bit rate frequency. The synchronizer output is a sine wave (approximately) whose zero crossings are the synchronizer estimates of the start times of each incoming pulse. (The output could be fed into a hard limiter followed by a differentiator to obtain timing pulses whose leading edges contain the synchronization information.) Note that the memory of the suboptimum synchronizer can be adjusted by changing the bandwidth of the bandpass filter.

The optimum system weighted the data from all past intervals equally, but the bandpass filter is characterized by an exponential weighting into the past. (The envelope of its impulse response decreases exponentially.) Therefore, in order to compare the memory times of the optimum and suboptimum systems an effective memory time (in symbol durations) was used as follows:

$$K = \frac{(\sum C_i)^2}{\sum C_i^2} \tag{12}$$

where C_i is the relative weighting assigned to the ith preceding time interval. (This definition of effective measurement time is due to Price [8]. See also [9].)

A mathematical analysis of the suboptimum synchronizer has been completed; it is quite complex and the computations very tedious. Hence we present here only a brief overview; details can be found in [7]. The input to the synchronizer is a sample function from a random process which is itself the sum of two random process, i.e., the random sequence of signals and the noise. The power spectrum of the signal process can be computed by drawing on results due to Huggins [10], Zadeh [11], and Barnard [12]. Since the low-pass filter is a linear device, the power spectrums at the filter output due to both signal and noise can be computed separately. Furthermore, they are simply related to the input processes. The signal spectrum and the noise spectrum (including signal-cross-noise terms) at the output of the square law device were computed by the direct method for nonlinear devices [13]. From these spectra the SNR at the input to the bandpass filter can be determined. Next, the probability density function for the phase of the output of the bandpass filter was assumed to have the same functional form as the density function of the phase of a sine wave in additive Gaussian noise, i.e.,

$$p(\theta) = \exp(-d)\{1 + [\sqrt{\pi d}\cos\theta][\exp(-\sqrt{d}\cos\theta)]^2 \times [1 + \text{erf}(\sqrt{d}\cos\theta)]\}. \tag{13}$$

(This assumption was verified by experiment.) This is a one-parameter density function depending only on the SNR parameter d; d is related by a constant to the SNR computed at the input to the bandpass filter.

This analysis was used to determine the dependence of the performance of the suboptimum synchronizer on the various system parameters and to optimize the suboptimuve synchronizer relative to these parameters. For example: What is the optimum bandwidth (3-dB frequency) of the low-pass filter, and how sensitive is the performance to variations in this parameter? Fig. 9 shows the jitter $\overline{|\epsilon|}$ for the raised cosine symbol and an equivalent memory time of $60(K = 60)$ symbol durations. f_c is the ratio of the filter cutoff frequency to the bit rate frequency. To interpret these curves we recall that the low-pass filter effects both the signal and the noise. (It eliminates noise, but also smears the signals.) Hence at high SNRs performance is limited by the signal smearing effect and the optimum bandwidth is $f_c = 1.0$ (or more). At low SNRs the noise effect predominates and performance is enhanced by accepting some smearing in order to eliminate more noise. However, since the effect of the noise can also be mitigated by increasing the memory time (decreasing the bandwidth of the bandpass filter), we conclude that the low-pass filter bandwidth should be set equal to the bit rate frequency ($f_c = 1.0$). We also conclude that the jitter is quite insensitive to moderate (±20 percent) variations in this parameter.

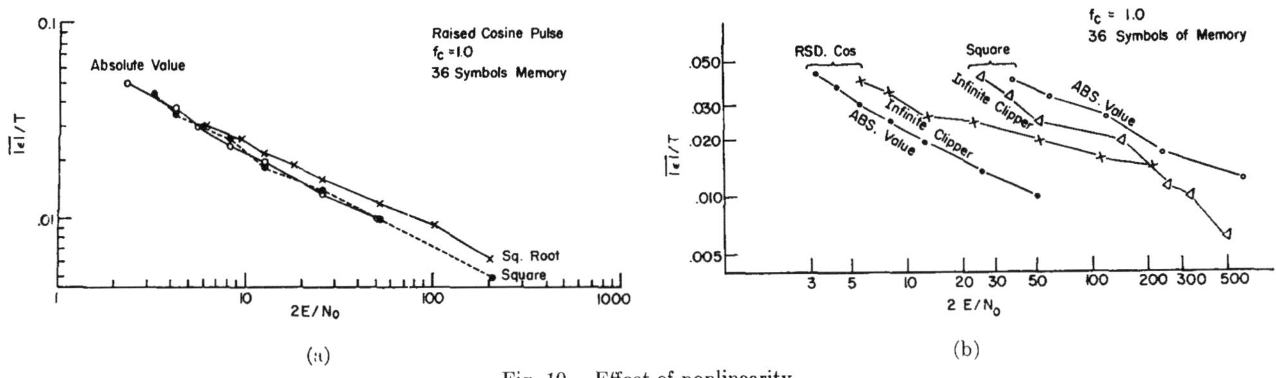

(a) (b)

Fig. 10. Effect of nonlinearity.

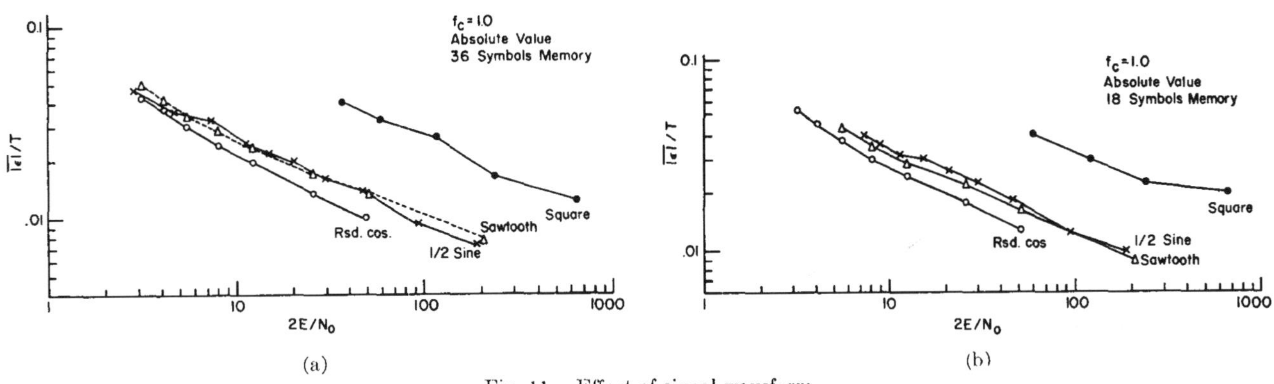

(a) (b)

Fig. 11. Effect of signal waveform.

The effect of low-pass filter bandwidth for the half-sine and square symbols was also investigated. The results for the half-sine symbol are similar to the results for the raised cosine symbol but not quite as pronounced at low SNRs. Hence the same conclusion is reached (set $f_c = 1.0$). For square pulses low-pass filtering is absolutely essential. Indeed, for the case of no noise and no low-pass filter, the output of the square law device would be a constant voltage. In order to generate energy at the bit rate frequency, some distortion must be introduced. We found that performance is essentially independent of low-pass filter bandwidth for sufficiently high SNRs; for low SNRs lower cutoff frequencies are better. Again, since the effect of the noise can be mitigated by increasing the memory time, $f_c = 1.0$ appears to be a reasonable setting. Detailed results for both cases (half-sine and square symbols) are given in [7].

A prototype of the suboptimum system was constructed in the laboratory, and the effect of the low-pass filter bandwidth on the jitter $\overline{|\epsilon|}$ was measured. These experimental results agreed quite well with those predicted by the analysis. Detailed results are given in [7].

The effect of the nonlinearity was also investigated by experiment since the square law nonlinearity is the only one amendable to analysis. Laboratory measurements were obtained for the absolute value and square root nonlinearities as well as for the square law device (see Fig. 7). Fig. 10(a) is typical of the results obtained for the rounded pulse shapes (raised cosine and half-sine). (Results for the other symbol waveforms can be found in [7].) We conclude that for rounded pulses the specific shape of the nonlinearity makes little difference so long as it is an nth law device with $1 < n < 2$.

We also took measurements with a hard limiter (infinite clipper) nonlinearity since it is commonly used, especially for square pulses. The performances for the hard limiter and absolute value nonlinearities are compared in Fig. 10(b) for the raised cosine and square symbol waveforms. For the raised cosine symbol the absolute value nonlinearity yields significantly better performance while the reverse is true for the square symbol. Note, however, that the shape of the symbol waveform is much more significant than the shape of the nonlinearity, especially at SNRs of practical interest.

Further experimental data were obtained to determine the effect of the symbol waveform on the performance of the suboptimum synchronizer. Performance for raised cosine, half-sine, ramp, and square pulses are presented in Fig. 11 for the system with the absolute value nonlinearity. Again we note that the square pulses are very inefficient (approximately one order of magnitude in SNR) compared to the other waveforms.

The effect of symbol waveform is even more pronounced when the synchronizer is incorporated into the detection system, as illustrated in Fig. 2(b). The memory times required to achieve any specified degradation is presented in Fig. 12 for the raised cosine, half-sine, and square symbols for SNRs $2E/N_0 = 4,9,16$. For 0.1-dB degration

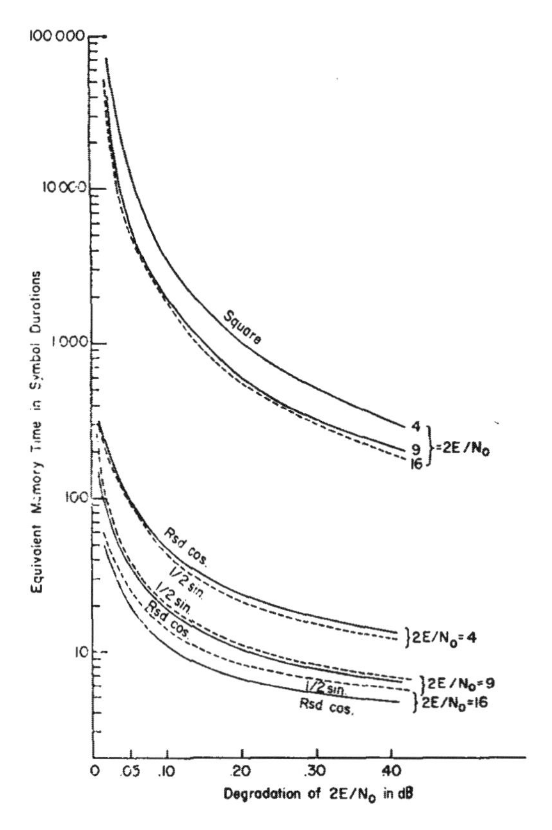

Fig. 12. Effect of symbol waveform.

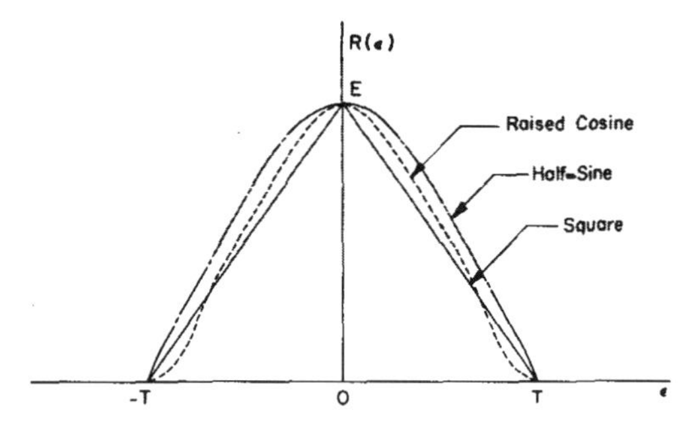

Fig. 13. Signal autocorrelation functions.

and an SNR of 9, the system requires two orders of magnitude more memory for square pulse than for rounded pulses (raised cosine or half-sine). Also note that this effect cannot be eradicated by increasing the SNR.

This drastic dependence of detection system performance on waveshape is due to two effects, one concerning the synchronizer and the other the correlation detector. We have already noted that the synchronizer performs significantly better with rounded pulses. (For the same amount of jitter $|\epsilon|$ rounded pulses outperform square pulses by an order of magnitude in SNR.) Hence we focus our attention on the correlation detector. Now, the probability of a detection error depends only on the SNR at the output of the correlation detector. The output of the correlation detector depends on the input signal, the input noise, and the timing error. Since it is a linear device, the outputs due to signal and noise can be considered separately. It is easy to show that the output due to noise is independent of the timing error. That is, the rms value of the noise depends only on the signal energy (not the waveform) and the correlation time, and for reasonable values of synchronizer memory the correlation times do not change significantly from bit to bit. Hence timing error effects only the correlator output due to signal. The correlator output due to signal depends on the timing error and the autocorrelation function of the signal waveform. Suppose, for the moment, that only a single signal is transmitted and that the timing error is ϵ. Then the correlator output due to signal is $R(\epsilon)$, the autocorrelation function evaluated at ϵ [see (6)]. When a sequence of signals is transmitted, the output due to signal also contains a component due to the preceding signal (for positive ϵ) or the succeeding signal (for negative ϵ), i.e., the correlator output is now given by $R(\epsilon) \pm R(T + \epsilon)$. If both signals have the same sign, the signal component at the correlator output is increased, while if one signal is a positive symbol and the other negative, the signal component is decreased. Since these two events occur with equal probability, and since the probability of error is nearly a linear function of SNR for small SNR perturbations, we conclude that this effect has a small effect on system performance. Hence the primary effect is due to the first term $R(\epsilon)$. In Fig. 13 we have sketched $R(\epsilon)$ as ϵ for the three symbol waveforms. Note that for the rounded pulses $R(\epsilon)$ is quite flat for small values of ϵ, while for the square pulse $R(\epsilon)$ decreases much faster with ϵ. Since $R(\epsilon)$ is essentially the SNR (except for a constant scale factor) at the correlator output, the dependence of detection system performance on symbol waveform is quite clear. For a given timing error ϵ (if the system is to be at all practical, ϵ is reasonably small) rounded pulses perform significantly better than square pulses. Furthermore, we recall that rounded pulses result in significantly smaller timing errors than square pulses. The combined effect yields the result illustrated in Fig. 12.

VI. Optimum and Suboptimum Synchronizers Compared

In Section IV we presented the optimum (maximum-likelihood) synchronizer and showed how its performance depends on the basic system parameters. In Section V we suggested a suboptimum synchronizer and showed how its performance depends on the basic system parameters. We have also pointed out that the suboptimum synchronizer is relatively simple to implement while the optimum synchronizer does not appear to be practical. In this section we compare the performances of the optimum and suboptimum synchronizers.

In Fig. 14(a) we compare the mean magnitude jitter performances of the optimum and suboptimum synchronizers [Fig. 2(a)] for $K = 8$ for the raised cosine, half-sine, and square symbol waveforms. Note first that for the raised cosine symbol the suboptimum system performs nearly as well as the optimum system (within 1 dB).

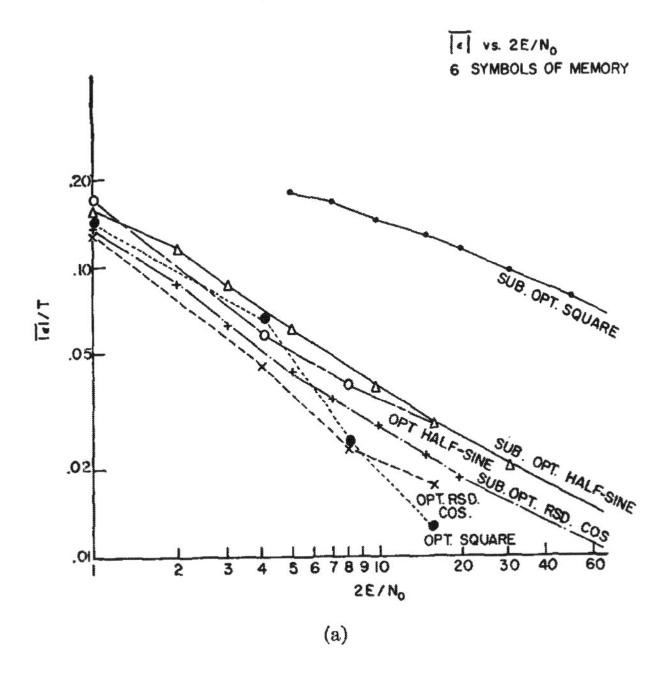

(a)

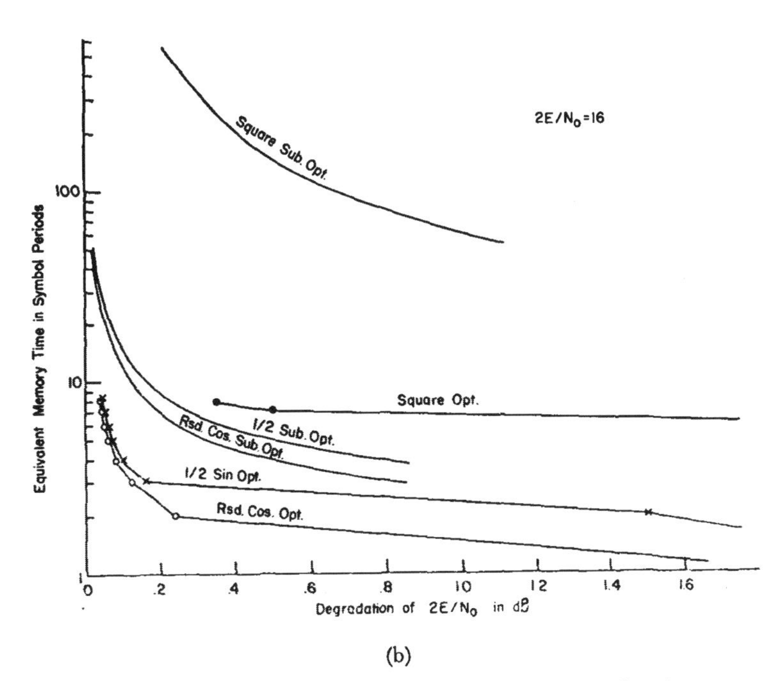

(b)

Fig. 14. Comparison of optimum and suboptimum synchronizers.

The same is true for the half-sine symbol. On the other hand, we note an order of magnitude difference between the performances of the optimum and suboptimum systems for the square pulse. From this and from additional curves given in [7] we conclude that for all practical purposes the suboptimum synchronizer performs as well as the optimum synchronizer for rounded pulses, but not for square pulses. As indicated in Fig. 10(b), the performance for square pulses can be improved somewhat by using the hard limiter nonlinearity.

The performances of the detection system [Fig. 2(b)] utilizing both the optimum and suboptimum synchronizers are compared on the basis of SNR degradation for the three symbol waveforms in Fig. 14(b). Note that for the rounded pulses the suboptimum detection system is inferior to the optimum system by approximately a factor of two in the memory time required for the same degradation. For the square symbol the discrepancy between the optimum and suboptimum detection systems is approximately a factor of 10 in equivalent memory time.

VII. Conclusions

We have determined the performances of the optimum (maximum-likelihood) and some suboptimum (Fig. 8) self-bit synchronization systems as functions of the pertinent system parameters. This was accomplished through a combined program of analysis, simulation, and laboratory experimentation. Two criteria were considered: synchronizer jitter and detection system SNR degradation. The most significant parameters (in terms of system performance) were found to be the input SNR, the shape of the signal waveform, and the memory time of the synchronizer. System performance is less sensitive to the bandwidth of the synchronizer input filter and, within limits, the synchronizer nonlinearity.

The dependence on input SNR and synchronizer memory are roughly equivalent in the sense that doubling the memory has the same effect on performance as doubling the input SNR. This is in agreement with some general conclusions on adaptive systems stated in [14]. Synchronizer jitter and detection system performance are very sensitive to the shape of the signal waveform for both the optimum and suboptimum synchronizers considered. Synchronizer performance can be improved by one order of magnitude, and detection system performance can be improved by two orders of magnitude by using rounded pulses rather than square pulses. That is, with square pulses the input SNR must be increased by a factor of 10 in order to achieve the same amount of synchronizer jitter incurred with rounded pulses. With square pulses the input SNR must be increased by a factor of 100 in order to achieve the same error rate obtained with rounded pulses. These comparisons were made with square and rounded pulses having equal energy or average power. On a peak power basis the difference in performance is not quite so great, but rounded pulses still yield significantly better performance.

We also found that the best nonlinearity for use with rounded pulses in the suboptimum system is an nth law device with $1 < n < 2$; for square pulses a hard limiter gives better performance.

Finally, we determined that the suboptimum synchronizer properly adjusted performs almost as well as the optimum synchronizer for rounded pulses; for square pulses this is not the case.

References

[1] S. W. Golomb, J. R. Davey, I. S. Reed, H. L. Van Trees, and J. J. Stiffler, "Synchronization," *IEEE Trans. Communications Systems*, vol. CS-11, pp. 481–491, December 1963.

[2] J. H. Van Horn, "A theoretical synchronization system for use with noisy digital signals," *IEEE Trans. Communication Technology*, vol. COM-12, pp. 82–90, September 1964.

[3] J. J. Stiffler, "Maximum likelihood symbol synchronization," *JPL Space Programs Summary*, vol. 4, pp. 349–357, October 1965.

[4] H. L. Van Trees, "Optimum power division in coherent communication systems," *IEEE Trans. Space Electronics and Telemetry*, vol. SET-10, pp. 1–9, March 1964.

[5] D. Chase, "Communication over noise channels with no a priori synchronization information," M.I.T. Res. Lab. of Electronics, Cambridge, Mass., Tech. Rept. 463, February 1968.

[6] P. A. Wintz and J. C. Hancock, "An adaptive receiver approach to the time synchronization problem," *IEEE Trans. Communication Technology*, vol. COM-13, pp. 90–96, March 1965.

[7] P. A. Wintz and E. J. Luecke, "Performances of self-bit synchronization systems," School of Electrical Engineering, Purdue University, Lafayette, Ind., Tech. Rept. TR-EE68-1, January 1968.

[8] R. Price, "Error probabilities for adaptive multichannel reception of binary signals," M.I.T. Lincoln Lab., Lexington, Mass., Tech. Rept. 258, July 1962.

[9] J. G. Proakis, P. R. Drouilhet, Jr., and R. Price, "Performance of coherent detection systems using decision-directed channel measurement," *IEEE Trans. Communications Systems*, vol. CS-12, pp. 54–63, March 1964.

[10] W. H. Huggins, "Signal-flow graphs and random signals," *Proc. IRE*, vol. 45, pp. 74–86, January 1957.

[11] L. A. Zadeh and W. H. Huggins, "Signal-flow graphs and random signals," *Proc. IRE* (Correspondence), vol. 45, pp. 1413–1414, October 1957.

[12] R. D. Barnard, "On the discrete spectral densities of Markov pulse trains," *Bell Sys. Tech. J.*, vol. 43, pp. 233–259, January 1964.

[13] W. Davenport and W. Root, *Random Signals and Noise*. New York: McGraw-Hill, 1958.

[14] J. C. Hancock and P. A. Wintz, *Signal Detection Theory*. New York: McGraw-Hill, 1966.

Paul A. Wintz (S'61–M'64), for a photograph and biography, please see page 290 of the April, 1969, issue of this Transactions.

PHOTO NOT AVAILABLE

Edgar J. Luecke (M'59) was born in Cleveland, Ohio, on October 6, 1933. He received the B.S.E.E. degree from Valparaiso University, Valparaiso, Ind., in 1955, the M.S.E.E. degree from the University of Notre Dame, Notre Dame, Ind., in 1957, and the Ph.D. degree in electrical engineering from Purdue University, Lafayette, Ind., in 1968.

From 1955 to 1963 he was an Instructor and Assistant Professor of Electrical Engineering at Valparaiso University. During 1963 he was an NSF Faculty Fellow at Purdue University and from 1964 to 1966 a Research Instructor. Since 1966 he has been Associate Professor of Electrical Engineering at Valparaiso University.

Dr. Luecke is a member of Tau Beta Pi, Eta Kappa Nu, and the American Society for Engineering Education.

Correlative Digital Communication Techniques

ADAM LENDER, MEMBER, IEEE

Abstract—**A new method for the transmission of intelligence by means of a signal having certain correlation properties has been evolved. The theoretical and practical aspects of this concept are presented. An important advantage of these techniques is that for a fixed performance criteria, considerably higher speeds are possible compared to the presently known methods. In addition, the implementation is simple and straightforward. An unusual property of these techniques is the capability of error detection without the introduction of redundancy into the original data. Finally, expressions for spectral distributions and error performance as well as methods for practical implementation, including the error-detection process, are presented.**

I. Review of Multilevel Techniques

According to the well-known postulate [1] of information theory, it is possible to attain virtually error-free transmissions up to the maximum rate of

$$C = W \log_2 \left(1 + \frac{S}{WN_0}\right) \text{ bits per second (b/s)} \qquad (1)$$

where W represents the channel bandwidth, S the average signal power, and N_0 the Gaussian noise power in a one-cycle band. The theory also states that to achieve such a capacity, an exceedingly complex signaling system is required. In existing systems only a small fraction of this maximum possible channel capacity has ever been obtained. This paper proposes methods and describes practical approaches which are felt to represent a step toward realizing this theoretical goal.

In high-speed digital communications, multilevel techniques have been known and used for some time; recently an experimental system with 32 discrete signaling levels was described [2]. Multilevel techniques employ b discrete signaling levels and represent $\log_2 b$ binary channels. For a fixed peak power, a b-level system has approximately 20 $\log_{10}(b - 1)$ dB penalty relative to binary, but at the same time has $\log_2 b$ times greater capacity. As a result, the net noise penalty of a b-level multilevel system relative to binary is

$$10 \log_{10} \frac{(b-1)^2}{\log_2 b} \text{ dB} \qquad (2)$$

Equation (2), however, does not take into account the noise impairment due to intersymbol interference which, in multilevel systems, increases rapidly with the number of signaling levels. Moreover, multilevel systems require complex equipment. Recognizing the shortcomings of multilevel systems, new techniques have been explored to obtain more efficient digital communications in terms of both performance and equipment.

The common characteristic of existing multilevel codes is the absence of correlation between the digits. Attention was directed to the possibility of utilizing discrete signaling levels which would be correlated in the process of generating such levels, yet be treated independently in the detection process. One such technique has been described in an earlier paper [3]. Generation of a time wave consisting of correlated signaling levels permits overall *spectrum* shaping in addition to individual *pulse* shaping. It is, for instance, possible to redistribute the spectral energy so as to concentrate most of it at low frequencies or to eliminate any power at low frequencies. Such techniques are termed correlative.

II. The Polybinary[1] Concept

Let a binary message with two signaling levels (MARK and SPACE) represented by NRZ (nonreturn to zero) digits be transformed into a signal with b signaling levels numbered consecutively from zero to $(b - 1)$ starting at the bottom. All even-numbered levels are identified as SPACE, and all odd-numbered ones as MARK, although this labeling can, of course, be easily reversed. Both the original message and the polybinary signal have an identical digit duration of T seconds. There are no restrictions on the number of levels, b. *Such a signal, termed polybinary, is capable of transmitting $(b - 1)$ binary channels over a bandwidth which normally accommodates only a single binary channel.*

The binary message is transformed into the polybinary signal in two steps. In the first step the original sequence a_n, consisting of MARKS and SPACES, is converted into another binary sequence d_n in such a manner that the present binary digit of sequence d_n represents the modulo 2 sum of $(b - 2)$ immediately preceding digits of sequence d_n and the present digit of a_n. The second step involves the transformation of the binary sequence d_n (in which 1 and 0 now no longer represent MARK and SPACE) into the polybinary pulse train p_n by adding algebraically (not modulo 2) the present digit of sequence d_n to the $(b - 2)$ preceding digits of sequence d_n. One result of the transformation from sequence a_n to p_n is that SPACE and MARK in sequence a_n are mapped into even- and odd-numbered levels respectively in sequence p_n. This is significant, since each digit in sequence p_n can be independently detected in spite of the strong correlation properties. The primary consequence of such properties is the redistribution of the spectral density of the original sequence a_n into energy compressed near low frequencies for the new sequence p_n.

Manuscript received July 21, 1964. Presented at 1964 IEEE Internat'l Convention and published originally in the *1964 IEEE Internat'l Convention Record,* vol 12, pt 5, pp 45–53.

The author is with Lenkurt Electric Co., Inc., San Carlos, Calif.

[1] i.e., a plurality of binary channels.

Reprinted from *IEEE Transactions on Communication Technology,* December 1964.

The Best of the Best. Edited by W. H. Tranter, D. P. Taylor, R. E. Ziemer, N. F. Maxemchuk, and J. W. Mark.

It is well known that the continuous component of the spectral density of the sequence a_n, consisting of uncorrelated binary digits, is

$$W_1(f) = \frac{1}{T}|G(f)|^2 pq \tag{3}$$

where p is the probability of MARK and $q = (1 - p)$ is the probability of SPACE, $G(f)$ is the Fourier transform of the pulse shape, and $1/T$ is the speed in b/s. Similarly, it is shown in Appendix I that the spectral density for the sequence d_n is

$$\left.\begin{aligned} W_2(f) &= W_1(f)Z_1 \quad \text{for } b = 3 \\ &= W_1(f)Z_2 \quad \text{for } b > 3 \end{aligned}\right\} \tag{4}$$

where

$$Z_1 = \frac{1}{(1 - 2p)^2 - 2(1 - 2p)\cos 2\pi fT + 1} \tag{5}$$

$$Z_2 = \frac{1 + (1 - 2p)^2}{(1 - 2p)^4 - 2(1 - 2p)^2 \cos (b - 1)2\pi fT + 1} \tag{6}$$

Finally, the spectral density of the sequence p_n as shown in Appendix I is

$$\left.\begin{aligned} W_3(f) &= \frac{1}{T}|G(f)|^2\left[\frac{\sin (b - 1)\pi fT}{\sin \pi fT}\right]^2 pqZ_1 \text{ for } b = 3 \\ &= \frac{1}{T}|G(f)|^2\left[\frac{\sin (b - 1)\pi fT}{\sin \pi fT}\right]^2 pqZ_2 \text{ for } b > 3 \end{aligned}\right\} \tag{7}$$

It is interesting to note that the weighting factor Z_2 of the spectral density in (6) and (7) is a symmetrical function of p and q, where $q = 1 - p$; when $p \neq 0.5$, the energy density is more concentrated near low frequencies than for $p = 0.5$. The case for $b = 3$ is an exception. Both weighting factors shown in (5) and (6) are plotted for representative values of p in Fig. 1.

That the spectrum of the polybinary sequence p_n is compressed by the factor of $(b - 1)$ relative to the binary sequence a_n can be shown for the simplest case of equal likelihood of MARK and SPACE and rectangular NRZ pulses of unit height, i.e., when

$$p = q = 1/2 \tag{8}$$

$$G(f) = T\frac{\sin \pi fT}{\pi fT} \tag{9}$$

Then $Z_1 = Z_2 = 1$, and substitution of (8) and (9) into (3) and (7) yields

$$W_1(f) = \frac{T}{4}\left(\frac{\sin \pi fT}{\pi fT}\right)^2 \text{(for binary)} \tag{10}$$

$$W_3(f) = \frac{(b - 1)^2 T}{4}\left[\frac{\sin (b - 1)\pi fT}{(b - 1)\pi fT}\right]^2 \tag{11}$$
(for polybinary)

In practical situations the $(b - 1)$ compression factor can be closely approached by judicious choice of network characteristics depending upon the value of b and the maximum permissible intersymbol interference.

The net noise penalty of the polybinary signal relative to binary is approximately

$$10 \log_{10}(b - 1) \text{ dB} \tag{12}$$

The intersymbol interference is inherently small, since the only possible transitions in two successive digit-time slots

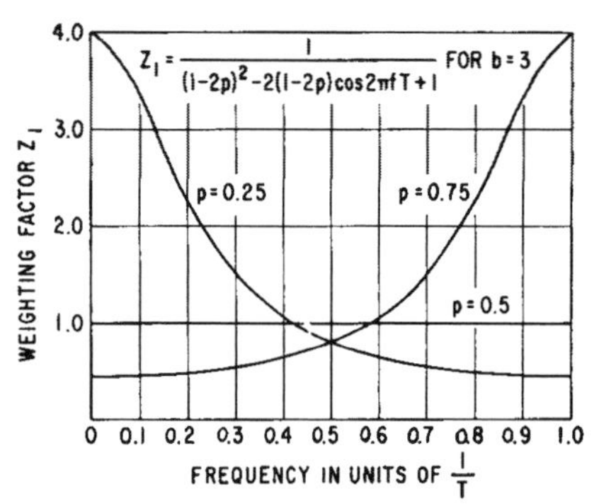

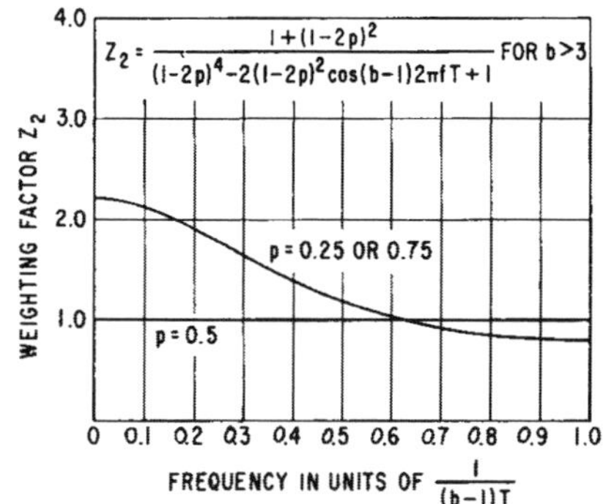

Fig. 1. Weighting factors of spectral densities.

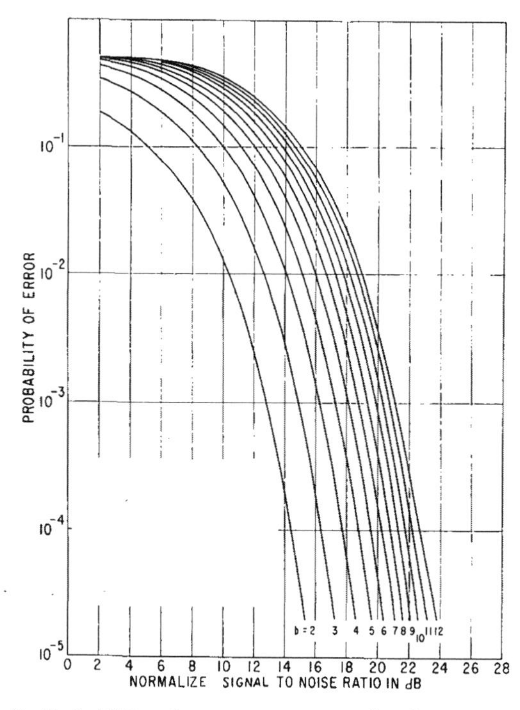

Fig. 2. Probability of error vs. average signal power per bit per second divided by noise power density in dB.

occur between adjacent signaling levels. Equation (12) should be compared with (2) for multilevel systems. However, (12) is pessimistic and does not take into account the peculiar property of the b-level polybinary sequence, namely, that errors occur only when an even-numbered level is changed to an odd-numbered level, or vice versa. An exact expression for the probability of error for any value of b in the presence of Gaussian noise for the polybinary system is developed in Appendix II and is

$$P_e = \frac{1}{2} + 2^{1-b} \sum_{k=1}^{b-1} \binom{b-1}{k} \sum_{i=1}^{k} (-1)^i erf \left\{ (2i-1) \left[\frac{E}{2bN_0} \right]^{1/2} \right\} \quad (13)$$

where E/N_0 is the normalized signal-to-noise ratio (SNR). Equation (13) is plotted for several values of b in Fig. 2.

III. The Polybipolar[2] Concept

In some applications it is desirable to transmit a baseband signal which has no dc component and only a small amount of energy at low frequencies. A well-known signal with this property is the bipolar [4]. It has three levels, occupies approximately the same bandwidth as binary and represents a single binary channel. However, when greater capacity is desired, the polybipolar concept is useful. As in the case of a bipolar signal, there is no dc component.

The polybipolar transformation process is similar to the polybinary transformation described in Section II. It maps the original binary sequence a_n into the polybipolar sequence g_n with b levels representing $(b - 1)/2$ binary channels. The only restriction is that b must be an odd number. Binary sequence a_n is converted into binary sequence d_n as before. In the second step, each digit of the sequence g_n is formed by adding algebraically to the present digit of the sequence d_n, $(b - 3)/2$ immediately preceding digits and subtracting the remaining $(b - 1)/2$ digits. As a result, the center level of the sequence g_n always represents SPACE of the original sequence a_n. The adjacent positive and negative levels are consecutively numbered ± 1, ± 2, etc., starting from the center level so that even-numbered levels again represent SPACES and odd-numbered MARKS. From Appendix I the spectrum of the polybipolar sequence g_n for $b > 3$ is

$$W_4(f) = \frac{1}{T} |G(f)|^2 \frac{\left[\sin \left(\frac{b-1}{2} \right) \pi f T \right]^4}{(\sin \pi f T)^2} pqZ_2 \quad (14)$$

The conventional bipolar spectrum is merely (14) with $b = 3$ and the weighting factor Z_1, i.e.,

$$W_5(f) = \frac{1}{T} |G(f)|^2 (\sin \pi f T)^2 \, pqZ_1 \quad (15)$$

where Z_1 and Z_2 are defined in (5) and (6). When the conditions in expressions (8) and (9) prevail, the polybipolar spectrum is compressed exactly by the factor $(b - 1)/2$ relative to bipolar. Such a compression is reasonably well approached in practice for more general conditions.

Polybipolar signals have greater intersymbol interference than polybinary because transitions in two successive digit slots involve one or two steps between signaling levels. All other characteristics of polybipolar signals such as independent detection of digits, orderly distribution of even and odd levels (SPACES and MARKS, respectively), as well as the probability of error developed in Appendix II and shown in Fig. 2 are similar to the properties of polybinary signals.

[2] i.e., a plurality of bipolar channels.

IV. Implementation of Polybinary and Polybipolar Systems and Experimental Results

Implementation of the correlative systems as shown in Fig. 3 is straightforward and simple. The transmission filter $H(\omega)$ is a passive LC filter with nominal half-amplitude point at f_1 Hz. It represents to a rough approximation the conversion characteristics combined with pulse shaping. The conversion characteristics correspond to the transformation rules for the sequence d_n to p_n for polybinary and the sequence d_n to g_n for polybipolar. The input data rate at A of Fig. 3 is $2(b - 1)f_1$ b/s for polybinary and $(b - 1)f_1$ b/s for polybipolar. At point B the coder transforms the original data sequence a_n to the binary sequence d_n which in turn is converted to the band-limited polybinary or polybipolar time function at C. In the reconversion at the receiving end, each signaling level is clearly associated with MARK or SPACE (1 or 0) regardless of previous decisions; this permits the use of one of the two methods shown in Fig. 3. Such an experimental system was implemented for specific values of b.

B of Fig. 4 shows the polybinary waveform for a 16-bit pattern with five levels representing four binary channels and corresponding to the original data input in A of Fig. 4 at 4800 b/s. These waveforms appear at points A and C of Fig. 3. To indicate the degree of intersymbol interference, an eye pattern is shown for random data input at the same speed. For comparison, eye patterns for $b = 3$ and 4 are also shown at speeds of 2400 and 3600 b/s with appropriate coding through the same filter $H(\omega)$ as in the system for $b = 5$.

Results of the polybipolar experimental system with $b = 5$ at 2400 b/s appear in Fig. 5. The bandwidth is approximately the same as for the polybinary system with $b = 5$ at 4800 b/s. No dc is present, and a higher intersymbol interference in the eye pattern is evident due to one- and two-step transitions. For comparison, the bipolar signal is also shown at 1200 b/s through the same filter.

V. Error Detection in Polybinary and Polybipolar Systems

Based on previous considerations, a b-level correlative time function has recurrent correlation properties extending over $(b - 1)$ digits. These properties are not utilized in the signal detection but can be employed to check pattern violations without introducing redundant digits into the original binary message. Such violations result

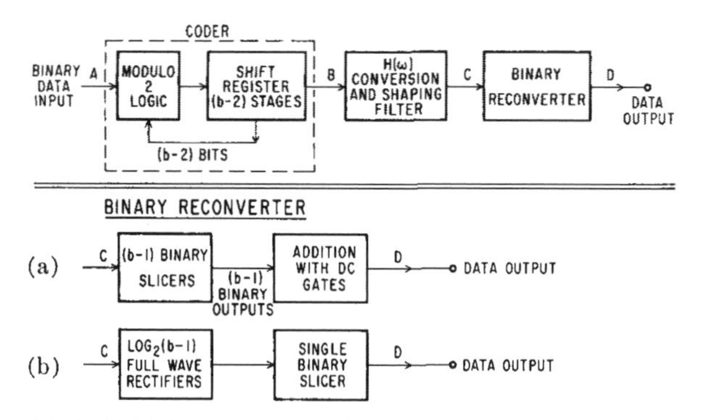

Fig. 3. Polybinary or polybipolar system (a) Applicable to any value of b. (b) Applicable when $b = 2^n + 1$ and n is an integer.

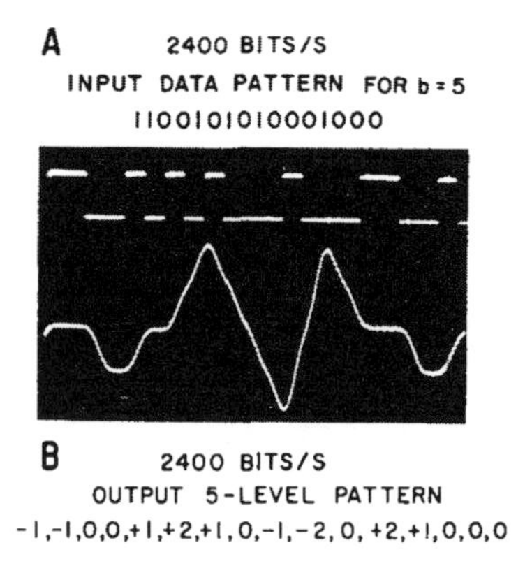

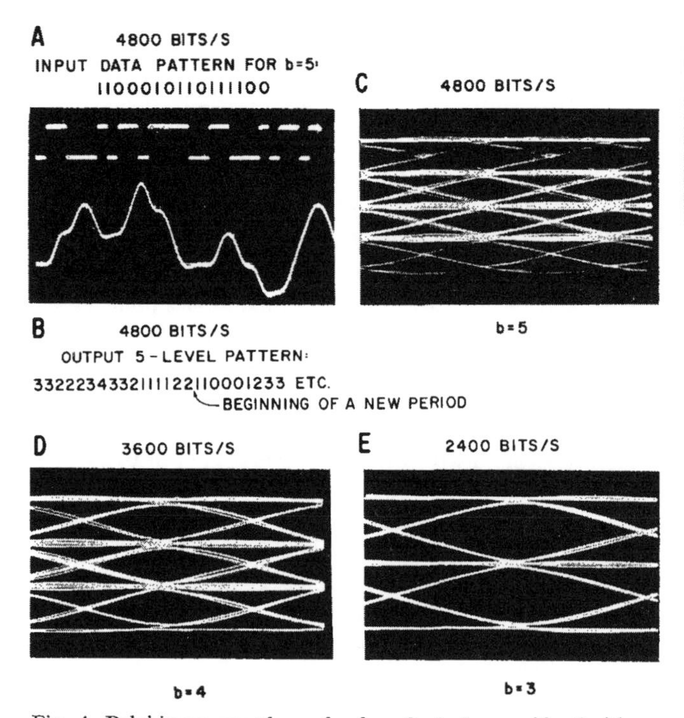

Fig. 4. Polybinary waveforms for $b = 3, 4, 5$, over identical low-pass filter.

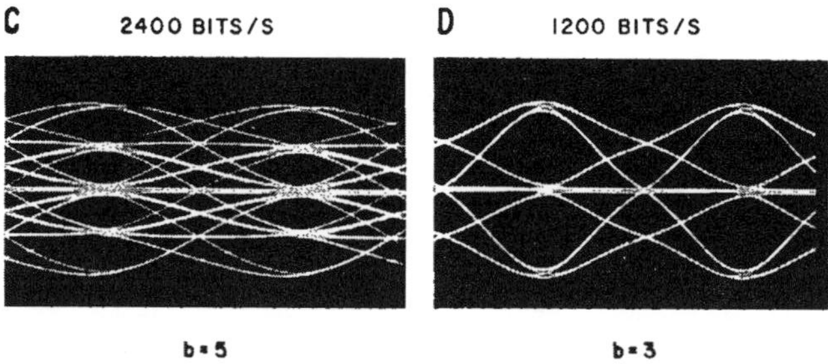

Fig. 5. Polybipolar waveforms for $b = 3, 5$, over identical band-pass filter.

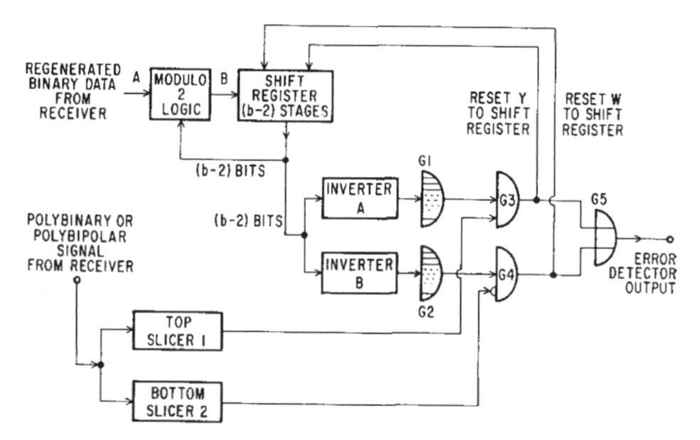

Fig. 6. Error-detection system for b-level signal

in errors which are easily detected. Error detection is performed at the receiving end by comparing the received polybinary or polybipolar waveform with the decoded binary data.

A simple, yet efficient method for implementation is shown in Fig. 6. The regenerated binary data sequence a_n at A is converted to the binary sequence d_n at B. As a result, the shift register always contains $(b - 2)$ past digits of sequence d_n. Every time the polybinary or polybipolar signal is in one of the extreme signaling levels, an error check is performed. The top and bottom slicers in Fig. 6 sense these extreme levels and provide respective indications to gates $G3$ or $G4$. When either of the extreme states occurs, the pattern of $(b - 2)$ digits in the shift register is examined. For example, when the top state is indicated for the polybipolar signal, the first $(b - 3)/2$ stages of the register should be in binary state 1 and the remaining $(b - 1)/2$ stages in binary state 0. If there is a disagreement, an error indication is provided, and the register is reset to the proper states. Consequently, for polybipolar signals inverter A inverts only the first $(b - 3)/2$ states of the shift register and inverter B the last $(b - 1)/2$ states. For polybinary signals inverter B is absent and inverter A inverts all the states of the shift register; at the same time, reset Y resets all stages to binary 1 and reset W to binary 0. In the case of polybipolar signals, reset Y resets the first $(b - 3)/2$ stages to binary 1 and the remaining to binary 0; reset W does exactly the opposite.

An interesting aspect of the error-detection process is demonstrated when a violation which does not result in an error occurs; e.g., when one even-numbered level is changed to another even-numbered level. In general, the type of error detector described will not give an indication except for the singular case when such violations result in a transition to one of the extreme signaling levels. It has been experimentally verified that such cases occur rarely and have relatively little bearing on the error-detection process.

VI. The Polybinary AM-PSK Process with Envelope Detection

The correlative baseband techniques already described are suitable for any type of carrier modulation. However, the most interesting signal characteristics result from the unqiue combination of the polybinary technique with AM-PSK modulation. The restriction on the number of levels is that b must be odd.

Suppose the center level of the polybinary signal is represented by the absence of carrier, the $(b - 1)/2$ levels above the center by equal increments of carrier and the $(b - 1)/2$ levels below the center also by equal increments but with carrier reversed by 180°. Remembering that b is odd and that inherent symmetry exists with respect to the center level, the original b amplitude levels of the polybinary signal are reduced to $(b + 1)/2$ amplitude levels of the AM-PSK carrier signal. The demodulator at the receiving end completely disregards phase reversals and detects only the envelope of the carrier. Since, however, the signal follows the polybinary rules, the bandwidth is compressed by a factor of $(b - 1)$ relative to the straight binary *AM*. For example, the system has only two carrier amplitude states for $b = 3$, yet it requires only half the bandwidth of conventional on-off AM. Conversely, for a fixed bandwidth, the polybinary AM-PSK for $b = 3$ has twice the bit capacity of binary AM. In fact, there is a 3-dB noise advantage for this case over binary AM, since AM-PSK still has two levels but requires half the bandwidth. The system should not be confused with AM-PSK multilevel systems [5]. For instance, in the case of four binary channels the AM-PSK multilevel system requires 16 states, all of which must be distinguished at the receiver; two separate detectors are needed to recover two amplitude levels and eight phases for a total of 16 possible conditions. In polybinary AM-PSK, to accommodate four binary channels, $b = 5$, but $(b + 1)/2$ calls for only three amplitude levels; the 180° phase reversal is ignored by the receiver.

Figure 7 shows the general implementation for the AM-PSK polybinary system. The coder again converts the original binary sequence a_n at A into the sequence d_n at B. Phase modulation is generated digitally to provide a binary two-phase modulated wave at C. The $H(\omega)$ bandpass filter performs the polybinary conversion and shaping so that at D the polybinary AM-PSK wave appears. The receiver is a conventional envelope detector followed by slicers. For such a digitally generated phase-modulated wave with one carrier cycle of unit amplitude per bit and 0° and 180° shift at C, the spectral density at point D can be represented by (7) with

$$G(f) = T \frac{(\sin \pi f T/2)^2}{\pi f T/2} \qquad (16)$$

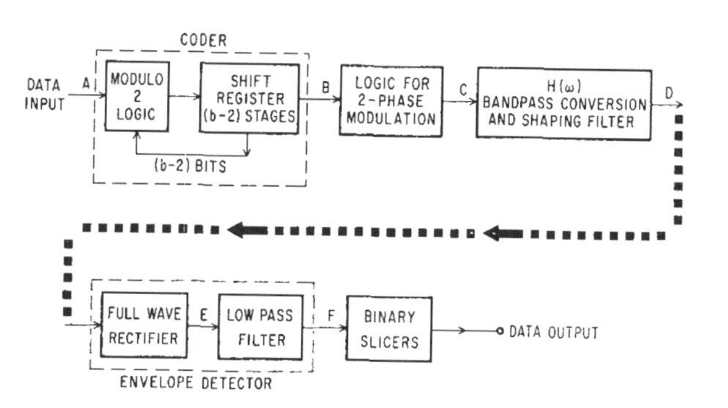

Fig. 7. AM-PSK polybinary system.

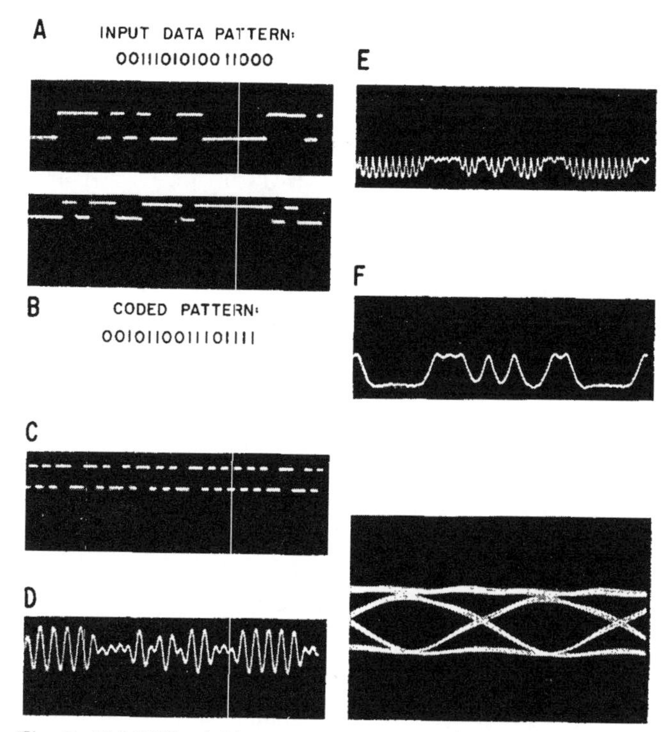

Fig. 8. AM-PSK polybinary waveforms with $b = 3$ at 2400 bits per second.

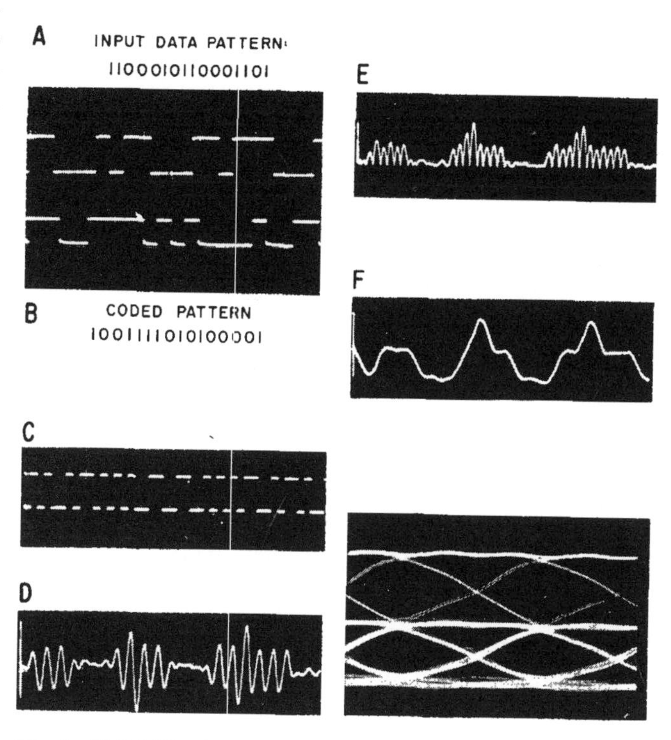

Fig. 9. AM-PSK polybinary waveforms with $b = 5$ at 2400 bits per second.

If a unit amplitude sine wave carrier with n-cycles per bit were employed in the generation of the phase modulation, the corresponding $G(f)$ in (7) would be

$$G(f) = \frac{T}{2\pi n} \frac{\sin \pi f T}{1 - (fT/n)^2} \tag{17}$$

The system was implemented for $b = 3$ and 5 at 2400 b/s with carrier at 2400 Hz. For $b = 3$ the band-pass filter attenuation increases rapidly above 3000 Hz and below 1800 Hz, and for $b = 5$ the bandwidth of this filter is reduced by approximately a factor of two. Figure 8 shows the AM-PSK waveforms for a 16-bit pattern and $b = 3$. The letter designations correspond to those in Fig. 7. The last waveform represents an eye pattern for a random input. Similarly, Fig. 9 shows the AM-PSK waveforms for $b = 5$ with the same designations. It should be noted that the 3-level pattern in F of Fig. 9 as well as the eye pattern for a random binary input are not duobinary [3] in spite of the fact that MARKS are at the center level and SPACES at the extreme levels.

VII. Conclusions

The correlative digital communication techniques described in this paper indicate that high data transmission rates per cycle of bandwidth can be achieved with much less SNR penalty compared to the present multilevel techniques. Moreover, the correlation properties of the coded wave, representing the binary message, can be used to detect errors without introducing redundant digits into the original data stream. Signals of this type can be easily generated with or without dc and low frequency components. A unique combination with AM-PSK results in a technique which is superior to other digital modulation techniques, such as, e.g., a multilevel AM-PSK combination. Finally, the circuit implementation is straightforward and simple.

Appendix I
Spectral Densities of Polybinary and Polybipolar Sequences

The continuous component of the spectral density [6] of a pulse train is

$$W(f) = \frac{1}{T} |G(f)|^2 \left\{ R(0) - m_1^2 + 2 \sum_{k=1}^{\infty} [R(k) - m_1^2] \cos 2\pi k f T \right\} \tag{18}$$

where $R(k) = \text{ave}(d_n\, d_{n+k}) =$ autocovariance of the binary sequence d_n. The binary sequences a_n and d_n, consisting of 1's and 0's, as well as their relationship are defined in Section II and can be expressed as

$$a_j = \text{mod } 2 \sum_{i=0}^{b-2} d_{j+i} \tag{19}$$

where a_j is the jth digit of sequence a_n. Also,

$$m_1 = \text{ave}(d_n) = 1/2 \tag{20}$$

$$R(0) = \text{ave}(d_n^2) = 1/2 \tag{21}$$

To evaluate $R(k)$ of sequence d_n, the probabilities of patterns with $(k + 1)$ digits will be considered. The probability of each pattern is 2^{2-b} multiplied by the product of the conditional probabilities of each of the remaining $(k + 3 - b)$ digits. This is in agreement with the physical process of generating sequence d_n, since the first $(b - 2)$ digits in the shift register of Fig. 3 are arbitrary with equal likelihood for 1 and 0. The product of the remaining $(k + 3 - b)$ conditional probabilities equals the product of the probabilities of MARKS and SPACES for the pattern in sequence a_n which corresponds to the pattern in sequence d_n. Such probabilities are denoted, respectively, by p and q, where $q = 1 - p$; and they are represented in sequence a_n by the symbols 1 and 0. The only binary patterns in sequence d_n which contribute to nonzero terms in $R(k)$ are those that start and end with binary 1. For such an arrangement we seek the relations, if any, which are imposed upon the sequence a_n by the value of the modulo 2 sum of the first and last digits of sequence d_n; i.e., by the value of $d_0 \oplus d_k$, where $\oplus$ indicates modulo 2 addition. When $b = 3$, a relation exists for any value of k and amounts merely to

$$\text{mod } 2 \sum_{i=0}^{k-1} a_i = d_0 \oplus d_k \tag{22}$$

When $b > 3$, a relation exists only when k is an integral multiple of $(b - 1)$, i.e., $k = n(b - 1)$, and is

$$\text{mod } 2 \sum_{i=0}^{n} [a_{i(b-1)} + a_{i(b-1)+1}] = d_0 \oplus d_k \tag{23}$$

Equation (23) is represented by the summation pattern in Fig. 10. Each horizontal line segment represents $(b - 1)$ digits of sequence d_n necessary to form a digit in sequence a_n in accordance with (19). The horizontal lines are summed up along the vertical axis corresponding to the right-hand side of (23). Since addition is modulo 2, all parallel line segments cancel leaving only single line segments corresponding to d_0 and $d_{n(b-1)}$. It follows from the pattern in Fig. 10 that unless the number of digits in a pattern of sequence d_n is a multiple of $(b - 1)$, it is not possible to obtain cancellation of all but the first and last digits. The summation of specific digits of sequence a_n, taken vertically, corresponds to the left-hand side of (23). The case of $b = 3$ in Fig. 10 is unique, since all, rather that the selected digits of sequence a_n, are summed up.

The case when $k = n(b - 1)$ is first considered. Since the only patterns of interest in sequence d_n are such that $d_0 = d_k = 1$, (23) is zero, and the positions of sequence a_n indicated by (23) have an even number of 1's. There are $2k/(b - 1)$ such positions. Moreover, sequence d_n has 2^{k-1} distinct patterns and sequence a_n only 2^{k+2-b} out of a possible 2^{k+3-b}. But each pattern in sequence d_n has a corresponding one in sequence a_n, therefore sequence a_n has $(2^{k-1})(2^{k+2-b})^{-1} = 2^{b-3}$ identical sets. Based on previous considerations, binary 1 in sequence a_n symbolizes the probability of MARK and binary 0 the probability of SPACE. Remembering that the product of the end digits in patterns of sequence d_n is 1, the autocovariance of sequence d_n for $k = n(b - 1)$ is

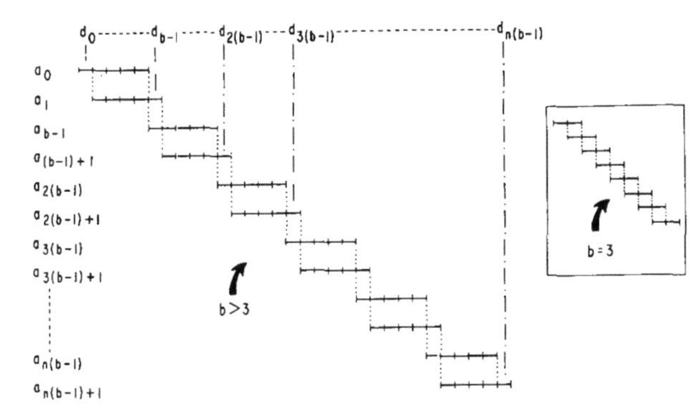

Fig. 10. Pattern diagrams for sequences a_n and d_n.

$$R(k) = (1/2)^{b-2}(2^{b-3}) \sum_{i=even}^{\frac{2k}{b-1}} \binom{\frac{2k}{b-1}}{i} p^i (q)^{\frac{2k}{b-1} - i} \tag{24}$$

The remaining $(k + 3 - b) - 2k/(b - 1)$ positions in sequence a_n form all possible binary patterns, and the corresponding summation yields unity in (24). When $k \neq n(b - 1)$and $b > 3$, all possible binary combinations of sequence a_n exist, and there are $(2^{k-1})(2^{k+3-b})^{-1} = 2^{b-4}$ identical sets each of which has all possible binary patterns. Hence, for this case the autocovariance of sequence d_n is

$$R(k) = (1/2)^{b-2}(2^{b-4}) = 1/4 \tag{25}$$

Based on (22), the case of $b = 3$ is included in (24).

Next, (20), (21), (24), and (25) are substituted into (18) to yield the spectral density of the binary sequence d_n

$$\left.\begin{aligned} W_2(f) &= \frac{1}{T} |G(f)|^2 pqZ_1 \text{ for } b = 3 \\ &= \frac{1}{T} |G(f)|^2 pqZ_2 \text{ for } b > 3 \end{aligned}\right\} \tag{26}$$

where p is the probability of MARK (binary 1) in sequence a_n, and Z_1 and Z_2 are expressed in (5) and (6).

The polybinary sequence p_n is obtained from sequence d_n by adding algebraically consecutive groups of $(b - 1)$ digits. This is equivalent to multiplying $W_2(f)$ in (26) by

$$\left| \sum_{m=0}^{b-2} e^{-jm\omega T} \right|^2 = \left[\frac{\sin (b-1)\pi f T}{\sin \pi f T} \right]^2 \tag{27}$$

so that the spectral density of sequence p_n is

$$W_3(f) = W_2(f) \left[\frac{\sin (b-1)\pi f T}{\sin \pi f T} \right]^2 \tag{28}$$

Similarly, the polybipolar spectral density is obtained by adding consecutively the first $(b - 1)/2$ digits of sequence d_n and subtracting the remaining $(b - 1)/2$ digits, so that $W_2(f)$ in (26) is multiplied by

$$\left| \sum_{m=0}^{\frac{b-3}{2}} e^{-jm\omega T} - \sum_{r=\frac{b-1}{2}}^{b-2} e^{-jr\omega T} \right|^2 = \frac{\left[\sin \left(\frac{b-1}{2} \right) \pi f T \right]^4}{(\sin \pi f T)^2} \tag{29}$$

and the spectral density of sequence g_n is

$$W_4(f) = W_2(f) \frac{\left[\sin \left(\frac{b-1}{2} \right) \pi f T \right]^4}{(\sin \pi f T)^2} \tag{30}$$

Appendix II
Probability of Error for Baseband Polybinary and Polybipolar Signals

In a b-level polybinary system with a peak voltage of A volts, the power dissipated in a one-ohm resistor for the kth level is

$$\left(\frac{Ak}{b-1} \right)^2 \tag{31}$$

Assuming MARK and SPACE being equally likely or $p = q = 1/2$, the probability p_k of the kth level in the polybinary system is

$$p_k = (1/2)^{b-1} \binom{b-1}{k} \tag{32}$$

where $k = 0, 1, 2, \ldots, b - 1$.

Similarly, in the polybipolar system the probability of the jth level is

$$p_j = (1/2)^{b-1} \sum_{i=j}^{\frac{b-1}{2}} \binom{\frac{b-1}{2}}{i} \binom{\frac{b-1}{2}}{i-j} \tag{33}$$

where $j = 0, \pm 1, \pm 2, \ldots, \pm (b - 1)/2$. Equation (33) reduces to (32) by using the identity (see [7])

$$\sum_{i=j}^{\frac{b-1}{2}} \binom{\frac{b-1}{2}}{i} \binom{\frac{b-1}{2}}{i-j} = \binom{b-1}{j + \frac{b-1}{2}} \tag{34}$$

and changing the variable $j + (b - 1)/2$ to k so that the results are applicable to both polybinary and polybipolar systems. From (31) and (32) the average signal power S in a b-level system is

$$S = \sum_{k=0}^{b-1} (1/2)^{b-1} \binom{b-1}{k} \left(\frac{Ak}{b-1} \right)^2 = \frac{A^2 b}{4(b-1)} \tag{35}$$

The key point in calculating the probability of error is that the only time an error occurs is when an even-numbered amplitude level is changed into an odd-numbered level, or vice versa. Hence the probability of error is the summation of the probabilities of noise being present in specific amplitude zones when the kth level is present. To illustrate this point, let the distance between any signaling level and the adjacent slicing level be α, where $\alpha = A/2(b - 1)$. If, for example, the zero level is present, only noise in the amplitude zones α to 3α, 5α to 7α, etc., will result in an error. Assuming additive Gaussian noise with noise power $N = \sigma^2$ and signal power given by (35), the probability of error P_e is

$$P_e = 1/2 + \sum_{k=0}^{b-1} p_k Y_k \tag{36}$$

where p_k is defined in (32) and

$$Y_k = 1/2 \sum_{i=1}^{b-k-1} (-1)^i \text{erf}\left\{(2i - 1)\left[\frac{S/N}{2b(b-1)}\right]^{1/2}\right\}$$

$$+ 1/2 \sum_{i=1}^{k} (-1)^i \text{erf}\left\{(2i - 1)\left[\frac{S/N}{2b(b-1)}\right]^{1/2}\right\} \quad (37)$$

$$\text{erf}(x) = \frac{2}{\sqrt{\pi}} \int_0^x e^{-t^2}\, dt \quad (38)$$

After simplification of (36) and (37)

$$P_e = 1/2 + 2^{1-b} \sum_{k=1}^{b-1} \binom{b-1}{k} \sum_{i=1}^{k} (-1)^i \text{erf}$$

$$\cdot \left\{(2i - 1)\left[\frac{S/N}{2b(b-1)}\right]^{1/2}\right\} \quad (39)$$

If E/N_0 represents the ratio of the average signal power per bit to the noise power density, then

$$E/N_0 = SWT/N\ (b - 1).$$

Here $1/T$ is the speed in b/s for the binary channel, and the noise bandwidth W Hz is assumed to equal the frequency of the first zero when signaling at this rate so that $WT = 1$ and

$$\frac{E}{N_0} = \frac{S}{N(b-1)} \quad (40)$$

Finally, combining (39) and (40), the probability of error vs. the normalized SNR is

$$P_e = 1/2 + 2^{1-b} \sum_{k=1}^{b-1} \binom{b-1}{k} \sum_{i=1}^{k} (-1)^i \text{erf}$$

$$\cdot \left\{(2i - 1)\left[\frac{E}{2bN_0}\right]^{1/2}\right\} \quad (41)$$

References

[1] Shannon, C. E., A mathematical theory of communication, *Bell Sys. Tech. J.*, vol 27, Oct 1948, pp 623–656.

[2] Lebow, I. L., et al., Application of sequential decoding to high-rate data communication on a telephone line, *IEEE Trans. on Information Theory (Correspondence)*, vol IT-9, Apr 1963, pp 124–126.

[3] Lender, A., The duobinary technique for high speed data transmission, *IEEE Trans. on Communication and Electronics*, vol 82, May 1963, pp 214–218.

[4] Aaron, M. R., PCM transmission in the exchange plant, *Bell Sys. Tech. J.*, vol 41, Jan 1962, pp 99–141.

[5] Hancock, J. C., and R. W. Lucky, Performance of combined amplitude and phase-modulated communication systems, *IRE Trans. on Communication Systems*, vol CS-8, Dec 1960, pp 232–236.

[6] Bennett, W. R., Statistics of regenerative digital transmission, *Bell Sys. Tech. J.*, vol 37, Nov 1958, pp 1501–1543.

[7] Ryshik, I. M., and I. S. Gradstein, Tables of series, products and integrals. Berlin: Veb Deutscher Verlag der Wissenschaften, 1963, p 4.

Characterization of Randomly Time-Variant Linear Channels

PHILIP A. BELLO, SENIOR MEMBER, IEEE

Summary—This paper is concerned with various aspects of the characterization of randomly time-variant linear channels. At the outset it is demonstrated that time-varying linear channels (or filters) may be characterized in an interesting symmetrical manner in time and frequency variables by arranging system functions in (time-frequency) dual pairs. Following this a statistical characterization of randomly time-variant linear channels is carried out in terms of correlation functions for the various system functions. These results are specialized by considering three classes of practically interesting channels. These are the wide-sense stationary (WSS) channel, the uncorrelated scattering (US) channel, and the wide-sense stationary uncorrelated scattering (WSSUS) channel. The WSS and US channels are shown to be (time-frequency) duals. Previous discussions of channel correlation functions and their relationships have dealt exclusively with the WSSUS channel. The point of view presented here of dealing with the dually related system functions and starting with the unrestricted linear channels is considerably more general and places in proper perspective previous results on the WSSUS channel. Some attention is given to the problem of characterizing radio channels. A model called the Quasi-WSSUS channel is presented to model the behavior of such channels.

All real-life channels and signals have an essentially finite number of degrees of freedom due to restrictions on time duration and bandwidth. This fact may be used to derive useful canonical channel models with the aid of sampling theorems and power series expansions. Several new canonical channel models are derived in this paper, some of which are dual to those of Kailath.

I. Introduction

During recent years there has been an increasing amount of attention given to the study of randomly time-variant linear channels. This attention has been motivated to a large extent by the advent of troposcatter, ionoscatter, chaff and moon communication links and radar astronomy systems. The determination of optimum modulation and demodulation techniques and the analytical determination of the efficacy of optimum and suboptimum communication (or radar) techniques for such channels depends heavily upon a satisfactory characterization of the transmission channel. Thus, the characterization of randomly time-variant linear channels is of some interest.

The characterization of time-variant linear filters (whether random or not) in terms of system functions received its first general analytical treatment by Zadeh,[1] who introduced the Time-Variant Transfer Function and the Bi-Frequency Function as frequency domain methods of characterizing time-variant linear filters to complement the time-variant impulse response which is a time domain method of characterization. Further interesting work on the characterization of time-varying linear filters in terms of system functions has been done by Kailath,[2] who has pointed out that a third type of impulse response may be defined in addition to the two already used for time-variant linear filters. He has defined single and double Fourier transforms of these impulse responses in order to demonstrate that certain variables may be identified with frequencies at the filter input and output and certain variables may be identified with the rate of variation of the filter. However, (excepting the Time-Variant Transfer Function) only the impulse responses and their double Fourier transforms were demonstrated to be system functions; *i.e.*, filter input-output relations were derived which used only impulse responses and their double Fourier transforms.

In Section II we demonstrate that time-varying linear channels (or filters) may be characterized in an interesting symmetrical manner in time and frequency variables by arranging system functions in (time-frequency) dual pairs.[3] Most of these system functions (which include, among others, those introduced by Zadeh and Kailath) are shown to imply circuit model interpretations or representations of the time-varying linear channels. The relationship between these system functions is demonstrated in a simple way with the aid of a graph involving duality and Fourier transformations.

When the filter becomes randomly time-variant the various system functions become random processes. An exact statistical characterization of a randomly time-variant linear channel in terms of multidimensional probability density distributions for system functions, while necessary for some theoretical investigations, presupposes more knowledge than is likely to be available in physical situations. A less ambitious but more practical goal involves a statistical characterization in terms of correlation functions for the various system functions, since knowledge of these correlation functions allows a

Received May 20, 1963. The work reported in this paper was performed, in part, under Subcontract 480.117D with ITT Communications Systems, Inc., Paramus, N. J.

The author is with ADCOM, Inc., Cambridge, Mass.

[1] L. A. Zadeh, "Frequency analysis of variable networks," Proc. IRE, vol. 38, pp. 291–299; March, 1950.

[2] T. Kailath, "Sampling Models for Linear Time-Variant Filters," M.I.T. Research Lab. of Electronics, Cambridge, Mass., Rept. No. 352; May 25, 1959.

[3] P. A. Bello, "Time-frequency duality," IEEE Trans. on Information Theory, vol. IT-10, pp. 18–33; January, 1964. Section V-D of this paper is essentially identical to Section II of the present one. *Note added in proof*: Since the present material has been accepted for publication the author has discovered that A. J. Gersho ["Characterization of time-varying linear systems," Proc. IEEE, (*Correspondence*), vol. 51, p. 238; January, 1963] also has determined a symmetrical formulation of system functions for time-variant linear channels. His formulation omits the Delay-Doppler Spread and Doppler-Delay Spread Functions and two Kernel System Functions.

Reprinted from *IEEE Transactions on Communications Systems,* December 1963.

The Best of the Best. Edited by W. H. Tranter, D. P. Taylor, R. E. Ziemer, N. F. Maxemchuk, and J. W. Mark.

determination of the autocorrelation function of the channel output.

In Section IV we define and determine relationships between the correlation functions of the various system functions for the general randomly time-variant linear channel. These results are specialized by considering three classes of practically interesting channels. These are the WSS channel, the US channel, and the WSSUS channel. The WSS and US channels are shown to be (time-frequency) duals.

Previous discussions[4-6] of channel correlation functions and their relationships have dealt exclusively with the WSSUS channel. Our point of view dealing with the dually related system functions and starting with the unrestricted linear channel is considerably more general and places in proper perspective previous results on the WSSUS channel.

Virtually all radio transmission media may be regarded as randomly time-variant linear channels. In the case of the transmission of digital signals over radio transmission media certain simplifications may be effected in channel characterization when the channel contains very slow fluctuations superimposed upon more rapid fluctuations, the latter of which exhibit an approximate statistical stationarity. In Section V we introduce the quasiwide-sense stationary uncorrelated scattering (QWSSUS) channel as a means for characterizing such channels.

All real-life channels and signals have an essentially finite number of degrees of freedom due to restrictions on time duration and bandwidth. This fact has been exploited by Kailath[2] to derive canonical channel models for the cases in which the channel band-limits signals at its input or output and in which the channel impulse response is time-limited. With the aid of the dual system functions derived in Section III we derive new canonic sampling models in Section V, some of which may be identified as dual to those of Kailath. As might be expected, these dual models are particularly useful under the dual time- frequency constraints, namely when the input or output time functions are time-limited or when the channel fading rate is band-limited. In addition we derive two new dually related canonic channel models, called f-power series and t-power series models. The f-power series model is of particular use in evaluating the effect of frequency selective fading on a signal whose bandwidth is less than the correlation bandwidth of a scatter channel. The t-power series model will be of use in the dual situation, *i.e.*, in evaluating the effect of (time-selective) fading on a pulse whose duration is less than the correlation time constant of a scatter channel.

II. Complex Envelopes

A process $x(t)$ whose spectral components cover a band of frequencies which is small compared to any frequency in the band may be expressed as

$$x(t) = Re\ \{\gamma(t)e^{j\omega_c t}\} \tag{1}$$

where $Re\{\ \}$ is the usual real part notation, ω_c is some (angular) frequency within the band and $\gamma(t)$ is the complex envelope of $x(t)$. This name for $\gamma(t)$ derives from the fact that the magnitude of $\gamma(t)$ is the conventional envelope of $x(t)$ while the angle of $\gamma(t)$ is the conventional phase of $x(t)$ measured with respect to carrier phase $\omega_c t$. The non-narrow-band case may be handled with the complex notation also by the use of Hilbert transforms.[7-10] However, the complex envelope will then no longer have the simple interpretation described above. Complex envelope notation will be used extensively for the remainder of this paper. However, it should be understood that there is always implied the existence of a center or reference frequency ω_c which via an equation such as (1) converts the complex time functions under discussion into physical narrow-band signals.

When dealing with problems in which there are wide-band filters (time-variant included) whose inputs and outputs are narrow-band (when expressed with reference to the same center frequency), it is possible to replace these filters with equivalent narrow-band filters which leave the input-output relations invariant. This fact becomes obvious when it is realized that by preceding and following a wide-band filter with narrow-band filters which have flat transfer functions over the range of input and output frequencies of interest, one produces a composite filter which is narrow-band and of course cannot change the input-output relations for the properly restricted class of input and output narrow-band signals. It is readily demonstrated that (except for an unimportant constant of one-half) the complex envelope of a narrow-band signal at the output of a narrow-band filter due to a narrow-band input may be obtained by passing the complex envelope of the input through an "equivalent" low-pass filter whose impulse response is just equal to the complex envelope of the narrow-band filter impulse response.

In defining the autocorrelation function of the complex envelope of a random process a certain difficulty appears that is not generally appreciated, namely, that *two* autocorrelation functions are needed in order to uniquely specify the autocorrelation function of the original real process. This fact is demonstrated by direct calculation

[4] T. Hagfors, "Some properties of radio waves reflected from the moon and their relation to the lunar surface," *J. Geophys. Res.*, vol. 66, pp. 777–785; March, 1961.

[5] R. Price and P. E. Green, Jr., "Signal Processing in Radar Astronomy," M.I.T. Lincoln Lab., Lexington, Mass., Rept. No. 234; October 6, 1960.

[6] P. E. Green, Jr., "Radar measurement of target characteristics," in "Radar Astronomy," J. V. Harrington and J. V. Evans, Eds., Chapter 9; to be published.

[7] P. M. Woodward, "Probability and Information Theory," McGraw-Hill Book Co., Inc., New York, N. Y.; 1953.

[8] J. Dugundji, "Envelopes and pre-envelopes of real waveforms," IRE Trans. on Information Theory, vol. IT-4, pp. 53–57; March, 1958.

[9] R. Arens, "Complex processes for envelopes of normal noise," IRE Trans. on Information Theory, vol. IT-3, pp. 204–207; September, 1957.

[10] D. Gabor, "Theory of communications," *J. IEE*, Part III, vol. 93, pp. 429–457; November, 1946.

of the autocorrelation function of $x(t)$, (1), as

$$\overline{x(t)x(s)} = \tfrac{1}{2} Re \{\overline{\gamma^*(t)\gamma(s)}\, e^{i\omega_c(s-t)}\} + \tfrac{1}{2} Re \{\overline{\gamma(t)\gamma(s)}\, e^{i\omega_c(s+t)}\}. \quad (2)$$

Thus the two autocorrelation functions

$$R_\gamma(t, s) = \overline{\gamma^*(t)\gamma(s)}$$
$$\tilde{\tilde{R}}_\gamma(t, s) = \overline{\gamma(t)\gamma(s)} \quad (3)$$

are needed to specify the autocorrelation function of the real process. Fortunately, in most applications the narrow-band process is so constituted that

$$\tilde{\tilde{R}}_\gamma(t, s) = 0. \quad (4)$$

In fact, from (2), one may readily deduce that (4) is necessary if $x(t)$ is to be wide-sense stationary.

A simple physical test of $x(t)$ (deterministic components removed) to determine whether (4) is satisfied is to multiply it by itself delayed and examine the sum frequency component for the presence of a deterministic component. According to (2) the complex amplitude of this component is $\frac{1}{2}\tilde{\tilde{R}}_\gamma(t, s)$, so that the presence of a deterministic component would mean that (4) is violated. In the subsequent discussion involving complex envelopes we shall deal only with that form of autocorrelation function which involves the conjugate under the expectation sign. It should be kept in mind, however, that an analogous discussion applies for $\tilde{\tilde{R}}_\gamma(t, s)$ in those cases where it is nonzero.

The above discussion of complex envelopes, equivalent noises and equivalent filters is supplied as a physical justification for our subsequent use of "low-pass" complex time functions, complex white noise and low-pass filters with complex impulse responses.

III. System Functions for Time-Variant Linear Filters

A. Dual Operators and Kernel System Functions

The concept of "time-frequency" duality is discussed at some length by Bello.[3] For the purposes of this section it will be sufficient to define the concept of dual operators.

A device which processes communication signals may be thought of in mathematical terms as an operator which transforms input signals into output signals. The inputs and outputs of such a device may be described in either the time or frequency domain according to convenience. Since either time or frequency domain descriptions may be used at the input and output, a two-terminal device (a single-input single-output device) may be described by any one of four operators. If we define time and frequency domain descriptions of processes as dual descriptions, then these four operators may be grouped into dual pairs with the aid of the following definition:

Two operators associated with a particular two-terminal-pair device are defined as duals when dual descriptions are used for corresponding inputs and outputs.

If $z(t)$, $Z(f)$ denote the input time function and spectrum, and $w(t)$, $W(f)$ denote the output time function and spectrum of a device, then the four possible operators are described by the equations

$$w(t) = O_{tt}[z(t)] \qquad W(f) = O_{ff}[Z(f)]$$
$$w(t) = O_{tf}[Z(f)] \qquad W(f) = O_{ft}[z(t)] \quad (5)$$

where the operator pairs O_{tt}, O_{ff} and O_{tf}, O_{ft} individually consist of dual operators.

In the case of a linear device, such as a linear time-variant channel, the four equations in (5) may be formally expressed[11] as linear integral operators with associated kernels; *i.e.*,

$$w(t) = \int z(s)K_1(t, s)\, ds \qquad W(f) = \int Z(l)K_2(f, l)\, dl \quad (6)$$

$$w(t) = \int Z(f)K_3(t, f)\, df \qquad W(f) = \int z(t)K_4(f, t)\, dt. \quad (7)$$

These kernels are, in effect, system functions and we shall call them *kernel* system functions to distinguish them from other classes of system functions to be described. It is clear that the system function pairs K_1, K_2 and K_3, K_4 may be considered as dual system function pairs.

The system functions $K_1(t, s)$ and $K_2(f, l)$ may be recognized as the Time-Variant Impulse Response and the Bi-Frequency Function respectively, used by Zadeh.[1] The system functions $K_3(t, f)$ and $K_4(f, t)$ have not been defined previously. Without difficulty it may be established that $K_1(t, s)$ and $K_2(f, l)$, besides being duals, are double Fourier transform pairs, and similarly that the dual pairs $K_3(t, f)$ and $K_4(f, t)$ are double Fourier transform pairs. Also $K_1(t, s)$ and $K_3(t, f)$ are single Fourier transform pairs with t considered as a parameter, while $K_2(f, l)$ and $K_4(f, t)$ are single Fourier transform pairs with f considered as a parameter. It is worth noting that K_1, K_2 and K_3, K_4 are the only dual pairs of system functions among those to be presented which are related directly as double Fourier transform pairs.

The kernel system functions have simple physical interpretations in terms of the response of the channel to impulses and cissoids. Thus, it is readily determined that if the channel is excited with a unit impulse at $t = s$, the resulting channel output is the time function $K_1(t, s)$ with spectrum $K_4(f, s)$, while if the channel is excited with the cissoid $e^{i2\pi lt}$ (*i.e.*, frequency impulse at $f = l$), the resulting channel output is the time function $K_3(t, l)$ with spectrum $K_2(f, l)$.

The present discussion of kernel system functions has been included primarily in the interest of making our discussion of system functions as complete as possible, and in clarifying our subsequent discussion of system functions. We shall actually make little use of the kernel system functions in the remainder of this paper, primarily because we are interested in circuit model descrip-

[11] In order to include linear differential operators one must assume that the kernels may include singularity functions.

tions of the time-variant linear channel and the kernel system functions do not lend themselves readily to such phenomenological descriptions.

B. Delay-Spread and Doppler-Spread Functions

From a strictly mathematical point of view the kernel system functions are sufficient to describe the time-frequency input-output relations for a time-variant linear channel. From a physical intuitive point of view they are not as satisfactory, since they do not readily allow one to grasp by inspection the way in which the time-variant filter affects input signals to produce output signals. Section IIIC we will be concerned with system functions which, via circuit model analogies, provide a somewhat more physical interpretation of the action of the linear time-variant channel.

Consider first the following input-output relationship for a linear time-variant channel obtained from the first equation in (6) by the transformation $s = t - \xi$:

$$w(t) = \int z(t - \xi)g(t, \xi)\, d\xi \tag{8}$$

where

$$g(t, \xi) = K_1(t, t - \xi). \tag{9}$$

Eq. (8) leads to a physical picture of the channel as a continuum of nonmoving scintillating scatterers, with with $g(t, \xi)d\xi$ equal to the (complex) modulation produced by hypothetical elemental "scatterers" that provide delays in the range $(\xi, \xi + d\xi)$. Fig. 1 illustrates such a physical picture with the aid of a densely tapped delay line. Note that the input signal is first delayed and then multiplied by the differential scattering gain. We shall call $g(t, \xi)$ the Input Delay-Spread Function to distinguish it from another system function called the Output Delay-Spread Function, to be described below, which leads to a channel representation similar to $g(t, \xi)$ except that the delay occurs on the output side of the channel (and the multiplication on the input).

If we consider $z(t)$ to be first multiplied by a differential gain function $h(t, \xi)d\xi$ and then delayed by ξ with a continuum of ξ values, we obtain the input-output relationship

$$w(t) = \int z(t - \xi)h(t - \xi, \xi)\, d\xi \tag{10}$$

and the circuit model representation of Fig. 2. By comparing (10) with (8) and (6) we quickly find that

$$h(t, \xi) = K_1(t + \xi, t) = g(t + \xi, \xi). \tag{11}$$

From the fact that $K_1(t, s)$ is the channel response at time t due to a unit impulse input at $t = s$, it is seen from (7) and (11) that $g(t, \xi)$ may be interpreted as the response at time t to a unit impulse input ξ seconds in the past, and $h(t, \xi)$ may be interpreted as the response ξ seconds in the future to a unit impulse input at time t. Since a physical channel (without internal sources) may not have an output before the input arrives, $K_1(t, s)$ must vanish for

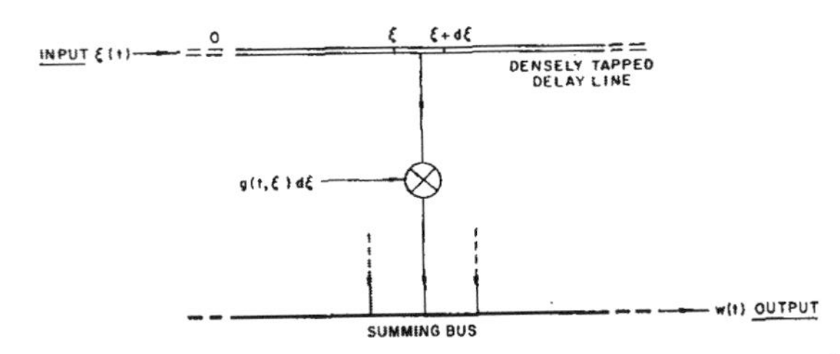

Fig. 1—A differential circuit model representation for linear time-variant channels using the Input Delay-Spread Function.

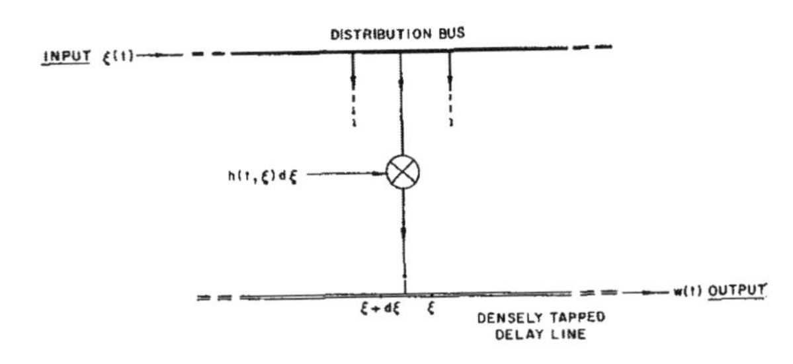

Fig. 2—A differential circuit model representation for linear time-variant channels using the Output Delay-Spread Function.

$t < s$ and $g(t, \xi)$, $h(t, \xi)$ must vanish for $\xi < 0$. These physical realizability conditions may be explicitly indicated by appropriate limits on the integrals defining input-output relations. However, for simplicity of presentation we will assume that the integral limits are $(-\infty, \infty)$, with the integrand being taken as zero in the appropriate intervals to assure physical realizability.

An entirely dual and just as general channel characterization exists in terms of frequency variables by employing the Input and Output Doppler-Spread Functions, the system functions which are the (time-frequency) duals of the Input and Output Delay-Spread Functions, respectively. Consider first the dual of the Input Delay-Spread Function. Such a system function must relate the channel output spectrum to the channel input spectrum in a manner identical in form to the way $g(t, \xi)$ relates the input and output time functions. This dual characterization involves a representation of the output spectrum $W(f)$ as a superposition of infinitesimal Doppler-shifted (the dual of delayed) and filtered (the dual of modulated) replicas of the input spectrum $Z(f)$. Thus we have

$$W(f) = \int Z(f - \nu)H(f, \nu)\, df \tag{12}$$

where $H(f, \nu)$ is the Input Doppler-Spread Function.

Eq. (12) may be interpreted physically with the aid of a model dual to that in Fig. 1. To construct such a dual it is necessary to note that the dual of a tapped delay line is a "frequency conversion chain," *i.e.*, a string of frequency converters arranged so that the output of one converter is not only the input to the next converter but is also the "local" frequency shifted output. Fig. 3 illustrates such an interpretation of (12) using a "dense" frequency conversion chain. Note that the quantity $H(f, \nu)d\nu$ is to be interpreted as the transfer function associated with hypothetical Doppler-shifting elements that provide frequency shifts in the range $(\nu, \nu + d\nu)$.

By comparing (12) and the last equation in (6) we find that the Input Doppler-Spread Function is related to Zadeh's Bi-Frequency Function by

$$H(f, \nu) = K_2(f, f - \nu) \tag{13}$$

which is an equation dual to (9).

From (10) we deduce that the dual of the Output Delay-Spread Function must provide the input spectrum-output spectrum relationship

$$W(f) = \int Z(f - \nu)G(f, \nu)\, d\nu \tag{14}$$

where $G(f, \nu)$ is defined as the Output Doppler-Spread Function. Whereas the Input Doppler-Spread Function leads to a cascaded Doppler shifter-filter realization as indicated in Fig. 3, the Output Doppler-Spread Function leads to a cascaded filter-Doppler shifter realization as shown in Fig. 4. The quantity $G(f, \nu)d\nu$ is the transfer function of a hypothetical differential filter at the input which is associated with a Doppler shift of ν cps at the channel output.

By comparing (14) with (13) and (6) we find that

$$G(f, \nu) = K_2(f + \nu, f) = H(f + \nu, \nu), \tag{15}$$

which is a set of equations dual to (11). Since $K_2(f, l)$ is the value of the spectral response of the channel at a frequency f due to a cissoidal excitation of frequency l cps, it is quickly seen from (13) and (15) that $H(f, \nu)$ may be interpreted as the spectral response of the channel at f cps due to a cissoidal input ν cycles below f, and $G(f, \nu)$ may be interpreted as the spectral response of the channel at a frequency ν cps above the cissoidal input at the frequency f cps.

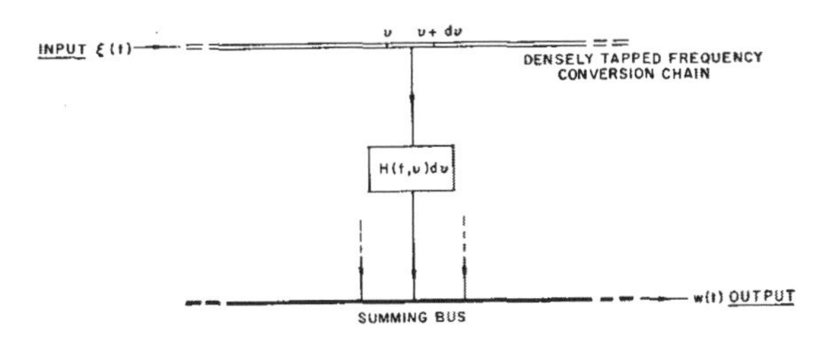

Fig. 3—A differential circuit model representation for linear time-variant channels using the Input Doppler-Spread Function.

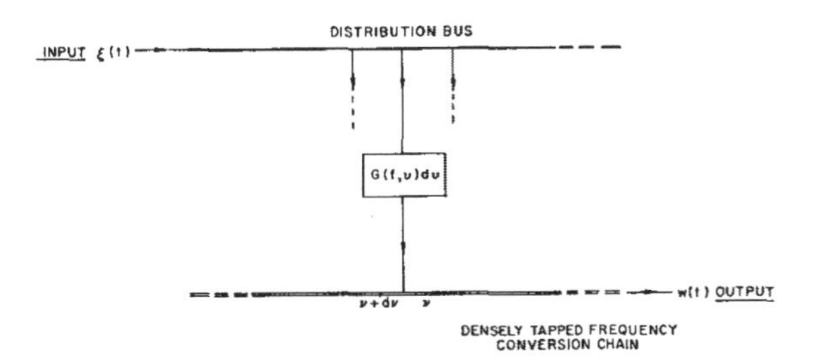

Fig. 4—A differential circuit model representation for linear time-variant channels using the Output Doppler-Spread Function.

C. Time-Variant Transfer Function and Frequency-Dependent Modulation Function

The characterizations of a time-variant channel in terms of the Delay-Spread Functions $g(t, \xi)$ and $h(t, \xi)$ and the Time-Variant Impulse Response $K_1(t, s)$ are strictly time domain approaches, while the characterizations in terms of the Doppler-Spread Functions $H(f, \nu)$ and $G(f, \nu)$ and the Bi-Frequency Function $K_2(f, l)$ are entirely frequency domain approaches. In the former cases the output time function is directly related to the input time function, while in the latter cases the output spectrum is directly related to the input spectrum. As discussed in Section III A and exemplified by the dual kernel system functions $K_3(t, f)$ and $K_4(f, t)$, two other approaches are possible. These involve an expression of the output time function directly in terms of the input spectrum in one case, and an expression of the output spectrum directly in terms of the input time function in the other. An example of the former approach was first introduced by Zadeh[1] with the aid of the Time-Variant Transfer Function.

In this section we will introduce a new system function called the Frequency-Dependent Modulation Function, which is the (time-frequency) dual of the Time-Variant Transfer Function. This system function relates the output spectrum to the input time function.

Assuming we have an input $z(t)$, which may be represented as a summation of infinitesimal cissoidal time functions, *i.e.*,

$$z(t) = \int Z(f)e^{j2\pi ft}\, df \tag{16}$$

where $Z(f)$ is the spectrum of $z(t)$, one may determine the channel output by superposing the separate responses to the infinitesimal cissoidal components. The response of the channel to the cissoidal time function exp $[j2\pi lt]$ (or spectral impulse $\delta(f - l)$) is given by [see (9)]

$$\int e^{j2\pi l(t-\xi)}g(t, \xi)\, d\xi = e^{j2\pi lt}T(l, t) \tag{17}$$

where

$$T(f, t) = \int e^{-j2\pi f\xi}g(t, \xi)\, d\xi \tag{18}$$

is the Fourier transform of the Input Delay-Spread Function with respect to the delay parameter. By superposition the network output is given by

$$w(t) = \int Z(f)T(f, t)e^{j2\pi ft}\, df. \tag{19}$$

Eq. (19) shows that even though the channel may be time-variant, one may determine the output by *exactly* the same frequency domain techniques as for time-variant (linear) channels. This involves, basically, a multiplication of the input spectrum by a system function followed by an inverse Fourier transformation with respect to the frequency variable. For time-variant channels, however, the system function is a function of the time variable. This explains use of the name Time-Variant Transfer Function to denote $T(f, t)$.

By using (14) to determine the spectrum of the response to the frequency impulse $\delta(f - l)$, and then inverse Fourier transforming to obtain the corresponding time response, it may be quickly determined that

$$T(f, t) = \int G(f, \nu)e^{i2\pi\nu t}\, d\nu, \qquad (20)$$

i.e., that the Time-Variant Transfer Function is the inverse Fourier transform of the Output Doppler-Spread Function with respect to the Doppler-shift variable.

Also, either by noting, as discussed in Section III A, that $K_3(t, l)$ may be interpreted as the channel response at time t to an excitation $e^{i2\pi lt}$, or by comparing (7) and (15), it is readily seen that

$$K_3(t, f) = e^{i2\pi ft}T(f, t). \qquad (21)$$

To develop the dual system function we assume that we have an input whose spectrum $Z(f)$ may be represented as a summation of infinitesimal cissoidal *frequency functions*; *i.e.*,

$$Z(f) = \int z(t)e^{-i2\pi ft}\, dt. \qquad (22)$$

The spectrum of the response of the channel to the cissoidal frequency function $\exp[-j2\pi fs]$ (*i.e.*, to the time function $\delta(t - s)$ whose spectrum is $\exp[-j2\pi fs]$) is given by [see (12)]

$$\int e^{-i2\pi s(f-\nu)}H(f, \nu)\, d\nu = e^{-i2\pi fs}M(s, f) \qquad (23)$$

where

$$M(t, f) = \int e^{i2\pi t\nu}H(f, \nu)\, d\nu \qquad (24)$$

is the Fourier transform of the Input Doppler-Spread Function with respect to the Doppler-shift variable. By superposition the network output spectrum is given by

$$W(f) = \int z(t)M(t, f)e^{-i2\pi ft}\, dt. \qquad (25)$$

Eq. (25) shows that, even though the channel may be a general time-variant linear filter, one may determine the output spectrum by exactly the same time domain techniques as for a channel which acts as a pure complex multiplier (or modulator). This involves, basically, a multiplication of the input time function by a complex time function characterizing the channel, followed by a Fourier transformation with respect to the time variable. For general time-variant channels, however, the complex multiplier is frequency-dependent. This explains our use of the name Frequency-Dependent Modulation Function to denote $M(t, f)$.

By using (6) to determine the time function response to the input $\delta(t - s)$ and then Fourier transforming to obtain the spectrum of the response, it may be quickly determined that

$$M(t, f) = \int e^{-i2\pi f\xi}h(t, \xi)\, d\xi; \qquad (26)$$

i.e., that the Frequency-Dependent Modulation Function is the Fourier transform of the Output Delay-Spread Function with respect to the delay variable.

Also, either by noting, as discussed in Section III A, that $K_4(f, s)$ is the spectrum of the channel response to an excitation $\delta(t - s)$, or by comparing (25) and (7), it is readily seen that

$$K_4(f, t) = e^{-i2\pi ft}M(t, f). \qquad (27)$$

In the case of time-invariant linear filters the transmission frequency characteristic of the filter can be determined by direct measurement as the cissoidal response or else indirectly as the spectrum of the impulse response. For time-variant linear filters these measurements procedures yield different results, as exemplified by the fact that $T(f, t)$, which corresponds to the cissoidal measurement, differs from $M(t, f)$ which corresponds to impulse response measurement followed by spectral analysis.

Moreover, as we have shown above, only $T(f, t)$ may properly be considered a transmission frequency characteristic, the proper interpretation of $M(t, f)$ being that of a channel "modulator."

D. Delay-Doppler-Spread and Doppler-Delay-Spread Functions

In Section III B it was demonstrated that any linear time-varying channel may be interpreted either as a continuum of nonmoving scintillating scatterers with the aid of the Delay-Spread Functions, or as a continuum of hypothetical Doppler-shifting elements with associated filters with the aid of the Doppler-Spread Functions. We demonstrate in this section that any linear time-varying channel may be represented as a continuum of elements which simultaneously provide both a corresponding delay and Doppler shift.

As in Section III B, we can consider system functions classified according to whether the corresponding phenomenological channel model has its delay operation or Doppler-shift operation at the channel input or output. Since delay and Doppler shift both occur in the model to be described, only two possibilities exist, *i.e.*, input-delay output-Doppler-shift and input-Doppler-shift output-delay, rather than the four possibilities of Section III B. To determine the system function corresponding to the input delay output Doppler-shift channel model, we express the Input Delay-Spread Function $g(t, \xi)$ as the inverse Fourier transform of its spectrum (where ξ is considered to be a fixed parameter), *i.e.*,

$$g(t, \xi) = \int U(\xi, \nu)e^{i2\pi\nu t}\, d\nu, \qquad (28)$$

and then use (28) in (8) to obtain the following input-output relationship:

$$w(t) = \iint z(t - \xi)e^{j2\pi\nu t}U(\xi, \nu)\, d\nu\, d\xi. \tag{29}$$

Examination of (29) shows that the output is represented as a sum of delayed and then Doppler-shifted elements, the element providing delays in the interval $(\xi, \xi + d\xi)$ and Doppler shifts in the interval $(\nu, \nu + d\nu)$ having a differential scattering amplitude $U(\xi, \nu)d\nu d\xi$. For this reason we call $U(\xi, \nu)$ the Delay-Doppler-Spread Function.

In an entirely analogous way, in order to determine the dual system function, *i.e.*, that corresponding to the input Doppler-shift output-delay channel model, we express the Input Doppler-Spread Function as a Fourier transform

$$H(f, \nu) = \int V(\nu, \xi)e^{-j2\pi\xi f}\, d\xi, \tag{30}$$

and then use (30) in (12) to obtain the input-output relationship

$$W(f) = \iint Z(f - \nu)e^{-j2\pi\xi f}V(\nu, \xi)\, d\xi\, d\nu. \tag{31}$$

Examination of (31) shows that the output is represented as a sum of Doppler-shifted and then delayed elements, the element providing Doppler shifts in the interval $(\nu, \nu + d\nu)$ and delays in the interval $(\xi, \xi + d\xi)$ having a differential scattering amplitude $V(\nu, \xi)d\xi\, d\nu$. For this reason we call $V(\nu, \xi)$ the Doppler-Delay-Spread Function.

If we Fourier transform both sides of (29) with respect to t and inverse Fourier transform both sides of (31) with respect to f we obtain the equations

$$W(f) = \iint Z(f - \nu)e^{-j2\pi\xi(f-\nu)}U(\xi, \nu)\, d\nu\, d\xi \tag{32}$$

and

$$w(t) = \iint z(t - \xi)e^{j2\pi(t-\xi)}V(\nu, \xi)\, d\xi\, d\nu. \tag{33}$$

A comparison of (31) and (32) or (29) and (33) reveals that $U(\xi, \nu)$ and $V(\nu, \xi)$ are simply related; *i.e.*,

$$U(\xi, \nu) = e^{-j2\pi\nu\xi}V(\nu, \xi). \tag{34}$$

If the integration with respect to ξ is carried out in (28) and the integration with respect to ν is carried out in (33), one finds that

$$h(t, \xi) = \int V(\nu, \xi)e^{-j2\pi\nu t}\, d\nu \tag{35}$$

and

$$G(f, \nu) = \int U(\xi, \nu)e^{-j2\pi\xi f}\, d\xi. \tag{36}$$

E. Relationship Between System Functions

At this point the reader may be somewhat bewildered by the variety of system functions that have been introduced. In addition to the four kernel system functions, we have discussed eight other system functions. The relationships between the kernel system functions are rather clearly outlined in Section III A. The relationships between the other eight system functions can be simply portrayed by grouping them according to duality and Fourier transform relationships. This grouping is illustrated in Fig. 5, in which the dashed line labeled D signifies that the system functions occupying mirror image positions with respect to the dashed line are duals, and the line labeled F signifies that the system functions at the terminals of the line are related by single Fourier transforms. Since the system functions involve two variables, any two system functions connected by an F must have a common variable which should be regarded as a fixed parameter in employing the Fourier transform relationships involving the other two variables. Note that one of these latter two variables is a time variable and the other is a frequency variable. To make the F notation unique we have employed the convention that in transforming from a time to a frequency variable a negative exponential is used in the Fourier integral, while in transforming from a frequency to a time variable a positive exponential is used.

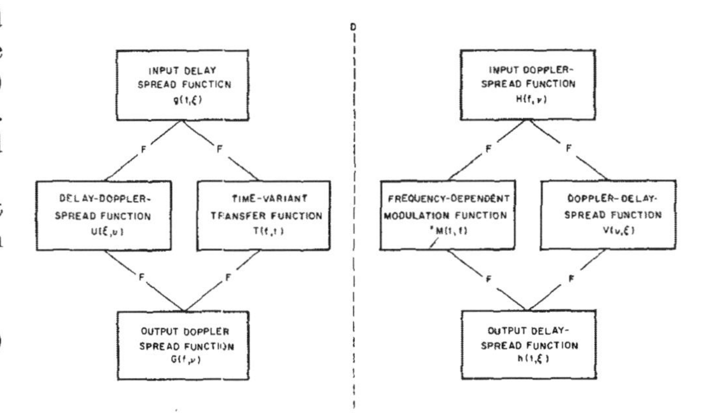

Fig. 5—Relationships between system functions for time-variant linear channels.

Examination of Fig. 5 reveals that the Time-Variant Transfer Function $T(f, t)$ and the Delay-Doppler-Spread Functions are double Fourier transforms; *i.e.*,

$$T(f, t) = \iint U(\xi, \nu)e^{-j2\pi\xi f}e^{j2\pi\nu t}\, d\xi\, d\nu. \tag{37}$$

Also the dual relationship exists. The Frequency-Dependent Modulation Function $M(t, f)$ and the Doppler-Delay-Spread Function are double Fourier transforms; *i.e.*,

$$M(t, f) = \iint V(\nu, \xi)e^{-j2\pi\xi f}e^{j2\pi\nu t}\, d\xi\, d\nu. \tag{38}$$

Since $T(f, t)$ and $M(t, f)$ are double Fourier transforms of system functions which differ only by the simple exponential factor $\exp[-j2\pi\nu\xi]$ [see (34)] it might be supposed that they also are related in a similar simple fashion. However, this is not the case. The analytic relationship between $T(f, t)$ and $M(t, f)$ is quickly obtained from (21) and (27) with the aid of the fact that $K_3(t, f)$ and $K_4(f, t)$ are double Fourier transform pairs. This relationship is

$$
\begin{aligned}
M(t, f) &= \iint T(f', t')e^{i2\pi(f-f')(t-t')}\, df'\, dt' \\
T(f, t) &= \iint M(t', f')e^{-i2\pi(f-f')(t-t')}\, df'\, dt'.
\end{aligned} \tag{39}
$$

$$
\begin{aligned}
\overline{g^*(t, \xi)g(s, \eta)} &= R_g(t, s; \xi, \eta) & \overline{H^*(f, \nu)H(l, \mu)} &= R_H(f, l; \nu, \mu) \\
\overline{T^*(f, t)T(l, s)} &= R_T(f, l; t, s) & \overline{M^*(t, f)M(s, l)} &= R_M(t, s; f, l) \\
\overline{G^*(f, \nu)G(l, \mu)} &= R_G(f, l; \nu, \mu) & \overline{h^*(t, \xi)h(s, \eta)} &= R_h(t, s; \xi, \eta) \\
\overline{U^*(\xi, \nu)U(\eta, \mu)} &= R_U(\xi, \eta; \nu, \mu) & \overline{V^*(\nu, \xi)V(\mu, \eta)} &= R_V(\nu, \mu; \xi, \eta)
\end{aligned} \tag{40}
$$

We also note from Fig. 5 that the Input Delay-Spread Function $g(t, \xi)$ and the Output Doppler-Spread Function $G(f, \nu)$ are double Fourier transforms. Also, the dual relationship exists; *i.e.*, the Output Delay-Spread Function $h(t, \xi)$ and the Input Doppler-Spread Function $H(f, \nu)$ are double Fourier transform pairs. Other analytic relationships between system functions are readily obtained by using (11) and (15) and the Fourier transform relationships indicated in Fig. 5.

IV. Channel Correlation Functions

When the channel is randomly time-variant the system functions discussed in Section III become stochastic processes. An exact statistical characterization of a randomly time-variant linear channel in terms of multidimensional probability density distributions for system functions, while necessary for some theoretical investigations, presupposes more knowledge than is likely to be available in physical situations. A less ambitious but more practical goal involves a statistical characterization in terms of correlation functions for the various system functions since (as will be shown below) these correlation functions allow a determination of the autocorrelation function of the channel output. In this section we will be concerned with defining correlation function for these system functions and showing their interrelationships. Special attention will be given to simplifications that result for certain classes of channels of practical interest.

In general, a randomly time-variant channel has a mixed deterministic and random behavior. Thus, for example, the Input Delay-Spread Function $g(t, \xi)$ may separated into the sum of a purely random part and a deterministic part [equal to the ensemble average of $g(t, \xi)$]. This separation implies a representation of the channel as the parallel combination of a deterministic channel and a purely random channel. For simplicity of discussion, we shall only be concerned in this paper with the correlation properties associated with the purely random part of the channel. Thus it should be understood in subsequent discussions of correlation functions that each of the system functions has a zero ensemble average.

A. General Case

We shall confine our discussion of channel correlation functions to the eight system functions shown in Fig. 5, since it is felt that these system functions provide a better picture of the operation of a time-varying linear channel than the kernel system functions.

The correlation functions for the system functions in Fig. 5 will be defined as follows: [see (40) above]

where correlation functions for dual system functions have been placed in the same row and correlation functions for Fourier-transform-related system functions have been placed in the same column.

It is readily appreciated that the relationships between correlation functions in any column are double and quadruple Fourier transform relationships since the corresponding system functions are related by single and double Fourier transforms, respectively. As an illustration, consider the derivation of the relationship between $R_g(t, s; \xi, \eta)$ and $R_T(f, l; t, s)$. The Fourier transform relationship between $g(t, \xi)$ and $T(f, t)$ is shown explicitly in (18). Using this equation we find that

$$T^*(f, t)T(l, s) = \iint g^*(t, \xi)g(s, \eta)e^{i2\pi(\xi f-\eta l)}\, d\xi\, d\eta. \tag{41}$$

Then, taking the ensemble average of both sides of (41) (and assuming the validity of interchanging the order of integration and ensemble averging), we find that

$$R_T(f, l; t, s) = \iint R_g(t, s; \xi, \eta)e^{i2\pi(\xi f-\eta l)}\, d\xi\, d\eta \tag{42}$$

and by inverting the Fourier transform relationship

$$R_g(t, s; \xi, \eta) = \iint R_T(f, l; t, s)e^{-i2\pi(\xi f-\eta l)}\, df dl. \tag{43}$$

If an identical procedure is followed to determine the other Fourier transform relationships between channel correlation functions, one finds that these relationships may be portrayed as shown in Fig. 6, wherein a double line labeled F indicates a double Fourier transform relationship between the connected correlation functions. The meaning of the dashed line labeled D is similar to the corresponding dashed line in Fig. 5, namely, the correlation functions which occupy mirror image positions with respect to the dashed line are correlation functions of dual system functions. Since the channel correlation functions involve four variables, any two correlation functions connected by an F must have two common variables [such as t, s in (42) and (43)] which should be regarded as fixed parameters in employing the double Fourier transform relationship involving the remaining variables. Note that the four variables of a correlation function are divided into two pairs separated by a semicolon. One of these pairs is involved directly in the Fourier transform relationship while the other pair is fixed. Note also that the double Fourier transform relationship always connects a pair of time (or delay) variables and a pair of frequency (or Doppler-shift) variables *e.g.*, pairs ξ, η and f, l in (42) and (43)]. In order to make the Fourier transform symbolism in Fig. 6 unique we have employed the convention that in Fourier transforming from a pair of time variables to a pair of frequency variables a positive exponential is to be used to connect the first variables in each pair and a negative exponential to connect the second variables [*e.g.*, $\exp[j2\pi\xi f]$ and $\exp[-j2\pi\eta l]$, respectively, in (42)], while in transforming from a pair of frequency variables to a pair of time variables the opposite signing procedure is to be used [*e.g.*, $\exp[-j2\pi\xi f]$ and $\exp[j2\pi\eta l]$ in (43)].

Examination of Fig. 6 reveals that the pairs of channel correlation functions (R_g, R_G), (R_U, R_T) and their duals (R_h, R_H), (R_V, R_M) are quadruple Fourier transform pairs. The actual fourfold integral relating any of these pairs is readily obtained by performing two successive double Fourier transforms as indicated in Fig. 6.

From (30) it is quickly determined that

$$R_V(\xi, \eta; \nu, \mu) = e^{j2\pi(\nu\xi - \eta\mu)} R_U(\nu, \mu; \xi, \eta) \tag{44}$$

is the relationship between the correlation functions of the Delay-Doppler-Spread and Doppler-Delay-Spread system functions.

Eq. (39) may be used to determine that the relationship between the correlation functions of the Time-Variant Transfer Function and the Frequency-Dependent Modulation Function is as shown below,

$$R_M(t, s; f, l) = \iiiint R_T(f', l'; t', s') \exp[j2\pi(l - l') \cdot (s - s') - j2\pi(f - f')(t - t')]\, df'\, dl'\, dt'\, ds' \tag{45}$$

$$R_T(f, l; t, s) = \iiiint R_M(t', s'; f', l') \exp[j2\pi(f - f') \cdot (t - t') - j2\pi(l - l')(s - s')]\, df'\, dl'\, dt'\, ds'.$$

From (11) and (15) we find the following relationships between the Delay-Spread and Doppler-Spread Functions:

$$\begin{aligned} R_h(t, s; \xi, \eta) &= R_g(t + \xi, s + \eta; \xi, \eta) \\ R_G(f, l; \nu, \mu) &= R_H(f + \nu, l + \mu; \nu, \mu). \end{aligned} \tag{46}$$

Other analytic relationships between channel correlation functions on either side of the dashed line in Fig. 6 are quickly obtained by using (40), (41) and (42) in conjunction with the Fourier transform relationships indicated in Fig. 6.

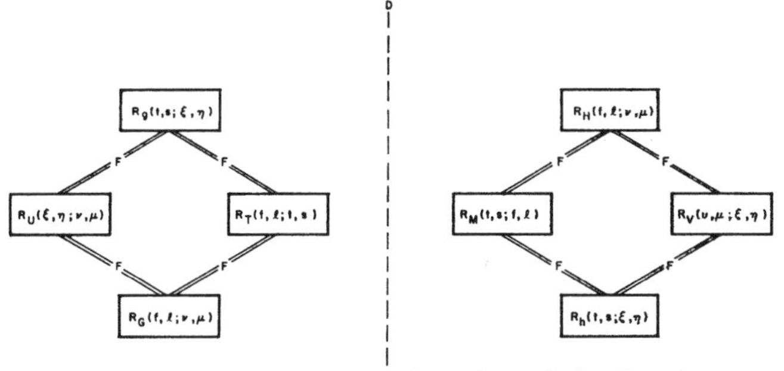

Fig. 6—Relationships between channel correlation functions.

With the aid of the input-output relationship and the correlation function associated with each system function one may readily determine a corresponding double integral relating the autocorrelation function of the output to the autocorrelation function of the system function. Thus consider the system function $g(t, \xi)$. Using (8) to form the product $w^*(t)\, w(s)$ as follows:

$$w^*(t)w(s) = \iint z^*(t - \xi)z(s - \eta)g^*(t, \xi)g(s, \eta)\, d\xi\, d\eta, \tag{47}$$

and then averaging, one finds that

$$R_w(t, s) = \iint z^*(t - \xi)z(s - \eta)R_g(t, s; \xi, \eta)\, d\xi\, d\eta \tag{48}$$

where we have defined

$$R_w(t, s) = \overline{w^*(t)w(s)} \tag{49}$$

as the autocorrelation function of the output time function.

When the input is random rather than deterministic, as in (48), the output autocorrelation function becomes

$$R_w(t, s) = \iint R_z(t - \xi, s - \eta)R_g(t, s; \xi, \eta)\, d\xi\, d\eta \tag{50}$$

where we have defined

$$R_z(t, s) = \overline{z^*(t)z(s)}. \tag{51}$$

The dual system function $H(f, \nu)$ leads to the following expression for the autocorrelation function of the output spectrum

$$\begin{aligned} R_W(f, l) &\equiv \overline{W^*(f)W(l)} \\ &= \iint Z^*(f - \nu)Z(l - \mu)R_H(f, l; \nu, \mu)\, d\nu\, d\mu \end{aligned} \tag{52}$$

when the input is deterministic and

$$R_W(f, l) = \iint R_Z(f - \nu, l - \mu)R_H(f, l; \nu, \mu)\, d\nu\, d\mu \tag{53}$$

when the input is random, where we have defined

$$R_Z(f, l) = \overline{Z^*(f)Z(l)} \tag{54}$$

as the autocorrelation function of the input spectrum.[12] The reader may readily form the input-output correlation function relationships corresponding to the remaining system functions.

B. The Wide-Sense Stationary Channel

In many physical channels the fading statistics may be assumed approximately stationary for time intervals sufficiently long to make it meaningful to define a subclass of channels, called Wide-Sense Stationary (WSS) Channels. A WSS channel has the property that the channel correlation functions $R_g(t, s; \xi, \eta)$, $R_h(t, s; \xi, \eta)$, $R_T(f, l; t, s)$ and $R_M(t, s; f, l)$ are invariant under a translation in time; *i.e.*, these correlation functions depend on the variables t, s only through the difference $\tau = s - t$. Thus for the WSS channel

$$\begin{aligned} R_g(t, t + \tau; \xi, \eta) &= R_g(\tau; \xi, \eta) \\ R_h(t, t + \tau; \xi, \eta) &= R_h(\tau; \xi, \eta) \\ R_T(f, l; t, t + \tau) &= R_T(f, l; \tau) \\ R_M(t, t + \tau; f, l) &= R_M(\tau; f, l). \end{aligned} \tag{55}$$

The restricted nature of the four channel correlation functions in (55) constrains the remaining four channel correlation functions in Fig. 6 to have a singular behavior in the Doppler-shift variables. As an example consider the double Fourier transform relationship between $R_g(t, s; \xi, \eta)$ and $R_U(\xi, \eta, \nu, \mu)$:

$$R_U(\xi, \eta; \nu, \mu) = \iint R_g(t, s; \xi, \eta) e^{i2\pi(\nu t - \mu s)}\, dt\, ds. \tag{56}$$

Upon making the change in variable $\tau = s - t$ in (56) and using the first equation in (55), one finds that

$$R_U(\xi, \eta; \nu, \mu) = \int e^{i2\pi t(\nu - \mu)}\, dt \int R_g(\tau; \xi, \eta) e^{-i2\pi\tau}\, d\tau. \tag{57}$$

The first integral in (57) is recognized as a unit impulse at $\nu = \mu$, *i.e.*, $\delta(\nu - \mu)$. It follows that $R_U(\xi, \eta; \nu, \mu)$ may be expressed in the form

$$R_U(\xi, \eta, \nu, \mu) = P_U(\xi, \eta; \nu)\delta(\nu - \mu) \tag{58}$$

where $P_U(\xi, \eta; \nu)$ is the Fourier transform of $R_g(\tau; \xi, \eta)$ with respect to the variable τ; *i.e.*,

$$P_U(\xi, \eta; \nu) = \int R_g(\tau; \xi, \eta) e^{-i2\pi\nu\tau}\, d\tau. \tag{59}$$

In an analogous fashion it is readily determined that

$$\begin{aligned} R_G(f, l; \nu, \mu) &= P_G(f, l; \nu)\delta(\nu - \mu) \\ R_V(\nu, \mu; \xi, \eta) &= P_V(\nu; \xi, \eta)\delta(\nu - \mu) \\ R_H(f, l; \nu, \mu) &= P_H(f, l; \nu)\delta(\nu - \mu) \end{aligned} \tag{60}$$

where $P_G(f, l; \nu)$, $P_V(\nu; \xi, \eta)$, and $P_H(f, l; \nu)$ are Fourier transforms of $R_T(f, l; \tau)$, $R_h(\tau; \xi, \eta)$, and $R_M(\tau; f, l)$, respectively, with respect to the delay variable τ.

[12] See Bello, *op. cit.*[3], for a discussion of the spectrum of a random process.

The singular behavior of the channel correlation functions in (58) and (60) has interesting implications with regard to the behavior of the associated circuit models. Thus the forms of R_H and R_G as shown in (60) imply that in Figs. 3 and 4 the transfer functions of the random filters associated with different Doppler shifts are uncorrelated. Similarly the forms of R_V and R_U in (60) and (58) imply that in the associated channel models consisting of a number of differential "scatterers" producing delay and Doppler shifts, the complex scattering amplitudes of two different elements are uncorrelated if these elements cause different Doppler shifts.

When the system functions are normally distributed stochastic processes, complete lack of correlation between two processes implies statistical independence. Then wide-sense stationarity implies strict-sense stationarity, and in the circuit models of Figs. 3 and 4 the transfer functions of random filters associated with different Doppler shifts are statistically independent. Similarly in the models consisting of a number of differential "scatterers" producing delay and Doppler shifts, the complex scattering amplitudes of two different elements are statistically independent if these elements cause different Doppler shifts.

The singular behavior of the correlation functions R_H, R_G, R_V and R_U might have been expected *a priori* from the observation that the corresponding system functions are interpretable as (complex) amplitude spectra of random processes and from the fact that the cross-correlation function between the amplitude spectra of two wide-sense stationary noises is an impulse located at zero frequency shift with a complex area equal to the cross-power spectral density between the original processes.[3] Thus, when considered as a function of the Doppler-shift variable ν, the functions $P_U(\xi, \eta; \nu)$, $P_G(t, l; \nu)$ $P_V(\nu; \xi, \eta)$ and $P_H(f, l; \nu)$ may be interpreted as cross-power spectral densities between the pairs of time functions $[g(t, \xi), g(t, \eta)]$, $[T(f, t), T(l, t)]$, $[h(t, \xi), h(t, \eta)]$ and $[M(t, f), M(t, l)]$, respectively. In the particular case that $\xi = \eta$, $P_V(\nu; \xi, \xi)$ and $P_U(\xi, \xi; \nu)$ may be interpreted as power spectral densities of the functions $g(t, \xi)$ and $h(t, \xi)$, respectively; while for $f = l$, $P_G(f, f; \nu)$ and $P_H(f, f; \nu)$ may be interpreted as power spectral densities of the functions $T(f, t)$ and $M(t, f)$, respectively. In view of the above it is clear that the system functions $U(\xi, \nu)$, $G(f, \nu)$, $V(\nu, \xi)$ and $H(f, \nu)$ will behave like nonstationary white noises in the Doppler-shift variable when the channel is WSS.

In Fig. 7 we have summarized the relationships between the channel correlation functions, using only the corresponding density function when the correlation function has an impulsive behavior. Note that the Fourier transform notations of Figs. 5 and 6 have been used.

Let us now consider some analytical relationships between functions on the opposite side of the dashed line in Fig. 7. From (44) and (60) we find that

$$P_V(\xi, \eta; \nu) = e^{-i2\pi\nu(\eta - \xi)} P_U(\mu; \xi, \eta) \tag{61}$$

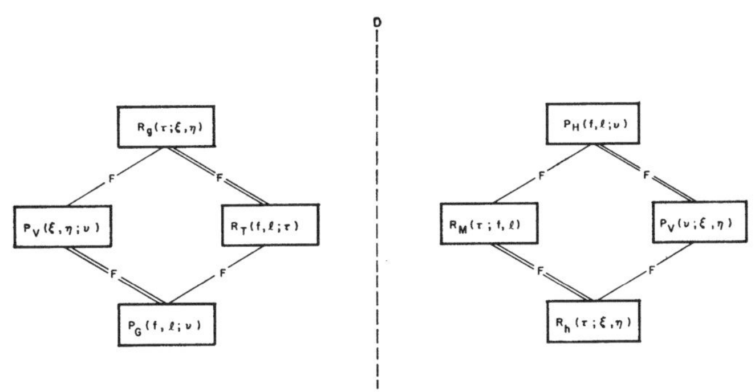

Fig. 7—Relationships between channel correlation functions for WSS channel.

while from (46), (55), (58) and (60) we find that

$$R_h(\tau;\xi,\eta) = R_g(\tau+\eta-\xi;\xi,\eta)$$
$$P_G(f,l;\nu) = P_H(f+\nu,l+\nu;\nu). \tag{62}$$

The relationship between the correlation functions of the Time-Variant System Function and the Frequency-Dependent Modulation Function is readily determined from (45) and (55) to be

$$R_M(\tau;f,l) = \iint R_T(f',f'+l-f;\tau') \cdot e^{j2\pi(f-f')(\tau-\tau')}\,df'\,d\tau'$$
$$R_T(f,l;\tau) = \iint R_M(\tau';f',f'+l-f) \cdot e^{-j2\pi(f-f')(\tau-\tau')}\,df'\,d\tau'. \tag{63}$$

Other analytic relationships between the channel correlation functions on either side of the dashed line in Fig. 7 are quickly obtained by using (61), (62) and (63) in conjunction with the Fourier transform relationships indicated in Fig. 7.

C. The Uncorrelated Scattering Channel

For several physical channels (*e.g.*, troposcatter, chaff, moon reflection) the channel may be modeled approximately as a continuum of uncorrelated scatterers. The mathematical counterpart of this statement is embodied in the following forms for the autocorrelation function of the Input and Output Delay-Spread Functions:

$$R_g(t,s;\xi,\eta) = P_g(t,s;\xi)\delta(\eta-\xi)$$
$$R_h(t,s;\xi,\eta) = P_h(t,s;\xi)\delta(\eta-\xi). \tag{64}$$

Because of the intimate relationship between the Input and Output Delay-Spread Functions, one of the equations in (64) implies the other. Moreover, these equations imply that the autocorrelation functions of the Doppler-Delay-Spread and Delay-Doppler-Spread Functions must have the form

$$R_U(\xi,\eta;\nu,\mu) = P_U(\xi,\nu,\mu)\delta(\eta-\xi)$$
$$R_V(\nu,\mu;\xi,\eta) = P_V(\nu,\mu;\xi)\delta(\eta-\xi). \tag{65}$$

The singular behavior of the channel correlation functions in (64) and (65) has implications with regard to the behavior of the associated circuit models. Thus the forms of R_g and R_h as shown in (64) imply that in Figs. 1 and 2 the complex gain functions associated with different path delays are uncorrelated. Similarly the forms of R_V and R_U in (65) imply that in the associated channel models consisting of a number of differential "scatterers" producing delay and Doppler shifts, the complex scattering amplitudes of two different scatterers are uncorrelated if these elements cause different delays. A channel whose system functions have correlation functions of the form shown in (64) and (65) will be called an Uncorrelated Scattering (US) channel.

When the system functions are normally distributed stochastic processes, the uncorrelatedness properties mentioned above for complex gain functions and scattering amplitudes become independence properties.

The forms of the correlation functions for the remaining four system functions which are readily determined from the Fourier transform relationships indicated in Fig. 6 are given by

$$R_t(f,f+\Omega;t,s) = R_T(\Omega;t,s)$$
$$R_M(t,s;f,f+\Omega) = R_M(t,s;\Omega)$$
$$R_G(f,f+\Omega;\nu,\mu) = R_G(\Omega;\nu,\mu)$$
$$R_H(f,f+\Omega;\nu,\mu) = R_H(\Omega;\nu,\mu). \tag{66}$$

A comparison of the channel correlation functions for the WSS and US channels reveals an interesting fact: the correlation function of a particular system function of the WSS channel and the correlation function of the dual system function of the US channel have identical analytical forms as a function of dual variables. For this reason one may consider the class of WSS channels to be the dual of class of US channels.[13]

As a consequence of this duality we note that the US channel may be regarded as a WSS channel in the frequency variable since, from (66), the channel correlation functions depend upon the frequency variables f, l only through the difference frequency $\Omega = l - f$. Similarly the WSS channel may be regarded as a form of US channel in the Doppler-shift variable.

While the Input and Output Doppler-Spread Function have the character of nonstationary white noise as a function of the Doppler-shift variable in the case of the WSS channel, the dual system functions, *i.e.*, the Input and Output Delay-Spread Functions, the Delay-Doppler-Spread and Doppler-Delay-Spread Functions, respectively, have the character of nonstationary white noise as a function of the dual variable, *i.e.*, the delay variable, in the case of the US channel.

By analogy with the dual functions in the WSS channel, the functions $P_g(t,s;\xi)$, $P_h(t,s;\xi)$, $PU(\xi;)$, ν, $\mu)$ and

[13] Using the definitions developed in Bello, *op. cit.* [3], one may state that the wide-sense dual of a WSS channel is a US channel and vice versa.

$P_V(\nu, \mu; \xi)$, when considered as a function of ξ, may be regarded as cross-power spectral densities while $P_g(t, t; \xi)$, $P_h(t, t; \xi)$, $P_U(\xi; \nu, \nu)$, $P_V(\nu, \nu; \xi)$ may be regarded as power spectral densities as a function of the delay variable. In Fig. 8 we have summarized the relationships between the channel correlation functions for the US channel using only the corresponding cross-power density function when the correlation function has an impulsive behavior.

We will now obtain some analytical relationships between functions on the opposite side of the dashed line in Fig. 8. First we have the relationship dual to (61) which may be obtained by using (65) in (44),

$$P_U(\nu, \mu; \xi) = e^{j2\pi\nu(\eta-\xi)} P_V(\xi; \nu, \mu). \tag{67}$$

Then, using (64) and the last two equations of (66) in (46) we obtain the dual to (62) as

$$\begin{aligned} R_G(\Omega; \nu, \mu) &= R_H(\Omega + \mu - \nu; \nu, \mu) \\ P_h(t, s; \xi) &= P_g(t + \xi, s + \xi; \xi). \end{aligned} \tag{68}$$

The relationships between the Time-Variant Transfer Function and the Frequency-Dependent Modulation Function dual to (63) are obtained by using the first two equations in (66) in (45) and carrying out two integrations of the appropriate quadruple integrals in (45),

$$\begin{aligned} R_T(\Omega; t, s) &= \iint R_M(t', t' + s - t; \Omega) \\ &\qquad \cdot e^{-j2\pi(t-t')(\Omega-\Omega')} \, dt' \, d\Omega' \\ R_M(t, s; \Omega) &= \iint R_T(\Omega'; t', t' + s - t) \\ &\qquad \cdot e^{j2\pi(t-t')(\Omega-\Omega')} \, dt' \, d\Omega'. \end{aligned} \tag{69}$$

As we have mentioned in a similar vein for the other classes of channels, further analytical relationships may be obtained between system functions on the opposite side of the dotted line in Fig. 8 by using (67), (68) and (69) and the Fourier transform relationships indicated in Fig. 8.

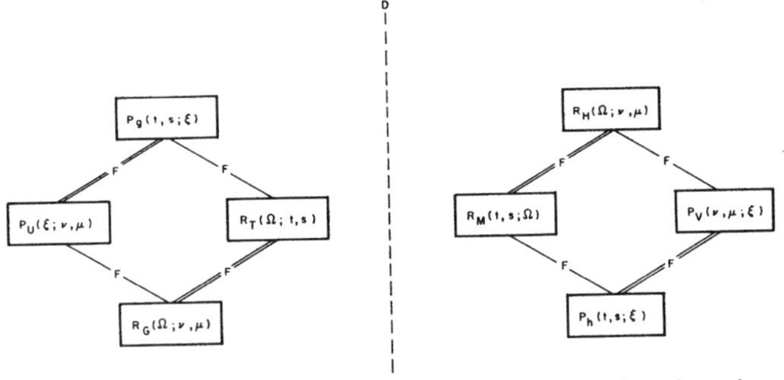

Fig. 8—Relationships between channel correlation functions for US channel.

D. *The Wide-Sense Stationary Uncorrelated Scattering Channel*

The simplest type of randomly time-variant linear channel to describe in terms of channel correlation functions, and one which, fortunately, is of practical interest is the WSSUS channel. As might be suspected from its name, the WSSUS channel is both a WSS and a US channel. Thus, the channel correlation functions of the WSSUS channel must have forms characteristic of both the WSS channel [(55), (58) and (60)] and the US channel [(64), (65) and (66)].

An examination of the correlation functions of the WSS and US channels reveals that for the WSSUS channel, the correlation functions of the Delay-Spread Functions must have the form

$$\begin{aligned} R_g(t, t + \tau; \xi, \eta) &= P_g(\tau, \xi)\delta(\eta - \xi) \\ R_h(t, t + \tau; \xi, \eta) &= P_h(\tau, \xi)\delta(\eta - \xi) \end{aligned} \tag{70}$$

while the correlation functions of the Doppler-Spread Functions must have the form

$$\begin{aligned} R_G(f, f + \Omega; \nu, \mu) &= P_G(\Omega, \nu)\delta(\mu - \nu) \\ R_H(f, f + \Omega; \nu, \mu) &= P_H(\Omega, \nu)\delta(\mu - \nu). \end{aligned} \tag{71}$$

The equations in (70) show that for the WSSUS channel, the system functions $g(t, \xi)$ and $h(t, \xi)$ have the character of nonstationary white noise in the delay variable and wide-sense stationary noise in the time variable. In terms of the channel models of Figs. 1 and 2, the WSSUS channel has a representation as a continuum of uncorrelated randomly scintillating scatterers with wide-sense stationary statistics.

The equations in (71) show that for the WSSUS channel the system functions $G(f, \nu)$ and $H(f, \nu)$ have the character of nonstationary white noise in the Doppler-shift variable and wide-sense stationary noise in the frequency variable. In terms of the channel models of Figs. 3 and 4, the WSSUS channel has a representation as a continuum of uncorrelated Doppler-shifting filtering (or filtering-Doppler shifting) elements with each filter having a transfer function with wide-sense stationary statistics in the frequency variable.

For the WSSUS channel the correlation functions of the Delay-Doppler-Spread and Doppler-Delay-Spread Functions simplify to

$$\begin{aligned} R_U(\xi, \eta; \nu, \mu) &= P_U(\xi, \nu)\delta(\mu - \nu)\delta(\eta - \xi) \\ R_V(\nu, \mu; \xi, \eta) &= P_V(\nu, \xi)\delta(\mu - \nu)\delta(\eta - \xi). \end{aligned} \tag{72}$$

Eq. (72) shows that for the WSSUS channel the system functions $U(\xi, \nu)$ and $V(\nu, \xi)$ have the character of nonstationary white noise in both the delay and Doppler-shift variables, *i.e.*, a form of two-dimensional nonstationary white noise. It follows that in terms of the corresponding channel models, the WSSUS channel may be represented as a collection of nonscintillating uncorrelated scatterers which cause both delays and Doppler shifts.

Finally, in the case of the WSSUS channel, the correlation functions for the Time-Variant Transfer Function and the Frequency-Dependent Modulation Function take the simple forms

$$\begin{aligned} R_M(t, t + \tau; f, f + \Omega) &= R_M(\tau, \Omega) \\ R_T(f, f + \Omega; t, t + \tau) &= R_T(\Omega, \tau); \end{aligned} \tag{73}$$

i.e., the system functions $T(f, t)$ and $M(t, f)$ are wide-sense stationary processes in both the time and frequency variables.

From previous discussions in Sections IV B and IV C, we know that when considered as a function of the Doppler-shift variable ν the functions $P_G(\Omega, \nu)$ and $P_H(\Omega, \nu)$ may be interpreted as the cross-power spectral densities of the pairs of time functions $T(f, t)$, $T(f + \Omega, t)$ and $M(t, f)$, $M(t, f + \Omega)$, respectively.

Similarly, when considered as a function of the delay variable ξ, the functions $P_g(\tau, \xi)$ and $P_h(\tau, \xi)$ may be interpreted as cross-power spectral densities of the frequency functions $T(f, t)$, $T(f, t + \tau)$ and $M(t, f)$, $M(t + \tau, f)$, respectively.

The functions $P_U(\xi, \nu)$ and $P_V(\nu, \xi)$ may be interpreted as a sort of two-dimensional power density spectrum as a function of delay and Doppler shift corresponding to the combined time and frequency functions $T(f, t)$ and $M(t, f)$, respectively.

In the case of the WSSUS channel the relationships between correlation functions on opposite sides of the dashed line in Fig. 6 become trivial. Thus, use of (73) in (44) shows that

$$P_U(\xi, \nu) = P_V(\nu, \xi) \equiv S(\xi, \nu); \tag{74}$$

i.e., that the two-dimensional power spectral densities in delay and Doppler shift associated with the Doppler-Delay-Spread and Delay-Doppler-Spread functions are identical. We have used the function $S(\xi, \nu)$ to denote this common function which is identical to the Target Scattering Function $\sigma(\xi, \nu)$ defined by Price and Green[5] in their work on radar astronomy. We shall call $S(\xi, \nu)$ the Scattering Function since it has more general applications than to radar problems.

If (72) is used in the last equation of (46) one finds immediately that

$$P_G(\Omega, \nu) = P_H(\Omega, \nu) \equiv P(\Omega, \nu). \tag{75}$$

Thus the Doppler cross-power spectral densities associated with the channel models of Figs. 3 and 4 become equal in the case of the WSSUS channel. We shall call this common function the Doppler Cross-Power Spectral Density and denote it by the function $P(\Omega, \nu)$. In the particular case that $\Omega = 0$, the cross-power spectral densities become simply power spectral densities. Thus we define

$$P(0, \nu) \equiv P(\nu) \tag{76}$$

where $P(\nu)$ is called the Doppler Power Density Spectrum. (This function is called the Echo Power Spectrum by Green.[6])

If (59) is used in the first equation of (42) one finds that

$$P_g(\tau, \xi) = P_h(\tau, \xi) \equiv Q(\tau, \xi). \tag{77}$$

Thus the delay cross-power spectral densities associated with the channel models of Figs. 1 and 2 become equal in the case of the WSSUS channel. We shall call this common function the Delay Cross-Power Spectral Density and denote it by the function $Q(\tau, \xi)$. In the particular case that $\tau = 0$, the cross-power spectral densities become simple power spectral densities. Thus we define

$$Q(0, \xi) \equiv Q(\xi) \tag{78}$$

where $Q(\xi)$ is called the Delay Power Density Spectrum. (This function has been called the Power Impulse Response by Green[6] and the Delay Spectrum by Hagfors.[4])

Finally, if (73) is used in (45) one finds that the quadruple integrals in (45) vanish leaving the interesting result

$$R_M(\tau, \Omega) = R_T(\Omega, \tau) \equiv R(\Omega, \tau). \tag{79}$$

Thus in the case of the WSSUS channel, the correlation functions of the Time-Variant System Function and the Frequency-Dependent Modulation Function become identical, *i.e.*,

$$\begin{aligned}&\overline{T^*(f, t)T(f + \Omega, t + \tau)}\\ &\quad = \overline{M^*(t, f)M(t + \tau, f + \Omega)} = R(\Omega, \tau).\end{aligned} \tag{80}$$

We shall call this common function the Time-Frequency Correlation Function and denote it by $R(\Omega, \tau)$. (This function has been called the Spaced-Time Spaced-Frequency Correlation Function by Green.[6])

Two correlation functions of practical interest derivable from $R(\Omega, \tau)$ are the Frequency Correlation Function $q(\Omega)$ (called the Spaced-Frequency Correlation Function by Green[6]) given by

$$\begin{aligned}&\overline{T^*(t, f)T(t, t + \Omega)}\\ &\quad = \overline{M^*(t, f)M(t, f + \Omega)} = R(\Omega, 0) \equiv q(\Omega)\end{aligned} \tag{81}$$

and the Time Correlation Function $p(\tau)$ (called the Echo Correlation Function by Green[6]) given by

$$\begin{aligned}&\overline{T^*(t, f)T(t + \tau, f)}\\ &\quad = \overline{M^*(t, f)M(t + \tau, f)} = R(0, \tau) \equiv p(\tau).\end{aligned} \tag{82}$$

The relationships between channel correlation functions are shown in Fig. 9. Note that $Q(\tau, \xi)$ and $P(\Omega, \nu)$ are double Fourier transform pairs as are $S(\xi, \nu)$ and $R(\Omega, \tau)$. The double Fourier transform relationship between the Scattering Function and the Time-Frequency Correlation Function appears to have been first pointed out by Hagfors.[4] It is readily determined from the Fourier transform relationships in Fig. 9 that $q(\Omega)$, $Q(\xi)$ and $p(\tau)$, $P(\nu)$ are single Fourier transform pairs.

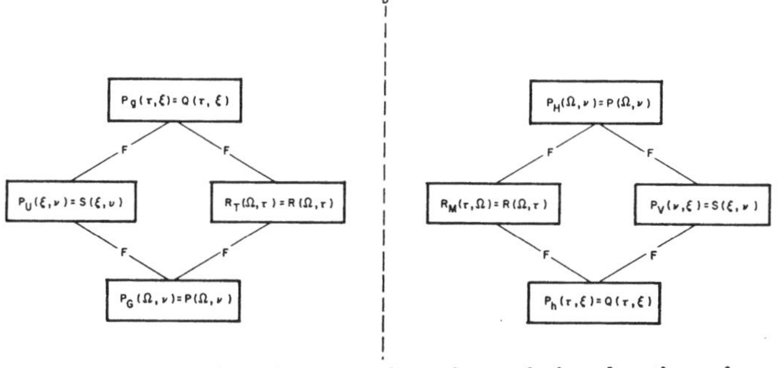

Fig. 9—Relationships between channel correlation functions for WSSUS channel.

V. Radio Channel Characterization

Virtually all radio transmission media may be characterized as linear in regard to their influence upon communication signals. Thus, from a phenomenological point of view, radio channels may be regarded as special cases of random time-variant linear filters. In the case of the transmission of digital signals over radio channels it appears that certain simplifications may be effected in the general characterization of randomly time-variant channels developed in the previous sections of this report. These simplifications arise when the time and frequency selective behavior of the channel may be regarded as wide-sense stationary for time and frequency intervals much greater than the durations and bandwidths, respectively, of the signaling waveforms of interest. Such a situation arises in practice when the channel contains very slow fluctuations superimposed upon more rapid fluctuations, the latter of which exhibit the desired statistical stationarity properties. Most radio channels do, in fact, appear to exhibit such "quasi-stationary" behavior. Moreover, the more rapid fluctuations often appear to be characterizable in terms of appropriately defined Gaussian statistics. Since a Gaussian process can be completely described statistically if its correlation function is known, it follows that a fairly complete statistical description of many quasi-stationary radio channels should be achievable by measuring the channel correlation functions for time and frequency intervals small compared to the fluctuation intervals of the slow channel variations, and then measuring the statistical behavior of these quasi-stationary channel correlation functions as caused by the slow channel variations. In this way one may compute quasi-stationary error probabilities for digital transmission which would accurately reflect the short-time error rate behavior of the channel.[14-16] The long-time error rate behavior of the channel could then be predicted by averaging the short-time error rate behavior over the long-time fading statistics of the channel.

To sum up, our thesis is that a useful way to perform measurements on radio channels is to determine the long-time statistics of short-time channel correlation functions. The resulting data should be sufficient to provide a fairly complete statistical description of some radio channels. To make these ideas more precise we shall now present a mathematical exposition of the above ideas.

The time and frequency selective behavior of a random time-variant linear channel may be described with the aid of several of the system functions described in Section III. For purposes of discussion it is sufficient to start with an examination of $T(f, t)$, the Time-Variant Transfer Function of the channel. It will be recalled that $T(f, t)$ is the complex envelope of the response of the channel to an excitation $\cos 2\pi(f_c + f)t$, where f_c is the "center" or "carrier" frequency at which the channel is being excited. Thus, the magnitude of $T(f, t)$ is the envelope of the channel response and the angle of $T(f, t)$ is the phase of the channel response measured with respect to the phase $2\pi(f_c + f)t$.

For a general input signal with complex envelope $z(t)$, the channel output complex envelope $w(t)$ is given by (19). Although the input time function $z(t)$ may be deterministic (nonrandom), the output time function $w(t)$ will be a random process since $T(f, t)$ is a random process. It is possible that for some radio channels $T(f, t)$ will contain a deterministic component so that $w(t)$ will also contain a deterministic component. However, for the purpose of the present discussion it is sufficient to confine our attention to the purely random part of $T(f, t)$ and $w(t)$. Thus, to avoid introducing unnecessary notation we shall assume in this section that $w(t)$ and $T(f, t)$ are purely random.

The time and frequency selective behavior of the channel is evidenced by the way $T(f, t)$ changes with changes in f and t. As far as conventional usage is concerned the concept of "fading" is associated only with the fact that $T(f, t)$ varies with the time variable t. However, it appears desirable to extend this definition to include variation of $T(f, t)$ with f, and thus talk of "frequency fading," the dual of conventional "time fading."

From a statistical point of view the simplest way to describe the sensitivity of $T(f, t)$ to changes in f and t is to form the correlation function

$$\overline{T^*\left(f - \frac{\Omega}{2}, t - \frac{\tau}{2}\right)T\left(f + \frac{\Omega}{2}, t + \frac{\tau}{2}\right)} \equiv R_{f,t}(\tau, \Omega). \quad (83)$$

In terms of the notation in Section IV,

$$R_{f,t}(\tau, \Omega) \equiv R_T\left(f - \frac{\Omega}{2}, f + \frac{\Omega}{2}; t - \frac{\tau}{2}, t + \frac{\tau}{2}\right). \quad (84)$$

When f and t are fixed, say at $f = f'$, $t = t'$, $R_{f',t'}(\tau, \Omega)$ describes the way in which the Time-Variant Transfer Function becomes decorrelated for a frequency interval Ω and a time interval τ centered on the "local" time-frequency coordinates f', t'. In the case of the WSSUS channel $R_{f,t}(\tau, \Omega)$ becomes independent of f, t; *i.e.*,

$$R_{f,t}(\tau, \Omega) = R(\tau, \Omega). \quad (85)$$

From an analytical point of view the WSSUS channel is the simplest nondegenerate channel that exhibits both time and frequency selective behavior. As discussed in Section IV, such a channel may be modeled as a continuum of uncorrelated scatterers such that each infinitesimal scatterer providing Doppler shifts in the range ν, $\nu + d\nu$ and delays in the range ξ, $\xi + d\xi$ has a scattering cross section of $S(\xi, \nu)d\xi, d\nu$, where

[14] P. A. Bello and B. D. Nelin, "The influence of fading spectrum on the binary error probabilities of incoherent and differentially-coherent matched filter receivers, "IRE Trans. on Communications Systems, vol. CS-10, pp. 160–168; June, 1962.

[15] P. A. Bello and B. D. Nelin, "The effect of frequency selective fading on the binary error probabilities of incoherent and differentially coherent matched filter receivers," IEEE Trans. on Communication Systems, vol. CS-11, pp. 170–186; June, 1963.

[16] P. A. Bello and B. D. Nelin, "The Effect of Combined Time and Frequency Selective Fading on the Binary Error Probabilities of Incoherent Matched Filter Receivers," ADCOM, Inc., Cambridge, Mass., Res. Rept. No. 7, March, 1963.

$$S(\xi, \nu) = \iint e^{i2\pi(\xi\Omega - \nu\tau)} R(\tau, \Omega)\, d\tau\, d\Omega \tag{86}$$

is the Fourier transform of $R(\tau, \Omega)$.

We will now demonstrate that the simplicity of the WSSUS channel can be transferred to a practically interesting class of channels which we shall call Quasi-WSSUS (or QWSSUS). This class contains two subclasses which are (time-frequency) duals. For the purposes of the present discussion we need only introduce that subclass which is based upon the properties of $R_{f,t}(\tau, \Omega)$. The dual subclass is based upon the Frequency Dependent Modulation Function and may be readily constructed by the reader.

To define the QWSSUS channel of interest here one must assume that the typical input signaling waveform has a constraint on bandwidth and that the resulting output waveform has a constraint on time duration. This bandwidth and time constraint can be centered anywhere, but for simplicity of discussion (and with no loss in generality) it is convenient to assume that the input bandwidth constraint is centered at $f = 0$ (*i.e.*, at the carrier frequency) and the output time constraint is centered at $t = 0$. Any other centering can be handled by a redefinition of carrier frequency and time origin.

The QWSSUS channel is defined as that subclass for which certain gross channel parameters have specified inequality relations with respect to the input-bandwidth and output-time constraints defined above. The gross channel parameters of concern here are measures of the maximum rate at which $R_{f,t}(\tau, \Omega)$ varies in the f and t directions. Let the maximum rate of fluctuation $R_{f,t}(\tau, \Omega)$ in the f and t directions be denoted by $\gamma_{\max}$ sec and $\theta_{\max}$ cps, respectively. Let W, T denote the bandwidth, and time duration of the input signal. Let Δ denote the multipath spread of the channel. Then the QWSSUS channel under discussion is a channel for which the following inequalities hold:

$$W \ll \frac{1}{\gamma_{\max}} \tag{87}$$

$$T + \Delta \ll \frac{1}{\theta_{\max}}, \tag{88}$$

or, in other words, one for which $R_{f,t}(\Omega, \tau)$ changes negligibly over "f" intervals equal to the input signal bandwidth (W) and over "t" intervals equal to the output signal time duration ($T + \Delta$).

It will now be demonstrated that if inequalities (87) and and (88) are satisfied, the actual channel may be replaced by a hypothetical WSSUS channel at least as far as the determination of the correlation function of the channel output is concerned. However, when the channel has Gaussian statistics, the actual channel may be replaced by a hypothetical WSSUS channel as far as the determination of any output statistics are concerned, since then the output will be a Gaussian process and thus completely determined statistically from knowledge of its correlation function.

From (19) we determine that the output correlation function is given by

$$\overline{w^*(t)w(s)} = \iint Z^*(f)Z(l)R_T(f, l; t, s)e^{-i2\pi(ft - ls)}\, df\, dl. \tag{89}$$

Upon making the transformations

$$t \to t - \frac{\tau}{2}, \qquad s \to t + \frac{\tau}{2}$$

$$f \to f - \frac{\Omega}{2}, \qquad l \to f + \frac{\Omega}{2}$$

in (89) one finds that

$$\overline{w^*\left(t - \frac{\tau}{2}\right)w\left(t + \frac{\tau}{2}\right)} = \iint Z^*\left(f - \frac{\Omega}{2}\right)Z\left(f + \frac{\Omega}{2}\right) \cdot R_{f,t}(\tau, \Omega)e^{-i2\pi f}e^{i2\pi\Omega t}\, df\, d\Omega. \tag{90}$$

If we consider the integration with respect to f first in (90) we note that the integrand is nonzero over an interval of f values of maximum width W centered on $f = 0$, because by hypothesis $Z(f)$ is zero for values f outside this interval and thus $Z^*(f - \Omega/2)Z(f + \Omega/2)$ must also be zero outside this interval. According to inequality (87), however, $R_{f,t}(\tau, \Omega)$ will vary negligibly for values of f in this interval and thus for values of f for which the integrand in (90) is nonzero. It follows that insignificant error will result in (90) if we use $R_{0,t}(\tau, \Omega)$ in place of $R_{f,t}(\tau, \Omega)$.

Furthermore we note that, since $w(t)$ is constrained by hypothesis to be nonzero only over an interval of t values of width $T + \Delta$ centered on $t = 0$, then the left-hand side of (90) must perforce exhibit the same property as a function of t. Since the double integral in (90) must vanish for values of t outside an integral of width $T + \Delta$ centered on $t = 0$, and since by inequality (88), $R_{f,t}(\tau, \Omega)$ in the integrand can vary negligibly in this interval, it follows that little error can result by replacing $R_{f,t}(\tau, \Omega)$ by $R_{f,0}(\tau, \Omega)$ in the integrand.

It then follows that if both inequalities (87) and (88) are valid, we have the close approximation

$$\overline{w^*(t)w(s)} = \int \chi(\Omega, \tau)R_{00}(\tau, \Omega)e^{i2\pi\Omega t}\, d\Omega \tag{91}$$

where

$$\chi(\Omega, \tau) = \int Z^*\left(f - \frac{\Omega}{2}\right)Z\left(f + \frac{\Omega}{2}\right)e^{-i2\pi\tau f}\, df \tag{92}$$

is the ambiguity function of the transmitted signal. The expression (91) for the output signal correlation function is identical in form to that for a WSSUS channel in which

$$R(\tau, \Omega) = R_{00}(\tau, \Omega). \tag{93}$$

It will be recalled that initially we had assumed the spectrum of the input and the output time function were centered at $f = 0$ and $t = 0$ respectively. If we had assumed instead that they were centered at $f = f'$ and $t = t'$, the satisfaction of the inequalities (87) and (88) would still have lead us to conclude that the output correlation function can be determined by replacing the actual channel by a hypothetical WSSUS channel. However, instead of (93) we must use

$$R(\tau, \Omega) = R_{f',t'}(\tau, \Omega). \tag{94}$$

We are now in a position to consider the application of the preceding analytical results to the characterization of radio channels. As discussed at the beginning of this section, many radio channels seem to exhibit a combination of fast fading of a nearly Gaussian nature[17] and very slow fading of a generally non-Gaussian nature. We shall assume this to be the case for radio channels using the extended concept of fading described previously. Thus we assume that $T(f, t)$ "fades" along the frequency axis with a "fast" frequency fading superimposed upon a "slow" frequency fading.

Let us momentarily consider that the very slow variations are deterministic by selecting a particular member function of the stochastic process defining the slow fading, and let us assume that the fast fading is Gaussian. Then all statistical information concerning the output of the channel may be obtained once the correlation function $R_{f,t}(\tau, \Omega)$ is known, since then (90) may be used to determine the correlation function of the output Gaussian process $w(t)$.

We may ascribe the variations of $R_{f,t}(\tau, \Omega)$ with f, t as due to the (temporarily deterministic) slow (time and frequency) fading of the channel. In practical channels it appears sufficient to consider

$$R_{f,t}(0, 0) = \overline{|T(f, t)|^2} \tag{95}$$

in order to obtain a feeling as to the degree to which $R_{f,t}(\tau, \Omega)$ varies with f and t. From the definition of $T(f, t)$ we see that $R_{f,t}(0, 0)$ is equal to twice the average power received from a unit amplitude sine wave at time t and of frequency f cps away from carrier frequency, as measured by averaging along an ensemble of channels *all with the same deterministic slow variations*. In practice we do not have available an ensemble of channels. However, we may obtain an approximate measurement of (95) with the determination of the time average;

$$\frac{1}{T_1}\int_{t-T_1/2}^{t+T_1/2} |T(f, t_1)|^2\, dt_1 \equiv \langle |T(f, t)|^2 \rangle_{T_1} \tag{96}$$

where the averaging T_1 is (hopefully) long enough to produce negligible measurement fluctuations due to the fast fading but yet short enough to reflect the long term fading behavior of the channel.

[17] Some confusion may exist in the reader's mind as to precisely what is meant by Gaussian fading, since fading is usually stated to be Rayleigh distributed. By Gaussian fading it is meant that the transmission of a sinusoid results in the reception of a narrow-band Gaussian process with a possible nonfading specular component present. It is the envelope which will be Rayleigh or Rice distributed depending upon the nonexistence or existence of the specular component.

Measurements such as (96) seem to indicate that the inequalities (87) and (88) will be satisfied for a large percentage of radio channels for operating frequencies and signaling element bandwidths and time durations of practical interest. Thus it appears that a useful model for several radio channels is the QWSSUS channel with Gaussian statistics. Measurement of the correlation function $R_{f,t}(\tau, \Omega)$ on a short-time basis will then provide the necessary statistical information to evaluate the short-time performance of a digital system, assuming the statistics of any additive interferences are known. This short-time performance will be a functional of $R_{f,t}(\tau, \Omega)$. Since $R_{f,t}(\tau, \Omega)$ is in effect a random process due to the slow channel fluctuations (we have removed the deterministic assumption), the performance index (say error probability) computed on a short time basis assuming a Gaussian QWSSUS channel must be averaged over the long term statistics of $R_{f,t}(\tau, \Omega)$ to determine a long-time basis performance index.

VI. Canonical Channel Models

All practical channels and signals have an essentially finite number of degrees of freedom due to restrictions on time duration, fading rate, bandwidth, etc. These restrictions allow a simplified representation of linear time-varying channels in terms of canonical elements or building blocks. Such channel representations, called canonical channel models, can simplify the analysis of the performance of communication systems which employ time-variant linear channels.

Two general classes of channel models, called Sampling Models and Power Series Models, will be considered in this paper. The Sampling Models apply when a system function vanishes for values of an independent variable (time t, frequency f, delay ξ, or Doppler shift ν) outside some interval or when the input or output time function is time-limited or band-limited. The conditions for the applicability of the Power Series Models are not so simply stated. Stated briefly it requires the existence of a power series expansion of a system function in an independent variable and, depending upon the channel model, the existence of derivatives of the input function or spectrum.

A. Sampling Models

In this section we will develop the various sampling canonical channel models referred to above. The models developed by Kailath[2] will also be included not only for completeness but because the significance of some of the new sampling models is enhanced since they are dual to those of Kailath.

It will be convenient to divide our discussion into two parts, one involving time and frequency constraints and the other involving delay and Doppler-shift constraints.

1) *Time and Frequency Constraints*: A quick understanding of the time and frequency limitations that are relevant in the case of the sampling channel models may be arrived at by examining the input-output relationships corresponding to the Time-Variant Transfer Function

$T(f, t)$ and the Frequency-Dependent Modulation Function $M(t, f)$. These relationships are repeated below for convenience:

$$w(t) = \int Z(f)T(f, t)e^{i2\pi ft}\, df$$
$$W(f) = \int z(t)M(t, f)e^{-i2\pi ft}\, dt. \tag{97}$$

If a time-variant linear filter is preceded by a time-invariant linear filter with transfer function $T_i(f)$ and is followed with a multiplication by a time function $M_0(t)$, a combination time-variant linear filter which includes the input filter and output multiplier has a Time-Variant Transfer Function $T'(f, t)$ given by

$$T'(f, t) = T_i(f)T(f, t)M_0(t). \tag{98}$$

Eq. (98) is quickly deduced from the first equation in (97) by noting that preceding the original filter by a time-invariant linear filter with transfer function $T_i(f)$ is equivalent to changing the input spectrum from $Z(f)$ to $Z(f)\ T_i(f)$, while following the original filter by a multiplication with $M_0(t)$ is equivalent to multiplying both sides of the first equation in (97) by $M_0(t)$.

Since a constraint on the bandwidth of the input signal to a frequency region of width W_i centered on f_i, *i.e.*, to $f_i - W_i/2 < f < f_i + W_i/2$ cps may be represented by means of a band-limiting filter at the channel input, it is clear from (98) that such a constraint can be handled conveniently by defining a hypothetical channel whose Time-Variant Transfer Function $T'(f, t)$ is given by

$$T'(f, t) = \text{Rect}\left(\frac{f - f_i}{W_i}\right)T(f, t) \tag{99}$$

where $T(f, t)$ is the actual system function and

$$\text{Rect}\,(x) = \begin{cases} 1 & |x| < \frac{1}{2}. \\ 0 & |x| \geq \frac{1}{2} \end{cases} \tag{100}$$

Since the Input Delay-Spread Function $g(t, \xi)$ is the inverse Fourier transform of the Time-Variant Transfer Function $T(f, t)$ with respect to the frequency variable, it follows that corresponding to the hypothetical system function $T'(f, t)$ in (99) there is a hypothetical Input Delay-Spread Function $g'(t, \xi)$ given by

$$g'(t, \xi) = \int e^{i2\pi f_i(\xi-\eta)}W_i \cdot \text{sinc}\,[W_i(\xi - \eta)]g(t, \eta)\, d\eta \tag{101}$$

where

$$\text{sinc}\ y = \frac{\sin \pi y}{\pi y}. \tag{102}$$

Eq. (101) is obtained from (99) by noting that multiplication corresponds to convolution in the transform domain and that the inverse transform of Rect $([f-f_i]/W_i)$ is $e^{i2\pi f_i \xi}W_i$ sinc $[W_i\xi]$.

Thus the case wherein the channel responds (*i.e.*, has a nonzero output) only to input frequencies in a given range, say $f_i - W_i/2 < f < f_i + W_i/2$, and the case wherein the channel responds to other frequencies but has an input signal limited to frequencies in the range $f_i - W_i/2 < f < f_iW_i/2$ may be handled by the same analytical approach. In deriving the sampling model relevant to an input signal bandwidth limitation or an input frequency response limitation it will be convenient to assume that $T(f, t)$ is nonzero only for values of f within the relevant frequency interval. It should be kept in mind, however, that when the input signal rather than the input frequency response of the channel is band-limited, one must eventually use an equation such as (99) or (101) in order to express the parameters of the canonical channel model in terms of the true channel system function.

If it is desired to observe the channel output for some finite time interval, say $t_0 - T_0/2 < t < t_0 + T_0/2$ or if, due to some gating operation in the receiver, only a finite time segment of the received waveform in the same interval is available, one has a constraint on the time duration of the channel output. It is clear from (98) that such a constraint may be handled analytically by defining a hypothetical channel whose Time-Variant Transfer Function $T'(f, t)$ is given by

$$T'(f, t) = T(f, t)\,\text{Rect}\left(\frac{t - t_0}{T_0}\right). \tag{103}$$

Since the Output Doppler-Spread Function $G(f, \nu)$ is the Fourier transform of the Time-Variant Transfer Function with respect to the time variable, it follows that corresponding to the hypothetical system function $T'(f, t)$ in (103) there is a hypothetical Output Doppler-Spread Function $G'(f, \nu)$ given by

$$G'(f, \nu) = \int e^{-i2\pi t_0(\nu-\mu)}T_0 \cdot \text{sinc}\,[T_0(\nu - \mu)]G(f, \mu)\, d\mu. \tag{104}$$

Eq. (104) follows from (103) for the same reasons that (101) followed from (99).

Thus the case wherein the channel output vanishes outside some time interval, say $t_0 - T_0/2 < t < t_0 + T_0/2$, and the case where the channel has outputs outside this range but the receiver observes the received waveform only in this range may be handled by the same analytical approach. In deriving the sampling model relevant to a limitation in a receiver observation time it will be convenient to assume that $T(f, t)$ is nonzero only for values of t within the relevant time interval. However, it should be kept in mind that when the output observation time is limited rather than the output time response of the channel, one must eventually use an equation such as (103) or (104) in order to express the parameters of the canonical channel model in terms of the true channel system function.

A discussion entirely dual to the one above concerning input frequency and output time constraints and dealing with $T(f, t)$ may be carried through for an input time and output frequency constraint by dealing with $M(t, f)$ the Frequency-Dependent Modulation Function. Thus, if a time-variant filter is preceded by a multiplier which multiplies the input by $M_i(t)$ and is followed with a time-invariant linear filter of transfer function $T_0(f)$, a combination time-variant linear filter which includes the input multiplier and output filter has a Frequency-Dependent Modulation Function given by

$$M'(t, f) = M_i(t)M(t, f)T_0(f). \tag{105}$$

Eq. (105) is quickly deduced from the second equation in (97) by noting that preceding the original filter with a multiplication by $M_i(t)$ is equivalent to changing the input time function from $z(t)$ to $z(t)M_i(t)$, while following the original filter by a filter with transfer function $T_0(f)$ is equivalent to multiplying both sides of the second equation in (97) by $T_0(f)$.

A constraint on the duration of the input waveform to, say, $t_i - T_i/2 < t < t_i + T_i/2$ can be handled analytically by using a hypothetical Frequency-Dependent Modulation Function $M'(t, f)$ given by

$$M'(t, f) = \text{Rect}\left(\frac{t - t_i}{T_i}\right)M(t, f), \tag{106}$$

or equivalently by using a hypothetical Input Doppler-Spread Function $H'(f, \nu)$ given by

$$H'(f, \nu) = \int e^{-i2\pi t_i(\nu-\mu)} T_i \operatorname{sinc}[T_i(\nu - \mu)]H(f, \mu)\, d\mu \tag{107}$$

in place of the actual system function and then assuming an internal input time constraint.

A constraint on the frequency interval over which the channel output is observed, say to $f_0 - W_0/2 < f < f_0 + W_0/2$ cps, may be handled analytically by using

$$M'(t, f) = M(t, f)\,\text{Rect}\left(\frac{f - f_0}{W_0}\right) \tag{108}$$

in place of the actual Frequency-Dependent Modulation Function $M(t, f)$, or equivalently by using

$$h'(t, \xi) = \int e^{i2\pi f_0(\xi-\eta)} W_0 \operatorname{sinc} W_0(\xi - \eta)h(t, \eta)\, d\eta \tag{109}$$

in place of the actual Output Delay-Spread Function $h(t, \xi)$ and then assuming an internal output bandwidth constraint.

Having completed our preliminary discussion of time and frequency constraints we may now proceed to the determination of the corresponding sampling canonical channel models.

All the sampling models are derived by application of the Sampling Theorem,[18] which states that if a function $h(x)$ is zero for values of x outside an interval $-X/2 < x < X/2$, then its Fourier transform $H(y)$ may be expressed as the following series

$$H(y) = \sum H\left(\frac{k}{X}\right) \operatorname{sinc}\left[X\left(y - \frac{k}{X}\right)\right] \tag{110}$$

where

$$\operatorname{sinc} y = \frac{\sin \pi y}{\pi y} \tag{111}$$

and

$$H(y) = \int h(x)e^{-i2\pi xy}\, dx \quad \text{or} \quad \int h(x)e^{i2\pi xy}\, dx. \tag{112}$$

When $h(x)$ vanishes outside an interval which is centered on x_1, *i.e.*, when $h(x)$ vanishes outside the interval $x_1 - X/2 < x < x_1 + X/2$, the Sampling Theorem becomes

$$H(y) = \sum H\left(\frac{k}{X}\right)e^{\pm i2\pi x_1(y-k/X)} \operatorname{sinc}\left[X\left(y - \frac{k}{X}\right)\right] \tag{113}$$

where the sign of the exponential used in (113) agrees with the sign of the exponential used in the Fourier transform definition of $H(y)$.

a) *Sampling Models for Input Time and Frequency Constraints*: In this subsection we will derive sampling models appropriate for input time and frequency constraints. Consider first the case of an input time constraint. As discussed above, such a case may be described by stating that $M(t, f)$ vanishes for values of t outside some interval, say, $t_i - T_i/2 < t < t_i + T_i/2$. Since the Input Doppler-Spread Function $H(f, \nu)$ is the Fourier transform of $M(t, f)$ with respect to t (see Fig. 5), it follows from (113) that

$$H(f, \nu) = \sum H\left(f, \frac{n}{T_i}\right)e^{-i2\pi t_i(\nu-n/T_i)} \cdot \operatorname{sinc}\left[T_i\left(\nu - \frac{n}{T_i}\right)\right]. \tag{114}$$

If the summation in (114) is used in place of $H(f, \nu)$ in the input-output relationship [(12)]

$$W(f) = \int Z(f - \nu)H(f, \nu)\, d\nu,$$

one finds that in the case of an input time constraint the output spectrum is given by

$$W(f) = \sum H_n(f) \int Z(f - \nu)T_i \cdot \operatorname{sinc}\left[T_i\left(\nu - \frac{n}{T_i}\right)\right]e^{-i2\pi t_i(\nu-n/T_i)}\, d\nu \tag{115}$$

where

$$H_n(f) = \frac{1}{T_i} H\left(f, \frac{n}{T_i}\right). \tag{116}$$

Note that the integral in (115) is just the convolution of the input spectrum $Z(f)$ with $e^{-i2\pi t_i(f-n/T_i)}T_i \operatorname{sinc}[T_i(f - n/T_i)]$. Since convolution becomes multiplication in the transform domain, and since the time function corresponding to $e^{-i2\pi t_i(f-n/T_i)}T_i \operatorname{sinc}[T_i(f - n/T_i)]$ is

[18] P. M. Woodward, *op. cit.*,[7] pp. 33–34.

$\exp [j2\pi n(t/T_i)]$ Rect $([t - t_i]/T_i)$, the time function corresponding to the convolution integral in (116) is just the product $z(t) \exp [j2\pi nt/T_i]$ Rect $([t - t_i]/T_i)$. Thus, (115) states that the channel output may be obtained by gating the input with the time function Rect $([t - t_i]/T_i)$, frequency shifting the gated input by harmonics of $1/T_i$ cps, filtering each of these gated frequency-shifted waveforms with an appropriate filter $H_n(f)$ for each harmonic and then summing the result. This series of operations immediately suggests the channel model shown in Fig. 10. Although, in theory, an infinite number of frequency converter-filtering elements would be required, in practice a finite number will suffice since for a physical channel the range of Doppler shifts is finite and thus $H(f, \nu)$ must effectively vanish for ν outside some interval.

When the channel is randomly time-variant the filters $H_n(f)$ become random filters. The correlation properties of these filters are defined by the average

$$\overline{H_n^*(f)H_m(l)} = \left(\frac{1}{T_i}\right)^2 R_H\left(f, l; \frac{n}{T_i}, \frac{m}{T_i}\right). \tag{117}$$

For the case of the US channel (90) simplifies to [see (66)]

$$\overline{H_n^*(f)H_m(f + \Omega)} = \left(\frac{1}{T_i}\right)^2 R_H\left(\Omega; \frac{n}{T_i}, \frac{m}{T_i}\right). \tag{118}$$

It is not difficult to see that a channel may not have an *internal* input (or output) time constraint and be a WSS (or WSSUS) channel since the existence of a time constraint is incompatible with the existence of stationarity. However, an *external* input (or output) time constraint may be associated with any type of channel.

The correlation properties of the random filters as expressed in (117) and (118) are pertinent to the case of an internal input time constraint. In order to obtain from (85) the corresponding cross-correlation function between the random filters for the case of an external input time constraint, *i.e.*, the case of a time-limited input waveform (*e.g.*, a pulse input), we replace the actual Input Doppler-Spread Function by a hypothetical one as indicated in (107). It is quickly seen that instead of (116) we have

$$H_n(f) = \int e^{-j2\pi t_i(\nu - n/T_i)} \operatorname{sinc}\left[T_i\left(\nu - \frac{n}{T_i}\right)\right] H(f, \nu)\, d\nu \tag{119}$$

as the expression for the transfer function of the random filter in the canonic channel model of Fig. 10 when the input time constraint is external to the channel proper. Also, instead of (117) we have

$$\overline{H_n^*(f)H_m(l)} = \iint e^{-j2\pi t_i(\nu - n/T_i - \mu + m/T_i)} \operatorname{sinc}\left[T_i\left(\nu - \frac{n}{T_i}\right)\right] \cdot \operatorname{sinc}\left[T\left(\mu - \frac{m}{T_i}\right)\right] R_H(f, l; \nu, \mu)\, d\nu\, d\mu \tag{120}$$

as the cross-correlation function between the random filters for the general channel. By using the appropriate form for the correlation functions, (120) may be specialized

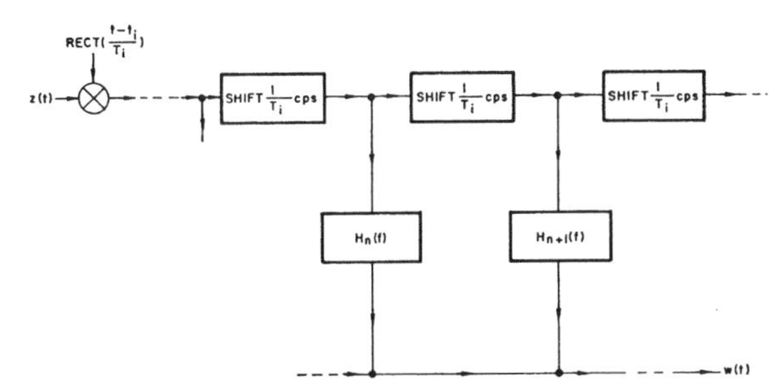

Fig. 10—Canonical channel model for input time constraint, output filter version.

for the US, WSS, and WSSUS channels. We mention only the simplified form it takes in the case of the WSSUS channel,

$$\overline{H_n^*(f)H_m(f + \Omega)} = \int e^{-j2\pi t_i(m-n)/T_i} \operatorname{sinc}\left[T_i\left(\nu - \frac{n}{T_i}\right)\right] \cdot \operatorname{sinc}\left[T_i\left(\nu - \frac{m}{T_i}\right)\right] P_H(\Omega, \nu)\, d\nu. \tag{121}$$

It will be recalled that $P_H(\Omega, \nu)$ is equal to the cross-power spectral density between the processes $M(f, t)$ and $M(f + \Omega, t)$. From (121) it is readily seen that if $P_H(\Omega, \nu)$ changes very little for changes in ν of the order of the reciprocal of the duration of the input waveform, $1/T_i$, we have the approximation

$$\overline{H_n^*(f)H_m(f + \Omega)} \approx \begin{cases} 0 & ; \quad m \neq n \\ \dfrac{1}{T_i} P_H\left(\Omega, \dfrac{n}{T_i}\right); & \quad m = n \end{cases}. \tag{122}$$

Thus, in the case of the WSSUS channel, when $P_H(\Omega, \nu)$ varies little for changes in ν of the order of the reciprocal of the duration T_i of the input time constraint, the various random filters become uncorrelated and the frequency correlation function of an individual filter transfer function $H_n(f)$ becomes proportional to the value of the density function $P_H(\Omega, \nu)$ at $\nu = n/T_i$.

We will now determine the channel model appropriate to the case of an input bandwidth restriction.[19] This restriction is dual to that of an input time limitation and, as will be seen, leads to a dual channel model.

Let us assume that the channel responds only to frequencies within the interval $f_i - W_i/2 < f < f_i + W_i/2$. From the discussion in Subsection VIA.1), we see that this is equivalent to stating that $T(f, t)$ is zero for values of f outside this interval. Since $g(t, \xi)$ is the inverse Fourier transform of $T(f, t)$, an application of the Sampling Theorem shows that $g(t, \xi)$ may be represented by the series

$$g(t, \xi) = \sum g\left(t, \frac{n}{W_i}\right) e^{j2\pi f_i(\xi - n/W_i)} \cdot \operatorname{sinc}\left[W_i\left(\xi - \frac{n}{W_i}\right)\right]. \tag{123}$$

[19] This channel model has previously been derived by Kailath, *op. cit.*[2]

If the summation in (123) is used in place of $g(t, \xi)$ in the input-output relationship

$$w(t) = \int z(t - \xi)g(t, \xi)\, d\xi,$$

one finds that in the case of an input frequency constraint the output time function may be represented in the form

$$w(t) = \sum g_n(t) \int z(t - \xi)W_i \cdot \text{sinc}\left[W_i\left(\xi - \frac{n}{W_i}\right)\right]e^{j2\pi f_i(\xi - n/W_i)}\, d\xi \tag{124}$$

where

$$g_n(t) = \frac{1}{W_i}\, g\left(t, \frac{n}{W_i}\right). \tag{125}$$

Note that the integral in (124) is just the convolution of the input with a time function $W_i \exp[j2\pi f_i(t - n/W_i)]$ sinc $[W_i(t - n/W_i)]$. Since the spectrum corresponding to this latter time function is $\exp[-j2\pi n(f/W_i)]$ rect $([f - f_i]/W_i)$, it follows that the spectrum corresponding to the convolution in (124) is just the product $Z(f)$ Rect $([f - f_i]/W_i) \exp[-j2\pi n(f/W_i)]$. Thus (124) states that the channel output may be obtained by passing the input through a band-limiting filter with transfer function Rect $([f - f_i]/W_i)$, delaying the resultant by multiples of a basic delay $1/W_i$, multiplying each of these delayed functions by a multiplier $g_n(t)$ appropriate to the delay n/W_i, and then summing the result. This series of operations immediately suggests the channel model shown in Fig. 11.

Although, in theory, an infinite number of taps would be required, in practice a finite number will suffice, since for a physical channel the spread of path delays is finite and thus $g(t, \xi)$ must effectively vanish for ξ outside some interval.

When the channel is randomly time variant the multipliers $g_n(t)$ become random processes. The correlation properties of these multipliers are defined by the average

$$\overline{g_n^*(t)g_m(s)} = \left(\frac{1}{W_i}\right)^2 R_g\left(t, s; \frac{n}{W_i}, \frac{m}{W_i}\right). \tag{126}$$

For the case of the WSS channel,

$$\overline{g_n^*(t)g_m(t, \tau)} = \left(\frac{1}{W_i}\right)^2 R_g\left(\tau; \frac{n}{W_i}, \frac{m}{W_i}\right). \tag{127}$$

It is not difficult to see that a channel may not have an *internal* input (or output) bandwidth constraint and be a US (or WSSUS) channel. To understand this fact recall [see discussion following (66)] that the US channel has a wide sense stationarity property in the frequency variable. Such a property is clearly incompatible with an internal input (or output) frequency constraint. However, an *external* frequency constraint may, of course, be associated with any type of channel.

The correlation properties of the random multipliers as expressed in (126) and (127) are pertinent to the case

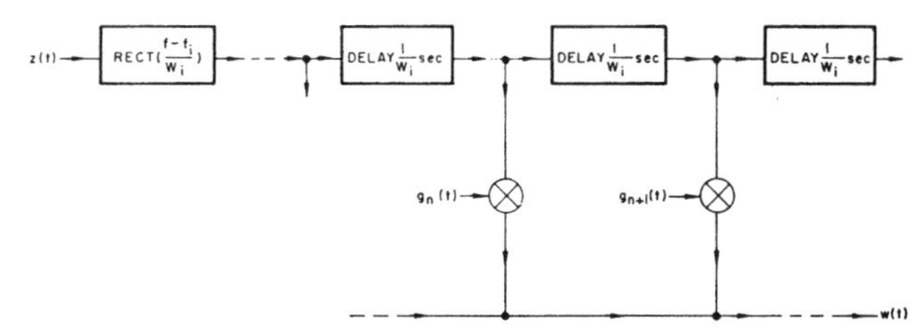

Fig. 11—Canonical channel model for input frequency constraint, output multiplier version.

of an internal input bandwidth constraint. In order to obtain the corresponding cross-correlation function for the case of an external input bandwidth constraint, *i.e.*, a band-limited input signal, we replace the actual Input Delay-Spread Function by a hypothetical one as indicated in (99). It is readily seen that instead of (125), we have

$$g_n(t) = \int e^{-j2\pi f_i(\xi - n/W_i)} \cdot \text{sinc}\left[W_i\left(\xi - \frac{n}{W_i}\right)\right]g(t, \xi)\, d\xi \tag{128}$$

as the expression for the multiplier associated with a delay of n/W_i sec in the canonic model of Fig. 11 when the input frequency constraint is external to the channel proper. Instead of (127), we have

$$\overline{g_n^*(t)g_m(s)} = \iint e^{j2\pi f_i(\xi - \eta - n/W_i + m/W_i)}\, \text{sinc}\left[W_i\left(\xi - \frac{n}{W_i}\right)\right] \cdot \text{sinc}\left[W_i\left(\eta - \frac{m}{W_i}\right)\right]R_g(t, s; \xi, \eta)\, d\xi\, d\eta. \tag{129}$$

By using the appropriate form for the correlation functions (129) may be specialized for the US, WSS, and WSSUS channels. We mention only the simplified form it takes in the case of the WSSUS channel,

$$\overline{g_n^*(t)g_m(t + \tau)} = \int e^{j2\pi f_i([n-m]/W_i)} \cdot \text{sinc}\left[W_i\left(\xi - \frac{m}{W_i}\right)\right]P_g(\tau, \xi)\, d\xi. \tag{130}$$

Examination of (130) shows that if $P_g(\tau, \xi)$ changes little for changes in ξ of the order of $1/W$, we have the approximation

$$\overline{g_n^*(t)g_m(t + \tau)} \approx \begin{cases} 0 & ; \quad m \neq n \\ \dfrac{1}{W_i} P_g\left(\tau, \dfrac{n}{W_i}\right) & ; \quad m = n \end{cases}. \tag{131}$$

There exist alternate channel models closely related to those of Figs. 10 and 11. Consider first the case of an input time constraint and the channel model of Fig. 10. If we define $Z'(f)$ as the spectrum of the time function, $z(t)$ Rect $([t - t_i]/T)$, which is the input to the frequency conversion chain, it is readily seen that the spectrum of the channel output $W(f)$ may be expressed as

$$W(f) = \sum_n H_n(f)Z'\left(f - \frac{n}{T_i}\right). \tag{132}$$

If we define the filter

$$G_n(f) = H_n\left(f + \frac{n}{T_i}\right) \tag{133}$$

then the output spectrum may be expressed as

$$W(f) = \sum G_n\left(f - \frac{n}{T_i}\right)Z'\left(f - \frac{n}{T_i}\right) \tag{134}$$

which states that the output may be obtained by gating, filtering, and frequency shifting as indicated in Fig. 12. With the aid of (133) and (116) and (119) one may determine expressions for $G_n(f)$ in terms of the system functions for the cases of internal and external time constraints. Also the correlation properties of $G_n(f)$ are readily determined from those of $H_n(f)$ with the aid of (133).

Similarly, if we define $z''(t)$ as the time function resulting from band-limiting the input as shown in Fig. 11, it is seen that the channel output $w(t)$ may be expressed as

$$w(t) = \sum_n g_n(t)z''\left(t - \frac{n}{W_i}\right). \tag{135}$$

If we define the time function

$$h_n(t) = g_n\left(t + \frac{n}{W_i}\right) \tag{136}$$

then the output time function may be expressed as

$$w(t) = \sum_n h_n\left(t - \frac{n}{W_i}\right)z''\left(t - \frac{n}{W_i}\right) \tag{137}$$

which states that the output may be obtained by band-limiting, multiplying, and delaying as indicated in Fig. 13. With the aid of (136) and (125) and (128) one may determine expressions for $h_n(t)$ in terms of the system functions for the cases of internal and external bandwidth constraints. Also the correlation properties of $h_n(t)$ are readily determined from those of $g_n(t)$ with the aid of (136).

b) *Sampling Models for Output Time and Frequency Constraints*: The development of channel models for output time and frequency constraints parallels the previous development of channel models for input time and frequency constraints. When considering an output time constraint one specifies that $T(f, t)$ vanishes for t outside some interval, say, $t_0 - T_0/2 < t < t_0 + T_0/2$ while for an output frequency constraint one specifies that $M(t, f)$ vanishes for f outside an interval $f_0 - W_0/2 < f < f_0 + W_0/2$. Then, for an output time constraint, one makes a sampling expansion of $G(f, \nu)$ [the Fourier transform of $T(f, t)$ with respect to t] in the ν variable while for an output frequency constraint one makes a sampling expansion of $h(t, \xi)$ [the inverse Fourier transform of $M(t, f)$ with respect to f] in the ξ variable. By using these expansions in the appropriate input-output relations and examining the resulting series expressions for the output (as was done for the case of input time and frequency constraints) one may obtain the appropriate canonical channel models for the case of output time and frequency constraints.

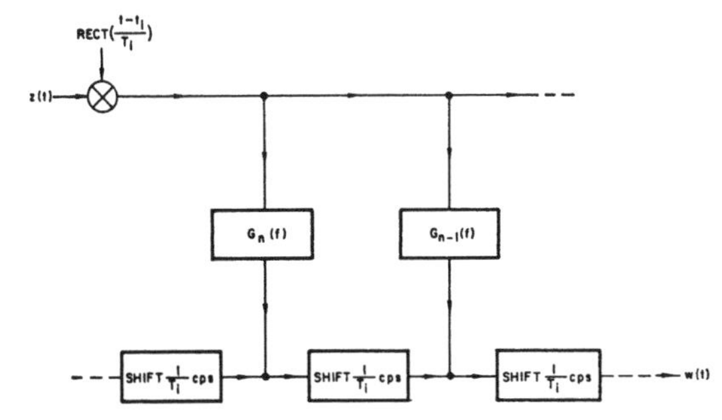

Fig. 12—Canonical channel model for input time constraint, input filter version.

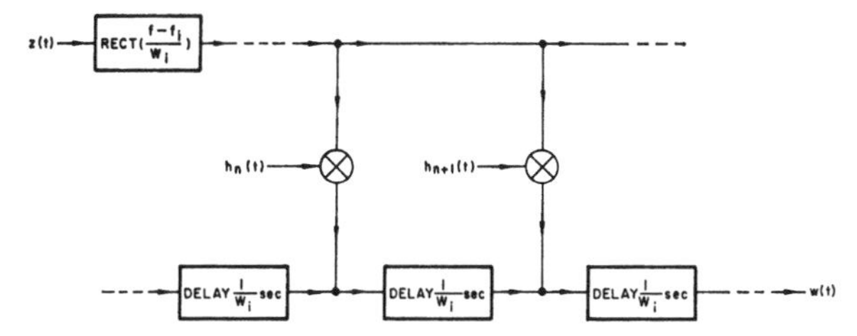

Fig. 13—Canonical channel model for input frequency constraint, input multiplier version.

We shall not give the details of the derivations because of their similarity to the derivations in the previous section. The resulting canonical models are shown in Figs. 14 to 17.[20] Note that the output time constraint models differ in form from the input time constraint models only in having an output time gate instead of an input time gate. Similarly, the output frequency constraint models differ from the input frequency constraint models only in having an output rather than input band-pass filter.

The filter transfer functions in Figs. 14 and 15 are given by

$$\hat{H}_n(f) = \frac{1}{T_0} H\left(f, \frac{n}{T_0}\right)$$
$$\hat{G}_n(f) = \frac{1}{T_0} G\left(f, \frac{n}{T_0}\right) = H_n\left(f + \frac{n}{T_0}\right) \tag{138}$$

in the case of internal output time constraints. When the output time constraint is external, however, one must use

$$\hat{G}_n(f) = \int e^{i2\pi t_0(\nu - n/T_0)} \operatorname{sinc}\left[T_0\left(\nu - \frac{n}{T_0}\right)\right]G(f, \nu)\, d\nu$$
$$\hat{H}_n(f) = G_n\left(f - \frac{n}{T_0}\right). \tag{139}$$

The gain functions in Figs. 16 and 17 are

$$\hat{h}_n(t) = \frac{1}{W_0} h\left(t, \frac{n}{W_0}\right)$$
$$\hat{g}_n(t) = \frac{1}{W_0} g\left(t, \frac{n}{W_0}\right) = h\left(t - \frac{n}{W_0}\right) \tag{140}$$

in the case of internal output bandwidth constraints.

[20]The model in Fig. 16 has been previously derived by Kailath, *op. cit.* [2]

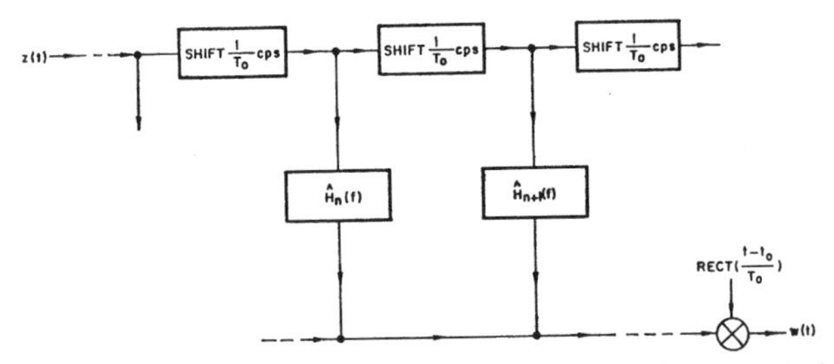

Fig. 14—Canonical channel model for output time constraint, output filter version.

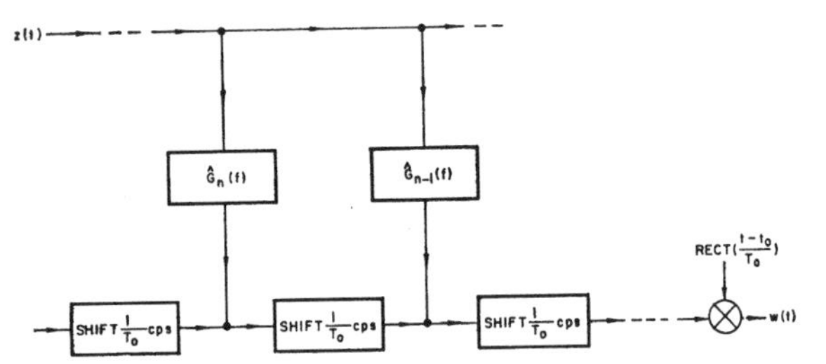

Fig. 15—Canonical model for output time constraint, input filter version.

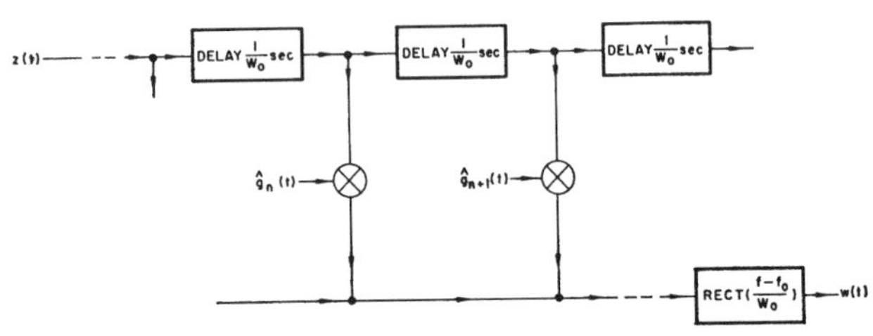

Fig. 16—Canonical channel model for output frequency constraint, output multiplier version.

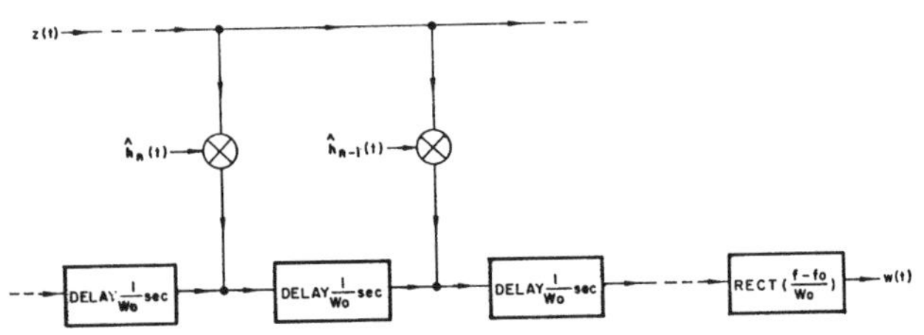

Fig. 17—Canonical channel model for output frequency constraint, input multiplier version.

For external output bandwidth constraints the gain functions become

$$\hat{h}_n(t) = \int e^{-j2\pi f_0(\xi - n/W_0)} \cdot \operatorname{sinc}\left[W_0\left(\xi - \frac{n}{W_0}\right)\right] h(t, \xi)\, d\xi$$
$$\hat{g}_n(t) = \hat{h}_n\left(t - \frac{n}{W_0}\right). \tag{141}$$

For randomly time-varying channels the correlation properties of the filter transfer functions in Figs. 14 and 15 and the gain functions in Figs. 16 and 17 are quickly determined in terms of the appropriate channel correlation functions, as was done for the case of input time and frequency constraints.

c) *Sampling Models for Combined Time and Frequency Constraints*: In Subsections VA. 1) a) and b), we have developed canonic channel models for the case of a single constraint on time or frequency at the input or output of the channel. However, it is possible for certain combinations of these constraints to exist for the same channel. Two interesting combinations, which are dually related, are the cases of a combined input time constraint and output frequency constraint and a combined input frequency constraint and output time constraint. Other combinations are either impossible or do not lead to new models. The impossible combinations are combined internal time and frequency constraints on the same end of the channel, since (as a study of the meaning of an internal constraint readily reveals) such combined constraints imply the existence of functions which are both time- and band-limited.

Consider now the case of an internal input time and output frequency constraint. Mathematically we can represent such a combined constraint by stating that the Frequency-Dependent Modulation Function $M(t, f)$ vanishes for values of t and f outside the rectangle $t_i - T_i/2 < t < t_i + T_i/2$, $f_0 - W_0/2 < f < f_0 + W_0/2$. Thus, $M(t, f)$ satisfies the equation

$$M(t, f) = \operatorname{Rect}\left(\frac{t - t_i}{T_i}\right) M(t, f) \operatorname{Rect}\left(\frac{f - f_0}{W_0}\right). \tag{142}$$

But it is readily seen that we can also write

$$M(t, f) = \operatorname{Rect}\left(\frac{t - t_i}{T_i}\right) \tilde{M}(t, f) \operatorname{Rect}\left(\frac{f - f_0}{W_0}\right) \tag{143}$$

where

$$\tilde{M}(t, f) = \sum_{m,n=-\infty}^{\infty} M(t - mT_i, f - mW_0) \tag{144}$$

since only the $m = 0$, $n = 0$ term in the sum (144) defining $\tilde{M}(t, f)$ contributes nonzero values to the left side of (143).

Eq. (143) states that the channel under discussion may be represented as the cascade of three operations. The first is an input gating operation with the function Rect $([t - t_i]/T_i)$, the second is a filtering operation by means of a time variant filter with Frequency-Dependent Modulation Function $\tilde{M}(t, f)$ and the last is a band-pass filtering operation with transfer function Rect $([f - f_0]/W_0)$. We will now develop a canonical channel model for the second operation and then obtain our desired canonic channel model for the combined input time and output frequency constraint by preceding the canonic channel model of the second operation by the time gate Rect $([t - t_i]/T_i)$ and following it by the band-pass filter Rect $([f - f_0]/W_0)$.

It will be recalled (see Fig. 5 for example) that the Doppler-Delay-Spread Function $V(\nu, \xi)$ is the double Fourier transform of $M(t, f)$,

$$V(\nu, \xi) = \iint M(t, f) e^{-j2\pi(t\nu - f\xi)}\, dt\, df.$$

Thus the Doppler-Delay-Spread Function corresponding to $\tilde{M}(t, f)$ is given by

$$\tilde{V}(\nu, \xi) = \sum_{m,n}\sum \iint M(t - mT_i, f - nW_0) \cdot e^{-i2\pi(t\nu - f\xi)}\, dt\, df$$
$$= V(\nu, \xi)\Big(\sum_m e^{-i2\pi mT_i\nu}\Big)\Big(\sum_n e^{i2\pi nW_0\xi}\Big) \tag{145}$$

where the second equation follows from the first by making the changes of variable $t - mT_i \to t$, $f - nW_0 \to f$ in the double integral in the first equation.

The sums in the second equation in (145) may be recognized as Fourier series expansions of periodic impulse trains, *i.e.*,

$$\sum_m e^{-i2\pi mT_i\nu} = \frac{1}{T_i}\sum_m \delta\Big(\nu - \frac{m}{T_i}\Big)$$
$$\sum_n e^{i2\pi nW_0\xi} = \frac{1}{W_0}\sum_n \delta\Big(\xi - \frac{n}{W_0}\Big) \tag{146}$$

so that

$$\tilde{V}(\nu, \xi) = \frac{1}{T_iW_0}\sum_{m,n}\sum V\Big(\frac{m}{T_i}, \frac{n}{W_0}\Big) \cdot \delta\Big(\nu - \frac{m}{T_i}\Big)\delta\Big(\xi - \frac{n}{W_0}\Big). \tag{147}$$

If we let $z'(t)$ and $w'(t)$ respectively denote the input and output of the channel with Doppler-Delay-Spread Function $\tilde{V}(\nu, \xi)$, we can express the channel output as

$$w'(t) = \sum_{m,n}\sum V_{mn} z'\Big(t - \frac{n}{W_0}\Big)e^{i2\pi(m/T_i)(t-n/W_0)} \tag{148}$$

where the complex amplitude

$$V_{mn} = \frac{1}{T_iW_0} V\Big(\frac{m}{T_i}, \frac{n}{W_0}\Big). \tag{149}$$

Thus, apart from the input gate and the output band-limiting filter, the channel representation corresponding to an input-time output frequency limitation is a discrete number of point "scatterers" each providing first a fixed Doppler shift which is some multiple of the reciprocal input time duration constraint and then a fixed delay which is some multiple of the reciprocal output bandwidth constraint. The complex amplitude of the reflection from the point scatterer is just $1/T_iW_0$ times the Doppler-Delay-Spread Function sampled at the same value of delay and Doppler shift provided by the scatterer. Fig. 18 demonstrates the realization of such a channel by tapped delay lines and frequency conversion chains.

The canonic model of Fig. 18 can be derived in a somewhat different manner than described above by making use of canonic channel models previously derived for single constraints on input time and output frequency. Thus, consider the canonic channel model for the case of an input time constraint, Fig. 10, and note that when an output frequency constraint exists the filters $H_n(f)$ must

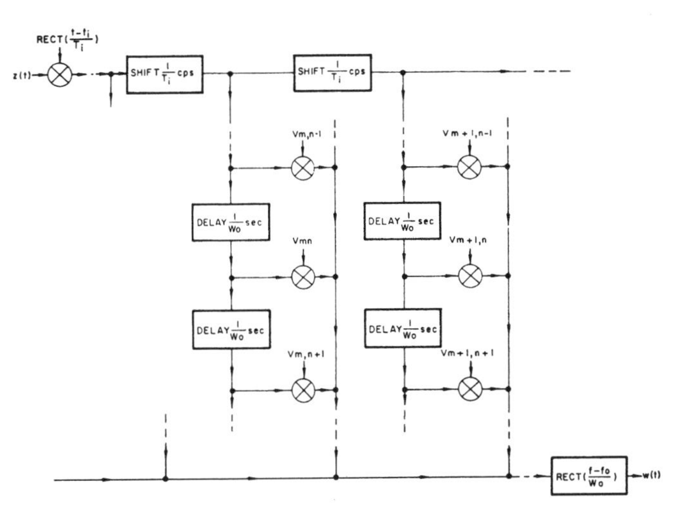

Fig. 18—Canonical channel model for input time-output frequency constraint.

have this same constraint, *i.e.*, $H_n(f)$ must vanish for values of f outside the interval $f_0 - W_0/2 < f < f_0 + W_0/2$. This latter fact becomes quite clear by examination of Fig. 10, since these filters are the only elements in the model which provide frequency selectivity. An analytical proof is readily obtained by noting that $H_n(f) = (1/T_i)H(f, n/T_i)$ and that the f variable in $H(f, \nu)$ and $M(t, f)$ are the same variables since $H(f, \nu)$ and $M(t, f)$ are Fourier transform pairs in the t, ν variables.

Since the filter $H_n(f)$ has an output frequency constraint (and also an input frequency constraint since it is time-invariant), one may represent this filter by means of a canonic channel model consisting of a tapped delay line as shown in Fig. 11, but with time-invariant gains. When this representation of $H_n(f)$ is made one arrives at the canonic channel model shown in Fig. 18.

When the channel is randomly varying the coefficients V_{mn} become random variables. The correlation between these random gains is given by

$$\overline{V_{mn}^* V_{rs}} = \frac{1}{(T_iW_0)^2} R_V\Big(\frac{m}{T_i}, \frac{r}{T_i}; \frac{n}{W_0}, \frac{s}{W_0}\Big). \tag{150}$$

The expressions for the gain (149) and its correlation properties (150) are applicable for the case of internal constraints on the input time and output frequency. For the case of external constraints the same canonical channel model applies, but in determining V_{mn} the actual Frequency-Dependent Modulation Function $M(t, f)$ should be replaced by a hypothetical one given by

$$M'(t, f) = \text{Rect}\Big(\frac{t - t_i}{T_i}\Big)M(t, f)\,\text{Rect}\Big(\frac{f - f_0}{W_0}\Big) \tag{151}$$

or, equivalently, the actual Doppler-Delay-Spread Function $V(\nu, \xi)$ should be replaced by a hypothetical one $V'(\nu, \xi)$ given by

$$V'(\nu, \xi) = \int \text{Rect}\Big(\frac{t - t_i}{T_i}\Big)e^{-i2\pi t\nu} \cdot \int \text{Rect}\Big(\frac{f - f_0}{W_0}\Big)e^{i2\pi f\xi}M(t, f)\, dt\, df. \tag{152}$$

With the aid of (108) and (109) the integral with respect to f can be expressed as

$$\int \text{Rect}\left(\frac{f-f_0}{W_0}\right)e^{i2\pi f\xi}M(t,f)\,df$$
$$= \int e^{i2\pi f_0(\xi-\eta)}W_0 \operatorname{sinc}[W_0(\xi-\eta)]h(t,\eta)\,d\eta. \quad (153)$$

Similarly the integration with respect to t may be expressed as

$$\int \text{Rect}\left(\frac{t-t_i}{T_i}\right)e^{-i2\pi t\nu}h(t,\eta)\,dt$$
$$= \int e^{-i2\pi t_i(\nu-\mu)}T_i \operatorname{sinc}[T_i(\nu-\mu)]V(\mu,\eta)\,d\mu \quad (154)$$

which results in the following expression for $V'(\nu, \xi)$:

$$V'(\nu,\xi) = \iint e^{-i2\pi t_i(\nu-\mu)}e^{i2\pi f_0(\xi-\eta)}T_iW_0$$
$$\cdot\operatorname{sinc}[T_i(\nu-\mu)]\operatorname{sinc}[W_0(\xi-\eta)]V(\mu,\eta)\,d\mu\,d\eta. \quad (155)$$

The gain V_{mn} in the case of an external input time-output frequency constraint is given by

$$V_{mn} = \frac{1}{T_iW_0}V'\left(\frac{m}{T_i},\frac{n}{W_0}\right) = \iint e^{i2\pi t_i(\nu-m/T_i)}$$
$$\cdot e^{-i2\pi f_0(\xi-n/W_0)}\operatorname{sinc}\left[T_i\left(\nu-\frac{m}{T_i}\right)\right]$$
$$\cdot\operatorname{sinc}\left[W_0\left(\xi-\frac{n}{W_0}\right)\right]V(\nu,\xi)\,d\nu\,d\xi. \quad (156)$$

The correlation $\overline{V_{mn}^*V_{rs}}$ is readily determined as a fourfold integral involving $R_V(\nu,\mu;\xi,\eta)$ by using the integral representation, (156), for V_{mn} and V_{rs} and then averaging under the integral sign. It does not appear desirable to take the space to present this fourfold integral. In the case of the WSSUS channel, for which [(72) and (74)]

$$R_V(\nu,\mu;\xi,\eta) = S(\xi,\nu)\delta(\mu-\nu)\delta(\eta-\xi),$$

the fourfold integral becomes the double integral

$$\overline{V_{mn}^*V_{rs}} = e^{-i2\pi(t_i/T_i)(m-r)}e^{i2\pi(f_0/W_0)(n-s)}$$
$$\cdot\iint\operatorname{sinc}\left[T_i\left(\nu-\frac{m}{T_i}\right)\right]\operatorname{sinc}\left[T_i\left(\nu-\frac{r}{T_i}\right)\right]$$
$$\cdot\operatorname{sinc}\left[W_0\left(\xi-\frac{n}{W_0}\right)\right]\operatorname{sinc}\left[W_0\left(\xi-\frac{s}{W_0}\right)\right]S(\xi,\nu)\,d\nu\,d\xi. \quad (157)$$

When the Scattering Function $S(\xi, \nu)$ varies very little for changes in ξ of the order of $1/W_0$ and changes in ν of the order of $1/T_i$, (157) simplifies to

$$\overline{V_{mn}^*V_{rs}} \approx \begin{cases} 0 & ;\ m\neq n,\ r\neq s. \\ \dfrac{1}{T_iW_0}S\left(\dfrac{n}{W_0},\dfrac{m}{T_i}\right) & ;\ m=n,\ r=s \end{cases} \quad (158)$$

Thus for the WSSUS channel and a sufficiently smooth Scattering Function, the gains of the point "scatterers" in the canonical channel model become uncorrelated and the strength of the reflection from a particular scatterer becomes proportional to the amplitude of the Scattering Function at the same value of delay and Doppler shift provided by the scatterer.

A somewhat different canonical channel model may be derived for the case of an input-time output-frequency constraint by using the relationship [see (34)]

$$U\left(\frac{n}{W_0},\frac{m}{T_i}\right) = V\left(\frac{m}{T_i},\frac{n}{W_0}\right)e^{-i2\pi(m/T_i)(n/W_0)} \quad (159)$$

in (121) to show that

$$w'(t) = \frac{1}{T_iW_0}\sum_{m,n}\sum U\left(\frac{n}{W_0},\frac{m}{T_i}\right)$$
$$\cdot z'\left(t-\frac{n}{W_0}\right)e^{i2\pi(m/T_i)t}. \quad (160)$$

Examination of (160) shows that $w'(t)$ is obtained by first delaying $z'(t)$ by multiplies of $1/W_0$ and then Doppler-shifting by multiples of $1/T_i$. Thus, this model will differ from the one shown in Fig. 18 only in that the order of delay and Doppler shift is reversed and the complex amplitude of the reflection from the point scatterer is equal to $1/T_iW_0\ U(n/W_0, m/T_i)$ rather than $1/T_iW_0\ V(m/T_i, n/W_0)$.

To derive the canonical channel model for the case of an input-frequency output-time constraint we may proceed in a manner entirely analogous to that in the case of the dual constraint, *i.e.*, the input time-output frequency constraint. We have omitted this derivation because of its similarity to that for the dual case. The resulting channel model is shown in Fig. 19, in which

$$U_{mn} = \frac{1}{T_0W_i}U\left(\frac{m}{W_i},\frac{n}{T_0}\right) \quad (161)$$

in the case of internal constraints and

$$U_{mn} = \iint e^{-i2\pi f_i(\xi-m/W_i)}e^{i2\pi t_0(\xi-n/T_0)}$$
$$\cdot\operatorname{sinc}\left[W_i\left(\xi-\frac{m}{W_i}\right)\right]\operatorname{sinc}\left[T_0\left(\nu-\frac{n}{T_0}\right)\right]U(\xi,\nu)\,d\nu\,d\xi \quad (162)$$

in the case of external constraints.

The correlation between the gains is given by

$$\overline{U_{mn}^*U_{rs}} = \frac{1}{(T_0W_i)^2}R_U\left(\frac{m}{W_i},\frac{r}{W_i};\frac{n}{T_0},\frac{s}{T_0}\right) \quad (163)$$

for the case of internal time and frequency constraints.

As in the dual situation, the correlation between the gains in the case of external constraint may be expressed in terms of a fourfold integral involving the Delay-Doppler-Spread Function. We present only the expression for the case of WSSUS channel,

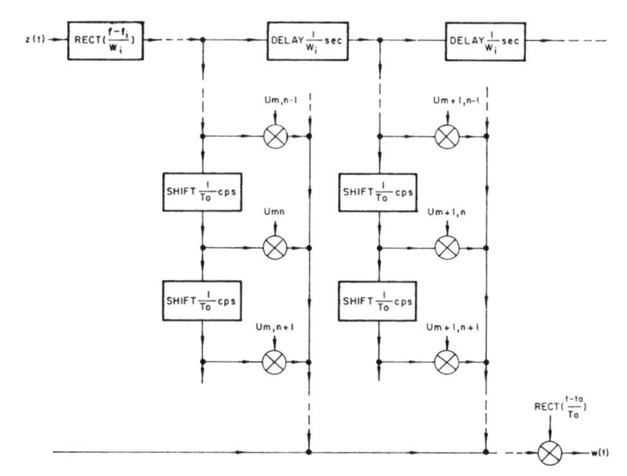

Fig. 19—Canonical model for input frequency-output time constraint.

$$\overline{U_{mn}^* U_{rs}} = e^{j2\pi(f_i/W_i)(m-r)} e^{-j2\pi(t_0/T_0)(n-s)}$$
$$\cdot \iint \operatorname{sinc}\left[W_i\left(\xi - \frac{m}{W_i}\right)\right] \operatorname{sinc}\left[W_i\left(\xi - \frac{m}{W_i}\right)\right]$$
$$\cdot \operatorname{sinc}\left[T_0\left(\nu - \frac{n}{T_0}\right)\right] \operatorname{sinc}\left[T_0\left(\nu - \frac{s}{T_0}\right)\right] S(\xi, \nu)\, d\nu\, d\xi \quad .(164)$$

When the Scattering Function varies very little for changes in ξ of the order of $1/W_i$ and changes in ν of the order of $1/T_0$, (164) simplifies to

$$\overline{U_{mn}^* U_{rs}} \approx \begin{cases} 0 & ; \quad m \neq n, \quad r \neq s \\ \dfrac{1}{W_i T_0} S\left(\dfrac{n}{W_i}, \dfrac{M}{T_0}\right) & ; \quad m = n, \quad r = s \end{cases} \tag{165}$$

yielding a collection of uncorrelated scatterers for the canonic channel model.

As in the dual case, a canonical channel model may be found for the case of an input-frequency output-time constraint which differs from that shown in Fig. 19 only in a reversal of the order of delay and Doppler shift. In this model the gain of a scatterer is set equal to

$$(1/T_0 W_i) V(n/T_0, m/W_i)$$

rather than $(1/T_0 W_i) U(m/W_i, n/T_0)$.

2) *Delay- and Doppler-Shift Constraints*: For physical channels the spread of path delays and the spread of Doppler shifts are effectively limited to finite values. According to our definition of system functions, a limitation in the spread of path delays means that *all* system functions containing the delay variable ξ vanish for values of ξ outside some specified interval. Similarly, a limitation in the spread of Doppler shifts means that *all* system functions containing the Doppler-shift variable ν vanish for values of ν outside some specified interval. This situation is somewhat different from the cases of time and frequency constraints discussed above where different physical interpretations (input as opposed to output constraints) might be associated with a specification that system functions vanish for values of the variables t or f outside specified ranges. The difference in behavior may be traced to the fact that the dual system functions $U(\xi, \nu)$ and $V(\nu, \xi)$ are so simply related both must vanish over the same intervals of ξ and ν, while the dual system functions $M(t, f)$ and $T(f, t)$, with their more complicated relationship, (39), need not vanish over the same intervals of f and t.

In the following subsections we shall develop canonical models for channels which are limited in either path delay spread or Doppler spread or both Doppler- and delay-spread.

a) *Sampling Models for Delay-Spread Constraint*:[21] If we assume that a channel provides path delays only in an interval Δ seconds wide centered at ξ_0 seconds, then $g(t, \xi)$, $U(\xi, \nu)$, $h(t, \xi)$, and $V(\nu, \xi)$ must vanish for values of ξ outside this interval. It follows that the Sampling Theorem, (86), may be applied to the Fourier transforms of these system functions with respect to the delay variable, *i.e.*, to the system functions $T(f, t)$, $G(f, \nu)$, $M(t, f)$, and $H(f, \nu)$, respectively. To derive the canonical channel models appropriate to a delay-spread limitation it is sufficient to deal with $T(f, t)$ and $M(f, t)$. Thus, according to (113), we find that $T(f, t)$ and $M(f, t)$ have the following expansions

$$T(f, t) = \sum_m T\left(\frac{m}{\Delta}, t\right) e^{-j2\pi\xi_0(f-m/\Delta)} \cdot \operatorname{sinc}\left[\Delta\left(f - \frac{m}{\Delta}\right)\right] \tag{166}$$

and

$$M(t, f) = \sum_m M\left(t, \frac{m}{\Delta}\right) e^{-j2\pi\xi_0(f-m/\Delta)} \cdot \operatorname{sinc}\left[\Delta\left(f - \frac{m}{\Delta}\right)\right]. \tag{167}$$

Upon using the expansion (166) to represent $T(f, t)$ in the input-output relationship (19), we find the series representation for the channel output to be

$$w(t) = \sum_m T\left(\frac{m}{\Delta}, t\right) \int Z(f)\, e^{-j2\pi\xi_0(f-m/\Delta)} \cdot \operatorname{sinc}\left[\Delta\left(f - \frac{m}{\Delta}\right)\right] df. \tag{168}$$

Examination of (168) shows that the channel output is represented as the sum of the outputs of a number of elementary parallel channels, each of which filters the input with a transfer function of the form

$$\exp\left[-j2\pi\xi_0\left(f - \frac{m}{\Delta}\right)\right] \operatorname{sinc} \Delta\left(f - \frac{m}{\Delta}\right)$$

for some value of m and then multiplies the resultant by a gain function $T(m/\Delta, t)$. The impulse response of the filter, *i.e.*, the inverse Fourier transform of

$$\exp\left[-j2\pi\xi_0\left(f - \frac{m}{\Delta}\right)\right] \operatorname{sinc} \Delta\left(f - \frac{m}{\Delta}\right)$$

[21] This case has been treated previously by Kailath, *op. cit.*[2]

is readily found to be

$$\exp\left[j2\pi \frac{m}{\Delta} t\right] \frac{1}{\Delta} \text{Rect}\left(\frac{t-\xi_0}{\Delta}\right),$$

which is the complex envelope of a rectangular RF pulse of frequency $f_c + m/\Delta$ (where f_c is the carrier frequency) and of width Δ seconds centered on $t = \xi_0$. Such a filter has frequently been called a band-pass integrator. If we let

$$I_m(f) = \exp\left[-j2\pi\xi_0\left(f - \frac{m}{\Delta}\right)\right] \text{sinc } \Delta\left(f - \frac{m}{\Delta}\right), \quad (169)$$

then we can represent the canonical channel model corresponding to (168) as shown in Fig. 20.

An alternate channel representation which involves multipliers on the input side rather than the output side may be derived by using the series (140) to represent $M(t, f)$ in the input-output relationship (21). The resulting series expression for the output spectrum is given by

$$W(f) = \sum_m I_m(f) \int z(t) M\left(t, \frac{m}{\Delta}\right) e^{j2\pi ft} \, dt. \quad (170)$$

Examination of (170) shows that the mth term in the sum involves a multiplication of the input by a (complex) gain function $M(t, m/\Delta)$ followed by a filtering operation with a filter having transfer function $I_m(f)$. Such a representation is shown in Fig. 21.

As with the previous channel models, although an infinite number of elements is involved, only a finite number is needed in practice. Thus, in Fig. 20, since the approximate bandwidth of each band-pass integrator is $1/\Delta$ cps, and since adjacent integrators are separated by $1/\Delta$ cps, an input signal of bandwidth W would require somewhat more than $W\Delta$, perhaps $10W\Delta$, judiciously selected adjacent multiplier-filter channels to produce a very close approximation to the channel output. More elementary channels may be needed in the model in Fig. 21 where the multipliers precede the filters, because the time varying gains $M(t, m/\Delta)$; $m = 0, \pm 1, \pm 2$, etc. spread the spectra of the inputs to the corresponding filters.

When the channel is randomly time-variant the gain functions $T(m/\Delta, t)$ and $M(t, m/\Delta)$ become random processes. It is clear that the correlation properties of these gain functions are completely determined from the correlation functions of the Time-Variant Transfer Function and Frequency-Dependent Modulation Function. Since these correlation functions have been discussed in detail in Section III, there is no need for further discussion here. However, it is interesting to note that only for the WSSUS channel do the correlation properties of the gain functions in Figs. 20 and 21 become identical, *i.e.*, for the WSSUS channel

$$\overline{M^*(t, m/\Delta) M(t + \tau, n/\Delta)}$$
$$= \overline{T^*(m/\Delta, t) T(n/\Delta, t + \tau)} = R\left(\frac{n-m}{\Delta}, \tau\right) \quad (171)$$

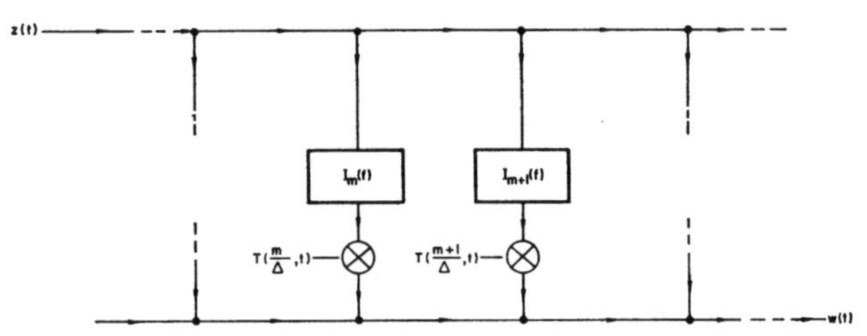

Fig. 20—Canonical channel model for delay-spread limited channel, output multiplier version.

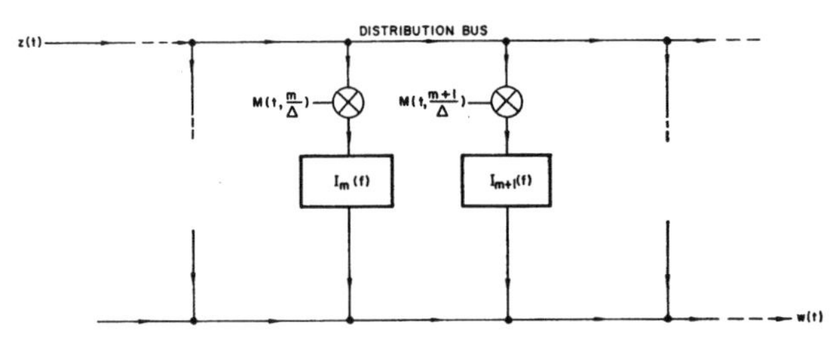

Fig. 21—Canonical channel model for delay-spread limited channel, input multiplier version.

where $R(\Omega, \tau)$ is the Time-Frequency Correlation Function defined in Section IV-D.

b) *Sampling Models for Doppler-Spread Constraint*: The Doppler-spread constraint is dual to the delay-spread constraint and the derivations and resulting canonic models follow a dual pattern. Thus, if we assume that a channel provides Doppler shifts only in an interval Γ cps wide centered on ν_0 cps, then $H(f, \nu)$, $V(\nu, \xi)$, $G(f, \nu)$, and $U(\xi, \nu)$ must vanish for values of ν outside this interval. Application of the sampling theorem then produces the expansions

$$T(f, t) = \sum_n T(f, n/\Gamma) e^{j2\pi\nu_0(t-n/\Gamma)} \text{sinc } [\Gamma(t - n/\Gamma)] \quad (172)$$

and

$$M(t, f) = \sum_n M(n/\Gamma, f) e^{j2\pi\nu_0(t-n/\Gamma)} \cdot \text{sinc } [\Gamma(t - n/\Gamma)]. \quad (173)$$

Upon using (172) in (15) and (173) in (21), we obtain the following expansions for the channel output:

$$w(t) = \sum_n e^{j2\pi\nu_0(t-n/\Gamma)} \text{sinc } [\Gamma(t - n/\Gamma)] \cdot \int Z(f) T(f, n/\Gamma) e^{j2\pi ft} \, dt \quad (174)$$

and

$$W(f) = \sum M(n/\Gamma, f) \int z(t) e^{j2\pi\nu_0(t-n/\Gamma)} \cdot \text{sinc } [\Gamma(t - n/\Gamma)] e^{j2\pi ft} \, dt. \quad (175)$$

The nth term of the series (174) may be interpreted as the result of a filtering operation with filter transfer function $T(f, n/\Gamma)$, followed by a multiplication with a gain function $\exp [j2\pi\nu_0(t - n/\Gamma)]$ sinc $\Gamma(t - n/\Gamma)$. If we let

$$p_n(t) = \exp [j2\pi\nu_0(t - n/\Gamma)] \text{ sinc } [\Gamma(t - n/\Gamma)], \quad (176)$$

then the canonical channel model which follows from the above interpretation of (174) is as shown in Fig. 22. In an entirely analogous fashion (175) leads to the model shown in Fig. 23.

Whereas in the dual cases described in the previous section a finite number of multiplier-filter combinations is satisfactory for representing the channel for a band-limited input, in the present cases a finite number of multiplier-filter combinations may be used when the input is time-limited. This fact may be appreciated by noting that the gain function $p_n(t)$ acts as a "gate" of duration $1/\Gamma$ and that the "gates" of adjacent elementary channels are separated by $1/\Gamma$ seconds. Thus, an input signal of duration T will require anywhere from, say $T\Gamma$ to $10T\Gamma$ judiciously selected adjacent elementary channels to characterize the channel.

When the channel is randomly time-variant the filters $T(f, n/\Gamma)$ and $M(n/\Gamma, f)$ become random processes in the frequency variable. The correlation properties of these filters may be determined from the results of Section IV, which deals with the correlation functions of the various system functions. It is interesting to note, as in the dual case, that only for the WSSUS channel do the correlation properties of the random filters in Figs. 22 and 23 become identical.

c) *Sampling Models for Combined Delay-Spread and Doppler-Spread Constraints*: When both a Delay-Spread and Doppler-Spread constraint exist, the sampling theorem may be applied twice to the system functions $T(f, t)$ and $M(t, f)$, *i.e.*, once when they are considered as time functions with the frequency variables fixed and once when they are considered as frequency functions with the time variables fixed. In this manner one may form sampling expansions and determine corresponding canonical channel models. However, this procedure is unnecessary since the desired models may be obtained by inspection of Figs. 20–23 by combining models appropriate to delay-spread and Doppler-spread constraints. To demonstrate this latter approach, examine the model of Fig. 20, which is appropriate to a delay-spread constraint. If we require that a Doppler-spread constraint also exist, the multiplication operation in each parallel channel is in effect a sub-channel with a Doppler-spread constraint and may be represented by the canonical model of Fig. 22 or 23 where the f variable is set equal to m/Δ for the mth branch in Fig. 20. If this procedure is followed using the model of Fig. 22 in Fig. 20, the model of Fig. 24 appears. In an entirely analogous fashion one may generate three additional models, one by using the model of Fig. 23 in Fig. 20 and two more by using the models of Figs. 22 and 23 in Fig. 21.

It is readily demonstrated that for waveforms which are effectively limited in time and frequency duration, *i.e.*, waveforms which have most of their energy located in a finite time-frequency interval, only a finite number of filters and multipliers may be used in the models of Figs. 24 and 25 to provide a close approximation to the actual channel output.

The complex gain constants $T(m/\Delta, n/\Gamma)$, $M(n/\Gamma, m/\Delta)$ become random variables when the channel is randomly time-variant. Their correlation properties are just sampled values of the correlation functions of $T(f, t)$ and $M(t, f)$ respectively.

3) *Combined Time and Delay-Spread or Frequency and Doppler-Spread Constraints*: In Section VIA. 1) we have developed canonical channel models for time and frequency constraints, *i.e.*, for situations in which system functions vanish for values of t and/or f outside specified intervals. In Section VIA. 2) we have developed canonical

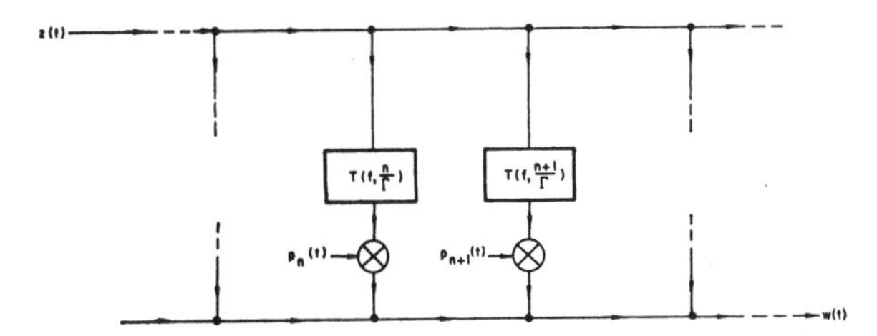

Fig 22—Canonical channel model for Doppler-spread limited channel, output multiplier version.

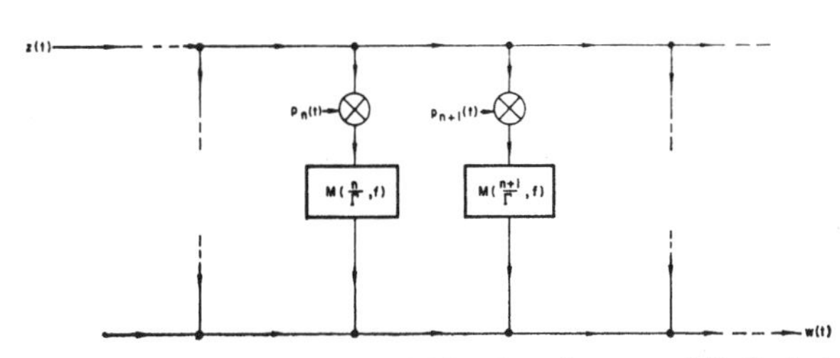

Fig. 23—Canonical channel model for Doppler-spread limited channel, input multiplier version.

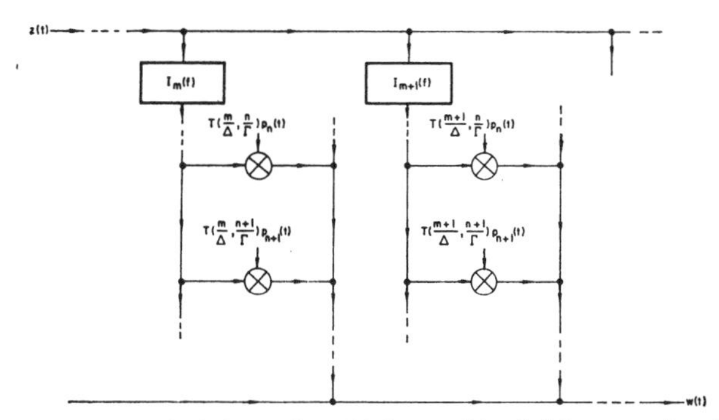

Fig. 24—A canonical channel model for combined delay-spread and Doppler-spread limited channel.

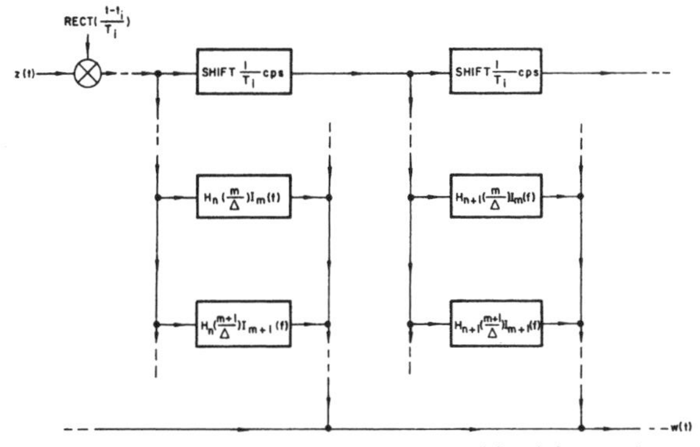

Fig. 25—A canonical channel model for combined input-time and delay-spread constraints.

channel models for delay-spread and Doppler-spread constraints, *i.e.*, for situations in which system functions vanish for values of ξ and/or ν outside specified intervals. Here we consider certain combinations of the constraints in Sections VIA. 1) and 2), namely, combined time and delay-spread constraints and combined frequency and Doppler-spread constraints. Other combinations are not possible since they are equivalent to requiring that a function be limited in both time and frequency.

The combined constraint models may be derived by combining the models appropriate to the individual constraints, as was done in Section VIA. 2) c). One may combine a delay-spread constraint with either an input or output frequency constraint. We shall present here only one model for a combined Doppler-spread and frequency constraint and one model for a combined delay-spread and time constraint. The remaining possible models may be quickly constructed by the reader.

To construct a model appropriate to an input time constraint and a delay-spread constraint we may make use of the models of Figs. 10 and 21. We note first that if the channel is delay-spread limited then the filters $H_n(f)$ in Fig. 10 are also delay-spread limited. Thus, each of these filters may be represented by means of a canonical model of the form of Fig. 21. Note, however, that since these filters are time-invariant, the gain functions in Fig. 21 are also time-invariant. To determine the value of the (time-invariant) gains it should be noted that a time-invariant filter with transfer function $H(f)$ has

$$M(t, f) = H(f). \tag{177}$$

Then, using the model of Fig. 21 to represent each filter in Fig. 10, we arrive at the model shown in Fig. 25.

To construct a model appropriate to an input frequency constraint and a Doppler-spread constraint we may use the Doppler-spread constraint model of Fig. 22 and input frequency constraint model of Fig. 11. In this connection one should note that a multiplier $g(t)$ is a degenerate time-variant linear filter with

$$T(f, t) = g(t). \tag{178}$$

It is readily determined that the model of Fig. 26 results when the model of Fig. 22 is used to represent each multiplier $g_n(t)$ in Fig. 11.

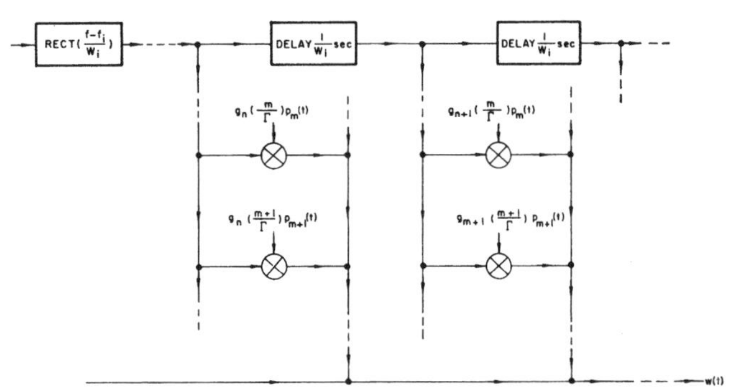

Fig. 26—A canonical channel model for combined input-frequency and Doppler-spread constraints.

B. Power Series Models

In this section we will derive certain canonical channel models which arise from power series expansions of $T(f, t)$ and $M(t, f)$ in either the time or frequency variables. The two models arising from expansions in the f variable will be called f-power series models, while those arising from expansions in the t variable will be called t-power series models. As might be expected, the f-power series models are dual to the t-power series models and are useful in dual situations.

1) *f-Power Series Models*: The starting point for our discussion is the input-output relationship corresponding to the Time-Variant Transfer Function $T(f, t)$ (19), which is repeated below:

$$w(t) = \int Z(f)T(f, t)e^{j2\pi ft}\, df.$$

If the input spectrum $Z(f)$ is confined primarily to a specified frequency interval over which $T(f, t)$ varies little in f, with a minimum fluctuation period much greater than the bandwidth of $Z(f)$, then a Taylor series representation of $T(f, t)$ in f will provide a rapidly convergent expansion of $Z(f)T(f, t)$ and $w(t)$.

Since the existence of a mean path delay ξ_o, *i.e.*, a value of ξ about which $g(t, \xi)$ may be considered centered, produces a factor $\exp[-j2\pi f\xi_o]$ in $T(f, t)$ which can fluctuate with f quite rapidly, it is desirable to expand only that portion of $T(f, t)$ which does not include this factor. To this end we may define a shifted Input Delay-Spread Function $g_0(t, \xi)$ in which the mean path delay ξ_o has been removed, *i.e.*,

$$g_0(t, \xi) = g(t, \xi + \xi_o) \tag{179}$$

where ξ_o is a mean multipath delay defined according to some convenient criterion. Then

$$T(f, t) = T_0(f, t)e^{-j2\pi f\xi_o} \tag{180}$$

where $T_0(f, t)$ is the Time-Variant Transfer Function of the medium after the mean path delay has been removed, *i.e.*,

$$T_0(f, t) = \int g_0(t, \xi)e^{-j2\pi f\xi}\, d\xi. \tag{181}$$

In the most general situation the input spectrum $Z(f)$ may not be centered at $f = 0$. Thus, assuming that $Z(f)$ is centered at $f = f_i$, the most rapid convergence of $Z(f)T_0(f, t)$ will be obtained by expanding $T_0(f, t)$ about $f = f_i$, *i.e.*,

$$T_0(f, t) = \sum_{n=0}^{\infty} T_n(t)(2\pi j)^n(f - f_i)^n \tag{182}$$

where

$$T_n(t) = \frac{1}{n!\,(2\pi j)^n}\left[\frac{\partial^n T_0(f, t)}{\partial f^n}\right]_{f=f_i}$$

$$= \frac{1}{n!}\int (-\xi)^n g_0(t, \xi)e^{-j2\pi f_i\xi}\, d\xi. \tag{183}$$

A filter with transfer function $(2\pi j)^n f^n$ is an nth order differentiator. We shall define a filter with transfer function $(2\pi j)^n (f - f_i)^n$ as an offset differentiator with an offset of f_i cps. If we let D_{f_i} be an operator denoting such an offset differentiation, it is quickly demonstrated that

$$D^n_{f_i}[f(t)] = e^{i2\pi f_i t} \frac{d^n}{dt^n} \{f(t)e^{-i2\pi f_i t}\}. \qquad (184)$$

Then use of (182) and (184) in (19) yields the following series representation of the channel output:

$$w(t) = \sum T_n(t) D^n_{f_i}[z(t - \xi_o)]$$
$$= e^{j2\pi f_i t} \sum T_n(t) \frac{d}{dt^n} \{z(t - \xi_o)e^{-i2\pi f_i t}\}. \qquad (185)$$

Examination of the last equation in (185) indicates that the channel output may be represented as the parallel combination of the outputs of an infinite number of elementary channels each consisting of a differentiation of some order followed by a time variant gain with all channels preceded by a delay ξ_o and then a frequency translation of $-f_i$ cps and all channels followed by a frequency translation of $+f_i$ cps. Study of the first equation in (185) shows that the channel output may also be represented as the parallel combination of elementary channels where now the typical channel consists of an offset differentiation of some order followed by the same time-variant gain, with all channels preceded by a delay ξ_o.

This latter channel representation is shown in Fig. 27, where the offset differentiators of different subchannels have been combined into a chain of offset differentiators.

For simplicity, in the following discussion of power series models, we shall present the particular forms that are simplest to diagram. It should be realized, however, that equations such as (185) may have a variety of interpretations in terms of channel models.

An understanding of the conditions leading to rapid convergence of the series (185) may be obtained by first defining a normalized shifted Input Delay Spread Function $\tilde{\tilde{g}}_0(t, \xi)$ whose "width" in the ξ direction is unity and a shifted normalized input time function $\tilde{\tilde{z}}_0(t)$ which is located at $f = 0$ and has unit bandwidth (using any convenient bandwidth criterion). These normalized functions are defined implicitly by the relations

$$z(t) = \tilde{\tilde{z}}_0(G_i t)e^{i2\pi f_i t}$$
$$g_0(t, \xi) = \frac{1}{\Delta_o} \tilde{\tilde{g}}_0\left(t, \frac{\xi}{\Delta_o}\right) \qquad (186)$$

where Δ_o is a measure of multipath spread given by the "width" of $g_0(t, \xi)$ in the ξ direction and B_i is the bandwidth of the input.

With the aid of the normalized functions one readily finds that the series (185) may be expressed in the form

$$w(t) = e^{i2\pi f_i(t-\xi_i)} \sum_{n=0}^{\infty} \frac{(j2\pi B_i \Delta_o)^n}{n!} \tilde{\tilde{T}}_n(t) \frac{\tilde{\tilde{z}}_0^{(n)}(B_i t - B_i \xi_o)}{(j2\pi)^n} \qquad (187)$$

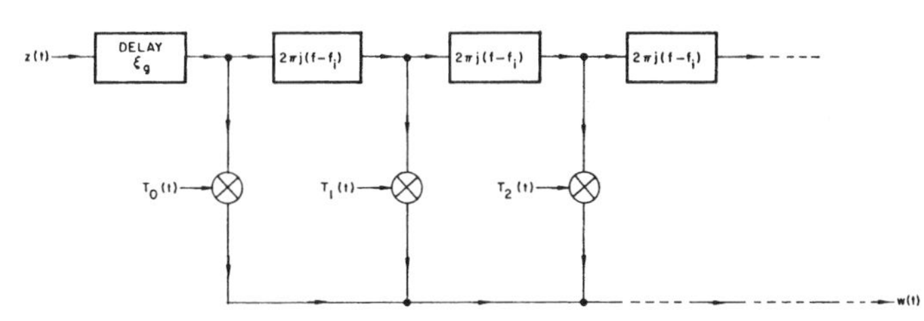

Fig. 27—f-power series channel model, output multiplier version.

where

$$\tilde{\tilde{T}}_n(t) = \int (-\xi)^n \tilde{\tilde{g}}_0(t, \xi)\, d\xi$$

$$\frac{\tilde{\tilde{z}}_0^{(n)}(t)}{(j2\pi)^n} = \frac{1}{(j2\pi)^n} \frac{d^n \tilde{\tilde{z}}_0(t)}{dt^n} = \int f^n \tilde{\tilde{Z}}_0(f) e^{i2\pi ft}\, df \qquad (188)$$

in which $\tilde{\tilde{Z}}_0(f)$ is the spectrum of $\tilde{\tilde{z}}_0(t)$.

Examination of (188) reveals that $T_n(t)$ and $[1/(j2\pi)^n] \cdot \tilde{\tilde{z}}_0^{(n)}(B_i t - B\xi_o)$ are both moments of functions having unit "duration." If, indeed, we assume $\tilde{\tilde{g}}_0(t, \xi)$ (as a function of ξ) and $\tilde{\tilde{Z}}_0(f)$ are zero outside the unit interval (centered at $\xi = 0$, and $f = 0$ respectively), then it is readily demonstrated that these moments may not increase and in practice will most likely decrease with increasing n. Examination of (187) then indicates that if

$$2\pi B_i \Delta_o \ll 1 \qquad (189)$$

the series will be rapidly convergent.

In the general case $\tilde{\tilde{g}}_0(t, \xi)$ and $\tilde{\tilde{Z}}_0(f)$ will have "tails" extending outside the unit interval. For (189) still to represent a useful convergence criterion, these tails must drop to zero sufficiently rapidly so that the moments do not increase too rapidly with increasing n.[22]

When the channel is randomly time-variant, the multiplier functions become random processes with correlation properties defined by

$$\overline{T_n^*(t)T_m(s)} = \frac{(-1)^{m+n}}{n!\, m!}$$
$$\cdot \iint \xi^m \eta^n R_o(t, s; \xi + \xi_o, \eta + \xi_o) e^{i2\pi f_i(\xi - \eta)}\, d\xi\, d\eta \qquad (190)$$

for the general channel. For the WSSUS channel, the cross-correlation function (190) specializes to (see Fig. 9)

$$\overline{T_n^*(t)T_n(t + \tau)} = \frac{(-1)^{m+n}}{n!\, m!} \int \xi^{m+n} Q(\tau, \xi + \xi_o)\, d\xi. \qquad (191)$$

In the case of the WSSUS channel, a desirable choice for ξ_o is given by

$$\xi_o = \frac{\int \xi Q(\xi)\, d\xi}{\int Q(\xi)\, d\xi} \qquad (192)$$

[22] The series will diverge if these tails fall too slowly. If the product of the two moments increases exponentially with n as α^n, one may modify (189) into $2\pi\alpha B_i \Delta_o \ll 1$ to obtain a suitable convergence criterion.

where $Q(\xi)$ is the Delay Power Density Spectrum, since such a choice not only minimizes $\overline{|T_1|^2}$ relative to $\overline{|T_0|^2}$ but also leads to $T_1(t)$ and $T_0(t)$ being uncorrelated. The ratio of the strength of T_1 relative to T_0 then takes the simple form

$$\frac{\overline{|T_1|^2}}{\overline{|T_0|^2}} = \frac{\int (\xi - \xi_o)^2 Q(\xi)\, d\xi}{\int Q(\xi)\, d\xi} = \Delta^2 \tag{193}$$

where Δ may be called the rms width of $Q(\xi)$.

When the frequency selective fading in the channel is sufficiently slow, only the first term in the series (185) will be sufficient to characterize the channel output, *i.e.*,

$$w(t) = T_0(t)z(t - \xi_o) \tag{194}$$

which may be recognized as a "flat-fading" or non-frequency-selective channel model. If the first two terms are used,

$$w(t) = T_0(t)z(t - \xi_o) + T_1(t)D_{f_i}[z(t - \xi_o)] \tag{195}$$

which may be called a "linearly frequency-selective fading" channel since it corresponds to approximating $T_0(f, t)$ by a linear term in the frequency variable. One may continue and define a "quadratically frequency-selective fading" channel, etc., depending upon the degree of approximation required.

We shall now investigate the error incurred in using a finite number of terms in the expansion (185) for the case of a WSSUS channel. If we assume the existence of derivatives of $T_0(f, t)$ with respect to f as high as Nth order, then we may expand $T_0(f, t)$ in a finite Taylor series expansion

$$T_0(f, t) = \sum_{n=0}^{N-1} T_n(t)(2\pi j)^n (f - f_i)^n + \frac{1}{N!}(f - f_i)^N \left[\frac{\partial^N T_0(f, t)}{\partial f^N}\right]_{f=f'} \tag{196}$$

where f' lies between f_i and f. Then using (196) in (15) and making use of (183), one obtains the following series expression for the channel output:

$$w(t) = \sum_{n=0}^{N-1} T_n(t)D_{f_i}^n[z(t - \xi_o)] + R_N(t) \tag{197}$$

where $R_N(t)$ is a remainder term given by

$$R_N(t) = \int (2\pi j)^N (f - f_i)^N Z(f) \frac{1}{N!} \frac{1}{(2\pi j)^N} \cdot \left[\frac{\partial^N T_0(f, t)}{\partial f N}\right]_{f=f'} e^{j2\pi f(t-\xi_o)}\, df$$
$$= \int (2\pi j)^N (f - f_i)^N Z(f) \frac{1}{N!} \cdot \int (-\xi)^N g_0(t, \xi) e^{-j2\pi f'\xi}\, d\xi\, e^{j2\pi f(t-\xi_o)}\, df. \tag{198}$$

It follows that

$$\overline{|R_N(t)|^2} = \iint (-2\pi j)^N (f - f_i)^N (2\pi j)^N (l - f_i)^N \frac{1}{(N!)^2} \overline{Z^*(f)Z(l)} \cdot \iint (-\xi)^N (-\eta)^N \overline{g_0^*(t, \xi) g_0(t, \eta)} \cdot e^{j2\pi(f'\xi - f''\eta)}\, d\xi\, d\eta\, e^{-j2\pi(f-l)(t-\xi_o)}\, df\, dl \tag{199}$$

where f'' lies between f_i and l.

For the WSSUS channel

$$\overline{g_0^*(t, \xi) g_0(t, \eta)} = Q(\xi + \xi_o)\delta(\eta - \xi), \tag{200}$$

and for a wide-sense stationary $z(t)$[12]

$$\overline{Z^*(f)Z(l)} = P_z(f)\delta(l - f) \tag{201}$$

where $P_z(f)$ is the power spectrum of $z(t)$.

Using (200) and (201) in (199) we readily find that

$$\overline{|R_N(t)|^2} = \frac{(2\pi)^{2N}}{(N!)^2} \int (f - f_i)^{2N} P_z(f)\, df \int (\xi - \xi_o)^{2N} Q(\xi)\, d\xi$$
$$= \frac{(2\pi B_i \Delta_o)^{2N}}{(N!)^2} \int f^{2N} \tilde{\tilde{P}}_z(f)\, df \int \xi^{2N} \tilde{\tilde{Q}}(\xi)\, d\xi \tag{202}$$

where $\tilde{\tilde{P}}_z(f)$ is the power spectrum of the normalized input signal $\tilde{\tilde{z}}(t)$ (unit bandwidth and centered at zero frequency) and $\tilde{\tilde{Q}}(\xi)$ is the Delay Power Density Spectrum associated with the normalized Delay-Spread Function $\tilde{\tilde{g}}_0(t, \xi)$.

One may show that the right-hand side of (202) is also just equal to the average magnitude squared of the $N + 1$th term in the series (185) for the case of a WSSUS channel. Thus, we have the simple error criterion that the average magnitude squared of the error incurred by using only a finite number of terms in (185) is just equal to the average magnitude squared of the first omitted when the channel is WSSUS and the input is wide-sense stationary.

We shall now derive a channel model which differs from Fig. 27 principally in a reversal of the order of the operations of differentiation and multiplication. This channel model is derived by making use of a Taylor series expansion of $M(t, f)$ in the frequency variable. The input-output relationship corresponding to $M(t, f)$ is given by (25), which is repeated below:

$$W(f) = \int z(t)M(t, f)e^{-j2\pi ft}\, dt.$$

If the output spectrum $W(f)$ is confined primarily to a specified frequency interval over which $M(t, f)$ varies little in f, with a minimum fluctuation period much greater than the bandwidth of $W(f)$, then a Taylor series representation of $M(t, f)$ in f can be used in (25) to obtain a rapidly convergent expansion of $W(f)$.

Since $M(t, f)$ is the Fourier transform of $h(t, \xi)$, the Input Delay-Spread Function, the presence of a nonzero value of ξ, say ξ_h, about which $h(t, \xi)$ is "centered" will result in a factor exp $[-j2\pi f\xi_h]$ in $M(t, f)$ which can fluctuate with f quite rapidly. Thus, it is desirable to expand only that portion of $M(t, f)$ which does not include this factor. If this portion of $M(t, f)$ is denoted by $M_0(t, f)$, we have

$$M_0(t, f) = \int h_0(t, \xi)e^{-i2\pi f\xi}\, d\xi \tag{203}$$

where

$$h_0(t, \xi) = h(t, \xi + \xi_h) \tag{204}$$

is a shifted Output Delay-Spread Function centered on $\xi = 0$. Using (204) and (203) we note that

$$M(t, f) = M_0(t, f)e^{-i2\pi f\xi_h}. \tag{205}$$

The most rapidly convergent expansion of $W(f)$ will be obtained by expanding $M_0(t, f)$ about the "center" frequency of $W(f)$, say, $f = f_0$, as follows:

$$M_0(t, f) = \sum_{n=0}^{\infty} M_n(t)(2\pi j)^n(f - f_0)^n \tag{206}$$

where

$$M_n(t) = \frac{1}{n!\,(2\pi j)^n}\left[\frac{\partial^n M_0(f, t)}{\partial f^n}\right]_{f=f_0}$$

$$= \frac{1}{n!}\int (-\xi)^n h_0(t, \xi)e^{-i2\pi f_0\xi}\, d\xi. \tag{207}$$

Upon using (205) and (206) in (25) we find that

$$W(f) = \sum_{n=0}^{\infty} (2\pi j)(f - f_0)^n e^{-i2\pi f\xi_h} \int z(t)M_n(t)e^{-i2\pi ft}\, dt. \tag{208}$$

By Fourier transforming (208) we obtain the following series expression for $w(t + \xi_h)$:

$$w(t + \xi_h) = \sum_{n=0}^{\infty} D^n_{f_0}[z(t)M_n(t)]. \tag{209}$$

Examination of (209) indicates that the channel output may be represented as the parallel combination of the outputs of an infinite number of elementary channels each consisting of a time-varying (complex) gain followed by an offset differentiation of some order with all channels followed by a delay of ξ_h seconds. Such a model is shown in Fig. 28, where the differentiators of different subchannels have been combined into a chain of differentiators.

In order to obtain a representation of (209) in terms of normalized functions analogous to (187) it is necessary to assume that the bandwidth of $z(t)M_n(t)$ does not exceed the bandwidth of $w(t)$. Then it may be shown that when $w(t)$ is band-limited to B_0 cps and $h(t, \xi)$ is zero outside a ξ interval of duration Δ_h, the condition for rapid convergence of (209) is given by

$$2\pi B_0\Delta_h \ll 1. \tag{210}$$

Even when $w(t)$ is not band-limited and $h(t, \xi)$ is not ξ-limited, (210) will still be a useful convergence criterion so long as the "tails" of $W(f)$ and $h(t, \xi)$ (as a function of ξ) drop to zero rapidly enough.

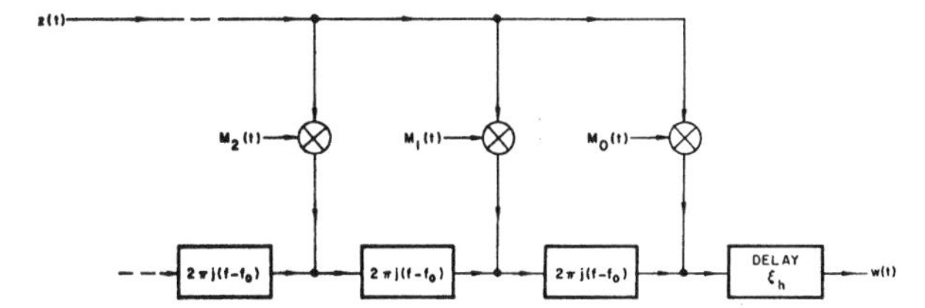

Fig. 28—f-power series channel model, input multiplier version.

It may also be shown in the same manner as was shown for the series (185) that the average magnitude squared error incurred by using a finite number of terms in the series (209) is just equal to the average magnitude squared of the first term omitted when the input is wide-sense stationary and the channel is WSSUS. As a final point we note that relationships analogous to (190) to (196) are readily constructed for the channel model of Fig. 28. Thus, it is readily shown that for the WSSUS channel the correlation properties of the multipliers in Fig. 28 are identical to those in Fig. 27 so that a choice of $\xi_h = \xi_g$ of (192) causes $M_0(t)$ and $M_1(t)$ to be uncorrelated and $\overline{|M_1|^2}$ to be minimized relative to $\overline{|M_0|^2}$.

2) *t-Power Series Models*: The t-Power Series Models are dual to the f-Power Series Models described in the previous section and thus involve power series expansions of $T(f, t)$ and $M(t, f)$ in the t variable. Whereas the f-Power series models are particularly useful when the frequency spread of the input (or output) spectrum and the delay spread of the channel are fairly sharply delimited and their product is much less than unity, the t-Power series models are particularly useful in the dual situation, *i.e.*, when the width of the input (or output) time function and the Doppler spread of the channel are fairly sharply delimited and their product is much less than unity.

Since the models presented in this section are dual to those in the previous section, the analytical part of their derivation is essentially identical to that in the previous section and thus it will be unnecessary to present as detailed derivations.

We will start our discussion by presenting the following expansion of $M(t, f)$:

$$M(t, f) = e^{i2\pi\nu_H t} \sum_{n=0}^{\infty} \hat{M}_n(f)(2\pi j)^n(t - t_i)^n \tag{211}$$

where t_i is a time instant about which the input may be assumed "centered" and ν_H is a value of ν about which $H(f, \nu)$ may be assumed "centered." The frequency function $\hat{M}_n(f)$ is given by

$$\hat{M}_n(f) = \frac{1}{n!\,(2\pi j)^n}\left[\frac{\partial^n}{\partial t^n}\{M(t, f)e^{-i2\pi\nu_H t}\}\right]_{t=t_i}$$
$$= \frac{1}{n!}\int \nu^n H(f, \nu + \nu_H)e^{i2\pi\nu t_i}\, d\nu. \tag{212}$$

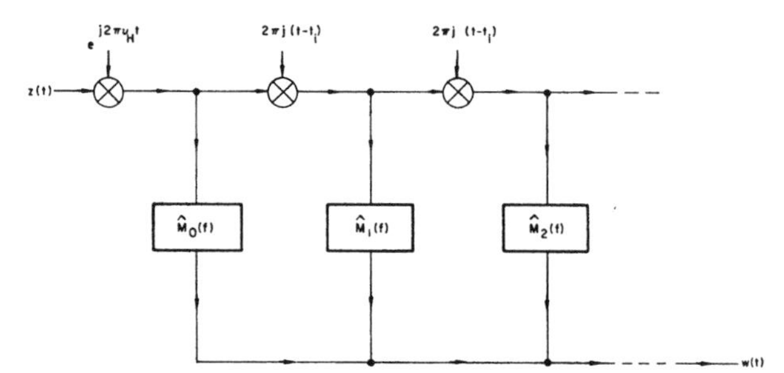

Fig. 29—t-power series model, output filter version.

Use of (211) in (25) leads to the following expansion for the output spectrum:

$$W(f) = \sum_{n=0}^{\infty} \hat{M}_n(f) \int z(t) e^{-j2\pi(f-\nu_H)t} (2\pi j)^n (t - t_i)^n \, dt$$

$$= \sum_{n=0}^{\infty} \hat{M}_n(f) D_{t_i}[Z(f - \nu_H)]$$

$$= e^{-j2\pi t_i f} \sum_{n=0}^{\infty} \hat{M}_n(f) \frac{d^n}{df^n} \{Z(f - \nu_H) e^{j2\pi t_i f}\}. \qquad (213)$$

Examination of the channel model in Fig. 29 readily reveals that the summed outputs of its elementary channels is identical to the series (186).

When the input time function is limited to a finite time interval T_i seconds and $H(f, \nu)$ is zero for values of ν outside an interval of duration β_H cps, one may show that the series (186) will be rapidly convergent if

$$2\pi T_i \beta_H \ll 1 \qquad (2.14)$$

[c.f. (189)]. Even if $z(t)$ is not time-limited and $H(f, \nu)$ is not ν-limited, (214) will still be satisfactory convergence criterion if the "tails" of $z(t)$ and $H(f, \nu)$ (as a function of ν) drop to zero sufficiently rapidly.

When the channel is randomly time-variant the filter transfer functions $\hat{M}_n(f)$ become random processes in the frequency variable with correlation properties defined by

$$\overline{\hat{M}_n^*(f)\hat{M}_m(l)} = \iint \frac{\nu^n \mu^m}{n!\, m!} R_H(f, l; \nu + \nu_H, \mu + \mu_H) \cdot e^{-j2\pi t_i(\nu-\mu)} \, d\nu \, d\mu \qquad (215)$$

for the general channel. In the case of the WSSUS channel, (215) simplifies to

$$\overline{\hat{M}_n^*(f)\hat{M}_m(f + \Omega)} = \frac{1}{n!\, m!} \int \nu^{m+n} P(\Omega, \nu) \, d\nu \qquad (216)$$

(see Fig. 9).

A desirable choice for ν_H in the case of the WSSUS channel is given by

$$\nu_H = \frac{\int \nu P(\nu) \, d\nu}{\int P(\nu) \, d\nu} \qquad (217)$$

where $P(\nu)$ is the Doppler Power Density Spectrum [see (76)], since in this case $\overline{|\hat{M}_1(f)|^2}$ is minimized relative to $\overline{|\hat{M}_0(f)|^2}$ and

$$\overline{\hat{M}_1^*(f)\hat{M}_0(f)} = 0. \qquad (218)$$

The ratio of the strength of $\hat{M}_1(f)$ relative to $\hat{M}_0(f)$ then takes the simple form

$$\frac{\overline{|\hat{M}_1(f)|^2}}{\overline{|\hat{M}_0(f)|^2}} = \frac{\int (\nu - \nu_H)^2 P(\nu) \, d\nu}{\int P(\nu) \, d\nu} \equiv \beta^2 \qquad (219)$$

where here β is the rms width of $P(\nu)$.

When the fading is sufficiently slow only the first term in (213) will be sufficient to characterize the channel output, *i.e.*, one may use

$$W(f) = \hat{M}_0(f) Z(f - \nu_H) \qquad (220)$$

which, apart from the frequency shift of ν_H cps, represents the channel as a time-invariant linear filter with transfer function $\hat{M}_0(f)$.

If the first two terms are used,

$$W(f) = \hat{M}_0(f) Z(f - \nu_H) + \hat{M}_1(f) D_{t_i}[Z(f - \nu_H)] \qquad (221)$$

which may be called a "linearly time-selective fading" channel since it corresponds to approximating $M(t, f)e^{-j2\pi\nu_H t}$ by a linear term in the time variable. One may continue and define a "quadratically time-selective fading channel," etc., depending upon the degree of approximation required.

An exact expression dual to (203) is readily formulated for the average magnitude squared error incurred by using only $N - 1$ terms in (213). However, this expression would be applicable in the dual situation, namely, when the input spectrum (rather than time function) is a wide-sense stationary process and the channel is WSSUS. Since such an input is not very common, the corresponding error expression may not be as useful as in the dual case. Thus, we present a different derivation which yields an upper bound on the average magnitude squared error for the case of arbitrarily specified $z(t)$ and a WSSUS channel.

We first express $M(f, t)e^{-j2\pi\nu_H t}$ in a finite Taylor series,

$$M(f, t)e^{-j2\pi\nu_H t} = \sum_{n=0}^{N-1} \hat{M}_n(f)(2\pi j)^n (t - t_i)^n + \frac{1}{N!}(t - t_i)^N \left[\frac{\partial^N}{\partial t^N}\{M(f, t)e^{-j2\pi\nu_H t}\}\right]_{t=t'} \qquad (222)$$

where t' lies between t_i and t. Then using (212) and (222) in (213) we obtain the finite series representation of the output spectrum

$$W(f) = \sum_{n=0}^{N-1} \hat{M}_n(f) D_{t_i}^n [Z(f - \nu_H)] + E_N(f): \qquad (223)$$

where the remainder term $E_N(f)$ is given by

$$E_N(f) = (2\pi j)^N \int (t - t_i)^N z(t) e^{j2\pi\nu_H t} \cdot \frac{1}{N!} \int \nu^N H(f, \nu + \nu_H) e^{j2\pi\nu t'} \, d\nu \, e^{-j2\pi f t} \, dt. \qquad (224)$$

It follows that

$$\overline{|E_N(f)|^2} = \frac{(2\pi)^{2N}}{(N!)^2} \iint (t - t_i)^N (s - t_i)^N z^*(t) z(s) e^{-i2\pi\nu_H(t-s)}$$

$$\cdot \iint \nu^N \mu^N \overline{H^*(f, \nu + \nu_H) H(f, \mu + \nu_H)}$$

$$\cdot e^{-i2\pi(\nu t' - \mu t'')} \, d\nu \, d\mu \, e^{i2\pi f(t-s)} \, dt \, ds \tag{225}$$

where t'' lies between t_i and s.

For the WSSUS channel,

$$\overline{H^*(f, \nu + \nu_H) H(f, \mu + \nu_H)} = P(\nu + \nu_H)\delta(\mu - \nu). \tag{226}$$

Using (226) in (225), we find

$$\overline{|E_N(f)|^2} = \frac{(2\pi)^{2N}}{(N!)^2} \iint (t - t_i)^N (s - t_i)^N z^*(t) z(s) e^{-i2\pi\nu_H(t-s)}$$

$$\cdot \int \nu^{2N} P(\nu + \nu_H) e^{-i2\pi\nu(t' - t'')} \, d\nu \, e^{i2\pi f(t-s)} \, dt \, ds. \tag{227}$$

Noting that the magnitude of an integral is less than the integral of the magnitude,

$$\overline{|E_N(f)|^2} \le \frac{(2\pi)^{2N}}{(N!)^2} \iiint |t - t_i|^N \, |s - t_i|^N$$

$$\cdot |z(t)| \, |z(s)| \, P(\nu + \nu_H) \nu^{2N} \, d\nu \, dt \, ds$$

$$= \frac{(2\pi)^{2N}}{(N!)^2} \int (\nu - \nu_H)^{2N} P(\nu) \, d\nu \left| \int |t - t_i|^N \, |z(t)| \, dt \right|^2$$

$$= \frac{(2\pi\beta_H T_i)^{2N}}{(N!)^2} \int \nu^{2N} \tilde{\tilde{P}}(\nu) \, d\nu \left| \int |t|^N \, |\hat{z}(t)| \, dt \right|^2 \tag{228}$$

where $\tilde{\tilde{P}}(\nu)$ is the Doppler Power Density Spectrum corresponding to a normalized Doppler-Spread Function $\tilde{\tilde{H}}(f, \nu)$ which differs from $H(f, \nu)$ in being translated and scaled along the ν axis so that it has $\beta_H = 1$ and $\nu_H = 0$. Similarly, $\hat{z}(t)$ is a shifted scaled version of the input which has unit duration and is located at $t = 0$.

The channel model dual to that in Fig. 28 may be arrived at with the aid of the following expansion:

$$T(f, t) = e^{i2\pi\nu_G t} \sum_{n=0}^{\infty} \hat{T}_n(f)(2\pi j)(t - t_0)^n \tag{229}$$

where t_0 is a time instant about which the output may be assumed "centered" and ν_G is the value of ν about which $G(f, \nu)$ may be assumed "centered." The frequency function $T_n(f)$ is given by

$$\hat{T}_n(f) = \frac{1}{n! \, (2\pi j)^n} \left[\frac{\partial^n}{\partial t^n} \{T(f, t) e^{-i2\pi\nu_G t}\} \right]_{t=t_0}$$

$$= \frac{1}{n!} \int \nu^n G(f, \nu + \nu_G) e^{i2\pi\nu t_0} \, d\nu. \tag{230}$$

Use of (229) in (19) leads to the following expansion for the output time function:

$$w(t) = e^{i2\pi\nu_G t} \sum_{n=0}^{\infty} (2\pi j)^n (t - t_0)^n \int \hat{T}_n(f) Z(f) e^{i2\pi f t} \, df \tag{231}$$

from which we readily infer the channel model shown in Fig. 30.

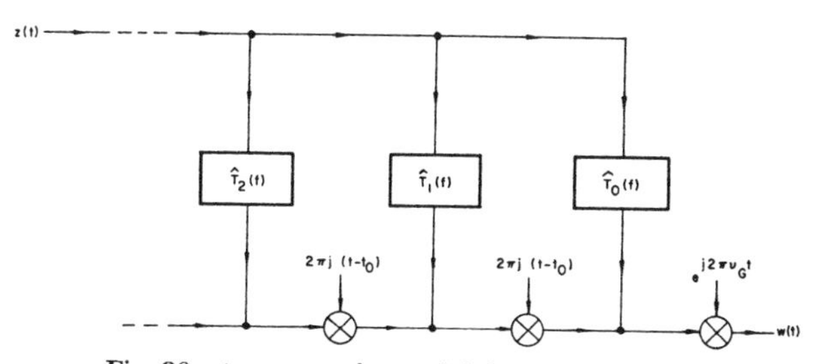

Fig. 30—t-power series model, input filter version.

When the output time function is limited to a finite interval of duration T_0 sec and $G(f, \nu)$ is zero for values of ν outside an interval of duration β_G cps, one may show that the series (231) will be rapidly convergent if

$$2\pi T_0 \beta_G \ll 1. \tag{232}$$

Even if $w(t)$ is not time-limited and $G(f, \nu)$ is not ν-limited, (232) will still be a satisfactory convergence criterion if the "tails" of $w(t)$ and $G(f, \nu)$ (as a function of ν) drop to zero sufficiently rapidly.

When the channel is randomly time-variant the filter transfer functions in Fig. 30, like those in Fig. 29, become random processes in the frequency variable. Relationships analogous to those in (215) to (228) are readily constructed for the model in Fig. 30. In particular, for the WSSUS channel the correlation properties of the random filters in Fig. 30 become identical to those of the random filters in Fig. 29.

3) *ft- and tf-Power Series Models*: In this section we present two channel models, one arising from an expansion of $T(f, t)$ and the other from an expansion of $M(t, f)$ in the t and f variables. As in the previous power series models, it is desirable to remove mean path delays and Doppler shifts before expanding these functions. Thus, we define

$$T_{00}(f, t) = T(f, t) e^{i2\pi(f\xi_0 - t\nu_0)} \tag{233}$$

where ξ_0, ν_0 are a mean delay and Doppler shift defined as the value of ξ and ν, about which the Delay-Doppler-Spread Function $U(\xi, \nu)$ may be assumed "centered", *i.e.*, $U(\xi + \xi_0, \nu + \nu_0)$ is "centered" at $\xi = \nu = 0$.

Since in $T(f, t)$ the variable f is associated directly with the spectrum of the input signal and the variable t is associated with the output time function, we expand $T_{00}(f, t)$ in the following double power series:

$$T_{00}(f, t) = \sum_0^{\infty} \sum_0^{\infty} T_{mn} (2\pi j)^{m+n} (f - f_i)^m (t - t_0)^n \tag{234}$$

where

$$T_{mn} = \frac{1}{m! \, n! \, (2\pi j)^{m+n}} \left[\frac{\partial^{m+n} T_{00}(f, t)}{\partial f^m \, \partial t^n} \right]_{\substack{f=f_i \\ t=t_0}}$$

$$= \frac{1}{m! \, n!} \iint (-\xi)^m (\nu)^n$$

$$\cdot U(\xi + \xi_0, \nu + \nu_0) e^{-i2\pi(f_i\xi - t_0\nu)} \, d\xi \, d\nu. \tag{235}$$

Using (234) in (19), we find that the output time function is represented by the series

$$w(t) = e^{j2\pi\nu_0 t} \sum_{m=0}^{\infty} \sum_{n=0}^{\infty} T_{mn}(2\pi j)^n (t - t_0)^n D_{f_i}^m [z(t - \xi_0)]. \quad (236)$$

It is readily seen by examination of (236) that an appropriate channel model whose output is given by (236) may be obtained by using the f-Power Series Model of Fig. 27 with each gain function represented by the t-Power Series Model of Fig. 30. We leave it to the reader to sketch this channel model, which we shall call the ft-Power Series Channel Model.

We may make statements concerning the convergence properties which are similar to those pertinent to the output series expansions of the t- and f-Power Series Models. However, since the tf-Power Series Model is, in essence, both a t- and an f-Power Series Model, the convergence requirements of both models need to be imposed. Thus, it is readily shown that when $U(\xi, \nu)$ is zero outside a rectangle whose sides are β_0 cps long in the ν direction and Δ_0 sec long in the ξ direction and when $z(t)$ is band-limited to a bandwidth B_i cps, a sufficient condition for convergence of (236) is the satisfaction of the inequalities

$$2\pi\ |t - t_0|\ \beta_0 \ll 1 \quad (237)$$

$$2\pi B_i \Delta_0 \ll 1. \quad (238)$$

It is clear from (237) that the series (237) may not converge for all values of t. However, if the significant values of the output are confined to an interval of duration T_0 one may change (237) to the inequality

$$2\pi T_0 \beta_0 \ll 1. \quad (239)$$

If the finite Taylor series

$$T_{00}(f, t) = \sum_{m=0}^{M} \sum_{n=0}^{N} T_{mn}(2\pi j)^{m+n} (f - f_i)^m (t - t_0)^n + \frac{(f - f_i)^M (t - t_0)^n}{M!\, N!} \left[\frac{\partial^{M+N} T_{00}(f, t)}{\partial f^M\, \partial t^N} \right]_{\substack{f=f' \\ t=t'}} \quad (240)$$

is used in (19), one readily finds that for the WSSUS channel the average magnitude squared error incurred by using terms up to $m = M - 1$ and $n = N - 1$ is bounded by

$$|E_{MN}|^2 \leq \frac{(2\pi T_0 \beta_0)^{2N} (2\pi B_i \Delta_0)^{2M}}{(N!)^2\, (M!)^2} \left| \int |f|^M\, |\tilde{Z}(f)|\, df \right|^2 \cdot \iint \xi^{2M} \nu^{2N} \tilde{\tilde{S}}(\xi, \nu)\, d\xi\, d\nu \quad (241)$$

where $\tilde{Z}(f)$ is a shifted scaled version of the input spectrum with unit bandwidth and located at zero frequency, and $\tilde{\tilde{S}}(\xi, \nu)$ is the Scattering Function [see (74)] associated with a shifted scaled version of $U(\xi, \nu)$ which is zero outside a unit square centered at $\xi = \nu = 0$. It is assumed in (241) that only those values of t are of interest for which $|t - t_0| \leq T_0$.

A discussion entirely dual to the one above may be formulated by expanding $M(t, f)$ rather than $T(f, t)$ in a Taylor series and using the resultant series to derive a series expansion for the output spectrum. Because the analytical procedure is identical to that above, except for a replacement of functions and variables by their duals, we shall not present these derivations. We note, however, that the resulting channel model, which we call the tf-Power Series Model may be obtained by using the t-Power Series Model of Fig. 29, with each filter represented by the f-Power Series Model of Fig. 28.

On the Optimum Detection of Digital Signals in the Presence of White Gaussian Noise—A Geometric Interpretation and a Study of Three Basic Data Transmission Systems*

E. ARTHURS†, MEMBER, IRE AND H. DYM‡, MEMBER, IRE

Summary—This paper considers the problem of optimally detecting digital waveforms in the presence of additive white Gaussian noise. A technique for representing the transmitted signals and the additive noise which leads to a geometric interpretation of the detection problem is presented on a tutorial level. Subsequently, this technique is used to derive the optimum detector for each of three basic data transmission systems: m-level Phase Shift Keyed, m-level Amplitude Shift Keyed and m-level Frequency Shift Keyed. Corresponding probability of error curves are derived, compared and discussed with reasonable detail.

I. Introduction

THIS PAPER WAS written to fulfill two objectives. The first objective is tutorial; the paper is intended to serve as an introduction to some of the ideas of modern statistical communication theory; in particular, the problem of detecting a known signal in a white noise background with minimum probability of error is introduced. The second objective is to analyze and compare in detail the performance of three basic data transmission systems, namely, m-level Phase Shift Keyed, m-level Amplitude Shift Keyed and m-level (orthogonal) Frequency Shift Keyed.

The approach adopted within this paper stresses the geometric viewpoint. Specifically, advantage is taken of the fact that in the white noise case it is possible to choose a convenient (orthonormal) representation for the transmitted signals and yet still be guaranteed that the noise can be decomposed suitably (*i.e.*, as described in Theorem II). This freedom of choice in signal representation leads in a natural way to a geometric interpretation of the detection problem which is both analytically correct and intuitively plausible. Furthermore, it is felt that the presented approach which, strictly speaking, is only applicable to the white noise case can serve as a useful introduction to the more general treatment wherein the transmitted signals are represented in terms of a Karhunen-Loeve expansion.[1,2]

The concepts developed in the early portions of the paper are used subsequently to derive the optimum detector for each of three systems mentioned above (both under the assumption of phase coherence and phase incoherence) and to derive the corresponding expressions for the probability of error. Several sets of curves are presented and discussed in reasonable detail in the final sections of the paper.

In the course of the presentation, references to additional articles and books on related subject matter are cited. However, appreciating the difficulty involved in sifting through many references each with its own peculiar notation, an effort has been made to write this paper as a complete unit. Accordingly, theoretical results which are utilized within the body of this paper without proof are discussed at length in the appendixes, which, with the exception of Appendix II, do not depend critically on outside source material. The treatment of Section V, which is concerned with deriving the probability of error for the systems under consideration, is somewhat different. The objective therein is to present a complete description of the calculations involved and the types of estimation which can be resorted to. It is believed that some of the presented results are new, although this is difficult to ascertain without an extensive search of literature. Principally, however, it is felt that the value of this paper lies in the presentation; considerable insight into the significant factors which contribute to error is gained by the geometric approach emphasized. References to some alternate techniques for calculating the probability of error are presented where thought to be of interest or where they have been of direct help to the authors. We remark that, as pointed out in the text, Section V-A to V-F may be skipped by the reader without loss of continuity.

* Received June 15, 1962. This work has been supported by the Mitre Corporation, Bedford, Mass., under Contract No. AF 33(600) 39852.

† Bell Telephone Laboratories, Murray Hill, N. J. On leave from Dept. of Elec. Engrg., M. I. T. Research Laboratory of Electronics, Cambridge, Mass. Formerly Consultant to the Mitre Corporation, Bedford, Mass.

‡ M. I. T. Dept. of Mathematics and Research Laboratory of Electronics, Cambridge, Mass. On leave from the Mitre Corporation, Bedford, Mass.

[1] W. B. Davenport and W. L. Root, "An Introduction to the Theory of Random Signals and Noise," McGraw-Hill Book Co., Inc., New York, N. Y., pp. 96–99, 338–345; 1958.

[2] C. W. Helstrom, "Statistical Theory of Signal Detection," Pergamon Press, Inc., New York, N. Y., pp. 95–109; 1960.

Reprinted from *IEEE Transactions on Communications Systems*, December, 1962.

II. Basic Geometric Concepts

A. Discussion of Assumed Model

The analysis of data transmission systems is commonly based on the following model. There is assumed to exist a message source generating a stream of equally likely messages, $M_1, M_2, \cdots, M_m$, into a waveform generator having available an alphabet of m distinct waveforms, $S_1(t), S_2(t), \cdots, S_m(t)$, each of duration T (and necessarily finite energy). One waveform is transmitted every T seconds, the choice of waveform depending in some fashion on the incoming message and possibly on the waveforms transmitted in preceding time slots. The medium coupling the transmitter to the receiver is assumed to add stationary-white-zero mean-Gaussian noise to the transmitted signal but otherwise is assumed to be distortion free. It is generally further assumed that the receiver is time synchronized with the transmitter (synchronous detection). Sometimes it is also assumed that the receiver is phase locked to the transmitter (coherent detection). In this report we shall always assume time synchronism but shall distinguish between coherent and incoherent detection.

The problem we are generally interested in solving, given this model (see Fig. 1), is how to design the receiver so that it makes as few errors as possible. Furthermore, assuming that an optimum receiver (optimum in the sense that it will make fewer errors in the long run than any other receiver) is constructed, we are interested in calculating its error rate.

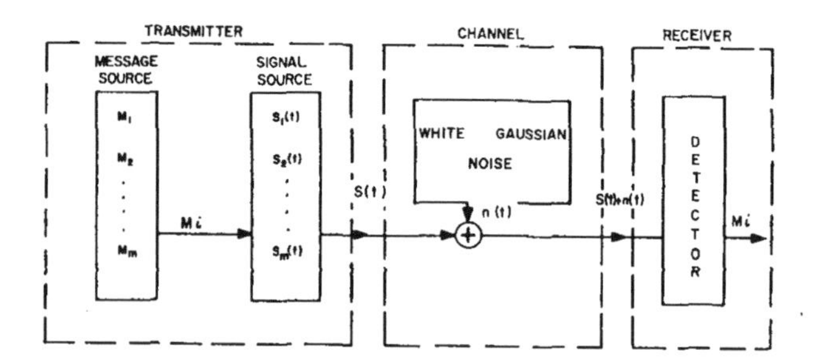

Fig. 1—Idealized model of data transmission system.

B. Geometric Representation of a Known Set of Waveforms

One purpose of this report is to point out that all problems of the type mentioned above may be transformed into geometric problems with considerable simplification of detail. Basic to the geometric viewpoint are two theorems, the first of which we shall now state.

Theorem I

Any finite set of physically realizable waveforms of duration T, say $S_1(t), S_2(t), \cdots, S_m(t)$, may be expressed as a linear combination of k orthonormal waveforms $\varphi_1(t), \varphi_2(t), \cdots, \varphi_k(t)$ $k \leq m$.

That is to say, we can rewrite the $S_i(t), i = 1, 2, \cdots, m$ in the form

$$\begin{aligned} S_1(t) &= a_{11}\varphi_1(t) + a_{12}\varphi_2(t) + \cdots + a_{1k}\varphi_k(t) \\ &\vdots \\ S_i(t) &= a_{i1}\varphi_1(t) + a_{i2}\varphi_2(t) + \cdots + a_{ik}\varphi_k(t) \qquad (1) \\ &\vdots \\ S_m(t) &= a_{m1}\varphi_1(t) + a_{m2}\varphi_2(t) + \cdots + a_{mk}\varphi_k(t) \end{aligned}$$

where the a_{ij} are real numbers given by the formula

$$a_{ij} = \int_0^T S_i(t)\varphi_j(t)\,dt \qquad (2)$$

and the $\varphi_j(t)$, $j = 1, 2, \cdots, k$ are waveforms having the property (definition of orthonormal) that

$$\int_0^T \varphi_i(t)\varphi_j(t)\,dt = \begin{cases} 0 & \text{if } i \neq j \\ 1 & \text{if } i = j \end{cases}. \qquad (3)$$

The proof of this theorem is presented in Appendix I. Note that the conventional Fourier Series expansion of a waveform of duration T is an example of a particular expansion of this type. There are, however, two very important distinctions we wish to make.

1) The form of the $\varphi_j(t)$ has not been specified. That is to say, we have not confined the expansion to be in terms of sinusoids and cosinusoids.
2) The expansion of $S_i(t)$ in terms of a finite number of terms is not an approximation wherein only the first k terms are significant but rather an exact expression where k and only k terms are significant. The number k, incidentally, is referred to as the dimension of the signal alphabet of waveforms.

The form of the $\varphi_j(t)$ is dependent upon the form of the message waveforms originally specified, $S_1(t), \cdots, S_m(t)$. The proof of Theorem I outlines a method of determining the $\varphi_j(t)$.

Accepting the fact that each signal waveform may be represented by a linear combination of $\varphi_j(t)$, $j = 1, \cdots, k$, namely,

$$S_i(t) = \sum_{j=1}^{k} a_{ij}\varphi_j(t), \qquad i = 1, 2, \cdots, m, \qquad (4)$$

it is apparent that each signal waveform may actually be specified uniquely in terms of the coefficients of the $\varphi_j(t)$, $(j = 1, \cdots k)$. Thus, we can represent $S_i(t)$ by the set of k-tuples $(a_{i1}, a_{i2}, \cdots, a_{ik})$. Furthermore, if we conceptually extend our conventional notion of 2 and 3 dimensional Euclidean spaces to a k-dimensional Euclidean space, we can think of the numbers

$$a_{i1}, a_{i2}, \cdots, a_{ik}$$

as the k coordinate projections of the signal point S_i on a k-dimensional Euclidean space.

Thus, for example, if $k = 3$ we may plot the point S_i corresponding to the waveform

$$S_i(t) = a_{i1}\varphi_1(t) + a_{i2}\varphi_2(t) + a_{i3}\varphi_3(t)$$

as a point in a 3-dimensional Euclidean space with coordinates (a_{i1}, a_{i2}, a_{i3}) as shown in Fig. 2.

We shall subsequently refer to the k-dimensional space on which S_i is plotted as the signal space.

There are some interesting relationships between the energy content of a signal and the distance between a signal point and the origin of the signal space. The distance between a pair of points, X, Y, with coordinates $(x_1, x_2, \cdots, x_k)$ and $(y_1, y_2, \cdots, y_k)$, respectively, is given by the formula

$$d(X, Y) = \sqrt{\sum_{j=1}^{k} (x_j - y_j)^2}. \tag{5}$$

It follows readily that the distance between a signal point S_i with coordinates $(a_{i1}, a_{i2}, \cdots, a_{ik})$ and the origin of the signal space is given by

$$d(S_i, 0) = \sqrt{\sum_{j=1}^{k} a_{ij}^2}. \tag{6}$$

Now, by (4), we may write

$$\int_0^T S_i^2(t)\, dt = \int_0^T \left[\sum_{j=1}^{k} a_{ij}\varphi_j(t)\right]\left[\sum_{h=1}^{k} a_{ih}\varphi_h(t)\right] dt$$

$$= \sum_{j=1}^{k} a_{ij} \int_0^T \varphi_j(t)[a_{i1}\varphi_1(t) + a_{i2}\varphi_2(t) + \cdots + a_{ik}\varphi_k(t)]\, dt$$

but since the $\varphi_j(t)$ are orthornormal [see (3)], this latter equation reduces simply to

$$\int_0^T S_i^2(t)\, dt = \sum_{j=1}^{k} a_{ij}^2. \tag{7}$$

That is to say, the energy content of $S_i(t)$, E_i is equal to

$$E_i = d^2(S_i, 0) = \sum_{j=1}^{k} (a_{ij})^2. \tag{8}$$

It may similarly be shown that

$$\int_0^T [S_i(t) - S_v(t)]^2\, dt = \sum_{j=1}^{k} (a_{ij} - a_{vj})^2. \tag{9}$$

Though not essential to the development we might point out an interesting sidelight. Namely, if we consider the plane formed by lines joining the signal points S_i, S_v, $(i \neq v)$ and the origin, then by the law of cosines we may express $\cos\theta$ (see Fig. 3) as

$$\cos\theta = \frac{d^2(0, S_i) + d^2(0, S_v) - d^2(S_i, S_v)}{2\, d(0, S_i)\, d(0, S_v)}. \tag{10}$$

If, in particular, $\theta = \pi/2$, then $\cos\theta = 0$ and (10) reduces to

$$d^2(0, S_i) + d^2(0, S_v) - d^2(S_i, S_v) = 0.$$

It follows, therefore, from (6) and (7) that

$$\int_0^T S_i^2(t)\, dt + \int_0^T S_v^2(t)\, dt - \int_0^T (S_i - S_v)^2\, dt = 0$$

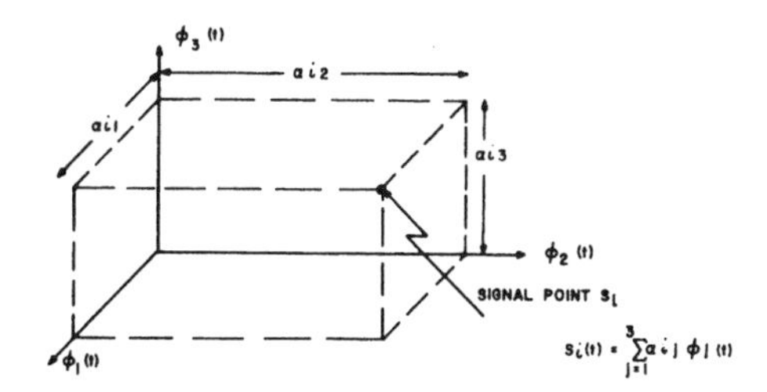

Fig. 2—Geometric representation of the signal waveform $S_i(t)$.

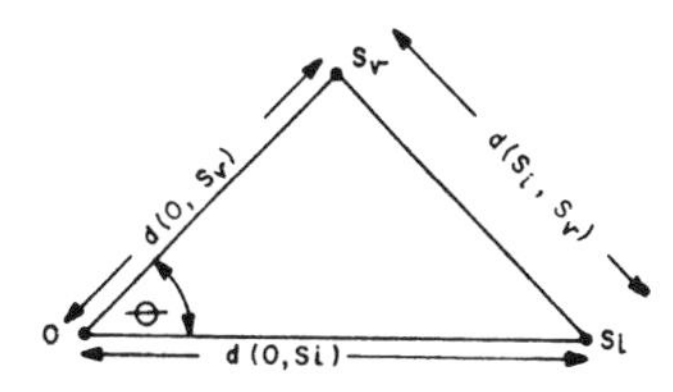

Fig. 3—Planar section in the signal space determined by the origin and the signal points S_i and S_v, $i \neq v$.

which, however, may be simplified to yield

$$\int_0^T S_i(t) S_v(t)\, dt = 0.$$

Thus, we can conclude that if the signal points corresponding to the pair of waveforms $S_i(t)$ and $S_v(t)$ are orthogonal to each other, the angle between a pair of signal points being defined as in Fig. 3, then

$$\int_0^T S_i(t) S_v(t)\, dt = 0 \qquad (i \neq v).$$

If, further, each of the waveforms $S_i(t)$, $i = 1, 2, \cdots, m$ is suitably scaled, that is, normalized, so that

$$\int_0^T S_i^2(t)\, dt = 1$$

for $i = 1, 2, \cdots, m$, then the set of waveforms $S_i(t)$, $i = 1, 2, \cdots, m$ is termed an *orthonormal* set. [See also (3)]. Thus, we see that there is a rather simple geometric interpretation which can be given to the notion of orthonormal waveforms.

C. Detection of Signals in the Presence of Noise

Now, returning to the main development, we wish to point out that the coefficients

$$a_{ij} = \int_0^T S_i(t)\varphi_j(t)\, dt, \tag{2}$$

may be calculated electrically by a series of product integrators (properly synchronized to the waveform generators) as shown in Fig. 4.

Such a series of product integrators can, in fact, be used as the first stage of a detector in a data transmission system. The function of the second stage or decision stage, as we shall term it, is then to decide, on the basis of the k outputs of the product integrators, what signal was actually sent.

The decision problem is complicated by the fact that the transmitted signal is perturbed by noise. (We are assuming, for the present, coherent detection.)

Typically, the noise is assumed to be additive white-stationary-zero mean-Gaussian, the reasons for this assumption being that

1) it makes calculations more tractable, and
2) it is a reasonable description of the type of noise present in many communication channels.

We shall now outline, briefly, the meaning of each of the terms used in the description of the noise.

Classifying the noise as additive implies simply that the received signal, which we shall designate as $z(t)$, consists of a noise term in addition to the originally transmitted signal. That is, if $S_i(t)$ was transmitted, the received signal

$$z(t) = S_i(t) + n(t). \tag{11}$$

Correspondingly, the output of the *jth* product integrator equals

$$\int_0^T z(t)\varphi_j(t)\,dt = a_{ij} + n_j \qquad j = 1, 2, \cdots, k \tag{12}$$

where

$$a_{ij} = \int_0^T S_i(t)\varphi_j(t)\,dt \tag{2}$$

$$n_j = \int_0^T n(t)\varphi_j(t)\,dt. \tag{13}$$

The amplitude of the term n_j will be dependent upon the particular noise sample which perturbed the transmitted signal waveform. Since there is an infinite number of such possible noise samples, each of which could have perturbed the transmitted signal (see Fig. 5), there is a correspondingly infinite number of values which the term n_j can take on. Accordingly, the amplitude of n_j cannot be specified in advance and can at best be described in a probabilistic sense.

The fact that the noise is stationary tells us that the statistics of the noise are independent of the particular time we choose to start transmitting data. In particular, referring to Fig. 5, the choice of the point $t = 0$ is arbitrary as far as the noise is concerned since all joint probability density functions will depend only upon time differences and not upon the actual values of time with respect to some absolute reference.

If, in particular, the noise model used is assumed to be stationary Gaussian with zero mean, it may be shown (Appendix II) that the probability density function of the noise perturbation n_j is Gaussian with zero mean. That is to say, the probability that

$$a \leq n_j < b$$

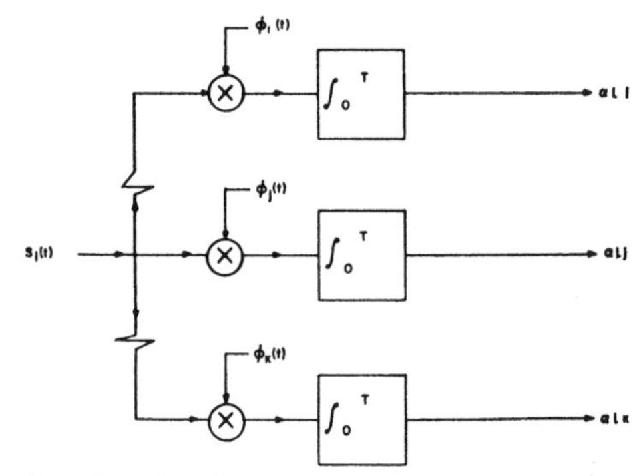

Fig. 4—Set of product integrators which may be used to calculate the signal space coordinates of the signal $S_i(t)$.

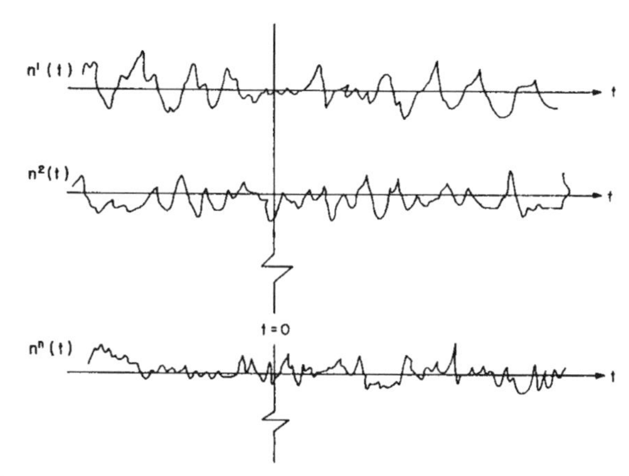

Fig. 5—Samples of possible noise voltages which might be superimposed on the transmitted signal.

is equal to

$$P(a \leq n_j < b) = \frac{1}{\sqrt{2\pi}\,\sigma_j}\int_a^b e^{-x^2/2\sigma_j^2}\,dx \qquad j = 1, 2, \cdots, k. \tag{14}$$

Evaluation of the integral described in (14) requires knowledge of the quantity σ_j^2, that is, the variance of the noise perturbation n_j. Since the noise is stationary, the variance of each noise perturbation is determined by the spectral density. In fact, if the noise is specified to be white which implies that the spectral density $W(f)$ equals

N_0 watts/cps for all f (positive and negative),

it may be shown (Appendix II) that the variance of n_j is

$$\sigma_j^2 = N_0 \qquad j = 1, 2, \cdots, k.$$

It may further be shown (Appendix II) that each of the perturbations are independent. That is the probability of the joint event that, say, $a_1 \leq n_1 < b_1$ and $a_2 \leq n_2 < b_2 \cdots$ and $a_k \leq n_k < b_k$ is equal to simply the product of the probabilities of the individual events.

$$P(a_1 \leq n_1 < b_1, a_2 \leq n_2 < b_2, \cdots a_k \leq n_k < b_k) = P(a_1 \leq n_1 < b_1)P(a_2 \leq n_2 < b_2) \cdots \cdot P(a_k \leq n_k < b_k). \tag{15}$$

We wish to point out that the $n_1, n_2, \cdots, n_k$ do not serve to completely characterize the noise but only that portion of the noise which interacts with the product integrators. That is, $n(t)$ cannot be expanded simply in terms of $\varphi_i(t)$ alone but, rather, must be expressed as

$$n(t) = n_1\varphi_1(t) + n_2\varphi_2(t) + \cdots + n_k\varphi_k(t) + h(t) \qquad (16)$$

where $h(t)$ is a sort of remainder term which must be included on the right to preserve the equality. [Contrast this with the expansion of $S_i(t)$, in (1).] Utilizing the fact that the $\phi_j(t)$ are orthonormal and that

$$n_j = \int_0^T n(t)\phi_j(t)\,dt \qquad j = 1, 2, \cdots, k \qquad (13)$$

it may be deduced from (16) that

$$\int_0^T h(t)\varphi_j(t)\,dt = 0 \qquad j = 1, 2, \cdots, k. \qquad (17)$$

This is, of course, no more than the statement that $h(t)$ does not have any components on the signal space.

The results of the preceding few pages may be summarized as the second basic theorem.

Theorem II

Given a set of orthonormal waveforms, $\varphi_1(t), \varphi_2(t), \cdots, \varphi_k(t)$, which characterize a signal space and a stationary-white-zero mean-Gaussian noise source, $n(t)$, with spectral density N_0, the noise may be decomposed into two portions, the first $n_1\varphi_1(t) + n_2\varphi_2(t) + \cdots + n_k\varphi_k(t)$ consisting of the projection of the noise on the signal space and the second consisting of that portion of the noise which is orthogonal to the signal space. [See (16) and (17).] The n_j $j = 1, 2, \cdots, k$, which are defined by (13), are independent Gaussian random variables with zero mean and variance N_0.

That is to say, the $n_1, n_2 \cdots, n_k$ represent the k coordinate projections of the noise on the signal space and represent that portion of the noise which will interfere with the detection process. The remaining portion of the noise $[h(t)]$ may be thought of as being effectively tuned out by the detector.

III. Coherent Detection

A. Statement of Detection Problem in Geometric Terms

Summarizing the results of Section II, we note that a received signal, $z(t)$, may be represented by a point in a Euclidean space of the appropriate dimension. The coordinates of the point are calculated by a series of product integrators which make up the first stage of our conceptual detector. Each coordinate, as may be deduced from (12), consists of two components—one due to the transmitted signal and the other due to the noise which has been superimposed on the signal in the channel coupling the transmitter to the receiver. The function of the decision stage of the detector is to guess which signal was transmitted from the position of the received "noisy" point. We emphasize the fact that the best the detector can do in the presence of a statistical perturbation such as the additive noise model assumed is to guess at the transmitted message. As a consequence, one reasonable measure for the performance of a detector is the number of times it guesses wrong in a long typical sequence of messages. Or, more precisely, since by assumption the *a priori* probability for transmission of each signal waveform $S_i(t)$ is known, we can calculate for each detector the probability of making an error.

B. Optimum Decision Rule

In the coherent case, the coordinates of each possible transmitted signal may be calculated by the detector. Thus, m points, each of which corresponds to a transmitted signal, may be plotted in the detector signal space. We shall subsequently refer to these m points as the message points or the transmitted signal points. Note that the received signal point will be displaced from the transmitted signal point due to the addition of noise. Since, as may be deduced from the bell shaped curve, small noise perturbations are much more likely than large ones in the Gaussian case, a reasonable decision rule to adopt is to assume that the signal whose message point lies closest to the received point was actually transmitted. In fact, in Appendix III the following theorem is verified.

Theorem III

If each signal waveform is transmitted with equal probability and if the received signal is perturbed by additive stationary-white-zero mean-Gaussian noise, then, for the case of coherent detection, that decision rule which selects the message point closest to the received point minimizes the probability of error.

A detector which embodies the decision rule of Theorem III is often referred to as a maximum likelihood Detector. It should be noted, however, that a maximum likelihood detector will only minimize the probability of error, when, as in this case, it is assumed that each possible signal waveform is transmitted with equal probability.[3]

We shall now illustrate this rule for three cases of practical interest—coherent Phase Shift Keyed, coherent Amplitude Shift Keyed and coherent Frequency Shift Keyed.

C. Coherent PSK

This modulation scheme is characterized by the fact that the information carried by the transmitted waveform is contained in the phase. A typical set of message waveforms is described by

$$S_i(t) = \begin{cases} \sqrt{\dfrac{2E}{T}} \cos\left(\omega_0 t + \dfrac{2\pi i}{m}\right) & 0 \le t \le T \\ 0 \quad \text{elsewhere} & i = 1, 2, \cdots, m \end{cases} \qquad (18)$$

[3] For further discussion, see Davenport and Root, *op. cit.*, pp. 317–324; R. M. Fano, "Transmission of Information," M. I. T. Press, Cambridge, Mass., John Wiley and Sons, Inc., New York, N. Y., p. 184; 1961.

where E is the energy content of $S_i(t)$ and

$$\omega_0 = \frac{2\pi n_0}{T} \quad \text{for some fixed integer } n_0.$$

Now, recognizing that each $S_i(t)$ may be written in terms of a sinusoid and cosinusoid, which are orthogonal, and then suitably scaling to fulfill the conditions of (3), we conclude that the appropriate form for the orthonormal waveforms $\varphi_1(t)$ and $\varphi_2(t)$ (alternately, we could have used the techniques described in Appendix I) to be used in the product integrators of Fig. 4 is

$$\varphi_1(t) = \sqrt{\frac{2}{T}} \cos \omega_0 t$$
$$\varphi_2(t) = \sqrt{\frac{2}{T}} \sin \omega_0 t. \tag{19}$$

The coordinates of the message points may be calculated by (2), (18) and (19).

$$a_{i1} = \int_0^T \sqrt{\frac{2E}{T}} \cos\left(\omega_0 t + \frac{2\pi i}{m}\right)\sqrt{\frac{2}{T}} \cos \omega_0 t \, dt = \sqrt{E} \cos \frac{2\pi i}{m} \tag{20}$$

$$a_{i2} = \int_0^T \sqrt{\frac{2E}{T}} \cos\left(\omega_0 t + \frac{2\pi i}{m}\right)\sqrt{\frac{2}{T}} \sin \omega_0 t \, dt = -\sqrt{E} \sin \frac{2\pi i}{m}.$$

Note that for the particular case $m = 2$ [often termed phase reversal since $S_i(t) = \sqrt{E/2T} \sin(\omega_0 t \pm \pi/2)$], $a_{i2} = 0$. Accordingly, we can dispense with $\varphi_2(t)$.

We shall now illustrate the decision rule embodied in Theorem III for the case $m = 4$. The four possibly transmitted signal points whose coordinates are given by (20) are shown in the diagram of the signal space displayed in Fig. 6. To realize the decision rule we must partition the signal space into four regions, namely, the set of points in the signal space closest to S_1, the set of points closest to S_2, the set of points closest to S_3 and the set of points closest to S_4. This is accomplished by constructing the perpendicular bisectors of the 4-sided polygon $S_1S_2S_3S_4$ and marking off the appropriate regions. It may in this way be deduced that the regions of interest are cones whose vertices coincide with the origin. These regions are marked zone 1, zone 2, zone 3 and zone 4 according to the transmitted signal point about which they are constructed.

The decision rule is now simply to guess $S_1(t)$ was transmitted if the received signal point falls in zone 1, guess $S_2(t)$ was transmitted if the received signal point falls in zone 2 and so on. An erroneous decision will be made if, for example, $S_4(t)$ is transmitted and the noise is such that the received signal point falls outside zone 4. The probability of error for the PSK coherent case is calculated for various values of m in Section V. Curves and discussions are presented in Section VI.

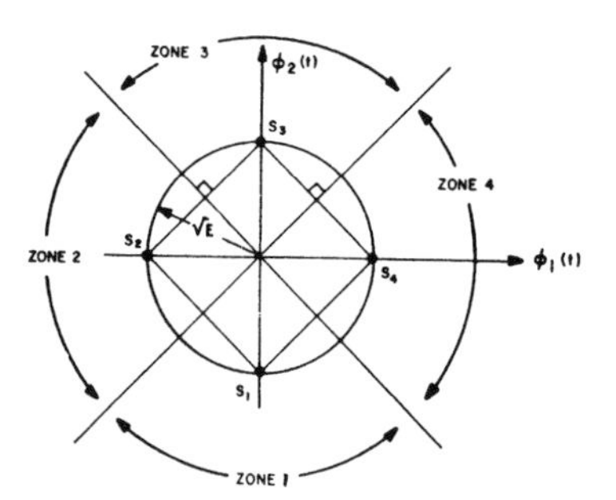

Fig. 6—Optimum partitioning of detector signal space for a 4-level PSK coherent system.

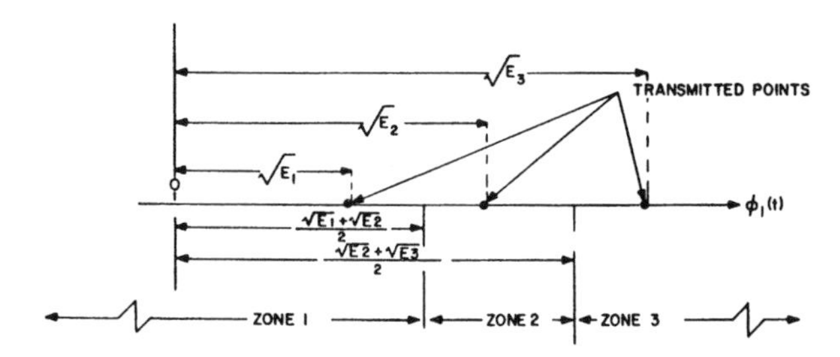

Fig. 7—Optimum partitioning of detector signal space for a 3-level ASK coherent system.

D. Coherent ASK

In this modulation scheme the information carried by the transmitted waveform is contained in the amplitude. A typical set of message waveforms is described by

$$S_i(t) = \begin{cases} \sqrt{\frac{2E_i}{T}} \cos \omega_0 t & 0 \le t \le T \\ 0 & \text{elsewhere} \quad i = 1, 2, \cdots, m \end{cases} \tag{21}$$

where E_i is the energy content of $S_i(t)$ and $\omega_0 = 2\pi n_0/T$ for some fixed integer n_0.

It should be clear that each transmitted waveform may be expanded in terms of the single orthonormal waveform

$$\varphi_1(t) = \sqrt{\frac{2}{T}} \cos \omega_0 t \tag{22}$$

and that

$$a_{i1} = \int_0^T \sqrt{\frac{2E_i}{T}} \cos \omega_0 t \sqrt{\frac{2}{T}} \cos \omega \, dt = \sqrt{E_i}. \tag{23}$$

The possible transmitted signal points are illustrated for the case $m = 3$ in Fig. 7. The signal space is partitioned into 3 distinct detection zones according to the techniques just discussed. Thus, for example, zone 2 consists of the set of points in the signal space which lie closer to S_2 than to S_1 or S_3.

Probability of error calculations for this case (under the further assumption of average power limitations and uniform amplitude spacing starting with zero) are presented in Section V. Curves and discussion appear in Section VI.

E. Coherent FSK

This modulation scheme is characterized by the fact that the information carried by the transmitted signal is contained in the frequency. A typical set of signal waveforms is described by

$$S_i(t) = \begin{cases} \sqrt{\frac{2E}{T}} \cos(\omega_i t) & 0 \le t \le T \\ 0 & \text{elsewhere} \end{cases} \quad (24)$$

where E is the energy content of $S_i(t)$,

$$\omega_i = 2\pi \frac{(n_0 + i)}{T} \quad \text{for some fixed integer } n_0 \quad i = 1, 2, \cdots, m.$$

Following the procedure of Appendix I or observing directly that the $S_i(t)$ are orthogonal (not orthonormal), it may be deduced that the most useful form for the orthonormal waveforms $\varphi_1(t), \varphi_2(t), \cdots, \varphi_k(t)$ is

$$\varphi_j(t) = \sqrt{\frac{2}{T}} \cos \omega_j t \qquad j = 1, 2, \cdots, k = m. \quad (25)$$

Correspondingly,

$$a_{ij} = \int_0^T \sqrt{\frac{2E}{T}} \cos \omega_i t \sqrt{\frac{2}{T}} \cos \omega_j t \, dt = \begin{cases} \sqrt{E} & \text{if } i = j \\ 0 & \text{otherwise} \end{cases}. \quad (26)$$

That is to say that *ith* signal point is located on the *ith* coordinate axis at a displacement of $\sqrt{E}$ from the origin of the signal space.

It should also be noted that in this modulation scheme the distance between any two signal points S_i and S_j is constant, since by (5) and (26) we have

$$d(S_i, S_j) = \sqrt{2E} \qquad i \neq j.$$

The detection rule is illustrated for the case $m = 3$ in Fig. 8.

Calculations for the probability of error of an m-level FSK coherent modulation scheme are presented in Section V. Curves and discussion appear in Section VI.

F. Remarks

The procedure we have followed in the last three examples is to partition the detector signal space into (m) distinct regions, each region containing one and only one message point and consisting of those points in the signal space which are closer to the contained message point than to any other message point. The received signal point will (with probability one) fall into one, and only one, of these regions.

The optimum decision rule (when the hypothesis of Theorem III is satisfied) is simply to identify the region in which the received signal point falls and assume that the signal corresponding to the contained message point was actually transmitted.

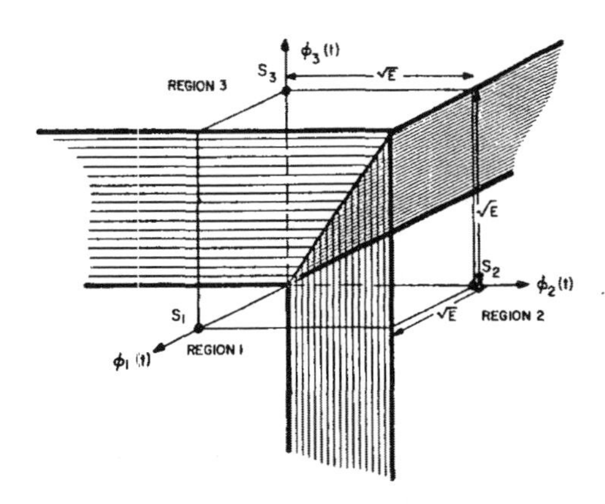

Fig. 8—Optimum partitioning of detector signal space for a 3-level FSK coherent system.

IV. Incoherent Detection

Incoherent systems differ from coherent systems in that no provisions have been made to phase synchronize the receiver with the transmitter. Accordingly, if the waveform

$$S(t) = \sqrt{\frac{2E}{T}} \cos(\omega_0 t + \phi) \quad (27)$$

is transmitted, the received signal $z(t)$ will be of the form

$$z(t) = \sqrt{\frac{2E}{T}} \cos(\omega_0 t + \phi + \alpha) + n(t) \quad (28)$$

where the angle α is unknown and is usually considered to be a random variable uniformly distributed between 0 and 2π.

It may readily be deduced that the detection schemes presented previously are inadequate for the incoherent case for if the received signal takes the form described by (28), the outputs of the product integrators will be functions of the unknown angle α. We shall now discuss in turn the modifications which must be introduced to the PSK, ASK and FSK systems discussed previously.

A. PSK Incoherent

It should be clear from (28) that the presence of the random phase angle in the argument of the cosine prevents the receiver from deriving any information from the phase of the incoming signal alone, namely, $(\phi + \alpha)$. If, however, α varies slowly (that is, slowly enough so that it may be considered constant over the period of time required to transmit two waveforms, $2T$), then the relative phase difference between two successive waveforms will be independent of α [*i.e.*, $(\phi_1 + \alpha) - (\phi_2 + \alpha) = \phi_1 - \phi_2$]. Thus, if the detector was equipped with storage, it could measure the phase difference between successive signals regardless of the value of α. This suggests that we modify the coding scheme at the transmitter as follows. To send the *ith* message ($i = 1, 2, \cdots, m$), phase advance the current signal waveform by $2\pi i/m$ radians over the previous waveform.

Correspondingly, the detector should (at least, conceptually) calculate the coordinates of the incoming signals by product integrating it with the locally generated waveforms $\sqrt{2/T}\cos\omega_0 t$ and $\sqrt{2/T}\sin\omega_0 t$. It should then plot the received signal points and measure the angle between the currently received signal point and the previously received signal point which has been stored. It may be shown (see Appendix IV) that the best rule for the detector to follow is to quantize the measured angle in steps of $2\pi/m$ and guess that the corresponding message was transmitted.

Thus, for example, if the ith message was transmitted, a pair of successively received signals $z_1(t)$ and $z_2(t-T)$ will be of the form

$$z_1(t) = \sqrt{\frac{2E}{T}}\cos(\omega_0 t + \alpha) + n(t) \tag{29a}$$

$$z_2(t-T) = \sqrt{\frac{2E}{T}}\cos\left(\omega_0 t + \alpha + \frac{2\pi i}{m}\right) + n(t-T) \tag{29b}$$

where the angle α is unknown and is assumed to be uniformly distributed over a 2π interval (symmetric with respect to some mean which may be unknown). The coordinates of the corresponding signal points z_1 and z_2, which we designate as (x_1, y_1) and (x_2, y_2), will be of the form

$$x_1 = \int_0^T z_1(t)\phi_1(t)\,dt = \sqrt{E}\cos\alpha + n_{11} \tag{30a}$$

$$y_1 = \int_0^T z_1(t)\phi_2(t)\,dt = -\sqrt{E}\sin\alpha + n_{12} \tag{30b}$$

$$x_2 = \int_T^{2T} z_2(t-T)\phi_1(t-T)\,dt = \sqrt{E}\cos\left(\alpha + \frac{2\pi i}{m}\right) + n_{21} \tag{30c}$$

$$y_2 = \int_T^{2T} z_2(t-T)\phi_2(t-T)\,dt = -\sqrt{E}\sin\left(\alpha + \frac{2\pi i}{m}\right) + n_{22} \tag{30d}$$

where n_{11}, n_{12}, n_{21} and n_{22} are independent-Gaussian-random variables, each having zero mean and variance N_0.

Suppose now that for some particular combination of received signals, the random variables x_1, y_1, x_2, y_2 take on the values a_1, b_1, a_2, b_2, respectively. The optimum rule for the detector to follow (see Appendix IV) is to measure the angle θ between the two points (a_1, b_1) and (a_2, b_2), which are shown plotted in Fig. 9, round off to the nearest integral multiple of $(2\pi/m)$ and guess that the signal corresponding to that phase rotation was transmitted.

A basic difference between coherent and incoherent PSK systems is that in the coherent case the received signal is being compared with a clean reference, that is, the known position of the transmitted point. In the incoherent case, however, two noisy signals are being compared with each other. Thus, we might, after a crude fashion, say that there is twice as much noise present in the incoherent case as in the coherent and, consequently, there will be a 3-db degradation in performance. This latter statement will in fact turn out to be approximately true under the appropriate restrictions (namely, high signal-to-noise ratio and $m > 2$) as will be discussed in Section V-D.

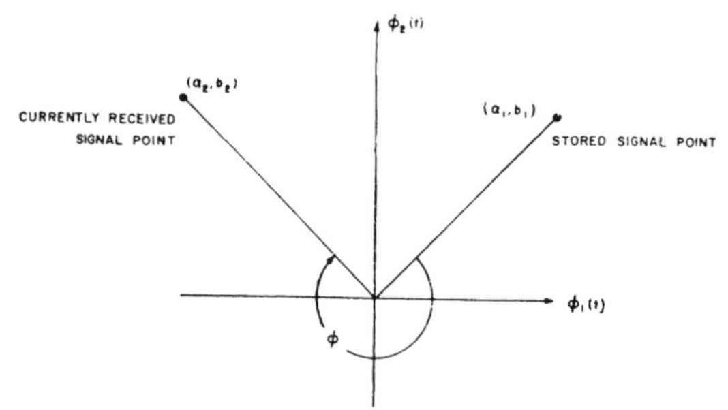

Fig. 9—Illustration of detection rule for PSK incoherent.

We wish also to point out that under the proposed detection scheme two product integrators will be necessary for the two-level case as well as for the multilevel case. Recall that only one product integrator is required for the coherent two-level case.

B. ASK Incoherent

In the ASK incoherent case the received signal $z(t)$ is of the form

$$z(t) = \sqrt{\frac{2E_i}{T}}\cos(\omega_0 t + \alpha) + n(t) \quad i = 1, 2, \cdots, m \tag{31}$$

where α is unknown and is assumed to be uniformly distributed over a 2π interval.

Consider for the present the signal portion of a particular received signal which we shall designate as $z^*(t)$, for which it is known that $\alpha = A$. That is,

$$z^*(t) = \sqrt{\frac{2E_i}{T}}\cos(\omega_0 t + A). \tag{32}$$

Any waveform of this form may be expressed as a linear combination of the pair of orthonormal waveforms $\sqrt{2/T}\cos\omega_0 t$ and $\sqrt{2/T}\sin\omega_0 t$. Product integrating $z^*(t)$ with $\sqrt{2/T}\cos\omega_0 t$ and $\sqrt{2/T}\sin\omega_0 t$, respectively, yields

$$a = \int_0^T z^*(t)\sqrt{\frac{2}{T}}\cos\omega_0 t\,dt = \sqrt{E_i}\cos A \tag{33a}$$

$$b = \int_0^T z^*(t)\sqrt{\frac{2}{T}}\sin\omega_0 t\,dt = -\sqrt{E_i}\sin A. \tag{33b}$$

In accordance with our previous discussion $z^*(t)$ may be represented by the point z^* with coordinates (a, b) shown plotted in Fig. 10.

It may readily be deduced that the line segment drawn from the origin to the point z^* has length $\sqrt{E_i}$ and is displaced A radians below the abscissa.

As far as recovery of the transmitted information is concerned, only the distance of the point from the origin,

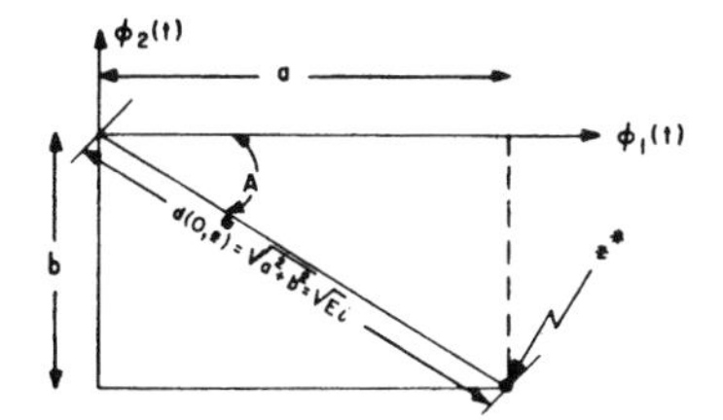

Fig. 10—Plot of the point z^* in the detector signal space.

namely, $\sqrt{E_i}$ is significant. We wish to point out, however, that Fig. 10 lends itself to a simple geometric interpretation of the difference between ASK coherent and ASK incoherent systems. In both cases the set of message points lies on a straight line (see, *e.g.*, Fig. 7). In the coherent case, however, the receiver knows the orientation of this line (*i.e.*, the angle A) and, thus, need only make one measurement (along the line) in order to deduce what message point was sent.

In the incoherent case the receiver does not know the position of the line and must, therefore, perform its analysis in the plane containing all possible rotations of the line. In particular, to calculate the distance of a point from the origin it must first measure the projections of the point on each of two perpendicular axes lying on the plane and passing through the origin and then take the square root of the sum of the squares of the two projections. Accordingly, it should be noted that the ASK coherent detector requires only one product integrator whereas the ASK incoherent detector requires two. (The logic following the product integrators will, of course, be different in the two cases.)

Owing to the presence of noise the actual outputs of the product integrators will be of the form

$$x = \int_0^T z(t)\phi_1(t)\,dt = \sqrt{E_i}\cos\alpha + n_1 \quad \text{(34a)}$$

$$y = \int_0^T z(t)\phi_2(t)\,dt = -\sqrt{E_i}\sin\alpha + n_2 \quad \text{(34b)}$$

where n_1 and n_2 are independent-Gaussian-random variables, each having zero mean and variance N_0 and α is assumed to be a uniformly distributed random variable over a 2π interval.

It may be shown (see Appendix IV) that a reasonable decision rule for the receiver to adopt is to measure the outputs of the two product integrators, calculate the rms amplitude, quantize in steps of $\sqrt{E_i}$ and guess that the corresponding signal was transmitted. Thus, for example, if, in particular, $x = a$ and $y = b$, then the receiver should guess that the message corresponding to the value of i which minimizes the quantity

$$|\sqrt{a^2 + b^2} - \sqrt{E_i}| \qquad i = 1, 2, \cdots, m$$

was sent. This is equivalent to saying that the two-dimensional space corresponding to all possible outputs of the two product integrators should be partitioned into m distinct regions (zones), each of which is associated with a particular message where the ith region corresponding to the *ith* message consists of those points with coordinates (x, y), for which

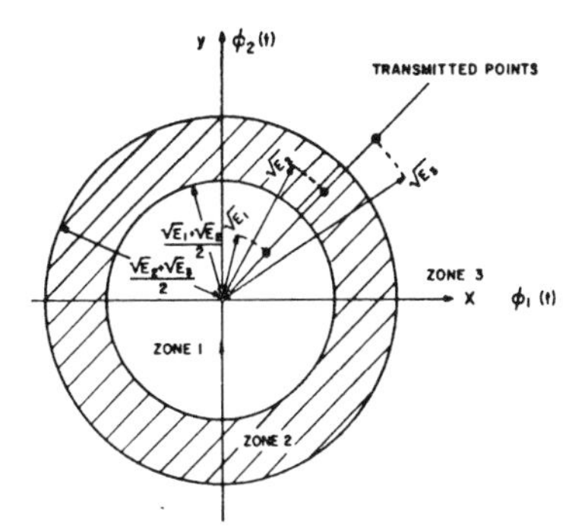

Fig. 11—Illustration of detection rule for 3-level ASK incoherent.

$$|\sqrt{x^2 + y^2} - \sqrt{E_i}| < |\sqrt{x^2 + y^2} - \sqrt{E_j}| \qquad \text{all } j \neq i. \quad (35)$$

Such a partitioning of the detector signal space is illustrated in Fig. 11 for the three-level case.

We wish to point out that this decision rule is not the optimum one to adopt but approaches the optimum rule in the case of high signal-to-noise ratio. The rationale for adopting this rule is that it is considerably easier to instrument than the optimum rule for which the decision regions are functions of the signal-to-noise ratio. The decision rule is discussed in Appendix IV. Probability of error calculations for the particular case of uniformly spaced signal amplitudes starting with zero and an average power limited transmitter are presented in Section V. Curves and discussion appear in Section VI.

C. FSK Incoherent

In the FSK incoherent case, if the *ith* message is transmitted, the received signal $z(t)$ will be of the form

$$z(t) = \sqrt{\frac{2E}{T}}\cos(\omega_i t + \alpha) + n(t) \quad i = 1, 2, \cdots, m \quad (36)$$

where the unknown angle α is assumed to be a random variable uniformly distributed over a 2π interval.

Although each of the transmitted signals may be represented by a point in an m-dimensional space, the presence of the unknown angle α makes it necessary to resolve the incoming signal in terms of the $2m$ orthonormal waveforms

$$\sqrt{\frac{2}{T}}\cos\omega_1 t, \sqrt{\frac{2}{T}}\cos\omega_2 t, \cdots, \sqrt{\frac{2}{T}}\cos\omega_m t$$

$$\sqrt{\frac{2}{T}}\sin\omega_1 t, \sqrt{\frac{2}{T}}\sin\omega_2 t, \cdots, \sqrt{\frac{2}{T}}\sin\omega_m t.$$

Correspondingly, the $2m$ product integrator outputs will be of the form

$$x_j = \int_0^T z(t)\sqrt{\frac{2}{T}}\cos\omega_j t\,dt = \begin{cases} n_{xj} & j \neq i \\ \sqrt{E}\cos\alpha + n_{xi} & j = i \end{cases} \quad \text{(37a)}$$

$$y_j = \int_0^T z(t)\sqrt{\frac{2}{T}}\sin\omega_j t\,dt = \begin{cases} n_{yj} & j \neq i \\ -\sqrt{E}\sin\alpha + n_{yi} & j = i \end{cases} \quad \text{(37b)}$$

where the $2m$ noise perturbations

$$n_{xj}, n_{yj} \qquad j = 1, 2, \cdots, m$$

are independent random variables with zero mean and variance N_0 and α is a uniformly distributed random variable over a 2π interval.

If, at some instant, the random variables $x_1, x_2, \cdots, x_m, y_1, y_2, \cdots, y_m$ take on the particular values $a_1, a_2, \cdots, a_m, b_1, b_2, \cdots, b_m$, respectively, it may be shown (Appendix IV) that the optimum decision rule for the receiver to follow is to find that value of j, $j = 1, 2, \cdots, m$ for which the quantity $\sqrt{a_j^2 + b_j^2}$ is a maximum and guess that the corresponding signal was transmitted. That is to say, the detector should calculate the rms amplitude associated with each possibly transmitted frequency and select the largest one.

The probability of error for an orthogonal FSK incoherent system is calculated in Section V. Curves and discussion appear in Section VI.

D. Interpretation of Decision Rules in Light of Appropriate Minimum Distance Criteria

In concluding our discussion of incoherent systems we wish to point out that the decision rule adopted in each case can be interpreted as a minimum distance type rule although the spaces in which the distances are measured are not simply related to the k-dimensional Euclidean spaces which characterize the k product-integrator outputs. Thus, in the PSK incoherent case the space of interest is that corresponding to the possible values of the relative phase difference between a pair of successively received signals, namely, an interval of length 2π. The possibly transmitted phase differences determine a set of m message points with coordinate displacements $2\pi i/m$ $i = 1, 2, \cdots, m$, respectively, and the optimum decision rule is equivalent to measuring the phase difference plotting the resultant number in the space and selecting the closest message point.

In the ASK incoherent case the space of interest is that corresponding to the possible values of the rms amplitude of the received signal. This space can be represented geometrically by a semi-infinite line running from zero through the positive real numbers to infinity. The set of possibly transmitted signals define a set of m messages points with coordinate displacements $\sqrt{E_i}$ $i = 1, 2, \cdots, m$, respectively, and the adopted decision rule (which is only asymptotically optimum) is equivalent to calculating the rms amplitude of the received signal, plotting the resultant value in the space and selecting the closest message point.

In the FSK incoherent case the space of interest is an m-dimensional Euclidean space (or, to be more exact, the positive "quadrant" of that space) wherein each direction in that space is associated with one of the possibly transmitted frequencies. The received signal may be considered to be an m-dimensional vector whose coordinate projection in each direction is equal to the rms amplitude of the outputs of the sine and cosine product integrators associated with that direction (frequency). The m distinct points, one on each coordinate axis, displaced $\sqrt{E}$ units from the origin, constitute the message points. Correspondingly, the optimum decision rule is equivalent to measuring the distance (in this space) from the received point to each message point and selecting the closest one.

V. Probability of Error Calculations

For the systems under consideration it is possible to obtain exact expressions for the probability of error in integral form. Unfortunately, however, in many cases the integrals in question are not simply integrable nor have they been tabulated over the ranges of interest. When this is the case it is sometimes possible to obtain upper and lower bounds on the probability of error which are usually adequate to predict the signal-to-noise ratio (within a decibel or so) required to maintain a prescribed error rate.

The approximations which can be made fall into two categories, namely, simplification of the integrand and simplification of the region of integration. The latter procedure is especially useful in the coherent case where the regions of integration are fixed relative to the signal space and the noise is symmetric Gaussian with zero mean. In fact, Theorem IV may be shown (see Appendix V).

Theorem IV

Given M message waveforms, each transmitted with equal probability and perturbed by additive stationary-white-zero mean-Gaussian noise with double-sided spectral density N_0 watts/cps, then the average probability of error for a maximum likelihood coherent detector is bounded by[4]

$$\frac{1}{\sqrt{2\pi}} \int_{\bar{\rho}/2\sqrt{N_0}}^{\infty} e^{-x^2/2}\, dx \leq P_e \leq \frac{(M-1)}{\sqrt{2\pi}} \int_{\rho^*/2\sqrt{N_0}}^{\infty} e^{-x^2/2}\, dx \qquad (38)$$

where ρ^* and $\bar{\rho}$ are defined in terms of ρ_i, the distance between message point i and its closest neighbor. That is,

$$\rho^* = \underset{i}{\text{minimum}}\,(\rho_i) \qquad i = 1, 2, \cdots, m$$

$$\bar{\rho} = \frac{1}{m} \sum_{i=1}^{m} \rho_i.$$

Actually, it is often possible to establish tighter bounds than those presented in Theorem IV. The results presented, however, are indicative of the type of bounds which may be achieved by overestimating and underestimating the regions of integration.

Similar reasoning may be employed to estimate the probability of error for the ASK incoherent and FSK

[4] The upper bound is similar to one presented by E. N. Gilbert, "A comparison of signaling alphabets," *Bell Sys. Tech. J.*, vol. 31, pp. 504–522 (Theorem 3); 1952. Gilbert's Lower Bound is incorrect.

incoherent systems where, again, we are dealing with fixed regions of integration independent of the incoming signal (though the probability of error for the latter system may be evaluated exactly). Unfortunately, however, these techniques are not readily applicable to the PSK incoherent case. Therein it was only possible, with the exception of the two-level case (for which the probability of error may be evaluated exactly), to obtain approximate expressions (in simple form) which represent neither an upper bound nor a lower bound to the probability of error.

Sections V-*A* to V-*F* are devoted to calculating the probability of error and approximations thereof for each of the six systems under consideration. They may be skipped without loss of continuity. Before proceeding to this section, however, we wish to point out that, in the ASK systems, the FSK systems and the coherent PSK system, an error occurs whenever the *ith* signal waveform $S_i(t)$ is transmitted and the received signal point does not land in the region associated with the message point S_i. Designating this region by R_i and the received signal point by z, the event z falling inside the region R_i will be written symbolically as $z \in R_i$ whereas the event z falling outside the region R_i will be denoted $z \notin R_i$. Averaging over all possibly transmitted signals it is readily seen that the average probability of error P_e equals

$$P_e = \sum_{i=1}^{m} P(S_i \text{ sent})\, P(z \notin R_i/S_i)$$

$$= \frac{1}{m} \sum_{i=1}^{m} P(z \notin R_i/S_i) \tag{39}$$

where we are using standard notation to denote the probability of an event and the conditional probability of an event.[5] Now let us consider each of the six systems individually.

A. Coherent PSK

In the coherent PSK case, if $S_i(t)$ is transmitted, the received signal point z has coordinates [see (11), (12) and (20)]

$$x = \int_0^T z(t)\varphi_1(t)\, dt = \sqrt{E} \cos \frac{2\pi i}{m} + n_1 \tag{40a}$$

$$y = \int_0^T z(t)\varphi_2(t)\, dt = -\sqrt{E} \sin \frac{2\pi i}{m} + n_2. \tag{40b}$$

Thus, x and y are independent-Gaussian-random variables with means $\sqrt{E} \cos 2\pi_i/m$ and $-\sqrt{E} \sin 2\pi i/m$, respectively, and common variance N_0. Consequently, the probability that z lands in R_i when $S_i(t)$ is transmitted is given by

$$P(z \in R_i/S_i) = \frac{1}{2\pi N_0} \iint_{R_i} \exp \left\{ -\frac{\left(x - \sqrt{E} \cos \frac{2\pi i}{m}\right)^2 + \left(y + \sqrt{E} \sin \frac{2\pi i}{m}\right)^2}{2N_0} \right\} dx\, dy$$

which, transforming to polar coordinates with $x = \rho \sqrt{N_0} \cos \theta$ and $y = \rho \sqrt{N_0} \sin \theta$, may be written as

$$P[z \in R_i/S_i] = \frac{1}{2\pi} \iint_{R_i} \exp \left\{ -\frac{1}{2} \left[\rho^2 - 2\rho \sqrt{\frac{E}{N_0}} \cdot \cos \left(\theta + \frac{2\pi i}{m} \right) + \frac{E}{N_0} \right] \right\} \rho d\, \rho d\, \theta. \tag{41}$$

But R_i, as may be deduced from Fig. 12, is simply the set of points satisfying the two conditions

$$0 \leq \rho < \infty$$

$$-\frac{2\pi i}{m} - \frac{\pi}{m} \leq \theta < -\frac{2\pi i}{m} + \frac{\pi}{m}.$$

Thus, substituting the appropriate limits of integration in (41), we get

$$P(z \in R_i/S_i) = \frac{1}{2\pi} \int_{-2\pi i/m - \pi/m}^{-2\pi i/m + \pi/m} d\theta \int_0^\infty d\rho \rho \cdot \exp \left\{ -\frac{1}{2} \left[\rho^2 - 2\rho \sqrt{\frac{E}{N_0}} \cos \left(\theta + \frac{2\pi i}{m} \right) + \frac{E}{N_0} \right] \right\}$$

$$= \frac{1}{2\pi} \int_{-\pi/m}^{+\pi/m} d\theta \int_0^\infty d\rho \rho \cdot \exp \left\{ -\frac{1}{2} \left[\rho^2 - 2\rho \sqrt{\frac{E}{N_0}} \cos \theta + \frac{E}{N_0} \right] \right\}. \tag{42}$$

Note that (42) is independent of the choice of i. That is to say, the probability of interpreting the received signal correctly is the same regardless of which particular signal was transmitted. Therefore, the probability of error for the m-level PSK coherent system is simply equal to one minus the right-hand side of (42). That is,

$$P_e = 1 - \frac{1}{2\pi} \int_{-\pi/m}^{+\pi/m} d\theta \int_0^\infty d\rho \rho \cdot \exp \left\{ -\frac{1}{2} \left[\rho^2 - 2\rho \sqrt{\frac{E}{N_0}} \cos \theta + \frac{E}{N_0} \right] \right\}. \tag{43}$$

Now, if we complete the square in exponent by writing

$$\left(\rho^2 - 2\rho \sqrt{\frac{E}{N_0}} \cos \theta + \frac{E}{N_0} \right) = \left(\rho - \sqrt{\frac{E}{N_0}} \cos \theta \right)^2 + \frac{E}{N_0} \sin^2 \theta$$

[5] See, *e.g.*, Davenport and Root, *op. cit.*, pp. 7–13.

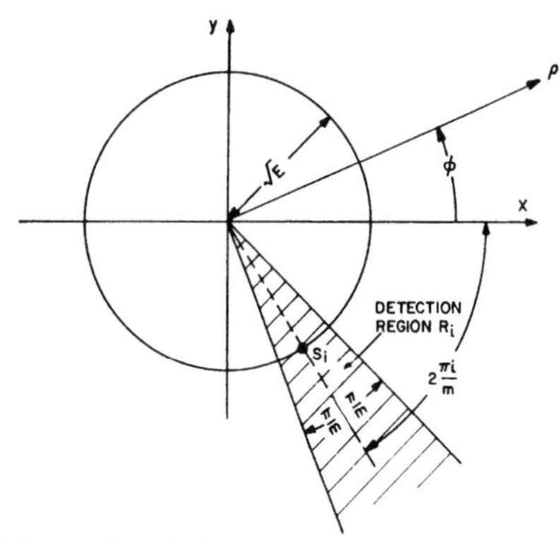

Fig. 12—Illustration of the detection region R_i corresponding to the signal $S_i(t)$ for the multilevel PSK coherent case.

and substitute this result into (43), we get

$$P_e = 1 - \frac{1}{2\pi}\int_{-\pi/m}^{+\pi/m} d\theta \exp\left\{-\frac{1}{2}\frac{E}{N_0}\sin^2\theta\right\} \cdot \int_0^\infty d\rho\rho \exp\left\{-\frac{1}{2}\left[\rho - \sqrt{\frac{E}{N_0}}\cos\theta\right]^2\right\}$$

$$= 1 - \frac{1}{2\pi}\int_{-\pi/m}^{\pi/m} d\theta \cdot \exp\left\{-\frac{E}{2N_0}\sin^2\theta\right\}\left\{\exp\left\{-\frac{E}{2N_0}\cos^2\theta\right\} + \sqrt{\frac{E}{N_0}}\cos\theta \int_{-\sqrt{E/N_0}\cos\theta}^\infty \exp\{-\tfrac{1}{2}t^2\}\, dt\right\}. \tag{44}$$

Eq. (44) may be bounded quite easily for $m > 2$. For if $m > 2$, then $-\pi/2 < \theta < \pi/2$ which implies that $\cos\theta > 0$ and, therefore, that[6]

$$\int_{-\sqrt{E/N_0}\cos\theta}^\infty \exp\{-\tfrac{1}{2}t^2\}\, dt > \sqrt{2\pi} - \frac{\exp\left\{-\frac{E}{2N_0}\cos^2\theta\right\}}{\sqrt{\frac{E}{N_0}}\cos\theta}. \tag{45}$$

Combining (44) and (45) results in the bound

$$P_e < 1 - \frac{1}{\sqrt{2\pi}} \cdot \int_{-\pi/m}^{\pi/m} \exp\left\{-\frac{E}{2N_0}\sin^2\theta\right\}\sqrt{\frac{E}{N_0}}\cos\theta\, d\theta. \tag{46}$$

If we now let $x = \sqrt{E/N_0}\sin\theta$, then the right-hand side of (46) may be written as

$$1 - \frac{1}{\sqrt{2\pi}}\int_{-\sqrt{E/N_0}\sin\pi/m}^{+\sqrt{E/N_0}\sin\pi/m} e^{-x^2/2}\, dx = \frac{2}{\sqrt{2\pi}}\int_{\sqrt{E/N_0}\sin\pi/m}^\infty e^{-x^2/2}\, dx$$

[6] This result follows from some elementary inequalities presented in W. Feller, "An Introduction to Probability Theory and its Applications," John Wiley and Sons, Inc., New York, N. Y., vol. 1, 2nd ed., pp. 164–166; 1959.

yielding finally that

$$P_e < \frac{2}{\sqrt{2\pi}}\int_{\sqrt{E/N_0}\sin\pi/m}^\infty e^{-x^2/2}\, dx. \tag{47}$$

Thus, we have established a simple upper bound to the probability of error for the case $m > 2$. We remark that if $\sqrt{E/N_0} \gg 1$, which is usually the case, the left-hand side of inequality, (47), represents a good approximation to P_e.

A lower bound for the probability of error for $m > 2$ may be established quite easily by geometric reasoning. In particular, it may be deduced from the symmetry of the message points [though it has been shown formally in the discussion following (42)] that the average probability of error is equal to the probability of landing outside detection region i when message point i is sent. The probability of landing outside R_i, however, is larger than the probability of landing in the shaded planar region of Fig. 13. But the received point z will only fall in the planar region if the component of noise perpendicular to the boundary line of the planar region exceeds $\sqrt{E}$ $\sin\pi/m$. That is, designating the planar region by the symbol B_1,

$$P_e > P[z \varepsilon B_1]$$

but

$$P[z \varepsilon B_1] = \frac{1}{\sqrt{2\pi N_0}}\int_{\sqrt{E}\sin\pi/m}^\infty e^{-x^2/2N_0}\, dx = \frac{1}{\sqrt{2\pi}}\int_{\sqrt{E/N_0}\sin\pi/m}^\infty e^{-x^2/2}\, dx.$$

Therefore,

$$P_e > \frac{1}{\sqrt{2\pi}}\int_{\sqrt{E/N_0}\sin\pi/m}^\infty e^{-x^2/2}\, dx. \tag{48}$$

Note that if we were to consider also the shaded planar region of Fig. 14, which we shall denote B_2, then it should be clear that

$$P_e < P[z \varepsilon B_1] + P[z \varepsilon B_2],$$

from which we can conclude that

$$P_e < \frac{2}{\sqrt{2\pi}}\int_{\sqrt{E/N_0}\sin\pi/m}^\infty e^{-x^2/2}\, dx.$$

The upper bound so calculated is identical to the one established previously [see (47)] with considerably more effort.

Now let us consider separately the cases $m = 2$ and $m = 4$, for which P_e may be evaluated exactly. If $m = 2$, it may be deduced from Fig. 15 that the probability of error is equal to the probability that a Gaussian random variable of mean zero and variance N_0 exceeds $\sqrt{E}$. That is, when $m = 2$,

$$P_e = \frac{1}{\sqrt{2\pi}}\int_{\sqrt{E/N_0}}^\infty e^{-x^2/2}\, dx. \tag{49}$$

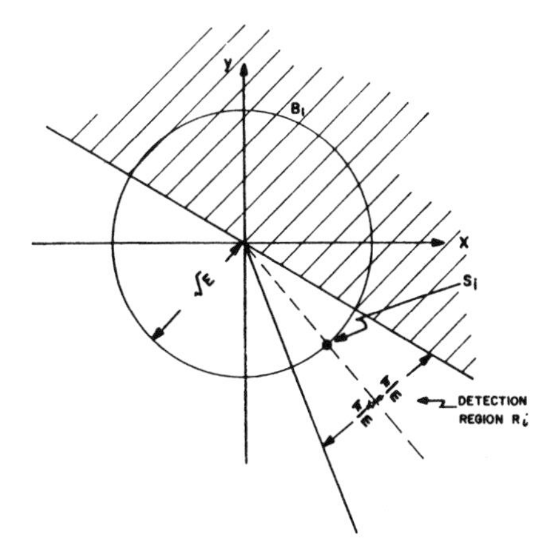

Fig. 13—Illustration of detection region R_i for the multilevel coherent PSK case and the planar region B_1.

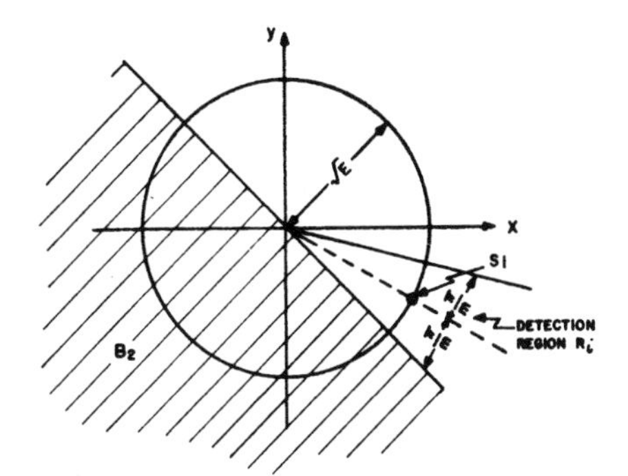

Fig. 14—Illustration of detection region R_i for the multilevel PSK coherent case and the planar region B_2.

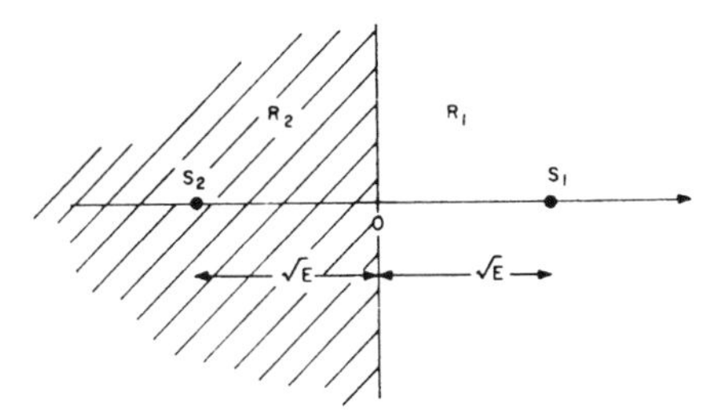

Fig. 15—Illustration of detection regions for 2-level PSK coherent system.

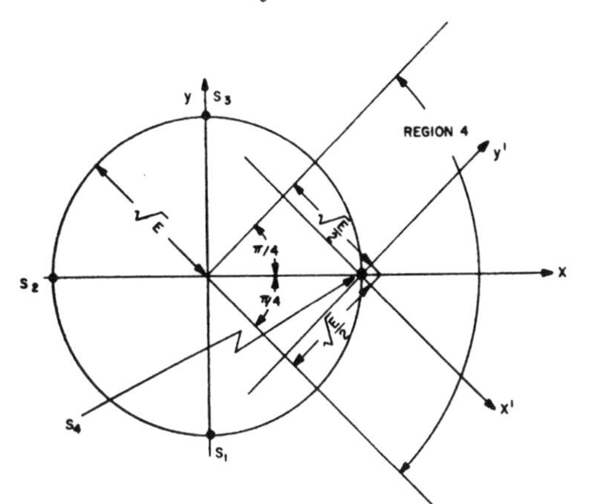

Fig. 16—Illustration of the detection region R_4 for a 4-level PSK coherent system.

The probability of error for the case $m = 4$ may be calculated most easily by resolving the noise into the two orthogonal directions x' and y' indicated on Fig. 16. It follows readily that an error will not occur if n_1' and n_2', the noise components in directions x' and y', satisfy the conditions

$$-\sqrt{\frac{E}{2}} \le n_1' < \infty \quad \text{and} \quad -\sqrt{\frac{E}{2}} \le n_2' < \infty$$

but since n_1' and n_2' are independent noise vectors with zero mean and variance N_0, the probability of this event is simply

$$\left[\frac{1}{\sqrt{2\pi N_0}} \int_{-\sqrt{E/2}}^{\infty} e^{-x^2/2N_0}\, dx\right]^2 = \left[1 - \frac{1}{\sqrt{2\pi}} \int_{\sqrt{E/2N_0}}^{\infty} e^{-x^2/2}\, dx\right]^2.$$

Thus, the probability of error for the 4-level case equals

$$P_e = 1 - \left[1 - \frac{1}{\sqrt{2\pi}} \int_{\sqrt{E/2N_0}}^{\infty} e^{-x^2/2}\, dx\right]^2 = \frac{2}{\sqrt{2\pi}} \int_{\sqrt{E/2N_0}}^{\infty} e^{-x^2/2}\, dx - \left[\frac{1}{\sqrt{2\pi}} \int_{\sqrt{E/2N_0}}^{\infty} e^{-x^2/2}\, dx\right]^2. \quad (50)$$

The results of the preceding calculations for the probability of error of an m-level PSK coherent system are summarized below.

An exact expression for the probability of error in integral form is, by (44),

$$P_e = 1 - \frac{1}{2\pi} \int_{-\pi/m}^{\pi/m} d\theta e^{-E/2N_0 \sin^2 \theta} \cdot \left\{e^{-E/2N_0 \cos^2 \theta} + \sqrt{\frac{E}{N_0}} \cos \theta \int_{-\sqrt{E/N_0} \cos \theta}^{\infty} e^{-t^2/2}\, dt\right\}.$$

If $m > 2$, simple bounds for the probability of error are given by (47) and (48). That is,

$$\frac{1}{\sqrt{2\pi}} \int_{\sqrt{E/N_0} \sin \pi/m}^{\infty} e^{-x^2/2}\, dx < P_e < \frac{2}{\sqrt{2\pi}} \int_{\sqrt{E/N_0} \sin \pi/m}^{\infty} e^{-x^2/2}\, dx.$$

The geometrical reasoning used to establish these bounds is similar to that used to establish Theorem IV (see Appendix V). Theorem IV, however, yields a set of slightly weaker bounds when applied to this case, namely,

$$\frac{1}{\sqrt{2\pi}} \int_{\sqrt{E/N_0} \sin \pi/m}^{\infty} e^{-x^2/2}\, dx \le P_e \le \frac{(m-1)}{\sqrt{2\pi}} \int_{\sqrt{E/N_0} \sin \pi/m}^{\infty} e^{-x^2/2}\, dx.$$

We might point out, however, that Theorem IV is valid even for $m = 2$, in which case both the upper and lower bounds coincide yielding a check for (49), namely,

$$P_e = \frac{1}{\sqrt{2\pi}} \int_{\sqrt{E/N_0}}^{\infty} e^{-x^2/2}\, dx.$$

For the particular case $m = 4$, the probability of error is given by (50) as

$$P_e = \frac{2}{\sqrt{2\pi}} \int_{\sqrt{E/2N_0}}^{\infty} e^{-x^2/2}\, dx - \left(\frac{1}{\sqrt{2\pi}} \int_{\sqrt{E/2N_0}}^{\infty} e^{-x^2/2}\, dx\right)^2.$$

As a final point it should be noted that both (49) and (50) could have been derived from (44).[7]

B. Coherent ASK

In the ASK coherent case, if $S_i(t)$ is transmitted, the received signal point z has a single coordinate [see (11), (12) and (23)],

$$x = \int_0^T z(t)\varphi(t)\, dt = \sqrt{E_i} + n. \tag{51}$$

Thus, x is a Gaussian random variable with mean $\sqrt{E_i}$ and variance N_0. Let us assume in particular that the message points are uniformly spaced, starting with $\sqrt{E_1} = 0$. That is,

$$\sqrt{E_i} = (i - 1)\,\Delta \qquad i = 1, 2, \cdots, m. \tag{52}$$

The corresponding detector signal space is shown in Fig. 17.

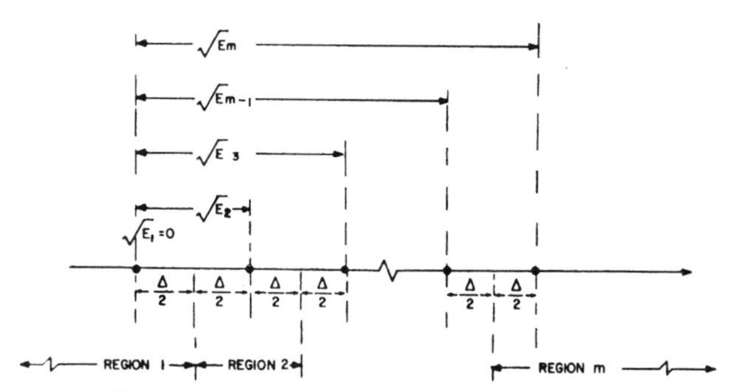

Fig. 17—Illustration of detector signal space for the coherent ASK case, assuming uniformly spaced message points starting with $\sqrt{E_1} = 0$.

It is readily deduced from Fig. 17 that if $1 < i < m$, then the probability of interpreting the transmitted signal incorrectly is simply

$$P[z \notin R_i/S_i] = \frac{2}{\sqrt{2\pi}} \int_{\Delta/2\sqrt{N_0}}^{\infty} e^{-x^2/2}\, dx \quad (1 < i < m). \tag{53}$$

On the other hand, if $i = 1$ or $i = m$, then the probability of interpreting the transmitted signal incorrectly is

$$P[z \notin R_1/S_1] = P[z \notin R_m/S_m]$$

$$= \frac{1}{\sqrt{2\pi}} \int_{\Delta/2\sqrt{N_0}}^{\infty} e^{-x^2/2}\, dx. \tag{54}$$

Now, combining (39), (53) and (54), we can compute the average probability of error P_e to equal

$$P_e = \frac{2(m-1)}{m} \frac{1}{\sqrt{2\pi}} \int_{\Delta/2\sqrt{N_0}}^{\infty} e^{-x^2/2}\, dx. \tag{55}$$

If we further assume that the transmitter is subject to an average power limitation of E/T (watts), it follows that

$$\sum_{i=1}^{m} \left(\frac{E_i}{T}\right) \frac{1}{m} = \frac{E}{T}. \tag{56}$$

Substituting (52) into (56) and making use of the equality[8]

$$\sum_{i=1}^{m} (i-1)^2 = \sum_{j=1}^{m-1} j^2 = \frac{(m-1)(m)(2m-1)}{6}, \tag{57}$$

it is readily seen that the quantity Δ is constrained to equal

$$\Delta = \sqrt{\frac{6E}{(m-1)(2m-1)}}. \tag{58}$$

Consequently, the average probability of error (valid for all m) may be written as

$$P_e = \frac{2(m-1)}{m} \frac{1}{\sqrt{2\pi}} \int_{\sqrt{6E/[4N_0(m-1)(2m-1)]}}^{\infty} e^{-x^2/2}\, dx. \tag{59}$$

C. Coherent FSK

In the coherent FSK case, when $S_i(t)$ is transmitted, the received signal point z has coordinates [see (11), (12) and (26)]

$$x_j = \int_0^T z(t)\varphi_j(t)\, dt = \begin{cases} n_j & j \neq i \\ \sqrt{E} + n_i & j = i \end{cases}$$

$$j = 1, 2, \cdots, m. \tag{60}$$

The x_j are independent-Gaussian-random variables with mean zero if $i \neq j$ with mean $\sqrt{E}$ if $i = j$, and each having variance N_0. The decision rule, namely, to choose the message point closest to the received signal point, is equivalent to choosing that value of j for which x_j is largest. This may be deduced by noting that if the received signal point z, with coordinates $x_1, x_2, \cdots, x_m$ is closer to, say, the signal point S_i than to any other signal point, then

$$d^2(z, S_i) < d^2(z, S_j) \quad \text{all} \quad j \neq i.$$

[7] An ingenious, though somewhat involved, derivation of (50) from a form of (44) has been presented by E. A. Trabka, "Embodiments of the Maximum Likelihood Receiver for Detection of Coherent Phase Shift Keyed Signals," Detect Memo. No. 5A (Appendix) in "Investigation of Digital Data Communication Systems," J. G. Lawton, Ed., Cornell Aeronautical Lab., Inc., Ithaca, N. Y., Rept. No. UA-1420-S-1; January, 1961.

[8] This equality may be verified by induction. See *e. g.*, G. Birkhoff and S. MacLane, "A Survey of Modern Algebra," The Macmillan Co., New York, N. Y., revised ed.; 1960. Note, in particular, example 5a, p. 13.

That is to say, [by (5), (26) and (60)]

$$x_1^2 + \cdots + x_j^2 + \cdots + (x_i - \sqrt{E})^2 + \cdots + x_m^2 < x_1^2 + \cdots + (x_j - \sqrt{E})^2 + \cdots + x_i^2 + \cdots + x_m^2$$

but this is true if, and only if,

$$-2\sqrt{E}\, x_i < -2\sqrt{E}\, x_j;$$

that is, if, and only if,

$$x_i > x_j.$$

Since, when $S_i(t)$ is sent, each of the x_j $(j \neq i)$ are independent-Gaussian-random variables with mean zero and variance N_0, the probability that each of the $(m-1)$ x_j is less than x_i is simply

$$P[x_j < x_i \quad \text{all} \quad j \neq i/x_i, S_i] = \left[\frac{1}{\sqrt{2\pi N_0}} \int_{-\infty}^{x_i} e^{-u^2/2N_0}\, du\right]^{m-1}. \tag{61}$$

Hence, the probability of a correct decision when S_i is sent to equal

$$P[x_j < x_i \quad \text{all} \quad j \neq i/S_i] = \frac{1}{\sqrt{2\pi N_0}} \int_{-\infty}^{\infty} \exp\left\{-\frac{(x_i - \sqrt{E})^2}{2N_0}\right\} \cdot \left[\frac{1}{\sqrt{2\pi N_0}} \int_{-\infty}^{x_i} \exp\left\{-\frac{u^2}{2N_0}\right\} du\right]^{m-1} dx_i. \tag{62}$$

The right-hand side of (62) is independent of the choice of i and is, in fact, equal to the average probability of a correct decision. It follows, therefore, that the average probability of error

$$P_e = 1 - \frac{1}{\sqrt{2\pi}} \int_{-\infty}^{\infty} \exp\left\{-\frac{\left(x - \sqrt{\frac{E}{N_0}}\right)^2}{2}\right\} \cdot \left[\frac{1}{\sqrt{2\pi}} \int_{-\infty}^{x} \exp\{-\tfrac{1}{2}u^2\}\, du\right]^{m-1} dx. \tag{63}$$

The integral appearing in (63) does not appear to be solvable in terms of standard functions for $m > 2$. It should be noted, however, that the integral has been tabulated for several values of m and $\sqrt{E/N_0}$ by Urbano,[9] although not over the ranges which are considered to be of interest within this report. Fortunately, simple bounds for the average probability of error may be found quite easily by applying Theorem IV. In particular, noting [by (5) and (26)] that the distance between any two distinct message points S_i, S_j is

$$d(S_i, S_j) = \sqrt{2E},$$

it follows that

$$\rho = \rho^* = \bar{\rho} = \sqrt{2E}$$

[9] R. H. Urbano, "Analysis and Tabulation of the M Positions Experiment Integral and Related Error Function Integrals," AF Cambridge Res. Ctr., Bedford, Mass., Tech. Rept. No. AFCRC TR-55-100; April, 1955.

and, therefore, that

$$\frac{1}{\sqrt{2\pi}} \int_{\sqrt{E/2N_0}}^{\infty} e^{-x^2/2}\, dx \leq P_e \leq \frac{(m-1)}{\sqrt{2\pi}} \int_{\sqrt{E/2N_0}}^{\infty} e^{-x^2/2}\, dx. \tag{64}$$

Actually a tighter upper bound for P_e has been derived by Fano[10] for use in a channel capacity argument. For our purposes, however, the considerably simpler, if less sophisticated, bounds presented above will be adequate.

Note that if, in particular, $m = 2$, the upper and lower bounds coincide. It follows, therefore, that in the two-level case,

$$P_e = \frac{1}{\sqrt{2\pi}} \int_{\sqrt{E/2N_0}}^{\infty} e^{-x^2/2}\, dx. \tag{65}$$

D. PSK Incoherent

A distinguishing feature of the PSK incoherent case is the fact that there are no pre-assigned detection regions in the signal space, each of which corresponds to a particular transmitted signal. The decision, rather, is based on the phase angle between successively received signals. If the *ith* message has been sent, such a pair of successively received signals will, by (29a) and (29b), be of the form

$$z_1(t) = \sqrt{\frac{2E}{T}} \cos(\omega_0 t + \alpha) + n(t)$$

$$z_2(t - T) = \sqrt{\frac{2E}{T}} \cos\left(\omega_0 t + \alpha + \frac{2\pi i}{m}\right) + n(t - T).$$

Correspondingly, the measurement will (at least conceptually) be based on the pair of signal points z_1 and z_2, having coordinates (x_1, y_1) and (x_2, y_2), respectively, where, by (30a)–(30d), it is known that x_1, y_1, x_2, y_2 are of the form

$$x_1 = \sqrt{E} \cos\alpha + n_{11}$$

$$y_1 = -\sqrt{E} \sin\alpha + n_{12}$$

$$x_2 = \sqrt{E} \cos\left(\alpha + \frac{2\pi i}{m}\right) + n_{21}$$

$$y_2 = -\sqrt{E} \sin\left(\alpha + \frac{2\pi i}{m}\right) + n_{22}$$

where $n_{11}, n_{12}, n_{21}, n_{22}$ are independent-Gaussian-random variables with zero mean and variance N_0 and α is a uniformly distributed random variable over a 2π interval.

A possible set of received signal points corresponding to the case $x_1 = a_1$, $y_1 = b_1$, $x_2 = a_2$, $y_2 = b_2$, $\alpha = A$ are shown plotted in Fig. 18 as vectors. Each vector is represented as the sum of a signal vector and a noise vector.

The decision will be based on the angle ψ which equals

$$\psi = \frac{2\pi i}{m} + \phi_2^* - \phi_1^*.$$

[10] R. M. Fano, *op. cit.*,[3] pp. 200–206.

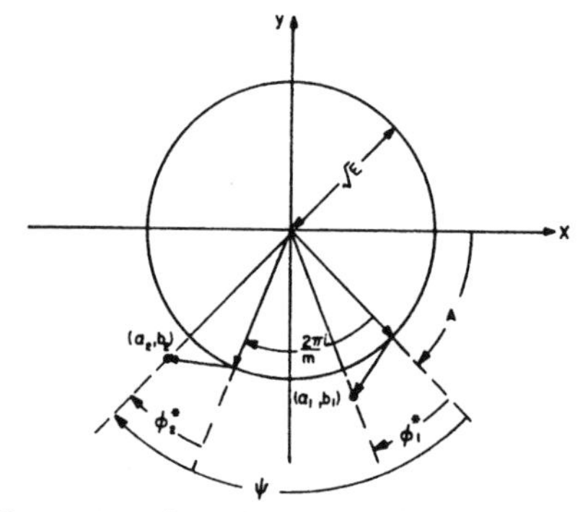

Fig. 18—Illustration of a pair of successively received signal points with coordinates (a_1, b_1) and (a_2, b_2), respectively.

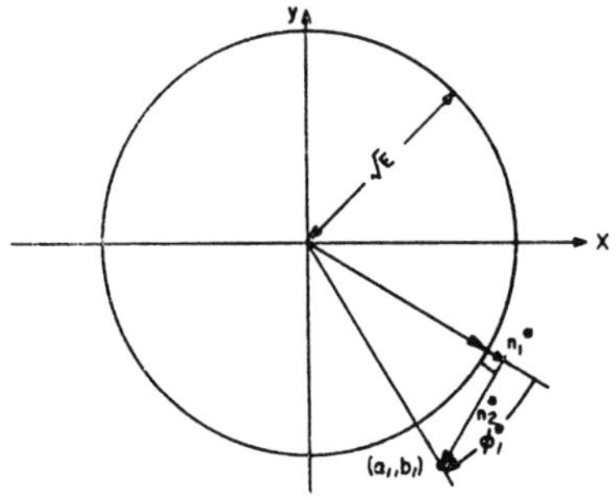

Fig. 19—Decomposition of received signal vector into components parallel and perpendicular to the signal component of the received signal.

An erroneous decision will be made if, and only if, the noise is such that

$$| \phi_2^* - \phi_1^* | > \frac{\pi}{m}. \tag{66}$$

Note that the angles ϕ_1^* and ϕ_2^* which are defined in Fig. 18 are samples of identically distributed random variables. Since the probability density of these random variables, which we shall denote by ϕ_1 and ϕ_2, may readily be determined, the probability density of the new random variable η,

$$\eta = | \phi_2 - \phi_1 |, \tag{67}$$

can be calculated. Designating this density function by $p(\eta)$, it follows that the average probability of error

$$P_e = \int_{\pi/m}^{\pi} p(\eta)\, d\eta. \tag{68}$$

The approach to calculating the probability of error which has just been outlined has, in fact, been used by Fleck and Trabka.[11] They have shown that[12]

$$p(\eta) = \frac{1}{\pi} \int_0^{\pi/2} \sin \psi \left[1 + \frac{E}{2N_0} (1 + \cos \eta \sin \psi) \right] \cdot \exp \left\{ -\frac{E}{2N_0} (1 - \cos \eta \sin \psi) \right\} d\psi \tag{69}$$

and, correspondingly, that

$$P_e = \int_{\pi/m}^{\pi} \frac{1}{\pi} \int_0^{\pi/2} \sin \psi \left[1 + \frac{E}{2N_0} (1 + \cos \eta \sin \psi) \right] \cdot \exp \left\{ -\frac{E}{2N_0} (1 - \cos \eta \sin \psi) \right\} d\psi\, d\eta. \tag{70}$$

Since the manipulations required to establish (69) are rather involved and since, furthermore, the derived expression for the probability of error, (70), is awkward to work with (except if $m = 2$, in which case it reduces to a more tractable form) unless some simplifying approximations are introduced, it is worthwhile to consider an alternate procedure for estimating the probability of error. This we shall now do.

Initially let us calculate the probability density of the random variable ϕ_1, corresponding to the angle ϕ_1^* defined in Fig. 18. The probability density of the random variable ϕ_2 is, of course, the same. It is convenient to resolve the noise component of the corresponding received signal vector into components which are parallel and perpendicular to the signal component as illustrated in Fig. 19.

Designating the projections of the received signal on the directions parallel to and perpendicular to the signal component of the received signal as $\bar{a}$ and $\bar{b}$, respectively, it follows readily from Fig. 19 that

$$\bar{a} = \sqrt{E} + n_1^* \tag{71a}$$

$$\bar{b} = n_2^* \tag{71b}$$

where n_1^* and n_2^* are observed samples of the independent-Gaussian-random variables n_1, n_2, each of which has zero mean and variance N_0. That is to say, $\bar{a}$ and $\bar{b}$ are sample values of a pair of independent random variables which we shall denote by $\bar{x}$ and $\bar{y}$. Since $\bar{x}$ is Gaussian with mean $\sqrt{E}$ and variance N_0 and $\bar{y}$ is Gaussian with mean zero and variance N_0, the joint probability density is given by

$$p(\bar{x}, \bar{y}) = \frac{1}{2\pi N_0} \exp \left\{ -\frac{[(\bar{x} - \sqrt{E})^2 + \bar{y}^2]}{2N_0} \right\}. \tag{72}$$

Now, transforming to polar coordinates by means of the relationships

$$\bar{x} = \nu \sqrt{N_0} \cos \phi_1 \tag{73a}$$

$$\bar{y} = \nu \sqrt{N_0} \sin \phi_1, \tag{73b}$$

we can express (72) in terms of the random variables ν and ϕ_1. Since the Jacobian of the transformation is equal to νN_0, the joint probability density of ν and ϕ_1 which we shall denote by $q(\nu, \phi_1)$ is equal to[13]

$$q(\nu, \phi_1) = p(\nu \sqrt{N_0} \cos \phi_1, \nu \sqrt{N_0} \sin \phi_1) \cdot \nu N_0.$$

That is, by (72),

$$q(\nu, \phi_1) = \frac{\nu}{2\pi} \exp \left\{ -\frac{1}{2} \left[\nu^2 - 2\nu \sqrt{\frac{E}{N_0}} \cos \phi_1 + \frac{E}{N_0} \right] \right\}. \tag{74}$$

[11] J. T. Fleck and E. A. Trabka, "Error Probabilities of Multiple-State Differentially Coherent Phase Shift Keyed Systems in the Presence of White Gaussian Noise," Detect Memo. No. 2A in "Investigation of Digital Data Communication Systems," J. G. Lawton, Ed., Cornell Aeronautical Lab., Inc., Ithaca, N. Y., Rept. No. UA-1420-S-1; January, 1961.

[12] *Ibid.*, see (32) and (33) from which (69) of this paper follows. It should be noted that $p(\eta) = 2h(\eta)$ and $R = E/2N_0$ since we are using a double-sided noise spectrum.

[13] Davenport and Root, *op. cit.*, pp. 37–38.

Integrating out the ν dependency, we get, finally, the probability density of ϕ_1, $q(\phi_1)$ equal to

$$q(\phi_1) = \frac{1}{2\pi}\int_0^\infty d\nu\nu \cdot \exp\left\{-\frac{1}{2}\left[\nu^2 - 2\nu\sqrt{\frac{E}{N_0}}\cos\phi_1 + \frac{E}{N_0}\right]\right\}. \tag{75}$$

The reader might find it interesting to compare the right-hand side of (75) with the integrand of (42) and to accordingly note that the probability of a correct decision in the PSK coherent case is equal to the probability that the angle between the transmitted signal vector and the received signal vector is less than π/m radians in magnitude.

Now, recall [see (66) and preceding discussion] that a correct decision will be made by the detector if, and only if, $|\phi_1 - \phi_2| \leq \pi/m$. The region in the $\phi_1 \times \phi_2$ space corresponding to a correct decision is illustrated by cross hatchings in Fig. 20 (we are assuming that ϕ_1 and ϕ_2 are restricted to lie between $-\pi$ and π modulo 2π).

Neglecting edge effects, it may be deduced from Fig. 20 that the correct decision region is characterized by the condition

$$|\phi_2'| \leq \frac{\pi}{\sqrt{2}\,m}. \tag{76}$$

Thus, the probability of a correct decision is approximately equal to the probability that inequality (76) is satisfied. Unfortunately, the exact probability density of ϕ_2' is not readily determinable. For small ϕ_2', however, (75) yields a reasonably good approximation to the probability density of ϕ_2', as we shall now demonstrate.

It is clear from (75) that $q(\phi_1)$ is an even function, ϕ_1, and that in the interval $|\phi_1| \leq \pi$, $q(\phi_1)$ takes on its maximum value at $\phi_1 = 0$ and decreases monotonically with $|\phi_1|$ to its minimum value at $|\phi_1| = \pi$.

Furthermore, following the procedure that was used to transform (43) into (44), (75) can be rewritten in the form

$$q(\phi_1) = \frac{1}{2\pi}e^{-E/2N_0 \sin^2\phi_1}\left\{e^{-E/2N_0\cos^2\phi_1} + \sqrt{\frac{E}{N_0}}\cos\phi_1 \int_{-\sqrt{E/N_0}\cos\phi_1}^{\infty} e^{-t^2/2}\,dt\right\}. \tag{77}$$

Substituting some particular values of ϕ_1 into (77) to gauge the way in which $q(\phi_1)$ decreases as ϕ_1 increases, we get[14]

$$q(0) = \frac{1}{2\pi}\left[e^{-E/2N_0} + \sqrt{\frac{E}{N_0}}\int_{-\sqrt{E/N_0}}^{\infty} e^{-t^2/2}\,dt\right] \tag{78a}$$

[14] Some plots of $q(\phi_1)$ for various values of $E/2N_0 (= S/N$ in his notation) have been presented by C. R. Cahn, "Performance of digital phase-modulation communication systems," IRE TRANS. ON COMMUNICATIONS SYSTEMS, vol. CS-7, pp. 3–6 (Fig. 2); May, 1959.

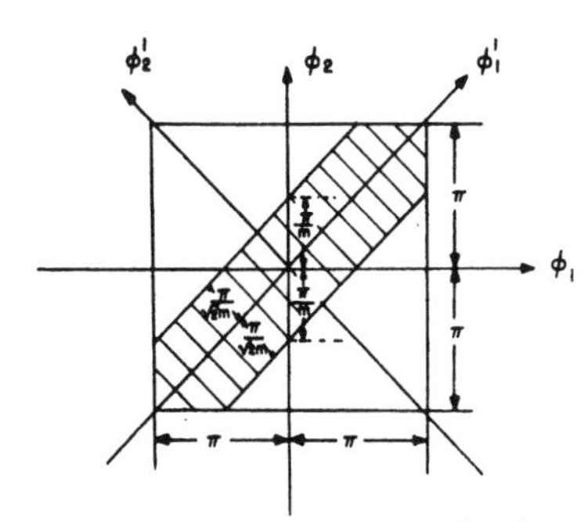

Fig. 20—Illustration of correct decision region in $\phi_1 \times \phi_2$ space.

$$q\left(\pm\frac{\pi}{4}\right) = \frac{1}{2\pi} \cdot\left[e^{-E/2N_0} + e^{-E/4N_0}\sqrt{\frac{E}{2N_0}}\int_{-\sqrt{E/2N_0}}^{\infty} e^{-t^2/2}\,dt\right] \tag{78b}$$

$$q\left(\pm\frac{\pi}{2}\right) = \frac{1}{2\pi}e^{-E/2N_0} \tag{78c}$$

$$q(\pm\pi) = \frac{1}{2\pi}\left[e^{-E/2N_0} - \sqrt{\frac{E}{N_0}}\int_{\sqrt{E/N_0}}^{\infty} e^{-t^2/2}\,dt\right]. \tag{78d}$$

The point we wish to make is that in the case of high signal-to-noise ratio, say $E/2N_0 \gg 20$, $q(\phi_1)$ falls off quite rapidly as $|\phi_1|$ departs from the origin. If, in particular, $|\phi_1| < \pi/2$, then inequality (45) is valid. Treating this inequality as an approximate equality (the approximation improves as $\sqrt{E/N_0}\cos\theta$ increases) and substituting into (77), we get

$$q(\phi_1) \approx \frac{1}{\sqrt{2\pi}}e^{-E/2N_0\sin^2\phi_1}\sqrt{\frac{E}{N_0}}\cos\phi_1 \quad (\text{if } |\phi_1| < \pi/2). \tag{79}$$

In the neighborhood of the origin

$$\sin\phi_1 \approx \phi_1$$

and

$$\cos\phi_1 \approx 1$$

and, correspondingly,

$$q(\phi_1) \approx \frac{1}{\sqrt{2\pi}}e^{-(E/2N_0)\phi_1^2}\sqrt{\frac{E}{N_0}}. \tag{80}$$

That is to say, the probability density of ϕ_1 is approximately Gaussian in the region where it has the most weight, namely, near the origin. Correspondingly, the joint distribution

$$q(\phi_1, \phi_2) = q(\phi_1)q(\phi_2)$$

is approximately circularly symmetric in the same region and, thus, for small ϕ_2' the probability density of ϕ_2' is approximately given by (79). Since the probability of error for the case $m = 2$ may be evaluated exactly as will be shown below, we shall only assume (79) to be a valid representation for the probability density of ϕ_2'

when $m \geq 4$ or, correspondingly, by (76) when

$$|\phi_2'| \leq \frac{\pi}{4\sqrt{2}}.$$

The probability of a correct decision when $m \geq 4$ is thus

$$1 - P_e \approx P\left[|\phi_2'| \leq \frac{\pi}{\sqrt{2}\,m}\right]$$

$$\approx \frac{2}{\sqrt{2\pi}} \int_0^{\pi/\sqrt{2}m} e^{-E/2N_0 \sin^2 \phi_2'} \sqrt{\frac{E}{N_0}} \cos \phi_2' \, d\phi_2'. \quad (81)$$

Transposing (81) and introducing a new variable

$$u = \sqrt{\frac{E}{N_0}} \sin \phi_2',$$

we get, finally, that for $m \geq 4$ the probability of error is approximately equal to

$$P_e \approx \frac{2}{\sqrt{2\pi}} \int_{\sqrt{E/N_0} \sin \pi/\sqrt{2}m}^{\infty} e^{-u^2/2} \, du. \quad (82)$$

We shall now consider the case $m = 2$. From our previous discussion it should be clear that the probability of error is dependent only on the angles ϕ_1 and ϕ_2 and not on the phase difference of $2\pi i/m$ introduced between successive signals at the transmitter. Accordingly, in examining the two-level case it is sufficient to calculate the probability of error for the particular case where the transmitter keeps sending the same waveform. That is to say, we shall consider the case for which the signal components of the successively received signal points z_1 and z_2 coincide. Assuming now that the angle ϕ_1 is known and is equal to, say, ϕ_1^* and that the signal point z_1 has coordinates (a_1, b_1), it may be deduced from Fig. 21 that an error will occur if, and only if, the noise component of z_2 in a direction parallel to the orientation of the stored vector z_1 exceeds $\sqrt{E} \cos \phi_1^*$.[15] That is,

$$P[\text{error}/\phi_1 = \phi_1^*] = \frac{1}{\sqrt{2\pi}} \int_{\sqrt{E/N_0} \cos \phi_1^*}^{\infty} e^{-x^2/2} \, dx. \quad (83)$$

The average probability of making an error is, however, equal to

$$P_e = \int_{-\pi}^{\pi} P[\text{error}/\phi_1 = \phi_1^*] q(\phi_1^*) \, d\phi_1^*. \quad (84)$$

Substituting (77) and (83) into (84) thus yields

$$P_e = \int_{-\pi}^{\pi} \left\{ \frac{1}{\sqrt{2\pi}} \int_{\sqrt{E/N_0} \cos \phi_1}^{\infty} e^{-x^2/2} \, dx \right\} \cdot \left\{ \frac{e^{-E/2N_0 \sin^2 \phi_1}}{2\pi} \left[e^{-E/2N_0 \cos^2 \phi_1} + \sqrt{\frac{E}{N_0}} \cos \phi_1 \int_{-\sqrt{E/N_0} \cos \phi_1}^{\infty} e^{-t^2/2} \, dt \right] \right\} d\phi_1. \quad (85)$$

[15] *Ibid.*, this approach to calculating the probability of error has been suggested here. An alternate technique involving a reduction of (70) is presented in Fleck and Trabka, *op. cit.*

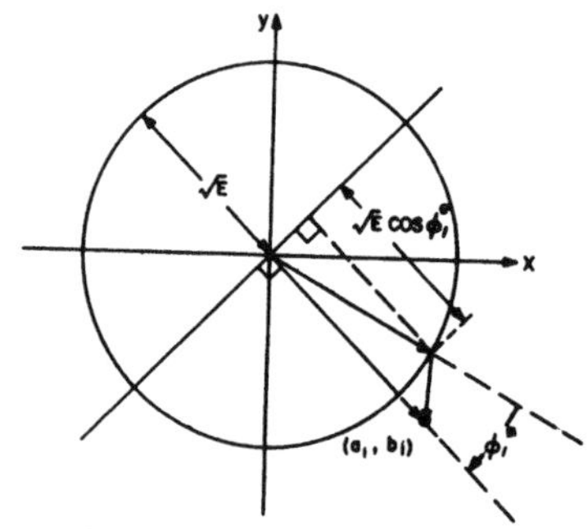

Fig. 21—Illustration of necessary and sufficient conditions for an error to occur in the 2-level PSK incoherent case, given that $\phi_1 = \phi_1^*$ and that $i = m$.

Defining the quantity

$$Q(\phi_1) = \frac{1}{\sqrt{2\pi}} \int_0^{\sqrt{E/N_0} \cos \phi_1} e^{-x^2/2} \, dx, \quad (86)$$

we can rewrite (85) as

$$P_e = \int_{-\pi}^{\pi} \{\tfrac{1}{2} - Q(\phi_1)\} \left\{ \frac{e^{-E/2N_0}}{2\pi} + \frac{e^{-(E/2N_0) \sin^2 \phi_1}}{2\pi} \sqrt{\frac{E}{N_0}} \cos \phi_1 [\tfrac{1}{2} + Q(\phi_1)] \right\} d\phi_1. \quad (87)$$

Now, by simple symmetry arguments it may be deduced that

$$\int_{-\pi}^{\pi} Q(\phi_1) \, d\phi_1 = 0 \quad (88)$$

$$\int_{-\pi}^{\pi} e^{-E/2N_0 \sin^2 \phi_1} \sqrt{\frac{E}{N_0}} \cos \phi_1 \, d\phi_1 = 0 \quad (89)$$

$$\int_{-\pi}^{\pi} e^{-E/2N_0 \sin^2 \phi_1} \sqrt{\frac{E}{N_0}} \cos \phi_1 Q^2(\phi_1) \, d\phi_1 = 0. \quad (90)$$

The expression for the average probability of error, (87), thus reduces to

$$P_e = \tfrac{1}{2} e^{-E/2N_0}. \quad (91)$$

E. ASK Incoherent

In the ASK incoherent case, if $S_i(t)$ is transmitted, the received signal point z has coordinates of the form [see (34a) and (34b)]

$$x = \int_0^T z(t)\varphi_1(t) \, dt = \sqrt{E_i} \cos \alpha + n_1$$

$$y = \int_0^T z(t)\varphi_2(t) \, dt = -\sqrt{E_i} \sin \alpha + n_2$$

where n_1 and n_2 are independent-Gaussian-random variables with zero mean and variance N_0 and α is a uniformly distributed random variable over a 2π interval.

The conditional joint density function of the random variables x and y, given that $\alpha = A$ and that $S_i(t)$ was transmitted, is thus equal to

$$p_i(x, y/\alpha = A) = \frac{1}{2\pi N_0} \cdot \exp\left\{-\frac{[(x - \sqrt{E_i}\cos A)^2 + (y + \sqrt{E_i}\sin A)^2]}{2N_0}\right\}. \quad (92)$$

Correspondingly, averaging over all possible values of A, we get the conditional joint density function of x and y, given only that $S_i(t)$ was transmitted, equal to

$$\begin{aligned} p_i(x, y) &= \int_\theta^{\theta+2\pi} p_i(x, y/\alpha = A)\frac{dA}{2\pi} \\ &= \frac{1}{2\pi N_0}\exp\left\{-\frac{x^2 + y^2 + E_i}{2N_0}\right\}\cdot\frac{1}{2\pi} \cdot \int_\theta^{\theta+2\pi}\exp\left\{\cdot\frac{x\sqrt{E_i}\cos A - y\sqrt{E_i}\sin A}{N_0}\right\}dA \\ &= \frac{1}{2\pi N_0}\exp\left\{-\frac{x^2 + y^2 + E_i}{2N_0}\right\}I_0\left(\frac{\sqrt{x^2 + y^2}\sqrt{E_i}}{N_0}\right) \end{aligned} \quad (93)$$

where I_0 is the modified Bessel function of the first kind of zero order.[16]

As previously noted [see discussion preceding and following (35)], the decision rule for this case is simply to round off the measured value of $|z| = \sqrt{x^2 + y^2}$ to the nearest value of $\sqrt{E_i}$ and guess that the corresponding signal was transmitted. Thus, if $S_i(t)$ is transmitted, the received signal will be interpreted correctly if, and only if,

$$\frac{\sqrt{E_{i-1}} + \sqrt{E_i}}{2} \leq \sqrt{x^2 + y^2} < \frac{\sqrt{E_i} + \sqrt{E_{i+1}}}{2} \qquad i = 1, 2, \cdots, m \quad (94)$$

where we define $\sqrt{E_0} = -\sqrt{E_1}$ and $\sqrt{E_{m+1}} = \infty$.

Now the probability that inequality (94) is satisfied, *i.e.*, the probability that the received signal point z falls into the detection zone R_i when $S_i(t)$ is transmitted, may be expressed as

$$P[z \in R_i/S_i] = \iint_{R_i} p_i(x, y)\,dx\,dy = \iint_{R_i}\frac{1}{2\pi N_0} \cdot \exp\left\{-\frac{x^2 + y^2 + E_i}{2N_0}\right\}I_0\left(\frac{\sqrt{x^2 + y^2}\sqrt{E_i}}{N_0}\right)dx\,dy. \quad (95)$$

Transforming to polar coordinates with

$$x = \nu\sqrt{N_0}\cos\phi \qquad y = \nu\sqrt{N_0}\sin\phi,$$

we get

$$P[z \in R_i/S_i] = \frac{1}{2\pi}\iint_{R_i}\exp\left\{-\left(\frac{\nu^2}{2} + \frac{E_i}{2N_0}\right)\right\}I_0\left(\nu\sqrt{\frac{E_i}{N_0}}\right)\nu d\nu d\phi. \quad (96)$$

[16] F. Bowman, "Introduction to Bessel Functions," Dover Publications, Inc., New York, N. Y., pp. 41–42; 1958.

Since the region R_i is defined by the equations

$$\frac{\sqrt{E_{i-1}} + \sqrt{E_i}}{2\sqrt{N_0}} \leq \nu < \frac{\sqrt{E_i} + \sqrt{E_{i+1}}}{2\sqrt{N_0}} \quad (97a)$$

$$0 \leq \phi < 2\pi, \quad (97b)$$

it follows readily that

$$P[z \in R_i/S_i] = \int_{(\sqrt{E_{i-1}}+\sqrt{E_i})/2\sqrt{N_0}}^{(\sqrt{E_i}+\sqrt{E_{i+1}})/2\sqrt{N_0}}\exp\left\{-\frac{1}{2}\left(\nu^2 + \frac{E_i}{N_0}\right)\right\} \cdot I_0\left(\nu\sqrt{\frac{E_i}{N_0}}\right)\nu\,d\nu. \quad (98)$$

Unfortunately, however, this integral is not simply solvable except for the particular case $\sqrt{E_i} = 0$ nor has it been tabulated over the regions of interest.[17]

It is possible, however, to bound integrals of this type from above and below by analyzing the integral geometrically and then modifying the regions of integration. Thus, for example, we might note that the integral in question represents the probability that a two-dimensional spherically symmetric Gaussian noise vector with mean zero and variance N_0, originating from some point lying on the circumference of a circle of radius $\sqrt{E_i}$, falls inside the ring determined by the circles (concentric with the first) of radius

$$\frac{\sqrt{E_{i-1}} + \sqrt{E_i}}{2} \quad \text{and} \quad \frac{\sqrt{E_i} + \sqrt{E_{i+1}}}{2},$$

respectively.

The probability of the noise vector falling inside the ring, however, is certainly larger than the probability of it falling inside the circle of radius $\mathcal{R}_4$ centered on the noise origin (see Fig. 22), where $\mathcal{R}_4$ is chosen as large as possible subject only to the constraint that the circle must lie in the ring.

On the other hand, the probability of landing in the ring is certainly less than the probability of landing in the shaded region of Fig. 22(c). Both these probabilities are easily evaluated.

The probability that the noise lands inside the circle of radius $\mathcal{R}_4$ [see Fig. 22(b)], which we shall denote by $\underline{P}(i)$ is equal to

$$\underline{P}(i) = \iint_{\sqrt{n_1^2+n_2^2}\leq\mathcal{R}_4}\frac{\exp\left\{-\frac{(n_1^2 + n_2^2)}{2N_0}\right\}}{2\pi N_0}dn_1\,dn_2 = 1 - \exp\left\{-\frac{\mathcal{R}_4^2}{2N_0}\right\}. \quad (99)$$

[17] Eq. (98) may be expressed as the difference of 2 Q functions. The Q function is defined in J. I. Marcum and P. Swerling, "Studies of Target Detection by Pulsed Radar," IRE TRANS. ON INFORMATION THEORY (*Special Monograph Issue*), vol. IT-6, p. 159; April, 1960. The Q function has been tabulated, though not within the ranges of interest of this report, by J. I. Marcum, "Table of Q Functions," Rand Corp., Santa Monica, Calif., Rept. RM-339; January 1, 1950.

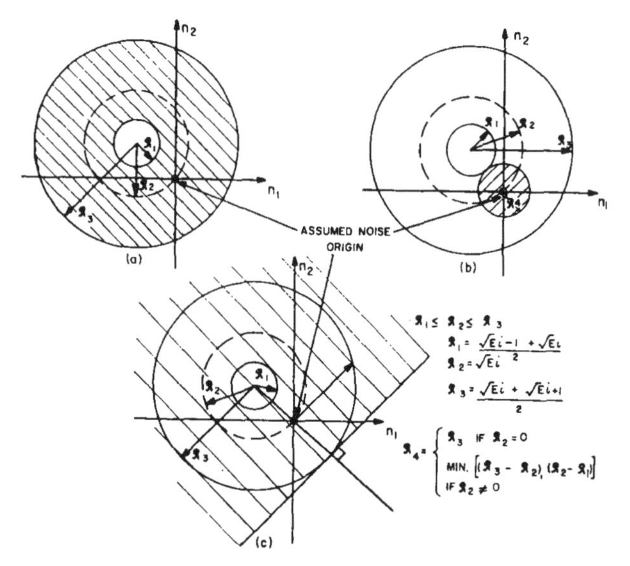

Fig. 22—Geometric interpretation of (98). (a) The right-hand side of (98) is equal to the probability that the two-dimensional spherically symmetric Gaussian noise vector, with coordinates n_1, n_2, each having zero mean and variance N_0, lands inside the shaded region. (b) The probability of this event is certainly larger than the probability of landing in this shaded region. (c) The probability of this event is certainly smaller than the probability of landing in this shaded region.

The probability that the noise lands in the shaded region of Fig. 22(c), $\bar{P}(i)$, is equal to

$$\bar{P}(i) = 1 - \frac{1}{\sqrt{2\pi}} \int_{(\mathfrak{R}_3 - \mathfrak{R}_2)/\sqrt{N_0}}^{\infty} e^{-x^2/2}\, dx. \tag{100}$$

Note that the quantities $\mathfrak{R}_1$, $\mathfrak{R}_2$, $\mathfrak{R}_3$, $\mathfrak{R}_4$, which are defined in Fig. 22 [see also the remark following (94)], are functions of i, the parameter used to index the possibly transmitted signals.

We have established in (99) and (100) lower and upper bounds, respectively, to the probability of the event of interest as expressed by (98). That is to say,

$$\underline{P}(i) \le P[z \,\varepsilon\, R_i/S_i] < \bar{P}(i). \tag{101}$$

Now if, in particular, we assume that the message points are uniformly spaced starting with zero, then, by (52),

$$\sqrt{E_i} = (i-1)\,\Delta \qquad i = 1, 2, \cdots, m$$

where, assuming the same average power constraints as in the ASK coherent case [see (58)],

$$\Delta = \sqrt{\frac{6E}{(m-1)(2m-1)}}.$$

It follows readily by direct substitution that $\mathfrak{R}_4 = \Delta/2$ independently of the choice of i and, therefore, that

$$\underline{P}(i) = 1 - e^{-\Delta^2/8N_0}. \tag{102}$$

The average probability of a correct decision is equal to

$$1 - P_e = \frac{1}{m} \sum_{i=1}^{m} P[z \,\varepsilon\, R_i/S_i] \tag{103}$$

and is, therefore, by (101) and (102), bounded from below by

$$1 - P_e \ge 1 - e^{-\Delta^2/8N_0}. \tag{104}$$

An upper bound to $1 - P_e$ can be obtained in a similar fashion by evaluating (100) for different values of i. The resultant expression is, however, rather cumbersome. A satisfactory upper bound can be obtained quite simply by noting that since

$$P[z \,\varepsilon\, R_i/S_i] > 0,$$

it is certainly true that [see (39)]

$$P_e = \frac{1}{m} \sum_{i=1}^{m} P[z \not\varepsilon\, R_i/S_i] > \frac{1}{m} [Pz \not\varepsilon\, R_1/S_1].$$

That is,

$$P_e > \frac{1 - P[z \,\varepsilon\, R_1/S_1]}{m} \tag{105}$$

but, for the case $i = 1$, (98) reduces to

$$P[z \,\varepsilon\, R_1/S_1] = \int_0^{\Delta/2\sqrt{N_0}} \nu e^{-\nu^2/2}\, d\nu = 1 - e^{-\Delta^2/8N_0}.$$

Substituting this latter result into (105) yields

$$P_e > \frac{1}{m} e^{-\Delta^2/8N_0}. \tag{106}$$

Finally combining (104), (106) and (58), we get the average probability of error bounded by

$$\frac{\exp\left\{-\frac{6E}{8N_0(m-1)(2m-1)}\right\}}{m} < P_e \le \exp\left\{-\frac{6E}{8N_0(m-1)(2m-1)}\right\}. \tag{107}$$

The bounds are valid for all $m \ge 2$.

F. FSK Incoherent

In the FSK incoherent case, if $S_i(t)$ is transmitted, the received signal point z has, by (37a) and (37b), coordinates of the form

$$x_j = \begin{cases} n_{xj} & j \neq i \\ \sqrt{E}\cos\alpha + n_{xi} & j = i \end{cases}$$

$$y_j = \begin{cases} n_{yj} & j \neq i \\ -\sqrt{E}\sin\alpha + n_{yi} & j = i \end{cases}$$

where the n_{xj} and n_{yj}, $j = 1, 2, \cdots, m$ are independent-Gaussian-random variables with mean zero and variance N_0 and α is a uniformly distributed random variable over a 2π interval.

A correct decision will be made by the detector when $S_i(t)$ is sent, if, and only if, for all $j \neq i$ the inequality

$$\sqrt{x_j^2 + y_j^2} < \sqrt{x_i^2 + y_i^2} \tag{108}$$

is satisfied. The probability density of the random variable $\sqrt{x_j^2 + y_j^2}$, $j = 1, 2, \cdots, m$ may be established readily by converting the joint density of x_j and y_j to polar form and then integrating out the angular dependency. In particular, if $j = i$, the joint conditional density of x_i, y_i, given that $S_i(t)$ was transmitted and that $\alpha = A$, is equal to

$$p_i(x_i, y_i/\alpha = A)$$

$$= \frac{1}{2\pi N_0} \exp\left\{-\frac{1}{2N_0}[(x_i - \sqrt{E}\cos A)^2 + (y_i + \sqrt{E}\sin A)^2]\right\}. \tag{109}$$

Averaging over all possible values of A to get the joint conditional density of x_i and y_i, given only that $S_i(t)$ was transmitted, yields

$$p_i(x_i, y_i) = \int_\theta^{\theta+2\pi} p_i(x_i, y_i/\alpha = A)\frac{dA}{2\pi}$$

$$= \frac{1}{2\pi N_0}\exp\left\{-\frac{1}{2N_0}[x_i^2 + y_i^2 + E]\right\} \cdot \int_\theta^{\theta+2\pi} \exp\left\{\frac{x_i\sqrt{E}\cos A}{N_0} - \frac{y_i\sqrt{E}\sin A}{N_0}\right\}\frac{dA}{2\pi}$$

$$= \frac{1}{2\pi N_0}\exp\left\{-\frac{1}{2N_0}[x_i^2 + y_i^2 + E]\right\} \cdot I_0\left(\frac{\sqrt{x_i^2 + y_i^2}\sqrt{E}}{N_0}\right) \tag{110}$$

where I_0 is the modified Bessel function of the first kind of zero order.[16]

Letting $x_i = \nu_i\sqrt{N_0}\cos\phi_i$, $y_i = \nu_i\sqrt{N_0}\sin\phi_i$ and noting that the Jacobian of the transformation is equal to $\nu_i N_0$, the joint probability density of the random variables ν_i and ϕ_i, $q(\nu_i, \phi_i)$ which is equal to[13]

$$q(\nu_i, \phi_i) = \nu_i N_0 \cdot p(\nu_i\sqrt{N_0}\cos\phi_i, \nu_i\sqrt{N_0}\sin\phi_i),$$

may be written as

$$q(\nu_i, \phi_i) = \frac{\nu_i}{2\pi}\exp\left\{-\frac{1}{2}\left[\nu_i^2 + \frac{E}{N_0}\right]\right\}I_0\left(\nu_i\sqrt{\frac{E}{N_0}}\right). \tag{111}$$

Therefore, the probability density of ν_i, $q(\nu_i)$ is equal to

$$q(\nu_i) = \int_0^{2\pi} q(\nu_i, \phi_i)\,d\phi_i$$

$$= \nu_i\exp\left\{-\frac{1}{2}\left[\nu_i^2 + \frac{E}{N_0}\right]\right\}I_0\left(\nu_i\sqrt{\frac{E}{N_0}}\right). \tag{112}$$

If $j \neq i$, then the joint density of the random variables x_j and y_j is equal to the right-hand side of (110) with E set equal to zero. Correspondingly, the probability density of the random variable ν_j is equal to the right-hand side of (112) with E set equal to zero. That is,

$$q(\nu_j) = \nu_j e^{-\nu_j^2/2} \qquad (j \neq i). \tag{113}$$

Condition (108) is equivalent to requiring that

$$\nu_j < \nu_i \tag{114}$$

for all $j \neq i$. Since the ν_j are independent random variables, the probability that inequality (114) is satisfied under the condition that ν_i is known is simply

$$P[\nu_j < \nu_i, \text{all } j \neq i/\nu_i] = \prod_{j\neq i} P[\nu_j < \nu_i/\nu_i]. \tag{115}$$

But the probability that for any particular $j \neq i$ that $\nu_j < \nu_i$, given ν_i, is equal to [by (113)]

$$P[\nu_j < \nu_i/\nu_i] = \int_0^{\nu_i} \nu_j e^{-\nu_j^2/2}\,d\nu_j$$

$$= 1 - e^{-\nu_i^2/2}. \tag{116}$$

Substituting (116) into (115) yields

$$P[\nu_j < \nu_i, \text{all } j \neq i/\nu_i] = (1 - e^{-\nu_i^2/2})^{m-1}. \tag{117}$$

Consequently, the probability of a correct decision when $S_i(t)$ is sent, which is given by

$$P[\nu_j < \nu_i, \text{all } j \neq i] = \int_0^\infty P[\nu_j < \nu_i \text{ all } j \neq i/\nu_i]q(\nu_i)\,d\nu_i,$$

is equal to [see (112) and (117)]

$$P[\nu_j < \nu_i, \text{all } j \neq i] = \int_0^\infty \nu_i\exp\left\{-\frac{1}{2}\left[\nu_i^2 + \frac{E}{N_0}\right]\right\}I_0\left(\nu_i\sqrt{\frac{E}{N_0}}\right) \cdot [1 - e^{-\nu_i^2/2}]^{m-1}\,d\nu_i. \tag{118}$$

Utilizing the binomial expansion, we can write

$$[1 - e^{-\nu_i^2/2}]^{m-1} = \sum_{k=0}^{m-1}\binom{m-1}{k}(-1)^k e^{-\nu_i^2 k/2}$$

which result, when combined with (118), yields

$$P[\nu_j < \nu_i, \text{all } j \neq i] = \exp\left\{-\frac{E}{2N_0}\right\}\sum_{k=0}^{m-1}\binom{m-1}{k}(-1)^k \cdot \int_0^\infty \nu_i\exp\left\{-\frac{\nu_i^2(k+1)}{2}\right\}I_0\left(\nu_i\sqrt{\frac{E}{N_0}}\right)d\nu_i. \tag{119}$$

The integral appearing in the right-hand side of (119) is a standard form whose solution is[18]

$$\int_0^\infty \nu_i\exp\left\{-\frac{\nu_i^2(k+1)}{2}\right\}I_0\left(\nu_i\sqrt{\frac{E}{N_0}}\right)d\nu_i = \frac{1}{k+1}\exp\left\{\frac{E}{2N_0(k+1)}\right\}. \tag{120}$$

[18] G. N. Watson, "A Treatise on the Theory of Bessel Functions," 2nd ed., Cambridge University Press, Cambridge, England, Sec. 13.3, p. 393, Eq. (1); 1958.

Combining (119) and (120), we get the probability of a correct decision when $S_i(t)$ is transmitted equal to

$$P[\nu_j < \nu_i, \text{ all } j \neq i] = \exp\left\{-\frac{E}{2N_0}\right\} \cdot \sum_{k=0}^{m-1} \binom{m-1}{k}(-1)^k \frac{\exp\left\{\frac{E}{2N_0(k+1)}\right\}}{(k+1)}. \quad (121)$$

The right-hand side of (121) is, however, independent of the choice of i and is, therefore, in fact, equal to the average probability of a correct decision. The average probability of error for an m-level incoherent (orthogonal) FSK system is, thus, equal to

$$P_e = 1 - \exp\left\{-\frac{E}{2N_0}\right\} \sum_{k=0}^{m-1} \binom{m-1}{k}(-1)^k \cdot \frac{\exp\left\{\frac{E}{2N_0(k+1)}\right\}}{k+1}. \quad (122)$$

Noting that (122) can be written as

$$P_e = -\exp\left\{-\frac{E}{2N_0}\right\} \sum_{k=1}^{m-1} \binom{m-1}{k}(-1)^k \frac{\exp\left\{\frac{E}{2N_0(k+1)}\right\}}{k+1}$$

and that

$$\binom{m-1}{k}\frac{1}{k+1} = \binom{m}{k+1}\frac{1}{m},$$

we can, introducing a new summation index, $q = k + 1$, rewrite the expression for the average probability of error in final form as

$$P_e = \frac{\exp\left\{-\frac{E}{4N_0}\right\}}{m} \sum_{q=2}^{m} \binom{m}{q}(-1)^q \exp\left\{\frac{E(2-q)}{4N_0 q}\right\}. \quad (123)$$

VI. Conclusions

A. Review of Basic Assumptions

In the preceding sections of this paper, an approach to the problem of optimally detecting a set of known waveforms in a stationary-white-Gaussian environment has been presented. Utilizing this approach, three basic data transmission systems have been studied and a set of error characteristics has been derived. Curves wherein the probability of error (actually $\log_{10} P_e$) is plotted as a function of signal-to-noise ratio (in decibels), the number of levels m appearing as a parameter, are presented in Figs. 23–28 (pages 358–359). Before proceeding to a detailed discussion of these curves, however, we wish to review explicitly the assumptions upon which they are based.

In particular we have assumed that:

1) Each signal waveform is transmitted with equal probability.
2) The transmitter is subject to an average power limitation, E/T (watts), where T is the duration of each transmitted signal waveform.
3) The received signal is the sum of the transmitted signal and a noise term, the noise being stationary-white-zero mean-Gaussian with double-sided spectral density N_0 (watts/cps). That is to say, the noise power passed by an ideal filter with unit gain and (positive) bandwidth W is $2N_0W$ watts.
4) The receiver is in time synchronism with the transmitter, by which we mean to say that the receiver knows when to sample and when to quench the product integrators. When, in addition, it is assumed that the receiver is phase locked to the transmitter, the system is referred to as coherent.
5) The received signal is processed by a maximum-likelihood detector except in the ASK incoherent case. For reasons of simplicity, an approximation to the maximum-likelihood detector which approaches the true maximum-likelihood detector in a high signal-to-noise ratio environment was chosen for this case.
6) In the FSK case and the PSK case, the transmitted signal waveforms, which are sinusoidal pulses, each contain equal energy E whereas in the ASK case the amplitudes (that is, the square root of the energy) of the transmitted pulses are uniformly spaced starting with zero. Furthermore, in the FSK cases the transmitted waveforms are orthogonal.

B. Physical Significance of P_e

It should be particularly noted that the calculated quantity designated as P_e represents the average probability of misinterpreting the transmitted waveforms. That is to say, if, in a long period of time KT, K waveforms are transmitted and, say, L of them are misinterpreted by the detector, then P_e will (almost always) be approximately equal to

$$P_e \approx \frac{L}{K},$$

the approximation becoming better as $K \to \infty$.

The point we wish to emphasize is that P_e, which is sometimes also referred to as the character error, is not, in general, equal to the probability that a single binary symbol is received incorrectly or that a binary sequence of some arbitrary length is received incorrectly. Thus, we might note, for example, that if, in an 8-level system the waveform corresponding to the binary sequence 001 is misinterpreted for the waveform corresponding to the binary sequence 011, a single character error has been made but two out of three of the binary digits have been received correctly. It is, of course, possible for a single

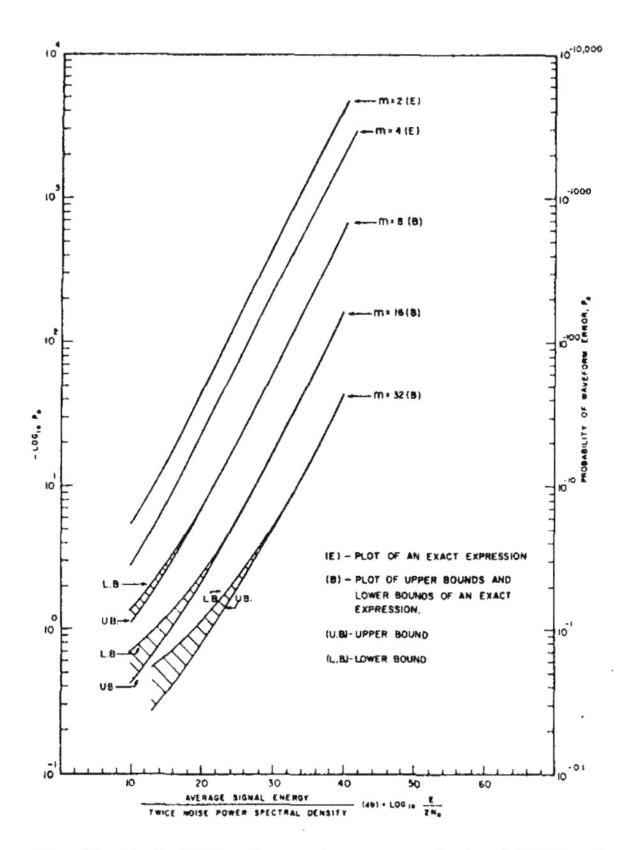

Fig. 23—Probability of waveform error (m-level PSK coherent), assuming that the duration of each signal is fixed independently of m.

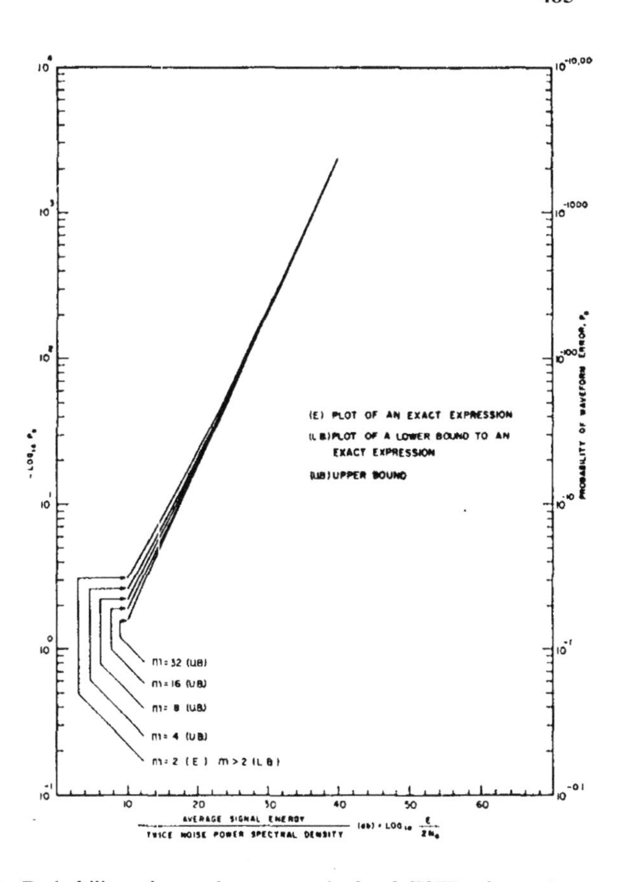

Fig. 25—Probability of waveform error (m-level FSK coherent), assuming that the duration of each signal is fixed independently of m.

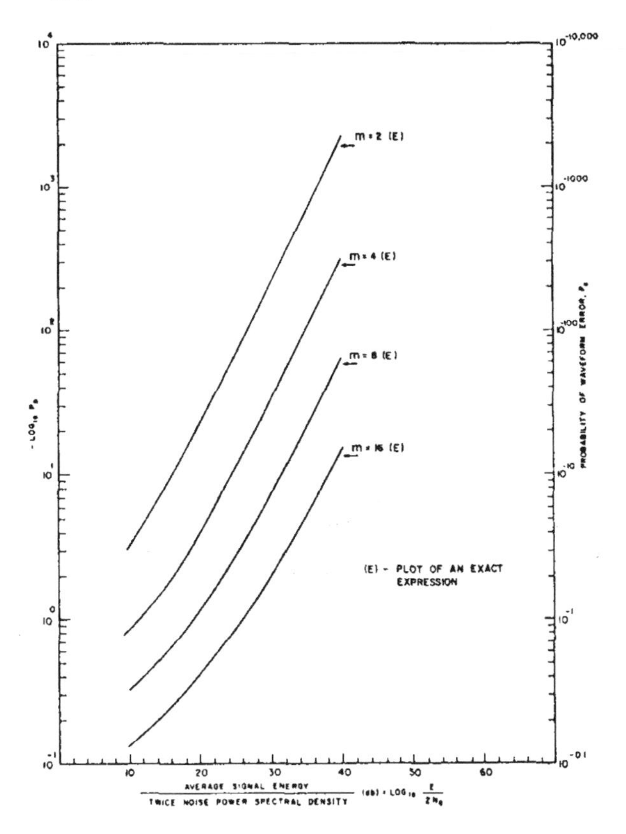

Fig. 24—Probability of waveform error (m-level ASK coherent), assuming that the duration of each signal is fixed independently of m.

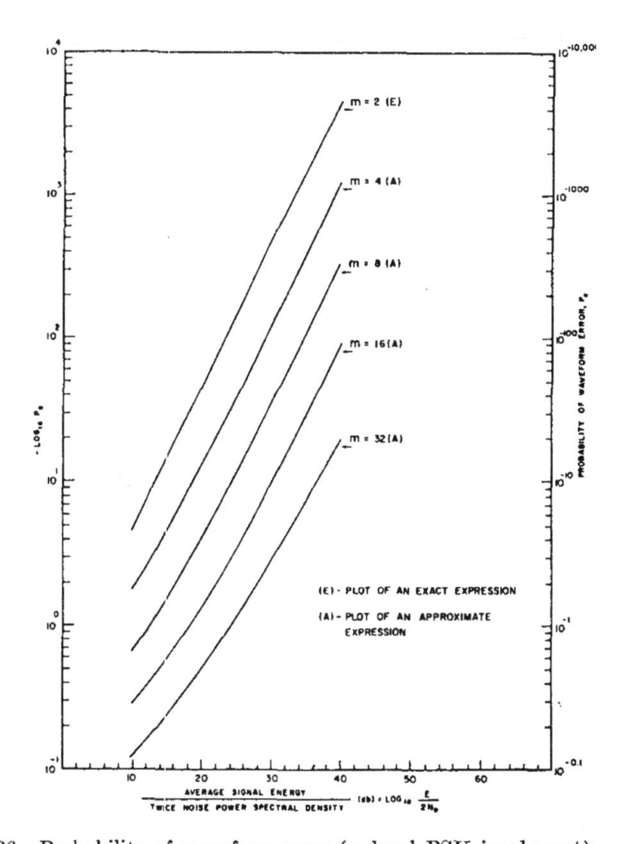

Fig. 26—Probability of waveform error (m-level PSK incoherent), assuming that the duration of each signal is fixed independently of m.

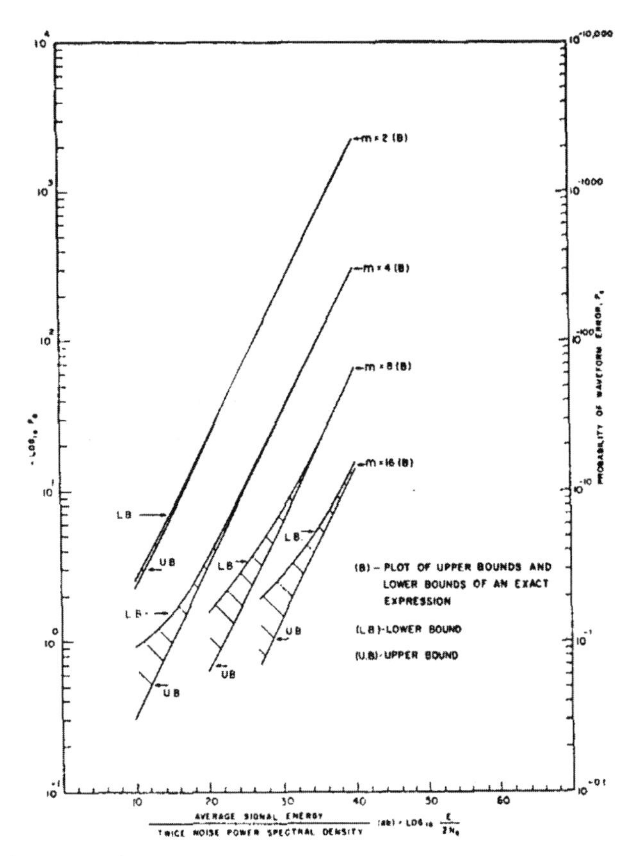

Fig. 27—Probability of waveform error (m-level ASK incoherent), assuming that the duration of each signal is fixed independently of m.

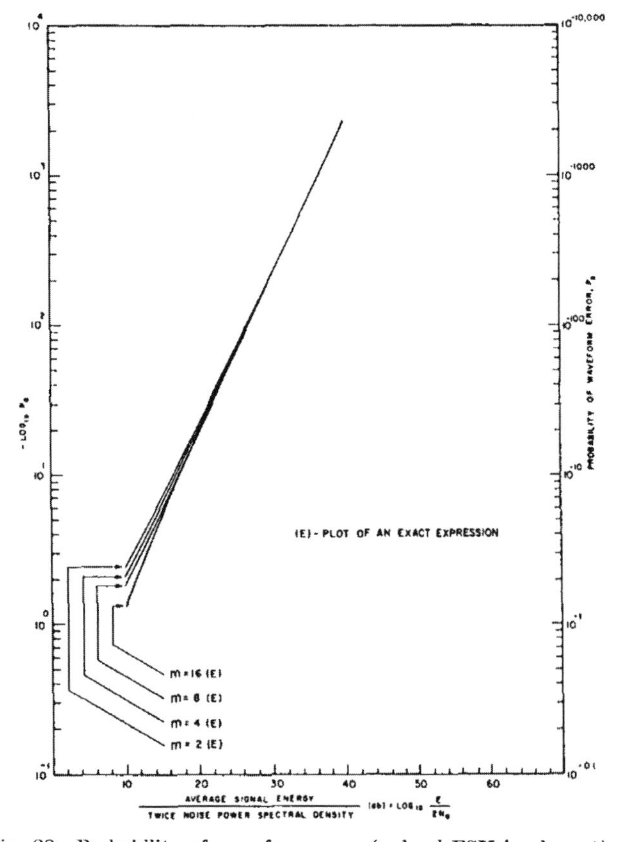

Fig. 28—Probability of waveform error (m-level FSK incoherent), assuming that the duration of each signal is fixed independently of m.

character error to correspond to two or three binary errors (in the 8-level case). Whether or not such a character error is more serious than the other kind depends upon the particular coding scheme used and cannot be predicted in advance. In one coding scheme, for example, we might be interested in the probability that binary sequences of length 6 are received correctly (alpha-numeric code). If an 8-level system is utilized in this situation, each binary sequence of length 6 must be associated with a pair of waveforms which will then be transmitted in succession. Correspondingly, the sequence will be received correctly if, and only if, both the associated waveforms are received correctly. The probability of receiving such a sequence incorrectly is, therefore,

$$P_{seq} = 1 - (1 - P_e)^2$$

which for small P_e is approximately equal to

$$P_{seq} \approx 2P_e.$$

In the last cited example, it was a relatively simple task to derive an expression for the probability of the event of interest, namely, that a binary sequence of length 6 is received incorrectly, in terms of P_e. In general, however, the events of interest will not be simply related to P_e. In fact, it may be impossible to calculate the probability of certain events such as the probability of a binary error without resorting to a detailed analysis of the probability with which various types of waveform errors can occur[19] (*e.g.*, what is the probability that if waveform number one is sent it is interpreted as waveform number two, number three, etc.?). Nevertheless, a set of inequalities which adequately relate the probability of the event of interest to the probability of a character error may frequently be established quite easily. Thus, if we are interested in the probability of a binary error we might note that if K waveforms of an m-level system are sent in a period of time KT, then $K \log_2 m$ (which is assumed to be an integer) binary symbols are sent in that time. If L out of K waveforms are misinterpreted by the receiver, then at least L but no more than $L \log_2 m$ binary symbols are received incorrectly. Consequently, the ratio r of the number of binary symbols received incorrectly to the number transmitted is bounded by

$$\frac{L}{K \log_2 m} \leq r \leq \frac{L}{K} \frac{\log_2 m}{\log_2 m}.$$

Now, as K becomes large L/K approaches P_e, the probability of a waveform error and r approaches P_b, the probability of a binary error. In the limit, therefore, as K approaches infinity we have

$$\frac{P_e}{\log_2 m} \leq P_b \leq P_e. \tag{124}$$

[19] Some special cases have been considered by J. K. Wolf, "Comparison of N-ary Transmission Systems," Rome Air Dev. Ctr., Rome, N. Y., Rept. No. RADC-TN-60-210; December, 1960.

C. Accuracy of Presented Curves

Regarding the accuracy of the curves presented, it should be noted that in all cases except PSK incoherent ($m > 2$) either an exact expression for P_e or an upper and lower bound to P_e is plotted. Unfortunately, for the PSK incoherent case simple bounds could not be found with the exception of the 2-level case which was evaluated exactly. Correspondingly, the curves presented for the PSK incoherent case ($m > 2$) represent simply an approximation to the actual probability of error, the approximation being quite good for large values of m.

We might point out that the plotted curves cannot be used to sharply determine the probability of error for a particular modulation scheme, given $E/2N_0$ and m. The reader can verify this himself simply by trial and error, noting in particular that a small uncertainty in log P_e results in a much larger uncertainty in P_e. The inverse problem, which is perhaps the more natural, can be handled much more satisfactorily. That is to say, given a value of P_e which it is desired to maintain and a value of m, the needed signal-to-noise ratio ($E/2N_0$) can be determined quite closely.

D. Comparisons Assuming Fixed Waveform Duration

Now, referring to the curves appearing on Figs. 23–28, there are a few general conclusions which can be drawn. In the first place, note that increasing m (the number of waveforms which may be transmitted in any time T) tends to increase the probability of error whereas increasing the energy content in each transmitted signal (*i.e.*, the signal-to-noise ratio) tends to decrease the probability of error. Furthermore, it should be noted that increasing m introduces the most degradation in the ASK case, somewhat less degradation in the PSK case and comparatively little degradation in the FSK case. Geometrically, the reason for this is clear. In all cases, increasing m increases the number of message points. In the ASK case, these points lie in a one-dimensional space; in the PSK case, they lie in a two-dimensional space whereas in the FSK case, the dimension of the space increases linearly with m. Correspondingly, in the first two cases the points (since we are subject to an average power limitation) become crowded together and the frequency of errors increase. In the FSK case, however, all message points may be maintained equidistant. Consequently, there is little deterioration in performance with increasing m. In fact, we might expect that if m is sufficiently large, the average probability of error of an FSK system should be smaller than that of a PSK or an ASK system (if it is not so already). That this is indeed the case may readily be deduced from Figs. 29–31 wherein some cross plots of the error rates for various systems are presented for given values of m. For the sake of clarity only four curves are presented in each diagram rather than the full six. The selected curves are, however, representative, as the following comparison of the per-

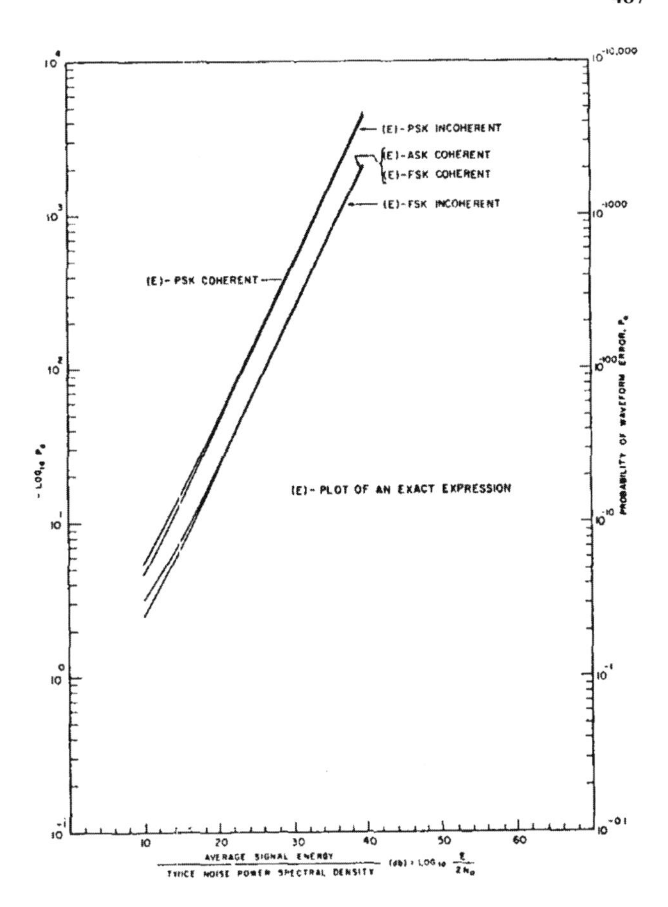

Fig. 29—Comparison of error performance (2-level systems).

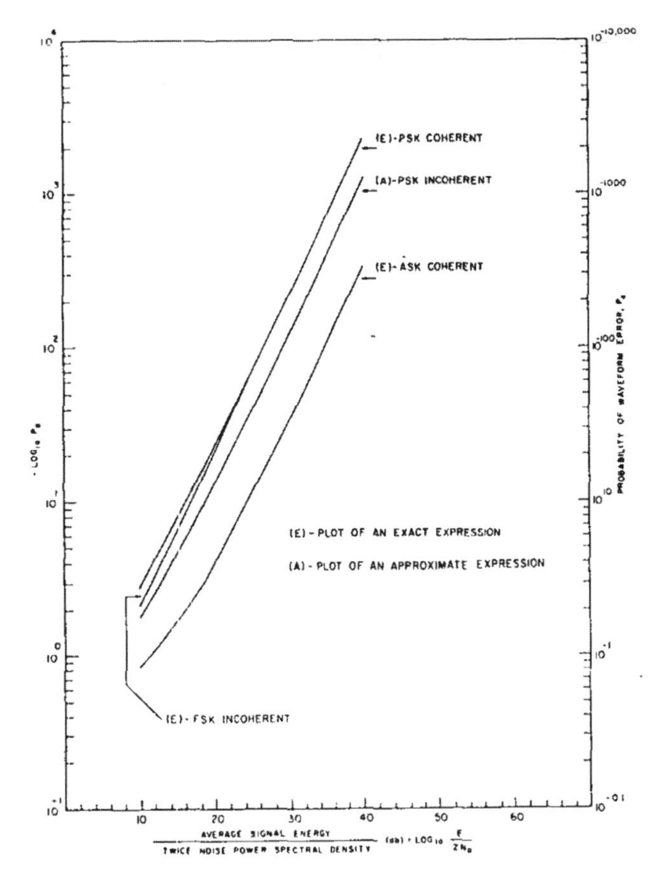

Fig. 30—Comparison of error performance (4-level systems).

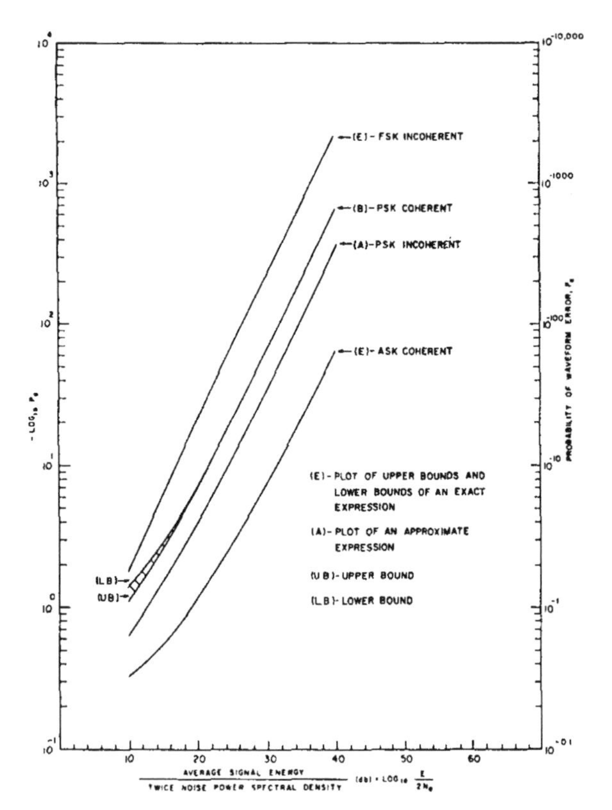

Fig. 31—Comparison of error performance (8-level systems).

formance of coherent systems vs incoherent systems will demonstrate.

E. Coherent vs Incoherent

Of the three basic systems under study, the PSK system suffers the most degradation in performance due to lack of coherence. The calculations performed in Section V show that for $m > 2$ the probability of error for a PSK coherent system and a PSK incoherent system are approximately given by [see inequality (47) and the remark following it and (82)]

$$P_{e\,\text{coherent}} \approx \frac{2}{\sqrt{2\pi}} \int_{\sqrt{E/N_0}\,\sin \pi/m}^{\infty} e^{-u^2/2}\, du$$

$$P_{e\,\text{incoherent}} \approx \frac{2}{\sqrt{2\pi}} \int_{\sqrt{E/N_0}\,\sin \pi/\sqrt{2}m}^{\infty} e^{-u^2/2}\, du.$$

From these equations it may be seen that for large m (*i.e.*, where $\sin \pi/m \approx \pi/m$) the cost of incoherence is a 3-db degradation in signal-to-noise ratio. For small m the degradation is somewhat less. In fact, for $m = 2$ the degradation approaches zero in the high signal-to-noise ratio case. This latter point may be verified by comparing the expressions for the probability of error when $m = 2$ [see (49) and (91)]. A graphical comparison is presented in Fig. 32.

In the ASK and FSK cases effective comparisons between the analytic expressions for the probability of error in the coherent case and the incoherent case do not seem feasible. Yet, examination of Figs. 33 and 34 indicate that in those regions where the probability of error is small enough to be of interest (say, $P_e < 10^{-5}$) the degradation between coherent and incoherent is of the order of a decibel or less. Consequently, either the coherent curves or the incoherent curves may be taken as a representative set of curves for either of these two systems.

F. Comparisons Assuming Fixed Signalling Rate

Returning now to the discussion of degradation in performance with increasing m, we wish to emphasize the fact that any conclusions drawn regarding the relative performance of various systems must take into account the constraints under which the comparisons are made. In particular, it should be noted that the curves presented in Figs. 23–28 are drawn under the assumption that the duration of each transmitted signal is maintained at a fixed value, T, independent of the choice of m. The relative positions of these curves might change considerably if, instead, we considered the equally valid constraint of fixed rate R. (We are still assuming that the transmitter is average-power limited.)

Under this constraint we can allow a longer time duration for each waveform in a multilevel system and, hence, increase the energy content of the transmitted signals. Thus, for example, if each waveform employed in a 2-level scheme has a duration T, the waveforms employed in the corresponding 4-level version of the scheme can have a duration $2T$. Doubling the allotted time duration doubles the energy content of the signal and, hence, is equivalent to a 3-db boost in signal-to-noise ratio $E/2N_0$. Now, in considering the general case let us attach the subscripts m to the parameters E and T to indicate the number of levels which are under discussion. Assuming that the rate at which data is being transmitted,

$$R = \frac{\log_2 m}{T_m} \quad \text{(bits/sec)}, \tag{125}$$

is maintained constant, it follows that

$$T_2 = \frac{1}{R} \tag{126}$$

and, therefore, that

$$T_m = T_2 \log_2 m. \tag{127}$$

Since the transmitter is average-power limited,

$$\frac{E_m}{T_m} = \frac{E_2}{T_2}. \tag{128}$$

Combining (127) and (128) yields

$$E_m = E_2 \log_2 m \tag{129}$$

from which it follows that

$$10 \log_{10} \left(\frac{E_m}{2N_0}\right) = 10 \log_{10} \frac{E_2}{2N_0} + 10 \log_{10} (\log_2 m). \tag{130}$$

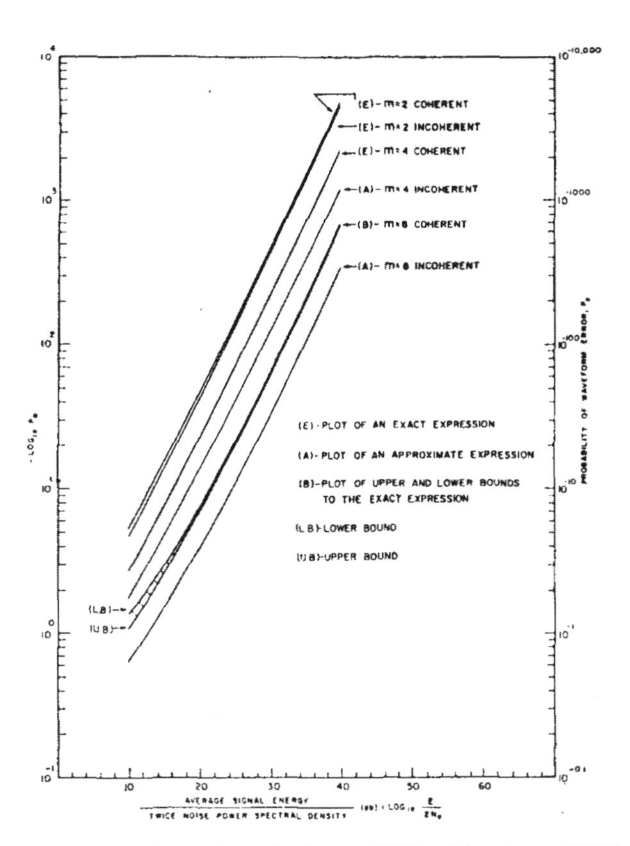

Fig. 32—Comparison of coherent PSK and incoherent PSK.

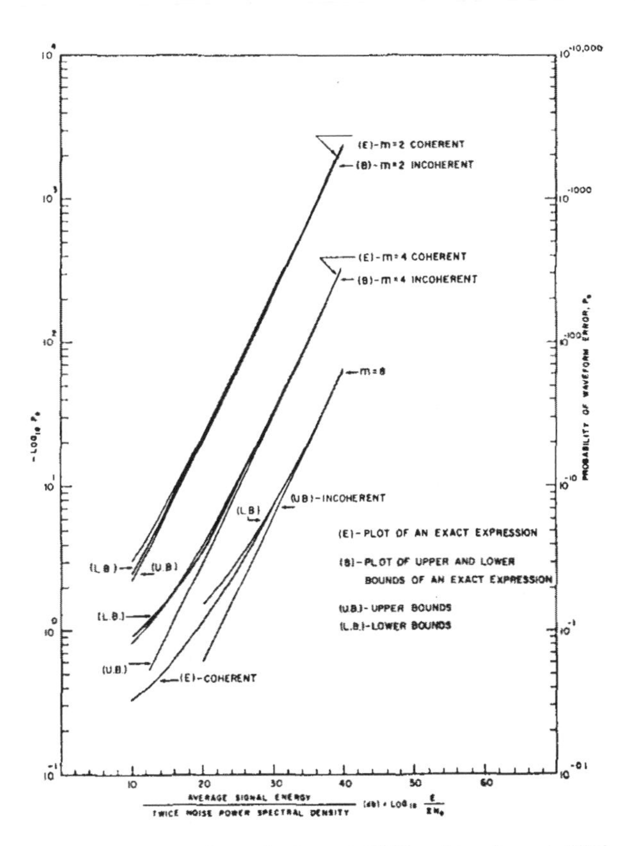

Fig. 33—Comparison of coherent ASK and incoherent ASK.

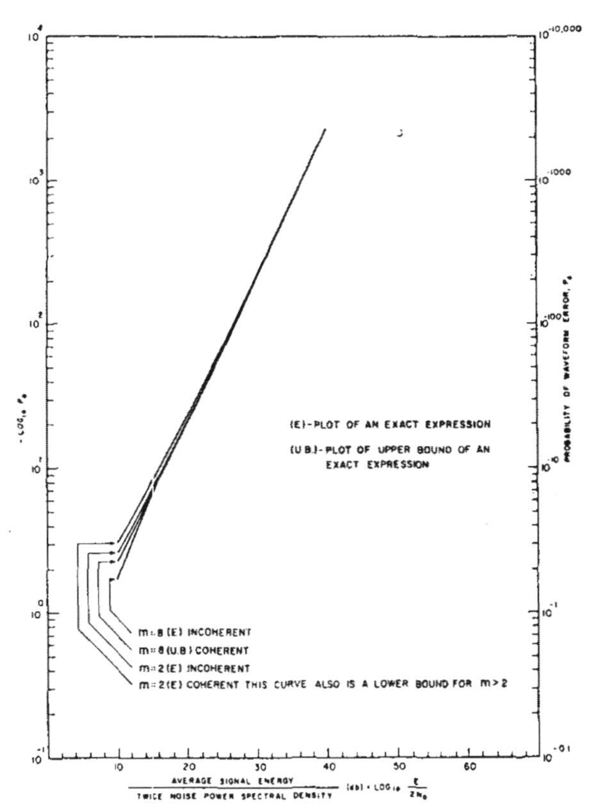

Fig. 34—Comparison of coherent FSK and incoherent FSK.

That is to say, under the assumptions of constant rate an m-level system has a signal-to-noise ratio advantage of $10 \log_{10} (\log_2 m)$ db over its 2-level counterpart. Making use of this fact a set of error characteristics for the assumption of constant rate may be derived from the curves presented in Figs. 23–28 (which were drawn under the assumption that the duration time of each transmitted waveform was fixed) simply by translating the m-level characteristic to the left by $\log_{10} (\log_2 m)$ db. A sample set of constant rate curves which were so constructed is presented in Figs. 35–38. It may be noted from these curves that multilevel PSK and ASK systems are inferior in performance to their two-level counterparts whereas FSK systems seem to improve in performance with increasing m. In both the PSK and the ASK cases, however, the difference in performance between the multilevel and the corresponding two-level schemes is smaller under the present assumption of constant rate than under the previous assumptions of constant time duration. These results are completely consistent with the geometric picture. In all cases, increasing m increases the number of message points. Under the restriction of constant rate, however, the energy content of the transmitted signals increases with m and, hence, the volume of the sphere within which the message points are constrained to lie also increases with m. This tends to partially counteract the crowding together of message points which occurs when m is increased in the PSK and ASK cases. In the FSK case, the message points are actually moved further apart as m increases, the result being an improvement in performance.

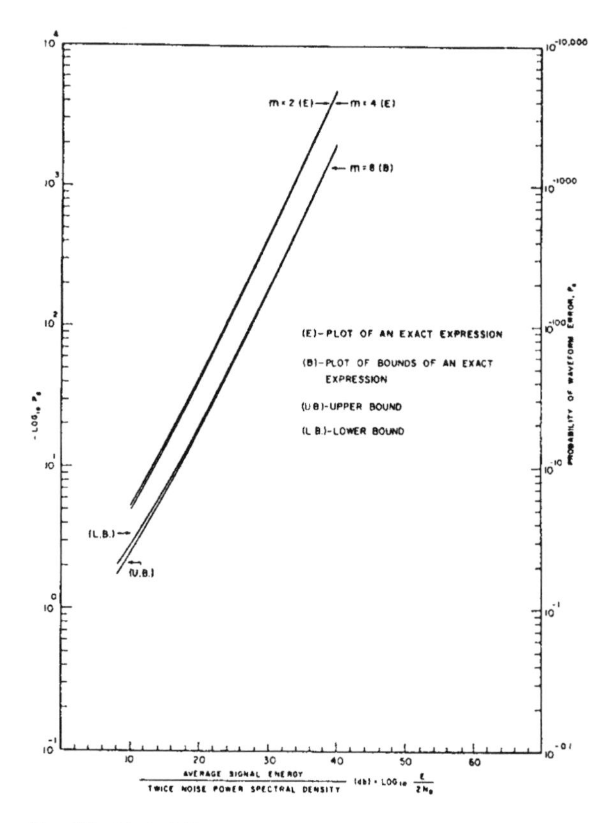

Fig. 35—Probability of waveform error (m-level PSK coherent), assuming that signal duration is adjusted to keep the data rate constant.

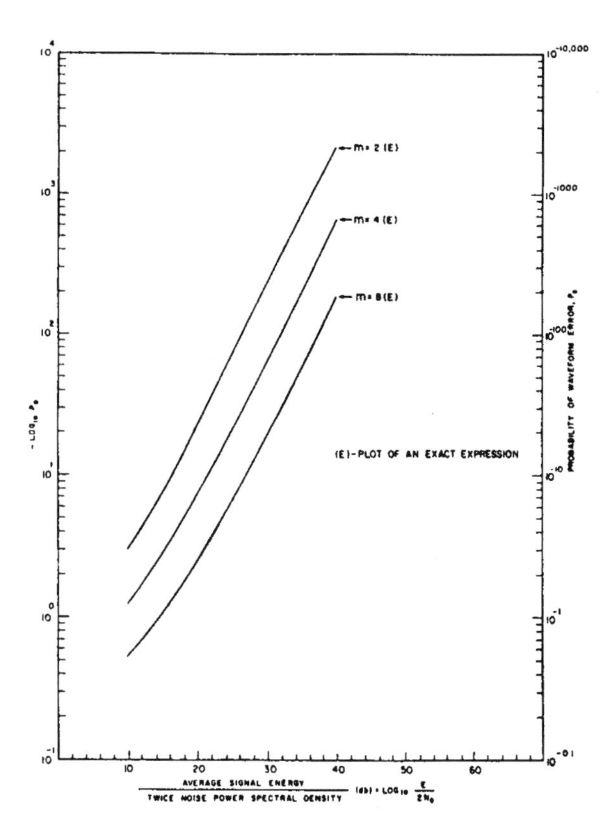

Fig. 37—Probability of waveform error (m-level ASK coherent), assuming that signal duration is adjusted to keep the data rate constant.

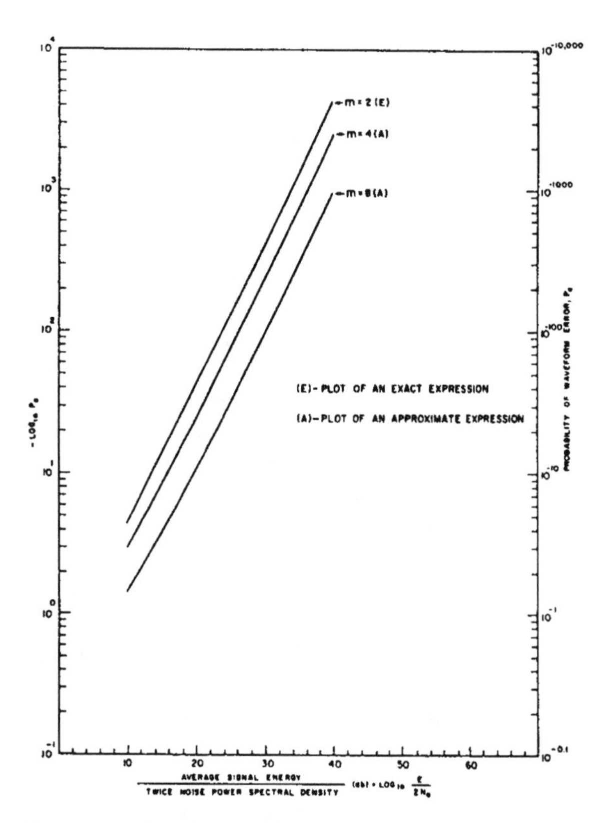

Fig. 36—Probability of waveform error (m-level PSK incoherent), assuming that signal duration is adjusted to keep the data rate constant.

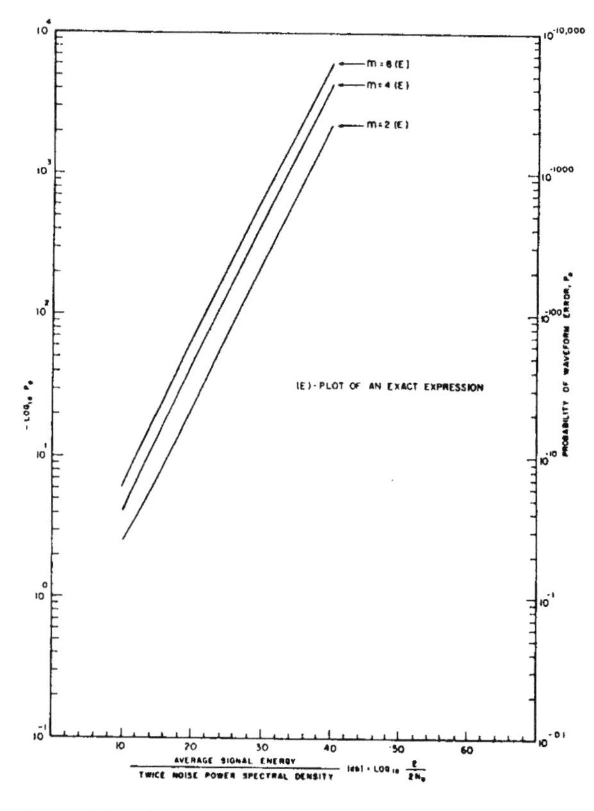

Fig. 38—Probability of waveform error (m-level FSK incoherent), assuming that signal duration is adjusted to keep the data rate constant.

G. Discussion of Bandwidth

A significant factor in the comparison of different communication systems is the quality of the channel required by the system to maintain a "satisfactory" flow of information between the transmitter and receiver sites. To this point in the discussion, all comparisons have been made under the assumption that a distortionless "wideband" Gaussian channel (which is completely specified by the spectral density of the noise N_0) is available. Unfortunately, channels of this type, while most amenable to analysis, are hard to come by in practice. In particular, the bandwidth allotted to any one transmitter is generally limited and, hence, the relative efficiency with which it uses the available bandwidth is of prime interest. As a measure of bandwidth efficiency we shall introduce the parameter r defined as the ratio of the rate at which information is being transmitted, $R = \log_2 m/T$ bits/sec, to the Nyquist rate of transmission, $2B$ bits/sec. That is,

$$r = \frac{R}{2B} = \frac{\log_2 m}{2BT} \tag{131}$$

where

m = the number of different waveforms which may be transmitted
T = the duration time of a waveform
B = bandwidth required to maintain satisfactory operation.

It is quite difficult to define B precisely. For the purposes of present discussion it will be adequate to assume simply that a sinusoid of duration T and frequency f_0 will be passed with negligible distortion by an ideal filter with pass band $1.5/T$ centered around f_0. It follows, therefore, that for the PSK and ASK modulation schemes wherein the frequency of the pulses sent in each time slot is fixed the required transmitter bandwidth $B \approx 1.5/T$ and, correspondingly, the bandwidth efficiency

$$r \approx \frac{\log_2 m}{3}. \tag{132}$$

In the FSK case, however, assuming a separation of $1/T$ between adjacent tones, an m-level transmitter requires a bandwidth (see Fig. 39)

$$B \approx \frac{m + 0.5}{T},$$

in which case

$$r \approx \frac{\log_2 m}{2m + 1}. \tag{133}$$

In the FSK coherent case, however, adjacent signals need only be separated by a frequency difference of $1/2T$ to maintain orthogonality and, hence, the required bandwidth may be reduced to

$$B \approx \frac{m + 2}{2T},$$

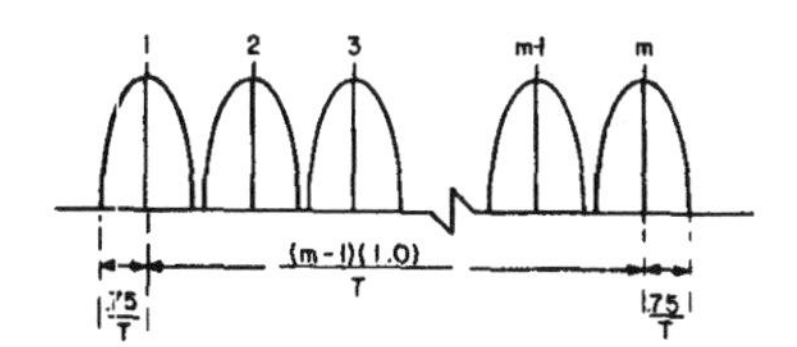

Fig. 39—Illustration of transmission bandwidth required by an m-level orthogonal FSK incoherent system.

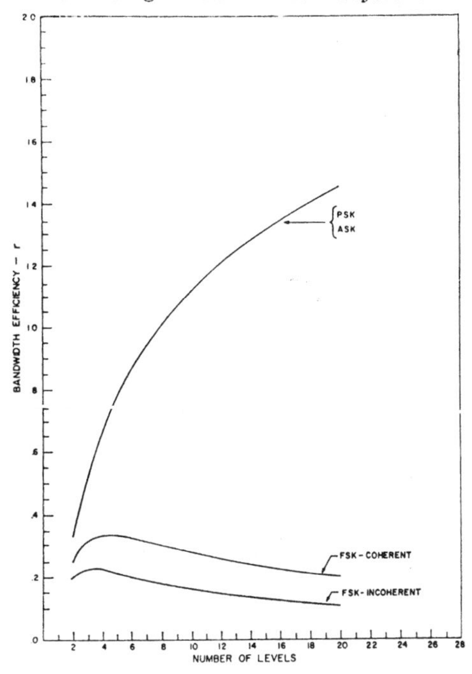

Fig. 40—Plot of bandwidth efficiency as a function of the number of levels.

resulting in a bandwidth efficiency of

$$r \approx \frac{\log_2 m}{m + 2}. \tag{134}$$

Eqs. (132), (133) and (134) are plotted for purposes of comparison in Fig. 40.

It is apparent from Fig. 40 that simple multilevel orthogonal FSK systems, even under idealized operating conditions, are inefficient users of bandwidth. Physically, the reason is clear; in any time period T only a fraction of the total system bandwidth, namely, that occupied by the particular tone transmitted, is utilized. Thus, we see that although the number of levels and, correspondingly, the rate (if we keep the duration time of each waveform T fixed) of such an FSK system may be increased with relatively little degradation in performance (see Figs. 25 and 28), there is, correspondingly, an increase in the bandwidth required by the system to operate. Consequently, the utility of simple multilevel orthogonal FSK systems is limited to situations where conservation of bandwidth is not of principle concern.

It is clear from Fig. 40 that multilevel PSK and ASK modulation schemes utilize bandwidth more efficiently than do the FSK modulation schemes. For these systems the bandwidth utilization factor r increases montonically with the number of levels m. The probability of error, however, as may be deduced from Figs. 23, 24, 26 and 27, also increases with m. There is, thus, an effective upper limit to r beyond which the error rate becomes intolerable. This upper limit will be a function of the signal-to-noise ratio on the channel.

H. Numerical Examples

A feel for the types of tradeoff involved can perhaps best be gained by considering a numerical example. Let us, therefore, investigate the feasibility of operating one of the modulation schemes discussed above at the Nyquist rate, that is, at $r = 1$. We assume for the sake of definiteness that the maximum acceptable error rate is 10^{-5} (characters per second) and that the signal-to-noise ratio available is 15 db. The curves of Fig. 40 indicate that a transmission rate corresponding to $r = 1$ limits our choice of systems to 8-level PSK or 8-level ASK. Furthermore, an examination of Fig. 23 reveals that if the signal-to-noise ratio is limited to 15 db, the error rate of an 8-level PSK coherent system is of the order of 10^{-3} character/sec, which is unacceptable whereas the error rates of PSK incoherent and ASK modulation schemes are even higher (see Fig. 31). It follows, therefore, that under the assumed constraints none of the systems considered will operate satisfactorily at the Nyquist rate. If, on the other hand, the available signal-to-noise ratio was increased to 22 db, both the PSK coherent and the PSK incoherent modulation schemes would meet the stated requirements.

As a second example let us consider in a semiqualitative way the type of performance we might expect from one particular modulation scheme, namely, 4-level PSK incoherent on a telephone channel. Assuming a useable bandwidth of about 2 kc, it follows from Fig. 40 that, for a 4-level PSK system, $r \approx .68$ and, thus, the rate at which information may be transmitted is of the order of $R \approx (4000)(0.68) \approx 2700$ bits/sec.

Correspondingly, from Fig. 26 we may deduce that this system will operate at a character error rate of 10^{-5} with a signal-to-noise ratio of about 15 ½ db and at a character error rate of about 10^{-20} with a signal-to-noise ratio of 22 db.

I. Final Remarks

Since typically the signal-to-noise ratio on a telephone channel is well in excess of 22 db and yet the error rates of existing 4-phase systems are more nearly on the order of 10^{-5}, one can only conclude that the principle sources of errors on a telephone channel are non-Gaussian. Indeed, Kelly and Mercurio[20] have noted that the major sources of errors on a telephone channel appear to be impulse noise and dropouts.

There are, of course, additional factors which tend to limit the performance of actual communication systems which have been ignored in the present analysis. Among these are distortion in the received signal and intersymbol interference due to the nonlinear delay, band-limiting and gain fluctuation of the medium coupling the transmitter to the receiver. Furthermore, the received signal is processed in less than ideal fashion by the detector due to imperfections in the hardware and timing recovery. It is to be noted that all these factors ultimately manifest themselves at the detector simply as a perturbation in the position of the transmitted message point. As such, the effects are similar to those produced by the noise and may largely be compensated for by an additional margin of signal-to-noise ratio at the detector.[21]

Conversely, we might, loosely speaking, say that some fraction of the total signal-to-noise ratio available at the detector is needed to compensate for effects of the type listed above which were not accounted for in the basic analysis. Consequently, only the remaining fraction of signal-to-noise ratio is available for combating Gaussian noise. It is to be expected, therefore, that any predictions of the probability of error based on estimates of the total signal-to-noise ratio available at the detector will be unduly optimistic.

Appendix I

The following is devoted to a proof of Theorem I. Within the course of the proof, a technique for calculating the orthonormal functions $\varphi_1(t), \varphi_2(t), \cdots, \varphi_k(t)$ will be outlined.

Initially, let us note that a set of functions $f_1(t), f_2(t), \cdots, f_q(t)$ are said to be linearly dependent if there exists a set of constants, $a_1, a_2, \cdots, a_q$, not all equal to zero such that

$$a_1f_1(t) + a_2f_2(t) + \cdots + a_qf_q(t) \equiv 0. \qquad (135)$$

If the set of functions is not linearly dependent, it is said to be linearly independent and vice versa.

Consider the given set of waveforms $S_1(t), S_2(t), \cdots S_m(t)$. Either this set is linearly independent or it is not linearly independent. If not, then (by definition) there exists a set of constants, $b_1, b_2, \cdots, b_m$, not all equal to zero such that

$$b_1S_1(t) + b_2S_2(t) + \cdots + b_mS_m(t) \equiv 0.$$

Suppose, in particular, that $b_m \neq 0$. Then,

$$S_m(t) = -\left(\frac{b_1}{b_m} S_1(t) + \frac{b_2}{b_m} S_2(t) + \cdots + \frac{b_{m-1}}{b_m} S_{m-1}(t)\right).$$

That is to say, $S_m(t)$ can be expressed in terms of the remaining $(m - 1)$ waveforms.

[20] J. P. Kelly, and J. F. Mercurio, "Comparative Performance of Digital Data Modems," Mitre Corp., Bedford, Mass., Tech. Memo., TM-3037, p. 4; April 14, 1961.

[21] An assessment of the effects of delay distortion in terms of a parameter related to signal-to-noise ratio for a particular data transmission scheme has been carried out by R. A. Gibby, "An evaluation of AM data system performance by computer simulation," *Bell Sys. Tech. J.*, vol. 39, pp. 675–704, Ref. 1; May, 1960.

Consider now the set of waveforms $S_1(t), S_2(t), \cdots, S_{m-1}(t)$. Either this set is linearly independent or it is not. If not, there exists a set of constants, $c_1, c_2, \cdots, c_{m-1}$, not all equal to zero such that

$$c_1 S_1(t) + c_2 S_2(t) + \cdots + c_{m-1} S_{m-1}(t) \equiv 0.$$

Suppose that $c_{m-1} \neq 0$. Then,

$$S_{m-1}(t) = -\left(\frac{c_1}{c_{m-1}} S_1(t) + \frac{c_2}{c_{m-1}} S_2(t) + \cdots + \frac{c_{m-2}}{c_{m-1}} S_{m-2}(t)\right)$$

which implies that $S_{m-1}(t)$ can be expressed as a linear combination of the remaining $(m-2)$ waveforms. Now, examining the set of waveforms $S_1(t), S_2(t), \cdots, S_{m-2}(t)$ for linear independence and continuing in this fashion, it is clear that we will eventually end up with a linearly independent subset of the original set of waveforms, say,

$$S_1(t), S_2(t), \cdots, S_k(t) \qquad k \leq m.$$

(The indicise of the given set of waveforms can always be permuted in such a fashion that the first k waveforms, $S_1(t), S_2(t), \cdots, S_k(t)$, will be linearly independent.) Note that each of the given waveforms $S_1(t), S_2(t), \cdots, S_m(t)$ may be expressed as a linear combination of these k waveforms.

We shall now, utilizing the Gram Schmidt process, show that if the given waveforms are physically realizable (or, to be more precise, L_2 functions), which condition guarantees the existence of the integrals in question, it is possible to construct a set of k orthonormal waveforms, $\varphi_1(t), \varphi_2(t), \cdots, \varphi_k(t)$, from the derived linearly independent waveforms $S_1(t), S_2(t), \cdots, S_k(t)$.

As a starting point set,

$$\varphi_1(t) = \frac{S_1(t)}{\sqrt{\int_0^T S_1^2(t)\, dt}}. \qquad (136)$$

It is clear that

$$\int_0^T \varphi_1^2(t)\, dt = 1. \qquad (137)$$

Now, define a new intermediate function,

$$h_2(t) = S_2(t) - \lambda\varphi_1(t), \qquad (138)$$

where λ is some constant which is yet to be determined. Since, by (137) and (138),

$$\int_0^T h_2(t)\varphi_1(t)\, dt = \int_0^T S_2(t)\varphi_1(t)\, dt - \lambda, \qquad (139)$$

it is clear that if we set

$$\lambda = \int_0^T S_2(t)\varphi_1(t)\, dt \qquad (140)$$

and

$$\varphi_2(t) = \frac{h_2(t)}{\sqrt{\int_0^T h_2^2(t)\, dt}} \qquad (141)$$

that

$$\int_0^T \varphi_2(t)\varphi_1(t)\, dt = 0$$

and

$$\int_0^T \varphi_2^2(t)\, dt = 1.$$

That is to say, $\varphi_1(t)$ and $\varphi_2(t)$ form an orthonormal set. Continuing in the same fashion, set

$$h_i(t) = S_i(t) - \gamma_1\varphi_1(t) - \gamma_2\varphi_2(t) - \cdots - \gamma_{i-1}\varphi_{i-1}(t) \qquad (142)$$

and the constants γ_j, $j = 1, 2, \cdots, i-1$ equal to

$$\gamma_j = \int_0^T S_i(t)\varphi_j(t)\, dt. \qquad (143)$$

It follows readily that the set of functions

$$\varphi_i(t) = \frac{h_i(t)}{\sqrt{\int_0^T h_i^2(t)\, dt}} \qquad i = 1, 2, \cdots, k \qquad (144)$$

form an orthonormal set.

Since each waveform of the derived subset $S_i(t)$ $i = 1, 2, \cdots, k$ may be expressed as a linear combination of the $\varphi_i(t)$ $i = 1, 2, \cdots, k$, it follows that each of the originally given waveforms $S_i(t)$ $i = 1, 2, \cdots, m$ may be expressed as a linear combination of the $\varphi_i(t)$ $i = 1, 2, \cdots, k$. That is to say, we can write

$$S_i(t) = \sum_{j=1}^{k} a_{ij}\varphi_j(t) \qquad i = 1, 2, \cdots, m \qquad (145)$$

where the a_{ij} are constants. Furthermore, multiplying both sides of (145) by $\varphi_j(t)$ and integrating from 0 to T, we can deduce the fact that

$$a_{ij} = \int_0^T S_i(t)\varphi_j(t)\, dt \qquad \begin{matrix} i = 1, 2, \cdots, m. \\ j = 1, 2, \cdots, k \end{matrix} \qquad (146)$$

We remark that the results of this appendix are abstracted from the general theory of vector spaces. The interested reader would do well to refer to some of the standard texts in the area.[22]

Appendix II

In this appendix we wish to investigate the properties of the quantities n_j, $j = 1, 2, \cdots, k$ defined in (13). In so doing, we shall use the notation of Davenport and Root[1] and shall also make use of some of the results derived therein.

Now, by (13),

$$n_j = \int_0^T n(t)\varphi_j(t)\, dt \qquad j = 1, 2, \cdots, k$$

where the $\varphi_j(t)$ form an orthonormal set.

[22] P. R. Halmos, "Finite Dimensional Vector Spaces," D. Van Nostrand Co., Inc., Princeton, N. J., 1958; G. Birkhoff and S. MacLane, "A Survey of Modern Algebra," The Macmillan Co., New York, N. Y.; 1960.

If $n(t)$ is a Gaussian random process, then n_i is a Gaussian random variable[23] and is thus characterized completely by its mean and variance. In particular, the mean of n_i is equal to

$$\bar{n}_i = E[n_i] = \int_0^T E[n(t)]\varphi_i(t)\,dt \tag{147}$$

and the variance of n_i is equal to

$$\begin{aligned}\sigma_i^2 &= E[n_i^2] - \bar{n}_i^2 \\ &= E\left[\int_0^T dt \int_0^T ds\, n(t)\varphi_i(t)n(s)\varphi_i(s)\right] - \bar{n}_i^2 \\ &= \int_0^T dt \int_0^T ds\, E[n(t)n(s)]\varphi_i(t)\varphi_i(s) - \bar{n}_i^2.\end{aligned} \tag{148}$$

If $n(t)$ is a zero mean process, then

$$E[n(t)] = 0 \tag{149}$$

which in turn, by (147), implies

$$\bar{n}_i = E[n_i] = 0. \tag{150}$$

By definition, the statistical autocorrelation function of the random process $n(t)$ is equal to

$$R(t, s) = E[n(t)n(s)]. \tag{151}$$

If $n(t)$ is stationary, then the autocorrelation function is a function of the time difference $t - s$ alone and not on the particular choice of t and s per se.

Summarizing the results to this point we note that if $n(t)$ is a stationary Gaussian random process with zero mean, then n_i is a Gaussian random variable with zero mean and variance σ_i^2 equal to

$$\sigma_i^2 = \int_0^T dt \int_0^T ds R(t - s)\varphi_i(t)\varphi_i(s). \tag{152}$$

For a stationary random process, the spectral density and the statistical autocorrelation function form a Fourier Transform pair.[24] In particular,

$$R(\tau) = \int_{-\infty}^{\infty} W(f)e^{+j2\pi f\tau}\,df \tag{153}$$

where $W(f)$ represents the spectral density of the stationary random process in question.

For white noise,

$$W(f) = N_0 \quad \text{for all } f$$

and, correspondingly,

$$R(\tau) = \int_{-\infty}^{\infty} N_0 e^{+j2\pi f\tau}\,df = N_0\,\delta(\tau) \tag{154}$$

where $\delta(\tau)$ is the unit impulse.[25]

Consequently, substituting this result into the expression for the variance, (152), we get

$$\begin{aligned}\sigma_i^2 &= N_0 \int_0^T dt \int_0^T ds\, \delta(t - s)\varphi_i(t)\varphi_i(s) \\ &= N_0 \int_0^T dt\varphi_i^2(t) = N_0\end{aligned} \tag{155}$$

where we have utilized the sifting integral[26] and the fact that the $\varphi_j(t)$: $j = 1, 2, \cdots, k$ form an orthonormal set.

It follows readily that if $j \neq k$,

$$\begin{aligned}E[n_j n_k] &= E\left[\int_0^T dt \int_0^T ds\, n(t)\varphi_j(t)n(s)\varphi_k(s)\right] \\ &= \int_0^T dt \int_0^T ds\, R(t - s)\varphi_j(t)\varphi_k(s) \\ &= N_0 \int_0^T dt\, \varphi_j(t)\varphi_k(t) = 0.\end{aligned} \tag{156}$$

This suffices to prove that the Gaussian random variable n_j and n_k are independent if $j \neq k$.[27]

We might further point out that successive noise outputs in time are independent. That is to say, if

$$n = \int_0^T n(t)\varphi(t)\,dt \tag{157}$$

and

$$n^* = \int_T^{2T} n(t)\varphi(t - T)\,dt \tag{158}$$

where it is assumed that $\varphi(t)$ is nonzero only for $0 \leq t \leq T$, then

$$E[nn^*] = 0. \tag{159}$$

An easy way to see this is to let

$$g_1(t) = \begin{cases}\varphi(t) & 0 \leq t < T \\ 0 & T \leq t < 2T\end{cases} \tag{160}$$

$$g_2(t) = \begin{cases}0 & 0 \leq t < T \\ \varphi(t - T) & T \leq t < 2T\end{cases} \tag{161}$$

and then to note that (157) and (158) can be rewritten in the form

$$n = \int_0^{2T} n(t)g_1(t)\,dt \tag{162}$$

$$n^* = \int_0^{2T} n(t)g_2(t)\,dt. \tag{163}$$

Now, noting further that

$$\int_0^{2T} g_1(t)g_2(t)\,dt = 0, \tag{164}$$

[23] Davenport and Root, *op. cit.*, pp. 155–156.
[24] Davenport and Root, *op. cit.*, p. 104.
[25] Davenport and Root, *op. cit.*, pp. 365–368.
[26] Davenport and Root, *op. cit.*, pp. 365–368.
[27] Davenport and Root, *op. cit.*, pp. 55–58.

it follows readily that

$$E[nn^*] = E\left[\int_0^{2T} dt \int_0^{2T} ds\, n(t)g_1(t)n^*(s)g_2(s)\right]$$

$$= \int_0^{2T} dt \int_0^{2T} ds\, N_0\, \delta(t-s)g_1(t)g_2(s)$$

$$= N_0 \int_0^{2T} dt\, g_1(t)g_2(t) = 0.$$

Appendix III

This appendix is devoted to a proof of Theorem III. The proof will proceed in two steps. Initially, we shall show that a maximum-likelihood detector minimizes the probability of error if each possible signal is transmitted with equal probability. Secondly, we shall show that if the transmitted signals are perturbed by additive stationary-white-zero mean-Gaussian noise, then, in the coherent case, maximum-likelihood detection is equivalent to picking the message point closest to the received signal point and guessing that the corresponding signal was transmitted.

Assume that in each time slot T one of the m possible signals $S_1(t)$, $S_2(t)$, $\cdots$, $S_m(t)$ is transmitted with equal probability, namely, $1/m$. Assume further that, whenever a signal is transmitted, a point (or vector) y is observed at the detector. Denoting the set of all possibly observed y by Y, the observation space, we suppose that the conditional probability density of y [under the condition that $S_i(t)$ is sent], $p_i(y)$ $i = 1, 2, \cdots, m$, is defined on Y. Our objective is to establish a rule for partitioning the space Y into a set of disjoint regions, Y_1, Y_2, $\cdots$, Y_m, such that if we guess that $S_i(t)$ was transmitted whenever y lands in the region Y_i, $i = 1, 2, \cdots, m$, the probability of error is at a minimum.

Note initially that the probability that y lands in the region Y_j when $S_k(t)$ is transmitted may be written in the following equivalent ways:

$$P[y \in Y_j/S_k] = \int_{Y_j} p_k(y)\, dy = 1 - \int_{Y-Y_j} p_k(y)\, dy$$

$$= 1 - P[y \notin Y_j/S_k]$$

Now, assuming that we do partition the observation space Y into a set of disjoint regions, Y_1, Y_2, $\cdots$, Y_m, and then guess that $S_i(t)$ was transmitted if y lands in the region Y_i, $i = 1, 2, \cdots, m$, the probability of incorrect decision P_e is equal to

$$P_e = \frac{1}{m}\sum_{i=1}^{m} P[y \notin Y_i/S_i]$$

$$= 1 - \frac{1}{m}\sum_{i=1}^{m} P[y \in Y_i/S_i]$$

$$= 1 - \frac{1}{m}\sum_{i=1}^{m} \int_{Y_i} p_i(y)\, dy.$$

It is clear that P_e will be minimized if $\sum_{i=1}^{m} \int_{Y_i} p_i(y)\, dy$ is maximized. A little reflection, however, indicates that this latter sum will take on its maximum value if we set Y_i equal to the set of points y in Y, for which

$$p_i(y) > p_j(y) \quad \text{all} \quad j \neq i.$$

(In case there is a point y_0 in Y for which, say,

$$p_1(y_0) = p_2(y_0) = p_3(y_0) > p_q(y_0), \qquad q = 4, 5, \cdots, m,$$

the point y_0 can be assigned to either Y_1 or Y_2 or Y_3.)

The decision rule embodied in this partitioning of the observation space is equivalent to taking the observed point, say y_0, finding that value of i for which $p_i(y_0)$ is a maximum and then guessing that the corresponding signal $S_i(t)$ was transmitted. That is to say, this decision rule is equivalent to maximum-likelihood detection. We conclude, therefore, that if each signal is transmitted with equal probability, a maximum-likelihood detector will minimize the probability of error.

In the text it has been shown that, in the case of coherent detection, if $S_i(t)$ is transmitted, the received signal can be characterized by a point in a k-dimensional Euclidean space with coordinates [see (12)]

$$a_{ij} + n_j \qquad j = 1, 2, \cdots, k$$

where the a_{ij} are the coordinates of the transmitted signal and the n_j are independent Gaussian random variables with zero mean and variance N_0.

The decision as to what signal was sent will be based on the coordinates of the received point. That is to say, in this case the observation space Y is a k-dimensional Euclidean space. Accordingly, let us designate the k-dimensional random vector corresponding to the received signal by $(\bar{y})$ (the symbol z was used in the paper). Since each coordinate of $\bar{y}$, namely, $y_j = a_{ij} + n_j; j = 1, 2, \cdots, k$, is an independent Gaussian random variable with mean a_{ij} when $S_i(t)$ is transmitted and variance N_0, the conditional probability density function $p_i(\bar{y})$ is equal to

$$p_i(y) = \left(\frac{1}{2\pi N_0}\right)^{k/2} \exp\left\{-\frac{1}{2N_0}\sum_{j=1}^{k}(y_j - a_{ij})^2\right\}.$$

But, by (5),

$$\sum_{j=1}^{k}(y_j - a_{ij})^2 = d^2(\bar{y}, S_i)$$

where $d(\bar{y}, S_i)$ is equal to the distance between the received point $\bar{y}$ and the transmitted point S_i.

Now, suppose a particular signal is observed at the receiver; that is to say, $\bar{y} = \bar{y}_0$. The maximum-likelihood detection rule is simply to choose that value of i for which

$$p_i(\bar{y}_0) = \left(\frac{1}{2N_0}\right)^{k/2} \exp\left\{-\frac{1}{2N_0} d^2(\bar{y}_0, S_i)\right\}$$

is a maximum and guess that the corresponding signal $S_i(t)$ was transmitted. This is, however, equivalent to selecting that value of i for which $d(\bar{y}_0, S_i)$ is a minimum or selecting the message point closest to the received signal point.

Appendix IV

The following is devoted to discussing the decision rules which have been adopted for the incoherent systems. As shown in Appendix III, a maximum-likelihood detector will minimize the average probability of error if each possible message is transmitted with equal probability. Accordingly, we shall, in each case derive the decision rule corresponding to maximum-likelihood detection but shall, for the purpose of simplicity, modify the decision rule derived for the ASK case.

A. PSK Incoherent

In the PSK incoherent case the decision as to what message was sent is based on the successive outputs of the two product integrators. Assuming, in particular, that the *ith* message has been sent, the decision will be based on the four quantities described by (30), namely,

$$x_1 = \sqrt{E}\cos\alpha + n_{11}$$
$$y_1 = -\sqrt{E}\sin\alpha + n_{12}$$
$$x_2 = \sqrt{E}\cos(\alpha + 2\pi i/m) + n_{21}$$
$$y_2 = -\sqrt{E}\sin(\alpha + 2\pi i/m) + n_{22}$$

where n_{11}, n_{12}, n_{21}, n_{22} are independent Gaussian random variables with zero mean and variance N_0 and α is assumed to be uniformly distributed over a 2π interval. That is to say,

$$p(\alpha) = \begin{cases} 1/2\pi & \theta \le \alpha < \theta + 2\pi \\ 0 & \text{elsewhere} \end{cases}. \tag{165}$$

The conditional joint density function of the random variables x_1, y_1, x_2, y_2, given, say, that $\alpha = A$ and that the *ith* message was transmitted, may be written as

$$\begin{aligned} p_i(x_1, y_1, x_2, y_2/\alpha = A) &= \left(\frac{1}{2\pi N_0}\right)^2 \exp\left\{-\frac{1}{2N_0}\left[(x_1 - \sqrt{E}\cos A)^2 \right.\right. \\ &\quad + (y_1 + \sqrt{E}\sin A)^2 \\ &\quad + \left(x_2 - \sqrt{E}\cos\left(A + \frac{2\pi i}{m}\right)\right)^2 \\ &\quad \left.\left. + \left(y_2 + \sqrt{E}\sin\left(A + \frac{2\pi i}{m}\right)\right)^2\right]\right\} \\ &= D\exp\{P\cos A + Q\sin A\} \end{aligned} \tag{166}$$

where

$$D = \left(\frac{1}{2\pi N_0}\right)^2 \cdot \exp\left\{-\frac{1}{2N_0}\left[x_1^2 + y_1^2 + x_2^2 + y_2^2 + 2E\right]\right\} \tag{167}$$

$$P = \frac{\sqrt{E}}{N_0}\left[x_1 + x_2\cos\frac{2\pi i}{m} - y_2\sin\frac{2\pi i}{m}\right] \tag{168}$$

$$Q = \frac{\sqrt{E}}{N_0}\left[-y_1 - x_2\sin\frac{2\pi i}{m} - y_2\cos\frac{2\pi i}{m}\right]. \tag{169}$$

The joint conditional density for x_1, y_1, x_2, y_2, assuming only that the *ith* message was sent, is equal to

$$p_i(x_1, y_1, x_2, y_2) = \int_\theta^{\theta+2\pi} p_i(x_1, y_1, x_2, y_2/\alpha = A)p(A)\,dA \tag{170}$$

where we are averaging over all possible A.

Substituting (165) and (166) into (170) and performing the indicated integration, we get

$$p_i(x_1, y_1, x_2, y_2) = DI_0(\sqrt{P^2 + Q^2}) \tag{171}$$

where I_0 is the modified Bessel function of the first kind of zero order.[16]

Now, in principle, to decide what signal was sent, the decision box of the detector should substitute the measured values of x_1, y_1, x_2, y_2 into the right-hand side of (171) and evaluate this expression for all values of i, $i = 1, 2, \cdots, m$. It should then select that value of i which yielded the largest value and assume that the corresponding signal was sent. However, since the quantity D is independent of the choice of i and the Bessel function increases monotonically with its argument, it is sufficient to find the value of i which maximizes the quantity

$$P^2 + Q^2.$$

By (168) and (169), we find that

$$P^2 + Q^2 = \frac{E}{N_0^2}\left\{x_1^2 + y_1^2 + x_2^2 + y_2^2 + 2(x_1x_2 + y_1y_2) \cdot \cos\frac{2\pi i}{m} + 2(y_1x_2 - x_1y_2)\sin\frac{2\pi i}{m}\right\}. \tag{172}$$

The right-hand side of (172) can be interpreted geometrically. Before doing so, however, we wish to point out that since [as can be deduced from (29a), (30a) and (30b)]

$$\int_0^T \phi_1(t)\sqrt{\frac{2E}{T}}\cos(\omega_0 t + \alpha)\,dt = \sqrt{E}\cos\alpha$$

$$\int_0^T \phi_2(t)\sqrt{\frac{2E}{T}}\cos(\omega_0 t + \alpha)\,dt = -\sqrt{E}\sin\alpha,$$

it is necessary to consider angular displacements in the clockwise direction to be positive if the output of the $\phi_1(t)$ product integrator is interpreted as the x coordinate projection of the received signal and the output of the $\phi_2(t)$ product integrator is interpreted as the y coordinate projection of the received signal. Under this convention, the transformation from polar coordinates to rectangular coordinates is given by

$$x = \rho\cos\theta$$
$$y = -\rho\sin\theta.$$

In Fig. 41, x, y, ρ and θ are defined.

Now, let us consider a pair of successively received signal points with coordinates (a_1, b_1) and (a_2, b_2), respectively. That is to say, we are assuming that the

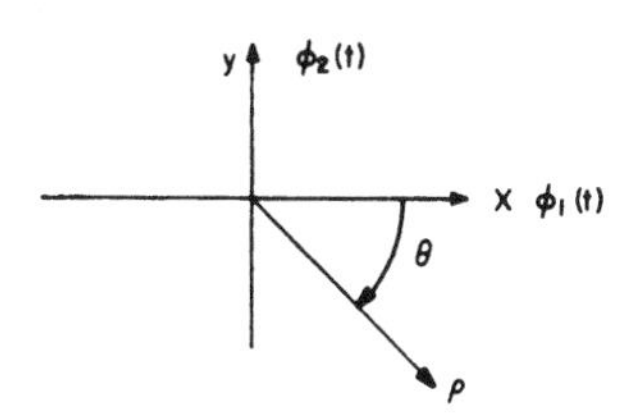

Fig. 41—Definition of polar and rectangular coordinate systems adopted in Section A, Appendix IV.

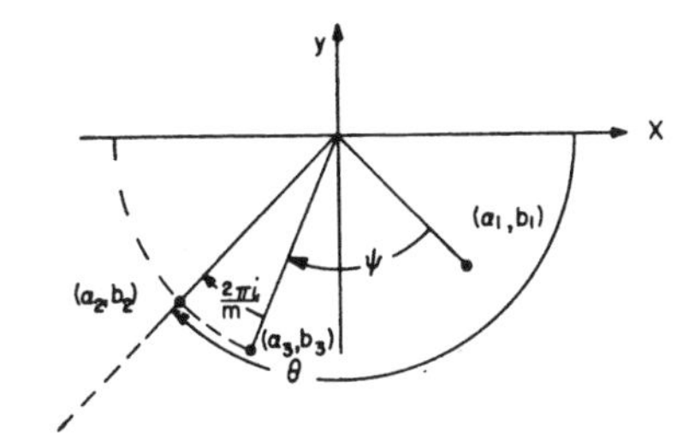

Fig. 42—Geometric interpretation of (172) and (177).

random variables x_1, y_1, x_2, y_2 take on the particular values a_1, b_1, a_2, b_2, respectively. The two signal points are shown plotted in Fig. 42.

If we denote the point (a_3, b_3) as the rotation of the point (a_2, b_2) in the counterclockwise direction by $2\pi i/m$ radians, it is apparent, since

$$a_2 = \sqrt{a_2^2 + b_2^2} \cos \theta$$
$$b_2 = -\sqrt{a_2^2 + b_2^2} \sin \theta,$$

that

$$a_3 = \sqrt{a_2^2 + b_2^2} \cos \left(\theta - \frac{2\pi i}{m} \right)$$
$$= a_2 \cos \frac{2\pi i}{m} - b_2 \sin \frac{2\pi i}{m} \tag{173}$$

and

$$b_3 = -\sqrt{a_2^2 + b_2^2} \sin \left(\theta - \frac{2\pi i}{m} \right)$$
$$= b_2 \cos \frac{2\pi i}{m} + a_2 \sin \frac{2\pi i}{m}. \tag{174}$$

Furthermore, applying the law of cosines to the angle ψ defined in Fig. 42, we get

$$\cos \psi = \frac{a_1 a_3 + b_1 b_3}{\sqrt{a_1^2 + b_1^2} \sqrt{a_3^2 + b_3^2}}. \tag{175}$$

Substituting (173) and (174) into (175) yields

$$\cos \psi = \frac{(a_1 a_2 + b_1 b_2) \cos \frac{2\pi i}{m} + (b_1 a_2 - a_1 b_2) \sin \frac{2\pi i}{m}}{\sqrt{a_1^2 + b_1^2} \sqrt{a_2^2 + b_2^2}}. \tag{176}$$

Substituting this result into (172) yields (recall we are considering that particular case where $x_1 = a_1$, $y_1 = b_1$, $x_2 = a_2$, $y_2 = b_2$)

$$P^2 + Q^2 = \frac{E}{N_0^2} \{a_1^2 + a_2^2 + b_1^2 + b_2^2 + 2\sqrt{a_1^2 + b_1^2} \sqrt{a_2^2 + b_2^2} \cos \psi\}. \tag{177}$$

The only term in the right-hand side of (177) which is dependent upon i, the index of the transmitted signal, is $\cos \psi$. Accordingly, that value of i, $i = 1, 2, \cdots, m$, which maximizes $\cos \psi$ will also maximize $(P^2 + Q^2)$. Equivalently, we can solve for that value of i which minimizes $|\psi|$. That is to say, the decision rule, as may be deduced from Fig. 42, reduces simply to measuring the phase difference between successively received signals, rounding off the measured value to the nearest value of $2\pi i/m$ $i = 1, 2, \cdots, m$ and guessing that the corresponding signal was transmitted.

B. *ASK Incoherent*

In the ASK incoherent case, if $S_i(t)$ is transmitted, the decision at the receiver as to what signal was transmitted is based on the random variables x and y described in (34a) and (34b), namely,

$$x = \sqrt{E_i} \cos \alpha + n_1$$
$$y = -\sqrt{E_i} \sin \alpha + n_2$$

where n_1, n_2 are independent Gaussian random variables with zero mean and variance N_0 and α is assumed to be uniformly distributed over a 2π interval.

The joint conditional probability density function of the random variables x and y, given that $\alpha = A$ and that $S_i(t)$ was transmitted, is equal to

$$p_i(x, y/\alpha = A) = \frac{1}{2\pi N_0} \exp \left\{ -\frac{1}{2N_0} \{(x - \sqrt{E_i} \cos A)^2 + (y + \sqrt{E_i} \sin A)^2\} \right\}. \tag{178}$$

Therefore, averaging over all possible values of A, the conditional joint density function of the random variables x and y, given only that $S_i(t)$ was transmitted, is equal to

$$p_i(x, y) = \int_\theta^{\theta + 2\pi} p_i(x, y/\alpha = A) \frac{dA}{2\pi}$$
$$= \frac{1}{2\pi N_0} \exp \left\{ -\frac{1}{2N_0} (x^2 + y^2 + E_i) \right\} \cdot I_0 \left(\frac{\sqrt{E_i} \sqrt{x^2 + y^2}}{N_0} \right) \tag{179}$$

where I_0 is the modified Bessel function of the first kind of zero order.[16]

The optimum decision rule for the detector to follow is to substitute the measured value of x and y into the right-hand side of (179), find the value of i which maximizes the resultant expression and assume that the corresponding signal was transmitted. Unfortunately, this decision rule is rather complicated to instrument and we shall adopt, instead, its asymptotic form although we shall not always be operating in ranges where the resultant rule is optimum.

The asymptotic expansion of the Bessel function is given by[28]

$$I_0(t) = \frac{e^t}{\sqrt{2\pi t}}\left\{1 + \frac{1}{8t} + \cdots\right\}. \tag{180}$$

Substituting the first term of the expansion into the right-hand side of (179), we get

$$p_i(x, y) \approx \frac{\exp\left\{-\frac{1}{2N_0}[\sqrt{x^2+y^2} - \sqrt{E_i}]^2\right\}}{2\pi\sqrt{2\pi N_0}\,[E_i(x^2+y^2)]^{1/4}}. \tag{181}$$

In the regions where this expansion is valid, the behavior of the right-hand side of (181) is dominated by the exponential term. It follows, therefore, that the set of (x, y) points for which $p_i(x, y) > p_j(x, y)$ all $j \neq i$ corresponds approximately to the set of (x, y) points for which

$$|\sqrt{x^2+y^2} - \sqrt{E_i}| < |\sqrt{x^2+y^2} - \sqrt{E_j}| \quad \text{all} \quad j \neq i. \tag{182}$$

That is to say, it is approximately true that if, in particular, $x = a$ and $y = b$, the value of i which maximizes $p_i(a, b)$ corresponds to the value of i which minimizes $|\sqrt{a^2+b^2} - \sqrt{E_i}|$. The indicated decision rule is, therefore, to calculate the rms amplitude of the received signal roundoff to the nearest value of $\sqrt{E_i}$, $i = 1, 2, \cdots, m$ and guess that the corresponding signal was transmitted. It should be noted that this decision rule, although seemingly a natural one to adopt, is only optimum in the region where it is legitimate to approximate the Bessel function by the first term of its asymptotic expansion—that is, for those values of x, y, $\sqrt{E_i}$ and N_0 for which $\sqrt{x^2+y^2}\sqrt{E_i}/N_0 \gg 1$.

C. FSK Incoherent

In the FSK incoherent case, if $S_i(t)$ is transmitted, the detector will base its decision as to what was sent on the $2m$ random variables [see (37a) and (37b)]

$$x_j = \begin{cases} n_{xj} & j \neq i \\ \sqrt{E}\cos\alpha + n_{xi} & j = i \end{cases}$$

$$y_j = \begin{cases} n_{yj} & j \neq i \\ -\sqrt{E}\sin\alpha + n_{yi} & j = i \end{cases}$$

where the n_{xj} and n_{yj} are independent Gaussian random variables, each having zero mean and variance N_0 and α is a random variable uniformly distributed over a 2π interval.

The conditional joint probability density of the random variables $x_1, x_2, \cdots, x_m, y_1, y_2, \cdots, y_m$, given that $S_i(t)$ was sent and that $\alpha = A$, is equal to

$$\begin{aligned} P_i(x_1, x_2, \cdots, x_m, y_1, y_2, \cdots, y_m/\alpha = A) &= \left(\frac{1}{2\pi N_0}\right)^m \exp\left[-\frac{1}{2N_0}\left\{\sum_{j\neq i}(x_j^2 + y_j^2)\right.\right. \\ &\quad \left.\left. + (x_i - \sqrt{E}\cos A)^2 + (y_i + \sqrt{E}\sin A)^2\right\}\right] \\ &= (L)\exp\left\{\frac{\sqrt{E}}{N_0}(x_i\cos A - y_i\sin A)\right\} \end{aligned} \tag{183}$$

where

$$L = \left(\frac{1}{2\pi N_0}\right)^m \exp\left\{-\frac{1}{2N_0}\left\{\sum_{j=1}^m (x_j^2 + y_j^2) + E\right\}\right\}. \tag{184}$$

Now, averaging over all possible values of A to find the joint conditional probability density of the random variables $x_1, x_2, \cdots, x_m, y_1, y_2, \cdots, y_m$, given only that $S_i(t)$ was sent, we get

$$\begin{aligned} p_i(x_1, x_2, \cdots, x_m, y_1, y_2, \cdots, y_m) &= \int_\theta^{\theta+2\pi} p_i(x_1, x_2, \cdots, x_m, y_1, y_2, \cdots, y_m/\alpha = A)\frac{dA}{2\pi} \\ &= LI_0\left(\frac{\sqrt{E}\sqrt{x_i^2+y_i^2}}{N_0}\right) \end{aligned} \tag{185}$$

where I_0 is the modified Bessel function of the first kind of zero order.[16]

Clearly, since L is independent of the choice of i and I_0 is a monotonically increasing function of its argument, the set of points $x_1, x_2, \cdots, x_m, y_1, y_2, \cdots, y_m$ for which

$$p_i(x_1, x_2, \cdots, x_m, y_1, y_2, \cdots, y_m) > p_j(x_1, x_2, \cdots, x_m, y_1, y_2, \cdots, y_m) \quad \text{for all} \quad j \neq i$$

is equal to the set of points for which

$$\sqrt{x_i^2 + y_i^2} > \sqrt{x_j^2 + y_j^2} \quad \text{all} \quad j \neq i. \tag{186}$$

Thus, if, at some instant, the random variables $x_1, x_2, \cdots, x_m, y_1, y_2, \cdots, y_m$ take on the particular values $a_1, a_2, \cdots, a_m, b_1, b_2, \cdots, b_m$, the receiver should calculate the m rms amplitudes $\sqrt{a_i^2 + b_i^2}$ $i = 1, 2, \cdots, m$ (each of which is associated with a particular frequency), select the largest one and guess that the corresponding signal was transmitted.

Appendix V

This appendix is devoted to a proof of Theorem IV. Recall that the decision rule for the maximum-likelihood detector in the coherent case is simply to choose the message point closest to the received signal point and to guess that the corresponding signal was transmitted. Accordingly, an error will occur if, and only if, when $S_i(t)$ is transmitted, the received signal point lies closer to one of the message points S_j $(j \neq i)$ than to the message point S_i. Denoting the distance between message

[28] Bowman, *op. cit.*, p. 84.

points S_i and S_j by ρ_{ij} and the noise components originating at the point S_i and directed towards the point S_j by n_{ij}, it follows that the probability of this event which equals the probability of an error, P_{ei}, when $S_i(t)$ is transmitted is equal to

$$P_{ei} = P\left[n_{ij} > \frac{\rho_{ij}}{2} \text{ for at least one } j \neq i\right] \quad (187)$$

where n_{ij} is a Gaussian random variable with mean zero and variance N_0. Since the events $n_{ij} > \rho_{ij}/2$ are not necessarily mutually exclusive, P_{ei} is certainly less than or equal to

$$P_{ei} \leq \sum_{j \neq i} P\left[n_{ij} > \frac{\rho_{ij}}{2}\right]. \quad (188)$$

Now, noting that

$$P\left[n_{ij} > \frac{\rho_{ij}}{2}\right] = \frac{1}{\sqrt{2\pi}} \int_{\rho_{ij}/2\sqrt{N_0}}^{\infty} e^{-x^2/2}\, dx \qquad j \neq i \quad (189)$$

and defining

$$H\left(\frac{\rho_{ij}}{2\sqrt{N_0}}\right) = \frac{1}{\sqrt{2\pi}} \int_{\rho_{ij}/2\sqrt{N_0}}^{\infty} e^{-x^2/2}\, dx, \quad (190)$$

it follows, since $H(\rho_{ij}/2\sqrt{N_0})$ is a monotonically decreasing function of its argument, that

$$P_{ei} \leq \sum_{j \neq i} H\left(\frac{\rho_{ij}}{2\sqrt{N_0}}\right) \leq (m-1)H\left(\frac{\rho_i}{2\sqrt{N_0}}\right) \quad (191)$$

where

$$\rho_i = \min_{j \neq i} [\rho_{ij}]. \quad (192)$$

The quantity ρ_i defined in (192) is equal to the distance between the signal point S_i and its nearest neighbor. It is clear [by (187) and (192)] that

$$P_{ei} \geq H\left(\frac{\rho_i}{2\sqrt{N_0}}\right) \quad (193)$$

which result, when combined with (191), yields

$$H\left(\frac{\rho_i}{2\sqrt{N_0}}\right) \leq P_{ei} \leq (m-1)H\left(\frac{\rho_i}{2\sqrt{N_0}}\right). \quad (194)$$

The average probability of error, P_e, which is equal to

$$P_e = \frac{1}{m} \sum_{i=1}^{m} P_{ei}, \quad (195)$$

is, therefore, by (194), bounded by

$$\frac{1}{m} \sum_{i=1}^{m} H\left(\frac{\rho_i}{2\sqrt{N_0}}\right) \leq P_e \leq \frac{m-1}{m} \sum_{i=1}^{m} H\left(\frac{\rho_i}{2\sqrt{N_0}}\right). \quad (196)$$

Since the second derivative of $H(\rho_i/2\sqrt{N_0})$ (with respect to its argument) exists and is greater than or equal to zero if $\rho_i \geq 0$, it follows that $H(\rho_i/2\sqrt{N_0})$ is a convex function[29] for $\rho_i \geq 0$. That is to say,

$$\frac{1}{m} \sum_{i=1}^{m} H\left(\frac{\rho_i}{2\sqrt{N_0}}\right) \geq H\left(\frac{1}{m} \sum_{i=1}^{m} \frac{\rho_i}{2\sqrt{N_0}}\right). \quad (197)$$

Defining the symbol $\bar{\rho}$ as the average of ρ_i, we have

$$\bar{\rho} = \frac{1}{m} \sum_{i=1}^{m} \rho_i \quad (198)$$

and combining (197) and (198) with the left-hand equality of (196), we get

$$P_e \geq H\left(\frac{\bar{\rho}}{2\sqrt{N_0}}\right). \quad (199)$$

Further, defining the symbol

$$\rho^* = \min_i [\rho_i] \quad (200)$$

and noting that

$$H\left(\frac{\rho_i}{2\sqrt{N_0}}\right) \leq H\left(\frac{\rho^*}{2\sqrt{N_0}}\right), \quad (201)$$

we get, by combining (201) with the right-hand inequality of (196),

$$P_e \leq (m-1)H\left(\frac{\rho^*}{2\sqrt{N_0}}\right). \quad (202)$$

Thus, by the inequalities of (199) and (202), we have

$$H\left(\frac{\bar{\rho}}{2\sqrt{N_0}}\right) \leq P_e \leq (m-1)H\left(\frac{\rho^*}{2\sqrt{N_0}}\right)$$

which, by (190), may be rewritten in the form presented in the text, namely,

$$\frac{1}{\sqrt{2\pi}} \int_{\bar{\rho}/2\sqrt{N_0}}^{\infty} e^{-x^2/2}\, dx \leq P_e \leq \frac{(m-1)}{\sqrt{2\pi}} \int_{\rho^*/2\sqrt{N_0}}^{\infty} e^{-x^2/2}\, dx. \quad (203)$$

[29] G. H. Hardy, J. E. Littlewood and G. Polya, "Inequalities," Cambridge University Press, Cambridge, England; 1959. In particular, note Secs. 3.5 and 3.10.

Performance of Combined Amplitude and Phase-Modulated Communication Systems*

J. C. HANCOCK†, MEMBER, IRE AND R. W. LUCKY†

Summary—**The performance of two types of digital phase- and amplitude-modulated systems is investigated for the high signal-to-noise ratio region. Approximate expressions for the probability of error and channel capacity of the more optimum of these two systems are compared with corresponding expressions for probability of error and channel capacity for a digital phase-modulated system. It is shown that the phase- and amplitude-modulated systems show a definite power advantage over the phase-only system when the information content per transmitted symbol must be greater than 3 bits. From a channel capacity standpoint, the phase- and amplitude-modulated systems make more efficient use of the channel for signal-to-noise ratios greater than 11 db. The more optimum of the two phase and amplitude systems has only a 3-db advantage over the less optimum and is considerably more difficult to instrument.**

INTRODUCTION

RECENT papers have investigated the performance of digital phase-modulation systems and have suggested the possibility of a digital system which is both phase and amplitude modulated [1]–[3].

Fig. 1 shows a typical transmitted signal in such a system. During each pulse length of T seconds the phase and amplitude of the transmitted signal assume values chosen from a discrete set of possible phases and amplitudes. Each combination of a particular amplitude level and phase position represents a transmitted symbol, and the totality of such combinations is the alphabet size.

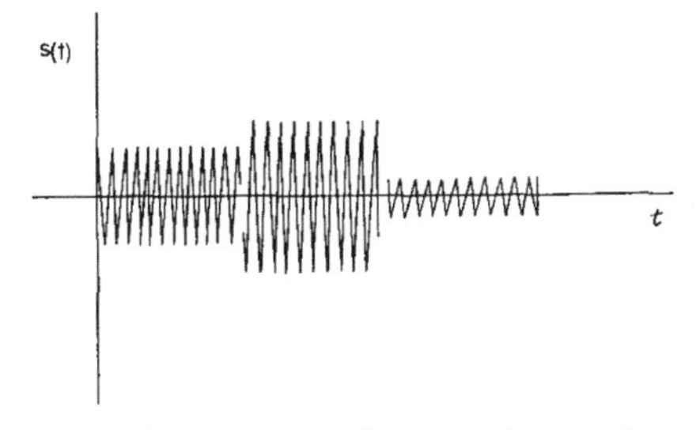

Fig. 1—Digital phase and amplitude modulation.

Since this transmitted signal is contaminated by noise in the channel, the received symbols cannot be detected with certainty. Fig. 2 shows the decision levels at the receiver for a system with two possible amplitude positions A_1 and A_2 and four possible phase positions $\phi_1 - \phi_4$. The phase and amplitude of an incoming pulse are detected, and the decision as to which symbol has been sent is made on the basis of which of the eight possible segments the resulting received phasor is in. In a system such as this, where there are the same number of phase positions available regardless of the amplitude level, the phase and amplitude channels are independent, and a

* Received by the PGCS, June 6, 1960.
† School of Elec. Engrg., Purdue University, Lafayette, Ind.

Reprinted from *IRE Transactions on Communications Systems,* December 1960.

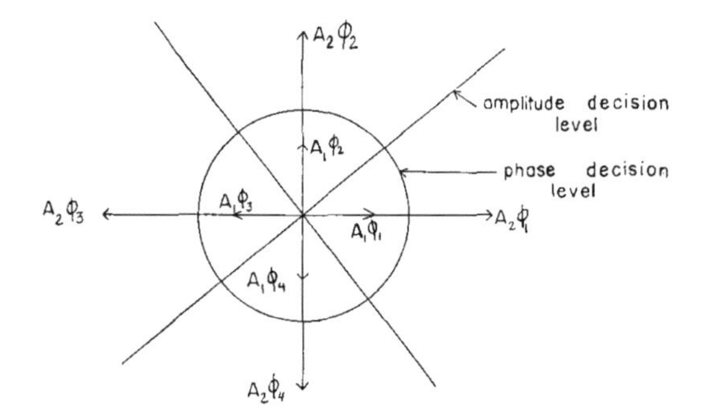

Fig. 2—Decision levels in a type *I* system.

transmitted symbol may be expressed as $A_i - \phi_j$, where A_i and ϕ_j may be chosen independently from separate discrete sets A and ϕ. Such a system will be designated a type *I* system.

It is possible to make a more efficient system by removing the restriction that the phase and amplitude channels be independent. Since the probability of an error in phase decreases with increasing amplitude levels, it is possible to increase the number of phase positions available with increasing amplitude levels, while maintaining a constant probability of phase error. Thus there are more symbols available for transmission for a given peak power. A typical system shown in Fig. 3 has three phase positions on the first amplitude level and nine on the second, for a total of twelve transmission symbols. This kind of system will be designated a type *II* system. While the type *II* system is theoretically superior, the type *I* system is much easier to instrument.

In the first section of this paper the structure of the type *II* system (*i.e.*, the placing of amplitude and phase levels) is derived so as to make the probability of error equal for each transmitted symbol. Such a system has its maximum capacity when all symbols are equally likely to be transmitted. Subsequent sections of the paper compare this system and a digital phase system with respect to probability of error and with respect to channel capacity. The latter comparison is particularly useful, since by using channel capacity both systems can be compared with the theoretical upper limit as given by Shannon [4].

As pointed out by Cahn [1], two types of phase demodulation are possible—coherent detection using a synchronized local reference, and phase comparison of two successive received samples. Only the use of coherent detection is considered in this paper. Similar results could be obtained for noncoherent detection and would show some degradation in performance. Cahn [2] has shown that in the case of digital phase systems this degradation approaches 3 db.

Approximations have been derived by Cahn for the probability of error in detecting the phase of a signal in the presence of Gaussian noise [2], [3]. These expressions are asymptotic in the high signal-to-noise ratio region and are useful when dealing with high accuracy systems where the probability of error must be small. Since these expressions have been used here, it should be

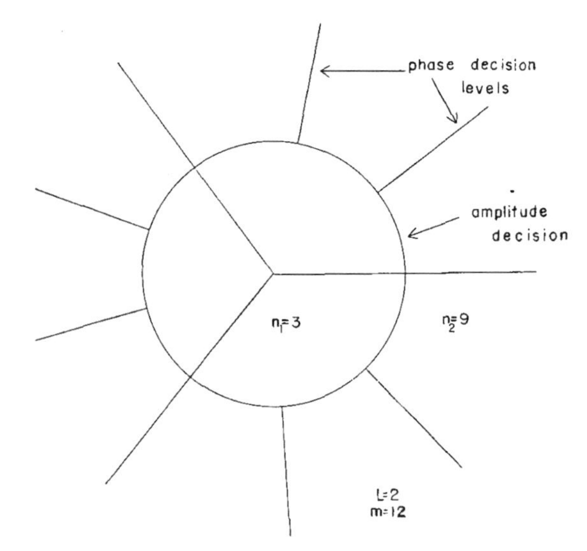

Fig. 3—Decision levels in a type *II* system.

noted that all results apply only to the high signal-to-noise ratio region. In most of the results given here it has also been necessary to employ other approximations rather freely. While there is some justification for each of these approximations, none of the stated results are intended to be exact.

Mathematical Description of System

We first consider a type *I* system. For large signal-to-noise ratios the probability of a phase error at the kth amplitude is given by the asymptotic approximation [1]–[3],

$$P_{e\phi} = \frac{e^{-(A_k^2/2)\sin^2 \pi/n_k}}{\sqrt{\pi}\,\dfrac{A_k}{\sqrt{2}}\sin \pi/n_k}, \tag{1}$$

where A_k is the normalized amplitude of the kth amplitude level based on a noise power of unity in the channel, and where n_k is the number of phase positions at this level. (For a type *I* system n_k is equal to n_1.) It should be noted that this expression is a function of the argument,

$$\frac{A_k}{\sqrt{2}}\sin\frac{\pi}{n_k},$$

i.e.,

$$P_{e\phi} = f\left(\frac{A_k}{\sqrt{2}}\sin\frac{\pi}{n_k}\right). \tag{2}$$

It can be assumed that the marginal amplitude distribution of the detected envelope is Gaussian for large signal-to-noise ratios. Using these distributions, Cahn has shown that for a type *I* system the placing of the amplitude levels is given by [3]

$$A_k = A_1\left[1 + 2(k-1)\sin\frac{\pi}{n_1}\right], \tag{3}$$

where there are n possible phase positions. When the amplitude levels are placed accordingly, the probability of an amplitude error at each amplitude level is equal to the probability of a phase error on the first amplitude level. The placing of A_1 is determined by the number of phase positions n and the desired probability of error.

For a type II system the number of phase positions on the higher amplitude levels is increased so as to keep the probability of a phase error constant and equal to the probability of an amplitude error. Thus the channel is symmetrical with respect to probability of error in a received symbol. By (2), $A_k \sin \pi/n_k$ must be kept constant for a constant probability of error. Thus we must have

$$\frac{A_1}{\sqrt{2}} \sin \frac{\pi}{n_1} = \frac{A_k}{\sqrt{2}} \sin \frac{\pi}{n_k}.$$

Eq. (3), relating A_k to A_1 for a type I system, may also be used for the type II system, since the amplitude levels are not changed.

Using (3) for A_k,

$$\sin \frac{\pi}{n_1} = \sin \frac{\pi}{n_k} \left[1 + 2(k-1) \sin \frac{\pi}{n_1} \right]. \quad (4)$$

Approximating the sine by its argument,

$$n_k \cong n_1 + 2\pi(k-1). \quad (5)$$

Since a fractional phase position is impossible,

$$n_{k+1} = n_k + 6. \quad (6)$$

Since most systems cannot have more than four or five amplitude levels without making the spacing between phase positions too close to be of practical value, (6) usually gives the integer closest to the true value of n_k as computed from (4). The alphabet size m is obtained by summing the number of phase positions on all amplitude levels:

$$m = \sum_{k=1}^{L} n_k. \quad (7)$$

Using (6) for n_k,

$$m = 3L(L-1) + n_1 L. \quad (8)$$

Thus the channel is determined by choice of the number of amplitude levels L, the number of phase positions on the first amplitude level n_1, and one additional factor which can be average or peak power, probability of error, or first amplitude level position A_1.

Probability of Error vs Signal-to-Noise Ratio

It is desired to find the minimum probability of error that can be attained for a given average power and alphabet size and the system parameters, A_1, n_1, and L, of the system which achieves this minimum error rate. The probability of error for a received symbol is the sum of the probability of a phase error and the probability of an amplitude error less the probability of both errors occurring. Neglecting the latter probability as small, the probability of error is just twice the probability of a phase error as given by (1). (All amplitude and phase errors are equally probable.) Thus,

$$P_e = \frac{2e^{-(A_1^2/2)\sin^2 \pi/n_1}}{\sqrt{\pi} \dfrac{A_1}{\sqrt{2}} \sin \pi/n_1}. \quad (9)$$

Since this error is a monotonically decreasing function of $A_1 \sin \pi/n_1$, the error may be minimized by maximizing $A_1 \sin \pi/n_1$ subject to the average power and alphabet size constraints.

Since all symbols are equally probable, the average power is given by

$$P = \sum_{k=1}^{L} \frac{n_k}{m} \frac{A_k^2}{2}. \quad (10)$$

Using (6) for n_k and (3) for A_k, and performing the summation,

$$\begin{aligned} P = \frac{A_1^2}{2m} \Big\{ & n_1 L + 2n_1 L(L-1) \sin \frac{\pi}{n_1} \\ & + \tfrac{2}{3} n_1 L(L-1)(2L-1) \sin^2 \frac{\pi}{n_1} \\ & + 3L(L-1) + 4L(L-1)(2L-1) \sin \frac{\pi}{n_1} \\ & + 6L^2(L-1)^2 \sin^2 \frac{\pi}{n_1} \Big\}. \end{aligned} \quad (11)$$

It should be noted that because the amplitudes have been normalized on the basis of unity noise power, this average power P is equivalent to the average signal-to-noise ratio.

Now A_1 may be solved in terms of P, m, and n_1 by use of (11) and (8) and the function to be maximized, $A_1 \sin \pi/n_1$, may be expressed as a rather lengthy function of n_1, P, and m; *i.e.*,

$$A_1 \sin \frac{\pi}{n_1} = g(n_1, P, m), \quad (12)$$

$$P, m = \text{given constraints}.$$

Setting $\partial g/\partial n_1 = 0$ to find the maximum, yields approximately

$$n_1 = 3. \quad (13)$$

Using this value for n_1 in (8) and (11) gives

$$L = \sqrt{\frac{m}{3}}, \qquad A_1^2/2 = P/(\tfrac{4}{3}m - 1). \quad (14)$$

With these optimum values, the minimum probability of error (9) becomes

$$P_{(e)\min} = \frac{2e^{-P/(8/9m-4/3)}}{\sqrt{\pi \dfrac{P}{8/9m - 4/3}}}. \quad (15)$$

Cahn has drawn error curves for phase modulation on a normalized signal-to-noise ratio abcissa, $P \sin^2 \pi/m$ [2], [3]. Fig. 4 shows (15) plotted together with Cahn's curve for phase modulation systems using this normalized coordinate. The signal-to-noise ratio improvement obtained

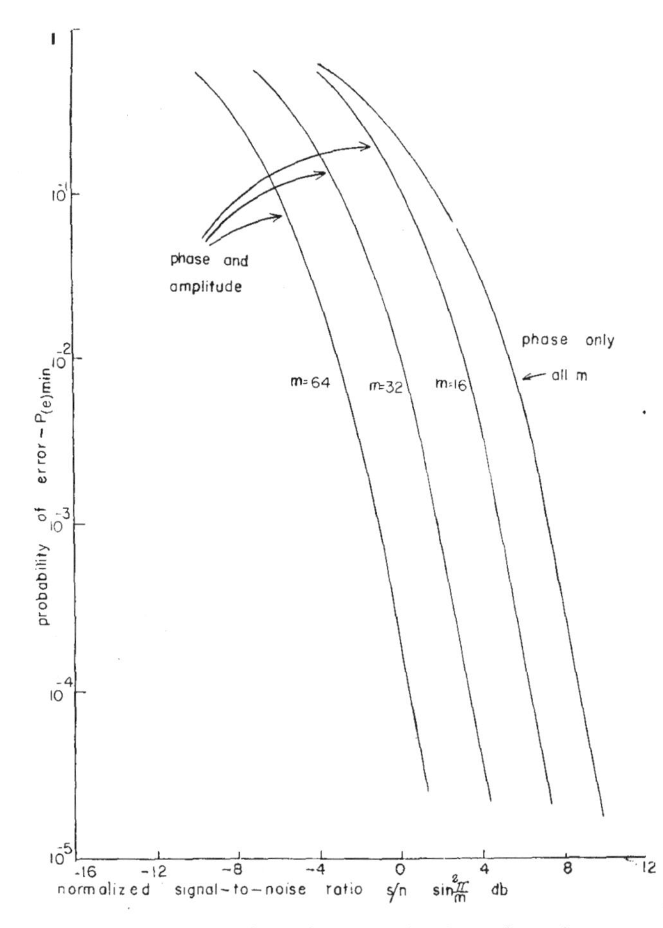

Fig. 4—Probability of error vs signal-to-noise ratio.

by the addition of amplitude modulation to phase-modulation systems is a function of the alphabet size. The improvement increases approximately 3 db every time the alphabet size is doubled.

It is interesting to note that the minimization of peak power for a given alphabet size and error rate also leads to the result that $n_1 = 3$ regardless of the power or alphabet size.

Channel Capacity

A. Phase Modulation Only

In order to provide a comparison with a type *II* phase and amplitude modulation system, the channel capacity will first be found for a digital phase-modulation system. In this system there is one amplitude level A, and m phase positions giving an alphabet size of m symbols.

In general the channel capacity as defined by Shannon can be found by

$$C = \max [H(y) - H_x(y)], \tag{16}$$

where the maximum is with respect to all possible information sources [4]. In a symmetrical system such as digital phase modulation, the maximum is obtained by making all input alphabet symbols equally probable; the expression reduces to [4]

$$C = \log m + \sum p_i \log p_i. \tag{17}$$

The p_i are the transitional probabilities from any one input symbol to each output symbol, and m is the total number of input symbols.

The probability of an error of more than one phase position is small, and it is assumed that the total probability of error is concentrated in the two neighboring phase positions with a transitional probability of

$$\beta = \tfrac{1}{2}P_{e\phi} \cong \frac{e^{-A^2/2\sin^2\pi/m}}{2\sqrt{\pi}\dfrac{A}{\sqrt{2}}\sin \pi/m} \tag{18}$$

to each. (See Fig. 5.) The probability of correct transmission is $(1 - 2\beta)$ and the channel capacity is

$$C = \log m + 2\beta \log \beta + (1 - 2\beta) \log (1 - 2\beta). \tag{19}$$

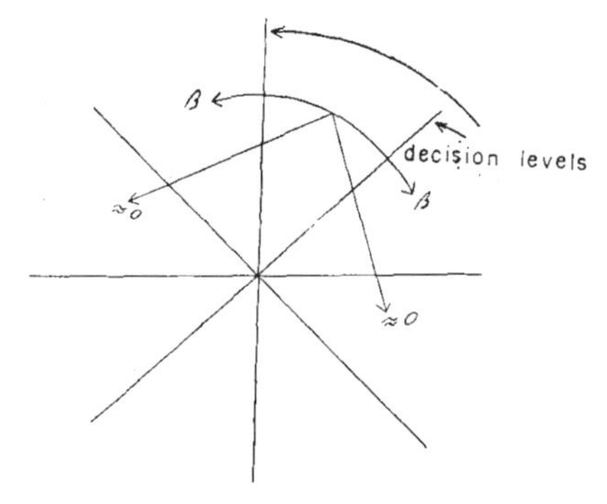

Fig. 5—Transitional probabilities in a phase-only system.

For small β, this can be approximated by

$$C \cong \log m + 2\beta(\log \beta - 1). \tag{20}$$

Substituting (18) for β gives C as a function of A and m. For a given average power A is a constant, and the expression may be differentiated with respect to m to give an optimum number of phase positions. Differentiation yields a transcendental equation of the form

$$h\left(\frac{A}{m}\right) = 0. \tag{21}$$

This equation may be solved by trial and error to give

$$\frac{A}{m} \cong 2/\pi, \tag{22}$$

$$M_{\text{opt}} \cong \pi \sqrt{P/2}. \tag{23}$$

Using this value for m in (20) gives an expression for channel capacity as a function of power (signal-to-noise ratio).[1]

$$C = \tfrac{1}{2} \log P + 0.814 \quad \text{bits/symbol}. \tag{24}$$

[1] The use of (18) as an approximation for the probability of error causes the channel capacity maximum to occur at the knee of the true C vs m curve instead of at $m = \infty$. As m becomes infinite $C = \frac{1}{2} \log P + 1.10$, but the probability of error approaches unity. Increasing m beyond the value given in (23) results in very little gain in channel capacity and only serves to complicate the coding problem.

B. Phase and Amplitude Modulation

Fig. 6 shows a typical portion of a phase and amplitude scheme. It can be seen that the channel is no longer symmetrical and that appreciable transitional probabilities exist to more than the four neighboring states of any transmitted symbol. Approximations have been found to take into consideration these asymmetries in calculating channel capacities of sample systems, and numerous channel capacities have been calculated using these approximations. However, these capacities do not differ appreciably from capacities calculated assuming that the channel is symmetrical with equal probability of error in each of four directions. To solve for the optimum system parameters, A_1, n_1, and L, for a given average power from the point of view of channel capacity, (8) and (9) are used in an expression entirely similar to (20). A Lagrangian multiplier is used to add (11) as a constraint. The resulting equation is differentiated with respect to A_1, n_1, L, and the Lagrangian multiplier. Unfortunately, the resulting simultaneous transcendental equations cannot be solved in closed form. If, however, it is assumed that $n_1 = 3$ (a not unreasonable assumption in view of the earlier results on probability of error and peak power minimization), new equations which may be solved may be constructed by the same method.

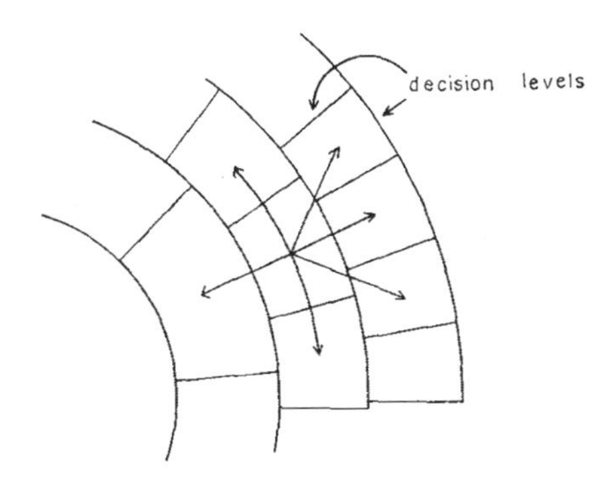

Fig. 6—Transitional probabilities in a phase and amplitude system.

For this case these results are obtained:

$$n_1 = 3 \quad \text{(assumption)}, \tag{25}$$

$$L = \tfrac{1}{2}\sqrt{P+2}. \tag{26}$$

(The result of these two equations is that $m = \frac{3}{4}P$.) (27)

$$A_1 = 2, \tag{28}$$

$$C = \log(P+2) - 1.19 \quad \text{bits/symbol}. \tag{29}$$

For the general case where no value is assumed for n_1, a digital computer was programmed to find the maximum of C over all integer combinations of n_1 and L while changing A_1 to hold average power constant. The computed maxima of C fall quite closely on the curve of (29) while the optimum values of m and L were those given by (27) and (26). The optimum value of n_1 is then found by using the integer values obtained for m and L. Thus a better specification of optimum system parameters is

$$m = \tfrac{3}{4}P \quad \text{(nearest integer)}, \tag{30}$$

$$L = \tfrac{1}{2}\sqrt{P+2} \quad \text{(nearest integer)}, \tag{31}$$

$$n_1 = \frac{m}{L} - 3(L-1). \tag{32}$$

It is now of interest to compare the channel capacities of these two systems with each other and with Shannon's upper limit [4]:

$$C = W \log(1 + S/N). \tag{33}$$

In order to compare these capacities with Shannon's result it is necessary to find the bandwidth associated with a given symbol rate in the case of phase modulation and combined phase and amplitude modulation. Both cases have spectral densities with a $(\sin^2 x)/x^2$ envelope. The width between first zeros is $2/\tau$ cps where τ is the pulse length. Thus, the bandwidth is twice the number of symbols transmitted per second. The three-channel capacities are plotted against signal-to-noise ratio for unit bandwidth in Fig. 7.[2]

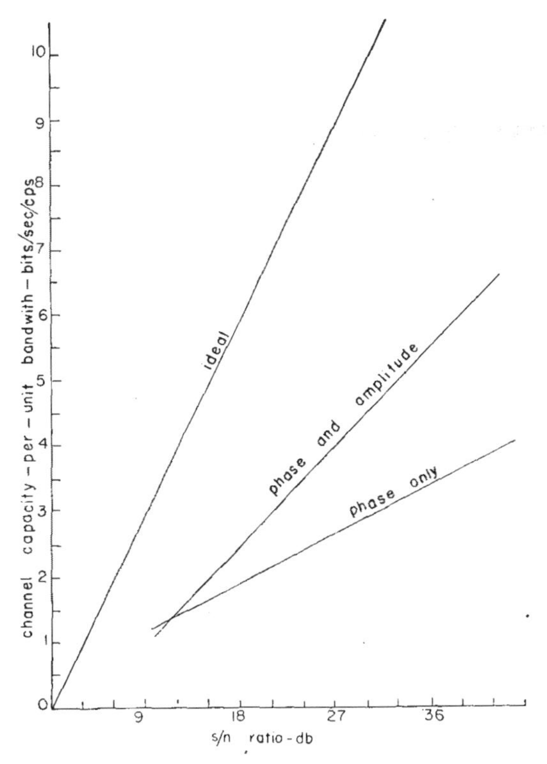

Fig. 7—Channel capacity vs signal-to-noise ratio.

[2] Although an additional improvement can be realized throug bandwidth reduction by sampling, the results of Fig. 7 demonstrat the relative capacities of the two systems. For the sampled systei the channel capacity scale for both the phase and amplitude systei and the phase only system in Fig. 7 would be doubled.

Conclusion

The structure, probability of error, and channel capacity of type *II* phase- and amplitude-modulated communication systems have been presented. In this type of system the phase and amplitude channels are dependent and cannot transmit separate information sources. A system in which the phase and amplitude channels are independent has been referred to as a type *I* system. Obviously the type *II* system is more difficult to instrument than the type *I* system, and both systems are in turn more complex than the phase-modulation-only system. The progressive complexity of the three systems must be balanced against the resulting savings in power.

The probability of error curves (Fig. 4) shows a 2.5-db advantage to the type *II* phase and amplitude system over the phase-only system for an alphabet size of 16 symbols (4 bits per symbol). This advantage increases to about 5.5 db for an alphabet size of 32 symbols (5 bits per symbol) and nearly 3 db for each bit increase thereafter.

The channel capacity curves (Fig. 7) show that the phase and amplitude system makes much more efficient use of the channel for high signal-to-noise ratios. This is because it is more efficient to use larger alphabet sizes at these powers and carry more bits of information per sample; this constitutes the power advantage for the phase and amplitude system. The channel capacity curves cross at a signal-to-noise ratio of about 11 db. Below this power there is no advantage to the phase and amplitude system over the phase-only system. At this crossover point of 11 db the optimum alphabet size for the two systems as given by (23) and (30) is about 8 symbols each.

The type *II* phase and amplitude system can be easily compared with the type *I* system on the basis of peak power required for a given probability of error and alphabet size. This comparison shows approximately a 1-db saving for an alphabet size of 8 symbols and a 2-db saving for an alphabet size of 16 symbols. The saving approaches 3 db asymptotically. Thus while both type *I* and type *II* systems increase their power advantage over the phase-only system for larger alphabet sizes, there is about a constant 3-db saving between the two.

In summary, combined phase- and amplitude-modulation systems are advantageous whenever an information capacity of more than 3 bits per symbol is used. For a signal-to-ratio of greater than 11 db, it is efficient from the point of view of channel capacity to use such information rates. If the channel bandwidth is not restricted, a phase-only system of nearly the same efficiency can be made by increasing the bandwidth and sending more symbols per second with each symbol carrying 3 bits or less of information.

Bibliography

[1] C. R. Cahn, "Performance of digital phase-modulation communication systems," IRE Trans. on Communication Systems, vol. CS-7, pp. 3–6; May, 1959.
[2] C. R. Cahn, "Performance of Digital Phase-Modulation Communications Systems," Ramo-Wooldridge, Corp., Los Angeles, Calif., Tech. Rept. No. M110-9U5; April, 1959.
[3] C. R. Cahn, "Combined digital phase- and amplitude-modulation communication systems," IRE Trans. on Communications Systems, vol. CS-8, pp. 150–154; September, 1960.
[4] C. E. Shannon, "The mathematical theory of communication," *Bell Sys. Tech. J.*, vol. 27, pp. 379–423, 623–656; July-October, 1948.

SYNCHRONOUS COMMUNICATIONS

J. P. Costas
General Electric Company
Syracuse, N.Y.

Summary

It can be shown that present usage of amplitude modulation does not permit the inherent capabilities of the modulation process to be realized. In order to achieve the ultimate performance of which AM is capable synchronous or coherent detection techniques must be used at the receiver and carrier suppression must be employed at the transmitter.

When a performance comparison is made between a synchronous AM system and a single-sideband system it is shown that many of the advantages normally attributed to single-sideband no longer exist. SSB has no power advantage over the synchronous AM (DSB) system and SSB is shown to be more susceptible to jamming. The performance of the two systems with regard to multipath or selective fading conditions is also discussed. The DSB system shows a decided advantage over SSB with regard to system complexity, especially at the transmitter. The bandwidth saving of SSB over DSB is considered and it is shown that factors other than signal bandwidth must be considered. The number of usable channels is not necessarily doubled by the use of SSB and in many practical situations no increase in the number of usable channels results from the use of SSB.

The transmitting and receiving equipment which has been developed under Air Force sponsorship is discussed. The receiving system design involves a local oscillator phase-control system which derives carrier phase information from the sidebands alone and does not require the use of a pilot carrier or synchronizing tone. The avoidance of superheterodyne techniques in this receiver is explained and the versatility of such a receiving system with regard to the reception of many different types of signals is pointed out.

System test results to date are presented and discussed.

Introduction

For a good many years a very large percentage of all military and commercial communications systems have employed amplitude modulation for the transmission of information. In spite of certain well-known shortcomings of conventional AM its use has been continued mainly due to the simplicity of this system as compared to other modulation methods which have been proposed. During the last few years, however, it has been felt by many responsible engineers that the increased demands being made on communications facilities could not be met by the use of conventional AM and that new modulation techniques would have to be employed in spite of the additional system complexity. Of these new techniques single-sideband has been singled out as the logical replacement for conventional AM and a great deal of publicity and financial support has been given SSB as a consequence.

Many technical reasons have been given to support the claim that SSB is better than AM and these points will be discussed in some detail later in this paper. In addition many experiments have been performed which also indicate a superiority for SSB over AM. Some care must be taken, however, in drawing conclusions from the above statements. We cannot conclude that SSB is superior to AM because we have no assurance whatever that conventional AM systems make efficient use of the modulation process employed. In other words AM as a modulation process may be capable of far better performance than that which is obtained in conventional AM systems. If an analysis is made of AM and SSB systems it will be found that existing SSB systems are very nearly optimum with respect to the modulation process employed whereas conventional AM systems fall far short of realizing the full potential of the modulation process employed. In fact it could honestly be said that we have been misusing rather than using AM in the past. Realization of the above situation raises some immediate questions: What are the equipment requirements of the optimum AM system? How does the performance of the optimum AM system compare with that of SSB? Which shows the greater promise of fulfilling future military and commercial communications requirements, optimum AM or SSB? The remainder of this paper will be devoted mainly to answering these questions.

Synchronous Communications - The Optimum AM System

Receiver

Conventional AM systems fail to obtain the full benefits of the modulation process for two main reasons: inefficient use of generated power at the transmitter and inefficient detection methods at the receiver. Starting with the receiver it can be shown that if maximum receiver performance is to be obtained the detection process must involve the use of a phase-locked oscillator and a synchronous or coherent detector. The basic synchronous receiver is shown in Figure 1. The incoming signal is mixed or multiplied with the coherent local oscillator signal in the detector and the demodulated audio output is thereby directly produced. The audio signal is then filtered and amplified. The local oscillator must be maintained at proper phase so that the audio output contributions of the upper and lower sidebands reinforce one another. If the oscillator phase is 90 degrees away from the optimum value a null in audio output will result which is typical of detectors of this type. The actual method of phase control will be explained shortly but for the pur-

Reprinted from *IRE Transactions on Communications Systems,* March 1957.

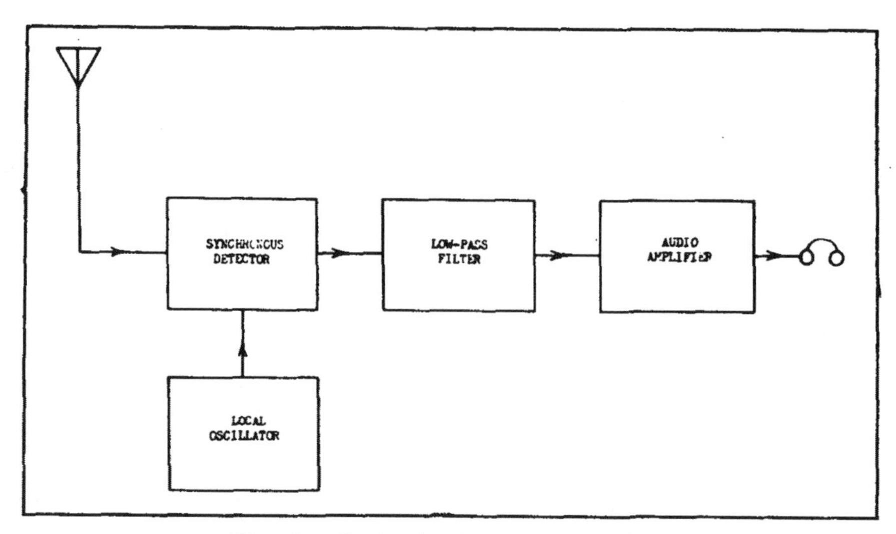

Fig. 1 - Basic synchronous receiver.

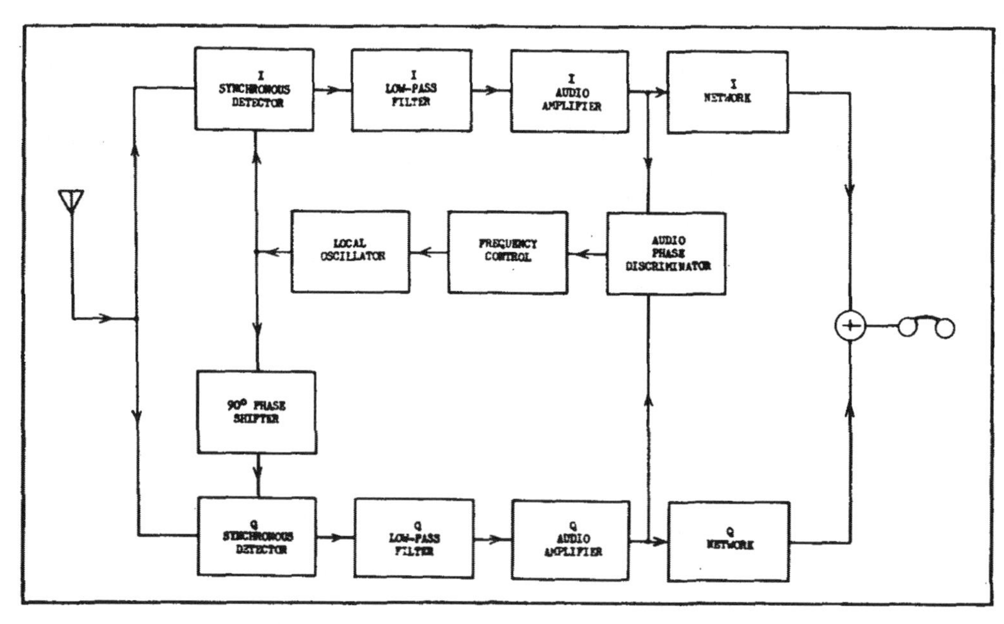

Fig. 2 - Two-phase synchronous receiver.

pose of this discussion maintenance of correct oscillator phase shall be assumed.

In spite of the simplicity of this type of receiver there are several important advantages worthy of note. To begin with no IF system is employed which eliminates completely the problem of image responses. The opportunity to use effectively post-detector filtering allows extreme selectivity to be obtained without difficulty. The selectivity curve of such a receiver will be found to be the low-pass filter characteristic mirror-imaged about the operating frequency. Not only is a high order of selectivity obtained in this manner but the selectivity of the receiver may be easily changed by low-pass filter switching. The carrier component of the AM signal is not in any way involved in the demodulation process and need not be transmitted when using such a receiver. Furthermore, detection may be accomplished at very low level and consequently the bulk of total receiver gain may be at audio frequencies. This permits an obvious application of transistors but more important it allows the selectivity determining low-pass filter to be inserted at a low-level point in the receiver which aids immeasurably in protecting against spurious responses from very strong undesired signals.

Phase Control To obtain a practical synchronous receiving system some additions to the basic receiver of Figure 1 are required. A more complete synchronous receiver is shown in Figure 2. The first thing to be noted about this diagram is that we have essentially two basic receivers with the same input signal but with local oscillator signals in phase quadrature to each other. To understand the operation of the phase control circuit consider that the local oscillator signal is of the same phase as the carrier component of the incoming AM signal. Under these conditions the in-phase or I audio amplifier output will contain the demodulated audio signal while the quadrature or Q audio amplifier will have no output due to the quadrature null effect of the Q synchronous detector. If now the local oscillator phase drifts from its proper value by a few degrees the I audio will remain essentially unaffected but there will now appear some audio output from the Q channel. This Q channel audio will have the same polarity as the I channel audio for one direction of local oscillator phase drift and opposite polarity for the opposite direction of local oscillator phase drift. The Q audio level is proportional to the magnitude of the local oscillator phase angle error for small errors. Thus by simply combining the I and Q audio signals in the audio phase discriminator a D.C. control signal is obtained which automatically corrects for local oscillator phase errors. It should be noted that phase control information is derived entirely from the sideband components of the AM signal and that the carrier if present is not used in any way. Thus since both synchronization and demodulation are accomplished in complete independence of carrier, suppressed-carrier transmissions may be employed.

It is unfortunate that many engineers tend to avoid phase-locked systems. It is true that a certain amount of stability is a prerequisite but it has been determined by experiment that for this application the stability requirements of single-sideband voice are more than adequate. Once a certain degree of stability is obtained the step to phase-lock is a simple one. It is interesting to note that this phase-control system can be modified quite readily to correct for large frequency errors when receiving AM due to doppler shift in air-to-air or ground-to-air links.

It is apparent that phase control ceases with modulation and that phase lock will have to be reestablished with the reappearance of modulation. This has not proved to be a serious problem since lock-up normally occurs so rapidly that no perceptible distortion results when receiving voice transmissions. It should be further noted that such a phase control system is inherently immune to carrier capture or jamming. In addition it has been found that due to the narrow noise bandwidth of the phase-control loop, synchronization is maintained at noise levels which render the channel useless for voice communications.

Interference Suppression The post-detector filters provide the sharp selectivity which of course contributes significantly to interference suppression. However, these filters cannot protect against interfering signal components which fall within the pass-band of the receiver. Such interference can be reduced and sometimes eliminated by proper combination of the I and Q channel audio signals. To understand this process consider that the receiver is properly locked to a desired AM signal and that an undesired signal appears, some of whose components fall within the receiver pass-band. Under these conditions the I channel will contain the desired audio signal plus an undesired component due to the interference. The Q channel will contain only an interference component also arising from the presence of the interfering signal. In general the interference component in the I channel and the interference component in the Q channel are related to one another or they may be said to be correlated. Advantage may be taken of this correlation by treating the I and Q voltages with the I and Q networks and adding these network outputs. If properly done this process will reduce and sometimes eliminate the interfering signal from the receiver output as a result of destructive addition of the I and Q interference voltages.

The design of these networks is determined by the spectrum of the interfering signal and the details of network design may be found in the literature.[1] Although such details cannot be given here it is interesting to consider one special interference case. If the interfering signal spectrum is confined entirely to one side of the desired signal carrier frequency the optimum I and Q networks become the familiar 90 degree phasing networks common in single-sideband work. Such operation does not however result in single-sideband reception of the desired signal since both desired signal sidebands contribute to receiver output at all times. This can be seen by noting that the Q channel contains no desired signal component so that network treatment and addition affects only the undesired audio signal components. The phasing networks are optimum only for the interference condition assumed above. If there is an overlap of the carrier frequency by the undesired signal spectrum the phasing networks are no longer optimum and a different network design is required for the greatest interference suppression.

This two-phase method of AM signal reception can aid materially in reducing interference. As a matter of fact it can be shown that the true anti-jam characteristics of AM cannot be realized unless a receiving system of the type discussed above is used. If we now compare the anti-jam characteristics of single-sideband and suppressed-carrier AM properly received it will be found that intelligent jamming of each type of signal will result in a two-to-one power advantage for AM. The bandwidth reduction obtained with single-sideband does not come without penalty. One of the penalties as we see here is that single-sideband is more easily jammed than double-sideband.

Transmitter

The synchronous receiver described above is capable of receiving suppressed-carrier AM transmissions. If a carrier is present as in standard AM this will cause no trouble but the receiver obviously makes no use whatever of the carrier component. The opportunity to employ carrier-suppressed AM transmissions can be used to good advantage in transmitter design. There are many ways in which to generate carrier-suppressed AM signals and one of the more successful methods is shown in Figure 3. A pair of class-C beam power amplifiers are screen-modulated by a push-pull audio signal and are driven in push-pull from an R.F. exciter. The screens are returned to ground or to some negative bias value by means of the driver transformer center-tap. Thus in the absence of modulation no R.F. output results and during modulation the tubes conduct alternately with audio polarity change. The circuit is extremely simple and a given pair of tubes used in such a transmitter can easily match the average R.F. power output of the same pair of tubes used in SSB-linear amplifier service. The circuit is self-neutralizing and the tune-up procedure is very much the same as in any other class-C R.F. power amplifier. The excitation requirements are modest and as an example the order of eight watts of audio are required to produce a sideband power output equivalent to a standard AM carrier output of one kilowatt. Modulation linearity is good and the circuit is amenable to various feedback techniques for obtaining very low distortion which may be required for multiplex transmissions.

This transmitter circuit is by no means new. The information is presented here to indicate the equipment simplicity which can be realized by use of synchronous AM communications.

Prototype Equipment

A synchronous receiver covering the frequency range of 2-32 mc. is shown in Figure 4. The theory of operation of this receiver is essentially that of the two-phase synchronous receiver discussed earlier. This is a direct conversion

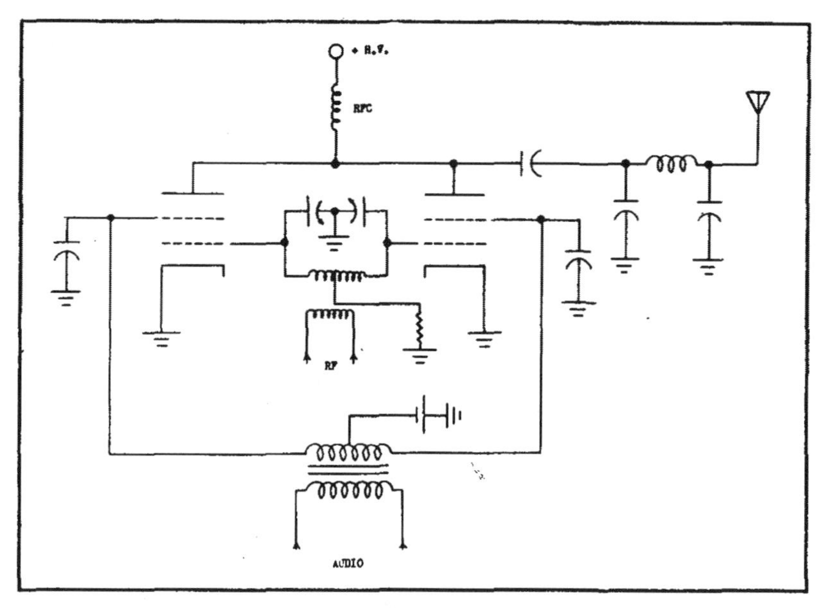

Fig. 3 - Suppressed-carrier AM transmitter.

Fig. 4 - The AN/FRR-48 (XW-1) synchronous receiver.

Fig. 5 - The AN/FRT-49 (XW-1) suppressed-carrier AM transmitter.

receiver and the superheterodyne principle is not used. A rather unusual frequency synthesis system is employed to give high stability with very low spurious response. Only one crystal is used and this is a 100 kc. oven-controlled unit.

This receiver will demodulate standard AM, suppressed-carrier AM, single-sideband, narrow-band FM, phase modulation, and CW signals in an optimum manner. This versatility is a natural by-product of the synchronous detection system and no great effort is required to obtain this performance.

Figure 5 shows a suppressed-carrier AM transmitter using a pair of 6146 tubes in the final. This unit is capable of 150 watts peak sideband power output for continuous sine-wave modulation. The modulator is a single 12BH7 miniature double triode. Figure 6 shows a transmitter capable of one-thousand watts peak sideband power output under continuous sine wave audio conditions. The final tubes are 4-250-A's and the modulator uses a pair of 6L6's. Both of these transmitters are continuously tuneable over 2 - 30 mc.

A Comparison of Synchronous AM and Single-Sideband

It is interesting at this point to compare the relative advantages and disadvantages of synchronous AM and single-sideband systems. Although single-sideband has a clear advantage over conventional AM this picture is radically changed when synchronous AM is considered.

Signal-To-Noise Ratio

If equal average powers are assumed for SSB and synchronous AM it can easily be shown that

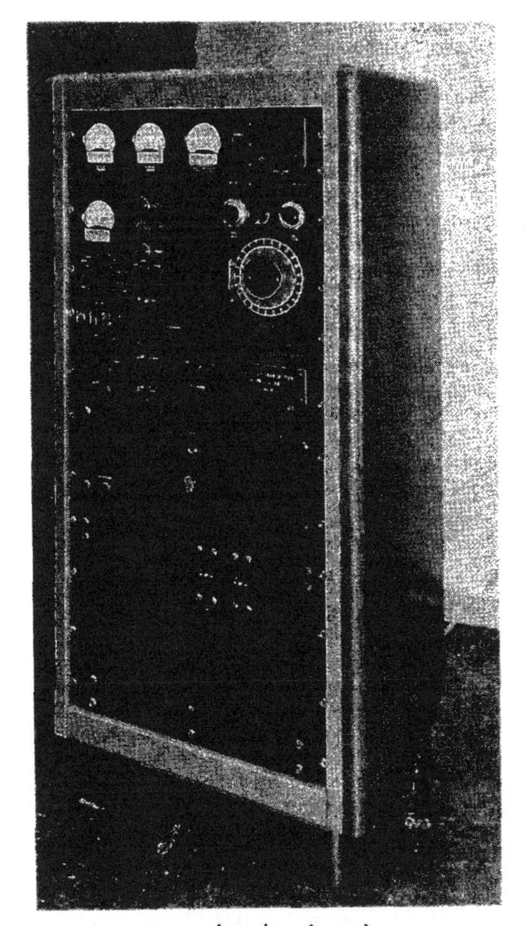

Fig. 6 - The AN/FRT/30 (XW-1) suppressed-carrier AM transmitter.

identical S/N ratios will result at the receiver. The additional noise involved from the reception of two sidebands is exactly compensated for by the coherent addition of these sidebands. The 9db advantage often quoted for SSB is based on a full AM carrier and a peak power comparison. Since we have eliminated the carrier and since a given pair of tubes will give the same average power in suppressed-carrier AM or SSB service there is actually no advantage either way. If intelligent jamming rather than noise is considered there exists a clear advantage of two-to-one in average power in favor of synchronous AM.

System Complexity

Since the receiver described is also capable of SSB reception it would appear that synchronous AM and SSB systems involve roughly the same receiver complexity. This is not altogether true since much tighter design specifications must be imposed if high quality SSB reception is to be obtained. If AM reception only is considered these specifications may be relaxed considerably without materially affecting performance. The synchronous receiver described earlier may possess important advantages over conventional superheterodyne receivers but this point is not an issue here.

The suppressed-carrier AM transmitter is actually simpler than a conventional AM transmitter. It is of course far simpler than any SSB transmitter. There are no linear amplifiers, filters, phasing networks, or frequency translators involved. Personnel capable of operating or maintaining standard AM equipment will have no difficulty in adapting to suppressed-carrier AM. The military and commercial significance of this situation is rather obvious and further discussion of this point is not warranted.

Long-Range Communications

The selective fading and multipath conditions encountered in long-range circuits tend to vary the amplitude and phase of one sideband component relative to the other. This would perhaps tend to indicate an advantage for SSB but tests to date do not confirm this. Synchronous AM reception of standard AM signals over long paths has been consistently as good as SSB reception of the same signal. In some cases it was noted that the SSB receiver output contained a serious flutter which was only slightly discernable in the synchronous receiver output. Some attempt has been made to explain these results but as yet no complete explanation is available. One interesting fact about the synchronous receiver is that the local oscillator phase changes as the sidebands are modified by the medium since phase control is derived directly from the sidebands. In a study of special cases of signal distortion it was found that the oscillator orients itself in phase in such a way as to attempt to compensate for the distortion caused by the medium. This may partially explain the good results which have been obtained. Perhaps another point of view would be that the synchronous receiver is taking advantage of the inherent diversity feature provided by the two AM sidebands.

Test results to date indicate that synchronous AM and single-sideband provide much the same performance for long-range communications. The AM system has been found on occasion to be better but since extensive tests have not been performed and since a complete explanation of these results is not yet available it would be unfair to claim any advantage at this time for AM.

Spectrum Utilization

In theory single-sideband transmissions require only half the bandwidth of equivalent AM transmissions and this fact has led to the popular belief that conversion to single-sideband will result in an increase in usable channels by a factor of two. If a complete conversion to single-sideband were made those who believe that twice the number of usable channels would be available might be in for a rather rude awakening. There are many factors which determine frequency allocation besides modulation bandwidth. Under many conditions it actually turns out that modulation bandwidth is not a consideration. This is a complicated problem and only a few of the more pertinent points can be discussed briefly here.

To begin with the elimination of one sideband is a complicated and delicate business. Any one of several misadjustments of the SSB transmitter will result in an empty sideband which is not actually empty. We are not thinking here of a telephone company point-to-point system staffed by career personnel but rather we have in mind the majority of military and commercial field installations. This is in no way meant to be a criticism but the technical personnel problem faced by the military especially in time of war is a serious one and this simple fact of life cannot be ignored in future system planning. Thus we must concede that single-sideband transmissions will in practice not always be confined to one sideband and that those who allocate frequencies must take this into consideration.

There may be those who would argue that SSB transmitting equipment can be designed for simple operation. This is probably true but in general operational simplicity can only be obtained at the expense of additional complexity in manufacture and maintenance. This of course trades one set of problems for another but if we assume ideal SSB transmission we are still faced with an even more serious allocation problem. We refer here to the problem of receiver non-linearity which becomes a dominant factor when trying to receive a weak signal in the presence of one or more near-frequency strong signals. Under such conditions the single-signal selectivity curves often shown by manufacturers are next to meaningless. This strong undesired-weak desired signal situation often arises in practice especially in the military where close physical spacing of equipment is mandatory as in the case of ships or aircraft and where signal environment changes due to changing locations of these vehicles. Because of this situation allocations to some extent must be made practically independent of modulation bandwidth and the theoretical spectrum conservation of single-sideband cannot always be advantageously used.

The problem of receiver non-linearity is especially serious in multiple conversion superheterodyne receivers for obvious reasons. This was the dominant factor in choosing a direct conversion scheme in the synchronous receiver described earlier. Although this approach has given good results and continued refinement has indicated that significant advances over prior art can be obtained, it cannot be said however that the receiver problem is solved. This problem will probably remain a serious one until new materials and components are made available. This is a relatively slow process and it is not at all absurd to consider that by the time this problem is eliminated new modulation processess will have appeared which will eclipse both of those now being considered.

In short the spectrum economies of SSB which exist in theory cannot always be realized in practice as there exist many important military and commercial communications situations in which no increase in usable channels will result from the adoption of single-sideband.

Jamming

The reduction of transmission bandwidth afforded by single-sideband must be paid for in one form or another. A system has yet to be proposed which offers nothing but advantages. One of the prices paid for this reduction in bandwidth is greater susceptibility to jamming as was previously mentioned. There is an understandable tendency at times to ignore jamming since the systems with which we are usually concerned provide us with ample worries without any outside aid. Jamming of course cannot be ignored and from a military point of view this raises a very serious question. If we concede for the moment that by proper frequency allocation single-sideband offers a normal channel capacity advantage over AM, what will happen to this advantage when we have the greatest need for communications? It is almost a certainty that at the time of greatest need jamming will have to be reckoned with. Under these conditions any channel capacity advantage of SSB could easily vanish. A definite statement to this effect cannot be made of course without additional study but this is a factor well worth considering.

Concluding Remarks

There is an undeniable need for improved communications and to date it appears that single-sideband has been almost exclusively considered to supplant conventional AM. It has been the main purpose of this paper to point out that the improved performance needed can be obtained in another way. The synchronous AM system can compete more than favorably with single-sideband when all factors are taken into account.

Acknowledgement

Much of the work reported here was sponsored by the Rome Air Development Center of the Air Research and Development Command under Air Force contract AF 30(602) 584.

The author wishes to acknowledge the support, cooperation, and encouragement which has been extended by the personnel of the Rome Air Development Center.

Bibliography

1. J. P. Costas, "Interference Filtering", Technical Report No. 185, Research Laboratory of Electronics, M.I.T.

NETWORKING

On the Self-Similar Nature of Ethernet Traffic (Extended Version)

Will E. Leland, *Member, IEEE*, Murad S. Taqqu, *Member, IEEE*, Walter Willinger, and Daniel V. Wilson, *Member, IEEE*

***Abstract*—We demonstrate that Ethernet LAN traffic is statistically *self-similar*, that none of the commonly used traffic models is able to capture this *fractal*-like behavior, that such behavior has serious implications for the design, control, and analysis of high-speed, cell-based networks, and that aggregating streams of such traffic typically intensifies the self-similarity ("burstiness") instead of smoothing it. Our conclusions are supported by a rigorous statistical analysis of hundreds of millions of high quality Ethernet traffic measurements collected between 1989 and 1992, coupled with a discussion of the underlying mathematical and statistical properties of self-similarity and their relationship with actual network behavior. We also present traffic models based on self-similar stochastic processes that provide simple, accurate, and realistic descriptions of traffic scenarios expected during B-ISDN deployment.**

I. Introduction

IN THIS PAPER[1], we use the LAN traffic data collected by Leland and Wilson [14] who were able to record hundreds of millions of Ethernet packets without loss (irrespective of the traffic load) and with recorded time-stamps accurate to within 100 μs. The data were collected between August 1989 and February 1992 on several Ethernet LAN's at the Bellcore Morristown Research and Engineering Center. Leland and Wilson [14] present a preliminary statistical analysis of this unique high-quality data and comment in detail on the presence of "burstiness" across an extremely wide range of time scales: traffic "spikes" ride on longer-term "ripples," that in turn ride on still longer term "swells," etc. This *self-similar* or *fractal*-like behavior of aggregate Ethernet LAN traffic is very different both from conventional telephone traffic and from currently considered formal models for packet traffic (e.g., pure Poisson or Poisson-related models such as Poisson-batch or Markov-Modulated Poisson processes (see [11]), packet-train models (see [13]), fluid flow models (see [1]), etc. and requires a new look at modeling traffic and performance of broadband networks.

The main objective of this paper is to establish in a statistically rigorous manner the *self-similarity* characteristic of the very high quality, high time-resolution Ethernet LAN traffic measurements presented in [14]. Moreover, we illustrate some of the most striking differences between self-similar models and the standard models for packet traffic currently considered in the literature. For example, our analysis of the Ethernet data shows that the generally accepted argument for the "Poisson-like" nature of aggregate traffic, namely, that aggregate traffic becomes smoother (less bursty) as the number of traffic sources increases, has very little to do with reality. In fact, using the degree of self-similarity (which typically depends on the utilization level of the Ethernet and can be defined via the *Hurst parameter*) as a measure of "burstiness," we show that the burstiness of LAN traffic typically intensifies as the number of active traffic sources increases, contrary to commonly held views.

The term "self-similar" was coined by Mandelbrot. He and his co-workers (e.g., see [21]–[23]) brought self-similar processes to the attention of statisticians, mainly through applications in such areas as hydrology and geophysics. For further applications and references on the probability theory of self-similar processes, see the extensive bibliography in [27]. For an early application of the self-similarity concept to communications systems, see the seminal paper by Mandelbrot [18].

The paper is organized as follows. In Section II, we describe the available Ethernet traffic measurements and comment on the changes of the Ethernet population, applications, and environment during the measurement period from August 1989 to February 1992. In Section III, we give the mathematical definition of self-similarity, identify classes of stochastic models which are capable of accurately describing the self-similar behavior of the traffic measurements at hand, and illustrate statistical methods for analyzing self-similar data sets. Section IV describes our statistical analysis of the Ethernet data, with emphasis on testing for self-similarity. Finally, in Section V we discuss the significance of self-similarity for traffic engineering, and for operation, design, and control of B-ISDN environments.

II. Traffic Measurements

2.1. The Traffic Monitor

The monitoring system used to collect the data for the present study was custom-built by one of the authors (Wilson) in 1987/88 and has been in use to the present day with one upgrade. For each packet seen on the Ethernet under study, the monitor records a timestamp (accurate to within 100μs—to within 20 μs in the updated version of the monitor), the packet

Manuscript received July 1, 1993; revised January 15, 1994; approved by IEEE/ACM Transactions on Networking Editor Jonathan Smith. This work was supported in part by Boston University, under ONR Grant N00014-90-J-1287.

W. E. Leland, W. Willinger, and D. V. Wilson are with Bellcore, Morristown, NJ 07962-1910 (email: wel@bellcore.com, walter@bellcore.com, dvw@bellcore.com),.

M. S. Taqqu is with the Dept. of Mathematics, Boston University, Boston, MA 02215-2411 (email: murad@bu-ma.bu.edu).

IEEE Log Number 9300098.

[1] An abbreviated version of this paper appeared in [15].

Reprinted from *IEEE /ACM Transactions on Networking*, vol. 2, no. 1, February 1994.

TABLE I
QUALITATIVE DESCRIPTION OF SETS OF ETHERNET TRAFFIC MEASUREMENTS USED IN THE ANALYSIS IN SECTION IV

Traces of Ethernet Traffic Measurements					
Measurement Period		Data Set	Total Number of Bytes	Total Number of Packets	Ethernet Utilization
	Total (27.45 h)		11 448 753 134	27 901 984	9.3%
AUGUST 1989	Low Hour	AUG89.LB	224 315 439		5.0%
Start of Trace:	(6:25 am–7:25 am)	AUG89.LP		652 909	
Aug. 29, 11:25 am	Normal Hour	AUG89.MB	380 889 404		8.5%
End of Trace:	(2:25 pm–3:25 pm)	AUG89.MP		968 631	
Aug. 30, 3:10 pm	Busy Hour	AUG89.HB	677 715 381		15.1%
	4:25 pm–5:25 pm)	AUG89.HP		1 404 444	
	Total (20.86 h)		14 774 694 236	27 915 376	15.7%
OCTOBER 1989	Low Hour	OCT89.LB	468 355 006		10.4%
Start of Trace:	(2:00 am–3:00 am)	OCT89.LP		978 911	
Oct. 5, 11:00 am	Normal Hour	OCT89.MB	827 287 174		18.4%
End of Trace:	(5:00 pm–6:00 pm)	OCT89.MP		1 359 656	
Oct. 6, 7:51 pm	Busy Hour	OCT89.HB	1 382 483 551		30.7%
	(11:00 am–12:00 am)	OCT89.HP		2 141 245	
	Total (40.16 h)		7 122 417 589	27 954 961	3.9%
JANUARY 1990	Low Hour	JAN90.LB	87 299 639		1.9%
Start of Trace:	(Jan. 11, 8:32 pm–9:32 pm)	JAN90.LP		310038	
Jan. 10, 6:07 am	Normal Hour	JAN90.MB	182 636 845		4.1%
End of Trace:	(Jan. 10, 9:32 am–10:32 am)	JAN90.MP		643 451	
Jan. 11, 10:17 pm	Busy Hour	JAN90.HB	711 529 370		15.8%
	(10:32 am–11:32 am)	JAN90.HP		1 391 718	
	Total (47.91 h)		6 585 355 731	27 674 814	3.1%
FEBRUARY 1992	Low Hour	FEB92.LB	56 811 435		1.3%
Start of Trace:	(Feb. 20, 1:21 am–2:21 am)	FEB92.LP		231 823	
Feb. 18, 5:22 am	Normal Hour	FEB92.MB	154 626 159		3.4%
End of Trace:	(Feb. 18, 8:21 pm–9:21 pm)	FEB92.MP		524 458	
Feb. 20, 5:16 am	Busy Hour	FEB92.HB	225 066 741		5.0%
	(Feb. 18, 11:21 am–12:21 am)	FEB92.HP		947 662	

length, the status of the Ethernet interface and the first 60 bytes of data in each packet (header information). As we will show in Section IV, the high-accuracy timestamps of the Ethernet packets produced by this monitor are crucial for our statistical analyses of the data. A detailed discussion of the capabilities of the original monitoring system, including extensive testing of its capacity and accuracy can be found in [14].

2.2. *The Network Environment at Bellcore*

The network environment at the Bellcore Morris Research and Engineering Center (MRE) where the traffic measurements used for the analysis presented later were collected is probably typical of a research or software development environment where workstations are the primary machines on people's desks. It is also typical in that much of the original installation was well thought out and planned but then grew haphazardly. For the purposes of this study, this haphazard growth is not necessarily a liability, as we are able to study the traffic on a network that is evolving over time. Table I gives a summary description of the traffic data analyzed later in the paper. We consider four sets of traffic measurements, each representing between 20 and 40 consecutive hours of Ethernet traffic and each consisting of tens of millions of Ethernet packets. The data were collected on different intracompany LAN networks at different times over the course of approximately four years (August 1989, October 1989, January 1990, and February 1992).

2.2.1. Workgroup Network Traffic Data: Four data sets will be considered in this paper. A summary description of these data sets is given in Table I. The first two sets of traffic measurements, taken in August and October of 1989 (see first two rows in Table I), were from an Ethernet network serving a laboratory of researchers engaged in everything from software development to prototyping new services for the telephone system. The traffic was mostly from services that used the Internet Protocol (IP) suite for such capabilities as remote login or electronic mail, and the Network File System (NFS) protocol for file service from servers to workstations. There were some unique services, though; for example, the audio of a local radio station was μ-law encoded and distributed over the network during portions of the day. While it is not our intent to provide here a detailed description of the particular MRE network segments under study, some words about the types of traffic on them are appropriate.

A snapshot of the network configuration at the time of collection of the earliest data set being used (August 1989) is given in Fig. 1: there were about 140 hosts and routers connected to this intra-laboratory network at that time, of which 121 spoke up during the 27 h monitoring period. This network consisted of two cable segments connected by a bridge, implying that not all the traffic on the network as a whole was visible from our monitoring point. During the period this data was collected, among the 25 most active hosts were two DEC 3100 fileservers, one Sun-4 fileserver, six Sun-3 fileservers, two VAX 8650 minicomputers, and one CCI Power 6 minicomputer. At that time, the less active hosts were mainly diskless Sun-3 machines and a smattering of Sun-4's, DEC 3100's, personal computers, and printers.

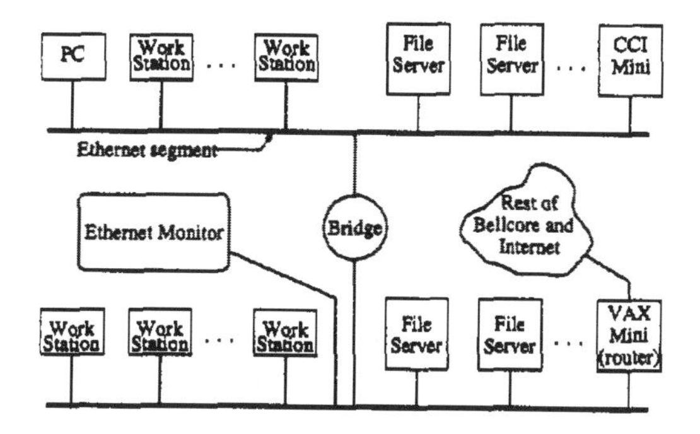

Fig. 1. Network from which the August 1989 and October 1989 measurements were taken.

During the latter part of 1989 when the first two data sets were collected, a revolution was taking place on this network. The older Sun-3 class workstations were rapidly replaced with RISC-based workstations such as the SPARC station-1 and DEC 3100. Many of the new workstations were "dataless" (where the operating system is stored on a local disk but user data on a server) instead of "diskless" (where all files for the user and for the operating system are stored on a remote server). Because of the increased computing power of the machines connected to this segment, the network load increased appreciably, in spite of the trend towards dataless workstations. Note, for example, that the "busy hour" from the October 1989 data set is indeed busy: 30.7% utilization as compared to 15.1% during the August 1989 busy hour; similar increases can also be observed for the low and normal hours. Not long after this data was taken, this logical Ethernet segment was again segmented by adding yet a third cable and a bridge, and moving some user workstations and their fileserver to that new cable. The above network has always been isolated from the rest of the Bellcore world by one or more routers. The other sides of these routers were connected to a large corporate internet consisting at that time of many Ethernet segments and T-1 point-to-point links connected together with bridges. Less than 5% of the total traffic on this workgroup network during either of the traces went out to either the rest of Bellcore or outside of the company.

2.2.2. Workgroup and External Traffic: The third data set, taken in January 1990 (row 3 in Table I), came from an Ethernet cable that linked the two wings of the MRE facility that were occupied by a second laboratory (see Fig. 2). At the time this data set was collected, this second laboratory comprised about 160 people, engaged in work similar to the first laboratory. This particular segment was unique in that it was also the segment serving Bellcore's link to the outside Internet world. Thus the traffic on this cable was from several sources: (i) two very active file servers directly connected to the segment; (ii) traffic (file service and remote login) between the two wings of this laboratory; (iii) traffic between the laboratory and the rest of Bellcore; and (iv) traffic between Bellcore as a whole and the larger Internet world. This last type of traffic we term *external* traffic, and in 1990 could come from conversations between machines in any part of Bellcore and the outside world. This Ethernet segment was specifically monitored to capture this external traffic. In Section IV, we

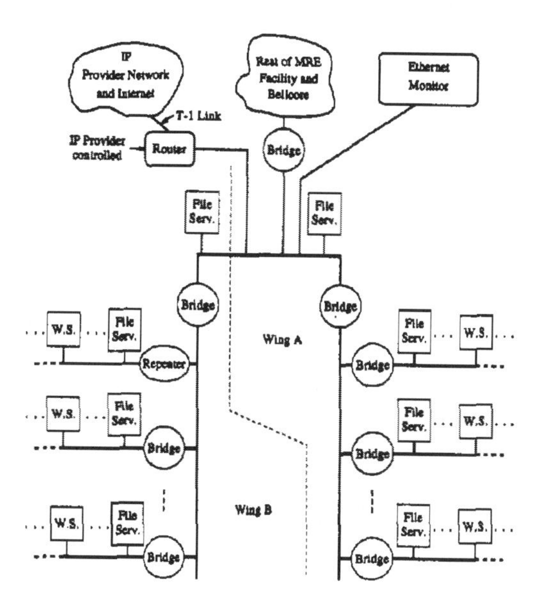

Fig. 2. Network for second laboratory from which the January 1990 measurements were taken.

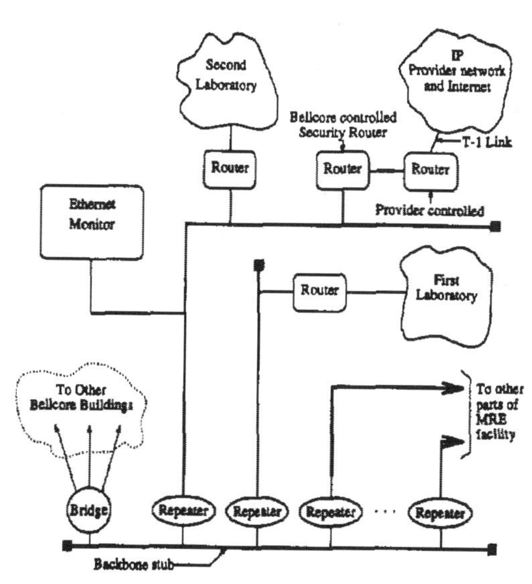

Fig. 3. Backbone network for MRE facility from which the February 1990 measurements were taken.

will be considering the aggregate and external traffic from this data set separately. This segment was separated from both the Bellcore internet and the two wings of the laboratory by bridges, and from the outside world by a vendor-controlled router programmed to pass anything with a Bellcore address as source or destination. In contrast to the two earlier data sets, over 1200 hosts spoke up during the 40 h monitoring period on this segment.

The last data set, from February 1992 (see row 4 in Table I), was taken from the building-wide Ethernet backbone in MRE after security measures mandated by the "Morris worm" (described in detail in [26]) had been put into place (see Fig. 3). This cable carried all traffic going between laboratories within MRE, traffic from other Bellcore buildings destined for MRE, and all traffic destined for locations outside of Bellcore.

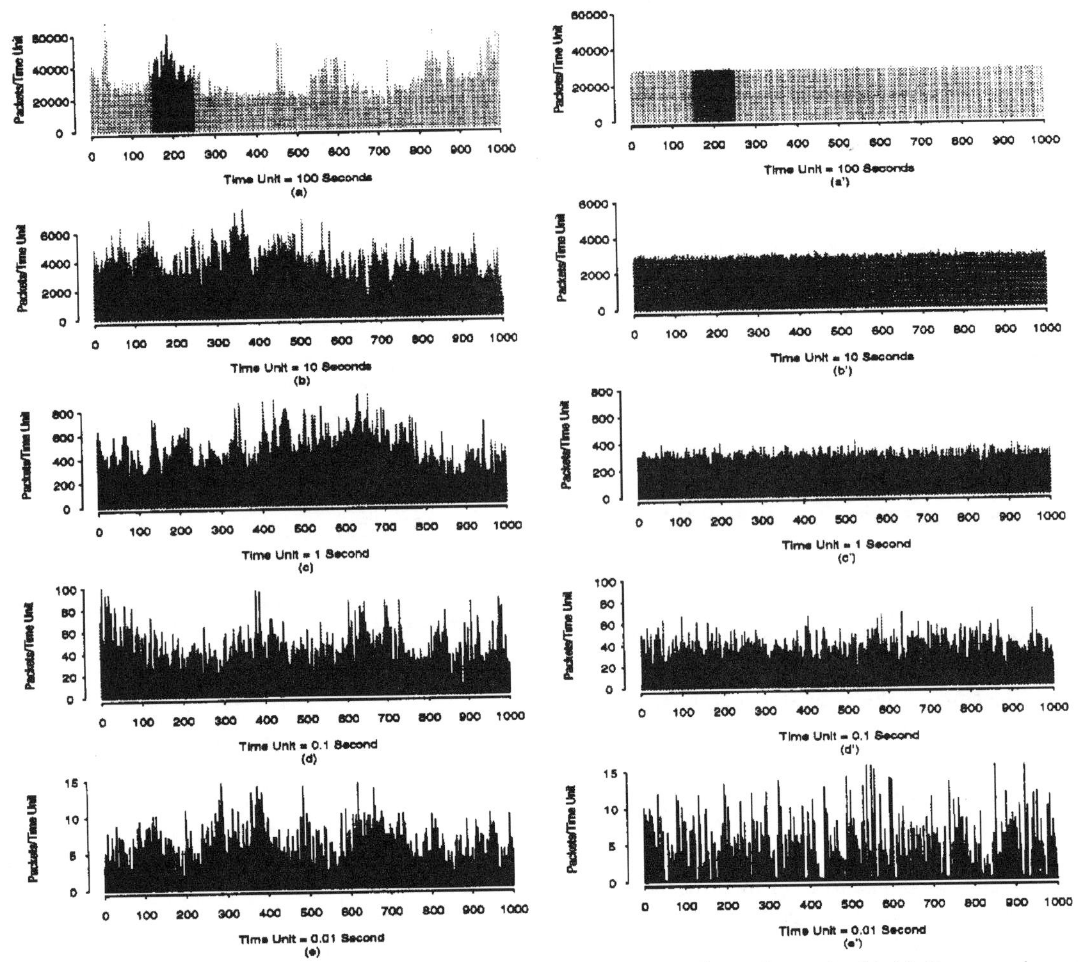

Fig. 4. Pictorial "proof" of self-similarity: Ethernet traffic (packets per time unit) on five different time scales (a)–(e). For comparison, synthetic traffic from an appropriately chosen compound Poisson model on the same five different time scales (a')–(e').

Some hosts were still directly connected to this company-wide network in early 1992, but the trend to move them from the Bellcore internet to workgroup cables connected to the Bellcore internet via routers continues to the present. Because this cable had very little host to file server traffic, the overall traffic levels were much lower than for the other three sets. On the other hand, the percentage of remote login and mail traffic was higher. This cable also carried the digitized radio traffic between the two laboratories under discussion. The most radical difference between this data set and the others is that the traffic is primarily router to router rather than host to host. In fact, about 600 hosts spoke up during the measurement period (down from about 1200 active hosts during the January '90 measurement period), and the five most active hosts were routers.

III. Self-similar Stochastic Processes

3.1. A Picture is Worth a Thousand Words

For 27 consecutive hours of monitored Ethernet traffic from the August 1989 measurements (first row in Table I), Fig. 4 (a)–(e) depicts a sequence of simple plots of the packet counts (i.e., number of packets per time unit) for five different choices of time units. Starting with a time unit of 100 s (Fig. 4(a)), each subsequent plot is obtained from the previous one by increasing the time resolution by a factor of 10 and by concentrating on a randomly chosen subinterval (indicated by a darker shade).

The time unit corresponding to the finest time scale (e) is 10 ms. In order to avoid the visually irritating quantization effect associated with the finest resolution level, plot (e) depicts a "jittered" version of the number of packets per 10 ms, i.e., a small amount of noise has been added to the actual arrival rate. Observe that with the possible exception of plot (a) which suggests the presence of a daily cycle, all plots are intuitively very "similar" to one another (in a distributional sense), that is, Ethernet traffic seems to look the same in the large (min, h) as in the small (s, ms). In particular, notice the absence of a natural length of a "burst:" at every time scale ranging from milliseconds to minutes and hours, bursts consist of bursty subperiods separated by less bursty subperiods. This scale-invariant or "self-similar" feature of Ethernet traffic is drastically different from both conventional telephone traffic and from stochastic models for packet traffic currently considered in the literature. The latter typically produce plots of packet counts which are indistinguishable from white noise after aggregating over a few hundred milliseconds, as illustrated in Fig. 4 with the sequence of plots (a')–(e'); this sequence was obtained in the same way as the sequence (a)–(e), except that it depicts synthetic traffic generated from a comparable (in terms of average packet size and arrival rate) compound Poisson process. (Note that while the choice of a compound Poisson process is admittedly not very sophisticated, even more complicated Markovian arrival processes would produce plots indistinguishable from Fig. 4(a')–(e').) Fig. 4 provides a surprisingly simple method for distinguishing clearly between our measured data and traffic generated by currently used models and strongly suggests the use of self-similar stochastic processes for traffic modeling purposes. Below, we give a brief description of the concept of self-similar processes, discuss their most important mathematical and statistical properties, mention some modeling approaches, and outline statistical methods for analyzing self-similar data. For a more detailed presentation and references, see [17], [4], or [2].

3.2. Definitions and Properties

Let $X = (X_t : t = 0, 1, 2, \dots)$ be a *covariance stationary* stochastic process with mean μ, variance σ^2 and autocorrelation function $r(k), k \geq 0$. In particular, we assume that X has an autocorrelation function of the form

$$r(k) \sim k^{-\beta} L(t), \text{ as } k \to \infty, \tag{1}$$

where $0 < \beta < 1$ and L is slowly varying at infinity, i.e., $\lim_{t\to\infty} L(tx)/L(t) = 1$, for all $x > 0$. (For our discussion below, we assume for simplicity that L is asymptotically constant.) For each $m = 1, 2, 3, \dots$, let $X^{(m)} = (X_k^{(m)} : k = 1, 2, 3, \dots)$ denote the new covariance stationary time series (with corresponding autocorrelation function $r^{(m)}$) obtained by averaging the original series X over non-overlapping blocks of size m. That is, for each $m = 1, 2, 3, \dots$, $X^{(m)}$ is given by $X_k^{(m)} = 1/m(X_{km-m+1} + \cdots + X_{km}), k \geq 1$. The process X is called *(exactly) second-order self-similar* with self-similarity parameter $H = 1 - \beta/2$ if for all $m = 1, 2, \dots$, $\text{var}(X^{(m)}) = \sigma^2 m^{-\beta}$ and

$$r^{(m)}(k) = r(k), k \geq 0. \tag{2}$$

X is called *(asymptotically) second-order self-similar* with self-similarity parameter $H = 1 - \beta/2$ if for all k large enough,

$$r^{(m)}(k) \to r(k), \text{ as } m \to \infty \tag{3}$$

with $r(k)$ given by (1). In other words, X is exactly or asymptotically second-order self-similar if the corresponding aggregated processes $X^{(m)}$ are the same as X or become indistinguishable from X—at least with respect to their autocorrelation functions.

Mathematically, self-similarity manifests itself in a number of equivalent ways: (i) the variance of the sample mean decreases more slowly than the reciprocal of the sample size (*slowly decaying variances*), i.e., $\text{var}(X^{(m)}) \sim a_2 m^{-\beta}$, as $m \to \infty$, with $0 < \beta < 1$ (here and below, $a_2, a_3, \dots$ denote finite positive constants); (ii) the autocorrelations decay hyperbolically rather than exponentially fast, implying a non-summable autocorrelation function $\sum_k r(k) = \infty$ (*long-range dependence*), i.e., $r(k)$ satisfies relation (1); and (iii) the spectral density $f(\cdot)$ obeys a power-law near the origin ($1/f$*–noise*), i.e., $f(\lambda) \sim a_3 \lambda^{-\gamma}$, as $\lambda \to 0$, with $0 < \gamma < 1$ and $\gamma = 1 - \beta$.

Intuitively, the most striking feature of (exactly or asymptotically) second-order self-similar processes is that their aggregated processes $X^{(m)}$ possess a nondegenerate correlation structure, as $m \to \infty$. This intuition is best illustrated with the sequence of plots in Fig. 4: if X represents the number of Ethernet packets per 10 ms (plot (e)), then plots (d)–(a) depict segments of the time series $mX^{(m)}, m = 10, 100, 1000, 10000$ (i.e., number of Ethernet packets per 0.1, 1, 10, 100 s), respectively. Note that all plots look "similar" and distinctively different from pure noise. The existence of a nondegenerate correlation structure for the processes $X^{(m)}$, as $m \to \infty$, is in stark contrast to typical packet traffic models currently considered in the literature, all of which have the property that their aggregated processes $X^{(m)}$ tend to second-order pure noise, i.e., for all $k \geq 1$,

$$r^{(m)}(k) \to 0, \text{ as } m \to \infty. \tag{4}$$

Equivalently, packet traffic models currently considered in the literature can be characterized by (i) a variance of the sample mean that decreases like the reciprocal of the sample mean, i.e., $\text{var}(X^{(m)}) \sim a_4 m^{-1}$, as $m \to \infty$, (ii) an autocorrelation function that decreases exponentially fast (i.e., $r(k) \sim \rho^k, 0 < \rho < 1$), implying a summable autocorrelation function $\sum_k r(k) < \infty$ (*short-range dependence*), or (iii) a spectral density that is bounded at the origin.

Historically, the importance of self-similar processes lies in the fact that they provide an elegant explanation and interpretation of an empirical law that is commonly referred to the *Hurst effect*. Briefly, for a given set of observations $(X_k : k = 1, 2, \dots, n)$ with sample mean $\overline{X}(n)$ and sample variance $S^2(n)$, the *rescaled adjusted range statistic* (or *R/S statistic*) is given by $R(n)/S(n) = 1/S(n)[\max(0, W_1, W_2, \dots, W_n) - \min(0, W_1, W_2, \dots, W_n)]$, with $W_k = (X_1 + X_2 + \cdots + X_k) - k\overline{X}(n) (k \geq 1)$. While many naturally occurring time series appear to be well represented by the relation $E[R(n)/S(n)] \sim a_5 n^H$, as $n \to \infty$, with *Hurst parameter* H "typically" about 0.7, observations X_k from a short-range dependent model are known to satisfy $E[R(n)/S(n)] \sim a_6 n^{0.5}$, as $n \to \infty$. This discrepancy is generally referred to as the *Hurst effect*.

3.3. Modeling of Self-Similar Phenomena

Since in practice we are always dealing with finite data sets, it is in principle not possible to decide whether the above asymptotic relationships (e.g., (1)–(4)) hold or not. For processes that are not self-similar in the sense that their aggregated series converge to second-order pure noise (see (4)), the correlations will eventually decrease exponentially, continuity of the spectral density function at the origin will eventually show up, the variances of the aggregated processes will eventually decrease as m^{-1}, and the rescaled adjusted range will eventually increase as $n^{0.5}$. For finite sample sizes, distinguishing between these asymptotics and the ones corresponding to self-similar processes is, in general, problematic. In the present context of Ethernet measurements, we typically deal with time series with hundreds of thousands of observations and are therefore able to employ statistical and data analytic techniques that are impractical for small data sets. Moreover, with such sample sizes, parsimonious modeling becomes a necessity due to the large number of parameters needed when trying to fit a conventional process to a "truly" self-similar model. Modeling, for example, long-range dependence with the help of short-range dependent processes is equivalent to approximating a hyperbolically decaying autocorrelation function by a sum of exponentials. Although always possible, the number of parameters needed will tend to infinity as the sample size increases, and giving physically meaningful interpretations for the parameters becomes more and more difficult. In contrast, the long-range dependence component of the process can be modeled (by a self-similar process) with only one parameter. Moreover, from a modeling perspective, it would be very unsatisfactory to use for a single empirical time series two different models, one for a short sequence, another one for a long sequence.

Two formal mathematical models that yield elegant representations of the self-similarity phenomenon but do not provide any physical explanation of self-similarity are *fractional Gaussian noise* and the class of *fractional autoregressive integrated moving-average (ARIMA) processes. Fractional Gaussian noise* $X = (X_k : k \geq 0)$ with parameter $H \in (0,1)$ has been introduced in [22] and is a stationary Gaussian process with mean μ, variance σ^2, and autocorrelation function $r(k) = 1/2(|k+1|^{2H} - |k|^{2H} + |k-1|^{2H}), k > 0$. Simple calculations show that fractional Gaussian noise is exactly second-order self-similar with self-similarity parameter H, as long as $1/2 < H < 1$. Methods for estimating the three unknown parameters μ, σ^2, and H are known and will be addressed below. *Fractional ARIMA(p, d, q) processes* are a natural generalization of the widely used class of Box–Jenkins models [3] by allowing the parameter d to take non-integer values. They were introduced by Granger and Joyeux [10] and Hosking [12] who showed that fractional ARIMA(p, d, q) processes are asymptotically second-order self-similar with self-similarity parameter $d + 1/2$, as long as $0 < d < 1/2$. Fractional ARIMA processes are much more flexible with regard to the simultaneous modeling of the short-term and long-term behavior of a time series than fractional Gaussian noise, mainly because the latter, having only the three parameter μ, σ^2, and H, has a very rigid correlation structure and is not capable of capturing the wide range of low-lag correlation structures encountered in practice. This flexibility can already be observed when considering the simplest processes of the fractional ARIMA(p, d, q) family, namely the two-parameter models ARIMA$(1, d, 0)$ and ARIMA$(0, d, 1)$.

Finally, we briefly mention a construction of self-similar processes (due to Mandelbrot [19] and later extended by Taqqu and Levy [28]), based on aggregating many simple renewal reward processes exhibiting inter-renewal times with infinite variances. Although the construction was originally cast in an economic framework involving commodity prices, it is particularly appealing in the context of high-speed packet traffic, and we will return to this construction in Section V when attempting to provide a "phenomenological" explanation for the observed self-similar nature of aggregate Ethernet traffic. In its simplest form, this construction requires a sequence of i.i.d. integer valued random variables $U_0, U_1, U_2, \ldots$ ("inter renewal times") with "heavy tails," i.e., with the property

$$P[U \geq u] \sim u^{-\alpha} h(u), \text{ as} u \to \infty, \quad (5)$$

where h is slowly varying at infinity and $0 < \alpha < 2$. For example, the stable (Pareto) distribution with parameter $1 < \alpha < 2$ satisfies the "heavy-tail" property (5). Furthermore, let $W_0, W_1, W_2, \ldots$ be an i.i.d. sequence ("rewards") with mean zero and finite variance, independent of the U's. Next, let $S_k = S_0 + \sum_{j=1}^{k} U_j, k \geq 0$ denote the delayed renewal sequence derived from $(U_j)_{j \geq 0}$ where S_0 is chosen such that the sequence $(S_k)_{k \geq 0}$ is stationary. The renewal reward process $W = (W(t) : t = 0, 1, 2, \ldots)$ is then defined by $W(t) = \sum_{k=0}^{t} W_k I_{(S_{k-1}, S_k]}(t)$, with $I_A(\cdot)$ denoting the indicator function of the set A. By aggregating M i.i.d. copies $W^{(1)}, W^{(2)}, \cdots, W^{(M)}$ of W, we obtain the model of interest, namely the process W^* given by $W^*(T, M) = \sum_{t=1}^{T} \sum_{m=1}^{M} W^{(m)}(t)$ with $W^*(0, M) = 0$. In [19] and [28] it is shown that for T and M both large with $T \ll M$, W^* behaves like *fractional Brownian motion*; in other words, properly normalized, $W^*(T, M)$ converges to the integrated version of fractional Gaussian noise, i.e., to a mean-zero Gaussian process $B_H = (B_H(s) : s \geq 0), 1/2 < H < 1$, with correlation function $R(s, t) = 1/2(s^{2H} + t^{2H} - |s - t|^{2H})$. For more details concerning fractional Brownian motion, see [22] and [21]. As an immediate consequence of Taqqu and Levy's result, we have that for T and M both large with $T \ll M$, the increment process of W^* behaves like fractional Gaussian noise.

3.4. Inference for Self-Similar Processes

Since slowly decaying variances, long-range dependence, and a spectral density obeying a power-law are different manifestations of one and the same property of the underlying covariance stationary process X, namely that X is asymptotically or exactly second-order self-similar, we can approach the problem of testing for and estimating the degree of self-similarity from three different angles: (1) time-domain analysis based on the R/S-statistic, (2) analysis of the variances of the aggregated processes $X^{(m)}$, and (3) periodogram-based analysis in the frequency-domain. The following gives a brief description of the corresponding statistical and graphical tools. For an engineering-based graphical tool that is related to the variance property of the aggregated processes, see Section 5.2.

The objective of the R/S analysis of an empirical record is to infer the degree of self-similarity H (Hurst parameter)—via the Hurst effect—for the self-similar process that presumably generated the record under consideration. Graphical R/S analysis consists of taking logarithmically spaced values of n (starting with $n \approx 10$), and plotting $\log(R(n)/S(n))$ versus $\log(n)$ results in the *rescaled adjusted range plot* (also called the *pox diagram of R/S*). When H is well defined, a typical rescaled adjusted range plot starts with a transient zone representing the nature of short-range dependence in the sample, but eventually settles down and fluctuates in a straight "street" of a certain slope. Graphical R/S analysis is used to determine whether such asymptotic behavior appears supported by the data. In the affirmative, an estimate $\hat{H}$ of H is given by the street's asymptotic slope which can take any value between 1/2 and 1. For practical purposes, the most useful and attractive feature of the R/S analysis is its relative robustness against changes of the marginal distribution. This feature allows for practically separate investigations of the self-similarity property of a given data set and of its distributional characteristics.

We have observed that for second-order self-similar processes, the variances of the aggregated processes $X^{(m)}, m \geq 1$, decrease linearly (for large m) in log-log plots against m with slopes arbitrarily flatter than -1. The so-called *variance-time plots* are obtained by plotting $\log(\text{var}(X^{(m)}))$ against $\log(m)$ ("time") and by fitting a simple least squares line through the resulting points in the plane, ignoring the small values for m. Values of the estimate $\hat{\beta}$ of the asymptotic slope between -1 and 0 suggest self-similarity, and an estimate for the degree of self-similarity is given by $\hat{H} = 1 - \hat{\beta}/2$.

The absence of any limit law results for the statistics corresponding to the R/S analysis or the variance-time plot makes them inadequate for a more refined data analysis (e.g., confidence intervals for H). In contrast, a more refined data analysis is possible for maximum likelihood-type estimates (MLE) and related methods based on the *periodogram* $I(x) = (2\pi n)^{-1}|\sum_{j=1}^{n} X_j e^{ijx}|^2, 0 \leq x \leq \pi$ of $X = (X_1, X_2, \ldots, X_n)$ and its distributional properties. In particular, for Gaussian or approximately Gaussian processes, Whittle's approximate MLE has been studied extensively and has been shown to have desirable statistical properties. Combined, Whittle's approximate MLE approach and the aggregation method discussed earlier give rise to an operational procedure for obtaining confidence intervals for the self-similarity parameter H. Briefly, for a given time series, consider the corresponding aggregated processes $X^{(m)}$ with $m = 100, 200, 300, \ldots$. For each of the aggregated series, estimate the self-similarity parameter $H^{(m)}$ via Whittle's method. This procedure results in point estimates $\hat{H}^{(m)}$ of $H^{(m)}$ and corresponding 95%-confidence intervals of the form $\hat{H}^{(m)} \pm 1.96\hat{\sigma}_{H^{(m)}}$, where $\hat{\sigma}^2_{H^{(m)}}$ is given by a known central limit theorem result (for references, see [17]). Plots of $\hat{H}^{(m)}$ (together with their 95%-confidence intervals) versus m will typically vary for small aggregation levels, but will stabilize after a while and fluctuate around a constant value, our final estimate of the self-similarity parameter H.

IV. Ethernet Traffic Is Self-similar

While Fig. 4 gives a pictorial "proof" of the self-similar nature of the traffic measurements described in Section II, using the statistical and graphical tools presented above, we establish in this section the self-similar nature of Ethernet traffic (and some of its major components, such as external traffic or external TCP traffic) in a statistically more rigorous manner. For each of the four measurement periods described in Table I, we identified typical low-, medium-, and high-activity hours. With the resulting data sets, we are able to investigate features of the observed traffic that persist across the network as well as across time, irrespective of the utilization level of the Ethernet. Only one LAN could be monitored at any one time (making it impossible to study correlations in the activity on different LAN's) and all data were collected from LAN's in the same company (making it not representative for all LAN traffic). For a similar analysis that uses different data sets from Table I, see [16].

4.1. Ethernet Traffic over a 27-Hour Period

In order to check for the possible self-similarity of the August 1989 Ethernet traffic data, we apply the graphical tools described in the previous section, namely, variance-time plots, pox plots of R/S, and periodogram plots, to the three subsets AUG89.LB, AUG89.MB, and AUG89.HB of the August '89 trace that correspond to a typical "low hour," "normal hour," and "busy hour" traffic scenario, respectively (see Table I). Each sequence contains 360 000 observations, and each observation represents the number of bytes sent over the Ethernet per 10 ms. As an illustration of the usefulness of the graphical tools for detecting self-similarity in an empirical record, Fig. 5 depicts the variance-time curve (a), the pox plot of R/S (b), and the periodogram plot (c) corresponding to the sequence AUG89.MB. The variance-time curve, which has been normalized by the corresponding sample variance, shows an asymptotic slope that is distinctly different from -1 (dotted line) and is easily estimated to be about $-.40$, resulting in an estimate $\hat{H}$ of the Hurst parameter H of about $\hat{H} \approx .80$. Estimating the Hurst parameter directly from the corresponding pox plot of R/S leads to a practically identical estimate; the value of the asymptotic slope of the R/S plot is clearly between 1/2 and 1 (lower and upper dotted line, respectively), with a simple least-squares fit resulting in $\hat{H} \approx .79$. Finally, looking at the periodogram plot, we observe that although there are some pronounced peaks in the high-frequency domain of the periodogram, the low-frequency part is characteristic for a power-law behavior of the spectral density around zero. In fact, by fitting a simple least-squares line using only the lowest 10% of all frequencies, we obtain a slope estimate $\hat{\gamma} \approx .64$ which results in a Hurst parameter estimate $\hat{H}$ of about .82. Thus, together the three graphical methods suggest that the sequence AUG89.MB is self-similar with self-similarity parameter $H \approx .80$. Moreover, Fig. 5(d) indicates that the normal hour Ethernet traffic of the August 1989 data is, for practical purposes, exactly self-similar: it shows the estimates of the Hurst parameter H for selected aggregated time series derived from the sequence AUG89.MB, as a function of the aggregation level m. For aggregation levels $m = 1, 5, 10, 50, 100, 500, 1000$, we plot the Hurst parameter estimate $\hat{H}^{(m)}$ (based on the pox plots of R/S ("$\star$"), the variance-time curves ("$\circ$"), and the periodogram plots ("$\square$")) for the aggregated time series $X^{(m)}$ against the logarithm of the aggregation level m. Notice that the estimates are extremely stable and practically constant over the depicted range of aggregation levels $1 \leq m \leq 1000$. Because the range includes small values of m, the sequence AUG89.MB

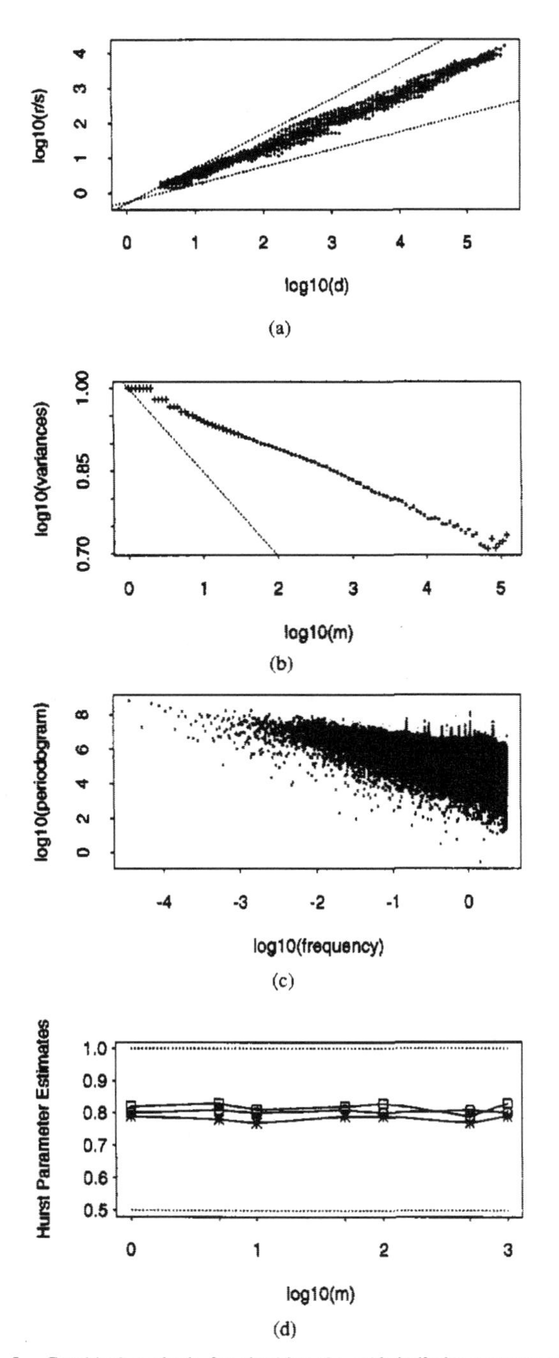

Fig. 5. Graphical methods for checking the self-similarity property of the sequence AUG89.MB.

can be regarded as exactly self-similar. Similar results are obtained for the sequences AUG89.LB and AUG89.HB, and for the corresponding packet count processes AUG89.LP, AUG89.MP, and AUG89.HP. Together, these observations show that Ethernet traffic over approximately a 24-hour period is self-similar, with the degree of self-similarity increasing as the utilization of the Ethernet increases.

4.2. Ethernet Traffic Over a Four-Year Period

In order to examine in detail the nature of Ethernet traffic across time as well as across the network under consideration, we now consider the remaining data sets described in Table I. In contrast to Section 4.1, our analysis below results in estimates of the self-similarity parameter H together with their respective 95%-confidence intervals. As discussed in Section 3.4, such a refined analysis is possible if maximum likelihood type estimates (MLE) or related estimates based on the periodogram are used instead of the mostly heuristic graphical estimation methods illustrated in the previous section. Plots (a)–(d) of Fig. 6 show the result of the MLE-based estimation method when combined with the method of aggregation. For each of the four sets of traffic measurements described in Table I, we use the time series representing the packet counts during normal traffic conditions (i.e., AUG89.MP in Fig. 6(a), OCT89.MP in (b), JAN90.MP in (c), and FEB92.MP in (d)), and consider the corresponding aggregated time series $X^{(m)}$ with $m = 100, 200, 300, \cdots, 1900, 2000$ (representing the packet counts per $1, 2, \cdots, 19, 20$ s, respectively). We plot the Hurst parameter estimates $\hat{H}^{(m)}$ of $H^{(m)}$ obtained from the aggregated series $X^{(m)}$, together with their 95%-confidence intervals, against the aggregation level m. Fig. 6 shows that for the packet counts during normal traffic loads (irrespective of the measurement period), the values of $\hat{H}^{(m)}$ are quite stable and fluctuate only slightly in the 0.85 to 0.95 range throughout the aggregation levels considered. The same holds for the 95%-confidence interval bands indicating strong statistical evidence for self-similarity of these four time series with degrees of self-similarity ranging from about 0.85 to about 0.95. The relatively stable behavior of the estimates $\hat{H}^{(m)}$ for the different aggregation levels m also confirms our earlier finding that Ethernet traffic during normal traffic hours can be considered to be exactly self-similar rather than asymptotically self-similar. For exactly self-similar time series, determining a single point estimate for H and the corresponding 95%-confidence interval is straightforward and can be done by visual inspection of plots such as the ones in Fig. 6 (see below). Notice that in each of the four plots in Fig. 6, we added two lines corresponding to the Hurst parameter estimates obtained from the pox diagrams of R/S and the variance-time plots, respectively. Typically, these lines fall well within the 95%-confidence interval bands which confirms our earlier argument that for these long time series considered here, graphical estimation methods based on R/S or variance-time plots can be expected to be very accurate.

In addition to the four normal hour packet data time series, we also appliedthe combined MLE/aggregation method to the other traffic data sets described in Table I. Fig. 7(a) depicts all Hurst parameter estimates (together with the 95%-confidence interval corresponding to the choice of m discussed earlier) for each of the 12 packet data time series, while Fig. 7(b) summarizes the same information for the time series representing the number of bytes. We also include in these summary plots the Hurst parameter estimates obtained via the variance-time plots ("∘") and R/S analysis ("⋆") in order to indicate the accuracy of these essentially heuristic estimators when compared to the statistically more rigorous Whittle estimator ("•").

Concentrating first on the packet data, i.e., Fig. 7(a),we see that despite the transition from mostly host-to-host workgroup traffic during the August 1989 and October 1989 measurement

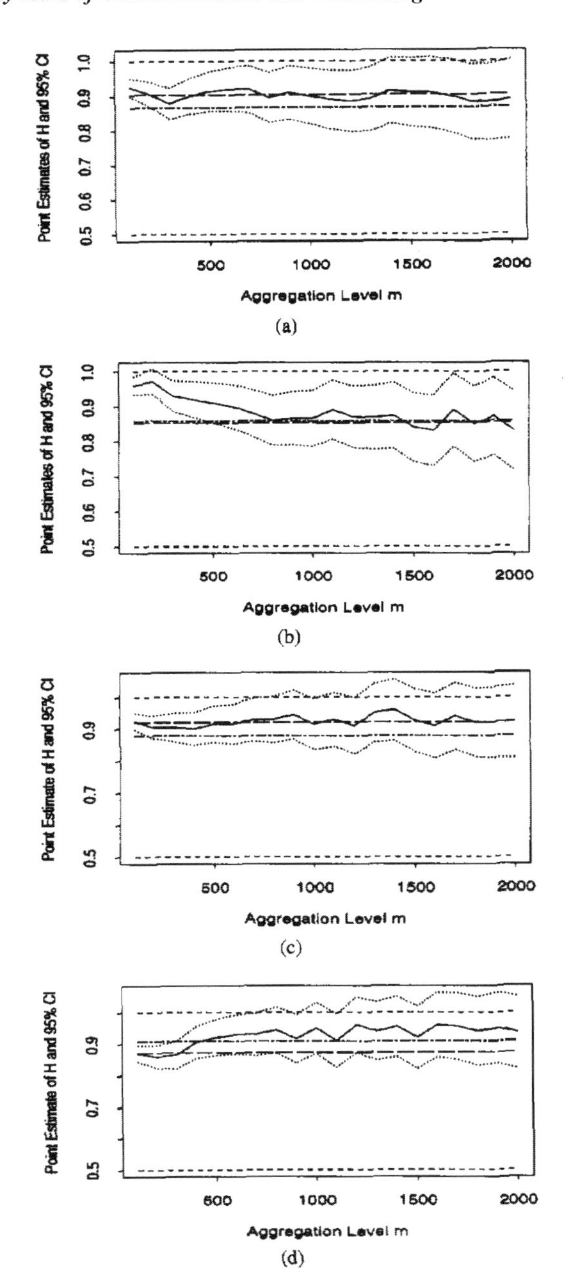

Fig. 6. Periodogram-based MLE/aggregation method for the sequences AUG89.MP, OCT89.MP, JAN90.MP, and FEB92.MP.

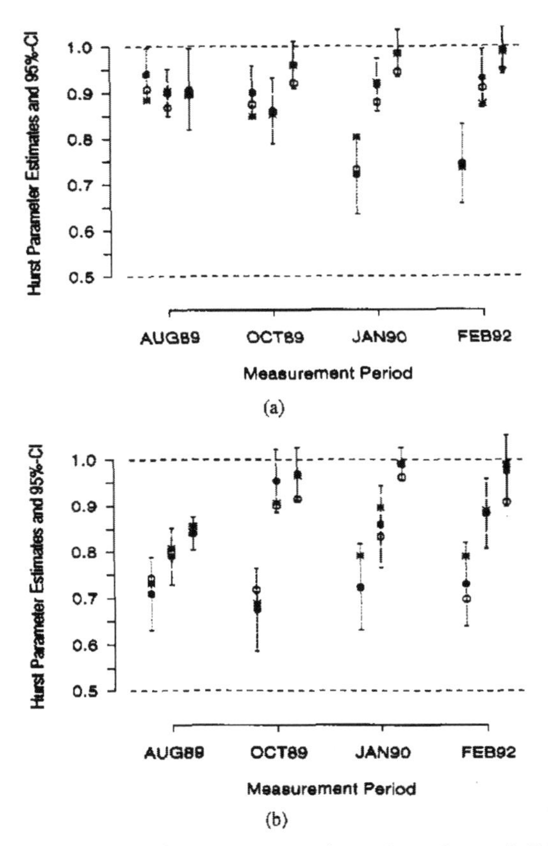

Fig. 7. Summary plot of Hurst parameter estimates for all data sets in Table 1.

periods, to a mixture of host-to-host and router-to-router traffic during the January 1990 measurement period, to the predominantly router-to-router traffic of the February 1992 data set, the Hurst parameter corresponding to the typical normal and busy hours, respectively, are comparable, with slightly higher H-values for the busy hours than for the normal traffic hours. This latter observation might be surprising in light of conventional traffic modeling where it is commonly assumed that as the number of sources (Ethernet users) increases, the resulting aggregate traffic becomes smoother and smoother. In contrast to this generally accepted argument for the "Poisson-like" nature of aggregate traffic, our analysis of the Ethernet data shows that, in fact, the aggregate traffic tends to become less smooth (or, more bursty) as the number of active sources increases (see also our discussion in Section 5.1). While there were about 120 hosts that spoke up during the August 1989 or October 1989 busy hour, we heard from an order of magnitude more hosts (about 1200) during the January 1990 high traffic hour; the comparable number of active hosts during the February '92 busy hour was around 600. The major difference between the early (pre-1990) measurements and the later ones (post-1990) can be seen during the low traffic hours. Intuitively, low period router-to-router traffic consists mostly of machine-generated packets which tend to form a much smoother arrival process than low period host-to-host traffic which is typically produced by a smaller than average number of actual Ethernet users, e.g., researchers working late hours. Next, turning our attention to Fig. 7(b), we observe that as in the case of the packet data, H increases as we move from low to normal to high traffic hours. Moreover, while there is practically no difference between the two post-1990 data sets, the two pre-1990 sets clearly differ from one another but follow a similar pattern as the post-1990 ones. The difference between the August 1989 and October 1989 measurements can be explained by the transition from diskless to "dataless" workstations that occurred during the latter part of 1989 (see Section 2.2). Except during the low hours, the increased computing power of many of the Ethernet hosts causes H to increase and gives rise to a bit rate that closely matches the self-similar feature of the corresponding packet process. Also note that the 95%-confidence intervals corresponding to the Hurst parameter estimates for the low traffic hours are typically wider than those corresponding to the estimates of H for the normal and high traffic hours. This widening indicates

TABLE II
QUALITATIVE DESCRIPTION OF THE SETS OF EXTERNAL ETHERNET TRAFFIC MEASUREMENTS USED IN THE ANALYSIS IN SECTION 4.3

Traces of Ethernet Traffic Measurements					
Measurement Period	Internal Traffic Data (see Table I)	Data Set	Total Number of Bytes	Total Number of Packets	Percentage of Internal Traffic
JANUARY 1990 Start of Trace: Jan. 10, 6:07 am End of Trace: Jan. 11, 10:17 pm	JAN90.LB	JAN90E.LB	1 105 876		1.27%
	JAN90.LP	JAN90E.LP		9369	3.02%
	JAN90.MB	JAN90E.MB	16 536 148		9.05%
	JAN90.MP	JAN90E.MP		87 307	13.57%
	JAN90.HB	JAN90E.HB	13 023 016		2.00%
	JAN90.HP	JAN90E.HP		68 405	4.96%
FEBRUARY 1992 Start of Trace: Feb. 18, 5:22 am End of Trace: Feb. 20, 5:16 am	FEB92.LB	FEB92E.LB	2 319 881		4.08%
	FEB92.LP	FEB92E.LP		25 247	10.89%
	FEB92.MB	FEB92E.MB	86 283 283		55.80%
	FEB92.MP	FEB92E.MP		270 636	51.60%
	FEB92.HB	FEB92E.HB	55 154 789		24.50%
	FEB92.HP	FEB92E.HP		202 367	21.35%

that Ethernet traffic during low traffic periods is asymptotically self-similar rather than exactly self-similar.

We also notice in Fig. 7 that some of the analyzed time series result in estimated Hurst parameters close to 1, i.e., their corresponding 95%-confidence intervals include the value $H = 1$. When finding an H-estimate close to 1, it is advisable to analyze the time series further to ensure that the observed high degree of self-similarity is genuine and cannot be explained by elementary arguments (see for example [21]). To illustrate, we consider the sequences JAN90.HP and FEB92.HP; visual inspection of both time series and comparisons with traces of fractional Gaussian noise with $H \approx 0.9$ (see, for example, the plots in [23] and [21]) show no obvious signs of non-stationarity; the mean seems to be changing with time but the overall mean appears constant and although, locally, there clearly exist spurious trends and cycles of varying frequencies, these "typical" features of nonstationarity are characteristic of stationary long-range dependent processes. Moreover, the variance-time plots as well as the pox diagrams of the adjusted range R (without rescaling by S) of the two time series yield slope estimates (not shown) that are consistent with the observed high H-values. As discussed in [2] this consistency is a strong indication that the given time series cannot be regarded as nonstationary due to a lack of differencing. Further tests for non-stationarity (e.g., due to nonhomogeneities of H) can be found in [17].

4.3. External Ethernet Traffic

The Ethernet traffic analyzed so far is also called *internal* traffic and consists of all packets on a LAN. An important component of internal Ethernet traffic is the so-called *remote* or *external* Ethernet traffic, consisting of all those Ethernet packets that originate on one LAN but are routed to another LAN. That is, for the traffic measurements at hand, an *external* packet is defined to be an IP (Internet protocol) packet with a source or destination address that is not on any of the Bellcore networks. This external traffic can be viewed as representative for LAN interconnection services, which are expected to contribute significantly to future broadband traffic.

Table II summarizes the external Ethernet traffic data analyzed in the process of this study. We consider the two most recent measurement traces i.e., the January 1990 and February 1992 data sets, and for ease of comparison, we analyze for both measurement periods the time series consisting of the number of external packets (bytes) per 10 ms during the same low-, normal-, and high-hours of (internal) Ethernet traffic as considered in Table I. The last column in Table II shows that external traffic (in terms of packets or bytes) makes up between 1–10% of the internal traffic during the low hours in January 1990 and February 1992, about 2–25% during the corresponding busy hours, and up to 56% during the February 1992 normal hour. As a result, it is reasonable to expect external traffic to behave very similarly to the overall traffic analyzed earlier in this section. Differences (if any) between the internal and external traffic can, in general, be attributed to NFS traffic between workstations and file servers which is missing completely in the external traffic.

Repeating the same laborious analysis of Section 4.2 for the data sets described in Table II, we find that in terms of its self-similar nature, external traffic does not differ from the internal traffic studied earlier. More specifically, the Hurst parameters for the external traffic during normal and high (internal) traffic hours (or during previously identified stationary parts of the corresponding data sets) are only slightly smaller than the ones depicted in Fig. 7. For instance, even though the portion of external packets during the high (internal) traffic hour of the January 1990 data is only 2% of all the packets seen during this period, the data set JAN90E.HP seems to be well described by an H-value that changes from $H = 0.82$ for the first 30 min to $H = 0.94$ for the second 30 min; recall that the corresponding data set of internal traffic, i.e., the sequence JAN90.HP, has an estimated Hurst parameter of 0.98. A more significant change in the Hurst parameter occurs during the low traffic hours. While the internal traffic data (JAN90.LB, JAN90.LP, FEB92.LB, and FEB92.LP) yield a Hurst parameter of about 0.70, the sequences JAN90E.LB, JAN90E.LP, FEB92E.LB, and FEB92E.LP have $H \approx 0.55$, and the corresponding 95 intervals contain the value $H = 0.5$. These are the only cases in all the data sets considered in this paper, where an H-value of 0.5 (i.e., conventionally used short-range dependent models such as Poisson, batch-Poisson, or Markov-Modulated Poisson Processes) seems to describe the data accurately. For all other data sets described in Tables I and II, the 95%-confidence intervals for the Hurst parameter estimates do not even come close to covering the value $H = 0.5$. As already mentioned in our discussion of Fig. 7, the low hour traffic in the January 1990 and February 1992 data is mostly machine-generated and produces traffic that is typically smoother (i.e., less bursty) than traffic that is generated during the normal and busy hours by humans using their workstations. This argument applies even more when considering low hour external traffic.

We also looked at the portion of external traffic using the Transmission Control Protocol (TCP) and IP. There were two main reasons for this. First, the traditional services offered by the Internet are for the most part based around TCP, which offers reliable delivery of data and protection against data loss due to lost or corrupted packets. These services include remote login, file transfer (including anonymous file transfer for making information and programs publicly available to any Internet user), electronic mail, and more recently the delivery of the electronic bulletin board known as Netnews. The second reason is that application programs using the TCP protocol have significantly *less* control over how their data is actually sent than do applications using the User Datagram Protocol (UDP) or their own protocol. The TCP protocol has significant control over how the user data is segmented and a great deal of control over the spacing of the packets as they are sent out. When investigating the external TCP traffic, we found that there was little point in doing a separate analysis. For instance, in the heavy traffic hour from the MRE backbone taken in 1992 (FEB92E.HP), 87% of the packets were TCP packets, and a plot of the external TCP traffic is practically indistinguishable from the corresponding plot of the entire external traffic. Of those TCP packets of the FEB92E.HP data set, about 66% of the packets were for file transfer, 9% for remote login/TELNET, 11% for electronic mail, and 13% for netnews delivery. The 12% of non-TCP traffic simply had no effect on the results of our analysis for this data set; external TCP traffic is practically identical to the external traffic, and our findings for the external traffic apply directly to external TCP traffic.

V. Engineering for Self-similar Network Traffic

The fact that one can distinguish clearly—with respect to second-order statistical properties—between the existing models for Ethernet traffic and our measured data is surprising and clearly challenges some of the modeling assumptions that have been made in the past. While this distinction is obvious from a statistical perspective, potential traffic engineering implications of this distinction are currently under intense scrutiny. Below, we concentrate on three implications of self-similar network traffic for traffic engineering purposes: modeling individual sources such as Ethernet hosts, inadequacy of conventional notions of "burstiness," and the generation of synthetic traces of self-similar traffic. For a simulation study of the effects of self-similar packet traffic on congestion control and management for B-ISDN, we refer to [7].

5.1. On the Nature of Traffic Generated by Individual Ethernet Hosts

In Section IV, we showed that irrespective of when and where the Ethernet measurements were collected, the traffic is self-similar, with different degrees of self-similarity depending on the load on the network. We did so without first studying and modeling the behavior of individual Ethernet users (sources). Although historically, accurate source modeling has been considered a prerequisite for successful modeling of *aggregate* traffic, we show here that in the case of self-similar packet traffic, knowledge of fundamental characteristics of the aggregate traffic can provide new insight into the nature of traffic generated by an individual user. Thus, in this section we attempt to give a phenomenological explanation for the visually obvious (see Fig. 4) and statistically significant (see Fig. 7) self-similarity property of aggregate Ethernet LAN traffic in terms of the behavior of individual Ethernet users.

To this end, we recall Mandelbrot's construction of fractional Brownian motion (see Section 3.3) and interpret the renewal reward process $W^{(m)} = (W^{(m)}(t) : t = 0, 1, 2, \ldots)$ introduced in Section 3.3 as the amount of information (in bits, bytes, or packets) generated by Ethernet host m at time t $(1 \leq m \leq M, t \geq 0)$. In fact, if bits or bytes are the preferred units, the renewal reward process source model resembles the popular class of fluid models (see [1]). On the other hand, if we think of packets as the underlying unit of information, the renewal reward process is basically a packet train model in the sense of [13]. For ease of presentation, we can assume that the "rewards" $W_0, W_1, W_2, \ldots$ take only the values 1 and 0 (or, to keep $E[W] = 0$, +1 and −1), with equal probabilities, where the value 1/0 during a renewal interval indicates an active/inactive period during which the source sends 1/0 unit(s) of information every time unit. The crucial property that distinguishes the renewal reward process source model from the above mentioned models is that the inter-renewal intervals (i.e., the lengths of the active/inactive periods) are *heavy-tailed* in the sense of (5) or, using Mandelbrot's terminology, exhibit the *infinite variance syndrome*. Intuitively, (5) states that with relatively high probability, the active/inactive periods are very long, i.e., each W_m can assume the same value for a long period of time. While this heavy-tailed property of the active/inactive periods seems plausible in light of the way a typical workstation user contributes to the overall traffic on the Ethernet, we have not yet analyzed the traffic generated by individual Ethernet users in order to validate the simple renewal reward source model assumption.

However, evidence in support of the infinite variance syndrome in packet traffic measurements already exists. For example, in a recent study of traffic measurements from an ISDN office automation application, Meier-Hellstern *et al.* [24] observed that the extreme variability in the data (e.g., interarrival times of packets, number of successive packet arrivals in certain states) cannot be adequately captured using traditional packet traffic models but, instead, seems to be best described with the help of heavy-tailed distributions of the form (5). These authors subsequently propose an elaborate and highly parameterized model for the measured traffic. In contrast, the renewal reward source model for the traffic generated by an individual workstation user is extremely simple; moreover, we have seen in Section 3.3 that when aggregating the traffic of many such source models, the resulting superposition process is a fractional Brownian motion with self-similarity parameter $H = (3 - \alpha)/2$, where α is given in (5), and that the time series representing, for example, the total number of bytes or Ethernet packets every 10 ms, behaves like fractional Gaussian noise with the same H-value. In this sense, our analysis in Section IV suggests that a simple renewal reward process is an adequate traffic source model for an individual Ethernet user and that often, a more detailed source modeling might not be needed since the convergence result in Section 3.3 shows that many of the details disappear during the process of aggregating the traffic of many sources and only property (5) is required for the fractional Brownian motion behavior of the superposition process to hold. Note that we have reached this conclusion by treating the Ethernet packets essentially as black boxes, i.e., we did not look into the packet header fields or distinguish packets based on their source or destination. Further work on extracting the relevant source-destination addresses from our measurements and on statistically validating the infinite variance property of the inter-renewal periods of a single source is currently in progress.

5.2. *On Measuring "Burstiness" for Self-Similar Network Traffic*

On an intuitive level, the results of our statistical analysis of the Ethernet traffic measurements in Section IV can be summarized by saying that typically, the higher the load on the Ethernet the higher the estimated Hurst parameter H, i.e., the degree of self-similarity in the arrival rate process (in terms of packets or bytes). Visual comparisons between the different traces also suggest that the larger H, the "burstier" the corresponding trace appears. Trying to capture the intuitive notion of "burstiness" with the help of the Hurst parameter H becomes particularly appealing in light of the relation $H = (3 - \alpha)/2$ mentioned in the previous section between the self-similarity parameter H and the parameter α that characterizes the "thickness" (see (5)) of the tail of the inter-renewal time distribution (i.e., of the lengths of the active/inactive periods). Clearly, the heavier the tail in (5) (i.e., the closer α gets to 1), the greater the variability of the active/inactive periods and hence, the burstier the traffic generated by an individual source. Going from α to H relates burstiness of an individual source to burstiness of the aggregate traffic: the higher the H, the burstier the aggregate traffic. The fact that the Hurst parameter H seems to capture the intuitive notion of burstiness through the concept of self-similarity and, at the same time, also seems to agree well with the visual assessment of bursty behavior challenges the feasibility of some of the most commonly used measures of "burstiness." The latter include the *index of dispersion (for counts)*, the *peak-to-mean ratio*, and the *coefficient of variation (of inter-renewal times)*.

A commonly used measure for capturing the variability of traffic over different time scales is provided by the *index of dispersion (for counts)* and has recently attracted considerable attention (see for example [11]). For a given time interval of length L, the index of dispersion for counts (IDC) is given by the variance of the number of arrivals during the interval of length L divided by the expected value of that same quantity. Fig. 8 depicts the IDC as a function of L in log-log

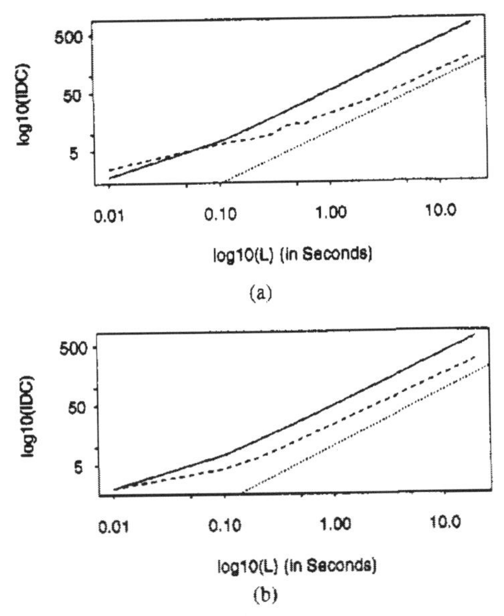

Fig. 8. Index of dispersion for counts (IDC) as a function of the length L of the time interval over which the IDC is calculated, for the high traffic hours of the January 1990 and February 1992 data sets.

coordinates; it shows the IDC for both internal (solid lines) and external (dashed lines) traffic from the high traffic hour of the January 1990 (Fig. 8(a)) and February 1992 data (b).

Note in particular that the IDC increases monotonically throughout a time span that covers 4–5 orders of magnitude. This behavior is in stark contrast to conventional traffic models such as Poisson or Poisson-like processes and the popular Markov-modulated Poisson processes where the IDC is either constant or converges to a fixed value quite rapidly. On the other hand, self-similar traffic models are easily shown to produce a monotonically increasing IDC. In fact, assume for simplicity that the process X representing the total number of packets seen in every 10 ms interval, is fractional Gaussian noise (with positive drift) with self-similarity parameter H. Then we have $IDC(L) = \mathrm{var}(\sum_{j=1}^{j=L} X_j)/E[\sum_{j=1}^{j=L} X_j] \sim cL^{2H-1}$ (where c is a finite positive constant that does not depend on L), and plotting $\log(IDC(L))$ against $\log(L)$ results in an asymptotic straight line with slope $2H - 1$. The dotted lines in Figure 5.1 represent the IDC curves predicted by self-similar traffic models with $H \approx 0.94$ (JAN90.HP) and $H \approx 0.96$ (FEB92.HP), respectively. Similarly striking agreement between the empirical and theoretical IDC curves can be observed for the corresponding external traffic data sets. Notice that plotting the IDC curve and estimating its slope provides a quick and simple engineering-based approach to testing for self-similarity of a set of traffic measurements.

Leland and Wilson [14] have pointed out the problem with using the *peak-to-mean ratio* as a measure for "burstiness" in the presence of self-similar traffic. The observed ratio of peak bandwidth (i.e., peak arrival rate of, say, bytes) to mean bandwidth depends critically on the time interval over which the peak and mean bandwidth is determined, i.e., essentially any peak-to-mean ratio is possible, depending on the length of the measurement interval. For a two-week long trace of the October 1989 measurements, they show that the peak rate in bytes for the external traffic observed in any 5 s interval is about 150 times the mean arrival rate, while the peak rate observed in any 5 ms interval is about 710 times the mean. The dependence of this burstiness measure on the choice of the time interval is clearly undesirable.

Finally, we remark that the use of the *coefficient of variation* (for interarrival times), i.e., the ratio of the standard deviation of the interarrival time to the expected number of the interarrival time, as a measure of "burstiness" becomes questionable because of the potential "heavy-tailedness" (in the sense of (5)) of the interarrival times and the implied infinite variance property. Although the empirical standard deviation can always be calculated, it will depend crucially on the sample size and can attain practically any value as the sample size increases.

5.3. On Generating Synthetic Traces of Self-Similar Traffic

As we have noted in Section IV, exactly self-similar models such as fractional Gaussian noise, or some nonlinear transformation of fractional Gaussian noise (in order to ensure for example that the process takes only positive values) or asymptotically self-similar models such as fractional ARIMA processes can be used to fit hour-long traces of Ethernet traffic very well. Parameter estimation techniques for these models are known but they often turn out to be computationally too intensive in order to work for large data sets. However, we have illustrated in Section IV how to estimate the Hurst parameter H for large data sets, and methods to adapt the existing parameter estimation techniques and to apply them to long time series are currently being studied (for references, see [17]). Notice also that our analysis of the measured data has shown that the Hurst parameter can be expected to change during a measurement period of an hour or more and that refinements such as modeling the change points of H may be needed in the future in order to produce more realistic traffic models. For other approaches to modeling self-similar packet traffic, see the recent articles by Erramilli and Singh [6] who use deterministic nonlinear chaotic maps in order to mimic the fractal-like properties of Ethernet traffic, and Veitch [29] whose work is motivated by the early paper of Mandelbrot [18].

An important requirement of practical traffic modeling is to generate synthetic data sequences that exhibit similar features as the measured traffic. While exact methods for generating synthetic traces from fractional Gaussian noise and fractional ARIMA models exist (see for example [12]), they are, in general, only appropriate for short traces (about 1000 observations). For longer time series, short memory approximations have been proposed such as the *fast fractional Gaussian noise* by Mandelbrot [20]. However, such approximations also become often inappropriate when the sample size becomes exceedingly large. Here, we briefly discuss two methods for generating asymptotically self-similar observations. The first method simulates the buffer occupancy in an $M/G/\infty$ queue, where the service time distribution G satisfies the heavy-tail condition (5), i.e., G has infinite variance. Cox [4] showed that an infinite variance service time distribution results in an asymptotically self-similar buffer occupancy process, and he relates the tail-behavior of the former to the degree of self-similarity of the latter. Generating a time series of length 100 000 this way requires about 2 h of CPU-time on a Sun SPARCstation 2. The second method exploits a convergence result obtained by Granger [9] who showed that when aggregating many simple AR(1)-processes, where the AR(1) parameters are chosen from a beta-distribution on $[0, 1]$ with shape parameters p and q, then the superposition process is asymptotically self-similar; Granger also showed that the Hurst parameter H depends linearly on the shape parameter q of the beta-distribution. This method is well-suited for parallel computers, and producing a synthetic trace of length 100 000 on a MasPar MP-1216, a massively parallel computer with 16 384 processors, takes only a few minutes. In contrast, Hosking's method to produce 100 000 observations from a fractional ARIMA$(0, d, 0)$ model requires about 10 h of CPU time on a Sun SPARCstation 2. Implementations of and experimentations with these and some other methods are currently under way.

VI. Discussion

Understanding the nature of traffic in high-speed, high-bandwidth communications systems such as B-ISDN is essential for engineering, operations, and performance evaluation of these networks. In a first step toward this goal, it is important to know the traffic behavior of some of the expected major contributors to future high-speed network traffic. In this paper, we analyze LAN traffic offered to a high-speed public network supporting LAN interconnection, an important and rapidly growing B-ISDN service. The main findings of our statistical analysis of hundreds of millions of high quality, high time-resolution Ethernet LAN traffic measurements are that (i) Ethernet LAN traffic is statistically self-similar, irrespective of when during the four-year data collection period 1989–1992 the data were collected and where they were collected in the network, (ii) the degree of self-similarity measured in terms of the Hurst parameter H is typically a function of the overall utilization of the Ethernet and can be used for measuring the "burstiness" of the traffic (namely, the burstier the traffic the higher H), (iii) major components of Ethernet LAN traffic such as external LAN traffic or external TCP traffic share the same self-similar characteristics as the overall LAN traffic, and (iv) the packet traffic models currently considered in the literature are not able to capture the self-similarity property and can therefore be clearly distinguished from our measured data.

For the purpose of modeling this self-similar or fractal-like nature of the Ethernet traffic data, we introduce novel methods based on self-similar stochastic processes. The motivation for these methods is the desire for an accurate and relatively simple (i.e., parsimonious) description of the complex packet traffic generation process. These modeling approaches typically yield a single parameter (i.e., the Hurst parameter) that describes the fractal nature of the measured traffic and appears to capture the intuitive notion of "burstiness" where conventional measures of burstiness no longer apply. From the point of view of queueing/performance analysis, the proposed modeling approaches pose new and challenging problems which are likely to require new sets of mathematical tools. Ultimately, in the context of traffic engineering, it is the predicted performance of appropriately chosen queueing systems that will decide the relevance of self-similar traffic models.

However, indications of the impact of the self-similar nature of packet traffic for engineering, operations, and performance evaluation of high-speed networks are already ample: (i) source models for individual Ethernet users are expected to show extreme variability in terms of interarrival times of packets (i.e., the infinite variance syndrome), (ii) commonly used measures for "burstiness" such as the index of dispersion (for counts), the peak-to-mean-ratio, or the coefficient of variation (for interarrival times) are no longer meaningful for self-similar traffic but can be replaced by the Hurst parameter, (iii) the nature of congestion produced by self-similar network traffic models differs drastically from that predicted by standard formal models and displays a far more complicated picture than has been typically assumed in the past, and (iv) first analytic results show a clear distinction between predicted performance of certain queueing models with traditional input streams and the same queueing models with self-similar inputs (see for example [25] and [5]). Finally, in light of the same fractal-like behavior recently observed in VBR video traffic (see [2] and [8])—another major contributor to future high-speed network traffic—the more complicated nature of congestion due to the self-similar traffic behavior can be expected to persist even when we move toward a more heterogeneous B-ISDN environment. Thus, we believe based on our measured traffic data that the success or failure of, for example, a proposed congestion control scheme for B-ISDN will depend on how well it performs under a self-similar rather than under one of the standard formal traffic scenarios.

ACKNOWLEDGMENT

This work could not have been done without the help of J. Beran and R. Sherman who provided the S-functions that made the statistical analysis of an abundance of data possible. The authors also acknowledge many helpful discussions with A. Erramilli about his dynamical systems approach to packet traffic modeling.

REFERENCES

[1] D. Anick, D. Mitra, and M. M. Sondhi, "Stochastic theory of a data-handling system with multiple sources," *Bell System Tech. J.* vol. 61, pp. 1871–1894, 1982.
[2] J. Beran, R. Sherman, M. S. Taqqu, and W. Willinger, "Variable-bit-rate video traffic and long-range dependence," accepted for publication in *IEEE Trans. Commun.*, 1993.
[3] G. E. P. Box and G. M. Jenkins, *Time Series Analysis: Forecasting and Control*, 2nd ed. San Francisco, CA: Holden Day, 1976.
[4] D. R. Cox, "Long-range dependence: A review," in *Statistics: An Appraisal*, H. A. David and H. T. David, eds. Ames, IA: The Iowa State University Press, 1984, pp. 55–74.
[5] N. G. Duffield and N. O'Connell, "Large deviations and overflow probabilities for the general single-server queue, with applications," preprint, 1993.
[6] A. Erramilli and R. P. Singh, "Chaotic maps as models of packet traffic," in *Proc. 14th ITC*, Antibes Juan-les-Pins, France, 1994 (to appear).
[7] H. J. Fowler and W. E. Leland, "Local area network traffic characteristics, with implications for broadband network congestion management," *IEEE J. Select. Areas Commun.* vol. 9, pp. 1139–1149, 1991.
[8] M. W. Garrett and W. Willinger, "Analysis, modelling, and generation of self-similar VBR video traffic," preprint, 1994.
[9] C. W. J. Granger, "Long memory relationships and the aggregation of dynamic models," *J. Econometr.* vol. 14, pp. 227–238, 1980.
[10] C. W. J. Granger and R. Joyeux, "An introduction to long-memory time series models and fractional differencing," *J. Time Series Anal.* vol. 1, pp. 15–29, 1980.
[11] H. Heffes and D. M. Lucantoni, "A Markov modulated characterization of packetized voice and data traffic and related statistical multiplexer performance," *IEEE J. Select. Areas Commun.*, vol. SAC-4, pp. 856–868, 1986.
[12] J. R. M. Hosking, "Fractional differencing," *Biometrika*, vol. 68, pp. 165–176, 1981.
[13] R. Jain and S. A. Routhier, "Packet trains: Measurements and a new model for computer network traffic," *IEEE J. Select. Areas Commun.*, vol. SAC-4, pp. 986–995, 1986.
[14] W. E. Leland and D. V. Wilson, "High time-resolution measurement and analysis of LAN traffic: Implications for LAN interconnection," in *Proc. IEEE INFOCOM '91*, Bal Harbour, FL, 1991, pp. 1360–1366.
[15] W. E. Leland, M. S. Taqqu, W. Willinger, and D. V. Wilson, "On the self-similar nature of Ethernet traffic," in *Proc. ACM Sigcomm '93*, San Francisco, CA, 1993, pp. 183–193.
[16] ——, "Statistical analysis of high time-resolution Ethernet LAN traffic measurements," in *Proc. 25th Interface*, San Diego, CA, 1993.
[17] ——, "Self-similarity in high-speed packet traffic: Analysis and modeling of Ethernet traffic measurements," *Statistical Science*, 1994 (to appear).
[18] B. B. Mandelbrot, "Self-similar error clusters in communication systems and the concept of conditional stationarity," *IEEE Trans. Commun. Techn.*, vol. COM-13, pp. 71–90, 1965.
[19] ——, "Long-run linearity, locally Gaussian processes, H-spectra and infinite variances," *Intern. Econom. Rev.*, vol. 10, pp. 82–113, 1969.
[20] ——, "A fast fractional Gaussian noise generator," *Water Resources Research*, vol. 7, pp. 543–553, 1971.
[21] B. B. Mandelbrot and M. S. Taqqu, "Robust R/S analysis of long run serial correlation," in *Proc. 42nd Session ISI*, 1979, pp. 69–99.
[22] B. B. Mandelbrot and J. W. Van Ness, "Fractional Brownian motions, fractional noises and applications," *SIAM Rev.*, vol. 10, pp. 422–437, 1968.
[23] B. B. Mandelbrot and J. R. Wallis, "Computer experiments with fractional Gaussian noises," *Water Resources Research*, vol. 5, pp. 228–267, 1969.
[24] K. Meier-Hellstern, P. E. Wirth, Y.-L. Yan, and D. A. Hoeflin, "Traffic models for ISDN data users: Office automation application," in *Teletraffic and Datatraffic in a Period of Change (Proc. 13th ITC, Copenhagen, 1991)*, A. Jensen and V. B. Iversen, eds. Amsterdam, The Netherlands: North-Holland, 1991, pp. 167–172.
[25] I. Norros, "Studies on a model for connectionless traffic, based on fractional Brownian motion," COST24TD(92)041, 1992.
[26] E. H. Spafford, "The Internet worm incident," in *Proc. ESEC 89* and *Lecture Notes in Computer Science* 87. New York: Springer-Verlag, 1989.
[27] M. S. Taqqu, "A bibliographical guide to self-similar processes and long-range dependence," in *Dependence in Probability and Statistics*, E. Eberlein and M. S. Taqqu, eds. Basel: Birkhauser, 1985, pp. 137–165.
[28] M. S. Taqqu and J. B. Levy, "Using renewal processes to generate long-range dependence and high variability," in *Dependence in Probability and Statistics*, E. Eberlein and M. S. Taqqu, eds. Boston, MA: Birkhauser, 1986, vol. 11, pp. 73–89.
[29] D. Veitch, "Novel models of broadband traffic," in *Proc. 7th Australian Teletraffic Research Seminar*, Murray River, Australia, 1992.

PHOTO NOT AVAILABLE

Will Leland (M'82/ACM'77) received the Ph. D. degree in computer science from the University of Wisconsin, Madison.

He is a Member of Technical Staff at Bellcore, where he works in the Network Systems Research Department.

PHOTO NOT AVAILABLE

Murad Taqqu (M'92) received the B. A. degree in mathematics and physics in 1966 from the Université de Lausanne-Ecole Polytechnique and the Ph. D. degree in statistics in 1972 from Columbia University, New York.

Since 1985, he has been Professor in the Department of Mathematics at Boston University.

Dr. Taqqu is a Guggenheim Fellow and a Fellow of the Institute of Mathematical Statistics. He is currently an Associate Editor for *Stochastic Processes and their Applications* and coauthor of the book *Stable Non-Gaussian Random Processes: Stochastic Models with Infinite Variance* (Chapman and Hall, 1994).

PHOTO NOT AVAILABLE

Walter Willinger received the Diplom (Dipl. Math.) in 1980 from the ETH Zurich, Switzerland, and the M. S. and Ph. D. degrees in 1984 and 1987, respectively, from the School of ORIE, Cornell University, Ithaca, NY.

He is a Member of Technical Staff at Bellcore, where he works in the Computing and Communications Research Department.

Dr. Willinger is currently an Associate Editor for *The Annals of Applied Probability*.

PHOTO NOT AVAILABLE

Daniel V. Wilson (M'85/ACM'85) received the M. S. degree in electrical engineering from Stanford University in 1983 and the B. S. degree in physics and mathematics from Southwest Missouri State University in 1977.

He is a Member of Technical Staff at Bellcore where he works on network monitoring and analysis.

A Generalized Processor Sharing Approach to Flow Control in Integrated Services Networks: The Single-Node Case

Abhay K. Parekh, *Member, IEEE,* and Robert G. Gallager, *Fellow, IEEE*

***Abstract*—The problem of allocating network resources to the users of an integrated services network is investigated in the context of rate-based flow control. The network is assumed to be a virtual circuit, connection-based packet network. We show that the use of Generalized Processor Sharing (GPS), when combined with Leaky Bucket admission control, allows the network to make a wide range of worst-case performance guarantees on throughput and delay. The scheme is flexible in that different users may be given widely different performance guarantees, and is efficient in that each of the servers is work conserving. We present a practical packet-by-packet service discipline, PGPS (first proposed by Demers, Shenker, and Keshav [7] under the name of Weighted Fair Queueing), that closely approximates GPS. This allows us to relate results for GPS to the packet-by-packet scheme in a precise manner.**

In this paper, the performance of a single-server GPS system is analyzed exactly from the standpoint of worst-case packet delay and burstiness when the sources are constrained by leaky buckets. The worst-case session backlogs are also determined. In the sequel to this paper, these results are extended to arbitrary topology networks with multiple nodes.

I. Introduction

This paper and its sequel [17] focus on a central problem in the control of congestion in high-speed integrated services networks. Traditionally, the flexibility of data networks has been traded off with the performance guarantees given to its users. For example, the telephone network provides good performance guarantees but poor flexibility, while packet switched networks are more flexible but only provide marginal performance guarantees. Integrated services networks must carry a wide range of traffic types and still be able to provide performance guarantees to real-time sessions such as voice and video. We will investigate an approach to reconcile these apparently conflicting demands when the short-term demand for link usage frequently exceeds the usable capacity.

We propose the combined use of a packet service discipline based on Generalized Processor Sharing and Leaky Bucket rate control to provide flexible, efficient, and fair use of the links. Neither Generalized Processing Sharing, nor its packet-based version, PGPS, are new. Generalized Processor Sharing is a natural generalization of uniform processor sharing [14], and the packet-based version (while developed independently by us) was first proposed in [7] under the name of Weighted Fair Queueing. Our contribution is to suggest the use of PGPS in the context of integrated services networks and to combine this mechanism with Leaky Bucket admission control in order to provide performance guarantees in a flexible environment.

A major part of our work is to analyze networks of arbitrary topology using these specialized servers, and to show how the analysis leads to implementable schemes for guaranteeing worst-case packet delay. In this paper, however, we will restrict our attention to sessions at a single node, and postpone the analysis of arbitrary topologies to the sequel.

Our approach can be described as a strategy for rate-based flow control. Under rate-based schemes, a source's traffic is parametrized by a set of statistics such as average rate, maximum rate, and burstiness, and is assigned a vector of values corresponding to these parameters. The user also requests a certain quality of service that might be characterized, for example, by tolerance to worst-case or average delay. The network checks to see if a new source can be accommodated and, if so, takes actions (such as reserving transmission links or switching capacity) to ensure the quality of service desired. Once a source begins sending traffic, the network ensures that the agreed-upon values of traffic parameters are not violated.

Our analysis will concentrate on providing guarantees on throughput and worst-case packet delay. While packet delay in the network can be expressed as the sum of the processing, queueing, transmission, and propagation delays, we will focus exclusively on how to limit *queueing* delay.

We will assume that rate admission control is done through *leaky buckets* [20]. An important advantage of using leaky buckets is that this allows us to separate the packet delay into two components: delay in the leaky bucket and delay in the network. The first of these components is *independent* of the other active sessions, and can be estimated by the user if the statistical characterization of the incoming data is sufficiently simple (see [1, Sect. 6.3] for an example). The traffic entering the network has been "shaped" by the leaky bucket in a manner that can be succinctly characterized (we will do this in Section V), and so the network can upper bound the second component of packet delay through this characterization. This

Manuscript received June 1992; revised February and April 1992; approved by IEEE/ACM Transactions on Networking Editor Moshe Sidi. This paper was presented in part at IEEE INFOCOM '92. The research of A. Parekh was partly funded by a Vinton Hayes Fellowship and a Center for Intelligent Control Systems Fellowship. The research of R. Gallager was funded by the National Science Foundation under 8802991-NCR and by the Army Research Office under DAAL03-86-K-0171.

A. K. Parekh is with the IBM T. J. Watson Research Center, Yorktown Heights, NY 10598.

R. G. Gallager's is with the Laboratory for Information and Decision Systems, Massachusetts Institute of Technology, Cambridge, MA.

IEEE Log Number 9211033.

Reprinted from *IEEE/ACM Transactions on Networking,* vol. 1, no. 3, June 1993.

upper bound is independent of the statistics of the incoming data, which is helpful in the usual case where these statistics are either complex or unknown. A similar approach to the analysis of interconnection networks has been taken by Cruz [5]. From this point on, we will not consider the delay in the leaky bucket.

Generalized Processor Sharing (GPS) is defined and explained in Section II. In Section III, we present the packet-based scheme, PGPS, and show that it closely approximates GPS. Results obtained in this section allow us to translate session delay and buffer requirement bounds derived for a GPS server system to a PGPS server system. We propose a virtual time implementation of PGPS in the next section. Then, PGPS is compared to weighted round robin, virtual clock multiplexing [21], and stop-and-go queueing [9]–[11].

Having established PGPS as a desirable multiplexing scheme, we turn our attention to the rate enforcement function in Section V. The Leaky Bucket is described and proposed as a desirable strategy for admission control. We then proceed with an analysis, in Sections VI–VIII, of a single GPS server system in which the sessions are constrained by leaky buckets. The results obtained here are crucial in the analysis of arbitrary topology and multiple node networks, which we will present in the sequel to this paper.

II. GPS Multiplexing

The choice of an appropriate service discipline at the nodes of the network is key to providing effective flow control. A good scheme should allow the network to treat users differently, in accordance with their desired quality of service. However, this *flexibility* should not compromise the *fairness* of the scheme, i.e., a few classes of users should not be able to degrade service to other classes, to the extent that performance guarantees are violated. Also, if one assumes that the demand for high bandwidth services is likely to keep pace with the increase in usable link bandwidth, time and frequency multiplexing are too wasteful of the network resources to be considered candidate multiplexing disciplines. Finally, the service discipline must be *analyzable* so that performance guarantees can be made in the first place. We now present a flow-based multiplexing discipline called Generalized Processor Sharing that is efficient, flexible, and analyzable, and that therefore seems very appropriate for integrated services networks. However, it has the significant drawback of not transmitting packets as entities. In Section III, we will present a packet-based multiplexing discipline that is an excellent approximation to GPS even when the packets are of variable length.

A Generalized Processor Sharing (GPS) server is work conserving and operates at a fixed rate r. By work conserving, we mean that the server must be busy if there are packets waiting in the system. It is characterized by positive real numbers $\phi_1, \phi_2, \ldots, \phi_N$. Let $S_i(\tau, t)$ be the amount of session i traffic served in an interval $(\tau, t]$. A session is backlogged at time t if a positive amount of that session's traffic is queued at time t. Then, a GPS server is defined as one for which

$$\frac{S_i(\tau, t)}{S_j(\tau, t)} \geq \frac{\phi_i}{\phi_j}, j = 1, 2, \ldots, N \qquad (1)$$

for any session i that is continuously backlogged in the interval $(\tau, t]$.

Summing over all sessions j:

$$S_i(\tau, t) \sum_j \phi_j \geq (t - \tau) r \phi_i$$

and session i is guaranteed a rate of

$$g_i = \frac{\phi_i}{\sum_j \phi_j} r. \qquad (2)$$

GPS is an attractive multiplexing scheme for a number of reasons:

- Define r_i to be the session i average rate. Then, as long as $r_i \leq g_i$, the session can be guaranteed a throughput of ρ_i independent of the demands of the other sessions. In addition to this throughput guarantee, a session i backlog will always be cleared at a rate $\geq g_i$.
- The delay of an arriving session i bit can be bounded as a function of the session i queue length, independent of the queues and arrivals of the other sessions. Schemes such as FCFS, LCFS, and Strict Priority do not have this property.
- By varying the ϕ_i's, we have the flexibility of treating the sessions in a variety of different ways. For example, when all ϕ_i's are equal, the system reduces to uniform processor sharing. As long as the combined average rate of the sessions is less than r, any assignment of positive ϕ_i's yields a stable system. For example, a high-bandwidth delay-insensitive session i can be assigned g_i much less than its average rate, thus allowing for better treatment of the other sessions.
- Most importantly, it is possible to make worst-case network queueing delay *guarantees* when the sources are constrained by leaky buckets. We will present our results on this later. Thus, GPS is particularly attractive for sessions sending real-time traffic such as voice and video.

Fig. 1 illustrates generalized processor sharing. Variable-length packets arrive from both sessions on infinite capacity links and appear as impulses to the system. For $i = 1, 2$, let $A_i(0, t)$ be the amount of session i traffic that arrives at the system in the interval $(0, t]$ and, similarly, let $S_i(0, t)$ be the amount of session i traffic that is served in the interval $(0, t]$. We assume that the server works at rate 1. When $\phi_1 = \phi_2$ and both sessions are backlogged, they are each served at rate $\frac{1}{2}$ (e.g., interval [1, 6]). When $2\phi_1 = \phi_2$ and both sessions are backlogged, session 1 is served at rate $\frac{1}{3}$ and session 2 at rate $\frac{2}{3}$. Notice how increasing the relative weight of ϕ_2 leads to better treatment of that session in terms of both backlog and delay. The delay to session 2 goes down by one time unit, and the delay to session 1 goes up by one time unit. Also, notice that under both choices of ϕ_i, the system is empty at time 13 since the server is work conserving under GPS.

It should be clear from the example that the delays experienced by a session's packets can be reduced by increasing the value of ϕ for that session. This reduction, though, may be at the expense of *a corresponding* increase in delay for packets from the other sessions. Fig. 2 demonstrates that this

TABLE I
HOW GPS AND PGPS COMPARE FOR THE EXAMPLE IN FIG. 1.

		Session 1				Session 2		
packet information	Arrival	1	2	3	11	0	5	9
	Size	1	1	2	2	3	2	2
$\phi_1 = \phi_2$	GPS	3	5	9	13	5	9	11
	PGPS	4	5	7	13	3	9	11
$2\phi_1 = \phi_2$	GPS	4	5	9	13	4	8	11
	PGPS	4	5	9	13	3	7	11

The lower portion of the table gives the packet departure times under both schemes.

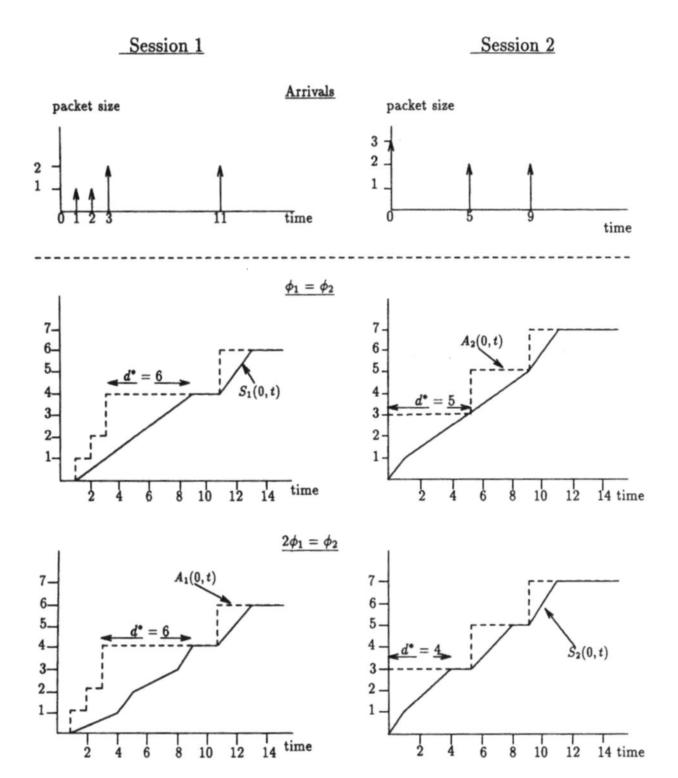

Fig. 1. An example of generalized processor sharing.

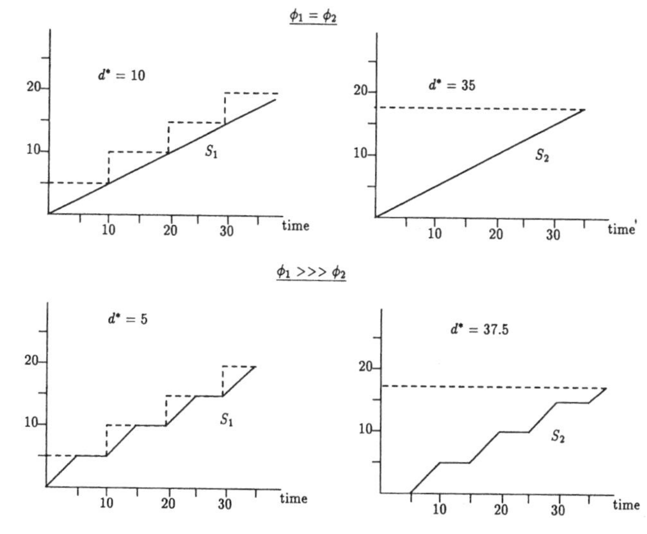

Fig. 2. The effect of increasing ϕ_i for a steady session i.

may not be the case when the better-treated session is steady. Thus, when combined with appropriate rate enforcement, the flexibility of GPS multiplexing can be used effectively to control packet delay.

III. A PACKET-BY-PACKET TRANSMISSION SCHEME–PGPS

A problem with GPS is that it is an idealized discipline that does not transmit packets as entities. It assumes that the server can serve multiple sessions simultaneously and that the traffic is infinitely divisible. In this section, we present a simple packet-by-packet transmission scheme that is an excellent approximation to GPS even when the packets are of variable length. Our idea is identical to the one used in [7]. *We will adopt the convention that a packet has arrived only after its last bit has arrived.*

Let F_p be the time at which packet p will depart (finish service) under Generalized Processor Sharing. Then, a very good approximation of GPS would be a work-conserving scheme that serves packets in increasing order of F_p. Now, suppose that the server becomes free at time τ. The next packet to depart under GPS *may not have arrived* at time τ and, since the server has no knowledge of when this packet will arrive, there is no way for the server to be *both* work conserving and serve the packets in increasing order of F_p. The server picks the first packet that would complete service in the GPS simulation if no additional packets were to arrive after time τ. Let us call this scheme PGPS for *packet-by-packet* Generalized Processor Sharing. As stated earlier, this mechanism was originally called Weighted Fair Queueing [7].

Table I shows how PGPS performs for the example in Fig. 1. Notice that when $\phi_1 = \phi_2$, the first packet to complete service under GPS is the session 1 packet that arrives at time 1. However, the PGPS server is forced to begin serving the long session 2 packet at time 0 since there are no other packets in the system at that time. Thus, the session 1 packet arriving at time 1 departs the system at time 4, i.e., 1 time unit later than it would depart under GPS.

A natural issue to examine at this point is how much later packets may depart the system under PGPS relative to GPS. First, we present a useful property of GPS systems.

Lemma 1: Let p and p' be packets in a GPS system at time τ, and suppose that packet p completes service before packet

p' if there are no arrivals after time τ. Then, packet p will also complete service before packet p' for any pattern of arrivals after time τ.

Proof: The sessions to which packets p and p' belong are both backlogged from time τ until one completes transmission. By (1), the ratio of the service received by these sessions is independent of future arrivals. □

A consequence of this lemma is that if PGPS schedules a packet p at time τ before another packet p' that is also backlogged at time τ, then in the simulated GPS system, packet p cannot leave later than packet p'. Thus, the only packets that are delayed more in PGPS are those that arrive too late to be transmitted in their GPS order. Intuitively, this means that only the packets that have a small delay under GPS are delayed more under PGPS.

Now let $\hat{F}_p$ be the time at which packet p departs under PGPS. We show that

Theorem 1: For all packets p,

$$\hat{F}_p - F_p \leq \frac{L_{\max}}{r},$$

where $L_{\max}$ is the maximum packet length and r is the rate of the server.

Proof: Since both GPS and PGPS are work-conserving disciplines, their busy periods coincide, i.e., the GPS server is in a busy period iff the PGPS server is in a busy period. Hence, it suffices to prove the result for each busy period. Consider any busy period and let the time that it begins be time zero. Let p_k be the kth packet in the busy period to depart under PGPS, and let its length be L_k. Also, let t_k be the time that p_k departs under PGPS and u_k be the time that p_k departs under GPS. Finally, let a_k be the time that p_k arrives. We now show that

$$t_k \leq u_k + \frac{L_{\max}}{r}$$

for $k = 1, 2, \ldots$. Let m be the largest integer that satisfies both $0 < m \leq k - 1$ and $u_m > u_k$. Thus,

$$u_m > u_k \geq u_i \quad \text{for } m < i < k. \tag{3}$$

Then, packet p_m is transmitted before packets $p_{m+1} \ldots, p_k$ under PGPS but after all these packets under GPS. If no such integer m exists, then set $m = 0$. Now, for the case $m > 0$, packet p_m begins transmission at $t_m - \frac{L_m}{r}$; so, from Lemma 1,

$$\min\{a_{m+1}, \ldots, a_k\} > t_m - \frac{L_m}{r}. \tag{4}$$

Since $p_{m+1}, \ldots, p_{k-1}$ arrive after $t_m - \frac{L_m}{r}$ and depart before p_k does under GPS,

$$u_k \geq \frac{1}{r}(L_k + L_{k-1} + L_{k-2} + \ldots + L_{m-1}) + t_m - \frac{L_m}{r}$$

$$\Rightarrow u_k \geq t_k - \frac{L_m}{r}.$$

If $m = 0$, then $p_{k-1}, \ldots, p_1$ all leave the GPS server before p_k does, and so

$$u_k \geq t_k.$$ □

Note that if N maximum-size packets leave simultaneously in the reference system, they can be served in arbitrary order in the packet-based system. Thus, $F_p - \hat{F}_p \geq (N-1)\frac{L_{\max}}{r}$ even if the reference system is tracked perfectly.

Let $S_i(\tau, t)$ and $\hat{S}_i(\tau, t)$ be the amount of session i traffic (in bits, not packets) served under GPS and PGPS in the interval $[\tau, t]$.

Theorem 2: For all times τ and sessions i:

$$S_i(0, \tau) - \hat{S}_i(0, \tau) \leq L_{\max}.$$

Proof: The slope of $\hat{S}_i$ alternates between r when a session i packet is being transmitted, and 0 when session i is not being served. Since the slope of S_i also obeys these limits, the difference $S_i(0, t) - \hat{S}_i(0, t)$ reaches its maximal value when session i packets begin transmission under PGPS. Let t be some such time, and let L be the length of the packet going into service. Then, the packet completes transmission at time $t + \frac{L}{r}$. Let τ be the time at which the given packet completes transmission under GPS. Then, since session i packets are served in the same order under both schemes,

$$S_i(0, \tau) = \hat{S}_i(0, t + \frac{L}{r}).$$

From Theorem 1,

$$\tau \geq (t + \frac{L}{r}) - \frac{L_{\max}}{r} \tag{5}$$

$$\Rightarrow S_i(0, t + \frac{L - L_{\max}}{r}) \leq \hat{S}_i(0, t + \frac{L}{r}) \tag{6}$$

$$= \hat{S}_i(0, t) + L. \tag{7}$$

Since the slope of S_i is at most r, the theorem follows. □

Let $\hat{Q}_i(\tau)$ and $Q_i(t)$ be the session i backlog (in units of traffic) at time τ under PGPS and GPS, respectively. Then, it immediately follows from Theorem 2 that

Corollary 1: For all times τ and sessions i

$$\hat{Q}_i(0, \tau) - Q_i(0, \tau) \leq L_{\max}.$$

Theorem 1 generalizes the result shown for the uniform processing case by Greenberg and Madras [12]. Notice that

- Theorem 1 and Corollary 1 can be used to translate bounds on GPS worst-case packet delay and backlog to corresponding bounds on PGPS.
- Variable packet lengths are easily handled by PGPS. This is not true of weighted round robin.
- The results derived so far can be applied to provide an alternative solution to a problem studied in [4],[19],[2],[8],[3]: There are N input links to a multiplexer; the peak rate of the ith link is C_i, and the rate of the multiplexer is $C \geq \sum_{i=1}^{N} C_i$. Since up to $L_{\max}$ bits from a packet may be queued from any link before the packet has "arrived," at least $L_{\max}$ bits of buffer must be allocated to each link. In fact, in [3] it is shown that at least $2L_{\max}$ bits are required, and that a class of buffer policies called Least Time to Reach Bound (LTRB) meets this bound. It is easy to design a PGPS policy that meets this bound as well: Setting $\phi_i = C_i$, it is clear that the resulting GPS server ensures

that no more than L_{max} bits are ever queued at any link. The bound of Corollary 1 guarantees that no more than $2L_{max}$ bits need to be allocated per link under PGPS. In fact, if L_i is the maximum allowable packet size for link i, then the bound on the link i buffer requirement is $L_i + L_{max}$. Further, various generalizations of the problem can be solved: For example, suppose the link speeds are arbitrary, but no more than $f_i(t) + r_i t$ bits can arrive on link i in any interval of length t (for each i). Then, if $\sum_i r_i \leq C$, setting $\phi_i = r_i$ for each i yields a PGPS service discipline for which the buffer requirement is $L_{max} + \max_{t\geq 0}(f_i(t) - r_i t)$ bits for each link i.

- There is no constant $c \geq 0$ such that

$$\hat{S}_i(0,t) - S_i(0,t) \leq cL_{max} \tag{8}$$

holds for all sessions i over all patterns of arrivals. To see this, let $K = \lfloor c+2 \rfloor$, $\phi_1 = K$, $\phi_2 = \ldots = \phi_N = 1$ and fix all packets sizes at L_{max}. At time zero, $K-1$ session 1 packets arrive and one packet arrives from each of the other sessions. No more packets arrive after time zero. Denote the $K-1^{st}$ session 1 packet to depart GPS (and PGPS) as packet p. Then, $F_p = \frac{K-1}{K}(N+K-1)\frac{L_{max}}{r}$, and $S_i(0,F_p) = \frac{K-1}{K}L_{max}$ for $i = 2,\ldots,N$. Thus, the first $K-1$ packets to depart the GPS system are the session 1 packets, and packet p leaves PGPS at time $(K-1)\frac{L_{max}}{r}$. Consequently,

$$\hat{S}_1(0,(K-1)\frac{L_{max}}{r}) = (K-1)L_{max}$$

and

$$S_1(0,(K-1)\frac{L_{max}}{r}) = \frac{K(K-1)L_{max}}{N-K+1}.$$

This yields

$$\hat{S}_1(0,(K-1)\frac{L_{max}}{r}) - S_1(0,(K-1)\frac{L_{max}}{r})$$
$$= (K-1)L_{max}(1 - \frac{K}{N-K+1}). \tag{9}$$

For any given K, the RHS of (9) can be made to approach $(K-1)L_{max}$ arbitrarily closely by increasing N.

A. *Virtual Time Implementation of PGPS*

In this section, we will use the concept of Virtual Time to track the progress of GPS that will lead to a practical implementation of PGPS. Our interpretation of virtual time generalizes the innovative one considered in [7] for uniform processor sharing. In the following, we assume that the server works at rate 1.

Denote as an event each arrival and departure from the GPS server, and let t_j be the time at which the jth event occurs (simultaneous events are ordered arbitrarily). Let the time of the first arrival of a busy period be denoted as $t_1 = 0$. Now observe that, for each $j = 2,3,\ldots$, the set of sessions that are busy in the interval (t_{j-1}, t_j) is fixed, and we may denote this set as B_j. Virtual time $V(t)$ is defined to be zero for all times when the server is idle. Consider any busy period, and let the time that it begins be time zero. Then, $V(t)$ evolves as follows:

$$V(0) = 0$$
$$V(t_{j-1}+\tau) = V(t_{j-1}) + \frac{\tau}{\sum_{i\in B_j}\phi_i},$$
$$\tau \leq t_j - t_{j-1}, j = 2,3,\ldots \tag{10}$$

The rate of change of V, namely $\frac{\partial V(t_j+\tau)}{\partial \tau}$, is $\frac{1}{\sum_{i\in B_j}\phi_i}$, and each backlogged session i receives service at rate $\phi_i \frac{\partial V(t_j+\tau)}{\partial \tau}$. Thus, V can be interpreted as increasing at the marginal rate at which backlogged sessions receive service.

Now suppose that the kth session i packet arrives at time a_i^k and has length L_i^k. Then, denote the virtual times at which this packet begins and completes service as S_i^k and F_i^k, respectively. Defining $F_i^0 = 0$ for all i, we have

$$S_i^k = \max\{F_i^{k-1}, V(a_i^k)\}$$
$$F_i^k = S_i^k + \frac{L_i^k}{\phi_i}. \tag{11}$$

There are three attractive properties of the virtual time interpretation from the standpoint of implementation. First, the virtual time finishing times can be determined at the packet arrival time. Second, the packets are served in order of virtual time finishing time. Finally, we need only update virtual time when there are events in the GPS system. However, the price to be paid for these advantages is some overhead in keeping track of sets B_j, which is essential in the updating of virtual time.

Define $\text{Next}(t)$ to be the *real* time at which the next packet will depart the GPS system after time t if there are no more arrivals after time t. Thus, the next virtual time update after t will be performed at $Next(t)$ if there are no arrivals in the interval $[t, \text{Next}(t)]$. Now, suppose a packet arrives at some time t (let it be the jth event) and that the time of the event just prior to t is τ (if there is no prior event, i.e., if the packet is the first arrival in a busy period, then set $\tau = 0$). Then, since the set of busy sessions is fixed between events, $V(t)$ may be computed from (10) and the packet stamped with its virtual time finishing time. $\text{Next}(t)$ is the real time corresponding to the smallest virtual time packet finishing time at time t. This real time may be computed from (10) since the set of busy sessions, B_j, remains fixed over the interval $[t, Next(t)]$: Let F_{min} be the smallest virtual time finishing time of a packet in the system at time t. Then, from (10)

$$F_{min} = V(t) + \frac{\text{Next}(t) - t}{\sum_{i\in B_j}\phi_i}$$

$$\Rightarrow \text{Next}(t) = t + (F_{min} - V(t))\sum_{i\in B_j}\phi_i.$$

Given this mechanism for updating virtual time, PGPS is defined as follows: When a packet arrives, virtual time is updated and the packet is stamped with its virtual time finishing time. The server is work conserving and serves packets in an increasing order of timestamp.

IV. Comparing PGPS to Other Schemes

Under weighted round robin, every session i has an integer weight w_i associated with it. The server polls the sessions according a *precomputed sequence* in an attempt to serve session i at a rate of $\frac{w_i}{\sum_j w_j}$. If an empty buffer is encountered, the server moves to the next session in the order instantaneously. When an arriving session i packet just misses its slot in a frame, it cannot be transmitted before the next session i slot. If the system is heavily loaded in the sense that almost every slot is utilized, the packet may have to wait almost N slot times to be served, where N is the number of sessions sharing the server. Since PGPS approximates GPS to within one packet transmission time regardless of the arrival patterns, it is immune to such effects. PGPS also handles variable-length packets in a much more systematic fashion than does weighted round robin. However, if N or the packet sizes are small, then it is possible to approximate GPS well by weighted round robin. Hahne [13] has analyzed round robin in the context of providing fair rates to users of networks that utilize hop-by-hop window flow control.

Zhang proposes an interesting scheme called virtual clock multiplexing [21]. Virtual clock multiplexing allows a guaranteed rate and (average) delay for each session, independent of the behavior of other sessions. However, if a session produces a large burst of data, even while the system is lightly loaded, that session can be "punished" much later when the other sessions become active. Under PGPS, the delay of a session i packet can be bounded in terms of the session i queue size seen by that packet upon arrival, even in the absence of any rate control. This enables sessions to take advantage of lightly loaded network conditions. We illustrate this difference with a numerical example:

Suppose there are two sessions that submit fixed-size packets of one unit each. The rate of the server is one, and the packet arrival rate is $\frac{1}{2}$ for each session. Starting at time zero, 1000 session 1 packets begin to arrive at a rate of 1 packet/second. No session 2 packets arrive in the interval [0900) but, at time 900, 450 session 2 packets begin to arrive at a rate of one packet/second. Now if the sessions are to be treated equally, the virtual clock for each session will tick at a rate of $\frac{1}{2}$, and the PGPS weight assignment will be $\phi_1 = \phi_2$. Since both disciplines are work conserving, they will serve session 1 continuously in the interval [0900).

At time 900^-, there are no packets in queue from either session; the session 1 virtual clock will read 1800 and the session 2 virtual clock will read 900. The 450 session 2 packets that begin arriving at this time will be stamped 900902904, ..., 1798, while the 100 session 1 packets that arrive after time 900 will be stamped 1800, 1804, ... 1998. Thus, **all** of the session 2 packets will be served under Virtual Clock before any of the session 1 packets are served. The session 1 packets are being punished since the session used the server exclusively in the interval [0900). Note, however, that this exclusive use of the server was *not at the expense of any session 2 packets.* Under PGPS, the sessions are served in round robin fashion from time 900 on, which results in much less delay to the session 1 packets.

The lack of a punishment feature is an attractive aspect of PGPS since, in our scheme, the admission of packets is regulated at the network periphery through leaky buckets and it does not seem necessary to punish users at the internal nodes as well. Note, however, that in this example PGPS guarantees a throughput of $\frac{1}{2}$ to each session even in the absence of access control.

Stop-and-Go Queueing is proposed in [9]-[11] and is based on a network-wide time slot structure. It has two advantages over our approach: it provides better jitter control and is probably easier to implement. A finite number of connection types are defined, where a type g connection is characterized by a fixed frame size of T_g. Since each connection must conform to a predefined connection type, the scheme is somewhat less flexible than PGPS. The admission policy under which delay and buffer size guarantees can be made is that no more than $r_i T_g$ bits may be submitted during any type g frame. If sessions $1, 2, \ldots, N$ are served by a server of capacity 1, it is stipulated that $\sum_{i=1}^{N} r_i \leq 1$, where the sum is only taken over the real-time sessions. The delay guarantees grow linearly with T_g, so in order to provide low delay one has to use a small slot size. The service discipline is not work conserving and is such that each packet may be delayed up to $2T_g$ time units, even when there is only one active session at the server. Observe that for a single-session PGPS system in which the peak rate does not exceed the rate of the server, each arriving packet is served *immediately* upon arrival. Also, since it is work conserving, PGPS will provide better average delay than stop-and-go for a given access control scheme.

It is clear that r_i is the average rate at which the source i can send data over a single slot. The relationship between delay and slot size may force Stop-and-Go to allocate bandwidth by peak to satisfy delay-senstive sessions. This may also happen under PGPS, but not to the same degree. To see this, consider an on/off periodic source that fluctuates between values $C - \epsilon$ and 0. (As usual, ϵ is small.) The on period is equal to the off period, say they are B seconds in duration. We assume that B is large. Clearly, the average rate of this session is $0.5(C - \epsilon)$. We are interested in providing this session low delay under Stop-and-Go and PGPS. To do this, one has to pick a slot size smaller than B, which forces $r = C - \epsilon$. The remaining capacity of the server that can be allocated is ϵ. Under PGPS, we allocate a large value of ϕ to the session to bring its delay down to the desired level; however, now the remaining capacity that can be allocated is $0.5(C + \epsilon)$. Now observe that if there is a *second* on/off session with identical on and off periods as the first sesision, but which is relatively less delay sensitive, then PGPS can carry both sessions (since the combined sustainable rate is less than C) whereas Stop-and-Go cannot.

V. Leaky Bucket

Fig. 3 depicts the Leaky Bucket scheme [20] that we will use to describe the traffic that enters the network. Tokens or permits are generated at a fixed rate, ρ, and packets can be released into the network only after removing the required number of tokens from the token bucket. There is no bound

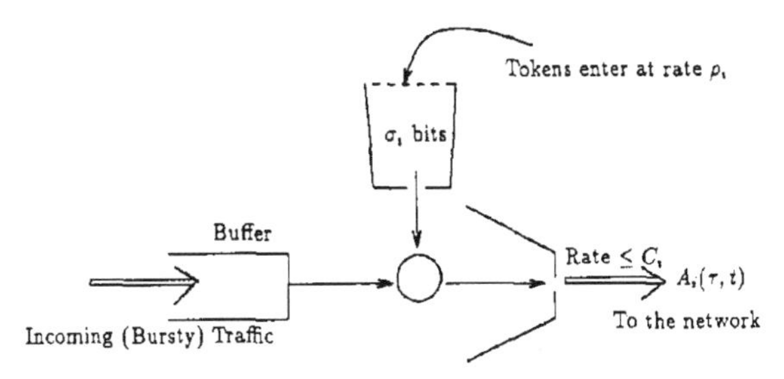

Fig. 3. A Leaky Bucket.

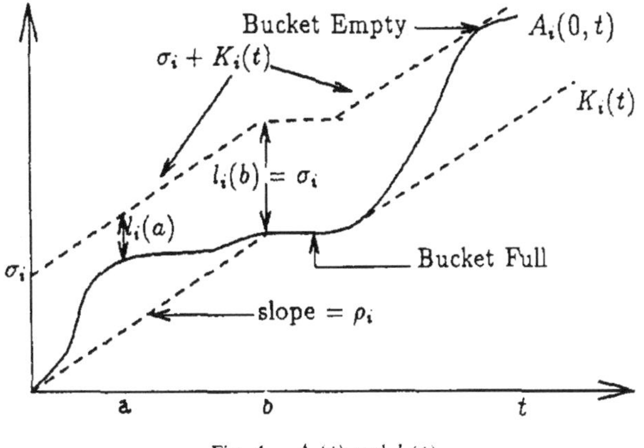

Fig. 4. $A_i(t)$ and $l_i(t)$.

on the number of packets that can be buffered, but the *token* bucket contains at most σ bits worth of tokens. In addition to securing the required number of tokens, the traffic is further constrained to leave the bucket at a maximum rate of $C > \rho$.

The constraint imposed by the leaky bucket is as follows: If $A_i(\tau, t)$ is the amount of session i flow that leaves the leaky bucket and enters the network in time interval $(\tau, t]$, then

$$A_i(\tau, t) \leq \min\{(t-\tau)C_i, \sigma_i + \rho_i(t-\tau)\}, \forall t \geq \tau \geq 0, \quad (12)$$

for every session i. We say that session i conforms to (σ_i, ρ_i, C_i), or $A_i \sim (\sigma_i, \rho_i, C_i)$.

This model for incoming traffic is essentially identical to the one recently proposed by Cruz [5], [6], and it has also been used in various forms to represent the inflow of parts into manufacturing systems by Kumar [18], [15]. The arrival constraint is attractive since it restricts the traffic in terms of average sustainable rate (ρ), peak rate (C), and burstiness (σ and C). Fig. 4 shows how a fairly bursty source might be characterized using the constraints.

Represent $A_i(0, t)$ as in Fig. 4. Let there be $l_i(t)$ bits worth of tokens in the session i token bucket at time t. We assume that the session starts out with a full bucket of tokens. If $K_i(t)$ is the total number of tokens accepted at the session i bucket in the interval $(0, t]$ (it does not include the full bucket of tokens that session i starts out with, and does not include arriving tokens that find the bucket full), then

$$K_i(t) = \min_{0 \leq \tau \leq t} \{A_i(0, \tau) + \rho_i(t - \tau)\}. \quad (13)$$

Thus, for all $\tau \leq t$

$$K_i(t) - K_i(\tau) \leq \rho_i(t - \tau). \quad (14)$$

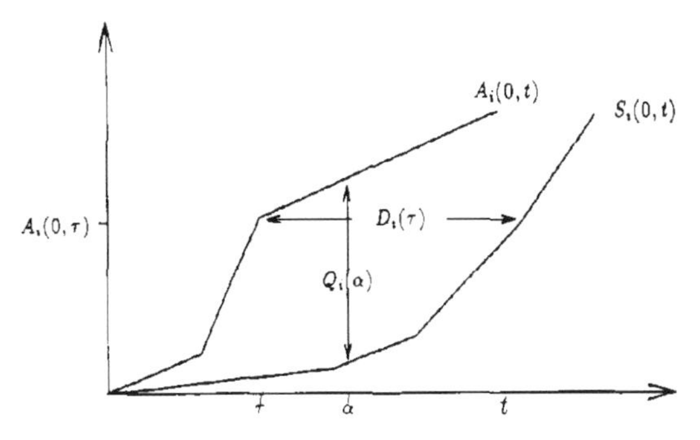

Fig. 5. $A_i(0, t)$, $S_i(0, t)$, $Q_i(t)$ and $D_i(t)$

We may now express $l_i(t)$ as

$$l_i(t) = \sigma_i + K_i(t) - A_i(0, t). \quad (15)$$

From (15) and (14), we obtain the useful inequality

$$A_i(\tau, t) \leq l_i(\tau) + \rho_i(t - \tau) - l_i(t). \quad (16)$$

VI. Analysis

In this section, we analyze the worst-case performance of single-node GPS systems for sessions that operate under Leaky Bucket constraints, i.e., the session traffic constrained as in (12).

There are N sessions, and the only assumptions we make about the incoming traffic are that $A_i \sim (\sigma_i, \rho_i, C_i)$ for $i = 1, 2, \ldots, N$ and that the system is empty before time zero. The server is work conserving (i.e., it is never idle if there is work in the system), and operates at the fixed rate of 1.

Let $S_i(\tau, t)$ be the amount of session i traffic served in the interval $(\tau, t]$. Note that $S_i(0, t)$ is continuous and nondecreasing for all t (see Fig. 5). The session i backlog at time τ is defined to be

$$Q_i(\tau) = A_i(0, \tau) - S_i(0, \tau).$$

The session i delay at time τ is denoted by $D_i(\tau)$, and is the amount of time that it would take for the session i backlog to clear if no session i bits were to arrive after time τ. Thus,

$$D_i(\tau) = \inf\{t \geq \tau : S_i(0, t) = A_i(0, \tau)\} - \tau. \quad (17)$$

From Fig. 5, we see that $D_i(\tau)$ is the horizontal distance between curves $A_i(0, t)$ and $S_i(0, t)$ at the ordinate value of $A_i(0, \tau)$.

Clearly, $D_i(\tau)$ depends on the arrival functions $A_1, \ldots, A_N$. We are interested in computing the maximum delay over all time, and over all arrival functions that are consistent with (12). Let D_i^* be the maximum delay for session i. Then,

$$D_i^* = \max_{(A_1, \ldots, A_N)} \max_{\tau \geq 0} D_i(\tau).$$

Similarly, we define the maximum backlog for session i, Q_i^*:

$$Q_i^* = \max_{(A_1, \ldots, A_N)} \max_{\tau \geq 0} Q_i(\tau).$$

The problem we will solve in the following sections is: Given $\phi_1, \ldots, \phi_N$ for a GPS server of rate 1 and given (σ_j, ρ_j, C_j), $j = 1, \ldots, N$, what are D_i^* and Q_i^* for every session i? We will also be able to characterize the burstiness of the output traffic for every session i, which will be especially useful in our analysis of GPS networks in the sequel.

A. Definitions and Preliminary Results

We introduce definitions and derive inequalities that are helpful in our analysis. Some of these notions are general enough to be used in the analysis of any work-conserving service discipline (that operates on sources that are Leaky Bucket constrained).

Given $A_1, \ldots, A_N$, let σ_i^τ be defined for each session i and time $\tau \geq 0$ as

$$\sigma_i^\tau = Q_i(\tau) + l_i(\tau) \quad (18)$$

where $l_i(\tau)$ is defined in (15). Thus, σ_i^τ is the sum of the number of tokens left in the bucket and the session backlog at the server at time τ. If $C_i = \infty$, we can think of σ_i^τ as the maximum amount of session i backlog at time τ^+ over all arrival functions that are identical to $A_1, \ldots, A_N$ up to time τ.

Observe that $\sigma_i^0 = \sigma_i$ and

$$Q_i(\tau) = 0 \Rightarrow \sigma_i^\tau \leq \sigma_i. \quad (19)$$

Recall (16)

$$A_i(\tau, t) \leq l_i(\tau) + \rho_i(t - \tau) - l_i(t).$$

Substituting for l_i^τ and l_i^t from (18)

$$Q_i(\tau) + A_i(\tau, t) - Q_i(t) \leq \sigma_i^\tau - \sigma_i^t + \rho_i(t - \tau). \quad (20)$$

Now notice that

$$S_i(\tau, t) = Q_i(\tau) + A_i(\tau, t) - Q_i(t). \quad (21)$$

Combining (20) and (21), we establish the following useful result:

Lemma 2: For every session i, $\tau \leq t$:

$$S_i(\tau, t) \leq \sigma_i^\tau - \sigma_i^t + \rho_i(t - \tau). \quad (22)$$

Define a **system busy period** to be a maximal interval B such that for any $\tau, t \in B$, $\tau \leq t$:

$$\sum_{i=1}^{N} S_i(\tau, t) = t - \tau.$$

Since the system is work conserving, if $B = [t_1, t_2]$, then $\sum_{i=1}^N Q_i(t_1) = \sum_{i=1}^N Q_i(t_2) = 0$.

Lemma 3: When $\sum_j \rho_j < 1$, the length of a system busy period is at most

$$\frac{\sum_{i=1}^N \sigma_i}{1 - \sum_{i=1}^N \rho_i}.$$

Proof: Suppose $[t_1, t_2]$ is a system busy period. By assumption,

$$\sum_{i=1}^{N} Q_i(t_1) = \sum_{i=1}^{N} Q_i(t_2) = 0.$$

Thus,

$$\sum_{i=1}^{N} A_i(t_1, t_2) = \sum_{i=1}^{N} S_i(t_1, t_2) = t_2 - t_1.$$

Substituting from (12) and rearranging terms:

$$t_2 - t_1 \leq \frac{\sum_{i=1}^N \sigma_i}{1 - \sum_{i=1}^N \rho_i}.$$

□

A simple consequence of this lemma is that all system busy periods are bounded. Since session delay is bounded by the length of the largest possible system busy period, the session delays are bounded as well. Thus, the interval B is finite whenever $\sum_{i=1}^N \rho_i < 1$ and may be infinite otherwise.

We end this section with some comments valid only for the GPS system: Let a **session** i **busy period** be a maximal interval B_i contained in a single system busy period, such that for all $\tau, t \in B_i$:

$$\frac{S_i(\tau, t)}{S_j(\tau, t)} \geq \frac{\phi_i}{\phi_j}, \quad j = 1, 2 \ldots, N. \quad (23)$$

Notice that it is possible for a session to have zero backlog during its busy period. However, if $Q_i(\tau) > 0$ then τ *must* be in a session i busy period at time τ. We have already shown in (2) that

Lemma: For every interval $[\tau, t]$ *that is in a session i busy period*

$$S_i(\tau, t) \geq (t - \tau)\frac{\phi_i}{\sum_{j=1}^N \phi_j}.$$

Notice that when $\phi = \phi_i$ for all i, the service guarantee reduces to

$$S_i(\tau, t) \geq \frac{t - \tau}{N}.$$

B. Greedy Sessions

Session i is defined to be greedy starting at time τ if

$$A_i(\tau, t) = \min\{C_i(t - \tau), l_i(\tau) + (t - \tau)\rho_i\}, \text{for all } t \geq \tau. \quad (24)$$

In terms of the Leaky Bucket, this means that the session uses as many tokens as possible (i.e., sends at maximum possible rate) for all times $\geq \tau$. At time τ, session i has $l_i(\tau)$ tokens left in the bucket, but it is constrained to send traffic at a maximum rate of C_i. Thus, it takes $\frac{l_i^\tau}{C_i - \rho_i}$ time units to deplete the tokens in the bucket. After this, the rate will be limited by the token arrival rate ρ_i.

Define A_i^τ as an arrival function that is greedy starting at time τ (see Fig. 6). From inspection of the figure [and from

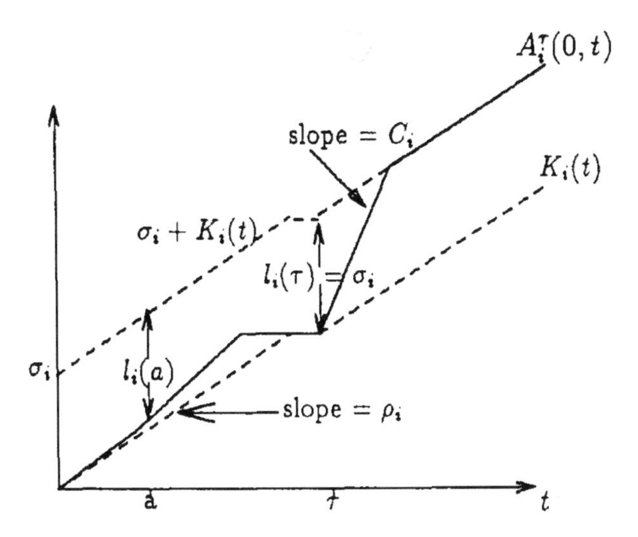

Fig. 6. A session i arrival function that is greedy from time τ.

(24)], we see that if a system busy period starts at time zero, then

$$A_i^0(0,t) \geq A(0,t), \forall A \sim (\sigma_i, \rho_i, C_i), t \geq 0.$$

The major result in this section is the following:

Theorem 3: Suppose that $C_j \geq r$ for every session j, where r is the rate of a GPS server. Then, for every session i, D_i^* and Q_i^* are achieved (not necessarily at the same time) when every session is greedy starting at time zero, the beginning of a system busy period.

This is an intuitively pleasing and satisfying result. It seems reasonable that if a session sends as much traffic as possible at all times, it is going to impede the progress of packets arriving from the other sessions. Notice, however, that we are claiming a worst-case result, which implies that it is never more harmful for a subset of the sessions to "save up" their bursts and to transmit them at a time greater than zero.

While there are many examples of service disciplines for which this "all-greedy regime" does not maximize delay, the amount of work required to establish Theorem 3 is still somewhat surprising. Our approach is to prove the theorem for the case when $C_i = \infty$ for all i—this implies that the links carrying traffic to the server have infinite capacity. This is the easiest case to visualize since we do not have to worry about the input links. Further, it bounds the performance of the finite link speed case since any session can "simulate" a finite speed input link by sending packets at a finite rate over the link. After we have understood the infinite capacity case, it will be shown that a simple extension in the analysis yields the result for finite link capacities as well.

C. Generalized Processor Sharing with Infinite Incoming Link Capacities

When all input link speeds are infinite, the arrival constraint (12) is modified to

$$A_i(\tau,t) \leq \sigma_i + \rho_i(t-\tau), \quad \forall 0 \leq \tau \leq t, \tag{25}$$

for every session i. We say that session i conforms to (σ_i, ρ_i) or $A_i \sim (\sigma_i, \rho_i)$. Further, we stipulate that $\sum_i \rho_i < 1$ to ensure stability.

By relaxing our constraint, we allow step or jump arrivals, which create discontinuities in the arrival functions A_i. Our convention will be to treat the A_i as *left-continuous* functions (i.e., continuous from the left). Thus, a session i impulse of size Δ at time 0 yields $Q_i(0) = 0$ and $Q_i(0^+) = \Delta$. Note also that $l_i(0) = \sigma_i$, where $l_i(\tau)$ is the maximum amount of session i traffic that could arrive at time τ^- without violating (25). When session i is greedy from time τ, the infinite capacity assumption ensures that $l_i(t) = 0$ for all $t > \tau$. Thus, (16) reduces to

$$A_i^\tau(\tau,t) = l_i(\tau) + (t-\tau)\rho_i, \text{ for all } t > \tau. \tag{26}$$

Note also that if the session is greedy after time τ, $l_i(t) = 0$ for any $t > \tau$.

Defining σ_i^τ as before (from 18), we see that it is equal to $Q_i(\tau^-)$ when session i is greedy starting at time τ.

An all-greedy GPS system: Theorem 3 suggests that we should examine the dynamics of a system in which all the sessions are greedy starting at time 0, the beginning of a system busy period. This is illustrated in Fig. 7.

From (26), we know that

$$A_i(0,\tau) = \sigma_i + \rho_i\tau, \quad \tau \geq 0$$

and let us assume, for clarity of exposition, that $\sigma_i > 0$ for all i.

Define e_1 as the first time at which one of the sessions, say $L(1)$, ends its busy period. Then, in the interval $[0, e_1]$, each session i is in a busy period (since we assumed that $\sigma_i > 0$ for all i) and is served at rate $\frac{\phi_i}{\sum_{k=1}^N \phi_k}$. Since session $L(1)$ is greedy after 0, it follows that

$$\rho_{L(1)} < \frac{\phi_i}{\sum_{k=1}^N \phi_k}$$

where $i = L(1)$. (We will show that such a session must exist in Lemma 5.) Now each session j still in a busy period will be served at rate

$$\frac{(1-\rho_{L(1)})\phi_j}{\sum_{k=1}^N \phi_k - \phi_{L(1)}}$$

until a time e_2 when another session, $L(2)$, ends its busy period. Similarly, for each k:

$$\rho_{L(k)} < \frac{(1-\sum_{j=1}^{k-1} \rho_{L(j)})\phi_i}{\sum_{j=1}^N \phi_j - \sum_{j=1}^{k-1} \phi_{L(j)}}, k = 1, 2, \ldots, N, i = L(k). \tag{27}$$

As shown in Fig. 7, the slopes of the various segments that comprise $S_i(0,t)$ are $s_1^i, s_2^i, \ldots$. From (27)

$$s_k^i = \frac{(1-\sum_{j=1}^{k-1} \rho_{L(j)})\phi_i}{\sum_{j=1}^N \phi_j - \sum_{j=1}^{k-1} \phi_{L(j)}}, k = 1, 2, \ldots, L(i).$$

It can be seen that $\{s_k^i\}, k = 1, 2, \ldots, L(i)$ forms an increasing sequence.

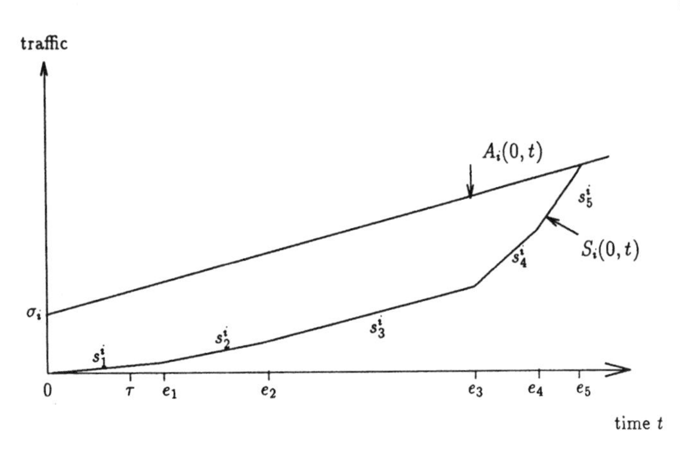

Fig. 7. Session i arrivals and departures after0, the beginning of a system busy period.

Note that:

- We only require that

$$0 \le e_1 \le e_2 \le \dots \le e_N.$$

allowing for several e_i to be equal.
- We only care about $t \le e_{L(i)}$ since the session i buffer is always empty after this time.
- Session $L(i)$ has exactly one busy period–the interval $[0, e_i]$.
- e_N is the maximum busy period length, i.e., it meets the bound of Lemma 3.

Any ordering of the sessions that meets (27) is known as a **feasible ordering**. Thus, sessions $1, \dots, N$ follow a feasible ordering if and only if:

$$\rho_k < \frac{(1 - \sum_{j=1}^{k-1} \rho_j)\phi_k}{\sum_{j=k}^{N} \phi_j}, k = 1, 2, \dots, N. \tag{28}$$

Lemma 5: At least one feasible ordering exists if $\sum_{i=1}^{N} \rho_i < 1$.

Proof: By contradiction, suppose there exists an index i, $1 \le i \le N$ such that we can label the first $i - 1$ sessions of a feasible ordering $\{1, \dots, i-1\}$ but (28) does not hold for any of the remaining sessions when $k = i$. Then, denoting $L_{i-1} = \{1, \dots, i-1\}$, we have for every session $k \notin L_{i-1}$:

$$\rho_k \ge (1 - \sum_{j \in L_{i-1}} \rho_j)(\frac{\phi_k}{\sum_{j \notin L_{i-1}} \phi_j}).$$

Summing over all such k, we have:

$$\sum_{k \notin L_{i-1}} \rho_k \ge 1 - \sum_{j \in L_{i-1}} \rho_j \Rightarrow \sum_{j=1}^{N} \rho_j \ge 1$$

which is a contradiction, since we assumed that $\sum_{j=1}^{N} \rho_j < 1$. Thus, no such index i can exist and the lemma is proven. □

In general, there are many feasible orderings possible, but the one that comes into play at time 0 depends on the σ_i's. For example, if $\rho = \rho_j$ and $\phi = \phi_j, j = 1, 2, \dots, N$, then there are $N!$ different feasible orderings. Similarly, there are $N!$ different feasible orderings if $\rho_i = \phi_i$ for all i. To simplify the notation, let us assume that the sessions are labeled so that

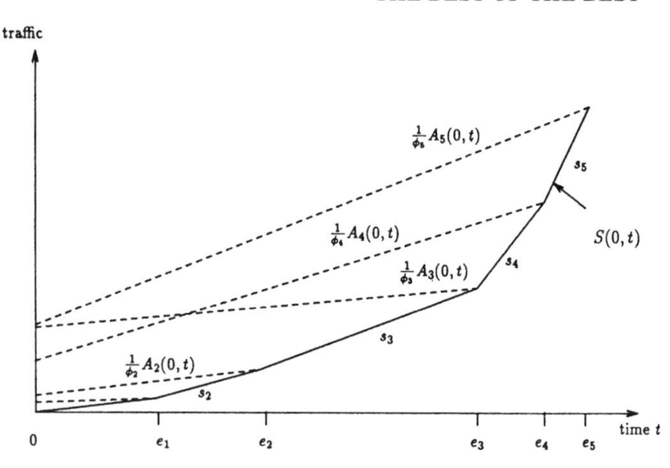

The arrival functions are scaled so that a universal service curve, $S(0,t)$, can be drawn. After time e_i, session i has a backlog of zero until the end of the system busy period, which is at time e_5. The vertical distance between the dashed curve corresponding to session i and $S(0,\tau)$ is $\frac{1}{\phi_i}Q_i(\tau)$, while the horizontal distance yields $D_i(\tau)$ just as it does in Figure 8.

Fig. 8. The dynamics of an all-greedy GPS system.

$j = L(j)$ for $j = 1, 2, \dots, N$. Then, for any two sessions i, j indexed greater than k we can define a "universal slope" s_k by:

$$s_k = \frac{s_k^i}{\phi_i} = \frac{s_k^j}{\phi_j} = \frac{1 - \sum_{j=1}^{k-1} \rho_j}{\sum_{j=k}^{N} \phi_k}, i, j > k, k = 1, 2, \dots, N.$$

This allows us to describe the behavior of **all** sessions in a single figure as is depicted in Fig. 8. Under the all-greedy regime, the function $V(t)$ (described in Section III-A) *corresponds exactly* to the universal service curve $S(0, t)$ shown in Fig. 8. It is worth noting that the virtual time function $V(t)$ captures this notion of generalized service for arbitrary arrival functions.

In the remainder of this section, we will prove a tight lower bound on the amount of service a session receives when it is in a busy period: Recall that, for a given set of arrival functions $A = \{A_1, \dots, A_N\}$, $A^\tau = \{A_1^\tau, \dots, A_N^\tau\}$ is the set such that for every session k, $A_k^\tau(0, s) = A_k(0, s)$ for $s \in [0, \tau)$ and session k is greedy starting at time τ.

Lemma 6: Assume that session i is in a busy period in the interval $[\tau, t]$. Then,

i) For any subset M of m sessions, $1 \le m \le N$ and any time $t \ge \tau$:

$$S_i(\tau, t) \ge \frac{(t - \tau - (\sum_{j \notin M} \sigma_j^\tau + \rho_j(t - \tau)))\phi_i}{\sum_{j \in M} \phi_j}. \tag{29}$$

ii) Under A^τ, there exists a subset of the sessions, M^t, for every $t \ge \tau$ such that equality holds in (29).

Proof: For compactness of notation, let $\phi_{ji} = \frac{\phi_j}{\phi_i}, \forall i, j$.

i) From (22),

$$S_j(\tau, t) \le \sigma_j^\tau + \rho_j(t - \tau)$$

for all j. Also, since the interval $[\tau, t]$ is in a session i busy period:

$$S_j(\tau, t) \le \phi_{ji} S_i(\tau, t).$$

Thus,

$$S_j(\tau, t) \le \min\{\sigma_j^\tau + \rho_j(t - \tau), \phi_{ji} S_i(\tau, t)\}.$$

Since the system is in a busy period, the server serves exactly $t-\tau$ units of traffic in the interval $[\tau, t]$. Thus,

$$t - \tau \leq \sum_{j=1}^{N} \min\{\sigma_j^\tau + \rho_j(t-\tau), \phi_{ji} S_i(\tau, t)\}$$

$$\Rightarrow t - \tau \leq \sum_{j \notin M} \sigma_j^\tau + \rho_j(t-\tau) + \sum_{j \in M} \phi_{ji} S_i(\tau, t)$$

for any subset of sessions M. Rearranging the terms yields (29).

ii) Since all sessions are greedy after τ under A^τ, every session j will have a session busy period that begins at τ and lasts up to some time e_j. As we showed in the discussion leading up to Fig. 8, $Q_j(t) = 0$ for all $t \geq e_j$. The system busy period ends at time $e^* = \max_j e_j$. Define

$$M^t = \{j : e_j \geq t\}.$$

By the definition of GPS, we know that session $j \in M^t$ receives exactly $\phi_{ji} S_i(\tau, t)$ units of service in the interval $(\tau, t]$. A session k is not in M^t only if $e_k < t$, so we must have $Q_k(t) = 0$. Thus, for $k \notin M^t$,

$$S_k(\tau, t) = \sigma_k^\tau + \rho_k(t-\tau).$$

and equality is achieved in (29).

D. An Important Inequality

In the previous section, we examined the behavior of the GPS system when the sessions are greedy. Here, we prove an important inequality that holds for any arrival functions that conform to the arrival constraints (25).

Theorem 4: Let $1, \ldots, N$ be a feasible ordering. Then, for any time t and session p:

$$\sum_{k=1}^{p} \sigma_k^t \leq \sum_{k=1}^{p} \sigma_k.$$

We want to show that at the beginning of a session p busy period, the collective burstiness of sessions $1, \ldots, p$ will never be more than what it was at time 0. The interesting aspect of this theorem is that it holds for **every** feasible ordering of the sessions. When $\rho_j = \rho$ and $\phi_j = \phi$ for every j, it says that the collective burstiness of *any* subset of sessions is no less than what it was at the beginning of the system busy period.

The following three lemmas are used to prove the theorem. The first says (essentially) that if session p is served at a rate *smaller* than its average rate ρ_p during a session p busy period, then the sessions indexed lower than p will be served correspondingly *higher* than their average rates. Note that this lemma is true even when the sessions are not greedy.

Lemma 7: Let $1, \ldots, N$ be a feasible ordering, and suppose that session p is busy in the interval $[\tau, t]$. Further, define x to satisfy

$$S_p(\tau, t) = \rho_p(t-\tau) - x \tag{30}$$

Then,

$$\sum_{k=1}^{p-1} S_k(\tau, t) > \sum_{k=1}^{p-1} (t-\tau)\rho_k + x\left(1 + \sum_{j=p+1}^{N} \frac{\phi_j}{\phi_p}\right). \tag{31}$$

Proof: For compactness of notation, let $\phi_{ij} = \frac{\phi_i}{\phi_j}, \forall i, j$. Now because of the feasible ordering,

$$\rho_p < \frac{1 - \sum_{j=1}^{p-1} \rho_i}{\sum_{i=p}^{N} \phi_{ip}}.$$

Thus,

$$S_p(\tau, t) < (t-\tau)\left(\frac{1 - \sum_{j=1}^{p-1} \rho_i}{\sum_{i=p}^{N} \phi_{ip}}\right) - x. \tag{32}$$

Also, $S_j(\tau, t) \leq \phi_{jp} S_p(\tau, t)$ for all j. Thus,

$$\sum_{j=p}^{N} S_j(\tau, t) \leq S_p(\tau, t) \sum_{j=p}^{N} \phi_{jp}.$$

Using (32),

$$\sum_{j=p}^{N} S_j(\tau, t) < (t-\tau)\left(1 - \sum_{j=1}^{p-1} \rho_j\right) - x \sum_{j=p}^{N} \phi_{jp}.$$

Since $[\tau, t]$ is in a system busy period,

$$\sum_{j=p}^{N} S_j(\tau, t) = (t-\tau) - \sum_{j=1}^{p-1} S_j(\tau, t).$$

Thus,

$$(t-\tau) - \sum_{j=1}^{p-1} S_j(\tau, t) < (t-\tau)\left(1 - \sum_{j=1}^{p-1} \rho_j\right) - x \sum_{j=p}^{N} \phi_{jp}$$

$$\Rightarrow \sum_{j=1}^{p-1} S_j(\tau, t) > (t-\tau) \sum_{j=1}^{p-1} \rho_j + x\left(1 + \sum_{j=p+1}^{N} \phi_{jp}\right)$$

since $\phi_{pp} = 1$. □

Lemma 8: Let $1, \ldots, N$ be a feasible ordering, and suppose that session p is busy in the interval $[\tau, t]$. Then, if $S_p(\tau, t) \leq \rho_p(t-\tau)$:

$$\sum_{k=1}^{p} S_k(\tau, t) > (t-\tau) \sum_{k=1}^{p} \rho_k \tag{33}$$

Proof: Let

$$S_p(\tau, t) = \rho_p(t-\tau) - x$$

$x \geq 0$. Then, from (31), we are done since $x \sum_{j=p+1}^{N} \frac{\phi_j}{\phi_p} \geq 0$. □

Lemma 9: Let $1, \ldots, N$ be a feasible ordering, and suppose that session p is busy in the interval $[\tau, t]$. Then, if $S_p(\tau, t) \leq \rho_p(t-\tau)$:

$$\sum_{k=1}^{p} \sigma_k^t \leq \sum_{k=1}^{p} \sigma_k^\tau.$$

Proof: From Lemma 2, for every k,

$$\sigma_k^\tau + \rho_k(t-\tau) - S_k(\tau,t) \geq \sigma_k^t.$$

Summing over k and substituting from (33), we have the result.

If we choose τ to be the beginning of a session p busy period, then Lemma 9 says that if $S_p(\tau,t) \leq \rho_p(t-\tau)$ then

$$\sigma_p^t + \sum_{k=1}^{p-1} \sigma_k^t \leq \sigma_p + \sum_{k=1}^{p-1} \sigma_k^\tau. \quad (34)$$

□

Now we will prove Theorem 4.

Proof (of Theorem 4): We proceed by induction on the index of session p.

Basis: $p = 1$. Define τ to be the last time at or before t such that $Q_1(\tau) = 0$. Then, session 1 is in a busy period in the interval $[\tau, t]$, and we have

$$S_1(\tau,t) \geq \frac{(t-\tau)\phi_1}{\sum_{k=1}^N \phi_k} > (t-\tau)\rho_1.$$

The second inequality follows since session 1 is first in a feasible order, implying that $\rho_1 < \frac{\phi_1}{\sum_{k=1}^N \phi_k}$. From Lemma 2,

$$\sigma_1^t \leq \sigma_1^\tau + \rho_1(t-\tau) - S_1(\tau,t) < \sigma_1^\tau \leq \sigma_1.$$

This shows the basis.

Inductive Step: Assume the hypothesis for $1, 2, \ldots, p-1$ and show it for p. Observe that if $Q_i(t) = 0$ for any session i then $\sigma_i^t \leq \sigma_i$. Now consider two cases:

Case 1: $\sigma_p^t \leq \sigma_p$: By the induction hypothesis:

$$\sum_{i=1}^{p-1} \sigma_i^t \leq \sum_{i=1}^{p-1} \sigma_i.$$

Thus,

$$\sum_{i=1}^{p} \sigma_i^t \leq \sum_{i=1}^{p} \sigma_i.$$

Case 2: $\sigma_p^t > \sigma_p$: Session p must be in a session p busy period at time t, so let τ be the time at which this busy period begins. Also, from (22): $S_p(\tau,t) < \rho_p(t-\tau)$. Applying (34):

$$\sigma_p^t + \sum_{k=1}^{p-1} \sigma_k^t \leq \sigma_p + \sum_{k=1}^{p-1} \sigma_k^\tau \leq \sum_{k=1}^{p} \sigma_k. \quad (35)$$

where, in the last inequality, we have used the induction hypothesis. □

Proof of the Main Result

In this section, we will use Lemma 6 and Theorem 4 to prove Theorem 3 for infinite capacity incoming links.

Let $\hat{A}_1, \ldots, \hat{A}_N$ be the set of arrival functions in which all the sessions are greedy from time 0, the beginning of a system busy period. For every session p, let $\hat{S}_p(\tau,t)$, and $\hat{D}_p(t)$ be the session p service and delay functions under $\hat{A}$. We first show

Lemma 10: Suppose that time t is contained in a session p busy period that begins at time τ: Then

$$\hat{S}_p(0, t-\tau) \leq S_p(\tau,t). \quad (36)$$

Proof: Define $\mathcal{B}$ as the set of sessions that are busy at time $t-\tau$ under $\hat{A}$. From Lemma 6:

$$S_p(\tau,t) \geq \frac{(t-\tau-\sum_{i\notin\mathcal{B}}(\sigma_i^\tau + \rho_i(t-\tau)))\phi_i}{\sum_{j\in\mathcal{B}}\phi_j}$$

Since the order in which the sessions become inactive is a feasible ordering, Theorem 4 asserts that:

$$S_p(\tau,t) \geq \frac{(t-\tau-\sum_{i\notin\mathcal{B}}(\sigma_i + \rho_i(t-\tau)))\phi_i}{\sum_{j\in\mathcal{B}}\phi_j}$$

$$= \hat{S}_i(0, t-\tau).$$

(from Lemma 6) and (36) is shown. □

Lemma 11: For every session i, D_i^* and Q_i^* are achieved (not necessarily at the same time) when every session is greedy starting at time zero, the beginning of a system busy period.

Proof: We first show that the session i backlog is maximized under $\hat{A}$: Consider any set of arrival functions $A = \{A_1, \ldots, A_N\}$ that conforms to (25), and suppose that for a session i busy period that begins at time τ:

$$Q_i(t^*) = \max_{t\geq\tau} Q_i(t).$$

From Lemma 10,

$$\hat{S}_i(0, t^*-\tau) \leq S_i(\tau, t^*).$$

Also,

$$A_i(\tau, t^*) \leq \sigma_i + \rho_i(t-\tau) = \hat{A}_i(0, t^*-\tau).$$

Thus,

$$\hat{A}_i(0, t^*-\tau) - \hat{S}_i(0, t^*-\tau) \geq A_i(\tau, t^*) - S_i(\tau, t^*)$$

i.e.,

$$\hat{Q}_i(t^*-\tau) \geq Q_i(t^*).$$

The case for delay is similar: Consider any set of arrival functions $A = \{A_1, \ldots, A_N\}$ that conforms to (25); for a session i busy period that begins at time τ, let t^* be the smallest time in that busy period such that:

$$D_i(t^*) = \max_{t\geq\tau} D_i(t).$$

From the definition of delay in (17):

$$A_i(\tau, t^*) - S_i(\tau, t^* + D_i(t^*)) = 0.$$

Let us denote $d_i^* = t^* - \tau$. From Lemma 10,

$$\hat{S}_i(0, d_i^* + D_i(t^*)) \leq S_i(\tau, t^* + D_i(t^*))$$

and, since $\sigma_i \geq \sigma_i^\tau$:

$$\hat{A}_i(0, d_i^*) \geq A_i(\tau, t^*).$$

Thus,

$$\hat{A}_i(0, d_i^*) - \hat{S}_i(0, d_i^* + D_i(t^*)) \geq$$

$$A_i(\tau, \tau + t^*) - S_i(\tau, t^* + D_i(t^*)) = 0$$

$$\Rightarrow \hat{D}_i(d_i^*) \geq D_i(t^*).$$

Thus, we have shown Theorem 3 for infinite capacity incoming links. □

VII. Generalized Processor Sharing with Finite Link Speeds

In the infinite link capacity case, we were able to take advantage of the fact that a session could use up all of its outstanding tokens instantaneously. In this section, we include the maximum rate constraint, i.e., for every session i, the incoming session traffic can arrive at a maximum rate of $C_i \geq 1$. Although this can be established rigorously [16], it is not hard to see that Theorem 3 still holds: Consider a given set of arrival functions for which there is no peak rate constraint. Now consider the intervals over which a particular session i is backlogged when the arrivals reach the server through (a) infinite capacity input links and (b) input links such that $1 \leq C_j$ for all j and $C_k < \infty$ for at least one session k. Since the server cannot serve any session at a rate of greater than 1, the set of intervals over which session i is backlogged is identical for the two cases. This argument holds for every session in the system, implying that the session service curves are identical for cases (a) and (b). Thus, Lemma 10 continues to hold, and Theorem 3 can be established easily from this fact. We have not been able to show that Theorem 3 holds when $C_j < 1$ for some sessions j, but delay bounds calculated for the case $C_j = 1$ (or $C_j = \infty$) apply to such systems since any link of capacity 1 (or ∞) can simulate a link of capacity less than 1.

VIII. The Output Burstiness σ_i^{out}

In this section, we focus on determining, for every session i, the least quantity σ_i^{out} such that

$$S_i \sim (\sigma_i^{out}, \rho_i, r)$$

where r is the rate of the server. This definition of output burstiness is due to Cruz [5]. (To see that this is the best possible characterization of the output process, consider the case in which session i is the only active session and is greedy from time zero. Then, a peak service rate of r and a maximum sustainable average rate of ρ_i are both achieved.) By characterizing S_i in this manner, we can begin to analyze networks of servers, which is the focus of the sequel to this paper. Fortunately, there is a convenient relationship between σ_i^{out} and Q_i^*:

Lemma 12: If $C_j \geq r$ for every session j, where r is the rate of the server, then for each session i:

$$\sigma_i^{out} = Q_i^*.$$

Proof: First, consider the case $C_i = \infty$. Suppose that Q_i^* is achieved at some time t^*, and session i continues to send traffic at rate ρ_i after t^*. Further, for each $j \neq i$, let t_j be the time of arrival of the last session j bit to be served before time t^*. Then, Q_i^* is also achieved at t^* when the arrival functions of all sessions $j \neq i$ are truncated at t_j, i.e., $A_j(t_j, t) = 0$, $j \neq i$. In this case, all other session queues are empty at time t^* and, beginning at time t^*, the server will exclusively serve session i at rate 1 for $\frac{Q_i^*}{1-\rho_i}$ units of time, after which session i will be served at rate ρ_i. Thus,

$$S_i(t^*, t) = \min\{t - t^*, Q_i^* + \rho_i(t^* - t)\}, \forall t \geq t^*.$$

From this, we have

$$\sigma_i^{out} \geq Q_i^*.$$

We now show that the reverse inequality holds as well: For any $\tau \leq t$:

$$\begin{aligned} S_i(\tau, t) &= A_i(\tau, t) + Q_i(\tau) - Q_i(t) \\ &\leq l_i^\tau + \rho_i(t - \tau) + Q_i(\tau) - Q_i(t) \\ &= \sigma_i^\tau - Q_i(t) + \rho_i(t - \tau) \end{aligned}$$

(since $C_i = \infty$.) This implies that

$$\sigma_i^{out} \leq \sigma_i^\tau - Q_i(t) \leq \sigma_i^\tau \leq Q_i^*.$$

Thus,

$$\sigma_i^{out} = Q_i^*.$$

Now suppose that $C_i \in [r, \infty)$. Since the traffic observed under the all-greedy regime is indistinguishable from a system in which all incoming links have infinite capacity, we must have $\sigma_i^{out} = Q_i^*$ in this case as well. □

References

[1] D. Bertsekas and R. Gallager, *Data Networks*. Englewood Cliffs, NJ: Prentice Hall, 1991.

[2] A. Birman, P. C. Chang, J. S. C. Chen, and R. Guerin, "Buffer sizing in an ISDN frame relay switch," Tech. Rep. RC14 386, IBM Res., Aug. 1989.

[3] A. Birman, H. R. Gail, S. L. Hantler, Z. Rosberg, and M. Sidi, "An optimal policy for buffer systems," Tech. Rep. RC16 641, IBM Res., Mar. 1991.

[4] I. Cidon, I. Gopal, G. Grover, and M. Sidi, "Real time packet switching: A performance analysis,"*IEEE J. Select. Areas Commun.*, vol. SAC-6, pp. 1576–1586, 1988.

[5] R. L. Cruz, "A calculus for network delay, Part I: Network elements in isolation,"*IEEE Trans. Inform. Theory*, vol. 37, pp. 114–131, 1991.

[6] ——, "A calculus for network delay, Part II: Network analysis," *IEEE Trans. Inform. Theory*, vol. 37, pp. 132–141, 1991.

[7] A. Demers, S. Keshav, and S. Shenkar, "Analysis and simulation of a fair queueing algorithm," *Internet. Res. and Exper.*, vol. 1, 1990.

[8] H. R. Gail, G. Grover, R. Guerin, S. L. Hantler, Z. Rosberg, and M. Sidi, "Buffer size requirements under longest queue first," Tech. Rep. RC14 486, IBM Res., Jan. 1991.

[9] S. J. Golestani, "Congestion-free transmission of real-time traffic in packet networks," in *Proc. IEEE INFOCOM '90*, San Fransisco, CA, 1990, pp. 527–536.

[10] ——, "A framing strategy for connection managment," in *Proc. SIGCOMM '90*, 1990.

[11] ——, "Duration-limited statistical multiplexing of delay sensitive traffic in packet networks," in *Proc. IEEE INFOCOM '91*, 1991.

[12] A. C. Greenberg and N. Madras, "How fair is fair queueing?," *J. ACM*, vol. 3, 1992.

[13] E. Hahne, "Round robin scheduling for fair flow control," Ph.D. thesis, Dept. Elect. Eng. and Comput. Sci., M.I.T., Dec. 1986.

[14] L. Kleinrock, *Queueing Systems Vol. 2: Computer Applications*. New York: Wiley, 1976.
[15] C. Lu and P. R. Kumar, "Distributed scheduling based on due dates and buffer prioritization," Tech. Rep., Univ. of Illinois, 1990.
[16] A. K. Parekh, "A generalized processor sharing aproach to flow control in integrated services networks," Ph.D. thesis, Dept. of Elect. Eng. and Comput. Sci., M.I.T., Feb. 1992.
[17] A. K. Parekh and R. G. Gallager, "A generalized processor sharing approach to flow control—The multiple node case," Tech. Rep. 2076, Lab. for Inform. and Decision Syst., M.I.T., 1991.
[18] J. R. Perkins and P. R. Kumar, "Stable distributed real-time scheduling of flexible manufacturing systems," *IEEE Trans. Aut. Contr.*, vol. AC-34, pp. 139–148, 1989.
[19] G. Sasaki, "Input buffer requirements for round robin polling systems," in *Proc. Allerton Conf. Commun., Contr., and Comput.*, 1989.
[20] J. Turner, "New directions in communications, or Which way to the information age?," *IEEE Commun. Mag.*, vol. 24, pp. 8–15, 1986.
[21] L. Zhang, "A new architecture for packet switching network protocols," Ph.D. thesis, Dept. Elect. Eng. and Comput. Sci., M.I.T., Aug. 1989.

Abhay K. Parekh (M'92) received the B.E.S. degree in mathematical sciences from Johns Hopkins University, the S.M. degree in operations research from the Sloan School of Management, and the Ph.D. degree in electrical engineering and computer science from the Massachusetts Institute of Technology in 1992.

He was involved in private network design as a Member of Technical Staff at AT&T Bell Laboratories from 1985 to 1987. From February to June 1992, he was a Postdoctoral Fellow at the Laboratory for Computer Science at M.I.T., where he was associated with the Advanced Network Architecture Group. In October 1992, he joined the High Performance Computing and Communications Group at IBM as a Scientific Staff Member. His current research interests are in application-driven quality of service for integrated services networks, and in distributed protocols for global client-server computing. While a student at M.I.T., he was a Vinton Hayes Fellow and a Center for Intelligent Control Fellow. A paper from his Ph.D. dissertation, jointly authored with Prof. Robert Gallager, won the INFOCOM '93 best paper award.

Robert G. Gallager (S'58–M'61–F'68) received the B.S.E.E. degree in electrical engineering from the University of Pennsylvania in 1953, and the S.M. and Sc.D. degrees in electrical engineering from the Massachusetts Institute of Technology in 1957 and 1960, respectively.

Following two years at Bell Telephone Laboratories and two years in the U.S. Signal Corps, he has been at M.I.T. since 1956. He is currently the Fujitsu Professor of Electrical Engineering and Co-Director of the Laboratory for Information and Decision Systems. His early work was on information theory, and his textbook *Information Theory and Reliable Communication* (New York: Wiley, 1968) is still widely used. Later research focused on data networks. *Data Networks* (Englewood Cliffs, NJ: Prentice Hall, 1992), coauthored with D. Bertsekas, helps provide a conceptual foundation for this field. Recent interests include multiaccess information theory, radio networks, and all-optical networks. He has been a consultant at Codex Motorola since its formation in 1962. He was on the IEEE Information Theory Society's Board of Governors from 1965 to 1970 and 1979 to 1988, and was its president in 1971. He was elected a member of the National Academy of Engineering in 1979 and a member of the National Academy of Sciences in 1992. He was the recipient of the IEEE Medal of Honor in 1990, awarded for fundamental contributions to communications coding techniques.

DQDB Networks with and without Bandwidth Balancing

Ellen L. Hahne, *Member, IEEE*, Abhijit K. Choudhury, *Member, IEEE*, and Nicholas F. Maxemchuk, *Fellow, IEEE*

Abstract— **This paper explains why long distributed queue dual bus (DQDB) networks without bandwidth balancing can have fairness problems when several nodes are performing large file transfers. The problems arise because the network control information is subject to propagation delays that are much longer than the transmission time of a data segment. Bandwidth balancing is then presented as a simple solution. By constraining each node to take only a certain fraction of the transmission opportunities offered to it by the basic DQDB protocol, bandwidth balancing gradually achieves a fair allocation of bandwidth among simultaneous file transfers. We also propose two ways to extend this procedure effectively to multipriority traffic.**

I. Introduction

The distributed queue dual bus (DQDB) [1], [2] is a metropolitan area network that has recently been standardized by the IEEE [3]. The dual-bus topology is identical to that used in Fasnet [4] and is depicted in Fig. 1. The two buses support unidirectional communications in opposite directions. Nodes are connected to both buses and communicate by selecting the proper bus. In both DQDB and Fasnet a special unit at the head-end of each bus generates slots; however, the protocols for acquiring slots differ significantly. Fasnet uses a protocol similar to that of a token ring, where each station is given an opportunity to transmit in order. DQDB resembles a slotted ring with free access, where stations transmit in every empty slot if they have data. Both token access and free access have performance drawbacks, so Fasnet and DQDB use the channel in the opposite direction from which they are sending data to derive performance improvements over the earlier networks. In a token-passing network, a significant fraction of the bandwidth can be wasted as the token circulates among a small number of active stations. Therefore, Fasnet includes several techniques for using the reverse channel to reduce the token circulation time and to let stations use slots that would otherwise have been wasted. In a slotted ring network with free access, one station may take all the slots and prevent the others from transmitting. This fairness problem is exacerbated when the ring topology is replaced by a bus because the station closest to the bus head-end always has first access to slots. Therefore DQDB uses the reverse channel to reserve slots for stations that are farther from the head-end, as explained in Section II. The aim of DQDB's reservations is to improve access fairness without the bandwidth wastage of a circulating token. Moreover, by associating priority levels with these reservations, DQDB can offer multipriority service.

Paper approved by the Editor for Wide Area Networks of the IEEE Communications Society. Manuscript received October 8, 1989; revised October 30, 1990 and August 6, 1991. This paper was presented in part at IEEE INFOCOM '90, San Francisco, CA, June 1990 and IEEE INFOCOM '91, Bal Harbour, FL, April 1991.

The authors are with AT&T Bell Laboratories, Murray Hill, NJ 07974.

IEEE Log Number 9201110.

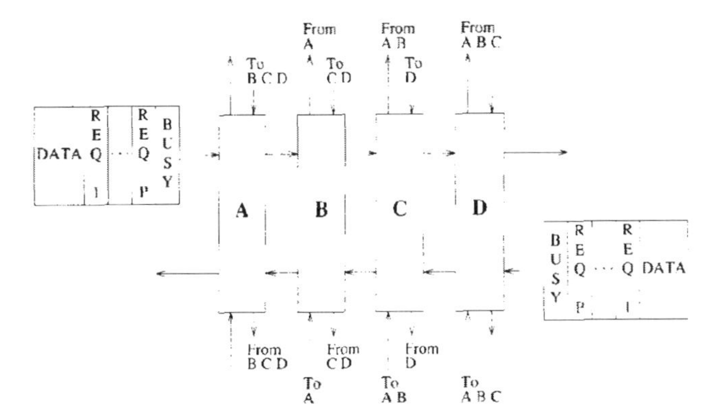

Fig. 1. DQDB network architecture.

Unfortunately, the DQDB reservation process is imperfect. The network span (up to 50 km), the transmission rate (assumed to be 150 Mbps in this paper), and the slot size (53 bytes) of DQDB allow many slots to be in transit between the nodes. Therefore, the nodes can have inconsistent views of the reservation process. If this happens, and if the access protocol is too efficient and tries never to waste a slot, then bandwidth can be divided unevenly among nodes simultaneously performing long file transfers. Since the network does not provide the same throughput to all of the nodes, in that sense it is unfair. This is the main problem to be addressed in this paper. In Section III we offer a novel analysis of a closed queueing network model for this scenario. (Other DQDB fairness studies appear in [5]–[31].)

In Section V we present a simple enhancement to the basic DQDB protocol, called "bandwidth balancing," that equalizes the throughput. (Other recent fairness proposals can be found in [28]–[37].) Bandwidth balancing intentionally wastes a small amount of bus bandwidth in order to facilitate coordination among the nodes currently using that bus, but it divides the remaining bandwidth equally among those nodes. The key idea (adapted from Jaffe [38]) is that the maximum permissible nodal throughput rate is proportional to the unused bus capacity; each node can determine this unused capacity by observing the volume of busy slots and reservations. The throughput is equalized gradually over an interval several times longer than the propagation delay between competing nodes. Bandwidth balancing is easy to implement: each node

Reprinted from *IEEE Transactions on Communications,* vol. 40, No. 7, July 1992.

permits itself to use only a certain fraction of the transmission opportunities offered to it by the basic DQDB protocol. If the traffic is all of one priority, then bandwidth balancing requires no additional control information to be transmitted, and it requires only one additional counter (per bus) in each node.

Bandwidth balancing has been incorporated into the DQDB standard as a required feature that is enabled by default; the ability to disable this feature is also required. In the standard, bandwidth balancing conforms to the priority structure of the basic DQDB protocol, which is explained in Section II below. In particular, a node is aware of the priority levels of incoming reservations, but not the priority levels of the data in busy slots. This asymmetry means that different nodes have different information about the traffic on a bus, making it difficult to control multipriority traffic. The version of bandwidth balancing specified in the standard guarantees equal allocations of bus bandwidth to nodes with traffic of the lowest priority level. A node with higher-priority traffic is guaranteed at least as much bandwidth as a lowest-priority node, but no further guarantees are possible. Furthermore, when there are nodes with different priorities, the throughputs achieved by nodes can depend on their relative positions on the bus [27], [39].

In Sections VI and VII of this paper we propose two better ways to extend the unipriority bandwidth balancing procedure of Section V to multipriority traffic. (Additional proposals appear in [40] and [41].) The first step is to correct the asymmetry of the priority information, either by adding priority information about the data in busy slots, or by removing priority information from the reservations. The former method we call the "global" approach, because the priority information for all traffic is available to all nodes by reading the data and reservation channels. The latter method we call the "local" approach because a node is only aware of the priority of its own locally generated data, and the node does not disseminate this information over the network. Section VI presents a version of bandwidth balancing based on local priority information. (Similar "local" versions of bandwidth balancing have been proposed by Damodaram [42], Spratt [43], and Hahne and Maxemchuk [44], [45]). Section VII presents a version of bandwidth balancing based on global priority information. (A cruder version of this scheme appears in [46].) Both multipriority schemes presented in this paper produce bandwidth allocations that are independent of the nodes' relative positions on the bus. Moreover, both schemes are fair, i.e., they allocate equal bandwidth shares to all nodes active at the same priority level. The schemes differ in the way they allocate bandwidth across the various priority levels. Either scheme could easily be included in a future version of the standard, and nodes satisfying the old standard could share the same network with the new nodes, provided that the old nodes only generate traffic of the lowest priority level.

II. DQDB Without Bandwidth Balancing

DQDB allows some network capacity to be set aside for synchronous services such as voice, but we will assume that no such traffic is present. DQDB supports asynchronous traffic of several priority levels, which for convenience we will number from 1 (least important) through P (most important).[1]

DQDB uses the dual-bus topology depicted in Fig. 1. The two buses support unidirectional communications in opposite directions. Nodes are connected to both buses and communicate by selecting the proper bus. The transmission format is slotted, and each bus is used to reserve slots on the other bus in order to make the access fairer. Each slot contains one request bit for each priority level and a single busy bit. The busy bit indicates whether another node has already inserted a segment of data into the slot. The request bits on one bus are used to notify nodes with prior access to the data slots on the other bus that a node is waiting. When a node wants to transmit a segment on a bus, it waits for an empty request bit of the appropriate priority on the opposite bus and sets it, and it waits for an empty slot on the desired bus to transmit the data.

The IEEE 802.6 Working Group is currently considering optional procedures for erasing the data from a slot once it passes the destination, so that the slot can be reused [47], [48], [49]. However, this paper assumes that once a data segment has been written into a slot, that data is never erased or overwritten. This paper only discusses how data segments are written onto a bus, since reading data from the bus is straightforward.

The operation for data transmission in both directions is identical. Therefore, for the remainder of this paper, the operation in only one direction is described. One bus will be considered the data bus. Slots on this bus contain a busy bit and a payload of one segment of data. These slots are transmitted from upstream nodes to downstream nodes. The other bus is considered the request bus. Each slot on this bus contains one request bit for each priority level, and slots are transmitted from downstream nodes to upstream nodes.

Fig. 2 shows how a DQDB node operates with bandwidth balancing disabled. We model each node as composed of P sections, one to manage the writing of requests and data for each priority level. We assume that the sections act like separate nodes: each section has its own attachments to the buses, and the data bus passes through the sections in order, starting with priority 1, while the request bus passes through the sections in the opposite order, starting with priority P. While this layout does not correspond to the actual physical implementation of DQDB,[2] for the purpose of this paper, they are functionally equivalent.[3]

[1] The DQDB standard calls for three priority levels, i.e., $P = 3$, but it labels them differently: 0, 1, 2.

[2] In an actual DQDB node, the sections of the various priority levels all read the request bus at the same place, before any of them has a chance to write. Write-conflicts on the request bus are impossible, though, because the priority-p request bit can only be set by the priority-p node section. There is some internal linkage among the sections: whenever a section generates a request, it notifies all lower-priority sections within the node as well as writing the request onto the bus. Similarly, all sections of the node read the data bus at the same place, before any of them can write. Nevertheless, because of the internal linkage just described, the protocol can guarantee that two node sections will not try to write data into the same empty slot.

[3] Figs. 2, 4, 8, and 10 and associated text are intended to be functional descriptions. The physical implementation of these ideas will not be discussed in this paper.

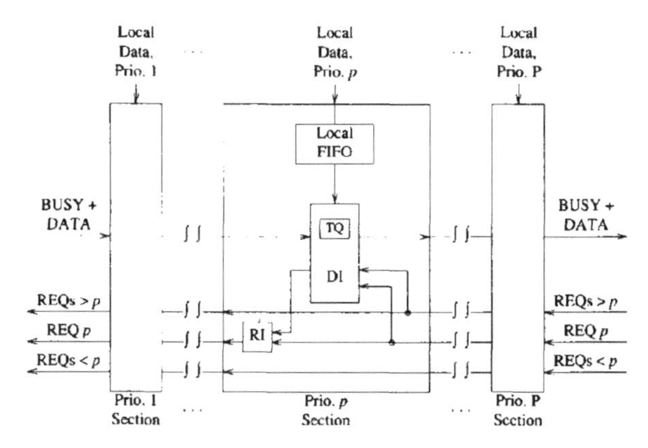

Fig. 2. DQDB node architecture without bandwidth balancing.

In Fig. 2 the details of the priority-p section are shown. This section has a local FIFO queue to store priority-p data segments generated by local users while these segments wait for the data inserter (DI) to find the appropriate empty slots for them on the data bus. The data inserter operates on one local data segment at a time; once the local FIFO queue forwards a segment to the data inserter, the local FIFO queue may not forward another segment until the data inserter has written the current segment onto the data bus. When the data inserter takes a segment from the local FIFO queue, first it orders the request inserter (RI) to send a priority-p request on the request bus. Then the data inserter determines the appropriate empty slot for the local segment by inserting the segment into the data inserter's transmit queue (TQ). All the other elements of this queue are requests of priority p or greater from downstream nodes. (The data inserter ignores all requests of priority less than p.) The transmit queue orders its elements according to their priority level, with elements of equal priority ordered by the times they arrived at the data inserter. The data inserter serves its transmit queue whenever an empty slot comes in on the data bus. If the element at the head of the queue is a request, then the data inserter lets the empty slot pass. If the head element is the local data segment, then the busy bit is set and the segment is transmitted in that slot. The transmit queue is implemented with two counters, called the request counter and the countdown counter. When there is no local data segment in the queue, the request counter keeps track of the number of unserved reservations from downstream nodes in the transmit queue. When the data inserter accepts a local data segment, the request counter value is moved to the countdown counter, which counts the number of reservations that are ahead of the local data segment in the transmit queue, and the request counter is then used to count reservations behind the local data segment. The request inserter sends one reservation of priority p for each data segment taken by the data inserter from the local FIFO queue. Since the incoming priority-p request bits may have been set already by downstream nodes, the request inserter sometimes needs to queue the internally generated reservations until vacant request bits arrive. Thus, it is possible for a data segment to be transmitted before its reservation is sent.

Perfect operation of the DQDB protocol without bandwidth balancing would occur if the system had no propagation or processing delays, and if it included an idealized reservation channel with no slotting, queueing or transmission delays. Under these conditions [50]:

- Slots are never wasted.
- The priority mechanism is absolute (i.e., a data segment can only be transmitted when there are no higher priority segments waiting anywhere in the network).
- Nodes with traffic at the current highest priority level are served one-segment-per-node in round-robin fashion.

However, if the propagation delay between nodes is much longer than the transmission time of a data segment, then performance deteriorates. This is the subject of the next section.

III. Throughput Fairness of DQDB Without Bandwidth Balancing

When the network propagation delay is larger than a slot transmission time, the DQDB access protocol without bandwidth balancing is unfair, in the sense that nodes simultaneously performing large file transfers can obtain different throughputs. The severity of the problem depends upon the propagation delay, the network utilization, and the lengths of the messages submitted to the network. In this section, we assume that users often submit messages consisting of a great many segments. This model (suggested to us by M. Rodrigues) seems to be increasingly appropriate as diskless workstations abound, because large files are typically transferred between a workstation and its file server. Network "overloads" are caused by as few as two users simultaneously performing file transfers and hence could be quite typical. We model an overloaded DQDB system as a closed network of queues and study its fairness through a novel approximate analysis.

We will examine scenarios similar to those explored in Wong's study [10] of an earlier version of DQDB. Consider two nodes that are transmitting very long messages of the same priority. Call the upstream node 1 and the downstream node 2. Ideally, each node should obtain half the bandwidth of the data channel, but this rarely happens.

Suppose that the propagation delay between the nodes equals D slot transmission times where D is an integer. Let Δ be the difference in the starting times of the two nodes, i.e., the time when node 2 wants to begin transmission minus the time when node 1 is ready; Δ is measured in slot times and is assumed to be an integer. Once both nodes are active, node 1 leaves slots idle only in response to requests from node 2. Therefore, once node 2 begins to receive segments transmitted by node 1, the only idle slots node 2 receives are in response to its earlier requests. Each idle slot received by node 2 results in a segment being transmitted, a new segment being queued, and a new reservation being transmitted. Therefore, the number X of requests plus idle slots circulating between the two nodes is fixed. (Some of these requests may be stored in node 1's transmit queue.) Let us call these conserved entities *permits*. This quantity X determines the throughput of the downstream node. Unfortunately, X depends strongly on D and Δ:[4]

$$X = 1 + D - c(\Delta) \tag{3.1}$$

where c is a function that clips its argument to the range $[-D, D]$. To clarify this claim, let us explain the system behavior for extreme values of Δ. If the first segment from node 1 has already been received at node 2 by the time node 2 becomes active, i.e., if $\Delta \geq D$, then node 2 inserts one data segment in its transmit queue and transmits one reservation. The segment will not be transmitted until the reservation is received by node 1 and an idle slot is returned. In this instance, there is one permit in the network. At the other extreme, consider $\Delta \leq -D$. Initially, only node 2 is active. It inserts its first segment in its transmit queue and sends its first reservation upstream. The first segment is transmitted immediately in the first slot. Then the second segment is queued, the second reservation sent, and the second segment is transmitted in the second slot, etc. The request channel is already carrying D requests when node 1 begins transmission, and in the D time slots that it takes for node 1's first segment to reach node 2, node 2 injects another D requests, so that $X \approx 2D$.

Now we will show the relationship between X and the nodal throughputs. Recall that permits can be stored in the request channel, in the data channel, and in the transmit queue of the upstream node. This transmit queue also includes a single data segment from node 1. When the second file transfer begins, there is a transient phase in which the transmit queue length moves to a steady-state average value Q. More precisely, we should distinguish between $Q(1)$, the average queue length observed by a data segment from node 1 just after it has been inserted into node 1's transmit queue and $Q(2)$, the average queue length observed by a request (permit) from node 2 just after it has been inserted into node 1's transmit queue. The difference between these two views of the queue will be explained shortly. The network's steady-state behavior can be determined approximately by simultaneously solving the following equations involving X, $Q(1)$, $Q(2)$, the nodal throughput rates $r(1)$ and $r(2)$, and the average round-trip delay T experienced by a permit. (Throughput rates are measured in segments per slot time, and round-trip delays are measured in slot times.)

$$r(1) + r(2) = 1 \tag{3.2}$$

$$r(1) = 1/Q(1) \tag{3.3}$$

$$r(2) = X/T \tag{3.4}$$

$$T = 2D + Q(2) \approx 2D + Q(1). \tag{3.5}$$

Before solving the equations above, let us discuss the approximation in the last equation. The difference between $Q(1)$ and $Q(2)$ is most pronounced when the internode distance D is large and only one permit circulates. In this case, the queue length observed by the permit is always two (itself plus node 1's data segment), so $Q(2) = 2$. The queue length observed by the data segment, however, is usually one (itself) and occasionally two (itself plus the permit), so $Q(1) \approx 1$. Even though $Q(1)$ and $Q(2)$ differ by almost a factor of two in this example, recall that D is large, so that the approximation shown above for the round-trip delay T is still justified.

[4]The analysis depends on some detailed timing assumptions. We have assumed that the bus synchronization is such that a node reads a busy bit on the data channel immediately after reading a request bit on the reservation channel. We also assume that there are no processing delays in the nodes, and that a node is permitted to insert a new data segment into the transmit queue and send the corresponding reservation as soon as it *begins* to transmit the previous data segment.

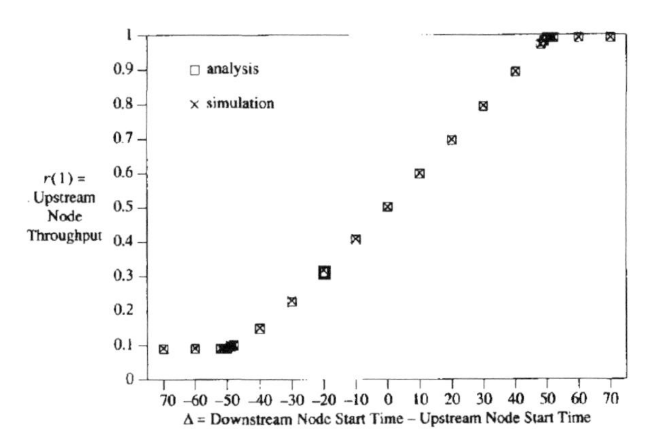

Fig. 3. Bandwidth division for two users without bandwidth balancing.

Solving (3.1)–(3.5) for the steady-state throughput rates yields:

$$r(1) \approx \frac{2}{2 - D - c(\Delta) + \sqrt{(D - c(\Delta) + 2)^2 + 4Dc(\Delta)}}$$

$$r(2) = 1 - r(1).$$

Note that if the nodes are very close together ($D = 0$) or if they become active at the same time ($\Delta = 0$), then each node gets half the bandwidth. However, if D is very large and the downstream node starts much later, its predicted throughput rate is only about $1/2D$. Node 1 is also penalized for starting late (though not as severely as node 2): the worst case upstream rate is roughly $1/\sqrt{2D}$.

The predicted throughputs match simulated values very well. Fig. 3 compares our approximate analysis with simulation results for an internode distance of $D = 50$ slots $\approx$ 29 km. The analysis can easily be generalized to multiple nodes, provided that these nodes are clustered at only two distinct bus locations.

The analysis and simulation studies in this section show perfect fairness when the propagation delay is negligible. However, moderate unfairness can be demonstrated even for $D = 0$ if the timing assumptions in the previous footnote are changed [51], [52]. Hence, bandwidth balancing may prove useful for fairness enhancement even on short networks.

IV. Definitions

This section gives various background assumptions and definitions to be used in the remainder of this paper. Recall that we are focusing on data transmission over one bus only. (Of course, the other bus is needed to carry requests for the use of the primary bus). The term *parcel* will be used to denote the traffic originating at one node at one priority level for transmission over one bus. All traffic rates will be measured in data segments per slot time. We will usually assume that the traffic demand of each node n has some fixed rate $\rho(n)$. This offered load may be stochastic, as long as it has a well-defined average rate. The offered load of the traffic parcel of priority level p at node n will be denoted $\rho_p(n)$. It is possible that not all this offered load can be carried. The actual long-term average throughput of node n will be denoted by $r(n)$, and that of its parcel p by $r_p(n)$. The unused bus capacity will be denoted by U:

$$U = 1 - \sum_m r(m) = 1 - \sum_m \sum_q r_q(m)$$

while U_{p+} will be the bus capacity left over by parcels of priority p *and greater:*

$$U_{p+} = 1 - \sum_m \sum_{q \geq p} r_q(m).$$

Of course, any individual node n at any instant t will not have direct knowledge of the long-term average rates defined above. All the node can see is: $B(n,t)$, the rate of busy slots coming into node n at time t from nodes upstream; $R(n,t)$, the rate of requests coming into node n at time t from nodes downstream; and $S(n,t)$, the rate at which node n serves its own data segments at time t. In one of our proposed protocols, the node can break these observations down by priority level p, in which case they are denoted $B_p(n,t)$, $R_p(n,t)$, and $S_p(n,t)$. These observations can be used to determine $U(n,t)$, the bus capacity *unallocated* by node n at time t. By "unallocated," we mean the capacity that is neither used by nodes upstream of n, nor requested by nodes downstream of n, nor taken by node n itself:

$$\begin{aligned} U(n,t) &= 1 - B(n,t) - R(n,t) - S(n,t) \\ &= 1 - \sum_q B_q(n,t) - \sum_q R_q(n,t) - \sum_q S_q(n,t) \end{aligned}$$

If node n can observe priority levels, then it can also measure $U_{p+}(n,t)$, the bus capacity not allocated by node n at time t to parcels of priority p *or greater*:

$$U_{p+}(n,t) = 1 - \sum_{q \geq p} B_q(n,t) - \sum_{q \geq p} R_q(n,t) - \sum_{q \geq p} S_q(n,t).$$

All the access control protocols described in this paper have a parameter M called the *bandwidth balancing modulus*;[5] in some schemes the modulus is different for each priority level p and is denoted M_p. For convenience we assume that the bandwidth balancing moduli are integers, though rational numbers could also be used.

Finally, let us define an alternative queueing discipline called *deference scheduling* for the data inserter's transmit queue. With deference scheduling, the transmit queue for the node section of priority p still holds at most one local data segment at a time, still accepts no requests of priority less than p, and still accepts all requests of priority p or greater. The difference is that all requests (of priority p or greater) are served *before* the local data segment—even those priority-p requests that entered the transmit queue *after* the local data segment. The local data segment is served only when there are no requests (of priority p or greater) in the transmit queue. Deference scheduling uses a request counter but needs no countdown counter. By itself, deference scheduling makes little sense because it offers no guarantee that the local data segment in the transmit queue will ever be served. This discipline does make sense, however, when used in conjunction with bandwidth balancing, as explained in the next section. To contrast with deference scheduling, we will call the two-counter discipline of Section II *distributed queueing*.

[5] In the DQDB standard the bandwidth balancing modulus is tunable, with a default value of 8.

V. Bandwidth Balancing for Unipriority Traffic

We contend that the unfairness problem discussed in Section III arises because the DQDB protocol pushes the system too hard. If it attempts to use every slot on the bus, DQDB can inadvertently lock the network into unfair configurations for the duration of a sustained overload. In this section, we introduce the concept of bandwidth balancing: the protocol of Section II is followed, except that a node takes only a fraction of the slots that are not reserved or busy. This rate control mechanism lets the system relax a bit so it can work as intended. In this section, we focus on traffic of one priority level only. In Sections VI and VII, we propose ways to extend bandwidth balancing effectively to multipriority traffic.

Section V-A presents our definition of fairness and the bandwidth balancing concept. In Section V-B, the implementation of unipriority bandwidth balancing is described. It requires only one extra counter, and either distributed queueing or deference scheduling may be used. The performance of bandwidth balancing is investigated in Section V-C through analysis and simulation. There we show the existence of a tradeoff between bus utilization and the rate of convergence to a fair operating point. In this paper, we consider only throughput performance during overloads involving two or three active nodes. Other simulation studies of bandwidth balancing using more nodes, different traffic models, and a variety of performance measures appear in [24]–[27].

A. Concept

When the bus is overloaded, we want to divide its bandwidth fairly. Our ideal of throughput fairness is that nodes with sufficient traffic to warrant rate control should all obtain the same throughput, called the *control rate*. As discussed in Section III, the performance of the DQDB protocol without bandwidth balancing can diverge from this ideal when the bus is long. One obvious solution is a centralized approach where nodes inform some controller node of their offered loads; the controller then computes the control rate and disseminates it. We will present an alternative way to do this–one that requires no controller node, no offered load measurements, and no explicit communication of the control rate. Our method intentionally wastes a small amount of bus bandwidth, but evenly divides the remaining bandwidth among the nodes.

The key idea is that the control rate is implicitly communicated through the idle bus capacity; since each node can determine this quantity by observing the passing busy bits from upstream and request bits from downstream, control coordination across the system can be achieved. This idea has also been suggested for congestion control in wide-area mesh-topology networks [38], but the problem there is more complex, since flow control must also be coordinated along multihop paths; in our case, the implementation is quite simple, adding very little to the complexity of DQDB.

More specifically, each node limits its throughput to some multiple M of the unused bus capacity; nodes with less demand than this may have all the bandwidth they desire:

$$r(n) = \min[\rho(n), M \cdot U] = \min\left[\rho(n), M \cdot \left(1 - \sum_m r(m)\right)\right]. \quad (5.1)$$

This scheme is fair in the sense that all rate-controlled nodes get the same bandwidth.

Given the offered loads $\rho(n)$ and the bandwidth balancing modulus M, (5.1) can be solved for the carried loads $r(n)$. If there are N rate-controlled nodes, then the throughput of each is

$$r(n) = \frac{M}{1 + M \cdot N} \cdot (1 - S) \quad (5.2)$$

and the total bus utilization is $\frac{S+M \cdot N}{1+M \cdot N}$ where S is the utilization due to the nodes that are not rate-controlled. (It takes some trial and error to determine which nodes are rate-controlled). The worst case bandwidth wastage is $1/(1 + M)$, which occurs when only one node is active.

For example, if there are three nodes whose average offered loads are 0.24, 0.40, and 0.50 segments per slot time and if $M = 9$, then only the last two nodes are rate-controlled, the carried loads are 0.24, 0.36, and 0.36, respectively, and the wasted bandwidth is 0.04.

One desirable feature of this scheme is that it automatically adapts to changes in network load.

B. Implementation

In order to implement unipriority bandwidth balancing, the slot header need only contain the busy bit and a single request bit. In theory, a node can determine the bus utilization by summing the rate of busies on one bus, the rate of requests on the other bus, and the node's own transmission rate. In the long run, this sum should be the same at every node (though the individual components will differ from node to node). In other words, each node n has enough information available to implement (5.1). Fortunately, it is not necessary for the node to measure the bus utilization rate over some lengthy interval. As the analysis and simulation of Section V-C will show, it is sufficient for node n to respond to arriving busy bits and request bits in such a way that

$$S(n,t) \leq M \cdot U(n,t) = M \cdot [1 - B(n,t) - R(n,t) - S(n,t)] \quad (5.3)$$

or equivalently:

$$S(n,t) \leq \frac{M}{1+M} \cdot [1 - B(n,t) - R(n,t)]. \quad (5.4)$$

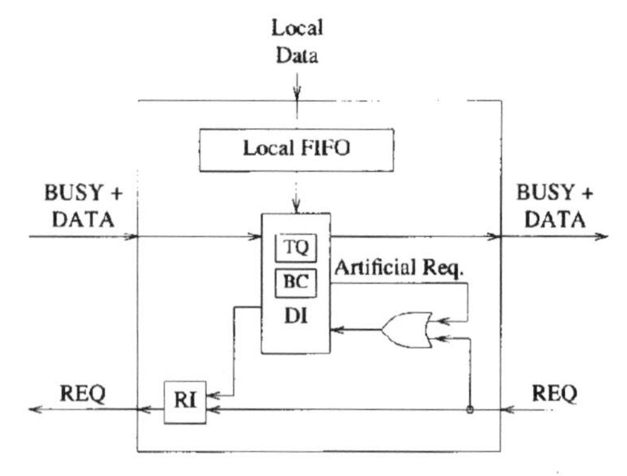

Fig. 4. Implementation of bandwidth balancing.

In other words, the node takes only a fraction $M/(1 + M)$ of the slots that are not reserved or busy at any point in time.

One simple way to implement (5.3) and (5.4) is to add a bandwidth balancing counter (BC) to the data inserter, as shown in Fig. 4. The bandwidth balancing counter counts local data segments transmitted on the bus. After M segments have been transmitted, the bandwidth balancing counter resets itself to zero and generates a signal that the data inserter treats exactly like a request from a downstream node. This artificial request causes the data inserter to let a slot go unallocated. (The request inserter is not aware of this signal; hence the node does not send any extra requests upstream corresponding to the extra idle slots it sends downstream.)

The data inserter may use either distributed queueing or deference scheduling in serving its transmit queue. (Since bandwidth balancing by all nodes ensures some spare system capacity, the local data segment in the transmit queue will eventually be served, regardless of the queue's scheduling discipline.) The advantages of deference scheduling are that it uses one counter (rather than two) and that it is easier to analyze, as we shall show shortly. However, we prefer distributed queueing for the following reasons. 1) While we will show that both versions of bandwidth balancing have the same throughput performance under sustained overload, the delay performance under moderate load is frequently better with distributed queueing [24]. 2) Many DQDB networks will have no significant fairness problems (e.g., if the buses are short, or if the transmission rate is low, or if the application is point-to-point rather than multi-access). In these cases, one would want to disable bandwidth balancing (because it wastes some bandwidth) and use the DQDB protocol as described in Section II, which works only with distributed queueing. It is convenient to build one data inserter that can be used with or without bandwidth balancing, and this would have to be the distributed queueing version.

C. Performance

1) *Analysis:* This section analyzes the transient behavior of bandwidth balancing. In preparation, we first present our modeling assumptions and a useful bound on the value of a node's request counter. Suppose the propagation delays between nodes are all integer numbers of slot transmission times. Let the propagation delay from the most upstream node to the most downstream node be D_{MAX} slot times. Assume

the bus synchronization is such that a node reads a busy bit on the data channel immediately after reading a request bit on the reservation channel. Let the nodes use deference scheduling. To simplify the counting, imagine that deference scheduling serves a node's transmit queue as follows: first all genuine requests (i.e., those from downstream nodes) are served in order of arrival, then the artificial request from the node's bandwidth balancing counter is served, then the node's local data segment is served. This viewpoint yields the following two upper bounds on the time when a genuine request r is served at a node n: (i) the time when r is served at the node immediately upstream from n, plus the one-way node-to-node propagation delay; (ii) the arrival time of r at n, plus the round-trip propagation delay between n and the network's upstream end. (Bound (i) can be proved by induction on the requests. Bound (ii) follows from (i) and induction on the nodes.) Bound (ii) shows that the number of genuine requests in a node's transmit queue can be at most $2D_{\text{MAX}}+1$. Adding in one possible artificial request from the bandwidth balancing counter bounds the request counter value at $2D_{\text{MAX}}+2$.

We will now offer an approximate analysis of bandwidth balancing during simultaneous file transfers by two nodes separated by a propagation delay of D slots. Call the upstream node 1 and the downstream node 2. We will show that the bandwidth balancing scheme converges to the steady-state throughputs given by (5.2), independent of the initial conditions created by the previous history of the system. We will also determine the rate of convergence. Although the analysis assumes deference scheduling, simulations will show that the distributed queueing implementation also achieves the desired steady-state throughputs.

First we show that the request counters of both active nodes drain rapidly. Suppose that both file transfers have started, and that all other nodes have been inactive for at least D_{MAX} slot times, so that the effects of these other nodes have disappeared from the buses (though not necessarily from the Request Counters). In every $M+1$ slot times, nodes 1 and 2 each transmit at most M data segments and at most M requests, i.e., node 1 leaves at least one idle data slot and node 2 leaves at least one vacant request bit. Each of these holes gives the other node a chance to decrement its request counter. Since the request counter values started at $2D_{\text{MAX}}+2$ or less, they will drain to zero within $(2D_{\text{MAX}}+2)\cdot(M+1)$ slot times, and thereafter they will never increase above one.

Now we can show how the throughput rates converge to a fair allocation. Assume that at time 0, the request counters have already drained. Also assume that a fraction f_B of the D busy bits in transit between nodes 1 and 2 on the data bus and a fraction f_R of the D request bits in transit between the nodes are set. For convenience, define

$$\alpha = \frac{M}{1+M}.$$

Each node transmits in a fraction α of the idle slots available to it for its own data transmission. Consequently, in the first D slot times, node 1 will transmit in $\alpha(1-f_R)D$ slots and node 2 will transmit in $\alpha(1-f_B)D$ slots. In the next D slot times, node 1 transmits in $\alpha[1-\alpha(1-f_B)]D$ slots, while node 2 transmits in $\alpha[1-\alpha(1-f_R)]D$ slots. The throughput of a node over half a round-trip time depends on the other node's throughput in the previous half round-trip time. (This analysis is approximate; in the interval D a node actually acquires an integer number of slots.)

Let $\gamma(1,k)$ and $\gamma(2,k)$ be the fraction of the bandwidth acquired by nodes 1 and 2, respectively, during slots kD to $(k+1)D$ where $k=0,1,2,\cdots$. The analyses for the two nodes are similar and we shall concentrate on the bandwidth acquired by node 1. Consider the sequence $\gamma(1,k)$ to be composed of two subsequences: a subsequence of even terms $\gamma^e(1,m)=\gamma(1,2m)$ and a subsequence of odd terms $\gamma^o(1,m)=\gamma(1,2m+1)$, for $m=0,1,2,\cdots$. Both subsequences $\gamma^e(1,m)$ and $\gamma^o(1,m)$ satisfy the same difference equation, for $m=1,2,3,\cdots$,

$$\gamma^e(1,m)=\alpha(1-\alpha)+\alpha^2\gamma^e(1,m-1)$$
$$\gamma^o(1,m)=\alpha(1-\alpha)+\alpha^2\gamma^o(1,m-1)$$

but they have different initial conditions:

$$\gamma^e(1,0)=\alpha(1-f_R)$$
$$\gamma^o(1,0)=\alpha[1-\alpha(1-f_B)].$$

The throughput of node 1 over half round-trip times can be found by separate Z-transform analyses of the even and odd subsequences:

$$\gamma(1,k)=\begin{cases}\frac{\alpha}{1+\alpha}-\left[f_R-\frac{\alpha}{1+\alpha}\right]\alpha^{k+1}, & k \text{ even}\\ \frac{\alpha}{1+\alpha}+\left[f_B-\frac{\alpha}{1+\alpha}\right]\alpha^{k+1}, & k \text{ odd}\end{cases} \quad (5.5)$$

Similarly, the throughput of node 2 over half round-trip times is given by

$$\gamma(2,k)=\begin{cases}\frac{\alpha}{1+\alpha}-\left[f_B-\frac{\alpha}{1+\alpha}\right]\alpha^{k+1}, & k \text{ even}\\ \frac{\alpha}{1+\alpha}+\left[f_R-\frac{\alpha}{1+\alpha}\right]\alpha^{k+1}, & k \text{ odd}\end{cases}. \quad (5.6)$$

We can use the model developed above to analyze various possible scenarios in the simultaneous transfer of two files, some of which are listed below.

- Both nodes turn on at the same time: $f_B=f_R=0$.
- The upstream node turns on at least half a round-trip time before the downstream node: $f_B=\alpha$, $f_R=0$.
- The downstream node turns on at least half a round-trip time before the upstream node: $f_B=0$, $f_R=\alpha$.

The approximate throughput expressions are found to match simulation results reasonably well. Fig. 5 compares the analysis with simulation results for the case where the two active nodes are separated by 38 slots ($\approx$ 22 km), the upstream node starts transmitting at least half a round-trip time before the downstream node, and $\alpha=0.9$. The plotted throughputs are measured over successive full round-trip times (i.e., successive 76 slot intervals). Simulation results are shown for both deference scheduling and distributed queueing.

Let us make a few remarks on (5.5) and (5.6). Note that in steady state the nodal throughputs are each $\frac{\alpha}{1+\alpha}=\frac{M}{1+2M}$ and the amount of system bandwidth wasted is $\frac{1-\alpha}{1+\alpha}=\frac{1}{1+2M}$, in accord with (5.2). For example, if $\alpha=0.9$, 5.3% of the

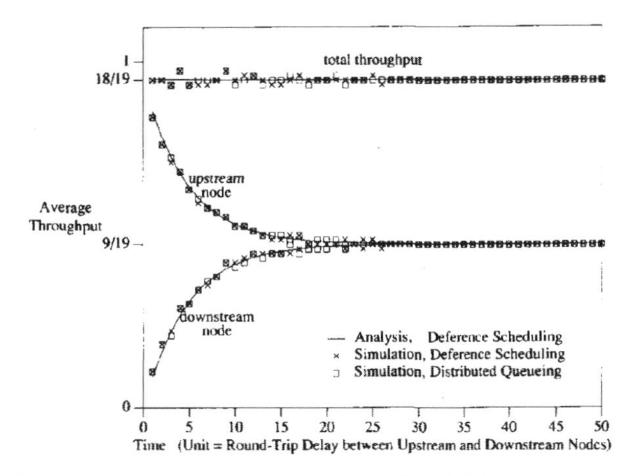

Fig. 5. Throughputs for two users with bandwidth balancing ($\alpha = 0.9$, $M = 9$).

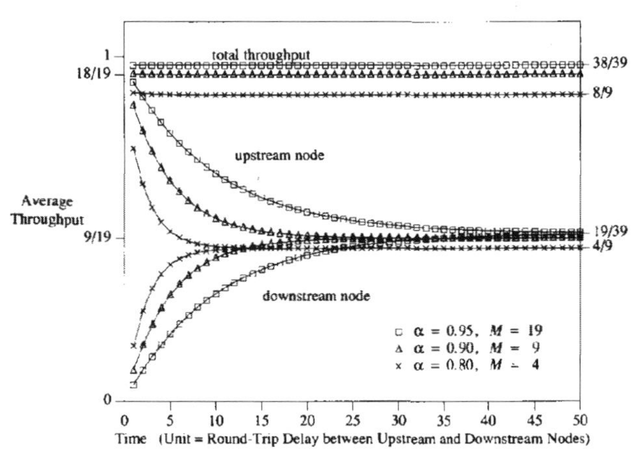

Fig. 6. Throughput performance of bandwidth balancing for various values of α.

bandwidth is wasted. Note, moreover, that the steady-state nodal throughputs are independent of the initial conditions f_B and f_R, in marked contrast to the behavior of DQDB without bandwidth balancing, shown in Fig. 3. Finally note that, while the exact transient depends on f_B and f_R, the rate of convergence depends only on α. For example, if $\alpha = 0.9$, then the error (i.e., the unfairness) in each nodal throughput $\gamma(n, k)$ shrinks by a factor of $0.9^{22} = 0.1$ every $22D$ slot times. In other words, each nodal throughput moves 90% of the way to its steady-state value every 11 round-trip times. A lower α results in faster convergence but more bandwidth wastage. The effect of different values of α on the convergence rate and on the steady-state throughputs is shown in Fig. 6, for the same scenario as Fig. 5.

2) *Simulation:* Fig. 7 depicts simultaneous file transfers by three nodes, with 28 slots ($\approx$ 16 km) between successive nodes. The plot shows the average nodal throughputs measured over successive 112 slot intervals. Bandwidth balancing is used, with $M = 9$. The system starts in an idle state. The most upstream node comes up first and immediately achieves a throughput of 9/10, in accord with (5.2). The most downstream node turns on next and contends with the upstream node for an equal share of the bandwidth, viz.,

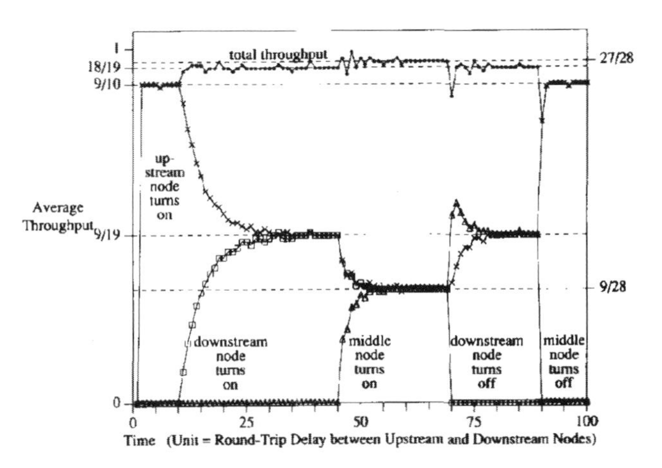

Fig. 7. Throughputs for three users with bandwidth balancing ($M = 9$).

9/19. The middle node turns on next, and the system again adjusts so that all three nodes achieve the throughput of 9/28 predicted by (5.2). The most downstream node and then the middle node complete their file transfers, and in each case the system adjusts rapidly and redistributes the available bandwidth equally. Note that the amount of wasted bandwidth decreases as the number of active nodes increases. (The simulation for Fig. 7 used distributed queueing; the simulation was also performed with deference scheduling and the results were virtually indistinguishable.)

An interesting feature of bandwidth balancing is that nodes whose offered load is less than the control rate are not rate-controlled. The remaining bandwidth is distributed equally among the rate-controlled nodes, in accord with (5.2). The simulation results in Table I show the distribution of bandwidth among three active nodes, when the upstream and downstream nodes are involved in long file transfers and the middle node is a low-rate user, with either Poisson or periodic segment arrivals. The results are the same whether distributed queueing or deference scheduling is used.

VI. Bandwidth Balancing Using Local Priority Information

A. Concept

Now let us introduce multipriority traffic. As before, our bandwidth balancing procedure will guarantee that there is some unused bus capacity and ask each parcel to limit its throughput to some multiple of that spare capacity. Now, however, the proportionality factor will depend on the priority level of the parcel. Specifically, the parcel of priority p is asked to limit its throughput to a multiple M_p of the spare bus capacity; parcels with less demand than this may have all the bandwidth they desire:

$$r_p(n) = \min\left[\rho_p(n), M_p \cdot U\right] = \min\left[\rho_p(n), M_p \cdot \left(1 - \sum_m \sum_q r_q(m)\right)\right]. \quad (6.1)$$

TABLE I
THROUGHPUTS FOR HETEROGENEOUS USERS WITH BANDWIDTH BALANCING ($M = 9$)

Offered load at the middle node		Throughputs			
		Upstream	Middle	Downstream	Total
deterministic	0.2	0.379	0.200	0.379	0.958
	0.25	0.355	0.250	0.355	0.961
	0.3	0.332	0.300	0.332	0.963
random	0.2	0.379	0.200	0.379	0.958
	0.25	0.355	0.250	0.355	0.961
	0.3	0.332	0.300	0.332	0.963

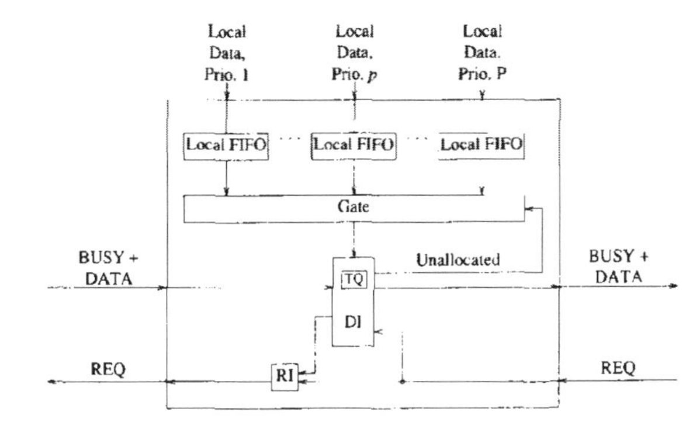

Fig. 8. Bandwidth balancing with local priority information.

Note that every active parcel in the network gets some bandwidth. This scheme is fair in the sense that all rate-controlled parcels of the same priority level get the same bandwidth. Parcels of different priority levels are offered bandwidth in proportion to their bandwidth balancing moduli M_p.

Given the offered loads $\rho_p(n)$ and the bandwidth balancing moduli M_p, (6.1) can be solved for the carried loads $r_p(n)$. In the special case where all N_p parcels of priority level p have heavy demand, the solution has an especially simple form:

$$r_p(n) = \frac{M_p}{1 + \sum_q M_q \cdot N_q}. \tag{6.2}$$

Suppose, for example, that there are three priority levels and $M_1 = 2$, $M_2 = 4$, and $M_3 = 8$. If there is one active parcel of each priority, then the parcels' throughput rates are 2/15, 4/15, and 8/15, and the unused bandwidth is 1/15 of the bus capacity.

B. Implementation

For this "local" version of bandwidth balancing, the slot header need only contain the busy bit and a single request bit. In order to implement (6.1), the node should respond to arriving busy bits and request bits in such a way that

$$S_p(n,t) \leq M_p \cdot U(n,t). \tag{6.3}$$

The most straightforward way to implement (6.3) is to construct a separate section for each priority level p, similar to Fig. 2, then add a bandwidth balancing counter with modulus M_p to that section.

A more compact implementation is shown in Fig. 8. Here the node has only one section with one data inserter and one request inserter to manage data of all priority levels, but a separate local FIFO queue for each priority is required. The data inserter may serve its transmit queue using either distributed queueing or deference scheduling. A gate controls the movement of local data segments from the local FIFO queues to the data inserter. A local data segment must be authorized (as explained below) before it may pass through the gate, and the data inserter may only accept and process one authorized segment at a time. Whenever the data inserter observes an unallocated slot, it authorizes M_p local data segments for each priority level p. (If fewer than M_p segments of priority p are available, then all these available segments are authorized and the extra authorizations expire.) The order in which authorized segments pass through the gate is unimportant, as long as FIFO order is preserved among segments of the same priority level. When all authorized segments of all priority levels have been transmitted, the data inserter is temporarily prevented from processing any more local data. Because the other nodes are following the same discipline, however, the data inserter will eventually detect an unallocated slot and create more authorizations.

C. Performance

Fig. 9 shows simulation results for the bandwidth balancing scheme with local priority information, using the compact implementation described above and using distributed queueing. As in the simulation of Fig. 7, the bus is shared by three nodes spaced apart by 28 slots ($\approx$ 16 km) for a round-trip delay of 112 slot times. Fig. 9 shows the nodal throughputs over successive round-trip times. There are three priority levels of traffic, and their bandwidth balancing moduli are 2, 4, and 8. First the node farthest upstream begins transmitting a long message of medium priority. As predicted by equation (6.2), this node acquires 4/5 of the bus bandwidth. Later the downstream node gets a high-priority message to transmit, and after several round-trip times it achieves a throughput rate of 8/13, while the medium-priority parcel is cut back to 4/13, again as predicted by (6.2). Finally, the middle node becomes active at low priority, and the nodal throughputs shift to 8/15, 4/15, and 2/15, in accord with (6.2).

VII. BANDWIDTH BALANCING USING GLOBAL PRIORITY INFORMATION

A. Concept

Now we assume that every node can determine the bus utilization due to traffic of *each* priority level. Each parcel is asked to limit its throughput to some multiple M of the spare bus capacity not used by parcels of equal or greater priority; parcels with less demand than this may have all the

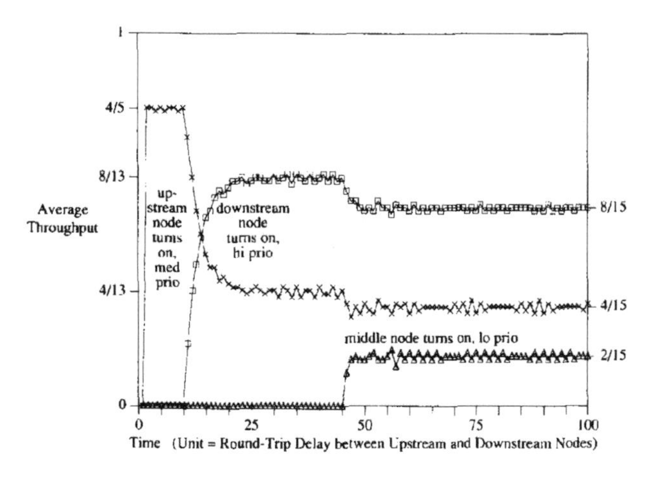

Fig. 9. Bandwidth balancing with local priority information ($M_1 = 2$, $M_2 = 4$, $M_3 = 8$).

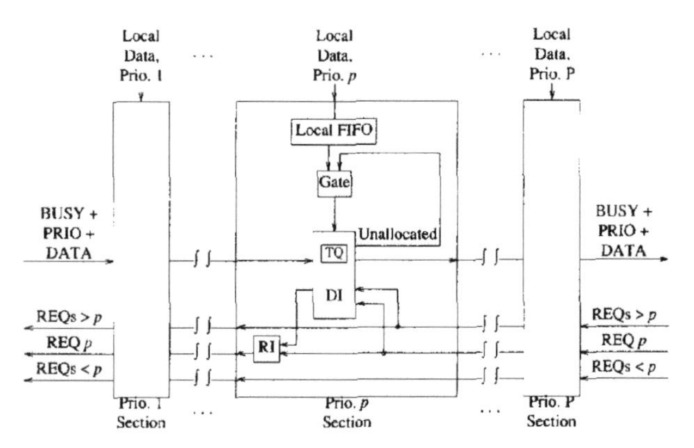

Fig. 10. Bandwidth balancing with global priority information.

bandwidth they desire:

$$r_p(n) = \min\left[\rho_p(n), M \cdot U_{p+}\right]$$
$$= \min\left[\rho_p(n), M \cdot \left(1 - \sum_m \sum_{q \geq p} r_q(m)\right)\right]. \tag{7.1}$$

This scheme is fair in the sense that all rate-controlled parcels of the same priority level get the same bandwidth. Allocation of bandwidth *across* the various priority levels is as follows. First, the entire bus capacity is bandwidth-balanced over the highest priority parcels, as though the lower priority parcels did not exist. Bandwidth balancing ensures that some bus capacity will be left unused by the highest priority parcels. This unused bandwidth is then bandwidth-balanced over the second highest priority parcels. The bandwidth left over after the two highest priorities have been processed is then bandwidth-balanced over the third highest priority parcels, etc. We emphasize that with this scheme, in contrast to the scheme of Section VI, the throughput attained by a parcel of a given priority is independent of the presence of lower priority parcels anywhere in the network.

Given the offered loads $\rho_p(n)$ and the bandwidth balancing modulus M, (7.1) can be solved for the carried loads $r_p(n)$. In the special case where all N_p parcels of priority level p have heavy demand, the solution has a simple form:

$$r_p(n) = \frac{M}{\prod_{q \geq p} (1 + M \cdot N_q)}. \tag{7.2}$$

For example, if $M = 4$ and there are three active parcels of three different priorities, then the parcels' throughput rates are 4/5, 4/25, and 4/125, and the unused bandwidth is 1/125 of the bus capacity.

B. Implementation

For this "global" version of bandwidth balancing, the slot header must contain the busy bit, an indication of the priority level of the data segment in a busy slot,[6] and one request bit for each priority level. By reading these fields, each node can determine the priority level of all traffic on the bus (i.e., there is "global" priority information). In order to implement (7.1), node n should respond to arriving busy and request information in such a way that

$$S_p(n,t) \leq M \cdot U_{p+}(n,t). \tag{7.3}$$

As shown in Fig. 10, the node needs a separate section to manage data for each priority level. Each section has its own data inserter, request inserter, local FIFO queue, and gate. Each data inserter may serve its transmit queue using either distributed queueing or deference scheduling. Inequality (7.3) can be implemented by the node section of priority p as follows. Only authorized segments may pass through the gate, from the local FIFO queue into the data inserter. The data inserter is still restricted to processing only one (authorized) local data segment at a time. Whenever the data inserter observes a slot that is not allocated to traffic of priority p or higher, it authorizes up to M additional local data segments that were not previously authorized. (If fewer than M unauthorized segments of priority p are available, then all these available segments are authorized and the extra authorizations expire.) Note that there are two circumstances under which the data inserter observes such a slot: a) the slot is already busy with a segment of priority less than p when the slot arrives at the data inserter, or b) the slot arrives empty and finds the transmit queue inside the data inserter also empty, holding no local data segment and holding no requests from downstream nodes.

C. Performance

Fig. 11 shows simulation results for this scheme with distributed queueing. As in the preceding simulation, the bus is shared by three nodes spaced apart by 28 slots ($\approx$ 16 km) for a round-trip delay of 112 slot times. There are three priority levels of traffic, and the bandwidth balancing modulus is 4. The arrival times and priority levels of the messages match those for Fig. 9. There are two key differences from the

[6] In the current DQDB standard, the access control field of the slot header does not include the data priority level. However, the field has enough spare bits that this information could be added in a future version of the standard.

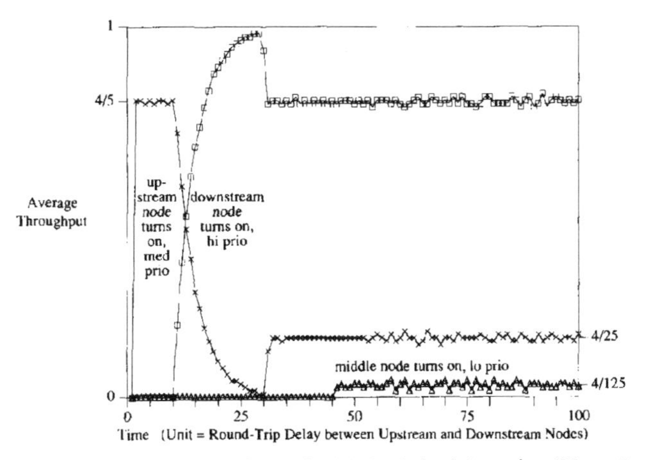

Fig. 11. Bandwidth balancing with global priority information ($M = 4$).

preceding simulation, however. First note that the throughput of the downstream node's high-priority parcel overshoots its target value of 4/5 before settling down at time 31. This happens because the parcel receives less than its correct bandwidth when it first becomes active; the authorizations it accumulates at that time allow it to compensate later. The second thing to notice is that a parcel's steady-state throughput is independent of the presence of lower-priority parcels: when the low-priority parcel turns on at time 45, the high-priority and medium-priority parcels surrender no bandwidth. Once again, for each interval over which the number of active parcels is constant, the steady-state parcel throughputs are close to the values predicted by (7.2).

VIII. Conclusion

If a dual-bus network uses the DQDB protocol with bandwidth balancing disabled, unfair operating conditions can occur during overloads. If the bus speed is 150 Mbps, then when two nodes separated by 30 km perform simultaneous file transfers, one can obtain a hundred times the throughput of the other.

Fortunately, bandwidth balancing is a simple technique that corrects this problem reasonably well. By having each node use the request and busy bits of the DQDB protocol to gauge competing traffic, and by constraining the node to acquire no more than a certain fraction of the remaining bandwidth, we can make the protocol converge to a fair operating point when several nodes are performing large file transfers. How rapidly the network converges to fair operation depends upon how much bandwidth we are willing to waste. The larger the waste, the faster the convergence. The bandwidth efficiency and convergence time of bandwidth balancing cannot match that attainable by a centralized scheduler. This difference suggests that there is still room for improvement in distributed schemes like bandwidth balancing, although there is likely to be a tradeoff between performance and complexity.

The implementation of bandwidth balancing is quite simple. For instance, a node can restrict itself to 90% of the available slots by letting one extra (i.e., unrequested) idle slot go by, every time it transmits nine of its own data segments. Bandwidth balancing somewhat reduces the importance of the discipline used by a node to schedule its own data segments among the requests from downstream nodes. The distributed queueing discipline of DQDB may be retained, or a streamlined discipline called deference scheduling may be substituted.

In this paper we also proposed two ways to balance the bandwidth of multipriority traffic more effectively than does the current standard. The two new methods differ in the priority information they use. In the "local" scheme, each node only knows the priority level of its own data; this scheme uses no priority information from the buses. In the "global" scheme, however, the slot headers need to show the priority of data segments in busy slots, as well as the request priorities. Both methods determine the carried loads of the nodes based only on their offered loads, not on their relative bus locations. (The version of multipriority bandwidth balancing in the current standard is not so robust.) Moreover, both methods are fair in the sense that traffic parcels of the same priority level get the same bandwidth in steady state. (The current standard lacks this level of fairness.) The methods differ in their allocation of bus bandwidth *across* the various priority levels. In the "local" scheme, each priority level has an associated bandwidth balancing modulus, and the scheme allocates bandwidth to all traffic parcels in proportion to their bandwidth balancing moduli. The steady-state throughput of a traffic parcel is affected by the presence of every other parcel for that bus. The "global" scheme, roughly speaking, allocates bandwidths first to the higher priority parcels, then allocates the leftovers to the lower priority parcels. Lower-priority parcels have no effect on the steady-state throughputs of higher priority parcels.

Either multipriority technique could be implemented through modest changes in the DQDB standard. The current slot header already contains more than enough control information to drive the "local" scheme, and while the header does not currently include all the information needed for the "global" scheme, there are spare header bits that could be used for that purpose in a future version of the standard. If either scheme were included in a future standard, then nodes satisfying the old standard could share the same network with the new nodes, provided that the old nodes only generate traffic of the lowest priority level. In fact, a crude variant of the "local" scheme (cf. [44] and the "local per-node" procedure in [45]) can be realized within the framework of the *current* standard as follows: each node executes the DQDB protocol and bandwidth balancing *as if* all its traffic were at the lowest priority level, while constantly varying its bandwidth balancing modulus according to the true priority level of its current traffic.

It also bears mention that all the bandwidth balancing schemes presented in this paper have an extra dimension of flexibility; different bandwidth balancing moduli may be assigned to different nodes [43]. By this means, extra bandwidth may be allocated to high-volume nodes such as file servers.

Acknowledgment

Our many discussions with U. Mukherji greatly increased our understanding of the performance of the DQDB protocol.

References

[1] R. M. Newman, Z. L. Budrikis, and J. L. Hullett, "The QPSX MAN," *IEEE Commun.*, vol. 26, pp. 20–28, Apr. 1988.

[2] Z. L. Budrikis, J. L. Hullett, R. M. Newman, D. Economou, F. M. Fozdar, and R. D. Jeffery, "QPSX: A queue packet and synchronous circuit exchange," *8th Int. Conf. Comput. Commun.*, Munich, Germany, Sept. 15–19, 1986, published by North-Holland, pp. 288–293.

[3] IEEE 802.6 Working Group, "IEEE standard 802.6: Distributed Queue Dual Bus (DQDB) subnetwork of a metropolitan area network (MAN)," Final Draft D15, approved by IEEE Standards Board on Dec. 6, 1990.

[4] J. O. Limb and C. Flores, "Description of Fasnet—A unidirectional local area communications network," *Bell Syst. Tech. J.*, vol. 61, no. 7, pp. 1413–1440, Sept. 1982.

[5] H. Kaur and G. Campbell, "DQDB—An access delay analysis," in *IEEE INFOCOM '90*, San Francisco, CA, June 5–7, 1990, pp. 630–635.

[6] R. M. Newman, "Distributed queueing: Performance characterisation," Contribution 802.6-88/11 to the IEEE 802.6 Working Group, 1988.

[7] M. Conti, E. Gregori, and L. Lenzini, "DQDB media access control protocol: Performance evaluation and unfairness analysis," *Third IEEE Workshop MAN's*, Mar. 28–30, 1989, San Diego, CA, pp. 375–408.

[8] M. W. Garrett, "Simulation study of dual bus MAN's," in *Third IEEE Workshop MAN's*, San Diego, CA, Mar. 28–30, 1989, pp. 409–419.

[9] P. Davids, T. Welzel, "Performance analysis of DQDB based on simulation," in *Third IEEE Workshop MAN's*, San Diego, CA, Mar. 28–30, 1989, pp. 431–445.

[10] J. W. Wong, "Throughput of DQDB networks under heavy load," EFOC/LAN-89, Amsterdam, The Netherlands, June 14–16, 1989, pp. 146–151.

[11] M. Zukerman and P. Potter, "The effect of eliminating the standby state on DQDB performance under overload," *Int. J. Digital Analog Cabled Syst.*, vol. 2, 1989, pp. 179–186.

[12] M. N. Huber, K. Sauer, and W. Schodl, "QPSX and FDDI-II: Performance study of high speed LAN's," EFOC/LAN'88, Amsterdam, June 29–July 1, 1988, pp. 316–321.

[13] R. M. Newman and J. L. Hullett, "Distributed queueing: A fast and efficient packet access protocol for QPSX," *8th Int. Conf. Comput. Commun.*, Munich, Germany, Sept. 15–19, 1986, published by North-Holland, pp. 294–299.

[14] H. R. van As, J. W. Wong, and P. Zafiropulo, "Fairness, priority and predictability of the DQDB MAC protocol under heavy load," in *Int. Zurich Sem. Digital Commun.*, Zurich, Switzerland, March 5–8, 1990, pp. 410–417.

[15] P. Tran-Gia and T. Stock, "Approximate performance analysis of the DQDB access protocol," Paper no. 16.1, ITC Specialist Seminar, Adelaide, Australia, 1989.

[16] B. T. Doshi and A. A. Fredericks, "Visual modeling and analysis with application to IEEE 802.6 DQDB protocol," Paper no. 16.2, ITC Specialist Seminar, Adelaide, Australia, 1989.

[17] K. Sauer and W. Schodl, "Performance aspects of the DQDB protocol," Paper no. 16.3, ITC Specialist Seminar, Adelaide, Australia, 1989.

[18] M. Zukerman and P. Potter, "The DQDB protocol and its performance under overload traffic conditions," Paper no. 16.4, ITC Specialist Seminar, Adelaide, Australia, 1989.

[19] O. Gihr, "The fairness issue in high speed local area networks," in *Fourth Australian Teletraffic Res. Sem.*, Bond Univ., Queensland, Australia, Dec. 4–5, 1989.

[20] P. Martini, "Fairness issues of the DQDB protocol," in *14th Conf. Local Comput. Networks*, Minneapolis, MN, Oct. 10–12, 1989, pp. 160–170.

[21] Y. Yaw, Y.-K. Yea, W.-D. Ju, and P. A. Ng, "Analysis on access fairness and a technique to extend distance for 802.6," in *14th Conf. Local Comput. Networks*, Minneapolis, MN, Oct. 10–12, 1989, pp. 171–176.

[22] C. Bisdikian, "Waiting time analysis in a single buffer DQDB (802.6) network," in *IEEE INFOCOM'90*, San Francisco, CA, June 5–7, 1990, pp. 610–616.

[23] H. Adiseshu and U. Mukherji, "An approximate performance analysis of the distributed queue dual bus metropolitan area network medium access control protocol," in *Workshop Signal Processing, Commun., Networking*, Indian Inst. Sci., Bangalore, India, July 23–26, 1990.

[24] E. L. Hahne, A. K. Choudhury and N. F. Maxemchuk, "Improving the fairness of distributed-queue dual-bus networks," in *IEEE INFOCOM '90*, San Francisco, CA, June 5–7, 1990, pp. 175–184.

[25] M. Conti, E. Gregori, and L. Lenzini, "DQDB under heavy load: Performance evaluation and fairness analysis," in *IEEE INFOCOM '90*, San Francisco, CA, June 5–7, 1990, pp. 313–320.

[26] S. Fdida and H. Santoso, "Simulation and approximate performance analysis of DQDB," Tech. Rep. 90-18, MASI Lab., Pierre and Marie Curie Univ., Paris, France, May 1990.

[27] H. R. van As, "Performance evaluation of bandwidth balancing in the DQDB MAC protocol," EFOC/LAN 90, Munich, Germany, June 27–29, 1990, pp. 231–239.

[28] A. Myles, "DQDB simulation and MAC protocol analysis," *Electron. Lett.*, vol. 25, no. 9, Apr. 27, 1989, pp. 616–618.

[29] A. Baiocchi, M. Carosi, M. Listanti, G. Pacifici, A. Roveri, and R. Winkler, "The ACCI access protocol for a twin bus ATM metropolitan area network," in *IEEE INFOCOM '90*, San Francisco, CA, June 5–7, 1990, pp. 165–174.

[30] K. M. Khalil and M. E. Koblentz, "A fair distributed queue dual bus access method," in *14th Conf. Local Comput. Networks*, Minneapolis, MN, Oct. 10–12, 1989, pp. 180–188.

[31] J. Filipiak, "Access protection for fairness in a distributed queue dual bus metropolitan area network," *IEEE ICC '89*, Boston, MA, June 1989, pp. 635–639.

[32] H. R. Muller, M. M. Nassehi, J. W. Wong, E. Zurfluh, W. Bux, and P. Zafiropulo, "DQMA and CRMA: New access schemes for Gbit/s LANs and MANs," in *IEEE INFOCOM '90*, San Francisco, CA, June 5–7, 1990, pp. 185–191.

[33] B. Mukherjee and J. S. Meditch, "The pi-persistent protocol for unidirectional broadcast bus networks," *IEEE Trans. Commun.*, vol. 36, pp. 1277–1295, Dec. 1988.

[34] J. O. Limb, "Load-controlled scheduling of traffic on high-speed metropolitan area networks," *IEEE Trans. Commun.*, vol. 37, pp. 1144-1150, Nov. 1989.

[35] M. A. Rodriques, "A fair MAC access scheme," Contribution 802.6-88/62 to the IEEE 802.6 Working Group, July 5, 1988.

[36] P. Potter and M. Zukerman, "Cyclic request control for provision of guaranteed bandwidth within the DQDB framework," *ISS '90*, Stockholm, Sweden, May 1990, Paper no. A4.1.

[37] Y.-S. Leu and D. H. C. Du, "Cycle compensation protocol: A completely fair protocol for the unidirectional twin bus architecture," in *15th Conf. Local Comput. Networks*, Minneapolis, MN, Sept. 30–Oct. 3, 1990, pp. 416–425.

[38] J. Jaffe, "Bottleneck flow control," *IEEE Trans. Commun.*, vol. COM-29, pp. 954–962, July 1981.

[39] M. Spratt, "A problem in the multi-priority implementation of the bandwidth balancing mechanism," Contribution to the IEEE 802.6 Working Group, Nov. 1989.

[40] V. Phung and R. Breault, "Enhancement to the bandwidth balancing mechanism (version 2)," Contribution 802.6-90/25 to the IEEE 802.6 Working Group, Mar. 1990.

[41] M. Spratt, "Implementing the priorities in the bandwidth balancing mechanism," Contribution 802.6-90/05 to the IEEE 802.6 Working Group, Jan. 1990.

[42] R. Damodaram, verbal proposal at the IEEE 802.6 Working Group

meeting of Sept. 1989.
[43] M. Spratt, "A non-unity ratio bandwidth allocation mechanism—A simple improvement to the bandwidth balancing mechanism," Contribution to the IEEE 802.6 Working Group, Nov. 1989.
[44] E. L. Hahne and N. F. Maxemchuk, "Bandwidth balancing with local priority information," Contribution 802.6-90/06 to the IEEE 802.6 Working Group, Jan. 22, 1990.
[45] ——, "Fair access of multi-priority traffic to distributed-queue dual-bus networks," in *IEEE INFOCOM '91*, Bal Harbour, FL, Apr. 9–11, 1991, pp. 889–900.
[46] ——, "Bandwidth balancing with global priority information," Contribution 802.6-90/07 to the IEEE 802.6 Working Group, Jan. 22, 1990.
[47] A. M. Perdikaris and M. A. Rodrigues, "Destination release of segments in the IEEE 802.6 protocol," Contribution 802.6-88/61 to the IEEE 802.6 Working Group, July 1988.
[48] R. Breault and V. Phung, "DQDB performance improvement with erasure nodes," Contribution 802.6-90/21 to the IEEE 802.6 Working Group, Mar. 1990.
[49] M. Zukerman and P. G. Potter, "A protocol for eraser node implementation within the DQDB framework," in *IEEE GLOBECOM '90*, San Diego, CA, Dec. 1990, pp. 1400–1404.
[50] P. Potter and M. Zukerman, "A discrete shared processor model for DQDB," ITC Specialist Sem., Adelaide, Australia, 1989, Paper No. 3.4.
[51] N. L. Golding, "DQDB D10 fairness analysis," Contribution 802.6-90/01 to the IEEE 802.6 Working Group, Jan. 1990.
[52] M. Zukerman and G. Shi, "Throughput analysis for a DQDB subnetwork with two close sources under overload traffic conditions," SNRB Branch Paper 205, Telecom Australia Res. Labs., Clayton, Australia, Jan. 1991.

PHOTO NOT AVAILABLE

Ellen L. Hahne (M'87) was born in 1956 in Pittsburgh, PA. She received the B.S. degree in electrical engineering from Rice University, Houston, TX, in 1978 and the S.M., E.E., and Ph.D. degrees in electrical engineering and computer science from the Massachusetts Institute of Technology, Cambridge, MA, in 1981, 1984 and 1987. Since 1978 she has worked for AT&T Bell Laboratories, Murray Hill, NJ.

Her primary research interest is the control of traffic in networks. Past projects have dealt with scheduling and flow control in wide-area data networks, access control in local- and metropolitan-area data networks, detection and control of focused overloads in telephone networks, and dynamic routing in automated factories.

Dr. Hahne was named a Presidential Scholar in 1974. She is a member of Phi Beta Kappa, Tau Beta Pi, and Sigma Xi.

PHOTO NOT AVAILABLE

Abhijit K. Choudhury (M'91) received the B. Tech. degree in electrical engineering from Indian Institute of Technology, Kanpur, India, in 1986 and the M.S. in electrical engineering from S.U.N.Y. at Stony Brook, New York, in 1987. He received the Ph.D. degree in electrical engineering from the University of Southern California in 1991.

During the summer of 1988, he worked at AT&T Bell Laboratories, Murray Hill, NJ, where he was part of a team that designed the bandwidth balancing technique for solving the unfairness problem in DQDB networks. In the summer of 1989, he was engaged in research at AT&T Bell Laboratories, Murray Hill, that lead to an understanding of the effect of a finite reassembly buffer on the performance of deflection routing. His dissertation work dealt with deflection routing in high-speed computer networks. He is currently working in the Distributed Systems Research Department of AT&T Bell Laboratories, Murray Hill. His current areas of interest are routing and flow control in high speed networks, performance evaluation of communication networks and computer systems, multiple access protocols and network design.

Dr. Choudhury is a member of IEEE Communications and Computer Societies and of Eta Kappa Nu.

PHOTO NOT AVAILABLE

Nicholas F. Maxemchuk (F'89) received the B.S.E.E. degree from the City College of New York, and the M.S.E.E. and Ph.D. degrees from the University of Pennsylvania.

He is currently the Head of the Distributed Systems Research Department at AT&T Bell Laboratories, where he has been since 1976. Prior to joining Bell Laboratories he was at the RCA David Sarnoff Research Center in Princeton, NJ for eight years.

Dr. Maxemchuk has been on the adjunct faculties of Columbia University and the University of Pennsylvania. He has served as the Editor for Data Communications for the IEEE TRANSACTIONS ON COMMUNICATIONS, as a Guest Editor for the IEEE JOURNAL ON SELECTED AREAS IN COMMUNICATIONS, and is currently on JSAC's Editorial Board. He has been on the program committee for numerous conferences and workshops, and was the Program Chairman for 1987 and 1990 workshops on Metropolitan Area Networks. He was awarded the RCA Laboratories Outstanding Achievement Award in 1970, the Bell Laboratories Distinguished Technical Staff Award in 1984, and the IEEE's 1985 and 1987 Leonard G. Abraham Prize Paper Award.

Input Versus Output Queueing on a Space-Division Packet Switch

MARK J. KAROL, MEMBER, IEEE, MICHAEL G. HLUCHYJ, MEMBER, IEEE, AND SAMUEL P. MORGAN, FELLOW, IEEE

Abstract—**Two simple models of queueing on an $N \times N$ space-division packet switch are examined. The switch operates synchronously with fixed-length packets; during each time slot, packets may arrive on any inputs addressed to any outputs. Because packet arrivals to the switch are unscheduled, more than one packet may arrive for the same output during the same time slot, making queueing unavoidable. Mean queue lengths are always greater for queueing on inputs than for queueing on outputs, and the output queues saturate only as the utilization approaches unity. Input queues, on the other hand, saturate at a utilization that depends on N, but is approximately $(2 - \sqrt{2}) = 0.586$ when N is large. If output trunk utilization is the primary consideration, it is possible to slightly increase utilization of the output trunks—up to $(1 - e^{-1}) = 0.632$ as $N \to \infty$—by dropping interfering packets at the end of each time slot, rather than storing them in the input queues. This improvement is possible, however, only when the utilization of the input trunks exceeds a second critical threshold—approximately $\ln(1 + \sqrt{2}) = 0.881$ for large N.**

I. Introduction

SPACE-DIVISION packet switching is emerging as a key component in the trend toward high-performance integrated communication networks for data, voice, image, and video [1], [2] and multiprocessor interconnects for building highly parallel computer systems [3], [4]. Unlike present-day packet switch architectures with throughputs measured in 1's or at most 10's of Mbits/s, a space-division packet switch can have throughputs measured in 1's, 10's, or even 100's of Gbits/s. These capacities are attained through the use of a highly parallel switch fabric coupled with simple per packet processing distributed among many high-speed VLSI circuits.

Conceptually, a space-division packet switch is a box with N inputs and N outputs that routes the packets arriving on its inputs to the appropriate outputs. At any given time, internal switch points can be set to establish certain paths from inputs to outputs; the routing information used to establish input–output paths is often contained in the header of each arriving packet. Packets may have to be buffered within the switch until appropriate connections are available; the location of the buffers and the amount of buffering required depend on the switch architecture and the statistics of the offered traffic.

Clearly, congestion can occur if the switch is a blocking network, that is, if there are not enough switch points to provide simultaneous, independent paths between arbitrary pairs of inputs and outputs. A Banyan switch [3]–[5], for example, is a blocking network. In a Banyan switch, even when every input is assigned to a different output, as many as $\sqrt{N}$ connections may be contending for use of the same center link. The use of a blocking network as a packet switch is feasible only under light loads or, alternatively, if it is possible to run the switch substantially faster than the input and output trunks.

In this paper, we consider only nonblocking networks. A simple example of a nonblocking switch fabric is the crossbar interconnect with N^2 switch points (Fig. 1). Here it is always possible to establish a connection between any idle input–output pair. Examples of other nonblocking switch fabrics are given in [3]. Even with a nonblocking interconnect, some queueing in a packet switch is unavoidable, simply because the switch acts as a statistical multiplexor; that is, packet arrivals to the switch are unscheduled. If more than one packet arrives for the same output at a given time, queueing is required. Depending on the speed of the switch fabric and its particular architecture, there may be a choice as to where the queueing is done: for example, on the input trunk, on the output trunk, or at an internal node.

We assume that the switch operates synchronously with fixed-length packets, and that during each time slot, packets may arrive on any inputs addressed to any outputs (Fig. 2). If the switch fabric runs N times as fast as the input and output trunks, all the packets that arrive during a particular input time slot can traverse the switch before the next input slot, but there will still be queueing at the outputs [Fig. 1(a)]. This queueing really has nothing to do with the switch architecture, but is due to the simultaneous arrival of more than one input packet for the same output. If, on the other hand, the switch fabric runs at the same speed as the inputs and outputs, only one packet can be accepted by any given output line during a time slot, and other packets addressed to the same output must queue on the input lines [Fig. 1(b)]. For simplicity, we do not consider the intermediate case where some packets can be queued at internal nodes, as in the Banyan topology.

It seems intuitively reasonable that the mean queue lengths, and hence the mean waiting times, will be greater for queueing on inputs than for queueing on outputs. When queueing is done on inputs, a packet that could traverse the switch to an idle output during the current time slot may have to wait in queue behind a packet whose output is currently busy. The intuition that, if possible, it is better to queue on the outputs than the inputs of a space-division packet switch also pertains to the following situation. Consider a single road leading to both a sports arena and a store [Fig. 3(a)]. Even if there are no customers waiting for service in the store, some shoppers might be stuck in stadium traffic. A simple bypass road around the stadium is the obvious solution [Fig. 3(b)].

This paper quantifies the performance improvements provided by output queueing for the following simple model. Independent, statistically identical traffic arrives on each input trunk. In any given time slot, the probability that a packet will arrive on a particular input is p. Thus, p represents the average utilization of each input. Each packet has equal probability $1/N$ of being addressed to any given output, and successive packets are independent.

With output queueing, all arriving packets in a time slot are

Paper approved by the Editor for Local Area Networks of the IEEE Communications Society. Manuscript received August 8, 1986; revised May 14, 1987. This paper was presented at GLOBECOM'86, Houston, TX, December 1986.

M. J. Karol is with AT&T Bell Laboratories, Holmdel, NJ 07733.

M. G. Hluchyj was with AT&T Bell Laboratories, Holmdel, NJ 07733. He is now with the Codex Corporation, Canton, MA 02021.

S. P. Morgan is with AT&T Bell Laboratories, Murray Hill, NJ 07974.

IEEE Log Number 8717486.

Reprinted from *IEEE Transactions on Communications,* vol. COM-35, no. 12, December 1987.

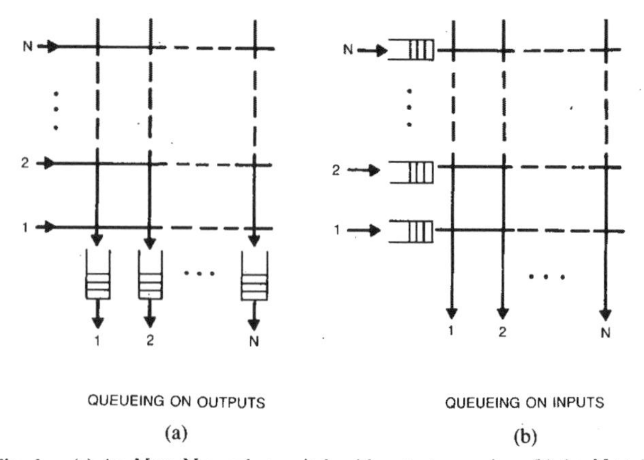

Fig. 1. (a) An $N \times N$ crossbar switch with output queueing. (b) An $N \times N$ crossbar switch with input queueing.

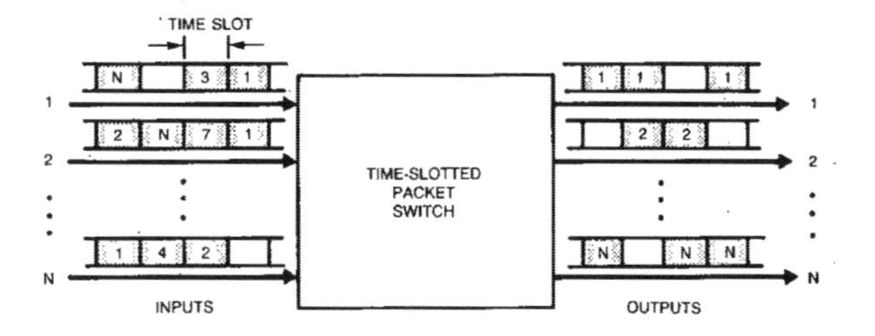

Fig. 2. Fixed-length packets arrive synchronously to a time-slotted packet switch.

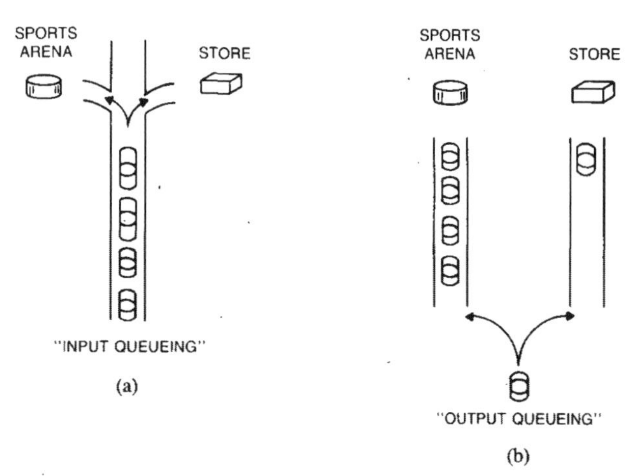

Fig. 3. "Output queueing" (b) is superior to "input queueing" (a). In (a), even if there are no customers waiting for service in the store, some shoppers might be stuck in stadium traffic. In (b), a bypass road around the stadium serves cars traveling to the store.

cleared before the beginning of the next time slot. For example, a crossbar switch fabric that runs N times as fast as the inputs and outputs can queue all packet arrivals according to their output addresses, even if all N inputs have packets destined for the same output [Fig. 1(a)]. If k packets arrive for one output during the current time slot, however, only one can be transmitted over the output trunk. The remaining $k - 1$ packets go into an output FIFO (first-in, first-out queue) for transmission during subsequent time slots. Since the average utilization of each output trunk is the same as the utilization of each input trunk, namely p, the system is stable and the mean queue lengths will be finite for $p < 1$, but they will be greater than zero if $p > 0$.

A crossbar interconnect with the switch fabric running at the same speed as the inputs and outputs exemplifies input queueing [Fig. 1(b)]. Each arriving packet goes, at least momentarily, into a FIFO on its input trunk. At the beginning of every time slot, the switch controller looks at the first packet in each FIFO. If every packet is addressed to a different output, the controller closes the proper crosspoints and all the packets go through. If k packets are addressed to a particular output, the controller picks one to send; the others wait until the next time slot, when a new selection is made among the packets that are then waiting. Three selection policies are discussed in Section III: one of the k packets is chosen at *random*, each selected with equal probability $1/k$, *longest queue* selection, in which the controller sends the packet from the longest input queue,[1] and *fixed priority* selection where the N inputs have fixed priority levels, and of the k packets, the controller sends the one with highest priority.

Solutions of these two queueing problems are given in Sections II and III. Curves showing mean waiting time as a function of p are plotted for various values of N. As expected, the mean waiting times are greater for queueing on inputs than for queueing on outputs. Furthermore, the output queues saturate only as $p \to 1$. Input queues, on the other hand, saturate at a value of p less than unity, depending weakly on N; for large N, the critical value of p is approximately $(2 - \sqrt{2}) = 0.586$ with the random selection policy. When the utilization p of the input trunks exceeds the critical value, the steady-state queue sizes are infinite, packets experience infinite waiting times, and the output trunk utilization is limited to approximately 0.586 (for large N). In the saturation region, however, it is possible to increase utilization of the output trunks—up to $(1 - e^{-1}) = 0.632$ as $N \to \infty$—by dropping packets, rather than storing them in the input queues. This improvement is possible, however, only when the utilization of the input trunks exceeds a second critical threshold—approximately $\ln(1 + \sqrt{2}) = 0.881$ for large N. Consequently, if the objective is maximum output utilization, rather than 100 percent packet delivery, then below the second threshold, it is better to queue packets until they are successful, whereas above the second threshold, it is better to reduce input queue blocking by dropping packets whenever there are conflicts. With high probability, new packets (with new destinations) will quickly arrive to replace the dropped packets.

Comparing the random and longest queue selection policies of input queueing, the mean waiting times are greater with random selection. This is expected because the longest queue selection policy reduces the expected number of packets blocked (behind other packets) from traversing the switch to idle outputs. For fairness, the fixed priority discipline should be avoided because the lowest priority input queue suffers large delays and is sometimes unstable, even when the other two selection policies guarantee stability.

II. Queues on Outputs

Much of the following analysis of the output queueing scheme involves well-known results for discrete-time queueing systems [6]. Communication systems have been modeled by discrete-time queues in the past (e.g., [7]); we sketch our analysis and present results for later comparison to the input queueing analysis.

We assume that packet arrivals on the N input trunks are governed by independent and identical Bernoulli processes. Specifically, in any given time slot, the probability that a packet will arrive on a particular input is p. Each packet has equal probability $1/N$ of being addressed to any given output, and successive packets are independent.

Fixing our attention on a particular output queue (the

[1] A random selection is made if, of the k input queues with packets addressed to a particular output, several queues have the same maximum length.

"tagged" queue), we define the random variable A as the number of packet arrivals at the tagged queue during a given time slot.[2] It follows that A has the binomial probabilities

$$a_i \triangleq \Pr[A=i]=\binom{N}{i}(p/N)^i(1-p/N)^{N-i} \qquad i=0, 1, \cdots, N \quad (1)$$

with probability generating function (PGF)

$$A(z) \triangleq \sum_{i=0}^{N} z^i \Pr[A=i]=\left(1-\frac{p}{N}+z\frac{p}{N}\right)^N. \quad (2)$$

As $N \to \infty$, the number of packet arrivals at the tagged queue during each time slot has the Poisson probabilities

$$a_i \triangleq \Pr[A=i]=\frac{p^i e^{-p}}{i!} \qquad i=0, 1, 2, \cdots \quad (3)$$

with probability generating function (PGF)

$$A(z) \triangleq \sum_{i=0}^{N} z^i \Pr[A=i]=e^{-p(1-z)}. \quad (4)$$

Letting Q_m denote the number of packets in the tagged queue at the end of the mth time slot, and A_m denote the number of packet arrivals during the mth time slot, we have

$$Q_m=\max(0, Q_{m-1}+A_m-1). \quad (5)$$

When $Q_{m-1} = 0$ and $A_m > 0$, one of the new packets is immediately transmitted during the mth time slot; that is, a packet flows through the switch without suffering any delay. The queue size Q_m is modeled by a discrete-time Markov chain; Fig. 4 illustrates the state transition diagram. Using (5) and following a standard approach in queueing analysis (see, for example, [8, sect. 5.6]), we obtain the PGF for the steady-state queue size:

$$Q(z)=\frac{(1-p)(1-z)}{A(z)-z}. \quad (6)$$

Finally, substituting the right-hand side of (2) into (6), we obtain

$$Q(z)=\frac{(1-p)(1-z)}{\left(1-\frac{p}{N}+z\frac{p}{N}\right)^N-z}. \quad (7)$$

Now, differentiating (7) with respect to z and taking the limit as $z \to 1$, we obtain the mean steady-state queue size $\bar{Q}$ given by

$$\bar{Q}=\frac{(N-1)}{N}\cdot\frac{p^2}{2(1-p)}=\frac{(N-1)}{N}\bar{Q}_{M/D/1} \quad (8)$$

where $\bar{Q}_{M/D/1}$ denotes the mean queue size for an $M/D/1$ queue. Hence, as $N \to \infty$, $\bar{Q} \to \bar{Q}_{M/D/1}$.

We can make the even stronger statement that the steady-state probabilities for the queue size converge to those of an $M/D/1$ queue. Taking the limit as $N \to \infty$ on both sides of (7) yields

$$\lim_{N\to\infty} Q(z)=\frac{(1-p)(1-z)}{e^{-p(1-z)}-z} \quad (9)$$

STATE TRANSITION PROBABILITIES

$a_i \equiv \Pr(A=i) \qquad i = 0, 1, 2, \ldots$

$$= \begin{cases} \frac{p^i e^{-p}}{i!} & \text{IF } N = \infty \\ \binom{N}{i}\left(\frac{p}{N}\right)^i\left(1-\frac{p}{N}\right)^{N-i} & \text{IF } N < \infty \end{cases}$$

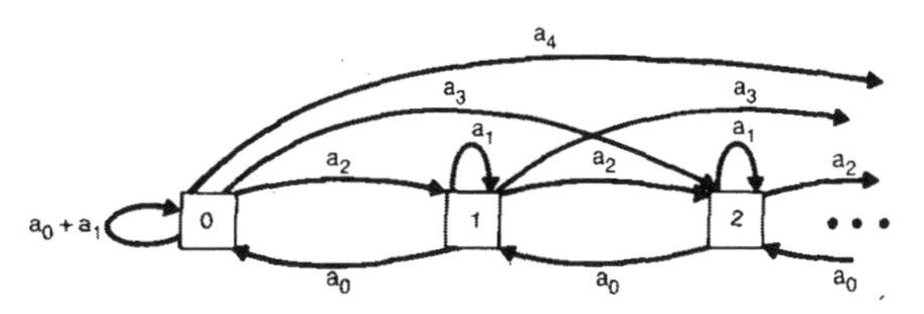

Fig. 4. The discrete-time Markov chain state transition diagram for the output queue size.

which corresponds to the PGF for the steady-state queue size of an $M/D/1$ queue. Expanding (9) in a Maclaurin series [9] yields the asymptotic (as $N \to \infty$) queue size probabilities[3]

$$\Pr(Q=0)=(1-p)e^p \quad (10)$$

$$\Pr(Q=1)=(1-p)e^p(e^p-1-p) \quad (11)$$

$$\vdots$$

$$\Pr(Q=n)=(1-p)\sum_{j=1}^{n+1}(-1)^{n+1-j}e^{jp} \cdot \left[\frac{(jp)^{n+1-j}}{(n+1-j)!}+\frac{(jp)^{n-j}}{(n-j)!}\right] \quad \text{for } n\geq 2 \quad (12)$$

where the second factor in (12) is ignored for $j = (n + 1)$.

Although it is mathematically pleasing to have closed-form expressions, directly using (12) to compute the steady-state probabilities leads to inaccurate results for large n. When n is large, the alternating series (12) expresses small steady-state probabilities as the difference between very large positive numbers. Accurate values are required if one is interested in the tail of the distribution; for example, to compute the probability that the queue size exceeds some value M. Numerically, a more accurate algorithm is obtained directly from the Markov chain (Fig. 4) balance equations. Equations (13)–(15) numerically provide the steady-state queue size probabilities.

$$q_0 \triangleq \Pr(Q=0)=\frac{(1-p)}{a_0} \quad (13)$$

$$q_1 \triangleq \Pr(Q=1)=\frac{(1-a_0-a_1)}{a_0}\cdot q_0 \quad (14)$$

$$\vdots$$

$$q_n \triangleq \Pr(Q=n)=\frac{(1-a_1)}{a_0}\cdot q_{n-1} - \sum_{i=2}^{n}\frac{a_i}{a_0}\cdot q_{n-i} \qquad n\geq 2 \quad (15)$$

[2] We use the phrase "arrivals at the tagged queue *during* a given time slot" to indicate that packets do not arrive instantaneously, in their entirety, at the output. Packets have a nonzero transmission time.

[3] The steady-state probabilities in [9, sect. 5.1.5] are for the *total number* of packets in an $M/D/1$ system. We are interested in *queue size;* hence, the modification to (10)–(12).

where the a_i are given by (1) and (3) for $N < \infty$ and $N = \infty$, respectively.

We are now interested in the waiting time for an arbitrary (tagged) packet that arrives at the tagged output FIFO during the mth time slot. We assume that packet arrivals to the output queue in the mth time slot are transmitted over the output trunk in random order. All packets arriving in earlier time slots, however, must be transmitted first.

The tagged packet's waiting time W has two components. First, the packet must wait W_1 time slots while packets that arrived in earlier time slots are transmitted. Second, it must wait an additional W_2 time slots until it is randomly selected out of the packet arrivals in the mth time slot.

Since packets require one time slot for transmission over the output trunk, W_1 equals Q_{m-1}. Consequently, from (6), the PGF for the steady-state value of W_1 is

$$W_1(z)=\frac{(1-p)(1-z)}{A(z)-z}. \tag{16}$$

We must be careful when we compute W_2, the delay due to the transmission of other packet arrivals in the mth time slot. Burke points out in [10] that many standard works on queueing theory are in error when they compute the delay of single-server queues with batch input. Instead of working with the size of the batch to which the tagged packet belongs, it is tempting to work with the size of an arbitrary batch. Errors result when the batches are not of constant size. The probability that our tagged packet arrives in a batch of size i is given by $ia_i/\bar{A}$; hence, the random variable W_2 has the probabilities

$$\Pr[W_2=k]=\sum_{i=k+1}^{\infty}\frac{1}{i}\,ia_i/\bar{A} \qquad k=0,1,2,\cdots$$

$$=\frac{1}{p}\sum_{i=k+1}^{\infty}a_i \tag{17}$$

where $\bar{A}$ $(=p)$ is the expected number of packet arrivals at the tagged output during each time slot, and the a_i are given by (1) and (3) for $N < \infty$ and $N = \infty$, respectively. The PGF for the steady-state value of W_2 follows directly from (17).

$$W_2(z)=\frac{1-A(z)}{p(1-z)}. \tag{18}$$

Finally, since W is the sum of the independent random variables W_1 and W_2, the PGF for the steady-state waiting time is

$$W(z)=Q(z)\cdot\frac{1-A(z)}{p(1-z)}. \tag{19}$$

$A(z)$ is given by (2) and (4) for $N < \infty$ and $N = \infty$, respectively.

Differentiating (19) with respect to z and taking the limit as $z \to 1$, we obtain the mean steady-state waiting time given by

$$\bar{W}=\bar{Q}+\frac{1}{2p}[\overline{A^2}-\bar{A}]. \tag{20}$$

Since $\bar{A} = p$ and $\overline{A^2} = p^2 + p(1 - p/N)$, substituting the right-hand side of (8) into (20) yields

$$\bar{W}=\frac{(N-1)}{N}\cdot\frac{p}{2(1-p)}=\frac{(N-1)}{N}\bar{W}_{M/D/1} \tag{21}$$

where $\bar{W}_{M/D/1}$ denotes the mean waiting time for an $M/D/1$ queue. The mean waiting time $\bar{W}$, as a function of p, is shown in Fig. 5 for several values of N. Notice that Little's result and (8) generate the same formula for $\bar{W}$.

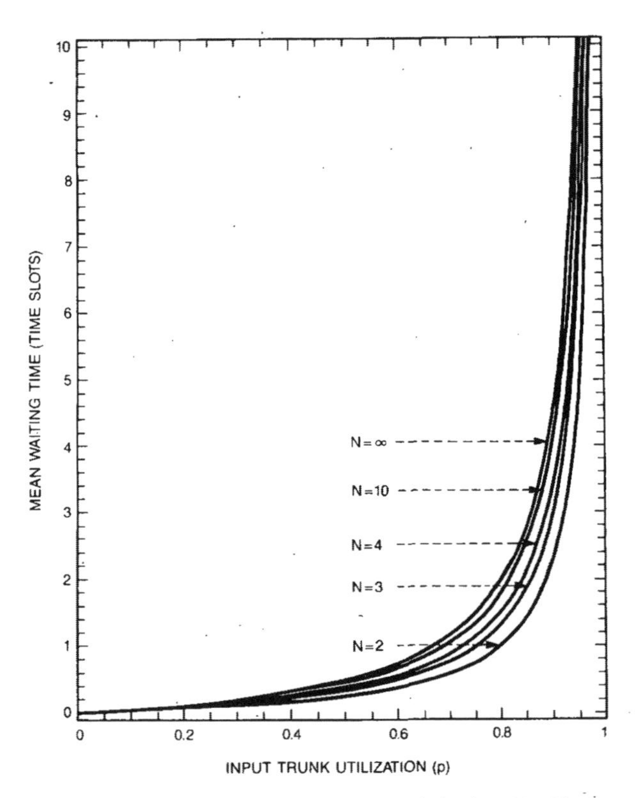

Fig. 5. The mean waiting time for several switch sizes N with output queueing.

Rather than take the inverse transform of $W(z)$, it is easier to compute the steady-state waiting time probabilities from

$$\Pr[W=k]=\Pr[W_1+W_2=k]$$

$$=\frac{1}{p}\sum_{n=0}^{k}q_n\cdot\sum_{i=k+1-n}^{\infty}a_i \qquad k=0,1,\cdots$$

$$=\frac{1}{p}\sum_{n=0}^{k}q_n\cdot\left[1-\sum_{i=0}^{k-n}a_i\right] \tag{22}$$

where the q_n are given by (13)–(15) and the a_i are given by (1) and (3) for $N < \infty$ and $N = \infty$, respectively.

III. Queues on Inputs

The interesting analysis occurs when the switch fabric runs at the same speed as the input and output trunks, and packets are queued at the inputs. How much traffic can the switch accommodate before it saturates, and how much does the mean waiting time increase when we queue packets at the inputs rather than at the outputs? As in the previous section, we assume that packet arrivals on the N input trunks are governed by independent and identical Bernoulli processes. In any given time slot, the probability that a packet will arrive on a particular input is p; each packet has equal probability $1/N$ of being addressed to any given output. Each arriving packet goes, at least momentarily, into a FIFO on its input trunk. At the beginning of every time slot, the switch controller looks at the first packet in each FIFO. If every packet is addressed to a different output, the controller closes the proper crosspoints and all the packets go through. If k packets are addressed to a particular output, one of the k packets is chosen at *random*, each selected with equal probability $1/k$. The others wait until the next time slot when a new selection is made among the packets that are then waiting.

A. Saturation Analysis—Random Selection Policy

Suppose the input queues are saturated so that packets are always waiting in every input queue. Whenever a packet is transmitted through the switch, a new packet immediately replaces it at the head of the input queue. We define B^i_m as the number of packets at the heads of input queues that are "blocked" for output i at the end of the mth time slot. In other words, B^i_m is the number of packets destined for output i, but not selected by the controller during the mth time slot. We also define A^i_m as the number of packets moving to the head of "free" input queues during the mth time slot and destined for output i. An input queue is "free" during the mth time slot if, and only if, a packet from it was transmitted during the $(m-1)$st time slot. The new packet "arrival" at the head of the queue has equal probability $1/N$ of being addressed to any given output. It follows that

$$B^i_m = \max(0, B^i_{m-1} + A^i_m - 1). \quad (23)$$

Although B^i_m does not represent the occupancy of any physical queue, notice that (23) has the same mathematical form as (5).

A^i_m, the number of packet arrivals during the mth time slot to free input queues and destined for output i, has the binomial probabilities

$$\Pr[A^i_m = k] = \binom{F_{m-1}}{k} (1/N)^k (1-1/N)^{F_{m-1}-k} \qquad k = 0, 1, \cdots, F_{m-1} \quad (24)$$

where

$$F_{m-1} \triangleq N - \sum_{i=1}^{N} B^i_{m-1}. \quad (25)$$

F_{m-1} is the number of free input queues at the end of the $(m-1)$st time slot, representing the total number of packets transmitted through the switching during the $(m-1)$st time slot. Therefore, F_{m-1} is also the total number of input queues with new packets at their heads during the mth time slot. That is,

$$F_{m-1} = \sum_{i=1}^{N} A^i_m. \quad (26)$$

Notice that $\bar{F}/N = \rho_0$ where $\bar{F}$ is the mean steady-state number of free input queues and ρ_0 is the utilization of the output trunks (i.e., the switch throughput). As $N \to \infty$, the steady-state number of packets moving to the head of free input queues each time slot, and destined for output i, (A^i) becomes Poisson at rate ρ_0 (see Appendix A). These observations and (23) together imply that we can use the results of Section II to obtain an expression for the mean steady-state value of B^i as $N \to \infty$. Modifying (8), we have

$$\overline{B^i} = \frac{\rho_0^2}{2(1-\rho_0)}. \quad (27)$$

However, using (25) and $\bar{F}/N = \rho_0$, we also have

$$\overline{B^i} = 1 - \rho_0. \quad (28)$$

It follows from (27) and (28) that $\rho_0 = (2 - \sqrt{2}) = 0.586$ when the switch is saturated and $N = \infty$.

It is interesting to note that this same asymptotic saturation throughput has also been obtained in an entirely different context. Consider the problem of memory interference in synchronous multiprocessor systems [11], [12] in which M memories are shared by N processors. Memory requests are presented at the beginning of memory cycles; a conflict occurs if more than one simultaneous request is made to a particular memory. In the event of a conflict, one request is accepted, and the other requests are held for future memory cycles. If $M = N$ and processors always make a new memory request in the cycle immediately following their own satisfied request, then our saturation model for input queueing is identical to the multiprocessor model. As $N \to \infty$, the expected number of busy memories (per cycle) is $(2 - \sqrt{2}) \cdot N$ [11].

When the input queues are saturated and $N < \infty$, the switch throughput is found by analyzing a Markov chain model. Under saturation, the model is identical to the Markov chain analysis of memory interference in [12]. Unfortunately, the number of states grows exponentially with N, making the model useful only for small N. The results presented in Table I,[4] however, illustrate the rapid convergence to the asymptotic throughput of 0.586. In addition, saturation throughputs obtained by simulation[5] (Fig. 6) agree with the analysis.

B. Increasing the Switch Throughput by Dropping Packets

Whenever k packets are addressed for a particular output in a time slot, only one can be transmitted over the output trunk. We have been assuming that the remaining $k-1$ packets wait in their input queues until the next time slot when a new selection is made among the packets that are then waiting. Unfortunately, a packet that could traverse the switch to an idle output during the current time slot may have to wait in queue behind a packet whose output is currently busy. As shown in Section III-A, input queue blocking limits the switch throughput to approximately 0.586 for large N.

Instead of storing the remaining $k-1$ packets in input queues, suppose we just drop them from the switch (i.e., we eliminate the input queues). Dropping packets obviously reduces the switch throughput when the input trunk utilization p is small; more time slots on the output trunks are empty because new packets do not arrive fast enough to replace dropped packets. Although dropping a significant number of packets (say, more than 1 out of 1000) may not be realistic for a packet switch, it is interesting to note that as the input utilization p increases, the reduction in input queue blocking when packets are dropped eventually outweighs the loss associated with dropping the packets.

We define A^i_m as the number of packet arrivals during the mth time slot that are addressed for output i. A^i_m has the binomial probabilities

$$\Pr[A^i_m = k] = \binom{N}{k} (p/N)^k (1-p/N)^{N-k} \qquad k = 0, 1, \cdots, N. \quad (29)$$

We also define the indicator function I^i_m as follows:

$$I^i_m = \begin{cases} 1 & \text{if output trunk } i \text{ transmits a packet during the } m\text{th time slot} \\ 0 & \text{otherwise.} \end{cases} \quad (30)$$

When we drop packets, only those that arrive during the mth time slot have a chance to be transmitted over output trunks during the mth time slot. If they are not selected in the slot in which they arrive, they are dropped. Consequently, for each output trunk i, the random variables I^i_r and $I^i_s (r \neq s)$ are independent and identically distributed, with probabilities

$$\Pr[I^i_m = 1] = \Pr[A^i_m > 0] = 1 - (1 - p/N)^N. \quad (31)$$

[4] The entries in Table I were obtained by normalizing (dividing by N) the values from [12, Table III].

[5] Rather than plot the simulation results as discrete points, the saturation throughputs obtained for N between 2 and 100 are simply connected by straight line segments. No smoothing is done on the data.

TABLE I
THE MAXIMUM THROUGHPUT ACHIEVABLE WITH INPUT QUEUEING

N	Saturation Throughput
1	1.0000
2	0.7500
3	0.6825
4	0.6553
5	0.6399
6	0.6302
7	0.6234
8	0.6184
∞	0.5858

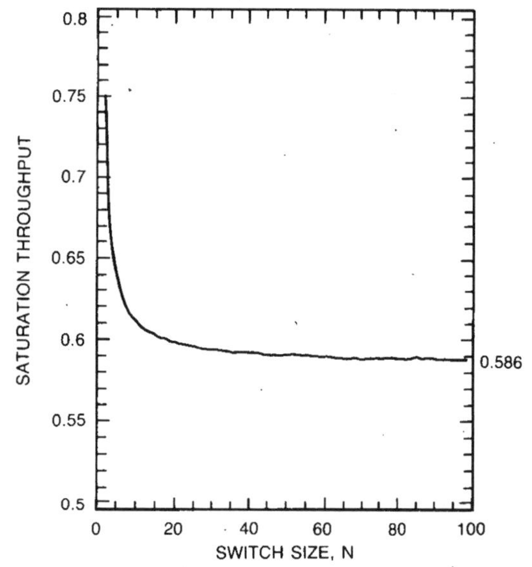

Fig. 6. The maximum throughput achievable with input queueing.

By symmetry, $1 - (1 - p/N)^N$ is the utilization of each output trunk; the switch throughput ρ_0 is given by

$$\rho_0 = 1 - (1 - p/N)^N. \tag{32}$$

As $N \to \infty$,

$$\rho_0 = 1 - e^{-p}. \tag{33}$$

The probability that an arbitrary packet will be dropped from the switch is simply $1 - \rho_0/p$.

The switch throughput ρ_0, as a function of p, is shown in Fig. 7 for several values of N. When the utilization p of the input trunks exceeds a critical threshold, the switch throughput ρ_0 is larger when we drop packets [(32) and (33)] than when we queue them on the input trunks (Table I). For example, when $N = \infty$ and $p > \ln(1 + \sqrt{2})$, the switch throughput when we drop packets is greater than $(2 - \sqrt{2})$—the throughput with input queues. Table II lists, as a function of p, which of the two strategies yields the larger switch throughput.

C. Waiting Time—Random Selection Policy

Below saturation, packet waiting time is a function of the service discipline the switch uses when two or more input queues are waiting to transmit packets to the same output. In this section, we derive an exact formula for the mean waiting

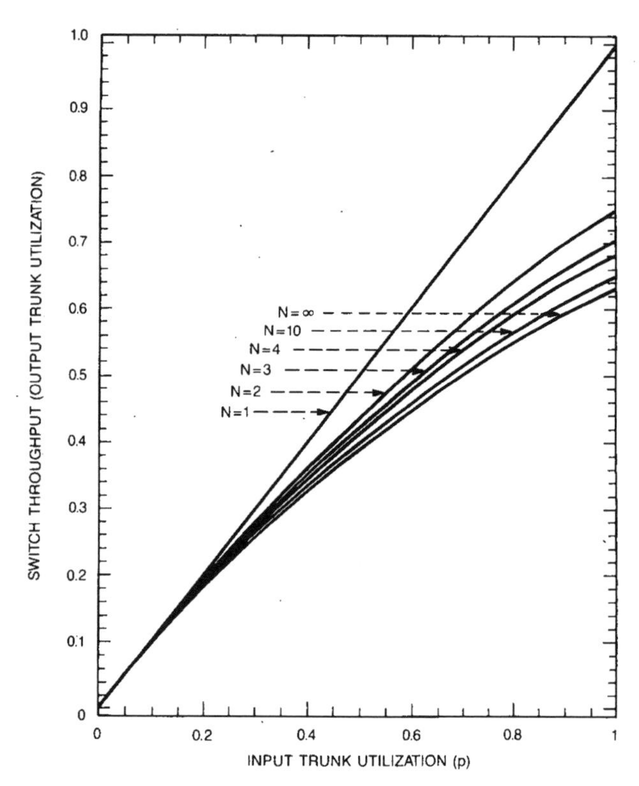

Fig. 7. The switch throughput when packets are dropped, rather than queued at the inputs.

TABLE II
THE STRATEGY (AS A FUNCTION OF p), INPUT QUEUEING, OR PACKET DROPPING THAT YIELDS THE LARGER SWITCH THROUGHPUT

N	Queues On Inputs - Finite Queue Sizes	Queues On Inputs - Saturated Queues	Drop Packets
1	$0 \leq p < 1$	$p = 1$	----------
2	$0 \leq p < 0.750$	$0.750 \leq p \leq 1$	----------
3	$0 \leq p \leq 0.682$	$0.683 \leq p \leq 0.953$	$0.954 \leq p \leq 1$
4	$0 \leq p \leq 0.655$	$0.656 \leq p \leq 0.935$	$0.936 \leq p \leq 1$
5	$0 \leq p \leq 0.639$	$0.640 \leq p \leq 0.923$	$0.924 \leq p \leq 1$
6	$0 \leq p \leq 0.630$	$0.631 \leq p \leq 0.916$	$0.917 \leq p \leq 1$
7	$0 \leq p \leq 0.623$	$0.624 \leq p \leq 0.911$	$0.912 \leq p \leq 1$
8	$0 \leq p \leq 0.618$	$0.619 \leq p \leq 0.907$	$0.908 \leq p \leq 1$
∞	$0 \leq p \leq 0.585$	$0.586 \leq p \leq 0.881$	$0.882 \leq p \leq 1$

time under the random selection policy for the limiting case of $N = \infty$. The waiting time is obtained by simulation for finite values of N. In Section III-D, numerical results are compared to the mean waiting time under the longest queue and fixed priority selection policies.

When the input queues are not saturated, there is a significant difference between our analysis of a packet switch with input queues and the analysis of memory interference in synchronous multiprocessor systems. The multiprocessor application assumes that new memory requests are generated only after a previous request has been satisfied. A processor never has more than one memory request waiting at any time. In our problem, however, packet queueing on the input trunks impacts the switch performance.

A discrete-time Geom/G/1 queueing model (Fig. 8) is used

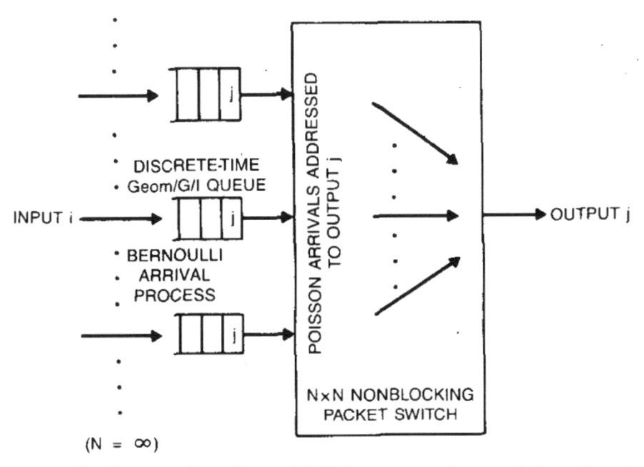

Fig. 8. The discrete-time Geom/G/1 input queueing model used to derive an exact formula for the mean waiting time for the limiting case of $N = \infty$.

to determine the expected packet waiting time for the limiting case of $N = \infty$. The arrival process is Bernoulli: in any given time slot, the probability that a packet will arrive on a particular input is p where $0 < p < 2 - \sqrt{2}$. Each packet has equal probability $1/N$ of being addressed to any given output, and successive packets are independent. To obtain the service distribution, suppose the packet at the head of input queue i is addressed for output j. The "service time" for the packet consists of the wait until it is randomly selected by the switch controller, plus one time slot for its transmission through the switch and onto output trunk j. As $N \to \infty$, successive packets in input queue i experience the same service distribution because their destination addresses are independent and distributed uniformly over all N outputs. Furthermore, the steady-state number of packet "arrivals" to the heads of input queues and addressed for output j becomes Poisson at rate p.[6] Consequently, the service distribution for the discrete-time Geom/G/1 model is itself the packet delay distribution of another queueing system: a discrete-time $M/D/1$ queue with customers served in random order. Analysis of the discrete-time $M/D/1$ queue, with packets served in random order, is given in Appendix B.

Using [6, eq. (39)], the mean packet delay for a discrete-time Geom/G/1 queue is

$$\bar{D} = \frac{p\overline{S(S-1)}}{2(1-p\bar{S})} + \bar{S} \tag{34}$$

where S is a random variable with the service time distribution given in Appendix B and mean value $\bar{S}$. The mean waiting time is $\bar{W} = \bar{D} - 1$

$$\bar{W} = \frac{p\overline{S(S-1)}}{2(1-p\bar{S})} + \bar{S} - 1. \tag{35}$$

$\overline{S(S-1)}$ and $\bar{S}$ are determined numerically using the method in Appendix B.

The mean waiting time $\bar{W}$, as a function of p, is shown in Fig. 9 for both input queueing and output queueing—in the limit as $N \to \infty$. As expected, waiting times are always greater for queueing on inputs than for queueing on outputs. Packet waiting times for input queueing and finite values of N, obtained by simulation,[7] agree with the asymptotic analytic results (Fig. 10).

[6] This follows from the proof in Appendix A.

[7] Rather than plot the simulation results as discrete points, the simulation results are simply connected by straight line segments; no smoothing is done on the data. The same comment applies to Figs. 11, 12, and 13.

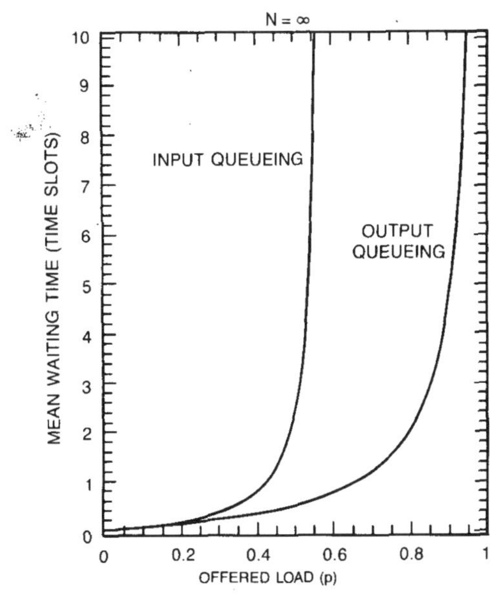

Fig. 9. A comparison of the mean waiting time for input queueing and output queueing for the limiting case of $N = \infty$.

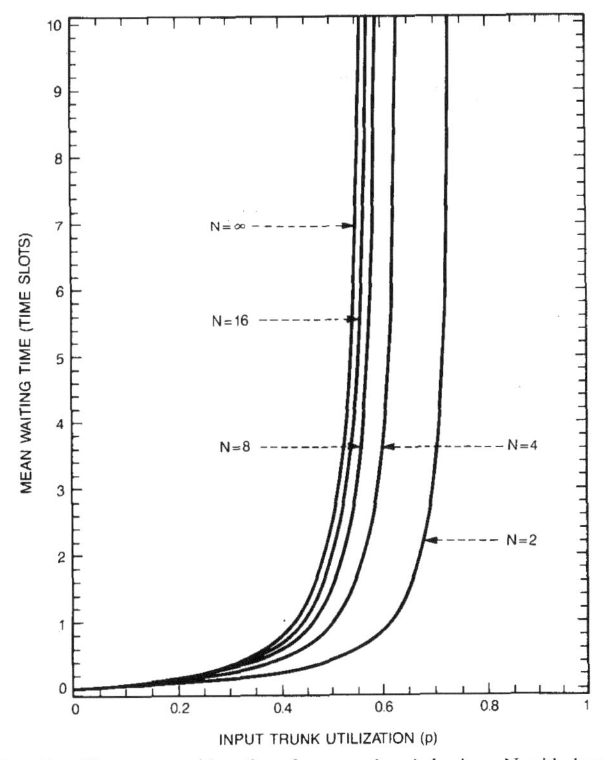

Fig. 10. The mean waiting time for several switch sizes N with input queueing.

D. Longest Queue and Fixed Priority Selection Policies

Until now, we have assumed that if k packets are addressed to a particular output, one of the k packets is chosen at random, each selected with equal probability $1/k$. In this section, we consider two other selection policies: longest queue selection, and fixed priority selection. Under the longest queue selection policy, the controller sends the packet from the longest input queue. A random selection is made if, of the k input queues with packets addressed to a particular output, several queues have the same maximum length. Under the

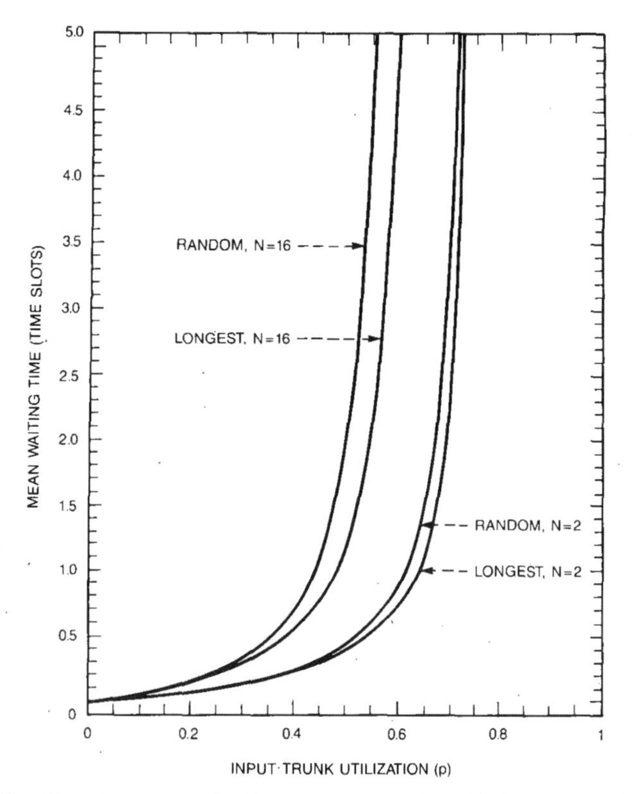

Fig. 11. The mean waiting time for input queueing with the random and longest queue selection policies.

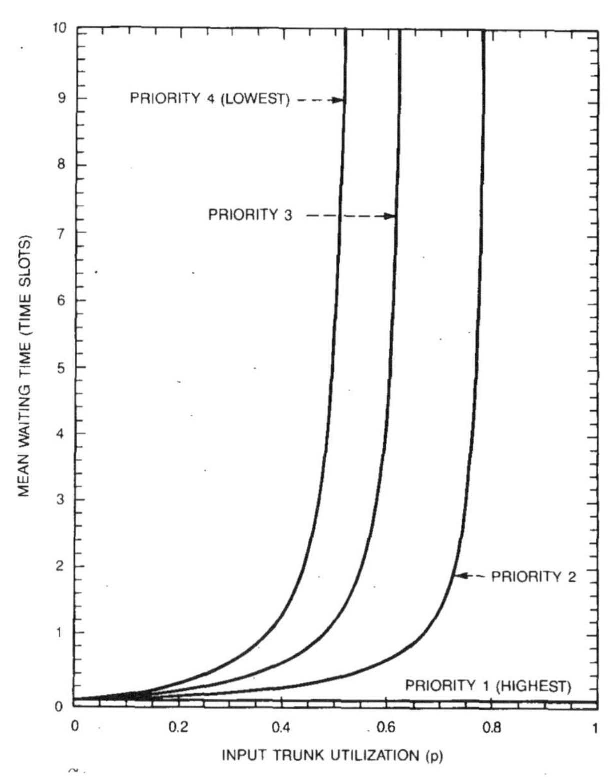

Fig. 12. The mean waiting time for input queueing with the fixed priority service discipline and $N = 4$.

fixed priority selection policy, the N inputs have fixed priority levels, and of the k packets, the controller sends the one with highest priority.

Simulation results for the longest queue policy indicate smaller packet waiting times than those expected with random service (Fig. 11). This is anticipated because the longest queue selection policy reduces the expected number of packets blocked (behind other packets) from traversing the switch to idle outputs.

For the fixed priority service discipline, our simulation results show that the lowest priority input queue suffers large delays and is sometimes saturated, even when the other two service disciplines guarantee stability. Although the saturation throughput is 0.6553 under the random selection policy when $N = 4$ (see Table I), it is shown in Fig. 12 that the lowest priority input queue saturates at approximately 0.55 under the fixed priority discipline. Fig. 13 illustrates the family of waiting time curves for $N = 8$.

These results are interesting because imposing a priority scheme on a single server queueing system usually does not affect its stability; the system remains work conserving. For the $N \times N$ packet switch, however, packet blocking at the low priority input queues does impact stability. More work remains to characterize the stability region.

IV. Conclusion

Using Markov chain models, queueing theory, and simulation, we have presented a thorough comparison of input versus output queueing on an $N \times N$ nonblocking space-division packet switch. What the present exercise has done, for a particular solvable example, is to quantify the intuition that better performance results with output queueing than with input queueing. Besides performance, of course, there are other issues, such as switch implementation, that must be considered in designing a space-division packet switch. The Knockout Switch [13] is an example of a space-division packet switch that places all buffers for queueing packets at the

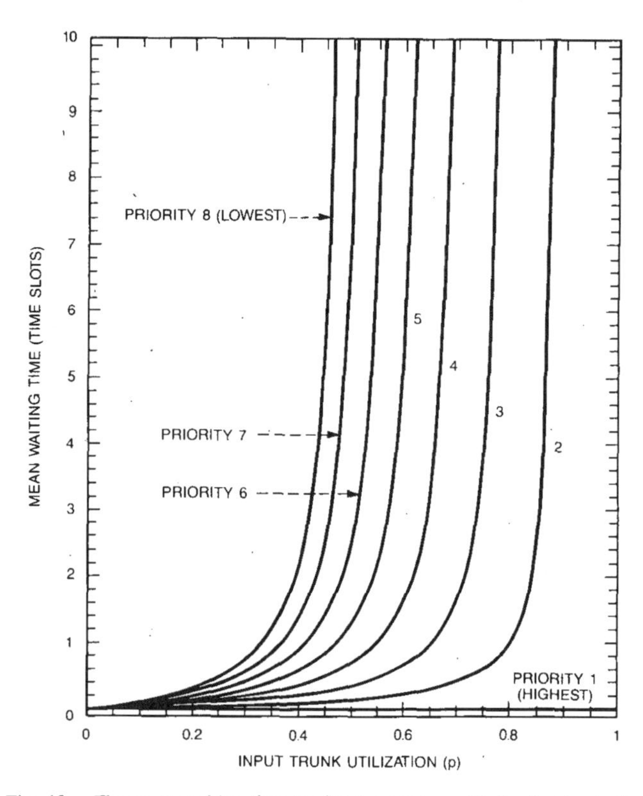

Fig. 13. The mean waiting time for input queueing with the fixed priority service discipline and $N = 8$.

outputs of the switch, thus enjoying the performance advantages of output queueing. Furthermore, the switch fabric runs at the same speed as the input and output trunks.

Appendix A

Poisson Limit of Packets Moving to the Head of Free Input Queues

For the input queueing saturation analysis presented in Section III-A, as $N \to \infty$, we show that the steady-state number of packets moving to the head of free input queues each time slot, and destined for output i, (A^i) becomes Poisson at rate $\rho_0 = \bar{F}/N$. To make clear the dependency of F and ρ_0 on the number of inputs N, we define $\bar{F}(N)$ as the steady-state number of free input queues and $\rho_0(N)$ as the output trunk utilization for a given N where $\rho_0 N = \bar{F}(N)/N$.

We can write

$$\operatorname{Var}\left\{\frac{F(N)}{N}\right\} = \frac{1}{N}\Pr[\text{input queue } r \text{ is free}] + \left(1-\frac{1}{N}\right)\Pr[\text{input queue } r \text{ is free, input queue } s(s\neq r) \text{ is free}] - (\Pr[\text{input queue } r \text{ is free}])^2. \tag{A1}$$

As $N \to \infty$, the events {input queue r is free} and {input queue s is free} become independent for $s \neq r$. Therefore, from (A1),

$$\lim_{N\to\infty} \operatorname{Var}\left\{\frac{F(N)}{N}\right\} = 0. \tag{A2}$$

Given $\epsilon > 0$, we define the set S_N by

$$S_N \triangleq \{L, L+1, \cdots, U-1, U\} \tag{A3}$$

where

$$L \triangleq \max\{1, \lfloor \bar{F}(N) - \epsilon N \rfloor\}, \tag{A4}$$

$$U \triangleq \min\{N, \lceil \bar{F}(N) + \epsilon N \rceil\}, \tag{A5}$$

and $\lfloor x \rfloor$ ($\lceil x \rceil$) denotes the greatest (smallest) integer less than (greater than) or equal to x. By the Chebyshev inequality,

$$\Pr[F(N) \in S_N] \geq 1 - \frac{\operatorname{Var}\left\{\frac{F(N)}{N}\right\}}{\epsilon^2}. \tag{A6}$$

Therefore,

$$\Pr[A^i = a] = \sum_{f=\max(a,1)}^{N} \Pr[F(N)=f] \cdot \binom{f}{a}(1/N)^a(1-1/N)^{f-a} \leq \sum_{f\in S_N} \Pr[F(N)=f] \cdot \binom{f}{a}(1/N)^a(1-1/N)^{f-a} + \frac{\operatorname{Var}\left\{\frac{F(N)}{N}\right\}}{\epsilon^2}. \tag{A7}$$

Case I ($a = 0$): If $f \in S_N$, then

$$(1-1/N)^U \leq \binom{f}{0}(1/N)^0(1-1/N)^{f-0} \leq (1-1/N)^L. \tag{A8}$$

Therefore, from (A6), (A7), and (A8), we obtain

$$\left[1 - \frac{\operatorname{Var}\left\{\frac{F(N)}{N}\right\}}{\epsilon^2}\right] \cdot (1-1/N)^U \leq \Pr[A^i = 0] \leq (1-1/N)^L + \frac{\operatorname{Var}\left\{\frac{F(N)}{N}\right\}}{\epsilon^2}. \tag{A9}$$

As $N \to \infty$,

$$e^{-(\rho_0+\epsilon)} \leq \Pr[A^i=0] \leq e^{-(\rho_0-\epsilon)}. \tag{A10}$$

Since this holds for arbitrarily small $\epsilon > 0$,

$$\lim_{N\to\infty} \Pr[A^i=0] = e^{-\rho_0}. \tag{A11}$$

Case II ($a > 0$): Since $\binom{f}{a}(1/N)^a(1 - 1/N)^{f-a}$ is a nondecreasing function of f for $1 \leq f \leq N$ and $a > 0$, for $f \in S_N$ we have

$$\binom{L}{a}(1/N)^a(1-1/N)^{L-a} \leq \binom{f}{a}(1/N)^a(1-1/N)^{f-a} \leq \binom{U}{a}(1/N)^a(1-1/N)^{U-a}. \tag{A12}$$

Therefore, from (A6), (A7), and (A12), we obtain

$$\left[1 - \frac{\operatorname{Var}\left\{\frac{F(N)}{N}\right\}}{\epsilon^2}\right] \cdot \binom{L}{a}(1/N)^a(1-1/N)^{L-a} \leq \Pr[A^i=a] \leq \binom{U}{a}(1/N)^a(1-1/N)^{U-a} + \frac{\operatorname{Var}\left\{\frac{F(N)}{N}\right\}}{\epsilon^2}. \tag{A13}$$

As $N \to \infty$,

$$e^{-(\rho_0-\epsilon)}\frac{(\rho_0-\epsilon)^a}{a!} \leq \Pr[A^i=a] \leq e^{-(\rho_0+\epsilon)}\frac{(\rho_0+\epsilon)^a}{a!}. \tag{A14}$$

Since this holds for arbitrarily small $\epsilon > 0$,

$$\lim_{N\to\infty} \Pr[A^i=a] = e^{-\rho_0}\frac{\rho_0^a}{a!}. \tag{A15}$$

Appendix B

Discrete-Time $M/D/1$ Queue—Packets Served in Random Order

In this Appendix, we present a simple numerical method for computing the delay distribution of a discrete-time $M/D/1$

queue, with packets served in random order. The number of packet arrivals at the beginning of each time slot is Poisson distributed with rate λ, and each packet requires one time slot for service. We fix our attention on a particular "tagged" packet in the system, during a given time slot. Let $p_{m,k}$ denote the probability, conditioned on there being a total of k packets in the system during the given time slot, that the remaining delay is m time slots until the tagged packet completes service. It is easy to obtain $p_{m,k}$ by recursion on m.

$$p_{1,1}=1 \tag{B1}$$

$$p_{m,1}=0 \qquad m\neq 1 \tag{B2}$$

$$p_{1,k}=\frac{1}{k} \qquad k\geq 1 \tag{B3}$$

$$p_{m,k}=\frac{k-1}{k}\cdot\sum_{j=0}^{\infty}p_{m-1,k-1+j}\cdot\frac{e^{-\lambda}\lambda^{j}}{j!} \qquad m>1,\ k>1. \tag{B4}$$

Averaging over k, the packet delay D has the probabilities

$$\Pr[D=m]=\sum_{k=1}^{\infty}p_{m,k}\cdot\Pr[k \text{ packets in system immediately after the tagged packet arrives}]$$

$$=\sum_{k=1}^{\infty}p_{m,k}\cdot\sum_{n=0}^{k-1}q_n\cdot\frac{e^{-\lambda}\lambda^{k-n-1}}{(k-n-1)!} \tag{B5}$$

where the q_n are the steady-state queue size probabilities given by (13)–(15).

The variance and mean of the packet delay distribution are determined numerically from the delay probabilities in (B5).

References

[1] J. S. Turner and L. F. Wyatt, "A packet network architecture for integrated services," in *GLOBECOM'83 Conf. Rec.*, Nov. 1983, pp. 45–50.

[2] J. J. Kulzer and W. A. Montgomery, "Statistical switching architectures for future services," in *Proc. Int. Switching Symp.*, May 1984.

[3] T.-Y. Feng, "A survey of interconnection networks," *Computer*, vol. 14, pp. 12–27, Dec. 1981.

[4] D. M. Dias and M. Kumar, "Packet switching in $N \log N$ multistage networks," in *GLOBECOM'84 Conf. Rec.*, Nov. 1984, pp. 114–120.

[5] Y.-C. Jenq, "Performance analysis of a packet switch based on a single-buffered Banyan network," *IEEE J. Select. Areas Commun.*, vol. SAC-1, pp. 1014–1021, Dec. 1983.

[6] T. Meisling, "Discrete-time queueing theory," *Oper. Res.*, vol. 6, pp. 96–105, Jan.–Feb. 1958.

[7] I. Rubin, "Access-control disciplines for multiaccess communication channels: Reservation and TDMA schemes," *IEEE Trans. Inform. Theory*, vol. IT-25, pp. 516–536, Sept. 1979.

[8] L. Kleinrock, *Queueing Systems, Vol. 1: Theory.* New York: Wiley, 1975.

[9] D. Gross and C. M. Harris, *Fundamentals of Queueing Theory.* New York: Wiley, 1974.

[10] P. J. Burke, "Delays in single-server queues with batch input," *Oper. Res.*, vol. 23, pp. 830–833, July–Aug. 1975.

[11] F. Baskett and A. J. Smith, "Interference in multiprocessor computer systems with interleaved memory," *Commun. ACM*, vol. 19, pp. 327–334, June 1976.

[12] D. P. Bhandarkar, "Analysis of memory interference in multiprocessors," *IEEE Trans. Comput.*, vol. C-24, pp. 897–908, Sept. 1975.

[13] Y. S. Yeh, M. G. Hluchyj, and A. S. Acampora, "The knockout switch: A simple, modular architecture for high-performance packet switching," in *Proc. Int. Switching Symp.*, Phoenix, AZ, Mar. 1987, pp. 801–808.

PHOTO NOT AVAILABLE

Mark J. Karol (S'79–M'85) was born in Jersey City, NJ, on February 28, 1959. He received the B.S. degree in mathematics and the B.S.E.E. degree in 1981 from Case Western Reserve University, Cleveland, OH, and the M.S.E., M.A., and Ph.D. degrees in electrical engineering from Princeton University, Princeton, NJ, in 1982, 1984, and 1985, respectively.

Since 1985 he has been a member of the Network Systems Research Department at AT&T Bell Laboratories, Holmdel, NJ. His current research interests include local and metropolitan area lightwave networks, and wide-band circuit- and packet-switching architectures.

PHOTO NOT AVAILABLE

Michael G. Hluchyj (S'75–M'82) was born in Erie, PA, on October 23, 1954. He received the B.S.E.E. degree in 1976 from the University of Massachusetts, Amherst, and the S.M., E.E., and Ph.D. degrees in electrical engineering from the Massachusetts Institute of Technology, Cambridge, in 1978, 1978, and 1981, respectively.

From 1977 to 1981 he was a Research Assistant in the Data Communication Networks Group at the M.I.T. Laboratory for Information and Decision Systems where he investigated fundamental problems in packet radio networks and multiple access communications. In 1981 he joined the Technical Staff at Bell Laboratories where he worked on the architectural design and performance analysis of local area networks. In 1984 he transferred to the Network Systems Research Department at AT&T Bell Laboratories, performing fundamental and applied research in the areas of high-performance, integrated communication networks and multiuser lightwave networks. In June 1987 he assumed his current position as Director of Networking Research and Advanced Development at Codex Corporation. His current research interests include wide-band circuit- and packet-switching architectures, integrated voice and data networks, and local area network interconnects.

Dr. Hluchyj is active in the IEEE Communications Society and is a member of the Technical Editorial Board for the IEEE NETWORK Magazine.

PHOTO NOT AVAILABLE

Samuel P. Morgan (SM'55–F'63) was born in San Diego, CA, on July 14, 1923. He received the B.S., M.S., and Ph.D. degrees, all in physics, from the California Institute of Technology, Pasadena, in 1943, 1944, and 1947, respectively.

He is a Distinguished Member of the Technical Staff at AT&T Bell Laboratories, Murray Hill, NJ. He joined Bell Laboratories in 1947, and for a number of years he was concerned with applications of electromagnetic theory to microwave antennas and to problems of waveguide and coaxial cable transmission. From 1959 to 1967 he was Head, Mathematical Physics Department, and from 1967 to 1982 he was Director, Computing Science Research Center. His current interests include queueing and congestion theory in computer-communication networks.

Dr. Morgan is a member of the American Physical Society, the Society for Industrial and Applied Mathematics, the Association for Computing Machinery, the American Association for the Advancement of Science, and Sigma Xi.

Routing in the Manhattan Street Network

NICHOLAS F. MAXEMCHUK, SENIOR MEMBER, IEEE

Abstract—**The Manhattan Street Network is a regular, two-connected network, designed for packet communications in a local or metropolitan area. It operates as a slotted system, similar to conventional loop networks. Unlike loop networks, routing decisions must be made at every node in this network. In this paper, several distributed routing rules are investigated that take advantage of the regular structure of the network.**

In an operational network, irregularities occur in the structure because of the addressing mechanisms, adding single nodes, and failures. A fractional addressing scheme is described that makes it possible to add new rows or columns to the network without changing the addresses of existing nodes. A technique is described for adding one node at a time to the network, while changing only two existing links. Finally, two procedures are described that allow the network to adapt to node or link failures. The effect that irregularities have on routing mechanisms designed for a regular structure is investigated.

I. Introduction

THE Manhattan Street Network (MSN) [1], Fig. 1, is a two-connected, regular network with unidirectional links. The links are arranged in a structure that resembles the streets and avenues in Manhattan. The MSN topology is being applied to a local or metropolitan area packet communication system.

The nodes in the MSN are described in Section II. The structure of the nodes and the access strategy are similar to those in a slotted loop system. The principle difference between the MSN and a loop network is that there are two links arriving at and leaving each node instead of a single link, and a routing decision must be made for each packet transmitted at each node. An experimental network is being constructed with a 50 Mbit/s transmission rate on each link and 128 bit fixed sized packets. More than 750 000 routing decisions per second may have to be made at each node in this network. In this type of network, the routing rule must be simple.

In this paper, distributed routing rules for the MSN are investigated. Simple routing rules that use the regular structure of the network are compared to shortest path algorithms and random routing strategies. In the MSN, shown in Fig. 1, the number of rows and columns completely defines the network, and if these numbers are known, the shortest path between any pair of nodes can be determined. In addition, because of the cyclic structure of the MSN, routing is only dependent upon the relative location of the current node with respect to the destination, as defined in Section III-B, and the same routing rule can be used at every node. In Section IV-A, a distributed rule is described that finds the shortest path. In Sections IV-B and IV-C, two simplifications of the shortest path rule are described. The simplified rules do not always find the shortest path, and the effect that these rules have on the average path length is investigated in Section IV-E.

In Section III-A, a fractional addressing scheme is described. This addressing scheme has two advantages over the integer addressing scheme in Fig. 1.

Paper approved by the Editor for Wide Area Networks of the IEEE Communications Society. Manuscript received November 25, 1985; revised December 11, 1986.

The author is with AT&T Bell Laboratories, Murray Hill, NJ 07974.

IEEE Log Number 8613925.

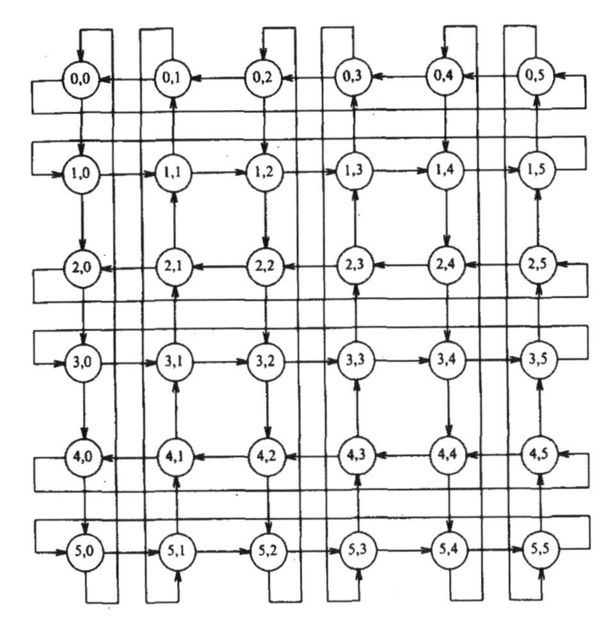

Fig. 1. 36-node MSN.

1) New rows or columns are added to the network without changing the addresses of existing nodes.

2) The distributed routing rules are independent of the number of rows or columns in the network.

A disadvantage of fractional addressing is that the distributed routing rules must operate without knowing the position of all of the rows and columns in the network and cannot always find the shortest path to the destination. The effect that this addressing scheme has on the average distance between nodes in investigated in Section IV-E.

The MSN is a regular structure with an even number of rows and columns and is not defined for an arbitrary number of nodes. A realistic network, in which nodes are added and other nodes or links fail, may be approximated by the MSN, but it is unlikely that it will exactly correspond to the regular structure. The routing rule that is selected for the MSN must operate in networks with irregularities. The effect that irregularities have on the routing rules depends upon the techniques used to add nodes and remove failed components. In Section V-A, a technique is described for adding one node at a time to the network. When this technique is used, only two links must be changed when a new node is added to the network. As nodes are added, a row or column may not have a full complement of nodes. The routing rules operate without knowing which rows or columns are incomplete. In Sections V-B and V-C, procedures are described that allow the network to adapt to node or link failures. The adaptations guarantee that the nodes continue to operate without losing packets at any of the surviving nodes. The routing rules operate without knowing which nodes or links have failed.

II. System Description

The MSN, Fig. 1, is a member of a class of multiply-connected, regular, mesh-configured networks. There is an

Reprinted from *IEEE Transactions on Communications,* vol. COM-35, no. 5, May 1987.

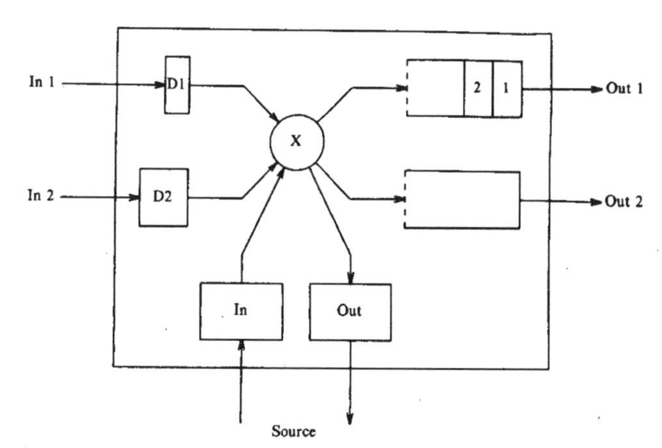

Fig. 2. The structure of a node in a two-connected network with fixed size packets.

even number of rows and columns with two links arriving at and two links leaving each node. Logically, the links form a grid on the surface of a torus, with links in adjacent rows or columns traveling in opposite directions.

In [2], it has been demonstrated that, because of the increased connectivity, mesh networks can achieve higher throughputs and support more sources than conventional loop [3]–[5] and bus [6], [7] networks. This occurs because of the following.

1) On the average, a smaller fraction of the links in the network are used to interconnect a source and destination.

2) Sources that communicate frequently can be clustered into communities of interest that do not interfere with one another.

In a network with several paths arriving at a node, messages from more than one incoming link may be destined for the same outgoing link. Data from several links can be concentrated onto one link by storing the data, forwarding them when the link is available, and establishing protocols to recover messages that are lost because of buffer overflows. In [2], a slotted system is described that does not require buffering on the output links and does not lose packets because of buffer overflows. The structure of a node in this network is shown in Fig. 2.

The packets in the slotted system are a fixed size. A node periodically transmits a packet from an input line, a packet from the source, or an empty packet on each output line. At each node, the packets from the input lines are delayed so that they arrive at the switch at the time that the node transmits a packet. The node switches each of the incoming packets not destined for the node to one of the output links. If the buffer for an output link is full, and two incoming packets are destined for this link, one of the packets is forced to take the other output link. This strategy guarantees that packets are never lost because of buffer overflow, even if the output buffer size at the nodes is reduced to zero; however, the larger the buffers at a node, the less likely it is that a packet must be misdirected. Packets from the source are only transmitted when there is an empty slot on an output link. The node controls the source so that packets do not arrive faster than they are transmitted, and the rate available to the source decreases when the network is busy. Packets that are misdirected take a longer path to their destination and prevent more new packets from entering the network. Therefore, there is a tradeoff between buffer size and the throughput of the network.

In the MSN, each time a packet is misdirected, the length of the path to the destination is increased by at most four links. In addition, there are many nodes for which either outgoing link provides the same path length to the destination, and when a packet may take either link, the probability any packet will have to be misdirected decreases. A recently published analysis and simulation of the MSN [8] indicates that the MSN operates reasonably efficiently without buffers on the outgoing links, and the experimental system that is being implemented does not have buffers.

III. Addressing Nodes in the Network

Each node in the network has a unique address that is called the node's absolute address. To simplify the routing rule, the absolute address of a node reflects the regular structure of the MSN. Because of the cyclic nature of the MSN, routing only depends on the relative position of the current node and the destination, called the current nodes relative address, and not on the absolute address of any node. Relative addresses allow the same routing rule to be used at each node.

A. Absolute Addresses

In Fig. 1, the rows in the MSN are sequentially numbered from 0 to $m - 1$, the columns are numbered from 0 to $n - 1$, and the absolute address of a node is its row and column. The odd-numbered rows have links in one direction and the even-numbered rows have links in the opposite direction. New rows or columns are added in pairs to preserve the alternating directions, and the address of an existing row or column changes when new elements are added in the middle of the network. To reduce the effect of changing addresses on the communications routines at the source, there must be a transformation between a logical address by which the source refers to the destination and a physical address that is the destination node's current row and column. The transformation need only be performed at the source node of a packet; however, a transformation table must be maintained at every node in the network, and a protocol must be developed to update the table as physical addresses change.

An alternative to changing the address of a node when new rows and columns are added is to plan for expansion by not using all of the addresses initially. For instance, in the initial implementation of the network, the rows may be numbered 0, 11, 22, $\cdots$ so that ten new rows can be added between each of the initial rows. By leaving an even number of integers between assigned rows or columns, the alternating direction can be retained as new rows are added. The spacing between rows can be decreased when nodes in the same community of interest are in adjacent rows, and it can be increased where new communities of interest may be inserted. This approach requires careful planning because the network growth must be predicted when the initial network is designed. The planning can be reduced by using fractional addresses rather than integer addresses. Fractional addresses allow an arbitrary number of pairs of rows to be added at any position in the network.

The fractional addressing scheme that has been selected is shown in Fig. 3. The first two rows or columns are labelel 0 and 1. Rows are added in pairs and are labeled as two fractions, 1/3 of the way between two other rows. For instance, two rows added between 0 and 1 are labeled 1/3 and 2/3 and two rows added between 2/3 and 1 are labeled 7/9 and 8/9. New rows that are added between 1 and 0 are considered to be between 1 and 2 so that they have different addresses from the rows between 0 and 1. For instance, two rows added between 1 and 0 are labeled 4/3 and 5/3 and two rows between 5/3 and 0 are labeled 16/9 and 17/9. Fractional addressing does not constrain the total number of rows that can be added to the network or the number of rows that can be added to a community of interest in a particular area of the network. In addition, the fractional addressing scheme selected guarantees that all rows with an even numerator have links in one direction and all rows with an odd numerator have links in the opposite direction, as in the integer addressed system.

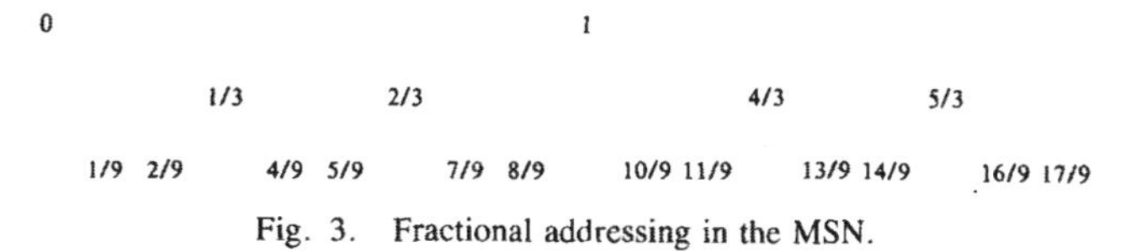

Fig. 3. Fractional addressing in the MSN.

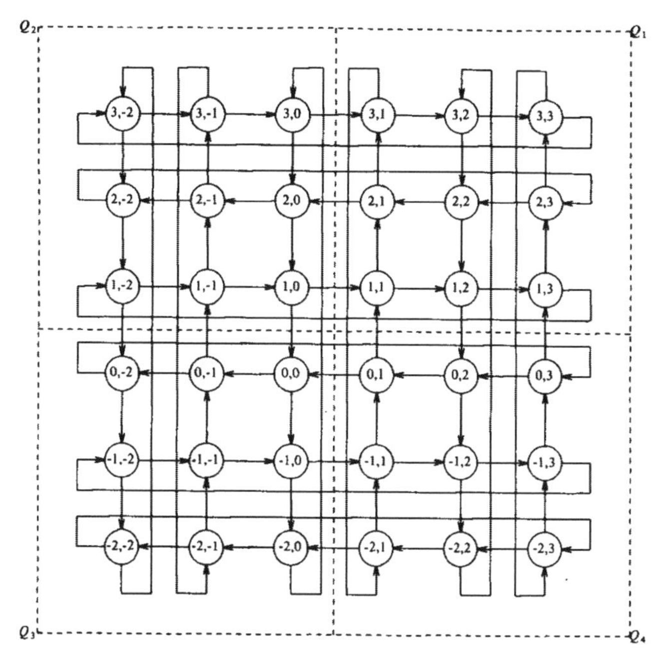

Fig. 4. Relative addresses in a 36-node MSN.

B. Relative Addresses

Because of the cyclic structure of the MSN, any node can be considered to be in the center of the network. The relative address (r, c) of a node with absolute address (r_{fr}, c_{fr}) with respect to the destination node with absolute address (r_{to}, c_{to}) is defined so that the destination node is approximately at the center of the network, has relative address (0, 0), and has both row and column links directed toward decreasing numbered nodes, as in Fig. 4.

The relative address in an $m \times n$ integer-addressed network is

$$r = \frac{m}{2} - \left\{\left(\frac{m}{2} - D_c(r_{fr} - r_{to})\right) \bmod m\right\}$$
$$c = \frac{n}{2} - \left\{\left(\frac{n}{2} - D_r(c_{fr} - c_{to})\right) \bmod n\right\}, \qquad (1)$$

and in a fractionally addressed network is

$$r = 1 - \{(1 - D_c(r_{fr} - r_{to})) \bmod 2\}$$
$$c = 1 - \{(1 - D_r(c_{fr} - c_{to})) \bmod 2\} \qquad (2)$$

where D_c and D_r are dependent upon the direction of the links at the destination node. In a network with the links in the even and odd rows and columns directed toward increasing or decreasing rows and columns, as in Fig. 1, $D_c = +1$ when c_{to} (the numerator of c_{to} in a fractionally addressed network) is even, $D_c = -1$ when c_{to} is odd, $D_r = +1$ when r_{to} is even, and $D_r = -1$ when r_{to} is odd.

The definition of the relative coordinates in (1) and (2) limits the relative address of the current node to $-(m/2) < r \leq m/2$, and $-(n/2) < c \leq n/2$ for an integer-addressed network and $-1 < r, c \leq 1$ for a fractionally addressed

Quadrant 3 or 4			Quadrant 1 or 2				
$-\frac{2}{3}$	$-\frac{1}{3}$	0	$\frac{1}{9}$	$\frac{2}{9}$	$\frac{1}{3}$	$\frac{2}{3}$	1

a) Actual assignment of rows to quadrants

Quadrant 3 or 4				Quadrant 1 or 2			
1	$-\frac{2}{3}$	$-\frac{1}{3}$	0	$\frac{1}{9}$	$\frac{2}{9}$	$\frac{1}{3}$	$\frac{2}{3}$

b) Expected assignment of rows to quadrants

Fig. 5. Assignment of rows to quadrants in a network with eight rows.

network. A node is in Q_1 when $r > 0$ and $c > 0$, Q_2 when $r > 0$ and $c \leq 0$, Q_3 when $r \leq 0$ and $c \leq 0$, and Q_4 when $r > 0$ and $c \leq 0$. The quadrant of the current node indicates the direction in which to proceed to get to the destination. Because the network has unidirectional links, this routing strategy must be modified when the current node is at the boundary of the quadrants, as discussed in Section IV. Fixing the orientation of the links at the destination allows the same routing decisions to apply at the boundaries.

An advantage of fractional addressing over integer addressing is that the relative addresses are independent of the number of rows or columns in the network. In an integer-addressed network, the arithmetic unit that calculates the relative address must be changed whenever the number of rows or columns changes. This arithmetic unit does not change in the fractionally addressed network.

A disadvantage of fractional addressing is that the destination is sometimes displaced from the center of the network when relative addresses are calculated. For instance, in Fig. 5(a), a possible assignment of rows to quadrants is shown for a network with eight rows and only two of a possible 12 rows in the 1/9th addressing level. Five rows are assigned to quadrants 1 or 2 and only three to quadrants 3 or 4, and as a result, packets routed from nodes with a relative address $(1, X)$ may take a longer path to the destination. In Fig. 5(b), the assignment of rows to quadrants that places the destination in the center is shown.

It is evident from this example that new rows should be added uniformly, when possible, in order to calculate the quadrant correctly. However, the quadrant is most likely to be calculated incorrectly for the nodes that are furthest apart. In large networks, with many small communities of interest, expanding the network with nonuniform addresses that keep nodes in their communities of interest is preferable to forcing nodes to join distant parts of the network. If most packets remain within the community of interest and are not directed to the nodes that are furthest away, the distance between nodes that communicate frequently is kept small and the effect of nonuniform addresses is less than in a network with uniform traffic requirements.

IV. Distributed Routing Rules

In Sections IV-A, B, and C, three distributed routing rules are described that use the regular structure of the MSN to select a path to the destination. Rule 1 determines all shortest paths to the destination for integer addressed MSN's. Rules 2 and 3 reduce the number of calculations that are performed at each node, but occasionally take longer paths. Rules 1 and 2 are dependent upon the addresses of the adjacent nodes to which a node is connnected; rule 3 is not.

In complete, integer-addressed networks, the address of adjacent nodes is known. In fractionally addressed networks or in networks with partially full rows or columns, as described in Section V-A, the address of adjacent nodes is not known. If rule 1 or 2 is used, a technique must be used to determine the node to which each node is connected. In the experimental network, the nodes to which a node is connected is stored locally and changed manually when the network connectivity

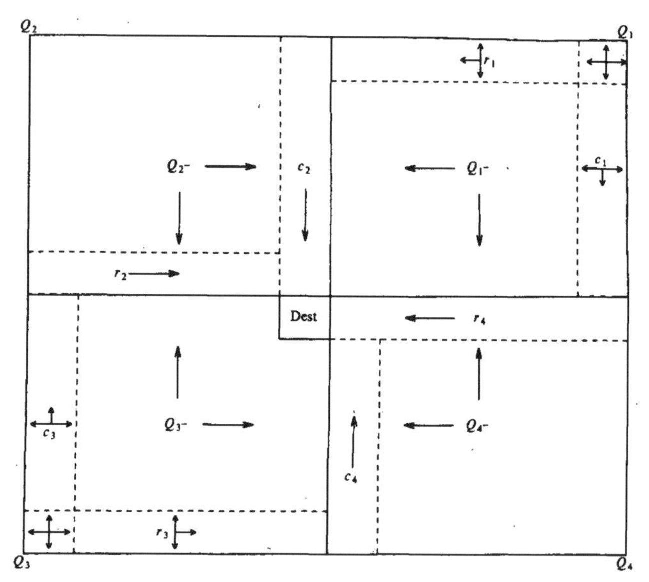

Fig. 6. Preferred paths in Rule 1.

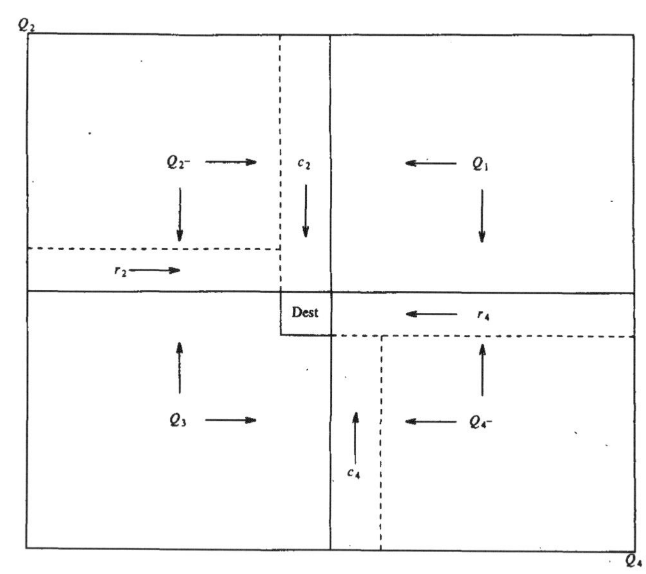

Fig. 7. Preferred paths in Rule 2.

changes. When the procedure described in Section V-A is used to add nodes, the connectivity for only two nodes changes when a new node is added to the network. Therefore, a manual rather than an automatic procedure is reasonable. When nodes or links fail and are bypassed automatically, as in Sections V-B and C, the connectivity information at a node is not changed, and it is incorrect.

These three rules are referred to as deterministic routing rules. In addition, two random routing rules are described in Section IV-D. Rule A is independent of the address of adjacent nodes, and rule B routes packets correctly when the destination is one node away. In Section IV-E, the path lengths resulting from the routing rules are compared in integer and fractionally addressed networks.

A. Deterministic Rule 1

The solid arrows in Fig. 6 show the preferred direction of travel from the relative positions in the network to the destination for the first routing rule. In this figure, r_1, r_2, r_3, and r_4 are rows at the edges of the quadrants, and c_1, c_2, c_3, and c_4 are columns at the edges of the quadrants. The first routing rule is as follows.

Rule 1:

- Select the preferred path if there is one preferred path from a node.
- Select either path if there are zero or two preferred paths from a node.

To implement the first rule, the relative addresses of the current node (r, c), the next node along the column (r_{nxt}, c), and the next node along the row (r, c_{nxt}) are calculated. The quadrant is determined from (r, c) as in Section III-B, the direction of the link along the row is determined from $c - c_{nxt}$, and the direction of the link along the column from $r - r_{nxt}$. A node in Q_4 is in row r_4 if $r = 0$, and a node in Q_2 is in column c_2 when $c = 0$. When a node in Q_2 is also in r_2, the link directed down is not preferred. These links can be determined because $r_{nxt} = 0$ and $c_{nxt} \neq 0$. Similarly, a link is directed to the left from c_4 if (r, c) is in Q_4, $c_{nxt} = 0$, and $r_{nxt} \neq 0$. Row r_1 is at the outside edge of the network. A node in Q_1 is in r_1 and has a preferred link that is not preferred in Q_1 if $c_{nxt} \neq 0$ and c_{nxt} is in Q_2. Similarly, when a node is in r_3, c_1, or c_3 and has a preferred link that is not preferred in the rest of the quadrant, an adjacent node is also in a different quadrant.

In the Appendix, it is shown that this routing rule selects the shortest path from any node to the destination in an integer addressed network. Furthermore, when there are several shortest paths, every path is selected as one of the alternatives; therefore, this rule has the maximum number of instances in which either link may be selected.

B. Deterministic Rule 2

Rule 2 is the same as Rule 1 except that the preferred paths are those shown in Fig. 7 instead of those in Fig. 6. Rule 2 has the advantage that there are fewer calculations than in Rule 1 because the special cases when nodes are in r_1, c_1, r_3, and c_3 are not determined. However, this rule has the disadvantage that nodes in these special rows and columns take a slightly longer path to the destination. In Rule 2, it is still necessary to know c_{nxt} and r_{nxt} in order to determine when a node is in r_2 or c_4.

The routing rule is simplified in this manner because it should have a relatively small effect on the average path length. The nodes that are affected are those that are furthest from the destination. Fewer packets are affected by changes in routing rules in these nodes than elsewhere in the network because packets from other nodes are not intentionally routed through these nodes to get to the destination, and in a network with communities of interest, nodes are more likely to communicate with nodes that are nearby. By contrast, a change in the routing rule in c_2 and r_4 would affect every packet headed for the destination. In addition, incorrect paths are not selected at all of the nodes in the affected rows and columns. From Fig. 6, preferred paths in the quadrants are also preferred paths in the special rows and columns at the edges of the network. Incorrect decisions may only be made at nodes where neither path is thought to be preferred and one of the paths is shorter. Furthermore, from Table V in the Appendix, the longer paths at the edge of the network are two greater than the preferred paths, while elsewhere in the network, they are four greater. The effect of longer paths on the average shortest path length is shown in Section IV-E.

C. Deterministic Rule 3

The solid arrows in Fig. 8 show the preferred paths and the dashed arrows show the alternate paths. The routing rule is as follows.

Rule 3:

- Select the preferred path if there is one preferred path from a node.
- Select the alternate path if there is no preferred path and one alternate path from the node.

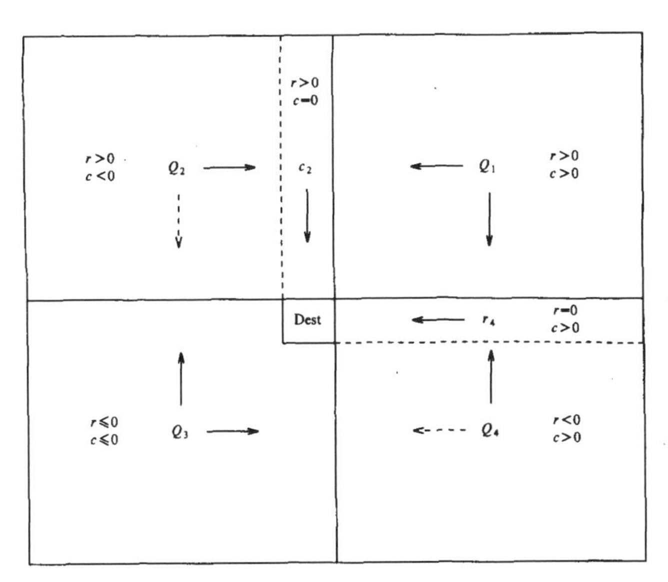

Fig. 8. Preferred and alternate paths in Rule 3.

• Select either path if neither path is a preferred or alternate path or if both paths are preferred.

The advantage of Rule 3 is that it uses fewer calculations than Rule 2 and is not dependent upon r_{nxt} or c_{nxt}. From Fig. 8, the regions of interest in Rule 3 depend only upon the relative address of the current node. The direction of the links at the current node is determined by assuming that a node has a link directed to the left when r is even and to the right when r is odd, and a link directed down when c is even and up when c is odd. The disadvantage of Rule 3 is that there are fewer instances in which either path from the node may be selected. In Q_2 and Q_4 where Rule 2 may select either path, Rule 3 is constrained to select one of the paths. This increases the number of times when two packets arriving at the node conflict. There are also instances in incomplete networks, in Section V-A, where Rule 3 cannot get to a specific destination while Rule 2 can.

In complete networks, Rules 2 and 3 result in the same distance from any node to the destination. Whenever there are one or two preferred paths in Rule 2, one of these paths is selected by Rule 3. When there are two preferred paths in Rule 3, both paths have the same distance to the destination. Therefore, the path length is the same for both rules.

D. Random Routing

Two random routing rules have been considered. Rule A is completely random, a packet selects either link with equal probability, and at each node, checks to see if it is at the destination. Rule B assumes that the two nodes to which a node is connected is known. At each node, if the destination is one node away, the packet is directed there; otherwise, a path is selected at random.

There are two advantages to using random routing rules rather than deterministic rules. First, they are extremely easy to implement. In the random routing rules, it is not necessary to calculate the relative address of a node, its quadrant, or the direction of the links emanating from the nodes. Second, random routing rules are extremely tolerant of network irregularities. If nodes are added or fail in a perverse manner, the network may bear little resemblance to the regular structure, and the deterministic rules may not work. The random rules provide an alternative to the deterministic rules when this occurs. These random routing rules were first investigated by Prosser [9] as a routing mechanism for survivable networks. The disadvantage of random routing rules is that they use more links to get between a source and destination, and this results in a smaller network throughput.

TABLE I
THE EFFICIENCY OF ROUTING RULES RELATIVE TO THE SHORTEST PATH ALGORITHM

Network	Short. Path	Efficiency of Routing Rules					
		Deterministic				Random	
		Integer Addr.		Fractional Addr.		A	B
		1	2, 3	1	2, 3		
4x4	2.93	1.00	1.00	.95	.94	.21	.79
4x6	3.30	1.00	.97	.97	.95	.14	.30
6x6	3.71	1.00	.97	1.00	.97	.10	.21
6x8	4.34	1.00	.98	.99	.97	.09	.17
8x8	5.02	1.00	1.00	1.00	.98	.07	.14
8x10	5.42	1.00	.99	1.00	.98	.06	.11
10x10	5.84	1.00	.99	1.00	.99	.05	.09
10x12	6.42	1.00	.99	1.00	.99	.05	.08
12x12	7.02	1.00	1.00	1.00	.99	.04	.07
12x14	7.45	1.00	1.00	1.00	.99	.04	.06
14x14	7.89	1.00	.99	1.00	.99	.03	.06

E. Comparison

A comparison of the deterministic routing rules in integer addressed and fractionally addressed networks is presented in Table I. The average distance between nodes for a routing rule is calculated by determining the average distance between each source and destination in the network. The efficiency of the routing rule is the average of the shortest distance between nodes over the average distance between nodes using the routing rule. In the comparisons, there is no contention, and a packet always takes the path specified by the routing rule. When the rule decides that both paths are equivalent, either path is selected with probability 0.5. Because of this random component, a packet does not always take the same length path from a source to the destination. To compensate for the random component, the efficiency is calculated by determining the average distance between each node several times and averaging the result. The number of times that the average distance is determined is varied so that the span of values representing a 95 percent confidence interval is less than 1 percent of the average value.

Table I shows that Rule 1 determines the shortest path in integer addressed networks, and that Rules 2 and 3 result in the same average distance between nodes. Rule 2 selects longer paths than Rule 1 when the relative location of a node is at the edge of the network, and this effect is also seen in the table. In the simulations, fractionally addressed rows and columns are added to the network two at a time in the order shown in Table II, which makes the depth or the row and column addresses as uniform as possible. It is evident that fractionally addressed rows can be added to large networks in a way that has a small effect on the average path length.

Random rules are inefficient. In the networks in Table I, the average path length using random routing can be 33 times longer than the path lengths resulting from the deterministic rules. It is inadvisable to use random routing when a network has some regularity to its structure. However, a hybrid random and deterministic rule can be used to obtain the efficiency of the deterministic rule in a regular network and the survivability of the random rule. For instance, a random component can be inserted in the routing rule after a packet has traversed a larger number of nodes than expected. The number of nodes a packet has traversed must be tracked in any practical network because when a node fails, packets destined for this node must be purged from the network.

V. Network Irregularities

In addition to getting packets quickly between nodes in complete, regular MSN's, the routing rules must continue to

TABLE II
THE ORDER IN WHICH ROWS AND COLUMNS ARE ADDED TO THE NETWORK

Number of Rows or Cols.	Address of Rows or Columns Added	
	0	1
4	1/3	2/3
6	4/3	5/3
8	1/9	2/9
10	10/9	11/9
12	4/9	5/9
14	13/9	14/9

function in irregular networks. The irregularities investigated in this section occur when the number of nodes that are added to the network are not sufficient to completely fill a row or column and when nodes or links fail and are deleted from the network.

The effect of the irregularities on the routing rules depends upon the procedures used to add and delete nodes and links from the network. In Section V-A, a procedure for adding one node at a time to a network is described. This procedure has the characteristic that only two existing links must be changed to add a new node. In Sections V-B and C, procedures for deleting failed nodes and failed links are described. These procedures are similar to those used in loop networks [10] and can be implemented automatically. The source is not informed when a packet cannot be delivered to the destination and not all failures are detected. Therefore, a higher level acknowledgment protocol is still required to guarantee that packets are delivered.

A. Adding Nodes One at a Time

A procedure is shown in Fig. 9 for adding one node at a time to an MSN. Each time a node is added, two links must be changed. The two links that will be changed when the next node is added are shown by dashed lines. When this procedure is followed, two complete new rows or columns are eventually added to the network.

Adding one node at a time makes the network less regular and affects the ability of the distributed routing rules to find the shortest path to a destination. The effect on a 6 × 6 network is shown in Table III. When 12 nodes are added, a 6 × 8 network is formed. In this table, the efficiency is calculated as in Section IV-E. The italicized numbers in parentheses indicate the fraction of source destination pairs that are unable to communicate.

There are several cases for which Rule 3 cannot find a path. The reason this routing rule fails is seen by examining Step 4 in Fig. 9. Assume that a packet at node A is destined for node B. Node A is an odd-numbered column in Q_4; therefore, in Rule 3, the link along the column is assumed to be directed upward and is selected. Unfortunately, the column is not complete and the packet ends up at node C. At node C, which is also in an odd-numbered column in Q_4, the upward-directed path is selected, and the packet arrives back at node A. At node A, the path to node C is again selected, and the packet is stuck in a loop. In Rules 1 and 2, at node A it is known that the next node along the column is node C and that both links are directed away from the destination. Therefore, at node C, either link is selected with probability 0.5 and the loop is avoided.

B. Node Failures

Loop systems have active components in the path at each node, and if one of these components fails, the loop is broken. When nodes fail, they are bypassed so that the remainder of the loop continues to operate. Loss of power at the node is a common failure because power is usually obtained from a local source. This type of failure is automatically corrected by using a relay to create a path around the node [10]. The relay is

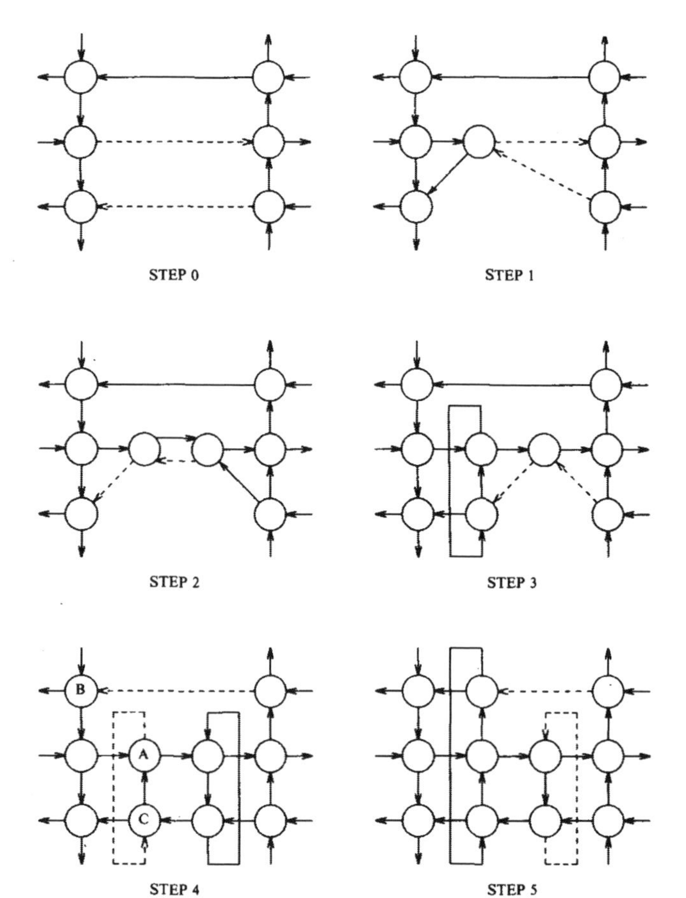

Fig. 9. Adding two columns to an existing network, one node at a time.

TABLE III
THE EFFICIENCY OF LOCAL ROUTING RULES RELATIVE TO THE SHORTEST PATH ALGORITHM AS SINGLE NODES ARE ADDED TO A 6 × 6 NETWORK

Network	Short. Path	Efficiency of Routing Rules				
		Deterministic			Random	
		1	2	3	A	B
6x6	3.71	1.00	.97	.97	.10	.20
add 1	3.77	.98	.95	.93	.10	.21
add 2	3.80	.95	.94	.93 *(.005)*	.10	.20
add 3	3.91	.92	.91	.91 *(.005)*	.10	.19
add 4	3.95	.92	.90	.91 *(.013)*	.10	.19
add 5	3.99	.89	.89	.88 *(.007)*	.09	.19
add 6	4.04	.88	.87	.87	.09	.18
add 7	4.07	.92	.90	.91	.09	.19
add 8	4.16	.95	.93	.93	.09	.18
add 9	4.18	.94	.93	.93	.09	.19
add 10	4.22	.94	.93	.93	.09	.18
add 11	4.26	.97	.95	.95	.09	.17
6x8	4.34	1.00	.97	.97	.09	.17

open when there is power and closes to bypass the node when power is lost. When nodes fail in the MSN, the system is not completely disabled as in a loop; however, packets that arrive at the node are lost and the node should be bypassed to prevent this from occurring. Loss of power at a node in the MSN can operate relays as in a loop system; however, there are two links entering and leaving each node and two relays must be used. The failure recovery procedure selected connects the row through and the column through, as shown in Fig. 10.

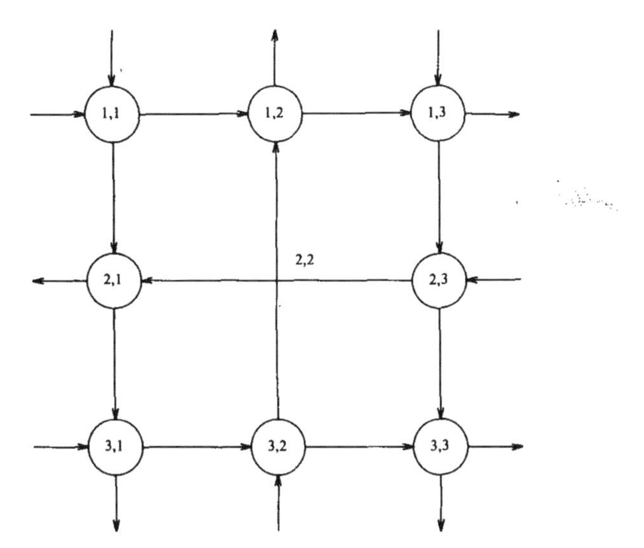

Fig. 10. Operation of the MSN when nodes fail.

C. Link Failures

In loop systems, a transmitter sends bits continuously, even when there is no information to send. This allows the receiver to retain bit synchronization between packets. Broken links and certain node failures are detected by the loss of signal. More subtle failures can be detected when the periodic start of slot does not occur in systems with fixed size slots or when there are violations of the pseudoternary modulation rules used in wire systems. Loop systems can be designed to bypass segments with failed links by constructing the loop as a series of subloops that start and end at a central location [10]. When the loss of signal is detected on a subloop, the subloop is bypassed and the signal from the previous subloop is switched to the next subloop. This allows a large part of the loop system to operate when links on one of the subloops fail.

The MSN is a slotted system with continuous transmission on each of the links; therefore, failures can be detected on the links arriving at a node as in a loop system. When a link has failed and is detected at the termination node, the origination node of the link must be informed. Otherwise, packets that are transmitted on the inoperable link will be lost, and if packets between a pair of nodes are always routed along that link, the pair of nodes will not be able to communicate. In addition, the implementation of the MSN described in Section II does not lose packets because the node can transmit as many packets as it receives. In order to preserve this characteristic, when a link leaving a node fails, data on one of the incoming links must stop so that the in-degree and out-degree of the node remains the same.

One way to prevent transmission on a link that has failed and to keep the in-degree equal to the out-degree at every node is to stop transmitting on a directed cycle of links that includes the link that has failed. The in-degree and out-degree of each node in the cycle is reduced by one and the link that has failed is not used. To minimize the effect that this strategy has on the throughput and connectivity of the network, the number of links in the cycle must be kept as small as possible and the cycle should not pass through any node twice.

A simple rule that meets these conditions most of the time is to stop transmitting on a row if a signal is not received on a column and to stop transmitting on a column when a signal is not received on a row. The operation of this rule is shown in Fig. 11. In this example, the dotted link from node 2,2 to node 2,3 fails. According to the rule, the dashed links are taken out of service. That is,

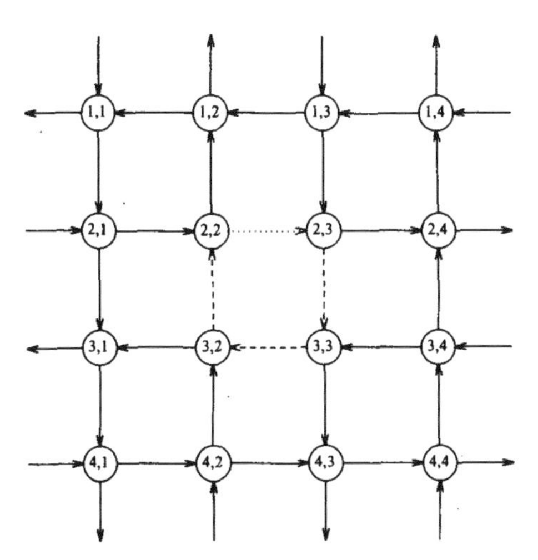

Fig. 11. Operation of the MSN when links fail.

- Node 2,3 receives no signal on the row; therefore, it stops transmitting on the column
- Node 3,3 receives no signal on the column; therefore, it stops transmitting on the row
- Node 3,2 receives no signal on the row; therefore, it stops transmitting on the column
- Node 2,2 receives no signal on the column; therefore, it does not try to transmit on the failed link on the row.

When the failed link is restored, the cycle is returned to service by forcing transmission on this link.

When a single failure occurs in a complete MSN, this procedure removes four links, which is the minimum number of links in a cycle. Also, at most one link is removed at each node. Therefore, this simple rule has the desirable characteristics in this instance. However, when there are mutliple link and node failures or if the network has partially full rows or columns, these characteristics are not always obtained. For instance, if the link from 4,4 to 3,4 also fails, both links to node 3,3 will stop, and an operable node is removed from the network. This removal rule is simple, and it works well when there are a few removals. Since a network will be repaired when removals occur, it is unlikely that there will be many removals, and this simple rule is adequate.

D. Effect on Routing Rules

Simulations were conducted to determine the effect that failures have on the distributed routing rules, and the results are presented in Table IV. The fraction of nodes that are in the network, but cannot communicate using the distributed routing rules, are italicized. In the simulations, a random selection of nodes or links fail in a 10×12 network and the bypass or link removal rules are applied. Each experiment is repeated ten times. This results in the span of values in the 95 percent confidence interval for the average path length being less than 1 percent of the mean for node failures. The span of values in the 95 percent confidence intervals for link failures ranges from 1 to 5 percent of the mean. The efficiency is calculated as in Section IV-E.

When nodes are added to each network in Section V-A, it is assumed that the routing rules that use information about the next node know about the changes. This is reasonable because adding nodes is a planned activity. When links or nodes fail, the network is modified automatically, without operator intervention. After failures occur, if the node that each node is connected to is to be known, a protocol must be developed to distribute this information. In the simulations, the change in network connectivity after failures is not known, and the routing rules operate with incorrect information.

TABLE IV
THE EFFICIENCY OF LOCAL ROUTING MECHANISMS RELATIVE TO THE SHORTEST PATH ALGORITHM WHEN NODES OR LINKS FAIL IN A 10 × 12 NETWORK

Failures Size	Short. Path	Efficiency of Routing Rules				
		Deterministic			Random	
		1	2	3	A	B
8 Nodes	5.94	.93	.91	.92	.05	.09
4 Nodes	6.15	.96	.94	.95	.05	.08
2 Nodes	6.28	.98	.96	.96	.05	.08
1 Node	6.34	.98	.97	.97	.05	.08
None	6.42	1.00	.99	.99	.05	.08
1 Link	6.51	.93	.93	.92	.05	.08
2 Links	6.59	.89 *(.007)*	.88 *(.001)*	.87 *(.001)*	.05	.08
4 Links	6.75	.81 *(.014)*	.80 *(.001)*	.79 *(.007)*	.04	.07

The simulations show that the distributed routing rules operate reasonably efficiently when up to eight nodes fail and are bypassed. When up to four links fail and up to 16 links are removed from the network, the distributed routing rules still operate reasonably efficiently. However, a small fraction of the nodes in the network, shown by italicized numbers, cannot communicate until the network is repaired. In the first deterministic rule, a greater fraction of the nodes cannot communicate compared to the other two rules. The first rule uses information about the edges of the network to improve routing decisions. When this information is incorrect, routing failures can occur.

VI. CONCLUSIONS

The objective of this paper is to study simple mechanisms for routing packets in the MSN. The routing rules must not only operate in complete rectangular networks, but must also operate when single nodes are added and when failures occur.

Three distributed routing rules are described in Section IV that use the regular structure of the MSN to simplify routing. The first rule provides the shortest path between any source and destination in an integer addressed network. The second is simpler to implement than the first rule, but results in slightly longer paths. The third rule is the simplest to implement; it has the same path length as the second rule in complete networks, but it does not determine equal length shortest paths as well as the second rule. The nodes with longer path lengths in the second and third rules are those that are furthest from the destination. Networks are divided into communities of interest and nodes communicate more frequently with nodes that are nearby than with nodes that are further away. Therefore, the simpler routing rules are preferable to the first rule.

In Section III-A, a fractional addressing scheme is described that does not change network addresses when new rows or columns are added. In addition, in a fractionally addressed network, the routing rules are independent of the number of rows or columns in the network. This makes the arithmetic unit that calculates relative addresses simpler to implement in a fractionally addressed network than in an integer-addressed network. The path selected between nodes in a fractionally addressed network may be longer than in an integer-addressed network; however, in Section IV-E, it is shown that new rows or columns can be added to a fractionally addressed network in a way that has very little effect on the average distance between nodes. In addition, the nodes that are most affected by fractional addressing are those that are furthest away from the destination. Therefore, fractional addressing is preferred to integer addressing.

In Section IV-A, a procedure is described for adding new nodes to a network. When the third routing rule is used in networks that use this procedure, there are nodes that cannot communicate. Since this is not a condition that can be repaired, and the network must operate for all combinations of nodes, the second routing rule is preferable to the third rule, even though the third rule is simpler to implement.

In Sections IV-B and C, procedures are described to automatically bypass nodes or links that fail. The procedure for bypassing links that fail guarantees that packets are not transmitted on the failed link as well as guaranteeing that all of the packets that arrive at a node can be transmitted. There is a small fraction of the nodes that cannot communicate when multiple link failures occur. Unlike adding nodes, multiple failures should be repaired, and it is unlikely that the network will operate in this mode frequently. Therefore, this condition does not preclude using these failure recovery mechanisms.

APPENDIX

Theorem: The first routing rule, Section IV-A, selects all possible shortest paths to the destination in a complete, integer addressed MSN.

To prove this theorem,

1) a hypothetical distance function $d_{sp}(r, c)$ from every node to the destination is defined,

2) a path from (r, c) to the destination that has this distance is found,

3) it is shown that a shorter path does not exist,

4) it is shown that the routing rule can select every path with this distance, and does not select any paths with a larger distance.

The distance function that will be tested is

$$d_{sp}(r, c) = |r| + |c| + x_{sp}(r, c)$$

where

$$\begin{aligned} x_{sp}(r, c) = {} & \left(2 * (r \bmod 2) * (c \bmod 2) * \left(1 - \delta\left(r - \frac{m}{2}\right)\right) * \left(1 - \delta\left(c - \frac{n}{2}\right)\right)\right) * ((1 - U(-r)) * (1 - U(-c))) \\ & + (2 * (1 - r \bmod 2) * (c \bmod 2)) * ((1 - U(-r)) * U(-c)) \\ & + \left(2 + 2 * (1 - r \bmod 2) * (1 - c \bmod 2) * \left(1 - \delta\left(r + \frac{m}{2} - 1\right)\right) * \left(1 - \delta\left(c + \frac{n}{2} - 1\right)\right) - 4 * \delta(r) * \delta(c)\right) \\ & * (U(-r) * U(-c)) + (2 * (r \bmod 2) * (1 - c \bmod 2)) * (U(-r) * (1 - U(-c))) \end{aligned}$$

and

$$U(x) = \begin{cases} 0 & \text{for } x < 0 \\ 1 & \text{for } x \geq 0 \end{cases} \quad \text{and} \quad \delta(x) = \begin{cases} 0 & \text{for } x \neq 0 \\ 1 & \text{for } x = 0. \end{cases}$$

The correction factor $x_{sp}(r, c)$ is included in the distance function to account for longer paths that must be taken because of the unidirectional links. It adds two to $d_{sp}(r, c)$ for nodes that only have links directed away from the destination. There are two additional links added to $d_{sp}(r, c)$ for all nodes in Q_3 because packets from this quadrant must pass the destination and return. In addition, there is a modification in $x_{sp}(r, c)$ at the edges of the first and third quadrant that is caused by the destination being displaced from the exact center of the network.

A node (r, c) is connected to node (r_{nxt}, c) along the column and (r, c_{nxt}) along the row where

$$r_{nxt}=r-(1-c \bmod 2) * \left(1-m\delta\left(r+\frac{m}{2}-1\right)\right) + (c \bmod 2) * \left(1-m\delta\left(r-\frac{m}{2}\right)\right)$$

and

$$c_{nxt}=c-(1-r \bmod 2) * \left(1-n\delta\left(c+\frac{n}{2}-1\right)\right) + (r \bmod 2) * \left(1-n\delta\left(c-\frac{n}{2}\right)\right).$$

The change in this distance function between the (r, c) and (r_{nxt}, c) is

$$\Delta_c(r, c)=d_{sp}(r_{nxt}, c)-d_{sp}(r, c)=(|r_{nxt}|-|r|) + (x_{sp}(r_{nxt}, c)-x_{sp}(r, c))$$

and between (r, c) and (r, c_{nxt}) is

$$\Delta_r(r, c)=d_{sp}(r, c_{nxt})-d_{sp}(r, c)=(|c_{nxt}|-|c|) + (x_{sp}(r, c_{nxt})-x_{sp}(r, c)).$$

In an $m \times n$, integer-addressed network, the regions in Fig. 6 are

$$Q_1=\left\{(r, c) : 1 \le r \le \frac{m}{2},\ 1 \le c \le \frac{n}{2}\right\}$$

$$Q_1-=\{Q_1 \cap \overline{(r_1 \cup c_1 \cup I_1)}\}$$

$$r_1=\left\{(r, c) : r=\frac{m}{2},\ 1 \le c < \frac{n}{2}\right\}$$

$$c_1=\left\{(r, c) : 1 \le r < \frac{m}{2},\ c=\frac{n}{2}\right\}$$

$$I_1=\left(\frac{m}{2}, \frac{n}{2}\right)$$

$$Q_2=\left\{(r, c) : 1 \le r \le \frac{m}{2},\ -\frac{n}{2}+1 \le c \le 0\right\}$$

$$Q_2-=\{Q_2 \cap \overline{(r_2 \cup c_2)}\}$$

$$r_2=\left\{(r, c) : r=1,\ -\frac{n}{2}+1 \le c < 0\right\}$$

$$c_2=\left\{(r, c) : 1 \le r \le \frac{m}{2},\ c=0\right\}$$

$$Q_3=\left\{(r, c) : -\frac{m}{2}+1 \le r \le 0,\ -\frac{n}{2}+1 \le c \le 0 \cap \overline{r=c=0}\right\}$$

$$Q_3-=\{Q_3 \cap \overline{(r_3 \cup c_3 \cup I_3)}\}$$

$$r_3=\left\{(r, c) : r=-\frac{m}{2}+1,\ -\frac{n}{2}+1 < c \le 0\right\}$$

TABLE V
COMPARISON OF ROUTING DISCUSSIONS FROM RULE 1 AND THE CHANGE IN DISTANCE FROM $d_{sp}(r, c)$

<table>
<tr><th colspan="3">Current Node (r,c)</th><th colspan="2">Column → (r_{nxt},c)</th><th colspan="2">Row → (r,c_{nxt})</th></tr>
<tr><th></th><th>mod 2
r c</th><th>Region</th><th>Δ_c</th><th>R_c</th><th>Δ_r</th><th>R_r</th></tr>
<tr><td rowspan="11">Q_1</td><td rowspan="3">0 1</td><td>Q_1-</td><td>$3-2\delta(r-(m/2-1))$</td><td></td><td rowspan="4">-1</td><td rowspan="4">X</td></tr>
<tr><td>c_1</td><td>+1</td><td></td></tr>
<tr><td>r_1+I_1</td><td rowspan="6">-1</td><td rowspan="6">X</td></tr>
<tr><td>0 0</td><td>Q_1</td></tr>
<tr><td rowspan="3">1 0</td><td>Q_1-</td><td>$3-2\delta(c-(n/2-1))$</td><td></td></tr>
<tr><td>r_1</td><td>+1</td><td></td></tr>
<tr><td>c_1+I_1</td><td rowspan="4">-1</td><td rowspan="4">X</td></tr>
<tr><td rowspan="4">1 1</td><td>Q_1-</td></tr>
<tr><td>c_1</td><td>+1</td><td></td></tr>
<tr><td>I_1</td><td rowspan="2">-1</td><td rowspan="2">X</td></tr>
<tr><td>r_1</td><td>+1</td><td></td></tr>
<tr><td rowspan="6">Q_2</td><td>1 1</td><td>Q_2-+r_2</td><td>$3-2\delta(r-m/2)$</td><td></td><td rowspan="3">-1</td><td rowspan="3">X</td></tr>
<tr><td rowspan="3">1 0</td><td>r_2</td><td>$3-2\delta(c+n/2-1)$</td><td></td></tr>
<tr><td>Q_2-</td><td rowspan="3">-1</td><td rowspan="3">X</td></tr>
<tr><td>c_2</td><td>$3-2\delta(r-m/2)$</td><td></td></tr>
<tr><td>0 0</td><td>Q_2-+c_2</td><td rowspan="2">$3-\delta(c+n/2-1)$</td><td rowspan="2"></td></tr>
<tr><td>0 1</td><td>Q_2-</td><td>-1</td><td>X</td></tr>
<tr><td rowspan="11">Q_3</td><td rowspan="3">0 1</td><td>Q_3-</td><td rowspan="4">-1</td><td rowspan="4">X</td><td>$3-2\delta(c+n/2-2)$</td><td></td></tr>
<tr><td>r_3</td><td>+1</td><td></td></tr>
<tr><td>c_3+I_3</td><td rowspan="6">-1</td><td rowspan="6">X</td></tr>
<tr><td>1 1</td><td>Q_3</td></tr>
<tr><td rowspan="3">1 0</td><td>Q_3-</td><td>$3-2\delta(r+m/2-2)$</td><td></td></tr>
<tr><td>c_3</td><td>+1</td><td></td></tr>
<tr><td>r_3+I_3</td><td rowspan="4">-1</td><td rowspan="4">X</td></tr>
<tr><td rowspan="4">0 0</td><td>Q_3-</td></tr>
<tr><td>r_3</td><td>+1</td><td></td></tr>
<tr><td>I_3</td><td rowspan="2">-1</td><td rowspan="2">X</td></tr>
<tr><td>c_3</td><td>+1</td><td></td></tr>
<tr><td rowspan="6">Q_4</td><td>1 1</td><td>Q_4-+c_4</td><td rowspan="3">-1</td><td rowspan="3">X</td><td>$3-2\delta(c-n/2)$</td><td></td></tr>
<tr><td rowspan="3">0 1</td><td>c_4</td><td>$3-2\delta(r+m/2-1)$</td><td></td></tr>
<tr><td>Q_4-</td><td rowspan="4">-1</td><td rowspan="4">X</td></tr>
<tr><td>r_4</td><td>$3-2\delta(c-n/2)$</td><td></td></tr>
<tr><td>0 0</td><td>Q_4-+r_4</td><td>$3-2\delta(r+m/2-1)$</td><td></td></tr>
<tr><td>1 0</td><td>Q_4-</td><td>-1</td><td>X</td></tr>
</table>

$$c_3=\left\{(r, c) : -\frac{m}{2}+1 < r \le 0,\ c=-\frac{n}{2}+1\right\}$$

$$I_3=\left(-\frac{m}{2}+1,\ -\frac{n}{2}+1\right)$$

$$Q_4=\left\{(r, c) : -\frac{m}{2}+1 \le r \le 0,\ 1 \le c \le \frac{n}{2}\right\}$$

$$Q_4-=\{Q_4 \cap \overline{(r_4 \cup c_4)}\}$$

$$r_4=\left\{(r, c) : r=0,\ 1 \le c \le \frac{n}{2}\right\}$$

$$c_4=\left\{(r, c) : -\frac{m}{2}+1 \le r < 0,\ c=1\right\}.$$

In Table V, $\Delta_c(r, c)$ and $\Delta_r(r, c)$ are listed for r and c even and odd in each of the regions. Those table entries that are not possible, such as r even in r_2, have been eliminated. An "X" in column R_c indicates that the first routing rule selects the link to (r_{nxt}, c) and an "X' in column R_r indicates that the routing rule selects the link to (r, c_{nxt}).

Table V shows the following.

1) $\Delta_r(r, c) = -1$ or $\Delta_c(r, c) = -1$ for all $(r, c) \neq (0, 0)$. Therefore, each node $(r, c) \neq (0, 0)$ is connected to at least one node that is one closer to the destination. It is possible to travel from any node to the destination in the number of steps specified by $d_{sp}(r, c)$ by selecting a link for which $\Delta_r(r, c) = -1$ or $\Delta_c(r, c) = -1$ at each node along the path. Therefore, $d_{sp}(r, c)$ is a valid distance function from every node (r, c) to the destination.

2) $\Delta_r(r, c) \ge -1$ and $\Delta_c(r, c) \ge -1$ for all $(r, c) \neq (0, 0)$. Therefore, $d_{sp}(r, c) =$ Minimum $(d_{sp}(r_{nxt}, c) + 1, d_{sp}(r,$

$c_{nxt}) + 1)$ for all $(r, c) \neq 0$. Since it is not possible to select a link from any node that results in a shorter path to the destination, $d_{sp}(r, c)$ is the shortest path to the destination. This is the basis for many shortest path algorithms and is proven in [11, pp. 193–195].

3) $R_c = X$ or $R_r = X$ for every entry in the table. Therefore, Rule 1 selects at least one outgoing link at every node in the network.

4) $R_c = X$ if and only if $\Delta_c(r, c) = -1$ and $R_r = X$ if and only if $\Delta_r(r, c) = -1$ for all $(r, c) \neq (0, 0)$. Since $R_c = X$ only if $\Delta_c(r, c) = -1$ and $R_r = X$ only if $\Delta_r(r, c) = -1$, routing Rule 1 selects a shortest path to the destination. Since $R_c = X$ whenever $\Delta_c(r, c) = -1$ and $R_r = X$ whenever $\Delta_r(r, c) = -1$, routing Rule 1 can find every shortest path to the destination.

REFERENCES

[1] N. F. Maxemchuk, "The Manhattan street network," in *Proc. GLOBECOM'85*, New Orleans, LA, Dec. 1986, pp. 255–261.

[2] ——, "Regular and mesh topologies in local and metropolitan area networks," *AT&T Tech. J.*, vol. 64, pp. 1659–1686, Sept. 1985.

[3] E. H. Steward, "A loop transmission system," in *Conf. Rec. Int. Conf. Commun.*, San Francisco, CA, June 1970, pp. 36-1, 36-9.

[4] B. K. Penney and A. A. Baghdadi, "Survey of computer communications loop networks: Part 1," *Comput. Commun.*, vol. 2, pp. 165–180, Aug. 1979.

[5] ——, "Survey of computer communications loop networks: Part 2," *Comput. Commun.*, vol. 2, pp. 224–241, Oct. 1979.

[6] N. Abramson, "The Aloha system—Another alternative for computer communications," in *Fall Joint Comput. Conf., AFIPS Conf. Proc.*, vol. 37, 1970, pp. 281–285.

[7] R. M. Metcalf and D. R. Boggs, "Ethernet: Distributed packet switching for local computer networks," *Commun. ACM*, vol. 19, pp. 395–404, July 1976.

[8] A. G. Greenberg and J. Goodman, "Sharp approximate analysis of adaptive routing in mesh networks," submitted to *Proc. Int. Sem. Teletraffic Anal. and Comput. Performance Eval.*, The Netherlands, June 1986.

[9] R. T. Prosser, "Routing procedures in communications networks—Part I: Random procedures," *IRE Trans. Commun. Syst.*, pp. 322–329, Dec. 1962.

[10] H. E. White and N. F. Maxemchuk, "An experimental TDM data loop exchange," in *Proc. ICC '74*.

[11] H. Frank and I. T. Frisch, *Communication, Transmission and Transportation Networks*. Reading, MA: Addison-Wesley, 1971.

PHOTO NOT AVAILABLE

Nicholas F. Maxemchuk (M'72–SM'85) received the B.S.E.E. degree from the City College of New York, New York, NY, and the M.S.E.E. and Ph.D. degrees from the University of Pennsylvania, Philadelphia.

He is the Head of the Distributed Systems Research Department at AT&T Bell Laboratories, Murray Hill, NJ. He has been at Bell Laboratories since 1976, and his research interests include local and metropolitan area networks, protocols, speech editing, and picture processing. Prior to joining Bell Laboratories, he was at RCA Labs, Princeton, NJ, for eight years where his research interests included local area networks, error-correcting codes, and graphics compression. He has been on the adjunct faculty at Columbia University, New York, NY, where he was associated with the Center for Telecommunications Research, and the University of Pennsylvania, where he taught courses in computer communications networks.

Dr. Maxemchuk has served as the Editor for Data Communications for the IEEE TRANSACTIONS ON COMMUNICATIONS, as Guest Editor for the IEEE JOURNAL ON SELECTED AREAS IN COMMUNICATIONS, and on the Program Committees of numerous conferences. He was awarded the RCA Laboratories Outstanding Achievement Award, the Bell Laboratories Distinguished Technical Staff Award, and the IEEE's Leonard G. Abraham Prize Paper Award.

Bottleneck Flow Control

JEFFREY M. JAFFE, MEMBER, IEEE

Abstract—**The problem of optimally choosing message rates for users of a store-and-forward network is analyzed. Multiple users sharing the links of the network each attempt to adjust their message rates to achieve an ideal network operating point or an "ideal tradeoff point between high throughput and low delay." Each user has a fixed path or virtual circuit.**

In this environment, a basic definition of "ideal delay-throughput tradeoff" is given and motivated. This definition concentrates on a fair allocation of network resources at network bottlenecks. This "ideal policy" is implemented via a *decentralized* algorithm that achieves the *unique* set of *optimal* throughputs. All sharers constrained by the same bottleneck are treated fairly by being assigned equal throughputs.

A generalized definition of ideal tradeoff is then introduced to provide more flexibility in the choice of message rates. With this definition, the network may accommodate users with different types of message traffic. A transformation technique reduces the problem of optimizing this performance measure to the problem of optimizing the basic measure.

I. INTRODUCTION

VARIOUS store-and-forward packet-switched computer networks have been developed in recent years. The primary function of these networks is to route messages or *packets* from one network location to another. Typically, the source of a message dispatches a packet to a neighboring location or node, which relays the message to another node and so forth, until the message arrives at the destination.

There are a number of disciplines used by networks to funnel a large number of packets from one source to a given destination. For example, ARPANET handles each packet individually [1], trying to find the shortest path for each packet based on changing network characteristics. In this paper we assume a fixed route approach whereby all messages from a given "session" are assigned to a fixed unique route. This approach is currently used in TYMNET [2], [3], IBM's network architecture [4], and various other networks (e.g., [5]). Many sessions may share a given route.

The total time required for transmission of a packet is called its delay. Assuming small nodal processing time, there are two major components to message delay. Since communication links take some time to transmit a message, there is a *transmission* delay component. Also, if a communication link needs to transmit too many packets at once, it temporarily buffers some of them, leading to a *queueing* delay component. The queueing delay clearly depends on the amount of network traffic, and roughly speaking, increases with greater traffic.

Paper appoved by the Editor for Computer Communication of the IEEE Communications Society for publication after presentation at the 5th International Conference on Computer Communication, Atlanta, GA, October 1980. Manuscript received April 25, 1980; revised January 6, 1981. This research was supported in part by the National Science Foundation under Grant EDS-79-25092.

The author is with the IBM Thomas J. Watson Research Center, Yorktown Heights, NY 10598.

Flow control regulates the amount of traffic to maintain good system performance. For example, if the buffers at a link are almost full, some mechanism is needed to slow down the rate of incoming traffic. Otherwise, the buffers would overflow, causing severe queueing delays or even deadlock. Another purpose of flow control is to maintain a good throughput delay tradeoff. If a user is sending a high average message rate (in our studies this is equated with throughput), the resulting delays may be intolerably long. On the other hand, the user would not want to sacrifice too much throughput in order to achieve low delay. Related to this is the notion of *fairly* dividing network resources between competing network users.

In this paper we discuss methods to achieve a well defined notion of system performance which results in fairness to users and a good delay-throughput tradeoff. We concentrate on network access means of flow control [6] where external inputs are throttled based on measurements of internal network congestion. The buffer depletion problem (see [7]) is ignored so that we may concentrate on delay and throughput. Formally, when our model is specified (in Section II), infinite buffers at each link are assumed.

This paper primarily concentrates on the fundamental questions of "what is optimum performance?" and "what notions of optimality are accomplishable in a decentralized environment?". No new method of constraining the input of messages is proposed; it is assumed that message rate is regulated by a simple rate mechanism, i.e., some "black box" at each route which chooses the message rate for that route.

Network access flow control schemes include the isarithmic scheme [8], input buffer limiting [9], and the choke packet scheme [10]. Other schemes are discussed in [6] and [11]. The isarithmic scheme limits the total number of packets allowable in the network. Input buffer limiting locally restricts input traffic in favor of transit traffic.

The "bottleneck flow control" presented here may be viewed as a generalization and abstraction of both the choke packet scheme and certain ideas presented in [9]. Common features with the choke packet scheme are that the decision to decrease message rate is a function of congestion in the bottleneck links. The relationship between the two is further developed throughout this paper. The main difference is that, while optimality is defined in a similar way, the control mechanisms are different. As a result, the choke packet scheme has no ***explicit*** way of ensuring a specified notion of fairness. On the other hand, bottleneck flow control uses fairness criteria related to those that are described in [9].

In Section III we define and motivate a notion of "optimal tradeoff." An adaptive algorithm is given in Section IV which attempts to achieve this tradeoff in a network that is experiencing changes in traffic patterns and numbers of users. Due

Reprinted from *IEEE Transactions on Communications,* vol. COM-29, no. 7, July 1981.

to the changing nature of such a network, it is difficult to state specific "steady-state" properties of the algorithm. We thus restate the problem somewhat to reflect a static network. In that environment it is easier to discuss properties of the "optimal tradeoff" and an algorithm that implements it. In particular, the following is achieved:

- A "decentralized" algorithm is given that always achieves the optimal tradeoff (Sections V and VII).
- The algorithm obtains the tradeoff in linear time [in the number of users (Section VII)].
- The "optimal tradeoff" defines a *unique* set of throughputs that the users of the network must achieve (Section VIII).
- The unique set of optimal throughputs has important "fairness" properties (Section IX).

Section X generalizes these results to the situation where different user classes have different network performance requirements. The main result of Section X is that the techniques developed earlier in the paper may be applied directly to the more general case by a simple transformation technique.

We briefly explain and motivate the notion of a "decentralized" algorithm for flow control. When a user chooses its throughput, the inputs to the process should consist of information locally available to it. The user might be permitted to use information about the interfering traffic on its path, but not about global topology. Basically, in a decentralized algorithm, information not readily available on a user's path should not be usable for throughput determination.

In [12] it is shown that a single user may optimize its power (ratio of throughput to delay) using only such local information. However, in [13] it is shown that, under certain conditions, no decentralized algorithm maximizes power in a multiple user system. Since certain optimality criteria are nondecentralizable, the importance of the decentralizable criterion discussed here is enhanced.

We further remark that the criterion expressed here has other advantages over the power concept. It is shown in [14] that, in some network configurations, optimizing power implies that certain users must choose zero throughput. A corollary of the fairness property of Section IX is that no users are required to have zero throughput at optimal performance. This fact is still true for the generalization of Section X where users are not handled identically in terms of throughput allotment.

II. NETWORK MODEL

We model a data network as a graph (N, L) with vertex (or node) set N and edge (or link) set L. Each link $l \in L$ has a *service rate* of $s(l)$ bits/s. A *path* p in the network is a sequence $p = (n_1, \cdots, n_k)$ with $n_i \in N$ such that for $i = 1, \cdots, k-1$, $l_i = (n_i, n_{i+1}) \in L$. The set $\{l_1, \cdots, l_{k-1}\}$ is denoted $l(p)$, the *links of* p. A path p models a fixed route that is used by one of the "users" of the network.

In order to evaluate the delays on the links, a queueing model is needed which relates throughputs to delay. We use a simple model ([15, Sect. 5.6]) which, as indicated above, has infinite buffers. Specifically, we assume that each link may be modeled as an $M/M/1$ queue, the average message length is b bits/message, there is no nodal processing time, and Kleinrock's independence assumption applies [15].

Define the *capacity* of link l, $c(l)$, by $c(l) = s(l)/b$. Assume that there are K users, all of whose fixed routes use a link l. Let γ_i denote the message rate of the ith user. In that case, the *average* steady-state *delay* for the packets (of each user) that traverse the link at l is $d_l(\gamma) = 1/(c(l) - (\gamma_1 + \cdots + \gamma_K))$. The *average total delay* of packets sent by user i, $D_i(\gamma)$ is the sum of the average delays experienced at the individual links.

III. OPTIMALITY CRITERION

In this section an optimality criterion is presented using several levels of description. First, optimum throughput is defined in terms of link capacity. We explain why our definition might be considered "the optimum operating point of a network." Next, the definition is reformulated to express a tradeoff between user throughput and delay. Section IV gives an adaptive algorithm for optimizing the criterion in a "dynamically changing" network. It is difficult, however, to present any concrete analysis for a rapidly changing network. Starting with Section V we analyze the optimality criterion in a "static" environment.

Recall that $c(l)$ is the capacity of the link l. Let $\gamma(l)$ denote the sum of the throughputs of all users of link l. The maximum value that $\gamma(l)$ can be is $c(l)$ or else messages are generated at a faster rate than they can be transmitted. Certainly, $\gamma(l) > c(l)$ is not a situation we would like to encourage for *any* link. In fact, it is probably not even desirable to have $\gamma(l) = c(l)$ for two reasons. First of all, if $\gamma(l) = c(l)$ the system "never reaches steady state"; the delays of the messages increase over time due to the fact that buffer occupancy approaches infinity. Also, choosing $\gamma(l) = c(l)$ leaves no room for fluctuations in the network. One user may be forced by certain considerations to increase his throughput or new users may attempt to open up new routes sharing link l. For that reason, optimum $\gamma(l)$ is chosen to be somewhat less than $c(l)$, as we proceed to describe. This distance is parameterized by a variable x. This variable permits designers of different systems to choose somewhat different notions of "ideal throughput-delay tradeoff." If they are throughput-oriented, they choose x large; if delay-oriented, then x should be small.

Define the *residual capacity of* l by $r(l) = c(l) - \gamma(l)$. Let γ denote the throughput of a user whose path includes link l. The user *saturates* l if $\gamma = x(r(l))$. The user *overloads* l if $\gamma > x(r(l))$. A user is *overloaded* if it overloads any link on its path. A user is *saturated* if it is not overloaded and it saturates at least one link on its path. These preliminaries prepare us for the following.

Definition: Given a data network (as modeled in Section II) with several paths through the network (corresponding to users of the network), and a rate assigned to each user, the rate assignment is *optimal* if all users are saturated.

Remarks: The way that we keep $\gamma(l)$ somewhat less than $c(l)$ is to guarantee that no user overloads any links. Thus, for each link l, $x(r(l)) \geqslant \gamma_{max}$ where γ_{max} is the largest throughput of any user of link l. In addition to keeping $\gamma(l)$ somewhat less than $c(l)$, we also desire a large measure of throughput in the network. Thus, each user must not only prevent overload—it also must be saturated. Each user would then have the largest possible throughput subject to x and the residual capacities.

To contrast this with the Cyclades choke packet proposal, we remind the reader that optimality in [10] basically requires that no link exceeds a certain threshold of utilization. For instance, $\gamma(l)$ should not exceed (0.8) $(c(l))$ if the threshold equals 0.8.

We feel that it is better to force *saturation* of each user and choose $\gamma(l)$ as a function of $\gamma_{\max}$ for a few reasons. The primary reason is that the choke packet scheme has no regard for the number or types of users of the link, and therefore loses the ability to fairly allocate resources. By fixing the requirement that no link should exceed a certain utilization, one loses the ability to predict transients in future utilization based on current utilization. This is developed further in Section IX. Also assume that $x(r(l)) = \gamma_{\max}$. Then, with our definition, if $x = 1$, we can accommodate one new user with throughput $\gamma_{\max}$ without causing $\gamma(l) > c(l)$. Similarly, choosing $r(l) = (\gamma_{\max})/x$ protects the network against percentage changes in each user's throughput due to transients. If a user increases his throughput by a factor of $1/x$, the inequality $c(l) \geqslant \gamma(l)$ still applies. Methods of obtaining an optimality criterion similar to "80 percent of utilization," as a limiting case of saturation, are discussed in Section XII.

Next, we motivate saturation as a means of expressing an "optimal delay-throughput tradeoff." Recall that the delay at l is given by $d_l = 1/(c(l) - \gamma(l))$. Thus, saturation for user p is equivalent to

$$\gamma = \min_{l:l \in l(p)} x/d_l(\gamma) \tag{1}$$

From (1) it is evident that saturation is a direct method of expressing a delay-throughput tradeoff for the users of the network. A user may increase its throughput until the delay on its "bottleneck" link is too large. As delay increases, γ is constrained by (1).

Note the role played by the parameter x in all viewpoints of the optimality criterion. From the network point of view, it indicates the amount of traffic fluctuation that is to be protected against. From the user viewpoint, it indicates the amount of effect that increased delay should have on throughput.

There is a third viewpoint of saturation. Using Little's theorem [16], the average number of messages waiting at a link l when the throughput of a user is γ, and the delay is d_l is $\gamma \cdot d_l$. Now if $\gamma \leqslant x/d_l$ for every link l in the path of a given user, the user is willing to tolerate x messages waiting at each link, and a total of x times # (user's links), messages waiting in the system. Thus, the average number of waiting messages that a user will tolerate varies linearly with the length of his path—if the path is longer, the user may have more messages in transit.

To review, the features of optimum network operation based on the use of the saturation measure are

1) protection for the network against changes in users' rates

2) protection for the network from arrivals of new users

3) establishment of delay/throughput tradeoff at the bottleneck link

4) use of the parameter (x) to permit flexibility in the definition of optimum performance

5) protection for the buffers in an average sense

6) fair allocation of resources (Section IX).

In addition to *stating* what optimal performance is (all users saturated), it might be helpful to evaluate how far suboptimal solutions are from optimal. To do this, it is useful to have an objective function which characterizes the quality of a set of throughput assignments. Assume that there are m users with throughputs $\gamma = (\gamma_1, \cdots, \gamma_m)$. Define

$$f(\gamma) = \sum_{i=1}^{m} \left| \gamma_i - \min_{l:l \in l(i)} \frac{x}{d_l(\gamma)} \right|. \tag{2}$$

If each user is saturated at γ, then for all i, $\gamma_i = \min_{l:l \in l(i)} x/d_l(\gamma)$ and $f(\gamma) = 0$. Conversely, if $f(\gamma) = 0$, all users are saturated. Thus, the goal of saturating all users may be conveniently restated as an attempt to minimize f.

IV. AN ADAPTIVE DISTRIBUTED ALGORITHM

An adaptive distributed algorithm which attempts to saturate all paths without overloading any is now given. Each user adjusts its message rate based on information sent to it by the links and nodes on its path. The information needed by a user with path p is

1) its current throughput γ

2) $\min_{l:l \in l(p)} r(l)$.

We do not specify the mechanics of when this information is made available and in what form the information arrives. Each link may know to dispatch information to all users of the link at regular intervals, or alternatively, information gathering may be prompted by a signal from the user. Each link may *compute* $r(l)$ or estimate it based on buffer occupancy. Also, the links may send the throughputs of the individual users of the links, and let user p calculate $r(l)$.

The algorithm executed by user p each time it desires to recalculate its message rate γ' from the old rate γ is

$$\gamma' \leftarrow \min_{l:l \in l(p)} \frac{x(r(l) + \gamma)}{x + 1}. \tag{3}$$

The following explains why we say that the above algorithm attempts to achieve saturation. First, note that after executing one step of the algorithm, the user is saturated. This can be seen as follows. For a link l, the new sum of throughputs $\gamma'(l) = \gamma(l) - \gamma + \gamma'$. Thus, $x(c(l) - \gamma'(l)) = xc(l) - x\gamma'(l) = xc(l) - x\gamma(l) + x\gamma - x\gamma' = \gamma'$ by (3) for the link at which $r(l)$ was minimized. Also, $x(c(l) - \gamma'(l)) \geqslant \gamma'$ for all other links, l, i.e., none is overloaded.

If there were no transients, such as no new users entering the system, *and* each user converged to a steady-state throughput, then those throughputs that are converged to will saturate all users. Any unsaturated or overloaded user must change its throughput! Unfortunately, we are unable to show, even without transients and new users, that each user does converge. To clearly express an algorithm that saturates all users, we spend the rest of this paper discussing a static case, i.e., no new users.

As a practical matter, the above algorithm would need to be modified in an adaptive situation. Choosing γ' by (3) may cause large deviations in certain user's message rates, leading to instabilities in the system. A better way is to have users slowly change rates in the direction (increase or decrease) implied by (3). The reader is referred to [14] for an algorithm to coordinate user updates, so that many users do not change their rates at once.

V. ALGORITHM TO SATURATE ALL USERS

In this section an algorithm is presented which saturates all users in a static network with a fixed set of users. It is assumed that if a user is assigned by the algorithm to send messages at a rate γ, that indeed its average throughput is γ. (Variations of this are described in Section XI.) The algorithm is decentralized in the sense described above. Each user chooses its throughput based on information provided from its links. In fact, the execution of the algorithm will be presented in a manner which distributes the computation even more—the links (or whatever controls the links) will do some computation in the algorithm. The link computation provides a concise description of the current traffic on the link.

There are a number of idealizations used in this section. It is assumed that each link may accurately calculate message rates of users that use the link. Also, in order to conveniently discuss the convergence time of the algorithm, a synchronous algorithm is assumed (i.e., a clock at each node permits all updates to occur at once). However, the main feature of using "local information," i.e., information accumulated along a user's path, is preserved. In practice, one would probably use a hybrid of the algorithm of Section IV and the algorithm that we proceed to present here.

The algorithm proceeds in iterations. Consider a link l which is shared by a number of users, exactly j of which are not saturated before the ith iteration. Let $\gamma_{sat}(l, i)$ denote the sum of the throughputs of the users of link l that are saturated before the ith iteration. Then the *saturation allocation of* l at i, denoted $\gamma(l, i)$, is

$$\gamma(l, i) = x\left(\frac{c(l) - \gamma_{sat}(l, i)}{1 + jx}\right). \tag{4}$$

Intuitively, if each unsaturated user of link l chooses the saturation allocation as its throughput, and each saturated user leaves its throughput unchanged, then all unsaturated users become saturated. This follows from the fact that $r(l)$ in that case would be $(c(l) - \gamma_{sat}(l, i))/1 + jx$.

The following is the algorithm for the ith iteration. Initially, all throughputs are 0 and each link knows how many users have paths which use it.

Saturation Algorithm (ith Iteration)

1) Each link l calculates $\gamma(l, i)$.

2) Each link sends the value $\gamma(l, i)$ to all users of l.

3) Each user sets its new throughput γ to the smallest value of $\gamma(l, i)$ among links l that it uses.

4) Each link l determines which of its users are now saturated at l and informs each such user.

5) Each user that is saturated at *any* link informs *all* of its links that it is saturated.

There are basically two computations done at each iteration. After receiving $\gamma(l, i)$ from each link l on its path, a user readjusts its throughput by taking the minimum allocation [step 3)]. Also, each link must calculate $\gamma(l, i)$. The information needed for this calculation is the number of saturated users [obtained in step 5)] and $\gamma_{sat}(l, i)$ (obtained in some way by measuring each saturated user's throughput).

One method whereby a link can determine $\gamma_{sat}(l, i)$ without explicitly finding out which user sent each message is briefly described. Let each saturated user set a bit in the message header to 1 and each unsaturated user to 0. Then $\gamma_{sat}(l, i)$ is just the average rate of messages arriving with header bit equal to 1. Further elaboration on implementation is omitted.

The key properties of the algorithm (proved in Section VII) follow.

- Any user that is saturated after iteration i, remains saturated after iteration $i + 1$.
- If not all users are saturated at the beginning of an iteration, then at least one becomes saturated at the iteration.

From the above two facts it is immediate that if there are m users, they are all saturated after no more m iterations.

VI. AN EXAMPLE

Consider the network of Fig. 1. The following is a trace of the iterations of the algorithm with $x = 1$. The labels of the links are the capacities.

	Iteration 1	Iteration 2	Iteration 3
γ_1	1/2 (from link *D*)	1/2	1/2
γ_2	1 1/2 (*F*)	7/4 (*E*)	7/4
γ_3	1 1/2 (*F*)	11/6 (*F*)	15/8 (*F*)
γ_4	10 (*A*)	10	10
γ_5	3 1/3 (*C*)	19/4 (*C*)	19/4

User 1 is saturated at link *D*, 2 at *E*, 3 at *F*, 4 at *A*, and 5 at *C*.

VII. PROOF OF CORRECTNESS

The main result of this section is the following.

Theorem 1: Fix a network with m paths. Define

$$f(\gamma) = \sum_{i=1}^{m} \left| \gamma_i - \min_{l: l \in l(i)} \frac{x}{d_l(\gamma)} \right|. \tag{5}$$

If the saturation algorithm is executed, then after at most m iterations, the resulting value of γ, satisfies $f(\gamma) = 0$. Furthermore, γ is unchanged by subsequent iterations of the algorithm.

Proof: As mentioned in Section IV, this is proved by showing that saturated users stay saturated—and each iteration produces at least one saturated user. (Recall that $f(\gamma) = 0$ if all users are saturated at γ.) The main technical result

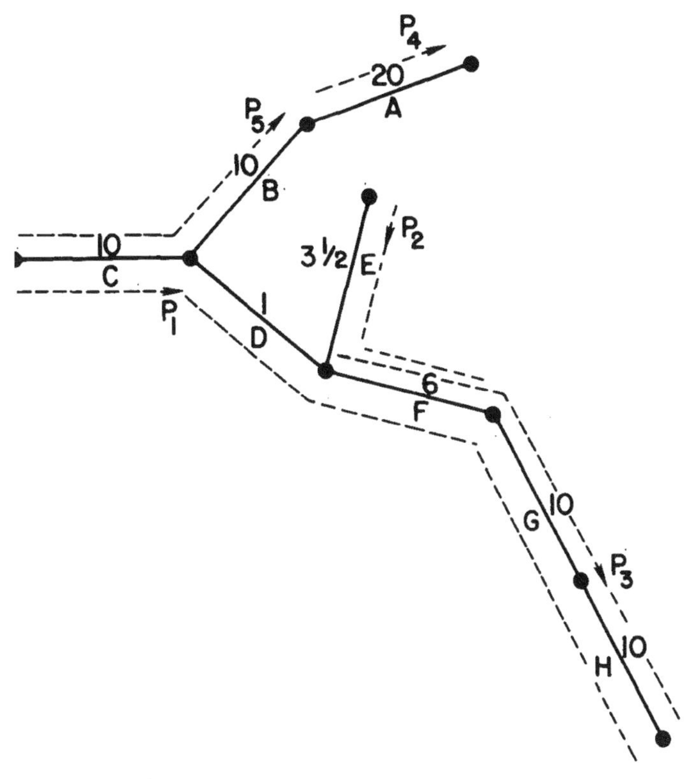

Fig. 1. Example network for execution of algorithm.

needed to prove Theorem 1 may be stated informally as "$\gamma(l, i)$ is a nondecreasing function of i." This fact, and the fact that saturated users stay saturated, are proved inductively in the following lemma.

Lemma 1:

1) For all $l \in L$, all $I \in \mathbb{Z}+$, $\gamma(l, i+1) \geqslant \gamma(l, i)$.

2) If any user becomes saturated at link l during the ith iteration, then all users of l that were not saturated before the ith iteration become saturated at l during the ith iteration.

3) If a user is saturated after the ith iteration with throughput γ, he remains saturated after the i + 1st iteration with throughput γ.

The proof of Lemma 1 is given in the Appendix. To complete the proof of Theorem 1 we prove the following.

Lemma 2: At each iteration which starts with some unsaturated users, at least one user becomes saturated.

Proof: For each link at which not all users are saturated at a given iteration, consider the saturation allocation of the link. Some link must have minimal allocation among all such links. All unsaturated users of that link choose that allocation. Since all saturated users of the link do not change their throughputs [3) of Lemma 1], all of the unsaturated users of that link become saturated. ■

Theorem 1 follows directly from Lemma 2 and 3) of Lemma 1. At each iteration at least one user becomes saturated--and saturated users stay saturated. ■

Corollary–(Existence): Given any network and set of users of the network, there is a throughput assignment γ, which saturates all of the users.

Note that the saturation algorithm determines the optimal throughputs exactly. In contrast, even when the adaptive

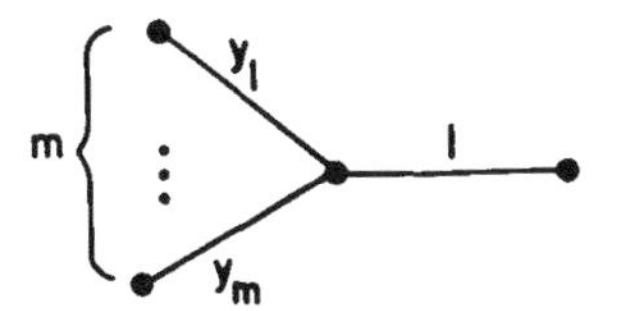

Fig. 2. Worst case network (in terms of number of steps).

algorithm of Section IV converges to an optimal solution, it does not converge exactly. Rather, the sequence of throughputs achieved by the users converge (in a Cauchy sense) to the optimal throughputs.

The fact that linear time is actually required by our algorithm in the worst case is proved by the example of Fig. 2. Basically, the y_i may be chosen so that each user converges at a different step. See [17] for details.

VIII. UNIQUENESS

In this section it is shown that for any network and any set of users there is a unique way to saturate all users. This is a "well-defined" result for the saturation measure: two different throughput assignments cannot both be optimal for the same network configuration. We first separate out a simple lemma which we refer to later.

Lemma 3: Assume user i is saturated at link l at an optimum solution γ, with throughput γ_i, and user j uses link l and has throughput γ_j. Then $\gamma_i \geqslant \gamma_j$.

Proof: Since user i is saturated, $\gamma_i = x(r(l))$. Since user j is not overloaded, $\gamma_j \leqslant x(r(l)) = \gamma_i$. ■

Theorem 2: The value γ obtained from the saturation algorithm *uniquely* minimizes the objective function f.

Proof: We prove by induction on the iteration number that all users saturated at step i must obtain the same throughput assignment in any optimal solution. The basis step is similar to the inductive step and is left to the reader.

Consider all users saturated at the ith step. By Lemma 1, part 2), a user may only be saturated if it takes the saturation allocation at some link l, and all other not previously saturated users also take their saturation allocations at l (and get saturated). Thus, we may study *all* users that are saturated at the ith step by looking at *all* links at which *all* nonsaturated users take the saturation allocation.

Assume, contrary to the hypothesis, that it is possible for the users saturated at step i to get different assignments in some optimal assignment γ^*. Consider a link l, which is saturated at the ith iteration and has some of its saturated users with different assignments in γ^*. By induction, recall that all users that share l, and are saturated before the ith iteration must receive the same throughputs in any optimal solution.

We first claim that at least one user saturated at l at the ith iteration must obtain less than $\gamma(l, i)$ in γ^*. For if all of them receive $\gamma(l, i)$ or more, and the users saturated before iteration i receive the same amounts, then $x(c(l) - \gamma^*(l)) < \gamma(l, i)$. But then, all those users that receive $\gamma(l, i)$ or more are overloaded at l in γ^*, and thus γ^* is not optimal.

Thus, one may consider a user which obtains throughput γ^* in γ^* where $\gamma^* < \gamma(l, i)$. Assume that the user is saturated in γ^* at link l'. Note that $\gamma(l', i) > \gamma^*$ since $\gamma(l', i) \geqslant \gamma = \gamma(l, i) > \gamma$. Consider the sharers of l'. Those saturated before iteration i may not change their throughput in γ^* by induction. The r other users must have throughputs in γ^* of at most γ^*, each by Lemma 3. Thus, $x(c(l') - \gamma^*(l')) \geqslant x(c(l') - \gamma_{sat}(l', i) - r\gamma^*)$. But $< \gamma(l', i)$ implies that $\gamma^* < x(c(l') - \gamma_{sat}(l', i))/(1 + rx)$. Thus,

$$x(c(l') - \gamma^*(l')) > x\,(c\,(l') - \gamma_{sat}(l', i) - \left(\frac{rx}{1 + rx}\right)(c(l') - \gamma_{sat}(l', i))) = \frac{x}{1 + rx}\,(c(l') - \gamma_{sat}(l', i)) > \gamma^*. \tag{6}$$

This contradicts the fact that the user is saturated at l' in γ^*.

IX. FAIRNESS

One aspect of a flow control optimality criterion which is difficult to evaluate is the elusive notion of fairness. One version of fairness is to insist that all users obtain equal throughputs. In a network with different users, using links of different capacities, it is unlikely that such a policy would be desirable.

Recall that flow control is instituted not only to protect a user against high delay due to traffic, but also to equitably divide network resources among competing users. The notion of fairness provided by saturation relates to the equitable division of resources. Briefly, saturation is "fair" because

- each user's throughput is at least as large as all other users that share its bottleneck link (Lemma 3)
- the only factor that prevents a user from obtaining higher throughput is the bottleneck link (which essentially divides resources equally).

X. GENERALIZATIONS

The fact that our algorithm saturates all users is interesting in a network with a homogenous user set, but suffers in that it provides too restrictive a notion of fairness. The property that "all users are treated equally" may not be desirable in practical networks. One user may be more important and thus deserving of a higher message rate. Alternatively, a user that interferes with *many* other users would probably deserve special treatment.

This is only one deficiency that results from the definition of saturation. A different problem arises if many (n) users share a single link. If the link is the bottleneck link for each, then (at $x = 1$) they each choose a message rate of $c(l)/(n + 1)$. As $n \to \infty$, the total rate approaches $c(l)$; thus, there is excessive throughput and disastrous delay. (This particular problem is dealt with both here and in Section XII.)

A final problem with the definition of saturation is that it may not be desirable to have a network-wide value of x as defined. Recall that one reason to choose $\gamma = x \min_l r(l)$ was to protect the network against transients in a user's message rate which were as large as a factor of $1/x$. Clearly, the variability in rates of different users is different. A user that has large variability would need a larger relative amount of residual capacity on its links.

This section solves the above problems by reformulating the definition of saturation. With user p, one associates a number x_p, the *throughput priority of user* p. User p's throughput priority expresses the desired message rate of user p as compared to the rates of interfering users. In particular, user p is saturated at l if $\gamma_p = x_p\, r(l)$. If users p and q are both saturated at l, the ratio of their throughputs is x_p/x_q. This generalization clearly treats users differently. Optimum performance is again equated with rate assignments that saturate all users.

In practice, some higher level protocol would decide what the relative values of x_p should be. If x_p were chosen as a function of the number of interfering users, some network manager could prevent the excessive use of an n user bottleneck. Similarly, a network manager could decide how to appropriately allocate relative priorities to competing users. In some network environments, each user might make a local decision choosing x_p based on the expected variability of its message rate to protect the network. A network manager is not needed if some convention is adopted by network users for determination of their throughput priorities.

We proceed to explain how the variable throughput priority case may be effectively reduced to the equal throughput priority case. In particular, the following questions are addressed:

- Is there a static algorithm to saturate all users?
- Is there a unique way to saturate all users?
- Is there an appropriate adaptive, distributed algorithm such as the one described in Section IV?
- What delay/throughput tradeoff is implied by the new definition of saturation?
- What fairness properties are implied?

First, consider the case that x_p is an integer for all p. We assume that each link knows the value of x_p for each user of the link. In this case, the variable x_p case is reduced to the $x = 1$ case as follows. A user with priority x_p is treated as x_p users each with $x = 1$ and identical paths. Initially, if there are j users of l with priorities $x_1, \cdots, x_j$, then

$$\gamma(l, 1) = \frac{c(l)}{1 + S} \tag{7}$$

where $S = \Sigma_{i=1}{}^{j}\, x_i$. If $\gamma(l, 1)$ is the minimal allocation for user k (with priority x_k), then user k chooses $\gamma = x_k\, \gamma(l, 1)$. In subsequent steps, γ_{sat} is measured as before, and

$$\gamma(l, i) = \frac{c(l) - \gamma_{sat}(l, i)}{1 + S(l, i)} \tag{8}$$

where $S(l, i)$ is the sum of the x_p's for users of link l that are not saturated before iteration i.

It can be shown that with this modified algorithm, the value of $\gamma(l, i)$ for every l and every i is identical here to the case where each user with priority x_p were replaced by x_p users with priority 1. Also, the message rate γ of a user with priority x_p after iteration i equals the sums of the rates of the x_p users with $x = 1$. These facts are proved trivially by induction on i. From this it follows that there is a static algorithm to saturate all users, and that saturation is unique.

Actually, using (7) and (8) uniquely saturates all users even if x_p is not an integer. The proof of this follows in a manner similar to the proof of Section VII.

Continuing with the aforementioned questions, the appropriate adaptive algorithm remains roughly the same as in Section IV; each user saturates itself based on current conditions (perhaps changing message rate slowly for stability reasons). The delay-throughput tradeoff defined for user p is

$$\gamma_p = \min_{l:l\in l(p)} \frac{x_p}{d_l(\gamma)}. \tag{9}$$

The relevant fairness statements are as follows.

• Each user's throughput is only constrained by its bottleneck link.

• At its bottleneck link a user gets at least "its share of capacity" based on its throughput priority. That is, the rate γ_p of user p satisfies $\gamma_p \geqslant (x_p/x_q)\,\gamma_q$ if q shares p's bottleneck link.

XI. LOW THROUGHPUT USERS

The saturation algorithm provides each user with an "optimum" throughput, but requires one special assumption to do so. It is assumed that each user has a throughput equal to that assigned in the algorithm. In practice, however, a user may not have enough data to send at the high rate. In this section we briefly discuss the required modifications to handle this case.

Assume that γ is the maximum possible rate for a user based on incoming data rate considerations. Then the user "pretends" that on its path is a "virtual link" of capacity $(\gamma)(1 + x)/x$, which is shared with no one. If all other links have saturation allocation larger than γ, then the rate chosen on the basis of the virtual link is γ. For example, if $x = 1$, the virtual link has capacity 2γ and the user is saturated if its rate is γ. Thus, by slightly modifying the network, the inherent throughput constraints of each user are taken into account, without changing the algorithms and their properties.

XII. RESTRICTING THE PERCENTAGE UTILIZATION OF A LINK

Assume that it was desired that no link exceed a fraction y of its capacity. This might be used to prevent $\gamma(l) \to c(l)$ as $n \to \infty$ in the case of n users sharing a bottleneck link. Section XI prevents $\gamma(l) \to c(l)$ by suggesting that the values x_p should be chosen as a function of n. In this section a more direct approach is used. This approach leads to a derivation of the "optimum Cyclades performance" as a limiting case of saturation.

Define the *effective capacity of* l, $e(l) = yc(l)$. This is the largest amount of capacity of l that should be used. If $e(l)$ is used instead of $c(l)$ in the algorithms to saturate all users, then the capacity of any link utilized is restricted to be at most $e(l)$.

This is not quite the Cyclades notion of optimality—they require that $e(l)$ not be exceeded, but place no other restrictions on the message rates (such as $\gamma \leqslant r(l)$). To effectively remove the restriction $\gamma \leqslant r(l)$, let $x \to \infty$; $\gamma \leqslant xr(l)$ is then trivially accomplished.

To review, a utilization of y at bottleneck links is accomplished by using $e(l)$ instead of $c(l)$, letting $x \to \infty$, and saturating all users. This accomplishes the desired utilization of bottlenecks, and also provides fairness not usually provided by just restricting link utilization. In this case, letting $x \to \infty$ does not strongly degrade delay at the cost of throughput, since the rates are all chosen based on $e(l)$, not $c(l)$.

XIII. CONCLUSIONS

We have presented a "fair" motivatable network performance criterion. Two algorithms have been presented to optimize performance, one of which is guaranteed to find the unique optimal throughput assignments in a static environment.

APPENDIX

PROOF OF LEMMA 1 (BY INDUCTION ON i)

$i = 1$:

1) $\gamma(l, 1) = x \cdot c(l)/(1 + jx)$ where j is the number of users that share l. $\gamma(l, 2) = x(c(l) - \gamma_{sat}(l, 2))/(1 + rx)$ where r is the number of users of l not saturated at the first iteration. By the way that throughputs are assigned, $\gamma_{sat}(l, 2) \leqslant (j - r)\,\gamma(l, 1) = (j - r)(x \cdot c(l))/(1 + jx)$. Thus,

$$\gamma(l, 2) = \frac{xc(l) - xy_{sat}(l, 2)}{1 + rx} \geqslant \frac{xc(l) - \dfrac{x(j - r)(x \cdot c(l))}{1 + jx}}{1 + rx}$$

$$= \frac{x(c(l))\left(1 - \dfrac{x(j - r)}{1 + jx}\right)}{1 + rx}$$

$$= \frac{xc(l)\left(\dfrac{1 + jx - jx + rx}{1 + rx}\right)}{1 + rx}$$

$$= \frac{xc(l)}{1 + jx}$$

$$= \gamma(l, 1). \tag{A1}$$

2) Recall that the saturation allocation is designed to guarantee saturation if all unsaturated users of a link choose the saturation allocation and all saturated users keep the same throughput. Before the first iteration, there are no saturated users, and each user chooses at most the saturation allocation. From this, 2) follows immediately.

3) Similar to the inductive step (below).

Inductive Step: Assume 1), 2), and 3) for $k < i$ and prove 1) and 2) for $k = i$. Then, using 1) and 2) for $k = i$ and 3) for $k < i$, prove 3) for $k = i$ as follows.

1) $\gamma(l, i+1) = x(c(l) - \gamma_{sat}(l, i+1))/(1+rx)$, $\gamma(l, i) = x(c(l) - \gamma_{sat}(l, i))/(1+sx)$ where there are r nonsaturated users of l before the $i+1$st iteration and s before the ith. By induction on 3), any user saturated before the ith iteration remains saturated before the $i+1$st (i.e., after the ith) with the same throughput. Thus, $\gamma_{sat}(l, i+1) = \gamma_{sat}(l, i) + \gamma_{new}$ where γ_{new} is the sum of the throughputs of the $s - r$ users that become saturated at the ith iteration. Note that $\gamma_{new} \leqslant (s-r)\,\gamma(l, i)$ since each newly saturated user has message rate at most $\gamma(l, i)$. Thus,

$$\begin{aligned}\gamma(l, i+1) &= \frac{x(c(l) - \gamma_{sat}(l, i+1))}{1+rx}\\ &= \frac{x(c(l) - \gamma_{sat}(l, i))}{1+rx} - \frac{x\gamma_{new}}{1+rx}\\ &\geqslant \left(\frac{sx+1}{rx+1}\right)\gamma(l, i) - \frac{x(s-r)}{1+rx}\gamma(l, i)\\ &= \gamma(l, i)\frac{sx+1-sx+rx}{1+rx}\\ &= \gamma(l, i)\end{aligned} \tag{A2}$$

2) By induction on 3), all users saturated before the ith iteration choose the same throughput at the ith iteration. Since each unsaturated user chooses, at most, the saturation allocation at l, by the definition of $\gamma(l, i)$, a user becomes saturated at l at iteration i only if *all* other unsaturated users choose $\gamma(l, i)$ and become saturated.

3) Fix a user that is saturated after the ith iteration with throughput γ. We must show that at the $i+1$st iteration, it chooses the same throughput and remains saturated. Consider a link l at which the user is saturated after the ith iteration. Using 2) for the iteration number k at which the user was first saturated at l, $(k \leqslant i)$, all users that share l are either saturated before the kth iteration or become saturated at the kth iteration. By induction on 3), it follows that all are saturated after the kth iteration. Also, the ones that were previously saturated use the same throughput as before the kth iteration. This continues through iteration i. Since the user is saturated at l, its throughput γ satisfies $\gamma = x(r(l))$. Also, since all users of l are saturated, $\gamma_{sat}(l, i+1) = \gamma(l)$ and $\gamma(l, i+1) = x(c(l) - \gamma(l)) = \gamma$. Thus, due to the saturation allocation at l, the user chooses a throughput of at most γ at iteration $i+1$. Since for every link l' in the user's path $\gamma(l', i+1) \geqslant \gamma(l', i)$ [by (1)], the user chooses exactly γ.

The above argument may be repeated for each user saturated after the ith iteration. Returning to the user fixed above, it is apparent that the user is saturated at l at the $i+1$st iteration, since all users that share l do not change their throughputs. Thus, $\gamma(l)$ is unchanged and $\gamma = x(r(l))$ still holds. To prove that the user is still saturated after the $i+1$st iteration, it suffices to show that it is not overloaded on any other link on its path.

To prove that the user is not overloaded at a link l', it suffices to show $\gamma \leqslant x(c(l') - \gamma(l'))$ where $\gamma(l')$ is the sum of throughputs of users of l' after the $i+1$st iteration. Consider the iteration (iteration k) at which the user became saturated (with rate γ). If l' is on its path, $\gamma(l', k) \geqslant \gamma$ by the way γ is chosen. By induction on (1), $\gamma(l', i+1) \geqslant \gamma$. Recall that $\gamma(l', i+1) = (x(c(l') - \gamma_{sat}(l', i+1)))/(1+jx)$ (if j users are not saturated before the $i+1$st iteration). Note, the value of $\gamma(l')$ after iteration $i+1$ is given by

$$\gamma(l') \leqslant \gamma_{sat}(l', i+1) + j\gamma(l', i+1).$$

Thus,

$$\begin{aligned}x(c(l') - \gamma(l')) &\geqslant x(c(l') - \gamma_{sat}(l', i+1) - j\gamma(l', i+1))\\ &= x(c(l') - \gamma_{sat}(l', i+1) - \frac{jx}{1+jx}(c(l')\\ &\quad - \gamma_{sat}(l', i+1)))\\ &= x(c(l') - \gamma_{sat}(l', i+1))\left(\frac{1}{1+jx}\right)\\ &= \gamma(l', i+1) \geqslant \gamma.\end{aligned} \tag{A3}$$

ACKNOWLEDGMENT

The author acknowledges helpful conversations with K. Bharath-Kumar, F. H. Moss, and M. Schwartz.

REFERENCES

[1] J. M. McQuillan, "Adaptive routing algorithms for distributed computer networks," Bolt Beranek and Newman Rep. 2831, NTISAD 781467, May 1974.
[2] L. Tymes, "TYMNET—A terminal-oriented communication network," in *AFIPS Conf. Proc., Spring Joint Comput. Conf.*, vol. 38, 1971, pp. 211–216.
[3] M. Schwartz, *Computer Communication Network Design and Analysis*. Englewood Cliffs, NJ: Prentice-Hall.
[4] J. P. Gray and T. B. McNeill, "SNA multiple-system networking," *IBM Syst. J.*, vol. 18, no. 2, 1979.
[5] A. Danet, R. Despres, A. LaRest, G. Pichon, and S. Ritzentheler, "The French public packet switching service: The transpac network," in *Proc. 3rd Int. Conf. Comput. Commun.*, Toronto, Ont., Canada, Aug. 1976, pp. 251–260.
[6] M. Gerla and L. Kleinrock, "Flow control: A comparative survey," *IEEE Trans. Commun.*, vol. COM-28, pp. 553–575, Apr. 1980.
[7] V. Ahuja, "Routing and flow control in systems network architecture," *IBM Syst. J.*, vol. 18, no. 2, 1979.
[8] D. W. Davies, "The control of congestion in packet switching networks," *IEEE Trans. Commun.*, vol. COM-20, June 1972.
[9] E. Raubold and J. Haenle, "A method of deadlock-free resource allocation and flow control in packet networks," in *Proc. 3rd Int. Conf. Comput. Commun.*, Toronto, Ont., Canada, Aug. 1976.
[10] J. C. Majithia *et al.*, "Experiments in congestion control techniques," in *Proc. Int. Symp. Flow Contr. Comput. Networks*, Versailles, France, Feb. 1979.
[11] *Proc. Int. Symp. Flow Contr. Comput. Networks*, Versailles, France, Feb. 1979.
[12] K. Bharath-Kumar, "Optimum end-to-end flow control in net-

works," in *Proc. Int. Conf. Commun.*, Seattle, WA, June 1980.
[13] J. M. Jaffe, "Flow control power is non-decentralizable," IBM Res. Rep. RC8343, July 1980; also to be published *IEEE Trans. Commun.*, 1981.
[14] K. Bharath-Kumar and J. M. Jaffe, "A new approach to performance oriented flow control," IBM Res. Rep. RC8307, May 1980; also, *IEEE Trans. Commun.*, vol. COM-29, pp. 427–435, Apr. 1981.
[15] L. Kleinrock, *Queueing Systems*, vol. 2. New York: Wiley, 1976.
[16] D. C. Little, "A proof of the queueing formula: $L = \lambda W$," *Oper. Res.*, vol. 9, pp. 383–387, 1961.
[17] J. M. Jaffe, "A decentralized, 'optimal,' multiple user flow control algorithm," in *Proc. 5th Int. Conf. Comput. Commun.*, Oct. 1980.

PHOTO NOT AVAILABLE

Jeffrey M. Jaffe (M'80) received the B.S. degree in mathematics and the M.S. and Ph.D. degrees in computer science from the Massachusetts Institute of Technology, Cambridge, in 1976, 1977, and 1979, respectively.

He is currently employed by IBM Research, Yorktown Heights, NY, where he is engaged in research on network algorithms and combinatorial optimization.

Dr. Jaffe is a member of ACM and Phi Beta Kappa, and was a National Science Foundation Fellow while a graduate student.

Routing and Flow Control in TYMNET

LA ROY W. TYMES, MEMBER, IEEE

(Invited Paper)

Abstract—**TYMNET uses two mechanisms for moving data: a tree structure for supervisory control of the original network and a virtual circuit approach for everything else. Each mechanism is described. The routing and flow control is contrasted with ideal routing and flow control, and also with conventional packet-switched networks. One of the mechanisms described, the virtual circuit as implemented in TYMNET, is compared to the ideal. This mechanism avoids several inefficiencies found in other packet networks. The tree structure is shown to have several problems which increase roughly with the square of the size of the network.**

INTRODUCTION

ROUTING and flow control are the two most important factors in determining the performance of a network. They determine the response time seen by the user, bandwidth available to the user, and, in part, the efficiency of node and link utilization. TYMNET has several years' experience with two different mechanisms of moving data through the network. The routing of one of them, the virtual circuit, is described in detail with emphasis on performance considerations. Flow control is then discussed with emphasis on heavy load conditions.

TYMNET is a commercial value added network (VAN) which has been in operation since 1971[1]. The original network, now called TYMNET I, was designed to interface low-speed (10-30 character/s) terminals to a few (less than 30) time-sharing computers. The data rate was expected to be low, the size of the network small (less than 100 nodes), and the log-on rate low (less than 10 new users/min). High efficiency of the 2400 bit/s lines interconnecting the nodes was required along with good response time for the user, who typically interacted with full duplex terminals on a character-by-character basis. Echo control with full duplex terminals was to be passed smoothly back and forth between host and network. Finally, memory in the nodes was expensive, and little money was available for development and deployment.

With these considerations, the nodes were made to be as small as possible, with very little buffering. All complexity that could be centralized, such as routing control, was put into a supervisor program which ran on a time-sharing system. A virtual circuit scheme was devised which allowed a smooth flow of data to the users with the small buffers and which allowed the multiplexing of data from several users into a single physical packet. This way a user could type one character at a time without generating a lot of overhead on the network lines.

When a supervisor took control of the net, it generated a tree structure with itself at the root. Through this tree, it could read out the tables which defined the existing circuits and make new entries in them to define new circuits. This kept the software in the nodes very simple at the cost of complexity in the supervisor.

As the years passed, all the design considerations except the need for efficiency and interactive response were completely reversed. In particular, with many terminals requiring much higher bandwidth, with the number of nodes approaching 1000 worldwide, and the log-on rate measured in log-ons per second, many of the design decisions for TYMNET I became inappropriate. In TYMNET II, which is displacing TYMNET I in high-density areas and new installations, the features which worked well in TYMNET I have been enhanced and generalized. The supervisor control tree, however, did not scale up very well, and TYMNET II uses virtual circuits exclusively.

TYMNET II is a more recent technology than TYMNET I, not really a different network. In fact, both technologies can and do exist in the same network. TYMNET I was implemented in Varian 16-bit minicomputers. TYMNET II was implemented in the TYMNET Engine [2], a 32-bit byte-addressable machine developed by Tymshare, Inc. specifically for network applications. A basic difference between the two technologies is that in TYMNET I the supervisor maintains an image of the internal routing tables of all the nodes and explicitly reads and writes the tables in the nodes, whereas in TYMNET II the nodes maintain their own tables, and there is much less interaction between node and supervisor.

IDEAL ROUTING AND FLOW CONTROL

Before getting into the details of the machinery, we should be clear about what it is supposed to do. In routing, the first consideration is to find the "optimum" path, where the definition of the word "optimum" can be rather complex. It should also be fast, so that the impatient human user is not kept waiting. The ideal routing algorithm should use little CPU time and network bandwidth, since these are both scarce resources. It must base its decisions on the current state of the network, not on the state of the network a minute ago, since many things can change in a minute. Finally, it must not spend a lot of network bandwidth to keep its data up to date. The practice of passing tables from node to node becomes expensive as the network, and hence the tables, grows large.

The ideal flow control mechanism must assure a smooth flow of data to the low-speed user. In TYMNET, it must do this with small buffers in the nodes. If the user wishes to abort his output (a common thing for an interactive user to do), the data buffered in the network must disappear quickly. An

Manuscript received April 23, 1980; revised August 15, 1980.

The author is with the Data Network Division, Tymshare, Inc., Cupertino, CA 95014.

Reprinted from *IEEE Transactions on Communications,* vol. COM-29, no. 4, April 1981.

The Best of the Best. Edited by W. H. Tranter, D. P. Taylor, R. E. Ziemer, N. F. Maxemchuk, and J. W. Mark.

additional consideration for a commercial network is that the network lines must be used efficiently. This means that all forms of overhead, such as end-to-end acknowledgment, deadlocks, discarding of packets, retransmission of data (other than for line errors), and so on, must be eliminated or greatly minimized, even if the user wishes to send just one byte at a time.

Interactive users will want fast response, so queuing delays must be kept short. Finally, when bandwidth on a link is oversubscribed, it must be partitioned among the various users in some satisfactory manner, without causing lockups, deadlocks, or other network malfunctions.

THE TYMNET VIRTUAL CIRCUIT

The TYMNET virtual circuit has been documented elsewhere [3], [4]. For the purposes of this paper, let us define it as a full duplex data path between two ports in the network. Data are always treated as a stream of 8-bit bytes. The two ports are usually, but not always, on two different nodes, and the circuit must therefore be routed through the network. The routing is normally done only once, when the user first requests that the circuit be built. If the circuit is rebuilt later (because of a node failure on the original path, for instance), the rebuild procedure is the same except for the recovery of data that may have been lost when the first circuit failed.

Each link between two adjacent nodes is divided into channels, and a circuit passing over that link is assigned to a channel. The communication between these nodes is concerned with channels, not circuits. Only the network supervisor knows about circuits. Data from various channels may be combined, or multiplexed, into one physical packet to share the overhead of checksums and packet headers among several low-speed channels. A high-speed channel with much data to send may use the whole physical packet by itself.

An example of a simple circuit is given in Fig. 1. Port 4 on node A is connected through node B to port 7 on node C. A pair of buffers (represented by a circle), one buffer for each direction of data flow, is associated with port 4 in node A and another pair with port 7 in node C. Another pair is used in node B for traffic passing through it. Each buffer is elastic, like a balloon, so it takes little memory when it is empty. The data are stored in a threaded list of bufferlets of 16 bytes each (14 data bytes plus a 2-byte pointer to the next bufferlet). When more data are put into a buffer, additional bufferlets are taken from a common pool. When data are removed from a buffer, the empty bufferlets are returned to the common pool. An empty buffer has no bufferlets associated with it. Buffers in TYMNET are empty almost all the time.

Each node has a table for each of its links, associating channels on that link with internal buffer pairs. Entry number 5 in a table in node A refers to the buffer pair for port 4, so data from port 4 are sent out on channel 5 on the link between A and B. Entry 5 in node B for that link refers to the passthrough buffer pair. Entry 12 in another table in node B also refers to the same passthrough buffer pair. This second table is for the link from B to C, so data from port 4 of node A will use channel 12 when going from B to C. Entry 12 in node C refers to the buffer pair associated with port 7, which completes the circuit.

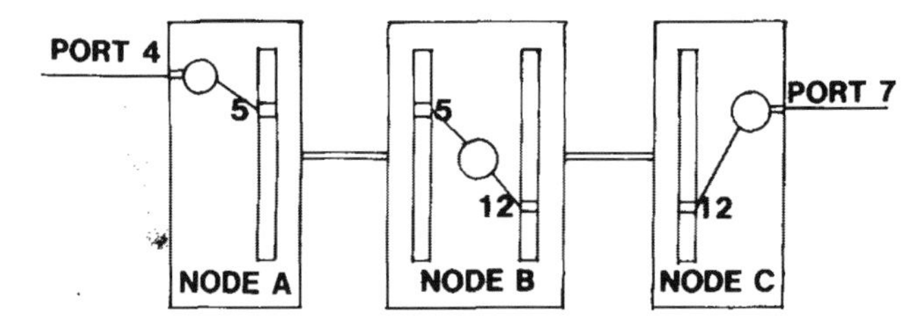

Fig. 1. A simple TYMNET virtual circuit.

All routing is done by the supervisor. It begins when the user identifies himself with a name and password, and presents a request to build a virtual circuit (log-in to a host). The supervisor hashes the user name into the MUD (master user directory) to get the attributes needed for access control, accounting information, and so on. Then the supervisor assigns a "cost" to each link in the net. This cost reflects the desirability of including that link in the circuit. This cost is based on link bandwidth, link load factor, satellite versus land-based lines, terminal type, and other factors. For instance, if the user has a low-speed interactive terminal, a 9600 bit/s land link will be assigned a lower cost than 56 000 bit/s satellite because satellites add a delay which is undesireable to interactive users. If the circuit is to be used to pass files between computers, however, the satellite will be assigned the lower cost because bandwidth is more important than response time.

A definition of link load factor will depend on what type of circuit is being routed. For instance, if a user wishes to pass files between background processes on two computers, response time is not required. High bandwidth is not needed either if the user is in no hurry. The objective is to move the data at the minimum cost to the network, which means giving the user the bandwidth left over from other users. A file moving from one computer to another can pass at a high enough rate to saturate any link. Such a link is not overloaded, however, since there is no reason not to build more circuits over it.

A second user may wish to run a line printer. He cannot saturate a link by himself because after the line printer is going full speed, it can use no more bandwidth. If several such users share the same link, they may saturate it and compete with each other for bandwidth. If the printers can no longer run at full speed, then the link has reached "high speed overload," and it is not desirable to route more such users over this link. A longer route which avoids the congested link may be prefered, even though it increases cost to the network and increases response time. Printer users do not care about response time. It is all right to send interactive users over a link with high-speed overload because they require so little bandwidth that they will not interfere very much with the printers. The printers, on the other hand, will not interfere with the response time of the interactive user.

The most severe form of overload is "low-speed overload." The formal definition of low-speed overload is that the average delay for a particular channel on a link to be serviced exceeds $\frac{1}{2}$ s several times in a 4-min period. When this happens, low-speed interactive users begin to experience degradation in response times. A high cost is given to such a link to avoid building more circuits on it.

Legal considerations also affect "cost." For instance, there may be prior agreements that the traffic between two countries be divided among the interconnecting links in a particular way. These rules can be enforced by assigning unacceptable costs to the unallowed options.

This assigning of costs is not compute intensive. It is mostly a matter of indexing into the correct table. Once the correct table has been selected, there remains the problem of finding the path of lowest cost through the net. This is complicated by the fact that some users may have more than one possible target. For instance, there may be several gateways to another network, and the cost of the gateways and the other network may have to be considered. The one path which produces the "best" load balancing is preferred.

The problem of finding the best path through a network has been investigated by many people [5]-[7]. The algorithm that TYMNET has been using is as follows.

T1: Initialize the cost of reaching the source node from the source node to 0. Initialize the cost of reaching all other nodes from the source node to an unacceptably large, finite number.

T2: Initialize a list of nodes to contain only the source node.

T3: If the list is empty, done. Otherwise, remove the next node from the list. For each neighbor of this node, consider the cost of going to that neighbor (the cost associated with this node plus the cost of the link to the neighbor). If this cost is less than the cost currently associated with that neighbor, then the newly computed cost becomes the cost associated with that neighbor and a path pointer for that neighbor is set to point to this node. If that neighbor is not on the list and the new cost is less than the cost associated with the target node, add the neighbor to the list. Repeat T3.

When the algorithm completes, the path of minimum cost is defined by the backward pointers. Furthermore, the minimum cost is precisely known. If the cost is high, the supervisor may elect to reject the user rather than tax the network to provide poor service.

A trivial example is given in Fig. 2. The problem is to find the path of least cost from node A to node D. The cost associated with node A is set to 0, and for all the other nodes it is set to 99. Node A has two neighbors, B and C, and the cost of each of them can be improved by going to them from node A, so both are added to the list. When node B is considered, the cost of neighbor D is reduced by reaching it from B, but it is not added to the list because the new cost is not less than the cost of reaching the target node (since D is the target node). Finally, when C is processed, the cost for node D is further reduced, redefining the best path from A to D, and the cost for node E is also reduced. Node E is not added to the list because its cost is not less than the cost already associated with the target node D. Nodes F and G were never considered. The best path is seen to be from A to C to D, and has a cost of 3.

There are other algorithms which require less CPU time

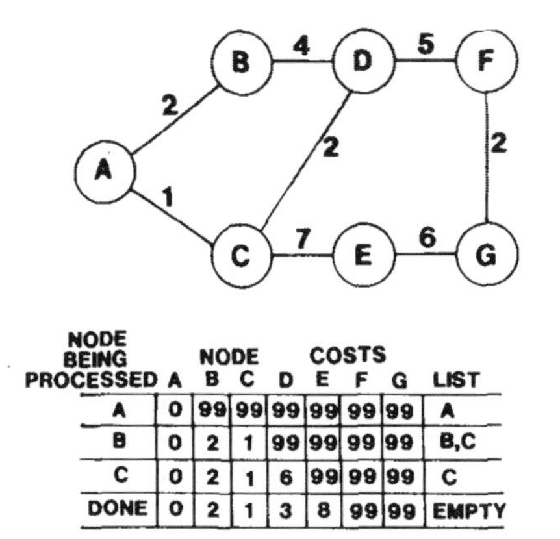

NODE BEING PROCESSED	NODE COSTS A	B	C	D	E	F	G	LIST
A	0	99	99	99	99	99	99	A
B	0	2	1	99	99	99	99	B,C
C	0	2	1	6	99	99	99	C
DONE	0	2	1	3	8	99	99	EMPTY

Fig. 2. Routing example. Find the best path from A to D.

than this one, but of those known to the author, all are more complex and require more memory. This algorithm has been satisfactory for TYMNET so far.

More important than the algorithm itself is the mechanism for correctly determining link costs. If the link costs are incorrect or inappropriate, the resulting circuit path will be unsatisfactory. Any time the network capacities change, e.g., link failure or overload, the supervisor is notified immediately and can use this information for the next circuit to be built. In TYMNET II, the next step is to send a needle to the originating port. This needle contains the routing information and threads its way through the net, building the circuit as it goes, with the user data following behind it. In TYMNET I, the entries in the routing tables are explicitly made by the supervisor for every node.

The TYMNET II needle is a list of node numbers, together with accounting information and some flags indicating circuit class, encoded as a string of 8-bit bytes. When data enter a TYMNET II node from a link on an unassigned channel, it is checked to see if it is a needle. If it is not, it is discarded. Otherwise, the channel is assigned and the needle checked to see what to do next. If the circuit terminates in this node, it is attached to a port. If the next node is a neighbor of this one, a channel on the link to that node is assigned and the needle, together with any other data behind it, is sent on its way. If the next neighbor is unknown (because of recent link failure or some other error), the data are destroyed and the circuit is zapped back to its origin.

Note that a needle followed by some data followed by a nongobbling zapper is similar to a datagram, one which contains explicit routing information. Thus, TYMNET II has a type of datagram capability, although it is not used.

VIRTUAL CIRCUIT FLOW CONTROL

The following analogy will help make clear the dynamics of TYMNET virtual circuit flow control. A building with many water faucets in it is supplied by one water pipe coming from a water main. If a faucet is turned on, water immediately flows out of it and water begins flowing in from the main. The main can supply water at a much faster rate, but the rate of flow is restricted by the faucet. When the faucet is turned off, the resulting backpressure stops the flow from the main instantly. If enough faucets are turned on simultaneously, the capacity of the pipe from the main may be oversubscribed. When this happens, faucets which are turned on only a little still get all the water they want, but faucets turned on all the way will not flow at their maximum rate. If any faucet is turned off, the water it was consuming is now available to the remaining faucets. Note that there is no machinery in the water pipe from the main to allocate the water. It is just a pipe and does not know or care where the water goes. Also note that when the capacity is oversubscribed, all faucets that want water still get some. Some faucets get less than they want, but they do not stop. Finally, note that no water is wasted. It does not have to be "retransmitted" to compensate for water spilled.

For each channel on a link between two nodes, there is a quota of 8-bit bytes which may be transmitted. This quota is assigned when the virtual circuit is built, and varies with the expected peak data rate of the circuit; in other words, the throughput class of the circuit. Once a node has satisfied the quota for a given channel, it may not send any more on that channel until the quote is refreshed from the other node. In TYMNET II, the receiving node remembers approximately how many characters it has received on a channel, so it knows when the quota is running low. It also knows how much data it is buffering for this circuit, and whether its total buffer space is ample. This last consideration is almost always true in TYMNET because at any given instant, most circuits are idle, at least in one direction, and no memory is needed for their empty buffers.

When the quota is low or exhausted and the receiving node does not have enough data buffered for the circuit to assure smooth data flow, it sends back permission to refresh the quota for this channel. This permission is highly encoded so as not to require much bandwidth. Note the passive nature of this backpressure scheme. Doing nothing is the way to stop the influx of data, so if a node is overloaded, one effect of the overload is to reduce the load. Also note that this mechanism does not know or care about the destination of the data on each channel. Backpressure propagates from node to node back to the source and effectively shuts it off. It does not matter whether the cause of the backpressure is inability of the destination to consume the data as fast as the source can supply it or congestion within the net. Either way, the source is quickly slowed down or shut off. Finally, note that only circuits which are actively moving data need attention. At any instant, this is a small percentage of the total number of circuits.

A complication arises when an interactive user realizes that what is printing on his terminal is of no interest to him and he wishes to stop it. He can type an abort command and the host computer may stop outputing, but there are still many characters buffered in the network. To clean out the circuit, the host can send a character gobbler. The character gobbler ignores backpressure and goes through the circuit at full speed, gobbling all characters in front of it.

Another exception to normal backpressure convention is the circuit zapper. When a session is over and one or both ports disconnect, a circuit zapper is released which not only gobbles up the characters, but releases the buffer pairs and clears the table entries as well to free up these resources for new circuits. A zapper must be able to override backpressure because some circuits may stay backpressured for a long time. Suppose, for instance, that the terminal is an IBM 2741 with the keyboard unlocked. It cannot accept output in this state, so output data remain buffered and backpressured in the net, waiting for the user to lock his keyboard. The zapper will not wait, but will clear his circuit and disconnect him.

TYMNET I has only one circuit zapper, but TYMNET II has a family of them. Each is generated by a different terminating condition, so that the port at the other end knows why the circuit is being zapped. There are hard zappers and soft zappers. A hard zapper disconnects the port, but a soft zapper allows the port to request a circuit rebuild. A soft zapper might be generated by a link failure, for instance, and a rebuilt circuit will allow the session to continue. A short history of characters sent may be kept at each end, so that when the circuit is rebuilt, data lost when the old circuit failed can be retransmitted. This is transparent to the user, and there is normally no indication that an outage has happened unless it is a special host that wishes to monitor such things. No user data are lost or allowed to get out of order. Note that there is no overhead on the links to provide this feature except briefly when the circuit is rebuilt.

In theory, TYMNET II could use this rebuild mechanism to redistribute network load, but in practice, it is not needed. Circuits come and go often enough that the supervisor has no difficulty redistributing load by proper routing of the new circuits. When the numbers of users are very large, as they are in TYMNET, a statistical approach to load leveling works well. The load on any small area of the net changes very little from one minute to the next.

PACKET TRANSMISSION AND BANDWIDTH ALLOCATION

To move data from one node to another, they must be assembled into packets. Since the greatest value of a value-added network is the sharing of line costs among many users, this must be done as efficiently as possible. The packet maker is a process which builds physical packets to send over a link. It will build a packet when there are data to send and the window of outstanding packets for the link is not full. The packet may contain data from several channels or from just one channel if only one channel has data to send or if a channel has so much data to send that it can fill a full length packet. A channel may send data if it has data to send, its backpressure quota is not zero, and if it has not had a turn recently. Once it is serviced, even if for only one byte, it is not serviced again until all other channels have had a chance (unless it is flagged as a priority channel, in which case it may be given extra turns to give it more bandwidth). On any particular turn, a channel is limited in the amount of data it may send by the amount of data in its buffer, its backpressure quota, and the amount of room left in the packet when it is serviced. Thus, when bandwidth is oversubscribed, channels which only need a little get what they want with little or no queuing delays, while channels that want all they can get share the remaining bandwidth equally (except for priority channels, which get extra bandwidth at the expense of nonpriority channels).

Packet making and teardown are link-related processes. Once a packet is made, it is handed over to the packet transmitting and receiving processes, which are line-related. A line is a physical connection between nodes, and is subject to noise and outages. There may be several lines on one link, for instance, three 9600-bit/s lines, all passing data in both directions simultaneously. There are several different packet transmitters and receivers because there are several different kinds of hardware for moving data between nodes. They differ in data rate, interface requirements, checksumming and formating, and in methods for getting data in and out of memory. Window size, which is the number of packets one may send before getting an acknowledgment, varies from 4 to 128, depending on the maximum number of outstanding unacknowledged packets likely in the absence of errors. When the window size is exhausted, the oldest packet is retransmitted (on all lines, if there is more than one line on this link) until it is acknowledged. Packets may be sent and received in any order, but are always built and torn down in sequence.

It is instructive to contrast TYMNET with ARPANET in an overload situation. In Fig. 3, assume that node *A* has high bandwidth access to sources of data bound for ports on nodes *B*, *D*, and *E*. Also assume that the links shown are of equal bandwidth and that nodes *B*, *D*, and *E* have equal appetites for data. In a packet-switched network, data from the sources to node *A* will be in packets to be distributed equally among nodes *B*, *D*, and *E*. Two thirds of these packets must be sent to node *C*, and one third to node *B*. However, the bandwidths to nodes *B* and *C* are the same, so node *A* can send only half as much data to *B* as to *C*. When *A* fills up with packets, it must reject all incoming packets, not just those bound for nodes *D* and *E*. Thus, the link to node *B* runs at half speed because of congestion on the link to node *C*.

Now suppose that the link from *C* to *D* goes out. *C* will wait a few seconds to be sure the link is out before discarding packets for *D*. In the meantime, it fills up with packets for node *D* and stops receiving packets from node *A*. Node *A* is already full of packets for nodes *D* and *E*, so it stops receiving packets from the sources. Now all traffic is stopped and the network is vulnerable to deadlocks which will require that packets be discarded at random and retransmitted. Even after node *C* starts discarding packets for node *D*, the sources for node *D* may try to retransmit their packets until they realize that node *D* can no longer be reached.

The reader may wish to construct more complex topologies which involve dynamic rerouting and bidirectional data flow. The general conclusion from these exercises is that *when ARPANET-like networks are oversubscribed in any area, they respond with poor efficiency in other areas.* In general, the problems become more severe as the size of the network and the volume of user traffic increases. In the author's opinion, an ARPANET-like network the size of TYMNET would give very poor performance. Of course, there are other considerations in the choice of one network architecture over another, but performance in the presence of overload is one of the most important for a commercial network.

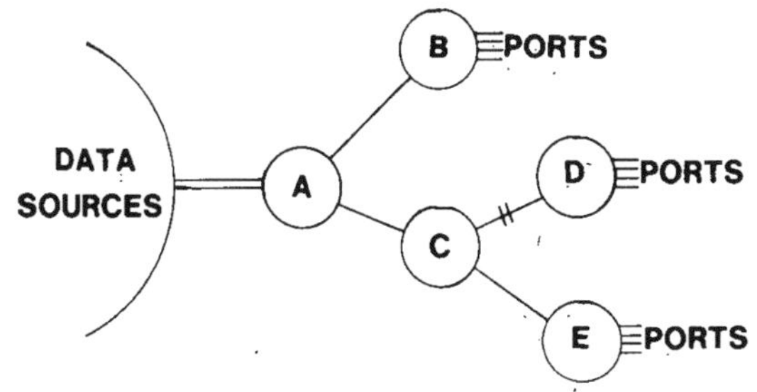

Fig. 3. Local section of loaded network.

In TYMNET, since the flow control applies to each direction of each channel of each link separately, the traffic bound for nodes *D* and *E* would be backpressured independently from the traffic bound for node *B*. The link from *A* to *B* would therefore run at full speed. When the link from *C* to *D* went out, the traffic for *D* would backpressure to the sources for *D* and the bandwidth on the link from *A* to *C* would be completely available for traffic bound for *E*. When *C* realized that the link to *D* was down, the circuits for *D* would be zapped back to the sources for *D*. These sources would then either give up or request a rebuild from the supervisor. The supervisor would reject the rebuild requests on the grounds that there is no longer a path to node *D*. At no point would any bandwidth be compromised. No deadlocks would occur. In fact, nothing like a deadlock has ever been observed in TYMNET circuits. Situations like this one occur several times a day in TYMNET, and the only harm done is that, when a link is oversubscribed, some users get less bandwidth than they would like.

After many years' experience with this method of routing and flow control, the only complaint we have with it is the amount of CPU time required to assemble and disassemble the physical packets. Packet-switched networks do not have this overhead since the packets merely pass through the node without being torn apart. We have implemented the more compute bound portions of these processes in microcode in our TYMNET Engine[2] and can sustain throughputs in excess of 25 000 bytes/s (25 000 in and 25 000 out), which has been adequate so far. A proposed enhancement, so far not implemented, is to allow a circuit to enter a "blast" mode in which it only uses full-sized physical packets. These packets would be so tagged and buffered separately to avoid the need to tear them down and reassemble them. Through any one node, only a few circuits could be in blast mode at a time because of limited buffer space, and a circuit would be in blast mode only while moving data at a high rate. Regulation of the buffer space would be handled by the standard flow control procedure. The primary application for this special mode would be file transfers between computers over very high bandwidth communication lines. In this mode, the throughput of an Engine might be about 200 000 bytes/s. The exact limit would depend on the method of getting the data in and out of memory.

FLOW CONTROL THROUGH GATEWAYS

TYMNET has many gateways, some to private or experimental networks using TYMNET technology, and some to networks quite alien to TYMNET. Fig. 4 illustrates the case where both networks use TYMNET technology. The node in the center is called "schizoid" and has two identities, one for each network. Within each network, each node number must be unique. The schizoid node has two numbers. In this case, it is known as node 12 to the supervisor of network *A* and node 2073 to the supervisor of network *B*. Each supervisor claims the node as its own and sees the other network as a host computer, which can originate and terminate circuits. It is possible for one supervisor to see the schizoid as a TYMNET I node, while the other sees it as a TYMNET II node. Each supervisor is responsible for circuit routing and resource allocation within its own network.

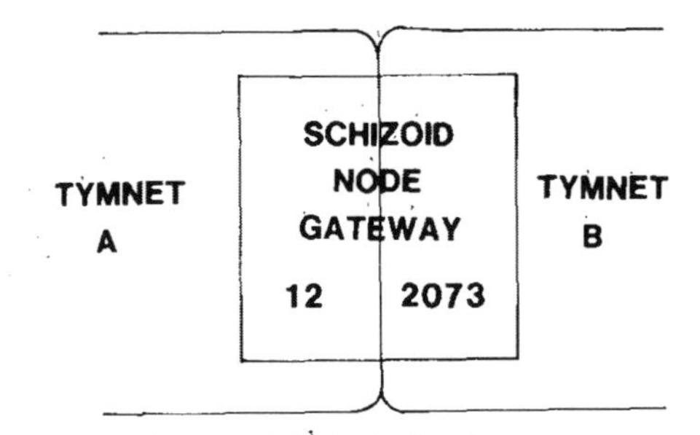

Fig. 4. TYMNET schizoid gateway.

Fig. 5 illustrates a TYMNET gateway to a different type of network. The actual interface is commonly an X.75 format on a high-speed synchronous line. Again, the TYMNET supervisor sees the other network as a host. The structure of any particular gateway is dictated by the sophistication of the other network and the availability of manpower to design the interface. In theory, it is certainly possible to build a schizoid node between TYMNET and any other network. This would be the most efficient interconnection, and would make the flow control work as well as the other network would allow, but so far the X.75 approach has been satisfactory. It has the advantage of keeping the two networks truly separated from each other, both organizationally and technically.

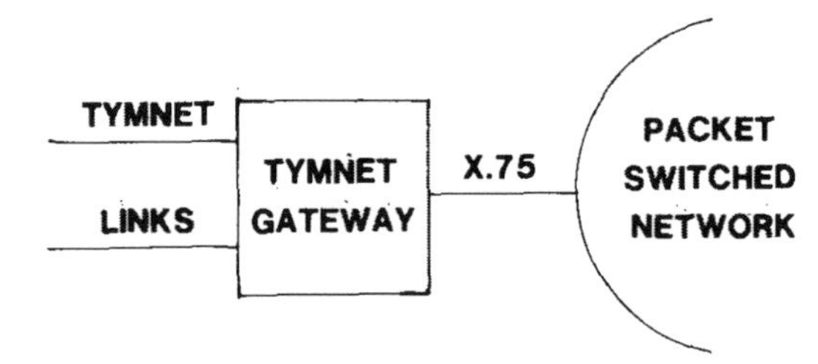

Fig. 5. TYMNET to alien network gateway.

TREE STRUCTURE FOR TYMNET I CONTROL

In the original TYMNET, the nodes were very small and the software primitive. They had no ability to process circuit-building needles. Instead, the supervisor maintained their internal tables directly. When the supervisor took control of the network, it first took control of the node nearest to itself, then the neighbors of those nodes, then the neighbors of those nodes, and so on. When a node was taken over on a link, that link became its upstream direction. A special channel on each link was dedicated to supervisory traffic. All such traffic consisted of 48-bit messages, either to the supervisor (upstream) or to the node (downstream). In this tree structure, upstream was always well defined. Any node wishing to send something to the supervisor sent it in the upstream direction, and it would proceed to the root of the tree. Downstream, however, had many branches. To get a message to a particular node, the supervisor had to first set switches in the intervening nodes so that each node would know which of its links was the downstream link.

When the net was small, the scheme worked fairly well. It was simple to implement and required little software. However, as the network grew beyond 200 nodes, problems occurred.

One obvious problem was that the routing was haphazard. It simply happened as a result of the order in which the nodes were taken over. If a major control link failed, the nodes which were downstream of the failure were retaken from another direction, perhaps a less optimum direction than before. The time required to get data to and from the nodes could be excessive.

The second problem was the flow control. At first there was no flow control. The volume of data was so small that none was needed. As the nodes grew in size, so did their tables. At the same time, the numbers of nodes grew. Because TYMNET I required the supervisor to know the contents of the routing tables in all the nodes, one thing the supervisor had to do when it took over the network was read all the tables. The data volume grew to hundreds of thousands of bytes. Obviously, something was needed to prevent nodes near the supervisor from running out of buffer space at takeover time.

The ad hoc solution that was applied was for a node to backpressure the upstream supervisor channel when the node had too much supervisory data in it. The node could still send data to the supervisor, but the supervisor could no longer send data to it. Since most supervisory traffic is either generated by the supervisor or generated in response to the supervisor, this had the effect of allowing the data to drain out of the congested area of the net. Temporarily, the supervisor was cut off from that portion of the net. What is worse is that the backpressure could easily back up into the supervisor itself, thus cutting it off from the entire network.

Another problem was the misbehaving node. If some node had a software bug which caused it to generate a lot of supervisory traffic, it could flood the treee structure, turn on backpressure at some points, and, in general, make it difficult for the supervisor to retain control. In a tree structure, there is no good way to selectively apply backpressure to a node of the tree without affecting other nodes downstream of it.

The most troublesome problem for TYMNET I was the "deadend circuit." To build a circuit, the supervisor had to send commands to each node involved with the circuit to set up the routing tables. If, because of loss of data in the supervisory tree, e.g., due to a link failure, some commands made it and some did not, the result was that only pieces of the circuit could get built. Not only would the user not be connected, but some unreachable (therefore unzappable) circuit fragments would remain to clutter the tables. Furthermore, the image in the supervisor of the routing tables would no longer agree with the tables in the nodes. For these reasons, the tree structure was abandoned for TYMNET II. The supervisor builds a separate virtual circuit to each node at takeover time and controls the node through that circuit.

SUMMARY

Since TYMNET was brought up in 1971, it has never been down. Nodes, lines, and even the supervisor have all had their failures, but there was never a moment when some part of the network was not running and able to do useful work. The TYMNET virtual circuit, with its associated routing and flow control, has stood the test of time in a commercial environment. Its efficiency and robustness when faced with a wide variety of overload and failure conditions closely matches the ideal. It has scaled very well from a rather small network to a very large one. The size of TYMNET provides abundant feedback to any changes in the machinery, which allows a quick check of theory against reality. The evidence so far is that the virtual circuit approach is entirely satisfactory.

The tree structure, on the other hand, did not scale very well. While quite acceptable in the original network, the absence of controlled routing and sloppy flow control proved inadequate as the volume of supervisory data grew and the average distance between node and supervisor increased. For this reason (and several others, such as the difficulty of maintaining synchronism between the tables in the nodes and the images of those tables in the supervisor), this type of control structure is not used in TYMNET II.

The routing takes into account all factors known at the time the circuit is to be built and attempts to find the "optimum" path, both for the user and for the network. The network overhead required to keep the supervisor informed about link outages and other factors that might affect routing is small. The overhead required to build the circuit is small, especially in TYMNET II. Normally, the response time to a circuit build request is less than two seconds, which is acceptable to most users. Load balancing is accomplished through proper routing of new circuits, according to the expected demands of the new circuits and the current network load. The centralized routing control, where all information required for optimum global strategy is available to a single program on a single computer (not counting the inactive backup supervisors) has proven to be fast, efficient, and versatile.

How large TYMNET can be and still be controlled by a single centralized supervisor is difficult to answer. For TYMNET I the answer appears to have been from 500 to 600 nodes because of the volume of data transferred between nodes and supervisor, combined with the damage done when any of that data was lost. For TYMNET II, however, it is probably at least 5000 nodes. When that limit is reached, regionalized networks connected with gateways are the most obvious answer and would require no new development. A possibly more efficient approach would be a "super" supervisor managing global strategy while delegating local routing to "sub" supervisors. The question may become academic before any limit is reached. Perhaps 5000 nodes is enough to cover the entire market. Also, value-added networks may soon be obsolete. Telephone networks of the future will be digital. Bandwidth between central offices will be very large and inexpensive, and the offices themselves will be automated. The cost of the telephone network will be largely in the local loops to the customer locations, since installation and maintenance of the local loops is labor-intensive and will not benefit from technology advances as much as the rest of the system. In such a situation, the "value" of the value-added network disappears, and straightforward digital circuit switching will be the most cost-effective way to connect the intelligent terminal with the sophisticated computer.

REFERENCES

[1] L. Tymes, "TYMNET—A terminal oriented communications network," in *Proc. AFIP Nat. Comput. Conf.*, Spring 1971, pp. 211–216.

[2] L. Tymes and J. Rinde, "The TYMNET II Engine," in *Proc. ICCC 78*, Int. Council Comput. Commun., Sept. 1978.

[3] J. Rinde, "TYMNET I: An alternative to packet switching," in *Proc. 3rd Int. Conf. Comput. Commun.*, 1976.

[4] M. Schwartz, *Computer Communication Network Design and Analysis*, ch. 2. Englewood Cliffs, NJ: Prentice-Hall, 1977.

[5] E. W. Dijkstra, "A note on two problems in connexion with graphs," *Numer. Math.*, vol. 1, pp. 269–271, 1959.

[6] J. Gilsinn and C. Witzgall, "A performance comparison of labeling algorithms for calculating shortest path trees," Nat. Bureau of Standards, Washington, DC, Tech. Note 772, 1973.

[7] D. R. Shier, "On algorithms for finding the *k* shortest paths in a network," *Networks*, vol. 9, pp. 195–214, 1979.

PHOTO NOT AVAILABLE

La Roy W. Tymes (M'79) received the B.S. degree in mathematics in 1966 and the M.A. degree, also in mathematics, in 1968, both from California State College at Hayward.

He joined Tymshare, Inc., in 1968 and has been involved in network development ever since. His duties have included design and implementation of the original TYMNET, supervision of network development, design and microcoding of the TYMNET Engine, and systems programming of Tymshare's timesharing systems. He has also been a consultant for numerically controlled machine tools and development of custom LSI. His research interests include subnanosecond LSI, supercomputers, and wide-band digital communication.

OSI Reference Model—The ISO Model of Architecture for Open Systems Interconnection

HUBERT ZIMMERMANN

(Invited Paper)

Abstract—**Considering the urgency of the need for standards which would allow constitution of heterogeneous computer networks, ISO created a new subcommittee for "Open Systems Interconnection" (ISO/TC97/SC16) in 1977. The first priority of subcommittee 16 was to develop an architecture for open systems interconnection which could serve as a framework for the definition of standard protocols. As a result of 18 months of studies and discussions, SC16 adopted a layered architecture comprising seven layers (Physical, Data Link, Network, Transport, Session, Presentation, and Application). In July 1979 the specifications of this architecture, established by SC16, were passed under the name of "OSI Reference Model" to Technical Committee 97 "Data Processing" along with recommendations to start officially, on this basis, a set of protocols standardization projects to cover the most urgent needs. These recommendations were adopted by TC97 at the end of 1979 as the basis for the following development of standards for Open Systems Interconnection within ISO. The OSI Reference Model was also recognized by CCITT Rapporteur's Group on "Layered Model for Public Data Network Services."**

This paper presents the model of architecture for Open Systems Interconnection developed by SC16. Some indications are also given on the initial set of protocols which will likely be developed in this OSI Reference Model.

I. INTRODUCTION

IN 1977, the International Organization for Standardization (ISO) recognized the special and urgent need for standards for heterogeneous informatic networks and decided to create a new subcommittee (SC16) for "Open Systems Interconnection."

The initial development of computer networks had been fostered by experimental networks such as ARPANET [1] or CYCLADES [2], immediately followed by computer manufacturers [3], [4]. While experimental networks were conceived as heterogeneous from the very beginning, each manufacturer developed his own set of conventions for interconnecting his own equipments, referring to these as his "network architecture."

The universal need for interconnecting systems from different manufacturers rapidly became apparent [5], leading ISO to decide for the creation of SC16 with the objective to come up with standards required for "Open Systems Interconnection." The term "open" was chosen to emphasize the fact that by conforming to those international standards, a system will be open to all other systems obeying the same standards throughout the world.

The first meeting of SC16 was held in March 1978, and initial discussions revealed [6] that a consensus could be reached rapidly on a layered architecture which would satisfy most requirements of Open Systems Interconnection with the capacity of being expanded later to meet new requirements. SC16 decided to give the highest priority to the development of a standard Model of Architecture which would constitute the framework for the development of standard protocols. After less than 18 months of discussions, this task was completed, and the ISO Model of Architecture called the Reference Model of Open Systems Interconnection [7] was transmitted by SC16 to its parent Technical Committee on "Data Processing" (TC97) along with recommendations to officially start a number of projects for developing on this basis an initial set of standard protocols for Open Systems Interconnection. These recommendations were adopted by TC97 at the end of 1979 as the basis for following development of standards for Open Systems Interconnection within ISO. The OSI Reference Model was also recognized by CCITT Rapporteur's Group on Public Data Network Services.

The present paper describes the OSI Architecture Model as it has been transmitted to TC97. Sections II–V introduce concepts of a layered architecture, along with the associated vocabulary defined by SC16. Specific use of those concepts in the OSI seven layers architecture are then presented in Section VI. Finally, some indications on the likely development of OSI standard protocols are given in Section VII.

Note on an "Interconnection Architecture"

The basic objective of SC16 is to standardize the rules of interaction between interconnected systems. Thus, only the external behavior of Open Systems must conform to OSI Architecture, while the internal organization and functioning of each individual Open System is out of the scope of OSI standards since these are not visible from other systems with which it is interconnected [8].

It should be noted that the same principle of restricted visibility is used in any manufacturer's network architecture in order to permit interconnection of systems with different structures within the same network.

These considerations lead SC16 to prefer the term of "Open Systems Interconnection Architecture" (OSIA) to the term of "Open Systems Architecture" which had been used previously and was felt to be possibly misleading. However, for unclear reasons, SC16 finally selected the title "Reference Model of Open Systems Interconnection" to refer to this Interconnection Architecture.

Manuscript received August 5, 1979; revised January 16, 1980.
The author is with IRIA/Laboria, Rocquencourt, France.

Reprinted from *IEEE Transactions on Communications,* vol. COM-28, no. 4, April 1980.

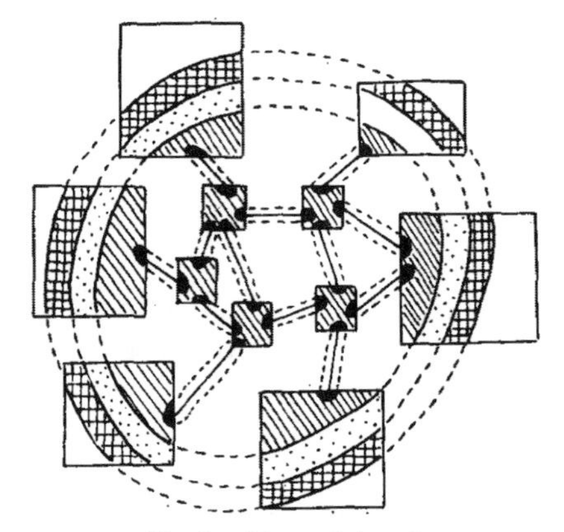

Fig. 1. Network layering.

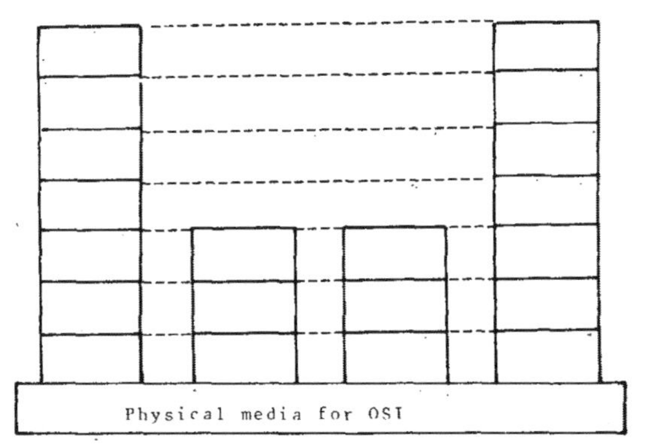

Fig. 2. An example of OSI representation of layering.

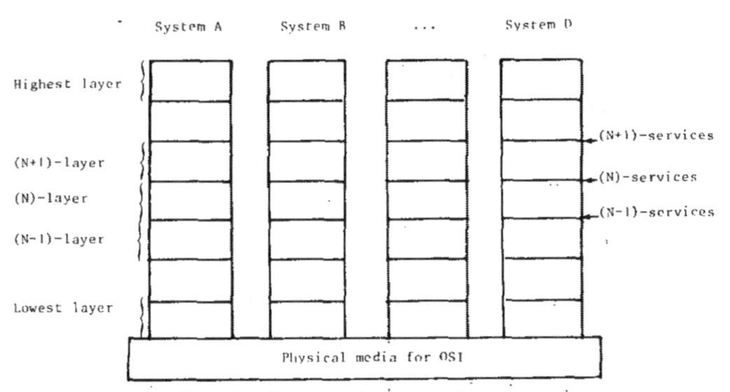

Fig. 3. Systems, layers, and services.

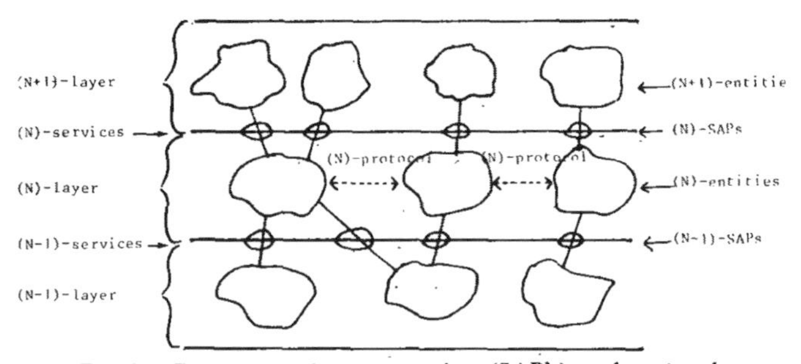

Fig. 4. Entities, service access points (SAP's), and protocols.

II. GENERAL PRINCIPLES OF LAYERING

Layering is a structuring technique which permits the network of Open Systems to be viewed as logically composed of a succession of layers, each wrapping the lower layers and isolating them from the higher layers, as exemplified in Fig. 1.

An alternative but equivalent illustration of layering, used in particular by SC16 is given in Fig. 2 where successive layers are represented in a vertical sequence, with the physical media for Open Systems Interconnection at the bottom.

Each individual system itself is viewed as being logically composed of a succession of subsystems, each corresponding to the intersection of the system with a layer. In other words, a layer is viewed as being logically composed of subsystems of the same rank of all interconnected systems. Each subsystem is, in turn, viewed as being made of one or several entities. In other words, each layer is made of entities, each of which belongs to one system. Entities in the same layer are termed *peer* entities.

For simplicity, any layer is referred to as the *(N) layer*, while its next lower and next higher layers are referred to as the $(N - 1)$ layer and the $(N + 1)$ layer, respectively. The same notation is used to designate all concepts relating to layers, e.g., entities in the (N) layer are termed *(N) entities*, as illustrated in Figs. 3 and 4.

The basic idea of layering is that each layer adds value to services provided by the set of lower layers in such a way that the highest layer is offered the set of services needed to run distributed applications. Layering thus divides the total problem into smaller pieces. Another basic principle of layering is to ensure independence of each layer by defining services provided by a layer to the next higher layer, independent of how these services are performed. This permits changes to be made in the way a layer or a set of layers operate, provided they still offer the same service to the next higher layer. (A more comprehensive list of criteria for layering is given in Section VI.) This technique is similar to the one used in structured programming where only the functions performed by a module (and not its internal functioning) are known by its users.

Except for the highest layer which operates for its own purpose, (N) entities distributed among the interconnected Open Systems work collectively to provide the *(N) service* to $(N + 1)$ entities as illustrated in Fig. 4. In other words, the (N) entities add value to the $(N - 1)$ service they get from the $(N - 1)$ layer and offer this value-added service, i.e., the (N) service to the $(N + 1)$ entities.

Communication between the $(N + 1)$ entities make exclusive use of the (N) services. In particular, direct communication between the $(N + 1)$ entities in the same system, e.g., for sharing resources, is not visible from outside of the system and thus is not covered by the OSI Architecture. Entities in the lowest layer communicate through the Physical Media for OSI, which could be considered as forming the (O) layer of the OSI Architecture. Cooperation between the (N) entities is ruled by the *(N) protocols* which precisely define how the (N) entities work together using the $(N - 1)$ services to perform the (N) functions which add value to the $(N - 1)$ service in order to offer the (N) service to the $(N + 1)$ entities.

The (N) services are offered to the $(N + 1)$ entities at the *(N) service access points,* or *(N) SAP's* for short, which represent the logical interfaces between the (N) entities and the $(N + 1)$ entities. An (N) SAP can be served by only one

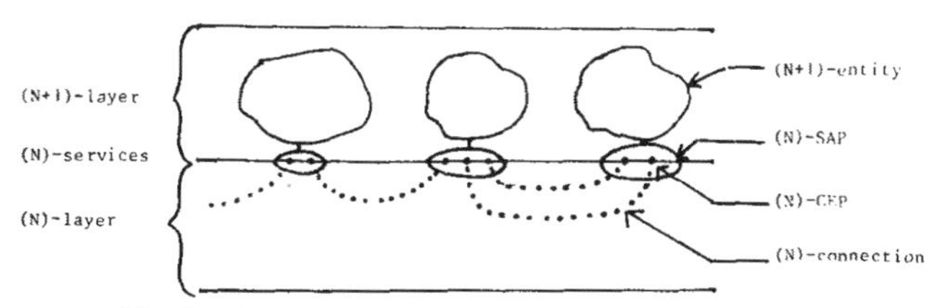

Fig. 5. Connections and connection endpoints (CEP's).

(*N*) entity and used by only one (*N* + 1) entity, but one (*N*) entity can serve several (*N*) SAP's and one (*N* + 1) entity can use several (*N*) SAP's.

A common service offered by all layers consists of providing associations between peer SAP's which can be used in particular to transfer data (it can, for instance, also be used to synchronize the served entities participating in the association). More precisely (see Fig. 5), the (*N*) layer offers (*N*) *connections* between (*N*) SAP's as part of the (*N*) services. The most usual type of connection is the *point-to-point* connection, but there are also *multiendpoint* connections which correspond to multiple associations between entities (e.g., broadcast communication). The end of an (*N*) connection at an (*N*) SAP is called an (*N*) *connection endpoint* or (*N*) *CEP* for short. Several connections may coexist between the same pair (or *n*-tuple) of SAP's.

Note: In the following, for the sake of simplicity, we will consider only point-to-point connections.

III. IDENTIFIERS

Objects within a layer or at the boundary between adjacent layers need to be uniquely identifiable, e.g., in order to establish a connection between two SAP's, one must be able to identify them uniquely. The OSI Architecture defines identifiers for entities, SAP's, and connections as well as relations between these identifiers, as briefly outlined below.

Each (*N*) entity is identified with a *global title*[1] which is unique and identifies the same (*N*) entity from anywhere in the network of Open Systems. Within more limited *domains*, an (*N*) entity can be identified with a *local title* which uniquely identifies the (*N*) entity only in the domain. For instance, within the domain corresponding to the (*N*) layer, (*N*) entities are identified with (*N*) *global titles* which are unique within the (*N*) layer.

Each (*N*) SAP is identified with an (*N*) *address* which uniquely identifies the (*N*)-SAP at the boundary between the (*N*) layer and the (*N* + 1) layer.

The concepts of titles and addresses are illustrated in Fig. 6.

Binding between (*N*) entities and the (*N* – 1) SAP's they use (i.e., SAP's through which they can access each other and communicate) are translated into the concept of (*N*) directory which indicates correspondence between global titles of (*N*) entities and (*N*) addresses through which they can be reached, as illustrated in Fig. 7.

[1] The term "title" has been preferred to the term "name" which is viewed as bearing a more general meaning. A title is equivalent to an entity name.

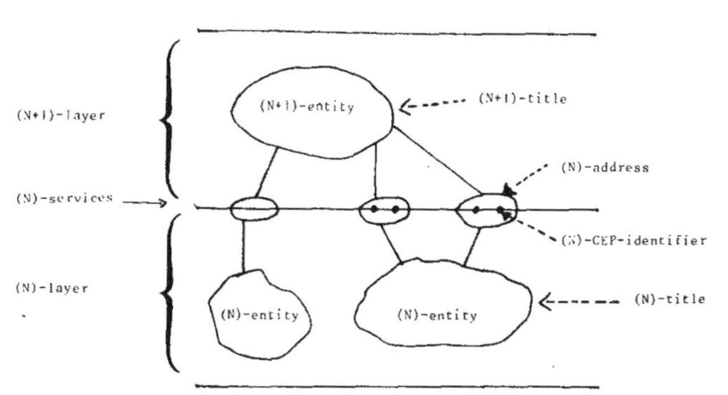

Fig. 6. Titles, addresses, and CEP-identifiers.

(N)-title	(N-1)-address
A	352
B	237
B	015
C	015

Fig. 7. Example of an (*N*)-directory.

Correspondence between (*N*) addresses served by an (*N*) entity and the (*N* – 1) addresses used for this purpose is performed by an (*N*) mapping function. In addition to the simplest case of one-to-one mapping, mapping may, in particular, be hierarchical with the (*N*) address being made of an (*N* – 1) address and an (*N*) suffix. Mapping may also be performed "by table." Those three types of mapping are illustrated in Fig. 8.

Each (*N*) CEP is uniquely identified within its (*N*) SAP by an (*N*) *CEP identifier* which is used by the (*N*) entity and the (*N* + 1) entity on both sides of the (*N*) SAP to identify the (*N*) connection as illustrated in Fig. 6. This is necessary since several (*N*) connections may end at the same (*N*) SAP.

IV. OPERATION OF CONNECTIONS

A. Establishment and Release

When an (*N* + 1) entity requests the establishment of an (*N*) connection from one of the (*N*) SAP's it uses to another (*N*) SAP, it must provide at the local (*N*) SAP the (*N*) address of the distant (*N*) SAP. When the (*N*) connection is established, both the (*N* + 1) entity and the (*N*) entity will use the (*N*)CEP identifier to designate the (*N*) connection.

(*N*) connections may be established and released dynamically on top of (*N* – 1) connections. Establishment of an (*N*) connection implies the availability of an (*N* – 1) connection between the two entities. If not available, the (*N* – 1) connection must be established. This requires the availability of an (*N* – 2) connection. The same consideration applies downwards until an available connection is encountered.

In some cases, the (*N*) connection may be established simultaneously with its supporting (*N* – 1) connection provided

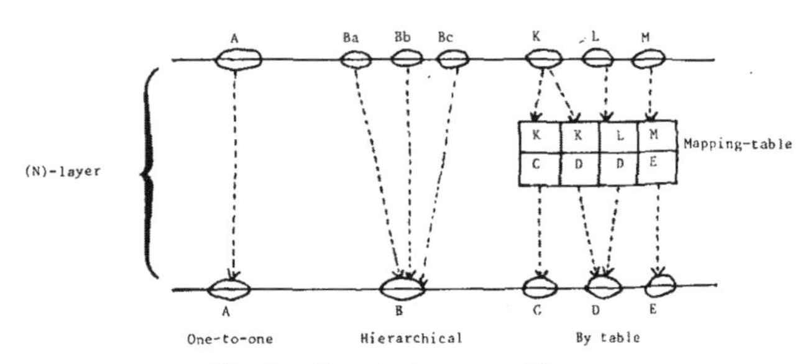

Fig. 8. Mapping between addresses.

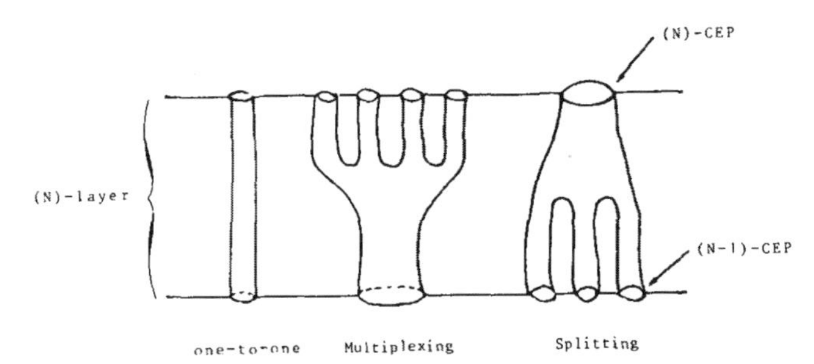

Fig. 9. Correspondence between connections.

the (N − 1) connection establishment service permits (N) entities to exchange information necessary to establish the (N) connection.

B. Multiplexing and Splitting

Three particular types of construction of (N) connections on top of (N − 1) connections are distinguished.

1) One-to-one correspondence where each (N) connection is built on one (N − 1) connection.

2) Multiplexing (referred to as "upward multiplexing" in [7]) where several (N) connections are multiplexed on one single (N − 1) connection.

3) Splitting (referred to as "downward multiplexing" in [7]) where one single (N) connection is built on top of several (N − 1) connection, the traffic on the (N) connection being divided between the various (N − 1) connections.

These three types of correspondence between connections in adjacent layers are illustrated in Fig. 9.

C. Data Transfer

Information is transferred in various types of data units between peer entities and between entities attached to a specific service access point. The data units are defined below and the interrelationship among several of them is illustrated in Fig. 10.

(N) *Protocol Control Information* is information exchanged between two (N) entities, using an (N − 1) connection, to coordinate their joint operation.

(N) *User Data* is the data transferred between two (N) entities on behalf of the (N + 1) entities for whom the (N) entities are providing services.

An (N) *Protocol Data Unit* is a unit of data which contains (N) Protocol Control Information and possibly (N) User Data.

(N) *Interface Control Information* is information exchanged between an (N + 1) entity and an (N) entity to coordinate their joint operation.

(N) *Interface Data* is information transferred from an (N + 1) entity to an (N) entity for transmission to a correspondent (N + 1) entity over an (N) connection, or conversely, information transferred from an (N) entity to an (N + 1) entity which has been received over an (N) connection from a correspondent (N + 1) entity.

(N) *Interface Data Unit* is the unit of information transferred across the service access point between an (N + 1) entity and an (N) entity in a single interaction. The size of (N)

	Control	Data	Combined
(N) - (N) Peer Entities	(N)-Protocol-Control-Information	(N)-User-Data	(N)-Protocol-Data-Units
(N)-(N-1) Adjacent layers	(N-1)-Interface-Control-Information	(N-1)-Interface Data	(N-1)-Interface-Data-Unit

Fig. 10. Interrelationship between data units.

interface data units is not necessarily the same at each end of the connection.

(N − 1) *Service Data Unit* is the amount of (N − 1) interface data whose identity is preserved from one end of an (N − 1) connection to the other. Data may be held within a connection until a complete service data unit is put into the connection.

Expedited (N − 1) service data unit is a small (N − 1) service data unit whose transfer is expedited. The (N − 1) layer ensures that an expedited data unit will not be delivered after any subsequent service data unit or expedited data unit sent on that connection. An expedited (N − 1) service data unit may also be referred to as an (N − 1) expedited data unit.

Note: An (N) protocol data unit may be mapped one-to-one onto an (N − 1) service data unit (see Fig. 11).

V. MANAGEMENT ASPECTS

Even though a number of resources are managed locally, i.e., without involving cooperation between distinct systems, some management functions do.

Examples of such management functions are

configuration information,
cold start/termination,
monitoring,
diagnostics,
reconfiguration, etc.

The OSI Architecture considers management functions as applications of a specific type. Management entities located in the highest layer of the architecture may use the complete set of services offered to all applications in order to perform

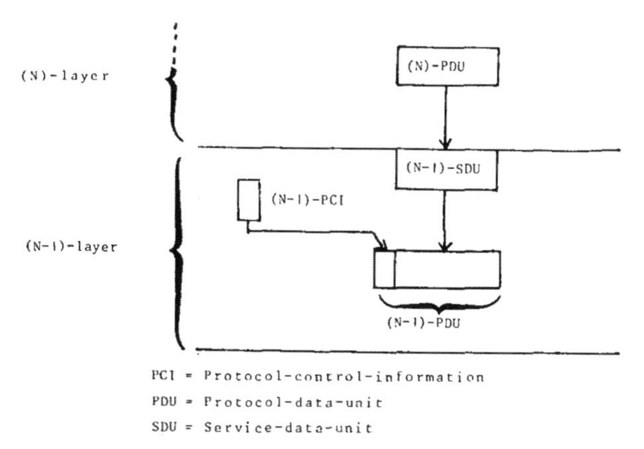

Fig. 11. Logical relationship between data units in adjacent layers.

management functions. This organization of management functions within the OSI Architecture is illustrated in Fig. 12.

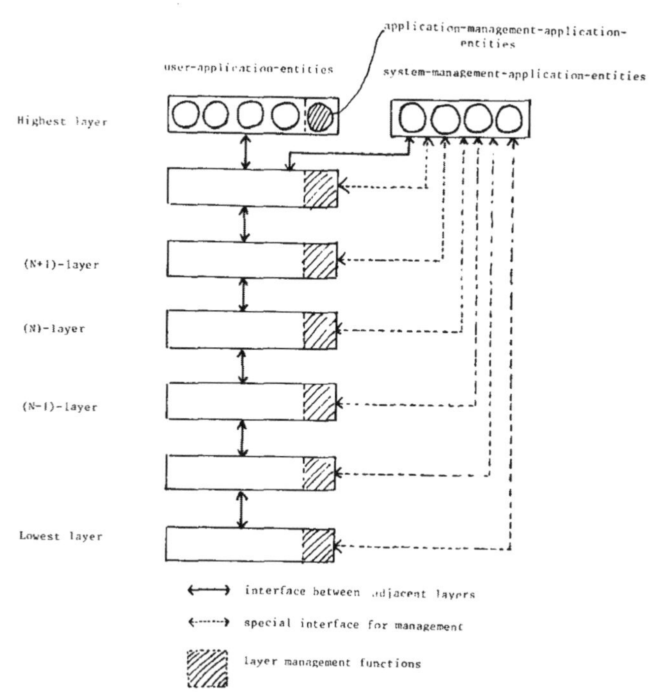

Fig. 12. A representation of management functions.

VI. THE SEVEN LAYERS OF THE OSI ARCHITECTURE

A. Justification of the Seven Layers

ISO determined a number of principles to be considered for defining the specific set of layers in the OSI architecture, and applied those principles to come up with the seven layers of the OSI Architecture.

Principles to be considered are as follows.

1) Do not create so many layers as to make difficult the system engineering task describing and integrating these layers.

2) Create a boundary at a point where the services description can be small and the number of interactions across the boundary is minimized.

3) Create separate layers to handle functions which are manifestly different in the process performed or the technology involved.

4) Collect similar functions into the same layer.

5) Select boundaries at a point which past experience has demonstrated to be successful.

6) Create a layer of easily localized functions so that the layer could be totaly redesigned and its protocols changed in a major way to take advantages of new advances in architectural, hardware, or software technology without changing the services and interfaces with the adjacent layers.

7) Create a boundary where it may be useful at some point in time to have the corresponding interface standardized.

8) Create a layer when there is a need for a different level of abstraction in the handling of data, e.g., morphology, syntax, semantics.

9) Enable changes of functions or protocols within a layer without affecting the other layers.

10) Create for each layer interfaces with its upper and lower layer only.

11) Create further subgrouping and organization of functions to form sublayers within a layer in cases where distinct communication services need it.

12) Create, where needed, two or more sublayers with a common, and therefore minimum, functionality to allow interface operation with adjacent layers.

13) Allow bypassing of sublayers.

B. Specific Layers

The following is a brief explanation of how the layers were chosen.

1) It is essential that the architecture permits usage of a realistic variety of physical media for interconnection with different control procedures (e.g., V.24, V.35, X.21, etc.). Application of principles 3, 5, and 8 leads to identification of a *Physical Layer* as the lowest layer in the architecture.

2) Some physical communications media (e.g., telephone line) require specific techniques to be used in order to transmit data between systems despite a relatively high error rate (i.e., an error rate not acceptable for the great majority of applications). These specific techniques are used in data-link control procedures which have been studied and standardized for a number of years. It must also be recognized that new physical communications media (e.g., fiber optics) will require different data-link control procedures. Application of principles 3, 5, and 8 leads to identification of a *Data link Layer* on top of the Physical Layer in the architecture.

3) In the Open Systems Architecture, some systems will act as final destination of data. Some systems may act only as intermediate nodes (forwarding data to other systems). Application of principles 3, 5, and 7 leads to identification of a *Network Layer* on top of the Data link Layer. Network-oriented protocols such as routing, for example, will be grouped in this layer. Thus, the Network Layer will provide a connection path (network connection) between a pair of transport entities (see Fig. 13).

4) Control of data transportation from source end system to destination end system (which need not be performed in

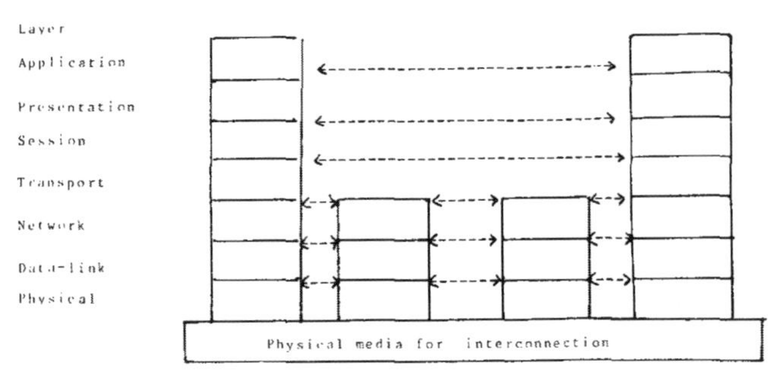

Fig. 13. The seven layers OSI architecture.

intermediate nodes) is the last function to be performed in order to provide the totality of the transport service. Thus, the upper layer in the transport-service part of the architecture is the *Transport Layer,* sitting on top of the Network Layer. This Transport Layer relieves higher layer entities from any concern with the transportation of data between them.

5) In order to bind/unbind distributed activities into a logical relationship that controls the data exchange with respect to synchronization and structure, the need for a dedicated layer has been identified. So the application of principles 3 and 4 leads to the establishment of the *Session Layer* which is on top of the Transport Layer.

6) The remaining set of general interest functions are those related to representation and manipulation of structured data for the benefit of application programs. Application of principles 3 and 4 leads to identification of a *Presentation Layer* on top of the Session Layer.

7) Finally, there are applications consisting of application processes which perform information processing. A portion of these application processes and the protocols by which they communicate comprise the *Application Layer* as the highest layer of the architecture.

The resulting architecture with seven layers, illustrated in Fig. 13 obeys principles 1 and 2.

A more detailed definition of each of the seven layers identified above is given in the following sections, starting from the top with the application layer described in Section VI-C1) down to the physical layer described in Section VI-C7).

C. Overview of the Seven Layers of the OSI Architecture

1) *The Application Layer*: This is the highest layer in the OSI Architecture. Protocols of this layer directly serve the end user by providing the distributed information service appropriate to an application, to its management, and to system management. Management of Open Systems Interconnection comprises those functions required to initiate, maintain, terminate, and record data concerning the establishment of connections for data transfer among application processes. The other layers exist only to support this layer.

An application is composed of cooperating *application processes* which intercommunicate according to application layer protocols. Application processes are the ultimate source and sink for data exchanged.

A portion of an application process is manifested in the application layer as the execution of application protocol (i.e., application entity). The rest of the application process is considered beyond the scope of the present layered model. Applications or application processes may be of any kind (manual, computerized, industrial, or physical).

2) *The Presentation Layer*: The purpose of the Presentation Layer is to provide the set of services which may be selected by the Application Layer to enable it to interpret the meaning of the data exchanged. These services are for the management of the entry exchange, display, and control of structured data.

The presentation service is location-independent and is considered to be on top of the Session Layer which provides the service of linking a pair of presentation entities.

It is through the use of services provided by the Presentation Layer that applications in an Open Systems Interconnection environment can communicate without unacceptable costs in interface variability, transformations, or application modification.

3) *The Session Layer*: The purpose of the Session Layer is to assist in the support of the interactions between cooperating presentation entities. To do this, the Session Layer provides services which are classified into the following two categories.

a) Binding two presentation entities into a relationship and unbinding them. This is called *session administration service.*

b) Control of data exchange, delimiting, and synchronizing data operations between two presentation entities. This is called *session dialogue service.*

To implement the transfer of data between presentation entities, the Session Layer may employ the services provided by the Transport Layer.

4) *The Transport Layer*: The Transport Layer exists to provide a universal transport service in association with the underlying services provided by lower layers.

The Transport Layer provides transparent transfer of data between session entities. The Transport Layer relieves these session entities from any concern with the detailed way in which reliable and cost-effective transfer of data is achieved.

The Transport Layer is required to optimize the use of available communications services to provide the performance required for each connection between session entities at a minimum cost.

5) *The Network Layer*: The Network Layer provides functional and procedural means to exchange network service data units between two transport entities over a network connection. It provides transport entities with independence from routing and switching considerations.

6) *The Data Link Layer*: The purpose of the Data link Layer is to provide the functional and procedural means to establish, maintain, and release data links between network entities.

7) *The Physical Layer*: The Physical Layer provides mechanical, electrical, functional, and procedural characteristics to establish, maintain, and release physical connections (e.g., data circuits) between data link entities.

VII. OSI PROTOCOLS DEVELOPMENTS

The model of OSI Architecture defines the services provided by each layer to the next higher layer, and offers concepts to be used to specify how each layer performs its specific functions.

Detailed functioning of each layer is defined by the protocols specific to the layer in the framework of the Architecture model.

Most of the initial effort within ISO has been placed on the model of OSI. The next step consists of the definition of standard protocols for each layer.

This section contains a brief description of a likely initial set of protocols, corresponding to specific standardization projects recommended by SC16.

A. Protocols in the Physical Layer

Standards already exist within CCITT defining:

1) interfaces with physical media for OSI, and

2) protocols for establishing, controlling, and releasing switched data circuits.

Such standards are described in other papers in this issue [9], [10], e.g., X.21, V.24, V.35, etc.

The only work to be done will consist of clearly relating those standards to the OSI Architecture model.

B. Protocols in the Data Link Layer

Standard protocols for the Data link Layer have already been developed within ISO, which are described in other papers within this issue [11], [12].

The most popular Data link Layer protocol is likely to be HDLC [13], without ruling out the possibility of using also other character-oriented standards.

Just as for the Physical Layer, the remaining work will consist mainly of clearly relating these existing standards to the OSI Architecture model.

C. Protocols in the Network Layer

An important basis for protocols in the network layer is level 3 of the X.25 interface [14] defined by CCITT and described in another paper in this issue. It will have to be enhanced in particular to permit interconnection of private and public networks.

Other types of protocols are likely to be standardized later in this layer, and in particular, protocols corresponding to Datagram networks [10].

D. Protocols in the Transport Layer

No standard exists at present for this layer; a large amount of experience has been accumulated in this area and several proposals are available.

The most widely known proposal is the Transport Protocol proposed by IFIP and known as INWG 96.1 [15], which could serve as a basis for defining an international standard.

E. Protocols for the Session Layer

No standard exists and no proposal has been currently available, since in most networks, session functions were often considered as part of higher layer functions such as Virtual Terminal and File Transfer.

A standard Session Layer Protocol can easily be extracted from existing higher layer protocols.

F. Presentation Layer Protocol

So far, Virtual Terminal Protocols and part of Virtual File are considered the most urgent protocols to be developed in the Presentation Layer.

A number of VTP's are available (e.g., [16], [17]), many of them being very similar, and it should be easy to derive a Standard VTP from these proposals, also making use of the ISO standard for "Extended Control Characters for I/O Imaging Devices" [18]. These protocols are reviewed in another paper in this issue [19].

The situation is similar for File Transfer Protocols.

G. Management Protocols

Most of the work within ISO has been done so far on the architecture of management functions, and very little work has been done on management protocols themselves. Therefore, it is too early to give indications on the likely results of the ISO work in this area.

VIII. CONCLUSION

The development of OSI Standards is a very big challenge, the result of which will impact all future computer communication developments. If standards come too late or are inadequate, interconnection of heterogeneous systems will not be possible or will be very costly.

The work collectively achieved so far by SC16 members is very promising, and additional efforts should be expended to capitalize on these initial results and come up rapidly with the most urgently needed set of standards which will support initial usage of OSI (mainly terminals accessing services and file transfers). The next set of standards, including OSI management and access to distributed data, will have to follow very soon.

Common standards between ISO and CCITT are also essential to the success of standardization, since new services announced by PTT's and common carriers are very similar to data processing services offered as computer manufacturer products, and duplication of now compatible standards could simply cause the standardization effort to fail. In this regard, acceptance of the OSI Reference Model by CCITT Rapporteur's Group on Layered Architecture for Public Data Networks Services is most promising.

It is essential that all partners in this standardization process expend their best effort so it will be successful, and the benefits can be shared by all users, manufacturers of terminals and computers, and the PTT's/common carriers.

ACKNOWLEDGMENT

The OSI Architecture model briefly described in this paper results from the work of more than 100 experts from many countries and international organizations. Participation in this collective work was really a fascinating experience for the author who acknowledges the numerous contributions from SC16 members which have been merged in the final version of the OSI Architecture briefly presented here.

REFERENCES

[1] L. G. Roberts and B. D. Wessler, "Computer network development to achieve resource sharing," in *Proc. SJCC*, 1970, pp. 543–549.

[2] L. Pouzin, "Presentation and major design aspects of the CYCLADES computer network," in *Proc. 3rd ACM-IEEE Commun. Symp.*, Tampa, FL, Nov. 1973, pp. 80–87.

[3] J. H. McFayden, "Systems network architecture: An overview," *IBM Syst. J.*, vol. 15, no. 1, pp. 4–23, 1976.

[4] G. E. Conant and S. Wecker, "DNA, An Architecture for heterogeneous computer networks," in *Proc., ICCC*, Toronto, Ont., Canada, A 1976, pp. 618–625.

[5] H. Zimmermann, "High level protocols standardization: Technical and political issues," in *Proc. ICCC*, Toronto, Ont., Canada, Aug. 1976, pp. 373–376.

[6] ISO/TC97/SC16, "Provisional model of open systems architecture," Doc. N34, Mar. 1978.

[7] ISO/TC97/SC16, "Reference model of open systems interconnection," Doc. N227, June 1979.

[8] H. Zimmermann and N. Naffah, "On open systems architecture," in *Proc. ICCC*, Kyoto, Japan, Sept. 1978, pp. 669–674.

[9] H. V. Bertine, "Physical level protocols," this issue pp. 433–444.

[10] H. C. Folts, "Procedures for circuit-switched service in synchronous public data networks," and "X.25 transaction-oriented features—Datagram and fast select," this issue, pp. 489–496.

[11] J. W. Conard, "Character oriented data link control protocols," this issue, pp. 445–454.

[12] D. E. Carlson, "Bit-oriented data link control procedures," this issue, pp. 455–467.

[13] ISO, "High level data link control-elements of procedure," IS 4335, 1977.

[14] CCITT, "X25," *Orange Book*, vol. VIII-2, 1977, pp. 70–108.

[15] IFIP-WG 6.1, "Proposal for an internetwork end-to-end transport protocol," INWG Note 96.1; also, doc. ISO/TC97 SC16/N24, 46 pp., Mar. 1978.

[16] IFIP-WG 6.1, "Proposal for a standard virtual terminal protocol," doc. ISO/TC97/SC16/N23, 56 pp., Feb. 1978.

[17] EURONET, "Data entry virtual terminal protocol for EURONET," VTP/D-Issue 4, doc. EEC/WGS/165.

[18] ISO, "Extended control characters for I/0 imaging devices," DP 6429.

[19] J. Day, "Terminal protocols," this issue, pp. 585–593.

Hubert Zimmermann received degrees in engineering from Ecole Polytechnique, Paris, France, in 1963, and from Ecole Nationale Superieure des Telecommunications, Paris, France, in 1966.

He is presently in charge of the computer communications group at IRIA, Rocquencourt, France. He was involved in development of command and control systems before joining IRIA in 1972 to start the CYCLADES project with L. Pouzin. Within CYCLADES, he was mainly responsible for design and implementation of host protocols.

Dr. Zimmermann is a member of IFIP WG 6.1 [International Network Working Group (INWG)]. He also chaired the Protocol Subgroup and coauthored several proposals for international protocols. He is an active participant in the development of standards for Open Systems Interconnection (OSI) within ISO, where he chairs the working group on OSI architecture.

Deadlock Avoidance in Store-and-Forward Networks—I: Store-and-Forward Deadlock

PHILIP M. MERLIN, MEMBER, IEEE, AND PAUL J. SCHWEITZER

Abstract—**Store-and-forward deadlock in store-and-forward networks may be avoided by forwarding messages from buffer to buffer in accordance with a loop-free directed buffer graph which accommodates all the desired message routes. Schemes for designing such buffer graphs are presented, together with methods for using them to forward the messages in an efficient and deadlock-free manner. These methods can be implemented by a set of counters at each node. Such an implementation increases the efficiency of buffer use, and simplifies jumping between normal low-overhead operation when deadlock is far and more careful operation when deadlock is near. The proposed deadlock avoidance mechanism works for any network topology and any finite routing algorithm.**

I. INTRODUCTION

IN most cases, the occurrence of network deadlock has a highly objectionable impact upon network users. When the deadlocks are discovered, they are frequently corrected in an ad hoc fashion [1]. Since changes in existing implementations can be costly, it is preferable to incorporate deadlock avoidance (or recognition-and-recovery) procedures into the design as early as possible and in an orderly fashion. A primary concern with any deadlock avoidance procedure is that it have minimal effect upon system performance during normal operating conditions.

In this paper, a general method is present for the design and implementation of store-and-forward networks [2] free from "store-and-forward-deadlock" (or circular deadlock) [3]. This method has minimal overhead during normal operations, and switches into a more careful mode of operation when deadlock is near. A companion paper [4] extends this method to other types of deadlock.

Store-and-forward deadlock refers to the situation in which there is a set of buffers, all of which hold messages waiting to be forwarded, and these messages can be forwarded *only* to other buffers of the set. The result is standstill. Figs. 1 and 2 show two examples of store-and-forward deadlock. In Fig. 1 there are two network nodes, each *full* with messages for the other and therefore neither node can receive or eject any messages. Fig. 2 shows a more complex example. All buffers marked x are occupied by messages waiting for other buffers marked x to become available. Since all x-buffers are full, the result is standstill. Notice there may be other empty buffers at these nodes, e.g., buffers y, but permanently allocated for other purposes.

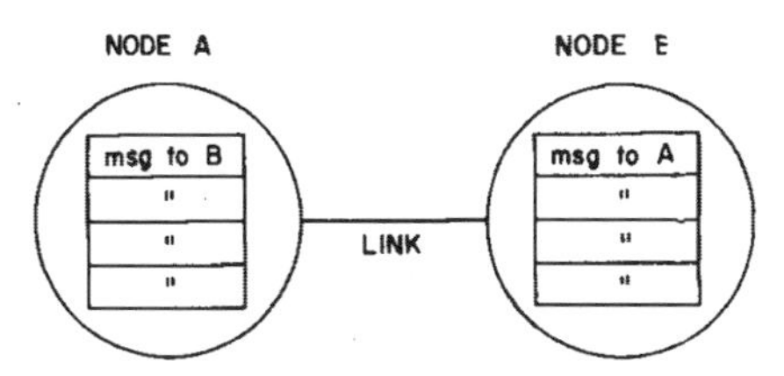

Fig. 1. Example of store-and-forward deadlock: all buffers deadlocked.

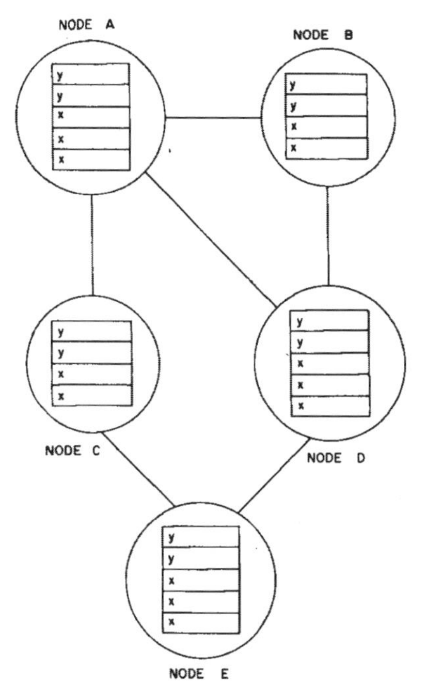

Fig. 2. Example of store-and-forward deadlock: smaller deadlocked set of buffers.

It can be shown that store-and-forward deadlock implies a cycle of buffer requests [5], [6]. Basically, our method is based on preventing cycles of requests, in an efficient way.

The proposed method generalizes schemes for deadlock avoidance proposed by [7] which are now incorporated in the GMD-net protocol, and which can also be used for congestion control [8]-[10]. Conceptually we use a directed graph

Paper approved by the Editor for Computer Communication of the IEEE Communications Society for publication after presentation at the Third Jerusalem Conference on Information Technology, Jerusalem, Israel, August 1978. Manuscript received November 7, 1977; revised June 11, 1979. This work was performed mainly at the IBM T. J. Watson Research Center, Yorktown Heights, NY, and was completed shortly before P. M. Merlin's premature death; by his request, it is dedicated to the medical staff at the B-Internal Disease Department and at the Institute of Oncology at the Rambam Hospital in Haifa, Israel and in particular to Dr. Yoram Cohen.

P. M. Merlin was with the Department of Electrical Engineering, Technion–Israel Institute of Technology, Haifa, Israel. He is now deceased.

P. J. Schweitzer is with the Graduate School of Management, University of Rochester, Rochester, NY 14627.

Reprinted from *IEEE Transactions on Communications,* vol. COM-28, no. 3, March 1980.

as in [11] to represent the possible forwarding of messages between buffers. However, we differentiate between "guaranteed paths" and "optional paths." This permits generation of deadlock-free schemes which would otherwise be impossible to generate or difficult to visualize.

Several assumptions are made in this paper.

a1) All nodes, communication channels, and destinations are operational, i.e., they will process a message in finite time. (This is relaxed later.)

a2) No messages are lost. (This is relaxed later.)

a3) Messages are transmitted as complete units, i.e., no packetizing or reassembly. (Reassembly deadlock is treated in [4].)

a4) No protocol prevents the destination node from turning messages over to the user as soon as they arrive. (This can be relaxed as shown for the pacing protocol in [4].)

a5) Any topology with a finite number of nodes is permitted. Any node-to-node routing algorithm is permitted which achieves message delivery within a bounded number of hops, e.g., a bounded number of repeat visits to a node is allowed. Adaptive routing can be employed provided route lengths are bounded.

a6) All message sizes are bounded; hence, no message requires more than, say, M buffers at a node. Throughout most of this paper, any message is assumed to fit into *one* buffer, and the extension to multibuffer messages is given in Section V.

The deadlock avoidance method described in this paper affects only the transfer of messages between adjacent network nodes. Provisions can be made such that no alterations are required in the protocols by which network nodes accept messages from attached users, except the ability to slow the admittance rate, if necessary, until the appropriate network buffers become available. The implementation can be arranged to preserve FIFO for all messages following a fixed route.

Section II presents the basic method for deadlock avoidance. Section III extends the method to allow for better utilization of the buffers as well as to provide more flexibility in the ways in which messages can be forwarded. Section IV presents several schemes for designing the directed graphs used by the proposed method. Section V presents an approach in which buffer "counters" implement the method efficiently.

II. GUARANTEED PATHS

Suppose one is given a store-and-forward network with prescribed nodes, prescribed communication channels between some pairs of nodes, a prescribed finite set of buffers at each node, and a prescribed set of all message routes permitted by the routing algorithm. Each route consists of a finite sequence of nodes and communication channels from the source node to the destination node. At each node the message is assumed to fit into one buffer. (Section V extends the method to multibuffer messages.)

A *buffer graph* (BG) is a directed graph whose nodes are a subset of the buffers, and where directed arcs connect some pairs of buffers, indicating permitted message flow from one buffer to the next. Arcs are permitted only between buffers in the same node, or between buffers in distinct nodes which are connected by communication channels.

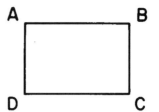

Fig. 3. Network for Example 2.1.

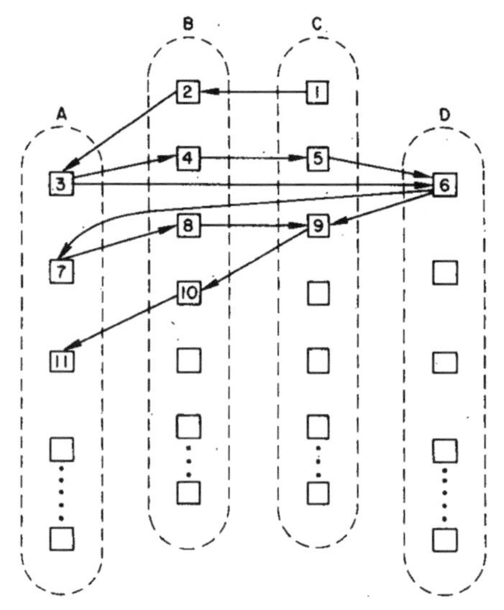

Fig. 4. BG for Example 2.1.

The BG is required to have the following properties.

1) There is at least one directed path, called a *guaranteed path,* in the BG corresponding to each route in the network; and

2) the BG contains no directed loops.

A fresh message accepted by the network is placed, at the source node, in a buffer from which there is at least one guaranteed path in the buffer graph to its destination node, which conforms to the given routing algorithm. The message makes its way from source to destination, using the buffers of such a guaranteed path. In this way, each message in the network is waiting for the next buffer in the path. However, since the BG is loop-free, waiting loops cannot form and, as proven below, deadlocks never occur.

Example 2.1: For illustration, the network of Fig. 3 is given. In this example, any route which does not visit a node more than once is permitted. Fig. 4 shows a corresponding BG in which the squares denote buffers (numbered for convenience). The reader can verify that, in both directions, any route visiting all four nodes is contained in some path in the buffer graph. Routes involving fewer than four nodes are clearly included within the routes which visit all four nodes.

If, for example, a message is to be transmitted from B to A to D to C, then it will be first placed in buffer 2 of B and moved in the path $2 \rightarrow 3 \rightarrow 6 \rightarrow 9$, and absorbed by the destination. A message in the reverse direction will travel in buffers $5 \rightarrow 6 \rightarrow 7 \rightarrow 8$, and a direct message from C to B will choose between the guaranteed paths $1 \rightarrow 2$ and $9 \rightarrow 10$. Note that the graph of Fig. 4 has no directed loops and therefore cannot have deadlocks.

In many cases it is difficult to graphically construct buffer graphs. The following schematic example from [7] will demonstrate that BG's can be constructed without actually drawing the directed graph. Furthermore, the example will demonstrate that for any given network, a corresponding BG can be constructed, provided that the routing algorithm puts a bound on the maximal number of hops a message may take.

Example 2.2—Hops-So-Far Scheme: Let K denote the maximal number of hops a message can make in the net. At each node, assign $K + 1$ buffers, say $[B0, B1, B2, \cdots, BK]$. At each node, from each buffer Bi there is an arc to the buffers of type $B(i + 1)$ located at nodes with communication channels to this node. A message is placed initially in $B0$ at the source node, transferred to buffer $B1$ at the next node on the route, buffer $B2$ at the following node, etc. At each node, buffer Bi holds messages which have made exactly i hops so far. Clearly, all possible routes involving at most K hops are accommodated and the monotonicity of the index i insures that loops are not formed.

The hops-so-far scheme has the convenient features of easy implementation and a simple criterion for admitting external messages to the network: whether or not buffer $B0$ is available.

Theorem 1: Any message accepted by the network will be delivered to the destination in a finite time, and all buffers occupied by the message will be released in a finite time, provided that the following conditions hold.

C1) Messages are accepted by the network only if there is an empty buffer in the source node from which there is at least one guaranteed path in the BG to its destination node.

C2) Any message in a buffer at other than its destination node is characterized by a nonempty set of buffers it can be transferred to. This set is called the *waiting set*, and each of its buffers lies on a guaranteed path from the buffer the message occupies to its destination. The waiting set is constant in time. If any buffer in the waiting set becomes vacant and grants permission for this message to enter, the message is transferred within a finite period of time.

C3) If a buffer becomes vacant, then within a finite period of time, one of the messages having this buffer in its waiting set is granted admission to the buffer.

C4) Consider any buffer in the waiting set for a given message. Then only a finite number of other messages can be admitted to this buffer before this message is granted admission ("fairness in admissions").

C5) When a message is transmitted from one buffer to another, the previous buffer is released within a finite time.

C6) A message occupying a buffer at its destination node is not forwarded any further; it is delivered to the user, and the buffer is released, within a finite time.

Schematic Proof:

1) Since the BG is loop-free, each buffer of the BG can be labeled by its *buffer class*, defined as the maximum number of arcs of all directed paths to this buffer from buffers lacking incoming arcs. (e.g., in Fig. 4, buffer j is in class $j - 1$, for $j = 1, \cdots, 11$). With this definition of buffer class, the forwarding of messages involves buffers having *strictly increasing* buffer classes.

2) Let the highest buffer class be class r. Any message occupying a buffer of class r is at buffer with no outgoing arcs; hence, it is a message at its destination. By C6), this message will be delivered within a finite time and its buffer released. By C5), within a finite time no copies of this message will exist anywhere in the BG.

3) An inductive proof proceeds downward on $j = r - 1, r - 2, \cdots, 0$. Any message in a buffer of class j is either at its destination [and will be delivered within a finite time with release of all buffers holding copies, via C6) and C5)] or is waiting for some buffer of class exceeding j [which will become vacant within a finite time, and will both become available to, and be used by, this message within a finite time, via C3), C4), and C2)]. Hence, any message not at its destination will be forwarded within a finite amount of time. Since the message always lies on a guaranteed path in the BG, all of which have finite length, it must be delivered within a finite time with release of all buffers containing copies.

Remark 1: Our proof carries over to some more complicated cases where the waiting set is not constant in time [12].

Remark 2: Assumptions a1), a2) can be relaxed. If messages are lost, the method still guarantees deadlock-free operation, but clearly, a lost message will not be delivered. The method can also cope with nonoperational nodes or links, provided that alternate routes are supplied or messages with unreachable destinations are discarded [4]. Under normal operations, however, deadlock is avoided *without* resorting to throwaway.

III. BUFFER GRAPH EXTENSIONS AND GENERALIZED USAGE

The proposed method restricts the set of messages which can use some of the buffers at a node. Therefore, at certain times there may be empty buffers at a node which cannot be used by messages waiting at adjacent nodes. In other words, the buffer usage is restricted for the sake of deadlock avoidance. There are several ways by which the frequency of such situations can be reduced to a very modest level by using the methods described in this section.

A. Additional Guaranteed Paths

One may add arcs to the BG, provided the loop-free property is maintained, and provided these new arcs connect either buffers in the same node or buffers in nodes having a communication channel between them. (The direction of such an arc is dictated if the original BG contained a path between these two buffers, and is otherwise arbitrary.) This creates alternative guaranteed paths in the BG while maintaining the deadlock-free property.

As an illustration of additional paths, consider Example 2.2. One may add arcs from buffer Bi at one node to every buffer Bj, $j > i$, at the same node or at nodes connected by communications channels to this node. This may achieve better performance than restricting j to $i + 1$.

The buffers lacking arcs can be put into use by adding arcs. This can be employed to augment the BG by parallel disjoint paths (e.g., for different traffic types) or replace a buffer by a set of interchangable buffers, all having the same incoming and outgoing arcs.

B. The Use of Path Switching

Suppose some buffer, say, $u1$, holds a message on a guaranteed path, and in the same or an adjacent node there is an empty buffer, say, $v1$, also lying on some guaranteed path for this message. Then, the message can be transferred from $u1$ to $v1$, even if there is no directed arc in the buffer graph from $u1$ to $v1$, provided that this transfer is allowed by the routing algorithm. This "path switching" does not compromise deadlock avoidance since from $v1$ there is also a guaranteed path in the BG to the destination. Path switching increases the number of possible buffers that a message can use at each node. For instance, a message one hop away from its destination may use *any* empty buffer at the destination node.

Example 3.1: Suppose that in the network of Figs. 3 and 4, a message is sent on route $C \to D \to A \to B$. The message will be accepted by the network only if buffer 5 is empty because the only path in the BG for this route is $5 \to 6 \to 7 \to 8$. If, when the message resides in 6, it happens that 7 is full but 3 is empty, then since 3 has a path in the same route to the destination (i.e., $3 \to 4$), it is possible to undertake a path switch by transferring the message to buffer 3. Although 7 is guaranteed to become available within a finite time, the use of 3 reduces the waiting time. Note that when the message arrives at 6, and 3 happens to be in use, it is not guaranteed that 3 will be available to the message in 6, for example, if 3 holds a message waiting for the release of buffer 6. Therefore, to avoid deadlock, *a message is not allowed to wait indefinitely only for buffers which are reachable exclusively via path switching*. That is, if in a finite time, buffers for path switching do not become available, the message should start waiting also for buffers which lie on a guaranteed path in the BG from the currently occupied buffer to the destination. (This is the motivation for Condition C8) of Theorem 2 below.)

Example 3.2: As another example, possible transfers also exist in the hops-so-far scheme in Example 2.2. For a given message, let t denote the total number of hops from source to destination. When the message has made $h < t$ hops it can make a transfer to any buffer Bg at an adjacent node (lying on the message route) such that

$$0 \leqslant g \leqslant K - (t - h) + 1.$$

From each of those Bg's there is a path to the destination:

$$Bg \to B(g+1) \to B(g+2) \to \cdots \to B(g+(t-h)-1).$$

and by definition of g, $g + (t-h) - 1 \leqslant K$.

Assuming that messages are forwarded from source to destination through a sequence of buffers, each of which has a guaranteed path in the BG from itself to the destination which satisfies the routing constraints, then the following theorem can be established.

Theorem 2: Any message accepted by the network will be delivered to the destination in a finite time, and all buffers occupied by the message will be released in a finite time, provided that: C1), C2) (without requiring the waiting sets to be constant in time), C3), C4), C5), and C6) hold as in Theorem 1, and also that the following hold.

C7) A message cannot wait indefinitely only for path switching. Within a finite time after the message enters the buffer, the waiting set must include buffers on guaranteed paths from that buffer.

C8) For each message in the current node, let U denote a subset of the set of buffers being waited for in *adjacent* nodes (i.e., buffers in a node connected to the current node by a communications channel with each such buffer having a guaranteed path in the BG to the destination). Within a finite time after the message enters the current node, from this time onwards until the message leaves the node:

a) U is nonempty

b) there exists a path in the BG from the buffer now holding the message at the current node to some buffer in U

c) elements of U are never deleted, even if the message is transferred among buffers in the current node.

This condition permits arbitrarily many path switchings within a node, but ensures departure from the node within finite time. If unbounded intranode transfers are not required, this condition can be simplified.

Schematic Proof:

1) The buffers of the BG can be labeled into classes $0, 1, 2, \cdots, r$ as in Theorem 1. Now messages can move into a buffer of either a lower or higher class. But because of C7), within a finite time after a message enters a buffer, either the message leaves the buffer or the waiting set for this message must include buffers of higher class.

2) As in Theorem 1, buffers of class r will empty in finite time and all copies of these messages will disappear in finite time.

3) The inductive proof of Theorem 1 shows each message must transfer buffers within finite time (since a higher class buffer will become available), but not always to a buffer of higher class if path switching occurs.

4) A message cannot stay forever in one node (via internal buffer transfers); it must leave the node within finite time. This holds because C8) ensures its waiting set eventually includes buffers (on guaranteed paths) in other nodes, and part 3) of this proof shows these will become available within a finite time. Then C2), C3), and C4) ensure that one of these buffers will receive the message in finite time.

5) A given message will travel only in buffers, each of which has a guaranteed path in the BG (from themselves to the destination) which satisfies the routing constraints. Part 4 shows that only a finite time is spent in each node and a5) limits the number of nodes visited. Hence the message must reach its destination in finite time.

Remark: If path switching is permitted, as above, into buffers lying on a guaranteed path to the destination, but relaxing the requirement that the original routing algorithm be satisfied, then new routes can be traversed but deadlock-free operation is still maintained. However, in this case, infinite looping is possible and additional precautions should be taken to ensure message delivery in a finite time.

C. The Common Buffer Pool

Via path switching, the buffers lacking arcs in the BG can be used more flexibly than as described in Section III-A. These buffers can be organized as a *common buffer pool* which can be used *arbitrarily*. Conceptually, this is done by adding an arc from each buffer in the common buffer pool to every buffer of the same node belonging to the original buffer graph and to every one of their immediate successors in the BG. These arcs ensure that from every buffer in the buffer pool, the message can be continued along any of the paths in the BG which pass through this node. Therefore, using the path switching mechanism, any buffer in the buffer pool can host any of the messages traveling through the node. Adding the common buffer pool to the BG maintains the loop-free property because those buffers have only exiting arcs.

Buffer availability is improved by reducing the number of buffers belonging only to special paths in the BG, and shifting these buffers into the common buffer pool. Section IV illustrates schemes for constructing BG's with a nearly minimal number of buffers at each node. This permits either shifting buffers into the buffer pool, or system design with small buffer requirements.

It is desirable for a node to move messages from the common buffer pool into the nonbuffer-pool buffers whenever possible, because this will increase the number of empty buffers which can be used arbitrarily. This shifting of messages is called "reclassification" and may be efficiently implemented via the logical buffer scheme of Section V.

D. Discussion

Design Freedom: Great flexibility in implementation exists because of design freedom in constructing the buffer graph (see Section IV), selecting among guaranteed paths, and selecting among possible path switchings.

Multiple Interpretations: The buffer graph concept requires only that the buffers and their allowed connections be known, and does not require that the buffers at each node be explicitly labeled in some special manner. Since buffers can be labeled in a variety of ways, multiple explanations can be given for how they are being used. For example, the *hops-so-far* scheme described in Example 2.2 has a BG which permits an i-hop message to travel either in buffers $B0 \rightarrow B1 \rightarrow B2 \rightarrow \cdots \rightarrow Bi$ or in, say, buffers $B(K-i) \rightarrow B(K-i+1) \rightarrow \cdots \rightarrow B(K-1) \rightarrow BK$. If one now renames the buffers so that Bj becomes $B(K-j)$, then the *same* use of the buffer graph may be newly interpreted as *remaining hops*, because the message now travels in buffers $BK \rightarrow B(K-1) \rightarrow B(K-2) \rightarrow \cdots \rightarrow B(K-i)$ or in buffers $Bi \rightarrow B(i-1) \rightarrow \cdots \rightarrow B1 \rightarrow B0$. Although an explanation (or interpretation) of the BG is unnecessary, it is sometimes convenient because it permits a straightforward implementation.

Reversed Routings: If the reversal of every permitted route is also a permitted route, then reversing the arc directions in a given BG produces another BG (e.g., the hops-so-far scheme can be converted into the hops-remaining scheme.)

Internode Transfer Implementation: Link-level protocols are described in [12] which implement the proposed deadlock avoidance method. These protocols can be implemented in a way which guarantees FIFO for all messages on the same fixed route.

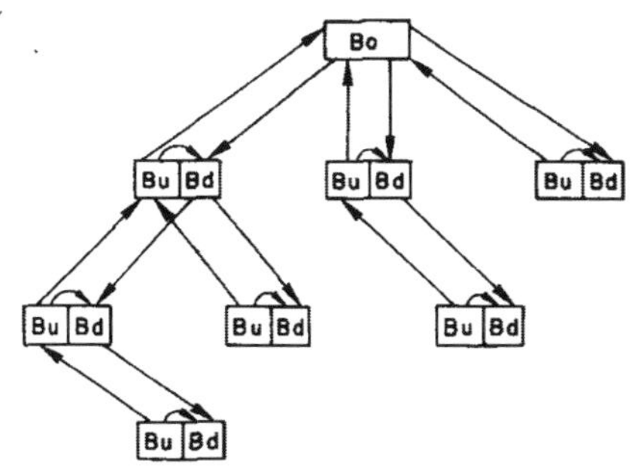

Fig. 5. BG for tree networks.

IV. SCHEMES FOR CONSTRUCTING BUFFER GRAPHS

This section describes several families of buffer graphs which are practical both regarding ease of implementation and having small buffer requirements. In [7], hop-type schemes which lead to practical buffer graphs are described. Below we generalize some of these schemes and describe new ones. Some of the schemes which follow exploit the network topology explicitly. Buffer graphs may also be constructed exploiting the specific given set of routes [13], but these may be more difficult to implement.

A. The Tree Scheme

Suppose a network with a tree topology is given, and the routing rules forbid a message to make repeated visits to any node. It is known [7] that deadlock-free operation in such a tree can be achieved using two buffers at each node. Let one node be arbitrarily designated as the root, and suppose "up" means toward the root while "down" means away from it. At each node two buffers *Bu* and *Bd*, are allocated for traffic going up or down, respectively. Because no node can be visited more than once, when a message starts going down it cannot turn up. On the other hand, a message going up may at certain nodes turn down to a different branch of the tree. Therefore, for each node, *Bd* is connected to the *Bd*'s of its sons, and *Bu* is connected to both *Bu* of its parent and *Bd* of the same node. Fig. 5 shows the BG for such a tree.

Note that only one buffer, named $B0$, is needed at the root node. On the other hand, in the special case in which messages never turn down after starting up the tree, and if the root node has two buffers *Bu* and *Bd*, then only *one* buffer *BL* is necessary at each leaf node (i.e., nodes without "sons"). This buffer is used for both down-moving messages, and, when empty, for accepting external traffic to the leaf node.

The BG of Fig. 5 will be used in following sections, as a basic building block of loop-free BG's for networks with more complex topologies.

B. The Mesh and Tree Scheme

In some practical cases [14], two conditions may hold: 1) the network decomposes into a mesh plus a collection of trees $T1$, $T2$, $\cdots$, such that each tree joins the mesh at one node (the node common to the mesh and tree is called the root), and any two trees are disjoint; 2) the routing of each message does not permit repeated visits to a node in a tree, and permits only a finite number of total visits to mesh nodes. In such cases, a BG can be constructed by connecting several independent BG's as follows.

1) A buffer graph BG0 is constructed for the mesh part (including the roots) in which the trees are disregarded (i.e., routes terminating (originating) in a tree are treated as terminating (originating) in the tree root). *Any* scheme can be used to construct BG0, provided it accommodates, in a loop-free manner, all of the desired routes within the mesh.

2) For each tree Ti, a buffer graph BGi is constructed as shown in Fig. 5, with the root node chosen to be the node in the mesh.

3) Each BGi is connected to BG0 as follows:

a) Let Su (Sd) denote the set of up (down) buffers in BGi at all nodes which are "sons" of the root node. Let So denote the set of buffers in BG0 at the root node for Ti.

b) in BGi, delete the root buffer (node) and all arcs attached to it.

c) Add an arc from each buffer in Su to each buffer in So. Add an arc from each buffer in So to each buffer in Sd.

The combined BG is loop-free because each buffer graph BG0, BG1, $\cdots$ is loop-free, the new arcs add loop-free paths from buffers Bu to buffers Bd (via buffers in BG0) but not the opposite, and a path from Bu to Bd does not close loops. The combined BG accommodates all permitted routes.

For illustration, if the mesh nodes are managed by the hop scheme of Example 2.2, then each mesh node has buffers $\langle B0, B1, \cdots Bq \rangle$ where q is the maximum number of hops within the mesh. The nodes in each tree, excluding the root, have buffers and arcs as in Fig. 5. The connections between the tree buffers and the mesh buffers are shown in Fig. 6.

Because the topology is exploited, the use of such mixed schemes reduces the buffer requirements. Such mixed schemes can be created not only for joining trees and mesh, but also for many other particular cases, as shown in Section IV-E.

C. The "Loop Breaking" Scheme

This scheme for the construction of loop-free buffer graphs can be applied to any network topology. It is especially attractive for cases in which there is a small set of nodes whose removal will leave the remaining network loop-free.

Given a network, suppose that a set of nodes $G = \langle G1, G2, \cdots, GL \rangle$ is marked. These nodes are chosen such that if all of them are removed along with all communication channels connected to them, then the remaining network has no loops. (It is immaterial whether the remaining network is connected.) The remainder of the network will consist of one or more unconnected trees, in each of which at most two buffers per node are enough to avoid deadlock (see Section IV-A) for messages moving exclusively *within* the tree.

The proposed buffer graph is defined as follows. We assume that no route of interest has repeated visits to any node, and that $q \leq L$ denotes the maximal number of nodes in G which can be visited on any route. Each node in G is allocated exactly q buffers labeled $\langle B1, B2, \cdots, Bq \rangle$ where Bi is reserved for messages which have made exactly i visits to G so far. Each of the trees is assigned a root node arbitrarily. Each

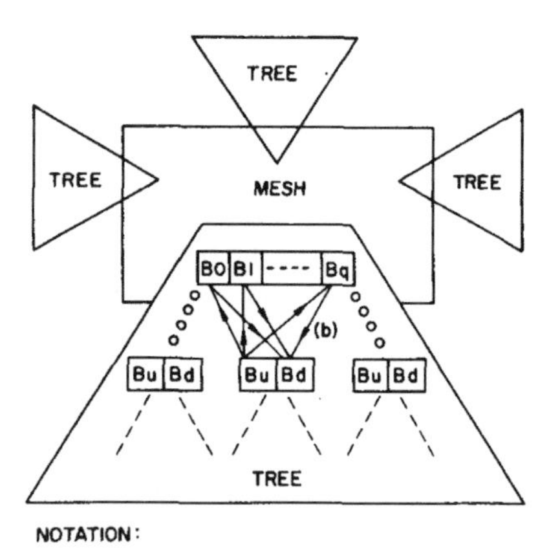

Fig. 6. BG for mesh-and-trees networks.

nonroot node in a tree is allocated $2(q + 1)$ buffers labeled $\langle Bu0, Bu1, \cdots, Buq, Bd0, Bd1, \cdots, Bdq \rangle$ where buffer Bui (Bdi) is reserved for messages currently heading up (down) the tree and which have already made i visits to nodes in G. The root node in a tree has $q + 1$ buffers labeled $\langle Bud0, Bud1, \cdots, Budq \rangle$ where buffer $Budi$ is reserved for messages which have already made i visits to G and which can be headed either up (i.e., out of) the tree or down the tree.

Fig. 7 schematically illustrates the buffer graph. The bottom plane holds the buffers for nodes in the trees which are reserved for messages which have not yet visited G: $\langle Bu0, Bd0, Bud0 \rangle$. The first visit to G is represented by the second plane, which holds buffers $B1$ for nodes in G. The third plane is essentially a replication of the first plane, and holds buffers $\langle Bu1, Bd1, Bud1 \rangle$ for messages in the tree nodes which have made exactly one visit to G. In general, copies of the first two planes are alternated, ending with a replication $\langle Buq, Bdq, Budq \rangle$ of the first plane.

Buffers in each plane for tree nodes have directed arcs only to buffers in the same plane or to the next higher plane of buffers for G nodes. The G buffers have arcs to the next higher plane for tree nodes and/or to the next level of plane of buffers for G nodes. Such a structure is loop-free because each plane is loop-free and the planes are connected only in one direction.

This scheme requires q buffers per node in G, $q + 1$ buffers per root node in a tree, and $2q + 2$ buffers per nonroot node in a tree. If q is small, the total number of buffers can be significantly smaller than the $K + 1$ buffers per node (K = maximum number of hops for any route, any given set of routes) required for the hop number scheme shown in Example 2.2. For illustration, Fig. 8 shows an example of a network with 9 nodes where a loop-free path exists which visits all the nodes; hence, $K = 8$. The hop number scheme requires 9 buffers per node or 81 buffers total. By taking $q = 2$ and $G = \langle G1, G2 \rangle$ as indicated, the removal of G leaves one tree consisting of $\langle A1, A2, \cdots, A7 \rangle$. The loop-breaking scheme requires 2 buffers each at $G1$ and $G2$, 3 buffers at the tree root, and 6 buffers at each of the 6 nonroot tree nodes. The total is 43 buffers contrasted with 81 for the hop number scheme.

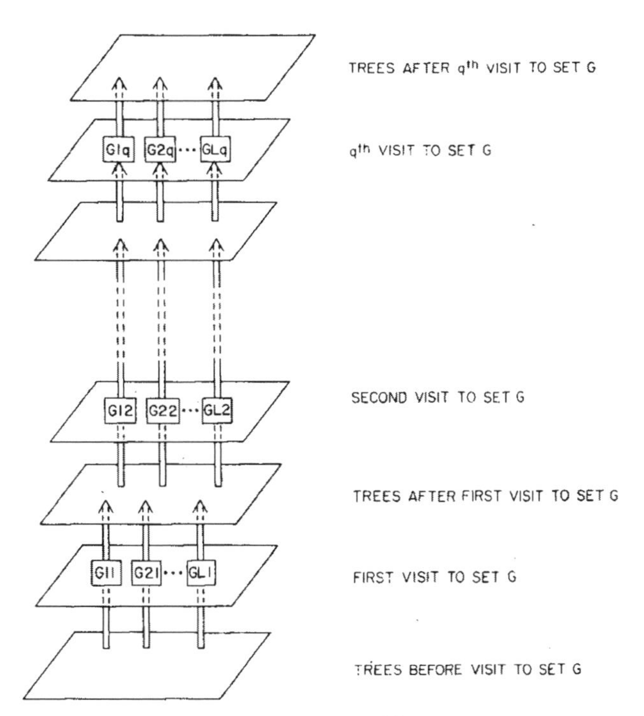

Fig. 7. BG for loop-breaking scheme.

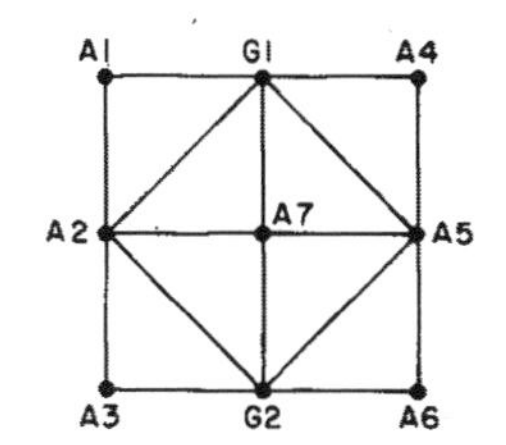

Fig. 8. Network illustrating loop-breaking scheme.

Another example consists of a ring of $R > 2$ nodes, with any routes permitted which do not visit a node more than once. The hop-number scheme requires R buffers per node, while the loop-breaking scheme with $q = L = 1$ requires 1 buffer at node $G1$, 2 buffers at the root node of the tree, and 4 buffers at each of the $R - 2$ nonroot tree nodes. For large rings, this is significantly better than the hop-number scheme.

D. The "Valley Counting" Scheme

This scheme can be applied to construct a buffer graph for any network topology. This scheme is based on valleys in the encountered node numbers, and requires half as many buffers as the "peaks and valleys" deadlock avoidance method given in [7]. A "peaks" scheme can be developed which completely parallels the "valley" scheme to be described below.

Given any N-node network, assign distinct static numbers $\langle n1, n2, n3, \cdots, nN\rangle$ to the nodes. Any static or adaptive routing algorithm is permitted which ensures message delivery within a bounded number of hops, i.e., a bounded number of repeated visits to any node is allowed.

A route which visits three consecutive nodes, say $a \to b \to c$, is said to have a *valley* at node b if nodes a and c have higher node numbers than node b. Since all messages depart after a bounded number of hops, there exists an integer $K \geqslant 0$ such that no route has more than K valleys. A loop-free buffer graph is constructed with $2K + 2$ buffers per node (the method in [7] requires $4K + 2$ buffers per node). The buffers at node are named $\langle u0, u1, \cdots, uK, d0, d1, \cdots, dK\rangle$ where the buffers up (down) will be occupied by messages which have just undergone an increase (decrease) in node number, or will undergo such an increase (decrease) when leaving this node.

Let ns and nt be two nodes connected by a communications channel; take $nt > ns$ without loss of generality. In the BG, the arcs between buffers are as follows (see Fig. 9).

1) At each node, buffer ui $(0 \leqslant i \leqslant K)$ is connected to buffer di at the same node.

2) At each node, buffer di $(0 \leqslant i \leqslant K - 1)$ is connected to buffer $u(i + 1)$ at the same node.

3) At node ns, buffer ui $(0 \leqslant i \leqslant K)$ is connected to buffer ui at node nt.

4) At node nt, buffer di $(0 \leqslant i \leqslant K)$ is connected to buffer di of node ns.

The BG is loop-free because of the following.

1) For fixed i, a path in the BG consisting exclusively of ui buffers involves monotonically increasing node numbers; hence, it cannot form a loop.

2) Similarly, for fixed i, a path in the BG consisting exclusively of di buffers cannot form a loop.

3) As shown in Fig. 9, the BG consists of alternating layers of buffers $\langle u0\rangle$, $\langle do\rangle$, $\langle u1\rangle$, $\langle d1\rangle$, $\cdots$, $\langle uK\rangle$, $\langle dK\rangle$ where each layer is itself loop-free and is connected only to the next higher layer. This prevents a loop involving multiple layers.

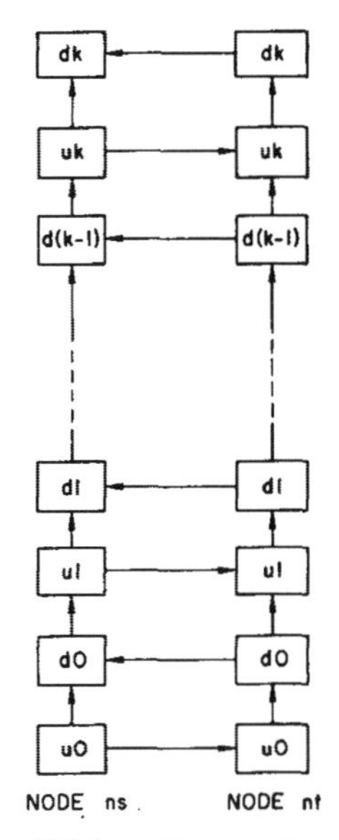

Fig. 9. BG for valley-counting scheme.

It remains to be shown that any route in the physical network has a corresponding path in the BG. Let the sequence of node numbers $m1$, $m2$, $m3$, $\cdots$, $m9$ denote any arbitrary route. The corresponding path in the buffer graph starts in buffer $u0$ of node $m1$, and continues in buffer $u0$ of successive nodes as long as the node numbers are increasing. If a node mp is encountered whose successor has a smaller node number, i.e., node mp is the first "peak" in the route, then the path shifts from buffer $u0$ to buffer $d0$ within node mp, and then to buffer $d0$ of node $m(p + 1)$. The path continues through $d0$ buffers of successive nodes as long as the node numbers are decreasing. If the path encounters a node mv whose successor node has a higher node number, i.e., node mv is a valley, then the path shifts from buffer $d0$ to buffer $u1$ within node mv, and continues to buffer $u1$ of the successor node $m(v + 1)$.

The path continues similarly to the end, with up-steps taken in buffers uj, with switches at a peak node from buffer uj to buffer dj, with down-steps taken in buffers dj, and with switches at a valley node from buffer dj to $u(j + 1)$. Since at most K valleys are encountered, each node requires at most $K + 1$ each of d-buffers and u-buffers.

E. Interconnection of Subnetworks

A hierarchical scheme may be used in many practical situations to generate efficient new buffer graphs from old ones. See also [7, Theorem 5].

Consider a network of N nodes, communication channels between some pairs of nodes, and suppose that a finite set R of permitted routes is given. Suppose that the nodes of the network are partitioned into $m < N$ nonempty subnetworks $s1$, $s2$, $\cdots$, sm. That is, the network may be abstractly viewed as consisting of "aggregate nodes" $s1$, $s2$, $\cdots$, sm with a "communication channel" between aggregate nodes si and sj if and only if a communication channel exists between some pair of the original nodes in si and sj.

Let $R1$ denote the set of all permitted routes *among* the subnetworks, i.e., a route in $R1$ is deduced from the corresponding route in R by identifying the sequence of subnetworks visited by a message following the route in R. Let $R2i$, $1 \leqslant i \leqslant m$, denote the set of routes *within* subnetwork si which are used by R, i.e., any list of consecutive nodes in a route of R, all of which lie in si, belongs to $R2i$.

The buffer graph BG for the given network can be constructed in the following three steps. First, construct a loop-free buffer graph BG1 which accommodates all routes in $R1$ for the abstract network of aggregate nodes. Second, construct BG as the collection, over all buffers b of BG1, of buffer graphs BG(b) defined next. Each BG(b), where b is at, say, aggregate node si, is a loop-free buffer graph which accommodates all the routes of $R2i$. Third, if in BG1, buffer b is connected to buffer $b' \neq b$, then in BG, an arc is added from a buffer in BG(b) to a buffer in BG(b') whenever those buffers lay in the same node or in nodes connected by a communications channel.

The combined BG accommodates all routes R because any route in R is a concatenation of routes of $R2i$, $1 \leqslant i \leqslant m$, according to one of the sequences of $R1$. The combined BG is also loop-free because each BG(b) is loop-free, and since BG1 is loop-free, the arcs added in the third step cannot close loops.

The mesh and tree scheme of Section IV-B can be constructed by an interconnection of subnetworks. In this case, the mesh, and each tree node (excluding the root node), are considered as subnetworks. Here BG1 will be as in Fig. 5, each BG(b) will be a single buffer, and BG will have the form of Fig. 6. The loop-breaking scheme of Section IV-C may also be interpreted as an interconnection of subnetworks. In this case, the nodes of the set G are taken as distinct subnetworks, and all of the remaining nodes are taken as a single subnetwork. Here BG1 agrees with Fig. 7 provided each plane labeled "trees" is regarded as a single buffer.

A simple deadlock-free connection of several subnetworks can be accomplished by providing buffers at each node with a two-part index (s, t) where s (for BG1) indicates, for instance, the valleys-so-far in subnetwork number, and t [for BG(b)] indicates, for instance, the hops-so-far within the current subnetwork.

The interconnection of subnetworks can be recursively applied, in the sense that while constructing BG(b) for some subnetwork, this subnetwork can itself be decomposed into subnetworks, applying the method again to each subnetwork. In other words, the original network can be hierarchically decomposed into subnetworks. Conversely, any given set of deadlock-free networks can be combined into a deadlock-free supernetwork while preserving its internal buffer scheme [the SB(b)] by replicating it as many times as BG1 demands.

V. LOGICAL BUFFERS

The schemes in previous sections assume that real buffers are actually earmarked (reserved) to match the buffers in the buffer graph, and that a message fits into one real buffer. Here we describe a way of managing logical buffers at each node so that real buffers are not physically earmarked, and so that multibuffer messages can be accommodated. The use of counters to manage logical buffers also permits rapid reclassification of messages from one buffer type to another, thereby improving performance within the network.

Suppose a loop-free BG is given which accommodates all the desired routes. For a given node in the network, let $\langle c1, c2, \cdots, ct\rangle$ denote the buffers of the BG which have restricted usage, i.e., are not in the common buffer pool for the BG. These will be called *logical buffers* of class $c1, c2, \cdots, ct$. Let $M \geqslant 1$ denote the maximal number of physical buffers that a message may occupy at this node.

If the node has B (equal-size) *physical buffers,* then at different times different physical buffers can be considered of class ci. The buffer management policy which avoids deadlock is to ensure that, *at all times, for each i, there are at least M physical buffers which can be considered to be logical buffer ci.* To meet this requirement, at least $B \geqslant M \cdot t$ physical buffers must be provided at the node. Furthermore, if TCi denotes the number of physical buffers currently containing messages assigned to logical buffer ci, then $RCi = \max(M - TCi; 0)$ empty physical buffers must be reserved for messages of logical class ci. These buffers are kept empty until either arrival of additional messages of class ci or reclassification of a current message into class ci.

Because of the possibilities of path switching and of multiple guaranteed paths, a message at a node might conceivably fit in any of several logical buffers. However, each message at a node must be actually associated with exactly *one* logical buffer ci, and *all* physical buffers actually occupied by this message will also be considered to be currently of class ci until this message departs or is reclassified. Consequently, $TCi = \Sigma \{xj | mj$ is assigned to logical buffer class $ci\}$ where $\langle m1, m2, \cdots, mp\rangle$ denotes the messages in the node and $\langle x1, x2, \cdots, xp\rangle$ denotes the number of physical buffers these messages occupy.

The total number of reserved empty buffers at the node is

$$R = \sum_{i=1}^{t} RCi.$$

If E denotes the actual number of empty physical buffers at the node, then in order to guarantee that the network is deadlock-free, we must ensure that $R \leqslant E$ holds at all times.

This implies that the node could now accept messages of a given logical class ci, $1 \leqslant i \leqslant t$, which occupies no more than $Ji = E - R + RCi$ physical buffers. Similarly, it could now accept messages which occupy up to $J0 = E - R + \min\{RCi \mid 1 \leqslant i \leqslant t\}$ physical buffers, irrespective of their logical classes.

A polling scheme which implements the deadlock avoidance method is described below. We assume also that the node is not allowed to discard incoming messages, and therefore it can issue a poll only when it can commit buffer space for the messages which can be received in response. A poll will be responded to within finite time, either with messages or with a null response if there is no message to send. For simplicity, we assume that a new poll can be sent only after the previous poll has been responded to, but the same ideas can be generalized to accommodate multiple outstanding polls. We assume also that the length of a message is not known by the polling node until the message is received.

When $J0/M$ is larger than the maximum number of messages which can be sent in response to a poll, then "send everything" polling is used. (Hopefully, this situation occurs most of the time.) If not, but $J0/M \geqslant 1$, then the node can issue a poll for at most $[J0/M]$ messages of arbitrary classes. If $J0 < M$, the node must resort to selective polling for a single message of any class ci such that $Ji \geqslant M$. Whenever every $Ji < M$, the node will suspend polling. For nonpolling implementation, see [12].

The buffer encounters TCi are incremented or decremented upon message arrivals, departures, or reclassifications, with recomputation of RCi, Ji, and R. As described in Section III-C, the main goal of reclassification is to increase the threshold for selective polling by increasing $J0$ (or decreasing R). Although this section has not explicitly treated the common buffer pool, all of its properties have been preserved in the implementation. The buffer counter scheme has the additional advantage of automatically putting as much as possible of an incoming (or already-present) class ci message into the reserved empty buffers RCi, and then putting the remainder of the message, if any, into the buffer pool. This is a simple form of reclassification which postpones the threshold for selective polling.

VI. Conclusions

The schemes for constructing buffer graphs, in conjunction with the different possibilities of using them for the actual forwarding of the messages, provide a large number of alternative ways of applying to actual networks the deadlock avoidance ideas presented in this paper. For conventional networks, many of these alternatives are a viable, practical solution to the store-and-forward-deadlock problem. The examples throughout the paper demonstrate that the buffer requirements for typical buffer graphs are modest, and that low overhead implementations can be designed.

References

[1] L. Kleinrock, "ARPANET lessons," in *Conf. Rec., Int. Conf. Commun.*, Philadelphia, PA, June 14–16, 1976.

[2] D. W. Davies and D. L. A. Barker, *Communication Networks for Computers*. New York: Wiley, 1973.

[3] R. E. Kahn and W. R. Crowther, "Flow control in a resource-sharing computer network," *IEEE Trans. Commun.*, vol. COM-20, pp. 539–546, 1972.

[4] P. M. Merlin and P. J. Schweitzer, "Deadlock avoidance in store-and-forward networks—II: Other deadlock types," this issue, pp. 355–360.

[5] E. G. Coffman, Jr., M. J. Elphick, and A. Shoshani, "System deadlocks," *ACM Comput. Surveys*, vol. 3, pp. 67–78, 1971.

[6] R. C. Holt, "Some deadlock properties of computer systems," *ACM Comput. Surveys*, vol. 4, pp. 179–196, 1972.

[7] K. D. Gunther, "Prevention of buffer deadlocks in packet-switching networks," rep. presented at the IFIP–IIASA Workshop on Data Commun., Laxenburg, Austria, Sept. 15–19, 1975.

[8] E. Raubold and J. Haenle, "A method of deadlock-free resource allocation and flow control in packet networks," in *Proc. 3rd Int. Conf. Comput. Commun.*, Toronto, Ont., Canada, Aug. 3–6, 1976.

[9] W. L. Price and J. D. Haenle, "Some comments on simulated datagram store-and-forward networks," *Comput. Networks*, vol. 2, pp. 70–73, 1978.

[10] A. Giessler, J. Hanle, A. Konig, and E. Dade, "Free buffer allocation—An investigation by simulation," *Comput. Networks*, vol. 2, pp. 191–208, 1978.

[11] R. C. Chen, "Bus communication systems," Ph.D. dissertation, Dep. Comput. Sci., Carnegie-Mellon Univ., Pittsburgh, PA, Jan. 1974, NTIS-PB-235 897.

[12] P. M. Merlin and P. J. Schweitzer, "Deadlock avoidance in store-and-forward networks I: Store-and-forward deadlock," IBM T. J. Watson Res. Cen., Yorktown Heights, NY, Rep. RC-6624, July 1977.

[13] B. Gavish, P. M. Merlin, and P. J. Schweitzer, "Minimal buffer requirements for deadlock avoidance in store-and-forward networks," *IBM J. Res. Develop.*, to be published.

[14] Program Product, "Introduction to advanced communication function. Multiple system data communication networks," IBM Corp., Dep. E01, P. O. Box 12195, Research Triangle Park, NC 27709, GC30-3033-0, Oct. 1976.

PHOTO NOT AVAILABLE

Philip M. Merlin (S'74–M'76) received the B.S. (cum laude) and M.S. degrees in electrical engineering from the Technion—Israel Institute of Technology, Haifa, and the Ph.D. degree in information and computer science from the University of California, Irvine, in 1971, 1973, and 1974, respectively.

From 1971 to 1973 he led several projects on digital systems and computer communications at the Technion. During 1974 he was employed by the University of California, Irvine; from 1974 to 1977 he was with the IBM T. J. Watson Research Center; and from 1977 to 1979 he was a faculty member with the Department of Electrical Engineering, Technion. Since 1974 he had been engaged in research activities on distributed computer systems, computer networks, communication protocols, Petri nets, and recoverability. He is now deceased.

Dr. Merlin was several times a Visiting Faculty Member at the IBM T. J. Watson Research Center; a British Science Research Council Senior Visiting Fellow at the University of Newcastle-upon-Tyne; and a consultant with Intel. He was a member of the Association for Computing Machinery and the IEEE Computer Society.

PHOTO NOT AVAILABLE

Paul J. Schweitzer was born in New York, NY, on January 16, 1941. He received B.Sc. degrees in physics and mathematics from M.I.T., Cambridge, in 1961, and the Sc.D. degree in physics from M.I.T. in 1965.

He was a member of the Institute for Defense Analyses, Arlington, VA, from 1965 to 1970, a Visiting Associate Professor of Operations Research in the Technion—Israel Institute of Technology, Haifa, from 1970 to 1972, and a Research Staff member in the Computer Science Department, IBM Watson Research Center, Yorktown Heights, NY, from 1972 until 1977. He has been a Professor in the Graduate School of Management at the University of Rochester, Rochester, NY, since 1977. His research interests include performance evaluation of telecommunications networks, computer protocols, queueing networks, optimization algorithms, and Markovian decision processes.

A Minimum Delay Routing Algorithm Using Distributed Computation

ROBERT G. GALLAGER, FELLOW, IEEE

Abstract—An algorithm is defined for establishing routing tables in the individual nodes of a data network. The routing table at a node i specifies, for each other node j, what fraction of the traffic destined for node j should leave node i on each of the links emanating from node i. The algorithm is applied independently at each node and successively updates the routing table at that node based on information communicated between adjacent nodes about the marginal delay to each destination. For stationary input traffic statistics, the average delay per message through the network converges, with successive updates of the routing tables, to the minimum average delay over all routing assignments. The algorithm has the additional property that the traffic to each destination is guaranteed to be loop free at each iteration of the algorithm. In addition, a new global convergence theorem for non-continuous iteration algorithms is developed.

Manuscript received March 16, 1976; revised September 15, 1976. This work was supported in part by the Advanced Research Projects Agency of the Department of Defense under Grant N00014-75-C-1183, in part by the National Science Foundation under Grant NSF-ENG75-14103, and in part by Codex Corporation, Newton, MA 02195. This paper was presented at the International Conference on Communications, Philadelphia, PA, June 14-16, 1976.

The author is with the Department of Electrical Engineering and Computer Science and the Electronic Systems Laboratory/Research Laboratory for Electronics, Massachusetts Institute of Technology, Cambridge, MA 02139.

INTRODUCTION

THE problem of routing assignments has been one of the most intensively studied areas in the field of data networks in recent years. These routing problems can be roughly classified as static routing, quasi-static routing, and dynamic routing. Static routing can be typified by the following type of problem. One wishes to establish a new data network and makes various assumptions about the node locations, the link locations, and the capacities of the links. Given the traffic between each source and destination, one can calculate the traffic on each link as a function of the routing of the traffic. If one approximates the queueing delays on each link as a function of the link traffic, one can calculate the expected delay per message in the network. The problem then is to choose routes in such a way as to minimize expected delay. This is a multicommodity flow problem, and the reader is referred to Cantor and Gerla [1] for a particularly elegant algorithm and for other references.

Quasi-static routing problems can be typified by the following situation. A data network is in operation, but over

Reprinted from *IEEE Transactions on Communications,* vol. COM-25, no. 1, January 1977.

time, new source-receiver pairs establish data transmission sessions and old sessions are terminated. It is necessary at the very least to establish routes for these new sessions and it might in addition be desirable to occasionally change routes for established sessions or to change the fraction of the traffic for a session that takes different routes. Over a longer range time scale, links or nodes fail, new links and nodes are added, and routings must be changed accordingly. The usual approach to this problem is to have a special node in the network that makes all decisions about routings. In principle such a node periodically gets information from all the other nodes about traffic requirements and uses this information to solve the current static routing problem. Such a strategy seems simple and straightforward, but in fact it is not. First there is the need for protocols for the nodes in the network to send updating information to the control node. Similarly protocols are required for the control node to send its routing decisions to the other nodes. There is also a serious problem about what to do when nodes or links in the network fail. The routes by which notification of such catastrophes are sent to the control node might in fact be destroyed by the catastrophe. Finally, there is the possibility that a failure of the control node may cause the whole network to fail. The point of this is not that central node routing is unworkable, but rather to convince the reader that the problems of communicating information about routing through a network is conceptually as difficult as making routing decisions once all the information is available.

Finally, dynamic routing refers to the kinds of problems that arise in a network when messages or packets are routed according to the instantaneous states of the queues at the links of the network. The routing of a particular message or packet is not determined when it enters the network; instead, each node that receives the message selects the next node to which the message is routed on its path to the destination. Here, in addition to the problem of determining an algorithm to make these decisions, there is also the problem of conveying information about queue lengths through the network and the problem of coping with lost messages and messages which arrive out of order at the destination node.

Our major interest here is in distributed algorithms for quasi-static routing, i.e., in algorithms in which each node constructs its own routing tables based on periodic updating information from neighboring nodes. We first develop a number of theoretical results that should be applicable to any such algorithm and then we develop a particular algorithm. The analysis is based on a static model with stationary traffic inputs and an unchanging network. We show that the average delay per message converges under these conditions to the minimum over all routing assignments. We have not addressed the problem of how well the algorithm adapts to variations in the input traffic or the network. Qualitatively, an algorithm's ability to adapt to variations is intimately connected with its speed of convergence in the static case and with its robustness. We feel that distributed algorithms have important advantages in both these areas. A distributed algorithm can react rapidly to a local disturbance at the point of the disturbance with slower "fine tuning" in the rest of the network. The robustness comes from lack of reliance on a central node that might fail and from avoiding the "chicken and egg" problem of centralized routing where one needs routes to transmit the routing information required to establish routes.

The algorithm here is quite similar to the algorithm used in the Advanced Research Projects Agency Network (ARPANET) [2]. The major difference is that the ARPANET attempts to send each packet over a route that minimizes that packet's delay with no regard to other packet's delays, whereas here packets are sent over routes to minimize the overall delay of all messages. This difference between "user optimization" and "system optimization" was evidently first noticed by Pigou [3], later used by Dafermos and Sparrow [4], and then by Agnew [5], [6]. Angew analyzed a network with a single source and destination and described an algorithm very similar to that described here. Kahn and Crowther [7] also developed a distributed algorithm which meters traffic so as to change routes slowly in response to quasi-static variations. Stern [8] developed another distributed algorithm based on an electrical network analogy of a communication network. Finally our algorithm has similarities to the centralized flow deviation strategy of Fratta *et al.* [8]. Their algorithm was the first to effectively exploit the marginal change in network delay with a change in link flow, a notion which we also use extensively.

One important characteristic of the algorithm, not possessed by any other routing algorithm to our knowledge, is its property of being loop free at every iteration. Aside from reducing delay, it appears that loop freedom can be important in simplifying higher level protocols. In fact, the major reason for building loop freedom into the algorithm was to prevent a potential deadlock in the protocol for communicating update information between the nodes.

FORMULATION OF THE MODEL

Let the nodes of an n-node network be represented by the integers $1, 2, \cdots, n$ and let a link from node i to node k be represented by (i,k). Let L be the set of links, $L = \{(i,k)$: a link goes from i to $k\}$. In order to discuss traffic flow, we distinguish link (i,k) from (k,i), but assume that if one exists the other does also.

Let $r_i(j) \geqslant 0$ be the expected traffic, in bits/s, entering the network at node i and destined for node j (see Fig. 1). We assume that this input traffic forms an ergodic process such as, for example, a Poisson process of message arrivals with a geometric distribution on message lengths. Let $t_i(j)$ be the total expected traffic (or node flow) at node i destined for node j. Thus $t_i(j)$ includes both $r_i(j)$ and the traffic from other nodes that is routed through i for destination j. Finally let $\phi_{ik}(j)$ be the fraction of the node flow $t_i(j)$ that is routed over link (i,k). We take $\phi_{i,k}(j) = 0$ for $(i,k) \notin L$ (i.e., no traffic is routed over nonexistent links). We also take $\phi_{ik}(j) = 0$ for $i = j$ (i.e., traffic which has reached its destination is not sent back into the network). Since the node flow $t_i(j)$ at node i is the sum of the input traffic and the traffic routed to i from other nodes,

$$t_i(j) = r_i(j) + \sum_l t_l(j)\phi_{li}(j), \qquad \text{all } i,j. \tag{1}$$

Equation (1) implicitly expresses the conservation of flow

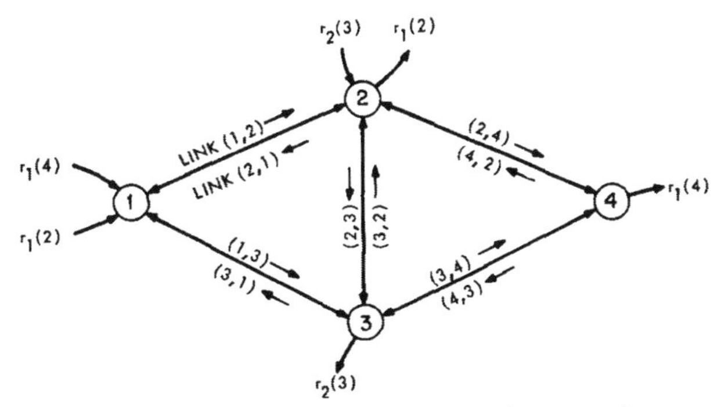

Fig. 1. Nodes, links, and inputs in a data network.

at each node; the expected traffic into a node for a given destination is equal to the expected traffic out of the node for that destination. Note that (1) deals with expected traffic and thus does not preclude the existence of traffic queues at the nodes.

Now let f_{ik} be the expected traffic, in bits/s, on link (i,k) (with $f_{ik} = 0$ if $(i,k) \notin L$). Since $t_i(j)\phi_{ik}(j)$ is the traffic destined for j on (i,k), we have

$$f_{ik} = \sum_j t_i(j)\phi_{ik}(j). \tag{2}$$

In what follows we refer to the set of expected inputs $\{r_i(j)\}$ as the *input* set r; the set of expected total node flows $\{t_i(j)\}$ as the *node flow* set t, the set of fractions $\{\phi_{ik}(j)\}$ as the *routing variable* set ϕ, and the set of expected link traffics $\{f_{ik}\}$ as the *link flow* set f. We have seen for an arbitrary strategy of routing (subject to the existence of the expectations $\{t_i(j)\}$ and the conservation of flow) that r, t, ϕ, and f all have meaning and satisfy (1) and (2). We are interested in distributed routing algorithms in which each node i chooses its own routing variables $\phi_{ik}(j)$ for each k, j. The question then arises whether the inputs r and the routing variables set ϕ uniquely specify t and f. Before answering this question, we define ϕ precisely, adding one additional constraint.

Definition: A routing variable set ϕ for an n-node network with links L is a set of nonnegative numbers $\phi_{ik}(j)$, $1 \leqslant i, k, j \leqslant n$, satisfying the following conditions.

1) $\phi_{ik}(j) = 0$ if $(i,k) \notin L$ or if $i = j$.

2) $\Sigma_k \phi_{ik}(j) = 1$.

3) For each i, j $(i \neq j)$ there is a routing path from i to j, which means there is a sequence of nodes $i, k, l, \cdots, m, j$ such that $\phi_{ik}(j) > 0, \phi_{kl}(j) > 0, \cdots, \phi_{mj}(j) > 0$.

Theorem 1: Let a network have input set r and routing variable set ϕ (according to the above definition). Then the set of equations (1) has a unique solution for t. Each component $t_i(j)$ is nonnegative and continuously differentiable as a function of r and ϕ.

This theorem is proved in Appendix A. It turns out that the constraint on the existence of routing paths in the definition of routing variables is necessary for this theorem. If this constraint were eliminated, one could still show, by the method in Appendix A, that $t_i(j)$ has a unique solution for each i, j for which a routing path from i to j exists. If no routing paths exist from some set of i to j, then there are two possible cases: 1) if no traffic for j comes into any of these nodes, either from inputs or from other nodes outside the set, then there are multiple solutions to (1); 2) otherwise there is no solution to (1). Physically, the first case above corresponds to a set of nodes which have no traffic for a given destination coming in or going out, but which might have some messages circulating around within the set. The second case corresponds to traffic coming into the set for a given destination, but none going out, leading to an infinite build-up of queues or lost traffic.

The more customary way to treat routing in a network is to regard it as a multicommodity flow problem (see, for example, Frank and Chou [9]). The traffic flow to each destination can be regarded as a commodity, and then (1) is equivalent to the multicommodity flow constraints. Our restrictions on the routing variables ϕ are somewhat more restrictive than the usual multicommodity flow constraints. In particular $\phi_{jk}(j) = 0$ prevents traffic at a destination j from looping back into the network, and the existence of routing paths prevents the isolated looping referred to in case 1) above.

We have seen that any routing policy, subject to the previously mentioned restrictions, leads to the sets t, ϕ, and f, and any distributed algorithm in which ϕ is selected by the individual nodes leads to a unique t, f. We now turn our attention to delay of messages in the network.

Let D_{ik} be the expected number of messages/s transmitted on link (i,k) times the expected delay/message (including queueing delays at the link input). Assume that D_{ik} is a function only of the link flow f_{ik}, i.e., that D_{ik} depends on the routing variables only through f_{ik}. We also make the assumption that messages are delayed only by the links of the network. This is reasonable if the processing time at an intermediate node is associated partly with the link on which the message arrives and partly with the link on which it departs.

It can now be seen with a little thought (or see Kleinrock [11]) that the total expected delay per message times the total expected number of message arrivals/s is given by

$$D_T = \sum_{i,k} D_{ik}(f_{ik}). \tag{3}$$

Since $f_{ik} = 0$ for $(i,k) \notin L$, we also take $D_{ik}(f_{ik}) = 0$ for $(i,k) \notin L$. Since the total message arrival rate is independent of the routing algorithm, we can minimize the expected delay/message on the network by minimizing D_T over all choices of routing variables (recall that f is a function of r and ϕ). The algorithm we describe subsequently will be an iterative algorithm for performing this minimization.

Before proceeding, however, we should point out some of the consequences of our assumption that D_{ik} is a function only of f_{ik}. Suppose that there are two paths from node i to j and that half the traffic is sent over each path, but the delay is greater on one path than the other. Then we could reduce the delay/message by sending the short messages over the small-delay path and the long messages over the long-delay path. Keeping the same traffic (in bits/s) on each path, we would have more messages on the short path than the long, and thus would reduce delay/message. The assumption that D_{ik} is a function only of f_{ik} restricts us from comparing such alternatives. Another consequence arises with dynamic routing, where one would hope to reduce the queueing delays on the links without reducing the long-term expected link flow. This, however, would change the functions $D_{ik}(f_{ik})$. Thus our assumption effectively masks the distinctions between dynamic and quasi-static routing (and for this reason makes the problem analytically tractable).

Kleinrock [11] showed that if queueing delays are the only nonnegligible source of delay in a network, and if each link traffic can be modeled as Poisson message arrivals with independent exponentially distributed lengths, then $D_{ik}(f_{ik}) = f_{ik}/(C_{ik} - f_{ik})$ where C_{ik} is the capacity of link (i,k). This formula has also been refined to account for overhead and propagation delays (Kleinrock [12]). For our purposes, it is immaterial what function D_{ik} is, although we shall make the reasonable assumption that D_{ik} is increasing and convex ∪ in f_{ik}. Before describing the algorithm, we develop necessary and sufficient conditions on ϕ to minimize D_T.

NECESSARY AND SUFFICIENT CONDITIONS FOR MINIMUM DELAY

First we calculate the partial derivatives of the total delay D_T with respect to the inputs r and the routing variables ϕ. Assume a small increment ϵ in the input $r_i(j)$. For each adjacent node k, an increment $\epsilon\phi_{ik}(j)$ of this new incoming traffic will flow over (i,k), and to first order, this will cause an incremental delay on that link of

$$\epsilon\phi_{ik}(j)D_{ik}'(f_{ik}), \qquad \text{where } D_{ik}'(f_{ik}) = \frac{dD_{ik}(f_{ik})}{df_{ik}}. \tag{4}$$

If node k is not the destination node, then the increment $\epsilon\phi_{ik}(j)$ of extra traffic at node k will cause the same increment in delay from node k onward as an increment $\epsilon\phi_{ik}(j)$ of new input traffic at node k. To first order this incremental delay will be $\epsilon\phi_{ik}(j)\partial D_T/\partial r_k(j)$. Summing over all adjacent nodes k, then, we find[1] that, for $i \neq j$,

$$\frac{\partial D_T}{\partial r_i(j)} = \sum_k \phi_{ik}(j)\left[D_{ik}'(f_{ik}) + \frac{\partial D_T}{\partial r_k(j)}\right]. \tag{5}$$

We take $\partial D_T/\partial r_j(j) = 0$ in this and subsequent equations and also take terms for which $(i,k) \notin L$ to be 0. Theorem 2, which follows, gives a rigorous justification of (5).

Next consider $\partial D_T/\partial\phi_{ik}(j)$. An increment ϵ in $\phi_{ik}(j)$ causes an increment $\epsilon t_i(j)$ in the portion of $t_i(j)$ flowing on link (i,k). If $k \neq j$, this causes an addition $\epsilon t_i(j)$ to the traffic at k destined for j. Thus for $(i,k) \in L$, $i \neq j$,

$$\frac{\partial D_T}{\partial\phi_{ik}(j)} = t_i(j)\left[D_{ik}'(f_{ik}) + \frac{\partial D_T}{\partial r_k(j)}\right]. \tag{6}$$

Theorem 2: Let a network have inputs r and routing variables ϕ, and let each marginal link delay $D_{ik}'(f_{ik})$ be continuous in f_{ik}, $(i,k) \in L$. Then the set of equations (5), $i \neq j$, has a unique (and correct) set of solutions for $\partial D_T/\partial r_i(j)$. Furthermore, (6) is valid and both $\partial D_T/\partial r_i(j)$ and $\partial D_T/\partial\phi_{ik}(j)$ for $i \neq j$, $(i,k) \in L$ are continuous in r and ϕ.

This theorem is proved in Appendix A. The appendix also gives explicit expressions for $\partial D_T/\partial r_i(j)$ and $\partial D_T/\partial\phi_{ik}(j)$, but it turns out that the implicit forms in (5) and (6) are needed in the algorithm to be presented.

One might now hope that all that is required to minimize D_T is to find a stationary point for D_T with respect to variations in ϕ. Using Lagrange multipliers for the constraint $\sum_k \phi_{ik}(j) = 1$, and taking into account the constraint $\phi_{ik}(j) \geq 0$, the necessary conditions for a minimum of D_T with respect to ϕ are, for all $i \neq j$, $(i,k) \in L$,

$$\frac{\partial D_T}{\partial\phi_{ik}(j)} \begin{cases} = \lambda_{ij}, & \phi_{ik}(j) > 0 \\ \geq \lambda_{ij}, & \phi_{ik}(j) = 0. \end{cases} \tag{7}$$

This states that for a given i, j, all links (i,k) for which $\phi_{ik}(j) > 0$ must have the same marginal delay $\partial D_T/\partial\phi_{ik}(j)$, and that this marginal delay must be less than or equal to $\partial D_T/\partial\phi_{ik}(j)$ for the links on which $\phi_{ik}(j) = 0$. Unfortunately, as Fig. 2 illustrates, (7) is not a sufficient condition to minimize D_T (i.e., D_T can have inflection points as a function of ϕ).

In Fig. 2, the only input traffic goes from node 1 to 4. It is easy to verify that (7) is satisfied at each node. The trouble is that the traffic at node 2, $t_2(4)$, is zero, which automatically satisfies (7); one does not get a better routing by decreasing $\phi_{2,3}(4)$, but one does move to a point, when $\phi_{2,3}(4) < 1/2$, where the routing can be improved by increasing $\phi_{1,2}(4)$. After studying this example, it is not difficult to hypothesize that (7) would be sufficient to minimize D_T if the factor $t_i(j)$ were removed from the condition.

Theorem 3: For each $(i,k) \in L$ assume that $D_{ik}(f_{ik})$ is convex ∪ and continuously differentiable for $0 \leq f_{ik} < C_{ik}$ where the capacity C_{ik} satisfies $0 < C_{ik} \leq \infty$. Let ψ be the set of ϕ for which the link flows satisfy $f_{ik} < C_{ik}$ for all $(i,k) \in L$. Then (7) is necessary for ϕ to minimize D_T over ψ and (8), for all $i \neq j$, $(i,k) \in L$ is sufficient.

$$D_{ik}'(f_{ik}) + \frac{\partial D_T}{\partial r_k(j)} \geq \frac{\partial D_T}{\partial r_i(j)}. \tag{8}$$

This theorem is proved in Appendix B. Note that the theorem does not assert the existence of a minimum; the conditions of the theorem do not even assert that ψ is nonempty. Note also that if we multiply both sides of (8) by $\phi_{ik}(j)$ and sum over k, then we see from (5) that (8) must be satisfied with equality for $\phi_{ik}(j) > 0$. Thus (8) is equivalent to

$$D_{ik}'(f_{ik}) + \frac{\partial D_T}{\partial r_k(j)} - \min_{m:(i,m)\in L}\left[D_{im}'(f_{im}) + \frac{\partial D_T}{\partial r_m(j)}\right] \geq 0 \tag{9}$$

for all $i \neq j$, $(i,k) \in L$ with equality for $\phi_{ik}(j)$ greater than 0.

THE ALGORITHM

The general structure of an algorithm to minimize D_T (assuming stationary traffic inputs) should now be clear. Each

[1] Agnew [5], [6] develops an equation similar to (5) but omits the final term $\partial D_T/\partial r_k(j)$; his algorithm, however, effectively includes the effect of this term.

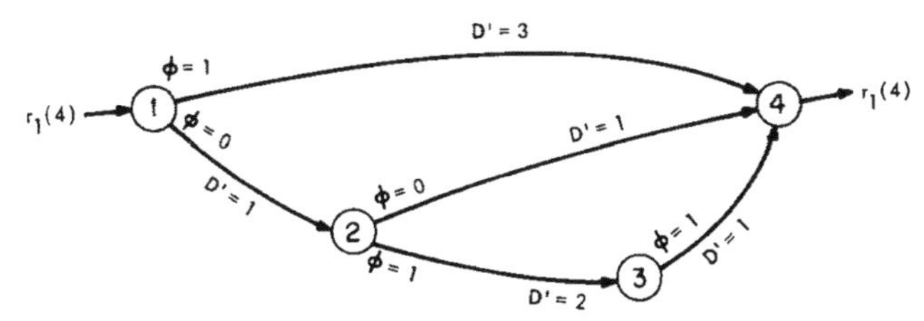

Fig. 2. Inflection point in $D_T(\phi)$.

node i must incrementally decrease those routing variables $\phi_{ik}(j)$ for which the marginal delay $D_{ik}'(f_{ik}) + \partial D_T/\partial r_k(j)$ is large, and increase those for which it is small. The algorithm breaks into two parts: a protocol between nodes to calculate the marginal delays and an algorithm for modifying the routing variables; we discuss the protocol part first.

Each node i can estimate, as a time average, the link traffic f_{ik} for each outgoing link. Thus with an appropriate formula for $D_{ik}(f_{ik})$, the node can also calculate $D_{ik}'(f_{ik})$. Since formulas for D_{ik} involve many assumptions which might be unwarranted, it might be preferable to estimate D_{ik}' directly; such estimation procedures are developed by Segall [13] and Bello [14].

In order to see how node i can calculate $\partial D_T/\partial r_j(k)$ for a neighboring node k, define node m to be *downstream* from node i (with respect to destination j) if there is a routing path from i to j passing through m (i.e., a path with positive routing variables on each link). Similarly, we define i as *upstream* from m if m is downstream from i. A routing variable set ϕ is *loop free* if for each destination j, there is no i, m $(i \neq m)$ such that i is both upstream and downstream from m. Note that if such an i, m pair existed, there would be a routing path from i to j that looped from i to m and back to i on its way to j. If ϕ is loop free, then for each destination j, the downstream (and the upstream) relation form a partial ordering on the set of nodes.

The protocol used for an update, now, is as follows: for each destination node j, each node i waits until it has received the value $\partial D_T/\partial r_k(j)$ from each of its downstream neighbors $k \neq j$ (i.e., nodes k with $\phi_{ik}(j) > 0$). The node i then calculates $\partial D_T/\partial r_i(j)$ from (5) (using the convention that $\partial D_T/\partial r_j(j) = 0$) and broadcasts this to all of its neighbors (except to the destination node j which has no need of the information). It is easy to see that this procedure is free of deadlocks (i.e., a node waiting forever for updating information from a downstream neighbor) if and only if ϕ is loop free. In fact, for a given j, the nodes can broadcast their values in any order consistent with the downstream partial ordering. For this reason we will be careful to ensure that the algorithm generates only loop free ϕ.

It can be seen that in an update, each link (i,k) must transmit $\partial D_T/\partial r_i(j)$ for each $j \neq i$, $j \neq k$. The same amount of updating information is used in the ARPANET strategy, but there delays rather than marginal delays are sent, and the transmissions are unordered so that many updates are required for changes to propagate through the network. Here, of course, changes propagate completely in one update, and the only inaccuracies come from inaccuracies in the estimates of the link marginal delays. One might object to sending each value $\partial D_T/\partial r_i(j)$ separately on a link, and indeed the inefficiency would be high if each such number required an individual packet. However, the routing update information could easily be piggy-backed on other packets, requiring very little overhead. One might also object to the time required for the updating to propagate through the network, but speed is relatively unimportant in a quasi-static algorithm.

We shall later define one small but important detail that has been omitted so far in the updating protocol between nodes; a small amount of additional information is necessary for the algorithm to maintain loop freedom. It turns out to be necessary, for each destination j and each node i, to specify a set $B_i(j)$ of blocked nodes k for which $\phi_{ik}(j) = 0$ and the algorithm is not permitted to increase $\phi_{ik}(j)$ from 0. For notational convenience we include k such that $(i,k) \notin L$ in the set $B_i(j)$. We first define and discuss the algorithm and then define the sets $B_i(j)$.

The algorithm A, on each iteration, maps the current routing variable set ϕ into a new set $\phi^1 = A(\phi)$. The mapping is defined as follows. For $k \in B_i(j)$,

$$\phi_{ik}{}^1(j) = 0, \qquad \Delta_{ik}(j) = 0. \tag{10}$$

For $k \notin B_i(j)$, define

$$a_{ik}(j) = D_{ik}'(f_{ik}) + \frac{\partial D_T}{\partial r_k(j)} - \min_{m \notin B_i(j)} \left[D_{im}'(f_{im}) + \frac{\partial D_T}{\partial r_m(j)} \right] \tag{11}$$

$$\Delta_{ik}(j) = \min\,[\phi_{ik}(j), \eta a_{ik}(j)/t_i(j)] \tag{12}$$

where η is a scale parameter of A to be discussed later. Let $k_{\min}(i,j)$ be a value of m that achieves the minimization in (11). Then

$$\phi_{ik}{}^1(j) = \begin{cases} \phi_{ik}(j) - \Delta_{ik}(j), & k \neq k_{\min}(i,j) \\ \phi_{ik}(j) + \sum\limits_{k \neq k_{\min}(i,j)} \Delta_{ik}(j), & k = k_{\min}(i,j). \end{cases} \tag{13}$$

The algorithm reduces the fraction of traffic sent on nonoptimal links and increases the fraction on the best link. The amount of reduction, given by $\Delta_{ik}(j)$, is proportional to $a_{ik}(j)$, with the restriction that $\phi_{ik}{}^1(j)$ cannot be negative. In turn $a_{ik}(j)$ is the difference between the marginal delay to node j using link (i,k) and using the best link. Note that as the sufficiency condition (9) is approached, the changes get small, as desired. The amount of reduction is also inversely proportional to $t_i(j)$. The reason for this is that the change in link traffic is related to $\Delta_{ik}(j)t_i(j)$. Thus when $t_i(j)$ is small, $\Delta_{ik}(j)$ can be changed by a large amount without greatly affecting the marginal link delays. Finally the changes depend on the scale factor η. For η very small, convergence of the algorithm is guaranteed, as shown in Theorem 5, but rather slow. As η

increases, the speed of convergence increases but the danger of no convergence also increases.

It is not difficult to develop heuristic improvements on this algorithm to speed up its convergence; we have settled on this particular version since it allows us to prove convergence.

We now must complete the definition of algorithm A by defining the sets $B_i(j)$. First define a routing variable $\phi_{ik}(j)$ to be *improper* if $\phi_{ik}(j) > 0$ and $\partial D_T/\partial r_i(j) \leqslant \partial D_T/\partial r_k(j)$. We have already said that $B_i(j)$ includes only k for which $\phi_{ik}(j) = 0$, and thus, from (5),

$$\min_{m \notin B_j(i)} D_{im}'(f_{im}) + \frac{\partial D_T}{\partial r_m(j)} \leqslant \frac{\partial D_T}{\partial r_i(j)}. \tag{14}$$

Assuming positive marginal link delays, $\partial D_T/\partial r_i(j) < \partial D_T/\partial r_k(j) + D_{ik}'(f_{ik})$ if $\phi_{ik}(j)$ is improper, and we see that the algorithm always reduces improper routing variables. In fact, since $\partial D_T/\partial r_i(j)$ is the marginal delay from i to j, we would expect marginal delay to decrease as we move downstream, and improper routing variables should be rather atypical.

For a given destination node j, the set of marginal delays $\partial D_T/\partial r_i(j)$ $(i \neq j)$ forms an ordering of the nodes i. Note that if there are no improper routing variables, this ordering is consistent with the downstream partial ordering. Fig. 3 illustrates these orderings. The horizontal axis represents marginal delay (for the given destination node $j = 5$) and the solid lines show the downstream partial ordering by denoting the links for which $\phi_{ik}(5) > 0$. The dotted lines are examples of links (i,k) for which loops would form if $\phi_{ik}(5)$ were increased from 0. We now see that if ϕ is loop free and $\phi^1 = A(\phi)$ contains a loop for some destination j, then the following two conditions must hold.

1) The loop contains some link (i,k) for which $\phi_{ik}(j) = 0$, $\phi_{ik}{}^1(j) > 0$, and $\partial D_T/\partial r_i(j) > \partial D_T/\partial r_k(j)$.

2) The loop contains some link (l,m) for which $\phi_{lm}(j)$ is improper and for which $\phi_{lm}{}^1(j) > 0$.

The first condition reiterates that some routing variables must be increased from 0 to form a loop and that the algorithm only increases routing variables on links to nodes with smaller marginal delay. The second makes use of the fact that if nodes i have numbers associated with them $(\partial D_T/\partial r_i(j))$, then it is impossible to move around a loop of nodes and have those numbers monotonically decrease.

Definition: The set $B_i(j)$ is the set of nodes k for which either $\phi_{ik}(j) = 0$ and k is blocked relative to j or $(i,k) \notin L$. A node k is blocked relative to j if k has a routing path to j containing some link (l,m) for which $\phi_{lm}(j)$ is improper and

$$\phi_{lm}(j) \geqslant \eta \left[D_{lm}'(f_{lm}) + \frac{\partial D_T}{\partial r_m(j)} - \frac{\partial D_T}{\partial r_l(j)} \right] \Big/ t_l(j). \tag{15}$$

Note that the definition permits k to be identical to l. The reason for (15) can be seen from (14) and (12). If (15) is not satisfied, then $\Delta_{lm}(j) = \phi_{lm}(j)$ and $\phi_{lm}{}^1(j) = 0$, so that (l,m) can not be part of a loop for destination j.

Theorem 4: If the marginal link delays D_{ik}' are positive and ϕ is loop free, then $\phi^1 = A(\phi)$ is loop free.

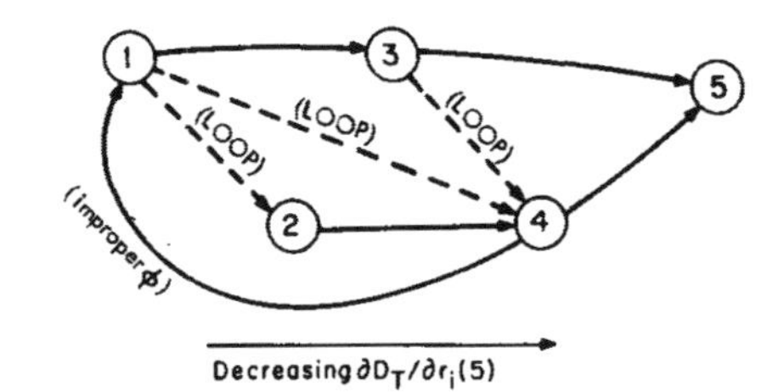

Fig. 3. Marginal delay ordering, downstream partial ordering, and possible loop formation.

Proof: Assume to the contrary that ϕ^1 has a loop, say with respect to destination j. Then from condition 2) above there is some link (l,m) on the loop for which $\phi_{lm}(j)$ is improper and $\phi_{lm}{}^1(j) > 0$. This implies that (15) is satisfied. Now move backward around the assumed loop to the first link (i,k) for which $\phi_j(i,k) = 0$; from condition 1), there must be such a link. Since (l,m) is on a routing path for ϕ from k to j, $k \in B_i(j)$. Thus, from the algorithm, $\phi_{ik}{}^1(j) = 0$, yielding the contradiction.

The protocol required for a node i to determine the set $B_i(j)$ is as follows. Each node l, when it calculates $\partial D_T/\partial r_l(j)$ determines, for each downstream m, if $\phi_{lm}(j)$ is improper and satisfies (15) (only the downstream neighbors could be improper). If any downstream neighbor satisfies these conditions, node l adds a special tag to its broadcast of $\partial D_T/\partial r_l(j)$. The node l also adds this special tag if the received value $\partial D_T/\partial r_m(j)$ from any downstream m contained a tag. In this way all nodes upstream of l also send the tag. The set $B_i(j)$ is then the set of nodes k for which either $(i,k) \notin L$ or the received $\partial D_T/\partial r_k(j)$ was tagged.

Theorem 5: Assume that for all $(i,k) \in L$, $D_{ik}(f_{ik})$ has a positive first derivative and nonnegative second derivative for $0 \leqslant f_{ik} < C_{ik}$ and that $\lim_{f_{ik} \uparrow C_{ik}} D_{ik}(f_{ik}) = \infty$. For every positive number D_0 there exists a scale factor η for A such that if ϕ^0 satisfies $D_T(\phi^0) \leqslant D_0$, then

$$\lim_{m \to \infty} D_T(\phi^m) = \min_{\phi} D_T(\phi) \tag{16}$$

where $\phi^m = A(\phi^{m-1})$ for all $m \geqslant 1$.

This is proved in Appendix C. Note that η depends on some upper bound D_0 to D_T; this is natural, since when the link flows are very close to capacity, small changes in the link flows cause large changes in marginal delay. The proof uses a ridiculously small value of η to guarantee convergence under all conditions and experimental work is necessary to determine practical values for η.

USE OF THE ALGORITHM FOR QUASI-STATIC ROUTING

We have shown in the last section that the algorithm A must eventually converge to the minimum average delay for a network with stationary inputs and links. The algorithm is really intended, however, for quasi-static applications where the input statistics are slowly changing with time and where occasionally links or nodes fail or are added to the network. Under these more general conditons, it is clear that the loop freedom of the algorithm is maintained since this is a mathematical property that is independent of the marginal link delays and node flows, which are the only inputs to the algorithm (note that the inputs r plays a role in the theoretical development, but do not appear in the algorithm and need not be estimated).

The question of whether the algorithm can adapt fast enough to keep up with changing statistics is difficult and requires more study. Clearly, the faster the statistics change, the more frequently the algorithm should be updated, but frequent updating has two undesirable effects. First, frequent updates require more updating protocol, thus reducing the effective link capacities available for data, and second, frequent updates will necessitate noisier measurements of marginal link delays and node flows. Experimentation would be helpful both in determining update rates and the scale parameter η.

Another open question is that of a starting rule for the algorithm (finding a loop free ϕ to start with). One possibility is to start with shortest paths; that is, set $\phi_{ik}(j) = 1$ for the link (i,k) that leads to j from i with the smallest number of links. Such a strategy might well lead to link flows which mathematically exceed capacity, but in this case a well designed flow control would limit the input to the network, thus yielding large but finite marginal delays on such links and allowing the routing algorithm to gradually adapt.

The problem of dropped links or nodes is somewhat more complicated. Some of the problems here must be solved by higher order protocols, since if the network becomes disconnected, there is no way to route data between disconnected parts of the network. However the routing algorithm should still adapt by finding routes for any data that can be sent. Each node i at the end of a link (i,k) that has failed or whose opposite node has failed should signal the fact that an update should start throughout the network. In addition, i should no longer regard k as being downstream with respect to any destination j, and if k was the *only* downstream neighbor, then i should broadcast $\partial D_T/\partial r_i(j) = \infty$. This latter broadcast prevents upstream nodes from waiting indefinitely for update information to propagate through a failed link or node. The exact details of updating protocols in the presence of link and node failures is a subject for futher research.

APPENDIX A

Proof of Theorem 1:

Without loss of generality, take the destination node j to be the nth of the n nodes and drop the argument j from (1),

$$t_i = r_i + \sum_{l=1}^{n-1} t_l \phi_{li}, \qquad 1 \leqslant i \leqslant n. \tag{A1}$$

Summing both sides over i, we see that any solution to (A1) satisfies

$$t_n = \sum_i r_i. \tag{A2}$$

Temporarily let $\phi_{ni} = r_i/t_n$ and substitute this in (A1).

$$t_i = \sum_{l=1}^{n} t_l \phi_{li}. \tag{A3}$$

Any solution to (A3) and (A2) satisfies (A1) and vice versa. Let $\hat{\Phi}$ be the $n \times n$ matrix with components ϕ_{li}. $\hat{\Phi}$ is stochastic (i.e., $\phi_{li} \geqslant 0$ for all l, i and $\Sigma_i \phi_{li} = 1$ for all i) and (A3) is just the formula for steady-state probabilities in a Markov chain.

It is well known (see, for example, Gantmacher [15]) that if $\hat{\Phi}$ is irreducible, then (A3) has a unique solution, aside from a scale factor determined by (A2), and $t_i > 0$, $1 \leqslant i \leqslant n$. The matrix $\hat{\Phi}$ is irreducible; however, if for each i, k there is a path $i, l, m, \cdots, p, k$ such that $\phi_{il} > 0, \phi_{lm} > 0, \cdots, \phi_{pk} > 0$. If $r_i > 0$ for $1 \leqslant i \leqslant n-1$, then node n has a path to each i, $1 \leqslant i \leqslant n-1$. By the definition of routing variables, each i has a path to n and consequently $\hat{\Phi}$ is irreducible. Thus (A1) has a unique solution, with positive t_i, if $r_i > 0$ for $1 \leqslant i \leqslant n-1$.

Now let $t = (t_1, \cdots, t_{n-1})$, $r = (r_1, \cdots, r_{n-1})$, and let Φ be the $n-1 \times n-1$ matrix with components ϕ_{li} $(1 \leqslant i, l \leqslant n-1)$. Equation (A1) for $1 \leqslant i \leqslant n-1$ is then $t(I - \Phi) = r$. Since this equation has a unique solution for $r_i > 0$, $I - \Phi$ must have an inverse, and

$$t = r(I - \Phi)^{-1}. \tag{A4}$$

Since the components of t are positive when the components of r are positive, components of t are nonnegative when the components of r are nonnegative. Differentiating (A4), we get the continuous function of Φ,

$$\frac{\partial t_i}{\partial r_l} = [(I - \Phi)^{-1}]_{li}. \tag{A5}$$

Using (A5) in (A4), the solution to (A1) is conveniently expressed, for any r, as

$$t_i = \sum_l \frac{\partial t_i}{\partial r_l} r_l. \tag{A6}$$

Finally, differentiating (A1) with respect to ϕ_{km}, we get

$$\frac{\partial t_i}{\partial \phi_{km}} = \sum_{l=1}^{n-1} \frac{\partial t_l}{\partial \phi_{km}} \phi_{li} + t_k \delta_{im}$$

where $\delta_{im} = 1$ for $i = m$ and 0 otherwise. For fixed k, m, this is the same set of equations as (A1), so that the solution, continuous in ϕ, is

$$\frac{\partial t_i}{\partial \phi_{km}} = \frac{\partial t_i}{\partial r_m} t_k. \tag{A7}$$

Proof of Theorem 2

First we show that (5), repeated below with the destination node again taken to be n, has a unique solution.

$$\frac{\partial D_T}{\partial r_i} = \sum_{k=1}^{n} \phi_{ik} D_{ik}'(f_{ik}) + \sum_{k=1}^{n} \phi_{ik} \frac{\partial D_T}{\partial r_k}. \tag{A8}$$

Let $b_i = \Sigma_k \phi_{ik} D_{ik}'(f_{ik})$ and let b be the column vector $(b_1, \cdots,$

b_{n-1}). Let $\nabla \cdot D_T$ be the column vector $(\partial D_T/\partial r_1, \cdots, \partial D_T/\partial r_{n-1})$. Then (A8) can be rewritten as

$$\nabla \cdot D_T = b + \Phi(\nabla \cdot D_T). \tag{A9}$$

We saw in the proof of Theorem 1 that $I - \Phi$ has a unique inverse with components given by (A5). Thus the unique solution to (A9) is

$$\frac{\partial D_T}{\partial r_i} = \sum_l \frac{\partial t_l}{\partial r_i} \sum_m \phi_{lm} D_{lm}'(f_{lm}) \tag{A10}$$

$$= \sum_{l,m} \frac{\partial f_{lm}}{\partial r_i} D_{lm}'(f_{lm}). \tag{A11}$$

Differentiating D_T directly with (2) and (3), we get the same unique solution, which, from Theorem 1, is continuous in ϕ.

Finally we calculate $\partial D_T/\partial \phi_{ik}$ directly using (3) and (2),

$$\frac{\partial D_T}{\partial \phi_{ik}} = \sum_{l,m} D_{lm}'(f_{lm})\phi_{lm} \frac{\partial t_l}{\partial \phi_{ik}} + D_{ik}'(f_{ik})t_i$$

$$= t_i \left[\sum_{l,m} D_{lm}'(f_{lm})\phi_{lm} \frac{\partial t_l}{\partial r_k} \right] + t_i D_{ik}'(f_{ik})$$

$$= t_i \left[\frac{\partial D_T}{\partial r_k} + D_{ik}'(f_{ik}) \right]. \tag{A12}$$

We have used (A7) and (A10) to derive (A12), which is the same as (6). This is clearly continuous in ϕ given the continuity of t_i and $\partial D_T/\partial r_i$, and the proof is complete.

APPENDIX B

Proof of Theorem 3

First we show that (7) is a necessary condition to minimize D_T by assuming that ϕ does not satisfy (7). This means that there is some i, j, k, and m such that

$$\phi_{ik}(j) > 0, \qquad \frac{\partial D_T(\phi)}{\partial \phi_{ik}(j)} > \frac{\partial D_T(\phi)}{\partial \phi_{im}(j)}. \tag{B1}$$

Since these derivatives are continuous, a sufficiently small increase in $\phi_{im}(j)$ and corresponding decrease in $\phi_{ik}(j)$ will decrease D_T, thus establishing that ϕ does not minimize D_T.

Next we show that (8), repeated below, is a sufficient condition to minimize D_T.

$$D_{ik}'(f_{ik}) + \frac{\partial D_T(\phi)}{\partial r_k(j)} \geqslant \frac{\partial D_T(\phi)}{\partial r_i(j)}, \qquad \text{all } i, j, k. \tag{B2}$$

Suppose that ϕ satisfies (B2) and has node flows t and link flows f. Let ϕ^* be any other set of routing variables with node flows t^* and link flows f^*. Define

$$f_{ik}(\lambda) = (1 - \lambda)f_{ik} + \lambda f_{ik}^* \tag{B3}$$

$$D_T(\lambda) = \sum_{i,k} D_{ik}(f_{ik}(\lambda)). \tag{B4}$$

There is a set of routing variables $\phi(\lambda)$ which gives rise to $f(\lambda)$, but they are *not* linear in λ and their existence is not relevant to our proof. Since each link delay D_{ik} is a convex ∪ function of the link flow, $D_T(\lambda)$ is convex ∪ in λ, and hence

$$\left.\frac{dD_T(\lambda)}{d\lambda}\right|_{\lambda=0} \leqslant D_T(\phi^*) - D_T(\phi). \tag{B5}$$

Since ϕ^* is arbitrary, proving that $dD_T(\lambda)/d\lambda \geqslant 0$ at $\lambda = 0$ will complete the proof. From (B4) and (B3),

$$\left.\frac{dD_T(\lambda)}{d\lambda}\right|_{\lambda=0} = \sum_{i,k} D_{ik}'(f_{ik})[f_{ik}^* - f_{ik}]. \tag{B6}$$

We now show that

$$\sum_{i,k} D_{ik}'(f_{ik})f_{ik}^* \geqslant \sum_{j,k} r_k(j) \frac{\partial D_T(\phi)}{\partial r_k(j)}. \tag{B7}$$

Note from (B2) that

$$\sum_k D_{ik}'(f_{ik})\phi_{ik}^*(j) \geqslant \frac{\partial D_T(\phi)}{\partial r_i(j)} - \sum_k \frac{\partial D_T(\phi)}{\partial r_k(j)} \phi_{ik}^*(j). \tag{B8}$$

Multiplying both sides of (B8) by $t_i^*(j)$, summing over i, j, and recalling that $f_{ik}^* = \Sigma_j t_j^*(j)\phi_{ik}^*(j)$, we obtain

$$\sum_{i,k} D_{ik}'(f_{ik})f_{ik}^* \geqslant \sum_{i,j} t_i^*(j) \frac{\partial D_T(\phi)}{\partial r_i(j)} - \sum_{i,j,k} t_i^*(j)\phi_{ik}^*(j) \frac{\partial D_T(\phi)}{\partial r_k(j)}. \tag{B9}$$

From (1), $\Sigma_i t_i^*(j)\phi_{ik}^*(j) = t_k^*(j) - r_k(j)$. Substituting this into the rightmost term of (B9) and canceling, we get (B7). Note that the only inequality used here was (B8), and that if ϕ is substituted for ϕ^*, this becomes an equality from the equation for $\partial D_T/\partial r_i(j)$ in (5). Thus

$$\sum_{i,k} D_{ik}'(f_{ik})f_{ik} = \sum_{j,k} r_k(j) \frac{\partial D_T(\phi)}{\partial r_k(j)}. \tag{B10}$$

Substituting (B10) and (B7) into (B6), we see that $dD_T(\lambda)/d\lambda \geqslant 0$ at $\lambda = 0$, completing the proof. We note in passing that (B10) is valid for any set of routing variables and appears to be a rather fundamental conservation equation.

APPENDIX C

We prove Theorem 4 through a sequence of seven lemmas. The first five establish the descent properties of the algorithm, the sixth establishes a type of continuity condition, showing that if ϕ does not minimize D_T, then for any ϕ^* in a neighborhood of ϕ, $D_T(A^m(\phi^*)) < D_T(\phi)$ for some m. The seventh lemma is a new global convergence theorem which does not require continuity in the algorithm A; Lemmas 6 and 7 together establish Theorem 4.

Let ϕ be an arbitrary set of routing variables satisfying $D_T(\phi) < D_0$ for some D_0. Let $\phi^1 = A(\phi)$ and let t, f, t^1, f^1 be the node and link flows corresponding to ϕ and ϕ^1, respectively. Let f^λ $(0 \leq \lambda \leq 1)$ be defined by $f_{ik}{}^\lambda = (1-\lambda)f_{ik} + \lambda f_{ik}{}^1$, and let

$$D_T(\lambda) = \sum_{i,k} D_{ik}(f_{ik}{}^\lambda). \tag{C1}$$

From the Taylor remainder theorm,

$$D_T(\phi^1) - D_T(\phi) = \left.\frac{dD_T(\lambda)}{d\lambda}\right|_{\lambda=0} + \frac{1}{2}\left.\frac{d^2D_T(\lambda)}{d\lambda^2}\right|_{\lambda=\lambda^*} \tag{C2}$$

where λ^* is some number, $0 \leq \lambda^* \leq 1$. The continuity of the second derivative above will be obvious from the proof of Lemma 4, which upper bounds that term. The first three lemmas deal with $dD_T(\lambda)/d\lambda\,|_{\lambda=0}$.

Lemma 1:

$$\left.\frac{dD_T(\lambda)}{d\lambda}\right|_{\lambda=0} = \sum_{i,j,k} -\Delta_{ik}(j)a_{ik}(j)t_i{}^1(j). \tag{C3}$$

Proof[2]*:* Using the definitions of $a_{ik}(j)$ and $\Delta_{ik}(j)$ in (11) and (12)

$$\sum_k \Delta_{ik}(j)a_{ik}(j)$$

$$= \sum_{k \neq k_{\min}(i,j)} [\phi_{ik}(j) - \phi_{ik}{}^1(j)] \left\{ D_{ik}'(f_{ik}) + \frac{\partial D_T}{\partial r_k(j)} - \min_{m \notin B_i(j)} \left[D_{im}'(f_{im}) + \frac{\partial D_T}{\partial r_m(j)} \right] \right\}$$

$$= \sum_k [\phi_{ik}(j) - \phi_{ik}{}^1(j)] \left[D_{ik}'(f_{ik}) + \frac{\partial D_T(\phi)}{\partial r_k(j)} \right] \tag{C4}$$

$$= \frac{\partial D_T(\phi)}{\partial r_i(j)} - \sum_k \phi_{ik}{}^1(j) \left[D_{ik}'(f_{ik}) + \frac{\partial D_T(\phi)}{\partial r_k(j)} \right] \tag{C5}$$

In (C4), we have used (13) to extend the sum over all k and in (C5), we have used (5). Multiplying both sides of (C5) by $t_i{}^1(j)$, summing, and using (1) and (2), we get

$$\sum_{i,j,k} \Delta_{ik}(j)a_{ik}(j)t_i{}^1(j)$$

$$= \sum_{i,j} t_i{}^1(j)\frac{\partial D_T(\phi)}{\partial r_i(j)} - \sum_{i,k} f_{ik}{}^1 D_{ik}'(f_{ik}) - \sum_{k,j} [t_k{}^1(j) - r_k(j)] \frac{\partial D_T(\phi)}{\partial r_k(j)}$$

$$= -\sum_{i,k} f_{ik}{}^1 D_{ik}'(f_{ik}) + \sum_{k,j} r_k(j) \frac{\partial D_T(\phi)}{\partial r_k(j)} \tag{C6}$$

$$= \sum_{i,k} (f_{ik} - f_{ik}{}^1) D_{ik}'(f_{ik}) \tag{C7}$$

$$= -\left.\frac{dD_T(\lambda)}{d\lambda}\right|_{\lambda=0} \tag{C8}$$

We have used (B10) to get (C7), and (C8) follows from (C1), completing the proof.

Lemma 2:

$$\left.\frac{dD_T(\lambda)}{d\lambda}\right|_{\lambda=0} \leq -\frac{1}{\eta(n-1)^3} \sum_{i,j} \Delta_i{}^2(j)t_i{}^2(j) \tag{C9}$$

where

$$\Delta_i(j) = \sum_k \Delta_{ik}(j). \tag{C10}$$

Proof: From the definition of $\Delta_{ik}(j)$ in (12), $-a_{ik}(j) \leq -t_i(j)\Delta_{ik}(j)/\eta$. Substituting this into (C3) yields

$$\left.\frac{dD_T(\lambda)}{d\lambda}\right|_{\lambda=0} \leq -\frac{1}{\eta} \sum_{i,j,k} \Delta_{ik}{}^2(j)t_i(j)t_i{}^1(j)$$

$$\leq -\frac{1}{(n-1)\eta} \sum_{i,j} \Delta_i{}^2(j)t_i(j)t_i{}^1(j) \tag{C11}$$

where (C11) follows from Cauchy's inequality, $(\Sigma_k \alpha_k \beta_k)^2 \leq (\Sigma \alpha_k{}^2)(\Sigma \beta_k{}^2)$, with $\alpha_k = 1, \beta_k = \Delta_{ik}(j)$, and the sum over $k \neq i$.

Now define $t_i{}^*(j)$ as the total flow at node i destined for j if the routing variables $\phi_{ik}(j)$ (for $k \neq k_{\min}(i,j)$) are reduced by $\Delta_{ik}(j)$ but $\phi_{ik}(j)$ for $k = k_{\min}(i,j)$ is not increased. Mathematically $t_i{}^*(j)$ satisfies

$$t_i{}^*(j) = \sum_l t_l{}^*(j)[\phi_{li}(j) - \Delta_{li}(j)] + r_i(j). \tag{C12}$$

This has a unique solution because of the loop freedom of ϕ. Subtracting (C12) from (1) results in

$$t_i(j) - t_i{}^*(j) = \sum_l [t_l(j) - t_l{}^*(j)]\phi_{li}(j) + \sum_l t_l{}^*(j)\Delta_{li}(j). \tag{C13}$$

[2] As mentioned before there is a ϕ^λ corresponding to f^λ, but it is nonlinear in λ, and $dD_T(\lambda)/\partial\lambda$ cannot be calculated in a straightforward way by differentiating with respect to ϕ^λ.

From (A6), using $\Sigma t_l{}^*(j)\Delta_{li}(j)$ for $r_l(j)$,

$$t_i(j) - t_i{}^*(j) = \sum_l \frac{\partial t_i(j)}{\partial r_l(j)} \sum_k t_k{}^*(j)\Delta_{kl}(j). \quad \text{(C14)}$$

Since ϕ is loop free, $\partial t_i(j)/\partial r_l(j) \leqslant 1$. Also if $\partial t_i(j)/\partial r_l(j) > 0$, then l is upstream of i for destination j and $\phi_{il}(j)$ (and hence $\Delta_{il}(j)$) is zero. Thus

$$t_i(j) - t_i{}^*(j) \leqslant \sum_l \sum_{k \neq i} t_k{}^*(j)\Delta_{kl}(j) = \sum_{k \neq i} t_k{}^*(j)\Delta_k(j). \quad \text{(C15)}$$

Multiplying the left side by $\Delta_i(j) \leqslant 1$ preserves the inequality, yielding

$$t_i(j)\Delta_i(j) \leqslant \sum_k t_k{}^*(j)\Delta_k(j). \quad \text{(C16)}$$

Since the right-hand side of (C14) is nonnegative, we also have $t_i(j)\Delta_i(j) \geqslant t_i{}^*(j)\Delta_i(j)$. We interrupt the proof now for a short technical lemma.

Lemma 3: Let α_i, β_i $(1 \leqslant i \leqslant m)$ be nonnegative numbers satisfying $\alpha_i \leqslant \Sigma_k \beta_k$; $\alpha_i \geqslant \beta_i$ for $1 \leqslant i \leqslant$ m. *Then*

$$\sum_{i=1}^m \alpha_i\beta_i \geqslant \frac{1}{m^2} \sum_i \alpha_i{}^2. \quad \text{(C17)}$$

Proof of Lemma 3:

$$\sum_i \alpha_i\beta_i \geqslant \sum_i \beta_i{}^2 \geqslant \frac{1}{m}\left(\sum \beta_i\right)^2 \quad \text{(C18)}$$

where we have used $\alpha_i \geqslant \beta_i$ and then Cauchy's inequality. Since $\Sigma\beta_i \geqslant \alpha_k$ for all k,

$$\sum_i \alpha_i\beta_i \geqslant \frac{1}{m} \alpha_k{}^2, \qquad \text{for all } k. \quad \text{(C19)}$$

This implies (C17), completing the proof of Lemma 3.

Now let $\alpha_i = t_i(j)\Delta_i(j)$ and $\beta_i = t_i{}^*(j)\Delta_i(j)$. Since these terms are nonzero only for $i \neq j$, we can take $m = n - 1$. Since the conditions of the lemma are satisfied for this choice,

$$\sum_i \Delta_i{}^2(j)t_i(j)t_i{}^*(j) \geqslant \frac{1}{(n-1)^2} \sum_i \Delta_i{}^2(j)t_i{}^2(j). \quad \text{(C20)}$$

Since $t_i{}^*(j) \leqslant t_i{}^1(j)$, we can substitute (C20) into (C11), getting (C9) and completing the proof of Lemma 2.

Lemma 4: Let M be an upper bound to $D_{ik}{}''(f_{ik}{}^\lambda)$ over all i, k and over $0 \leqslant \lambda \leqslant 1$. Then for any λ, $0 \leqslant \lambda \leqslant 1$,

$$\frac{d^2D_T(\lambda)}{d\lambda^2} \leqslant M(n+2)(n-1)n \sum_{j,k} \Delta_k{}^2(j)t_k{}^2(j). \quad \text{(C21)}$$

Proof: The bound M must exist because $D_{ik}{}''(f_{ik}{}^\lambda)$ is a continuous function of λ over the compact region $0 \leqslant \lambda \leqslant 1$. Taking the second derivative, we get

$$\frac{d^2D_T(\lambda)}{d\lambda^2} = \sum_{i,k} D_{ik}{}''(f_{ik}{}^\lambda)[f_{ik}{}^1 - f_{ik}]^2 \leqslant \sum_{i,k} M[f_{ik}{}^1 - f_{ik}]^2. \quad \text{(C22)}$$

We now upper bound $|f_{ik}{}^1 - f_{ik}|$ by first upper bounding $|t_i{}^1(j) - t_i(j)|$. As in the proof of Lemma 2, we have

$$\begin{aligned} t_i{}^1(j) - t_i(j) &= \sum_l [t_l{}^1(j) - t_l(j)]\phi_{li}{}^1(j) \\ &\quad + \sum_l t_l(j)[\phi_{li}{}^1(j) - \phi_{li}(j)] \\ &= \sum_l \frac{\partial t_i{}^1(j)}{\partial r_l(j)} \sum_k t_k(j)[\phi_{kl}{}^1(j) - \phi_{kl}(j)]. \end{aligned} \quad \text{(C23)}$$

Since $0 \leqslant \partial t_i{}^1(j)/\partial r_l(j) \leqslant 1$, we can upper bound this by

$$t_i{}^1(j) - t_i(j) \leqslant \sum_k t_k(j)\Delta_k(j).$$

We can lower bound (C23) in the same way, considering only terms in which $\phi_{kl}{}^1(j) - \phi_{kl}(j) < 0$, and this leads to

$$|t_i{}^1(j) - t_i(j)| \leqslant \sum_k t_k(j)\Delta_k(j) \quad \text{(C24)}$$

$$\begin{aligned} f_{ik}{}^1 - f_{ik} &= \sum_j [t_i{}^1(j) - t_i(j)]\phi_{ik}{}^1(j) \\ &\quad + t_i(j)[\phi_{ik}{}^1(j) - \phi_{ik}(j)] \\ |f_{ik}{}^1 - f_{ik}| &\leqslant \sum_j \sum_l t_l(j)\Delta_l(j)\phi_{ik}{}^1(j) \\ &\quad + \sum_j t_i(j)\,|\phi_{ik}{}^1(j) - \phi_{ik}(j)|. \end{aligned} \quad \text{(C25)}$$

The double sum in (C25) has at most $(n-1)^2$ nonzero terms $(j \neq i,\ l \neq j)$ and the second sum at most $n - 1$ terms. Using Cauchy's inequality on both terms together, we get

$$|f_{ik}{}^1 - f_{ik}|^2 \leqslant n(n-1)\left\{\sum_{j,l} t_l{}^2(j)\Delta_l{}^2(j)[\phi_{ik}{}^1(j)]^2 + \sum_j t_i{}^2(j)[\phi_{ik}{}^1(j) - \phi_{ik}(j)]^2\right\}$$

$$\sum_k |f_{ik}{}^1 - f_{ik}|^2 \leqslant n(n-1)\left\{\sum_{j,l} t_l{}^2(j)\Delta_l{}^2(j) + 2\sum_i t_i{}^2(j)\Delta_i{}^2(j)\right\}. \quad \text{(C26)}$$

Summing over i and substituting the result in (C22), we get (C21), completing the proof.

Lemma 5: For given D_0, define

$$M = \max_{i,k} \max_{f: D_{ik}(f) \leqslant D_0} D_{ik}''(f) \tag{C27}$$

$$\eta = [Mn^6]^{-1}. \tag{C28}$$

Then for all ϕ such that $D_T(\phi) \leqslant D_0$,

$$D_T(\phi^1) - D_T(\phi) \leqslant -\frac{1}{2\eta(n-1)^3} \sum_{i,j} \Delta_i^2(j) t_i^2(j). \tag{C29}$$

Proof: Temporarily let M be as defined in Lemma 4. Combining Lemmas 2 and 4,

$$D_T(\phi^1) - D_T(\phi) \leqslant \left[-\frac{1}{\eta(n-1)^3} + \frac{Mn(n-1)(n+2)}{2} \right] \sum_{i,j} \Delta_i^2(j) t_i^2(j). \tag{C30}$$

For $\eta = [\mathrm{Mn}^6]^{-1}$, the second term in brackets above is less than half the magnitude of the first term, yielding (C29). It follows that $D_T(\phi^1) \leqslant D_T(\phi) \leqslant D_0$. By convexity then $D_{ik}(f^\lambda) \leqslant D_0$ for $0 \leqslant \lambda \leqslant 1$. Thus M as given in (C27) satisfies the condition on M in Lemma 4, compleing the proof.

Lemma 6: Let the scale factor η satisfy (C28) for a given D_0 and let ϕ be an arbitrary set of routing variables that does not minimize D_T and that satisfies $D_T(\phi) \leqslant D_0$. Given this ϕ, there exists an $\epsilon > 0$ and an m, $1 \leqslant m \leqslant n$, such that for all ϕ^* satisfying $|\phi - \phi^*| < \epsilon$,

$$D_T(A^m(\phi^*)) < D_T(\phi). \tag{C31}$$

Proof: We consider three cases. The first is the typical case in which no blocking occurs and $D_T(A(\phi)) < D_T(\phi)$, the second is the case in which blocking occurs, and the third is the situation typified by Fig. 2 in which $D_T(A(\phi)) = D_T(\phi)$.

Case 1: No blocking; $\Delta_i(j)t_i(j) > 0$ for some i, j. If no nodes are blocked for ϕ, then by the definition of blocking (15), there is a neighborhood of ϕ^* around ϕ for which no blocking occurs. In this neighborhood,

$$a_{ik}(j) = \left[D_{ik}'(f_{ik}) + \frac{\partial D_T}{\partial r_k(j)} \right] - \min_{1 \leqslant m \leqslant n} \left[D_{im}'(f_{im}) + \frac{\partial D_T}{\partial r_m(j)} \right] \tag{C32}$$

which is continuous in ϕ. It follows from (12) that $\Delta_{ik}(j)$ is continuous in ϕ, and the upper bound to $D_T(A(\phi)) - D_T(\phi)$ in (C29) is continuous[3] in ϕ. Since by assumption the bound in (C29) is strictly negative, there is a neighborhood of ϕ^* around ϕ for which

[3]As a precaution against being too casual about these arguments, one should note that if the minimizing m in (C32) is not unique, then $A(\phi)$ is *not* continuous in ϕ.

$$D_T(A(\phi^*)) - D_T(\phi^*) < -\frac{1}{4\eta(n-1)^3} \sum_{i,j} \Delta_i^2(j) t_i^2(j) \tag{C33}$$

where $\Delta_i(j)$ and $t_i(j)$ correspond to the given ϕ. Choose ϵ small enough so that (C33) is satisfied for $|\phi - \phi^*| < \epsilon$ and also so that

$$|D_T(\phi^*) - D_T(\phi)| < \frac{1}{4\eta(n-1)^3} \sum_{i,j} \Delta_i^2(j) t_i^2(j).$$

Combining this with (C33), we have (C31) for $m = 1$.

Case 2: Blocking occurs. For any ϕ, we can use (5) to lower bound $a_{ik}(j)$ by

$$a_{ik}(j) \geqslant D_{ik}'(f_{ik}) + \partial D_T/\partial r_k(j) - \partial D_T/\partial r_i(j) \tag{C34}$$

$$\Delta_{ik}(j)t_i(j) \geqslant \min \left\{ \phi_{ik}(j)t_i(j),\ \eta \left[D_{ik}'(f_{ik}) + \frac{\partial D_T}{\partial r_k(j)} - \frac{\partial D_T}{\partial r_i(j)} \right] \right\} \tag{C35}$$

The lower bounds above are continuous functions of ϕ. Since blocking occurs in ϕ, there is some i, j, k such that both

$$\frac{\partial D_T}{\partial r_k(j)} - \frac{\partial D_T}{\partial r_i(j)} \geqslant 0 \tag{C36}$$

and

$$\phi_{ik}(j)t_i(j) \geqslant \eta \left[D_{ik}'(f_{ik}) + \frac{\partial D_T}{\partial r_k(j)} - \frac{\partial D_T}{\partial r_i(j)} \right] \tag{C37}$$

Combining (C35) to (C37),

$$\Delta_{ik}(j)t_i(j) \geqslant \eta D_{ik}'(f_{ik}). \tag{C38}$$

Since the right-hand side of (C35) is continuous in ϕ, there is a neighborhood of ϕ^* around ϕ for which

$$\Delta_{ik}{}^*(j)t_i{}^*(j) \geqslant \frac{\eta}{2} D_{ik}'(t_{ik}). \tag{C39}$$

Equation (C31), for $m = 1$, now follows in the same way as in case 1.

Case 3: $\Delta_{ik}(j)t_i(j) = 0$ for all i, j, k. Let Φ_3 be the set of ϕ for which $\Delta_{ik}(j)t_i(j) = 0$ for all i, j, k. Let $\phi^{(l)} = A^l(\phi)$ for the given ϕ and let $m \geqslant 2$ be the smallest integer such that $\phi^{(m-1)} \notin \Phi_3$. We first show that $m \leqslant n$. Note first that for any $\phi \in \Phi_3$, A changes $\phi_j(i,k)$ only for i, j such that $t_i(j) = 0$ and thus the node flows and link flows cannot change. $\partial D_T/\partial r_i(j)$ can change, however, and as we shall see later, must change for some i, j if ϕ does not minimize D_T.

Now consider $\phi^{(l)}$ $(0 \leqslant l \leqslant m-2$, where $\phi^{(0)}$ denotes the original $\phi)$. Since $\phi^{(l)} \in \Phi_3$, $\Delta_{ik}{}^{(l)}(j) > 0$ implies that $t_i(j) = 0$. From (12), $\phi_{ik}{}^{(l)}(j) = \Delta_{ik}{}^{(l)}(j)$ and $\phi_{ik}{}^{(l+1)}(j) = 0$. For a given i, j all $\phi_{ik}{}^{(l)}(j)$ are reduced to 0 except for the k which minimizes $D_{ik}'(f_{ik}) + \partial D_T(\phi^{(l)})/\partial r_k(j)$. Thus, using (5),

$$\frac{\partial D_T(\phi^{(l+1)})}{\partial r_i(j)} = \min_k \left[D_{ik}'(f_{ik}) + \frac{\partial D_T(\phi^{(l)})}{\partial r_k(j)} \right] \leqslant \frac{\partial D_T(\phi^{(l)})}{\partial r_i(j)}. \quad \text{(C40)}$$

Since this equation is satisfied for all l, $0 \leqslant l \leqslant m-2$, we see that $\partial D_T(\phi^{(l)})/\partial r_i(j)$ can be reduced on iteration l only if $\partial D_T(\phi^{(l-1)})/\partial r_k(j)$ is reduced on iteration $l-1$ for some k such that $\partial D_T(\phi^{(l-1)})/\partial r_k(j) < \partial D_T(\phi^{(l)})/\partial r_i(j)$. This reduction at node k however implies a reduction at some node k' of smaller differential delay at iteration $l-2$ and so forth. Since this sequence of differential delays is decreasing with decreasing l and since (from (C40)) the differential delay at a given node is nondecreasing with decreasing l, each node in the sequence must be distinct. Since there are $n-1$ nodes other than the given destination available for such a sequence, the initial l in such a sequence satisfies $l \leqslant n-2$. On the other hand, if $D_T(\phi^{(l)})/\partial r_i(j)$ is unchanged for all i, j, we see from (C40) that $\phi^{(l)}$ satisfies the sufficient conditions to minimize D_T' and then ϕ also minimizes D_T' contrary to our hypothesis; thus we must have $m \leqslant n$.[4]

Now observe that the middle expression in (C40), for $l=0$, is a continuous function of ϕ and consequently $\partial D_T(\phi^{(1)})/\partial r_i(j)$ is a continuous function of ϕ for all i, j. It follows by induction that $\partial D_T(\phi^{(l)})/\partial r_i(j)$ is a continuous function of ϕ for all i, j and for $l \leqslant m-1$. Finally $\phi^{(m-1)} \notin \Phi_3$, so it must satsify the conditions of case 1 or 2; it will be observed that the analyses there apply equally to $\phi^{(m-1)}$ because of the continuity of $\partial D_T(\phi^{(m-1)})/\partial r_i(j)$ as a function ϕ. This completes the proof.

Our last lemma will be stated in greater generality than required since it is a global convergence theorem for algorithms that avoids the usual continuity constraint on the algorithm (see Luenberger [16] for a good discussion of global convergence).

Lemma 7: Let Φ be a compact region of Euclidean N space. Let A be a mapping from Φ into Φ and let D_T be a continuous real valued function in Φ. Assume that $D_T(A(\phi)) \leqslant D_T(\phi)$ for all $\phi \in \Phi$. Let $D_{\min}$ be the minimum of D_T over Φ and let $\Phi_{\min}$ be the set of $\phi \in \Phi$ such that $D_T(\phi) = D_{\min}$. Assume that for every $\phi \in \Phi - \Phi_{\min}$, there is an $\epsilon > 0$ and an integer $m \geqslant 1$ such that for all $\phi^* \in \Phi$ satisfying $|\phi - \phi^*| < \epsilon$, we have $D_T(A^m(\phi^*)) < D_T(\phi)$. Then for all $\phi \in \Phi$,

$$\lim_{m \to \infty} D_T(A^m(\phi)) = D_{\min}. \quad \text{(C41)}$$

Proof: Since Φ is compact, the sequence $\{A^m(\phi)\}$ has a convergent subsequence, say $\{\phi^l\}$, with

$$\phi' = \lim_{l \to \infty} \phi^l, \qquad \phi' \in \Phi. \quad \text{(C42)}$$

Since D_T is continuous,

$$D_T(\phi') = \lim_{l \to \infty} D_T(\phi^l). \quad \text{(C43)}$$

Furthermore, by assumption, $D_T(A^m(\phi))$ is nonincreasing in m, so that

$$D_T(\phi') = \lim_{m \to \infty} D_T(A^m(\phi)) \quad \text{(C44)}$$

$$D_T(\phi') = D_T(A^m(\phi)), \qquad \text{all } m \geqslant 1. \quad \text{(C45)}$$

To complete the proof, we must show that $\phi' \in \Phi_{\min}$; we assume the contrary and demonstrate a contradiction. By assumption then, there is an $\epsilon > 0$ and an m' associated with ϕ' such that $D_T(A^{m'}(\phi^*)) < D_T(\phi')$ for all $\phi^* \in \Phi$, $|\phi^* - \phi'| < \epsilon$. By (C42) there is an l such that $|\phi^l - \phi'| < \epsilon$, and thus $D_T(A^{m'}(\phi^l)) < D_T(\phi')$. Since $\phi^l = A^m(\phi)$ for some m, $D_T(A^{m'+m}(\phi)) < D_T(\phi')$, contradicting (C45) and completing the proof.

Proof of Theorem 4: Let Φ be the set of loop free routing variables ϕ such that $D_T(\phi) \leqslant D_0$. We have verified that A maps loop free routing variables into loop free routing variables, and from Lemma 5, $D_T(A(\phi)) \leqslant D_T(\phi)$ for $\phi \in \Phi$. Thus A is a mapping from Φ into Φ. It is obvious that Φ is bounded and easy to verify that any limit of loop free variables with $D_T(\phi) \leqslant D_0$ is also loop free with $D_T(\phi) \leqslant D_0$. Thus ϕ is compact. The final assumption of Lemma 7 is established by Lemma 6. Thus Lemma 7 asserts the conclusion of Theorem 4.

[4] It can be seen from this that the algorithm converges in at most n steps to a ϕ satisfying the sufficient conditions (8) if D_{ik} is linear in f_{ik} for each i, k (in this case, from (C28), $\eta = \infty$).

ACKNOWLEDGMENT

The author would like to thank L. Kleinrock, A. Segall, J. Wozencraft, and two anonymous reviewers for a number of helpful comments on an earlier version of this paper.

REFERENCES

[1] D. G. Cantor and M. Gerla, "Optimal routing in a packet switched computer network," *IEEE Trans. Comput.*, vol. C-23, pp. 1062-1069, Oct. 1974.
[2] F. E. Heart, R. E. Kahn, S. M. Ornstein, W. R. Crowther, and D. C. Walden, "The interface message processor for the ARPA Computer Network," in *Conf. Rec. 1970 Spring Joint Comput. Conf., AFIPS Conf. Proc.*, 1970, pp. 551-566.
[3] A. C. Pigou, *The Economics of Welfare.* London, England: MacMillan, 1920.
[4] S. C. Dafermos and F. T. Sparrow, "The traffic assignment problem for a general network," *J. Res. Nat. Bureau of Standards-B Math. Sci.*, vol. 73B, no. 2, pp. 91-118, 1969.
[5] C. Agnew, "On the optimality of adaptive routing algorithms," in *Conf. Rec. Nat. Telecommun. Conf.*, 1974, pp. 1021-1025.
[6] —, "On quadratic adaptive routing algorithms," *Commun. Ass. Comput. Mach.*, vol. 19, no. 1, pp. 18-22, 1976.
[7] R. E. Kahn and W. R. Crowther, "A study of the ARPA Network design and performance," BBN rep. 2161, Aug. 1971.
[8] T. E. Stern, "A class of decentralized routing algorithms using relaxation," to be published.
[9] L. Fratta, M. Gerla, and L. Kleinrock, "The flow deviation method: An approach to store-and-forward communication network design," *Networks*, vol. 3, pp. 97-133, 1973.
[10] H. Frank and W. Chou, "Routing in computer networks," *Networks*, vol. 1, pp. 99-122, 1971.
[11] L. Kleinrock, *Communication Nets: Stochastic Message Flow and Delay.* New York: McGraw-Hill, 1964.
[12] —, "Analytic and simulation methods in computer network design," in *Conf. Rec., Spring Joint Comput. Conf., AFIPS Conf. Proc.*, 1970, pp. 569-579.

[13] A. Segall, "The modeling of adaptive routing in data communication networks," this issue, pp. 85–95.
[14] M. Bello, S. M. thesis, Dep. Elec. Eng. and Comput. Sci., Massachusetts Inst. Technol., Cambridge, Sept. 1976.
[15] F. R. Gantmacher, *Matrix Theory*, vol. 2. New York: Chelsea, 1959.
[16] D. G. Luenberger, *Introduction to Linear and Nonlinear Programming*. Reading, MA: Addison Wesley, 1973.

PHOTO NOT AVAILABLE

Robert G. Gallager (S'58–M'61–F'68) was born in Philadelphia, PA on May 29, 1931. He received the S.B. degree in electrical engineering from the University of Pennsylvania, Philadelphia, in 1953 and the S.M. and Sc.D. degrees in electrical engineering from the Massachusetts Institute of Technology, Cambridge, in 1957 and 1960, respectively.

From 1953 to 1954 he was a member of the technical staff at Bell Laboratories and from 1954 to 1956 was in the signal corps of the U.S. Army. He has been at the Massachusetts Institute of Technology since 1956 and was Associate Chairman of the Faculty from 1973 to 1975. He is currently a Professor of Electrical Engineering and Computer Science and is the Associate Director of the Electronic Systems Laboratory. He is also a consultant to Codex Corporation, Newton, MA. He is the author of the text book *Information Theory and Reliable Communication* (New York: Wiley, 1968), and was awarded the IEEE Baker Prize Paper Award in 1966 for the paper "A Simple Derivation of the Coding Theorem and Some Applications."

Mr. Gallager was a member of the Administrative Committee of the IEEE group on Information Theory, from 1965 to 1970 and was Chairman of the group in 1971. His major research interests are data networks, information theory, and computer architecture.

Packet Switching in Radio Channels: Part I—Carrier Sense Multiple-Access Modes and Their Throughput-Delay Characteristics

LEONARD KLEINROCK, FELLOW, IEEE, AND FOUAD A. TOBAGI

Abstract—**Radio communication is considered as a method for providing remote terminal access to computers. Digital byte streams from each terminal are partitioned into packets (blocks) and transmitted in a burst mode over a shared radio channel. When many terminals operate in this fashion, transmissions may conflict with and destroy each other. A means for controlling this is for the terminal to sense the presence of other transmissions; this leads to a new method for multiplexing in a packet radio environment: carrier sense multiple access (CSMA). Two protocols are described for CSMA and their throughput-delay characteristics are given. These results show the large advantage CSMA provides as compared to the random ALOHA access modes.**

I. INTRODUCTION

LARGE COMPUTER installations, enormous data banks, and extensive national computer networks are now becoming available. They constitute large expensive resources which must be utilized in a cost/effective fashion. The constantly growing number of computer applications and their diversity render the problem of *accessing* these large resources a rather fundamental one. Prior to 1970, wire connections were the principal means for communication among computers and between users and computers. The reasons were simple: dial-up and leased telephone lines were available and could provide inexpensive and reasonably reliable communications for short distances, using a readily available and widespread technology. It was long recognized that this technology was inadequate for the needs of a computer-communication system which is required to handle bursty traffic (i.e., large peak to average data rates). For example, the inadequacies included the long dial-up and connect time, the minimum three-minute tariff structure, the fixed and limited data rates, etc. However, it was not until 1969 that the cost to switch communication bandwidth dropped below the cost of the bandwidth being switched [1]. At that time, the new technology of packet-switched computer networks emerged and developed a cost/effective means for connecting computers together over long-distance high-speed lines. However, these networks did not solve the *local* interconnection problem, namely, how can one efficiently provide access from the user to the network itself? Certainly, one solution is to use wire connections here also. An alternate solution is the subject of this paper, namely, ground radio packet switching.

We wish to consider broadcast radio communications as an alternative for computer and user communications. The ALOHA System [2] appears to have been the first such system to employ wireless communications. The advantages in using broadcast radio communications are many: easy access to central computer installations and computer networks; collection and dissemination of data over large distributed geographical areas independent of the availability of preexisting (telephone) wire networks; the suitability of wireless connections for communications with and among mobile users (a constantly growing area of interest and applications); easily bypassed hostile terrain; etc. Perhaps, this broadcast property is the key feature in radio communication.

The Advanced Research Projects Agency (ARPA) of the Department of Defense recently undertook a new effort whose goal is to develop new techniques for packet radio communication among geographically distributed, fixed or mobile, user terminals and to provide improved frequency management strategies to meet the critical shortage of RF spectrum. The research presented in this paper is an integral part of the total design effort of this system which encompasses many other research topics [3]–[9].

Consider an environment consisting of a number of (possibly mobile) users in line-of-sight and within range of each other, all communicating over a (broadcast) radio channel in a common frequency band. The classical approach for satisfying the requirement of two users who need to communicate is to provide a communication channel for their use so long as their need continues (line-switching). However, the measurements of Jackson and Stubbs [10] show that such allocation of scarce communication resources is extremely wasteful. Rather than providing channels on a user-pair basis, we much prefer to provide a single high-speed channel to a large number of users which can be shared in some fashion. This, then, allows us to take advantage of the powerful "large number laws" which state that with very high probability, the demand at any instant will be approximately equal to

Paper approved by the Associate Editor for Computer Communication of the IEEE Communications Society for publication after presentation at the National Computer Conference, Anaheim, Calif., 1975. Manuscript received December 5, 1974; revised June 11, 1975. This work was supported in part by the Advanced Research Projects Agency, Department of Defense under Contract DAHC15-73-C-0368.

The authors are with the Computer Science Department, University of California, Los Angeles, Calif. 90024.

Reprinted from *IEEE Transactions on Communications,* vol. COM-23, no. 12, December 1975.

the sum of the average demands of that population. We wish to take advantage of these gains due to resource sharing.

Of interest to this paper is the consideration of radio channels for packet switching (also called packet radio channels). A packet is merely a package of data prepared by one user for transmission to some other user in the system. As soon as we deal with shared channels in a packet-switching mode, then we must be prepared to resolve conflicts which arise when more than one demand is simultaneously placed upon the channel. In packet radio channels, whenever a portion of one user's transmission overlaps with another user's transmission, the two collide and "destroy" each other. The existence of some acknowledgment scheme permits the transmitter to determine if his transmission was successful or not. The problem we are faced with is how to control the access to the channel in a fashion which produces, under the physical constraints of simplicity and hardware implementation, an acceptable level of performance. The difficulty in controlling a channel which must carry its own control information gives rise to the so-called random-access modes. A simple scheme, known as "pure ALOHA," permits users to transmit any time they desire. If, within some appropriate time-out period, they receive an acknowledgment from the destination, then they know that no conflicts occurred. Otherwise, they assume a collision occurred and they must retransmit. To avoid continuously repeated conflicts, some scheme must be devised for introducing a *random* retransmission delay, spreading the conflicting packets over time. A second method for using the radio channel is to modify the completely unsynchronized use of the ALOHA channel by "slotting" time into segments whose duration is exactly equal to the transmission time of a single packet (assuming constant-length packets). If we require each user to start his packets only at the beginning of a slot, then when two packets conflict, they will overlap completely rather than partially, providing an increase in channel efficiency. This method is referred to as "slotted ALOHA" [11]–[13].

The radio channel as considered in this paper is characterized as a wide-band channel with a propagation delay between any source-destination pair which is very small compared to the packet transmission time.[1] This suggests a third approach for using the channel; namely, the carrier sense multiple-access (CSMA) mode. In this scheme one attempts to avoid collisions by listening to (i.e., "sensing") the carrier due to another user's transmission.[2] Based on this information about the state of the channel, one may think of various actions to be taken by the terminal. Two protocols will be described and analyzed which we call "persistent" CSMA protocols: the nonpersistent and the p-persistent CSMA. Below, we present the protocols, discuss the assumptions, and finally establish and display the throughput-delay performance for each.

II. CSMA TRANSMISSION PROTOCOLS AND SYSTEM ASSUMPTIONS

The various protocols considered below differ by the action (pertaining to packet transmission) that a terminal takes after sensing[3] the channel. However, in all cases, when a terminal learns that its transmission was unsuccessful, it reschedules the transmission of the packet according to a randomly distributed retransmission delay. At this new point in time, the transmitter senses the channel and repeats the algorithm dictated by the protocol. At any instant a terminal is called a *ready terminal* if it has a packet ready for transmission at this instant (either a new packet just generated or a previously conflicted packet rescheduled for transmission at this instant).

A terminal may, at any one time, either be transmitting or receiving (but not both simultaneously). However, the delay incurred to switch from one mode to the other is negligible. Furthermore, the time required to detect the carrier due to packet transmissions is negligible (that is a zero detection time is assumed).[4] All packets are of constant length and are transmitted over an assumed noiseless channel (i.e., the errors in packet reception caused by random noise are not considered to be a serious problem and are neglected in comparison with errors caused by overlap interference). The system assumes noncapture (i.e., the overlap of any fraction of two packets results in destructive interference and both packets must be retransmitted). We further simplify the problem by assuming the propagation delay (small compared to the packet transmission time) to be identical[5] for all source-destination pairs.

We first consider the *nonpersistent CSMA*. The idea here is to limit the interference among packets by always rescheduling a packet which finds the channel busy upon arrival. More precisely, a ready terminal senses the channel and operates as follows.

1) If the channel is sensed idle, it transmits the packet.

2) If the channel is sensed busy, then the terminal schedules the retransmission of the packet to some later time according to the retransmission delay distribution. At this new point in time, it senses the channel and repeats the algorithm described.

A slotted version of the nonpersistent CSMA can be

[1] Consider, for example, 1000-bit packets transmitted over a channel operating at a speed of 100 kbits/s. The transmission time of a packet is then 10 ms. If the maximum distance between the source and the destination is 10 mi, then the (speed of light) packet propagation delay is of the order of 54 μs. Thus the propagation delay constitutes only a very small fraction ($a = 0.005$) of the transmission time of a packet. On the contrary, when one considers satellite channels [13] the propagation delay is a relatively large multiple of the packet transmission time ($a \gg 1$).

[2] Sensing carrier prior to transmission is a well-known concept in use for (voice) aircraft communication. In the context of packet radio channels, it was originally suggested by D.Wax of the University of Hawaii in an internal memorandum dated Mar. 4, 1971.

[3] Each terminal has the capability of sensing carrier on the channel. The practical problems of feasibility and implementation of sensing, however, are not addressed here.

[4] The detection time is considered negligible for relatively wide-band channels (100 kHz). In Part II [19] the detection time on the "busy-tone" narrow-band channels (on the order of 2 kHz) will be accounted for in the analysis.

[5] By considering this constant propagation delay equal to the largest possible, one gets lower (i.e., pessimistic) bounds on performance.

considered in which the time axis is slotted and the slot size is τ seconds (the propagation delay). All terminals are synchronized[6] and are forced to start transmission only at the beginning of a slot. When a packet's arrival occurs during a slot, the terminal senses the channel at the beginning of the next slot and operates according to the protocol described above.

We next consider the *p-persistent CSMA* protocol. However, before treating the general case (arbitrary p), we introduce the special case of $p = 1$.

The *1-persistent CSMA* protocol is devised in order to (presumably) achieve acceptable throughput by never letting the channel go idle if some ready terminal is available. More precisely, a ready terminal senses the channel and operates as follows.

1) If the channel is sensed idle, it transmits the packet with probability one.

2) If the channel is sensed busy, it waits until the channel goes idle (i.e., persisting on transmitting) and only then transmits the packet (with probability one—hence, the name of 1-persistent).

A slotted version of this 1-persistent CSMA can also be considered by slotting the time axis and synchronizing the transmission of packets in much the same way as for the previous protocol.

The above 1-persistent and nonpersistent protocols differ by the probability (one or zero) of not rescheduling a packet which upon arrival finds the channel busy. In the case of a 1-persistent CSMA, we note that whenever two or more terminals become ready during a transmission period (TP), they wait for the channel to become idle (at the end of that transmission) and then they all transmit with probability one. A conflict will also occur with probability one! The idea of randomizing the starting time of transmission of packets accumulating at the end of a TP suggests itself for interference reduction and throughput improvement. The scheme consists of including an additional parameter p, the probability that a ready packet persists ($1 - p$ being the probability of delaying transmission by τ seconds). The parameter p will be chosen so as to reduce the level of interference while keeping the idle periods between any two consecutive nonoverlapped transmissions as small as possible. This gives rise to the *p-persistent CSMA*, which is a generalization of the 1-persistent CSMA.

More precisely, the protocol consists of the following: the time axis is finely slotted where the (mini) slot size is τ seconds. For simplicity of analysis, we consider the system to be synchronized such that all packets begin their transmission at the beginning of a (mini) slot.

Consider a ready terminal. If the channel is sensed idle, then: with probability p, the terminal transmits the packet; or with probability $1 - p$, the terminal delays the transmission of the packet by τ seconds (i.e., one slot). If at this new point in time, the channel is still detected idle, the same process is repeated. Otherwise, some packet must have started transmission, and our terminal schedules the retransmission of the packet according to the retransmission delay distribution (i.e., acts as if it had conflicted and learned about the conflict).

If the ready terminal senses the channel busy, it waits until it becomes idle (at the end of the current transmission) and then operates as above.

III. TRAFFIC MODEL: ASSUMPTIONS AND NOTATION

In the previous section, we identified the system protocols, operating procedures, and assumptions. Here we characterize the traffic source and its underlying assumptions.

We assume that our traffic source consists of an infinite number of users who collectively form an independent Poisson source with an aggregate mean packet generation rate of λ packets/s. This is an approximation to a large but finite population in which each user generates packets infrequently and each packet can be successfully transmitted in a time interval much less than the average time between successive packets generated by a given user. Each user in the infinite population is assumed to have at most one packet requiring transmission at any time (including any previously blocked packet).

In addition, we characterize the traffic as follows. We have assumed that each packet is of constant length requiring T seconds for transmission. Let $S = \lambda T$. S is the average number of packets generated per transmission time, i.e., it is the input rate normalized with respect to T. Under steady-state conditions, S can also be referred to as the channel throughput rate. Now, if we were able to perfectly schedule the packets into the available channel space with absolutely no overlap or gaps between the packets, we could achieve a maximum throughput equal to 1; therefore we also refer to S as the *channel utilization*. Because of the interference problem inherent in the random nature of the access modes, the achievable throughput will always be less than 1. The maximum achievable throughput for an access mode is called the *capacity* of the channel under that mode.

Since conflicts can occur, some acknowledgment scheme is necessary to inform the transmitter of its success or failure. We assume a positive acknowledgment scheme[7]: if within some specified delay (an appropriate time-out period) after the transmission of a packet, a user does not receive an acknowledgment, he knows he has conflicted. If he now retransmits immediately, and if all users behave likewise, then he will definitely be interfered with again (and forever!). Consequently, as mentioned above, each user delays the transmission of a previously collided packet by some random time whose mean is $\bar{X}$ (chosen, for example, uniformly between 0 and $X_{max} = 2\bar{X}$). The traffic

[6] In this paper, the practical problems involved in synchronizing terminals are not addressed.

[7] The channel for acknowledgment is assumed to be separate from the channel we are studying (i.e., acknowledgments arrive reliably and at no cost).

offered to the channel from our collection of users consists not only of new packets but also of previously collided packets: this increases the mean *offered* traffic rate which we denote by G (packets per transmission time T) where $G \geq S$.

Our two further assumptions are the following.

Assumption 1: The average retransmission delay $\bar{X}$ is large compared to T.

Assumption 2: The interarrival times of the point process defined by the start times of all the packets plus retransmissions are independent and exponentially distributed.

It is clear that Assumption 2 is violated in the protocols we consider. (We have introduced it for analytic simplicity.) However, in Section V, some simulation results are discussed which show that performance results based on this assumption are excellent approximations, particularly when the average retransmission delay $\bar{X}$ is large compared to T. Moreover, in the context of slotted ALOHA it was analytically shown [14] in the limit as $\bar{X} \rightarrow \infty$, that Assumption 2 is satisfied; furthermore, simulation results showed that only the first moment of the retransmission delay distribution had a noticeable effect on the average throughput-delay performance.

So far, we have defined the following important system variables: S (throughput), G (offered channel traffic rate), T (packet transmission time), $\bar{X}$ (average retransmission delay), τ (propagation delay), and p (p-persistent parameter). Without loss of generality, we choose $T = 1$. This is equivalent to expressing time in units of T. We express $\bar{X}$ and τ in these normalized time units as $\delta = \bar{X}/T$ and $a = \tau/T$.

IV. THROUGHPUT ANALYSIS

We wish to solve for the channel capacity of the system for all of the access protocols described above. This we do by solving for S in terms of G (as well as the other system parameters). The channel capacity is then found by maximizing S with respect to G. S/G is merely the probability of a successful transmission and G/S is the average number of times a packet must be transmitted (or scheduled) until success. In Section V, we discuss delay and give the throughput-delay tradeoff for these protocols.

This analysis is based on renewal theory and probabilistic arguments requiring independence of random variables provided by Assumption 2. Moreover steady-state conditions are assumed to exist. However from the (S,G) relationships found below one can see that steady state may not exist because of inherent instability of these random-access techniques. This instability is simply explained by the fact that when statistical fluctuations in G increase the level of mutual interference among transmissions, then the positive feedback causes the throughput to decrease to 0. Nevertheless, the results are useful for the following reasons.

1) They are meaningful for a finite (and possibly long) period of time. (Simulations supporting these analytic results showed no saturation over the simulated period of time when $\bar{X}$ was large enough; see Section V.)

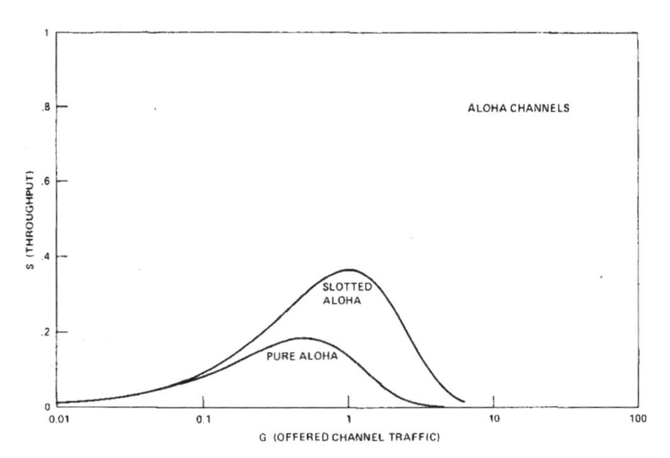

Fig. 1. Throughput in ALOHA channels.

2) In finite population cases, stable situations are possible for which steady-state results prevail over an infinite time horizon. (See [14] and [16].)

3) Control procedures have been prescribed for the slotted ALOHA random access [14] which stabilize unstable channels, achieving performance very close to the equilibrium results.

A. ALOHA Channels

In the *pure ALOHA* access mode, each terminal transmits its packet over the data channel in a completely unsynchronized manner. Under the system and model assumptions (mainly Assumption 2), we have

$$S = GP_s$$

where P_s is the probability that an arbitrary offered packet is successful. A given packet will overlap with another packet if there exists at least one start of transmission within T seconds before or after the start time of the given packet (i.e., over a "vulnerable" interval of length $2T$). Using the Poisson traffic assumption, Abramson [2] first showed that

$$S = Ge^{-2G}. \tag{1}$$

Thus, we see that pure ALOHA achieves a maximum throughput of $1/(2e) = 0.184$ (at $G = 1/2$).

In the *slotted ALOHA*, if two packets conflict, they will overlap completely rather than partially (i.e., a vulnerable interval only of length T). The throughput equation then becomes

$$S = Ge^{-G} \tag{2}$$

and was first obtained by Roberts [12] who extended Abramson's result in (1). With this simple change, the maximum throughput is increased by a factor of two to $1/e = 0.368$ (at $G = 1$). In Fig. 1, we plot the throughput S versus the offered traffic G for these two systems. From these results, it is all too evident that a significant fraction of the channel's ultimate capacity ($C = 1$) is not utilized with the ALOHA access modes; we recover a major portion of this loss with the CSMA protocols, as we now show.

B. Nonpersistent CSMA

The basic equation for the throughput S is expressed in terms of a (the ratio of propagation delay to packet transmission time) and G (the offered traffic rate) as follows:

$$S = \frac{Ge^{-aG}}{G(1 + 2a) + e^{-aG}}. \tag{3}$$

Proof: G denotes the arrival rate of new and rescheduled packets. All arrivals, in this case, do not necessarily result in actual transmissions (a packet which finds the channel in a busy state is rescheduled without being transmitted). Thus, G constitutes the "offered" channel traffic and only a fraction of it constitutes the channel traffic itself. Consider the time axis[8] (See Fig. 2)[9] and let t be the time of arrival of a packet which senses the channel idle and such that no other packet is in the process of transmission. Any other packet arriving between t and $t + a$ will find (sense) the channel as unused, will transmit, and hence will cause a conflict. If no other terminal transmits a packet during these a seconds (the "vulnerable" period), then the first packet will be successful.

Let $t + Y$ be the time of occurrence of the last packet arriving between t and $t + a$. The transmission of all packets arriving in $(t, t + Y)$ will be completed at $t + Y + 1$. Only a seconds later will the channel be sensed unused. Now, any terminal becoming ready between $t + a$ and $t + Y + 1 + a$ will sense the channel busy and hence will reschedule its packet. The interval between t and $t + Y + 1 + a$ is called a *transmission period* (TP). Note that there can be at most one successul transmission during a TP. Define an *idle period* to be the period of time between two consecutive TP's (also called busy periods in this simple case). A busy period plus the following idle period constitute a cycle. Let $\bar{B}$ be the expected duration of the busy period, $\bar{I}$ the expected duration of the idle period, and $\bar{B} + \bar{I}$ the expected length of a cycle. Let U denote the time during a cycle that the channel is used without conflicts. Using renewal theory arguments, the average channel utilization is simply given by

$$S = \frac{\bar{U}}{\bar{B} + \bar{I}}. \tag{4}$$

The probability that a TP is successful is simply the probability that no terminal transmits during the first a seconds of the period and is equal to e^{-aG}. Therefore

$$\bar{U} = e^{-aG}. \tag{5}$$

The average duration of an idle period is simply $1/G$. The average duration of a busy interval is $1 + \bar{Y} + a$, where $\bar{Y}$ is the expected value of Y.

[8] The reference time axis considered in this and subsequent proofs is the transmitter's time. Shifting all transmissions by τ seconds will give a description of events on the station's time axis. Any time overlap in transmission on the station's time axis results in packet interference.

[9] In this and other figures, a vertical arrow represents a terminal becoming ready.

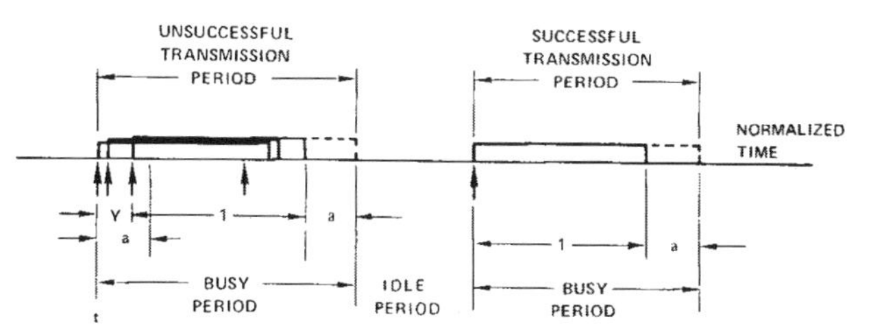

Fig. 2. Nonpersistent CSMA: Busy and idle periods.

The distribution function for Y is

$$F_Y(y) \triangleq \Pr\{Y \le y\} = \Pr\{\text{no arrival occurs in an interval of length } a - y\} = \exp\{-G(a - y)\}, \quad (y \le a). \tag{6}$$

The average of Y is therefore given by

$$\bar{Y} = a - \frac{1}{G}(1 - e^{-aG}). \tag{7}$$

Applying (4) and using the expressions found for $\bar{U}$, $\bar{B}$, and $\bar{I}$, we get (3). Q.E.D.

It is easy to prove that the throughput equation for the *slotted* nonpersistent CSMA is given by

$$S = \frac{aGe^{-aG}}{(1 - e^{-aG}) + a}. \tag{8}$$

Note that for both cases we have

$$\lim_{a \to 0} S = G/(1 + G). \tag{9}$$

This shows that when $a = 0$, a throughput of 1 can theoretically be attained for an offered channel traffic equal to infinity. S versus G for various values of a is plotted in Fig. 3.

C. 1-Persistent CSMA

The throughput equation for this protocol is given by

$$S = \frac{G[1 + G + aG(1 + G + aG/2)]e^{-G(1+2a)}}{G(1 + 2a) - (1 - e^{-aG}) + (1 + aG)e^{-G(1+a)}}. \tag{10}$$

Proof: Consider Fig. 4 and again let t be the time of arrival of a packet which senses the channel to be idle with no other packet in the process of transmission. In this protocol, any packet arriving in the interval $[t + a, t + Y + 1 + a]$ will sense the channel busy and hence must wait until the channel is sensed idle (at time $t + 1 + Y + a$) at which time they will *all* transmit! The number of packets accumulated at the end of TP is the number of arrivals in $1 + Y$ seconds. If this total is equal to or greater than two, then a conflict occurs in the next TP with probability 1.

Define a *busy period* to be the time between t and the end of that TP during which no packets accumulate. De-

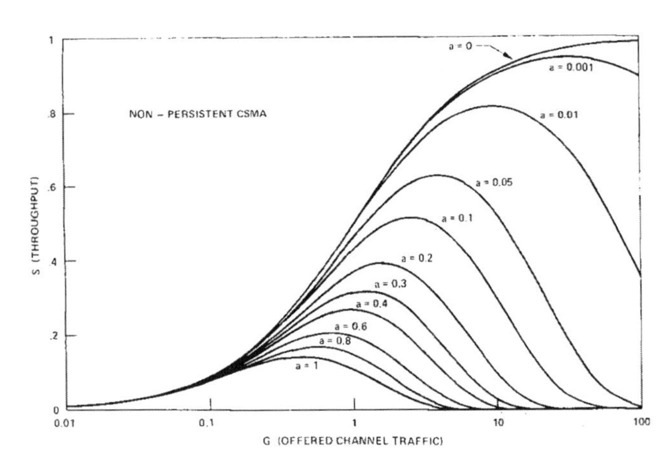

Fig. 3. Throughput in nonpersistent CSMA.

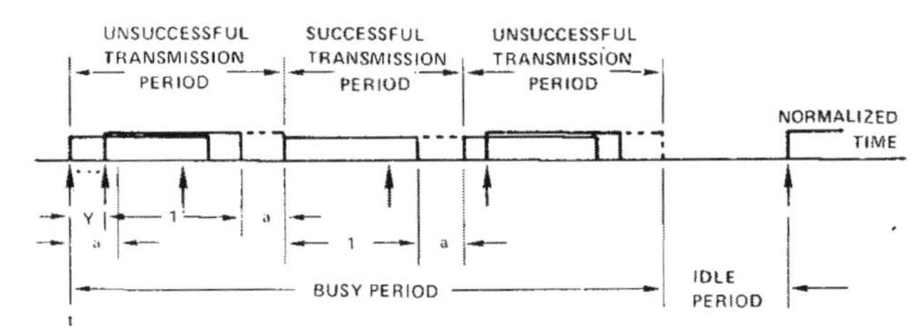

Fig. 4. 1-persistent CSMA: TP's, busy periods, and idle periods.

fine an *idle period* to be the period of time in which the channel is idle and no packets are present awaiting transmission. A busy period plus the following idle period constitute a *cycle*.

Let $\bar{B}$ be the expected duration of the busy period, $\bar{I}$ the expected duration of the idle period, and $\bar{B} + \bar{I}$ the expected length of a cycle.

Let us now consider the transmission of an arbitrary packet. Three situations must be considered.

1) If the packet arrives to an idle system, then its transmission is successful if and only if no packets arrive during its first a seconds; its probability of success is therefore e^{-aG}.

2) If the packet arrives during the first a seconds of a TP, then its probability of success is 0.

3) If the packet arrives during the channel busy period (excluding the first a seconds of the TP), then it is successful (in the next TP) if and only if it is the only packet to arrive during this TP and no packets arrive during *its* first a seconds. To calculate its probability of success, we observe that a TP is of random length equal to $1 + a + Y$ where Y is a random variable. Let B' denote the time during a cycle that the channel is in its busy period excluding the first a seconds of each TP. B' is a sequence of segments of random length $1 + Y \triangleq Z$ separated by periods of a seconds. Knowing that a packet arrives in B', this packet is more likely to arrive in a longer segment Z than in a shorter one (due to the "paradox of residual life" [17]). Let $\hat{Z}$ denote the segment in which the arrival occurred, and $\hat{q}_0$ (derived below) be the probability that no arrival occurs in $\hat{Z}$; the probability of success of the packet is therefore $\hat{q}_0 e^{-aG}$.

Only cases 1) and 3) contribute to a successful transmission. Let $\bar{B}'$ be the expected value of B'. From renewal theory arguments, the probability that an arrival finds the channel idle [case (1)] is given by $\bar{I}/(\bar{B} + \bar{I})$, and the probability that an arrival finds the channel in situation 3) is $\bar{B}'/(\bar{B} + \bar{I})$; then the probability of success of the packet is given by

$$P_s \triangleq \Pr\{\text{success}\} = \frac{\bar{I}}{\bar{B} + \bar{I}} e^{-aG} + \frac{\bar{B}'}{\bar{B} + \bar{I}} \hat{q}_0 e^{-aG}. \quad (11)$$

The determination of $\bar{I}, \bar{B}, \bar{B}'$, and $\hat{q}_0$ follows.

Since the traffic is Poisson, it is clear that the average idle period is given by

$$\bar{I} = 1/G. \quad (12)$$

For $\bar{B}, \bar{B}'$, and $\hat{q}_0$ we must first obtain some intermediate results as follows. The distribution function for Y and its average are given in (6) and (7), respectively. The Laplace transform of the probability density function of Y, defined as

$$F_Y^*(s) \triangleq \int_0^\infty e^{-sy}\, dF_Y(y),$$

is given by

$$F_Y^*(s) = e^{-aG} + \frac{G(e^{-as} - e^{-aG})}{G - s}. \quad (13)$$

Let us now find the distribution of the number of packets accumulated at the end of a TP.

Let

$$q_m(y) \triangleq \Pr\{m \text{ packets accumulated at end of TP} \mid Y = y\}$$

and

$$q_m = \int_0^a q_m(y)\, dF_Y(y).$$

Let $Q(z)$ denote the generating function of q_m defined by

$$Q(z) \triangleq \sum_{m=0}^{\infty} q_m z^m.$$

The number of packets accumulated at the end of a TP is equal to the number of packets arriving during a period of time equal to $1 + Y$. Let m_1 denote the number of packets arriving in $T = 1$, and m_2 the number of packets arriving in Y. Let $Q_1(z)$ and $Q_2(z)$ denote the generating functions of the probability distributions for m_1 and m_2, respectively. Since the arrival process is Poisson, the random variables m_1 and m_2 are independent and the generating function $Q(z)$ of q_m, where $m = m_1 + m_2$, is given by

$$Q(z) = Q_1(z) Q_2(z).$$

We have [17]

$$Q_1(z) = \exp\{G(z - 1)\}$$

and

$$Q_2(z) = F_Y^*(G(1 - z)).$$

From (13) we get

$$Q(z) = \exp\{G(z-1)\}\exp\{-aG\}\left[1+\frac{\exp\{aGz\}-1}{z}\right]. \quad (14)$$

We may invert this explicit expression for $Q(z)$; in particular we find that the probability of zero packets accumulated at the end of a TP is

$$q_0 = Q(z)|_{z=0} = \exp\{-G(1+a)\}[1+aG]. \quad (15)$$

To find the average busy period, we let Y_i denote the random variable Y defined above corresponding to the ith TP in a busy period. All Y_i, $i = 1,2,\cdots$, are independent and identically distributed. It is easy to see that the number of TP's in a busy period is geometrically distributed with mean $1/q_0$. Conditioned on the fact that we have exactly k TP's in the busy period and that $Y_i = y_i$ for $i = 1,2,\cdots,k$, the average busy period is

$$\bar{B}(y_1,y_2,\cdots,y_k) = k(1+a) + y_1 + y_2 + \cdots + y_k.$$

Therefore, by removing the conditions on k and Y_i, we get $\bar{B}$ as

$$\bar{B} = \cdots\int_{y_i=0}^{a}\cdots\int_{y_1=0}^{a}\sum_{k=1}^{\infty}[k(1+a)+y_1+\cdots+y_k]$$

$$\cdot\, q_0(y_k)\prod_{i=1}^{k-1}(1-q_0(y_i))\, dF_{Y_1}(y_1)\cdots dF_{Y_i}(y_i)\cdots.$$

It is easy to see that by inverting the order of summation and integration, the contribution of the term $k(1+a)$ reduces to $(1+a)/q_0$ and the contribution of the generic term y_j simply reduces to $\bar{Y}(1-q_0)^{j-1}$. Finally, we have

$$\bar{B} = \frac{1+a}{q_0} + \sum_{j=1}^{\infty}\bar{Y}(1-q_0)^{j-1} = \frac{1+a+\bar{Y}}{q_0}. \quad (16)$$

Since the average number of TP's is $1/q_0$, from the distribution of B' we have

$$\bar{B}' = \frac{1+\bar{Y}}{q_0}. \quad (17)$$

In (11) for P_s, it remains only to compute $\hat{q}_0$. The probability density function of $Z = 1 + Y$ is easily obtained from the distribution of Y. From (6), the probability density function of Y can be expressed as

$$f_Y(y) = \exp\{-aG\}u_0(y) + G\exp\{-aG\}\exp\{Gy\}, \qquad 0 \le y \le a$$

where $u_0(y)$ is the unit impulse at $y = 0$. Thus we have

$$f_Z(x) = \exp\{-aG\}u_0(x-1) + G\exp\{-aG\} \cdot \exp\{G(x-1)\}, \qquad 1 \le x \le 1+a.$$

The probability density function of $\hat{Z}$ is given by [17]

$$f_{\hat{Z}}(x) = \frac{xf_Z(x)}{\bar{Z}} = \frac{e^{aG}}{1+\bar{Y}}u_0(x-1) + \frac{Gxe^{-aG}e^{G(x-1)}}{1+\bar{Y}}, \qquad 1 \le x \le 1+a.$$

Finally, the probability that no arrival occurs (from our Poisson source) in the interval $\hat{Z}$ is simply

$$\hat{q}_0 = \int_{x=1}^{1+a}\exp\{-Gx\}f_{\hat{Z}}(x)\,dx = \frac{\exp\{-G(1+a)\}}{1+\bar{Y}}[1+aG(1+a/2)]. \quad (18)$$

Using our expressions for $\bar{I}$, $\bar{B}$, $\bar{B}'$, and $\hat{q}_0$ in (12), (16), (17), and (18), respectively, we immediately obtain from (11)

$$P_s = \frac{\dfrac{1+\bar{Y}}{q_0}e^{-aG}\hat{q}_0 + \dfrac{1}{G}e^{-aG}}{\dfrac{1+a+\bar{Y}}{q_0} + \dfrac{1}{G}}.$$

Substituting the expressions obtained for q_0, $\hat{q}_0$, and $\bar{Y}$, and recalling that $S = GP_s$, we have finally established (10).

Q.E.D.

Slotted 1-persistent CSMA: Let us now consider the *slotted* version of 1-persistent CSMA. The throughput equation for this case is given by

$$S = \frac{G\exp\{-G(1+a)\}[1+a-\exp\{-aG\}]}{(1+a)(1-\exp\{-aG\}) + a\exp\{-G(1+a)\}} \quad (19)$$

Proof: In this slotted version, as in slotted ALOHA, if two packets conflict, they will overlap completely. The length of a TP is always equal to $1 + a$. (We have assumed that the packet transmission time is an integer multiple of the propagation delay.)

Since the traffic process is an independent one (Assumption 2), the number of slots in an idle period is geometrically distributed with a mean equal to $1/(1-e^{-aG})$. Thus the average idle period is given by

$$\bar{I} = \frac{a}{1-e^{-aG}}. \quad (20)$$

Using a similar argument, we find that the average busy period is given by

$$\bar{B} = \frac{1+a}{\exp\{-G(1+a)\}}. \quad (21)$$

Let $\bar{U}$ again denote the expected time during a cycle that the channel is used without conflicts. In order to find $\bar{U}$ we need to determine the probability of success over each

TP in the busy period. The probability of success over the first TP is given by

$$\Pr\{\text{success over first TP}\} = \Pr\{\text{only one packet arrives during the last slot of the preceding idle period/some arrival occurred}\} = \frac{aGe^{-aG}}{1 - e^{-aG}}.$$

Similarly we have:

$$\Pr\{\text{success over any other TP}\} = \frac{G(1+a)\exp\{-G(1+a)\}}{1 - \exp\{-G(1+a)\}}.$$

The number of TP's in a busy period is geometrically distributed with a mean equal to

$$\exp\{G(1+a) \triangleq 1/q_0,$$

thus

$$\bar{U} = \frac{aG\exp\{-aG\}}{1 - \exp\{-aG\}} + \left(\frac{1}{q_0} - 1\right) \cdot \frac{G(1+a)\exp\{-G(1+a)\}}{1 - \exp\{-G(1+a)\}}. \quad (22)$$

Applying (4) and using the expressions found for $\bar{U}$, $\bar{I}$, and $\bar{B}$, we get (19). Q.E.D.

The ultimate performance in the ideal case ($a = 0$), for both slotted and unslotted versions, is

$$S = \frac{Ge^{-G}(1+G)}{G + e^{-G}}. \quad (23)$$

For any value of a, the maximum throughput S will occur at an optimum value of G. In Fig. 5 we show S versus G for the nonslotted version of 1-persistent CSMA for various values of a.

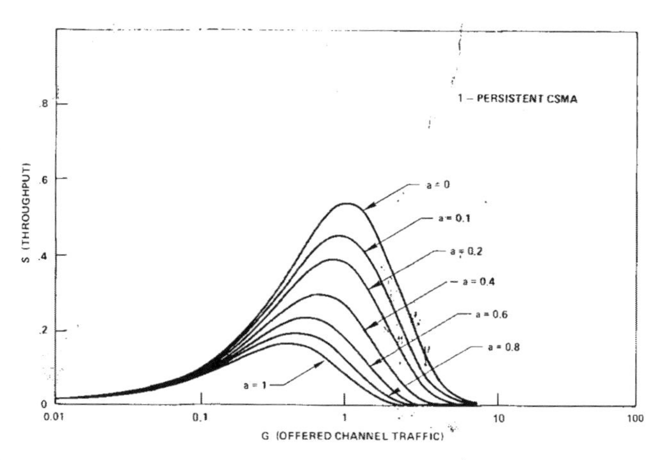

Fig. 5. Throughput in 1-persistent CSMA.

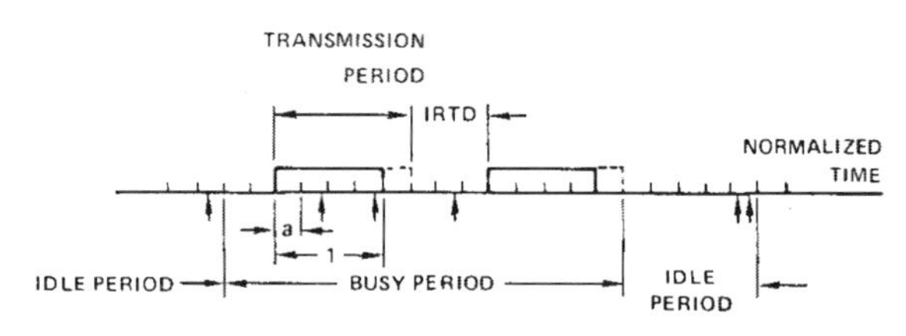

Fig. 6. p-persistent CSMA: TP's, busy periods, and idle periods.

D. *p-Persistent CSMA*

For a given offered traffic G and a given value of the parameter p, we can determine the throughput S as

$$S(G,p,a) = \frac{(1 - e^{-aG})[P_s'\pi_0 + P_s(1 - \pi_0)]}{(1 - e^{-aG})[a\bar{t}'\pi_0 + a\bar{t}(1 - \pi_0) + 1 + a] + a\pi_0} \quad (24)$$

where P_s', P_s, $\bar{t}'$, $\bar{t}$, and π_0 are defined in the following proof in (37), (34), (36), (30), and (25), respectively.

Proof: Consider a TP and assume that some packets arrive during the period as shown in Fig. 6. These packets sense the channel busy and accumulate at the end of the TP, at which point they randomize the starting times of their transmission according to the randomizing process described in Section II. This randomization creates a random delay before a TP starts, called the initial random transmission delay (IRTD), during which time the channel is "wasted." If, at the start of a new TP, two or more terminals decide to transmit, then a conflict will certainly occur. All other packets which have delayed their transmission by τ seconds will then sense the channel busy and will have to be rescheduled for transmission by incurring a retransmission delay δ. Thus, at the expense of creating this IRTD, we greatly improve the probability of success over a TP.

Consider Fig. 6 in which we observe two TP's separated by an IRTD. One can also define busy periods and idle periods in much the same way as before. An idle period is that period of time during which the channel is idle and no packets are ready for transmission. A busy period consists of a sequence of transmission periods such that some packets arrive during each transmission period *except* the last one. Let $\mathfrak{I}_i$ denote the ith TP of a busy period. In order to find the channel utilization, we once again apply (4), which requires identifying and determining the average busy and idle periods, the gaps between TP's, as well as the condition for success over each TP. This we do as follows.

Recall that we require the system to be (mini-) slotted (the slot size equal to a, the normalized propagation delay) and all transmissions to start at the beginning of a slot. Here again we consider the transmission time of a packet to be an integer number $1/a$ slots (recall $T = 1$). Let $g = aG$; g is the average arrival rate of new and rescheduled packets during a (mini) slot.

We first determine the distribution of the number of packets accumulated at the end of a TP. Let N denote this number and let $\pi_n \triangleq \Pr\{N = n\}$. According to the protocol described in Section II, only those packets arriving during a TP will accumulate at the end of that TP. Therefore, by Assumption 1, we have

$$\pi_n = \frac{[(1+a)G]^n}{n!} \exp\{-(1+a)G\}, \qquad n \geq 0. \quad (25)$$

To find the distribution of the IRTD between two successive TP's in the same busy period, we condition $N = n$ and we let t_n be the number of slots elapsed until some packet is transmitted. Let $q = 1 - p$. It is easy to see that

$$\Pr\{t_n > k\} = q^{(k+1)n} \prod_{j=1}^{k} \left[\sum_{m=0}^{\infty} \exp\{-g\} \frac{g^m}{m!} q^{m(k-j+1)} \right]$$

$$= q^{(k+1)n} \prod_{j=1}^{k} \exp\{g(q^{k-j+1} - 1)\}$$

$$= q^{(k+1)n} \exp\left\{g\left(\frac{q(1-q^k)}{p} - k\right)\right\} \quad (26)$$

and, therefore, for $k > 0$ we have

$$\Pr\{t_n = k\} = \Pr\{t_n > k-1\} - \Pr\{t_n > k\}$$

$$= q^{kn}[1 - q^n \exp\{-g(1-q^k)\}]$$

$$\cdot \exp\left\{g\left(\frac{q(1-q^{k-1})}{p} - (k-1)\right)\right\} \quad (27)$$

and for $k = 0$,

$$\Pr\{t_n = 0\} = 1 - q^n. \quad (28)$$

The average IRTD is given by

$$\bar{t}_n = \sum_{k=0}^{\infty} \Pr\{t_n > k\}. \quad (29)$$

Removing the condition on N, we get

$$\bar{t} = \sum_{n=1}^{\infty} \bar{t}_n \frac{\pi_n}{1 - \pi_0}. \quad (30)$$

$\bar{t}$ is the average gap between two consecutive TP's in a busy period.

In order to find the probability of success over a TP $\mathfrak{I}_i$ one has to distinguish two cases: $i = 1$ and $i \neq 1$. We first treat the second case, $i \neq 1$. Given $N = n$, define[10]:

- $P_s(n)$ probability of success over $\mathfrak{I}_i$
- L_n the number of packets present at the starting time of $\mathfrak{I}_i$
- $L_n - n$ merely the number of packets arriving during the gap t_n.

By the Poisson assumption we have

$$\Pr\{L_n = l/t_n = k\} = \frac{(kg)^{l-n}}{(l-n)!} e^{-kg}, \qquad l \geq n. \quad (31)$$

[10] The quantities $P_s(n)$, P_s, and L_n need no index i since they are identical for all $\mathfrak{I}_i$, $i \neq 1$.

Removing the condition on t_n,

$$\Pr\{L_n = l\} = \sum_{k=1}^{\infty} \frac{(kg)^{l-n}}{(l-n)!} e^{-kg} \Pr\{t_n = k\} + (1 - q^n)\delta_{l,n}, \qquad l \geq n \quad (32)$$

where $\delta_{i,j}$ is the Kronecker delta. The probability of success over $\mathfrak{I}_i$ is equal to the probability that none of the L_n transmit over $\mathfrak{I}_i$:

$$P_s(n) = \sum_{l=n}^{\infty} \frac{lpq^{l-1}}{1 - q^l} \Pr\{L_n = l\}. \quad (33)$$

Removing the condition on N, we get

$$P_s = \sum_{n=1}^{\infty} P_s(n) \frac{\pi_n}{1 - \pi_0}. \quad (34)$$

For the probability of success over $\mathfrak{I}_1$ we note that the number of packets present at the beginning of a busy period, denoted by N', is the number of packets arriving in the last slot of the previous idle period. We then have

$$\pi_n \triangleq \Pr\{N' = n\}$$

$$= \frac{g^n}{n!} \frac{e^{-g}}{1 - e^{-g}}, \qquad n \geq 1. \quad (35)$$

Given $N' = n$, let t_n' denote the first initial random transmission delay of the busy period, and $P_s'(n)$ denote the probability of success over $\mathfrak{I}_1$. The distribution of t_n' and its average $\bar{t}_n'$ are the same as for t_n [(27) and (29)]. $P_s'(n)$ is the same as $P_s(n)$ [see (33)]. Removing the condition on N', we get

$$\bar{t}' = \sum_{n=1}^{\infty} t_n' \pi_n' \quad (36)$$

$$P_s' = \sum_{n=1}^{\infty} P_s'(n) \pi_n'. \quad (37)$$

It remains to compute $\bar{B}$, $\bar{U}$, and $\bar{I}$. It is clear that the number of TP's in a busy period is equal to m with probability $\pi_0(1 - \pi_0)^{m-1}$.

Consider a busy period with m TP's. Let N_i denote the number of packets accumulated at the end of the ith TP. We know that $N_m = 0$, and that all other $N_i \geq 1$ are independent and identically distributed random variables. Conditioned on the fact that $N_i = n_i, i = 1,\cdots,m-1$, the average busy period is given by

$$\bar{B}_m(n_1,\cdots,n_{m-1}) = a\bar{t}' + \sum_{i=1}^{m-1} a\bar{t}_{n_i} + m(1+a). \quad (38)$$

The expected time, during the busy period, that the channel is used without conflicts is given by

$$\bar{U}_m(n_1,\cdots,n_{m-1}) = P_s' + \sum_{i=1}^{m-1} P_s(n_i). \quad (39)$$

On the other hand, we know that

$$\Pr\{N_i = n_i\} = \frac{\pi_{n_i}}{1-\pi_0}, \qquad n_i \geq 1,\ i = 1,2,\cdots,m-1. \tag{40}$$

Therefore, removing the conditions $N_i = n_i$ in (38) and (39), we get

$$\bar{B}_m = a\bar{t}' + (m-1)a\bar{t} + m(1+a) \tag{41}$$

$$\bar{U}_m = P_s' + (m-1)P_s \tag{42}$$

and removing the condition on m we get

$$\bar{B} = \sum_{m=1}^{\infty} \bar{B}_m \pi_0 (1-\pi_0)^{m-1} = a\bar{t}' + \frac{a\bar{t}(1-\pi_0) + 1 + a}{\pi_0} \tag{43}$$

$$\bar{U} = P_s' + \frac{1-\pi_0}{\pi_0} P_s. \tag{44}$$

The idle period is geometrically distributed with mean $1/(1-e^{-g})$; its average is:

$$\bar{I} = \frac{a}{1-e^{-g}}. \tag{45}$$

Finally, using (4) and substituting for $\bar{B}$, $\bar{U}$, and $\bar{I}$ the expressions found in (43), (44), and (45), respectively, we get the throughput S; it is a function of G, p, and $a = 1/T$ and is expressed as

$$S(G,p,a) = \frac{P_s' + \dfrac{1-\pi_0}{\pi_0} P_s}{a\bar{t}' + a\bar{t}\dfrac{1-\pi_0}{\pi_0} + \dfrac{1+a}{\pi_0} + \dfrac{a}{1-e^{-g}}} \tag{46}$$

which reduces to (24). Q.E.D.

In order to evaluate $S(G,p,a)$, a PL/1 program was written and run on the IBM 360/91 of the Campus Computing Network at UCLA. For small values of p ($0.01 \leq p \leq 0.1$), the numerical computation as suggested by (24) becomes time consuming and requires an extremely large amount of storage. Fortunately some approximations have been found useful which lead to a closed-form solution for the throughput (see the derivations of $S'(G,p,a)$ in Appendix A).

Special case $a = 0$: Let us now consider the special case $a = 0$. For finite G, $g = aG = 0$. Equation (26) becomes

$$\Pr\{t_n > k\} = q^{(k+1)n}.$$

The average IRTD is then given by (29), and is expressed as

$$\bar{t}_n = \sum_{k=0}^{\infty} \Pr\{t_n > k\} = \frac{q^n}{1-q^n}.$$

It is important to note that $\bar{t}_n$ is finite, so is $\bar{t}$. On the other hand the idle period given in (45) becomes

$$\bar{I} = \frac{1}{G}.$$

Since $\bar{t}$ and $\bar{t}'$ are finite, by letting $a \to 0$ in (46) we get

$$S(G,p,a=0) = \frac{P_s' + \dfrac{1-\pi_0}{\pi_0} P_s}{\dfrac{1}{\pi_0} + \dfrac{1}{G}}. \tag{47}$$

To compute P_s we have to get back to (31) through (34). With $a = 0$ we have

$$\Pr\{L_n = l/t_n = k\} = 1, \qquad l = n$$

and

$$\Pr\{L_n = n\} = 1.$$

Therefore

$$P_s(n) = \frac{npq^{n-1}}{1-q^n} \tag{48}$$

and

$$P_s = \sum_{n=1}^{\infty} \frac{npq^{n-1}}{1-q^n} \frac{\pi_n}{1-\pi_0} \tag{49}$$

where

$$\pi_n = \frac{G^n}{n!} e^{-G}. \tag{50}$$

By the same token, we see from (35) that

$$\pi_1' = \frac{ge^{-g}}{1-e^{-g}} \xrightarrow[g \to 0]{} 1$$

and that

$$P_s' = P_s'(1) = 1.$$

With these considerations, the throughput is given by

$$S(G,p,a=0) = \frac{G[\pi_0 + (1-\pi_0)P_s]}{G+\pi_0} \tag{51}$$

where P_s and π_n are given in (49) and (50), respectively. When $p = 1$, we have, from (48),

$$P_s(1) = 1$$

$$P_s(n) = 0, \qquad n > 1$$

and therefore

$$P_s = \frac{Ge^{-G}}{1-e^{-G}}.$$

Equation (51) then becomes

$$S(G,p=1,a=0) = \frac{G(1+G)e^{-G}}{G+e^{-G}}$$

which is (and should be) identical to the 1-persistent CSMA when $a = 0$ [see (23)]. Let us now consider $p \to 0$. Since $1 - q_n \approx np$, (48) then becomes

$$P_s(n) = q^{n-1}$$

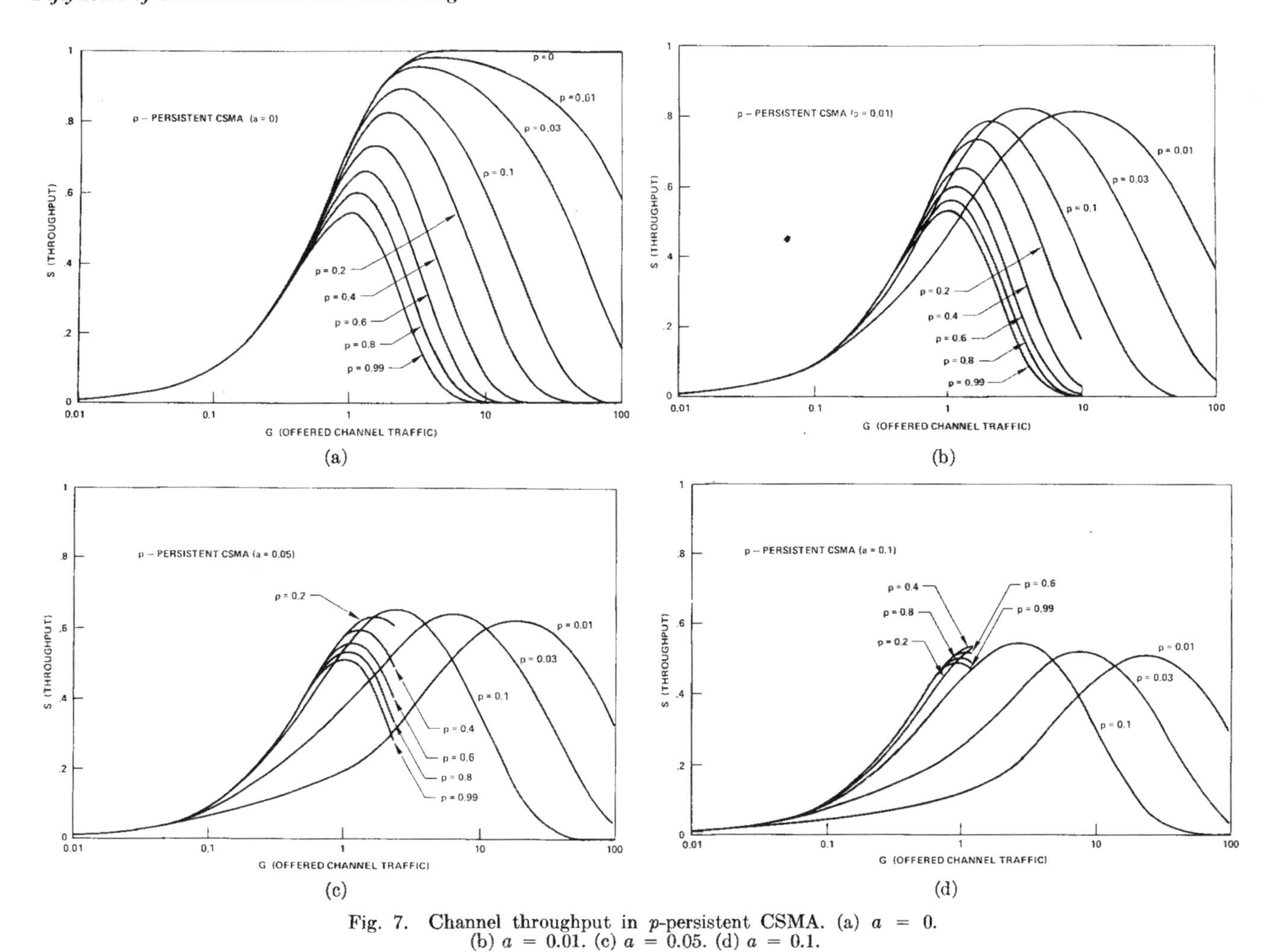

Fig. 7. Channel throughput in p-persistent CSMA. (a) $a = 0$. (b) $a = 0.01$. (c) $a = 0.05$. (d) $a = 0.1$.

and

$$P_s = \sum_{n=1}^{\infty} \frac{q^{n-1}G^n e^{-G}}{n!(1-e^{-G})} = \frac{qe^{-G}(e^{qG}-1)}{1-e^{-G}}.$$

In particular, $p \to 0$ gives $P_s(n) \to 1$, for all n, and $P_s \to 1$. In this limit the throughput is then given by

$$S(G,p \to 0, a = 0) \to \frac{G}{G+e^{-G}} \tag{52}$$

which shows that a channel capacity of 1 can be achieved when $G \to \infty$.

For each value of a, one can plot a family of curves S versus G with parameter p [as shown in Fig. 7 (a)–(d)]. The channel capacity for each value of p can be numerically determined at an optimum value of G. In Fig. 8 we show the channel capacity as a function of p, for $a = 0$, 0.01, 0.05, and 0.1. We note that the capacity is not very sensitive to small variations of p; for $a = 0.01$, it reaches its highest value (i.e., the channel capacity for this protocol) at a value $p = 0.03$. When $p = 1$, the (slotted) p-persistent CSMA reduces to the slotted 1-persistent CSMA. Indeed we can check that, when $p = 1$, (24) reduces to (19), since P_s, $\bar{I}$, and $\bar{I}'$ then become

$$P_s = \frac{aGe^{-G}}{1-e^{-aG}}$$

$$\bar{I}' = \bar{I} = 0.$$

E. Performance Comparison and Sensitivity of Capacity to the Parameter a

To summarize, we plot in Fig. 9 for $a = 0.01$, S versus G for the various access modes introduced so far and thus show the relative performance of each, as also indicated in Table I.

While the capacity of ALOHA channels does not depend on the propagation delay, the capacity of a CSMA channel does. An increase in a increases the vulnerable period of a packet. This also results in "older" channel state information from sensing. In Fig. 10 we plot, versus a, the channel capacity for all of the above random-access modes. We note that the capacities for nonpersistent and p-persistent CSMA are more sensitive to increases in a, as compared to the 1-persistent scheme. Nonpersistent CSMA drops below 1-persistent for larger a. Also, for large a,

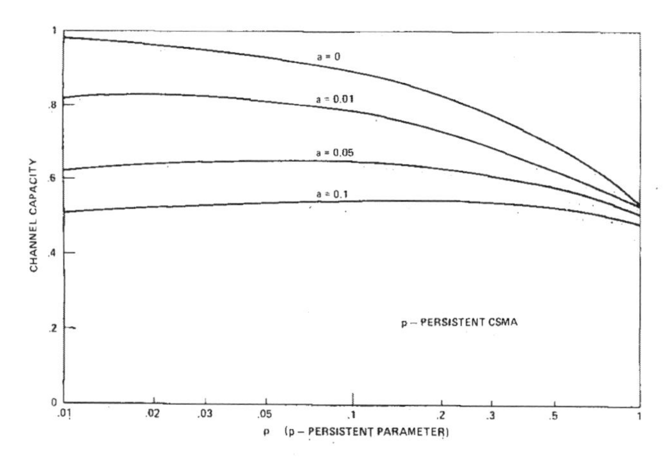

Fig. 8. p-persistent CSMA: effect of p on channel capacity.

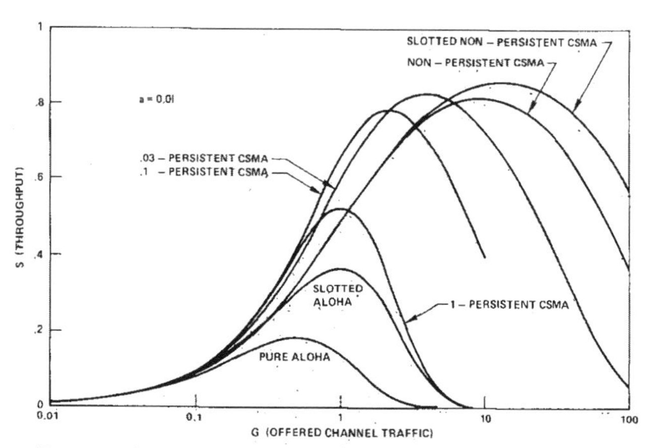

Fig. 9. Throughput for the various access modes (a = 0.01).

TABLE I
CAPACITY C FOR THE VARIOUS PROTOCOLS CONSIDERED (a = 0.01)

Protocol	Capacity C
Pure ALOHA	0.184
Slotted ALOHA	0.368
1-Persistent CSMA	0.529
Slotted 1-Persistent CSMA	0.531
0.1-Persistent CSMA	0.791
Nonpersistent CSMA	0.815
0.03-Persistent CSMA	0.827
Slotted Nonpersistent CSMA	0.857
Perfect Scheduling	1.000

slotted ALOHA (and even "pure" ALOHA) is superior to any CSMA mode since decisions based on partially obsolete data are deleterious: this effect is due in part to our assumption about the constant propagation delay. (For p-persistent, numerical results are shown only for $a \leq 0.1$. Clearly, for larger a, optimum p-persistent is lower-bounded by 1-persistent.)

V. DELAY CONSIDERATIONS

A. Delay Model

In the previous section, we analyzed the performance of CSMA modes in terms of maximum achievable throughput. We now introduce the expected packet delay D de-

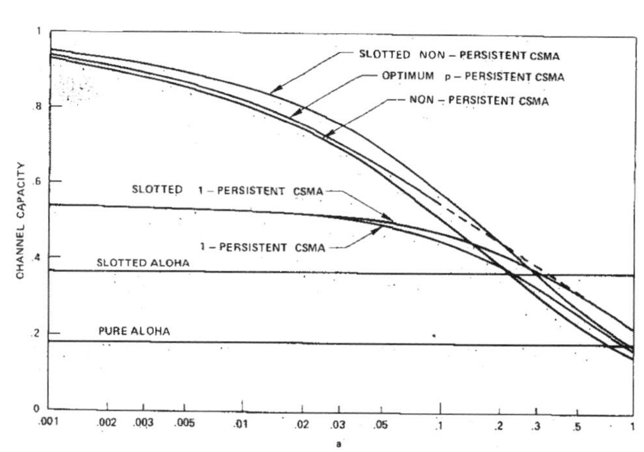

Fig. 10. Effect of propagation delay on channel capacity.

fined as the average time from when a packet is generated until it is successfully received.

Our principal concern in this section is to investigate the tradeoff between the average delay D and the throughput S.

As we have already stated, for the correct operation of the system, a positive acknowledgment scheme is needed. If an acknowledgment is not received by the sender of a packet within a specified time-out period, then the packet is retransmitted (incurring the random retransmission delay X, introduced to avoid repeated conflicts). For the present study, we assume the following.

Assumption 3: The acknowledgment packets are always correctly received with probability one.

The simplest way to accomplish this is to create a separate channel[11] (assumed to be available) to handle acknowledgment traffic. If sufficient bandwidth is provided to this channel overlaps between acknowledgment packets are avoided, since a positive acknowledgment packet is created only when a packet is correctly received, and there will be at most one such packet at any given time. Thus, if T_a denotes the transmission time of the acknowledgment packet on the separate channel, then the time-out for receiving a positive acknowledgment is $T + \tau + T_a + \tau$, provided that we make the following assumption.

Assumption 4: The processing time needed to perform the sumcheck and to generate the acknowledgment packet is negligible.

Assumption 2 further simplifies our delay model by implicitly assuming that the probability of a packet's success is the same whether the packet is new or has been blocked, or interfered with any number of times before; this probability is simply given by the throughput equation, i.e.,

$$P_s = \frac{S}{G} = \frac{\text{throughput}}{\text{offered traffic}}.$$

Bearing these assumptions in mind, we can write the delay equations for each of the previous access modes.

[11] The reader is referred to [16] for a study of the effect of acknowledgment traffic on channel throughput when acknowledgment packets are carried by the same channel.

As an example let us consider the ALOHA mode. Let R be the average delay between two consecutive transmissions (i.e., a retransmission) of a given packet. R consists of the transmission time of the packet, the transmission time of the acknowledgment packet, the round-trip propagation delay, and the average retransmission delay, that is

$$R = T + \tau + T_a + \tau + \bar{X}.$$

Using our normalized time units, we have

$$R = 1 + 2a + \alpha + \delta \tag{53}$$

where $\alpha = T_a/T$. Since $(G/S - 1)$ is the average number of retransmissions required, the average delay is given by

$$D = \left(\frac{G}{S} - 1\right)R + 1 + a. \tag{54}$$

(Special attention must be devoted to the CSMA modes in which packets may incur pretransmission delays, and in which all arrivals do not necessarily correspond to actual transmissions. The delay equations and their derivations are given in Appendix B.)

Let us begin with some comments concerning the above delay equations. First, G/S as obtained from the throughput equations rests on two important and strong Assumptions 1 and 2; namely, that we have an independent Poisson point process and that δ is infinite, or large compared to the transmission time (in which case delays are also large and unacceptable). On the other hand, δ cannot be arbitrarily small. It is intuitively clear that when a certain backlog of packets is present, the smaller δ is, the higher is the level of interference and hence the larger is the offered channel traffic G. Thus, $G = G(S,\delta)$ is a decreasing function of δ such that the average number of transmissions per packet, $[G(S,\delta)]/S$, decreases with increasing values of δ, and reaches the asymptotic value predicted by the throughput equation. Thus, for each S, a minimum delay can be achieved by choosing an optimal δ. Such an optimization problem is difficult to solve analytically, and simulation techniques have been employed in our evaluations below.[12]

Before we proceed with the discussion of simulation results, we compare the various access modes in terms simply of the average number of transmissions (or average number of schedulings[13]) G/S. For this purpose, we plot G/S versus S in Fig. 11 for the ALOHA and CSMA modes, when $a = 0.01$. Note that CSMA modes are superior in that they provide lower values for G/S than the ALOHA modes. Furthermore, for each value of the throughput, there exists a value of p such that p-persistent is optimal. For small values of S, $p = 1$ (i.e., 1-persistent) is optimal. As S increases, the optimum p decreases.

[12] We have been able to solve the problem analytically in the case of the nonpersistent CSMA when we are in presence of a large population but with a finite number of users; all conclusions obtained from simulation in Section V-B have been verified by the analysis. For this the reader is referred to reference [16].

[13] For the nonpersistent and p-persistent CSMA, G measures the offered channel traffic and not the actual channel traffic. G/S represents, then, the average number of times a packet was scheduled for transmission before success.

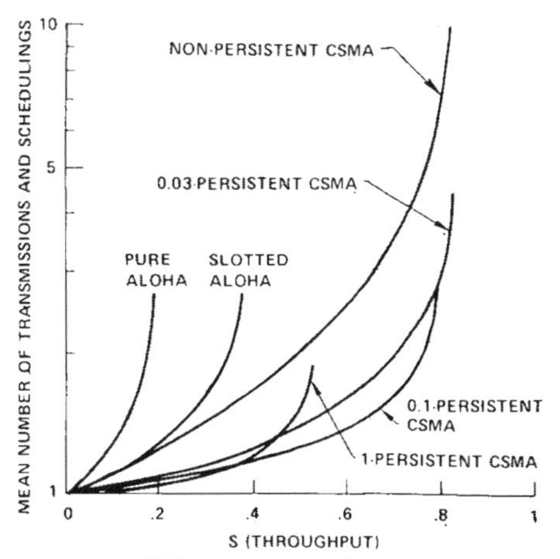

Fig. 11. G/S versus throughput ($a = 0.01$).

B. Simulation Results

The simulation model is based on all system assumptions presented in Section II. However, we relax Assumptions 1 and 2 concerning the retransmission delay and the independence of arrivals for the offered channel traffic. That is, in the simulation model, only the newly generated packets are derived independently from a Poisson distribution; collisions and uniformly distributed random retransmissions are accounted for without further assumptions.

In general, our simulation results indicate the following.

1) For each value of the input rate S, there is a minimum value δ for the average retransmission delay variable, such that below that value it is impossible to achieve a throughput equal to the input rate.[14] The higher S is, the larger δ must be to prevent a constantly increasing backlog, i.e., to prevent the channel from saturating. In other words, the maximum achievable throughput (under assumed stable conditions) is a function of δ, and the larger δ is, the higher is the maximum throughput.

2) Recall that the throughput equations were based on the assumption that $\bar{X}/T = \delta \gg 1$. Simulation shows that for finite values of δ, $\delta > \delta_0$, but not too large compared to 1, the system already "reaches" the asymptotic results ($\delta \rightarrow \infty$). That is, for some finite values of δ, Assumption 2 is excellent and delays are acceptable. Moreover, the comparison of the (S,G) relationship as obtained from simulation and the results obtained from the analytic model exhibits an excellent match. Simulation experiments were also conducted to find the optimal delay; that is, the value of $\delta(S)$ which allows one to achieve the indicated throughput with the minimum delay.

Finally, in Fig. 12[15] we give the throughput-minimum delay tradeoff for the ALOHA and CSMA modes (when $a = 0.01$). *This is the basic performance curve.* We conclude

[14] Such behavior is characteristic of random multiple-access modes. Similar results were already encountered by Kleinrock and Lam [13] when studying slotted ALOHA in the context of a satellite channel.

[15] In Fig. 12, the curve corresponding to slotted ALOHA is obtained from the analytical model developed in [13] successfully verified by simulation.

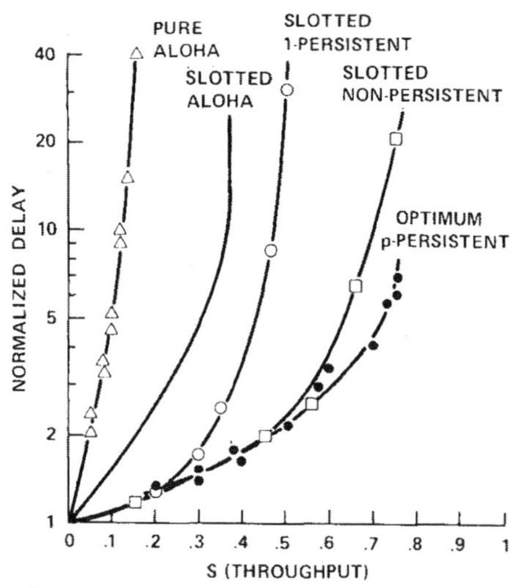

Fig. 12. Throughput-delay tradeoffs from simulation ($a = 0.01$).

that the optimum p-persistent CSMA provides us with the best performance; on the other hand the performance of the (simple) nonpersistent CSMA is quite comparable.

VI. SUMMARY AND DISCUSSION

We have introduced and evaluated the new CSMA mode and have shown it to be an efficient means for randomly accessing packet switched radio channels which have a small ratio of propagation delay to packet transmission time. Just as with most "contention" systems, these random multi-access broadcast channels (ALOHA, CSMA) are characterized by the fact that the throughput goes to zero for large values of channel traffic. At an optimum traffic level, we achieve a maximum throughput which we define to be the system capacity. This and the throughput-delay performance were obtained by a steady-state analysis under the assumption of equilibrium conditions.

However, these channels exhibit unstable behavior at most input loads as shown by Kleinrock and Lam [18]. In this last reference, the dynamic behavior and stability of an ALOHA channel are considered; quantitative estimates for the relative stability of the channel are given, which indicate the need for special control procedures to avoid a collapse. Optimal control procedures have been found [14], [15] and similar procedures are necessary for CSMA as well, since it can be shown [16] that CSMA exhibits similar unstable behavior.

Throughout the paper, it was assumed that all terminals are within range and in line-of-sight of each other. A common situation consists of a population of terminals, all within range and communicating with a single "station" (computer center, gate to a network, etc.) in line-of-sight of all terminals. Each terminal, however, may *not* be able to hear all the other terminals' traffic. This gives rise to what is called the *"hidden-terminals"* problem. The latter badly degrades the performance of CSMA as shown in Part II of this paper [19]. Fortunately, in a single-station environment, the hidden-terminal problem can be eliminated by dividing the available bandwidth into two separate channels: a busy tone channel and a message channel. As long as the station is receiving a signal on the message channel, it transmits a busy tone signal on the busy tone channel (which terminals sense for channel state information). The CSMA with a busy tone under a nonpersistent protocol has been analyzed. It is shown to provide a maximum channel capacity of approximately 0.65 when $a = 0.01$ for a channel bandwidth W of 100 kHz (modulated at 1 bit/Hz); when $W = 1$ MHz and $a = 0.01$, the channel capacity is 0.71 [19]. These values compare favorably with the capacity of 0.815 for nonpersistent CSMA with no hidden terminals.

APPENDIX A

SMALL p APPROXIMATIONS IN p-PERSISTENT CSMA

We claim, for small p, that $S(G,p,a)$ may be approximated by

$$S'(G,p,a) = \frac{(1-e^{-aG})[\hat{P}_s'\pi_0 + \hat{P}_s(1-\pi_0)]}{(1-e^{-aG})[a\hat{\hat{t}}'\pi_0 + a\hat{\hat{t}}(1-\pi_0) + 1 + a] + a\pi_0} \tag{A1}$$

where $\hat{P}_s'$, $\hat{P}_s$, $\hat{\hat{t}}$, and $\hat{\hat{t}}'$ are defined hereafter in the proof.

Proof: We show here that, with some approximations, we can get a closed-form solution for the throughput when p has *small values* ($p < 0.1$). These approximations are validated by comparing the results obtained in this section with those obtained from Section IV-D for $p = 0.1$.

For the distribution of idle time between two TP's, we have from (26)

$$\Pr\{t_n > k\} = q^{(k+1)n} \exp\left\{g\left(\frac{q(1-q^k)}{p} - k\right)\right\}. \tag{A2}$$

When p is small, we may make the following approximation (actually a lower bound):

$$q^k = (1-p)^k \simeq 1 - kp \tag{A3}$$

and therefore we may rewrite (A2) as

$$\Pr\{t_n > k\} \simeq q^{(k+1)n}e^{-kpg} = q^n[q^ne^{-pg}]^k. \tag{A4}$$

Let $t_{n>}^*(z)$ and $t_n^*(z)$ be the generating functions defined by

$$t_{n>}^*(z) \triangleq \sum_{k=0}^{\infty} \Pr\{t_n > k\}z^k \tag{A5}$$

$$t_n^*(z) \triangleq \sum_{k=0}^{\infty} \Pr\{t_n = k\}z^k. \tag{A6}$$

We have

$$t_{n>}^*(z) = q^n \sum_{k=0}^{\infty} (q^ne^{-pg}z)^k = \frac{q^n}{1 - q^ne^{-pg}z}. \tag{A7}$$

Since

$$\Pr\{t_n = k\} = \Pr\{t_n > k-1\} - \Pr\{t_n > k\}, \qquad k > 0$$

and

$$\Pr\{t_n = 0\} = 1 - \Pr\{t_n > 0\},$$

we have

$$t_n^*(z) = 1 + (z-1)t_{n>}^*(z) = 1 + \frac{q^n(z-1)}{1 - q^n e^{-pg}z}. \tag{A8}$$

The averages defined in (29) can now be written as

$$\bar{t}_n = \left.\frac{\partial t_n^*(z)}{\partial z}\right|_{z=1} = \frac{q^n}{1 - q^n e^{-pg}}. \tag{A9}$$

Equation (30), which defines $\bar{t}$ as $\sum_{n=1}^{\infty} \bar{t}_n \pi_n/(1-\pi_0)$, does not lead to a closed-form expression. Instead, we replace $\bar{t}$ by $\hat{\bar{t}}$, which is defined as

$$\hat{\bar{t}} = \frac{C}{1 - Ce^{-pg}} \tag{A10}$$

where $C = \sum_{n=1}^{\infty} q^n \pi_n/(1-\pi_0)$. ($\hat{\bar{t}}$ is smaller than $\bar{t}$ since $\bar{t}_n = q^n/(1 - q^n e^{-pg})$ is a convex function of q^n.)

C can be expressed as

$$C = \frac{\exp\{-(1+a)pG\} - \pi_0}{1 - \pi_0} = \frac{\pi_0^p - \pi_0}{1 - \pi_0} \tag{A11}$$

and therefore,

$$\hat{\bar{t}} = \frac{\pi_0^p - \pi_0}{1 - \pi_0 - (\pi_0^p - \pi_0)e^{-pg}}. \tag{A12}$$

To find the probability of success over TP $\mathfrak{I}_i$, $i \neq 1$, we first define the following generating functions:

$$L_n^*(z) \triangleq \sum_{l=n-1}^{\infty} \Pr\{L_n = l\} z^l \tag{A13}$$

$$L_n^*(z/k) \triangleq \sum_{l=n-1}^{\infty} \Pr\{L_n = l / t_n = k\} z^l. \tag{A14}$$

It is clear that

$$L_n^*(z/k) = \exp\{kg(z-1)\} z^{n-1}. \tag{A15}$$

Removing the condition on k, we get

$$\begin{aligned} L_n^*(z) &= \sum_{k=0}^{\infty} L_n^*(z/k) \cdot \Pr\{t_n = k\} \\ &= z^{n-1} \sum_{k=0}^{\infty} \exp\{kg(z-1)\} \cdot \Pr\{t_n = k\} \\ &= z^{n-1} t_n^*(\exp\{g(z-1)\}) \\ &= \frac{q^n(\exp\{g(z-1)\} - 1)z^{n-1}}{1 - q^n \exp\{-pg\} \exp\{g(z-1)\}} + z^{n-1}. \end{aligned} \tag{A16}$$

The probability of success $P_s(n)$, defined in (33), is now simply expressed (since $1 - q^l \approx lp$) as

$$\begin{aligned} P_s(n) &= L_n^*(q) \\ &= q^{n-1} - \frac{(1 - e^{-gp})q^{2n-1}}{1 - q^n e^{-2gp}}. \end{aligned} \tag{A17}$$

Here again, (34) defines

$$P_s = \sum_{n=1}^{\infty} P_s(n) \cdot \frac{\pi_n}{1 - \pi_0}$$

which does not lead to a closed-form expression. Instead, we replace P_s by $\hat{P}_s$, which is defined as

$$\hat{P}_s = \frac{C}{q} - \frac{(1 - e^{-gp})C'}{q(1 - Ce^{-gp})} \tag{A18}$$

where C is as expressed in (A11), and

$$C' = \sum_{n=1}^{\infty} q^{2n} \frac{\pi_n}{1 - \pi_0} = \frac{\pi_0^{1-q^2} - \pi_0}{1 - \pi_0}. \tag{A19}$$

Finally, $\hat{P}_s$ can be expressed as shown in the following equation:

$$\hat{P}_s = \frac{\pi_0^p - \pi_0}{q(1-\pi_0)} - \frac{(1 - \exp\{-gp\})(\pi_0^{1-q^2} - \pi_0)}{q(1-\pi_0) - q\exp\{-2gp\}(\pi_0^p - \pi_0)}. \tag{A20}$$

The quantities $\hat{\bar{t}}'$ and $\hat{P}_s'$ are readily obtained from (A12), and (A20), respectively, by replacing

$$\pi_0 \triangleq \exp\{-G(1+a)\}$$

by the quantity e^{-g}. The substitution of $\hat{P}_s$, $\hat{\bar{t}}$, $\hat{P}_s'$, and $\hat{\bar{t}}'$ for P_s, $\bar{t}$, P_s', and $\bar{t}'$, respectively, in (46) provides us with a closed-form solution for $S(G,p,a)$ when p is small.

In Table II, we compare for $p = 0.1$ the "exact" results obtained from Section IV-D to those obtained by the approximation; note that the closed-form solution is quite satisfactory for $p < 0.1$.

APPENDIX B

DELAY EQUATIONS

A. *Nonpersistent CSMA*

In this case, the average delay R between two successive sense points of the same packet is

$$R = \begin{cases} 1 + \alpha + 2a + \delta, & \text{if the packet is transmitted} \\ \delta, & \text{if the packet is blocked.} \end{cases} \tag{B1}$$

Let P_b be the probability that an arrival gets blocked (i.e., senses the channel busy). We have

$$\begin{aligned} 1 - P_b &= \frac{a + 1/G}{\bar{C}} \\ &= \frac{1 + aG}{1 + G(1 + a + \bar{Y})}. \end{aligned} \tag{B2}$$

Under the traffic independence assumption, the rate of

TABLE II
COMPARISON OF RESULTS FOR THROUGHPUT S OBTAINED FROM THE EXACT ANALYSIS (24) AND RESULTS OBTAINED FROM THE APPROXIMATION (APPENDIX A) WHEN $p = 0.1$

	$a = 0.01$		$a = 0.05$	
G	Exact	Approximate	Exact	Approximate
0.1	0.098	0.098	0.095	0.094
0.2	0.192	0.192	0.179	0.178
0.3	0.279	0.279	0.252	0.251
0.4	0.358	0.358	0.316	0.314
0.5	0.428	0.428	0.370	0.367
0.6	0.490	0.490	0.417	0.413
0.7	0.544	0.544	0.457	0.453
0.8	0.589	0.590	0.490	0.486
0.9	0.628	0.630	0.519	0.515
1.0	0.661	0.663	0.543	0.539
1.1	0.689	0.691	0.563	0.560
1.2	0.711	0.714	0.580	0.578
1.3	0.730	0.733	0.594	0.593
1.4	0.745	0.749	0.606	0.605
1.5	0.757	0.761	0.616	0.616
1.6	0.766	0.771	0.624	0.625
1.7	0.773	0.778	0.630	0.632
1.8	0.778	0.784	0.635	0.638
1.9	0.781	0.787	0.639	0.643
2.0	0.783	0.790	0.642	0.647
2.1	0.784	0.791	0.644	0.649
2.2	0.784	0.791	0.645	0.651
2.3	0.783	0.790	0.646	0.653

actual transmissions is given by

$$H = G(1 - P_b).$$

Since $(H/S) - 1$ represents the average number of actual retransmissions per packet, the average delay D is therefore

$$D = (H/S - 1)[1 + \alpha + 2a + \delta][(G - H)/S]\delta + 1 + a \tag{B3}$$

where G/S is given by the nonpersistent CSMA throughput equation (3).

If we choose to treat all packet arrivals in a uniform manner, we may assume that when a packet is blocked, it behaves as if it could transmit, and learned about its blocking only T_a seconds after the end of its "virtual" transmission. With this simplification, the delay equation is

$$D = (G/S - 1)(1 + 2a + \alpha + \delta) + 1 + a \tag{B4}$$

thus introducing an additional delay equal to $(GP_b/S)[1 + \alpha + 2a]$.

B. 1-Persistent CSMA

Unlike the ALOHA channel, a packet on a CSMA channel incurs an additional pretransmission delay r, if upon its arrival, that packet detects the channel busy. Recall that the probability of finding the channel busy is given by (see Section IV-C)

Pr {a packet finds the channel busy}

$$= \frac{\bar{B} - a/q_0}{\bar{B} + \bar{I}} = \frac{1 + \bar{Y}}{q_0(\bar{B} + \bar{I})} \tag{B5}$$

where $\bar{B}$, $\bar{I}$, $\bar{Y}$, and q_0 are given in (16), (12), (7), and (15), respectively.

Under the condition that the packet found the channel busy, the average waiting time until the channel is detected idle (i.e., until the end of the TP) is simply equal [17] to $\overline{Z^2}/2\bar{Z}$ by the Poisson assumption. The second moment of Z is simply given by

$$\overline{Z^2} = \overline{(1 + Y)^2} = 1 + 2\bar{Y} + \overline{Y^2}.$$

From the distribution of Y given in (6) we then have

$$\overline{Z^2} = 1 + a^2 + 2(1 - 1/G)\bar{Y}. \tag{B6}$$

Therefore the average pretransmission delay $\bar{r}_1$ can be easily expressed as

$$\bar{r}_1 = \frac{1 + a^2 + 2(1 - 1/G)\bar{Y}}{2(1 + \bar{Y})} \cdot \Pr\{\text{the packet finds the channel busy}\}$$

$$= \frac{1 + a^2 + 2(1 - 1/G)\bar{Y}}{2q_0(\bar{B} + \bar{I})}. \tag{B7}$$

Finally, the expected packet delay is

$$D = (G/S - 1)(1 + 2a + \alpha + \delta + \bar{r}_1) + \bar{r}_1 + 1 + a \tag{B8}$$

where G/S is given by the 1-persistent CSMA throughput equation (10).

C. p-Persistent CSMA

Similar to the special case of 1-persistent CSMA, a packet in this general scheme incurs an initial delay which we denote by r_p. In order to compute its expected value $\bar{r}_p$, one must consider the following situations.

1) An arbitrary packet, upon arrival, will find the channel idle with probability $\bar{I}/(\bar{B} + \bar{I})$, in which case its average initial wait is $a\bar{t}'$.

2) An arbitrary packet, upon arrival, will find the channel in the first IRTD (first t' seconds) of a busy period with probability $a\bar{t}'/(\bar{B} + \bar{I})$. In this case, its average initial delay is $a\overline{t'^2}/2\bar{t}'$.

3) An arbitrary packet, upon arrival, will find the channel in the remaining part of a busy period with probability $(B - a\bar{t}')/(\bar{B} + \bar{I})$, in which case the average initial wait is $\overline{(1 + a + at)^2}/2(1 + a + a\bar{t})$.

Therefore

$$\bar{r}_p = \frac{\bar{I}}{\bar{B} + \bar{I}} a\bar{t}' + \frac{a\bar{t}'}{\bar{B} + \bar{I}} \cdot \frac{a\overline{t'^2}}{2\bar{t}'} + \frac{B - a\bar{t}'}{\bar{B} + \bar{I}} \cdot \frac{\overline{(1 + a + at)^2}}{2(1 + a + a\bar{t})}. \tag{B9}$$

Treating all transmissions and schedulings uniformly (by introducing artificial delays due to "virtual" transmissions and acknowledgment), the expected delay can simply be expressed as

$$D = (G/S - 1)[1 + 2a + \delta + \bar{r}_p] + 1 + a + \bar{r}_p \tag{B10}$$

where G/S is given by the p-persistent CSMA throughput equation (24).

REFERENCES

[1] L. G. Roberts, "Data by the packet," *IEEE Spectrum*, vol. 11, pp. 46–51, Feb. 1974.

[2] N. Abramson, "The ALOHA System—Another alternative for computer communications," in *1970 Fall Joint Comput. Conf., AFIPS Conf. Proc.*, vol. 37. Montvale, N. J.: AFIPS Press, 1970, pp. 281–285.

[3] R. E. Kahn, "The organization of computer resources into a packet radio network," in *Nat. Comput. Conf., AFIPS Conf. Proc.*, vol. 44. Montvale, N. J.: AFIPS Press, 1975, pp. 177–186.

[4] L. Kleinrock and F. Tobagi, "Random access techniques for data transmission over packet-switched radio channels," in *Nat. Comput. Conf., AFIPS Conf. Proc.*, vol. 44. Montvale, N. J.: AFIPS Press, 1975, pp. 187–201.

[5] R. Binder, N. Abramson, F. Kuo, A. Okinaka, and D. Wax, "ALOHA packet broadcasting—A retrospect," in *Nat. Comput. Conf., AFIPS Conf. Proc.*, vol. 44. Montvale, N. J.: AFIPS Press, 1975, pp. 203–215.

[6] H. Frank, I. Gitman, and R. Van Slyke, "Packet radio system—Network considerations," in *Nat. Comput. Conf., AFIPS Conf. Proc.*, vol. 44. Montvale, N. J.: AFIPS Press, 1975, pp. 217–231.

[7] S. Fralick and J. Garrett, "Technological considerations for packet radio networks," in *Nat. Comput. Conf., AFIPS Conf. Proc.*, vol. 44. Montvale, N. J.: AFIPS Press, 1975, pp. 233–243.

[8] J. Burchfiel, R. Tomlinson, and M. Beeler, "Functions and structure of a packet radio station," in *Nat. Comput. Conf., AFIPS Conf. Proc.*, vol. 44. Montvale, N. J.: AFIPS Press, 1975, pp. 245–251.

[9] S. Fralick, D. Brandin, F. Kuo, and C. Harrison, "Digital terminals for packet broadcasting," in *Nat. Comput. Conf., AFIPS Conf. Proc.*, vol. 44. Montvale, N. J.: AFIPS Press, 1975, pp. 253–261.

[10] P. E. Jackson and C. D. Stubbs, "A study of multi-access computer communications," in *1969 Spring Joint Comput. Conf., AFIPS Conf. Proc.*, vol. 34. Montvale, N. J.: AFIPS Press, 1969, pp. 491–504.

[11] N. Abramson, "Packet switching with satellites," in *Nat. Comput. Conf., AFIPS Conf. Proc.*, vol. 42. Montvale, N. J.: AFIPS Press, 1973, pp. 695–702.

[12] L. Roberts, "ARPANET Satellite System," Notes 8 (NIC Document 11290) and 9 (NIC Document 11291), available from the ARPA Network Information Center, Stanford Research Institute, Menlo Park, Calif.

[13] L. Kleinrock and S. Lam, "Packet-switching in a slotted satellite channel," in *Nat. Comput. Conf., AFIPS Conf. Proc.*, vol. 42. Montvale, N. J.: AFIPS Press, 1973, pp. 703–710.

[14] S. Lam, "Packet switching in a multi-access broadcast channel with application to satellite communications in a computer network," School of Eng. and Appl. Sci., Univ. of California, Los Angeles, rep. UCLA-ENG 7429, Apr. 1974.

[15] S. Lam and L. Kleinrock, "Dynamic control schemes for a packet switched multi-access broadcast channel," in *Nat. Comput. Conf., AFIPS Conf. Proc.*, vol. 44. Montvale, N. J.: AFIPS Press, 1975, pp. 143–153.

[16] F. Tobagi, "Random access techniques for data transmission over packet switched radio networks," Ph.D. dissertation, Comput. Sci. Dep., School of Eng. and Appl. Sci., Univ. of California, Los Angeles, rep. UCLA-ENG 7499, Dec. 1974.

[17] L. Kleinrock, *Queueing Systems, Vol. I, Theory; Vol. II, Computer Applications.* New York: Wiley Interscience, 1975.

[18] L. Kleinrock and S. S. Lam, "Packet switching in a multi-access broadcast channel: Performance evaluation," *IEEE Trans. Commun.*, vol. COM-23, pp. 410–423, Apr. 1975.

[19] F. A. Tobagi and L. Kleinrock, "Packet switching in radio channels: Part II—The hidden terminal problem in carrier sense multiple access and the busy tone solution," *IEEE Trans. Commun.*, this issue, pp. 1417–1433.

Leonard Kleinrock (S'55–M'64–SM'71–F'73), for a biography, see page 662.

PHOTO NOT AVAILABLE

Fouad A. Tobagi was born in Beirut, Lebanon on July 18, 1947. He received the Engineering Degree from Ecole Centrale des Arts of Manufactures, Paris, France, in 1970 and the M.S. and Ph.D. degrees in computer science from the University of California, Los Angeles, in 1971 and 1974, respectively.

From 1971 to 1974 he was with the University of California, Los Angeles, where he participated in the ARPA Network Project as a Postgraduate Research Engineer and did research on packet radio communication. During the summer of 1972 he was with the Communications Systems Evaluation and Synthesis Group, IBM J. Watson Research Center, Yorktown Heights, N.Y. Since December 1974 he has been a Research Staff Project Manager with the ARPA project, Computer Science Department, University of California, Los Angeles. His current research interests include computer communication networks, packet switching over radio, and satellite networks.

From 1967 to 1970 he held a scholarship from the Ministry of Foreign Affairs of the French government. During the academic year 1972–1973 he held an Earl Anthony Fellowship.

Packet Switching in a Multiaccess Broadcast Channel: Performance Evaluation

LEONARD KLEINROCK, FELLOW, IEEE, AND SIMON S. LAM, MEMBER, IEEE

Abstract—**In this paper, the rationale and some advantages for multiaccess broadcast packet communication using satellite and ground radio channels are discussed. A mathematical model is formulated for a "slotted ALOHA" random access system. Using this model, a theory is put forth which gives a coherent qualitative interpretation of the system stability behavior which leads to the definition of a stability measure. Quantitative estimates for the relative instability of unstable channels are obtained. Numerical results are shown illustrating the trading relations among channel stability, throughput, and delay. These results provide tools for the performance evaluation and design of an uncontrolled slotted ALOHA system. Adaptive channel control schemes are studied in a companion paper.**

INTRODUCTION

IN THIS and a forthcoming paper [1], a packet switching technique based upon the random access concept of the ALOHA System [2] will be studied in detail. This technique, referred to as slotted ALOHA random access, enables efficient sharing of a data communication channel by a large population of users, each with a bursty data stream. This packet switching technique may be applied to the use of satellite and ground radio channels for computer–computer and terminal–computer communications, respectively [3]–[10]. The multiaccess broadcast capabilities of these channels render them attractive solutions to two problems: 1) large computer–communication networks with nodes distributed over wide geographic areas, and 2) large terminal access networks with potentially mobile terminals.

The objective of this study is to develop analytic models and methods for the evaluation and optimization of the channel performance of a slotted ALOHA system. The problem of performance evaluation is addressed in this paper. In [1], we present dynamic channel control procedures as solutions to some of the issues considered herein.

In this paper, the rationale for multiaccess broadcast packet communication is first discussed. The mathematical model to be considered is then described. Following that, a theory is proposed which explains the dynamic and stochastic channel behavior. In particular, we display the delay-throughput performance curves obtained under the assumption of equilibrium conditions [6]. We then demonstrate that a slotted ALOHA channel often exhibits "unstable behavior." A stability definition is proposed which characterizes stable and unstable channels. A stability measure (FET) is then defined which quantifies the relative instability of unstable channels. An algorithm is given for the calculation of FET. Finally, numerical results are shown which illustrate the trading relations among channel stability, channel throughput, and average packet delay. Our main concern in this paper is the consideration of the stability issue and its effect on the channel throughput-delay performance.

MULTIACCESS BROADCAST PACKET COMMUNICATION

Rationale

For almost a century, circuit switching dominated the design of communication networks. Only with the higher speed and lower cost of modern computers did packet communication become competitive. It was not until approximately 1970 that the computer (switching) cost dropped below the communication (bandwidth) cost in a packet switching network [11]. This also marked the first appearance of packet switched computer–communication networks [2], [12].

Circuit switching is relatively inefficient for computer

Paper approved by the Associate Editor for Computer Communication of the IEEE Communications Society for publication after presentation at the 7th Hawaii International Conference on System Sciences, Honolulu, Hawaii, January 8–10, 1974. Manuscript received June 30, 1974; revised September 30, 1974. This research was supported by the Advanced Research Projects Agency of the Department of Defense under Contract DAHC 15-73-C-0368.

L. Kleinrock is with the Department of Computer Science, University of California, Los Angeles, Calif. 90024.
S. S. Lam is with the IBM Thomas J. Watson Research Center, Yorktown Heights, N. Y. 10598.

Reprinted from *IEEE Transactions on Communications,* April 1975.

communications, especially over long distances. Measurement studies [13] conducted on time-sharing systems indicate that both computer and terminal data streams are *bursty*. Depending on the channel speed, the ratio between the peak and the average data rates may be as high as 2000 to 1 [5]. Consequently, if a high-speed point-to-point channel is used, the channel utilization may be extremely low since the channel is idle most of the time. On the other hand, if a low-speed channel is used, the transmission delay is large.

The above dilemma is caused by channel users imposing bursty random demands on their communication channels. By the law of large numbers in probability theory, the total demand at any instant from a large population of independent users is, with high probability, approximately equal to the sum of their average demands (i.e., a nearly deterministic quantity). Thus, if a channel is dynamically shared in some fashion among many users, the required channel bandwidth to satisfy a given delay constraint may be much less than if the users are given dedicated channels. This concept is known as *statistical load averaging* and has been applied in many computer–communication schemes to various degrees of success. These schemes include: polling systems [14], loop systems [15], asynchronous time division multiplexing (ATDM) [16], and the store-and-forward packet switching concepts [17]–[19] implemented in the ARPA network [12].

We are currently facing an enormous growth in computer networks [20]. To design cost-effective computer–communication networks for the future, new techniques are needed which are capable of providing efficient high-speed computer–computer and terminal–computer communications in a large network environment. The application of packet switching techniques to radio communication (both satellite and ground radio channels) appears to provide a solution.

Radio is a multiaccess broadcast medium. That is, a signal generated by a radio transmitter may be received over a wide area by any number of receivers. This is referred to as the *broadcast* capability. Furthermore, any number of users may transmit signals over the same channel. This is referred to as the *multiaccess* capability. (However, if two signals at the same carrier frequency overlap in time at a radio receiver[1], we assume that neither is received correctly. This destructive interference is the key issue in studying the multiaccess radio channel used in a packet switching mode.) Thus, a single ground radio channel provides a completely connected network topology for a large number of nodes within range of each other. Similarly, a satellite transponder in a geostationary orbit above the earth acts as a radio repeater. Any number of earth stations may transmit signals up to the satellite at one carrier frequency (the multiaccess channel). Any signal received by the satellite transponder is beamed back to earth at another frequency (the broadcast channel). This broadcasted signal may be received by all earth stations covered by the transponder beam. Thus, a satellite channel (consisting of both carrier frequencies) provides a completely connected network topology for all earth stations covered by the transponder beam.

[1] This event will be referred to as a *channel collision*.

Consider the use of packet communication in a computer–communication network environment to support large populatons of (bursty) users over a wide area. We can then identify and summarize the following advantages of satellite and ground radio channels over conventional wire communications.

1) Elimination of Complex Topological Design and Routing Problems: Topological design and routing problems are very complex in networks with a large population of users. Existing implementations suitable for a (say) 50 node network may become totally inappropriate for a 500 node network required to perform the same functions [21]. On the other hand, ground radio and satellite channels used in the multiaccess broadcast mode provide a completely connected network topology, since every user may access any other user covered by the broadcast.

2) Wide Geographical Areas: Wire communications become expensive over long distances (e.g., transcontinental, transoceanic). Even on a local level, the communication cost for an interactive user on an alphanumeric console over distances of over 100 miles may easily exceed the cost of computation [2]. On the other hand, satellite and radio communications are relatively distance independent, and are especially suitable for geographically scattered users.

3) Mobility of Users: Since radio is a multiaccess broadcast medium, it is possible for users to move around freely. This consideration will soon become important in the development of personal terminals in future telecommunication systems [22] as well as in aeronautical and maritime applications [23].

4) Large Population of Active and Inactive Users: In wire communications, the system overhead usually increases with the number of users (e.g., polling schemes). The maximum number of users is often bounded by some hardware limitation (e.g., the fan-in of a communications processor). In radio communication, since each user is merely represented by an ID number, the number of active users is bounded only by the channel capacity and there is no limitation to the number of inactive (but potentially active) users beyond that of a finite address space.

5) Flexibility in System Design: A radio packet communication system can become operational with two or three users. The size of the user population can be increased up to the channel capacity. More users can be accommodated by increasing the radio channel bandwidth. In other words, the communication system can be expanded or contracted without major changes in the basic system design and operational schemes.

6) Statistical Load Averaging: Wire communication links are more efficiently utilized in a store-and-forward packet switched network than in a circuit switched network. However, at any instant, there may be unused channel capacity in some parts while congestion exists in other parts of the network. The application of packet switching techniques to a single high-speed satellite or radio channel permits the total demand of all user input sources to be statistically averaged at the channel. Note also that each user transmits data at the wide-band channel rate.

7) *Multiaccess Broadcast Capability:* This capability in radio communication may be useful for certain multipoint-to-multipoint communication applications.

The Multiaccess Channel Model

Consider a radio communication system such as a packet switched satellite system [5]-[10] or the ALOHA System [2]. In each case, there is a *broadcast* channel for point-to-multipoint communication and a *multiaccess* channel shared by a large number of users. Since the broadcast channel is used by a single transmitter, no transmission conflict will arise. All nodes covered by the radio broadcast can receive on the same frequency, picking out packets addressed to themselves and discarding packets addressed to others.

The problem we are faced with is how to effect time-sharing of the multiaccess channel among all users in a fashion which produces an acceptable level of performance. As soon as we introduce the notion of sharing in a packet switching mode, we must be prepared to resolve conflicts which arise when simultaneous demands are placed upon the channel. There are two obvious solutions to this problem: the first is to form a queue of conflicting demands and serve them in some order; the second is to "lose" any demands which are made while the channel is in use. The former approach is taken in ATDM and in store-and-forward networks assuming that storage may be provided economically at the point of conflict. The latter approach is adopted in the ALOHA System random access scheme; in this system, in fact, *all* simultaneous demands made on the radio channel are lost.

Let us define *channel throughput rate* S_{out} to be the average number of correctly received packet transmissions per packet transmission time (assuming stationary conditions). We also define *channel capacity* S_{max} to be the maximum possible channel throughput rate. The channel capacity of a pure ALOHA multiaccess channel was shown by Abramson to be $1/2e \simeq 18$ percent for a fixed packet size [2]. Under similar assumptions, Gaarder showed that a pure ALOHA channel with a fixed packet size is always superior (in terms of channel capacity) to one with different packet sizes [24].

Roberts suggested that the channel may be slotted by requiring all users to synchronize[2] the leading edges of their packet transmissions to coincide with an imaginary time slot boundary at the multiaccessed radio receiver [25]. The duration of a channel time slot is chosen to be equal to a packet transmission time. The resulting scheme will be referred to as "slotted ALOHA random access" or "slotted ALOHA." In this scheme, the users transmit newly generated packets into channel time slots independently. In the event of a channel collision, the collided packets are retransmitted after *random* retransmission delays. (See Fig. 1.) The channel capacity of a slotted ALOHA channel was shown to be $1/e \simeq 36$ percent [25].

To achieve a channel throughput rate larger than the 36 percent limitation, various other multiaccess broadcast packet swiching schemes have been proposed to take advantage of special system and traffic characteristics. The reader is referred to the references [3], [7], [26] for description of these schemes.

Consider a slotted ALOHA channel. The *channel input* in a time slot is defined to be a random variable representing the total number of *new* packets transmitted by all users in that time slot. Assuming stationary conditions, the channel input rate S is the average number of new packet transmissions per time slot. The *channel traffic* in a time slot is defined to be a random variable representing the total number of packet transmissions (both *new and previously collided* packets) by all users in that time slot. Assuming stationary conditions, the channel traffic rate G is the average number of packet transmissions per time slot. The *channel throughput* (or output) in a time slot is defined to be a random variable representing the number (0 or 1) of successful packet transmissions in that time slot. Assuming stationary conditions, the channel throughput (output) rate S_{out} is the probability of exactly one packet transmission in a channel time slot.

The retransmission delay (RD) incurred by an unsuccessful packet transmission may be regarded as the sum of a deterministic component (R) and a random component. The random component is necessary since if collided packets are retransmitted after the same deterministic delay, they will collide again for sure. In a ground radio system, RD corresponds to the positive acknowledgment time-out interval [2]. In a satellite system, since each channel user listens to the satellite broadcast, one round-trip propagation time after transmitting a packet he knows whether he was successful or if a channel collision occurred. In this case, the deterministic component corresponds to a round-trip satellite propagation delay. We shall assume a noise-free channel such that a packet is received incorrectly if and only if it suffered a channel collision. In [6], a uniform probability distribution is assumed for the random component of RD such that each user retransmits a previously collided packet at random during one of the next K slots (each such slot being chosen with probability $1/K$). Thus, retransmission will take place either $R + 1, R + 2, \cdots$ or $R + K$ slots after the previous transmission. This is said to be the uniform retransmission randomization scheme. Under this scheme, equilibrium throughput-delay tradeoffs have been obtained for a slotted ALOHA channel with a Poisson input source (the infinite population model). Such throughput-delay contours are shown here in Fig. 2 for different values of K. Note that the minimum envelope of these contours defines the optimum channel perform-

[2] The problem of synchronizing channel users is a nontrivial one. It will not be addressed in this paper.

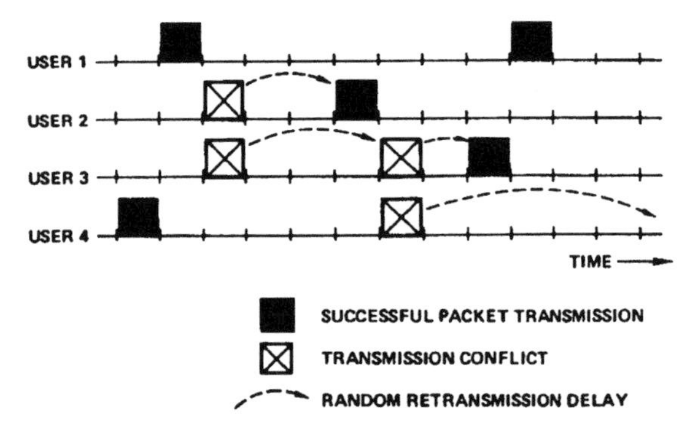

Fig. 1. Slotted ALOHA random access.

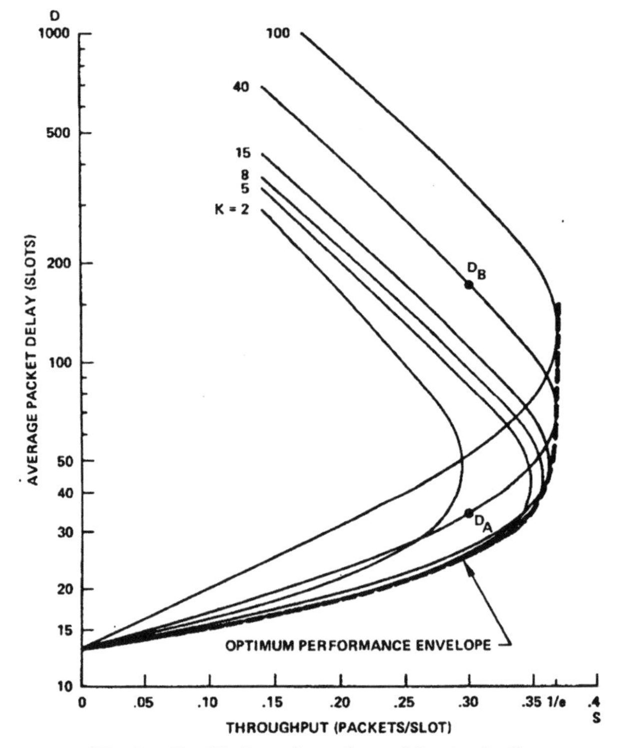

Fig. 2. Equilibrium throughput-delay tradeoff.

ance. These results correspond to the use of a 50 KPBS satellite channel, 1125 bits per packet, and a satellite round-trip propagation delay of 0.27 s for all users. Thus R is equal to 12 slots and there are 44.4 slots in one second. (These numbers will be assumed throughout this paper.) In Fig. 2, D represents the average packet delay in slots. Note that the channel input rate S is equal to the channel throughput rate S_{out} under the assumption of channel equilibrium. The channel capacity S_{max} approaches $1/e$ in the limit as $K \rightarrow \infty$. For $K = 15$, it is almost there. For values of K between 8 and 15, the equilibrium throughput-delay tradeoffs are very close to the optimum performance envelope over a wide range of S.

The analytic results presented so far are based upon the assumption that the channel is in equilibrium. Referring to Fig. 2, we see that given S and K (say $K = 40$), there are two possible equilibrium solutions for D! They correspond to a small delay value D_A and a much larger delay value D_B. (We shall refer to the equilibrium point given by S, K, and D_A as the *channel operating point*, since this is the desired channel performance given S and K.) This observation suggests that the assumption of equilibrium conditions adopted in most previous analytic models [4]–[7] may not be valid.

In order to study the dynamic behavior of these channels, simulations were performed for the infinite population model [10]. Each simulation run was observed to behave in the following manner. Starting from an initially empty system, the channel stays in equilibrium at the channel operating point for a finite period of time until stochastic fluctuations give rise to some high channel traffic rate which reduces the channel throughput rate which in turn further increases the channel traffic rate. As this vicious cycle continues, the channel becomes inundated with collisions and retransmissions. At the same time, the channel throughput rate vanishes rapidly to zero. This phenomenon will be referred to as *channel saturation.* Thus, we realize that the equilibrium throughput-delay tradeoffs are not sufficient to characterize the performance of the infinite population model. A more accurate measure of channel performance must reflect the trading relations among channel stability, throughput and delay. A mathematical model with a simpler structure than that used in [6] will be defined below. This model is similar to the one studied by Metcalfe [4]. Using this model, the concepts of channel saturation and stability in a slotted ALOHA random access channel have been characterized [8], [10].

STABILITY-THROUGHPUT-DELAY TRADEOFF PERFORMANCE

In this section, a Markovian model is first formulated for a population of M channel users. The variable M is assumed to be large and may be either finite or infinite. A theory is then proposed which characterizes the instability phenomenon in the following ways.

1) Stable and unstable channels are defined.

2) In a stable channel, equilibrium throughput-delay results (as shown in Fig. 2) are achievable over an infinite time horizon. In an unstable channel, such channel performance is achievable only for some finite time period before the channel goes into saturation.

3) For unstable channels, a stability measure is defined and an efficient computational procedure for its calculation is given.

4) Using the above stability measure, the stability-throughput-delay tradeoff for unstable channels is examined.

The Markovian Model

We consider a slotted ALOHA channel with a user population consisting of M users. Each such user can be in one of two states: *blocked* or *thinking*. In the thinking state, a user generates and transmits a new packet in a time slot with probability σ. A packet which had a channel collision and is waiting for retransmission is said to be *backlogged.* The retransmission delay RD of each backlogged packet is assumed to be geometrically distributed, i.e., each backlogged packet retransmits in the current time slot with probability p. Assuming bursty users, we must have $p \gg \sigma$. From the time a user generates a packet until that packet is successfully received, the user is blocked in the sense that he cannot generate (or accept from his input source) a new packet for transmission.

Let N^t be a random variable (called the *channel backlog*) representing the total number of backlogged packets at time t. The channel input rate at time t is $S^t = (M - N^t)\sigma$. Note that S^t decreases linearly as N^t increases. The vector (N^t, S^t) will be denoted as the *channel state vector*. In this context, both M and σ may be functions of time. We shall assume M and σ to be time-invariant unless stated otherwise. In this case, N^t is a Markov process (chain) with stationary transition probabilities and serves as the state description for the system. The discrete *state space* will now consist of the set of integers $\{0,1,2,\cdots,M\}$. The *one-step state transition probabilities* of N^t are, for $i = 0,1,2,\cdots,M$,

$$p_{ij} = \text{Prob}\,[N^{t+1} = j \mid N^t = i] = \begin{cases} 0 & j \le i-2 \\ ip(1-p)^{i-1}(1-\sigma)^{M-i} & j = i-1 \\ (1-p)^i(M-i)\sigma(1-\sigma)^{M-i-1} + [1 - ip(1-p)^{i-1}](1-\sigma)^{M-i} & j = i \\ (M-i)\sigma(1-\sigma)^{M-i-1}[1-(1-p)^i] & j = i+1 \\ \binom{M-i}{j-i}\sigma^{j-i}(1-\sigma)^{M-j} & j \ge i+2. \end{cases} \tag{1}$$

For the infinite population model in which $M \to \infty$ and $\sigma \to 0$ such that $M\sigma = S$ which is constant and finite, the above equation becomes

$$p_{ij} = \begin{cases} 0 & j \le i-2 \\ ip(1-p)^{i-1}\exp(-S) & j = i-1 \\ (1-p)^i S\exp(-S) + [1 - ip(1-p)^{i-1}]\exp(-S) & j = i \\ S\exp(-S)[1-(1-p)^i] & j = i+1 \\ \dfrac{S^{j-i}}{(j-i)!}\exp(-S) & j \ge i+2. \end{cases} \tag{2}$$

The assumption that RD has a memoryless geometric distribution permits a simple state description for the mathematical model. However, this assumption implies that RD has a zero deterministic component ($R = 0$). In a satellite channel this obviously represents an approximation. (However, it may be physically realizable in radio communications over short distances in which channel propagation delays are negligible compared to a packet transmission time.) A (geostationary) satellite channel has a round-trip propagation delay of 0.27 s, which necessitates a state description consisting of the channel history for at least R consecutive time slots. The difficulty in mathematical analysis using such a state description was illustrated in [10]. However, simulation results have shown that the slotted ALOHA channel performance (in terms of average throughput and delay) is dependent primarily upon the *average* retransmission delay ($\overline{\text{RD}}$) and quite insensitive to the exact probability distributions considered [10]. In order to use the analytic results of the Markovian model here to predict the throughput-delay performance of a slotted ALOHA channel with nonzero R, it is necessary to use a value of p in the Markovian model which gives the same $\overline{\text{RD}}$. For example, to approximate a slotted ALOHA channel with uniform retransmission randomization, we must let

$$p = \frac{1}{R + (K+1)/2} \tag{3}$$

such that $\overline{\text{RD}} = R + (K+1)/2$ in both cases.

We define the length of time for which a packet is backlogged to be the backlog time of the packet and denote the *average backlog time* by D_b. To obtain the average packet delay (as defined in [6]), we must add to D_b, $R + 1$ time slots, which represent the delay incurred by each successful transmission. Thus, we have

$$D = D_b + R + 1. \tag{4}$$

Numerical results in this paper will be expressed in terms of K (rather than p) through use of (3) and (4) for comparison with previous results for channel performance [6].

The Theory

Conditioning on $N^t = n$, the expected channel throughput $S_{out}(n,\sigma)$ is the probability of exactly one packet transmission in the tth time slot. Thus,

$$S_{out}(n,\sigma) = (1-p)^n(M-n)\sigma(1-\sigma)^{M-n-1} + np(1-p)^{n-1}(1-\sigma)^{M-n}. \quad (5)$$

For the infinite population model, i.e., in the limit as $M \uparrow \infty$ and $\sigma \downarrow 0$ such that $M\sigma = S$ is finite and the channel input is Poisson distributed at the constant rate S, the above equation reduces to

$$S_{out}(n,S) = (1-p)^n S \exp(-S) + np(1-p)^{n-1}\exp(-S). \quad (6)$$

This expression is very accurate even for finite M if $\sigma \ll 1$ and if we replace $S = M\sigma$ by $S = (M-n)\sigma$. We assume that the condition $\sigma \ll 1$ (which implies bursty users) is always satisfied in problems of interest to us.

In Fig. 3, for a fixed K we sketch $S_{out}(n,S)$ as a three-dimensional surface above the (n,S) plane. Note that there is an *equilibrium contour* in the (n,S) plane defined as the locus of points on which the channel input rate S is equal to the expected channel throughput $S_{out}(n,S)$ given by (6). In the crosshatched region enclosed by the equilibrium contour, $S_{out}(n,S)$ exceeds S; elsewhere, S is greater than $S_{out}(n,S)$. In Fig. 4, a family of equilibrium contours for various K are displayed. We see that if we increase the average retransmission delay (by increasing K or equivalently decreasing p), the equilibrium contour moves upwards. We show below that these equilibrium contours play a crucial role in determining the stability behavior of the channel.

Given an equilibrium contour in the (n,S) plane, we first consider the dynamic behavior of the channel subject to *time-varying* inputs using a *fluid approximation* interpretation. The following example serves to illustrate the underlying concepts.

Consider the case in which σ is constant while $M = M(t)$ is a function of time as shown in Fig. 5. We use the fluid approximation for the trajectory of the channel state vector (N^t,S^t) in the (n,S) plane as sketched in Fig. 6. Recall that $S^t = (M - N^t)\sigma$. The arrows indicate the "fluid" flow direction which depends on the relative magnitudes of the instantaneous channel throughput rate $S_{out}(n,S)$ and the channel input rate S. Two possible cases are shown corresponding to different values of the amplitude M_3, of the input pulse in Fig. 5. The solid line (Case 1) represents a trajectory which returns to the original state on the equilibrium contour despite the input pulse. The dashed line (Case 2) represents a less fortunate situation in which the decrease in the channel input rate at time t_2 is not sufficient to bring the trajectory back into the "safe" region (shown shaded) in which $S < S_{out}(n,S)$; eventually, the channel "fails" as a result of an increasing backlog and a vanishing channel throughput.

The above example demonstrates channel saturation due to a time-varying input. Let us now study the con-

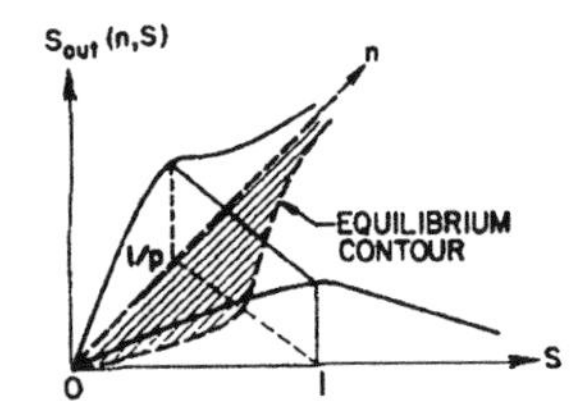

Fig. 3. Throughput surface above the (n,S) plane.

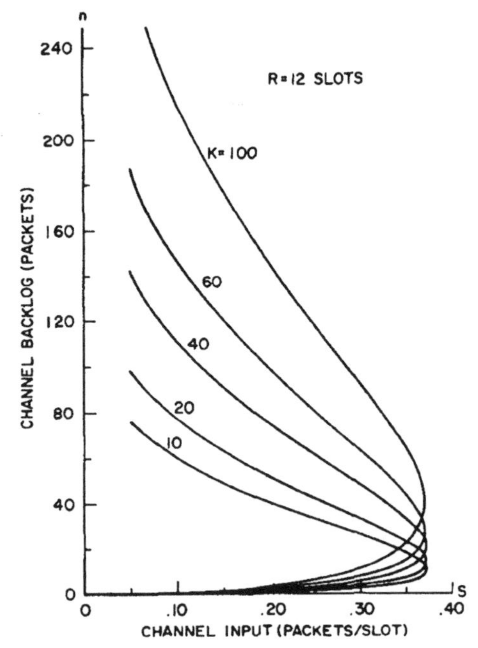

Fig. 4. Equilibrium contours in the (n,S) plane.

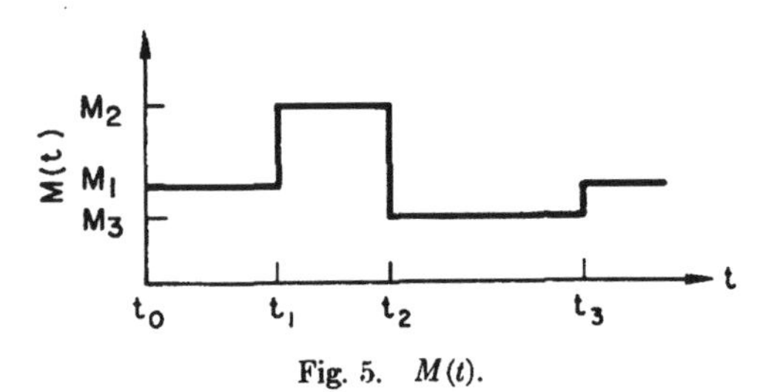

Fig. 5. $M(t)$.

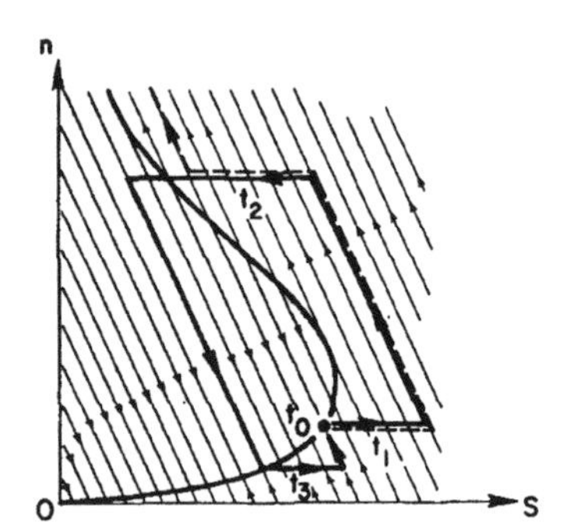

Fig. 6. Fluid approximation trajectories.

ditions under which the slotted ALOHA channel with a *stationary* input (constant M and σ) can go into saturation as a result of statistical fluctuations.

Assume that M and σ are constant. The trajectory of (N^t,S^t) is constrained to lie on the straight line $S = (M - n)\sigma$ called the *channel load line* which intercepts the n-axis at $n = M$ and has a slope equal to $-1/\sigma$. We now propose the following definition for characterizing stable and unstable channels.

The Stability Definition: A slotted ALOHA channel is said to be *stable* if its load line intersects (nontangentially) the equilibrium contour in exactly one place. Otherwise, the channel is said to be *unstable.*

Examples of stable and unstable channels are shown in Fig. 7. Arrows on the channel load lines indicate directions of fluid flow given by the fluid approximation. In other words, the arrows point in the direction of increasing backlog size if $S > S_{out}(n,S)$ and in the direction of decreasing backlog size if $S_{out}(n,S) > S$.

Each channel load line may have one or more equilibrium points. A point on the load line is said to be a *stable equilibrium point* if it acts as a "sink" with respect to fluid flow. It is a *globally stable equilibrium point* if it is the only stable equilibrium point on the channel load line. Otherwise, it is a *locally stable equilibrium point.* (Each stable equilibrium point is identified by a dot on channel load lines in Fig. 7 except in Fig. 7(c), where one of the stable equilibrium points is at $n = \infty$.) An equilibrium point is said to be an *unstable equilibrium point* if fluid flow emanates from it. Thus, the channel state N^t sitting on such a point will drift away from it given the slightest perturbation. The stability definition given above is equivalent to defining a stable channel to be one whose channel load line has a globally stable equilibrium point.

In Fig. 7(a), we show the channel load line of a stable channel. The globally stable equilibrium point on the load line, (n_o,S_o), will be referred to as the *channel operating point.* If M is finite, a stable channel can always be achieved by using a sufficiently large K (see Fig. 4). Of course, a large K implies that the equilibrium backlog size n_o is large; the corresponding average packet delay may be too large to be acceptable. Since the Markov chain N^t has a finite state space and is irreducible (assuming $p,\sigma > 0$), a stationary probability distribution always exists [27], [28]. The stationary probability distribution $\{P_n\}_{n=0}^M$ of N^t can be computed by solving the following set of linear simultaneous equations

$$P_j = \sum_{i=0}^{M} P_i p_{ij} \qquad j = 0,1,\cdots,M$$

and

$$\sum_{i=0}^{M} P_i = 1$$

where the state transition probabilities p_{ij} are given by (1). The steady-state channel throughput rate S_{out} and

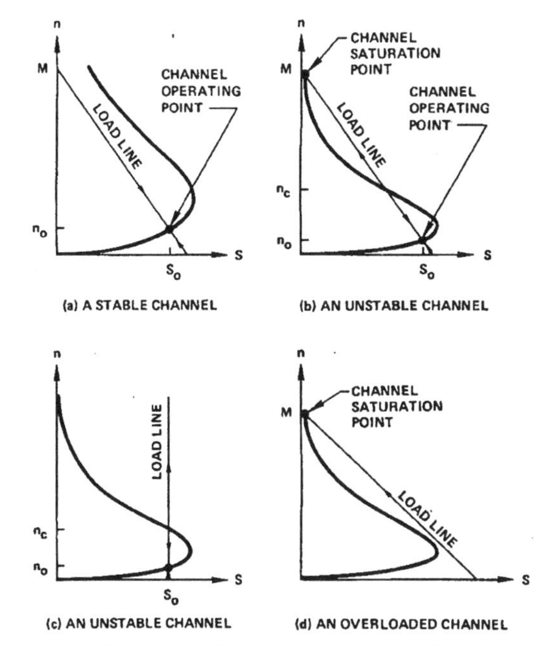

Fig. 7. Stable and unstable channels.

expected channel backlog $\bar{N}$ can then be obtained from

$$S_{out} = \sum_{n=0}^{M} S_{out}(n,\sigma)P_n \tag{7}$$

and

$$\bar{N} = \sum_{n=1}^{M} nP_n. \tag{8}$$

Numerical results have shown that these values of S_{out} and $\bar{N}$ for a stable channel are closely approximated by the equilibrium S_o and n_o at the channel operating point, and also by the equilibrium throughput-delay values in Fig. 2 for the infinite population model. For example, suppose $K = 60$, $M = 200$, and $1/\sigma = 536.1$; the equilibrium channel throughput rate at the channel operating point is $S_o = 0.346$. In Fig. 9 below (to be described later), we see that the steady-state channel throughput rate computed by using (7) is $S_{out} = 0.344$. For the same example, $\bar{N}$ is calculated to be 15.4 slots. By Little's result [27], the average backlog time is

$$D_b = \frac{\bar{N}}{S_{out}} = \frac{15.4}{0.344} = 44.8 \text{ slots}.$$

Applying (4), we get $D = 44.8 + 13 = 57.8$ slots. Now given $S_o = 0.346$, the $K = 60$ equilibrium throughput-delay contour for the infinite population model [6] gives $D = 56.5$ slots.

In Fig. 7(b), we show the channel load line of an unstable channel. The point (n_o,S_o) is again the desired channel operating point since it yields the larger channel throughput and smaller average packet delay between the two locally stable equilibrium points on the load line. In fact, the other locally stable equilibrium point, having a huge backlog and virtually zero throughput, corresponds to the channel saturation state; it will be referred to as the *channel saturation point*. Although it has a stationary probability distribution, N^t will "flip-flop" between the two locally stable equilibrium points in the following manner. Starting from an empty channel ($N^0 = 0$) quasi-stationary conditions will prevail at the operating point (n_o, S_o). The channel, however, cannot maintain equilibrium at this point indefinitely since N^t is a random process; that is, with probability one, the channel backlog N^t crosses the unstable equilibrium point n_c in a finite time, and as soon as it does, the channel input rate S exceeds $S_{out}(n,S)$. Under this condition, N^t will drift toward the saturation point. Although there is a nonzero probability that N^t may return below n_c, all our simulations show that the channel state N^t accelerates up the channel load line producing an increasing backlog and a vanishing throughput rate. Since the saturation point is a locally stable equilibrium point, quasi-stationary conditions will prevail there for some finite (but probably very long) time period. In this state, the communication channel can be regarded as having failed. (In a practical system, external control should be applied at this point to restore proper channel operation.) Thus, the two locally stable equilibrium points on the load line of an unstable channel correspond to the channel being "up" or "down". An unstable channel may be acceptable if the average channel up time is large and external control is available to bring the channel back up whenever it goes down.

In Figs. 8 and 9, we see how, as the number of channel users M increases, an originally stable channel becomes unstable although the channel input rate S_o at the operating point remains constant (by reducing σ). (These results are obtained by first solving for the stationary probability distribution of N^t and then applying (7) and (8).) For $S_o = 0.36$ and $K = 10$, we see that as M exceeds 80, the stationary channel throughput rate decreases and the average packet delay increases very rapidly with M. Using the $K = 10$ equilibrium contour in Fig. 4, the maximum value of M that is possible without making the channel load line intersect the equilibrium contour more than once is determined (graphically) to be $M_{max} = 79$, which exactly gives the knees of the curves in Fig. 8. This excellent agreement provides the motivation for the stability definition proposed above. In Fig. 9, by using a larger value of K ($=60$), a larger M_{max} is possible. Note, however, that the average packet delay ($\simeq 56$ slots) for $K = 60$ is much larger than the average packet delay ($\simeq 36$ slots) for $K = 10$.

Given K and S_o, M_{max} can be obtained graphically from the equilibrium contours such as shown in Fig. 4. In Fig. 10 we show M_{max} as a function of K with S_o fixed at the maximum possible value given K. Note the linear relationship between M_{max} and K for the values shown. In Fig. 11, we illustrate how an originally unstable channel can be rendered stable by using a sufficiently large K.

In Fig. 7(c), we show the channel load line of an infinite population model. This is an unstable channel since

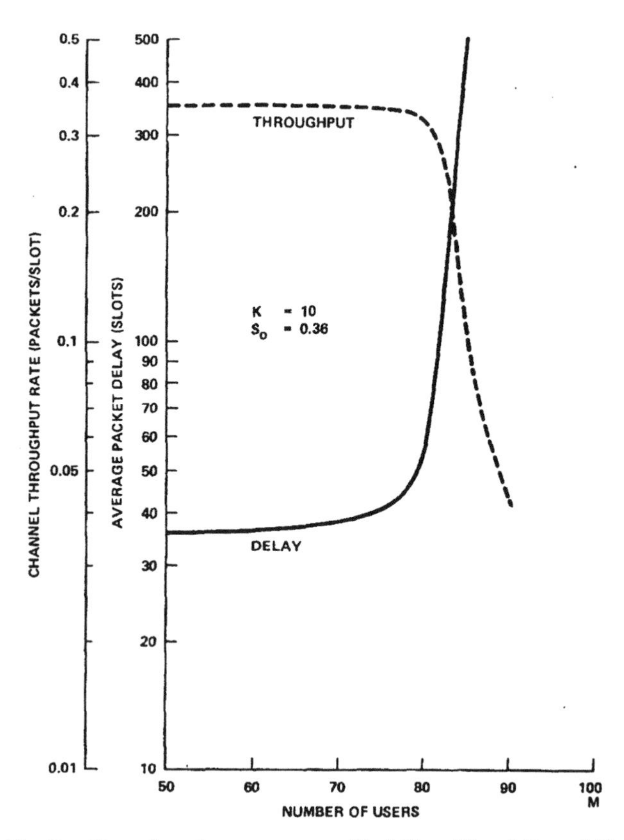

Fig. 8. Channel performance versus M at $K = 10$ and $S_o = 0.36$.

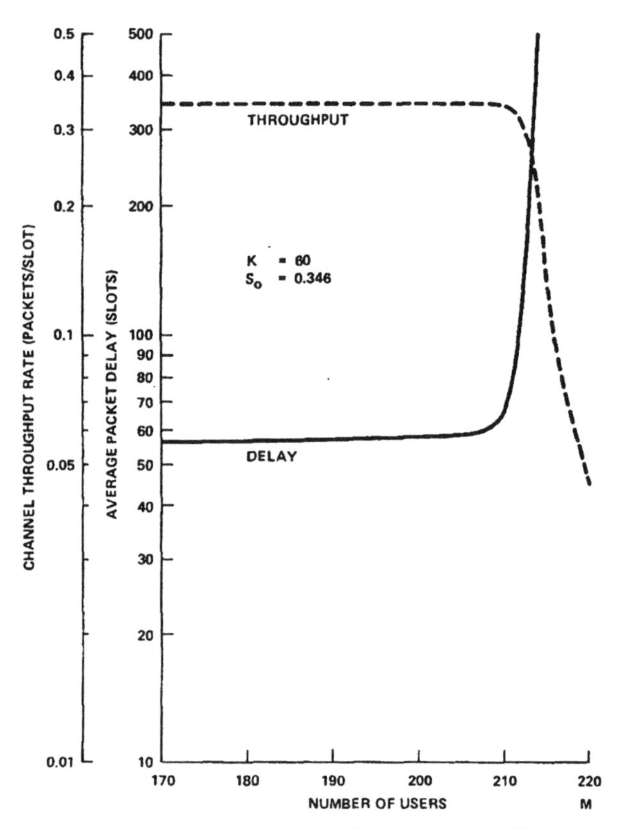

Fig. 9. Channel performance versus M at $K = 60$ and $S_o = 0.346$.

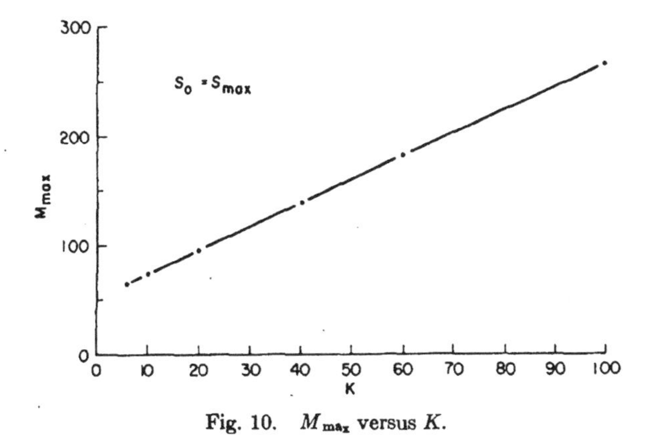

Fig. 10. M_{max} versus K.

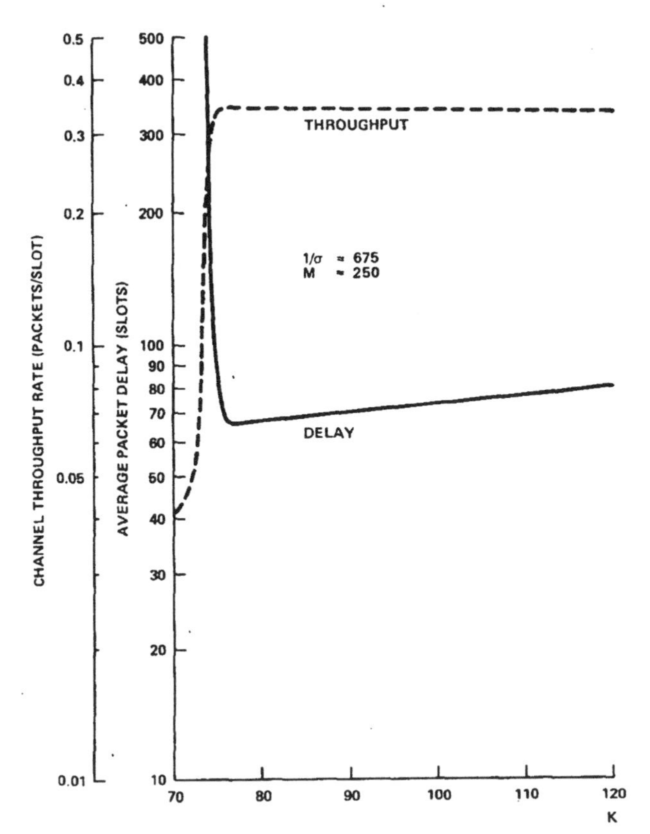

Fig. 11. Channel performance versus K at $M = 250$ and $1/\sigma = 675$.

stable equilibrium point in this case is the channel saturation point! Thus, this represents an "overloaded" channel as a result of bad system design. To correct this situation, the number of active users M supported by the channel should be reduced. From now on, a stable channel will always refer to the load line depicted in Fig. 7(a) instead of Fig. 7(d).

Let us summarize the major *conclusions* in the above discussion.

1) The steady-state throughput-delay performance of a stable channel is closely approximated by its globally stable equilibrium point and by the equilibrium throughput-delay results for the infinite population model.

2) In an unstable channel, the throughput-delay performance at a locally stable equilibrium point can be achieved only for some finite time period.

A Stability Measure

From the above discussion and referring to Fig. 7(b), the load line of an unstable channel can be partitioned into two regions. The *safe* region consisting of the channel states $\{0,1,2,\cdots,n_c\}$ and the *unsafe* region consisting of the channel states $\{n_c + 1,\cdots,M\}$. A good stability measure (for these unstable channels!) is the average time to exit into the unsafe region starting from a safe channel state. To be exact, we define FET to be the *average first exit time* into the unsafe region starting from an initially empty channel ($N^0 = 0$). Thus, FET gives an approximate measure of the average up time of an unstable channel. Below we derive the probability distributions and expected values of such first exit times. The derivations are based upon well-known results of first entrance times in the theory of Markov chains with stationary transition probabilities [28], [30].

Consider the Markovian model with constant M and σ, where M may be infinite. N^t is a Markov process (chain) with stationary transition probabilities $\{p_{ij}\}$ given by (1) or (2). Define the random variable T_{ij} to be the number of transitions which N^t goes through until it enters state j for the first time starting from state i. The probability distribution of T_{ij} (called the *first entrance probabilities* from state i to state j) may be defined as

$$f_{ij}(m) = \text{Prob}\,[T_{ij} = m] = \begin{cases} 0 & m = 0 \\ p_{ij} & m = 1 \\ \text{Prob}\,[N^{t+m} = j,\ N^{t+h} \neq j,\ h = 1,\cdots,m-1 \mid N^t = i] & m \geq 2. \end{cases} \quad (9)$$

$n = \infty$ is a stable equilibrium point. In fact, since N^t has an infinite state space and $S > S_{out}(n,S)$ for $n > n_c$, a stationary probability distribution does not exist for N^t. (See, for example, [29, pp. 543–546] for such a proof in a queueing context.)

The channel load line shown in Fig. 7(d) is stable according to the stability definition. However, the globally

The state space $\mathcal{S}$ for N^t consists of the set of nonnegative integers $\{0,1,2,\cdots,n_c,\ n_c + 1,\cdots,M\}$ which is partitioned into the safe region $\{0,1,2,\cdots,n_c\}$ and the unsafe region $\{n_c + 1,\cdots,M\}$. Now consider the modified state space $\mathcal{S}' = \{0,1,2,\cdots,n_c,n_u\}$ where n_u is an absorbing state such that N^t is now characterized by the transition probabilities

$$p_{ij}' = \begin{cases} p_{ij} & i,j = 0,1,\cdots,n_c \\ \sum_{l=n_c+1}^{M} p_{il} & i = 0,1,\cdots,n_c; j = n_u \\ 0 & i = n_u; j = 0,\cdots,n_c \\ 1 & i = j = n_u. \end{cases} \quad (10)$$

Define the random variable T_i to be the number of transitions which N^t goes through before it enters the unsafe region for the first time starting from state i in the safe region. T_i is called the *first exit time from state* i. The probability distribution of T_i is defined to be $\{f_i(m)\}_{m=1}^{\infty}$ which are called the *first exit probabilities*. It is trivial to show that starting from state i $(0 \leq i \leq n_c)$, the first entrance probabilities into the absorbing state n_u in the modified state space $\mathcal{S}'$ are the same as the first exit probabilities into the unsafe region of $\mathcal{S}$. Using (9), such probabilities are given by the following recursive equation [30],

$$f_{in_u}(m) = p_{in_u}'\delta(m-1) + \sum_{j=0}^{n_c} p_{ij}'f_{jn_u}(m-1) \qquad m \geq 1; i \neq n_u$$

where

$$\delta(m) = \begin{cases} 1 & m = 1 \\ 0 & \text{otherwise.} \end{cases}$$

The above equation can be rewritten in terms of the first exit probabilities as

$$f_i(m) = \sum_{j=n_c+1}^{M} p_{ij}\delta(m-1) + \sum_{j=0}^{n_c} p_{ij}f_j(m-1) \qquad m \geq 1; 0 \leq i \leq n_c \quad (11)$$

where $f_i(m)$ can be solved recursively for $m \geq 1$ starting with $f_i(0) = 0$ for all i.

The probability distribution $\{f_i(m)\}_{m=1}^{\infty}$ for the random variable T_i typically has a very long tail and cannot be easily computed. We had defined earlier FET as a stability measure for an unstable channel. By our definition, FET is the same as the expected value of the random variable T_0. Let $\bar{T}_i$ be the expected value and $\overline{T_i^2}$ be the second moment of T_i. These moments can be obtained by solving a set of linear simultaneous equations. It can easily be shown[30] that

$$T_i = \begin{cases} 1 & \text{with probability } p_{in_u}' \\ 1 + T_j & \text{with probability } p_{ij} \end{cases}$$

from which we obtain [28], [30]

$$\bar{T}_i = 1 + \sum_{j=0}^{n_c} p_{ij}\bar{T}_j \qquad i = 0, 1,\cdots,n_c \quad (12)$$

$$\overline{T_i^2} = 2\bar{T}_i - 1 + \sum_{j=0}^{n_c} p_{ij}\overline{T_j^2} \qquad i = 0, 1,\cdots,n_c. \quad (13)$$

Equation (12) forms a set of $n_c + 1$ linear simultaneous equations from which $\{\bar{T}_i\}_{i=0}^{n_c}$ can be solved and the stability measure FET $(=\bar{T}_0)$ determined. After $\{\bar{T}_i\}_{i=0}^{n_c}$ have been found, (13) can then be solved in a similar manner for $\{\overline{T_i^2}\}_{i=0}^{n_c}$.

Numerical Results

With the stability measure defined above, we are now in a position to examine quantitatively the tradeoff among channel stability, throughput and delay for *unstable* channels. Below we first give a computational procedure to solve for $\bar{T}_i$ and hence, FET. We then compute these quantities for various values of K, S_o, and M (corresponding to different channel load lines). The trading relations among channel stability, throughput, and delay are then illustrated.

The solution of the set of simultaneous equations in either (12) or (13) requires inverting the $(n_c + 1)$ by $(n_c + 1)$ matrix of p_{ij} for $i, j = 0, 1,\cdots,n_c$. When n_c is large, this becomes a nontrivial task because of the large number of computational steps and large computer storage requirement for the $[p_{ij}]$ matrix. The fact that $p_{ij} = 0$ for $j \leq i - 2$ in (1) and (2) enables us to use an algorithm given in the Appendix which is very efficient in terms of both computer time and space requirements. For our purposes, this algorithm is superior to conventional methods such as Gauss elimination [31] for solving linear simultaneous equations. In this algorithm, each p_{ij} is used exactly once and can be computed using (1) or (2) only when it is needed in the algorithm. This eliminates the need for storing the $[p_{ij}]$ matrix and practically eliminates any computer storage constraint on the dimensionality of the problem. The number of arithmetic operations $(+ - \times \div)$ required by the above algorithm is in the order of $2n_c^2$ which is comparable to that of Gauss elimination.

In Fig. 12, we show FET as a function of K for the infinite population model and for fixed values of the channel throughput rate S_o (at the channel operating point). We see that FET can be improved by either decreasing the channel throughput rate S_o or by increasing K (which in turn increases the average packet delay). The infinite population model results give the worst case estimates for channel stability as demonstrated in Fig. 13 in which we show FET as a function of M for $K = 10$ and four values of S_o. Note that FET increases as M decreases and there is a critical value of M below which the channel is always stable in the sense of Fig. 7(a). As M increases to infinity, FET reaches a limiting value corresponding to the infinite population model with a Poisson channel input. Fig. 14 is similar to Fig. 12 except now the number of users

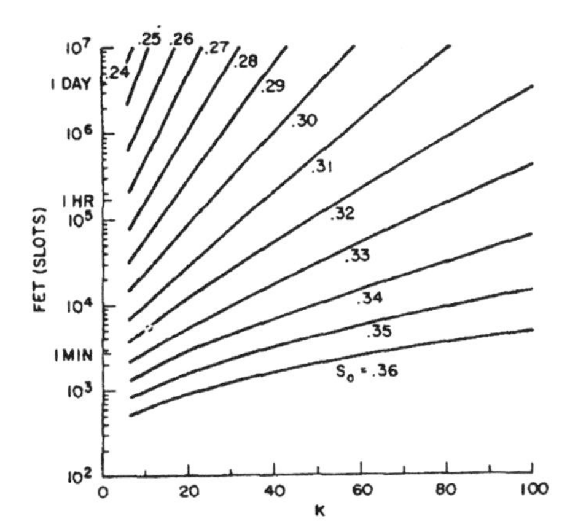

Fig. 12. FET values for the infinite population model.

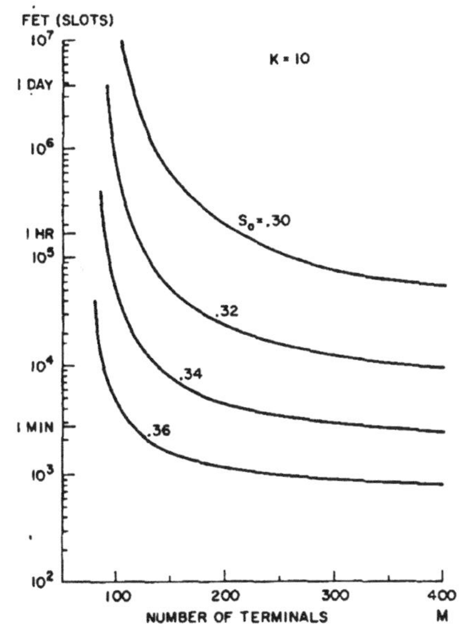

Fig. 13. FET versus M.

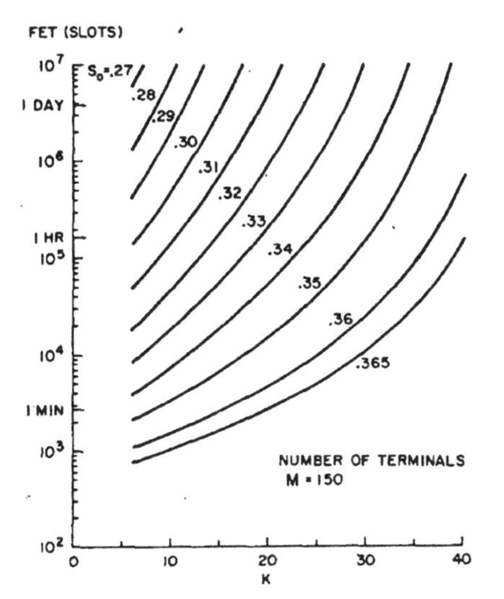

Fig. 14. FET values for a finite user population ($M = 150$).

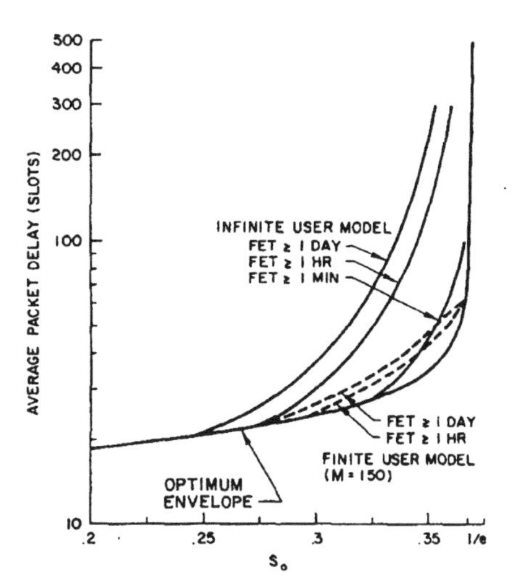

Fig. 15. Stability-throughput-delay tradeoff.

M is 150. Recall that if M is finite, the channel will become stable when K is sufficiently large.

As an example, we see that in Fig. 14 for $M = 150$, if the channel throughput rate S_o is kept at approximately 0.28 and $K = 10$ is used, the channel is estimated to fail once every two days on the average. If this is an acceptable level of channel reliability, then no other channel control procedure is necessary except to restart the channel whenever it goes into saturation. However, if absolute channel reliability is required at the same throughput-delay performance, then dynamic channel control strategies should be adopted. Channel control schemes have been studied [10] and the results will be published in a forthcoming paper [1].

In Fig. 15, we show the optimum performance envelope in Fig. 2 as a lower bound for the throughput-delay tradeoff of the infinite population model. This corresponds to the performance of the channel at the channel operating point. However, from Fig. 7, we see that *the channel operating point* (n_o,S_o) *provides no information regarding the stability behavior of the channel.* The equilibrium performance given by (n_o,S_o) is achievable in the long run if M is small enough such that the channel is stable; elsewhere it is achievable only for some random time period whose average is estimated by our stability measure FET.

In addition to the infinite population model optimum envelope, we also show in Fig. 15 two sets of equilibrium throughput-delay performance curves with guaranteed FET values. The first set consists of three solid curves corresponding to an infinite population model with the stability measure FET $\geq$ 1 day, 1 hour, and 1 minute. Again, these results represent worst case estimates if M is actually finite. The second set consists of two dashed curves corresponding to $M = 150$ with FET $\geq$ 1 day and 1 hour. These results were obtained by looking up the values of K and S_o in Fig. 12 or Fig. 14 corresponding to a fixed FET. The average packet delay was then obtained from Fig. 2. This figure illustrates the *fundamental tradeoff* among channel stability, throughput and delay. In [1], [10], control strategies are devised to dynamically regulate the channel usage to achieve truly stable throughput-delay performance close to the optimum performance envelope.

A Design Example

The designer of a slotted ALOHA channel is faced with the problem of deciding whether he wants 1) a stable channel by limiting its use to a small population of users and sacrificing channel utilization, or 2) an unstable channel which supports a large population of users operating at a certain level of reliability (some value of FET). For example, suppose K is chosen to be 10. (Note in Fig. 2 that $K = 10$ gives close to optimum equilibrium throughput-delay performance over a wide range of channel throughput rate.) Also, suppose that the channel users have an average think time of 20 s which, for our channel numerical constants, correspond to 888 time slots. Now if we draw channel load lines in Fig. 4 with a slope equal to -888, the channel is stable up to approximately 110 channel users. For $M = 110$, the channel throughput rate S_o is about 0.125 packet/slot. From Fig. 2, the average packet delay is roughly 16.5 time slots (=0.37 s). The same channel can be used (in an unstable mode) to support 220 users at a channel throughput rate of $S_o = 0.25$ packet/slot. The average packet delay is 21 time slots (=0.47 s). From Fig. 12, for $K = 10$ and $S_o = 0.25$, the average up time (FET) of the channel is approximately two days for the infinite population model. Note that this value represents a lower bound for the FET of $M = 220$. Thus, we see that if a channel failure rate of once every two days on the average is an acceptable level of reliability, the second channel design is much more attractive than the first since the number of channel users is more than doubled at a modest increase in delay.

CONCLUSIONS

In this paper, the rationale and some advantages for broadcast packet communication have been discussed. A mathematical model was then formulated for a slotted ALOHA random access system. Using this model, a theory was put forth which gives a coherent qualitative interpretation of the system stability behavior. Quantitative estimates for the relative instability of unstable channels were obtained through definition of the stability measure FET. Numerical results were shown illustrating the trading relations among channel stability, throughput and average packet delay. These results establish tools for the performance evaluation and design of an uncontrolled slotted ALOHA system. Further improvement in the system performance may be accomplished through adaptive control techniques studied in [1], [10].

APPENDIX

The algorithm below solves for the variables $\{t_i\}_{i=0}^{I}$ in the following set of $(I + 1)$ linear simultaneous equations,

$$t_0 = h_0 + \sum_{j=0}^{I} p_{0j} t_j \qquad \text{(A1)}$$

$$t_i = h_i + \sum_{j=i-1}^{I} p_{ij} t_j \qquad i = 1, 2, \cdots, I. \qquad \text{(A2)}$$

The Algorithm

1) Define

$$e_I = 1$$

$$f_I = 0$$

$$e_{I-1} = \frac{1 - p_{II}}{p_{I,I-1}}$$

$$f_{I-1} = -\frac{h_I}{p_{I,I-1}}.$$

2) For $i = I - 1, I - 2, \cdots, 1$ solve recursively

$$e_{i-1} = \frac{1}{p_{i,i-1}} [e_i - \sum_{j=i}^{I} p_{ij} e_j]$$

$$f_{i-1} = \frac{1}{p_{i,i-1}} [f_i - h_i - \sum_{j=i}^{I} p_{ij} f_j].$$

3) Let

$$t_I = \frac{f_0 - h_0 - \sum_{j=0}^{I} p_{0j} f_j}{\sum_{j=0}^{I} p_{0j} e_j - e_0}$$

$$t_i = e_i t_I + f_i \qquad i = 0, 1, 2, \cdots, I - 1.$$

Derivation of the Algorithm

Define

$$t_i = e_i t_I + f_i \qquad i = 0, 1, 2, \cdots, I - 1 \qquad \text{(A3)}$$

and

$$e_I = 1$$

$$f_I = 0. \qquad \text{(A4)}$$

The last equation in (A2) is

$$t_I = h_I + p_{I,I-1} t_{I-1} + p_{II} t_I.$$

Substituting $t_{I-1} = e_{I-1} t_I + f_{I-1}$ into the above equation, we get

$$t_I = h_I + p_{I,I-1} e_{I-1} t_I + p_{I,I-1} f_{I-1} + p_{II} t_I.$$

Equating the coefficients of t_I and the constant terms, we have

$$e_{I-1} = \frac{1 - p_{II}}{p_{I,I-1}}$$

$$f_{I-1} = -\frac{h_I}{p_{I,I-1}}. \tag{A5}$$

Equation (A2) can be rewritten as follows,

$$t_{i-1} = \frac{1}{p_{i,i-1}}\left[t_i - h_i - \sum_{j=i}^{I} p_{ij}t_j\right]. \tag{A6}$$

In each of the above equations, use (A3) to substitute for t_i. We then have

$$e_{i-1}t_I + f_{i-1} = \frac{1}{p_{i,i-1}}\left[e_it_I + f_i - h_i - \left(\sum_{j=i}^{I} p_{ij}e_j\right)t_I - \sum_{j=i}^{I} p_{ij}f_j\right].$$

Equating the coefficients of t_I and the constant terms, we get

$$e_{i-1} = \frac{1}{p_{i,i-1}}\left[e_i - \sum_{j=i}^{I} p_{ij}e_j\right]$$

$$f_{i-1} = \frac{1}{p_{i,i-1}}\left[f_i - h_i - \sum_{j=i}^{I} p_{ij}f_j\right]. \tag{A7}$$

From (A4), (A5), and (A7), e_i and f_i $(i = I-2, I-3, \cdots, 1, 0)$ can then be determined recursively.

We next solve for t_I. Equation (A3) is used to substitute for t_i in (A1), which then becomes

$$e_0t_I + f_0 = h_0 + \left(\sum_{j=0}^{I} p_{0j}e_j\right)t_I + \sum_{j=0}^{I} p_{0j}f_j.$$

Solving for t_I in the above equation, we have

$$t_I = \frac{f_0 - h_0 - \sum_{j=0}^{I} p_{0j}f_j}{\sum_{j=0}^{I} p_{0j}e_j - e_0}. \tag{A8}$$

Finally, $t_i (i = 0, 1, 2, \cdots, I-1)$ can be obtained from (A3), since e_i, f_i, and t_I are all known. The derivation of the algorithm is now complete.

REFERENCES

[1] S. S. Lam and L. Kleinrock, "Packet switching in a multiaccess broadcast channel: dynamic control procedures," *IEEE Trans. Commun.*, to be published; also in IBM Corp., Yorktown Heights, N. Y., Res. Rep. RC-5062, Oct. 1974.
[2] N. Abramson, "The ALOHA system—another alternative for computer communications," in *1970 Fall Joint Comput. Conf., AFIPS Conf. Proc.*, vol. 37. Montvale, N. J.: AFIPS Press, 1970, pp. 281–285.
[3] W. Crowther, R. Rettberg, D. Walden, S. Ornstein, and F. Heart, "A system for broadcast communication: reservation—ALOHA," in *Proc. 6th Hawaii Int. Conf. System Sciences*, Univ. Hawaii, Honolulu, Jan. 1973.
[4] R. M. Metcalfe, "Steady-state analysis of a slotted and controlled ALOHA system with blocking," in *Proc. 6th Hawaii Int. Conf. System Sciences*, Univ. Hawaii, Honolulu, Jan. 1973.
[5] N. Abramson, "Packet switching with satellites," in *1973 Nat. Comput. Conf., AFIPS Conf. Proc.*, vol. 42. New York: AFIPS Press, 1973, pp. 695–702.
[6] L. Kleinrock and S. S. Lam, "Packet-switching in a slotted satellite channel," in *1973 Nat. Comput. Conf., AFIPS Conf. Proc.*, vol. 42. New York: AFIPS Press, 1973, pp. 703–710.
[7] L. G. Roberts, "Dynamic allocation of satellite capacity through packet reservation," in *1973 Nat. Comput. Conf., AFIPS Conf. Proc.*, vol. 42. New York: AFIPS Press, 1973, pp. 711–716.
[8] L. Kleinrock and S. S. Lam, "On stability of packet switching in a random multi-access broadcast channel," in *Proc. 7th Hawaii Int. Conf. System Sciences (Special Subconf. Computer Nets)*, Univ. Hawaii, Honolulu, Jan. 8–10, 1974.
[9] S. Butterfield, R. Rettberg, and D. Walden, "The satellite IMP for the ARPA network," in *Proc. 7th Hawaii Int. Conf. System Sciences (Special Subconf. Computer Nets)*, Univ. Hawaii, Honolulu, Jan. 8–10, 1974.
[10] S. S. Lam, "Packet switching in a multi-access broadcast channel with application to satellite communication in a computer network," Ph.D. dissertation, Dep. Comput. Sci., Univ. Calif., Los Angeles, Mar. 1974; also in Univ. of Calif., Los Angeles, Tech. Rep. UCLA-ENG-7429, Apr. 1974.
[11] L. G. Roberts, "Data by the packet," *IEEE Spectrum*, vol. 11, pp. 46–51, Feb. 1974.
[12] L. G. Roberts and B. D. Wessler, "Computer network development to achieve resource sharing," in *1970 Spring Joint Comput. Conf., AFIPS Conf. Proc.*, vol. 36. Montvale, N. J.: AFIPS Press, 1970, pp. 543–549.
[13] P. E. Jackson and C. D. Stubbs, "A study of multiaccess computer communications," in *1969 Spring Joint Comput. Conf., AFIPS Conf. Proc.*, vol. 34. Montvale, N. J.: AFIPS Press, 1969, pp. 491–504.
[14] J. Martin, *Systems Analysis for Data Transmission.* Englewood Cliffs, N. J.: Prentice-Hall, 1972.
[15] J. R. Pierce, "Network for block switching of data," in *IEEE Conv. Rec.*, New York, Mar. 1971.
[16] W. W. Chu, "A study of asynchronous time division multiplexing for time-sharing computer systems," in *1969 Fall Joint Comput. Conf., AFIPS Conf. Proc.*, vol. 35. Montvale, N. J.: AFIPS Press, 1969, pp. 669–678.
[17] P. Baran, "On distributed communications XI. Summary overview," Rand Corp., Santa Monica, Calif., Memo. RM-3767-PR, Aug. 1964.
[18] L. Kleinrock, *Communication Nets: Stochastic Message Flow and Delay.* New York: McGraw-Hill, 1964 (out of print); reprinted by New York: Dover, 1972.
[19] D. W. Davies, "The principles of a data communication network for computers and remote peripherals," in *Proc. Int. Fed. Information Processing Congr.*, Edinburgh, Scotland, 1968, pp. D11–D15.
[20] P. Wright, "Facing a booming demand for networks," *Datamation*, vol. 19, pp. 138–139, Nov. 1973.
[21] H. Frank, M. Gerla, and W. Chou, "Issues in the design of large distributed computer communication networks," in *Proc. Nat. Telecommunications Conf.*, Atlanta, Ga., Nov. 26–28, 1973.
[22] L. G. Roberts, "Extensions of packet communication technology to a hand held personal terminal.," in *1972 Spring Joint Comput. Conf., AFIPS Conf. Proc.*, vol. 40. Montvale, N. J.: AFIPS Press, 1972, pp. 295–298.
[23] In *Inst. Elec. Eng. (London) Proc. Int. Conf. Satellite Systems for Mobile Communications and Surveillance*, Mar. 13–15, 1973.
[24] N.T . Gaarder, "ARPANET satellite system," ARPA Network Inform. Center, Stanford Res. Inst., Menlo Park, Calif., ASS Note 3 (NIC 11285), Apr. 1972.
[25] L. G. Roberts, "ALOHA packet system with and without slots and capture," ARPA Network Inform. Center, Stanford Res. Inst., Menlo Park, Calif., ASS Note 8 (NIC 11290), June 1972.
[26] L. Kleinrock and F. A. Tobagi, "Carrier-sense multiple access for packet switched radio channels," in *Proc. Int. Conf. Communications*, Minneapolis, Minn., June 1974.
[27] L. Kleinrock, *Queueing Systems, Vol. I, Theory, Vol. II, Computer Applications.* New York: Wiley-Interscience, 1975.
[28] E. Parzen, *Stochastic Processes.* San Francisco, Calif.: Holden-Day, 1962.
[29] J. W. Cohen, *The Single Server Queue.* New York: Wiley, 1969.
[30] R. Howard, *Dynamic Probabilistic Systems, Vol. 1: Markov Models and Vol. 2: Semi-Markov and Decision Processes.* New York: Wiley, 1971.
[31] E. J. Craig, *Laplace and Fourier Transforms for Electrical Engineers.* New York: Holt, Rinehart, and Winston, 1964.

PHOTO NOT AVAILABLE

Leonard Kleinrock (S'55–M'64–SM'71–F'73) was born in New York, N. Y., on June 13, 1934. He received the B.E.E. degree from the City College of New York, N. Y., in 1957, and the S.M.E.E. and Ph.D. degrees in electrical engineering from the Massachusetts Institute of Technology, Cambridge, in 1959 and 1963, respectively, while participating in the Lincoln Laboratory Staff Associate Program.

From 1951 to 1957, he was employed at the Photobell Company, Inc., New York, N. Y., an industrial electronics firm. He spent the summers from 1957 to 1961 at the M.I.T. Lincoln Laboratory, Lexington, Mass., first in the Digital Computer Group and later in the Systems Analysis Group. At M.I.T. he was a Research Assistant, initially with the Electronic Systems Laboratory, and later with the Research Laboratory for Electronics, where he worked on communication nets in the Information Processing and Transmission Group. After completing his graduate work at the end of 1962, he worked at Lincoln Laboratory on communication nets and on signal detection. In 1963 he accepted a position on the faculty at the University of California, Los Angeles, where he is now Professor of Computer Science. He is a referee for numerous scholarly publications, book reviewer for several publishers, and a consultant for various aerospace, research, and governmental organizations. He is principal investigator of a large contract with the Advanced Research Projects Agency (ARPA) of the Department of Defense. He has published over 60 papers and is the author of *Communication Nets; Stochastic Message Flow and Delay* (New York: McGraw-Hill, 1964), *Queueing Systems, Vol. 1: Theory* and *Vol. 2: Computer Applications* (New York: Wiley-Interscience, 1975). His main interests are in communication nets, computer nets, data compression, priority queueing theory, and theoretical studies of time-shared systems.

Dr. Kleinrock is a member of Tau Beta Pi, Eta Kappa Nu, Sigma Xi, the Operations Research Society of America, and the Association for Computing Machinery. He was awarded a Guggenheim Fellowship in 1971.

PHOTO NOT AVAILABLE

Simon S. Lam (S'69–M'74) was born in Macao on July 31, 1947. He received the B.S.E.E. degree in electrical engineering from Washington State University, Pullman, in 1969, and the M.S. and Ph.D. degrees in engineering from the University of California, Los Angeles, in 1970 and 1974, respectively.

At the University of California, Los Angeles, he held a Phi Kappa Phi Fellowship from 1969 to 1970, and a Chancellor's Teaching Fellowship from 1969 to 1973. He also participated in the ARPA Network project at UCLA as a postgraduate research engineer from 1972 to 1974 and did research on satellite packet communication. Since June 1974 he has been a research staff member with the IBM Thomas J. Watson Research Center, Yorktown Heights, N. Y. His current research interests include computer-communication networks and queueing theory.

Dr. Lam is a member of Tau Beta Pi, Sigma Tau, Phi Kappa Phi, Pi Mu Epsilon, and the Association for Computing Machinery.

A Protocol for Packet Network Intercommunication

VINTON G. CERF AND ROBERT E. KAHN, MEMBER, IEEE

Abstract—**A protocol that supports the sharing of resources that exist in different packet switching networks is presented. The protocol provides for variation in individual network packet sizes, transmission failures, sequencing, flow control, end-to-end error checking, and the creation and destruction of logical process-to-process connections. Some implementation issues are considered, and problems such as internetwork routing, accounting, and timeouts are exposed.**

INTRODUCTION

IN THE LAST few years considerable effort has been expended on the design and implementation of packet switching networks [1]–[7],[14],[17]. A principle reason for developing such networks has been to facilitate the sharing of computer resources. A packet communication network includes a transportation mechanism for delivering data between computers or between computers and terminals. To make the data meaningful, computers and terminals share a common protocol (i.e., a set of agreed upon conventions). Several protocols have already been developed for this purpose [8]–[12],[16]. However, these protocols have addressed only the problem of communication on the same network. In this paper we present a protocol design and philosophy that supports the sharing of resources that exist in different packet switching networks.

After a brief introduction to internetwork protocol issues, we describe the function of a GATEWAY as an interface between networks and discuss its role in the protocol. We then consider the various details of the protocol, including addressing, formatting, buffering, sequencing, flow control, error control, and so forth. We close with a description of an interprocess communication mechanism and show how it can be supported by the internetwork protocol.

Even though many different and complex problems must be solved in the design of an individual packet switching network, these problems are manifestly compounded when dissimilar networks are interconnected. Issues arise which may have no direct counterpart in an individual network and which strongly influence the way in which internetwork communication can take place.

A typical packet switching network is composed of a set of computer resources called HOSTS, a set of one or more *packet switches*, and a collection of communication media that interconnect the packet switches. Within each HOST, we assume that there exist *processes* which must communicate with processes in their own or other HOSTS. Any current definition of a process will be adequate for our purposes [13]. These processes are generally the ultimate source and destination of data in the network. Typically, within an individual network, there exists a protocol for communication between any source and destination process. Only the source and destination processes require knowledge of this convention for communication to take place. Processes in two distinct networks would ordinarily use different protocols for this purpose. The ensemble of packet switches and communication media is called the *packet switching subnet*. Fig. 1 illustrates these ideas.

In a typical packet switching subnet, data of a fixed maximum size are accepted from a source HOST, together with a formatted destination address which is used to route the data in a store and forward fashion. The transmit time for this data is usually dependent upon internal network parameters such as communication media data rates, buffering and signaling strategies, routing, propagation delays, etc. In addition, some mechanism is generally present for error handling and determination of status of the networks components.

Individual packet switching networks may differ in their implementations as follows.

1) Each network may have distinct ways of addressing the receiver, thus requiring that a uniform addressing scheme be created which can be understood by each individual network.

2) Each network may accept data of different maximum size, thus requiring networks to deal in units of the smallest maximum size (which may be impractically small) or requiring procedures which allow data crossing a network boundary to be reformatted into smaller pieces.

3) The success or failure of a transmission and its performance in each network is governed by different time delays in accepting, delivering, and transporting the data. This requires careful development of internetwork timing procedures to insure that data can be successfully delivered through the various networks.

4) Within each network, communication may be disrupted due to unrecoverable mutation of the data or missing data. End-to-end restoration procedures are desirable to allow complete recovery from these conditions.

Paper approved by the Associate Editor for Data Communications of the IEEE Communications Society for publication without oral presentation. Manuscript received November 5, 1973. The research reported in this paper was supported in part by the Advanced Research Projects Agency of the Department of Defense under Contract DAHC 15-73-C-0370.

V. G. Cerf is with the Department of Computer Science and Electrical Engineering, Stanford University, Stanford, Calif.

R. E. Kahn is with the Information Processing Technology Office, Advanced Research Projects Agency, Department of Defense, Arlington, Va.

Reprinted from *IEEE Transactions on Communications*, vol. COM-22, no. 5, May 1974.

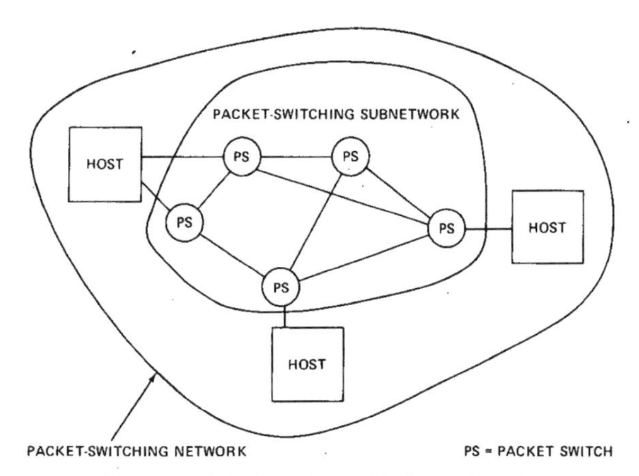

Fig. 1. Typical packet switching network.

5) Status information, routing, fault detection, and isolation are typically different in each network. Thus, to obtain verification of certain conditions, such as an inaccessible or dead destination, various kinds of coordination must be invoked between the communicating networks.

It would be extremely convenient if all the differences between networks could be economically resolved by suitable interfacing at the network boundaries. For many of the differences, this objective can be achieved. However, both economic and technical considerations lead us to prefer that the interface be as simple and reliable as possible and deal primarily with passing data between networks that use different packet switching strategies.

The question now arises as to whether the interface ought to account for differences in HOST or process level protocols by transforming the source conventions into the corresponding destination conventions. We obviously want to allow conversion between packet switching strategies at the interface, to permit interconnection of existing and planned networks. However, the complexity and dissimilarity of the HOST or process level protocols makes it desirable to avoid having to transform between them at the interface, even if this transformation were always possible. Rather, compatible HOST and process level protocols must be developed to achieve effective internetwork resource sharing. The unacceptable alternative is for every HOST or process to implement every protocol (a potentially unbounded number) that may be needed to communicate with other networks. We therefore assume that a common protocol is to be used between HOST's or processes in different networks and that the interface between networks should take as small a role as possible in this protocol.

To allow networks under different ownership to interconnect, some accounting will undoubtedly be needed for traffic that passes across the interface. In its simplest terms, this involves an accounting of packets handled by each net for which charges are passed from net to net until the buck finally stops at the user or his representative. Furthermore, the interconnection must preserve intact the internal operation of each individual network. This is easily achieved if two networks interconnect as if each were a HOST to the other network, but without utilizing or indeed incorporating any elaborate HOST protocol transformations.

It is thus apparent that the interface between networks must play a central role in the development of any network interconnection strategy. We give a special name to this interface that performs these functions and call it a GATEWAY.

THE GATEWAY NOTION

In Fig. 2 we illustrate three individual networks labeled A, B, and C which are joined by GATEWAYS M and N. GATEWAY M interfaces network A with network B, and GATEWAY N interfaces network B to network C. We assume that an individual network may have more than one GATEWAY (e.g., network B) and that there may be more than one GATEWAY path to use in going between a pair of networks. The responsibility for properly routing data resides in the GATEWAY.

In practice, a GATEWAY between two networks may be composed of two halves, each associated with its own network. It is possible to implement each half of a GATEWAY so it need only embed internetwork packets in local packet format or extract them. We propose that the GATEWAYS handle internetwork packets in a standard format, but we are not proposing any particular transmission procedure between GATEWAY halves.

Let us now trace the flow of data through the interconnected networks. We assume a packet of data from process X enters network A destined for process Y in network C. The address of Y is initially specified by process X and the address of GATEWAY M is derived from the address of process Y. We make no attempt to specify whether the choice of GATEWAY is made by process X, its HOST, or one of the packet switches in network A. The packet traverses network A until it reaches GATEWAY M. At the GATEWAY, the packet is reformatted to meet the requirements of network B, account is taken of this unit of flow between A and B, and the GATEWAY delivers the packet to network B. Again the derivation of the next GATEWAY address is accomplished based on the address of the destination Y. In this case, GATEWAY N is the next one. The packet traverses network B until it finally reaches GATEWAY N where it is formatted to meet the requirements of network C. Account is again taken of this unit of flow between networks B and C. Upon entering network C the packet is routed to the HOST in which process Y resides and there it is delivered to its ultimate destination.

Since the GATEWAY must understand the address of the source and destination HOSTS, this information must be available in a standard format in every packet which arrives at the GATEWAY. This information is contained in an *internetwork header* prefixed to the packet by the source HOST. The packet format, including the internet

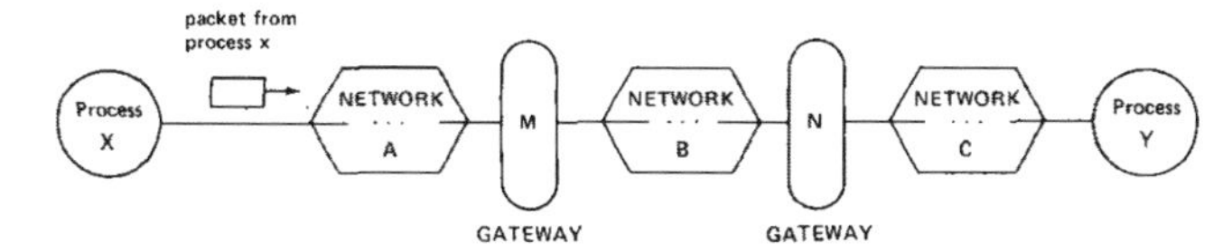

Fig. 2. Three networks interconnected by two GATEWAYS.

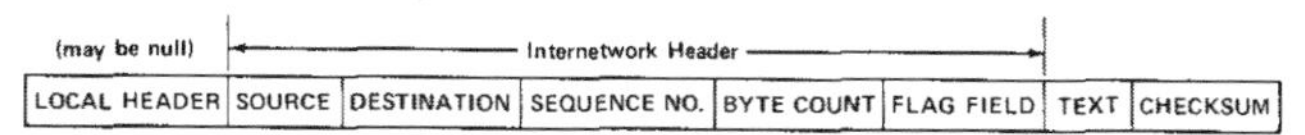

Fig. 3. Internetwork packet format (fields not shown to scale).

work header, is illustrated in Fig. 3. The source and destination entries uniformly and uniquely identify the address of every HOST in the composite network. Addressing is a subject of considerable complexity which is discussed in greater detail in the next section. The next two entries in the header provide a sequence number and a byte count that may be used to properly sequence the packets upon delivery to the destination and may also enable the GATEWAYS to detect fault conditions affecting the packet. The flag field is used to convey specific control information and is discussed in the section on retransmission and duplicate detection later. The remainder of the packet consists of text for delivery to the destination and a trailing check sum used for end-to-end software verification. The GATEWAY does *not* modify the text and merely forwards the check sum along without computing or recomputing it.

Each network may need to augment the packet format before it can pass through the individual network. We have indicated a *local header* in the figure which is prefixed to the beginning of the packet. This local header is introduced merely to illustrate the concept of embedding an internetwork packet in the format of the individual network through which the packet must pass. It will obviously vary in its exact form from network to network and may even be unnecessary in some cases. Although not explicitly indicated in the figure, it is also possible that a local trailer may be appended to the end of the packet.

Unless all transmitted packets are legislatively restricted to be small enough to be accepted by every individual network, the GATEWAY may be forced to split a packet into two or more smaller packets. This action is called fragmentation and must be done in such a way that the destination is able to piece together the fragmented packet. It is clear that the internetwork header format imposes a minimum packet size which all networks must carry (obviously all networks will want to carry packets larger than this minimum). We believe the long range growth and development of internetwork communication would be seriously inhibited by specifying how much larger than the minimum a packet size can be, for the following reasons.

1) If a maximum permitted packet size is specified then it becomes impossible to completely isolate the internal packet size parameters of one network from the internal packet size parameters of all other networks.

2) It would be very difficult to increase the maximum permitted packet size in response to new technology (e.g., large memory systems, higher data rate communication facilities, etc.) since this would require the agreement and then implementation by all participating networks.

3) Associative addressing and packet encryption may require the size of a particular packet to expand during transit for incorporation of new information.

Provision for fragmentation (regardless of where it is performed) permits packet size variations to be handled on an individual network basis without global administration and also permits HOSTS and processes to be insulated from changes in the packet sizes permitted in any networks through which their data must pass.

If fragmentation must be done, it appears best to do it upon entering the next network at the GATEWAY since only this GATEWAY (and not the other networks) must be aware of the internal packet size parameters which made the fragmentation necessary.

If a GATEWAY fragments an incoming packet into two or more packets, they must eventually be passed along to the destination HOST as fragments or reassembled for the HOST. It is conceivable that one might desire the GATEWAY to perform the reassembly to simplify the task of the destination HOST (or process) and/or to take advantage of a larger packet size. We take the position that GATEWAYS should not perform this function since GATEWAY reassembly can lead to serious buffering problems, potential deadlocks, the necessity for all fragments of a packet to pass through the same GATEWAY, and increased delay in transmission. Furthermore, it is not sufficient for the GATEWAYS to provide this function since the final GATEWAY may also have to fragment a packet for transmission. Thus the destination HOST must be prepared to do this task.

Let us now turn briefly to the somewhat unusual accounting effect which arises when a packet may be fragmented by one or more GATEWAYS. We assume, for simplicity, that each network initially charges a fixed rate per packet transmitted, regardless of distance, and if one network can handle a larger packet size than another, it charges a proportionally larger price per packet. We also assume that a subsequent increase in any network's packet size does not result in additional cost per packet to its users. The charge to a user thus remains basically constant through any net which must fragment a packet. The unusual effect occurs when a packet is fragmented into smaller packets which must individually pass through a subsequent network with a larger packet size than the original unfragmented packet. We expect that most networks will naturally select packet sizes close to one another, but in any case, an increase in packet size in one net, even when it causes fragmentation, will not increase the cost of transmission and may actually decrease it. In the event that any other packet charging policies (than the one we suggest) are adopted, differences in cost can be used as an economic lever toward optimization of individual network performance.

PROCESS LEVEL COMMUNICATION

We suppose that processes wish to communicate in full duplex with their correspondents using unbounded but finite length messages. A single character might constitute the text of a message from a process to a terminal or vice versa. An entire page of characters might constitute the text of a message from a file to a process. A data stream (e.g., a continuously generated bit string) can be represented as a sequence of finite length messages.

Within a HOST we assume the existence of a transmission control program (TCP) which handles the transmission and acceptance of messages on behalf of the processes it serves. The TCP is in turn served by one or more packet switches connected to the HOST in which the TCP resides. Processes that want to communicate present messages to the TCP for transmission, and TCP's deliver incoming messages to the appropriate destination processes. We allow the TCP to break up messages into segments because the destination may restrict the amount of data that may arrive, because the local network may limit the maximum transmission size, or because the TCP may need to share its resources among many processes concurrently. Furthermore, we constrain the length of a segment to an integral number of 8-bit bytes. This uniformity is most helpful in simplifying the software needed with HOST machines of different natural word lengths. Provision at the process level can be made for padding a message that is not an integral number of bytes and for identifying which of the arriving bytes of text contain information of interest to the receiving process.

Multiplexing and demultiplexing of segments among processes are fundamental tasks of the TCP. On transmission, a TCP must multiplex together segments from different source processes and produce internetwork packets for delivery to one of its serving packet switches. On reception, a TCP will accept a sequence of packets from its serving packet switch(es). From this sequence of arriving packets (generally from different HOSTS), the TCP must be able to reconstruct and deliver messages to the proper destination processes.

We assume that every segment is augmented with additional information that allows transmitting and receiving TCP's to identify destination and source processes, respectively. At this point, we must face a major issue. How should the source TCP format segments destined for the same destination TCP? We consider two cases.

Case 1): If we take the position that segment boundaries are immaterial and that a byte stream can be formed of segments destined for the same TCP, then we may gain improved transmission efficiency and resource sharing by arbitrarily parceling the stream into packets, permitting many segments to share a single internetwork packet header. However, this position results in the need to reconstruct exactly, and in order, the stream of text bytes produced by the source TCP. At the destination, this stream must first be parsed into segments and these in turn must be used to reconstruct messages for delivery to the appropriate processes.

There are fundamental problems associated with this strategy due to the possible arrival of packets out of order at the destination. The most critical problem appears to be the amount of interference that processes sharing the same TCP–TCP byte stream may cause among themselves. This is especially so at the receiving end. First, the TCP may be put to some trouble to parse the stream back into segments and then distribute them to buffers where messages are reassembled. If it is not readily apparent that all of a segment has arrived (remember, it may come as several packets), the receiving TCP may have to suspend parsing temporarily until more packets have arrived. Second, if a packet is missing, it may not be clear whether succeeding segments, even if they are identifiable, can be passed on to the receiving process, unless the TCP has knowledge of some process level sequencing scheme. Such knowledge would permit the TCP to decide whether a succeeding segment could be delivered to its waiting process. Finding the beginning of a segment when there are gaps in the byte stream may also be hard.

Case 2): Alternatively, we might take the position that the destination TCP should be able to determine, upon its arrival and without additional information, for which process or processes a received packet is intended, and if so, whether it should be delivered then.

If the TCP is to determine for which process an arriving packet is intended, every packet must contain a *process header* (distinct from the internetwork header) that completely identifies the destination process. For simplicity, we assume that each packet contains text from a single process which is destined for a single process. Thus each packet need contain only one process header. To decide whether the arriving data is deliverable to the destination process, the TCP must be able to determine whether the data is in the proper sequence (we can make provision for the destination process to instruct its TCP to ignore sequencing, but this is considered a special case). With the assumption that each arriving packet contains a process header, the necessary sequencing and destination process identification is immediately available to the destination TCP.

Both Cases 1) and 2) provide for the demultiplexing and delivery of segments to destination processes, but only Case 2) does so without the introduction of potential interprocess interference. Furthermore, Case 1) introduces extra machinery to handle flow control on a HOST-to-HOST basis, since there must also be some provision for process level control, and this machinery is little used since the probability is small that within a given HOST, two processes will be coincidentally scheduled to send messages to the same destination HOST. For this reason, we select the method of Case 2) as a part of the *internetwork transmission protocol.*

ADDRESS FORMATS

The selection of address formats is a problem between networks because the local network addresses of TCP's may vary substantially in format and size. A uniform internetwork TCP address space, understood by each GATEWAY and TCP, is essential to routing and delivery of internetwork packets.

Similar troubles are encountered when we deal with process addressing and, more generally, port addressing. We introduce the notion of *ports* in order to permit a process to distinguish between multiple message streams. The port is simply a designator of one such message stream associated with a process. The means for identifying a port are generally different in different operating systems, and therefore, to obtain uniform addressing, a standard port address format is also required. A port address designates a full duplex message stream.

TCP ADDRESSING

TCP addressing is intimately bound up in routing issues, since a HOST or GATEWAY must choose a suitable destination HOST or GATEWAY for an outgoing internetwork packet. Let us postulate the following address format for the TCP address (Fig. 4). The choice for network identification (8 bits) allows up to 256 distinct networks. This size seems sufficient for the forseeable future. Similarly, the TCP identifier field permits up to 65 536 distinct TCP's to be addressed, which seems more than sufficient for any given network.

As each packet passes through a GATEWAY, the GATEWAY observes the destination network ID to determine how to route the packet. If the destination network is connected to the GATEWAY, the lower 16 bits of the TCP address are used to produce a local TCP address in the destination network. If the destination network is not connected to the GATEWAY, the upper 8 bits are used to select a subsequent GATEWAY. We make no effort to specify how each individual network shall associate the internetwork TCP identifier with its local TCP address. We also do not rule out the possibility that the local network understands the internetwork addressing scheme and thus alleviates the GATEWAY of the routing responsibility.

PORT ADDRESSING

A receiving TCP is faced with the task of demultiplexing the stream of internetwork packets it receives and reconstructing the original messages for each destination process. Each operating system has its own internal means of identifying processes and ports. We assume that 16 bits are sufficient to serve as internetwork port identifiers. A sending process need not know how the destination port identification will be used. The destination TCP will be able to parse this number appropriately to find the proper buffer into which it will place arriving packets. We permit a large port number field to support processes which want to distinguish between many different messages streams concurrently. In reality, we do not care how the 16 bits are sliced up by the TCP's involved.

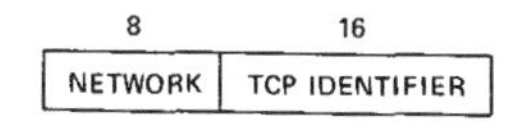

Fig. 4. TCP address.

Even though the transmitted port name field is large, it is still a compact external name for the internal representation of the port. The use of short names for port identifiers is often desirable to reduce transmission overhead and possibly reduce packet processing time at the destination TCP. Assigning short names to each port, however, requires an initial negotiation between source and destination to agree on a suitable short name assignment, the subsequent maintenance of conversion tables at both the source and the destination, and a final transaction to release the short name. For dynamic assignment of port names, this negotiation is generally necessary in any case.

SEGMENT AND PACKET FORMATS

As shown in Fig. 5, messages are broken by the TCP into segments whose format is shown in more detail in Fig. 6. The field lengths illustrated are merely suggestive. The first two fields (source port and destination port in the figure) have already been discussed in the preceding section on addressing. The uses of the third and fourth fields (window and acknowledgment in the figure) will be discussed later in the section on retransmission and duplicate detection.

We recall from Fig. 3 that an internetwork header contains both a sequence number and a byte count, as well as a flag field and a check sum. The uses of these fields are explained in the following section.

REASSEMBLY AND SEQUENCING

The reconstruction of a message at the receiving TCP clearly requires[1] that each internetwork packet carry a sequence number which is unique to its particular destination port message stream. The sequence numbers must be monotonic increasing (or decreasing) since they are used to reorder and reassemble arriving packets into a message. If the space of sequence numbers were infinite, we could simply assign the next one to each new packet. Clearly, this space cannot be infinite, and we will consider what problems a finite sequence number space will cause when we discuss retransmission and duplicate detection in the next section. We propose the following scheme for performing the sequencing of packets and hence the reconstruction of messages by the destination TCP.

A pair of ports will exchange one or more messages over a period of time. We could view the sequence of messages produced by one port as if it were embedded in an infinitely long stream of bytes. Each byte of the message has a unique sequence number which we take to be its byte location relative to the beginning of the stream. When a

[1] In the case of encrypted packets, a preliminary stage of reassembly may be required prior to decryption.

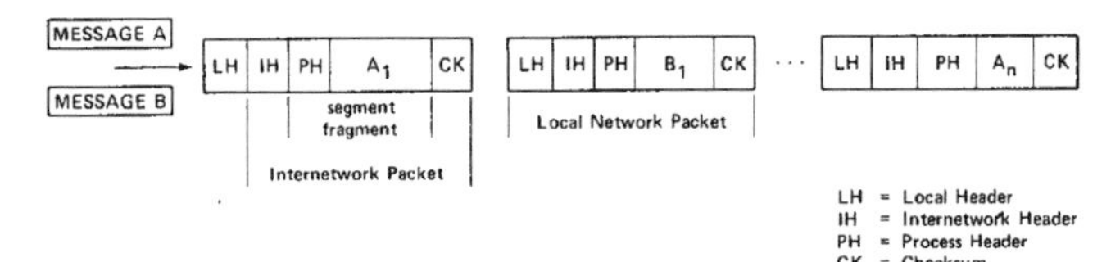

Fig. 5. Creation of segments and packets from messages.

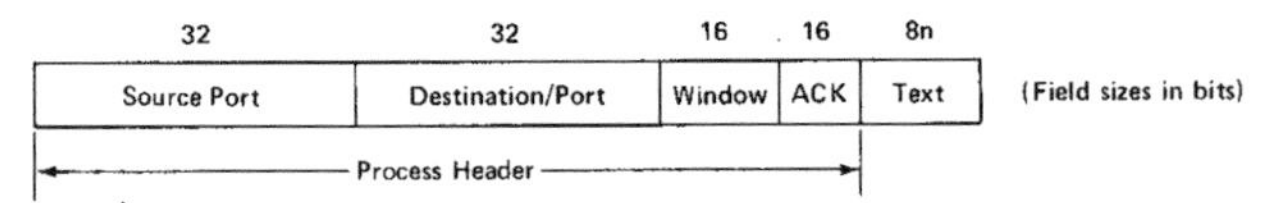

Fig. 6. Segment format (process header and text).

byte identification → sequence number

0	1	2	. . .	k	. . .		
First Message				Second Message		Third Message	. . .
Segment		Segment		Segment	Segment	. . .	

(SEQ = k)

Fig. 7. Assignment of sequence numbers.

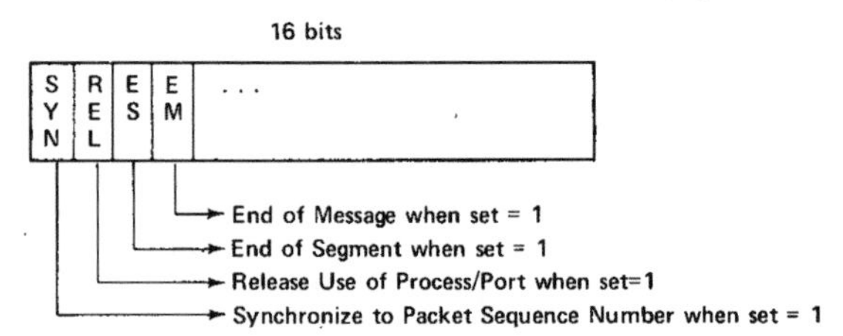

Fig. 8. Internetwork header flag field.

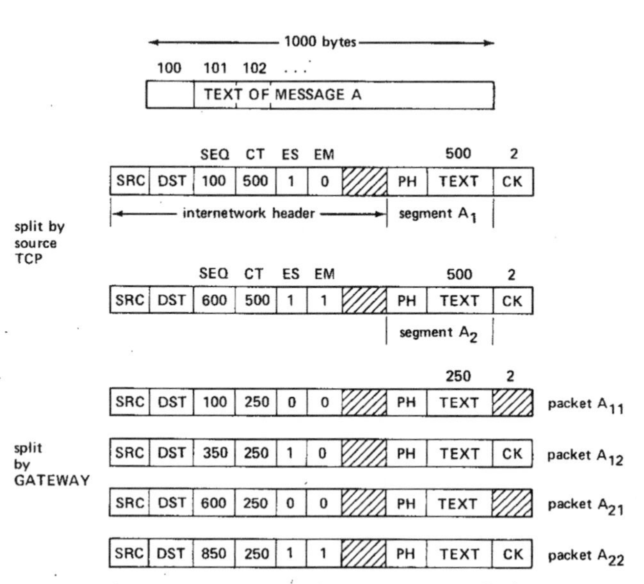

Fig. 9. Message splitting and packet splitting.

segment is extracted from the message by the source TCP and formatted for internetwork transmission, the relative location of the first byte of segment text is used as the sequence number for the packet. The byte count field in the internetwork header accounts for all the text in the segment (but does not include the check-sum bytes or the bytes in either internetwork or process header). We emphasize that the sequence number associated with a given packet is unique only to the pair of ports that are communicating (see Fig. 7). Arriving packets are examined to determine for which port they are intended. The sequence numbers on each arriving packet are then used to determine the relative location of the packet text in the messages under reconstruction. We note that this allows the exact position of the data in the reconstructed message to be determined even when pieces are still missing.

Every segment produced by a source TCP is packaged in a single internetwork packet and a check sum is computed over the text and process header associated with the segment.

The splitting of messages into segments by the TCP and the potential splitting of segments into smaller pieces by GATEWAYS creates the necessity for indicating to the destination TCP when the end of a segment (ES) has arrived and when the end of a message (EM) has arrived. The flag field of the internetwork header is used for this purpose (see Fig. 8).

The ES flag is set by the source TCP each time it prepares a segment for transmission. If it should happen that the message is completely contained in the segment, then the EM flag would also be set. The EM flag is also set on the last segment of a message, if the message could not be contained in one segment. These two flags are used by the destination TCP, respectively, to discover the presence of a check sum for a given segment and to discover that a complete message has arrived.

The ES and EM flags in the internetwork header are known to the GATEWAY and are of special importance when packets must be split apart for propagation through the next local network. We illustrate their use with an example in Fig. 9.

The original message A in Fig. 9 is shown split into two segments A_1 and A_2 and formatted by the TCP into a pair of internetwork packets. Packets A_1 and A_2 have the ES bits set, and A_2 has its EM bit set as well. Whe packet A_1 passes through the GATEWAY, it is split into tw pieces: packet A_{11} for which neither EM nor ES bits a set, and packet A_{12} whose ES bit is set. Similarly, pack A_2 is split such that the first piece, packet A_{21}, has neith bit set, but packet A_{22} has both bits set. The sequen number field (SEQ) and the byte count field (CT) of eac packet is modified by the GATEWAY to properly identi the text bytes of each packet. The GATEWAY need onl examine the internetwork header to do fragmentation.

The destination TCP, upon reassembling segment A will detect the ES flag and will verify the check sum knows is contained in packet A_{12}. Upon receipt of pack A_{22}, assuming all other packets have arrived, the dest nation TCP detects that it has reassembled a comple message and can now advise the destination process of i receipt.

RETRANSMISSION AND DUPLICATE DETECTION

No transmission can be 100 percent reliable. We propose a timeout and positive acknowledgment mechanism which will allow TCP's to recover from packet losses from one HOST to another. A TCP transmits packets and waits for replies (acknowledgements) that are carried in the reverse packet stream. If no acknowledgment for a particular packet is received, the TCP will retransmit. It is our expectation that the HOST level retransmission mechanism, which is described in the following paragraphs, will not be called upon very often in practice. Evidence already exists[2] that individual networks can be effectively constructed without this feature. However, the inclusion of a HOST retransmission capability makes it possible to recover from occasional network problems and allows a wide range of HOST protocol strategies to be incorporated. We envision it will occasionally be invoked to allow HOST accommodation to infrequent overdemands for limited buffer resources, and otherwise not used much.

Any retransmission policy requires some means by which the receiver can detect duplicate arrivals. Even if an infinite number of distinct packet sequence numbers were available, the receiver would still have the problem of knowing how long to remember previously received packets in order to detect duplicates. Matters are complicated by the fact that only a finite number of distinct sequence numbers are in fact available, and if they are reused, the receiver must be able to distinguish between new transmissions and retransmissions.

A *window* strategy, similar to that used by the French CYCLADES system (voie virtuelle transmission mode [8]) and the ARPANET very distant HOST connection [18], is proposed here (see Fig. 10).

Suppose that the sequence number field in the internetwork header permits sequence numbers to range from 0 to $n - 1$. We assume that the sender will not transmit more than w bytes without receiving an acknowledgment. The w bytes serve as the window (see Fig. 11). Clearly, w must be less than n. The rules for sender and receiver are as follows.

Sender: Let L be the sequence number associated with the left window edge.

1) The sender transmits bytes from segments whose text lies between L and up to $L + w - 1$.

2) On timeout (duration unspecified), the sender retransmits unacknowledged bytes.

3) On receipt of acknowledgment consisting of the receiver's current left window edge, the sender's left window edge is advanced over the acknowledged bytes (advancing the right window edge implicitly).

Receiver:

1) Arriving packets whose sequence numbers coincide with the receiver's current left window edge are acknowledged by sending to the source the next sequence number expected. This effectively acknowledges bytes in between. The left window edge is advanced to the next sequence number expected.

2) Packets arriving with a sequence number to the left of the window edge (or, in fact, outside of the window) are discarded, and the current left window edge is returned as acknowledgment.

3) Packets whose sequence numbers lie within the receiver's window but do not coincide with the receiver's left window edge are optionally kept or discarded, but are not acknowledged. This is the case when packets arrive out of order.

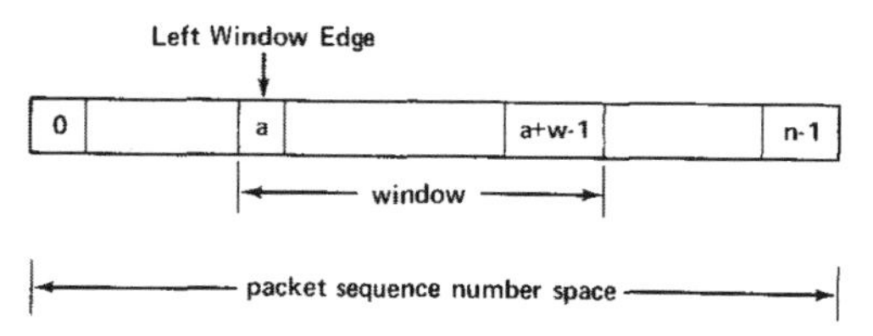

Fig. 10. The window concept.

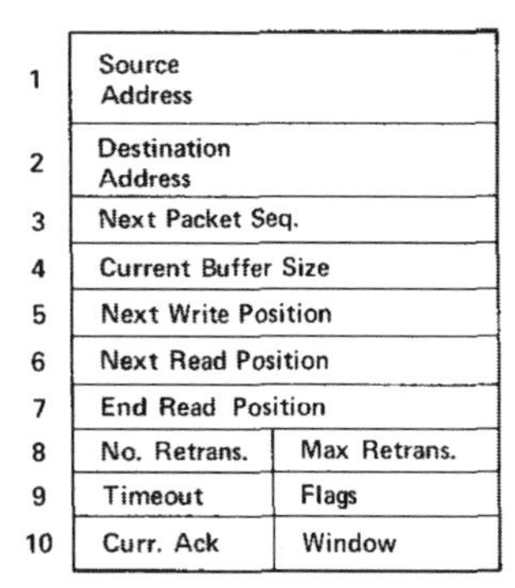

Fig. 11. Conceptual TCB format.

We make some observations on this strategy. First, all computations with sequence numbers and window edges must be made modulo n (e.g., byte 0 follows byte $n - 1$). Second, w must be less than $n/2$[3]; otherwise a retransmission may appear to the receiver to be a new transmission in the case that the receiver has accepted a window's worth of incoming packets, but all acknowledgments have been lost. Third, the receiver can either save or discard arriving packets whose sequence numbers do not coincide with the receiver's left window. Thus, in the simplest implementation, the receiver need not buffer more than one packet per message stream if space is critical. Fourth, multiple packets can be acknowledged simultaneously. Fifth, the receiver is able to deliver messages to processes in their proper order as a natural result of the reassembly mechanism. Sixth, when duplicates are detected, the acknowledgment method used naturally works to resynchronize sender and receiver. Furthermore, if the receiver accepts packets whose sequence numbers lie within the current window but

[2] The ARPANET is one such example.

[3] Actually $n/2$ is merely a convenient number to use; it is only required that a retransmission not appear to be a new transmission.

which are not coincident with the left window edge, an acknowledgment consisting of the current left window edge would act as a stimulus to cause retransmission of the unacknowledged bytes. Finally, we mention an overlap problem which results from retransmission, packet splitting, and alternate routing of packets through different GATEWAYS.

A 600-byte packet might pass through one GATEWAY and be broken into two 300-byte packets. On retransmission, the same packet might be broken into three 200-byte packets going through a different GATEWAY. Since each byte has a sequence number, there is no confusion at the receiving TCP. We leave for later the issue of initially synchronizing the sender and receiver left window edges and the window size.

FLOW CONTROL

Every segment that arrives at the destination TCP is ultimately acknowledged by returning the sequence number of the next segment which must be passed to the process (it may not yet have arrived).

Earlier we described the use of a sequence number space and window to aid in duplicate detection. Acknowledgments are carried in the process header (see Fig. 6) and along with them there is provision for a "suggested window" which the receiver can use to control the flow of data from the sender. This is intended to be the main component of the process flow control mechanism. The receiver is free to vary the window size according to any algorithm it desires so long as the window size never exceeds half the sequence number space.[3]

This flow control mechanism is exceedingly powerful and flexible and does not suffer from synchronization troubles that may be encountered by incremental buffer allocation schemes [9],[10]. However, it relies heavily on an effective retransmission strategy. The receiver can reduce the window even while packets are en route from the sender whose window is presently larger. The net effect of this reduction will be that the receiver may discard incoming packets (they may be outside the window) and reiterate the current window size along with a current window edge as acknowledgment. By the same token, the sender can, upon occasion, choose to send more than a window's worth of data on the possibility that the receiver will expand the window to accept it (of course, the sender must not send more than half the sequence number space at any time). Normally, we would expect the sender to abide by the window limitation. Expansion of the window by the receiver merely allows more data to be accepted. For the receiving HOST with a small amount of buffer space, a strategy of discarding all packets whose sequence numbers do not coincide with the current left edge of the window is probably necessary, but it will incur the expense of extra delay and overhead for retransmission.

TCP INPUT/OUTPUT HANDLING

The TCP has a component which handles input/output (I/O) to and from the network.[4] When a packet has arrived, it validates the addresses and places the packet on a queue. A pool of buffers can be set up to handle arrivals, and if all available buffers are used up, succeeding arrivals can be discarded since unacknowledged packets will be retransmitted.

On output, a smaller amount of buffering is needed, since process buffers can hold the data to be transmitted. Perhaps double buffering will be adequate. We make no attempt to specify how the buffering should be done, except to require that it be able to service the network with as little overhead as possible. Packet sized buffers, one or more ring buffers, or any other combination are possible candidates.

When a packet arrives at the destination TCP, it is placed on a queue which the TCP services frequently. For example, the TCP could be interrupted when a queue placement occurs. The TCP then attempts to place the packet text into the proper place in the appropriate process receive buffer. If the packet terminates a segment, then it can be checksummed and acknowledged. Placement may fail for several reasons.

1) The destination process may not be prepared to receive from the stated source, or the destination port ID may not exist.

2) There may be insufficient buffer space for the text.

3) The beginning sequence number of the text may not coincide with the next sequence number to be delivered to the process (e.g., the packet has arrived out of order).

In the first case, the TCP should simply discard the packet (thus far, no provision has been made for error acknowledgments). In the second and third cases, the packet sequence number can be inspected to determine whether the packet text lies within the legitimate window for reception. If it does, the TCP may optionally keep the packet queued for later processing. If not, the TCP can discard the packet. In either case the TCP can optionally acknowledge with the current left window edge.

It may happen that the process receive buffer is not present in the active memory of the HOST, but is stored on secondary storage. If this is the case, the TCP can prompt the scheduler to bring in the appropriate buffer and the packet can be queued for later processing.

If there are no more input buffers available to the TCP for temporary queueing of incoming packets, and if the TCP cannot quickly use the arriving data (e.g., a TCP to TCP message), then the packet is discarded. Assuming a sensibly functioning system, no other processes than the one for which the packet was intended should be affected by this discarding. If the delayed processing queue grows

[4] This component can serve to handle other protocols whose associated control programs are designated by internetwork destination address.

excessively long, any packets in it can be safely discarded since none of them have yet been acknowledged. Congestion at the TCP level is flexibly handled owing to the robust retransmission and duplicate detection strategy.

TCP/PROCESS COMMUNICATION

In order to send a message, a process sets up its text in a buffer region in its own address space, inserts the requisite control information (described in the following list) in a transmit control block (TCB) and passes control to the TCP. The exact form of a TCB is not specified here, but it might take the form of a passed pointer, a pseudointerrupt, or various other forms. To receive a message in its address space, a process sets up a receive buffer, inserts the requisite control information in a receive control block (RCB) and again passes control to the TCP.

In some simple systems, the buffer space may in fact be provided by the TCP. For simplicity we assume that a ring buffer is used by each process, but other structures (e.g., buffer chaining) are not ruled out.

A possible format for the TCB is shown in Fig. 11. The TCB contains information necessary to allow the TCP to extract and send the process data. Some of the information might be implicitly known, but we are not concerned with that level of detail. The various fields in the TCB are described as follows.

1) *Source Address*: This is the full net/HOST/TCP/port address of the transmitter.

2) *Destination Address*: This is the full net/HOST/TCP/port of the receiver.

3) *Next Packet Sequence Number*: This is the sequence number to be used for the next packet the TCP will transmit from this port.

4) *Current Buffer Size*: This is the present size of the process transmit buffer.

5) *Next Write Position*: This is the address of the next position in the buffer at which the process can place new data for transmission.

6) *Next Read Position*: This is the address at which the TCP should begin reading to build the next segment for output.

7) *End Read Position*: This is the address at which the TCP should halt transmission. Initially 6) and 7) bound the message which the process wishes to transmit.

8) *Number of Retransmissions/Maximum Retransmissions*: These fields enable the TCP to keep track of the number of times it has retransmitted the data and could be omitted if the TCP is not to give up.

9) *Timeout/Flags*: The timeout field specifies the delay after which unacknowledged data should be retransmitted. The flag field is used for semaphores and other TCP/process synchronization, status reporting, etc.

10) *Current Acknowledgment/Window*: The current acknowledgment field identifies the first byte of data still unacknowledged by the destination TCP.

The read and write positions move circularly around the transmit buffer, with the write position always to the left (module the buffer size) of the read position.

The next packet sequence number should be constrained to be less than or equal to the sum of the current acknowledgment and the window fields. In any event, the next sequence number should not exceed the sum of the current acknowledgment and half of the maximum possible sequence number (to avoid confusing the receiver's duplicate detection algorithm). A possible buffer layout is shown in Fig. 12.

The RCB is substantially the same, except that the end read field is replaced by a partial segment check-sum register which permits the receiving TCP to compute and remember partial check sums in the event that a segment arrives in several packets. When the final packet of the segment arrives, the TCP can verify the check sum and if successful, acknowledge the segment.

CONNECTIONS AND ASSOCIATIONS

Much of the thinking about process-to-process communication in packet switched networks has been influenced by the ubiquitous telephone system. The HOST–HOST protocol for the ARPANET deals explicitly with the opening and closing of simplex connections between processes [9],[10]. Evidence has been presented that message-based "connection-free" protocols can be constructed [12], and this leads us to carefully examine the notion of a connection.

The term *connection* has a wide variety of meanings. It can refer to a physical or logical path between two entities, it can refer to the flow over the path, it can inferentially refer to an action associated with the setting up of a path, or it can refer to an association between two or more entities, with or without regard to any path between them. In this paper, we do not explicitly reject the term connection, since it is in such widespread use, and does connote a meaningful relation, but consider it exclusively in the sense of an association between two or more entities without regard to a path. To be more precise about our intent, we shall define the relationship between two or more ports that are in communication, or are prepared to communicate to be an *association*. Ports that are associated with each other are called *associates*.

It is clear that for any communication to take place between two processes, one must be able to address the other. The two important cases here are that the destination port may have a global and unchanging address or that it may be globally unique but dynamically reassigned. While in either case the sender may have to learn the destination address, given the destination name, only in the second instance is there a requirement for learning the address from the destination (or its representative) each time an association is desired. Only after the source has learned how to address the destination can an association be said to have occurred. But this is not yet sufficient. If

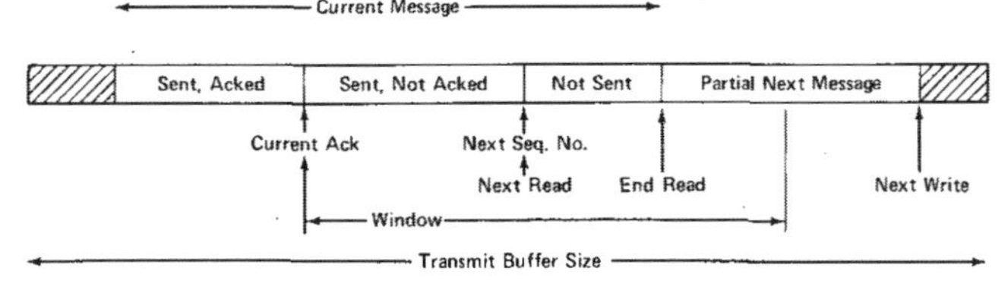

Fig. 12. Transmit buffer layout.

ordering of delivered messages is also desired, both TCP's must maintain sufficient information to allow proper sequencing. When this information is also present at both ends, then an association is said to have occurred.

Note that we have not said anything about a path, nor anything which implies that either end be aware of the condition of the other. Only when both partners are prepared to communicate with each other has an association occurred, and it is possible that neither partner may be able to verify that an association exists until some data flows between them.

CONNECTION-FREE PROTOCOLS WITH ASSOCIATIONS

In the ARPANET, the interface message processors (IMP's) do not have to open and close connections from source to destination. The reason for this is that connections are, in effect, always open, since the address of every source and destination is never[5] reassigned. When the name and the place are static and unchanging, it is only necessary to label a packet with source and destination to transmit it through the network. In our parlance, every source and destination forms an association.

In the case of processes, however, we find that port addresses are continually being used and reused. Some ever-present processes could be assigned fixed addresses which do not change (e.g., the logger process). If we supposed, however, that every TCP had an infinite supply of port addresses so that no old address would ever be reused, then any dynamically created port would be assigned the next unused address. In such an environment, there could never be any confusion by source and destination TCP as to the intended recipient or implied source of each message, and all ports would be associates.

Unfortunately, TCP's (or more properly, operating systems) tend not to have an infinite supply of internal port addresses. These internal addresses are reassigned after the demise of each port. Walden [12] suggests that a set of unique uniform external port addresses could be supplied by a central registry. A newly created port could apply to the central registry for an address which the central registry would guarantee to be unused by any HOST system in the network. Each TCP could maintain tables matching external names with internal ones, and use the external ones for communication with other processes. This idea violates the premise that interprocess communication should not require centralized control. One would have to extend the central registry service to include all HOST's in all the interconnected networks to apply this idea to our situation, and we therefore do not attempt to adopt it.

Let us consider the situation from the standpoint of the TCP. In order to send or receive data for a given port, the TCP needs to set up a TCB and RCB and initialize the window size and left window edge for both. On the receive side, this task might even be delayed until the first packet destined for a given port arrives. By convention, the first packet should be marked so that the receiver will synchronize to the received sequence number.

On the send side, the first request to transmit could cause a TCB to be set up with some initial sequence number (say, zero) and an assumed window size. The receiving TCP can reject the packet if it wishes and notify the sending TCP of the correct window size via the acknowledgment mechanism, but only if either

1) we insist that the first packet be a complete segment
2) an acknowledgment can be sent for the first packet (even if not a segment, as long as the acknowledgment specifies the next sequence number such that the source also understands that no bytes have been accepted).

It is apparent, therefore, that the synchronizing of window size and left window edge can be accomplished without what would ordinarily be called a connection setup.

The first packet referencing a newly created RCB sent from one associate to another can be marked with a bit which requests that the receiver synchronize his left window edge with the sequence number of the arriving packet (see SYN bit in Fig. 8). The TCP can examine the source and destination port addresses in the packet and in the RCB to decide whether to accept or ignore the request.

Provision should be made for a destination process to specify that it is willing to LISTEN to a specific port or "any" port. This last idea permits processes such as the logger process to accept data arriving from unspecified sources. This is purely a HOST matter, however.

The initial packet may contain data which can be stored or discarded by the destination, depending on the availability of destination buffer space at the time. In the other direction, acknowledgment is returned for receipt of data which also specifies the receiver's window size.

If the receiving TCP should want to reject the synchronization request, it merely transmits an acknowledgment carrying a release (REL) bit (see Fig. 8) indicating that the destination port address is unknown or inaccessible. The sending HOST waits for the acknowledgment (after accepting or rejecting the synchronization request) before sending the next message or segment. This rejection is quite different from a negative data acknowledgment. We do not have explicit negative acknowledgments. If no acknowledgment is returned, the sending HOST may

[5] Unless the IMP is physically moved to another site, or the HOST is connected to a different IMP.

retransmit without introducing confusion if, for example, the left window edge is not changed on the retransmission.

Because messages may be broken up into many packets for transmission or during transmission, it will be necessary to ignore the REL flag except in the case that the EM flag is also set. This could be accomplished either by the TCP or by the GATEWAY which could reset the flag on all but the packet containing the set EM flag (see Fig. 9).

At the end of an association, the TCP sends a packet with ES, EM, and REL flags set. The packet sequence number scheme will alert the receiving TCP if there are still outstanding packets in transit which have not yet arrived, so a premature dissociation cannot occur.

To assure that both TCP's are aware that the association has ended, we insist that the receiving TCP respond to the REL by sending a REL acknowledgment of its own.

Suppose now that a process sends a single message to an associate including an REL along with the data. Assuming an RCB has been prepared for the receiving TCP to accept the data, the TCP will accumulate the incoming packets until the one marked ES, EM, REL arrives, at which point a REL is returned to the sender. The association is thereby terminated and the appropriate TCB and RCB are destroyed. If the first packet of a message contains a SYN request bit and the last packet contains ES, EM, and REL bits, then data will flow "one message at a time." This mode is very similar to the scheme described by Walden [12], since each succeeding message can only be accepted at the receiver after a new LISTEN (like Walden's RECEIVE) command is issued by the receiving process to its serving TCP. Note that only if the acknowledgment is received by the sender can the association be terminated properly. It has been pointed out[6] that the receiver may erroneously accept duplicate transmissions if the sender does not receive the acknowledgment. This may happen if the sender transmits a duplicate message with the SYN and REL bits set and the destination has already destroyed any record of the previous transmission. One way of preventing this problem is to destroy the record of the association at the destination only after some known and suitably chosen timeout. However, this implies that a new association with the same source and destination port identifiers could not be established until this timeout had expired. This problem can occur even with sequences of messages whose SYN and REL bits are separated into different internetwork packets. We recognize that this problem must be solved, but do not go into further detail here.

Alternatively, both processes can send one message, causing the respective TCP's to allocate RCB/TCB pairs at both ends which rendezvous with the exchanged data and then disappear. If the overhead of creating and destroying RCB's and TCB's is small, such a protocol might be adequate for most low-bandwidth uses. This idea might also form the basis for a relatively secure transmission system. If the communicating processes agree to change their external port addresses in some way known only to each other (i.e., pseudorandom), then each message will appear to the outside world as if it is part of a different association message stream. Even if the data is intercepted by a third party, he will have no way of knowing that the data should in fact be considered part of a sequence of messages.

We have described the way in which processes develop associations with each other, thereby becoming associates for possible exchange of data. These associations need not involve the transmission of data prior to their formation and indeed two associates need not be able to determine that they are associates until they attempt to communicate.

[6] S. Crocker of ARPA/IPT.

CONCLUSIONS

We have discussed some fundamental issues related to the interconnection of packet switching networks. In particular, we have described a simple but very powerful and flexible protocol which provides for variation in individual network packet sizes, transmission failures, sequencing, flow control, and the creation and destruction of process-to-process associations. We have considered some of the implementation issues that arise and found that the proposed protocol is implementable by HOST's of widely varying capacity.

The next important step is to produce a detailed specification of the protocol so that some initial experiments with it can be performed. These experiments are needed to determine some of the operational parameters (e.g., how often and how far out of order do packets actually arrive; what sort of delay is there between segment acknowledgments; what should be retransmission timeouts be?) of the proposed protocol.

ACKNOWLEDGMENT

The authors wish to thank a number of colleagues for helpful comments during early discussions of international network protocols, especially R. Metcalfe, R. Scantlebury, D. Walden, and H. Zimmerman; D. Davies and L. Pouzin who constructively commented on the fragmentation and accounting issues; and S. Crocker who commented on the creation and destruction of associations.

REFERENCES

[1] L. Roberts and B. Wessler, "Computer network development to achieve resource sharing," in *1970 Spring Joint Computer Conf., AFIPS Conf. Proc.*, vol. 36. Montvale, N. J.: AFIPS Press, 1970, pp. 543–549.

[2] L. Pouzin, "Presentation and major design aspects of the CYCLADES computer network," in *Proc. 3rd Data Communications Symp.*, 1973.

[3] F. R. E. Dell, "Features of a proposed synchronous data network," in *Proc. 2nd Symp. Problems in the Optimization of Data Communications Systems*, 1971, pp. 50–57.

[4] R. A. Scantlebury and P. T. Wilkinson, "The design of a switching system to allow remote access to computer services by other computers and terminal devices," in *Proc. 2nd Symp. Problems in the Optimization of Data Communications Systems,* 1971, pp. 160–167.
[5] D. L. A. Barber, "The European computer network project," in *Computer Communications: Impacts and Implications,* S. Winkler, Ed. Washington, D. C., 1972, pp. 192–200.
[6] R. Despres, "A packet switching network with graceful saturated operation," in *Computer Communications: Impacts and Implications,* S. Winkler, Ed. Washington, D. C., 1972, pp. 345–351.
[7] R. E. Kahn and W. R. Crowther, "Flow control in a resource-sharing computer network," *IEEE Trans. Commun.,* vol. COM-20, pp. 539–546, June 1972.
[8] J. F. Chambon, M. Elie, J. Le Bihan, G. LeLann, and H. Zimmerman, "Functional specification of transmission station in the CYCLADES network. ST-ST protocol" (in French), I.R.I.A. Tech. Rep. SCH502.3, May 1973.
[9] S. Carr, S. Crocker, and V. Cerf, "HOST-HOST Communication Protocol In the ARPA Network," in *Spring Joint Computer Conf., AFIPS Conf. Proc.,* vol. 36. Montvale, N. J.: AFIPS Press, 1970, pp. 589–597.
[10] A. McKenzie, "HOST/HOST protocol for the ARPA network," in *Current Network Protocols,* Network Information Cen., Menlo Park, Calif., NIC 8246, Jan. 1972.
[11] L. Pouzin, "Address format in Mitranet," NIC 14497, INWG 20, Jan. 1973.
[12] D. Walden, "A system for interprocess communication in a resource sharing computer network," *Commun. Ass. Comput. Mach.,* vol. 15, pp. 221–230, Apr. 1972.
[13] B. Lampson, "A scheduling philosophy for multiprocessing systems," *Commun. Ass. Comput. Mach.,* vol. 11, pp. 347–360, May 1968.
[14] F. E. Heart, R. E. Kahn, S. Ornstein, W. Crowther, and D. Walden, "The interface message processor for the ARPA computer network," in *Proc. Spring Joint Computer Conf., AFIPS Conf. Proc.,* vol. 36. Montvale, N. J.: AFIPS Press, 1970, pp. 551–567.
[15] N. G. Anslow and J. Hanscoff, "Implementation of international data exchange networks," in *Computer Communications: Impacts and Implications,* S. Winkler, Ed. Washington, D. C., 1972, pp. 181–184
[16] A. McKenzie, "HOST/HOST protocol design considerations," INWG Note 16, NIC 13879, Jan. 1973.
[17] R. E. Kahn, "Resource-sharing computer communication networks," *Proc. IEEE,* vol. 60, pp. 1397–1407, Nov. 1972.
[18] Bolt, Beranek, and Newman, "Specification for the interconnection of a host and an IMP," Bolt Beranek and Newman, Inc., Cambridge, Mass., BBN Rep. 1822 (revised), Apr. 1973.

PHOTO NOT AVAILABLE

Vinton G. Cerf was born in New Haver Conn., in 1943. He did undergraduate wor in mathematics at Stanford University Stanford, Calif., and received the Ph.D. de gree in computer science from the Universit of California at Los Angeles, Los Angele Calif., in 1972.

He was with IBM in Los Angeles fror 1965 through 1967 and consulted and/c worked part time at UCLA from 1967 throug 1972. Currently he is Assistant Professor c Computer Science and Electrical Engineering at Stanford Universit and consultant to Cabledata Associates. Most of his current researc is supported by the Defense Advanced Research Projects Agency an by the National Science Foundation on the technology and economic of computer networking. He is Chairman of IFIP TC6.1, an inte national network working group which is studying the proble of packet network interconnection.

★

PHOTO NOT AVAILABLE

Robert E. Kahn (M'65) was born in Brookly N. Y., on December 23, 1938. He received th B.E.E. degree from the City College of Ne York, New York, in 1960, and the M.A and Ph.D. degrees from Princeton Universit Princeton, N. J., in 1962 and 1964, re spectively.

From 1960 to 1962 he was a Member of th Technical Staff of Bell Telephone Labora tories, Murray Hill, N. J., engaged in traffi and communication studies. From 1964 t 1966 he was a Ford Postdoctoral Fellow and an Assistant Professo of Electrical Engineering at the Massachusetts Institute of Tech nology, Cambridge, where he worked on communications and in formation theory. From 1966 to 1972 he was a Senior Scientist a Bolt Beranek and Newman, Inc., Cambridge, Mass., where h worked on computer communications network design and technique for distributed computation. Since 1972 he has been with the Ad vanced Research Projects Agency, Department of Defens Arlington, Va.

Dr. Kahn is a member of Tau Beta Pi, Sigma Xi, Eta Kappa Nu the Institute of Mathematical Statistics, and the Mathematica Association of America. He was selected to serve as a Nation Lecturer for the Association for Computing Machinery in 1972.

On Distributed Communications Networks

PAUL BARAN, SENIOR MEMBER, IEEE

Summary—This paper[1] briefly reviews the distributed communication network concept in which each station is connected to all adjacent stations rather than to a few switching points, as in a centralized system. The payoff for a distributed configuration in terms of survivability in the cases of enemy attack directed against nodes, links or combinations of nodes and links is demonstrated.

A comparison is made between diversity of assignment and perfect switching in distributed networks, and the feasibility of using low-cost unreliable communication links, even links so unreliable as to be unusable in present type networks, to form highly reliable networks is discussed.

The requirements for a future all-digital data distributed network which provides common user service for a wide range of users having different requirements is considered. The use of a standard format message block permits building relatively simple switching mechanisms using an adaptive store-and-forward routing policy to handle all forms of digital data including digital voice. This network rapidly responds to changes in the network status. Recent history of measured network traffic is used to modify path selection. Simulation results are shown to indicate that highly efficient routing can be performed by local control without the necessity for any central, and therefore vulnerable, control point.

INTRODUCTION

LET US CONSIDER the synthesis of a communication network which will allow several hundred major communications stations to talk with one another after an enemy attack. As a criterion of survivability we elect to use the percentage of stations both surviving the physical attack and remaining in electrical connection with the largest single group of surviving stations. This criterion is chosen as a conservative measure of the ability of the surviving stations to operate together as a coherent entity after the attack. This means that small groups of stations isolated from the single largest group are considered to be ineffective.

Although one can draw a wide variety of networks, they all factor into two components: centralized (or star) and distributed (or grid or mesh). (See types (a) and (c), respectively, in Fig. 1.)

The centralized network is obviously vulnerable as destruction of a single central node destroys communication between the end stations. In practice, a mixture of star and mesh components is used to form communications networks. For example, type (b) in Fig. 1 shows the hierarchical structure of a set of stars connected in the form of a larger star with an additional link forming a loop. Such a network is sometimes called a "decentralized" network, because complete reliance upon a single point is not always required.

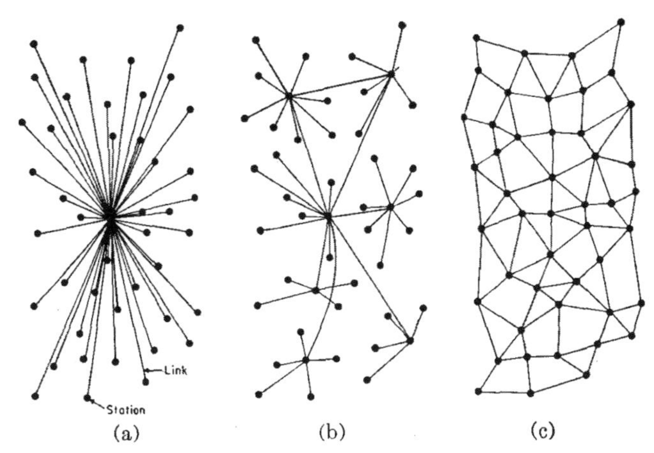

Fig. 1—(a) Centralized. (b) Decentralized. (c) Distributed networks.

EXAMINATION OF A DISTRIBUTED NETWORK

Since destruction of a small number of nodes in a decentralized network can destroy communications, the properties, problems, and hopes of building "distributed" communications networks are of paramount interest.

The term "redundancy level" is used as a measure of connectivity, as defined in Fig. 2. A minimum span network, one formed with the smallest number of links possible, is chosen as a reference point and is called "a network of redundancy level one." If two times as many links are used in a gridded network than in a minimum span network, the network is said to have a redundancy level of two. Fig. 2 defines connectivity of levels 1, $1\frac{1}{2}$, 2, 3, 4, 6 and 8. Redundancy level is equivalent to link-to-node ratio in an infinite size array of stations. Obviously, at levels above three there are alternate methods of constructing the network. However, it was found that there is little difference regardless of which method is used. Such an alternate method is shown for levels three and four, labelled R'. This specific alternate mode is also used for levels six and eight.[2]

Each node and link in the array of Fig. 2 has the capacity and the switching flexibility to allow transmission between any ith station and any jth station, provided a path can be drawn from the ith to the jth station.

Starting with a network composed of an array of stations connected as in Fig. 3, an assigned percentage of nodes and links is destroyed. If, after this operation,

Manuscript received October 9, 1963. This paper was presented at the First Congress of the Information Systems Sciences, sponsored by the MITRE Corporation, Bedford, Mass., and the USAF Electronic Systems Division, Hot Springs, Va., November, 1962.

The author is with The RAND Corporation, Santa Monica, Calif.

[1] Any views expressed in this paper are those of the author. They should not be interpreted as reflecting the views of The RAND Corporation or the official opinion or policy of any of its governmental or private research sponsors.

[2] See L. J. Craig, and I. S. Reed, "Overlapping Tessellated Communications Networks," The RAND Corporation, Santa Monica, Calif., paper P-2359; July 5, 1961.

Reprinted from *IEEE Transactions on Communications Systems,* March 1964.

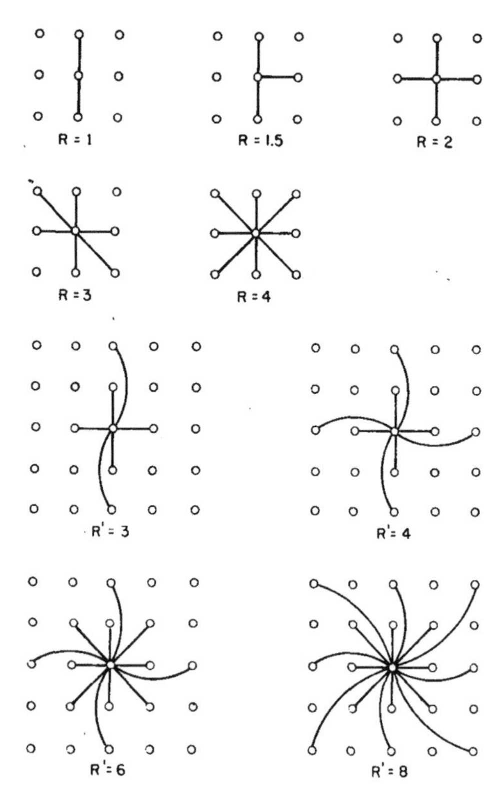

Fig. 2—Definition of redundancy level.

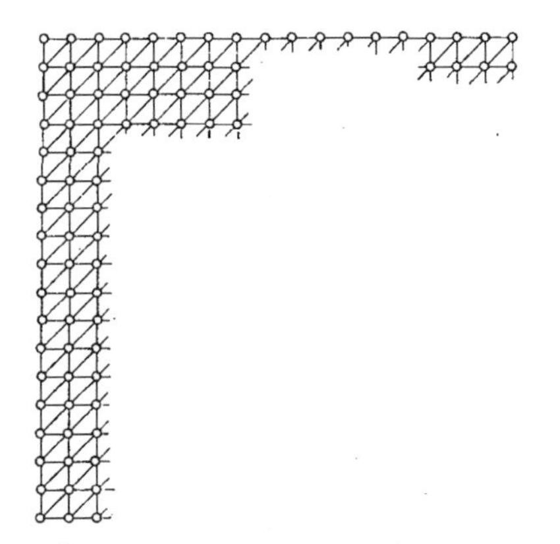

Fig. 3—An array of stations.

it is still possible to draw a line to connect the ith station to the jth station, the ith and jth stations are said to be connected.

Node Destruction

Fig. 4 indicates network performance as a function of the probability of destruction for each separate node. If the expected "noise" was destruction caused by conventional hardware failure, the failures would be randomly distributed through the network. But if the disturbance were caused by enemy attack, the possible "worst cases" must be considered.

To bisect a 32-link network requires direction of 288 weapons each with a probability of kill, $p_k = 0.5$, or 160

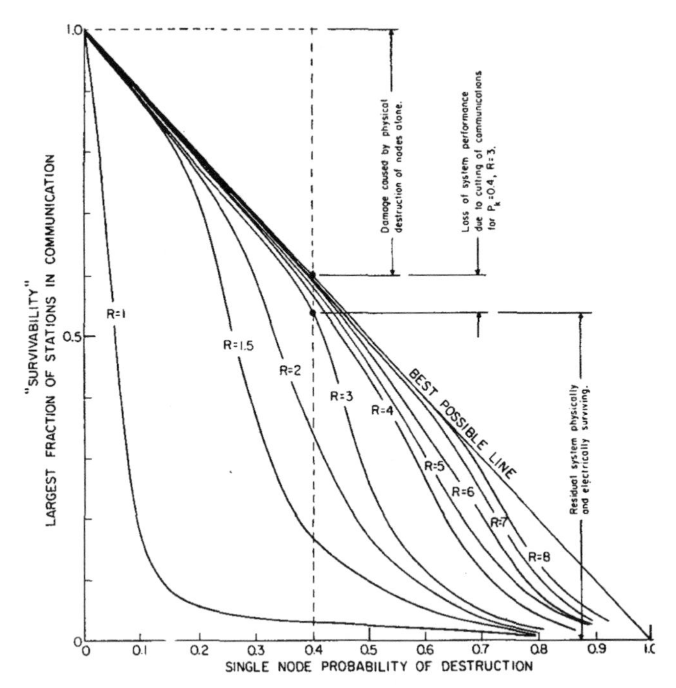

Fig. 4—Perfect switching in a distributed network: sensitivity to node destruction, 100 per cent of links operative.

with a $p_k = 0.7$, to produce over an 0.9 probability of successfully bisecting the network. If hidden alternative command is allowed, then the largest single group would still have an expected value of almost 50 per cent of the initial stations surviving intact. If this raid misjudges complete availability of weapons, complete knowledge of all links in the cross section, or the effects of the weapons against each and every link, the raid fails. The high risk of such raids against highly parallel structures causes examination of alternative attack policies. Consider the following uniform raid example. Assume that 2000 weapons are deployed against a 1000-station network. The stations are so spaced that destruction of two stations with a single weapon is unlikely. Divide the 2000 weapons into two equal 1000-weapon salvos. Assume any probability of destruction of a single node from a single weapon less than 1.0; for example, 0.5. Each weapon on the first salvo has a 0.5 probability of destroying its target. But, each weapon of the second salvo has only a 0.25 probability, since one half the targets have already been destroyed. Thus, the uniform attack is felt to represent a worst-case configuration.

Such worst-case attacks have been directed against an 18×18-array network model of 324 nodes with varying probability of kill and redundancy level, with results shown in Fig. 4. The probability of kill was varied from zero to unity along the abscissa, while the ordinate marks survivability. The criterion of survivability used is the percentage of stations not physically destroyed and remaining in communication with the largest single group of surviving stations. The curves of Fig. 4 demonstrate survivability as a function of attack level for networks of

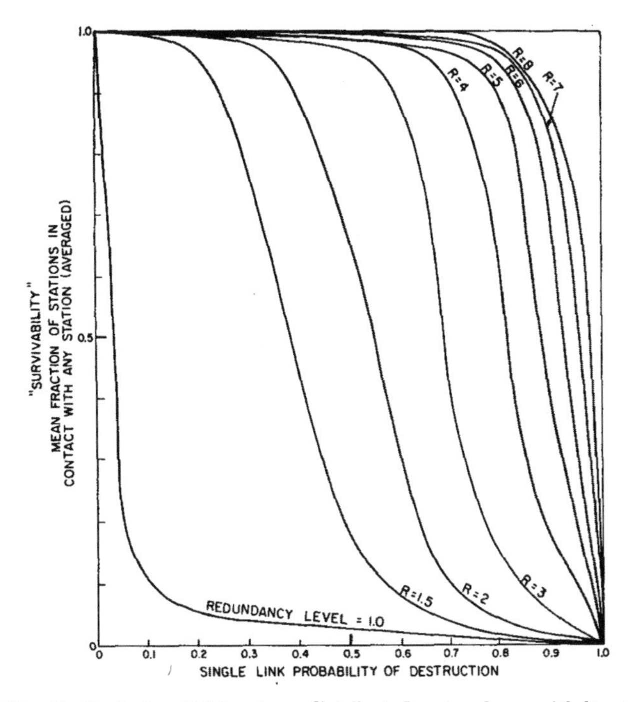

Fig. 5—Perfect switching in a distributed network: sensitivity to link destruction, 100 per cent of nodes operative.

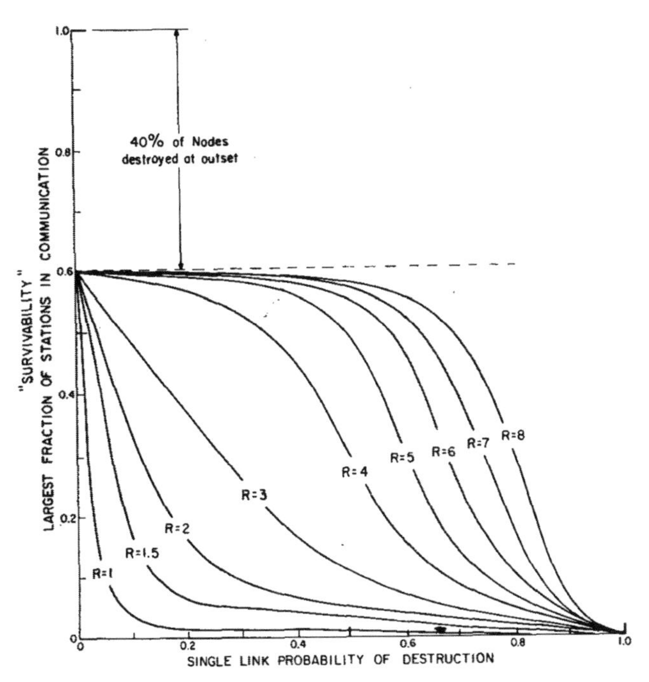

Fig. 6—Perfect switching in a distributed network: sensitivity to link destruction after 40 per cent nodes are destroyed.

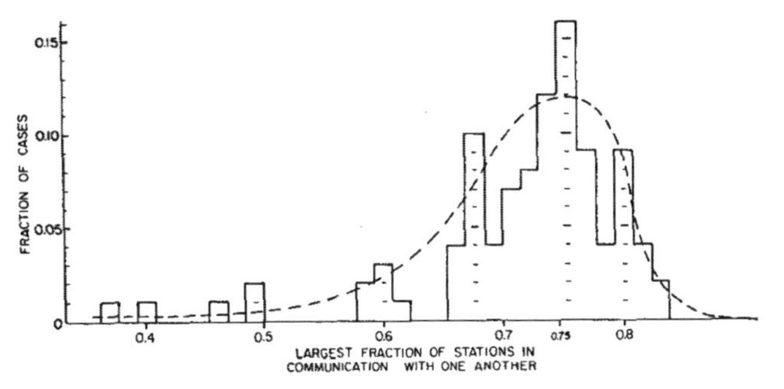

Fig. 7—Probability density distribution of largest fraction of stations in communication: perfect switching, $R = 3$, 100 cases, 80 per cent node survival, 65 per cent link survival.

varying degrees of redundancy. The line labeled "best possible line" marks the upper bound of loss due to the physical failure component alone. For example, if a network underwent an attack of 0.5 probability destruction of each of its nodes, then only 50 per cent of its nodes would be expected to survive, regardless of how perfect its communications. We are primarily interested in the additional system degradation caused by failure of communications. Two key points are to be noticed in the curves of Fig. 4. First, extremely survivable networks can be built using a moderately low redundancy of connectivity level. Redundancy levels on the order of only three permit the withstanding of extremely heavy level attacks with negligible additional loss to communications. Secondly, the survivability curves have sharp break points. A network of this type will withstand an increasing attack level until a certain point is reached, beyond which the network, rapidly deteriorates. Thus, the optimum degree of redundancy can be chosen as a function of the expected level of attack. Further redundancy gains little. The redundancy level required to survive even very heavy attacks is not great; it is on the order of only three or four times that of the minimum span network.

Link Destruction

In the previous example we have examined network performance as a function of the destruction of the nodes (which are better targets than links). We shall now reexamine the same network, but using unreliable links. In particular, we want to know how unreliable the links may be without further degrading the performance of the network.

Fig. 5 shows the results for the case of perfect nodes; only the links fail. There is little system degradation caused even using extremely unreliable links, on the order of 50 per cent down time, assuming all nodes are working.

Combination Link and Node Destruction

The worst case is the composite effect of failures of both the links and the nodes. Fig. 6 shows the effect of link failure upon a network having 40 per cent of its nodes destroyed. It appears that what would today be regarded as an unreliable link can be used in a distributed network almost as effectively as perfectly reliable links. Fig. 7 examines the result of 100 trial cases in order to estimate the probability density distribution of system performance for a mixture of node and link failures. This is the distribution of cases for 20 per cent nodal damage and 35 per cent link damage.

Diversity of Assignment

There is another and more common technique for using redundancy than in the method described above in which each station is assumed to have perfect switching ability. This alternative approach is called "diversity of assignment." In diversity of assignment, switching is not required. Instead, a number of independent paths are selected between each pair of stations in a network which requires reliable communications. However, there are marked differences in performance between distributed switching and redundancy of assignment as revealed by the following Monte Carlo simulation.

Simulation

In the matrix of N separate stations, each ith station is connected to every jth station by three shortest but totally separate independent paths ($i = 1, 2, 3, \cdots, N$; $j = 1, 2, 3, \cdots, N$; $i \neq j$). A raid is laid against the network. Each of the *preassigned* separate paths from the ith station to the jth station is examined. If one or more of the preassigned paths survive, communication is said to exist between the ith and the jth station. The criterion of survivability used is the mean number of stations connected to each station, averaged over all stations.

Unlike the distributed perfect switching case, Fig. 8 shows that there is a marked loss in communications capability with even slightly unreliable nodes or links. The difference can be visualized by remembering that fully flexible switching permits the communicator the privilege of *ex post facto* decision of paths. Fig. 8 emphasizes a key difference between some present-day networks and the fully flexible distributed network we are discussing.

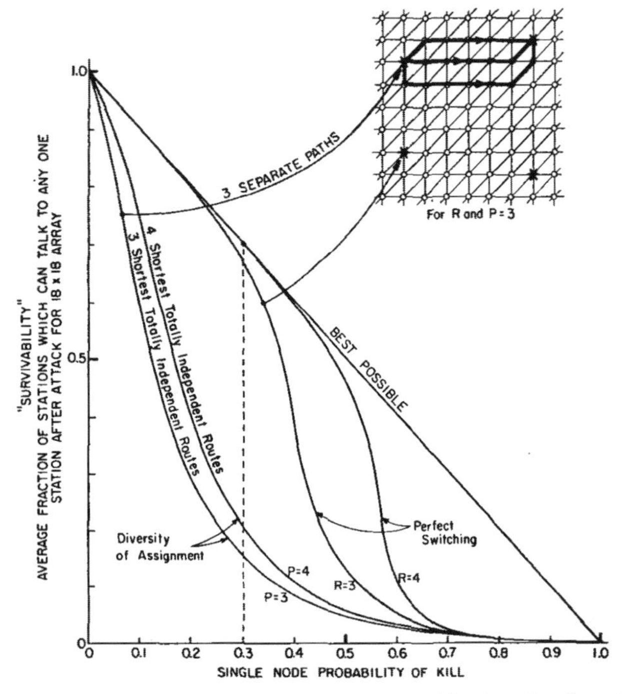

Fig. 8—Diversity of assignment vs perfect switching in a distributed network.

Comparison with Present Systems

Present conventional switching systems try only a small subset of the potential paths that can be drawn on a gridded network. The greater the percentage of potential paths tested, the closer one approaches the performance of perfect switching. Thus, perfect switching provides an upper bound of expected system performance for a gridded network; the diversity of assignment case provides a lower bound. Between these two limits lie systems composed of a mixture of switched routes and diversity of assignment.

Diversity of assignment is useful for short paths, eliminating the need for switching, but requires survivability and reliability for each tandem element in long-haul circuits passing through many nodes. As every component in at least one out of a *small* number of possible paths must be simultaneously operative, high reliability margins and full standby equipment are usual.

On Future Systems

We will soon be living in an era in which we cannot guarantee survivability of any single point. However, we can still design systems in which system destruction requires the enemy to pay the price of destroying n of n stations. If n is made sufficiently large, it can be shown that highly survivable system structures can be built, even in the thermonuclear era. In order to build such networks and systems we will have to use a large number of elements. We are interested in knowing how inexpensive these *elements* may be and still permit the *system* to operate reliably. There is a strong relationship between element cost and element reliability. To design a system that must anticipate a worst-case destruction of both enemy attack and normal system failures, one can combine the failures expected by enemy attack together with the failures caused by normal reliability problems, provided the enemy does not know which elements are inoperative. Our future systems design problem is that of building at lowest cost very reliable systems out of the described set of unreliable elements. In choosing the communications links of the future, digital links appear increasingly attractive by permitting low-cost switching and low-cost links. For example, if "perfect switching" is used, digital links are mandatory to permit tandem connection of many separately connected links without cumulative errors reaching an irreducible magnitude. Further, the signaling measures to implement highly flexible switching doctrines always require digits.

Future Low-Cost All-Digital Communications Links

When one designs an entire system optimized for digits and high redundancy, certain new communications link techniques appear more attractive than those common today. A key attribute of the new media is that it permits cheap formation of *new routes*, yet allows transmission on the order of a million or so bits per second, high enough to be economic yet low enough to be inexpensively processed with existing digital computer techniques at the relay station nodes. Reliability and raw error rates are secondary. The network must be built with the expectation of heavy damage anyway. Powerful error removal methods exist.

Some of the communication construction methods that look attractive for the near future include pulse regenerative repeater line, minimum-cost or "mini-cost" microwave, TV broadcast station digital transmission and satellites.

Pulse Regenerative Repeater Line: S. F. B. Morse's regenerative repeater invention for amplifying weak telegraphic signals has recently been resurrected and transistorized. Morse's electrical relay permits amplification of weak binary telegraphic signals above a fixed threshold. Experiments by various organizations (primarily the Bell Telephone Laboratories) have shown that digital data rates on the order of 1.5 million bits per second can be transmitted over ordinary telephone line at repeater spacings on the order of 6000 feet for 22-gage pulp paper insulated copper pairs. At present, more than 20 tandemly connected amplifiers have been used without retiming synchronization problems. There appears to be no fundamental reason why either lines of lower loss, with corresponding further repeater spacing, or more powerful resynchronization methods cannot be used to extend link distances to in excess of 200 miles. Such distances would be desired for a possible national distributed network. Power to energize the miniature transistor amplifier is transmitted over the copper circuit itself.

"Mini-Cost" Microwave: While the price of microwave equipment has been declining, there are still untapped major savings. In an analog signal network we require a high degree of reliability and very low distortion for each tandem repeater. However, using digital modulation together with perfect switching we minimize these two expensive considerations from our planning. We would envision the use of low-power, mass-produced microwave receiver/transmitter units mounted on low-cost, short, guyed towers. Relay station spacing would probably be on the order of 20 miles. Further economies can be obtained by only a minimal use of standby equipment and reduction of fading margins. The ability to use alternate paths permits consideration of frequencies normally troubled by rain attenuation problems reducing the spectrum availability problem. Preliminary indications suggest that this approach appears to be the cheapest way of building large networks of the type to be described.

TV Stations: With proper siting of receiving antennas, broadcast television stations might be used to form additional high data rate links in emergencies.

Satellites: The problem of building a reliable network using satellites is somewhat similar to that of building a communications network with unreliable links. When a satellite is overhead, the link is operative. When a satellite is not overhead, the link is out of service. Thus, such links are highly compatible with the type of system to be described.

Variable Data Rate Links

In a conventional circuit-switched system each of the tandem links requires matched transmission bandwidths. In order to make fullest use of a digital link, the post-error-removal data rate would have to vary, as it is a function of noise level. The problem then is to build a communication network made up of links of variable data rate to use the communication resource most efficiently.

Variable Data Rate Users

We can view both the links and the entry point nodes of a multiple-user all-digital communications system as elements operating at an ever-changing data rate. From instant to instant the demand for transmission will vary. We would like to take advantage of the average demand over all users instead of having to allocate a full peak demand channel to each. Bits can become a common denominator of loading and we would like to efficiently handle both those users who make highly intermittent bit demands on the network and those who make long-term continuous, low-bit demands.

Common User

In communications, as in transportation, it is most economic for many users to share a common resource rather than each to build his own system, particularly when supplying intermittent or occasional service. This intermittency of service is highly characteristic of digital communication requirements. Therefore, we would like to consider one day the interconnection, of many *all-digital* links to provide a resource optimized for the handling of data for many potential intermittent users: a new common-user system.

Fig. 9 demonstrates the basic notion. A wide mixture of different digital transmission links is combined to form a common resource divided among many potential users. But each of these communications links could possibly have a different data rate. How can links of different data rates be interconnected?

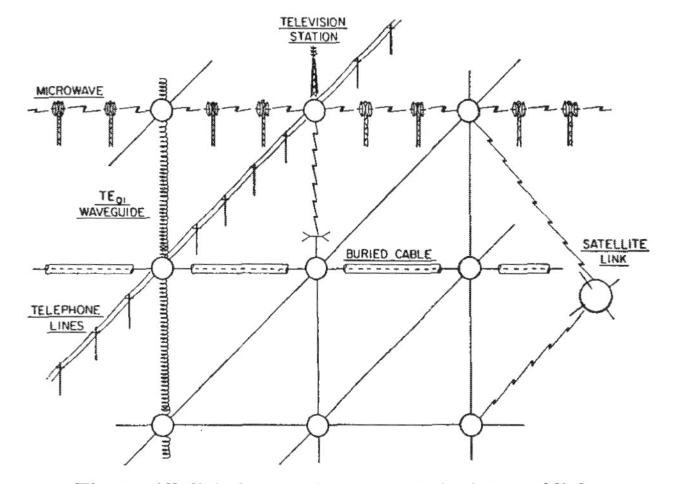

Fig. 9—All-digital network composed of mixture of links.

A Model All-Digital Distributed System

A future system, incorporating the features outlined in the preceding section, has been modeled and simulated. The key attribute of the system is in its switching scheme. But prior to considering the way in which the system would work, some thought must be given to message format standardization.

Standard Message Block

Present common carrier communications networks, used for digital transmission, use links and concepts originally designed for another purpose—voice. These systems are built around a frequency division multiplexing link-to-link interface standard. The standard between links is that of data rate. Time division multiplexing appears so natural to data transmission that we might wish to consider an alternative approach, a standardized message block as a network interface standard. While a standardized message block is common in many computer-communications applications, no serious attempt has ever been made to use it as a universal standard. A universally standardized message block would be composed of perhaps 1024 bits. Most of the message block would be reserved for whatever type data is to be transmitted, while the remainder would contain housekeeping information such as error detection and routing data, as in Fig. 10.

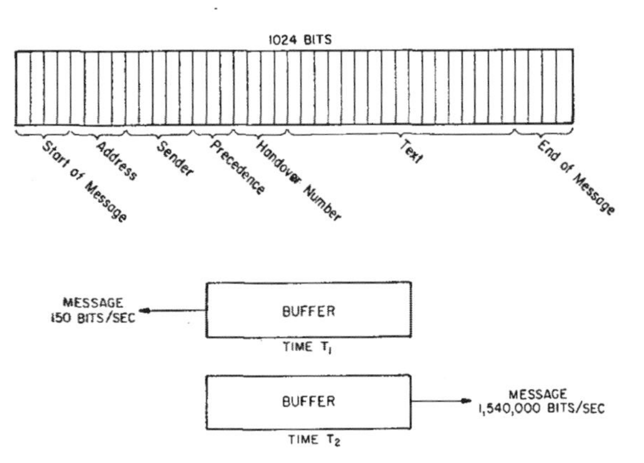

Fig. 10—Message block.

As we move to the future, there appears to be an increasing need for a standardized message block for our all-digital communications networks. As data rates increase, the velocity of propagation over long links becomes an increasingly important consideration.[3] We soon reach a point where more time is spent setting the switches in a conventional circuit-switched system for short holding-time messages than is required for actual transmission of the data.

Most importantly, standardized data blocks permit many simultaneous users, each with widely different bandwidth requirements to economically share a broad-band network made up of varied data rate links. The standardized message block simplifies construction of very high speed switches. Every user connected to the network can feed data at any rate up to a maximum value. The user's traffic is stored until a full data block is received by the first station. This block is rubber stamped with a heading and return address, plus additional housekeeping information. Then it is transmitted into the network.

[3] 3000 miles at $\simeq$150,000 miles/sec $\simeq$50 msec transmission time, T. 1024-bit message at 1,500,000 bits/sec $\simeq$2/3 msec message time, M. Therefore, $T \gg M$.

Switching

In order to build a network with the survivability properties shown in Fig. 4, we must use a switching scheme able to find any possible path that might exist after heavy damage. The routing doctrine should find the shortest possible path and avoid self-oscillatory or "ring-around-the-rosey" switching.

We shall explore the possibilities of building a "real-time" data transmission system using store-and-forward techniques. The high data rates of the future carry us into a hybrid zone between store-and-forward and circuit switching. The system to be described is clearly store and forward if one examines the operations at each node singularly. But, the network user who has called up a "virtual connection" to an end station and has transmitted messages across the United States in a fraction of a second might also view the system as a *black box providing an apparent circuit connection* across the country. There are two requirements that must be met to build such a quasi-real-time system. First, the in-transit storage at each node should be minimized to prevent undesirable time delays. Secondly, the shortest instantaneously available path through the network should be found with the expectation that the status of the network will be rapidly changing. Microwave will be subject to fading interruptions and there will be rapid moment-to-moment variations in input loading. These problems place difficult requirements upon the switching. However, the development of digital computer technology has advanced so rapidly that it now appears possible to satisfy these requirements by a moderate amount of digital equipment. What is envisioned is a network of unmanned digital switches implementing a self-learning policy at each node, without need for a central and possibly vulnerable control point, so that over-all traffic is effectively routed in a changing environment. One particularly simple routing scheme examined is called the "hot-potato" heuristic routing doctrine and will be described in detail.

Torn-tape telegraph repeater stations and our mail system provide examples of conventional store-and-forward switching systems. In these systems, messages are relayed from station to station and stacked until the "best" outgoing link is free. The key feature of store-and-forward transmission is that it allows a high line occupancy factor by storing so many messages at each node that there is a backlog of traffic awaiting transmission. But the price for link efficiency is the price paid in storage capacity and time delay. However, it was found that *most of the advantages of store-and-forward switching could be obtained with extremely little storage* at the nodes.

Thus, in the system to be described, each node will attempt to get rid of its messages by choosing alternate routes if its preferred route is busy or destroyed. Each message is regarded as a "hot potato," and rather than hold the hot potato, the node tosses the message to its neighbor who will now try to get rid of the message.

The Postman Analogy: The switching process in any store-and-forward system is analogous to a postman sorting mail. A postman sits at each switching node. Messages arrive simultaneously from all links. The postman records bulletins describing the traffic loading status for each of the outgoing links. With proper status information, the postman is able to determine the best direction to send any letters. So far, this mechanism is general and applicable to all store-and-forward communication systems.

Assuming symmetrical bidirectional links, the postman can infer the "best" paths to transmit mail to any station merely by looking at the cancellation time or the equivalent handover number tag. If the postman sitting in the center of the United States received letters from San Francisco, he would find that letters from San Francisco arriving from channels to the west would come in with later cancellation dates than if such letters had arrived in a roundabout manner from the east. Each letter carries an implicit indication of its length of transmission path. The astute postman can then deduce that the best channel to send a message *to* San Francisco is probably the link associated with the latest cancellation dates of messages *from* San Francisco. By observing the cancellation dates for all letters in transit, information is derived to route *future* traffic. The return address and cancellation date of recent letters is sufficient to determine the best direction in which to *send* subsequent letters.

Hot-Potato Heuristic Routing Doctrine: To achieve real-time operation it is desirable to respond to change in network status as quickly as possible, so we shall seek to derive the network status information directly from each message block.

Each standardized message block contains a "to" address, a "from" address, a handover number tag and error detecting bits together with other housekeeping data. The message block is analogous to a letter. The "from" address is equivalent to the return address of the letter.

The handover number is a tag in each message block set to zero upon initial transmission of the message block into the network. Every time the message block is passed on, the handover number is incremented. The handover number tag on each message block indicates the length of time in the network or path length. This tag is somewhat analogous to the cancellation date of a conventional letter.

The Handover Number Table: While cancellation dates could conceivably be used on digital messages, it is more convenient to think in terms of a simpler digital analogy; a tag affixed to each message and incremented every time the message is relayed. Fig. 11 shows the handover table located in the memory of a single node. A row is reserved for each major station of the network allowed to generate traffic. A column is assigned to each separate link connected to a node. As it was shown that redundancy levels on the order of four can create extremely "tough" networks and that additional redundancy can bring little, only about eight columns are really needed.

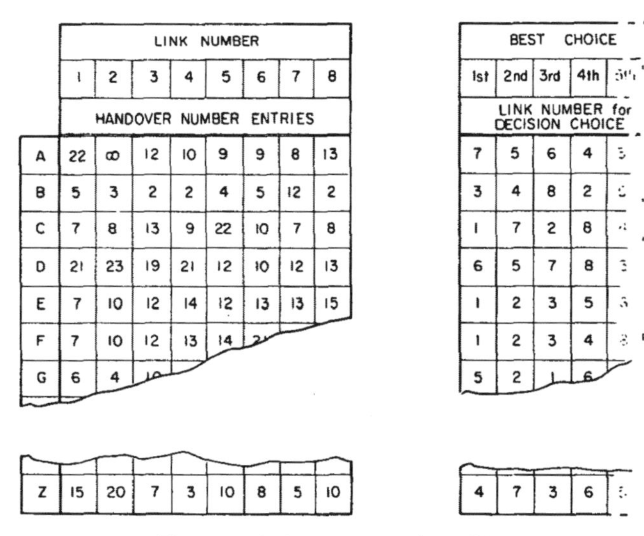

	LINK NUMBER							
	1	2	3	4	5	6	7	8
	HANDOVER NUMBER ENTRIES							
A	22	∞	12	10	9	9	8	13
B	5	3	2	2	4	5	12	2
C	7	8	13	9	22	10	7	8
D	21	23	19	21	12	10	12	13
E	7	10	12	14	12	13	13	15
F	7	10	12	13	14			
G	6	4						
Z	15	20	7	3	10	8	5	10

BEST CHOICE			
1st	2nd	3rd	4th
LINK NUMBER for DECISION CHOICE			
7	5	6	4
3	4	8	2
1	7	2	8
6	5	7	8
1	2	3	5
1	2	3	4
5	2		
4	7	3	6

Fig. 11—The handover number table.

Perfect learning: If the network used perfectly reliable, error-free links, we might fill out our table in the following manner. Initially, set entries on the table to high values. Examine the handover number of each message arriving on each line for each station. If the observed handover number is less than the value already entered on the handover number table, change the value to that of the observed handover number. If the handover number of the message is greater than the value on the table, do nothing. After a short time this procedure will shake down the table to indicate the path length to each of the stations over each of the links connected to neighboring stations. This table can now be used to route new traffic. For example, if one wished to send traffic *to* station *C*, he would examine the entries for the row listed for station *C* based on traffic *from C*, and select the link corresponding to the column with the lowest handover number. This is the shortest path to *C*. If this preferred link is busy, do not wait, choose the next best link that is free.

Digital Simulation: This basic routing procedure was tested by a Monte Carlo simulation of a 7 × 7 array of stations. All tables were started completely blank to simulate a worst-case starting condition where no station knew the location of any other station. Within one-half second of simulated real-world time, the network had learned the locations of all connected stations and was routing traffic in an efficient manner. The mean measured path length compared very favorably to the absolute shortest possible path length under various traffic loading conditions. Preliminary results indicate that network loadings on the order of 50 per cent of link capacity could be inserted without undue increase of path length. When local busy spots occur in the network, locally generated traffic is intermittently restrained from entering the busy points while the potential traffic jams clear. Thus, to the node the network appears to be a variable data rate system, which will limit the number of local subscribers that can be handled. If the network were carrying light traffic, any new input line into the network would accept full traffic, perhaps 1.5 million bits per second. But, if every station had heavy traffic and the network became heavily loaded, the total allowable input data rate from any single station in the network might drop to perhaps 0.5 million bits per second. The absolute minimum guaranteed data capacity of the network from any station is a function of the location of the station in the network, the redundancy level and the mean path length of transmitted traffic in the network. The "choking" of input procedure has been simulated in the network and no signs of instability under overload noted. It was found that most of the advantage of store-and-forward transmission can be provided in a system having relatively little memory capacity. The network "guarantees" very rapid delivery of all traffic that it has accepted from a user.

Forgetting and Imperfect Learning

We have briefly considered network behavior when all links are working. But we are also interested in determining network behavior with real-world links, some destroyed, while others are being repaired. The network can be made rapidly responsive to the effects of destruction, repair and transmission fades by a slight modification of the rules for computing the values on the handover number table.

Learning: In the previous example, the lowest handover number ever encountered for a given origination, or "from" station, and over each link was the value recorded in the handover number table. But if some links had failed, our table would not have responded to the change. Thus, we must be more responsive to recent measurements than old ones. This effect can be included in our calculation by the following policy. Take the most recently measured value of handover number; subtract the previous value found in the handover table; if the difference is positive, add a fractional part of this difference to the table value to form the updated table value. This procedure merely implements a "forgetting" procedure: placing more belief upon more recent measurements and less on old measurements. In the case of network damage, this device would automatically modify the handover number table entry to exponentially and asymptotically approach the true shortest path value. If the difference between measured value minus the table value is negative, the new table value would change by only a fractional portion of the recently measured difference.

This implements a form of skeptical learning. Learning will take place even with occasional errors. Thus, by the simple device of using only two separate "learning constants," depending on whether the measured value is greater or less than the table value, we can provide a mechanism that permits the network routing to be responsive to varying loads, breaks and repairs. This learning and forgetting technique has been simulated for a few limited cases and was found to work well.

Adaptation to Environment: This simple simultaneous learning and forgetting mechanism implemented independently at each node causes the entire network to suggest the appearance of an adaptive system responding to gross changes of environment in several respects, without human intervention. For example, consider self-adaptation to station location. A station, Able, normally transmitted from one location in the network, as shown in Fig. 12(a). If Able moved to the location shown in Fig. 12(b), all it need do to announce its new location is to transmit a few seconds of dummy traffic. The network will quickly learn the new location and direct traffic toward Able at its new location. The links could also be cut and altered, yet the network would relearn. Each node sees its environment through myopic eyes by only having links and link-status information to a few neighbors. There is no central control; only a simple local routing policy is performed at each node, yet the over-all system adapts.

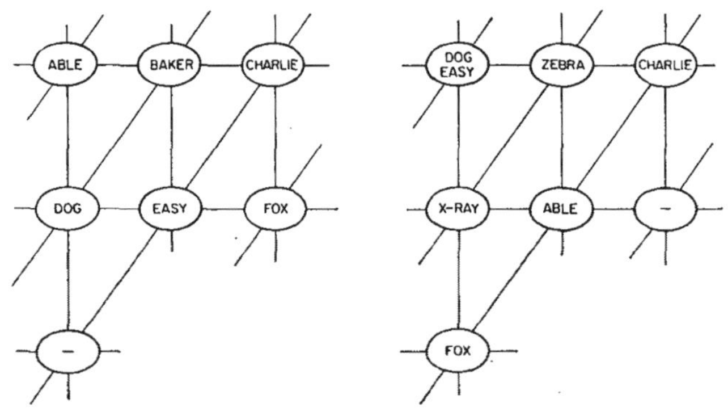

Fig. 12—Adaptability to change of user location. (a) Time "T_1." (b) Time "T_2."

Lowest Cost Path

We seek to provide the lowest cost path for the data to be transmitted between users. When we consider complex networks, perhaps spanning continents, we encounter the problem of building networks with links of widely different data rates. How can paths be taken to encourage most use of the least expensive links? The fundamentally simple adaptation technique can again be used. Instead of incrementing the handover by a fixed amount, each time a message is relayed, set the increment to correspond to the link cost/bit of the transmission link. Thus, instead of the "instantaneously shortest nonbusy path" criterion, the path taken will be that offering the cheapest transportation cost from user to user that is available. The technique can be further extended by placing priority and cost bounds in the message block itself, permitting certain users more of the communication resource during periods of heavy network use.

Where We Stand Today

Although it is premature at this time to know all the problems involved in such a network and understand all costs, there are reasons to suspect that we may not wish to build future digital communication networks exactly the same way the nation has built its analog telephone plant.

There is an increasingly repeated statement made that one day we will require more capacity for data transmission than needed for analog voice transmission. If this statement is correct, then it would appear prudent to broaden our planning consideration to include new concepts for future data network directions. Otherwise, we may stumble into being boxed in with the uncomfortable restraints of communications links and switches originally designed for high-quality analog transmission. New digital computer techniques using redundancy make cheap unreliable links potentially usable. Some sort of switched network compatible with these links appears appropriate to meet this new upcoming demand for digital service.

Of course, we could use our existing circuit switching techniques, but a system with greater capacity than the long lines of telephone plants might best be designed for such data transmission and survivability at the outset. Such a system should economically permit switching of very short blocks of data from a large number of users simultaneously with intermittent large volumes among a smaller set of points. Considering the size of the market, there appears to be an incommensurately small amount of thinking about a national data plant, designed primarily around bit transportation.

Postscript

This paper was, in essence, written about 18 months ago. Since that time the most critical aspects of the system have been examined and developed in detail, and a series of amplifying RAND Memoranda is in preparation. An idea of the subjects covered can be gained from the following list of tentative titles:

[1] Paul Baran, "Introduction to Distributed Communications Networks."[4]
[2] S. Boehm and P. Baran, "Digital Simulation of Hot-Potato Routing in a Broadband Distributed Communications Network."
[3] J. W. Smith, "Determination of Path-Lengths in a Distributed Network."
[4] P. Baran, "Priority, Precedence, and Overload."
[5] ——, "History, Alternative Approaches, and Comparisons."
[6] ——, Mini-Cost Microwave."
[7] ——, "Tentative Engineering Specifications and Preliminary Design for a High Data Rate Distributed Network Switching Node."
[8] ——, "The Multiplexing Station."
[9] ——, "Security, Secrecy, and Tamper-Free Considerations."
[10] ——, "Cost Analysis."
[11] ——, "Summary Overview."

Because of the dependence of each of these Memoranda (vols. 2–11) upon one another, we have elected to release the volumes as a set as an aid to the reader.

Acknowledgment

In discussing this work, I received a number of helpful ideas and suggestions. Wherever possible, these acknowledgments are included within the detailed papers amplifying the subject.

Specific acknowledgments for the present paper include the excellent programming assistance provided by Sharla Boehm, J. Derr, and J. W. Smith. I am also indebted to J. Bower for his suggestions that switching in any store-and-forward system can be described by a model of a postmaster and a blackboard.

[4] This is essentially the present paper.

Routing Procedures in Communications Networks—Part I: Random Procedures*

REESE T. PROSSER†

Summary—**A study is made of possible routing procedures in military communications networks in order to evaluate these procedures in terms of future tactical requirements. In Part I this study is devoted to procedures involving random choices. In such networks each message path is essentially a random walk. Estimates of the average traverse time of each message and average traffic flow through each node are derived by statistical methods under reasonable assumptions on the operating characteristics of the network for various typical random routing procedures.**

This paper does not purport to present a complete system design. Many design questions, common to all network routing problems—response to temporary loss of links or nodes, rules for handling of message priorities, etc.—are not considered here.

It is shown that random routing procedures are highly inefficient but extremely stable. A comparison of these theoretical results with the results of an extended computer simulation effort lends support to their reliability, discrepancies being accounted for by the simplifying nature of the statistical assumptions. It is suggested that in circumstances where the need for stability outweighs the need for efficiency, this type of network might be advantageously employed.

A. Introduction

THIS REPORT grew out of an attempt to describe a military communications system suitable for combat units operating in a hostile environment. The requirements for such a system differ in various respects from those of a civilian system, such as the telephone or telegraph systems, operating in normally favorable environments, and these differences are reflected not only in the operating characteristics of the components of the system but also in the organization of the system itself. This report is devoted to a study of the statistical properties of different kinds of communications system organizations and an evaluation of their relative effectiveness under different types of environment.

The central problem, stated in general terms, appears to be one of efficiency vs reliability: In an entirely favorable environment, the reliability of a communications system is limited only by the characteristics of the components themselves, and the system should be organized for maximum efficiency. In an extremely hostile environment, on the other hand, the system must be organized for maximum reliability, and the efficiency is limited by the characteristics of this organization. Thus in a civilian system, messages normally get through quickly and accurately, but if fire sweeps a central office, the communications failure may be absolute. In a military system, however, degradation must be only gradual: messages must still be delivered—possibly delayed and distorted—even after a loss of half the system.

A closer analysis reveals that these general considerations can be given a theoretical basis, based on the following observations.

Any large-scale civilian communications system is essentially a directory system: Every operating station in the system has a directory, or has access to a directory, which contains complete information on how to reach every other station in the network. In most cases there are also "central" stations, whose primary function is to handle, or to assist in handling, the routing of all messages through the system. This type of organization has the obvious advantage that efficient routing procedures can be obtained by all stations from the directory, and the obvious disadvantage that it depends heavily upon this directory. In a hostile environment the disadvantage is likely to outweigh the advantage: any change in the system requires a revision of the directory of every station, and the removal of a central station is likely to paralyze a large segment of the system.

In contrast, any military communications system operating in an extremely hostile environment tends to operate as a random system: there are no directories, and each station sends messages to whatever stations he can, based on whatever information he has, and hopes for the best. This type of organization has the obvious disadvantage that it is likely to be extremely inefficient, but has the advantage that it will continue to operate in the presence of serious degradation. In a favorable environment the disadvantage surely outweighs the advantages: No military establishment would tolerate so high a degree of inefficiency in its communications unless driven to it by truly desperate circumstances.

These observations can be restated in terms of the information available to each station about the rest of the system. On the one hand, the directory system supposes that each station has access to complete and correct information about the system and routes messages on the basis of this information. Such a system can be made efficient, but is faced with the problem of devising optimum routing procedures from this information, and keeping this information up to date. On the other hand, the random system supposes that no station has access to any information about the system and each station routes message on the basis of random choices. Such a system can be made reliable, but must determine how to organize itself to ensure best possible efficiency. It is likely that any military communications system suitable

* Received September 25, 1962. This work was supported by the U. S. Continental Army Command under Air Force Contract AF 19 (604)-5200.

† M.I.T. Lincoln Laboratory, Lexington, Mass.

Reprinted from *IEEE Transactions on Communications Systems*, December 1962.

for operating under a wide variety of conditions will necessarily lie somewhere between these extremes, incorporating elements of both in a proportion which might well change with circumstances.

In this report we shall attempt to reinforce these qualitative considerations with quantitative information on the statistical properties and behavior of various kinds of communications systems organizations. In Part I we consider essentially random systems in which the routing procedures depend to a large extent on random choices made at each station. Theoretical estimates of average traffic rates, average traverse times, and average storage requirements are derived for various routing procedures and compared with the results of an extensive simulation experiment made on the IBM 709 computer. The behavior of such systems under degradation is then investigated in terms of these parameters. In Part II we consider essentially deterministic systems, in which the routing procedures are determined from directory information at the initial station. Here algorithms for optimum efficiency routing procedures and possible updating schemes are investigated, average traffic rates, traverse times and storage requirements are derived, and behavior under degradation investigated. The report concludes with a table of comparisons and summary of results.

B. Operating Characteristics

An accurate analysis of an arbitrary communications system operating under a random routing procedure is exceedingly difficult to achieve. For this reason we begin by introducing a series of broad assumptions designed to simplify the analysis without adulterating the problem. Specifically, all of the communications systems considered here are assumed to have the following features in common.

Each network may be considered as a connected *graph* [1] consisting of a fixed number n of *nodes* of which certain pairs ("neighbors") are joined by *links*. Since we are dealing with systems with random routing procedures, we shall assume that the network is (approximately) homogeneous in the sense that each node is linked to (approximately) k others (*i.e.*, has k "neighbors").

Messages are supposed to originate at each node of the network with a Poisson distribution in time of mean λ and with an exponential distribution in length of mean $1/\mu$. Both λ and μ are to be the same for all nodes. Each message is addressed to a node in the network chosen at random. It is convenient to include the (unlikely) possibility that the sender and receiver nodes for a message are identical. Each message is relayed from node to node along existing links in the network subject to the following general rules.

1) Messages arriving at a node, whether relayed or originating there, are placed in storage in order of arrival. Any message arriving while the storage is full is dropped from the network (*i.e.*, "lost").
2) Messages are taken one at a time from storage on a first-come, first-serve basis and handled according to a prescribed procedure (the routing procedure) which varies with the system under study, but always includes the following:
   a) If the message is addressed to the node where it is located, it is dropped from the net (*i.e.*, "received").
   b) If the node has (partial) information on the location of the addressee, it relays the message to one of its k neighbors selected on the basis of that information.
   c) Otherwise the node relays the message to one of its k neighbors selected at random.
3) Each node handles messages one at a time only, and the time required to handle a message[1] is directly proportional to the length of the message. For convenience we take the constant of proportionality to be 1.

It is desirable to avoid specifying the unit of time, and so we shall refer to it as a *cycle*. The length of time required to send a message from one station to a neighbor station is called a *relay*. Note that this length depends on the message, but not on the station. The *average* relay is then $1/\mu$ cycles. We assume $1/\mu > 1$.

C. Mean Total Traffic

We now proceed to make a series of rough estimates of the operating characteristics of the system described in the previous section. These estimates are statistical in nature, and are based on assumptions, stated explicitly below, about the statistical properties of the system which are gross over-simplifications of the actual state of affairs. The estimates are consequently to be regarded as indicative, but not descriptive of the actual characteristics of the system. A comparison of these estimates with the results of the simulation experiment described in Section I. will give some idea of their reliability.

The first step in our analysis consists in estimating the average total number $N(t)$ of messages in the network at the end of a given cycle t. Roughly speaking, this average is given by 1), the average total number of messages in the network at the end of the previous cycle, plus 2), the average total number of messages added to the network during a single cycle, minus 3), the average total number of messages dropped from the network during a single cycle. Now 1) is simply $N(t - 1)$ and 2) is just λn. 3) is the sum of two terms, one arising from messages lost because of insufficient storage facilities, and the other from messages delivered to their destinations. We shall assume here that the storage facilities at each node are sufficient to justify neglecting the first of these terms. (Estimates of these storage requirements are derived below.) The second term is the product of the average *fraction* f of $N(t)$ actually handled during a cycle (*i.e.*, not waiting in storage), times the probability p that one of these messages will be relayed to its destina-

[1] *i.e.*, the time required to process and transmit a message. Time spent in storage is not included.

tion, divided by the expected number of cycles $1/\mu$ required to relay such a message to its destination. Thus we are led to the following equation for $N(t)$:

$$N(t) = N(t-1) + \lambda n - \mu p f N(t-1). \quad (1)$$

This is an ordinary first-order difference equation whose solution is determined once we settle on an initial condition. Since we are principally interested in steady-state estimates, it is convenient to assume that the network is empty at $t = 0$ and investigate the behavior of $N(t)$ as $t \to \infty$. This leads to

$$N(0) = 0. \quad (2)$$

The solution now depends on assumptions made on p and f. Now p (and essentially only p) depends on the routing procedure of the system, and is investigated in Section E. Tentatively we shall assume here that p is a constant, independent of t or N, and postpone its evaluation until Section E. On the other hand, f is the fraction of N not waiting in storage, and in general depends on N. If we assume that the messages are independently randomly distributed throughout the network, then the methods of Riordan [6] may be used to show that the expected value of f is given by

$$f(N) = \frac{n}{n + N - 1}. \quad (3)$$

This value leads to a troublesome equation for (1). But two cases of interest are immediately accessible.

We first note that

$$f(N) \sim \begin{cases} \dfrac{n}{N} & \text{if } N \gg n \\ 1 & \text{if } N \ll n; \end{cases} \quad (4)$$

if $N \gg n$, we get

$$N(t) = N(t-1) + (\lambda - \mu p)n \quad (5)$$

whose solution is easily seen to be

$$N(t) = (\lambda - \mu p)nt. \quad (6)$$

In this case $N(t) \to \infty$ as $t \to \infty$, and the system has no steady state.

If $N \ll n$, we get

$$N(t) = N(t-1) + \lambda n - \mu p N(t-1) \quad (7)$$

whose solution is given by (*cf.* [4])

$$N(t) = \frac{\lambda n}{\mu p}(1 - (1 - \mu p)^t). \quad (8)$$

In this case $N(t) \to \lambda n/\mu p$ as $t \to \infty$, and we may take this as the expected value for N when the network is operating in steady state. Formally,

$$N(\infty) = \frac{\lambda n}{\mu p}. \quad (9)$$

The requirement that $N \ll n$ places a condition on the parameters of the system; namely, that $\lambda n/\mu p \ll n$. This leads to

$$\frac{\lambda}{\mu} \ll p \quad (10)$$

We conclude that when (10) is satisfied, the network will operate at a steady state with (9) as the expected value for N. As the ratio λ/μ increases past p, however, then messages originate in the system faster than they can be delivered, and N may be expected to increase without limit, and ultimately behave like (6). These conclusions are borne out reasonably well by the simulation results (*cf.* Section I).

We have assumed above that the number of messages lost because of insufficient storage is negligible. This assumption places a requirement on the storage facilities at each node which we shall estimate here.

The storage facilities at each node may be considered as forming a queue, with messages arriving with a Poisson distribution of mean which because of the homogeneous character of the net we may take to be $n\lambda$, and leaving with an exponential distribution of mean $1/\mu$. Thus the methods of queueing theory apply (*cf.* [5], chapter 2). According to this theory, the probability that the queue is empty at any cycle is given by

$$P_0 = \frac{1 - \rho}{1 - \rho^{s+1}} \quad (11)$$

where $\rho = n\lambda/\mu$, and s is the length of the queue. The probability that the queue contains b messages is

$$P_b = \rho^b P_0 \qquad (0 \le b \le s). \quad (12)$$

The mean of this distribution over b when $\rho \ll 1$ is approximately ρ (see [5], page 18).

Now the probability that a message is lost at a given node during a cycle is certainly no greater than the probability P_s that the storage queue at that node is full. The requirement that this is negligible relative to the probability that the message is delivered during this cycle is simply (assuming $f \sim 1$)

$$P_s < \alpha\mu p \quad (13)$$

where α is a preassigned fraction (for instance $\alpha = 0.01$). Using (11) and (12) we get

$$\rho^s \frac{1 - \rho}{1 - \rho^{s+1}} < \alpha\mu p. \quad (14)$$

This is surely satisfied if

$$\rho^s < \alpha\mu p \quad (15)$$

and this implies

$$s > \frac{\log(\alpha\mu p)}{\log \rho}. \quad (16)$$

The condition expressed in (16) places a lower bound on the size of the storage facilities at each node needed to ensure that message loss is negligible relative to message delivery. Thus we must have storage for at least s messages at each node, where s satisfies (16).

D. Mean Traverse Times

The next step in our analysis consists of estimating the average traverse time of a single message, *i.e.*, the average number of cycles required to relay the message from sender to receiver by routing procedures of the type described in Section B.

As in Section C, let p be the probability that a given message, placed at random in the network, is relayed to its destination on the next relay. Again we shall assume tentatively that p is independent of time. Then the probability that the message is still in the network after s relays is simple $(1 - p)^s$. This gives a familiar distribution over s, whose mean m and variance σ^2 are readily computed [3]:

$$m = \frac{1-p}{p} \sim \frac{1}{p} \tag{17}$$

$$\sigma^2 = \frac{1-p}{p^2} \sim m^2. \tag{18}$$

Note that σ^2 is larger than m. This may be interpreted as saying that while the mean traverse time is of the order of $1/p$, the variance is very large, and some messages may be expected to have very large traverse times.

It remains to express m and σ^2 in terms of cycles. The number z of cycles required for the average relay is just 1), the average service time in cycles, plus 2), the average time in cycles spent waiting in storage. Now 1) is just $1/\mu$, and 2) is equal to $1/\mu$ times the average number of messages waiting in storage. The average number of messages waiting in storage, assuming adequate storage facilities, is approximately $\rho/(1 - \rho)$ (*cf.* [5], page 22). Combining, we get

$$z = \frac{1}{\mu}\left(1 + \frac{\rho}{1-\rho}\right) = \frac{1}{\mu}\frac{1}{(1-\rho)}. \tag{19}$$

Thus the mean traverse time in cycles, from (18) and (19), is

$$m = \frac{(1-p)}{\mu p}\cdot\frac{1}{1-\rho}. \tag{20}$$

When $p \ll 1$ this becomes

$$m \sim \frac{1}{\mu p}\frac{1}{(1-\rho)}. \tag{21}$$

When ρ is negligibly small, the network is essentially empty, and (21) becomes

$$m \sim \frac{1}{\mu p}. \tag{22}$$

Thus the delaying effect caused by the presence of other messages is given by the factor $(1/[1 - \rho])$.

E. Effects of Various Routing Procedures

It remains to give estimates for the value of p, the probability that a message will be relayed to its destination in a single relay. This probability obviously depends directly upon the choice of routing procedure; and in this section we estimate the value of p for various typical procedures, each of them consistent with the general description given in Section B, and each of them involving random choices.

It is convenient for this purpose to regard a message as having reached its destination as soon as its subsequent route is completely deterministic. Thus a message is regarded as "in the network" only as long as its subsequent path is not completely determined by the routing procedure. In all cases considered here this convention does not materially affect the results of Sections C and D.

With this in mind, we see that the value of p has the following form: it is given by 1), the probability that the message is located at a neighbor of its destination, taken in the sense described above, times 2), the probability that this neighbor will relay it to its destination. The value of 1) is computed assuming that the message is located at random in the network, and 2) is computed on the basis of the routing procedure.

We now examine various cases:

a) *pure random*. In this case we assume each node knows only its own identity. Thus each node relays every message to one of its k neighbors selected at random unless the message is addressed to it. Here the "destination" of the message consists of the addressee, and there are k neighbors of this destination. The probability that message is located at such a neighbor is k/n, and we get

$$p = \frac{k}{n}\cdot\frac{1}{k} = \frac{1}{n}. \tag{23}$$

b) *first-order neighbors*. In this case we assume each node knows the identity of itself and each of its neighbors. Messages are relayed at random unless addressed to a neighbor, in which case they are relayed to that neighbor. The "destination" of a message now consists of $k + 1$ nodes (addressee plus k neighbors) and the number of neighbors of this destination depends somewhat on the details of the network graph. Two cases are accessible when n is large. If neighbors of each node are located at random through the network, then they are relatively independent of each other, and there are approximately $k(k - 1)$ neighbors of the destination (*i.e.*, neighbors of neighbors of the addressee). The probability that a message is located at one of these neighbors is then $k(k - 1)/n$, and we get

$$p = \frac{k(k-1)}{n}\cdot\frac{1}{k} = \frac{k-1}{n}. \tag{24}$$

If the graph is planar, however, and if neighbors are located among the geographical neighbors of a node then they are no longer independent of each other; and a study of various regular lattice configurations in the plane shows that when k is small there are approximately $2k$ neighbors of the destination, each of which is connected to the destination in two different ways. This gives us

$$p = \frac{2k}{n}\cdot\frac{2}{k} = \frac{4}{n}. \tag{25}$$

c) *second-order neighbors*. In this case each node knows the identity of itself, its neighbors and their neighbors. This case is an extension of the previous one, and the dependence on the details of the graph, already noted there, also applies here.

If neighbors are located at random through the network, then the "destination" consists of $k^2 + 1$ nodes and there are approximately $k(k - 1)^2$ neighbors of the destination. In this case we get

$$p = \frac{k(k-1)^2}{n}\cdot\frac{1}{k} = \frac{(k-1)^2}{n}. \tag{26}$$

If neighbors are located only among geographical neighbors then the "destination" consists of approximately $3k + 1$ nodes and there are approximately $3k$ neighbors each of which is connected to the destination in two different ways. Thus we get

$$p = \frac{3k}{n}\cdot\frac{2}{k} = \frac{6}{n}. \tag{27}$$

It is evident that we can continue this extension until each node has complete information about the net, though the approximations made here become less and less valid as we proceed. When k is large, however, not many steps are required. (Thus it has been said that any pair of persons in the United States are fourth-order neighbors via acquaintance links.)

We may summarize the preceding arguments in the following way: if the "destination" of a message is defined as the set of nodes which know the location of the addressee node and can assure a deterministic path to the addressee, and if there are h neighbors of this destination, each of them connected to the destination in j different ways, then we may estimate the value of p as

$$p = \frac{h}{n}\cdot\frac{j}{k}. \tag{28}$$

We assemble our results in Table I.

TABLE I

Routing Procedure	*Value of p*
pure random	$1/n$
first-order neighbors (arbitrary)	$(k - 1)/n$
first-order neighbors (geographic)	$4/n$
second-order neighbors (arbitrary)	$(k - 1)^2/n$
second-order neighbors (geographic)	$6/n$

Recall that for large n we have

$$m \sim \frac{1}{p} \tag{17}$$

$$\sigma^2 \sim \frac{1}{p^2} \tag{18}$$

$$N \sim \frac{1}{p}\cdot\rho \quad \text{where} \quad \rho = \frac{\lambda n}{\mu}. \tag{9}$$

From Table I we see that in general, the larger the value of p, the smaller the value of m, σ^2 and N, and the more efficient the behavior of the network. In this sense the more information a node has, the more efficiently it will perform as a relay station in this type of network.

F. Critique of Assumptions

In deriving the formulas of the preceeding sections we have relied heavily upon several simplifying assumptions, which may well be expected to introduce errors into the results. In this section we examine briefly the nature of these errors.

In the first place, we have assumed that the network is homogeneous, *i.e.*, connected in such a way that, statistically speaking, the nodes are essentially identical, yet statistically independent, in their behavior. In point of fact, such a network is hard to realize. Any graph with a high degree of homogeneity in its connectivity is likely to have a high degree of symmetry as well, and the statistical independence of the different nodes is not easy to justify. This is particularly true for planar graphs, where neighbors are chosen geographically.

It is difficult to get a clear view of the situation without becoming hopelessly entangled in advanced combinatorics. Nevertheless, it seems reasonable to suppose that the dependence of our results upon these assumptions is not critical when n, the number of nodes, is large ($n \sim 100$) and deviations from homogeneity are not too pronounced ($\Delta k \leq 2$). Thus we shall expect the behavior of such networks to follow at least qualitatively the predictions of our results and actual mean traverse times and mean total traffic to approximate the values we have predicted with at least the right order of magnitude. Excessive reliance on the predicted values, however, is to be avoided.

In the second place, we have assumed throughout that the value of p, the probability that a message reaches its destination on the next relay, is constant in time. Actually, it can be shown that p *decreases* somewhat in time. This effect tends to make the mean traverse times somewhat longer, and the mean total traffic somewhat larger, than predicted in Section D.

The actual behavior of p is, like everything else, shrouded in combinatorics. In the pure random case a rough idea may be obtained in the following way: the process of relaying a single message through a network may be regarded as a finite Markov process, with the nodes as states, and the routing doctrine determining the transition matrix (*cf.* [3]). The behavior of the network is then determined by the rules for finite Markov processes.

As an example we consider the pure random case with n large and $k = 2$. Here the graph consists of a single closed loop. Suppose a message originates somewhere at random in the network, addressed for node n. Its probable position at time $t = 0$ is described by the one-column matrix

$$\mathbf{v} = \begin{bmatrix} 1/n \\ 1/n \\ \vdots \\ 1/n \end{bmatrix} \tag{29}$$

and its position after one relay is described by $\mathbf{A}\cdot\mathbf{v}$, where $\mathbf{A}$ is the transition matrix.

$$\mathbf{A} = \begin{bmatrix} 0 & \frac{1}{2} & 0 & 0 & \cdots\cdots & & & 0 \\ \frac{1}{2} & 0 & \frac{1}{2} & 0 & \cdots\cdots & & & 0 \\ 0 & \frac{1}{2} & 0 & \frac{1}{2} & \cdots\cdots & & & 0 \\ 0 & 0 & \frac{1}{2} & 0 & \cdots\cdots & & & 0 \\ \cdots & \cdots & \cdots & \cdots & \cdots & \cdots & \cdots & \cdots \\ 0 & \cdots\cdots & & & \frac{1}{2} & 0 & \frac{1}{2} & 0 \\ 0 & \cdots\cdots & & & 0 & \frac{1}{2} & 0 & 0 \\ \frac{1}{2} & \cdots\cdots & & & 0 & 0 & \frac{1}{2} & 1 \end{bmatrix}. \tag{30}$$

It is easy to see that after several relays the probability that the message is located at node $i \neq n$ is not the same for all i, being least at the nodes closest to node n, and greatest at the nodes farthest away from node n. Nevertheless, the decay of the probability that the message is not at n is still exponential in time, being determined essentially by the largest eigenvalue of $\mathbf{A}$ different from 1 ([3]). This eigenvalue depends on n but is always somewhat larger than $(1 - p)$, which is the value used in Section E. Similar arguments can be given for other values of k and other routing logics. The problem in each case is to find the largest eigenvalue (different from 1) of the associated transition matrix, and this depends in general on the details of the network graph. The net result, however, is as stated above: the mean traverse times are increased, and the mean total traffic increased, by the inclusion of this effect.

These observations are borne out by the simulation results described in Section I.

G. Effect of Unreliable Links and Unreliable Nodes

Efforts to evaluate the behavior of a random routing system operating in a hostile environment will depend in general on assumptions made on the effect of the environment upon the system.

Perhaps the simplest way to learn something about this behavior is to assume that the graph of the network is unaffected, but that the links are "unreliable", *i.e.*, that each link is operational only a part of the time. To make this precise, we suppose that the probability that any link is operational at any moment is r_l, r_l being the same for all links.

The effect of this assumption is twofold: in the first place, it obviously reduces the probability that a message is delivered to its destination (in the sense of Section E) in the next relay by the probability that the appropriate links are operational. Thus we have

$$p_{\text{unreliable}} = r_l p_{\text{reliable}} \tag{31}$$

In the second place, our assumption increases the number of cycles in a relay and hence the mean traverse time by a factor depending on the (small) probability that a message is delayed at a relay node because no link leading from that node is operational. This probability is just $(1 - r_l)^k$, and we have

$$z_{\text{unreliable}} = (1 + (1 - r_l)^k) z_{\text{reliable}} \tag{32}$$

This effect is negligible when $r_l \sim 1$.

Another type of effect of environment upon the system results from supposing that links are unaffected, but that *nodes* are "unreliable." In this case we suppose that each node is operational with probability r_n, and further, that each node knows his neighbors, and relays messages only to operational nodes. Again the effect is two-fold: in the first place the probability that a message is delivered to its destination in the next relay is reduced by a factor of r_n,

$$p_{\text{unreliable}} = r_n p_{\text{reliable}}, \tag{33}$$

and the number of cycles in a relay is increased by a factor of the form $1 + (1 - r_n)^k$, so that

$$z_{\text{unreliable}} = (1 + (1 - r_n)^k) z_{\text{reliable}} \tag{34}$$

which is negligible when $r_n \sim 1$.

From these arguments we see that this type of network is extremely stable in the presence of a hostile environment, in the sense that the principle effect of the environment on the over-all operation of the network is a moderate attenuation of efficiency; in no sense is this operation destroyed.

H. Effects of Sequential Reception

The results of the preceding sections have been derived under the assumption that each node of the network sends messages sequentially (*i.e.*, one at a time) but receives messages simultaneously, being limited only by its storage facilities. In practice it is more likely that each node both sends and receives messages sequentially. This is certainly the case, for example, in any military communications system organized around microwave relay links connecting mobile stations with a single directional antenna. At any given instant, such a station may send or receive a simple message, or neither, but not both. The situation is not appreciably changed if each link consists of several channels. For such a system the previous derivations are no longer valid, but a slight modification makes it possible to include this more "realistic" case within our study.

Let us assume, then, that each node of the network operates sequentially, in the sense that it can perform only one operation (send or receive a single message) at a time. The principle effect of this assumption is to render the nodes "unreliable"—each node is operational only during that fraction of the time that it is not busy. Thus the results of Section G apply. The value of p must be reduced by a factor r_n and the value of z increased by a factor $1 + (1 - r_n)^k$. Here r_n is the probability that a node is operational, *i.e.*, not busy.

A crude estimate of r_n may be given along the following lines. The fraction of the time that a node is busy is roughly equal to the average density of messages in the network, *i.e.*, to the average number of messages located at each node. As in Section D we take for this average the value $\rho/(1 - \rho)$. If we neglect ρ^2 relative to ρ, we find that $r_n = 1 - \rho$ and hence

$$p_{\text{sequential}} = (1 - \rho)p_{\text{simultaneous}} \tag{35}$$

$$z_{\text{sequential}} = (1 + \rho^k)z_{\text{simultaneous}}. \tag{36}$$

The factor modifying z may certainly be neglected. Other effects of sequential operation are negligible relative to these.

I. Results of a Simulation Experiment

In an effort to verify quantitatively the conclusions described in this report, an extensive simulation experiment was performed by D. F. Clapp at Lincoln on the IBM 709. Details of this effort are described elsewhere [2]. Here it suffices to say that the simulation involved typical homogeneous networks with typical values of n and k, and routing logics including the pure random, first-order neighbors, and second-order neighbors cases as described in Section E. The operating characteristics of these networks were made consistent with the general principles laid down in Section B, except that it was felt desirable to include provisions for the sequential reception feature described in Section G, and the actual simulation was run in this mode.

Each node was provided with storage facilities for seven messages. The values of the ratio λ/μ were chosen by trial and error to be as large as possible subject to the requirement that the message loss through storage overflows be negligible. In this sense the networks were operated at the largest possible average traffic load. Smaller values of λ/μ are expected to result in a better over-all performance.

The data recorded in this simulation included values of the mean total traffic and mean traverse time as described in Sections C and D. From the point of view of this study, these are the factors necessary to evaluate the performance of this type of network. A comparison of their values as determined by analysis and by simulation for typical situations is assembled in Tables II and III. We feel that the results of the simulation tend to bear out the results of the previous sections within the broad range of uncertainty introduced by the complexities of the problem.

TABLE II
MEAN TOTAL TRAFFIC—ANALYSIS VS SIMULATION
$N = \lambda n/\mu p$

n	k	$\lambda/\mu \times 10^3$	A		B		C	
50	3	2.60	7.48	**28.9**	3.74		1.87	
		3.89	12.1	**38.3**	6.06		3.04	
		4.55	14.8	**52.4**	7.38	**40.3**	3.69	**35.8**
		5.20	17.6	**60.3**	8.78		4.39	
50	9	4.55	14.8	**11.7**	1.85		0.231	
		6.50	30.1	**35.0**	3.73		0.471	
		9.09	41.9	**52.2**	5.23	**5.60**	0.653	**1.70**
100	3	0.250	2.70	**11.4**	1.35		0.675	
		1.30	14.9	**61.7**	7.45	**60.8**	3.73	**54.7**
		1.82	22.3	**91.2**	11.2		5.60	
100	9	1.30	15.0	**23.7**	1.88		0.234	
		2.60	35.2	**74.1**	4.40	**14.6**	0.550	**6.30**

Column A: pure random case
Column B: 1st-order neighbors case
Column C: 2nd-order neighbors case
Results of Analysis: 0.000
Results of Simulation: **0.000**

TABLE III
MEAN TRAVERSE TIMES—ANALYSIS VS SIMULATION
$m = 1/\mu p \times 1/(1 - \rho)^2$

n	k	$\lambda/\mu \times 10^3$	A		B		C	
50	3	2.60	106	**164**	53.0		26.5	
		3.89	124	**187**	62.0		31.0	
		4.55	135	**191**	67.5	**131**	33.8	**126**
		5.20	147	**161**	73.5		36.8	
50	9	4.55	135	**51.3**	16.9		2.11	
		6.50	206	**94.9**	25.6		3.20	
		9.09	272	**95.0**	34.0	**12.1**	4.25	**4.0**
100	3	0.250	170	**311**	85.0		47.5	
		1.30	213	**252**	107	**194**	53.5	**206**
		1.82	241	**250**	121		60.5	
100	9	1.30	213	**128**	26.6		3.33	
		2.60	294	**141**	36.7	**50.1**	4.60	**22.6**

Column A: pure random case
Column B: 1st-order neighbors case
Column C: 2nd-order neighbors case
Results of Analysis: 0.000
Results of Simulation: **0.000**

J. Summary and Conclusions

The quantitative results established in this report tend to bear out and to emphasize the qualitative conclusions drawn in the introduction on the performance characteristics of communications systems depending upon random routing procedures as defined in Section B. From the point of view of military communications requirements, these conclusions may be summarized as follows:

1) The system provides access to the whole network for each station.
2) It does not require the use of directories.
3) It does not require the use of complex routing doctrines.
4) It does not require the use of central offices or trunk lines.

5) Its performance is essentially independent of the actual structure of the network.
6) Its performance is extremely stable in the presence of a hostile environment, in the sense of Section G.

Disadvantages:

1) Average traverse times are long, and variances large.
2) Traffic rates must be kept low.
3) The system is not suitable for operation in real time. In particular, it cannot be employed as a telephone system.
4) The system is not suitable for any but the simplest types of messages. In particular, it cannot be expected to deliver "all points" messages.
5) The system is extremely vulnerable to overloading. Consequently, it cannot be expected to handle nonessential or duplicate messages.

It is evident that from a military point of view the advantages are extremely attractive but that the price to be paid for them is high. Perhaps the severest disadvantage is the limitation on traffic rates. This can be alleviated somewhat by radically increasing the transmission rates, thus reducing the effective message lengths. If transmission rates of the order of a millisecond[2] can be achieved by high-speed microwave components, then messages may originate at the rate of one per second at each station without overloading a system of one hundred stations. Mean traverse times are then of the order of a tenth of a second. In this case it might be possible to provide a simple, workable, non-real-time communications system which could assume all of the advantages listed above, and which could be supplemented with whatever additional facilities seem desirable.

In Part II we will consider essentially deterministic systems, in which the routing procedures are determined from directory information at the initial station.

Bibliography

[1] C. Berge, "Theorie des Graphes et ses Applications," Dunod, Paris, France; 1958.
[2] D. Clapp, "On A Communications Network Simulation Program," Lincoln Lab., Lexington, Mass., Group Rept. No. 22G-0016; May, 1960.
[3] W. Feller, "An Introduction to Probability Theory and Its Applications," John Wiley and Sons, Inc., New York, N. Y.; 1950.
[4] F. Hildebrand, "Methods of Applied Mathematics," Prentice-Hall Inc., New York, N. Y.; 1952.
[5] P. Morse, "Queues, Inventories and Maintenance," John Wiley and Sons, Inc., New York, N. Y.; 1958.
[6] J. Riordan, "An Introduction to Combinatorial Analysis," John Wiley and Sons, Inc., New York, N. Y.; 1958.

[2] Per relay.